COMMUNICATION AND COMPUTING SYSTEMS

PROCEEDINGS OF THE INTERNATIONAL CONFERENCE ON COMMUNICATION AND COMPUTING SYSTEMS (ICCCS-2016), DRONACHARYA COLLEGE OF ENGINEERING, GURGAON, INDIA, 9–11 SEPTEMBER 2016

Communication and Computing Systems

Editors

B.M.K. Prasad
Principal, Dronacharya College of Engineering, Gurgaon, India

Krishna Kant Singh
Department of Electrical and Electronics Engineering, Dronacharya College of Engineering, Gurgaon, India

Neelam Ruhil
Department of Electronics and Computer Engineering, Dronacharya College of Engineering, Gurgaon, India

Karan Singh
Department of Computer Science and Engineering, Jawaharlal Nehru University, New Delhi, India

Richard O'Kennedy
Biomedical Diagnostics Institute (BDI), Dublin City University, Ireland

CRC Press
Taylor & Francis Group
Boca Raton   London   New York   Leiden

CRC Press is an imprint of the
Taylor & Francis Group, an **informa** business

A BALKEMA BOOK

CRC Press/Balkema is an imprint of the Taylor & Francis Group, an informa business

© 2017 Taylor & Francis Group, London, UK

Typeset by V Publishing Solutions Pvt Ltd., Chennai, India

Published by: CRC Press/Balkema
    P.O. Box 11320, 2301 EH Leiden, The Netherlands
    e-mail: Pub.NL@taylorandfrancis.com
    www.crcpress.com – www.taylorandfrancis.com

ISBN: 978-1-138-02952-1 (Hbk)
ISBN: 978-1-315-61650-6 (eBook)

Communication and Computing Systems – Prasad et al. (Eds)
© 2017 Taylor & Francis Group, London, ISBN 978-1-138-02952-1

Table of contents

Communication and computer networks

Soft computing, intelligent system, machine vision and artificial neural network

VLSI and embedded system

Software engineering and emerging technologies

Preface

The International Conference on Communication and Computing Systems (ICCCS-2016) took place at Dronacharya College of Engineering, Gurgaon on September 9–11, 2016. The purpose of the conference was to establish a platform for interaction among the knowledge holders belonging to industry, academia and various areas of society to discuss the current scenario of the advancements in the field of communication and computing systems. The Conference theme was chosen to facilitate discussions and personal interaction between academics involved in engineering and technology from different cultural backgrounds. The theme allowed the participants to identify and present best collaborative research and innovative ideas, as well as examples relevant to the main theme.

This book is a collection of accepted papers. The papers presented in the proceedings were peer-reviewed by 2–3 expert referees. This volume contains 5 main subject areas: 1. Signal and Image Processing, 2. Communication & Computer Networks, 3. Soft Computing, Intelligent System, Machine Vision and Artificial Neural Network, 4. VLSI & Embedded System, 5. Software Engineering and Emerging Technologies. The committee of ICCCS-2016 would like to express their sincere thanks to all authors for their high quality research papers and presentations. Also, we would like to thank the reviewers for their valuable comments and advises. Finally, thanks are expressed to CRC Press/Balkema as well for producing this book.

Organizing Committee of ICCCS-2016
Conference Chair Dr. K.K. Singh & Dr. Neelam Ruhil
Dronacharya College of Engineering, Gurgaon

Communication and Computing Systems – Prasad et al. (Eds)
© 2017 Taylor & Francis Group, London, ISBN 978-1-138-02952-1

Organization

Dronacharya College of Engineering, Gurgaon was established in 1998. The college has accepted education as a highway to achieve its long cherished goals. It is one of the best technical institutions in the state of Haryana. Founded by prominent educationalists and industrialist, Dronacharya College of Engineering strives to maintain high standards in providing quality education. All through the years of its existence, we have been making remarkable contributions in moulding the young minds socially responsible and productive human resource serving the needs of the country. Our college has the proud privilege of the following accreditations, affiliations and certificates: Approved by All India Council for Technical Education (AICTE), Permanent Affiliation to Maharashi Dayanand University, Rohtak, Accreditation by National Board of Accreditation (NBA) and National Assessment & Accreditation council (NAAC), Certification by ISO-14001-2004 Environment Management System & Certification by ISO-9001-2008 Quality Management System.

Signal and image processing

Communication and Computing Systems – Prasad et al. (Eds)
© 2017 Taylor & Francis Group, London, ISBN 978-1-138-02952-1

DCT, DWT, SVD & integrated DCT-DWT based robust digital image watermarking algorithm

Kamini Negi & Rahul Chauhan

ECE Department, Graphic Era Hill University, Dehradun, Uttrakhand, India

ABSTRACT: Paper presents a comparative analysis of various digital image watermarking techniques like DCT, DWT, Integrated approach of DCT-DWT and SVD. The experimental results demonstrate the superiority of SVD based watermarking over DCT, DWT and Integrated approach of DCT-DWT watermarking techniques on the basis of visual quality of watermarked images and also on the basis of PSNR and MSE by changing the value of embedding factor. The performance of proposed techniques are compared on the basis of two performance metrics one is imperceptibility another is robustness. The imperceptibility is analyzed through PSNR and MSE calculation and robustness of techniques is checked by various signal processing attacks.

1 INTRODUCTION

Digital watermark is a technique which allows an individual to add hidden copyright information or other verification message to digital media. It is inserted visibly or invisibly into another image so that it can be extracted later as an evidence of authentic owner. The watermark may be a label, logo or a random number (sequential). A typical good watermarking scheme should extract the embedded watermark efficiently under various watermark attack in real domain (Podilchuk & Delp, 2001, p. 346).

Basically the imperceptibility and robustness are defined as fundamental properties of embedded watermark. If host image and watermarked image is not distinguishable then it called imperceptibility. Similarity between host image and watermarked image defines the imperceptibility. If it difficult to remove or destroy watermark from watermarked image then it said to be robustness. Digital image watermarking schemes can be placed under two categories one is blind watermarking techniques which do not require original image and non-blind watermarking that requires original image to exist for detection. Robustness, Perceptual transparency, capacity and blind watermarking are four essential factors to determine quality of watermarking scheme. Two major applications of watermarking are copyright protection (proving ownership of data) and data authentication (Baisa & Mali, 2011). There are two main classes in watermarking techniques, namely the spatial domain, embed the watermark by directly modifying the pixel values of the original image. The transform domain technique, data is embedded by modulating the transform domain signal coefficients. Frequency based schemes spread the watermark over the whole spatial extent of the image, and thus frequency domain has an advantage over the spatial domain and is therefore less likely to be affected by attacks. Hence the transform domain techniques are most successful and popular for image watermarking (Barni & Bovid, 2001).

In this work a comparative analysis of various digital image watermarking algorithms based on two-dimensional Discrete Cosine Transform (DCT2), 2-dimensional Discrete Wavelet Transform (DWT2), Singular Value Decomposition (SVD) and integrated approach of DCT-DWT are evaluated and analyzed. Various performance indices like Peak Signal to Noise Ratio (PSNR) and Mean Square Error (MSE) (Lin, 2009) are used for evaluation. And further quality of watermarked image is analyzed for various signal processing attacks and variation in PSNR is evaluated.

2 PROPOSED WATERMARKING ALGORITHM

This paper study and analyzes four different types of watermarking algorithms in transform domain which are:

2.1 Discrete Cosine Transform (DCT) algorithm

The Fundamental property of Discrete Cosine Transform (DCT) is that it converts a signal into elementary frequency components. 'Energy compaction Property' is the special property that most of the visually significant information of the image is concentrated in just a few coefficients of the DCT (Tsai & Hung, 2005). Techniques based on DCT watermarking are more robust as compared to spatial domain watermarking techniques (Harish, Kumar, Kusagur (2013)). Joint Photographic Experts Group (JPEG) standard, which is a file format that is widely used on the Internet utilize DCT as a compaction algorithm. And thus ensures that the watermark is robust against JPEG compression. The 2-D DCT and 2-D IDCT transforms is given by equation (1) and (2).

$$F(u,v) = \alpha(u)\alpha(v)\sum_{m=0}^{M-1}\sum_{n=0}^{N-1}f(m,n)$$
$$\cos\left[\frac{(2m+1)u\pi}{2N}\right]\cos\left[\frac{(2n+1)v\pi}{2N}\right] \quad (1)$$

2-D inverse DCT (*IDCT*) is obtained as

$$f(m,n) = \sum_{m=0}^{M-1}\sum_{n=0}^{N-1}\alpha(u)\alpha(v)F(u,v)$$
$$\cos\left[\frac{(2m+1)u\pi}{2N}\right]\cos\left[\frac{(2n+1)v\pi}{2N}\right] \quad (2)$$

where,

$$\alpha(u) = \begin{cases} \frac{1}{\sqrt{2}} \\ 1 \end{cases}$$

$$\alpha(v) = \begin{cases} \frac{1}{\sqrt{2}} & x = 0 \\ 1 & x = 1,2,\ldots.N-1 \end{cases}$$

where, $x = (u, v)$ Frequency components.

Watermark insertion algorithm for DCT

Step 1. Read original image or host image H (to be watermarked) and the watermark image W.
Step 2. 2-DCT is applied on original image and watermark image.
Step 3. The value of embedding factor defined to be suitable for invisible watermarking.
Step 4. The DCT coefficient of the original image and watermark image is modified using the following equation.

$$H_{wi,j} = H_{i,j} + \alpha W_{ij}, \quad i,j = 1,\ldots,n \quad (3)$$

The IDCT of modified coefficients give the watermarked image.

To extract the watermark applying the following equation:

$$W_{ij} = \left(H_{w,ij} - H_{ij}\right)/\alpha \quad (4)$$

2.2 Discrete Wavelet Transforms (2-Dwt)

Discrete wavelet transform, decomposed the image into different space and in different frequency sub images (Wang, Su & Kuo, (1998)). The digital image through a two-dimensional discrete wavelet transform, the sub image can get four space of equal size: high-frequency component sub images of horizontal HF1, low-frequency approximation sub graph LF1, vertical high-frequency sub images VF1 and diagonal high-frequency DF1.

High-frequency detail sub images of DF1 obtained by wavelet transform in digital image of the 3 sub—HF, LF and DF. And the diagonal contains the main image texture background and other details of the original image; generally the human perception is low in this region. Better invisibility can achieve in this part of the information by watermark embedding operations. But this part of the information in image processing operations is subjected to noise interference, and therefore cannot guarantee the robustness of algorithm; most of the energy of the original image lies inside the low-frequency sub image LF And the image processing operations in general will not be disturbed, so the watermark is embedded into the low-frequency approximation sub image in to obtain better robustness, but because the human visual perception is sensitive to this part of the information, thus the watermark information directly in the region of the embedded operation cannot guarantee the imperceptibility of the watermark (Zhu & Lin, 2015).

Watermark insertion Algorithm for DWT

Step 1. Read Cover image (*Img*) and watermark image (*W*).
Step 2. 2-DWT is applied to obtain the first level decomposition matrix of the cover image (*Img*).
Step 3. Modify the DWT coefficients in the LL band:

$$LL_{wi,j} = LL_{i,j} + \alpha_k w_{ij}, \quad i,j = 1,\ldots,n \quad (5)$$

Step 4. Watermarked cover Image Ic_W is obtain by Inverse DWT to watermarked image.

Extraction Algorithm

Step 1. Read the input Watermarked image.

Step 2. First level decomposition of the water-marked cover image Ic_w^* is obtain after applying *DCT.*

Step 3. Extract the binary visual watermark from the LF band:

$$w_{ij} = \left(LL_{w,ij} - LL_{ij}\right)/\alpha \qquad (6)$$

2.3 *Integrated approach of Dct-Dwt*

Watermark Insertion Algorithm

Step 1. Read host image H and the watermark image W_m.

Step 2. 2-DCT is applied on host image and then 2-DWT.

Step 3. 2-DCT is applied on watermark image.

Step 4. Modify the DWT coefficients in the LL band of host image.

$$LL_1 = LL_h + \alpha W_m \qquad (7)$$

Step 5. Inverse DWT then IDCT is applied to obtain the watermarked image W_{md} (Rahman, 2009).

Extraction Algorithm

Step 1. Following equation is applied to obtain the watermark image (Jiansheng1, Sukang1 & Xiaomei, 2009)

$$W_m = \left(LL_1 - LL_h\right)/\alpha \qquad (8)$$

2.4 *Singular Value Decomposition (Svd)*

Imperceptibility, perfect reconstruction and a robustness is achieved if watermark is embedded

(a)　　　　　(b)

(c)　　　　　(d)

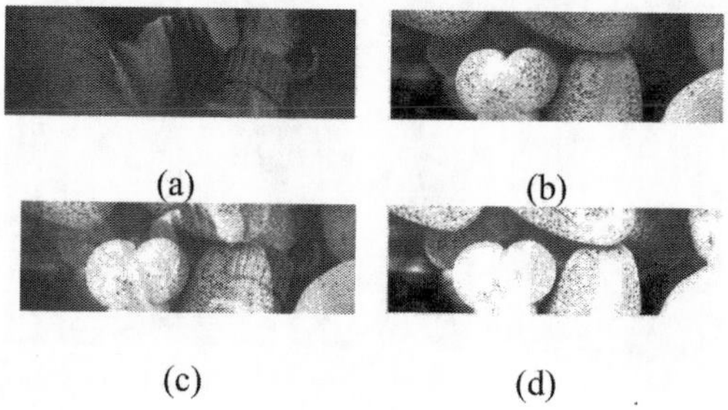

Figure 1. (a) Original image (Gonzalez & Woods, 3rd edition) (b) Watermark image (c) Watermarked response evaluated at embedding factor (α) = 0.8 (d) Extracted watermark.

into a singular value of selected sub band. SVD is a linear algebraic numerical technique and used to form diagonalizable matrices in numerical analysis. Characteristic equation is formed in Singular value matrix and the degree of characteristic equation is equal to the dimension of column or row of the image. The SVD decompose *A* matrix into three matrices i.e. *R, S* and *V*.

$$A = RSV^T \qquad (9)$$

where *A* is a $m \times n$ matrix

S is a $m \times n$ diagonal matrix in which the entries along the diagonal of *S* are singular values of *A*. Singular value of a matrix is calculated by taking square root of its Eigen value.

S in matrix form

$$\begin{bmatrix} s_1 & 0 & . & 0 \\ 0 & s_2 & 0 & 0 \\ . & 0 & s_{n-1} & 0 \\ 0 & 0 & 0 & s_n \end{bmatrix} \qquad (10)$$

R is a $m \times m$ matrix containing left singular vectors of *A* and

V is a $n \times n$ matrix containing right singular vectors of *A*.

R and *V* are orthonormal matrices which means

$$RR^T = I \text{ and } VV^T = I \qquad (11)$$

The singular values are arranged along the diagonal of *S* in such a way that

$$s_1 \geq s_2 \geq s_3 \dots \dots \geq s_n \geq 0 \qquad (12)$$

And $s_i = \sqrt{\lambda_i}$
where λ_i for i = 1, 2,.........., N are the eigen values of image matrix *A*.

Insertion Algorithm for SVD

Step 1. Read host image *H* (to be watermarked) and watermark image *W*.

Step 2. SVD is applied on host image.

$$H = R\_h\,S\_h\,V\_h$$

Step 3. SVD is applied on watermark image.

$$W = R\_w\,S\_w\,V\_w$$

Step 4. Modify the singular values of the host image with the singular values of watermark image.

$$S\_h' = S\_h + \alpha S\_w$$

5

where α is embedding factor, S_h and S_w are the diagonal matrices of singular values of the host and watermark images, respectively.

Step 5. Inverse SVD is applied on the transformed host image with modified singular values to obtain the watermarked image (Wang1 et al., 2009).

Extraction Algorithm

Step 1. SVD is applied on watermarked image

$$W\_ex = W_R W_S W_V \qquad (13)$$

Step 2. SVD is applied on host image

$$H_{ex} = H_R H_S H_v \qquad (14)$$

Step 3. Extract the singular values from watermarked and host image

$$S\_ex = (W_S - H_S)/\alpha \qquad (15)$$

Step 4. Inverse SVD is applied to obtain watermark image (Zhang & Li, 2009).

3 RESULT ANALYSIS

Simulation of watermarking algorithms is carried out on digital image processing toolbox in MATLAB R2013a. Results are analyzed on the basis of visual similarity & dissimilarity between watermarked image and original image. Further robustness is checked by various watermark attacks.

3.1 Discrete Cosine Transform (DCT)

Result shows that when the value of α is taken as 0.002 there is no visual degradation on image that watermarked by DCT based techniques or the embedding watermark is totally invisible in watermarked image, hence good imperceptibility achieved in Fig. 2(c). And Fig. 2(d) shows extracted watermark similar to original.

3.2 Discrete Wavelet Transforms (2-DWT)

Simulation results of DWT show that when the value of α is taken as 0.002 the embedding watermark is totally invisible in watermarked image

(a)   (b)

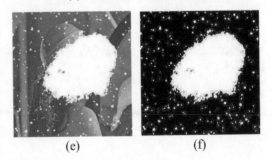

(c)   (d)

(e)   (f)

Figure 3. (a) Original image (b) Watermark image (c) Watermarked response evaluated at $(\alpha) = 0.002$ (d) Extracted watermark at $(\alpha) = 0.002$ (e) Watermarked response evaluated at $(\alpha) = 1.5$ (f) Extracted watermark at $(\alpha) = 1.5$.

(a)   (b)

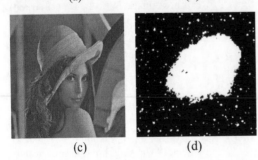

(c)   (d)

Figure 2. (a) Original image (b) Watermark image (c) Watermarked response evaluated at $(\alpha) = 0.002$ (d) Extracted watermark at $(\alpha) = 0.002$.

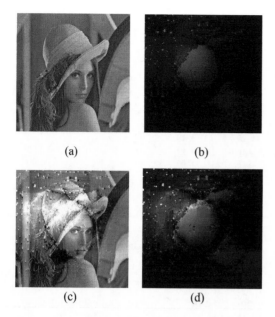

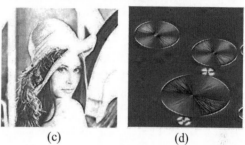

| (a) | (b) |
| (c) | (d) |

Figure 4. (a) Watermarked response evaluated at (α) = 0.002 (b) Extracted watermark at (α) = 0.002 (c) Watermarked response evaluated at (α) = 1.5 (d) Extracted watermark at (α) = 1.5.

Figure 5. (a) Original image (b) Watermark image (c) Watermarked response evaluated at (α) = 1.5 (d) Extracted watermark at (α) = 1.5.

as same in case of DCT and the image is similar to original image hence good imperceptibility achieved in Fig. 3(c). And Fig. 3(d) shows extracted watermark similar to original.

By increasing the value of α = 1.5 the embedding watermark is somewhat visible in watermarked image as visualize in Fig. 3(e) due to this impercep-tibility is degrade but the extracted watermark is similar to original watermark as in Fig. 3(f).

3.3 Analysis of integrated DCT-DWT

When the value of α is taken as 0.002 the embed-ding watermark is totally invisible in watermarked image as same and hence good imperceptibility achieved in Fig. 4(a) and Fig. 4(b) shows extracted watermark which is distorted and totally different from the original one.

By increasing the value of α = 1.5 the embedding watermark is visible in watermarked image as visualize in Fig. 4 (c) hence in this case both watermarked and extracted watermark visible performance is poor.

3.4 Singular Value Decomposition (SVD)

When value of α = 1.5 the embedding watermark is not visible in watermarked image but image is over-saturated as visualize in Fig. 5 (c) due to this imper-ceptibility is degrade but the extracted watermark is similar to original watermark as in Fig. 5 (d).

Table 1. Comparison between PSNR (in db) Values of DCT, DWT, DCT-DWT and SVD.

Value of Embedding factor (α)	PSNR (peak signal to noise ratio)			
	DCT	DWT	DCT-DWT	SVD
0.0005	26.34	29.44	29.44	INF
0.002	26.32	29.43	29.44	INF
0.003	26.32	29.38	29.44	57.53
0.005	26.30	29.37	29.43	46.06
0.02	25.99	28.74	29.39	39.66
0.5	10.78	15.27	24.44	14.84
1.5	8.57	8.21	16.88	8.01

4 PERFORMANCE ANALYSIS OF WATERMARKING TECHNIQUES

The performance evaluation of proposed water-marking techniques are based on two performance metrics one is imperceptibility another is robustness.

4.1 Measuring imperceptibility

Imperceptibility represents a measure of similarity between the original and watermarked image. For analyzing the performance of watermarked image PSNR (Peak Signal to Noise Ratio) and MSE (Mean Square Error) is chosen as a performance

Table 2. Comparisons between MSE (mean square error) Values of DCT, DWT, DCT-DWT and SVD.

Value of Embedding factor (α)	MSE (Mean square error)			
	DCT	DWT	DCT-DWT	SVD
0.0005	150.86	73.90	73.9040	0
0.002	151.44	74.13	73.9053	0
0.003	151.48	74.34	73.9086	0.11
0.02	163.65	78.11	74.8313	7.025
0.5	5.4311e +03	1.9317e +03	233.60	2.1334e +03
1.5	9.0377e +03	8.9479e +03	1.3335e +03	1.0284e +03

Table 3. PSNR (in db) for various Attacks on watermarked image.

Watermark attacks	PSNR (in db) Embedding factor (α) = 0.002			
	DCT	DWT	DCT-DWT	SVD
Crop	8.03	9.91	7.50	3.90
Average	8.19	28.97	29	3.90
Median	8.11	37.01	37.02	3.90
Noise	8.03	25.49	25.39	3.99
Rotate	8.09	18.68	18.69	3.92
Shear	4.89	4.3	4.3	2.46

criteria. It is calculated for various values of embedding factor, α (ranges from 0.0005 to 1.5).

When PSNR approaches infinity, the original image and output image are totally the same. Hence, large PSNR and small MSE is desirable

$$PSNR = 10 * \log_{10} \left[\frac{max(H(i,j))^2}{MSE} \right] \quad (16)$$

$$MSE = \frac{1}{m*n} \sum_{i=0}^{m-1} \sum_{j=0}^{m-1} \left[H(i,j) - H_W(i,j) \right]^2 \quad (17)$$

Watermark results are analyzed on the basis of PSNR and MSE values for different digital watermarking techniques i.e. Discrete Cosine Transform, Discrete Wavelet Transform, SVD and combined DCT-DWT, algorithm were implemented using MATLAB In the procedure of watermarking the value of embedding factor (α) is taken in such a way so as to produce almost visually similar watermarked image to the original image without appearance of watermark behind the original image and also analyze results of watermarked images for different values of α. As the value of embedding factor is scaled down a good visual quality of watermarked image is obtained. When α is taken as 0.002 best result is achieved in terms of PSNR and MSE values as well as in terms of imperceptibility.

The simulation results of combined DCT-DWT are analyzed and observe that the extracted watermark is degrade and different from the original one. And in SVD based technique it is observed that the visual quality of watermarked image is not degraded by watermark in fact watermark is not visible in the host image for the value of α taken same in case of other techniques.

In the Table 1 for different values of embedding factor PSNR is evaluate for all proposed techniques and observed that the watermarking by proposed SVD based method have superior performance in

terms of PSNR value and also in terms of MSE as well as in terms of imperceptibility.

4.2 Measuring robustness

Robustness test of watermarking techniques are carried out by applying different signal processing attacks like crop, average filtering, median filtering, noise attack etc. The different values of PSNR for different attacks are listed in Table 3.

After different attacks watermarked image is affected visually and also PSNR is degrade. In case of SVD the watermarked image is not much affected by attacks which is clear from PSNR Table 3.

5 CONCLUSION

The experimental results show that better imperceptibility and robustness is achieved in SVD based techniques against various watermarking attacks. And also by comparing the values of PSNR at different values of embedding factor (α), it is concluded that the SVD technique is much better than other proposed techniques. For various values of embedding factor, and also from the analysis it has been found that for the value of α=0.002 and below same PSNR and MSE is obtained i.e. INF and 0 respectively.

REFERENCES

Al-Haj Ali. 2007. Combined DWT-DCT Digital Image Watermarking, Journal of Computer Science 3 (9): 740–746, ISSN 1549-3636 © 2007 Science Publications.

Barni M. and Bovid B. 2001. Digital Watermarking for Copyright Protection: A Communication Perspective, IEEE Communication Magazine, vol. 39, no. 8, pp. 90–91.

Gonzalez and woods, Digital image processing:Database of Digital Images, 3rd edition.

Gunjal Baisa L. and N. Mali Suresh. 2011. Secured Color Image Watermarking Technique in DWT-DCT

Domain, International Journal of Computer Science, Engineering and Information Technology (IJCSEIT), Vol.1, No.3, August.

Jiansheng1 Mei, Sukang1 Li and Xiaomei Tan. 2009. A Digital Watermarking Algorithm Based on DCT and DWT, 104–107, International Symposium on Web Information Systems and Applications (WISA'09).

Lue Lin. 2009. A Survey of Digital Watermarking Technologies, 1–12, WISA.

Harish, N J, Kumar B B S and Kusagur Ashok. 2013. Hybrid Robust Watermarking Technique Based on DWT, DCT and SVD, International Journal of Advanced Electrical and Electronics Engineering, (IJAEEE) ISSN (Print): 2278–8948, Volume-2, Issue–5.

Podilchuk C.I. and Delp E.J. 2001. Digital watermarking: algorithm and application, IEEE Signal Processing Magazine, vol. 18, no. 4, pp. 346.

Rahman Md. Maklachur. A DWT, DCT and SVD based watermarking technique to protect the image piracy, International Journal of Managing Public Sector Information and Communication Technologies (IJMPICT).

Tsai M and Hung H. 2005 DCT and DWT-based Image Watermarking Using Sub sampling, Proceeding of the IEEE Fourth Int. Conf. on Machine Learning and Cybernetics, China, pp: 5308–5313.

Wang. H.J.M. Su. P.C., and Kuo. C.C.J. 1998. Wavelet—base digital image watermarking, Opt. Express, Vol.3, No.12, pp.491496.

Wang1 Ben, Ding2 Jinkou, Wen1 Qiaoyan, Liao1 Xin, Liu Cuixiang. 2009. "An image watermarking algorithm based on DWT DCT AND SVD" 1034–1038, IEEE.

Yuefeng Zhu and Li Lin. 2015. Digital image watermarking algorithms based on dual transform domain and self-recovery, International journal on smart sensing and intelligent systems vol. 8. No. 1. March.

Zhang Lijing, Li Aihua. 2009. Robust watermarking scheme based on singular value of decomposition in DWT domain, 19–22, Asia-Pacific Conference on Information Processing IEEE.

Communication and Computing Systems – Prasad et al. (Eds)
© 2017 Taylor & Francis Group, London, ISBN 978-1-138-02952-1

Evaluating of file systems, applications and MapReduce logs to support functional analysis

Chitresh Verma & Rajiv Pandey
Amity University, UP, India

Devesh Katiyar
D.S.M.N.R.U., UP, India

ABSTRACT: The distributed computation for problem statement requires a programming model. MapReduce is one of these computation models. Google MapReduce and Hadoop MapReduce are popular MapReduce implementations on the global market. This paper provides in-depth logs analysis for both implementations of MapReduce programming model.

1 INTRODUCTION

The Big Data is set of large data in various Vs. There are 4 Vs in the Big Data according to IBM data scientists. (IBM Big Data & Analytics Hub, 2016). These 4 Vs are volume, variety, velocity and veracity. The volume describes the scale of the data, the variety describes different forms of data, the velocity is an analysis of streaming the data and veracity is regarding the uncertainty of data. The MapReduce is one of top 10 hot Big Data technologies in the list prepared by Forbes Magazine. (Forbes Magazine, 2016) MapReduce is the programming model for distributed environment. It was first implemented by Google Inc and later on Yahoo under Apache Foundation built Hadoop MapReduce as an open source framework.

This paper is divided into six sections. The section I provide an introduction of Big Data, MapReduce and other related terms and technologies. The section II describes the distributed programming in detail. The section III provides insight of the Google MapReduce similarly the section IV provides for Hadoop MapReduce. The section V compares the Google MapReduce and Hadoop MapReduce. The section VI makes a conclusion of the log analysis of both implementations of MapReduce.

2 DISTRIBUTED PROGRAMMING

The distributed programming is the programming where data is stored different system and computation is done on these distributed systems rather than a single piece of hardware. The MapReduce programming model takes input key-value pairs and computes them to produce the output key-value pairs. The MapReduce consists of two stages and they are mapped stage and reduce stage. The input is taken in map stage and the output is produced on reduce stage.

3 GOOGLE MAPREDUCE

The Google MapReduce was designed by Google Inc. (IBM Big Data & Analytics Hub, 2016) It has used various Google products like Google Search, Gmail, Google Plus and other Google products.

Features

The Google MapReduce is programming model for distributed data processing on the commodity hardware. It provides the Big Data computations for various usages.

Architecture

The architecture of the Google MapReduce consists of two functions. These functions are Map stage and Reduce stage.

Map stage

The map stage reads the spited files as an input from Google Distributed File System (GFS). The worker in map stage converts the raw data into key value pairs.

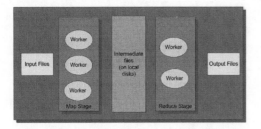

Figure 1. Diagram showing Google MapReduce (Ghemawat, Sanjay et al. 2003).

Reduce stage

The reduce stage merges the larger set of key-value pairs from intermediate files to a smaller set. The usual mirage results in one key-value pair.

Performance

The Google MapReduce can be implemented Google App Engine using Java or Python language. The Google MapReduce is not open source product, so we implemented the same in the App Engine for performance.

Figure 2. Screenshot showing logs-based metrics for MapReduce project on App Engine.

The MapReduce project is created on App Engine which uses Google MapReduce. The logs of the MapReduce project are mentioned below.

```
{
"@type":"type.googleapis.com/google.clou
d.audit.AuditLog",
  "authenticationInfo":{
    "principalE-
mail":"chitreshsrmcem@gmail.com"
```

```
},
"service-
Name":"appengine.googleapis.com",
  "method-
Name":"com.google.appengine.legacy.app_
created",
  "resourceName":"apps/mapreduce-1261",
  "serviceData":{

"@type":"type.googleapis.com/google.appe
ngine.legacy.AuditData",
    "eventMessage":"Created the applica-
tion",
    "eventData":{
      "title":"MapReduce"
    }
}
```

The log provides the details of MapReduce project. The principal email, is mentioned which is chitreshsrmcem@gmail.com. The service name is appehgine.googleapis.com. The resource name is mentioned, which is apps/mapreduce–1261.

However this log does not return info like file reads and writes MapReduce job details. Thus may not be suitably considered we have explored in the subsequent section about the Hadoop logs that retrieve in substantial info and the same are deliberated.

Figure 3. Screenshot showing the Google MapReduce for App Engine.

4 HADOOP MAREDUCE

The Hadoop MapReduce is the programming model in Hadoop framework. It also consists of two functions and they are map and reduce.

Features

The 80% of the contribution of HDFS and Hadoop MapReduce was by Yahoo Inc. (Ricky Ho., 2016) Although Hadoop is open source license product.

Architecture

The architecture of Hadoop MapReduce also consists of two functions. These functions are Map stage and Reduce stage.

Map stage

The map stage reads the raw data from the Hadoop Distributed File System (HDFS). (Shvachko, Konstantin et al. 2010).

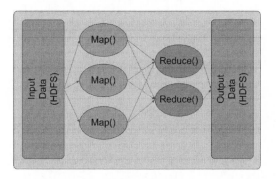

Figure 4. Diagram showing Hadoop MapReduce (Hadoop Wiki, 2016).

Reduce stage

The reduce stage is second stage in Hadoop MapReduce. It combines the resulted key-value pairs for output data which is stored on the HDFS.

Performance

The Hadoop MapReduce can be used by downloading the Hadoop from Apache Hadoop website. The Hadoop version 2.4.1 with Ubuntu Linux version 14.04 LTS are used for performance measurement of the Hadoop MapReduce. The Hadoop MapReduce job ID and counters provide below as logs.

> INFO mapreduce.Job: Job job_1458909437881_0001 completed successfully
> INFO mapreduce.Job: Counters: 49

Figure 5. Screenshot showing the Hadoop MapReduce runs on Ubuntu Linux.

Application Overview of word count application on Hadoop MapReduce:

Table 1

User:	Chitresh
Name:	WordCountNewAPI.jar
Application Type:	MAPREDUCE
State:	FINISHED
Final Status:	SUCCEEDED
Started:	25-Mar-2016 18:13:50
Elapsed:	4 mins, 33 sec
Tracking URL:	History

The logs related to the file system are given below.

> File System Counters
> FILE: Number of bytes read=111177
> FILE: Number of bytes written=3195140
> FILE: Number of read operations=0
> FILE: Number of large read operations=0
> FILE: Number of write operations=0
> HDFS: Number of bytes read=73421
> HDFS: Number of bytes written=81665
> HDFS: Number of read operations=96
> HDFS: Number of large read operations=0
> HDFS: Number of write operations=2

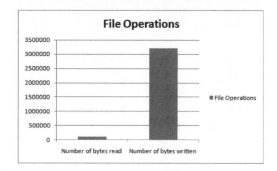

Figure 6. Column chart showing comparisons of file operations.

The output logs of the application are provided below.

```
Job Counters
    Launched map tasks=31
    Launched reduce tasks=1
    Data-local map tasks=31


    Total time spent by all maps in occupied slots
(ms) =1355869
    Total time spent by all reduces in occupied
slots (ms) =59079
    Total time spent by all map tasks (ms)
=1355869
    Total time spent by all reduce tasks (ms)
=59079
    Total vcore-seconds taken by all map
tasks=1355869
    Total vcore-seconds taken by all reduce
tasks=59079
    Total megabyte-seconds taken by all map
tasks=1388409856
    Total megabyte-seconds taken by all reduce
tasks=60496896


    Map-Reduce Framework
    Map input records=1796
    Map output records=7376
    Map output bytes=96418
    Map output materialized bytes=111357
    Input split bytes=3932
    Combine input records=0
    Combine output records=0
    Reduce input groups=1263
    Reduce shuffle bytes=111357
    Reduce input records=7376
    Reduce output records=7376
    Spilled Records=14752
    Shuffled Maps =31
    Failed Shuffles=0
    Merged Map outputs=31
    GC time elapsed (ms) =38206
    CPU time spent (ms) =33820
    Physical    memory    (bytes)    snap-
shot=4967350272
    Virtual    memory    (bytes)    snap-
shot=11721990144
    Total committed heap usage (bytes)
=3772837888


    Shuffle Errors
    BAD_ID=0
    CONNECTION=0
    IO_ERROR=0
    WRONG_LENGTH=0
    WRONG_MAP=0
    WRONG_REDUCE=0


    File Input Format Counters
    Bytes Read=69489
    File Output Format Counters
    Bytes Written=81665
```

The logs related to MapReduce Framework shows the number of input and output records that are mapped as well as same kind of data for reduce function is provided.

The shuffle errors are also provided in the logs. They can be used for tests and improvement of the algorithm. The file format counter for both input as read and output as written bytes are also available for analysis.

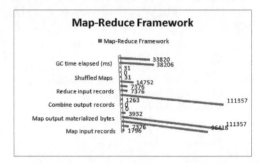

Figure 7. Bar chart showing the Map-Reduce Framework.

Table 2. Processor configuration used for performance measurement.

Processor: 2	Vendor ID: Genuine Intel
CPU family: 6	model: 37
Model name: Intel(R) Core(TM) i3 CPU 540 @ 3.07GHz	stepping: 5
microcode: 0 × 616	CPU MHz: 3058.238
cache size: 6144 KB	physical id: 0
siblings: 2	core id: 0
CPU cores: 2	apicid: 0
initial apicid: 0	fdiv_bug: no
f00f_bug: no	coma_bug: no
fpu: yes	fpu_exception: yes
CPU id level: 5	wp: yes
bogomips: 6116.47	clflush size: 64
cache alignment: 64	address sizes: 36 bits physical, 48 bits virtual

5 COMPARSION OF GOOGLE MAPREDUCE AND HADOOP MAPREDUCE

The various terms related to MapReduce programming model describe below in detail.

Key-Value pair

The key-value pair sets of data. Each data entity is stored in key-value pairs during the mapping

14

stage. Key is used for reference and value stores the actual value.

Mapper stage

The mapper stage is an input processing stage. The input file is taken from the user and split into small chucks of data. These small data sets are stored on the Hadoop distributed file system. This data set is called intermediate data.

Shuffling and sorting

Shuffling is the first phase on the intermediate data sets. It combines all values into a collection associated with same key. Sorting is the second phase on the intermediate data sets. It sorts the pairs (key-value). Due to shuffling and sorting all the unique keys will able to compare with each other which result in output in same sorting order.

Figure 8. Screenshot showing results of the Hadoop MapReduce.

Reducer stage

The reducer stage processes the data after shuffling and sorting. It generates new output data set. It stores the output data set in the Hadoop distributed file system.

In Reducer stage, the data are processed with intermediate values for a particular key generated in Mapper Stage and the processed data is stored as output.

Figure 9. Screenshot showing the Dashboard of App Engine which uses the Google MapReduce.

The comparison of both implementation of MapReduce is provided below in the context of various parameters.

Table 3.

Feature	Google MapReduce	Hadoop MapReduce	Observation
Developer	Google Inc.	Apache Software Foundation	Hadoop MapReduce has 80% contribution by Yahoo. (Ricky Ho 2016)
Year	2003	2006	First Google MapReduce came
Language	C++	Java	Google MapReduce has C++ implementation in their paper. (Ghemawat, Sanjay et al. 2003)
License	Proprietary	Apache License 2.0	
DFS	Google DFS	HDFS	DFS stands for distributed file system.
Usage	Google products mainly	Used by many companies like Adobe, EBay, Facebook, IBM, LinkedIn, The New York Times, Twitter and Yahoo.	Google has stopped using the Google MapReduce in 2014. (Sverdlik, Yevgeniy 2014)

6 CONCLUSION AND FUTURE SCOPE

This paper compared the Google MapReduce and Hadoop MapReduce in context of feature, architecture and performance. The Google Inc has stopped using the Google MapReduce in favor of a new system called Cloud Dataflow. (Sverdlik, Yevgeniy 2014) While Hadoop MapReduce continue to power many big names like Adobe, EBay, Facebook, IBM, LinkedIn, The New York Times, Twitter and Yahoo. (Kadirvel, S. et al. 2015). Thus, the Hadoop MapReduce being an open source under the Hadoop framework continues as a popular solution while Google MapReduce has been shelved by Google itself. This paper can used for log based analysis of any algorithmic behavior in context of MapReduce. The analysis may be related to security, forensics and optimization of the algorithm.

REFERENCES

Agnivesh and Dr. Rajiv Pandey. "Shiny Based Elective Recommendation Web App through K Means Clustering" Proc. of *IEEE International Conference on Communication Systems and Network Technologies (CSNT-2016)*. Web < http://www.csnt.in/Proceedings.html>. ISBN 978-1-4673-9950-0.

Agnivesh and Dr. Rajiv Pandey. "Elective Recommendation Support through K Means Clustering Using R Tool" Proc. of *IEEE International Conference on Computational Intelligence and Communication Networks (CICN-2015)*. Web <http://www.cicn.in/Proceedings.html>. ISBN 978-1-5090-0076-0.

Chitresh Verma and Dr. Rajiv Pandey. "Big Data Representation for Grade Analysis Through Hadoop Framework." Proc. *of Confluence-2016–Cloud System and Big Data Engineering.* ISBN: 978-1-4673-8202-1.

Chitresh Verma and Dr. Rajiv Pandey. "Comparative Analysis of GFS and HDFS: Technology and Architectural Landscape." Proc. *of International Conference on Communication Systems and Network Technologies (CSNT-2016)*. Web. <http://www.csnt.in/Proceedings.html>. ISBN 978-1-4673-9950-0.

Chitresh Verma and Dr. Rajiv Pandey. "An Implementation Approach of Big Data Computation by Mapping Java Classes to MapReduce." IndiaCom 2016. Proc. of *INDIACom–2016: Computing For Sustainable Global Development*, India, at New Delhi. IEEE Delhi Section. Web. 30 Mar. 2016. <http://bvicam.ac.in/news/INDIACom 2016 Proceedings/Main/index.html>. ISSN 0973-7529; ISBN 978-93-80544-20-5.

Forbes. Forbes Magazine. Web. 25 Mar. 2016. <http://www.forbes.com/sites/gilpress/2016/03/14/top-10-hot-big-data-technologies/#7a0588897f26>.

Ghemawat, Sanjay, Howard Gobioff, and Shun-Tak Leung. "The Google File System". *ACM SIGOPS Operating Systems Review SIGOPS Oper. Syst.* Rev. (2003): 29.

Kadirvel, S., & Fortes, J. A. (2015). Towards self caring MapReduce: a study of performance penalties under faults. *Concurrency and Computation: Practice and Experience*, 27(9), 2310–2328.

"PoweredBy."—Hadoop Wiki. Web. 25 Mar. 2016. <https://wiki.apache.org/hadoop/PoweredBy>.

Ricky Ho. "How Hadoop Map/Reduce Works—DZone Big Data." Dzone.com. Web. 25 Mar. 2016. <https://dzone.com/articles/how-hadoop-mapreduce-works>.

Shvachko, Konstantin, HairongKuang, Sanjay Radia, and Robert Chansler. "The Hadoop Distributed File System." 2010 *IEEE 26th Symposium on Mass Storage Systems and Technologies (MSST)*.

Shvachko, Konstantin, et al. "The hadoop distributed file system." Mass Storage Systems and Technologies (MSST), 2010 *IEEE 26th Symposium on. IEEE*, 2010.

Sverdlik, Yevgeniy (2014-06-25). "Google Dumps MapReduce in Favor of New Hyper-Scale Analytics System". Data Center Knowledge. Retrieved 2015-10-25. "We don't really use MapReduce anymore" [UrsHölzle, senior vice president of technical infrastructure at Google]".

"The Four V's of Big Data." IBM Big Data & Analytics Hub. Web. 22 Mar. 2016. <http://www.ibmbigdatahub.com/infographic/four-vs-big-data>.

Communication and Computing Systems – Prasad et al. (Eds)
© 2017 Taylor & Francis Group, London, ISBN 978-1-138-02952-1

Temporal effects in collaborative filtering for recommendation systems

R. Aggarwal & V. Kumar

School of ICT, Gautam Buddha University, Greater Noida, Uttar Pradesh, India

ABSTRACT: This paper attempt to explore the various techniques used to build a recommendation systems. Since the need of an efficient recommendation systems is rising with respect to the increment in the volume of database on e-commerce websites. These database consists of millions of data about different items and it becomes difficult for users to browse through it to find a desired item. A lot of research had been done to build a more accurate recommendation system till now, but very rare have considered the time-dependent collaborative filtering approach. In this paper we highlight some previous research work done in this field and discuss about the importance of time-dependent collaborative filtering.

1 INTRODUCTION

With the constant rise in e-commerce websites, people are getting attracted more towards browsing and purchasing products online such as mobiles, books, clothes etc. The Ecommerce websites provides each user a variety of products to choose from with just one click. These websites maintains extremely large database having information about their products which get updated with time. Often this causes difficulty for a user to browse for the products of his/her interest from various different items. To deal with this issue recommendation systems were build which make recommendations to the user with respect to its past browsing behavior.

Recommendation Systems (RS) are the web applications which are basically used to deal with the excess of data maintained by the e-commerce websites related to their items/products. These application aims to make recommendations for an active user based on his browsing behavior or its interest. It helps user to avoid grossing through various items and suggest products which a user may find interesting.

The recommendation system allows different users to rate different items on the website according to their interest. These ratings are recorded in the database and based on these ratings a recommendation is made for a specific user. Some recommendation may not use the rating criteria rather they focus on the visits or browsing done by a particular user, in that case it record the links which the user has viewed.

A recommendation may be designed to, either search a set of similar users with respect to an active user and recommend the products liked by similar users to the active user or it may find the items similar to the items liked by the active user and recommend those items. Both of the approaches has been used by many recommendation systems. These based on these approaches recommendation systems are broadly classified in 3 major types:-

1.1 *Content based filtering*

Content-based filtering is concerned with the detailed information about the item that is various features and characteristics of an item associated with it. This approach attempts to search for the items which are similar to those liked by a user previously. It exploits the different features of the item by breaking down them in a set of keywords in the form of linguistic grammar. These keywords of an item are compared with others to calculate the similarity among them. Depending upon the items previously liked by the user and the keywords related to these items, a user profile is made which is used to determine the interests of a user which further helps in searching for items which user may find interesting.

They keywords formed by various features of items are quite specific to a domain. This means if the recommendation system is developed to recommend mobile phones, the features of each item will be related to the basic features of mobile phones like camera quality, storage, battery backup etc. It is necessary so as to make more accurate comparisons among different products and to find the appropriate recommendation. This property of content-based filtering is also its biggest drawback as this makes a content based filtering domain restricted. Since it will be very difficult to match a book with a mobile phone based on their features so a book recommendation system can work effectively in the domain of books and the mobile phones can only be used to recommend the mobile phones only.

1.2 *Collaborative filtering*

In collaborative filtering instead of finding similar items as in content-based filtering, a set of similar users is identified. In this approach ratings of different users on various items is considered, and a set of similar users is determined based on these ratings. It works on the assumption that users having similar behavior in past will have similar behavior in future. This means if two users had rated similar items in the past will tend to like similar items in the future.

The collaborative filtering first find the similar users with respect to an active users. The users which rated or viewed similar items as compared to an active user are considered to be similar. The similarity between the users can be determined by various operations like using correlation coefficients and the users with highest similarity with the active user are considered. Then at last the items which the most similar users have rated highly are recommended to the active user.

It is a simple technique which doesn't require any prerequisite information about an item or to make any used profile to make recommendations. It only considers the ratings provided by different users on various items. This feature of collaborative filtering overcomes the limitation of content-based filtering for being domain restricted since collaborative filtering can give effective results regardless of the domain.

Major issues with collaborative filtering:-

a. Cold start: This issues occurs when there is a new user to the system and it has not rated many items yet. It becomes difficult to make comparisons between the new user and existing user due to lack of ratings, thus recommendation becomes a challenging task.
b. Scalability: The recommendation systems are designed to work on database consisting of millions of data so a large amount of intensive computations is required.
c. Sparsity: Since the database on websites gets added with more and more new items everyday which causes the isolation of old items. This means due to the introduction of new items sometimes old items may get fewer ratings this may lead to isolation of those items and they may not be considered for recommendation to any user.

1.3 *Hybrid recommender system*

As the name suggests the hybrid recommender system uses both content-based and collaborative filtering. It exploits the best features of both the pure techniques used for making recommendation systems. Hybrid approach can either calculate the results using both collaborative filtering and content based filtering and then combining their results to achieve recommendations for a user or it can combine the features of both the methods and form a new model for making recommendations.

Recent studies have indicated that hybrid recommender systems can provide much better results as compared to both pure approaches of collaborative filtering and content based filtering but these models are found to be more complicated to build as compared to models using a single approach.

Traditionally many recommendation systems have been build, each using different approaches, using different statistical coefficients like, Jaccard's coefficient, Pearson's coefficient etc., also considering various factors like demographic data, multi-criteria etc., but very few have considered time as factor for computing the similarity. Since this paper focuses on collaborative filtering, we will comment on the importance of time in collaborative filtering.

Time-dependent behavior of users can also be an effective factor while using collaborative filtering. A profile for each user can be maintained keeping record of each time when the user rate a product. Based on the timeline related to an active user, a set of similar users can be identified having similar behavior in that timeline. This can improve the comparisons between the users as. It considers the timely behavior of each user and it can be useful to compare a new user (rated few products) and old user (rated many products).

Our paper will first highlight the various research been done on different types of recommendation systems. Then a small discussion about an issue in recommendation systems when timely behavior of users are not considered while calculating similarity between different users while using user based collaborative filtering method for making predictions about items.

2 RELATED WORK

Recommendation systems became an important area of research in mid-1990's. A Personalized elearning material Recommender System (PLRS) was proposed by Lu (2004) which was designed to identify the best studying material for the students. A record of student's personal information was maintained which was used to determine the requirements of the student and based on the material were suggested to the individual student. PLRS dealt with major problems of recommender system like sparsity and was quite accurate in identifying the required materials.

2.1 *Content based recommendation*

A photo recommendation system was built by Peng (2010) which used a content based filtering approach and used the concept of weighted uni-directional graph. The graph represented the similarity between the items which was computed with the use of content-based filtering with the use of semantic relations. A ranking list for each item in the neighbor node was made and based on which the recommendations was made.

Adnan & Chowdury (2014) implemented a fuzzy based approach to find the similarity between the articles. They used fuzzy logic based system to compare various articles on the basis of their attributes. The attributes of articles were identified and based on the set of fuzzy rules and content based filtering similarity among these articles was computed and hence the recommendations were made for a user.

Blanco-Fernandez & Pazos-arias (2008) applied content-based fileting and semantic reasoning to perform recommendations. They collected a large information about the user's preference based on the semantic reasoning and compared it with the details of an item to find the appropriate match for the user. They overcome the overspecialization caused by the syntactic metrics.

A news recommendation system by Moerland et al. (2013), where news was represented as key-word vector, where each keyword is associated with a corresponding TF-IDF value. Removing of stop words, tokenizing and stemming etc. were performed before computing the value of TF-IDF. A user profile is created based on his browsing behavior. The user profile is then compared with the news articles using Jaccard similarity or Cosine similarity. The news articles were represented as a list of words which lack the semantic relations. Goossen et al. (2011) extended the TF-IDF measure by adding various ontological concepts.

2.2 Collaborative filtering recommendation

Collaborative filtering makes use of the ratings provided by the users to identify the accurate item for the user. A set of similar users is made and items which are highly rated by them as recommended to the browsing user. Resnick et al. (1994) collected the ratings of different users to determine the rating of the news article for a specific user. It assumed that the browsing behavior of the users remains constant that is the interest of similar people always remains the same. Jiang et al. (2015) proposed a Bayesian model which considered both user ratings and review exhaustively which linked collaborative filtering with a topic model. By employing a topic model with the review text and aligning user review topics with user attitudes (i.e., abstract rating patterns) over the same distribution they show great results and in addition solved the cold start problem was alleviated to a large extend.

Kaushik & Tomar (2015) evaluated the similarity functions using user based collaborative filtering. A rating matrix was taken as input, similarity between the users was determined by use of Euclidean distance, city-block distance, Pearson's correlation and adjusted cosine relation.

Then the nearest neighbor were selected based on the above similarity functions and hence the recommendations were generated. This paper also conducted a comparative analysis between the similarity functions to identify the best function.

Sneha & Mahadevan (2015) integrated the concept of multi-criteria rating in recommendation system where every item has ratings for different characteristics. They used two different approaches that is a clustering approach and weighted correlation approach. The former one uses Spearman's rank correlation, Euclidean distance metrics and Karl Pearson's correlation. This will provide the list of similar users from clustering on the bases of mentioned metrics. The second approach assigns certain weights to the criterions depending on the time stamp. And finally the best recommendation were made for users.

Hu et al. (2014) proposed a cluster based collaborative filtering which aimed to recruit similar services in several groups called the clusters. The method executed in two steps, firstly the services were divided into small-clusters according to their similarity by using agglomerative hierarchal clustering algorithm also known as the bottom-up clustering and in the second step collaborative filtering was computed by Pearson's correlation among these clusters.

The approach instead of considering an item as a whole entity it considers the multiple parameters of an item and based on the ratings of these parameters the average rating of the item is calculated. This method lead to a more efficient and accurate result about recommending an item. Hwang (2010) proposed a system which integrates multi-criteria into collaborative filtering algorithm with the help of genetic algorithm for optimal feature weighting. This technique used collaborative filtering and genetic algorithm to compute the prediction for individual criterion and aggregates the final prediction by weighting values respectively.

Parveen et al. (2015) proposed a multi-criteria recommendation system using fuzzy integrated meta-heuristics which recommends the relevant item to the user with the help of fuzzy multi-criteria decision making approach. This approach was based on the domain of movies which has multiple characteristic as Actor, Actress and director. These multiple characteristics are assigned some linguistic variables which the users uses for providing their preferences. These variables are then fuzzified using triangular membership functions. Users are then allowed to provide their ratings for each characteristic and a user-item matrix is maintained. Then to rank the items for a user defuzzification is done. Based on the normalized weights ranks are provided to the list of items on the basis of ranks provided by other users. Finally the best possible item is selected and is recommended to the user.

Ge & Ge (2012) presented a model to deal with the most common problem of collaborative filtering by using Singe Value Decomposition (SVD) based collaborative filtering. They used the lower-rank approximation to eliminate the noisy data which is caused due to the irregularities in the browsing behavior of users. They achieved good results of recommendations with their model.

2.3 *Hybrid recommender system*

It combines best features of both content-based filtering and collaborative filtering to make more accurate suggestions to the user. It overcomes the disadvantages of both the techniques and can be more effective and efficient in recommending items to the users. For example Liu et al. (2010) proposed a Bayesian framework which uses browsing behavior of the user and the news trends of all the users to predict the user's current news interest. It used both content based and collaborative filtering for creating user profile and generating new recommendations respectively.

Reddy et al. (2014) proposed a novel algorithm based a hybrid approach where clustering is used to cluster the similar users on the basis of their listening history and then the items which are closely related are identified by the use of association rule mining. At last the items are recommended using strong association rules.

Wang & Blei (2011) proposed a recommendation system based Collaborative Topic Regression (CRT). They combined the use of both collaborative filtering and content based filtering. The best features of collaborative filtering and LDA were used to make interpretable user profiles and latent structures for items to make recommendations for scientific articles.

McAuley & Leskovec (2013) considered the review text along with user ratings for building a hybrid recommendation system. They combined the latent ratings dimensions and latent review topics with a help of a transformation function, to obtain interpretable textual labels to compare the user ratings with their reviews and thus make predictions for products. Their results were better than those approaches which ignores the review text and considers ratings only for prediction.

Ling et al. (2014) proposed a hybrid model of recommendations system. They emphasizes on both review text and the ratings provided by the

users. They overcome the problem of cold start by applying topic modelling methods and using the information present in the review text. To improve their predictions they aligned the topics and associated them with latent item factors.

3 COLLABORATIVE FILTERING BASED ON TIME

Traditionally many research has been proposed for building a more accurate and more efficient recommendation system as we have discussed above by highlighting various different types of method used to develop a recommendation system. Some of them have used collaborative filtering, some applied content-based and also hybrid approach, many have focused on association rules, fuzzification/defuzzification, clustering etc. Many factors have been exploited previously such as review text, ratings, sentiments, demographic details etc. to compare different users and items for measuring the similarity to generate the most appropriate recommendations for a user.

Time as a factor has been rarely used in computing the similarity among the users or items. In earlier approaches specially in the case of collaborative filtering a set of similar users is identified based on their browsing behavior. This means the ratings provided by the users are considered for computing the similarity among different users.

The collaborative filtering pre-assumes that the users having similar interest in the past tends to have similar interest in future. But what happens when any two users having similar behavior in past, in future tends to have a slightly deviated browsing behavior? The predictions will be made for them on the basis of their past behavior which will not generate accurate results.

Since we cannot ignore the similar behavior of users in the past and considering the question raised above, we can use time as factor for determining the similar users. As discussed the browsing behavior of user may change with time so it highlights the importance of considering time as factor calculating similarity among the users. Time factor can also be useful in comparing a new user (have liked less items) and an old user (have liked much more items).

Yingyuan et al. (2015) raised a problem in their paper that suppose there are two users 'a' and 'b' and they have to be compared for similarity. Here 'a' is a new user and have liked Na items and user 'b' being an old user have liked Nb items, such that Na << Nb. To compare the similarity among the both the user we use a simple metrics called the Jaccard's correlation coefficient. It is given by:

$$\frac{|Na \cap Nb|}{|Na \cap Nb|}$$

According to the formula above the similarity is calculated by dividing the items common to both of the users by the total number of items they have liked. Suppose user a and user b have similar interests in past and have liked similar items. Since b is an old user and have liked much more items as compared to the new user a. Due to this the set of similar items is much smaller than the set of items which are liked by both the users. The result of Jaccard's coefficient will now result in a very small similarity between the two users. This shows that despite of having similar behavior in the past the similarity may not be always constant. What if we have considered the time factor of comparing these two users? What will happen if we check for the similarity between the users based on the time news user have been active on the system?

To answer these questions a news recommendation was built by Yingyuan et al. (2015), which calculates the time-dependent similarity among the users and uses a Time-Ordered Collaborative Filtering (TOCF) to predict the news articles for the users. They made a profile for each user having the record of at what time a specific user read an article. Based on this profile a time interval is made for each user and the similarity is computed on the basis of user's behavior during that time interval. Using above example, for determining the time-dependent similarity between users a and b, the time when they both first read a news article and the time they last read a news article is considered. The lower value of the time interval is determined by the maximum (first time of a, first time of b) and the upper value of interval is determined by minimum (last time of a, last time of b). Now to compute the similarity Jaccard's correlation is used but only the articles read between the time intervals defined are considered. This provides the time-dependent similarity between the users which can compare a new user to a very old user and it can even deal with timely changing browsing behavior of the users.

After determining the time-dependent similarity are Time Ordered Collaborative Filtering (TOCF) is executed in which the set of time-dependent similar users is taken as input and the articles which are highly rated by these similar users are recommended to a user considering the fact the articles must have been read after the upper interval of the timeline.

The method discussed is the one recently made approach for developing a recommendation system which uses time as an important factor for determining the similar users and recommending articles

to the user. Since they have use the Jaccard's correlation coefficient which only considers the number of articles which are similar between the users and ignore the ratings which the users have provided on those articles. This highlights a point what if the users have read similar articles but have rated them with extremely different rating? What if, on a similar article one user has provided very low rating and other provided a very high rating? The Jaccard's correlation will simply considered the news articles that are common to both the user despite of what ratings have been provided to those articles which can hamper the accuracy of recommendation. So there is need to use some other correlation metrics which can be used to for calculating the similarity between the users by considering the ratings provided by them, this could lead to more accurate results and together with the time-dependent collaborative filtering can result in much better recommendations for the products as compared to other traditional approaches used. Some good result has been displayed by Time Ordered Collaborative Filtering (TOCF) developed by Yingyuan et al. (2015), as compared with simple collaborative filtering approaches and has the capability of improving by considering the preferences provided by different users on the news articles while computing the similarity between users instead of their viewing behavior on different news articles.

4 CONCLUSION

We concluded that, previously many research work has been done to build more accurate recommendation systems using various methods and metrics and considering various factors to achieve better performance but very rare have given importance to the time as factor for making recommendations. The use of time factor can be very useful along with the collaborative filtering method. Since the browsing behavior may deviate for different users with time which makes it very difficult to make accurate comparisons between the users. But if we consider timelines for each user and use it while comparing them we can obtain much better results.

5 FUTURE WORK

Recommendation systems can be built by exploiting time as an important factor for making better recommendations for users. Some other metrics can be used to calculate the similarity between the user which not just dependent on the number of items liked by users as in case of Jaccard's correlation coefficient, but also consider the ratings provided by different users on different items. This can improve the recommendation results as it may deal with issue of users browsing similar items but rating them in different ways.

REFERENCES

Adnan, M.N.M. & Chowdury, M.R. 2014. Content based news recommendation system based on fuzzy logic, Informatics, Electronics & Vision (ICIEV). In International Conference, IEEE 978-1-4799-5179-6.

Blanco-Fernandez, Y. & Pazos-arias, J.J. 2008. Providing entertainment by content-based filtering and semantic reasoning in intelligent recommender systems. IEEE Transactions on Consumer Electronics (Volume: 54, Issue: 2).

Ge, S. & Ge, X. 2012. An SVD-based Collaborative Filtering approach to alleviate cold-start problems. Fuzzy Systems and Knowledge Discovery (FSKD), 9th International Conference. 978-1-46730025-4.

Goossen, F. et al. 2011. News personalization using the CFIDF semantic recommender. In Proceedings of WIMS '11. ACM, New York, NY, USA, No. 10.

Hu, R. et al. 2014. ClubCF: A Clustering Based Collaborative Filtering Approach for Big Data Application. IEEE Transactions on Emerging Topics in Computing (Volume: 2, Issue: 3). IEEE, 2168–6750.

Hwang, C. 2010. Genetic Algorithms for Feature Weighting in Multi-criteria Recommender Systems. Dept. of Information Management Chinese Culture University Taipei, Taiwan cshwang@faculty.pccu.edu.tw, doi: 10.4156/jcit, vol5, issue8.13.

Jiang, M. et al. 2015. A Bayesian Recommender Model for User Rating and Review Profiling. ISSN 1007–0214 11/13 pp634–643. Volume 20, Number 6.

Kaushik, S. & Tomar, P. 2015. Evaluation of Similarity Functions by using User based Collaborative Filtering approach in Recommendation Systems. International Journal of Engineering Trends and Technology (IJETT), Volume 21, Number 4.

Ling, G. et al. 2014. Ratings meet reviews, a combined approach to recommend. In Proc. 8th ACM Conference on Recommender Systems, Foster, Silicon Valley, CA, USA, pp. 105–112.

Liu, J. et al. 2010. Personalized news recommendation based on click behavior. In Proceedings of IUI '10. ACM, New York, NY, USA: 31–40.

Lu, J. 2004. Personalized e-learning material recommender system. In Proceedings of International Conference on International Conference Information Technology for Application, pp. 374–379.

McAuley, J. & Leskovec, J. 2013. Hidden factors and hidden topics: Understanding rating dimensions with review text. In Proc. 7th ACM Conference on Recommender Systems, Hong Kong, China, pp. 165–172.

Moerland, M. et al. 2013. Semantics-based news recommendation with SF-IDF+. In Proceedings of WIMS '13. ACM, New York, NY, USA, No. 22.

Parveen, H. et al. 2015. Improving The Performance of Multi-Criteria Recommendation System Using Fuzzy Integrated Meta heuristic. International Conference on Computing, Communication and Automation (ICCCA2015), ISBN: 978-1-4799-8890-7/15R.

Peng, T. 2010. A Graph Indexing Approach for Content-Based Recommendation System. Multimedia and Information Technology (MMIT), Second International Conference. IEEE, 978-076954008-5.

Reddy, M.S. et al. 2014. A Novel Association Rule Mining And Clustering Based Hybrid Method For Music Recommendation System. International Journal of Research in Engineering and Technology. eISSN: 2319–1163 | pISSN: 2321–7308. Volume: 03 Special Issue: 05.

Resnick, P. et al. 1994. GroupLens: an open architecture for collaborative filtering of netnews. In Proceedings of CSCW '94. ACM, New York, NY, USA: 175–186.

Sneha, Y.S. & Mahadevan, G. 2015. A Novel Approach to Personalized Recommender Systems Based on Multi Criteria Ratings. ISSN: 2040–7459; eISSN: 2040–7467, Maxwell Scientific Organization.

Wang, C. & Blei, D.M. 2011. Collaborative Topic Modelling for Recommending Scientific Articles. Proceedings of the 17th ACM SIGKDD international conference on knowledge discovery and data mining, pg. 448–456, ACM, New York, USA. ISBN: 978-1-4503-0813-7.

Yingyuan, X. et al. 2015. Time-Ordered Collaborative Filtering for News Recommendation", China Communications.

Communication and Computing Systems – Prasad et al. (Eds)
© 2017 Taylor & Francis Group, London, ISBN 978-1-138-02952-1

Study on the formation of wrinkles in the panel drawing operation using FEM

J. Kaur, S.S. Dhami & B.S. Pabla
Department of Mechanical Engineering, NITTTR, Chandigarh, India

ABSTRACT: Sheet metal forming process is widely used technique in deep drawing for producing parts from sheet metal blanks in a variety of fields such as aerospace and automobile. Wrinkling and tearing are common surface defects in sheet metal forming. In experimental methodology, three basic steps were carried out to find out the cause of defect in the drawn component. In the first step, the process was studied to explore the various causes for the wrinkle formation. In the second step, Taguchi method was used to design the set of experiments that were to be carried out for defect analysis of the drawn component. In the third step, simulation of the experiments gave results in the form of thickness distribution, effective stress distribution, effective plastic strain distribution and the yield stress distribution. Using signal to noise ratio, the optimum level of each parameter was decided. Further, the contribution of each parameter to the quality of the drawn component was studied.

1 INTRODUCTION

Sheet metal forming is one of the most widely used manufacturing processes in many industries, but, the improper design of process parameters can lead to defects such as wrinkles and tearing. Various parameters that affect the drawing operation are blankholder force, lubricity, drawbeads contour and location, material properties of the blank, friction, die radius, punch die clearance and sheet thickness. In the present paper, these parameters were varied to find out the actual reason that caused wrinkling. The modeling of dies was done by using Pro/E and simulation was done using Simufact application software. Drawing is a technological process during which a flat piece of sheet-metal material (i.e., blank) is transformed into a hollow, three-dimensional object. Such transformation can be produced either in a single step, or in a sequence of operations, each of them changing the shape but partially. During the process of drawing, the material is forced to follow the movement of a punch, which pulls it along, on its way through the die. At first, the drawn material has to overcome its own elastic limit, succumbing to plastic deformation right afterwards. Drawing of irregular shapes is a complex process that always demands many special considerations. Irregular shapes, such as covers, car fenders, and many other such elements, demand a tight control of the dimensional stability of the die, coupled with a proper material flow during drawing. Since production of many of these parts is further complicated by cosmetic requirements pertaining to their exposed surfaces, no defects of any kind are often acceptable on such parts. For these reasons, the metal flow must be tightly controlled. Where unrestricted flow of material will tend to create wrinkles, draw beads have to be added selectively, where needed. The degree of bead restriction may be altered by varying the bead contour and height (Ivana Suchy, 2006).

2 LITERATURE SURVEY

Feng et al. (Yang Feng, Xiaochun Lu and Bing Gao, 2011) investigated the deep drawing of irregular square cup by studying different blankholder forces and blank shapes in the numerical model and accurately predicted the wrinkle formation through experimentation of blankholder force and forming limit diagram. Murali et al. (A. Murali G., B. Gopal M. and C. Rajadurai A, 2010) used finite element method to optimize the location of circular and rectangular drawbeads and analyzed the strain and thickness variations during the cup drawing process using Dynaform and Ls-Dyna. Watanbe et al. (Akihro Watanabe, Hisaki Watari, Takehiro Shimizu, Yuji Kotani and Takanori Yamazaki, 2010) investigated the effects of variable blank holding force on press formability and compared it to the use of conventional press machine using FEM and observed wrinkling for different loading conditions. Aye and Xing (Winn Wah Wah Aye and Li Xiao Xing, 2010) optimized the forming process for transiting part of combustion chamber by using ABAQUS software and investigated the influences of friction coefficient and blankholder force in wrinkling. The values were optimized to eliminate wrinkles. Raju et al. (S. Raju, G. Ganesan & R. Karthikeyan, 2010) discussed the cup drawing and studied the effect of equipment and tooling parameters and discussed the resulting complex deformation mechanism in sheet metal formability using the Taguchi's signal-to-noise ratio. Ayari et al. (F. Ayari, T. Lazghab and E. Bayraktar, 2009) developed a parametric study to predict accurately the final geometry of the sheet blank and the distribution of strains and stresses and also to control various forming defects, such as thinning as well as parameters affecting strongly the final form of the sheet after forming process using ABAQUS/Explicit standard code. Ji et al. (Lianqing Ji, Zhiyuan Zhang and Qian Li, 2009) suggested that that unbalanced compressive forces resulted in wrinkling. The results described that the reasonable selection of mould clearance, blank holder force, drawing speed, etc. could effectively control the distortion during drawing of components. Păunoiu and Nicoară (Viorel Păunoiu & Dumitru Nicoară, 2003) presented the technological test used for characterizing the friction coefficient in deep drawing. The numerical simulation was used and the influences of clearance and friction coefficient value on the deep drawing force were explained. Hu et al. (Z. Hu, H. Schulze Niehoff and F. Vollertsen, 2003) investigated the sheet metal forming process in detail and described about the influence of friction between the workpiece and tool on the process. Liu et al. (Gang Liu, Zhongqin Lin and Youxia Bao, 2002) optimized the design of the drawbead which affected the strain and thickness distributions in the formed part. Wong et al. (C. Wong, R.H. Wagoner and M.L. Wener, 1997) investigated the influence of die corner geometry on the attainable draw depth of rectangular parts using 3D FEM and optimum rectangular blanks. In the present paper, the drawing simulation of a dashboard top was done using Simufact application software. The modeling of the dies was done in Pro/Engineer. The paper aimed at the elimination of wrinkles that appeared in the corners of the dashboard after the drawing operation.

3 METHODOLOGY

3.1 Design of experiments

Design of experiments is a structured, organized method that is used to determine the relationship between the different factors (Xs) affecting a process and the output of that process (Y). The parameters that were selected for analysis were corner radius, clearance between punch and die, die

entry radius, drawbead height, friction coefficient between punch and workpiece, and blankholder force. The low, medium and high levels of the control factors are given in Table 1.

The L18 array was selected based upon the number of design factors and their levels, as designed by Taguchi. The workpiece was a sheet of size 500 mm × 470 mm × 1.2 mm. The material of sheet was IS: 513 EDD grade (Extra Deep Draw) Cold Rolled Close Annealed steel. The cad models of the dies were imported into Simufact in stereolithographic (.stl) format. The meshing of the sheet was done using sheetmesher with 5 mm element size. A stroke of 100 mm was defined. Multifrontal Sparse solver was used to carry out the simulation. The outputs for the various results obtained with actual process parameters are shown in Figure 1.

The thickness distribution pattern shows that the thickness after the drawing operation varies from about 0.139 mm to 1.25 mm. So, the maximum thickening of metal is in the flange region and the region near the corner radius. It can be observed that the region where the value of effective plastic strain is maximum is the same region at the side walls where the maximum thinning of the sheet has occurred. It can be noticed that the value of effective stress is maximum at the side wall region where the maximum thinning of the sheet has occurred. It can be observed that the region where the yield stress is maximum is the same region of the side wall where the maximum thinning of the sheet has occurred. The thickness distribution obtained by simulation and photo-graph of the actual component with wrinkles are shown in Figure 2.

It can be observed from the Figure 2 (a) that the thickness distribution of the metal is fairly gradual at most of the regions. But, the local thickening of metal at the corners shows that the region would be susceptible to wrinkles. As explained earlier, the region where local thickening occurs is the same region where the effective stress is maximum. So, the corner regions can be clearly defined as the problematic area. The same results are shown in Figure 2(b), which shows wrinkles in the corners of the component which actually appear after the

Table 1. Levels of control factors.

Design Factors	Levels		
	1	2	3
A (Corner Radius) (mm)	15	20	25
B (Die Punch Clearance) (mm)	1.2	1.35	1.5
C (DieEntry Radius) (mm)	6	7	8
D (Drawbead Height) (mm)	2.2	2.6	3
E (Friction Coefficient)	0.01	0.08	0.15
F (Blankholder Force) (kN)	100	125	150

(a)   (b)

(c)   (d)

Figure 1.   Results obtained in terms of (actual case) (a) Thickness (b) Effective stress (c) Effective plastic strain and (d) Yield stress variations.

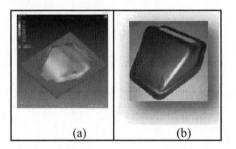

(a)   (b)

Figure 2.   Comparison of (a) Simulated result (actual case) and (b) Actual component.

drawing operation. So, it may be concluded that the simulation results are valid. For comparing the results obtained using optimum parameter values with those obtained with current parameter values, in terms of thickness and thickness variation of the region prone to wrinkles, a grid of points was identified, shown in Figure 3(a). This grid was used for measuring the thickness of the part after the simulation at specified points for obtaining comparative results. The thickness measurements were obtained at two corner regions, one on the narrow side (p) and the other on the broader side (q) of the drawn component, as shown in Figure 3(b). Since the component is sym-metrical about the section XX', the thickness distribution in the corners opposite to p and q will be similar.

The thickness and thickness variation of the wrinkled regions at the corners was plotted using

thickness values measured at 300 grid points, which are given in Figure 4.

Figure 4(a) and (b) show thickness in corner regions p and q, respectively. The surface of corner q is more uneven than that of corner p which shows that the variation in thickness is more in case of corner q than corner p. The figures give values of the thickness at different points on the corners. To obtain a clearer picture of the thickness variation, the plots (c) and (d) were plotted about the mean thickness of the component. The unevenness of the surface reduces in both plots, but, the variation in corner q is still more than in corner p. Thus, the thickness obtained at the grid points gives a quantitative measure of the wrinkles at the corners.

3.2 *Signal-To-Noise ratio and response of process*

One of the quality criteria in sheet metal formed parts is thickness distribution. The objective is to reduce thickness variation in deep drawn part as well as minimize thinning. Therefore, in this study,

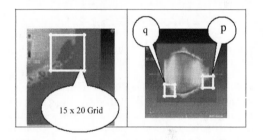

15 x 20 Grid

Figure 3.   (a) Grid used for thickness measurement (b) Corners for thickness measurement.

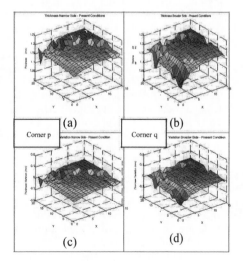

(a)   (b)

Corner p   Corner q

(c)   (d)

Figure 4.   (a), (b) Thickness and (c), (d) Thickness variation of actual case.

the response selected from the experiments is the thickness distribution. It analyzes the joint effects of control and noise factors, and for this purpose, proposed a performance criterion called signal-to-noise ratio (S/N). Defectives due to wrinkles and excessive localized thinning alter the product geometry from the designed one, causing difficulties in joining and assembly of sheet products, and limiting the product serviceability. Therefore, thickness of the deep drawn component should be as uniform as possible, i.e. the nominal values are preferred throughout the section. If the nominal value for a characteristic is the best, then the designer should maximize the S/N ratio, accordingly the S/N ratio chosen was given below (Winn Wah Wah Aye and Li Xiao Xing, 2010):

$$S/N = 10\log\left(\frac{y^2}{s^2}\right)....$$ (1)

$$s^2 = \sum_{i=1}^{n}(y_i - y_m)^2/(n-1)....$$ (2)

$$y^2 = \sum_{i=1}^{n} y_i/n.....$$ (3)

where y is the measured value of thickness, and n is the number of experiments.

The thickness of the drawn component varies in a different manner along different contours. It thins along the side walls, while, it thickens along the corner edge. The optimum value of parameters would depend on the thickness values considered. So, along different profiles, there would be different results. Therefore, the S/N ratios were calculated along two profiles on the component as shown in Figure 5.

For Computing The Signal To Noise Ratio The Thickness Values Were Measured At 12 Points Along The Profile I. The S/N ratios that were calculated from equations 1 to 3 are given in Table 2.

The percent contribution measures proposed by TAGUCHI were used for the interpretation of experimental results. The calculated values of level average for all the parameters are given in Table 3.

The mean value of S/N ratio in all experiments is given as S/N$_m$ 9.947374. The optimum levels for the significant factors, for the most even wall thickness distribution are given in Table 4.

The contribution of the control factors is given in Figure 6.

Finally, the comparison of the actual values of control factors and the optimum values as described from the results is shown in Table 5.

Simulation of the Drawing Operation using Optimized Parameters (Case 1)

The setup of simulation was done using the optimum parameters as defined by Table 5. The result of case 1 is shown in Figure 7.

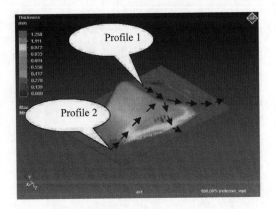

Figure 5. Thickness values taken along two profiles case-1: signal-to-noise ratio along profile I.

Table 2. S/N ratios for experiments (case 1).

Expt. No.	S/N Ratio	Expt. No.	S/N Ratio
1	11.91202	10	8.132855
2	8.067483	11	10.13141
3	9.804496	12	9.855615
4	8.131779	13	9.730056
5	12.68915	14	10.40492
6	9.928137	15	7.623429
7	11.46781	16	7.659717
8	11.24959	17	11.12385
9	10.68021	18	9.046293

Table 3. Level averages of S/N ratios (case 1).

Control factors	Level	S/N$_{mn}$
A (mm)	15	9.650647
	20	9.751245
	25	10.22125
B (mm)	1.2	9.522373
	1.35	10.61107
	1.5	9.489697
C (mm)	6	9.933783
	7	10.14273
	8	9.546619
D (mm)	2.2	10.48177
	2.6	8.781269
	3	10.3601
E	0.01	10.63321
	0.08	9.077257
	0.15	9.912669
F (kN)	100	10.55027
	125	9.39059
	150	9.68227

On comparison of the actual case and case 1, it was found that the minimum thickness in case 1 was 0.417 mm which was more than the actual case where the minimum thickness was 0.139 mm. So,

Table 4. Optimum parameter values (case 1).

Parameter	Optimum value
Corner Radius	25 mm
Clearance	1.35 mm
Die Entry Radius	7 mm
Drawbead Height	2.2 mm
Friction Coefficient	0.01
Blankholder Force	100 kN

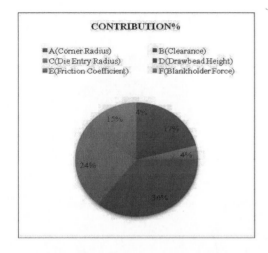

Figure 6. Contributions of parameters (case 1).

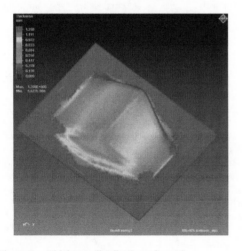

Figure 7. Thickness distribution of case 1.

the thinning was reduced along the side wall. But, the wrinkled region at the corners was almost the same. The thinning is not giving any problem at pre-sent, so, only the wrinkled region was focused. The thickness and thickness variation measured at grid points in the wrinkle prone region were plotted, as shown in Figure 8.

From the comparison of thickness distribution and thickness variation (Figure 8) with those obtained in the actual case (Figure 4) it is observed that the minimum and maximum thickness values obtained with optimized parameters are 1.05 mm and 1.22 mm, respectively whereas, those obtained with actual case are 1.08 mm and 1.20 mm, respectively in the wrinkle prone region on the narrow side. This indicates an improvement in the thickness distribution. However, the thickness variation about mean does not show any improvement as compared to actual case. On the broader side, the minimum and maximum thickness values obtained with optimized parameters are 1.05 mm and 1.25 mm, respectively whereas, those obtained with actual case are 1.06 mm and 1.20 mm, respectively in the wrinkle prone region on the broader side. Thus, the thickness values as well as the thickness variation about the mean indicate that the results have not improved with the new parameters. For computing the Signal to Noise ratio the thickness values were measured at 15 points along profile II. The S/N ratios that were calculated from equations 1 to 3 are given in Table 6.

The calculated values of level average for all the parameters are given in Table 7.

The mean value of S/N ratio in all experiments is given as S/N_m 14.05351. The optimum levels for the significant factors, for the most even wall thickness distribution are given in Table 8.

The contribution of the control factors is given in Figure 9.

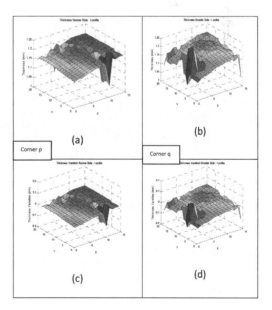

Figure 8. (a), (b) thickness and (c), (d) thickness variation of corners in case 1.

Table 5. Comparison of actual values with optimum values (case 1).

Parameter	Actual value	Optimum value
Drawbead Height (mm)	2.2	2.2
Friction Coefficient	0.01	0.01
Clearance (mm)	1.2	1.35
BHF (kN)	100	100
Corner Radius (mm)	20	25
Die Entry Radius (mm)	8	7

Table 6. S/N ratios for experiments (case 2).

Expt. No.	S/N Ratio	Expt. No.	S/N Ratio
1	12.74322	10	13.11521
2	13.86883	11	16.0617
3	16.3681	12	12.44249
4	13.29376	13	12.19386
5	16.84759	14	13.03901
6	13.5431	15	14.99935
7	16.25178	16	14.16467
8	13.24656	17	13.45652
9	13.60767	18	13.71979

Table 7. Level averages of S/N ratios (case 2).

Control factors	Level	S/N_{mn}
A (mm)	15	14.09993
	20	13.98611
	25	14.0745
B (mm)	1.2	13.62708
	1.35	14.42004
	1.5	14.11342
C (mm)	6	14.02704
	7	14.22072
	8	13.91278
D (mm)	2.2	14.22643
	2.6	13.26483
	3	14.26483
E	0.01	12.93763
	0.08	13.44071
	0.15	15.7822
F (kN)	100	13.80744
	125	14.2058
	150	14.1473

Finally, the comparison of the actual values of control factors and the optimum values as described from the results is shown in Table 9.

The setup of simulation was done in Simufact using the optimum parameters as defined by Table 9. The result of case 2 is shown in Figure 10.

Table 8. Optimum parameter values (case 2).

Parameter	Optimum value
Corner Radius	15 mm
Clearance	1.35 mm
Die Entry Radius	7 mm
Drawbead Height	3 mm
Friction Coefficient	0.15
Blankholder Force	125 kN

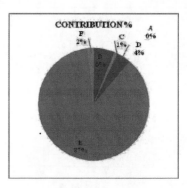

Figure 9. Contribution of parameters (case 2).

Table 9. Comparison of actual values with optimum values (case 2).

Parameter	Actual value	Optimum value
Friction Coefficient	0.01	0.15
Clearance (mm)	1.2	1.35
Drawbead height (mm)	2.2	3
BHF (kN)	100	125
Die Entry Radius (mm)	8	7
Corner Radius (mm)	20	15

On comparison of the actual case and case 2, it was found that the wrinkled region at the corners had decreased. Though, the thinning region increased, but still, the minimum thickness was 0.287 mm which was more than the actual case where the minimum thickness was 0.139 mm. So, it was concluded that the wrinkles could be reduced by using the optimized parameters of case 2. The thickness and thickness variation at the wrinkled region were plotted, as shown in Figure 11 below.

From the comparison of thickness distribution and thickness variation (Figure 11) with those obtained in the actual case (Figure 4) it is observed that the minimum and maximum thickness values obtained with optimized parameters are 1.12 mm and 1.20 mm, respectively whereas, those obtained with actual case are 1.08 mm and 1.20 mm, respectively in the wrinkle prone region on the narrow

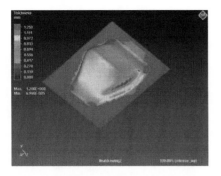

Figure 10.   Thickness distributions in case 2.

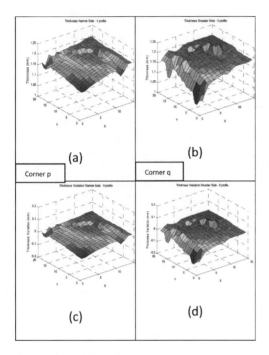

(a)   (b)

Corner p   Corner q

(c)   (d)

Figure 11.   (a), (b) thickness and (c), (d) thickness variation of corners in case 2.

side. This indicates an improvement in the thickness distribution. However, the thickness variation about mean shows significant improvement when compared to actual case. On the broader side, the minimum and maximum thickness values obtained with optimized parameters are 1.08 mm and 1.20 mm, respectively whereas, those obtained with actual case are 1.06 mm and 1.20 mm, respectively in the wrinkle prone region on the broader side. The thickness variation about mean on the broader side is marginal.

Thus, the thickness values as well as the thickness variation about the mean indicate that the results have improved with the new parameters.

Since, the dial gauge panel is fitted on the the broader side, the wrinkle prone regions are covered by the frame of the panel. Therefore, the rework, which is carried out on the narrow side of the dashboard, will not be required if the dashboard is drawn with the optimized parameters of case 2.

From the analysis of the results, it may be concluded that the optimized parameters obtained using S/N ratios computed for points along profile II are suitable for reducing wrinkles. The results were experimentally verified.

REFERENCES

Murali A.G., B. Gopal M. and C. Rajadurai A, "Analysis of Influence of Drawbead Location and Profile in Hemispherical Cup Forming", *IACSIT International Journal of Engineering and Technology*, Vol. 2, No. 4, August 2010, pp. 356–360.

Akihro Watanabe, Hisaki Watari, Takehiro Shimizu, Yuji Kotani and Takanori Yamazaki, "Improvement of Formability by servo die cushion in deep drawing", *SICE Annual Conference*, August 18–21, 2010, pp. 3174–3178.

Wong, C., R.H. Wagoner and M.L.Wener, "Corner Design in Deep Drawn Rectangular Parts", *International Congress and Exposition, Detroit, Michigan*, February 24–27, 1997.

Ayari, F., T. Lazghab and E. Bayraktar, " Parametric Finite Element Analysis of square cup deep drawing", *Archives of Computational Materials Science and Surface Engineering*, 2009, pp. 106–111.

Gang Liu, Zhongqin Lin and YouxiaBao, "Optimization Design of Drawbead in Drawing Tools of Autobody CoverPanel",*Journal of Engineering Materials and Technology*, Vol.124,April 2002, pp.278–285.

Ivana Suchy, "Handbook of Die Design", *Tata McGraw Hill Company Limited*, second edition, 2006

Lianqing Ji, Zhiyuan Zhang and Qian Li, "Study on the Drawforming Process of the Complex Shaping Part", *IEEE*, 2009, pp. 929–932.

Raju, S., G. Ganesan & R. Karthikeyan, "Influence of variables in deep drawing of AA 6061 sheet", *Tranactions of Nonferrous Metal Society*, China, 2010, pp. 1856–1862.

Viorel Păunoiu & Dumitru Nicoară, "Simulation of Friction Phenomenon in Deep Drawing Process", *National Tribology Conference, Galaţi*, Romania, September 24–26, 2003, pp. 407–412.

Winn Wah Wah Aye and Li Xiao Xing, "Optimization of Forming Process for Transiting Part of Combustion Chamber", *International Conference on Mechanical and Electrical Technology*, 2010, pp. 327–332.

Yang Feng, Xiaochun Lu and Bing Gao, "Numerical Damage prediction and Experiments in Deep Drawing of Irregular Square Cup", *Fourth International Conference on Intelligent Computation Technology and Automation*, 2011, pp. 853–856.

Hu, Z., H. Schulze Niehoff and F. Vollertsen, "Determination of the Friction Coefficient in Deep Drawing", *1st Colloquium Processscaling, Bremen*, October 28–29, 2003.

Communication and Computing Systems – Prasad et al. (Eds)
© 2017 Taylor & Francis Group, London, ISBN 978-1-138-02952-1

Metaheuristics: Modeling variant of ant colony optimization for image edge detection using self adaptive approach

Jyotika Pruthi & Gaurav Gupta

Department of Computer Science, The NorthCap University, Gurgaon, India

ABSTRACT: Ant Colony Optimization (ACO) algorithm is nature inspired that relies on the way the ants make a route to reach the destination with food. The principle lies in the fact that as ants make a way they release a chemical substance known as pheromone that helps other ants to follow the shortest path. ACO also holds an application in the area of edge detection in an image by creating the pheromone matrix that represents the intensities of the pixels of the image. The manner in which the ants move is analogous to the way the variation is seen in the intensity values of image. In this paper a variant of ant colony has been proposed that modifies the way the global path is updated thereby making the edge detection to be more precise and accurate. Experimental results have been shown and been compared with the previously existing techniques.

1 INTRODUCTION

As metaheuristics algorithms came into the domain with a view to obtain global optimization, ant colony optimization was one of these algorithms that mimics the traits of the natural ants (MaulikUjjwal, 2009) (Glover F. and Kochenberger G. A, 2003). Ants find their route to reach the food by randomly searching the different paths. As each moves, it deposits the substance called pheromone along the path. On the basis of the length of the path, the concentration of pheromone varies. The other ants choose the path that has the highest concentration of pheromone. The varying randomness determines the optimal solution. In this paper, ACO has been used to detect and deal with the edge detection problem, with a view to detect the edges in the image as they provide really important information (MaulikUjjwal, 2009). It considers the ants that make a move on the image as the intensity of the pixels vary thereby generating a pheromone matrix. There has not been much work done in this particular direction except some of it being shown in (Collet Pierre & Jean-Philippe, 2006). The difference that has been shown in our approach in the way the pheromone matrix is updated. In this updation has been done combining it with the self-adaptive approach thereby detecting the edges more clearly.

The following Figure 2 shows the various possibilities encountered by the ants while foraging.

1.1 Mathematical model of ant colony algorithm

ACO focusses on finding the global optimum solution through a guided search mechanism i.e. with the movement of the ants by generating the pheromone matrix.

At the nth construction-step of ACO, the kth ant shows movement from the node i to the node j as per the probabilistic action rule. The transition probability will be as follows in equation 1. When the tour of the ants is completed, the pheromone values are updated as per the following equation 2.

1.2 Proposed approach

In the following four equations, lambda value determines curve of the function used for calculating heuristic value. To determine best result, 4 lambda values are considered: 5, 10, 15, 20.

When new heuristic value is found which is best, then updation for single ant is carried out. It includes "memory" factor (rho) which means for how long the path will remain in ant's movement path. This local updation is carried out by ant in program using:

$$\tau_{i,j}^{(n-1)} = \begin{cases} (1-\rho) \cdot \tau_{i,j}^{(n-1)} + \rho \cdot \Delta_{i,j}^{(k)}, & \text{if } (i,j) \text{ is visited by} \\ & \text{the current } k\text{-th ant;} \\ \tau_{i,j}^{(n-1)}, & \text{otherwise.} \end{cases}$$

where right side of eqn. is existing pheromone path and after updation for each ant, overall update takes place, which determines the final extension in existing path. This new extension is corrected using genetic algorithm in program using:

$$p' = \left(1 + \frac{1-p}{p} \exp(-\gamma \cdot N(0,1))\right)^{-1}$$

After this extension is merged with existing path in program using:

```
delta_p = (delta_p + (delta_p_current > 0))>0;
p = (1-phi).*p;
```

2 RESULTS AND ANALYSIS

2.1 Parameters for comparison

2.1.1 Pratts figure of merit

$$R = 1/I_n \sum 1/ad^2$$

where $I = \max\{Ii,Ia\}$ where Ii and Ia are expected and actual edge map points, d is the distance between actual edge point and ideal edge points. a is the scaling factor. Basically the quality of Edge detector is analyzed

2.1.2 Shannon's entropy
Image entropy is a quantity that refers to the amount of information.

$$Entropy = -\sum PLogP$$

In the above expression, P i is the probability that the difference between 2 adjacent pixels and Log 2 is the base 2 logarithm. It compares the Ideal image and our output image.

$$p_{ij}^k(t) = \frac{[\tau_{ij}(t)]^\alpha [n_{ij}]^\beta}{\sum_{k \in allowed} [\tau_{ij}(t)]^\alpha (n)_{ij}^\beta}$$

if j belongs to allowed otherwise 0

$$\tau_{ij}(t+n) = \rho.\tau_{ij}(t) + \Delta\tau_{ij}$$

Where rho is the evaporation rate as pheromone evaporates over a period of time.

$$\Delta\tau_{ij} = \sum_{k=1}^{m} \Delta\tau_{ij}^k$$

Where left hand side shows the quantity of pheromone on edge (i,j) and is calculated as:

$$\Delta\tau_{ij}^k = \frac{q}{L_k}$$

if kth ant uses (i,j) in tour or 0 otherwise. Now to calculate heuristic value for that location, Vi,j is calculated as:

$$V_c = \left(\left| I_{i-2,j-1} - I_{i+2,j+1} \right| + \left| I_{i-2,j+1} - I_{i+2,j-1} \right| \right)$$

The function f(.) is determined using the four functions mentioned below:

$$f(x) = \Lambda x \text{ for } x \geq 0$$

$$f(x) = \Lambda x^2 \text{ for } x \geq 0$$

$$f(x) = \sin\Lambda x/2 \Lambda$$

$$f(x) = \prod x \sin \prod x / \Lambda$$

Here I is the intensity and whole equation inside f() in 1st equation, is calculated using intensity of neighborhood positions.

Table 1. Four functions considered.

Image	Lena(128*128)
Func 1	Λx
Func 2	Λx^2
Func 3	$\sin\Lambda x/2 \Lambda$
Func 4	$\prod x \sin \prod x / \Lambda$

Table 2. Basic algorithm.

Basic (Λ =)	5	10	15	20
Func 1	0.662	0.682	0.681	0.667
Func 2	0.430	0.482	0.399	0.427
Func 3	0.665	0.707	0.685	0.680
Func 4	0.470	0.450	0.418	0.501

Table 3. Modified algorithm.

Modified (Λ =)	5	10	15	20
Func 1	0.648	0.701	0.693	0.664
Func 2	0.430	0.442	0.442	0.443
Func 3	0.699	0.695	0.718	0.659
Func 4	0.462	0.456	0.402	0.453

Table 4. Comparison with other algoithms.

Algorithm	Pratts FOM	Entropy
Edison	0.5461	0.6777
Rothwell	0.5184	0.7438
LoG	0.4721	0.8114
SUSAN	0.5115	0.7928
Roberts	0.3592	0.5124
Sobel	0.3761	0.5303
Genetic Ant	0.7180	0.4546

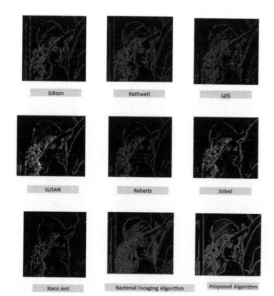

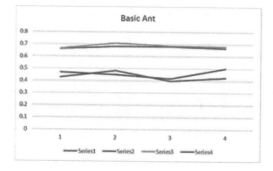

Figure 1.   Comparison of algorithms.

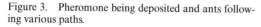

Figure 2.   Pheromone being deposited and ants following various paths.

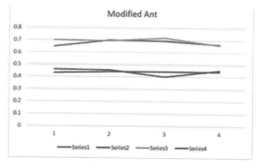

Basic Ant

Figure 3.   Pheromone being deposited and ants following various paths.

Modified Ant

Figure 4.   Maximum value of Pratt's figure of merit is when using func3 with $\lambda = 15$ using modified ACO Algorithm: 0.718042.

REFERENCES

Collet Pierre, Jean-Philippe (2006), "Stochastic Optimization Algorithms", published in Handbook of Research on Nature Inspired Computing for Economics and Management, Hershey.

Glover F. and Kochenberger G. A., Handbook of Metaheuristics, Springer, (2003).

MaulikUjjwal, Senior Member, IEEE, (2009)," Medical Image Segmentation Using Genetic Algorithms", IEEE Transactions on Information Technology In Biomedicine, vol. 13, no. 2.

Otsu, N., (1979), A threshold selection method from gray-level histogram, IEEE Transactions on Systems, Man and Cybernatics., vol. 9, pp. 62–66.

Communication and Computing Systems – Prasad et al. (Eds)
© 2017 Taylor & Francis Group, London, ISBN 978-1-138-02952-1

Improving medical image classification model using data mining technique

Ujjwal Gupta, Vimlesh Kumar & Sandeep Sharma
School of ICT, Gautam Buddha University, Greater Noida, Uttar Pradesh, India

ABSTRACT: A huge increase in medical image database volume has set new challenges to clinical routine for patient's record about diagnosis, treatment & follow-up, and with help of data & image mining it is possible to assist or automate the radiologist for diagnosis. Therefore research should be done in this area of data mining and its methods can help as remedy in this circumstances. This paper discusses about an improved method of classification using glcm features to classify medical images especially retinal fundus images. Here for various image pre-processing techniques and for the purpose of calculating and extracting various intensity related features MATLAB tool is used. Pre-processed retinal fundus images are used to model the classifier and classification accuracy and area under the ROC curve are used as performance metrics in order to compare the existing kNN classifier with proposed classifier.

1 INTRODUCTION

Data Mining is the process where one tries to retrieve information or find out the hidden patterns in a data set in order to predict future. So many advancements has been done in the field of image processing that can help in analyzing anomalies/irregularities in images using their pixel values. Numerical data can be extracted from images by using statistical methods, which can be further used for classifying images. One such tool is KNN classifier, where k nearest neighbors are considered while predicting the class of a new test instance. But the problem arises when we have even value of k and also the class labels of the nearest neighbors are equal in number. In such a case random class label is chosen by the kNN classifier. In this paper an improved kNN model for classifying retinal fundus images has been used as proposed by Satej (2013) along with some additional features. Retinal fundus images are specific class of images used in medical studies that provides the details of the inner lining of a human eye. The model proposed will help to identify a patient affected by diabetic retinopathy.

2 RELATED WORK

Medical images can be classified using various techniques and these are categorized into Neural Network classification, Texture-based classification, and Data Mining as mentioned by Smitha (2011). Neural network classifier has a shortcoming that they don't perform well when we have large datasets as mentioned by Hosseini (2012). Data mining is a better technique among all as mentioned by Gandhi (2015). The idea of using statistical features was derived from Satej (2013). kNN algorithm is simple to implement & execute and also gives better accuracy, therefore kNN method was pursued. The concept of adding a weighting technique for the kNN algorithm was derived from Satej (2013) which also considered weights of the training instance while classifying test instance. This method is effective but lack the selection of better features for classification. The idea of using some new features came from Kumar (2015) and Aggarwal (2012).

3 PROPOSED WORK

In the below figure overall classification model for classifying medical images as proposed in this paper has been mentioned. The initial step of classification includes image pre-processing followed by feature extraction and feature selection (feature scaling (optional)). Once all this has been done, the mentioned k nearest neighbor algorithm is then applied on the data set which we obtained as a result set containing reduced set of features. The process of image processing, feature extraction and feature selection is performed on test images as well before feeding to classifier.

3.1 Image pre-processing

For this paper retinal fundus images are used where there is a high variation in the pigment colors of the eye therefore image pre-processing step is applied in order to counter the non-uniformity in color

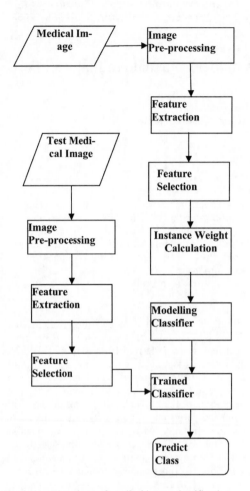

Figure 1. Flowchart of medical image classification.

features which include mean, median, mode, variance, Skewness, kurtosis and second order statistical features i.e. glcm (gray level co-occurrence matrix) features which include contrast, energy, correlation and homogeneity. While extracting glcm features the image was divided into sub-images of size 36×36 pixels leading to 320 sub-images and an offset of 17 pixels is considered in four directions only, also images were divided into sub-images of size 36×90 pixels leading to 120 sub-images. Glcm features were calculated and extracted using MATLAB.

Features extracted are:

1. Mean $= \mu = \sum_{i=1}^{n} X_i$ (1)

2. Variance $= \dfrac{\sum_{i=1}^{n}(X_i - \bar{X})^2}{N}$ (2)

3. Skewness $= \dfrac{1}{n} \dfrac{(x-\mu)^2}{\sigma^3}$ (3)

4. Kurtosis $= \left(\dfrac{1}{N} \dfrac{(x-\mu)^4}{\sigma^4} \right) - 3$ (4)

5. Energy $= \sum_{i,j} p(i,j)^2$ (5)

6. Contrast $= \sum_{i,j} |i-j|^2 \, p(i,j)^2$ (6)

7. Median   (7)
8. Mode   (8)

9. Correlation $= \sum_{i,j} \dfrac{(i-\mu i)(j-\mu j)p(i,j)}{\sigma_i \sigma_j}$ (9)

10. Homogeneity $= \sum_{i,j} \dfrac{p(i,j)}{1+|i-j|}$ (10)

distribution. So gray scaling of the image is done and then for adjusting the variation in contrast an adaptive histogram equalization process is applied. Gray scaling is a process where instead of converting a color image to black/white, image is converted to various shades of gray with black being darkest and white being lightest. Adaptive Histogram equalization is a process of contrast adjustment of smaller regions in the image.

3.2 Feature extraction

The data set consisted of images of size 1500×1152 pixels. So, images were resized to 576×720 pixels and these were used for building the classification model. Then images were further divided into smaller 128 sub-images and the size of the sub-image being 36×90 pixels. After this from each block/sub-image features are extracted as mentioned in equations 1–10 below. Two types of features were extracted from the images, first order statistical

3.3 Algorithm

Algorithm 1: Calculating weights of all instances in training set

Requirements: X_train, temp_weight, total_weight, Weight, k
Output: Addition of Weight column in X_train
/ * Steps * /

```
1: for each instance i ∈ X_train do
2:     find k neighbors of i and put it in a list N
3:     for each neighbor b ∈ N do
4:         if (class (b) == class (i)) then
5:             add 1 to temp_weight
6:         else
7:             add 0 to temp_weight
8:         end if
9:     end for
10:    total_weight (i) = temp_weight * (1/k)
11:    add total_weight (i) to Weight
12: end for
```

Algorithm 2: Improved KNN

Requirements: X_test, X_train, Neighbor_weightage, class0, class1, Q, D
Output: Predict class of test instance
/ *Steps * /

```
1: for each instance q ∈ X_test do
2:       find k neighbors of q with their Euclidean distances in X_train and
         put neighbors in Q & their respective distances in D
3:       for each neighbor z ∈ Q do
4:               Neighbor_weightage (z) = Weight (z) * (1/D (z) +0.1)
5:               if (class (z) == 0) then
6:                       append Neighbor_weightage (z) to class0
7:               else
8:                       append Neighbor_weightage (z) to class1
9:               end if
10:      end for
11:      if (sum (class0) > sum (class1)) then
12:              return 0
13:      else
14:              return 1
15:      end if
16: end for
```

Here first algorithm assigns a weight to each training instance and second algorithm uses those weights in order to predict the class of its test instance. While applying the second algorithm 10 fold cross—validation is used in order to have unbiased results. For first algorithm 10% neighborhood size is considered. While applying Monte Carlo simulation only second algorithm is considered.

3.4 Feature selection

Feature Selection is a major step that helps out in improving the quality of data by selecting the most valuable/distinguishing features for classification. For this various feature selection techniques were applied in order to obtain the best features and it was observed that they resulted out mostly same set of features.

Various techniques applied are:

1. "CfsSubsetEval" with "Best First Search" in Forward direction which resulted in 19 features.
2. "CfsSubsetEval" with "Greedy Search Forward Selection" which resulted in 19 features.
3. "CfsSubsetEval" with "Greedy Search Forward Selection" on the result of "Information Gain" with "Rank Search" (zero threshold) which resulted in 19 features.

4 PERFORMANCE METRICS

The Analysis of the performance of the proposed classifier is done using following performance metric tests, as mentioned below:

1. Accuracy: The classification accuracy is calculated by comparing the predicting class with the actual test class. The formula is given below

$$Accuracy(\%) = \frac{T_{CP}}{N} * 100 \qquad (11)$$

where T_{cp} denotes total number of correct predictions and N is the total number of instances in the test dataset.

2. Precision: It is defined as the number of true positives (T_p) over the number of true positives plus the number of false positives (F_p).

$$P = \frac{T_P}{T_P + F_P} \qquad (12)$$

3. Recall: It is defined as the number of true positives (T_p) over the number of true positives plus the number of false negatives (F_N).

$$R = \frac{T_P}{T_P + F_N} \qquad (13)$$

4. Receiver Operating Characteristic (ROC) and Area Under the Curve (AUC): ROC is a graphical plot with FPR on X axis and TPR on Y axis as mentioned in Fawcett (2006). FPR is False Positive Rate and TPR is the True Positive Rate of a classifier. By calculating the Area Under this Curve (AUC) one can measure the classifier performance. The case where AUC $< = 0.5$, denotes poor classification. For an AUC > 0.5 the classifier is considered to be good. When we have a TPR equal to 1 and FPR equal to 0, resulting in AUC being 1 means perfect classifier.

5 EXPERIMENTATION RESULTS

The experiment was done using 38 pre-processed images where 21 images were of 'Normal' category and rest of 'Severe' category. Dataset was acquired from DIARETDB0 which is a diabetic retinopathy image database available at http://www.it.lut.fi/project/imageret/diaretdb0/.

Dataset consists of 130 images out of which only 21 are Normal images. A sample of the raw images of each category is shown below.

Image 120, 121 are Normal whereas 45, 46 are Severe conditions. Here only proliferative diabetic retinopathy stage has been considered as Severe class. Data about the state of retina condition was acquired from "groudtruth" file which was provided with the dataset.

Using feature extraction techniques proposed in section 3, seven types of datasets were prepared where first four and sixth one resulted in 512 features, fifth one with 1280 features and last one with 768 features in total:

1. One containing <Mean, Variance, Skewness, Kurtosis> as features with each sub-image of size 36×90
2. One containing <Mean, Standard Deviation, Skewness, Kurtosis> as features with each sub-image of size 36×90
3. Another one containing <Median, Standard Deviation, Skewness, Kurtosis> as featureswith each sub-image of size 36×90
4. Another one containing <Mode, Standard Deviation, Skewness, Kurtosis> as features with each sub-image of size 36×90
5. Another one containing <Contrast, Correlation, Energy, Homogeneity> as features with each sub-image of size 36×36
6. Another one containing <Contrast, Correlation, Energy, Homogeneity> as features with each sub-image of size 36×90
7. Last one containing <Contrast, Correlation, Energy, Homogeneity, Mean, Standard Deviation, Skewness, Kurtosis> as features

From each of the thirty eight images, features were extracted and the existing features set was updated by adding the class label of each image. The features were then brought down to minimum features (including the class label) by using various feature selection algorithm as mentioned in section 3. Then it was observed that scale of standard deviation was larger in comparison to other features therefore, MinMax feature scaling was applied on the dataset. The table columns represents the features and the rows represent feature vectors of an image. The feature value is the intersection of a column and row. The data set was then divided on 60%-40% basis i.e. 60% of the data will be treated as training data and 40% will be used as test data. Also 10 fold cross validation was applied in order to have better results. Then the weighted K nearest neighbor was modelled as mentioned in section 3 to predict the class of each image that are present in the test set. And all this process was simulated using Monte Carlo Simulation, where 500 simulations were run and plotted as well.

Firstly it was observed that replacing variance with standard deviation and applying feature scaling improved the classification performance a little bit. See Figures 3 & 4.

It was also observed that while using the second, third and fourth data sets i.e. feature vector consisting only statistical features replacing mean with median or mode does not enhance the accuracy of classifier.

But when feature vector comprises of only glcm features i.e. the fifth data set then accuracy increases. See Figures 5, 6 & 7.

Another interesting thing observed was that when the data set was shuffled before splitting into test and train datasets and 500 Monte Carlo simulations were performed then accuracy spread over an interval of +/– 5%. One can observe Figures 8, 9.

It was observed that glcm features are so powerful that when each sub-image is of size 36×90 pixels (i.e. 6th dataset) then also accuracy of the classifier increases in comparison to first order statistical features. Also noticeable difference in accuracy can be observed between simulation of fifth and sixth datasets. See Figure 10.

Figure 11 shows that when both first order and second order statistical features were used together for classification, performance degrades.

Naïve Bayes when applies on fifth dataset also resulted in 92.1% accuracy with 0.2 RMSE and 0.989 as ROC area (using WEKA). A comparison chart is shown in Figure 12.

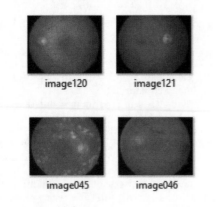

Figure 2. Sample images.

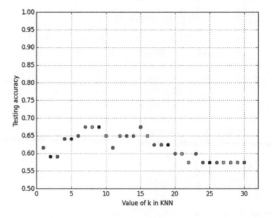

Figure 3. Variation of Accuracy with k value in first dataset.

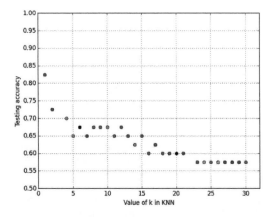

Figure 4. Variation of Accuracy with k value in second dataset.

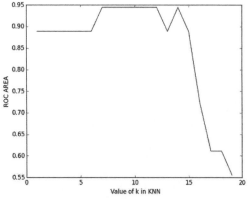

Figure 7. Variation of AUC with k value in fifth dataset.

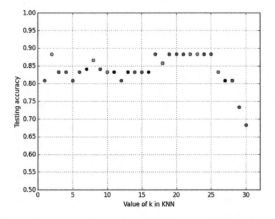

Figure 5. Variation of Accuracy with k value in fifth dataset.

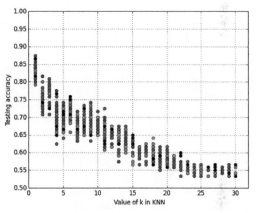

Figure 8. Monte Carlo of Accuracy second dataset.

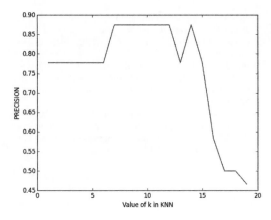

Figure 6. Variation of Precision with k value in fifth dataset.

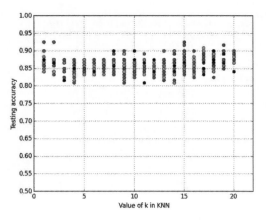

Figure 9. Monte Carlo of Accuracy fifth dataset.

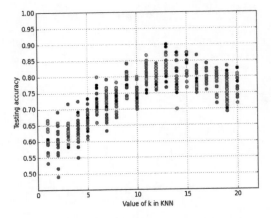

Figure 10.   Variation of Accuracy with k value in sixth dataset.

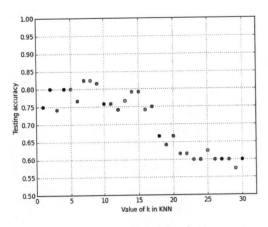

Figure 11.   Variation of Accuracy last dataset.

Dataset	k	Classification Accuracy	RMSE	Precision	Recall	ROC AUC	False Positive	True negative
I	4	78.9	0.37	0.81	0.79	0.96	1	7
II	4	81.5	0.37	0.83	0.81	0.94	1	6
III	4	81.5	0.37	0.83	0.81	0.94	1	6
IV	4	81.5	0.37	0.83	0.81	0.94	1	6
V	4	92.1	0.26	0.92	0.92	0.97	1	2
VII	4	81.5	0.37	0.82	0.82	0.88	2	5

Figure 12.   Comparison chart of datasets.

6   CONCLUSIONS AND FUTURE WORK

Here in this paper an effort has been made to improve image classification of retinal fundus medical images. The model uses an improved kNN algorithm. The model was applied to differentiate between "Severe" and "Normal" class of images. Classification accuracy and AUC helped in evaluating the model's performance. The results with the use of glcm features extracted have shown better performance with the improved kNN model leading to classification accuracy of around 90% while classifying retinal fundus images.

Novelty of the work is that in previous papers mostly statistical features were extracted from retinal fundus images for classification purpose but here Monte Carlo simulation and second order statistical features viz. contrast, correlation, energy and contrast along with weighted kNN algorithm are used for improving the classifier performance.

One can improve the model by improving the feature set or improving feature weighting schemes or both.

REFERENCES

Aggarwal Namita, Agrawal R.K., 2012. "First and Second Order Statistics Features for Classification of Magnetic Resonance Brain Images", *Journal of Signal and Information Processing*, Vol.3: 146–153.

Fawcett T. 2006 "An Introduction to ROC Analysis", *Pattern Recognition Letters*, pp.: 861–874.

Gandhi M., Singh S.N., 2015 "Predictions in heart disease using techniques of data mining", *Futuristic Trends on Computational Analysis and Knowledge Management (ABLAZE) International Conference*.

Hosseini Monireh Sheikh, Zekri M., 2012. "Review of Medical Image Classification using the Adaptive Neuro-Fuzzy Inference System" *Journal of Medical Signals and Sensors*, Volume 2, Issue 1:49–60.

Kumar Sathees, Selvi R. Anbu, 2015. "Feature extraction using image mining techniques to identify brain tumors", *Innovations in Information, Embedded and Communication Systems (ICIIECS) International Conference*.

Smitha, P. 2011. "A Review of Medical Image Classification Techniques" *JCA Proceedings on International Conference on VLSI, Communications and Instrumentation (ICVCI)* (11):34–38.

Wagle Satej, Mangai J. Alamelu and Kumar V. Santosh, 2013. "An Improved Medical Image Classification Model using Data Mining Techniques", *IEEE GCC Conference and exhibition*, Doha, Qatar.

Communication and Computing Systems – Prasad et al. (Eds)
© 2017 Taylor & Francis Group, London, ISBN 978-1-138-02952-1

Iris recognition using hybrid transform

Bhawna Kaliraman & Rekha Vig
Department of EECE, The NorthCap University, Gurgaon, India

ABSTRACT: Biometrics provides more secure method for identification as compared to old traditional methods. Iris is found to be most secure and accurate biometric among all biometrics. This paper includes method of recognition of individuals by their unique iris patterns. In this paper, transform domain technique is used to extract features of transformed iris image. Hybrid transform are used which are generated using the Kronecker product of any two existing orthogonal transform matrices. Here we use various combinations of Walsh, DCT and Haar transform matrices to extract features of iris image. Combination of two transforms is generally gives better result than when used alone as properties of both transforms are applied together. Energy compaction technique is used where threshold value of energy reduces the size of feature vector. By varying the threshold values of cut-off energy, we get up to 100% efficiency with the help of hybrid transform using Haar-Walsh transform.

1 INTRODUCTION

In today's world security has become most important issue. We can bring more secure methods with the help of biometrics because these are difficult to replace or steal; in fact their duplication is virtually impossible. Traditional identification methods were not so secure these can be divided into two parts: one of them is based on something the user has i.e. smart card and another is based on something the user known such as passwords. But these traditional methods suffer from several drawbacks such as password may be hacked or key may be stolen. In order to get rid of these kinds of problems we move to identification based on biometric information. This is unique for each individual. Authentication systems based on bio-metrics uses physiological characteristics such as finger print, iris, face or ear shape or, behavioral characteristics such as voice gait or signature to identify the person. Human body parts can be used as biometric information such as fingerprint, iris, speech, face, retina, teeth etc. which provide reliable performance.

Human eye can provide a huge amount of information to identify the individual. Iris can be a better option to identify an individual because its pattern is unique for each individual. Also it remains unchanged with age and it is different for twins also. It can act as a living password. It is also known as eye iris pattern recognition technology. It can act as most promising biometric identification technologies with high uniqueness and stability, non-invasiveness, anti-falsification and many other qualities.

In earlier methods of iris recognition there are three main stages which are image preprocessing, feature extraction and template matching. Many systems utilize the patented Daugman's algorithm. Iris recognition consists of image acquisition, image preprocessing, iris segmentation, normalization, iris feature extraction, comparison and matching process.

Image is acquired for preprocessing, which includes iris localization, normalization etc. Iris localization detects the inner and outer boundaries of iris. The region covered by eyelids and eyelashes is detected and removed. Iris localization include different methods such as Integro-Differential operator, Hough transform, Discrete circular active contour, Bisection method and Black hole search method. Iris segmentation is the process of locating iris region in acquired input image. Normalization is transformation of image from Cartesian coordinates to polar coordinates. Image enhancement process is generally used to improve the results of normalization process because iris image has low contrast and non-uniform illumination which can be caused by the position of light source which are compensated by enhancement process. Normalization includes Homogeneous rubber sheet model. In feature extraction features are extracted from normalized image which extract the significant features of the iris for identification purpose. Feature extraction can be done with the help of Gabor filters and wavelet transform, Laplacian of Gaussian filter, Hilbert transform and DCT. And then template matching compares the input image with database images. Matching results tells whether input image is accepted or rejected.

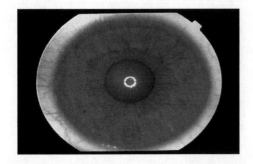

Figure 1.   Iris image.

Template matching can be done with help of Hamming distance, weighted Euclidean distance, Normalized correlation, nearest feature line.

In this paper we design an iris recognition algorithm based on hybrid transforms. Hybrid transform [10] is formed by using two different transforms so that we can use properties of both the transforms to get better results. In this paper we use different combinations of Walsh, DCT and Haar. Walsh is preferred because it has only real values while others such as DFT consists of complex values which can increase the complexity of computation. DCT is also a real valued function, its basis functions are only cosines and it gives maximum compression among other transforms. Haar transform gives lossless compression.

The first step is to represent the captured iris image in frequency domain. Using energy compaction analysis, features of all the database images are extracted and saved as energy map then feature of test image is extracted and then these test image features are compared with features of database images. A decision is then made whether input test image is accepted or not. (i.e features of test image are matched with database image or not.)

This paper contains following sections: section II describes the generation of hybrid transform, Methodology of feature vector generation, feature extraction, feature matching is explained in section III. Section IV evaluates the results and V concludes the paper.

2   GENERATION OF HYBRID TRANSFORM

Hybrid transform is generated using Kronecker product i.e. we apply Kronecker product on two different transform matrices to form single hybrid transform matrix. If A and B are existing orthogonal transform matrices, Kronecker product is given by H where,

$$H = A \otimes B = [aij\ B]$$

$$[H] = \begin{pmatrix} a_{11}B & a_{12}B & a_{1m}B \\ a_{21}B & a_{22}B & a_{2m}B \\ \cdot & \cdot & \cdot \\ \cdot & \cdot & \cdot \\ a_{1m}B & a_{m2}B & a_{mm}B \end{pmatrix}$$

If matrix A and B are orthogonal then matrix H is also orthogonal.

Here we are using DCT, WALSH and HAAR matrices as different input transform matrices. Say T is hybrid transform matrix and T1 is Walsh matrix (2 × 2) and T2 is DCT matrix (3 × 3) then Kronecker product of T1 and T2 is given by:

$$T1 = \begin{pmatrix} 1 & 1 \\ 1 & -1 \end{pmatrix}$$

$$T2 = \begin{pmatrix} 0.5574 & 0.5574 & 0.5574 \\ 0.7071 & 0 & -0.7071 \\ 0.4082 & -0.8165 & 0.4082 \end{pmatrix}$$

$$T = T_1 \otimes T_2 =$$
$$\begin{pmatrix} 0.5574 & 0.5574 & 0.5574 & 0.5574 & 0.5574 & 0.5574 \\ 0.7071 & 0 & -0.7071 & 0.7071 & 0 & -0.7071 \\ 0.4082 & -0.8165 & 0.4082 & 0.4082 & -0.8165 & 0.4082 \\ 0.5574 & 0.5574 & 0.5574 & -0.5574 & -0.5574 & -0.5574 \\ 0.7071 & 0 & -0.071 & -0.071 & 0 & 0.071 \\ 0.4082 & -0.8165 & 0.4082 & -0.4082 & 0.8165 & -0.4082 \end{pmatrix}$$

The row and column size of T is given by the product of row and column sizes of T1 & T2. In this case the size of T is (6 × 6).

Since the size of iris images are (512 × 512). In this paper we are using combinations of Walsh (16 × 16) and DCT (32 × 32) and Haar (32 × 32).

3   METHODOLOGY

3.1   Input training image

These are database images stored in the system for valid users.

3.2   Input test image

These images are those images which we want to authenticate.

3.3   Transform matrix

Here transform matrix is hybrid transform matrix formed by Kronecker product of two different transform matrices (DCT or WALSH or HAAR).

Here we use Walsh (16×16) and DCT (32×32), Walsh (16×16) and Haar (32×32), DCT (32×32) and Walsh (16×16), Haar (32×32) and Walsh (16×16) to form hybrid transform matrix.

3.4 *Transformed image*

Transformed image is obtained by using this equation:

$$I = T * i * T^t$$

where
i = image in spatial domain
T = transform matrix
T^t = transpose of transform matrix
Each image is converted in transform domain using this equation and then further processing is done on transformed image.
The Figure 2 explains the flow of method used.

3.5 *Average energy matrix*

Energy matrix of transformed image is generated for each of database image. Energy matrix is formed by squaring the transformed image. Averages of energy matrices of all database images are calculated.

3.6 *Average energy block matrix*

Energy block matrix is formed by using energy matrix. Here blocks of 8×8 are taken form

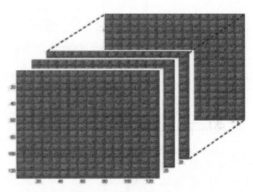

Figure 3.   Energy matrix for all the database images.

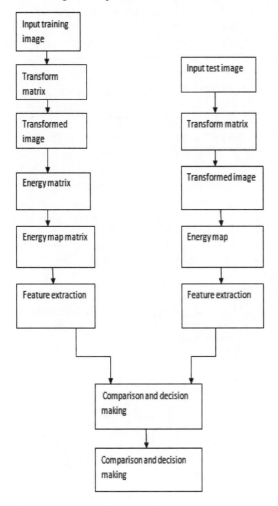

Figure 2.   Process flow of method.

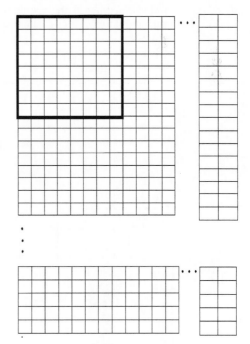

Figure 4. 512×512 energy matrix with 8×8 block shown.

43

512 × 512 and average of each block is computed. The average energy block matrix of size 64 × 64 is generated.

One 8 × 8 block is shown in Fig. 4 with solid outlines.

The average energy block matrix is converted into column matrix (4096 × 1). The second column for this matrix consists of the sequence no. from 1 to 4096 for ordering purpose.

Then we got 4096 × 2 matrix which is shown in Figure 5.

The values of first column are arranged in descending order, their cumulative sum was taken and then thresholding was applied by choosing a threshold value. Threshold values are varied to get better results.

Due to this thresholding we get zero at some places whose value doesn't contains much information (shown in Figure 6. with dark cells). This matrix is converted back into 64 × 64 matrix to obtain energy map matrix (64 × 64). All energy coefficients below the threshold value are made zero to reduce the size of feature vector.

3.7 Feature extraction

Features of each database images are selected according to the energy map. Based on threshold value only some blocks of energy map are selected (i.e. which are non zero). Again convert each database image into transformed image and standard deviation of the same 8 × 8 block is calculated to

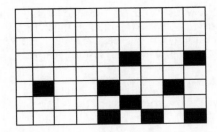

Figure 6.   Energy map matrix (64 × 64).

convert database image into 64 × 64 matrix. Features are then standard deviation values of the same block which are selected from energy map. These features are stored as. mat file. Similarly this analysis is done for all the database images.

3.8 Generate the transform and energy matrix for test image

Transform matrix is same for test image also as in database image. Energy map of test image is calculated in the same way as in database image to find the features of test image.

3.9 Feature extraction of test image

Features of each test image are selected as features of database image as explained above.

3.10 Comparing and decision making

Features of test image are compared with each of database image by finding the Euclidian distance between two images i.e. test and all database images. The image which is having least Euclidian distance is selected as authenticated image.

3.11 Accept reject

Image is accepted or rejected based on result of Euclidian distance calculated between test images and training images.

4 RESULTS

The database consists of iris samples for both left and right eye of around 90 samples (of 15 persons 3–3 samples each for their left and right eye image) one of the sample is considered as test image which is used to test the efficiency of proposed method.

To analyze performance of proposed method energy threshold is varied i.e. the percentage energy of energy matrix which is considered as threshold

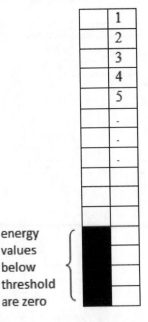

Figure 5.   4096 × 2 energy column matrix.

Table 1. Comparison between efficiencies using Walsh, DCT and Haar transforms.

Threshold (%)	Efficiency values for different combinations			
	Walsh, DCT	DCT, Walsh	Walsh, Haar	Haar, Walsh
99	95.4%	90.9%	95.4%	95.4%
98	95.4%	90.9%	95.4%	100%
95	95.4%	90.9%	95.4%	90.9%
91	90.9%	86.3%	90.9%	86.3%

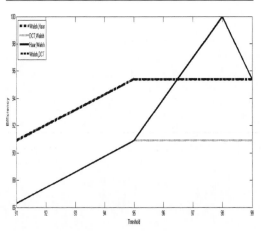

Figure 7. Threshold vs. efficiency plot for different transforms.

for selection of feature vector. Along with percentage energy we analyze performance of system by using different transforms combinations in hybrid transform technique such as (Walsh and DCT), (DCT and Walsh), (Haar and Walsh), (Walsh and Haar) etc. threshold is varied between 91% to 99% and efficiencies up to 100% is obtained by using Haar and Walsh combination when its threshold value is 98%.

For each combination we take different threshold values and get different values of efficiency as shown in Table 1. We got same efficiency values for (Walsh, DCT) and (Walsh, Haar) combination so their graph coincides with each other.

5 CONCLUSION

In this paper we have used transform domain technique to extract features of iris images.

Hybrid transform is used as a transform matrix which is Kronecker product of any two existing transform matrices. We have used three transform matrices Walsh, DCT, Haar and use their different combinations to generate hybrid transform matrix. Such as Walsh and DCT, Walsh and Haar, DCT and Walsh, Haar and Walsh. No preprocessing of iris image is required in this method. By varying the energy threshold values in different combinations of transform matrices we get different efficiencies. Efficiency as high as 100% is obtained by using Walsh and Haar transform combination.

REFERENCES

Daugman, J. 2004. How iris recognition works, IEEE Trans. CSVT, vol. 14, no. 1, pp. 21–30.
Grabowski, K. Sankowski, W. zubert, M & Napieralska, M 2006. Reliable iris localization method with application to Iris recognition in Near Infrared Light, MIXDES 2006. pp. 684–687.
Kekre, H.B. Sarode, T. Vig, R. Arya, P. Irani, A. Bisani, S. 2011. Palm print identification using Kronecker Product of DCT and Walsh transform for multi spectral images, International Conference on Hand-Based Biometrices.
Li, S.Z. & Lu, J. 1999. Face recognition using nearest feature line method, IEEE Transactions on neural network, 10(2), pp. 439–443.
Lim, S. Lee, K. Byeon. O. & Kim, T. 2001. Efficient iris recognition through improvement of feature vector and classifier, ETRI Journal, vol. 23, no.2, pp. 61–70.
Monro, D.M. Rakshit, S & Zhang, D. 2007. DCT based iris recognition, IEEE Trans pattern Analysis and Machine Intelligence, vol. 29, no. 4.
Ritter, N 1999. Location of Pupil-Iris Border in slit Lamp images of a cornea, Proceeding of international conference on Image Analysis and Processing. pp. 740–745.
Tisse, C. Martin, L. Torres, L. & Robert, M. 2002. person identification technique using human iris recognition, International conference on vision interface. vol. 4, pp. 249–299.
Wildes, R.P. 1997. Iris Recognition an Emerging Biometric technology, Proceedings' of IEEE, vol. 85, pp. 1348–1363.
Zhu, Y., Tan, T. & Wang, Y. 2000. Biometric personal identification based on Iris patterns, National Laboratory Of Pattern Recognition (NLPR), Institute of automation, Chinese academy of sciences, China, Proceeding of the 15th international conference on pattern recognition, vol. 2, pp. 2801–2804.

Communication and Computing Systems – Prasad et al. (Eds)
© 2017 Taylor & Francis Group, London, ISBN 978-1-138-02952-1

SIMO and MISO universal filters employing OTRA

Ujjwal Chadha & Tajinder Singh Arora
Maharaja Surajmal Institute of Technology, New Delhi, India

ABSTRACT: This paper introduces two universal filter employing operational trans-resistance amplifier. One proposed filter configuration is of single-input and multiple-output type whereas other one is multiple-input and single-output type filter. Both the biquadratic filters operate in trans-resistance mode and all the employed passive components are virtually grounded. The availability of output voltage at the buffered port makes the circuit more reliable. Circuits offer independent tunability of quality factor and bandwiidth. PSPICE simulations have been performed to ensure its workability. Non-ideal analysis and sensitivity analysis also have been included.

1 INTRODUCTION

The analog world around us necessitates the use and creation of analog devices in the electronics industry. Various devices have been realised that are capable of processing the analog electrical signals and modifying them according to their properties. These devices can be utilised to obtain circuits having different transfer functions corresponding to useful applications like amplifiers, summers, filters, oscillators, wave generators etc. (See Allen 1995, Salama & Soliman 1999, Elwan, & Soliman 1997, Kerwin, Huelsman & Newcomb, 1966, Altuntaş & Toker 2002, Senani, Singh & Singh 2013, Soliman 1996, Arora & Sharma (2016) Salama & Soliman 2000 and references cited therein)

Electronic filter is one of the important applications of analog signal processing. Afilter allows electrical signals having certain frequencies to pass through depending upon their type, which can be Low Pass (LP), High Pass (HP), Band Pass (BP), Band Reject (BR) or All Pass (AP) filter. A filter that is capable of performing two or more types of filtering is called a multifunction filter whereas the one that can perform all five types of filtering is called a universal filter. Further, a filter can be of Single-Input-Single-Output (SISO), Single Input-Multiple-Output (SIMO), Multiple-Input-Single-Output (MISO) or Multiple-Input-Multiple-Output (MIMO) type. One way of realising a SIMO or MISO multi function filter is using the KHN (KerwinHuelsman Newcomb) filter technique [4]. Various filters have been realised that utilises the properties of KHN state variable technique and use different active building blocks such as Current Feedback Operational Amplifier (CFOA), Current Conveyorsto achieve it. Some of the eminent references are Altuntaş & Toker (2002), Senani, Singh & Singh (2013), Soliman (1996), Arora & Sharma (2016) and their cited in. Any filter circuitthat is based on the KHN filter Signal Flow Graph (SFG) contains two integrators and one summer. An active device like OTRA (Operational transresistance Amplifier) (Salama & Soliman 1999) can be combined with some passive components like linear resistors and capacitances to make a summer or an integrator (Salama & Soliman 2000). These can be then used to design a multi function filter or universal filter. The good stability, low sensitivity, low requirement, of components and independent tunability of quality factor and simultaneous availability of different outputs is what makes such a design better than some other filter designs.

Various second order filter functions have been realized using OTRA but most of them suffer from one or the other problem. Kılınç, Keskin & Çam (2007), for example, have proposed a multifunction filter that requires a hardware change every time a different filter response is required and although single OTRA has been used to design the multifunction filter, each filter response requires certain conditions to be satisfied in terms of passive component values and node potentials. Gökçen, KilincÇam (2011), and Salama & Soliman (1999) have proposed universal filters using two and three OTRAs respectively but different conditions need to be satisfied to get a different filter response. Pandey, Pandey, Paul, Singh, Sriram & Trivedi (2012) have proposed a circuit where all the five response can be achieved by employing as many as five OTRAs but is suffers from the drawback of using a large number of passive components.

Here, we propose two universal filters of SIMO and MISO type that utilises the properties of classical KHN signal flow graph and its adjoint form.

Both filters have been designed using Operational Transresistance Amplifier (OTRA) as the active building block. SIMO filter can give all five filter responses (LP, HP, BP, BR and AP) at different points on the circuit. MISO filter also gives all filter responses without changing any hardware configuration at a single output port depending upon the node(s) at which input current is given. The proposed circuits do not require any hardware change every time while obtaining a different filter response, work in trans-resistance mode, employ virtually grounded passive components and provide independent tunability of quality factor. The output is available at buffered port.

The next section gives a brief introduction to OTRA. Section 3 gives the proposed circuit designs of both filters and the equations describing the filter properties. Section 4 shows the non ideal analysis of the proposed circuits. The sensitivity analysis is provided in section 5. Section 6 gives the simulation results of the proposed circuits which is done using PSPICE. Conclusion is provided in the last section.

2 OTRA—AN INTRODUCTION

OTRA (Operational Transresistance Amplifier) is a three-terminal device that provides high transresistance gain and makes the output potential proportional to the differential input current (Salama & Soliman 1999). Symbolic representation of OTRA is shown in Figure 1. Vo is the output potential and I+ and I− non-inverting and inverting input currents respectively.

The characteristic equations of the device and the input-output relationship is shown by (1)

$$\begin{bmatrix} V_+ \\ V_- \\ V_o \end{bmatrix} = \begin{bmatrix} 0 & 0 & 0 \\ 0 & 0 & 0 \\ R_m & -R_m & 0 \end{bmatrix} \begin{bmatrix} I_+ \\ I_- \\ I_o \end{bmatrix} \qquad (1)$$

The equations contained by the above matrix, when written individually give the equations as given in (2). The input potentials V+ and V− are virtual grounds and the output potential Vois dependent on the differential input current.

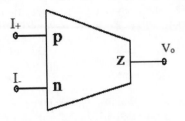

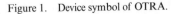

Figure 1.   Device symbol of OTRA.

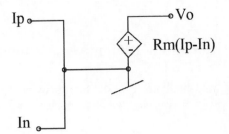

Figure 2.   Circuit of ideal OTRA.

$$V+ = V- = 0 \ \& \ Vo = Rm(I+ - I-) \qquad (2)$$

Ideally, the transresistance gain Rm approaches infinity and thus the input currents I+ and I− are equal. This type of representation of the device is shown in Figure 2 where $R_m \to \infty$.

The characteristics of the device can be obtained using MOS transistors connected in certain configurations. One such implementation of OTRA is discussed in section 5.

3 PROPOSED CIRCUITS

We propose two biquadratic universal filters operating in Transresistance mode. First one is a universal filter of Single-Input-Multiple-Output (SIMO) type. It is based on properties of classical KHN signal flow graph (Kerwin, Huelsman & Newcomb 1966) which is redrawn in Figure 3. The proposed filter circuit is shown in Figure 5.

The second one isa universal filter having multiple inputs and single output and is based on properties of adjoint KHN signal flow graph redrawn in Figure 4. The proposed MISO filter circuit shown in Figure 6 has been realised using OTRA as the active building block.

In the above single input, multiple output filter shown in Figure 5(a), the HP, BP and LP filter responses can be obtained as VHP, VBP, and VLP respectively. The transfer functions of this filter can be obtained by using the characteristic equations of the device given in (1) and (2). The obtained transfer functions of HP, BP and LP filter are given by (3a), (3b) and (3c) respectively. The BR response can be obtained by the summation of the HP and LP responses as given by (3d). Similarly, all pass response can be obtained by the summing the HP and LP response while subtracting the BP response as given by (3e). A simple voltage summer circuit can achieve this. One such possibility is shown in Figure 5(b). Here, Vo is the BR filter output when the BP input key K is closed and AP filter response when K is open.

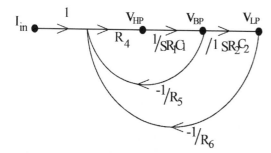

$$\frac{V_{HP}}{I_{in}} = \frac{s^2 \cdot R_4}{D(s)} \quad (3a)$$

$$\frac{V_{BP}}{I_{in}} = \frac{\dfrac{s}{R_1 C_1} R_4}{D(s)} \quad (3b)$$

Figure 3. Classical KHN signal flow graph.

$$\frac{V_{LP}}{I_{in}} = \frac{\dfrac{1}{R_1 C_1 R_2 C_2} \cdot R_4}{D(s)} \quad (3c)$$

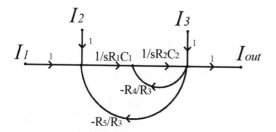

$$\frac{V_{BR}}{I_{in}} = \frac{s^2 R_4 + \dfrac{R_4}{R_1 C_1 R_2 C_2}}{D(s)} \quad (3d)$$

Figure 4. Adjoint KHN signal flow graph.

$$\frac{V_{AP}}{I_{in}} = \frac{s^2 R_4 - \dfrac{s R_4}{R_1 C_1} + \dfrac{R_4}{R_1 C_1 R_2 C_2}}{D(s)} \quad (3e)$$

$$D(s) = s^2 + \frac{s}{R_1 C_1}\frac{R_4}{R_5} + \frac{1}{R_1 C_1 R_2 C_2}\frac{R_4}{R_6} \quad (3f)$$

The angular frequency (ω) and quality factor (Q) are given by (4).

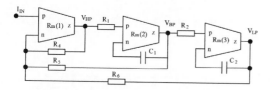

Figure 5(a). Proposed SIMO filter with 3 basic filter outputs.

$$\omega = \sqrt{\frac{\dfrac{R_4}{R_6}}{R_1 C_1 R_2 C_2}} \quad \& \quad Q = R_5\sqrt{\frac{R_1 C_1}{R_2 C_2 \cdot R_4 R_6}} \quad (4)$$

It can be seen that the quality factor Q is independently tunable. Varying the value of R5 will vary the quality factor proportionally. This makes the bandwidth (ω/Q) to be independently tunable as well.

In the above Multiple-Input-Single-Output (MISO) universal filter, the HP, BP and LP responses are obtained at output terminal when input current is given at I1, I2 and I3 respectively. BR output will be obtained when equal input current is given at both I1 and I3 (I1 = I3 = Ia) whereas AP response would be obtained when input current is given at all three inputs such that I1 = −I2 = I3 (= Ib).

The obtained transfer functions of HP, BP, LP, BR and AP filter are given by (5a), (5b), (5c), (5d) and (5e) respectively.

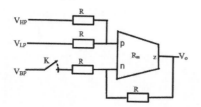

Figure 5(b). Extension of 5(a) to get all universal filter outputs.

$$\frac{V_o}{I_1} = \frac{s^2 \cdot R_3}{D(s)} \quad (5a)$$

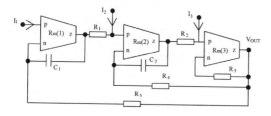

Figure 6. Proposed MISO universal filter circuit.

$$\frac{V_o}{I_2} = \frac{\dfrac{s}{R_2 C_2} R_3}{D(s)} \quad (5b)$$

$$\frac{V_o}{I_3} = \frac{\dfrac{1}{R_1 C_1 R_2 C_2}.R_3}{D(s)} \tag{5c}$$

$$\frac{V_o}{I_a} = \frac{s^2 R_3 + \dfrac{R_3}{R_1 C_1 R_2 C_2}}{D(s)} \tag{5d}$$

$$\frac{V_o}{I_b} = \frac{s^2 R_3 - \dfrac{s R_3}{R_2 C_2} + \dfrac{R_3}{R_1 C_1 R_2 C_2}}{D(s)} \tag{5e}$$

$$D(s) = s^2 + \frac{s}{R_2 C_2}\frac{R_3}{R_4} + \frac{1}{R_1 C_1 R_2 C_2}\frac{R_3}{R_5} \tag{5f}$$

The expressions of angular frequency (ω) and quality factor (Q) for MISO filter are given by (6).

$$\omega = \sqrt{\frac{\dfrac{R_3}{R_5}}{R_1 C_1 R_2 C_2}} \ \& \ Q = R_4 \sqrt{\frac{R_2 C_2}{R_1 C_1.R_3 R_5}} \tag{6}$$

Similar to the previous filter, the quality factor and bandwidth of this filter are independently tunable and can be changed by changing the value of R4.

3.1 Non-ideality analysis

When non-idealities are considered, the transresistance gain of OTRA is considered to be finite (here Rm) instead of infinity. The individual gains of different parts of the signal flow graph are multiplied by the following factors.

$$\text{Integrator non ideal error} = \alpha = \frac{1}{1 + \dfrac{1}{R_m S c}} \tag{7a}$$

$$\text{Summer non ideal error} = \beta = \frac{1}{1 + \dfrac{R_f}{R_m}} \tag{7b}$$

Here, Rf is the feedback resistance in summer and C is the integrating capacitance. The signal flow graph of the filter, considering these errors is shown in Figure 7.

The non-ideal analysis of the SIMO filter represented by signal flow graph in Figure 7 (and its OTRA circuit equivalent shown in Figure 5) leads to the value of angular frequency and quality factor given by (8a) and (8b).

$$\omega_{non-ideal} = \sqrt{\frac{\dfrac{R_4}{R_6}(\beta_2 \alpha_1 \alpha_2)}{R_1 C_1 R_2 C_2}} = \varepsilon_1 \omega_{ideal} \tag{8a}$$

(where, $\varepsilon_1 = \sqrt{\beta_2 \alpha_1 \alpha_2}$)

$$Q_{non-ideal} = \left(\frac{1}{\beta_1}\sqrt{\frac{\alpha_2 \beta_2}{\alpha_1}}\right).R_5\sqrt{\frac{R_1 C_1}{R_2 C_2.R_4 R_6}} = \varepsilon_2 Q_{ideal} \tag{8b}$$

(where, $\varepsilon_2 = \frac{1}{\beta_1}\sqrt{\frac{\alpha_2 \beta_2}{\alpha_1}}$)

Both error factors ε_1 and ε_2 (corresponding to angular frequency and quality factor) are same for the proposed MISO filter shown in Figure 6. These factors are close to unity, thus making the errors extremely small.

3.2 Sensitivity analysis

The sensitivity analysis of the SIMO proposed circuit shown in Figure 5 gives the following results.

$$S^\omega_{R_1,R_2,R_6,C_1,C_2} = -S^\omega_{R_4} = -\frac{1}{2} \tag{9a}$$

$$S^Q_{R_5} = 1 \tag{9b}$$

$$S^Q_{R_2,C_2,R_4,R_6} = -S^Q_{R_1,C_1} = -\frac{1}{2} \tag{9c}$$

Sensitivity analysis of MISO proposed circuit shown in Figure 6 gives the following results.

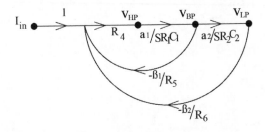

Figure 7. Adjoint KHN signal flow graph with non ideal gains.

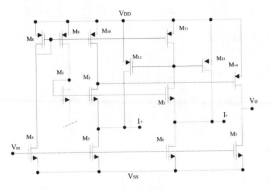

Figure 8. CMOS implementation of OTRA.

$$S^{\omega}_{R_1,R_2,R_5,C_1,C_2} = -S^{\omega}_{R_3} = -\frac{1}{2} \qquad (10a)$$

$$S^{Q}_{R_4} = 1 \qquad (10b)$$

$$S^{Q}_{R_1,C_1,R_3,R_5} = -S^{Q}_{R_2,C_2} = -\frac{1}{2} \qquad (10c)$$

For the non ideal case, sensitivity analysis of the both SIMO and MISO circuits gives the following results.

$$S^{\omega_{non-ideal}}_{\alpha_1,\alpha_2,\beta_2} = \frac{1}{2} \qquad (11a)$$

$$S^{Q_{non-ideal}}_{\alpha_2,\beta_2} = \frac{1}{2} \qquad (11b)$$

$$S^{Q_{non-ideal}}_{\alpha_1} = -\frac{1}{2} \qquad (11c)$$

$$S^{Q_{non-ideal}}_{\beta_1} = -1 \qquad (11d)$$

It can be noted form the above equations that that the values of passive sensitivities are all less than or equal to 1 in magnitude

4 SIMULATION RESULTS

The proposed circuits shown in Figure 5 and Figure 6 were tested for their responses using PSPICE simulation. OTRA was designed using CMOS transistors as proposed by Mostafa & Soliman (2006).The 0.5 μm process parameters provided by MOSIS (AGILENT) for CMOS were used during the simulation. The CMOS circuit defining OTRA has been redrawn in Figure 8. The transistor aspect ratios and the process parameters for the circuit used were same as used in proposed modified OTRA by Mostafa & Soliman (2006). Responses of filters using the non ideal OTRA are plotted along with their ideal responses. Both of the proposed filters were tested at 1MHz centre frequency. For the Single-Input-Multiple-Output (SIMO) filter shown in Figure 5, the obtained responses for HP, LP, BP, BR and AP are shown in Figure 9. The component values used during simulation were: R1 = 28KΩ, R2 = 56KΩ, R4 = R5 = R6 = 50KΩ, C1 = 4pF and C2 = 4pF The supply voltages of OTRA used were: $+V_{DD} = -V_{SS} = 1.5V$. The constant bias voltage was −0.5V. The continuous line shows the non-ideal filter response whereas the ideal filter response is shown by dashed lines.

The difference between the ideal and non-ideal curves occur due to non ideal errors discussed in section 4. Any small difference between ideal and

non ideal response in centre frequency and quality factor can be explained by non ideal expressions of angular frequency and quality factor given by (8a) and (8b) respectively. The Q-factortunability of the circuit was also tested by varying the value of R5. The variation of quality factor with R5 is shown in Figure 10.

For the Multiple Input Single Output (MISO) universal filter shown in Figure 6 the components values used during the simulation were: R1 = 56KΩ, R2 = 28KΩ, R3 = R4 = R5 = 50KΩ, C1 = 4pF and C2 = 4pF. The filter response for HP, LP, BP and

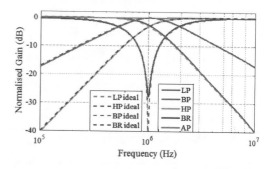

Figure 9. Ideal and non ideal responses of SIMO filter.

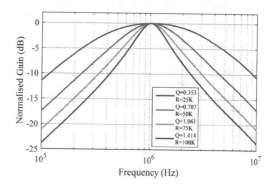

Figure 10. Variation of Q factor with R5.

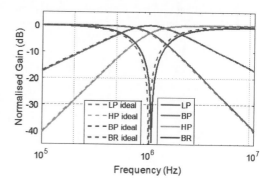

Figure 11. Ideal and non ideal responses of MISO filter.

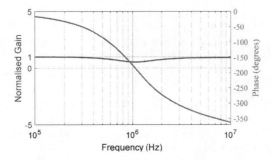

Figure 12. All pass magnitude and phase response of MISO filter.

BR filter are shown in Figure 11. The non ideal response is given by continuous line where as ideal response is given by dashed lines. Figure 12 shows the magnitude as well as phase response for the all pass filter.

5 CONCLUSION

This paper proposes two new universal filters employing OTRA and virtually grounded passive components. The filters have been implemented in PSPICE program to verify its theoretical properties. The proposed filters show various useful characteristics like independent tunability of quality factor and bandwidth. The availability of single-input-multiple-output design as well as multiple-input-single-output design and the simultaneous availability of all filter responses in the proposed SIMO filter ensure the ease of use and the circuits' capability to work in different situations without changing their hardware configuration. The operation of circuits' in trans-resistance makes them current to voltage convertor. Thus, these can be used to interconnect two circuits working in current mode and voltage mode. Moreover, the circuits show low sensitivity value. The simulated as well as the theoretical results are agreement with each other to a high degree.

REFERENCES

Allen, B. W. (1995). Applications of Op-amps. In *Analogue Electronics for Higher Studies* (pp. 91–108). Macmillan Education UK.

Altuntaş, E., & Toker, A. (2002). Realization of voltage and current mode KHN biquads using CCCIIs. *AEU-International Journal of Electronics and Communications*, *56*(1), 45–49

Arora T.S. & Sharma R.K. (2016). Realization of Current Mode Universal Filter Employing Third Generation Current Conveyor. *Indian Journal of Science and Technology,* Vol. 9, issue 13, pages1–7

Elwan, H. O., & Soliman, A. M. (1997, June). Novel CMOS differential voltage current conveyor and its applications. In *Circuits, Devices and Systems, IEE Proceedings-* (Vol. 144, No. 3, pp. 195–200). IET.

Gökçen, A., Kilinc, S., & Çam, U. (2011). Fully integrated universal biquads using operational transresistance amplifiers with MOS-C realization. *Turkish Journal of Electrical Engineering & Computer Sciences*, *19*(3), 363–372.

Kerwin, W. J., Huelsman, L. P., & Newcomb, R. W. (1966). *State-variable synthesis for insensitive integrated circuit transfer functions* (no. Tr-6560–10). Stanford univ ca stanford electronics labs.

Kılınç, S., Keskin, A. Ü., & Çam, U. (2007). Cascadable voltage-mode multifunction biquad employing single OTRA. *Frequenz*, *61*(3–4), 84–86.

Mostafa, H., & Soliman, A. M. (2006). A modified CMOS realization of the operational transresistance amplifier (OTRA). *Frequenz*, *60*(3–4), 70–77.

Pandey, R., Pandey, N., Paul, S. K., Singh, A., Sriram, B., & Trivedi, K. (2012). Voltage mode OTRA MOS-C single input multi output biquadratic universal filter. *Advances in Electrical and Electronic Engineering*, *10*(5), 337.

Salama, K. N., & Soliman, A. M. (1999). CMOS operational transresistance amplifier for analog signal processing. *Microelectronics Journal*, *30*(3), 235–245.

Salama, K. N., & Soliman, A. M. (2000). Active RC applications of the operational transresistance amplifier. *Frequenz*, *54*(7–8), 171–176.

Senani, R., Singh, A. K., & Singh, V. K. (2013). *Current feedback operational amplifiers and their applications*. Springer Science & Business Media.

Soliman, A. M. (1996). Applications of the current feedback operational amplifiers. *Analog Integrated Circuits and Signal Processing*, *11*(3), 265–302.

Communication and Computing Systems – Prasad et al. (Eds)
© 2017 Taylor & Francis Group, London, ISBN 978-1-138-02952-1

Modeling of brain tumor detection using image processing

Pooja Dang
Dronacharya College of Engineering, Gurgaon, Haryana, India

Jyotika Pruthi
The Northcap University, Gurgaon, Haryana, India

ABSTRACT: In this paper, a technique for Detection and analysis of tumor in human brain from MRI-Magnetic Resonance Imaging has been proposed without any manual intervention. MRI technique is best suited for diagnosis and treatment planning. In this, first stage is acquisition of an image followed by preprocessing and post processing steps. Segmentation is basically used to extract the features. Image segmentation helps in simplification or modification of an input image. For the initial segmentation, we have used thresholding technique. Others segmentations such as watershed and texture are used to further produce the richer results. Set of morphological operations are been applied to extract out the degenerated tissue. Our detection and segmentation methodology plays an important role in detection and performance analysis.

1 INTRODUCTION

Medical Image Processing is the standard method to detect the abnormal tissue. The key benefit of this field is reduction of manual work and early detection of the disease. Medical Imaging provides the facility of analyze the inner parts of the body and to diagnose them. The main goal of medical imaging is to advance the quality of images of inner body parts for interpretation by humans.

Human body consists of various cells. Brain is the most important and highly sensitive organ. Tumor is very harmful disease that leds to diminished speech, hearing and various other problems. Tumor is basically any abnormal growth of the tissue in any organ. It can be classified as benign or malignant type. The benign tumor is non-cancerous and somewhat less harmful and malignant is cancerous one. Malignant tumor can be further divided into two categories as primary and secondary tumor. The Benign type tumor is less dangerous than malignant. The malignant tumor spreads so easily entering nearby cells of the brain hence, worsening the condition of the patient. Due to the complex structure of the brain, detection of abnormal tissue is very difficult task in medical image processing. There are various types of medical imaging modalities such as X-Ray, Magnetic Resonance Imaging (MRI), and Computer Tomography Scan (CT Scan) are present to capture the images. MRI is best suited technology for collecting the internal data of the human body for diagnosing.

In this paper we propose a methodology for detecting the region of interest i.e. the tumor and then performing segmentation to extract it. The proposed methodology is described in three phases: First-phase is Preprocessing for the enhancement of the image. Second-phase is Post Processing which includes division of an image into top and bottom, histogram plotting to analyze the plot and location of tumor. Then various segmentation techniques and morphology operators such as erosion and dilation are used to extract the tumor region.

2 METHODOLOGY

This algorithm for the detection and analysis of tumor region is been proposed. Preprocessing and Postprocessing are been further categorized. After the tumor detection area and volume of the abnormal tissue is calculated using MATLAB.

2.1 Preprocessing

In this module, series of steps are followed such as skull stripping in which non brain regions are been removed. Secondly input image is converted into gray scale to change its original form. Then various enhancement techniques are applied such as contrast stretching techniques known as Histogram equalization that stretch out the intensity values. In this module, we are basically concerned with the location of the tumor.

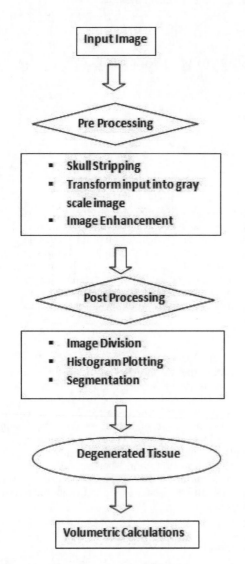

Figure 1.   Flow chart of methodology.

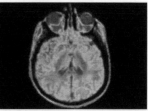

Original Image

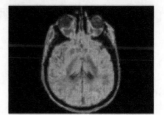

Grayscale image

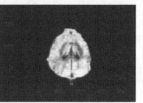

Removed skull image

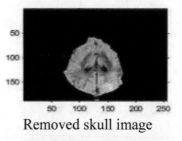

Contrast Streched Image

Figure 2.   Preprocessing steps.

3   POST PROCESSING

In this section, we have divided the image into two halves top and bottom so as to categorize and extract features from each half.

The algorithm followed to accomplish this task is:

i. Take two variables as A and B for storing the number of rows and columns.
ii. Take some other variable C that contains the number of columns divided by 2.
iii. Using the command "imcrop()", output image will convert into half of the input image.

3.1   *Histogram plotting*

In medical imaging, Histogram Plotting is a graphical depiction between scale of pixels and the intensity of pixels of the image. This histogram is a chart showing the number of pixels in an image at each different intensity value found in the particular image. In the figure, the horizontal axis shows the scale of pixels and vertical axis shows the intensity of pixel present in the image. Figure 3 (a), (b), (c) shows the histogram graph of the original image and the plot of both halves. It is clearly visible that the bottom half has more deviation around 200. Figure 4 indicates that abnormal tissue is present in the bottom part of the image.

3.2 Threshold segmentation

Threshold segmentation is the common and easiest segmentation which is widely used. In this we are basically concerned with images which have lesser objects of interest. This method is based on getting binary image out of gray scale image.

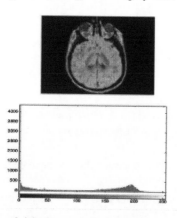

Figure 3. Original Image and its Histogram.

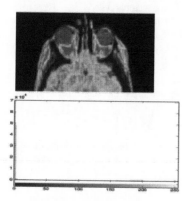

Figure 4. Top half and its histogram.

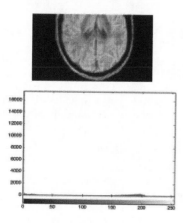

Figure 5. Bottom half and its histogram.

To segment abnormal tissue we are using thresholding as segmentation algorithm. In this first of all, select a threshold value such as T for initial segmentation. Various methods are present for this task such as K-means clustering, Otsu's method. Using K-means clustering, representation of an image can be made which is somewhat easier to analyze the results. This is basically used to locate boundaries and objects. As in Figure 7(b), objects in the image are very less, therefore thresholding segmentation is best suited. Then we have "grad-mag" matlab toolbox function to create boundaries of an image.

3.3 Compute watershed segmentation

In this transformation, image is considered as a topographic surface and the gray level of the image represents the altitudes. Regions with a constant gray level constitute the flat area of an image

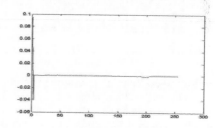

Figure 6. Difference of two histograms.

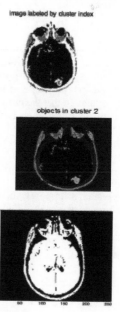

Figure 7. (a) Object in cluster index. (b) Objects in cluster. (c) After thresholding technique.

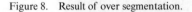

Watershed transform of gradient magnitude (Lrgb)

Figure 8.   Result of over segmentation.

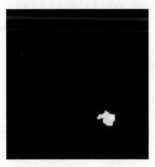

Figure 9.   Tumor region as final output.

and region edges depict high watersheds and low rise region depicts catchment basis. Every part has their own minima, and even if single drop of water falls on one plane of the watershed, minima of that region will be reached. Watershed lines represent the boundaries of the object.

3.4   *Morphological operators*

Morphological operations such as dilation and erosion are used as an image processing tools for sharpening the regions. Morphological image processing relates to the shape or morphology of features in an image. In this we have used two functions as erosion operator to shrink the objects in the image and dilation function for thickening the objects in the image. Size and shape of the structuring element is been controlled by this operation. Structuring element plays an important role in both of the operations. Then we have applied the operations such as opening and closing formed from combining these two (dilation and erosion) operations. After applying all these operations, we can clearly figure out the tumor (abnormal) region present in the brain.

3.5   *Performance analysis*

Performance analysis is the measure of accuracy of the proposed work. Various performance measures are true positive, true negative, false positive and

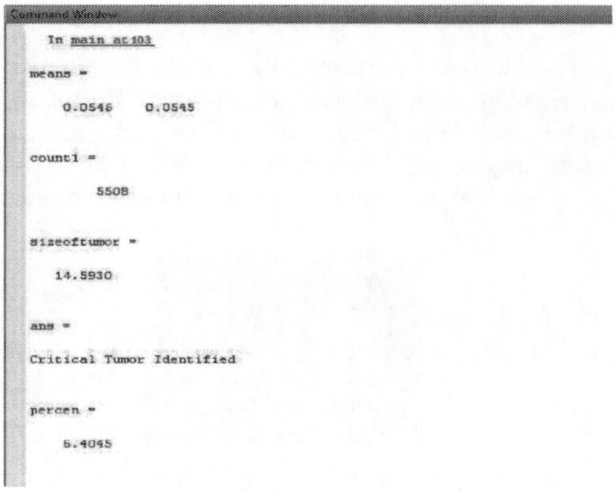

Figure 10.   Calculated size or area of detected tumor.

false negative. Based on these parameters accuracy, sensitivity and specificity can be calculated.

- True Positive (TP) – Count of correct cases identified correctly.
- True Negative (TN) – Count of correct cases identified as incorrect.
- False Positive (FP) – Count of incorrect cases identified as correct.
- False Negative (FN) – Count of incorrect cases identified as incorrect.

Sensitivity– is the quantity that a diagnostic test is positive when patient has tumor.

Sensitivity = TP / (TP + FN)

Specificity – is the amount that a diagnostic test is negative when patient does not have tumor.

Specificity = TN / (TN + FP)

Accuracy – is the quantity that a diagnostic test is correctly tested.

Accuracy = (TP + TN) / (TP + TN + FN + FP)

The proposed work has been performed on 20 Brain MRI images, in which 10 images are abnormal (tumor is present) and others are normal.

The proposed system provides an accuracy of 94.4%, sensitivity of 100% and specificity of 90%.

3.6   *Volume calculation*

For volume calculation first we need to calculate the area of the degenerated tissue. Then volume is calculated using the area and thickness of the image slice by both. We have taken the thickness of image slice as 5 mm. After detection of tumor, its

Table 1. Shows the value of sensitivity, specificity and accuracy.

Normal	Abnormal	TP	TN	FP	FN	Sensitivity	Specificity	Accuracy
10	10	8	9	1	0	100%	90%	94.4%

size or area is calculated. And on the basis of size it is classified as either it is in critical or not. The size of tumor is calculated using the formula:

$$Size\_of\_tumor, S = \left[\left(\sqrt{P}\right)^* 0.264\right] mm^2$$

where, P = no-of white pixels; 1 Pixel = 0.264 mm

The size or area of the tumor in this case is 14.5930 mm^2. If the area or size or tumor is greater than 6 mm^2 then it is considered in critical stage. So the detected tumor is critical.

For volume calculation,

Volume of abnormal tissue = Area * 0.5

Therefore, Volume = 14.5930 * 0.5 = 7.2965 mm³.

4 FUTURE SCOPE

In this work the image processing approach has been used to detect the abnormal tissue or tumor from the human brain image. We can extend the work in different ways as detecting the degenerated tissue for some other image formats. And second some other segmentation approach can be used to detect brain tumor with better efficiency.

REFERENCES

Akram, M. Usman & Usman2, Anam "Computer Aided System for Brain Tumor Detection and Segmentation" IEEE 2011.

Bandyopadhyay, Dr. Samir Kumar "Detection of Brain Tumor-A Proposed Method" Journal of Global Research in Computer Science Volume 2, No. 1, January 2011.

Elmoataz, Li. & Fadili & Ruan, S. "Segmentation based on enhanced morphological watershed algorithm" Journal of Global Research in Computer Science.

Patil, Rajesh C. & Bhalchandra, Dr. A. S. Bhalchandra "Brain Tumour Extraction from MRI Images Using MATLAB" International Journal of Electronics, Communication & Soft Computing Science and Engineering ISSN: 2277–9477, Volume 2, Issue 1.

Rahnamayan, S & Tizhoosh, H.R. & Salama, M.M.A. "Automatic Acquisition of Image Filtering and Object Extraction Procedures from Ground-Truth Samples". Journal of Advanced Computational Intelligence and Intelligent Informatics, Volume 13, No. 2, March 2009, pp. 115–127.

Selvanayaki, K. & Dr. Karnan, M. "CAD System for Automatic Detection of Tumor through Magnetic Resonance Image - A Review". International Journal of Engineering Science and Technology Vol. 2(10), 2010, 5890–5901.

Communication and Computing Systems – Prasad et al. (Eds)
© 2017 Taylor & Francis Group, London, ISBN 978-1-138-02952-1

Improved advanced modified decision based unsymmetric trimmed median filter for removal of salt and pepper noise

K. Sharma & S.K. Malik
SRM University, Delhi NCR, Sonepat, Haryana

ABSTRACT: Noise removal from images is an interesting field of research. IAMDBUTMF (proposed) by means of global trimmed mean approach removes the salt and pepper ensuring that the mean and median for the image is calculated from the non noisy pixels. The proposed algorithm shows better results when compared with standard Median Filter (MF), Adaptive Median Filter (AMF), Decision Based Algorithm (DBA), Decision Based Unsymmetric Trimmed Median Filter (DBUTMF) and Advanced Modified Decision Based Unsymmetric Trimmed Median Filter (AMDBUTMF) on the basis of IEF v/s noise density and PSNR v/s noise density values.

Keywords: mean filter, image filtering, image denoising, median filter, salt and pepper noise

1 INTRODUCTION

Also familiar as black and white noise, the salt and pepper noise replaces the image pixels with black (as 0 - minimum pixel intensity) and white (as 255 - maximum pixel intensity) dots. Salt and pepper noise can occur in the image at any step of image capturing and transferring which makes it important to process an image. There are many filters like gaussian filter, mean filter, median filter, adaptive filter, wavelets, wiener filter which are used to get rid of the noise in image but median filter is the most popular filter. In the literature many techniques are described to deal with salt and pepper noise. Standard Median Filter (MF) (Astola et al. 1997) which is effective only for low density. Better than MF, Adaptive Median Filter (AMF) (Hwang et al. 1995) works efficiently but only at low level. The Switching Median Filter (SMF) (Karim et al. 2002) (Ng et al. 2011) works with the help of pre defined threshold value but the robustness is hard to find. The solution is given by Decision Based Algorithm (DBA) (Ebenezer et al. 2007) but repeated replacement of neighboring pixels used in this algorithm produces streaking effect. To surmount the streaking effect difficulty Decision Based Unsymmetric Trimmed Median Filter (DBUTMF) (Ebenezer et al. 2010) is proposed but it doesn't provides better visual quality. To avoid this problem Modified Decision Based Unsymmetric Trimmed Median Filter (MDBUTMF) (Esakkirajan et al. 2011) as well as Advanced Modified Decision Based Unsymmetric Trimmed Median Filter (AMDBUTMF) (Sreenivasulu

et al. 2014) is proposed which again does not produces much better performance. IAMDBUTMF (proposed algorithm) by means of global trimmed mean approach provides better results as compared others.

The rest of paper is organised as follows: a brief description of advanced modified decision based unsymmetric median filter is given in section 2, explanation of global trimmed mean is discussed in section 3, IAMDBUTMF (proposed algorithm) is described in section 4. Illustration of IAMDBUTMF is described in section 5. Analysis and simulation results are shown in section 6. Finally conclusion is made in section 7.

2 ADVANCED MODIFIED DECISION BASED UNSYMMETRIC TRIMMED MEDIAN FILTER

This algorithm describes the preservation of edge details by increasing the size of image by adding zeros to the all the four sides of the image (Sreenivasulu et al. 2014). Hence the image size increases from 256×256 to 258×258. Now the image is made through filtering process by selecting a window matrix of size 3×3 with the central pixel as processing pixel P. If P = 0 or P = 255 then it is considered to be noisy else noise free. If the matrix consists of all the elements as 0's as well as 255's then swap the processing pixel with mean value of the matrix window. If there are some of the non noisy elements in the matrix then remove all 0's as well as 255's from the matrix and interchange the central pixel with the median of the remaining non noisy pixels. For entire image the process is repeated.

3 GLOBAL TRIMMED MEAN APPROACH

When the elements in the selected matrix consist of all the 0's as well as 255's then the processing pixel is interchanged by the mean value of the matrix. The mean value is obtained from the 0's and 255's only which are noisy elements and hence proves that the noise impact is not reduced (Veerakumar et al. 2012). So the global trimmed mean is used to decrease the effect of noise which describes that when entire matrix is noisy then instead of considering the selected matrix to measure the mean consider the whole noisy image and remove all the noisy elements in the corrupted image and interchange the processing central pixel with the mean value of the left over non noisy elements of the image.

4 PROPOSED ALGORITHM

The IAMDBUTMF filters the image by initially increasing the image size and then by processing the image. It then detects the noisy pixels followed by filtering only the noisy pixels. Interchange the central pixel of the matrix with the global trimmed mean of image if the selected matrix consists of all the noisy pixels. Also swap the central pixel with trimmed median of the selected matrix if it contains some of the non noisy pixels.

Algorithm

Step 1: Add additional zeros around the sides of image.

Step 2: Assume P as the processing pixel in the selected 3×3 matrix 2-D window.

Step 3: If $0 < P < 255$, the pixel is non noisy and is kept unaffected.

Step 4: P is considered noisy if $P = 0$ or 255 and there may be following cases:

    a. If entire matrix is mix of 0's as well as 255's, interchange the processing pixel with global trimmed mean of image.

    b. If entire matrix is either only 0's or only 255's, interchange the processing pixel with the global trimmed mean of image.

    c. If there are some of the non noisy pixels in the window, interchange the central pixel with the trimmed median of the selected matrix.

Step 5: Repeat steps 2 to 4 until entire image is processed.

Step 6: Remove the additionally added 0's from the image.

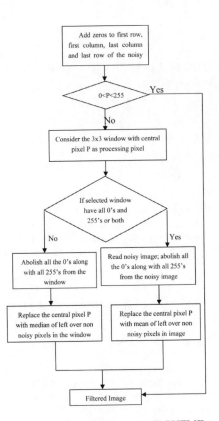

Figure 1. Flowchart of improved AMDBUTMF.

5 ILLUSTRATION

To preserve details as described in the algorithm, the image is surrounded by adding extra zeros to the image, thus making the image size from 256×256 to 258×258 after which noisy pixels are detected.

Case1: If the central element is neither 0 nor 255 then P is non noisy and is kept unaffected.

Case2: If the central pixel P is noisy (0 or 255) and matrix is mix of noisy and non noisy pixels then P is interchanged by trimmed median of the selected matrix. To measure the trimmed median, for instance, consider the array in the selected window as M [100 0 255 150 0 200 10 50 50] with central pixel 0 as processing pixel. Eliminate all 0's as well as 255's. Rearrange the matrix array as [100 150 200 10 50 50] which contains only the non noisy pixels and calculate the median. The median value of this array is 75. Replace the noisy pixel 0 with 75.

Case3: (a) If the central pixel P is noisy (0 or 255) and matrix contains combination of 255's as well as 0's. Interchange the processing pixel by global trimmed mean i.e. the trimmed mean of image.

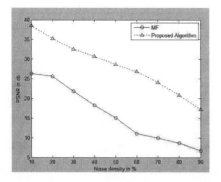

Figure 2.   PSNR values of IAMDBUTMF with MF.

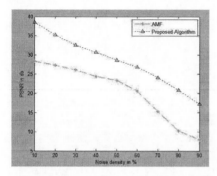

Figure 3.   PSNR values of IAMDBUTMF with AMF.

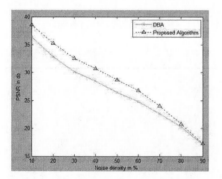

Figure 4.   PSNR values of IAMDBUTMF with DBA.

b. If the matrix contains either only 0's or 255's then; interchange the central pixel by trimmed mean of image.

To measure the trimmed mean of image, for instance, consider the array in the selected window as M [0 0 255 0 0 255 0 255 255] with central pixel 0 as processing pixel. Now as the entire matrix contains noisy elements, instead of measuring the mean of this matrix, the noisy image is considered for filtering and all the noisy elements in the image

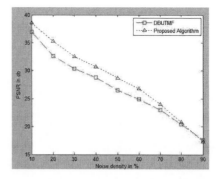

Figure 5.   PSNR   values   of   IAMDBUTMF   with DBUTMF.

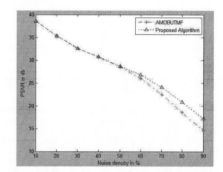

Figure 6. PSNR values of IAMDBUTMF with AMDBUTMF.

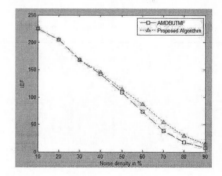

Figure 7. IEF values of IAMDBUTMF with AMDBUTMF.

are removed. For instance consider the noisy image array as N [100 0 255 150 200 110 0 0 255 50 60 0 0 255 20 85 0 255 255 40 0 0 255 255 0]. Remove all the noisy elements from the image array and rearrange to form a new array as [100 150 200 110 50 60 20 85 40] and calculate the trimmed mean which comes out to be 90. Replace the noisy element 0 as 90.

Table 1. PSNR measure at varied noise density for lena image.

Noise varied in %	MF	AMF	DBA	DBUTMF	AMDBUTMF	PA
10	26.34	28.43	36.4	36.94	38.54	38.54
20	25.66	27.40	32.9	32.69	35.28	35.28
30	21.86	26.11	30.15	30.41	32.60	32.61
40	18.21	24.40	28.49	28.79	30.68	30.76
50	15.04	23.36	26.41	26.52	28.53	28.68
60	11.08	20.60	24.83	24.91	26.03	26.80
70	9.93	15.25	22.64	22.97	22.50	24.05
80	8.68	10.31	20.32	20.44	18.57	20.81
90	6.65	7.93	17.14	17.56	14.55	17.28

Table 2. IEF measure at varied noise density for lena image.

Noise varied in %	AMDBUTMF[8]	PA
10	225.23	225.23
20	205.38	205.38
30	167.93	168.15
40	142.53	145.07
50	108.64	114.28
60	73.12	87.44
70	38.10	54.42
80	17.68	29.68
90	7.87	14.76

Figure 9. (a) Uncorrupted image, Image restored by AMDBUTMF at (b) 50% (c) 70% (d) 80%.

Figure 8. a) Uncorrupted image (b) 50% corrupted image (c) 70% corrupted image (d) 80% corrupted image.

6 SIMULATION RESULTS AND ANALYSIS

The IAMDBUTMF is tested only for salt and pepper noise by 256×256 grayscale image and the performance is calculated in terms of PSNR (Peak Signal to Noise Ratio) in decibel(dB) which is shown in Table 1 and IEF (Image Enhancement Factor) shown in Table 2 which describes the

Figure 10. (a) Uncorrupted image, Image restored by IAMDBUTMF at (b) 50% (c) 70% (d) 80%.

visual quality of image. Visual quality improved due to the proposed algorithm is shown through the images (Figures 8, 9 and 10). The mean and median measured only from the non noisy elements ensure the decrease of noise impact which improves the visual quality and increase in PSNR and IEF values. The experiment is performed in MATLAB. Density of noise is assorted from 10% to 90% and is inspected with some of the existing methods. Results show better performance than some of the existing algorithms.

7 CONCLUSION AND FUTURE SCOPE

IAMDBUTMF proposed in the paper reduces the noise impact by replacing the processing pixel with the trimmed mean and trimmed median measured only from non noisy elements hence reducing the noise impact. Because of the global trimmed mean approach, when noise density increases from 60% it becomes time consuming as the mean is calculated from the image and not from the selected window and hence further studies can be carried out by future researchers.

REFERENCES

Aiswarya, K. Jayaraj, V. & Ebenezer, D. January 2010. A new and efficient algorithm for removal of high density salt and pepper noise in images and videos. *Second International Conference on Computer Modeling and Simulation* 4: 409–413.

Astola, J. & Kuosmaneen, P. 1997. Fundamentals of Nonlinear Digital Filtering. Boca Raton: FL:CRC.

Chaitanya, N.K. & Sreenivasulu, P. February 2014. Removal of salt and pepper noise using advanced modified decision based unsymmetric trimmed median filter. *International Conference on Electronics and Communication Systems*: 1–4.

Esakkirajan S. Veerakumar, T. Subramanyam, A.N. & PremChand, C.H. May 2011. Removal of high density salt and pepper noise through modified decision based unsymmetric trimmed median filter. *IEEE Signal Processing Letters* 18(5): 287–290.

Hwang, H. & Hadded, R.A. April 1995. Adaptive median filter: New algorithms and results. *IEEE Transactions on Image Processing* 4(4): 495–502.

Ng, P.E. & Ma, K.K. June 2006. A switching median filter with boundary discriminative noise detection for extremely corrupted images. *IEEE Transactions on Image Processing* 15(6): 1506–1516.

Srinivasan, K.S. & Ebenezer, D. March 2007. A new fast and efficient decision based algorithm for removal of high density impulse noise. *IEEE Signal Processing Letters* 14(3): 189–192.

Veerakumar, T. Esakkirajan, S. & Vennila, IIa. 2012. An approach to minimize very high density salt and pepper noise through trimmed global mean. *International journal of computer applications* 39: 29–33.

Zhang, Shuqun & Karim, M.A. November 2002. A new impulse detector for switching median filters. *IEEE Signal Processing Letters* 9(11): 360–363.

Communication and Computing Systems – Prasad et al. (Eds)
© *2017 Taylor & Francis Group, London, ISBN 978-1-138-02952-1*

Threshold segmentation technique for tumor detection using morphological operator

Manorama Sharma
Air Force Vocational College, Delhi, India

G.N. Purohit & Saurabh Mukherjee
Banasthali University, Rajasthan, India

ABSTRACT: Medical imaging plays a revolutionary role in the field of medical science. In present scenario Image processing is frequently used in diagnosis. The main reason for this is that it extract abnormal tissues from normal tissues clearly and consuming less time. Image quality is enhanced and it provides a great help to the doctor to identify the diseases. A Tumor may lead to cancer, which is major leading cause of death and responsible for around 13% deaths worldwide. Automation of tumor detection is required for detecting tumor on right stage. This paper reviews the process and techniques used in detecting tumor based on medical images such as CT scan, MRI with the help of gray scale image. We applied morphological operator, edge detection and threshold segmentation approach for detecting tumor. We use MatLab R2013a to show the result.

1 INTRODUCTION

Image processing is used to extract the edges of affected part of human body for better diagnosis. Brain is responsible for controlling all over functioning such as memory, learning, emotions, and blood vessels. Sometime unnatural growth in the form of lump is found and this growth may be benign or malignant. Identifying tumor affected part within brain is called brain tumor detection. With the help of image processing doctors can identify tumor shape and size which is used for diagnosis. Image processing is used to enhance image quality to extract information from acquired image. Brain tumor is a dangerous disease commonly found in human being. Unwanted cells are detected through image processing for diagnoses purpose. By image processing death ratio is decreased. For finding tumor size, shape, and type MRI and CT scans are commonly used. In this paper we use MRI images for finding affected area in brain. MRI is used to show the internal structure of the body. MRI shows the difference between normal tissues and abnormal tissues. In this paper we used threshold segmentation for tumor detection and morphological operator.

1.1 *Tumor*

Abnormal cell in brain are called tumor. There are three stage of tumor -

1.1.1 *Benign*
In this type normal tissues are not affected by abnormal tissues. When it detect it can be diagnosis.

1.1.2 *Pre-malignant*
If it not diagnosis properly. It can convert into cancer.

1.1.3 *Malignant*
It is cancer and cause to death.

1.2 *Magnetic Resonance Imaging (MRI)*

It is used to show the internal structure of human body. It is used to detect abnormal tissues in brain. It gives better result than CT and commonly used to take images of brain.

1.3 *Threshold segmentation*

It is commonly used for image segmentation. It is used to discriminate foreground from the background. First of all an image is converted into gray scale and then into binary image. In this technique a threshold value T is selected from binary image. Histogram is frequently used to select T value from binary image. Threshold technique is global and local threshold. In global threshold a single value is selected and multiple values are selected in local threshold.

1.4 *Morphological operator*

➢ It is used on binary image for background subtraction.
➢ It is also used on gray value images.
➢ It is good for noise removal in background.
➢ It is suitable for removal of holes in foreground and background.

Common Morphological Operations— Shrinking the foreground ("erosion"), Expanding the foreground ("dilation"), Removing holes in the foreground ("closing"), Removing stray foreground pixels in background ("opening"), Finding the outline of the foreground, Finding the skeleton of the foreground.

2 LITERATURE REVIEW

Hiran and Doshi, Authors were developed a technique for image enhanced for brain tumor detection. Their algorithm was based on digital image segmentation. This algorithm was used to present edge pattern and segment of brain tumor through MRI images. Using this technique they were successful to find the size and region of tumor. They were used preprocessing, image enhancement, Thresholding and Morphological operation. They were used color image and then it was converted into gray for processing. (Kamal Kant Hiran & Ruchi Doshi, 2013).

Syed and Narayanan proposed a method for Brain Tumor Detection based on artificial neural network categorized into Multi-layer perceptron neural network. They were used segmentation for feature extraction. They developed this method to discriminate normal and abnormal tissues through MRI scan images. It was helpful to doctor to analyze stage of cancer and less time consuming. They were used preprocessing, histogram, binarization, thresholding, Morphological operation, GLCM based feature extraction and BPN based classifier. (Aqhsa Q. Syed & K. Narayanan, 2014)

Viji et al. developed an effective modified region growing technique. Comparative analyses were made for the normal and the modified region growing using both the Feed Forward Neural Network (FFNN) and Radial Basis Function (RBF) neural network. The results were better than normal technique. Technique was applied on MRI images for tumor detection. For evaluation of the proposed method the sensitivity, specificity and accuracy values were used.

Malakooti et al. proposed a method which combines both Neural Network and fuzzy clustering method. Using the proposed method they classified tumor region from non tumor candidate areas. They used morphological operation for extracting candidate abnormal areas and used this technique for brain tumor MRI images. They found better result compared to existing methods. The proposed technique then increased correctness for brain tumor MRI image for diagnosis.

Hoseynia et al. proposed a method to minimize the error in the process of image segmentation and for improving edge detection in MRI (brain tumor) images, combining fuzzy c-means algorithm with watershed algorithm. They found better results by using this method and accuracy helped them to improve images, edge detection and noise reduction in brain tumor MRI images. The results indicated that using this combination method, presented more accuracy helped them to improve images, edge detection. They applied fuzzy algorithm before applying watershed makers and they proposed high accuracy in images.

3 PROPOSED METHOD

Input MRI images which includes range of brain, a dark black background and signs. Convert this image into grey scale image for this segmentation is used. Segmentation is used to provide division between regions and categories. Type pixels are presented in similar grey scale and different pixels are presented with different value. For segmentation threshold is used. Threshold is commonly used in segmentation. It provides discrimination between foreground and background. Here images are converted into grey scale and then into binary image. Binary image is required to reduce the complexity of data. It provides actual shape and position of the object. Using threshold segmentation feature extraction is done. Tumor segmentation is done on the basis of affected cells. This process is done on the basis of different behavior of pixels in brain image. Different behavior is found according to their shape, brightness and color. Feature extraction is done for brightness, shape and texture. Threshold segmentation is one of the simplest segmentation methods. The input gray scale image

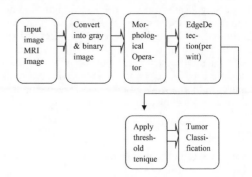

Figure 1. Flow chart for proposed method.

is converted into a binary format. The method is based on a threshold value which will convert gray scale image into a binary image format. The main logic is the selection of a threshold value.

There are no individual techniques for detecting tumor. For different body part different segmentation techniques are used. A combination of two techniques always shows a better result than other. We used threshold with morphological operator. Steps which are carried out for detection of tumor are enlisted below:-

Step 1: Consider MRI scan image of brain of patients.
Step 2: Convert image into gray image and then binary for segmentation.
Step 3: Apply morphological operator.
Step 4: Apply edge detection technique.
Step 5: Apply threshold technique on binary image and find the results.
Step 6: Tumor is classified.

4 EXPERIMENTAL RESULT

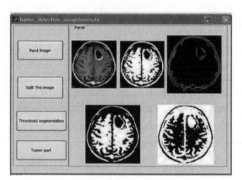

Figure 2. Raw image, binary, edge detection, morphological operator.

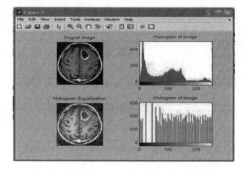

Figure 3. Histogram of image.

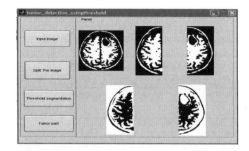

Figure 4. Split image.

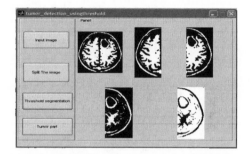

Figure 5. Split image with tumor.

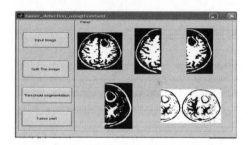

Figure 6. Image with tumor.

Raw image is input for process then histogram and histogram equalization is applied as follows:-
MATLAB: A = imhist(I); B = histeq(I);

4.1 *Morphological erosion*

This is used to fill the gap of binarized image. Four morphological operator s are used for sharpen region 9dilation, erosion, opening, closing. In our research wok we used erosion. This is used to convert pixel into background pixels which is near to background pixels. After applying this objects become smaller. Mathematically representation is as follows:-

$$(A\theta B)(x) = \{x \in X, x = a + b: a \in A\ b \in B\}$$

A–Matrix of binary image
B–Mask MATLAB: = imerode (BW1, SE);
where SE is structuring element

Edge Detection reduces the unwanted data from image. Four types are used for edge detection (sobel, prewitt, canny, Roberts). We used prewitt edge detection algorithm for image processing as follows:-

MATLAB: BW1 = edge(gray, 'prewitt');

After that tumor is extracted using thresholding technique, which is very clear image.

5 RESULT AND DISCUSSION

In this section we are showing are result find by proposed technique:-

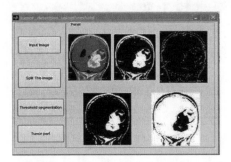

Figure 7. Raw image, binary, edge detection, morphological operator.

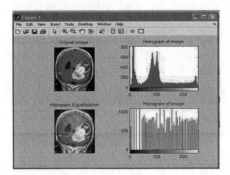

Figure 8. Histogram of image.

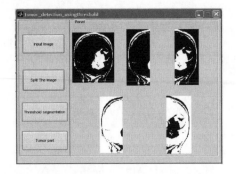

Figure 9. Split image.

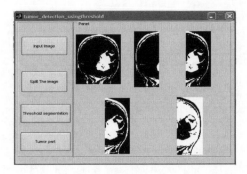

Figure 10. Split image with tumor.

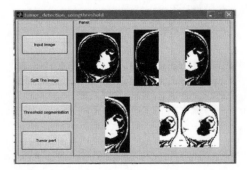

Figure 11. Image with tumor.

6 CONCLUSION

Segmentation techniques are used to detect brain tumor. Ii is very helpful in patient treatment. Brain tumor detection and classification is successfully implemented by using the image processing tool box in MAT Lab. We present an automated recognition system for the MRI image using the threshold and morphological operator. It is observed that results are improved than the individual technique. In image processing thresholding is very important technique. By this objects are extracted from background. It provides fast processing, smaller storage space, and easy to manipulate. The considerable iteration time and the accuracy level are improved as compared to individual technique.

REFERENCES

Aqhsa Q. Syed, K. Narayanan. "Detection of Tumor in MRI Images Using Artificial Neural Networks" IJAREEIE Vol. 3, Issue 9, September 2014.

Brain Tumor Detection Using Digital Image Segmentation" IJETTCS Volume 2, Issue 5, Sep-October 2013.

Dr Mohammad V. Malakooti, Seyed Ali Mousavi, and Dr Navid Hashemi Taba, "MRI Brain Image

Segmentation Using Combined Fuzzy Logic and Neural Networks for Tumor Detection", May 2013.

Farnaz Hoseyn, Siamak Haghipourb and Amirhoseyndaei Sorkhabic, "Improvement of Segmentation on MRI Image Using Fuzzy Clustering C-means and watershed Marker control Algorithm".

Gopal, N. Nandha, and M. Karnan. "Diagnose brain tumor through MRI using image processing clustering algorithms such as Fuzzy C Means along with intelligent optimization techniques." Computational Intelligence and Computing Research (ICCIC), 2010 IEEE International Conference on. IEEE, 2010.

Kamal Kant Hiran and Ruchi Doshi. "An Artificial Neural Network Approach for brain tumor detection using digital image segmentation." IJETTCS, Volume 2, Issue 5, September–October 2013.

Li, Gang, and Youchuan Wan. "Improved watershed segmentation with optimal scale based on ordered dither halftone and mutual information." Computer Science and Information Technology (ICCSIT), 2010 3rd IEEE International Conference on. Vol. 9. IEEE, 2010.

Minakshi Sharma and Saurabh Mukherjee. "Fuzzy c-means, anfis and genetic algorithm for segmenting astrocytoma-a type of brain tumor." IAES International Journal of Artificial Intelligence 3.1 (2014): 16.

Roshan G. Selkar, Prof. M. N. Thakare "Brain tumor detection and segmentation by using thresholding and watershed algorithm" IJAICT Volume 1, Issue 3, July 2014.

Siva Sankari S., Sindhu M, Sangeetha R., Shenbaga Rajan A. "Feature Extraction of Brain Tumor Using MRI", IJIRSET, Vol. 3, Issue 3, March 2014.

Viji, K.S.A.; JayaKumari, J. "Modified texture based region growing segmentation of MR brain images", Information & Communication Technologies (ICT), 2013 IEEE Conference.

Communication and Computing Systems – Prasad et al. (Eds)
© 2017 Taylor & Francis Group, London, ISBN 978-1-138-02952-1

FPGA implementation of low power & high speed accurate ECG signal detection using DABLMS algorithm

M.B. Dembrani, K.B. Khanchandani & Anita Zurani

SSGMCE, Shegaon, SGBAU Amravati, India

ABSTRACT: Biomedical Signals are such kind of signals which play a vital role in the analysis of the human health diagnosis. This Signal includes such as ECG Signal for analysis of heart rate, PCG signal for analysis of heart and blood, VAG signal for analysis of Knee Joint, EMG Signal for analysis of muscle fibers, EEG signal for analysis of brain activities. The paper analyzes the Distributed Arithmetic (DA) based Block LMS (BLMS) based on that we propose iteration of LUT sharing reduces the energy consumption and provides higher speed to detect the accurate ECG signal. The FPGA implementation is carried out on virtex board having 5 grade speed xc2vpx70-5ff1704. The result focuses on the low power and low memory is required for implementation which provides higher speed and accuracy.

1 INTRODUCTION

It is known that in real world there are different kinds of the signals. All these signals consist of Varieties of information which are ease to diagnosis. The number of different techniques is available for analyzing, interpreting, manipulating and processing this signal so as to obtain the faithful information from it. ECG signal analysis is the easiest way to check the condition of heart beat rate. The signals generated from human certainly carry the information about the human health activities. This information helps to understand the condition of human health where the processing of this signal is to be required to obtain the accurate health monitoring.

1.1 *Distributed Arithmetic (DA)*

Distributed Arithmetic is an efficient technique for calculation of inner product or Multiplies and Accumulates (MAC). The MAC operation is common in Digital Signal Processing Algorithms the direct method involves using dedicated multipliers. Multipliers are fast but they consume considerable hardware. DA is the bit level rearrangement of the inner products which reduces the computational requirements of multipliers.

The mathematical formulation for DA is given as

$$y = \sum_{K=1}^{k} A_k * x_k$$

where x_k is an N-bit scaled twos complement number

$$x_k = -b_{k0} + \sum_{n=1}^{N-1} b_{kn} 2^{-n}$$

By having the bit level rearrangement in the equation. The Final DA formulation can be obtains as

$$y = -\sum_{k=1}^{K} A_k *(b_{k0}) + \sum_{n=1}^{N-1} \left[\sum_{k=1}^{K} A_k *b_{kn} \right] 2^{-n}$$

This final equation shows that how the bit level rearrangement reduces the computational mathematical operation thereby increasing the speed of performances.

1.2 *Block LMS (BLMS)*

BLMS algorithm permits processing of data over a wide range at a processor complexity and cost as low as that of a fixed-point processor. BLMS algorithm is based on appropriate representation of the filter coefficients and the data. The filter coefficients and the filter output can be evaluated to obtain the faster processing.

The mathematical formulation for BLMS is given as

For updating of filter weights at k^{th} iteration in block LMS filter is

$$w_{k+1} = w_k + \Delta w_k$$

where Δw_k is given as

$$\Delta w_k = \mu X_k^T * e^k$$

e_k and w_k are error vector and weight vector for k^{th} iteration it is given as

$$e_k = \left[e(kL), e(kL-1), ..., e\left(L(k-1)+1\right)\right]^T$$

$$w_k = \left[w_k(0), w_k(1), ..., w_k(N-1)\right]^T$$

where μ is the step size and X_k is the input matrix which is derived from the current input block $\left[x(kL), x(kL-1), ..., x(kL-L+1)\right]$ of length L and (N − 1) past samples is given by

$$\begin{bmatrix} x(kL) & x(kL-1)\cdots\cdots & x(kL-N+1) \\ x(kL-1) & x(kL-2)\cdots\cdots & x(kL-N) \\ \vdots & \vdots & \vdots \\ x(kL-N+1) & x(kL-L)\cdots\cdots & x(kL-L-N+2) \end{bmatrix}$$

The error vector is calculated as

$$e_k = d_k - y_k$$

where the desired response vector d_k is given as

$$d_k = \left[d(kL), d(kL-1), ..., d\left(L(k-1)+1\right)\right]^T$$

The final iteration k^{th} block of filter output y_k is calculated by matrix vector product as

$$y_k = X_k * w_k$$

1.3 Distributed Arithmetic Block LMS (DABLMS)

DA techniques require less multipliers which reduce the hardware requirement thereby reduces the power requirement. Thus the computational analysis is reduces which increase the higher

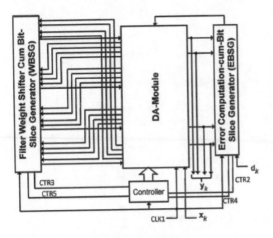

Figure 1. DA based BLMS architecture.

performance and higher efficiency is obtained using DA. BLMS adaptive filter reduces the iteration period, which reduces the hardware resources requirement. As hardware resources are reduces it also reduces the energy consumption for the same.

The figure shows the block diagram for the DA based BLMS adaptive filter. Since if we combine both the DA and BLMS it increase the speed, reduces the power consumption which gives higher accuracy for the detection of the signal. in this paper we have considered in ECG signal. this technique can be applied to different medical, communication application which can widely increase the performance as per the application.

2 LITERATURE SURVEY

The signal processing involves algorithms implementation to analyze the biomedical signals. Since there are many problems in acquisition, analysis and processing of biomedical signals, this paper (Rangayyan R. M. 2002) resembles on this various difficulties approach. The number of different methods is available to solve this problem. Segmentation method is use to solve nonstationary problem and parametric modeling method is used to solve the stochastic problem. The segmentation is efficient method to analyze with nonstationary signals (Rangayyan R. M. 2002) (Ahmadi B. et al. 2008) (Moussavi Z. M. K. et al. 1996) and parametric modeling method to analyze with random time series (Rangayyan R. M. et al. 1997) (Tavathia S. et al. 1992) (Akay M. et al. 1990).

FPGA (Field Programmable Gate Array) technology is developed significantly and obtained more preference due to the functional advantage of FPGA. FPGA application is widely spread in different areas such as digital signal processing, computer vision, ASIC prototyping, aerospace, bioinformatics, medical imaging. EDA tools are flexible and highly efficient design methodology applied C to FPGA (Moussavi Z. M. K. et al. 1996), Matlab to FPGA (Camera K. 2002) (Haldar M. et al. 2001), State flow to FPGA (Diaby M. et al. 2004), Simulink to FPGA (Haldar M. et al. 2001) (Sun X. et al. 2006). This paper focuses on Simulink to FPGA design is used to implement two biomedical algorithms for autoregressive modeling and adaptive segmentation algorithms. The Simulink to FPGA is the combination of Xilinx system generator and implementation tools to implement the design in graphical and flexible.

3 IMPLEMENTATION AND RESULTS

In this paper DA based BLMS adaptive filter is presented for medical application for ECG signal monitoring. The filter gets the input ECG signal from the database having a finite length. This signal is added with the noise signal and applied to the input to filter to recover the original signal. The signal is converted into values using the Matlab. The response of the filter signal is obtained for various normal, abnormal and severe conditions. Based on it the performance measure response of the proposed method is calculated.

The Figures 2 and 3 shows the input signal for two conditions normal and severe conditions of the ECG signal respectively. These signals are applied DA based BLMS adaptive filter for the exact analysis of the ECG signal. The response of proposed

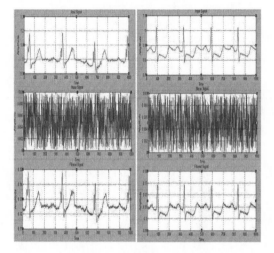

Figure 2. Normal condition a) ECG Signal b) Noise signal c) Filtered signal.

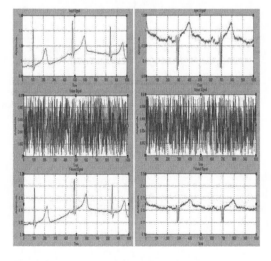

Figure 3. Severe condition a) ECG Signal b) Noise signal c) Filtered signal.

method provides the various results to obtain the physical condition of the health. The performance measures shows the result for Accuracy, Specificity, Sensitivity, false negative rate FNR, false discovery rate FDR, detection error rate DER.

3.1 Performance measures

To Evaluate the Performance of the proposed algorithm, several terms are considered as FP (False Positive) which means false heart beat detection and FN (False Negative) which means failed to detect true heart beat rate.

3.1.1 Accuracy

Accuracy is used as a statistical measure of how a classifier and filtering techniques identifies the condition. The accuracy is the proportion of true results both true positives and true negatives among the total number of cases examined.

$$Accuracy = \frac{TP + TN}{TP + FN + TN + FP} \times 100\%$$

3.1.2 Specificity

Specificity is related to the ECG signal condition is normal (no disease). High Specificity shows that the Monitoring System obtains the Normal Condition as Normal.

$$Specificity = \frac{TN}{TN + FP} \times 100\%$$

3.1.3 Sensitivity

Sensitivity is related to the ECG signal condition is abnormal (disease). High Sensitivity shows that the Monitoring System obtains the Abnormal Condition as Abnormal.

$$Sensitivity = \frac{TP}{TP + FN} \times 100\%$$

3.1.4 False Negative Rate

The false negative rate is the proportion of events that are being tested for which yield negative test outcomes with the test, i.e., the conditional probability of a negative test result given that the event being looked for has taken place.

$$False\ Negative\ Rate = FN/(TP + FN) = 1 - TPR$$

3.1.5 False Discovery Rate

The False Discovery Rate (FDR) is one way of conceptualizing the rate of type I errors in null hypothesis testing when conducting multiple comparisons. FDR-controlling procedures are designed to control the expected proportion of rejected null hypotheses that were incorrect rejections.

$$False\ Discovery\ Rate = FP / (TP + FP) = 1 - PPV$$

3.1.6 *Detection Error Rate*

Detection error rate gives the error response of the system where it detects the error in calculating the heart beat rate

$$Detection\ Error\ Rate = FP + FN / CompleteQRS$$

4 RESOURCE UTILIZATION AND FPGA IMPLEMENTATION

The Table 1 shows the minimum period and frequency requirement for implementation of the design on virtex 2pro. The design is implemented on virtex 2p board with a grade speed xc2vpx70-5ff1704. Alongwith the detail timing analysis results are also shown which give the time required for the performance after and before the clock signal. It also gives the combinational path delay time.

The Device Utilization Summary includes various logic utilization factors obtained by exploring the Synthesis report of the Xilinx ISE tool. Device utilization summary provides complete

Table 1. Time utilization for implementation on Virtex board.

Parameter	Estimated values
Frequency	25.643 MHz
Minimum period	38.997 ns
Minimum input arrival time before clock	30.302 ns
Maximum output required time after clock	17.144 ns
Maximum combinational path delay	8.499 ns

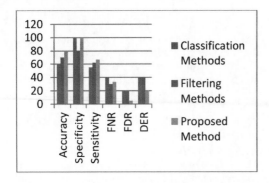

Figure 4. Graphical representation of the performance measures.

information of 4 input LUTs, Slices, IOBs, MULT18X18SIOs, GCLKs utilization. The graphical representation of the synthesis report is shown.

Since the utilization of synthesis is under the requirement utilization means not overflowed we can implement the design on board and generate obtain the complete design consisting of RTL schematic and complete place and route ant generation of files.

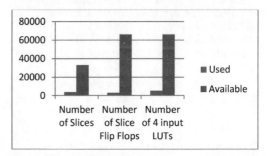

Figure 5. Graphical representation of 4 input LUTs, Slices & Slice flip flops synthesis on virtex board.

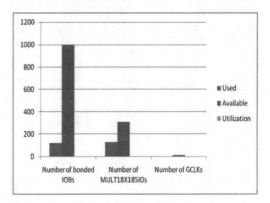

Figure 6. Graphical representation of IOBs, MULT18X18SIOs, & GCLKs synthesis on virtex board.

Figure 7a. Internal view of synthesis results at Register Transfer Level (RTL).

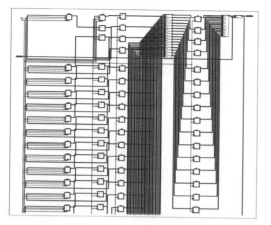

Figure 7b. Internal flow of the placing and routing on the board.

The Figure 7a and b shows the RTL Schematic of the DA based BLMS adaptive filter obtained using Xilinx ISE tool. The figure also shows the complete internal flow of the placing and routing on the board.

5 CONCLUSION

The DA based BLMS algorithm is presented that ensures simple Functional and statistical operations in most of the computations while maintaining wide dynamic range via a block exponent. Overflow by a new upper bond in the step size should be prevented. A analysis of DA based BLMS adaptive filter to reduce the LUT size. The block sizes can be varied which reduces the LUT size leads to the saving of LUT, respectively. FPGA implementation on virtex board show s the utilization of resources and performance measures for the accurate detection of the ECG Signal A faster realization of the proposed scheme can be extended by having suitable modification of the FFT-based FBLMS algorithm can be used.

REFERENCES

Ahmadi B., Aimrfattahi R., Negahbani E., Mansouri M., and Taheri M., 2008, Comparison of adaptive and fixed segmentation in different calculation methods of electroencephalogram time series entropy of estimating depth of anesthesia, 6th International special topic conference on TAB, Tokyo, pp. 265–268.

Akay M., Semmlow J. L., Welkowitz W., Bauer M. D., and Kostis J. B., 1990, Detection of Coronary occlusions using autoregressive modeling of diastolic heart sounds, IEEE transactions on biomedical engineering, vol. 37, pp. 366–373.

Camera K., 2001, SF2VHD: A stateflow to VHDL translator, Master thesis, UC Berkeley.

Diaby M., Tuna M., Desbarbieux J., and Wajsburt F., 2004, High level synthesis methodology from C to FPGA used for a network protocol communication, 15th IEEE International Workshop on Rapid System Prototyping, pp. 103–108.

Haldar M., Nayak A., Chaudhary A., and Banerjee P., 2001, A system for synthesizing optimized FPGA hardware from Matlab, IEEE/ACM International Conference on Computer Aided Design, pp. 314–319.

Haldar M., Nayak A., Chaudhary A., and Banerjee P., 2001, Automated synthesis of pipelined designs on FPGAs for signal and image processing applications described in MATLAB, Proceedings of the ASP-DAC 2001. Asia and South Pacific Design Automation Conference, 2001, pp. 645–648.

Moussavi Z. M. K., Rangayyan R. M., Bell G. D., Frank C. B., Ladly K. O. and Zhang Y. T., 1996, Screening of vibroathrographic signals via adaptive segmentation and linear prediction modeling, IEEE transactions on biomedical engineering, vol. 43, pp. 15–23.

Rangayyan R. M., 2002, Biomedical signal analysis: a case-study approach. New York, N.Y. Wiley-Interscience.

Rangayyan R. M., Krishnan S., Bell G. D., Frank C. B. and Ladly K. O.,1997, Parametric representation and Screening of knee joint vibroarthrographic signals, IEEE Transactions on biomedical engineering, vol. 44, pp. 1068–1074.

Sun X. Li, F. and Wu E., 2006, A Simulink-to-FPGA Co-Design of encryption module, IEEE Asia Pacific Conference on Circuits and Systems, 2006, pp. 2008–2011.

Tavathia S., Rangayyan R., Frank C., Bell G., Ladly K. and Zhang Y., 1992, Analysis of knee vibration signals using linear prediction, IEEE transactions on biomedical engineering, vol. 39, no. 9, pp. 959–970.

Communication and Computing Systems – Prasad et al. (Eds)
© 2017 Taylor & Francis Group, London, ISBN 978-1-138-02952-1

Robust vehicle detection and localization using normalized color illumination

Sandeep Singh & Bikrampal Kaur
CGC, Landran, Mohali, Punjab, India

ABSTRACT: The vehicle detection and license plate localization are the techniques to extract the number from the number plate image containing a multiple vehicles on the road. The traffic surveillance cameras are utilized to obtain the vehicular image data from the roads in the multiple lanes and the vehicles travelling in both directions. The traffic surveillance includes the various vehicle object localization and number plate localization. In this paper, the optimized solution has been proposed to improve the number plate localization accuracy over the traffic surveillance data obtained from the various sources. The proposed model is designed along multiple factor analysis in a sequence and extract a vehicle object in the day and night images.

1 INTRODUCTION

It basically defines the significance of number plate Recognition system. It also defines the structures of Number Plate Recognition System. It defines that what has been described for the type it inherits since its development in the recent years. With increase number of vehicles on route, it is unmanageable manually for control the traffic flow. Numbers of toll-plaza are manufactured on routes & in some parking areas, at there we stop a vehicle and give some fees for safe parking, and also check the speed of vehicles etc. All these problems having a solution to improve this. To improve such a scheme, very important to discover the particular vehicle. The Question is that can we realize the any vehicle? And its simple Answer by Number Plate or License Plate.

In each Country, every vehicle has their own number plate. That License Plate Number distinguish it from others i.e useful to identify if two of these are same make and model. By extracting Number Plate from vehicle it is easy to collect the information about the owner of vehicle & by using information model should be little in size, portability form and ability to access the data at particular rate. Simply in this we use binary process concept i.e 0 and 1. One (1) is for white where zero (0) is for Black. The segmentation process used to simply extract the particular character or number from the Number Plate.

2 RELATED STUDY

Jingyu Dun and Sanyuan Zhang et al. (2015) proposed a system called China license plate localization at multilane along difficult background based on attendant colors. A color plays very important role on number plate. In this system we mainly follows blue and yellow plates. It follow the basics steps as described below-First to convert the color image into grayscale form and then highlight the yellowish and bluish areas of the actual image, and at the end using color attendant property and transition between the backdrop and words, duplicate plates will be abstracted and original plates will be restrained. By using this system the chances of accuracy to get good results and more & we can remove unoriginal plates by using kinship b/w width and size of words on number plate. This system use two basic techniques-globally. Threshold technique and adaptive block threshold technique. This algorithm decide which method has to be used by it. This algorithm with difficult backdrop as on images along uneven illumination. In this we test the proposed technique having approx 1315 daytime images and approx 894 nighttime and (ALR) average location rate is 96.0% and it surpasses the other techniques.

Hitesh Rajput and Tanmoy Som et al. (2015) proposed a model which overcome the limitations of existing model. In this we follow mainly 3 steps first capture the image, second to detect the characters and third is segmentation i.e Recognition of characters. In this, we follow the processing-Binarized image processing, Gray level processing and third one is colored processing. In this we take different-different models of vehicles having different structures, sizes and colors of plate under different circumstances it works & give accuracy about 87% result.

Abdul Mutholib and Teddy Surya Gunawan et al. (2012) proposed a system using android application which works on different steps. Firstly capture the image using good quality camera, which may capture the image static or dynamic form then to recognize the characters using Neural Network Algorithm and atlast using contrast enhancement, filtering and straightening recognition of characters takes place. It gives result 88% accurate and store the data i.e characters on number plate online in database. This system is used for Malaysia cars in that country, The model is figure in IDE eclipse and SDK android using API7 for realization of plates.

Kamalaruban P, Ranga Rodrigo et al. (2013) proposed an automatically number plate pickout system that uses settlement camera to acquire (detect) the better caliber images to pass through it. In this 3 steps are mainly follows, first to localized the plates second characters are segmented and last the recognition of characters to identify the vehicle. This system produces very less accurate result due to the quality of images. In this we follow the two techniques to improve the result as well as to improve the quality of image. These techniques are image processing and domain knowledge technique. Best thing of this system is to operate in real time. In this we get 90% accurate result to detect the image from the car and results as well. Apart from these techniques another algorithm also used for video taking with regular cameras. These algorithms are—Viola Jones and Kalman filter. Another algorithm used i.e SVM also use for multiple detection of plates for character recognition.

Jian Yang, Bin Hu et al. (2013) proposed a system called LPR license plate realize based on machinary vision which is popularly used in smart parking system & ITS intelligent Transportation System. The main steps followed by this system is to detect the image, segmentation of characters and to recognize the characters. For better result we improve the effect on binarized images processing technique. In this we solved distinctive features about character segmentation & for character realize to amend the condition or truth. At last we only adopted image realizable to realize the eccentric(characters). To clear the result we take a plate having blue background & white characters on it. First we enlarge the number plate area, which decrease the difficulty of localization and increase of pixels happen while we do enlarge. For this we use about 97 samples of plate from which the number of accurate segmentation gives 73 and number of precise recognition give 64 from 97. Christos-Nikolaos E. Anagnostopoulos et al. proposed, A license plate realizable system which use an algorithm to detect the images or still images or video. It follows the following 3 steps—first to extract of license plate of a vehicle second the segmentation of plate characters of a vehicle and third the recognition of character one by one. This task is difficult sometimes because of using different background or the stylish form of characters or color illumination on characters due to certain conditions. Number of techniques are used to get the results accurate from still image or videos the main purpose of this system is to categrise or access the result. It overcomes such problem like Accessing time, computation ability and recognizable rank. This supports mainly the system performance valuation in the scientific profession worldwide & allows the system developers to know what techniques are helpful in particular application.

3 COMPARISON TABLE

Author publication and year	Problem addressed	Techniques proposed	Merits	Demerits
Jingyu Dun et al., IEEE, 2012	To situate the license plates in multi-lane with complicated background	Global threshold method & adaptive block threshold method	This algorithm is used to made a full use of the color attendant property & Modulation b/w the characters and backdrop.	One problem is that the bounding box of plate is a much larger, so it is not so precise.
Hitesh Rajput and Tanmoy Som, et al., WWW.COMPUTER. ORG/COMPUTER, 2015	To keyout vehicles and gives a denotation for further Vehicle in tracking and activity analysation.	Use formula matching or erudition-based classification algorithm.	Method which gives higher accuracy and indicate technique's effectiveness in real-life uses.	It is focus on single-level wavelet translate, and have proven it to be efficient.

Abdul Mutholib, Teddy Surya Gunawan et al., (ICCCE 2012)	To draw out the numbers or alphabets from number plate.	ANPR algorithm, Android SDK, Neural network.	System is made for keyout Malaysia number plates & the system is checked for multiple images and gives accuracy 84.1% for the partition of the characters & 92.1% for the realizable unit, giving an complete system performance of 89% realizable rank.	An algorithm or techniques used in it work for only Malaysian number plates only.
Kamalaruban P., Ranga Rodrigo and Ajanthan T., et al. IEEE, 2013.	High declaration cameras to get good caliber images of the vehicles passing from.	Viola-Jones for number plate localizable with the Kalman filter.	Getting 90.1% realizable rate in data set that is unbelievable for us.	Equipment cost is more.
Jian Yang, Bin Hu et al. IEEE, 2013.	To draw out the preciseness of the license plate position.	Template matching method	Sweep up the image recognition to realize the words, along not significant vulnerability.	The number plate must be bluish with backdrop and white words. Also image should atleast in the object area.
Christos-Nikolaos E. Anagnostopoulos, Member et al. IEEE, 2008	To improve the variety of plate formatting and the non uniform outside illumination situations during visualize accusive.	Morphological operation called "top-hat, and edge statistics and morphology	LPR may also hold intelligence vehicle manageable.	Problems such as accessing time, computation ability, and recognition rate are also addressed.

4 SYSTEM DESIGN

In this research, at very initial stage a detailed literature survey is being carried out with which all the earlier described methods analyzed in detail and an algorithm with better result can be developed. In this, first an image is captured, it may be colored also or we can say complex background, firstly convert this color background into gray color or then into binary i.e. 0 and 1. This can be used as double fashion purpose. First we can find or see the size & location of the character, it can be done using Morphological Algorithm and second Adaptive color illumination with boundary detection, in this the color histogram, histogram equilization, Canny edge and Sobel edge detection technique are used. More than two or more frames are detected per second. At last performance of proposed Algorithm is measured on some Parameters like location rate, false positive rate, elapsed time, accuracy, & Precision. Automatic vehicle designation is an necessary stage in Intelligence traffic system. As we know that now a days, a huge traffic is present on roads everywhere, to control

such a flow, we design a system to control these traffic problems. Automatic License plate system helps for this effective controls. This system use image processing techniques to detect the plates from the vehicle. This system maintains some laws for public on roads also. Uniquely every vehicle has their own personal number plate, no other thing is used by which it is recognize by the system. In the existing model, the complex background based on concomitant colors method has been realized for the vehicle detection in the multi-lane system.

Initial step to that side of research is the detail study of the already existing algorithms for Automatic License Plate Recognition Systems (ALPR). It comprises of Segmentation (Number Plates Extraction), Optical Characters Recognizable (OCR) from the extracted number plate and Decision Logic (Process Successful or Failed). Literature study will help us for the development of an ALPR system. The Project would be carried out in MATLAB Simulator. A proper performance will check including its security, if any failure occurs, first to detect it and then to recover that failure. N number of Algorithms were developed in Previous

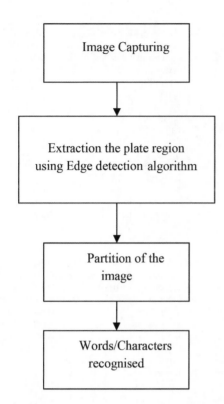

Image Capturing

↓

Extraction the plate region
using Edge detection algorithm

↓

Partition of the
image

↓

Words/Characters
recognised

Flowchart 1. Working of traditional number plate recognition systems.

few years. Every algorithm has their own pros and cons. Arth *et al.* (1) describes a method in which license plate is detected very confidentially. As we know we can detect this multiple times detection for a single plate, a model is used called Pre-Processing to combine all these detected regions, with the help of trackers, a particular area can be detected in the image. Kwasnicka *et al.* (2) gives a method for detecting the unuseful region from an image. In this approach, Firstly image is detected then convert it into binarized form based on contrast between words and backdrops in license plate. When binarized the image happen, it is divided into 2 regions black and white n(0 and 1). Based on these two colors, actual output take place.

Algorithm 1: Neural network based number plate character recognition method

1. Load the input test image from given dataset
2. Load the training data into the training matrix
3. Find the ROI containing the number plate region with morphological operations
4. Calculate the rationale component analyzation over training image
5. Calculate the rationale portion analysation over test image

6. Compute the total offset from the PCA feature descriptor for training and testing data.
7. Apply feature preprocessing over the Test and Training image features
8. Compute the early stage similarity between the testing and training samples and return the early pruning result matrix
9. Perform fuzzy logic method to early eliminate the non-matching training elements than the minimum matching threshold
10. Run the neural network feed forward back propagation classification
11. Return the matching sample according to decision logic
12. Return the matching text samples

5 RESULT ANALYSIS

In this research project, the work has been performed using the proposed model for the character recognition in the natural scene. The natural scene text extraction procedure has used several morphological methods amalgamated with the principle component analysis for character recognition. The text region is extracted using the proposed model's centrally weighted morphological operations. The natural scene text recognition can assist the vehicle drivers, and make their ride easy and informative to read all of the traffic messages flashed on the road. This can provide the critical information to the drivers and hence can assist them to avoid the traffic hours, road blockage due to accidental hazards, etc.

We have changed over the picture into grayscale from its hued particular. At that point we have utilized middle channel to de-clamor the auto picture in the event of any commotion created in the picture securing procedure. Subsequently, we have

Figure 1. The figure from the dataset for number plate recognition.

Table 1. Showing performance of proposed system on per sample basis.

File index	Total characters	Correctly detected characters	Character level accuracy in percentage
1	10	10	100
2	19	16	84.21
3	38	25	65.78
4	15	15	100
5	4	4	100
6	9	1	11.11
7	13	8	61.53
8	13	13	100
9	16	10	62.5
10	9	9	100
11	12	12	100
12	10	10	100
13	10	0	0
14	8	8	100
15	5	0	0
16	5	5	100
17	8	4	50
18	16	8	50
19	5	4	80
20	4	4	100
Average accuracy			73.25

Table 2. Overall performance parameters of proposed system.

Overall character level accuracy mean	Total characters	Correctly detected characters
73.25%	229	166

utilized picture widen morphological strategy to improve the splendor of the picture to highlight the number plate district. Later, we have utilized article sketching out determination strategy with the end goal of extraction of the different protests in the picture. In the last step, we have utilized the number plate extraction area utilizing the most noticeable articles chose in the recent step and utilizing it with littler article evacuation strategy for morphological picture handling.

The proposed calculation is intended to give back the slipped by time per revolution and aggregate reaction time for all pivots. The precision of the project is figured on the premise of sort 1 and 2 factual mistakes. Sort 1 & 2 factual slips incorporate the genuine positive, genuine negative, false positive and false negative qualities. On the premise of the recent parameters, specificity and affectability are likewise figured.

The above results table has demonstrated aggregate characters, effectively recognized characters, character level precision rate. The character recognition in the scene text model has been evaluated in the details in this result analysis module.

6 CONCLUSION

In this paper, the proposed model has been designed in order to overcome the shortcomings of the existing models. The proposed model has been aimed at solving the problem of false detection or no detection cases over the vehicular image obtained from the multiple lanes camera. The proposed model has been designed to improve the issue of uneven color illumination in order to improve the number plate region localization. The existing model has been designed to work on the basis of the uneven illumination for the purpose of multiple number plate detection on the multiple vehicles captured in the single frame. In each Country, every vehicle has their own number plate. That License Plate Number distinguish it from others i.e useful to identify if two of these are same make and model. By extracting Number Plate from vehicle it is easy to collect the information about the owner of vehicle & by using information model should be little in size, portability form and ability to access the data at particular rate. Simply in this we use binary process concept i.e 0 and 1. One (1) is for white where zero (0) is for Black. The segmentation process used to simply extract the particular character or number from the Number Plate.

The existing model has been designed to work in the multiple rounds to compute the number plate region and then uses the various mathematical and morphological operations to extract the vehicle number on the detected license plate. The location rate of almost 94% has been recorded on an average for the detection of vehicles during day and night. But the existing model is not very capable in the recognizing the number plate regions where it posts false positive rate of almost 48% during day and 35% during the night. Also the elapsed time of the existing model is little higher as it takes nearly 1 second to process the one frame, where it can be applicable over the camera capturing more than 1 frame every second, which is generously required to achieve the best of the recognition.

REFERENCES

Benjapa Ratchatasriprasert, Kittawee Kongpan, Paruhat Punyarprateep, "License Plate detection Based on Template Matching Algorithm", vol. 1, pp. 139–143, ICCCT, 2012.

Ching-Liang Su, "Car Plate recognition by whole 2-D image", Expert Systems with Applications, vol. 38, pp. 7195–7200, Science Direct, 2011.

Clemens Arth, Florian Limberger and Horst Bischof, "Real-Time License Plate Recognition on an Embedded DSP-Platform", Proceedings of IEEE conference on Computer Vision and Pattern Recognition, pp. 1–8, June 2007.

Dening Jiang, Tulu Muluneh Mekonnen, Tiruneh Embiale Merkebu, Ashenafi Gebrehiwot, "Car Plate Recognition System", ICINIS, vol. 1, pp. 9–12, IEEE, 2012.

Du, Shan, Mohammad Ibrahim, Mohamed Shehata, and WaelBadawy. "Automatic license plate recognition (ALPR): A state-of-the-art review." Circuits and Systems for Video Technology, IEEE Transactions on 23, no. 2 (2013): 311–325.

Dun, Jingyu, Sanyuan Zhang, Xiuzi Ye, and Yin Zhang. "Chinese License Plate Localization in Multi-Lane with Complex Background Based on Concomitant Colors." Intelligent Transportation Systems Magazine, IEEE 7, no. 3 (2015): 51–61.

Halina Kwasnicka and Bartosz Wawrzyniak, "License plate localization and recognition in camera pictures", AI-METH 2002, November 13–15, 2002.

Hsu, Gee-Sern, Jiun-Chang Chen, and Yu-Zu Chung. "Application-oriented license plate recognition." Vehicular Technology, IEEE Transactions on 62, no. 2 (2013): 552–561.

Janowski, Lucjan, Piotr Kozłowski, Remigiusz Baran, Piotr Romaniak, Andrzej Glowacz, and Tomasz Rusc. "Quality assessment for a visual and automatic license plate recognition." Multimedia Tools and Applications 68, no. 1 (2014): 23–40.

Kaur, ErKavneet, and Vijay Kumar Banga. "Number Plate Recognition Using OCR Technique." International Journal of Research in Engineering and Technology 2, no. 09 (2013).

Massoud, M. A., M. Sabee, M. Gergais, and R. Bakhit. "Automated new license plate recognition in Egypt." Alexandria Engineering Journal 52, no. 3 (2013): 319–326.

Patel, Chirag, Dipti Shah, and Atul Patel. "Automatic number plate recognition system (anpr): A survey." International Journal of Computer Applications 69, no. 9 (2013).

Pei-Chen Tseng, Jiun-KueiShiung, Chun-Ting Huang, Shih-Mine Guo, Wen-Shyang Hwang, "Adaptive Car Plate Recognition in QoS-Aware Security Network", SSIRI, vol. 1, pp. 120–127, IEEE, 2008.

Ping Wang, Wei Zhang, "Research and Realization of Improved Pattern Matching in License Plate Recognition", ISIITAW, vol. 1, pp. 1089–1092, IEEE, 2008.

Rajput, Hitesh, TanmoySom, and SoumitraKar. "An Automated Vehicle License Plate Recognition System." Computer 8 (2015): 56–61.

Ratree Juntanasub, Nidapan Sureerattanan, "Car License Plate Recognition through Hausdorff Distance Technique", IICTAI, vol. 1, pp. 645–651, IEEE, 2005.

Roy, Sourav, AmitavaChoudhury, and Joydeep Mukherjee. "An approach towards detection of indian number plate from vehicle." International Journal of Innovative Technology and Exploring Engineering (IJITEE) 2, no. 4 (2013): 241–244.

Sarfraz, M. Saquib, AtifShahzad, Muhammad A. Elahi, Muhammad Fraz, Iffat Zafar, and Eran A. Edirisinghe. "Real-time automatic license plate recognition for CCTV forensic applications." Journal of real-time image processing 8, no. 3 (2013): 285–295.

Setumin, U.U. Sheikh, S.A.R Abu-Bakar, "Car Plate Character Extraction and Recognition using Stroke Analysis", SITIBS, vol. 1, pp. 30–34, IEEE, 2010.

Communication and Computing Systems – Prasad et al. (Eds)
© 2017 Taylor & Francis Group, London, ISBN 978-1-138-02952-1

Digital image forgery detection using the wavelet decomposition and low level feature points

Pahulpreet Kaur & Bikrampal Kaur
CGC, Landran, Mohali, Punjab, India

ABSTRACT: The image copyright protection is the important issue to protect the digital ownership of the image or video data. The copyright bypassing by using the several data hijacking techniques, such as copy-move forgery, splicing forgery, etc, are utilized primarily to forge the copyright of the digital data. The object are moved, cloned or the new objects are inserted in the image matrix to edit the original image. The image forgery detection techniques utilize the various color based or low-level feature based methods to detect the digital image forgery. The image forgery detection techniques must be capable of detecting the objects on the whole image by generating the local feature vectors from everywhere in the image. In this paper, the robust low level image forgery detection method has been proposed by using the low-level feature points over the decomposed component with the wavelet decomposition. The proposed model design is expected to improve the forgery detection results of the existing models by improving its ability to detect the higher level of feature.

1 INTRODUCTION

Image forgery detection is the technique of developing or matching objects with the objective to receive the for the sake of transformed or to make some profit. Concept of forgery is necessarily related with a distributed or exercised object. In this technically leading world, capturing image instructions is very common. Instructions of an image are captured in form of digitalized picture of forgery detection, it completely emphasize on the authenticity and originality of an image. There are plenty of image editing software's that can modify an image according to its own malicious intentions. Such software's are like MS-Paint, Photoshop etc. Digitalized images plays vital role in our daily life. These are being used for capturing the information that is used in many fields like magazine, newspaper, medical etc. So the authenticity to be achieved is must.

Forgery is of two types i.e. copy move is one in which area of an image is copied from a particular point or place and is pasted on the other point of similar image and other one is image splicing forgery also known as Image splicing. This is very similar to Copy move. The only difference lies in that here one of the region of an image is copied and pasted into some another distinct image.

The two basic approaches for forgery detection are

i. Active approach
ii. Passive approach.

In active approach a small piece of information is interspersed into digital signatures. This is also titled as Reference based or intrusive technique. The Passive approach discovers the passive ways to attest the integrity of digital images. Also Entitled as Non-intrusive or blind approach. This document describes the introduction and related work about the image forgery detection techniques. The introduction section describes the basic idea of the image forgery detection along with its popular methods. The financial losses have also been discussed due the image forgery based copyright material tampering and hijacking.

Nowadays, the era of the technically advanced world has been reached, where capturing pictorial information of any event in the form of digital images has become very simple. Currently, digital images play significant role in the everyday life, where they are being used as means for capturing pictorial information and are being employed in various domains such as medical diagnosis, daily newspapers, magazines, and as the evidence at court or for insurance claims. Because of the widespread applications of digital images, very powerful and easy-to-use image editing tools like Photoshop are available. Using these tools even a novice can alter the digital contents of a digital image without leaving any visible traces, which can be noticed by human eyes. The digital contents are often altered with illicit designs in mind by hiding or adding important information to an image. Therefore, the authenticity of digital images cannot be taken for granted, it needs verification and is an object for research. Copy-Move Forgery (CMF) is the most common type of image forgery; in this case one region is copied from one place and pasted to another place of the same image in order to conceal important information. Sometimes, the copied region is modified by pre-processing operations like scaling, rotation, adding noise, etc. to make it matching with the surrounding region so that the tempering is not visible. In another similar kind of forgery, a part is copied from one image and is pasted to a different image. This type of forgery is called image splicing.

Authenticating digital images is a very serious issue and so far the researchers developed many methods, which can mainly be classified into (1) intrusive (active) and (2) non-intrusive (blind or passive) techniques. Further, intrusive methods can be divided into two classes based on (1) embedding a watermark and (2) incorporating digital signature in an image. In each of these techniques, a piece of information is integrated into digital images as an aid for authenticating digital contents and security rights. Once the digital contents of an image are changed, the incorporated information is also modified. The authenticity of an image is validated by ensuring that the embedded information is unaltered. Though these methods are robust, their domain of application is restricted because all digital cameras are not equipped with the feature of embedding digital signature. These limitations and constraints of active methods motivated the research to propose non-intrusive methods for authenticating digital images. This class of methods do not take into consideration any kind of embedded information (such as watermarks or signatures) to validate the authenticity of a digital image. Instead, these methods draw their conclusions about the originality of the digital content of images using its structural changes, which take place due to tempering.

2 RELATED STUDY

Jian Li, Xialong Li Proposed a model for the detection of copy-move forgery in the images by obtaining some major key points for their comparison. It is typically different from the already existing methods, because in this firstly the image is segmented to the semantically independent patches and giving priority to its key point extraction. In this matching is done in two stages: first, the pairs of mistrustful patches are matched then second, the E-M based algorithm is proposed to affirm the presence of copy-move forgery and it produces a better results with better performance.

Abhishek Kashyap, B. suresh described the concept of detection of image splicing forgery. As the Authentication of an image is a major issue, so it is important to secure the images, a powerful methods or models are required to detect forgery from the damaged images. In this paper the authors has proposed a new model which uses the method of block matching and gives the accuracy of 87.5%.

Andrea Costanzo, Irene Amerini has worked upon the attacks of the removal of the SIFT, key points in an image, with which one has to compromise with the correctness of the image. In this paper, three novel forensic detectors have been followed to recognize the image. Firstly, the validation of methods is done, second their robustness is assessed and then the detectors are applied and gives a better efficiency and exposure to the new upcoming tools.

David Cozzolino, Giovanni Poggi has proposed a modified and unique model for the correct detection and place of copy-move forgery. In this paper the detection is based on rotation invariant features, which has been computed densely on an image, which guarantees the remarkable performance and this has been done by analyzing on the online available databases and gives very accurate, robust and faster technique.

Chi-Man Pun, seni proposed an integration of booth key-point based and block based forgery detection methods. Firstly, the segmentation of the host image is done then all its feature points has been extracted from blocks as features of block. For accurate results the forgery region extraction algorithm is used, which gives the correct results.

E. Ardizzone, A. Bruno has given the concept of copy-move forgery detection by matching triangles of key points, it is done in case of alteration of digital images. In this paper a hybrid approach is followed in which there is a comparison between triangles is done in single points. These methods are being proposed for the robust geometric transformations and produces better and accurate results.

3   COMPARISON TABLE

Author, publication and year	Problem addressed	Techniques proposed	Merits	Demerits
Jian Li et al., IEE, 2014	To detect the copy-move forgery in an image.	Matching process and E-M based algorithm.	This algorithm segments the test image and gives prior to the key point extraction, leads to good performance.	Computational complexity of the combination of matching and algorithm is very high and detection speed is low.
Abhishek kashyap, et al., ICCCA, 2015	Detection of splicing-forgery in an image.	Wavelet decomposition and block matching.	Efficiency is increased with block matching and gives up to 87.75% accuracy.	This doesn't works on the larger sized images and cannot find all the little discrepancies and cannot work on scaled or rotated spliced images.
Andrea Costanzo et al., IEEE, 2014.	Analysis of SIFT key point removal and injection.	Three novel forensic detectors.	Detectors are effective for both key point removal and injection of fake key points.	In some attacks, it would allow to counter those applications that, unlike copy-move detection, rely on less numerous but more robust key points.
Davide cozzolino et al., IEEE, 2015	Dense-field copy-ove forgery detection.	Nearest-neighbor search algorithm and patch matching.	This scheme efficiently deals with invariant features, high robustness and also works on rotated or scaled images.	Its features fail to guarantee the invariance properties and are ineffective in case of larger images.
Chi-Man Pun et al., IEEE, 2015	Image forgery detection.	Adaptive over-segmentation and feature point matching.	This proposed scheme achieves better detection results even under challenging and complex conditions like geometric transformations.	This scheme is not efficient in other types of forgery such as splicing or other type of media like video and audio.
E. Aridzzone et al., IEEE, 2015.	Copy-move forgery detection.	Matching triangles of key points	This method robust the geometric transformations as compared to other matching/point based methods.	It does not work on complex scenes where the number of detected triangles and key points are higher and results into worse performances.

4   SYSTEM DESIGN

In this research, at the first stage a whole literature survey is reviewed from the existing system so that problems of existing system can be mitigated. In this paper, contrast between two images is done for forgery detection i.e. one is the original image and other is the suspected image. On both image 'Dyadic Wavelet Transformation (DWT)' is applied for decomposition of both images and with the help of "Canny Edge Detection Technique" objects and shapes are recognized from the decomposed parts of both images. After that a robust SURF feature technique is used for feature description and after that KNN technique is applied on that which will perform 2-level classification i.e. to classify the similarity and injection probability between both the original and suspected image. Sat last performance of proposed algorithm is measured on some parameters like precision, true position, recall false positive, false negative and false positive rate etc.

To mitigate the problems of the existing system, we propose the use of pixel based image region localization and neighbor pixel and neighbor region analysis to find the abnormalities in the pixel based pattern in order to detect the image forgery along with Speeded-Up Robust Features (SURF) and Discrete Wavelet Transform (DWT). The k-nearest neighbor algorithm can be used to match and find the matching pixel groups based upon the region localization. k-NN (k-Nearest Neighbor) algorithm has been used between the SUFT and DWT. The DWT has been used to decompose the image into multiple coefficients. All of the image coefficient will be examined using the k-NN and SURF, and feature descriptor vectors have been generated on the basis of k-NN over SURF. These descriptor vectors have been analyzed on the final stage and compare with the feature descriptor vectors obtained from the target image. Then the results have undergone the final matching and return the forged regions in case any forged (matching region or similar region) regions are found in the scanned image. The performance of the proposed algorithm has been measured on the basis of Precision, Recall, False positive rate, true positive, true negative, false positive and false negatives.

Algorithm 1: SURF points based image forgery detection

1. Load the original image
2. Get the size of original image
3. If original image depth is 3
   a. Convert the image to grayscale
4. Apply SURF over the original image
5. Get the SURF data matrix from the SURF function
6. Load the suspected image
7. Get the size of suspected image
8. If suspected image depth is 3
   a. Convert the image to grayscale
9. Apply SURF over the suspected image
10. Get the SURF data matrix from the SURF function
11. Evaluate the point locations between the point location arrays obtained from original and suspected image using the k-Nearest Neighbor (kNN) algorithm.
12. The Euclidean distance is calculated using the kNN algorithm
13. Evaluate the point descriptors on the location where it matches.
14. Return the matching points
15. Return the non-matching points in original image (Points of Removal)
16. Return the non-matching points in suspected image (Points of Injection)

5 RESULT ANALYSIS

The above table (Table 1) is demonstrating the general execution of the proposed framework. The proposed framework has been ended up being more exact than the current framework. The proposed framework is 73.25% exact match ratio. The proposed framework is turned out to be equipped for accurately recognizing the 166 images out of 229 aggregate images.

Sensitivity (RECALL): True Positive/(True Positive + False Negative)

Positive: True Positive + False Positive

Negative: False Negative + True Negative

Precision or Positive Predictive Value: True Positive /(True Positive + False Positive)

In the Table 2 and Table 3, the measurable mistakes have been gotten from the after effects of proposed calculation. The aggregate number of images, accurately and dishonestly recognized images, not distinguished numbers, genuine positives, genuine negatives, false positive and false negative properties have been gotten from the

Table 1. Overall performance parameters of proposed system.

Overall image level accuracy mean	Total images	Correctly detected images
73.25%	229	166

Table 2. Type 1 and 2 statistical errors.

Category	Result
Total images	229
Correctly detected images	166
False detected images	41
Not detected images	22
True positive	166
True negative	0
False positive	41
False negative	22

Table 3. Statistical indices calculated on the basis of Type 1 & 2 errors.

Statistical error type	Value
Positive	207
Negative	22
Sensitivity (recall)	88.30%
Precision	80.19%
Positive likelihood ratio	0.88
Negative likelihood ratio	0
Result prevalence	82.10%

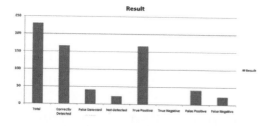

Figure 1. The result chart of the image forgery detection model.

consequences of the proposed calculation. The Total positive messages and aggregate negative messages have been acquired from the outcomes. Likewise the specificity and affectability has been computed on the got results. Additionally the positive probability proportion, negative probability proportion and ailment pervasiveness has been computed. From the got results, the precision of the proposed calculation has been demonstrated.

6 CONCLUSION

The existing system is based upon the two level analyses of the images to find the copy paste forgery. The SIFT has been used after the image segmentation in patches in the first stage. The existing model has used to EM clustering in the second level to reevaluate the copy move forgery again. The authors have combined the EM-segmentation with SIFT matching points for the purpose of forgery detection. The authors have used these two techniques in the series defined as the difference of Gaussian, SIFT points, k-NN and Expectation Maximum (EM). The existing system is time consuming as well as less accurate due to the use of image segmentation techniques. The image segmentation techniques are based upon the pixel to pixel variation or similarity in the images, but the problem lies in the cluster center selection, which are usually selected randomly. The randomly extracted clusters always produce the different results on the same image, which makes the system inefficient in terms of accuracy. A variety of the experiments have been performed over the proposed model to evaluate its performance. The results have proved the efficiency of the proposed model while evaluated on the basis of various performance parameters.

REFERENCES

Al-Qershi, Osamah M., and Bee EeKhoo. "Passive detection of copymove forgery in digital images: State-of-the-art." Forensic Science International, vol. 231, no. 1, pp. 284–295, 2013.

Amerini, Irene, Lamberto Ballan, Roberto Caldelli, Alberto Del Bimbo, and Giuseppe Serra. "A sift-based forensic method for copy–move attack detection and transformation recovery." IEEE Transactions on Information Forensics and Security, vol. 6, no. 3, pp. 1099–1110, 2011.

Anand, Vijay, Mohammad Farukh Hashmi, and Avinash G. Keskar. "A Copy Move Forgery Detection to Overcome Sustained Attacks Using Dyadic Wavelet Transform and SIFT Methods." In Proceedings of the 6th Asian Conference on Intelligent Information and Database Systems (ACIIDS 2014), Springer International Publishing, pp. 530–542, 2014.

Ardizzone, E. A. Bruno, and G. Mazzola, "Detecting multiple copies in tampered images." In Proceedings of the 17th IEEE International Conference on Image Processing (ICIP -10), pp. 2117–2120, September 2010.

Ayalneh, Dessalegn Atnafu, Hyoung Joong Kim, and Yong Soo Choi. "JPEG copy paste forgery detection using BAG optimized for complex images." In Advanced Communication Technology (ICACT), 2014 16th International Conference on, pp. 181–185. IEEE, 2014.

Bo, Xu, Wang Junwen, Liu Guangjie, and Dai Yuewei. "Image copymove forgery detection based on SURF." In Proceedings of IEEE International Conference on Multimedia Information Networking and Security (MINES-2010), pp. 889–892, 2010.

Cheng, Ming, Niloy J. Mitra, Xumin Huang, Philip HS Torr, and Song Hu. "Global contrast based salient region detection." Pattern Analysis and Machine Intelligence, IEEE Transactions on 37, no. 3 (2015): 569–582.

Hashmi, Mohammad Farukh, Aaditya R. Hambarde, and Avinash G. Keskar. "Copy Move Forgery Detection using DWT and SIFT Features." In Proceedings of 13th IEEE International Conference on Intelligent Systems Design and Applications (ISDA-2013), pp. 188–193, December 2013.

Hashmi, Mohammad Farukh, Vijay Anand, and Avinash G. Keskar. "A copy-move image forgery detection based on speeded up robust feature transform and Wavelet Transforms." In Computer and Communication Technology (ICCCT), 2014 International Conference on, pp. 147–152. IEEE, 2014.

Hussain, Muhammad, Sahar Q. Saleh, Hatim Aboalsamh, Ghulam Muhammad, and George Bebis. "Comparison between WLD and LBP descriptors for non-intrusive image forgery detection." In Innovations in Intelligent Systems and Applications (INISTA) Proceedings, 2014 IEEE International Symposium on, pp. 197–204. IEEE, 2014.

Jaberi, Maryam, George Bebis, Muhammad Hussain, and Ghulam Muhammad. "Accurate and robust localization of duplicated region in copy–move image forgery." Machine vision and applications 25, no. 2 (2014): 451–475.

Lei, Jie, Mingli Song, Ze-Nian Li, and Chun Chen. "Whole-body humanoid robot imitation with pose similarity evaluation." Signal Processing 108 (2015): 136–146.

Li, Jian, Xiaolong Li, Bin Yang, and Xingming Sun. "Segmentation-based Image Copy-move Forgery Detection Scheme.", Information Forensics and Security, IEEE Journals, 2014.

Li, Leida, Shushang Li, Hancheng Zhu, Shu-Chuan Chu, John F. Roddick, and Jeng-Shyang Pan. "An Efficient Scheme for Detecting Copy-move Forged Images by Local Binary Patterns", Journal of Information Hiding and Multimedia Signal Processing, vol. 4, no. 1, pp. 46–56, January 2013.

Lin, Liang, Xiaolong Wang, Wei Yang, and Jian-Huang Lai. "Discriminatively trained and-or graph models for object shape detection." Pattern Analysis and Machine Intelligence, IEEE Transactions on 37, no. 5 (2015): 959–972.

Mishra, Parul, Nishchol Mishra, Sanjeev Sharma, and Ravindra Patel. "Region Duplication Forgery Detection Technique Based on SURF and HAC." The Scientific World Journal, vol. 2013, Article ID 267691, pages 8, 2013.

Muhammad, Ghulam, Munner H. Al-Hammadi, Muhammad Hussain, and George Bebis. "Image forgery detection using steerable pyramid transform and local binary pattern." Machine Vision and Applications 25, no. 4 (2014): 985–995.

Shivakumar, B. L., and S. Santhosh Baboo. "Detection of Region Duplication Forgery in Digital Images Using SURF." International Journal of Computer Science Issues (IJCSI), vol. 8, no. 4, pp.199–205, 2011.

Zhang, Chenyang, Xiaojie Guo, and Xiaochun Cao. "Duplication localization and segmentation." In Proceedings of Advances in Multimedia Information Processing (PCM 2010), pp. 578–589, Springer Berlin Heidelberg, 2010.

Zhang, Ruimao, Liang Lin, Rui Zhang, Wangmeng Zuo, and Lei Zhang. "Bit-scalable deep hashing with regularized similarity learning for image retrieval and person re-identification." Image Processing, IEEE Transactions on 24, no. 12 (2015): 4766–4779.

Zoccolan, Davide. "Invariant visual object recognition and shape processing in rats." Behavioural brain research 285 (2015): 10–33.

Communication and Computing Systems – Prasad et al. (Eds)
© 2017 Taylor & Francis Group, London, ISBN 978-1-138-02952-1

Binary image reconstruction using the adaptive hybrid mechanism with swarm optimization

Ishu Garg & Bikrampal Kaur
CGC, Landran, Mohali, Punjab, India

ABSTRACT: The image reconstruction is the procedure to reproduce the image from the image contents provided in the form of projections or damaged image matrix. The image matrix damages are produced due to many internal or external factors which cause the image degradation. The image compression is also one the primary reasons behind the image reconstruction. The image reconstruction from the projection data is incorporated using the iterative reconstruction mechanisms. The data matrix processing and matrix regeneration is the iterative process to regenerate the matrix by estimating the data values from the provided projections. The projection data in obtained in the multi-dimensional format which includes the 2-D, 3-D or N-D projection data obtained from the various types of images. The proposed model is aimed at solving the problem of the binary matrix reconstruction by estimating the pixel values from the projection data. The swarm optimization techniques would be incorporated to effectively regenerate the image matrix. The proposed model design has been aimed at improving the performance of the existing models by eliminating the shortcomings of the existing schemes.

1 INTRODUCTION

Binary Image Reconstruction outlines the technique of reproducing an image from the traced image. This traced image is acquired in the structure of projections such as diagonal, vertical and horizontal. These projections are absolutely based on the type of an image that may be 2-D or 3-D image. This involves the concept of binary image reconstruction. A binary image is a digitalized image containing on white and black pixels where white pixels corresponds to binary value 1 & black pixels corresponds to binary value 0.

In binary images, Image reconstruction is a mathematical process that produces images from Horizontal-Vertical projection data derived from the different angles. Image reconstruction has a major influence on image quality also on radiation noise. For a radiation dose it is necessary to reconstruct images with the lower noise without sacrificing spatial resolution and image quality. Reconstructions methods that improve image quality can be converted into a decrease of radiation dose as images of acceptable quality can be reconstructed at lower dose.

There are plenty of techniques of image reconstruction. Mostly used are:

i. Analytical Reconstruction or Filtered Back Projections (FBP).
ii. Iterative reconstruction.

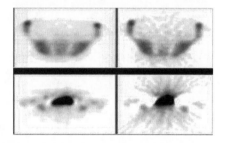

Figure 1. Example showing differences between filtered back projection (right half) and iterative reconstruction method (left half).

Analytical Reconstruction constructs an image in a single step. The widest use of this projection is in the field of clinical binary image scanners. The major reason behind its use is its numerical stability and computation efficiency whereas Iterative Reconstruction constructs an image in multiple steps but provides a complete solution which leads to efficient reconstruction of an image as compared to filtered back projection. Also it decreases the image artifacts like beam hardening and metal artifacts.

The existing work is done on simulated annealing that uses the two level solution for reproducing an image. The simulated annealing inhibits the concept of metallurgy where it is enforced on glass or metal. This provides the excellent strength

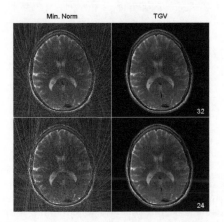

Figure 2. Reconstruction example over the MRI images.

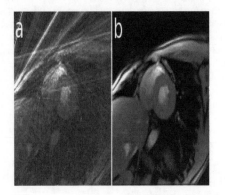

Figure 3. A single frame from a Real-time MRI movie of a human heart. a) Direct reconstruction b) iterative (nonlinear inverse) reconstruction.

in glass with accurate flexibility. In this, there are more chances of occurring error while reconstruction phase. As a result of which there is a need of single efficient initial solution algorithm that will produce more accurate results.

This paper emphasizes on iterative reconstruction of binary images from HV projections analyzed from different angles. It also emphasizes on obtaining the more efficient results as the solution of optimization problem or to a set of equations which are concluded in a iterative loop.

2 RELATED STUDY

Patel, Divyesh et al. (2015) describes a technique for the reconstruction of h-convex binary images form its horizontal and vertical projections using the idea of simulated hardening. In this research,

the use of convexity property in case of binary images is represented and it also downside the reproduction of h-convex binary image from its vertical and horizon ions. This downside is altered into two completely different optimization problems by the concept of two acceptable objective features. Then the two simulated hardening algorithms to untangle the two atomization problems are constructed by this algorithm. The sturmarbeiteilung algorithm are verified on infinite aimlessly generated test images. These algorithm verifies the weedy pictures.

Shiv Kumar Verma et al. (2013) The Branch and Bound method has been carried efficiently and effectively over the convex binary images in the discrete tomography. The authors laid emphasis on the diagonal and anti-diagonal projections and afterwards the results are compared with the traditional horizontal-vertical projections. With the use of Branch and Bound method, if number of elements increases, then the trees grows exponentially but this works efficiently well for images having matrix order below 10. Also the bound is computed at each branching point and also on intermediary node points to limit the unbound expansion of tree. Only if the diagonal and anti-diagonal vectors sums are consistent, only then reconstruction of unique binary images is possible. The authors analyzed that in HV-convex images, the number of nodes and execution time increases exponentially but while using diagonal and anti-diagonal convex images, execution time decreases. No doubt the height of tree is larger as compared to HV-projections. This method has slower convergence rate as compared to the other methods but it works efficiently on 50×50 matrices.

Abris Nagy et al. (2014) The authors recommended an algorithm for the reconstruction of convex bodies from booming X-Ray computation with a complete evidence of convergence. He presented some advanced walk into the direction of reconstruction. The continuity features are used for transmitting the connected dense HV-convex sets possessing the similar axis parallel bounded box to the correlated generalized conic function. These results are used to schedule an algorithm for the reconstruction of densely connected bodies. Greedy and Anti greedy variants are also presented with quota system strategy. The co-ordinates of X-Ray are determined by generalized conic function that are related to a dense planar set. The generalized methods of reconstruction are established on the Fourier transformation while the algebraic reconstruction are carried out in case of pixel based objects.

Shiv Kumar Verma et al. (2012) discussed the difficulties encountered while reconstructing a binary image only from the sums of line disposed in two orthogonal projections. These two projections may be diagonal (45°) and anti-diagonal (135°) or horizontal and vertical projections. Using diagonal and anti-Diagonal projections provides better and accurate information about the image which further lessens the number of solutions. The authors suggested this algorithm and further test this on sample images of varying size and patterns. After that the obtained results are contrast with the horizontal-vertical projections using Chang's algorithm in terms of misclassifications. No doubt both the algorithms works and performs in identical manner, the only variation can be seen in misclassification parameter which is because of only particular image patterns are selected leaving the others, therefore the tie breaks and weights generates the variations.

Norbert Hantos et al. (2013). In this, the reconstruction of binary images using Morphological skeleton over Horizontal-Vertical projections is proposed. If only the two directions are given, to satisfy these two directions there may be many exponentially Horizontal-Vertical convex 4-connected images. Also if the morphological skeleton is known in advance, one can decrease the number of possible solutions. According to the author, the reconstruction solution can only be find out in polynomial time and in last and final step one needs to validate if the morphological structure of the reconstructed images lies in the set. The author defines horizontal-vertical convex images in terms of object co-ordinates. If in binary image (F), the object co-ordinates are chronological in every column (row), then F is known as V-convex (H-convex) image. And the binary image is anti-diagonal (diagonal) convex, if the object co-ordinates are chronological in every anti-diagonal (diagonal). The authors concluded that even if the morphological skeleton is known the reconstruction of HV convex is not unique. No doubt for a certain parameter value this can be possible.

Zoltan Ozsvar et al. (2013) The authors proved the problem of reconstruction of binary images from Horizontal-Vertical projection is NP-Hard. The authors examined the various heuristic algorithms of reconstruction and also contrast the results from the perspective of reconstruction quality and execution time. The quality of reconstruction not only depends on the size of binary image but also on the positions and number of its components. The authors also revealed that the time of reconstruction is also concerned with the number of switching components which are present in a binary image. A binary image contains of only black and white regions. The authors assumed the object which is examined is unvarying i.e. homogenous then the image must consists of only white (background) pixels and black (objects). This can generate images of superior quality even from little quantity of data. Many experiments are performed to contrast various algorithms of reconstruction of binary images. The authors wind up with the result that for fast and efficient reconstruction algorithm of Horizontal-Vertical convex images, the various feature such as localization of components, core-shell operator and switching operators to be combined.

3 COMPARITIVE ANALYSIS

Table 1. The comparison of various techniques has been shown in the table below.

Author, publication and year	Problem addressed	Techniques proposed	Merits	Demerits
Divyesh Patel et al., IEEE, 2015	To reconstruct the binary images from the image traces i.e. its Vertical and Horizontal Projections	Simulated annealing	The simulated annealing provides the apt. strength in metal or glass with appropriate flexibility	More probability of occurring error during the reconstruction process and also the two optimization solutions mess up with each other.
Shiv Kumar Verma et al., Research Gate, 2013	Reconstruction of convex binary images from Diagonal and Anti-Diagonal projections.	Branch and bound technique	Execution time and Number of nodes in a tree decreases to a great extent. Works well for up to 50×50 matrices size.	Height of tree is large as well as the converges rate is slow.

(Continued)

Table 1. (*Continued*).

Author, publication and year	Problem addressed	Techniques proposed	Merits	Demerits
Abris Nagy et al., Journal of mathematical imaging and vision, 2014.	Reconstruction of HV-convex sets with the help of their co-ordinate X-ray functions.	Use of Continuity properties and Greedy and Anti-Greedy algorithms of reconstruction.	Work well on Noisy data as well as.	The proposed algorithm considers the measurements of only in a finite points of sets.
Shiv Kumar Verma et al., International Journal of Tomography and Statistics, 2012	Reconstruction of convex binary images from Diagonal and Anti-Diagonal projections.	Reconstruction projection algorithms	More Information about the image from these two projections that lessens the no. of possible solutions. Gives Better reconstruction when contrast with other algorithms.	The proposed solutions has been found lacking in the ability to accurately reconstruct the non-centrally weighted image.
Norbert Hantos et al., IEEE, 2013	Reconstruction of the binary images Vertical and Horizontal Projections and morphological skeleton	Polynomial Time calculation method	For Finding the solutions, There exists a polynomial time technique.	If the morphological skeleton is known is addition even then the Reconstruction of HV-convex polynomies is not unique
Zoltan Ozsvar et al., Acta Cybern, 2013.	Reconstruction of the HV-Convex binary matrices using Horizontal and Vertical Projections.	Kernel-Shell method, simulated annealing	The efficiency of reconstruction is determined by Size, number and location of components. Hence, a fast and efficient algorithm is proposed.	Objects to be examined must be unvarying i.e. they must be homogenous.

4 SYSTEM DESIGN

In this research, at very initial stage a detailed literature survey is being carried out with which all the earlier descriptor methods analyzed in detail so that an algorithm with better precision can be developed. A new approach has been proposed for image reconstruction having unique optimization and initial solution design techniques for the binary images. The image will be regenerated using the image traces. These traces are acquired in the form of horizontal and vertical projections. The proposed algorithm is a hybrid method of image reconstruction that will be utilizing the pixel labelling and pre-shape analysis based on initial solution which derives the initial combinations to satisfy the projection data. The projection data is then further used to recreate the projected binary image in the iterative context which then runs the whole program and finds the various combinations to recreate the initial image matrix. This will be done using the "Iterative Reconstruction method" and "Chang or Reyser working".

The results obtained from this initial solution are then passed onto optimization stage. The optimization will be performed using either the "Genetic Algorithm" or the "PSO (Particle Swarm Optimization)". At last the results are compared with the existing work (stimulated annealing) using the performance parameters such as accuracy, elapsed time, Correction factor etc.

Algorithm 1: Genetic Algorithm with Noise Purification

  i. Input binary image
 ii. Read the binary image and convert into image matrix data
iii. Compute the projection vectors
 iv. Acquire the projection vector
  v. Apply the condition of the horizontal and vertical summation difference to zero
 vi. If iteration count is 1
vii. Compute the binary compositions in the given sized matrix (N × M)
viii. Otherwise

92

ix. Update the binary composition in the given sized matrix (N × M)

x. If condition satisfies

xi. Return from the Iteration

xii. Apply the genetic programming over the constructed binary matrix

xiii. Perform the new population selection

xiv. Compute the solution

xv. Calculate the solution fitness

xvi. Optimize the matrix components by applying the computed solution

xvii. Return the Optimized image matrix

xviii. Check the pixel conditions in the given window of 3 × 3.

xix. Check the density of the surrounding pixels

xx. Update the center value according to the condition received condition

xxi. Return the finally regenerated matrix

Table 2. Results obtained from the proposed model simulation.

Image size	Reconstruction accuracy (%)	Elapsed time (sec)
20 × 20	79.54	5.31 seconds
30 × 30	81.32	7.59 seconds
40 × 40	78.36	9.42 seconds
50 × 50	84.65	11.95 seconds
60 × 60	90.75	14.23 seconds
80 × 80	73.17	17.19 seconds
100 × 100	82.77	20.22 seconds

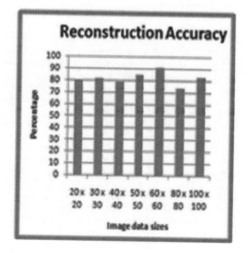

Figure 4. Reconstruction accuracy measured in the percentage.

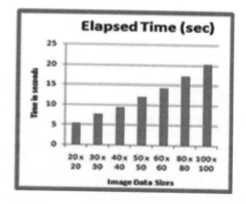

Figure 5. Elapsed time for the reconstruction accuracy.

5 RESULT ANALYSIS

We have implemented our algorithms using the MATLAB simulator over the windows 7. The experiments were run on an Intel Core-i3 CPU based PC with 6 GBs of memory. The program has been specifically built for the requirement of the binary image reconstruction using the iterative pixel guessing solution with data matrix reconstruction. The following Table 2 describes the results obtained from the proposed model in the form of the accuracy of reconstruction and elapsed time.

The reconstruction accuracy has been measured over each reconstructed image matrix against the original image. The image reconstruction accuracy depicts the accuracy of the results obtained from the proposed solution. The proposed solution has been evaluated in the graphical form in the Figure 4, which describes the overall accuracy of all 7 reconstructed images.

The elapsed time graph has been obtained from the proposed model. The elapsed time has been measured in the seconds for the entire solution over the input image data.

6 CONCLUSION

The horizontal, vertical and diagonal projections can be obtained from the image data. The projection data contains the sum-ups of the rows and columns in the 2-D or 3-D formation according to the type of images. The binary images are the 1-bit images which contain only two types of combinations or 1 and 0. The binary image reconstruction is based upon the initial solution which computes the initial combinations to satisfy the projection data. The projection data is utilized to recreate the projected binary image in the iterative context which runs the whole programs and computes the various combinations to recreate the initial image matrix. The matrix result obtained from the initial solution is then passed on to the genetic algorithm for the matrix optimization. The GA is responsible for returning the optimal solution with best combination available according to the projection data. The results of the proposed model have been analyzed in the form of various performance parameters such of reconstruction accuracy and elapsed time.

In this paper, the use of swarm intelligent solution has been suggested for the image matrix reconstruction. The image reconstruction process has been aimed at producing the image matrix from the projection data obtained from the binary images. The hybrid image matrix regeneration sequence will be utilized to improve the overall quality of the image matrix reconstruction. The proposed model is aimed at solving the problem of improving the stability of the iterative solution in the lowest possible regeneration process time.

REFERENCES

Alessio, Adam M., Charles W. Stearns, Shan Tong, Steven G. Ross, Steve Kohlmyer, Alex Ganin, and Paul E. Kinahan. "Application and evaluation of a measured spatially variant system model for PET image reconstruction." Medical Imaging, IEEE.

Bittencourt, Márcio Sommer, Bernhard Schmidt, Martin Seltmann, GerdMuschiol, Dieter Ropers, Werner Günther Daniel, and Stephan Achenbach. "Iterative reconstruction in image space (IRIS) in cardiac computed tomography: initial experience." The international journal of cardiovascular imaging 27, no. 7 (2011): 1081–1087.

Hantos, Norbert, and Péter Balázs. "A Fast Algorithm for Reconstructinghv-Convex Binary Images from Their Horizontal Projection." In Advances in Visual Computing, pp. 789–798. Springer International Publishing, 2014.

Hantos, Norbert, and Péter Balázs. "A uniqueness result for reconstructing hv-convex polyominoes from horizontal and vertical projections and morphological skeleton." In Image and Signal Processing and Analysis (ISPA), 2013 8th International Symposium on, pp. 795–800. IEEE, 2013.

Huang, Junzhou, Shaoting Zhang, and Dimitris Metaxas. "Efficient MR image reconstruction for compressed MR imaging." Medical Image Analysis 15, no. 5 (2011): 670–679.

Kashuk, Sina, and MaguedIskander. "Reconstruction of three dimensional convex zones using images at model boundaries." Computers & Geosciences 78 (2015): 96–109.

Leipsic, Jonathon, Giang Nguyen, Jaqueline Brown, Don Sin, and John R. Mayo. "A prospective evaluation of dose reduction and image quality in chest CT using adaptive statistical iterative reconstruction." American Journal of Roentgenology 195, no. 5 (2010): 1095–1099.

Mohamed, Hadded, and Hasni Hamadi. "Combining Genetic Algorithm and Simulated Annealing Methods for Reconstructing HV-Convex Binary Matrices." In Hybrid Metaheuristics, pp. 78–91. Springer Berlin Heidelberg, 2013.

Moscariello, Antonio, Richard AP Takx, U. Joseph Schoepf, Matthias Renker, Peter L. Zwerner, Terrence X. O'Brien, Thomas Allmendinger et al. "Coronary CT angiography: image quality, diagnostic accuracy, and potential for radiation dose reduction using a novel iterative image reconstruction technique—comparison with traditional filtered back projection." European radiology 21, no. 10 (2011): 2130–2138.

Nagy, Ábris, and CsabaVincze. "Reconstruction of hv-convex sets by their coordinate X-ray functions." Journal of mathematical imaging and vision 49, no. 3 (2014): 569–582.

Ozsvár, Zoltán, and PéterBalázs. "An Empirical Study of Reconstructing hv-Convex Binary Matrices from Horizontal and Vertical Projections." Acta Cybern. 21, no. 1 (2013): 149–163.

Patel, Divyesh, and Tanuja Srivastava. "Reconstructing h-convex binary images from its horizontal and vertical projections by simulated annealing." In Contemporary Computing (IC3), 2015 Eighth International Conference on, pp. 117–121. IEEE, 2015.

Patel, Divyesh, and Tanuja Srivastava. "Reconstruction of binary matrices satisfying neighborhood constraints by simulated annealing." World Academy of Science, Engineering and Technology 8, no. 5 (2014): 760–763.

Pontana, François, Alain Duhamel, Julien Pagniez, Thomas Flohr, Jean-Baptiste Faivre, Anne-LiseHachulla, Jacques Remy, and Martine Remy-Jardin. "Chest computed tomography using iterative reconstruction vs filtered back projection (Part 2): image quality of low-dose CT examinations in 80 patients." European radiology 21, no. 3 (2011): 636–643.

Ravishankar, Saiprasad, and Yoram Bresler. "MR image reconstruction from highly undersampled k-space data by dictionary learning." Medical Imaging, IEEE Transactions on 30, no. 5 (2011): 1028–1041.

Renker, Matthias, John W. Nance Jr, U. Joseph Schoepf, Terrence X. O'Brien, Peter L. Zwerner, Mathias Meyer, J. Matthias Kerl et al. "Evaluation of heavily calcified vessels with coronary CT angiography: comparison of iterative and filtered back projection image reconstruction." Radiology 260, no. 2 (2011): 390–399.

Ritschl, Ludwig, Frank Bergner, Christof Fleischmann, and Marc Kachelrieß. "Improved total variation-based CT image reconstruction applied to clinical data." Physics in medicine and biology 56, no. 6 (2011): 1545.

Srivastava, Tanuja, Shiv Kumar Verma, and Divyesh Patel. "Reconstruction of binary images from two orthogonal projections." IJTS 21, no. 2 (2012): 105–114.

Verma, Shiv Kumar, Tanuja Shrivastava, and Divyesh Patel. "Efficient Approach for Reconstruction of Convex Binary Images Branch and Bound Method." In Proceedings of the Third International Conference on Soft Computing for Problem Solving, pp. 183–193. Springer India, 2014.

Zhang, Xinpeng. "Lossy compression and iterative reconstruction for encrypted image." Information Forensics and Security, IEEE Transactions on 6, no. 1 (2011): 53–58.

Communication and Computing Systems – Prasad et al. (Eds)
© 2017 Taylor & Francis Group, London, ISBN 978-1-138-02952-1

Application of signal processing to optimize EM attack on AES

A.K. Singh, S.P. Mishra, B.M. Suri & A. Khosla
SAG, DRDO, Metcalfe House, Civil Lines, Delhi, India

ABSTRACT: EM attack is one of the most effective Side Channel Attack techniques. It does not require the close proximity of the adversary to the crypto systems to acquire the EM signals radiated. This paper contains brief implementation detail of AES on FPGA and illustrates the EM signals measurement process. This explains briefly about the philosophy of correlation EM analysis attack technique and presents totally algorithmic approach to reduce the requirement of EM signals acquired by placing EM probe in the vicinity of the FPGA device executing AES algorithm to extract the key. This cost effective approach includes different signal processing techniques i.e. moving average, low pass filtering, harmonics removal, spectral band pass filtering, combination of more than one of them, etc. The application of these techniques on EM signals eliminates/suppresses noise contents. Effectiveness of these techniques depends on the clock frequency of the board. The requirement of EM signals has been reduced up to 43% with lesser computational overheads and without use of additional complex equipments.

1   INTRODUCTION

Side Channel Attack is a newer and fast evolving technological field used to extract the secret information of crypto systems with lesser computational efforts and time as compared to classical cryptanalysis techniques. It deduces information by acquiring and analyzing data dependent side channel leakages like power consumption (Kocher et al., 1999), EM radiation (Gandolfi et al., 2001), execution time (Kocher, 1996), etc. of a crypto algorithm running on electronic circuitry.

Most of the electronic circuits are based on CMOS technology (Kang & Leblebici, 2002) where state of the circuit changes with the data being processed. The change of the states causes variation in the transient current drawn. This results in variation of power consumptions and EM radiations, which are exploited to extract the secret key of the crypto algorithms. AES (Stallings, 2003) algorithm is being widely used due to its robustness towards crypt analysis and its ease of implementation either in S/W or H/W. FPGA based platforms are becoming natural choice of crypto systems because of their reconfigurable feature. But FPGA based crypto systems have been found leaking information through power consumption (Berna et al., 2003) and EM radiation (Carlier et al., 2004.). This enables adversary to mount the power/EM attack and extract the secret key.

EM Analysis (EMA) (Quisquater & Samyde, 2001), (Gandolfi et al., 2001), (Agrawal et al., 2002) or EM attack extracts information by analyzing EM signals using statistical tool. EM attack is of two types. First is known as Simple EM Analysis

(SEMA) attack, which deduces some secret information of crypto system by visual inspection of few EM signals with detail knowledge of the system. Second is Differential EM Analysis (DEMA) attack, which is analogous to Differential Power Analysis (DPA) (Kocher et al., 1999) attack technique. DEMA (Mathews, 2006) attack uses statistical tool to deduce the secret key by analyzing EM signals. It needs less details of crypto system. It forms two groups of EM signals based on intermediate data values and computes average of each group. Finally, it generates differential signal by subtracting the averaged EM signals. In the case of correct guessed key, dominant peaks are found in the differential signal. An improved version of DEMA attack is known as Correlation EM Analysis (CEMA) attack which is similar to the Correlation Power Analysis (CPA) (Brier et al., 2004) attack. CEMA attack compares measured EM signals with the EM signals computed using a hypothetical model.

EM attack is now being considered more efficient attacks method because EM emissions are very rich source of information (Maistri et al., 2013.), (Rao & Rohatgi, 2001) and measurement of EM emissions is contactless process. EM attack does not require the close proximity of the attacker to the crypto systems as required in the case of power attack to measure the power consumptions. EM signals of a crypto system can be acquired from distance (Mangard, 2003) with the help of appropriate antenna and data acquisition card or oscilloscope. Thus monitoring of EM emissions is easier and less detectable (Maistri et al., 2013).

But problem with EM attack is the channel noise. With the increase of distance, EM radiations are buried in a lot of noise due to interferences of radio signals and radiations of other electronic equipments put in the reception area of the receiver used by the adversary. The more problem is posed by dominant emissions from power supply grid of a cryptographic device (Mangard, 2003). This inhibits the extraction of information from emissions of specific part of a device. Combined effect of these noises make difficult task to detect compromising EM signals of a device and hence retrieval of information by mounting EM attack becomes highly tedious job. This necessitates further actions to be taken to address this problem. Towards this, some ideas have already been proposed in the literature (Mangard, 2003), (Rao & Rohatgi, 2001) including the use of heterodyne receiver of variable bandwidth and tunable with large frequency values. In order to exploit these noisy emissions, it was proposed to tune the receiver to clock frequency or to one of its harmonics, down convert the side-bands to some intermediate frequency and analyze down converted signal like in power analysis attacks.

This paper proposes software oriented approach to reduce the requirement of EM signals measured from close proximity (near field) of the cryptographic device running the crypto algorithm without employing additional complex equipments. Various signal processing techniques like Moving Average, Low Pass Filtering, Harmonics Removal, Spectral Band Pass Filtering and combination of more than one of them are applied on EM signals emitted from FPGA board executing AES.

Remaining part of the paper is structured as follows. Next section of the paper contains brief implementation details of the AES on FPGA (Spartan 3E) board. Measurement process of EM signals is given in section 3. Section 4 contains the philosophy of CEMA attack. Effect of different signal processing techniques on EM signals captured at different clock frequencies is presented in section 5. Section 6 includes conclusions of the paper.

2 IMPLEMENTATION OF AES ON FPGA

AES algorithm is a symmetric block cipher used to encrypt electronic data. Its execution starts with add key operation followed by nine similar rounds of processing (Singh et al., 2015). Each round contains substitute byte, shift rows, mix column and add round key operations. Mix column operation is not used in final (tenth) round processing.

FPGA is configured for 128 bits AES algorithm using VHDL and Xilinx tool. Ten round keys are generated in ten clocks before the encryption starts. One clock is used in processing of each round and full encryption process consumes ten clocks.

Byte wise XORing of key with plaintext (Singh et al., 2015) is the first step of encryption process

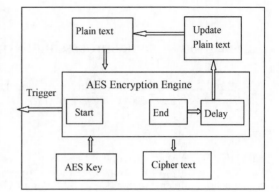

Figure 1. AES encryption core on FPGA (Singh et al., 2015).

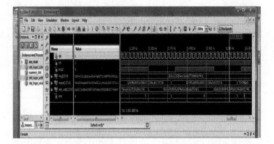

Figure 2. Simulation window of AES encryption core.

followed by substitute byte, shift rows, mix column operations. Bytes resulting from mix column operation are XORed with the corresponding bytes of round keys in add round key operation. The output bytes of shift rows operation are byte wise XORed with the respective bytes of the tenth round key in last round processing.

Figure 1 shows the AES encryption core on FPGA (Singh et al., 2015). This core performs encryption in continuous manner. Appropriate waiting time has been inserted between encryptions. Each encryption process starts with generation of a trigger signal which remains at high logic value during the whole encryption time. This is illustrated in Figure 2 with the help simulation window. Simulation window shows clock, reset, trigger, key, input, output and encryption count signals starting from the top. This core produces plaintext of random nature during the waiting time to serve as plaintext for next encryption.

3 MEASUREMENT OF EM SIGNALS

EM radiations (leakages) are captured using appropriate magnetic sensors (Masoumi & Rezayati, 2015) and they are digitized with the help of suitable ADC or oscilloscope. These digitized data of leakages are stored for further processing/analysis purposes. Care is required during measurement process. Representation of each intermediate result requires sufficient numbers of samples. Use of optimum number of samples improves the analysis results (Singh et al., 2015). Using either too large or too less numbers of samples increases analysis time and computations.

Figure 3 depicts the measurement setup (Singh et al., 2015) to measure EM signals emitted from FPGA board. It consists of workstation, oscilloscope (DPO 7254), FPGA board and EM probe fixed on the top of FPGA with the help of EM probe station. Specific part of the FPGA board, where radiations are strong and possesses unique patterns of AES processing, is identified for picking the EM signals. PC communicates with FPGA board using JTAG port. Xilinx tool at PC generates application (bit) file, transfers it to the board and executes it. EM signals radiated during encryptions time are captured with the help of probe and oscilloscope. The start of algorithm's execution and capturing time of EM signals is matched by employing a trigger signal. Trigger signal is produced in FPGA at the moment when execution starts. It is sent to one of the inputs of oscilloscope to activate it for acquisition of EM signals. Tektronix make Oscope tool is used to transfer the data acquired by oscilloscope to the PC. In order to avoid the overwriting of present encryption data (EM signals) of oscilloscope with the data of next encryption, a appropriate waiting time (1.5 s) is introduced between encryptions.

Figure 4 below depicts screenshot of the oscilloscope containing trigger and EM signals for single encryption of AES. It shows that the trigger signal remains high for full encryption time.

Figure 5 shows the plot of the EM signals (Singh et al., 2015) radiated at various clock frequencies of the FPGA board. Improvement in radiation quality with the increase of clock frequency is observed. It exhibits that the clarity and visibility of distinct patterns of AES processing enhances with the increase of clock frequency. Improvement

Figure 3. Measurement setup for acquisition of EM signals (Singh et al., 2015).

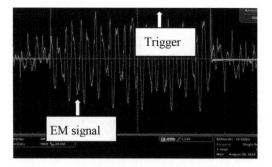

Figure 4. Trigger and EM signals for single encryption of AES.

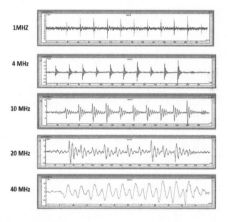

Figure 5. EM signals of different clock frequencies (Singh et al., 2015).

in radiation reduces the requirement of EM signals obtained at higher clock frequency to deduce the key used in the crypto device.

4 CORRELATION EM ANALYSIS ATTACK PHILOSOPHY

The correlation EM analysis attack philosophy is based on existence of statistical correlation of leakages with operations and data being used by the cryptographic devices. Measurement and analysis of leakages i.e. power consumptions and EM signals, captured when execution of cryptographic algorithms takes place, results in revealing the secret information. CEMA attack performs comparison between measured and predicted EM signals. Development of a suitable model is needed to predict the leakages. Quality of a model is dependent on skills and knowledge of the adversary. Hamming Weight and Hamming Distance (HD) models (Brier et al., 2004), (Peeters et al., 2007) are well known. Among them, the HD model is mostly used for devices based on CMOS technology. HD model assumes a kind of linear relation between the leakages and changes of states taking place in the devices. According to the HD model, the number of change of states ($0 \rightarrow 1$ and $1 \rightarrow 0$) are proportional to leakages. This model proposes that the same amounts of leakages are resulted from the both type of change in states. The HD model Predicts Leakage (PL) using equation 1 for two successive intermediate values (let Z_1 and Z_2) of an algorithm under consideration.

$$PL = HD(Z_1, Z_2) = HW(Z_1 \oplus Z_2) \qquad (1)$$

where HW stand for Hamming Weight, $\oplus$ indicates exclusive—OR operation. Part of key (Ks) and plaintext (PT) are involved in producing the intermediate values Z_1 and Z_2.

Equation 1 is used to predict leakage PL for different plaintexts and guessed keys. Real EM signal, RLj, is measured for each plaintext at different time j. These real EM signals are correlated to the predicted leakages. The correlation is measured using Pearson's correlation coefficient (Clarke & Cooke, 1998) between predicted leakage PL and measured EM signal RLj using the formula given as:

$$Cs = \frac{E(PL, RLj) - E(PL).E(RLj)}{\sqrt{Var(PL).Var(RLj)}} \qquad (2)$$

where E and *var* represent average and variance operation respectively. RLj has little correlation with the corresponding PL when Ks is not the correct key guess. When Ks is the correct key guess, the RLj shows highest correlation with the corresponding PL.

Part of EM signals representing tenth round of AES processing is used for CEMA attack. Guessed key byte which shows highest correlation value is considered as actual key byte. All 16 key bytes of AES algorithm are extracted by analysing 8450 EM signals which are acquired at 5 GS sampling rate of the oscilloscope when 1 MHz was the clock frequency of the FPGA board. Table 1 contains the results of EM attack mounted on EM signals of several clock frequencies of FPGA board (Singh et al., 2015) and sampling rates of the oscilloscope. This shows that the need of EM signals at 4 MHz is larger than at 10 MHz. This is the result of improved radiation quality.

Table 1. Result of CEMA attack at different clock frequencies and sampling rates (Singh et al., 2015).

S. No.	Clock Frequency/ Sampling rate	No. of samples/ Clock	No. of Traces
1	1 Mhz/5.0GS	5000	8000
2	4 MHz/2.5 GS	625	17500
3	10 MHz/2.5GS	250	7500

5 SIGNAL PROCESSING OF EM SIGNALS

The CEMA attack needs substantially large amount of EM signals as compared to the amount of power waveforms required by CPA attack (Singh et al., 2015) to extract the AES key. The reason for this is the presence of surrounding noises. This inspired to explore some cost effective techniques to reduce the requirement of EM signals without employing additional equipments and significant increase in computations. Hence, the use of software oriented signal processing approach is finalized to achieve the goal by removing/suppressing the noise contents.

Based on spectrum and spectrogram analysis of EM signals acquired at several clock frequencies of the FPGA board, some techniques namely Moving Average (MA), Low Pass Filtering (LPF), Harmonics Removal (HR), Spectral Band Pass Filtering (SBPF) and combination of more than one of them have been tried. These techniques have considerably reduced the amount of EM signals required to reveal the secret key.

One thing needs to be mentioned here that all these techniques are not equally effective in reducing the requirement of EM signals of all clock frequencies. Different techniques are found effective for EM signals of different clock frequencies. The effective techniques for EM signals of 1, 4 and 20 Mhz frequencies are discussed below in case 1, case 2 and case 3 respectively.

Case 1: 1 Mhz frequency.
In this case, processing is performed at 1 MHz which is derived from 4 MHz clock frequency through software means. HR technique followed by LPF technique has enabled in reducing the amount of signals needed to deduce the full key. HR technique is used to remove the 4 MHz clock frequency component and low pass filtering removes the high frequencies noise components. EM signal along with signal resulting from HR+LPF technique for last three rounds of AES processing are shown in Figure 6. EM signal of only last three rounds of AES is chosen to improve the clarity. Second reason is the way of mounting the attack. CEMA attack has been mounted on last round. Figure 6 shows that the signal resulting from HR+LPF technique has become smooth due to removal of unwanted noise signals. This technique has reduced the amount of EM signals from 8450 to 7000, which indicates that the requirement of signals has reduced by 17.2%.

Case 2: 4 MHz frequency.
Here, both processing and clock frequency are same (4 Mhz). In this case, LPF technique is found effective in reducing the requirement of EM signals by removing/suppressing the high frequencies noise signals. The plots of the EM signal for last 3 rounds of AES, average spectrum of EM signals, signal resulting from the LPF technique

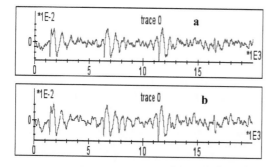

Figure 6. a) EM signal, b) Signal resulting from HR + LPF technique.

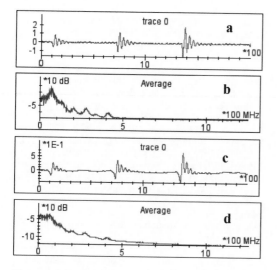

Figure 7. a) EM signal, b) Average spectrum of a, c) After applying LPF technique on a, d) Average spectrum of c.

and its average spectrum are shown in Figure 7. From this figure, it is obvious that the signal resulting from LPF technique and its average spectrum have become smooth due to elimination of noise contents. Therefore, the number of EM signals required has reduced from 17725 to 12000. It implies that 32.3% less amount of EM signals is required, which is significant quantity.

Case 3: 20 MHz frequency.

In this case also, both processing and clock frequencies are same. Only difference is the clock frequency value. In place of 4 MHz, 20 MHz clock frequency has been used. The plots of EM signal of 20 MHz clock frequency, spectrum of EM signal and the average spectrum of EM signals are depicted in Figure 8. Average spectrum of EM signals (100) exhibits that the signal is concentrated in the frequency range from 0 to 350 MHz. Therefore, application of SBPF technique with

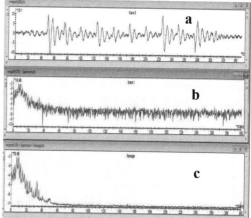

Figure 8. a) EM signal (20 MHz), b) Spectrum of EM signal c) Average of Spectrum.

Table 2. Effect of signal processing techniques.

S. No.	Clock/ Sampling Frequency	Techniques	No of Traces
1	1 MHz/5GS	NIL	8450
		MA	8000
		LPF	8000
		SBPF	8000
		HR (4 MHz) + LPF	7000
2	4 MHz/2.5GS	NIL	17725
		MA	17500
		SBPF	17500
		LPF	12000
3	10 MHz/2.5GS	NIL	12850
		MA	12500
		HR+LPF	10500
		SBPF (0–450 MHz)	8000
4	20 MHz/5GS	NIL	11350
		MA	11000
		LPF	11000
		SBPF(20–250 MHz)	10000
		SBPF (0 to 350 MHz)	6500
5	40 MHz/20GS	NIL	9500
		MA	9000
		Others	No improv.

frequency range 0 to 350 MHz on EM signals has greatly reduced their requirement to extract the key. The requirement of number of EM signals has decreased from 11350 to 6500. Here, maximum 43% of reduction in the requirement of number of EM signals is noted down.

The effect of different signal processing techniques on EM signals of different clock frequencies is shown in Table 2. From this table, it is obvious that LPF technique is effective for EM signals of low clock frequency values (1 and 4 MHz). For the clock frequency having medium values (10 and 20 MHz), SBPF technique becomes more effective in terms of reducing the amount of EM signals to reveal the secret key. In the case of high value of clock frequency (40 MHz), none of the techniques (except MA) is able to reduce the requirement of EM signals. MA technique has offered insignificant amount of improvement.

6 CONCLUSIONS

Compromising EM signals are radiated from the FPGA chip executing AES algorithm. The secret key is revealed by mounting CEMA attack on EM signals representing tenth round of AES processing. Huge amount of EM signals are needed to deduce the key successfully by mounting CEMA attack due to presence of surrounding noises. To reduce the amount of EM signals required to extract the key with minimal computational overhead, cost effective approach is explored. This completely software oriented approach includes signal processing techniques namely moving average, low pass filtering, spectral band pass filtering and combination of more than one of them. The application of these techniques on EM signals of FPGA board executing AES algorithm enables in minimizing their requirement to break the system. The requirement of EM signals has been reduced up to 43% without employing additional complex equipments and increasing significant amount of computations. Different techniques are found effective in reducing the requirement of EM signals of different clock frequencies by eliminating/suppressing the noise contents. LPF and SBPF techniques are found more effective for low (1 & 4 MHz) and medium (10 & 20 MHz) clock frequency values respectively. At high clock frequency value (40 MHz), none of the techniques has emerged as a winner.

ACKNOWLEDGEMENTS

The authors would like to thank Dr. G. Athithan, DS, CCR&D, SAM for providing invaluable guidance and motivation during this work. Gratitude is extended to Sh. A. K. Sharma, Sc 'G', SAG for his encouraging and informative inputs during the preparation of this paper. Sincere thanks goes to both Sh. Akash Gupta, Sc 'C' and Ms Amita Malik, To 'D' of SAG for their invaluable suggestions and supports.

REFERENCES

Agrawal, D., Archambeault, B., Rao, J. & Rohatgi, P. 2002. The EM Side–Channel(s): Attacks and Assessment Methodologies. Proceedings of Cryptographic Hardware and Embedded Systems-2002. LNCS, vol. 2523: pp. 29–45.

Berna Ors, S., Oswald, E. & Preneel, B. 2003. Power- Analysis Attacks on an FPGA – First Experimental Results. CHES 2003, LNCS 2779: pp. 35–50, Springer-Verlag Berlin Heidelberg.

Brier, E., Clavier, C. & Olivier, F. 2004. Correlation power analysis with a leakage model. In the proceeding of CHES 2004. Springer, Hiedelberg, LNCS, vol. 3156: pp. 16–29.

Carlier, V., Chabanne H., Dottax, E. & Pelletier, H. 2004. Electromagnetic side channels of an FPGA implementation of AES. Cryptology ePrint Archive-2004/145, http://eprint.iacr.org/.

Clarke, G. M. & Cooke, D. 1998. A basic course in statistics, Arnold London, 4th edition.

Gandolfi, K., Mourtel, C., Oliver F. 2001. Electromagnetic Analysis: Concrete Results. In the proceedings of the workshop on Cryptographic Hardware and Embedded Systems 2001, LNCS 2162 Paris, France: pp. 251–261.

Kang, S.-M. & Leblebici, Y. 2002. CMOS Digital Integrated Circuits: Analysis and Design. McGraw Hill.

Kocher, P. 1996. Timing attacks on implementations of Diffie-Hellman, RSA, DSS and other systems. in Advances in Cryptology: Proceedings of CRYPTO'96, N. Koblitz, Ed., vol. 1109 of LNCS, pp. 104–113.

Kocher, P.C. & Jaffe, J., Jun, B. 1999. Differential Power Analysis. Springer, Heidelberg In: Wiener, M. (ed.) CRYPTO 99. LNCS, vol. 1666: pp. 388–397.

Maistri, P., Tiran, S., Maurine, P., Koren, I., Leveugle, R. 2013. Countermeasures against EM analysis for a secured FPGA based AES implementation. ReConFig, pp. 1–6.

Mangard, S. 2003. Exploiting radiated emissions -EM attacks on cryptographic ICs. In Proceedings of Austrochip.

Masoumi, M. and Rezayati, M. H. 2015. Novel Approach to Protect Advanced Encryption Standard Algorithm Implementation Against Differential Electromagnetic and Power Analysis, IEEE Transactions on Information Forensic and Security, Vol. 10, No. 2, pp. 256–265.

Mathews, A. 2006 Low cost attacks on smart cards: the electromagnetic side-channel. Technical report, NGSSoftware Insight Security Research. http://www.ngssoftware.com/research/papers/EMA.pdf

Peeters, E., Standaert, F-X. & Quisquater, J-J. 2007. Power and Electromagnetic Analysis: Improved Model, Consequences and Comparisons. Integration, the VLSI Journal, Elsevier, vol. 40, pp. 52–60.

Quisquater, J.-J. & Samyde, D. 2001. ElectroMagnetic Analysis (EMA): Measures and Countermeasures for Smart Cards. International Conference on Research in Smart Cards –E-smart 2001. Springer. LNCS, vol. 2140: pp.200–210.

Rao, J.R. & Rohatgi, P. 2001. Empowering Side-Channels Attacks. https://eprints.iacr.org/2001/037.pdf

Singh, A.K., Mishra, S.P., Suri, B.M., & Khosla, Anu. 2015. "Investigations of Power and EM Attacks on AES Implemented in FPGA", 2nd international conference on Soft computing in Problem Solving, SocProS-2015, held at IIT Roorkee.

Stallings, W. 2003. Cryptography and Network Security: Principles and Practice. 3rd edition: Prentice Hall.

Communication and Computing Systems – Prasad et al. (Eds)
© 2017 Taylor & Francis Group, London, ISBN 978-1-138-02952-1

Resistorless realization of universal filter employing CCIII and OTA

Tajinder Singh Arora
Maharaja Surajmal Institute of Technology, New Delhi, India

Deepali Srivastava
Noida Institute of Engineering and Technology, Greater Noida, Uttar Pradesh, India

Manish Gupta
Greater Noida Institute of Technology, Greater Noida, Uttar Pradesh, India

ABSTRACT: A new current mode universal filter is being proposed in this paper. The operational transconductance amplifier and third generation current conveyor are used here as an active building blocks. This second order biquadratic filter has used only two capacitors as passive components and they both are grounded, which is good for integrated circuit implementation. The current input is at low impedance port and current output is at high impedance port, making this circuit a better composition. The circuit is able to produce all the five basic filtering function i.e. low pass, high pass, band pass, band reject and all pass responses, without changing any hardware configuration as well as any component matching condition. PSPICE simulation has been done to test and verify the theoretical results.

1 INTRODUCTION

Despite the advancement in digital signal processing and their applications in real world, analog signal processing still cannot be replaced with it, because of its certain advantages. There are various processing of natural signals that should be carried-out with analog signal processing only, such as amplification, rectification etc. (Senani et al. 2014). Various devices have been realized that are capable of processing the analog electrical signals and modifying them according to their properties. These devices can be utilized to obtain circuits having different transfer functions corresponding to useful applications like amplifiers, summers, filters, oscillators, wave generators etc. (Senani et al. 2014). Electronic filter is one of the important application of analog signal processing. A filter allows electrical signals having certain frequencies to pass through it depending upon their type, which can be "low pass" (LP), "high pass" (HP), "band pass" (BP), "band reject" (BR) or "all pass" (AP) filter. "Current–mode (CM)", "voltage-mode (VM)", "trans-conductance-mode (TC)" and "trans-resistance-mode (TR) filters" are the various available modes of the filter operation. A filter that is capable of performing two or more types of filtering functions is called a multifunction filter whereas the one that can perform all five types of filtering functions is called a universal filter. Further, a filter can be of "Single-Input-Single-Output" (SISO), "Single-Input-Multiple-Output" (SIMO),

"Multiple-Input-Single-Output" (MISO) or "Multiple-Input-Multiple-Output" (MIMO) type.

Various universal filters based on Operational Transconductance Amplifier (OTA) and third generation current conveyor (CCIII) are available in literature, for instance (Horng & Jiun-Wei 2003, Kumngern et al. 2013, Siripruchyanun & Jaikla 2009, Tsukutani et al. 2007) and they are cited in. In (Horng & Jiun-Wei 2003) only 3 active devices has been utilized but suffers from the drawback of using floating capacitors. In (Tsukutani et al. 2007) proposes a universal filter using OTA and Current Conveyor (CC), but they employed three OTA and one CC, to get all the five responses. In (Kumngern et al. 2013, Siripruchyanun & Jaikla 2009, Tsukutani et al. 2007) simply OTA has been employed to get a second order standard filtering function, but uses a large number of active devices.

The authors propose one such proposition that eliminates all the above mention criteria. This circuit proposes a universal filter that gives all the five basic responses i.e. "LP", "HP", "BP", "BR" and "AP", simultaneously. This CM circuit offers electronic tunability of quality factor and frequency of operation. All the utilized capacitors are grounded in nature, which makes this circuit useful for their implementation and use in the form of integrated circuit. The current mode outputs are available explicitly, which makes the circuit good for the use of making higher order filter.

Introduction of the employed active building block, utilized in the designed filter, is given in section 2. Section 3 shows the "block diagram" and "transfer functions" of the proposed filter circuit. Section 4 provide the "sensitivity analysis" and section 5 depicts the graphical "simulation results" of the proposed resistor-less filter circuit that has been carried out using the software "PSPICE". The conclusion is presented in section 6.

2 INTRODUCTION TO OTA AND CCIII

The Operational Transconductance Amplifier (OTA), symbolically shown in Figure 1, is characterized by

$$I_o = \pm g_m (V_+ - V_-) \tag{1}$$

The transconductance factor is defined by g_m and it is controlled by the bias current inside the circuitry of the OTA. The polarity symbol + or − used before the transconductance parameter represents the direction of the current at the output.

A "third generation current conveyor" (CCIII), symbolically shown in Figure 2, is characterized by

$$\begin{bmatrix} I_y \\ V_x \\ I_{z\pm} \end{bmatrix} = \begin{bmatrix} 0 & -\alpha & 0 \\ \beta & 0 & 0 \\ 0 & \pm\gamma & 0 \end{bmatrix} \begin{bmatrix} V_y \\ I_x \\ V_{z\pm} \end{bmatrix} \tag{2}$$

where α, β and γ represent non-ideal port transfer ratios of X, Y and Z terminals respectively and ideally $\alpha = \beta = \gamma = 1$.

3 CIRCUIT CONFIGURATION

The proposed second order universal filter using OTA and CCIII, by utilizing all grounded passive

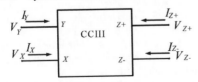

Figure 1. Symbolic notation of OTA.

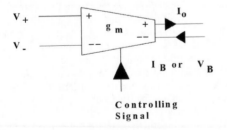

Figure 2. Symbolic notation of CCIII.

components is shown in Figure 3. By using the port's relationships given in (1) and (2), routine analysis yields derivation of all the "transfer functions" as given in (3–6). The "natural angular frequency" ω_0 and the "quality factor" Q of the filter are presented in (7).

$$\frac{I_{o1}}{I_{in}} = \frac{\left(\dfrac{g_{m1}g_{m2}}{C_1 C_2}\right)}{D(s)} \tag{3}$$

$$\frac{I_{o2}}{I_{in}} = -\frac{\dfrac{s.g_{m1}}{C_1}}{D(s)} \tag{4}$$

$$\frac{I_{o3}}{I_{in}} = \frac{s^2}{D(s)} \tag{5}$$

where $\quad D(s) = s^2 + \dfrac{s.g_{m1}}{C_1} + \dfrac{g_{m1}g_{m2}}{C_1 C_2} \tag{6}$

$$\omega_0 = \sqrt{\frac{g_{m1}g_{m2}}{C_1 C_2}} \quad \text{and} \quad Q = \sqrt{\frac{g_{m2}C_1}{g_{m1}C_2}} \tag{7}$$

The summation of I_{o1} and I_{o3} will give the BR response whereas summation of currents I_{o1}, I_{o2} and I_{o3} will provide an AP response. It is very much evident from the given equations (3–7) that all the filter responses, namely "high-pass" (HP), "band-pass" (BP), "low-pass" (LP), "band reject" (BR) and "all pass" (AP) can be obtained in the proposed filter circuit, with an ease. Any proposed universal filter must give all the said five basic filter responses. It is very much evident from (7) that ω_0 and Q have electronic tunability and their value can be changed by varying the value of g_{m1} or g_{m2}. In section 5 we have presented simulation results depending upon the proposed filter configuration.

4 SENSITIVITY ANALYSIS

By the sensitivity analysis of the realized network it was found that the proposed configuration has a very low sensitivity for all the active and passive components employed. Therefore, this analysis justified the performance of our network.

$$S_{C_1}^{\omega_0} = S_{C_2}^{\omega_0} = -\frac{1}{2} \tag{8(a)}$$

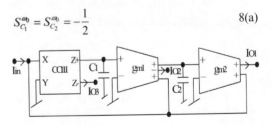

Figure 3. Proposed universal filter configuration.

$$S_{g_{m1}}^{\omega_0} = S_{g_{m2}}^{\omega_0} = \frac{1}{2} \qquad \text{8(b)}$$

$$S_{C_1}^{Q} = S_{g_{m2}}^{Q} = \frac{1}{2} \qquad \text{8(c)}$$

$$S_{C_2}^{Q} = S_{g_{m1}}^{Q} = -\frac{1}{2} \qquad \text{8(d)}$$

5 SIMULATION RESULTS

The testing and the workability of the proposed design have been checked, for the "current-mode" response of the proposed filter circuit of Figure 3, on "PSPICE" software. The CMOS versions of OTA (Senani et al. 2014) and CCIII (Arora & Sharma 2016) are shown in Figure 4 and Figure 5 respectively, for the use of making a universal filter. The mode of operation of MOS transistors utilized in Figure 4 and Figure 5 is saturation –mode. The "aspect ratios" of all the MOS transistors, utilized in Figure 4 and Figure 5, are given in Table 1 and Table 2 respectively. The 0.18 μm CMOS model parameters used for the PSPICE simulations have been taken from (Arora & Sharma 2016). The selected supply voltage has been $V_{DD} = -V_{SS} = 1.0$ V for OTA and $V_{DD} = -V_{SS} = 1.45$ V for the CCIII. The bias current is taken as $4\,\mu$A for the OTA to get the values of $g_{m1} = g_{m2}$ as $42.8\,\mu$. The ω_0 chosen for the design was 1MegaHz and the passive component values chosen were $C_1 = 3.41$ pF and $C_2 = 13.46$

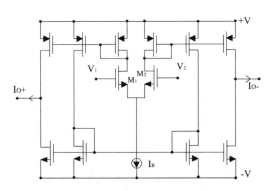

Figure 4.   CMOS realization of the OTA (Senani et al. 2014).

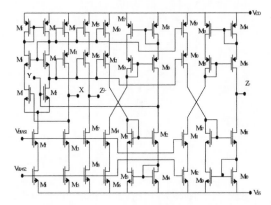

Figure 5.   Third generation current conveyor (CCIII).

Table 1.   Aspect ratios of CMOS utilized in OTA.

MOSFET	W(μm)	L(μm)
M1, M2	5.76	0.72
M3, M4, M5, M6, M7, M8	2.16	0.72
M9, M10, M11, M12	1.44	0.72

Table 2.   Aspect ratios of the MOSFETs of CCIII.

CMOS Transistors	W(μm)	L(μm)
M1-M4	5.714	0.43
M5, M6, M7, M8, M13-M16, M21-M24, M27, M28, M31, M32, M37-M40	9.285	0.43
M9-M12, M17-M20, M25, M26, M29, M30, M33-M36	18.571	0.43

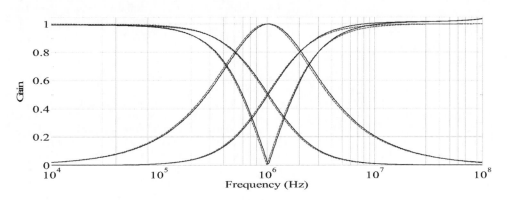

Figure 6.   Frequency response curves of the proposed filter circuits.

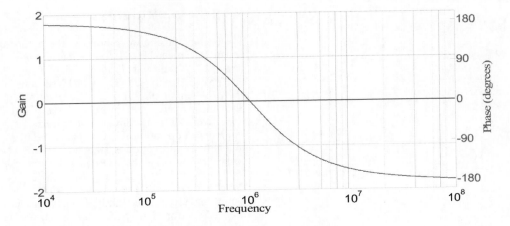

Figure 7. Magnitude and Phase response of an all pass filter configuration.

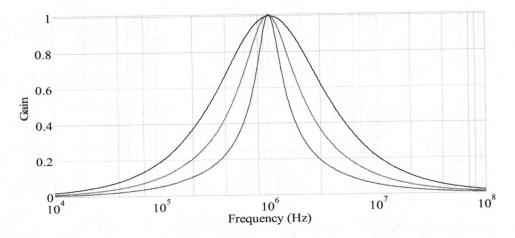

Figure 8. Tunability of Q_0 on the band pass filter response.

Table 3. Different component values for Q_0 tunability.

S.No.	Conditions	Q	ω_0
1	$g_{m1} = g_{m2} = g_m$ $C_1 = C/2, C_2 = 2C$	½	1 MHz
2	$g_{m1} = 2\,g_m, g_{m2} = g_m$ $C_1 = 2C, C_2 = C$	1	1 MHz
3	$g_{m1} = g_m, g_{m2} = 2\,g_m$ $C_1 = 2C, C_2 = C$	2	1 MHz

pF. Simulation results based on the Figure 3 of the proposed "current mode" filter for "LP", "BP", "HP" and "BR"are shown in Figure 6. Dotted line in Figure 6 shows the ideal or theoretical outputs whereas continuous line represents the simulated values of the outputs of the proposed filter configuration. For an "all-pass" (AP) filter, the gain

i.e. magnitude and phase response are shown in Figure 7. One can inspect in Figure 6 and find that ideal responses and simulated are very close to each other. The tunability of Q while keeping the value of ω_0 fixed is shown in Figure 8, by taking the different combination values of g_{m2}, g_{m1}, C_1 and C_2 as shown in Table 3.

6 CONCLUSION

The paper proposes a new and novel current-mode universal filter based on OTA and a CCIII. It has been simulated using Orcad PSPICE software. The circuit offers several advantages such as, 1) The circuit does not have any resistor as a passive component, 2) requirement of minimum number of capacitors i.e. only two, 3) to achieve all 5 responses it does not require any matching conditions for passive components, 4) all the five responses can be achieved without changing any hardware configuration, 5) the use of all grounded passive components, 6) this current mode circuit has an electronic tunability of Q_0, 7) explicit outputs available at high impedance port makes it a better proposition for cascading, 8) The proposed second order filter operates on extensively high range of frequencies. The theoretical results and the obtained practical results are very much close to each other and thus justified the design.

REFERENCES

Arora, T.S. and Sharma, R.K. 2016. Adjoint-KHN equivalent realization of current mode universal biquad employing third generation current conveyor. *Indian Journal of Science and Technology,* 9, no. 7: 1–7.

Arora, T.S. and Sharma, R.K. 2016. Realization of current mode KHN-equivalent biquad employing third generation current conveyor. *IEEEICCTICT'2016, pp. 20–24.*

Horng, Jiun-Wei. 2003. High input impedance voltage-mode universal biquadratic filter using two OTAs and one CCII. *International Journal of Electronics* 90, no. 3: 185–191.

Kumngern, Montree, Peerawut Suwanjan, and Kobchai Dejhan. 2013. Electronically tunable voltage-mode universal filter with single-input five-output using simple OTAs. *International Journal of Electronics* 100, no. 8: 1118–1133.

Senani, Raj, Gupta, Manish, Bhaskar, D.R. and Singh, A.K. 2014. Generation of equivalent forms of operational trans-conductance amplifier-RC sinusoidal oscillators: the nullor approach. *The Journal of Engineering* 1, no. 1.

Senani, Raj, Bhaskar, D.R. and Singh, A.K. 2014. Current conveyors: variants, applications and hardware implementations. *Springer.*

Siripruchyanun, M. and Jaikla, W. 2009. "Cascadable current-mode biquad filter and quadrature oscillator using DO-CCCIIs and OTA", *Circuits, Systems & Signal Processing, vol. 28, no. 1, pp. 99–110.*

Tsukutani, T., Sumi, Y. and Fukui, Y. 2007. Novel current-mode biquad filter using OTAs and DO-CCII. *International journal of electronics* 94, no. 2: 99–105.

Communication and Computing Systems – Prasad et al. (Eds)
© 2017 Taylor & Francis Group, London, ISBN 978-1-138-02952-1

Face detection from digital images: A comparative study

Garima Sharma, Akansha Singh & A.K. Yadav
The Northcap University, Gurgaon, India

Krishna Kant Singh
Dronacharya College of Engineering, Gurgaon, India

ABSTRACT: Detecting human faces from digital images are of great significance as faces represent important and meaningful information which play a vital role in all face analysis systems. Face detection is the first step in any face processing system. This field has gained enormous attention by the researchers due to its wide variety of applications. Performing face detection is a challenging task due to certain problems associated with it. This paper presents, a survey and comparison of various approaches available for the process of face detection along with the challenges associated with it.

1 INTRODUCTION

The field of face detection from digital images has gained enormous interest from researchers in recent areas. It has became a popular area of research in computer vision due to its wide range of applications such as human face recognition, surveillance systems, human computer interfaces, video conferencing, lip tracking, gender classification [8] and so on.

Face Detection as defined by Yang M.H et al. (2002) can be defined as the process to determine the presence of any faces in a given image and if present, return the location and extent of each face. It is the first step to build any face processing system such as face recognition, face tracking, pose estimation and expression classification. Although it appears to be an easy task for humans but it is a very challenging task for computers due to various difficulties associated to it. The challenges associated with automatic facial detection can be listed as follows:

a. Pose: Relative camera-face position can cause part of face to be occluded partially or even wholly.
b. Occlusion: Occlussion is obstructions to the faces in images. Facial features can be occluded by beard, spectacles etc.
c. Facial expression: Different expressions can directly affect the facial appearance.
d. Imaging Conditions: Factors like camera characteristics and lighting conditions affect the facial apperances in images.
e. Illumination: The lighting and the angle of light may cause faces to appear different in different light conditions.

This paper contains a survey on different approaches to face detection i.e. knowledge based approach, feature invariant approach, template based approach and appearance based. Section 2 contains a literature review on different approaches and methods for face detection. Section 3 provides a comparison between the approaches listing their merits and demerits.

2 REVIEW OF APPROACHES FOR FACE DETECTION

2.1 Knowledge based approach

Knowledge based approach for face detection as discussed by by Yang M.H et al. (2002) is a rule based top-down approach. Face detection using this approach is done based on certain rules which are derived from researcher's knowledge of human faces. These rules describe the facial features and relationship between them in terms of their relative distances and positions. For instance, a facial image can be described as a combination of two eyes that are symmetrical to each other, a nose and a mouth. Thus, an image which satisfies these rules can be considered as human face. It is a top-down approach as it works by first extracting the features from the face followed by the complete facial detection based on the derived rules. Advantage of using this approach is that it is easy to come up with simple rules to define human face. But on the other hand this approach also suffers certain setbacks i.e. if the derived rules are too general, it may result in false positives and if the rules are too strict, it may fail to detect faces that don't pass all the rules.

A face detection method is proposed by] Nam M et al. (2006), which used a context based face detection method to identify facial regions from an image. The context-based face detection method works by first reducing the search space of the image by finding regions of face color. Later, the remaining regions are searched upon using the multiple Bayesian classifiers under varying illuminations. Image is organized into sub-bands for processing. As the facial images have variant illuminations, the proposed model used multiple classifiers which carry the face illumination information. It was observed that multiple classifiers yielded better results as compared to single classifier.

Jung J & Horace H.S (2007) presented a scale independent technique to detect the locations of human faces in images. It is a hierarchical approach which is performed in three different levels. At level 1, certain rules are derived using the eye model to find out the presence of human face in the image. Level 2 uses an improved Yang's mosaic image model to check the consistency of features with respect to human face. Finally, SVM based face model is applied at the third level to eliminate false positives obtained from second level.

Khanum. A et al. (2008) proposed a method to detect multiple faces in image invariant to scale and position. It combined human skin color detection technique with knowledge based approach. The search space of the input image is reduced by extracting the skin coloured clusters from the image. The resultant dataset is processed by neural networks which are trained to detect features like: left eye, right eye, nose and mouth. Finally when both colour and human features match the input image, facial region is detected.

Another method presented by Wang D, Ren J et al. (2008) works by firstly detecting the skin pixels in an image using supervised learning. After the skin regions are detected, they are labeled to obtain the outer boundary rectangle and no. of pixels in each region. Regions are subjected to a threshold condition i.e. the regions having pixels less than the threshold value are removed and are then filtered based on SR parameter which defines rules based on width and height of the regions. These regions are then fed as an input to an algorithm which contains simple rules based on relationships of the facial features present in human face.

An efficient method for facial detection is presented by Devadethan S et al. (2014) which works by detecting the eye region from the image. After detecting the facial region, feature points are extracted and distance between them is used to find a possible face candidate. At first eye regions are processed to verify the possible eye regions in the image and then the further processing is done in order to find out the face candidate.

2.2 Feature based approach

Yang M.H et al. (2002) discussed feature invariant approach as a bottom up approach i.e. based on the assumption that humans can effortlessly detect faces in varying poses and lighting conditions, so there must be certain properties that are independent of these variabilities. These methods aim to extract facial features like eyes, nose, mouth and hairline using edge detection and corner detection methods and then, building a statistical model based on them to describe their relationships in order to judge the presence of a human face. One disadvantage of using this approach is that image features can be severely corrupted due to conditions like illumination, noise and occlusion.

Skin color is widely used as an efficient feature for face detection although different people have different skin colours. But studies have shown that the difference is based on their intensity rather than their chrominance. Multiple feature method uses a combination of facial features to detect faces.

A feature-based approach by Gizatdinova Y & Surakka V. (2006) is used for detection of facial landmarks which successfully detected eyes from facial images regardless of the expressions. The method extracts the oriented edges and constructs edge maps at two resolution levels. Landmark candidates are then obtained by selecting regions with characteristic edge pattern.

Face detection by Miry L.D.A.H. (2014) is performed by first segmenting the skin region from the image using three different colorspaces RGB, YCbCr and HIS and then in order to extract features, high information segments of facial images are selected by comparing it with ORL image using Euclidean distances. It also incorporated fuzzy systems to determine presence of face or not. This method showed a detection accuracy of 94.74%.

Yan X, Chen X.W. (2009) used skin color and wavelet express of images along with PCA to obtain Eigen vectors. These eigen vectors distinguish the face and non-face regions. In order to detect multiple faces from input image, modified Bayesian classifier is used.

An algorithm for robust and efficient face detection is proposed by Abdellatif Hajraoui & Mohamed Sabri (2014). The algorithm contains 2 modules. The first module uses watershed segmentation method to detect skin pixels and the second module determines the presence of face in each region of skin with a cascade of Gabor filters.

Another high level skin detection technique is presented by Mohd Zamri Osman et al. (2016) based on online sampling where offline training is not required. It uses a dynamic skin detector which is a combination of multicolor spaces in order to reduce false positives and increase the precision rate of detection as compared to single color space.

Taylor M. J. & Morris T (2014) adopted a Viola-Jones feature based face detector in high precision configuration to eliminate false positive results. The segmentation step begins with the detection of faces and then defining sub region within the obtained regions. These pixel regions are then filtered according to their luma values in order to obtain skin pixels within the set.

2.3 Template based approach

In Template based approach as mentioned by Yang M.H et al. (2002), certain standard patterns of face are stored which describe the face as a whole or facial feature separately. Face detection is performed by calculating the correlation values between the input image and the stored patterns. Advantage of using this approach is that it is very simple to implement and easily determines facial regions using the correlation values. However it is inefficient to detect faces in the conditions of variation in scale, pose and shape. Multiscale, sub templates and deformable templates have been proposed where the templates are deformed by translation, rotation and scaling in order to achieve scale and shape invariance.

A hierarchical model is proposed by Wang J & Yang H (2008) for face detection which uses template matching algorithm along with 2DPCA algorithm. At the first level, a rough classifier is used to filter out the non-face regions and the output of this step is fed into the second level which uses a core classifier based on 2DPCA algorithm in order to detect the facial regions the image.

Tripathi Smitha et al. (2011) combined the skin colour detection and template matching method. After the skin areas are detected, sobel edge detector is used to detect edges and finally the trained images are used in template matching to find out the location of eyes, nose, mouth and discard the non-face regions.

T. Venugopal & T. Archana (2015) compared template based approach with a statistical based approach i.e PCA over the images of FERET database. It was observed that the recognition rate of PCA was only 70–75% and it failed to detect faces in changes in illumination, pose, in plane rotation whereas template matching is a very simple approach and gave results 20% better and even recognized faces efficiently invariant to all factors.

A system is proposed by Sharma S (2013) where users can backup and compress their database servers remotely. The system works by capturing the image of user and exhaustively matching it by comparing each pixel of the captured image to the template saved on server. The exhaustive matching procedure utilizes the neural networks algorithm and template matching.

Another method propose by Nikan Soodeh & Ahmadi Majid (2015) works by localizing the partial template face image based on the similarities between the intensities of partial face and gallery samples. ZNCC and NSSD techniques are proposed for template matching. Then, feature extrac-

tion and classification approaches are incorporated to find the best match.

2.4 Appearance based approach

Yang M.H et al. (2002) explains appearance based methods are similar to template matching but instead of using predefined templates, they adopt machine learning techniques to extract faces. Image vector is considered as a random variable having some probability of being a face or not a face. Another approach is to define a discriminant function between facial and non-facial regions. Some of the most relevant methods of this approach are:

a. Eigen faces: A. R. Mohan et al. (2009) built a classifier to filter the non-face regions by boosting a set of weak classifiers constructed by projection into eigen vectors of face space. An efficient algorithm is proposed by Tayal Yogesh et al. (2013), where weight and Euclidean distance of input image is considered to compute the features under varying lighting and backgrounds.

b. Neural Networks: Neural Networks is successfully used for face detection due to its feasibility to train a system to capture complex class conditional density of face pattern. A.Bouzalmat et al. (2011) proposed BPNN with gabor wavelet for face detection. Feature vector is extracted based on Fourier Gabor filter and is then fed in as input to BPNN for recognition. Another method proposed by Bailing Zhang et al. (2015), detects the occluded facial features based on convolution NN with multitask learning. This approach operates directly on the image pixels and is based on end to end principle and is thus extremely effective as well as optimal from system point of view.

c. Support Vector Machine: A method is used by Ruan J & J. Yin (2009) which filters the skin regions from the image and then uses linear SVM to exclude the non-facial regions. Face candidates are finally selected by detecting features like eyes and mouth. Deep feature face detection for mobile network is performed in [16] where DFFDM algorithm extracts deep features from facial images using Alexnet layers. Faces are detected by feeding different sized sliding window into SVM i.e trained for each window size.

d. SNoW classifier: Nilsson M et al. (2007) and Somashekar K et al. (2011) performed face detection in three steps. In the first step, pixel info of an image is obtained. Second step uses local SMQT features as feature extraction for object detection. Finally the task of face detection is accomplished by combining the feature and the fast SNoW classifier.

3 COMPARATIVE STUDY

Table 1 lists the discussed approaches along with their merits and demerits.

Table 1. A comparative study of face detection approaches.

Approach	Merits	Demerits
1. Knowledge based approach	(i) Based on user's interpretation of human face geometry. (ii) Simple rules. (iii) It is an efficient method to detect frontal faces. (iv) It is a Top-down approach.	(i) If the rules are too detailed, it may fail to detect faces. (ii) If the rules are too general, it may result in false positives. (iii) It is unable to detect faces in different poses.
2. Feature Invariant approach	(i) Features are invariant of vari-ability's of poses and lighting conditions.	(i) Features can be severely corrupted due to illumina-tion, noise and occlusion. (ii) Shadows can cause strong edges and thus weaken feature boundaries. (iii) It is difficult to detect features in complex background.
3. Template Based approach	(i) Simple method. (ii) Requires less data points for face detection.	(i) It is not suitable to detect faces if there is obstruc-tion in front of face. (ii) More templates are required to cover more views of face thus increasing detec-tion time. (iii) Depends on size, scale and rotation.
4. Appearance Based approach	(i) Fast, efficient and robust. (ii) Uses powerful machine learning techniques. (iii) Works effi-ciently in differ-ent poses and orientation. (iv) Provides good empirical results.	(i) Learning requires a lot of positive and neg-ative examples. (ii) The architecture has to be exten-sively tuned to get exceptional performance.

4 RESULTS

This section shows some results of the different approaches discussed in this paper:

Table 2. Results of using different approaches.

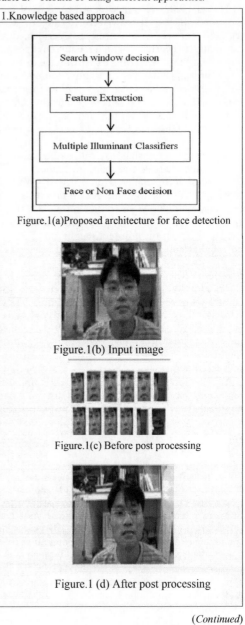

1.Knowledge based approach

Figure.1(a)Proposed architecture for face detection

Figure.1(b) Input image

Figure.1(c) Before post processing

Figure.1 (d) After post processing

(Continued)

Table 2. (*Continued*)

2.Feature invariant approach
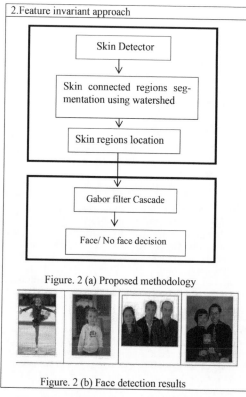
Figure. 2 (a) Proposed methodology
Figure. 2 (b) Face detection results

3.Template based approach
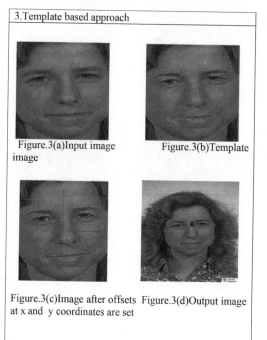
Figure.3(a)Input image image   Figure.3(b)Template
Figure.3(c)Image after offsets at x and y coordinates are set   Figure.3(d)Output image

(Continued)

Table 2. (*Continued*)

4.Apperance based approach
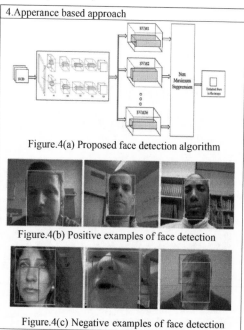
Figure.4(a) Proposed face detection algorithm
Figure.4(b) Positive examples of face detection
Figure.4(c) Negative examples of face detection

5 CONCLUSION

This paper shows a survey of different techniques for face detection. Face detection is the process of identifying faces in digital images and returning its position. It is an interesting and active area of research. Various methods are discussed in this paper. A comparative study of different approaches is done to list out the merits and demerits of using each approach. The future scope of this area is to make an efficient and robust face detection system which can provide accurate results in any circumstances.

REFERENCES

Abdellatif Hajraoui & Mohamed Sabri. 2014. Face Detection Based On Skin Detection, Watershed Method and Gabor Filters. *International Journal of Computer Applications.* 94(6):.33–39.

Bailing Zhang et al. 2015. Face occlusion detection based on multi-task convolution neural network. *In Proceedings of 12th International Conference on Fuzzy Systems and Knowledge Discovery (FSKD).* Pp. 375–379. IEEE.

Bouzalmat, Anissa et al. 2011. Face Detection And Recognition Using Back Propagation Neural Network And Fourier Gabor Filters. *Signal & Image Processing: An International Journal (SIPIJ).* 2(3):15.

Devadethan S et al. 2014. Face detection and facial feature extraction based on a fusion of knowledge based method and morphological image processing. *In Pro-*

ceedings of Annual International Conference on Emerging Research Areas: Magnetics, Machines and Drives (AICERA/iCMMD). pp. 1–5. IEEE.

Gizatdinova Y & Surakka V. 2006. Feature-based detection of facial landmarks from neutral and expressive facial images. *IEEE Transactions on Pattern Analysis and Machine Intelligence.* 28(1):135–139.

Jung J & Horace H.S Ip. 2007. A real time hierarchical rule based approach for scale Independent human face detection. In *Proceedings of Electronic Imaging, International Society of Photo-Optimal Instrumentation Engineers* (SPIE) Conference Series 2007. pp.64960p-64960p.

Kaushal A & Raina JPS. 2010. Face Detection using Neural Network and Gabor Wavelet Transform. *International Journal of Computer Science and Technology.* 1(1).

Khanum A et al. 2008. A Heuristically Guided Hybrid Approach to Face Detection for Context Based Image Retrieval in Internet Images. In *Proceedings of 4th International Conference on Emerging Technologies.* p.237–41.

Miry L.D.A.H. 2014. Face Detection Based on Multi Facial Feature using Fuzzy Logic. *Al-Mansour Journal. Issue(21).*

Mohan, A. R. & N. Sudha. 2009. Fast Face Detection Using Boosted Eigenfaces. In *Proceedings of IEEE Symposium on Industrial Electronics and Applications (ISIEA 2009).* Vol. 2:1002–1006.

Mohd Zamri Osman et al. 2016. Improved Dynamic Threshold Method for Skin Colour Detection Using Multi-Colour Space. *American Journal of Applied Sciences.* 13(2): 135–144.

Nam M et al. 2006. An efficient face and eye detector modeling in external environment. *Artificial Intelligence and Soft Computing-ICAISC.* 4029:841–9. Springer Berlin Heidelberg.

Nikan Soodeh & Ahmadi Majid. 2015. Partial Face Recognition Based on Template Matching. In *proceedings of 11th International Conference on Signal-Image Technology & Internet-Based Systems* (SITIS). pp.160–163.

Nilsson M et al. 2007. Face Detection using Local SMQT Features and Split Up SNOW Classifier. In *Proceedings of IEEE International Conference on Acoustics, Speech and Signal Processing (ICASSP),* Vol.2. pp.II – 589.

Ruan J & J. Yin. 2009. Face detection based on facial features and linear support vector machines. In *Proceedings of the International Conference on Communication Software and Networks ICCSN'09.* pp: 371–375. IEEE.

Sarkar, S., Patel, V.M. and Chellappa, R., 2016. Deep Feature-based Face Detection on Mobile Devices. *arXiv preprint arXiv:1602.04868.*

Sharma S. 2013. Template Matching Approach for Face Recognition System. *International Journal of Signal Processing Systems.* Vol. 1(2).

Somashekar K et al. 2011. Face detection by smqt features and snow classifier using color information. *International Journal of Engineering Science and Technology.* 3(2):1266–1272.

Tayal Yogesh et al. 2013. Face Recognition using Eigenface. *International Journal of Emerging Technologies in Computational and Applied Sciences (IJETCAS).* Vol. 3 (1).pp: 50–55.

Taylor M. J. & Morris T. 2014. Adaptive skin segmentation via feature-based face detection. In *Proceedings of SPIE Photonics Europe– Real-time Image and Video Processing.* pp: 91390p-91390p.

Tripathi Smitha et al. 2011. Face Detection using Combined Skin Color Detectorand Template Matching Method. *International Journal of Computer Applications.* 26(7): 5–8.

Venugopal, T. & T. Archana. 2015. Face recognition:A template based approach. In *Proceedings of International Conference on Green Computing and Internet of Things (ICGCIoT).* pp.966–969. IEEE.

Wang D, Ren J et al. 2008. Skin Detection from Different Color Spaces for Model Based Face detection. *Advanced Intelligent Computing Theories and Applications: With Aspects of Contemporary Intelligent Computing Techniques.* Vol.15:487–94.

Wang J. & Yang H. 2008. Face Detection Based on Template Matching and 2DPCA Algorithm. *In Proceedings of Congress on Image and Signal Processing(CISP).* Vol. 4. pp.575–579.

Wong K.W. et al. 2001. An Efficient Algorithm for Human Face Detection and Facial Feature Extraction under Different Conditions. *Pattern Recognition,* Vol. 34:1993–2004.

Yan X, Chen X.W. 2009. Multiple faces detection through facial features and modified Bayesian classifier. In *Proceedings of International Conference on Multimedia Information Networking and Security.* Vol.1. p.73–7.

Yang M.H et al. 2002. Detecting Faces in images: A Survey. *In IEEE Trans. On Pattern Analysis and Machine Intelligence.* Vol. 24(1):34–58.

Communication and Computing Systems – Prasad et al. (Eds)
© 2017 Taylor & Francis Group, London, ISBN 978-1-138-02952-1

Improved iris biometric system by exploring iris texture features

N. Rathee
Maharaja Surajmal Institute of Technology, New Delhi, India

A. Rathee
IIT, Delhi, India

ABSTRACT: In the present era of automation, biometrics has drawn the interest of researchers working in the field of computer vision and robotics. The features, which has been popularly accepted for biometrics, include facial features (iris, ear), palm, thumb and gait. Among the above mentioned features, iris is the most reliable feature for biometrics as the texture features of eyes are consistent and independent of age. In the presented approach, texture features of iris have been explored for iris biometrics. The KAZE features have been used for extracting the iris features. The extracted multiple features from the same eye are checked for consistency. The inconsistent features are removed using the mask and the consistent features are fused together to generate a feature vector. The extracted features are applied to support vector machine for iris recognition. The proposed approach is evaluated on the Chinese Academy of Sciences—Institute of Automation database and the results are compared with the recently proposed methods to prove its efficacy.

1 INTRODUCTION

Biometric has drawn the interest of researchers since last many decades and had achieved great success. The biometric techniques, which were earlier based on finger print recognition, have been developed now by using other features also. The commonly used features can be classified as physiological features and behavioral features. The physiological features such as palm print, face, lips and iris are capable of representing static characteristics and are hereditary in nature. The behavioral features represent dynamic characteristic and are produced by a person unconsciously. The most commonly adopted behavioral features are handwriting, speech and gait. The behavioral methods can mislead as they can be copied and are less reliable.

Recently, iris, which is one of the most reliable physiological features, has been increasingly used for biometric (Jain et al. 2004) due to the consistency of iris pattern of a person. Moreover, it needs a small region of face to be processed and hence reduces the computation in comparison with complete face.

Recently, iris, which is one of the most reliable physiological features, has been increasingly used for biometric (Jain et al. 2004) due to the consistency of iris pattern of a person. Moreover, it needs a small region of face to be processed and hence reduces the computation in comparison with complete face.

The pioneering work in iris biometric area is done by Daugman (Daugman 1993) by extracting phase information of iris texture at various scales using quadrature wavelets (Daugman 2001 & Daugman 2004). For iris recognition, he adopted template matching based on Hamming distance. Later, many researchers have worked in this area mainly for performing two tasks: iris feature extraction and iris classification. The iris feature extraction is done in literature by Laplacian Pyramid (Wildes 1997), wavelet transform (Boles & Boashash 1998), circular symmetric filters (Ma et al. 2002), log Gabor filters (Yao et al. 2006) and by using the random and uniqueness properties (Daugman 2006). Other research in this area include improvement in image acquisition (Matey et al. 2006 & Park & Jim 2005), iris segmentation algorithm [Sung et al. 2004 & Proenca & Alexandre 2006).

For iris recognition, template matching is widely adopted. The similarity between the features is measured by computing the distance between the base template and the input image; and score of matching for the classes is computed. The average of the matched scores is used for iris recognition [Ma et al. 2003, 2004). Another approach is suggested by Krinchen et al. (Krichen et al. 2005) by adopting minimum value of score fusion. In contradiction to this, Schmid et al. (Schmid et al. 2006) computed average of distance for matching. To further improve the performance of the system, Hollingsworth et al. (Hollingsworth et al. 2006) suggested to acquire multiple codes from the same eye and to remove the redundant data by masking the inconsistent bit.

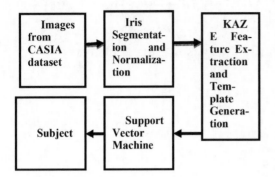

Images from CASIA dataset	Iris Segmentation and Normalization	KAZE Feature Extraction and Template Generation
Subject	Support Vector Machine	

Figure 1. Block diagram of the proposed iris biometric system.

From above mentioned brief literature survey, it is clear that for an efficient iris recognition system, three criteria should be taken care of: computation cost, computation time and accuracy. In the proposed approach, iris features have been extracted using KAZE features. The dimension of the extracted features is reduced by removing the inconsistent features and taking the consistent features as representative of iris features. The processed features have been applied to support vector machine for iris recognition. The proposed approach is evaluated on Chinese Academy of Sciences—Institute of Automation (CASIA) dataset for biometric purposes.

The approach presented in this paper is organized as follows: section 2 describes the preprocessing step. The KAZE feature extraction and generation of base template by post processing the extracted features is presented in section 3. The brief of SVM classifier used for biometric is explained in section 4. The CASIA database, used to evaluate the proposed approach is mentioned in section 5. Experimental results are discussed in detail in section 6 and finally conclusion are drawn in section 7. The overview of the proposed approach is well depicted in Figure 1.

2 PREPROCESSING

The pre-processing step is the primary step, which is taken prior to the feature extraction. For the proposed approach, pre-processing involves two main tasks: iris segmentation and iris normalization. The images from the CASIA dataset include pupil, eyelashes, eyelids and sclera. The iris is localized in the region between the iris and sclera (outer boundary) and iris and pupil (inner boundary).

So, iris segmentation needs localization of the iris as well as pupil. In literature, Hough transform has been frequently adopted to accomplish the same. The major challenge in iris localization

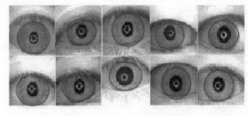

Figure 2. Result of iris segmentation using Hough Transform.

is specular reflection caused by occlusion of upper and lower parts of the iris by eyelids and eyelashes, respectively. This results in distortion of eye pattern, which may deteriorate the performance of iris biometric system. To handle the above issue, linear Hough Transform with threshold is applied. The results for iris segmentation are shown in Figure 2.

The next step in pre-processing is normalization in terms of size and illumination. The inconsistency in size is due to varying distance between camera and variation in illumination. The varying illumination may cause dilation or contraction of pupil, thus affecting the size and pattern of iris. The size normalization is needed for faithful comparison of iris templates.

To avoid the inconsistency in iris size, iris is modelled as a rubber-sheet (flexible) and this rubber sheet is unwrapped into a rectangular block with constant polar dimension with the assumption that variation in size of iris is linear during dilation and contraction. The mathematical formulation for mapping cartesian coordinates (x,y) of the iris features to polar coordinates (r, θ) is given as

$$x(r,\theta) = (1-r)x\_r(\theta) + rx\_i(\theta) \qquad (1)$$

$$y(r,\theta) = (1-r)y_r(\theta) + ry_i(\theta) \qquad (2)$$

where (xr(θ), yr(θ)) are the set of pupillary boundary points, and (xi(θ), yi(θ)) are the set of limbus boundary points and the boundary condition on polar coordinates are: $0 < r \leq 1$ and $0 \leq \theta \leq 360°$.

3 FEATURE EXTRACTION AND POSTPROCESSING

3.1 KAZE feature extraction

The previous feature descriptors such as SIFT and SURF depicted features at various scale levels by constructing or rounding off the Gaussian scale space of an image. But the borderline of the objects are smoothened to the same extent as that of both details and noise thereby severely affecting localization accuracy and distinctiveness. The object boundaries can be preserved by making blurring locally adaptive to the image data without severely affecting the noise reducing capability of the Gaussian blurring.

From the previous research, (Weickert et al. 1998 & Romeny 2008) it is duly known that non-linear diffusion approaches yield better results than linear ones and astonishing results have been obtained in the applications such as image segmentation (Weickert 2001) and denoising (Qiu et al. 2011). Therefore, 2D features are detected in a nonlinear scale space by the means of nonlinear diffusion filtering. The expansive computations and high number of iterations because of small step size limited the usage of the nonlinear diffusion filtering from practical computer vision components such as feature detection and description. This limitation is overcome by the use of Additive Operator Splitting (AOS) techniques (Alcantarilla et al. 2012). By means of AOS schemes we can obtain stable nonlinear scale spaces for any step size in a very efficient way.

The classic nonlinear diffusion formulation is depicted in equation 3.

$$\partial K / \partial t = div\left(o(x,y,t).\nabla K\right) \tag{3}$$

where div and ∇ are respectively the divergence and gradient operators. The diffusion is made adaptive to the local image structure due to conductivity function (o) in the diffusion equation. The function c is dependent on the local image differential structure. The scale parameter is represented by t.

The scale space in logarithmic steps that are arranged in a series of C Octaves and L sub-levels are discretized. Downsampling is not performed at every new octave as was the case with SIFT, thus the original image resolution is maintained. The sub-level indexes are usually identified by a discrete index c and a sub level l and are mapped to their corresponding scale f by:

$$f_j(c,l) = f_0 2^{\frac{c+l}{L}}, \\ c \in [0...C-1], \quad l \in [0...L-1], j \in [0...M] \tag{4}$$

where f0 represents the base scale level and N gives the total number of the filtered images. The non-linear diffusion filtering is defined for time domain thus the set of discrete set levels in pixel units fj is converted to time terms. Convolving an image with Gaussian of standard deviation f is similar to filtering the image for time = f^2/2. A set of evolution times is obtained and the scale space is transformed to time units through the mapping given below $f_j \to t_j$

$$t_j = \frac{1}{2} f_j^2, j = \{0...M\}, \tag{5}$$

A given image is convolved with a Gaussian Kernel having a standard deviation f_0 for reduction of noise. Image gradient histogram is computed and a contrast parameter is obtained. Finally, the position of key points is estimated. The non-linear scale space is built in an iterative way using the AOS schemes as following

$$K^{j+1} = \left(J - \left(t_{j+1} - t_j\right).\sum_{k=1}^{n} A_k\left(K^j\right)\right)^{-1}.K^j \tag{6}$$

The response of scale normalized determinant of Hessian is calculated at multiple scale levels. The set of differential operators has to be normalized with respect to scale for the purpose of multiscale feature detection by:

$$K_{Hessian} = f^2\left(K_{xx}K_{yy} - K_{xy}^2\right) \tag{7}$$

where K_{yy} and K_{xx} are second order vertical and horizontal derivatives where as K_xy is the second order cross derivative. From the set of filtered images from nonlinear scale space Kj the detector response is analyzed at different scale levels f_j. The maxima is searched in all images except where j = 0 and j = M. The output of the filter is checked over a window of 3 × 3 and consecutive Schurr filters have been adopted to approximate the second order derivatives at desired co-ordinates.

The features from iris region possess high degree of freedom, so the features from different individuals are quite different and of the same person are highly correlated. This property is explored in the proposed approach for iris biometric.

3.2 Post Processing of extracted features

The KAZE features extracted from iris images are large in dimension and contain redundant information. The motive of the post processing step is to extract the consistent features from the various iris images of the same eye and to remove the inconsistent features.

The Euclidian distance is computed between all the KAZE features extracted from the various images of an iris. A threshold is taken to select the consistent features and rest of the features are assumed to be inconsistent and discarded. By fusing the consistent features of an image a base template is generated that represent the characteristic feature of the iris of a person. Similarly base template for all the subjects are computed and used to train the classifier.

4 SUPPORT VECTOR MACHINE

Support Vector Machine (SVM) is the maximum margin classifier. The SVM classifier applies the

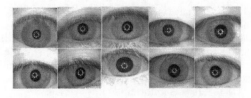

Figure 3. Sample images from CASIA-Iris-Interval.

Table 1. Comparison of various feature extraction methods.

Sr.No.	Features	DEC	AUC	EER (%)
1	KAZE	1.01	0.80	25.87
2	LBP	0.99	0.76	31.87
3	DWT	0.90	0.75	31.20
4	SIFT	0.87	0.74	32.09

kernel trick, which uses dot product, to keep computational loads reasonable. The kernel functions enable the SVM algorithm to fit a hyperplane with a maximum margin into the transformed high-dimensional feature space. SVM is a unique discriminating approach that models the boundary between classes of training data in higher dimensional space. The data are classified by finding a hyperplane with a maximum margin between the classes. Several kernel functions have been proposed. Some of the most popular kernels are linear kernel, polynomial kernel, Gaussian Radial Basis Function (RBF) kernel, and sigmoid kernel. Since, it is assumed the variation in iris size is linear so linear kernel of SVM is adopted for subject recognition.

Since the size of data set is very small so we have computed leave-one-sequence-out cross validation accuracy for evaluating the proposed approach. In leave one sequence out approach, one image is used for testing and rest of the data is used for training. The accuracy of the SVM is computing by taking average of the test accuracy of all the images. An SVM is originally designed for binary classification; however, it is applied for multiclass SVM in the proposed approach by adopting one against one technique of LIBSVM toolkit (Chang & Lin 2012).

5 CASIA DATASET

To evaluate the performance of the proposed approach, CASIA-Iris-Interval dataset (Set 2007) is used. The dataset is recorded by using a special camera having circular NIR LED array and is having sufficient luminous flux. Due to the above mentioned feature the iris images are suitable for texture analysis of iris images.

The contribution towards database is made by 249 volunteers and a total of 2651 iris images were captured. The size of the acquired images is 320×280 and each pixel is represented by 8 bits. A few samples from the iris database are represented in Figure 3.

6 EXPERIMENTAL RESULTS

The images from the CASIA dataset are pre-processed to segment the iris and to normalize their size so that variation due to different size may not affect the performance of the iris biometric system. The pre-processed images are used for KAZE feature extraction. The dimension of the KAZE features is reduced by generating a template of the extracted features from the various images of the same iris. The final feature vector is applied to support vector machine for evaluating the discriminative ability of the KAZE features. The cross validation accuracy achieved by the proposed approach is 92%, which represent the recognition accuracy of the KAZE features.

The efficacy of the proposed approach is further measured by using three parameters: Area Under the Curve (AUC), DECidability (DEC) and Equal Error Rate (EER). The area under the curve is the integration of area between x-axis and y-axis. The decidability is represented by taking the difference between mean of inter class and intra class and dividing the same by the square root of the difference of square of standard deviation of inter and intra class iris images. The EER represent the value where false positive and false negative are equal.

All the above mentioned values have been computed for KAZE features and represented in Table 1. To prove the improvement using KAZE features, all the above parametric measures have been computed using Local Binary Pattern (LBP), Scale Invariant Feature Extraction (SIFT), Discrete Wavelet Transform (DWT) and are represented in Table 1 for comparison. Though the KAZE features are better in terms of all the three performance parameters but improvement is significantly high in terms of EER.

7 DISCUSSION AND CONCLUSION

In this paper, KAZE features have been explored for iris biometrics. The KAZE feature are extracted by building a nonlinear scale space with a variable conductance diffusion up to a maximum evolution time. The features are more robust as compared to the other popular features. In most of the iris biometric systems, mask is used to remove the noise caused due to occlusion and it increases the computation complexity. The KAZE features are robust to these occlusion and avoid this masking step.

Another concern of biometric systems is accuracy. The SVM classifier results in 92% for biometric purpose. The discriminative ability of the KAZE features is measured in terms of AUC, EER and DEC and compared with the other features also.

The presented approach for iris biometric is based on KAZE features. In future, KAZE features will be fused with other popular descriptors so as to increase the recognition accuracy. Moreover, the evaluation of the proposed approach may be extended to the other database also.

REFERENCES

Alcantarilla, P. F., Bartoli, A., & Davison, A. J. (2012, October). KAZE features. In *European Conference on Computer Vision* (pp. 214–227). Springer Berlin Heidelberg.

Boles, W. W., & Boashash, B. (1998). A human identification technique using images of the iris and wavelet transform. *IEEE transactions on signal processing*, 46(4), 1185–1188.

Chang, C. C., & Lin, C. J. (2011). LIBSVM: a library for support vector machines. *ACM Transactions on Intelligent Systems and Technology (TIST)*, 2(3), 27.

Daugman, J. (1993). High confidence visual recognition of persons by a test of statistical independence. *IEEE transactions on pattern analysis and machine intelligence*, 15(11), 1148–1161.

Daugman, J. (2001). Statistical richness of visual phase information: update on recognizing persons by iris patterns. *International Journal of computer vision*, 45(1), 25–38.

Daugman, J. (2004). How iris recognition works. *IEEE Transactions on circuits and systems for video technology*, 14(1), 21–30.

Daugman, J. (2006). Probing the uniqueness and randomness of IrisCodes: Results from 200 billion iris pair comparisons. *Proceedings of the IEEE, 94*(11), 1927–1935.

Hollingsworth, K. P., Bowyer, K. W., & Flynn, P. J. (2008, September). All iris filters are not created equal. In *Biometrics: Theory, Applications and Systems, 2008. BTAS 2008. 2nd IEEE International Conference on* (pp. 1–6). IEEE.

Jain, A. K., Ross, A., & Prabhakar, S. (2004). An introduction to biometric recognition. *IEEE Transactions on circuits and systems for video technology, 14*(1), 4–20.

Krichen, E., Allano, L., Garcia-Salicetti, S., & Dorizzi, B. (2005, July). Specific texture analysis for iris recognition. In *International Conference on Audio- and Video-Based Biometric Person Authentication* (pp. 23–30). Springer Berlin Heidelberg.

Ma, L., Tan, T., Wang, Y., & Zhang, D. (2003). Personal identification based on iris texture analysis. *IEEE Transactions on Pattern Analysis and Machine Intelligence*, 25(12), 1519–1533.

Ma, L., Tan, T., Wang, Y., & Zhang, D. (2004). Efficient iris recognition by characterizing key local variations. *IEEE Transactions on image processing*, 13(6), 739–750.

Ma, L., Wang, Y., & Tan, T. (2002). Iris recognition using circular symmetric filters. In *Pattern Recognition, 2002. Proceedings. 16th International Conference on* (Vol. 2, pp. 414–417). IEEE.

Matey, J. R., Naroditsky, O., Hanna, K., Kolczynski, R., LoIacono, D. J., Mangru, S., ... & Zhao, W. Y. (2006). Iris on the move: Acquisition of images for iris recognition in less constrained environments. *Proceedings of the IEEE, 94*(11), 1936–1947.

Park, K. R., & Kim, J. (2005). A real-time focusing algorithm for iris recognition camera. *IEEE Transactions on Systems, Man, and Cybernetics, Part C (Applications and Reviews), 35*(3), 441–444.

Proença, H., & Alexandre, L. A. (2006). Iris segmentation methodology for non-cooperative recognition. *IEE Proceedings-Vision, Image and Signal Processing*, 153(2), 199–205.

Qiu, Z., Yang, L., & Lu, W. (2011). A New Feature-preserving Nonlinear Anisotropic Diffusion Method for Image Denoising. In *BMVC* (pp. 1–11).

Romeny, B. M. H. (2008). Front-end vision and multi-scale image analysis: multi-scale computer vision theory and applications, *written in mathematica* (Vol. 27). Springer Science & Business Media.

Schmid, N. A., Ketkar, M. V., Singh, H., & Cukic, B. (2006). Performance analysis of iris-based identification system at the matching score level, *IEEE Transactions on Information Forensics and Security, 1*(2), 154–168.

Set, I. D. (2007). Comments on the CASIA version 1.0 iris data set. *IEEE Transactions on Pattern Analysis and Machine Intelligence*, 29(10), 1869.

Sung, H., Lim, J., Park, J. H., & Lee, Y. (2004, August). Iris recognition using collarette boundary localization. In *Pattern Recognition, 2004. ICPR 2004. Proceedings of the 17th International Conference on* (Vol. 4, pp. 857–860). IEEE.

Weickert, J. (2001). Efficient image segmentation using partial differential equations and morphology. *Pattern Recognition*, 34(9), 1813–1824.

Weickert, J., Romeny, B. T. H., & Viergever, M. A. (1998). Efficient and reliable schemes for nonlinear diffusion filtering. *IEEE transactions on image processing*, 7(3), 398–410.

Wildes, R. P. (1997). Iris recognition: an emerging biometric technology. *Proceedings of the IEEE, 85*(9), 1348–1363.

Yao, P., Li, J., Ye, X., Zhuang, Z., & Li, B. (2006, August). Iris recognition algorithm using modified log-gabor filters. In *18th International Conference on Pattern Recognition (ICPR'06)* (Vol. 4, pp. 461–464). IEEE.

Communication and Computing Systems – Prasad et al. (Eds)
© 2017 Taylor & Francis Group, London, ISBN 978-1-138-02952-1

Feature extraction of hyperspectral face images using PCA in NIR

N. Pratap & S. Arya
GKV, Haridwar, India

V. Kumar
NGI, Meerut, India

ABSTRACT: Face identification or Face Recognition (FR) is emergent research area because of wide-ranging of applications in the fields of business and regulation enforcement. Conventional FR methods are facing different challenges of types like object lighting, position dissimilarity, appearance variations, and lead to decrease in performance of object identification and verification. To prevail over all these challenges, Hyperspectral Image Set (HIS) may be used in human FR. The HIS minimize the several limitations because the skin spectra derived with these cubic dataset depicts the unique features for an individual. In this paper a modest and effective technique is discussed to take out the set of Attributes Vectors (AVs) with HIS and correspondingly to decrease the size of HIS. PCA is applied as FR method to extract the AVs in these large data sets. PCA has been proved as a capable tool in Hyperspectral Image Processing (HIP) as well as to minimize the dimensions of data set. Research is conducted by means of CMU hyperspectral dataset by considering the image wavelength in NIR region of Infrared Spectrum (IRS). A successful attributes extraction technique using PCA is studied intended for HIS and investigational outcomes are presented with AVs.

1 INTRODUCTION

FR is a challenging task in which the face image is recognized by analyzing and comparing patterns. Generally the appearance is our key attention of consideration, in community association, having main part in carrying uniqueness and feeling (M. Turk, 1991). Frequently there are three steps in FR. One is acquisition, in which the facial images are captured from various sights. Another is normalization, in which the segmentation, arrangement and consistency of facial images are performed. Third step is Facial Recognition, in which illustrations, modeling of unfamiliar facial descriptions are performed and correspondingly links them with well-known models to offer the identities. Attributes extraction is main step in FR for which a face picture need to be listed in a regular extent prior the processing is performed. HIP practices has intricacy that hinge on straight on a total of spectral layers in the acquired HIS. Because HIS contains a number of layers, so it is always a primary objective to use such type of technique which converts the HIS into low dimensions without the loss of information. These methods are familiar with the common name of attributes extraction (Pan, 2003). Attribute withdrawal is through by whichever choosing some layers by means of an alter that yields the attributes by way of groupings of layers. PCA provides a straightforward, nonparametric scheme of taking out pertinent data with huge HIS.

This scheme can be well-defined in relations of the preliminary processing and some more phases. In preliminary processing, the HIS are kept and signified by means of a vector of picture element values. PCA is the method to discover the maximum difference in the unique space. The linear conversion maps the unique space on a multi-dimensional feature space to identify the AVs as well as storing them in face recognition module. PCA is also useful to compress the data by dropping a number of layers deprived of any loss of data and similarly can be applied in reduction of the size of an image. The detailed features of PCA are depicted through Table 1. Other popular methods such as Eigenface method (Sharma 2015) and Fisherface method can

Table 1. Detailed features of principal component analysis.

Features	Principal component analysis
Perception among classes	Accomplishes the whole data for the primary components study deprived of taking into attention the major class.
Calculation for huge data	Not require large computations
Focus	Studies the components having extensive deviations
Applications	In the major arenas of culprit analysis

also be used to take out the attributes, but the significance of these methods are not much effective as compared to PCA. In this paper, the core emphasis is to take out AVs from CMU hyperspectral face image data set using PCA.

2 METHODOLOGY FOR FEATURE EXTRACTION USING HYPERSPECTRAL IMAGES

The AVs taking out method is the significant phase in FR. Because the vector space is much high, it is not easy process to find the Covariance Matrix (COVM) for these HIS. PCA technique shows a great significance to achieve this which is based on image matrices. The image matrices and its transformation have vital significance in image AVs extraction as well as in the reduction of dimensions. PCA is used to take out AVs as well as decrease the sizes of HIS (Arya, 2015). Figure 1 points the methodology of planned practice. Suppose the HIS comprising of total P images taking a dimensions of p × q picture element. At this point, the major objective to calculate AVs with fewer calculations. Here the suggested technique is demonstrated to take out the AVs for verification of face images by calculating EVALs and EVs. To find the match, the selected sample HIS is compared with hyperspectral

data base images. The selected set of input images having n pixels in each image can be pointed as a t-dimensional region or appearance region. The specific vertical value in every appearance depicts strength of every picture element and also make a row direction. The row direction is formed by combining the pixels of each and every row to a size of 128 by 128. This is a type of linear mapping which selects a novel axis position intended for the HIS by a way with the purpose that first difference by mapping of the HIS occur on principal axis i.e. 1st Principal Elements (PEs). Similarly 2nd difference on the next axis i.e. 2nd PEs and so on. Dimensions reduction using PCA can be performed by removing the PEs which explore the less information (Pandey, 2011). Generally the later PEs are removed because of having the less information. PCA provides a methodology to diminish a HIS to the lesser appearance size (Uzair, 2013). A sample of HIS is considered with the CMU HIS (Denes, 2002), TDS = (image1, image2, image3 ... image P) with P imageries and every have q picture elements.

We have represented HIS as p × q matrix where every row signifies to a single image. Because the hyperspectral images are mapped into a subset of area, the identification and verification accuracy of HIS may be obtained with mapping the set of query HIS into subset of area and the resulting imageries closer to the query HIS may be selected.

2.1 Computing Primary Elements (PEs)

PEs is calculated with the help of EV and EVAL of concerned COVM. It is similar as to discover the coordinate position where the COVM is slanting. Though generally the HIS have a number of layers (Robila, 2008), and objective of analyzing these HIS is to check the relationship between these dimensions. Covariance represents association among these layers in the HIS (Uzair, 2013) and is calculated among two dimensions. Each row in the matrix represents each dimensions of a specific kind W_i as well as every single column points to a number of layers at a time. Because of the positive value of covariance, the resulting COVM is a slanting matrix or also called as diagonal matrix. First of all PCA pick out a stabilized path in t-dimension area such that the discrepancy in W is exploited. Again it selects one more direction, in which the variance W is maximized. Similarly in this way, p directions can be designated and the resulting set yields to the PEs.

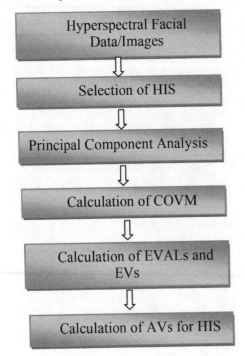

Figure 1. Flow Chart for PCA based AVs Extraction using Hyperspectral Images.

$$\text{Covariance (W, Z)} = \sum \frac{(wi - W_{avg})(zi - Z_{avg})}{(q-1)} \quad (1)$$

Generally the COVM signify interdependence structure and basically in the form of a matrix of square type. The EV having the biggest EVALs is the path of difference and is measured 1st PE. Similarly the 2nd biggest EVALs with the next maximum difference is considered as the 2nd PE and so on. Suppose P is matrix of size n × n. The EVs e1, e2 ... en span on to the eigen space and also called as orthonormal vectors. The EVs e1 is the EVs of P having EVALs λ1. Similarly the EVs e2 is the EVs of P having EVALs λ2 and so on. The next step is to arrange them on the basis of EVALs from highest to lowest.

2.2 Brief summary about HIS

In this study, CMU hyperspectral image set is used in which the image wavelength vary from 450 nm to 1100 nm and having sixty five spectral layers (Denes, 2002). The available portion of the CMU database contain imageries for fifty four diverse subjects. The thirty six subject images were captured in two different slots on the different days. Moreover, a subgroup of images are available for several sessions i.e. twenty eight subjects for three sessions, twenty two subjects for four sessions and sixteen subjects for five sessions (Denes, 2002).

2.3 Experimental outcomes

Herein study, CMU Hyperspectral Image Set (Denes 2002) is selected to pull out AVs. The images are selected at a wavelength interval of 10 nm. The experiment is carried on two phases. In Phase 1, four images of CMU data base in NIR region of Electromagnetic Spectrum is selected. Here in suggested scheme, the sample HIS are considered in the form of a matrix and each particular hyperspectral image is saved in a particular row in the matrix. So it results that the four hyperspectral images will be saved in four different rows. Here the single hyperspectral image is nearly of 301 Kb having 640 rows, 480 columns (640 × 480). Subsequently accumulation for four hyperspectral images would become 4 × 640 × 480. By using PCA, the COVM is calculated.

In phase 2, six images are considered. Similarly according to the suggested scheme, the hyperspectral images are retrieved in the form of row and column and one hyperspectral image is saved in a particular row.

2.4 Developing attributes vectors by selecting principal elements

PCA is a promising tool to calculate the AVs as well as in the area of FR. The Main objective of PCA is to reduce the number of layers in HIS which introduce the simplicity in computations. To take out AVs is an essential phase in FR (Pratap, 2014). The size of all face images are same and each pixel is treated as a variable and analyzed with PCA. Here by using the suggested scheme, the EVs having biggest EVALs is calculated and points to the association in the middle of the data sizes. After calculating EVs from the COVM, subsequent step is arrange the AVs on the basis of their EVALs in non-increasing order. This provide the components by their significance. Now the AVs is formed in the form of a matrix called as vector matrix. This is through by selecting the EVs with the help of list and constitute a AVs matrix.

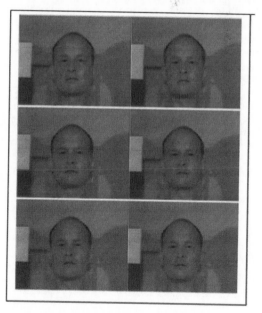

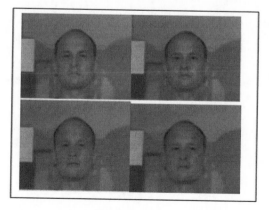

Figure 2. Four Face Image of CMU Hyperspectral database in NIR (750 nm–780 nm) (Denes, 2002).

Figure 3. Six Face Image of CMU Hyperspectral database in NIR (750 nm–780 nm) (Denes, 2002).

AVs = (eigen 1, eigen 2, eigen 3 ... eigen n)
Result of Phase 1 using four images in NIR of Electromagnetic Spectrum:
The EVALs are = [5.4391, 3.0067, 2.290, .7567]
The AVs matrix:

$$
\begin{bmatrix}
.3013 & .9974 & -.0868 & .0672 \\
.5624 & -.2589 & -.2107 & .9849 \\
.7487 & -.0714 & .8238 & -.3665 \\
.6942 & -.2709 & -.7885 & -.5091
\end{bmatrix}
$$

Result of Phase 2 using Six Images in NIR of Electromagnetic Spectrum:
The EVALs are = [6.5099, 3.1190, 3.3361, 0.8189, 0.7928, 0.4629]
The AVs matrix:

$$
\begin{bmatrix}
.2828 & -.9968 & -.2269 & .0799 & -.0053 & .0499 \\
.4961 & .2589 & -.3546 & .7793 & -.6779 & .0627 \\
.6917 & .1699 & .6185 & .4199 & .6622 & -.0628 \\
.6125 & .2597 & -.8369 & -.4372 & .4155 & -.0354 \\
.2919 & -.0448 & .2478 & -.3344 & -.3942 & -.9132 \\
.4442 & .0177 & .3528 & -.5452 & -.4494 & .4228
\end{bmatrix}
$$

3 CONCLUSIONS

Through this study, a straightforward method to take out the Attributes Vector is explored, centered on PCA. PCA is one of the utmost important approach, widely used in matching applications and is applied to collect attributes as well as decrease the sizes of hyperspectral face image set for the face recognition. This study successfully uses the PCA to take out the set of attributes vector by selecting and analyzing a number of layers from the HIS. The investigational outcomes on CMU hyperspectral database in two phases is described to demonstrate the attributes vector matrix.

REFERENCES

Arya S, Pratap N. Kumar V. 2015. Enhancement in Security by Reducing Dimensions of Hyperspectral Face Images for Face Recognition. *African Journal of Computing & ICTs*. Volume 8, No. 2. 35–40.

Belhumeur, PN., JP, Hespanha. DJ, Kriegman. 1997. Eigenfaces vs Fisherfaces Recognition using class specific linear projection. *IEEE Transactions and Pattern Analysis Machine Intelligence*. 19 (7): 711–20.

Brunelli R, and Poggio T. 1993. Face recognition: features versus templates. *IEEE Transactions and Pattern Analysis and Machine Intelligence*, Volume 15, pp. 1042–1052.

Cevikalp H, 2010. Face recognition based on image sets. *In Proceeding of IEEE International Conference of Computer Vision and Pattern Recognition*. pp. 2567–2573.

Denes L. J, Metes P. Liu, Y. 2002. Hyperspectral face database. *Robot. Inst., Carnegie Mellon University Pittsburgh, PA, USA*, Tech. Rep. CMU-RI-TR-02-25.

Jayasri T, Hemalatha. 2013. Categorization of respiratory signal using ANN and SVM based feature extraction algorithm. *Indian Journal of Science and Technology*. 6 (9): 5195–200.

Muhammad, U. 2013. Hyperspectral Face Recognition using 3D-DCT and Partial Least Squares. *In Proceedings British Machine Vision Conference*. Pages 57.1–57.10.

Mukherjee Amit, Reyes-Miguel. 2009. Interest Points for Hyperspectral Image Data, *IEEE Transactions on Geoscience and Remote Sensing*. Volume 47, No. 3, pp. 748–760

Pan Z. Healey G. Prasad M. 2003. Face recognition in hyperspectral images. *IEEE Transactions Pattern Analysis and Machine Intelligence*. Volume. 25, No. 12, pp. 1552–1560.

Pandey P. 2011. Image Processing using Principle Component Analysis, *International Journal of Computer Applications*. 0975–8887, Volume 15, No.4.

Pratap N., Arya S. 2014. Classification of Imagery Data and Face Recognition Techniques. *International Journal of Computer Applications* (0975–8887). Volume 85, No. 10.

Robila S.A. 2008. Toward hyperspectral face recognition. *Proceeding SPIE, Image Processing, Algorithms Syst. VI*, Volume 6812, pp. 1–9.

Sharma, UM. 2015. Hybrid feature based face verification and recognition system using principal component analysis. *Indian Journal of Science and Technology*. 8(S1): 115–20.

Siddharth, G., Face Recognition Using Eigen Faces and Dimensionality Reduction by PCA, *International Journal of Emerging Research in Management & Technology*. ISSN: 2278–9359, Volume-2, Issue-5.

Sivaram KS, et al. 2015. Object recognition by feature weighted matrix - A novel approach. *Indian Journal of Science and Technology*. 8(S7): 278–91.

Turk, M. A. 1991. Face recognition using Eigenfaces. In Proceeding of IEEE Computer Society Conference. *Computer Vision Pattern Recognition*. pp. 586–591.

Turk, M., A. Pentland. 1991. Eigenfaces for recognition, *Journal of Cognitive Neuroscience*. 3 (1): 71–86.

Yang J, and Zhang D. 2004. Two-dimensional PCA: a new approach to appearance-based face representation and recognition. *IEEE Transactions Pattern Analysis and Machine Intelligence*, Volume 28, pp. 131–137.

Communication and Computing Systems – Prasad et al. (Eds)
© 2017 Taylor & Francis Group, London, ISBN 978-1-138-02952-1

Design and analysis of high gain-low power operational-amplifier for bio-medical applications

Rohit Singh Toliya, Nitin Dhaka, Mohit Kumar & Ashutosh Pranav
Graphic Era Deemed University, Dehradun, Uttrakhand, India

ABSTRACT: This paper represents designing of Operational Amplifier using a composite cascode stage which operating in subthreshold region and gives gain of 97 dB at 20 μW power consumption. The designing is done on Cadence Virtuoso 6.1.5 180 nm process technology with supply voltage 1.8 volt. The circuit eliminated the need of compensation capacitor. The main benefits of the proposed design include high voltage gain, low quiescent current with low power, low noise and reduced hardware. As the proposed architecture use very low power and provide high gain with lower bandwidth range, it is beneficial for bio medical application which required high gain at low power consumption with frequency range of mHz-kHz.

1 INTRODUCTION

In past few years, area optimization of any design is of increasing interest in low-power applications like biomedical instrumentation amplifiers, sensors, comparators and ADC. Amplifiers are the most basic building blocks for these instruments or applications. Biomedical applications required high gain, low power consumption, low bandwidth with reduced chip area (Rajput et al, 2002), (Comer et al, 2010). In this work the amplifiers are in sub threshold region operation which provides high gain and low noise with less power consumption and low chip area. This design eliminates the need of compensation capacitor which reduced the area overhead. As CMOS circuits grew in popularity, BJT architecture was applied to CMOS designs. Transconductance of the MOS devices is low as compared to the BJT. The fabrications are much easier of the MOS devices compared to BJT. Generally two stages of voltage amplification are used, voltage gain of the two stage op amp is relatively low. To increase the overall gain we need to add more gain stages, if do this the circuit are more complex and other factors are effected on our designs.

2 CONVENTIONAL OPERATIONAL AMPLIFIER ARCHITECTURE

The earlier operational amplifier was designed in 1964. The conventional operational amplifier has a high voltage gain differential input stage with a second stage having high voltage gain with a last stage called buffer stage. Buffer stage provides stumpy voltage gain and large current gain. The

first two stages are responsible for providing the overall voltage gain of the operational amplifier circuit. This operational amplifier circuit sometimes called as the Widlar architecture (Johns et al, 1997). Compensating capacitor is placed between the output of the first stage and the input of the second stage (Allen et al, 2002). The conventional type of operational amplifier is shown in Figure 1.

CMOS circuit is basic building block for designing operational amplifier because CMOS circuits have various advantages over BJT including power consumption and noise reduction. The .CMOS required less fabrication steps then BJT. Two stage operational amplifiers have low voltage gain. To increase the overall gain we need to add more gain stages, if do this the circuit are more complex and other factors are effected on our designs Proposed architecture

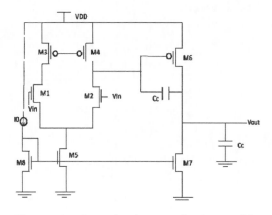

Figure 1. Conventional operational amplifier architecture.

provides large CMOS op amp gain without adding gain stages. Proposed design works in sub threshold region (Comer et al, 2004), (Comer et al, 2010).

The proposed design eliminates the need of compensation capacitor by virtue of which chip area reduced. Proposed design required low dc current for differential input stage so it has less dc power consumption. Operation in the subthreshold region of MOS devices can also lead to lower harmonic distortion than saturation region.

3 THE SINGLE ENDED, COMPOSITE CASCODE LOAD, COMPOSITE CASCODE STAGE

In conventional cascode structure, the gain is increased by increasing the output impedance. But the conventional cascode structure face limitations in low voltage applications and low power due to limited output swing & bias voltage requirement. The composite cascode circuit proposed by (Comer et al., 2010) is shown in Figure 2.

MN1 and MN2 shows the n type of transistors and MP1 and MP2 shows the p type of transistors. V_{dd} is external operating voltage and Vin is signal source. The transistors MN1 and MN2 are driven by the same input and bias voltage. The transistors MP1 and MP2 are load and biased by a voltage source Vbias2.

3.1 Operation in subthreshold region

The high gain is obtained by the composite cascode stage operates in subthreshold region. The

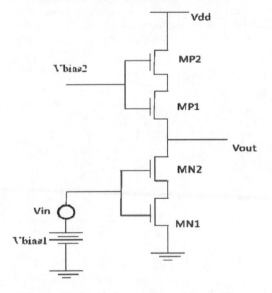

Figure 2. Single ended composite cascode stage.

first stage of differential pair is design by using composite cascode stage. The subthreshold region has become an important aspect of low power circuit design. High voltage gain and low current draw are some of the most important features of the MOSFET operating in weak inversion regions (Comer et al, 2004), (Comer et al, 2010). The Inversion Coefficient (IC) of a device is calculated by using

$$IC = I_D/I_0$$

Where I_D is drain current through device, I_0 is technology current, W is width, L is length of device. The technology current is calculated by

$$I_0 = 2\mu n \, C_{ox} U_T^2$$

where μ is carrier mobility, C_{ox} is the capacitance per unit area, n is substrate factor; U_T^2 is the thermal voltage of silicon (26 mV at room temperature).

If the value of Inversion Coefficient (IC) is larger than 10 the device will be in saturation (strong inversion) region. The value of an IC is less than 0.1 the device will be in subthreshold region. If the value of IC falls even lower value, the device may enter the sub threshold region that is a subset of the weak inversion region where the channel has barely left the depletion region.

4 PROPOSED ARCHITECTURE

In proposed op amp architecture the gain is most important features so design of op amp to achieve 97 dB gain. The aspect ratio chose such a way that some of the transistors are in subthreshold region and some are in strong inversion (saturation region) and drain current ID set to appropriate value (Comer et al, 2004), (Rajput et al, 2002). Aspect ratio for subthreshold region;

$$\frac{W}{L} = \frac{I_D}{I_o IC}$$

aspect ratio for weak inversion region.

The current equation of n-MOS for saturation region;

$$I_D = \frac{1}{2}\mu_n C_{ox}\frac{W}{L}(V_{gs}-V_{th})^2,$$
$$V_{gs} \geq 0, V_{ds} \geq (V_{gs}-V_{th})$$

The transistor MN1, MN2, MN3 and MN4 are the differential pair and MP1, MP2, MP3,

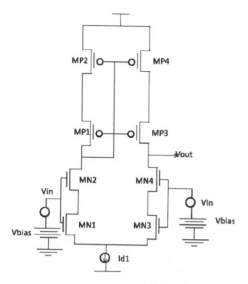

Figure 3.   Differential stage using composite cascade.

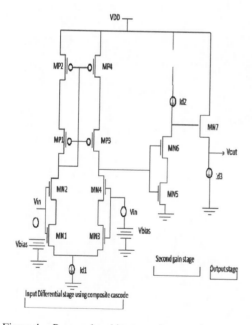

Figure 4.   Proposed architecture of composite cascode stage op-amp.

MP4 are the active current mirror load and I_{D1} is tail current source. The second stage consist of current source I_{D2} and transistor MN5 and MN6. The second stage is followed by a buffer stage MN7. This proposed architecture MN1, MN3, MP1, MP3 and MN6 are in subthreshold region. Rests of the transistors are in saturation region.

The drain current 200 nA flow through tail current source Id1 and the value of current source Id3 is 10 uA.

5   SIMULATION RESULTS

The operation of composite cascade stage operational amplifier simulated with Cadence Virtuoso 6.5.1. The overall gain of this op amp is 97 dB and the bandwidth is 495 kHz. The frequency response graph includes magnitude and phase plot is shown in Figure 5 that shows the Phase Margin (PM) of 60° for supply voltage of 1.8 volt.

Voltage gain of proposed amplifier is 97 dB as shown in Figure 5. The power consumption is 20 μW and bandwidth is quite low. PM is 60° that is sufficient for stability of the op amp. The voltage gain of proposed amplifier varies from 95 dB to 105 dB over the temperature variation from −10°C to 50°C.

Table 1.   Result analysis.

Open Loop Gain, dB	97
Phase Margin, degree	60°
Unity Gain Bandwidth, KHz	495
Power Dissipation, μW	20
Temperature variation, C°	−10 to 50

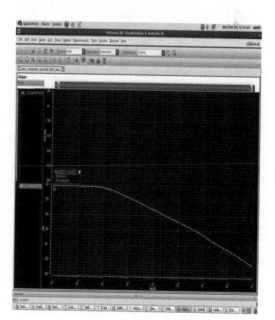

Figure 5.   Output.

The chip area of this operational amplifier is minimized by removing the Compensating capacitor (Cc).

6 CONCLUSION

The proposed work represents the possibility of composite cascode stage, working in the sub-threshold region. The proposed design provide gain 97dB, low power 20 uW and low bandwidth requirement. It removes the need of the Compensation capacitor (Cc). It has also low nonlinear distortion, low noise and reduced chip area. It is very efficient in terms of power consumption and high gain requirement, so it can be used in bio-medical application which requires high gain with very small power consumption.

REFERENCES

Allen, P. E. & Holberg, D. R. (2nd Edition) 2002. CMOS Analog Circuit Design. New York: Oxford University Press.

Comer, D. T., Comer, D. J. & Li, L. 2010. A high-gain CMOS op amp using composite cascode stages. International Journal of Electronics 97: 637–646.

Comer, D. J. & Comer, D. T. 2004. Using the weak inversion region to optimize input stage design of CMOS op amps. IEEE Transactions on Circuits and Systems-Part 51: 8–14.

Comer, D. J., Comer, D. T. & Singh, Rishi Pratap 2010. A High-Gain, Low-Power CMOS Op Amp Using Composite Cascode Stages. IEEE Transactions on Circuits and Systems-Part: 5–11.

Fay, Misra, V. & Sarpeshkar, R. 2009. A micro power electrocardiogram amplifier. IEEE Transactions on Biomedical Circuits and Systems 3: 312–320.

Johns, D. A. & Martin, K. 1997. Analog Integrated Circuit Design. New York: John Wiley and Sons.

Mollazadeh, M., Murari, K., Cauwenberghs, G. & Thakor, N. 2009. Micro power CMOS integrated low-noise amplification, filtering, and digitization of multimodal neuropotentials. IEEE Transactions on Biomedical Circuits and Systems 3: 1–10, 2009.

Rajput, S. S. & Jamuar, S. S. 2002. Low voltage, low power, high performance current mirror for portable analogue and mixed mode applications. in Proc. IEEE Circuits, Devices and Systems 148: 273–278.

Communication and Computing Systems – Prasad et al. (Eds)
© 2017 Taylor & Francis Group, London, ISBN 978-1-138-02952-1

Early detection of symptoms in tongue cancer through image processing

Palak Goyal, Ritu Yadav & Sonal A. Singh
Banasthali Vidyapeeth, Jaipur, India

ABSTRACT: Oral Cancer is one of the most common diseases in the world. It has its effect on various parts of mouth such as tongue, lips, inner chicks & throat. Tongue cancer is common among all to occur and spread easily. It increases due to insufficient detection of symptoms at the initial levels. In this paper various image processing has been done on different images at different angle to detect the early stage in tongue cancer.

1 INTRODUCTION

Cancer is an abnormal division of cells in tissues. Cancer can have impact on any part of human body with endless limitations and conditions such as oral cancer, urine cancer etc. Tongue cancer is common in Oral cancers. Early stage of Tongue cancer, it's hard to identify. They are painless and symptoms are not very visible. The first sign is developing a red and white patch on the tongue surface. Along with it a sore throat that does not go away, a sore spot (ulcer) or lump on the tongue that does not go away, pain when swallowing, numbness in the mouth that will not go away, unexplained bleeding from the tongue (that is not caused by biting your tongue or other injury), pain in the ear (rare) diagnosis in the later stages. Basically these symptoms are common for any diagnosis so normally a person does not go for the X-ray and other higher treatment until it occurs in cancer. Medical image processing helps to identify these symptoms as the turning point of cancer at the earlier stage. The various processing done on the samples of diagnosis helps a person to take the correct treatment at correct time.

2 MOTIVATION AND SCOPE

Cancer being a common disease has the higher death rates due to failure in detection of early stages. In this case medical image processing helps to identify the early stages of the cancer at the possible extents so that right treatment can be taken at the right time. This helps to extent the livelihood of a person though it is not possible to avoid the death.

3 LITERATURE REVIEW

3.1 *A survey on the imaging systems, pattern recognition and classification techniques used in the detection of oral cancer [Rajesh et al., 2014]*

It has been found out that highest average sensitivity is seen in fusion of PET and CT imaging whereas highest average specificity is in fusion of MRI and CT imaging. Applications of various image processing techniques, pattern recognition and classification algorithms to detect and classify cancers are being reviewed.

3.2 *Oral cancer staging established by magnetic resonance imaging [Rogerio et al, 2011]*

This paper is all about comparison of clinical staging and Magnetic Resonance Imaging (MRI) staging for oral cancer, and to assess inter-observer agreement between oral and medical radiologists.

3.3 *Oral cancer early detection and stages using various methods (G. Visalaxi, 2007)*

This paper emphasizes on various aids indiagnosing potentially malignant lesions or early oral cancer along with their critical evaluation. The visual detection of premalignant oral lesions remains problematic. Recent advances have shown that the risk of malignant transformation is associated with chromosomal aberrations. The proposed work has explored different enhancement techniques to improve the quality of images capturing devices.

3.4 Oral cancers at an earlier stage—A literature survey (K Anuradha & K Sankarnarayan, 2012)

In this paper, a comparison has done between various methods for the identification and classification of cancers. The proposed work has explored techniques to improve the quality of images capturing devices.

3.5 Prevalence of oral cancer in India (Varshitha, 2015)

This paper gives an idea on burden, prevalence, regional variation in India, cause, symptoms and prevention of oral cancer. The lifetime risk for mortality from cancer in India for both males and females is 61%. People less than 40 years who are habitual cigarette smokers, alcohol consumers, and betel quid chewers must undergo oral mucosa screening regularly so that oral cancer can be identified as early as possible.

3.6 Research gap

In the reviewed papers mentioned above, the work related to pattern recognition and comparison of different stages of oral cancer using different modules has been done. Basically the whole work is about the types & pattern of oral cancer could occur. In this paper we are dealing with early stage detection so that we could tackle the cancer cells at initially stage itself.

3.7 Research objectives

The sample images taken on which the various experiments such as MRI, CT scan, X-Ray have been done does not give the good result image in terms of brightness and visibility. In context of image processing, it helps to enhance the given sample image into an effective image by removing the various disturbances like blurring of image, noise presence etc. for better future references in terms of treatment.

Figure 1.   Original sample.

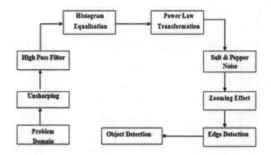

Figure 2.   Technical used.

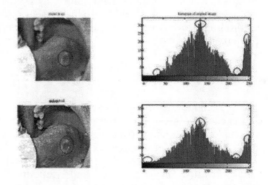

Figure 3.   Unsharp masking.

3.8 Research methodology

The methodology used for visibility and contrast enhancement of the sample taken of the affected person such as unsharp masking, power law transformation, histogram equalization, edge detection, salt & peeper noise, zooming effect with RGB, high pass filtering.

4   TECHNIQUES USED

4.1 Unsharp masking consists simply of generating a sharp image by subtracting from an image a blurred version of itself using frequency domain terminology this means obtaining a high pass filtered image by subtracting from the image a low pas filtered version of itself (R Gonzalez & R Wood, 2007).

4.2 A high-pass filter is a filter that passes high frequencies well, but attenuates frequencies lower than the cut-off frequency. Sharpening is fundamentally a high pass operation in the frequency domain.

4.3 Histogram Equalization is a simple and effective image enhancement technique. Image enhancement is a process of changing the pixels' intensity of the input image to make

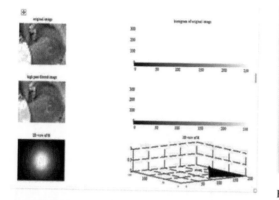

Figure 4.   High pass filtering.

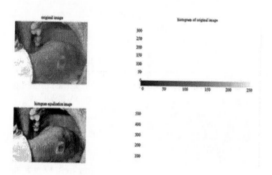

Figure 5.   Histogram equalization.

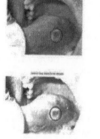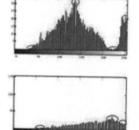

Figure 6.   Power law transformation.

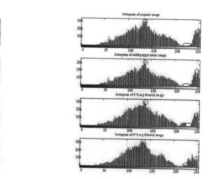

Figure 7.   Salt & pepper noise.

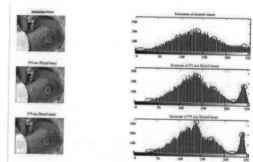

Figure 8.   Gaussian image.

Figure 9.   Zooming factor in RGB.

the output image subjectively look better. Contrast enhancement is an important area in image processing for both human and computer vision.

4.4 In poor contrast images, the adjacent character's merge during binarization. We have to reduce the spread of the characters before applying a threshold to the word image. Hence, we introduce power-law transformation which increases the contrast of the characters and helps in better segmentation.

4.5 Salt & Pepper noise is also known as impulse noise. This noise can be caused by sharp and sudden disturbances in the image signal.

4.6 Gaussian filtering is used to blur images and remove noise and detailed. It defines a probability distribution for noise or data. It is a smoothing operator (S. Jayaraman et al, 2010).

4.7 The zooming operation helps to view fine details in the image. Zooming is equivalent to holding a magnifying class in front of the screen. The simplest zooming operation is through replication of pixels (S Jayaraman et al, 2010).

4.8 Edge detection is a process of locating an edge of an image. Detection of edges in an image is a very important step towards understanding

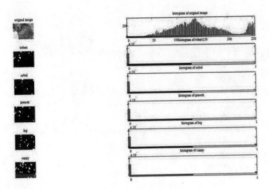

Figure 10. Edge detection.

Table 1. Experiment observations.

Experiments Performed	Observations
1. Unsharp Masking	Amount of visibility increased (60%)
2. High Pass Filter	Low frequency noise in the resultant image removed (80%)
3. Histogram Equalization	Enhancing imaging by increasing pixel intensity (70%)
4. Power Law Transformation	Increase displaying high contrast (55%)
5. Salt & Pepper Noise	Noise disturbance removed in the resultant image (40%)
6. Zooming Effect with RGB	Every pixel of the resultant image zoomed in RGB format highly in green factor of image (Factor x)
7. Edge Detection	Image brightness changes sharply (or) more (85%)

image features. Edges consist of meaningful features and contain significant information. It significantly reduces the image size and filters out information that may be regarded as less relevant, thus preserving the important structural properties of an image.

5 RESULTS AND DISCUSSIONS

Number of experiments are being done with the sampled image and the various observation corresponding to each experiment has been taken. The observation of the respective experiments is shown in Table 1 given above.

6 CONCLUSION

In this paper, we have analyzed the efficiency various medical imaging systems withrespect to the detection of tongue cancer. Applications of various image processing techniques to detect early stage of tongue cancer are reviewed. Various pattern recognition and studies are being done with the enhanced resultant sample of the affected person by the specialist. This helps him to decide whether the person has to undergo the X-ray, MRI or any other required treatment to conquer the diagnosis at the early stage itself.

REFERENCES

Anuradha K., K. Sankaranarayanan. Oral cancers at an earlier stage–A Literature Vol. 3, Issue 1, pp. 84–91. International Journal of Advances in Engineering & Technology, March 2012. ©IJAET ISSN: 2231–1963.

Gonzalez R. & R. Wood, "Digital Image Processing," 3rd ed, Englewood Cliffs, NJ: Prentice Hall, 2007.

Jayaraman S., S. Esakkirajan and T. Veerakumar Digital Image Processing. Edition: 3 Year: 2010 ISBN: 978-0-07-014479-8.

Rajashekhargouda C. Patil, Dr. Mahesh P. K. DOI: © Association of Computer Electronics and Electrical Engineers, 2014. A Survey on the Imaging Systems, Pattern Recognition and Classification Techniques used in the Detectionof Oral Cancer, 03.AETS.2014.5.349

Rogério Ribeiro de Paiva, Paulo Tadeu deSouza, Figueiredo, André Ferreira Leite, Maria Alves Garcia Silva, ElieteNeves Silva Guerra. Braz, Oral cancer staging established by magnetic resonance imaging, Oral Res. 2011 Nov-Dec; 25(6):512–518.

Varshitha, A., Prevalence of oral cancer in India A/J. Pharm. Sci. & Res. Vol. 7(10), 2015, 845–848.

Visalaxi G., Oral Cancer Early Detection and Stages Using Various Methods. (An ISO 3297: 2007 Certified Organization)Vol. 2, Issue 11, November 2014.

http://www.cancerresearchuk.org/about-cancer/cancers-in-general/cancer-questions/tongue-cancer.

http://www.deakmedicaldentistry.com/blog/uncategorized/april-is-oral-cancer-awareness-month/.

http://research.ijcaonline.org/volume109/number14/pxc3900999.pdf.

http://www.sersc.org/journals/IJSIP/vol6_no5/31.pdf.

http://mile.ee.iisc.ernet.in/mile/publications/softCopy/DocumentAnalysis/deepak_SPCOM2012.pdf.

http://mnikolova.perso.math.cnrs.fr/ChanHoNikoIP05.pdf.

https://www.math.washington.edu/~morrow/336_13/papers/debosmit.pdf.

Communication and Computing Systems – Prasad et al. (Eds)
© 2017 Taylor & Francis Group, London, ISBN 978-1-138-02952-1

A study on adaptive wavelet technique for speckle noise removal

Shallu
ECE Department, National Institute of Technical Teachers Training and Research, Chandigarh, India

Sumit Kumar
ECE Department, Dronacharya College of Engineering, Gurgaon, India

Ekta Aggarwal
ECE Department, IIT Roorkee, Uttarakhand, India

ABSTRACT: This paper presents the comparative study of adaptive wavelet technique using Global and Baye's Thresholding method to de-noise an image corrupted with speckle noise of different variance values. Speckle noise is of great concern for a large community of researchers & academia's because it causes degradation in Medical Ultrasound Images, Synthetic Aperture Radar (SAR) Images and Optical Coherence Tomography Images etc. So, it is essential to provide a best solution for the problem. Adaptive Wavelet Technique remove speckle noise by distinguishing it from the image and the results are obtained after the implication of adaptive wavelet technique on test image for different noise variance. Comparison of PSNR values for different noise variance shows that the 'dmey' wavelet with Baye's Thresholding method provides best Peak Signal to Noise Ratio as compared to Mean and Median filter.

1 INTRODUCTION

Digital images play a vital role in daily life applications such as satellite television, ultrasound image, magnetic resonance imaging, radar images and computer tomography as well as in areas of research and technology such as astronomy, geographical information systems and many more. Images collected by camera sensors are generally corrupted by noise and degrade the data & information of interest. So, de-noising is often an essential. It is the first step to be taken before the evaluation of image data. So, it is requisite to apply an efficient denoising algorithm to compensate for such data contamination. The main objective of image de-noising algorithms is to remove the noise by distinguishing it from the signal. In any noisy image, Speckle Noise comes into existence when a returning wave suffered from interference at the transducer aperture. This interference may be constructive or destructive which depends on relative phases of scattered waveform, result in bright and dark spots in the image. There are numerous advantages of image de-noising; one of it is improvement in quality of medical image that may provide help in medical diagnosis.

1.1 *Previous work*

Recently, many DWT based de-noising algorithm have been proposed to de-noise an image contaminated with different types of noises. Some previous works are presented here that are related to process of image de-noising by implementing wavelet transform. In 1998, Zhaohui in collaboration with colleagues proposed a de-noising algorithm for SAR images corrupted with Speckle noise which can be reduced by using discrete wavelet transform, Bayesian estimator (Zeng, Zhaohui., & Cumming, Ian. 1998). The evaluation of SAR images was done by considering two measurements: (1) Standard Deviation to Mean Ratio and, (2) Target to Clutter Ratio. They concluded that the estimator depends on two key factors: sub band decomposition & the statistical model of wavelet coefficients. Also, the noise from the image can be removed efficiently by the careful choice of mother wavelet function. Mohideen et al. 2008 implemented the image de-noising algorithm on natural images contaminated with Gaussian noise using wavelet technique with modified Neigh shrink thresholding method (Mohideen, S. Kother., Perumal, Dr. S. Arumuga., & Sathik, Dr. M. Mohamed. 2008). The experimental results showed that modified Neigh shrink provide better result than Neigh shrink, Weiner filter and Visu shrink. In 2003, X. H. Wang and his colleagues presented a new approach for microarray image de-noising which was Stationary Wavelet Transform approach (Wang, X. H., Istepanian, Robert S. H., & Song, Yong Hua. 2003). According to him, microarray imaging is considered as an important tool in analysis of gene expression to easily identify the diseased gene for diagnosing critical diseases. But Stationary Wavelet Transform approach due to its time invariant property plays a wonderful role in image de-noising. His results showed that the stationary wavelet de-noising has 16% better performance than Wiener filter which is widely used in commercial de-noising software system.

In this work, Adaptive Wavelet Technique is used along with Baye's and Global thresholding method to calculate the threshold for each sub band. Then, original image is reconstructed after applying Inverse transform. Adaptive Wavelet Technique is adaptive in behaviour which is decided by the thresholding method as some methods of choosing thresholds are adaptive to different spatial characteristics and found to be more efficient than the global one. Effectiveness of an algorithm is also depends more on choice of thresholding method. Adaptive behaviour of wavelet makes it more effective in process of image de-noising. Adaptive Wavelet Technique performs remarkably in process of de-noising due to its energy compactness property. Due to this property a major portion of signal energy is incorporated in a few large wavelet coefficients whereas small portion of energy is incorporated in large number of small wavelet coefficients. These coefficients of wavelet are threshold by applying appropriate shrinkage method for de-noising of image.

1.2 Wavelet transform

A wavelet is a wave like structure or oscillation that has amplitude. Its energy is concentrated in time and provides a tool for the analysis of non stationary, time varying or transient phenomenon. It has an oscillating nature but still it has capability to allow frequency and time analysis simultaneously. It gives complete three dimensional information of any signal i.e. what different frequency components present in a signal and what are their respective amplitude and at time axis where these different frequency components exists. In wavelet transform, there will be use of wavelet for series expansion of a function or signal like Fourier transform and a discrete time version is developed by this series expansion that is similar to discrete Fourier transform. But the wavelet transform has the zooming property which is not supported by Fourier transform and Short Time Fourier Transform (STFT); as the window width in these transform are constant and non-adaptive (Hans-George Stark. 2005).

1.3 Image decomposition by wavelet

In image processing, Discrete Wavelet Transform (DWT) perform the multi-differentiated decomposition of image. After performing it on image, sub-images are produced. DWT transform decomposes the image into four frequency band, among these one with low frequency components is termed as LL and other three with high frequency components are LH, HL and HH (Sendur, Levent & Selesnick, Ivan W. 2002) and illustrated below in Figure 1.

These three high frequency sub-images also known as detailed sub-bands are corresponding to orientation in three different directions such as LH in Horizontal, HL in Vertical and HH in Diagonal direction. The low frequency sub-band LL is termed as approximation. Multilevel decomposition can be achieved by continuous application of decomposition on LL band in a similar manner as shown in Figure 2.

Due to the addition of noise into the image; the information present in the detail coefficients gets destroyed as a result the edges of the objects

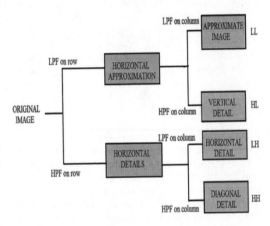

Figure 1. First level decomposition of image by DWT.

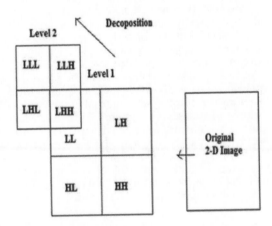

Figure 2. Multilevel decomposition hierarchy (Levent Sendur et al., 2002).

present in the image get destroyed or become invisible. There is no such major effect can be seen in approximate image, this is due to the fact that high frequency component much more effected by the noise and approximate image is a sub band of low frequency.

1.4 Thresholding

Thresholding is a nonlinear process which helps in removing the noise present in the signal. Thresholding is simple process because it operates on one wavelet coefficients at a time (Mahalakshmi, B.V., & Anand, M.J. 2014). In wavelet thresholding by discarding insignificant coefficients noise is removed from the image. The selection of threshold method determines the efficiency of de-noising to a great level.

1.4.1 Global thresholding
It is a uniform and simple thresholding method. The threshold

$$\lambda_{UNIV} = \sqrt{2\ln N}\sigma \qquad (1)$$

(N being the signal length, σ being the noise variance) is well known in wavelet literature as the Universal threshold.

1.4.2 Baye's thresholding
This method was proposed by Chang, Yu and Vetteli. In this a threshold can be calculated using Bayesian framework and suggested a simple and approximation formula to calculate the threshold [10] as defined below:

$$T_B = \frac{\sigma_n^2}{\sigma_s} \qquad (2)$$

where, σ_n is the noise variance and σ_s is the variance of original signal without noise. The estimation of noise variance is done by the median estimator from the sub-band such as HH sub band—

$$\sigma_n = \frac{median(|HH|)}{0.6745} \qquad (3)$$

Further, the additive noise in the signal can be represented as:

$$w(a,b) = s(a,b) + n(a,b) \qquad (4)$$

But, the signal and noise are independent of each other, it can be expressed as:

$$\sigma_w^2 = \sigma_s^2 + \sigma_n^2 \qquad (5)$$

σ_w^2 can be calculated as depicted below:

$$\sigma_w^2 = \frac{1}{n^2}\sum_{a,b-1}^{n} w^2(a,b) \qquad (6)$$

Now, signal variance can be calculated as:

$$\sigma_s = \sqrt{Max(\sigma_w^2 - \sigma_n^2),0} \qquad (7)$$

2 PROCESS OF IMAGE DE-NOISING USING BAYE'S & GLOBAL THRESHOLDING

The image de-noising subroutine which is used in this work is shown in form of flow chart in Figure 3. Removal of speckle noise from the image contaminated with different noise variances has been carried out using image denoising algorithm. In this algorithm, a mathematical transformation i.e. Discrete Wavelet Transform (DWT) is applied for image analysis and it also preserve the important features of image effectively. The Discrete Wavelet Transformation of image has been performed using different wavelets such as: Haar and Dmey. Two thresholding methods (Global and Baye's) are selected and implemented to shrink the noisy coefficients. Later, Inverse Discrete Wavelet Transform (i.e. IDWT) is applied to obtain the original image.

Stepwise methodology of the Image De-noising algorithm is described as follows:

1. Corrupt original image by speckle noise with different noise variance.
2. Perform Discrete Wavelet Transformation of the noisy image using Haar and Dmey wavelets.

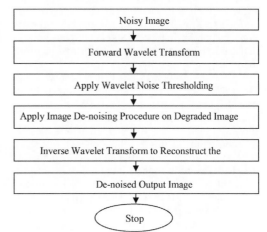

Figure 3. Flow chart for De-noising of image.

133

3. Calculate threshold for each sub band with Baye's and Global thresholding method using above equations.
4. Apply denoising procedure on degraded image to shrink the noisy coefficients.
5. Perform the inverse DWT to reconstruct the denoised image.

The evaluation of Image quality is carried out on the basis of two measurements—Peak Signal to Noise Ratio (PSNR) which is calculated using formula—

$$PSNR = 10 * log_{10} \frac{255^2}{MSE} \qquad (8)$$

& Mean Squared Error as—

$$MSE = \frac{1}{MN} \sum_{b=1}^{M} \sum_{a=1}^{N} \left[X(a,b) - X'(a,b) \right]^2 \qquad (9)$$

where, $X'(a,b)$ is the original signal without any noise and $X(a,b)$ is the estimate of the signal (Sendur, Levent., & Selesnick, Ivan W. 2002).

The conclusion is based on the PSNR value obtained for all the wavelets which are used. All the results are compared and best wavelet with best thresholding method is selected. After obtaining the best wavelet technique with optimum thresholding method, a comparison is made with Mean and Median filter.

3   RESULTS AND DISCUSSION

The simulation of speckle de-noising code is performed in MATLAB tool. Here, original image corrupted by Speckle noise with different noise variance 0.01, 0.02, 0.03 and 0.04 is used. The performance analysis of denoising algorithm is done by measuring Peak Signal to Noise Ratio and comparison of PSNR is performed as given in Table 1.

From the results in Table 1, it is observed that Dmey Wavelet with Baye's thresholding method provides best result. The best PSNR value is achieved in Dmey wavelet while using Baye's threshold method for denoising of image because adaptive nature of this thresholding method makes the Dmey wavelet more efficient in performance as clearly observed from the Figure 4.

In Figure 5, the comparison is depicted among Dmey wavelet, Mean and Median filter. Here, it can be clearly analysed that Dmey wavelet is much efficient in image de-noising as compared to Mean and Median filter because these filters introduce blurring in edges of image.

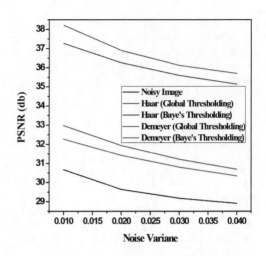

Figure 4.   PSNR vs variance for Haar and Demeyer wavelet with Baye's Thresholding and Global Thresholding.

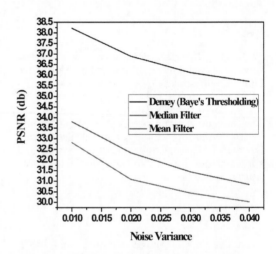

Figure 5.   PSNR vs variance for Dmey wavelet (Baye's Thresholding), Median and Mean filter.

4   CONCLUSION

It is concluded from the comparative study among the filters that the use of the wavelets with Global Threshold method gives poor results as, it is fixed & non-adaptive in behavior, on the other hand use of Bayes Threshold method improve the performance of wavelets to a great extent as it has adaptive nature. Dmey wavelet provides better result as compared to other wavelets because wavelet is capable to differentiate infinitely and also adequate for non-stationary signals. The wavelet technique provides 16% better result as compared to mean filter. In near future research will be carried in the

Table 1. PSNR values for different noise variance by using MATLAB.

Wavelet	Filtering technique	Variance (0.01)	Variance (0.02)	Variance (0.03)	Variance (0.04)
Noisy Image PSNR (Speckle noise)		30.67	29.64	29.18	28.91
Haar	Global Thresholding	32.27	31.42	30.8	30.34
	Bayes Thresholding	37.27	36.26	35.6	35.14
Dmey	Global Thresholding	32. 98	31. 96	31.21	30.69
	Bayes Thresholding	38.22	36.9	36.12	35.7
Non-Linear	Median	32.82	31.09	30.44	30.02
Linear	Mean	33.81	32.34	31.44	30.84

Visual comparison

(a) Original Image

(b) Noisy Image

(c) Haar (Global)

(d) Haar (Baye's)

(e) Demeyer (Global)

(f) Demeyer (Baye's)

(g) Median

(h) Mean

area of image de-noising via implementation of Wiener Filter in Wavelet Domain, especially for two conditions: (a) When noise variance is known; (b) When noise variance is unknown.

For further enhancing the performance of de-noising algorithms through wavelet domain with various other types of shrinkage methods and appropriate threshold techniques would be employed in image processing so that more detail information can be extracted from the noisy image.

REFERENCES

Achim, A. & Tsakalides, P. 2003. SAR image de-noising via Bayesian wavelet shrinkage based on heavy-tailed modelling. IEEE Transaction on International Geoscience and Remote Sensing. 41(8): 1773–1784.

Aggrawal, Ekta. & Kumar, Nishant. 2014. High throughput pipelined 2-D Discrete cosine transform for video compression. IEEE International Conference on Issues and Challenges in Intelligent Computing Techniques (ISBN: 978-1-4799-2899-6): 708–710.

Antonini, M., Barlaud, M., Mathieu, P., & Daubechies, I. 1992. Image coding using wavelet transform. IEEE Transaction on Image Processing 1(2): 69–87.

Ben Hamza, A., & Krim, Hamid. 2001. Image De-noising: A Nonlinear Robust Statistical Approach. IEEE Transaction on Signal Processing. 49(12): 3045–3054.

Chang, S. Grace., Yu, Bin., & Vetterli, Martin. 2000. Adaptive Wavelet Thresholding for Image De-noising and Compression. IEEE Transaction on Image Processing. 9(9): 1532–1546.

Dass, Ashish Kumar., Shial, Rabindra Kumar., & Gouda, B.S. 2012. Improvising MSN and PSNR for Finger-Print Image noised by Gaussian and Salt & Pepper. The International Journal of Multimedia & Its Applications (IJMA). 4(4): 59–72.

Gonzalez, R.C., Steven, R.E. Woods., & Eddins, L. 2009. Digital Image Processing using MATLAB. Published by Gatesmark.

Gupta, Shweta., & Kumar, Sumit. 2012. Variational Level Set Formulation and Filtering Techniques on CT Images. International Journal of Engineering Science and Technology (IJEST) (ISSN: 0975-5462) 4(7): 3509–3513.

Hans-George Stark. 2005. Wavelets and Signal Processing. Published by Springer.

Mahalakshmi, B.V., & Anand, M.J. 2014. Adaptive Wavelet Packet Decomposition for Efficient Image De noising By Using Neigh Sure Shrink Method. The International Journal of Computer Science and Information Technologies. 5(4): 5003–5009.

Mohideen, S. Kother., Perumal, Dr. S. Arumuga., & Sathik, Dr. M. Mohamed. 2008. Image De-noising using Discrete Wavelet Transform. International Journal of Computer Science and Network Security. 8(1): 213–216.

Pizurica, A., Philips, W., Lemahieu, I., & Acheroy, M. 2003. A Versatile Wavelet Domain Noise Filtration Technique for Medical Imaging. IEEE Transaction on Medical Imaging. 22(3): 323–331.

Sendur, Levent., & Selesnick, Ivan W. 2002. Bivariate Shrinkage Functions for Wavelet-Based De-noising Exploiting Inter scale Dependency. IEEE Transaction on Signal Processing. 50(11): 2744–2756.

Shallu., Narayan, Yogendera., & Nanglia, Pankaj. 2016. A Comparative Analysis for Haar Wavelet Efficiency to Remove Gaussian and Speckle Noise from Image. IEEE International Conference on Computing for Sustainable Global Development (ISSN 0973-7529; ISBN 978-93-80544-20-5): 2361–2365.

Wang, X. H., Istepanian, Robert S. H., & Song, Yong Hua. 2003. Microarray Image Enhancement by Denoising Using Stationary Wavelet Transform. IEEE Transaction on Nano Bioscience. 2 (4):184–189.

Zeng, Zhaohui., & Cumming, Ian. 1998. Bayesian Speckle Noise Reduction Using the Discrete Wavelet Transform. International Geo-science and Remote Sensing Symposium, (IGARSS-98), Seattle: 1–3.

Communication and Computing Systems – Prasad et al. (Eds)
© 2017 Taylor & Francis Group, London, ISBN 978-1-138-02952-1

Image enciphering using modified AES with secure key transmission

Shahina Anwarul & Suneeta Agarwal
Motilal Nehru National Institute of Technology Allahabad, Allahabad, U.P., India

ABSTRACT: In this era of Internet and communication technology, sharing information through digital images has become the importunate technique. But transmission over public network requires high level of security to protect the digital images from unauthorized access. Image enciphering is a widely used technique to provide the security. Advanced Encryption Standard (AES) is a well-known block cipher encryption algorithm. National Institute of Standards and Technology (NIST) has declared AES as one of the most secure algorithm. AES gives best results for text encryption but it has some drawbacks for image encryption because of the presence of high correlation among the pixel values in it. AES suffers from pattern appearance problem and requires high computation. In order to remove these drawbacks some modifications have been proposed in this paper. Firstly, modification of mix columns step of AES algorithm to decrease the encryption time is made. Second modification is to scramble the image before encryption to avoid pattern appearance problem. Third modification is done by decreasing some rounds of AES algorithm to reduce computations and hence enciphering time and security is enhanced by encrypting the hash value of key instead of using the original key. *Ruth Rendell said that AES is very simple algorithm. But if you don't know the key is it is virtually indecipherable* (Stallings 2005). So, the focus of the work is not only on the secure encryption but also on the secure key exchange. Cipher Block Chaining (CBC) mode of encryption algorithms is used in the proposed work to enhance the security. Experimental results proved that the proposed method is highly secure and faster than the existing encryption algorithms.

Keywords: Advanced Encryption Standard (AES), ciphering modes, Electronic Code Book (ECB), hashing, image enciphering, Modified Advanced Encryption Standard (MAES), steganography

1 INTRODUCTION

With the rapid growth of internet and communication technology, people share information through digital images. A digital image is a group of pixels that are arranged in the form of matrix. Image encryption is used in various fields e.g., in military communication, data transfer should be highly secure so that no one can breach the security. In the field of medical, a small change in patient's report can misdiagnose the patient and this can lead to the patient's death. Transmission over public network requires high level security and encryption is the most efficient technique to provide the security. Encryption algorithms make the information indiscernible.

Various encryption algorithms have been emerged like DES, 3DES, RSA etc. (Gupta et al. 2014). But these algorithms suffer from some drawbacks. Security of DES has already been cracked because of its small key length. 3DES has replaced DES but it is not efficient in terms of encryption time. It takes too long to encrypt the data. Then, AES has replaced 3DES. AES is a

symmetric key based block cipher algorithm. AES gives best results for text encryption but does not give optimum results for image encryption because of high correlation present among the pixels. AES suffers from some drawbacks like pattern appearance problem and high computations (Wadi & Zainal 2013). Figure 1 shows the pattern appearance problem after encrypting the image using AES (ECB mode). By looking the encrypted image we can easily predict the content of image. So, in order to remove these drawbacks several attempts have been made but proposed work is more efficient in terms of security and encryption time.

Kamali et al. (2010) proposed the modification of shift rows transformation of AES algorithm but the focus of the work was only on one problem of AES. Their work reduced the pattern appearance problem but require too much time and high computations. Wadi & Zainal (2013) used single S-box for both encryption and decryption process to reduce memory requirement. They used only one round of AES but theoretically it has been found that AES algorithm can be cracked by the reduced-round adversary in 2^{120} iterations if number of

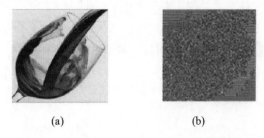

(a)	(b)

Figure 1.   (a) Original image (b) Encrypted image.

rounds reduced to 7. Therefore, reducing number of rounds can make this algorithm non-resilient to this type of attack (Wadi & Zainal 2014). Tran et al. (2008) modified the S-box to improve the security level by applying the gray code conversion to S-box. But this improvement increased the computation cost of algorithm. Huang et al. (2011) and Telagarapu et al. (2011) modified the algorithm to reduce pattern appearance problem using compression. The focus of all the previous work is on secure and efficient encryption only but as the key plays a crucial role in encryption algorithm so present work considers the secure key transmission also. As we know AES is the most secure algorithm but after some modifications the complexity of algorithm reduces and it affects the security level of the algorithm. In order to maintain the security, security is provided at multiple levels and achieved time efficient encryption algorithm.

The organization of this paper is as follows. Section 2 and section 3 give brief description of initial existing AES algorithm and modified AES algorithm respectively. The detailed discussion of the proposed system is done in section 4. Experimental results and evaluation of security parameters is discussed in section 5. The conclusions of this paper are in section 6.

2   EARLIER VERSION OF AES

Before starting the discussion on proposed modifications, the working of earlier version of AES is stated first.

2.1   Salient features of AES

- AES is a symmetric key based algorithm with block length of 128 bits developed by Joan Daemen and Vincent Rijmen.
- It consists of 10, 12 or 14 rounds with different key lengths. 128-bit key is used for 10 rounds, 192-bit key is used for 12 rounds and 256-bit key is used for 14 rounds.
- Each round includes 4 operations—substitute bytes, shift rows, mix columns and add round key.

2.2   The overall structure of AES

Figure 2 (Kak 2015) represents the overall working of AES algorithm for encryption and decryption. 128-bit input is passed through 10, 12 or 14 rounds to generate the ciphertext. On the other side ciphertext is passed to generate the original plaintext.

2.3   Four steps in each round of processing

2.3.1   Substitute bytes

A simple substitution of each byte is performed using S-box of 16×16 bytes which is constructed using defined transformation of values in GF (2^8). Each byte of state array is replaced by the value of S-box indexed by the row (first four bits of byte of state array) and column (next 4 bits of byte of state array) e.g., byte {82} is replaced by the 8th row and 2nd column of S-box whose value is {13}.

2.3.2   Shift rows

Since state is processed by columns, this step permutes bytes between the columns (Kak 2015). First row remains unchanged. Each byte of second row shifts left one position. Each byte of third row shifts two positions in left direction. Similarly, each byte of fourth row shifts three positions in left.

2.3.3   Mix columns

The value of each byte of a column is replaced by a new value which is a function of all four bytes in that column (Kak 2015).The columns are considered as polynomials over GF (2^8) and multiplied by a fixed polynomial a(x) modulo x^4 +1 given by (1):

$$\alpha(x) = \{03\}x^3 + \{01\}x^2 + \{01\}x + \{02\} \qquad (1)$$

The mix columns operation is defined by the following matrix multiplication (Stallings 2005).

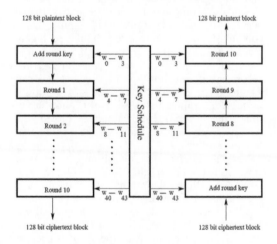

Figure 2.   The overall description of AES algorithm.

$$\begin{bmatrix} 02 & 03 & 01 & 01 \\ 01 & 02 & 03 & 01 \\ 01 & 01 & 02 & 03 \\ 03 & 01 & 01 & 02 \end{bmatrix} \begin{bmatrix} s_{0,0} & s_{0,1} & s_{0,2} & s_{0,3} \\ s_{1,0} & s_{1,1} & s_{1,2} & s_{1,3} \\ s_{2,0} & s_{2,1} & s_{2,2} & s_{2,3} \\ s_{3,0} & s_{3,1} & s_{3,2} & s_{3,3} \end{bmatrix}$$

$$= \begin{bmatrix} s'_{0,0} & s'_{0,1} & s'_{0,2} & s'_{0,3} \\ s'_{1,0} & s'_{1,1} & s'_{1,2} & s'_{1,3} \\ s'_{2,0} & s'_{2,1} & s'_{2,2} & s'_{2,3} \\ s'_{3,0} & s_{3,1} & s'_{3,2} & s'_{3,3} \end{bmatrix}$$

where, value of each byte can be calculated using (2):

$$
\begin{aligned}
s_{0,j}^{G} &= (2 \bullet s_{0,j}) \oplus (3 \bullet s_{1,j}) \oplus s_{2,j} \oplus s_{3,j} \\
s_{1,j}^{G} &= s_{0,j} \oplus (2 \bullet s_{1,j}) \oplus (3 \bullet s_{2,j}) \oplus s_{3,j} \\
s_{2,j}^{G} &= s_{0,j} \oplus s_{1,j} \oplus (2 \bullet s_{2,j}) \oplus (3 \bullet s_{3,j}) \\
s_{3,j}^{G} &= (3 \bullet s_{0,j}) \oplus s_{1,j} \oplus s_{2,j} \oplus (2 \bullet s_{3,j})
\end{aligned}
\tag{2}
$$

In inverse mix columns step each column is multiplied by a fixed polynomial b(x) modulo x^4 +1 given by (3):

$$b(x) = \{0B\}x^3 + \{0D\}x^2 + \{09\}x + \{0E\} \tag{3}$$

The inverse mix columns operation is defined by the following matrix multiplication (Stallings 2005).

$$\begin{bmatrix} 0E & 0B & 0D & 09 \\ 09 & 0E & 0B & 0D \\ 0D & 09 & 0E & 0B \\ 0B & 0D & 09 & 0E \end{bmatrix} \begin{bmatrix} s_{0,0} & s_{0,1} & s_{0,2} & s_{0,3} \\ s_{1,0} & s_{1,1} & s_{1,2} & s_{1,3} \\ s_{2,0} & s_{2,1} & s_{2,2} & s_{2,3} \\ s_{3,0} & s_{3,1} & s_{3,2} & s_{3,3} \end{bmatrix}$$

$$= \begin{bmatrix} s'_{0,0} & s'_{0,1} & s'_{0,2} & s'_{0,3} \\ s'_{1,0} & s'_{1,1} & s'_{1,2} & s'_{1,3} \\ s'_{2,0} & s'_{2,1} & s'_{2,2} & s'_{2,3} \\ s'_{3,0} & s'_{3,1} & s'_{3,2} & s'_{3,3} \end{bmatrix}$$

2.3.4 *Add round key*

The state array is XORed with 128-bits of expanded key.

2.4 *Key expansion*

In key expansion, a master key of 128-bit is converted into 10, 12 or 14 128-bit keys for each round of AES algorithm. Each column of 32-bit is represented as a word w and four words are expanded into 44 words. The key expansion operation is performed to ensure that a change in one bit of the encryption key affects the round keys for several rounds (Kak 2015). Master key can be expanded using (4):

$$w_{i+4} = w_i \oplus g(w_{i+3}) \tag{4}$$

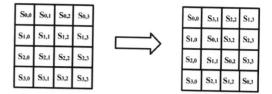

Figure 3. Modified mix columns step.

3 A MODIFIED VERSION OF AES

Modification is done in mix columns step and reduces two rounds of AES algorithm to make the algorithm efficient in terms of encryption time and computations. A lot of computations are required in mix columns step which makes the algorithm slow. In order to make it time efficient algorithm mix columns step is modified in a similar way as shift rows step in AES algorithm. First column remains unchanged. Second column shifts one step downwards cyclically and so on. At the time of decryption inverse mix columns step from second column onwards involves shifting upwards cyclically. Figure 3 shows the modification in mix columns step. This modification simply mixes the values of the column but reduces the security of AES algorithm. In order to recompense security is provided at multiple levels in efficient time.

4 PROPOSED SYSTEM

Figure 4 shows the proposed system for image encryption and secure key transmission. In this paper, several modifications have been proposed to provide fast and high level security to image transmission over public network.

The following steps will be performed for secure transmission of digital images.

Sender side

Step 1: Select the input image.

Step 2: Scramble the image pixels using the property of XOR given in (5). Original image is XORed with random image that is already stored in receiver's system.

$$original \oplus random = result \tag{5}$$

So, this step will diffuse all the pixels of image to remove pattern appearance in the standard AES caused by the same intensity pixel values.

Step 3: Now encrypt the scrambled image using modified AES as discussed in above section.

{Encryption is done using the hash value of key instead of original key and this information is only known to the receiver so that this technique can

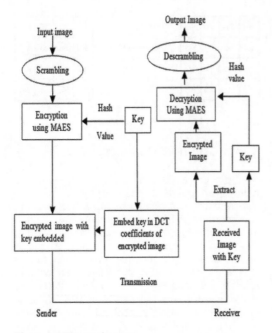

Figure 4. Proposed system.

swindled the attacker. Here, SHA-256 algorithm is used to calculate hash value because collision attack has been found in other hashing algorithms like MD5, SHA-1 etc. First 128-bits of hash value are used as a key for encryption and next 128-bits are used as initialization vector (iv) in CBC mode of AES algorithm.}

Step 4: Convert the encrypted image from spatial domain to frequency domain because processing in frequency domain is more secure as compared to spatial domain.

Step 5: Embed the original key into the DCT coefficients of encrypted image for secure transmission of key.

{Steganography is used to embed the key into the image. Several attacks can be possible on embedded data in spatial domain like noising, image compression, smoothing, sharpening, rotation and other type of attacks to disturb the embedded content. That is why frequency domain is used for embedding.}

Step 6: Transmit encrypted image over public network.

Receiver side

Step 1: Receive the encrypted image with key embedded in the image.

Step 2: Convert the received image from spatial domain to frequency domain and extract the embedded key.

Step 3: Calculate the hash value of extracted key using SHA-256 algorithm and use

initial 128-bits as a key and next 128-bits are used as initialization vector (iv).

Step 4: Decrypt the encrypted image using modified AES as discussed in section 3.

Step 5: Descramble the encrypted image using (6) and get the original image sent by the sender.

$$random \oplus result = original \qquad (6)$$

5 EXPERIMENTAL RESULTS AND SECURITY ANALYSIS

It is proved here experimentally that the proposed method is more efficient and secure than the initial version of AES algorithm and other State-of-the-Art by evaluating various parameters like histogram analysis, entropy, simulation time, pattern appearance and correlation between original image and encrypted image.

Various attacks can be possible while transmitting the images (Sun et al. 2008). The effects of some of these attacks on the embedded content at the time of decryption are shown in Figure 5. Results show that embedded content is not disturbed however some noise is observed present in the results.

5.1 Pattern appearance

This problem arises due to the pixels of same intensity values. It can be reduced by using the CBC

(a) Original image

(b) Data for embedding

(c) Salt-pepper attack on embedded image

(d) Extracted data

(e) Cropping attack

(f) Extracted data

Figure 5. The effects of attacks on the embedded content.

140

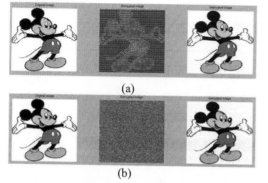

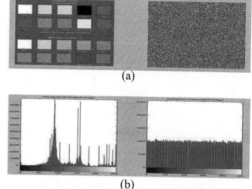

(a)

(a)

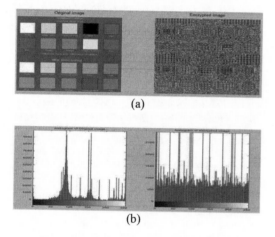

(b)

(b)

Figure 6. (a) Encryption by AES with pattern appearance problem (b) Encryption by proposed method.

Figure 8. Histogram analysis of results obtained from modified AES (a) Original and encrypted image (b) Histogram of original and encrypted image.

(a)

Table 1. Entropy values.

(b)

	Image 1	Image 2	Image 3	Image 4
Original image	2.6797	7.8362	7.2226	5.9484
Encrypted image Using initial AES	6.5719	7.9988	7.9992	7.8325
Encrypted image using method by Kamali et al. (2010)	7.9993	7.9993	7.9994	7.9993
Encrypted image using proposed method	7.9993	7.9994	7.9994	7.9994

Figure 7. Histogram analysis of results obtained from earlier version of AES (a) Original and encrypted image with pattern appearance (b) Histogram of original and encrypted image.

the ideal value is 8. Better the encryption tends the entropy value towards 8.

$$E(x) = -\sum_{i=1}^{K} P(x_i) \log_2 P(x_i) \qquad (7)$$

mode of AES but takes long strands of time that's why we have used scrambling with modified AES.

where $P(x_i)$ is the probability of the ith pixel and K is the number of pixels in an image. Results in Table 1. show that the calculated entropy values using proposed method and proposed method by Kamali et al. (2010) are near to 8.

5.2 Histogram analysis

Statistical attacks can be possible by analyzing the histogram values. Uniform distribution of values in histogram represents that the algorithm is resistant to statistical attacks. Results in Figure 7. and Figure 8. proved that proposed system is more secure than earlier version of AES.

5.4 Correlation between original and encrypted image

A good encrypted image should have very low correlation with the original image. Equation (8) is being used to calculate correlation between original and encrypted image (MathWorks).

5.3 Entropy

It can be defined as the degree of randomness which can be used to characterize the texture of image. The number of gray levels for grey scale image is 256 or 2^8 (Wadi & Zainal 2013). Using (7),

$$r = \frac{\sum_m \sum_n (A_{mn} - \bar{A})(B_{mn} - \bar{B})}{\sqrt{(\sum_m \sum_n (A_{mn} - \bar{A})^2)(\sum_m \sum_n (B_{mn} - \bar{B})^2)}} \qquad (8)$$

Table 2. Correlation coefficient between original and encrypted image.

	Image 1	Image 2	Image 3	Image 4
Encrypted Image using initial AES	−0.2048	−0.0004202	0.0017145	−0.015924
Encrypted image using method by Kamali et al. (2010)	−0.000249	0.0005630	0.001404	0.0034004
Encrypted Image using proposed method	0.0014188	−0.0030691	0.0017545	−0.000366

Table 3. Encryption and decryption time (in seconds).

	Image 1	Image 2	Image 3	Image 4
Encrypted Image using initial AES	469.76892	472.98429	490.40811	466.770302
Encrypted image using method by Kamali et al. (2010)	518.96447	495.43535	497.47588	507.205127
Encrypted Image using proposed method	73.482489	76.926034	87.096232	83.769738

where $\bar{A}$, $\bar{B}$ are the mean of gray values of original image and encrypted image respectively, m and n refer to the pixel location in original/encrypted image.

5.5 Simulation time analysis

A good algorithm is one which takes less time and gives optimum results. Proposed work takes approximately 85% lesser time in comparison to initial version of AES.

All the simulations have been done using MATLAB R2010a on a laptop with specifications: Intel Core i3-4010U 1.70GHz processor with 2GB random access memory, 500GB hard disk and Windows 7 Home Premium 64-bit. All the images for simulation have been taken from Berkeley standard database of images.

6 CONCLUSIONS

Several attempts have been made to modify AES algorithm. But all these algorithms take too long for the encryption process. In this paper, modifications are made to remove the mentioned drawbacks and make the algorithm time efficient. Scrambling is done to remove the existing pattern appearance problem. SHA-256 algorithm is used to calculate the hash value of key for the encryption process. Steganography is used to hide the key in the encrypted image so that the key can be transferred securely.

REFERENCES

Gupta, A., Tiwari, N., Chawla, M., & Shandilya, M. 2014. An Image Encryption using Block based Transformation and Bit Rotation Technique. *International Journal of Computer Applications (0975–8887)*, 98(6).

Huang, C.W., Tu, Y.H., Yeh, H.C., Liu, S.H., & Chang, C. J. Image Observation on the Modified ECB Operations in Advanced Encryption Standard. *International Conference on Information Society (i-Society 2011)*, *IEEE*: p. 6.

Kak, A. 2015. AES: The Advanced Encryption Standard [Lecture notes on computer and network security]. *https://engineering.purdue.edu/kak/compsec/NewLectures/Lecture8.pdf*

Kamali, S.H., Shakerian, R., Hedayati, M., & Rahmani, M. 2010. A New Modified Version of Advanced Encryption Standard Based Algorithm for Image Encryption. *International Conference on Electronics and Information Engineering (ICEIE 2010)*, IEEE: p.1.

MathWorks (http://in.mathworks.com/help/images/ref/corr2.html)

Stallings, W. 2005. *Cryptography and Network Security: Principles and Practices*. New Jersey: Prentice Hall.

Sun, T., Shao, X., & Wang, X. 2008. A Novel Binary Image Digital Watermarking Algorithm Based on DWT and Chaotic Encryption. *The 9th International Conference for Young Computer Scientists, IEEE*: 2797–2802.

Telagarapu, P., Biswal, B., & Guntuku, V.S. 2011. Security of Image in Multimedia Applications. *International Conference On Energy, Automation And Signal, IEEE*, p.5.

Tran, M.T., Bui, D.K., & Duong, A.D. 2008. Gray S-box for Advanced Encryption Standard. *In IEEE computer society meeting*, Suzhou: China.

Wadi, S.M. & Zainal, N. 2013. Rapid Encryption Method Based on AES Algorithm for Grey Scale HD Image Encryption. *The 4th international conference on Electrical Engineering and Informatics(ICEEI 2013)*, Procedia Technology (2013): 51–56.

Wadi, S.M. & Zainal, N. 2014. High Definition Image Encryption Algorithm Based on AES Modification. *Wireless Personal Communications An International Journal (0929–6212)*, *Springer*, 77(3).

Communication and Computing Systems – Prasad et al. (Eds)
© *2017 Taylor & Francis Group, London, ISBN 978-1-138-02952-1*

Trust version of AASR

Pragya & Vidushi Sharma
Gautam Buddha University, Greater Noida, India

ABSTRACT: Mobile Ad Hoc Network is a network which does not contain any infrastructure and have wireless communications. Every device in the network can freely move in any direction. By reason of dynamic nature of this network, it is very essential to secure network from adversaries. Anonymity is an issue which must be considered while securing communication in MANET. Anonymous on demand routing protocols secure the network from both active and passive attacks. But these protocols contain some vulnerability such as modification of packets, DOS attacks which can be removed by providing Authentication with the help of Group Signature. AASR introduces authentication as well as maintains anonymity but it experiences the problem of packet delay. A trust model with AASR is helpful to resolve this problem.

Keywords: MANET, on demand routing, anonymity, onion routing, group signature

1 INTRODUCTION

Making the communication secure in networking environment is an important aspect. In MANET maintaining security is quite difficult as compared to wired network because MANET is configured automatically and exist without any infrastructure. Another important feature of this type of network is that it contains mobility nature of the various nodes and dynamic topology. In MANET, nodes itself works as the server as well as the client. This nature helps it to be used in military battlefield, emergency operations such as fire, flood. Due to the range limitation of the nodes in MANET, it requires routing to communicate with different nodes. Many active and passive attacks could occur from external environment by malicious hosts or from compromised hosts inside the network (M. Yu et al. 2009).

So many On-demand routing protocols have been anticipated for securing routing and data content. Including anonymity factor makes these protocols much better for the security of the network. Though anonymity introduces unidentifiability and unlinkability over the network but it is vulnerable to modification of data packet illegally and DOS attacks. For this authentication can be provided by group signature scheme. The fine points of the concepts are as follows:

1.1 *Onion routing*

Onion (M.G Reed et al. 1998) is a kind of data structure which is based on layered approach. The source node setup the root of an onion with

a precise route message. While communicating in MANET environment during the route request phase an encrypted layer is added to RREQ message by the forwarding node. Due to this, the actual source and destination remain disclosed which establish an anonymous route.

1.2 *Anonymous routing*

Anonymous routing provides various anonymity such as (a) Identity Anonymity: In this, identity of the nodes which are taking part in the communication remains hidden from forwarded packets. (b) Route Anonymity: The nodes cannot be mapped to the route. The source intermediate and destination nodes contain information regarding the pseudonyms only. Even participating nodes don't know about the information of entire route. (c) Topology Anonymity: Even in the case of network traffic analysis number of participating nodes, hop count can keep of sight by making location anonymous.

1.3 *Trapdoor*

A trapdoor is an ordinary idea in cryptographic functions which describes a one-way function between two sets. A global trapdoor reflects a method to collect information in which intermediate nodes may append information, like node ID's with the trapdoor. Only some of the nodes like the source and destination can decipher and recover the elements by means of pre-established secret keys. The procedure of trapdoor needs an anonymous end-to-end key acknowledgment between the source and destination.

1.4 *Group signature*

Group trusted authority issues a pair of group public and private keys which every member of the cluster may contain. Group Signature scheme allows a member of a group to produce its signature with the help of its private key which can be verified by other members of the group by keeping identity of signer unrevealed.

Though numerous anonymous routing protocols exist such as ANODR (J. Kong et al. 2003), ASR (B. Zhu et al. 2005), ARM (S. Seys et al. 2009) have been introduced. But most of them do not add authentication to the protocols. In the lack of authentication, an adversary can behave maliciously without any kind of restrictions during the route discovery. To grant anonymity & authentication simultaneously is difficult for authentication identity of signer is required during verification process. Group signature scheme provides authentication without revealing identity of nodes.

2 RELATED WORK

In existing On Demand Routing protocols AODV (C. Perking et al. 2003) and DSR (D. Jhonson, 2007) adversary can easily detect the identities of the nodes. DSR is observable by a single eavesdropper as it adds routing information in packet headers. While in AODV information is stored in routing tables as an alternative of packet headers. ANODR (J. Kong & H. Kong, 2003) was proposed to avoid adversaries from following a packet flow to the source or destination. It is an anonymous routing protocol which is based on route pseudonymity approach. This approach allows unlinking the node's location and identity.

Another secure distributed anonymous routing protocol SDAR (A. Boukerche et al. 2004) is introduced in which public key algorithm is used which makes the system very unscaleable as each intermediate forwarding node need to attempt decryption of the trapdoor during route discovery. In AnonDSR (R. Song et al. 2005), three protocols are introduced i.e. security parameter establishment, anonymous route discovery and anonymous data transfer in which a cryptographic onion is created for anonymous communication data protection. In MASK (Y. Zhang et al. 2006) a clear plain node ID is used for route discovery. It provides neighbourhood authentication but it cannot sign the routing packets. This protocol is based on new cryptographic concept known as 'pairing' which allows authenticating neighbouring nodes without disclosing their identity.

ASR (B. Zhu et al) provides supplementary properties on anonymity i.e. Identity anonymity and

Strong location privacy by using shared secrets between any two consecutive nodes. Some small size information known as TAG is sent combined with the data packet so that forwarding node can verify that information instead of the whole packet. No node identity involved in ASR, as well as external & internal nodes, cannot detect the location of source and destination. Another protocol proposed as ARM (S. Seys & B. Preneel, 2009), in which random padding and time to live values for RREQ messages are used. Padding is applied to stop an adversary about the number of hops to source or destination of RREQ or REEP messages. The source randomly chooses a padding length as per the probability distribution proposed by author. Distribution function is used with high probability that a node will select a padding length. Another scheme used is TTL field in which source adds a random value to the required TTL size to reach destination. The TTL remains randomized not to reveal the distance between source and destination.

Some of the vulnerabilities such as modification of data packet data & DOS attacks may occur which is reduced by introducing authentication with anonymity proposed in A3RP (J. Paik et al. 2008). The routing and data packets are preserved by Group Signature scheme (D. Boneh et al. 2004). Secure hash functions are also implemented for mapping keys and node pseudonyms along a route. Session key between source and destination is established for secure data transmission after route discovery. In Group signature scheme, every member of the group can generate its own signature by its own private key which is issued by trust authority.

Previous protocols use hash function which is not that much scalable. In AASR (Liu Wei & Yu Ming, 2014) encrypted onion mechanism is used which records discovered route. It creates an encrypted secret message to authenticate RREQ-RREP linkage. Group Signature scheme is used for authenticating the RREQ packet per hop as well as prevent the intermediate nodes from modifying any routing packet.

3 EXISTING SYSTEM

The Authenticated Anonymous Secure Routing (AASR) protocol is the extended version of AODV element which maintains some cryptographic operations (William et al. 2006). AASR provides improved hold up for secure communications with high throughput and lower packet loss. AASR can be further modified by combing trust-based routing (M. Yu & K. Leung, 2009) which will reduce end to end packet delay.

The routing algorithm of the existing system can be achieved on the basis of current on-demand ad-hoc routing protocol like AODV or DSR. The fundamental strategy of routing procedure can be described step by step as follows:

- A source node transmits an RREQ packet during route discovery phase.
- In the case of receiving the RREQ packet by intermediate nodes, it cross checks the RREQ by utilizing its group public key and integrates one coat on top of the key-encrypted onion. Same procedure is rechecked until the RREQ packet is not received by the destination or gets expired.
- After receiving RREQ by the destination node, it verifies the RREQ and then collects RREP packet to broadcast it back to the source node.
- Validation of the RREP packet is done by each intermediate node on the path from destination to source and updates its routing and forwarding tables. Then it peel off one layer on the top of the key-encrypted onion and pursues broadcasting the updated RREP.
- At the time of receiving the RREP packet by source node, it authenticates the packet and rejuvenates its routing and forwarding tables which results the completion of route discovery phase.
- The source node starts transmitting data in the conventional route. The route pseudonym is utilized for forwarding the data packets by each intermediate node.

4 PROPOSED WORK

4.1 Assumptions

The network for the proposed work is assumed as symmetric i.e. if node A comes in range of node B then node B also comes in the range of node A. Each group as well as group in the network contains its public/private keys issued by PKI (Public Key Infrastructure) or different Certificate Authority (CA).

4.2 Model

- Source node S has information in the format of <Network Size, Traffic, Message>
- While Destination node D will have following details:
  <Node, PKR, PKS, T, $R_{DataTrf}$, Intermediate, Node>
  Where PKR, PKS are packets received and packets send respectively.
  $R_{DataTrf}$ is the rate of data transfer.
- Node's pseudonym have been generated with the help of node's private key.

- Next hop is selected as per the trust value in the terms of packet loss. Trust calculation is done as per indirect observation i.e. on the basis of neighbour nodes.

$$t = pow(e, f) \qquad (1)$$

where, t is trust value, e is packet loss & f = 0.5
e can be calculated as a = dest (X_1)-source (X_1)

$b = a*a$
$c = dest (Y_1) - source (Y_1)$
$d = c*c$
$e = b + d$

4.3 Protocol design

The neighbour node which will have minimum t value will be selected as next hop as it results into more trustworthiness.

- Selected node will send ACK (acknowledgement) to the source node.
- Routing table has various attributes in the form of <Source, RREQ_ID, pseudonym, Dest_address, hop count, route record>
- Route Reply will flow over the network with details i.e. <source_address, RREP_Id, route record>

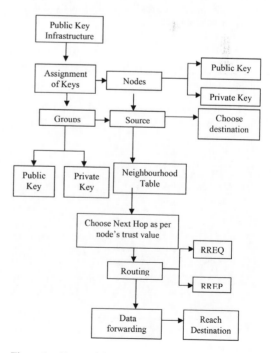

Figure 1. Protocol design of proposed scheme.

145

5 SIMULATION

5.1 Network configuration

To simulate our protocol, 1000 m × 1000 m network space has been mediated having thirty-five nodes which are uniformly distributed. Maximum numbers of packets in Interface Queue (IfQ) are fifty. Nodes can check packet integrity with the help of Link Layer (LL). MAC Layer is used to tell node how to share broadcast medium.

5.2 Evaluation metrics

- End To End Delay:
End to End delay can be defined as the average time taken by the data packet to arrive at the destination. This can be caused due to buffering during route discovery latency, queuing at the interface queue. Average End to End delay can be calculated with the help of following formula i.e.

$$\text{Avg EED} = S/N \tag{2}$$

where, Avg EED: Average End to End delay
  S: sum of the time spend to deliver packets to each destination i.e. ∑ (Arrive time-Spend time).
  N: Number of packets received by all destination nodes. Low End to End delay results better performance.
- Throughput:
Throughput can be defined as the total number of packets delivered over the total simulation time. It can be calculated as:

$$\text{Throughput} = N/1000 \tag{3}$$

where, N: Number of bits received successfully by all destinations.
- Packet Loss:
The total number of packets dropped during the simulation is known as packet loss. It can be calculated as:

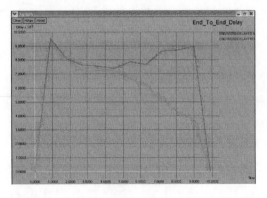

Figure 2.   End to end packet delay.

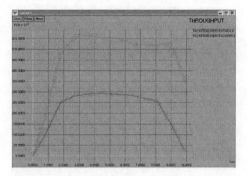

Figure 3.   Throughput.

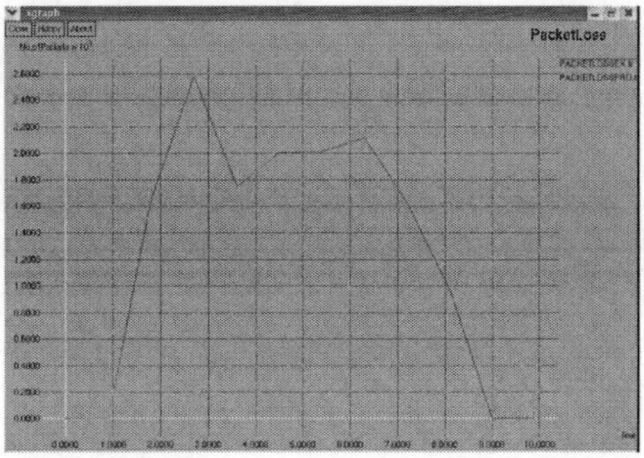

Figure 4.   Packet Loss.

Packet Loss = No. of packet send–No.
                of packet received   (4)

5.3 Results

Simulation results are shown in the following graph by comparing metrics of proposed scheme with AASR (existing scheme).

In graph (a), End to End delay is represented between delay and time. Graph (b) shows high throughput of proposed method shown between Packet delivery ratio and time which results into better connectivity over the network. Packet Delivery Ratio (PDR) can be calculated as:

PDR = ∑ Number of packet received/∑
Number of packet send   (5)

where as graph(c) denotes lower packet loss which increase the efficiency.

6 CONCLUSION

In this paper, we incline to approach associated genuine and suggest routing protocol for. The route request packets are authenticated by cluster signatures, which might protect the potential active innominate attacks while not unveiling the node identities. The key-encrypted onion routing with a route secret verification message is proposed to not solely record the innominate routes however conjointly prevent the intermediary nodes from inferring the consequential destination. By combining trust model in AASR utilizes trust values to favour packet forwarding by maintaining a trust breach for every node. In this scheme, we compare trust value with other node's trust value and finding the best node for the route which reduces End to End delay and gives high throughput with lower packet loss.

REFERENCES

Anonymizer. http://www.anonymizer.com.

Boneh, D., X. Boyen, and H. Shacham, Aug. 2004, "Short group signatures," in Proc. Int. Cryptology Conf. (CRYPTO'04).

Boukerche, A., K. El-Khatib, L. Xu, and L. Korba, Nov. 2004, "SDAR: a Secure Distributed Anonymous Routing Protocol for Wireless and Mobile Ad hoc Networks," in Proc. IEEE Int'l Conf. Local Computer Networks (LCN'04), pp. 618–624.

Johnson, D., Y. Hu, and D. Maltz, 2007, "RFC 4728 - The Dynamic Source Routing Protocol (DSR) for Mobile Ad Hoc Networks for IPv4," Internet RFCs.

Kong J. and X. Hong, Jun. 2003, "ANODR: ANonymous On Demand Routing with Untraceable Routes for Mobile Ad hoc networks," in Proc. ACM Mobi-Hoc'03 pp. 291–302.

Liu Wei and Yu Ming., March 2014, "AASR: Authenticated anonymous secure routing for manets in adversarial environments", IEEE Transactions on vehicular technology.

Paik, J., B. Kim, and D. Lee, Apr. 2008, "A3RP: Anonymous and Authenticated Ad hoc Routing protocol," in Proc. International Conf. on Information Security and Assurance (ISA'08).

Perkins, B., E. Belding-Royer, S. Das, et al., 2003, "RFC 3561 - Ad hoc On—Demand Distance Vector (AODV) Routing," Internet RFCs.

Reed, M. G., P. F. Syverson, and D. M. Goldschlag, May 1998,"Anonymous Connections and Onion Routing," IEEE Journal on Selcted Area in Comm., vol. 16, no. 4, pp. 482–494.

Seys S. and B. Preneel, Oct. 2009, "ARM: Anonymous Routing protocol for mobile ad hoc networks," Int. Journal of Wireless and Mobile Computing, vol. 3, no. 3, pp. 145–155.

Song, R., L. Korba, and G. Yee, Nov. 2005, "AnonDSR: efficient anonymous dynamic source routing for mobile ad hoc networks," in Proc. ACM Workshop Security of Ad Hoc and Sensor Networks (SASN'05).

William S. and W. Stallings, 2006 Cryptography and Network Security, 4th Edition. Pearson Education India.

Yu M. and K. Leung, Apr. 2009."A Trustworthiness-based QoS routing protocol for ad hoc networks," IEEE Trans. on Wireless Comms., vol. 8, no. 4, pp. 1888–1898.

Yu, M., M. C. Zhou, and W. Su, Jan. 2009 "A secure routing protocol against Byzantine attacks for MANETs in adversarial environment," IEEE Trans. on Vehicular Tech., vol. 58, no. 1, pp. 449–460.

Zhang, Y., W. Liu, W. Lou, and Y. G. Fang, Sept. 2006, "MASK: Anonymous On-Demand Routing in Mobile Ad hoc Networks," IEEE Trans. on Wireless Comms., vol. 5, no. 9, pp. 2376–2386.

Zhu, B., Z.Wan, M.S. Kankanhalli, F. Bao and R.H. Deng, "Anonymous Secure Routing in Mobile Ad-Hoc Networks".

Communication and Computing Systems – Prasad et al. (Eds)
© *2017 Taylor & Francis Group, London, ISBN 978-1-138-02952-1*

Sparse filter design techniques: A review

P. Dalal & S. Dhull
Guru Jambheshwar University of Science and Technology, Hisar, India

ABSTRACT: Digital filters have a wide variety of applications in signal processing field and communication field. Research is being focused now a day, to improve the implementation efficiency of digital filters. To achieve this, designing sparse filters i.e. filters having small number of non-zero coefficients in its impulse response, is a good solution, where the adders and multipliers corresponding to the zero value coefficients are simply omitted. In this paper an extensive literature survey is being done on design of sparse digital filters. Various techniques and algorithms that are used in the literature to design sparse filters are investigated and discussed.

Keywords: FIR filter; sparse filter; compressed sampling; implementation efficiency

1 INTRODUCTION

Digital filters are being used these days in various signal processing applications and communication applications like in radar, sonar, biomedical signal processing, image processing, consumer electronics and many more (Mitra & Kaiser 1993). Digital filter design has remained an ongoing research field. Researchers aim to design digital filters with desired level of performance along with minimizing complexity. Computation time, power consumption, hardware required and circuit area utilized are some of the factors that must be taken care of while designing a digital filter.

To design Finite Impulse Response (FIR) digital filters a number of algorithms have been proposed. The traditional or conventional FIR filter designing methods take total number of coefficients also called as order of the filter as a measure of complexity. Minimizing order of the filter minimizes the computation time and memory required. Thus, all traditional filter designing algorithms such as Parks-McClellan aim to obtain an optimal filter that minimizes error of the designed filter with respect to desired frequency response under some criteria with a specified filter order (McClellan et al. 1973). With the recent advancements and demands in the field of signal processing, there is a need to focus on improving the implementation efficiency, which can be achieved by reducing number of arithmetic operations, especially multiplications. To achieve this, some authors have tried to eliminate multipliers by restricting impulse response coefficient to take either binary values (Lim & Parker 1983) or composed of power-of-two terms (Chen & Willson 1999, Lim et al. 1999, Li et al. 2002).

Another way to improve the implementation efficiency is by designing FIR filters with a least number of non-zero coefficients, as the number of additions and multiplications depend on the number of non-zero coefficients and the adders and multipliers corresponding to the zeroed value coefficients may simply be omitted from the design. This leads to a desire for sparse designs i.e design with fewer non-zero coefficients. Sparse FIR filters tend to obtain better performance compared to non-sparse optimum filters by designing a higher order filter but setting some of the coefficients to zero value. Sparse filters thus improve the implementation efficiency and hence possess real time application potential while being cost effective also. Sparse filter designs have proved to be efficient in terms of cost of computation, hardware utilized and power consumed. Many authors have worked upon this target area to find the most efficient solution for the sparse FIR filter design problem.

The remainder of the paper is presented as follows. In Section II, we describe the problem of interest. In Section III, a brief literature review is presented. In Section IV, applications of sparse filters are discussed. Finally, in Section V conclusion is being presented.

2 PROBLEM FORMULATION

Given a desired frequency response $H_d(j\omega)$, optimal FIR filters are aimed to minimize error between the designed response $H(j\omega)$, and the desired response in either least-squares or minimax (Chebyshev) sense. Some researchers have also worked to minimize the

L_p norm of the error (Grossmann & Eldar 2007). But to design a sparse filter objective function must be set to minimize, the error and the number of non-zero filter coefficients.

Consider an order 'N' linear-phase FIR filter with impulse response coefficient vector denoted as $\mathbf{h} = h_i$ {for i = 0,1...N}. The transfer function $H(z)$ of the filter is defined as follows:

$$H(z) = \sum_{i=0}^{i=N} h_i z^{-i} \tag{1}$$

The impulse response vector $\mathbf{h}$ for a FIR filter is called as sparse impulse response if significant number of $h_i's$ exactly equals to zero. The impulse response vector is understood as K-sparse if it consists of only K number of nonzero coefficients. The p-norm of vector $\mathbf{h}$ denoted as $\|\mathbf{h}\|_p$ or L_p for $p \geq 1$, is defined as:

$$\|\mathbf{h}\|_p = \left(\sum_{i=0}^{N} |h_n|^p \right)^{1/p} \tag{2}$$

Also the 0-norm, written as L_0 can be defined as

$$\|\mathbf{h}\|_0 = \lim_{p \to 0} \|\mathbf{h}\|_p^p \tag{3}$$

The L_0 norm of filter coefficients is generally used for the sparsity measure of filter impulse response coefficients, which actually counts the number of nonzero coefficients. The objective function for designing a sparse FIR filter can be defined in a manner to minimize the L_0 norm of filter coefficient vector. But sparse FIR filter design problems defined using L_0 norm can be highly non-convex, as L_0 norm is non-differentiable (Xie & Hu 2013). So, it would be difficult to find the optimum solution using the conventional optimization algorithms. By doing exhaustive search optimal solution may be obtained, but this could lead to large amount of computation. Thus, better way is to look for locally optimal solutions. A sparse filter design is also inspired by the advancement in the field of sparse representation of signals (Donoho 2006) and compressive sensing (Elad 2010, Zibulevsky & Elad 2010), so the techniques and algorithms developed for sparse coding of signals can be utilized to design sparse filters. So some of the techniques like hard-thresholding, Orthogonal Matching Pursuit, Basis Pursuit, Iterative shrinkage and thresholding algorithms, Branch and bound algorithms, Iterative Reweighted Least Square algorithms, which were utilized in sparse and redundant representation of signal are also being utilized in the literature to design sparse filters.

In general, to solve the objective problem, the authors in the literature have proposed a two step process, in the first step the those coefficients are identified that could be set to zero while minimally affecting the optimal filter performance, using an appropriate optimization algorithm and in the second step the sparse FIR filter is designed that minimizes the error while constraining the identified coefficients to be zero.

3 LITERATURE REVIEW

(Mattera et al. 2002) defined the sparse FIR filter designing problem as a Linear Programming Problem. The problem aims at designing a FIR filter that tends to minimize the weighted error in Chebyshev sense with imposed constraints of setting K0 number of filter coefficients as zero. For choosing the set of filter coefficients that need to be set to zero OMP (Orthogonal Matching Pursuit) procedure is employed. This algorithm is a greedy algorithm that sequentially keeps on adding one additional coefficient to a set of zero-valued coefficients, while minimizing the lingering approximation error. Moreover, there is no guarantee of achieving convergence. This paper solves the optimization problem with the help of MINOS, which is a commercial software package to solve linear and non-linear problems.

(Gustafsson et al. 2007) have defined the problem of sparse filter designing as a Mixed Integer Linear Programming problem. The objective function is defined that tends to minimize the weighted error in Chebyshev sense subject to constraints that tends to minimize the filter order and the number of non-zero coefficients. These two constraints are weighted, with more weight given to obtain sparse impulse response. Different weighting may be used, to obtain a different trade-off between sparsity and order of the filter. To solve the MILP problem either branch-and-bound or branch-and-cut algorithm is being used. Simulation experiments have been performed by varying passband ripple. Results show that filter order is larger than the optimal non-sparse filter, but the number of non-zero valued coefficients has significantly reduced. It can also be concluded from the results that as the passband ripple is increased, the number of zero value coefficients also increases.

(Tseng & Lee 2009), utilized the constrained L_1 minimization method used in compressed sampling for designing the sparse filter. The filter design problem is specified in the form of standard compressed sensing equation. It is demonstrated in the paper that minimizing the L_1 norm of the filter coefficients leads to a sparse filter design with the reduced range of filter coefficients and reduced maximum value of the filter output. So the objective function is defined that tends to minimize the L_1 norm of the filter coefficients for obtaining a sparse impulse response subject to a condition that L_2 norm of the error between the desired response and designed response is less than a specified error 'ϵ'. Design examples are presented in the paper which demonstrate the flexibility of the new design method compared to the conventional Least Squares design. In a similar paper (Tseng & Lee 2012), the Orthogonal Matching Pursuit (OMP) algorithm is being utilized to design a constrained sparse FIR filter. The authors try to obtain trade-off between sparsity of filter coefficients and magnitude response errors by defining an error bound that can be suitably chosen. Simulation examples are given for sparse FIR notch filter. (Tseng & Lee 2011) again utilized the OMP algorithm to design a digital differentiator. In a similar paper (Tseng & Lee 2013), authors have worked to design digital differentiator using improved Iterative Hard-Thresholding (IHT) by defining a smooth thresholding curve. In the improved IHT the number of elements thresholded to zero value increases gradually as compared to conventional IHT that is greedy. Simulation results have proven that filter designed using this technique has less error than standard OMP or IHT based designs.

(Lu & Hinamoto 2010) defined the objective function for the sparse filter design as to minimize the weighted summation of either the L_∞ or L_0 norm of the error between the desired response and designed response and L_1 norm of the filter coefficients. The L_1 norm term help produce a sparse impulse response. This technique is inspired by the Basis Pursuit used in compressive sensing, where because of non-convexity of L_0 norm, we can use L_1 norm. The defined objective function is a convex problem and solved using SeDuMi. After solving the objective function hard-thresholding is applied on the obtained coefficients with appropriate threshold to obtain sparse impulse response. Design examples are presented in the paper that proves the improved sparsity of the designed filter. In a similar paper (Lu & Hinamoto 2011), authors have worked to design a sparse IIR filter. The objective function minimizes the L_1 norm of the coefficients of numerator to promote sparsity subject to stability of transfer function. The objective function is a standard second-order cone program is solved using efficient convex program solver SeDuMi. After solving the objective function hard-thresholding is applied to obtain zero-valued coefficients.

(Wei 2009) adopts a method to design sparse FIR filter that sequentially minimizes the p-norm of coefficient vector with p gradually decreasing from 1 toward 0. Since the p-norm of a vector is highly non-convex for p < 1, this problem is overcome by appropriately initializing each subproblem. Design examples demonstrate that the filters obtained using method for a given set of specifications posses an optimal level of sparsity. (Wei & Oppenheim 2010) defined the objective function for sparse filter design to obey two constraints, first minimizes the weighted least squares error of the frequency response and second maximizes the signal-to-noise ratio so that detection of signal in the noisy environment is possible. Both the problems are combined to form a single quadratic constraint. Simultaneously to maximize the sparsity, the L_0 norm of the filter coefficient vector is minimized. In a similar paper (Wei et al. 2013) have defined the objective function to obey three constraints; first minimizes weighted least-squares error of frequency response, second constraint minimizes the mean squared error for estimation, and third maximizes the signal-to-noise ratio for detection. The design problem is thus defined that maximizes the sparsity while following the three above defined performance constraints, which can be reduced to a single quadratic constraint. Authors in this paper try to develop efficient and exact solution. Examples are presented for wireless channel equalization and distortion less response beamforming. In a similar paper, (Wei & Oppenheim 2013) proposed a method based on Branch and Bound algorithm to solve a combinatorial optimization problem. Design experiments demonstrate that techniques used in this paper decrease the complexity of designed filter dramatically.

(Baran et al. 2010) proposed two approaches in this paper. In the first approach, the impulse response of non-sparse filter is iteratively thinned. Two rules called as minimum increase rule and smallest coefficient rule are defined by the authors that will in each iterative step, set the value of one or more filter coefficients to be zero. In the first rule, the coefficient with the smallest magnitude is set to zero, while second rule searches for a coefficient whose value if set to zero results in the minimum increase of error. The second approach, firstly intends to obtain smaller magnitude filter coefficients by minimizing the L_1-norm of impulse response, and then re-optimization is done by setting the K smallest magnitude coefficients to zero. The algorithms are evaluated by the authors in the context of array design and acoustic equalization. In both the approaches filter is being designed using direct form structure and compared with standard Parks McClellan algorithm. Because of the iterative nature of both the approaches, the drawback is that, for designing higher order filters, there is a need of high computational effort.

(Jiang et. al 2012a) have defined sparse filter design problem that tends to minimize the L_0 norm of the coefficients subject to minimizing the approximation error in minimax sense. To solve the design problem Iterative Shrinkage and Thresholding (IST) algorithm is employed. In IST algorithms optimization problem is decomposed into set of independent scalar optimization problems. Subproblems in a simpler form are defined in each iterative step and respective dual problems of these nonconvex subproblems are constructed using Lagrangian function. This dual problem now can be easily solved. The design procedure is successively run for several times to achieve better results and is continued until more sparsity can't be further obtained. Simulation results of filter design examples are presented and results are compared. The results show that the sparsity of the designed filter is improved but at a cost of slight increase in the filter order. A similar approach is being used in paper (Jiang et. al 2012b). In paper (Jiang & Kwan 2012) defined the objective function to minimize the L_0 norm of the coefficient vector subject to a WLS approximation error constraint. To solve the objective function IST algorithm is being employed. An iterative algorithm called successive activation algorithm is proposed that constructs a simplified subproblem. The design method can be used to design linear phase, as well as nonlinear-phase FIR filter designs. Authors have proposed a similar design strategy in paper (Jiang & Kwan 2013) as of the previous paper. In paper (Jiang et al. 2014) authors aim to optimize the order and the sparsity of an FIR filter simultaneously, so that both the group delay and complexity is being reduced. The basic idea to reduce the filter order is to force some initial coefficients of the impulse response to zero and to achieve this a regularization term is being integrated in the objective function. So the objective function acquires the form of a weighted L_0 norm optimization problem. To solve the objective function an efficient numerical method is proposed which is based on the Iterative-Reweighted-Least-Squares (IRLS) algorithms. In the IRLS algorithms the L_0 norm of a vector is replaced its L_1 norm, which is its convex relaxation form, and can be more easily solved. Example designs are presented and the results are compared. It is proven that the proposed method is jointly optimizes both the filter order and sparsity of the filter as well. In paper (Jiang et al. 2015), the authors share the similar idea as of previous paper but with a slight modification in the objective function.

(Matsuoka et al. 2014) proposes design of sparse filter that minimizes the error in the least square sense. But according to well know Gibbs phenomenon, a filter designed using Least-squares criteria posses large error near the cut-off frequencies. To minimize this error, authors have imposed constraints on the peak value of error in the frequency response of the designed filter, without increasing the transition band. So, a two-step approach is being proposed by the authors to design constrained sparse FIR filter. In the first step, optimal sparse coefficients are obtained using the Graduate Non-Convexity algorithm. The second step, while restricting the zero valued coefficients obtained in first step to zero value, aims to reduce the peak error of the filter by optimizing the values of left over coefficients. Experimental results demonstrate that the proposed design approach jointly obtains a sparse impulse response and minimizes the peak error as compared to a conventional least squares design.

4 APPLICATIONS OF SPARSE FILTERS

In the literature, numerous applications of sparse filters have been explored by various researchers. The frameworks in which these applications are explored include beam forming, frequency response approximation, speech coding, channel equalization, signal detection, estimation and linear phase acoustic equalizer. The reason behind using a sparse filter depends on the application area where the filter is being used. For example, in situations where large amount of computation needs to done, the use of sparse filters can reduce the computation complexity. Similarly, if a filter is required for a wireless application, where power consumption is a primary concern, the use of sparse filters can reduce the power consumed, as fewer computations are done. In integrated circuit implementations, circuit area is of prime concern, the use of sparse filters decrease the circuit area and hardware utilized. In sensor array design, where the fabrication and operation of the individual sensors determines the cost and capabilities of the array, a maximally sparse array design can be very cost effective. So, based on above discussion, it can be concluded that the sparse design can be utilized in variety of application areas, being efficient and cost effective.

5 CONCLUSION

Sparse FIR filters can dramatically improve the implementation efficiency of digital filters. Various techniques for sparsity maximization of FIR digital filter are discussed. Based on the discussion, there is a need to design optimized sparse filters that maximize sparsity, minimize complexity with an optimum filter order that maintains a negotiable group delay. Future work can be done in the direction of improving the sparsity of IIR Digital filter, notch filters, 2-dimensional filters and adaptive filters.

REFERENCES

Baran, T., Wei, D. & Oppenheim, A.V. 2010. Linear Programming Algorithms for Sparse Filter Design. IEEE Transactions on Signal Processing 58(3): 1605–1617.

Chen, C.L. & Willson, A.N. 1999. A trellis search algorithm for the design of FIR filters with signed-powers-of-two coefficients. IEEE Transactions on Circuits Systems II 46(1): 29–39.

Donoho, D. 2006. Compressed sensing. IEEE Transactions on Information Theory 52(4): 1289–1306.

Elad, M., Figueiredo, M. A. T. & Ma, Y. 2010. On the Role of Sparse and Redundant Representations in Image Processing. *Proceedings of the IEEE 98(6):* 972–982.

Grossmann, L.D. & Eldar, Y.C. 2007. An L1-Method for the Design of Linear-Phase FIR Digital Filters. IEEE Transactions on Signal Processing 55(11): 5253–5266.

Gustafsson, O., DeBrunner, L.S., DeBrunner, V. & Johansson, H. 2007. On the Design of Sparse Half-Band Like FIR Filters. Conference Record of the Forty-First Asilomar Conference on Signals, Systems and Computers, 4–7 Nov 2007: 1098–1102.

Jiang A., Kwan H.K., Zhu, Y. & Liu, X. 2012b. Minimax design of sparse FIR digital filters. IEEE International Conference on Acoustics, Speech and Signal Processing, 25–30 March 2012: 3497–3500.

Jiang, A. & Kwan, H. K. 2012. Efficient design of sparse FIR filters in WLS sense. IEEE International Symposium on Circuits and Systems, Seoul: 41–44.

Jiang, A. & Kwan, H.K. 2013. WLS Design of Sparse FIR Digital Filters. IEEE Transactions on Circuits and Systems 60(1): 125–135.

Jiang, A., Kwan, H.K. & Zhu, Y. 2012a. Peak-Error-Constrained Sparse FIR Filter Design Using Iterative SOCP. IEEE Transactions on Signal Processing 60(8): 4035–4044.

Jiang, A., Kwan, H.K., Tang, Y. & Zhu, Y. 2014. Efficient design of sparse FIR filters with optimized filter length. IEEE International Symposium on Circuits and Systems, 1–5 June 2014: 966–969.

Jiang, A., Kwan, H.K., Zhu, Y., Liu, X., Xu, N. & Tang, Y. 2015. Design of Sparse FIR Filters with Joint Optimization of Sparsity and Filter Order. IEEE Transactions on Circuits and Systems I: Regular Papers 62(1):195–204.

Li, D., Lim, Y.C., Lian, Y. & Song, J. 2002 A polynomial-time algorithm for designing FIR filters with power-of-two coefficients. IEEE Transactions on Signal Processing 50(8): 1935–1941.

Lim, Y. & Parker, S. 1983. FIR filter design over a discrete powers-of-two coefficient space. IEEE Transactions on Acoustics, Speech and Signal Processing 31(3): 583–591.

Lim, Y.C., Yang, R., Li, D. & Song, J. 1999. Signed power-of-two term allocation scheme for the design of digital filters. IEEE Transactions on Circuits and Systems II: Analog and Digital Signal Processing 46(5): 577–584.

Lu, W.S. & Hinamoto, T. 2010. Digital filters with sparse coefficients. Proceedings of 2010 IEEE International Symposium on Circuits and Systems, 30 May-2 June 2010: 169–172.

Lu, W.S. & T. Hinamoto. 2011. Minimax design of stable IIR filters with sparse coefficients. IEEE International Symposium on Circuits and Systems, 15–18 May 2011: 398–401.

Matsuoka, R., Baba, T. & Okuda, M. 2014. Constrained design of FIR filters with sparse coefficients. Annual Summit and Conference Asia-Pacific Signal and Information Processing Association, 2014 Annual Summit and Conference, 9–12 December 2014.

Mattera, D., Palmierl, F., & Haykin, S. 2002. Efficient sparse FIR filter design. IEEE International Conference on Acoustics, Speech, and Signal Processing (ICASSP) 2: 1537–1540.

McClellan, J.H., Parks, T.W. & Rabiner, L.R. 1973. A computer program for designing optimum FIR linear phase digital filters. IEEE Transactions on Audio Electroacoustics 21: 506–526.

Mitra, S.K. & Kaiser, J.F. 1993. Handbook for Digital Signal Processing. New York: Wiley.

Tseng, C. C. & Lee, S. L. 2011. Design of sparse digital differentiator using orthogonal matching pursuit method. IEEE 54th International Midwest Symposium on Circuits and Systems (MWSCAS), Seoul: 1–4.

Tseng, C.C. & Lee, S.L. 2009. Design of FIR filter using constrained L1 minimization method. IEEE Region 10 Conference TENCON, 23–26 Jan 2009: 1–5.

Tseng, C.C. & Lee, S.L. 2012. Design of sparse constrained FIR filter using orthogonal matching pursuit method. International Symposium on Intelligent Signal Processing and Communications Systems (ISPACS): 308–313.

Tseng, C.C. & Lee, S.L. 2013. Design of sparse low-pass differentiator using iterative hard thresholding method. European Conference on Circuit Theory and Design 2013, 8–12 Sept. 2013: 1–4.

Wei, D. & Oppenheim, A.V. 2010. Sparsity maximization under a quadratic constraint with applications in filter design. IEEE International Conference on Acoustics Speech and Signal Processing, 14–19 March 2010: 3686–3689.

Wei, D. & Oppenheim, A.V. 2013. A Branch-and-Bound Algorithm for Quadratically-Constrained Sparse Filter Design. IEEE Transactions on Signal Processing 61(4): 1006–1018.

Wei, D. 2009. Non-convex optimization for the design of sparse fir filters. IEEE/SP 15th Workshop on Statistical Signal Processing, 31 Aug-3 Sept 2009: 117–120.

Wei, D., Sestok, C.K. & Oppenheim, A.V. 2013. Sparse Filter Design under a Quadratic Constraint: Low-Complexity Algorithms. IEEE Transactions on Signal Processing 61(4): 857–870.

Xie, Z. & Hu, J. 2013. Reweighted L1-minimization for sparse solutions to underdetermined linear systems. 6th International Congress on Image and Signal Processing (CISP) Hangzhou: 1660–1664.

Zibulevsky, M. & Elad, M. 2010. L1-L2 optimization in signal and image processing. IEEE Signal Processing Magazine 27: 76–88.

Communication and Computing Systems – Prasad et al. (Eds)
© 2017 Taylor & Francis Group, London, ISBN 978-1-138-02952-1

Biomedical image mosaicing: A review

Kshitija Pol, Diksha Chaudhary & Parul Bansal
Dronacharya College of Engineering, Gurgaon, India

ABSTRACT: One of key research area in the field of Image Processing & Computer Vision is Image mosaicing. When image mosaicing used for medical diagnosis purpose it is called as Biomedical Image Mosaicing. This paper takes an overview of how image mosaicing is used in medical application such as X-Ray stitching, Visualization of Tissue structure, Robotically-assisted Biomedical image mosaicing system, Blood Vessel image, whole Body MRI.

1 INTRODUCTION

Image Stitching is the technique to stitch various images having overlapped fields of view to construct a panoramic image. Process of Medical Image stitching is similar to creation of panorama image of a scene by using several images of that scene. Creation of Panorama Image is done in two stages (Banarase 2013). First stage is to combine two or more images, identified and registered. In second stage the corresponding pixels of the images are blended. The central step is image registration. It is done to align the images in a precise way and this can be achieved through a combination of different techniques (Brown 1992), (Vercauterene T. Pennec 2007), (Vercauteren,. Pennec, Perchant & Ayache (2008). Biomedical image stitching is mainly used in clinical diagnosis, such as diagnosis of cardiac, retinal, pelvic, renal, abdomen, liver, tissue and other disorders. As advances in technology in the field of computer science have led to reliable and efficient image processing methods useful in medical diagnosis, treatment planning, and medical research. In clinical diagnosis, Subheading integration of useful data obtained from separate images is often desired.

2 BIOMEDICAL MOSAICING

2.1 *Visualization of tissue structure*

It is done in three steps namely Pair wise Mosaicing, Global Optimization & Local Optimization.

In Pair wise mosaicing, we assume that two image is having 80% overlap section. Correct the scan distortion, then track each frame. Select the image only if minimum motion exceeds or max number frame elapsed. Now use fine tune registration using template matching & pyramidal blending. In Global Optimization, first apply pairwise mosaicing. Template matching is used to check nearby images for cumulative error. After enough error accumulations perform optimization & updates mosaic. Local Optimization starts with global optimization. In local optimization, next step is partition images into matches & register overlapping patches. Here intermediate step is run local optimization. In a last step, wrap the images using radial basis function & update mosaic. Fig. 1 shows image wrapping with radial basis function. Local optimization algorithm is applied toward in vivo imaging with a miniature confocal microendoscope (Lowke 2011).

This technique is implemented to improve physician confidence during in vivo pathology, it will be essential to visualize tissue at micro-meter scale resolution across centimeter-sized fields of view for larger tissue coverage.

2.2 *Stitching of X-Ray images*

X-Ray systems cannot cover all the parts of body. Here we take several X-Ray images of body part and

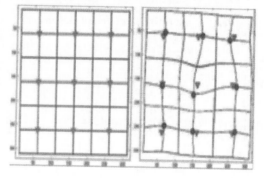

Figure 1. Image wrapping with radial basis function.

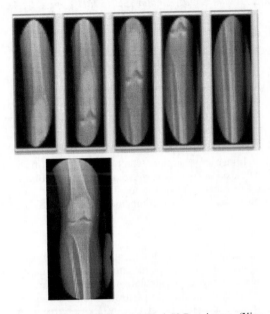

Figure 2. Stitching of Medical X-Ray image (Xing 2007).

generate a single high resolution image. It is shown in Figure 2. It has two main stage namely Image matching & Image Blending. Two commonly used image matching algorithm is SIFT & SURF. SIFT is slow process, suffers from illumination changes but scale invariant, rotation & affine transformation. While SURF requires high computation time but invariant to illumination change. Combination of SURF & SIFT gives better panoramic image with good features. Stitching of Medical X-Ray image is as shown in Figure 2.

Stitching of Medical X-ray image is done as follows.

Find out features in both image using SIFT & SURF, correlates those features. Now correct the features using RANSAC from each image. RANSAC removes unwanted features point. Correlate correct feature point. Apply RANSAC on these feature points to obtain the stitched image of input images. In last step apply the image blending process to remove the seam between the stitched images. Image blending techniques removes the visible seam between the stitched images.

2.3 Robotically-assisted Biomedical image mosaicing system (Vercauteren 2008) (Xing 2007)

It involves fields of projective geometry, camera calibration, sensor-based robot kinematics, hand eye calibration, and image mosaicing algorithms. Here camera takes picture of planner space in 3D

space. The camera is allowed any arbitrary movement with respect to the scene as long as it stays in focus and there are no major artifacts that would cause motion parallax. Now obtain image point using word point through perspective projection and rigid transformation. Then find out homographic matrix. In camera calibration method, first task is to find out the homography between image pairs which gives intrinsic camera parameter. Now crop the image to remove blurred edges caused by the large focal length at near-field. In Robot kinematic step we use the Phantom forward kinematics to measure the rotation and translation of the point where the 3 gimbal axes to obtain robot's reference frame Hand eye calibration is used to find out rigid transformation between the end-effector and the camera's optical center, which is the same for all views. It is denoted by 4×4 matrix composed of rotation & translation. Hand-eye calibration is solved during camera calibration to find out hand eye equation. After solving this equation we get hand eye transformation (H).

Image mosaicing algorithm used is Levenberg- Marquardt (LM) (Szeliski 1996). The LM algorithm requires an initial estimate of the homography in order to find a locally optimal solution, making it an ideal candidate for integrating our position sensing. The initial estimate is often obtained using optical flow, feature detection, or correlation based techniques in the spatial or frequency domain. With robotic position sensing, we get an accurate estimate of the homography that requires relatively few iterations of the LM algorithm for optimization. Position sensing eliminates cumulative error. If each new image is aligned to the previous image, alignment errors will propagate through the image chain (Fleischer 1997), (Lowke 2011, Szeliski 1996, Vercauterene T. Pennec 2007) becoming most prominent when the path closes a loop or traces back upon itself.

2.4 Blood vessel image

It has two main stages namely image stitching & image blending. In image, stitching stage, the first step is generation of relative position of acquired image and creation of empty array where his image is stored. Then the point of best correlation is searched, by sliding adjacent image edges in both the directions until the best match of edge features is found. It is required to choose an optimum search space where search is performed for the best correlation. If images are of equal dimension then to stitch it is adequate to extract a thin, narrow strip from the edge of one image and to correlate this with a larger rectangle from the other image. In Image Blending stage feathering algorithm is used. It is used to improve visual qual-

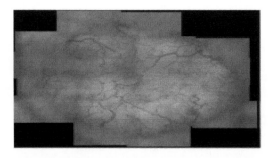

Figure 3.   Image after cross relation & blending.

ity of the composite image and making the edges invisible. Position of image is determined by cross-correlation between the new image and the composite image. The blending algorithm is then applied and the process is repeated for all the other images. Now, blending algorithm is repetitively applied on all the other images. For overlapped area the image blending algorithm calculates the contribution of the new image and the composite image at every pixel. It creates a look up table for each new image. This look up table has the size and shape of the overlap. Look up table is normalized to define proportion of intensities of two overlapped region. This overlapped region is used to generate a new composite image. This approach is used on three types of images namely RGB images, and 8bit gray scale images. RGB images is first decomposed into three-band planes, the blending algorithm is applied to each band and lastly image planes is recombined again Fig. 3 shows result.

2.5   Whole body MRI

Speeding up the acquisition for WB-MRI with large FOV images leads to significant distortions towards the boundaries. The use of simultaneous deformable registration in a mosaicing scenario is useful. Key for the simultaneous registration is the creation of a linear weighted average, each of the two images is registered to. The unaltered clinical workflow integrated into further MR scanner generations.

3   CHALLENGES OF BIOMEDICAL IMAGE MOSAICING & SOLUTION

The main challenges associated with image mosaicing in medical applications: Cumulative image registration errors To deal with cumulative errors a global alignment algorithm that draws upon techniques commonly used in probabilistic robotic Scene deformation. To accommodate scene deformation, a local alignment algorithm is used to corporates deformable surface models into the mosaicing framework.

4   CONCLUSION

Image mosaicing is useful for a variety of tasks in Computer vision, Image processing and computer graphics. Due to the wide range of applications, image mosaicing is one of the important research area in the field of Biomedical. Then it is Biomedical Image mosaicing Here we have presented some of the very fundamental and basic techniques used in Biomedical image mosaicing. This paper presents a various process for Biomedical image mosaicking. We also discuss some of challenges of Biomedical Image Mosaicing. One of Biomedical Image Mosaicing application is Clinical diagnosis. A lot of research needs to be done in this area.

REFERENCES

Banarase, S.J. Banaswakar & Banswaka, M.R. 2013. "Multiresolution Panorama Image Applications." *International Journal of Advanced Research in Computer Science and Software Engineering*, 2013.

Brown, L., "A Survey of Image Registration Techniques." *ACM Comput Surv*, 1992.

Fleischer, S.D., Rock, S.M. & Burton, S.M. "Global Position Determination and Vehicle Path Estimation from a Vision Sensor for Real Time Videao Mosaicing and Navigation." *OCEANS' 97*, 1997.

Lowke, K.E. Camarillo, D.B. Piyawattanametha, W. Mandella, M.J. Catag, C.H. Thrun S. & Salsibury, J.K. "Vivo Micro-Image Mosaicing." *IEEE Transactions on Biomedical Engineering*, 2011.

Szeliski, R. "Video Mosaics for Virtual Environments." *IEEE Computer Applications*. 1996.

Vercauteren, T. Pennec, X. Perchant, A and Ayache. "Symmetric ldgmain diffemorphic registration: A demons based approach." *MICCAI*. 2008.

Vercauterene T. Pennec, X. Perchant, A. & Ayache, N. "Non parametric diffeormorphic image registration with the demons algorithms." *MICCAI*. 2007.

Xing, J. and Miao, Z. "An Improved Algorithm on Image Stitching based on SIFT features." *IEEE 2007*. 2007.

Communication and Computing Systems – Prasad et al. (Eds)
© 2017 Taylor & Francis Group, London, ISBN 978-1-138-02952-1

A novel PSO based algorithm to find initial seeds for the k-means clustering algorithm

Lavika Goel, Nilay Jain & Shivin Srivastava
Department of Computer Science and Information Systems, BITS Pilani, Pilani Campus, India

ABSTRACT: Clustering is a very fundamental problem in machine learning. Many algorithms have been proposed but none is more popular than the k-means algorithm developed by Lloyd around 50 years ago. K-means is a fast and simple algorithm but often gives a sub-optimal clustering. This is due to the initialization that is employed in the simple K-means algorithm. K-means++ provides a new way to seed the k-means algorithm that is O(log k)-competitive with optimal clustering. In this paper we use Particle Swarm Optimization to find the initial seeds of k-means algorithm. Our approach basically tries to select the k centroid points as far away as possible from each other so that good clusters are generated.

1   INTRODUCTION

Swarm Intelligence (SI) is the collective behavior of decentralized, self-organized systems, natural or artificial. This concept is employed in work in artificial intelligence. SI systems consist typically of a population of simple agents or bodies interacting locally with one another and with their environment. The inspiration often comes from nature, especially biological systems. (Kennedy et al. 1995, 2002, F van den Bergh, 2002). Swarm Algorithms can be applied to a variety of problems like optimization problems, scheduling problems, clustering problems etc. In this paper we propose to apply swarm algorithms to data clustering and cluster analysis problems.

Cluster analysis falls into unsupervised learning ategory of machine learning. Here we do not have any prior information about the category labels of the data points. The task is to cluster the data points in such a manner that, data points falling into the same cluster are related to each other. One metric that we use to specify good clustering is that the intra cluster distance between the data points should be small and the inter cluster distance must be large. Clustering has a lot of applications in the scientific fields ranging from Computational Biology and Medical Imaging to Market Research, etc.

One of the most popular clustering algorithms is K-means. This algorithm is still widely used despite being proposed more than 50 years ago. There have been many more clustering algorithms invented since then, but the simplicity and scalability of k-means algorithm makes it one of the most widely used clustering algorithm even today.

K-means++ algorithm is an addition to the k-means algorithm to give us an even better clustering on the data. More precisely it gives a (logk)

competitive bound with the optimal clustering. The strategy that k-means++ algorithm employs is in the initialization of the clusters. k-means++ tries to select k initial cluster points in such a manner that each cluster point i is chosen at random from the weighted probability distribution of the distances of point i from the other i-1 points. This initialization strategy makes k-means++ give a (logk) competitive bound with the optimal clustering (Arthur, David & Vassilvitskii, Sergei, 2007).

2   RELATED WORK

The clustering problem and k-means algorithm have a very rich history. Because k-means algorithm is very simple and has a good observed speed, it is one of the most widely used clustering algorithms ever since being introduced by Lloyd (Lloyd, 1955). The simplicity and scalability of k-means algorithm implied that it was very widely adopted peer reviewed in the computer science community.

K-means begins with k centers that are chosen randomly from the data points. Each point is assigned to a centroid, and then we compute the centroids again, to the mean of all the points assigned to that centroid. This process is done repeatedly until the algorithm converges. (Lloyd, 1982 & Jain, 2010). The problems that came to be associated with k-means was the random initialization step. The optimal clustering is given by k-means algorithm only when each initial point was part of one cluster. This means the number of iterations for k-means initialization should be increased when the number of clusters were more (Arthur et al. 2006 & Dasgupta, 2003). But the probability that each initial cluster centroid was part of one cluster in the final clustering was still small. Hence increasing the number of iterations still not helped and we were left with suboptimal clustering.

K-means++ algorithm solved the problem of suboptimal clustering by focusing on the initialization technique of k-means. Instead of randomly initializing the k-cluster points, David Arthur et al. proposed in their remarkable paper that initializing k particles based on the probability of distance of the new cluster from the existing cluster points gives a more "optimal" clustering. The larger the distance of particle from currently chosen points, the greater is its probability of getting selected as a cluster point (Arthur, David & Vassilvitskii, Sergei, 2007). This paper analyses mathematically the benefits of using such an initialization, and they arrive at the conclusion that using such an initialization gives us an O(log k) competitive algorithm to optimal clustering.

The k-means++ initialization technique, is an NP hard problem known as the k-center problem in the computer science literature. The k-center problem is defined as following: Given n cities with specified distances, we want to build k warehouses in different cities such that the sum of maximum distance of a city to a warehouse is minimum. (Dasgupta, 2013) analyses the approximate methods to solve the k-center problem using Farthest First traversal and Covering Numbers. The initialization technique used by David Arthur et al. is a probabilistic variant of the greedy method Farthest first traversal. K-means++ has a further advantage in that it can be made extremely scalable by parallelizing it as referenced in the paper by Sergei Vassilvitskii et al. (Bahmani 2011, Zhang 1996).

We plan to apply Particle Swarm Optimization technique, to find the solution of the k-center problem and use it to initialize the k-means algorithm (Cui, 2005). Particle Swarm Optimization (PSO) is a novel algorithm developed by (Kennedy, 1995). PSO was introduced to optimize continuous nonlinear functions. Though now there are many variants of PSO that can work in different settings like dynamic environment of PSO, Multi-objective optimization and Discrete PSO. PSO was discovered by simulating behaviours in a social setting. In PSO we allow particles to wander in a search space. PSO tries to optimize the given fitness function.

Given this function, particles remember their personal best and global best position values, on the basis of which a converging point for all particles is obtained which is the solution to the equation which was described by the fitness function.

PSO has been applied in data clustering by Van Der Merwe et al. (Merwe, 2003). They use PSO to initialize the k-centroids and then extend this method to apply k-means.

3 A BRIEF REVIEW OF K-MEANS++ AND HYBRID K-MEANS/PSO TECHNIQUES

This section provides a brief review of the k-means++ algorithm, Hybrid PSO with k-means and other techniques that have been used in literature up until now.

3.1 k-means++ algorithm

The k-means++ algorithm starts by choosing the first centroid arbitrarily. (Arthur, David & Vassilvitskii, Sergei, 2007). Let $D(x)$ denote the shortest distance of a data point to the Cluster we have already chosen, then k-means++ initialization is defined as follows:

1. Choose an initial center c uniformly at random from X.
2. Choose the next center c_1, selecting $C_i = \frac{D(x)^2}{\sum_{x \in X} D(x)^2}$

   $x \in X$ with probability
3. Repeat Step 1b until we have chosen a total of k centers.

3.2 Hybrid k-means PSO

Swarm Intelligence algorithms have been applied to wide range of algorithms in the literature and hybrid techniques have been developed. (Kennedy 2002, Goel et al. 2012) In the Hybrid k-means PSO technique, the strategy adopted while performing PSO is to concatenate the dimensions of the k centroid points to get into a k*d dimensional space. Then the optimal centroid point is found using PSO with the sum of squared distance error as the fitness function metric which represents the optimal solution. Here is the algorithm (Merwe, 2003):

1. Initialize each particle to contain N_e randomly selected cluster centroids.
2. For t = 1 to t_{max} do

   (a) For each particle x do
   (b) For each data vector z_p do

   (1) Calculate the Euclidean Distance $d(z_p, m_{ij})$ to all the cluster centroids C_{ij}
   (2) assign z_p to cluster C_{ij} such that $d(z_p, m_{ij}) = \min_{c=1 \text{ to } Ne}\{d(z_p, m_{ic})\}$
3. Calculate fitness using eqn 3.

   (c) Update the global best and local best positions
   (d) Update the cluster centroids using equations (1) and (2).

$$v_{i,k}(t+1) = w^*v_{i,k}(t) + c_1 r_{1,k}(t)(y_{i,k}(t) - x_{i,k}(t)) + c_2 r_{2,k}(t)(\hat{y}_k(t) - x_{i,k}(t)) \qquad (1)$$
$$x_i(t+1) = x_i(t) + v_i(t+1) \qquad (2)$$

$$J_e = \frac{\sum_{j=1}^{N_e}\left[\sum_{\forall Z_p \in C_{ij}} d\left(z_p, m_j\right) / \left|C_{ij}\right|\right] -}{N_c} \qquad (3)$$

But the idea of concatenating so many dimensions gives rise to an inefficient algorithm and we plan to improve on this strategy with our method.

4 PROPOSED METHOD

In this section we describe our main algorithm.

4.1 Challenges

Since the standard PSO works on the continuous space and our problem involves choosing k points from the set of n points, we needed to find a suitable mapping from the continuous space to the discrete space. We decided to map the n points which were originally in the continuous space to a binary ceil (lg n) dimensional hypercube whose vertices encode the data points in the original euclidean space. This restricts the movement of search agents to the n points only.

Next was the problem of adapting the PSO algorithm for this modified search space. The Optimum point to which the PSO search agents must converge to represents the configuration of the k particles. So ideally this required us to move to another higher dimensional search space where each point represents a configuration of the k points, as seen in [8]. Since this would be a costly affair both in terms of time and space, we decided to use the hypercube as search space only.

In our modified approach the k particles play a dual role. They are search agents of the PSO algorithm which not only explore the search space but whose final configuration in the hypercube represents the position the k selected points. This deviation from the original PSO required us to modify the optimizing function so that we could make sense of the particles positions. Two different optimising functions were used i.e. one for calculating global best and one for calculating local best of the particles. Our selection of the functions is crucial as the algorithm finally needs the local and global best positions to converge.

Also, the meaning of velocity and distance between particles needs to redefined for particles in a hypercube because the particle is constrained to move only at the corners of the hypercube and nowhere else. We define the distance or separation between the particles as their edit distance, as described in the next subsection.

4.2 Approach

We aim to use PSO to disperse the 'k' centroids as far away from each as possible. The standard PSO algorithm (Kennedy, 1995) leaves a fixed number of agents in the search space which follow a particular heuristic rule to explore and exploit the search space. Since we want to select k points as initial centroids our metric of choosing a good solution is the sum of inter particle distances.

To find the optimum configuration of the particles, we modify our PSO algorithm so that instead of the particles converging at a single point representing the optimum configuration in a higher dimensional space (as seen in (Merwe, 2003)), the configuration of the particles in the search space itself becomes the optimum configuration. The PSO is performed in the discretized space in the hypercube, where each data point is at the corner of the hypercube. This constraints the PSO particles to move only inside the hypercube. The changed "interpretations" of position and velocity inside the hypercube were described as follows: Let key of any data point be denoted by k. Position of particle inside the hypercube is defined as the binary representation of key k. Velocity of particle inside the hypercube is an integer number n. To move the particle at position p with velocity v, we flip randomly v bits of the particle from the binary position vector p.

The global fitness function maximizes the distance of PSO agents from each other. The local fitness function maximizes the fitness of one particular particle.

4.3 Pseudocode

Below is the pseudocode of our algorithm followed by a comprehensive flowchart.

Algorithm 1. K-means PSO.

```
1:  procedure KMEANS PS(X, k)
2:    iterations ← 1
3:    n ← number of data points
4:    Map the data points randomly to a hypercube
5:    of ceil(lg(n)) dimentions
6:    Define the global and local fitness functions as:-
7:      • J(X) = Σ_{i=1}^{k} Σ_{j=i+1}^{k} ||x_i − x_j||²
        • H(x_p) = Σ_{i=1}^{k} ||x_i − x_p||²
8:    while iterations < max_iterations do
9:      Update the position of particle using the PSO equations
10:       for all particles i do
11:         v_i = v_i + c1 ∗ rand(−1, 1) ∗ (pbest_i − position_i)
12:         +c2 ∗ rand(−1, 1) ∗ (gbest_i − position_i)
13:         X_i = X_i + v_i
14:         if H(position_i) < J(pbest_i) then
15:           pbest_i = position_i
16:         else if J(X) < H(gbest) then
17:           gbest_i = position_i
18:         end if
19:       end for
20:       iterations ← iterations + 1
21:    end while
22:    return gbest as the seeds to initialize kmeans algorithm
23:  end procedure
```

161

5 RSEULTS

We performed k-means clustering using random initialization (number of iterations = 300), k-means++ initialization and PSO initialization. Following SSE (sum of squared errors) scores were obtained on the standard data sets (Blake, 1998) as shown in Tables 1–3. The first and second datasets are 12 dimensional. The third is the iris dataset is 4 dimensional. We perform the above algorithms on 3 values of k. In each case the first value is the number of natural clusters that are there in a dataset. Then we increase the number of clusters in the other 2 experiments and compare the SSE errors. We can see that our algorithm per forms comparable to state-of-the-art on real world datasets. A lower SSE implies better clustering (Jain, 2010). The datasets are described as follows below:

Iris plants database: This is a well-understood database with 4 dimensions, 3 natural clusters and 150 data vectors.

White-Wine: This is a classification problem with well behaved class structures. There are 12 dimensions, 6 natural clusters and around 5000 data vectors.

Wine dataset: This dataset has 2 natural clusters redwine and whitewine. It has 13 dimensions and around 7000 data vectors.

We show the clustering visualization [Figure 3] using our algorithm on the iris dataset along with the ground truth [Figure 2]. The algorithm gives optimal clustering.

In Figure 4 and Figure 5 we compare the performance of k-means-random and k-means++ with k-means-pso (the algorithm that we propose in this work). Relative SSE is plotted on the y-axis and number of clusters is varied on x-axis. Relative SSE of algorithm a w.r.t algorithm b is defined as: Relative $SSE = SSE_a - SSE_b$.

Positive Relative SSE of k-means-random and k-means++ w.r.t k-means-pso implies that k-means-pso gives better clustering (lower SSE) than k-means++ and k-means-random.

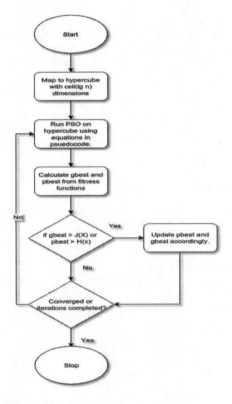

Figure 1.   Flowchart.

Table 1.   Comparative results on whitewine-dataset.

k	k-means-random	k-means++	k-means-pso
6	1544344.6209	1544327.8491	1544327.8491
9	1146811.0615	1141490.8237	1139974.6475
12	941364.00364	899171.39676	897445.64313

Table 2.   Comparative results on whitewine-redwine-dataset.

k	k-means-random	k-means++	k-means-pso
6	8589514.6373	8589514.6373	8589514.6373
9	2041453.5289	2041432.8925	2041680.3178
12	1365612.9213	1370097.5526	1364864.6436

Table 3.   Comparative results on iris-dataset.

k	k-means-random	k-means++	k-means-pso
6	78.918808773	78.918808773	78.918808773
9	38.882367282	38.856319175	38.856232051
12	28.294123636	28.150368578	28.383452381

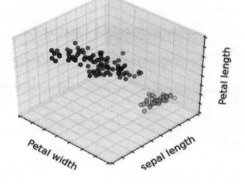

Figure 2.   Ground truth for Iris dataset.

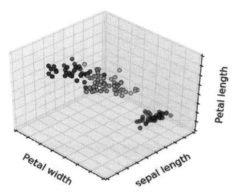

Figure 3. Clustering result using k-means-pso.

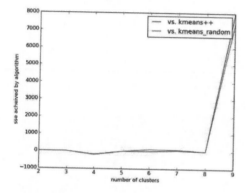

Figure 4. Relative performance of K means PSO on whitewine dataset.

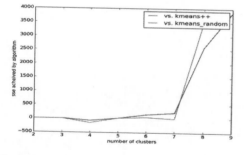

Figure 5. Relative performance of k-means PSO on wine dataset.

6 CONCLUSIONS AND FUTURE WORK

In this work we have introduced a novel algorithm to solve the k center problem and applied it for selecting the k centroids for initializing the k means algorithm. We have presented a modified PSO algorithm to select k initial points to feed into the k-means means algorithm as the seed points. In future we would like to make this approach more scalable and robust by providing parallel implementations of our approach. Also we would like to analyze different methods that we can use in our hypercube space by which we can have greater coverage of particles in lesser number of iterations.

REFERENCES

Arthur, David & Vassilvitskii, Sergei, 2007, k-means++ test code. http://www.stanford.edu/darthur/kMeansppTest.zip.

Arthur, David & Vassilvitskii, Sergei, 2006, How slow is the k-means method? In SCG 06: Proceedings of the twenty-second annual symposium on computational geometry. ACM Press.

Arthur, David & Vassilvitskii, Sergei, 2007 k-means++: The Advantages of Careful Seeding.

Bahmani, Moseley, Vattani, Kumar and Vassilvit skii, 2011, Scalable k-means++, Stanford University and Yahoo Research.

Blake, Merz C.J., 1998. UCI repository of ma chine learning databases.

Cui X., Potok T. E., 2005. Document Clustering using Particle Swarm Optimization, IEEE Swarm Intelligence Symposium 2005, Pasadena, Califor nia.

Dasgupta, 2013, k center problem, CS291 Geometric Algorithms, lecture 1. http://cseweb.ucsd.edu/dasgupta/291geom/kcenter.pdf

Dasgupta, Sanjoy, 2003, How fast is k-means? In Bernhard Scholkopf and Manfred K. Warmuth, editors, COLT, volume 2777 of Lecture Notes in Computer Science, page 735. Springer

Goel, Lavika, Gupta, Daya, and Panchal V.K., 2012 "Hybrid bioinspired techniques for land cover feature extraction: A remote sensing perspective", Applied Soft Computing Volume 12, Issue 2. Pages 832–849.

Jain, Anil K., 2010 Data clustering: 50 years beyond K-means.

Kennedy J, Eberhart RC, 1995, Particle Swarm Optimization, Proceedings of the IEEE International Joint Conference on Neural Networks, Vol. 4, pp 1942–1948.

Kennedy J, RC Eherhart, Y Shi, 2002, Swarm Intelligence, Morgan Kaulrnann.

Lloyd, Stuart P, 1982, Least squares quantization in pcm, IEEE Transactions on Information Theory, 28(2):129 136.

Merwe, DW van der & Engelbrecht, 2003, A.W, Data-Clustering using particle swarm Optimization, Evolutionary Computation, CEC.

van den Bergh, F., 2002, An Analysis of Particle Swarm Optimizers, PhD Thesis, Department of Computer Science, University of Pretoria, Pretoria, South Africa.

Zhang T., Ramakrishnan R., and Birch, M. Livny., 1996: An efficient data clustering method for very large databases, SIGMOD Record, 25:103114.

Communication and Computing Systems – Prasad et al. (Eds)
© *2017 Taylor & Francis Group, London, ISBN 978-1-138-02952-1*

Color image segmentation using MST based approach

Arpita Kumari & Suneeta Agarwal

Motilal Nehru National Institute of Technology, Allahabad, India

ABSTRACT: In this paper we have proposed a novel graph based approach for segmentation of color images. Our algorithm adapts a minimum spanning tree based clustering approach which results in dividing the image into various regions on the basis of color and spatial distance. Key feature of the approach used is that the shape and size of segmentation depends on the surrounding color variation in the image. It is very fast and can be applied to megapixel sized images in a matter of seconds. The performance of proposed algorithm is tested on the Berkeley dataset and compared with state-of-the-art segmentation algorithm. The proposed algorithm yields better segmentation with respect to compactness of segments. The new approach offers a significant speedup over Simple Linear Iterative Clustering (SLIC) based segmentation.

Keywords: Cluster Difference, Clustering, Image Segmentation, Minimum spanning tree

1 INTRODUCTION

Image segmentation is division of an image into meaningful disjoint regions. It is a challenging problem in computer vision and it is normally used for finding objects and their boundaries in images. Each pixel of the image is assigned a label such that pixels with same label share certain properties. It is used as principal step in object recognition, image analysis, image mining and many other image processing tasks. A wide range of segmentation approaches exists; each has its own advantages and disadvantages, which are the deciding factor for its suitability in a particular application.

Everyday vision spans a wide range of abilities including motion detection, shape identification, learning the cues to the size of distant objects, measuring distance and speed etc. Vision is more than just scanning what meets the eye; it's the ability to understand, what we see, almost instantly; and this happens in brain. A human brain reads processes and analyzes enormous amount of information as fast as it can and simultaneously segments or divides the visual image into related objects. The same process becomes extremely difficult when it has to be performed by a machine using an algorithm. An effective image segmentation algorithm efficiently reduces the amount of visual information that is to be processed and hence reduces object identification time in a visual scan.

The field of image segmentation has a promising role in multifarious applications ranging from medical image processing (where doctors use modalities such as CT Scan and MRI to measure the size and condition of organs and tissues using scans obtained from their patients) to multimedia applications (where robust segmentation results benefits immensely in image and video encoding, leading to more efficient image transfer and storage) and even encompasses fields like machine learning, physics, engineering and pattern recognition, where image segmentation, as a tool, provides the proving ground for realization of ideas. It is well accepted that there exists no general algorithm which caters to all the needs effectively. Since image segmentation is domain specific hence algorithms which are highly successful for a particular scenario might not work at all for the other. Thus, these algorithms may have to be highly altered or adapted to serve a particular scenario in order to achieve better performance with respect to specific application.

Goal of the approach proposed is to develop a graph based segmentation approach that should capture visually important components or regions, while reflecting global aspects of an image. Graph-based segmentation is known to consider the global image properties as well as local spatial relationships. Now the problems that arise are to provide description of what is visually important, to define what a developed segmentation approach does and accurate definition of the features of a resulting segmentation, so as to better understand the approach as well as compare it with different approaches. The segmentation approach should run in nearly linear time and with low constant factors. It should run at speed closer to edge detection or other visual processing techniques so that the new approach is of practical importance. Also, the visual quality of segmentation needs to be preserved.

This paper presents a graph based image segmentation using efficient Minimum Spanning Tree (MST) based clustering approach. The proposed approach has adapted (Jana et al. 2009) approach for the purpose of segmentation. The problem of segmenting the image is modeled in terms of clustering the nodes of the graph into various subgraphs such that each represents a significant region in the image. The proposed approach, segments an image by collecting the similar pixels together. The results have been tested on a number of natural images from Berkeley Image Database as well as some synthetic images from online sources. This method is very fast and more memory efficient. Apart from being faster, the approach is easy to use and it can be easily extended for higher dimensions as well.

The organization of this paper is as follows. Section II includes the previous work. Section III focuses on the proposed approach where the working of the MST based algorithm for segmenting an image is described along with its pseudo code. Section IV emphasize on the experimental results along with comparison of the obtained results followed by the discussion about the experimental results. Section V addresses the conclusions.

2 PREVIOUS WORK

A lot of work has been done on segmentation and clustering, having its application in tasks that require analysis of images such as object detection, object recognition, Optical Character Recognition (OCR), medical imaging and so on. In this section the work that is most related to the proposed approach is described: (i) Efficient Graph based Image Segmentation proposed by Felzenszwalb et al. (ii) improvements to it (Zang & Alhajj 2006, Riemersma 2012) and (iii) an approach which focuses on conversion of pixels into superpixels (Achanta et al. 2012).

The image segmentation method in (Felzenszwalb & Huttenlocher 2004) uses a simple adaptation of Kruskal's algorithm. A predicate is defined for evaluating whether there is an evidence of a boundary between two components in segmentation. The defined predicate measures the dissimilarity between pixels that are along the boundary relative to dissimilarity among pixels that are within each of the two regions. An important property of the method is its ability to preserve details in low variability region while ignoring details in high variability regions in the image. However, the method (Felzenszwalb & Huttenlocher 2004) has various drawbacks: (i) the internal difference is calculated based on the extreme values of edge weights in the components which does not give accurate description of the components; (ii) It is very difficult to choose an appropriate threshold value k for expected size of segment. This value specifies minimum number of pixels in a region.

Zang et al. proposed some significant refinements for method in (Felzenszwalb & Huttenlocher 2004). The method was proposed for sensor devices which are used for monitoring purposes. The internal difference, which previously was used to define the property of the component, is redefined to give more accurate description of the components. The threshold function which in (Felzenszwalb & Huttenlocher 2004) was used to determine the size of the component is also redefined. The threshold value is adapted during the segmentation thereby making the process independent of edge weight scale. This method overcomes all the drawbacks of method in (Felzenszwalb & Huttenlocher 2004) and also increases its efficiency and effectiveness.

Edge Weight is mainstay in the construction of graph, which governs the segmentation result. The methods in (Felzenszwalb & Huttenlocher 2004) & (Zang & Alhajj 2006) use Euclidean metric to calculate weight of the edge connecting any two pixels. The article on color metric (Riemersma 2012) proposed the use of weighted Euclidean Distance in RGB color space since, if color images are taken into consideration, it is not just enough to take the spatial distance between the two pixels. RGB image contains three components viz. red, green and blue. Thus it is needed to assign some weight to each component in the RGB color space. The edge weight is calculated as in (1) (Riemersma 2012):

$$|\Delta C| = \sqrt{2 \times \Delta R^2 + 4 \times \Delta G^2 + 3 \times \Delta B^2} \qquad (1)$$

A new superpixel algorithm, SLIC is proposed in (Achanta et al. 2012) which adapt k-means clustering to produce superpixels. Two distinctive characteristics of SLIC are, firstly, the number of distance calculations is greatly reduced by reducing the search space and secondly, the edge weight is a combined measure of both spatial and color proximity so as to control the size and compactness of the superpixels. It also has a user specified parameter k as the count of number of superpixels that should be formed. In order to produce approximately equal sized superpixels, a grid interval $S = \sqrt{(N/k)}$ is used. Like k-means, k initial cluster centers are sampled S pixels apart. The similar pixels are searched in the region $2S \times 2S$ around the superpixel center. Each pixel gets associated to the nearest cluster center whose search space spans the pixel. After all the pixels are associated to nearest cluster center an update step adjusts the cluster center as the mean vector $[l \ a \ b \ x \ y]^T$ of all the pixels in the cluster. In the final postprocesing step, all the disjoint pixels are assigned to the closest superpixel.

3 PROPOSED ALGORITHM

We propose a new MST based approach for segmenting an image into various regions. The proposed method uses an adaptation of an efficient MST based clustering method (Jana & Naik 2009). The clustering approach used in this algorithm has following major differences from efficient MST based clustering method:

- The number of distance calculation between pixels is decreased to a great extent. The distance of a pixel is calculated with its neighbors within a radius of r pixels.
- The weighted distance measure is calculated on the basis of color distance and spatial distance.
- The step size is removed in order to reduce the number of iterations required. Therefore, next threshold is based on the minimum weight edge from the removed edges.

Graph based approach is easy to use and has been proved to be more efficient for segmenting an image. Since it is a clustering based approach, therefore it clusters the similar pixels together. The pixels are considered similar when they are of almost the same color and are also close to each other.

3.1 Algorithm

For color images, first the image is converted in CIELAB color space. Although RGB color space is most commonly preferred, it does not cover the entire visible color spectrum and also it is device dependent color space. A color space is device dependent when the color produced depends on both, the device that is used for display and the parameters used. CIELAB color space has been used for this approach as it is perceptually uniform as well as device independent. Also CIELAB has the advantage of a gamut that encompasses all the colors the human eye can perceive.

Next, a proximity graph G (V, E) is constructed where each node $v_i \in V$ represents a pixel; each edge $(v_i, v_j) \in E$ is the set of weighted edges connecting the neighboring pixels. Each edge $(v_i, v_j) \in E$ has a weight $w(v_i, v_j)$ which is the measure of dissimilarity between the neighboring pixels v_i and v_j. The distance measure defined in (Achanta et al. 2012) has been modified in this approach, where instead of using the count of superpixels in the image, a handy variable r is used (2), which defines the maximum acceptable distance between any two pixels. Therefore the weight (2) is calculated as the

distance of a pixel with only those pixels which are at less than or equal to r distance from it.

$$w_{ij} = \begin{cases} \sqrt{\left(dc_{ij}\right)^2 + \left(\dfrac{ds_{ij}}{r}\right)^2} \, c^2 & \text{if } ds_{ij} \leq r \\ 0 & \text{otherwise} \end{cases} \quad (2)$$

Each pixel in CIELAB is represented as $[l \ a \ b \ x \ y]^T$. dc_{ij}(3) is Euclidean color distance between pixels v_i and v_j. ds_{ij} (4) is Euclidean spatial distance between pixels v_i and v_j.

$$dc_{ij} = \sqrt{\left(l_j - l_i\right)^2 + \left(a_j - a_i\right)^2 + \left(b_j - b_i\right)^2} \quad (3)$$

$$ds_{ij} = \sqrt{\left(x_j - x_i\right)^2 + \left(y_j - y_i\right)^2} \quad (4)$$

By using handy variable 'c' in distance measure (2), helps us to weigh the relative importance of color similarity and spatial proximity. When c is large, spatial distance gains importance over color difference. However, if c is small, only similar color pixels fall together.

By using this distance measure, smaller weight values are assigned to edges between pixels which are close to each other as well as have similar color value. Each pixel is connected with pixels that are at distance less than or equal to r from it. Therefore, the maximum number of edges in the graph is $|V| \times r^2$. Therefore, the complexity to construct graph is equal to O (V) because of the sparseness of graph.

After the graph is constructed, the minimum spanning tree T is obtained for the graph for identifying the inconsistent edges i.e. edges whose weight is larger than weight of nearby edges in the tree. Here Prim's algorithm is used for finding MST. However any other algorithm for constructing MST can also be used. The cost of constructing MST using traditional algorithms is O (E log V) where E is the count of edges in the graph and V is the count of vertices in the graph. Some efficient algorithms (Gabow et al. 1986) for constructing MSTs, have already been researched which promise an almost linear execution time under different assumptions.

A threshold value t is calculated as the sum of mean weight $\hat{W}$ of all the edge weights in the minimum spanning tree and standard deviation σ of all the edge weights in the minimum spanning tree. All the edges in T having weight greater than the threshold value t are removed. This results in a set S of edges and disjoint sets of subtrees $T_1, T_2, \ldots$ generating various clusters $C_1, C_2, \ldots$.

The Efficient MST based Clustering Algorithm is a nearest-centroid based approach. Let z_i be arbitrarly selected cluster center for cluster C_i. Intra cluster distance (5) is calculated as spatial distance of the pixels in a cluster from the cluster center. Inter cluster distance (6) is spatial distance between two cluster centers. The cluster difference (7) is calculated as the ratio between the maximum intra cluster distance and minimum inter cluster distance. It is used to finally determine best number of clusters/regions in the image. After the clusters are generated, cluster difference is calculated.

$$\text{Intra} = \frac{1}{N} \sum_{i=1}^{k} \sum_{x \in Cl} (x - z_i)^2 \qquad (5)$$

$$\text{Inter} = \min(z_i - z_j)^2 \qquad (6)$$

$$\text{Cluster Difference (CD)} = \frac{\text{Intra}}{\text{Inter}} \qquad (7)$$

For cluster validation, find the minimum weight edge from removed set S' of edges i.e. the edges with $w > \bar{W} + \sigma$ and use this edge weight as the threshold value for next iteration. Using this edges weight as the next threshold value, add all the edges with weight equal to threshold to the S, this produces clusters $C'_1, C'_2, \ldots$ Again the cluster difference is calculated using the inter cluster and intra cluster distances of the new clusters formed after adding the edges. The addition of new edges to S is done so as to reduce the number of iterations and thereby reduce the overall time to O (V) because the edge count is $|V| \times r^2$. A low value of cluster difference means that the clusters are too close to each other, and thus clustering is not proper. A high value of cluster difference means that the clusters so formed are more likely to be valid. Thus the threshold value which leads to maximum cluster difference is found. This procedure is performed until either no edge remains in S' or the cluster difference is more than the threshold t (Peter J. S. 2011).

The steps involved are as follows:

1. Read the color image.
2. Convert the RGB image to LAB space.
3. Create a graph G (V, E) such that each pixel v_i is connected to most similar pixels that are at most r Euclidean distance from it.
4. Find MST for the proximity graph obtained in step 3.
5. The edges that are part of MST are denoted as E'.
6. Calculate threshold using the average weight and standard deviation of E'.
7. Repeat Steps 8 to 19 until CD ≤ threshold
8. Add all edges from E' with w ≤ threshold to set S.
9. Store the removed edges in S'.
10. if S' is empty {
11. go to step 20
    }

12. The MST is broken into disjoint subtrees T_1, T_2,…. Each subtree represents a cluster.
13. Find maximum Intra cluster Distance
14. Find minimum Inter cluster Distance
15. Calculate CD
16. If CD > previous CD value{
17. Store the clustering obtained using this threshold value.
    }
18. Threshold = min-edge-weight(S')
19. Now set E' equal to edges removed from MST i.e. E' = S'
20. Exit

4 EXPERIMENTAL RESULTS

We have performed comparison of proposed method with the method in (Achanta et al. 2012). The presented approach focuses on improving the quality of segmented images in less time. The methods are compared in terms of time as well as segmentation results.

Test images from Berkeley Segmentation dataset (Berkeley EECS) and also images available online have been used for performance evaluation.

From the results in Fig. 1., it can observed that results obtained using the proposed approach is closer to ground truth than the results obtained from SLIC based method.

4.1 Performance evaluation

Results obtained from (Achanta et al. 2012) have been compared with results obtained from the proposed approach. The objective evaluation parameters described in (Monteiro & Campilho 2006) and (Shivani & Agarwal 2016) have been used here for performance evaluation of segmentation approach. Precision and recall values (Monteiro & Campilho 2006) are to identify the agreement between the pixels of region boundaries of two segmentations.

Let S_1 be the result of any segmentation algorithm and S_2 is the ground truth. Precision is directly proportional to the fraction of boundary elements from S_1 that matches with the boundary elements of ground truth S_2, and recall is proportional to the fraction of the boundary pixels from S_2 for which a acceptable match was found in S_1.

$$Presicion(P) = \frac{Matched(S_1, S_2)}{|S_1|} \qquad (8)$$

$$Recall \, / \, Sensitivity(R) = \frac{Matched(S_2, S_1)}{|S_2|} \qquad (9)$$

(a)

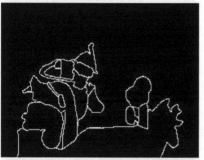

(b)

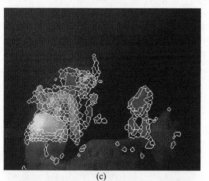

(c)

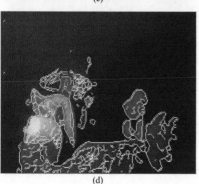

(d)

Figure 1. (a) Input Image "45096. jpg" (b) Ground truth for input (c) Segmentation result using SLIC superpixel based segmentation (d) Segmentation result using the proposed approach.

$$F - measure = \frac{2 \times R \times P}{R + P} \qquad (10)$$

Precision (8) is the probability that the result is valid while recall (9) is the probability that ground truth data is detected. A low precision value indicates the over-segmentation. Similarly, a low recall value indicates under-segmentation. The two values are combined into one (10) as F-measure.

The advantage of using precision and recall values for the evaluation of segmentation results is that the segmentation produced by different methods can be compared. Also comparison of segmentation produced from the same algorithm using different input parameters is easier.

Some of the error metrics (Shivani & Agarwal 2016) are used for objective evaluation. Let N_{TP}, N_{TN}, N_{FP} and N_{FN} are the total number of true positive, true negative, false positive and false negative with respect to corresponding pixels in the two images that are to be compared respectively. The following parameters (11), (12), (13) are used for performance evaluation and comparison of approaches.

$$Specificity = \frac{N_{TN}}{N_{TN} + N_{FP}} \qquad (11)$$

$$BCR = 0.5 \times (Specificity + Sensitivity) \qquad (12)$$

$$BER = 100 \times (1 - BCR) \qquad (13)$$

The Negative Rate Matrix (NRM) (14) is based on pixel wise mismatch between the corresponding pixels of the ground truth and the result of segmented image.

Table 1. Comparison of the methods on the evaluation parameters.

	Ideal values	SLIC based segmentation	Segmentation using proposed approach
Precision	1	0.807	0.841
Recall/Sensitivity	1	0.844	0.704
Specificity	1	0.382	0.660
F-measure of Sens/Spec (%)	100	52.45	66.96
BCR	1	0.614	0.682
BER (%)	0	38.64	31.85
NRM	0	0.386	0.318
Geometric Accuracy	1	0.567	0.680
Segmentation Time	–	18.83	11.58

$$NRM = \frac{NR_{fn} + NR_{fp}}{2} \qquad (14)$$

where,

$$NR_{fn} = \frac{N_{FN}}{N_{FN} + N_{TP}} \qquad (15)$$

$$NR_{fp} = \frac{N_{FP}}{N_{FP} + N_{TN}} \qquad (16)$$

Table 1 shows the results calculated on images of size 200×250 on Intel Core 1.87 GHz processor with 2 GB RAM using MATLAB R2014a. The average values obtained by testing on a large number of images have been used in the table.

5 CONCLUSIONS

In this paper, efficient MST based clustering method (Jana & Naik 2009) is adapted to obtain segmentation of color images. An empirical comparison of the proposed approach with SLIC based segmentation method in terms of time and evaluation parameters has been done. The better values of precision, F-measure, specificity, BCR and NRM for proposed approach in Table 1 indicate that the new approach gives comparatively better segmentation. Unlike the other segmentation algorithms, this approach does not require any prior count of number of segments. This algorithm creates best number of clusters. It needs less computational work because it uses MST

therefore requires lesser time. Thus, the approach gives better segmentation in lesser time.

REFERENCES

Achanta R., Shaji A., Smith K., Lucchi A., Fua P. & Susstrunk S. 2012. SLIC Compared to State-of-the-Art Superpixel Method. *IEEE Transactions on Pattern Analysis and Machine Intelligence* 34(11): 2274–2281 *Child Development* 65(1): 13–16.

Berkeley EECS (https://www.eecs.berkeley.edu).

Felzenszwalb P. & Huttenlocher D. 2004. Efficient Graph Based Image Segmentation. *International Journal of Computer Vision* 59(2):167–181.

Gabow H., Spencer T. & Naik A. 2009. Efficient Algorithm for Minimum Spanning Trees in Undirected and Directed Graphs. *Combinatorica* 6(2): 109–122.

Jana P. K. & Naik A. 2009. An Efficient Minimum Spanning Tree based Clustering Algorithm. *Methods and Models in Computer Science:* 1–5.

Monteiro F.C. & Campilho A.C. 2006. Performance Evaluation of Image Segmentation. *Springer* 4141:248–259.

Peter J.S. 2011. A novel dynamic minimum spanning tree based clustering method for image mining. *Journal of Mathematical Sciences and Cyrptography* 14(5): 405–419.

Riemersma T. 2012. Color Metric. Retrieved from http://www.compuphase.com/cmetric.htm.

Shivani S. & Agarwal S. 2016. Novel basis matrix creation and preprocessing algorithm for friendly progressive visual secret sharing with space efficient share. *Multimedia Tools and Application Springer* 75(8): 1–34.

Zhang M. & Alhajj R. 2006. Improving the Graph Based Image Segmentation Method. 18th IEEE International Conference on Tools with Artificial Intelligence: 617–624.

Communication and Computing Systems – Prasad et al. (Eds)
© 2017 Taylor & Francis Group, London, ISBN 978-1-138-02952-1

Morphology based segmentation of optic cup and disc on retinal fundus images

R.K. Sidhu & J. Sachdeva

Department of Electrical and Instrumentation Engineering, Thapar University, Patiala, India

D. Katoch

Department of Ophthalmology, Advanced Eye Centre, PGIMER, Chandigarh, India

ABSTRACT: Glaucoma is persistent eye disease leading to permanent loss of vision. This paper presents a methodology to extract optic cup and optic disc for fundus retinal images of healthy and glaucomatous subjects. The automatic segmentation is performed that includes preprocessing and morphological operations. The database of fundus retinal images of 50 subjects is collected from internet repository. This database is divided into healthy and diseased class each consisting of 25 subjects. In the initial stage, extraction of the green and red plane followed by morphological operations is performed on retinal images for higher accuracy. The Cup To Disk Ratio (CDR) is the key indicator of this disease and is compared with the threshold value for detecting glaucoma. CDR is evaluated for the segmented images for better visual and clinical diagnosis.

1 INTRODUCTION

Medical imaging is the visual diagnosis of various body organs for treatment. The techniques of imaging include the fields of nuclear medicine, radiology and optical imaging. The fundus imaging is the technique of capturing the picture of the inside of the eye consisting of optic disc, posterior pole, fovea, macula and retina (Yu et al. 2012). Glaucoma and diabetic retinopathy are amongst the leading causes of blindness in today's world (Welfer et al. 2010). It is estimated that there are around 60 million suspected cases of glaucoma and 93 million people suffer from Diabetic Retinopathy (DR) worldwide.

Glaucoma is a chronic neurodegenerative disease of the optic nerve. It can be potentially blinding if left untreated (Maldhure & Dixit 2015). It is a condition in which the increase in Intra-Ocular Pressure (IOP) results in optic nerve damage (Mahalakshmi & Karthikeyan 2014a, b, Whardana & Suciati 2014). Glaucoma is characterized by apoptosis of the retinal ganglion cells. One of the major causes of this is a rise in the IOP. The increase in pressure occurs when aqueous humor doesn't circulate normally in the front part of eye. Normally, this fluid flows out of eye through a mesh like channel known as trabecular meshwork. If this channel becomes blocked, fluid builds up causing glaucoma. The disease can be treated either by medication or surgery (Heijl et al. 2002).

The risk of visual loss due to glaucoma can be minimized by careful monitoring and early diagnosis of progression of disease. The ocular examination for glaucoma involves measurement of the intraocular pressure, evaluation of anterior chamber angle by Gonioscopy and structural assessment of the optic disc. One of the key components of optic disc assessment involves measurement of the vertical cup to disc ratio.

The optic disk morphology is a structural indicator for evaluating the existence and severity of retinal disease like Glaucoma. The optic disc is pinkish orange in color and marks the point of exit for the axons of the ganglion cells from the eye. The normal disc is vertically oval in shape. There is a central depression, of variable size devoid of neuroretinal tissue called the optic cup (Tielsch et al. 1998). In glaucomatous eye, the optic cup grows larger and deeper. The CDR is defined as ratio of area of optic cup to that of optic disk and is the clinical indicator of glaucoma. In healthy eye images, CDR< = 0.3 and it exceeds 0.3 in patients suffering from glaucoma (Kwon et al. 2009).

The ancillary investigations for glaucoma include visual fields, stereo-disc photographs, optic nerve head tomography (Heidelberg Retina Tomography) and the Optical Coherence Tomography (OCT) based peripapillary nerve fibre layer thickness (Chauhan et al. 2001a, b, Schuman et al. 1995). Most of these require individual and costly equipment which may not be readily available. The ophthalmologists perform the examination manually in mass screening of subjects limiting its use in early detection of glaucoma (Joshi et al. 2011). The fundus retinal images are routinely obtained for both healthy and diseased persons to enable documentation of disease as well as to allow comparisons over time.

Thus, in modern ophthalmology, segmentation of retinal image structures is of interest as a non-invasive diagnostic method (Gonzalez et al. 2014). There are methods that help in automatic detection of glaucoma. However, these are generally unavailable at local clinics due to high cost (Welfer et al. 2010). Hence, such equipments do not provide a solution for mass screening program. The aim of this study is to develop a method for automatically detecting glaucoma.

The algorithm discussed in this paper for segmenting optic disk and cup is basically carried out in three steps viz. preprocessing, morphological operations and optic disc/cup detection.

2 LITERATURE SURVEY

In the previous papers, various authors have discussed CDR computation with different methods. Few of them are discussed below.

Wong et al. (Wong et al. 2008) worked on Glaucoma diseased images in which images of optic disc and cup were segmented by applying variation level-set approach. The initial contour images were preprocessed to reduce noise in the boundary detected. The presence of retinal vasculature traversing the disc and cup boundaries which can cause inaccuracies in contour was marked. The study was conducted on 104 images and Cup to Disk Ratio (CDR) was evaluated. It was observed that 0.2 CDR units were attained near to manually marked samples.

Similar study was done by Liu et al. (Liu et al. 2008). A vibration level set method was used with the initiative of marking the cup by evaluating two methods namely color intensity and threshold level. The dataset of 73 retinal images was taken and accuracy of 97% by thresholding and 18% improvement results were attained as compared to the state of art methods.

Another related study for segmentation of optic disc and cup was proposed by Kavita et al. (Kavita et al. 2010) in which the Region of Interest (ROI) and the component analysis methods were used. The final extraction was done by active contour for plotting the boundary. The glaucomatous and healthy images were differentiated by calculation of CDR.

However, the major drawback for the above methods was that the manual CDR calculation was marked by a single ophthalmologist. However, it should have been marked by 2–3 ophthalmologists and then the average should have been taken.

Aquino et al. (Aquino et al. 2010) proposed a template based method using morphological, edge detection techniques and circular hough transform for calculation of boundary with accurate measurement of CDR in nearly 99% of cases with a minimum average time of 1.67 s and deviation of 0.14 s. The circular methodology gave better results as compared to elliptical based methods. However, this method failed to deform on certain images where CDR was quite high.

Joshi et al. (Joshi et al. 2011) proposed automatic optic disk segmentation based on parameterization technique. A multi-dimensional space of features was evaluated on each point of retina to minimize changes in the optic disk region. A very important feature of vessel bending named as R bending was also included in this research. A wide dataset consisting of 33 normal and 105 glaucomatous images was used and cup-to-disk diameter ratio of 0.09/0.08 and cup-to-disk area ratio of 0.12/0.10 was attained. However, small error rate existed as compared to the average results by three ophthalmologists.

Cheng et al. (Cheng et al. 2013) proposed a sub-pixel based methodology followed by histograms and center surround calculations to classify each super pixel as disc or non-disc for segmentation of optic cup and disc. Total of 650 images with optic disc and optic cup boundaries were manually marked by trained professionals. An overlapping error of 9.5% and 24.1% in optic disc and optic cup segmentation existed respectively. The achieved areas calculated were in the range of 0.800–0.822 which is much higher as compared to other state of art methods. However, this method used the prior information of location of cup which is always not possible.

In this paper, automated method of segmentation based on morphological operations is used. It segments optic cup and disk by extracting green plane and red plane respectively from colored retinal fundus image giving better results for CDR evaluation. The methodology for the proposed work is discussed in next section.

3 METHODOLOGY

This paper presents a methodology (as shown in Fig. 1) in which certain operations are performed on fundus images to obtain the desired results.

3.1 Image acquisition

Image data is collected from internet repository. It includes retinal fundus images of healthy subjects and Glaucoma patients. These images are acquired by trained ophthalmic technician/ophthalmologist on a fundus camera.

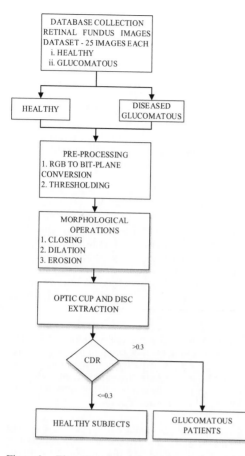

Figure 1. Flow chart for methodology.

3.2 Pre-processing

The pre processing stage is used to enhance the retinal images by contrast enhancement for improved visual quality (Choukikar et al. 2014). The Optic Disk (OD) segmentation is normally carried out in red plane as it shows better contrast between optic disk and the non-disk region. The optic cup segmentation is more difficult as the cup is interwoven with retinal blood vessels and tissues surrounding it. Thus, it is carried out in green channel plane as it shows least visibility of blood vessels.

a. RGB to bit plane conversion

The red channel and green channel are extracted from colored RGB image for optic cup and disc segmentation respectively. The contrast enhancement process is applied to respective bit channels of fundus images.

b. Thresholding

It is simple way of partitioning an image into background and foreground pixels. This technique isolates object by converting grayscale images into binary

images. The level of threshold is set and the gray scale or bit plane image is converted into binary image.

3.3 Morphological operations

The binary image from step 3.2 consists of some irregular gaps. The gaps can be filled by using morphological operations like dilation, erosion, opening and closing. In these operations, an image is searched with a structuring element, i.e., disk with a different value for every operation.

The erosion of A by structuring element B is denoted by $A \odot B$ and is given by equation

$$A \odot B = \{z \mid (B)_z \subseteq A\} \tag{1}$$

It states that erosion of A by B is a set of all points z, so that B, displaced by z, is contained in A. Since B is not sharing any element common with the background pixels, another expression for erosion is given by

$$A \odot B = \{z \mid (B_z) \cap A^c = \varnothing\} \tag{2}$$

where, A^c is the compliment of set A and $\varnothing$ is null set.

The dilation of A by B is denoted by $A \oplus B$, with A and B being sets of Z^2, and is defined as

$$A \oplus B = \{z \mid (\hat{B}) \cap A \neq \varnothing\} \tag{3}$$

where, A is the set to be dilated and B is the structuring element. B element is reflected about its origin and this reflection is shifted by z. The dilation of A by B is the set of all displacements, such that B and A overlap by at least 1 element. Mathematically,

$$A \oplus B = \{z \mid [(\hat{B})_z \cap A] \subseteq A\} \tag{4}$$

The closing of set A by structuring element B, denoted by $A \bullet B$, is given by

$$A \bullet B = (A \oplus B) \odot B \tag{5}$$

which states the dilation of A by B and further the erosion of result by B.

3.4 Cup to Disk Ratio (CDR)

Cup to Disk Ratio (CDR) is a key identifier for confirming glaucoma in a subject (Yin et al. 2012). In this paper, CDR is calculated as ratio of area (in pixels) of optic cup to that of optic disk. The CDR value obtained is thus compared with certain threshold value to detect whether the subject is healthy or diseased.

4 EXPERIMENTAL SET UP

The three set of experiments are performed to analyze the proposed methodology.

Experiment 1: The dataset of 25 healthy subjects consisting of 25 images is considered for this set up. Morphology based segmentation is performed to segment optic cup and disc on given dataset.

Experiment 2: The dataset of 25 images of 25 patients suffering from glaucoma is used in this experiment. The segmentation based on morphological operations is applied on the available database to extract the optic cup and disc.

Experiment 3: Comparison Analysis of cup to disc ratio (CDR) is performed.

4.1 Database details

The fundus retinal images of size 250*250 are collected from internet repository (1). Total database consists of 50 images of which 25 are of healthy subjects and 25 are of glaucoma patients.

4.2 Software development

An Image Processing Toolbox integrated in MAT-LAB software package R2014a running on 64 bit operating system (Windows 8.1) is used for segregating optic cup and disc for evaluation of CDR.

5 RESULTS AND DISCUSSIONS

The following section includes the results of segmentation specified under section 4.

Experiment 1: Segmentation of optic cup and disc for healthy subject is shown by images obtained at each step of algorithm. These are depicted in Figure 2 and 3.

Figure 2 depicts the steps of optic cup segmentation. The colored retinal fundus image is shown in Figure 2(a). It is preprocessed in which green plane is extracted for cup segmentation as shown in Figure 2(b). Figure 2(c) depicts the thresholded binary image. The final stage includes the segmented optic cup after the application of morphological operations shown by Figure 2(d).

Similarly, Figure 3 shows the various stages of optic disc extraction. Figure 3(a) shows colored fundus image, Figure 3(b) shows the red plane extracted out of colored image. This contrast of this red plane is enhanced using histogram equalization and is depicted in Figure 3(c). Figure 3(d) shows the thresholded binary image. Figure 3(e) shows the segmented optic disc of healthy image. Similar steps are applied on remaining 24 healthy subjects.

Experiment 2: Glaucomatous subjects are taken and segregation of optic cup and disc is done. It is observed from the experiment that the size of optic cup in glaucoma patients is larger than the size of optic cup in healthy subjects. The results are shown by Figure 4 and Figure 5.

Figure 4 demonstrates the different stages involved in optic cup extraction of glaucomatous image. Figure 4(a) shows the colored retinal fundus image. The green plane is extracted from the RGB image in preprocessing step as depicted by Figure 4(b). This green plane is thresholded to binary image as shown in Figure 4(c). Figure 4(d)

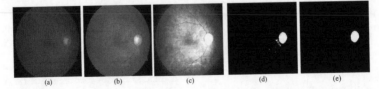

Figure 2.   Healthy image (a) normal fundus image (b) green plane (c) thresholded image (d) optic cup extracted.

Figure 3.   Healthy image (a) normal fundus image (b) red plane (c) histogram equalized image (d) thresholded image (e) optic disc extracted.

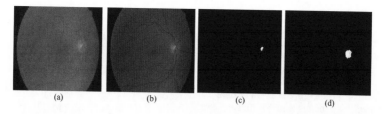

Figure 4. Glaucomatous image (a) colored fundus image (b) green plane (c) thresholded image (d) optic cup extracted.

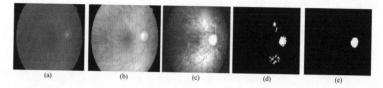
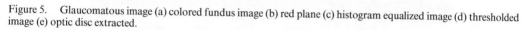

Figure 5. Glaucomatous image (a) colored fundus image (b) red plane (c) histogram equalized image (d) thresholded image (e) optic disc extracted.

shows the optic cup extracted by applying morphological operations.

Figure 5 displays the results of various steps in optic disc extraction of images of glaucoma patients. Figure 5(a) shows the colored fundus image of eye. Figure 5(b) shows the preprocessed image in which the red plane is extracted from the colored fundus image. The contrast of the red plane is enhanced using histogram equalization technique as depicted in Figure 5(c). The histogram equalized image is thresholded and converted into binary image as shown in Figure 5(d). Finally, Figure 5(e) displays the result of segmented optic disc after application of morphological operations that refine the output and ignore certain portions that are not required. Similar results are obtained for other 24 glaucoma patients.

Experiment 3: The CDR is calculated which is given by ratio of area of optic cup to that of disc. The number of pixels is calculated for both optic cup and disc.

Among 25 subjects, CDR of 10 subjects for each healthy and glaucoma patients is shown in Table 1 and Table 2. The following parameters are calculated for 25 healthy subjects and 25 glaucoma patients.

It is observed from the above calculations that CDR for healthy subjects is less than 0.3 and CDR for glaucoma patients is more than 0.3 on an average. Thus, the results evaluated in this section for both healthy and glaucoma subjects are in accordance with the standards acknowledged globally.

Table 1. CDR of healthy subjects.

	Cup Area	Disk Area	
S. No.	No. of pixels	No. of pixels	CDR
1	261	920	0.283
2	216	924	0.233
3	260	1013	0.256
4	217	672	0.322
5	250	665	0.375
6	211	675	0.312
7	222	765	0.290
8	193	669	0.288
9	168	695	0.241
10	143	492	0.290

For 25 healthy subjects
Average CDR = 0.2905
Standard deviation = 0.0363.

Table 2. CDR of Glaucoma patients.

	Cup Area	Disk Area	
S. No.	No. of pixels	No. of pixels	CDR
1	351	835	0.420
2	214	547	0.391
3	405	991	0.408
4	343	905	0.379
5	438	1007	0.434
6	193	470	0.410
7	235	540	0.435
8	151	416	0.362
9	273	788	0.346
10	357	709	0.503

For 25 glaucoma patients
Average CDR = 0.4129
Standard deviation = 0.0471.

6 CONCLUSION

In this paper, methodology based on morphological operations was developed to extract optic cup and disc from retinal fundus images of healthy and glaucomatous subjects. The database of 50 images was collected from internet repository of which 25 images were of healthy subjects and 25 images were of glaucoma patients. This method made use of bit plane extraction followed by morphology for segmentation of optic cup and disc. It is observed from the experimentation that clear cut peripheries of cup and disc were extracted. Further, calculation of Cup to Disc Ratio (CDR) with mean 0.2905 and standard deviation 0.0363 was obtained for healthy subjects. It is demonstrated that CDR mean is 0.4129 and standard deviation is 0.0471 for glaucomatous subjects. The gold standards (i.e. CDR< = 0.3 for healthy eye and CDR > 0.3 for glaucomatous eye) confirm the results. The results were further confirmed by experienced ophthalmologist of PGIMER. The outcomes of the methodology may assist medical fraternity for better visual and clinical diagnosis.

REFERENCES

Aquino, A. Arias, M. E. G. & Marín, D. 2010. Detecting the Optic Disc Boundary in Digital Fundus Images using Morphological, Edge Detection, and Feature Extraction Techniques. *IEEE Transactions on Medical Imaging* 29(11): 1860–1869.

Cheng, J. Liu, J. Xu, Y. Yin, F. Wong, D. W. K. Tan, N. M. Tao, D. Cheng, C. Y. Aung, T. & Wong, T. Y. 2013. Super-pixel Classification Based Optic Disc and Optic Cup Segmentation for Glaucoma Screening. *IEEE Transaction on Medical Imaging* 32(6):1019–1032.

Chauhan, B. McCormick, T. Nicolela, M. & LeBlanc, R. 2001. Optic Disc and Visual Field Changes in a Prospective Longitudinal Study of Patients with Glaucoma: Comparison of Scanning Laser Tomography with Conventional Perimetry and Optic Disc Photography. *Archives of Ophthalmology* 119: 1492–1499.

Choukikar, P. Patel, A. K. & Mishra, R. S. 2014. Segmenting the Optic Disc in Retinal Images using Thresholding. *International Journal of Computer Applications* 94(11): 6–10.

Gonzalez, A. S. Kaba, D. Li, Y. & Liu, X. 2014. Segmentation of Blood Vessels and Optic Disk in Retinal Images. *IEEE Journal of Biomedical and Health Information* 18(6): 1874–1886.

Heijl, A. Leske, M. C. Bengtsson, B. Hyman, L. & Hussein, M. 2002. Reduction of Intraocular Pressure and Glaucoma Progression: Results from the Early Manifest Glaucoma Trial. *Archives of Ophthalmology* 120: 1268–1279.

Joshi, G. Sivaswamy, J. & Krishnadas, S. R. 2011. Optic Disk and Cup Segmentation From Monocular Color Retinal Images for Glaucoma Assessment. *IEEE Transactions on Medical Imaging* 30(6): 1192–1205.

Kavitha, S. Karthikeyan, S. & Duraiswamy, K. 2010. Early Detection of Glaucoma in Retinal Images Using Cup to Disc Ratio. *Second International conference on Computing, Communication and Networking Technologies*: 1–5.

Kwon, Y. Adix, M. Zimmerman, M. B. Piette, S. Greenlee, E. C. & Abramoff, M. D. 2009. Variance owing to Observer, Repeat Imaging, and Fundus Camera Type on Cup-to-disc Ratio estimates by Stereo Planimetry. *Journal of Glaucoma* 18: 305–310.

Liu, J. Wong, D. W. K. Lim, J. H. Jia, H. Yin, F. Li, H. Xiong, W. & Wong, T. Y. 2008. Optic Cup and Disk Extraction from Retinal Fundus Images for Determination of Cup-to-Disc Ratio. *3rd IEEE Conference on Industrial Electronics and Applications*: 1828–1838, Singapore.

Mahalakshmi, V. & Karthikeyan, S. 2014. Clustering Based Optic Disc and Optic Cup Segmentation for Glaucoma Detection. *International Journal of Innovative Research in Computer and Communication Engineering* 2: 3756–37561.

Maldhure, P. N. & Dixit, V. V. 2015. Glaucoma Detection Using Optic Cup and Optic Disc Segmentation. *International Journal of Engineering Trends and Technology (IJETT)* 20(2): 52–55.

Schuman, J. Hee, M. Puliafito, C. Wong, C. Pedut-Kloizman, T. Lin, C. P. Hertzmark, E. Izatt, J. A. Swanson, E. A. & Fujimoto, J. G. 1995. Quantification of Nerve Fiber Layer Thickness in Normal and Glaucomatous Eyes using Optical Coherence Tomography. *Archives of Ophthalmology* 113: 586–596.

Tielsch, J. M. Katz, J. Quigley, H. A. Miller, N. R. & Sommer, A. 1998. Intraobserver and Interobserver Agreement in Measurement of Optic Disc Characteristics. *Ophthalmology* 95: 350–356.

Welfer, D. Scharcanski, J. Kitamura, C. Pizzol, M. Ludwig, L. & Marhino, D. 2010. Segmentation of Optic Disc in Color Eye Fundus Images using Adaptive Morphological Approach. *Computers in Biology and Medicine* 40: 124–137.

Whardana, A. K. & Suciati, N. 2014. A Simple Method for Optic Disk Segmentation from Retinal Fundus Image. *Graphics and Signal Processing* 11: 36–42.

Wong, D. Liu J. Lim, J. H. Jia, X. Yin, F. Li, H. & Wong, T. Y. 2008. Level-set based Automatic Cup-to-disc Ratio Determination using Retinal Fundus Images in ARGALI. *30th Annual International IEEE EMBS Conference*: 2266–2269, Vancouver: Canada.

Yu, H. Barriga, E. S. Agurto, C. Echegaray, S. Pattichis, M. S. Bauman, W. & Soliz, P. 2012. Fast Localization and Segmentation of Optic Disk in Retinal Images using Directional Matched Filtering and Level Sets. *IEEE Transactions on Information Technology in Biomedicine* 16(4): 644–657.

Yin, F. Liu J. Wong, D. W. K. Tan, N. M. Cheung, C. Baskaran, M. Aung, T. & Wong, T. Y. 2012. Automated Segmentation of Optic Disc and Optic Cup in Fundus Images for Glaucoma Diagnosis. *25th International Symposium on Computer-Based Medical Systems (CBMS)*.

Communication and Computing Systems – Prasad et al. (Eds)
© 2017 Taylor & Francis Group, London, ISBN 978-1-138-02952-1

Recent trends in arrhythmia beat detection: A review

Manisha Jangra & Sanjeev Kumar Dhull
Guru Jambheshwar University of Science and Technology, Hisar, India

Krishna Kant Singh
Dronacharya College of Engineering, Gurgaon, India

ABSTRACT: Electrocardiogram (ECG), a typical non-stationary signal, is used for diagnosis of cardiac disorders. The electrical activities such as depolarization and re-polarization of heart are reflected in P, QRS complex and T waveforms. The minute changes in duration, rhythm and amplitudes of these waves indicates pathological alterations of heart viz arrhythmia. A number of sophisticated ECG arrhythmic beat detection algorithms have been proposed so far aimed at high classification accuracy. However, recent advancements are more focused at algorithms suited to mobile, battery operated ECG analyzing systems. With the change in application, now the requirements are to develop not only accurate, but also fast and less complex algorithms. The purpose of this paper is to study the recent trends and asses the performance of existing algorithms.

1 INTRODUCTION

One of the most investigated fields in biomedical engineering is the automatic analysis of the Electrocardiogram (ECG) and, inside that, the detection of arrhythmic heart beats (Silipo& Berthold 2000). An ECG Signal shown in Figure 1 is characterized by mainly P, QRS complex, T and sometimes U waveform. P, Q, R, S, & T waves represent the atrial depolarization, septal depolarization, early ventricular depolarization, late ventricular depolarization, and repolarization of the ventricles respectively.

Cardiovascular Diseases has drawn attention worldwide due to its increased incidence and prevalence. Arrhythmias commonly occur due to cardiac rhythm disturbances. These cardiac arrhythmias can be non-invasively diagnosed using ECG signal. Cardiac arrhythmia is a collective term for heterogeneous group of conditions in which there is abnormal cardiac electrical activity. In case of fetal and life threatening arrhythmias like ventricular techy-arrhythmia, accurate and early detection is must (Martis et al. 2012).

The automatic beat detection systems are useful not only at diagnostic phase where long term ECG is analyzed for early recognition of cardiac disease but also in intensive care units where real time monitoring is required for treatment of patients. Ability to process real time data accurately is an attribute of automatic beat detection system. Other than that, in today's scenario, it is expected that holter devices will be replaced by portable,

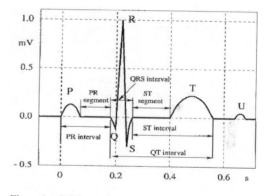

Figure 1. ECG waveform.

and wireless ECG systems such as mobile phones in the near future (Elgendi et al. 2014).

A plenty of research has been done over ECG beat detection systems. ECG beat analysis consists of following modules also shown in Figure 2: (i) Preprocessing (ii) Feature Extraction (iii) Feature selection (iv) Classification. The objective of this paper is to get an insight into these module's state-of–the art methods and techniques.

This paper is organized as follows: We discuss the availability of ECG signals in terms of standardized database in section 2. In section 3 recent trends in the pre-processing techniques are discussed. Section 4 covers the feature extraction and feature selection module. Classification module is discussed in section 5. Section 6 describes the

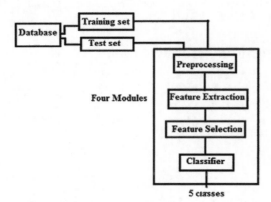

Figure 2.   ECG beat detection system modules.

performance measures. Section 7 gives a quick discussion over literature. Finally paper is concluded in section 8.

2   ECG DATABASE

Human heart can be exposed to three pathological alterations: Hypertension, Cardiac ischemia and cardiac arrhythmia. Plenty of standard Databases exist to train and test the automatic system. These database signals are well annotated and validated with patient specific information, type of beat, and other sampling rate related information. MIT-BIH database, AHA database, CSE database, PTB database are few of benchmark databases reviewed well in (Kohler et al. 2002).

The MIT-BIH Database (Mark & Moody 1997) provided by MIT and Boston's Beth Israel Hospital, is the most used database by researchers for cardiac arrhythmia detection. It contains 48 recordings, each of 30 min, of annotated ECG of two leads. Signals are sampled at 360 Hz and having 11 bit resolution. Each recording is supported with header file having detailed information about source lead, sampling frequency & patient details etc. Other than that each recording is annotated by experts for validation of automatic system performance. There are over 110157 labeled arrhythmic beats from 15 different heartbeat types. The database is not uniformly distributed. There is huge difference between number of examples belonging to different classes causes biased results from classifier.

3   PRE-PROCESSING

While recording, the ECG signal gets interfered by various types of noises. Mainly ECG signal noises can be classified in three types: (i) Baseline drift/ wander (ii) Power Line Interference (PLI) (iii) EMG (Electromyogram) Interference also known as Muscle Artifacts.

3.1   Baseline drift/wander

This is a low frequency noise generated by human breathing and motion at the time of recording. It mainly exists in frequency range 0–0.5 Hz. This noise causes drift in baseline of original ECG signal. In presence of Baseline wander, detection of low amplitude fiducial points such as P, S & T wave is affected. Studies reported miss and false detection of true R-peak with less effective baseline removal in ECG signals. Thus suitable pre-processing methods are needed for removal of Baseline wander.

3.2   Power Line Interference (PLI)

This kind of noise signal is caused by coupling of human body distribution capacitance with power lines connected with the ECG recording instrument. American Heart Association (AHA) suggests operating range from 0.67 Hz–150 Hz for the ECG recorders (Butt, Akram& Khan 2015). PLI noise lies in range of 50–60 Hz depending on the power line frequency used in that region. PLI generates impulses of said frequency in ECG signal and thus affects the ECG analysis.

3.3   EMG (Electromyogram) interference/muscle artifacts

This high frequency noise is caused by electrical activity of muscles or muscle tension in contact with electrodes. The amplitude of EMG noise depends upon rate of the muscle movements (Butt et al. 2015). It exists in frequency range 20 Hz-100 Hz. Several pre-processing techniques exist in literature based on filtering (Butt et al. 2015, Mar et al. 2011, Qiao et al. 2014), Neural Network (Singh et al. 2015), Wavelet transform (Su & Zhao 2005, Zhang et al. 2010, Martis et al. 2013) for elimination of noise.

In automatic arrhythmia detection systems, the pattern is to use filtering based pre-processing. (Mar et al. 2011, Chazel et al. 2004) used two median filters for removal of baseline wander. High frequency noise and power line interference is removed using a 12-tap FIR low pass filter with cut-off frequency 35 Hz. A cascade of filters is used for ECG signal preprocessing. 1) a high-pass filter with 3 db frequency of 1 Hz to suppress residual baseline wander (2) a 2nd order Butterworth low-pass filter with cutoff frequency 30 Hz to overcome high frequency noise, and (3) a notch filter to eliminate PLI (Qiao et al. 2014).

In today's world, when ECG analyzing system has emerged from ICU to mobile handsets for easy monitoring by the patients themselves, the Pre-processing should be efficient and should offer less computational load. Wavelet transform based de-noising techniques seems more promising as the coefficients can be reused for feature extraction exerting less computational load on system.

Su & Zhao (2005) discussed two kinds of wavelet denoising techniques. One technique based on Modulus Maxima method and other is Threshold Dependent technique. Modulus Maxima method have tradeoff of efficiency and complexity. Theoretically there are two thresholding techniques: hard thresholding illustrated by Equation 1 and soft thresholding illustrated by Equation 2.

$$x_{j,k}^{\wedge} = \begin{cases} x_{j,k} & |x_{j,k}| \geq h \\ 0 & |x_{j,k}| < h \end{cases} \tag{1}$$

$$x_{j,k}^{\wedge} = \begin{cases} \text{sgn}(x_{j,k})(|x_{j,k}| - h) & |x_{j,k}| \geq h \\ 0 & |x_{j,k}| < h \end{cases} \tag{2}$$

where x^j,k, and xj, k are wavelet coefficients after and before threshold comparison. Limitation of hard-thresholding is oscillatory nature of re-constructed signal whereas soft-thresholding causes reduction in amplitude of ECG waveform especially R peak (Su & Zhao 2005). An improved thresholding technique is proposed in (Su & Zhao 2005) given by Equation 3.

$$x_{j,k}^{\wedge} = \begin{cases} \text{sgn}(x_{j,k})(|x_{j,k}| - \beta^{(|x_{j,k}| - h_j)}.h) & |x_{j,k}| \geq h_j \\ 0 & |x_{j,k}| < h_j \end{cases} \tag{3}$$

$h_j = \sigma\sqrt{2\log N}/\log(j+1); \beta > 1$ choosen as 15.

(Zhang et al. 2010) pointed out the importance of choice of wavelet basis function and decomposition scale. The selective approach towards choice of wavelet basis function directly affects the accuracy and efficiency. The knowledge of frequency range of ECG signal and noise signal helps in deciding the decomposition level. Coif3 (Zhang et al. 2010), Doubechies 6(db6) (Martis et al. 2013), and Symlet (Su & Zhao 2005) are the mother wavelets used frequently. (Su & Zhao 2005) used 4 scale of decomposition whereas 8–9 levels of decompositions are suggested in (Su & Zhao 2005, Zhang et al. 2010). 9th level approximation of frequency range 0–0.351 Hz refers to baseline wander. First two level detailed coefficients show main line interference and high frequency noise (Martis et al. 2013).

Combinations of wavelet transform with filters do exist in literature. (Zhang et al. 2010) proposed wavelet transform with notch filter to remove PLI at 60 Hz having transfer function given by Equation 4. (Qiao et al. 2014) proposed combination of DWT with wiener filter

$$H(z) = \frac{z^2 - z + 1}{z^2 - 0.9z + 0.81} \tag{4}$$

4 FEATURE EXTRACTION AND FEATURE SELECTION

4.1 *Feature extraction*

An expert cardiologist analyzes the ECG signal by its shape, amplitude and duration of PQRST complex. The amplitudes of all the samples of a beat are raw features of an ECG signal. Such raw features can be cost effective in terms of complexity but they tend to scatter in classification space if R-positions are not correctly detected (Emanet 2009). For an automatic system, a category of most effective feature extraction methods is to decompose the signal into various components such that valuable features can be uncovered from the otherwise hidden details (Yu & Chen 2009). The feature extraction should represent the data in such a way, that the differences among data samples of the same class are reduced and the differences among the classes enhanced (Osowski & Hoai 2004). Features can be categorized into morphological features, temporal features, features based on transformation such as wavelet transform, Hilbert transform, statistical features, Hermite polynomial, and vector cardiogram features.

Temporal features include interval based features like RR interval, QRS duration etc. Morphological features are related to shape of waveforms measured in terms of area, power, or extrema (Mar et al. 2011). Morphological variations among beats distinguish them from normal beats. e.g. VPC affected beat have prolonged QRS duration than normal beat. (Mar et al. 2011) have proved the significance of temporal and morphological features in classification. Detection of R peak or one can say QRS complex is first milestone for feature extraction and segmentation. Most beat detection algorithms proposed by researcher uses Pan-Tompkins algorithm (Pan & Tompkins 1985) for QRS complex detection. New QRS complex detection detection algorithm proposed in (Arefin et al. 2015, Ramakrishna et al. 2014) offer less computational load than (Pan & Tompkins, 1985) (Pan & Tompkins, 1985). Sometime QRS complex detection, segmentation and fiducial point detection are considered as part of pre-processing module only.

179

Frequency domain features are obtained by transformation of time-domain signal. Transformed features do not scatter in classification space. Their limitation is complexity. Wavelet Transform represents the time domain signal in both time & Frequency plane. Wavelet analysis inherits and develops the localization idea of short-time fourier and changes the window with the variant of frequency (Guangying & Yue 2010).

Wavelet Transform of function f, with respect to a given mother wavelet, φ is defined as:

$$wf(a, b) = \int_{\infty}^{-\infty} f(t)\varphi^*_{a,b}(t)\, dt \qquad (5)$$

with

$$\varphi_{a,b} = \frac{1}{\sqrt{a}}\,\varphi\!\left(\frac{t - b}{a}\right) \qquad (6)$$

where a and b are the scale (dilation) and translation parameters, respectively (Shyu et al. 2007). The ability of WT to extract ECG features has been demonstrated by several studies (Martis et al. 2012, Emanet 2009, Guangying & Yue 2010, Shyu et al. 2004, Banerjee & Mitra 2014). Mother wavelet should have morphological similarity with ECG signal. Researches' findings prove superiority of Doubenchis and Spline wavelet for their smoothness and performance (Shyu et al. 2007). Importance of WT features can be infer from abundant use of DWT at various stages whether it is pre-processing, QRS complex detection or feature extraction. Computational cost of using DWT gets balanced by its coefficient reuse throughout classification.

All types of beats occupy similar range of amplitudes and frequencies. It is, thus, difficult to separate one from the other on the basis of only time or frequency representations (Osowaski et al. 2004). Another important feature is Statistical features include mean, variance, entropy etc. of higher order cumulants. The cumulants are the coefficients of the Taylor expansion of the cumulant generating function (Osowaski et al. 2004). By comparing HOS characteristics of different rhythm types, it was observed in (Osowaski et al. 2004) that the differences among them have been increased and they are easier to distinguish.

Features extracted from Vector cardiogram does seem promising keeping in mind today's requirements. Vector cardiogram is nothing but X-Y mode representation of signal from two leads of ECG. So information of two leads can be analyzed from one graphical signal. Features from ECG complex network along with temporal and morphological features are utilized in (Queiroz et al. 2015). Results catch the attention of researchers due to performance and cost effectiveness.

4.2 Feature selection

This module controls the dimensionality of feature set. Generally combination of feature extraction methods are applied creating very large feature set. Further a classifier complexity increases with feature set. It is said that lower dimensional systems have better generalization capability. To overcome dimensionality curse feature selection or detection methods are used. These methods involve a process wherein a number of subsets of the available features are evaluated, and the best one is selected for application on the learning algorithm. "The best subset contains the least number of dimensions that most contribute to the application's performance; the remaining, unimportant dimensions are discarded" (Mar et al. 2011).

Various feature selection techniques are Principle Component Analysis (PCA) (Martis et al. 2013), Independent Component Analysis (ICA), Linear Correlation Based Filter (LCBF), Linear Discriminant Analysis (LDA) (Yu & Chen 2009), & Quality measures (Osowski & Hoai 2004). Investigators find optimization techniques such as Ant Colony Optimization (ACO) (Dogan & Korürek 2012), Particle Swarm Optimization (PSO) (Melgani & Bazi 2008), Genetic Algorithms (GA) promising for feature selection and to find optimized parameters for modeling.

5 CLASSIFIER

This module generates a model of system based on the training data feature set. A number of classifiers have been reported in literature. These methods include Feed Forward Neural Network (FFNN) (Martis et al. 2012, 2013), Probabilistic Neural Networks (PNN) (Martis et al. 2012), Support Vector Machine (SVM), Least Square Support Vector Machine (LS-SVM) (Silipo & Berthold 2000, Martis et al. 2012), Self-Organizing Map networks (SOM) (Lagerholm et al. 2000), Extreme learning machine (Karpagachelvi et al. 2011), Fuzzy Neural Network (FNN) (Shyu et al. 2004), Radial Basis Function Neural Network (RBF NN) (Guangying & Yue 2010), random Forest algorithm (Emanet 2009), multiple SVM (Zellmer et al., 2009), and the combination of different Neural based systems, called hybrid systems (Shyu et al. 2007).

Fuzzy Neural Network is nothing but a hybrid of Neural Network and Fuzzy System. The repeated use of ANN systems justifies the generalization capability of neural networks. Support Vector Machine (SVM) is another frequently used classifier by the researchers. Random Forest Algorithm (Emanet 2009) is from family of Decision tree based algorithms.

Another popular Neural Network, RBF Neural Network is characterized by its simple topological structure and fast learning rate as compared to multilayer feed-forward neural networks. The RBF neural network has one input layer, one hidden layer followed by output layer. Gaussian density function is used as its activation function.

The network works on basis of the following Equation 6–7 (Guangying & Yue 2010):

$$y_k(x) = \sum w_{kj}h_j(x) + w_{ko} \tag{7}$$

where

$$h_j(x) = \exp\left\{\frac{-\|x - \mu_j\|^2}{2\sigma^2}\right\} \tag{8}$$

Usually the number of hidden nodes is equals the number of classes.

Another Simple but fast classifier suitable for low power devices is Linear Discriminant Classifier (LDC). The model parameters are determined by maximum likelihood estimates from the training data. The relative proportions of the classes of the available training examples influence the performance of an LD classifier. SVM is considered as state-of-the-art solution for classifier using different kernel functions viz linear, quadratic, polynomial and RBF kernel. (Martis et al. 2012, 2013) compared Least-square SVM, a modified SVM using different kernel functions, with FFNN and PNN.

Different sensitivities and specificities for different type of beats like Normal (N), Ventricular (VC), Super Ventricular (SVAB) beats for the same classifier proves the no free lunch theorem.

6 PERFORMANCE MEASURES

The performance of the classifier is evaluated using confusion matrix. Sensitivity (Se), Specificity (Sp), Positive Predictivity Value (PPV) and classification Accuracy (Ac) are used as performance measures of automatic ECG beat classifier. These are

$$S_e = \frac{TP}{TP + FN} \tag{9}$$

$$PPV = \frac{TP}{TP + FP} \tag{10}$$

$$S_p = \frac{TN}{TN + FP} \tag{11}$$

$$A_c = \sum \frac{TPi}{T} \tag{12}$$

T = total no of beats used for testing
where TP is true positive, No. of normal beats classified as normal correctly. TN is true negative, no. of abnormal beat not classified as normal (i.e. LBBB, RBBB, PVC etc not classified as NORM). FN is false negative, denotes the no. of beat misclassified i.e. normal beat classified as abnormal. FP is false positive, represents the no of abnormal beats classified as normal.

7 DISCUSSION

The objective of automatic ECG arrhythmia beat classifier is to categories long term test signal into 5 classes as given by AAMI/ANSI:57 standard. Overall performance of beat classifier depends on performance of pre-processing, feature extraction and classification algorithms. Table 1 lists some of work reported by researchers. It is possible to infer from Table 1 that Probabilistic neural network classifers report highest classification accuracy (Martis et al. 2013). Applicability of WT is evinced from Table 1 at both pre-processing and feature extraction stage.

Table 2 is zest of Work done by (Martis et al. 2012, 2013). Accuracy of LS-SVM with RBF kernel varies with the feature extraction method chosen. The classification performance reported to degrade when principal components of DWT or DCT coefficients are used as features as compared to principal components of segmented beats only. Thus the same classifier performs differently for different feature sets. Highest accuracy is reported for PNN classifier using principal components of DCT coefficients as features. Complexity constrained of the methods is not reported.

(Ghahremani 2010) reported 99.5% classification accuracy using PNN classifier. Katz fractal algorithm at feature extraction stage provide noise robustness, thus no pre-processing is required. This technique not only provides noise robustness but also offers less computational load.

Recent work by Queiroz et al. 2015 reports only 84.1% classification accuracy. Degradation in accuracy is compensated by simple pre-processing method and feature set. Thus complexity of system reduces, making it suitable for mobile, battery-operated ECG systems.

To overcome the biasing of result due to different percentage of examples for different beats in the training and testing set, cross-validation methodology is adopted. 22 fold cross-validation adopted by (Chazel et al. 2004) is also implemented by (Queiroz et al. 2015). 10 fold cross-validations is adopted by (Martis et al. 2012, 2013).

Table 1.

Reference	type of beat	pre-processing	feature extraction	feature selection	classifier	sensitivity	specificity	accuracy	ppv
(Ghahremani 2010)	4 type	no need	DWT+katz' fractal (noise robust)also pqrst	–	PNN	–	–	99.5	–
(chi 2012)	6 type	–	COMPLEX FEATURE	PCA	Fuzzy logic	–	–	94.03	–
(Martis et al. 2012)	5 type AAMI standars	Wavelet Transform db6, 6 level, pan-tompkins qrs detection	PCA of segmented ecg beats	–	LS-SVM (RBF kernel)	99.9	99.1	98.11	99.61
(Dogan 2012)	6 type	median filter+notch filter	kernalized fuzzy c-mean	HACO	Supervised classifier (NN)	93.766	–	–	–
(karpagachelvi et al. 2011)	5 type AAMI standars	–	third order commulant, AR model, variance of DWT (db5)	–	ELM	–	–	89.74	–
(Martis et al. 2013)	5 type AAMI standars	WT db6 9 level decomposition+pan-tompkins QRS detection	DCT+PCA	ANNOVA +Fisher index	PNN	98.69	99.91	99.52	99.58
(Martis et al. 2013)	5 type AAMI standars	WT db6 9 level decomposition+pan-tompkins QRS detection	DCT+PCA	ANNOVA +Fisher index	LS-SVM (RBF kernel)	99.58	97.97	96.61	91.5
(Banerzee & Mitra 2014)	N and abnormal IMI	DWT	XWT, WCS &WCOH	–	Threshold based classifier	97.3	98.8	97.6	–
(Queiroz et al. 2015)	5 type AAMI standars	common filter	VCG+temporal+morphological	–	SVM	–	–	86.2	–
(Queiroz et al. 2015)	5 type AAMI standars	no filter	VCG+temporal+morphological	–	SVM	–	–	84.1	–

Table 2.

Reference	Features	Classifier	Class	Accuracy in %
(Martis et al. 2012)	DWT+PCA	FFNN	AAMI standardised 5 type	94.9
(Martis et al. 2013)	DCT+PCA	LS-SVM with RBF kernel	AAMI standardised 5 type	96.61
(Martis et al. 2012)	DWT+PCA	LS-SVM with RBF kernel	AAMI standardised 5 type	96.88
(Martis et al. 2012)	PCA	FFNN	AAMI standardised 5 type	97.58
(Martis et al. 2012)	PCA	LS-SVM with RBF kernel	AAMI standardised 5 type	98.11
(Martis et al. 2013)	DCT+PCA	FFNN	AAMI standardised 5 type	99.12
(Martis et al. 2013)	DCT+PCA	PNN	AAMI standardised 5 type	99.52

8 CONCLUSION

ECG Signal processing have moved a long way. However, requirements of each system changes with time. This paper gives a detailed understanding of ECG beat classification problem in context of recent trends and requirements. The ECG beat classification is a cascade of three modules viz. pre-processing, feature extraction and classifier. Recent advancements and algorithms used at each stage are reviewed giving a clear bird eye view to someone new to this field.

REFERENCES

ANSI/AAMI EC57: Testing and Reporting Performance Results of Cardiac Rhythm and ST Segment Measurement Algorithms (AAMI Recommended Practice/ American National Standard), Order Code: EC57-293, 1998. <http://www.aami.org>

Arefin, M.R. et al. 2015. QRS complex detection in ECG signal for wearable devices. Annual International Conference of the *IEEE Engineering in Medicine and Biology Society (EMBC)*. 37: 5940–5943.

Banerjee, S. & Mitra, M. Feb 2014. Application of Cross Wavelet Transform for ECG Pattern Analysis and Classification. *IEEE Transactions on Instrumentation and Measurement*. 63(2): 326–333.

Butt, M.M., Akram, U., & Khan, S.A. April 2015. Denoising practices for electrocardiographic (ECG) Signals: A Survey. *International conference on Computer, Communication, and control technology*. 264–268.

Chazel, P.D et al. July 2004. Automatic Classification of Heartbeats Using ECG Morphology and Heartbeat Interval Features. *IEEE Transactions on Biomedical Engineering*. 51(7): 1196–1206.

Dogan, B. & Korürek, M. 2012. A New ECG Beat Clustering Method Based On Kernelized Fuzzy C—mean And Hybrid Ant Colony Optimization For Continuous Domains. *Applied Soft Computing, Elsevier*: 3442–3451.

Elgendi, M. et al. Jan 2014. Revisiting QRS detection methodologies for portable, Wearable, battery operated, and wireless *ECG* systems. *PLOS* 1.9(1): e84018.

Emanet, N. 2009. ECG beat classification by using Discrete Wavelet Transform and Random Forest Algorithm. Fifth International Conference on Soft Computing, Computing with Words and Perceptions in System Analysis, Decision and Control, ICSCCW: 1–4.

Guangying, Y. & Yue, C. 2010. The study of electrocardiograph based on radial basis function neural network. Third International Symposium on Intelligent Information Technology and Security Informatics IEEE. DOI 10.1109/IITSL.2010. 85: 143–145.

Karpagachelvi, S., Arthanari, M. & Sivakumar, M. 2011. Classification of Electrocardiogram Signals with Support Vector Machines and Extreme Learning Machines. *Neural Computing & Applications, Springer*. 21(6): 1331–1339.

Kohler, B.U. et al. Feb 2002. The principles of software detection. *IEEE Engineering in Medicine and Biology Engineering*. 21(1): 42–57.

Lagerholm, M. et al. July 2000. Clustering ECG Complexes Using Hermite Functions and Self Organizing Maps. *IEEE Transaction on Biomedical Engineering*. 47: 838–848.

Mar T. et al. August 2011. Optimization of ECG Classification by Means of Feature Selection", *IEEE Transactions on Biomedical Engineering*. 58(8): 2168–2177.

Moody, G.B., Mark, R.G. 2001. The impact of the MIT-BIH Arrhythmia Database. *IEEE Engineering in Medicine and Biology*. 20(3): 45–50.

Martis, R.J. et al. 2012. Application of principle component analysis to ECG signals for automated diagnosis of cardiac health. *Expert System With Applications*. 39(14): 11792–11800.

Martis, R.J., Acharya, U.R., Lim, C.M. & Suri, J.S. June 2013. Characterization of ECG Beats From Cardiac Arrhythmia Using Discrete Cosine Transform In PCA Framework. *Knowledge-Based Systems, Elsevier*. 45: 76–82.

Melgani, F. & Bazi, Y. Sept. 2008. Classification of Electrocardiogram Signals with Support Vector Machine and Particle Swarm Optimization. *IEEE Transactions on Information Technology in Biomedicine*. 12(5): 667–677.

Osowaski, S. et al. April 2004. Support Vector Machine Based Expert System For Reliable Heartbeat Recognition. *IEEE Transactions on Biomedical Engineering*. 51(4): 582–589.

Osowski, S. & Hoai, L.T. 2004. Analysis of Features for Efficient ECG Signal Classification Using Neuro-Fuzzy Network. *IEEE International Joint Conference on Neural Networks*. 3: 2443–2448.

Osowski, S. & Linh, T.H. 2000. Fuzzy clustering neural network for classification of ECG beats. *IEEE-INNS-ENNS International Joint Conference on Neural Network*. 5: 26–30.

Pan, J. & Tompkins, W.J. March 1985. A real time QRS detection Algorithm. *IEEE Transactions on Biomedical Engineering*. BME-32(3): 230–236.

Qiao, L., Cadathur, R., & Gari, D. C. June 2014. Ventricular Fibrillation and Tachycardia Classification Using a Machine Learning Approach. *IEEE Transactions on Biomedical Engineering*. 61(6): 1607–1613.

Queiroz, V. et al. 2015. Automatic cardiac arrhythmia detection and classification using vectorcardiograms and complex networks. Annual International Conference of the IEEE Engineering in Medicine and Biology Society (EMBC). 37: 5203–5206

Ramakrishna, A.G. et al. May 2014. Threshold-independent QRS detection using the Dynamic plosion-index. *IEEE Signal Processing Letter*. 21(5): 554–558.

Shyu, L.Y et al. July 2004. Using wavelet transform and fuzzy neural network for VPC detection from the holter ECG. *IEEE Transaction on Biomedical Engineering*. 51(7): 1269–1273.

Shyu, L.Y. et al. 2007. Intelligent Hybrid Methods for ECG classification-A review. *Journal of Medical and Biological Engineering*. 28(1): 1–10.

Silipo, R. & Berthold, M.B. Dec, 2000. Input features' impact on fuzzy decision processes. *IEEE transactions on systems, man and cybernetics-part b: cybernetics*. 30(6): 821–834.

Singh, B., Singh, P. & Budhiraja, S. 2015. Various Approaches To Minimize Noises In ECG Signal: A Survey. *IEEE International Conference On Advanced Computing & Communication Technologies*. 5: 131–137.

Su L. & Zhao G. Sept 2005. De-noising of ECG signal using translation-invariant wavelet denoising method with improved thresholding. *Annual Conference on Engineering In Medicine And Biology*. 25: 5946–49.

Yu, S.N. & Chen, Y.H. 2009. ECG Beat Classification Based on Signal Decomposition: A Comparative Study. *International Symposium on Circuits & Systems, (IEEE ISCAS)*: 3090–3093.

Zellmer, E., Shang, F. & Zhang, H. 2009. Highly Accurate ECG Beat Classification Based On Continuous Wavelet Transformation and Multiple Support Vector Machine Classifiers. *International Conference on Biomedical Engineering & Informatics (IEEE BMEI)*. 2: 1–5.

Zhang, W., Li, M., Zhao, J. 11–14 July 2010. Research on electrocardiogram signal noise reduction based on wavelet multi-resolution analysis. *International Conference on Wavelet Analysis and Pattern Recognition*: 351–354.

Communication and Computing Systems – Prasad et al. (Eds)
© *2017 Taylor & Francis Group, London, ISBN 978-1-138-02952-1*

Mobile biometrics using face recognition: A survey

Alpa Choudhary & RekhaVig
The Northcap University, Gurgaon, India

ABSTRACT: Security management on mobile devices is of great significance and challenging task as they contain lot of private information such as photographs, payment information and other important details. In this paper a survey on security management on mobile phones using face recognition techniques is presented. An overview of developed face recognition methods implemented and tested on mobile devices is also studied. The effectiveness of the methods used in terms of the database experimented on, algorithms applied for face recognition and the result obtained is also summarized in this paper. Methods implemented for providing online security management on mobile phones and integration of different biometric methods have also been presented in this paper.

Keywords: face recognition, security management, mobile biometrics

1 INTRODUCTION

Biometric technology came to existence since 1998 but faced many challenges in order to achieve significant adoption rates. Nowadays Mobile phones provides access to various services and applications in daily life for people. Security management should be necessary on mobile phones extending from traditional Personal Identification Number (PIN) to user authentication level. Various online applications are provided on mobiles which can solve our daily problems like online billing services, internet banking, e commerce etc. These applications are accessed through internet. So online security management should also be taken into consideration to make the network secure to avoid frauds or any other criminal offense. The most significant biometric technologies which can provide security on mobile phones presented by Gofman, M. I., et al. S. (2016) are Face recognition, Voice recognition, Iris recognition, Fingerprint recognition, Keystroke recognition, Hand, Signature and Gait recognition. Using these characteristics of individual mobile phones can be accessed by recognizing the identity of that individual. Among all these recognition technologies this paper focuses mainly on face recognition biometrics. As almost all the mobile phones are provided with cameras which helps in image acquisition, faces can be captured in form of images or videos and can be saved on these mobiles. To meet security, usability and costs requirements face recognition is most effective biometrics for now. Face Recognition is a process in which we locate human face and extract features for matching it in database of human faces. The

reason for choosing facial biometric method is because this trait is most acceptable by users and hardware resources provided by mobile devices are better suited in case of facial biometrics. Wide applications using facial recognition biometric methods have been developed and implemented so far. Due to this aspect this paper presents a survey on recent biometric technologies and methods used for implementing face recognition techniques. A simple mobile biometric system consists of four basics components presented by Delac & Grgic (2004):

1. Sensor Module: Mobile sensors like cameras perform image acquisition.
2. Feature Extraction: The features are extracted from acquired images.
3. Template Matching: The extracted features are compared with the stored templates.
4. Decision making: The claimed identity is accepted or rejected.

The block diagram illustrates the proposed framework: There are two modes, in the first step one to one identification is performed and in second one to many verification is performed. The mobile sensors acquire images which are further pre processed and features are extracted for template matching wherein the extracted features are compared with stored templates which tests the user's identity and verifies the correct individual.

Based on number of sensors used we can categorize our system into two types:

1. Unimodal Biometrics: It uses single sensor to capture the human characteristic for recognition

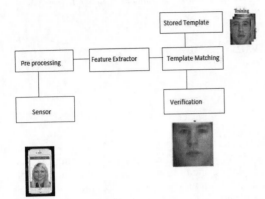

Figure 1. Block diagram of mobile biometrics using face recognition.

for example: face, iris, and fingerprint. Single human characteristic is used to verify the correct individual.

2. Multimodal Biometrics: To overcome the limitations of unimodal system multimodal system was developed which uses multiple sensors for image acquisition. It is the combination of biometrics like voice recognition combined with face recognition or any other combination of biometrics.

2 WHY USE FACE RECOGNITION FOR MOBILE BIOMETRICS

Biometric based techniques are prominently used for recognizing individuals. To provide access to virtual domains using passwords, PIN's, smart cards and so forth methods have been adopted. It is a difficult task to remember these passwords, PINs also they can be hacked, or misplaced so physiological characteristics and behavioural traits are better to be used. Face recognition offers several advantages over other recognition techniques such as: It is a passive technique since it does not require users presence during action since faces can be acquired at a distance from cameras which is beneficial in security and surveillance applications. Data acquisition is simpler and efficient as compared to other biometric techniques as for example hands and fingers can be useless as they can be damaged in some way, iris and retina require expensive equipments, voice recognition is affected by background noises, and signatures can be modified or forged. Face recognition requires inexpensive cameras also they are non intrusive and free from health risks presented in the work by Jafri & Arabnia (2009).

3 RELEVANT WORKS ON FACE RECOGNITION

With the improvement of mobile devices capabilities, security of data stored on them has become very important. Face recognition schemes provide user with flexible security as there is no need to remember passwords. In this section works done so far to provide mobile security using face recognition is presented.

An open source face recognition system named XFace was designed and implemented by Jiawei Hu et al. (2015). It was designed and implemented on Android platform. It was based on open CV (open source computer vision) SDK to provide security. Haar based and local binary patterns (LBP) techniques were used for face detection and features were extracted using Eigen faces and Principle component analysis (PCA). Finally the faces were recognised using Linear Discriminant Analysis (LDA). Satisfactory performance was achieved in providing security using XFace method on Android platforms. They used 20 face images of 2 persons under different emotions as shown:

A method using sparse representation for face recognition on mobile devices was proposed by Chou et al. (2014). Sparse representation method used for information generated from cameras achieved high performance. They also developed applications for Android platforms. The Viola Jones technique was used for face detection and features were extracted using Eigen faces and sparse representation technique was used for recognition of faces.

Powell et al. (2014) in their paper provided online security on mobile devices using CAPTCHA (Completely Automated Public Turing test to tell Computers and Human Apart) technique. In this technique they combined touch based input methods with genetically optimized face detection tests. It provided high level of security in applications like email, instant messaging, text message spam. They also prevent password attacks.

Barra et al. (2013) proposed a method for Face Authentication for Mobile Encounter (FAME). FAME was embedded on mobile devices to provide face verification and identification. It also

(a)　　　　　(b)

Figure 2. Face images of 2 persons under different environments.

supported social activities like finding doubles in social network like facebook. The proposed method performs image acquisition, spoofing, face detection, face segmentation, feature extraction and matching. It was implemented real time on Android platform. An example screenshot of FAME and finding a double is shown in figure 3.

Kremic et al. (2012) proposed a client server model for providing network security on mobile devices using face recognition. The proposed model was also compared with recent client server models and was tested on Android OS platform and DROID emulator. They used eigen faces and Principle Component Analysis (PCA) technique in their paper for feature extraction and recognition of faces. This methodology aimed to reduce fraud with cell phone authentication.

Chen et al. (2012) implemented a fast face recognition system on mobile phones. Initially the pictures taken using mobile phone camera faces can be detected using adaboost algorithm with some pre processing techniques and Local Binary Patterns (LBP) are used for face recognition. Haar—like features are also used in the system. It was implemented on Nokia 5230. The main interface of the system can be as shown:

Xi et al. (2010) proposed a hierarchical correlation based face verification (HFV) with its usability on low end mobile phones with memory less than 1 Mb. This method performed better than conventional correlation schemes. This method consumed less resources, performs partial correlation output peak analysis (analyse relation among multiple cross relation peaks) which are generated from selected regions of faces. This method was implemented on nokia S60 CLDC emulator using Java ME programming.

Tao et al. (2006) in their method use face detection, registration and authentication process to provide security on mobile phone device. They used Viola Jones face detection method for face detection and registration and subspace metrics process was used for face authentication. This method

Figure 3. Screenshot of FAME + finding double in a network.

Figure 4. Interface of face recognition system on mobile phone.

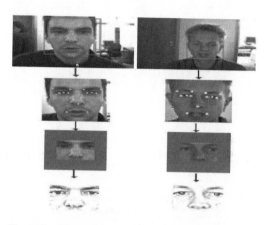

Figure 5. Shows feature extraction procedure.

proved efficient using simple rectangular binary features also eigen faces using single value decomposition was used. It provided secure, convenient and efficient connection in personal network.

Figure 5 Feature Extraction procedure (top to down) face detection, feature extraction, registration and masking.

Ijri et al. (2006) used face recognition in two different steps of verification and identification but used only verification step for mobile devices. They proposed their OKAO vision algorithm for providing security on mobile devices. This method had advantage that users don't have to adjust their face to predefined points. It contributes to usability and supports 4 major environments: symbian OS, embedded Linux, ITRON, BREW.

Al. Baker et al. (2005) proposed a system which uses General Packet Radio System (GPRS) used in mobile phones or PDA for providing security. In this method a mobile phone with GPRS activated acquires images from camera, user sends these images for automatic face recognition. Template matching was used for required recognition. This method can be applied to real time criminal identification, wireless home surveillance, imaging security services. An efficient way to provide GPRS based online security can be as shown:

Tresader et al. (2012) proposed integrated real time face and voice verification for better security of personal data stored or accessible from mobile. It provides a software that uses both voice and face captured from mobile to ensure your identiy. The integrated performance was better than individual recognition performances. Figure 7. Demonstrates

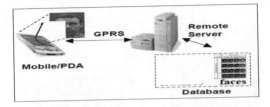

Figure 6.   GPRS based security management.

Figure 7.   Mobile biometrics interface demonstration.

Table 1.   Summary of proposed algorithms for mobile biometrics using face recognition.

Author	Database	Modes for recognition	Recognition algorithm
Jiawei Hu et al.	Face dataset captured from Google Nexus 5.	Unimodal	Haar—Based, LBP, Eigen faces, PCA, LDA.
Kuan–Yu Chou et al.	Sfeffield Face database, Cohn—Kanade database, ORL database.	Unimodal	Viola Jones, Eigen faces, Sparse representation of L2 norm minimisation.
Brian M. Powell et al.	LFW face database, photobucket.com	Unimodal	Genetic Algorithm Optmization technique used.
Silvio Barra et al.	Face dataset captured on mobile.	Unimodal	Viola Jones, Self Quotient Image Algorithm (SQI), Spatial Correlation.
Emir Kremic et al.	Small database developed consisting of face images.	Unimodal	Eigen faces, Principle Component analysis (PCA).
Bin Chen et al.	100 pictures of 10 people in different illumination and background, each with 10 pictures were used.	Unimodal	Adaboost face detection algorithm, Local Binary Patterns (LBP) for face recognition.
Kai Xi et al.	Yale Face Database.	Unimodal	Hierarchical Correlation based Face Verification (HCFV) Algorithm.
Qian Tao et al.	BioID Face database.	Unimodal	Viola Jones, Subspace metrics, eigen faces.
Yoshihisa Ijri et al.	Images of wide variation of genders and ages were used.	Unimodal	Fast rotation invariant Multiviewadaboost face detection algorithm, ScoreNormaliastion, OKAO vision.
O. Al. Baker et al.	Human face images captured using images at AT&T laboratory Cambridge.	Unimodal	Template Matching, Dual Mode recognition schemes.
Phil Tresader et al.	BANCA, XM2VTS dataset.	Bimodal	Active appearance mode (AAM), Mel Frequency cepstral coefficient (MFCC), Gaussian Mixture mode (GMM), fusion was done using boosted slice classifier.
Chris McCool et al.	MOBIO Database.	Bimodal	Probabilistic Linear Discriminant Analysis (PDLA), Local Binary Patterns.
Dong-Ju King et al.	Database collected via a smart-phone.	Multimodal	2D-DCT, Embedded Hidden Markov model (EHMM), Adaboostalgorithm, MFCC, GMM algorithm. Fusion: K-NN fisher, weighted summation.

Table 2. System performance of implemented methods.

Author	System performance
Jiawei Hu et al.	High recognition rates achieved. Training set—100% accuracy. Test set—96% to 93.8% accuracy.
Kuan–Yu Chou et al.	High recognition rates. Frontal view—90 to 95%. Varying illumination—70 to 72%. Different poses—95 to 99%. Different expression—99%.
Brian M. Powell et al.	Recognition studied under distortions such as height scaling and rotation. 88% human success rate was achieved with 0% attack rate.
Silvio Barra et al.	For overall accuracy of system values obtained were FAR = 0.055 & FRR = 0.177. Also doubles can be found on social networks using FAME.
Emir Kremic et al.	Identity authentication provided for access control and prevention of unauthorised mobile phone 88.88% of accuracy was achieved.
Bin Chen et al.	87% recognition rate was obtained using this method. It can handle only frontal and near frontal images.
Kai Xi et al.	This method consumed 1/10 memory, 1/6 storage space, 3.5% processing time. An equal error rate of 1.2% was reported.
Qian Tao et al.	Reported equal error rate for face authentication test was abot 1.2% and almost 0 for validation test.
Yoshihisa Ijri et al.	On basis of threshold set, if threshold is increased proposed method accepts more number of users. False rejection rate decreases and false acceptance rate increases.
O. Al. Baker et al.	System was sensible to head rotation greater than 20 degrees along x and y axis.
Phil Tresader et al.	97% of true faces were recognised.
Chris McCool et al.	Face and speaker recognition can be performed in a mobile environment and using score fusion can improve the performance by more than 25% in terms of error rates.
Dong- Ju King et al.	The results obtained showed that fusion error rates outperforms single modality error rate.

face detection, face feature localization, user interface with automated login and logout for social networking.

McCool et al. (2012) proposed a real time face and speaker recognition on mobile phone. The score fusion technique was used to combine individual face and speaker recognition results which improved performance by more than 25% in terms of error rate. The method was implemented real time on NokiaN900.

Kim et al. (2010) proposed an enhanced multimodal personal authentication system for mobile device [11]. The information's obtained from teeth, face and voice were fused together to give better recognition accuracy. The overall performance of the system was improved. The table summarises the standard database for mobile recognition, algorithms used for face identification, verification and fusion of biometric methods and also highlights the result obtained using each methodology.

4 PERFORMANCE ANALYSIS

The geometry changes, imperfect imaging conditions, noise and environmental changes, processor used in mobile can affect the performance of the system. System errors can be measured in terms of: False Acceptance Rate (FAR) and False Rejection Rate (FRR). Based on these factors the table shows performance of developed systems studied so far.

5 CONCLUSION

Biometric based face recognition is gaining importance in our daily business applications. Data on mobile is secured using mobile biometrics by extracting facial characteristics. Various face recognition algorithms for unimodal, bimodal and multimodal have been studied in this paper. A comparative study of these algorithms reveals that multimodal approach outperforms other approaches. So multimodal approach is the future research area wherein there is scope for mobile biometrics.

REFERENCES

Al-Baker, O., Benlamri, R., & Al-Qayedi, A. 2005, March. A GPRS-based remote human face identification system for handheld devices. In *Wireless* and *Optical Communications Networks, Second IFIP International Conference on* (pp. 367–371). IEEE.

Barra, S., De Marsico, M., Galdi, C., Riccio, D., & Wechsler, H. 2013, September. Fame: face authentication for mobile encounter. In *Biometric Measurements and Systems for Security and Medical Applications (BIOMS), IEEE Workshop on* (pp. 1–7). IEEE.

Chen, B., Shen, J., & Sun, H. 2012, May. A fast face recognition system on mobile phone. In *Systems and Informatics (ICSAI), 2012 International Conference on* (pp. 1783–1786). IEEE.

Chou, K. Y., Huang, G. M., Tseng, H. C., & Chen, Y. P. 2014, November. Face recognition based on sparse representation applied to mobile device. In *Automatic Control Conference (CACS), 2014 CACS International* (pp. 81–86). IEEE.

De Marsico, M., Nappi, M., & Riccio, D. 2010, September. Face: face analysis for Commercial Entities. In *ICIP* (pp. 1597–1600).

Delac, K., & Grgic, M. 2004, June. A survey of biometric recognition methods. In *Electronics in Marine, 2004. Proceedings Elmar 46th International Symposium* (pp. 184–193). IEEE.

Gofman, M. I., & Mitra, S. 2016, Multimodal biometrics for enhanced mobile device security. *Communications of the ACM, 59*(4), 58–65.

Hu, J., Peng, L., & Zheng, L. 2015, August. XFace: A Face Recognition System for Android Mobile Phones. In *Cyber-Physical Systems, Networks, and Applications (CPSNA), IEEE 3rd International Conference on* (pp. 13–18). IEEE.

Ijiri, Y., Sakuragi, M., & Lao, S. 2006, May. Security management for mobile devices by face recognition. In *Mobile Data Management,. 7th International Conference on* (pp. 49–49). IEEE.

Jafri, R., & Arabnia, H. R. 2009, A Survey of Face Recognition Techniques. *JIPS, 5*(2), 41–68.

Kim, D. J., Chung, K. W., & Hong, K. S. 2010. Person authentication using face, teeth and voice modalities for mobile device security. *Consumer Electronics, IEEE Transactions on, 56*(4), 2678–2685.

Kremic, E., Subasi, A., & Hajdarevic, K. 2012, June. Face recognition implementation for client server mobile application using PCA. In *Information Technology Interfaces (ITI), Proceedings of the ITI 2012 34th International Conference on* (pp. 435–440). IEEE.

McCool, C., Marcel, S., Hadid, A., Pietikainen, M., Matejka, P., Cernocky, J., & Matrouf, D. 2012, July. Bi-modal person recognition on a mobile phone: using mobile phone data. In *Multimedia and Expo Workshops (ICMEW), 2012 IEEE International Conference on* (pp. 635–640). IEEE.

Powell, B. M., Goswami, G., Vatsa, M., Singh, R., & Noore, A. 2014. fgCAPTCHA: Genetically Optimized Face Image CAPTCHA 5. *Access, IEEE, 2*, 473–484.

Tao, Q., & Veldhuis, R. N. 2006, July. Biometric authentication for a mobile personal device. In *Mobile and Ubiquitous Systems-Workshops, 3rd Annual International Conference on* (pp. 1–3). IEEE.

Tresadern, P. A., McCool, C., Poh, N., Matejka, P., Hadid, A., Levy, C., & Marcel, S. 2012, Mobile biometrics (mobio): Joint face and voice verification for a mobile platform. *IEEE pervasive computing, 99*.

Xi, K., Tang, Y., Hu, J., & Han, F. 2010, June. A correlation based face verification scheme designed for mobile device access control: From algorithm to Java ME implementation. In *Industrial Electronics and Applications (ICIEA), the 5th IEEE Conference on* (pp. 317–322). IEEE.

Communication and computer networks

Communication and Computing Systems – Prasad et al. (Eds)
© 2017 Taylor & Francis Group, London, ISBN 978-1-138-02952-1

Coverage of underwater sensor networks in shadowing environment

Anvesha Katti & D.K. Lobiyal
Jawaharlal Nehru University, New Delhi, India

ABSTRACT: Coverage being one of the indicators of Quality of service refers to the extent to which a given area is tracked by sensors. Objective of the paper is to determine the coverage of underwater sensor networks in shadowing environment. We come up with a channel model which considers the impact of shadowing on the coverage of underwater sensor networks.

1 INTRODUCTION

A Wireless Sensor Network (WSN) is a collection of sensor nodes that monitor physical or environmental conditions, such as temperature, pressure, etc. Nodes in this network cooperatively pass their data through multiple hops to a main location (Akyildiz, Su, Sankarasubramaniam, & Cayirci 2002) (Hossain, Biswas, & Chakrabarti 2008). An Underwater Sensor Network (UWSN) consists of a number of underwater sensor nodes and Autonomous Underwater Vehicles (AUVs) with sensors that perform monitoring and exploration over a given area in the ocean (Ghosh & Das 2008).

Coverage is a confronting issue in designing a WSN and refers to how well sensors monitor the events of interest (Mao, Anderson, & Fidan 2007), (Zhu, Zheng, Shu, & Han 2012), (Wang, Xing, Zhang, Lu, Pless, & Gill 2003), and (Xing, Wang, Zhang, Lu, Pless, & Gill 2005).

There are many kinds of signals that can be used for underwater communication such as acoustic signal, seismic signal, radio waves, and optical waves (Tsai 2008). These signals are affected by various factors like noise, interference, reflection of signals, obstructions in the propagation path, and movement of other objects. The power loss due to the above factors leads to large deviations in the received signal strength. The deviation in the received signal strength due to obstructions in propagation path is known as shadowing.

Efficient communication underwater among UWSN is one of the critical issues. Underwater communication systems can use sound, electromagnetic (EM), or optical waves to transmit information. Acoustic communication is the most versatile and widely used technique in underwater environments due to low attenuation (signal reduction) of sound wave in the water.

Almost all of the known research works have considered coverage problem for terrestrial sensor networks. To the best of our knowledge, no study has focused on the impact of shadowing on the network coverage in underwater sensor networks.

We derive coverage probability in the presence of shadowing using mathematical model based on received signal strength. The rest of the paper is organised as follows: Section 2 presents the research work related to coverage problem already reported in the literature. In Section 3, the details of proposed sensing channel model are presented. In Section 4 analytical and simulation results for the proposed model are explained. The conclusion of the work is given in Section 5.

2 RELATED WORK

The research work in the area of sensing coverage for underwater sensor networks under shadowing fading has received almost no attention till now. However, significant work has been done in the area of terrestrial sensor networks. In (Lanbo, Shengli, & Jun-Hong 2008), the authors derive sensing coverage under shadowing-fading environment with asymmetric sensing ability of sensors in a wireless sensor network. Some work relating to electromagnetic waves through soil and water is presented in (Daniels 1996) where D. Daniels introduces the empirical attenuation and relative permittivity values for soil at 100 MHz frequency range. The Propagation of electromagnetic waves l of frequency range from 1 to 2 GHz through soil is also studied in (Weldon & Rathore 1999). In (Tsai 2008), the author describes sensing coverage under shadowing-fading environment with asymmetric sensing ability of sensors in a randomly deployed wireless sensor network. The coverage probability has been determined using lognormal shadowing fading for variation in the received signal power.

3 CHANNEL MODEL

We assume that each sensor has a constant sensing range and the sensing region of a sensor is a sphere of volume $v = 4/3\pi r^3$ and the sensor nodes are randomly deployed and static. A sensor can sense and detect the event within its sensing range. A target is said to be covered if it is within the sensing range of a sensor. Figure 1 describes the sensing area. The probability of target detection by arbitrary sensor is defined as the ratio of sensing volume to network volume. Therefore, it can be expressed as $d = v/V$, where V is network volume in which sensor nodes are deployed uniformly. The probability (P_c) of target detection (Coverage Probability) by at least one of the sensors can be expressed as

$$P_c = 1 - (1 - P_d)^N \qquad (1)$$

As shown in the below figure, a sensor is positioned at distance r from the target placed at the origin of the circle. Rmax is the maximum sensing range of a sensor. R is an average sensing radius of the sensor.

As we know from the equality approximation $[1 - x]^n \approx e^{-nx}$ for large n, the above equation can be rewritten as

$$P_c = 1 - \exp\left(-\frac{N\pi r^3}{3V}\right) \qquad (2)$$

We can get the profile of the received signal from the transmitted signal by a channel model as it captures the characteristics of the medium. We assume uniformly deployed sensors in a sphere with sensing field of volume V having non-uniform sensing range. Sensors are homogenous having the same sensing threshold power λ (in dB). The sensing threshold is defined as that least strength of

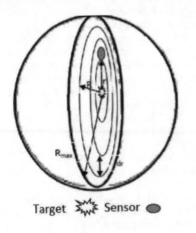

Target ⚡ Sensor ●

Figure 1.   Sensing area.

the signal that can be justly decoded at the receiver. The sensing field or range of a sensor is determined by the transmit power of a sensing signal, sensing threshold power, and power attenuation. The sensing signal power (transmitted power) generated by an event is assumed to be P_t (in dB). In this work, we have adopted log-normal shadowing fading model for the proposed sensing channel model. We present the derivation for Probability detection and coverage for acoustic signals. The received signal power $P_r(r)$ (in dB) at a sensor according to log-normal shadowing can be expressed as in (Rappaport et al. 1996) (Joshy & Babu 2010).

$$P_r(r) = P_t - 10slog_{10}\left(\frac{r}{d_0}\right) - 10rlog_{10}a(f) + y_\sigma \qquad (3)$$

where s is a path loss exponent that represents the rate at which path loss increases with the distance and $a(f)$ denotes absorption losses. d_0 is the reference distance and r is the distance between a sensor and its target. y_σ is a Gaussian random variable (in dB) with zero mean and variance σ^2 and represents lognormal shadowing effects. P_t represents transmitted power. The received signal power $P_r(r)$ (in dB) usually demonstrates a Gaussian distribution. The reference distance d_0 in wireless sensor network is generally assumed to be 1 meter, and therefore d_0 becomes a constant. The parameters s and σ depend on environmental factors. The path loss exponent (spreading factor) s is very important parameter for coverage and assumed to be constant. The spreading factor s typically takes values of 1, 1.5 and 2 depending on whether it is spherical, cylindrical or practical spreading. Threshold is assumed to be power threshold and not SNR threshold as noise is not being taken into consideration.

The probability that a sensor S detects an event occurring at r can be expressed as

$$P_d(r) = P(P_r(t) > \delta) \qquad (4)$$

where δ denotes the power threshold value.

$$P_d(r) = P(y_o > \delta - P_t + 10s\ log_{10}r + 10rlog_{10}a(f)) \qquad (5)$$

where P denote the probability function. y_σ is a Gaussian random variable with zero mean and variance σ^2, and represents shadowing effect in the propagation path. Therefore, the detection probability $P_d(r)$ can be expressed as

$$P_d(r) = \int_{10slog_{10}r + 10(r)log_{10}a(f)}^{\infty} \frac{1}{\sqrt{2\pi\sigma^2}} e^{[-(x^2/2\sigma^2)]}dx$$

$$= Q\left(\frac{10slog_{10}r + (r)10log_{10}a(f)}{\sigma}\right) \qquad (6)$$

$$P_d(r) = Q\left(\frac{10slog_{10}r - (r)10log_{10}a(f)}{\sigma}\right) \qquad (7)$$

Assumption includes that a sensor is positioned at distance r from the target which is located at the origin of the sphere. R_{max} is the maximum sensing range of the sensor. The distance r is continuous. When dr approaches 0, the chance that the target is found by a random sensor placed in the stated volume $4\pi r^2 dr$ of network area A can be calculated as follows

$$P_d = \frac{1}{A}\int_{r=0}^{R_{max}} P_d(r) \times 4\pi r^2 \ dr \qquad (8)$$

$$P_d = \frac{1}{A}\int_{r=0}^{R_{max}} Q\left(\frac{10slog_{10}r - (r)10log_{10}a(f)}{\sigma}\right) \times 4\pi r^2 \ dr \qquad (9)$$

The coverage probability P_c can be expressed as

$$P_c = 1 - \exp$$
$$\times\left(-\frac{N}{A}\int_0^{R_{max}} 4\pi r^2 Q\left(\frac{10slog_{10}(r)-(r)10log_{10}a(f)}{\sigma}\right)dr\right) \qquad (10)$$

4 RESULTS AND ANALYSIS

In this section, we present the simulation and numerical results to show the impact of sensing channel model under shadowing environment on the network coverage. All the simulations are done using Matlab. In the simulation, the entire sensing field is assumed to be a sphere with volume $V = 50 \times 50 \times 50$ m^3. The maximum sensing radius R_{max} is assumed to be 10 m. The parameters along with their values as used in simulation are shown in Table 1.

Figure 2 shows coverage probability and number of sensor nodes for variable shadowing parameter. We can clearly interpret for decreasing shadowing the number of nodes required to achieve a coverage probability of 1 also decreases. As shadowing decreases, therefore the number of nodes to provide full coverage also decreases.

Figure 3 highlights the outcome of coverage probability for normalized sensing radius node deployment is obtained using derivation given in (10). It is observed that coverage probability is directly proportional to normalized sensing radius and thus increases with the increase in normalized sensing radius r/Rmax. The maximum sensing coverage ≈ 1 can be achieved when sensing radius maximum or $r/R_{max} = 1$.

Table 1. Parameters.

Parameter	Value
No of sensors	0 to 550
speed of light	3×10^8 m/sec
s (Path loss exponent/spreading factor)	2
a(f) absorption coefficient	3.4

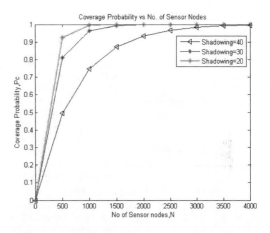

Figure 2. Coverage probability vs no. of senor nodes.

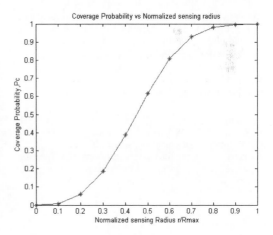

Figure 3. Coverage probability vs normalized sensing radius.

Figure 4 shows coverage probability and sensing radius for fixed shadowing. We interpret that in shadowing environment the coverage probability increases for increase in sensing radius and reaches a maximum of 1 for sensing radius of approx. 23 m.

195

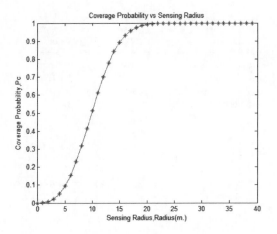

Figure 4. Coverage probability vs sensing radius.

5 CONCLUSIONS

We propose a channel model which considers the impact of fading on underwater sensor networks. We derive a mathematical model for calculating coverage probability. We observe that in shadowing environment the coverage probability increases for increase in sensing radius. We also observe that as shadowing becomes less, less number of sensor nodes are required for the full coverage. As the proposed sensing model provides good coverage for shadowing environment, it will be more useful to evaluate the performance of wireless sensor networks in realistic environment with interference.

REFERENCES

Akyildiz, I. F., W. Su, Y. Sankarasubramaniam, & E. Cayirci 2002. A survey on sensor networks. IEEE communications magazine 40 (8), 102–114.

Daniels, D. J. 1996. Surface-penetrating radar. Electronics & Communication Engineering Journal 8 (4), 165–182.

Ghosh, A. & Das S. K. 2008. Coverage and connectivity issues in wireless sensor networks: A survey. Pervasive and Mobile Computing 4 (3), 303–334.

Hossain, A., Biswas P., & Chakrabarti S. 2008. Sensing models and its impact on network coverage in wireless sensor network. In 2008 IEEE Region 10 and the Third international Conference on Industrial and Information Systems, pp. 1–5.

Joshy, S. & Babu A. 2010. Capacity of underwater wireless communication channel with different acoustic propagation loss models. International Journal of Computer Networks & Communications (IJCNC).

Lanbo, L., Shengli Z., & Jun-Hong C. 2008. Prospects and problems of wireless communication for underwater sensor networks. Wireless Communications and Mobile Computing 8 (8), 977–994.

Mao, G., Anderson B. D., & Fidan B. 2007. Path loss exponent estimation for wireless sensor network localization. Computer Networks 51 (10), 2467–2483.

Rappaport, T. S. et al. 1996. Wireless communications: principles and practice, Volume 2. Prentice Hall PTR New Jersey.

Tsai, Y.R. 2008. Sensing coverage for randomly distributed wireless sensor networks in shadowed environments. IEEE Transactions on Vehicular Technology 57 (1), 556–564.

Wang, X., Xing G., Zhang Y., Lu C., Pless R., & Gill C. 2003. Integrated coverage and connectivity configuration in wireless sensor networks. In Proceedings of the 1st international conference on Embedded networked sensor systems, pp. 28–39. ACM.

Weldon, T. P. & Rathore A. Y. 1999. Wave propagation model and simulations for landmine detection. Department of Electrical & Computer Engineering University of North Carolina-Charlotte.

Xing, G., Wang X., Zhang Y., Lu C., Pless R., & Gill C. 2005. Integrated coverage and connectivity configuration for energy conservation in sensor networks. ACM Transactions on Sensor Networks (TOSN) 1 (1), 36–72.

Zhu, C., Zheng C., Shu L., & Han G. 2012. A survey on coverage and connectivity issues in wireless sensor networks. Journal of Network and Computer Applications 35 (2), 619–632.

Communication and Computing Systems – Prasad et al. (Eds)
© 2017 Taylor & Francis Group, London, ISBN 978-1-138-02952-1

Technological advances in computing and it's coherence with business needs

Amit Mishra, Pradeep Tomar & Anurag Singh Baghel
Gautam Buddha University, Greater Noida, UP, India

ABSTRACT: Today's business scenarios are made up of global needs, complex designs, complex integrations and products, and very fast changing trends and customer requirements for different market segments. In this scenario when technology is changing so fast, it is inevitable for everyone to work and collaborate with each other. If different stakeholders are not communicating to each other the situation may land up into a situation where technology has advanced as thought and assumed by researchers and engineers but of not much useful for the business or end users. All systems, technological innovations must respond appropriately to the business needs continuously. Convergence of business and technology is must to benefit all stakeholders.

1 INTRODUCTION

Technological advances have a real meaning for the business world only if it is helping in economic value creation in some sense. New business models must ultimately return value of their technology investments. On one hand if software technology evolution is completely based on customer needs it may become counterproductive and response of technology to business would be reactive and is not an evolution in real sense. On the other hand if evolution in technology is independent of user needs it may become incoherent. Degree of incoherency may be of two natures. One is incoherent as per todays need but coherency level is improved with time and finally it matches with business evolution. Another is incoherent and becomes further incoherent with time and never meeting business evolution and expectations. The study presented here would focus on coherence of software development w.r.t. business needs and Return on Investment (ROI) correlation w.r.t. level of coherence factor.

2 REVIEW OF LITERATURE

Formally the software evolution was first proposed by Dr. Meir Lehman in 1980. Fundamentally Lehman stated that Software requirements change continuously as they are never static in nature. Lehman's laws describe "the software evolution as a force responsible for driving new and revising developments in a system" (Lehman, M. M. 1980). Lehman did the classification of all software systems into three types. First classification is static (S-Type) where solutions are well understood and

less likely to change, Second classification is practical systems (P-Type) where solution is not immediately clear. Third classification is E-type, which is embedded software systems (Lehman, M. M. 1980 & Lehman, M. M. et al. 1997). According to study done in (Rajesh Vasa et al. ICSM 2009) to explore the role of software matrices, it was derived that developers prefer to organize their solutions around a limited set of design options. Several studies have been done around innovation patterns and co-evolution with business strategies. Long back, Ettlie et al. discussed with an analogy with "jigsaw puzzle" for these two types of innovation models. More conventional model is the Incremental innovation model where the changes are limited to the individual jigsaw pieces that actually not transforming the total picture, while the radical innovation is innovation in real sense which is the act of reconfiguring the whole jigsaw and its pieces. Natural evolution of systems happens when the business Change Requests (CRs) are carried out to meet immediate demands. These CRs along with current technology develops islands of automation and a complex mosaic of bridges with lot of redundancies. Many other authors focus on organizational strategy. More fine-grained framework proposed by Henderson and Clark, 1990 is discussed here below, who distinguish four types of innovations. Incremental when the design is reinforced by improving only at the individual component level. Architectural is core design is untouched and linked modules are changed together. Modular innovation is when only the core design concepts of a technology are changed, not the linkages. Radical is a completely new design establishment with a new set of core design concepts. The large and complex classes do not get bigger and morecomplex (internally) purely due to the process of evolution, rather there are other contributing factors that determine which classes gain complexity and volume (Vasa & Rajesh, 2010). Evolution models have organizational specificities as well. Normally designers and architects face a situation in which they have to consider the desired functional requirements as well as project-specific and organizational principles governing local development practices and styles (Vasa et al. ICSM 2009 & Yasir et al. 2016). Focuses if evolution has to base on open source software.

3 PROPOSED WORK

The study presented in this paper would focus on coherence of software evolution models w.r.t. business needs and ROI correlation w.r.t. level of coherence factor. This paper would illustrate the fact on case study of software products evolution used in hi-tech industry. Based on this study the software technology and product evolution is divided in three main categories:

3.1 Reactive Evolution Model (REM)

This is the typical software product/technology development model which simply starts answering what is needed today. So this model has perfect coherence on day 1, but along with evolution of business it becomes less and less useful and finally the level of coherence with business needs goes down and after certain time it finally goes for end of life. In most of the cases ROI is achieved in this model but due to early end of life this model is not able to pay back much. However ROI is fastest in this model and this is the reason why most of the companies and management teams get attracted to this model.

3.2 Proactive Evolution Model (PEM)

This is a software product/technology development model which is developed with most visionary thoughts. It does best speculations in business evolution over a future horizon and technology evolution follows the similar path and trend. In this model when it starts it's life looks less coherent with business needs but it pulls/guides the business evolution in it's own direction and with time the level of coherence increases. It requires more money and resources to develop and evolve and hence ROI is reached in longer horizon. However it does not reach to end of life so fast. In long term it pays off the organization.

3.3 Corrective Evolution Model (CEM)

This is a mixed model. It starts with a Proactive Evolution Model (PEM) but during course of

time it is realized that business needs are divergent to the predictions of evolution. A correction feedback is supplied but still the generic implementation remains the key. Normally it results due to mistakes in understanding the future business evolution. In this case the technology development is started with certain assumptions on business evolution but they don't follow later. Finally business evolution goes in different direction. Hence it starts with incoherence and the level of incoherence increases with time. It need to applied a correction soon else it goes below the threshold of 'end of life' line. ROI is normally achieved very late as it requires duplicate effort for correction.

This study is based on the application evolution and roadmap on actual projects executed in world class hi-tech electronics company. Consider the fact that the applications under study are not commercial applications being sold to customers but are used for internal world-wide operations. Following are the typical custom definitions of certain terms used in the organization. Remedy data of past 5 years is used to conclude this study.

3.4 Degree of coherence

Degree of coherence is indicator for application behavior vs. business needs. It is normalized to highest value 1. Degree of coherence one indicates that application under consideration is in full coherence with the then business needs. It is computed on monthly basis based on the remedy data. It's basically function of Change Requests (CRs) by business, incident tickets assigned by business to ICT support and number of question tickets assigned by business to business champion and ICT support.

$$DoC \ (degree \ of \ coherence) = Function$$
$$(CRs, \ Incident \ ticket \ count, \ Question \ ticket \ count)$$

Actual computation considers the increase of above three parameters over past one month data and normalized to one.

3.5 Point of ROI

With respect to the roadmap on set of applications, certain applications are chosen for evolutionary development or new application development with allocation of budget for each application. Beyond this development budget, there is a certain maintenance budget for each application for next few years (which is x% of development budget). In the half yearly exercise product owner (business owner) confirms if ROI has met or not based on level of increased automation and usage of application as foreseen at

Table 1. Degree of coherence data for 60 months for applications under REM, PEM and CEM.

Degree of coherence (Software evolution vs. Biz evol.)

Month	REM	PEM	CEM	Grand total
1	1	0.5	0.65	2.15
2	1	0.5	0.65	2.15
3	1	0.5	0.65	2.15
4	1	0.5	0.65	2.15
5	1	0.5	0.65	2.15
6	1	0.5	0.65	2.15
7	0.9	0.5	0.65	2.05
8	0.9	0.5	0.65	2.05
9	0.9	0.55	0.65	2.1
10	0.9	0.55	0.65	2.1
11	0.8	0.6	0.65	2.05
12	0.8	0.6	0.65	2.05
13	0.8	0.6	0.65	2.05
14	0.8	0.6	0.6	2
15	0.8	0.6	0.6	2
16	0.8	0.6	0.5	1.9
17	0.8	0.65	0.5	1.95
18	0.7	0.65	0.5	1.85
19	0.7	0.65	0.5	1.85
20	0.7	0.65	0.4	1.75
21	0.7	0.65	0.4	1.75
22	0.7	0.65	0.4	1.75
23	0.7	0.7	0.4	1.8
24	0.6	0.7	0.4	1.7
25	0.6	0.7	0.4	1.7
26	0.6	0.7	0.4	1.7
27	0.6	0.7	0.3	1.6
28	0.6	0.75	0.3	1.65
29	0.6	0.75	0.3	1.65
30	0.6	0.75	0.3	1.65
31	0.6	0.75	0.3	1.65
32	0.6	0.75	0.3	1.65
33	0.5	0.8	0.3	1.6
34	0.5	0.8	0.3	1.6
35	0.5	0.8	0.3	1.6
36	0.5	0.8	0.3	1.6
37	0.4	0.85	0.3	1.55
38	0.4	0.85	0.3	1.55
39	0.4	0.85	0.3	1.55
40	0.4	0.85	0.3	1.55
41	0.4	0.85	0.3	1.55
42	0.4	0.85	0.3	1.55
43	0.4	0.85	0.3	1.55
44	0.4	0.85	0.3	1.55
45	0.4	0.85	0.3	1.55
46	0.4	0.9	0.3	1.6
47	0.4	0.9	0.3	1.6
48	0.4	0.9	0.2	1.5
49	0.3	0.9	0.2	1.4
50	0.3	0.9	0.2	1.4
51	0.3	0.9	0.2	1.4

(Continued)

Table 1. (Continued)

Degree of coherence (Software evolution vs. Biz evol.)

Month	REM	PEM	CEM	Grand total
52	0.3	0.9	0.2	1.4
53	0.3	0.9	0.2	1.4
54	0.2	0.9	0.1	1.2
55	0.2	0.9	0.1	1.2
56	0.2	0.9	0.1	1.2
57	0.2	0.9	0.1	1.2
58	0.2	0.9	0.1	1.2
59	0.2	0.9	0.1	1.2
60	0.2	0.9	0.1	1.2

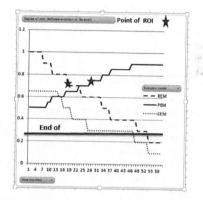

Figure 1. REM, PEM and CEM s/w technology evolution graph vs. ROI and 'End of Life.

the time of release definition. At the same time ICT confirms if the maintenance cost is under the budget for developed application release. If above both parameters confirmed by business and ICT, this is called a point of ROI.

4 RESULTS AND VALIDATION

Based on the analysis presented in this study, it is very evident that level of coherence in software technology is crucial and plays a big role in achieving the ROI with greater sustenance. Proactive and visionary evolution model is found to be the best. This study tells us the fact that even the highest degree of technological advances won't help if they are not in line with what is the benefit for the purpose it is to be used for? REM is best when it is known that full technology is to be replaced after certain period of time and is given preference even over PEM in such scenarios. As cost of following the path of REM evolution model is lowest but less sustainable. Phasing out applications must target to reduce cost and stay sustained.

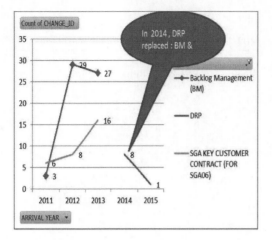

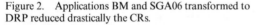

Figure 2. Applications BM and SGA06 transformed to DRP reduced drastically the CRs.

Case studies have been performed under this work related to all three models: REM, PEM and CEM. It is proven by the study that products falling under CEM were requiring often the correction else they are scrapped very fast. REM is the most usual model and every team by default goes into this safe trap. However few products categorized as CEM were STAR products even if many of them were recognized much later than they were deployed and adopted. The products developed during the recession periods were mostly fallen in the category of REM due to pressures on ROI. Two applications Backlog Management (BM) and Key Customer Contract (SGA06) were resulting into several change requests (CRs) and development activities every year for the evolution in more reactive way. Typically in 2012 these two applications had 37 change requests and in year 2013 they met with 43 CRs. But after transforming the solution with PEM it met only 8 CRs in 2014 while in 2015 just met 1. This shows that if proactive evolutionary model is developed with all considerations and parameters with well understanding of business needs and trends it results to better model which suits the need.

However the work done here is for internal software used in more predictable environment and predictable business scenarios. Going in PEM model in general software development models is much more difficult and study to be reviewed for those complex scenarios and results to be proven.

5 CONCLUSION

Based on the study done it's clear that the requirement management is key to any technology

evolution model. Anything developed which is incoherent to the current business needs and also not converging in future with business evolution it is bound to DIE. It reaches end of life without giving it's ROI. For making the investments viable for technology evolution partnering with non-technical or business stakeholders is must. Technology roadmap definition team must constitute practitioners, technical experts, architects and visionary people. Even if future coherence is not perfectly predictable yet it can be improved if degree of coherence is thought when building the roadmap. In summary it can be concluded –

- Nothing can move in isolation, coherence is required in full ecosystem
- Incremental innovations does not result incoherency
- Time to respond and showing the sensitivity is less and less in current situations
- Thought process of innovation must have the parameters of full ecosystem.

REFERENCES

Ettlie, J.E., Bridges W.P., R.D. O'Keefe, Organization strategy and structural differences for radical vs. incremental innovation, Management Science 30 (6) (1984).

Henderson R.M., Clark K.B., Architectural innovation: the reconfiguration of existing product technologies and the failure of established firms, Administrative Science Quarterly 35, 1990, pp. 9–30.

Lehman, M.M., Programs, Life Cycles, and Laws of Software Evolution. Proceedings of IEEE, Special Issue on S/W Evolution, 68(9):1060–1076, Sept. 1980.

Lehman, M.M., D.E. Perry, J.C.F. Ramil, W.M. Turski & P. Wernik. Metrics and Laws of Software Evolution—The Nineties View. In Proceedings of the Fourth International Symposium on Software Metrics (Metrics '97), pages 20–32, Albuquerque, New Mexico, Nov. 1997.

Rajesh Vasa, Markus Lumpe, Philip Branch, Comparative Analysis of Evolving Software Systems Using the Gini Coefficient, 25th International Conference on Software Maintenance (ICSM 2009), pp. 179–188, IEEE Computer Society.

Vasa, Rajesh., Growth and change dynamics in open source software systems., Faculty of Information and Communication Technologies (2010): 254.

Vasa, R., Lumpe, M., Branch, P. & Nierstrasz, O. "Comparative analysis of evolving software systems using the Gini coefficient. In Software Maintenance", ICSM 2009. IEEE International Conference on (pp. 179–188).

YasirJaved, MamdouhAlenezi, Defectiveness Evolution in Open Source Software Systems, Symposium on Data Mining Applications, SDMA2016, 30 March 2016, Riyadh, Saudi Arabia, Procedia Computer Science 82 (2016) 107–114.

Communication and Computing Systems – Prasad et al. (Eds)
© 2017 Taylor & Francis Group, London, ISBN 978-1-138-02952-1

Auction mechanism using SJF scheduling for SLA based resource provisioning in a multi-cloud environment

Naela Rizvi, Prashant Pranav & Bibhav Raj
Birla Institute of Technology, Mesra, Ranchi, Jharkhand, India

Sanchita Paul
Department of Computer Science and Engineering, Birla Institute of Technology, Mesra, Ranchi, Jharkhand, India

ABSTRACT: Cloud computing provides users a place required to store their large and valuable data. Cloud computing is a utility aware service which is based on pay as you use methodology. The stored user's data and information must be kept secured and the customers must have privilege to retrieve and process their data as and when required from anywhere. So, besides security, cloud computing has many features. Flexibility, works from anywhere, disaster recovery are some to name from others. Resource provisioning is very important in context of clouds. Allocating as well as managing of resources is the responsibility of cloud service providers. While allocating resources to a already waiting job, the service provider must keep into consideration the predefined SLA promised with customers. SLA, a Service Level Agreement which is signed initially between the customer and service provider to ensure QoS to customers. So, completing a request within the defined SLA is a big challenge in cloud computing. This paper aims at reducing SLA violation while also decreasing the waiting time of jobs. This is fulfilled through an auction based policy with SJF.

1 INTRODUCTION

Cloud computing is a new computing paradigm which gained momentum in late of 2007. Cloud computing refers to delivery of resources on internet. Because of utility based nature, cloud offers computational services whenever users need it. Thus transferring to more commoditized utilities like water, gas, electricity etc. Users pay for whatever they use there is no need to purchase whole infrastructure. Computing resources are distributed worldwide so that companies are able to access their applications and services anytime from anywhere. Due to reduced computational cost, high degree of reliability and flexibility cloud has become one of the exciting technologies.

In cloud computing resource provisioning is the process of assigning resources to the needed applications over the internet. Resource Provisioning means the selection, deployment and runtime management of software and hardware resources. Resource provisioning should be done in an efficient manner such that it avoid Service Level Agreement (SLA) violation by meeting Quality of Service (QoS) parameters such as availability, response time, throughput, security, etc.

In grid environment, resources are accessed based on a queuing model which provides best-effort QoS. Jobs are put in the queue till the resources the jobs require are freed. As jobs have to wait for resources to become available, this approach results in long delay. These delays vary according to the number of jobs any application have. To improve QoS for workflow application (more number of jobs) a model for resource allocation based on provisioning can be used which allows a single user to gain total control of the resources for a given period. This minimizes queuing delay. Provisioning is more complex than queuing in the way it requires more sophisticated resource allocation decisions.

2 RELATED WORK

There are many resource provisioning techniques detailed so far. These were analyzed on various aspects. (Buyya et al. 2011, Garg et al. 2012, Dhingra et al. 2013, Vecchiola et al. 2011, Munteanu et al. 2013, Chang et al. 2013) are more SLA centric. The techniques of (Vecchiola et al. 2012, Dhingra et al. 2013 and Munteanu et al. 2013) focus on reducing the application and execution time while those discussed in (Dhingra et al. 2014 and Dhingra et al. 2014) reduce energy consumption. (Li et al. 2012 and Chang et al. 2013) explains better ways for optimizing QoS while (Buyya et al. 2011, Xu et al. 2007 and Dhingra et al. 2014) explains cost optimization techniques. Profit is the main concern of techniques in (Xu et al. 2007, You et al. 2009 & Chang et al. 2013). Resource utilization in a better way is discussed in (Xu et al. 2007 and Chang et al. 2013).

3 PROPOSED WORK

3.1 System architecture

Resource provisioning is done here through an auction mechanism. This allows jobs to not wait for longer in the for resources. If the resources are not available with the service provider, then through the use of auction policy, a service provider can lease resources from other cloud vendors and in turn pay rent to the vendors. The system architecture of the auction model consists of many cloud vendors. Among the available vendors some are taught to be bidders. So, auction model comes into play when a request arrives. Different components like Billing system, Monitor, Clients SLA Decision, Local VM Scheduling and Coordinator. Coordinator performs the main role as it provide communications between various cloud, status of resources at particular time and migration of jobs to other cloud vendors.

When a request arrives to a particular cloud, the cloud checks if the available resources are enough to fulfill the request or not. Job is scheduled using SJF policy if resources are enough otherwise an auction policy is called upon. Bidder checks for the amount of resources required by a job and accordingly calls the desired vendors to participate in auction. The vendor which gives it resources at minimum possible price wins the bid and henceforth it leases its resources to the job and in turn the service provider will have to pay a rent to the vendor.

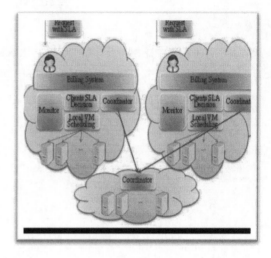

Figure 1. System architecture of auction mechanism.

3.2 Flow chart of proposed model

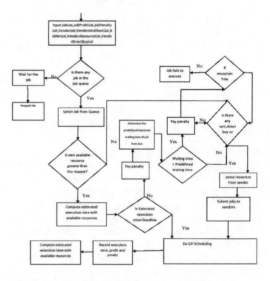

Figure 2. Flow chart of the proposed model.

3.3 Proposed Job Allocation Algorithm with SLA (JAS Algorithm)

Input:CloudletList,CloudletProfitList,CloudPenalty List,VendorsList,VendorsInitialPricingList, BidderList,VendorsResourcesList,VendorsDirectBuy

202

List

Return:Profit, Penalty, Execution time, Rent, Directbuyprice, Vendors

1. CloudletList   {j1, j2, ..., jN};

2. CloudletProfitList   {p1, p2, ..., pN};

3. CloudletPenaltyList   {l1, l2, ..., lN};

4. Vendors List   {V1, V2, ..., VN};

5. VendorsIPList   {VIR1, VIR2, ..., VIRN};

6. VendorsResourceList   {VR1, VR2, ..., VRN};

7. Bidder List   {B1, B2, ..., BN};

8. VendorsDirectBuyList   {VDB1, VDB2, ..., VDBN};

9. **for Cloudlets from j1 to jN in Cloudlet List do {**

10. **if (own available resources >=   request) then**

11. **Calculate estimated execution time**

12. **If (estimated execution time =< deadline)**

13. SJF(Scheduling);

14. record execution time;

15. profit = CloudletProfit

16.   else

17. **Pay Penalty**

18. **Determine the predefined maximum waiting time of job from SLA.**

19. **while (waiting time > predefined waiting time)**

20. **do {**

21. **Pay Penalty**

22. **}**

23. **else**

24. call x = {rent, direct buy price, vendors}

25.   if (x == null) then

26.   penalty = CloudletPenalty;

27. end if

28. end if }

29. return profit, penalty, execution time, rent, direct buy price, vendors

3.4 *Performance analysis using a case study*

We tried to simulate our work in CloudSim software. For this we created seven clouds with two working as bidders (Cloud 2 and Cloud 3). Suppose that all the clouds posses specific characteristic as CPU, memory and storage. The seven clouds are listed below:

There are 10 jobs to be executed on both the bidders. Besides deadline, profit, penalty and others among the variables, the jobs have specific characteristics and requirement. All the cloud vendors have their initial rents and resource utilization and they host one or more virtual machines on them. Now when the jobs come, the bidders execute them on their specific VMs. If the bidders fall sort of resources, they go for auction among the available vendors. First the bidders check for the availability of resources among them, this condition satisfies the SLA with respect to availability. Second if resources are not available with the bidders, they

Table 1. Seven clouds with their specific characteristics.

Clouds	CPU	Memory	Storage
Cloud 1	1	4 GB	150 G
Cloud 2	1	2 GB	100 G
Cloud 3	2	2 GB	120 G
Cloud 4	1	2 GB	110 G
Cloud 5	1	2 GB	90 G
Cloud 6	2	4 GB	100 G
Cloud 7	2	4 GB	200 G

lease resources from vendors and as such takes into consideration the waiting time of a job and thus satisfying the reduced waiting time SLA constraint. Deadline SLA parameter is also met as by leasing resources, the jobs are taught to be executed before their respective deadlines. The requirements of 10 jobs are shown below:

The proposed model considers three SLA parameters namely deadline, waiting time and availability and so is better than the existing auction model. Further we are trying to implement it using SJF scheduling. This ensures that shorter jobs will never face starvation.

To elaborate it further, let us define the constraints associated with each job. We define the deadline, total execution time, predefined waiting time, actual waiting time and penalty. Deadline is the time in which a job is taught to be executed successfully. Total execution time is the actual time of execution of a job including the time for which the job remains idle. Predefined waiting time is the waiting time confronts with the customer in SLA and actual waiting time is the time for

which the job actually waits for resources to be available. Penalty is the cost which a cloud provider will have to pay if it does not complete the job in time. We have taken to pay Rs 10 as a penalty, if the job exceeds the deadline by 1 ms. Similarly, if the job exceeds the predefined waiting time by 1 ms, the provider will have to pay a penalty of Rs 10. All time are in Millisecond (ms) and. We have also done the calculation of FCFS scheduling and have compared it with our model.

3.5 Comparison graph

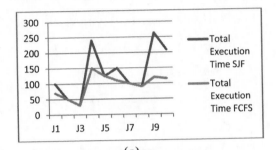

(a)

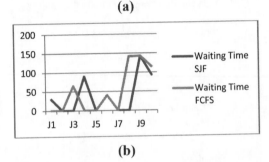

(b)

Figure 3. (Continued).

Table 2. Ten jobs with different requirements.

Job id	Processing unit required	Ram	Storage
Job 1	1	2 GB	70 G
Job 2	1	2 GB	50 G
Job 3	1	2 GB	30 G
Job 4	1	4 GB	150 G
Job 5	2	4 GB	125 G
Job 6	2	2 GB	110 G
Job 7	2	4 GB	100 G
Job 8	1	4 GB	90 G
Job 9	1	2 GB	120 G
Job 10	2	2 GB	115 G

Table 3. Comparison of SJF and FCFS scheduling.

Job id	Arrival time	Deadline	Execution time		Predefined waiting time	Actual waiting time		Penalty	
			SJF	FCFC		SJF	FCFC	SJF	FCFC
Job 1	0	95	100	70	40	30	0	50	0
Job 2	2	110	50	50	65	0	0	0	0
Job 3	5	70	30	95	30	0	65	0	600
Job 4	8	240	240	150	100	90	0	1000	100
Job 5	10	130	125	125	40	0	0	0	0
Job 6	12	200	150	150	130	40	40	0	0
Job 7	15	200	100	100	50	0	0	0	0
Job 8	16	200	90	232	80	0	142	0	940
Job 9	19	250	263	263	150	143	143	130	130
Job 10	22	210	208	208	100	93	115	0	150
						Total penalty		1180	1920

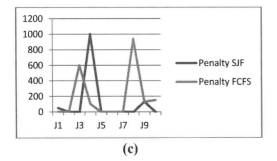

(c)

Figure 3. Graphs showing the difference between the two approaches. a) shows the relation between total execution time for both SJF and FCFS scheduling; b) depicts the relation between both in terms of waiting time while; c) shows that in terms of penalty & Figure 4 shows the overall relation between these two policies.

4 CONCLUSION AND FUTURE WORK

By comparing the results of the two scheduling approaches, we conclude that SJF and FCFS policies go hand by hand as far as total execution time is considered. But, SJF gives slightly better result in terms of waiting time, Further, it can also be concluded that SJF is far better than FCFS when penalty is concerned.

In the future we would like to implement the above mechanism using other scheduling policies such as priority scheduling, credit based scheduling.

REFERENCES

Buyya, R., Garg, S.K., Calheiros, R. N.: SLA-Oriented Resource Provisioning for Cloud Computing: Challenges, Architecture, and Solutions. International Conference on Cloud and Service Computing, 1–10 (2011).

Bhavani, B.H., Guruprasad, H. S.: Rresource Provisioning Techniques in Cloud Computing Environment: A Survey. International Journal of Research in Computer and communication Technology, Vol. 3, Issue 3, 395–401 (2014).

Chang, C., Lai, K., Yang, C.: Auction Based Resource Provisioning with SLA Consideration on Multi-Cloud Systems. IEEE 37th Annual Computer Software and Applications Conference Workshops. 445–450 (2013).

Dhingra, M., Lakshmi, J., Nandy, S., Bhattacharyya, C., Gopinath, K.: Elastic Resource Framework in IaaS, Preserving Performance SLA. IEEE 6th International Conference on Cloud Computing, 430–437 (2013).

Dhingra, A., Paul, S.: A Survey of Energy Efficient Data Centres in a Cloud Computing Environment. International Journal of Advanced Research in Computer and Communication Engineering, Vol. 2, Issue 10, 4033–4040 (2013).

Dhingra, A., Paul, S.: Green Cloud: Smart Resource Allocation and Optimization using Simulated Annealing Technique. Indian Journal of Computer Science and Engineering (IJCSE), Vol. 5 No. 2 (2014).

Dhingra, A., Paul, S.: Green Cloud: Heuristic based BFO Technique to Optimize Resource Allocation. Indian Journal of Science and Technology, Vol. 7(5): 685–691 (2014).

Garg, S.K., Srinivasa, K.G., Buyya, R.: SLA Based Resource Provisioning for Heterogeneous Workload in Virtualized Cloud Data Centres. Journal of Network and Computer Applications. (2012).

Li, C., La, Y.L.: Optimal Resource Provisioning for Cloud Computing. Springer Science + Business Media, LLC, 62: 989–1022 (2012).

Munteanu, V., Fortis, T., Negru, V.: Evolutionary approach for SLA based cloud resource provisioning. IEEE 27th International Conference on Advanced Information Networking and Applications, 506–513 (2013).

Pandi. D., Chattopadhya, S., Chattopadhya, M., Chaki, N.: Resource Allocation in Cloud Using Simulated Annealing. Applications and Innovations in Mobile Computing (AIMoC), 21–27 (2014).

Suri, P.K., Mittal, S.: A Comparative study of Various Computing Processing Environments: A Review. International Journal of Computer Science and Information Technologies, Vol. 3(5), 5215–5218 (2012).

Sharma, M., Bansal, H., Sharma, A.K.: Cloud Computing: Different Approach & Security Challenge. International Journal of Soft Computing and Engineering, Vol. 2, Issue 1, 421–424 (2012).

Vecchiola, C., Calheiros, R., Karunamoorthy, D., Buyya, R.: Deadline Driven Provisioning of Resources for Scientific Applications in Hybrid Cloud with Aneka. Future Generation Systems, 58–65 (2012).

Vecchiola, C., Calheiros, R., Karunamoorthy, D., Buyya, R.: The Aneka Platform and QoS Driven Resource Provisioning for Elastic Applications on Hybrid Cloud. Future Generation System 28 (2012) 861–870 (2011).

Xu, J., Zhao, M., Fortes, J., Carpenter, R., Yousif, M.: On The Use of Fuzzy Modelling in Virtualized Data Centre Management. Fourth International Conference on Autonomic Computing (ICAC'07). (2007).

You, X., Xu, X., Wan, J., Yu, D., RAS-M: Resource Allocation Strategy Based on Market Mechanism in Cloud Computing. Fourth China Grid International Conference, 256–263 (2009).

Communication and Computing Systems – Prasad et al. (Eds)

Energy efficient routing algorithm for wireless sensor networks: A distributed approach

Suneet Kumar Gupta
Department of Information Technology, ABES Engineering College, Ghaziabad, Uttar Pradesh, India

Pratyay Kuila
Department of Computer Science and Engineering, National Institute of Technology, Sikkim, Gangtok, India

Prasanta K. Jana
Department of Computer Science and Engineering, Indian School of Mines, Dhanbad, Jharkhand, India

ABSTRACT: In this article, we present an energy efficient routing algorithm for Wireless Sensor Networks (WSNs). Proposed algorithm finds the next hop in each round which is based on the residual energy. The proposed algorithm selects the next hop based on residual energy in each round. The parameters residual energy and distance make the algorithm energy efficient and other parameters i.e. number of predecessor and successor also improve the chances that each node equally act as the next hop. Experimentally, it is demonstrated that the performance of proposed algorithm is better than existing algorithms in terms of inactive relay nodes, data packet send to BS and consumption of energy in each round.

1 INTRODUCTION

In the last decade wireless sensor networks had gained massive attention due to varied applications in the area of health-care, military, environment monitoring, underground mining etc. (Akyildiz & Vuran 2010, Akyildiz, Su, Sankarasubramaniam, & Cayirci 2002, Yick, Mukherjee, & Ghosal 2008). WSN is a network of interconnected small sensor nodes which are equipped with limited computational power, sensing capabilities and limited power source. Therefore, energy consideration of nodes in one of the most important challenges. Generally sensors are randomly deployed in such region, where human intervention is almost negligible. Due to less human intervention, it is almost impossible to replace the battery of sensors. Therefore, energy of sensor node is one of the most important challenge. To reduce the energy consumption, clustering is one of the important method (Akyildiz & Vuran 2010, Anastasi, Conti, Di Francesco, & Passarella 2009). In clustering sensors are grouped and formed a cluster, in each cluster there is a Cluster Head (CH) and every sensor forwards the sensed data to respective CH. An example of clustered WSN is pictorially represented in Figure 1. Generally CHs are selected amongst the sensor nodes, but due to extra workload, e.g. aggregating the data, these CHs die quickly. So in past few years, many researchers have proposed the use of relay nodes as CHs and these relay nodes are provisioned

with extra energy as compared to the normal sensor nodes. The role of CHs is to collect the sensed data from the member sensor nodes, aggregate the data and forward it to BS directly or via multi hop communication (Akkaya & Younis 2005a, Prabhu & Sophia 2011, Jiang, Yuan, & Zhao 2009). However, energy of relay nodes is also one of the most important constraint, because relay nodes are also battery operated and failure of relay nodes due to complete depletion of energy may partition the network even if other relay nodes have sufficient energy for forwarding the data. So, with the help of energy efficient routing algorithm we

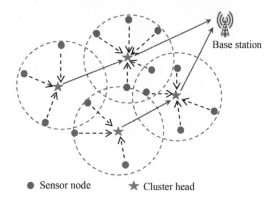

Figure 1. An example of clustered wireless sensor network.

207

may conserve the energy of relay nodes as in WSN energy consumption is directly proportional to square of the distance between sender and receiver. So, during routing selection of proper next hop is the most important.

In this research article, an energy efficient routing algorithm for wireless sensor networks has been proposed by us using a distributed approach. In proposed work, for selecting the next hop we mainly consider four parameters 1) residual energy of next hop 2) distance between sender and receiver 3) distance between next hop and BS 4) number of predecessor and successor nodes. First three parameters makes the algorithm energy efficient and last two parameters try to minimize the participation of that node which has higher probability to act as CH and having minimum number of possible next hop.

The rest of the paper is organized as follows. Related work is discussed in section 2. System model and used terminologies are explained in section 3. The detailed description of proposed algorithm is presented in section 4, followed by discussion about the experimental results in section 5 and paper is concluded in section 6.

2 RELATED WORK

Many researchers have proposed various routing algorithms for WSN that are based on centralized and distributed approach (Akkaya & Younis 2005b, Al-Karaki & Kamal 2004, Heinzelman, Chandrakasan, & Balakrishnan 2000, Zungeru, Ang, & Seng 2012, Chakraborty, Das, & Abbott 2012, Shukla, Chandel, Gupta, Jain, & Bhansali 2015, Gupta & Jana 2015, Gupta, Kuila, & Jana 2013). In centralized approach, all decisions about routing are taken care by the base station and base station is able to take the decision only when the geographical coordinates of all the nodes is well known in advance. However, this approach is not providing the scalability of the network. In distributed approach, the decisions for selection of next hop is taken by the nodes in each round and proposed work is based on distributed approach, so we only discuss the existing routing algorithms which are based on distributed approach. In (Heinzelman, Chandrakasan, & Balakrishnan 2000), authors have presented cluster based routing algorithm named as LEACH (Low-energy adaptive clustering hierarchy). LEACH dynamically assigns the responsibility of cluster heads amongst the sensor nodes. The drawback of this algorithm is that it assigns the role of CH to a sensor node which has very low energy, due to this sensor may die quickly. In Minimum Hope Routing Model (MHRM) (Chiang, Huang, & Chang 2007), authors presented an algorithm where individual

relay nodes find the path with minimum number of intermediate relay nodes. The disadvantage of this method is that entire relay nodes communicate with the farthest relay nodes, due to this long haul communication relay nodes die quickly. In DFCR (Azharuddin, Kuila, & Jana 2013), authors discussed about clustering and routing algorithm simultaneously using distributed approach. In routing every relay node selects the next hop using a cost function. This cost function is based on residual energy, distance between sender and possible next hop with distance between next hop to BS and also consider hop counts. However they did not consider the energy balancing or avoiding the bottleneck situation.

The authors of (Ok, Lee, Mitra, & Kumara 2009) have discussed a data forwarding algorithm known as DEBR (Distributed Energy Balanced Routing). This algorithm provides the balanced energy consumption in the network. However the disadvantage of this method is that a node selects that next hop which does not have any node with in communication range of it. Liu et al. have presented a energy aware routing algorithm known as ESCFR (Exponential and Sine Cost Function based Route) in (Liu, Ren, Li, Chen, & Shen 2012). The cost function of ESCFR is designed in such a manner that it reflects the large changes if there is a small changes in nodal remaining energy. The function is as follows:

$$EC_{ij} = e_{ij} * exp(1 \setminus sin(\pi - (\pi \setminus 2) * (E_i^r \setminus E_0)$$

However, the working of ESCFR very similar to DEBR (Ok, Lee, Mitra, & Kumara 2009), so in ESCFR same problem persists as in DEBR.

3 SYSTEM MODEL AND TERMINOLOGIES

It is assumed that all the sensor nodes are randomly deployed with some relay nodes in $2D$ area, after deployment it is assumed that the sensors and relay nodes are stationary. After deployment of these nodes clusters are formed after execution of algorithm as discussed in (Gupta, Kuila, & Jana 2014). The data gathering operation is divided in round as similar to LEACH (Heinzelman, Chandrakasan, & Balakrishnan 2000). One round is completed only when entire sensor nodes transmit the data to BS, in proposed work network lifetime is considered in terms of rounds when first relay node dies. In proposed work, we use same energy model as discussed in LEACH (Heinzelman, Chandrakasan, & Balakrishnan 2000).

The following terminologies is used in proposed work.

1. $R = \{r_1, r_2, \ldots r_n\}$ represents set of relay nodes and BS is represented by r_{n+1}.
2. The maximum communication range is denoted by CR.
3. $Dist(r_i, r_j)$ represents the distance between relay nodes r_i and r_j.
4. The residual energy of relay r_i is represented by $REn(r_i)$.
5. $Next(r_i)$ represents the set of all those relay node(s), which may act as next hop of r_i as well as within the communication range of r_i. It can be also defined as follows:

$$Next(r_i) = \begin{cases} BS & \text{If} \quad Dist(r_i, BS) \leq CR \\ r_j & \text{If} \quad Dist(r_i, r_j) \leq CR; \forall r_j \in R \end{cases}$$

6. $NextHop(r_i) = \{r_j\}$; represents that r_j is the next hop of r_i; $r_j \in Next(r_i)$ and $r_i, r_j \in R$ with hop count to reach BS from r_j is less than hop count to reach BS from r_i. Also BS may act as the next hop of any relay node.
7. $HCount(r_i)$: It represents the number of intermediate nodes required to forward the data the BS from r_i. If distance between r_i and BS is less than the communication range of r_i then $Hcount(r_i)$ is one, otherwise it is recursively defined as follows:

$$HCount(r_i) = \begin{cases} 1, & NextHop(r_i) = r_{m+1} \\ 1 + HCount(r_j), & NextHop(r_i) = r_j \end{cases} \quad (1)$$

8. $Pred(r_j)$: It denotes the set of relay nodes, which may select r_j as next hop. In other words, we can define $Pred(r_j)$ as follows:

$$Pred(r_j) = \{r_k \mid j \neq k \,\&\,\& Dist(r_j, r_k) \leq CR \,\&\, \& HCount(r_k) > HCount(r_j)\}$$

9. $Succ(r_j)$: It denotes the set of that relay nodes, which may act as next hop of r_j. In other words, we can define $Succ(r_j)$ as follows:

$$Succ(r_j) = \{r_k \mid j \neq k \,\&\,\& Dist(r_j, r_k) \leq CR \,\&\, \& HCount(r_k) < HCount(r_j)\}$$

4 PROPOSED WORK

4.1 Cost function

$cost(r_i, r_j)$: It represents the cost involved when r_i selects r_j as next hop for forwarding the data to sink. The cost function is based on residual energy, distance, number of predecessor and successor

node of possible next hop. By considering possible number of predecessor and successor node we try to minimize participation of such node which has maximum number of predecessor node and minimum number of successor node. Higher predecessor value of a node indicates the higher probability to act as a next hop. Higher successor value indicates that node has maximum option for next hop. Moreover it also helps during the failure of some node(s). The cost function is calculated with following equation (refer eq. 2).

$$cost(r_i, r_j) = \frac{REn(r_j)}{X} \quad (2)$$

where $X = Dist(r_i, r_j) \times Dist(r_j, BS) \times |Pred(r_j)| \times |Succ(r_j)|$

4.2 Boot strapping

The process of bootstrapping, start with flooding the hop packet by BS. Each hop packet contains a counter which represents the hop value, initially the value of this counter is one. When a node receives the packet then node increments the counter value by one and again broadcast the packet. The said procedure repeats until entire nodes do not aware about their hop values. This hop value plays a vital role when a node selects the next hop for sending the data to BS. However, a node may receive multiple copies of hop packet, in this situation a node discards the packet if counter value is greater than the count value stored at a node. If the counter value is less than the count value stored at a node then node increment the new value by one and store it. After a predefined time this process is over and hop value stored in each node is final. After calculation of the hop count, the routing process will be taken place. The detailed discussion about routing is discussed in following section.

4.3 Next hop selection

In routing setup phase every node finds its next hop for forwarding the data. The selection of next hop depends on cost between sender and receiver and the cost function depends on four parameters, namely residual energy of possible next hop, distance between sender and receiver, number of successor and predecessor nodes of next hop. The proposed algorithm is distributed in nature, so in every round next hop may be changed. The detailed algorithm is presented as follows (refer algorithm 1).

Lemma 1. *Next hop selection by proposed algorithm is energy efficient.*

Proof. The *cost* function of proposed algorithm (refer section 2) is based on parameters, namely residual energy, distance, number of predecessor and successor nodes. Moreover, the *cost* is ratio of residual energy and distance, number of successor and predecessor nodes. Therefore, the value of variable *cost* is maximum only when either residual energy of next hop is maximum or distance between sender and receiver is minimum. It means a node select the next hop in such a manner that next hop is either having higher residual energy or distance between both node is comparatively less. Moreover, other two parameters namely number of successor and predecessor also makes the algorithm energy efficient by minimizing the participation of node having morepredecessor and successor nodes. Suppose node *i* has comparatively higher number of predecessor nodes then there is a higher probability that node *i* act as next hop, such node quickly deplete the energy due to receiving and forwarding the data. So, by considering this parameter we try to minimize the participation of such node, which has higher number of predecessor nodes. □

Algorithm 1 : Next hop selection by proposed algorithm.

Initialization: $m = 0; count = 0; max = -\infty$
Input: relay node r_i
Output: NextHop of relay node r_i

1: **procedure** :
2:   **if** $(Dist(r_i, BS) \leq CR)$ **then**
3:     $NextHop(r_i) = \{BS\}$
4:   **else**
5:     **for** $(\forall\ r_j \in Next(r_i))$
6:       $X[m] = cost(r_i, r_j); Y[m] = j; m = m + 1; count + +;$
7:     **end for**
8:   **end if**
9:   **for** $(P = 1\ \text{to}\ count)$
10:     **if** $(max \leq X[P])$ **then**
11:       $NextHop(r_i) = r_{Y[P]}; max = X[P];$
12:     **end if**
13:   **end for**
14:   $NextHop(r_i) = r_{Y[P]};$
15: **end procedure**

Lemma 2. *The runtime complexity of proposed algorithm is linear.*

Proof. The run time complexity of above algorithm (refer algorithm 1) is linear. As in algorithm there are two *for* loops started at line number 5, 9 and ended at line number 7, 13. Both the loops are executed individually and execution of both loops is up to number of nodes in worst case. So, the worst case complexity for individual loop is $\theta(n)$, *n* represents the number of relay nodes in network. Therefore, the worst case run time complexity

of proposed algorithm is $\theta(n) + \theta(n) = 2 \times \theta(n) = \theta(n)$.   □

5 EXPERIMENTAL RESULTS

For the simulation purpose we have used all the parameters, same as described in (Heinzelman, Chandrakasan, & Balakrishnan 2000). For the simulation purpose, we select two different scenarios WSN#1, WSN#2 with size of 300×300 and 500×500 met^2 respectively. In WSN#1 scenario, sink is placed at (200, 100) and in scenario WSN#2 the BS is placed at corner of the scenario i.e. (0, 0). For the results comparison purpose, DEBR (Ok, Lee, Mitra, & Kumara 2009) and ESCFR (Liu, Ren, Li, Chen, & Shen 2012) algorithms are also executed.

Figures 2, 3 and Figures 4, 5 demonstrate the comparison of algorithms in terms of data packet sent in a round and from the figures it is clearly represented that the performance of proposed algorithm is better than others, because we consider residual energy and distance to make algorithm energy efficient in contrast to DEBR and ESCFR as the cost function of these algorithms only consider remaining energy of node as a parameter.

Figures 6 and 7 demonstrate the energy consumption and from the figures, it is noted that the proposed algorithm performs better than DEBR and ESCFR because to make the algorithm energy

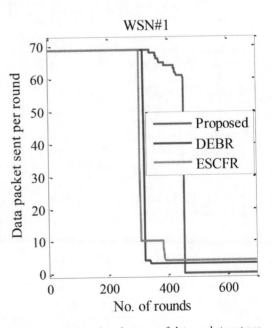

Figure 2.   Comparison in terms of data packet sent per round in WSN#1.

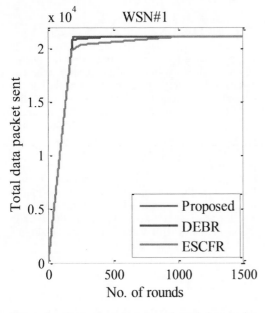

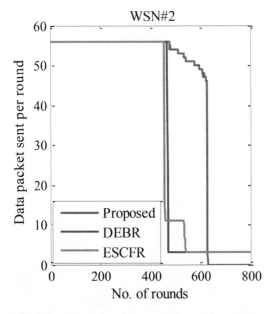

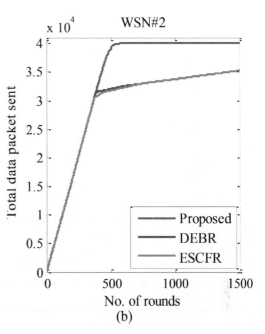

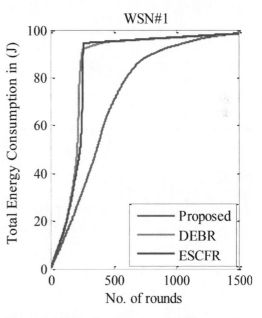

Figure 3. Comparison in terms of data packet sent per round in WSN#2.

Figure 4. Comparison in terms of total data packet sent in WSN#1.

Figure 5. Comparison in terms of total data packet sent in WSN#2.

Figure 6. Comparison in terms of total energy consumption in WSN#1.

efficient we consider distance and residual energy of possible next hop. However both DEBR and ESCFR only consider the residual energy of possible next hop as parameter due to which a node transmits the data to such node which compara-

tively far away and possibly drains the energy quickly due to long haul communication.

Figures 8 and 9 demonstrate the comparison of EEBR, DFBR and ESCFR in terms of inactive sensor nodes. The proposed algorithm performs

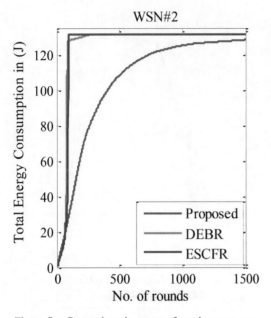

Figure 7. Comparison in terms of total energy consumption in WSN#2.

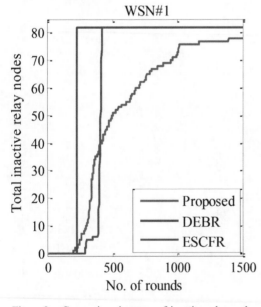

Figure 8. Comparison in terms of inactive relay nodes in WSN#1.

better than other because EEBR always selects such next hop which has less chances to act as next hop as well as has maximum number of possible next hop.

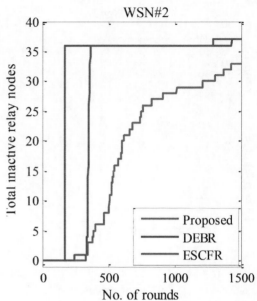

Figure 9. Comparison in terms of inactive relay nodes in WSN#2.

6 CONCLUSION

In proposed work, we have presented the energy efficient routing using distributed approach. The algorithm is energy efficient by considering the parameter e.g. residual energy, distance and also tried to select that next hop which has minimum number of predecessor node with maximum number of successor node. Moreover, the run time complexity of proposed algorithm is linear. The presented routing technique is also tolerate the failure of next hop, as it is distributed in nature. Experimentally, it is shown that the performance of presented algorithm is better than DEBR and ESCFR. As a future work, we extend the work in mobile scenario with consideration of Quality of Service (QoS) parameter.

REFERENCES

Akkaya, K. & M. Younis (2005a). A survey on routing protocols for wireless sensor networks. *Ad hoc networks* 3(3), 325–349.

Akkaya, K. & M. Younis (2005b). A survey on routing protocols for wireless sensor networks. *Ad hoc networks* 3(3), 325–349.

Akyildiz, I. F. & M. C. Vuran (2010). *Wireless sensor networks*, Volume 4. John Wiley & Sons.

Akyildiz, I. F., W. Su, Y. Sankarasubramaniam, & E. Cayirci (2002). Wireless sensor networks: A survey. *Computer networks* 38(4), 393–422.

Al-Karaki, J. N. & A. E. Kamal (2004). Routing techniques in wireless sensor networks: A survey. *Wireless communications, IEEE* 11(6), 6–28.

Anastasi, G., M. Conti, M. Di Francesco, & A. Passarella (2009). Energy conservation in wireless sensor networks: A survey. *Ad Hoc Networks* 7(3), 537–568.

Azharuddin, M., P. Kuila, & P. K. Jana (2013). A distributed fault-tolerant clustering algorithm for wireless sensor networks. In *Advances in Computing, Communications and Informatics (ICACCI), 2013 International Conference on*, pp. 997–1002. IEEE.

Chakraborty, U. K., S. K. Das, & T. E. Abbott (2012). Energy efficient routing in hierarchical wireless sensor networks using differential-evolution-based memetic algorithm. In *Evolutionary Computation (CEC), 2012 IEEE Congress on*, pp. 1–8. IEEE.

Chiang, S.-S., C.-H. Huang, & K.-C. Chang (2007). A minimum hop routing protocol for home security systems using wireless sensor networks. *Consumer Electronics, IEEE Transactions on* 53(4), 1483–1489.

Gupta, S. K. & P. K. Jana (2015). Energy efficient clustering and routing algorithms for wireless sensor networks: Ga based approach. *Wireless Personal Communications* 83(3), 2403–2423.

Gupta, S. K., P. Kuila, & P. K. Jana (2013). GAR: An energy efficient GA-based routing for wireless sensor networks. In *International Conference on Distributed Computing and Internet Technology 2013, LNCS (Springer)*, Volume 7753, pp. 267–277. Springer.

Gupta, S., P. Kuila, & P. Jana (2014). E³BFT: Energy efficient and energy balanced fault tolerance clustering in wireless sensor networks. In *Contemporary Computing and Informatics (IC3I), 2014 International Conference on*, pp. 714–719.

Heinzelman, W. R., A. Chandrakasan, & H. Balakrishnan (2000). Energy-efficient communication protocol for wireless microsensor networks. In *System Sciences, 2000. Proceedings of the 33rd Annual Hawaii International Conference on*, pp. 10–pp. IEEE.

Jiang, C., D. Yuan, & Y. Zhao (2009). Towards clustering algorithms in wireless sensor networks—a survey. In *Wireless Communications and Networking Conference, 2009. WCNC 2009*. IEEE, pp. 1–6. IEEE.

Liu, A., J. Ren, X. Li, Z. Chen, & X. S. Shen (2012). Design principles and improvement of cost function based energy aware routing algorithms for wireless sensor networks. *Computer Networks* 56(7), 1951–1967.

Ok, C.-S., S. Lee, P. Mitra, & S. Kumara (2009). Distributed energy balanced routing for wireless sensor networks. *Computers & Industrial Engineering* 57(1), 125–135.

Prabhu, S. B. & S. Sophia (2011). A survey of adaptive distributed clustering algorithms for wireless sensor networks. *International Journal of Computer Science and Engineering Survey* 2(4), 165–176.

Shukla, R., A. Chandel, S. Gupta, J. Jain, & A. Bhansali (2015). GaE³br: Genetic algorithm based energy efficient and energy balanced routing algorithm for wireless sensor networks. In *Advances in Computing, Communications and Informatics (ICACCI), 2015 International Conference on*, pp. 942–947.

Yick, J., B. Mukherjee, & D. Ghosal (2008). Wireless sensor network survey. *Computer networks* 52(12), 2292–2330.

Zungeru, A. M., L.-M. Ang, & K. P. Seng (2012). Classical and swarm intelligence based routing protocols for wireless sensor networks: A survey and comparison. *Journal of Network and Computer Applications* 35(5), 1508–1536.

Communication and Computing Systems – Prasad et al. (Eds)
© 2017 Taylor & Francis Group, London, ISBN 978-1-138-02952-1

Review of WiMax routing protocols

Vidhu Kiran, Shaveta Rani & Paramjeet Singh
GZS CCET, Bathinda, India

ABSTRACT: In this wireless world, there is a challenge for wireless services to give high data rate, quality of services and security. Worldwide interoperability for microwave access provides these features but the selection of proper routing as per avail environment is necessary. A lot of routing: Proactive, reactive and hybrid are there. In this paper, there is an analysis of different routing with possible parameter at national and international level for future use.

1 INTRODUCTION

WiMax stands for Worldwide Interoperability for Microwave Access can support both fixed and mobile wireless broadband having two types: Non line of sight and Line of sight. Line of sight services is just like a transceiver and receiver structure using antenna, tower etc. Non line of sight services is like a WI-Fi services having no fixed resources. LOS is stronger than NLOS for providing higher frequency.

2 WIMAX REFERENCE MODEL

WiMax reference model is divided into three parts:

2.1 Mobile Stations (MS)

It is used by end user for accessing service from BSs.

2.2 Access Service Network (ASN)

Access service network contain one or more than one BSs. BS is required for providing interface to all MSs and also manage mobility function, hand-off (within or out of network), traffic classification, session management, resource management etc.

2.3 Connectivity Service Network (CSN)

It can provide connectivity with internet and other networks such as PSTN or corporate network.

3 WIMAX ARCHITECTURE

WiMAX has layered architecture as:

3.1 Physical layer

This layer allows FDD and TDD for system implementation. FDD (Frequency Division Duplexing) uses lot of frequency and cover unlimited range. TDD (Time Division Duplexing) is required for short range and less costly. Physical layer support both half and full duplex (TDD and FDD). Instead of this, it can encrypt data (scrambling), reordering of data (interleaving), addition of information to electronic and optical carrier (modulation) etc.

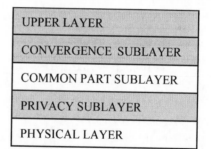

3.2 Mac layer

It is a package of three sub layers: Privacy sub layer, Common part sub layer and Convergence sub layer.

3.2.1 Privacy sub layer
It can perform authentication of two nodes, encryption of message to be sent and function of management of keys.

3.2.2 Common part sub layer
This layer can setup a connection, allocate bandwidth, Automatic Retransmission Request (ARQ) and functioning for Quality of services.

3.2.3 Convergence sub layer
This layer provide compatibility with upper layer for map address from upper layer address to WiMAX protocols address and maintain TCP/IP based traffic regulations.

3.3 Upper layer
Upper layer functioning is similar to TCP/IP layer.

4 ROUTING PROTOCOLS IN WIMAX

Many types of routing protocols can be used by WiMAX network such as Proactive, Reactive and Hybrid.

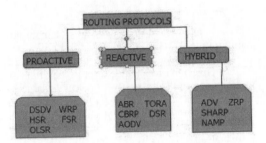

4.1 Proactive routing protocols

4.1.1 Destination sequence distance vector
Mobile Node (MN) has table containing entries (Dstn, hop, Sq no sent by destination node). Updation occurs after modification periodically or event drive. Each mobile node broadcast packet to neighbor node. Updation can be in full dump or incrementally.

4.1.2 Wireless routing protocol
It is based on shortest path algorithm. Each MN has 2 to last of shortest path to each destination. Routing contain distance, route, cost, Retransmission table. Update occurs, all nodes wait for best path.

4.1.3 Hierarchal state routing
Use location management scheme, divide network into cluster. Only Cluster head communicate with each other.

4.1.4 Fisheye state routing
Use fisheye structure. Basic idea is to update only nearer node instead of far node.

4.1.5 Optimized link state routing
Can deliver packet to a movable node having predictable speed that neighbor can analyze. It is more optimized.

4.2 Reactive routing protocols

4.2.1 Associatively Based Routing (ABR)
It is based on association stability of node. MN send beacon message to Neighbor Node (NN) for show existence. NN increment associative tick with each beacon. Increment in tick shows node is stable. If MN moves from its location, Tick value will be reset.

4.2.2 Temporarily ordered routing algorithm
This is similar of proactive method for searching destination. Use "height" parameter as a distance b/w responding node and destination. Response of query provides route table with height to destination comes from multiple nodes.

4.2.3 Cluster based routing
New MN send hello to Cluster Head (CH). CH send 'hello' to MN. CH maintain all nodes information. MN can be a CH if it has bidirectional links with 1 or more nodes.

4.2.4 Dynamic source routing
Each MN checks route cache before route discovery. Broadcast discovery message, get response.

4.2.5 Ad hoc On demand Distance Vector(AODV)
AODV is an improvement of DSDV. Match destination address, If not same, broadcast it to next. Reply with similar path.

4.3 Hybrid routing protocol

4.3.1 Adaptive distance vector
Use adaptive mechanism to mitigate the effect of periodic transmission during changes. MN will ON receiver flag. Source send packet to active MN, get receiver alert packet in reverse.

4.3.2 Zone Routing Protocol (ZRP)
ZRP is useful for large span. Each node having HOP = 1 comes under a zone. Proactively maintain routes table in local region. No blind broadcasting, use control mechanisms with "cover nodes".

4.3.3 Sharp hybrid adaptive routing protocol
All MN in zone radius are member of zone, store their data in central node. Use Proactive method in zone. If MN not avails, use query-reply method.

4.3.4 Neighbor aware multicast protocol
Use tree structure. It can perform three operations: tree creation, maintenance, joining and leaving. Every MN has two hop information. Secondary Forward List (SFL) having intermediate node detail for reverse path.

5 PERFORMANCE ANALYSIS IN WIMAX

Performance can be done into two categories: National and international level.

5.1 At national level

All routing protocols with various performance matrices like packet delivery, delay, energy, throughput etc can be analyzed on various simulators by varying nodes, time, and speed/mobility. It can be concluded as:-

5.1.1 On the basis of nodes

When topology changes from low to high node densities, AODV is more favorable for throughput, packet delivery and end to end delay instead of DSDV [1]. OLSR is better than DSDV for packet

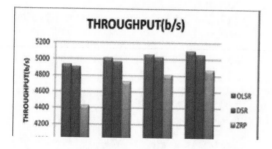

Figure 1. Throughput for OLSR, DSR, ZRP.

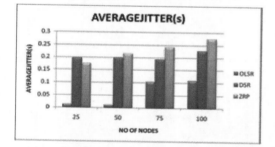

Figure 2. Avg Jitter for OLSR, DSR, ZRP.

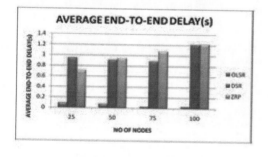

Figure 3. End to End delay b/w OLSR, DSR, ZRP.

delivery, end to end delay and throughput [15]. When node increase with CBR then DSR gave more throughputs, less jitter and end to end delay [17]. Similarly, OLSR is better than DSR and ZRP for throughput, end to end delay, and jitter and packet delivery (Suresh et al. 2013) (Figs. 1, 2, and 3). OLSR is also better than AODV and DSDV when network is large for successful delivery of packets (Dubey and Patel 2012). DSR provides better throughputs, delay and jitter than FSR. In integration of Wi-Fi and Wimax, DYMO's throughput is greater than Bellman Ford, AODV, DYMO, OLSRv2-NIIGATA, OLSRv2, RIP (Fig. 4) but more end to end delay (Fig. 6) and jitter (Fig. 5).

5.1.2 On the basis of time

For high load application FTP, BGP is more favorable for throughput than OSPF and IGRP.OSPF has less end to end delay than BGP (Kaur and Kaur 2015). AODV provides maximum through-

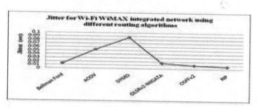

Figure 4. Throughput for Bellman ford, AODV, DYMO, OLSRv2, RIP.

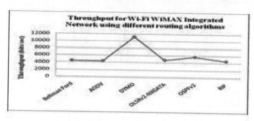

Figure 5. Jitter for Bellman ford, AODV, DYMO, OLSRv2, RIP.

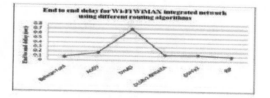

Figure 6. Delay for Bellman ford, AODV, DYMO, OLSRv2, RIP.

put than DSDV and DSR. AOMDV provide maximum packet delivery (Fig. 7) than AODV, DSR and DSDV, TORA and less end to end delay (Fig. 8) with respect to time in on demand multi path routing (Seethalakshmi and Kumar 2013).

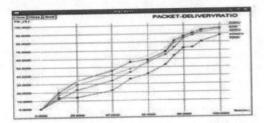

Figure 7. Packet delivery in DSDV, DSR, AODV, AOMDV, TORA.

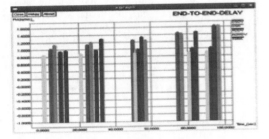

Figure 8. DSDV, DSR, AODV, AOMDV, TORA End to End delay.

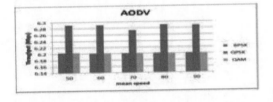

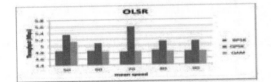

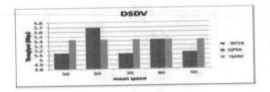

Figure 9. Throughput for AODV, DSDV, OLSR according to BPSK, QPSK, QAM.

5.1.3 *On the basis of speed/mobility*

According to change in speed, AODV have less variation than DSDV and OLSR when different modulation is applied on them. QPSK modulation is better than BPSK and QAM (Abdullah and Singh 2015) (Fig. 9). Similarly when mobile node has different speed, path and class, result for throughput, end to end delay and jitter vary with various modulation (Hamodi et al. 2014). For high mobility, DSR is better than AODV and DSDV but for low mobility, DSDV is not good. DSDV is best for end to end delay and throughput with varying mobility model (Rahel and Shukla 2013) (Figs. 10, 11, 12).

Other parameters that can be analyzed are:-

1. Network Load: In a computer network the load is a measure of the amount of computational work that a computer system performs (Mishra et al. 2015).
2. Media Access Delay: It is the time taken by nodes in routing to access resources and to complete services required by them includes all delays: transmission delay, queuing delays and

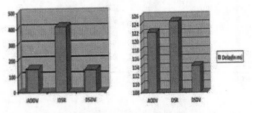

Figure 10. Comparison with Low/High Mobility b/w AODV, DSDV, DSR in term of End to end delay.

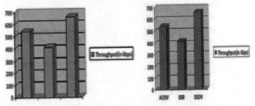

Figure 11. Comparison with Low/High Mobility b/w AODV, DSDV, DSR in term of Throughput.

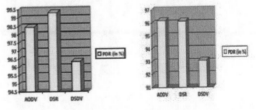

Figure 12. Comparison with Low/High Mobility b/w AODV, DSDV, DSR in Packet delivery ratio.

delays due to contentions and back offs (Kaur and Kaur 2015).

3. Retransmission Attempts: It is a total attempts done by network for retransmission of packets until it will reach successfully or rejected after long delay (Kaur and Kaur 2015).

4. Packet loss: In this, sum of corrupted, lost, or excessively delayed packets received at receiver end divided by the total number of packets send by sender side (Hamodi et al. 2014).

5. Energy consumption: It can be calculated by dividing consumed energy by individual node to total energy (Seethalakshmi and Kumar 2013).

6. Number of Packets dropped: It is a calculation of failed packets that was not reached at destination within a fixed time (Seethalakshmi and Kumar 2013).

7. Channel Utilization: It is the ratio of received bandwidth out of total bandwidth available at network (Prasad et al. 2013).

8. Blocking Probability: It is the ratio of number of rejected request to the total requests done by nodes in routing (Prasad et al. 2013).

5.2 At international level

5.2.1 On the basis of nodes
When nodes increase, DSDV is better than AODV and DSR for packet delivery and less end to end delay (Figs. 13,14) (Rasheed et al. 2010). OHLAR performs outstanding than AODV and LAR because it can provides maximum packet delivery and less end to end delay when density of node is high (Liao and Jung 2008).

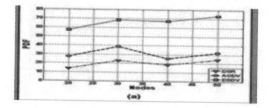

Figure 13.   PDF for DSR, AODV and DSDV.

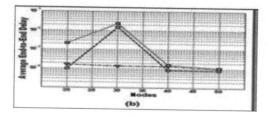

Figure 14.   Average delay for DSR, DSDV, AODV.

5.2.2 On the basis of time
DSDV performs better than AODV and DSR for packet delivery and throughputs but with more end to end delay (Fig. 15) (Rasheed et al. 2010).

5.2.3 On the basis of mobility
OHLAR perform outstanding than AODV and LAR for maximum packet delivery and less end to end delay and control overhead (Liao and Jung 2008). ZRP gave better results than AODV, DSR and OLSR for packet delivery ratio (Fig. 16). OLSR is best for routing overhead and end to end delay than other but DSR is good in throughput than AODV and ZRP (Figs. 17, 18) (Anwar et al. 2008).

Other parameters that can be analyzed are:

1. Wireless User Background Work Traffic: It is a calculation of data sent by server node for

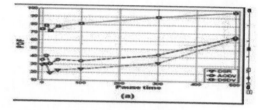

Figure 15.   PDF for DSR, DSDV, AODV.

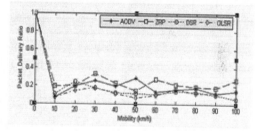

Figure 16.   Packet delivery for AODV, ZRP, DSR, OLSR.

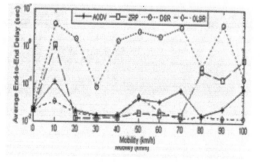

Figure 17.   End to End delay for AODV, OLSR, ZRP, DSR.

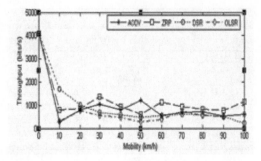

Figure 18. AODV, ZRP, DSR, OLSR Throughput (bits/sec).

background process only instead of any meaning task for example for checking network, node, and resources availability before routing (Anouari and Haqiq 2014).

2. Delay: Delay at client as well as at server side (Anouari and Haqiq 2014).

3. Routing Load: It can be calculated by determining overhead bytes to delivered data bytes (Dubey and Patel 2012).

6 CONCLUSIONS

This paper should be very useful for beginners in the selection of parameter and their effect on different Wimax routing protocols. The most important propositions of the paper are detail of routing protocols with their graph on the basis of three parameter: Increment in nodes, time and mobility. This paper also shows work done on wimax routing done at national as well as international level.

REFERENCES

Abdullah, N. & Singh H. 2015. International Journal of Emerging Research in Management & Technology. *Performance Based Comparison of Different Routing Protocol under Different Modulation over WiMAX* Volume 4 Issue 2.

Alam, D. et al. 2014. International Journal of Advanced Science & Technology. *Performance Evaluation of Different Frequency Bands of WiMAX and their selection procedure* Volume 62 pp.1–18.

Anouari, T. & Haqiq, A. 2014. International Journal of Soft Computing and Engineering. *A QoE-Based Scheduling Algorithm for UGS Service Class in WiMAX Network* Volume-4 Issue-1.

Anwar, F. et al. 2008. International Journal of Computer Science. *Performance analysis of Ad hoc routing protocols in mobile WiMax Environment* Volume 35 Issue 3.

Chaniwal, O. & Haldar, M. 2015. International Journal of Technology Research and Management. *Analysis of Wimax Module for Reactive and Proactive Routing Protocols under ns2* Volume 2 Issue 2.

Das, P. et al. 2012. National Conference on Communication, Measurement and Control. *Performance Analysis of Different Routing Algorithms in WiMAX Network caring VoIP traffic using Qualnet* Volume 1.

Dubey, P. & Patel, B. 2012. International Journal of Engineering Research and Applications. *An Improved routing efficiency for WiMAX Network using Random 2D direction model in NS3 simulator* Volume 2, Issue 5 pp: 420–424.

Hamodi, J. et al. 2014. International Journal of Communications, Network and System Sciences. *Performance Study of Mobile TV over Mobile WiMAX considering Different Modulation and Coding Techniques* Volume 7 pp: 10–21.

Kaur, H. & Saini, J. 2013. International Journal of Multi disciplinary and Current Research. *Comparative Analysis of Routing protocols in Wimax environment.*

Kaur, J. & Kaur, M. 2015. International Journal of Electronics & Communication. *Simulation based analysis of IEEE 802.16 Wimax routing protocols for high load applications* Volume 3 Issue 7.

Liao, H.C. & Jung, C. 2008. Asian Network for Scientific Information. *A Position-Based Connectionless Routing Algorithm for MANET and WiMAX under High Mobility and Various Node Densities* Volume 7 (3) pp: 458–465.

Mishra, A. et al. 2015. International Journal of Application or Innovation in Engineering & Management. *Performance Evaluation of WIMAX Network with High QOS Services Incorporating different Physical & Mac Layer Standards* Volume 4, Issue 4.

Mohammed, H. et al. 2014. International Journal of Current Engineering and Technology *Performance Comparison of a Reactive and Hierarchical Routing Protocol in 802.16e using Qualnet 6.1.*

Prakash, S. et al. 2015. International Conference of Advance Research and Innovation. *Quality of Service Based Handoff Schemes for WiMAX/WLAN Networks.*

Prasad, R. M. et al. 2013. International Arab Journal of Information Technology. *Joint Routing, Scheduling and Admission Control Protocol for WiMAX Networks* Volume 10.

Rahel, S. & Shukla, D. 2013. International Journal of Engineering Research and Application. *Implementation and Analysis of Wimax Module under Ns2 with Varying Mobility Model* Volume 3, Issue 6 pp: 2181–2185.

Rasheed, M.R. et al. 2010. International Journal of Engineering & Technology. *Performance of Routing Protocols in WiMAX Networks* Volume 2.

Seethalakshmi, V. & kumar, G.M. 2013. International Journal of Advanced Research in Computer Science and Software Engineering. *A Survey of Routing Protocols in Mobile Ad Hoc Network* Volume 3, Issue 8.

Sharma, M. & Shrimal, V.M. 2012. International Journal of Computer Networks and Wireless Communications. *Comparison of different routing protocols (DSR & AODV) on behalf of evaluation of different routing parameters with constraints* Volume 2.

Sharma, R. & Gupta, A. 2014. International Journal of Advance Research in Computer Science and Management Studies. *Review of QOS Parameters for Wimax Environment* Volume 2 Issue 5.

Shrivastava, P. et al. 2012. International Journal of Engineering Research & Applications. *Vehicle-to-Road-Side-Unit Communication Using Wimax* Volume 2 Issue4 pp: 1653–1655.

Shrivastava, S. et al. 2014. International Journal of Emerging Trends & Technology in Computer Science. *A Study and comparison of various routing protocol in WiMAX networks* Volume 3 Issue 3.

Singh, T.P. & Dua, S. 2012. International Journal of Advanced Research in Computer Science and Software Engineering, *Energy efficient Routing Protocols in Mobile Ad-Hoc Networks* Volume 2 Issue 1.

Song, S. & Issac, B. 2014. International Journal of Computer Networks & Communications. *Analysis of Wi-Fi and Wimax and wireless network coexistence* Volume 6.

Suresh, M. et al. 2013. International Journal of Advanced Research in Computer Science and Software Engineering. *Improving the Performance of WIMAX Using Various Routing Protocols* Volume 3 Issue 10.

Yasien, S. et al. 2015. International Journal of Advanced Research in Computer Science and Software Engineering. *Assessment of Worldwide Interoperability for Microwave Access Technology (WiMAX) Quality of Services (QoS)* Volume 5 Issue 7.

Communication and Computing Systems – Prasad et al. (Eds)
© 2017 Taylor & Francis Group, London, ISBN 978-1-138-02952-1

Cross layer based data delivery mechanism over heterogeneous wireless networks

A.S. Dhama & S. Sharma

Gautam Buddha University, Greater Noida, India

ABSTRACT: As the number of wireless network user are increasing so do the different type of wireless network, as a result whenever we receive some data it has to travel through different wireless network. Therefore a Heterogeneous Wireless Network (HWN) supporting video data will be considered. A Cross Layer (CR) between data link and transport layer and another CR between network and transport layer will be suggested to increase the overall throughput of wireless network and also to decrease the overall packet loss and end to end delay. This paper features Stream Control Transmission Protocol (SCTP) based video delivery over network. A noticeable improvement in the throughput and in the end to end delay was achieved while simulating over mechanism over two different HWN. The simulator used here to implement the mechanism is NS2.

Keywords: Heterogeneous Wireless Networks; NS2; Cross Layer

1 INTRODUCTION

The increase in the number of user using wireless technologies has pushed over existing networks to its limit, therefore whenever we receive data in over mobile it had travelled through different network using different wireless technologies. The data that we receive can be voice, text or video. Video data require more bandwidth than any other type of data that we receive and it is also delay sensitive. So to transfer video over a network the throughput should be high and end to end delay must be low and also our bit error rate should be low. The SCTP has been considered to be a better transport layer protocol than the existing protocol to transfer video over a HWN (XuC). The route through which our data travel are not very reliable because packet get dropped very often even in ideal condition also the route are unstable because the receiver is most of the time in motion. So to counter these many difficulties a CR mechanism is proposed. There are two CR between the data link and transport layer and second CR between the network and the transport layer. In first CR (XuC) we will get the frame error rate from over data link layer and from there we will calculate bit error rate. Now according to over bit error rate we will select or deselect the route. If bit error rate of that particular route get increased due to any reason a queue will be activated at the transport layer to store the packets. If the bit error rate crossed the threshold than existing route will be changed and a new route will be considered. The

second CR is between the network layer and the transport layer (Dhama S). We propose a queue at the transport layer that will be activated in case of congestion and by using CR the packet of network layer will be buffered in the queue. This will result in the decrease of the packet dropping rate and energy consumption. In our proposed design the received packets are being sent to CR queue at the transport layer and after the congestion or when the node finds a new path they are sending back to network layer and restart the transfer of data. Apart from these problems there are security issues which can be overcome by techniques like authentication with CR approach (Sharma S). Also when a packet get dropped even in the ideal condition its frame is transported to the higher layer while its parent in the lower layer is missing (Eddine N), so it is useless to carry such frames and these frames will not be included because they increase the burden of network. So by not including these packets we will decrease the load of network and increasing band width efficiency.

2 RELATED WORK

As we have discussed in our earlier paper (Dhama S) on the basis of past research the following route failure and congestion can be categorized into flowing types based on their working principle. One of the types uses a backup route for all active routes. When primary route fails backup route is used.

The example of this type of protocol is AODV-BR (Wahi C), but this category suffers from problem like the maintenance of multipath is difficult, costly and energy efficiency is less as a result the performance get effected. Other type uses secondary route to send the packets (Akintola A. A.). For example in DSR protocol when congestion occurs, it split traffic and mitigates congestion; this will increase QoS for the network and as a result the overall maintains of network increases decreasing the throughput. The third type focuses on local recovery. For example, in case of (Harshada A), a node check the signal coming from other nodes if signal are low, it starts a local recovery and finds a new route instead of sending error message for the transmitting node. Here neighbor node stores information as a backup node, due to this the adjacent node keeps the packet without any reason and at the end more than one adjacent node can transmit packets after finding the new route. This results in decrease of network efficiency. The last type is most recent that focuses on the CR approach. For example a cross-layer design for resource allocation in (Alima P.K.) and RCECD proposed in (Hassan M. M.). We are proposing a design using this CR approach which will have a queue at the transport layer as proposed in (Dhama S). There are other future aspects of CR (Sharma S) also by the help of which users can access the information over different network. We can also work in field of security (Sharma S) where we can provide security to our wireless network using CR (Sharma S). Research in the field of transport layer protocol (SCTP) has increased in recent year (Stewart R). Some have researched in the field of congestion window management (Wallace) and concurrent multipath transfer (Huang) of the data in HWN. The research is also being done on SCTP concurrent multipath transfer and showing its capability for video transfer as transport layer. Other techniques like round robin scheduling (Perotto) are used to transfer the data with lowest transmission time. But these all focus on the transport layer QoS parameter they does not make any noticeable difference with the throughput, end to end delay or packet loss. So it is hard to use them with wireless network.

3 THE CROSS LAYER TECHNIQUE

The proposed CR approach uses two CR for transfer of packets in HWN. Here we combine the data link layer and transport layer and second CR combines the network layer and transport layer with each other, hence named CR. At the transport layer we are going to buffer our packet during route failure, congestion and when bit error rate is high of a particular route. So in these case nodes will generate a message that will indicate the prob-

lem and all respective node will start buffering the packet in their respective transport layer.

3.1 *Architecture overview*

Figure 1 shows the architecture overview of the CR design. The sender and receiver are connected with each other through wirelesslink.

3.2 *Buffering of packets at the transport layer*

In our simulation model nodes will act as a receiver as well as the transmitter, so when it is going to receive the data, in our CR, the nodes will have a Transport Layer Queue (TLQ) to store the packets. Now this TLQ will only be used in case of congestion and it case of high BER when there is no congestion the normal operation will take place, this can be seen clearly in Figure 2. But when a node detects congestion or node detects that the intermediate link between two nodes is broken, nodes will start using their TLQ to buffer the packet until a new route is formed or there is no more congestion.

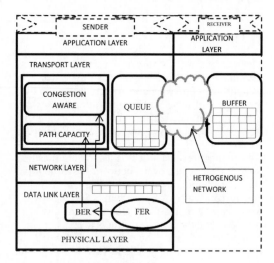

Figure 1. Architecture Design.

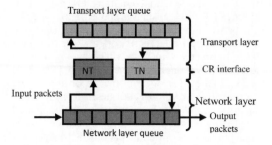

Figure 2. Interface between transport queue and network queue.

In the above Figure 2 we can see the mechanism of CR. Our interface will have two components, receiver interface NT (network to transport layer) and TN (transport layer to network layer). Our first interface NT receives packet from the network layer queue and transfers the incoming data to TLQ, NT only works in case of route failure, congestion and high BER. In second case the packet from transport layer is transferred from transport layer queue through TN interface that will on getting the information that congestion is over or in case of rote failure that a new route is found, TN transfer will send the data to network layer queue. Let us take an example that there is a route which has two intermediate nodes C and D. Now take a condition that the link between both the nodes is broken. Now C will detect that the link between both of the nodes is broken, so it will start buffering the packets from the network layer to transport layer queue, and C will send a message to other nodes that the route is failed and a RF (route failure) message will be sent. Now as soon as other nodes get the RF message they will also start buffering the packet to their transport layer queue. Now all the source will stop there transition and wait for the new route notification message. Now if node C finds a new local path it will send a message New Route (NR) to all the nodes. After getting this message all the nodes will send data through TN interface to the network layer queue from their respective transport layer queue and after that normal operation will begin. This mechanism will decreases packet dropping rate. If no new route is found node C will send a No New Route message (NNR). The source node on receiving this will start a new route discovery. The procedure to start a new route will be same as finding the route first time, the source node floods the RREQ (Route Request) packet all over the network in case of not finding the destination node. After that from the single RREQ it may receive many (RREP) route reply. A route reply carries the source identifier, the destination identifier, the source sequence number, the destination sequence number, the broadcast identifier and the time to live field. Based on the number of hop count and time to live the source node will reply with ACK (Acknowledgement) packet and then data packets will be sent. After all packets will be sent the destination node will send an ACK packet.

3.3 BER calculation at data-link layer

The calculation of BER at the physical layer require bit by bit comparisons of all the frames which is very complex so we will use the FER at the data link layer instead. The frame error rate can be calculated as follows:

$$FER = (error\ frame/total\ frame) \tag{1}$$

where error frame are lost frame and total frame are total frame send by sender. Now BER can be calculated [13] as:

$$FER = 1 - (1 - BER)^{8L} \tag{2}$$

Here L is defined as the number of bytes in a data frame. From equation two we can find the value of BER. From here we can get the bit error rate of a particular path and according to the BER we can change the path of that route. If the BER of a particular route has increased due to known or unknown reasons we can activate the queue at the transport layer by comparing the value of BER to a threshold value or a pre-defined value. By doing so we can save the packet loss and thereby increase the throughput.

3.4 Detection of congestion level

Our CR design detects the congestion so that at time of congestion the packet can be buffered in the transport layer queue. At each intermediate node we measure the amount of data that is in queue of the network layer, on the basis of this information we measure the level of congestion and according to this the action are taken. A two bit flag is used in both the packet; the packets are data packets and the acknowledgement of data packet. This is known as congestion level CL flag. The value of our flag is measured on the basis of the Table 1.

To determine the congestion we set minimum and maximum threshold, Qmin and Qmax, respectively, for the data present in queue by as follows:

$$Qmin = 0.4 \times Qsize; \tag{3}$$
$$Qmax = 0.9 \times Qsize; \tag{4}$$

If the current size of queue is less than Qmin, then we can say that there is no congestion, if the queue size is greater than Qmin but less than Qmax, then it is light congestion, and if packet length exceeds Qmax, then there is congestion. We introduce another parameter Qwarn, for warning stated below:

Table 1.   Congestion notification (Cl).

Value of CL	Congestion level
00	No congestion
01	Light congested
10	Heavy congested
11	congested

225

$$Qwarn = w \times Qsize; \tag{5}$$

where w is a weight factor, we choose w = 0.8. We then calculate average queue occupancy of a node after every certain interval using exponentially weighted moving average formula as follows:

$$Qavg = (BER) \times Qavg + Qcurr \times \alpha \tag{6}$$

where α is a weight factor and Qcurr is the current queue size. Now, on the basis of the value of Qavg, the value of CN flags is as follows:

- if Qavg < Qmin then CN = 00
- if Qavg> = Qmin and Qavg < Qwarn then CN = 01
- if Qavg> = Qwarn and Qavg< = Qmax then CN = 10
- if Qavg > Qmax then CN = 11

There can be a one more case where the queue size of transport layer and network layer queue is full and in the meantime if No New Route is found node will send a no new route message (NNR). The source node on receiving this will start a new route discovery. So we can say that the threshold to determine failure of the route will be the queue size of transport layer and when TLQ queue size is equal to packet queued during congestion then route is fail. On the basis of above calculation we can make a congestion control mechanism as given below in Table 2.

4 PERFORMANCE EVALUATIONS

We have evaluated our performance on three parameters that are the throughput, end to end delay and packet loss by our network. To evaluate result we have shown the graphs against the protocol that have the proposed mechanism to the one that does not have the mechanism.

4.1 Simulation setup

The simulator used here is NS2. To simulate a heterogeneous network we have chosen two different wireless technologies. The two different technologies are the IEEE 802.11 and IEEE 802.16. The receiver trav-

els from the first network to the other network while sender and the receiver remain constant. So there will be over all two different paths by which data will travel. The path parameter is given in Table 3.

After getting the results we can see the difference between the throughput of our CR approach and traditional approach, the one that does not include the CR approach. In the simulation scenario, we have considered square area of size 500* 500 m², where 16 random moving nodes. The time for our simulation is 80 s. Nodes with transmission range of 250 m. Our source node will send random message at constant interval of time. By this we can check the connectivity of link. Table 4 gives the simulation parameters.

4.2 Simulation results

From the analysis, discussion and the simulation results shows that our proposed approach perform better than the one without CR.

In Figure 3 we can see that end to end delay of proposed solution in low than that of normal solution. We can see that there is a spike in the delay at the starting of the simulation but as the queue starts buffering the packets the delay decreases because our mechanism not only stores the packet but also finds the new route if BER is high of that route.

In Figure 4 we can see that packet loss of proposed solution in nearly zero whereas the packet

Table 2. Action taken by node in case of congestion.

Congestion level	Action by node
No congestion	Normal operation
Light congested	Do no send RREQ
Heavy congested	Warning message to sources
Congested	Enable TLQ

Table 3. Path_parameter_in_the_simulation.

Parameters	Path_A	Path_B
Wireless __technology	IEEE_802.11	IEEE_802.16
Access__bandwidth	2 Mbps	10 Mbps
Access_link delay	10–20 _ms	10–20 _ms
Access_queue_limt	50	50

Table 4. Simulation parameters.

Parameter	Value
Network area	500 m*500 m
Number of nodes	21
Number of sources	1
Transmission range	250_m
Transport layer protocol	SCTP
MAC layer protocol	IEEE_802.11/802.16
Control packet size	100_bits
Bandwidth	10, 2 Mbps
Data burst size	2_Mbytes/flow
Packet size	512_bytes
Propagation model	Free space
Weight factor	0.20
Simulation time	80 s

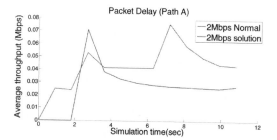

Figure 3. Packet delay at the receive comparession between solution and normal protocol with Access bandwidth of 2 Mbps.

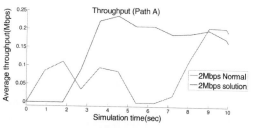

Figure 5. Throughput at the receive comparession between solution and normal protocol with Access bandwidth of 2 Mbps.

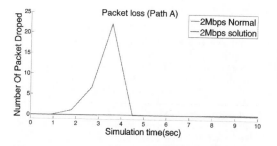

Figure 4. Packet loss at the receive comparession between solution and normal protocol with Access bandwidth of 2 Mbps.

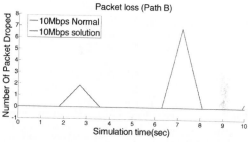

Figure 6. Packet loss at the receive comparession between solution and normal protocol with Access bandwidth of 10 Mbps.

loss of normal protocol is not nearly zero. We can see that there is a approx. no packet loss in our simulation because we have super cross layer here which activate as network layer starts having congestion or BER of increases. The decrease in the packet loss helps the overall increase in the throughput.

In Figure 5 we can see that throughput of proposed solution is higher than the normal solution. Our throughput is high because our packet loss is very less as well as the delay. The queue in the transport layer plays a very important role in this because of the buffering of the packet which gets delivered for sure thereby increasing the throughput.

In Figure 6 we can see that packet loss of proposed solution in nearly zero whereas the packet loss of normal protocol is not nearly zero. We can see that there is a approx. no packet loss in our simulation because we have super cross layer here which activate as network layer starts having congestion or BER of increases. The decrease in the packet loss helps the overall increase in the throughput.

In Figure 7 we can see that end to end delay of proposed solution in low than that of normal solution. We can see that there is a spike in the delay at the starting of the simulation but as the queue starts buffering the packets the delay decreases because our mechanism not only stores the packet but also finds the new route if BER is high of that route.

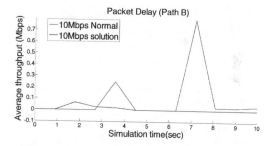

Figure 7. Packet delay at the receive comparession between solution and normal protocol with Access bandwidth of 10 Mbps.

In Figure 8 we can see that throughput of proposed solution is higher than the normal solution. Our throughput is high because our packet loss is very less as well as the delay. The queue in the transport layer plays a very important role in this because of the buffering of the packet which gets delivered for sure thereby increasing the throughput.

In the above given results we have divided the result in two main part. Fig. 3, Fig. 4 and Fig. 5 are of path A where we have taken Wireless technology IEEE 802.11 and access bandwidth of 2 Mbps on the hand path B uses Wireless technology IEEE 802.16 and result are shown by figure Fig. 6, Fig. 7

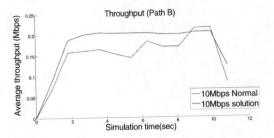

Figure 8. Throughput at the receive comparession between solution and normal protocol with Access bandwidth of 10 Mbps.

and Fig. 8. In the above results we have compared the normal protocol with the modified one.

5 CONCLUSION AND FUTURE WORK

Our paper proposed a congestion aware reliable data delivery mechanism over heterogeneous wireless network using two CR between transport layer and network layer and second between the data link layer and transport layer. The transport layer protocol used here to transfer video data is SCTP. This paper focus on the network layer where we have proposed a queue that stores the packets from network layer at the time of congestion or failure of route. So a CR design is introduced between these layers. Our design get activated only at the time of congestion or failure of route, upon the discovery of any of these two factor the packet that are being received by node are transferred to the TLQ and after resolving the problem the data in queue in transferred to network layer queue and normal operation begin. This has increased our throughput as shown by result at different bandwidth. For future work we can transfer the data using concurrent multipath transfer SCTP where same data is sent through different paths at the same time and we can also employ the CR design at the receiver.

REFERENCES

Akintola A. A., Aderounmu G. A., and Akanbi L. A., Adigun M. O. Modeling and Performance Analysis of Dynamic Random Early Detection (DRED) Gateway for Congestion Avoidance. Issues in Informing Science and Information Technology.

Alima Beebi P.K., Singha S, Mane R. A Study on CR MAC design for performance optimization of routing protocols in MANETs. (IJACSA) International Journal of Advanced Computer Science and Applications, Vol. 2, No. 2, February 2011.

Dhama S, Sharma S. Reliable Data Delivery Mechanism for Mobile Ad hoc Network Using CR Approach accepted and presented for publication in the proceeding of 50th Golden Jubilee Annual Convention of Computer Society of India (CSI-2015), December 02–05, 2015. The proceeding will be published by Springer.

Harshada A, Kokate S.R. A Study of Congestion Aware Adaptive Routing Protocols in Manet. International Journal of Advanced Technology & Engineering Research (IJATER) Volume 2, Issue 2, May 2012.

Hassan M. M., Kamruzzaman S. M., Alamril A. Design of an energy-efficient and reliable data delivery mechanism for mobile ad hoc networks: a cross-layer approach. 2014 Published online in Wiley Online Library (wileyonlinelibrary.com). DOI: 10.1002/cpe.3309.

Huang C.-M., Lin M.S. Multimedia streaming using partially reliable concurrent multipath transfer for multihomed networks. IET Communications, vol.5, no.5, pp.587–597, Apr. 2011.

Perotto F, Casetti C, Galante G. SCTP-based Transport Protocols for Concurrent Multipath Transfer. in Proc. IEEE WCNC, Mar. 2007.

Rikli N, Alabdulkarim M. Cross-Layer-Based Adaptive Video Transport Over Low Bit-Rate Multihop WSNs. in Canadian Journal of Electrical and Computer Engineering, Vol. 37, No. 4, Fall 2014.

Sharma S, Mishra R, Singh K. A review on wireless network security.in Springer Berlin Heidelberg, Nov 2013.

Sharma S, Mishra R, Singh K. Current Trends and Future Aspects in Cross-Layer Design for the Wireless Networks in Advances in Computer Science and Information Technology. Computer Science and Engineering: Second International Conference 2012.

Sharma S, Mishra R. A CR Approach for Intrusion Detection in MANETs in International Journal of Computer Applications, Jan 2014.

Sharma S, Mishra R. Authentication in wireless network. IEEE Conference Publications, 2015.

Stewart R. Stream Control Transmission Protocol. RFC 4960, IETF, Sep. 2007.

Wahi C, Kumar S. Mobile Ad Hoc Network Routing Protocols: A Comparative Study. DOI: 10.5121/ijasuc.2012.3203.

Wallace T. D. and Shami A. Concurrent Multipath Transfer using SCTP: Modelling and Congestion Window Management. IEEE Transactions on Mobile Computing, vol. PP, no. 99, Feb. 2014.

Xu C, Li Z. Cross-layer Fairness-driven Concurrent Multipath Video Delivery over Heterogenous Wireless Networks. 10.1109/TCSVT.2014.2376138, IEEE Transactions, 2013.

Communication and Computing Systems – Prasad et al. (Eds)
© 2017 Taylor & Francis Group, London, ISBN 978-1-138-02952-1

Simulation and analysis of OCDMA system using fiber bragg grating technique

Sandeep Sharma & Aarti Bhardwaj
Gautam Buddha University, Greater Noida, India

ABSTRACT: Tremendous growth of communication accelerated the access network capacity by widely deployment of broadband solution and the consequent increase in internet traffic which has motivated the need of new technology. This trend is instrumental to the high increment of number of users in network communication. In this paper, we present a design of optical spectral coding scheme for OCDMA communication. These OCDMA codes have been designed in the Fiber Bragg Grating (FBG) with appropriate encoder/decoder scheme. With this scheme, an intended receiver will discard all interfering signal from intending users and obtain orthogonality between users with OCDMA. However, there may be degradation in the network by Multiple Access Interference (MAI) effects which is induces by non-ideal FBG coders that lead to reduction in average error probability in the system performance. Simulated system shows the signature codes for six users with data rate of 200 Mbps. The performance is analyzed by comparing transmitter and receiver signal for the intended user.

1 INTRODUCTION

Data access security, bursty data transmission and connectivity to abundant bandwidth are main forces that generate interest in OCDMA techniques to transport information from one place to another. Multiple users can access asynchronously on same optical fiber with no delay with OCDMA technique (J. A. Salehi. 1989). There are many methods to achieve passive OCDMA which include optical orthogonal codes and modified prime sequence codes through Phase/amplitude masks employing spectral coding OCDMA system has an important role in high speed communication which not only allow increased number of users and simultaneous communication subscribers in a network, improving the performance of the network, but also simplify the network control and management, reducing processing time and alleviating the complexity and cost of hardware implementation (Vincent J. Hernandez et al. 2005)-(J. Faucher et al.). A unique signature code is assigned to intended user and that code will be modulated by the data of corresponding user. All the users of the network combine signals and transmitted on a single optical fiber and broadcast to each user. A block diagram of OCDMA system that most commonly used, is shown in Figure 1 and a basic diagram of multi-user shown in Figure 2.

The proposed OFDMA system is based on fiber Bragg grating to control the incoherent optical broadband signals by using On-Off shift key-

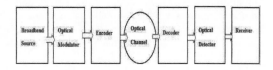

Figure 1. Block diagram of OCDMA system.

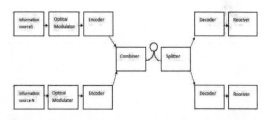

Figure 2. OCDMA transmission for N number of users.

ing with data (M. Kavehrad and D. Zaccarin. 1995)-(K.O. Hill and G. Meltz, 1997). Decision of reflected and transmitted components from the —high or —low frequency chip band is done by accurate spectral filtering of FBG. Here, FBG decoder is configured to overcome the MAI effect on by pseudo orthogonal codes. We establish an optical network by assigning N number of cycles of M number of sequence signature codes to N users. The spectral frequency pattern are determined by Signature code for the intended user that are centered about the grating frequencies. The ability to

Table 1. Signature codes for three users.

Users Number	Signature code
1	FBG1 + FBG3
2	FBG2 + FBG3
3	FBG1 + FBG2

decode the signal with decreasing MAI depends on number of users.

2 DESIGN OF CODE

Simulation of OCDMA network has been done for three users at 200 Mbps and 1550.1 nm frequency for FBG. If data is send by user no 1, then user1 and user2 are ON and user3 is OFF.

All the transmitted signals are multiplexed by the combiner and send through 10 Km optical fiber. Decoding is at receiving end, photo detectors are used to decode the signal with low pass filter by keeping user1 and user2 off. Similarly, signature code for 7 users can be given as in Table 2.

The desired input signal can be received by obtaining autocorrelation and cross correlation of the received OCDMA signals which reduced the MAI effect. The autocorrelation of the code should high and the cross correlation value should low as possible.

The signature code given by code vector Z_k,

$$Z_k = (Z_{k,0}, Z_{k,1}, \dots, Z_{k,N-1}) \quad (1)$$

where N = length of the code

$$Z_{k,N} \in \{0,1\}$$

for $0 \leq n \leq N-1$

The spectra of received signal (r) can be express as

$$r = \sum_{k=1}^{k} a_k Z_k \quad (2)$$

where $a_k \in \{0,1\}$ for $k = 1,2,\dots$

Table 2. Signature codes for seven users.

Users Number	Address Sequence	Data Bit	Transmitting Bit
1	1010101	0	10101010
2	1100110	0	11001100
3	1001100	1	10011001
4	1111000	0	11110000
5	1010010	1	10100101
6	1100001	1	11000011
7	1001011	0	10010110

3 SYSTEM SIMULATION

Figure 3 shows schematic of FBG based OCDMA system at 200 Mbps data rate and 10 Km fiber

length for seven users. Figure 4 shows the block diagram of encoder for user1 in which four FBGs at different wavelengths, spacing of 0.5 nm are used to extract the signature code for each user. Similarly, Figure 5 shows the decoder for user1 in which four FBGs are used to achieve the transmitted signal.

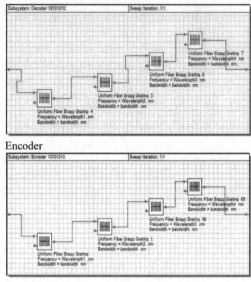

Encoder

Decoder

Optical power splitter is used on each output port which equally splits the incoming signal on each receiver's end. Seven users multiplexed their signature code into the fiber and decoder matched the received code to any of the seven transmitters. The received code intended to particular user is selected and Multiple User Interference (MUI) is rejected. Hence, degrading of MAI is achieved.

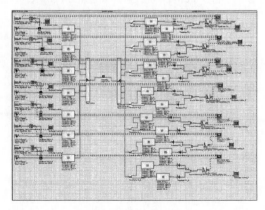

Figure 3. FBG based OCDMA system.

4 RESULT

FBG based encoding and decoding OCDMA system has been revealed for seven users at 200 Mbps data rate. Figures 4–10 shows the simulation results of transmitting signal, signature code and received signal for each user. The intended user can match the output signal with the input one by achieving high autocorrelation value. Hence MAI and transmission impairments can be degraded at a satisfactory level. The results show that the performance of the OCDMA system depends on number of users, data rate and length of the optical fiber. As the number of users increases, the received signal deteriorates and the matching of the signal become difficult which further results in increment of MAI.

The comparison of Quality Factor vs. length of the Fiber is shown in Figure 11. It has been analyzed that Quality factor decreases with the increment of the fiber length. Here, we also observed that Quality factor at 200 Mbps data rate is more than at 300 Mbps which shows that as the data rate increases, quality of the signal decreases.

The comparative results between BER and fiber length are shown in Figure 12 and Figure 13 for data rates of 200 Mbps and 300 Mbps. The graphs can be analyzed that User2 is affected much than User1,

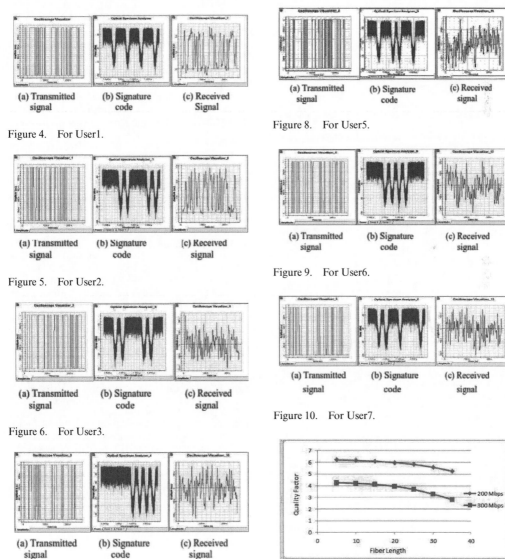

(a) Transmitted signal   (b) Signature code   (c) Received Signal

Figure 4. For User1.

(a) Transmitted signal   (b) Signature code   (c) Received signal

Figure 5. For User2.

(a) Transmitted signal   (b) Signature code   (c) Received signal

Figure 6. For User3.

(a) Transmitted signal   (b) Signature code   (c) Received signal

Figure 7. For User4.

(a) Transmitted signal   (b) Signature code   (c) Received signal

Figure 8. For User5.

(a) Transmitted signal   (b) Signature code   (c) Received signal

Figure 9. For User6.

(a) Transmitted signal   (b) Signature code   (c) Received signal

Figure 10. For User7.

Figure 11. Quality factor vs. Fiber length.

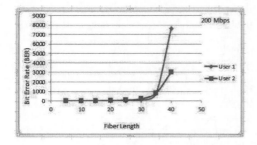

Figure 12.   BER vs. Fiber length (200 Mbps).

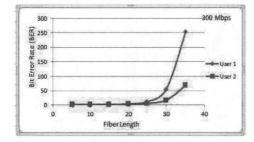

Figure 13.   BER vs. Fiber length (300 Mbps).

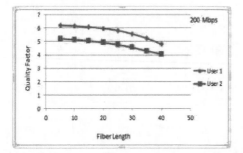

Figure 14.   Quality factor vs. Fiber length.

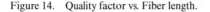

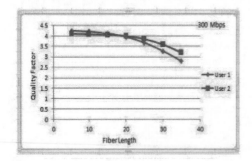

Figure 15.   Quality factor vs. Fiber length.

when data rate is increases from 200 Mbps to 300 Mbps. Also, it can be analyzed that BER is much more in case of data rate of 300 Mbps. When data rate further increased, BER becomes dominant and the data may distorted. Therefore, MAI increases as the data rate and number of users increases.

A comparative difference can be seen from Figure 14 and Figure 15 between Quality factor and number of users for data rates of 200 Mpbs and 300 Mbps. As the number of users increases, received power weakens and continuously deteriorates as the fiber length increases. Moreover, the quality factor diverted more at data rate of 300 Mbps than at 200 Mbps.

5 CONCLUSION

An OCDMA system based on FBG is presented in this paper for asynchronous transmission. The design and analysis of OCDMA system shows significant stability and scalability improvement with vulnerable degradation in MAI. These interference problems are produced because each user occupies different wavelength distribution. This system supports seven simultaneous users and gives acceptable signature codes. The transmitted code can be recovered with less MAI effect through proper encoder and decoder configurations. This system can be used for low data transmission with at a very low expense.

REFERENCES

Faucher, J., Adams, R., Chen, L. R. and Plant, D.V. Multiuser OCDMA system demonstrator with full CDR using a novel OCDMA receiver, *IEEE photon technology letter*: 1115–1117.

Griffin, R. A., Sampson, D. D. and Jackson, D. A. 1995. Coherence coding for photonic code-division multiple-access networks. *J. Lightwave Technology*: 1826–1837.

Hill, K. O. and Meltz, G. 1997. Fiber Bragg grating technology fundamentals and overview. *J. Lightwave Technology*: 1263–1276.

Hunter, D. B. and Minasian, R. A. 1999. Programmable high-speed optical code recognition using fiber Bragg gratings arrays. *Electron. Letter*: 412–414.

Kavehrad, M. and Zaccarin, D. 1995. Optical code-division-multiplexed systems based on spectral encoding of noncoherent sources. *J. Lightwave Technology*: 534–545.

Salehi, J. A. 1989. Code-division multiple-access techniques in optical fiber networks Part I: Fundamental principles. *IEEE Trans. Communication*: 824–833.

Salehi, J. A., Weiner, A. M., and Heritage, J. P. 1990. Coherent ultrashort light pulse code-division multiple-access communication systems, *J. Lightwave Technology*: 478–491.

Vincent J. Hernandez, Antonio J. Mendez, Bennett, V., Robert M. Gagliardi and William J. Lennon. 2005. Bit-Error-Rate Analysis of a 16-User Gigabit Ethernet Optical-CDMA (OCDMA) technology Demonstrator Using Wavelength/Time Codes, *IEEE Photonics Technology Letters*: 2784–2786.

Yu, K. and Park, N. 1999. Design of new family of two-dimensional wavelength-time spreading codes for optical code-division multiple-access networks. *Electronic Letter*: 830–831.

Communication and Computing Systems – Prasad et al. (Eds)
© 2017 Taylor & Francis Group, London, ISBN 978-1-138-02952-1

Application of MIMO in satellite communication subsystem

Gopesh Sharma, Praval Kumar Udaniya & Lokesh Tharani
Department of Electronics Engineering, Rajasthan Technical University, Kota, Rajasthan, India

ABSTRACT: Satellite communication subsystem is one of most important subsystem which is responsible to earn the revenue for the satellite designer. The revenue obtained by selling the total channel capacity is used to pay the whole cost of satellite. So it is one of important research field to improve the communication channel capacity within available limited resources (limited power). If the satellite communication subsystem adopts the Multiple Input Multiple Output (MIMO) technology's features it is sure that the communication channel capacity as well as energy efficiency will improve. The satellite MIMO communication subsystem will have high data transmission rate, high channel capacity, more energy efficiency, diversity.

1 INTRODUCTION

A satellite communication subsystem provides channel for data transmission the data may be audio, video or any other signal format, to users thus it generates revenue for satellite designer. So one should think about to increase channel capacity without increasing the power consumption. To improve the channel capacity many successive satellites are introduced with large capacity, lower cost but the number of users are increasing day by day need more channel bandwidth and higher data transmission rate. So transponders are used with communication satellite to full fill user demand. Early transponders of 250 or 500 MHz were used to provide bandwidth at 6GHz uplink and 4GHz downlink. Further satellites use the transponder which provides more bandwidth at 14/11GHz. For more improvement in channel capacity, bandwidth and signal to noise ratio there may be combination of MIMO with satellite communication. The satellite designer may inherit the properties of MIMO technology in the design of transponder. Thus it may provide more bandwidth within limited resources. In satellite MIMO communication subsystem the available bandwidth can be used very efficiently without any more additional power requirement and with high signal to noise ratio. Because the MIMO supports very important feature that is Diversity. Diversity means 'link reliability'. There are multiple paths between earth station and the satellite transponder uplink antennas. By using MIMO the system will able to choose most efficient (bandwidth, power) path to transmit/receive the data due to link reliability. Thus the path having highest SNR will be chosen. So there will be maximum utilization of available

bandwidth at minimum power consumption. Basically MIMO in based on Beam forming, spatial multiplexing and Diversity coding.

2 MIMO SATELLITE COMMUNICATION SUBSYSTEM DESIGN

MIMO satellite communication subsystem design is based on design of transponder. The transponder design (Fig. 1) contains uplink antenna, LNA (Low Noise Amplifier), BPF, Down converter, BPF, IF (Intermediate Frequency), up converter and HPA (High Power Amplifier). The LNA gives high output power and removes noise from upcoming data stream. BPF is used to remove unwanted signals. Further down conversion process is done. An IF amplifier is used for further operation. A up converter is used to upgrade the output of IF amplifier. HPA is used to provide redundancy. This is major component to provide life to satellite. Most HPAs have very large bandwidth than the allocated frequency band so it able to amplify the signal. And the last component is the downlink antenna. In MIMO satellite communication subsystem the both ends may adopt the properties of MIMO to increase the channel capacity, data rate and high SNR.

3 MIMO CHANNEL CAPACITY

This fig shows the relationship between the channel capacity and signal to noise ratio. The black line shows the relationship in capacity and SNR for a SIMO system which is very less as compare to MIMO system. There are three different cases of

MIMO in order of number of antennas are used. The red line shows MIMO system having 2 × 2 antennas. The green and yellow line show MIMO system with 3 × 3 and 4 × 4 respectively.

Thus it can be easily say that increment in the number of antennas will improve the capacity as well as SNR. One can use these properties in satellite transponder to provide much larger bandwidth.

In this Fig. the black line shows the results of SISO system which is very low. Blue and purple shows the results of MISO for 2 × 1 and 3 × 1 antennas. It shows that 3 × 1 MISO has high capacity and SNR. The red and green line shows the results of MIMO system MIMO capacity is increasing by increasing the number of antennas and there is also increment in the average SNR.

Fig. 3 shows the relation between SNR and the data transmission rate. It shows that the increasing number of antennas (MIMO) gives higher signal to noise ratio which provides very low fading environment and thus gives very high speed data transmission rate. The MIMO system supports the unique feature called diversity. In which the data is to be transmitted from transmitter chooses the most efficient path (high SNR, low fading). Thus it increases the over all channel capacity which is the main task of satellite communication subsystem.

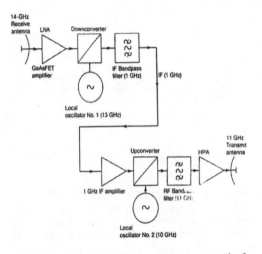

Figure 1. Simplified double conversion transponder for 14/11 GHz band.

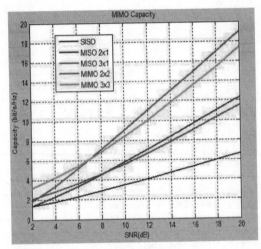

Figure 3. SISO Vs MISO Vs MIMO capacity.

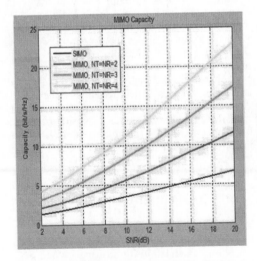

Figure 2. Capacity of SIMO and MIMO with increasing number of antennas.

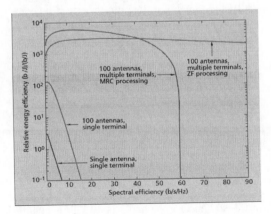

Figure 4. Energy efficiency and Spectral efficiency of MIMO system.

4 ENERGY EFFICIENCY AND SPECTRAL EFFICIENCY OF MIMO SYSTEM.

This Fig shows the results that the increment in the number of antennas always gives higher spectral efficiency as well as higher energy efficiency. In this Fig the purple line shows the results when single antenna used which is very low. The green line shows the value of EE (Energy Efficiency) and SE (Spectral Efficiency) when 100 antennas are used. And red and blue shows the high EE and SE.

5 FUTURE SCOPE

Traditional satellite communication subsystem can be replaced by MIMO satellite communication subsystem. This will provide high data transmission rate, diversity, high channel capacity and low power consumption. And thus there will be increment in revenue generated by selling the channel.

6 CONCLUSION

This paper highlights that the all cost of satellite earned by the selling of channel capacity which is generated by the transponders. Transponders are the major part of the communication subsystem. This paper shows if the designer implements the features of MIMO in communication subsystem then the channel capacity and data rate will improve which will generate more revenue.

REFERENCES

Erik G Larsson, Ove Edfors, Fredrik Tufvesson and Thomas L. Marzetta, Massive MIMO for Next Generation Wireless Systems, 2014, IEEE Communications Magazine, (52), 2, 186–195.
Kritika Sengar Nishu Rani, & Dolly Sharma2 Study and Capacity Evaluation of SISO, MISO and MIMO RF Wireless Communication Systems: Amity Institute of Information and Technology University, Noida, India.
Pratt, T., Bostian, C. W. and Allnutt, J. Satellite Communications, 2/e. Wiley, 2003.

Communication and Computing Systems – Prasad et al. (Eds)
© 2017 Taylor & Francis Group, London, ISBN 978-1-138-02952-1

Energy efficient heterogeneity overlap sensing ratio

Sandeep Sharma & Amit Shandilya
Gautam Buddha University, Greater Noida, India

ABSTRACT: In wireless sensor network directional and Omni-directional mobility have the ability to sense the region of the directional sensor node. It also provides a better efficiency effect on the coverage enhancement. Mobility or movable ability can overcome the coincided field of views and void areas occurring for the duration of the initial positioning. These are created on the random wave point measuring device or sensor model, used to estimate the performance of sensor nodes. Through the experimental results, in projected algorithm we will merge the overlap-sense ratio (OSRCEA) with enhance coverage overlap-sense ratio in Monte Carlo simulation (EEHOSR—MCS) in a direction to improve the network connectivity and the coverage ratio of the sensored area. Our lower energy shifting propagation loss model comprises of path loss function with randomly distributed shadowing, independent across with AWGN channel, which is simulated in MATLAB 2012a.

1 INTRODUCTION

In the field of wireless sensor network, the presentation of an MSN (Multimedia Sensor Network) is strongly associated with the position of separate combination sensors. In specific, alignment of a discrete multimedia sensor (way of its identifying element) is most significant for the sensor network used in the instruction to capture the complete network of the field. This is done to achieve the allocated detecting tasks positively. Nodes, by virtue of their position, have to cover and confide the network area to specific point locations, also known as the complete mark sensing zone. The method is used by an associated network over multi-hop wireless infrastructures. In a system, the sensors can be classified as directional and omnidirectional sensors. Omnidirectional sensors can correspondingly identify the nearby situation with their omnidirectional projections. On the other hand, the Omni-directional sensors in the field of the directional sensor have a restricted collection of sensing and communication abilities; meanwhile, it can perceive only a particular field of an idea in a narrow way. A decent amount of applied directional sensor nodes is now accessible in the real time, including cameras, electromagnetic and ultrasonic sensors (Wang, Z et al. 2009). In the field of DSN, the communication zone of a measuring device is a segment rather a disk. Directional sensors progress the feature of sensing and scale the interfering and disappearing, which in turn, improve the system shown, by way of the network period.

Demonstration of the directional sensors & their characteristic of:

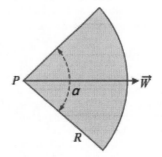

Figure 1. The directional sensing model of maximum directional area coverage.

1. In wireless sensor network each parameter of sensor is uniform. These include sensing radius, communication radius and sensing angle.
2. Every indicator sensor can intellect one fractional of Omni-direction angle only.
3. All directional sensors are secure and can exchange random viewpoint regarding the sensing area.
4. The message or communication range is double than the identifying ranges such that sensing neighbors can dependably connect.

1.1 *The sensing coverage*

As sensors yield their places on wireless nodes, omnidirectional identifying range senses the supposition losses processed meaningfully. A characteristic coverage sensor has is a sectoral insight and gets strongly affected by nearby obstacles. Multisensors, such as a controlling multi-dimensional.

In this work, the assumption of sensors nodes have a *secure* lenses only if field of vision with angle Θ, and they can only regulate their FoV. The use in terms of *"sensors"* for easiness to signify wireless sensors including MCS and OSR sensors consuming directional view. Each node is prepared to study its position evidence via any weak node localization method for WSN (M. A. Guvensan and A. G. Yavuz, 2011).

1.2 *Network lifetime*

In the wireless network, we identified that most Sensor nodes suffer due to its partial battery power volume. If sensor nodes are smaller in size than present batteries, nodes of sensor do not last as extensive as preferred. Then maneuvering sensor nodes have various occupied instructions, rotatable powered project is compulsory to use all engaged directions. Indeed, physical program consumes considerable extra power than other actions (Kansa et al. 2007). So, tangible accomplishments like revolving a sensor node about it reduce or affecting it to an alternative location should be well deliberate to reduce the energy feeding.

1.3 *Additive White Gaussian Noise (AWGN)*

The Additive white Gaussian noise is a sound frequency. It is a simple model of the imperfections that a communication channel consists of the disturbance caused by the Thermal noise is modeled as Additive White Gaussian Noise.

This sound frequency is decent for cable and profound space communication but not for experienced communication in multipath, land obstructive and interfering.

AWGN is used to pretend contextual noise (sound) of channel. In time domain the noise gets added in the transmitted signal so the established signal can be represented as—r (t) = s(t) + n(t), wherever s(t) is conveyed signal and n(t) is contextual noise (sound). C AWGN = W log2 (1+P/N0 W) bits/Hz.

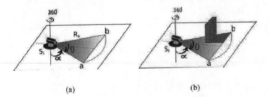

(a)   (b)

Figure 2. Design of 2 (a) and (b) dimensional field of view of a software sensor node, where α is the perpendicular angle to the border edge of FoV, Θ is the FOV top angle, and Rs is the supreme software sensing range. Sensors that can sense a directional view, typically called *Field of View (FoV)*.

2 PROBLEM STATEMENT

In WSN, the analysis of coverage issue is the foremost concern of the entire network. The problem of Omni-directional sensor networks has intensively been deliberated in the previous decade (Friis, H.T. 1946) In this part, we discuss the primary approaches for coverage development in omnidirectional sensor grids. Similarly, coverage difficulties in DSNs need more precise explanations since maneuvering sensor nodes prepared through ultrasound, electromagnetic and audiovisual sensors may affect in various instructions. Manipulating mobility ability of those nodes is one of the simple solutions. A casually organized Directional sensor network; the power can be overlay zones and blocked areas after the initial positioning. A DSN through motility corrects its occupied position beside x, y, and z-axis. Therefore, the node delivers them through a new FoV deprived of affecting to a unique position. This novel direction points towards an entirely improved coverage by (i) minimizing overlapped regions and (ii) if an occlusion-free field of view.

The accessible explanation for the coverage issues movement is slightly better than motility because of its insignificant rate and less energy feeding.

(Y. Mohamed and K. Akkaya, 2008) Subsequently, a directional sensing node preserves its environmental location whereas correcting its Field of View; the node does not need another driving device or GPS device. This efficiently reduces the entire manufacture price of the sensing node. Simulation results display that the best point for the positioning price and the coverage development relation can be attained by making use of a certain number of inactive, motile, and mobile nodes. Therefore, it proposes an EEHOR-MCS to enhance the area coverage. This research aims to exploit the space coverage of a randomly located DSN. The issue of occupied direction forecast to cover best areas, also known as Maximum Directional Area Coverage (MDAC) is complex. We define the directional sensing ideal and schemes for the Maximum Directional Area Coverage (MDAC) problem. In DSN model, every directional sensor cannot sense the complete spherical zone.

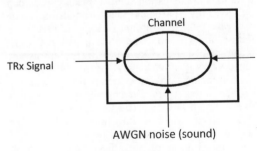

Figure 3.   AWGN channel (frequency).

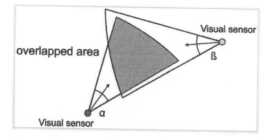

Figure 4. The overlapped are of the 2 sensor nodes.

3 SYSTEM MODEL

Viewing targets or sensing area grid of weak nodes, the FoV (field of view) of sensor nodes might overlay. In other words, when the field of view of two or multiple nodes crosses, the same object observed by more than one visual sensor, frequently from diverse directions and standpoints. In such a process, requests can describe severance built on the field of view overlaying. In the meantime measuring device with overlay identifying areas considers the same objective, but from different viewpoints. Therefore, they don't save same information.

4 PROPOSED IMPLEMENTATION

1. The segments for lower energy node consumption formulation & 2 categories of important things to estimate the k-coverage probability in a network with log-normal investigation (however the shadowing distribution can be slightly random and without coverage sensing area of higher energy shifting nodes.
2. Sensors node uses quadrature technique or a modest analytic formula for coverage sense with advance node energy.
3. The new composite high-dimensional uses inquadrature approaches for low dimensions and quasi-random integration for higher (n > 2) computes coverage probability for a network with Rayleigh fading (exponentially circulated with unit mean) and log-normal shadowing.
4. We present a modification of variables motivated by the dimensional spherical coordinates.

4.1 Algorithm for EEHOR-MCS

1. Find neighboring sensors nodes in grid;
2. If dead node occurrences;
3. Shift advance node due to dead node in another grid or area;
4. Set parameter for state = 1
5. While (overlap == 1)
6. Calculate OSR;
7. The Nodes are overlap optimal angle according to the rotation angle function
8. If (network is equilibrium)
9. Overlap = 0;
10. End
11. End
12. Calculate OSR;
13. While (OSR> = predefined threshold)
14. Calculate priority;
15. If (priority is highest)
16. State = 0;
17. Send state information to its neighboring sensors
18. Else
19. Calculate OSR;
20. End

5 RESULTS AND DISCUSSIONS

In this part we show the outcome of the proposed (EEHOR-MCS) algorithm, with the using of Random approach in terms of each sensor being excellent in sensing track arbitrarily.

In Figure 6 sensor area predict the minimum energy node which is executing towards low or dead.

In Figure 7, higher energy node is shifting towards weak node in the void grid or single sensor node in a network.

In Figure 8, when higher energy node shifts towards weak node then network updates the shifted position.

In Figure 9, after updating entire network, it views like hexagonal grid. The position of node changes to avoid any kind of intrusion.

In Figure 10, average energy of nodes in EEHOSR-MCS is higher in comparison to Energy of nodes in OSRCEA for the same round numbers.

Table 1. Simulation parameter.

Simulation parameters	Values
Network of field size (area)	200*200
Number of sensor nodes (N)	100
Number of advanced nodes (an)	0.2
Number of normal nodes(nn)	0.8
Energy of a normal node (E_0)	0.5
Location of the dual base station	X, Y
Sensor network deployment type	Random
Simulator software Version	2012a
Mobility model	Random wave-point
Sensing range	100
Grid radius	3.5
Fading	AWGN

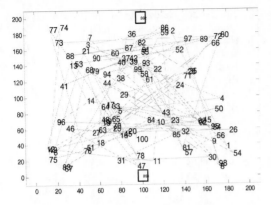

Figure 5. 100 Nodes with node shifting to void grid in 100 coverage range.

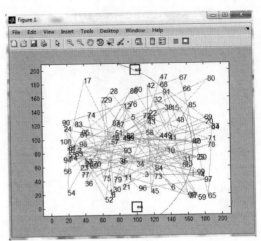

Figure 8. After shifting the nodes network update the node position.

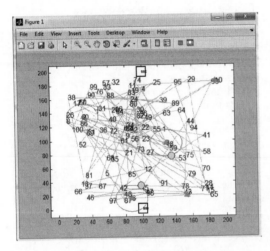

Figure 6. Node sense the weak node position in grid.

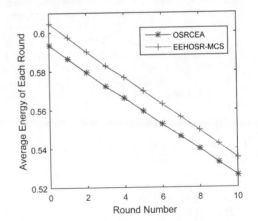

Figure 9. Hexagonal grid for 100 nodes.

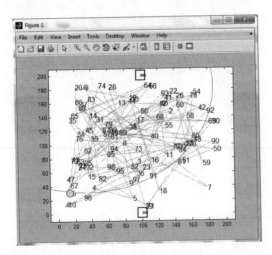

Figure 7. Higher energy node shifting towards the weak node.

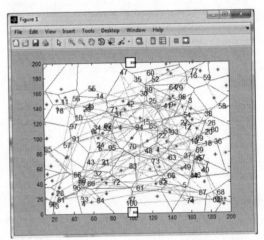

Figure 10. Average energy of each node w.r.t round number.

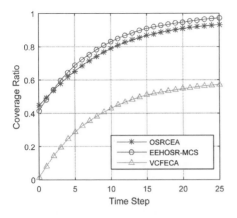

Figure 11. Showing Improvement Coverage ratio for EEHOR-MCS comparatively to Energy Coverage ratio.

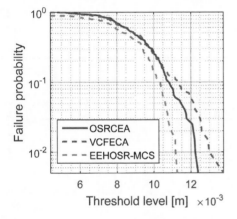

Figure 12. Failure probability of OSRCEA, VCFECA and EEHOSR-MCS.

In Figure 11, the coverage ratio is getting improved with respect to time-step at initial stage to gain better improvement when coverage increases in response to time as compare to Grid Based Data Aggregation (GBDAS).

In Figure 12, as per the graph, the EEHOSR-MCS shows the less failure probabilty as compared to existing techniques (OSRCEA,VCFECA).

6 CONCLUSIONS

In this paper, we worked on the grid based enhance coverage ratio with overlap sensing connected to associated coverage in directional WSNs. Specifically, we investigated the dead node problem. Furthermore, we proposed the Enhanced Coverage Overlap Sensing Ratio (EEHOR-MCS) method, created on the dimension of an overlay area among directional sensors. The experimental outcomes prove that our planned procedures outperformed

for the coverage enhancement and overcame the dead node occurrence.

Impending work will focus on merging the algorithm with an energy consumption model to give deliberation to both coverage and lifetime performance in mobile and direction-rotatable directional sensor networks.

REFERENCES

Adler, L., R., Buonadonna, P., Chhabra, J., Flanigan, M., Kushalnagar, N., Nachman, L., and Yarvis, M. (2014): Design and deployment of industrial sensor networks: experiences from a semiconductor plant and the North Sea. In Proceedings of the 3rd international conference on Embedded networked sensor systems. 64–75.

Friis, H.T.1946: A Note on a Simple Transmission Formula. In Proceedings of the IRE, Vol. 34, No. 5, 254–256 2. He, T., Huang, C., Blum, B. M., Stankovic, J. A., and Abdelzaher, T.: Range-free localization schemes for large scale sensor networks.

Girik Pachouri and Sandeep Sharma (2015). Anomaly Detection in Medical Wireless Sensor Networks using Machine Learning Algorithms. Elsevier Procedia Computer Science proceeding of 4th International Conference on Eco-friendly Computing and Communication Systems (ICECCS), NIT Kurukshetra, vol.70, pp.325–333.

Guvensan, M. A. and Yavuz, A. G. February 2011. "On coverage issues in directional sensor networks: A survey," Elsevier AdHoc Networks.

Jaiprakash Nagar and Sandeep Sharma (2015).K Barrier Coverage Based Intrusion Detection System for Wireless Networks. 50th Golden Jubilee Annual Convention of Computer Society of India (CSI-2015).

Kansal, W. Kaiser, G. Pottie, M. Srivastava, and G. Sukhatme, 2007. "Reconfiguration methods for mobile sensor networks," ACM Transactions on Sensor Networks, vol. 3, no. 4.

Ma, H., Zhang, X. and Ming, A. April 2009 "A coverage-enhancing method for 3d directional sensor networks," in Proc. Of IEEE Intl. Conf. on Computer Communications(INFOCOM'09), Rio de Janerio, Brazil,, pp. 2791–2795.

Mohamed, Y. and Akkaya, K. 2008. "Strategies and techniques for node placement in wireless sensor networks: A survey," AdHoc Networks, vol. 6, no. 4, pp. 621–655.

Sandeep Sharma and Rajesh Mishra (2014). A Cross Layer Approach for Intrusion Detection in MANETs. International Journal of Computer Applications, 93(9):34–41.

Tao, D., Ma, H. and Liu, L. November 2006, "Coverage-enhancing algorithm for directional sensor networks," in Lecture Notes in Computer Science: Mobile Ad-hoc and Sensor Networks, vol. 4325, pp. 256–267.

Wang, Z., Bulut, E., Szymanski, B. K. 30 November–4 December 2009. Distributed Target Tracking with Directional Binary Sensor Networks. In Proceedings of the 28th IEEE Conference on Global Telecommunications, Honolulu, HI, USA,

Zhao, J. and Zeng, J.-C. September 2009 "An electrostatic field-based coverage-enhancing algorithm for wireless n multimedia sensor networks," in Proc. of IEEE Intl. Conf. on Wireless Communications, Networking and Mobile Computing (WiCom'09), Beijing, China, pp. 1–5.

Communication and Computing Systems – Prasad et al. (Eds)
© 2017 Taylor & Francis Group, London, ISBN 978-1-138-02952-1

On energy-spectrum efficiency trade-off in 5G cognitive radio system

Sandeep Sharma & Anupam Kumar Yadav
Gautam Buddha University, Greater Noida, India

ABSTRACT: Cognitive Radio (CR) is a technique which makes use of radio spectrum in a more efficient way by intelligently exploiting licensed spectrum. In cognitive radio cellular networks, we combine cellular radio and cognitive radio into a system for fifth generation cellular systems. Cellular radio resources also known as licensed band, they have small bandwidth, high transmission power and high reliability. Whereas cognitive radio also known as unlicensed band and have broad bandwidth, low transmitting power and low reliability. In this research paper our challenge is to jointly utilize both the bands to increase the system performance. We will also perform EE and SE trade-off in CR Cellular networks at different architecture levels. For future cognitive cellular networks, we have illustrated four different examples for SE and EE trade-off using Matlab R2015a.

1 INTRODUCTION

There is increase in number of smart-phones, laptops and PDAs every year, which have different Quality of service requirement like web browsing, faster internet, multimedia downloads etc. So we have started research on fifth generation technology for wireless communication in order to meet there increasing demand. This fifth generation cellular technology is expected to be deployed beyond 2020 C.X Wang (2014). According to a survey economy of the world can be improved by proper utilization of Information and Communication Technology (ICT) advancement. It will also allow easy and fast access to the technology (2012). Advancement in the field of wireless communication technology has improved ability of people to communicate over long distance in both commercial and social aspect. 1G cellular network was analog system, invented during 1980s with working frequency of 150 MHz. Whereas 2G cellular network was launched on GSM standard in Finland. 2G supports technologies like GPRS, CDMA and GSM. It also supports digital encryption of data. 3G is the third generation in cellular network, which uses wideband wireless network and for data transmission it uses packet switching. 3G also enables devices like PDAs, mobile phones and mobile dongles to deliver broadband speed internet. This mobile broad band allows its customers to download files, browse internet, download music, videos and email with extremely fast speed. 3G is the third generation in cellular network, which uses wideband wireless network and for data transmission it uses packet switching. 3G also enables devices like PDAs, mobile phones and mobile dongles to deliver broad-band speed internet.

This mobile broad band allows its customers to download files, browse internet, and download music, videos and email with extremely fast speed.

4G cellular network was launch in 2008. This system not only provides voice and other 3G services but also provides ultra-broadband network access to various mobile devices. Their applications vary from IP telephony, HD Mobile Television, live video conferencing to gaming services and other cloud computing operations.

As we know energy and spectrum both are natural and limited resources. So the challenge for 5G cognitive cellular network is to utilize both the resources judiciously, in order to meet future requirements. Cognitive cellular network utilizes both licensed and unlicensed band for data transmission and reception, to increase the system performance. Here SE means how efficiently bandwidth of the system's network can be utilized, while EE measures how efficiently it consumes energy F. Heliot (2012). We also know that SE and EE are inversely proportional to each other, increasing either of them means degrading the other resource. So a trade-off between them is taken.

2 CAPACITY OF COGNITIVE CELLULAR NETWORK

In cognitive cellular network we have two types of natural resources spectrum (Bandwidth) and energy (Power). For SE unit is bits per second per hertz and for EE we make use of bit per joule capacity Hong (2014). In this paper centralized fifth generation cellular network capacity is classified into three different levels.

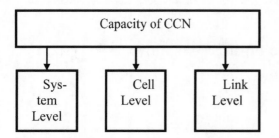

Figure 1.   Levels in cognitive cellular network.

In cognitive cellular network we have two types of natural resources spectrum (Bandwidth) and energy (Power). For SE unit is bits per second per hertz and for EE we make use of bit per joule capacity Hong (2014). In this paper centralized fifth generation cellular network capacity is classified into three different levels.

3   ENERGY EFFICIENCY

Here trade-off between Energy efficiency and Spectrum efficiency is studied with the help of different architecture combinations, levels and capacity, which is still an open research field. This EE and SE has become one of the key performances to be used as evaluation criteria in a communication system. These two are conflicting criteria, but with the help of trade-off they can be linked. Here improvement of one of them means degradation of other. So in order to get better performance a balance is to be achieved between two desirable but incompatible features like spectrum efficiency and energy efficiency. We have considered three examples of these combinations.

Architecture for Cognitive radio in Cellular Networks (CCN) can be categorized broadly into two types I. Humor (2011). Non-Cooperative cellular network and Cooperative cellular network.

4   NON-COOPERATIVE ARCHITECTURE IN CCN

This Architecture makes use of two different radio interfaces, licensed band and cognitive radio resource. Here base station can utilize both or single interface. It mainly supports communication from short-to-medium range and capacity of this CCN is the combination of two networks I. Ku (2013). We have assumed different scenarios for different users. For those users which are closer to a macro-cell base station uses power limited cognitive radio resources. But for users which are further away from macro-cell base station make use of

licensed radio resources. Similarly, this architecture provides services with strict Quality of Service requirement to the licensed RR while services with relaxed Quality of Service requirement are given to the cognitive radio interfaces.

5   COOPERATIVE ARCHITECTURE IN CCN

This architecture also uses both licensed and unlicensed radio resources, the only difference is that it allows processing and relay information of distributed users in a coordinated fashion. As compared to non cooperative architecture this has got more system capacity for longer range of communications. In this architecture we make use of relay in order to communicate to base station. Cognitive radio resources are used for local coverage. It also makes use of Virtual Antenna Array (VAA), where a base station array consists of several antennas placed together to form several independent VAA groups. Here useful information is extracted by individual antenna from the group of signals and then forwarded to other antennas.

6   EE-SE TRADE OFF IN CCN

To study further about cognitive cellular networks we have combined different architectures, levels and capacities which is still an open research field CX Wang (2013). These combinations not only give us information about EE-SE Trade-off, it also reflects the latest research progress in the field of CCN J. G. Andrews (2011). Following are the combinations:

1. System level + Non Cooperative Cognitive
2. Cell Level + Cooperative Cognitive
3. Link level + Cooperative Cognitive

7   SIMULATIONS AND RESULTS

For the simulation purpose MATLAB R2015a is used. In order to detect signal and estimate performance we find Probability of detection (Pd) and Probability of false alarm (Pfa). Input given to the detector is combination of signal + noise.

For better performance we add Additive White Gaussian Noise (AWGN) to the channel. With number of observations N = 100 and SNR = −10 dB.We obtain a graph between probability of detection is taken on y axis and probability of false alarm on x axis as shown in Fig. 3, here performance is defined by Receiver Operating Characteristics (ROC). As the value of Pd and Pfa increases same time threshold value decreases.

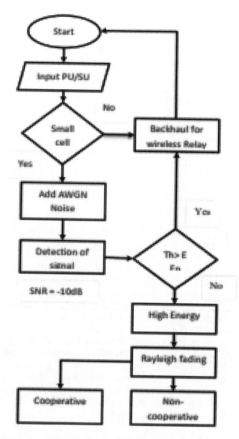

Figure 2.   Flow chart.

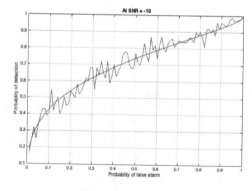

Figure 3.   Pd Vs Pfa.

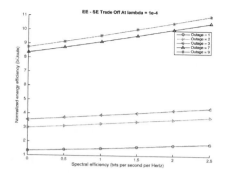

Figure 4.   Normalized EE vs SE.

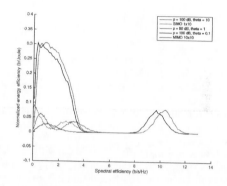

Figure 5.   Normalized EE vs SE.

8   NON-COOPERATIVE SYSTEM LEVEL

Here we consider non-cooperative system level outage capacity of small cells. Outage Capacity is the highest rate of reliable communication for some fixed outage probability (p), it is find out by $(1 - p)$. Our interest is to study EE-SE trade off in non cooperative architecture. We consider density of cognitive Base stations as lambda under Rayleigh fading channel. Here EE-SE is derived from outage capacity. Three factors that affect EE-SE are Base station density (lambda), outage probability (p) and reliability (k = 1) K. Xue (2013).

We observe that BS density (lambda) has got little impact on Normalized energy efficiency. With the increase in the value of outage probability (p) SE-EE also increases linearly. So more outage probability better EE-SE trade-off.

9   COOPERATIVE CELL LEVEL

Here we consider Cooperative cognitive VAA Cell level system. Reliability is assumed to be (k = 1), in MIMO network we have multiple mobile users and BSs. We consider that all BSs are having n number of antennas and each mobile user is having a single antenna. They use Rayleigh fading channel for transmission of signals. EE-SE Trade off depends upon the Theta = bandwidth ratio and p = transmitting power L. Musavian (2010).

There is no inter cell interference. We observe from the graph that when we increase the value of power (p) and bandwidth ratio (theta) we get EE-SE trade off for a small region i.e. when power = 100 dB and bandwidth ratio is 0.1.

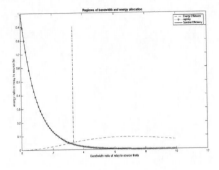

Figure 6.   Power ratio vs Bandwidth ratio.

10   COOPERATIVE LINK LEVEL

Here we consider cooperative cognitive relay link level system. In this example we consider the role of relay and three nodes. Node1 known as source node broadcasts its message to node2 which is relay node using licensed radio resources. Then node2 (relay node) forwards information to the destination node using unlicensed radio resources. Rayleigh fading channel is used S. Sridharan (2008).

Here we calculate EE-SE trade off with different values Cognitive bandwidth ratio (theta) and power ratio (phi) where reliability constant k is assumed to be 1. We observe that capacity of relay channel is quite limited, we can only achieve EE and SE trade off region when values of theta and phi are relatively small X. Hong (2010) and X. Hong (2013). On combining all three parameters EE, SE and Capacity we get required trade-off region for better system performance. This region would contribute increase in cell through put and also increase in data rate.

11   CONCLUSION AND FUTURE WORK

From our above results we can see that EE-SE trade off study is a powerful tool, for getting us know about the fundamental limits, pros and cons of cellular cognitive radio. It also offers theoretical knowledge about various CCN levels and architectures. Study tells us how our natural resources like energy and spectrum can be properly utilized and managed in fifth Generation (5G) cognitive cellular networks. This will allow accommodating more number of mobile users in the same spectrum at the same time. In future this fifth generation technology will contribute to increase data rate, throughput and further improvements in spectral and energy efficiency. In future for better performance we can use Nakagami fading channel and Rician fading channel which not been discussed in this paper. These channels can be used in different CCN architectures in order to study system performance. With the help of this 5G technology everything in our life becomes more connected. From our mobiles to our fitness trackers from cars to robotic manufacturing plants and banking. 5G will enable capabilities that will revolutionize how we live our life.

REFERENCES

Andrews J. G., F. Bacccelli, and R. K. Ganti. Nov. 2011. "A Tractable Approach to Coverage and Rate in Cellular Networks," *IEEE Trans. Commun.*, vol. 59, no. 11, pp. 3122–3134.

Anju Singh, Karan Singh, Sandeep Sharma. Sept. 2013, "Capacity Based Multicast Channel Assignment in Wireless Mesh Network", *International Journal of Communication and Networks*, vol. 5, no. 3B, pp. 673–679.

Apr. 2012. Commission of the European Communities, Staff Working Document, "Exploiting the Employment Potential of ICTs".

Heliot F., M. A. Imran, and R. Tafazolli. May 2012 "On the Energy-Efficiency Spectral-Efficiency Trade-Off over the MIMO Rayleigh Fading Channel," *IEEE Trans. Commun.*, vol. 60, no. 5, pp. 1345–1356.

Hong X. *et al.* Sept. 2010. "Capacity of Hybrid Cognitive Radio Networks with Distributed VAAs," *IEEE Trans. Vehic. Tech.*, vol. 59, no. 7, pp. 4201–4213.

Hong X. *et al.* Oct. 2013. "Energy-Spectral Efficiency Trade-Off in Virtual MIMO Cellular Systems," *IEEE JSAC*, vol. 31, no. 10, pp. 2128–2140.

Hong, Xuemin, Jing Wang, Cheng-Xiang Wang, and Jianghong Shi. 2014. "Cognitive radio in 5G: a perspective on energy-spectral efficiency trade-off", IEEE Communications Magazine.

Humar I. *et al.* Mar. 2011. "Rethinking Energy-Efficiency Models of Cellular Networks with Embodied Energy," *IEEE Network*, vol. 25, no. 3, pp. 40–49.

Ku I., C.-X. Wang, and J. S. Thompson. Oct. 2013. "Spectral-Energy Efficiency Trade-Off in Relay-Aided Cellular Networks," *IEEE Trans. Wireless Commun.*, vol. 12, no. 10, pp. 4970–4982.

Musavian L. and S. Aissa. Mar. 2010. "Effective Capacity of Delay-Constrained Cognitive Radio in Nakagami Fading Channels," *IEEE Trans. Wireless Commun.*, vol. 9, no. 3, pp. 1054–1062.

Sandeep Sharma and Rajesh Mishra. Oct. 2015, "A Simulation Model for Nakagami-m Fading Channel for m > 1", in *International Journal of Advanced Computer Science and Applications*, vol. 6, no. 10, pp. 298–305.

Sandeep Sharma, Jaiprakash Nagar, Poonam Singh. March 2016, "Modeling and Performance Analysis of Wireless Channel", *in the proceeding of 2016 IEEE 3rd International Conference Computing for Sustainable Global Development (InciaCom-2016)*.

Sridharan S. and S. Vishwanath. Feb. 2008. "On the Capacity of A Class of MIMO Cognitive Radios," *IEEE J. Sel. Signal Proc.*, vol. 2, no. 1, pp. 103–117.

Wang C. X. *et al.* Feb. 2014. "Cellular Architecture and Key Technologies for 5G Wireless Communication Networks," *IEEE Commun. Mag.*, vol. 52, no. 2, pp. 122–130.

Xue K. *et al.* Dec. 2013. "Performance Analysis and Resource Allocation of Heterogeneous Cognitive Gaussian Relay Channels," *Proc. IEEE GLOBECOM '13*, Atalanta, GA.

Communication and Computing Systems – Prasad et al. (Eds)
© 2017 Taylor & Francis Group, London, ISBN 978-1-138-02952-1

Energy minimization using NP-completeness Lower Energy Adaptive Aware Sink Relocation (NP-LEASR) in wireless sensor network

Sandeep Sharma & Jitendra Kumar Sonkar
Gautam Buddha University, Greater Noida, Uttar Pradesh, India

ABSTRACT: WSN has numerous application like traffic monitoring, temperature monitoring, health-care monitoring and so on. In WSN, Sink mobility is important term for relocation scheme in different sensing node application. Sink creates hot-spot to distribute load from one to another node, in this way energy consumption in sensing node application is less and increases the lifetime of the sensor network. As we know that in WSN energy and bandwidth resources are limited and these are the main parameter used for efficient increasing Sink mobility Scheme and lifetime of the Sensor Network. In the previous EASR method there is gradually increasing and decreasing energy which comes under the transmission range. In this paper we proposed method i.e. LEASR (LOWER ENERGY SINK RELOCATION) for minimize energy consumption, comparison of the existing EASR method with proposed method LEASR and increases the performance of the network. Thereby we further discuss numerical methods to extend the lifetime and result plotted in MATLAB software.

Keywords: WSN, Energy, Network lifetime, Sink mobility

1 INTRODUCTION

Sink Mobility Node scheme are the best solutions for the improving effective the lifetime of the Wireless sensor network only when sink is in moving position (J. Luo and J. P. Hubaux, 2005). Base station (mobile agents) aims to solve the problems hot—spot of the WSN where first hop neighbor node of the moving sink node is depletion due to the fast forwarding message. In order to create distributed forwarding load of the traffic through all the sensor nodes sink must be mobile through all the region of the transmission range, information is stored in moving node. Researcher said for the last few years many of the energy efficient protocol and increases the performance of the networks in terms of lifetime, energy and so on. Although the mobile sink when its moves according to load balancing. Shifted to other nodes is the eight half quadrant for specific movement of the sink (L. Sun et al., 2006). Multiple Sink mobility is major term for sink relocation and used in the many high gathering data collection application and also small application in WSN (Chu-Fu Wang et al. 2014). Sink mainly depend upon the residual battery and for the forbid placement of nodes check nearest to the low energy nodes. To our best knowledge work under sink relocation mainly focus on optimal location of the sink according to the initial battery energy and area of the network setup.

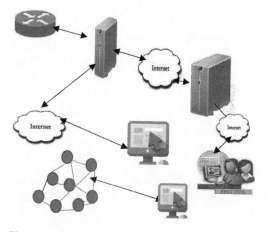

Figure 1. Operating Scheme for WSN.

The paper is explained as follows. In section 2, we describe about works related to Sink Relocation method. Section 3 describe related work. Section 4 presents proposed work. Section 5 describe the Numerical Analysis and finally, we discuss about Conclusion part in 6.

2 RELATED WORK FOR SINK DISPLACEMENT

In EASR scheme or technique focus on Sink Displacement Mechanism and Energy Aware transmission range and with 8 fan shaped sector decide when to move Sink with half quadrant based method. In this first we discuss Energy Consumption model and further discuss about the Maximum Capacity Graph

2.1 The energy consumption modelwsn

Firstly we consider Energy Consumption model for the proposed method LEASR and for this first order ratio model and by using this model we perform various simulation and performance parameter. In this two terms are used for transmitting and receiving of total energy sensed in the sensor node with the separation of a distance d respectively according to the proposed method LEASR i.e. E_{TX} and E_{RX}. E_{TX} can be divided into two parts, first one is energy that is consumed for the message transmitting is equal for the electronic component of transmit and receiving circuit, second one is used for the component of amplifier loss and is equal to the $\varepsilon_{amp} \times k \times d^n$. Summary of the above theory part as below in the mathematical equations as shown below:

$$E_{TX}(k, d) = E_{elec} * k + \varepsilon_{amp} + k * d^n \qquad (1)$$

$$E_{Rx}(K) = E_{elec} * k * k \qquad (2)$$

2.2 Load balancing model

As we know that for the increases the lifetime of the sensor network we should talk about energy saving mode and this is the main key issue of the WSN network. In WSN, routing protocol for the sensing and reporting message to the node generally classified in two technique static routing and dynamic routing. In the static routing reporting message to the different node is fixed and sensed according to the routing fixed in the loop of the network. On the other side dynamic routing transmission can be done through no of rounds and current position of the sensor nodes. Dynamic routing is used in this proposed method LEASR and energy consumption rate is much achievable amount. Dynamic routing i.e. Maximum capacity graph Jan (Sandeep Sharma and Rajesh Mishra, 2015) is taken as load balancing and this is the main underlying scheme for the proposed method LEASR. The description of MCP graph briefly discuss in this paper through illustration diagram as show in the Figure 2 and explained the procedure one by one of the MCP Method. According to the MCP (Maximum C, actual residual battery energy represent in the form of set equation G = (U, Y) where set U represents sensing nodes heap of the network and Y represent desirable forthright transmission through other node of the network. Suppose that r: U → Y⁺ may be function actual extra battery energy. For the ex the node with sink S with infinity energy for the

large capacity in the network The MCP mainly consist of three procedure steps. (1) The graph G should be mapped in the network N. (2) Determining MCP for every sensor node network (3) Loop performance and actual residual battery energy should be up-to-date. Above all these steps repeated again and again reporting each round of the message of the neighbor nodes. In the Fig. (1) The mapped graph for MCP operations are explains as follows. Suppose that level no be L_v with every sensor node v ε V express minimal path length form v up to Sink S. As seen in the first figure $L_g = L_h = 4$ and La = $L_B = 1$, these analysis done through layered network graph N from G. So we delete that edges (b, a) and (h, g) from the mapped layered network graph G. As we goes through procedure and find minimal path we must removes edges and obtained final result in the 3 figure of MCP routing. The procedure repeat until each round on the sensed node will exhaust its battery energy.

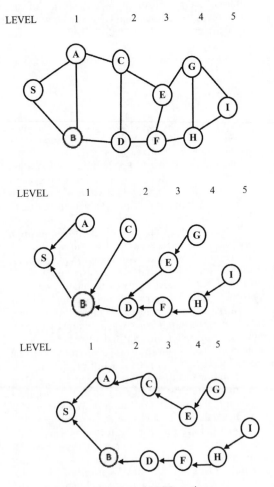

Figure 2. An Explanation of MCP routing.

248

3 RELATED WORK

J. Luo and J. P. Hubaux (2005) Joint Mobility Routing work on the circular trajectory with utilization of the WSN. It mainly work on the constant velocity with respect to circular trajectory movement. In this JMR two approaches are used fast mobility and slow mobility to increase the performance of the network. Distinguished based on the speed moving sink and delay of distributed data. This mainly work on the mobility of the Sink with avoidance traffic and decrease the load of the network.

L. Sun et al. (2006) Moving scheme for the sink is describe in this paper. It mainly discuss about the sink position and condition related to when sink actually move with the half quadrant Based moving scheme. Half quadrant based moving work on the 6 fan shaped sector with miry (one sector must be quasi-hotspot) and clean sector (none sector be quasi-hotspot).

I. S. AlShawi et al. (2012) discussed about life span of the sensor network by utilizing combination Fuzzy approach and A-star algorithm. In this paper approaches is to determine the optimal routing path with the maximum battery capacity, less number of nodes and jam loads on network. In this every sensor nodes are distribution random nodes in the given area and suppose all the sensor nodes its own position as well as that of its neighbor and the sink as well as the nodes lies with the same transmission range and additionally nodes queue application launch to wait for the forward transmission. Overall this paper to achieve network lifetime elongate of the network and distribution of energy consumption equal through limiting energy cost.

KOC and korpeoglu (2015) discussed about two main method i.e MSMA (Multiple Sink Movement Algorithm) & PMA (Prevent & Move Away) This paper explains about the much improvement in irregular and immobile sink of different cases. These two methods decrease the load and traffic load forwarding to the next level of node.

4 PROPOSED WORK

Our proposed technique mainly resolve problems of the sink relocation with a new technique or method NP-LEASR algorithm. NP-LEASR algorithm moves the sink approaches to the maximum area respectively in the overall network system. The battery exhaust out nodes may cause several problems such as, coverage hole and communication hole problems which can be validate by higher energy cluster node from the sink which improve the network life span, the sink is moved towards the last-hop relays which are the most involved in packet transmitting by using NP-Completeness problem. In this proposed paper polynomial time algorithms t give better performance for packet transmission. This NP-LEASR improve the average jitter as well as other parameter like end to end delay, energy consumption and so on.

4.1 Alogrthim for proposed NP-LEASR

Procedure for NP-LEASR
γ =Actual Communication range
Z = Startup battery energy
t = Communication range
i = No of iteration per nodes
$k(v)$ = Actual residual battery energy
n = Total no of nodes
N_n = No of specific nodes
V = Collection of sensor node
N = the neighbor collection of s with γ
Now, we suppose have a N_n in r(u)
Set nodes of with γ with two conditions
* Communication Range adapt */
1. While
2. {
3. Case 0:
4. {
5. Case 1:
6. $(0 \le k(v) < Z/3)$
7. $t = \gamma^i/E_n^i$;
8. Case 2:
9. $(Z/3 \le k(v) Z/2)$
10. $t = \gamma^i/E_n^i$;
11. Case 3:
12. $(Z/2 \le k(v) < Z)$
13. $t = \gamma^i$
14. End Case
15. {
16. While (true)
17. {
/ data collecting /

Find the mapped communication graph G & calculated communication range for every node of the network and inspection MCP p^*_{us} and its best capacity amount $c(P^*_{us}) \forall \varepsilon V$;

/* group residual energy
17. $r(u) \forall_u \varepsilon v$;
18. if $((\exists_u \varepsilon N, C(P^*_{avg}) * 10 \log (P_{random}) *10 \log (P_{random}) < B/2$
19. or
20. $\Sigma_{n \varepsilon N} r(u)/N < B/2)$
21. Or
22. $\Sigma_{n \varepsilon N} r(C_{HEAD}) * 100$
23. Then
24. Compute the neighbor set
25. $N_1, N2 .... N$
26. Subset $N_i (1 \le i \le N_n)$

Table 1. Sink mobility management comparison.

Reference	Description	Technique	Routing strategy	Metrics	Centralized/ Distributed	Advantage
3	6 fan shaped sector when sink to move according to various condition	HUMS (half quadrant moving strategy)	Multi hop	Energy/Delay	Distributed	Determination of sink when to move according to condition of half quadrant strategy
5	Lifetime enhancement by using A-Star and fuzzy approach	Fuzzy approach and A-Star algorithm	Shortest path	Energy/Delay	Distributed	Improve lifetime enhancement in wsn
2	Sink relocation with multiple predetermined hexagon trajectory	Distributed localized solutions & hexagon trajectory technique	Multi-hop	Delay/Energy	Distributed	Increases lifetime of the sensor with the 4.86% efficiency rate as the static sink
8	Optimal energy strategy	Energy saving opportunistic	Multi-hop	Delay/Energy	Distributed	More no of packet transmission for source to destination in minimum energy
6	Relocation technique description	EASR METHOD	Multi-hop	Delay/Energy	Distributed	Increases lifetime of the Network as well as rate of efficiency is increased

For each assignment $n \in \{1, ..., Nn\}$ for all $j \in XP^*$

With $\in$ Eo do
27. {
28. Store r (u) in the table of signatures for N_n
29. }
30. If
31. {
32. n u(i) = 0 or i = t then {
33. Complete a designation ui = (1 ... N_n)}}
34. End while (true) loop
35. }

5 RESULT ANALYSIS

Analysis of LEASR method performance different simulations methods are conducted by various Non-polynomial value. In the previous method EASR energy consumption rate is high as seen in the graph transmission range Vs. Network lifetime rapidly increasing and decreasing energy level as some point value on x-axis and y-axis plotting. In this paper mainly focuses on the energy consumption rate for both previous and proposed work i.e. EASR AND LEASR. It also focus on the end to end delay of the nodes and different parameter like average jitter, bit error rate, Sink load minimization and throughput through graph for both method EASR and LEASR as shown in the graph. Analysis is done through for both method and compared which one is better for increasing the lifetime of the network. As shown in the next page different comparison graph with different parameter and show that the NP-LEASR is more efficient for increasing the life span of the network as compared through the previous work EASR. In the first scenario energy consumption rate is low in the proposed work i.e. NP-LEASR take average of the x-axis (50) and see there is difference for both NP-LEASR AND EASR, for EASR (0.51 Jbit/s) & NP-LEASR (0.43 Jbit/s). Assume comparison always take form the average of the graph. In scenario 1 BER for NP-LEASR (1.4) & EASR (1.9). Similar scenario comparison 3, 4, 5, 6 take average and show that NP-LEASR is more efficient than EASR for increasing the lifetime of the network. According to energy level we gain 0.82 ratio in the form of percentage and show that 82% improvement and throughput we can say that 16% more improvement approx. as per NP-LEASR method.

250

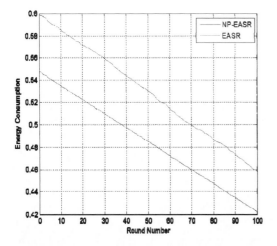

Figure 3.   Energy consumption scenario 1 with a round number.

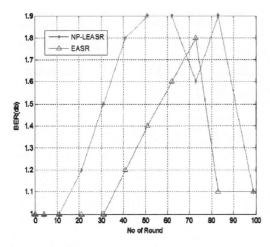

Figure 4.   Bit error rate simulation scenario with number of round.

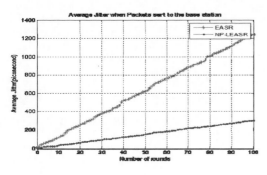

Figure 5.   Average jitter simulation scenario with rounds.

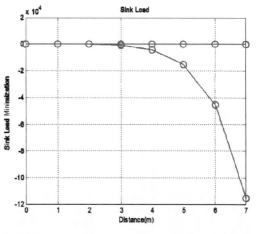

Figure 6.   Sink load minimization scenario with a distance.

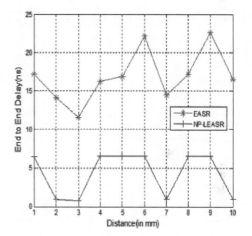

Figure 7.   End to end to delay simulation scenario with distance.

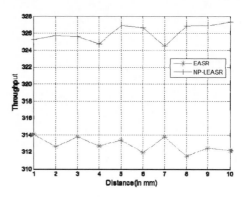

Figure 8.   Throughput simulation scenario with distance.

251

6 CONCLUSION

Depletion speed of battery exhaust of every sensed nodes affect the life span of the sensor network. Researcher main aimed focus on the energy aware routing to increase the life span if save the battery energy for long duration. A NP-LEASR is another method of approach for increasing the life span of the sensor network with move of multiple sinks of optimal position. This is not generally create problem of hotspot in the NP-LEASR. Comparison shows with different parameter like end to end delay, energy consumption rate, and average jitter and so on give better performance of the network. For the NP-LEASR, Reliability and Efficiency is increased in the proposed method as compared to the existing method EASR. As discuss in terms of Reliability overall increment in Energy rate, BER and throughputs, the proposed method is much more reliable and efficiency is increased as above graph in terms BER, throughputs, Average Jitter. So we say that the proposed method increases the life span of the network and minimize energy consumption as compared to the existing method EASR.

REFERENCES

AlShawi I.S., Y. Lianshan, P. Wei, and L. Bin, "Lifetime enhancement in wireless sensor network using fuzzy approach and A-star algorithm,"*IEEE Sensors J.*, vol. 12, no. 10, pp. 3010–3018, Oct. 2012.

Anitha S., N. Janakiraman, "Network Lifetime Augmentation by EASR METHOD in wireless Sensor Network" International Conference on Circuit, Power and Computing Technologies [ICCPCT], 2015.

Chu-Fu Wang, Jau-Der Shih, Bo-Han Pan, and Tin-Yu "A Network Lifetime Enhancement Method For Sink Relo-cation and Its Analysis in Wireless Sensor Networks", IEEE Sensors Journal, vol. 14, no. 6, June 2014.

Darsh Bhagwat, Ashish B. Girapur, "Optimistic Technique of Sink Relocation for a Network Lifetime Enhancement WSN" In International Journal of Innovative Research in Computer and Communication Engineering vol. 3, Issue 6, June 2015.

Juan Luo, Jinyu Hu, Di Wu, Renfa Li, "Opportunistic Routing Algorithm for Relay Node Selection in Wireless Sensor Networks" In IEEE Transactions on Industrial Informatics, vol. 11, no. 1, February 2015.

Koc and korpeoglu "Coordinated movement of multiple mobile Sink in Wireless sensor network for the improved lifetime" in EURASIP Journal on Wireless communications and networking, Sep. 2015.

Luo J. and J. P. Hubaux, "Joint mobility and routing for lifetime Elongation in wireless Sensor network", in proc IEEE Inf. Commun. vol. 3. Mar. 2005, pp. 1735–1746.

Marta M. and M. Cardei, "Improved sensor network lifetime with multiple mobile sinks," *J. Pervas. Mobile Comput.*, vol. 5, no. 5, pp. 542–555, Oct. 2009.

Sandeep Sharma and Rajesh Mishra, "A Simulation Model for Nakagami-m Fading Channel for m >1", in International Journal of Advanced Computer Science and Applications, vol. 6, no. 10, pp. 298–305, October 2015.

Sara G. S. and D. Sridharan, "Routing in mobile wireless sensor network: A survey," *Telecomm. Syst.*, Aug. 2013.

Sun L., Y. Bi, and J. Ma, "A moving strategy for mobile Sinks in wireless sensor networks," in *Proc. 2nd IEEE Workshop Wireless Mesh Netw.*, Sep. 2006, pp. 151–153.

Wei-Chieh Ke, Bing-Hong Liu, and Ming-Jer Tsai, "Constructing a Wireless Sensor Network to Fully Cover Critical Grids by Deploying Minimum Sensors on Grid Points is NP-Complete" In Ministry of Economic Affairs of ROC under grant 94-EC-17-A-04-S1–044.

Zytoune O., M. EL-Aroussi and D. Aboutajdine, "A Uniform balancing routing protocol for wireless sensor network" in WIRELESS PERSONAL Commun. vol. 55 no. 2, pp. 147–161, 2010.

Communication and Computing Systems – Prasad et al. (Eds)
© 2017 Taylor & Francis Group, London, ISBN 978-1-138-02952-1

Security mechanism in cognitive radio ad hoc networks

Sandeep Sharma & Shreyansh Singh
Gautam Buddha University, Greater Noida, India

ABSTRACT: Cognitive Radio (CR) technology is designed to solve the problems arising from limited available spectrum in wireless networks. It can ease the problem of spectrum shortage by allowing unlicensed users (secondary users) to use the same spectrum with imperative users in licensed spectrum bands without creating any interference in imperative communications. A lot of research has been done in Spectrum Sensing of CRs but not much research has been done regarding its security against selfish attacks. In CR network a selfish node is a serious issue which occupies a part or full resources (pre-occupation problem) in the multiple channels, and forbids other legitimate cognitive nodes from consuming the resources. The performance of CR network is degraded by these selfish nodes by capturing the licensed resources reserved for licensed users. In this article we will study about the various security threats to a CR network and will study the relation between the no. of nodes and detection rate of a cognitive radio network by using mechanism of cooperative neighboring cognitive radio nodes.

1 INTRODUCTION

COGNITIVE RADIO is a wireless technology designed to meet the ever increasing demand of spectrum. Unlike the conventional radio it is an intelligent radio which allows a Secondary User (SU) to use unused spectrum unoccupied by a primary (licensed) user by detecting the presence of a spectrum hole. Most of the spectrum is reserved for licensed users in traditional spectrum management. CR technology search for available or unused spectrum for unlicensed users by spectrum sensing technology. When any licensed spectrum is not being used, it is considered to be available. Now the available unused bands are given to secondary users (unlicensed) by *Dynamic Spectrum Access* (DSA).

In *Dynamic Spectrum Access* unlicensed users or secondary users use the available spectrum and *Dynamic Spectrum Access* allows CR to use the best available spectrum through its two main features: Cognitive Capability and Reconfigurability.

1.1 Cognitive capability

In radio technology it refers to the ability to sense and capture the information from the environment. Unused spectrum can be identified and best available band can then be used.

1.2 Reconfigurability

While above capability is used for spectrum detection and selecting best available spectrum, this technique enables the cognitive radio to be programmed dynamically. Because of this a cognitive radio is used to transmit and receive on various frequencies thus increasing the efficiency.

2 SECURITY CHALLENGES IN CR NETWORK

Cognitive radio are network are prone to several security challenges which acts as a hindrance in its functioning. This hindrances acts as a barrier in its performance and leads to the loss of valuable spectrum. So additional functions must be implemented in CR network like proper sensing protocol, correct decision making and appropriate switching to face these security challenges.

2.1 Security challenges in spectrum sensing

Spectrum sensing is main aspect in a CR network. Because of spectrum sensing unlicensed secondary makes use of unused licensed bands by not causing any interference with licensed primary users. Effective spectrum sensing ensures a CR network detects and differentiates between a secondary user and a primary user otherwise the attackers may emulate the signals of PUs thus selfishly occupying the spectrum.

2.2 Challenges in spectrum mobility

Spectrum mobility ensures seamless and smooth handoff from one channel to another.

When attackers launches this attack, the waiting time involved in achieving a proper handoff increases, thus leading to loss of channels.

2.3 Challenges in spectrum sharing

CR Ad hoc networks are infrastructure- less networks, without any central entity. This central infrastructure provides security among users. This decreases the probability of successful communication among SUs. SUs are the unlicensed users prone to attackers, which might leads to loss of spectrum of PU as well. This issue becomes more critical without any central entity and PUs have to give unused spectrum to SUs without any authenticity. If any SU turns out to be selfish, then it may occupy the spectrum reserved for PU.

A lot of research has been going on for developing security mechanism in centralized Cognitive radio network. But the main issue is that no research has been carried out in developing security mechanism for decentralized Cognitive radio network.

3 TYPES OF SELFISH ATTACKS IN COGNITIVE RADIO NETWORK

Selfish attack is an attack of MAC layer in wireless network in which nodes launch a series of false behaviors to degrade the performance of the network. These selfish attacks are launched to gain an unfair advantage to maximize channel usage and minimize their energy consumption. In CR network there is always a competition among nodes to sense and acquire the available channel. But some unlicensed users (secondary users) turns out to be selfish and then will try to acquire channels at the cost of other channels. These attacks are done by broadcasting fake channel information or fake signals. If any PU is using the channel, the secondary user will not use the channel. But a selfish secondary user will prohibit other potential SUs from sensing and using the channels. Second attack is carried out by selfish SU sharing fake channel information among other SU. Thus legitimate SU senses the channel and when it finds no available channel it eventually gives up sensing channel, thus giving selfish SU hold of all the available channels. All the selfish attacks are discusses in detail below.

3.1 Type 1 attack (primary user emulation attack)

In this type of attack a secondary user gets prohibited from making use available channels. This is carried out by selfish SU acting as a PU or technically emulating primary user signal characteristics. Upon sensing the faked signals, a legal SU gets to know that PU is active and is using the channels, thus doesn't sense channels.

This PUE attack is possible due to highly flexible software-based air interface of CR network which makes the CRs highly reconfigurable. An attackers makes use of this and modifies the cognitive radio interface to mimic the signal characteristics of a primary user. Thus causing the secondary user to identify attacker as a primary user. Therefore this attack is known as *Primary User Emulation (PUE) attack*. So this attack can be overcome by a scheme that can distinguish between legitimate PU and fake primary users.

The main challenge in spectrum sensing is to make difference between PU signal and SU signals. The current schemes use technique based on energy detectors. But still these techniques are not strong enough to prevent these attacks. Soto thwart this attack a transmitter verification scheme is used which can be integrated into spectrum sensing scheme to detect PUE attacks.

3.2 Type 2 attack

This attack is similar to PUE attack in which selfish secondary user emulates the characteristics of a PU signal. But dynamic spectrum access is used to carry out this attack. In normal Dynamic Signal Access (DSA) process, secondary user sense the current band and check whether if primary user is using the channel or not. But in this attack which is carried out by dynamic multiple channel access, instead of periodically sensing the channels, the channels are sensed continuously by using a chain of fake signal attack on various channels like in continuous fashion. Thus an attacker can restrict legitimate secondary user from sensing and identifying available channels and limiting it to use available spectrum channels.

3.3 Type 3 attack

This type of attack occurs in Common Control Channel (CCC). This channel is used for handling information related to exchange of information among the CR users. Spectrum sensing coordination, neighbor discovery and exchange of local measurements between the CR users is done with help of Common Control Channel (CCC).This attack is known as channel pre-occupation attack. In this attack a selfish SU broadcasts fake available lists of channels to legitimate secondary users which are in line to occupy a part or all of the free available channel. Thus selfish users restricts legitimate users from using available channels. Suppose there are 10 channels in a network out of which primary user is using 4 channels. This means that there are 6 available channels which can be occupied by secondary user. But a selfish user can broadcast fake information regarding free available channels. Even though it will use only 3 channels out of the remaining 6 channels but it will broadcast fake list of all the 6 occupied channels. Therefore a legitimate secondary user will think that all the free available free channels are occupied and gives up sensing channels. Thus a selfish SU gets to use all the available 6 channels. This attack reduces the efficiency of any cognitive radio network as the selfish user gets to occupied all the available channels thus compromising the performance of the CRN network.

4 DETECTION MECHANISM AND ALGORITHM OF *TYPE 3 ATTACK*

The detection mechanism works on the exchanged channel allocation among neighboring secondary users. Cooperative behavior among nodes is the core of this mechanism. Suppose a network consists of 5 nodes referred to as NU_Node. Out of these nodes a node is selected as a target node referred to as TU_Node. Thus remaining 4 nodes becomes neighboring nodes as NU_Node 1, NU_Node 2, NU_Node 3 and NU_Node 4. Now this target node and all other 4 neighboring nodes will exchange the channel allocation information list. All the neighboring nodes will send individual information regarding how many channels it is currently using. In return the target node, TU_Node will send channel list to each individual neighboring node about how many channel it is currently using. If there is any discrepancy between the information shared by target node, TU_Node and neighboring nodes, NU_Node then all the neighboring nodes will reach the conclusion that this node is the selfish node. This behavior is carried out using cooperative behavior among nodes. Thus all the neighboring nodes will come to know about the selfish behavior of the target node. Once a decision is made regarding target node of being selfish or not, another neighboring node is made target node and the same mechanism will be applied on it. Thus one by one all the nodes will be tested of selfish behavior and each node gets to become the target node.

But this detection mechanism has one major drawback. Since this technique is based on cooperative behavior of nodes, so it will be less reliable for more than 1 neighboring selfish SU. This is because 2 neighboring selfish nodes can exchange false information of channel allocation among them. It would become tough to know from where the fake information is coming from. Thus it will cause disparity in the network and reduce reliability. But it can be overcome by increasing the no. of legitimate neighboring nodes because in that case more genuine information will be broadcasted and good detection rate can be expected.

Detection Algorithm - Let Channel T_Node denotes sum of channels used as reported by the targetnode to each node around it and Channel N_Node denotes the sum of currently used channels reported by neighboring node to the target node. Suppose these four neighboring nodes are currently using 2, 2, 3, 2 channels respectively such that

NU_Node1 = 2
NU_Node2 = 2
NU_Node3 = 3
NU_Node4 = 2

So ChannelN_Node = (NU_Node1 + NU_Node2 + NU_Node3 + NU_Node4). Therefore ChannelN_Node = 2 + 2 + 3 + 2 = 9

And let the no. channels used reported by target node to each neighboring node be 2, 2, 5, 2 respectively.

Therefore ChannelT_Node = 2 + 2 + 5 + 2 = 11

Since ChannelT_Node>ChannelN_Node as 11 > 9, therefore the selected target secondary is detected as a selfish node. After this the next neighboring node will be made as a target node and the same algorithm will be made to run on that node. Every time a node gets checked, the loop increments by 1. After the selfish node is identified all the communication with that node is halted. This detection procedure continues till all nodes of the following network are verified and checked.

5 SIMULATION RESULT AND ANALYSIS

In order to study the effect of SUs on detection rate of network, simulation is conducted on MATLAB to check this mechanism's efficiency. This is calculated by detection rate. This is the ratio of no. of SU detected to the no. of actual SSU. The number of SU used in the experiment were 50, 100 and 150 and detection rate is calculated for each of these SUs. The effect on detection rate with the increasing no. of secondary users is negative. As the no. of nodes increases, detection rate of this mechanism falls rapidly. This is because the probability of having more than one selfish secondary user in the neighbor increases with the increasing SU. If more than one selfish secondary user exists in the network, wrong channel allocation information can be exchanged among them. Thus forcing the network to make wrong decision due to faked exchanged information. As observed from the Figure 1 network with 50 nodes have less selfish secondary user density as compared to network with 100 and 150 nodes. Total no. of nodes in a neighbor also effect selfish user detection accuracy. Detection accuracy will be high if there are more no. of neighboring nodes in the network, as genuine nodes can share correct channel

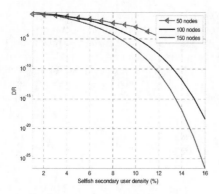

Figure 1. DR vs selfish secondary user density (%).

Table 1. Simulation parameters.

Parameters	Settings
Type of antenna	Omni antenna
Size of network	3000 m * 3000 m
Selfish secondary users	2%, 4%, 6%, 8%, 10%, 12%, 14%, 16%
CCC	1
Data rate of channel	11 Mbits/S
Secondary users	50, 100, 150

allocation information among them. Thus increasing the detection rate of the CR network. Thus we did the simulation with 2 to 5 neighboring nodes. Network with five neighboring node achieve highest accuracy as probability of sharing more correct channel allocation information increases. 4 neighboring SU achieve very good accuracy. But CR network with two neighboring nodes gets seriously affected by the number of actual selfish secondary users. Therefore for avoiding selfish attack and to increase the detection rate more than 3 secondary users are recommended.

Detection rate = Detected selfish secondary user/Actual selfish secondary users

6 CONCLUSION AND FUTUREWORK

In this paper we studied a selfish attack in which selfish user broadcasts fake channel allocation information to other neighboring nodes who are waiting to occupy the available channels. It detection mechanism is based on cooperative behavior among neighboring nodes in which all the neighboring nodes share channel allocation information with the target node and in return receives the same from target node and we proposed a detection algorithm for it. Results of this algorithm are very high and reliable. But this mechanism has some limitations as it gives less detection

accuracy as no. of nodes increases. Also if there are more than selfish users in the network the accuracy of this cooperative behavior mechanism decreases. So for future work a proper authentication system should be provided in which primary user should deliver information securely to secondary users. In this method a cryptographic link signature to its signal to authenticate primary users. This scheme will give better results and a better detection rate and accuracy as the malicious secondary users will not be able to decode the cryptographic signature modulated with the signal. Thus protecting primary users from selfish secondary user attacks.

REFERENCES

Anju Singh, Karan Singh, Sandeep Sharma, "Capacity Based Multicast Channel Assignment in Wireless Mesh Network", in the *Int. Journal of Communication and Networks*, vol. 5, no. 3B, pp. 673–679, 2013.

Chen. R, Park. J-M, Reed. J-H. 2008. Defense against Primary User Emulation Attacks in Cognitive Radio Networks. *IEEEJSAC(1)*: 25–36.

Cheng H.K., Maa. M., Qin. Y. 2012. An Altruistic Differentiated Service Protocol in Dynamic Cognitive Radio Networks Against Selfish Behaviors. *Computer Networks(56)*: 2068–2079.

Ganesan. G. Li. Y. 2005. Cooperative Spectrum sensing in cognitive radio networks. *IEEE DySPAN*: 137–143.

Haykin. S. 2005. Cognitive radio: brain-empowered wireless communications(23). *IEEE J. Select. Areas Commun*: 201–220.

Jaiprakash Nagar and Sandeep Sharma, "K Barrier Coverage Based Intrusion Detection System for Wireless Networks", accepted in the *proceeding of 50th Golden Jubilee Annual Convention of Computer Society of India (CSI-2015)*, December 2015.

Liang, Y.C., Y. Zeng, A.T. Hoang. 2008. Sensing-Throughput Tradeoff for Cognitive Radio Networks. *IEEE Transactions on Wireless Communication(7)*: 1326–1337.

Minho, J., H. Longzhe, K. Dohoon, P. Hoh. 2012. Selfish Attacks and Detection In CogntiveRadio Ad-Hoc Networks.*IEEE Network(27):* 46–50.

Sandeep Sharma, Rajesh Mishra, Karan Singh, "A Review on Wireless Network Security", in the *Lecture Notes of the Institute for Computer Sciences, Social Informatics and Telecommunications Engineering, Springer*, vol. 115, pp. 668–681, 2013.

Sandeep Sharma, Rajesh Mishra, Pratik Singh, "Authentication in Wireless Network", in the proceeding of *2015 IEEE 2nd International Conference Computing for Sustainable Global Development (InciaCom-2015)*, pp. 2031–2035, 2015.

Yan. M. 2011. Game-Theoretic Approach against Selfish Attacks in Cognitive Radio Networks. *IEEE/ASIC 10th Int'l Conf. Computer and Information Science (ICIS)*: 58–61.

Zhayou, H.G., S.Z. Haizin, Li. Shuai. 2012. Security and Privacy of Collaborative Spectrum Sensing In Cognitive Radio Networks. *IEEE Wireless Communications(12)*: 106–112.

Communication and Computing Systems – Prasad et al. (Eds)
© 2017 Taylor & Francis Group, London, ISBN 978-1-138-02952-1

Optimization of physical layer parameters for energy efficient wireless sensor networks

Karuna Babber & Rajneesh Randhawa
Department of Computer Science, Punjabi University, Patiala, Punjab, India

ABSTRACT: Wireless sensor networks can be seen as an effective base for pervasive intelligent computing but due to their unique characteristics, these networks face many implementation challenges. One of the major design constraints is its shorter network lifetime. This paper explores different parameters of OSI protocol stack. An algorithm to uniformly cluster the sensor nodes and thereby appropriate selection of modulation scheme for different members of the clusters has been proposed in this paper. Matlab Simulation results indicate the proposed scheme not only enhances energy efficiency of WSNs but also helps in expanding the network lifetime to a desirable level.

1 INTRODUCTION

Wireless Sensor Networks (WSNs) are composed of large number of tiny intelligent low-cost, low-power multi functional sensor nodes deployed over a wide geographical area in an arbitrary fashion. Each sensor node in WSN is equipped with a small processor, radio transceiver and a battery as an energy source. To replace or recharge the batteries of these sensor nodes is almost infeasible.

The various factors contributing to energy dissipation of sensor nodes can be channel loss, hop distance, large number of hops, coding/modulation scheme, packet size and fixed energy cost to run transmission and reception circuitry (Rapparport 1996). In order to make best possible use of available energy of sensor nodes lower layers of OSI protocol stack can be explored.

2 RELATED WORK

Authors (Heinzelman et al. 2002) suggested that with the increase of hop distance channel loss increases and with the increase of number of hops, the circuit's energy cost also increases.

(Kun Yang et al. 2012) explored different coding and modulation techniques like BPSK, 16QAM, QPSK and 64QAM to achieve better system performance.

(Sami et al. 2011) advocated selection of location aware modulation with error control codes for energy efficient WSNs. Authors used an Adaptive Modulation and Coding (AMC) in WiMAX to achieve higher throughputs when covering long distances.

(Ammer et al. 2006) defined Energy Per Useful Bit (EPUB) metric where in impact of preamble on the effectiveness of system is taken into consideration. But a more complex MAC can outweigh the gains of energy conservation.

Authors (Cui et al. 2005) suggested that for short transmission distances un-coded linear modulation schemes can be beneficial but for long transmission distances coded non-linear modulation schemes prove to be energy efficient.

(Shih 2001) explored various trade-off between modulation scheme and complexity of circuit, power consumption & transmission time and suggested that binary modulation scheme with an effective startup power dominant condition is more energy-efficient for WSNs.

(Korhonen et al. 2005) recommended optimization of packet size for adapting radio parameters to not only improve link performance but also minimizing energy consumption.

3 PROPOSED MODEL

The wireless sensor network comprises of some sensor nodes and base station at every half quarter. The entire sensing area is partitioned into different clusters with ordinary sensor nodes, cluster heads and gate-way nodes as its members.

To form uniform clusters of the entire sensing area, the base station dissects the sensing area into a defined angle and the sensor nodes falling under this defined angle can be grouped into number of clusters. The average distance of all the sensor nodes is taken into consideration for the selection of cluster heads and the selection of gate-way nodes is based on set theory.

By following the steps provided in flow chart, the sensing area can be split into uniform clusters. The cluster architecture is depicted in Figure 3. The

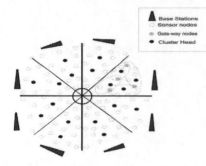

Figure 1.   System model.

base station broadcasts the routing information of the clusters to all the sensor nodes including cluster heads and gate-way nodes. Based on the number of sensor nodes within the cluster, the cluster head allocates the time schedule in the form of TDMA to all its members. Hence each sensor node has its own routing table and TDMA schedule, to follow.

The sensor nodes of the cluster transmit data to their cluster heads during their allocated TDMA schedule. The cluster heads after collecting data from all the member sensor nodes, compress & aggregate the data into a single packet and forward it to gate-way node. In a similar manner cluster head of adjacent cluster forward its data packet to this particular gate-way node. With this way one gate-way node (which is supposed to be member of at least two neighboring clusters) receives two data packets each from two clusters and send it to base station. This algorithm helps in reducing the over-head of selecting cluster heads after every round and also saves cluster head's energy by providing gate-way nodes within the clusters.

Optimization of Physical Layer Parameters

i. *Proposed Physical layer packet structure:* Traditional physical layer data packets of wireless networks contain 4 bytes preamble for symbol synchronization (Proakis et al. 2007), 1 byte as delimiter for frame synchronization, 1 byte as PHY header for length of actual data packet and up to 127 bytes of actual data. To reduce packet size, bytes of preamble can be removed as there is no need for symbol synchronization in case we apply pre-determined hop distance (using proposed cluster formation algorithm). Instead of preamble, a pattern sequence (11 or 00) for the estimation of channel condition can be added for better synchronization between transmitter and receiver. Figure 4 depicts the proposed packet structure.

ii. *Energy Model:* Energy consumption during single data packet transmission (Holland et al. 2009) can be given as:

$$E = \alpha E_{rx}d^n + E_{fixed} \qquad (1)$$

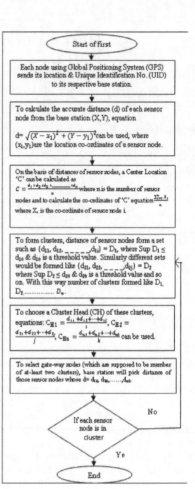

Figure 2.   Flow chart of the initial cluster formation.

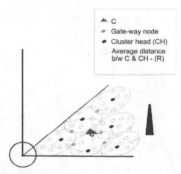

Figure 3.   Cluster architecture.

where α is the amplifier co-efficient, E_{rx} is the energy consumption during the reception of per data packet, d is the hop distance (distance covered by data packet) and E_{fixed} is the fixed energy required by transmitter/receiver to transmit or receive one data packet. Time taken to transmit a data packet of r bits is given by:

$$\text{Tr} = \frac{r_1}{bB} + \frac{r_0}{B} \tag{2}$$

where r_1 are actual data bits and r_0 are overhead (delimiter/header and pattern sequence) bits and B is the signal bandwidth. Total energy consumed during transmission/reception per data packet in terms of bits can be given as:

$$E = (E_{tx} + E_{rx})T_r \tag{3}$$

$$E = (E_{tx} + E_{rx})(\frac{r_1}{bB} + \frac{r_0}{B}) \tag{4}$$

From equations (1) and (4) it is clear that the total energy consumption during transmission/reception per data packet is directly proportional to the number of bits contained in data packet and the hop distance. So, the energy consumption of WSNs can be reduced by reducing packet sizes and hop distance among sensor nodes.

iii. *Proposed modulation schemes:* From equation (1) it is clear that energy efficiency is directly proportional to the distance covered by data packet. So, picking one modulation scheme for the entire network may result in lots of energy wastage, as the sensor nodes within the sensing area might be at different distances from their respective cluster heads and gate-way nodes. Modulation schemes (Kazem et al. 2007) such as MSK, BPSK, 16PSK and 64QAM were analyzed in two scenarios:

Scenario-I

• Each modulation technique was used on all sensor nodes of the sensing area.
• Constant Bit Error Rate (BER) value was maintained in all the modulation techniques.
• Each modulation scheme was used in White Gaussian Noise Channel (AWGN) environment.

In this scenario, MSK and BPSK modulation techniques being very simple to generate and demodulate performed good in terms of energy efficiency but proved fairly in-effective in terms of channel throughput whereas 16PSK and 64QAM modulation techniques being complex to design provided high channel throughput but at the cost of energy consumption of the network. This way it is highly recommended that modulation schemes

should be picked according to the optimal hop distance and the applicability of the network.

Scenario-II

• Dissimilar modulation techniques were applied on different sensor nodes (ordinary sensor nodes, cluster heads & gate-way nodes).
• Constant Bit Error Rate (BER) value was maintained in all the modulation techniques.
• Each modulation scheme was used in White Gaussian Noise Channel (AWGN) environment.

In this scenario, it was observed that when the information transmitted with in a distance of up to 40 m i.e. between ordinary sensor nodes to their cluster heads and then packets (with optimized packet-sizes as is discussed above) transmitted to the gate-way nodes the energy consumption by all the modulation schemes was almost same but when the data packets were sent to the base station, the distance covered was more than 100 m, the conventional modulation schemes i.e. MSK & BPSK performed well in terms of energy efficiency but at the cost of channel throughput whereas advanced modulation schemes such as 16PSK & 64QAM performed well in terms of channel throughput but with degraded energy efficiency.

4 PERFORMANCE ANALYSIS

In this study, 1000 sensor nodes were distributed randomly in a sensing area. The base station controlled uniform clustering of the sensing area was performed in accordance to the algorithm given above with the selection of cluster heads and gate-way nodes for each cluster. The initial battery energy of each sensor node is taken to be 2 J. BER to SNR ratio have been calculated for all the modulation techniques and it can be seen in the Figure 5 given below that MSK & BPSK modulation techniques are about 6db better than 64QAM or 16PSK. Hence MSK or BPSK can be applied for short distance transmissions.

It is evident from equation 1 that distance is directly proportional to energy, Figure 6 gives us the fair idea about energy consumption and transmission time for short distance. Figure 7 illustrates energy efficiency of different modulation schemes during changing distance.

Above figures clearly shows that during short transmission distance, consumption of energy by any of the modulation scheme is very less, though it may rise complexity of circuitry design in case of 16PSK or 64QAM. Therefore, it is highly recommended that MSK or BPSK modulation techniques being simple to design and demodulate can be picked during data transmission of distance less than 50 meters i.e. from ordinary sensor nodes

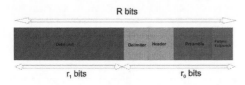

Figure 4.   Physical layer packet structure.

to their cluster heads and from cluster heads to gate-way nodes of the cluster. But as the transmission distance from gate-way node to base station exceeds 50 meters say more than 100 meters, 16PSK or 64QAM modulation techniques can be preferred.

No doubt there is a trade-off between bandwidth efficiency and energy efficiency. Modulation techniques like 64QAM, 16PSK provided higher channel throughput but at the cost of increased power consumption whereas modulation techniques such as MSK and BPSK proved to be energy efficient but at the cost of channel throughput. Now it is up to the applicability and resources of WSNs that if the target is high channel throughput, power consumption of the network increases by using 64QAM or 16PSK but if the target is energy efficiency then MSK or especially BPSK modulation techniques can be used.

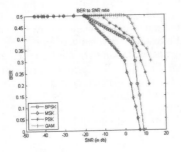

Figure 5.   BER to SNR ration.

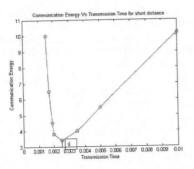

Figure 6.   Communication energy Vs Transmission time.

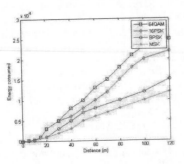

Figure 7.   Energy consumption of modulation schemes w.r.t to distance covered.

5 CONCLUSION

This paper explored different parameters of physical layer to enhance the network lifetime. Firstly the hop distance to transmit data between different nodes has been reduced by splitting the sensing area into uniform clusters. With the role of each node defined within the cluster, the need for preamble of packet doesn't arise here instead the provision of pattern sequence for the estimation of channel condition with lesser bits has been included in the packet structure. This way our packet size reduces and so is the energy consumption. Matlab simulation results indicate that if different modulation schemes are used by different members of the clusters it will enhance the network lifetime to a considerable length though there is trade-off between energy efficiency and high channel throughput and the same can be decided on the basis of applicability of WSNs.

REFERENCES

Ammer, J. and Rabaey, J. 2006. The energy-per-useful-bit metric for evaluating and optimizing sensor network physical layers. In Proceedings of the IEEE International Workshop on Wireless Ad-hoc and Sensor Networks: 695–700.

Cui, S., Goldsmith, A. J. and Bahai, A. 2005. Energy-contrained modulation optimization. In IEEE Transactions on Wireless Communications:2349–2360.

Heinzelman, W, Chandrakasan, A and Balakrishnan, H. 2002. An application specific protocol architecture for wireless micro sensor networks. In IEEE Transactions on Wireless Communication: 660–670.

Holland, Matthew., TianqiWang, Tavli, Bulent., Seyedi, Alireza. and Heinzelman, Wendi. 2009. Optimizing physical layer parameters for wireless sensor networks. ACM:107–130.

Kazem, Sohraby., Daniel, Minoli. And Taieb, Znati. 2007. Wireless Sensor Networks—Technology, Protocols and Applications. Wiley-Interscience, A John Wiley & Sons Publication.

Korhonen, J. and Wang, Y. 2005. Effect of packet size on loss rate and delay in wireless links. In proceedings of the IEEE Wireless Communications and Networking Conference: 1608–1613.

Kun Yang and Ling Yang. 2012. Adaptive Modulation and Coding for Two-Way Amplify-and-Forward Relay Networks. In proceedings of IEEE International Conference on Communications, China.

Proakis, J.G. 2007. Digital Communications. McGraw-Hill Inc.

Rapparport, T S. 1996. Wireless Communications, Principles and Practice. The Prentice Hall.

Sami H. O. Salih and Mamoun M. A. Suliman. 2011. Implementation of Adaptive Modulation and Coding Technique. In International Journal of Scientific and Engineering Research 2(5).

Shih, E. 2001. Physical layer driven protocol and algorithm design for energy efficient wireless sensor networks. In the proceedings of the 7th ACM International Conference on Mobile Computing and Networking (MobiCom'01):272–287, Rome, Italy.

Communication and Computing Systems – Prasad et al. (Eds)
© 2017 Taylor & Francis Group, London, ISBN 978-1-138-02952-1

Security and reliability trade-off in cognitive radio networks

Charchit Singh & Sandeep Sharma
Gautam Buddha University, Greater Noida, India

ABSTRACT: Secondary transmitter, numerous secondary relays secondary destination are the main elements of cognitive radio in the occurrence of an unlicensed invader. By the help of secondary relays, secondary transmitter is transmit the data to secondary destination. So the secondary transmission is interrupted by an unlicensed invader in the cognitive radio. By the help of single-relay and multi-relay selection, we need an accurate relay selection to give the secure transmission from secondary transmitter to secondary destination in the occurrence of an unlicensed invader. To sending the secondary transmission from secondary transmitter to secondary destination, we need only a specific relay after the particular transmit collection scheme and similarly to sending the secondary transmission from secondary transmitter to secondary destination, we need numerous secondary relays from multiple relay selection. During the secondary transmission, the intercept and outage probability will be analyzed on behalf of the particular transmit and numerous transmit collection scheme in realistic spectrum sensing channel. Uninterrupted communication plays as a benchmark scheme and artificial noise based methods and will do the comparison between uninterrupted communication and particular transmit and numerous transmit collection scheme. So in the numerical analysis when the intercept probability undisturbed so the outage presentation increases of uninterrupted communication, noise based schemes and particular and numerous transmit collection scheme and when the intercept probability is disturbed so the outage presentation of shortest communication, noise based schemes and the particular and numerous transmit collection scheme decreases. So in the secondary transmission the security and the reliability tells us about a trade-off which is also called as security reliability trade-off in the occurrence of an unlicensed invader. Basically it proves that the particular transmit and numerous transmit collection scheme have good security reliability trade-off in comparison of classic direct transmission. In the occurrence of an unlicensed invader, during the secondary communication particular transmit and numerous transmit collection scheme gives the good security. Because of the particular transmit and numerous transmit have good security reliability trade-off if we increased the number of secondary relays. By the numerical analysis, we have observed that numerous transmit collection have good security reliability trade off in comparison of particular transmit collection.

1 INTRODUCTION

In cognitive radio networks, Security problems Plays an important role. Primary user emulation attack plays an important role in cognitive radio, this attack was introduced in {S. Anand, 2008, K. Bian, Nov. 2008, R. Chen, 2008, Z. Jin, 2009}. Cognitive radio networks are unprotected because of the internal and external attackers, due to this reason, cognitive radio networks are highly accessible to malignant nature. An unlicesnsed invader user may deliberately impose interference for the sake of artificial tarnishing the CR environment. So the unlicesnsed invader try to hack the confidential information in CR networks.

So in spectrum sensing, Cognitive radio network face security threats. From the earlier studies that they are basically supported that we can secure the transmission in cognitive radio from primary user emulation attack and denial of service.

But snooping is also a major problem in front of protecting the data Privacy. Secured transmission can be achieved by cryptographic techniques in contrast to an unlicensed invader. By the easily decryption of cryptographic techniques they can't provide the secure and reliable communication in the occurrence of an unlicensed invader. Because of the broadcast nature of wireless channels, the Problems of security and privacy or reliability have taken on an increasingly important role in wire-less networks, mainly in Military and homeland security applications. Legitimate

Physical layer security using an information-theoretic point of view is considerable recent attention in this context. By the use of wiretap channel spanning, we can get the secure and reliable transmission at the logical destination, the maximal secrecy rate can be obtained, this is to referred as the secrecy capacity. Variance among the size of the central network and the spy network is called secrecy capacity. And previous research studies are dedicated to improve the security in wireless network. So we will analyze the tradeoff between security and the reliability in CR network, where the secondary transmitter, secondary destination, numerous secondary relays are the main elements of Cognitive radio in the occurrence of an unlicensed invader, by the help of secondary relays, secondary transmitter is transmit the data to secondary destination. So in the occurrence of accurate band detecting we will focus on the investigation of band detecting dependability in cognitive radio. Security and reliability was obtained in the term of intercept and outage probability and also non cognitive wireless networks. Security and dependability tradeoff analysis of cognitive radio networks revealed from this work. There are some main points which plays an important Characterstics in this paper.

- To secure the secondary transmission in the occurrence of an unlicensed invader, there are basically two schemes first one is single-relay transmission and second is known as multi-relay selection scheme.
- To sending the secondary transmission from secondary transmitter to secondary destination, we need only a specific relay from particular relay selection and similarly we need numerous secondary relays from the multiple relay selection to sending the secondary transmission from secondary transmitter to secondary destination.
- The occurrence of accurate band detecting we will analyze the scientific security reliability trade-off of the particular transmit collection and numerous transmit collection schemes.
- Uninterrupted communication and artificial noise based schemes are considered as the benchmark schemes and basically used for the comparison between classic direct trans-mission and particular transmit and numerous transmit collection scheme, security and reliability are measured in term of interrupt possibility and outage possibility to the communication over Rayleigh fading channel.
- By improving the security reliability trade-off of particular transmits collection and numerous transmit collection so the band detecting reliability is increased.
- We have seen from the numerical results that the security reliability trade-off of particular transmit collection and numerous transmit collection scheme have good results in comparison of benchmark schemes.

Basically the paper is described as follows.
Section 2 will discuss about physical layer security structure exemplary, single relay and multiple relay selection schemes and uninterrupted communication scheme in Cognitive radio.

Section 3 shows the related work of security reliability trade-off of particular transmit and numerous transmit collection schemes with accurate band detecting above Rayleigh fading channel.

Section 4 shows the numerical results part, where we will check the SRT analysis comparison between the shortest communication, the SRS and MRS schemes

2 AN UNLICENSED INVADER AND RELAYSELECTION IN CR NETWORK

This part shows the whole structure exemplary, physical layer security, direct transmission scheme which will serve as a benchmark scheme and also will discuss about the particular and numerous relay selection scheme in the occurrence of an unlicensed invader.

2.1 Structure exemplary

From the Figure 1, basically we have a structure exemplary. The main elements of this structure exemplary are primary network and secondary network. So in the primary network, multiple primary user and a primary base station where the multiple primary user are communicate with the help of primary base station over a licensed spectrum. And in the secondary network where we have secondary transmitter, secondary destination and an unauthorized attacker, with the help of spectrum sensing, secondary transmitter is transmit the data to secondary destination. If secondary transmitter will detect that licensed spectrum is engaged or not, if licensed spectrum is engaged by primary base station so secondary transmitter will not transmit the data and if licensed spectrum is not engaged by primary base

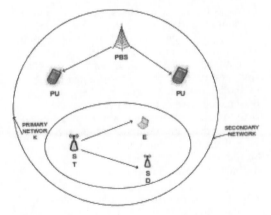

Figure 1. The coexistence of primary and secondary network.

station so secondary transmitter will transmit the data to secondary destination, in the occurrence of an authorized attacker where the secondary transmission may be interrupted by an unauthorized attacker.

Let us assume that R_0 and R_1 present the event. During an exact time slot, R_0 and R_1 represent that the licensed spectrum is engaged and free by primary base station

If the permitted band is free by main base station so it shows that

$$R = R_0$$

If the permitted band is engaged by main base station so it shows that

$$R = R_1$$

2.2 Direct communication scheme

An uninterrupted communication scheme is basically considered as a benchmark scheme. Where x_p and x_s are the random symbols transmitted by the primary base station. From the theory, as we know that

The licensed spectrum is free by primary base station so it shows that

$$R = R_0$$

It means that secondary transmitter will transmit the data to secondary destination and the signal received at destination point in cognitive radio, so in the numeric term we can write it as

$$z_d = t_{sd}\sqrt{p_s}x_s + t_{pd}\sqrt{\alpha}p_{px_p} + n_d \qquad (1)$$

From the channel spanning between source to destination where t_{sd} and t_{pd} are the fading coefficients of the channel and nd, It shows that AWGN received at secondary destination and α is a random variable.

If the licensed spectrum is engaged by primary base station so it shows that

$$R = R_1$$

By transmitting the signal of primary base station which is ended with a licensed spectrum so the secondary transmitter signal is heard by an unauthorized attacker because of the broadcast nature and in the numeric term we can write it as

$$z_e = t_{se}\sqrt{p_s}x_s + t_{pe}\sqrt{\alpha}p_px_p + n_e \qquad (2)$$

From the channel spanning where t_{se} and t_{pe} are the fading coefficients of the channel secondary transmitter to an authorized attacker, ne it shows that the AWGN received at at-tacker node.

2.3 Single-relay selection

To sending the secondary transmission from secondary transmitter to secondary destination, we need only a specific relay from particular transmit collection Let us assume as secondary destination and an authorized attacker behind the secondary transmitter.

To be specific, only a specific relay is selected from particular transmit to frontward the message to secondary destination. Where the licensed spectrum is free by primary base station so it shows that

$$R = R_0$$

So we can write the equation as

$$z_i = t_{si}\sqrt{p_s}x_s + t_{pi}\sqrt{\alpha}p_px_p + n_i \qquad (3)$$

From the channel spanning, Where t_{si} and t_{pi} are the fading Coefficients of the channel secondary transmitter to single relay selection ST and ni it shows that the AWGN received at single relay.

2.4 Multi-relay selection

We need numerous secondary relays from the multiple-relay selection to sending the secondary transmission from secondary transmitter to secondary destination.

So we can write it as equation (4)

$$z_d^{multi} = \sqrt{p_s}w^tH_dx_s + \sqrt{\alpha}p_ph_{pd}x_p + n_d$$

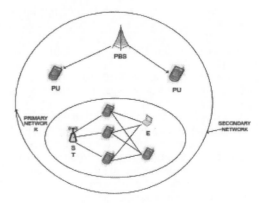

Figure 2. Multiple primary user, primary base station and secondary transmitter, secondary destination elements in CR.

And the signal, which is received at attacker node, we can write it as equation (5)

$$z_e^{multi} = \sqrt{p_s}\,w^t H_e x_e + \sqrt{\alpha}\,p_p h_{pe} x_p + n_e$$

3 RELATED WORK

3.1 *To check the craft-off among security and dependability*

This part will analyze the results of security dependability craft-off of particular transmit and numerous transmit collection scheme in the occurrence of ac-curate band detecting above Rayleigh failing channel. And the security and dependability are measured in the form of interrupt possibility and outage possibility and in this case when secondary transmitter will detect a spectrum hole so it will start the trans-mission in Cognitive radio network.

4 MATHEMATICAL OUTCOMES AND DISCUSSION

The section shows the SRT analysis comparison between the shortest communication, the SRS and MRS schemes. And we will also analyze the comparison between artificial noise based methods and relay selection scheme. Relay selection schemes and the equal power allocation method are used for numerical analysis. Figure 3 shows Graph between intercept and outage probability of SRS and MRS scheme with different probability density and false alarm probability.

Band detecting reliability's where

$$(P_d, P_f) = (0.9, 0.1), (P_d, P_f) = (0.99, 0.01)$$

By enhancing the band detecting reliability from

$$(P_d, P_f) = (0.9, 0.1) \text{ to } (P_d, P_f) = (0.99, 0.01)$$

So we have observed that spectrum sensing reliability is increased because of SRS and MRS schemes have good SRT.

Figure 4 shows that, to increase the number of secondary relays after N = 2 to 8 we have obtained that single relay and multi relay selection have good SRT. So we have ob-served from the numerical results, to increase the number of secondary relays the security reliability trade-off of single relay selection and multiple relay selection schemes are improved and the multiple re-lay selection scheme have good SRT in comparison of single relay selection.

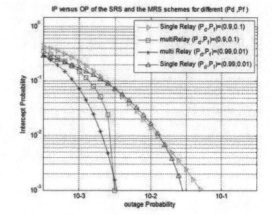

Figure 3. Graph between intercept and outage probability of SRS and MRS scheme.

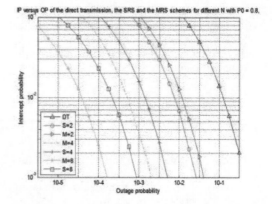

Figure 4. Graph between intercept and outage probability of SRS and MRS scheme with Po = 0.8. *DT = Direct Transmission.

5 CONCLUSION

So we have observed from the numerical results, to increase the number of secondary relays the security reliability trade-off of single relay selection and multiple relay selection schemes are improved and we have also observed that multiple relay selection scheme have good SRT in comparison of particular single relay selection. Basically the secondary transmitter, numerous secondary relays, secondary destination are the main elements in the occurrence of an authorized invader, where the intercept probability and outage probability are measured in the form of security and reliability.

As we have seen that outage probability is high of direct transmission so we can take direct transmission to as our future work where we will not manually define the value of SRS and MRS. We will identify outage probability for multiple-relay selection.

REFERENCES

Anand, S., Z. Jim and K.P. Subbalakshmi, 2008 "An analytical model for primary user emulation attacks in cognitive radio networks," in Proc. IEEE International Symposium of New Frontiers in Dynamic Spectrum Access Network (DySPAN).

Bian K. and J.M Park, Nov. 2008 "Security vulnerabilities in IEEE 802.22," in Proc. Fourth International Wireless Conference (WICON).

Chen, R., J.M Park and J.H. Reed, Jan. 2008 "Defense against primary user emulation attacks in cognitive radio network," IEEE J. Sel. Areas Commun, vol. 26, no. 1.

Chen, R., J.M Park and K. Bian, 2008, "Robust distributed spectrum sensing in cognitive radio networks," iProc. IEEE Conference on Computer Communication (Infocom).

Jeong, C., I. Kim and K. Dong, Jan. 2012 "Joint secure beam forming design at the source and the relay for an amplify and forward MIMO untrusted relay system," IEEE Trans. Signal Process, vol. 60, no. 1, pp. 310–325.

Jin, Z., S. Anand and K.P Subbalakshmi, 2009 "Detecting primary user emulation attacks in dynamic spectrum access networks," in Proc. IEEE International Conference on Communication (ICC).

Jin, Z., S. Anand and K.P. Subbakshmi, "Mitigting pri-mary user emulation attacks in dynamic spectrum access network using hypothesis testing," ACM Mobile Computing and Commun. Review to appear.

Leung-Yan-Cheong S.K. and M.E. Hellman, 1978 "The Gaussian wiretap channel," IEEE Trans Inf. Theory, vol. 24, no., pp. 451–456.

Oggier F. and B. Hassibi, Aug. 2007 "The secrecy capacity of the MIMO wiretap channel," IEEE Trasn. Inf. Theory, vol. 57, no. 8, pp. 4961–4972.

Yuksel M. and E. Erkip, Sep. 2007 "Secure communication with a relay helping the wire tapper," in Proc. IEEE Inf. Theory Workshop, Lake Tahoe, CA, USA, pp. 595–600.

Communication and Computing Systems – Prasad et al. (Eds)
© 2017 Taylor & Francis Group, London, ISBN 978-1-138-02952-1

MIMO based routing and optimization in ad-hoc networks

Siddhant Kumar & Sandeep Sharma
Department of Electronics and Communication, Gautam Buddha University Greater, Noida, India

ABSTRACT: Ad-hoc networks are one of the most subtle networks. To deal with them u have to create a balance of power consumption, bit error rate, delay and efficiency. So in this paper we will be using power control technique and MIMO based routing. In present scenario the nodes are extensively using the high power while transmitting the signal (which is basically a waste). So we can rather conserve the power by dynamically regulating the amount of transmission power according to the neighboring nodes and average power of network. And then we need to introduce the MIMO routing algorithm to further increase the performance of our system.

1 INTRODUCTION

MIMO is one of the most advanced technologies in field of wireless networking. It is high time that we introduce this technology in field of ad-hoc networks. But the major issue is the power consumption, which we are also going to handle in our work. There has been an issue of link breakages in Ad-hoc network and our route is most vulnerable when only one route is selected as there is motion all the time we can get changes in routes. But when implementing MIMO we introduce multiple routes to our destination. So even if route changes we have information coming from multi routes. There has been a paper on power control technique in Ad-hoc networks, which we are going to implement in our project but with slight modification.

2 MIMO BASED ROUTING AND CLPC

2.1 Concept

MIMO is a concept which includes various concepts. Spatial diversity, time diversity, frequency diversity and spatial multiplexing are few concepts of that are used in MIMO. We will use few of these concepts in Ad-hoc networks to improve the performance of our network. By using this technology we are basically increasing the spectral efficiency. When we talk about ad-hoc network we are using antennas which are omnidirectional in nature. We can use this to our advantage to make improvements in our routing protocol technology. We will be using special multiplexing in our protocol. This is basically the making use of channels more efficiently to create a more reliable data transmission. This decreases the bit error rate in our

channel very effectively. This exploits the multipath nature of wave propagation in scattered environment to create a high reliability model. This will be the main focus of our research. MIMO based routing includes the concept of sending the Data through multiple routes at the same time. In this using various routes if we get an error at anyone of the routes, the holes is filled up by other routes. There are a lot of channels in our network that go unused or we can say wasted but if we can make use of these unused channels we can improve the efficiency of our network such that we make use of every possible thing. This technique uses help of nodes ability to broadcast data to multiple nodes at the same time. So we can use up the free nodes to formulate a route to the destination and create a highly efficient network. This technique will not only create a dynamic use of nodes according to the traffic present in our network but also create a more reliable approach towards the network. We are going to use reduces the bit error rate by send message through multi routes and reconcile them at the end node, which will eliminate the retransmission of the message. Our objective will be mainly:

- To explore the tradeoff between the MIMO based routes and power consumption.
- Reduce effect of route breakages.
- Increase the throughput to its maximum.
- Reduce the Bit error rate.

2.2 CLPC modified concept

We will be using the modified version of CLPC in this research paper. CLPC is cross layer power control concept as given in research paper (A. Sarfaraz Ahmed et al. 2015). Here we will be sending hello packets throughout the network and

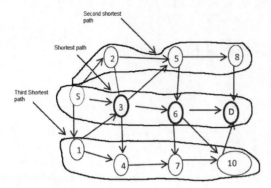

Figure 1. MIMO based routesHere.

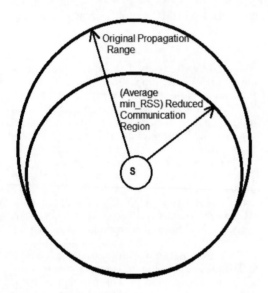

Figure 2. Power control.

check the RSSI value of each node. On basis of Lowest RSSI value of node we are going to set the power of each node in the network as the lowest value of RSSI was achieved.

Then what we will do is we will see what all transmission power of nodes were near the value of Max_avg power and according to that we will increase the transmission power of that node. This thing will help us in getting a better bit error rate performance over the previous protocol that we were using. The major difference will be that earlier we had aa few nodes which had a borderline range and it lead to node becoming unusable in some cases. This was because of very low RSSI values especially due to the mobility in our network this caused some reliability of link issues. The power control we are using will be used hand in hand with MIMO based routing. The usage of MIMO

complements the power control concept because it will reduce the overall power usage and at same time increase the efficiency of network in terms of power consumption and reliability of network.

2.3 LDPC

Low Density Parity Check Protocol (LDPC) (H. Futaki & T. Ohtsuki 2003) is used to encode and decode the message. It is a low power consuming protocol that we are using. In ad-hoc networks when using multi channels it has shown great results while implementing LDPC with turbo codes giving a very high performance, low bit error rate. The LDPC codes can be decoded using a probabilistic propagation algorithm. These codes are especially useful when being used with MIMO because they are highly efficient.

3 IMPLEMENTATION

3.1 LDPC

First we are going to encode the signal using LDPC (H. Futaki & T. Ohtsuki 2003). In this we are going to interleave the data with the first encoder and the second encoder. Then the output of both the encoders are mixed using a multiplexer and modulated. Here S1, P1, P2 are the information bits and parity bits.

At the receiver side the data is decoded by demodulating the data and then demultiplexing it into ys, yp1, yp2 where the ys ⌐⌐ is the information bit and yp1 and yp2 is the parity bits.

Then they are passed through the first decoder where we get the Ps(t) which is the likelihood of received bits corresponding to information and

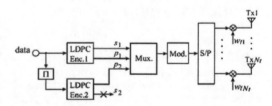

Figure 3. Transmitter.

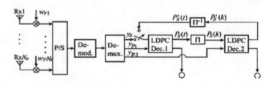

Figure 4. Receiver.

Ps(k) is the likelihood of the interleaved bits which will be decoded by second decoder. From the output of second decoder we get the message signal that was originally transmitted by the transmitter.

3.2 *Power control*

In this section we will be adjusting the power for all the nodes in our network. For that we are going to use following algorithm. In this algorithm power control is implemented.

Algorithm 1

1. Node initiates Hello packet for gathering Neighbors RSS value
2. Get number of neighbors and their RSS value
3. Receiving node checks its routing table neighbors RSS value
4. If value present
   Update routing table
5. Else
   Store as a new value
6. Calculate Average of RSS
7. If neighbors RSS value < Average RSS
   Find Lower Average value say AMin_RSS
8. Else if neighbors RSS value > Average RSS
   Find Upper Average value say AMax_RSS.
9. Segregate three Transmission regions
   a. Average of RSS region
   b. Amax_RSS region
   c. Amin_RSS region
10. Every node adjusts its transmission Power based on Amax_RSS value

Algorithm 2

1. Initialize all nodes by sending hello packet gathering node routes
2. Initialize source and destination.
3. Calculate various routes from Source to destination.
4. Select the shortest 3 paths for transmission.
5. Start encoding the Data for transmission with LDPC-TD encoder.
6. Create a routing overhead and include the information of all the nodes on above route.
7. Start packet transmission to the nodes calculated above.
8. Keep broadcasting to nodes as specified in above routing overhead.
9. Reject the Packet if already received.
10. If packet reaches destination
    Initiate an acknowledgement.
    Start decoding the packets using LDPC-TU decoder
11. Use the shortest path to send the Acknowledgement.
12. If the acknowledgement received Stop.
13. Data transmission is complete

4 PERFORMANCE ANALYSIS

When listing facts use either the style tag List signs or the style tag List numbers.

4.1 *NS-2 simulator configuration*

We will be including the impact of Rayleigh fading. This will be conducted using collision model as specified in NS-2. This will help determine the effects on the packet error rate. This is necessary in calculating the packet delivery ratio with high accuracy with respect to realtime environment. We are here describing a method to evaluate the reliability of our network in a harsh environment. This is basically a kind of stress test on the protocol. We will generate the scenario using a method called as random waypoint model. We will take different readings varying different parameters. First of all we will vary the speed of node and take reading based off that. Then we will vary the number of nodes and take the reading with as we slowly increase the number of nodes.

4.2 *Mobility in network*

When we introduced mobility in nodes increasing speed from 0–30 m/s there were high link breakages. But as we had a multipath propagation of data it will not impact our transfer of data as much as other protocols. This simulation will be done by randomly moving the nodes and checking parameters like throughput. Here we will see what kind of impact mobility has on parameters like throughput. This evaluation is important to see if our protocol it is worth switching to our or not. This will give results with which we can get results which will

Table 1. Parameters of the network.

Network parameters	Range
Speed	0–30 m/s
Load	20% network size
Packet rate	1–10 Packets/s
Terrain size	1000*1000
Max propagation range	300 m
Receiver sensitivity	−90 dBm
Mac protocol	IEEE 802.11
Routing protocol	MIMO, AODV, CLPC
Packet size	512 bytes
Transport layer protocol	TCP/UDP
Application	CBR
Simulation time	900 s
Node density	10–120
Channel propagation	Two ray model
Mobility	Contunues mobility

represent a practical behavior of our model. As this is a multi-hop network we get a lot of routes. These routes change when there is mobility and the node goes out of range and comes in range of another node. When we vary the size of network meaning increase the number of nodes we get different kind of results. As the number of node increases the complexity of network also increases. The size of network and mobility together gives us results of a real time scenario. The greater the network the higher the probability of breakage is there. The networks packet delivery ratio is ideally calculated this way. The reason why the MIMO-PC will give better results than CLPC and AODV is because as the network size increases it becomes more favorable condition for MIMO network because it is designed to utilize such kind of environment more effectively. The scattering effect of model is going to send signals to different nodes at different times and this is effectively going to improve the spatial diversity of our network. So in short we have the technology required to implement the MIMO network. This will also test the decoding capability of the code that we are using. Turbo codes LDPC is going to kick in when the network size increases and improve the quality of received message signal and decode is successfully.

4.3 *Varying the size of network*

When we vary the size of network meaning increase the number of nodes we get different kind of results. As the number of node increases the complexity of network also increases. The size of network and mobility together gives us results of a real time scenario. The greater the network the higher the probability of breakage is there. The networks packet delivery ratio is ideally calculated this way. The reason why the MIMO-PC will give better results than CLPC and AODV is because as the network size increases it becomes more favorable condition for MIMO network because it is designed to utilize such kind of environment more effectively. The scattering effect of model is going to send signals to different nodes at different times and this are effectively going to improve the spatial diversity of our network. So in short we have the technology required to implement the MIMO network. This will also test the decoding capability of the code that we are using. Turbo codes LDPC is going to kick in when the network size increases and improve the quality of received message signal and decode is successfully.

4.4 *Results*

These results are based on simulation done in Network Simulator-2.34.

From Figure 5 we can clearly see that MIMO and MIMO-PC are giving better results as compared to CLPC and AODV. In Figure 5 we compare the throughput of various protocols with respect to number of nodes. We vary the number of nodes from 5–25 and check what throughput we are getting for different values. Initially we started with 20 nodes then we increased the number to 40, 60, 80, and 100. We observed that we are getting very close results for MIMO and MIMO-PC but we are getting a huge difference from AODV. We are getting a close values between CLPC and MIMO when at 100 nodes but still MIMO is able to outperform CLPC.

In Figure 6 Throughput vs mobility we are able to see the performance improvement of MIMO and MIMO-PC. Here we can see that even when there are a lot of breakages it is having a less impact on MIMO and MIMO-PC as compared to AODV and CLPC. Due the multiple routes that we have implemented in our protocol it is more immune to breakages because even if one path breaks then second route still remains unharmed by it.

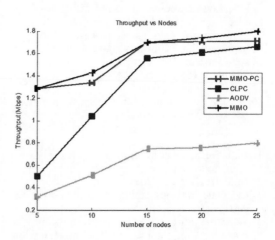

Figure 5.   Throughput vs Number of nodes.

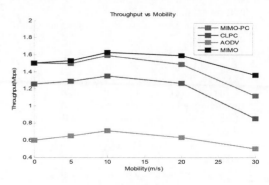

Figure 6.   Throughput vs mobility.

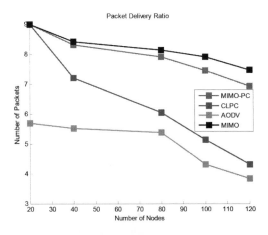

Figure 7. Packet delivery ratio.

Packet Delivery Ratio: It is calculated by number of packets received by the receiver to the number of packets that were transmitted by the receiver.

In Figure 7 we are comparing the packet delivery ratio. Here we observe how the number of nodes is going to affect the number of packets received. In this scenario we can get a very close result when there is less number of nodes. But as the number of nodes increase the performance of CLPC and AODV drastically goes down but the MIMO and MIMO-PC still show consistent packet delivery ratio even when number of node are high. This shows the performance difference between the newly introduced protocol and conventional ones.

5 CONCLUSIONS

It has been observed that when the number of nodes increases the effectiveness of our protocol also increases. So when having nodes >20 the true purpose of our protocol can be seen. It has also been seen that it gives are constant throughput as we reach the 1.8 Mbps mark. When mobility was introduced our protocol was able to give us surprising results. We were able to achieve 1.2 mbps of Data rate at speed of 30 m/s. That is a .3 mbps

of improvement over the CLPC protocol. We have also observed in Figure 7 that we are getting pretty close results at 10 nodes in CLPC and MIMO. But the story changes as the number of nodes significantly increase the packet delivery ratio shows better result over AODV and CLPC. This is a step towards MIMO based Ad-hoc networks. There is still room for improvement in power saving. The protocol can further be improved by reducing the routing overhead. There can be further improvement in how the rerouting is done in case of breakages as of right now traditional methods were used.

REFERENCES

Futaki, H. and Ohtsuki, T. 2003, Low-Density Parity-Check (LDPC) coded MIMO systems with iterative turbo decoding, Vehicular Technology Conference, 2003. VTC 2003-Fall. 2003 IEEE 58th, Year: 2003, Volume:1, Pages:342–346 vol. 1, DOI: 10.1109/VET-ECF. 2003.1285036.

Han C., Green radio 2010, Radio techniques to enable energy-efficient wireless networks, IEEE Commun. Mag., vol. 49, no. 6, pp. 46–54, 2011.

Jun Chen 2014, Energy Efficiency of Space and Polarization MIMO Communications With Packet Erasures Over Wireless Fading Channels, Thomas G. Pratt, IEEE Transactions on Wireless Communications, Year: 2014, Volume: 13, Issue: 12, Pages: 6557–6569, DOI: 10.1109/TWC.2014.2322055.

Jung S. Hundewale, N. and Zelikovsky, A. Energy efficiency of load balancing in MANET routing protocols, Proc. 6th Int. Conf. Softw. Eng., Artif. Intell., Netw. Parallel/Distrib. Comput., 1st ACIS Int. Workshop SNPD/SAWN, pp. 476–483, 2005.

Miao, G. Himayat, N. Li, Y. G. and Swami, A. 2009, Cross-layer optimization for energy-efficient wireless communications: A survey, Wireless Commun. Mobile Comput., vol. 9, no. 4, pp. 529–542, 2009.

Miao, G., Himayat, N. and Li, G. Y. 2010, Energy-efficient link adaptation in frequency-selective channels, IEEE Trans. Commun., vol. 58, no. 2, pp. 545–554, 2010.

Sarfaraz Ahmed, A. Senthil Kumaran, T. Himayat, and Syed Abdul Syed, S. 2015, Cross-Layer Design Approach for Power Controlin Mobile Ad Hoc Networks, Egyptian Informatics Journal (2015) 16, 1–7.

Communication and Computing Systems – Prasad et al. (Eds)
© 2017 Taylor & Francis Group, London, ISBN 978-1-138-02952-1

Intermediate node selection by using energy saving—opportunistic routing algorithm

Sandeep Sharma, Deepak Kumar Katiyar & Pranjal Katiyar
Gautam Buddha University, Greater Noida, India

ABSTRACT: Wireless Sensor Network are the network in which a lot of sensor nodes are deployed with a limited and a non-rechargeable battery power in the regions of consideration. So how to utilize available energy for the sensor node up to the maximum extent had always been a major concern for the design and practical implementation of the Wireless Sensor Network. So in this paper, we have focused on a routing protocol i.e. Energy Saving Opportunistic Routing Protocol for optimization of energy consumption of sensor node which focuses on minimizing energy consumption and maximizing network lifetime for a sensor node acting as data relay in one-Dimensional (1-D) queue network which is used for comparatively dense network. Selection of intermediate nodes is a crucial part for optimizing the energy efficiency of network. Opportunistic routing improve all over performance of network on the grounds of energy saving with a large margin on comparison to other protocols which are used in Wireless Sensor Networks.

1 INTRODUCTION

Sensor nodes are designed with the basic operations like sensing the data, processing it and transmitting the collected data to intermediate node which is selected by the designated routing algorithm to the sensor node. Intermediate nodes collaborate with each other on the basis of the packet forwarding maintaining the uniformity in the process of sending the data to the other intermediate nodes by using the desired routing protocol which is designated to them in order maintain the optimal characteristics of the particular network. The main aim of using opportunistic routing protocol is the selection offorwarding candidate node and coordination among the node to deliver the data packet to their desired and intended destination which is based on the calculations performed by the sensor nodes. Consequently, these protocols are efficient for multiple path selections and intermediate node calculations, so that it achieve larger transmitting range with efficient energy consumption.

Opportunistic routing protocol which is used after the various comparisons with the protocols like MTE, GeRaF on the various ground and then overall comparison is made between the protocols which proved Opportunistic Routing protocol as the best. This protocol broadcasts a data packet to all intended forwarder candidate, then one of forwarder candidates among the ones present successfully receives the data packet. Further these forwarder candidates declare a next forwarder for data transmission using this routing protocol only.

2 PRIMARY OPERATIONS OF OPPORTUNISTIC ROUTING

2.1 Candidate selection

The source select a group of nodes for transmission of data packet from source to destination. These group of sensor nodesor relays are known as the forwarding candidates. Forwarder candidate creates a list with the residual energy and distance for the other nodes present in the network for finding the next-hop forwarder.

2.2 Candidate priority assignment

Selection of nodes is consideredwhen next intermediate node (next-hop forwarder) is not found. Selection of next-hop forwarder depends on priority of candidate which is decided by the list which was created in previous step. If forwarder candidate is of higher priority than forwarding candidate is selected as next-hop forwarder else if forwarder candidate is of lower priority then forwarding candidate is not selected as next-hop forwarder.

2.3 Data transmission

In opportunistic routing protocol, transmission of packets is based on the packet distribution technique which is broadcasting in nature. This packet distribution technique focusses on the reception of packet from different neighboring node. However some opportunistic routing protocol are helped by channeling of unicast packet so that they can received from selected node. Data packet is transmit by higher priority of node that is called next-hop forwarder.

2.4 *Receiver coordination*

There will only be just one relaying node for the particular packet among all the forwarding candidates. That relay node will also authorized for data reception at the MAC layer. The selection of the intermediate node is allotted with the assistance of the distributed procedure within the nodes. The purpose of this selection is choose a highest-priority relay to receive the packet successfully. Thus some modification are tired, for currently (S. Jain & S. Das, 2005) a listing of four fields are enclosed within the RTS messages. The candidates then respond with one CTS message respectively. Then, the supply elect a node to work as forwarder to transmit the information to the chosen node. Normally this arrangement is followed however there also are alternative ways that don't use this arrangement. As (M. Zorzi and R. R. Rao, Oct/Dec. 2003) refers that every node can decide about work as candidate node according to their location and the final destination of the package. After that selection of nodes can be done without supervision of the source node As shown in Figure 1, an illustration of an information gathering system of traffic pattern which is widely spreading among the group of people or is covering a particular area

which is based on 1-D queue network platform, in which the nodes acting as the intermediate nodes are deployed linearly and along the road. As far as we know that most of the existing traditional traffic gathering system are deployed by not considering the power-saving mechanism into consideration and are basically designed for a specific purpose only. With the time, where the development of the cities and town is of prime importance an smart city is the requirement and for which the energy optimization is a prime requirement and the task of information gathering must be done with minimum energy consumption. In the analysis performed further, when the primary node responsible for detection of the vehicle in its range, acquires the relevant information such as velocity of the vehicle, volume of the traffic, density of the traffic and etc. information, it sends the collected traffic information to the intermediate nodes and then the intermediate nodes forwards this data along the path which is more efficient in terms of the energy towards the sink node which is further sent to the traffic management center. In the meantime the Traffic Management Center will select the information required for offering that information to the purchaser with the help of the network present at the sink.

The model discussed above also finds its relevance of extending the lifetime of the network because of saving the energy of the sensor nodes in 1-D queue network and can be implemented in WSN based infrastructure.

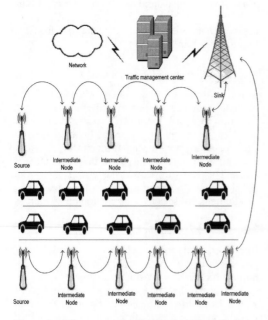

Figure 1. Diagram explaining the designing of cluster of intermediate nodes in 1-D Queue fashion.

3 RELATED WORK

As far as the advancements in recent years, various studies of the 1-D queue networks various parameters which are related to routing, like parameters related to connectivity and density of the distributed nodes can be studied easily. Also, more studies (P. Piret, Sep. 1991) and (E. P. Santi and D. M. Blough, Jan/Mar. 2003) on uniform and independent distribution based on the assumption of fixed transmission range among nodes of the sensor.

Few energy-efficient approaches are analyzed in the abstract (V. Ramaiyan et al., 2012), and (Y. Keshtkarjahromi et al., 2013) As major part of the energy of the sensor node is consumed in transmitting data when compared to the other processes performed by the sensor nodes, thus a mechanism of saving of energy is realized by the calculation of the path which is consuming minimum energy between the sources and sink in the considered WSN's. In (V. Ramaiyan et al., 2012) the discussion of the optimal distance for forwarding and also the power control for single hop is done after analyzing that there is a regular pattern variation between the nodes using the high power and hop lengths which are longer and the nodes which using the low power and having hop lengths which are shorter in regards of the sensor network size considered. The minimum energy consumption is achieved when the optimal distance for transmission of data is calculated by the sensor node from the other nodes in the dense multi-hop WSN's. In (S. Dulman et al., 2006) the approach of for 1-D queues network is also considered which is named as MFR, which deals with the concept of choosing the node which is the farthest as the next forwarder among all the neighboring nodes which ultimately resulted in the lesser delay and also lesser energy consumption because of multi-hop. In (H. Liu et al., 2009) another concept was discussed which focused on reduction of the total energy consumed based on the two objectives i.e. allocation of bits and selection of path. Because of networks nature of being wireless, major problem faced in the wireless network is the unreliability of the link. In (S. Biswas and R. Morris, 2005) the solution for this problem using the opportunistic routing is explained. In (M. Zorzi and R. R. Rao, Oct/Dec. 2003), (L. Cheng et al. 2014) (X. Mao et al., 2011) the opportunistic routing is further improved by the concept of Ex-OR, GeRaF, and EQ-GOR which are of more advantage when compared to the traditional best path routing as it takes into consideration the broadcast nature of the sensor networks and also allows multiple neighbors, which can enhance the problem in forwarding of the packet in the network and can result in overheat in transmission of the packets further. Further the issues like selecting the particular forwarding node or rejecting it to minimize the consumption of energy and optimizing the design of opportunistic routing protocol for wireless networks was not addressed by protocols discussed above. Mao et al. & Trevino-Cabrera et al. 2011 introduced EE-OR which was strategically more energy efficient which focused on the concept of selecting the forwarder set of nodes and prioritizing the particular set of nodes by using the solution of forwarding data to sink nodes by energy saving optimization solution.

Along with finding the best possible way of routing methods which emphasizes on improving the efficiency of the individual node which can be implemented in improving the efficiency of the whole network, it is also necessary to keep an eye on other objectives such as lifetime of the network and the energy left out in the intermediate node. Thus, the concept of the left out energy after performing the particular task or residual energy of the sensor nodes which is also acting as primary parameter is also considered for analysis.

4 EFFECTIVENESS OF OPPORTUNISTIC ROUTING IN MULTIHOP WIRELESS NETWORK

4.1 Achievement of optimal energy strategy

Optimal Energy strategy can be achieved by designating optimal hops for determining the optimal transmission distance. While choosing the unoccupied next hop forwarder, additional factors like energy-balancing of a system and remaining energy of nodes also are considered.

4.2 Assignment of each candidate priority

To maximize the progress of every transmission following priority based mostly forwarding is being employed. The packet transmitted by the sender specifies an inventory of forwarding nodes in associate increasing order of ETX towards the destination. Each packed detected by the node is initial checked whether or not it's present in the forwarding list or not. If it's not present then it discarded otherwise packet sets the forwarding timer proportional to its position in the forwarding list. The results of that is that the node with lower ETX towards the destination forwards the packet previous and therefore the alternative nodes, that were antecedently hearing the forwarding, cancels their forwarding timer and therefore the packet is being aloof from the queues, avoiding the duplicate forwarding.

4.3 Formation of farwarding candidate list

The forwarder list is fixed by the supply supported the priority order of the expected price of delivering a packet from every node in the list to the destination. Price metric is outlined because the range of transmissions needed to maneuver a packet on the most effective traditional route from node to destination by reckoning hops and retransmissions furthermore. The only distinction between ExOR (Opportunistic MultiHop Routing for Wireless) and ETX (Estimated transmission count) is that ExOR only uses the forward delivery chance. Complete set of inter-node loss rates are used for scheming ETX values by ExOR.

If in an exceedingly network the range of nodes are too many then the expected number of a batch's packet that any given node is to blame for forwarding could also be near zero also. If this is the case, then ExOR's agreement and programming protocols can have high overhead because the price is proportional to the quantity of nodes. This can be the explanation why the ExOR supply includes only a sub set of the nodes in the forwarder list. ExOR simulation that relies on the link loss possibilities is employed for running the supply and it selects only the nodes that transmit minimum of 10 percent of the full transmissions in the batch.

5 NETWORK MODEL, ENERGY MODEL AND OPTIMAL TRANSMISSION TECHNIQUE

5.1 Network model

In a 1-D queue model, We suppose that our strategy goes to work for comparatively dense network i.e., all intermediate node have a lot of surrounding nodes and that they all have some information concerning the situation information of their surrounding nodes and site of source and sink node. Each wireless device node has fastened most transmission vary r and minimum transmission vary D min. The 1-D queue network is then build by an attached graph G = (T, N), wherever T could be a bunch of nodes placed linearly and N is about of links between nodes used for communication.

5.2 Energy model

The energy used may be written as below:

$$E_T = \left(E_{elec} + \varepsilon_{amp} d^\tau \right) B \qquad (1)$$

where,

E_{elec} - Fundamental energy used of sensing node to power up the circuit for transmitter device or receiver device,

ε_{amp} is energy vanished in the transmit amplifier,
d is the space between transmitter device and receiver device,

τ is the channel path-loss exponent of the antenna,

Referred as energy used to transfer a B-bit message in a space d.

Energy used by receiver E_R may be evaluated as:

$$E_R = E_{elec} B \qquad (2)$$

5.3 Optimal distance

In the network of WSNs considered above the nodes are uniformly and independently distributed and optimal distance ( ) for any node "h" is defined as

$$d_{op} = M - x_h = \sum_{i=1}^{n} (x_i - x_{i-1}) \qquad (3)$$

where is defined as the position of the sensor node "h"

6 PERFORMANCE METRICS

6.1 First dead node

FDN is connectivity of the network defined for evaluating the influence of the connectivity of the network.

The connectivity of the network becomes lousy once the primary energy drained node seems and also the chance of the network partition will increase.

6.2 Network lifetime

When the receiver becomes unable to receive the packet sent by the supply, it is called Network

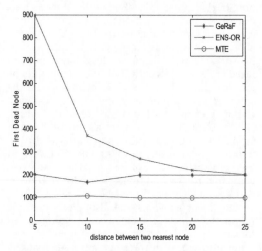

Figure 2. Performance comparison of FDN according to theminimum distance between two nodes.

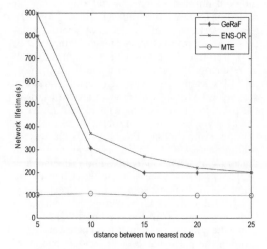

Figure 3. Performance comparison of NL according to minimum distance between nearest nodes.

276

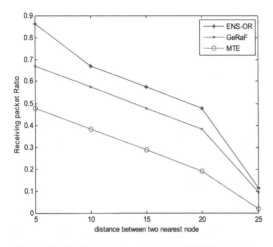

Figure 4. Performance comparison of RPR according to the minimum distance between two nearest nodes.

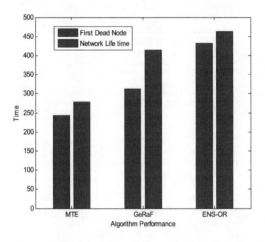

Figure 5. Overall performance comparison of all three protocols on the basis of time.

lifetime of a 1-D queue network. High balance of energy consumption and network partition can be achieved by higher network lifetime. Thus, network time period is strictly associated with the consumption of energy and network partition.

6.3 *Receiving Packets Ratio (RPR)*

The quantitative relation of variety of packet received by the receiver to the total packets sent by the supply is well-known as Receiving Packet Ratio (RPR). For avoiding the network partition adequately, most of the packets sent by the source should be received by sink node which ends up in good connectivity of the network. Overall performance comparison of all three protocols on the basis of time in Fig. 5.

7 CONCLUSION

Because of its efficient characteristics when compared to other protocols like simple implementation and low maintenance within WSN this protocol is mainly used for observance and management application in our lifestyle. But because of the limitation of the battery life of the sensor nodes, it is only the reason why Energy saving optimization becomes one among the main considerations in WSN routing protocol style. In our research, we mainly focused on reducing the energy used and increasing the network time period of 1-D queue network with already allocated sensor nodes. For doing this we tend to used timeserving routing protocol tomaximize the network energy efficiency with the help of distance (called as optimal distance above) between the sensor nodes acting as relay nodes and sink and also the energy remaining in each node is also considered. We focused this scheme after considering that the relay node cannot move and hence the optimal transmission distance is not changed and this indeed resulted in increased network lifetime. So, our aim is to work on developing an energy efficient opportunistic routing scheme that consumes the minimum power and hence the relay nodes with could save lot of energy which can increase the lifetime of the network. Various graphs and results shown above concludes that ENS_OR can help us in increasing the level of power saving and network partition in a better way with respects to other existing routing protocols considered for comparison.

8 FUTURE WORK

This routing protocol can further be modified by introducing various modifications in the routing protocol and the way in which the data is processed such as:

8.1 *Replacing the source nodes used to sense the data with VANET's*

Since the designing of the specialized sensor nodes requires a lot hard work for finding its efficient and practical implementation and after that also the sensor nodes designed are mainly capable of doing single task. Considering this fact only Vehicular Ad-Hoc Networks (VANET's) can be used which are based on the fact that vehicles which normally run on the roads can act as sensor nodes and can get equipped with the sensors acting as intermediate nodes to send the data towards the sink. This implementation can increase the practical consideration of this protocol and its design.

8.2 Replacing the 1-D approach with multihopping and efficient path calculation

The major consideration which we did for this protocol is considering that the network is a 1-dimensional network and which is used for sending the data towards the sink node. This approach of 1-D network can be replaced by considering the more efficient way of sending the data and go for the multi-hop approach in the network.

This approach can be included in out technique in the process of maintaining the forwarder table in which the count of minimum number of hops to sink can be included which can help in deciding the best possible path to the sink and also path with going to sink with passing through the multiple hops with least energy consumption.

8.3 Development of most efficient traffic system

With the data which is stored on the network which was processed with the minimum possible energy consumption with protocol suggested above can be used in various ways such as broadcasting the information in a particular area, connecting the display devices which can display the traffic density of a particular area, and etc. which can be further be used in designing the efficient traffic system which is the mandatory requirement as there is an exponential increase in the population and hence the traffic on the road.

REFERENCES

Biswas, S. and R. Morris, "Exor: Opportunistic multi-hop routing for wireless networks," 2005, in Assoc. Comput. Mach. SIGCOMM Comput. Commun. Rev, vol. 35, no. 4, pp. 133–144.

Cheng, L., J. Niu, J. Cao, S. Das, and Y. Gu, Jul. 2014, "Qos aware geographic opportunistic routing in wireless sensor networks," IEEE Trans. Parallel Distrib. Syst, vol. 25, no. 7, pp. 1864–1875.

Dulman, S., M. Rossi, P. Havinga, and M. Zorzi, 2006 "On the hop count statistics for randomly deployed wireless sensor networks", *International Journal Sensor Network*, vol. 1, no. 1, pp. 89–102.

Eric Rozner, Jayesh Seshadri, Yogita Mehta, and Lili Qiu, 2000 "Simple Opportunistic Routing Protocol for Wireless Mesh Networks." The universit of Texas at Austin.

Juan Luo, Jinyu Hu, Di Wu, and Renfa Li, February 2015, "Opportunistic Routing Algorithm for Relay Node Selection in Wireless Sensor Networks", IEEE Transactions on Industrial Informatics, vol. 11, no. 1.

Jain, S. and S. Das, 2005, "Exploiting Path Diversity in the Link Layer in Wireless Ad Hoc Networks", in the Proc. Of the 6th IEEE WoWMoM Symposium.

Keshtkarjahromi, Y., R. Ansari, and A. Khokhar, 2013 "Energy efficient decentralized detection based on bit-optimal multi-hop transmission in one dimensional wireless sensor networks," in Proc. International Federation of Information Process. Wireless Days (WD), pp. 1–8.

Liu, H. B., Zhang, H. T. Mouftah, X. Shen, and J. Ma, Dec. 2009, "Opportunistic routing for wireless ad hoc and sensor networks: Present and future directions" IEEE Communicatio Mag, vol. 47, no. 12, pp. 103–109.

Mao, X., S. Tang, X. Xu, X. Li, and H. Ma, Nov. 2011, "Energy efficient opportunistic routing in wireless sensor networks," IEEE Trans. Parallel Distrib. Syst., vol. 22, no. 11, pp. 1934–1942.

Piret, P. Sep. 1991 "On the connectivity of radio networks", IEEE Trans. Inf. Theory, vol. 37, no. 5, pp. 1490–1492.

Ramaiyan, V., A. Kumar, and E. Altman, Nov. 2012, "Optimal hop distance and power control for a single cell, dense, ad hoc wireless network," IEEE Transection Mobile Computing, vol. 11, no. 11, pp. 1601–1612.

Santi, E. P. and D. M. Blough Jan/Mar. 2003, "The critical transmitting range for connectivity in sparse wireless ad hoc networks," IEEE Transection Mobile Computing.

Trevino-Cabrera, S. Canadas-Hurtado, August 2011 "Survey on Opportunistic Routing in Multihop Wireless Networks", International Journal of Communication Networks and Information Security (IJCNIS).

Wendi Rabiner Heinzelman, Anantha Chandrakasan and Hari Balakrishnan, 2000 "Energy-Efficient Communication Protocol for Wireless Microsensor Networks", Proceedings of the 33rd Hawaii International Conference on System Sciences.

Zorzi, M. and R. R. Rao, Oct/Dec. 2003 "Geographic random forwarding (GeRaF) for ad-hoc and sensor networks: Multihop performance" IEEE Transection Mobile Computing, vol. 2, no. 4, pp. 337–348.

Communication and Computing Systems – Prasad et al. (Eds)
© 2017 Taylor & Francis Group, London, ISBN 978-1-138-02952-1

Spectrum sharing protocol for ad-hoc device-to-device users

Kshitiz Sekhri & Sandeep Sharma
School of ICT, Gautam Buddha University, Greater Noida, India

ABSTRACT: In this paper a technique is shown in which the resources of the spectrum has been efficiently used by using a protocol that dynamically allocates the channel to Ad-hoc users in order to increase the overall capacity of the channel. In order to maintain the quality of the cellular network, management of the interference plays a vital role. We design an on-demand spectrum allocation protocol for the ad-hoc user, which helps them to access licensed spectrum without causing interference in the channel. Firstly, we must estimates the gain over the channel so that the level of the interference has been calculated for ad-hoc users. Then information of the network has been distributed by using route finding packets either by single-hop manner or multi-hop manner in a variable access manner. Failure rate has been decrease while finding a route with the help of network information. With the help of route finding packets, the ad-hoc users can communicate over a licensed spectrum with minimum possibility of outage while the cellular network has not been affecting. In this method, we can save the total transmission power of the ad-hoc user which is directly connected to each other without using the cellular main station.

1 INTRODUCTION

Device-to-Device (D2D) networking allows explicit communication link between cellular networks, thus it bypasses the base station. As compared to the new spectrum sources obtained by the service providers, the number of wireless customers is on the peak. So we adopt new techniques in order to increase their performance. IMT and LTE are some new techniques in order to fulfill the increasing demand of the users but still it require more efforts. In this paper, we use Dynamic spectrum access techniques to fulfill the high requirement for service. For improving the whole performance of the customer network, different ways has been used in (Middleton G.B, 2006). In a process to reorganize surplus users over frequency channel with excess capacity has been given in (Jha S, Aug 2011). A similar work is done in (Lien S.Y, July 2011) where femtocell unit have been formed when we placed fixed relay station in a cell. A protocol is designed in (Middleton G.B, 2007) for assigning the basic resources to licensed user. By using this protocol cellular users can connect straight to one another without passing through Base Station (BS). In order to make a direct link between the device-to-device users, a protocol has been developed by RI.

We can also use a different method in which unlicensed user can dynamically access licensed spectrum in (Satapathy D.P, 1996). A simple model of cellular network is formed in (Liang Y, 2007) that help us to understand the phenomena of frequency reuse via different channel through near-by base stations. In ad-hoc network, transmission capacity plays a vital role to study performance metric and it shows the area spectral efficiency constrained by the outage probability (Weber S.P, Dec 2005) A method is recognize in that will allocate spectrum resources to number of frequencies in order to increase the rate of transmission and also decreases interference throughout the channel. Moreover, a similar technique is developed in which the device-to-device user can pick particular frequency over which it passes the information. In order to specify the physical region for the licensed users, a method has been used in where unlicensed users can also access the spectrum for licensed users.

We are trying to understand how ad-hoc device-to-device communication is possible over a particular frequency when the licensed spectrum channel is idle. The unlicensed users broadcast the discovery packets for the route mapping, hence it is possible to access the channel and restrict the total power of ad-hoc devices. But our first priority is, how to manage interference that cellular users receive so that the level of interference should be minimum while unlicensed user also access the same channel according to the availability. For interweave spectrum sharing, the impacts of spectrum sensing in the interweave spectrum sharing to the primary transmission were acknowledged by evaluating the characteristics of the aggregate interference. Interference is often controlled by providing the unlicensed users with frequency resources that are dislocate to those of the licensed network via completely different frequency band or a subset of the same band that is presently not in use.

In our technique, we make a dynamically allocating spectrum protocol to facilitate direct communication. We examine the scenario for the whole frame in which the path condition is fixed for a system. Furthermore, we use specific coding methods to grant us the power level of the network. D2D user will try to find either a single-hop route or multiple hop routes, and according to the requirement it provides power for transmission. In order to increase both efficiency and success rate of route discovery, network information has been used. If the path is establish between the device-to-device users, and then we are able to calculate outage probability and saving in power consumption of D2D network.

Our results conclude that the count of necessary transmissions is reduced while discovering a route and also decrease the possibility of failure in ending a route while using the information of the route. To determine the lower bound on the probability of outage, we are acknowledging perfect route inversion method. Moreover, ad-hoc users are trying to communicate over more concise route that consumes lesser power rather than using the base station.

Now the rest part of our paper is standardized as given. We will explain about the channel model and its structure in the next section. Later we describe about the device-to-device communication and power management in Section III. The analytical implementation of the outage probability and will study about the performance of the discovery packets and its simulation in Section IV. Conclusion and future work shows in Section V.

2 NETWORK STRUCTURE

2.1 Framework of the channel

We comprises a networking structure that have seven round microcells and a main station that is equipped with Omni-directional antenna that is consider at the centre of the microcell. Here we mainly think about the uplink frequency of the network and distributed the channel in N_C Orthogonal channels. In order to organize a cellular link with respect to the main station, we must have the minimal S1 NR of μm. We assume that a minimum margin of interference κ is present at each main station. This level of margin helps us to check the maximum interference that can channel handle without any distortion.

Macro user is the basic category of users that creates a network path between the main stations and accordingly the data has been send to the prior receiver. The basic users can utilize the main

station by signaling that is also in use by the cellular system. The next type of user is ad-hoc user that creates the route by broadcasting the discovery packets and accordingly mapped the network for single-hop or multi-hop in a given range of cellular dimensions. When the ad-hoc users utilizing the same channel that is also used by the macro user, then we must check the interference should be in allowed range and it will not harm the S1 NR of active licensed users.

2.2 Infrastructure of the network

Structure of the network also remains same for our channel model. We are assuming three different users, in which transmitter is referring as i, receiver is referring as j and accordingly interferer by k. When the channel is influenced by multiple fading and Gaussian white noises, we must calculate the pathloss of the data. If we want to estimate the large scale fading, we must use distance formula of Euclidian for d_{ij} for source and receiver and also for exponential of interferer i.e. α. Here our main purpose is to calculate the minimum power required by the user and accordingly the S1 NR estimation for the network connections. Therefore, we define jth user's S1 NR as

$$\Gamma_j = P_i d_{ij}^{-\alpha} h_{ij} \qquad (1)$$

where $d_{ij}^{-\alpha}$ is the pathloss for the signal, P_i is the power used by transmitter and h_{ij} is the channel gain. Furthermore, interferer's power consumption is denoted by P_i and $d_{ij}^{-\alpha}$ is loss in path and h_{ij} is denoted for gain over the channel, which is the link between the destination and interferer. In order to manage the interference of the signal, we use fading control techniques and check the condition of the channel by applying random environment for the channel.

3 RELATED WORK

3.1 Device-to-Device (D2D) communication

In this type of communication, we are trying to use on-demand spectrum allocation protocol so that the ad-hoc device can form a direct link with the licensed user over a similar frequency channel. By using this protocol, ad-hoc devices can transmit data only if the interference remains in the allowed margin while using cellular spectrum. There are two primarily steps that are discussed in our paper. First step is to control power consumption of the unlicensed user so that they can manage to transmit packets over device-to-device the channel when it works on the same bandwidth as of the macro user. But to compute the performance of the protocol, we must look upon the power consumption by each D2D user. If we calculate the transmission energy of the source user, we will further work on the route discovery and different methods of hopping to reach their expected destination. Later we will study about the working of the channel component.

But to compute the performance of the protocol, we must look upon the power consumption by each D2D user. If we calculate the transmission energy of the source user, we will further work on the route discovery and different methods of hopping to reach their expected destination. Later we will study about the working of the channel component.

3.2 *Power management for unlicensed user*

To establish a link between the main station and macro-unit user, a minimal S1 NR is required that is denoted as μ_m.

But in this process of link establishment, there is an interference change noted, called interference margin κ is introduced in the network. To get the required S1 NR of μ_m by an ideal power source of the macrocell, ad-hoc user must control the interference. When the macrocell user increases the transmitting power by interference margin, then only this type of outcome has been obtained. Hence, there is a condition when there is no interference margin κ, and then the ad-hoc user can link a path with other users and will achieve the required S1 NR. In order to analyze the effect of interference margin over a link that is established by the macrocell user, we must observe on the power used for the transmission and accordingly find out the S1 NR value of the network. It helps us to virtually notice the power consumption of microcell users as:

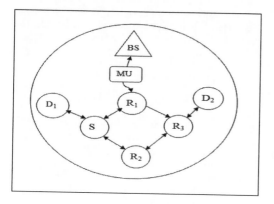

Figure 1.   Realization of Ad-hoc network with cellular infrastructure.

$$P_i \, d_{MB}^{-\alpha} \, h_{MB} \, \sigma \geq \kappa \mu_m \qquad (2)$$

$$P_i \geq \kappa \mu_m \, d_{MB}^{\alpha} \, h_{MB}^{-1} \, \sigma^{-2} \qquad (3)$$

where $d_{MB}^{-\alpha}$ is denoted as the loss over the path and h_{MB} is denoted as gain in channel between the ad-hoc user and cellular users. In this part, we have to calculate the gain in the channel that is exhibit through the main station with minimum error cause over the ad-hoc user. But in practical, estimation of the route may cause some sort of interference in the channel, therefore the macrocell user burst into outage. Hence the ad-hoc network must have autonomous level of outage value.

In a present scenario, we have to check the value of S1 NR of an ad-hoc connection for any variable unlicensed user. Let us assume that ad-hoc user as a source.

If we take P_i or P_s (i.e. source power) that should be minimal in (2), then re-ordering the equation, we will get

$$P_s \, d_{SB}^{-\alpha} \, h_{SB} + \sigma^2 \geq \mu_m \qquad (4)$$

where $d_{SB}^{-\alpha}$ is denoted as the loss over the path in between source and base station and h_{SB} is the total gain over the channel.

3.3 *Duplex channel allocation for ad-hoc users*

In our network, we are using on-demand channel allocation routing protocol that is used to dynamically access the channel and make a direct link between two D2D users.

DSR is a route on-demand packet based discovery protocol which broadcasts the network with discovery packets and hence, it swaps the destination address of every node in the system so that the receiver must have a brief idea about the path. There are number of route discovery protocol like DSR in which the source node broadcast the packet in order to reach the destination and must have the address of each node so that it is possible for source to reach at the right destination. The similar results showed when route detail is passed over one node to other node with shortest path distance through the link that helps us to tell about the route existence.

3.4 *Total power gain estimation*

In order to access the bounded dimension of the network and do not travel again and again over a same path, DSR protocol has been used which avoid the formation of loop. Moreover, D2Ds operate CSMA/CA to assure that only one D2D approaches the channel at a given time. If we want to access only one device at a time, CSMA/CA

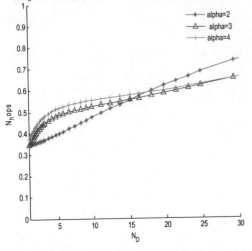

Figure 2. Total no. of hops (N_{hops}) for a multi-route ad-hoc network.

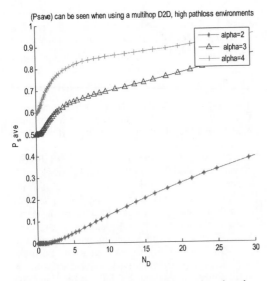

Figure 3. Power saving in ad-hoc network rather than using cellular network using discovery packets.

technique has been used by the ad-hoc device-to-device user. In order to decrease the level of interference and overhead, these methods should be apply and the total numbers of discovery packets reduces through the source.

Our priority is to calculate the power transmit by the source and also power dissipated by the interferer that manages the node address given by the route discovery packet. In the starting of the process, a packet has been broadcast to the destination which contains the transmission power Ps and interference 1s that is measured by the each node. Node j is used to resemble the received packet of the jth ad-hoc node user.

For analyzing the impact of these two power, we look at

$$\Gamma j = Ij + \sigma 2 \tag{5}$$

where Γj is denoted as S1 NR and Pj is denoted as received power at the node j. The calculated power of interference has been represented as at node j with Ij.

$$dsj\text{-}\alpha \; hsj = \Gamma j \; (I + \sigma 2) \tag{6}$$

which gives the solution for both path-loss and gain of the channel. In some topologies and channel conditions, the value of Pj is null so that a full duplex channel between ad-hoc users can be possible.

If node j is the unlicensed user's destination address, then single hop route has been laid between the ad-hoc users. The discovery packets broadcast the route request until a path has been found or go for multi-hop process. Whenever, there is a possibility for a full duplex channel, the protocol has been started for searching for near-by node. Both, power of transmission and power of interference has been included whenever discovery packets has been sent.

3.5 Results and discussions

In MATLAB, around 106 variable topologies have been simulated by using our channel scheme. There are different parameter that has been used for simulation process is mentioned in Table 1. In order to see the work of the central macro unit that is kept between the other 6 macro units has been show via number of simulation results. For a mutual connection between the two end users, not all the nodes participate for any value that is given for N_D. Number of failure has been decreased when the value of N_D increased. The efficiency of finding

Table 1. Network parameters.

Channel parameter	Value
Radius of a microcell (R)	2000 m
Radius of cluster (r)	500 m
Total channels (N_C)	30
Total MUs (N_M)	30
Noise in the network (σ^2)	−104 dBm
S1NR of base station (β_B)	10 dB
S1NR of D2D user (β_D)	5 dB
Interference (κ)	3 dB

a duplex path can be significantly improves when we have the information about the network with the help of route-request and route-reply packets that is shown in the Figure. We can also calculate the time saving during the transmission of route-request packet as

$$T_{save} = T - T_{NI} \qquad (7)$$

where T is the time interval in which the route-request packet is not present in the Network Information (NI) and T_{NI} is the time interval in which route request is present in the NI. These values are used to find T_{save} for a given node.

4 CONCLUSIONS

We present in our paper that how is it possible for an ad-hoc user that linked with other ad-hoc devices, can opportunistically access the same source of frequency as a fully weighted licensed user network. We develop a practical protocol for D2Ds to use this scheme in a distributed manner and with no coordination from the base station.

In order to use Ad-hoc device to device network structure in a shared manner, this protocol has been used. There is no need for the main station to establish a link between the ad-hoc users over a cellular spectrum. If we want to control the interference at the main station, then we should minimize the consumption of power used for the transmission of data packets. For establishing a route between the source and the destination, a protocol has been used which uses packet for finding the route with the minimum power for transmission. Route-request and route-reply packets significantly reduces the probability of failure in order to find a route and also lower the total power needed for transmission of packet, which includes the information about the network structure. We can also calculate the outage over a channel when the ad-hoc network has been linked with other devices that already provide us the route between source and destination. Hence we display

in our results that the unlicensed user can fully access the spectrum, which improves the trade-off between the networks and also increase the overall efficiency of the spectrum. Moreover, simulation results show that significant power savings can he gained using D2D mutes rather than connecting to the cellular base station. Moreover, the saving in the power can be achieved when we are using ad-hoc's node to node connection rather than using the main station via licensed users that has been showed in our simulation results.

REFERENCES

Doppler. K, Yu. C.H, Ribeiro. C, and Janis. P, in 2010. "Mode selection for device-to-device communication underlaying an LTE-advanced network." IEEE WCNC.

Jha. S, Rashid. M, Bhargava. V, Aug. 2011. "Medium access control in distributed cognitive radio network," IEEE Wireless Comm. vol. 18. no. 4. pp 41–51.

Kaufman. B, Aazhang. B, and Litlehers. J, in 2009. "Interference aware link discovery for device to device communication," Asilomar Conf. Signals, Syst. Comput.

Liang. Y, and Goldsmith. A, in 2007. "Adaptive channel reuse in cellular systems." IEEE Int. Conf. Commun.

Lien. S.Y, Lin. Y.Y, and Chen. K.C, July 2011. "Cognitive and game-theoretical radio resource management for autonomous femtocell with QoS guarantee," IEEE trans.Wireless Comm.

Middleton. G.B, and Hoch. K, in 2007. "An algorithm for efficient resource allocation in realistic wide area cellular networks." Svmp Wireless Personal Multimedia Comm.

Middleton. G.B, Hoch. K, Toth. A, and Lilleherg. J, 2006. "Inter-operator spectrum sharing in a broadband network." IEEE Int. Symp. Spread Spectrum Techniques Appl.

Saraperhy. D.P, and Peha. J.M, in 1996. "Spectrum sharing without licensing opportunities and dangers"- in Interconnection Internet in Telecommun. Policy Research Conf.

Weber. S.P, Yang. X and Andrews. G, in Dec 2005. "Transmission capacity of wireless ad-hoc networks" IEEE Trans. Inf. Theory, vol. 51, no.12, pp. 409.

Communication and Computing Systems – Prasad et al. (Eds)
© 2017 Taylor & Francis Group, London, ISBN 978-1-138-02952-1

Performance analysis of multichannel OCDMA-FSO network under different pervasive conditions

S. Arora, A. Sharma & H. Singh
CT Institute of Technology and Research, Jalandhar, Punjab

ABSTRACT: To meet the growing need of high data rate and bandwidth, various efforts has been made nowadays for the efficient communication systems. Optical Code Division Multiple Access over Free space optics communication system seems an effective role to provide transmission at high data rate with low bit error rate and low amount of multiple access interference. This paper demonstrates the OCDMA over FSO communication system upto the range of 7000 m at data rate of 5 Gbps. Initially, the 8 user OCDMA-FSO system is simulated and pseudo orthogonal codes are used for encoding. Also, the simulative analysis of various performance parameters like power and core effective area that are having effect on the Bit Error Rate (BER) of system is carried out. The simulative analysis reveal that the length of transmission is limited by the Multi-Access Interference (MAI) effect which arise when the number of users increases in the system.

1 INTRODUCTION

FSO is an optic technology that propagates light in free space to provide data transmission between two points. The technology is very useful where the physical connectivity by means of fiber cables is not practicable (Kumar et al. 2012). FSO communication has major advantages over fiber communication and terrestrial microwave communication because of high carrier frequencies, high data rate, high capacity, secure communication (Kumble et al. 2011; Sangma et al. 2012). FSO communication can be used in many applications such as communication between various buildings, ships and aircrafts. CDMA has become a very popular system for wireless mobile communication for the reason of its high bit rate, security, very low BER and system capacity increases (Kennedy et al. 2011). The very intelligent integration of FSO and optical CDMA is proposed and this is OCDM-FSO communication system (Singh et al. 2011). OCDMA-FSO communication plays an effective role to provide high data transmission very low amount of multiple access interference (Menninger et al. 2013).

(Mahajan et al. 2013) propesed a work that 2.5 Gbps signal is generated with the NRZ modulation.

(Nadhila et al. 2013) reviewed a concepts related to challenges a system designer has to consider while implementing an FSO system and atmospheric effects are also studied. (Phao et al. 2011) presenting a demonstration of fading resistant FSO system using simulative bed test employing

FSO link with acceptable BER with the highest stream rate of 2.5 Gbps. (Rang et al. 2015) presented a view to evaluate the quality of data transmission using Wavelength Division Multiplexing.

The FSO transmits invisible eye safe light beams from sender to the receiver using very low power lasers in the terahertz spectrum. FSO can function over kilometers (Rhag et al. 2014).

FSO communications provides cost effective optical connectivity (Smphawamb et al. 2014). It provides the economical valuable optical solutions systems to fulfill the demand of telecom users (Singh et al. 2015; Wang et al. 2013). The FSO system has more flexibility in providing intra campus connectivity and connection between various LANs (Vang et al. 2015; Vakaya et al. 2015). FSO system has high accessibility to provide high capacity links to businesses (Zazumdar et al. 2015).

2 SYSTEM MODEL AND WORKING

The system design shown in Figure 1 representing the block diagram for the 8 users OCDMA over FSO systems. The system design consists of sine generator, summer, electrical generator, DM lasers, pre-amplifier and optical Multiplexers. The sine wave generator produces the rectangular pulses of the light. The electrical generator produces electrical pulses of short amplitude, discrete mode laser used to provide narrow beam width optical frequency pulses, pre amplifier boost the small electrical pulses for future process of

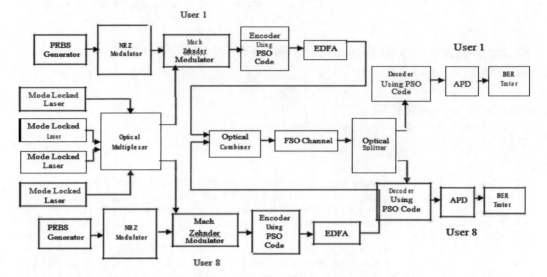

Figure 1.   Block diagram of 8 users OCDMA over FSO communication system.

amplification. In the designed OCDMA-FSO communication system (Figure 1), 5Gbps signal is generated with NRZ modulation format. The signal is then further modulated by means of the Mach-Zender Modulator (MZM) used to modulate the signal and then provide the transmission over FSO by providing means of a four mode-locked lasers used for creating a dense WDM multi-frequency source of light of 3 mW operating at 1550–1551.2 nm.

Then OCDMA signal is transmitted over FSO. The 8 different signature codes for the 8 OC-48 users are required. Pseudo Orthogonal (PSO) matrix codes (Singh et al. 2015) are very popular for OCDMA applications primarily as they also retains the correlation advantages of PSO linear sequences (Wang et al. 2013) while reducing the need for the bandwidth expansion. PSO codes also generates the larger code sets of various signature codes efficiently. In this simulation model, an extensive time-slot reuse sequences are used for user 1 to user 8. There are four time slots used without guard-band which gives the chip period of 100 ps. At the base station, the PSO codes are retrievable using decoder and the optical signal gets converted into electrical signal and data is recovered successfully.

3   RESULTS AND DISCUSSIONS

The analysis are performed by observing two cases: case (I) demonstrates an effect of Multiple Access Interference (MAI) in designed system under different data rates, case (II) investigated the

optimized design with different parameters under different pervasive conditions.

The various parameters used in the designed system are data rate = 5 Gb per second, core effective area = 300 cm², transmitted power = 9 mW, sigma add = 1.9 divergence angle = 0.25 mrad. These systems are implemented with the OPT-SIM software.

Case I: To compare and analyze the impact of Multiple Access Interference (MAI) in designed system:

The crosstalk among various users sharing the common FSO channel is particularly known as Multiple Access Interference (MAI). It is the significant source of bit errors in the proposed OCDMA over FSO system (Phao et al. 2011; Rang et al. 2015). Initially, (Figures 2a, b) and (Figure 3a, b) indicates the graph of Q values and BER versus distance with and without effect of MAI in OCDMA over FSO communication system.

In first case of without MAI effect, from (Figure 2a), it has been observed that that there is significant decrease in the value of Q factor which lies with values 18 to 12 at 5 Gbps in without MAI at one user. (Figure 2b) depicts that there is significant increase in the value of BER which lies within 10^{-17} to 10^{-7} at 5 Gbps in without MAI at single user users for transmission distance of 1000–7000 m. Further from (Figure 3a), which also indicates the graph between Q factor value versus distance with MAI effect at eight users in optical CDMA over FSO system, it has been observed that Q-factor lies between 4 and 1. From (Figure 3b) shows the BER analysis which depicts that with the increase

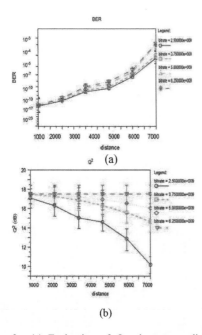

(a)

(b)

Figure 2. (a) Evaluation of Q value versus distance without MAI and (b) Evaluation of BER versus distance without MAI.

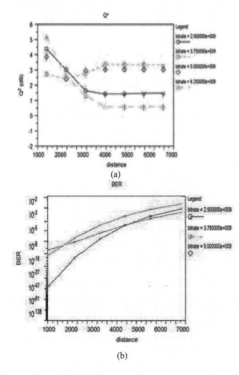

(a)

(b)

Figure 3. (a) Evaluation of Q value versus distance with MAI and (b) Evaluation of BER versus distance with MAI.

in number of users, BER lies within 10^{-12} to 10^{-3} at 5 Gbps which depicts sharply increase in the BER value in the plot. It shows that the Multi Access Interference (MAI) has an impact on OCDMA-FSO network. The (Figures 4a, b) depicts the eye diagram for data and voice in case of with and without MAI. Eye diagram shows that the width and height of the eye opening is large which means that the received signal is very much clear. If we increase the number of users then our data and voice distorted and become error full.

So, the results have shown that there is less distortion in the system with less number of users and it will further increase with the increased MAI effect.

Case II: To investigate the performance of an optimized design with various parameters under different pervasive conditions:

In this case, we are analyzing the designed system by varying the various parameters such as various power levels and attenuation levels.

From (Figure 5a), it has been predicted that there is significant increase in the value of BER with the increase in power transmission, which lies within range 10^{-10} to 10^{-14} and 10^{-1} to 10^{-7} for transmission distance from 1000 to 7000 m in case of power levels 1 mW from 9 mW respectively.

(Figure 5b) depicts the graph between Q-factor versus distance at different transmitter power.

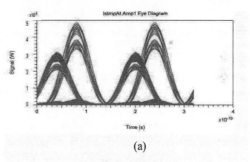

(a)

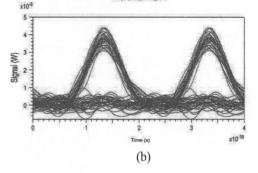

(b)

Figure 4. (a, b) Eye diagram analysis of of received signal for single user taking NRZ signal without MAI and with MAI at eight users.

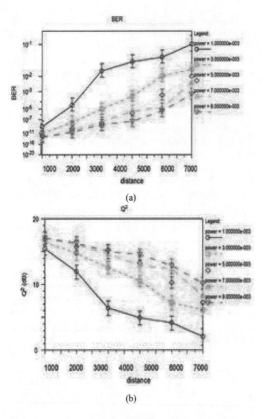

(a)

(b)

Figure 5.   (a) Evaluation of BER value versus distance and (b) Evaluation of Q-value versus distance at different power levels.

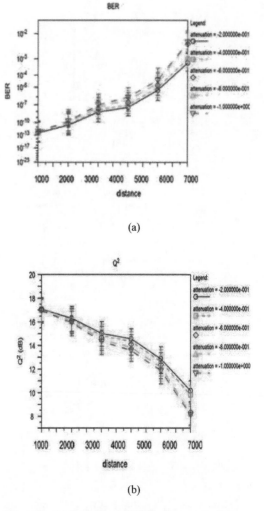

(a)

(b)

Figure 6.   (a) Evaluation of BER value versus distance and (b) Evaluation of Q-value versus distance at different attenuation levels.

From results, it has been observed that the Q factor value decreases which lies within 18 to14 and 13 to 3 for transmission distance of 1000–7000 m in case of 9 mw to 1 mw respectively.

(Figure 6a) indicates that there is valuable increase in the BER, which lies within range 10^{-15} to 10^{-11} and 10^{-5} to 10^{-1} for transmission distance from 1000 to 7000 m in case of attenuation –0.2 dbm to –1.0 dbm.

(Figure 6b) indicates the graph between Q factor value versus distance at different power levels. From results, it has been observed that the Q factor value decreases sharply which lies within range of 17 to16 and 12 to 8 for transmission distance of 1000 to 7000 m in case of attenuation –0.2 dbm to –1.0 dbm respectively.

So, the analysis of Bit Error Rate (BER) and Q-factor value with respect to the distance reveals that the significance decrease in the value of the Q-factor level and there is an increase in the value of the BER as far as transmission distance varies upto 7000 m.

4   CONCLUSION

The design of an OCDMA over FSO communication system at data rate of 5 Gbps for 7000 m length is presented. Performance investigation of this designed OCDMA over FSO system was carried out using NRZ modulation with different parameters and with and without MAI for comparative study. It is concluded that NRZ give us better performance in optical CDMA over FSO communication system. Further transmission range increases with the increase in transmitter power.

The eye diagrams for single and eight users are shown in (Figure 4a, b). It has been observed that the efficiency of the system is degraded with the increase of users. Thus it is determined that MAI play the key role in OCDMA over FSO communication system.

ACKNOWLEDGEMENT

The author would like to express his sincere gratitude and appreciation to the supervisor Er. Anurag Sharma (Assistant Professor, Department of Electronics and Communication for his invaluable guidance, motivation and discussions through all the stages of this research work which enabled to complete the research work on time.

The author would also like to thank all the faculty members, the staff of Electronics and Communication and my classmates, CT Institute of Technology and Research, Jalandhar. They have made the institute a wonderful place to gain knowledge and an enjoyable place to work.

REFERENCES

Kennedy, P. & Bakay, R. 2011. Performance Evaluation of the Free space Optical communication with the effects of the atmospheric turbulences. International Journal of Optical Communication. California University: USA.

Kumar, B. & Singh, N. 2011. 1.25Gbit/s optical CDMA over FSO comm. System. Industrial Electronics (ISIE), IEEE symposium: 911–934.

Kumar, N. & Singh, T. 2012. 2.50 Gbit/s optical CDMA over FSO communication system. Industrial Electronics (ISIE). IEEE symposium: 957–962.

Kumar, P. & Sangma, A. 2012. Performance evaluation of free space optics communication system in the presence of forward error correction techniques. Industrial Electronics (ISIE), IEEE symposium: 911–922.

Kumble, P. & Sharma, H. 2011. Performance of the bit error rate of light wave systems with optical amplifiers. Interna—tional Journal of Optical Communication. Vanderbilt University: USA.

Mahajan, V. & Kaur, G. 2013. High speed, long reach OFDM-FSO transmission link incorporating OSSB and OTSB schemes. Advances in Optics. Elsevier 34: 140–144.

Menninger, H. & Willert, O. 2013. An Introduction to Free-space Optical Communication. Radio Engineering Communications 19(48): 200–204.

Nadhila, H. & Amphawanb, A. 2013. Optimization of free space optics parameters. International Journal Institute Optical Communication. Vanderbilt University: USA.

Phao, L. & Peng, L. 2011. A MPR optimization algorithm for FSO communication system with star topology. International Journal of Optics Communications 4(8): 147–154.

Rang, P. & Lixin, B. 2015. Performance analysis of MPPM FSO system with three decision thresholds over exponentiated bull fading channels. International Journal of Optics Communications 7(9): pp 173–185.

Rhag, V. & Lumba, M. 2014. Severe climate sway in coherent OCDMA-OSSB-FSO carrying transmission system. Advances in Optics. Elsevier 12: 145–148.

Singh, T. & Lamba, S. 2015. Optical CDMA over FSO communication system. International Journal of Optics Communications 4(7): 173–185.

Smphawanb, A. & Shamsuddina, B. 2014. Optimization of free space optics parameters: An optimum solution for bad weather conditions. Elsevier 11: 152–155.

Vakaya, K. & Inagaki, K. 2015. Applications of the FSO communication systems. International Journal of Optics Communications 14(8): 173–185.

Vang, J. & Zhao, G. 2015. Free-space laser communication system with rapid acquisition based on astronomical telescopes. Elsevier 33: 814–816.

Wang, N. & Sharma, A. 2013. Performance evaluation of free space optics communication system in the presence of forward error correction techniques, International Journal of Optics Communications 15: 133–155.

Zajumder, S. & Azhari, A. 2015. Impact of chromatic dispersion on the BER performance of an optical CDMA IM/DD transmission system. IEEE Photon. Technol:150–154.

Communication and Computing Systems – Prasad et al. (Eds)
© 2017 Taylor & Francis Group, London, ISBN 978-1-138-02952-1

On improving performance of wireless sensor networks using mobile sink

Vineet Kamal & Sandeep Sharma
Gautam Buddha University, Greater Noida, India

ABSTRACT: A network of sensors is a collection of sensor nodes in a particular region. The sensor nodes collect data and send it to the sink. Initially data transmission takes place smoothly but after some time the problem of energy hole occurs. The energy of sensor nodes near the sink depletes faster as compared to the sensor nodes away from the sink as they have to send their data as well as the data collected from far off nodes also to the sink and this is known as the Energy Hole problem. Hence in order to get rid of this problem this paper proposes the concept of Mobile Sink. This concept divides the entire sensor network into smaller regions and the Mobile Sink goes into each region, collects data and thus saves energy. This concept is collectively known as Stable Election Adaptive Immune Algorithm. This algorithm increases both network lifetime as well as stability period of WSNs.

1 INTRODUCTION

A network of sensors is a wireless infrastructure consisting of devices that sense, compute and communicate information to an administrator by recording data in a specific environment. Sensor nodes are sometimes also called motes. The power efficiency of WSNs is being increased by employing following measures: low duty cycle operation, reduction in data volume by in network processing and using multi-hop networking. The working operation of a node is given step by step—after sensing physical changes in the environment electrical signals are produced in the transducer. These are analyzed and kept in a microcomputer. As soon as the microcomputer gets command it transfers this data to the centralized computer where this information is processed.

The proper placement of nodes and the problem of energy depletion can be solved with the help of clustering protocols. In this paper, a Stable Election Adaptive Immune Algorithm has been given to enhance the working life of WSNs and to remove the energy-depletion problem. SEAIA uses the Adaptive Immune Algorithm (AIA) to get the proper placement of the moving sink and the required number of CHs. This is achieved by decreasing the amount of energy required in performing the communication process and also to manage the overhead control packets. Further we have done literature survey in section (II). Advantages of mobile sink and proposed mobility path pattern are described in

section (III) and (IV). Network model and proposed protocol is given in (V) and (VI). Section (VII) and (VIII) tell about performance measures and simulation results. Conclusion and future work along with references in (IX) and (X) finally end the paper.

2 RELATED WORK

Various clustering methods in which both the nodes as well as the sink remain stationary are already present. Starting of clustering methods is done by Low-Energy Adaptive Clustering Hierarchy (LEACH). Information is gathered by the CHs and then directly given to the sink and thus completing the mechanism of LEACH. Random selection of CHs is the main drawback of LEACH protocol. Whether a node becomes Cluster Head or not is determined by the probability given by the user. However, the optimal probability of CHs (p_{opt}) is found by applying Genetic Algorithm based LEACH. Even though LEACH-GA enhances the CHs threshold function but it is not able to pick CHs by following a proper defined mechanism. The amount of energy left in a node at a particular time is not considered while picking CHs. Election of CHs by using a well-established mechanism and improvement in the alive time of multi-level hierarchical heterogeneous WSNs is done in the Amend LEACH protocol. To reduce signal overhead and efficient data delivery to the sink Intelligent Agent-based Routing was proposed.

3 ADVANTAGES OF MOBILE SINK

This section will list some of the advantages of having a mobile sink. A mobile sink does not stays at a place for a long time. It keeps on changing its position over time and this makes almost impossible for the attacker node to extract confidential information from it. Another usefulness of mobile sink is that it goes close to the location where the sensor nodes are present. As a result less number of hops are required for packet transfer between sensor nodes and the sink.

4 PROPOSED MOBILITY PATH PATTERN

A moving sink is used in this algorithm that decreases the amount of energy lost by each node. To prevent unnecessary wastage of energy information is transferred in lesser number of hops by dividing the given area into smaller regions. This mechanism not only decreases the number of packets lost but also the time required to send data from nodes to the sink. Now sensor nodes with small transmission range are required in comparison to the situation in which the whole sensor area is present where sensor nodes with extremely large transmission range are required to have a link between the nodes and sink. Here, we take three types of division of regions. The first and second type are moving sink with 4 parallel regions and 4 square regions and the last one is 8 boxes regions.

5 ASSUMPTIONS

To get proper functioning from this algorithm these assumptions are being made:

- The location of each sensor node is provided by a GPS module or a certain localization technique and each one of them is stationary.
- Data aggregation techniques are used to join various correlated signals into one fixed length packet and this makes intra-cluster sensed data is extremely symmetric.
- Same energy is required to send a message between any two nodes which makes the communication channel symmetric.

Parameter	Notation
Length of overhead control packet (bits)	k_{CP}
Residual energy of the sensor node	E_{RS}
Number of member nodes for CH	m_i
Size of the sensing field	$M \times P$
Path loss	p
Total number of sensor nodes in the network	N
Initial energy of each sensor node	E_o

- The location of the sink keeps on changing and it has the highest energy among the various sensor nodes.

6 STABLE ELECTION ADAPTIVE IMMUNE ALGORITHM

The right movement of the sink along with the correct election of CHs are the main two criterion by which the amount of energy lost by a node can be minimized along with enhancement in the alive time of the nodes. Hence Stable Election Adaptive Immune Algorithm is used to get rid of these problems. Various rounds are run to find the functioning of the proposed protocol. One single round consists of three parts-

- Prepare Phase: To have uniform distribution of CHs in the sensor field and to achieve energy efficiency this step is used.
- Set-up Phase: Proper placement of CHs and sensor nodes is done
- Steady State Phase: Framing of data packets and then transmission to the sink is performed.

Algorithm for one round of SEAIA protocol Notations:

N: Number of sensor nodes in the sensor field
E_o: Initial energy
k: Number of bits in the message
k_{CP}: Number of bits in control packet
$M \times P$: Size of sensor field

Algorithm for SEAIA
1: //**Initialize the network**//
2: xm=100;
3: ym=100;
4: sink.x=0.5*xm;
5: sink.y=0.5*ym;
6: N=100;
7://**Divide the sensor field in R equal regions**//
8: rectangle('Position',[0 0 50 50]);
9: rectangle('Position',[50 50 50 50]);
10: rectangle('Position',[0 50 50 50]);
11: rectangle('Position',[50 0 50 50]);
12 //**Creation of random sensor network**//
13: **for** i=1:1:n
14:    S(i).xd=rand(1,1)*xm;
15:    XR(i)=S(i).xd;
16:    S(i).yd=rand(1,1)*ym;
17:    YR(i)=S(i).yd;
18: **end for**
19://**Election of Normal Nodes**//
20: **if** (temp_rnd0>=m*n+1)
21:    S(i).E=Eo;
22:    S(i).ENERGY=0;

(Continued)

Algorithm for SEAIA (*Continued*)

23:    plot(S(i).xd,S(i).yd,'o');
24: **end if**
25: //**Election of Cluster Heads**//
26 :**if** (temp_rnd0<m*n+1)
27:    S(i).E=Eo*(1+a)
28:    S(i).ENERGY=1;
29:    plot(S(i).xd,S(i).yd, '+';);
30: **end if**
31: //**Data transmission**//
32: **if** (Cluster Head=1) //**Create TDMA**
33:    { packets_TO_BS=0;
34:      packets_TO_CH=0; }
35: **end if**
36: **elseif** (Cluster Head=0)
37:        S(i).min_dis=min_dis;   //**Wait for info. about TDMA**
38: **end if**
39: //**Checking the coverage capacity of mobile sink**//
40: **if** (min_dis<do) //**If all regions not visited**
41:    **Go to Step 7**
42: **end if**
43:    **elseif** (min_dis>do) //**If all regions visited**
44:    **Go to Step 46**
45: **end if**
46: //**Analyzing energy profile of nodes**//
47:**while** (S(i).ENERGY==1)
48:        **Go to Step 7**
49:**elseif** (S(i).ENERGY==0)
50:**end if**
51: //**Dividing the whole sensor network into dy-namic area regions**//
52: [vx,vy]=voronoi(X,Y);
53: plot(X,Y,'r*',vx,vy,'b-');
54: //**End**//

---

7 PARAMETER ANALYSIS

The following parameters are being checked:
   Network lifetime
   Stability Period
   Instability Period
   First Dead Node (FDN)
   Half Dead Node (HDN)
   Last Dead Node (LDN)
   Number of Cluster Heads per Round
   Number of Alive Nodes per Round
   Throughput
   Packet Lost Ratio
   Packet Transmission Ratio
   Packet Delay

8 SIMULATIONS AND RESULT

The efficiency of the proposed algorithm and its comparison with the other protocols is done in this section with the help of MATLAB software. In these tests, we assume that all the sensors are of the same type and every sensor is making one data packet per round to be sent to the sink.

A *Analysis of dividing in regions*

Here, we analyse the movement of the sink to find the performance of this protocol. So we consider four cases:

   i. stationary sink
   ii. moving sink with 4 parallel regions
   iii. moving sink with 4 square regions
   iv. moving sink with 8 box regions

The simulation consists of 100 nodes (with starting energy of 0.5 J) placed within a 100×100 m^2 sensor field. 2000 message bits along with 100 control bits are being sent.

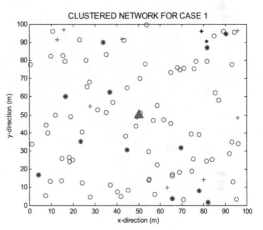

Figure 1.   Stationary sink.

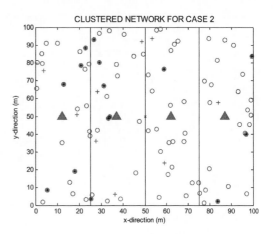

Figure 2.   Moving sink with 4 parallel regions.

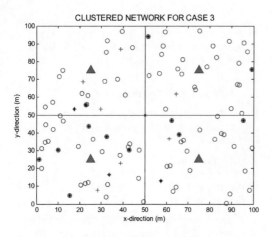

Figure 3.   Moving sink with 4 square regions.

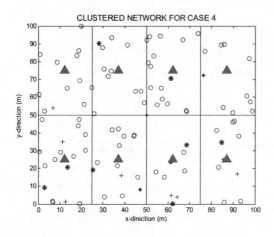

Figure 4.   Moving sink with 8 box regions.

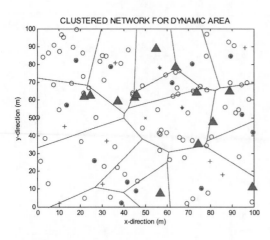

Figure 5.   Dynamic area graph.

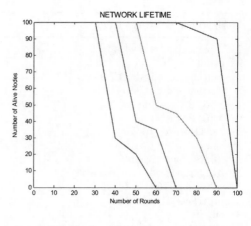

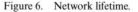

——— Stable Election Adaptive Immune Algorithm
——— Mobile Sink using 8 boxes region
——— Mobile Sink using 4 square region
——— Mobile Sink using 4 parallel region

Figure 6.   Network lifetime.

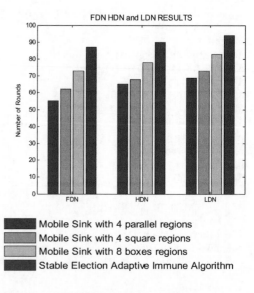

■ Mobile Sink with 4 parallel regions
■ Mobile Sink with 4 square regions
■ Mobile Sink with 8 boxes regions
■ Stable Election Adaptive Immune Algorithm

Figure 7.   FDN, HDN and LDN results.

B   *Alive time and stability period*

The time period till which the nodes remain alive as well as they give effective performance is checked in this test. 100 sensor nodes of the same type with starting energy of 0.3 J are placed randomly within a 100×100 m² sensor field. 3000 bits of message was transmitted.

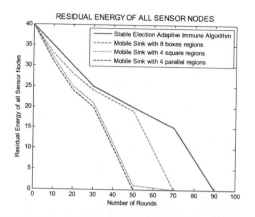

Figure 8. Residual energy of all sensor nodes.

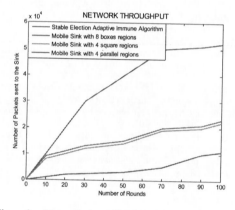

Figure 9. Network throughput.

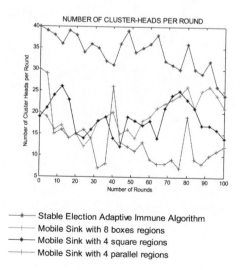

Figure 10. Number of Cluster heads per Round.

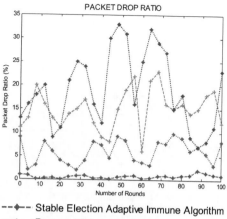

Figure 11. Packet drop ratio.

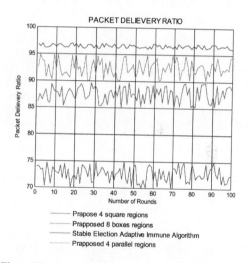

Figure 12. Packet delivery ratio.

C Moving sink vs Stationary sink

In this test, the data transmission capabilities of LEACH, A-LEACH and LEACH-GA is compared with the proposed protocol. With starting energy of 0.5 J, 100 nodes are being kept in an area of 50×50 m². 60 bits of control packets together with 2500 data packets are being transmitted.

D Delivery of packets and delay

With stating energy of 0.2 J, 150 nodes are being placed in an area of 100×100 m² sensor field. 3000

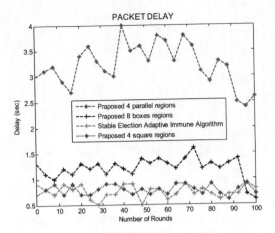

Figure 13.   Packet delay.

data packets are being sent. The number of packets delivered and the number of packets lost is being checked in this test.

9   CONCLUSION AND FUTURE WORK

Enhancement in the packet transmission rate along with a considerable decrease in the number of packets lost is achieved with this protocol as compared to the other protocols. This is basically due to the reduction in the distance between the sink and nodes. For future work we would like to have both the sensor motes as well the base station with travelling capabilities so that their effect on the protocol can be analyzed.

REFERENCES

Abbasi A. A. and M. Younis, ‖A survey on clustering algorithms forwireless sensor networks, ‖ *Int. J. of Computer Communications*, vol. 30, no. 14, pp. 2826–2841, Oct. 2007.

Girik Pachouri and Sandeep Sharma, "Anomaly Detection in Medical Wireless Sensor Networks using Machine Learning Algorithms", in Elsevier Procedia Computer Science proceeding of *4th International Conference on Eco-friendly Computing and Communication Systems (ICECCS)*, NIT Kurukshetra, vol. 70, pp. 325–333.

Hamida E. B. and G. Chelius, ‖Strategies for data dissemination to mobile sinks in wireless sensor networks, *IEEE Trans. Wireless Communications*, vol. 15, no. 6, pp. 31–37, February 2008.

Heinzelman, W., A. Chandrakasan, and H. Balakrishnan, "An application-specific protocol architecture for wireless microsensor networks," *IEEE Tran. Wireless Comm.*, vol. 1, no. 4, pp. 660–670, 2002.

Singh S. K., M. P. Singh and D. K. Singh, ‖Routing protocols in wireless sensor networks–a survey, ‖ *Int. J. of Computer Science and Engineering Survey*, vol. 1, no. 2, pp. 63–83, Nov. 2010.

Sandeep Sharma and Rajesh Mishra, "A Cross Layer Approach for Intrusion Detection in MANETs", in *International Journal of Computer Applications*, Vol. 93, No. 9, pp. 34–41, May 2014.

Communication and Computing Systems – Prasad et al. (Eds)
© *2017 Taylor & Francis Group, London, ISBN 978-1-138-02952-1*

Optimization of communication parameters in wireless sensor network

Abhishek Raman & Sandeep Sharma
Gautam Buddha University, Greater Noida, India

ABSTRACT: Nowadays, opportunistic routing is being used as an important parameter. It improves the network parameters which are used to forward the data packets from source to sink in more efficient and optimized way through Opportunistic Routing (OR), the data can be forwarded from one node to the second and, selection of more prioritized node is possible. Here, in this paper the focus is on optimization of network parameters for better results. The protocol used is for Energy Optimization Opportunistic Routing (ENO-OR). The protocol evaluates the best test results with the proposed routing scheme. The constraint parameters have been used are; lifetime of the network, Residual Energy, Packet Received, dead nodes received. Here we compared wireless connectivity with the other existing wireless sensor network routing schemes.

Keywords: opportunistic routing, ENO-OR, Network lifetime, residual energy

1 INTRODUCTION

Nowadays, WSN is very commonly used in commercialized and industrial areas because their application used in biological observation, habitat observation, health care, process observation, and surveillance. The military area, we are capable of use WSN to check a human action. When an event is occurred the source node i.e. transmitter sense it and throw the data to the initial place to sink by communication through connections present in the scenario. The utilization of WSN is growing every day as well as at this moment it faces some difficulties main crisis is force constraints with restricted battery lifetime. Since every connection going on force to the actions, it is becoming the leading problem in WSN. Not a success process to connection could be communication the absolute system or application. Each detection connection could dynamic form the half transmittal actions sleep along with idle mode. In the active form of connections consume force when receiving or transmission data. Here idle manner connections devour approximately to the similar sum of money of force within the dynamic mode, although in sleep mode; connections closure to the radio of keeping the physical phenomenon.

In adjustment to cooperatively adviser concrete or ecological conditions, the most important function of sensors which are deployed in a particular area or region is to collect and address the data. It is well acknowledged that sending data from the transmitters end consumes lot of energy as compared to receiving data at the receiver end (Pottie et al. 2000). To modified the performance of energy efficiency for sending the data from the transmitter end, a number of former presented energy-efficient protocols are used to acquisition very less energy route between transmitter to receiver to the accomplish most advantageous energy utilization (Hoang, Duc Chinh, et al. 2014, LoBello, Lucia, and Emanuele Toscano 2009, V 2014). However, the layout of the process an energy-efficient routing protocol, since it involves not alone award the minimum consumption of energy route from the single sensor node to the destination, but as well acclimation the administration of balance activity of the accomplished arrangement (Ren, Fengyuan, et al. 2011).

Additionally, undependable wireless network could be loss of packet and assorted retransmissions in a preselected acceptable aisle (Behnad, Aydin, and Said Nader-Esfahani 2011).

The network can be optimized by applying the Opportunistic routing theory using the one-Dimensional (1-D) network that used should be kept focused to the destination in order to forward the data that is listed and should not concentrate on the extra data traffic in Figure 1. Priorities are set by the nodes and if the same packet is transmitted by different nodes the one with lower priority will be ignored and higher priority node will be accepted. While making the forward list is also a concern, the issue is raised in a selection and setting the priority on the transmission list that would enable to save energy over all the sensors.

The Figure 1 shown above is vehicle information system through relay nodes. In this network

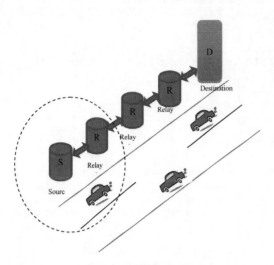

Figure 1. A relay network that showing the communication between source to destination.

a source node S which carries the information like information about vehicle average speed, vehicle density etc. and forward the data by using relay nodes R placing between source and destination D so that the destination can proceed it for further to the traffic management system. This study is concentrated on how to select and arrange the forward node list so the energy could be saved.

1.1 *Related work*

There are lots of methods in the literature dealing with routing in WSN. Some papers are also proposing energy efficient routing protocols. The sensor will sense the traffic information in their range, and pass it on to the sink. When it is passed, it should be sent through the energy efficient path. (Luo, Juan, et al. 2015) proposes an energy efficient routing scheme ENSOR for such 1-D networks.

It introduces a concept called EEN (Energy Equivalent Node). A forwarder set list is prepared based on the residual energy of the neighbors. The nodes in the forwarder set rank themselves on the basis of distance from the EEN. Specifically, the Energy Saving through the Opportunistic Routing (ENS_OR) algorithm is created to protect minimum power consumption during data transmission and also protect the nodes with low residual energy. Extensive simulations and authentic test results show that solution of ENO_OR can considerably improve the performance on energy and lifetime of network in comparison with the other existing routing scheme. Opportunistic routing for WSN is dealt in (Biswas et al. 2005, Chachulski, Szymon, et al. 2007). In ExOR (Biswas et al. 2005) the opportunistic routing concept is used for reduc-

ing the total cost of the route. The forwarder is adopted based on following a stringent scheduling mechanism so that only one node is forwarding the data and others will discard it. Packets are mixed randomly before transmission in MORE (Chachulski, Szymon, et al. 2007), so that same packet is not forwarded by all. In (Mao, Xufei, et al. 2011) opportunistic routing is used for minimizing the energy cost by appropriately selecting a forwarding list. It also explains the usage of opportunistic routing for adjustable and non-adjustable power models. The proposed approach of this paper is used opportunistic routing for selecting the relay nodes, which the total cost of the route is reduced and hence network lifetime is increased.

Some energy-efficient approaches have been prospected in the literature (Ramaiyan, Venkatesh, Anurag Kumar, and Eitan Altman. 2012, Keshtkarjahromi, Yasaman, Rashid Ansari, and Ashfaq Khokhar. 2013). The Most Forward within Range (MFR) (Dulman, Stefan, et al. 2006) routing protocol can select and choose the distant neighboring node as the next node and results in minimizing the delay and power consumption. Another approach ENO_OR in (Keshtkarjahromi, Yasaman, Rashid Ansari, and Ashfaq Khokhar. 2013) which minimizes total energy consumption based on path selection and bit allotment.

An Extremely Opportunistic Routing (ExOR), a new uni-cast routing technique for multi-hop wireless networks. ExOR specify which node from all the nodes that opportunely received that transmission is the node most proximate to the sink. That most proximate node transmits the packet (Biswas, Sanjit, and Robert Morris. 2004). Simulations based on measured radio characteristics (De Couto, Douglas SJ, et al. 2003) suggest that ExOR minimize the total number of transmissions by approximately a factor of two with best possible pre-determined path.

2 PROPOSED ALGORITHM

Here build a contrast of ENS_OR and ENO_OR Model idea to assorted data.

Algorithm: ENO_OR Algorithm

| Event: Node n contains data to send data packet to the destination. |
| Step 1: set the coordinates for the node transmission |
| Step 2: cost function calculation to select forwarder |
| Step 3: set transmission for forwarder node between the neighboring nodes |
| Step 3: *for* i = 1: length(x0) where i = 1:1:node, x0 is the energy of the node |

[min_node, I] = min (cost_function);
Node_num = I;
node_sel (r) = node_num;
end

Step 4: check energy of node and compare with
the threshold energy
if node energy >threshold energy
Then node forwarder
End

Step 5: set direct transmission if nodes energy
decreases below the threshold.

Step 6: set the priority to forwarder node by
using Optimal Energy;

Step 7: set the broadcast of data packet;

Step 8: check the node priority
if highest priority node n received the
data packet successfully then
reply an ACK signal to inform the sender
to get the data packet
else
if priority timer expires then
set n = lower-priority;
back to the start of the loop step 8

Step 9: if no forward node has received the
packet successfully then
if the retransmission timer expires then
Drop the packet;
Back to step 2

Step 10: *return*

---

3 ENERGY MODEL AND SIMULATION PARAMETERS

We utilize the parallel radio model concerned
in (Rappaport, Theodore S. 1996) that is initial
order radio model. The radio has power control
and can spend the lowly necessary energy to get
to the ENO_OR recipients. The radios can be

Table 1. Relative table for ENS_OR and ENO_OR
model.

Parameters	ENS_OR	ENO_OR
Average of Residual Energy (ARE)	High	Low
Standard Deviation of Residual Energy (SRE)	Minimum	Minimum
Receiving Packet Ratio (RPR)	High	Very high
First Dead Node (FDN)	Low	Low
Network Lifetime (NL)	High	Very high

twisted off to avoid in receipt of accidental
transmissions.

The equations that we have used here for calculating transmission costs and in receipt of cost the
consumption of energy at transmitter end is given
by E_{TX} and for receiver end E_{RX}.

$$E_{tx}(l,D) = (E_{\alpha} + E_{\beta}D^{\varphi})\, l \qquad (1)$$
$$E_{rx}(l) = E_{\alpha} l \qquad (2)$$

$E_{\alpha} = E_{elec}$ is the energy required for the functionality of transmitter and receiver circuit board,
$E\beta = \in_{amp}$ is the energy required for transmitter
amplifier, D is the distance from transmitter to the
receiver, φ is the path loss due to transmission channel [Efficiency, Aperture. 1994] value of and satisfies the condition $2 \le \varphi \le 4$. l is size of the packet.

The above Figure 2 shows the data transmission, where the next optimal node is selecting the
next node for data transfer. EEN (Energy Equivalent Node) is used to find the minimum energy
consumption of the nodes by replacing them.

Corresponding optimal transmission distance
d_{op} for node n is given by

$$D_{op} = \frac{m - x_n}{n_{op}} = \left\{ \frac{2E_{\alpha}}{[(\varphi - 1)E_{\beta}]} \right\}^{\frac{1}{\varphi}} \qquad (3)$$

$$D_{min} < D_{op} \le R$$

here, x_n is position of the sensor node ($x_{n} \ll m$).
where m is last sensor position.

For simulation, the software which is being used
is MATLAB. The experiment is for 8 nodes uniformly distributed in one dimension. Each node
has firmware character E_{α} and E_{β} in (1) is set as
5*0.000000001 J/bit and 1.97*0.000000001 J/bit/m²,
respectively. Hence, the value of optimal transmission distance D_{op} in (3) is approximately 0.1 m. Since
E_{α} and E_{β} are stable and permanent, no matter if the
distance between two most proximate node changes,
D_{op} still will be 0.1 m, without change. In this single

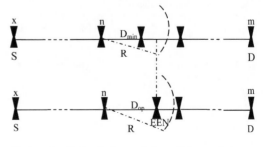

Figure 2. Optimal transmission distance between
nodes.

dimensional network, a node between the transmitter and receiver only acting as relay node. To analyze the performance of our proposed protocol i.e. ENO_OR, we compared it with the existing one that is ENS_OR, which represent the power transmission strategy with minimum transmission power, to ensure the Quality of Service (QOS).

4 SIMULATION METRICS AND DOMAIN

The simulation is of 2.5*6 feet. Here, we define five main quantifiable metrics to estimate the efficacy of ENO_OR algorithm for packet forwarding in one dimensional network. Table (2) shows the different parameters through which the performance metrics can be evaluated. In according to the above simulation domain parameters we, thus optimize the results of our performance metrics.

4.1 Residual Energy (RE)

The sensor nodes that have moreover residual energy show that all nodes are alive and in working condition for more time period, that increase the life of the network. Figure 3 presents the average saved power in terms of time when all the elements in the system are fully utilized. The figure shows that when the time in simulation increases the saved energy decreases. (1) & (2) interprets the explanation, when the size of the data packet increases gradually over the time, it can transmit data in a specific distance with more time period. As compared to ENS_OR this technique can achieve optimized residual energy, because of its opportunistic routing method and its optimized energy using the technique. These customized ENS_OR techniques keep the power usage at the lowest possible level. A higher lifetime can be achieved by reducing the energy usage by using this technique.

4.2 Standard deviation of Residual Energy (SRE)

To quantify the power consumption standard deviation is used as a metric for the routing protocol.

Table 2. Simulation domain.

Parameters	Parameter's values
Bit rate	2048 bit/sec.
Deployment distribution	Uniform
Frequency	2.4GHz
No. of nods	8
Source node	1
Sink node	1
Channel	AWGN
Tolerance factor	1.67 mJ
Packet size	1024 bit

It's been observed that high standard deviation in calculating saved energy shows that sensor nodes are using the unbalanced energy and it's important to reduce the SRE for the routing protocol. In Figure 4 it is observed that the standard deviation of ENO_OR technique and ENS_OR is nearly same.

4.3 Receiving Packets Ratio (RPR)

It is the proportion of the number of data packets received to the packets sent by the source. The sink must receive the most number of the packets from the source, presents the good connection of the network and also escape from the network division. Figure 5 shows the different time period of

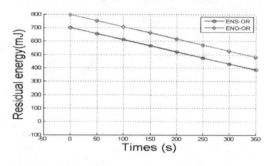

Figure 3. RE comparison with time.

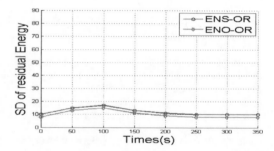

Figure 4. SRE comparison with time.

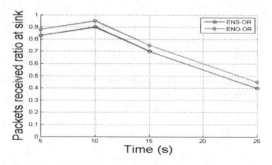

Figure 5. Comparison of packets received at sink with time between two proximate nodes.

receiving packets. While when the 5 seconds time increases it displays the average at sink increases in next 10 seconds and then gradually starts decreasing in time period, it presents the better receiving ratio with respect to the ENS_OR method, so that the ENO_OR method is better in receiving the data packets as compared to ENS_OR, and it can more efficiently reduce the network partition and also connectivity of the network increases. As more packets are received it shows that more energy should be consumed. It shows the direct relationship between the data transmitted and the consumption of energy.

4.4 First Dead Node received (FDN)

It is defined to measure the effect of the network connections. When the energy drained node appears the chances of the division of network increases, there is bad connectivity of the network. As in Figure 6, presents that energy exhausted node appearance in discussed method is later than ENS_OR.

4.5 Network Lifetime (NL)

It determined the time period when transmitter and receiver cannot establish connections anymore. A lifetime of the network is dependent on the power usage and network division. To increase the lifetime, we need to be at optimized energy usages point and the network partition is also going to occur. The relationship among the FDN and NL is very strong. As we prolong the network lifetime the appearance of the first dead, not id also going to be slowed. Figure 7 shows that the discussed method's lifetime is longer than ENS_OR. As the optimized use of energy will rescue the low energy sensors, the discussed method performs better. So the ENO_OR method ensures the extended lifetime as well as the optimal use of energy.

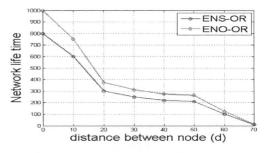

Figure 7. NL comparison according to the least distance between two proximate nodes.

5 CONCLUSION

In this paper, the concentration was on optimized energy usage and increase network life of one dimensional network, where the areas of sensors are fixed and cannot be changed. The optimization through opportunistic routing (ENO_OR) was used in simulation to suppose the sensors, as in real world node location cannot be changed to the optimal level. So by using this method the lifetime of the network can be increased extend. However, the focus was to make an optimization of opportunistic routing plan that guarantees the less energy is used and also saves sensors with low power. Different simulation figures and test results present that the discussed ENO_OR method can reduce the energy usage and network division also exists.

REFERENCES

Behnad, A., & Nader-Esfahani, S. (2011). On the statistics of MFR routing in one-dimensional ad hoc networks. *IEEE Transactions on Vehicular Technology*, *60*(7), 3276–3289.

Biswas, S., & Morris, R. (2004). Opportunistic routing in multi-hop wireless networks. *ACM SIGCOMM Computer Communication Review*, *34*(1), 69–74.

Biswas, S., & Morris, R. (2005). ExOR: opportunistic multi-hop routing for wireless networks. *ACM SIGCOMM Computer Communication Review*, *35*(4), 133–144.

Brenner, M. J., Ellder, A. J., & Zarghamee, M. S. (1994). Upgrade of a large millimeter-wavelength radio telescope for improved performance at 115 GHz. *Proceedings of the IEEE*, *82*(5), 734–741.

Chachulski, S., Jennings, M., Katti, S., & Katabi, D. (2007). *Trading structure for randomness in wireless opportunistic routing* (Vol. 37, No. 4, pp. 169–180). ACM.

De Couto, D. S., Aguayo, D., Chambers, B. A., & Morris, R. (2003). Performance of multihop wireless networks: Shortest path is not enough. *ACM SIGCOMM Computer Communication Review*, *33*(1), 83–88.

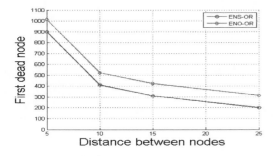

Figure 6. FDN comparison with minimum time between two nodes.

Dulman, S., Rossi, M., Havinga, P., & Zorzi, M. (2006). On the hop count statistics for randomly deployed wireless sensor networks. *International Journal of Sensor Networks*, *1*(1–2), 89–102.

Hoang, D. C., Yadav, P., Kumar, R., & Panda, S. K. (2014). Real-time implementation of a harmony search algorithm-based clustering protocol for energy-efficient wireless sensor networks. *IEEE Transactions on Industrial Informatics*, *10*(1), 774–783.

Keshtkarjahromi, Y., Ansari, R., & Khokhar, A. (2013, November). Energy efficient decentralized detection based on bit-optimal multi-hop transmission in one-dimensional Wireless Sensor Networks. In *Wireless Days (WD), 2013 IFIP* (pp. 1–8). IEEE.

LoBello, L., & Toscano, E. (2009). An adaptive approach to topology management in large and dense real-time wireless sensor networks. *IEEE Transactions on Industrial Informatics*, *5*(3), 314–324.

Luo, J., Hu, J., Wu, D., & Li, R. (2015). Opportunistic routing algorithm for relay node selection in wireless sensor networks. *IEEE Transactions on Industrial Informatics*, *11*(1), 112–121.

Mao, X., Tang, S., Xu, X., Li, X. Y., & Ma, H. (2011). Energy-efficient opportunistic routing in wireless sensor networks. *IEEE Transactions on Parallel and Distributed Systems*, *22*(11), 1934–1942.

Pottie, G. J., & Kaiser, W. J. (2000). Wireless integrated network sensors. *Communications of the ACM*, *43*(5), 51–58.

Ramaiyan, V., Kumar, A., & Altman, E. (2012). Optimal hop distance and power control for a single cell, dense, ad hoc wireless network. *IEEE Transactions on Mobile Computing*, *11*(11), 1601–1612.

Rappaport, T. S. (1996). *Wireless communications: principles and practice* (Vol. 2). New Jersey: Prentice Hall PTR.

Ren, F., Zhang, J., He, T., Lin, C., & Ren, S. K. D. (2011). EBRP: energy-balanced routing protocol for data gathering in wireless sensor networks. *IEEE Transactions on Parallel and Distributed Systems*, *22*(12), 2108–2125.

Zhang, D., Li, G., Zheng, K., Ming, X., & Pan, Z. H. (2014). An energy-balanced routing method based on forward-aware factor for wireless sensor networks. *IEEE transactions on industrial informatics*, *10*(1), 766–773.

Communication and Computing Systems – Prasad et al. (Eds)
© 2017 Taylor & Francis Group, London, ISBN 978-1-138-02952-1

Analysis of compression techniques to improve QoS in MANET

Sumitra Ranjan Sinha & Pallavi Khatri
ITM University, Gwalior (Madhya Pradesh), India

ABSTRACT: Improving Quality of Service (QoS) in MANET is very complicated due to its constantly changing topology and unpredictable network behaviours. This work proposes a QoS-aware compression technique for improving the performance and efficiency of a Mobile Ad-hoc Network (MANET). Proposed system implement Mpeg traffic compression over routing protocol that used for multimedia data transmission. Network parameters effecting on QoS like network delay; throughput; Packet Delivery Ratio (PDR) are being taken in to consideration for this study. This work suggested the comparative analysis over different compression techniques and provides the best effort performance in terms of QoS in MANET. Simulations are carried out on Network Simulator (NS-2) tool and the result matrix of our proposed compression based protocol clearly outperforms the existing protocols with respect to QoS parameters of the network.

1 INTRODUCTION

Mobile-ad-hoc networks are formed using group of mobile routers that are capable of developing a self-configuring infrastructure less network in which nodes are linked via wireless channel and they move autonomously in any direction. One of the demanding tasks in MANET is to offer a relevant QoS supported network, due to lack of network resources and frequent changing topology. QoS is a term that widely used in the area of ad-hoc networks as well as multimedia streaming networks. QoS defines the combined effect of service performance which determines the scale of contentment of a user of that specific service. For the enhancement of the performance in MANET, Quality of Services (QoS) must be increased in order to construct a faster and fluent real time communication network. To formulate the data transmission more effective and QoS radiant, the network allows different data compression techniques for effectual enhancement of parametric performance and network effectiveness.

1.1 Data compression

Any compression algorithm converts the actual data from an accessible arrangement to one optimized identical data format for compactness. Due to compression, the number of bits can be reduced to maximum extend so that the memory and bandwidth of the entire network remain less and that tends the network to an improvised network in terms of performance parameters.

Figure 1 elaborates the data compression techniques that are applicable in scalable multimedia

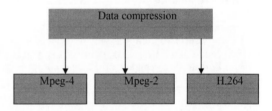

Figure 1. Categorization of video compression.

data traffic. These techniques are applied at the time of traffic generation on any network and provide a improve network traffic with respect to different network parameters.

1.2 Mpeg-4

Mpeg-4 is a compression technique that defines compression of multimedia data of a multimedia network in an effective manner. It compresses any audio visual data of a network and also take a major role for transmit it in network as a traffic generator. It supports better reliability as compare to the preceding techniques of Mpeg for enhancement of QoS in multimedia streaming. MPEG-4 data can be transmit and stream into other signal types according to compatible with available network. Different levels and profiles are categorized for performing compression of respective data.

1.3 Mpeg-2

Mpeg-2 is implemented by "the generic coding of moving pictures and ISO/IEC 13818 Mpeg-2 supported modern ISO standard. It incorporates with the lossy video compression which allow storage and transmission of data within a lower transmission bandwidth. MPEG-2 compression is very less effective for low bit-rates, especially less than 1 Mbit/s at standard definition resolutions.

1.4 *H.264 coding*

The intent of the H.264/AVC project was to create a standard capable of providing good video quality at substantially lower bit rates than previous standards (i.e., half or less the bit rate of MPEG-2, H.263, or MPEG-4 Part 2), without increasing the complexity of design so much that it would be impractical or excessively expensive to implement. An additional goal was to provide enough flexibility to allow the standard to be applied to a wide variety of applications on a wide variety of networks and systems.

Table 1. demonstrate the existing compression techniques that are used for video traffic according to their respective bit rate. ISO standards demonstrate the configuring details of the each existing protocol.

Figure 2 shows various compression techniques and their respective compression ratio according to their individual bit-rate. The comparative graph exhales that the performance of Mpeg-4 compression is better in bit rate of 6–8 mbps as compare to Mpeg-2 and H.264 which are works on the bit-rate of 2–4 mbps and 1–2 mbps. Improvement of bit rate as well as compression ratio precedes the remarkable perfection on QoS parameters in multimedia MANET.

Table 1. Overview table.

Compression techniques	Description	Standard	Bit-rate
Mpeg-4	Coding of audio-visual objects	ISO/IEC 14496	>2 mb/sec
Mpeg-2	Generic coding of moving pictures and associated audio information	ISO/IEC 13818	Variable bit rate
H.264	Covers all forms of digital compressed video from low bit-rate Internet streaming applications to with nearly lossless coding	ISO/IEC JTC1 Moving Picture Experts Group (MPEG)	10–100 kb/sec

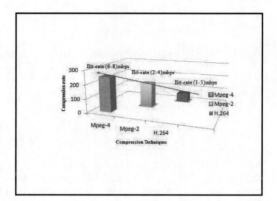

Figure 2. Compression techniques.

The remnants of this paper are prearranged respectively according section 2 gives the literature review of different mechanisms that are used for supporting QoS. In section 3 proposed protocol and compression methodology with relevant algorithm are to be discussed. Implementation and simulation results are exhibited in section 4 by using network simulator-2(NS2); the paper ends up with conclusion and future recommendations of researchers.

2 RELATED WORK

In the recent past lot of researches had been done for improving QoS in MANET and also many compression techniques were applied for data compression individually. This proposed research mainly focuses on combined study of QOS parameters as well as the invoking Mpeg4 lossless compression on video data to enhance the performance of data transmission to improve the QoS.

A review of all types of QoS parameters that are supported in MANET are demonstrated by M. Abbas. Different types of QoS constraints of different network layer like adaptive, multiplicative, concave that are effecting on network are discussed in this work.

Different constraints to enhance quality of service in term of video streams over wireless networks are discussed by M. A. Malgi. Cross layer designing and Integrated Cross-Layer Optimization Algorithm and Adaptive Dynamic Mapping Algorithm optimally used the control parameters at different protocol layers.

T.Vishnu Priya discusses the block coding interference of Mpeg-2 compression over scalable network in variable transmission conditions. Mpeg-2 compression works successfully on lower bandwidth as well as lower bit-rate compression from 2–10 mbps. The author exhibits the profiles and levels categorization of MPEG-2. MPEG 2 supposes compression of the raw frames into three kinds of frames: Intra-coded frames (I-frame), Predictive-coded frames (P-frames), and Bidirectional-predictive-coded frames (B-frames). This format defines 4 profiles and 4 levels that correspond to the quality and resolution.

R.B. Mamata proposed some recently computed video compression techniques that are applicable for Mpeg format. H.264 technique is applied on video data in which bit rate is less than the existing protocol like mpeg. H.264 also provides significant improvements in coding efficiency, latency, complexity and robustness. The paper also provides new possibilities for creating better video encoders and decoders that provide higher quality video streams at maintained bit-rates T. Samanchuen exhibits the proposed video data transmission and sends the transmission back to the predefined server and stream through the different frames. It also proposes the Mpeg-4 technique and other H.263 algorithms that are frequently used on client server architecture.

3 ANALYSIS OF EXISTING ALGORITHMS

Several researches have been done over different compression techniques according to the network behavior and network scenario.

3.1 Mpeg-4

Mpeg-4 is one of the important and new featured compression techniques that supported lower bandwidth and reduced delay rate Data Communication between different scalable nodes is set Mpeg-4 traffic that Improve the network performance from source to destination.

Figure 3 illustrate how actually Mpeg-4 compression works in case of any video input. Different video frame (I, B and P) first converted from RGB to YUV color coding. Any error in video frame are displayed and simplified on error block. Motion estimation block finds the best matching block for all the reference frames for enhancing the performance. Entropy coding is finally performed for getting the actual compressed data.

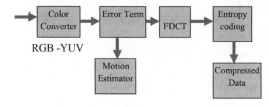

Figure 3.   Mpeg-4 compression.

Convert video frames From RGB to YUV conversion

$$Y = 0.299R + 0.587G + 0.144B \qquad (1)$$
$$U = 0.492(B - Y) \qquad (2)$$
$$V = 0.877(R - Y) \qquad (3)$$

It can also be represented as:

$$Y = 0.299R + 0.578G + 0.114B \qquad (4)$$
$$U = -0.147R - 0.289G + 0.436B \qquad (5)$$
$$V = 0.615R - 0.515G - 0.100B \qquad (6)$$

3.2 MPEG-2

The MPEG-2 video compression algorithm introduces efficient rates of compression at the minimum bit-stream rate of 1.5–6 mb/s by reducing the redundancy of respective frames in video information. MPEG-2 removes both the temporal redundancy and spatial redundancy which are present during the video transmission.

Figure 4 exhibits the characteristics of redundancy that are affecting on the video streaming as well as QoS parameters. Redundancy degrades the performance of QoS in case of loss rate and delay rate.

Block diagram on Figure 5 demonstrate the overall compression techniques of Mpeg-2 compression. Video samples are first simplified in respective frames and then spatial and temporal redundancy is removed.

Table 2 gives the comparative study of both compression techniques on different parameters and exhibits the resulting parameters of existing protocols.

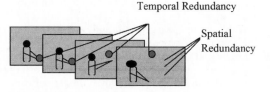

Figure 4.   Demonstration of redundancy.

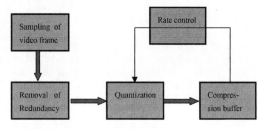

Figure 5.   Mpeg-2 compression.

Table 2. Comparative parameters.

Parameter	Mpeg-4	Mpeg-2
Bit-rate error	>1 in 10 mb/sec	>1.5 in10 mb/sec
Packet delay	Less than 500 ms	Less than 300 ms
Bandwidth	10 kb/s	5 kb/sect
PSNR	18.4%	25.1%
Transmission time	60 sec	20 sec
Bit-rate	2 mb/sec	1.5 mb/sec
Partitioning	16 × 16 micro blocks	4 × 4 blocks
Scalability	100 nodes	100 nodes
mobility	3.5	10.5

4 OVERVIEW OF PROPOSED ALGORITHM COMPRESSION

As per previous extensions and discussion of different compression mechanism Mpeg-4 and Mpeg-2 traffic compression on multimedia data over large range of scalability and different QoS parameters for improvement of network performance.

Figure 6 exhibits the proposed flowchart of the entire system. The overall mechanism of the system performed according to this flowchart for the effective improvement of QoS parameters of the network.

5 SIMULATION SETUP

Network simulator (NS2) is simulation tool which is used to study the performance of different communication networks. This paper evaluates the performance matrices of Mpeg-4 and Mpeg-2 compressions over MANET. Different network parameters that are effecting on QOS improvement are evaluated.

Table 3 give the corresponding network parameters and their preceding values that are used for effective and successful fulfillment of simulation.

Figure 7 shows the simulation scenario having 200 nodes in the entire network. In the network scenario, node 0 having green color indicates the source node.

6 RESULTS AND DISCUSSIONS

The simulation result of the entire network is generated by routing protocol AOMDVc and AOMDV. This work is based on Mpeg4 and Mpeg-2 compression traffic simulation in NS2. Different QoS effecting network parameters are calculated over scalable nodes.

Parametric results show that AOMDVc protocol with the invocation of Mpeg-4 compression

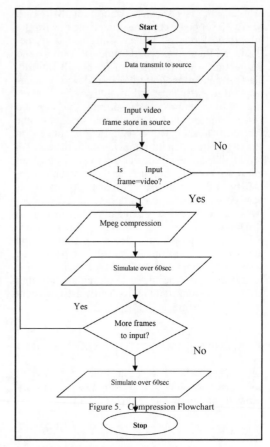

Figure 5. Compression Flowchart

Figure 6. Compression flowchart.

Table 3. Simulation table.

Parameter	Value
Topology	1000 * 1000
Number of nodes	5, 10, 50, 100, 150, 200
MAC protocol	MAC/802_11
Propagation model	Random way point
Routing protocol	AODV, AOMDV, AOMDVc
Antenna type	Omni-Antenna
Data traffic	CBR, FTP
Packet size	1500
Compression	Mpeg4, Mpeg-2
Mobility	2.3
Connection	UDP
Frame rate	5
Initial seed	0.4
Bandwidth	11 mb

improves the throughput of the large multimedia network as compare to the Mpeg-2 compression in the same protocol.

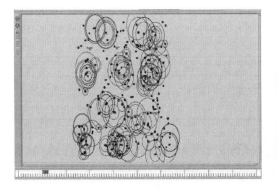

Figure 7.   Simulation scenario.

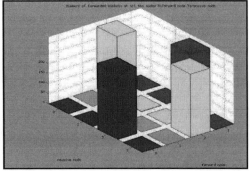

Figure 8.   Packet drop histogram.

6.1   *Packet drop*

Packet drop of any network make the network less effective and less reliable. In case of any large network the amount of packet drop is increases with respect to the different network parameters like network delay, network congestion.

Packet drop in the entire network in different scalable nodes are shown by the color histogram at the time of invocation of the of Mpeg-4 and Mpeg-2 traffic.

6.2   *Throughput*

In the above experimental result, Figure 9 elaborates the throughput of the network in respective nodes with the successful implementation of Mpeg-4 and Mpeg-2 compression technique. Resulting graph and parametric results show that AOMDVc protocol with the invocation of Mpeg-4 compression improves the throughput of the large multimedia network as compare to the Mpeg-2 compression in the same protocol.

6.3   *Packet Delivery Ratio (PDR)*

High degree of PDR in any network is necessery for improvement of network effectiveness and QoS enhancement. Packet Delivery Ratio (PDR) of the network in the proposed system is shown in Figure 10. The result of PDR is very satisfectory in both he cases of Mpeg-4 and Mpeg-2 compression. Performance of Mpeg-4 providing a better result in some extend as compare to the other ones.

6.4   *End-To-End delay*

Figure 11 defines the avg. delay of the network after implementing Mpeg-4 and Mpeg-2 compression. Graphical result exhibits the betterment of result in case of Mpeg-4 as compare to the Mpeg-2.

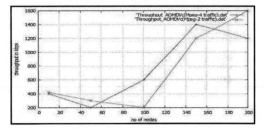

Figure 9.   Throughput graph.

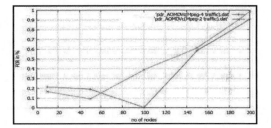

Figure 10.   PDR graph.

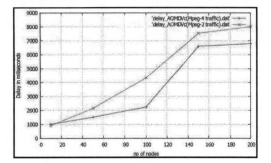

Figure 11.   Average delay graph.

All the evaluated parameters shows the performance and efficiency of mpeg4 compression that resulting better than Mpeg-2 compression when it implemented video traffic for providing QOS radiant network. After calculate all the network parameters results and graphs clearly exhibits that proposed work over Mpeg-4 is better than the previous conventional counterparts.

7 CONCLUSION

Invocation of Mpeg traffic compression on the proposed AOMDVc protocol provides a QoS attentive network platform over multimedia video transmission. Advancement and improve compression techniques enhanced the performance and efficiency of network by providing the enhanced results on parameters like jitter, packet drop,delay,throughput and PDR.The performance of network reagents depend on the different scalability. All the simulation result and statistical studies pretends relatively improved performance of Mpeg-4 over Mpeg-2 compression in case of different QoS parameters.

REFERENCES

Abbas A.M, June, 2010 "Quality of Service In Mobile Ad Hoc Networks: A Survey". International journal of ad hoc and ubiquitous computing.

Ahmad I., U. Ashraf, S. Anum, and H .Tahir. March 2014 "Enhanced AODV Route Discovery and Route Establishment for QOS Provision ForReal Time Transmission In MANET". International Journal of Computer Networks & Communications (IJCNC) Vol. 6, No. 2.

Baldaniya M, Prof. N. Sisodiya. 2014. "Improving Multipath AODV For Video Transmission". (IJCSIT) International Journal of Computer Science and Information Technologies, Vol. 5 (3), pg. 3590–3594.

Carlos T. Calafate, M. P. Malumbres, P. Manzoni, "Performance of H.264 compressed video streams over 802.11b based MANETs". Circuits and Systems (ISCAS), Officinal de Ciencia y Tecnologia de la Generalitat Valenciana, Spain, under grants CTIDIB.

Durdi, V.B., Chiariglione L. (ed.), 2012. "The MPEG Representation of Digital Media", Springer Science Business 7, DOI 10.1007/978-1-4419-6184-LLC.

Fakhar G. Y., S. R. Chaudhry. August 2013 "MANET Routing Protocols for Real-Time Multimedia Applications". SEAS Transactions on Communications, E-ISSN: 2224–2864, Issue 8, Volume 12.

Hung L. E., Hsiao Hsu-Feng, May 2015 "QoS-driven optimization for video streaming using layer-aligned multipriority rateless codes". Circuits and Systems (ISCAS), 2015 IEEE International Symposium.

Kaur G, Kad S, October 2015. "A Compressive Sensing based BEEIP Protocol for WANETS," International Journal of Computer Applications. (0975–8887) Volume 128, No. 6.

Li & Drew Prentice Hall. 2003. "Fundamentals of Multimedia", Chapter 7.

Li. S, Neelisetti R, Liu C and Lim A, June 2014. "Delay-Constrained High Throughput Protocol for Multi-Path Transmission over Wireless Multimedia Sensor Networks" 978-1-4244-2100-82008.

Md. Hossain A, Rashid A, N, M, B. 2014, "Performance Evaluation of MPEG-4 Video Transmission over IP-Networks Best-Effort and Quality-of-Service". IEEE. ISSN 2222–2863, Vol. 3.

Malgi M. A., Gaikwad G. N. 2015. "A Study on QoS Enhancement of MPEG-4 Video Transmission over Wireless Mesh Network." International Conference on Pervasive Computing (ICPC), 1-4799-62723/15/.00(c).

Mamata R.B, November 2014 "A Comparative Study of Video Compression Techniques-H. 264/Avc," International Journal of Advanced Research in Computer Science and Software Engineering Volume 4, Issue 11, ISSN: 2277 128X.

Samanchuen T, Pathom N, 9–12 Dec 2014. "Implementation and quality evaluation of video telephony using Session Initiation Protocol," IEEE Asia-Pacific Signal and Information Processing Association, 2014 Annual Summit and Conference (APSIPA).

Sheganaz A. January 2014, "An Efficient Approach for Multimedia Transmission Using Multipath Routing In MANET".International Journal of Computer Science and Mobile Computing, Vol. 3 Issue 1, pg. 447–451.

Sedrati M, Benmohamed M. January 2011, "M-AODV variant to Improve Quality of service in MANETs". IJCSI international Journal of computer science issues, January vol. 8, issue 1 ISSSN (online): 1694-0814.

Communication and Computing Systems – Prasad et al. (Eds)
© 2017 Taylor & Francis Group, London, ISBN 978-1-138-02952-1

Low complexity service optimization reliable routing algorithm for Vehicle Ad-hoc Network (VANETs)

Deepak Kumar & Praveen Pandey
Gautam Buddha University, Greater Noida, India

ABSTRACT: Vehicular specially appointed Systems (VANETs) are a specific type of remote system made by vehicles imparting among themselves and with roadside base stations. An extensive variety of administrations has been created for VANETs going from security to infotainment applications. A key necessity for such administrations is that they are offered with Nature of Administration (QoS) ensures as far as administration unwavering quality and accessibility. QoS steering assumes a crucial part in distinguishing courses that meet the QoS necessities of the offered administration over VANETs. For scanning the achievable routes it subject to different QoS imperatives is all in all a NP-hard problem. Besides, the dependability of the routing should be given extraordinary consideration as correspondence connections as often as possible in VANETs. In present scenario, the most existing QoS routing algorithms are been intended for stable systems without considering the security and safety of the routing process. In this way, they are not suitable for applications in VANETs. The above issues are tended to firstly by building up a connection unwavering quality model taking into account the topological and numerical properties of vehicular developments and speeds. The VANET communication graph in evolving graph theory is utilized to model and coordinate the formed connection dependability model into it. In light of the subsequent amplified advancing chart demonstrate, the most dependable course in the system is picked. However course determination does not guarantee great transmission too. Thusly low complexity optimized service reliable routing algorithm is created to guarantee to lessen transmission repetition with utilization of Subterranean insect State calculation with better after effects of route choice and transmission taking into account better inquiry and sorting algorithms in the proposed work algorithm for very dynamic systems with less many-sided quality in assessment with preferred information transmission over alternate QoS calculations as exhibited by the outcomes.

Keywords: VANETs, graph theory, Qos, low complexity optimization

1 INTRODUCTION

Wireless network which are a particular form of network made by communicating among the base station and among themselves an extensive variety of administrations has been created for VANETs it is ranging from safety to various infotainment applications. A key necessity for such administrations is that they are offered with Quality of Service (Qos) it ensures the administration unwavering quality and accessibility. In distinguishing routes that meet the QoS necessities of the offered administration service over VANETs Qos routing plays an important role.

Extensive variety of work has been done for organization in VANETs to guarantee productive information transmission and unwavering quality of the information. To guarantee wellbeing different information must be exchanged. It is important to guarantee better administration accessibility and dependability that is Nature of Administration. The extraordinary qualities of VANETs raise imperative specialized difficulties that should be considered with a specific end goal to bolster the transmission of various information sorts.

The most difficult issue is possibly the high portability and the successive changes in the system topology. The topology of vehicular systems could fluctuate when vehicles change their speeds and/or paths. These progressions rely on upon the drivers, street circumstances, and activity status and are not booked ahead of time. Thusly, asset reservation can't be utilized to give QoS ensures. The steering calculations that might be utilized in VANETs ought to have the capacity to build up courses that have the properties required to meet the QoS prerequisites characterized by the offered administration. Directing unwavering quality should be given unique consideration if a dependable information transmission ought to be accomplished. Be that as it may, it is a confounded assignment to procurement dependable courses in VANETs on the grounds that it is affected by numerous elements, for example, the vehicular versatility design and the vehicular movement circulation the routing unwavering quality, directing calculations ought to likewise give a conclusion to-end delay compelled information conveyance, particularly for postponement bigoted information. The current QoS routing they are generally intended for stable systems, for example, MANETs and remote sensor systems are not suitable for applications in VANETs. Without loss of all inclusive statement, distinguishing a doable course in a multi-jump VANETs environment subject to various added substance and free QoS requirements is a Multi-Compelled Way (MCP) issue. The MCP issue is turned out to be a NP-hard. Besides, it is regularly sought to distinguish the ideal course among the possible courses found by the directing calculation as per a particular rule, e.g., the briefest way. This case is known as the Multi-Compelled Ideal Way (MCOP) issue, which is additionally a NP-hard and for the MCOP issue is likewise an answer for the MCP issue however not as a matter of course the other way around. Building up a Multi-Obliged QoS (MCQ) algorithm it encourages the transmission of various information sorts as per different QoS limitations is one of the essential worries to send VANETs viably. Moreover, because of the absence of insurance of VANETs' remote channels, outer and inward security assaults on the directing procedure could essentially debase the execution of the whole system. Hence, the configuration of the created MCQ routing ought not add additional security dangers to manage but rather gives focal points while actualizing security instruments.

In VANETs, security instruments are required to ensure the MCQ directing process and give a powerful and dependable steering administration. These key difficulties rouse us to propose a novel secure multi-obliged QoS dependable steering calculation that locations them. In this exploration, we concentrate on vehicle-to-vehicle correspondences on roadways, i.e., the main system hubs are the vehicles. Thruways are relied upon to be the principle focus for the organization of vehicular correspondence systems to give security, help with movement administration, and offer Web availability to vehicles by means of portable passages. We expect that vehicles move at variation speeds for long separations along a thruway and are permitted to quicken, decelerate, stop, turn, and leave the roadway as in a genuine circumstance.

2 RELATED WORK

For systems are information driven in nature the routing traditional algorithms are given the unattended and un-fastened nature of systems, information driven steering must be community oriented over the entire system. In this system worldview, the system gadgets work together to accomplish regular system wide objectives, for example, course unwavering quality and way length while minimizing singular expenses.

The model can be utilized to characterize the nature of directing ways in the system which portray results on getting ways with limited shortcoming alongside a few heuristics for acquiring solid ways the advancement of calculations for around ideal solid directing is an open issue that can be investigated further. Three forms–uncast, telecast and multicast–of the steering issue have gotten huge consideration in the writing. The general target of these calculations is to either expand the information unwavering quality or the limit of the system (measure of information movement conveyed by the system over some settled time of time) (Akyildiz et al. 2002).

The optimizer algorithm is being used, here to maximize simulation parameters.

ACO Algorithm (Zhang et al. 2004) is used for the local search is binary algorithm to reduce the complexity (Haanpää et al. 2007)

Step 1: [Initialize] the initial parameter Eliminate from graph theory G every edge (u, v) here u and v are used for edge and vertex. Let L is list of possible values for maximum data reliability.

Step 2: In this step we used Binary Search. The parameter of L is find the maximum value, which is used for max pheromone ant, path P from source to destination that uses at most z*Pmin energy. Here test a value q from L, and then find a shortest path from source to destination path.

Step 3: [Wrap Up] here no path is found in Step 2, and then uncast is not possible. Otherwise, use the path P corresponding to max.

Algorithm start

Step: 1 set the antcoly function based on capacity number of Ants

Step: 2 Set parameter {lambda=0.1; phero= [ ]; [m, n]=size(A)}

for i=1: m   clients(i)=i; end

Step 3; calculate the best feasible path
      best_feasible = pfih_sol;

Step 4: calculate best_feasible_cost=
totalCost(best_feasible, B, 0);

Step 5 make a path from source to destination
[best_route, phero]=make_way(capacity, clients, A, B, phero, tauZero);

Step 6: estimate current_cost=
totalCost(current_route, B, truck_value);

Step 7: compare the current cost and best feasible cost

      if current_cost<best_feasible_cost
      best_feasible=current_route;
      best_feasible_cost=current_cost;
      a=0;
      end
      end

step 7: store the first cost and best feasible cost.
first_cost = totalCost(pfih_sol, B, 0)
best_feasible_cost = totalCost(best_feasible, B, 0)

Step 8: find the best optimal path
Route=best_feasible;

2.1 Solution construction

The rules followed by ANTS for computation of probabilities during ants' solution construction have different form that used in other ACO algorithms. In ANTS, an ant k that is positioned at city i choose the next city j with a probability given by

$$p_{ik}^{A_n} = \begin{cases} \dfrac{[\tau_{ik}(t)]^\beta [T_{ik}(t)]^\gamma}{\sum_{r_c \in N(r_i)} [\tau_{ic}(t)]^\beta [T_{ic}(t)]^\gamma} & \text{if } C_k \in N(C_i) \\ 0 & \text{otherwise} \end{cases} \quad (1)$$

While searching for possible routes, ants select next hop. They arrive at intermediate nodes based on the stochastic mechanism called the state transition rule. The probability function is given as:

$$p_{ij}^{A_k} = \begin{cases} \dfrac{[\tau_{ij}(t)]^\alpha [T_{ij}(t)]^\beta}{\sum_{C_c \in N(C_i)} [\tau_{ic}(t)]^\alpha [T_{ic}(t)]^\beta}, & \text{if } C_j \in N(C_i) \\ 0 & \text{otherwise} \end{cases} \quad (2)$$

In the hyper-cube framework the pheromone trails are forced to stay in the interval ½0; 1. This is achieved by adapting the standard pheromone update rule of ACO algorithms. Let us explain the necessary change considering the pheromone

update rule of AS [Equations (1) and (2)]. The modified rule is given by

$$\tau_{ij} \leftarrow (1-\rho)\tau_{ij} + \rho \sum_{k=1}^{m} \Delta \tau_{ij}^k,$$

$$\Delta \tau_{ij}^k = \begin{cases} \dfrac{1/C^k}{\sum_{h=1}^{m}(1/C^h)}, & \text{if are } (i, j) \text{ is used by ant } k; \\ 0, & \text{otherwise.} \end{cases}$$

This process update rule guarantees that shows trails remain smaller than 1 and new pheromone vector can be interpreted as a shift of the old pheromone vector toward the vector given by the weighted average of the solutions used in the pheromone update.

2.2 Service optimized reliable routing algorithm

This algorithm comprises of 2 main stages:

1. Selection stage optimized by ant colony algorithm and link stage.
   Basically computes feasible routes according the constrained parameters to maximize the data efficiency. The possibility of some route terminate while data transmission then algorithm process next best route for data transmission.
2. The algorithm also checks if a better link is established then primary link due to moving vehicle then the primary link is updated by the new link for data accuracy and minimum delay design.

3 SIMULATION RESULTS

The performance of our result is shown by given Table 1 in which we have considered some parameters.

3.1 Simulation A—effects of network density

Figure 1 demonstrates the recreation results for the directing calculations analyzed in this re-enactment.

Table 1. Simulation metrics.

Simulator	MATLAB
Routing protocol	Reliable SAMQ
Optimization protocol	Acs (with binary search)
Data	Random generation
Packet size	1024 bytes
Communication range	500 m
Simulation time	200 s
MAC layer	IEEE812.11P

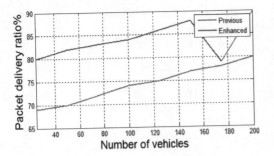

Figure 1. Packet delivery ratio verus number of vehicles.

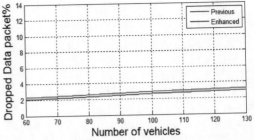

Figure 3. Dropped data packet versus number of vehicles.

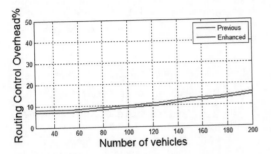

Figure 2. Routing control overhead versus number of vehicles.

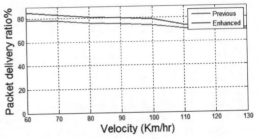

Figure 4. Packet delivery ration versus velocity (Km/hr).

By and large, higher system thickness adds to a superior PDR of the directing calculations since more vehicles infer more potential connections; consequently, there are more alternatives from which to register achievable courses to the destination. Besides, higher system thickness enhances the association length the other way for vehicles, which expands the PDR too. It can be seen in Fig. 1 that the PDR of the proposed directing calculation is higher than past calculation of SAMQ.

Figure 2 demonstrates the directing control overhead proportion produced by each steering calculation analyzed in this reproduction. It is foreseen that the steering control overhead would increment when the system thickness increments since more hubs are accessible for directing solicitations to navigate. The proposed routing keeps up the minimum directing control overhead in contrast with past algorithms in the base paper.

At last, Figure 3 obviously demonstrates the benefit of SAMQ in keeping away from dropped information bundles in contrast with Blemish DYMO and VACO. To keep away from higher rates of dropped information bundles at the destination hub, they chose plausible course ought to have the capacity to convey information parcels as indicated by their QoS necessities. The execution of SAMQ in this figure demonstrates the effectiveness of its ACS rules in distinguishing attainable courses.

3.2 Simulation B—effects of dynamic changes in network topology

The point of this re-enactment is to research the effect of expanded elements of the vehicular system on the directing calculation execution when higher speeds are presented. It is appeared in Figure 4 that the normal PDR diminishes detectably for all directing calculations when the normal speed in the third path begins to surpass. This decrease originates from the way that the system topology turns out to be more dynamic, and consequently, interfaces/courses are more helpless against disengagement.

In Figure 5, it can be watched that all the directing calculations are influenced by the dynamic changes in the system topology. Be that as it may, proposed calculation keeps up the minimum directing control overhead among the inspected steering calculations in this figure. In Figure 6 higher Re-enactment Transmission delays is shown. SAMQ excellently keep the minimum delay of transmission.

In Figure 7, Reproduction Dropped information parcels. Vehicular speeds build the elements of the system topology and expansion the likelihood of disengagement of built up connections/courses. Along these lines, it is key for a directing calculation to be set up to react to such element changes in

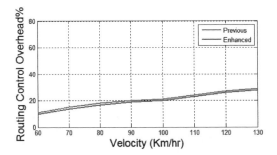

Figure 5. Routing control overhead versus velocity (Km/hr).

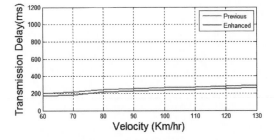

Figure 6. Transmission Delay (9 ms) versus velocity (Km/hr).

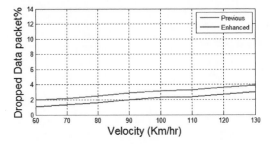

Figure 7. Dropped data packet rate versus velocity (Km/hr).

the system and pick the most solid courses between the conveying vehicles.

4 CONCLUSION

In this paper, we have explored the highlighted obliged QoS routing issue in VANETs. All the more particularly, we proposed the low unpredictability and solid directing calculation for VANETs taking into account the enhanced and les complex ACS calculation instruments. This calculation expects to choose the best course over numerous processed plausible courses between the conveying vehicles subject to various QoS requirements, if such a course exists. However its low multifaceted nature conduct displays that gives a system to enhancing QoS directing. The execution of the proposed directing calculation is assessed through broad reenactments of irregular information transmission on virtual urban interstate and contrasted and past proposed SAMQs. The calculation demonstrates promising results as far as accomplishing a higher PDR and, subsequently, ensuring more solid directing of information bundles. It has been apparent that using better ACS because of sort and hunt show down VANETs is exceptionally encouraging to accomplish a ceaseless and stable information transmission.

REFERENCES

Akyildiz, I.F., Su, W., Sankarasubramaniam, Y. Cayirci, E.: A survey on sensor networks. IEEE Communication Magazine 40(8) (2002) 102–114.

Cerpa, A., Estrin, D.: ASCENT: adaptive self-configuring sensor networks topologies. IEEE Transactions on Mobile Computing 3(3) (2004) 272–285.

Guo, S., Yang, O.W.W., Leung, V.C.M.: Tree-based distributed multicast algorithms for directional communications and lifetime optimization in wireless ad hoc networks. EURASIP Journal on Wireless Communications and Networking 2007 (2007) Article ID 98938, 10 pages.

Haanpää, H., Schumacher, A., Thaler, T., Orponen, P.: Distributed computation of maximum lifetime spanning subgraphs in sensor networks. In: Proceedings of The 3rd International Conference on Mobile Ad-hoc and Sensor Networks (MSN2007) (to appear).

Lloyd, E.L., Liu, R., Marathe, M.V., Ramanathan, R., Ravi, S.: Algorithmic aspects of topology control problems for ad hoc networks. Mobile Networks and Applications 10(1–2) (2005) 19–23.

McCanne, S., Floyd, S., Fall, K., Varadhan, K.: The network simulator ns2 (1995)The VINT project, available for download at http://www.isi.edu/nsnam/ns/.

Communication and Computing Systems – Prasad et al. (Eds)
© 2017 Taylor & Francis Group, London, ISBN 978-1-138-02952-1

Energy analysis of efficient power aware AODV routing protocol in MANET

C. Mafirabadza & P. Khatri

Department of Computer Science and Engineering, ITM University Gwalior, MP, India

ABSTRACT: Mobile ad-hoc networks is an area in the communication networks which is mostly referred as MANETs. In this network, there are independent nodes which are self organizing and they are able to send messages between themselves. The main challenge faced in this type of network is limited battery energy. The nodes rely on the limited energy, meaning as soon as the battery energy is depleted, the nodes will die. All communication is brought to a halt. The authors in this paper have come up with a proposed protocol which we name Efficient Power Aware Ad-hoc on-Demand Distance Vector (EPAAODV) protocol. This paper presents an energy consumption analysis of the proposed protocol. This proposed protocol is the modification of the well known Ad-hoc on-Demand Distance Vector protocol. Packet Delivery Ratio (PDR) and throughput will also be analyzed on the proposed EPAAODV protocol and AODV protocol using network simulator (NS2).

1 INTRODUCTION

Nodes in MANET are able to setup connection and route messages between themselves. Due to their ease of deployment, MANETs have found use in many applications which include but not limited to: battlefield, natural disaster rescue operations and sensor networks. Mohapatra & Kanungo (2012) suggest that two main routing protocols are implemented in MANET; Reactive and Proactive. To get the good traits of the two types, both reactive and proactive protocols combine to form a Hybrid protocol. Example of hybrid protocol is Zone Routing Protocol (ZRP) described in (Buruhanudeen et al. 2007, Giannoulis et al. 2005).

Many routing protocols only concern with latency; they put less importance on energy consumption. As highlighted in (Safa et al. 2014, Sumathi & Priyadharshini 2015), once node drains all its energy, it virtually becomes dead, and cannot send or receive any messages. However various techniques that are concerned with energy consumption as well as making sure that lifetime of a node is increased, have been proposed.

Load distribution and Transmission power are two main approaches which are used for minimizing energy consumption during active communication. In the Transmission power approach, stronger transmission results in an increase in transmission range and a reduction in the hop count. Weaker transmission power may result in partitioning of network and also high end-to-end delay due to increase in hop count. (Sofi et al. 2012, Yu et al. 2003) lists examples such as: Online Max-Min Routing (OMM), Smallest Common Power (COMPOW) and Power Aware Localised Routing (PLR). In Load Distribution Approach, the aim will be balancing of energy usage by mobile nodes through selection of route with nodes which are underutilized rather than shortest route. (Sofi et al. 2012, Yu et al. 2003) lists examples which include: Localized Energy Aware Routing (LEAR), Floor Augmented Routing (FAR) and Conditional Max-Min Battery Capacity Routing (CMMBCR). The method of Sleep/Power down is used for energy consumption minimization during inactivity. Spanning Tree (SPAN) and Geographic Adaptive Fidelity (GAF) are examples of this approach given in (Sofi et al. 2012, Yu et al. 2003).

The paper will be organized as follows; section (2) presents related work; while section (3) presents the proposed work; section (4) presents the simulation set up; section (5) details the results and analysis from the simulation, whilst section (6) contains the conclusion.

2 RELATED WORK

Lots of progress has been achieved before in this area, in this section, we take a look at some past algorithms that have already been proposed.

(AL-Gabri et al. 2012) proposed an energy model that ensured all nodes were balanced in their energy consumption for the purpose of prolonging the network's lifetime. The node would compare its battery level with a predefined threshold (θ). If (θ) <power_level, it would respond to the RREQ

as usual, else, it would simply drop the request. This approach is local since each node makes a decision basing on local information only. Results indicated improved performance with better network lifetime as a result of low energy consumption.

(Halder et al. 2014) proposed a protocol employing use of "Rank" parameter

Rank = Residual capacity/Rate.

Route Requests are rerouted on basis of RREQ with high "Rank" parameter. Results revealed comparable residual capacity differences between nodes, thereby indicating a balanced distribution of energy consumption resulting in a longer network lifetime.

The method proposed by (Maleki et al. 2003) was a prediction algorithm which took its basis from past activity and use of the Simple Moving Average; each node tried estimation of battery lifetime. Results revealed an increase in network lifetime by 32%.

(Maleki et al. 2002) proposed protocol that extended service life of a MANET. The link cost of a RREQ which is lowest is forwarded. Similarly at the destination node, link with lowest cost is used to send a RREP. The result showed that network lifetime by using PSR was increased by 30%.

(Othmen et al. 2014) developed protocol to select many paths that had longest lifetime in network without performance degradation on delay time. The Cost (C) of each route was found by the equation: $C = \frac{ML}{NH}$; where: ML is minimum remaining life of route and NH is number of hops. The source node selected primary route with the greatest C value and other routes are saved as secondary routes which can only be used in case of fault in the primary path. The results when compared with MAODV showed better throughput and low end to end delay.

(Pappa et al. 2007) proposed a modified protocol for ad-hoc sensor networks. Link cost was mainly used for selecting a path; destination node sends a RREP using route having minimum link cost. Results had a decrease in energy consumption without affecting other network parameters.

(Poongkuzhali et al. 2011) proposed a protocol that consisted of two phases: Finding Node Stability using Relative Mobility and also finding node which has more power and minimum hop count. The results selected a shortest path with longer overall battery lifetime.

Tarique & Tepe (2009) proposed a combination of two protocols. The duty of MEDSR protocol was energy consumption reduction in MANET while maintaining connectivity in the network whilst HMEDSR is responsible for MEDSR overhead reduction. The overall result is that energy spend in transmitting overhead packets is reduced.

The results obtained showed 25% energy consumption per data packet.

Vijayakumar & Poongkuzhali (2012) proposed a protocol whereby a node will rebroadcast messages according to high probability of not having more nearby nodes. Rebroadcast probability is high in sparser areas, whilst it is low in denser areas. Results indicated a 95% or higher packet delivery fraction using EPAB in all scenarios, giving a suggestion that EPAB is effective in route discovery and maintenance.

(Yitayal et al. 2014) proposed an algorithm which combined threshold, summation of residual energy, minimum residual energy and hop count cost metric. These metrics were integrated into AODV in an efficient way. They added fields: Minimum Residual Energy (MRE) and Sum of Residual Energy (SRE) into the RREQ header. An Energy Difference (D) field, which stores difference between either Average Minimum residual Energy (AME) and Threshold (Th) or Average Sum of residual Energy (ASE) and Threshold (Th), was also added on the routing table of the destination node. The result was network lifetime increase, due to increase in time taken for death of first node which is greater compared with other protocols.

3 PROPOSED WORK

The processing of an incoming RREQ packet by an intermediate node in proposed EPAAODV

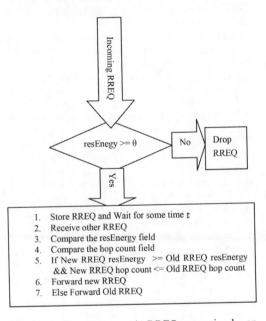

Figure 1. Proposed epaaodv RREQ processing by an intermediate node.

routing protocol differs from the normal AODV protocol. The flowchart below shows the proposed model algorithm for an intermediate node when it has received a RREQ.

In the proposed scheme, there is an addition of the residual energy (resEnergy) field packet header. When an intermediate node receives a RREQ packet, it will check the resEnergy field of RREQ packet header. If the resEnergy value is > = threshold energy (θ), then the request is stored and considered as eligible candidate for forwarding, else the RREQ is dropped. The RREQ is then stored in cache of intermediate node for some time (t). Intermediate node will then receive other RREQs during waiting time period. The intermediate node then compares the resEnergy field of the new incoming requests with that RREQ already in cache routing table. If the resEnergy of new RREQ is higher and hop count is low compared to stored old RREQ, then the new RREQ is forwarded. Otherwise, if the conditions mentioned above are not met, and the waiting period expires, then the old RREQ stored in routing table is forwarded. In this way, a RREQ with high resEnergy value and low hop count is always selected, thereby creating stable routes which are also fast. This result in not only longer network lifetime but also shortest path is selected.

4 SIMULATION SETUP

NS2.35 simulator is used to simulate the proposed EPAAODV protocol and the conventional AODV protocol and also to analyze energy consumption of the above mentioned protocols. Between source and destination node, we applied constant bit rate. It is applied over a User Datagram Protocol (UDP). Nodes in sizes of 5; 50; 100 and 300 nodes are used.

Table 1. Simulation parameters.

Number of nodes	5; 50; 100; 300
Speed of nodes	5 m/s; 10 m/s; 15 m/s
Area	(1000 * 1000) m
Protocol	epaaodv; aodv
Packet size	1000
Initial energy	100 J
TxPower	1.4 W
RxPower	1.0 W
Idle power	0.7 W
Sleep power	0.01 W
Mobility model for movement	Random waypoint
Simulation time	100 sec
Antenna model	Omni-directional
Path loss model	Two ray point
MAC Protocol	IEEE802.11
Physical layer radio type	IEEE802.11b

The nodes are given different movement speeds of 5; 10; and 15 m/s in square field (1000*1000) m. The whole simulation parameters are shown below.

5 SIMULATION RESULTS & ANALYSIS

This section outlines results obtained from the simulation of proposed EPAAODV and AODV protocol with 5, 50, 100 and 300 nodes.

In Figure 2, proposed protocol has a low energy consumption compared to the conventional AODV protocol. When 5 nodes are simulated, the proposed EPAAODV consumes an average of 80 J compared to 100 J consumed by the AODV protocol. This shows a reduction of 20% in consumption.

In all other simulated scenarios of 50; 100 and 300 nodes; EPAAODV consumes less energy compared to AODV. This means proposed protocol, uses less energy, thereby increasing network lifetime.

Our proposal also shows a reduction in percentage of dead nodes, as illustrated in Figure 3. With increase in number of simulated nodes, the dead nodes are reduced. In all situations, the proposed shows less percentage of dead nodes. Since less node number use up all their energy in the proposed work, it means network lifetime is greatly increased.

Packet delivery ratio is higher compared to the conventional protocol as illustrated in Figure 4. This could be due to stability of path in our proposed protocol. Since our proposal prolongs network lifetime, it means there are less link breakages, and our protocol guarantees that most of the send packets are delivered. In all scenarios simulated, our proposed EPAAODV has high packet delivery ratio compared to conventional AODV.

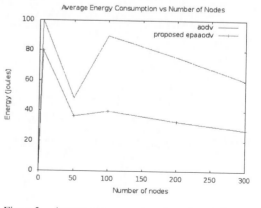

Figure 2. Average energy consumption of epaaodv and aodv.

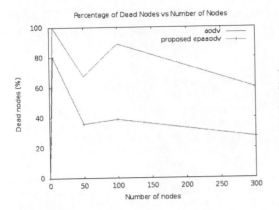

Figure 3. Percentage of dead nodes of epaaodv and aodv.

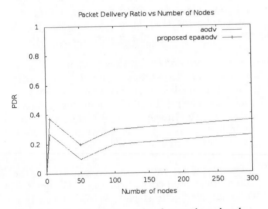

Figure 4. Packet delivery ratio of epaaodv and aodv.

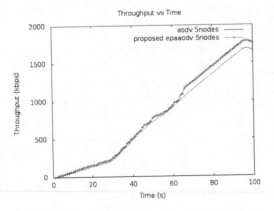

Figure 5. Throughput of epaaodv and aodv using 5 nodes.

In Figure 5, throughput of proposed work is slightly higher compared to conventional AODV protocol. When 5 nodes are simulated, during first 30 seconds, throughput is low, and then it starts

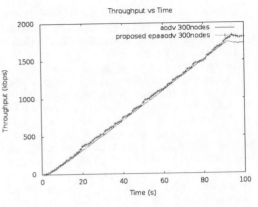

Figure 6. Throughput of epaaodv and aodv using 300 nodes.

to increase with time. In Figure 6, throughput of proposed work when 300 nodes are simulated is always better compared to conventional protocol. Reason for our proposed work having a slightly higher throughput is the increased network lifetime, meaning in our network, there is no need for reconstruction of new path.

6 CONCLUSION

From the results it is deduced that our proposed EPAAODV outperforms the conventional AODV protocol in many aspects. Proposed protocol saves more energy; has less dead nodes; better packet delivery ratio and better throughput; compared to conventional AODV. It is also concluded that both proposed EPAAODV and conventional AODV, are suitable in high dense networks, as performance in node scenarios of 100 and 300 nodes is better compared to scenarios of 5 and 50 nodes. The purpose of prolonging network life time is achieved by the proposed EPAAODV protocol because the protocol consumes less energy.

REFERENCES

AL-Gabri, M., L. Chunlin, Y. Zhiyong, A. Hasan., and Z. Xiaoqing. 2012. Improved The Energy Of Ad Hoc On-Demand Distance Vector Routing Protocol. IERI Procedia 2: 355–361. Elsevier BV.

Alslaim, M., H. Alaqel, and S. Zaghloul. 2014. A Comparative Study Of MANET Routing Protocols. The Third International Conference On E-Technologies And Networks For Development (Icend2014). Institute of Electrical & Electronics Engineers (IEEE).

Giannoulis, S., C. Antonopoulos, E. Topalis, and S. Koubias. 2016. ZRP Versus DSR And TORA: A Comprehensive Survey On ZRP Performance. 2005

IEEE Conference On Emerging Technologies And Factory Automation. Institute of Electrical & Electronics Engineers (IEEE).

Halder, T., C. Chowdhury, and S. Neogy. 2014. Power Aware AODV Routing Protocol For MANET.2014 Fourth International Conference On Advances In Computing And Communications. Institute of Electrical & Electronics Engineers (IEEE).

Maleki, M., K. Dantu, and M. Pedram. 2016. Lifetime Prediction Routing In Mobile Ad Hoc Networks.2003 IEEE Wireless Communications And Networking, 2003. WCNC 2003. Institute of Electrical & Electronics Engineers (IEEE).

Maleki, M., K. Dantu, and M. Pedram. 2002. Power-Aware Source Routing Protocol For Mobile Ad Hoc Networks. Proceedings Of The International Symposium On Low Power Electronics And Design. Institute of Electrical & Electronics Engineers (IEEE).

Mohapatra, S., and P. Kanungo. 2012. Performance Analysis Of AODV, DSR, OLSR And DSDV Routing Protocols Using NS2 Simulator. Procedia Engineering 30: 69–76. Elsevier BV.

Othmen, S., A. Belghith, F. Zarai, M. Obaidat, and L. Kamoun. 2014. Power And Delay-Aware Multi-Path Routing Protocol For Ad Hoc Networks. 2014 International Conference On Computer, Information And Telecommunication Systems (CITS). Institute of Electrical & Electronics Engineers (IEEE).

Pappa, K., A. Athanasopoulos, E. Topalis, and S. Koubias. 2007. Implementation Of Power Aware Features In AODV For Ad Hoc Sensor Networks. A Simulation Study. 2007 IEEE Conference On Emerging Technologies & Factory Automation (EFTA 2007). Institute of Electrical & Electronics Engineers (IEEE).

Poongkuzhali, T., V. Bharathi, and P. Vijayakumar. 2011. An Optimized Power Reactive Routing Based On AODV Protocol For Mobile Ad-Hoc Network. 2011 International Conference On Recent Trends In Information Technology (ICRTIT). Institute of Electrical & Electronics Engineers (IEEE).

Safa, H., M. Karam, and B. Moussa. 2014. PHAODV: Power Aware Heterogeneous Routing Protocol For Manets. Journal Of Network And Computer Applications 46: 60–71. Elsevier BV.

Shafinaz Buruhanudeen, Mohamed Othman, Mazliza Othman, and Borhanuddin Mohd Ali. 2007. Existing MANET Routing Protocols And Metrics Used Towards The Efficiency And Reliability—An Overview. 2007 IEEE International Conference On Telecommunications And Malaysia International Conference On Communications. Institute of Electrical & Electronics Engineers (IEEE).

Sofi, S., I. Ashraf, F. ud-Din, A. Ayub, and R. Mir. 2012. An Analysis Of Control Packets, Packet Delivery Ratio & Residual Energy In Power Aware Dynamic Source Routing In Mobile Adhoc Networks (Manets). Fourth International Conference On Advances In Recent Technologies In Communication And Computing (Artcom2012). Institution of Engineering and Technology (IET).

Sumathi, K., and A. Priyadharshini. 2015. Energy Optimization In Manets Using On-Demand Routing Protocol. Procedia Computer Science 47: 460–470. Elsevier BV.

Tarique, M., and K. Tepe. 2009. Minimum Energy Hierarchical Dynamic Source Routing For Mobile Ad Hoc Networks. Ad Hoc Networks 7, no. 6: 1125–1135. Elsevier BV.

Vijayakumar, P., and T. Poongkuzhali. 2012. Efficient Power Aware Broadcasting Technique For Mobile Ad Hoc Network. Procedia Engineering 30: 782–789. Elsevier BV.

Yitayal, E., J. Pierson, and D. Ejigu. 2014. A Balanced Battery Usage Routing Protocol To Maximize Network Lifetime Of MANET Based On AODV. Lecture Notes In Computer Science: 266–279. Springer Science + Business Media.

Yu, C., B. Lee, and H. Yong Youn. 2003. Energy Efficient Routing Protocols For Mobile Ad Hoc Networks. Wirel. Commun. Mob. Comput. 3, no. 8: 959–973. Wiley-Blackwell.

Communication and Computing Systems – Prasad et al. (Eds)
© 2017 Taylor & Francis Group, London, ISBN 978-1-138-02952-1

Strengthening of MAC layer security to prevent jamming attacks in wireless lans

A. Sharma & A. Solanki
Gautam Buddha University, Greater Noida, India

ABSTRACT: The mobility and ease of deployment of Wireless Local Area Networks (WLANs) make them widely useful in present time. However, due to their open nature and undefined boundaries, their users themselves are challenge in insuring complete security to them. Under detailed exploration into the subject, jamming attacks involving 'Media Access Control' (MAC) Layer Denial of Services (DoS) and De-authentication and Disassociation emerge as the main concern. The management frames carrying non encrypted MAC addresses of the Client or Access Point (AP) being vulnerable facilitate launch of such attacks. To prevent them there is need of such mechanism which does not change in the standard protocol and the hardware of the system. In the subject research it is attempted to strengthen the MAC layer against such attacks using an innovative strategy, titled "Anti-spoof Passkey-MAC Switching Strategy", the hybrid of two strategies identified through problem analysis and literature survey. The first one uses an algorithm to identify the intruder and gives the solution for MAC spoofing DoS attacks with using the passkey values in the communication. And the second involves use of dynamically changing MAC addresses without any change in 802.11 protocols. The hybrid strategy developed under the subject research utilizes only the advantageous features of the two base strategies. The results have been successfully validated through simulation.

1 INTRODUCTION

A wireless network is a type of computer network that uses wireless data connections for connecting network nodes. Wireless telecommunication networks are generally implemented and administered using radio communication. Their implementation takes place at the physical level (layer) of the Open System Interconnection (OSI) Model Network-structure. The users under a Wireless Local Area Network.

A user is still connect to the network while he/she is moving around the coverage area of the WLANs. IEEE 802.11 is the standard protocol, mostly used by the various Wi-Fi brands in the market.

WLANs being built upon a shared medium become easier for adversaries to let them conduct radio interference or jamming attacks towards them that effectively cause a Denial of Attack (DoS) of either transmission or reception functionalities. Such attacks are generally accomplished by either bypassing Media Access Control (MAC) Layer Protocol, or emitting a radio signal targeted at a particular channel for jamming.

OSI (open system interconnection) model have 7 layers: Application layer, Presentation layer, Session layer, Transport layer, Network layer, Data link layer, and Physical layer. MAC layer is the sub layer of the Data Link layer. A multiple access, shared medium is used to communicate by network nodes. This medium is provided due to MAC sub layer addressing system e.g. an Ethernet network. The hardware that implements the MAC is referred to as a Media Access Controller.

The MAC sub-layer acts as an interface between the Logical Link Control (LLC) sub-layer and the Network's Physical Layer. The MAC layer emulates a full-duplex logical communication channel in a multi-point network. This channel may provide unicast, multicast or broadcast communication services.

The problem targeted under the subject research was founded by going through a vast literature review about WLANs. After analyzing the problems therein, the objective of the research was mapped. According to which strengthening of MAC layer security to prevent jamming attacks in WLANs is a task of utmost requirement. But the traditional or the two base strategies merged in the subject research into "Anti-spoof Passkey-MAC Switching Strategy" have their own disadvantages and compromises with their limited benefits when used separately and alone.

The strategy that uses exchange of passkey to protect WLAN from any of such attack is useful only till the attacker becomes aware about it and hacks the passkey. Because after that the attacker can also use that passkey to send the bunch of extra data packets to jam the network. This way, the lone passkey strategy fails. The second strategy of using a dynamic MAC for each session, though works better, it has its own disadvantage of compromising with the network performance.

Thus, by all and both such ways, prevention of jamming attacks in WLANs poses as a very difficult task resulting in conceptually compromising problem in strengthening of MAC layer security. It leads toward the requirement of a hybrid strategy, the Anti-spoof Passkey-MAC Switching Strategy, that overcome the disadvantageous phases of the two identified strategies and works with them only till their use is advantageous by switching from the first to second.

2 THE FIRST BASE STRATEGY

This strategy (Arockiam & Vani 2013) secure the WLAN from the attacks basically attempted on *MAC Layer* such as de-authentication attack and dis-association attacks. Due to the openness of the system these attacks are launched to spoof the management frames and the MAC address of the client or *Access Point (AP)* which are not encrypted with security system. The algorithm developed under this strategy detects and prevent MAC spoofing DoS attacks by exchanging passkey values.

It uses an authentication method for each de-authentication request and response. An integer generated randomly and consider as a passkey, which is used for identify the genuine client, while communicate between the legitimate client and AP. The passkey is random integer so that the intruder cannot easily predict it. Though the MAC address of the client is spoofed by the intruder, it is impossible to make dummy node of the same client. Every time when the intruder makes a de-authentication request, every time it is it has to exchange the passkey.

This way, it not only protects the WLAN communication, but also improves the performance of WLAN by increasing the throughput and reducing the packet resend rates to a greater extend. The recovery time also gets reduced as compared with the existing method.

3 THE SECOND BASE STRATEGY

In WLANs, MAC address authentication of nodes is not completely secure because the intruder can make the dummy node by knowing the authorized MAC address of any node connected to the network. Under this strategy (Bicakci & Uzunay 2008), MAC address spoofing attacks are prevented using dynamically changing MAC addresses that are usable for only one session. There is no need for any changes in 802.11 protocol but it compromise with the performance a bit. This way, no third person can connect in any communication sessions of the same client by using MAC addresses. So, the strategy is preferable also with respect to the user privacy.

To protect the MAC address encryption is not the best option if viewed in depth. The secret used should be secure. If not so, its security is in doubt. In such a scenario, there is only one alternative left: only one-time usable secret. So, if somebody somehow gets it, it cannot be used for another time. The One-time Password Scheme is its example.

In the design of the algorithm developed under this strategy, the size limit of the MAC addresses field is 48 bits. It is so due to the requirement of conforming to 802.11 standards. It has been observed that the hash chain based solutions are not enough secured, for their only 48-bits security. But in their capability to make the server free of any secret and hold only public information to verify the secrets, they have the security advantage over the secret-key alternatives. In the design, it is assumed that the servers in the system can be kept secure so that the secret information remains safe.

4 LIMITATIONS OF THE BASE STRATEGIES

Although, the two base strategies provide significant protection against a variety of jamming attack on WLAN, they have their own disadvantage and compromise, too, when used separately and alone.

The first strategy that uses exchange of passkey to protect WLAN from any of such attack is useful only till the attacker becomes aware about it and hacks the passkey. Because after that the attacker can also use that passkey to send the bunch of extra data packets to jam the network. This way, the lone passkey strategy fails.

The second strategy of using a dynamically changing MAC for each session, though works better in terms of protection, it compromises a bit with the network performance.

5 ANTI-SPOOF PASSKEY-MAC SWITCHING STRATEGY

In a wireless network, the client communicates with the AP by authorization, so that only the authorized users can communicate. If in between the communication, any intruder tries to do de-authentication or disassociation attack, it can create a big problem as the DoS attack. The hybrid strategy works by utilizing the advantageous features of the both base strategies in optimum phases during the WLAN communication. The hybrid algorithm developed under this strategy has been tested on NS2, a network simulation tool.

Initially, under the first strategy, the Passkey Exchange Strategy, a passkey is attached with the every frame at the sending time and checked on the receiving time. If the frame and packet have the passkey (both should have the same passkey), the data packets are real, otherwise those are considered as fake. The passkey can be any randomly generated static key, or a timestamp, or a numeric or alphanumeric key. This way, the authentication of passkey is done. It is shown in Figure '1'.

The finding of the fake packets conveys that someone is trying to do spoofing in the network. Thereon the data packets without passkey would not be processed in the network communication. After three without or with wrong passkey attempts from a node, the situation would be declared as an attack. Figure '2' shows a Node's 3 attempts without passkey under the Passkey Exchange Strategy.

From this point of time, the hybrid algorithm would switch the task of security to the second selected strategy, Dynamically Changing MAC Strategy. The intruder node is isolated and WLAN communication continues under this second strategy as shown in Figure '3'.

In case of the algorithm under the first strategy, the process starts with the passkey between user and AP, and it sets the parameters zero as the de-authentication and the number of attacks, and then initializes the process. If the initial authentication is not successful for the security, the process will go for the next one and after successful authentication the communication will start with applying this strategy. The process runs continuously till the communication is on.

After its switching to the second strategy, upon declaration of an attack, the communication

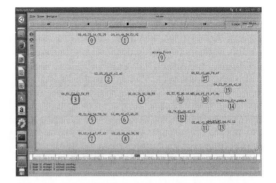

Figure 2. Node's 3 attempts without passkey under the Passkey Exchange Strategy.

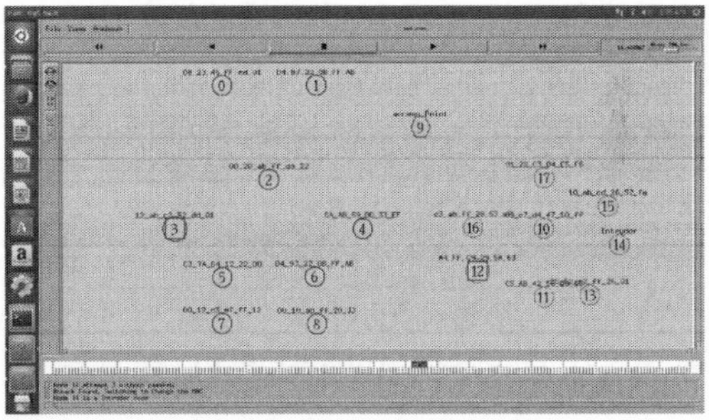

Figure 3. Intruder node isolation and beginning of communication under the Dynamically changing MAC Strategy.

remains on till manual switching back of the task of attack prevention to the first strategy.

Through Anti-spoof Passkey-MAC Switching Strategy, the task of prevention of jamming attacks is initially done under the first strategy, which does so without compromising with the network performance. Later, after detection of an attack, the second strategy provides better protection, though with a bit overload on the network's performance. It can be overcome by manually switching back the WLAN under the protection of the first strategy.

6 RESULTS

The Anti-spoof Passkey-MAC Switching Strategy has been tested on network simulator tool NS2. Under the implementation of the strategy, total 16 nodes were generated. One among these nodes was the AP, and 12 other nodes that were pre-connected were denoted as 'clients'.

Upon initialization of connectivity, 2 nodes among the rest 3 nodes sent frames for getting

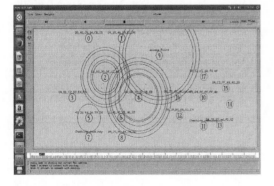

Figure 1. Passkey Authentication.

connected. Both nodes, with valid passkey got connected to the network. The then MAC address of each node was displayed in the simulation.

The remaining one node attempted to get authenticated without passkey, but got response: 'authentication failure_1'. Its request was dumped. The node tried again without correct passkey and 'authentication failure_2' message sent by AP with again dumping its packet. When the node attempted third time without passkey or with invalid passkey, AP dumped its request packet again and sent response: 'authentication failure_3'. At that point of event, the network declared it an attack, and the node that tried getting connected thrice without a valid passkey was denoted as 'intruder'. This attack and intruder information was sent to all nodes.

At this point, the strategy that used dynamic changing MAC address became active. All genuine nodes in the network started changing their MAC address after each session. Thereon the MAC address of each node was checked by the AP. Upon knowing whether a node had valid MAC or not, the authentication/association 'success' or 'failure' responses were sent and further communication took place.

During the communication, each node changed its MAC address after completion of each session successfully.

It is, thus, successfully resulted that the Antispoof Passkey-MAC Switching Strategy can work according to expectation for strengthening of MAC layer security to prevent jamming attacks in WLANs.

7 CONCLUSION

The hybrid algorithm developed under the Antispoof Passkey-MAC Switching Strategy has been implemented on NS2.

The utility of the subject hybrid algorithm may prove its place wherever WLANs require smart protection against the jamming attacks target at MAC layer security.

8 FUTURE DIRECTION

The hybrid algorithm developed under the subject strategy is itself capable of providing enhanced security to WLANs against the jamming attacks targeted through MAC layer. Being an innovative solution in the particular field, feasibility of improving it for a few left limitations will be worked out.

A research may be initiated in future to do so by introducing few measures in the code to automatically bring the network back on the Passkey Exchange Strategy from the Dynamically Changing MAC Strategy after a certain time, so that the network could overcome the performance overhead imposed by dynamically changing MACs. In case of further attacks, the network can again switch to the Dynamically Changing MAC Strategy, and return back after a defined time.

ACKNOWLEDGEMENT

We deeply acknowledge Mr Vala Kiritkumar, Scientist, R & D E (Engineers), DRDO, Pune, for his precious guidance to complete this research, and also the authors of following papers for their valuable research work:

"Medium Access Control Spoof Detection and Prevention Algorithm (MAC SDPDoS) for Spoofing Attacks in WLAN", by L. Arockiam, and B. Vani, in "IRACST—International Journal of Computer Science and Information Technology & Security (IJCSITS), ISSN: 2249–9555, Vol. 3, No.2, April 2013"; and "Pushing the Limits of Address Based Authentication: How to Avoid MAC Address Spoofing in Wireless LANs", by Kemal Bicakci, and Yusuf Uzunay, in "International Journal of Electrical, Computer, Energetic, Electronic and Communication Engineering Vol: 2, No: 6, 2008".

REFERENCES

AUSCERT. "Denial of Service Vulnerability in IEEE 802.11 Wireless Devices". At 'http://www.auscert.org.au/render.html?it = 4091'.

Alnie, G and R Simon: "A multi-channel defense against jamming attacks in wireless sensor networks". In "ACM International Workshop on Modeling Analysis and Simulation of Wireless and Mobile Systems", pages 95–104, 2007.

Arockiam, L and B Vani: "Medium Access Control Spoof Detection and Prevention Algorithm (MAC SDP DoS) for Spoofing Attacks in WLAN". In "IRACST - International Journal of Computer Science and Information Technology & Security (IJCSITS), ISSN: 2249–9555, Vol. 3, No. 2, April 2013".

Bellardo J, Savage S.: "802.11 Denial-of-service attacks: Real Vulnerabilities and Practical Solutions". In USENIX Security Symposium, 2003.

Bosiljka Tadíc, and Stefan Thurner: "Transport on Complex Networks: Flow, Jamming and Optimization". In "International Journal of Bifurcation and Chaos, Vol. 17, No. 7 (2007) 2363–2385".

Clayton W. Commander, Panos M. Pardalos, Valeriy Ryabchenko, Oleg Shylo, Stan Uryasev, and Grigoriy Zrazhevsky: "Eavesdropping and Jamming Communication Networks".

David Slater, Patrick Tague, Radha Poovendran, and Mingyan Li: "A Game-Theoretic Framework for

Jamming Attacks and Mitigation in Commercial Aircraft Wireless Networks".

Guevara Noubir: "On Connectivity in Ad Hoc Networks under Jamming Using Directional Antennas and Mobility".

Haller N. M.: "The S/KEY One-time Password System". In proceedings of the 'ISOC Symposium on Network and Distributed System Security, 1994'.

Kartik Siddhabathula, Qi Dong, Donggang Liu, and Matthew Wright: "Fast Jamming Detection in Sensor Networks".

Kemal Bicakci, and Yusuf Uzunay: "Pushing the Limits of Address Based Authentication: How to Avoid MAC Address Spoofing in Wireless LANs". In "International Journal of Electrical, Computer, Energetic, Electronic and Communication Engineering Vol: 2, No: 6, 2008".

Kyasanur, P and N Vaidya: "Selfish MAC layer misbehavior in wireless networks". In IEEE Trans. Mobile Computing, Vol. 4, No. 5, Sept./Oct. 2005.

Mina Malekzadeh, Abdul Azim Abdul Ghani, and Shamala Subramaniam: "Protected control packets to prevent denial of services attacks in IEEE 802.11 wireless networks". In "Malekzadeh et al. EURASIP Journal on Information Security 2011, 2011:4".

Mingyan Li, Iordanis Koutsopoulos, and Radha Poovendran: "Optimal Jamming Attacks and Network Defense Policies in Wireless Sensor Networks".

Patrick Tague, David Slater, and Radha Poovendran: "Linear Programming Models for Jamming Attacks on Network Traffic Flows".

Radosavac, S. I. Koutsopoulos and J.S. Baras: "A framework for MAC protocol misbehavior detection in wireless networks," in Proc. ACM Workshop on Wireless Security (WiSe), 2005.

Sampigethaya K, Poovendran R, and Bushnell L: "Secure Operation, Control, and Maintenance of Future e-Enabled Airplanes". Proc. IEEE, Vol. 96, No. 12, Dec. 2008, pp. 1992–2007.

Shanshan Jiang, and Yuan Xue: "Attack for Multi-Radio Multi-Channel Wireless Mesh Networks".

Timothy X, Brown Jesse E James, and Amita Sethi: "Jamming and Sensing of Encrypted Wireless Ad Hoc Networks".

Wenyuan Xu, Ke Ma, Wade Trappe, and Yanyong Zhang: "Jamming Sensor Networks: Attack and Defense Strategies".

Wright J.: "Detecting Wireless LAN MAC Address Spoofing". In white paper, available at 'http://www.logisense.com/docs/wlan-mac-spoof.pdf'.

Communication and Computing Systems – Prasad et al. (Eds)
© *2017 Taylor & Francis Group, London, ISBN 978-1-138-02952-1*

Analysis of energy detection and cyclostationary feature detection under fading channels in CRNs

Raman Kaur, Paras Chawla, Pooja Sahni & Sukhdeep Kaur
Department of Electronics and Communication Engineering, Chandigarh Group of College, Landran, Greater Mohali, Punjab, India

Kusum Yadav
Department of Information Systems, College of Computer Engineering and Science, Prince Sattam Bin Abdulaziz University, Kingdom of Saudi Arabia

ABSTRACT: Cognitive Radio (CR) is a modernistic approach of wireless communication which permits frequency spectrum to be used more efficiently. This becomes feasible by using dynamic spectrum access in which the spectrum is allocated to different users dynamically rather than statistically. Within the area of cognitive radio, in order to attain the most favorable system different characteristics of the system must be considered. This involves the techniques for spectrum sensing i.e. the spectrum is available or not, sensing the best channel to use, determining the time for data transmission. This paper reviews various spectrum sensing methods (cooperative and non-cooperative methods) and the challenges relating to spectrum sensing. The energy detection and cyclostationary feature detection methods under Rayleigh fading are mainly discussed in this paper with experimental setup based results in Matlab.

Keywords: cognitive radio, cooperative sensing, fading, non cooperative sensing, spectrum sensing

1 INTRODUCTION

Radio spectrum is a section of electromagnetic spectrum and it lies 3 KHz and 300 GHz frequencies. Spectrum is divided into bands and different bands are used for different applications. It is an essential medium for wireless communication. Presently wireless communication is highly in demand and this demand leads to constraint on accessible radio spectrum. The significant amount of the spectrum is not fully utilized as illustrate in Figure 1. It has been ascertained, at frequencies less than 3 GHz the use of spectrum is more enormous and harsh as compared to the frequencies between 3 to 6 GHz bands i.e. in these bands the spectrum is underutilized. The traditional approach used for the allocation of radio spectrum is static or fixed approach. It does not let an unlicensed or secondary user to utilize scarcely used spectrum assigned to licensed or primary user. This results in underutilization of radio spectrum and reduces system flexibility. To make substantial use of spectrum, the spectrum should be allocated dynamically rather than statistically. To make use of dynamic spectrum allocation, a technology called as Cognitive Radio (CR) is used. Cognitive radio is a technology which provides effective utilization of unused radio spectrums or bands. It makes an

unlicensed user to use license frequency band without interrupting licensed user.

In terms of cognitive radios, Primary User (PU) is a licensee of a particular frequency band and has higher priority to use that band. Secondary User

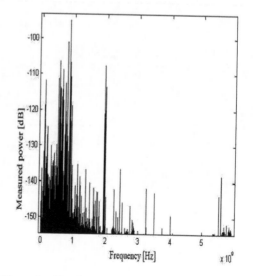

Figure 1. The Spectrum utilization (Carbic D et al. 2004).

(SU) is an unlicensed user which opportunistically utilize spectrum without causing interference to or from the primary user and has lower priority. Therefore, SU requires cognitive capabilities and reconfigurability to sense a specific geographical area for the presence of primary user. The cognitive radio is a radio for wireless communication in which transmission and reception parameters of a network or a node can be changed to communicate efficaciously without interfering licensed user (Wang & Ray Liu 2011).

The main challenge for cognitive radio network is to identify the licensed or primary users over a considerable range of spectrum at a specific time. The free bands or generally called spectrum holes are then allocated to secondary or unlicensed user for proper utilization of spectrum as shown in Figure 2.

For this purpose various spectrum sensing techniques are considered. The cognitive radio is an emerging technology which provides an idea for flexible radio system, multiband, reprogrammable and reconfigurable by software for Personal Communication Services (PCS).

1.1 The cognitive radio has a few basic functions which are as follows (Shabham & Mahajan 2015)

- **Spectrum Sensing**: it the way to recognize the existence of free spectrum and licensed user.
- **Spectrum management:** The use of frequency bands is managed by spectrum management i.e.

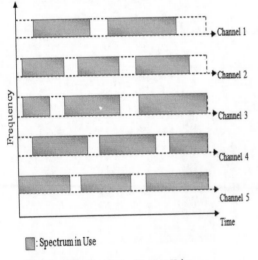

Legend:
▨ : Spectrum in Use

▢ : Under-utilized spectrum or Spectrum Holes

Figure 2. Spectrum Holes.

how long the unlicensed user can use the free frequency band.
- **Spectrum sharing:** it the process of sharing frequency bands of primary users with the secondary users. So the spectrum can be used more efficiently.
- **Spectrum Mobility:** The working frequency of an unlicensed user can be different while transmission.

2 SPECTRUM SENSING

Spectrum sensing is sensing the environment for underutilized spectrum or spectrum holes in time, frequency, space and code dimensions (Ghasemi & Sousa 2008). Spectrum sensing alludes to learn, measure and be aware of availability of spectrum, noise, temperature, power and interference in the operating environment. It is an important feature of cognitive radio, as spectrum will be efficiently used only if spectrum is sensed well. This leads to high Quality of Service (QOS) for the Primary User (PU).

2.1 Challenges in Spectrum sensing

Before getting into the complete detail of spectrum sensing techniques, there are few challenges in spectrum sensing that must be considered. Which are described in this section.

2.1.1 Hidden primary user problem
This problem is almost identical to the hidden node problem in Carrier Sense Multiple Access (CSMA) (Yucek & Arslan 2009). While examining the transmission of primary user, many factors such as multipath fading and shadowing are responsible for this problem. This may affect the wireless communication.

Hence the existence of primary user is not detected by secondary users. This leads to decrease the accuracy of the system. Hidden primary user problem happens when a node at access point is visible to other nodes but cannot communicate within the distance of communication (Reddy 2012).

2.1.2 Hardware requirement
In cognitive radio applications, spectrum sensing requires Analog to Digital Converters (ADCs) with resolution and high energy range, fast signal processors and high sampling rate. Unlike noise/interference estimation problem in which receivers have potential to process transmission over narrow bands with low power processors and low complexity, CRs requires terminals which should process transmission over much wider band for proper utilization if spectrum holes. Hence CRs

must have capability of capturing and analyzing a wider band for spectrum holes. This requires additional radio frequency components. Moreover, for executing signal processing task with relatively low delay, high speed processing units (DSPs or FPGAs) are used. A few available hardware and software platforms for cognitive radio are GNU radios (Blossom 2004), Universal Software Radio Peripherals (USRP) and shared spectrum XG radios (McHenry et al. 2007).

2.1.3 *Sensing duration and frequency*
As the cognitive radio make use of frequency bands of Primary User (PU) for the transmission. But these can by asserted by PU at any time for their transmission. The CR must be capable for vacating that frequency bands to circumvent the interference to or from the licensed users. This becomes sensible if CR is that much intelligent to discern the actuality of PU as fast as possible. Despite the facts that this condition places some complexity and challenges in cognitive radio design. Selection of sensing parameters carries a tradeoff between the reliability of sensing and speed i.e. sensing time. Sensing frequency is an essential parameter which should be selected inventively. It describes how often sensing should be achieved by cognitive radio. The slow change in the status of primary user makes sensing frequency requirement flexible. For example, TV channels detection. Generally, the existence of station is not often changes unless a present station goes offline or new station starts broadcasting. In IEEE 802.22 WRAN (wireless regional area network) standard, a sensing period of 30 seconds is chosen. In this standard, the channel detection time, channel moving time and other timing parameters are considered along with sensing frequency (Cordeiro et al. 2006). Interference tolerance is another factor that affects sensing frequency. Sensing time affects the performance of secondary user or CR. In non-adaptive sensing methods, sensing duration is reduced.

2.1.4 *Detecting spread spectrum primary users*
The two major types of wireless technologies are fixed frequency transmission and spread spectrum. Spread spectrum is the combination of Frequency-Hopping Spread Spectrum (FHSS), Direct-Sequence Spread Spectrum (DHSS), Time-Hopping Spread Spectrum (THSS). Narrowband or fixed frequency transmission is easy to implement and has a fixed carrier frequency. Whereas, in Frequency Hopping Spread Spectrum (FHSS) the signal's frequency changes periodically and FHSS devices distribute their energy by using single band. Licensed users that utilizes spread spectrum signaling, spreads the power over a wide frequency range. This becomes challenge for secondary user

to detect the tenancy of PU. By knowing the perfect hopping pattern and proper synchronization this problem can be avoided.

2.1.5 *Occurrence of false detection*
The signal strength of received signal is changes or decreased by multipath fading, shadowing and noise. Due to this change in signal strength, spectrum sensing become difficult and causes interference to or from the primary license holder.

2.1.6 *Decision fusion in cooperative sensing*
It is a challenging task in cooperative sensing to share data among cognitive radios and merging results from different measurements The shared information by each cognitive device can be soft or hard decisions. Soft information decision has better performance than hard information decision in terms of missed opportunity probability. But the results are in favor of hard decision if the numbers of users are increased. The decision of each cognitive radio is considered for the information fusion at Access Point (AP).

2.1.7 *Security*
In CRNs, a signal with high power can be transmitted or some features of primary user can be imitated by an attacker to delude the primary user detection. Therefore, an incorrect identification of primary user is made a secondary user and the frequency band remains underutilized. Such a attack is termed as Primary User Emulation (PUE) attack. It has harmful effect on CRs. To prevent this attack, the primary user must be authenticated i.e. using primary user identification. The PU needs to transmit a public key or a signature along with transmitted signal which is produced by private key. This key is used for formalize the primary user. Furthermore, the secondary user must be capable of demodulate for synchronize primary users signal.

3 CLASSIFICATION OF SPECTRUM SENSING METHODS IN CRN

Sensing techniques are the fundamental element of Cognitive Radio (CR). Sensing techniques can be classified into two main categories: cooperative spectrum sensing and non-cooperative spectrum sensing. In cooperative sensing, a prior knowledge of primary user is compared against received signal for consistent detection of primary user. In cooperative sensing priori knowledge is not required for detection purpose. There are many other sensing techniques such as interference based detection, wavelet based detection, filter bank based spectrum sensing, radio identification based detection, multi taper spectrum sensing and estimation are developed. This paper mainly focuses on cooperative and non cooperative sensing techniques.

3.1 Non cooperative spectrum sensing

In this technique, a decision on channel is made by each cognitive radio by its sensed data and this is achieved by sensing itself (Sharma & Katoch 2015). Non cooperative sensing includes:

Primary transmitter detection

Cognitive radio ought to specify that the signal from primary detector is present locally in definite spectrum. This approach involves matched filter, cyclostationary feature and energy detection.

3.1.1 Energy detection

Energy detection is a signal detection in which radio frequency energy in a media is required to calculate the absence and presence of licensed users. Then the calculated energy or received signal is estimated against a threshold value and therefore occupancy of channel is made. The existence of Primary User (PU) is indicated if obtained signal is larger than the threshold value. There is no preliminary knowledge of PU hence this method is known as non coherent detection. The evaluation of energy detection is given as in equation 1.

$$E = \frac{1}{N} \sum_{k=0}^{n} |y(k)|^2 \tag{1}$$

where, the size of observation vector is represented by N. By comparing the decision metric E with the threshold factor λ, the occupancy of the frequency band is determined. This is the simplest and most popular method among all the non cooperative methods. However, energy detection coincides with some disadvantages (i) performance of this method depends upon the noise power uncertainty. (ii) High sensing time is required to attain a given probability of detection. (iii) Unable to differentiate primary user from secondary user and noise thus secondary user must be properly synchronized. (iv) Detection of spread spectrum primary users is difficult (Akyildiz et al. 2011). (v) Performance of energy detection degrades under low Signal to Noise Ratio (SNR). Regardless of all these problems, energy detection is most commonly used detection method in non cooperative sensing. SNR is the major factor that influences the decision threshold. If the noise level in the channel is high, the measurements of energy detection can be distorted. Thus false detection may occur. Energy detection is basically used in time and frequency domain (Jiang et al. 2012). The block diagram below shows the basic functions in energy detection method.

Where, the Band Pass Filter (BPF) is subject to remove the out-of-band noise. The received energy signal is measured by the squaring device and integrator is therefore decides the observational interval of time. The output of the integrator is called as test stats which are compared to a threshold value to determine the PU's presence and absence.

3.1.2 Cyclostationary feature detection

In this detection method, by exploring the cyclostationary features i.e. periodicity of received signal, the detection of primary license holder or free spectrum is done. This can be achieved by examining the cyclic autocorrelation function of received signal z (t) as shown in equation 2.

$$R_z^\alpha(t) = E\left[z(t).z^*(t - \tau).e^{-j2\pi\alpha} \right] \tag{2}$$

where z (t) is received signal, [.] is expectation operator, * is complex conjugate and a cyclic frequency is α. Fourier transform expansion i.e. Cyclic Spectral Density (CSD) function is utilized to characterize Cyclic Autocorrelation Function (CAF) which is given in equation 3.

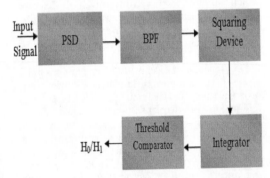

Figure 3. Block diagram of energy detection (Jiang et al. 2012).

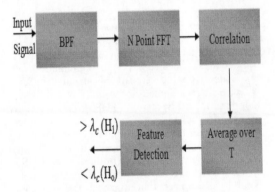

Figure 4. Block diagram of cyclostationary feature detection (Rebeiz et al. 2013).

$$S(f,a)=\sum_{\tau=-\infty}^{\infty}R_\tau^\alpha(\tau).e^{-j2\pi\alpha i} \qquad (3)$$

The fundamental frequency of transmitted signal is equivalent to cyclic frequency. CSD function demonstrates the peaks. There is no peak under H_0 hypothesis (Rebeiz et al. 2013). The modulated signal is incorporating with sine wave, cyclic prefix, pulse train etc. Because of this periodicity feature the periodic statics i.e. mean and auto correlation function of cyclostationary signal are revealed which cannot be found in interference and stationary noise (Lunden et al. 2009). These features are deliberately introduced to give assistance to spectrum sensing. The block diagram for cyclostationary feature detection is shown below.

In spite of using Power Spectral Density (PSD), thus method uses Cyclic Correlation Function (CCF) for detection of signal in given frequency band. It distinguishes noise from the primary signals hence it can detect weak signals at low SNR. This method is highly immune to noise uncertainty. The performance of this method is highly reliable when no information of primary user is there. However, this method has some cons: slow sensing, implementation and computation is complex as compared to energy detection method.

3.1.3 Matched filter detection

Matched filter technique is a well known technique of range detection for primary user signals. In other words, it is an ideal scheme if the secondary user has information about the primary user. In order to detect the primary user, a know PU signal is correlated with the received signal. For maximizing the SNR in presence of noise channel, match filter is an optimal linear. Matched filter detection is convolving received signal $q(t)$ with unknown signal of time-reversed version (Ejaz et al. 2013) as expressed in equation 4.

$$q(t)*r(T-t+\tau) \qquad (4)$$

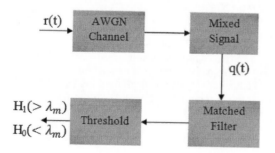

Figure 5. Block diagram of Matched Filter Detection (Ejaz et al. 2013).

where $q(t)$ convolving with $r(t)$, T is time duration and shift. The estimation of matched filter detection is shown is equation 5.

$$y(n)=\sum_{k=0}^{\infty}h(n-k)r(k) \qquad (5)$$

The unknown signal r is convolved with impulse function h, this matches the impulse response of matched filter with reference signal for increasing SNR.

The above figure shows the block diagram of matched filter detection. Where there are two main functions of convolution in equation 5. Firstly, a single value output is produced by placing one function over another function. Secondly, it carries first function at extremely small distance and establishes another value. Finally the comparison between the output M of matched filter with a threshold value λ is made to determine whether the PU is present or not as shown in Figure 5. In this technique secondary users must demodulate the received signal. Thus, an appropriate knowledge such as operating frequency, modulation type, frame format, pulse shaping and bandwidth of PU is required (Jaglan et al. 2015). This makes it inappropriate detection.

3.2 Cooperative spectrum sensing

Cooperative Spectrum Sensing (CSS) is generally deals with: Sensing and reporting (Sun et al. 2007). In sensing, by using some detection methods each cognitive user severally performs spectrum sensing. In reporting session, a common channel or central user gathers all local sensing observations and then the decision about the presence (H_1) and absence (H_0) of the licensed user is made. For each cognitive user CSS needs a control channel to describe its sensing observations. The ultimate goal of cooperative sensing is to identify the unused portion of spectrum temporally and spatially. Cooperative sensing can be classified as centralized and distributed.

3.2.1 Centralized cooperative sensing

In centralized cooperative sensing, there is a Fusion Center (FC) called common node or base station that mainly performs three tasks: sensing, reporting and decision as illustrated in Figure 7. In centralized sensing, CR1, CR2, CR2, CR3, CR4 and CR5 are the cooperating cognitive users which are connected to a common receiver CR0 (FC) as shown in Figure 6 (a).

CR1-CR5 performs sensing and result is reported to CR0. A sensing channel in which all the secondary users are tuned to a selected licensed

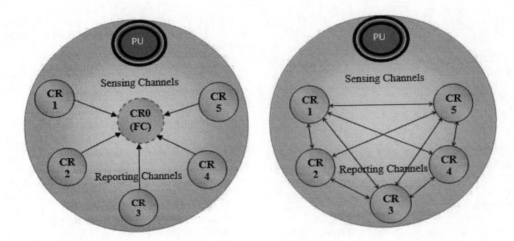

Figure 6. (a) Centralized cooperative sensing (b) distributed cooperative sensing.

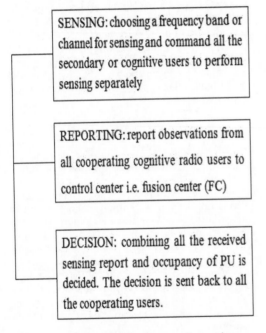

SENSING: choosing a frequency band or channel for sensing and command all the secondary or cognitive users to perform sensing separately

REPORTING: report observations from all cooperating cognitive radio users to control center i.e. fusion center (FC)

DECISION: combining all the received sensing report and occupancy of PU is decided. The decision is sent back to all the cooperating users.

Figure 7. Tasks of centralized cooperative sensing.

channel is used for local cooperating sensing. For detection of primary signal, a point to point link is established between the primary users. A control channel is used for reporting purpose where all the CR users are connected via a point to point configuration. A reporting channel which send the decision back to all the cooperating users.

3.2.2 *Distributed cooperative sensing*

In distributed sensing, there is no need to relay on FC for decision making. In this case, all users are connected to one another. A unite decision is made on the existence of primary user or nit. A sensing result is shared between CR1-CR5 and other users within their range of transmission as shown in Figure 6 (b). On the basis of distributed algorithm, each user exchanges their data with each other and combine that information with their own data then decides whether a PU is present or not. If the combined data is not matched, other iteration is takes place until the algorithm coincides and a decision is attained. Hence distributed sensing takes some iterations to achieve a consentient cooperated decision.

4 FADING

The transmission of radio signals is affected by channel or medium through which it is transmitted. Some of the channels are affected by fading: slow fading and fast fading

- *Slow fading:* the signal is faded by shadowing effects which are caused by building, mountains and many other obstacles in range of transmission.
- *Fast fading:* the amplitude of the signal at output is fluctuated rapidly and it is caused by multipath propagation of the signal.

Fading causes distortion in amplitude and phase of the signal in wireless medium. This may cause the high error rates by degrading the performance of the communication system. Some fadings are described below.

4.1 *Rayleigh fading model*

Due to the scattered nature of signal transmission over a channel, there are random amplitudes and angles of the received signal. This occurs when there is no Line-of-Sight (LOS) between receiver and transmitter. The Probability Density Function (PSD) of

Rayleigh fading is given in equation 6.

$$p_y(y) = \frac{y}{\sigma^2} \exp\left(\frac{-y^2}{2\sigma^2}\right) \qquad (6)$$

where, is σ variance of quadrature and in phase element and y represents the amplitude of the signal.

4.2 Racian fading model

The distribution models when there is a line of sight with rayleigh distribution. The PDF of racian distribution is given in equation 7.

$$p_y(y) = \frac{y}{\sigma^2} \exp\left(\frac{-y^2 + A^2}{2\sigma^2}\right) I_0\left(\frac{Ay}{\sigma^2}\right) \qquad (7)$$

4.3 Nakagami-m fading model

This distribution model covers a wireless environment in a wide range and often used to specify the transmitted signal statics via multipath fading environment (Carvalho et al. 2015). The PDF of nakagami-m distribution is given in equation 8.

$$p_y(y) = \frac{2m^m y^{2m-1}}{|(m)\Omega^m} \exp\left(\frac{-my^2}{\Omega}\right) \qquad (8)$$

where, m is the fading figure which is given as follows.

$$m = \frac{\Omega^2}{E[y^2 - \Omega^2]} \text{ And } \Omega = E[y^2]$$

5 SIMULATION RESULTS

The effect of fading over cognitive transmission is discussed in this section. In this paper, energy detection and cyclostationary feature detection techniques are considered under rayleigh fading.

Firstly the gain of the antenna is considered and then set the samples accordingly. Simulate the detection under Monte Carlo simulation rule by initiating a specific sample space in the spectrum and apply gaussian channel. Furthermore, simulate the effect of fading under rayleigh rule. Simulation

results of energy detection and cyclostationary feature detection under rayleigh fading is presented below in Figure 8. This shows the curve for probability of detection and false alarm detection. Initially the gain is set to be −10dB and 200 samples. From the graph below it is shown that the probability of false alarm at given probability of detection is more in case of energy detection as compared to CDF.

For the value of P_d i.e. 0.4, 0.6, 0.8 and 1, the P_{fa} for CFD is 0.02, 0.07, 0.29 and 0.97 and For ED P_{fa} is 0.03, 0.2, 0.5 and 1. This shows that the Pd is more relevant in case of Cyclostationary Feature Detection (CFD) as compared to Energy Detection (ED) and Pfa is more in ED.

In Figure 9, the effect of fading is considered with 200 samples and SNR at −15dB. The probability of detection of cyclostationary detection is more as compare to base technique and for the values of Pd i.e. 0.2, 0.4, 0.6, 0.8 and 1, the P_{fa} for CDF is 0.02, 0.15, 0.38, 0.68 and 0.98 and for ED it is 0.05, 0.22, 0.51, 0.8 and 1. From these observations it is it is shown that the probability of false alarm in energy detection is higher in comparison with cyclostationary feature detection.

The Figure 10 shows the probability of detection and probability of false alarm in ED and CFD. The P_{fa} for CDF is 0.05, 0.2, 0.42, 0.75 and 1 whereas for ED theses values are 0.1, 0.3, 0.6, 0.85 and 1. These values are for the Pd i.e. 0.2, 0.4, 0.6, 0.8 and 1. Hence rayleigh fading effects the performance of energy detection more.

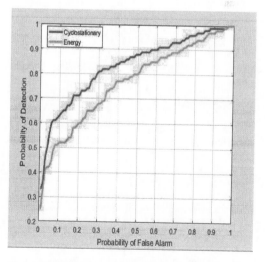

Figure 8. The probability of detection (P_d) as a function of probability of false alarm (P_{fa}) for ED and CFD subject to Rayleigh fading with −10 dB SNR and 200 samples.

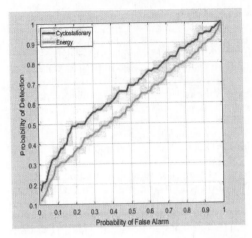

Figure 9.   Rayleigh fading with 200 samples and −15 dB.

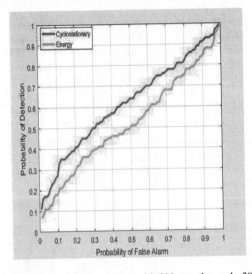

Figure 10.   Rayleigh fading with 200 samples and −20 SNR.

6   CONCLUSIONS

Cyclostationary feature detection and energy detection spectrum sensing techniques are analyzed for fading channel i.e. rayleigh fading channel. A simulation result shows that energy detector measurements are degraded with fading channel. The simulations are done with different SNR. The performance evaluation is made in all scenarios while the best performance has been ascertained at value of −10db SNR. It has been concluded that cyclostationary feature detection outperforms the energy detection if fadings are considered and in case of detection of spectrum or spectrum space. The probability of detection is more in case of CDF and the probability of false alarm is less in case of CDF as compared to ED.

REFERENCES

Akyildiz, I. F. et al. 2011. Cooperative spectrum sensing in cognitive radio networks: A survey. Physical Communication, Elsevier: 40–62.

Blossom, E. 2004. GNU radio: Tools for exploring the radio frequency spectrum. Linux Journal.

Cabric, D. et al. 2004. Implementation issues in spectrum sensing for cognitive radios. In signals, systems and computers IEEE, Vol. 1: 772–776

Cordeiro, C. et al. 2006. IEEE 802.22: An introduction to the first wireless standard based on cognitive radios. Journal of communications, Vol. 1: 38–47.

de Carvalho, F. B. S. et al. 2015. Performance of Cognitive Spectrum Sensing Based on Energy Detector in Fading Channels. International Conference on Communication, Management and Information Technology (ICCMIT): 140–147.

Ejaz, W. et al. 2013. Intelligent spectrum sensing scheme for cognitive radio networks. EURASIP Journal on Wireless Communications and Networking: 1–10.

Ghasemi, A & Sousa, E. S. 2008. Spectrum sensing in cognitive radio networks: requirements, challenges and design Trade-offs. IEEE Communications Magazine: 32–39.

Jiang, C. et al. 2012. Statistical matched filter based robust spectrum sensing in noise uncertainty environment. IEEE 14th International Conference on Communication Technology (ICCT): 1209–1213, Chengdu.

Jaglan, R. R. et al. 2015. Comparative study of single-user spectrum sensing techniques in cognitive radio networks. Second International Symposium on Computer Vision and the Internet (VisionNet'15) Procedia Computer Science 58: 121–128.

Lunden, J. et al. 2009. Collaborative cyclostationary spectrum sensing for cognitive radio systems. IEEE Transaction On Signal Processing, Vol. 57: 4182–4195.

McHenry, M. et al. 2007. XG dynamic spectrum sharing field test results. In Proceeding IEEE International Symposium onNew Frontiers in Dynamic Spectrum Access Networks: 676–684, Dublin, Ireland.

Reddy, Y. B. 2012. Solving hidden terminal problem in cognitive networks using cloud technologies. Sensorcomm, the Sixth International Conference on Sensor Technologies and Applications: 235–240.

Rebeiz, E. et al. 2013. Optimizing wideband cyclostationary spectrum sensing under receiver impairments. IEEE Transaction On Signal Processing, Vol. 61: 3931–3943.

Shabnam & Mahajan. R. 2015. Performance analysis of cyclostationary and energy detection spectrum sensing techniques. International Conference on Signal Processing, Computing and Control (ISPCC): 247–251.

Sharma, A. & Katoch, M. 2015. Analysis of various spectrum sensing techniques in cognitive radio. International Journal of Advanced Research in Computer Science and Software Engineering, Vol. 5: 1100–1104.

Sun, C. et al. 2007. Cooperative Spectrum Sensing for Cognitive Radios under Bandwidth Constraints. IEEE Wireless Communication and Networking Conference: 1–5, Kowloon.

Wang, B. & Ray Liu, K. J. 2011. Advances in cognitive radio networks: A survey. IEEE Journal of Selected Topics in Signal Processing, Vol. 5: 5–23.

Yücek, T & Arslan, H. 2009. A survey of spectrum sensing algorithms for cognitive radio application. IEEE Communications Surveys & Tutorials, Vol. 11: 116–130.

Communication and Computing Systems – Prasad et al. (Eds)
© 2017 Taylor & Francis Group, London, ISBN 978-1-138-02952-1

SKIP: A novel self-guided adaptive clustering approach to prolong lifetime of wireless sensor networks

Ankit Thakkar
Computer Science and Engineering Department, Institute of Technology, Nirma University, Ahmedabad, Gujarat, India

ABSTRACT: Wireless Sensor Network (WSN) consists of energy constraint sensor nodes, and it is difficult to replace or recharge batteries of these nodes when they operate in hostile environments. Hence, prolonging the lifetime of WSN nodes is an important issue for any WSN. Cluster based routing techniques improve the lifetime of WSNs, wherein longer stability and shorter instability periods are important aspects to measure network lifetime. Stable period ensures reliability of the data received from the network, as all nodes are alive during this period. This paper discusses a novel distributed adaptive clustering approach, *Skip* that guides every node to take part in the Cluster Head (CH) election process or not for the current epoch, based on residual energy of a node. Extensive simulations have been carried out to compare the proposed approach with prominent clustering techniques such as Low Energy Adaptive Clustering Hierarchy (LEACH), Stable Election Protocol (SEP) and enhanced Stable Election Protocol (SEP-E). Simulation results show that the proposed approach outperforms to LEACH, SEP and SEP-E protocols. The proposed approach *Skip* provides longer stability and does not deteriorate instability period. The performance of *Skip* is also validated using statistical test.

1 INTRODUCTION

Now days, Wireless Sensor Network (WSN) is gaining popularity among researchers because of the improvement in communication and sensor technology that makes sensor nodes inexpensive (Thakkar and Kotecha 2014b). This encourages use of sensor networks in a wide variety of applications such as military applications, environment monitoring, inventory control, human-centric applications, agriculture applications, structural health monitoring, and many more (Arampatzis et al. 2005). Depending upon the requirement of the underlying application, WSN nodes can be deployed either in a pre-planned manner or randomly. The first approach is called structured WSN, while the other is known as unstructured WSN (Yick et al. 2008). This paper focuses on unstructured WSN. In unstructured WSN, sensor nodes are randomly placed in the Region of Interest (ROI) which is hostile in nature in most cases. In addition, nodes are densely deployed in the ROI to increase reliability of the required information provided by the network. Because of the dense deployment of the WSN nodes in a hostile environment, it is difficult to replace or recharge batteries of the WSN nodes. This makes prolonging the lifetime of the WSN as the prime focus irrespective of the application being served by the network. There are many techniques to prolong

the lifetime of the WSN such as sleep scheduling, MAC protocols, routing protocols, data aggregation, topological control, etc. (Li et al. 2010). This paper mainly focuses on routing techniques to prolong the lifetime of WSNs.

Routing protocols can be classified as flat-based routing, hierarchical-based routing and location-based routing depending upon the underlying network structure used by the routing protocol (Al-Karaki and Kamal 2004). This paper focuses on hierarchical routing schemes as it is scalable and provides energy efficient communications (Al-Karaki and Kamal 2004, Thakkar and Kotecha 2015). In a hierarchical routing scheme, protocol operation is divided into rounds. During a particular round, few nodes elect themselves as Cluster Heads (CHs) and the remaining nodes act as member nodes. The member nodes transmit their data to either a selected CH or Base Station (BS), whichever is energy efficient. This reduces a long distance communication and provides pronged network lifetime (Thakkar and Kotecha 2015, Karkvandi et al. 2011). A group of few rounds makes an epoch. During every epoch, each alive node elects itself as CH once per epoch (Heinzelman et al. 2000, Smaragdakis et al. 2004, Thakkar and Kotecha 2014a, Aderohunmu et al. 2011).

In this paper, a novel self-guided adaptive clustering scheme *Skip* is proposed, which allows/disallows nodes to become CH during the current epoch based on their own energy level and number of members served by it when it was elected as CH in the recent past. The proposed approach *Skip* improves the stability period (the time between network setup and death of the first node (Smaragdakis et al. 2004)) and does not deteriorate instability period (the time between death of the first node and death of the last node (Smaragdakis et al. 2004)). The proposed approach is compared with prominent clustering schemes such as LEACH (Heinzelman et al. 2000), SEP (Smaragdakis et al. 2004) and SEP-E (Aderohunmu et al. 2011).

The major contributions of the paper are as follows:

- A novel self-guided adaptive clustering scheme *Skip* is proposed in the paper, which reduces the probability of electing less energy nodes as CH
- The proposed approach *Skip* is compared with prominent clustering schemes such as LEACH, SEP and SEP-E by doing extensive simulations
- The result have been statistically verified

The rest of the paper organized as follows: Existing cluster-based routing protocols are discussed in section 2, a novel self-guided adaptive clustering scheme *Skip* is discussed in section 3, Simulation parameters and result discussions are given in section 4, and Concluding remarks are given in section 5.

2 RELATED WORK

Low Energy Adaptive Clustering Hierarchy (LEACH) (Heinzelman et al. 2000) is a pioneer cluster-based routing protocol that improves the network lifetime of WSNs. LEACH works well for the homogeneous network[1], but fails to provide sufficiently prolonged network lifetime when works in heterogeneous network environments[2]. This is because of the same CH election probability assigned to all the nodes of a network.

The problem of LEACH is overcome by Stable Election Protocol (SEP) (Smaragdakis et al. 2004) by assigning different probability to the different group of nodes as per their initial energy. SEP has defined new parameter to measure the lifetime of the WSN named stability period and instability period. Stability period refers to the time between the network setup to the death of the first node and instability period refers to the time between the death of the first node and death of a last node. This is mainly important for the applications wherein feedback from the network must be reliable.

Many protocols have been proposed by extending LEACH protocol to enhance the lifetime of the

WSNs using different metrics such as use of coverage preservation (CVLEACH (Thakkar and Kotecha 2012a), CPCHSA (Tsai 2007), WCALEACH (Thakkar and Kotecha 2012c)), use of two level clustering scheme (TL-LEACH (Loscri et al. 2005)), considering energy while electing CHs (ALEACH (Ali et al. 2008), LEACH with deterministic Cluster Head Selection (Handy et al. 2002)), considering number of alive nodes in the network (AL-LEACH (Thakkar and Kotecha 2014a)), use of weighted clustering scheme (WEEC (Behboudi and Abhari 2011), WALEACH (Thakkar and Kotecha 2012b), WCALEACH (Thakkar and Kotecha 2012c)), etc. Also, many protocols have been proposed to extend stability period and to minimize instability period such as E-SEP (Aderohunmu et al. 2009), SEP-E (Aderohunmu et al. 2011), T-SEP (Kashaf et al. 2012), Extended SEP protocol with three levels of clustering (Islam et al. 2012), zonal based stable election protocol (Z-SEP) (Faisal et al. 2013), to name a few.

The proposed approach *Skip* is compared with LEACH (Heinzelman et al. 2000), SEP (Smaragdakis et al. 2004) and SEP-E (Aderohunmu et al. 2011) protocols, and hence, they are discussed in detail in this section.

2.1 Low Energy Adaptive Clustering Hierarchy (LEACH)

LEACH works in rounds, and each round is divided into cluster setup phase and steady state phase. During cluster setup phase, each node assumes a random number *rnd*, between 0 and 1, and it is compared with the threshold value $T(n)$ given by the equation 1. A node elects itself as CH, if *rnd* is less than $T(n)$, otherwise it works as a member node. Once a node elected as CH during a particular round of an epoch, it works as a member node for the remaining rounds of the same epoch.

$$T(n) = \begin{cases} \dfrac{p}{1 - p\left(r \bmod \dfrac{1}{p}\right)} & if \ n \in G \\ 0 & Otherwise \end{cases} \qquad (1)$$

Here G is the set of nodes, which are yet to be elected as CH for the current epoch, r is the current round, p is the optimal cluster head selection probability, and it is 5% for LEACH. During the first round the value of $T(n)$ is 0.05, and it increases with increase in the round number. $T(n)$ attains the value of 1 during the last round of every epoch. Once a node is elected as CH, it informs its status to other nodes. Each non-CH (member) node sends a join message to the selected CH if the selected CH is nearer to the member node as compared to BS;

---

[1]All nodes in the network have same initial energy.
[2]A fraction of the total nodes has higher initial energy than the remaining nodes.

otherwise, node transmits its data to BS directly during steady state phase. After receiving join messages, each CH node prepares a TDMA schedule, and informs to its member nodes. During the steady state phase, each node sends its data either to BS or the selected CH, whichever is energy efficient. Each CH node aggregates the data and sends it to BS, after receiving data from its member nodes.

2.2 Stable Election Protocol (SEP)

Like LEACH, SEP also works in rounds. As stated earlier, LEACH does not work well in heterogeneous network environments because the same optimal cluster head election probability (p) is assigned to all the nodes. SEP assigns α times higher energy to m fraction of total nodes (n) called advanced nodes and the remaining nodes are called as normal nodes. SEP defines different optimal cluster head election probabilities for advanced nodes and normal nodes, which is given by equation 2 and 3 respectively.

$$p_{adv} = \frac{p_{opt}(1+\alpha)}{1+\alpha m} \tag{2}$$

$$p_{nrm} = \frac{p_{opt}}{1+\alpha m} \tag{3}$$

Here, p_{adv} and p_{nrm} denote the optimal CH election probability for the advanced nodes and normal nodes respectively. The threshold value for advanced nodes and normal nodes are also different, and it is given by equation 4 and 5 respectively.

$$T(S_{adv}) = \begin{cases} \dfrac{p_{adv}}{1-p_{adv}\left(r\bmod\dfrac{1}{p_{adv}}\right)} & \text{if } n \in G'' \\ 0 & \text{Otherwise} \end{cases} \tag{4}$$

$$T(S_{nrm}) = \begin{cases} \dfrac{p_{nrm}}{1-p_{nrm}\left(r\bmod\dfrac{1}{p_{nrm}}\right)} & \text{if } n \in G'' \\ 0 & \text{Otherwise} \end{cases} \tag{5}$$

Here, $T(S_{adv})$ and $T(S_{nrm})$ denote the threshold value for advanced nodes and normal nodes respectively. G'' is the set of nodes, which are yet to be elected as CH for the current epoch.

2.3 Enhanced Stable Election Protocol (SEP-E)

SEP-E defines three level of nodes' heterogeneity by extending SEP protocol. In SEP-E, nodes are classified in three groups named normal nodes, advanced nodes and intermediate nodes depending upon the initial energy given to the nodes. Intermediate nodes have initial energy higher than

normal nodes, but lower energy than advanced nodes. SEP-E defines three different optimal clustering probabilities for advanced, normal and intermediate nodes.

SEP-E assigns α times higher energy to m fraction of total nodes (n) called advanced nodes, β times higher energy to μ fraction of total nodes (n) called intermediate nodes, and the remaining nodes are called as normal nodes. SEP-E defines different optimal cluster head election probabilities for advanced nodes, intermediate nodes and normal nodes, which is given by equation 6, 7 and 8 respectively.

$$p_{adv} = \frac{p_{opt}(1+\alpha)}{1+\alpha m+\beta\mu} \tag{6}$$

$$p_{int} = \frac{p_{opt}(1+\mu)}{1+\alpha m+\beta\mu} \tag{7}$$

$$p_{nrm} = \frac{p_{opt}}{1+\alpha m+\beta\mu} \tag{8}$$

Here, p_{adv}, p_{int} and p_{nrm} denote the optimal CH election probability for the advanced nodes, intermediate nodes and normal nodes respectively and p_{opt} is the optimal CH election probability, which is 10% for SEP and SEP-E. The threshold value for advanced nodes, intermediate nodes and normal nodes are given by equation 4, 9 and 5 respectively. SEP-E takes the value of p_{adv}, p_{int}, and p_{nrm} defined by the equation 6, 7 and 8 respectively.

$$T(S_{int}) = \begin{cases} \dfrac{p_{int}}{1-p_{int}\left(r\bmod\dfrac{1}{p_{int}}\right)} & \text{if } n \in G'' \\ 0 & \text{Otherwise} \end{cases} \tag{9}$$

Here, $T(S_{int})$ denotes the threshold value for intermediate nodes.

3 SKIP PROTOCOL

Skip protocol works in rounds like prominent clustering schemes. However, Skip has proposed a new threshold value by extending LEACH, SEP and SEP-E, and it is discussed later in this section. Apart from this modification, Skip works similar to LEACH, SEP and SEP-E.

LEACH, SEP and SEP-E allows each alive node to become CH once per epoch. This results in insufficient improvement in the network lifetime as residual energy of a node is not considered while electing the node as CH. This results in the early death of a node. To overcome this issue of LEACH, SEP and SEP-E protocols, threshold value(s) of these protocols are modified, and it is given by equation 10.

$$T(n*_{skip}) = \begin{cases} T(n*).Er_i & if\ n \in G \\ 0 & Otherwise \end{cases} \quad (10)$$

Here $T(n*)$ refers to the threshold value of nodes defined by LEACH, SEP and SEP-E, and Er_i is given by the equation 11.

$$Er_i = \left(\frac{E_{icurrent}}{E_{iinit}} \right)^{\frac{1}{1+ESC_i}} \quad (11)$$

Here, $E_{icurrent}$ and E_{iinit} denote the current energy and initial energy of a node i, ESC_i is the controlling parameter for a node i that increases/reduces threshold value of a node i over a period of time, and it is given by equation 12.

$$ESC_i = \begin{cases} ESCLast_i + \dfrac{E_{iinit}}{E_{icurrent}} & \begin{array}{l} if\ ((i \in CH_{last} \\ And\ Mmbr_i < Th) \\ OR(i \notin CH_{last})) \end{array} \\ 0 & \begin{array}{l} if\ i \in CH_{last} \\ And\ Mmbr_i \geq Th \end{array} \end{cases} \quad (12)$$

Here, $ESCLast_i$ denotes the maximum value of ESC_i when a node i is not elected as CH. The value of $ESCLast_i$ and ESC_i is calculated once per epoch. It should be noted that $ESCLast_i$ is updated and takes the value of ESC_i only when the node i is not elected as CH in the last epoch as shown in the equation 12 by the second part of the "**OR**" condition. The initial value for ESC_i and $ESCLast_i$ is set to 0 for all the nodes at the time of network setup. The value of Th is same for all nodes, and it is given by $p_{min} * 100$. Here, p_{min} is the minimum value of the optimal CH election probability. For LEACH, p_{min} gets the value of p; and for SEP and SEP-E, p_{min} gets the value of p_{nrm}. This ensures that each CH must be having minimum required number of member nodes associated with it, failing to which may indicate that a few nodes remains alive in the network. Hence, the CH node cannot be disallowed to skip itself for taking part in CH election process for the next epoch.

The threshold value (see equation 10) reduces as energy of a node drains out over a period. Hence, after some epoch, a node i is not able to become CH for the first time. This is because its threshold value is very small, and rnd generated during the epoch is not less than the threshold. Hence, the value of ESC_i and $ESCLast_i$ increased by $E_{iinit}/E_{icurrent}$. This will increase the threshold value of node i as compared to the previous value. If node i is not selected as CH in the next epoch, ESC_i is further updated, which results in increase of the threshold value of

node i. The value of ESC_i is set to zero when node i is elected as CH during the last epoch, and has required number of member nodes associated with it. It guides to node i that the sufficient number of nodes is still alive in the network, and the node i is allowed to escape from the CH election process. If a node i is elected as CH and number of member nodes associated with the node i are less than Th during the last epoch, then it guides to node i that a few number of nodes are alive in the network, and hence, it has to update its ESC_i value so that it can be elected as CH in the next epoch.

4 SIMULATION PARAMETERS AND RESULT DISCUSSIONS

4.1 Assumptions about the network model

The proposed approach *Skip* is tested under following assumptions as these assumptions are also used to test LEACH (Heinzelman et al. 2000) and SEP (Smaragdakis et al. 2004).
- Nodes are randomly placed in a two-dimensional space with uniform distribution
- BS is placed at the center of the node deployment area
- During each round, each node has some data that should be sent to the BS
- Each node is aware about the location of the BS
- Sensor nodes are energy constrained
- There is no energy constraint for the BS
- Each node is able to transmit at different power levels, depending upon the communication distance between sender and receiver
- *Skip* is tested using the first order radio energy model, which is presented in (Heinzelman et al. 2000, Smaragdakis et al. 2004). A radio dissipates the energy E_{Tx} to transmit l-bit message over a distance d meter is given by equation *13*.

$$E_{Tx} = \begin{cases} l.E_{elect} + l.\varepsilon_{fs}.d^2 & if\ d < d_0 \\ l.E_{elect} + l.\varepsilon_{mp}.d^4 & Otherwise \end{cases} \quad (13)$$

Here, ε_{fs} and ε_{mp} are the free space and multipath channel constants, E_{elect} denotes the energy required per bit to run transmitter or receiver, and d_0 is a cross-over distance. A radio spends energy E_{Rx} to receive l-bit message is given by $l.E_{elect}$.

4.2 Simulation parameters

Skip is tested using MATLAB. Parameters used for simulations are shown in the Table 1. The optimal CH selection probability, p is set to 5% for LEACH and p_{opt} is set to 10% for both SEP and SEP-E.

338

4.3 Result discussions

The proposed approach *Skip* is integrated with LEACH, SEP and SEP-E protocol, and this integrated approach is called as LEACH_Skip, SEP_Skip and SEP-E_Skip respectively. As discussed in the section 3, the proposed approach *Skip* improves the stability period and does not deteriorate instability period. It can be evident through Figures 1–4 for an initial energy of 0.5 J and 1 J. The controlling parameter ESC_i plays a key role to increase/decrease threshold value of the node i depending upon ratio of the current and initial energy levels and number of member nodes associated with it when node i was elected as CH.

4.4 Statistical significance

Wilcoxon signed-rank test is used to verify statistical significance of the empirical results. It is a

Table 1. Parameters used for simulations.

Parameter name	Value
Node deployment area	100 m × 100 m
Number of Nodes (n)	100, 200 and 500
Initial Energy/Node (E_0)	0.5 J and 1 J
Simulation Time	Till death of a last node
Sink position	at (50 m, 50 m)
Packet Size	4000 bits
Additional bits for $T(n^*_{Skip})$	64 bits (Hollasch)
E_{elect}	50 nJ/bit
ε_{fs}	10 pJ/bit/m²
ε_{mp}	0.0013 pJ/bit/m^4
Optimal CH election probability	5% (for LEACH) and 10% (for both SEP-E and SEP)
Simulation runs	Each protocol runs six times for a given n and E_o

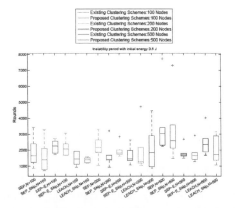

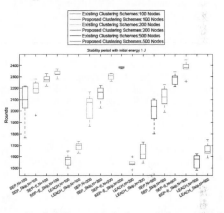

Figure 2. Instability period for existing clustering schemes (LEACH, SEP, SEP-E) with the integration of proposed scheme *Skip* for different node densities with initial energy 0.5 J.

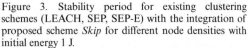

Figure 3. Stability period for existing clustering schemes (LEACH, SEP, SEP-E) with the integration of proposed scheme *Skip* for different node densities with initial energy 1 J.

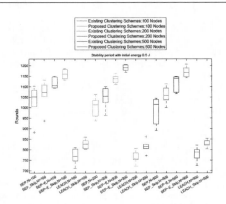

Figure 1. Stability period for existing clustering schemes (LEACH, SEP, SEP-E) with the integration of proposed scheme *Skip* for different node densities with initial energy 0.5 J.

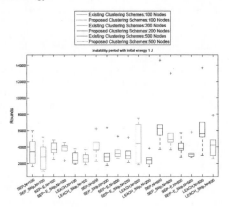

Figure 4. Instability period for existing clustering schemes (LEACH, SEP, SEP-E) with the integration of proposed scheme *Skip* for different node densities with initial energy 1 J.

nonparametric test to test for the median differences for two populations when the observations paired. The results of this test conclude that the median of stability period improves with the proposed approach with 95% probability.

5 CONCLUDING REMARKS

A self-guided distributed adaptive clustering approach *Skip* is proposed in this paper. The proposed scheme enhances stability period and does not deteriorate instability period, which are important parameters for applications where data reliability is a key requirement. The proposed approach has been tested by doing extensive simulations. Empirical results are also verified through statistical test. *Skip* works for homogeneous as well as heterogeneous networks. The simulation results provide sufficient insights to the researchers before doing real test-bed implementations.

REFERENCES

Aderohunmu, F. A., J. D. Deng, et al. (2009). An enhanced Stable Election Protocol (SEP) for clustered heterogeneous WSN. Department of Information Science, University of Otago, New Zealand.

Aderohunmu, F. A., J. D. Deng, & M. Purvis (2011). Enhancing clustering in wireless sensor networks with energy heterogeneity. IGI Global, 18–31.

Al-Karaki, J. N. & A. E. Kamal (2004). Routing techniques in wireless sensor networks: a survey. Wireless communications, IEEE 11(6), 6–28.

Ali, M. S., T. Dey, & R. Biswas (2008). ALEACH: Advanced LEACH routing protocol for wireless microsensor networks. In Electrical and Computer Engineering, 2008. ICECE 2008. International Conference on, pp. 909–914. IEEE.

Arampatzis, T., J. Lygeros, & S. Manesis (2005). A survey of applications of wireless sensors and wireless sensor networks. In Intelligent Control, 2005. Proceedings of the 2005 IEEE International Symposium on, Mediterrean Conference on Control and Automation, pp. 719–724. IEEE.

Behboudi, N. & A. Abhari (2011). A Weighted Energy Efficient Clustering (WEEC) forWireless Sensor Networks. In Mobile Ad-hoc and Sensor Networks (MSN), 2011 Seventh International Conference on, pp. 146–151. IEEE.

Faisal, S., N. Javaid, A. Javaid, M. Khan, S. H. Bouk, & Z. Khan (2013). Z-SEP: Zonal-stable election protocol for wireless sensor networks. arXiv preprint arXiv:1303.5364.

Handy, M., M. Haase, & D. Timmermann (2002). Low energy adaptive clustering hierarchy with deterministic cluster-head selection. In Mobile and Wireless Communications Network, 2002. 4th International Workshop on, pp. 368–372. IEEE.

Heinzelman, W. R., A. Chandrakasan, & H. Balakrishnan (2000). Energy-efficient communication protocol for wireless microsensor networks. In System sciences, 2000. Proceedings of the 33rd annual Hawaii international conference on, pp. 10p. IEEE.

Hollasch, S. IEEE Standard 754 Floating Point Numbers. http://steve.hollasch.net/cgindex/coding/ieee-float.html/. [Online; accessed 25-March-2016].

Islam, M. M., M. A. Matin, & T. Mondol (2012). Extended Stable Election Protocol (SEP) for three-level hierarchical clustered heterogeneous WSN. In Wireless Sensor Systems (WSS2012), IET Conference on, pp. 1–4. IET.

Karkvandi, H. R., E. Pecht, & O. Yadid-Pecht (2011). Effective lifetime-aware routing in wireless sensor networks. Sensors Journal, IEEE 11(12), 3359–3367.

Kashaf, A., N. Javaid, Z. A. Khan, & I. A. Khan (2012). TSEP: Threshold-sensitive stable election protocol for WSNs. In Frontiers of Information Technology (FIT), 2012 10th International Conference on, pp. 164–168. IEEE.

Li, W., M. Bandai, & T. Watanabe (2010). Tradeoffs among delay, energy and accuracy of partial data aggregation in wireless sensor networks. In Advanced Information Networking and Applications (AINA), 2010 24th IEEE International Conference on, pp. 917–924. IEEE.

Loscri, V., G. Morabito, & S. Marano (2005). A two-levels hierarchy for low-energy adaptive clustering hierarchy (TLLEACH). In IEEE Vehicular Technology Conference, Volume 62, pp. 1809. IEEE; 1999.

Smaragdakis, G., I. Matta, A. Bestavros, et al. (2004). SEP: A stable election protocol for clustered heterogeneous wireless sensor networks. In Second international workshop on sensor and actor network protocols and applications (SANPA 2004), Volume 3.

Thakkar, A. & K. Kotecha (2012a). CVLEACH: Coverage based energy efficient LEACH Algorithm. International Journal of Computer Science and Network (IJCSN) Volume 1.

Thakkar, A. & K. Kotecha (2012b). WALEACH: Weight based energy efficient Advanced LEACH algorithm. In 4th International Workshop on Wireless and Mobile Networks (WiMo- 2012), Volume 2.

Thakkar, A. & K. Kotecha (2012c). WCVALEACH:Weight and Coverage based energy efficient Advanced LEACH algorithm. Computer Science & Engineering 2(6), 51.

Thakkar, A. & K. Kotecha (2014a). Alive nodes based improved low energy adaptive clustering hierarchy for wireless sensor network. In Advanced Computing, Networking and Informatics-Volume 2, pp. 51–58. Springer.

Thakkar, A. & K. Kotecha (2014b). Cluster head election for energy and delay constraint applications of wireless sensor network. Sensors Journal, IEEE 14(8), 2658–2664.

Thakkar, A. & K. Kotecha (2015). A new bollinger band based energy efficient routing for clustered wireless sensor network. Applied Soft Computing 32, 144–153.

Tsai, Y.-R. (2007). Coverage-preserving routing protocols for randomly distributed wireless sensor networks. Wireless Communications, IEEE Transactions on 6(4), 1240–1245.

Yick, J., B. Mukherjee, & D. Ghosal (2008). Wireless sensor network survey. Computer networks 52(12), 2292–2330.

Communication and Computing Systems – Prasad et al. (Eds)
© 2017 Taylor & Francis Group, London, ISBN 978-1-138-02952-1

Impact of CBIR journey in satellite imaging

Chandani Joshi, G.N. Purohit & Saurabh Mukherjee
Banasthali University, Banasthali, India

ABSTRACT: Content-based image retrieval system has shown a massive growth in the past decades. Content Based Image Retrieval (CBIR) is a technique which uses visual features of image such as color, shape, texture, etc., to retrieve the relevant images from the database. Earlier TBIR system was used for the retrieval of relevant images but its ineffectiveness has raised the CBIR system. A lots of research works had been completed in the past decade to design efficient image retrieval techniques from the image or multimedia databases. This paper focuses on the major feature extraction techniques and methods since from the inception of the CBIR.

Keywords: TBIR, CBIR, ANN, FNN

1 INTRODUCTION

The collection of images has increased tremendously during the last decade owing to the rapid growth of the Internet. Image capturing has become easy in the areas such as geography, medicine, architecture, advertising, design, and satellite which led to the creation of huge databases of the images. Now there is a need of a system to retrieve those images from the massive database. Due to this the study of image retrieval became more interesting and challenging to retrieve the desired images from the huge and varied database. Image retrieval is concerned with techniques for storing and retrieving images both efficiently and effectively.

Earlier the images were retrieved by matching the keywords that were assigned to each image manually. However, manual processing became impractical due to the large collection of the images; every user had a different perception of images, so it becomes difficult to retrieve the correct image by a Keyword Search. Sometimes a typing mistake in assigning the keywords for images also retrieves undesired images. The National Science Foundation of the United States organized a workshop on Visual Information Management Systems to identify new directions in image database management systems in 1992. It was widely recognized that the visual content would be helpful for the problems associated. The concept of Content Based Image Retrieval came into existence in the workshop.

Content Based Image Retrieval (CBIR) is a technique to retrieve images on the basis of visual content such as color, shapes and textures (Saad 2008). "With the help of single content let's say

color we would not be able to get the optimized result because of the similarity of the color of different objects. For this along with the color, the texture or shape or both should be combined to retrieve similar result. Color, textures, shapes are still low level feature and they should be used along with the high level features like text annotation for the optimized results" (Mallick et al. 2014). "By extracting the, descriptors, histograms, colors, shapes, textures, etc., the content of the image is analyzed. The performance or the accuracy of the retrieved images can be calculated by some of the methods available such as by the Precision and recall or by Length of String to Recover All Relevant Images i.e. LSRR" (Pal et al. 2006, Kekre et al. 2010). The problem faced with the low level feature extraction was the semantic gap. This was due to the difference between the low level features connected with the high level user semantics. It was difficult to convert the user need for the image in a complete manner to a Content Based Image Retrieval (CBIR). Due to this problem, the images retrieved would not be more effective and efficient. Through extensive research, Content Based Image Retrieval came into existence in 1992 and since then many systems have been developed for Content Based Image Retrieval for uses in commercial fields. Some of these systems are briefly mentioned in the following paragraph.

This field of image retrieval is attracting substantial number of researchers, however till now the use of Content Based Image Retrieval is not very common and still in infancy. Many big organizations's search engine such as Yahoo Image Search and Google Image Search still depends heavily on the metadata (Banda et al. 2013). Due to which

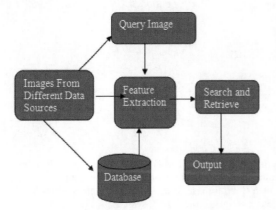

Figure 1. Existing model of content based image retrieval.

their performance is not very efficient, particularly when we come for the complicated searches such as a particular region within the images.

The basic Content Based Image Retrieval (CBIR) Model is shown below:

2 LITERATURE SURVEY

2.1 1995–2000

Since the inception of CBIR system the following best known systems were developed such as Query by Image Content (QBIC) (Flickner et al. 1995) and Photo-book (Rui et al. 1997) and its new version Four-Eyes. Other well-known systems are the searchengine family Visual-SEEk, Meta-SEEk and Web-SEEk (Bach et al., 1996), NETRA, Multimedia Analysis and Retrieval System (MARS) (Honkela et al., 1997) (Hui et al. 2009).

Stefanidis in 1998 proposed a new method, Intelligent Query (IQ) to retrieve the images from the database. The images were obtained on the basis of the semantic method rather than the metadata. The sketch created by the human was provided as the query image and similar images to that of a sketch were extracted from the database. The extracted images were ranked according to their similarity. If the result obtained was not so accurate than the sketch was again modified and the same process was repeated.

The results shown by the IQ were very accurate, but it could have been more effective if the images were retrieved with some more semantic features.

2.2 2001–2005

(Zhang 2002) have proposed a system to generate the mosaic of the system, which is called Mosaicture. The system is implemented using the color feature of the database image, using the color histogram values followed by binary signature.

In this work all the features are not addressed. So accuracy of the system is not so high.

(Srikrishna et al. 2004) have retrieved the images based on the concept of Fuzzy-Neural hybrid system. A Feed Forward Back Propagation Neural Network (FNN) was used for the image classification. The images were retrieved using the low level feature extraction. The attempt was made to show the maximum number of details of an image as provided by its histogram. For reorganization, Fuzzy Logic Approach (FLA) and Artificial Neural Network Approach (ANN) were used. They established that the histogram of Discrete Cosine coefficients, rather than the conventional intensity histogram, is a better measure for image details.

The obtained result was accurate, but the study was done on the basis of the low level feature extraction, but it would have been better if the high level feature extraction have been used along with the low level.

(Hafiane et al. 2005) retrieved the satellite images on the basis of the region, using the Motif Co-occurrence Matrix (MCM) in conjunction with spatial relationships. The image was decomposed into segments and then the MCM was computed for each region. The spatial relationship was evaluated using the a* algorithm. The images were represented by the Attribute Relational Graph (ARG). At the time of retrieval each image in the database was divided into different regions by the defined methods. Then the similarity measures were done on the basis of visual and the spatial characteristics. The performance of the result obtained was measured by the precision and recall method.

The obtained results would have been more effective if the combination of the semantic feature have been used.

2.3 2006–2010

(Hiremath et al. 2006) have retrieved the images using the texture and color analysis, Four different approaches of color texture are used for the image classification from the VisTex database. In this paper the experimental results were acquired in different color spaces such as RGB, RGB, HSV and YCbCr and are compared with that in gray scale. The results show that RGB provides better classification results. For texture classification Haar Wavelet are used as it had shown the encouraging results.

All the features are not addressed in the classification of the result. The precision and recall value is not so high.

(Ning et al. 2006) developed a web based application through which the user can find the images by the query being raised. The domain-dependent concept for the image retrieval was used using ontology approach. The user queries raised were based on the satellite images attributes such as sensor type, the time of the image captured, etc. The experiment was performed on the LANDSAT TM imagery for the water body. The result was prioritized using the prioritizing algorithm.

So far, the images were retrieved using the low level and the high level feature extraction method. But the author (Gao et al. 2007) have used measurements for the image retrieval. The three measurements are the interval scale, ordinal scale, and ratio scale. The method used to find the performance of semantic features with different measurement scales was Average Normalized Modified Retrieval Rank (ANMR). The result showed that ordinal measurement scale was effective for image retrieval among the three measurement scales.

A new technique was proposed by (Miguel et al. 2008) for the supervised classified images. In case of supervised classification, through the known categories and through the ground truth data the correct result could be validated, but if we have lack of ground truth data than the technique used in this paper could be used. The obtained result was validated on the unsupervised image classification using the Data-Driven Quality Map Assessment (DAMA) technique.

(Maheshwary et al. 2009) have used the 3 LISS III + multi-spectral satellite images with 23.5 m resolution. With the features like color and texture the four semantic categories such as mountain, vegetation, water bodies, and residential area, were used for the retrieval of the similar images from the database. The near-IR (infra-red), red and green bands as the three spectral channels. The following techniques were used for the image extraction and matching: the Color Moment was used for the Feature Extraction and GLCM was used for Texture Feature Extraction followed by K-means clustering to form index and then the images were retrieved using the query image and the images stored in the database. The entropy was small for the GLCM method used for textually uniform images.

The experiment was performed on images obtained by LISS III sensor and multi-spectral images, the wide variety of images were not used.

(Maheshwary 2010) have tried to bridge the semantic gap using the semantic category, such as field, water and vegetation for image retrieval. In this paper the two LISS III images in addition with multi-spectral satellite images with 23.5 m resolution have been used. The similarity measures have been performed by the color moments, GLSM, NDVI and using the combined features. The performance was validated by using the Precision & Recall.

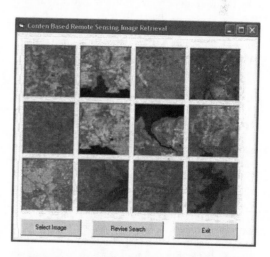

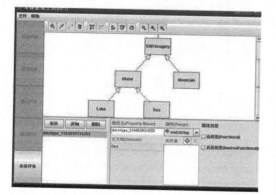

Figure 2. Interface for ontology building tool (Ning 2006).

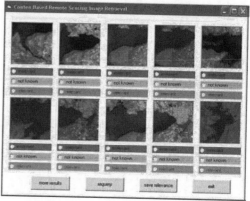

Figure 3(a), (b). CBIR system (Maheshwary & Srivastava 2009).

(Mamatha et al. 2010) have obtained the images with the low level and along with the high level feature extraction techniques. The two types of satellite images were taken, one was of the urbanized area and another one was of rural area. The color based content processing technique and the histogram technique was applied on the images. It was found that the color based content processing technique was useful for the urbanized area, however not for the rural area because there is no variation in the features in the rural area. So, for the rural area histogram provided the better result.

2.4 2011–till date

In another research paper, (Mamatha et al. 2011) used the color feature for the similarity measure for the image retrieval. A low resolution satellite image of the rural area had been taken for the experiment. The complete images were divided into the four quadrants. Initially two quadrants were selected and the content based retrieval of the image feature was applied on both the quadrants. Than the color classification were done, through this the four semantic features were identified, such as Natural Vegetation, Water Bodies, Land, and Housing. The identified features were estimated using L*a*b color spectral distribution and histogram techniques. At the end the same process was applied for all the four quadrants. The research was useful for Resource Development and Management in the Rural Regions by estimating natural resources available in the rural area.

(Bhandari et al. 2013) decomposed the images based on spatial and spectral heterogeneity. The features were extracted based on the visual feature, spatial relationship, semantic, scene semantic and object semantic basis. Then the images were retrieved based on the mapping and the SS modeling. In the end the computed result was validated with the traditional SBRSIR technique. The performance was evaluated using the methods like precision, recall and F-score. The result showed that the Remote Sensing Image Retrieval System (RSIR) has provided more precise images as compared to the conventional SBRSIR technique.

(Fernández et al. 2014) have work related to Oceanography. The objective was the retrieval of the mesoscale structures in the oceans through CBIR technique. The requirement for the study has been raised due to the occurrence of the natural hazards in the oceans. Through the detection of the similar images the researchers would be able to detect the hazard prone area in the ocean. The classifications of the input images were done using the neurofuzzy logic. The comparative study of the classification techniques was done and found that the runtime executions for the classification using neorfuzzy and image

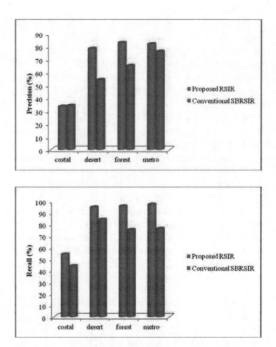

Figure 4.   Graph based on visual feature comparison.

retrieval of regions of interest was less than 0.001 sec. The images were gathered from the Advanced Very High Resolution Radiometer (AVHRR) sensors from the National Oceanic and Atmospheric Administration (NOAA) were classified.

The study was performed on the images acquired from the single sensor. The results would have been more effective if it would have been validated by the different sensors.

(Bouteldja et al. 2014) have designed a system using texture feature for high resolution satellite images. They have used a local binary feature and a block based scheme for the image retrieval. The obtained results were very accurate. But the high resolution satellite images contained the texture as well as the structure. If structural features would have been included then the results could have been much more accurate.

3 CONCLUSION

Research in Content-Based Image Retrieval (CBIR) in the past has been focused on image processing, extraction of low-level features, etc. Extensive experiments on CBIR systems demonstrate that low-level image features cannot always describe high-level semantic concepts in the users' mind. It is believed that CBIR systems should provide maximum support for bridging the 'semantic gap' between low-level visual features and the richness of human semantics. This paper provides a comprehensive survey of recentwork towards narrowing down the 'semantic gap'.

REFERENCES

Danish M., Rawat R., Sharma R., [2013], "A Survey: Content Based Image Retrieval Based On Color, Texture, Shape & Neuro Fuzzy" Journal of Engineering Research and Applications, Volume 3.

Ding Hui, Pan Wei, Guan Yong [2009], "Image Acquisition, Storage and Retrieval", Image Processing, Book edited by: Yung-Sheng Chen, ISBN 978-953-307-026-1, pp. 5.

Dr. Kekre H. B., Mishra D., Kariwala A., [2010], "A Survey of CBIR Techniques and Semantics", International Journal of Engineering Science and Technology (IJEST).

Hafiane A., Chaudhuri S., Seetharaman G., Zavidovique B., [2005], "Region-based CBIR in GIS with local space filling curves to spatial representation", Pattern Recognition Letters 27.

Hiremath, P.S., S. Shivashankar, and Jagadeesh Pujari, [2006] "Wavelet Based Features for Color Texture ClassificationwithApplication to CBIR", IJCSNS International Journal of Computer Science and Network Security, Vol. 6 No. 9 A.

Maheshwary P., Sricastava N., [2009] "Prototype System for Retrieval of Remote Sensing Images based on Color Moment and Gray Level Co-Occurrence Matrix" International Journal of Computer Science Issues, Voume 3.

Mallick A., Kapgate D., Vaidya N. [2014], "A Review on Feature Extraction Techniques for CBIR" IJCAT (International Journal of Computing and Technology), Volume 1.

Mallick A., Kapgate D., Vaidya N. [2014], "A Review on Feature Extraction Techniques for CBIR" IJCAT (International Journal of Computing and Technology), Volume 1.

Mamatha Y. N., Ananth A. G., [2011], "Feature Extraction from Rural Satellite Imagery Using Color Based CBIR Techniques", International Journal of Software Engineering & Applications (IJSEA), Volume 2.

Neha, [2014], "Content Based Image Retrieval: A Review", International Journal of Science, Engineering and Technology Research (IJSETR), Volume 3.

Ning R., Ning H., Hong W., [2006], "Semantic-Based Image Retrieval in Remote Sensing Archive: An Ontology Approach", IEEE.

Pal M.S., Dr. Garg S.K., [2006], "Image Retrieval: A Literature Review", International Journal of Advanced Research in Computer Engineering and Technology (IJARCET), Volume 2.

Saad M., [2008], "Content Based Image Retrieval Literature Survey", EE 381 K: Multi Dimensional Digital Signal Processing.

Yue Zhang, [2002] "On the use of CBIR in Image Mosaic Generation", Technical Report TR 02-17.

Communication and Computing Systems – Prasad et al. (Eds)
© 2017 Taylor & Francis Group, London, ISBN 978-1-138-02952-1

Generation of spectral efficient modulation format for fiber transmission by hybridization of duo binary and Manchester

Sarita
National Institute of Technical Teachers Training and Research, Chandigarh, India

Sandeep Vishwakarma
Computer Science and Engineering, Raj Kumar Goel Institute of Technology, India

Maitreyee Dutta
National Institute of Technical Teachers Training and Research Chandigarh, India

ABSTRACT: In the coming generation of optical fiber communication systems, very high data rate like 20 Gb/s, 40 Gb/s up to 100 Gb/s per channel is attractive. In addition to send more channels into a single fiber, channel spacing has to be decreased from 250GHz to 50GHz or smaller. The result of that linear and nonlinear degrading effect will be severe in such high-data rate optical systems. Several different modulation formats are compared and spectral efficient modulation format in high speed data rate lightwave systems has been generated. The modulation formats: NRZ, RZ, Duobinary, Manchester, and Manchester Duobinaryare under research. First of all, system performances of NRZ, RZ, Duobinary and Manchester over several existing transmission fibers are compared at different data rate, length and laser power. We found that the dominant degrading effect is dependent on laser power, data rate andlength. Then we will discuss about Manchester Duobinary. After that we will judge the effect of different parameter on receiver sensitivity (Q) and spectrum width. Manchester Duobinary is found to be the most spectral efficient among the investigated modulation formats. And finally, we will discuss how we can improve the Quality factor of Manchester Duobinary when data rate increase for future research.

1 INTRODUCTION

Optical fiber communication has an important role in the progress of high-speedand high quality communication systems. At the present time optical fiber are not only used in telecommunication but also used in the internet and local area networks to achieve high signaling rates. For transmitting data or information from one place to another place the use of light is an old technique. In 790 BC., the Greeks used smoke and fire signals for sending information like call for help, alert against enemy, victory in a war, etc. Mostly only one type of signal was conveyed. There was very slow development in optical communication up to the starting of the 19th century. In the old optical communication system, the speed of the optical communication link was limited due to unreliable nature of transmission paths affected by atmospheric effects such as fog and rain (M. Arumugam 2001). The bit rate distance product is only about 1 (bit/s)-km due to enormous transmission loss (105 to 107 dB/km).

As we have studied in survey that narrow bandwidth optical filter at the receiver side can greatly improve the receiver sensitivity of a non return-to-zero Duobinary system. We also studied that when we generate hybrid modulation format by using two existing formats, then hybrid modulation format show some interesting result ex-improve dispersion tolerance, improve spectral efficiency or reduced in chirping effect or many more effect. Some hybrid modulation formats are NRZ-DPSK, PSK-RZ-DPSK, RZ-Duobinary, and NRZ-QPSK. So our aim is to generate spectral efficient modulation format so that bandwidth requirement to transmit the data is reduced or we can also say that we can transmit more data into available bandwidth. Because the demand of the bandwidth requirement is increasing day by day, so we have requirement of such type of modulation format which required less bandwidth to transmit the same data as compare to available formats.

2 MODULATION FORMATS

2.1 Non-Return-to-Zero (NRZ)

NRZ line code is a binary code in which 1's are represented by one significant condition (commonly a positive voltage) and 0 s are represented by some other significant condition (commonly a negative voltage), with no other rest condition. The NRZ pulses contain more energy than a return-to-zero RZ code (Chongjin Xie et al. 2004- Hoon Kim et al. 2002).

Figure 1 shows block diagram of Non-Return-to-Zero (NRZ) has been the dominant modulation format in IM/DD fiber optical communication systems for a long time. There are various reasons for using Non-Return-to-Zero (NRZ) in the early days of fiber-optical communication. The simulation setup of NRZ shown in Figure 2.

- NRZ requires a relatively low electrical bandwidth as compared to RZ for the transmitters and receivers.
- NRZ is not sensitive to laser phase noise compared to PSK.
- NRZ has the simplest configuration of transceivers [9].

The spectrum of NRZ with 10GB/s of data rate at 4 mw laser power and 120 km transmission distance. From above Figure 3 we observed that main lobe is

between 193.09 to 193.11 THz. So the bandwidth requirement for NRZ is equal to.02 THz or 20 GHz. In general, NRZ modulated optical signal has the most compact spectrum compared with other modulation formats. Though, this does not mean that NRZ has best resistance to residual chromatic dispersion in an amplified fiber system with dispersion compensation (Lee Jaehoon et al. 2004).

2.2 Return to Zero (RZ)

Return-to-Zero (RZ) describes a line code used in telecommunication signals in which the signal drops (returns) to zero between each pulse. RZ have half energy required compared to NRZ. Usually a clock signal with the same data rate as electrical signal is used to generate RZ shape of optical signals. Figure 4 shows the block diagram of RZ transmitter and Figure 5 shows simulation of RZ transmission. Initially NRZ optical signal is generated by an external intensity modulator as we described in above discussion. Then, it is modulated by a pulse train with the same data rate as the electrical signal using another intensity modulator. RZ optical signal has been found to be more tolerant to nonlin-

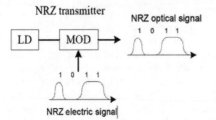

Figure 1.   Block diagram of NRZ transmitter.

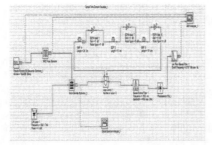

Figure 2.   Simulation set up of NRZ transmitter.

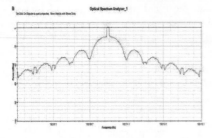

Figure 3.   NRZ transmitter spectrum.

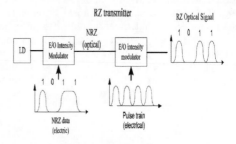

Figure 4.   Block diagram of RZ transmitter.

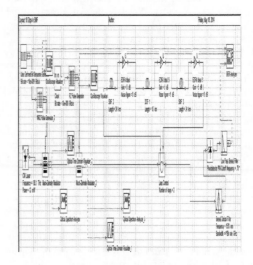

Figure 5.   Simulation set up of NRZ transmitter.

earity than NRZ light wave signal. The reason for RZ superior resistance to nonlinearity than above discussed NRZ is possibly due to its regular data pattern of optical signal. (Yi-Hsiang Wang 2011) (Ilya Lyubomirsky et al. 2005).

The spectrum of RZ with 10GB/s of data rate at 4 mw laser power and 120 km transmission distance. From above Figure 6 we observed that main lobe lies between 193.08 to 193.12 THz centre about 193.10 THz. So the bandwidth requirement for RZ is equal to.04 THz or 40 GHz. At the same parameters bandwidth of NRZ was 20 GHz In general, NRZ modulated optical signal has the most compact spectrum compared to that with other modulation formats. But power requirement to transmit the same data in RZ is approximately half to the NRZ. Since spectral width of RZ is two times to NRZ so bandwidth requirement to transmit the sane data is double as compare to NRZ (Rahul Chhilar 2011- Gilad Goldfarb 2006).

2.3 Duobinary

Duobinary has been introducing into Optical transmission since mid–90. The main feature of Duobinary modulation is a strong correlation between consecutive bits, which results in a more compressed spectrum. It can either be generated as a 3-level amplitude signal ('0', '0.5', '1') or as a 3-level signal that combines phase and amplitude signaling ('–1', '0', '+1'), which is referred to as PSK Duobinary. The use of Duobinary generation

is advantageous in optical communication as it has ideally the same or an even better optical signal to noise ration requirement than NRZ-OOK modulation. At the same time the narrower optical spectrum and strong correlation between consecutive bits result in an increased dispersion tolerance (MaikeWichers 2002). Figure 7 shows the block diagram of Duobinary modulation and Figure 8 shows the simulation setup of Duobinary.

Duobinary is about a factor of two spectral efficient as compared to RZ. Thus, it is natural to assume that the gain in Duobinary spectral efficiency must carry with it a price in performance (Xie Chongjin 2004) shown in Figure 9. However, we show below that such thinking is defective. It is well known that the Duobinary technique is used to double the transmission capacity at a given channel bandwidth. In the context of 10 GB/S optical systems, an effective channel (fiber) bandwidth limitation results from

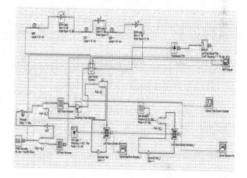

Figure 8.   Simulation setup for Duobinary.

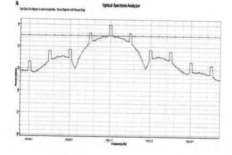

Figure 6.   Spectrum of RZ.

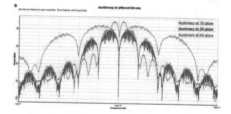

Figure 9.   Spectrum of Duobinary.

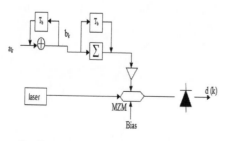

Figure 7.   Duobinary modulation.

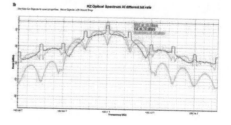

Figure 10.   Comparative spectral NRZ, RZ and Duobinary.

fiber chromatic dispersion which limits the maximum achievable distance. The dispersion limited transmission distance is usually defined as that giving rise to 1dB penalty shown in Figure 10.

2.4 *Manchester transmitter*

Manchester is an excellent modulation format for a variety of applications in optical fiber communication systems. As compared with the NRZ has rich clock component and enables simple level recovery and clock recovery. Figure 11 shows the block diagram of Manchester and the simulation shown in Figure 12. In addition, Manchester has the characteristic of zero dc components, which makes Manchester highly tolerant to signal intensity variation when differential detection is used. With these advantages, Manchester code was broadly studied in high-speed mode communication system (Nataša B. Pavlović 2005- Ilya Lyubomirsky 2005). Manchester has also found application in public optical network (Ex-WDM-PON). Manchester also has equal power in every transmitted bit shown in Figure 13. In frequency domain the main lobe of the Manchester coded signal is lies at the frequency which equals PRBS data rate. So, such intrinsic property could effectively lessen the optical beat noise between the optical line terminal. In particular, one of its variant, the PSK-Manchester signal has been shown in Figure 14 and Figure 15 to have more tolerance to the BIN (beat interference noise), compared with other modulation formats shown in Figure 16.

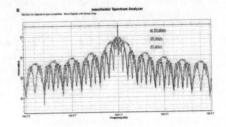

Figure 13. Spectrum at different bit rate with 120 km fiber length and 4 mW laser power.

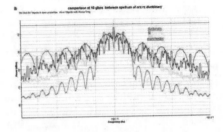

Figure 14. Spectrum width of NRZ, RZ, Duobinary and Manchester.

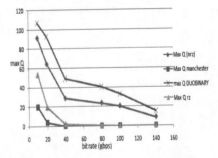

Figure 15. Effect of bit rate on Q OF NRZ, RZ, Duobinary and Manchester.

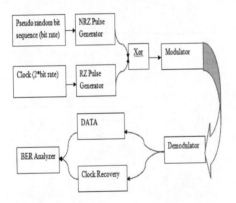

Figure 11. Block diagram for Manchester signal.

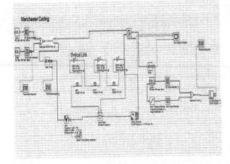

Figure 12. Simulation setup for Manchester signal.

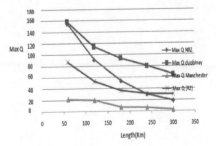

Figure 16. Comparative quality factors of different formats.

3 MANCHESTER-DUOBINARY TRANSMITTER

If we combine the Manchester and Duobinary together, a new format with better dispersion tolerance is projected and demonstrated, name as Manchester-Duobinary. With the advantage of improved spectral efficiency, Manchester-Duobinary has been found many useful applications as compare to conventional Manchester in optical access networks WDM.

We shall discuss the principle of Manchester-Duobinary shown in Figure 17, generation of Manchester-Duobinary signal shown in Figure 18, our contributions include a Manchester-Duobinary transmitter 128-km-Reach with 4 mw power our proposed Manchester-Duobinary transmitter.

3.1 Comparative spectrum of Manchester Duobinary, Duobinary, and Manchester

We observe that the Manchester Duobinary has narrower spectral width in these three formats. In Figure 19 for the 10 GB/s Manchester signal spectral width of main lobe (main lobe exist) from 93.08 THz to 93.12 THz. So the bandwidth requirement for the Manchester signal is 40 GHz. For the 10 GB/s Duobinary signal spectral width of main lobe

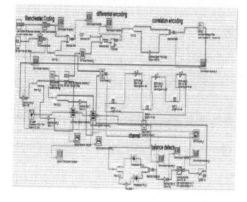

Figure 17. Block diagram of Manchester Duobinary.

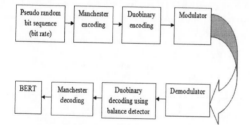

Figure 18. Simulation setup Manchester-Duobinary transmitter and corresponding receiver system.

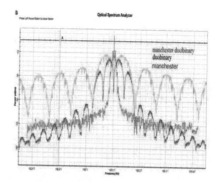

Figure 19. Spectrum width of NRZ, RZ, Duobinary and Manchester.

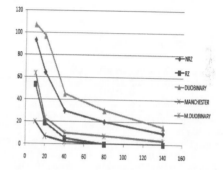

Figure 20. Comparative quality factor of different formats with bit rate variation at 120 km and 4 mw.

Table 1. Quality factor of different parameter at different data rate.

Pulse format	10 GB/s	20 GB/s	40 GB/s	80 GB/s	140 B/s
NRZ	93.72	64.32	29.76	20.00	12.05
RZ	56.32	19.36	6.12	.02	0.0
Duobinary	106.73	96.78	43.94	20	15.32
Manchester	20.00	8.93	4.56	0.00	0.00
D. Manchester	63.24	22.73	11.83	0.00	0.00

(main lobe exist) from 93.09 THz to 93.11 THz. So the bandwidth requirement for the Manchester signal is 20 GHz. Sensitivity of the Duobinary is better than RZ and NRZ. Comparative Sensitivity of RZ and NRZ with respect to Duobinary with fiber length shown in Figure 20 and Table 1.

4 CONCLUSION AND FUTURE WORK

In this paper, we have proposed Duobinary, Manchester and Manchester-Duobinary formats. We focused on the format spectral efficiency, transmitter and receiver design and receiver sensitivity. We have projected and studied these formats with an enthusiasm of fulfilling various kinds of requirements for next-generation networks so that we can transmit more data in available bandwidth In every chapter we have first explained the proposed modulation format, and after that we had studied the property of the each modulation format. we have proposed hybrid Manchester-Duobinary transmitter. The spectrum characteristics of the hybrid Manchester-Duobinary signal had been investigated and compared with that of Manchester and Duobinary signal. Experimental results show that Manchester-Duobinary format show more compact spectrum compared with Duobinary and Manchester. In future the study can be to improve the quality factor of signal and how can be receiving sensitive.

REFERENCES

Arumugam M. 2001. "Optical fiber communication—An overview", *Journal of physics*, Vol. 57, pp. 849–869.

Chhilar Rahul 2011."Modulation Formats in Optical Communication System", IJCEM International *Journal of Computational Engineering & Management*, Vol. 13 pp. 110–115.

Chongjin Xie et al. 2004. "Improvement of Optical NRZ— and RZ-Duobinary Transmission Systems With Narrow Bandwidth Optical Filters", *IEEE Photonics Technology Letters*, Vol. 16, No. 9, pp. 2162–2164.

Djordjevic 2010. "Coding for Optical Channels", Springer Publishers chapter 2, pp. 49–53.

Dong Yi 2004. "Pulsewidth-Tunable CS-RZ Signal Format With Better Tolerance to Dispersion and Nonlinear Degradation in Optical Transmission System", IEEE Photonics Technology Letters, Vol. 16, No. 5, pp. 1409–1411.

Eugen Lach & Wilfried Idle 2011. "Modulation formats for 100G and beyond", *Journal of Optical Fiber Technology*, Vol. 17, No. 5, pp. 377–386.

Ezra Ip & Joseph M. Kahn 2006. "Power Spectra of Return-to-Zero Optical Signals", *Journal of Lightwave Technology*, Vol. 24, No. 3, pp. 1610–1618.

Fonseca D. & A. 2005. Cartaxo, "On the transition to pseudo-linear regime in dispersion managed systems with NRZ, RZ andduobinary signal formats", IEEE Proceedings on Optoelectron., Vol. 152, No. 3, pp. 181–187.

García-Pérez A. 2006. "Efficient Modulation Formats for High Bit-Rate Fiber Transmission", Vol. 16 No. 2, pp. 17–26, 2006.

Goldfarb Gilad 2006. "Improved Chromatic Dispersion Tolerance for Optical Duo binary Transmission Using Coherent Detection", IEEE Photonics Technology Letters, Vol. 18, No. 3, pp. 517–519.

Hitoshi Murai 2007. "EA-Modulator-Based Optical Time Division Multiplexing/ Demultiplexing Techniques for 160-Gb/s Optical Signal Transmission", *IEEE Journal of Selected Topics in Quantum Electronics*, Vol. 13, No. 1, pp. 70–77.

Hoon Kim et al. 2002. "Demonstration of Optical Duo binary Transmission System Using Phase Modulator and Optical Filter", IEEE Photonics Technology Letters, Vol. 14, No. 7, pp. 1010–1012.

Jaehoon Lee et al. 2004. "Chromatic Dispersion Tolerance of New Duobinary Transmitters Based on Two Intensity ModulatorsWithout Using Electrical Low-Pass Filters", Journal Lightwave Technology, Vol. 22, No. 10, pp. 2264–2270.

Lyubomirsky Ilya & Cheng-Chung Chien 2005. "Optical Duo binary Spectral Efficiency versus Transmission Performance: Is There a Trade off?", IEEE Conference on Quantum Electronics and Laser Science (QELS), pp. 1774–1776.

Majid Moghaddasi 2011. "Comparison between NRZ and RZ OOK Modulation Format in Chromatic Dispersion Compensation in Both Electrical and Optical Compensator", 2011 *IEEE Symposium on Business, Engineering and Industrial Applications* (ISBEIA), pp. 494–496.

Maike Wichers 2002. "Improved dispersion tolerance of duo binary optical transmission considering the influence of duo binary filters and optical input power", IEEE International Conference, pp. 5–8.

Nataša B. Pavlović 2005. "Optimized Bandwidth-Limited Duo binary Coding Format for Ultra Dense WDM System", 7th IEEE International Conference on Tranceparent Optical Networks, Portugal, pp. 385–388.

Noriaki Kaneda & Xiang Liu 2003. "Improved Polarization-Mode-Dispersion Tolerance in Duo binary Transmission", IEEE Photonics Technology Letters, Vol. 15, No. 7, pp. 1005–1007.

Pachnicke S. 2012. "Fiber Optical Transmission Systems", *Springer Publishers*, pp. 11–29.

Shaoliang Zhang 2011. "Pulse Shaping on Quadrature Duo binary Format", IEEE Photonics Society Summer Topical Meeting Series, pp. 149–150.

Sheldon Walklin & Jan Conradi 1997."On the Relationship Between Chromatic Dispersion and Transmitter Filter Response in Duobinary Optical Communication System," IEEE Photonics Technology Letters, Vol. 9, No. 7, pp. 1005–1007.

Wang Yi-Hsiang, & Ilya Lyubomirsky 2011."Balanced Detection Schemes for Optical Duobinary Communication System,"*Journal Lightwave Technology*, Vol. 29, No. 12, pp. 1739–1745.

Zhixin Liu & Chun-Kit Chan 2011. "Generation of Dispersion Tolerant Manchester-Duo binary Signal Using Directly Modulated Chirp Managed Laser Generation of Dispersion Tolerant Manchester-Duo binary Signal Using Directly Modulated Chirp Managed Laser", IEEE Photonics Technology Letters, Vol. 23, No. 15, pp. 1043–1045.

Communication and Computing Systems – Prasad et al. (Eds)
© 2017 Taylor & Francis Group, London, ISBN 978-1-138-02952-1

Energy and coverage aware algorithm for data gathering in densely distributed sensor networks

Rajesh Kumar Singh & Sandeep Sharma
School of ICT, Gautam Buddha University, Greater Noida, India

ABSTRACT: These days the magnification rate of a multimedia data is increasing continuously in cyberspace technology. So gathering and analysis of Energy-efficient big data is very difficult within the densely distributed sensor networks. One among the foremost effective solution to deal with this problem is clustering and it uses the mobile sink to gather the information. This paper proposes a replacement movable sink routing and data collection technique through network cluster Expectation-Maximization algorithm with Gaussian Mixture Model (EMGMM) and applies DBSCAN for residual sensor nodes which are not a cluster member of EMGMM algorithm. It works in such a way that coverage and efficiency of sensor network is improved without increasing number of cluster heads so that data request flooding problem is also minimized. Proposed DBEMGM Algorithm considers both energy consumption and efficiency of network in densely distributed WSN.

Keywords: WSNs (Wireless Sensor Networks), Big Data, Energy Efficiency, DBSCAN, Clustering, Data gathering, EMGMM (Expectation Maximization algorithm with Gaussian mixture model), DBEMGM (Density Based Expectation Maximization algorithm with Gaussian Mixture Model), Efficiency of network

1 INTRODUCTION

In dense WSNs, sensor nodes are closely packed and the life time of sensors is very less since every sensor mote has to send huge amount of information collected by it from the neighboring sensor motes. For solving this type of problem, an energy-efficient technique is required to collect large amount of information from densely distributed sensor nodes in WSNs. Distributed sensor network is better because in centralized sensor network if central node fails then entire network collapses, so by using distributed controlled architecture such problem can be mitigated.

So an effectual solution is proposed to lessen the energy consumption in sensor networks and to use movable sink node to make easy the accumulation of data. Here, an advanced mobile sink routing and data collecting technique through network cluster is proposed which is based on customized EM technique.

To obtain energy-efficient information amassment in densely distributed sensor network, there are several subsisting approaches such as LEACH protocol, ECDC algorithm and other data compressing techniques which reduce the amount of transmitted data.

LEACH algorithm can also work for minimization of energy consumption of data but it is not favorable for the WSNs deployed in large area in which it will have node's communication range problem. k-CONID (k-hop Connectivity Identity) algorithm is based on ID number and it is a distributed clustering algorithm. In k-CONID the sensor motes give their unique identification code to one another and hence the mote with the least value of identification code in a k hop gets elected as the forward mote or CH. In this clustering algorithm energy consumption is not reduced to a great extent so this paper approaches to DBEMGM algorithm.

1.1 Overview of DBSCAN

DBSCAN stands for density based spatial clustering of application with noise, which aggregates nodes that are proximately packed. By analysis of the various motes that have been placed in a very small area increasing the density of the motes in a particular area, DBSCAN tries to divide the motes into groups by choosing a particular property of closeness between the motes.

This algorithmic guideline wants 3 input parameters:

The neighbor list size, Eps-the radius that delimitate the neighborhood space of a point and MinPts- the minimum range of points that has to exist within the Eps-neighborhood.

This algorithm starts by going to each of the sensor motes and asks it for its unique cluster ID. If the mote provides it with the unique ID then the algorithm goes to the next mote and again performs the same process. If the mote is not able to provide a unique ID then its Eps neighbourhood is studied and if suitable number of motes (greater than Minpts) is present around it then a new group is formed. Otherwise, the node is tagged as noise. DBSCAN doesn't need one to tell the exact number of groups that have to be formed which is a compulsion in some algorithms and DBSCAN can form random shape of the clusters.

2 NETWORK MODEL

A network model is considered in which a travelling base station (sink) and various sensor motes are placed in a fixed region.

Every sensor mote is able to find its position in the sensor network by utilizing the concept of localization and also the travelling base station has complete information regarding the exact placement of each and every mote.

All sensor nodes and sink also have their fixed area in which they can sense the data and if any disturbance happens in this particular region they sense the activity and send the information for further processing.

There is a specialized sensor mote which is present in the network that has extra capabilities as compared to the other motes and this mote is known as the 'sink mote' as it has travelling power that enables it to move from one location to another. Only this is the node that changes its position during the entire working process of the algorithm. When this sink comes near to sensing area

of the CH, then the CH sends information to the sink. Thus, energy consumption of transmission is minimized by decreasing the amount of relays within the sensor network.

The movable sink node travels surrounding the CH of the clusters that have been selected to reduce the resource consumption during the information transfer process and to collect information from sensor motes.

Sensor nodes have buffer memory which is used to keep detected data until mobile sink reaches near the CH of the clusters.

Data detected by one mote is sent to another mote which then sends it to another mote and the process keeps on going until it finally reaches the base station. This process is collectively known as multi hop communication process.

Let the nodes be distributed in L X L area by following the Gaussian Mixture Distribution rule.

$$p(x) = \sum_{k=1}^{k} \pi_k N(x / \mu_k, E_k)$$

$$N(x|\mu, E) = \frac{1}{(2\pi)|E|^{\frac{1}{2}}}$$
$$\exp\left\{ -\frac{1}{2}(x-\mu)^T E^{-1}(x-\mu) \right\}$$

where
x = position vector of all motes
E = 2 × 2 covariance matrix
μ = position vector of centroids
k = total number of clusters
π = mixing coefficient
These parameters are defined for k_{th} cluster in given equations.

3 PROPOSED ALGORITHM

1. First initialize the cluster centroids at random locations.
2. Calculate communication distance of each node from centroids then mixing coefficient and covariance matrix.
3. Calculate responsibility (R) and Log Likelihood (LL).
4. Until the distinction between the newly calculated LL_{new} and formerly calculated LL becomes less, calculations are repeatedly executed with the help of updated responsibility (R) in clusters.
5. Centroids and the matrix of covariance are recalculated and also the numbers of motes that fall in the k_{th} group are calculated.
6. The number of motes that fall in each of the group, the matrix of covariance and the centroid of the clusters are returned.

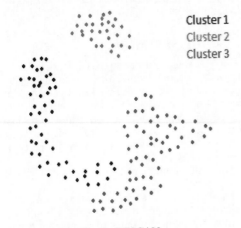

Cluster 1
Cluster 2
Cluster 3

Figure 1.   Clustering by DBSCAN.

7. Calculate optimal number of clusters each with single cluster-head.
8. Initialize Eps, MinPts
9. Check each node if it has any cluster id or it is a cluster member. Then mark the node as visited and continue for next node.
10. If number of neighbor nodes is less than number of MinPts, mark node as noise otherwise form cluster and expand the cluster.
11. So clusters without CH are formed of arbitrary shape of residual nodes (which are not cluster member in EMGMM).
12. If residual sensor nodes have information then it finds nearer cluster member of EMGMM by using localization technique and then sends the data to that node
13. This node transfers the data to sink node through cluster head which is chosen in EMGMM.

First step of EM algorithm is expectation in which the current estimate parameter is fixed and the expected value of the log likelihood function and responsibility is calculated.

Second step is maximization in which the model parameters are re-estimated using expected log-likelihood of the data.

In EM algorithm its steps are repeated until value of log likelihood converge.

Log likelihood is given by:

$$LL = \ln p\left(X|\mu, E, \pi\right) = \sum_{n=1}^{n} \ln\left\{\sum_{k=1}^{K} \pi_k N(x_n|\mu_k, E_k)\right\}$$

This evaluated value of LL is repeatedly decrementing and the value of EM algorithm comes to an end. This is because the EM algorithm perpetually changes the group centroid's position vector and the nodes responsibility to kth cluster. The responsibility describes dependency of node on a cluster. Responsibility calculates with the help of Baye's theorem and it has range from 0 to 1. The responsibility of nodes to kth cluster is given by

$$R_k = \frac{\pi_k N(x_n|\mu_k, E_k)}{\sum_{j=1}^{k} \pi_j N(x_n|\mu_j, E_j)}$$

When optimal numbers of clusters to reduce energy consumption are obtained by using EMGMM algorithm then DBSCAN with EM algorithm is applied for improving coverage and efficiency of WSN in terms of number of nodes that can participate in transmission and receiving of the data.

The parameters given in Table 1 are used in simulation and analysis of the results of DBEMGM algorithm.

Fig. 2 shows that efficiency of EM with GMM is higher than that of k-CONID and EM algorithm.

Fig. 3 shows that as the number of nodes increases the energy consumption also increases but EM algorithm with GMM has lesser energy consumption than that of k-CONID and EM

Table 1. Parameter used in simulation of DBEMGM.

Parameters of the simulation	
No. of nodes	100
Communication Range	200 m
No. of Clusters	4–50
Clustering Algorithm	DBEMGM
Area	200×200 m^2
Initial energy of nodes	3–6 J

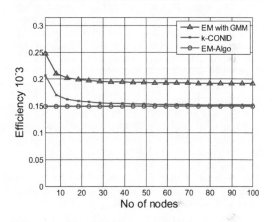

Figure 2. Efficiency of k-CONID, EM algorithm and EM algorithm with GMM.

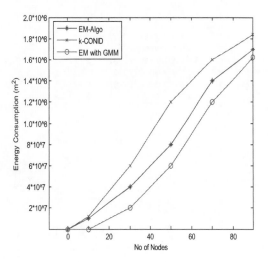

Figure 3. Energy consumption of k-CONID, EM algorithm and EM algorithm with GMM.

algorithm. EM algorithm lessens the distance from source mote to sink and energy consumption is proportional to square of communication distance so here it is taken in m^2.

Mostly anterior researches over energy efficient clustering say that incrementing the number of clusters leads to the decrementing the resource exhaustion for transmission of data. But these researches do not consider the amount of energy that was consumed in transmission of the data request messages. In this paper this quandary is considered and an analysis is done to obtain optimal number of clusters by using EMGMM algorithm.

When the numbers of clusters increase there is a considerable enhancement in the number of requests for the data packets to be transmitted.

The effect of flooding of data request packets increases highly as there is an enhancement in the connectivity. This happens because the data request messages are repeatedly sent to different nodes in a cluster so that the resource exhaustion in the transmission of data request messages increase which is shown in Figure 4.

As there is an increase in the number of clusters there is proportionate reduction in the energy that is used for transfer of data and data request messages but at higher connectivity when number of clusters increases then the resource consumption also increases. This is the reason why it is very essential to find the right number of groups that have to be made by keeping the packet transfer and resource exhaustion issues in mind and this same work is done by the EMGMM algorithm.

When optimal number of clusters are used then there are number of sensor nodes which remain outside the cluster in densely distributed network.

These nodes can't participate in communication process so DBEMGM provides the effective solution to this problem.

$$network\ efficiency_{EMGMM} = \frac{n_{EMGMM}*100}{N}$$

$$network\ efficiency_{DBEMGM}$$
$$= \frac{(n_{EMGMM} + n_{DBSCAN})*100}{N}$$

n_{EMGMM} = no. of nodes paricipate through EMGMM cluster

n_{DBSCAN} = no. of nodes paricipate through DBSCAN cluster

N = total number of nodes in the network

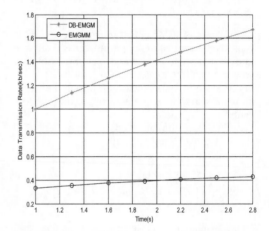

Figure 5. Data transmission rate of EMGMM and DBEMGM.

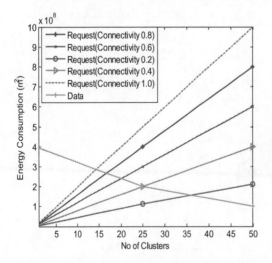

Figure 4. Energy consumption for data transmissions and data request messages with different connectivity.

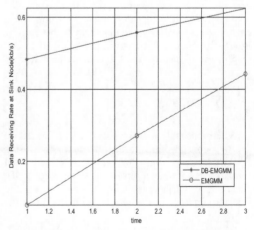

Figure 6. Data receiving rate of EMGMM and DBEMGM at sink node.

Thus it can be easily understood that the network efficiency of DBEMGM is higher than that of EMGMM algorithm

In Figure 5 and Figure 6 both the data transmission rate and the data receiving rate in DBEMGM are higher than EMGMM because in DBEMGM more number of sensor nodes participates in data transmission process in any particular time interval.

4 CONCLUSION

In this paper a study was conducted on EMGMM algorithm and we proposed DBEMGM algorithm which reduced energy consumption of total transmitted data, increased the coverage and number of node participation in data transmission thus the efficiency of the network improved. This algorithm gives better performance than EMGMM algorithm in densely distributed sensor network.

REFERENCES

Baum. D and CIO Information Matters, in 2013, Big Data, Big Opportunity.

Dimokas. N, Katsaros. D, Manolopoulos.Y, in 2010 "Energy-efficient distributed clustering in wireless sensor networks", J. Parall. Distrib. Comput. 371–383.

Rai, Santosh Kumar, and Nishchol Mishra, in 2012 "DBCSVM: Density Based Clustering Using Support Vector Machines",Vol. 9, Issue 4, No 2.

Ramaswamy. L, Lawson. V, and Gogineni. S, in Jul. 2013, "Towards a quality-centric big data architecture for federated sensor services," in Proc. IEEE Int. Big Data Congr. 86–93.

Taksai1,Hiroki Nikshiyama1,Nei Kato1, Ryu Miura,in April 2014,"Toward Energy Efficient Big Data Gathering in Densely Distributed Sensor Networks", Daisuke, *IEEE Transactions on emerging topic in computing* 2, no. 3: 388–397.

Xin Gu, Jiguo Yu, Dongxiao Yu, Guanghui Wang, Yuhua Lv, in 2014," ECDC: An energy and coverage-aware distributed clustering protocol for wireless sensor networks", *Computers & Electrical Engineering* 40, no. 2: 384–398.

Youssef. M, Younis. M, Youssef. A, in Dec. 2009 "Overlapping multihop clustering for wireless sensor networks," *IEEE Trans. Parallel Distrib. Syst.*, vol. 20, no. 12, pp. 1844–1856.

Communication and Computing Systems – Prasad et al. (Eds)
© 2017 Taylor & Francis Group, London, ISBN 978-1-138-02952-1

Mobile Cloud Computing Energy-aware Task Offloading (MCC: ETO)

A. Ahmed, A. Abdul Hanan, K. Omprakash, M.J. Usman & O.A. Syed
Faculty of Computing, Universiti Teknologi Malaysia, Skudai Johor Bahru, Malaysia
Department of Mathematics, Bauchi State University Gadau, Itas/Gadau, Bauchi State, Nigeria

Aanchal
School of Computer and Systems Sciences, Jawaharlal Nehru University (JNU), New Delhi, India

ABSTRACT: Cloud Computing (CC) holds a new dawn of computing where multiple services are offered to the cloud users through internet. CC has a qualitative, flexible and cost effective delivery platform for providing services to IT users with the aid of internet. Due to advent of CC, capabilities of Mobile devices' system has been improved. Mobiles devices can now depend on CC and information storage resources, to carryout complex computational operations which includes multimedia, searching and data mining. Also when considering local cloud, computation service is provided by enhancing operation of mobile cloud which handles mobile devices as a service nodes for example sensing services and computation of task. The task that need to be computed has to be offloaded to local or conventional cloud to ease battery power. The energy consumption during offloading of task is highly significant. Hence, characterization of some of the recent research works on battery energy in MCC are grouped into dynamic and non-dynamic energy-aware task offloading and a comparative analysis table are presented and finally open research issues are qualitatively discussed based on critical assessment of the literature

1 INTRODUCTION

Mobile cloud computing has witnessed more rapid growth in terms of research due to the fact that mobile phones has become a vital part of human life as it is portable to move around with, which is very effective and suitable for communication irrespective of time and place (Fernando et al., 2013, Khan et al., 2013). The birth of MCC is a significant turnaround for computer science technology and also phone developers. Meanwhile, MDs are becoming more sophisticated due to development of large and complex applications. Consequently, MDs are constrained with challenges of battery power, memory space and computation power, for these reasons the idea of offloading task to the cloud has been integrated into mobile devices. When offloading task to cloud so many issues suffices such as security, quality of service and mobile application development (Kumar and Lu, 2010, Zissis and Lekkas, 2012). Figure 1. Represent system view of Mobile cloud computing.

Applications which requires complex computations such as real time computing, image and voice processing, online game and video streaming, language and wearable computing demands high processing capabilities. These complex applications are challenge for MDs application developers in implementing application for MDs. The

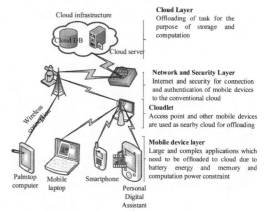

Figure 1. System view of mobile cloud computing.

problem of battery power and memory space are increasing due to high demand for smaller sized MDs, therefore, these challenges of battery power, memory space and computation power will be a lingering issue (Satyanarayanan, 1993). Hence, it has become paramount to adopt cloud computing solutions for mobile devices. Recently, many researches has been directed towards CC to address these challenges. CC can provides Infrastructure as a Services (IaaS) for MDs storage and Platform as a Services (PaaS) (Fox et al., 2009, Vogels, 2008) for MDs computation to alleviate their limitations. The basic idea of CC is to offload large and complex task to remote resource providers. Thus, security is needed for offloaded task confidentiality and integrity.

The remaining part of this paper is structured as follows: Section 2 describes MCC energy-aware task offloading taxonomy and comparative analysis of literature reviewed. Section 3 show outlined research challenges and open issues and finally section 4 is the concluding section.

1.1 MCC energy-aware task offloading

Energy task offloading is the most challenging issue in Mobile Computing (MC), due to this challenging issue a new idea has been presented dubbed Mobile Cloud Computing (MCC). MCC has been adopted so as to address the battery energy issue in MC. Despite the effort, MCC also faced battery energy issue, when trying to offload task to the cloud, energy could be caused due to distance between mobile and the network base station or due to quality of network of the mobile device.

In this section, a qualitative review of energy-aware mobile offloading is presented. We categorize them into two, namely dynamic and non-dynamic energy-aware task offloading. Figure 2 shows MCC energy-aware task offloading taxonomy.

1.2 Dynamic energy-aware task offloading

Dynamic resource allocation and parallel execution in the cloud for mobile code offloading (ThinkAir) addressed difficulty of computational power and energy due to increasing complex applications (Kosta et al., 2012). A framework has been developed and result analysis were benchmarked. ThinkAir framework uses concept of smartphone virtualization in cloud, which dynamically migrates application using parallelization with multiple Virtual Machine (VM) clone, when task

are offloaded. Two benchmark approaches are considered which includes micro-benchmark and application benchmark to evaluate two performance metric namely; energy and time consumption for four different scenario including WIFI, 3G, WIFI local and Phone.

Partial Mobile Application Offloading to the Cloud for energy-efficiency with security measures (PMAOC) addressed challenges of low battery life and systematic mechanism to evaluate effect of offloading an application onto the cloud (Saab et al., 2015). A mathematical model, dynamic algorithm, Free Sequence Protocol (FSP) and experiment were carried-out. Mathematical model has been deduced to save energy while preserving security using variables such as time and energy. Algorithm called dynamic minimum-cut has been used to preserve and enhance security in the system. FSP permits mobile phone applications to run in several possible combination in client and server sides. A complete system consisting of profiling, decision, offloading engine and an FSP-based mobile app has been developed on an Android platform. Figure 3 shows the system module for parallel execution and resources allocation.

Tradeoff between Performance Improvement and Energy saving in Mobile Cloud Offloading systems (PIEMCO) handle the challenges of low computation and battery power of mobile devices (Wu et al., 2013). A novel adaptive offloading scheme has been suggested and examined based on tradeoff analysis. The tradeoff analysis can be achieved due to the elasticity of cloud computing, in such a way that resources can be supplied on demand. Furthermore, the authors present a tradeoff analysis of performance enhancement and energy saving during offloading decisions. The execution time was divided into three intervals, namely, tradeoff, always offload and never offload to enhance energy saving. Figure 4 represent cloud offloading system architecture.

Dynamic Energy-aware Cloudlet-based Mobile cloud computing model for green computing (DECM) addressed issue of energy waste and latency delay due to restrictions of wireless bandwidth and device capacity (Gai et al., 2016). A DECM aimed

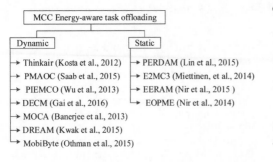

Figure 2. MCC energy-aware task offloading taxonomy.

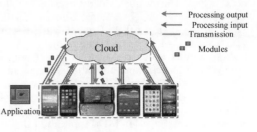

Figure 3. System module.

360

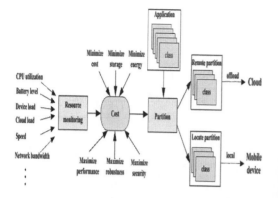

Figure 4. Architecture of cloud offloading systems.

at bringing solution to the extra energy consumptions during wireless communications by leveraging Dynamic CloudLets-based (DCL) model. The model was examined by simulating practical scenario and providing solid results for the evaluations. Actual contributions of this work is solving energy waste issue and suggesting future research direction within the networking environment.

MOCA: A lightweight MObile Cloud offloading Architecture addressed low battery and computation power of mobile devices, by designing an offloading architecture to the cloud (Banerjee et al., 2013). The architecture uses an in-network cloud platform to offer offloading resources. MOCA incorporates with present mobile network architectures without demanding significant variations, and utilizes software based networking techniques in the data plane to redirect appropriate traffic to and from the cloud platform. Viability of MOCA has been depicted by a prototype implementation using a LTE/EPC mobile testbed.

DREAM: Dynamic Resource and task allocation for Energy minimization in Mobile cloud systems handles complicated resource interference issue due to the resource sharing in dynamic mobile cloud offloading, in other words DREAM addresses issues of minimizing high CPU and network energy usage of current smartphones (Kwak et al., 2015). A joint CPU, network resource and task allocation, bearing in mind all types of tasks such as networking, processing and cloud offload-able task has been suggested. In essence, an architecture and algorithm has been designed to minimize mobile resource consumption. The architecture establishes applicability of the idea in practice and its implementation on android platform. The algorithms has been designed to use Lyapunov optimization and also was mathematically proven that it reduces CPU and network energy usage for given delay constraints. From the investigation conducted using trace-driven simulation based on

actual measurements shows that DREAM saves more than 35% of the total energy usage than the existing algorithms with the same delay.

MobiByte: An application development model for Mobile cloud computing to addressed issue of traditional mobile development model, which do not support offloading of task to the cloud (Othman et al., 2015). Mobile application development model called MobiByte has been suggested to enhance execution support, performance and energy efficiency for mobile device applications. MobiByte outperforms the present application models in several aspect such as energy efficiency, context awareness, privacy and performance. Meanwhile communication aspects of the applications offloading not considered. Figure 5 shows Factors affecting computation offloading decision.

2.2 Non-dynamic energy-aware task offloading

Performance Evaluation of Remote Display Access for Mobile cloud computing (PERDAM) addressed low computation, battery power remote display access as a means for mobile cloud computing, focusing on power utilization of mobile devices (Lin et al., 2015). Putting into considerations the existing literatures, experiment approach based on real client sessions, using different remote access protocol and different kinds of applications, for example gaming. Via numerous experiments, characterization of effects of different protocols and their features on network utilization and power consumption were performed. User experience and usability has also been analyzed. Model was not suggested for the energy consumption of remote access on mobile devices in a specific context provided.

Energy Efficiency of Mobile Clients in Cloud Computing (E2MC3) addresses cost of energy during offloading of task to the cloud, cost of offloading not to exceed the mobile device's computation energy consumption (Miettinen and

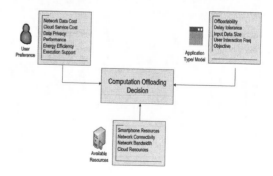

Figure 5. Factors affecting computation offloading decision.

361

Table 1. Comparative analysis for different research work on dynamic energy-aware task offloading.

Characteristics & protocol	Issues	Contribution	Techniques	Environment	Metrics	Remarks
ThinkAir (Kosta et al., 2012)	– Increase in complex mobile application – Low energy and computational power	– Virtualizations – Parallelization of VM clone	– Framework – Virtualization	Cloud platform	Energy and time consumption	Implementation support for parallelizable application is required
PMAOC (Saab et al., 2015)	– Battery lifetime – Lack of systematic mechanism to evaluate the effect of offloading an app onto the cloud.	– Mathematical model – Dynamic algorithm and free sequencing protocol (FSP)	Model, Algorithm and Protocol (FSP)	An Amazon EC2 Windows cloud instance	– Workload size – Network type-Computation cost – Signal strength – Call graph structure	More time and complexity can be experienced due to encryption/decryption
PIEMCO (Wu et al., 2013)	Low battery power and low computation power for mobile devices	Analysis and Architecture for Mobile Cloud offloading	Analysis and Architecture	Nil	Nil	No detail experiments was implemented in this work
DECM (Gai et al., 2016)	Restrictions of wireless bandwidth and device capacity have caused energy waste and latency delay	Energy waste was minimized	Model	User developed simulator called DECM-Sim	– Energy consumption – Time constraint – Service quality	Applicability of DECM in multiple industries with different service requirements, has not been investigated
MOCA (Banerjee et al., 2013)	Low battery and computation power of mobile devices	Architecture of the in-network cloud offloading	Architecture	AlterNet Mobile testbed, Openstack cloud platform with laptop with WIFI	Nil	No standardized performance metric was considered to evaluate the performance of the MOCA
DREAM (Kwak et al., 2015)	To handle complicated resource interference issue due to the resource sharing in dynamic mobile cloud	Resource and task allocation architecture and algorithm	Architecture and algorithm	Phone, WIFI and virtual cloud server on the Windows Azure cloud system	– Average energy consumption – Average delay	Security issues not looked into in this architecture
MobiByte (Othman et al., 2015)	Traditional mobile development model do not support offloading of task to the cloud	Application development model	Context aware model	Prototype application using Android platform and virtual public cloud and smartphone	Computation time – Energy consumption – Memory utilization	Communication aspects of the applications offloading not considered

Table 2. Summary of comparative analysis for different research on dynamic energy-aware task offloading.

Characteristics & protocol	Technique			Approach		Implementation		Metrics								
	MD	FW/AT	AL	DM	N-DM	CP	U-DT	EG	CT	DL	SQ	WS	NT	BU	MU	TP
ThinkAir (Kosta et al., 2012)	x	√	x	√	x	√	x	√	√	x	x	x	x	x	x	x
PMAOC (Saab et al., 2015)	√	x	x	√	x	√	x	√	√	x	√	√	√	x	x	x
PIEMCO (Wu et al., 2013)	√	x	√	√	x	x	√	√	√	x	x	x	x	x	x	x
DECM (Gai et al., 2016)	√	x	x	√	x	x	√	√	√	x	√	x	x	x	x	x
MOCA (Banerjee et al., 2013)	x	√	x	√	x	x	√	x	x	√	x	x	x	x	x	x
DREAM (Kwak et al., 2015)	√	√	√	√	x	√	x	√	x	x	x	x	x	x	x	x
MobiByte (Othman et al., 2015)	√	x	x	√	x	x	x	√	√	x	x	x	x	x	√	x

Note: √ = considered, x = not considered, MD = Model, FW/AT = Framework/Architecture, AL = AnaLysis, DM = DynaMic, N-DM = Non-DynaMic, CP = CloudPlatform, U-DT = User Developed Testbed, EG = Energy, CT = Computation Time, DL = Delay, SQ = Service Quality, WS = Workload Size NT = Network Time, BU = Bandwidth Usage.

Table 3. Comparative analysis of different research on non-dynamic energy-aware task offloading.

Characteristics & protocols	Issues	Contribution	Techniques	Environment	Metrics	Remarks
PERDAM (Lin et al., 2015)	To tackle low computation, battery power and Issue of remote display access	Evaluation of performance of remote display access in MCC	Performance evaluation.	Experimental testbed: Android phone, Amazon Computer server WIFI and LTE	Bandwidth utilization Transfer time – Power consumption	No model of the energy consumption of remote access on mobile devices in a specific context was considered
E2MC3 (Miettinen and Nurminen, 2010)	The issue of energy cost during offloading of task to the cloud	Analysis of power consumption in some Handheld mobile devices	Analysis	Mobile as thin client, 3G and Wi-Fi	– Energy Usage – Throughput	An offloading decision making model not considered
EERAM (Nir et al., 2015)	Task scheduling to minimize energy and monetary cost	Task scheduling	Mathematical model	Server machine and Mobile phone.	– Energy – Delay	The scheduler model does not handle network congestion and task priority.
EOPME (Nir et al., 2014)	Task scheduling in MCC environment	Task scheduling	Mathematical model	Server machine and Mobile phone.	– Energy consumed	Task priority not considered

Table 4. Summary of comparative analysis for different research on non-dynamic energy-aware task offloading.

Reference	Technique			Approach		Implementation		Metrics								
	MD	FW/AT	AL	DM	N-DM	CP	U-DT	EG	CT	DL	SQ	WS	NT	BU	MU	TP
PERDAM (Lin et al., 2015)	x	x	√	x	√	√	x	√	x	x	x	x	x	√	x	x
E2MC3 (Miettinen and Nurminen, 2010)	x	x	√	x	√	x	√	√	x	x	x	x	x	x	x	√
EERAM (Nir et al., 2015)	√	x	√	x	√	x	√	√	x	√	x	x	x	x	x	x
EOPME (Nir et al., 2014)	√	x	x	x	√	x	√	√	x	x	x	x	x	x	x	x

Note: √ = considered, x = not considered, MD = Model, FW/AT = Framework/Architecture, AL = AnaLysis, DM = DynaMic, N-DM = Non-DynaMic, CP = CloudPlatform, U-DT = User Developed Testbed, EG = Energy, CT = Computation Time, DL = Delay, SQ = Service Quality, WS = Workload Size NT = Network Time, BU = Bandwidth Usage.

Nurminen, 2010). An analysis of main factors affecting energy usage of mobile client in cloud computing has been outlined. Measurement of some essential characteristics of recent mobile phones and trade-off sensitivity between two scenarios local (mobile device) and cloud were presented.

Economic and Energy Considerations for Resource Augmentation in Mobile Cloud Computing (EERAM) addressed task scheduling for increasing resources of multiple mobile devices to minimize energy and computational cost (Nir et al., 2015). A model called mathematical task scheduler has been suggested which optimized both energy and economic parameters. The model handles mobile devices to offload more than one tasks to the cloud. Analysis of task offloading has been carried-out, by using two resource augmentation environment for MCC namely; public cloud and local private cloud. Evaluation of total energy consumption in relation the two environment scenarios has been tested to solve task scheduling issues.

An Energy Optimizing Scheduler for Mobile Cloud Computing Environments (EOPME) tackles challenges of task scheduling in MCC environment, so as to reduce energy issue of mobile devices (Nir et al., 2014). A task scheduler model has been suggested which schedules task based on user defined constraints. The task scheduler model is situated at centralized broker, which optimally offload task and offers reasonable reduction in energy consumption of MDs in when related consumption of energy when task offloaded from centralized scheduler without optimization.

2 RESEARCH CHALLENGES AND OPEN ISSUES

I. The scheduler model need to be explored for efficient energy minimization when offloading task to the cloud.

II. To achieve efficient energy minimization, network congestion and task priority has to be critically explored.

III. Standard and intelligent offloading mobile application need to be developed for effective offloading decision to the cloud, so as to avoid unnecessary offloading. And to consider criteria for decision making on offloading.

IV. Fine grained task offloading to cloud need to be also explored, so as to avoid offloading of non-required data to the cloud and this will reduces overhead.

V. Cloudlet in MCC needs more exploration to reduce latency in communication due to long distant network connectivity between MDs and mobile station or access point etc.

3 CONCLUSION

We have extensively reviewed related work and outlined most recent ideas on energy issue in MCC. MCC suffices from the integration of Wireless Area Networks (WAN), CC, and MDs and it applications. Highly computation required application are been developed for MDs, which consumes energy of these devices. Offloading in cloud has become the option for energy minimization during complex processing MDs. Consequently, offloading also has its own shortcoming since it depends on network connectivity, once the offloading devices (MDs) are distant from the network, more energy is consumed or whenever the network quality is low, also more energy is tend to be consumed.

In this work, we proposed a systematic review of MCC energy-aware issues, open research issues are outlined and parameters of MCC are discussed. Energy issue in MCC need to be further explored extensively. This work has proof its significance both in academia and industrial.

ACKNOWLEDGMENT

The research is supported by Ministry of Education Malaysia (MOE) and conducted in collaboration with Research Management Center (RMC) at University Teknologi Malaysia (UTM) under VOT NUMBER: Q.J130000.2628.11 J31.

REFERENCES

Banerjee, A., Chen, X., Erman, J., Gopalakrishnan, V., Lee, S. & Van Der Merwe, J. Year. MOCA: a lightweight mobile cloud offloading architecture. *In*: Proceedings of the eighth ACM international workshop on Mobility in the evolving internet architecture, 2013. ACM, 11–16.

Fernando, N., Loke, S. W. & Rahayu, W. 2013. Mobile cloud computing: A survey. *Future Generation Computer Systems*, 29, 84–106.

Fox, A., Griffith, R., Joseph, A., Katz, R., Konwinski, A., Lee, G., Patterson, D., Rabkin, A. & Stoica, I. 2009. Above the clouds: A Berkeley view of cloud computing. *Dept. Electrical Eng. and Comput. Sciences, University of California, Berkeley, Rep. UCB/EECS*, 28, 2009.

Gai, K., Qiu, M., Zhao, H., Tao, L. & Zong, Z. 2016. Dynamic energy-aware cloudlet-based mobile cloud computing model for green computing. *Journal of Network and Computer Applications*, 59, 46–54.

Khan, A. N., Kiah, M. M., Khan, S. U. & Madani, S. A. 2013. Towards secure mobile cloud computing: A survey. *Future Generation Computer Systems*, 29, 1278–1299.

Kosta, S., Aucinas, A., Hui, P., Mortier, R. & Zhang, X. Year. Thinkair: Dynamic resource allocation and parallel execution in the cloud for mobile code offloading. *In:* INFOCOM, 2012 Proceedings IEEE, 2012. IEEE, 945–953.

Kumar, K. & Lu, Y.-H. 2010. Cloud computing for mobile users: Can offloading computation save energy? *Computer*, 51–56.

Kwak, J., Kim, Y., Lee, J. & Chong, S. 2015. DREAM: Dynamic Resource and Task Allocation for Energy Minimization in Mobile Cloud Systems. *Selected Areas in Communications, IEEE Journal on*, 33, 2510–2523.

Lin, Y., Kämäräinen, T., Di Francesco, M. & Ylä-Jääski, A. 2015. Performance evaluation of remote display access for mobile cloud computing. *Computer Communications*, 72, 17–25.

Miettinen, A. P. & Nurminen, J. K. 2010. Energy Efficiency of Mobile Clients in Cloud Computing. *HotCloud*, 10, 4–4.

Nir, M., Matrawy, A. & St-Hilaire, M., 2015. Economic and Energy Considerations for Resource Augmentation in Mobile Cloud Computing. *IEEE Transactions on Cloud Computing*. DOI 10.1109/TCC.2015.2469665, IEEE Transactions on Cloud Computing

Nir, M., Matrawy, A. & St-Hilaire, M. Year. An energy optimizing scheduler for mobile cloud computing environments. *In:* Computer Communications Workshops (INFOCOM WKSHPS), 2014 IEEE Conference on, 2014. IEEE, 404–409.

Othman, M., Khan, A. N., Abid, S. A. & Madani, S. A. 2015. MobiByte: an application development model for mobile cloud computing. *Journal of Grid Computing*, 13, 605–628.

Saab, S. A., Saab, F., Kayssi, A., Chehab, A. & Elhajj, I. H. 2015. Partial mobile application offloading to the cloud for energy-efficiency with security measures. *Sustainable Computing: Informatics and Systems*, 8, 38–46.

Satyanarayanan, M. 1993. Mobile computing. *Computer*, 26, 81–82.

Vogels, W. Year. Head in the Clouds—The Power of Infrastructure as a Service. *In:* First workshop on Cloud Computing and in Applications (CCA'08) (October 2008), 2008.

Wu, H., Wang, Q. & Wolter, K. Year. Tradeoff between performance improvement and energy saving in mobile cloud offloading systems. *In:* Communications Workshops (ICC), 2013 IEEE International Conference on, 2013. IEEE, 728–732.

Zissis, D. & Lekkas, D. 2012. Addressing cloud computing security issues. *Future Generation Computer Systems*, 28, 583–592.

Communication and Computing Systems – Prasad et al. (Eds)
© 2017 Taylor & Francis Group, London, ISBN 978-1-138-02952-1

Optimizing energy consumption and inequality in wireless sensor networks using NSGA-II

Aanchal & Sushil Kumar
School of Computer and Systems Sciences, Jawaharlal Nehru University, New Delhi, India

Omprakash Kaiwartya & Abdul Hanan Abdullah
Faculty of Computing, Universiti Teknologi Malaysia (UTM), Johor Bahru, Malaysia

ABSTRACT: Due to the widespread growing usage of Wireless Sensor Network (WSNs) applications in almost every areas including medical, battle field, surveillance, industrial production, etc., lifetime maximization of the WSNs using energy optimization has attracted a lot of attention of researchers. Various techniques for energy optimization has been suggested based on single objective optimization or conversion of multiple objectives into single objective optimization and thus, enhances the lifetime of WSNs but led to increase standard deviation of energy among sensors. Due to the higher standard deviation, the possibility of optimal usage of energy hinders and many sensors retain higher energy when a network goes down because of out of energy in some sensors of the network. In this context, this paper proposes a technique for optimizing both energy consumption and inequality Non-dominated Sorting Algorithm (NSGA) which is a well know two objective optimization method. The optimization problem is mathematically formulated and an adapted NSGA-II is proposed to solve the problem. Major components of adapted NSGA-II include representation of chromosome, computation of energy consumption and energy inequality and sorting using NSGA-II. The analysis of simulation results have clearly shown that the proposed techniques effectively optimizes both the objectives.

1   INTRODUCTION

In the past decade Wireless Sensor Networks (WSNs) have witnessed much development and due to recent technological and manufacturing improvements, which make it economically feasible. The availability of low cost sensors resulted in the wide range deployment of WSNs in every possible area of usage staring from accessible areas to non-accessible or remote areas. Its applications ranges from monitoring, such as battlefield investigation, building inspection, security surveillance, to civil applications for example weather monitoring, disaster management, etc. The sensor nodes of the networks perform sensing task and report the data gathered while sensing to the sink node using the underlying infrastructure. Exhaustive research and development are undergoing in WSNs to succeed in dealing with the issues including memory constraint (Pathak et al., 2015), inadequate computation and communication capability (Chen et al., 2013), bandwidth limitation (Azizi et al., 2014) and most important limited energy.

Energy is one of the most important issue in the sensor network as the nodes are having a limited energy. That's why a lot of research is going on and literature contains a lot of popular energy conservation techniques that include duty cycling, data aggregation, load balancing and many others Optimization of network parameters is crucial for effective lifetime maximization which is lacking in traditional approached (Elhoseny et al., 2015). Optimization of forwarding path from the available multiple paths based on optimized value of the parameters including energy inequality, energy consumption, etc. has the potential to address the issues of the traditional approaches.

This paper proposes the technique for maximizing lifetime and minimizing energy deviation among nodes in wireless sensor network using Non dominating Sorting Genetic Algorithm (NSGA-II). The optimization problem is mathematically formulated by defining maximizing function for lifetime and minimization function for energy inequality. To solve the optimization function the adapted NSGA-II is proposed. Various components of NSGA-II are Chromosome representation energy consumption with respect to maximum energy, energy inequality sorting using non-domination, mutation and crossover operations. Two new functions are proposed for lifetime maximization and energy inequality of the nodes.

The simulator developed using C# under VISUAL studio 2010 IDE. The results are further analyzed using optimized value of chromosomes of last generation.

The rest of the paper is organized as follows. Related literature is reviewed in section 2. The optimization problem is formulated in section 3. Adapted NSGA-II is proposed in section 4. Analysis of empirical results are discussed in section 5. Section 6 concludes the paper.

2 LITERATURE REVIEW

In this section, single objective optimization based lifetime maximization in WSNs is qualitatively reviewed. Ant Colony Optimization (ACO) based energy optimization technique has been suggested in (Zhong et al., 2012). The technique finds the optimal Mobile Sink Schedule (MSS) for reducing energy consumption. Various parameters such as forbidden region, moving distance have been considered but these parameters are reduced as a single objective which optimize energy consumption. Therefore, in spite of enhancing life time using optimal use of energy, energy inequality increases which reduces the overall network life time. Another Energy Aware Evolutionary Routing (EAER) technique has been presented in (Enan et al., 2011) for balancing two conflicting objectives; namely, lifetime and network stability. Here also, both the objectives are not optimized separately. Motivated from social welfare approach, Maximum Energy Welfare (MaxEW) technique is suggested which consider number of network parameters including energy consumption, distributed and robust network operations (Changsoo et al, 2010). Social welfare functions have been utilized for defining parameters. All these aforementioned techniques use either singe objective optimization or converted the multiple objectives into single objective for optimization. Therefore, there is a need of optimizing both energy consumption and energy inequality using effective double objective optimization technique which is the aim of the proposed work.

3 THE OPTIMIZATION PROBLEM

The proposed technique maximizes the network lifetime by normalizing the energy consumption with minimized energy deviation of nodes. as the transmission proceed the network connections are updated with residual energy of the nodes and by considering the deviation or inequality of energy of nodes, it avoids using the same path frequently so to prevent quick depletion of energy of frequently used sensors.

Lifetime of the network is defined as the time when first node runs out of energy. Let initially all the nodes are having the same energy E_i where $i = 2,3,4....,N$ and $\bar{E}$ represent the average energy of sensors in the network. Consider the source node v_s wants to send data to destination node that is called as sink, the energy utilized in each communication is calculated using well known radio model which is defined by given below equations

$$E_{TE}(i,j) = k\left(E_e + E_{ampl} \times d^\partial\right) \quad (1)$$

$$E_{RE}(i,j) = kE_e \quad (2)$$

Where E_{TE} and E_{RE} represent energy of transmitting and receiving $k-bit$ of energy with d being the distance between the pair of sensor nodes and ∂ as path loss exponent of the medium. Using the total number of paths of equation, maximization function for lifetime can be expressed as

$$Maximize\ lifetime = minimize\left(E^{path}\right) \quad (3)$$

$$E^{path} = \sum_{i=S,j=1}^{i=n,j=D}\left(\frac{E_{i,j}^{link}}{E_{max}^{link}}\right) \quad (4)$$

Where E^{path} is the energy consumption of the path with respect to maximimum energy of the network. The constraint for the minimization function are as follows $\frac{E_{i,j}^{link}}{E_{max}^{link}} > 0, \frac{E_{i,j}^{link}}{E_{max}^{link}} \leq 1, E^{path} > 0, E^{path} \leq 1$.

For normalizing, energy consumption energy is divided by maximum energy consumption with respect to distance. Similarly the minimization function for energy inequality can be expressed as

$$minimize\ EI = \min_{a=1,2,...N^p}\left(EI_a^{path}\right) \quad (5)$$

Where EI^{path} represent inequality of energy of nodes along the path from source to destination. Normalized inequality of energy for the path can be calculated as

$$EI = \frac{1}{n}\left[\sum_{i=S}^{j=D}\left(\frac{E_i - E_{avg}}{E_{avg}}\right)\right] \quad (6)$$

Where, E_i is the residual energy of the node along the path, and $i,j \in opt_a^{path}$. E_{avg} represent the average residual energy of the network. The constraints for the minimization function are as follows: $EI \leq 1$, $EI > 0$, $\left(\frac{E_i - E_{avg}}{E_{avg}}\right) > 0$ and $\left(\frac{E_i - E_{avg}}{E_{avg}}\right) < 1$.

4 THE ADAPTED NSGA-II

In this section the major components of original NSGA-II are modified to fit for our proposed optimization problem. The modified components are described below.

4.1 Representation of chromosome

The ordered set of nodes as the path p_i with life-time maximization for energy efficiency and energy inequality values are represented as a chromosome of the modified NSGA-II. Each node of the set shows its corresponding genes. The chromosome is represented below.

4.2 Computation of energy consumption and energy inequality

Energy consumption of the chromosome is calculated using the energy consumption of individual links of the chromosome. Energy consumption of the link is equal to amount of energy utilized to transfer 1 unit of data from one node to another. As defined, initially all the nodes are having the same amount of energy. So to transfer a unit of data from node i to node i is represented as

$$E_{i,j}^{link} = \begin{cases} E \times d^\alpha \times \sin\theta^\beta & when \quad 0 < \theta < 180 \\ -E \times d^\alpha \times \sin\theta^\beta & when \quad 180 < \theta < 360 \end{cases} \quad (7)$$

Where E is defined as the unit of energy, d as the distance between node and the sink and θ represent the angle of the node from the x-axis if $0 \le \theta \le 180$ then the transmission is towards sink and if $180 \le \theta \le 360$ then transmission is away from the sink. So all those transmission which are away from the sink are rejected.

$$E_{max}^{link} = E \times d_{max}^\alpha \times 1 \quad when \quad \theta = 90$$

Where d_{max} represent the maximum distance between source and the sink. α and β represent the parameters to control the weights of distance and angle.

Energy inequality is calculated using residual energy of the ith node at time $t+1$ as

$$E_{i,t+1} = E_t - E_{i,j}^{link}$$

S(source)	Node-1	Node-2	...	Node-n	D(S\sink)
EC				EI	
(normalized Energy Consumption)				(normalized Energy Inequality)	

Figure 1. Chromosome representation.

Where E_t represent energy of the node at time t and $E_{i,j}^{link}$ represent energy consumption by the node to transfer a unit of data from node i to j.

4.3 Sorting using non-dominated approach

The population of the chromosome is sorted using non domination sorting method. The sorting is done based on domination. Let a and b are the two chromosomes of a population. The chromosome a is said to dominate chromosome b if atleast one of the objective function value of a is better than that of b. The ranking of chromosome in the respective population is done on basis of objective functions. The chromosome which are superioir most and not dominated by any other chromosome are ranked as first and further the one which is dominated by only one chromosome are ranked as second and so on. After ranking, crowding distance is calculated for each chromosome of the population. The population of next generation is selected using tournament selection approach with crowded comparison operator. For more details, non-dominated sorting the readers are advised to consult the advice.

4.4 Crossover and mutation for offspring operation

To perform crossover operation two chromosomes are randomly selected from the population and group of nodes are changed between these chromosomes in the same order. To perform mutation operation one chromosomes is randomly selected from the population and position of randomly selected pair of nodes is exchanged in the chromosome. Example of crossover and mutation operations is shown in Fig. 2 and 3 respectively.

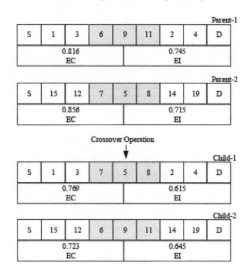

Figure 2. Crossover operation.

369

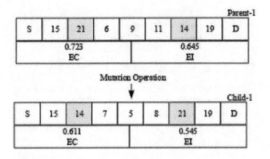

Figure 3.   Mutation operation.

5   EMPIRICAL RESULTS AND ANALYSIS

The proposed optimization of energy consumption and energy inequality in WSN is implemented in a simulator developed using C# programming language under Visual Studio 2010 Integrated Development Environment (IDE). The simulator has four classes mainly: *mainwindow.cs, RandomProvider. cs, pathSearchNSGA-11.cs, Networknodes.cs*. All the characteristics of the node in the network like its energy consumption, energy residual, distance from the sink, lifetime, etc. are implemented in *Networknodes.cs* for different simulation different set of nodes are randomly generated by *Random-Provider.cs* the optimization of energy consumption and energy inequality are implemented by P*athSearchNSGA-11.cs* and the result of the optimization are represented pictorially on *main-window.cs*.

Figure. 4(a) show the value of energy consumption and energy inequality after 500 generations with 100 nodes. It is depicted in the figure that the optimized value of energy consumption is approx. 0.99 whereas the optimized value for energy inequality is approx. 0.12. Initially we took less number of nodes and generations so chromosome have higher energy consumption and inequality. The result in Fig. 4(b) shows the value of two functions for the optimized chromosomes after 1000 generations with 500 nodes. It can be clearly shown that the optimize value of energy consumption is reduced to around 0.95 whereas optimized value for energy inequality is approx. 0.05. Hence by increasing the number of chromosomes it has reduced the energy consumption and energy inequality compared to network with 500 generations. The results in Fig. 4(c) show the value of energy consumption and energy inequality of optimized chromosome after 1500 generations with 1000 network nodes. It can be clearly observed that the optimized value of energy consumption is approx. 0.76 and correspondingly the optimized value of energy inequality is approx. 0.02 respectively. Hence by increasing

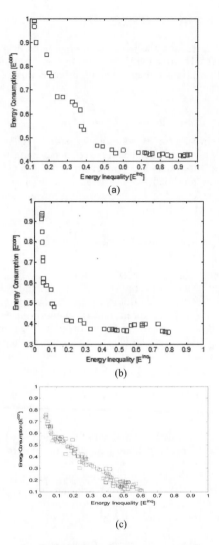

Figure 4.   Optimization of energy consumption and inequality: (a) with 500 generations, (b) with 1000 generations, (c) with 1500 generations.

the number of chromosomes and number of generations the chromosomes are tending towards the objective if the paper.

6   CONCLUSION

In this paper the recent approach for maximizing the network lifetime and minimizing the energy inequality in Wireless sensors network has been proposed using NSGA-II. The optimization problem is mathematically derived. To solve the optimization problem using NSGA-II, major components of the solution are representation of chromosome, computation of energy consumption and energy inequality, cross over operation, mutation have been clearly described. Two optimization functions for energy consumption and energy inequality have been derived. The simulation results of the optimization approach indicates that the two objectives considered in the paper optimizes significantly with increasing number of generations. In future research the authors will explore each component of NSGA-II based solution in detail by providing the pseudocodes and flowcharts. Including more objectives are considered in future.

REFERENCES

Azizi, T. and Beghdad, R., August. Maximizing bandwidth in wireless sensor networks using TDMA protocol. In Science and Information Conference (SAI), pp. 678–684, IEEE, 2014.

Changsoo Oka, Seokcheon Leeb. and Prasenjit Mitrac, Soundar Kumarad Distributed routing in wireless sensor networks using energy welfare metric Information Sciences Volume 180, Issue 9, pp. 1656–1670, May 2010.

Chen, Y.T., Horng, M.F., Lo, C.C., Chu, S.C., Pan, J.S. and Liao, B.Y., 2013. A transmission power optimization with a minimum node degree for energy-efficient wireless sensor networks with full-reachability. Sensors, 13(3), pp. 3951–3974, 2013.

Elhoseny, M., Yuan, X., Yu, Z., Mao, C., El-Minir, H.K. and Riad, A.M., 2015. Balancing energy consumption in heterogeneous wireless sensor networks using genetic algorithm. *Communications Letters, IEEE*, *19*(12), pp. 2194–2197, 2015.

Enan A. Khalil. and Bara'a A. Attea Energy-aware evolutionary routing protocol for dynamic clustering of wireless sensor networks, Swarm and Evolutionary Computation Volume 1, Issue 4, December 2011, pp. 195–203, 2011.

Pathak, A.A. and Deshpande, V.S., 2015, January. Buffer management for improving QoS in WSN. In Pervasive Computing (ICPC), International Conference on pp. 1–4. IEEE, 2015.

Zhong, J.H. and Zhang, J., 2012, July. Ant colony optimization algorithm for lifetime maximization in wireless sensor network with mobile sink. In Proceedings of the 14th annual conference on Genetic and evolutionary computation pp. 1199–1204. ACM, 2012.

Communication and Computing Systems – Prasad et al. (Eds)
© *2017 Taylor & Francis Group, London, ISBN 978-1-138-02952-1*

A comparative analysis of load balancing technique in cloud computing

Anvita Dixit & Arun Kumar Yadav
ITM University, Gwalior, Madhya Pradesh, India

Ram Shringar Raw
Indira Gandhi National Tribal University, Amarkantak, MP, India

ABSTRACT: Now days the cloud computing is one of the growing study zone its involves storage, software virtualization, networking, and on request web services. The main fundamental components of cloud infrastructure data centers, distributed server, Clients. Every day cloud computing users are increases. There is too much load on cloud server when vast number of client requests and tries to run application and send request on server this request occupied so many resources. To plummeting the heavy work load on cloud server. The Resource demand increases day by day for load balancing in cloud computing it is typical challenge to be faced. This paper discussed diverse load balancing techniques are used to clarify the problem with different parameter. Several methods given by the researchers using the load balancing method and comparison between several algorithms is done in this review paper.

1   INTRODUCTION

Cloud computing is the make use of the shared computing resources available over Internet. Computing resources can be hardware or software. It delivers services as per requirement and offers services as per payment. It allows user to customize, organize, and deploy cloud services. Cloud delivers resources above Internet with using various technologies like as virtualization, multi-tenancy, web services, etc. Applications interrelate above the Internet using web services.

Load balancing is solitary of the vital concerns in cloud computing. It is a system that allows the active local workload equally transversely each and every node and to avoid a condition where certain nodes are over loaded at the equivalent time as others are idle or liability little work. It helps to get a more user satisfaction, resource consumption ratio since improve the overall presentation and resource effectiveness of the system.

The main object is refining the performance with the help of balancing the load among these numerous resources like as network links, cpu, disk drives etc. to get optimum resource consumption, maximum throughput, maximum response time, avoid overload and performances.

Load balancing is a method that is use to partitioning a sequence of task into small task and executes each task simultaneously by assigning them to cloud assets like processors, memory, applications etc.

1.1   *Goals of load balancing*

• Optimum resource utilization.
• Maximum throughputs.
• Maximum response times.
• Avoiding overload.
• Performances.
• Throughputs.

Load balancing are of two types static or dynamic.

Static load balancing does not maintain the occurrence of the previous request while transferring the load to server.

Dynamic load balancing each request is maintain by load balancer so that work is equally distributed to various nodes on server and execute concurrently.

Load balancing technique uses various techniques to distribute these tasks to processor and other resources and manage the balance on server. Load balancing balance out the load, overall transforming time and improve the processor utilization.

It is a process of reallocating the absolute load to the separate nodes of the shared system to make resource consumption effectively and to pull through the response time interval of the job. Concurrently eliminating a state in which such nodes are over loaded while others are lightly overloaded.

Static load balancing does not maintain the case of the earlier request while transferring the load to

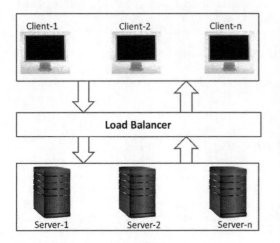

Figure 1. Load balancing.

server. Each request is maintain by load balancer in dynamic load balancing so that work is equally distributed to various nodes on server and execute concurrently.

2 RELATED WORK

Amandeep et al. (2014) presented algorithm is entirely contemptible on virtual machine. It will find the node for handover the new job and user first insist the load controller to search the suitable virtual machine, which access that easily load and perform the tasks and that is given by the client. The job supervisor will maintain a list of node using index list; that it will assign the job to rigorous node. And also in this algorithm the client initial request to the load balancer and find a relevant Virtual Machine to perform the appropriate task.

Dong, Bin et al. (2012) proposed an algorithm base on distributed architecture for dynamic and adaptive approach. It is also aware of network transmission rate which it generally avoids to take load decision as only single network transmission is considered. Self Active Load Balancing (SALB) ability to figure out future load help other servers to take decisions thus reducing decision delay a network transmission. Its threshold value can also be adjusted dynamically and also SALB works silently and whole system provides service does not interrupt by SALB.

Gupta, Ruhi (2014) presented a Role Based Access Control (RBAC) method that is use to decrease the load of the cloud. The role is designate to each user so that inadequate applications of the cloud can be access by their individual number of users. So this approach, the resources are restricted to the users.

Kumar, Suresh (2015) suggested a pheromone table designed that is efficient by ants as per the source utilization and node selection methods. Ants shift in forward direction even if the node is lightly overloaded or overloaded. At the time of movement, if finds the overloaded node, its movement in backward direction else it ants finds a few loaded node once more it shift in a forward way and replaces were the node found before. The overloaded node is negotiated, then ants move back to fill the recently faced loaded node, so a distinct table is updated every time.

L.D, Dhinesh Babu et al. (2013) introduces with honey bee behavior algorithm process the task on cloud server using priorities which balances the workload of nodes and it reduce the delay time interval of the tasks in queue. This introduces a major improvement in a time execution and decrease the delay time interval of tasks in queue. The algorithm which is motivated from the foraging behavior of honey bees. Search the food and startling others to go and eat the food. Honey bees sent to search the appropriate source of food is called forager bees. Forager bees firstly go and find their food. When they found the food, after they coming reverse to their own bee and they advertise it in the form of dance. After seeing the power of dance, the leader bees chase the forager bees and grow the food. So this entire procedure is generalized to overloaded or lightly weighted virtual servers. The server evolution the requests of the users like as to the food of the bees. As the server gets overloaded, the bees search for alternative place i.e. client is rearranged to any other virtual server.

Priya, Mohana S. et al. (2013) proposed Resource Allocation Scheduling Algorithm (RASA). This algorithm is the grouping of the Max-Min and Min-Min algorithms. In this algorithm, virtual nodes generated first and have no time consuming instruction. The probable response time interval of all virtual node is found. According to the minimum loaded node norms, efficient virtual node is found and ID of that node is return to the client. In this, Min-Min and Max-Min approaches are followed. If available numbers of resources are even, and then Max-Min approach is applied else Min-Min approach is applied. The time density of RASA is given as $O\ (mn2)$ where m and n are the numbers of resources and tasks.

Rajan, George Rajesh et al (2013) proposed Active load balancing in spread virtual atmosphere using heat diffusion. In spread virtual atmosphere various number of users and they load access simultaneous it can cause problem, that is based on selecting best efficient cell in conjugation with two heat based diffusion algorithm called Local and Global diffusion. Heat diffusion concept heat run from high temperature to low temperature. Simi-

Table 1. Comparison of existing algorithm.

Load balancing methods	Parameters	Merits	Demerits
Throttled load Balancing Algorithm	Load Movement Factor, Communication Cost, Network Delay	High Load Movement Factor	High Communication Cost, High
A dynamic and adaptive load balancing strategy	Scalability, availability, throughput, speed processing, elasticity, response time, resource utilization	High speed, good response time, good resource utilization	Due to continuous file migration the system performance get degrade.
RBAC	Performance, Resource Utilization, Energy, CPU Burst Time	Performance and Resource Utilization are high, CPU Burst time decreases	Response Time increases
Ant Colony Optimization	Performance, Resource Utilization, Fault tolerance, Scalability	Performance improved, Resource utilization high, Fault tolerance tremendous, Scalability Good	Complex Network
Honey Bee foraging algorithm	Resource utilization, Execution Time, Overheads, Throughput	Good, Maximize throughput, Low overheads	Low Priority task
RASA	Performance, Execution Time	Performance increases, Execution time decreases	Less Fault Tolerance
Dynamic load Balancing	Communication, speed	Communication overhead less, speed high	Network delay is high
Task Scheduling Algorithm based on Load Balancing	Response Utilization, Performance, Response to Request Ratio	Maximize Response Utilization, Increased Performance	Doesn't improve response to request ratio
Fast Adaptive Load Balancing	Communication, speed, efficiency	Lower Communication, faster speed, high efficiency	Maintain network topologies

larly transfer of the load from heavy Overloaded to Low Over loaded Virtual Machine. In local diffusion algorithm, it is local decision making and able to cell selection pattern are usage. In narrow diffusion they are basically compare the adjacent node loads. If load is lesser then the transfer of load becomes possible. Once global diffusion algorithm deliberated, it has two phase that is global scheduling and local load migration stage.

Sidhu, Kaur Amandeep et al. (2013) suggested a two-level task scheduling mechanism. It is established on load balancing to meet vigorous fulfillment of users and get high resource consumption. It accomplish load balancing by primary mapping jobs to virtual machines and then to host resources. Task scheduling approaches only focus on efficiency will increase the rate of time, throughput, space, the job response time, source utilization and on the entire performance in environment of the cloud computing at the equivalent time.

Verma, Manisha et al. (2015) studied parallel practical based simulations use a fast adaptive method. The main characteristic of this algorithm is to transfer or adjust the load between numerous processors from local domain to global domain through reducing and extending the cells according to net difference of workload among adjacent cells.

3 COMPARATIVE ANALYSIS

3.1 Roundrobin algorithm

Round robin algorithm is random selection based. And also in this algorithm load is equally distributed to all given nodes in circular order without any priority. It means it selects the load arbitrarily in case that some server is deeply loaded or some are without due consideration loaded. Each one node creates its own load table independent of provisions from remote node. It is easy to implement and simple. It does not need inter process communication and provides superlative performance for distinct purpose applications. It cannot give conventional result in general case and when the jobs are extreme processing time.

Equally spread current execution algorithm
It latest execution algorithm techniques control with priorities. It allot the load randomly in terms of size and transfer the load to virtual machine which is lightly loaded node that handle job easily and proceeds minimum time, and give maximize throughput. It is spread spectrum method in which the load balancer spread the load of the task in hand into various virtual machines.

3.2   *Throttled load balancing algorithm*

It is effectively based on virtual machine and firstly user request the load balancer to analysis the reasonable virtual machine that access load and execute the tasks which is given by the user. And also the client first requests the load balancer to

Table 2.   Comparative analysis of different algorithm on the basis of parameter.

Algorithm/ Parameter	Job scheduling	Time	Cost
Round Robin	R1	50.37	16.09
	R10	50.45	20.81
	R25	78.9	28.36
ECSE	EC1	50.09	0.99
	EC10	50.07	5.5
	EC25	50.17	13.03
Throttled	TH1	50.1	0.99
	TH10	50.09	5.5
	TH25	50.19	13.03

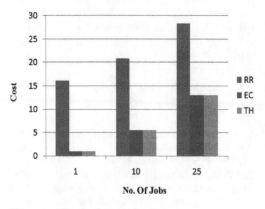

Figure 2.   Graph of cost estimation.

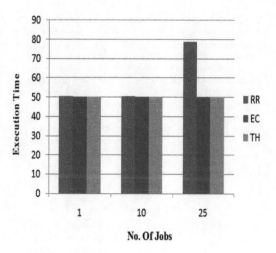

Figure 3.   Graph of time execution.

treasure an accurately Virtual Machine to perform the required operation.

4   CONCLUSION

In this paper we review the load balancing concern and studied various techniques used for load balancing in cloud computing. Load balancing is the typical problem to be faced and it uses various techniques to distribute these tasks to storage, processor and other resources and manage the balance on server. Load balancing balance out the load, overall processing time and improve the processor utilization. It also certifies that all computing resource is shared well and equally. Many researchers used the load balancing techniques to propose new approaches. Relative analysis of these above algorithms with respect to scalability, resource utilization, performance, response time, communication, throughput and overhead associated. But still the load balancing issue is open for research work so that high user agreement and resource utilization achieved.

REFERENCES

Amandeep, Yadav, Vandana Mohammad, Faz 2014 Different Strategies for Load Balancing in Cloud Computing Environment: a critical Study, International Journal of Scientific Research Engineering & Technology (IJSRET), ISSN 2278-0882 Volume 3 Issue 1, April.

Beniwal, Payal & Garg, Atul 2014a comparative study of static and dynamic Load Balancing Algorithms, International Journal of Advance Research in Computer Science and Management Studies Volume 2, Issue 12, December.

Dillon, Tharam and Wu, Chen and Chang, Elizabeth 2010 "Cloud Computing: Issues and Challenges" 24th IEEE International Conference on Advanced Information Networking and Applications.

Dong, Bin Li, Xiuqiao Wu, Qimeng & Xiao, Limin & Ruan, Li 2012 "A dynamic and adaptive load balancing strategy for parallel file system with large-scale I/O servers", J. Parallel Distribution Computing. 72 () 1254–1268.

Gupta, Ruhi 2014 Review on Existing Load Balancing Techniques of Cloud Computing, International Journal of Advanced Research in Computer Science and Software Engineering, Volume 4, Issue 2, February.

Kumar, Suresh 2015 Review on Existing Load Balancing Techniques of Cloud Computing INDIA/International Journal of Research and Computational Technology, Vol.7 Issue 1 ISSN: 0975-5662, March.

L.D, DhineshBabu & Krishna, Venkata P. 2013 Honey bee behavior inspired Load Balancing of tasks in cloud computing environments, ELSEVIER, vol.13, pp. 2292–2303.

NITIKA, Ms. Comparative Analysis of Load Balancing Algorithms in Cloud Computing, International Journal of Engineering and Science.

Priya, Mohana S. & Subramani, B. 2013 A New Approach for Load Balancing in Cloud Computing, International Journal of Engineering and Computer Science ISSN: 2319-7242 Volume 2 Issue 5 May.

Rajan, George Rajesh & Jeyakrishnan, V. 2013 A Survey on Load Balancing in Cloud Computing Environments, International Journal of Advanced Research in Computer and Communication Engineering Vol. 2, Issue 12, December.

Sidhu, kaurAmandeep & Kinger, Supriya 2013 Analysis of Load Balancing Techniques in Cloud Computing, International Journal of Computers & Technology www.cirworld.com Volume 4 No. 2, March–April, ISSN 2277-3061.

Verma, Manisha, Bhardawaj, Neelam & Yadav, Kumar Arun 2015 An architecture for Load Balancing Techniques for Fog Computing Environment, Vol 6 • Number 2 April–Sep pp. 269–274 DOI: 10.090592/IJCSC.2015.627.

Communication and Computing Systems – Prasad et al. (Eds)
© 2017 Taylor & Francis Group, London, ISBN 978-1-138-02952-1

Implementation of security & challenges on vehicular cloud networks

Raj Kamal, Ram Shringar Raw, Nidhi Gulati Saxena & Sachin Kr. Kaushal
Gautam Buddha University, Greater Noida, India

ABSTRACT: Basically Vehicular Cloud Network is combination of Vehicular Cloud (VC) and Vehicular Cloud Computing (VCC). In Vehicular Cloud Networks (VCN) the vehicles are communicated through two ways: primary is Vehicle to Vehicle (V2V) and secondary is Infrastructure to Vehicle (I2V). We apply Vehicular to Vehicular (V2V) security in cloud network to secure the privacy, authentication, protection, Security challenges and issues of V2V and I2V cloud infrastructure. In this paper, we have proposed a Key Generation & Distribution Protocol (KGDP) to select the maximum Key size and apply authenticate the secure vehicle communication. We are used maximum group key (G_{max}) with adding a random number for the vehicle's password and vehicle id as user name to secure the encrypted message with maximum key size with adding a random number which are sent by one vehicle to another vehicle. In this paper, we have shown the comparison of Throughput/Overhead for key sizes that are used in secure vehicle communication.

1 INTRODUCTION

This work is to facilitate their vehicles proficient in the market; the vehicles manufacturers are offering more and more potent onboard devices, including influential computers, a large link of wireless transceivers. These are all devices provide a set of clientele that except their vehicles to give unified addition of their home surroundings settled by refined amusement centers, access to internet, and other similar requirements and needs. Vehicular Cloud Networks (VCN) is basically combination of Vehicular Cloud (VC) and Vehicular Cloud Computing (VCC). Security and privacy problems want to be addressed if the VC (Vehicular Cloud) thought is to be broadly adopted. In VC (Vehicular Cloud), each and every one the users, including the attackers, are equal. The attackers can be bodily located on one machine. VANET is a vehicular communication network which has transformed the transportation systems. VANET (Vehicular Ad-hoc Network) possesses different characteristics such as mobility, go-ahead in nature, actual time processing, self-organizing of the nodes etc.

The figure of vehicular cloud network is shown below. In this figure the vehicles have communicated through Vehicle-to-Vehicle (V2V) and Infrastructure-to-Vehicle (I2V). When vehicles are in road then they are also connected through Road Side Units (RSUs) and On Board Units (OBUs) with cloud computing network.

The main purpose of VANET is to give secure, at ease drive and propose safety measures in traffic.

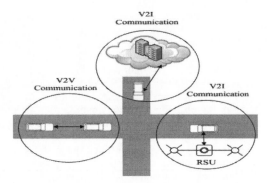

Figure 1. Vehicles communication in vehicular cloud computing.

Vehicle-to-Vehicle (V2V) and Infrastructure-to-Vehicle (I2V) allows the OBU's to communicate with each other and with RSU's using wireless technology. In this paper, we are proposed a Key Generation & Distribution Protocol (KGDP) designed for maximum group key (G_{max}) used for all group vehicles and communicate the G_{max} as vehicles password to communicate the message to other vehicles.

2 RELATED WORKS

There are many works pertaining to the secure Group Communication and that have been carried out, but some of the important works has been surveyed and cited here. Among them, Mike

Burmester and Yvo Desmedt (D. Huang, 2011) obtainable a Group Key Exchange protocol which extends the Diffie-Hellman protocol (Bohli, 2006). Bohli developed a framework for robust group key agreement that provides security against malicious insiders and active adversaries in an unauthenticated point-to-point network. Bresson et al. (Hellman, 1976) constructed a general authenticated group Diffie-Hellman key exchange algorithm which is more secure. Katz and Yung (E. Bresson, 2000) proposed the initial constant-round and completely scalable group Diffie-Hellman protocol which is provably secure. There are many other works related to group key management protocols based on non-DH key accord approaches. Now a days, vehicular ad hoc network (VANET) security and privacy have been addressed by a enormous number of papers. Yan et al. (J. Sun, 2010)proposed active and passive location security algorithms. Public Key Infrastructure (PKI) and digital signature-based methods have been fine out in VANETs (Olariu, 2009). A Certificate Authority (CA) generates public and private keys for nodes. The use of digital signature is to prove and authenticate the sender.

3 PROPOSED WORKS

In this paper, we are going to put forward a Key Generation & Distribution Protocol to calculate maximum group key (G_{max}) and adding with a random number on this maximum group key (G_{max}) for use as password for sending message between group vehicles which are connected through cloud Key Distribution Center (KDC). We generate maximum group key (G_{max}) such as 16 byte because maximum length of group key is very hard to detect from any system or other unauthorized system. We also compare the throughput/overhead for 16 byte using authentication mechanism for group vehicles and symmetric vehicles (i.e. means between two vehicles).

3.1 Research objectives

- To study the various security issues and challenges in vehicular cloud network and VANETs.
- To generate the Key Generation & Distribution Protocols (KGDP) for protected group communication.
- To comparison the authentication mechanism of throughput or overhead between digital signature and group key.

3.2 Assumptions

1. This system uses SHA-128 Hash mechanism that is very secure to send the authenticated message.

2. Throughput/Overhead, depend upon the number of vehicles, so we take less number of vehicles to minimize the throughput/overhead.
3. The throughput/overhead of group keys count less than of signature key but overhead of signature is best.
4. For Key Generation & Distribution Protocol (KGDP) algorithm, we select the random group key. After comparison random key we choose a maximum random group key Kn and apply a random password with max group key (G_{max}) to verify for different-2 group vehicles.

3.3 Process for generating maximum key (G_{max})

Key Generation & Distribution Protocol (KGDP) is the most appropriate method to generate the key and distribute to all vehicle in that group. In the KGDP, first of all vehicles registered with their permanent is Ai and secret id Bi with Key Generation Centre (KGC). When registration complete KGC gives a random key for each vehicle in that group.

1. Key generation
Too much secure the vehicles System, we will generate a group key which is very hard will not detect any other system such as man-in the center attack and Brute Force attack. There are groups of vehicles participating grouping communication. In the KGDP (Key Generation Distribution Protocol), there are n vehicles of keys such as K1, K2, and K3....K_n. We have to select maximum number of key to secure and efficient for group communication. Key Generation Distribution Protocol (KGDP) selects a random group key $K > (A_i \otimes B_i)$ compute the message (M_i, N_i) for all i of group G_i.

$$Mi = \frac{K}{Ai \oplus Bi} \tag{1}$$

$$Ni = K \bmod (Ai \oplus Bi) \tag{2}$$

When pair generates KGP published (M_i, N_i). From this public information each group vehicles V_i can be able to retrieve the key group computing.

$$Kn = Mi * (Ai \oplus Bi) + Ni \tag{3}$$

Only authorized vehicle can be access the hidden key using (M_i, N_i).

2. Group Re-Keying
The group keys needs to be modernized to maintain the frontward and rearward secrecy. To achieve this, the two important tasks namely vehicles join and vehicles leave process is performed.

A. Vehicle Join:

When a Vehicle wants to join the system, the new vehicle will register with the KGP. KGP will share a prime number A_{n+1} and the vehicle will provide the secret id B_{n+1} where $K' > (A_{n+1} \oplus B_{n+1})$. KGP generates the new pair of values (M_i, N_i) by using equation (1) and (2). After receiving the (M_i, N_i) pairs, the newly joined member can use the prime number along with his secret id to derive the key K_n from equation (3).

B. Vehicle Leave:

When a Vehicle leaves the system, the vehicle should inform to the KGP. Now KGP generates a new group key as follows.

Step 1. KGC selects a original prime maximum number K' (where $K' > (A1 \oplus B1)$ for all i).

Step 2. New pair of values (M_i, N_i) generates with the new Key K_n and distributes it to all the group vehicles.

3.4 Proposed algorithm for KGDP & authentication with group vehicles

Step (1):- Every vehicle registered with Key Distribution Center (KDC) with their random secret id (Bi) and permanent id (Ai).

Step (2):- Key Distribution Center (KDC) distribute a random key K to all registered vehicles.

Step (3):- Vehicles generate a message pair (Mi, Ni) and Kn from equation number (1), (2) and (3) for authenticate to KDC.

Step (4):- Generate Kn for all vehicles and select them maximum group key (G_{max}).

Step (5):- Now add a random number with the maximum group key (G_{max}) as a password for all vehicles to connected and sending the information.

Step (6):- Choose vehicles and random distance then apply the password. (Random distance = 50 meters and password = maximum group key (G_{max}) + A random number).

Step (7):- If random distance greater than 50 meters and enter, the password wrong then vehicles are not in the range e and not connected of lead vehicle.

Step (8):- Then exit or wrong result.

Step (9):- Repeat step 5–7.

Step (10):- End.

3.5 Methodology to calculate the throughput/ overhead

We use the two type's methods to calculate the throughput/overhead. These are given as:

3.5.1 Pair wise keys

In the pair wise key, we will communicate between two vehicles P and Q because this method is based on symmetric cryptography, which is additional efficient rather than asymmetric technique (G. Yan S. O., 2009). Vehicles P sends the session key K to Q encrypted with B's public key:

$$P \rightarrow Q : Q|K|TpubQ, SigQ|K|T_{priP} \qquad (4)$$

where P and Q are the identities of Vehicles P and Q. K is the secret key common by the both vehicles; m is the sensitive message; t is the time stamp; pub_Q is the public key of Q and pri_P is the private key of P.

3.5.2 Secure group communication

The make use of symmetric keys for authentication would minimize the security throughput. Vehicle X allocate the group key K to other vehicles Y, A, B, C, d and others. Vehicle X is the lead car Group leader.

$$X \rightarrow * : Hy, \{K\} \, PuKY, Ha, \{K\} \, PuKa, Hb, \{K\} \\ PuKb, Hc, \{K\} \, PuKc, SigprK_X \qquad (5)$$

Only HMAC secure message will broadcast itself:

$$X \rightarrow * : m, HMACkm.$$

When a new vehicle D enters the area, it receives the group key from the current group leader:

$$X \rightarrow D : \{K\} PuKd, Sig_{Pr}K_L \, [\{K\} \, PuKd] \qquad (6)$$

3.6 Comparison of authentication mechanism for throughput

In this section, we almost evaluate the performance of the dissimilar authentication mechanisms discussed, namely ElGamal based digital signatures, symmetric pair wise keys and symmetry group keys. We will spotlight on the message size and message number throughput corresponding to each mechanism. We suppose that Public Key Infrastructure (PKI) is ElGamal algorithm with a key size is 128 bits or 16 bytes; the signature and cipher text size just twice of key size (32 byte each)

Table 1. Notations of KGDP.

Parameters	Description	Parameters	Description
K	Random Group key	Vi	Vehicles for i^{th} group
Ai	Permanent Secret Id	I, $\oplus$	No. of groups, XOR Operation
Bi	Random Private Value	Gi	I^{th} Vehicles group
Mi	Message Pair	Ni	Message Pair

Table 2. Notation for authentication mechanism.

Parameters	Description	Parameters	Descriptions
M	Message	K	Group or Shared Key
K, *	Secret key, Whole message	T	Timestamp
Pubq	Vehicle Q public key	Prip	Private key for vehicle P
X	Lead Car	Y, A, B, C, D	Group Car

and certificate mainly as public key and CA's Signature over it (i.e., 48 bytes total). The computation of the overheads follows:

3.6.1 Digital signature
Using the message set-up in equation 1, the size of the throughput/overhead is the same for each message that is:

$$\left(Certificate \left(Key\ Size + Signature\ Size \right) + ciphertext\ or\ message\ lenght \right)$$

3.6.2 Group keys
Using the message set-up in equation 2, key organization for N vehicles requires the leader to send $(N-1)$ cipher texts in adding to one signature, which expenses $(32(N-1)+32)/N = 32$ bytes/vehicle. Assuming cells overlap over 100 m (for reliable handover), 3 s is the time a vehicle wants to switch between two cells while driving at a speed of 120 km/h. Moreover, let us assume the leader adds 3 new members every 3 s (assuming a highway with 3 lanes per direction), which expenses one encryption and one signature, i.e., **(32 + 32) * 3 = 192 bytes.**

The total throughput/overhead for group keys given below:

$$\frac{Signature\ size}{Message\ lenght\ (coiphertext)} + Key\ size$$
$$+ \frac{total\ overhead\ of\ digital\ signature}{10}$$
$$+ (one\ encryption + one\ signature)$$
$$* Highway\ lanes\ per\ direction\ /$$
$$10\ Bytes\ *\ /\ Message\ *\ Vehicles$$

4 RESULTS AND ANALYSIS

In simulation we take the random key for every group vehicles from Key Distribution Center (KDC). In this section of this paper, we generate the maximum group key (G_{max}) adding a random number to generate a password to sending the message other vehicles which are in random range less than 50 meter. Every vehicle is in range of 50 meter then vehicles receive the message from lead vehicles.

4.1 Maximum key generation (G_{max}) with password

The Key Generation & Distribution Protocol (KGDP) provides the G_{max} (Maximum group key) in five groups of vehicles. Each vehicles have own registered user id and their password for authentication. Every vehicle registered in Key Distribution Center (KDC) and finds their random key $K > (Ai \oplus Bi)$ and vehicles generate with random private value B_i and permanent private id A_i then generate a message pair (M_i, N_i) above formula. The table is given below for vehicles keys and their permanent private id (Ai), random secret id (Bi) and also generated message pair for all group vehicles.

Hence the value of maximum group key (G_{max}) is 99901. For vehicle-to-vehicle, authentication we use adding random number password with G_{max} and vehicles use this password to communicate the message each other. The random vehicle V_A acts as the lead vehicle and distributes the message to vehicles V_B and V_C, which are in range 24 and 19 meters. These vehicles have connected with lead vehicles use the password. The figure for this scenario is shown below.

Table 3. Group vehicles information for results.

SR. No	VA		VB		VC		VD		VE	
1	K1	73133	K2	60041	K3	63617	K4	99901	K5	15973
2	A1	51061	A2	86813	A3	79903	A4	91453	A5	83663
3	B1	59399	B2	65599	B3	97553	B4	87403	B5	90007
4	(M1,N1)	(6,257)	(M2,N2)	(2,17477)	(M3, N3)	(3,105 83)	(M4,N4)	(8,90 9)	(M5,N5)	(2,299 7)
5	EXTRACT KEY	73133	EXTRACT KEY	60041	EXTRACT KEY	63617	EXTRACT KEY	99901	EXTRACT KEY	15973

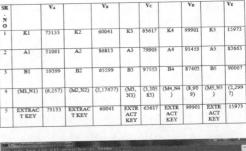

Figure 2. For Gmax 99901, the VA acts as lead vehicle and VB, VC receive message.

Figure 3. For Gmax 76871, the V_E acts as lead vehicle and V_B receive message.

Table 4. Comparison of throughput for key size 16 byte.

S. No.	Key Size	Ciphertext	Signature Size	CA	Throughput/overhead for Signature	Throughput/overhead for Group Key
1	16	32	32	48	80	44.2

Figure 4. Throughput/overhead for group key when key size 16 byte.

Figure 5. Throughput/overhead for digital signature when key size 16 byte.

4.2 Compare results of overhead/throughput for digital signature & group key

We compare the result of throughput/overhead for digital signature and group key from above formula, which is given in above. For this system, we take key size 16 byte. The comparison result for this key looks at below table.

The result of above table is given below:

For key size 16 byte, the throughput/overhead of group key and digital signature such as:

In above figures, we find that key size different throughput for one approach in bytes. The overhead for group vehicles save the 45% message throughput/overhead.

5 CONCLUSION

In this paper, we have proposed a technique to generate the maximum group key (G_{max}) with adding a random number to calculate the password for every vehicle which are a random range less than 50 meter. If range greater than 50 meter then lead vehicle could (G. Yan, 2008) (G. Yan S. O., 2009) (J. Sun, 2010) (Olariu, 2009)not communicate through the other neighboring vehicles. We also compare the throughput/overhead for maximum group key size for digital signature and group key. Our proposed algorithm is very capable in conditions of key extraction and message generation. In addition, we wish to compare our algorithm with the dual-level key management for secure group communication.

REFERENCES

Bohli. (2006). A Framework for Robust Group Key Agreement. *In Computational Science and Its applications ICCSA Lecture Notes in Computer Science in Springer.*

Bresson, E. O. C. (2000). Provably Authenticated Group Diffie-Hellman Key Exchange. *Proc. ACM Conf. Computer and Comm. Security (CCS '01).*

Desmed, M. B. (n.d.). *A Secure and Scalable Group Key Exchange System.* 2005: Information Processing Letters.

Hellman, W. D. (1976). New Directions in Cryptography. *IEEE Trans. Information Theory.*

Huang, D. S. M. (2011). PACP: An Efficient Pseudonymous Authentication Based Conditional Privacy Protocol for VANETs. *IEEE Trans. Intell. Transp. System.*

Olariu, G. Y. (2009). An efficient geographic location-based security mechanism for vehicular ad hoc networks. *Proc. IEEE Int. Symp. TSP, Macau SAR, China,.*

Sun, J. C. Z. (2010). An identity-based security system for user privacy in vehicular ad hoc networks. *IEEE Trans. Parallel Distrib., Syst.*

Tzeng. (2002). Secure Fault-Tolerant Conference-Key Agreement Protocol. *IEEE Trans. Computers.*

Xie, H. L. K. (2010). Privacy-aware traffic monitoring. *IEEE emTrans. Intell. Transp. Syst.*

Yan, G. S. O. (2008). Providing VANET Security Through Active Position Detection. *Comput. Commun.*

Yan, G. S. O. (2009). Providing Location Security in Vehicular ad hoc Networks. *IEEE Wireless Communnication.*

Yung, J. K. (2007). Scalable Protocols for Authenticated Group Key Exchange. *J Cryptology.*

Communication and Computing Systems – Prasad et al. (Eds)
© 2017 Taylor & Francis Group, London, ISBN 978-1-138-02952-1

Ad hoc network in MANET routing using swarm intelligence

Anshika Shukla
Banasthali University, Rajasthan, India

Neelam Ruhil
Dronacharya College of Engineering, Gurgaon, Haryana, India

ABSTRACT: This Paper shows research related to Manet routing using Swarm Intelligence. Basically the joint actions of, self-organized systems, natural or un-natural called as Swarm intelligence. The thought is being used in works in the field of AI (Artificial Intelligence). Networks are flattering extra and additional composite and this is advantageous that these networks can categorize on their own as well as configure themselves, adjusting according to latest circumstances with respect to movement, services, system connectivity etc. In Order to back the latest model, upcoming network algorithms should be stout, able to perform in a distributed manner, should detect variations in the network, and adjust accordingly. Natural self-organizing systems such as insect colonies demonstrate exactly the desired behavior.

1 INTRODUCTION

The motivation often come from natural surroundings, particularly from natural systems. The representatives go after simpler and straightforward principles, and even though there's no central regulating arrangement ordering in what way discrete actors are supposed to act, confined, and to a some point arbitrary, collaborations amongst these actors leads to rise of "intelligent" overall conduct, anonymous to each discrete actor. Instances from biological coordination of SI comprise ant societies, bird clustering, animal herding, and bacteriological evolution, fish shoaling and microbial intelligence. The impression is like networks are turning even more complicated and thus there is a need such these networks can categorize on their own as well as configure themselves, adjusting according to new circumstances with respect to movement, services, system connectivity etc. For backing the latest model, upcoming network algorithms should be stout, able to perform in a distributed manner, should detect variations in the network, and adjust accordingly. Natural self-organizing systems such as insect colonies demonstrate exactly the desired behavior. Employing no. of comparatively simple natural actors (e.g., ants) various dissimilar structured conducts are created on system level because of localized collaborations amongst actors and surroundings. Strength and efficiency of such combined actions in context with differences of environmental circumstances are some vital reasons for their

biological victory. These kinds of arrangements are stated as Swarm Intelligence. Swarm networks have lately emerge as basis for motivation for designing distributed and adaptive systems, and specifically routing procedures. Routing is a job of giving directions to data that streams from origin node to terminal node enhancing system efficiency. Numerous effective routing algorithms have been projected that have been inspired by ants society conduct and linked structure of Ant Colony Optimization (ACO) (V. Maniezzo, 2004). Instances of ACO based algorithms are AntNet (Blum, 2005) and ABC (A. Colorni, 1991). An Example of network where necessity for autonomous regulation stays fundamentally essential is Mobile Ad Hoc Networks (MANETs). They are systems where every node is moveable and interact within themselves through wireless mode. Nodes can connect and disconnect any point of time. No defined structure exists. Every node is considered equal, also there's no central controller. Selected routers are also absent: nodes function as router for each other, and data packets are transmitted within nodes in multiple hop manner. The ACO routing algorithms stated earlier have been proposed to be employed in wired systems. They operate in distributive and confined manner. They are also capable of observing and adapting to variations, in movement configurations. Though, variations in MANETs are even extra extreme. In addition to changes in movement, network topography as well as no. of nodes changes continuously.

2 SWARM INTELLIGENCE BASED MANET ROUTING

The elementary notion behind ant based routing algorithm has been adopted from ants' food searching tactic. Group of ants start looking for food from their Anthill and move in the direction of food, creating dissimilar paths/routes. Whenever an ant come across a diversion it needs to choose which route to move on. Furthermore when moving on the route (to and fro from the food source), they leave pheromone, a chemical substance, that marks the path they followed. Rest of the ants can smell the pheromone and are able to differentiate its concentration as well, this provides an idea to them about the usage of that particular route hence impacts their decision. With time this accumulation of pheromone reduces due to dissemination. This characteristic is significant to know which path is being rarely engaged, maybe because of degradation. Figure 2 demonstrates a situation when the best route amongst two options is taken by ants after a while. Bee-inspired algorithms are primarily established on exploring principle of bees. Two different kinds of bees being used for implementing routing in MANETS are scouts and foragers. Scouts determine new nodes from source node to target node. Once a scout reaches the target, it commences a backward trip to the source. After returning to the source, a scout employs the foragers for its route by the metaphor of dance (A. Colorni, 1991).

2.1 Ant colony optimization

Here we will see what ant colony optimization is and what all the hype is around it. Suppose you are an ant, living alone far away from other ants, looking for food. One approach could be that you walk to each and every point in each and every direction around your starting point by yourself and then see where you can get food from. That approach would no doubt be tedious not to mention time

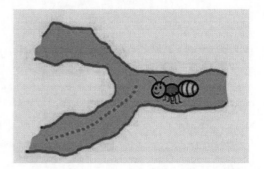

Figure 2. Ant choose the path where maximum pheromone occur and move towards the path.

consuming. Now consider this: you live in a colony of ants. You want food. You have say six ant friends willing to help you. You send each of these ants to six points in different directions. They send you their positions. Now from their respective positions each of them moves towards the food source choosing any random path they feel like- and then back towards their initial starting points.

At this point, note that ants secrete a substance called pheromone on the paths that they take from food to back home. The quantity of pheromone secreted along a route is inversely proportionate to the distance from food source using that path. This means that the shortest path to the food source will get maximum pheromone deposited on it. Now, you can see which path has the maximum pheromone and move towards the starting point of that path.

3 SOLVING TRAVELLING SALESMAN PROBLEM USING ANG COLONY OPTIMIZATION

3.1 Travelling salesman problem

Instinctively, the travelling salesman problem is a salesman dilemma. He starts from home wishes to discover a shorter path which will lead him via specified collection of consumer towns there after return back to the place he started from, going to every consumer town precisely once. The TSP is symbolized using fully weighted graph G = (N, E) with (N) representing collection of vertices depicting towns, & (E) representing collection of edges/paths (Kangshun Li, 2008).

Every path (i, j) ∈ E is allocated some cost (length) dij, which depicts the distance amidst towns i and j, with (i, j) ∈ N. The aim of TSP is to search for a minimal span Hamiltonian circuit for the graph. A Hamiltonian circuit represents a closed route consisting each one of the n = |N| vertices of the G precisely once [10]. The main reason behind the choice of TSP for ACO are.

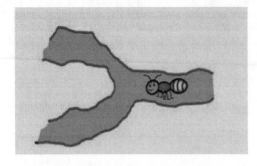

Figure 1. Sending ants for finding the path.

- TSP is a significant NP hard optimization problem which rises in various implementations. To this problem ACO algorithm can be implemented without much difficulty.
- It's easily comprehensible, such that algorithms behavior is not disguised by lots of technical details.
- It is typical trial ground for latest algorithmic thoughts- a decent efficiency on the TSP is usually considered as evidence of algorithm's effectiveness.
- Travelling Salesman Problem is a classic combinatorial optimization problem. It's being frequently employed for authenticating some algorithms, turning the evaluation with other algorithms a simpler task.

Every TSP examples used have been taken from the TSPLIB Benchmark library (Kangshun Li, 2008) that comprises a huge examples assembly; they have been employed for lot of other researches.

4 ANT COLONY ALGORITHM FOR TSP

In Ant Colony Optimization based systems ants are actors who, in the TSP instance, create routes by means of travelling from town to town on the problem graph (Reinelt, 1991). It makes a choice of next town to visit by some probabilistic function based on amount collected on arcs and of an experimental value that is calculated as a function of arcs length. Simulated ants probabilistically select cities which are linked by edges having more amount of pheromone concentration and also located nearby. To begin with, m simulated ants are positioned on arbitrarily chosen cities. When each and every ant has concluded a trip the ant which makes the minimal length path adjusts the arcs consisting its loop— called global trail updation—by addition of some quantity of pheromone trail which is inversely proportional to length of the path (V. Maniezzo, 2004).

Tour Construction: At first, ants are placed at arbitrarily chosen nodes/towns. For every creation stage, ant k uses a probabilistic decision rule, known as random proportional rule, to choose which town to visit next. Specifically, the possibility with which ant k, presently at town i, decides to visit town j is

$$p_{ij}^k = \frac{\left(\tau^\alpha_{ij}\right)\left(\eta^\beta_{ij}\right)}{\Sigma\left(\tau^\alpha_{ij}\right)\left(\eta^\beta_{ij}\right)} \quad , \quad \text{If } J \in N^i k$$

where $\eta_{ij} = 1/d_{ij}$ is an experimental value which is accessible a priori, α & β being factors that regulates comparative influence of pheromone track and N^i_k the experimental info, and N the possible neighborhood of ant k at town i, i.e. the collection of towns which ant k hasn't visited yet (the possibility of selecting a town out of N^i_k is

Nil). Every ant keeps a memory M^k that comprises the cities previously visited, in exact sequence they were visited. This memory is utilized for describing possible neighborhood N^i_k in the creation rule mentioned above. Apart from this, this memory M^k lets ant k to compute its tour length T^k it created as well as to go back over the route for depositing pheromone. Pheromone Update: When every ant is done with creation of its tour, the pheromone traces are updated which is accomplished via: firstly reducing the pheromone concentration on all edges by some fixed value, and then addition of pheromones on edges they had traversed in their trip. Pheromone disappearance is given as

$$\tau_{ij} \leftarrow (1-\rho)\,\tau_{ij}\, o\,(i,j) \in L$$

Where $0 < \rho \le 1$ degree of pheromone disappearance. The factor ρ being used for avoiding infinite accrual of the pheromone traces. After disappearance, every ant deposits pheromone on the edges which it had traversed during its trip:

$$\tau_{ij} \leftarrow \tau_{ij} + \Sigma^m_{k=1}\,\Delta\,\tau_{ij}^k,\ \forall(i,j) \in L$$

Where $\Delta\tau$ is the quantity of pheromone ant k deposits on the edges it had visited. It is given as below:

$$\Delta\tau_{ij}^k = \begin{cases} \dfrac{1}{C^k}, & \text{if edge}(i,j)\text{ belongs to } T^k; \\ 0, & \text{Or else} \end{cases}$$

Where C, the distance of tour T made by the Kth ant, is calculated as addition of the weight/length of the edges defining T. Using the immediate above equation, the better some ant's route is, the more pheromone the edges present in the tour receives [8]. Higher level interpretation of an Ant Colony Optimization algorithm for the TSP is given below (Figure 3):

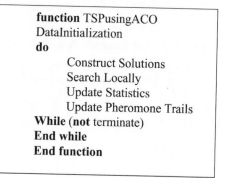

function TSPusingACO
DataInitialization
do
    Construct Solutions
    Search Locally
    Update Statistics
    Update Pheromone Trails
While (**not** terminate)
End while
End function

Figure 3. High level overview of an ACO algorithm for the TSP procedure ACO for TSP.

Generally, arches which are taken by most of the ants and are also fragments of shorter routes, have extra pheromone thus have greater chances of being preferred by ants during forthcoming repetitions.

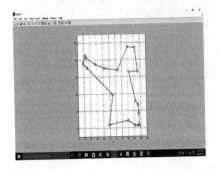

Figure 4. Shortest path.

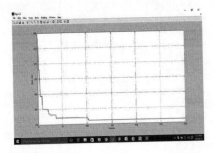

Figure 5. Iterative best cost.

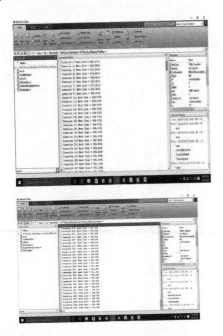

Figure 6. Simulation results in MATLAB.

5 RESULTS

When executed ACO on a well-known problem instance of classical city dataset from TSPLIB, which is att40 (40 capitals of the US), under the MATLAB tool. We set the parameters according to many experiments, which include $\alpha = 1$, $\beta = 1$, $\rho = 0.05$ & $Q = 10$. The maximum no. of iterations is 300. Running the program & calculating, we can see the result path (Shortest Path) in Figure 4 while Figure 5, show the iterative best cost over the 300 iterations.

The path length computed by ACO for att40 dataset of TSPLIB is 34266.15 which is very close to its optimal solution. We also check how the efficiency of ACO varies with the no. of ants and with pheromone importance parameter. All the values obtained from experimental results are rounded off to nearest integer value for better understandability and readability.

6 CONCLUSION

This paper is success to solve the Travelling Salesman Problem by ACO under the MATLAB environment. The results of the experiment depicts the decent ability of enhancement in ACO. The performance of ACO for TSP has been analyzed under various parameters which shows that most optimal paths are obtained when the total ants count is same as no. of cities as well as value of α should be close to 1. In addition, we also find that the ACO has a problem of stagnation, how to combat this problem is worthful to study.

REFERENCES

Blum, D. a. (2005). Ant colony optimization theory: A survey. *Theoretical Computer*.

Blum. (2005). Ant colony optimization: Introduction and recent trends. *hysics of Life Reviews*.

Colorni, A., M. D. (1991). Distributed optimization by ant colonies. *First European conference on artificial life*.

Dorigo, T. S. (1999). ACO algorithms for the traveling salesman problem. *Evolutionary Algorithms in Engineering and Computer Science*.

Kangshun Li, L. K. (2008). Comparative Analysis of Genetic Algorithm and Ant Colony Algorithm on solving Travelling Salesman Problem. *International Workshop on Semantic Computing and Systems.* IEEE.

Lanlan Kang, W. C. (2010). An Improved Genetic & Ant Colony Optimization Algorithm for Travelling Salesman Problem. *International Symposium on Information Science and Engineering (ISISE)*.

Maniezzo, V., L. M. (2004). Ant Colony Optimization. *Optimization Techniques in Engineering, Springer-Verlag*.

Reinelt, G. (1991). Traveling Salesman Problem Library. *ORSA Journal on Computing*.

Communication and Computing Systems – Prasad et al. (Eds)
© 2017 Taylor & Francis Group, London, ISBN 978-1-138-02952-1

Coalition game theory based multi-metric routing in wireless adhoc networks

Upasana Dohare & D.K. Lobiyal
School of Computer and Systems Sciences, Jawaharlal Nehru University (JNU), New Delhi, India

ABSTRACT: Fairly and efficiently power utilization is being a major constraint in operation of a mobile ad-hoc network. In this paper, we focus on how coalition game theory is applied in multipath routing for promising the quality of service. Incorporating the coalition game with multipath routing not only enforce cooperation among the nodes but also improve the performance of the network. The payoff function of coalition and utility function to assign benefit to individual node have been proposed. AODV has been used for evolution of the proposed coalition game based routing. The simulation results show that the performance of the modified AODV is comparably better that of AODV in terms of delay, energy and packet delivery ratio.

1 INTRODUCTION

Ad hoc networks are formed by a collection of mobile hosts, in self-organizing manner without any pre-established infrastructure support. Due to the lack of central administration it forces nodes to implement all networking tasks by themselves. These networks have no fixed routers, thus, every node could be router. Nodes are capable of moving randomly and can be connected dynamically in arbitrary manner. In this type of networks, some pairs of nodes may not be able to communicate directly with each other and have to rely on some nodes so that the messages are delivered to their destinations. Multihop transmission is used to increase the energy efficiency of the network by allowing packets to be delivered over several short links rather than one long link (Aggelou, 2009). The design objectives of ad hoc networks include the speedup of connection setup, energy efficient, the ease of removal of services and users, and the any-time, anywhere network services access on handheld devices. The ad hoc networking offers unique benefits and flexibility for certain environments and applications. They can be created and used anytime, anywhere (Marco, & Lui, 2003).

Ad hoc wireless networks inherit the traditional problems of wireless communications and wireless networking such as the Link layer design, Channel access and frequency reuse, Reliability, Routing, Resource Allocation, Network Capacity, Cross Layer Design, Power/energy management, Internet connectivity, security and node cooperation problems. Besides these problems and complexities, the multi-hop nature, and the lack of fixed infrastructure add a number of characteristics, complexities, and design constraints that are specific to ad hoc networking.

Besides all the perceived benefits of Ad Hoc Network, it is still in its infantry phase. Most of the research in Ad Hoc Network has focused on simulation results without forming a strong theoretical framework. This framework for Ad Hoc Network still needs to be explored, though few researchers have used probabilistic, geometric, number theoretic, game theoretic approaches to lay foundation for theoretical models in Ad Hoc Network. In this paper, we focus on how game theory is applied in multipath routing for promising the Quality of Service (QoS). In the multipath routing, a numbers of routes searched by the routing protocols, having different values of QoS parameters such as residual energy, and delay etc. The examples of such multipath protocols are MAODV and MDSR (Renke, et al., 2015). The performances of these multipath routing protocols have been improved by modifying the discovery and maintenance mechanisms of the route in terms of energy, delay, bandwidth, and reliability etc. All the nodes in a route can be considered as a coalition. If a node has multiple choices for joining a route, the node computes the payoff for every route and joins with highest payoff coalition.

In this paper, we proposed Coalition Game inspired Multi-metric Routing (CGMR) for use in ad hoc networks. We also proposed a coalition game model which selects a higher payoff coalition for a node among multiple coalitions to achieve QoS efficiency. The Shapely value has been used in the proposed game based routing to provide a solution for all the nodes to choose the better coalition. The proposed game based routing is implemented by modifying the existing AODV protocol to discover the multiple coalitions. The rest of the paper is organized as follows: Section 2 describes the existing related works. In Section 3, the proposed coalition game based routing algorithm is presented. In Section 4, the simulation and results are described. Finally, Section 5 concludes the paper.

2 RELATED WORK

Game theory is discipline of applied mathematics that models and analyses interactive decision situations. The main areas of application of game theory are mathematics, economics, political science, biology, and sociology. It was founded by the great mathematician John von Neumann. Games are models for the collaboration among individual rational decision makers. The rational decision makers are called to as players of the games. These players choose a single action from a set of possible actions. Each player get the resulting outcome after performing their chosen actions, the outcomes influence the players is called interaction. Each player evaluates the resulting outcome through a payoff or "utility" function representing their objectives (Nisan, et al., 2007). Game theory offers a suite of tools that may be used effectively in modeling and analyzing the interaction among independent nodes in an ad hoc network. Therefore, game theoretic models developed for an ad hoc networks focus on distributed systems.

Recently there has been a sequence of research papers (Xiao, et al., 2005, Jose, et al., 2010, Srinivasan, et al., 2003, Yang, et al., 2009, Song & Zhang, 2010, Flegyhagi, et al., 2006, Renke, et al., 2015) published in the area of ad hoc networks that made efforts to solve problems of selfish nodes, packet forwarding, and channel access using game theory. In (Xiao, et al., 2005) a game model is proposed to interpret the IEEE 802.11 distributed coordination function mechanism. In (Jose, et al., 2010) a game theoretic reputation mechanism is introduced to incentivize nodes which forward the packet for others, where cooperation is induced by the threat of partial or total network disconnection if a node acts selfishly. It is shown that a node which is perceived as selfish node due to the problem of packet collisions and interference can be avoided. A method named as DARWIN (Distributed and Adaptive Reputation mechanism for Wireless ad hoc Networks) has been introduced to avoid retaliation situations after a node is falsely perceived as selfish to help restore cooperation quickly. In (Srinivasan et al., 2003), wireless nodes are considered with the energy constraints.

Nodes are assumed to be rational. A rational node means that its actions are strictly determined by self-interest. Each node is associated with a minimum lifetime constraint. The throughput of each node is measured in terms of the ratio of the number of successful rely requests generated by the node. The optimal trade-off between the throughput and lifetime of nodes are studied using the game theory. A distributed Generous TFT (Tit For Tat) algorithms was introduce which decides whether to accept or reject a rely request. Authors in (Yang et al., 2009) provided an introduction to neutral cooperation in the ad hoc network which is based on game theoretic analysis of selfish nodes with a focus on the packet forwarding and relaying scenarios. Authors explained two-player packet forwarding scenario as well as three-player packet forwarding scenario. In (Song & Zhang, 2010) a game theoretic framework has been introduced to identify the condition under which equilibrium based cooperation among nodes takes place. In (Flegyhagi et al., 2006), a game theoretic model to investigate the conditions for cooperation in wireless ad hoc networks without incentive mechanisms has been presented. Several theorems for the strategy always defects are stated and proved for cooperation, considering the topology of the network and the existing communication routes. It is concluded that with a very high probability, there will be some nodes that have always defects as their best strategy.

3 SYSTEM MODEL

3.1 Coalition form game

The coalition form of a n-person game is defined as an ordered pair of players, and a real valued function say (N, f) where N is the set of players and f is a real valued function. This real valued function is known as characteristic function of the game, defined as $f : 2^N \rightarrow R$ on the all coalitions (subsets of N), and satisfying

$f(\phi) = 0$, and

If X and Y are disjoint coalitions $(X \cap Y = \phi)$, then $f(X) + f(Y) \leq f(X \cup Y)$.

3.2 Network model for coalition form game

Let us consider N is a set of nodes in an ad hoc network whose cardinality is n represents the number of players in the coalitional game. Two nodes are said to be neighbour nodes if they lie within the power range of each other. Let n nodes of network are formed connected, undirected graph G(V, E), where $V = \{v_1, v_2, \ldots, v_n\}$ denotes the set of network nodes and $E = \{e_{12}, e_{13}, \ldots, e_{mn}\}$ denotes the set of undirected communication links between nodes. Where, e_{12} is representing a link between v_1 and v_2.

In ad hoc networks, multi-hopes transmission is used when a source s and destination d nodes are not in their direct communication range. Routing is used to make transmission easier but there are many challenges in making networks stable and long lasting. Some of them are energy constraints, bandwidth allocation, delay, packet loss etc. all the paths between the nodes s and d knows as coalitions. In this paper we focus on finding delay efficient and less energy consumable coalition for a node.

Delay of route: Let the path between s and d is $p = \{V_m, \ldots\ldots, V_n\}$ and delay of the path p is D so

$$D_{ij} = d_{max} - d_{ij},\tag{1}$$

Where, d_{ij} is delay of link e_{ij}, d_{max} is maximum allowed delay of link e_{ij} and it can set artificially according to need. D_{ij} should be positive for node selection of route.

$$D = \sum_{e_{ij} \in p} d_{ij}\tag{2}$$

Pay out Energy: it is assumed that all nodes having the same initial energies. Time is divided into cycles, at the end of every cycle $E_{rem}(i)$ is remaining energy of the i'th node of path p. A node should have E_{min} as minimum required energy for node selection in a path. E_{min} can be set ourselves according to networks requirement.

$$E(i) = E_{rem}(i) - E_{min},$$

In above equation $E(i)$ should be positive, because it decides that the node be a part of route or not.

$$E_r = \sum_{i=m}^{n} E(i)\tag{3}$$

Where, E_r is total remaining energy of path $p = \{V_m, \ldots\ldots, V_n\}$. According to our prioritise, we should decide payoff function with the consideration of the following facts-

- The expected payoff will be higher if delay between links of the path p will become lesser.
- The remaining energy of the nodes of the path p should be higher for the better payoff.

Payoff: The players of the game are nodes and function of game is to compute the payoff for a node joining a coalition. For computing the payoff, we need a utility function which provides the credits to a node joining a coalition. According to our requirement mentioned above, the payoff of a coalition is defined as;

$$f(p) = \frac{\prod_{i=m}^{n}[E_{rem}(i) - E_{min}]}{\sum_{e_{ij} \in p}[d_{max} - D_{ij}]},\tag{4}$$

Cooperative game theory focus on how coalitions should earn the benefits for cooperating. How the benefits should be assigned to an individual node? The best known solution is the Shapely value provides a unique way to assign single benefit allocation among the nodes in such a way as to satisfy anonymity, dummy, and additivity. There exactly one mapping $\varphi_i : R^{2n-1} \to R^n$ such that satisfy anonymity, dummy, and additivity. The mapping is known as Shapely value is given by

$$\varphi_i(f) = \sum_{C \subseteq N-i} \frac{|C|!(n - |C| - 1)!}{n!}(f(C \cup \{i\}) - f(C))\tag{5}$$

Where, $N - i$ represents the set $N \setminus \{i\}$ that is not containing the node i. The term $\dfrac{|C|!(n - |C| - 1)!}{n!}$ is the probability that in any permutation, the nodes of C are ahead of a node i. The term $f(C \cup \{i\}) - f(C)$ represents the contribution of a node i to a coalition C. Thus, the equation (5) gives the expected contribution $\varphi_i(f)$ of node i to the worth of any coalition.

4 SIMULATION AND DISCUSSION

In this, we present the simulation results for the proposed Coalition Game inspired Multi-metric Routing (CGMR). The performance of the proposed work and AODV were evaluated using network simulator ns–2. In the simulation, one source node and one destination node were placed. Others have been placed with the different metric values between the source and destination nodes. The number of nodes were varies to show the performance of both the protocols. The following metrics used in the simulation are defined as follows: end-to-end delay per packet is the time required for successful delivery of the packet. Packet delivery ratio is the ratio of number of packets received at destination to number of packet transmitted by the source. Energy consumption balance factor is defined as the standard deviation of all nodes residual energy $E_f = (1/n) \sum_{i \in n} (E_{rem}(i) - E_a)$, where E_a is the average energy of the all nodes. The network area of 200×200 m2 has been taken for ad hoc network. The initial energy of the all the nodes are assumed to be 5 joule. The number of nodes between the source and destination are varied from 20 to 100.

Figure 1 shows the result obtained for the average end-to-end delay versus different network

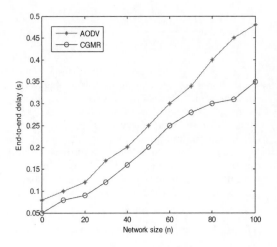

Figure 1. Average end-to-end delay versus different network sizes.

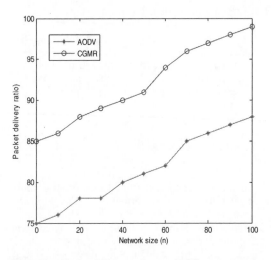

Figure 2. Packet delivery ratio versus different network sizes.

sizes. It is observed that the packet delay for both the protocols increase with the increase of network size, whereas the increment in delay is low for the proposed routing, however it is suddenly increases for the AODV. Figure 2 shows the packet delivery ratio versus different network sizes for both the modified AODV and classical AODV. It is observed that the packet delivery ratio for the AODV with proposed coalition game is improved as compared with classical AODV. Figure 3 show the energy consumption balance factor versus different network sizes. It is seen that the nodes consume energy in balance way for the proposed coalition game routing as compare to AODV. This

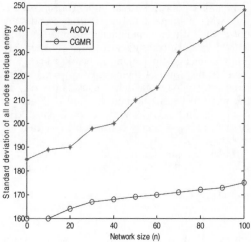

Figure 3. The energy consumption balance factor versus different network sizes.

due the fact that node is instruct to follows the effective guideline of the coalition game and use the energy fairly.

5 CONCLUSION

In this paper, we have presented a coalition game theoretic approach for multi-metric routing for ad hoc networks. In the proposed routing, nodes in all the paths are denoted as coalitions. The Shapely value has been introduced to assign the benefit to a node for cooperating in a coalition. The coalition game is merged with AODV and simulated in ns-2. The simulation results show that the performance of the modified AODV is comparably better that of AODV in terms of delay, energy and packet delivery ratio. In future, the proposed work will be extended core coalition game.

REFERENCES

Aggelou, G. 2009, ed. Mobile Ad Hoc Networks: from wireless LANs to 4G networks. Tata McGraw-Hills Publications.

Felegyhazi, M., Hubaux, J. P., & Buttyan, L. 2006. Nash equilibria of packet forwarding strategies in wireless ad hoc networks. IEEE Trans. Mobile Computing, 5, 463–476.

José, J., Jaramillo & Srikant, R. 2010. A game theory based reputation mechanism to incentivize cooperation in wireless ad hoc networks. Ad Hoc Networks, 8., 416–429.

Marco, I., & Lui, J. 2003. Mobile ad hoc networking: imperatives and challenges, Ad Hoc Networks, 1, 13–64.

Nisan, N., Tim R, EVA, & Vijaya, V.V. 2007. Algorithmic Game theory. Cambridge university press, 2007.

Renke, S., Enjie, D., Haifeng, J., Runtong, G., & Wei, C. 2015. Game theoretic Approch in adapting QoS routing protocol for wirelesss multimedia sensor networks. Internation journal of distributed sensor networks, 15, 1–12.

Song, C. & Zhang, Q. 2010. Protocol for stimulating packet forwarding in wireless ad hoc networks. IEEE wireless communication, 24, pp. 50–55.

Srinivasan, V., Nuggehalli, P., Carla, F. C., Rao, R. R. 2003. Cooperation in Wireless Ad Hoc Networks. IEEE INFOCOM, pp. 1–10.

Vivesk, S, James, N., Allen, B., Rekha, M., Luiz, A., James, E., Jeferey, H., & Robert, P. 2005. Using game theory to analyze wireless ad hoc networks. IEEE communication surveys and tutorials, 12, 46–56.

Xiao, Y., Shan, X., & Ren, Y. 2005. Game Theory models for IEEE 802.11 DCF in wireless ad hoc networks. IEEE radio communication, 14, 22–26.

Yang, J., Klein, A. G., & Brown, D. R. 2009. Natural Cooperation in Wireless Networks. IEEE signal processing magazine, 26(5), 98–106.

Communication and Computing Systems – Prasad et al. (Eds)
© 2017 Taylor & Francis Group, London, ISBN 978-1-138-02952-1

Adaptive compressive sensing based routing algorithm for internet of things and wireless sensor networks

Ahmed Aziz & Karan Singh

School of Computer and System Science, Jawaharlal Nehru University India, New Delhi, India

ABSTRACT: The simplicity of Compressive sensing scheme in reducing the data dimension without go through a lot of complex mathematical computation, make it a preferable method to compress the data traffic transmitted through Wireless sensor networks and Internet of Things. In this paper, we explain the definition of Compressive sensing method, propose an efficient technique that uses Compressive sensing to compress the sensors data then combine it with efficient routing algorithm to prolong the wireless Sensor Networks and Internet of Things life time. Our proposed research work firstly, describes how to use Compressive sensing from the Compressive sensing matrix selection step, creates data aggregation path, then applies compressive sensing method at each node and finally sends the compressed samples to the base station. We evaluate our work in term of network life time and average power consumption.

1 INTRODUCTION

Recently Internet of Things (IoT) has become one of the most important and interesting field of researches through which the researchers hope to control all everyday usages via the Internet (Xu 2011). IoT elements include objects different from the technological environment (clothes or foods), or even living organisms (plants, woods or domestic animal)as well as devices that are already deeply embedded in the technological environments (such as smart phones or cars). By integrating computational abilities in all kinds of things and living organisms, it will be possible providing a big leap in many sectors: Health, military, home, entertainment and so on (Palopoli, Passerone, & Rizano 2011).

It is expected that IoT will be a global network of interdependent Objects. Actually, Wireless Sensor Networks (WSNs) is considered the most important element in the IoT model. Utilization of wireless sensors devices and other IoT technologies in green applications and environmental conservation are one of the most promising market segments in the future. There will be an increased usage of wireless identifiable devices in environmentally friendly programs worldwide. For example, IoT offers solutions for fare collection and toll systems, screening of passengers and bags boarding commercial carriers and the goods moved by the international cargo system that support the security policies of the governments and the transportation industry, to meet the increasing demand for security in the globe. Monitoring traffic jams

through cell phones of the users and deployment of Intelligent Transport Systems (ITS) will make the transportation of goods and people more efficient (Bandyopadhyay & Sen 2011).

The main task for IoT and WSNs sensors nodes is to sense the data then send them to the base station; otherwise how to reduce the energy that consumed by the sensors during the transmission is still a challenge.

For this, it is highly required to find creative techniques that reduce the energy consumption which will lead to increase network lifetime. In order to solve this problem a lot of techniques have been proposed, such as routing protocols or data compression scheme. The data compression methods can be used to decrease transmitted data over wireless channels. The inter-node communication, which is the main power consumer in WSNs and IoT, can be reduced by using this technique.

In the context of the compression algorithms, Compressive Sensing (CS) has been proposed as a novel concept of signal sensing and compression. In CS, the signal can be successfully sampled less than the rate of Nyquist theory, if the signal is sparsed by natural or by transformer. In CS, the signal is sampled with compress simultaneously, rather than sample it then compress like the other traditional compression techniques, and it can be reconstructed again without any significant losing in the information.

In this paper, we adapt Compressive Sensing method with routing algorithm for information acquisition in IoT and WSNs to solve the previous problems. The basic idea of our work is to: firstly, construct a data aggregation path then apply a CS compression method at each node to reduce the local data traffic so that reduces the energy consumption prolonging the life of the whole network remarkably.

The remainder of the paper is organized as follows: The related work is briefly reviewed in Section. 2. Section. 3 presents Compressive Sensing background. We introduce our approach to solve the proposed problem in Sect. 4. In Section. 5, we present a simulation of our approach. Section. 6 is the conclusion to our research paper.

2 RELATED WORK

During the last few years, WSNs applications such as surveillance and monitoring applications have attracted number of researchers. The massive energy constraints of large number of densely deployed sensor nodes led to several routing protocols for wireless sensor network are discussed. For example, Low Energy Adaptive Clustering Hierarchy (LEACH) (Heinzelman, Chandrakasan, & Balakrishnan 2000) is a hierarchical protocol in which WSNs nodes are divided into clusters with specific Cluster Head (CH) to each one; each node in each cluster has to transmit its data to the CH which will collect the data and then send them to Base Station (BS). LEACH protocol improves the performance of WSNs in comparison to the direct transmission method.

Power-Efficient Gathering in Sensor Information Systems (PEGASIS) (Heinzelman, Chandrakasan, & Balakrishnan 2000) is the best protocol for high rate data collecting applications in WSNs. The basic idea of the PEGASIS algorithm is the creation of the chain list among WSNs nodes where, for each two neighbours in this list are the nearest nodes to each others. Thus, PEGASIS reduces the consumed energy from each node. The PEGASIS protocol has achieved improvement in comparison to the LEACH protocol.

But all routing algorithms cannot satisfy the huge data traffic of wireless sensor networks. So, it is effective to apply compression before transmitting data to reduce total power consumption by a sensor node (Aziz, Osamy, & Salim 2013).

For these reasons, latest researches have explained that, using Compressed Sensing (CS) scheme in WSNs can significantly decrease the total number of data gathered and improve WSNs performance (Jun, Liu, & Catherine 2010, Chong, Feng, Jun, & Chang 2009, Xiang, Jun, & Athanasios 2011, Haupt, Bajwa, & Rabbat. 2008). Compressive Data Gathering (CDG) (Chong, Feng, Jun, & Chang 2009), is the first research that introduced

CS to WSNs. It combines routing technique with CS in order to reduce the total energy consumption. The main problem of CDG that it explains the CS idea, without analyzing. The Efficient CS based routing Technique (ECST) which proposed in (Aziz, Osamy, & Salim 2013) prolongs the network life time using the compressive method to compress sensors reading before sending them to the base station, but this work did not take in consideration the effect of the compression matrix on the sensors data. (Xiang, Jun, & Athanasios 2011) aims to reduce the energy consumption by combining between routing algorithm and compression technique. Waheed et al. (Haupt, Bajwa, & Rabbat. 2008), proposed CS scheme for collecting data problem in large-scale WSNs. Shriram et al. utilized the signals data correlation to reduce the sampling ratio (Duarte, Sarvotham, Wakin, Baron, & Baraniuk 2005). Tang et al. proposed a method to reconstruct specially the losing data by exploiting the correlation between different nodes data (Jin, ShaoJie, Baocai, & Yang 2012).

In this paper, we propose an efficient work, which takes advantage of the correlations between sensors readings to compress them by using CS method. And also, we combine the CS method with suitable routing technique in order to reduce the overall power consumption and increase the network life time.

3 COMPRESSIVE SENSING BACKGROUND

The CS provides a direct method in which the data is compressed and sampled in one step in stead of sampling and then compressing such as conventional compression (Donoho 2006), as shown in Fig. 1. In addition, the CS reconstruction algorithm can successfully reconstruct the original data

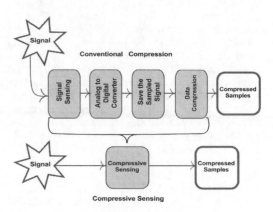

Figure 1. Fundamental idea behind CS.

from the compressed samples without any prior knowledge (Donoho 2006).

3.1 Mathematical definition

Let $x[n], n = 1, 2, ..., N$. the collected set of sensors readings is vector in R^N, where N represent number of sensors. Any signal in R^N can be represented in terms of a basis of $N \times 1$ vectors $\{\Psi_i\}_{i=1}^N$. For simplicity, assume that the basis is orthonormal. Using the $N \times N$ basis matrix $\Psi = [\Psi_1 | \Psi_2 | \Psi_3 | ... | \Psi_N]$ with the vectors Ψ_i as columns, a signal x can be expressed as (Donoho 2006)

$$x = \sum_{i=1}^N g_i \Psi_i \ or \ x = \Psi g \qquad (1)$$

where g is the $N \times 1$ sparse presentation of x. The transform matrix $\Psi \in N \times N$ is an orthonormal basis. CS will focus on signals that have a sparse representation, i.e x has just S basis vectors, with $S \ll N$. That is, $(N - S)$ are zero and only S of g are nonzero. The used notations through this paper are given in Table 1.

By using The Eq. (1), the compressed samples y (compressive measurements) can be obtained from Eq. (2):

$$y = \Phi x = \Phi \Psi g = \Theta g, \qquad (2)$$

where the compressed samples vector is $y \in R^M$, with $M \ll N$ and Θ is an $M \times N$ matrix.

3.2 Selecting suitable CS matrix

CS technique faces the problem of finding a compression matrix Θ which must allows the reconstruction of the original signal x of length-N from the compressed signal y of length M where $M < N$

Table 1. Notions description.

Notation	Description
x	Sensors readings
Ψ	Transform matrix
Φ	Measurement matrix
Θ	$M \times N$ matrix such that $\Theta = \Phi \Psi$
y	Measurement vector (compressed samples)
g	Sparse presentation of x
S	Sparse level (number of non zeros values)
ξ	The global seed
r	Number of round
$n_j . \alpha$	The coefficient vector for node n_j
$n_j . y$	The compressed vector for node n_j

and since $M < N$, the solution becomes ill-posed in general. to solve this problem the measurement matrix must satisfy the following conditions:

1. Restricted Isometry Property (RIP) (Candes & Tao 2006), the salient feature of CS that it focuses on the spares signal in which the measurement vector y is just a linear combination of the S columns of Θ whose corresponding $g_i \neq 0$ (see Fig. 2). Now, CS changes the problem from recovering N from M to recover S from M, where $S \ll M$, which make the solution well-posed. the prime perquisite of the matrix Θ to ensure that the solution will be well-posed must satisfy the RIP condition:

$$1 - \varepsilon \leq \frac{\|\Theta v\|_2}{\|v\|_2} \leq 1 + \varepsilon \qquad (3)$$

for some $\varepsilon > 0$ and any vector v sharing the same G nonzero entries as g. In words, the matrix Θ must preserve the lengths of these particular S-sparse vectors. In fact, it is very difficult to locate the nonzero values in s. Fortunately, to maintain a stable inverse for both S-sparse and compressible signals, only one condition is needed which is Θ to satisfy Eq. (3).

2. Incoherent: the matrix ϕ must be incoherence with the matrix Ψ.

Mostly CS methods use random matrices like Gaussian or Bernoulli distribution matrix, which fulfil all the previous conditions.

3.3 Signal reconstruction

For many years, the problem of finding the solution to an unspecified set of linear equations has grabbed the attention in the literature. Different practical applications has been carried out for this problem, for example compressed sensing. In CS scenario, few measurements coefficients are available and the task is to recover a larger, sparse signal. How the signals are reconstructed from this incomplete set of measurements, depending on the fact that the signal has sparse representation. One of the simplest solutions to recover such a vector from its measurements Eq. 2 is to solve $\|L\|_0$ minimization problem that counts the number of non-zeros entries, the reconstruction problem turns to be:

$$x = arg \ min \|x\|_0 \ subject \ to \ y = \Phi x \qquad (4)$$

Thus, this $\|L\|_0$ minimization problem works perfectly theoretically. However, it is computationally NP-Hard in general (Mallat 1999). It is computationally intractable to solve Eq. (4) for any matrix and vector. Fortunately, in the framework

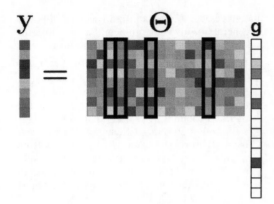

Figure 2. Columns which create the measurement vector y.

of CS, There are two families can be alternatively used to solve Eq. (2) with computationally efficient for this computationally NP-Hard problem. One is the basic pursuit that is a convex relaxation leading to $\| L \|_1$ norm minimization (Venkataramani & Bresler. 1998) and the other is greedy pursuit such as Orthogonal Matching Pursuit (OMP) (Tropp & Gilbert 2007), Stagewise Orthogonal Matching Pursuit (StOMP) (Donoho, Yaakov, Drori, & Jean. 2012) and Regularized Orthogonal Matching Pursuit (ROMP) (Deanna & Roman 2009). In the next section, we propose our work to solve the exist problem.

4   OUR PROPOSED WORK

Recently, IoT technologies has attracted many researchers as a new model in the area of wireless networks. This is because IoT changes the human vision for the world where they can control any physical objects through the internet. However, sensors energy constrain, designing effective sensor data aggregation techniques and managing large amount of information are considered the major challenges facing IoT technologies. To address these problems, in this section we introduce our propose work which aims tosolve these problems by using CS frame integrated with efficient routing technique. Our work consists of three stages:Seed estimation stage, Sensors organizing stage and Data Acquisition and Compression stage. am of ECST symbol:ecst technique.

4.1   Seed estimation stage

As discussed perviously, CS method uses the random matrix such as gaussian matrix as a measurement

matrix which is used in the compression and reconstruction process. This random matrix is generated by using seed so the aim of this stage is to allow the BS to estimate the best seed which will be used by the sensors to compress or by BS to reconstruct. To achieve this aim we proposed the following algorithm:

4.1.1   Seed estimation algorithm

In order to create sensing matrix Φ, a global seed ξ is generated and broadcasted to the network nodes. When each node (n_j) in the network receive the global seed ξ, it begins to generate the corresponding series of coefficients $n_j.\alpha = \begin{pmatrix} \Phi_{1j} \\ \vdots \\ \Phi_{Mj} \end{pmatrix}$,

those will be used to compute its measurement $n_j.y$, using its unique identification ($n_j.id$) and the global seed ξ. These coefficients can be regenerate at the BS in order to be used in the reconstruction process given that the BS knows the identifications of all nodes and the global seed ξ which BS generate it before.

Therefore, the global seed selection is a very important process because of its effect in sensors' data, where this seed is used to construct the measurement matrix which used by the sensors nodes to compress their data and also used by the BS to reconstruct sensors' data.

So, incorrect selection leads to loss lots of data. For that reason, we propose an adaptive technique given in algorithm 1, in which the base station decides whether it needs to dynamically change the global seed ξ or not according to the following algorithm:

For each time t, the BS generates a global random seed ξ and then transmits this seed to a set of closest nodes denoted by P. In order to find an empirically optimal rule for choosing P, we test various P choices among the experiments. It turns out that choosing $P >= 16\% \ of \ N$, where N is number of nodes in the network, leads to the optimal reconstruction performance in practice. Each node $i \in P$ generates its own seed using ξ and its unique identification (P_i, id) and then generates the corresponding series of coefficients $\alpha_i = \begin{pmatrix} \Phi_{1i} \\ \vdots \\ \Phi_{Mi} \end{pmatrix}$.

Then each node $i \in P$ computes the measurement $n_i.y = n_i.\alpha \ n_i.d$ and transmits the measurement $n_i.y$ to the BS. After that, the BS collects these measurements and generates the measurement matrix Φ by using global seed ξ which is used by CS reconstruction algorithm to reconstruct the original data from the compressed data. The BS will test this measurement matrix Φ either it gives the minimal reconstruction error, or generate another measurement matrix by using different seed to get the best seed that satisfy the minimal reconstruction error.

4.2 Sensors organization stage

Through this Stage, the BS uses the proposed Algorithm which called Chain Construction Algorithm to organize the network sensors into chain list then sends the chain list information to the entire network. The proposed algorithm goes as the following:

4.2.1 Chain construction algorithm

The Base Station (BS) begins to create the chain list by adding the nearest node c_0 to it and then updates the chain list with the nearest unselected node (c_1) to node c_0. After that, the BS holds the nearest unselected neighbor node c_i to node c_1 in waiting to decide wherever it will be placed in the chain list depending on the following scenario:

1. The BS adds it to the end of the chain list if the distance between c_1 and c_i is less than the distance between c_i and any consecutive nodes in the chain list.
2. Otherwise c_i will be added between the consecutive nodes that have minimum distance to c_i, for example if c_j and c_k are consecutive nodes in the chain list and if $dis(c_i,c_{last}) > dis(c_i,c_j)$ and $dis(c_i,c_{last}) > dis(c_i,c_k)$ then node c_i will be added between c_j and c_k; otherwise node c_i will

---

Algorithm 1: Seed Estimation Algorithm.

1: The BS multi-cast a new random seed to a set of closest nodes denoted by P.
2: **for** each node $i \in P$ **do**
3:     generates its seed using ξ and its unique identification (i, id).
4:     generates the corresponding series of coefficients $n_i.\alpha$.
5:     computes the measurement $n_i.y = n_i.\alpha n_i.d_i$.
6:     transmits the measurement $n_i.y$ to the BS.
7: **end for**
8: The BS collects the measurements $n_k.y \ \forall k \in P$.
9: The BS generates the measurement matrix Φ by using global seed ξ.
10: The BS calculates the reconstruction error.
11: **if** reconstruction error is not satisfied **then**
12:     GOTO Step 1.
13: **else**
14:     The BS starts the aggregation process based on this seed.
15: **end if**

---

be added to the end of the chain list after node c_{last}, where c_{last} is the last node added to the chain list and $dis(c_i,c_k)$ is the distance between node c_i and node c_k).

By following this scenario, our algorithm rearranges the chain list each time it adds anew node to the chain list so it saves power for each node rather than adds only the nearest node to the last node in the chain list without taking into consideration the other nodes in the chain list like ECST (Aziz, Osamy, & Salim 2013). The BS The repeating the previous update process to include all nodes in the chain list. By this way the overall power consumption required to transmit data per round is reduced.

4.3 Data acquisition and compression stage

This stage consists of two steps: firstly, header selection step. Then, adapt the CS method at each sensor node in the chain to compress its data while sending them (CS step).

4.3.1 Header selection step

At each round r, chain list header is selected randomly at any a random position in the chain list. Any node in the chain list can be chain list header, if its identification number equals $r \ mod \ N$, where N represents the total number of nodes in the network.

4.3.2 CS step

In this Step, we aim to reduce the total data traffic and decrease the energy consumption for each node by using CS. In order to make combination between sensors reading while sending them, every node c_i uses the global seed ξ which was sent by theBS to generate $c_i.\alpha$ then computes its compress vector (measurement) $c_i.y = c_i.\alpha c_i.d$ and then transmits the measurement $c_i.y$ to its neighbour node in the chain list c_{i+1}. After that, node c_{i+1} uses the same global seed ξ to generate $c_{i+1}.\alpha$ and compute its measurement $c_{i+1}.y = c_{i+1}.\alpha c_{i+1}.d$ and then transmits the summation vector $\Sigma_{i=1}^{2} c_i.y$ to the next node c_{i+2}. Once c_{i+2} received the last values, it computes its measurement $c_{i+2}.y$, adds it to the sum of receiving values and sends the joined value to next node in chain list and so on till the chain list header. Finally, the chain list header sends the packet containing the collection of all reading in the chain list to the BS. To cover all positions of the header node in the chain list, this step has three cases:

Case 1: If chain list header located at the end of chain list, then the step will going as follow: First of all, the chain list header sends an identification message to the first node c_1. second of all, node c_1 uses ξ to generate $c_1.\alpha$ and compute measurement $c_1.y$ and then transmit the measurement $c_1.y$ to c_2. Then, c_2 uses ξ to generate $c_2.\alpha$ and compute the measurement $c_2.y$ then transmit $c_1.y + c_2.y$ to c_3 and so on till chain header. Finally, the chain list header node computes $\sum_{i=1}^{N} c_i.y$ by using the data received from c_{N-1} and sends it to the BS.

Case 2: If chain list header located at the beginning of chain list list, then the step will going as follow: First of all, the chain list header passes an identification message to the last node of the chain list c_N. Second of all, node c_N uses ξ to generate $c_N.\alpha$ and compute measurement $c_N.y$ and then transmit the measurement $c_N.y$ to c_{N-1}. Then node c_{N-1} uses ξ to generate $c_N - 1.\alpha$ and compute the measurement $c_{N-1}.y$ and then transmit $c_N.y + c_{N-1}.y$ to c_{N-2} and so on till chain list header. Finally, the chain list header node computes $\sum_{i=N}^{1} c_i.y$ by using the data received from c_2 and sends it to the BS.

Case 3: If chain list header located at position $(j: 1 < j < N)$ of chain list, then the step will going as follow: First of all, the chain list header passes an identification message to the first node of the chain list c_1 as well as to the last node of the chain list c_N. Second of all, node c_1 uses ξ to generate $c_1.\alpha$ and compute measurement $c_1.y$ and then transmit the measurement $c_1.y$ to c_2. Third of all, node c_2 uses ξ to generate $c_2.\alpha$ and compute the measurement $c_2.y$ and transmits $c_1.y + c_2.y$ to c_3 and so on till chain list header at position j. At the same time Node c_N computes measurement $c_N.y$ and transmits the measurement $c_N.y$ to c_{N-1}. Then node c_{N-1} computes the measurement $c_{N-1}.y$ and transmits $c_N.y + c_{N-1}.y$ to c_{N-2} and so on till chain list header at position j. Finally, the chain list header node computes $\sum_{j=1}^{N} c_j.y$ by using the data received from c_{j-1} and c_{j+1} and sends it to the BS.

The compressed data from the chain list header will be received by the BS, then the BS reconstructs the original data from the compressed data by using any CS reconstruction algorithms such as OMP (Tropp & Gilbert 2007).

5 SIMULATION RESULTS

The network has been simulated in MATLAB. In our simulation, network region size is 100 m × 100 m, and the number of sensor nodes ranges from 50 to 200 nodes in the increments of 50 nodes and the base station is located at (x = 50, y = 50).

We used the same energy parameters that used in (Heinzelman, Chandrakasan, & Balakrishnan 2000) where to transmit an l-bit message over a distance d, the radio power consumption will be,

$$E_{Tx}(l,d) = \begin{cases} l\,E_{elec} + l\,\varepsilon_{fs}\,d^2 & d < d_0 \\ l\,E_{elec} + l\,\varepsilon_{mp}\,d^4 & d \geq d_0 \end{cases} \tag{5}$$

and to receive this message, the radio expends will be

$$E_{Rx}(l) = l\,E_{elec} \tag{6}$$

Simulated model parameters are set as: $E_{elec} = 50\,nJ/bit$, $\varepsilon_{fs} = 10\,pJ/bit/m^2$, $\varepsilon_{mp} = \frac{13}{10000}\,pJ/bit/m^4$, $d_0 = \sqrt{\varepsilon_{fs}/\varepsilon_{mp}}$, and the initial energy per node = $2J$.

5.1 Performance metrics

We compare the proposed algorithm performance with PEGASIS (Lindsey & Raghavendra. 2002) and ECST (Aziz, Osamy, & Salim 2013), in term of the following:

Average Energy Consumption
It is the average energy consumed by all the nodes during their operations like forwarding operations, sending and reeving. The average energy consumption per round can be estimated as:

$$E = \frac{\sum_{i=1}^{N} E_i(r)}{r} \tag{7}$$

Where N is the number considered WSNs sensor nodes and r is referred to rounds number.

Network lifetime
We measured the network life time according to the death of the First node.

5.2 Effect of seed selection on data compression and reconstruction process

In order to explain the effect of global seed selection on data compression and reconstruction process, consider the following:

1. Given the parameters $M = 25$, $N = 50$.
2. Draw the elements of the vector d (sensors reading data) from the standard Gaussian distribution; we refer to this type of signal as a Gaussian signal.
3. Let Φ_i is the measurement matrix which generate by using the corresponding seed (S_i), $i = 1,2,...,10$, from the standard i.i.d. Gaussian ensemble with mean 0 and standard deviation $1/N$.
4. For each Φ_i compute the measurement y (such that y is an $M \times 1$ vector):

$$y_i = \Phi_i d$$

5. Finally, we used OMP algorithm to reconstruct the original data from y_i and then compute reconstruction accuracy which is given in terms the Average Normalized Mean Squared Error (ANMSE) (Burak & Erdogan 2013) defined as:

$$ANMSE = \frac{1}{10} \sum_{i=1}^{10} = \frac{\| d_i - \hat{d}_i \|_2^2}{\| d_i \|_2^2} \qquad (8)$$

Table 2 show the effect of seed on reconstruction process and it is clear that seed S_8 is the best seed which satisfies the minimal ANMSE.

Another interesting test case is the reconstruction ability when the numbers of nodes N changes. This test is designed to determine the best number of nodes denoted by P that can be used by the BS to evaluate the effectiveness of the selected seed in the reconstruction process for each time t. Fig. 3 depicts the reconstruction performance over N where we take the Φ_8 as measurement matrix which is generated by using seed S_8 and N is changed from 1 to 150 with step size 2 to observe the variation of the average reconstruction error. We observe that the curve oscillating between

increasing or decreasing when $N \in [1,7]$ and up to the lowest level in the reconstruction error and settle this situation when $N \geq 8$. Foregoing we can conclude that choosing $P >= 16\%$ of N leads to the optimal reconstruction performance. We got the same results when we repeat the last scenario with different N sizes, where $N = 100$, 150 and 256. Table 3, show the ANMSE for this test when $N = 150$ observe that the ANMSE up to the lowest level when $N \geq 24$.

5.3 Evaluate the performance of the proposed algorithm

Fig. 3, shows the average energy consumption per round where the number of network nodes are ranging from 50 to 200 nodes with step size 50 sensor nodes. In Fig. 3, its clear that, our algorithm has less energy consumption compared with standard PEGASIS and ECST.

Fig. 4, shows the lifetime and the average energy consumption per round for our protocol. And also the same figure illustrates the effectiveness of the proposed algorithm in prolonging network lifetime than its counterparts.

In summary, the results prove that our research work success to prolong the lifetime of the IoT and WSNs by reducing the total energy.

Table 2. Seed and ANMSE.

S_i	ANMSE
S_1	0.0026
S_2	2.478e-16
S_3	0.0019
S_4	0.0322
S_5	0.001
S_6	2.454e-16
S_7	0.026
S_8	2.176e-16
S_9	0.0027
S_{10}	0.1029

Table 3. Variation of ANMSE VS number of nodes.

Number of Nodes	ANMSE
1–23	0.1947
24–47	8.3187e-016
48–71	2.9695e-016
72–95	2.5927e-016
96–119	2.5611e-016
120–144	2.6678e-016
145–150	2.9452e-016

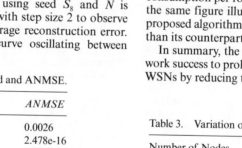

Figure 3. Average consumed energy per round as a function of number of sensor nodes.

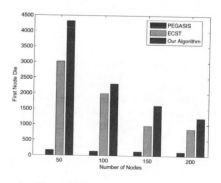

Figure 4. Network life time as a function of number of sensor nodes.

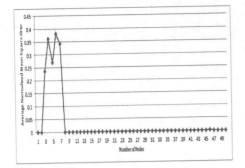

Figure 5. ANMSE as a function of number of sensor nodes.

6 CONCLUSIONS

The main goal of IoT and WSNs is to obtain the infirmations about any event correctly. However, there are some problems which impede the way to achieve this goal such as sensor batteries constrain and managing large amount data acquisition. To solve those problems, in this research paper we were able successfully to recruit CS scheme for WSNs and IoT and explained how this scheme could be used to compress and reduce the total data traffic through the network. Our proposed work consists of three stages. The first stage allows the BS to select the best seed of CS random matrix by introducing seed estimation algorithm. In the second stage, the sensor nodes are organized in chain list depending on Chain construction proposed algorithm. Through the third stage, the sensor nodes select the chain header and then apply CS method to compress their data. The proposed work achieved our goal to prolong the network life time by combining between routing protocols and CS method. Simulation results proved that our proposed algorithm was effective data collection tool for decreasing the energy consumption in networks.

REFERENCES

Ali, S. & S. Refaay (2011). Chain-chain based routing protocol. *IJCSI International Journal of Computer Science Issues 8*(2), 1694–0814.

Aziz, A., W. Osamy, & A. Salim (2013). Efficient compressive sensing based technique for routing in wireless sensor networks. *INFOCOMP Journal of Computer Science 12*(1), 01–09.

Bandyopadhyay, D. & J. Sen (2011, May). Internet of things: Applications and challenges in technology and standardization. *Wirel. Pers. Commun. 58*(1), 49–69.

Burak, N. & H. Erdogan (2013). Compressed sensing signal recovery via forward-backward pursuit. *Digital Signal Processing 23*(5), 1539–1548.

Candes, E. & T. Tao (Feb. 2006). Robust uncertainty principles: Exact signal reconstruction from highly incomplete frequency information. *IEEE Trans. Inform. Theory 52*(2), 489509.

Chong, L., W. Feng, S. Jun, & C. Chang (2009). Compressive data gathering for large-scale wireless sensor networks. In *Proceedings of the 15th Annual International Conference on Mobile Computing and Networking*, MobiCom '09, New York, NY, USA, pp. 145–156. ACM.

Deanna, N. & V. Roman (2009, April). Uniform uncertainty principle and signal recovery via regularized orthogonal matching pursuit. *Found. Comput. Math. 9*(3), 317–334.

Donoho, D. (Apr. 2006). Compressed sensing. *IEEE Trans. Inform. Theory 52*(4), 1289–1306.

Donoho, D., T. Yaakov, I. Drori, & S. Jean. (2012, February). Sparse solution of underdetermined systems of linear equations by stagewise orthogonal matching pursuit. *IEEE Trans. Inf. Theor. 58*(2), 1094–1121.

Duarte, M., S. Sarvotham, M. Wakin, D. Baron, & R. Baraniuk (2005). Joint sparsity models for distributed compressed sensing. In *Online Proceedings of the Workshop on Signal Processing with Adaptive Sparse Structured Representations (SPARS)*, Rennes, France.

Haupt, J., W. Bajwa, & M. Rabbat. (2008, March). Compressed Sensing for Networked Data. *Signal Processing Magazine, IEEE 25*(2), 92–101.

Heinzelman, W., A. Chandrakasan, & H. Balakrishnan (2000). Energy efficient communication protocol for wireless microsensor networks. In *Proceedings of the 33rd Annual Hawaii International Conference*, pp. 3005–3014.

Jin, W., T. ShaoJie, Y. Baocai, & X. Yang (2012). Data gathering in wireless sensor networks through intelligent compressive sensing. In A. G. Greenberg and K. Sohraby (Eds.), *INFOCOM*, pp. 603–611. IEEE.

Jun, L., X. Liu, & R. Catherine (2010). Does compressed sensing improve the throughput of wireless sensor networks. In *ICC*, pp. 1–6. IEEE.

Lindsey, S. & C. Raghavendra. (2002). Pegasis: Power-efficient gathering in sensor information systems. In *Aerospace Conference Proceedings, 2002. IEEE*, Volume 3, pp. 3–1125–3–1130 vol.3.

Mallat, S. (1999). *A Wavelet Tour of Signal Processing*. Academic Press.

Palopoli, L., R. Passerone, & T. Rizano (2011). Scalable offline optimization of industrial wireless sensor networks. *IEEE Trans. Industrial Informatics 7*(2), 328–339.

Tropp, J. & A. Gilbert (Dec. 2007). Signal recovery from random measurements via orthogonal matching pursuit. *IEEE Trans. Inform. Theory 53*(12), 4655–4666.

Venkataramani, R. & Y. Bresler. (1998). Sub-nyquist sampling of multiband signals: perfect reconstruction and bounds on aliasing error. In *IEEE International Conference on Acoustics, Speech and Signal Processing (ICASSP)*, pp. 1633–1636.

Xiang, L., L. Jun, & V. Athanasios (2011). Compressed data aggregation for energy efficient wireless sensor networks. In *SECON*, pp. 46–54. IEEE.

Xu, L. D. (2011, Nov). Enterprise systems: State-of-the-art and future trends. *IEEE Transactions on Industrial Informatics 7*(4), 630–640.

Communication and Computing Systems – Prasad et al. (Eds)
© 2017 Taylor & Francis Group, London, ISBN 978-1-138-02952-1

Energy efficient power allocation for OFDMA based cognitive radio

Chandra Shekhar Singh & Krishna Kant Singh
Department of EEE, Dronacharya College of Engineering, Gurgaon, India

Debjani Mitra
Department of E&I, IIT(ISM), Dhanbad, India

Akansha Singh
Department of CSE, The NorthCap University, Gurgaon, India

ABSTRACT: Cognitive Radio (CR) is a novel wireless technology which can be used as a solution to crisis of spectrum resources. CR is a technology to increase spectral efficiency by exploiting frequency holes in dynamically varying environments. Energy efficient design in wireless network is very important. In this paper, we investigate an optical power allocation procedure that maximizes the Energy Efficiency (EE) of transmissions from a cognitive base station. The objective function of Energy efficient problem is neither concave nor convex. Therefore we have to use a concave fractional program for optimizing the objective function. Dinkelbach parametric method is used to transform the objective function into a concave program. Simulation results exhibit that performance of the proposed method is much more effective as compared to the state of arts methods.

1 INTRODUCTION

The frequency spectrum is becoming a rare resource due to explosive growth of wireless technology and increase in the number of wireless users (Ian et al, 2006). The scarcity of spectrum resources becomes more severe in the next generation wireless communication due to rapid increase of multimedia traffic such as TV broadcasting, radio, navigation, etc. (Islam et al 2008).

Cognitive Radio (CR) is a highly promising wireless technology which can be used as a solution of crisis of spectrum resources. CR has been proposed as a technology for higher energy efficiency in wireless communication system. Dynamic Spectrum Access (DSA) through cognitive radio technology is extensively used to mitigate the scarcity of spectral resources by replacing the existing policy of assigning fixed bands only to licensed users. (Ian et al 2006) & (Beibei et al 2011).

In a Cognitive Radio Network (CRN), there is usually a licensed Primary User (PU) and many unlicensed Secondary Users (SU) sharing the common spectrum. In general, the transmission modes in cognitive radio network can be classified into two types: underlay and overlay modes. However, in underlay mode, Secondary users transmit concurrently and share the same spectrum band with primary network. The interference at the receiver of primary network is controlled carefully in this mode.

In overlay mode, cognitive network focuses on the spectrum hole via spectrum sensing (Song et al 2012).

In cognitive radio networks, efficient wireless resource allocation is a primary issue to be addressed. Also, it is necessary to minimize the impact of interference due to secondary users on the primary user that share the spectrum (Hisham et al 2009).

Literature projects Orthogonal Frequency Division Multiplexing (OFDM) is a popular transmission technology for cognitive radio network (Tuan et al 2014). Xu et al are interested in the resource allocation problem of the underlay scenario for cognitive network and select OFDM as an air interface due to the inherent significant advantages of flexibly re-source allocation (Xu et al 2015).

A two-step sub channel and power allocation protocol is presented. Firstly a near-optimal but low-complexity sub channel assignment is done and then optimal power is allocated (Illanko et al 2015).

A power allocation framework for spectrum sharing cognitive radio systems is presented. A non-convex problem is solved and the optimal power is derived under either a peak or an average power constraint. When the instantaneous channel is not available, a necessary and sufficient condition for the optimal power is provided and a simple sub-optimal power is presented (Sboui et al 2015).

Another scheme presents a subcarrier power allocation scheme for Orthogonal Frequency Division Multiplexing (OFDM) based cognitive users. Based on the quality of channel gains and spectral distance from PU bands, the powers of subcarriers adjacent to PU bands are suitably controlled till the interference constraint of PUs are satisfied (Bepari & Mitra, 2015).

This paper proposes a concave fractional program for optimizing the objective function. Dinkelbach parametric method is used to transform the objective function into a concave program.

2 SYSTEM MODEL

We consider OFDM based wireless cognitive radio network that consists of Primary User (PUs) and Secondary Users (SUs) geographically distributed in a cognitive cell. PUs are assumed as a licensed user for the allocated spectrum whereas SUs are unlicensed one, but has a capability to opportunistically access the available unused spectrum both in frequent and time. Figure 1 shows the generalized wireless cognitive network consists of cognitive PUs and SUs communicates with each other at any instant of time.

When SU communicate on a spectrum, it causes interference to the PUs as well as other SUs. For geographically far located PUs the accurate information of the signal is not known, so underlay spectrum access for secondary users can be setup by thresholding the signal to interference noise ratio in a certain limit or QoS defined in terms of outrage probability of PU does not affected. For example, in Figure 2, spectrum allocated to PU₁ is opportunistically sensed and partitioned into three sub carriers. SU₁ uses two subcarriers and SU₂ uses one subcarrier to transmit signal.

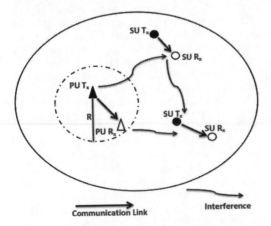

Figure 1. Generalized wireless cognitive radio network.

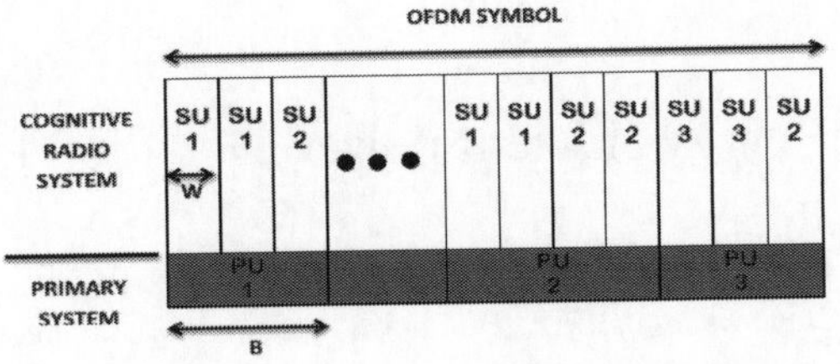

Figure 2. Spectrum allocation to secondary cognitive user.

The existence of SU_1 and SU_2 in the same channel causes interference to the PU_1, PU_2 and PU_3 as well as mutual interference to SU_1 and SU_2. In order to SU_1 transmit; the interference power received by PU_1 should be below the threshold value. The Power Spectral Density (PSD) of the kth subcarrier signal is given by (Weiss et al 2004).

$$\varnothing_k(f) = p_{tk}T_s\left(\frac{\sin\pi f T_s}{\pi f T_s}\right)^2 \tag{1}$$

p_{tk} = transmit power of kth subcarrier
T_s = Symbol duration

The interference introduced to PU by subcarrier k's transmission is

$$
\begin{aligned}
I(k) &= \int_{f\in F}|g(k)|^2 \varphi_k(f)df \\
&= p_{tk}\int_{f\in F}|g(k)|^2 T_s\left(\frac{\sin\pi f Ts}{\pi f T_s}\right)^2 df \\
&= p_{tk}I_k \le I_{th} = p_{tk} \le \frac{I_{th}}{I_k} = p_{tk}^*
\end{aligned} \tag{2}
$$

$g(k)$ = channel gain from secondary base station to PU_1 over subcarrier k. F is the frequency band allotted to PU_1.

p_{tk}^* is the carrier power limit imposed on the cognitive secondary user SU_1 to limit the interference level to the PU_1.

The expression for the maximum achievable data rate(bits/second) for the user mth transmitting over kth subcarrier is given by

$$dr_{m,k} = Wlog_2\left(1 + \gamma_{m,k}\right) \tag{3}$$

Where W is the bandwidth allocated to each subcarriers by distributing total bandwidth B equally. Hence $W = B/N$. $\gamma_{m,k}$ is the Signal to noise ratio for the mth user over kth subcarrier. If we assume the ideal modulation and coding scheme as given by [Thomas & Cover, 2006], then $\gamma_{m,k}$ defined by

$$\gamma_{m,k} = \frac{p_{tm,k}|g_{m,k}|^2}{\sigma^2} \tag{4}$$

So the aggregate rate for the mth cognitive user

$$R_m = \sum_{k=1}^{N} dr_{m,k} \tag{5}$$

Overall throughput

$$T_{overall} = \sum_{m=1}^{M} \sum_{k=1}^{N} dr_{m,k} \tag{6}$$

3 PROBLEM FORMULATION

3.1 Problem definition

In this paper, we have considered an optimization of the energy efficient resource allocation in general downlink OFDMA cognitive networks. For the given static circuit power Pc and ω is the reciprocal of the power efficiency of the amplifier, EE (Energy Efficient) in the downlink transmission expressed as bit per joule per hertz can be given by

$$U_{EE}(P) = \frac{R_m}{p_c + \omega \sum_k p_{tk}} \tag{7}$$

Our objective is to maximize the $U_{EE}(P)$ stated above with the constraints on interference, throughput and allocated power over all subcarriers.

The power allocation among the N number of subcarriers the maximizes above equation can be determined by solving the below optimization problem

$$\max_P U_{EE}(P) = \frac{R_m}{p_c + \omega \sum_k p_{tk}} \tag{8}$$

s.t

C1: $p_{tk} \geq 0$

C2: $\sum_{k=1}^{N} p_{tk} \leq P_{Total}$

C3: $\sum_{k=1}^{N} I_k p_{tk} \leq I_{th}$

C4: $R_m = \sum_{k=1}^{N} dr_{m,k} \geq R_{minimum}$

P_{Total} – Total transmitted power
$R_{minimum}$ – Minimum Data rate
Constraint C1 ensures that the power level is always non negative, C2 ensures that total allocated power is below total transmitted power, C3

provides the limit on the interference and it must be below threshold value whereas C4 always maintain data rate above minimum required data rate.

3.2 Solution to the optimization problem

Above optimization problem is very complex to solve, to reduce the complexity to some extend it is required to implement a low complexity channel assignment protocol. The channel assignment protocol proposed here will assign every sub channel to the user having largest channel gain on that sub channel. The channel assignment protocol starts with ordering the secondary users according to the decreasing order of their channel gain. After the channel assignment, we can determine the optimal power allocation that maximizes the EE by solving the optimization problem (8) given in previous section.

3.2.1 Converting EE to tractable form

Problem defined in equation (8) is neither convex nor concave with respect to power. Thus obtaining a global optimum solution is difficult. To make optimization problem tractable, we modify the problem stated above.

$$R_{modified} = \sum_{k=1}^{N} \rho_{m,k} dr_{m,k} \tag{9}$$

Here $\rho_{m,k}$ is real and [0, 1] with following properties

$$\rho_{m,k} = \begin{cases} 1 \, if \, subcarrier \, k \, is \, allocated \, to \, m \\ 0 \, otherwise \end{cases} \tag{10}$$

So the modified optimization problem is

$$\max_P U_{EE}(P) = \frac{R_{modified}}{p_c + \omega \sum_k p_{tk}} \tag{11}$$

s.t

C1: $P_{Total} - \sum_{k=1}^{N} p_{tk} \geq 0$

C2: $I_{th} - \sum_{k=1}^{N} I_k p_{tk} \geq 0$

C3: $R_{modified} - R_{minimum} \geq 0$

3.2.2 Optimal solution using parametric optimization

The optimization problem broadly termed as concave-convex fractional programming problem having the general representation

$$\max_{x} \frac{f_0(x)}{h(x)}$$

s.t

$$f_i(x) \le 0$$

where $f_0, f_1, f_3, f_4 \ldots \ldots \ldots \ldots, f_m$ are concave. $f_0(x), f_1$ and $f_i(x)$ are real valued functions. These types of problem are quasiconcave in nature and it gives local optima solution. But if $f_0(x)$ is strictly concave or $h(x)$ is strictly convex then the local optima solution is global optimum solution. The above problem can be solved by various techniques such as Charles-Cooper transformation (Charles & Cooper, 1962), nonlinear optimization technique (Bonami, P. et al, 2008) and Dinkelbach parametric method (Dinkelbach W., 1967).

In this paper, we propose an algorithm having less complexity, based on Dinkelbach parametric method, the fractional programming problem is converted to parametric by using Dinkelbach technique. The solution is obtained in iterative manner by defining the tolerance of convergence.

Dinkelbach method for the optimization problem is defined in (11) is given by

$$N(p) = R_{modified} = \sum_{k=1}^{N} \rho_{m,k} dr_{m,k} \qquad (12)$$

$$D(p) = p_c + \omega \sum_{k} p_{tk} \qquad (13)$$

$$Q(p) = \frac{N(p)}{D(p)} \qquad (14)$$

The feasible region for p given by

$$R: \{C1 \text{ to } C3 \text{ of equation } (); D(p) > 0 \qquad (15)$$

$$F(q_k) = max\{N(p) - q_k D(p) | p \in R\} \qquad (16)$$

The solution of the algorithm is

$$\max Q(p) | p \in R$$

More formal description of the proposed algorithm is given in Table 1

Algorithm: EE Power Allocation for Cognitive Radio

Channel Assignment:
Rearrange secondary users according to the decreasing order of their channel gain.

Optimization:

Step 1: Parameters initialization
Choose $p_1 \in R$ δ (*tolerence of convergence*)
Let k = 0

Set $q_k = \dfrac{N(p_{k+1})}{D(p_{k+1})}$

Step 2: Solve the problem

$$F(q_k) = max\{N(p) - q_k D(p) | p \in R\}$$

Obtain optimal solution as p_k

Step 3: Convergence of iterations

If $F(q_k) \ge \delta$
Update

$$q_{k+1} = \frac{N(p_k)}{D(p_k)}$$

Go to Step 2
If $F(q_k) < \delta$, Stop the iterations
Denote p_K as optimal solution

---

4 RESULTS AND DISCUSSION

To implement the simulation of proposed method, we consider a traditional OFDM based cognitive radio system with 256 subcarriers opportunistically accessed by SUs in underlay propagation system. The channel is assumed to be Rayleigh fading channel with an average power gain of −10 dB. The other system parameters listed below in Table 2.

Figure 3 and 4, displays the variation of EE with increase in number of users for fixed number of subcarrier channels (C). Both results in figure shows that EE is increasing with increase in number of C. With increase in C, both numerator and denominator of equation (8) increase. Numerator increases by factor 1/C. The $\omega \sum_k P_{tk}$ term in the denominator also increases by same factor 1/C, but the presence of static circuit power (P_c), the overall increase in the denominator will be less than numerator. Hence the overall EE of the system increases.

In above Figure 3 and 4, we see that with increase in number of users, there is slight increase in the value of EE, this is due to the optimization of the power in the cognitive radio networks.

If we increase the interference threshold I from 10^{-13} to 10^{-16}, there is significant drop in the EE of the system. This is due to the assumption of Interference constraint in the optimization problem.

The above results shows that the system tries to optimize the value of EE by optimizing power

Table 2. System parameters for OFDM CR network.

No of Channel Varied from (C)	64–256
Efficiency of power amplifier (ω)	1
No of PUs varied from	6–18
Bandwidth of each PU	0.96 MHz
Duration of OFDM symbol	1 μs
Static Circuit Power	2.5 W
Noise power variance	0.1 mW
Rate threshold	100 kbps
Interference Threshold varied from	10^{-13} to 10^{-16}

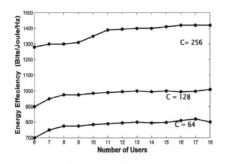

Figure 3. EE vs number of users and channels for interference $I = 10^{-13}$.

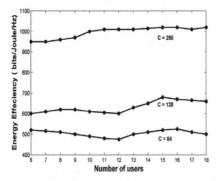

Figure 4. EE vs number of users and channels for interference $I = 10^{-16}$.

allocation to the cognitive radio secondary users in such a manner that the overall EE of the system remains constant for fixed value of subcarrier channels. As we see in both above figures that the EE is almost constant for increase in number of users.

5 CONCLUSION

We presented the energy efficient resource allocation in a downlink OFDMA cognitive radio that maximizes the energy efficiency of the network. The optimization problem developed has low complexity and practically implementable. The proposed iterative algorithm is based on Dinkelbach method. The proposed method is found to be effective and simple in solving concave convex fractional complex optimization problem. The results show that the method has good performance and can be effectively used for power allocation.

REFERENCES

Beibei, W., K., J.R.L.: 'Advances in cognitive networks: a survey', IEEE J. Sel. Topics Signal Process., 2011, 5, (1), pp. 5–23.

Bonami, P., Biegler, L. T., Conn, A. R., Cornuéjols, G., Grossmann, I. E., Laird, C. D., Lee, J., Lodi, A., Margot, F., Sawaya, N. & Waechter, A. (2008). An Algorithmic Framework for Convex Mixed Integer Nonlinear Programs. Discrete Optimization, 5(2), 186–204.

Charnes, A. & Cooper, W.W. Programming with linear fractional functionals. Naval Res. Logist. 1962, 9, 181–186.

Dinkelbach, W. (1967). On Nonlinear Fractional Programming. Management Science, 13, (7), 492–498.

Dipen Bepari & Debjani Mitra "Improved power loading scheme for orthogonal frequency division multiplexing based cognitive radio" IET Communication, 2015.

Federal Communications Commission: "Spectrum policy task force". Report ET Docket no. 02–135, November 2002.

Ian, F.A., Won-Yeol, L., Mehmet, C.V., et al. 'NeXt generation/dynamic spectrum access/cognitive radio wireless networks: A survey', Comput. Netw., 2006, 50,(13), pp. 2127–2159.

Islam, M. "spectrum survey in Singapore: Occupancy measurements and analysis," in Proc. 3rd Int conf. cognitive radio orient. Wireless netw. Commun. Singapore, may 13–17, 2008, pp. 1–7.

Kandasamy Illanko, Muhammad Naeem, Alagan Anpalagan, and Dimitrios Androutsos. "Energy-Efficient Frequency and Power Allocation for Cognitive Radios in Television Systems" IEEE System Journal March 10(1), 313–324, 2016.

Lokman Sboui, Zouheir Rezki, and Mohamed-Slim Alouini "Energy-Efficient Power Allocation for Underlay Cognitive Radio Systems" IEEE Transaction on cognitive and Networking 1.3, 2015.

Song, M., C. Xin, Y. Zhao & Cheng, X. Dynamic spectrum access: from cognitive radio to network radio. IEEE Trans. Wireless Commun. 19(1), 23–29 (2012).

Thomas M., Cover, J.T, "Elements of Information Theory" (John Wiley and Sons, 2006, 2nd Edition).

Weiss, T., Hillenbrand, J., Krohn, A., & Jondral, F.K. (2004). Mutual interference in OFDM-based spectrum pooling systems. In Proc. of IEEE vehicular technology conference (Vol. 4, pp. 1873–1877), Milan, Italy, May 2004.

Xu, Lei, Jun Wang, Ya-ping Li, Qianmu Li, and Xiaofei Zhang. "Resource allocation algorithm based on hybrid particle swarm optimization for multiuser cognitive OFDM network", Expert Systems with Applications, 2015.

*Soft computing, intelligent system, machine vision
and artificial neural network*

Communication and Computing Systems – Prasad et al. (Eds)
© 2017 Taylor & Francis Group, London, ISBN 978-1-138-02952-1

Advance diagnosis of MSME turning into NPA through intelligent techniques: An Indian perspective

Agnivesh & Rajiv Pandey
AIIT, Amity University, Lucknow, India

ABSTRACT: Prevention is better than cure, a very well said proverb, is applicable to all types of ailments. Any system, before it starts malfunctioning, gives specific symptoms and by those symptoms a diagnosis can be inferred to prevent the malfunctioning and bring the system to normal without much effort. Unchecked malfunctioning results in further decay and then it may require extra efforts involving extra cost to rejuvenate the system back. This fact is not only applicable to human and animal anatomy but extendable to all systems, industrial units being no exception. An industrial unit has its own system of functioning, growth and survival, giving symptoms not beyond System Analysis. Only what is required is proper monitoring and analysis. Recurring slippages by industries in their financial performance result in NPA, Non-Performing Assets. Today NPAs are one of the biggest challenges faced by Indian Banks. This problem of NPA will further deteriorate because of the global economic slowdown and continued complacence in monitoring periodical financial figures of industries in absence of any intelligent technique. The paper presents our approach to use intelligent technique for advance diagnosis of Micro, Small and Medium Enterprise segment termed as MSME, turning into NPA category.

1 INTRODUCTION

Over 13 million MSMEs (Micro, Small and Medium Enterprises) employing over 42 million persons have a very effective impact on the growth of Indian Economy. The contribution of this sector is so enormous that to give protection and encouragement to MSMEs, a special law has been enacted "The Micro, Small and Medium Enterprises Development Act 2006". As per provisions of the Act, the policies and practices in respect of financial credit to the micro, small and medium enterprises to be Credit Progressive to ensure timely and smooth flow of credit to such enterprises, minimize the incidence of sickness among and enhances the competitiveness of such enterprise.

The years of survival of such enterprises are larger compared to other enterprises for the reasons of annual burden of investment in Plant & Machinery being low, the strength of workforce being small and manageable and because of inherent resilience of the MSME's to cope up other requirements and problems from their own resources.

Being not immune to slippages, MSMEs also default financially which results in non-payment of interest and installments to the Banks and Financial Institutions affecting their assets.

The remainder of this paper is organized as follows. The section "Non-Performing Asset" defines non-performing asset as per Reserve Bank of India (RBI) prudential norms. "Working Capital" is describing the finance made available to MSME for purchase of raw materials, consumables and other elements required for production. The "Monthly Stock Statement" section highlights a mandatory provision to be furnished to the Bank by the MSME Unit. "Current Asset and Current Liability" is depicting the elements that make the foundation of our model experiment. In "Intelligent Technique" intelligent technique and its importance are described. The "K Means Algorithm Explored" and "R-Analytics" highlights K Means algorithm and features of R respectively. The proposed method is explained in "Methodology". "Result and Discussion" and "Conclusion and Future Work" are designed for experimental results of the proposed method and conclusion and future work respectively.

1.1 Non-performing asset

An asset becomes non-performing when it ceases to generate income for the Bank. A Non-Performing Asset (NPA) has been defined as a credit facility in respect of which the interest and/or installment of principle has remained due for a specified period of time. As per prudential norms of Reserve Bank of India (RBI) for income recognition and assets classification this period has been fixed as 90 days. The instructions of RBI to the Banks and financial institutions are to establish appropriate internal systems for quick identification of NPAs.

The Banks and financial institutions after classifying an asset non performing proceed to take legal action for recovery after issuing legal notices of recall of the loan and demanding the entire loan amount due with pendentelite interest till the amount is recovered in full within a specified period, failing which the recovery proceedings continue as per provisions of the laws of the land (RBI, 2014).

Huge amount of public money is thus locked in litigation which prevents proper utilization and recycling of the funds for development of the country. The figures of NPAs are mind boggling and increasing unchecked year after year. Though the major share of these NPAs is of large and corporate businesses but defaulting MSMEs also contribute to this huge amount of public money being locked.

Preventive mechanism for saving these MSMEs turning into NPA category may be searched out as their financial transactions are not so complicated as compared to large industries and corporate businesses but very simple financial operations.

1.2 Working capital

94.94% Micro units, 4.89% of Small units and 0.17% Medium units presently contribute the entire MSME segment in India as per a recent survey conducted by Ministry of MSME, Government of India. Proper and timely Working Capital Loan has been found to be their main requirement for thriving and survival.

Working Capital facilities by the Banks and financial institutions are provided to MSMEs by fixing an annually reviewable limit on the basis of projected business by the MSME. The annual review is based on the performance and this limit can be enhanced or reduced depending on the growth or decline of the business (Tannan, 1995).

1.3 Monthly stock statement

It is mandatory as per RBI guidelines for every MSME segment units availing working capital facility limit to submit Stock statement of the month at the month end. Based on this stock statement the Drawing Power (DP) of the unit for the next month financial operations is computed by the Bank. The unit can avail the so allowed funds only as per the DP. Stocks constitute raw materials, consumables, semi-finished goods and finished goods, finished goods dispatched but sales realization due. A unit in its premises may have at any time both paid stocks and non-paid stocks. Paid Stocks are those which are unit's absolute prop-

erties fully paid against and the Bank considers only Paid Stocks hypothecated to it and that is the Block Asset of the Bank against which the working capital facility is allowed. The Bank is at liberty to verify quantity, quality and rates to check the evaluation of the stocks (RBI, 2014).

Modification Suggested

- The stock statement must also have the declaration of non-paid stocks separately
- MSME should also declare values of sundry credits and sundry liabilities along-with stock statement

1.4 Current asset and current liability

Financial operations of MSMEs being simple, the current asset of an MSME at the month end may conveniently be depicted by the following expression (Tannan, 1995):-

$$\text{Current Asset} = \sum \text{Cr.} + \text{PS} + \text{US} + \text{SC} \qquad (1)$$

where $\sum$ Cr. = sum of all credit proceeds of the month; PS = value of paid stocks; US = value of unrealized sales; SC = value of sundry credits.

Similarly current liability at the month end may be depicted by the following expression:-

$$\text{Current Liability} = \sum \text{Dr.} + \text{UPS} + \text{INTT} + \text{INST} + \text{SL} \qquad (2)$$

where $\sum$ Dr. = sum of all withdrawals during the month; UPS = value of unpaid stocks; INTT = value of interest against Bank finance; INST = value of installment against any term loan of the Bank; SL = value of sundry liability.

It is an obligation on the part of MSME to deposit all sales realization with the financing Bank exclusively hence monthly $\sum$Cr and $\sum$Dr are available on the Banker's system, paid stocks, unpaid stocks and sold stocks (unrealized sales) are reflected in the stock statement. The modified statement shall also provide sundry credit and sundry liability. Now all key figures are available to compute current asset and current liability of the month. Using intelligent techniques the current asset and current liability can be compared to assess in advance the tending growth or decline of the MSME and the growing number of NPAs may be controlled.

2 INTELLIGENT TECHNIQUE

The ability to reason, perceive relationships and analogies, calculating and learning is known as intelligence. Definitions of intelligent technique may be many but they have a common meaning. It is an approach to develop new and effective system/ systems for problem solving and decision making with respect to a group or a combination of inter related, inter dependent or interacting elements or facts forming a collective entity, by applying system analysis. The system can manage knowledge and information which is known as the crucial issue in smart decision making support. Such a system can function in information rich environments, deal with complex and imprecise information, rely on its knowledge and learn from experience (Adla & Zarate, 2006).

In today's competitive environment, data is no longer labeled as simple and there is a need of intelligent techniques to mine these data and extract meaningful information from it. K means algorithm is one such technique of data mining and to knowledge discovery technique.

2.1 K means algorithm explored

There is a set of objects. Some of the objects may be similar to each other and some may vary on the basis of certain attributes. Group of objects similar to each other is known as a cluster. This approach of partitioning objects is known as clustering. There are various intelligent techniques for problem solving and knowledge discovery. Clustering is one of them. Clustering tries to maximize similarity among objects in the same cluster and minimize the similarity among objects in different clusters. Data clustering is the process of mining hidden patterns from large data sets (Kumar, Tan & Steinbach, 2004). The technique of clustering is most widely used in future prediction. Many different clustering techniques have been defined in order to solve the problem from different points of view. K means clustering is widely used method for clustering (Jain & Dubes, 1988).

K means is an algorithm to classify or to group objects based on attributes/features into K number of groups or clusters. K is a positive integer number. It is a user input to the algorithm. The most common algorithm has following steps:-

1. Enter the number of clusters to which data set is grouped and the data set to cluster.
2. Initialize the first k clusters-take first k instances or take random sampling of k elements.
3. Calculate the arithmetic means of each clusters.
4. K means assigns each record in the dataset to one of the clusters in the beginning—each record is assigned to the nearest cluster using a measure of distance (e.g. Euclidean distance).
5. K means re-assigns each record in the dataset to the most similar cluster and re-calculates the arithmetic means of all the clusters in the dataset.

The dissimilarity or similarity between objects is computed on the basis of the distance between each pair of objects. The most popular distance measure is Euclidean distance, which is defined in (3):

$$d(i,j) = ((x_{i1} - x_{j1})^2 + (x_{i2} - x_{j2})^2 + ... + (x_{in} - x_{jn})^2)^{1/2} \quad (3)$$

where $i = (x_{i1}, x_{i2}, ..., x_{in})$ and $j = (x_{j1}, x_{j2}, ..., x_{jn})$ are two-dimensional data objects (Han & Kamber, 2010).

2.2 R-analytics

There are multiple tools that provide statistical analysis. R is our choice for under-mentioned analysis because of the following reasons (Underwood, 2015):-

- R is being used for data analysis
- R libraries implement linear and non-linear modeling, classical statistical tests, time-series analysis, classification, clustering and others.
- R compiles and runs on a wide variety of UNIX platforms, Windows and Mac OS.
- For computationally heavy tasks, C, C++, and FORTRAN code can be linked and called at run time.
- R is easily extensible through functions and extensions.

3 METHODOLOGY

There may be number of units availing a limit of say Rs. 2.5 lacks as we have considered in our data set. Yet another number of units availing a limit of 4.0 lacks, 5 lacks, 6 lacks and so on so forth. Our methodology compares actual monthly performance or current asset as given by expression (1) with current liability as given by expression (2). The model data set has the following features:-

- Set of 24 MSMEs availing a limit of Rs 2.5 lacks.
- The data set is in Excel's CSV (Comma delimited) format consisting of 2 columns as shown in Fig. 1.
- Column 1 consists of MSMEs monthly current asset. Column 2 is their current liability at the month end corresponding to their current asset.
- The input file containing the data set is named as msme1a.

The model data set is imported from excel to R Studio as shown in the Fig. 1. Performance of the MSME units is analyzed by applying K means clustering algorithm using R. Two clusters will be created in the present experiment. One cluster is for the MSME units having higher current asset and the other cluster is for the MSME units having

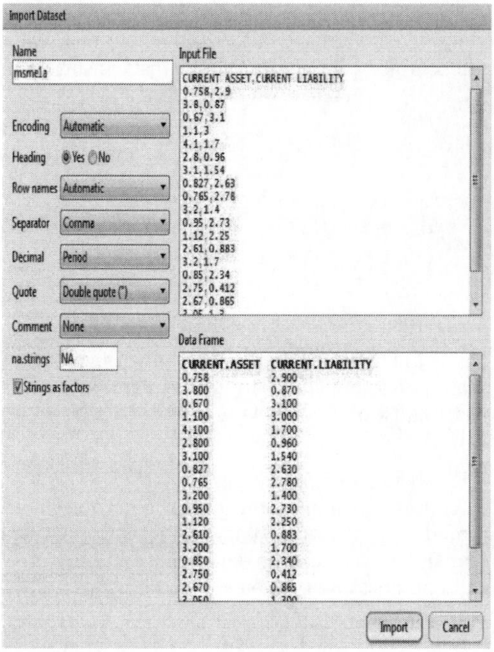

Figure 1. Data Set being imported from Excel to R Studio.

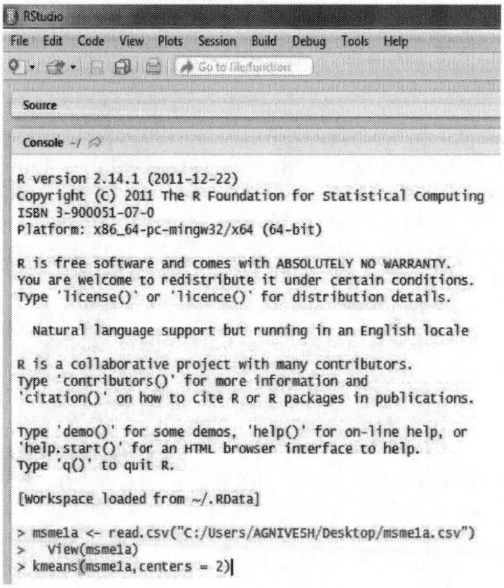

Figure 2. Creation of clusters using K Means.

higher current liability. The procedure for importing data set from Excel to R Studio and creation of cluster using K Means algorithm in R Studio is shown in Fig. 2.

4 RESULT AND DISCUSSION

Figure 3 is showing the analysis of our data set by R. The figure is showing 2 clusters of sizes 12 each. Mean value of cluster 1 is (0.9058333, 2.8125) and the mean value of cluster 2 is (3.0525000, 1.2350). Centroid of Cluster 1 has low current asset value and high current liability value whereas cluster 2's centroid has high current asset value and low current liability value. Therefore cluster 1 contains the MSME units whose performance is unsatisfactory. Cluster 2 contains the MSME units whose performance is satisfactory.

In Fig. 3 the clustering vector is indicating the cluster to which each data point is allocated. Value 1 is the cluster whose current asset is low and value 2 is the cluster whose current asset value is high. Similarity between objects within a cluster is measured by "within cluster sum of squares" and dissimilarity between clusters is measured by "between cluster sum of squares". R analyzed the data set MSME1a and the value of "within cluster sum of squares by cluster" comes out to be (1.433413, 5.894583) as shown in Fig. 4. K means minimizes within cluster dispersion and maximizes between cluster dispersion. The ratio of "between cluster sum of squares" and "total sum of squares" comes out to be 85.3 percent. Available comments are "cluster", "centers", "totss", "withinss", "betweenss", "size"," iter" and" ifault".

The graph of the clusters is shown in Fig. 4.

```
> kmeans(msme1a,centers = 2)
K-means clustering with 2 clusters of sizes 12, 12

Cluster means:
  CURRENT.ASSET CURRENT.LIABILITY
1    0.9058333            2.8125
2    3.0525000            1.2350

Clustering vector:
 [1] 1 2 1 1 2 2 2 1 1 2 1 1 2 2 1 2 2 2 1 2 1 2 1 1

Within cluster sum of squares by cluster:
[1] 1.433413 5.894583
 (between_SS / total_SS =  85.3 %)

Available components:

[1] "cluster"      "centers"     "totss"      "withinss"
[5] "tot.withinss" "betweenss"   "size"
> |
```

Figure 3. Clustering details.

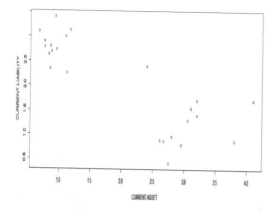

Figure 4.   Graph of clusters.

5   CONCLUSION AND FUTURE WORK

The MSME units which are in cluster 1 are lacking in their performance and therefore they need proper assistance to overcome what is lacking in their business. A subsequent month's performance shall also be checked. If the performance is still lacking, an adequate assistance should be provided before they turn into NPA category in the third month that is before 90 days. This model will definitely be suitable into multidimensional data set analysis as well. The future work also involves designing this model into an app. The app can be hosted in the cloud so that the online use of this app by the Bankers does not need the knowledge of R and K means.

REFERENCES

Adla, A., & Zarate, P. (2006). *A Cooperative Intelligent Decision Support System*. Paper presented at the International Conference on Service Systems and Service Management.

Han, J. & Kamber, M., Data Mining Concepts and Techniques, 2nd Edition, Elsevier, 2010.

Horton, N.J., Kaplan D.T. & Pruim, R. (2015). The Five College Guide to Statistics with R, Project MOSAIC.

Hughes G. & Dobbins C., 2015. The utilization of data analytics technique in predicting student's performance in massive open online courses (MOOCs), Research and Practice in Technology Enhanced learning, a Springer Open Journal.

Jain, A.K. & Dubes R.C.,1988. Algorithm for Clustering Data, Prentice Hall.

Kanungo, T., Mount, D.M. & Netanyahu, N.S. (July 2002). An Efficient k Means Clustering Algorithm: Analysis and Implementation, Vol. 24, No. 7.

Kumar, Tan & Steinbach, 2004. Introduction to Data Mining, Data Mining Cluster Analysis: Basic Concepts and Algorithms.

Madhu Yedla, Srinivasa Rao Pathakota, T.M. Srinivasa, Enhancing K-means Clustering Algorithm with Improved Initial Center, Madhu Yedla et al./(IJCSIT) International Journal of Computer Science and Information Technologies, Vol 1(2), 2010, 121–125.

Marakas, G. M. (1999). *Decision Support System in the 21st Century*. New Jersey: Prentice Hall, Inc.

Master Circular-Prudential Norms on Income Recognition, Asset Classification and Provisioning Pertaining to Advances, Retrieved May 9, 2016 from https://rbi.org.in

Tannan,M.L. (1999). Banking Law and Practice in India, Indian Law House, New Delhi.

The Securitization and Reconstruction of Financial Assets and Enforcement of Security Interest Act, 2002, BARE ACT with Short Comments. Professional Book Publishers, Delhi. 2009.

Underwood, J. (2015). Part 1: Integrating R with Web Applications. Retrieved May 10, 2016 from http://www.jenunderwood.com/2015/01/12/part-1-integrating-r/.

Verma C. & Pandey R., Big Data Representation for Grade Analysis Through Hadoop Framework, Proc. of Confluence-2016—Cloud System and Big Data Engineering, ISBN: 978-1-4673-8202-1.

Yedla, M., Pathakota, S.R. & Srinivasa, T.M. (2010). Enhancing K-means Clustering Algorithm with Improved Initial Center, Madhu Yedla et al./(IJCSIT) International Journal of Computer Science and Information Technologies, Vol 1(2), 2010, 121–125.

Communication and Computing Systems – Prasad et al. (Eds)
© 2017 Taylor & Francis Group, London, ISBN 978-1-138-02952-1

Relay node placement in constrained environment

A. Verma & V. Ranga
Department of Computer Engineering, National Institute of Technology, Kurukshetra, India

ABSTRACT: Relay node placement in wireless sensor network for constrained environment is a critical task due to various unavoidable constraints. Unpredictable obstacles are one of the most important constraints. Handling obstacles during relay node placement is complicated because of complexity involved in shape and size estimation of obstacles. This paper presents an Obstacle-Resistant Relay Node Placement strategy (ORRNP) which handles obstacles as well estimates best locations for relay node placement in the network. The approach does not require any additional hardware to estimate node locations thus can significantly reduce the deployment costs. Simulations are presented to show effectiveness of our proposed approach.

1 INTRODUCTION

In the recent years, there has been a rise in the applications of Wireless Sensor Networks (WSNs) (Ranga, Dave, & Verma, 2013). In applications such as forest fire detection, combat field reconnaissance, and machine health monitoring, to cooperatively observe the area and keep an eye on certain activities, some sensor nodes will be deployed in the network. By obtaining these sensors to control unattended in astute surroundings, it might be attainable to avoid the danger to human life and reduce the value of the appliance. These applications use Sensor Nodes (SNs) which are battery driven and have limited computational power and communication capabilities. After deployment, the sensor nodes set up a network with the target of sharing the data and synchronizing the actions performed. To assist such collaboration, nodes must be reachable to each other. Long distance transmission for sensor nodes would be expensive and will exhaust them very rapidly (as energy transmission is proportional to the distance). Thus, Relay Nodes (RNs) are deployed to perform transmission of sensed data from sensor nodes and forward that data to Base Station (BS). The problem of placing minimum relay nodes in the environment so that the entire network is interconnected is shown to be an NP-hard problem (Ranga, Dave, & Verma 2015) and is called Relay Node Placement (RNP) problem. RNP is classified into constrained or unconstrained. In unconstrained RNP, the relay nodes can be deployed anyplace in the network, while in constrained RNP, there are physical constraints (Karaki & Kamal, 2004) on where relay nodes are deployed. This is done to consider the physical limitations of a geographical area, for example, a volcano is present, a huge rock or even mountains or water bodies. These practical assumptions are closer to the real world problem in idealistic situations.

This paper presents a strategy to solve RNP problem in constrained environment. We also discuss about the state of the present research and review a mixture of already published techniques in this area stating their features and limitations.

2 RECENT WORK

In (Yu, Song, & Mah, 2011) an approach named RNIndoor is proposed which targets to provide at least one path among deployed sensor nodes by deploying few additional relay nodes. This work considers deployment area as floor which contains various obstacles like walls and restricted areas. Approach handles obstacles using structural information of floor and path loss generated from propagation models. The complexity involved in this approach limits it to simple obstacle handling only. In (Wang, Hu, & Tseng, 2005) presented a scheme for efficiently deploying sensor nodes. Sensing field is subdivided into smaller sub-regions based on the shape of field followed by deployment of sensor nodes in these sub-regions. Sensing field is assumed to be an arbitrary shaped region. This work considered obstacles while deploying sensors. The work is limited to a special case which is sensing range is same as transmission range. This problem is limited to a special case and hence cannot be applied to complex obstacles. There are few robot-deployment techniques present in literature. Like, In (Chang, Chang, Chen, & Chang, 2009) proposed a robot-deployment algorithm which is capable of handling unpredictable obstacles. The approach achieved full coverage deployment with minimum sensor nodes without using location information of nodes. Movement of robot is guided by some policies which include node placement policy, spiral movement policy and for obstacle a surround movement policy is proposed. In addition to node placement techniques we review some shape approximation techniques present in literature which we have referred for obstacle estimation and detection. In (Ramer U., 1972) an algorithm is proposed which iteratively approximated the two-dimensional arbitrary curves to polygons. The ideology behind this work is use small number of vertices lying on plane curve to produce polygons. Fit criterion is chosen as the maximum distance between any curve and approximating polygon. (Edelsbrunner, Kirkpatrick, & Seidel, 1983) presents a computational geometry based scheme that generalizes straight line graphs named "alpha-shapes". These shapes show much better notion of real complex shapes. Using point set data these shapes are produced. Obstacle estimation and detection problem is well mapped with this approach. Obstacle detection and estimation scheme is showed in (Wang & Ssu, 2013). In this scheme sensor nodes are marked around the boundary of obstacle to approximate the shape of obstacle. Geometrical rules have been used as key criterion to select the node to be marked for boundary estimation.

3 ASSUMPTIONS AND SYSTEM MODEL

A two-dimensional sensing field has been considered for the obstacle resistant relay node placement approach. Sensor and relay nodes are assumed to be static in nature after they have been deployed in the constrained environment. Absolute location of sensor nodes is the last known GPS location of the sensors. A GPS locater is assumed to be equipped with each sensor and relay nodes. Sensing Range (SR_s) and Transmission Range (TR_s) of each sensor node is in ratio ($SR\_s:SR\_s\sqrt{3}$) respectively. Sensors whose sensing range overlaps each other can communicate directly. Relay node can communicate with each node which is in its transmission range. In our system we assume that sensor to sensor, sensor to relay (vice-versa) and relay to relay connectivity for minimizing the relay node count and network deployment cost. Obstacles of arbitrary complex polygonal shapes are assumed to completely block the signal when they fall in transmission way. Proposed approach primarily considers connectivity of nodes.

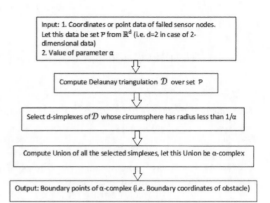

Figure 1.   Procedure to estimate obstacles.

3.1 Concept

Based on the idea of (Edelsbrunner et al., 1983) shape of any obstacle is approximated. We use the location of the failed sensor nodes to approximate the obstacle shape. The coordinates of the failed sensor node is its last known GPS location. These coordinates are the input to obstacle estimation procedure which approximates shape of obstacles in present given deployment area.

3.2 Mechanism

Any shape can be generalized by Alpha-shapes. These shapes are a family of straight Line graphs. Alpha shapes can be defined as a polytypic generated over a point set P, which depends on point p of point set P and a parameter α. This α parameter restricts the detail level of corresponding alpha shape. Radius $1/\alpha$ is assumed as the sensing range of any sensor node in case of WSN. Figure 1 shows the methodology used in obstacle estimation.

The underlying space of α-complex is the area which obstacle is present. Placement of relay nodes must be avoided in this underlying space. As mentioned above, the parameter plays an important role in estimation of obstacle shape thus a proper input of its value is required for efficient computation and better approximation of obstacle shape. α value 0 outputs convex hull and 1 outputs Delaunay triangulation of coordinate set. Figure 2 shows examples of some estimated obstacles. The output boundary coordinates of this procedure are used in ORRNP procedure.

4 ORRNP: OBSTACLE-RESISTANT RELAY NODE PLACEMENT STRATEGY

In this work we have solved the RNP problem in constrained environment using Steiner Minimum

Tree (SMT) approach (Hwang & Richards, 1992) for both rectilinear and Euclidean case. Steiner minimum tree is the shortest possible interconnection of terminals (vertices) using some extra points (Steiner points) which minimize the tree length. Every Steiner point has degree three and these points are the locations representing relay nodes. SMT are further classified as Rectilinear Steiner Minimum Tree (RSMT) (Garey & Johnson, 1977) and Euclidean Steiner Minimum Tree (ESMT) (Winter & Zachariasen, 1997). In Figure 3 procedure to compute RSMT and ESMT is presented.

We have used simple heuristics to reduce the candidate locations set for Steiner point in order to reduce unnecessary computation and minimize the time complexity. The procedure is divided into parts which include COMPUT-RSMT which computes rectilinear SMT similarly COMPUTE-

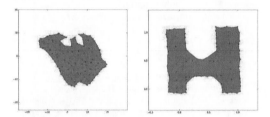

Figure 2. Estimation of various obstacle shapes.

```
Initialization: Input1: Obstacles: Boundary points of α-complex
shape (i.e. boundary co-ordinates of obstacles)
            Input2: Co-ordinates of working sensor nodes (i.e.
nodes outside the obstacles covered area)
/* Computes Euclidean Minimum Steiner tree with Obstacle(s) */
COMPUTE-ESMT(Input1, Input2)
    CANDIDATE_SET ← COMPUTE-CANDIDATE-
SET(Input1, Input2)
    Initialize MAX_POINT to a positive number
    while MAX_POINT ≥ 0
        COST = 0
        Assign MAX_POINT a negative value
        for each point p ∈ CANDIDATE_SET
            DeltaCost ← DELTA-MST(Input2+ SteinerPoints, p)
            if DeltaCost > COST
                COST ← DeltaCost
        if MAX_POINT ≥ 0
            SteinerPoints = SteinerPoints ∪ {MAX_POINT}
        for each point p ∈ SteinerPoints
            if Degree ≤ 2
                SteinerPoints = SteinerPoints - {p}
    FINAL_ESMT ← KRUSKAL(Input2 + SteinerPoints)
    DISTANCE_ESMT ← Σ Edges.distance of FINAL_ESMT
    return FINAL_ESMT
/* Computes Candidate set of points which can be used as
Steiner Points required for COMPUTE-ESMT */
COMPUTE-CANDIDATE-SET(Input1, Input2)
    CANDIDATE_SET ← {}
    ConvexHull = Convex_Hull(Input2)
    for each point p ∈ Points inside ConvexHull
        if point p ∉ Points within Obstacles
            CANDIDATE_SET = CANDIDATE_SET ∪ {p}
    return CANDIDATE_SET
```

Figure 3. (Continued)

```
/* Computes difference between costs of two MST's  one before
and one after adding SteinerPoint */
DELTA-MST(SetOfPoints, TestPoint)
    MST_ONE ← KRUSKAL-MODIFIED(SetofPoints,
Obstacles)
    MST_TWO ← KRUSKAL-MODIFIED(SetOfPoints +
TestPoints, Obstacles)
    COST1 ← Σ Edges.distance of MST_ONE
    COST2 ← Σ Edges.distance of MST_TWO
    return COST1 – COST2
/*Computes cost of MST; if edge intersects Obstacle(s) then
distance is set to infinity */
KRUSKAL-MODIFIED(SetOfPoints, Obstacles)
    Edges ← {}
    for each point p ∈ SetOfPoints
        for each point p' ∈ SetOfPoints
            if edge(p, p') intersects Obstacles
                Distance ← ∞
            else
                Distance ← CALCULATE-DISTANCE (p, p')
            Edges.add(p, p', Distance)
    Sort(Edges, key = Distance)
        MST ← {}
        while MST does not contains (SetOfPoints – 1) edges
        and MST does not connect all SetOfPoints
        for each edge ∈ Edges
            if edge(p, p') does not create cycle
                MST = MST ∪ {edge(p, p')}
    return MST
/* Computes Rectilinear Minimum Steiner Tree with Obstacle(s)
*/
COMPUTE-RSMT(Input1, Input2)
    CANDIDATE_SET ← HANAN-GRID-SET(Input1,
Input2)
    Initialize MAX_POINT to a positive number
    while MAX_POINT ≥ 0
        COST = 0
        Assign MAX_POINT a negative value
        for each point p ∈ CANDIDATE_SET
            DeltaCost ← DELTA-MST(Input1+
SteinerPoints, p)
            if DeltaCost > COST
                COST ← DeltaCost
        If MAX_POINT ≥ 0
            SteinerPoints = SteinerPoints ∪ {MAX_POINT}
        for each point p ∈ SteinerPoints
            if Degree ≤ 2
                SteinerPoints = SteinerPoints – {p}
    FINAL_RSMT ← KRUSKAL(Input2 + SteinerPoints)
    DISTANCE_RSMT ← Σ Edges.distance of
FINAL_RSMT
    return FINAL_RSMT
/*Computes set of Hanan Points excluding points in
Obstacle(s); which can be used as Steiner Points in
COMPUTE-RSMT */
HANAN-GRID-SET(Input1, Input2)
    CANDIDATE_SET ← {}
    HananPoints ← HANAN-GRID(Input2)
    for each point p ∈ HananPoints
        if point p ∉ Points within Obstacles
            CANDIDATE_SET = CANDIDATE_SET ∪ {p}
    return CANDIDATE_SET
```

Figure 3. ORRNP procedure.

ESMT which computes Euclidean SMT, MODIFIED-KRUSKAL part computes Minimum spanning tree with obstacle avoidance, DELTA-MST computes cost of tree when any new node

419

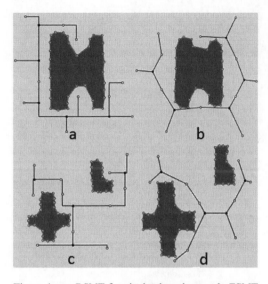

Figure 4. a. RSMT for single obstacle case, b. ESMT for single obstacle case, c. RSMT for multiple obstacle case, d. ESMT for multiple obstacle case.

is added, HANAN-GRID-SET generates a set of Hanan points which are used when computing RSMT.

5 SIMULATION AND RESULTS

This section presents the detail of simulation environment used for investigating the performance of proposed approach ORRNP. All Procedures have been coded in Python programming language. The computational experiments have been executed on computer with Intel i5 II gen 64×2, 2.5 GHz and 4 GB RAM. The parameters α is changed according to the sensor node range while computing obstacles. 800×800 grid is considered for node deployment simulation environment.

Figure 4a and Figure 4b show RSMT and ESMT respectively when single obstacle is considered in deployment area. Figure 4c and Figure 4d show relay node deployment in multiple obstacle case.

6 CONCLUSION AND FUTURE SCOPE

In this paper, the RNP in constrained environment is examined and possible solution has been for-

mulated. The objective of this work is to perform the relay node placement in WSN when the obstacles are present in deployment area. Since RNP in constrained environment is an NP-hard problem, a strategy to find a good solution in a reasonable amount of time is formulated. In future we will try to minimize the running time of ORRNP, integrate coverage factor and fault tolerance.

REFERENCES

Chang, C.Y., Chang, Y.C., Chen & Chang, H.R. 2009. Obstacle-resistant deployment algorithms for wireless sensor networks. *IEEE Transactions Vehicular Technology* 58(6): 2925–2941.

Edelsbrunner, H., Kirkpatrick, D.C., & Seidel, R. 1983. On the shape of a set of points in the plane. *IEEE Transactions on Information Theory* 29(4): 551–559.

Garey, M.R. & Johnson, D.S. 1977. The rectilinear Steiner tree problem is NP-complete. *SIAM Journal of Applied Mathematics* 32(4): 826–834.

Hwang, F.K. & Richards, D.S. 1992. Steiner Tree Problems. *Networks* 22: 55–89.

Karaki, J. & Kamal, A. 2004. Routing techniques in wireless sensor networks: a survey. *Journal of IEEE Wireless Communications* 11(6): 6–28.

Ramer, U. 1972. An iterative procedure for the polygonal approximation of plane curves. *Computer Graphics and Image Processing* 1(3): 244–256.

Ranga, V., Dave, M., & Verma, A.K. 2013. Network Partitioning Recovery Mechanisms in WSANs: a Survey. *Wireless Personal Communications* 72(2): 857–917.

Ranga, V., Dave, M., & Verma, A.K. 2015. Relay Node Placement for Lost Connectivity Restoration in Partitioned Wireless Sensor Networks. *Proceedings of International Conference on Electronics and Communication Systems*: 170–175. Barcelona, Spain: IEEE.

Wang, W. & Ssu, K. 2013. Obstacle Detection and Estimation in Wireless Sensor Networks. *Journal of Computer Networks* 57(4): 858–868.

Wang, Y., Hu, C., & Tseng, Y. 2005. Efficient Deployment Algorithms for Ensuring Coverage and Connectivity of Wireless Sensor Networks. *1st International Conference on Wireless Internet:* 114–121, Budapest, Hungary: IEEE.

Winter, P. & Zachariasen, M. 1997. Euclidean Steiner minimum trees: An improved exact algorithm. *Networks* 30: 149–166.

Yu, M., Song, J., & Mah, P. 2011. Rnindoor: A relay node deployment method for disconnected wireless sensor networks in indoor environment. *Proceedings of 3rd International Conference on Ubiquitous and Future Networks:* 19–24, Dalian, China: IEEE.

Communication and Computing Systems – Prasad et al. (Eds)
© 2017 Taylor & Francis Group, London, ISBN 978-1-138-02952-1

Evolving programs using genetic programming to minimize single objective problem

Anurag Upadhyay & Sandeep Sharma
School of ICT, Gautam Buddha University, Greater Noida, Uttar Pradesh, India

ABSTRACT: This paper discusses the use of Genetic Programming to minimize objective of a single objective problem. Genetic Programming is part of Evolutionary Computing Paradigm. It is partly evolved from Genetic algorithm concepts, out of the need to represent solution to the problem in a more complex manner. In this paper, a problem is considered for which solution can be represented in form of If-Then and Condition rules, wherein Genetic Programming is used to automatically formulate and evolve these rules to arrive at an effective solution of the problem.

1 INTRODUCTION

1.1 Topics covered

Genetic Programming is used to minimize objective of the given problem. The scenario is as follows, in which we need multiple or single repairers to fix the solar panels. To let them accomplish their tasks efficiently, rules or algorithms are written. We can come up with a solution in which we write multiple rules/programs and then select which one is better according to their performance with which it solves the problem. A more efficient approach is, to form-late these rules with a computer, and then evaluate it on the problem instance, and chose best rule. This can be repeated to get to rules/programs that solve the given problem really efficiently.

1.2 Organization

The organization of paper is as follows. First Genetic Programming (GP) field is introduced, afterwards the problem solving technique using GP tools is discussed. Which is followed by a brief description of the objective optimization problem. Next, is a discussion about the GP framework that have been built in C++ and how it has been applied to this specific problem. Finally, the results are discussed on the various runs of GP that were conducted. Later conclusions and future work are presented.

2 INTRODUCTION TO GENETIC PROGRAMMING

2.1 Genetic algorithms

The primary difference between Genetic Algorithm and Genetic Programming is the way in which the solution to the problem is represented. In GA, problem is represented in terms of chromosomes, i.e. in bits of string, different combination of characters of string represent different solution, which are then crossed over and mutated to form new bits of string. The genetic programming paradigm continues the trend of dealing with the problem of representation in genetic algorithms by increasing the complexity of the structures undergoing adaptation Koza (1992).

2.2 Genetic programming

Genetic programming is part of evolutionary computing paradigm of problem solving in which rather than writing explicit programs/rules to solve a problem, we define a basic set of hierarchical computer programs of various shape and sizes which undergo genetic procedure such as crossover and mutation to form better rules/programs by themselves.

A large number of population of programs are geneticily bred in genetic programming. Programs in genetic programming are represented as trees. The idea of representing programs in trees comes from the compiler because, internally the compiler translates written code to trees for evaluation and execution.

2.3 Representation

The tree representation of the genetic programming is also called s-expression (after LISP programs). It consists of two different types of nodes: function nodes and terminal nodes. The internal nodes of the tree which take an argument are called function nodes and the external nodes are called terminal nodes. In standard GP all the function nodes should be compatible with each other, i.e. in any argument of the function node any other function or terminal node can be added. Another approach to construct GP trees, is to specify a grammar which explicitly states which function/terminal can be coupled with which function's argument. The function sets and terminal sets are devised from the problem itself.

2.4 *Procedure of evolution*

The procedure of evolving programs by genetic programming is as follows. First, an initial populations of computer programs are generated from the function and terminal sets. For the problem in consideration, the programs are generated according to a specified grammar. Then, these programs are run on the instances of the problem, and based on how they performed on the given objective, each program is assigned a fitness value. Different fitness values are raw fitness, standardized fitness, adjusted fitness, and normalized fitness. Next, the programs accord-in to the selection criterion, e.g. tournament selection, uniform selection, or fitness proportionate selection are selected. Genetic operations are performed in which two separate programs are combined (crossover) or one program is changed individually (mutation). After creation of next population, each individual of population is evaluated on instance of the problem. The process is repeated until a termination criterion of the GP run is satisfied. Termination criterion can be maximum number of generations or getting a specified number of very accurate results on the problem (hits).

3 PROBLEM IN CONSIDERATION

3.1 *Problem overview*

The problem in consideration is as follows. The scenario is that, there are 'n' number of solar panels, a collection of panels close to each other are depicted as nodes. Out of these 'n' nodes, 'k' of them are defected nodes, i.e. in 'k' either some of the panels have been defected due to some malfunction in them or they just simply stopped working due to some constraint. These are also addressed as jobs. There are 'm' number of repairers available who have to finish 'k' jobs. For now, there is only single objective which is being considered to minimize: i.e. time. The repairers focus on minimizing the time (objective) in finishing all the jobs. For this problem, currently single repairer is considered.

The amount of time that it will take for them to travel-el between one node to another is known, and the amount of time it will take for the repairer to complete the job is also known. On an abstract level, the problem can be seen as: There are 'n'

number of nodes, out of which 'k' are defective. There are agents, which will transverse all the 'k' nodes with goal of minimizing a given objective. Objectives can be single or multiple.

3.2 *Representation of the problem*

One solar panel or a collection of solar panels that are close enough are depicted as a single node. That is, in a node, all the solar panels in that node are at best next to each other or at worst at a minimal distance, All the nodes are connected to each other, i.e. from any node it is possible to travel to any other node, and a direct distance to a node is shorter than starting from the same node and going through different nodes and then reaching the destination node.

Out of the grid of all the nodes, the defected nodes are also connected to each other.

The representation of the problem in the program is done as follows. The solar panel farm or the solar panel scenario is randomly generating from the inputs such as the max-min number of nodes to put in, the max-min number of nodes to make defected (i.e. jobs to make). Since the solar panels are arranged in a grid fashion, the max-min distance between the rows of solar panels is inputted, and max-min-time to complete a job, i.e. to fix a node, is also taken as an input.

Out of all the range of max-min values, several instances of same problem is generated, each having different site, different number of solar panels, different number of defective solar panels and different distance between row and column of solar panels.

Once the arrangement of solar panel has been generated, the distance between all the jobs to every other job is calculated and put in a matrix called job graph. It tells the time it would take to reach from one job to another and also the time it would take to complete the job. This matrix is the one, which is provided as an input of the problem to the Genetic Programming framework.

4 EXPERIMENTAL SETUP

The GP framework in consideration here, is based on the libgp framework provided by Larry I Gritz. The capabilities of the libgp framework has been extended by adding in details such as functionality of grammar. With the current addition, in order to evolve code, a specific grammar can be placed and the rules that are formed, i.e. the function set and the terminal set that are attached to form a rule are arranged in strict fashion described by grammar. Crossover and mutation also follow grammar rules; a crossover point is chosen on the basis of compatibility of one crossover point to another.

The crossover point in the first tree is randomly chosen and then the crossover point in next tree is chosen on the basis of grammar, if any point in the tree is grammatically incompatible, a new point is chosen in the first tree, else two trees are crossed over.

In creating programs/rules, function sets and terminal sets are provided, grammar is defined, and then the initial population is randomly generated from the sets. After each generation, different fitness measures are calculated e.g. raw fitness, standardized fitness, adjusted fitness and normalized fitness. These different fitness measures are used internally to select of parents for crossover, mutate and reproduce. The objective that is considered to minimize is raw fitness. Normalized fitness is used when using Roulette Wheel selection or Tournament selection is used in selecting parents.

5 RESULTS

The experiment was run with following parameters. Population size was 500, number of generation was 51, initial and created depth of computer programs generated were 10. One, single repairer was considered, on first a single instance and then on a multiple instance of the problem. Function set used were, If-Statement and Condition-statement.

The terminal set used were, comparison operator, threshold values ranging from 0.0 to 1.0 with 0.1 steps. Three ratio type values were used, which were dependent upon distances between the nodes, processing time of jobs, and travelling time between jobs. The algorithms that were chosen by the programs to move to next job after computing values of terminals were greedy, 2-greedy, 3-greedy, and random select.

Figure 1 is the graph plot of raw fitness of best of generation individual (program) against each generation. As can be clearly observed, better programs/rules are evolved and programs/rules evolved in the later generation minimize the objective much better than the programs evolved in earlier generation.

6 CONCLUSIONS AND FUTURE WORK

As evident from Figure 1, the programs evolved are certainly optimized for the objective, they can be further enhanced by choosing better functions and terminals which reveal more characteristics about the problem. Also further enhancements can be made if a greater collection of algorithms and different types of algorithms were chosen to apply on the problem. Future work will consist of multiple agents (repair members) optimizing the same objective on an instance, and also optimizing multiple objectives on instances.

REFERENCES

Dorigo, M., & Stutzle, T. 2004. Ant colony optimization. *Bradford Company*, Scituate, MA, USA.

Fraser. Adam P. 1994. "Genetic Programming in C++", a manual in progress for gpc++ a public domain genetic programming system. University of Salford Cybernetics Research Institute, Technical Report 040.

Gritz, I. Larry. 1993. A C++ implementation of genetic programming. Department of EE&CS. George Washington University.

Jabeen, Hajira & Baig, Abdul Rauf. 2010. Review of Classification Using Genetic Programming". *International Journal of Engineering Sciences and Technology*. Vol. 2. 94–103.

Koza, J.R. 1992. Genetic Programming: On the Programming of computers by means of Natural Selection. MIT Press.

McKay, R. I., Hoai, N. X., Whigham, P. A. Shan, Y., & O'Neill, M. 2010. Grammar-based genetic programming: a survey, *Genetic Programming and Evolvable Machines* 11, 3–4, 365–396.

Nguyen, S., Zhang, M., Johnston, M. & Tan, K.C. 2012. A computational study of representations in genetic programming to evolve dispatching rules for the job shop scheduling problem. *IEEE Transactions on Evolutionary Computation*. DOI: 10.1109/TEVC.2012. 22273261965.

Oltean, M., Grasan Crina, Diosan L. & Mihaila Cristina. 2008. Genetic Programming with linear representation, a survey. *International Journal on Artificial Intelligence Tools*. World Scientific Publishing Company.

Poli, R. & Langdon, W. B. and Mcphee, N.F. 2008.A field guide to Genetic Programming. *Published via http://lulu.com, freely available at http://www.gp-field-guide. org.uk.*

Poli, R., Langdon, William B., McPhee, Nicholas F. & Koza, John R. 2007. Genetic Programming: An Introductory Tutorial and a Survey of Techniques and Applications. *Technical Report CES-475.ISSN 17448050.*

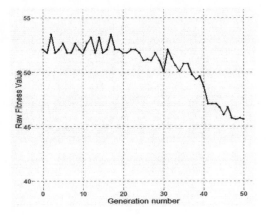

Figure 1. Graph plot of best of generation Raw Fitness Value vs. Number of Generations, showing programs optimizing on objective at later generations.

Communication and Computing Systems – Prasad et al. (Eds)
© 2017 Taylor & Francis Group, London, ISBN 978-1-138-02952-1

Modified GA approach for cloud based job scheduling

Sandeep Sharma & Sunpreet Singh Virk
Gautam Buddha University, Greater Noida, India

ABSTRACT: Cloud Computing is a very fast emerging and important field. It helps us in storing large chunks of data. An important work in this field is related to scheduling of tasks to various workers with the help of a job scheduling mechanism. This paper discusses some conventional methods in comparison to a new proposed algorithm. This was done by adding a fitness evaluation method which helped in faster convergence of solution. Since cloud computing promotes parallel computing, maximum make span of jobs was reduced to just time taken by the largest worker. We were able to establish that the proposed algorithm works better than the conventional methods.

1 INTRODUCTION

Popularity of cloud computing is growing day by day in distributed computing environment. The rising trend of using cloud environments for storage and data processing needs is a sign of its popularity. Providing applications, platforms and infrastructure over the internet is the basic requirement of cloud computing. Integrating wireless sensor network and cloud computing has been one of the most vital developments in recent times. Sensor network enables us to collect information even from remote areas. Cloud computing is widely used in distributed/mobile computing environment. This aids in resolving data access and security problems with ease. This is possible due to efficiency of communication technology. Scheduling of jobs is a tricky task in cloud computing. Various job scheduling algorithms have been deployed for the same. Fair scheduling of different type of jobs can be accomplished by choosing an appropriate algorithm. Clients seek out services from the organization. Sensor nodes are used to collect raw data from the environment. Cloud provides an online service to provide storage area to clients. Cloud computing platform dynamically provides services through the workers as in when needed by end users. These workers have the ability to parallel execute tasks. Cloud computing provides:

1. Efficiency, which is achieved through the highly scalable hardware and software resources, and
2. Agility, which is achieved through parallel batch processing. The main advantage of cloud computing is that exact location of sensor nodes shall not to be known by the end-user. They can access these sensor nodes by virtue of virtualization.

Cloud computing is a roaring area in the information technology domain. However its integration with sensor network has not been exploited fully.

There are still some areas that are needed to be focused on

1. Resource Management.
2. Task Scheduling.

Task scheduling and resource management are two tricky situations faced while integrating cloud computing. Cloud computing is emerging technology in IT domain. Scheduling is the process of assigning tasks to available resources on the basis of tasks' qualities and need. Increased utilization of the resources without affecting the services provided by cloud is the main goal of cloud computing. There are two types of scheduling i.e. resource scheduling and job scheduling. In this paper, we concentrate our attention to job scheduling. Job scheduling is one of the tough problems faced in cloud computing. Minimizing cost and attaining the objective are the main goals of job scheduling. Selection of the best sequence of job is the main purpose of scheduler in Cloud environment. Different algorithms have different characteristics A good scheduling algorithm must possess following characteristics

Minimum context switches
Maximum CPU utilization
Maximum throughput
Minimum turnaround time
Minimum waiting time

1.1 Scheduling criteria

Various job scheduling algorithms have different characteristics as mentioned above. The selection

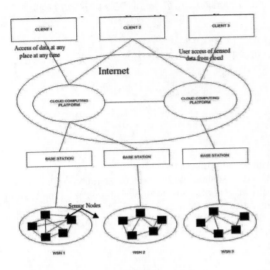

Figure 1. Sensor-cloud Integration.

of an algorithm may be critical to give priority to certain type of jobs. In other words, different algorithms work differently depending upon the data set given to them. For instance, small number of jobs may be scheduled using a simple algorithm such as FIFO, Round Robin; however, for large number of jobs they may not be sufficient and we need to use a different type of algorithm such as GA, GEP. Different characteristics used for evaluation can prove to be a considerable difference in which algorithm is judged to be the best. The criteria include the following –.

1. Context Switch: It deals with storage and restoration of state of a pre-empted process, which enables to resume the execution from the same point. Being computationally intensive, causes waste of time and memory, increasing the overhead of scheduler, scheming of operating system is done to optimize only these switches, intend being to diminish it.
2. Throughput: Throughput is defined as number of tasks finished in unit time. It is inversely related to context switching.
3. CPU Utilization: It describes how busy a processor is. Usually, the aim is always to maximize utilization of the CPU.
4. Turnaround Time: This refers to total duration of completion of task and time taken by the CPU to execute that task. The time period between the submission of task and completion of that task is known as turnaround time.
5. Waiting Time: It is the total waiting time of a task in ready queue. The CPU scheduling algorithm is not concerned with the execution time of the task or we can say input = output; it only affects the waiting time of a task.

6. Response Time: It is the time taken to produce a response to a submitted task. So this means that it is best if the response time is low.

2   LITERATURE SURVEY

Job scheduling is one of the tough problems faced in cloud computing. Minimizing cost and attaining the objective are the main goals of job scheduling. Selection of the best sequence of job is the main objective of scheduler in Cloud environment and it has static and dynamic conditions. The scheduler is responsible to select the best suitable resources in a Cloud Computing. The performance of the parallel system is affected by many significant factors out of which scheduling is one. The user submitted jobs are stored in Virtual Machine. Virtual Machine consists of many Processing Elements (PE's) and jobs are scheduled by the scheduler.

David S. L. Wei et al., 2013 proposed various issues in merging cloud computing with WSN. It included requirement to improve the throughput and reduce the processing time by the servers. The most simple CPU scheduling algorithm is First-Come First-Serve (FCFS). This algorithm is easily executed and managed with help of First In-First-Out (FIFO) row. With entry of the task in ready queue, task goes onto end of the ready queue and with the idle CPU, the successive task is removed from the leading position of the row. The task is served by the CPU based of process arrival. Simplicity and low overhead are its key characteristics. But long CPU bound tasks may govern the CPU and may compel shorter task to linger for long period, that in turn curtail the average CPU consumption and throughput.

Ranjan Kumar Mondal et al.,2015 proposed a model for job scheduling with Load Balancing Scheduling with Shortest Load First. This logic links every task length and successive CPU burst time. On availability of CPU, it is assigned the process that has the smallest next CPU burst time. If the next bursts of two tasks are identical, FCFS scheduling decides the priority of the task to be implemented first. It results in calculation of minimum average waiting time and minimum average turnaround time for the group of task. Difficulty in knowing the length of the next CPU burst it cannot be implemented. Due to the constant flow of short running jobs to the CPU, it may result in starvation of the long running jobs. Priority Scheduling (PS) is a more general case of SJF. In this assignation of priority to every task and the task which has the highest priority gets the first allocation of the CPU. The tasks having the same priority gets served on FCFS basis. The response to the highest priority processes has been really good. But, a major drawback that is known as indefinite blocking, or starvation, in which the low-priority task have to wait forever because the possibility that there can be other jobs round the corner having higher priority.

Round Robin Scheduling (RR) is another such algorithm (Shahram Behzad et al. 2013) (Abdulrazaq Abdulrahim et al.). It is used in time-sharing systems. It's same as FCFS scheduling, but preemption is the new feature that helps in switching between processes. A small time unit that is known as time quantum or time slice is defined. Circular queue is the form in which the ready queue is kept. The CPU scheduler gets through the queue that is ready, allocating the CPU to every task for time period of up to 1 time quantum. Implementation of the Round Robin scheduling involves, keeping ready queue as a First-In-First-Out (FIFO) queue of task. Addition of new tasks takes place at the tail of the ready queue. The first task from the ready queue gets picked by CPU scheduler, sets timer which interrupts the CPU after 1 time quantum, and dispatches the process. There are two possibilities now. The task can be having CPU burst which is lower than 1 time quantum. Under such situation CPU is released voluntarily by the task. The scheduler will then takes the successive

Table 1. Comparison of various algorithms.

Algorithm	Complexity	Allocation	Waiting Time
First-Come First Serve Algorithm	Simplest Scheduling Algorithm	CPU is allocated in the order in which the processes arrive	More
Shortest Job First Algorithm	Difficult to understand and code	CPU is allocated to the process with least CPU burst time	Lesser than FCFS
Priority Algorithm	Difficult to understand	Based on priority, So the higher priority job can run first	Lesser
Round Robin Algorithm	Performance heavily Depends upon the size of time quantum	The preemption take place after a fixed interval of time	More than all
Genetic Algorithm	Complexity depends on the task to be scheduled	This is a greedy algorithm and pick the best job to allocate the CPU	Waiting time is less

task in the ready queue. Otherwise, if the CPU burst of any task under process if burst is greater than 1 time quantum, the timer goes off, and causing interrupt to the operating system. Execution of a context switch takes place, and the process is put at the tail of the ready queue. Being effective in general-purpose, this algorithm is also efficient in times-sharing system or transaction-processing system. Being fairly efficient in handling all the tasks, CPU overhead is low. Choice of quantum value is important as efficiency becomes low when time quantum is too high or too low.

AV. Karthick et al., 2013 proposed a tri-queue scheduling algorithm which parallel queues are classified as small, medium and long based on sorted order of the processing time and needed processor. It improves the performance of system. It involves use of dynamic quantum time. Fragmentation is avoided as equal opportunity is available to all queues. Some researchers dedicate to heuristic scheduling algorithms, including GA, GEP and so on. The advantage of heuristic algorithms over exact methods such as Critical Path Method is that exact methods assume infinite resources to attain minimum make span. Introducing too many constraints makes it tough for exact methods to provide an optimal solution. On the other hand, heuristic methods concentrate on finding a "good solution". All the above mentioned work cannot efficiently acclimatize to scheduling large chunks of data. We propose an algorithm for incorporating fitness function in GA. This helps in increasing task execution and resource utilization efficiency.

3 TECHNIQUES USED

In job scheduling, we have a set of n tasks (T1, T2, T3, ..., Tn) that need to be scheduled onto m available resources (R1, R2, R3, ..., Rm). Following methods were adopted to allocate tasks

3.1 Priority min-min

Min-Min is a basic and conventional job scheduling algorithm. It has the ability to give good performance when it operates on limited number pf jobs. Explicitly, it starts with a set T of all unmapped tasks. A matrix containing Early Completion Time (ECT) of all tasks corresponding to each resource is created. Then the minimum of the matrix is found. Next, the task Tk is allocated to the corresponding resource Rj. Lastly, the task Tk is removed from set T and the same process is reiterated until all tasks are assigned (i.e., set T is empty). Priority min-min works in two phases where in tasks are divided into types-high priority WSN related tasks and normal tasks. In 1st phase, min-min algorithm is applied to high priority tasks. This is done till there are WSN related tasks left. Following this is the 2nd phase which involves implementation of min-min algorithm on non-priority tasks. The following flowchart depicts the same.

3.2 *Genetic algorithm*

Genetic algorithm is heuristic search method which means it tries to find an optimal solution rather

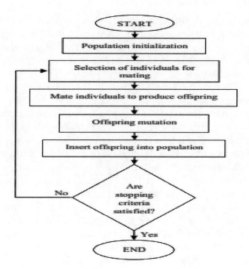

Figure 3. Flowchart of GA.

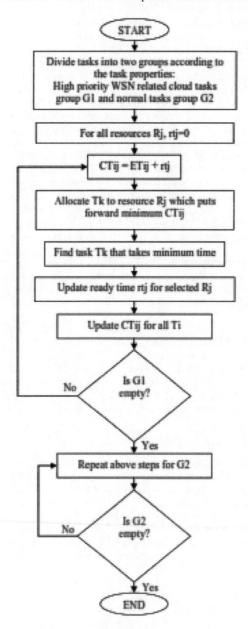

Figure 2. Flowchart for priority min-min.

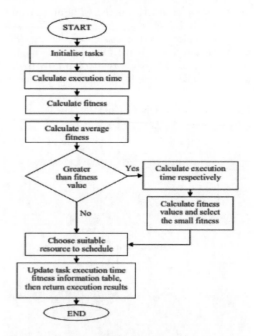

Figure 4. Flowchart for fitness evaluation.

than an exact solution. Its mechanism is based upon evolutionary methods observed in natural world. Genetic algorithms are usually good at both the exploration and exploitation of the search space because they operate on more than one key at once.

The algorithm begins by population initialization of individuals. Individual keys, or genomes, are selected from the population, and new solutions are created by mating them. The mating process is normally implemented by fusing. This is followed by genetic material cross over from two parents to form the genetic material for new solutions. Random mutation is applied at times to promote variety. The data is given to the next generation; if it is found to be better, it replaces individuals in the population. In general, the genetic algorithm only needs to make out how good a schedule is and how to fuse two schedules to form another schedule.

3.3 Proposed algorithm

Fitness function is very important as it helps to determine the convergence speed of the algorithm. We propose a different method of fitness evaluation in this paper.

Fitness function is calculated using the following formula:

$$f_i = 100 \times \frac{1}{1+U_i} \qquad (1)$$

$$U_i = 1/N \sum_{j=1}^{N} (C_{i,j} - T_{avg})^2 \qquad (2)$$

where f is fitness,
U is an intermediate variable,
$C_{i,j}$ is the execution time of task i in worker j,
T_{avg} is the average time of the same task in all workers,
N is the worker number
The maximum fitness occurs when $C_{i,j} = T_{avg}$. The algorithm for fitness function evaluation is given below:

4 RESULTS AND DISCUSSION

For evaluation of performance and efficiency of the algorithms, they were made to operate on a set of jobs of random jobs (which contains task objects). These are then scheduled to various workers which can work on independent tasks as in when they get free. For different scheduling analysis, worker number was set to 8, 16, 32 and 64 respectively.

Execution time, Throughput and Resource utilization are used to evaluate the system for 8, 16, 32 and 64 workers respectively. For evaluation of performance of different algorithms, it is necessary to implement the algorithms in the same environment.

The following parameters have been chosen to implement the algorithm.

4.1 Execution time comparison

When the number of tasks is small, the algorithms do no show great variance in execution time. However, as number of tasks is increased, the proposed algorithm shows better performance.

Table 2. Parameters used for implementing GA.

Parameter	Value
Population Size	100
Max. Generations	100
Crossover probability	0.7
Mutation probability	0.05
Children from elite parents	2

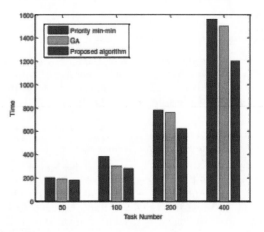

Figure 5. Execution time comparison.

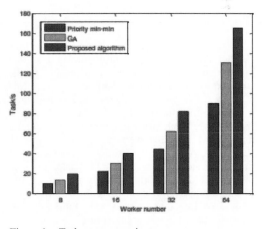

Figure 6. Task rate comparison.

4.2 Throughput comparison

Again like the previous comparison, not much difference in throughput exists when 8 workers are set

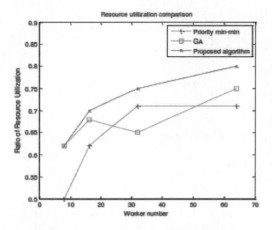

Figure 7.   Resource utilization comparison.

to work on the jobs. However, the proposed algorithm again shows promise when number of workers is increased. There is a significant difference in throughput when the worker number is increased to 64.

4.3 *Resource utilization comparison*

Resource utilization is compared by initializing 400 tasks for 8, 16, 32 and 64 workers respectively. Resource utilization of the algorithms is roughly located between 0.5 and 0.65 for 8 workers. This increases with increase in number of workers. The most significance increase in resource utilization was observed in proposed algorithm.

5 CONCLUSION

The priority min-min algorithm takes in account the task number, task size, task priority, resource number and their respective processing speeds. The results are adequately satisfactory. But this algorithm may prove to be less effective for some other type of data sets. For instance, if some of the tasks are much larger in comparison to others, it may slow down the average completion time of the system. Also, the algorithms consider only the processing speed of resources. GA helps in finding global optimal solution. Hence, it improves the process of task allocation to resources when job set is large. Introduction of a new fitness function improves GA and aids in faster and better convergence of solution.

Since these algorithms have been able to efficiently reduce time and there is no problem with storage area of cloud, we can focus more on load balancing to increase the efficiency of task allocation to resources. Also, we need to focus on introducing different data types and allocate them to corresponding type of resources. This will make it more useful for practical use where we can encounter large and different types of data.

The efficiency of resource allocation can be further enhanced by improving load balance and fully exploit parallel computing in cloud by dispatching tasks to an idle resource.

REFERENCES

Abbas Noon, Ali Kalakech and Seifedine Kadry in March 2014, Analysis of Job Scheduling Algorithms in Cloud Computing, International Journal of Computer Trends and Technology (IJCTT)—volume 9, number 7.

Abdulrazaq Abdulrahim, Ahmadu Bello, Nigeria and Saleh E Abdullahi, Department of Mathematics, Ahmadu Bello University, Zaria, Nigeria Salisu Aliyu Department of Mathematics, Ahmadu Bello University, Zaria, Nigeria Ahmad M Mustapha Department of Computer Engineering, University of Maiduguri, Nigeria Saleh E Abdullahi.

David S. L. Wei, Sarit Mukherjee, Kshirasagar Naik, Amiya Nayak, Yu-Chee Tseng, and Li-Chun Wang in December 2013, IEEE journal on selected areas in communication, VOL. 31, NO. 12.

Karthick, AV., Dr. E. Ramaraj and R. Kannan in 2013, An Efficient Tri Queue Job Scheduling using Dynamic Quantum Time for Cloud Environment, 978-1-4673-6126-2 IEEE.

Kaur, Shaminder, and Amandeep Verma in 2012, "An Efficient Approach to Genetic Algorithm for Task Scheduling in Cloud Computing Environment", International Journal of Information Technology and Computer Science.

Ranjan Kumar Mondal, Enakshmi Nandi and Debabrata Sarddar in 2015, Load Balancing Scheduling with Shortest Load First, International Journal of Grid Distribution Computing Vol. 8, No. 4, pp. 171–178.

Shahram Behzad, Reza Fotohi & Mehdi Effatparvar in 2013, Queue based Job Scheduling algorithm for Cloud computing, International Research Journal of Applied and Basic Sciences.

Sourabh Songar, Mrs. R. Annie Uthra in April 2015, Data Gathering through Wireless Sensor Network in Cloud, International Journal of Innovative Science, Engineering & Technology, Vol. 2, Issue 4.

Zhao, Yong, Liang Chen, Youfu Li, and Wenhong Tian in 2014, "Efficient task scheduling for Many Task Computing with resource attribute selection", China Communications.

Zhu, Chunsheng, Xiuhua Li, Victor C.M.Leung, Xiping Hu, and Laurence T. Yang in 2014, "Job Scheduling for Cloud Computing Integrated with Wireless Sensor Network", IEEE 6th International Conference on Cloud Computing Technology and Science.

Communication and Computing Systems – Prasad et al. (Eds)
© 2017 Taylor & Francis Group, London, ISBN 978-1-138-02952-1

A stochastic model for queueing CAMs in vehicular adhoc networks: Simulation and performance analysis

P. Verma & N. Singh
School of Information and Communication Technology (ICT), Gautam Buddha University, Greater Noida, Uttar Pradesh, India

ABSTRACT: One of the factors governing the tendency of broadcasting Cooperative Awareness Messages (CAMs) messages is the mobility of vehicles within the vehicular network. As Vehicular Adhoc Networks (VANETs) possesses the characteristic of high mobility, the distribution of CAMs should be managed precisely to mitigate channel load as a result of ample CAM messages. To cope up with this issue, this paper presents a stochastic model for queuing CAMs within each vehicle contributing to the network. The main feature of the model is to enable the CAMs to depart deterministically from the queue, allowing precise management of messages utilizing the Control Channel (CCH). The model is incorporated with test-drive sample data of vehicle to provide a proximity to a realistic scenario. The performance of the network is evaluated in the terms of channel utilization, average waiting time and queue length using MATLAB-Simulink.

1 INTRODUCTION

VANETs are proving to be one of the promising approach, ensuring safety alongwith providing other infotainment applications to the drivers as well as passengers. For this, VANET supports a wide range of applications from single-hop information dissemination of CAMs (containing information about vehicle id, speed, position, etc.) to multi-hop dissemination of messages supporting infotainment applications. The US Federal Communications Commission (FCC) has allocated a frequency band of 75 MHz comprising seven channels for message dissemination, each of 10 MHz, for VANETs (ASTM Standard, 2003). Among these seven channels, only one channel, i.e., CCH is dedicated to the dissemination of CAM and Event Driven Messages (EDMs). As both type of messages shares same CCH, in opaque traffic, periodic beacons may consume the entire channel bandwidth leading to a drenched/congested channel (Poonam et al. 2015). Therefore, the distribution of CAMs should be managed precisely to mitigate the channel load.

As CAMs are the building block of VANETs, we are confining our research to CAM broadcast only. In order to manage the distribution of CAMs, allowing their deterministic departure from the vehicle's queue can be an effective approach towards appropriate channel utilization. Taking into consideration only CAM broadcast, vehicular network can be considered as a network in which each vehicle is serving a single server i.e., single CCH (Poonam et al. 2015). In this paper, the proposed stochastic model deals with the deterministic departure of CAM from the queue that are apparently served by the CCH for evaluating the performance of the model.

The rest of the paper is outlined as follows: Section 2 gives an overview of some of the existing queueing models. Section 3 provides an insight to the proposed stochastic queueing model for vehicular network. Section 4 presents the performance evaluating metrics for the model. Section 5 provides the simulation details of the model. Section 6 presents the simulation results and provide the performance evaluation of the model. Section 7 concludes the paper.

2 RELATED WORK

In recent years, there has been a great emphasis on managing the CAM broadcast for optimizing the performance of vehicular network. As CAM are broadcasted through the network, there is no provision of receiving acknowledgement from the receiver. Also, for enhancing the performance of vehicular networks, merely dropping or delaying CAM's transmission during congestion is not an acceptable solution, as CAM messages are the base for a vehicular network to establish.

Some of the recent works are discussed in this section. Yang et al. 2011 proposes a Markov model for analysing the performance of periodic broadcast in VANETs. Daneshgaran et al. 2007 provides a multi-dimensional Markovian state transition model characterizing the behaviour at the MAC layer. These authors consider the situation when there is no packet waiting for transmission in the buffer of a node practically not possible due to the highly mobile nature of vehicular network. Yadumurthy et al. 2005, proposes a Batch Mode Multicast MAC (BMMM) protocol, which uses control frames for broadcast transmissions. Si et al. 2004, presents a new MAC protocol called RMAC that supports reliable multicast for wireless ad hoc networks. These researches uses acknowledgement information from receivers to reduce unnecessary forwarding transmission, which is not feasible in vehicular networks.

Yin et al. 2013, proposes a generalized M/G/1 queueing model for the performance evaluation of vehicular safety related services. Xaiomin et al. 2007, uses a discrete time M/G/l queue to model occasional occurrences of safety related message in each vehicle. Chen et al. 2007, presents a quantitative approach to evaluate the quality of service that the inter-vehicle communication can provide to the inter-vehicle safety application in the context of highway scenarios. All these researches evaluate the network performance in terms of packet delivery ratio, which is not one of the suitable performance parameters for VANETs (Poonam et al. 2016).

3 STOCHASTIC CAM QUEUEING MODEL

This section provides a detailed description of the stochastic queueing model developed for the vehicular network. The paper concentrates only on the MAC layer of the VANETs architecture (Yunxin et al. 2010). For modelling purpose the following assumptions are made:

- As the model is confined to CAM broadcast only, the CCH is considered to be fully utilized by CAM messages.
- λ_{CAM} denotes the poison arrival of CAM messages and μ_{CAM} denotes the deterministic departure of CAM messages.
- The model considers the mean arrival rate of CAM messages as the ratio of CAM queue length to the average waiting time for CAM.

Figure 1 depicts the layered behavior of the model. The model consist of four blocks, namely, CAM generation/broadcast, arrival at the Queue, CAM departure, service to the CCH. These blocks are discussed as follows:

- CAM Generation/Broadcast: Each vehicle requires to continuous generate/broadcast received CAMs from other vehicles to participate in the network.

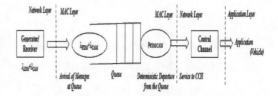

Figure 1. Layered Behavior of Model.

- Arrival at the Queue: Entering the MAC layer, the CAM reaches the queue within the vehicle, where there are prioritized on the basis of LIFO i.e., the most recent information is broadcasted first to keep the network up-to-date.
- CAM Departure: The model allows deterministic departure of CAMs from the queue so as to follow a systematic pattern of departure and also reduce the load of the CCH to optimize the performance of the network.
- Service to the CCH: Entering the Network Layer, the departed CAMs are served to the control channel for evaluating the performance of the network in terms of various performance evaluating metrics.

4 PERFORMANCE EVALUATING METRICS FOR THE MODEL

Using the assumptions of previous section, we have established various performance metrics as follows:

4.1 Channel utilization (ρ_{CAM})

Channel utilization is the ratio of CAM arrival rate to their service rate at any instant of time within the network.

$$\rho_{CAM} = \frac{\lambda_{CAM}}{\mu_{CAM}}$$

4.2 Average Waiting Time (AWT_{CAM})

Average Waiting Time of CAM is the average waiting time elapsed between arrivals of CAM messages in the queue to their deterministic departure from the queue within the vehicle.

$$AWT_{CAM} = \rho_{CAM}/2\mu_{CAM}(1-\rho_{CAM})$$

4.3 Queue Length (L_{Queue})

Queue Length of the system is defined as the total number of CAM messages within the queue at any instant of time.

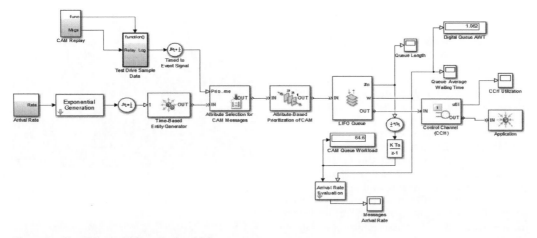

Figure 2. Simulink model for a single vehicle.

$$L_{Queue} = {\rho_{CAM}}^2 \Big/ 2(1 - \rho_{CAM}) + \rho_{CAM}$$

4.4 Arrival rate (λ_{CAM})

According to Little's Law,

$$Mean\ Arrival\ Rate = {L_{Queue}} \Big/ AWT_{CAM}$$

5 MODEL'S SIMULATION

This section discussed details of the simulation of the model. We have conducted the simulation using MATLAB's Simulink (MATLAB, R2015a). Figure 3 shows the general block diagram for simulating the model.

The block diagram consists of five phases of the simulation; message arrival to service of message to the application through the CCH. The setting of the parameters is summarized in the Table 1.

The queue within each vehicle is considered to have a buffer of infinite length so as to accommodate the maximum arrival of CAM messages. The arrival pattern of queue is considered to be in Poisson distribution and the service rate is deterministically distributed. Upper and lower bound on arrival rate are set as <0.999 and >0 respectively. The bandwidth of the control channel is taken of 10 MHz (Amendment 7: WAVE, 2009). A test-driven sample data of vehicle is taken to provide a proximity to a realistic scenario.

The Simulink Model for a single vehicle is illustrated in Figure 2. The test-drive subsystem provides the vehicles raw data for simulating the model which is further applied to attribution selection block before being fed to the LIFO queue.

Figure 3. Block diagram of simulation.

Table 1. Parameters setting.

Sr. no.	Parameter	Value
1.	Channel bandwidth	10 MHz
2.	No. of channel	1
3.	Queue length	∞ (Infinite)
4.	Data set	Test-drive sample data
5.	Arrival pattern	$\lambda^k e^{-\lambda}/k!$
6.	Lower bound on arrival rate	>0
7.	Queue discipline	LIFO
8.	Upper bound on arrival rate	<0.9999
9.	Service rate	$(1 - \lambda)(e^{\lambda} - 1)$
10.	Simulation time	100 units

The Arrival rate evaluation subsystem computes the ratio of CAM queue length to average waiting time, alongwith the ratio of mean service time to mean arrival time. Various scopes are employed to provide graphical visualization of the simulation results.

6 PERFORMANCE EVALUATION OF THE MODEL

This section provides the simulation results of the developed queueing model. For evaluating the performance of the model, performance metrics, namely, channel utilization, queue's average waiting time, message arrival rate and queue content are measured. Simulation result showing the average number of CAMs in the queue at any time instant is illustrated in Figure 4. The distribution of CAM is in accordance to the test-drive driven sample data, and hence, is random in nature, to provide a realistic environment for simulation.

Figure 5 shows the average waiting time of CAMs in the queue. AWT is proportional to the queue length, i.e., the waiting time increases with increase in queue length and vice versa. Figure 5 clearly depicts maximum waiting time corresponding to the highest peak of the CAM queue length. Figure 6 depicts the channel utilization of the network. As the channel utilization is given by λ_{CAM}/μ_{CAM}, the more is the arrival rate, the more is the channel utilization as shown in the Figure 6.

Figure 7 provides a comparative view between the theoretical and simulated arrival rate. Theoretically, arrival is considered to be constant throughout the simulation. On the contrary, the simulation's arrival rate depends on the test driven CAMs, and therefore, varies according to the queue content as shown in the figure.

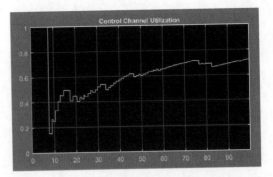

Figure 6. ρ_{CAM} at any instant of time.

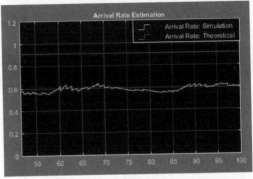

Figure 7. Arrival rate evaluation: simulation vs theoretical.

Figure 4. L_{CAM} at any instant of time.

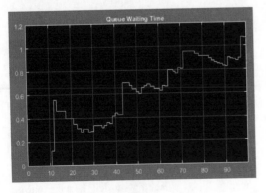

Figure 5. AWT_{CAM} at any instant of time.

7 CONCLUSION

As CAMs are the building blocks of vehicular network, their broadcast is one of the important factor to be considered in proposing any model. For this, a deterministic departure queueing model is introduced in this paper to precisely manage the load of the CCH. The simulation results elaborates the performance of the network in terms of channel utilization, average waiting time and queue length. The working principle of the model has been evaluated through simulation and the obtained result will be used to provide a platform for developing a model for a large number of vehicles.

REFERENCES

Daneshgaran, F., Laddomada, M., Mesiti, F., & Mondin, M. 2007. A model of the IEEE 802.11 DCF in presence of non-ideal transmission channel and capture effects. In *IEEE GLOBECOM 2007-IEEE Global Telecommunications Conference*: 5112–5116. IEEE.

Chen, X., Refai, H. H., & Ma, X. 2007. A quantitative approach to evaluate DSRC highway inter-vehicle safety communication. In *IEEE GLOBECOM 2007- IEEE Global Telecommunications Conference*: 151–155. IEEE.

IEEE P802.11p/D8.0, Draft Standard for Information Technology—Telecommunications and information exchange between systems—Local and metropolitan area networks—Specific requirements; Part 11: Wireless LAN Medium Access Control (MAC) and Physical Layer (PHY) specifications; *Amendment 7: Wireless Access in Vehicular Environments, 2009*.

Li, Y. J. 2010. An overview of the DSRC/WAVE technology. In *International Conference on Heterogeneous Networking for Quality, Reliability, Security and Robustness*: 544–558. Springer: Berlin Heidelberg.

MATLAB & Simulink. 2015. matlab R2015a, "http://in.mathworks.com/".

Ma, X., Chen, X., & Refai, H. H. 2007. Unsaturated performance of IEEE 802.11 broadcast service in vehicle-to-vehicle networks. In *2007 IEEE 66th vehicular technology conference*: 1957–1961. IEEE.

Si, W., & Li, C. 2004. RMAC: a reliable multicast MAC protocol for wireless ad hoc networks. In *Parallel Processing, 2004. ICPP 2004*: 494–501. IEEE.

Standard Specification for Telecommunications and Information Exchange Between Roadside and Vehicle Systems—5 GHz Band Dedicated Short Range Communications (DSRC) Medium Access Control (MAC) and Physical Layer (PHY) Specifications, *ASTM Standard E2213-03, September 2003*.

Verma, P., & Singh, N. 2015. An Analytical Review of the Algorithms Controlling Congestion in Vehicular Networks, *In IOSR Journal of Computer Engineering (IOSR-JCE), volume 17, issue 3, March 2015*: 32–41.

Verma, P., & Singh, N. 2015. Statistical behavior analyses of CAM queue model in Vehicular Adhoc Networks. In *2015 2nd International Conference on Recent Advances in Engineering & Computational Sciences (RAECS)*: 1–5. IEEE.

Verma, P., & Singh, N. 2016. Performance Evaluating Metrics for CAM Collision over Control Channel in VANETs, In *International Conference on Recent Trends in Engineering and Material Sciences (ICEMS), March, 2016 (in press)*.

Yang, Q., Zheng, J., & Shen, L. 2011. Modeling and performance analysis of periodic broadcast in vehicular ad hoc networks. In *Global Telecommunications Conference (GLOBECOM 2011), 2011 IEEE*: 1–5. IEEE.

Yadumurthy, R. M., Sadashivaiah, M., & Makanaboyina, R. 2005. Reliable MAC broadcast protocol in directional and omni-directional transmissions for vehicular ad hoc networks. In *Proceedings of the 2nd ACM international workshop on Vehicular ad hoc networks*: 10–19). ACM.

Yin, X., Ma, X., & Trivedi, K. S. 2013. An interacting stochastic models approach for the performance evaluation of DSRC vehicular safety communication. *IEEE Transactions on Computers*, 62(5): 873–885.

Communication and Computing Systems – Prasad et al. (Eds)
© 2017 Taylor & Francis Group, London, ISBN 978-1-138-02952-1

A novel band pass filter using parallel and direct-end coupling

L.S. Purohit

Department of Electronics Engineering, Rajasthan Technical University, Kota, Rajasthan, India

ABSTRACT: A Band Pass Filter (BPF) is proposed using parallel and direct-end coupling. A circular dot and circular half ring structure is utilized to do parallel coupling because this providing longer coupling length. Symmetric and asymmetric structure is analyzed with different spacing parameters. Tuning of resonant frequency can be done by some parameters like, circumference of outer fractional ring, center position of outer ring. Symmetric and asymmetric structures are compared. By using Asymmetric filter structure insertion loss and return loss performance in pass band is improved. Radio location, space research etc. are some applications of concern band. Filter has some commendable features like miniature size, low loss in pass band and long and deep stop band. Alumina material having epsilon 9.9 is used as substrate for microstrip filter design. Simulation of design is performed on commercial software CST microwave studio 2013.

1 INTRODUCTION

The Filter is most commonly used component in microwave systems for selecting desired frequencies. Band pass filters are made with different resonator structures. Length and width of structure utilized for variation in L and C parameter so resonant frequency can be tune to desired range (Haoran and Junfa 2012). Defected Ground Structure (DGS) is also utilized tuning resonate frequency to desired range, DGS can also provide some advantageous effects for good in-band characteristics (Kufa and Raida 2013).

In this paper, a novel planer structure for bandpass filter design is proposed. Parallel and direct-end coupling is utilized. Parallel coupling is significant out of two. Arrangement made for parallel coupling is of a parallel ring and a circular dot structure in the patch. Design is made using microstrip structure of size 75 mm² (length and width respectively are 15 mm and 5 mm). Alumina dielectric material of epsilon 9.9 (of height 0.5 mm) is used as substrate material of filter design. Designing and simulation is performed on commercial software CST microwave studio. Result covers frequency range of radio location, space research applications and result can also tune to other application frequencies by tuning its designing parameters.

2 DESIGN OF PROPOSED FILTER

Filter is designed using microstrip structure. Ground and patch layer made of copper conductor and middle substrate layer made of Alumina material. Schematic view of filter design (patch layer) with different dimensions is shown in Figure 1. Table 1 covers values of different parameters of Figure 1. Circle of radius R_1 and outer ring of

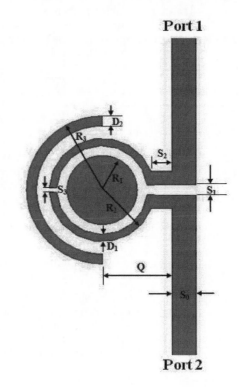

Figure 1. Patch of proposed symmetric filter structure.

Table 1. Values of parameters used in Figure 1.

Parameters	S_0	S_1	S_2	S_3	S_4	R_1	R_2
Calculated values (mm)	0.48	0.21	0.42	0.05	0.27	0.7	1.05
Parameters	R_3	D_1	D_2	Q	L	W	H
Calculated values (mm)	1.5	0.15	0.2	1.4	15	5	0.5

Table 2. Resonant frequencies of symmetric proposed filter (with $R_3 = 1.4$ mm constant) with variation in fraction of circumference.

Different lengths of fractional circumference (in mm)	Center frequency (GHz) (Return loss/Insertion loss in dB)
$C_1 = 5.65$	10.818 (25.45/0.44)
$C_2 = 6.59$	9.684 (22.85/0.62)
$C_3 = 7.54$	8.604 (21.40/0.771)
$C_4 = 8.17$	8.168 (22.78/0.656)

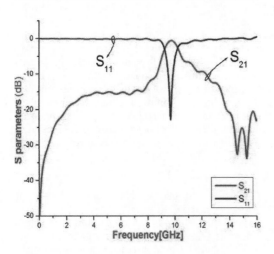

Figure 2. S parameters of symmetric proposed filter.

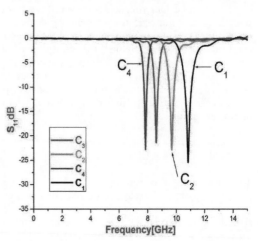

Figure 3. S_{11} Vs Frequency (Using different fractional circumference C_n) for proposed symmetric filter structure.

radius R_3 is placed for providing parallel coupling. Direct-end coupling can be seen at end of input feed line (at S_3). Coupling varies as coupling length and distance between coupled part is varies. Input and output feed line made of width S_0 for impedance match to 50 Ω.

For a particular values of design parameter (values as in Table 1) results are shown in Figure 2. Result shows a pass-band of frequency 9.506–9.859 GHz centered at frequency 9.684 GHz. Insertion loss and return loss at center frequency are respectively 0.621 dB and 22.855 dB. Application of this range is radio location.

3 ANALYSIS

3.1 Effect of length of Outer Partial Ring (OPR) of radius R_3

As we vary the confinement by outer ring, the resonating frequency varies and filter resonates to different frequency. Different fractional circumference ($\pi\theta$) results in different center frequency tabulated in Table 2 and shown in Figure 3 diagrammatically. By this figure can be concluded that as confinement increases, resonating frequency lowers.

For fixed $R_3 = 1.4$ mm, $C = 2\pi R_3 = 8.79$ mm, Fractional circumference shown by $C_n = \pi\theta$.

3.2 Effect of shifting center of outer ring (asymmetric structure)

If center of outer fractional ring is shifted right, center frequency shifted right with good insertion and return loss performance. Mean asymmetric like shown in figure can be utilized for good pass band performance. By Table 3 and Figure 4, effect can be analyzed. Shifting centre of inner circle not effect very much, that's why no results shown here.

3.3 Effect of spacing between outer ring, feed ring structure and inner circle

Variation in Spacing between Outer Fractional Ring (OFR) and Feed Line Ring Structure (FLRS) (for symmetric structure all rings having

Table 3. Center of Outer Fractional Ring (OFR) shifted by 0.2 mm and analyzed for different fractional circumference.

Different lengths of fractional circumference	Center frequency (GHz) (Return loss / Insertion loss in dB)
C_2 = 6.59 mm	11.977 (54.70/ 0.052)
C_3 = 7.54 mm	10.556 (47.79/0.001)
C_5 = 8.48 mm	9.396 (36.84 / 0.017)
C_6 = 8.796 mm = C	8.323 (25.96 / 0.46)

Table 4. Various values P and resonant frequency.

P in mm	Resonant frequency GHz (Return loss/ Insertion loss in dB)
P_1 = 0.25	9.684 (22.85/0.62)
P_2 = 0.15	9.684 (33.53/0.12)
P_3 = 0.10	9.684 (41.31/0.02)

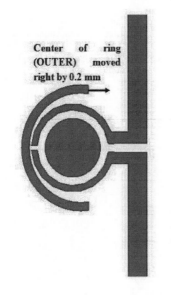

(a)

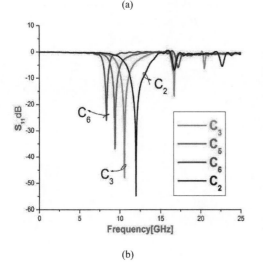

(b)

Figure 4. Asymmetric filter structure and its S_{11} Vs frequency graph (for different fractional circumferences).

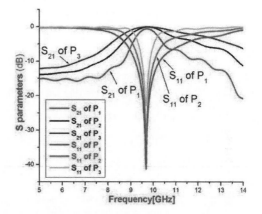

Figure 5. S parameters Vs frequency for different value of Spacing P (for symmetric filter structure, all other parameters kept constant).

same center, R_3 = 1.5, R_2 = 1.05 kept constant and only D_2 varied) is analyzed first.

Spacing between OFR and FLRS is can be calculated by

$$P = [(R_3 - D_2) - R_2] \qquad (1)$$

where R_3 = outer radius of outer fractional ring, D_2 = Width of outer fractional ring and R_2 = outer radius of feed line ring as shown in Figure 1.

By decreasing spacing P, insertion loss is decrease and return loss is increased, Bandwidth is also improved but sharpness of edges of stopband is reduced. By seeing caparison Table 4 and Figure 5 can be conclude that resonant frequency not vary with changing spacing P.

4 CONCLUSIONS

There are two controlling parameters for tuning resonant frequency one is Length of fractional circumference of outer ring, and another is variation in center position of OFR discussed. Change fractional circumference is give variation in resonant frequency with almost same return loss and

439

insertion loss if structure is symmetric. If structure is asymmetric then variation in fractional circumference results in improved insertion and return loss characteristics. Spacing between outer ring and input feed line structure can be utilized for bandwidth, insertion loss, return loss improvement but it may generate some small spurious-passband in stopband. Filter structure is designed and simulated on CST microwave studio.

REFERENCES

Bahl, I. J. 1989. Capacitively compensated performance parallel coupled microstrip filter. *IEEE MTT-S Int. Microwave Symp*. Dig., 679–682.

Cheng, C. M. & Yang, C. F. 2010. Develop Quad-Band (1.57/ 2.45 / 3.5 /5.2 GHz) Bandpass Filters on the Ceramic Substrate with defected ground structure. *IEEE Microwave And Wireless Component Letters* 20(5):268–270.

Haoran, Z. & Junfa, M. 2012. Miniaturized Tapered EBG Structure With Wide Stopband and Flat Passband, *IEEE Antennas And Wireless Propagation Letters*, 11: 314–317.

Hong, J. S. & Lancaster, M. J. 2001. Microstrip Filter for RF/Microwave Applications, *A Wiley Interscience Publication*, Canada.

Kufa, M. & Raida, Z. 2013. Lowpass filter with reduced fractal defected ground structure. *Electronics Letters* 49(3).

Kuo, J. T. Hsu, W. & Huang, W. 2002. Parallel-coupled microstrip filters with suppression of harmonic response. *IEEE Microwave and Wireless Components Letters* 13:383–385.

Purohit, L. S. & Tharani, L. 2015 A miniature quad-band bandpass filter using modified impedance resonator and parallel resonating stubs for satellite communication. *IEEE International conference on Next Generation Computing Technologies*, 527–530.

Communication and Computing Systems – Prasad et al. (Eds)
© *2017 Taylor & Francis Group, London, ISBN 978-1-138-02952-1*

Temperature and power efficient scheduling for cloud data centers: Green approach

Premkumar Ch, Rohit Joshi, Preeti Pathak & Naveen Garg
CIT Department, University of Petroleum and Energy Studies, Dehradun, India

ABSTRACT: Retrieval from the internet through web-based tools and applications, instead of connecting to a server gives rise to Cloud Computing. It indeed, is a model where data is stored in servers. Cloud computing structure allows access to information as long as an electronic device has access to the web. It provides us with hosting of applications from consumer, scientific and business domains. However, data centers hosting cloud computing applications consumes huge amount of energy, leading to high operational costs and carbon footprints to the environment. As because of energy shortages and global climate change leading our concerns these days, the power consumption of data centers has become a key issue. In this paper, we aim to solve the problem by using green cloud computing solutions which saves energy as well as reduces operational costs. The vision for energy efficient management of cloud computing environments is presented here.

1 INTRODUCTION

Cloud computing has emerged as one of the biggest revolution for the Information and Communication Technology (ICT) industry (Owusu, Francis & Colin Pattinson 2012). According to National Institute of Standards and Technology (NIST) (www.nist.gov/itl/csd/cloud–102511.cfm), cloud computing is a model for enabling convenient, on-demand network access to a shared pool of configurable computing resources networks, servers, storage, applications, and services) that can be rapidly provisioned and released with minimal management effort or service provider interaction. An organization can build a build a private Cloud data centre to improve the resource management and provisioning processes or can either outsource its computational needs to the Cloud (Andrew J. Younge 2010). Cloud Computing may be defined as IT resources, services, data and information abstracted from underlying infrastructure and provides on demand and at scale in a multitenant environment (L. Minas & B. Elliso et al. 2009). With the proliferation of Cloud computing, many large scale nodes, called data centers have established around the world (Mehta, Avinash, Mukesh Menaria, Sanket Dangi, & Shrisha Rao 2011). These data centers contain thousands of computer nodes for provision of data resources. However, Cloud data centers causes high operating costs and carbon dioxide (CO_2) emissions to the environment due to their consumption of huge amounts of electrical energy (Gartner 2007). A six-month study led by Lawrence Berkeley National Laboratory (Berkeley Lab) with funding from Google has found that moving common software applications used by 86 million U. S. workers to the cloud could save enough electricity annually to power Los Angeles for a year (Berkeley News 2013). Energy consumption in data centers will continue to grow rapidly unless advanced energy efficient resource management solutions are developed and applied (Anton Beloglazov 2013). As reported by the Open Compute project, Facebook's Oregon data centre achieved a Power Usage Effectiveness (PUE) of 1.09 (Open Compute Project 2015), which represents that about 91% of the energy consumption is consumed by the computing resources in the data centre. There are extra expenses on over-provisioning because most of the data centre servers operate at 10–50% of their full capacity (Anton Beloglazov 2010). Moreover, due to narrow dynamic power ranges of servers, the problem of low server utilization is also aggravated. This is because of the result that even completely Subheading completely idle servers consume up to 70% of their peak power (Anton Beloglazov 2010). Therefore, keeping servers underutilized is highly inefficient from the energy consumption perspective. The reduction in CO_2 emission and power consumption can be achieved with the aid of green computing.

2 SYSTEM MODEL

In the Green Cloud architecture, users submit their Cloud service requests through a new middleware Green Broker that manages the selection of the

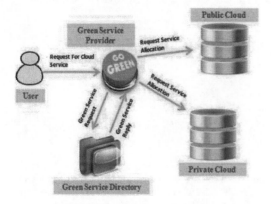

Figure 1. Carbon Aware Proposed Framework for Green Clouds.

greenest Cloud provider to serve the user's request. A user service request can be of three types i.e. software, platform or infrastructure. The Cloud providers can register their services in the form of "green offers" to a public directory which is accessed by Green Broker. The green offers consist of green services, pricing and time when it should be accessed for least carbon emission.

The Green Cloud framework is relies on two main components, Carbon Emission Directory and Green Cloud offers. From user side, the Green Broker plays a crucial role in monitoring and selecting the Cloud services based on the user QOS requirements, and ensuring minimum carbon emission for serving a user (R. Joshi & P. Pathak 2015). Green Cloud Framework proposed by R. Buyya (Garg, S., Yeo, C., Buyya, R 2011) illustrate the effectiveness of architecture in reducing the energy and CO_2 emissions across the entire Cloud infrastructure, while this framework present a simple scenario focused on IaaS.

3 TEMPERATURE AND ENERGY AWARE TASK SCHEDULING ALGORITHM

The technique consists of two levels. In the first level, a function is generated to compare system temperature with the threshold temperature, in the second level for each system with system temperature less then threshold; the increment rate of power is calculated after the task allocation. After the task is generated it is allocated to the system with lowest power increment rate. Each task has some requirements like; temperature, energy consumption, time, system requirements (CPU utilization, Memory).

The system with less power consumption is chosen, means the task will be performed on that system. The systems generated by the administrator have their own specifications. Researches specify

system feature are: Threshold Temperature, Power Consumption, Power increment rate (ΔP).

4 ALGORITHM

/* Tsys is system current temperature; Tth threshold temperature; Taski is current task; SYSj current system; ΔPj rate of change of power in SYSj */

IF (event (Task$_i$==True)) do

    FOREACH System in SYSTEM queue do

        IF ($T_{sys} < T_{th}$)

            Add System to SYSTEMP queue

        End IF

    End FOREACH

    FOREACH system in SYSTEMP queue

        Calculate Power (SYS$_j$, ΔP)

        P_b: Power before task allocation

        P_a: Power after task allocation

        $\Delta P = Pa - Pb$

    End FOREACH
    Sort system in ascending order of ΔP
    Allocate Task to system with least ΔP
End IF

By the arrival of each job request the scheduler compares all available systems temperature with the threshold temperature (Tsys < Tth) and maintain a queue i.e. SYSTEMP queue. If the temperature of a system is less then threshold temperature scheduler will add it to the SYSTEMP queue which means the system is eligible for further processing otherwise it will discard the system. With the completion of final SYSTEMP queue level one of the algorithm is completed. In second level for all system in queue power increment rate will be calculated. Here scheduler will check the power of system before task allocation (Pb) and after task allocation (Pa) and then calculate power increment rate (ΔP) as ($\Delta P = Pa-Pb$). After calculating ΔP scheduler will sort the system in the ascending order of ΔP. Finally the task will be assign to the system with least ΔP. The complexity of this algorithm is dominated by the process request and the system sorting function. Therefore, the time complexity of the algorithm is O (pm log (m)).

5 IMPLEMENTATION

From the point of view of experiment this algorithm is implemented in Object Oriented Language and C++ platform is used with Windows XP 32 bit Operating System.

5.1 Experimental results and analysis

The program is executed three times and then it is compared by already existing Round Robin algorithm. Here are the parameters for execution of Temperature and Power aware Algorithm:

Case I. We have taken the 10 virtual machines with initial temperature and Power and the

Table 1. Parameters for the execution of proposed algorithm.

VM name	Temperature (Dynamic)	Power (ΔP) KW
VM0	201	50
VM1	198	50
VM2	204	55
VM3	195	44
VM4	195	44
VM5	200	56
VM6	203	52
VM7	202	51
VM8	195	50
VM9	200	51

Figure 2. Screen shot for the Temperature and Power aware algorithm result.

program is executed as: Average time taken by the scheduler to sort virtual machines for load balancing—7.087912 Seconds

Case II. Same program is executed with different random values and the results are: Average time taken by the scheduler to sort virtual machines for load balancing—4.5249854 Seconds

Case III. Result after executing the program third time with different random values: Average time taken by the scheduler to sort virtual machines for load balancing—5.0203045 Seconds

So, the average time from the above 3 cases—5.5480183 Seconds.

5.2 Comparison with round robin algorithm

Case I. changing the parameters of burst time and arrival time we got the result—8.0 Seconds.

Case II. Changing the parameters of burst time and arrival time we got the result—8.928571 Seconds.

Case III. Changing the parameters of burst time and arrival time we got the result—12 Seconds.

So, the average time is **9.6434 Seconds.**

Figure 3. Screen shot for the Round Robin algorithm result.

Table 2. Parameters for the execution of Round Robin algorithm.

Virtual machine	Burst time	Arrival time
VM0	1	1
VM1	2	2
VM2	3	3
VM3	4	4
VM4	5	5
VM5	6	6
VM6	7	7
VM7	8	8
VM8	9	9
VM9	10	10

Table 3. Analysis of experimental result.

Algorithm/cases	Proposed algorithm (Time in sec.)	Round robin (Time in sec.)
Case I	7.087912	8
Case II	4.5249854	8.928571
Case III	5.0203045	12.00
Average time	5.5480183	9.6434

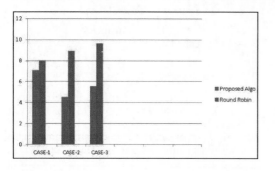

Figure 4. Comparison between algorithms using bar graph.

5.3 Analysis of experiment

During the analysis of result we found that the average load balancing time of our proposed algorithm is less than the existing Round Robin algorithm.

Average load balancing value of our proposed algorithm is lesser than Round robin load balancing algorithm.

This comparison graph depicts that blue bar column are taking less time for execution than the Red column. Therefore, we can say temperature and power aware algorithm is better than round robin algorithm for this scenario.

6 CONCLUSION

Proposed technique "Temperature and Power Aware Task scheduling for Green Cloud Framework" aims to maintain the temperature of the virtual machines below threshold temperature by scheduling task according to the temperature of VMs to avoid risk of system failure due to high temperature. As a conclusion, Green Cloud effectively saves energy by dynamically adapting to workload leveraging live task allocation, at the same time meeting system SLAs.

ACKNOWLEDGEMENT

The authors thank Mr. Mukesh Kestwal, Junior Research Fellow from UCOST, Dehradun and Mr. Mukesh Kumar, Corporate trainer, who provided insight and expertise that greatly, assisted the research. And we also thank to all those researchers whose research work has been used as a reference in this paper and all people who directly or indirectly helped in this research work. This research work is funded by University of Petroleum and Energy Studies, Dehradun.

REFERENCES

Andrew J. Younge 2010. Efficient Resource Management for Cloud Computing Environments. IEEE.

Anton Beloglazov 2010. Energy Efficient Allocation of Virtual Machines in Cloud Data Centers. In 10th IEEE/ACM International Conference on Cluster, Cloud and Grid Computing.

Anton Beloglazov 2013. Energy-Efficient Management of Virtual Machines in Data Centers for Cloud Computing. PhD thesis.

Berkeley News 2013. Berkeley Lab Study Finds Moving Select Computer Services to the Cloud Promises Significant Energy Savings. http://newscenter.lbl.gov/

Cloud Computing—Cloud Primer Introductory Course [online] Available: http://honim.typepad.com/biasc/2014/02/cloud-computing-cloud-primer-introductory-course.html

Garg, S., Yeo, C., Buyya, R 2011. Green cloud framework for improving carbon efficiency of clouds. In 17th International Conference on Parallel Processing: 491–502.

Gartner 2007, Inc. Gartner estimates ICT industry accounts for 2 percent of global CO2 emissions. Available at: http://www.gartner.com/it/page.jsp?id = 503867{6}

Joshi R. and P. Pathak 2015. Carbon aware Architecture for Green Clouds: A Proposed Framework. In Uttarakhand State Science & Technology Congress Conducted by Uttarakhand State Council for Science and Technology, at 9th USSTC 2014–15, during 26–28 February 2015.

Mehta, Avinash, Mukesh Menaria, Sanket Dangi, and Shrisha Rao 2011. Energy conservation in cloud infrastructures. In IEEE International Systems Conference.

Minas L. and B. Elliso et al. 2009. Energy Efficiency for Information Technology: How to Reduce Power Consumption in Servers and Data Centers. Intel Press.

Open Compute Project 2015. Prineville Energy-Efficient Data Centre-Facbook:http://www.opencompute.org/about/energy-efficiency/

Owusu, Francis, and Colin Pattinson 2012. The Current State of Understanding of the Energy Efficiency of Cloud Computing. In IEEE 11th International Conference on Trust Security and Privacy in Computing and Communications.

Communication and Computing Systems – Prasad et al. (Eds)
© *2017 Taylor & Francis Group, London, ISBN 978-1-138-02952-1*

QoS aware VM allocation policy in cloud using a credit based scheduling algorithm

Prashant Pranav, Naela Rizvi & Ritesh Jha
Birla Institute of Technology, Mesra, Ranchi, Jharkhand, India

ABSTRACT: With the emergence of cloud computing as a powerful computing storage off late, it has become easier for businesses and users to store and share their large data and application over the cloud. Besides providing space for storing large and valuable data of the users, cloud computing provides security to these stored valuable data as well. Jobs coming for getting executed on the clouds are executed on the virtual machines created for the jobs. When more and more jobs start coming, it becomes difficult for them to get the VMs and as such they are rested in a queue. From the queue these jobs are scheduled by using different scheduling algorithms. Now, it is very necessary to ensure customers of not waiting too longer for VMs i.e. ensuring QoS to the customers. In the proposed model, QoS is ensured by recycling the VMs created for specific type of requests and allocating them to other jobs needing those requests. In this way, the time wasted in creating and destroying VMs is saved reasonably. Further we have used a credit based scheduling algorithm which does not let jobs in the queue to wait for longer. QoS, is further ensured by not allowing jobs to enter in the system if those cannot be met in time defined.

1 INTRODUCTION

Cloud computing means using a collection of remote servers hosted over the internet. These remote servers can in turn store, manage and process the user's data. Cloud computing has a much of benefits. Flexibility, disaster recovery, security, work from anywhere are some to name among others. Clouds allow companies as well as individuals to store their large and valuable data over the internet despite of their personal computers. Cloud service providers in turn ensure security to these stored data at the same time giving customers a QoS. VM allocation is very important in aspect of cloud computing. VM are the actual location where a job is executed. So, when the number of jobs arriving into the system increases, it becomes difficult to allocate VMs to all of them. So, here we have proposed a comparatively better VM allocation policy. The scheduling policy used is credit based scheduling policy which works by giving a job a specific credit on the basis of their length and priority.

2 RELATED WORK

Many works have been done till date on VM allocation policy for incoming jobs. Focuses on an adoptive QoS aware VM allocation policy through a priority factor based scheduling approach. It ensures efficient resource utilization. In a provisioning approach that changes as the workloads changes and thus offering end users a Quality of Service is proposed. A dynamic provisioning technique which adopts itself in peek to peek workload changes is given in. Based on some analytical results, the provisioning of VMs is given by authors in while in QoS modelling based on FCM (Fuzzy Cognitive Mapping) is proposed by the authors. The use of queuing networks in the provisioning of multitier applications in clouds has been detailed by authors. Authors gives a reactive algorithm for dynamic VM provisioning.

3 PROPOSED METHODOLOGY

A self adaptive QoS based VM provisioning using a credit based scheduling algorithm is proposed here to control and manage uncertain behavior with regards to dynamic workload changing in a cloud environment. The system architecture of the proposed model is shown in Figure 1 below. The model has three basic components viz. Admission Controller/Requirement Analyzer, Resource Manager and VM Server.

Whether to accept or reject request is decided by Admission Controller. Admission of new jobs is allowed only if the system is not congested and QoS can be promised to the customers. The task

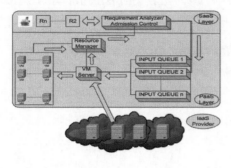

Figure 1. System model.

of requirement analyzer on the other hand is to predict the amount of resources of specific type required by each job and also the specific queue where the jobs will have to wait. Resource manager make the Admission Controller/ Requirement Analyzer to decide whether to let the jobs enter or not by analyzing the resources currently in use and the resources available to fulfil the incoming jobs. VM server allocates VM to time critical jobs and rests are put in a queue.

An Adaptive QoS Aware VM Allocation Using Credit Based Scheduling Algorithm

In order to ensure QoS jobs are entered in a controlled fashion. Before letting a job to enter the system the negotiated time for the jobs in service and all the jobs in queue is calculated denoted by $T_{negotiated}$. After that total time to complete all the jobs in service and in queue is estimated which is denoted by T_{total}. Now T_{total} is added with a reserved value of time which is the value taken for coping up with uncertain network behavior. When new jobs arrive its actual working time denoted by T_{new_act} is predicted by the Admission Controller/Requirement Analyzer. Also the negotiated time T_{new_neg} of the new job is calculated and it is checked by combining it with $T_{negotiated}$. that if it is greater than the sum of T_{total}, T_{new_act} and threshold time and if the total affordable service time $T_{service}$ becomes greater than the sum of $T_{negotiated}$. and T_{new_act}, then only the job is allowed to enter in the system. The entered job is then put into the queue according to its requirement. Once a VM gets free, credit based scheduling is performed in every queue and the job with highest credit is scheduled first.

3.1 Assigning credit to each job

Credit of a job is assigned based on two factors viz. Job Length and User Priority

Job Length Credit: Jobs having different lengths are executed in a cloud. The credit system based on job length first finds the length of each job ($Jlen_i$) and next average of job length is calculated as len_{avg}. The difference between each length with the length average is then done by the equation

$|JLD_i = len_{avg} - Jlen_i|$, where JLD_i is the job length difference of job i

Four different values is calculated from the length array on the basis of which credits are assigned. These four values must lie in the region of task length and are calculated as:

value_1 = high_len/5

value_2 = high_len/4

value_3 = value_2 + value_1

value_4 = value_3 + value_2,

where high_len is the highest value of job lenghth.

The algorithm adds credit to each job as (Credit_Lengthi)

Job Priority Credit: Each job has different priority which is denoted by value assigned to each job. Two or more jobs can have same value and hence same priority. So, dealing with jobs of equal priority is a concerned tackled well in this approach as it does not just rely on job priority, the final scheduling is done based on both task length and task priority In job priority credit algorithm, the highest priority number is found first and next choosing a division factor for finding Priority Factor for each job. The division factor is taken based on the highest priority of a job. If the highest priority is two digit number then division factor is taken as 100 and if it is a three digit number than division factor is chosen as 1000. Both the algorithms are given below

Job Length Credit	Job Priority Credit
For all submitted jobs in the set ;Ji $\| JLD_i = len_{avg} - Jlen_i \|$, If JLDi ≤ value_1 then credit =5 else if value_1 < JLDi ≤ value_2 then credit =4 else if value_2 < JLDi ≤ value_3 then credit =3 else if value_3 < JLDi ≤ value_4 then credit =2 else value_4 > JLDi then credit =1 End For	For all submitted tasks in the set ;Ji Find out task with highest priority(Priority Number) Choose division_part For each task with priority Jpri find Pri_frac(i)=Jpri /division_factor set credit as Pri_frac End For

Finally the total credit is assigned to each job by the below equation:

Total_Krediti = Credit_Lengthi*Credit_Priorityi

VM servers provide virtual machines to the job which has the highest total credit. The algorithm of the whole process is given below:

446

INPUT: Tmean :	else if value_1 < JLDi
Monitored mean execution time,	≤ value_2 then credit =4
Tservice : Total affordable service time by the service provider,	else if value_2 < JLDi ≤ value_3 then credit =3
ReservedTime : Time for secured QoS provisioning,	else if value_3 < JLDi ≤ value_4 then credit =2
n : Number of jobs,	else value_4 > JLDi
j : number of queues	then credit =1
i : number of jobs in queue j	End For
OUTPUT: QoS aware VM provisioning	16. Assign priority to each task and do{
1. Tnegotiated = ΣTreq;	For all submitted tasks
2. Ttotal = n *Tmean;	in the set ;Ji
3. Testimate =Tnew act + Ttotal + Reserved Time;	Find out task with highest
4. Tmax limit = Tnegotiated + Tnew neg;	priority(Priority Number)
5. if Tmax limit >= Testimate&&Tservice >= Tmax limit	Choose division_part
then	For each task with
6. Allow job to enter into input queue;	priority Jpri Find
7. else	Pri_frac(i)=Jpri
8. Reject job to enter into input queue;	/division_factor
9. end if	Set credit as Pri_frac
10. Calculate task length and priority;	End For
11. if resources available for creating VM then	17. Total Crediti = Credit_lengthi* Credit_priorityi
12. Repeat for all queuc;	18. Find maximum
13. Calculate |JLDi = lenavg; - Jleni|	crediti for creating new VM of type j;
14. Assign credit to each job;	15. else
15. For all submitted tasks in the set ;Ji	16. Wait for VM of same type for completing current
	JLDi = lenavg - Jleni		job;
If JLDi ≤ value_1	17. end if
then credit =5	

3.2 Performance analysis using a case study

For the purpose of looking for the feasibility of the proposed work we created a list of 21 jobs which were of different requirements type. Some require CPU while others RAM or storage as resources. Three VMs were created with specific feature. VM1 was to handle the CPU type of resource requests, VM2 was for that of RAM type and VM3 was to fulfil storage requests of jobs. Also three different queues were created to store jobs of specific type. Queue1 was to store CPU requested jobs, Queue2 was to handle RAM requested jobs and Queue3 for that of storage required. We compared the existing priority factor based approach with that of ours credit based approach on the basis of waiting time and found that for individual jobs in some cases priority factor is good and in some credit based gives better result. Also, when the credit factor based approach comes into play, it reduces the waiting time of some jobs to a great extent as compared to priority based. Average waiting time of all the 21 jobs is comparatively low in credit based scheduling. Queue wise average waiting time is also low for all the three queues in credit based scheduling as compared to that of priority factor based scheduling.

Table 1. List of jobs with different requirements.

Job Id	Requirement type	Quantity
Job 1	CPU	1
Job 2	RAM	4
Job 3	CPU	1
Job 4	Storage	150
Job 5	RAM	4
Job 6	CPU	2
Job 7	Storage	200
Job 8	CPU	2
Job 9	CPU	1
Job 10	RAM	2
Job 11	RAM	1
Job 12	Storage	180
Job 13	Storage	120
Job 14	RAM	2
Job 15	CPU	1
Job 16	Storage	100
Job 17	Stoarge	80
Job 18	RAM	4
Job 19	CPU	2
Job 20	RAM	2
Job 21	Storage	50

3.3 Comparison graph

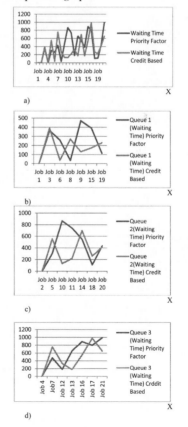

a)

b)

c)

d)

Figure 2. Figure a depicts the overall difference in waiting time for the two methods while Figure b, c and d shows the queue-wise differences for Queue 1, Queue 2 and Queue 3 respectively.

447

Table 2. Comparison of waiting time of priority factor based and credit based algorithms.

Job Id	Arrival time	Execution time	Negotiation time	Actual time	Queue	Waiting time Priority factor	Credit based
Job 1	0	50	70	50	1	0	0
Job 2	2	150	180	150	2	0	0
Job 3	5	55	150	100	1	365	390
Job 4	8	200	250	200	3	0	0
Job 5	9	170	350	313	2	303	553
Job 6	11	100	145	139	1	259	39
Job 7	13	250	470	445	3	425	755
Job 8	15	110	170	145	1	35	270
Job 9	18	60	100	92	1	472	132
Job 10	21	190	235	229	2	865	131
Job 11	24	60	195	188	2	743	218
Job 12	28	230	400	390	3	180	330
Job 13	32	150	340	326	3	656	176
Job 14	35	95	210	212	2	547	697
Job 15	36	65	100	79	1	389	174
Job 16	39	110	300	279	3	889	549
Job 17	41	90	300	279	3	797	977
Job 18	42	160	300	270	2	110	260
Job 19	45	110	140	115	1	115	230
Job 20	47	100	210	205	2	435	415
Job 21	49	70	240	229	3	989	649
Average waiting time						408.2	330.7

Table 3. Queue-wise comparison of the two algorithms.

Job Id	Queue 1 Waiting time Priority	Credit	Queue 2 Waiting time Priority	Credit	Queue 3 Waiting time Priority	Credit
Job 1	0	0				
Job 2			0	0		
Job 3	365	390				
Job 4					0	0
Job 5			303	553		
Job 6	259	39				
Job 7					425	755
Job 8	35	270				
Job 9	472	132				
Job 10			865	131		
Job 11			743	218		
Job 12					180	330
Job 13					656	176
Job 14			547	697		
Job 15	389	174				
Job 16					889	549
Job 17					797	977
Job 18			110	260		
Job 19	115	230				
Job 20			435	415		
Job 21					989	649
AWT	225.5	205.8	499	379	656	572.6

4   CONCLUSION AND FUTURE WORK

To conclude with it can be said that the results given shows the superiority of credit based scheduling over priority factor based scheduling. The average waiting time reduces reasonably for both the summation of all the 21 jobs and summation of 6 jobs each in the queue in the case of credit based approach as compared to that of priority factor based approach.

In the future, other scheduling approaches will be considered such as SJF, FCFS and round robin. Also, this methodology can be applied to resource allocation policy in the VMs.

REFERENCES

Bi, J., Z. Zhu, R. Tian, And Q. Wang, "Dynamic provisioning modeling for virtualized multi-tier applications in cloud data center," in *Proceedings of the 3rd International Conference on Cloud Computing (CLOUD10)*, 2010.

Calheiros, R. N., R. Ranjan and R. Buyya, "Virtual Machine Provisioning Based on Analytical Performance and QoS in Cloud Computing Environments," in *Parallel Processing (ICPP), 2011 International Conference*, 2011.

Chieu, T. C., A. Mohindra, A. A. Karve, and A. Segal, "Dynamic scaling of web applications in a virtualized cloud computing environment," in *Proceedings of the 6th International Conference on e-Business Engineering (ICEBE09)*, 2009.

Das, A., T. Adihkari, C. Hong, "An Intelligent Approach for Virtual Machine and QoS Provisioning in Cloud Computing," in ICOIN 2013, 978-1-4673-5742-5/13/, pp. 462–468.

Hitesh Beda, Jignesh Lakhani, "QoS and Performance Optimization with VM Provisioning Approach in Cloud Computing Environment," 2012 Nirma University International Conference On Engineering, Nuicone-2012, 06–08 december, 2012, Pp. 1–5.

Rodrigo. N, R. Ranjan, R. Buyya, "Virtual Machine Provisioning Based on Analytical Performance and QoS in Cloud Computing Environments," 2011 International Conference on Parallel Processing, 0190-3918/11, pp. 295–304.

Thomas, A., Krishanlal G, Jagathy Raj, "Credit Based Scheduling Algorithm in Cloud Computing Environment," Procedia Computer Science 46 (2015), 913–920.

Zhang And P., Z. Yan, "A Qos-Aware System For Mobile Cloud Computing," In *Proceedings Of IEEE CCIS2011*, 2011.

Communication and Computing Systems – Prasad et al. (Eds)
© 2017 Taylor & Francis Group, London, ISBN 978-1-138-02952-1

Classification of Pima indian diabetes dataset using naive bayes with genetic algorithm as an attribute selection

Dilip Kumar Choubey, Sanchita Paul & Santosh Kumar
CSE, BIT, Mesra, Ranchi, India

Shankar Kumar
Polytechnic, BIT, Mesra, Ranchi, India

ABSTRACT: Diabetes means blood sugar is above desired level on a sustained basis. The prime objective of this research work is to provide a better classification of diabetes. There are already several existing method, which have been implemented for the classification of diabetes dataset. In medical sector, the classifications systems have been widely used to exploit the patient's data and make the predictive models or build set of rules. In this manuscript firstly NBs used for the classification on all the attributes and then GA used as an attribute selection and NBs used on that selected attribute for classification. The experimental results show the performance of this work on PIDD and provide better classification for diagnosis.

1 INTRODUCTION

Diabetes is a problem and a major public health challenge worldwide. This is one of the most widespread disease, now a day's very common. In this manuscript, Genetic Algorithm (GA) has been used as an attribute (feature) selection method by which four attributes have been selected from eight attributes. Naive Bayes (NBs) are statistical, supervised learning method for classification. Here, NBs has been used for the classification of the diabetes diagnosis.

The paper is organized as follows: Proposed methodology is discussed in section 2, Results and Discussion are devoted to section 3, Conclusion and Future Direction are discussed in section 4.

2 PROPOSED METHODOLOGY

Here, the proposed methodology is implemented by GA as an Attribute Selection and NBs for Classification on PIDD which has been taken from UCI machine learning repository.

The block diagram of proposed approach is shown above and next proposed approach is as follows:

1. The PIDD has been taken from UCI machine learning repository.
2. Apply GA as an Attribute Selection on PIDD.
3. Do the Classification by using NBs on selected attributes and all the attributes in PIDD.

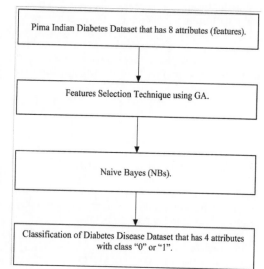

Figure 1. Proposed system.

2.1 Used diabetes disease dataset

The Pima Indian Diabetes Dataset (PIDD) has been taken from the UCI Machine Learning repository. The same dataset used in the reference (Polat and Gunes 2007; Seera and Lim 2014; Lukka 2011; Gajni and Abadeh 2011; Choubey and Paul 2016; Ephzibah 2011; Choubey and Paul 2015).

2.2 GA for attribute selection

Either the algorithms stops forming new iterations when a maximum number of iterations have been formed or a satisfactory fitness value is achieved for the problem.

The standard pseudo code of GA is given in below algorithm:

Algorithm 1
Begin
$q = 0$

Randomly initialize individual members of population P(q)
Evaluate fitness of each individual of population P (q)
while termination condition is not satisfied do
$q = q+1$

       selection (of better fit solutions)
    crossover (mating between parents to generate off-springs)
       mutation (random change in off-springs)
end while
Return best individual in population;

In Algorithm 1, q represents the iteration counter, initialization is done randomly in search space, and corresponding fitness is evaluated based on the function. After that, GA algorithm requires a cycle of three phases: selection, crossover, and mutation which is briefly explained in (Choubey and Paul, 2016).

2.3 NBs

The pseudo code of NBs are discussed below. The same pseudo code (Siddique and Hossain, 2013) of NBs are used for predicting Heart disease.

2.3.1 Pseudo code

Calculate diagnosis = "yes", diagnosis = "no" probabilities Pyes, Pno from training input
    For each test input samples
    For each feature
    Calculate of category of feature based on categorical
Division
    Calculate probabilities of diagnosis = "yes", diagnosis = "no" corresponds to that category P(feat, yes), P(feat, no) from training input
For each feature
    Calculate the resultyes = P(feat, yes), resultno = P(feat, no);
    Calculate the resultyes = resultyes *Pyes, resultno = resultno*Pno;
    If (resultyes > resultno) then diagnosis = "yes", else then diagnosis = "no".
Where,
    Pyes = Total number of yes/Total number of samples or instances;

Pno = Total number of no/Total number of samples;
P(attr, yes) = Total number of yes in corresponding category/Total number of yes;
P(attr, no) = Total number of no in corresponding category/Total number of no;

3 RESULTS AND DISCUSSION

In Experimental studies the dataset have been partitioned between 70–30% (538–230) for training & test of NBs, GA_NBs. It has been performed on PIDD and the results compared with several existing method which is noted in Table 5.

It may be seen that in Table 2, by applying the GA method, four attributes have been selected from eight attributes. This means the cost have reduced to s(x) = 4/8 = 0.5 from 1 and an improvement on the training and classification by a factor of 2.

It is well known that diagnostic performance is usually evaluated in terms of Accuracy, Precision, Recall, Fallout and F–Measure, ROC, Confusion Matrix, Kappa Statistics, Mean Absolute Error (MAE), Root Mean Square Error (RMSE), Relative Absolute Error (RAE), Root Relative Squared Error (RRSE) which is shown below.

The time taken to build model training set evaluation = 1.22 seconds, and time taken to build model testing set evaluation = 1.14 seconds for NBs Performance. The Table 1 shows the results of both the training set and testing set evaluation by using NBs method for PIDD based on some parameters, which are noted below:

The Table 2 shows the Attribute selection by using GA on PIDD, which is noted below.

Table 1. Results of NBs performance for PIDD.

Measure	Training set evaluation	Testing set evaluation
Precision	0.769	0.766
Recall	0.773	0.77
F–Measure	0.769	0.767
Accuracy	77.3234%	76.9565%
ROC	0.816	0.846
Kappa statistics	0.4875	0.478
Mean Absolute Error	0.2868	0.2768
Root Mean–Squared Error	0.4157	0.3973
Relative Absolute Error	62.9039%	61.0206%
Root Relative Squared Error	87.075%	83.654%

Cofusion Matrix for Training set
a	b	<--classified as
300	49	a = tested_ negative
73	116	b = tested_ positive

Cofusion Matrix for Testing set.
a	b	<--classified as
128	23	a = tested_ negative
30	49	b = tested_ positive

452

Table 2. GA for attributes selection.

Data set	No. of attributes	Name of attributes	No. of instances	No. of classes
PIDD (Without GA)	8	1. Number of times pregnant 2. Plasma glucose concentration a 2 hours in an oral glucose tolerance test 3. Diastolic blood pressure 4. Triceps skin fold thickness 5. 2 – hour serum insulin 6. Body mass index 7. Diabetes pedigree function 8. Age (years)	768	2
PIDD (With GA)	4	2. Plasma glucose concentration a 2 hours in an oral glucose tolerance test 5. 2 – hour serum insulin 6. Body mass index 8. Age (years)	768	2

The time taken to build model training set evaluation = 1.18 seconds, and time taken to build model testing set evaluation = 1.09 seconds for GA_NBs methodology. The Table 3 shows the results of both the training set and testing set evaluation by using NBs method for PIDD on the selected attributes by using GA based on some parameters, which is noted below.

The above figure is the ROC graph for tested_positive class by using GA_NBs on PIDD, achieved 0.844 ROC. It generates very less error rate, and ratio between FPR vs TPR is also good.

The Table 4 shows the analysis of comparison result with and without GA on NBs for PIDD by several measures along with several methods i.e., noted in table.

In the Table 4 it may be seen that with GA the improvement has occurred in every measure except ROC in the case of NBs. The only ROC measure achieved slightly very less, may be by applying this method only on this particular dataset but mostly in any cases by applying attribute selection method improvement occur in every measure.

In the above table, mentioned methods i.e. J48 graft DT, GA_J48 graft DT, MLP NN, GA_MLP NN implemented by Dilip Kumar Choubey et al. have mentioned Precision, Recall, F–Measure, Accuracy, ROC value results in the publication not the Kappa statistics, MAE, RMSE, RAE, RRSE. So once again went to implement the above-mentioned methods to find the not available value results i.e. Kappa statistics, MAE, RMSE, RAE, and RRSE.

Table 3. Results of GA_NBs for PIDD.

Measure	Training set evaluation	Testing set evaluation
Precision	0.75	0.782
Recall	0.757	0.787
F–Measure	0.748	0.78
Accuracy	75.6506%	78.6957%
ROC	0.811	0.844
Kappa statistics	0.4364	0.5021
Mean Absolute Error	0.3105	0.295
Root Mean–Squared Error	0.4158	0.3919
Relative Absolute Error	68.0976%	65.0317%
Root Relative Squared Error	87.1025%	82.5241%

Cofusion Matrix for Training set

```
a     b    <--classified as
305   44 |  a = tested_ negative
87   102 |  b = tested_ positive
```

Cofusion Matrix for Testing set

```
a     b    <--classified as
135   16 |  a = tested_ negative
33    46 |  b = tested_ positive
```

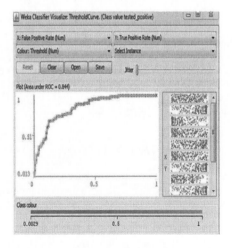

Figure 2. ROC graph for tested_positive class by using GA_NBs methodology on PIDD.

Table 4. Evaluation of NBs & GA _ NBs, along with several method performance for PIDD.

Measure	J48 graft DT Dilip Kumar Choubey et al. (2015)	GA_J48 graft DT Dilip Kumar Choubey et al. (2015)	MLP NN Dilip Kumar Choubey et al. (2016)	GA_MLP NN Dilip Kumar Choubey et al. (2016)	NBs	GA _ NBs
Precision	0.761	0.789	0.781	0.79	0.766	0.782
Recall	0.765	0.748	0.783	0.791	0.77	0.787
F–Measure	0.762	0.754	0.77	0.78	0.767	0.78
Accuracy	76.5217% (0.765217)	74.7826% (0.747826)	78.2609% (0.782609)	79.1304% (0.791304)	76.9565%	78.6957%
ROC	0.765	0.786	0.853	0.842	0.846	0.844
Kappa statistics	0.4665	0.4901	0.4769	0.5011	0.478	0.5021
MAE	0.3353	0.3117	0.2716	0.2984	0.2768	0.295
RMSE	0.4292	0.4114	0.387	0.387	0.3973	0.3919
RAE	73.9186%	68.7038%	59.8716%	65.7734%	61.0206%	65.0317%
RRSE	90.3686%	86.6146%	81.4912%	81.4774%	83.654%	82.5241%

Table 5. Results and comparison with other methods for the PIDD.

Source	Method	Accuracy	ROC
Pasi Luukka (2011)	Sim	75.29%	0.762
	Sim + F1	75.84%	0.703
	Sim + F2	75.97%	0.667
H. Hasan Orkcu et al. (2011)	Binary– coded GA	74.80%	....
	BP	73.80%	....
	Real–coded GA	77.60%	....
Manjeevan Seera et al. (2014)	FMM	69.28%	0.661
	FMM– CART	71.35%	0.683
	FMM-CART– RF	78.39%	0.732
Dilip Kumar Choubey et al. (2015)	J48 graft DT	76.5217%	0.765
	GA_J48 grft DT	74.7826%	0.786
Dilip Kumar Choubey et al. (2016)	MLP NN	78.2609%	0.853
	GA_MLP NN	79.1304%	0.842
Our Study	NBs	76.9565%	0.846
	GA _NBs	78.6957%	0.844

In the Table 5, It may be seen that there are already several existed method for PIDD. The Table 4 shows the result comparison in terms of accuracy and ROC on PIDD for the diagnosis of diabetes. The proposed method i.e., GA_NBs provides the almost highest accuracy and better ROC from all other existing method.

4 CONCLUSION AND FUTURE DIRECTION

Diabetes is a problem with your body that causes blood sugar levels to rise higher than normal. Diabetes can cause serious health complications including blindness, blood pressure, heart disease, kidney disease and nerve damage, etc. which is hazardous to health. The PIDD obtained from UCI repository of machine learning databases on which NBs, GA_NBs method have been applied. In this manuscript, firstly the classification has been done on PIDD by using NBs, and then using GA for Attributes selection, and there by performed classification on the selected attributes. The proposed method minimizes the computation cost, computation time and maximizes the ROC and classification accuracy than several other existing methods. With GA, the improvement has been occurred in every measure except ROC as we may compare from Table 1 and 3, may be by applying this method only on this dataset achieved little less ROC but mostly in any cases by applying attribute selection method the ROC also improved however the classification accuracy and several measure has been improved.

For the future research work, we suggest to develop an expert system of diabetes, which will provide good ROC, accuracy and this is possible to achieve only by using different Attribute selection and classification method which, could significantly decrease healthcare costs via early prediction and diagnosis of diabetes. The proposed method can also be used for other kinds of diseases but not sure that in all the medical diseases either same or greater than the existing results.

REFERENCES

Choubey, Dilip Kumar., Paul, Sanchita. (2016) 'GA_MLP NN: A Hybrid Intelligent System for Diabetes Disease Diagnosis', International Journal of Intelligent Systems and Applications (IJISA), MECS, ISSN: 2074–904X (Print), ISSN: 2074–9058. (Online), Vol. 8, No. 1, pp. 49–59.

Choubey, Dilip Kumar., Paul, Sanchita. (2015) 'GA_J48 graft DT: A Hybrid Intelligent System for Diabetes Disease Diagnosis', International Journal of Bio-Science and Bio-Technology (IJBSBT), SERSC, ISSN: 2233–7849, Vol. 7, No. 5, pp. 135–150.

Ephzibah, E.P. (2011) 'Cost Effective Approach on Feature Selection using Genetic Algorithms and Fuzzy Logic for Diabetes Diagnosis' International Journal on Soft Computing (IJSC), Vol. 2, No. 1.

Ganji, Mostafa Fathi., Abadeh, Mohammad Saniee. (2011) 'A fuzzy classification system based on Ant Colony Optimization for diabetes disease diagnosis' Expert Systems with Applications, Elsevier, Vol. 38, pp. 14650–14659.

Luukka, Pasi. (2011) 'Feature selection using fuzzy entropy measures with similarity classifier', Expert Systems with Applications, Elsevier, Vol. 38, pp. 4600–4607.

Polat, Kemal., Gunes, Salih. (2007) 'An expert system approach based on principal component analysis and adaptive neuro-fuzzy inference system to diagnosis of diabetes disease', Digital Signal Processing, Elsevier, Vol. 17, pp. 702–710.

Seera, Manjeevan., Lim, Chee Peng. (2014) 'A hybrid intelligent system for medical data classification', Expert Systems with Applications, Elsevier, Vol. 41 pp. 2239–2249.

Siddique, Aieman Quadir., Hossain, Md. Saddam. (2013) 'Predicting Heart-Disease from Medical Data by Applying Naive Bayes and Apriori Algorithm', International Journal of Scientific and Engineering Research (IJSER), Vol. 4, Issue 10.

UCI Repository of Bioinformatics Databases [online] Available: http://www.ics.uci.edu./~mlearn/ML Repository.html.

Communication and Computing Systems – Prasad et al. (Eds)
© 2017 Taylor & Francis Group, London, ISBN 978-1-138-02952-1

Semantically ontology merging with intelligent information retrieval using SPARQL on education domain

Abhishek Tiwari & Ankit Vidyarthi
Department of Information Technology, ABES Engineering College Ghaziabad, Ghaziabad, India

ABSTRACT: Retrieving information from the web based on semantics is a challenging task for Intelligent Information Retrieval (IIR) system. Various domain specific retrieval systems uses semantic web for specifying and retrieving user specific queries through ontology's. In past, various ontology merging algorithms had been proposed which was based on the concept of removing overlaps. In this paper a new concept for ontology merging is introduced for education domain which is based on semantics. Further, user specific natural queries are translated to SPARQL queries which are used for the retrieval of information semantically. Proposed merging algorithm is using the class semantics and class property semantics by using word net.

Keywords: Semantic Web, Ontology Merging Algorithm, Semantic, SPARQL Queries, Education ontologies

1 INTRODUCTION

Semantic web is an extension of current World Wide Web (WWW) and it is understood by computers, and performs operations like finding semantic information, combining and aggregating information. The main purpose of the Semantic web is to improvement in Intelligent Information Retrieval (IIR) techniques with well defined meaning. Ontology plays an important role in knowledge extraction and representing knowledge of the semantic web. The main purpose of ontology creation are defines concepts, defines relationship between concept nodes and defines properties of instances and concepts. Domain specific ontologies are used to describe a specific knowledge domain, Reuse of domain knowledge and analysis of domain knowledge. Ontologies have become more and more popular in Intelligent Information Retrieval systems and techniques such as Natural Language Processing, Web Technologies, Information management systems and Database Integration etc. The idea Ontology Merging arrives when there are more than one same domain ontology exist and to create a common repository of Knowledge Base (KB). The main purpose of Ontology Mapping and Merging removes inconsistencies in merged ontology, creation of new extended knowledge and may lead to loss of some valuable information. In the recent time ontology merging becomes a complex task for researchers and many issues are occur like creation of same domain ontology, Ontology Alignment, Ontology Mapping, Ontology Merging of two ontologies must be belong with same domain, Creating a single coherent merged ontology and reuse of existing ontology's.

Maintenance of existing ontology's efficiently, Maintenance of merged ontology, Use of ontology for applications, Selection of correct ontology, Execution of SPARQL query, Evolution and storage of ontology etc. In the recent web search engines Google, Yahoo, Bing, GoodSearch and mywebsearch etc. are not able to provide the relevant information according to the user's search query. SPARQL Protocol and RDF Query Language (SPARQL) is a metadata description language and used to retrieve and manipulate data stored in Resource Description Framework (RDF) and Web Ontology Language (OWL). SPARQL engines process different type of queries given by user and execute them on available RDF and OWL ontology data sets, thereby extracting the information that is relevant for the user.

In the recent web the unambiguous and uniform expression of various domains are available like online education, Medical, Agriculture and Industries. Proposed approach initially considers the domain Ontologies of Education domain. Secondly, ontology merging takes two source ontologies from the same domain and finds the semantic correspondence between two ontologies then it provides a single coherent merged ontology. In this domain specific ontology merging algorithm aim to find the semantic correspondences between two different ontologies and then it provides single coherent merged ontology. Finally, execution of SPARQL queries on a merged ontology. The Proposed Domain specific Semantic Ontology Merging algorithm have been developed which merges the important features of ontologies that will help in exchanging and searching information.

The rest of this paper is organized as follows: Section 2 discusses proposed approach for creation of Education domain Ontologies, proposed semantic ontology merging algorithm, SPARQL Queries for merged education ontology. Section 3, gives the result when applying Ontology Merging Algorithm on education ontologies and SPARQL queries execution. Section 4, discusses conclusion and future work of the paper.

2 PROPOSED APPROACH

The purpose of this approach is to provide semantic search results according to the query. The proposed System is divided into different modules firstly creation of ontology for an education domain, Ontology mapping finds the relationship between two ontologies, Ontology merging combines the two ontologies, Execution of SPARQL queries and finally provides the results. Figure 1 gives the detail block diagram for proposed Intelligent Information Retrieval System.

2.1 Creation of ontology for education domain

The purpose of creating the education ontologies is to provide a knowledge model for the university domain. In these ontologies includes the concept

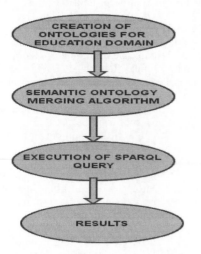

Figure 1. Proposed approach for intelligent information retrieval system.

of Departments, Courses are delivering by institute, Faculty information, Students, Research and publications, distribution of campus according to their location and semester information. In the first ontology, i.e., session distribution ontology, divide the classes according to their sessions fall semester and spring semester. Similarly in the second ontology (course ontology) describes what type of courses are delivered by university and degree awarded by university. In the Education Person ontology describe the person (Student and Staff) that all are belongs to University. In the Event ontology describes the related events which are occur in the organization like Workshop, Conference Activity, Meeting and Special Issue Event. In the department ontology Organization can be divided into departments which all are belongs to an academic field. In Research work ontology describe research group in which developing a project, developing new models and theories, advising and supervising students at academic levels and collaborating with research colleagues. In Location ontology divide the campus according to their location.

2.2 Domain specific semantic ontology merging algorithm

The proposed Domain specific Semantic Ontology Merging algorithm is fully automatic ontology mapping and merging algorithm and divided into two sections initially merge classes then perform merge operation on properties of two ontologies. Domain specific Semantic Ontology Merging interface designed such that any novice user can easily submit their ontology files but must be related to the same domain. In This interface user job is only to give the two .owl ontologies related to same domain as input and merged .owl file will be produced as output. In the first process individual classes are evaluated and punctuations are separated from classes then perform comparison between classes by using wordnet. When performing merging between two same domain .owl files they will have at least few similar properties. In Semantic Merging algorithm for properties Firstly, divide the ontologies into object and data properties. In the First phase find the common properties between two classes with the help of wordnet and then it performs the matching between properties. In the Overlaid Matching phase include the remaining properties of classes. In this phase considered Properties which are in owl1 but not in owl2 and Properties which are in owl2 but not in owl1.

2.3 Execution of SPARQL query

To retrieve the intelligent results the translation of natural language questions into SPARQL queries is done. SPARQL queries Structure are very similar to SQL queries and provides RDF graph as triples [10]. It follows various steps: Firstly Creating question related to domain ontologies according to their expected answers, determining the expected ideal results, Extracted entities from queries and the resultant answers, Identifying answer entity like Classes, Object Properties, Data Properties, Annotation, Instances, Axioms etc., Developing appropriate SPARQL query that gives the closest answer to correct answer.

3 EXPERIMENTAL RESULTS

The proposed domain specific semantic ontology merging algorithm takes two same domain ontology files. For result analysis we created Education domain ontologies. The domain specific semantic ontology merging algorithm applied on Education

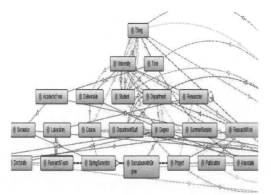

Figure 2. Course ontology.

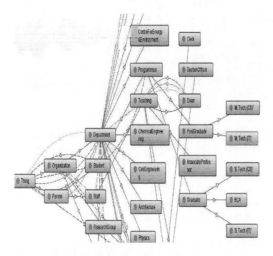

Figure 3. Department ontology.

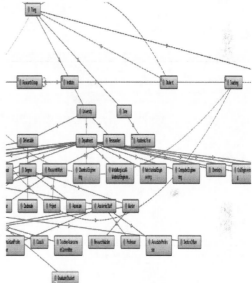

Figure 4. Result of merged ontology.

Table 1. Natural Language Query to SPARQL conversion.

Natural language query	Query identifier	SPARQL query
Describe the definition of a Course?	SP-1	DESCRIBE http://www.semanticweb.org/ankitasharma/ ontologies/2013/10/30/Deliverable#Course>
Select domain and range of the Object Property Appointed to?	SP-2	SELECT ?domain ?range WHERE { http://www.semanticweb.org/ ankitasharma/ontologies/2013/10/30/Deliverable#IsHiredBy> rdfs:domain ?domain; rdfs:range ?range.}
What is the Definition of Department Id?	SP-3	DESCRIBE http://www.semanticweb.org/ankitasharma/ ontologies/2013/10/30/Deliverable#DepartmentId>
What is the definition of a professor or full professor?	SP-4	DESCRIBE http://www.semanticweb.org/ankitasharma/ ontologies/2013/10/30/Deliverable#Professor>
What is a campus?	SP-5	SELECT ?comment WHERE { http://www.semanticweb.org/ ankitasharma/ontologies/2013/10/30/Deliverable#Campus> rdfs:comment ?comment}

Table 2. Execution of SPARQL query on merged Ontology.

SP-1

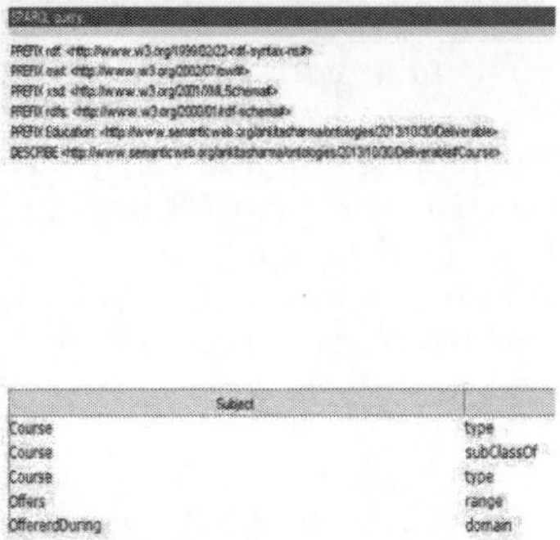

Subject	
Course	type
Course	subClassOf
Course	type
Offers	range
OfferedDuring	domain

SP-2

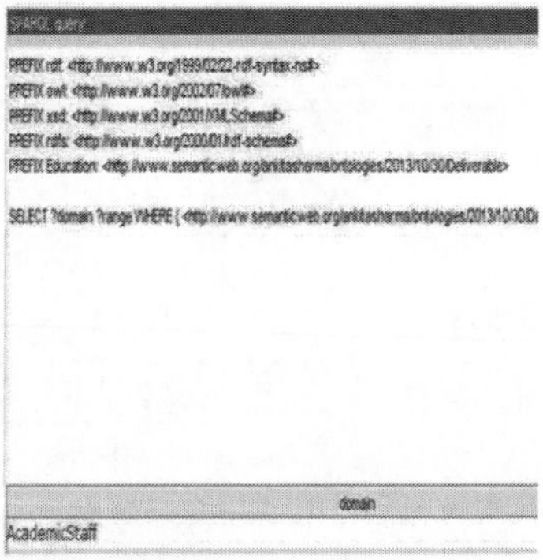

domain
AcademicStaff

SP-3

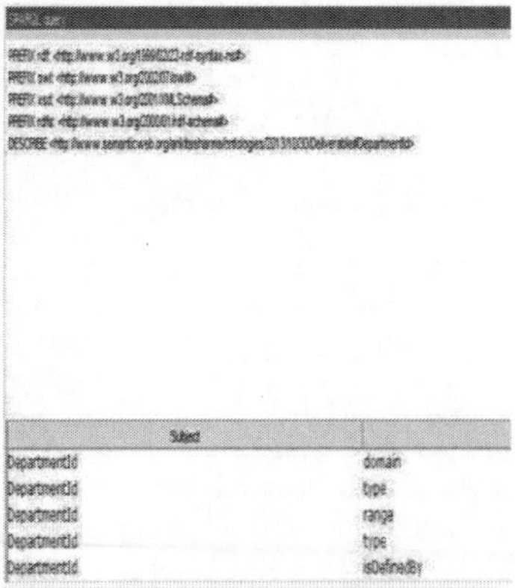

Subject	
DepartmentId	domain
DepartmentId	type
DepartmentId	range
DepartmentId	type
DepartmentId	isDefinedBy

(Continued)

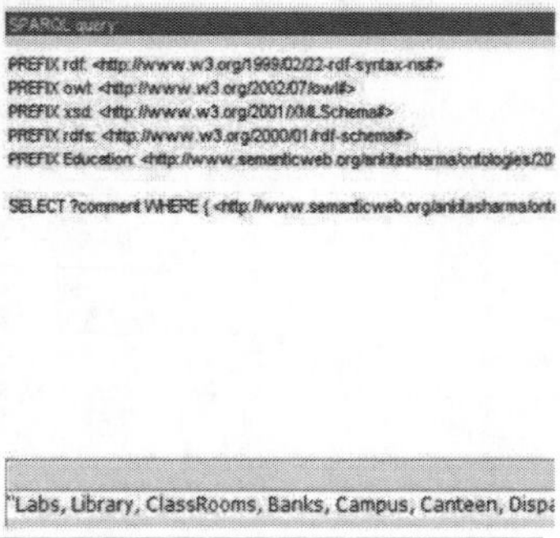

Domain Ontology's for testing purpose. For this purpose firstly it requires two same domain ontologies Owl1 and owl2 as shown in Figure 2 and Figure 3, and finally merged owl file will be produced as output as shown in Figure 4. In the Figure 2,

Table 2. *(Continued)*.

SP-4

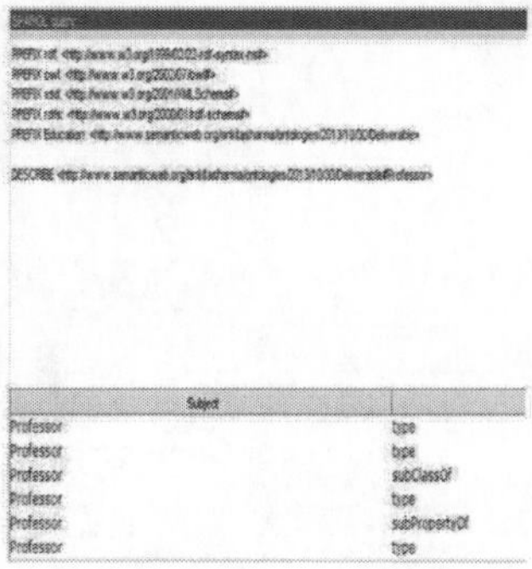

Subject	
Professor	type
Professor	type
Professor	subClassOf
Professor	type
Professor	subPropertyOf
Professor	type

SP-5

"Labs, Library, ClassRooms, Banks, Campus, Canteen, Dispa

ontology describes type of courses deliverable by the university, Classes information according to their sessions and a degree is awarded by University. In figure 3, Ontology describes the Department information in the Organization and maintains the information related to Staff and student. In these ontologies many similar name classes available like administrative staff—Non Teaching Staff. In figure 4, explains the merged result by combining two same domain ontologies and created single coherent merged ontology.

This Intelligent Information Retrieval System is mainly concerned with retrieving information from a merged Education ontology using the SPARQL query language. A Simple demonstration of sample natural language to its corresponding SPARQL query generation in shown in Table 1. Each of the SPARQL query is identified by the corresponding identifier whose execution results is shown in Table 2.

4 CONCLUSION AND FUTURE WORK

In this paper we have introduced Intelligent Information Retrieval Techniques apply on same domain ontology and provides final merged ontologies. In this paper firstly creation of ontologies for Education Domain, then apply fully auto-mated domain specific semantic ontology merging algorithm and Execution of SPARQL queries. The proposed Ontology merging algorithm defines a new approach that finds the relationship between classes, Object properties and data properties by using wordnet. It follows two steps merging algorithm for classes and merging algorithm for properties and finally it gives a single coherent merged ontology. The main aim of this proposed merging approach is to reduce human interaction. This Domain Specific semantic ontology merging algorithm applies on education domain ontolo-gies and provides the efficient result as output single merged education ontology. This Merging algorithm can be applied on other domains also like Agricultural, Medical and Industry also. To retrieve the intelligent results on a single coher-ent merged ontology use a SPARQL queries and provides the result that gives the closest answer to correct answer. This paper is useful for researchers to find an idea of ontology mapping and merging ontology's and proceeds for further research in Ontology merging. Ontology merging individually can be taken as topic of fully research. Proposed algorithm is not suitable for merging two different domain ontology's.

REFERENCES

Bhadgale, A M. el at, 2013, "Natural Language to SQL Conversion System", International Journal of Com-puter Science Engineering and Information Technol-ogy Research (IJCSEITR), Vol. 3(2), pp. 161–166.

Borsje, J., 2006 "Graphical Query Composition and Natural Language Processing in an RDF Visualiza-tion Interface", Informatics and Economics Erasmus School of Economics and Business Economics Eras-mus University Rotterdam.

Chandrasekaran B., Richard V., 1999 "What are ontol-ogy's, and Why do we need them?", Intelligent system and their applications, Vol. 14(1), pp. 20–26.

Choi N., Song Y., and Han H. 2006 "A Survey on Ontol-ogy Mapping", SIGMOD Record, Vol. 35(3), pp. 34–41.

Fikes R., Rice J., and S. Wilder, 2000 "The Chimaera Ontology Environment", In Proceedings of the 17 National Conference on Artificial Intelligence.

Fridman N. Noy and Musen A., 2000 "PROMPT: Algo-rithm and Tool for Automated Ontology Merging and Alignment", Seventeenth National Conference on Intelligent (AAAI), pp. 450–455.

Fridman, N. Noy and Musen A., 1999 "SMART: Auto-mated Support for Ontology Merging and Alignment", Workshop on Ontology Management at the Sixteenth National Conference on Artificial Intelligence.

Jain, P., 2005 "Intelligent Information Retrieval" 3rd International Conference: Sciences of Electronic, Technologies of Information and Telecommunica-tions, pp. 27–31.

Kannanl P., Shanthi P. Bala and Aghila, G., 2012 "A Comparative Study of Multimedia Retrieval Using Ontology for Semantic Web", IEEE-International Conference On Advances In Engineering, Science And Management (ICAESM), pp. 400–405.

Maree, M., S. Alhashmi, M., and Belkhatir M., 2011 "A Unified Ontology Merging and Enrichment Frame-work", 23rd IEEE International Conference on Tools with Artificial Intelligence (ICTAI), pp. 669–674.

Stumme G. and Maedche A., 2001 "FCA-MERGE: Bottom-Up Merging of Ontology's", 17th Interna-tional joint conference on Artificial intelligence – Vol. 1, pp. 225–230.

Zhai J., Zhou K., 2010 "Semantic Retrieval for Sports Information Based on Ontology and SPARQL", International Conference of Information Science and Management Engineering (ICME), Xian, Vol. 19 no. 2 pp: 315–323.

Communication and Computing Systems – Prasad et al. (Eds)
© 2017 Taylor & Francis Group, London, ISBN 978-1-138-02952-1

Probabilistic neural networks for hindi speech recognition

Poonam Sharma & Akansha Singh
The Northcap University, Gurgaon, India

Krishna Kant Singh
Dronacharya College of Engineering, Gurgaon, India

ABSTRACT: Automatic Speech Recognition System has been a challenging and interesting area of research in last decades. Only a few researchers have worked on Hindi and other Indian languages. In this paper, a Speech Recognition System for Hindi language based on Mel frequency Cepstral Coefficients (MFCC), its time derivatives and Linear Predictive Code (LPC) using Probabilistic Neural Networks is proposed. The proposed method works in two phases. In the first phase the Delta and Delta-Delta features from MFCC are calculated. In the second phase a Probabilistic Neural Network (PNN) is used to classify the input signal. The results obtained from the proposed method are compared with other state of art methods. Results show that the proposed method has better accuracy as compared to the other methods.

Keywords: Hidden Markov Model (HMM), Guassian Mixture Model (GMM), Deep Neural Networks (DNN), Mel Frequency Cepstral Coefficients (MFCC), Linear Predictive Code (LPC)

1 INTRODUCTION

The purpose of Automatic Speech Recognition is to develop a system that can transcribe the human speech into readable text despite of noise, speaking styles, reverberations and other different environmental conditions. Research on designing such systems is going on since decades and has resulted into development of many commercial products. Still in real usage scenarios the performance lags far behind human level performance (Baker et al. 2009). Many different techniques like feature enhancement and transformations, acoustic model adaptations has been used to overcome all the problems associated with the current systems, but most of them are using Hidden Markov Models (HMMs) combined with Gaussian Mixture Models (GMMs) which are conventional methodologies having problems like limited modeling capability and inability to make the use of unlabeled data.

Recently, research on HMM models combined with Artificial Neural Networks (ANNs) have shown a great possibility of improvement in designing the acoustic models for Speech Recognition (X. Li 2013, Sinha et al. 2013, Mohamed et al. 2011). Deep Neural Networks (DNNs) using both kind of approaches whether hybrid (HMM-DNN) approach or Tandem approach (Yoshioka & Gale 2014) have shown much improvement in both, speaker independent and Speaker adaptive

scenarios. Convolution Neural Networks (CNNs) are also helpful in further reducing the error rate as these networks use limited weight sharing scheme that can better model the speech features (Hamid et al. 2014). But most of the systems which are developed till now are for English or other foreign languages. Very less work has been done in Hindi and Other Indian languages.

In this paper we first review the various techniques available for speech recognition and their use within hybrid architecture. We then propose a speech recognition system for Hindi language based on PNN.

2 BACKGROUND

The main components of any speech recognition system are feature extraction, acoustic modeling, language modeling and searching algorithm. Among all these acoustic modeling is the most important component on which the success of any speech recognition system majorly depends. Modeling the acoustics focuses on defining the relationship between a speech signal and its linguistic units which can be phonemes, words or sentences. HMM was a dominant choice for evaluating the feature vectors obtained after preliminary processing and feature extraction from a speech signal. Several variants of HMM like CDHMM (Continuous density

Hidden Markov models), HMM-GMM were used for the purpose but main drawbacks involved were language characteristics, optimal number of Gaussian mixtures (X. Li. 2013). Therefore as more complex computational distributions and data resources are used in developing the ASR its accuracy is decreasing.

Machine Learning (ML) techniques like discriminative, active learning and Bayesian learning started being used for developing ASR systems also (X. Li. 2013). The relationship between the two came from the fact that ML is basically used to develop the systems after generalizing the relationships between previously taken data, inputs and expected outcomes to the system. Thus ML can be effectively used in ASR also.

Probabilistic neural networks have gained prominence in the area of pattern recognition, has many properties that makes it useful to be used for speech recognition as well. Speech recognition basically follows a statistical model in which highest probability word/sentence/phoneme has to be chosen given observation or speech signal (Rabiner & Juang 1986).

$$Outputword = \max_{W \in L} P(W \mid O) \qquad (1)$$

As we cannot estimate $P(W|O)$ Bayes rule can be used in above equation and finally it can be converted as

$$Outputword = \max_{W \in L} P(O \mid W)P(W) \qquad (2)$$

The probabilistic neural networks unlike other ANN is basically a statistical neural networks which is designed on Bayes decision strategy. Since PNN is based on Bayes decision strategy gives optimal decisions specially in case of Gaussian distributuions therefore it can be used for speech recognition.

3 PROPOSED METHOD

In this paper, we propose a method for speech recognition for Hindi language using PNN and MFCC, LPC for feature extraction. The flow of the proposed method is shown in Fig 1. The various steps of the method are discussed in the next sections.

3.1 Preprocessing

The speech signals acquired from different sources often contain some disturbances. These disturbances are modeled as noise and needs to be

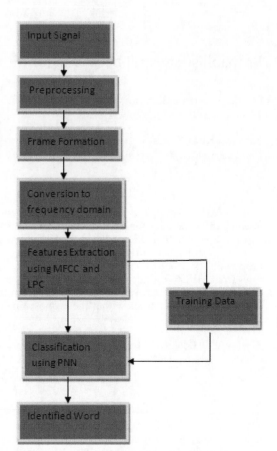

Figure 1. Basic methodology.

removed before further processing. In this paper, we applied the guassian noise removal filter to remove unwanted parts or noise.

3.2 Feature extraction

The speech is highly random in nature and thus temporal features alone cannot be used for recognition. Hence we select MFCC and LPC as they contain significant spectral features of the input speech signal.

These features are extracted as follows:

Step I: Take the input signal $X(n)$ Input Signal for "ek" is shown in Figure 2.

Step II: Divide the signal into K frames of duration of t ms with a framing shift of μms to get $X_i(n)$.

Step III: Compute the Fourier transform of the input signal using eq. 3.

$$X_f(K) = \sum_{n=0}^{N-1} X_i(n)h(n)e^{-\frac{2\pi kn}{N}} \quad k = 0,1,2 \dots\dots, N \ (3)$$

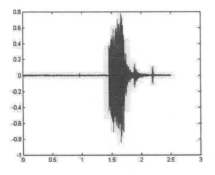

Figure 2.   Time vs Speech Signal for word "ek".

the output layer consists of K nodes where K is equal to number of resultant classes. A Gaussian function is associated with each hidden node corresponding to a particular class or pattern which is to be classified. At the output node the sum of all these Gaussian values is taken to identify the particular pattern. The likelihood of an observation belonging to a particular class in case of PNN can be calculated using eq 7.

$$O_i = \left[\frac{1}{(\sqrt{2\pi\sigma^2})^P}\right]\left(\frac{1}{Q}\right)\sum_{q=1}^{Q} exp\left\{-\frac{\|V-X\|^2}{2\sigma^2}\right\} \quad (7)$$

Here V is input feature vector, σ is standard deviation for Class i, P is dimension of input feature vector which is 24 in this case, Q is the number

where N is the size of input signal vector and $h(n)$ is hamming window.

Step IV: Apply m filter banks spaced according to Mel scale and calculate log of the resultant. This will give m filter bank energies.

Step V: Calculate DCT of the filter bank energies using eq 4. This will result in vector of size m. Only lower 6 are kept for recognition as they contain the major information of the signal.

$$V_j = \sum_{c=1}^{m} log(c)\cos\left[\frac{j\left(m-\frac{1}{2}\right)\pi}{m}\right] \quad j=1,2\ldots m \quad (4)$$

Step VI: To retain the temporal information in the signal calculate time derivatives of the signal i.e. delta and delta-delta MFCCs Using eq 5 (Arora 2013).

$$\Delta V(j) = \frac{\sum_{l=1}^{2} l(C[j+l]-C[j-l])}{2\sum_{l=1}^{2} l^2} \quad (5)$$

Here $V(j)$ is the jth MFCC Coefficient calculated earlier. Similarly delta-delta MFCC $\Delta^2 V_j$ vectors are also calculated for every input utternece.

Step VII: In the next step we compute six LPC of the input signal V_L.

Step VIII: The features obtained from step 4, 5 and 6 are combined to obtain the feature vector of size 24 features per frame using eq 6.

$$V = [V_j \Delta V_j \Delta^2 V_j V_L] \quad (6)$$

The resultant feature vector is used for training and classification purpose.

3.3   Recognition Using PNN

A probabilistic neural network has 3 layers of nodes as shown in Figure 4. Input layer consists of N nodes representing the input features and

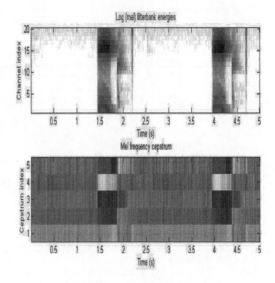

Figure 3.   Mel frequency cepstrum for word "ek".

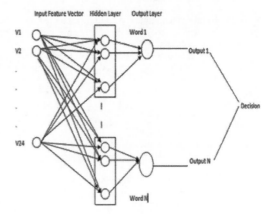

Figure   4.   Architecture   of   probabilistic   Neural   networks.

of Hidden nodes, X^q is the feature vector on which the system is trained.

In case of speech, utterances are series of observation vectors as described above. The vectors in the training set are labeled with the class of the particular word being spoken. Recognition is done based on output probabilities belonging to each class. The word which will be having higher probability among all the classes will be taken as the output word.

$$X \in \{\max O_i\} \, i = 1 \ldots \ldots K \tag{8}$$

where O_i is the output for each class.

4 RESULTS AND DISCUSSION

The proposed method was implemented in Matlab R2013a. Standard Wav files of Hindi speech from Linguistic Data Consortium for Indian Languages (LDC-IL) (Online available). Two speech signals of duration 1 minute with twenty spoken words were recorded and used for the experiments. The performance of the proposed method was tested in different scenarios. The accuracy is calculated using equation 9.

$$Accuracy = \frac{R}{S} * 100 \tag{9}$$

Where R = Number of correctly recognized words
S = Total number of Spoken Words

Case 1: The input signal and training signal were recorded from the same speaker. Results are shown in Table 1.
Case 2: The input signal was taken from different speaker. Results are shown in Table 2.

Table 1.   Results when input and training data are same.

Speakers	Number of spoken words	Recognized words	Accuracy
Speaker-1	20	18	90%
Speaker-2	20	19	95%

Table 2.   Recognition accuracy using different speakers.

Speakers	Number of spoken words	Recognized words	Accuracy
Speaker-1	20	15	75%
Speaker-2	20	16	80%

Table 3.   Recognition accuracy using MFCC and LPC only.

Speakers	Number of spoken words	Recognized words	Accuracy
Speaker-1	20	14	70%
Speaker-2	20	14	70%

Table 4.   Recognition accuracy using Different Methods.

Database	Method	Number of spoken words	Recognized words	Accuracy
LDC-IL	HMM	20	16	80%
Databse	PM	20	19	95%
	SVM	20	18	90%
Speaker 1	HMM	20	15	75%
	PM	20	16	80%
	SVM	20	16	80%
Speaker 2	HMM	20	14	70%
	PM	20	17	85%
	SVM	20	16	80%

Case 3: Accuracy was calculated only using MFCC features with LPC. Results are shown in Table 3.

The results obtained from the Proposed Method (PM) were compared with the other existing methods namely Hidden Markov Models (HMM). Table 4 shows the results obtained.

It is observed that the results obtained from the PM have higher accuracy as compared to others.

5 CONCLUSION AND FUTURE WORK

In this paper we proposed a method for Hindi Speech Recognition using PNN. MFCC, first and second order derivatives of MFCC and LPC features were used for the recognition. It was observed that results which were obtained using the proposed method were better as compared to other available methods like HMM. In this paper the focus was only on word detection so the work can be extended to recognize continuous speech.

REFERENCES

Arora, Sanghmitra, V. 2013. Effect of time Derivatives of MFCC Features on HMM Based Speech Recognition System, *ACEEE Internation Journal on Signal and Image Processing*, vol. 4, no. 3, pp: 50–55.

Baker, J.M., Deng, L., Glass, J., Khudanpur, S., Lee, C., Morgan, N. & O'Shaugnessy, D. 2009. Research developments and directions in speech recognition and understanding, part 1, *IEEE Signal Process. Mag.*, vol. 26, no. 3, pp: 75–80.

Baker, J.M., Deng, L., Glass, J., Khudanpur, S., Lee, C., Morgan, N. & O'Shaugnessy, D. 2009. Research developments and directions in speech recognition and understanding, part 2, *IEEE Signal Process. Mag.*, vol. 26, no. 4, pp: 78–85.

Dengand, L., Li, X. 2013. Machine learning paradigms for speech recognition: An overview, *IEEE Trans. Audio, Speech, Lang. Process.*, vol. 21, no. 5, pp: 1060–1089.

Hamid, O.A., Mohamed, A.R. & Jiang, H. 2014. Convolution Neural Networks for speech recognition, *IEEE/ACM transactions on Audio Speech and Language Processing*, vol. 22, no. 10, pp: 1533–1545.

Mohamed, A., Sainath, T., Dahl, B. Ramabhadran, G. Hinton & Picheny, M. 2011. Deep belief networks using discriminative features for phone recognition, *in Proc. IEEE Int. Conf. Acoust., Speech, Signal Process. (ICASSP)*, pp: 5060–5063. Online Availaible: http://www.ldcil.org/resourceSpeechCorp.aspx accessed on 2–2–2016.

Rabiner, L.R. & Juang, B.H. 1986. An introduction to Hidden Markov Models, *IEEE Signal Process. Mag.*, pp: 4–16.

Sinha, S., Aggarwal, S.S., Jain, Aruna., Continuous Density Hidden Markov Models for Context Dependent Hind Speech Recognition. 2013, *ICACCI*, pp: 1953–1958.

Yoshioka, T., Gales, M.J.F. 2014. Environmentally robust ASR for deep neural network acoustic models, *Computer Speech and Language*, pp: 65–86.

Communication and Computing Systems – Prasad et al. (Eds)
© *2017 Taylor & Francis Group, London, ISBN 978-1-138-02952-1*

Application of artificial neural network in traffic noise pollution modeling

Rajeev Kumar Mishra & Anuj Kumar
Delhi Technological University, Delhi, India

Kranti Kumar
School of Liberal Studies, Ambedkar University Delhi, Kashmere Gate, India

ABSTRACT: Noise is a global problem due to several factors: increase in demographic density and the number of per capita vehicles, appliances and vehicles capable of generating loud noise, and also the fact that society is getting used to higher noise levels. In this study Artificial Neural Network (ANN) has been applied to predict noise pollution level in Delhi, capital city of India. Factors that predominantly influence noise pollution level in a traffic noise model framework were classified into two categories: traffic volume and traffic speed. Volume, speed and noise level data of traffic were collected at six identified locations in the city. Structure of the proposed model consists of input variables as two wheelers (2W), three wheelers (3W), Car, Jeep, Van, Bus, Truck, corresponding traffic speed on both sides of the road and L_{eq} as output variable. The best model was selected on the basis of Mean Square Error (MSE). Results of the study reveal that artificial neural networks can be a useful tool for the prediction of noise level with sufficient accuracy.

1 INTRODUCTION

This study has been motivated by the fact that present road traffic noise prediction models have not been improved significantly since their development in the 1970s and 1980s, although traffic noise nuisance is a significant and growing issue in India and abroad as the number of vehicles on roads are increasing day by day.

Due to increasing motorization, construction of flyovers and growth in transport network, noise level has exceeded the prescribed limits in many Indian cities (CPCB, 2000). Adverse effects of high noise levels on human health are being identified as hypertension, sleeplessness, mental stress, etc. because of these physiological/psychological effects of noise level, it is important to assess the impact of traffic noise on residents and road users. Noise from roads depends on type of traffic, number of vehicles, number of heavy vehicles, location and the height of buildings and other noise barriers and the condition of surface of the roads. Excess noise pollution commonly occurs in predictable settings like industry, near traffic or transportation systems and impacts individuals living in such areas. In this study noise pollution has been chosen because it is believed that low income populations, accepts to live in less desirable housing conditions and can compromise living in areas where noise pollution is a prominent issue due to vehicular traffic.

The present work is an effort to quantify and analyze the traffic noise emissions in various locations in Delhi and to investigate, whether a neural network can be used for the prediction noise level in the case of Indian conditions, where most of the traffic is heterogeneous in nature. Field measurements were carried out to understand and assess various aspects of the impact of noise on the social lives of residents and road users in Delhi.

2 LITERATURE SURVEY

Literature survey was carried out to investigate the use of ANN for traffic noise modelling cited in literature. Dougherty (1995) presented a review on ANNs use in the field of transportation. Use of ANN for the first time in the field of traffic noise modeling was carried out by Cammarata et al. (1995). They proposed a neural architecture made up of two cascading levels. At first level a supervised classifying network, named Learning Vector Quantization (LVQ) network, filters the data, discarding all wrong measurements, while on the second level Back Propagation Network (BPN) predicts the sound pressure level. Results obtained by the comparison of the BPN approach with those provided by selected relationships found in relevant literature, it was concluded that ANN based BPN model was capable of predicting traffic noise more accurately and effectively as compared to empirical and regression models. Gupta et al. (1981) analyzed the traffic noise pollution for mixed traffic flow for various land uses in Roorkee city. A computer program was developed in FORTRAN IV to evaluate the noise parameters. From the study it was found the value of 'K' reported by Robinson does not match for mixed traffic flow (highway heterogeneous flow) and varies between 2.37 to 3.54 as against value of 2.56.

Srivastava et al. (1995) studied the environmental noise pollution level on the NH-45 between Ghaziabad and Roorkee in continuation of their earlier study. For this purpose they developed a computer software package named "Traffic Noise Analysis Package" (TNAP). This package has got three different alternative options. Using first option the various noise parameters like L_{10}, L_{50}, L_{90} and L_{eq} etc. may be calculated. Second option is for predicting the noise level L_{eq}. L_{eq} can be predicted for a given classified volume per hour and at a desire distance (in m) from the centre of traffic flow. The third option provides facility to obtain the combine noise level of a mixed traffic flow stream by giving their individual noise levels as input data.

Wu and Zhang (2000) concluded that the artificial neural network provides a new method for traffic noise prediction. Based on the measured traffic noise data near a highway in China, an artificial neural network model was formed to predict further traffic noise levels. It was found that the predicted data from ANN model were in good agreement with measured data.

ANNs have been used as predictors for many regression problems. Knowing how well predictions match the real world is crucial to some of them, and so many research groups have developed strategies to tackle this problem (Webera et al., 2003). A statistical model of road traffic noise in an urban setting which is based on the fact that percentage of heavy vehicles plays an important role over road traffic noise emission was developed by Calixto et al. (2003). Genaro et al. (2010) developed a neural network based model for urban noise prediction. In this study 12 street locations with different characteristics of the city of Granada (Spain) were selected to obtain a representative sample of the complexity of urban streets with presence of road traffic. A set of 289 data vectors, each one with 26 components, was obtained. A total of 25 input variables were used (Torija et al., 2010a), being the only output variable the A-weighted energy-equivalent sound pressure level (L_{Aeq}). Results obtained from ANN modelling were compared to those obtained with mathematical models. It was found that the proposed ANN system was able to predict noise level with greater accuracy.

Yıldırım et al., (2008) presented a noise analysis of passenger cars using Artificial Neural Networks (ANNs). The research comprises the experimental analysis of car's engines and simulation analysis of noise parameters using ANNs. Taghavifar et al. (2014) showed that the ANN advantages are fast, precise and reliable computation of multi-variable, non-linear and complex problems as compared to the conventional mathematical and numerical methods. Mishra et al. (2010) made an effort to quantify and analyze the traffic noise emissions along bus rapid transit corridor in Delhi.

Kumar et al. (2011) summarize the findings of research concerning the application of neural networks in traffic noise prediction and critically reviewed various neural network based models developed for road traffic noise prediction cited in the literature. They concluded that ANN based models were capable of predicting traffic noise more accurately and effectively as compared to deterministic and statistical models reported in various studies. Recent study by Sharma et al. (2014) have also tried to model the traffic noise by using ANN and regression approach. Garg and Maji (2014) presented an exhaustive review of principal traffic noise models with critical discussion and comparison of the traffic noise models developed and adopted in various countries.

3 DATA COLLECTION

To measure the traffic noise pollution level, first task was site selection. So, after performing several surveys of different areas and nature of noise problem in Delhi, six sites were selected where continuous uninterrupted flow of vehicles was occurring. Sites selected were Karol bagh, Uttam nagar, Punjabi bagh, Janpath, Laxmi nagar, Madhuban chowk. The map of Delhi showing the selected locations for field studies is presented in Figure 1. Following parameters were collected during Data collection.

1. Traffic volume studies were carried out at all the locations. At all the sites, traffic volume studies were conducted continuously for a period of twelve hours. Traffic volume data were manually recorded for every hour.
2. Traffic speed study was carried out continuously for a period of twelve hours at all the identified locations. Classified traffic speed study was carried out for both the directions using radar gun.
3. Meteorological parameters which have a major effect on noise pollution are wind speed, wind direction, mixing height, ambient temperature, and humidity. Data about these parameters on each study day was obtained from Indian Meteorological Department, New Delhi.

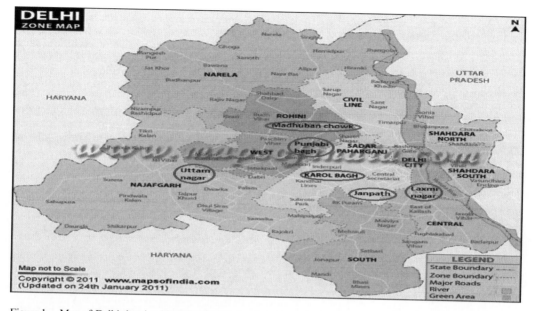

Figure 1. Map of Delhi showing identified locations of study.

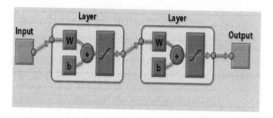

Figure 2. Structure of the proposed neural network.

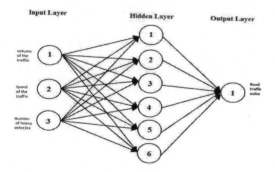

Figure 3. Structure of single layer ANN.

4. Ambient noise pollution data was collected continuously for a period of twelve hours at all the identified locations using Noise Level Meter.

4 ANN MODELLING OF NOISE POLLUTION

4.1 MATLAB programming

MATLAB was used to build neural network architectures. Various networks were created by varying the parameters like number of neurons, transfer function, number of epochs. We have used feed forward back propagation network (Sivanandam et al. 2006) as it is a systematic method for training multi-layer network. Levenberg Marquardt learning technique has been used in the current study. The structure of proposed neural network built by using MATLAB is shown in Figure 2.

Road traffic noise was calculated in the batch experiments as a function of traffic volume and speed. ANN model based on single layer recurrent back propagation algorithm for the data, generated from the above experiments was applied to train the Neural Network. During training, the output is computed by a forward pass in which the input is propagated forward through the network to compute the output value of each unit. The output is then compared with the desired values which gives errors for each output unit. In order to minimize the error, appropriate adjustments were made using different weights for the network. After several such iterations, network was trained to give the desired output for a given input. The single layer network structure having six hidden neurons, describing the dynamics of road traffic noise is shown in Fig. 3. The performance of network simulation was evaluated in terms of Mean Square Error (MSE), Mean Absolute Error (MAE) and coefficient of correlation.

4.2 Data parameters

Two inputs i.e. classified traffic volume and classified traffic speed has been used as input parameters to model the network, whereas L_{eq} has been considered as an output for all networks. The details of all the locations, the total data sets in the zone and the division of the total data sets into training and testing are given in Table 1.

Two inputs i.e. classified traffic volume and classified traffic speed has been used as input parameters to model the network, whereas L_{eq} has been considered as an output for all networks. The details of all the locations, the total data sets in the zone and the division of the total data sets into training and testing are given in Table 1.

4.3 Model development

Vehicles were divided in five categories (car/jeep/van, three wheelers, scooter/motor cycle, Mini bus/bus, Mini truck/truck of both the sides of roads) and the average hourly spot speed of each vehicle was calculated.

Table 1. List of complete data parameters used in ANN model.

No. of study locations	6
Total no. of data sets	66
No. of data sets used for training	54 (81%)
No. of data sets used for testing	12 (18%)

Traffic volume of both the sides was considered. The number of inputs parameters was ten (five on each side) when taking only classified traffic volume as input, and twenty (ten on each side) when taking classified traffic volume and classified traffic speed of each vehicle individually as input.

The number of obtained output was only one, i.e. L_{eq}. Details of all the networks made with Mean Square Error (MSE), Mean Absolute Error (MAE), coefficient of correlation (r) for classified traffic volume as input is given in Table 2 and for the classified traffic volume and classified traffic speed both as input is depicted in Table 3. The final mean square error is the one obtained after the network stops training based on the specified termination criterion.

4.4 Results and discussion

Networks formed are able to show good results when both the classified volume and classified speed were considered as an input.

It is observed from the analysis that different land use pattern has a different effect on the traffic noise generation with the same inputs.

This would be due to the different characteristics of traffic passing through different areas.

In case of classified traffic volume and classified speed the best results was of network 1 with 3 numbers of hidden neurons: with 6 epochs.

The value of r was 0.89279 and MSE was 0.03349. In case of classified volume as input the best result was of network 4 with 6 numbers of hidden neurons: with 10 epochs. The r value was found to be 0.87042 and MSE value was 0.0260.

It is also clear that results obtained in the case when both classified traffic volume and classified traffic speed were used as inputs were better than

Table 2. Details of networks with classified traffic volume as input parameter.

Trial No	Train 1	Train 2	Train 3	Train 4	Train 5	Train 6	Train 7	Train 8
No. of hidden layers	1	1	1	1	1	1	1	1
No. of hidden neurons	3	4	5	6	7	8	9	10
Transfer Function	Sigmoid	Sigmoid	Sigmoid	Sigmoid	Sigmoid	Sigmoid	Sigmoid	Sigmoid
Number of epochs	13	11	9	10	11	7	24	10
Learning	Levenberg	Levenberg	Levenberg	Levenberg	Levenberg	Levenberg	Levenberg	Levenberg
MSE	0.0371	0.0305	0.0273	0.0260	0.0819	0.0357	0.0473	0.04319
MAE	0.0168	0.0187	0.0243	0.0154	0.0268	0.0162	0.0189	0.02295
R	0.838816	0.86037	0.754848	0.87042	0.819903	0.816697	0.849937	0.85354

Notes: MSE – Mean Square Error; MAE – Mean Absolute Error; r – coefficient of correlation

Table 3. Details of networks with classified traffic volume and classified traffic speed as input parameters.

Trial No	Train 1	Train 2	Train 3	Train 4	Train 5	Train 6	Train 7	Train 8	Train 9
No. of hidden layers	1	1	1	1	1	1	1	1	1
No. of hidden neurons	3	4	5	6	7	8	9	10	11
Transfer Function	Sigmoid	Sigmoid	Sigmoid	Sigmoid	Sigmoid	Sigmoid	Sigmoid	Sigmoid	Sigmoid
No. of epochs	6	6	9	11	7	6	6	11	8
Learning	Levenberg	Levenberg	Levenberg	Levenberg	Levenberg	Levenberg	Levenberg	Levenberg	Levenberg
MSE	0.03349	0.04036	0.04342	0.05716	0.03500	0.04268	0.03935	0.03946	0.03503
MAE	0.01709	0.01845	0.02536	0.02525	0.02320	0.02227	0.01857	0.02288	0.01849
R	0.892792	0.757996	0.819661	0.829879	0.85949	0.87959	0.847829	0.883784	0.843735

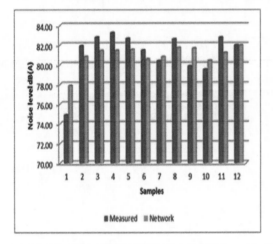

Figure 4. Measured and predicted (by ANN model) noise levels.

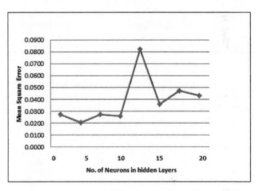

Figure 5. Variation of MSE with nunber of hidden layer neurons.

the results of only classified traffic volume alone. Figure 4 shows comparative analysis of measured and predicted (by ANN model) noise levels.

From figure it is clear that noise levels calculated by ANN are in good agreement with the measured noise level data.

Variation of MSE with no. of hidden layer neurons has been shown in Figure 5, from where it is clear that initially increasing number of hidden layer neurons MSE decreases but after the number of hidden layer neurons increases more than 5 MSE increases always whether sometimes abruptly and sometimes moderately.

5 CONCLUSION

From the study conducted at different locations in Delhi, it is observed that different locations have different effects on the traffic noise generated as shown in the results with the same inputs in each location. It is clear that the networks were able to produce better results when both classified speed and classified traffic volume were used as input instead of classified volume alone, thus indicating that speed of each category of vehicles plays a significant role in noise generation. Moreover current study reveals that road transportation noise prediction model can be used to reduce the traffic noise level by designing noise absorbent systems on the basis of noise prediction.

REFERENCES

Calixto A., Diniz, F.B. & Zannin, P.H.T. 2003. The statistical modeling of road traffic noise in an urban setting. *Cities* 20 (1): 23–29.

Cammarata, G., Cavalieri, S. & Fichera, A. 1995. A neural network architecture for noise prediction. *Neural Networks* 8: 963–973.

CPCB 2000. Noise Standards—Central Pollution Control Board. Available at: http://www.cpcb.nic.in/Noise_Standards.php

Dougherty, M. 1995. A review of neural networks applied to transport. *Transportation Research Part C* 3(4): 247–260.

Garg, N. & Maji, S. 2014. A critical review of some principle traffic noise models: Strategies and Implications. *Environmental Impact and Assessment Review* 46: 68–81.

Genaro, N., Torija, A., Ramos-Ridao, A., Requena, I., Ruiz, D.P. & Zamorano, M. 2010. A neural network based model for urban noise prediction. *Journal of Acoustical Society of America* 128: 1738–1746.

Gupta, A.K., Nigam & Hansi 1981. A study on traffic noise for various land use for mixed traffic flow. *Indian Highways,* Indian Road Congress, New Delhi.

Kumar, K., Parida, M. & Katiyar, V.K. 2012. Road Traffic Noise Prediction with Neural Networks—A Review. *An International Journal of Optimization and Control: Theories and Applications* 2(1): 29–37.

Kumar, P, Nigam, S. P. & Kumar, N. 2014. Vehicluar traffic noise modelling using artificial neural network approach. *Transportation Research Part C* 40: 111–122.

Mishra, R.K., Parida. M. & Rangnekar, S. 2010. Evaluation and analysis of traffic noise along bus rapid transit system corridor. *International Journal of Environmental Science and Technology* 7(4):737–750.

Sharma, A., Bodhe, G. L. & Schimak, G. 2014. Development of a traffic noise prediction model for an urban environment. *Noise Health* 16: 63–67.

Sivanandam, S. N., Sumathi, S. & Deepa, S. N. 2006. *Introduction to Neural Networks Using Matlab 6.0.* Tata McGraw-Hill Publishing Company.

Srivastava, J. B., Gupta, A. K. & Khanna, S. K. 1995. Air pollution modeling of highway traffic. Proc., ICORT-95, *University of Roorkee, Roorkee* 1264–1273.

Taghavifar, H. & Mardari, A. 2014. Application of artificial neural networks for the prediction of traction parameters. *Journal of Saudi Society of Agricultural Sciences* 13 (1): 35–43.

Torija, A. J., Genaro, N., Ruiz, D. P., Ramos-Ridao, A., Zamorano, M. & Requena, I. 2010a. Priorization of acoustic variables environmental decision support for the physical characterization of urban sound environments. *Building and Environment* 45 (6): 1477–1489.

Webera, K. E., Schlagnerb, W. & Schweierc, K. 2003. Estimating Regional Noise on Neural Network Predictions. *The Journal of the Pattern Recognition Society* 36(10): 2333–2337.

Wu, S. & Zhang, J. 2000. The application of neural network to the prediction of traffic noise. *International Journal of Acoustics and Vibration* 5(4): 179–182.

Yıldırım, S., Erkaya, S., Eski, I. & Uzmay, I. 2008. Design of neural predictor for noise analysis of passenger car's engines. *Journal of Scientific & Industrial Research* 67(5): 340–347.

Communication and Computing Systems – Prasad et al. (Eds)
© 2017 Taylor & Francis Group, London, ISBN 978-1-138-02952-1

A survey on the applicability of fourier transform and fractional fourier transform on various problems of image encryption

Priyanka Maan & Hukum Singh
The Northcap University, Gurugram, India

ABSTRACT: As data security is of utmost importance to every individual in some way or other. This letter presents the innovation in the image encryption techniques form the past time incorporating the use of Fourier transform and fractional Fourier transform in various ways by different authors. Many aspects of the usage of these transforms are explored here in this work.

1 INTRODUCTION

Security is the basic requirement of every human being in day to day life in various tasks. Information in modern world deals with electronic devices and the internet in multimedia form. The safe and secure storage and communication of data over the internet is essential. Most of the data is transmitted in the form of images so the image encryption algorithms come in view. These algorithms convert the image to be stored or transmitted into unintelligible form so that the unauthorised users will not be able to gain any knowledge from them. The conversion of the original image into the unreadable image known as encrypted image by implementing an algorithm and special set of characters known as key to the input image. The decryption process is normally the reverse of encryption process. There are number of digital and optical encryption techniques and cryptosystem developed in the past decades for the purpose of protected and unharmed data storage and transmission. Digital techniques uses various cryptographic algorithms based on stream and block cipher concepts. In the optical field of encryption the classic approach used relies on Double Random Phase Encoding (DRPE) system in which two random phase masks are deployed as the keys for both encryption and decryption process. One of them is applied in the input plane and the second one in the Fourier plane proposed by Refregier & Javidi (1995). A large number of optical techniques for image encryption developed after the DRPE come into existence are energized by the architecture of DRPE. Some of them are discussed in the literature section.

In the field of optical information processing for providing reliable security of data, the optical encryption techniques played a crucial and vital role. In this letter we mainly focus on the use of Fourier and fractional Fourier transforms for the image encryption.

Fourier Transform (FT) comes from the concept of Fourier series. FT is used to convert the original signal in time or spatial domain to the frequency domain. According to FT any signal can be said to be a combination of sine and cosine signals. FT is considered as an important image processing tool that attempts to break down the given image into its sine and cosine components and the output transformed image is in frequency domain. It is linear and reversible transform. Fractional Fourier Transform (FrFT) is derived from the FT by the addition of a scaling parameter say 'a'. When the order a = 1, FrFT is FT itself. Rotation is done in time-frequency domain by the use of FrFT. It has found to have several applications in the field of optical image processing. It adds one more parameter of security that is the fractional order in the encryption schemes.

The rest of the paper is constituted as the literature demonstrating the use of FT and FrFT in the context of image encryption. Many techniques and approaches developed in the last two decades are surveyed in this paper. Their related pros and cons are discussed. The security provided by the proposed works is examined in this work and the comparison of some approaches differing in the key space for the encryption algorithm is also presented here. The results shows which techniques are more resistant to attacks and hence offers more security.

2 REVIEW OF ENCRYPTION TECHNIQUES BASED ON FOURIER AND FRACTIONAL FOURIER TRANSFORMS

The authors Refregier & Javidi (1995) designed a new optical method for encoding of images for various security applications. The encoding is done using two random phase masks, one in the input plane that makes the input image a white noise and the other in the Fourier plane that makes the image stationary and perform encoding. The reconstruction of primary image from the stationary white noise is robust and the optical efficiency is quite high against blind deconvolution. The method proposed is one of the novel method at that time and is good for high security applications.

An algorithm is proposed for Fractional Fourier Transform (FrFT) calculation using Fast Fourier Transform (FFT) algorithm for one dimensional signals in the paper authored by Garcia et al. (1996). The authors also discussed about DFT, Fresnel – Diffraction calculation, FrFT numerical and digital computations. The proposed algorithm has complexity [Log (N)] as it uses two FFT's in cascade and three quadratic phase factors multiplied by it. The method work for fractional orders in range −1 to 1. FRFT and Fresnel diffraction scaling factors have been presented. Results for two dimensional signals can also be calculated using the mentioned algorithm.

Bahram Javidi (1997) discussed the need for securing information and communication over computers and achieving the security using optical techniques in the letter. These techniques has several advantages over the digital one for processing information like parallel processing of data, hiding information in phase, frequency, wavelength and other dimensions which are detailed in this paper. Optical encryption techniques are more preferred over the electronic processors as they provide more secure and robust encrypted image using phase encoding for security in different domains using Fourier transform is also discussed here. Future scope for fulfilling more security needs using optical processing method is presented here.

A new method using double random phase encoding (DRPE) in fractional Fourier domain is proposed by Unnikrishnan & Singh (1999). The method adds more security by addition of fractional order parameter 'a' along with the key for decryption procedure. For two dimensional input data anamorphic orders based fractional Fourier transform is also discussed. Robustness and fault tolerance properties of the techniques are discussed with respect to distortions in encrypted image like occlusion, noise and quantization.

The authors Unnikrishnan & Singh (2000) presented a technique using fractional Fourier transform to optically decrypt the input data using two random phase masks into white noise. Different orders fractional transform is taken in the input, encryption and output plane which needs to be correctly specified along with the key during decryption otherwise the Mean Square Error (MSE) increases. Both numerical simulation and optical implementation are carried out.

The new technique for data security by encoding the data using Quadratic Phase System (QPS) distribution in the input, encryption and output plane in the fractional Fourier domain based double random phase encoding is presented by Unnikrishnan et al. (2000). In addition with the two random masks, six other FrFT parameters must be used for the encrypted image which enhances the security of the system as guessing the QPS parameters values correctly has quite low probability. This method can be used to provide security in holographic storage.

The optical encryption scheme proposed earlier which uses two random phase masks as keys is prone to Known—Plaintext Attack (KPA) is shown by Peng et al. (2006). The counterfeiters can access the keys using phase retrieval techniques as a plaintext-cipher text pair is already known in KPA. The authors use Hybrid Input-Output algorithm (HIO) and the Sum-squared-error concept and proved the vulnerability to KPA.

The use of fractional Fourier transform for color digital images performing phase encryption is demonstrated in the letter by Vilardy et al. (2006). The method is much more complex due to addition of more parameters to work as key i.e. three fractional orders and two random keys which are 5 in total and three times fractional transformed is applied. To properly recover the image all keys must be used exactly and incurs difficulty for the unauthorised cryptanalyst.

A new scheme for encoding two different images to one stationary white noise like encoded image by employing 2-D fractional Fourier transform and random phase encoding has been put forward by Tao et al. (2007). The key space is increased by introduction of pixel scrambling operation to serve as a key in addition to random masks and fractional order. During encryption process, one image is taken as amplitude based image and another as phase based image. The proposed method is applied on combination of these images to create a complex combined signal. As pixel scrambling is only applied to phase-based image, so the one among two input image which requires higher security can be converted into pure phase function. The results of numerical simulation shows that the phase based image is more effected by noise perturbation and the susceptibility of the varying fractional orders along with robustness against data loss during encryption is also proved.

The authors Pei and Ding (2007) derives the eigen functions for the Fourier Transform (FT), the Fractional Fourier Transform (FRFT) and the Linear Canonical Transform (LCT) in this letter with both complex and offset parameters. Non offset FT and FrFT eigen functions are Hermite-Gaussian functions. Firstly for the offset Discrete Fourier Transform the eigen vectors are defined and then generalisation is done. The examples show that "smoothed" Hermite-Gaussian functions represent the shifted and modulated eigenfunctions of offset FT. With certain parameters offset LCT represents Fresnel Transform. The derived eigenfunctions have applications in mode selection, resonance analysis etc are also discussed here.

The various fields where the Fractional Fourier Transform (FrFT) can be applied are initially mentioned in the work by Lai, Liang & Cui (2010). As the need of secret data for secure transmission over internet is increased, here the Discrete Fractional Fourier Transform (DFrFT) is calculated using matrix algorithm and application of FrFT on a digital image is shown along with comparison of results by using both correct and incorrect decryption parameters.

In the proposed letter by Du et al. (2009) a notion of mixed Fourier transform is presented. It says that differences between the time and frequency domain are vanished when the images and signals are transformed in time-frequency domain. The idea is presented for both discrete and continuous time signals. The approach of discrete Mixed Fourier transform allows us to determine the square roots and high degree roots of various transformations like Fourier, Cosine, and Hadamard transformations etc which have special applications in image encryption. A more secure and complicated image of original image can be formed by practicing different parameters of 1-D mixed transform as the key in the encryption process.

A digital cryptographic algorithm for digital images that rest on the concepts of fractional Fourier transform and chaotic system is invented by Zhang & Zhao (2009). For obtaining the secret image the application of double random phase encoding based on fractional Fourier transform on the input image succeeded by encryption using confusion matrix originated using chaotic system is done. The security parameters constitute four fractional orders varying from 0 to 4 and two random phase masks. The total key space of the proposed scheme is quite large i.e. 10^6 which makes it robust against brute force attack.

The manifesto presented by Tao et al. (2010) is an innovative scheme for image encryption by using multi orders of FrFT. The input image is divided into sub images of equal size and the summation of 2-D Inverse Discrete Fractional Fourier Transform (IDFrFT) of the interpolated images results in the cipher image. Due to non orthogonal nature of Kernel functions of multi order FrFT, the whole transform orders are required for decoding of each sub image. The key used are the transform orders of adopted IDFrFT and even the slight deviation in them causes large errors. The unique feature of the scheme is that it can work for the encoding of more than two images and provide more security due to very large key space. Fast Fourier Transform (FFT) realisation can be done for the discussed one.

In the study of digital images importance and demand for the protection of images during transmission from attackers a technique that compares the performance of application of Discrete Fractional Fourier Transform (DFrFT) and Discrete Fractional Cosine Transform (DFrCT) for image encryption and decryption is explored in the work presented by Jindal & Singh (2010) and shows that DFrCT provides better results. In every stage different fractional orders are used which serves as additional parameter for encryption in addition with random masks. PSNR value is also calculated and shows the effect of usage of even one incorrect key during decryption.

A new cryptographic scheme that depend upon Phase-Truncated Fourier Transform (PTFT) and uses different keys for encryption and decryption is designed by Qin & Peng (2010) which produces real valued encrypted text which is a stationary white noise. The asymmetric scheme uses public keys (two random phase masks) for encryption and private keys (phase keys) for decryption depicting the non linearity. The phase reservation operation preserves the decryption keys. Contrasting feature of PTFT with respect to DRPE is studied here. PTFT is safe against certain types of attack possible on DRPE scheme, so asymmetric cryptosystems are more secure and powerful.

The authors Feng et al. (2011) explored an advanced cryptographic algorithm which performs double image encryption by employing Discrete Fractional Fourier Transform (DFrFT), chaotic sequences and magic cube based scrambled rotated function in the time-frequency domain. The key space is enlarged by using 4 keys: 2 RPM's and 2 scrambling keys each of which appropriate values are required for correct decryption of the ciphered image. The performance of the proposed algorithm is proved better in contrast to only DFrFT based encryption through results analysis.

The outline of digital watermarking and the properties and characteristics of Fractional Fourier Transform (FrFT) are given in introductory part of the paper written by Bo et al. (2011). To prevent the unauthorised access and extraction of watermark a watermarking algorithm is presented here using FrFT. The fractional orders work as encryption parameters to get the watermark embedded digital image. This encrypted image exploits the advantage of time and frequency domain characteristics of FrFT and hence yields more secure and robust technique than the conventional ones.

A new scheme to produce two independent keys used for decryption namely universal and special key employing asymmetric phase-truncated Fourier transform method is proposed by Qin et al. (2011). The universal key is generated by applying phase reservation operation on the Fourier transform of the encryption key and therefore can be used for decrypting any cipher text encoded by the encryption key but with poor discernability. The special key can be obtained during the encryption process from the phase reservation operation on the Fourier transform of the plaintext, hence can perform decryption of only the one cipher text related to the corresponding plain text legibly. Anyone of them can be separately used for decryption process.

In this paper proposed by Alam & Pandey (2011), an analogy of the comparison of the random phase masks and chaotic functions based masks for color images is carried out build on fractional Fourier transform. The technique presented in the work take advantage of 16 parameters (12 fractional orders, 4 seed values for random phase masks generated using chaos function) to work as decryption key. The results shows that the MSE values of chaos phase masks is much greater than the random masks for all three channels for color images. Hence, worthwhile the random masks deliver better results for color images.

The authors Elhosany et al. (2012) schemed a new cryptographic algorithm that apply 2-D chaotic Baker map in fractional Fourier domain with three variations of operation mode each having own power i.e. Cipher Block Chain which offers advantage of added security, noise resistant and fast operation; Cipher Feed-Back which gave best results in every term and the Output Feed-Back mode which exhibit preference over processing time and not subject to noise. The proposed scheme passes all tests of security and comes out to be a solid, robust and secure method.

A new methodology for confidential transmission of information is detected which relies on the decorrelation property and the real valueness of the reality-preserving Multiple-Parameter Fractional Fourier Transform (MPFrFT) is presented by Lang (2012). The method is resistant to statistical attack, follows the Shannon's principle of Confusion and Diffusion and implementation on real valued input provides real valued output itself which are easy to store and display. Hence the presented method is efficient and reliable.

The authors Cui et al. (2013) originated a new approach in for encryption that hinge on block based transformation on the input image by dividing the input image into blocks of small size and then rearrangement or shuffling of the positions of blocks are done by transformation algorithm. The output image is than fractionally transformed and cipher image is produced. The algorithm support more key space than traditional one, increases the entropy and limits the correlation among pixels of an image.

A new and simple method of encryption based on phase-truncated Fourier transform that use Random Amplitude Mask (RAM) to prevent it from amplitude-phase retrieval attack is detailed in the paper authored by Wang & Zhao (2013). The encryption process is non linear in contrast to decryption process which is linear. The two private keys are highly secure and provide reliability and are based on RAM.

An interesting technique to encrypt two images in fractional Fourier domain by converting the first image into phase only image by application of modified Gerchberg -Saxton algorithm which than act as key for the fractional domain for encrypting the second image using the DRPE scheme is searched out in the work authored by Rajput & Nishchal (2013). The unique key generation process adds more security. The concept of Known Plaintext Attack is used during decoding.

Extracting the original image by employing attack on DRPE based image hiding technique is detected in the letter written by Xu et al. (2014). The stego-image is obtained by superimposing the host image with the DRPE encoded image. The super imposition coefficient is chosen at random to find the original one and chosen-plaintext attack to find the spatial and frequency domain secret key for retrieving the original image. The results show the image retrieved is the correct one.

The presented work by Dahiya et al. (2014) exercise the combination of Fourier and fractional Fourier transform and uses the 4 random quad phase masks, 2 fractional order as the secret key and ensures the goals of security. A case study is done by dividing the image into four parts and applying different algorithms on them and finds the one which executes in minimum time and provides best results.

3 COMAPRISON OF ENCRYPTION APPROACHES

Table 1. List the discussed approaches along with advantages of each.

Publication/Year	Title	Technique used	Advantages
1995, Optics letters	Optical image encryption based on input plane and Fourier plane random encoding	DRPE in Fourier domain, two Random Phase Masks (RPM) used in input and Fourier plane	The method is tough against Blind deconvolution. Key space constitutes 2 RPM's.
2000, Optical engineering,	Double random fractional Fourier-domain encoding for optical security.	DRPE in fractional Fourier domain	Key space is extended by addition of fractional order as a key with 2 RPM's. So security extended.
2000, Optics letters	Optical encryption by double-random phase encoding in the fractional Fourier domain.	DRPE in fractional Fourier domain with optical distribution in Quadratic Phase System (QPS).	3 planes of QPS, 3 FrFT parameters and 2 random masks serves as key. All keys are necessary for successful decryption. So more secure.
2006, IEEE	Digital images phase encryption using fractional Fourier transform.	Phase Encryption using fractional Fourier Transform (FrFT) for color images.	Decryption becomes complicated because 5 keys musts be used i.e. (2 RPM's and 3 fractional orders)
2007, Optics express	Double image encryption based on random phase encoding in the fractional Fourier domain.	Encryption of 2 input images into one by application of DRPE using FrFT.	Robustness against loss of data. Extra degree of freedom included as key space constitutes 2 FrFT order, 2 RPM's, Pixel scrambling operator.
2010, IEEE	A novel image encryption algorithm based on fractional Fourier transform and chaotic system.	Fractional Fourier transform and Chaotic system	Enhance the security due to sensitivity of transform orders and use of logistic maps. 6 keys (4 FrFT orders, 2 RPM's)
2010, IEEE	Image encryption with multi orders of fractional Fourier transforms.	Multi orders FrFT based random encoding	Enlarged key space as number of keys can be twice the number of pixels in input image.
2010, Optics letter	Asymmetric cryptosystem based on phase-truncated Fourier transforms.	Phase-truncated Fourier Transform	Phase key pair is different during encryption and decryption owing to non linear nature and hence prevention against existing attacks on DRPE.
2011, IEEE	A novel Image encryption algorithm based on fractional Fourier transform and magic cube rotation.	Discrete fractional Fourier transform and algorithm of magic cube rotation based scrambling.	Addition of scrambling operators as key with transform orders.
2014, IEEE	Image encryption using quad phase masks in fractional Fourier domain and case study.	Fourier transform (FT) and fractional Fourier Transform (FrFT)	Better encryption quality by taking in account 4 random quad phase keys, 2 fractional orders and combination of FT and FrFT.

4 CONCLUSION

In the modern world dependent on internet, the security of images is an important concern. In this letter I have reviewed various image encryption and decryption techniques and methods based on Fourier and fractional Fourier transforms in the span of 18 years (1995–2014). As transmission of digital images occurs very often on the open network. The proposed techniques are analysed and studied thoroughly. To epitomize, every encryption technique is unique in its own way of providing security and prevention to known attacks and are useful. As the novel techniques for encryption are evolving steadily so the fast and more secure conventional techniques will always come out in better and advanced way.

REFERENCES

Alam, M. and Pandey, S., 2011, April. Color image security based on random phase masks in fractional Fourier transform domains. In Emerging Trends in Networks and Computer Communications (ETNCC), 2011 International Conference on (pp. 434–437). IEEE.

Bo, W., Cui, X. and Zhang, C., 2011, August. Realization of digital image watermarking encryption algorithm using fractional Fourier transforms. In Strategic Technology (IFOST), 2011 6th International Forum on (Vol. 2, pp. 822–825). IEEE.

Cui, D., Shu, L., Chen, Y. and Wu, X., 2013, August. Image encryption using block based transformation with fractional Fourier transform. In Communications and Networking in China (CHINACOM), 2013 8th International ICST Conference on (pp. 552–556). IEEE.

Dahiya, M., Sukhija, S. and Singh, H., 2014, February. Image encryption using quad phase masks in fractional Fourier domain and case study. In Advance Computing Conference (IACC), 2014 IEEE International (pp. 1048–1053). IEEE.

Du, N., Devineni, S. and Grigoryan, A.M., 2009, October. Mixed Fourier transforms and image encryption. In Systems, Man and Cybernetics, 2009. SMC 2009. IEEE International Conference on (pp. 547–552). IEEE.

Elhosany, H.M., Hossin, H.E., Kazemian, H.B. and Faragallah, O.S., 2012, April. C9. Chaotic encryption of images in the Fractional Fourier transform domain using different modes of operation. In Radio Science Conference (NRSC), 2012 29th National (pp. 223–235). IEEE.

Feng, X., Tian, X. and Xia, S., 2011, October. A novel image encryption algorithm based on fractional Fourier transform and magic cube rotation. In 2011 4th International Congress on Image and Signal Processing.

Garcia, J., Mas, D. and Dorsch, R.G., 1996. Fractional-Fourier-transform calculation through the fast-Fourier-transform algorithm. Applied optics, 35(35), pp.7013–7018.

Javidi, B., 1997. Securing information with optical technologies. Physics Today, 50, pp. 27–32.

Jindal, N. and Singh, K., 2010. Image encryption using discrete fractional transforms. In Advances in Recent Technologies in Communication and Computing (ARTCom), 2010 International Conference on (pp. 165–167). IEEE.

Lai, J., Liang, S. and Cui, D., 2010, August. A novel image encryption algorithm based on fractional Fourier transform and chaotic system. In Multimedia Communications (Mediacom), 2010 International Conference on (pp. 24–27). IEEE.

Lang, J., 2012, October. The reality-preserving multiple-parameter fractional Fourier transform and its application to image encryption. In Image and Signal Processing (CISP), 2012 5th International Congress on (pp. 1153–1157). IEEE.

Pei, S.C. and Ding, J.J., 2007. Eigenfunctions of Fourier and fractional Fourier transforms with complex offsets and parameters. Circuits and Systems I: Regular Papers, IEEE Transactions on, 54(7), pp. 1599–1611.

Peng, X., Zhang, P., Wei, H. and Yu, B., 2006. Known-plaintext attack on optical encryption based on double random phase keys. optics letters, 31(8), pp. 1044–1046.

Qin, W. and Peng, X., 2010. Asymmetric cryptosystem based on phase-truncated Fourier transforms. Optics letters, 35(2), pp. 118–120.

Qin, W., Peng, X., Meng, X. and Gao, B., 2011. Universal and special keys based on phase-truncated Fourier transform. Optical Engineering, 50(8), pp. 080501–080501.

Rajput, S.K. and Nishchal, N.K., 2013, December. Double image encryption scheme based on known-plaintext attack in fractional Fourier transform domain. In Recent Advances in Photonics (WRAP), 2013 Workshop on (pp. 1–2). IEEE.

Refregier, P. and Javidi, B., 1995. Optical image encryption based on input plane and Fourier plane random encoding. Optics Letters, 20(7), pp. 767–769.

Tao, R., Meng, X.Y. and Wang, Y., 2010. Image encryption with multi orders of fractional Fourier transforms. Information Forensics and Security, IEEE Transactions on, 5(4), pp. 734–738.

Tao, R., Xin, Y. and Wang, Y., 2007. Double image encryption based on random phase encoding in the fractional Fourier domain. Optics Express, 15(24), pp. 16067–16079.

Unnikrishnan, G. and Singh, K., 1999. Double random fractional Fourier domain encoding: fault tolerance properties. Asian Journal of Physics, 8(4), pp. 507–514.

Unnikrishnan, G. and Singh, K., 2000. Double random fractional Fourier-domain encoding for optical security. Optical Engineering, 39(11), pp. 2853–2859.

Unnikrishnan, G., Joseph, J. and Singh, K., 2000. Optical encryption by double-random phase encoding in the fractional Fourier domain. Optics letters, 25(12), pp. 887–889.

Vilardy, J.M., Calderon, J.E., Torres, C.O. and Mattos, L., 2006, September. Digital images phase encryption using fractional Fourier transform. In Electronics, Robotics and Automotive Mechanics Conference, 2006 (Vol. 1, pp. 15–18). IEEE.

Wang, X. and Zhao, D., 2013. Amplitude-phase retrieval attack free cryptosystem based on direct attack to phase-truncated Fourier-transform-based encryption using a random amplitude mask. Optics letters, 38(18), pp. 3684–3686.

Xu, H., Sang, N., Zhang, B. and Sang, J., 2014. Attack on double-random-phase-encoding-based image hiding method. Optik-International Journal for Light and Electron Optics, 125(13), pp. 3043–3050.

Zhang, Y. and Zhao, F., 2009. The algorithm of fractional Fourier transform and application in digital image encryption. In 2009 International Conference on Information Engineering and Computer Science.

Communication and Computing Systems – Prasad et al. (Eds)
© 2017 Taylor & Francis Group, London, ISBN 978-1-138-02952-1

Short-term load forecasting using ANN technique

Ramij Raza, M.A. Ansari & Rupendra Kumar Pachauri
Electrical Engineering Department, Gautam Buddha University, Greater Noida, India

ABSTRACT: Load forecasting technique is used for the prediction of electrical load. The generation utility company must know about the consumer's demand so that it can generate close to accurate power. If generation is not sufficient to fulfill the consumer demand, there would be problem of irregular supply and if there is excess generation the generating company will have to bear losses. Most of the researchers have been recently recommended the neural network techniques for short-term load forecasting. This paper presents a novel approach for 1 to 24 hours ahead load forecasting using Multilayer Perceptron (MLP) also referred to as multilayer feed-forward Artificial Neural Network (ANN) of a 220/132 kv substation. The inputs to the ANN model are hourly load of the day and daily average minimum temperature. The output to the model is 24 hours forecasted load. The Levenberg- marquadt optimization technique was used for training and testing the model using MATLAB R2010b. A Mean Square Error (MSE) of 3.46e-06 was obtained. This result shows that for modeling of short term load forecasting system the MLP artificial neural network may be considered as a good method

1 INTRODUCTION

Short term load forecasting is the hour to hour forecasting and important for daily maintenance of power plant. A STLF forecaster calculates the estimated load for each hour of the day, the daily maximum load or daily or weekly energy generation. The Levenberg-marquadt optimization technique is used as a back propagation algorithm for a multilayer feed forward ANN model. The output obtained from the ANN model shows the high accuracy in forecasting the future load demand. The Taxono my of load forecasting may be considered as Spatial forecasting and Temporal forecasting. The spatial forecasting includes the estimation of future load distribution in a special region such as state, a region or a whole country. The temporal forecasting deals with the particular supplier or total consumers in future hours, days, month or even years. Proper operation and planning of an electrical generation company needs the almost accurate forecasting of electrical demand of its consumers. According to time interval load forecasting may be categorized as short term, medium term and l0 ng term forecasting. The Short term load forecasting is done for one hour to one weak, medium term load forecasting done for one month to one year and long term load forecasting done for more than one year. Medium and long term load forecasting depends upon customer class and growth rate of population.There are various techniques for these three types of forecasting which can be classified as follows-

Linear methods

1. Linear regression
2. Time series approach

Nonlinear methods

1. Artificial Neural Networks
2. Nonlinear Regression
3. Fuzzy approach
4. Bayesian network technique

In this paper artificial neural network technique is used for the forecasting of future load demand.

2 LITERATURE REVIEW

Various researchers have used different methods to address a load forecasting. In 1992 Ho and Hsu designed a multilayer ANN. They used new adaptive learning algorithm for short term Load forecasting. In 2002 Chen et al. analyzed that how the load forecasting model gets affected by the electricity prices. The forecasting efficiency depends upon the electricity tariff increment hence it is suitable for this type of areas. In 2004 Satish et al. proposed a method for load forecasting which was based on ANN that shows the effect of temperature on it. It was observed that there was very less error in load forecasting using this method because the temperature was used in the model with the other environmental factors. In 2005 Sharif et al. compared the exactness of an ANN-based model and time series method. They proposed a multilayer feed-forward neural network model for improved results. In 2005 Rashid et al. presented the realistic phenomenon for load forecasting. They proposed the feed forward and feedback multi-context artificial neural network (FFFB–MCANN) for load forecasting. To obtain good efficiency they have used the rate values. In 2006 Topalli et al. have used recurrent neural network method to predict Turkey's total load one day in advance. They have used hybrid learning strategy for offline learning. Nearly 1.6% of error was found in load forecasting. The accuracy of load forecasting can be obtained by employing good network training. In 2006 Kandil et al. have proposed the method of load forecasting without use of the historical load demand. They only considered temperature values. They have seen that there was greater error in the forecasting when estimated load was used. Hence at input, temperature was used. In 2007 Adepoju et al. proposed a model which was based on supervised neural network. This model was used to forecast the load values in the Nigerian power system. The exactness in forecasting was improved because it did not consider the weather conditions influences. In the hidden layer they have used 5 to 11 neurons. They observed that when 11 neurons were utilized, it gave better model characteristics. In 2007 Lauret et al. have designed a network for short term load forecasting which was based on Bayesian neural network optimization. The Bayesian neural network model requires the uncertainties contemplations and superior noise model derivation. In 2007 Xiao et al. have developed the rough model and its ability to study and remember the input and output relationship. In this study a multi-layer back propagation neural network was used and to decrease the sensitivity of local parts of error curve surface momentum method was used.

3 LOAD FORECASTING

Energy is the most used and valuable thing in present world. This energy can be used in different forms in our daily life such as, electricity, LPG, chemical energies in form of batteries, solar energy, refined oils, wind energy, and many other forms. Sometimes we are watchful and sometimes we are improvident. But in order to provide regular and uninterrupted power supply to the user, present day and future power demand must be suitably assessed. Hence we need a suitable technique which has the ability to tell us about the consumer's demand. For this load forecasting technique is used. Power sector companies uses these forecasting techniques to foretell the amount of power needed to fulfill the power demand. It also tells about the present and future load demand scenario. There are many

applications such as switching of load, development of infrastructure, energy purchasing and its generation and contract evaluation. In electrical engineering the technique of load forecasting has become the major area of research.

Load forecasting is quite difficult work, because the series of load is complicated and it contains various levels of seasonality. Secondly, load at particular hour is dependent on the previous day load and there some important exogenous variables that must be considered.

4 PROPOSED ANN MODEL

In designing of ANN model matlab toolbox was utilized. A multilayer perceptron was chosen which basically consist of two layers that are hidden layer and output layer. The l0 g-sigmoid function is used as a transfer function for hidden layers is and for the output layer purelin function is used. To handle the nonlinearities in the input and output these functions are used. This would be able to give the wide range of values. The input consists of 24 hour load data for the 12 months of the year and daily average maximum temperature. The output layer will be a day's 24 hours load forecasted data for the substation. To enumerate the actual number of hidden layer nodes is quite difficult because there is no conceptual approach. By calculating the Mean Squared Error (MSE) over a validation data set for a varying number of hidden layer nodes, total number of hidden layer nodes was determined. The particular number of nodes in the hidden layer were selected which gives the lowest error.

4.1 Neural network training

The inputs are given to the input layer then these inputs are multiplied by interconnecting weights when they passed through the input layer to hidden layer. Within the hidden layer summing process is done by sigmoid function (non- linear). After processing of data in hidden layer the interconnecting weights are multiplied by these inputs and processed within the output layer which is performed by Purelin function.

The above Figure 2 shows the graphical interface for the neural network training process. The

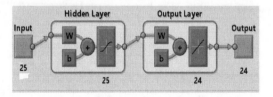

Figure 1. Model of ANN architecture.

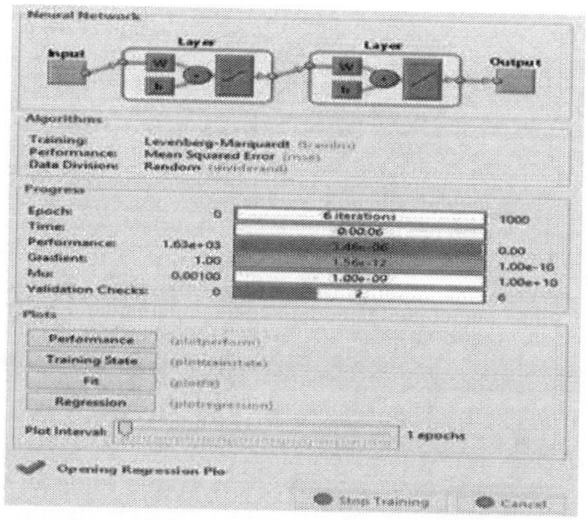

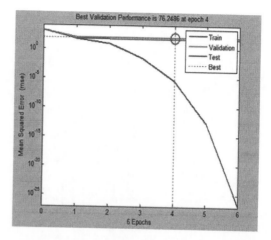

Figure 2. Neural Network training graphical interface.

Figure 5. The performance plot.

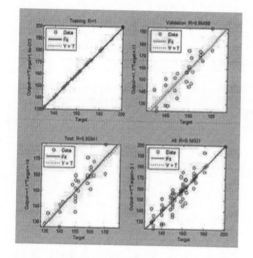

Figure 3. The Regression plot.

Table 1. 220/132 KV substation daily load profile data set.

Actual data (sample) in MVA

Time	Mon	Tue	Wed	Thu	Fri	Sat	Sun
	1st	2nd	3rd	4th	5th	6th	7th
0.00	156	160	163	140	130	160	160
1.00	140	150	140	132	130	158	148
2.00	150	150	130	140	158	142	148
3.00	130	130	140	132	148	150	132
4.00	160	140	150	132	152	150	150
5.00	140	150	160	150	130	142	150
6.00	162	148	148	132	140	142	148
7.00	170	160	150	160	148	170	150
8.00	180	170	160	150	150	140	150
9.00	162	160	150	160	152	142	160
10.00	200	150	152	160	150	160	160
11.00	170	160	140	170	150	170	148
12.00	150	140	150	142	150	152	148
13.00	148	152	150	142	148	150	150
14.00	152	162	150	142	150	152	150
15.00	160	162	142	132	150	148	140
16.00	170	160	150	150	160	140	158
17.00	160	170	162	158	162	158	152
18.00	168	170	162	152	160	158	148
19.00	170	163	172	158	160	158	160
20.00	150	160	180	170	162	162	162
21.00	180	160	170	180	162	170	172
22.00	200	180	172	162	160	150	180
23.00	182	200	160	152	152	162	170
Avg. Temp.	20	20	18	20	20	20	18

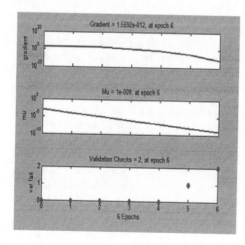

Figure 4. The training state plot.

Levenberg-Marquardt optimization technique is used for the training process. This algorithm basically consists of six basic steps which can be referred in Amjady 2001.To insure the zero toler-

Table 2. Simulation results table.

Forecasted data in MVA

	Mon	Tue	Wed	Thu	Fri	Sat	Sun
Time	1st	2nd	3rd	4th	5th	6th	7th
0.00	156.0161	160.0051	160.0155	140.6859	130.0018	159.0006	159.0109
1.00	139.0002	150.0184	139.9594	133.1022	130.7687	158.5101	148.6605
2.00	150.5039	149.9125	130.2578	140.0025	158.0094	142.0003	147.9991
3.00	129.7572	131.0025	139.9965	131.9995	148.0014	150.0100	132.0009
4.00	159.6661	140.3012	150.0012	132.0017	152.9916	149.9963	150.0100
5.00	140.5615	150.1201	160.1035	150.0010	130.0008	141.9998	149.9993
6.00	162.4541	147.9951	147.8889	131.9990	140.0004	141.9999	148.0003
7.00	169.7851	160.0014	150.0006	160.0012	147.9894	170.0004	150.0010
8.00	179.8803	170.8951	161.0025	149.9989	150.0100	140.0013	150.0031
9.00	162.2228	160.0213	150.0001	160.0013	151.9999	141.9996	159.9994
10.00	200.7373	150.1400	151.9996	160.0578	149.9891	161.9999	160.9999
11.00	169.8512	161.0325	140.9678	170.0030	150.0014	170.0002	148.0007
12.00	149.5351	140.3012	150.0010	141.9995	150.0102	151.9699	147.9999
13.00	147.8741	151.9875	150.1023	141.9991	147.8999	150.0001	150.0002
14.00	151.6176	161.9930	149.8999	142.0013	150.0001	152.0012	150.0019
15.00	159.3724	162.0058	142.9897	132.0049	150.0100	147.9990	139.9999
16.00	169.9287	161.0002	150.0013	149.9987	160.0009	139.9989	158.0004
17.00	160.0635	170.1012	162.1000	158.0008	161.9994	158.0014	152.0031
18.00	168.0021	169.8998	162.1009	152.0001	160.0002	158.0024	148.0002
19.00	170.0100	162.9961	171.9990	158.0013	161.0500	158.0009	160.0001
20.00	149.9982	160.0009	179.9699	169.0025	162.0004	162.0001	161.9997
21.00	180.0025	161.0045	169.9878	180.0014	161.9993	169.9991	172.0001
22.00	199.9962	180.1001	172.0009	161.9996	160.0101	150.0009	180.0010
23.00	182.0014	200.0005	160.0100	152.0006	152.0040	161.9991	169.9997

ance to the computational error the training goal was set at '0'. The maximum number of epoch is set to 1000. The steepest gradient descent function was used as a training function. The learning rate was set to its default value. As training progress it made adjustment automatically.

5 RESULTS AND DISCUSSION

The result was obtained from the trained ANN model. These results can be represented in term of following three plots-

1. Regression plot
2. Training state plot
3. Performance plot

Regression plot
There are four different plots in the regression analysis as shown in Figure.3.The first plot shows the graph between network output of training data and the target output. The second plot is the graph between validation data output and target output. The third is that of the Test data output against the target output. The last plot is the graph between overall network output and the target data. These graphs shows the relation between the target data

and the output data. This also gives an idea that how well network has learned the complicated relationship of input data.

Regression analysis is a statistical process for estimating the relationships among variables. It basically specifies how the typical value of the dependent variable changes when any one of the independent variables is varied, while the other independent variables are held fixed. There are some techniques that carry out regression analysis such as linear regression and ordinary least squares regression. The regression function is defined in terms of a finite number of unknown parameters that are estimated from the data. It basically refers to the estimation of continuous response variables. It opposes the discrete response variable used in classification.

Training state plot
The training state plot basically consists of three different plots as shown in Figure 4. The first plot shows the graph between gradient values against the number of epochs. It describes the manner how the training progress. The second plot is the graph between learning rate (mu) against increasing number of epochs. This plot shows the rate at which network error reduces as the process of training progresses. The third plot shows the

variation of val fail with respect to the number of epochs. This plot basically describes the function of validation.

Performance plot

The performance plot graph describes the co-relation between mean squared error versus the number of training epochs. As the number of iterations increases the computational error also improves as shown in the graph below It can also be stated that the training process is done until the zero error condition is reached.

5.1 Comparision between simulated and actual result

Table 1 shows the one week load profile data set for 220/132 Kv substation.

Now the simulation results (forecasted load) can be obtained using artificial neural network which can be shown as-

6 CONCLUSION

The load forecasting model was designed using Matlab R2010b ANN Toolbox. The training of the Neural Network and simulation of test results were all are satisfactory with a high degree of accuracy. This can be resulted into almost accurate 24 hour load data output. It can be verified by comparing the simulated outputs from the network with obtained results from the utility company. Before arriving at the best Mean squared error performance of 3.46e-06, various neural network architecture were trained and simulated.

REFERENCES

Adepoju, G. A. et al. (2007), "Application of Neural Network to Load Forecasting in Nigerian Electrical Power System". The Pacific Journal of Science and Technology, Vol. 8, No. 1, May.

Amjady, N. (2001) Short-Term Hourly Load Forecasting Using Time Series Modeling With Peak Load Estimation Capability. IEEE Trans. on Power Syst.16, 798–805.

Buhari, Muhhamad 2012 "Short-Term Load Forecasting Using Artificial Neural Network", Proceedings of the International multi conference of Engineers and Computer scientist, Vol. 1.March 14–16, Hong Kong.

Chen, H. et al. (2001), "ANN Based Short-Term Load Forecasting in Electricity Markets," *IEEE Proceedings of Power Engineering Society Winter Meeting*, Columbus, pp. 411–415. doi:10.1109/PESW.2001.916876.

Ho. et al. (1992), "Short-term load forecasting using a multilayer neural network with an adaptive learning algorithm", IEEE Transactions on Power Systems, Vol. 7, No. 1, pp. 141–148.

Kalaitzakis, K. et al. (2002), "Forecasting based on artificial neural networks parallel implementation", Electric Power Systems Research 63 185-/19.

Kandil, Nahi et al. (2006), "An efficient approach for short term load forecasting using artificial neural networks", International Journal of Electrical Power and Energy Systems, Volume 28. issue 8, pp. 525–530.

"Load forecasting" chapter 12, E.A Feinberg and Dora Genethlio, Page 269–285, from links: www.ams. sunysb.edu nd, www.usda.gov

Lauret, Philippe et al. (2008) "Bayesian Neural Network Approach to ShortTime Load Forecasting", Energy Conversion and Management, Vol. 49, No. 5, pp.1156–1167.

Rashid, T. & T. Kechadi, T. (2005). 'A practical approach for electricity load forecasting', In the proceeding WEC'05, The 3rdWorld Enformatika, volume 5, pp. 201–205, Isanbul, Turky, ACTA Pres.

Satish, B. et al (2004). "Effect of temperature on short term load forecasting using an integrated ANN", Electric Power System Research, vol. 72, pp. 95–101.

Sharif, S. & Taylor, J.H. (2000). Short term Load Forecasting by Feed Forward Neural Networks Proc. IEEE ASME First Internal Energy Conference (IEC), Al Ain, United Arab Emirate.

Topall et al. (2006). Intelligent short term load forecasting in turkey. Int .J. Electric Power Energy system.28; 437–447.

Xiao, Z. et al. (2009) "BP neural network with rough set for short term load forecasting". Expert Systems with Applications, 36(1): 273–279.

Communication and Computing Systems – Prasad et al. (Eds)
© 2017 Taylor & Francis Group, London, ISBN 978-1-138-02952-1

Solar radiation forecasting using artificial neural network with different meteorological variables

Puneet Kumar, Nidhi Singh Pal & M.A. Ansari
School of Engineering, Gautam Buddha University, Greater Noida, India

ABSTRACT: Renewable energy sources quickly replenish themselves and can be used again and again therefore it is now becomes an alternative source of energy. Solar energy has now gained a momentum to use it as a great source of alternative energy and is an important parameter in design of photovoltaic systems. An accurate knowledge of the location of solar radiation is required for the design and proper utilization of PV systems. Solar radiation is varies with time, weather conditions and geographical locations. This paper presents an annual prediction of solar radiation using an Artificial Neural Network (ANN). Neural fitting tool (nftool) is used for the prediction of clear sky diffuse insolation. The different meteorological variables like humidity, air temperature, atmospheric pressure and number of days of the year are used as input to the ANN and mean square error and regression values are calculated for the proposed ANN model.

1 INTRODUCTION

Renewable energy sources are best alternative source of energy as compared to any other type of conventional energy resources. The problem with the conventional energy sources is that it is responsible for the emission of carbon dioxide (CO_2) and many other harmful gases which pollutes the atmosphere results in various severe health issues to the human being. Major factor of causing the emission of CO_2 is electricity generation. Large amount of CO_2 is emits by the combustion of fossils fuel in generation of electricity using conventional sources of energy. Then transportation is the second main cause of CO_2 emission and last but not least the industries are also responsible for the CO_2 emission. Renewable energy resource is the clean source of energy without any CO_2 emission and does not make any harm to the environment. That's why RES draws the attention of government, power generating companies and consumer to use the renewable energy resources as an alternative source up to the great extent so that the use of conventional energy resources and fossils fuel could get minimized.

Recently, India adopted a new program called Intended Nationally Determined Contribution (INDCs) where India has announced that 40% of the country's electrical energy would come from non-conventional energy resources or non-fossil fuel based sources such as solar and wind by 2030. Now India has come forward as one of the last major global economics to release its goal to tackle the changes in climate. India has added 3,790 MW of solar power capacity by March 2016 and has taken the total installed capacity to 9,038 MW in India. Now India is moving towards the goal of achieving the target of 100,000 MW solar power capacity by 2022.

In India, heavy fluctuations and variations are found in the generation of solar power with respect to different time zones and places. Uncertainty and intermittency are the following terms which characterized the solar energy. The solar energy is greatly influenced by the different weather parameters and meteorological variables. Solar power forecasting has various advantages such as planning and set-up of solar plant, power reserve, unit commitment and integration of grid connection etc. Artificial Neural Network (ANN) technique is used for the purpose of solar radiation forecasting of the capital New Delhi. New Delhi has a total commissioned capacity (MW) of 6.172 MW as on March 7th, 2016.

Vikas Pratap Singh et al. (2013) used solar radiation, ambient temperature, wind velocity and module temperature as the input variables to the proposed ANN model. This model was used for short term solar power forecasting. In proposed model the Root Mean Square Error (RMSE) was 0.1019. Then Adaptive Neuro-Fuzzy Inference (ANFIS) Model and Generalized Neural Network (GNN) were used and the calculated RMSE error was 0.0965 and 0.093. Amit Kumar Yadav et al. (2014) used latitude, longitude, height above mean sea level and sunshine hours as the inputs to the two different models i.e.

ANN and Radial Basis Function Neural Network (RBFN) which were used in the forecasting of solar radiation for the purpose of power generation. For RBFN, the input has the Mean Absolute Percentage Error (MAPE) of 4.94% and absolute fraction variance (R2) of 96.18%. B. M. Alluhaidah et al. (2014) used the best combinations of different meteorological variables such as day, cloud cover, temperature, humidity, vapor, wind direction, wind velocity for the global solar radiation forecasting in Riyadh, Saudi Arabia. The best combinations of the inputs that provide low RMSE and MAPE and high values of regression were day, temperature, cloud cover, humidity and vapor. The RMSE and r were calculated as 3.4764% and 0.99208. Jose G. Hernandez et al. (2014) used humidity, meteor, cloudiness, precipitation, radiation, temperature and wind speed as input to the proposed ANN multilayer perceptron model for the solar radiation forecasting. The data was collected form the meteorological stations in Grain Canaria and Tenerife. The study of the solar radiation forecasting achieves a mean average error of 0.04 KWh/square meter. Xingyu Yan et al. (2014) used temperature, cloud cover, humidity and measured solar radiation as the inputs to an ANN model for the next 24-hours solar radiation prediction. Back propagation neural network is used to train the neural network. The RMSE and MSE for training set were 9.359% and 4.964%. For validation set RMSE and MSE were 11.849% and 6.714% and for test set RMSE and MSE were calculated as 9.811% and 5.057%. Jun Liu et al. (2015) used temperature, humidity, wind speed data with an additional input i.e. Aerosol Index (AI) data to an Back Propagation (BP) neural network to forecast the next 24-hours photovoltaic power output. The average relative error MAPE was about 7.65%.

In this paper four meteorological variables of the capital New Delhi (India) will be used as a case study to predict the annual solar radiation for the year 2003. The four meteorological variables with the historical data of year 2002 which are applied as input to the ANN model is number of days of the year, maximum air temperature, humidity and average atmospheric pressure. Clear sky diffuse insolation (solar radiation) is given to the ANN model as target vector which in turn gives the output solar radiation for the year 2003.

2 ARTIFICIAL NEURAL NETWORK

The concept of artificial neural network is totally inspired by the functioning of human brain. The artificial neural network is information processing model which consists with the highly interconnected element called neurons. These neurons are connected with the other neurons through a connection link.

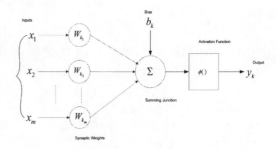

Figure 1. Schematic diagram of ANN.

These connections links are consist with the synaptic weights and these weights carry the information about the input signal. The neurons with the synaptic weights are then followed by the summing junction and bias. The bias is added to the neural network in order to have the impact on the output of neural network. There are two types of bias one is positive and another is negative. If the positive bias is applied to the network then it helps in increasing the net input of the neural network and if negative bias is applied then it decreases the net input of the neural network. Next is the activation function that is applied over the net input to calculate the output. The threshold value is used in the activation function. The threshold value is a fixed value and on the basis of this fixed value the output is calculated.

The ability of learning by example makes it very flexible and powerful. Artificial neural network exhibits adaptive learning which is the ability of learning based on the data given for training. It is also used for real time operation. During the training process in order to obtain the desired output the network weights and parameters are adjusted and minimize the error function. The network is trained by epochs. The epoch may be defined as it is the complete run of all training data applied to the network and are processed using learning algorithm.

The learning methods of ANN are divided into three categories: supervised learning, unsupervised learning and reinforcement learning. Basically the ANN is learned here by the supervised learning. Supervised learning is a method of learning in which each input vector requires a corresponding target vector. Then a training pair is formed by the input vector along with the target vector. The network has knowledge about the output. During the training process of the network the input vectors are applied to the network which results in output vector and this output vector is the actual output of the network. Then the actual output is compared with the desired output vector. If there is any difference between the two outputs then the network generates an error signal. Here the error

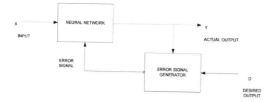

Figure 2.   Block diagram of supervised learning.

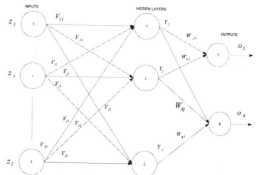

Figure 3.   Structure of back propagation network.

signal is used to adjust the network weights until and unless the actual output matches the desired one. The supervised learning block diagram is shown in Figure 2.

The error measure is defined by Mean Squared Error (MSE) and is given by:

$$MSE = \frac{1}{N}\sum_{K=1}^{N}\left|Y_K - \hat{Y}_K\right|^2 \qquad (1)$$

where N is the number of pattern pairs, Y_k is the output of K_{th} pattern pair, $\hat{Y}_k$ is the network output corresponding to the pattern pair K.

3   BACK PROPAGATION NEURAL NETWORK

The back propagation learning algorithm is developed by Bryson and Ho in 1969. The networks associated with the back propagation learning algorithm are also called back propagation networks. A back propagation neural network is a multilayer, feed-forward neural network which consists with an input layer, a hidden layer and an output layer. In back propagation learning the error is propagates backward from the output layer to the hidden layer in order to calculate the update of the weights in hidden layer units. The error is calculated from the difference between the actual output and desired output at each of the output units. Before applying the back propagation learning law for a given training set it is important to initialize the weights properly. The performance of the back propagation trained network is totally depends on the knowledge which is present to train the network in the form of initial weights. The back propagation neural network is the most widely used artificial feed forward neural network structure for the purpose of solar radiation forecasting.

4   LEVENBERG MARQUARDT ALGORITHIM

Levenberg Marquardt Algorithm is also known as Damped least square method and is used to solve the non-linear least square problems. In

MATLAB, Levenberg Marquardt Algorithm is known as trainlm which is a network training function which is used for the weight updates and biases. It is the fastest training algorithm in neural network toolbox and is recommended as the first choice to use it as a supervised learning method.

5   IMPLEMEMTATION OF ANN MODEL

In the present study, the four meteorological variables are applied to the artificial neural network. The ANN model with the four inputs is shown below.

The above ANN model is considered for the purpose of solar radiation forecasting with the four input meteorological variables namely as the days of the year, maximum air temperature, relative humidity and average atmospheric pressure. The data of the year 2002–2003 of the city New Delhi is taken from the surface meteorology and solar radiation website. The geographical description of the location is given in Table 1.

The data which is applied to the ANN model is divided in to three parts:

a. Training set- in this data set neural network is trained by the supervised learning method. Input data along with the target vectors is applied to the neural network to find the optimal weights and bias.
b. Validation set-this data set is used to determine the optimal number of hidden units. The validation of the neural networks stops when the further training of the neural network is not possible.
c. Test set-.in this data set the performance of the ANN is measured during and after the training.

In the neural fitting tool, 70% of the data is utilized for training of network, 15% of data is used

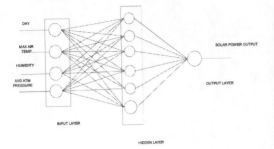

Figure 4.   Solar radiation forecasting model of ANN.

Table 1.   Geographical location of New Delhi.

City	Latitude	Longitude
New Delhi	28.350 N	77.120 E

for validation and another 15% data is utilized for testing of the neural network. The training set is used to teach the neural network continuously as long as the network continues to improve on the validation sets. When the network stops improving on validation set the training also stops. The neural network begins with arbitrarily chosen initial weights, in this way the results of neural network fitting tool comes marginally distinctive each time it runs. Best performance is decided on the basis of Mean Squared Error (MSE) which is the average difference between the output and target values. Low mean square error results in better output values with minimum errors.

The performance plot is shown in Figure 5 which indicates that the mean square error is decreasing by increasing the number of epochs. The best validation performance is obtained at epoch 13.

Next is the regression plot for training shown in Figure 6a and for total response in Figure 6b, between the output and target value. Regression plot shows the variation in output values with respect to the target. For the perfect fit the data should nearly fall along the 45 degree line. The regression value for the training of the network is obtained R = 0.98978 or 98.97%.

The comparison between the forecasted solar radiation and actual solar radiation plot is shown in Fig. 7. The forecasted solar radiation graph follows the symmetry as of the actual solar radiation graph of the year 2003.

The mean square error and regression values of an ANN model for training, validation and testing are shown below in Table 2.

The mean square error and regression value for training set are 0.13 and 98.97%. For validation set mean square error and regression value are 0.14

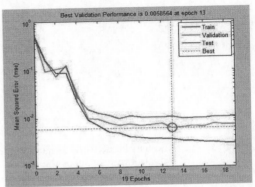

Figure 5.   Validation performance plot.

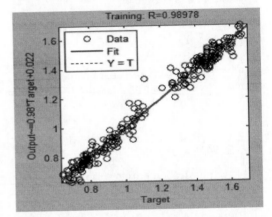

Figure 6a.   Regression plot for training.

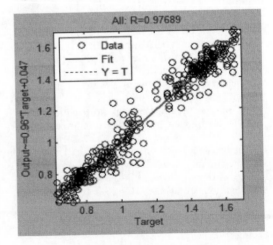

Figure 6b.   Regression plot for total response.

and 94.428% and for test data set the MSE and regression value are 0.13 and 95.66%. In the solar radiation forecasting the minimum error is −0.2905 and maximum error is obtained as 0.2583.

490

Table 2. MSE and R value for ANN model.

Data	Mean Square Error	Regression value
Training	0.13	0.98978
Validation	0.14	0.94428
Testing	0.13	0.95668

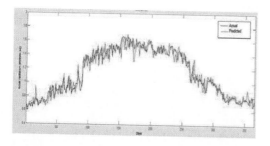

Figure 7. Plot between predicted and actual solar radiation.

6 CONCLUSIONS

The artificial neural network model is developed for solar power forecasting with the four inputs namely, days of the year, maximum air temperature, humidity, average atmospheric pressure. The MSE and R values for training, validation and testing samples are shown in Table 2. The regression for training of the neural network is obtained as 98.97%. It is concluded that for minimum mean square error higher regression values are obtained which is responsible for the perfect fit of the data nearby the target value. If the mean square error is high then regression values would be low. The comparison between the predicted and actual solar radiation plot is shown in Figure 7 and the difference between the measured values and predicted values of solar radiation was calculated in terms of minimum error and maximum. The minimum error and maximum error was calculated as −0.29 and 0.25. The results show that the prediction accuracy is enough and the input variables used for the forecasting of solar radiation are suitable for the prediction and provides enough accuracy.

In future, other weather parameters such as maximum ambient temperature, specific humidity, earth skin temperature, aerosol index can be used to forecast the solar radiation through ANN. Future work is focused on various intelligent techniques like Particle Swarm Optimization, Genetic Algorithm and Bees algorithm etc. to reduce the Mean Square Error.

REFERENCES

Alluhaidah, B.M., S.H. Shehadeh and M.E. El-Hawary 2014. Most influential variables for solar radiation forecasting using artificial neural network, *Electrical power and energy conference*, IEEE.

Amit Kumar Yadav and Hasmat Malik 2014. Comparison of different artificial neural network techniques in prediction of solar radiation for power generation using different combinations of meteorological variables, *International conference on power electronics, drives and energy systems*, PEDES, IEEE.

Hamid Oudjana, S., A. Hellal, and I. Hadj Mahammed 2014. Neural network based photovoltaic electrical forecasting in south algeria, *Applied solar energy*, vol. 50, n0. 4, pp. 273–277.

Haykin. S., Neural Networks 1999. A Comprehensive Foundation, Second Edition: Prentice Hall International, Inc.

Jiahao Kou, JunLiu, Qifan Li, W. Fang and Zhenhuan Chen 2013. Photovoltaic power forecasting based on artificial neural network and meteorological data, *IEEE Region 10th Confrence*, pp. 1–4, IEEE.

Jose G. Hernandez, Carlos M. Travieso, Travieso and M.K Dutta 2014. Solar Radiation modelling for the estimation of the solar energy generation, *IEEE Seventh International Confrence on Contemporary Computing (IC3)*, pp. 536–541, IEEE.

Jun Liu, Xudong Zhang, and Chunxiang Yang 2015. An improved photovoltaic power forecasting model with the assistance of aerosol index, *IEEE Transactions on Sustainable Enerrgy*, vol. 6, no.2, pp. 1949–3029.

Leva, S., A. Dolara, F.Grimaccia, M. Mussetta and E. Ogliari 2015. Analysis and Validation of 24 hours ahead neural network forecasting of photovolatic output power, *Mathematics and Computer in Simulation*, Elsevier.

Rojas, R. 1996. Neural Networks, Springer-Verlag, Berlin

Vikas Pratap Singh and M. Siddharta Bhatt 2013. Generalized Neural Network Methodology for Short Term Solar Power Forecasting, *The 13th IEEE Conference on Environment and Electrical Engineering (EEIC, pp. 58–62, IEEE.

Xingyu Yan, Dhaker Abbes and Bruno Francois 2014. Solar radiation forecasting using artificial neural network for local power reserve, *IEEE Electrical Sciences and Technologies in Magherb (CISTEM)*, pp. 1–6, IEEE.

Yadav A.K. and S.S. Chandel 2012. Artificial Neural Network based Prediction of Solar Radiation for Indian Stations, *International Journal of Computer Applications*, vol. 50, pp. 1–4.

Communication and Computing Systems – Prasad et al. (Eds)
© 2017 Taylor & Francis Group, London, ISBN 978-1-138-02952-1

Integrated approach for performance scrutiny of cloud resource provisioning algorithms

S. Dalal, S. Kukreja & Vishakha
Department of CSE, SRM University, Sonepat, Haryana, India

ABSTRACT: Cloud Computing is an emerging technology which is extending business day by day. Cloud provides innovative services on rent and the user has to pay as per the usage of the service. Cloud services are available for 24/7, accessed according to the requirement over the internet from anywhere in the world. The customer has to pay for the renting resources. So cloud computing is referred to as utility-oriented model through which users use the resources only when they need. Cloud infrastructure manages and leases the resources and price of the leased resource is charged according to the consumption. That is why cloud is a famous technological trend and is growing rapidly. The amazing features of cloud lets the user to use the cloud services again and again. In this paper, a review of the various resource provisioning techniques is covered and analyzed on the basis of their performance and parameters.

Keyword: Cloud computing, resource provisioning and availability, load balancing, QoS

1 INTRODUCTION

The computing word has travelled from parallel to distributed computing to cluster computing to grid computing and now to cloud computing. Cloud Computing appears as a buzzword in the modern era (Jadeja et al. 2012). Most of the scientific and engineering applications, computational finances, social network as well as industries and academicians are paying attention to cloud computing because of its pay-as-you-go nature. Every person can store a broad range of data like confidential data, statistical data, audios, videos and movies on the cloud (Marinescu et al. 2013). Google Drive is a common storage cloud. The user has to log in with a Google account and select the new option to upload the file or folder.

Before Cloud Computing came into existence, there were lots of websites on the internet having their own infrastructure. They thought that why not give our infrastructure on rent. These large websites started their business by renting the infrastructure, termed it as a cloud. Such an infrastructure is made from thousands of computing nodes that are interconnected by the huge network fabrics (architecture). Each node has their own hardware, operating system, protocols, scheduling and application programs. The cloud infrastructure provisions the resources at low power, low cost with high availability (Naeimi et al. 2013). According to the definition of the National Institute of Standards and Technology (NIST), cloud computing is defined as:

NIST Definition of Cloud Computing: "Cloud Computing is a model for enabling convenient, On-demand network access to a shared pool of configurable computing resources (e.g. Networks, servers, storage, applications, and services) that can be rapidly provisioned and released with minimal management effort or service provider interaction."

Cloud computing is built with deployment models, service models, characteristics, resources and organizational infrastructure. The deployment of the cloud services such as Infrastructure-as-a-Service (IaaS), Platform-as-a-Service (PaaS), and Software-as-a-Service (SaaS) is accomplished through public cloud, private cloud, hybrid cloud, and community cloud (Marinescu et al. 2013). These service models helps to deploy the applications into the cloud.

The above diagram shows different kind of services offered by cloud. IaaS provides the infrastructure such as number of servers, storage devices and the network resources for the specified time period on minimum cost. Platform-as-a-Service(PaaS) delivers all the resources required to build the application. The customer need not to purchase or install any software. SaaS enables the customer to use the hosted applications through the internet. It is the task of provider to manage the software.

Features that makes cloud famous are given here:

• Economically, Cloud Computing is effective as the user need not to purchase the resource and

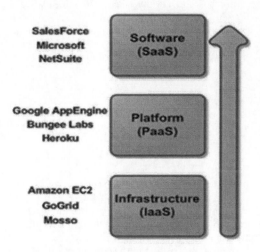

Figure 1.   Cloud service models and providers.

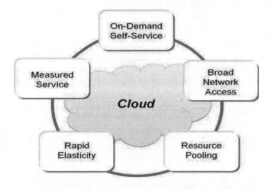

Figure 2.   Five Important characteristics of cloud.

the overhead of maintaining the infrastructure is also done by the provider.

• The resources can be added or removed dynamically. This shows the elasticity of cloud as the user can demand for the resource any time and release when task is finished.

• SLA is the initial agreement between the service provider and the customer in which the provider promises to give say 95% of the service and if failed to deliver 95% service, then the customer need not to worry about the money paid for the service because the provider will pay the penalty to the customer. This is a beneficial point for the customer.

2   TOP CLOUD VENDORS

There are a lot of vendors in the market depending upon the kind of service they provide. These services are infrastructure-as-a-Service, platform-as-a-Service and Software-as-a-Service. Every vendor has their pricing model. Some of the vendors who offer these services are google, yahoo, Rackspace, Salesforce, Microsoft, Amazon, IBM, etc. we are giving brief knowledge of some of the vendors:

2.1   *Amazon*

Amazon is one of the biggest player in the market. Amazon offers a number of cloud services to the customer that includes Elastic Cloud Compute (EC2), Simple Storage Service (S3), Simple DB, AMI, Elastic Block Storage (EBS) etc. Famous services provided by amazon are EC2 and S3. EC2 allows the customers to create, launch, start, stop, and terminate the instances. It enables the scalable infrastructure as the users can create a number of instances, use them and can terminate when not required. EC2 also has the capability to save the images of running instances in S3. S3 consists of buckets that store the objects of any size in binary form. These objects can be a file or image that can be accessible from anywhere in the world (Velte et al. 2010).

2.2   *Google*

Google App engine is a powerful and secured platform to build scalable web applications from simple to complex. App engine automatically scales your apps based on the need, from zero to millions of users. We just pay for the resources thereby reducing the overhead to maintain or upgrade the servers. Developers can build apps on their machine using the App engine SDK. Google App Engine is built with lots of features and services such as user authentication, NoSQL Datastore, Memcache, security scanner, traffic splitting, job Queues etc. Popular languages supported by Google are Java, Python, PHP (Rajkumar et al. 2013).

2.3   *Microsoft*

Microsoft cloud ensures high reliability through robustness, environmental sustainability, cost-effectiveness, security, efficiency and trustworthiness. The infrastructure is comprised of more than 100 datacenters, one million servers, network edges, compute nodes and fiber optic cable. The team of experts builds and manages the infrastructure to support the service for more than one billion customers. Microsoft's software-defined infrastructure model can immediately move workloads to the alternate server or even to the other datacenters (Velte et al. 2010).

Microsoft offers 200 plus online services including office 365, MSN, Skype and Microsoft Azure. Microsoft's Azure provides operating system,

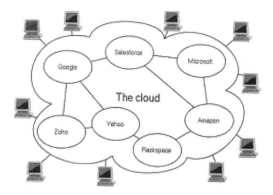

Figure 3. Diagram of vendors of cloud computing.

programming languages, frameworks, tools, database and devices to build and deploy the application. Customers can build apps with JavaScript, Python, PHP, .NET, Java, which are supported by microsoft. Moreover, Azure service can easily integrate with your existing IT environment through the secured private network along with data residency and encryption features. Azure's pay-as-you-go service can quickly scale-up and down the infrastructure on demand and user pays only for the using services.

One of the major challenges faced by the cloud providers is to efficiently allocate the resources and maximize the revenue. Cloud infrastructure allocates heterogeneous resources such as CPU, memory and storage in the form of virtual machine instances. The elastic nature of cloud provisions the resources which can be added or removed as per the need. As resource allocation is an important part of cloud computing technology.

3 NEED OF RESOURCE PROVISIONING IN CLOUD COMPUTING

The demand for the cloud resources is growing day by day. Cloud provisions the computing resources from the shared pool over the internet. Cloud computing offers resource provisioning in two different ways: first is advance reservation and second is on-demand resource provisioning. Economically, the advance reservation plan is cheaper than on-demand resource provisioning, but if we think from the effectiveness point of view, on-demand plan is more effective than the former one. On-demand plan lets the user to acquire the resource any time and utilize when required. Since resource utilization and provisioning are dynamic, the user can demand for the additional resource and the provider has to optimize the old request to cope with new demands. Due to the high resource availability, on-demand plan improves the performance of the system,

customer satisfaction, reduces the SLA violations meeting the QoS objectives. Since the plan offers on-demand provisioning of resources that meets the actual demand of the customer, the customer has to pay high due to the expensive service of the plan. In case of advance reservation, the resource can be idle until it is used or can be under provisioned to meet the consumer's requirement (Chiang et al. 2014). So efficient resource provisioning is one of the major issues faced by the cloud provider.

4 PRIORWORK

Kaveri et al. (2015) have discussed clustered shared resources in cloud. The authors used horizontal clustering to analyze the result. On the basis of the result, it has been found that the overall response time to complete the job has been reduced and the resources which were free will get available to the next job. This approach increased the availability of resources due to the fast response time, balanced the load between the virtual machines and decreased the total cost of resource utilization.

Hosseini et al. (2011) proposed a work to maintain the degree of resource availability and reliability of long-term applications. The technique in this work focused on the improvement of MTTF factor and hence the resource reliability increased and failure rate decreased. In order to maintain the quality of service, the goal is to minimize the resource consumption and cost of service provided.

Niu et al. (2013) have presented the computing model Semi-Elastic Cluster (SEC) for High Performance Computing (HPC) to reserve and manage the variable-size cloud-based clusters to be shared within the organization. The result obtained from this model has shown that SEC delivered better cost-effectiveness when performing with the load aggregation and simultaneous instance reuse.

Wang et al. (2013) addressed the problem of service availability to support the long-term applications on minimum lease cost in SpotCloud. An optimal resource provisioning algorithm has been provided to overcome the mentioned problem. The algorithm also found the minimum cost for dynamic resource availability. On the basis of results, it has been examined that the algorithm truly presented performance gain in service availability at minimum cost.

The authors Chiang et al. (2015) have provided two scheduling strategies Optimal Cost-efficient Provisioning policy (OCP) and OCP-Flexible (OCPF) to overcome the problem of overprovisioning and underprovisioning. Yi-Ju Chiang et al. developed a cost function that takes so many parameters into consideration to reduce the provisioning cost of the resources and also met the hard waiting time simultaneously. The performance of these two algorithms was satisfactory and have achieved the optimal quality of servers and service rate at minimum cost.

Yousefyan et al. (2013) developed a Cost-Effective Cloud Resource Provisioning with Imperialist Competitive Algorithm Optimization (CECRPI-CAO) that predicted future demands accurately so as to achieve the best of reservation plan when the user could not pay so much in the on-demand plan. This algorithm allocated the virtual machines on the reservation and minimized the provisioning cost. The suggested model MLP-EBP have predicted demands more accurately.

Authors Dong et al. (2013) invented QoS oriented availability monitoring model to monitor the availability of cloud computing resources. A QoS model is created to determine the quality of service that a computing resource could provide and according to the QoS attributes, it became easier to select the suitable cloud resource to meet the customer's demand.

Table 1. Resource provisioning techniques.

Year	Algorithms	Parameters considered	Cons
2010	Robust cloud resource provisioning	Customers demand, price, availability	1. Computing complexity is high.
2011	Resource Monitoring System	Availability, Reliability, Cost	1. Estimated deadline is hard to achieve. 2. GRMS spends a lot of time to allocate dulicate resources.
2012	Dynamic provisioning and Scheduling Approach	Deadline, budget	1. Public cloud executes the task which could not met their deadline with in-house resources.
2013	Optimal Resource Provsioning Algorithm	Service Availability, minimized lease cost	1. higher cost has to be paid by the buyers to acquire more VMs.
2013	Simulated-Annealling Load Balancing Algorithm	Resource availability, utilization	1. Migration cost is not considered.
2013	Semi-Elastic Cluster Computing Model	Waiting time, cost	1. Total cost is very high. 2. Overhead of acquiring and releasing the redundant nodes.
2013	QoS Oriented Availability	Availability, service	1. The local maximization algorithm failed to choose thecloud resources to fulfill th ovrall demand of the customer.
2014	Priority based VM Allocation Policy	Priority, resource utilization, availability, response time	1. Lot of time is consumed in calculating the parametric value. 2. process is long.
2014	Semantic Cloud Resource Broker	Availability	1. Semantic search delays in finding the exact resource. 2. Semantic based resource discovery takes more time to compare the conventional keywords.
2014	Optimal Cost-Efficient Resource Provsioning & OCP-Flexible	Service waiting, time, cost	1. OCPF algorithm is costly.
2014	Clustered Approach for High Avaiability	Scalability, availability	1. Resources are not properly utilized.
2015	Cluster based Load Balancing	Waiting time,execution time, turnaround time, throughput	1. Proposed algorithm is complex.
2015	Shared Resource Clustering	Availability, Scalability, elasticity	1. users suffered from unexpected expenses in case of bundled clustering.

The authors Fan et al. (2013) proposed simulated-Annealing load balancing algorithm for resource allocation and scheduling problem in cloud computing. The Load balancing algorithm has been developed under the SLA constraints and improved the utilization rate of resources. The evaluation was done using CloudSim toolkit and found that the proposed algorithm balanced the load effectively.

Girase et al. (2014) have presented the design and implementation of priority based resource management system. The scheduling heuristic mechanism has been constructed by creating new virtual machines on physical machine based the availability of the resources, load balancer preemption mechanism. The scheduling request on VMs had to ensure agreed SLA. The evaluated result showed that the heuristic method helped to select the request from waiting queue and optimized the performance.

Semantic Cloud Resource Broker (SCRB) based mechanism has been introduced by Kavitha et al. (2014) to manage the private cloud datacenter resources efficiently and to increase the resource availability. In this paper, the proposed mechanism managed the private clouds by using Open Nebula and Eucalyptus as middleware. SCRB used two different middleware both had their own mechanisms and protocol. So to avoid the conflict that has been raised by using different mechanisms and protocols, SCRB was used. As a result, system increased the success rate of resources and enhanced the performance.

Calheiros et al. (2012) have considered the whole organization's workload and proposed dynamic provisioning and scheduling architecture that enabled the applications to complete within the deadline in a cost-effective way. The accounting mechanism was introduced to know the cost utilized by the public cloud resources assigned to the users. The results showed that the approach made an efficient use of the public cloud resources and enabled deadlines to be met on reduced cost.

Cluster-based load balancing algorithm was developed by Kapoor et al. (2015) to serve the heterogeneous environment. The proposed algorithm was dynamic and considered the resource specific task. Also the overhead of distributing the load was managed by clustering of machines. The best output could be obtained by considering the various time parameters.

Robust Cloud Resource Provisioning algorithm proposed by Chaisiri et al. (2010) to minimize the cost of overprovisioning and underprovisioning. The proposed algorithm found the optimal solution for the cost and resulted in minimizing the on-demand cost.

The clustered virtual machine architecture has been proposed by Chavan et al. (2014) and provided higher availability of resources and improved scalability. The clustered approach helped in scaling the architecture, ease the scheduling and saved the time. The model maximized the usability and resource utilization.

5 CONCLUSION

We have studied various algorithms based on resource provisioning, cost and load balancing. We also have noted that the previous algorithms used so many parameters to measure the performance of the proposed work. As we know that every mechanism has their pros and cons, we are providing a tabular view to examine the performance of the system on the basis of cons.

REFERENCES

Amshavalli, R.S. and Kavitha, G. Increasing the Availability of Cloud Resources using Broker with Semantic Technology. in proceedings of IEEE International Conference on Advance Communication Control and Computing Technologies, Ramanathapuram, 2014, pp. 1578–1582.

Buyya, R., Vecchiola, C. and Selvi S.T. 2013. Mastering Cloud Computing. ISBN 978-1-25-902995-0: McGraw Hill Education Pvt. Ltd.

Calheiros, R.N. and Buyya, R. Cost-effective provisioning and scheduling of deadline-constrained Applications in Hybrid Cloud., in proceedings of 13th International Conference on Web Information System Engineering, Springer, Heidelberg, 2012, pp. 171–184.

Chaisiri, S., Niyato, D. and Lee, B.S. Robust Cloud Resource Provisioning for Cloud Computing Environment. in proceedings of IEEE International Conference on Service-Oriented Computing and Appications, Perth, 2010, pp. 1–8.

Chavan, V., Kaveri, P.R. Shared Resource Clusteing for Load Balancing and Availability in Cloud. in proceedings of 2nd IEEE international Conference on Computing for Sustainable Global Development, New Delhi, 2015, pp 1004–1007.

Chavan, V., and Kaveri, P.R. Clustered Virtual Machines for Higher Availability of Resources with Improved Scalability in Cloud Computing", in proceedings of Ist IEEE International Conference on Networks and Soft Computing (ICNSC), Guntur, 2014, pp. 221–225.

Chiang, Y.J., Ouyang, Y.C., Hsu, C.H. An Optimal Cost-Efficient Resource Provisioning for Multi-Servers Cloud Computing. in proceedings of IEEE International Conference on Cloud Computing and Big Data, Fuzhou, 2014, pp. 225–231.

Dong, W.E., Nan, W. and Xu, L. QoS-oriented Monitoring Model of Cloud Computing Resource Availability. in proceedings of 5th IEEE International Conference on Computational and Informational Services, Shiyang, 2013, pp. 1537–1540.

Fan, Z., Shen, H., Wu, Y. and Li, Y. Simulated-Annealing Load Balancing for Resource Allocation in Cloud Environments. in proceedings of IEEE International Conference on Parallel and Distributed Computing, Applications and Technologies, Taipei, 2013, pp. 1–6.

Girase, S.D., Sohani, M. and Patil, S. Dynamic Resource Provisioning in Cloud Computing Environment using Priority based Virtual Machines. in proceedings of IEEE International Conference on Advanced Communication Control and Computing Technologies (ICACCCT), Ramanathapuram, 2014, pp. 1777–1782.

Google App Engine https://cloud.google.com/appengine/.

Hadley, B., Hume, A., Lindberg, R. and Obraczka, K. Phantom of the Cloud: Towards Improved Cloud Availability and Dependability. in proceeding of International Conference on Cloud Networking (CloudNet), IEEE, Niagara Falls, 2015, pp. 14–19.

Hosseini, M.J., Arasteh, B. An Efiicient and Low Cost Monitoring System to Improve Availability and Reliability of Grid Services. in proceedings of International Conference on Computer Science and Network Technology, Harbin, 2011, pp 2597–2601.

Jadeja, Y. and Modi, K. Cloud Computing:Concepts, Architecture and Challenges", in Proceeding of IEEE transaction on Computing, Electronics and Electrical Technologies, Kumarc Oil, 2012, pp. 877–880.

Kapoor, S. and Dabas, C. Cluster Based Load Balancing in Cloud Computing. in proceedings of 8th International Conference on Contemporary Computing, Noida, 2015, pp. 76–81.

Marinescu, D.C. 2013. Cloud Computing: Theory and Practice. ISBN 978-0-12404-627-6: Elsevier.

Microsoft Azure Service https://azure.microsoft.com/en-in/overview/what-is-azure/.

Microsoft Cloud Platform https://www.microsoft.com/india/datacenter/about.aspx.

NIST Definition of Cloud Computing http://www.nist.gov/itl/csd/cloud–102511.cfm.

Naeimi, H., Natarajan, S., Vaid, K., Kudva, P. and Natu, M. Cloud Atlas-Unreliability through Massive Connectivity. VLSI Test Symposium, IEEE, Berkeley, 2013, pp. 1.

Niu, S., Zhai, J., Ma, X., Tang, X. and Chen, W. Cost-Effective Cloud HPC Resource Provisioning by Building Semi-Elastic Virtual Clusters. in proceedings of International Conference on High Performance Computing, Networking, Storage and Analysis, Denver, CO, 2013, pp. 1–12.

Phatak, M. and Kamalesh V.N. On Cloud Computing Deployment Architecture.in proceeding of International Conference on Advances in ICT for Emerging Regions, Bangalore, 2010, pp. 11–14.

Rittinghouse, J.W. and Ransome, J.F. 2010. Cloud Computing: Implementation, Management and Security. ISSN 978-1-4398-0680-7: CRC Press Taylor and Francis group.

Velte, A.T., Velte, T.J. and Elsenpeter, R. 2010. Cloud Computing: A Practical Approach. ISBN 978-0-07-162695-8. Mc Graw Hill.

Wang, H., Wang, F., Liu, J., Xu, K., Wu, D. and Lin, Q. Resource Provisioning on Customer-Provided Clouds: Optimization of Service Availability. in proceedings of IEEE International Conference on Communication(ICC), Budapest, 2013, pp. 2954–2958. www.google.co.in/cloud-computing-images.

Yousefyan, S., Dastjerdi, A.V. and Salehnamadi, M.R. Cost-Effective Cloud Resource Provisioning with Imperialist Competitive Algorithm Optimization. in proceedings of 5th IEEE Conference on Information and Knowledge Technology (IKT), Shiraz, 2013, pp. 55–60.

Zhang, Q., Cheng, L. and Boutaba, R. 2010. Cloud Computing: State-of-the Art and Research Challenges. Journal of Internet Services and Applications. 1(1): 7–18.

Communication and Computing Systems – Prasad et al. (Eds)
© 2017 Taylor & Francis Group, London, ISBN 978-1-138-02952-1

Intelligent tool wear monitoring in machining TI6AL4V alloy using support vector machines

Abhineet Saini, Vanraj, Deepam Goyal, B.S. Pabla & S.S. Dhami
Department of Mechanical Engineering, National Institute of Technical Teachers' Training and Research, Chandigarh, India

ABSTRACT: The ever increasing demand of product quality with economics, in manufacturing industries has attracted the researchers to develop real time monitoring systems for enhanced tool life in machining processes. The tool wear prediction in face milling operation has been presented in this paper to enhance the machinability of Ti6Al4V alloy. Experiments were performed in dry cutting conditions using carbide tools, which were evaluated for flank wear at constant length of cuts. The vibration responses were acquired for these lengths to correlate with the corresponding flank wear values, providing information about the useful cutting tool life. The vibration data was analyzed using support vector machines for prediction of tool failure based on various statistical parameters. The results showed the potential of soft computing techniques in the anticipation of tool life in conventional machining.

Keywords: Tool life; vibration signatures; soft computing; Ti6Al4V alloy; face milling

1 INTRODUCTION

Fiercely international competition has forced manufacturing industries to develop an intelligent system for providing greater value to the process and/or product that have culminated in the creation of minimally manned factories. Due to presence of uncertainty and inaccuracy in manufacturing processes, the various soft computing algorithms have been applied for anticipating the performance of the metal cutting processes and optimizing them. Several monitoring techniques have come into existence to detect tool damage and to measure tool wear by evaluating the key characteristics of the signals acquired from sensors such as, thermocouples, accelerometer(s), acoustic emission sensors, dynamometer, and microphones. Vibration is considered as a diagnostic tool for real time monitoring of the machine as it is non-destructive, authentic, and an uninterrupted monitoring technique. The various vibration monitoring techniques for analyzing the vibration parameter has been reviewed (Goyal et al. 2015).

The desire for maximum tool life for enhanced productivity is today's need. So, the input parameters i.e. feed, speed and depth of cut plays a significant role in predicting the tool life which directly affects response parameters viz. SR, Material Removal Rate (MRR), cutting forces, vibration signature, power requirement during machining. Thus, the input parameters need to be optimized for desired response parameters, or response parameters could be predicted for the selection of input parameters. Both these ways can be applied for the optimization of tool wear, hence, tool life.

At present era of CNC machine tools, it is a critical task to specify optimal machining parameters for achieving machining cost and efficiency have been hot areas of research. In order to meet this, there is a need to introduce artificial intelligence techniques for the prediction and optimization of multi-response parameters. The emerging of Soft Computing (SC) has produced novel opportunities for the manufacturers, which is undergoing a major transformation named as cloud manufacturing. The various condition monitoring indicators for fault diagnosis of dynamic machines has been reviewed (Goyal et al. 2016). Thus, the present paper focuses on the application of SC techniques for the anticipation of tool life in material removal processes and optimization is restricted to so-called traditional machining processes.

Ti6Al4V alloy finds application in number of demanding fields including aerospace, automobile and biomedical industries due to its exceptional properties viz. high specific strength, high corrosion resistance, biocompatibility and potential to retain its mechanical properties at elevated temperatures (Donachie 2000). Thus, the need for machining arises for achieving high dimensional tolerances and quality in components employed in these fields. However, the poor machinability of this alloy is a challenge and results in high cost of manufacturing. This is due to low thermal conductivity, high hot hardness, low chemical inertness and low elastic modulus of this alloy, which makes it a difficult material to cut (Ribeiro et al. 2003). This restricts the application of the alloy in limited fields despite of their unique properties. So, it is essential to develop techniques for economical machining of this alloy (Boyer 2003). Researchers have suggested some specialized techniques for enhanced machinability of Ti6Al4V alloy in terms of cutting tool life, but these techniques have resulted in further increased costs (Bermingham et al. 2011). Hence, there is a need to obtain balance between economics and quality achieved in machining which has been presented using carbide cutting tools (Saini et al. 2016). An experimental study has been attempted to optimize tool life in conventional milling of Ti6Al4V alloy using vibration signatures of cutting tool.

The tool life is generally considered as the measure of flank wear value and is taken to be maximum of 0.3 mm average value for uniform wear. For non-uniform wear, chipping as well as a maximum of 0.6 mm flank wear may be considered as a tool wear criterion. The generalized tool wear growth in a single point cutting tool corresponding to different cutting speeds in shown in Figure 1 (Shaw 2005).

2 EXPERIMENTAL METHODOLOGY

The experimental methodology for the work was completed in two phases. *Firstly*, experimental setup containing description of material, equipment and methodology is presented. *Secondly*, the vibration data of cutting tool, obtained during

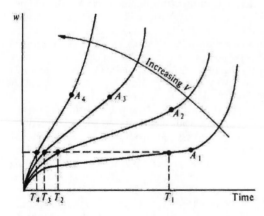

Figure 1. Variation in wear land with time for various cutting speeds.

machining of Ti6Al4V alloy using Support Vector Machine (SVM) was analyzed.

2.1 *Experimental setup*

In the present work, face milling of Ti6Al4V alloy using tungsten carbide tool (WC) was performed in dry conditions on Vertical Milling Centre (VMC). The Tool wear (Tw) and Tool vibration (Tv) data was collected and analyzed for prediction of tool life. Fig. 2 shows the schematic diagram of the machining setup used for face milling. The workpiece in form of plates of annealed Ti6Al4V alloy, size 110 mm × 115 mm × 65 mm, were utilized for the milling operation.

The cutting tools were uncoated tungsten carbide inserts of triangular shape having 3 cutting edges of length 22 mm, 4.5 mm thickness and axial rake angle of 110. Two cutting edges per cut were utilized for the face mill operation. The output parameters were noted after each 110 mm of cut for a maximum up to the point of failure of tool i.e. either tool flank wear (VB) value of 300 μm or tool premature failure stage. The VB of tool was measured for each length of cut for various experiments, using a machine vision system, and an average of maximum, middle and minimum VB values was considered. The images for VB measurement, Built-Up Edge (BUE) and crater face are shown in Fig. 3. For tool vibration, a single axis accelerometer was attached to the spindle of machine, in order to measure acceleration vibration amplitude of cutting tool in the feed direction of cut.

The average of the vibration amplitude data for each length of cut was considered as the output for that specified machined length. The data was recorded and analyzed using NI cRIO and NI LabVIEW software respectively, for single axis accelerometer. The use of Central Composite Design (CCD) and regression technique in Response Surface Methodology (RSM) was utilized for design of experiments. A full CCD model was selected to generate a total of 20 design experiments covering 5 levels for 3 input variables viz. cutting speed (s), feed (f) and depth of cut (d). The vibration ampli-

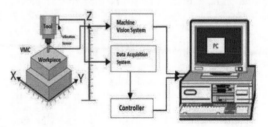

Figure 2. Schematic diagram of the experimental setup.

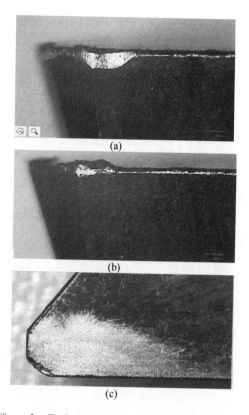

(a)

(b)

(c)

Figure 3. Tool wear pattern from machine vision system showing (a) VB (b) BUE (c) Crater Face.

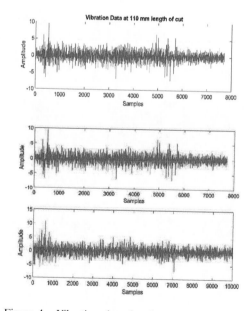

Figure 4. Vibration data for three successive length of cuts (top to bottom) at optimal conditions.

tude data at optimal conditions, thus obtained by RSM, is shown in Fig. 4. The data show an increasing trend of vibration amplitude with successive cuts of same length indicating effect of tool wear on vibration signatures.

2.2 Soft computing technique

Cortes and Vapnik (Cortes & Vapnik 1995) developed a Support Vector Machine (SVM), a learning machine, based on the concept of decision planes for categorization of two group problems.

SVMs are termed as supervised learning models in machine learning with associated learning algorithms that was implemented to classify the training data without errors. This technique involves the representation of the mapped data points in space, so that the separate categories could be formed from the data which are divided by a clear gap as wide as possible. Generally, the hyperplane is used for achieving good separation that has largest distance to the nearest training-data point of any category. The Lagrangian multipliers as well as matrix fundamentals are prerequisites in optimizing the data in SVM. This technique is employed for machine fault diagnosis to maximize the distance of the closest point to the boundary curve (Guo et al. 2003).

Where (w,b) defines the hyperplane separating the two classes of data. 1 and 2 are support vectors, is normal to the plane, b (scalar constant) is the minimum distance from the origin to the plane, is a vector. In order to make each decision surface (w,b) unique, we normalize the perpendicular distance from the origin to the separating hyperplane by dividing it by 'w', giving the distance as b/'w', as shown in Fig. 5.

The statistical parameters of vibration data were incorporated to SVM model and the

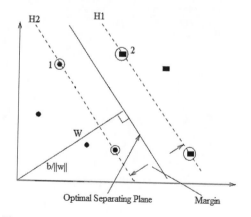

Figure 5. Optimal separating plane and support vectors.

501

results obtained were represented graphically. The 70% of the data was utilized for training and the remaining 30% for validation. Three classes were defined for flank wear to be predicted from the responses of the statistical data. The flank wear value up to 100 μm is considered as the *low flank wear region*, 100–200 μm is considered as the intermediate flank wear region and beyond that up to a maximum of 300 μm average flank wear is considered as the high flank wear region.

3 RESULTS AND DISCUSSION

The scatter plots of Root Mean Square (RMS) vs. all other vibration statistical parameters i.e. kurtosis, skewness, mean, variance and standard deviation (Figures 6–10) showed a liaison with all these parameters representing well distributed data within the domain. The time domain results of vibration amplitude data represented a non-predictable behavior with some higher amplitudes of vibration visible in initial stages, in case of new

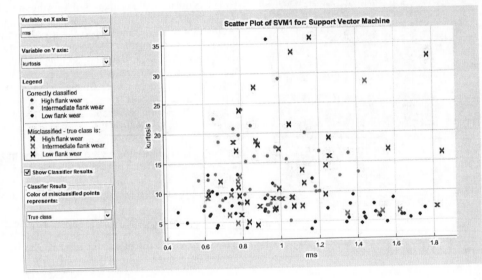

Figure 6. Scatter plot of RMS vs Kurtosis.

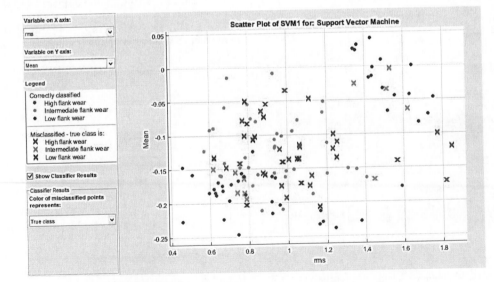

Figure 7. Scatter plot of RMS vs Mean.

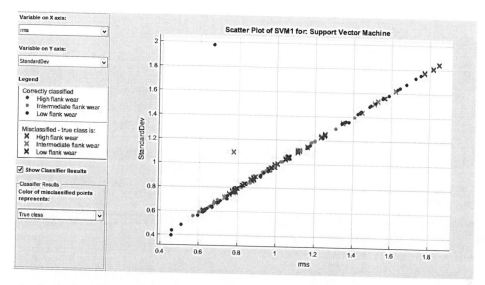

Figure 8.   Scatter plot of RMS vs Standard Deviation.

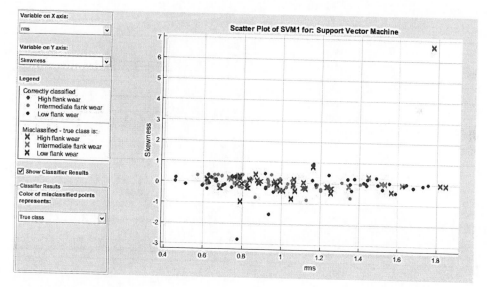

Figure 9.   Scatter plot of MS vs Skewness.

inserts having sharp cutting edges. These vibrations were found to be lesser in case of higher cutting speeds as well as for blunt or wore out edges. Moreover, the abrupt failures due to chipping and BUE had resulted in increased vibration amplitude.

Thus, for the time domain amplitude data, the increase in vibration amplitude was visible due to unpredictable behaviour for certain cutting conditions. The overall prediction accuracy of 63.3% was achieved for fine Gaussian SVM model with low and intermediate levels having high predictability and almost none for high flank wear values. The resulting confusion matrix for the percentage accuracy and error in predicted results is represented in Figure 11.

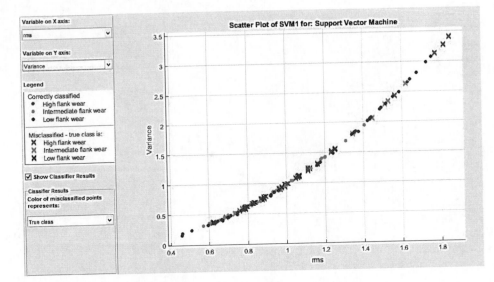

Figure 10.   Scatter plot for RMS vs Variance.

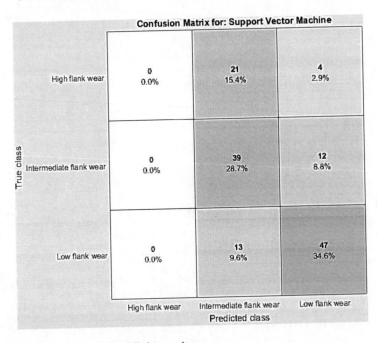

Figure 11.   Confusion matrix of overall prediction results.

The confusion matrix of the predicted data showed a maximum accuracy of 34.6% in predicting accurate class with misclassified class at 9.6%. Further, the SVM model successfully predicted the intermediate flank wear at 28.7% accuracy and wrong prediction as low flank wear at 8.8%. The model failed to predict high flank wear, with a total of 18.3% error as misclassified data. Thus, the overall success rate of the proposed SVM model was for low and intermediate flank wear values which accounted the 82% data. Moreover, the results analyzed for time domain represented statistical parameters in very close vicinity to each other resulting in overlapped data points which further makes it difficult for the classifier to identify and evaluate them. Thus, an improved prediction could be achieved with frequency domain statistical features as well as time-frequency domain features by identifying the unique frequency features for a non-uniformly distributed response data. In nutshell, the prediction of tool wear for difficult to cut materials, resulting from unpredictable cutting behaviors, may be analyzed using soft computing techniques, provided the sensory data be collected precisely and processed using advanced time-frequency signal processing techniques rather than individual time domain or frequency domain features.

4 CONCLUSION

The proposed SVM model shows a good confidence for the predicted results with minimal input data for learning. The conclusions drawn from the results are mentioned and measures for improved prediction are recommended.

1. The SVM model was used for prediction of tool wear from tool vibration data by considering the time domain amplitude data. The vibration amplitude for constant length of cuts was compared with flank wear values. Three classes were defined for flank wear values for classification.
2. The vibration amplitude for constant length of cuts was compared with flank wear values.
3. The classifier was incorporated with RSM, kurtosis, skewness, mean, variance and standard deviation values. The model presented an overall accuracy of 63.3%. The maximum accuracy of 78% was achieved for low flank wear values. Future research may focus to consider frequency domain and time-frequency domain data for improved accuracy. Further, sound signal features in combination with vibration can be utilized to improve the overall accuracy of the proposed model.

REFERENCES

Bermingham, M. J., Kirsch, J., Sun, S., Palanisamy, S., & Dargusch, M. S. (2011). New observations on tool life, cutting forces and chip morphology in cryogenic machining Ti-6Al-4V. International Journal of Machine Tools and Manufacture, 51(6), 500–511.

Boyer, R. R. (1996). An overview on the use of titanium in the aerospace industry. Materials Science and Engineering: A, 213(1), 103–114.

Cortes, C., & Vapnik, V. (1995). Support-vector networks. Machine learning, 20(3), 273–297.

Donachie, M. J. (2000). Titanium: a technical guide. ASM international, ISBN 0-87170-309-2 (1998), pp. 13.

Goyal, D., and Pabla, B. S. (2015). "Condition based maintenance of machine tools: a review", CIRP Journal of Manufacturing Science and Technology, Vol. 10, pp. 24–35.

Goyal, D., and Pabla, B. S. (2015). "The vibration monitoring methods and signal processing techniques for structural health monitoring: a review", Archives of Computational Methods in Engineering, doi 10.1007/s11831–015–9145–0.

Goyal, D., Vanraj, Pabla, B.S., & Dhami, S.S. (2016). "Condition Monitoring Parameters for Fault Diagnosis of Fixed Axis Gearbox - A Review", Archives of Computational Methods in Engineering, 2016, DOI: 10.1007/s11831-016-9176-1

Guo, M., Xie, L., Wang, S., Zhang, JM. (2003) "Research on an integrated ICA-SVM based framework for fault diagnosis", in: Proceedings of the 2003 IEEE International Conference on Systems, Man and Cybernetics, vol. 3, Washington, DC, USA, pp. 2710–2715.

Ribeiro, M. V., Moreira, M. R. V., & Ferreira, J. R. (2003). Optimization of titanium alloy (6 Al–4V) machining. Journal of Materials Processing Technology, 143, 458–463.

Saini, A., Pabla, B. S., & Dhami, S. S. (2016). Developments in cutting tool technology in improving machinability of Ti6Al4V alloy: A review. Proc IMechE Part B: J Engineering Manufacture, DOI: 10.1177/0954405416640176.

Shaw MC. Metal cutting principles. Tempe, AZ: Oxford University Press, 2005.

Communication and Computing Systems – Prasad et al. (Eds)
© *2017 Taylor & Francis Group, London, ISBN 978-1-138-02952-1*

Availability and performance aspects for mainframe consolidated servers

Manish Bhatnagar & Jayant Shekhar

S.I.T.E, Swami Vivekanand Subharti University, Meerut, India

ABSTRACT: Many people think that the mainframe platform now a days suffering from slow but a definite death. But this is a common misconception. The mainframe is alive and going well. This paper will help to recognize the mainframe platform's main qualities and its new driving forces like Server strengthening which is based on the z/VM operating system and Linux. Some of efforts are firstly made to provide a fundamental understanding of the IBM mainframe architecture in order to setup a common ground. z/VM based virtualization provides an insight that how resources can be shared between processes. This report also introduces some of the most essential issues, which are set separated mainframe Linux from Linux on other mainframe platforms. Some Performance tests have been carried out, and through which an insight into the systems behavior can be achieved. This paper also includes some initial considerations on monitoring of the software for the virtual server environment on the mainframe platform.

Keywords: Linux, mainframe, virtualization, partitioning reliable, availability, services, monitoring, memory

1 INTRODUCTION

In educational world of mainframe, and mainly mainframe virtualization technology, this is generally overlooked for the years. The concept is sparsely researched platform which is mostly ignored from an academic point of view. This is natural effect of common perception which mainframe has dead and bare memory of past. As mentioned above, reality is something different, and absolutely there is no indications that situation is about to change. Financial Services Sector has already started the server consolidation process: A handful of perspective application service provider customers are now running Linux on the mainframe. (Keth Adams & Ole, 2006) (Keth et al, 2006).

Application's migration successfully operates on new z/VM-Linux environment, and with promising results. It is nevertheless still a young platform from industry perspective. Industries wishes a deeper understanding of new environment, e.g. of interaction between the virtualization layer (z/VM), hardware and Linux.

The ability to control and predict the distribution of resources is also of interest. Unfortunately the existing review tend to be very "how-to oriented", unnecessary detailed, or unfeasibly extensive (IBM's "z/VM library" alone include over 50 books). To sum up, the report is expected to:

- Provide an introduction to the mainframe architecture.
- In details, describe virtual environment and explain the interaction between, and mutual support of:
  The hardware and hardware partitioning
  Software virtualization layer (z/VM)
  Linux as operating system in a virtual machine
- Show the resource sharing/virtualization options, and in particular the possibility to control and predict processor resource distribution.

Finally, gained platform insight should be used to address matter of availability: "monitoring" in particular. Its goal is to ensure optimum and flawless 24 × 7 operation of the mainframe based perspective ASP (Application Service Provider) solution.

2 RELATED WORK

Virtualization principles and concepts—Virtualization is a huge term to cover several different concepts and techniques. Virtualization is in general terms, ability to abstract computer resources into virtual or logical resources and/or computing environments. Virtualization can cover technique that can be obtained using resource aggregation, resource sharing, insulation, emulation. Resource sharing is ability to make multiple virtual resources based upon physical resource e.g. by partitioning and time sharing. Resource aggregation is ability to combine multiple physical resources into fewer virtual resources. With emulation available physical resources can be used to imitate or simulate resource types and features, which are physically unavailable. The last cornerstone of virtualization is insulation: the ability to segregate resources and/or environments and render it impossible for them to influence each other.

Together are four fundamental capabilities which are enabling various kinds of virtualization. The area of resource virtualization can cover like an example several techniques, which are not immediately associated with virtualization:

- Combine disk into huge logical disks (RAID and Volume managers)
- Combine several discrete computers into a whole (grids, clusters)
- Creating virtual networks within another network (VLAN, VPN) (Amit Singh, 2004).

System virtualization: Create virtual machines. In their context the focus is on "system virtualization" or "platform virtualization"—the ability to make several "virtual machines", which in every detail architecturally resemble hardware platform—in this case 'the mainframe'. It is an ability to transparently share the resources without consumer's knowledge to archive positive resource utilization.

Two major approaches for system virtualization exist:

"Hardware partitioning" and "hypervisor based" technology.

2.1 Hardware partitioning

Few computer platforms can support "hardware partitioning". They will allow for hardware to partition while using coarse grained units. Each partition can run separate operating system. This approach can allow for hardware consolidation but lacks ability to share the resources in order to prolong utilization. "HP nPartitions" and "Sun Domains" are examples of this technique. (Damian et al, 1991).

2.2 Hypervisor based partitioning

Virtualization by "hypervisors", on other side, allow for the fine-grained, and dynamic sharing of resources. A hypervisor constitutes a shallow software layer used to make and overcome virtual machines. Hypervisors are called "Virtual Machine Monitors" (VMM) (John et al, 2003) (Amit Singh, 2004).

2.3 Variant types of hypervisor based virtualization

Multiple techniques are used within each of these main categories. The differences arise from varying levels of virtualization support within hardware and from varying levels of interaction between hypervisor and guest operating system. The following sections are important hypervisor techniques, but first few low level computation fundamentals are recapped.

Multitasking operating systems can run several processes and shield from each other. To ensure operative system remains in control CPUs which can be typically supports two modes of operation: privileged (kernel or supervisor) mode and unprivileged (or user) mode. Only operating system can run in the privileged mode is allowed to run privileged instructions, which are instructions that can change the overall state of the system. When an unprivileged process executes a privileged instruction it causes a trap (an exception), which forces the CPU back into the operating system code. The operating system handles the situation (in privileged CPU mode) and returns control to application process afterwards. There are other situations, which drive a CPU from user mode to kernel mode: for instance (timer) interrupts and page faults. Under normal conditions privileged instructions are avoided in applications programs. Instead they depend on syscalls (operating system functions), which deliberately switches into kernel mode and handles the operating in a more efficient manner avoiding traps (Andrew Whitaker et al, 2002).

2.4 Paravirtualization (Hypervisor call method)

Sometimes a hypervisor creates virtual machines with an architecture, which differs slightly from the physical machine architecture. As a consequence parts of the operating system running in such a virtual machine have to be rewritten to run in the special virtual environment. The changes typically include explicit hypervisor calls for I/O and memory management. This kind of virtualization is known as para virtualization. In other words, special code segments that explicitly call hypervisor functions are introduced to source code of guest operating system. This allows for better optimization of interaction between hypervisor and guest. The operating system has become aware of virtual environment. Therefore virtualization strategy may improve scalability; it can reduce system complexity, allows for high efficiency. Unfortunately this method cannot be applied to proprietary operating systems, if the vendor can choose not to support virtualization platform.

The method has already applied in VM/370 and which is still used in z/VM today. A newer example is Xen, which among other operating system runs OpenBSD, Linux, FreeBSD, and OpenSolaris (Andrew et al, 2002).

2.5 Direct hardware support

The hypervisor variant to mention here which relies on direct virtualization support in the hardware. In this case of architecture includes hardware constructs and special instruction to run virtual machines very efficiently. This can be run for example include special "guest" CPU mode. In this mode guest operating system can run most instructions. The hardware can assist when processors are dispatched to the virtual processors and when execution can be handed back to hypervisor. This can provide the hypervisor with better measures to accommodate the conditions, which causes exceptions and hand back control to the hypervisor. Other examples of virtualization mechanism in hardware include the ability to route I/O interrupts directly to the guests; optimizations for multiple levels of virtual memory (multiple levels of dynamic address translation); and hardware measures to keep the hypervisor immune for critical errors (including I/O related errors) within the guest. The mainframe is the classical example having provided hardware "assist" since the "370 days". z/VM (and PR/SM) utilizes all these capabilities. The last few years of increased focus on virtualization have also made Intel and AMD integrate (incompatible) virtualization technology in newer x86 chipsets and processors. (Keith Rochord et al, 2006).

3 OUR NEW APPROACH

3.1 Performance tests and optimization

In this, multiple monitoring tools, scripts and test programs have been developed in order to observe the values of special interest and to be able to influence and stress the system in particular ways.

TEST: Number of processors per Linux guest. This test can be designed to search the optimum processor configuration for Linux guests in company specific z/VM environment. The test can display how number of virtual processors can be influenced the duration of typical, real and demanding jobs/workloads. The diagram x'44 instruction counts is determined to verify the 2-CPU guests actually call instruction.

3.2 Measurement methods and tools

qMonitor: Specialized monitoring in general, Linux provides many features and tools, which can be used when monitoring or testing the system.

That is tools like free and top, not to mention "raw data" provided by kernel in /proc/ file system e.g. meminfo, stat, and vmstat. Unfortunately output can be hard to comprehend and interpret in a "live" situation and output format is seldom suited for subsequent data analysis. Naturally, dedicated monitoring programs and products exist to counter these difficulties. But in other cases such monitors products are completely overkill, their sample rate is too low or they miss customization capabilities to allow for highly specialized measurements. Fortunately Linux systems typically come with many small tools and capabilities, which make it possible to create a customized monitoring service with a relatively small amount of work. This particular approach has been adopted here. 'qMonitor' A monitor solution, named qMonitor, has been developed for this project with following in mind:

- Extreme customization abilities
- Near real time data delivery to allow real time graphing
- Data gathering option for later data analysis
- Simple data format allowing easy processing in standard tool (e.g. Excel)
- Connectivity to let remote data graphing and gathering
- Simplicity is important than avoid communication overhead etc.

qMonitor utilizes a mix of xinetd (the eXtended InterNET Daemon), pipes, scripts and values from /proc and programs like top as shown in Figure 1.

The real monitoring process can be controlled and instantiated by client. It connects to configure TCP port and xinetd invokes mon.sh script, if connection can be successfully open. Now the client sends a character represents the wanted monitoring

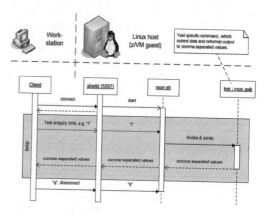

Figure 1. The monitoring setup constructed to deliver one of multiple sets of customized measurements to an arbitrary workstation via a TCP/IP socket.

data, for instance a 't' represents information from top. The mon.sh script parses input and invoked a corresponding command. This command can perform reformats and necessary measurements the output into comma separated string, which is written to stdout and thereby returned to the client.

Each time client sends a new code; a corresponding data string can be returned and generated. The client can easily parse comma separated string and potentially graph values in near real-time. Write their unaltered data directly to text file which is automatically generates "CSV" file, which is easily readable for most spreadsheet programs for their further analysis and later graphing. In this section, a simple client with graphing and logging capabilities have been developed in Java, but any type of program or script could be used. The GUI of the Java client program is presented in Figure 2.

When the client decides to stop the monitoring process it sends a stop code "q". This stops the mon.sh script by breaking an internal loop. Xinetd automatically closes the connection to the client as the mon.sh script terminates.

3.3 Data gathering from within Linux

Test data for this particular test is mainly derived from the standard UNIX tool "top". The program provides a real time view of the running system, including both system summary information as well as information on the individual processes. Especially the latter can make top useful in particular case, since it can provide easy access to exact information needed.

The mon.sh-script can invokes "top" in batch mode with one repetition, which disables the interactive interface and sends single output set to stdout (stand-

ard out). The output can be piped into an awk script (ora_mr.awk), that parses it and generates comma separated string including the following values:

a. Regarding processes owned by application user (mrdata):
- number of processes can run, sleep, in uninterruptible sleep
- memory usage
- % CPU usage (inaccurate "time tick based" value)

b. Regarding processes owned by the DBMS user (oracle):
- number of processes can run, sleep, in uninterruptible sleep o memory usage
- % CPU usage (inaccurate "time tick based" value)

c. Number of other processes (not oracle) running.

d. CPU usage related to other (not oracle) processes.

3.4 Data gathering outside Linux

The "IBM Performance Toolkit" can be used to verify CPU readings above mentioned that is basically cannot be trusted, since present Linux version is unaware of time VM spends servicing guests. IBM Performance Toolkit is licensed product that can be purchased separately, although it is shipped as component of z/VM. The Toolkit can provide variety of performance measurements, system information, and logging options (see Figure 3). Like all things it runs in virtual machine, but it depends upon special data provided by CP that enables it to present picture of complete virtual environment from "hypervisor view" (see Figure 4).

The toolkit CPU readings are used to confirm, that workloads are single threaded and virtual machines are generally occupy single processor regardless of number of virtual processors. The toolkit can use here

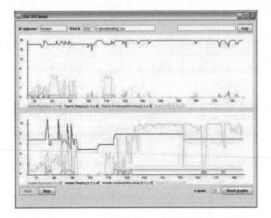

Figure 2. The GUI of the developed Java client program, which is used together with the qMonitor monitor. (Graphs are based on a class from the JChart2D library from http://jchart2d.sourceforge.net).

Figure 3. The selection menu from the "IBM Performance Toolkit" web interface.

Figure 4. The "IBM Performance Toolkit for VM" runs in its own virtual machine (typically "perfsvm") but the data is provided from CP.

to verify the tests which are completed in a "stable" environment without any big fluctuations which is caused by other guests. Finally Performance Toolkit can be used to verify the number of CPUs influences number of diagnose x'44 instructions. Apparently is not possible to monitor the number per user.

3.5 Workloads

The test can be performed to use workload definition WL1 and WL2 that have following characteristics.

3.5.1 Workload 1
Workload 1 is "complete payroll processing" (payslip creation excluded): Several independent application executables are called during the payroll processing that results in birth and termination of various newest processes. Several programs are individually access and "manipulate" the database.

The workload is of "mixed nature": A combination of CPU usage (application calculation + database queries) and disk usage. The CPU usage can typically fluctuate and rarely hits 100% (equivalent to one virtual CPU) for very long.

3.5.2 Workload 2
(WL2) Workload type 2 consists of Payslip PDF and PS file creation. It is dominated by some sequentially running and very CPU intensive processes. The processing can include creation of PDF files from XML, and P00DF to PS conversion. WL2 provides a more steady state / long run / high CPU load situation.

3.6 Test description

The goal is to display, if number of virtual processors influences the duration of workloads above. To eliminate this influence from other z/VM guests (resource availability), durations for workloads can be evaluated relatively: By running same workloads in 1-CPU guest and a 2-CPU guest concurrently. This allows for tests to be run in production environment. There are two guests which are running concurrently and completely identical except for number of CPUs. As safety precaution test can be repeated with their opposite CPU configuration

on same two guests in order to eliminate any unintended configuration differences.

As supplement to these "relative" tests with their concurrently running guests, workloads are run separately in one guest at time: first in 1-CPU guest and afterward in a 2-CPU guest. This is necessary to get Diagnose X'44 instruction count, since is impossible to monitor for individual guest. These separate tests can ensure that concurrent processing doesn't distort the results notably.

3.7 Test results and analysis

The CPU readings received from the Linux guests clearly confirm the mentioned workload characteristics. Figure 5 and Figure 6 give typical CPU readings for workload 1 and 2 respectively. Generally the readings for 2-CPU guests fluctuate more than the case for 1-CPU guest.

It should be mentioned that the data illustrated have not been normalized using values from Performance Toolkit. It has, however, been confirmed, that CPU usage does not exceed 100% (~1 real CPU). An example of the Performance toolkit readings are

Table 1. Test: Combinations of CPU and "relative" share settings together with workload name, the number of repetitions and monitor settings.

Test #	LINUX GUEST1 CPUs Ref share Workload			LINUX GUEST2 CPUs Ref share Workload			Rep
T1.1	2	200	WL1	1	100	WL1	4
T1.2	2	200	WL2	1	100	WL2	4
T1.3	2	200	WL1	–	–	–	2
T1.4	–	–	–	1	100	WL1	2
T1.5	2	200	WL2	–	–	–	2
T1.6	–	–	–	1	100	WL2	2
T1.7	1	100	WL1	2	200	WL1	4
T1.8	1	100	WL2	2	200	WL2	4
T1.9	1	100	WL2	–	–	–	2
T1.10	–	–	–	1	100	WL2	2
T1.11	1	100	WL2	–			2
T1.12	–	–	–	2	200	WL2	2

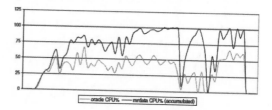

Figure 5. CPU usage characteristics for WL1. Contributions from oracle owned and application (mrdata) owned processes in accumulated (stacked) view. Performance toolkit readings confirm maximum CPU readings ~1 whole CPU.

511

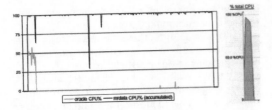

Figure 6. CPU usage characteristics for WL2. Contributions from oracle owned and application (mrdata) owned processes in accumulated (stacked) view. (Not normalized Linux readings, T1.2, 1-CPU-guest, workload during/execution time: 200 s).

Table 2. Execution time savings in percent gained by using 1-CPU guests (in the test setups with two concurrently running guests) generally no effect on workload 2.

T1.1	14%	T1.2	−4%	T1.7	6%	T1.8	1%
WL1	14%	WL2	2%	WL1	6%	WL2	0%
	22%		0%		6%		−1%
	17%		−1%		10%		1%

given in Figure 6. Furthermore Performance Toolkit readings also confirm the general picture of a less intensive CPU usage for workload 1.

Performance toolkit readings, shown in the graph to the right, confirm maximum CPU readings ~1 whole CPU. All tests based on workload 1 show longer execution times on the guest running with 2 CPUs. The tests show execution time savings on the 1-CPU guest between 6% and 22% (see Table 2 and Figure 7).

Workload 2 is on the other hand not affected. Actually two of the tests does show slightly longer execution times on the 1-CPU guests (2–4%), but the general picture is that WL2 is unaffected by the number of CPUs (most values vary ± 1%).

Furthermore there is no reason to believe that the effect is related to the two guests running concurrently—influenced by the test condition that is. Although less trustworthy the measurement from the tests with one guest running give the exact same picture. One exception is test T1.5 and T1.6 (Appendix A.1.4.1). These tests, however, were carried out an hour apart and seem to be influenced by VM environmental matters.

The Diagnose x'44 readings show, as expected, that guests with two CPUs issues many of these privileged instructions: The rate explodes with more than 5000 instructions per second for WL1 and nearly 3000 instructions per second for WL2.

Naturally a CPU reduction restricts the amount of work a multiprocessing guest can do. It does however also make the scheduling job easier for the hypervisor. It will always leave one CPU to achieve acceptable response time for others, while a guest generates pdf pay slips, a Linux guest otherwise gets out of control or similar.

3.8 Availability monitoring

Monitoring of hardware and network devices as well as software services, are important concepts when managing an IT infrastructure. Larges enterprises and Companies are dependent upon IT or

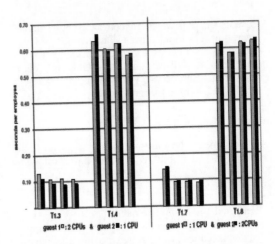

Figure 7. Workload duration measures. 1-CPU systems are generally faster on the database intensive workload which involves several sequential processes (T1.3 + T1.7). The difference disappears on the "pure" CPU intensive workload (T1.4 + T1.8).

profit on providing them, often have dedicated monitoring centre. Here people can supervise the operation of complete IT environment 24 × 7. In this way it is possible to react quickly when an issue arises—and hopefully resolve them before they become real problems. Regardless of size and complexity of system(s) being monitored, it requires some kind of programs or systems to do the job.

Monitoring methods: Monitoring systems fall into two groups depends upon technique used to gather data. The first group relies on centralized approach. Here single program or a few processes in single location collects data. The central monitoring system actively can try to determine the status of monitored entities. This is accomplished simply to establish some kind of connection to them or by performing operations involving them.

The second group is based upon distributed approach. Here special monitoring programs can be placed on or close to monitored entities. These programs can often called "agents" and perform necessary tests and report outcome to relevant entities. These agents may range from simpler programs to complex intelligent autonomous mobile soft2.

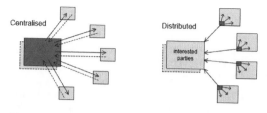

Figure 8. Centralized versus distributed monitor approach.

3.9 Components to monitor

3.9.1 Standard warning / error conditions

This section represents typical conditions that result in warning or errors on normal servers, e.g. UNIX servers and Microsoft Windows servers. It does not take number of repetition, duration, the severity of event into account, although thresholds/levels typically are specified.

Processor: Processor usage is monitored in order to track system irregularities, or general resource shortages. Concrete warning condition might be included:

- High system (kernel) CPU usage, e.g. above 80%
- Generally low idle time (might indicate CPU overload/resource shortage).

Memory: Memory is naturally an important resource. This is a severe problem, if system runs out of the memory. It can have a negative impact on the performance if system constantly requires swapping because the active working set is bigger than physical memory. Examples of error/alert conditions:

- The system is low on free swapping space and thereby on memory.
- The page rate to/from swap is too high.

Disk, file system, and files: Disks can be monitored in order to ensure the space for persistent as well as temporary data. In other situation monitoring include physical wellbeing of disk devices, in order to ensure the data integrity. Unexpected periods with large amounts of writes/ reads might indicate an I/O bottleneck or fail application out of control. Individual files might be monitored for changes to reveal security breaches or because files can be important for system operations. Concrete examples of warning/alarms in this category:

- Disk less on available space, e.g. below threshold in MB or a percentage.
- Low disk on i-nodes, e.g. below specific percentage of available i-nodes.
- High disk utilization, "too many" reads or write.
- Any modification of files, e.g. /etc/passwd and /etc/group.

Network interfaces and connectivity: A general well-functioning server does not worth much, if services are running on it which cannot be reached, because network is malfunctioning. Examples, this could raise warning or errors:

- Higher Packet loss.
- Multiple packet checksum errors or packet data collisions
- Network interface down. Ware elements. (Linus et al, 1999)

Programs, services, processes, and logs: The last category focuses on the individual programs. If key process, program, or service is not found on process list as expected, this is clear indication of error. In some situations it can be useful to follow (filter) log files to reveal the error conditions. Sometime this can even be important to call actual service and ensure it provides expected results with adequate response time.

- Unexpected large processor utilization of individual programs/daemons
- High number of zombie processes
- Special entries appear in log files
- A service does not respond as expected within acceptable time.

3.9.2 Mainframe/virtualization considerations

Running servers in virtual environment such as z/VM need some consideration of what to monitor, how to do it, and from where to do it.

Processor: Processor usage readings are good example of to monitor the data, which cannot be delivered from the Linux guest. Although never Linux kernels include "steal time". This is definitely not case with older kernels. In virtual environments dependent on older kernels, is therefore necessary to trustworthy CPU usage readings from the CP layer.

Hypervisor data: Monitoring from within individual guests (Linux or not), will generally not be able to reveal much useful information about virtual environment as whole. Only data from hypervisor level can offer a trustworthy picture of all virtual machines together. Important values include CP paging activity, I/O rates, and the overall CPU utilization. Luckily z/VM provides general method to access CP monitor data for programs such as the "IBM Performance Toolkit" and other monitoring products.

Instrumentation overhead: When running virtual servers, focus on the resource sharing and consumption typically increases, and it becomes much more apparent, if monitoring software squanders processor cycles and memory away. If instrumentation overhead for single Linux instance come close to feeble 1% of a processor, monitoring system may alone will quickly consume half or processors depends upon the number of guests. Adding some processors can increase the license costs of remaining software portfolio, and monitoring software suddenly becomes incredible expensive. Monitoring solutions should therefore be small, fast and efficient and preferably have a small memory footprint.

4 CONCLUSIONS

In this, beginning from scratch, with limited academic understanding of mainframe architecture, analysis can provide basic insight in, and appreciation of, IBM mainframe hardware and architecture; in particular when it comes to topic of the system virtualization while using z/VM operating system. According to paper goals and aspects the report defines how the z/VM operating system will be able to establish the virtual machines that behave such as real machines in every detail. The theoretical assessment of the virtual environment is supported empirically, with a number of performance tests. These tests have provided hands-on experience, and a practical comprehension of the system. The performance tests given in report illustrate the behavior of mainframe hardware, z/VM, Linux, and the combination of the three. These tests should generally contribute to a better understanding of running Linux under z/VM on the mainframe. Additionally the tests contribute with a few concrete performance optimization recommendations for the company specific setup. The paper finally touches upon the topic of availability of mainframe Linux services—solely focusing on monitoring software according to company wishes. The monitoring chapter should only be regarded as an offset for further investigation, since it has not been possible to make thorough study of this topic in due time.

All in all the main goals for this paper have been achieved, except when it comes to finding most suitable monitoring software solution for Industries. The work does however still provide the detailed platform knowledge, and initial monitoring considerations, which can useful in order to fulfill the goal.

REFERENCES

Amit Singh. An introduction to virtualization.

Andrew Whitaker, Marianne Shaw, and Steven D. Gribble. Denali: Lightweight virtual machines for distributed and networked applications. University of Washington Technical Report, February 2002.

Charlie Burns. The mainframe is dead: Long live the mainframe. Saugatuck Technology, Research Alert: http://research.saugatech.com/fr/researchalerts/364RA.pdf, July 2007. RA364

Damian L. Osisek, Kathryn M. Jackson, and Peter H. Gum. Esa/390 interpretive-execution architecture, foundation for vm/esa. IBM Systems Journal, Vol 30, No 1, 1991.

Edi Lopes Alves, Eravimangalath P. Naveen, Manoj S Pattabhiraman, and Kyle Smith. Introduction to the New Mainframe: z/VM Basics. IBM Redbooks, 1th, draft edition, August 2007. SG24-7316-00.

Gregory Geiselhart, Laurent Dupin, Deon George, Rob van der Heij, John Langer, Graham Norris, Don Robbins, Barton Robinson, Gregory Sansoni, and Steffen Thoss. Linux on IBM eServer

http://h20338.www2.hp.com/hpux11i/cache/323751-0-0-0121.html, August 2007.

http://www.kernelthread.com/publications/virtualization/, February 2004.

http://publib.boulder.ibm.com/infocenter/eserver/v1r2/topic/eicay/eicay% .pdf, September 2006.

http://linas.org/linux/i370why.html, 2007-09-16, November 1999.

IBM systems virtualization (r2v1).

John Langer, Graham Norris, Don Robbins, Barton Robinson, Gregory Sansoni and Steffen "Linux on IBM server"

Keith Adams and Ole Agesen. A comparison of software and hard ware techniques for x86 visualization, 2–13, 2006.

Keith Rochford, Brian Coghlan, and John Walsh. An agent-based approach to grid service monitoring, ISPDC, 10–109, 2006.

Linus Vepstas. Homepage: Why port linux to the mainframe?

Linus Vepstas. Homepage: Linux on the ibm esa/390 mainframe architecture. http://www.linas.org/linux/i370-bigfoot.html, 2007-09-16, February 2000.

Lydia Parziale, Eli Dow, Klaus Egeler, Jason Herne, Clive Jordan,

Michael Steil. Inside vmware: How vmware, virtualpc and parallels.

Communication and Computing Systems – Prasad et al. (Eds)
© 2017 Taylor & Francis Group, London, ISBN 978-1-138-02952-1

Software quality assurance using firefly optimization algorithm

D. Panwar & M.H. Siddique
Amity School of Engineering and Technology, Amity University Rajasthan, India

P. Tomar
Department of Computer Science and Engineering, School of ICT, Gautam Buddha University, India

ABSTRACT: Software quality assurance means the stakeholder satisfaction. When the producer and consumer both are satisfied from a given product than it is a quality product either the product is from mechanical industry, electrical industry or from software industry. Current scenario of today's software development market is dependent upon Commercial-Off-The-Self components (COTS) and the Open source software. So it is crucial thing to predict the quality of Component-Based Software (CBS). This study proposed an objective function for the prediction of software quality with the help of firefly optimization technique of Computational Intelligence and gives the better result on MATLAB with real data.

1 INTRODUCTION

The quality of software is fundamental for success of a software product (Kitchenham et al. 1989). The examination of meanings, views and quality definitions and denotes the quality of product as hard to measure and define whereas easily recognizable. Till now, various definitions define a quality (Kitchenham et al. 1989), (Wong 2002) and also various models have been introduced. Figure 1 describes the timeline for various software quality models introduced. According to IEEE, quality

is "the degree to which a system, component, or process meets specified requirements and customer or user needs or expectations" (IEEE 1998) and ISO has defined that the quality is "the totality of features and characteristics of a product or service that bear on its ability to satisfy specified or implied needs" (ISO 1986). The dependency of quality view is based upon various product contexts, the meanings that are assigned to the quality attributes as well as the relationships among them. The International Standard Organization (ISO) and Electrical Technical Commission (IEC) introduced ISO/IEC 9126 Standards for Product Quality for Software Engineering in order to provide evaluation model and comprehensive specification for the evaluation of software products quality evaluation. The focus of this paper is to enhance the quality of a software product by using Firefly Algorithm (FA) (Yang 2008) on six major characteristics of ISO/IEC 9126. The idea behind this research is that optimization algorithms have become powerful in solving various modern global optimization problems as well as combinatorial problems and have shown positive result in optimizing the solution. Thus FA has been used to optimize the solution and enhance the quality of the software product. Our research is mainly aimed to analyze the internal model quality metrics of ISO/IEC 9126 Standard Software Quality Assurance model.

The paper takes the following format: In section 2, the researchers have discussed ISO/IEC 9126 briefly. In section 3, related work in context of software quality assurance models have been discussed. In section 4, traditional Firefly Algorithm is discussed. In section 5, the proposed work

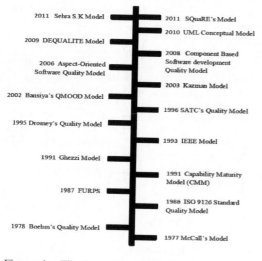

Figure 1.   Timeline of various Software Quality Models.

done by researchers has been discussed and the validated by an example. In section 6, the paper has provided some conclusion.

2 ISO/IEC 9126

ISO 9126 is an international standard for software product evaluation from the International Organization for Standardization which defines six characteristics that describes software quality with minimal overlap. ISO 9126 consists of four parts:

- ISO/IEC 9126-1: Quality Model
- ISO/IEC TR 9126-2: External Metrics
- ISO/IEC TR 9126-3: Internal Metrics
- ISO/IEC TR 9126-4: Quality in use Metrics

The ISO/IEC 9126 defines a two-part quality model for software product that is internal/external quality and quality in use. Six characteristics are specified for internal and external qualities which are further subdivided into sub-characteristics. The six characteristics are *functionality, reliability, usability, efficiency, maintainability* and *portability*.

The first three parts of the ISO/IEC 9126 is related to the measurement and description of the quality of software product while the fourth part describes the evaluation of the product from the user/customer point of view. There are four substantial documents related to ISO/IEC 9126 Standard but this paper provides a brief overview due to paper length restrictions. The main intention for introduction of ISO/IEC 9126 software quality assurance model is:

- To provide comprehensive evaluation as well as specification model for quality of software product.
- To address user needs of the software product by applying common language for the specification of user requirement that is supposed to be understandable by customers, developers and evaluators.
- To evaluate the quality of software on the basis of observation rather than opinion.
- To make the evaluation of quality reproducible (Punter et al. 1997).

The two-part quality model for software product that is internal/external quality and quality in use are the main parts of ISO/IEC 90126 standard model for software quality evaluation.

Part one describes the quality of the software product by the use of six defined characteristics of quality i.e. functionality, reliability, usability, efficiency, maintainability and portability. Each particular quality characteristic has related sub-characteristics which are described by appropriate internal as well as external software quality

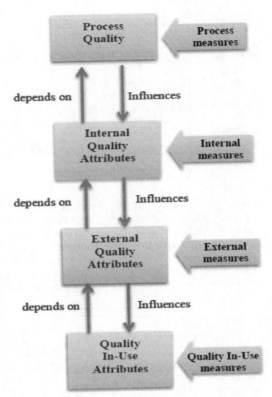

Figure 2. ISO/IEC 9126 Quality Model Framework Lifecycle (ISO 2001).

attributes which are based on measurement by specified metrics. All the metrics defined in (ISO 2002), are described in terms of formula, purpose for the software development according to the customer's requirement, method of application used, computation of data elements, measurement, scale and measure type, value interpretation as well as source of input and output audience.

In second part of quality model, "Quality in Use", four different quality characteristics are defined namely, productivity, effectiveness, satisfaction and safety, which according to the suggestion of ISO/IEC 9126, are indicative of the user's view of quality that are based on the attribute's combined effects which are specified in standard's part 1 (ISO 2002). Figure 2 describes the Quality Model Framework Lifecycle defined by ISO/IEC 9126 (ISO 2001).

3 RELATED WORK

Since last three decades various quality models were introduced for modeling the quality product in order to fulfill the customer requirement specification as described in Figure 1. Various combinatorial problem solving optimization algorithms have been applied on ISO/IEC 9126 Standard software quality model. A number of studies (Behkamal et al. 2009), (Chang et al. 2008), (Srivastava et al. 2011), (Yuen 2011), (Punter et al. 1997) integrated the quality factors of ISO/IEC 9126 to the approaches for decision making using fuzzy multi-criteria. To evaluate B2B e-commerce applications, Analytic Hierarchy Process was applied (Behkamal et al. 2009) to ISO/IEC 9126. Fuzzy Analytical Hierarchy Process was applied by Yuen and Lau (Yuen et al. 2011), in which its fuzzy proportion had used the modified form of traditional fuzzy logic i.e. fuzzy logarithmic least square method (Wang et al. 2006) (Wang et al. 2008), to ISO/IEC 9126 for selection of software vendors on the basis of relative measurement. During last few decades nature inspired meta-heuristic algorithms have become powerful in order to solve combinatorial problems, especially problems like NP-Hard such as travelling salesman problem. The main advantage of using such meta-heuristic algorithms is their simplicity as well as easiness in implementation. These meta-heuristics are inspired by natural and social behaviors of some species. Some of them are Differential Evolution (Price et al. 1995), Particle Swarm Optimization (Kennedy et al. 1995), Genetic algorithm (Goldberg 1989), Cuckoo Search Algorithm (Yang et al. 2009), Shuffled Frog Leaping Algorithm [8], Firefly Algorithm [10] etc. A significant number of researchers have done work on software quality assurance using artificial intelligence by examining the effective use of artificial intelligence and software engineering related activities that are inherently human-centered and knowledge intensive. These types of issues lead to the need for the investigation of the suitability of above mentioned search algorithms. In this paper, FA has been used to measure the quality of the software product by evaluating the six main characteristics (i.e., *functionality, reliability, usability, efficiency, maintainability* and *portability*) of ISO/IEC 9126 software quality assurance model and one additional characteristic i.e. *Reusability* (Panwar et al. 2012).

4 FIREFLY ALGORITHM

The flashing light patterns of fireflies are unique for a particular species. Bioluminescence is the process due to which the flashing light is produced. There are two fundamental functions of these flashing patterns that is to attract mating partners and also attract potential prey. Females respond to the unique flashing pattern of males of same species for mating, while in some species like *photuris*, females are capable of generating other species pattern in order to lure and eat male fireflies of other species attracted by those patterns. The flashing light phenomena is described in such a manner that the associativity of this phenomena with objective function in order to be optimized, due to which FA can be used for enhancement of quality of software products. For simplicity in description of FA, the researchers have used three idealized assumptions.

1. All the fireflies are supposed to be unisex for the fireflies to be attracted by other fireflies regardless of their sex.
2. The more the brightness the more is attractiveness. Therefore, less bright firefly will be attracted towards much bright firefly. As the distance between two fireflies increase, attractiveness as well as brightness will be decreased. The firefly with highest intensity (i.e. brightest firefly) will move randomly.
3. The brightness of firefly is determined by the objective function for FA.

Formation of attractiveness and the variation in light intensity are the two main issues in FA. So, for simplicity of the algorithm, this paper assumes that brightness B of firefly is responsible for its attractiveness with other fireflies. For maximum optimization problem, attractiveness β is varied with distance d_{ij} between firefly i and firefly j whereas the brightness B of a firefly can be chosen as $B(x) \propto f(x)$ where x is the particular location of a firefly. As the intensity of light decreases by increasing the distance from its source as well as some proportion of light is absorbed by media in which light is traversing, therefore this algorithm allows the attractiveness factor to be varied with the degree of absorption. So, by inverse square law, the intensity of light denoted by $\tau(d)$ is varied which is given by:

$$\tau(d) = \tau_s/d^2 \tag{1}$$

Where τ_s is the source intensity. For the given medium having γ as the fixed light absorption coefficient, the light intensity has to be varied with respect to distance d, given by:

$$\tau = \tau_0 \, e^{-\gamma d} \tag{2}$$

Where τ_0 describes the original intensity of light. Attractiveness of fireflies is proportional to the intensity of light that has to be seen by other fireflies. Therefore, attractiveness is given by:

$$\beta = \beta_0 e^{-\gamma d^2} \tag{3}$$

where β_0 is attractiveness of fireflies at distance $d = 0$.

5 PROPOSED WORK AND EXAMPLE VALIDATION

5.1 Firefly algorithm pseudo-code

```
Begin
  Initialize with the objective function f(x).
  Initialize by generating the initial firefly population x_i,
  i=(1,2...n)
    Define the light intensity using equation 2.
    Describe the coefficient light absorptionγ.
    while(i<MaxGeneration)
    for i = 1
      for j= i all n fireflies
        if(τ_j > τ_i)
        move i^th firefly for firefly j.
        end if
      end
    end
    provide the rank to each firefly and find one with cur-
rent best.
    end while
    post-process the result obtained by objective function.
  end //end of algorithm
```

For the implementation of algorithm, any of the monotonically decreasing functions can be implemented for the actual parametric form of attractiveness function as given below:

$$\beta(d) = \beta_0 e^{-\gamma d^m}, (m \geq 1) \qquad (4)$$

For a fixed absorption coefficient γ, the characteristic length Γ, becomes:

$$\Gamma = \gamma^{-1/m} \to 1 \quad as \; m \to \infty \qquad (5)$$

Cartesian distance is the actual distance between i^{th} firefly at positions x_i and j^{th} firefly positions x_j and is determined by:

$$d_{ij} = \|x_i - x_j\| = \sqrt{\sum_{k=1}^{d}\left(x_{ik} - x_{jk}\right)^2} \qquad (6)$$

where x_{ik} denotes the kth component of ith firefly's spatial coordinate x_i. The next position of the ith firefly is based on its attraction to more brighter j^{th} firefly, and is determined by:

$$x_i = x_i + \beta_0 e^{-\gamma d_{ij}^2}\left(x_j - x_i\right) + \alpha \; sign\left(rand - \frac{1}{2}\right) \qquad (7)$$

where the second term is because of attraction and the third one is randomization having α as the randomization parameter.

5.2 Example validation

The firefly algorithm is applied on the objective function. In order to validate the FA algorithm, the algorithm has been implemented in Matlab. The proposed objective function $f(x)$ is used for calculation of software quality of a product given by:

$$f(x) = \sum_{i=1}^{m}\sum_{j=1}^{n}k_{ij}w_{ij} \qquad (8)$$

where k_{ij} is the priority given to each characteristic of ISO/IEC 9126 by the customer and w_{ij} describes the weight of each characteristic obtained by summation of all the sub-characteristics of each characteristic according to ISO/IEC 9126 Standard. i.e. Suppose that the first characteristic of IOS/IEC 9126 functionality has five sub-characteristics (i.e. Suitability, Accuracy, interoperability, Security and Functionality compliance) having weights w1, w2, w3, w4 and w5 respectively and the priority given by the customer to functionality is 0.9 then the value for functionality characteristic will be (0.9*w1) + (0.9*w2) + (0.9*w3) + (0.9*w4) + (0.9*w5) and so on for the other characteristics. Similarly, after calculation of all other characteristics, all the values obtained are added. The resultant value will define the quality of software product enhanced by using FA algorithm.

5.3 Project description

The values for the objective function defined in equation 8, was obtained by using metric table described by ISO/IEC 9126-3: Internal Metrics for six main characteristics for software quality assurance model (ISO 2002) and Reusability characteristic (Panwar et al. 2012) which was applied on a live project namely, "Pay-roll Application System".

Pay-roll Application System was a software developed in Java (core) programming language having the description given below:

- The software keeps datasheet of employees stored in Oracle Database.
- Both admin as well as employee can login to the portal.
- Employee can view detail of their own after login.
- Details contain date, hour wages, rate per unit hour, and total salary of the day and week.

The software was developed as per the customer's requirement specification as well as validations of the project described below:

- The admin and employee have to login first to view data of the respective employee.
- Admin has the authority to add data to the database for respective employee.

- Only admin can add or edit data of employees and add details about new employee to the database.
- Employees have no authority to edit or add data to the database.
- For validation of software, login user id and password are made case sensitive.
- Random Employee ID is generated by Oracle DB auto increment facility. Employee ID acts as the primary key for retrieval of data from the database by Employee as well as admin.

5.4 *Parameter selection*

In the implementation of example validation of FA with proposed objective function, we have taken light absorption coefficient, $\gamma = 1$, attractiveness coefficient, $\beta_0 = 2$ and randomization coefficient $\alpha \in [0, 1]$. The scale of the variables used for solution varies between 0–1. The priority provided to the characteristics by the customer also lies between 0 and 1. The light absorption coefficient γ is responsible for the variation in attractiveness. Therefore its value is important to determine the behavior of FA algorithm and the speed of convergence.

The quality measurement of software is based on the priority given by the customer as well as the weights of each sub-characteristic obtained by metric table introduced in ISO/IEC 9126 Standard. The weight of the sub-characteristics is obtained by a live project described in section 5.3. The seventh additional characteristic *Reusability* is very important requirement for component based software development (Panwar et al. 2011).

Figure 3 describes the plot solution graph of the objective function applied on FA. At first iteration, value obtained is 2.7387 and then gradually decreases after each iteration. This shows that after each iteration, the heuristic information increases i.e. the developer understands customer requirement more appropriately after each iteration and the quality of software increases after each iteration. The decrement in best cost shows decrement in requirement specification and increment in quality of the software. The result shows positive

effect on measurement of quality as expected by the researchers.

6 CONCLUSION

Firefly optimization technique played a crucial role to optimize the proposed objective function of software quality prediction. The positive effect on measurement according to the iteration executed with the help of MATLAB. Other optimization techniques could also applied on the proposed objective function to the better result in future.

REFERENCES

Behkamal B., Kahani M. & Akbari M.K. (2009), Customizing ISO 9126 quality model for evaluation of B2B applications, Information and Software Technology, 51, pp. 599–609.

Chang C.W., Wu C.R. & Lin H. L. (2008), Integrating fuzzy theory and hierarchy concepts to evaluate software quality. Software Quality Journal, 16 (2), pp. 263–276.

Eusuff, M., Lansey, K.E. (2003). Optimization of water distribution network design using the shuffled frog leaping algorithm. Water Resources Planning and Management 129(3), 210–225.

Goldberg D (1989), Genetic Algorithms in Search, Optimization, and Machine Learning. Addison Wesley, Reading, MA.

IEEE, "IEEE Std 1074–1997—Standard for Software Life Cycle Processes," 1998.

ISO, "ISO 8402 Quality Vocabulary," in International Organisation for Standardization. Geneva, 1986.

ISO/IEC, "ISO/IEC 9126-3 Software engineering-Product quality-part3: Internal metrics," 2002.

ISO/IEC, "ISO/IEC 9126-2 Software engineering-Product quality-part2: External metrics," 2002.

ISO/IEC, "ISO/IEC 9126-4 Software engineering-Product quality-part4: Quality In Use metrics," 2002.

ISO/IEC, "ISO/IEC 9126-1 Software engineering-Product quality-Part 1: Quality model," 2001.

Kennedy J. & Eberhart R. C. (1995) Particle Swarm Optimization, Proceeding of IEEE International Conference on Neural Networks, Perth, Australia, IEEE Service Center, Piscataway, NJ, pp. 1942–1948.

Kitchenham B. & Walker J. (1989), "A Quantitative Approach to Monitoring Software Development," Software Engineering Journal, pp. 1–13.

Panwar D., Tomar P. (2011), "Investigation of Methods to Make the CBSE Process More Effective to Develop Quality Software" proceedings of National Conference on Advancements in Communication, Computing & Signal Processing jointly organized by COMMUN, Noida.

Price K. & Storn R. (1995), Differential Evolution—a Simple and Efficient Adaptive Scheme for Global Optimization Over Continuous Spaces, Technical Report, International Computer Science Institute, Berkley.

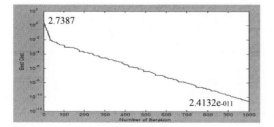

Figure 3. Plot solution graph describing best solution for software quality measurement.

Punter T., Solingen R., & Trienekens J. (1997), "Software Product Evaluation," presented at 4th IT Evaluation Conference (EVIT-97), Netherlands, Delft.

Srivastava1 P. R., Kumari Sirish, Singh A.P. & Raghurama G. (2011), Software Testing Effort: An Assessment Through Fuzzy Criteria Approach, Journal of Uncertain Systems, 5 (3), pp.183–201.

Wang Y.M., Elhag T.M.S. & Hua Z. (2006), A modified fuzzy logarithmic least squares method of fuzzy analytic hierarchy process. Fuzzy Sets and Systems, 157, pp. 3055–3071.

Wang Y.M., Luo Y. & Hua Z. (2008), On the extent analysis method for fuzzy AHP and its application. European Journal of Operational Research, 186, pp. 735–747.

Wong B. (2002), "An Investigation of the Cognitive Structures used in Software Quality Evaluation; PhD Thesis," in Information Systems, Technology and Management. Sydney: University of New South Wales.

Yang, X. S. (2008),: Nature Inspired Metaheuristic Algorithms, Luniver Press.

Yang, X. S., Deb. S. (2009), "Cuckoo Search via L`evy Flights", in proceedings of World Congress on Nature and Biologically inspired Computing (NaBIC), India, IEEE publications, USA, pp. 210–214.

Yuen K.K.F. & Lau H.C.W., (2008), Software Vendor Selection using Fuzzy Analytic Hierarchy Process with ISO/IEC 9126, IAENG International Journal of Computer Science, 35(3), 2008, pp. 267–274.

Yuen K.K.F. & Lau H.C.W. (2011), A Fuzzy Group Analytical Hierarchy Process Approach for Software Quality Assurance Management: Fuzzy Logarithmic Least Squares Method, Expert Systems with Applications, 38, pp. 10292–10302.

Communication and Computing Systems – Prasad et al. (Eds)
© 2017 Taylor & Francis Group, London, ISBN 978-1-138-02952-1

Data encryption in cloud computing: A review

Anuradha & Suman Sangwan
DCRUST, Murthal, Sonepat, India

ABSTRACT: Cloud computing is an architecture which is a combination of cluster and grid computing. Cloud computing follows pay-per-use scheme. Now a day's cloud computing having a great requirement in the field of IT industries. Cloud computing provides easy access to the computing resources without software installation. For providing security and fast communication between clients and the server many researches and the developments are taking place. Data encryption is a process which is used to provide on data for the protection purpose. Data protection in a cloud, encryption and decryption keys are the great solution and for providing maximum security, at the source point data should be encrypted and after this transmission process can take place in the cloud. The cloud computing become a more and more adaptive technology now a days.

1 INTRODUCTION

Data encryption used for protecting the data. Encryption of data done before it is moved to the cloud (Al-Sakran & Omar 2015). If any unknown user try to access the critical data, encryption makes it much more hard for unknown user to do anything with wrong inception (Arapinis et al. 2012). Public clouds basically provides an external services to the clients. For protecting data in a cloud, encryption keys are the best solution and for the maximum security, data should be encrypted at the source point and after that on transmission and then in the cloud. Intrusion detection system plays an important role in a computer security approach as the number of security-breaking attempts originating inside organizations is increasing (Boopathy & Sundaresan 2014). The cloud owner's relation with the client is depends upon how efficiently the clients are able to use the cloud resources, which later depends upon the effectiveness in cloud management (Gupta et al. 2015). AODV is a routing protocol used for MANET which uses hop count as a path selection process. Performance analysis of simulation shows that the proposed protocol shows better performance than AODV protocol (Gupta & Sanghwan 2015).

1.1 *Cloud service model*

Infrastructure as a service: In IaaS clients gets access to infrastructure for deploying their stuff in the cloud. This service does not control or manages the infrastructure. In fact this manages or controls the Operating system, storage, applications.

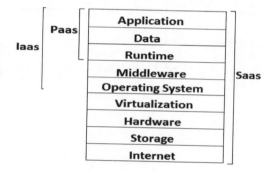

Figure 1.   Cloud service models.

Here client is not control the cloud infrastructure but controls the operating system, storage, limited users and controls of host firewalls.

Platform as a service: In platform as a service generally user deploys and controls their applications in a cloud. In this user never ever manages the servers and the storage. Here client is not able to control the cloud infrastructure for example servers, operating systems, data storage, network connections but controls the deployed cloud applications and hosting configurations.

Software as a service: In software as a service we basically use the provider applications. In this user never ever manage or controls the network, operating system and applications. Here client does not control the cloud infrastructure for example operating system, storage, servers and the limited users cloud based application setting. It generally provides services to the users or the service providers provide virtualization abilities. Multiple types of services are provided by the service interface.

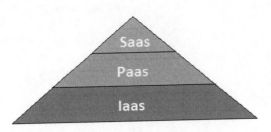

Figure 2.   Cloud Models or Layers [2].

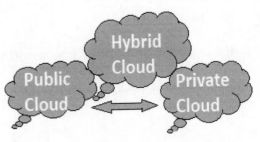

Figure 3.   Deployment of cloud.

1.2   *Deployment of clouds*

Public Cloud: Public clouds are the clouds in which service providers provides their resources or services to the public. Public clouds are unrestricted clouds. In public cloud the clients using shared infrastructure information just because of this the cost of services and resources are spread among all the clients. They are cheaper. In public cloud the shared resources are more efficient and scalable. Example of public cloud is Amazon web services, Google App Engine, Microsoft Azure.

Private Cloud: Private clouds are generally developed for a group, company or an organization and its limits are within that group or organization. These private clouds have great performance, security and reliability. Private clouds are more expensive as compared to public cloud. It uses high degree of sensitive applications. And it also has higher level of security. Example of a private cloud is Eucalyptus Systems.

Hybrid Cloud: Generally hybrid cloud is a combination of two clouds that is public cloud and private cloud. This hybrid cloud is so reliable whenever it required then the private cloud is burst into the public cloud. It also provides benefits of multiple deployment models and maintains the cloud as recovery of data. It is more flexible than public and private clouds. Example of hybrid cloud is Akamai.

1.3   *Benefits*

Cloud computing is a hottest area nowadays just because of its tendencies to decrease the costs related with the cloud computing. Virtualization is a major benefit in the public cloud computing. Services that use multiple sites can support reliability. Security also provides an advantages to the public cloud computing. As we know that in cloud data is stored and provides data and resources in open way too so due to this its maintenance is also a benefit in the computing. Software Delivery can easily done in cloud computing. At the starting companies usually start with the small deployment and as the grow to a higher deployment rapidly, and can scale back if required. Elasticity is an advantage of computing. Flexibility in cloud computing offers companies to use the extra resources many times and also satisfy the client's demands. Device and Location Independent also offers a client high degree of mobility of data in cloud computing. It also provides Multi-tenancy in the public cloud. In cloud computing no need to download or install any particular software. If in any case a system crashes so no data of a client is lost because all data of a client is stored in a cloud. Client has not to update his/her system for new launched software or the packs.

1.4   *Applications*

Now a day's many applications of cloud are available in the market for helping any individual or the companies for getting any data or the resources in the cloud. And the first approach of the cloud computing application is the Amazon and its AWS (Amazon Web Services). This AWS is a remote service which is launched in 2002 due to which any client can use it. Here clients pay some fees to AWS and gets virtual machines. And the most used approach of cloud computing is Google. By using Google apps Google cloud can be used. In Google, data can be viewed without paying any charges or installing or download any software in a system and have to just connect the system or the devices with the internet. For Infrastructure as a service, Platform as a service and Software as services have different- different trust boundaries are developed. Also provides a provider web application security. The most and important fact is to secure inter host communication channel.

2   CHALLENGES AND ISSUES

The main challenges of the research are to provide more security on a data in the public cloud and also to improve performance of data encryption and decryption in the public cloud on the behalf of existing some of algorithms. Also to propose a new algorithm which is more secure as compared to existing some of algorithms? The challenging task of the business organization is to provide higher degree of security to their data since the data are sensible related to their business. Some more challenges which are related with the cloud computing are lack of standards, continuously evolving and privacy.

The major problem associated with cloud computing is data privacy, security, data stealing, etc. Cloud computing basically reduces the operating cost and also increases the efficiency of computing in the clouds. The most associated issue with the cloud computing are services availability, confidentiality and audit ability of data, unpredictability of performance, storage scalability, bugs and software licensing. The major issues which can be faced by cloud computing are malicious insider, insecure interfaces of application programming, nefarious use of cloud computing, data leakage, information security, vulnerabilities in shared technologies, services and traffic hijacking.

3 LITERATURE SURVEY

In this review paper following literature is used which mainly on data encryption in public cloud is computing:-

Jasleen Kaur et al. 2014 represented the encryption algorithms which do not provide any authentication check for the clients but the survey security is the weakest area of this research paper. It also increases the ease of use of data in a public cloud.

K. Sudha et al. 2015 presented that the encryption of original data before outsource it using advanced encryption techniques and retrieve the original data by using coordinate matching algorithm. It also provides more security to the original data.

Shakeeba S. Khan et al. 2015 proposed a cryptographic algorithms for security in cloud computing. The proposed plan is to eliminate the concerns in data privacy for enhancing cloud's security in different cloud customers. In this research paper if some intruder gets a valid key that he/she can easily gets the data intentionally.

Ramalingam Sugumar et al. 2015 proposed a symmetric encryption algorithm which is used to secure a outsource data in a public cloud storage. This proposed algorithm minimizes time for the encryption and decryption. And the cloud should not access data stored in cloud storage server.

Al-Sakran 2015 represented a scheme which minimize communication overhead and computations on client and server sides. And the issues of this research paper of an untrusted client and another one is the outsourced data and the data owner's application software to a cloud provider.

Sengupta 2015 presented a hybrid RSA Encryption algorithm for the cloud security in which a proposed hybrid RSA algorithm provide a higher level of security but cannot maintains a integrity of data in a public cloud.

Masthanamma et al. 2015 proposed a RSA encryption algorithm which is used to enhance the security of cloud and it also increases the security of data and consumes less time and the less cost. But the disadvantages of this work is a fake public key algorithm, key generation complexity, security needs and the low speed.

Sebastian et al. represented data security issues in public cloud. The data in cloud, identify and discuss security risks associated with it and analyzes its solution strategies. For maintaining a secure environment it requires shared responsibilities of cloud providers and the customers.

Gupta et al. 2015 proposed a encryption techniques which are used to provide different issues of security to cloud and different cryptographic algorithms which are adoptable to provide better security for the cloud.

Sharmila 2014 presented a multi-keyword searching using homomorphic encryption which is an algorithm that provides confidentiality for user querying patterns and privacy for user data. It reduces time taken to retrieve files.

Boopathy et al. 2014 proposed a data encryption framework model which required a maximum levels of security in a cloud storage. The algorithms for secure cloud storage are in not developed up till now in the real world.

Khachatrian et al. 2014 presented a new public key encryption system related to permutation polynomials in which security of the system depends upon the difficulty to finding out the polynomials over the field. The presented system will be significantly. Security level is very low.

Hu et al. 2015 represented a public key encryption for protecting data in cloud system in where public key encryption is used for protecting the sensitive data in a cloud. And an adaptive leakage resilient and RRK secure IBE approach.

Luo et al. 2012 presented a scheme which is based on the factoring in public key encryption which secure and based on the bilinear map. But the shortcoming of this approach is that this approach cannot hide the access patter which allows the server to infer some information about the queries.

Long et al. 2012 proposed a approach which improves the performance of public encryption. A constant parameter is used to constrain the document type. But this cannot deal with most documents of the daily life.

Arapinis 2012 represented privacy support in cloud computing in ConfiChair which is generally based on the key translation and mixes in web browsers. This Confichair is a conference management system which is used sensitive data only in a encrypted form.

Renu et al. 2015 prosposed a biometric based approach for data sharing which can easily identifies the authorized user. This scheme combines a digital key and the biometric image to create bioscrypt. The proposed approach solves the problem of escrow and revocation. But there is a lot of noise inside the cipher text.

Sur et al. 2013 presented a certificate based proxy re-encryption for data confidentiality. This proposed scheme provides precise definitions for constructing secure certificate based proxy re-encryption approach and also proposed a concrete approach based on bilinear pairing.

Liang et al. 2014 proposed a DFA based functional proxy re encryption approach. This scheme motivates a problem which is used to convert DFA based FPRE in prime order bilinear group. Fu-Kuo et al. 2013 represented a fast and secure cloud data retrieval scheme which improves the efficiency. It cannot exploits iPEKS to support richer predicates. This scheme is much faster and more secure for retrieving data in a cloud.

Prakash et al. 2014 proposed a protect outsourced sensitive data in cloud. Data owner utilizes the benefits of file splitting to reduce storage and overheads. It also reduces the burden of data owner.

Sangwan et al. 2012 presented the heterogeneous wireless networks, seamless services continuity in mobile devices. This paper represents User Specific Intelligent Vertical Handoff which is multi-criteria algorithm in HWS.

Table 1.

Title	Strength	Weakness
Access to secure data in cloud computing Environment	Secure and efficient access to data in a cloud, Minimizes communication overhead	Untrusted Client.
Security in Cloud Compuitng Using Cryptographic Algorithm	Multi level encryption and decryption, Only authorized users can access the data, Enhanced the security in cloud.	If an intruder gets a valid key then he/she can easily gets data accidententally or intentionally.
Designing of Hybrid RSA Encryption Algorithm for Cloud security.	High Security level, Secure protocols transactions.	Implementation is difficult.
Public Key Encryption with keyword Search	Cipher Text Security Searchable Encryption	Requires support for gateway.
Survey Paper on Basics of Cloud Computing and data security	Increases the ease of use, Provides authenticity of the user.	Survey security is a weakest area, Encryption algorithms not allow to provide authentication check for user.
A Study: Data Security Issues In Public Cloud	Obtained a ownership of the data, Use and maintained own data, Having shared responsibilities of cloud providers.	Securing data in transit and at the rest.
Secure retrieval of files using homomorphic encryption for cloud	Fast retrieval of desired files, Provides Confidentiality and privacy for user querying patterns and user data, Reduces time taken to retrieve files.	Larger database is used.
Approach of Biometric Based Data Sharing in Public Cloud	Improve the security of data sharing, Solves key escrow problem and revocation problem.	Noise inside the cipher text.
Privacy-Preserving Public Auditing for Secure Cloud Storage	Security, Performance, Storage, Communication tradeoff	Batch auditing, It is not possible to download data.
The Framework Model of Data Encryption with Watermark Security for Data Storage in Cloud Model	It provides maximum level of secure cloud storage in cloud environment.	Less secure, Less Data authentication, Less data authorization.

4 CONCLUSION

From the customers' perspective, cloud computing security concerns on data security and privacy protection issues, remaining the primary inhibitor for adoption of cloud computing services. The advantages of cloud computing are when organizations or companies are supposed to move their services in the cloud, analyses the benefits that the improvements that they can get.

If the consumers decide to incorporate their business or part of it to the cloud, they need to take into account of risks and threats that arise, the possible solutions which can be carried out to provide protection to their applications and services from those risks, and some best recommendations may be helpful when the clients want to integrate their applications in the Cloud. Cloud computing is a technology which is used to deliver on-demand storage and capabilities of cloud computing. Security is a basic need in the public cloud computing whenever we are talking about the storage of data.

REFERENCES

Al-Sakran, Hasan Omar. "ACCESSING SECURED DATA IN CLOUD COMPUTING ENVIRONMENT." *International Journal of Network Security & Its Applications* 7.1 (2015): 19.

Arapinis, Myrto, Sergiu Bursuc, and Mark Ryan. "Privacy supporting cloud computing: Confichair, a case study." Principles of Security and Trust. Springer Berlin Heidelberg, 2012. 89–108.

Boopathy, D., and M. Sundaresan. "Data encryption framework model with watermark security for Data Storage in public cloud model." Computing for Sustainable Global Development (INDIACom), 2014 International Conference on IEEE, 2014.

Gupta Diksha., Chakraborty Partha Sarathi., Rajput Pragya., "Cloud Security using Encryption Techniques", "International Journal of Advanced Research in Computer Engineering & Technology (IJARCET)", Volume 5, Issue 2, February 2015.

Gupta, Shikha, and Suman Sanghwan. "Load Balancing in Cloud Computing: A Review." International Journal of Science, Engineering and Technology Research (IJSETR), June 2015.

Hu, Chengyu, et al. "Public-key encryption for protecting data in cloud system with intelligent agents against side-channel attacks." Soft Computing (2015):1–14.

Kaur, Jasleen, Ms Anupma Sehrawat, and Ms Neha Bishnoi, "Survey Paper on Basics of Cloud Computing and Data Security." *International Journal of Computer Science Trends and Technology (IJCSTT)* (2014).

Khachatrian, Gurgen and Melsik Kyureghyan. "A New Public Key Encryption System Based on Permutation Polynomials." Cloud Engineering(IC2E), 2014 IEEE International Conference on IEEE, 2014.

Khan, Miss Shakeeba S., Prof. R. R. Tuteja, "International Journal of Innovative Research in Computer and Communication Engineering", Volume 3, Issue 1, (2015).

Liang. Kaitai, et al. "DFA-based functional proxy re-encryption scheme for secure public cloud data sharing." Information Forensics and Security, IEEE Transactions on 9.10 (2014): 1667–1680.

Long, Bin, et al. "On Improving the Performance of Public Key Encryption with Keyword Search" Cloud and Service Computing (CSC), 2012 International Conference on IEEE, 2012.

Luo Wenjum, Tan Jianming., "IEEE CCIS", 2012.

Masthanamma V., Preya G. Lakshmi, "International Journal of Innovati ve Research in Science Engineering and Technology", Volume 4, Issue 3, (2015).

Prakash G L, Dr. Manish Prateek, Dr. Inder Singh, "International Journal of Engineering and Computer Science", Volume 3, Issue 4, Pg. 5215–5223, April 2014.

Rani, Amita, and Mayank Dave. "Weighted load balanced routing protocol for MANET." Networks, 2008. ICON 2008. 16th IEEE International Conference on. IEEE, 2008.

Renu S, Hasna Parveen O H, "International Journal of Advanced Research in Computer and Communication Engineering", Volume 4, Issue 2, February 2015.

Sangwan, Suman, Parvinder Singh, and R. B. Patel, "Uivh-algorithm for semless mobility in heterogeneous wireless network." Proceedings of the CUBE International Information Technology Conference. ACM, 2012.

Sebastian, Alycia, and L. Arockiam. "A Study On Data Security Issues In Public Cloud." *International Journal of scientific and technology research"*, Volume 3.

Sengupta Dr. Nandita, "International Journal of Innovative Research in Computer and Communication Engineering, Volume 3, Issue 5, (2015).

Sharmila, R., "Secure retrieval of files using homomorphic encryption for cloud computing." International Journal of Research in Engineering and Technology 3.7 (2014).

Singhrova, Anita. "A host based intrusion detection system for DDoS attack in WLAN." Computer and Communication Technology (ICCCT), 2011 2nd International Conference on. IEEE, 2011.

Singla, Sanjoli, and Jasmeet Singh, "Cloud data security using authentication and encryption technique." *Global Journal of Computer Science and Technology* 13.3 (2013).

Sudha K., B. Anusuya, P. Nivedha, A. kokila, "International Journal of Advanced Research in Computer Science and Software Engineering." Volume 5, Issue 1, (2015).

Sugumar, Ramalingam, and Sharmila Banu Sheik Imam. "Symmetric encryption algorithm to secure outsourced data in public cloud storage."*Indian Journal of Science and Technology* 8.23 (2015):1.

Sur, Chul, et al. "Certificate-based proxy re-encryption for public cloud storage." Innovative Mobile and Internet Services in Ubiquitous Computing (IMIS), 2013 Seventh International Conference on. IEEE, 2013.

Tseng Fu-Kuo., Chen Rong-Jaye., Lin Bao-Shuh Paul., "IEEE International Conference on Trust, Security and Privacy in Computing and Communication", 2013.

Communication and Computing Systems – Prasad et al. (Eds)
© 2017 Taylor & Francis Group, London, ISBN 978-1-138-02952-1

Big data medical analytics with hadoop to analyze various types of image data—A gentle review

M. Pareek, C.K. Jha & S. Mukherjee
Banasthali Vidhyapith, Tonk, Rajasthan, India

ABSTRACT: With the term 'Big Data' everyone thought about the huge data which could be structured, unstructured and semi-structured. Due to web world, the unstructured data grow rapidly. The unstructured data is generated from various sources like facebook, whatsapp, fliker, twiter, instagram, hospitals, etc. Every day 1.8 billion images are uploaded, so it is a very tedious task to store, process and analyzed the petabyte or large sized data in the unstructured form. Apart from above areas, in medical field Electronic Health Record (EHR) increasing day by day, so that it's challenging task to store and analyzed these E-data of patients. Because of traditional database tools don't support the unstructured data. To overcome these problems a new and emerging tool is widely used named 'Big Data Hadoop'. Recently Hadoop provided parallel and distributed computing environment for large scale images. The paper is mainly focused on, how Big Data is helpful in health care analytic, techniques how medical images are processed and analyzed by using the techniques of Hadoop Map-Reduce. For this purpose Hadoop Image Processing Framework has used. The frame work provides the different types of libraries to support large scale image processing for analysis purpose.

1 INTRODUCTION

"Big data" as the name implies the data is large in size. The definition of Big data is not defined, because of uncertain and inconsistence nature of data. When the data is growing day by day, it becomes cumbersome task to manage, store, process and analyzes the large scale data. Many other relational databases are used to store data but when they didn't support unstructured and inconsistent data. In general terms, Big data is a collection of data whose basic features are Volume, Variability, and velocity. The V3 terms feature of Big Data Volume, Variety and Velocity. Harshwardhan give the system architecture of Big data and divide the architecture into the layers, to store and retrieved the useful information. (Harshwardhan et al. 2014) Figure 1 shows the Layered Architecture of Big Data System. The layers can be divided in 3 Categories, such that Application Layer, Computing Layer and Infrastructure Layer.

In recent time data are increased rapidly, so the size of data is not defined and concise. Data are continuously changes over the times GB (Gigabytes) to TB (Terabytes) and TB to PB (Patabyets). For these scenario James gives the statistical approach to analyze the how the data would be reached in year the 2020 so rapidly (James 2013).

The term, Big Data is still not defined and has defined the V4 Characteristic of big data. Rahul

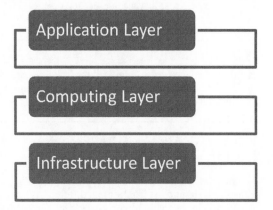

Figure 1. Layers architecture of big data system.

Beakta introduced the new term Veracity. This term has used because of uncertain and inconsistent nature of data (Beakta 2015). The term Veracity defined the accuracy of data. Ravikumaran depict the variety and volume of data are increased so rapidly from different sources so how to analyzed the worth of data. Because it is necessary when we have to use large size data it must be meaningful so they have depicted V5 feature of BIG Data and define Value as a new term with V4 (Ravikumaran et al. 2015). Javier and Merrifield has managed the V6 of BIG DATA and discussed how big data are

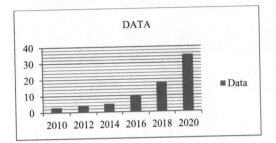

DATA

Figure 2.   Source:- International data composition.

Table 1.   V6 are managed clinical data in health industries.

Value	Information should be worth it.
Volume	The size of data.
Velocity	Increasing Rate of growth of clinical data with high-speed health care tools and equipment.
Variety	Increasing clinical data could be structured and unstructured.
Veracity	Data comes from different types of resources, and clinical data are very sensitive, so the reliability of data is a big question.
Variability	Seasonal health effect and disease evolution. on-deterministic models of illness and health.

useful to manage health-related data (Javier et al. 2015). The above table specified how V6 are managed clinical data in health industries.

2   BIG DATA ANALYTIC IN HEALTH CARE

Patient related information is very sensitive. The patient data could be containing different types of records. In medical industries as well as health industries, the term "Big data" is most relevant. All the main characteristics of Big data like volume, velocity and variety are fitted to healthcare industries. To obtain the best outcomes of particular patient and population, deep analysis is required. With the variety of data, analysis becomes easier and physician/doctors use the previous different types of records of the patient to determine the appropriate treatment. They can easily diagnose the disease to analyze the existing records.

Mr. Barak Obama, President of USA has taken initiative and proposed BRAIN project with new technology using BIG DATA (Sun et al. 2013). On April 2, 2013, the White House proposed a major national project to unlock the mysteries of

the brain—the "Brain Research through Advancing Innovative Neurotechnologies" (BRAIN) Initiative. advances in big data processing for health informatics, bioinformatics, sensing, and imaging will have a great impact on future clinical research. Another important factor to consider is rapid and seamless health data acquisition, which will contribute the success of Big Data in medicine.

Javier and Merrifield depict the ability of big data to deal with large and voluminous data; they focused on how big data are useful to store and analyzed the clinical data. In this paper, they derived how traditional medical services different from e-medical services. They have integrated the patient's EHR (Electronic Health Record) with social media. They realized a problem when we have used tradition services the question is raised how to analyzed the patients, those who have similar kind of disease and symptoms. Regarding this problem, EHR is used to store the patient's details for analytics purpose. Social media also play an important role in improving medical health, because it provide the platform where patient share their experiences and discussed about their disease (Javier et al. 2015).

Mathew Divide the clinical health issues into different types of levels such that-Bioinformatics, Nueroinformetics, Genomic, real-time health prediction and last but not the least Social media. In this paper they correlated all levels with Bid Data's feature "V6" (Mathew et al. 2014). These following fields are fulfilled the V6 features. They noted that Big Data play an important role to analyze the clinical data. In this paper they mentioned the, Brain tumor example for analytics purpose (Sun et al. 2013). The medical images are a large size and it is difficult to extract valuable data. X-ray, MRI, fMRI, CT-Scan images etc are imaging modalities. If patient's long time health records are managed then real time prediction is possible with the help of Big Data Clinical Analytics. Other section is social media that provide the platform to sharing experiences.

In health care organization and large hospitals, network realizes the significant benefits of Big data. (Ravikumaran et al. 2016) With the help of Big Data, we can easily detect any disease at an earlier stage. Big data could helpful in a different clinical field to reduce the waste and inefficiency. The following areas are where Big Data are useful:-

Clinical operations: In this field research determines relevant and cost-effective methods to detect disease and treat patient very well (Ravikumaran et al. 2016).

Research & development: Beneficial in R & D to manage Drugs and devices (Ravikumaran et al. 2016).

Public health: To improve public health, analyze the complex disease pattern to surveillance of public health (Ravikumaran et al. 2016).

Evidence-based medicine: With the help of EHR, analysis of structured and unstructured data, clinical data, and genomic data used to match the result so that they predict the patient health condition and diagnose particular disease on time and care about the risk factor of patient (Ravikumaran et al. 2016).

Genomic analytics: Big Data are very effective in this field because for analytics purpose the genes sequences are stored and process for a long time, so that medical decision takes place easily (Ravikumaran et al. 2016).

Device/remote monitoring: In hospitals and labs real time images are captured every day, voluminous image data are growing very fast on a daily basis, in order to manage data it also necessary to monitor devices (Ravikumaran et al. 2016).

Patient profile analytics: Profile analyses are used to identify the patient's details and life style. Patients life style interpret possibilities to detects specific disease like diabetes, Blood Pressure etc, that would be benefited to patient those who those types of disease to cure at early stage (Ravikumaran et al. 2016).

3 HADOOO: PROBLEM SOLVING FOR BIG DATA PROCESSING

Big data processing and analytics is a very cumbersome task. For this purpose, Apache provides a tool named "Hadoop" that is work with HDFS (Hadoop Distributed File System) environment. Map and Reduce functions are used to provide parallel and distributed environment. With the help of mappers, the task is distributed to different nodes and all nodes are process the task parallel. In medical field, there can be large amount data which could be structured and unstructured. Different types of images are used to diagnostic purpose like X-ray, MRI, CT-scan, fMRI etc, the size of these images are very large and they are increased rapidly. To storing and processing of these types of images is very complex. "Hadoop" provides the best platform to store, process and analyze those images in distributed and parallel environment.

There are following basic building blocks of Hadoop (Chuk Lam. 2013):-

I. Name Node
II. Data Node
III. Job tracker
IV. Task tracker

Name Node:- In the Hadoop Architecture Name Node plays an important role. For distributed computation and distributed storage Hadoop uses Master/Slave Architecture. This distributed system is known as HDFS (Hadoop Distributed File System). Name node is the master node in the system. In distributed system data has been distributed to the nodes. How files are distributed into blocks, which node store these blocks all information regarding controlling trace by the Master Node (Name Node). Master Node guides the Slave Node (Data Node).

Data Node:- Data Node is hosted by slave machine. Data node is used to execute the task of distributed file system. To perform any operation (read and write) HDFS divide the file into blocks, the Name Node decides which block resides in which Data Node. There different Data Node (Slave Node) communicates with each other and have a duplicate copy of the file so that content of files are distributed. If any Data Node crash or damaged, after that we can also read a file.) (Vemula et al. 2015) For computation each Data Node report to the Name Node (Master Node).

Job Tracker:- The client connects to the job tracker to perform data processing task. Every Hadoop cluster has only one Job tracker. Job tracker is liable to divide the files and assign the (map) tasks to different nodes and have control on all node Suppose when a task is failed due to any reason job tracker immediately restarts the task) (Vemula et al. 2015).

Task Tracker:- Job tracker is Master for Task tracker. Every task is completed by Task tracker that is assigned by Job Tracker. In Hadoop Cluster, every slave node has individual task tracker. Task tracker individually performs the task of Job tracker. The duty of the Task tracker is to regularly report to Job tracker. If Task tracker does not communicate to Job tracker in specified duration will be supposed Task tracker has crashed, and will assign the task to another node in a cluster. This mechanism provides the parallel computing and map reduce also) (Vemula et al. 2015).

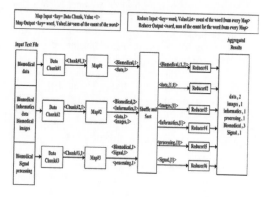

Figure 3.   (Emad et al. 2014).

4 METHODOLOGY FOR IMAGE PROCESSING WITH HADOOP MAP REDUCE

Apache HADOOP provides the suitable platform to handle and process huge data set. One of another easy to use tool is developed by Apache is Apache HIPI (Hadoop Image Processing Interface) is means to provide different users to process and access the large sized images. The basic purpose of HIPI (Hadoop Image Processing Interface) is to develop large-scale images, store images in various file formats to analyze those images. HIPI is tool that was developed to use by the researchers so that they would process as well as analyze large-scale data (Sweeney et al. 2011). Following are basic feature of HIPI:-

1. For image processing task provides open library using MapReduce framework.
2. Store the images in an efficient manner.
3. Only suitable for image-based operation.
4. With the help of MapReduce, HIPI uses parallel and distributed environment.
5. Hiding the other unnecessary information by the reducer.

Following are some steps for image processing using Hadoop image Processing Framework:-

Step 1: The first step specifies to collect all images. We can get images from query sting and also use the URL to download the images. To provide the parallel computing the image data sets are distributed o different nodes to run the task, to get maximum efficiency. At next step, the mapper does their task and all downloaded images are stored at HIPI image bundle that is generated by Map task, after that reducer will merge all the images at HIPI image Bundle for further processing (Sweeney et al. 2011).

Step 2: when the image URL are stored then next step to downloading of images with a list of URLs that will be done by mappers. To perform parallel computation first of deciding no of maximum nodes that perform a downloading task. After that the URL are sent to different nodes parallel downloading will start by nodes (Map function). Downloading will begin for each image is according to the URL list (Sweeney et al. 2011).

Step 3: In the next step the downloaded images are stored in HIPI image Bundle. To retrieve downloaded images from the data base a connection will be created in JAVA using URLConnection class. When the connection is established the file format is

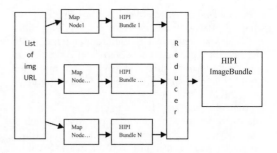

Figure 4. HIPI framework (Sweeney et al. 2011).

checked, images are valid or not. To add images in image Bundle the Input Stream class is used.

Step 4: From the URLConnection class we can add images using IputStream to HIPI image bundle. When all the nodes download the images amp task will generate separate image bundle where their respective images are stored. After that, reducer will perform their task. In reducer phase, all images are merged in single Large Image Bundle using addImage method (Sweeney et al. 2011).

HIPI provide the best method to store images as well as process the images with using Hadoop Map-Reduce framework.

5 CHALLENGES AND LIMITATION OF USE OF CLINICAL BIG DATA

As we had discusses above, about how the Hadoop and HIPI deal and process the large-scale clinical data very well, but when we deal with huge amount data it becomes complex, and sometimes we have to face many problems. Every technology faces some issues and challenges. Bhawana and Jyoti focused on two main issues first security and reliability of data, so that these are challenging issues of bid data to provide Valuable data among giant data with reliability and security (Bhawana et al. 2014). Mukesh also found it's a tedious job to manage large data sets because most of the data are noisy and few of data are valuable (Mukesh et al. 2014). The tools and technology are not so user-friendly so they required data scientist and skill experts. Macro and Hunter basically emphasis on Electronic Health Records of a patient to analyze their disease and improved the treatment. For analytical purpose they have used Big Data Hadoop Technology, but they found some shortcoming in big data platform. According to them when we talk about huge data, it does not deal with an individual so that there are lots of chances to lost patients details means there is no confidentiality of data another risk factor of social media to exchange their experiences and details (Macro V. et al. 2015). Wang and Krishnan found there are many other different challenges for big data that derived such that size, selection of data, missing information interpretation of data, data handling methodologies to how to extract, retrieve and access the information, which types of techniques are suitable. Whenever the data size is growing so fast it becomes complex to handle the uncountable data (Weigiwang et al. 2014). The main issues of Big Data processing are to maintain the reliability and confidentiality.

6 CONCLUSION

We have entered in "Big Data" world that has many features. Big Data Concept is started with V3 to V6. In this paper, we also focused on V6. Due to the social sites and other resources, the volume of data is increased so fast, to handle and manage the data, above we discussed new technology "Big Data" that is using to handle large scale medical images. The paper also focuses on how Big data is helpful for health industries, with the help of this technique experts cure the disease at the initial stage and EHR also helpful to analyze the patient's life style and previous records that could be helpful to interpret their upcoming condition. Big Data provide the distributed and parallel computing environment. Apache HIPI (Hadoop Image Processing Interface) tool used with Map Reduce programming to process and analyze the large-scale image data of health industries. But the Big Data have many advantages apart all this technique have some shortcomings also. We had also discussed in the paper the challenges of Big Data like reliability, security and extraction relevant information.

REFERENCES

Andreu-Perez, J., Poon, C. C., Merrifield, R. D., Wong, S. T., & Yang, G. Z. (2015). Big data for health. Biomedical and Health Informatics, IEEE Journal of, 19 (4), 1193–1208.

Bhosale, H. S., & Gadekar, D. P. (2014). A Review Paper On Big Data And Hadoop. International Journal of Scientific and Research Publications, 756.

Beakta, R. Big Data and Hadoop: A Review Paper. International journal of Computer Science. 2 (2).

Chuck Lam [2013], "Hadoop In Action", Chapter 2.

Gupta, B., & Jyoti, K. Big Data Analytics with Hadoop to analyze Targeted Attacks on Enterprise Data.

Herland, M., Khoshgoftaar, T. M., & Wald, R. (2014). A review of data mining using big data in health informatics. Journal of Big Data, 1(1), 1–35.

Kataria, M., Mittal, P., (2014). Big Data: A Review. International Journal of Computer Science and Mobile Computing, 2(1), 106–109.

Ravikumaran, P., & Devi, K. V. (2016). A review: big data and analytics in health care. Indian Journal of Engineering, 13(31), 1–10.

Sun, J., & Reddy, C. K. (2013). Big data analytics for healthcare. In Proceedings of the 19th ACM SIGKDD international conference on knowledge discovery and data mining (pp. 1525).

Sweeney, C., Liu, L., Arietta, S., & Lawrence, J. (2011). HIPI: A Hadoop Image Processing Interface for Image-based mapreduce tasks. Chris. University of Virginia.

Tien, J. M. (2013). Big data: Unleashing information. Journal of Systems Science and Systems Engineering, 22(2), 127–151.

Vemula, S., & Crick, C. (2015, June). Hadoop Image Processing Framework. In Big Data (BigData Congress), 2015 IEEE International Congress on (pp. 506–513). IEEE.

Viceconti, M., Hunter, P., & Hose, R. (2015). Big Data, Big Knowledge: Big Data for Personalized Healthcare. Biomedical and Health Informatics, IEEE Journal of, 19(4), 1209–1215.

Wang, W., & Krishnan, E. (2014). Big data and clinicians: a review on the state of the science. JMIR medical informatics, 2(1).

Communication and Computing Systems – Prasad et al. (Eds)
© *2017 Taylor & Francis Group, London, ISBN 978-1-138-02952-1*

Mitigation of EDoS attacks in cloud: A review

Preeti Daffu & Amanpreet Kaur
Chandigarh Engineering College, Landran, Punjab, India

ABSTRACT: Cloud computing has provided a platform to its users where they are charged on the basis of usage of the cloud resources; this is known as "pay-as-you-use". Providing the desired level of assurance and safeguarding the stored data on the cloud platform is a crucial issue these days. Issue of protecting the cloud from the attackers and hackers cannot be underestimated. In cloud environment it is very difficult to detect and filter the attack packets because everything is virtualized there. EDoS (Economic Denial of Sustainability) is a cloud-specific attack which is a form of a DDoS attack; intended to hurt the user financially or economically. Such attacks do not exhaust the bandwidth of the user; their main aim is to put a huge financial loss or burden on the user. In this research article various techniques of mitigation of EDoS attacks have been thoroughly discussed and compared.

Keywords: Distributed Denial of Service (DDoS), Economic denial of sustainability (EDoS), Cloud Service Provider (CSP)

1   INTRODUCTION

Cloud computing is a paradigm that enables access to computing and data storage resources that can be configured to meet unique constraints with minimal data overhead. Cloud computing model has five characteristics, three service models and four deployment models. These characteristics are: on demand service, resource pooling, broad network access, rapid elasticity and measured service (A. Zubair et al. 2013).

Cloud platform has provided its users on demand network access to shared pool of resources. In cloud environment large pool of resources are available and these are allocated dynamically among its users. Cloud computing has gained an immense popularity by allowing its users to lease the computer resources when they run out of them, a pay-based usage of resources and this is known as "pay-as-you-use". It allows its users to run the applications directly from the cloud (A. Bakshi et al. 2010). Today protecting the stored data on the cloud platform is an important issue that cannot be understated. Protecting the cloud from various attacks and their adversarial effects is a major concern that has stalled many users from shifting their data to the cloud.

The cloud infrastructure is fully virtualized and it supports all type of hardware architectures. Thus providing security to cloud is an important issue these days. The papers (Alzamil et al.) (A. Zubair et al. 2016) (A. Bakshi et al. 2010) give a clear idea

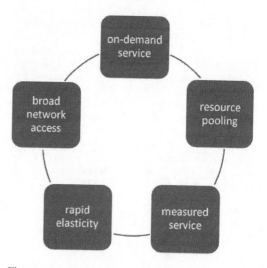

Figure 1. Five main characteristics of cloud computing.

about the security issues related to the cloud environment. Cloud is prone to a variety of attacks such as malware injection attack, Spoofing attack, flooding attack, wrapping attack, DoS, DDoS and EDoS attack. It is very difficult to detect and filter the attack packets. Denial of Service (DoS) attack is an intended attempt to exhaust the computing resources and the network bandwidth of the target user. Distributed Denial of Service (DDoS) attack is attempted to multiple users by flooding the packets. Economic Denial of Sustainability (EDoS) attack is a new breed of DDoS attack and it is a cloud specific attack which is intended to put a huge financial loss on its users.

1.1 DDoS attacks

DDoS is a killer application for cloud computing environments on Internet today. It is a Distributed Denial of Service. The main aim of DDoS attacks is to exhaust the resources of victim machines such as network bandwidth, computing power. DDoS attacks can be carried out in various forms, such as flooding packets or synchronization attacks or by creating zombies towards the victim machine. Flooding packets is the most common and effective DDoS attack strategy among all the strategies.

DDoS attacks have always been a threat to individual cloud customers as they have fewer resources to beat such attacks. To prevent DDoS attacks, employ the idle resources of cloud to filter out the packets and guarantee the Quality-of-Service (QoS). The essential issue of DDoS attacks and defense is competition for resources. The side (attacker or user) which have more resources will win the battle. The side (attacker or user) which have more resources will win the battle. Most of DDoS attacks are nowadays carried out by Botnets and the issue behind this is resource management problem. We can beat the DDoS attacks if we have the sufficient resources. But the client-server and peer to peer system don't have the sufficient resources to beat them. Dynamic resource allocation strategy is used to counter the DDoS attacks. However, Individual cloud hosted server are still vulnerable to DDoS attacks because they run in a traditional way.

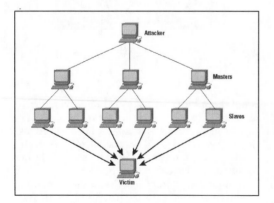

Figure 2.   Typical DDoS attack.

1.2 EDoS attacks

Economic Denial of Sustainability (EDoS) attacks targets the financial component of the victim. Such attacks are a form of traditional DDoS attacks on the cloud platform these days. Many organizations move their business on cloud because there is no need to buy the entire infrastructure and the maintenance cost is null. The cloud services provided to its users are in the form of SLA (Service Level Agreement) and it provides the facility of using the resources as you utilize them. Some of the organizations still don't run their business on cloud due to the threats and security issues related to the cloud environments. EDoS attacks are not intended to exhaust the bandwidth of the users; main aim of such attacks is to hurt the financial component of the user. The following actors are involved in attempting an EDoS attack:

1. Cloud Service Provider (CSP): a user who rents its resources and performs billing
2. Cloud Service Consumer: it uses cloud resources to host its web application
3. Legitimate Client: access the services provided by cloud consumer.
4. Attacker: intentionally generates fraudulent traffic to hit the economy of cloud consumer.

In literature while studying about cloud computing security various security issues have been found; concerning CSP (Cloud service provider) and Cloud Consumers. Liu, Huan (H. Liu 2010), Chi-Chun Lo (Lo Chi-chu et al. 2010) discussed about security and privacy in cloud computing. Zubair Baig discussed some security along with some of their solution using the parameters that affect the cloud security like network, resource allocation and de-allocation, operating system, resource scheduling, management of transactions, load balancing, concurrency control, virtualization, and memory management (G.K Gupta et al. 2010). The author classifies the security issues into four categories Data issues, Privacy Issues, Infected Application and other Security issues.

1.3 Difference between DDoS and EDoS attacks

There is a major difference between the DDoS (Distributed Denial of Service) attack and EDoS (Economic Denial of Sustainability) attack. First, EDoS attacks are aimed to put a huge financial burden on the user but DDoS attacks aims to consume the bandwidth of the user and to make the resources unavailable.

Secondly, EDoS attacks aim to make cloud resources unsustainable to the victim and DDoS attacks are attempted to block the cloud services.

EDoS attacks are carried out for a long period of time whereas DDoS attacks are short lived type of attacks. Therefore, these attacks are different from each other but EDoS attacks are considered as a next form of DDoS attacks. The methodologies and techniques that have been implemented to mitigate the DDoS attacks cannot be used for the mitigation of the EDoS attacks.

2 RELATED WORK

Morein, William G. et al. (2003) has proposed the method to counter the ddos attacks using graphic turing tests against web servers. The authors have discussed novel overlay-based architecture that provides guaranteed access to web servers that have been targeted by the denial of service attack. This approach has exploited two key characteristics of the web environment. The first characteristic is its design around a human-centric interface, and the other one is its extensibility inherent in many browsers through downloaded "applets." Khor and Nakao (Sqalli 2011) has implemented a scheme to make a client's commitment to the resources of cloud on demand by solving the crypto-puzzles. The access to resources was granted only to those genuine users who were intended to pay for the resources they use or utilized. Clients first need to define the difficulty level k of the crypto puzzles and then request for the resources they need to utilize. If the initial request of the client is not accepted in the given interval of time due to resource constraint then client may request for a more difficult puzzle to solve. After solving the complex crypto puzzle a secure communication channel is established between client and the server so that they can exchange the messages. This secure channel is established by the server always. This proposed technique has several shortcomings such as the correlation between puzzle difficulties, problem of asymmetric power consumption, its vulnerability to puzzle accumulation attacks, and associated false positives.

Zunnurhain, Kazi. et al. has analyzed the security attacks and their effects on the cloud. The authors have also proposed the solutions to such attacks. Cloud platform offers great potential and reduce costs but at the same time it is vulnerable to various threats and security risks. The authors have identified the possible security attacks on clouds including wrapping attacks, Flooding attacks, Browser attacks, malware injection attacks and also accountability checking problems. The authors have identified the root causes of these attacks and proposed the specific solutions. In Sqalli, Mohammed H. et al. has implemented two step mitigation technique i.e. EDoS-shield against

EDoS attacks on cloud platform. In this paper, the authors have discussed a novel solution as an enhancement to prior work, named it EDoS-Shield, for mitigation of EDoS attacks that have been originated from spoofed IP addresses. The authors have done simulation experiment to evaluate the performance as its solution to mitigate the EDoS attacks originating from spoofed IP. In this technique a virtual firewall and a verifier node perform the EDoS mitigation task. The firewall filters the incoming requests from the users on the basis of two lists i.e. white list and black list. White list contains the genuine users while black list contains the attacker nodes. Whenever a client request for gaining the access, the verifier node verifies it through a turing test. If a client passes the turing test then its IP is held in white list and if it fails the test then verifier node held its IP address in black list. The requests from black listed IP will be blocked. The shortcomings of this technique are its vulnerability to IP spoofing. The attack that used a spoofed IP address belonging to white list of verifier node will remain undetected. Secondly, it has high number of false positives identified. Many legitimate users are identified as attackers and get blocked.

Al-Haidari, Fahd et al. has discussed the Enhanced EDoS-Shield for mitigation of EDoS attacks those are generated from the spoofed IP addresses. A traditional DDoS attack has been transformed into an EDoS (Economic-Denial-of-Sustainability) attack which is a new breed of cloud attacks and targets the economic resources of the users. The authors have proposed the discrete simulation technique for the mitigation of EDoS attack. This technique has a drawback that it do not have the auto scaling cloud feature of the cloud.

In (G. Booth et al. 2013) the author has conducted a survey on DDoS attacks, its detection and the defense approaches for such attacks. Nowadays DDoS attacks have become a serious threat to the cloud users and the buzzing technology. Distributed Denial-of-Service (DDoS) attack is a serious problem because such an attack is very hard to detect, there is no comprehensive solution and it can shut an organization business off from the Internet. The primary goal of an attack is to deny the victim's access to a particular resource. The authors have reviewed the current DoS and DDoS detection and defense mechanism.

In (A. Chonka et al. 2011) authors have discussed about the cloud services divided into two categories: one is cloud service provider and other is cloud service consumer. As the security in the cloud service is the biggest concern; the control measures for the threats and the attacks prevailing the cloud services have been used. A framework is used to experiment the XML based and HTTP

Table 1. Summary of the techniques surveyed.

Authors and year	Problem addressed	Techniques proposed	Drawbacks
Morein, William G. et al. (2003)	Counter the DDoS attacks against web servers.	Overlay based architecture using graphic turing tests.	Low performance and requires asymmetric traffic routing.
Khor and Nakao (2009)	Packet matching mechanism and crypto puzzle.	SPoW (Self Verifying Proof of Work).	Prevents EDoS attacks only at network level. IP Spoofing and asymmetric power consumption.
Zunnurhain, Kazi. et al. (2010)	Analysis of the security attacks on cloud and their solution.	FAT system architecture, Hypervisor.	Solutions of Security issues need to be improved.
Sqalli, Mohammed H. et al. (2011)	Mitigation of Edos attacks generated from spoofed IP.	Two step mitigation technique i.e. EDoS-Shield.	False Positive Problem and false negative problem.
Al-Haidari, Fahd. et al. (2012)	Simulation technique for the mitigation of EDoS attacks.	Enhanced EDoS-Shield.	Methods need more clarification.
Baig, Zubair A. et al. (2013)	Controlled virtual resource access.	Normal and suspicious lists.	Normal users get blocked sometimes.
Al-Haidari, Fahd. et al. (2015)	Evaluated the impact of EDoS attacks.	Queuing model.	Attacks were not completely mitigated.
Baig, Zubair A. et al. (2016)	Detection and mitigation of EDoS attacks.	Rate Limit technique and duration factor.	Pattern attacks not determined.

based DDoS attacks for the protection of EDoS attacks. The transformation of DDoS attacks into EDoS attacks is also explored.

Baig, Zubair A. et al. has analyzed the controlled virtual resources access to mitigate the EDoS attacks against cloud infrastructure. Through this analysis the authors classified the incoming requests into two categories: normal list and suspect list. It ensures that the priority to cloud service access is given to the legitimate end-users whereas the suspected users are given the lesser priority to access the cloud services. Suspected users cannot access cloud services until they are present in the suspect list.

Al-Haidari, F. et al. has evaluated the impact of EDoS attacks against cloud computing services. A cloud introduces resource-rich computing platforms, where adopters are charged based on the usage of the cloud's resources or utility computing. However, traditional Distributed Denial-of-Service (DDoS) attacks on server consume the resources of the users and make the resources unavailable for them. The authors have developed a simulation model to evaluate the impact of the EDoS attacks. This analytical model is based upon the queuing model that captures the cloud services. Baig, Zubair A. et al. has proposed a novel approach based upon the rate limit technique and data with low overhead for the detection and mitigation of the EDoS attacks. It is a model which relies upon the incoming request rate from one source (a client

of cloud network) based upon the fixed threshold value. The duration factor also lies as the major factor while evaluating the nodes sending the request rate more than threshold. It is unable to detect the pattern-based controlled EDoS attacks where the attacker nodes work in the group. Multiple nodes might attack the target cloud by beating the threshold for request rate and duration. It is not capable of analyzing the proposed scheme to optimize the overall performance while looking at the service provider and network-level variations.

2.1 Summary of the techniques surveyed

The techniques that have been studied and discussed in the related work are addressed in the following table with each including its drawbacks.

3 RESEARCH CHALLENGES IN EXISTING MODELS

The existing models basically rely upon differentiating the legitimate users and the malicious users. The current approaches lack to distinguish whether the incoming request is generated from a bot or it is from another website. The incoming request rate from one source (a client of cloud network) based upon threshold value. The duration factor also lies as the major factor while evaluating the nodes sending the request rate more than threshold. The existing model has not been found capable of doing the following while protecting against the EDoS attacks:

- The existing models are unable to detect the EDoS attacks by recognizing their behavior. The attacker nodes work in the group. Multiple nodes might attack the target cloud by beating the threshold for request rate and duration. These kinds of attacks can beat the EDoS attacks by detecting the attacks by evaluating the whole reception group in the aggregate.
- The existing model is not capable of analyzing the proposed scheme to optimize the overall performance while looking at the service provider and network-level variations.
- The black and while list keeping is always a problematic issue. The existing model does evaluate the variety of factors for the selection of the node in black-list or while-list. The black-listed nodes are again marked in the white list while it shows right behavior. In this form, the malicious nodes can again gain the access to the cloud network. The nodes sending the request rate more than threshold gets blocked.

4 CONCLUSIONS

Unlike DDoS attacks, EDoS attacks span over a long period of time. Hence, traditional DDoS mitigation techniques are not sufficient to defend EDoS attack. Cloud is still vulnerable to various threats and attacks and EDoS attacks have been the most powerful and subtle attacks as they are the cloud-specific attacks. Various techniques have been implemented to mitigate the EDoS attacks but still such attacks exist. So it is necessary to implement such a model that can mitigate EDoS attacks effectively.

REFERENCES

Al-Haidari, F., M. Sqalli, and K. Salah. 2015. "Evaluation of the Impact of EDoS Attacks Against Cloud Computing Services." *Arabian Journal for Science and Engineering*, 40(3): 773–785.

Al-Haidari, F., Sqalli, M.H. and Salah, K., 2012. June. Enhanced edos-shield for mitigating edos attacks originating from spoofed ip addresses. In Trust, Security and Privacy in Computing and Communications (TrustCom), 2012 IEEE 11th International Conference on: 1167–1174. IEEE.

Alzamil, Ibrahim. "Simulation of Cloud Computing Eco-Efficient Data Centre."

Bala, A., Bansal, M. and Singh, J., 2009. December. Performance analysis of MANET under blackhole attack. In Networks and Communications, 2009.

NETCOM'09. First International Conference on 1:141–145 IEEE.

Baig, Zubair A., and Binbeshr Farid 2013. "Controlled Virtual Resource Access to Mitigate Economic Denial of Sustainability (EDoS) Attacks against Cloud Infrastructures." Cloud Computing and Big Data (CloudCom-Asia), International Conference: 346–353, IEEE.

Baig, Zubair A., Sait S.M. and Binbeshr F. 2016. "Controlled Access to Cloud Resources for Mitigating Economic Denial of Sustainability (EDoS) Attacks." Computer Networks. 97: 31–47.

Bakshi, A. and Yogesh, B., 2010. February. "Securing cloud from ddos attacks using intrusion detection system in virtual machine." In Communication Software and Networks, ICCSN'10. Second International Conference on: 260–264, IEEE.

Bhandari, N.H. 2013. "Survey on DDoS Attacks and its Detection & Defence Approaches." *International Journal of Science and Modern Engineering* (IJISME) ISSN, 2(7): 2319–6386.

Booth, G., Soknacki, A. and Somayaji, A., 2013. June. "Cloud Security: Attacks and Current Defenses." *8th Annual Symposium on Information Assurance* (ASIA'13): 56.

Chaudhary, A., Kumar, A. and Tiwari, V.N., 2014. February. A reliable solution against Packet dropping attack due to malicious nodes using fuzzy Logic in MANETs. In Optimization, Reliabilty, and Information Technology (ICROIT), 2014 International Conference on: 178–181. IEEE.

Chonka, A., Xiang, Y., Zhou, W. and Bonti, A., 2011. "Cloud security defence to protect cloud computing against HTTP-DoS and XML-DoS attacks." *Journal of Network and Computer Applications*, 34(4): 1097–1107.

Gupta, G.K. and Singh, M.J., 2010. Truth of D-DoS Attacks in MANET. *Global Journal of Computer Science and Technology*, 10(15): 15–22.

Harb, L.M., Tantawy, M. and Elsoudani, M., 2013. January. Performance of mobile ad hoc networks under attack. In Computer Applications Technology (ICCAT), 2013 International Conference on IEEE, 27(12): 1201–1206.

Liu, H., 2010. October. A new form of DOS attack in a cloud and its avoidance mechanism. In Proceedings of the 2010 ACM workshop on Cloud computing security workshop: 65–76, ACM.

Lo, C.C., Huang, C.C. and Ku, J., 2010. September. A cooperative intrusion detection system framework for cloud computing networks. In Parallel processing workshops (ICPPW), 2010 39th international conference on: 280–284. IEEE.

Morein, W.G., Stavrou, A., Cook, D.L., Keromytis, A.D., Misra, V. and Rubenstein, D., 2003, October. Using graphic turing tests to counter automated ddos attacks against web servers. In Proceedings of the 10th ACM conference on Computer and communications security: 8–19. ACM.

Naresh Kumar, M., Sujatha, P., Kalva, V., Nagori, R., Katukojwala, A.K. and Kumar, M., 2012, November. Mitigating economic denial of sustainability (edos) in cloud computing using in-cloud scrubber service. In Computational Intelligence and Communication

Networks (CICN), 2012 Fourth International Conference on: 535–539. IEEE.

Shakshuki, E.M., Kang, N. and Sheltami, T.R., 2013. EAACK—a secure intrusion-detection system for MANETs. Industrial Electronics, IEEE Transactions on, 60(3): 1089–1098.

Sqalli, M.H., Al-Haidari, F. and Salah, K., 2011. December. Edos-shield-a two-steps mitigation technique against edos attacks in cloud computing. In Utility and Cloud Computing (UCC), 2011 Fourth IEEE International Conference on: 49–56. IEEE.

Zunnurhain, K. and Vrbsky, S.V., 2010. December. Security attacks and solutions in clouds. In Proceedings of the 1st international conference on cloud computing: 145–156.

Communication and Computing Systems – Prasad et al. (Eds)
© 2017 Taylor & Francis Group, London, ISBN 978-1-138-02952-1

Performance evaluation of large scale cloud computing environment based on service broker policies

Apoorva Tripathi & Deepak Arora
Department of Computer Science and Engineering, Amity University, Lucknow, India

Varun Kumar Manik
Prospus Consulting Pvt. Ltd., India

ABSTRACT: Cloud computing is an opportune platform for executing sizeable applications with the availability of colossal computational resources on demand. Infrastructure costs can be decimated by dynamically leasing and releasing resources on demand. It is paramount to leverage existing simulation methodologies to evaluate and study the behavior of applications and algorithms before the actual development of these massive internet applications and distributed systems. Thus, Cloud analyst is a useful tool for researchers to simulate cloud applications in a timely and repeatable manner. Cloud Analyst provides value added services such as a Service Broker Policy whose basic functionality is to route the traffic between data centers and user bases and to choose the most appropriate data center to fulfill each user request. In this paper, authors have performed cost based analysis of the most desirable traits to select the most appropriate data centre for the optimization of application performance and the cost incurred.

1 INTRODUCTION

Cloud Computing known popularly as 'on demand computing', or 'pay as you go computing', is a state-of-the-art technology to offload computation to remote resource providers. Cloud computing fosters collaboration amongst diffuse communities by making it viable for them to partake in an interaction virtually. Furthermore, the resources as well as the information can be divvied up, amongst various users spread globally, in real time and via shared storage using cloud computing. This capability improves product development and customer service and brings about a reduction in time-to-market. Other merits of cloud computing include the curtailment of the size of the data centers by the respective companies and decimation of the data center footprint. As the number of data centers worldwide lessen and overall efficiency increases, it proves to be less detrimental to the environment. Therefore, the 'green' credentials of the companies using shared resources are also boosted. Other advantages of cloud adoption include quick deployment and ease of integration (Buyya, R. et al. 2010). However, there are certain issues that yet to be mitigated, such as security, confidentiality, outage and downtime as well as the tacit dependency on the cloud service provider which makes switching between the service providers complicated, a phenomenon known as

"vendor lock-in" (Zhang, Q. et al. 2010). Despite these drawbacks, its user base is burgeoning and more fine-tuned services and solutions are being offered.

Cloud computing is characterized by a feature known as virtualization. In virtualization, virtual machines are run over the available hardware to meet the requirements of the user (Sahoo, J. et al. 2010). Virtual machine management is an integral part of the concept, and is performed by a piece of computer software called the hypervisor or virtual machine manager. There is a load balancer which performs the selection of virtual machines whenever the workload is encountered. Load balancing intents to optimize resource use by apportioning the work-load in a manner that no node is overloaded, under loaded or idle (Kaur, R. & Luthra, P. 2014). This is entailed by maximization of throughput and minimization of response time as well as resource consumption. It helps in implementing fail over, scalability, resource availability and avoiding bottlenecks. Another abstraction lies above this level called the service broker. A service broker acts as an agent between the users and cloud service providers. The service broker routes the user request to the most appropriate data centre. Hence, the selection of the best policy is paramount as it decides the response time of a particular request as well as the efficiency of data centre utilization.

In this paper, a performance analysis of the current broker policies has been provided. We present a brief overview of these policies and compare these policies based on the average response time by region, average data centre request servicing time and overall cost through an efficient simulation tool known as Cloud Analyst.

2 CLOUD ANALYST

Cloud Analyst is a comprehensive simulation tool based on GUI (Wickremasinghe, B. & Calheiros, R.N. 2010). It emanated from CloudSim and extends some of its functionalities. High degree of configurability is a requisite while modeling sizeable applications with intricate details. In such cases fine-tuning the parameters through repeated simulations is essential. Graphical representation through charts and tables helps abridge the copious statistics accumulated during the simulation. It is useful in comparing and deducing logical patterns contained in the output parameters. Cloud Analyst gives a vivid description of application workloads. It includes the information about the user base where the traffic is emanating from and data centers as well as their location. This information is then processed by Cloud Analyst to generate the values of average response time by region. It also generates the values for other parameters such as overall data center request servicing time, total expenses involved, etc. Cloud Analyst aids in devising the best strategy for optimal selection of data centers, resource allocation amongst the available data centers, and the expenses involved with each of these operations. Cloud Analyst, is efficacious in performing a number of simulation experiments by tweaking the parameters.

Cloud Analyst is comprised of these basic entities (Buyya, R. et al. 2009):

GUI—Cloud analyst contains an exhaustive Graphical User Interface. Its salient attributes include a high degree of configurability and repeatability during simulation.

Region—Cloud analyst performs a geographical segmentation by fragmenting the world into six 'Regions'. These six regions coincide with the six major continents of the globe.

Internet—Cloud Analyst Internet implements the attributes which are crucial for the simulation while abstracting the other details of the physical or real world internet. It emulates physical world Internet traffic routing by injecting data transfer delays as well as transmission latency analogous to the real world internet.

User Base—A group of users constitute a 'user base'. A user base is deemed a single unit. Its task is to generate traffic. All the user bases involved in the simulation are listed in the User Bases Table. Each user base has several configurable fields in the user base table such as name, region, peak hours, average users during peak and off-peak hours, data size per request and average users during peak hours.

Cloud Application Service Broker—A Cloud Application Service Broker routes the traffic amongst the user bases where the traffic is emanating from and the various data centers fulfilling the user's requests.

Data Center Controller—A Data Center Controller administers the management activities of a data center e.g. addition of a new data center, creation and destruction of virtual machines and routing of client requests to the virtual machines, etc.

VM Load Balancer—VM Load balancer is efficacious in determining that which virtual machine should be allocated to the next request for processing.

3 EXISTING SERVICE BROKER POLICIES

In this paper, the three popular broker policies will be analysed.

3.1 *Service proximity based routing*

Service Proximity Based Routing uses the concept of region proximity by choosing the data center closest to the user base (Wickremasinghe, B. 2009). Data Center Controller keeps a region wise index table of all the data centres in the zone. As soon as a user base sends a message, a query is sent to the 'Service Proximity Service Broker' to know the targeted Data Center Controller. The sender's location is retrieved by the 'Service Proximity Service Broker'. A proximity list is maintained region wise for all the regions by the 'Internet Characteristics'. The remaining data centers are sorted in the order of increasing network latency criteria for that particular zone. The 'Service Proximity Service Broker' picks the data centre that occurs first/earliest in the proximity list, to fulfill the request. The service broker selects the data center arbitrarily, in case there are more than one data centers having the same network latency. This policy proves to be the most beneficial in cases where the request can be fulfilled by a data center that is quite close or within the same region.

3.2 *Performance optimized routing*

Best Response Time Service Broker implements Performance Optimized Routing (Wickremasinghe, B. 2009). 'Service Proximity Service Broker' emanated from it. 'Best Response Time Service Broker' always keeps an updated index of all data centers in that zone. As soon as the internet receives a user request from any of the user bases, a query is sent to the 'Best Response Time Service Broker' to get the information about the targeted Data Center Controller. 'Best Response Time Service Broker' has the responsibility of finding the closest data center i.e. the one with minimum latency.

3.3 Dynamically reconfigurable routing

On the basis of the current load that it is facing, Dynamically Reconfigurable Router (Wickremasinghe, B. 2009) performs many functions such as scaling the application deployment, increasing or decreasing the number of virtual machines. Here, scaling is done on the basis of the current processing time as well as the best processing time ever attained.

4 SIMULATION RESULTS AND DISCUSSIONS

To analyse the performance of various Service Broker Policies, configurations of various parameters have to be defined. In this case, the parameter that needs to be defined includes data centre configuration. We also need to set the values of parameters such as user base configuration as well as application deployment configuration. We have defined six user bases located in six different region of the world. We have created 5 user bases and 4 data centres. The location of the first data centre is in region 'zero', the second data center is located in region 'one', the third one is located in region 'three' and the fourth data centre lies in the region 'two'. Number of VMs in DC1 is 100, in DC2 is 80, in DC3 is 50 and DC4 it is 25.

Other parameters are set as follows:
We have set the duration of the simulation as 24 hours. Round Robin is the policy that is used for load balancing. For each request, there are 100 bytes of 'executable instruction length'. In a user base, the 'request grouping factor' and 'user grouping factor' have been set as 100 and 250 respectively.

4.1 Service proximity based routing

For Service Proximity Based Routing policy, the 'response time by region', the 'overall response time', the 'data center request servicing time', 'overall data center processing time', 'cost by data centres' and the 'overall cost' have been calculated by Cloud Analyst and is shown in the Tables 1, 2, 3, 4, 5, and 6, respectively.

Table 1. Response time by region.

User base	Average ms	Minimum ms	Maximum ms
UB1	79.20	45.86	115.47
UB2	59.83	43.07	86.46
UB3	62.08	43.94	84.23
UB4	52.30	38.36	66.50
UB5	313.13	226.96	409.64
UB6	230.98	173.29	291.01

Table 2. Overall response time.

	Average ms	Minimum ms	Maximum ms
Overall Response time	161.11	38.36	409.64

Table 3. Data center request servicing time.

Data centre	Average ms	Minimum ms	Maximum ms
DC1	30.42	1.83	71.56
DC2	10.14	0.77	32.14
DC3	12.74	0.65	29.52
DC4	2.60	0.25	7.89

Table 4. Overall data center processing time.

	Average ms	Minimum ms	Maximum ms
Data center Processing Time	16.30	0.25	71.56

Table 5. Cost by data centers.

Data centre	VM cost $	Data transfer cost $	Total $
DC1	1440.02	37.32	1477.34
DC2	960.02	16.81	976.82
DC3	600.01	40.73	640.74
DC4	300.01	20.27	320.27

Table 6. Overall cost.

	Total VM cost $	Total data transfer cost $	Grand total $
Overall cost	3300.06	115.13	3415.18

4.2 Performance optimized routing

For Performance Optimized Routing policy, the 'response time by region', the 'overall response time', the 'data center request servicing time', 'overall data center processing time', 'cost by data centres' and the 'overall cost' have been calculated by Cloud Analyst and is shown in the Tables 7, 8, 9, 10, 11, and 12, respectively.

Table 7. Response time by region.

User base	Average ms	Minimum ms	Maximum ms
UB1	78.36	43.33	113.07
UB2	58.87	44.33	79.41
UB3	61.33	45.52	77.87
UB4	52.07	39.03	65.16
UB5	311.14	242.15	392.60
UB6	226.86	172.22	414.50

Table 8. Overall response time.

	Average ms	Minimum ms	Maximum ms
Overall Response time	159.32	39.03	414.50

Table 9. Data center request servicing time.

Data centre	Average ms	Minimum ms	Maximum ms
DC1	27.49	1.57	59.77
DC2	9.17	0.75	26.33
DC3	11.35	1.33	24.99
DC4	2.36	0.33	7.82

Table 10. Overall data center processing time.

	Average ms	Minimum ms	Maximum ms
Data center Processing Time	14.68	0.33	59.77

Table 11. Cost by data centers.

Data centre	VM cost $	Data transfer cost $	Total $
DC1	1440.02	37.34	1477.36
DC2	960.02	16.82	976.84
DC3	600.01	40.76	640.77
DC4	300.01	20.28	320.29

4.3 Dynamically reconfigurable router

For Dynamically Reconfigurable Router policy, the 'response time by region', the 'overall response time', the 'data center request servicing time', 'overall data center processing time', 'cost by data centres' and the 'overall cost' have been calculated by Cloud Analyst and is shown in the Tables 13,14,15, 16,17, and 18, respectively.

Table 12. Overall cost.

	Total VM cost $	Total data transfer cost $	Grand total $
Overall cost	3300.06	115.13	3415.18

Table 13. Response time by region.

User base	Average ms	Minimum ms	Maximum ms
UB1	79.20	45.86	115.47
UB2	59.83	43.07	86.46
UB3	62.08	43.94	84.23
UB4	52.30	38.36	66.50
UB5	313.13	226.96	409.64
UB6	230.98	173.29	291.01

Table 14. Overall response time.

	Average ms	Minimum ms	Maximum ms
Overall Response time	1091.34	44.23	76847.01

Table 15. Data center request servicing time.

Data centre	Average ms	Minimum ms	Maximum ms
DC1	27.50	1.57	59.74
DC2	266.84	0.98	71998.28
DC3	290.46	2.60	70657.75
DC4	4468.08	0.93	76794.14

Table 16. Overall data center processing time.

	Average ms	Minimum ms	Maximum ms
Data center Processing Time	937.35	0.93	76794.14

Table 17. Cost by data centers.

Data centre	VM cost $	Data transfer cost $	Total $
DC1	1440.00	37.34	1477.34
DC2	1211.58	16.82	1228.40
DC3	1209.54	40.76	1250.30
DC4	1206.75	20.28	1227.03

Table 18. Overall cost.

	Total VM cost $	Total data transfer cost $	Grand total $
Overall cost	5067.87	115.20	5183.07

5 ANALYSIS

Post analysis of Figure 1 and Figure 2, it can be explicitly stated that there is a perceptible difference in the values of 'Service Proximity Based Routing' and 'Performance Optimized Routing' policies whereas the large disparity in the values of 'Dynamically Reconfigurable Routing', when juxtaposed with the former two policies, is conspicuous. It can be apprehended by looking at Figure 1 that the Performance Optimized Routing policy gives a better average response time than the Service Proximity Based Routing policy. Figure 2 shows that the average data center servicing time of Performance Optimized Routing policy less as compared to the Service Proximity Based Routing policy. A close inspection of Figure 3 depicts that the overall cost is the same in the case of Service Proximity Based Routing and Performance Optimized Routing policies whereas it is remarkably higher in case of Dynamically Reconfigurable Router.

One of the most prominent issues causing the degradation in the performance of Service Proximity Based Routing is that if there happen to be many data centers within the same zone, then the Service Proximity Based Routing performs the selection of data centers arbitrarily.

Also, there is a likelihood of choosing a data center with a higher cost. In addition, there is a possibility that the results may vary even if the configuration is kept unaltered.

The disparate values and the significant difference in the performance level of Dynamically Reconfigurable Router in Fig. 1, Fig. 2 and Fig. 3 provides a definitive conclusion that this policy falls short in terms of giving a good average response time as compared to Service Proximity Based

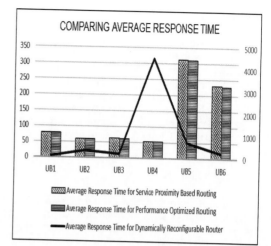

Figure 1. Comparing average response time.

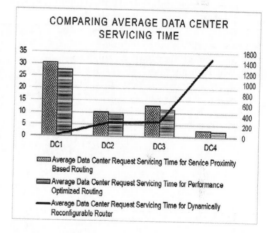

Figure 2. Comparing average data center servicing time.

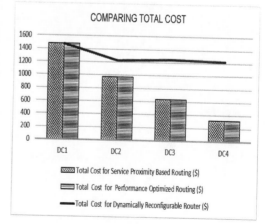

Figure 3. Comparing total cost.

Routing and Performance Optimized Routing policies. Moreover, the average data center servicing time as well as the overall cost of the Dynamically Reconfigurable Router is considerably high as compared to Service Proximity Based Routing and Performance Optimized Routing policies.

The above data and charts provide a conclusive evidence that the Performance Optimized Routing policy is the best approach amongst the three broker policies in the Cloud Analyst framework.

6 CONCLUSION

Cloud based sector is a lucrative business area but various objectives such as minimization of average response time, average data center servicing time and overall expense need to be attained to achieve customer satisfaction. Repeated simulations can give a new insight into ameliorating the quality of service in the cloud environment. In this paper, authors have found that in the case of the parameters such as the average response time and the average data center servicing time, Optimize Response Time policy outperforms Service Proximity Based Routing and Dynamically Reconfigurable Router. Authors have observed that the overall cost is same in case of Service Proximity Based Routing and Performance Optimized Routing. The performance level of Dynamically Reconfigurable Router for all these three parameters is a quite poor as compared to the Service Proximity Based Routing and Dynamically Reconfigurable Router. Hence, we have come to the conclusion that the Performance Optimized Routing is the best policy amongst these three policies.

Further, this research work can be extended towards the inclusion of various dynamic scheduling algorithms and related parameters.

REFERENCES

Buyya, R; Broberg, J. & Goscinski, A. 2011. Cloud Computing: Principles and Paradigms. John Wiley & Sons.

Buyya, R; Ranjan, R. & Calheiros, R. N. 2009. Modeling and Simulation of Scalable Cloud Computing Environments and the Cloudsim Toolkit: Challenges and Opportunities. 7th High Performance Computing and Simulation Conference; IEEE Computer Society.

Kaur, R. & Luthra, P. 2014 Load Balancing in Cloud Computing. ACEEE.

Sahoo, J; Mohaptra, S. & Lath, R. 2010. Virtualization: A survey on concepts, taxonomy and associate security issues. ICCNT, IEEE: 222–226.

Wickremasinghe, B. 2009. Cloud Analyst: A Cloud-Sim-based tool for Modeling and Analysis of Large Scale Cloud Computing Environments.

Zhang, Q.; Cheng, L. & Boutaba, R. 2010. Cloud computing: state-of the-art and research challenges. J Internet ServAppl. Springer: 7–18.

Communication and Computing Systems – Prasad et al. (Eds)
© *2017 Taylor & Francis Group, London, ISBN 978-1-138-02952-1*

Enhancing big data security in cloud using Hadoop

Ayushi Pathak & Deepak Motwani
ITM University, Gwalior, Madhya Pradesh, India

ABSTRACT: Now a day's big data is one of the key problems that researchers endeavor to get rid of and focusing their researches over it to solve the problem of how big data could be tackled and managed with the cloud of computing, and the main issue is how to gain a flawless security for BigData in the cloud computing. In this paper, we will propose a model for securing BigData using Hadoop. The main emphasis is on how to resolve security issues trigger in cloud environment that are related to big data. Big Data applications are of great advantage to companies, organizations, business and many small scale and large scale industries. We will also discuss several provisions for the complication in cloud computing security opting as a fast speed that encompasses possible network security, improper communication, leakage of information, and data privacy. Furthermore, cloud computing, bigdata and its appliance, benefits are likely to signify the most favorable opportunities in data science boundary.

1 INTRODUCTION

In current era to examine intricate information and to categorize facts which required to securely store, organize and share too big and too large data. Cloud comes with several complications, i.e. the data proprietor might not have any command about where the information is placed. The motive behind this control issue is that the resource of cloud computing are organized in a distributive manner. Hence it is insist to guard the information in the middle of unreliable job. As cloud provide infinite service number of complexity is triggered, we presume that instead of providing a holistic solution to securing the big data, it would be better to make remarkable improvements in securing the cloud that will eventually provide us with a safe and secure cloud. Map Reduce framework has been initiated by Google for processing bulk of data and information on hardware. Apache's Hadoop introduce circulated file system (HDFS) which is developing as a surpassing ingredient for cloud computing united along with integrated pieces of software such as Map Reduce. Hadoop is an open-source implementation of Google Map Reduce, including distributed directories, which provide the abstraction of the map to the application programmer. In the paper, we will proposed a model which make BigData secure at different level in Cloud computing

Cloud Computing rely on sharing of computing resources i.e., distributed servers instead of centralized servers or devices to manage the resource. In Cloud Computing, the word "Cloud"

is other name for "The Internet", thus Cloud Computing means a solution in which service are offered by internet. Cloud Computing aim to get benefit of growing computing power so that multiple of instructions get executed in a parallel fashion. This environment is being used to utilize the usage amount of computing resources. The cloud environment involves clusters of computers which take care of load. The cost of software and hardware is to minimize at user end. The process could be done at the user's end by software interface which is linked to cloud. Cloud Computing comprises of a front end and back-end. In the cloud computing framework the user can access applications from anywhere and anytime by connecting to the Cloud servers using the Internet.

1.1 Big-Data

BigData is the term for detail too large, too big amount of structured and unstructured data and it is very difficult to process this data using traditional databases and software technologies. The word "BigData" is an industry that had to query loosely structured too large distributed data. Following are three main key to brief the properties Big Data:

Volume: BigData have the volume, many factors are responsible increasing Volume streaming data and data gathered from sensors etc.

Variety: Now a day the data which is generated in different formats such as: text, graph, images etc.

Velocity: The data rates are rapidly therefore the data needs to be handle to meet the demand and how quick the data value is processed.

Variability: Including speed, the data flow can be highly inconsistence with respect to time.

1.2 *Hadoop*

Hadoop, is an open source framework which allow parallel processing of data. Apache Software Foundation sponsored an Apache project and it is a piece of this project. Master/Slave structure is used by its clusters. Bulky data items can be processed across a group of servers and applications can run on systems with huge numbers of nodes involving thousands of bytes, using Hadoop. Hadoop framework is reliable in case of node failure because it offers master/slave node. Even in the case of node failures this approach decreases the probability of a complete system failure. Hadoop gives a computing result which is cost effective, scalable, fault tolerant and flexible. Its Framework is used by world famous companies like Yahoo, Google, IBM and Amazon etc., to support their software which has a huge amount of data. Two main core components of Hadoop are—Map Reduce and HDFS.

1.3 *Map Reduce*

Map Reduce is a framework which is used to write software that processes bulky amounts of data process together on clusters of commodity hardware resources in a fault-tolerant and reliable manner. A Map Reduce splits information into various parts and they are processed parallel by Map jobs. The outcomes of the mapper stage are then put into the reducer phase which is sorted by the framework. Usually the input and the output, both are store in a directory. Framework Organize, Examine and re-executing of failed tasks.

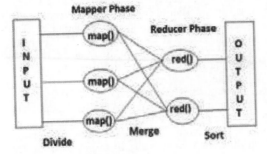

Figure 1.   Map Reduce framework.

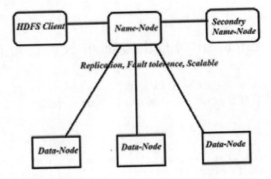

Figure 2.   HDFS.

1.4 *HDFS*

Hadoop distributed file system is a file system which is used for data storage and traverse all the nodes in a Hadoop cluster. To make file system as a single broad file system, it group together file systems on local nodes. The reliability of HDFS is improving by making the replica of nodes across multiple sources to overcome node failures.

2   CLOUD COMPUTING IN BIGDATA

With the emergence of BigData, the rises of cloud computing and cloud data stores have proved a precursor and facilitator. It is the commoditization of data storage along with computing time through standardized technologies. It has noteworthy advantages over conventional physical distribution. However, Cloud sometimes needs to be conjoined with traditional architectures. This makes decision makers confused, leading to a question which is the best option for their computing needs, particularly if it is a big data project? These projects frequently display immense computing power and storage needs or unpredictable bursting. But the organizations expect inexpensive, swift, and trustworthy products and project outcomes.

3   LITERATURE SURVEY

Santanu Chatterjee et al., (2015) proposed an efficient fine grained access control scheme for cloud environment, which allow user to authenticate with the cloud server to access the cloud object. The communication between user/client is done through insecure channel where no third party involves security protocol like secure shell without elevating any extra cost. The symmetric key encryption is used in these schemes which supports both user and attribute revocation functionalities.

The author compare existing scheme with proposed scheme and observe that proposed scheme incurs less computational cost. Sunny Singh et al., (2015). The motive of this paper is to tackle privacy issues and draft a framework pertaining to privacy and confidentiality concerns in adopting cloud computing. The author study about different framework such as: ISO/IEC29100, NSTIC2 FIPPs, U.S-EU Safe Harbor extract some of the privacy principle such as: Compliance, Access, Storage, Retention, Audit and monitoring, Destruction, Privacy breaches, Law. By covering all of these issues cloud service provider may serve their service to customers in a better way. Manjeet Singh and Jasbeer Singh et al., (2015) proposed virtualized architecture combines the properties of both parallel and Map Reduce DBMSs with data encryption using Hadoop framework with RSA encryption scheme using Java Cryptography Architecture (JCA). RSA is an scheme for public key cryptography that is based on the presumed difficulty of factoring large integers, the factoring problem The architecture extends the Map Reduce functionality to work in a secure environment with homogeneous data. Priya P. Sharma et al., (2015) focus at studying "BigData" security at the environmental level, including examination of built-in protections and the Achilles heel of these systems, and also boundary on a journey to assess a few issues that we are dealing with today in procuring contemporary BigData and proceeds to propose security solutions and commercially accessible

techniques to address the same. The author analysis different component and the need of security control for them. Muhtaroglu Pembee et al., (2013) conclude that data exploration and analysis became difficult problems in many sectors in the span of big data. With bulky and intricate data, computation turned difficult to be handled by the conventional data processing applications which initiate the advance of bigdata applications. Yi Jiang et al., (2012) Bioinformatics is a paramount application for Hadoop, which covers the biological domains and next generation sequencing. Huge scale data analysis is required by Bioinformatics, which uses Hadoop. Cloud computing gets computer clusters and web inter face together with parallel distributed computing framework.

4 ADVANTAGE OF BIGDATA

In BigData, the software packages supply a rich package of tools additionally options where the individuals can plan the whole of the data setting across the company, hence allowing the individual to examine the threats he/she faces within. Big data keeps the data safe and this is considered as the main advantages With this a person is capable to tackle the potentially sensitive information which was not protected in a proper manner and makes certain that it is saved according to the regulatory necessities. All the business and organizations will be benefited from scalability, capacity and speed,

Table 1. Comparative study of existing approach.

Author	Objective	Approach	Summary
Santanu Chatterjee, Amit Kumar Gupta & G.V. Sudhakar IEEE 2015	An Efficient Dynamic Fine Grained Access Control Scheme for Secure Data Access in Cloud Networks	Symmetric key encryption and decryption	Authenticate user to access cloud object
Sunny Singh & Nitin Goel IEEE 2015	Framework Approach to Extract Privacy Issues in Cloud Computing	Tackle privacy issues	Study of different framework and extract some privacy principle
B. Saraladevi, N. Pazhaniraja, P. Victer Paul, M.S. Saleem Basha, P. Dhavachelvan ISBCC 2015	Big Data and Hadoop-A Study in security Perspective	Kerberos, Bull eye Algorithm and NameNode	Issues arises in HDFS are discussed by author
Manjeet Singh & Jasbeer singh Springer Proceeding 2015	Virtualized Architecture for Secure Database Management in Cloud Computing	RSA encryption scheme with MapReduce	Parallel & MapReduce DBMS with RSA encryption scheme using Hadoop for secure Data Management
Venkata Narasimha Inukollu, Sailaja Arsi and Srinivasa Rao Ravuri, IJNSA 2014	Security Issues Associated With Big Data in cloud computing	Big Data, MapReduce and Hadoop	Discuss various security issues associated with big data

of cloud storage. Furthermore, companies can give creation to several business opportunities and end users can envision the data. Another remarkable gain of big-data is, data analytics, which permits the individual to change the look or content of the website in actual time so that it is liked by each customer accessing the website. The group in gleams to exploration of these four areas:

1. Analyze the risks on large portfolios.
2. Notice, check, and re-audit monetary fraud.
3. Make delinquent collections better.
4. Carry out lofty value marketing campaigns.

5 ISSUES AND CHALLENGES

Thus security issues occurs in these technologies and systems are need to be handled at different level of cloud computing. It is vital for the network which inter-connects the systems in a cloud framework to be safe and secure. Also, several security concerns are the result of virtualization paradigm in cloud computing. For example, mapping of the virtual node to the local node has to be done with full safety. Encryption of the data is not the only things involve in Data security, but it also gives surety that suitable policies are imposed for data sharing. As well as, memory management algorithms and resource allocation also have to be safe and secure. Problems related to big data are most intensely arises in certain industries, such as e-commerce, telecoms, advertising and retail and banking services, and also in several activities of government. This data explosion is a main topic for achieving reliability and scalability in many industries, and those companies will achieve significant benefit that are competent of adapting nicely and expand the ability to examine the reason for such data explosions in other companies. Finally, in the malware detection in clouds, data mining approaches can be used. User authentication level, generic issues, network level, and data level are the challenges of security which can be label in cloud computing environment.

5.1 Network level

The challenges which manage the network protocols can be fixing out under a network level, such as apportion of data, Inter node communication and distributed nodes.

5.2 Authentication level

The challenges at authentication level is that some person may hack your login detail make use with them, therefore number of challenges may occur while authenticating user.

5.3 Data level

If there is an delay in request and response of data client so this kind of issues are treated at data level.

5.4 Generic types

The issues which are related to software up gradation, tool is compatible or not, system which is to be use is free from virus or not is categorizes under generic level.

6 PROPOSED WORK

To improve the security of Hadoop ecosystem various security measures are presented here. As Hadoop environment is a fusion of many components, various solutions are proposed which will jointly make the environment safe and secure. The utility of multifarious technologies/tools to lessen the security problem specified in the previous section is stimulated by proposed solutions. Security recommendations are planned so as not to lessen the efficiency and scaling of cloud systems. To confirm the security for Hadoop environment these security solution should be followed.

6.1 Authentication

Hadoop should be configured with an authentication mechanism to check user identity. The main aim of authentication in Hadoop, as in other systems, is to prove that a user or service is who he or she claims to be. Authentication in enterprises is performed by a single distributed system, such as a Lightweight Directory Access Protocol (LDAP) directory. Some of the common enterprise-rank authentication systems is Kerberos. The threat of impersonation is virtually eliminates because it never sends a user's log in detail in plain text through the network.

6.2 Encryption

Just processing of data is not sufficient, some alteration is ought to be performed so that data become unintelligible for security purpose. The purpose of encryption is to guard that only authorized users can use, view, or contribute to a data set. The encryption adds another covering of protection against potential threats by administrators, endusers and other malicious actors on the network. Data protection can be performed at different levels within Hadoop.

OS File system-level: To guard files in a volume encryption at the Linux operating system file system level can be applied. Encryption should be performed in such a way that it should protect data inside or outside the cluster.

HDFS-level—Encryption can be performed to a Hadoop file system (HDFS). Each block is split each file into fixed size block and encrypts it by client, before uploading to Hadoop file system. The encryption is performed by the client on the CPU and transfer encrypted block to HDFS (Data Node). Then receiver Data Node (First DataNode where block store) replicate block into two other DataNodes.

Network-level-Encryption of all the data communication should be done while transferring the data over a network. Encryption can be performed to encrypt data when the data is being sent through a network and get decrypt it as soon as it is received. The RPC process calls should happen over SSL so that even if a hacker gets a chance to tap into network communication packets, he will not be successful in extracting useful information or manipulating packets.

6.3 *Authorization*

Authorization is different from authentication. Authorization describes us what any user can or cannot perform within a Hadoop cluster, after the user has been successfully authenticated. In the context of MapReduce the user and group and groups are used to determine who is allowed to submit or modify the job. Data security within Hadoop needs to support multiple use case for data access, while also providing a framework for central administration of security policies and monitoring of user access. The monitoring and management of data security across the Hadoop platform can be enabled by using Apache Ranger is a framework. Apache Ranger provides enhanced support for different authorization methods-Role based access control, Attribute Access Control.

6.4 *Software node maintenance*

Nodes running the software should be structured regularly to remove any virus existing. To make the system more secure all the application software and Hadoop software should be updated. For resolving, these problem secure version of Apache Hadoop should be used.

7 RESULT AND FUTURE WORK

Big Data refers to a huge collection of data which may contain sensitive information. Therefore it is duty of service provider to guard this sensitive information. By analyzing this situation we adopt a Hadoop framework which can process huge collection of data more efficiently and implement an encryption technique which protects this sensitive information from unauthorized user. By default Hadoop HTTP web-consoles allow access without any form of authentication, therefore we configured Hadoop HTTP web consoles with Kerberos authentication using HTTP SPENGO protocol supported by browser like Firefox and Internet Explorer. In future we will focus on image and video encryption methods.

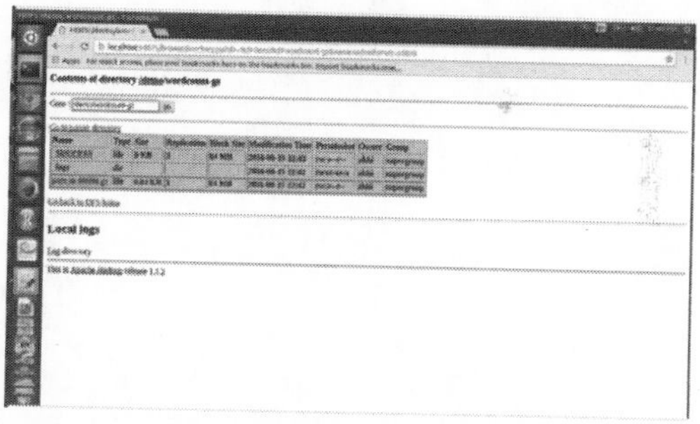

Figure 4.   Permission for accessing the file.

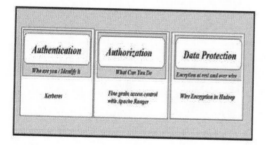

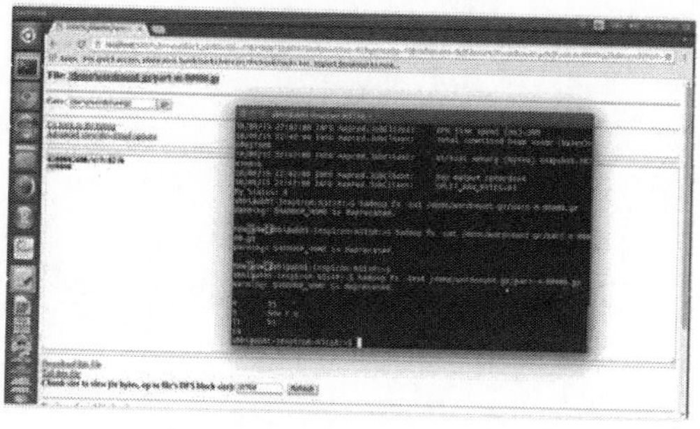

Figure 3.   Proposed model for securing BigData.

Figure 5.   Encryption with compression.

8 CONCLUSION

Data is growing at a very high speed in a cloud environment. Each and every data has its own importance. Therefore, the major demand of an organization that data should be kept as secure as possible. Many security issues may arise while taking the benefit of a cloud environment. In this paper, we proposed a model, which help to data owner for securing Big Data.

REFERENCES

Chatterjee Santanu, 2015, An Efficient Dynamic Fine Grained Access Control Scheme for Secure Data Access in Cloud Networks Electrical. *Computer and Communication Technologies (ICECCT), 2015 IEEE International Conference* DOI: 10.1109/ICECCT.2015.7226107.

Inukollu Narasimha Venkata, Arsi Sailaja and Ravuri Rao Srinivasa, 2014, Security issues associated with big data, *International Journal of Network Security & Its Applications (IJNSA), Vol. 6*, DOI: 10.5121/ijnsa.2014.6304: 45–56.

Li Xubin, Yi Jiang, Wenrui Jiang Yi and ZouQuan, 2012, Hadoop Applications in Bioinformatics, *Open Cirrus Summit (OCS), 2012 Seventh*: 48–52.

Muhtaroglu Pembe, Canari F., Demir Seniz, Obali Murat and GirginCanan, 2013, *Business model canvas perspective on big data applications: Big Data, IEEE International Conference* DOI: 10.1109/BigData.2013.6691684.

Sarala devi B., Pazhaniraja N., Paul Victer P., Basha Saleem M.S. and Dhavachelvan P., 2015, Big Data and Hadoop-A Study in Security Perspective, *2nd International Symposium on Big Data and Cloud Computing (ISBCC'15)*: 596–601.

Sharma P. Priya, 2014, Securing Big Data Hadoop, *A Review of Security Issues, Threats and Solution (IJCSIT) International Journal of Computer Science and Information Technologies* Vol. 5 (2): 2126–2131.

Singh Manjeet and Singh Jasbeer, 2015, A Virtualized Architecture for Secure Database Management in Cloud Computing, Proceedings of Fourth International Conference on Soft Computing for Problem Solving, Advances in Intelligent Systems and Computing, DOI: 10.1007/978-81-322-2220-0_25: 336.

Singh Sunny and Goel Nitin, 2015, Efficient Framework Approach to Extract Privacy Issues in Cloud Computing *Fifth International Conference on Communication Systems and Network Technologies* DOI: 10.1109/CSNT.2015.141: 698. http://ranger.apache.org/

Communication and Computing Systems – Prasad et al. (Eds)
© 2017 Taylor & Francis Group, London, ISBN 978-1-138-02952-1

Route Detection using Segmented Path Vector (RD-SPV) based connectivity-aware geocast routing in VANETs

D. Durga Prasada & K. Sushil
School of Computer and Systems Sciences, Jawaharlal Nehru University, New Delhi, India

K. Omprakash & A. Abdul Hanan
Faculty of Computing, Universiti Teknologi Malaysia (UTM), Johor Bahru, Malaysia

ABSTRACT: Emergency signalling, geographical advertising, information sharing through traffic networks etc. are some Intelligent Transport System (ITS) applications for which geocast routing is pre-requisite. Also, through geocast routing the genesis and realization of virtual warning system in vehicular communication is possible for enhancing road safety and security. Due to high dynamicity of network topology, large scale vehicular system and highly mobile vehicular nodes, relaying of geocast packets through multi-hop relay nodes become quite cumbersome in vehicular traffic environment. To address the issues, this paper purposes Route Detection using Segmented Path Vector (RD-SPV) based Connectivity Aware Geocast Routing (CAGR). SPV is used to determine optimal route for geospatial packet dissemi-nation which enhances packet delivery ratio of CAGR significantly. On-the-Fly Segment Density (OFSD) based Adaptive Beaconing System (ABS) is used to adjusts rate of beaconing with density of nodes in specific path segment which reduces congestion in the network by minimizing beacon overflow. Simula-tion results show that the purposed approach minimizes packet loss, reduce end-to-end delay as compared to Greedy Forwarding (GF) and Greedy Perimeter Stateless Routing (GPSR).

1 INTRODUCTION

The process of accessing, communicating and navigating information has completely changed due to the co-evolution of telecommunication, wireless sensor networks and emerging transpor-tation technologies. (Fan et al., 2007). Transporta-tion safety, security and efficiency are three most essential objectives of vehicular communication (Al-sultan et al., 2014). VANETs and Mobile Ad-hoc Networks (MANETs) have certain simi-larities such as ad-hoc infrastructure facility, multi hop node communication. Higher mobility, dynamic topology and self-organization character-istics of VANETs distinguish it from MANETs.

Many scientific projects funded by Trans-National Organization (TNO) with the help of automobile, Information and Communication Technology (ICT) and educational institutions are exploring different possibilities and opportunities regarding VANETs (Khatri et al, 2011). The most fundamental challenges of VANETS are reliable, speedy and error free dissemination of urgent and crucial information (Panichpapiboon et al., 2012). Vehicular Trajectory Information (VTI) obtained through Navigation Systems (NS) embedded in each vehicle which offers to the vehicular node

drivers the avenues to choose route. But, relating vehicular trajectories information to geocast rout-ing (Imielinski et al., 2009) is a challenging issue for researchers (Kaiwartya et al., 2015).

Dissemination of data packets from source vehicle to targeted region, called Zone of Interest (ZOI) is the aim of geocast routing in VANETS (Kaiwartya et al., 2014). Blind corner, wrong turn, "danger ahead" signals, mud slides, natural dev-astation such as heavy downpour, cloud burst are such road as well as environment conditions which are attached to a geocast message and disseminate in Intelligent Transportation System (Maihofer, 2004). Urban VANETs is different from others due to its peculiar characteristics such as constrained movement area, multiple traffic signals, size and number of vehicular nodes is large, random vehic-ular distributions, unlimited transmitting power, and mammoth hindrances such as big buildings, tower, tree, and skyscraper (Sum et al., 2000). Hence, connectivity and communication quality are the most challenging issues which need to be resolved for smooth data dissemination in urban VANETs (Rao et al., 2014).

Availability of efficient connectivity-aware geo-cast routing protocols are the basis for the work-ability of different applications of ITS (Wang et al.,

2008). But designing of such protocols are cumbersome process due to random node mobility and dynamic disconnected network topology (Luong et al, 2007). Position based routing instead of topology based routing is used for the enhancement of reliability and easy functioning. GPS enabled vehicles generate geospatial data which is used for data dissemination to the ZOI (Jiang et al., 2003).

Our proposed connectivity-aware geocast routing protocol based on Route Detection using Segmented Path Vector (RD-SPV) where relay node is selected by analyzing the Coverage Area Potential (CAP) calculated from SPV urban environment by checking credibility of the sender from TMS. We organize the paper as follows. Section 2 gives the required information regarding prior and current research work. Section 3 gives details regarding the proposed scheme and corresponding algorithms. Section 4 gives details regarding simulation results and performance analysis and in section 5 the conclusion part is presented.

2 LITERATURE REVIEW

Landmark Overlays for Urban Vehicular Routing Environments (LOUVRE) (Lee et al., 2008) builds landmark overlay network by creating the overlay links if the vehicle density is higher than threshold to solve frequent network disconnection problem. The packet delivery ratio and hop count of LOUVRE are better than GPSR. But, threshold is calculated by considering that all vehicles are characterized by uniform probability distribution function. Intersection based Geocast Routing Protocol (IGRP) (Saleet et al., 2011) choses path having maximum connectivity probability. Computational complexity, convergence speed of algorithm, centralized architecture, and assumption that vehicles follow Poisson's distribution are certain flaws which leads to Junction-Based geographic Routing (JBR) (Tsiachris et al., 2012). It uses selective greedy forwarding up to a junction node which is closer to the junction.

Coordinator, located at junction who is closest to the destination is chosen as NHV by using minimum angle method. As multiple paths are generated due to broadcasting to multiple coordinators, packet drop increases and to alleviate Junction-based multipath source routing is proposed (Sermpezis et al., 2013). For medium or long source-destination distance or for medium to high traffic loads multipath is beneficial. But, local optimum problem is not considered. Shortest path based Traffic Aware Routing (STAR) (Change et al., 2012) enhances delivery ratio and throughput by considering effect of traffic light on designing

of routing protocols and assuming high density on green light segment. A Connectivity-Aware Intersection-based Routing in VANETs (CAIR) (Chen et al., 2014) chooses intersected connected routes having higher connectivity and lower transmission delay based on the fly real time traffic density collection scheme. Searching area limitation strategy is used to reduce routing overhead and transmission delay.

For prediction of possible link breakage event (i.e. local disconnection problem) before it happens the author purposes uses information contained on vehicle heading and avoid forwarding to the disconnected neighbor (Taleb et al., 2006). Expected Disconnection Degree (EDD) is used to find the probability of path would be broken. In (Naumov et al., 2007), the concepts of guards is used which adjusts and maintains the connectivity path in commensurate with the change in speed or direction of the end nodes. Network overhead is very high in this approach because to determine the route it broadcasts the route discovery request. Mobile node trajectory is modified to address the frequent network disconnection so that data can be disseminated and spread in the disconnected area (Zhao et al., 2004). Packets are buffered when network is disconnected and forwarded when NHV is within the transmission range (Chen et al., 2001). In (Lochert et al., 2005), optimum route is calculated by using Dijkstra shortest path algorithm based on road maps. In VANETs the existing routing protocols works efficiently either in a fully connected network or packer buffering mechanism should be incorporated without considering the network disconnections. In Adaptive Connectivity Aware geocast Routing protocols (ACAR), the route is selected having best link quality adaptively. Real time statistical data which gives density information is generated during data dissemination. Carry-and-forward mechanism is used during network partition due to low vehicular density. It chooses the next hop by considering the minimum packet loss rate (Qing et al., 2008). Adaptive Traffic Beacon (ATB) is used for efficient message dissemination which has distributed architecture (Sommer et al., 2011). It shows that adoptive beaconing enhances the penetration rate by broadening dissemination of message than flooding based approaches.

3 PROTOCOLS DETAILS

In any geocast routing algorithm availability of suitable, registered and efficient relay node is most essential. It means less data dissemination delay, less network overload, more packet delivery ratio. But the main problem in greedy forwarding is that the relay node generally lies in the boundary zone

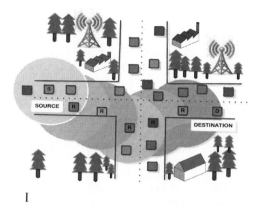

Figure 1.   Greedy forwarding in VANETs.

of sender and during transmission the relay node goes out of the transmission range which causes to retransmit the data packet and ultimately enhances traffic congestion, consumes bandwidth and takes more propagation (Karp et al., 2000).

In Fig. 1, the source s sends data packet to the relay node R. In greedy forwarding relay node is chosen among the neighboring node as that node which is farthest from the sender and nearer to the destination. Hence, from the figure it is clearly seen as all relay nodes R lies on the boundary zone of the respective sender. During data dissemination it may also happen that the relay node is out of transmission range and causes network congestion. Local optimum problem occurs in greedy forwarding where the sender is itself the router and not able to forward the data packet. GPSR somehow eliminates the local maximum problem by combining greedy routing with face routing approach but it is unable to reduce traffic density.

Location aware information based on geospatial data generated from navigation system such as GPS (Global Positioning system) which is incorporated in each vehicular node in a high speedy vehicular environment used by geocast routing protocol for route selection sometimes leads to selection of wrong route such as in a blind corner. In our purposed protocol the route selection is based on vehicular path segment which gives right shortest route to locate relay node which ultimately reduces the transmission cost and hence reduce propagation time.

Route Detection using Segmented Path Vector (RD-SPV) based connectivity-aware geocast routing works in two phases.

A-Node Trajectory Table Maintenance (NTTM) phase.

B-Node Trajectory Table Look-up (NTTL) phase.

A-Node Trajectory Table Maintenance (NTTM) phase

Segmented Path Vector (SPV) is collection of all Path Segments (PS) that the vehicular node is intent to travel from source to destination. Path segments are calculated by considering the intersection points lie in the square zone formed by considering the shortest route from source to destination. PS is the joining path of two consecutive intersection points. Which PS lies on which zone is calculated by PS selection algorithm. If the positional value of both the corner points of the path segment lies inside the area then PS is accepted. If the positional value of both the corner points of the path segment lies outside then the PS is rejected. If only one corner point of the PS lies inside then that point is chosen unless all other available PS are being rejected. In case no suitable PS is found Store-Carry-Forward (SCF) approach is used where it stores the data packet in its cache and forward when it reaches at another node's transmission range.

Each node constructed SPV by extracting information from GPS and beacon SPV with Time Stamp (TS) value. Adaptive beaconing system (ABS) is used which beacon on demand by calculating the node density of its neighbors. On-the-Fly Segment Density (OFSD) collection scheme is used where density of each segment is calculated by the ABS which sends node density information, which is the number of nodes that is within the transmission range of the sender along with SPV. If node density is higher than the Threshold Density (TD) then it increments the Beaconing Interval (BI) as at least one node can get the SPV. When the node density is lower it decrements the BI so that ABS can beacon regularly.

Node Trajectory Table (NTT) is constructed in a distributed manner and maintained by each node by executing Node Trajectory Table Maintenance (NTTM) algorithm which receives the beaconed information from the neighboring node as its input. When a source wants to disseminate the geocast packet it execute a Node Trajectory Table Lookup (NTTL) algorithm which extracts value of SPV and TS for each neighboring node stored in NTT and check the Coverage Area Potential (CAP) from SPV to get the relay node which is the node having highest CAP in the intended direction and lowest TS.

Ip is the collection of intersection points of roads.

Path Segment $(PS) = PS_{i,j}, \ i,j \in I_p$

$SPV(v) = \{PS_{i,j}, \ PS_{j,k}, \ PS_{k,l}, \ PS_{l,m}, \ PS_{m,n} \}, \ v \in V$

$NTT(v) = \langle SPV(v), TS(v) \rangle, \ v \in V$

In Fig. 2, it is shown that

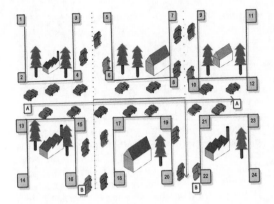

Figure 2. Selection of segmented path from intersection points.

$$I_p = \{1,2,3,4,5,6,7,8,9,10,11,12,13,14,15,16,17,$$
$$18,19,20,21,22,23,24\}$$

$$SPV(A) = \{PS_{2,4}, PS_{4,6}, PS_{6,8}, PS_{8,10}, PS_{10,12}\}$$

$$SPV(B) = \{PS_{16,15}, PS_{15,17}, PS_{17,19}, PS_{19,21}, PS_{21,22}\}$$

$$CAP(A) = 5, CAP(B) = 5$$

Algorithm 1: (RD-SPV).

1. FIND$_{relay}$();//*to find relay node from NHV*
2. NTTM()//*node trajectory table maintenance algorithm*
3. createSPV();//*calculate threshold time and wait;*
4. While(i≤n)
5. inf=extractINFO(PS,TS);
6. SPV[i]=buildSPV(inf);
7. Beacon(SPV[i],TS[i]);
8. createNTT();
9. While(i≤n)
10. inf= Beacon(SPV[i],TS[i]);
11. NTT[i]=setINFO(inf);
12. NTTL()//*node trajectory table look up algorithm*
13. While(i≤n)
14. info=getINFO(NTT);
15. CAP[i]=(info(PS[i]));
16. NODE$_{relay}$=max(CAP[i]);//*find relay node*

B-Node Trajectory Table Look-up (NTTL) phase

Each node is registered itself with the Traffic Monitoring System (TMS) which is a Road Side Unit (RSU). TMS maintains IP-to-Rid table by linking unique registration id (Rid) with IP address and allocated Rid during vehicle registration to the new vehicle. Each node has IP-to-Rid table beaconed by TMS and it extracts Rid from beaconed information and match it with table to check whether registered or not. If the receiver confirmed that the sender is registered then only it updates its NTT. Otherwise it rejects the data packet. When an accident occurred, the vehicles following that vehicle form the geocast region. The vehicles in the opposite lane which are moving in opposite direction constitutes forwarding region. The vehicle moving in boundary zone and forwarding region act as a relay node. As shown in the figure 3 accident occurred at intersection point 17. Hence, the first lane having intersection points17, 19, 21, 23 and second lane having intersection points15, 16 constitutes geocast region. The lane having intersection points 6, 8, 10, 12 constitute forwarding region. The vehicle at intersection point 12 acts as relay node.

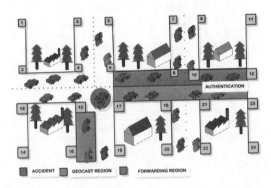

Figure 3. Identification of geocast region and forwarding region.

Algorithm 2: Node Trajectory Table Lookup (NTTL).

1. NTTL()//*node trajectory table look up algorithm*
2. Call V$_{regst}$()//*check vehicle's registration*
3. V$_{sender}$(sendIP);//*node send IP to TMS*;
4. TMS(getIP);//*TMS receive IP*
5. TMS(IP-to-Rid table)//*TMS matches with table*
6. TMS(sendRid)//*send registration id of vehicle*
7. V$_{sender}$ (getRid)//*node gets the id*
8. If(getRid(beconedINFO)==getRid(IP-to-Ridtable))
9. Then print("vehicle is registered");
10. Else
11. Print("reject")
12. V$_{sender}$(IP-to-Rid table); //*update its table*
13. Beacon(Rid)//*beacon id*

Algorithm 3: On-the-Fly Segment Density (OFSD) based Adaptive Beaconing System (ABS).

1. Beacon$_{ABS}$()
2. send(INFO,N$_{NBR}$)//*send data with number of neighbor*
3. receive(INFO,N$_{NBR}$);
4. While(N$_{NBR}$!=NULL)
5.   If(CN$_{NBR}$>receive(N$_{NBR}$))//*receiver's neighbors are more.*
6.   Then N$_{NBR}$=CN$_{NBR}$//*update number of neighbor*
7.   Else
8.   send(INFO,N$_{NBR}$)//*beacon*
9.   If(N$_{THRS}$<N$_{NBR}$)
10.   I$_{BCON}$++;//*increment beaconing interval*
11.   Else
12.   I$_{BCON}$--;//*decrement beaconing interval*
13. Beacon();

4 SIMULATION RESULTS AND PERFORMANCE ANALYSIS

Simulation is done using network simulator, ns-2.34. For the generation of realistic mobility model and realistic urban based vehicular traffic model MOVE (mobility model generator for vehicular networks), top level of an open source micro traffic simulator, SUMO (simulation of urban mobility) has been used. Network size and node density are taken as the setup parameters. Quadratic network size ranges from 500 m to 5000 m side length and 150 is the node density. Transmission range is taken as 500 m. In VANETs the vehicular speed is quite high. For simulation the speed between 150 km/h to 30 km/h is taken. Nodes are uniformly distributed and for each phase of simulation 250 messages are disseminated. For performance evaluation, delivery ratio, end-to-end delay and network overhead are taken into account. 802.11 layer-2 ACK is used in simulation for implicit acknowledgement, when data packet is delivered successfully.

In Fig. 4, it clearly shows that Route Detection using Segmented Path Vector (RD-SPV) based connectivity-aware geocast routing enhances the delivery ratio (percentage of delivered packets successfully to the total packets transmitted) significantly by maintain the link between nodes. As OFSD based ABS is used the vehicular density of each path segment is known before transmitting the data packet. The connectivity is maintained based on node density and it chooses the best NHV among the available nodes. GPSR performs better that simply greedy routing and our purposed scheme performs better as compared to other.

In Figure. 5, the purposed scheme shows least end to end delays especially when networks having high node density. In dense networks the purposed scheme selects the next hop which has higher Coverage Area Potential (CAP). As the CAP is known to the source before forwarding the data packet, it doesn't hold the data packet and disseminate immediately to the router which reduces the end-to-end delay up to a large extent. It ensures that the relaying node should not leave the transmission range before receiving the data packet. In sparse network also it chooses the same relaying node.

In Fig. 6, the network overload is reduced significantly in the purposed scheme due to the

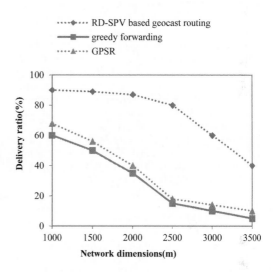

Figure 4. Delivery ratio (%) vs network dimensions (m).

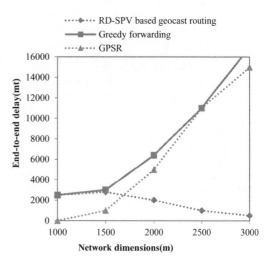

Figure 5. End-to-end delay (mt) vs network dimensions (m).

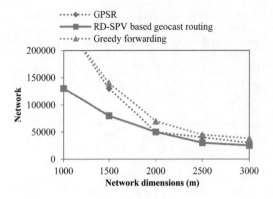

Figure 6. Network overhead (n) vs network dimensions (m).

incorporation of On-the-Fly Segment Density (OFSD) based Adaptive Beaconing System (ABS) which is almost reduced to half in comparison with other regular scheme. Optimal route is chosen based on Route Detection using Segmented Path Vector (RD-SPV) where the path segments are selected based on CAP. The receiver checks the registration of the sender from TMS before updating its NTT and 802.11 layer-2 ACK scheme avoids retransmission due to loss of acknowledgement by incorporating implicit retransmission.

5 CONCLUSION

An improved and efficient connectivity-aware geocast routing based on Route Detection using Segmented Path Vector (RD-SPV) has been proposed. Simulation result shows that the purposed approach improved significantly the dissemination of data packet to the Zone Of Relevance (ZOR) by appropriately choosing the optimal route based on Route Detection using Segmented Path Vector (RD-SPV) where the path segments are selected based on CAP and deserving relay node among the available forwarding nodes. The proposed scheme chooses only those neighborhood nodes which are having maximum coverage area potential (CAP). Hence, next hop will not leave the transmission range before receiving the data packet. In future, we explore different possibilities regarding the effect of node speed, node density and transmission range on connectivity of the geocast routing protocol in detail and also give mathematical models for our proposed scheme.

REFERENCES

Abdalla, G.M.T., Abu-Rgheff, M.A., Senouci, S.M. Current Trends in Vehicular Ad-hoc Networks. IEEE Global Information Infrastructure Symposium, Morocco July, 2007.

Al-Sultan, S., Al-Doori, M., Al-Bayatti, A.H. and Zedan, H. A comprehensive survey on vehicular Ad Hoc network. Journal of Network and Computer Applications, Elsevier, 37(1), pp. 380–392, 2014.

Chang, J.J., L YH, Liao, W.C I–C. Intersection-based routing for urban vehicular communications with traffic-light considerations. J Wireless Commun IEEE 19, pp. 82–88, 2012.

Chen, C., Jin, Y. Pei, Q. and Zhang, N.A connectivity-aware intersection-based routing in VANETs. Chen et al. EURASIP Journal on Wireless Communications and Networking 2014.

Chen, Z.D., Kung, H. and Vlah, D. Ad hoc relay wireless networks over moving vehicles on highways. In Mobi-Hoc '01: Proceedings of the 2nd ACM international symposium on Mobile ad hoc networking & computing, pp. 247–250, 2001.

Fan Li and Ya Wang. Routing in Vehicular Ad-Hoc Networks: Survey. IEEE Vehicular Technology Magazine-I, June 2007.

Imielinski, T., Navas, J. GPS-Based Addressing and Routing. Internet Engineering Task Force (IETF), Network Working Group, Internet Draft, RFC, 2009.

Jiang, Y., Li, S., Shamo, D.E. Development of vehicle platoon distribution models and simulation of platoon movements on Indian rural corridors. Joint Transport Res. Program, 2003.

Kaiwartya, O. and Kumar, S. Geocasting in vehicular adhoc networks using particle swarm optimization. In Proceedings of the International Conference on Information Systems and Design of Communication, pp. 62–66. ACM, Lisbon Portugal, May 2014.

Kaiwartya, O. and Kumar, S. Cache Agent based Geocasting (CAG) in VANETs. International Journal of Information Communication Technology, 2015.

Karp, B. and Kung, H.T. GPSR: greedy perimeter stateless routing for wireless networks. In Proc. ACM/IEEE MOBICOM'00, Boston, Massachusetts, USA, pp. 243–254, 2000.

Khatri, M., Malhotra, S. An insight overview of issues and challenges in Vehicular Ad Hoc Networks. Journal of Global Research in Computer Science, Volume 2, No. 12, pp. 47–50, December 2011.

Lee, K.C., Le, M., Harri, J., Gerla, M. Louvre: landmark overlays for urban vehicular routing environments. In IEEE 68th Vehicular Technology Conference, VTC -Fall (IEEE, Piscataway, 2008), pp. 1–5, 2008.

Lochert, C., Mauve, M., Füssler, H. and Hartenstein, H. Geographic routing in city scenarios. SIGMOBILE Mob. Comput. Commun. Rev vol. 9, no. 1, pp. 69–72, 2005.

Luong, T.T, Seet, B.C, Lee, B.S. Local maximum avoidance with correlated street blocking for map-based geographic routing in VANETs. In 2nd International Symposium on Wireless Pervasive Computing, 2007. ISWPC'07 (IEEE, Piscataway, 2007).

Maihofer, Christian. A survey of geocast routing protocols. IEEE Communications Surveys & Tutorials, Volume 6, No. 2, pp. 32–42, 2004.

Naumov, V. and Gross, T.R. Connectivity-Aware Routing (CAR) in vehicular ad-hoc networks. In INFOCOM 2007. 26th IEEE International Conference on Computer Communications. IEEE, pp. 1919–1927, 2007.

Panichpapiboon, S. and Pattara-Atikom, W.A Review of Information Dissemination Protocols for Vehicular Ad Hoc Networks.Communications Surveys & Tutorials, IEEE, 14(3), pp. 784–798, 2012.

Qing, Yang., Lim, A., Shuang, Li. Jian, Fang., Agrawal, P., ACAR: Adaptive Connectivity Aware Routing Protocol for Vehicular Ad Hoc Networks. Computer Communications and Networks, 2008. ICCCN '08. Proceedings of 17th International Conference on, vol., no., pp. 1,6, 3–7 Aug. 2008.

Rao, R.S., Soni, S.K., Singh, N. and Kaiwartya, O. A probabilistic analysis of path duration using routing protocol in VANETs. International Journal of Vehicular Technology, vol. 14, no. 1, pp. 1–10, 2014.

Saleet, H., Langar, R., Naik, K., Boutaba, R., Nayak, A., Goel, N. Intersection-based geographical routing protocol for VANETs: a proposal and analysis. J Vehicular Technology IEEE Trans 60(9), pp. 4560–4574, 2011.

Sermpezis, P., Koltsidas, G., Pavlidou, F.N. Investigating a junction-based multipath source routing algorithm for VANETs. Wireless Commun. IEEE17, pp. 600–603, 2013.

Sommer, C., Tonguz, O.K., Dressler, F. Traffic information systems: efficient message dissemination via adaptive beaconing. Communications Magazine, IEEE, vol.49, no.5, pp. 173–179, May 2011.

Sum, M., Feng, W., Lai,T.H., Yamada, K.H., Okada, and Jimura, K. Fu. GPS-based message broadcasting for Inter-vehicle Communication. Proc. of 2000 Int. Conf on Parallel Processing, 2000.

Taleb, T., Ochi, M, Jamalipour, A., Kato, N. and Nemoto, Y. An efficient vehicle-heading based routing protocol for VANET networks. In Wireless Communications and Networking Conference, 2006. WCNC 2006. IEEE, vol. 4, pp. 2199–2204, 2006.

Tsiachris, S., Koltsidas, G., Pavlidou, F.N. Junction-based geographic routing algorithm for vehicular ad hoc networks. Wirel. Pers. Commun. 71, 955–973 (2012).

Wang, Z., Liang, Y. and Wang, L. Multicast in Mobile Ad Hoc Networks. Computer And Computing Technologies In Agriculture, Springer Publication, pp. 151–164, 2008.

Zhao, W., Ammar, M. and Zegura, E. A message ferrying approach for data delivery in sparse mobile ad hoc networks. In MobiHoc '04: Proceedings of the 5th ACM international symposium on Mobile ad hoc networking and computing, 2004, pp. 187–198, 2004.

Communication and Computing Systems – Prasad et al. (Eds)
© *2017 Taylor & Francis Group, London, ISBN 978-1-138-02952-1*

Security over cloud computing using different encryption techniques–review

Rama Mishra & Arun Kumar Yadav
ITM University, Gwalior, Madhya Pradesh, India

Ram Shrinagar Rao
Indra Ghandhi National Tribal University, Amarkantak, Madhya Pradesh, India

ABSTRACT: Cloud is new computing model and technique which provides a huge benefit towards Information and Communication Technology (ICT) and it's capable of manufacture. The word Cloud emerges from the graphical scnario that was always utilized to present difficult infrastructure and heterogeneous networks. For describing some properties of Cloud Computing (CC), this graphic is adopted. Security is the main concern of CC because data on cloud distributed where all the user accessing the data so securely accessing the information over cloud it's a tough task, there were many security related algorithm present. Here presenting a survey on CC, that highlights particular issues relating CC. The aim of our paper is identifying the major problems in cloud security.

1 INTRODUCTION

Cloud Computing (CC) is a form of on-demand system were easily available and used via shared pool of computing resources which will swiftly provide with few service efforts (Peter Mell & Timothy France). The definition of CC is "Cloud is hugely dispersed computing model which have a group of interconnected computer or virtualized which might be provided as one and more than one combined resources, headquartered on Service Level Agreements (SLA) headquartered through negotiation between the carrier supplier and purchasers" (R. Buyya et al., 2009). It is a kind of computing where IT-related abilities were given to clients as service" in place of an item utilizing the gateway that is internet.

The Main motive of CC is that it provides us an affordable and high or scalable on-demand computing infrastructures, or we can say it a infrastructure which is always available when needed and it provide a high quality level of services. For applying high security into applications of cloud there were a huge amount of inventories are struggling for it. And in some other cases, the developers can't provide a real security so easily by the technologies which are early available (Tackle your client).

The design of CC includes at least two and more cloud parts associating with each other concerning the different data they are holding on as well, therefore providing to the customer to get to the required information on a rapid expense. In the case of cloud it's extra centered upon the front end and the again finish. The front end user wants data, and backend users need numerous information storage devices, and the sever on which cloud is made (Lord CrusAd3r).

Users connect the cloud they apparently cloud by single application, gadget, or report. All things in the cloud process like hardware with cloud and OS which provides hardware connections undetectable. CC begins with the person interface obvious by means of individual users. This is how customers give their request then gets passed to the procedure service, which finds the right resources and then calls the process's appropriate provisioning offerings. CC is not only used for knowledge storage.

Right here the data is put away on numerous outsider servers. The individual sees an effective server; which shows up as though the data is secured in a specific position with an exact identifier, when storing the data. This doesn't exist simply. It's simply used to reference the digital aspects of the cloud. Virtually, the client's information should be saved on any one entity and many computer systems are utilized to create the cloud.

2 CLOUD COMPUTING SERVICE MODELS

The providers or suppliers of cloud service were incorporating with three major services based on their capabilities likewise SaaS, PaaS, and IaaS (R. Buyya et al., 2009).

3 CLOUD COMPUTING DEPLOYMENT MODEL

The initiation for the concern about security starts with the cloud arrangement models. Depending on framework possession, there are 4 arrangement things of CC (Tharam Dillon et al., 2010).

- *Public Cloud*
- *Private Cloud*
- *Community Cloud*
- *Hybrid Cloud*

4 SECURITY REQUIREMENTS OF CLOUD COMPUTING

Various security requirements are depicted in (M. Armbrust et al., 2009).

4.1 *Access control*

It describes the "degree at which system limits access to its resources as it were authorized externals". Authorized users is social users, program parts, devices, other kinds of frameworks which have any kind of identity and conformation to access any program, but also services. Access control may be refined as a combo of Identification, Authentication and Authorization. All three have the common objective to control and supervise a certain range of permissions granted to those users who can claim their identity and be allowed for their defined privileges (e.g. read, modify) over assigned resources. After granting an external permission over collection of resources, access control furthermore has to hold and guarantee this state until a clean and successful termination of the temporary established access takes place.

4.2 *Non-repudiation*

Non-repudiation is referred as "the point at which party to a interaction (e.g., message, transaction, transmission of information) is maintained a strategic distance from adequately revoking (i.e., denying) any side of the interaction."

4.3 *Integrity*

The integrity describes necessities being deployed to protect add-ons of the method from purposeful and unapproved harm or debasement Integrity requirements may also be wonderful for data

integrity, hardware integrity, personnel integrity and software integrity.

4.4 *Privacy and confidentiality*

Privacy and confidentiality refer back to the reach at which unauthorized parties are prevented from obtaining touchy understanding. Many papers are focusing on information protection, implying the presence of correct access controls to assurance confidentiality of confidential information within main place. Anonymity closer to CSP poses another security requirement measures. Privacy used to be in most cases located to be instantly related to entry control standards. Pushing or enforcing strict access control mechanisms or method were relates to a high measure of privacy & confidentiality.

5 SECURITY ISSUES RELATED TO CLOUD

Cloud enables customers to pick up the influence of computing which bangs up their particular real zone. It prompts numerous security issues. The provider of cloud service, it ought to be essential that cloud constructs a distinct infrastructure in which purchaser does not confront any type of block, for example, absence of data or data inconsistency. CC foundations utilize creative technologies and services; most which haven't been totally assessed with acknowledge to insurance. There might be additionally probability where a malicious individual can penetrate the cloud by impersonating a reliable user, there with the guide of tainting the whole cloud. This prompts affects various purchasers who're sharing the polluted cloud. The assurance issues confronted with the guide of CC are specified under (Anitha Y, 2013).

5.1 *Data access control*

Commonly private information can be illegally accessed as a result of absence of secured knowledge access control. Private data in CC climate rise as vital concerns with reference to security in a cloud focused procedure Information exists for extended time in a cloud, the grater in hazard of unauthorized entry.

5.2 *Data integrity*

It has these circumstances, at the time any errors happen when knowledge is introduce. Errors are emerge when data will be forwarded to one to some other; in any other case fault be able to arise from some hardware faults, like disk crashes. Application fails and bugs may additionally make viruses. So in same time, many CC carrier customer and supplier accesses and modify information. Hence some necessary information is formulated to provide data integrity in CC.

5.3 Data theft

CC makes use of external information server for fee affective and bendy for operation. That's why here is danger of information may also be theft from the external server.

5.4 Data loss

Data loss is a remarkably serious hindrance into CC. In the field of bank or industrial transactions, research, advance procedures are all occurring online, unapproved persons may be prepared to passage the data shared. Despite the facts that everything is comfortable imagine a scenario where a server may be damaged or affected by method for a virulent disease, the entire framework would go down and viable knowledge loss could emerge. On the off chance that the seller closes because of financial or legal issues there are loss of data for the shoppers. The purchasers won't be prepared to get to these knowledge's thinking about that data is no additional available for the patron as the vendor close down.

5.5 Data location

Customer doesn't normally recognize the place of their knowledge. The vendor does not disclose where all of the knowledge's are saved. CC presents a excessive measure of data mobility. The data's gained even be in the identical country of client, it possibly placed anywhere on the earth. They wish to specify a desired place (e.g. Knowledge to be kept in US) then requires a contractual contract between suppliers of Cloud service and the patron that knowledge will have to keep in a specified place or reside on a given recognized server.

5.6 Privacy issues

Privacy at the customer side is very much important in case of personal as well as exceptional use of CC. security level at external server is highly essential for performing different computing operations from different operators.

5.7 Security concerns in provider level

A Cloud infrastructure is most excellent when it have recently secure and outfitted by the provider to the buyers. Supplier must make a decent security layer for the client and individual and need to affirm that server is efficiently protected from all the outside dangers it will come all through cloud computing bearer supplier.

5.8 User level issues

Client must establish a model that deals with truth of its individual movement; there shouldn't be any absence of data or altering of data for various clients who are using the equivalent Cloud.

5.9 Infected application

Service suppliers will need to have the entire access of server with all the permission and method of reasoning of checking and insurance of server. So this will probably block any malevolent individual from importing any impure function onto cloud having perspective to extremely affect the customer and CC service.

6   LITERATURE SURVEY

Wang et al (Cong Wang et al., 2009), investigated the situation of knowledge security in cloud knowledge storage, that is very nearly an allotted storage procedure. To be certain the exactness of customers' data present in cloud storage, invoked a potent and bendy allotted scheme with specific dynamic information help, together with block replace, delete, and append. the author of paper define that it depends on relies on upon eradication adjusting code which was available in the record dissemination practice for outfitting repetition equality vectors, assention the data constancy. By means of utilizing homomorphic token by assigning plan check for erasuring coded learning, this phenomenon achieves the combo of capacity accuracy protection and information blunder confinement.

(Jung et al., 2010), implement DAA protocol for security of the individual devices under the CC environment. It also defines TPM module using JAVA. And we have tested authentication and communication of each node through DAA protocol. And this model determines dynamically protection stage and entry manipulate for the original resources. Thus, it is purported to provide suitable protection offerings consistent with the dynamic changes of the original resources. Here expecting this model to solve security faults in the static environment utilizing MAUT, simple heuristics. The static security service system can't manage dynamic changes of the numerous natural variables. however, security managements flexible in this model based on RBAC may be expected for resolving that problems effectively. Also these modelsolved problems of requiring unreasonable frameworks resources and longer waiting time. And it protects resources safely from mal-intentional users.

(Li et al., 2010) exhibited novel cloud trust model furthermore it introduced a novel security structure reliant on module of trust service which is autonomous of cloud infrastructure. Utilizing of our proposed model, trust-based security mechanism were proposed to guarantee the security of both cloud clients and suppliers. While for suppliers, these explanations will withdraw from serving clients or coordinating with destructive different suppliers.

(Mohammed et al., 2011) proposed another model known as MCDB which utilize Shamir's mystery imparting calculations to multi-mists suppliers as an option of single cloud. Also, it's said its structure with its extras and layers. The objective of this model is to eliminate security dangers happen in CC and locations the issues that including information integrity, and repair accessibility.

(Rewagad et al., 2013) chosen for utilizing a mixture of authentication framework and key trade and encryption algorithm will merge together. This combo is known as "Three means mechanism" on the grounds that it ensures the entire three security

Table 1. Literature survey.

Author	Paper title	Journal	Proposed Solution	Applying algorithm	Type of service
Cong Wang, et al., 2009	Ensuring Data Storage Security in CC	IEEE 2009	1. Examine the problem offline-grained data fault localization	Homomorphism token	Multi Tire Security
			2. Cloud data storage security identification of misbehaving server(s)		
Youngmin Jung & Mokdong Chung, 2010	Adaptive Security Management Model in the CC Environment	International Conference on Advanced Communication Technology (ICACT). IEEE 2010	DAA protocol implement for security on CC environment	JAVA and MAUT	Single Tire Security
Mr. Prashant Rewagad et al., 2013	Use of Digital Signature with Diffie Hellman Key Exchange and AES Encryption Algorithm to Enhance Data Security in CC	International Conference on Communication Systems and Network Technologies, IEEE, 2013	1. Enhance data security in CC using Diffie Hellman and Advanced Encryption Standard encryption algorithm	Diffie Hellman key replace blended with (AES) Advanced Encryption Standard encryption algorithm	Double Tire Security
Ranjit Kaur et al., 2014	Enhanced CC Security and Integrity Verification via Novel Encryption Techniques	IEEE, 2014	1. Provide security on various attacks like brute force attack.	SAES, IDEA AES and Blow—Fish	Multi Tire Security
Meena Kumari, Rajender Nath, 2015	Security Concerns and Counter measures in CC Paradigm	Fifth International Conference on Advanced Computing & communication Technologies, 2015.	1. Detect the top security concerns of CC, these concerns are Data leakage, Governance and compliance, Trust, Virtualization vulnerabilities and various attacks.	Trust management module	Doule Tire Security

scheme of authentication, information protection and verification, equally. The author proposed about the utilization of digital sign and AES (Advanced Encryption Standard encryption algorithm) which is associated by Diffie Hellman key trade for ensuring classification of data. The essential thing in communication is hacked; the capacity of Diffie Hellman key exchange render it futile, considering key in travel is of no utilization without purchaser's selective key, which is limited best to the respectable customer.

(Kaur et al., 2014) introduce cloud storage security model provides an ethically secure environment of cloud by explaining the three areas with storage client data base on the security limitations are to be maintain specific confirmation categorization,, non repudiation, reliability, sanctuary, convenience and security. It confines unapproved substances to get switch of user's data by executing double authentication frameworks. It likewise offers wellbeing on premise of different security breaks, for example, masquerade attack,, brute force attack, data altering, and cryptanalysis of trustworthiness key.

(Shahzad et al., 2014), introduced the five vital characteristics of CC, three cloud supplier units, and four cloud usage units. Research in the comfortable cloud storage is aggravated by method for reality that clients data may be put away at a couple of zones for both redundancy/fault to internal failure or in light of the fact that the service is supplied through arrangement of service vendors.

(Lija et al., 2014) implemented a secure environment of cloud the place owner can add data, via making use of any symmetric encryption strategy and must furnish GAC (great grained access control) to records without using any figure concentrated operations. Right here utilizing Blundo Scheme1, a t-variety polynomial established mystery sharing for offering first-rate grained section control and Lagrange addition for client access Revocation.

(Kumari et al., 2015) speaks to a development for the day where CC security issues can be recorded in one far reaching report jointly with its answers. This paper distinguishes best security points of CC, these worries are Data misfortune, domination and consistence, Trust, Virtualization vulnerabilities and different attacks. It additionally compresses different countermeasures which can be received to overcome security issues.

(Kaur et al., 2015) considering difficulties of security in CC and approaches to overcome the information privacy obstacle. it invent the cloud security issues and primary issue confronted through cloud service supplier. Accordingly, defining the Pixel key sample and picture Steganography strategies that will be used to prevent the problem of data security.

(Rao et al., 2015) surveyed various data security problems, features and solutions for overcome those challenges in CC.

Researches and purpose of data security concern of cloud system in which it is crucial to control the capacity of distributed data system. Server talked about third grouping examiners and hash code implementers, utilizing these techniques clients can achieve data authentication in cloud storage system. Creator has exploit SHA-1 algorithms. Whenever data degradation distinguished crosswise over scattered server, we can give just about affirmation of recognizable proof of making trouble server. We demonstrate that in oue plan it unfathomably viable for malicious information change, server plotting attack with least calculation overhead. Table 1 show the review of the research work done in past.

7 COMPARISON ANALYSIS

The given below Fig.1 presents a comparison between different encryption algorithm, basically here comparing RSE, AES and DES encryption algorithm execution time with different data sizes. Blue represents RSA, red presents DES and green is for AES. X axis presents size in Kbytes and Y axis presents time in mili seconds.

The given below Fig. 2 presents a comparison between different decryption algorithms.

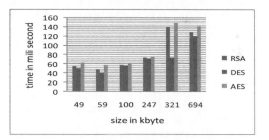

Figure 1. Comparison graph between encryption algorithms.

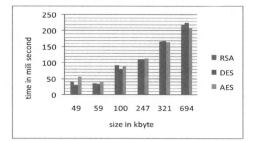

Figure 2. Comparison graph between decryption algorithms.

The comparison of execution time of various algorithms like RSE, AES and DES with different data sizes is shown in graph. Blue represents RSA, red presents DES and green is for AES. X axis presents size in Kbytes and Y axis presents time in mili seconds.

8 CONCLUSION AND FUTURE DIRECTIONS

The popularity of CC is by and large because it consists of many business applications and information are moving into cloud organizations; however, be short of security is the main obstacle for cloud acceptance. A few of the threats observed in current systems. In our review paper we compare a lots of security algorithm, in brief we study that if we provide security over cloud using data classification method in which data are transferred depend on data type so that cloud security become high. Out of them, the security threat is viewed to be of excessive threat.

REFERENCES

Anitha, Y.; "2013 Security Issues in Cloud Computing—A Review". International Journal of Thesis Projects and Dissertations (IJTPD) 2013.

Armbrust, M., A. Fox, R. Griffith, A. Joseph, R. Katz, A. Konwinski, G. Lee, D. Patterson, A. Rabkin, and I. Stoica; "Above the clouds: A 32Berkeley view of cloud computing," EECS Department, University of California, Berkeley, Tech. Rep. UCB/EECS-2009-28, 2009.

Buyya, R., C. S. Yeo, S. Venugopal, J. Broberg, and I. Brandic; Cloud computing and emerging IT platforms: Vision, hype, and reality for delivering computing as the 5th utility, Future Generation Computer Systems, 25:599616, 2009.

Buyya, R., S. Pandey, and C. Vecchiola; Cloudbus toolkit for market-oriented cloud computing, in Proceedings 1st International Conference on Cloud Computing (CloudCom 09), Beijing, 2009, pp. 3–27.

Cong Wang, Qian Wang, Kui Ren, and Wenjing Lou; "Ensuring Data Storage Security in Cloud Computing". IEEE, 2009.

Farrukh Shahzad; "State-of-the-art Survey on Cloud Computing Security Challenges, Approaches and Solutions". The 6th International Symposium on Applications of Ad hoc and Sensor Networks (AASNET'14), Elsevier.

Lija Mohan, and Sudheep Elayidom M; "Fine Grained Access Control and Revocation for Secure Cloud Environment—a polynomial based approach". International Conference on Information and Communication Technologies (ICICT 2014), Elsevier, 2014.

Lord CrusAd3r, "Problems Faced by Cloud Computing", dl.packetstormsecurity.net/.../ProblemsFacedbyCloudComputing.pdf.

Meena Kumari Rajender Nath; "Security Concerns and Countermeasures in Cloud Computing Paradigm". Fifth International Conference on Advanced Computing & Communication Technologies, 2015.

Mohammed, A., AlZain, Ben Soh, and Eric Pardede; "MCDB: Using Multi-Clouds to Ensure Security in Cloud Computing". Ninth IEEE International Conference on Dependable, Autonomic and Secure Computing, 2011.

Mr. Prashant Rewagad, and Ms. Yogita Pawar; "Use of Digital Signature with Diffie Hellman Key Exchange and AES Encryption Algorithm to Enhance Data Security in Cloud Computing". International Conference on Communication Systems and Network Technologies, IEEE, 2013.

Patrick Höner; "Cloud Computing Security Requirements and Solutions: a Systematic Literature Review" 2013.

Peter Mell, and Timothy France; "The NIST Definition of Cloud Computing" NIST Special Publication 800-145.

Randeep Kaur, and Jagroop Kaur; "Cloud Computing Security Issues and its Solution: A Review". IEEE, 2015.

Ranjit Kaur, and Raminder Pal Singh; "Enhanced Cloud Computing Security and Integrity Verification via Novel Encryption Techniques". IEEE, 2014.

Tackle your client's security issues with cloud computing in 10 steps, http://searchsecuritychannel.techtarget.com/tip/Tackleyour-clients-security-issues-withcloudcomputing-in-10-steps.

Tharam Dillon, Chen Wu, and Elizabeth Chang; Cloud Computing: Issues and Challenges, 2010 24th IEEE International Conference on Advanced Information Networking and Applications, 1550-445X/10.

Velumadhava Rao R. and K. Selvamani; "Data Security Challenges and Its Solutions in Cloud Computing". International Conference on Intelligent Computing, Communication & Convergence (ICCC-2015), Elsevier.

Wenjuan Li, Lingdi Ping, and Xuezeng Pan; "Use Trust Management Module to Achieve Effective Security Mechanisms in Cloud Environment". International Conference on Electronics and Information Engineering, IEEE, 2010.

Youngmin Jung, and Mokdong Chung; "Adaptive Security Management Model in the Cloud Computing Environment". The 12th International Conference on Advanced Communication Technology (ICACT). IEEE 2010.

Communication and Computing Systems – Prasad et al. (Eds)
© *2017 Taylor & Francis Group, London, ISBN 978-1-138-02952-1*

Information security: In android expedients

Neelam Ruhil & Jitender Kumar
Dronacharya College of Engineering, Gurgaon, Haryana, India

Rajesh Kumar
GlobalLogic, Noida, UP, India

ABSTRACT: This paper discuss about different technologies to secure sensitive information in user's Android device. It also explains about advantages and disadvantage of storing information in Android device as well as different method and technologies to save data and safety level of data stored in Android. This paper also discuss about storing data using shared preferences, SQLite db, internal storage and external storage. It also explains use of different tools like android proguard, android keystore etc.

1 INTRODUCTION

Smart phones have become an integral part of today's society, and Android is playing a big role in these smart phones. Most of the banks, email clients and other companies have provided there apps to Android users. As per a report by Android, Android devices are used in 190 + countries and almost one million new Android devices are activated every day (Makan, 2011). This is indeed a big number that have attracted hackers to target Android devices.

Hackers can hack user's sensitive information from their devices. Hackers can use this confidential or sensitive data to hack user's bank account information, email information, and other important information that needs to be secure (Kaur S., 2010). Android being an open source have provided lots of flexibility, hackers normally take advantage of this flexibility to steal sensitive data from user's Android phone.

Android provide frequent updates of the operating systems that help Android developers to make their application more secure and hack proof. But few small mistakes made by developers can allow hackers to attack (Makan, 2011) the device and steal sensitive information. For Android users this is very important to know about the developer of the application and do not grant unnecessary permissions to the app those are not supposed to be required for the application. Android user should always install apps from the authorized store only. While installing app it always ask user what all permission that Android app require. It also explains use of different tools like android proguard, android keystore etc.

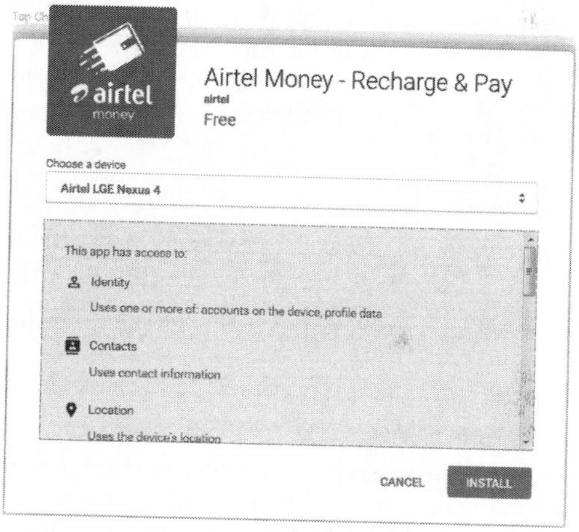

Figure 1. An app asking to grant permission that can be used by the application.

Figure 1 demonstrates an app asking for permission to user. After getting the permission app can access the granted areas. To quickly access the user information and other sensitive information these apps store user's confidential data on the local storage of the device. This locally saved information can be accessed by the hackers.

1.1 Background

Android apps take advantage of saving data locally on user's device. It helps developers to access data quickly and makes their app to work fast.

For example few Android app provide a feature to remember user name and password on the device, so that user is not required to enter credential every time he/she uses the application. To accomplish this developer some time save this information on device itself. Other use case is sometime developers need to access WEB API's, these WEB API's require to provide *application id* and *application secret*. Sometime developer hardcode this information in code files itself. Other example is storing user's name, email address and other personal information that is required for the app. To quickly access these information, instead of making web call developer store these information on the device itself. Any information that is stored on the device can be accessed by the hackers. This is very important to store information that is absolutely must to be stored. If developer can avoid storing information on device then it is perfect otherwise developer need to make sure that the information that is being stored is secured and is not easy to be accessed by the hackers. Stored information should be properly encrypted. Android provides several ways to store information. Below are few of the way that developer normally uses to store information locally on Android device (Anmol Mishra, 2008).

1.2 Store using database

Every Android device is equipped with inbuilt database SQLite. Every app can use this database to store information in this DB. Developer can use SQLiteOpenHelper class to store data in the DB.

1.3 Store in shared preferences

Shared Preferences provide a way to store data in key-value pair. It provides a space in app memory to store data. Only app can access the stored information. It provides several functions to store and retrieve the data. Below are few of the functions that can be used with Shared Preferences:

```
public class DictionaryOpenHelper extends SQLiteOpenHelper {

    private static final int DATABASE_VERSION = 2;
    private static final String DICTIONARY_TABLE_NAME = "dictionary";
    private static final String DICTIONARY_TABLE_CREATE =
            "CREATE TABLE " + DICTIONARY_TABLE_NAME + " (" +
            KEY_WORD + " TEXT, " +
            KEY_DEFINITION + " TEXT);";

    DictionaryOpenHelper(Context context) {
        super(context, DATABASE_NAME, null, DATABASE_VERSION);
    }

    @Override
    public void onCreate(SQLiteDatabase db) {
        db.execSQL(DICTIONARY_TABLE_CREATE);
    }
}
```

Figure 2.   Code snippet to store data in SQLite DB.

1.3.1. *getSharedPreferences( )*

1.3.2. *getPreferences( )*

1.3.3. *edit( )*

1.3.4. *putString( ), putBoolean( )*

1.3.5. *getString( ), getBoolean( )*

1.3.5. *commit( )*

1.4 Store in internal storage

Files can be stored in Internal Storage of Android device; by default this storage is private to that app. Developer need to write content of the data by converting content into bytes and then writing those bytes into FileOutputStream type.

1.5 Store in external storage

Android device also support external storage, this storage can be a removable storage or internal storage. To store data in this storage, developer first needs to ask for permission in this storage.

```
public class Calc extends Activity {
    public static final String PREFS_NAME = "MyPrefsFile";

    @Override
    protected void onCreate(Bundle state){
        super.onCreate(state);
        . . .

        // Restore preferences
        SharedPreferences settings = getSharedPreferences(PREFS_NAME, 0);
        boolean silent = settings.getBoolean("silentMode", false);
        setSilent(silent);
    }

    @Override
    protected void onStop(){
        super.onStop();

        // We need an Editor object to make preference changes.
        // All objects are from android.context.Context
        SharedPreferences settings = getSharedPreferences(PREFS_NAME, 0);
        SharedPreferences.Editor editor = settings.edit();
        editor.putBoolean("silentMode", mSilentMode);

        // Commit the edits!
        editor.commit();
    }
}
```

Figure 3.   Code snippet to store and retrieve data using share preferences.

```
String FILENAME = "hello_file";
String string = "hello world!";

FileOutputStream fos = openFileOutput(FILENAME, Context.MODE_PRIVATE);
fos.write(string.getBytes());
fos.close();
```

Figure 4.   Code snippet to store file in internal storage.

```
<manifest ...>
    <uses-permission android:name="android.permission.WRITE_EXTERNAL_STORAGE" />
    ...
</manifest>
```

Figure 5.   Permission request in manifest file.

```
/* Checks if external storage is available for read and write */
public boolean isExternalStorageWritable() {
    String state = Environment.getExternalStorageState();
    if (Environment.MEDIA_MOUNTED.equals(state)) {
        return true;
    }
    return false;
}

/* Checks if external storage is available to at least read */
public boolean isExternalStorageReadable() {
    String state = Environment.getExternalStorageState();
    if (Environment.MEDIA_MOUNTED.equals(state) ||
        Environment.MEDIA_MOUNTED_READ_ONLY.equals(state)) {
        return true;
    }
    return false;
}
```

Figure 6.   Code snippet to check if external media is available or not.

Then write code to check if the external storage of device is available or not. Application can crash if developer didn't make a check. Once verified save file in the external device.

2   PROPOSED METHOD

In this paper we propose few methods to store data securely in Android device locally.

2.1   Set allow backup flag to false

In real scenario hackers can take backup of any android application. This backup allows them to take backup of all the data and files stored by that app. Once the hackers have backup of this data, they convert it into the desired format and then can fetch all the information that was stored in the device. It is a good idea to not allow hackers to take backup of the application. By default allow Backup flag is true for all the application, this can be turned off by changing its value to false in AndroidManifest.xml file.

2.2   Encrypt stored information

By setting allowBackup to false will surely create difficulty for hackers to hack stored information, but if we need to allow to take backup of the application, in that case it is better to encrypt information before storing in Android device. By using this approach even if hackers hack the encrypted it would be of no use for them, because the

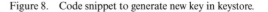

Figure 7.   AllowBackup value set to True | False.

```
/*
 * Generate a new EC key pair entry in the Android Keystore by
 * using the KeyPairGenerator API. The private key can only be
 * used for signing or verification and only with SHA-256 or
 * SHA-512 as the message digest.
 */
KeyPairGenerator kpg = KeyPairGenerator.getInstance(
        KeyProperties.KEY_ALGORITHM_EC, "AndroidKeyStore");
kpg.initialize(new KeyGenParameterSpec.Builder(
        alias,
        KeyProperties.PURPOSE_SIGN | KeyProperties.PURPOSE_VERIFY)
        .setDigests(KeyProperties.DIGEST_SHA256,
            KeyProperties.DIGEST_SHA512)
        .build());

KeyPair kp = kpg.generateKeyPair();
```

Figure 8.   Code snippet to generate new key in keystore.

information is in encrypted form. Hackers cannot decrypt this information till the time they do not have encryption key of that information.

2.3   Use of proguard

Developer usually store encryption key or any other secret key in the code itself. But this code is not safe. Hackers can fetch an apk installed on device. Once then have apk with them this apk can be converted to readable form. By using reverse engineering hackers can get the key or any other secret key then can help then in decrypting stored information. To avoid such situation developer should obfuscate there code. Android provide proguard to obfuscate the code. Proguard also help to shrink the code [3]. Proguard generate four additional files to make it work. Below are the four files those are generated after using the proguard.

2.3.1.   *Dump.txt*

2.3.2.   *Mapping.txt*

2.3.3.   *Seeds.txt*

2.3.4.   *Usage.txt*

2.4   Use of android keystore

Android Keystore can be used to store cryptographic key, it is very difficult to hack this key from the device. Device should be encrypted (Nina, 2009) before using the keystore. This key can be used for cryptographic operations. Encryption/ Decryption can be performed using this key.

```
/*
 * Use a PrivateKey in the KeyStore to create a signature over
 * some data.
 */
KeyStore ks = KeyStore.getInstance("AndroidKeyStore");
ks.load(null);
KeyStore.Entry entry = ks.getEntry(alias, null);
if (!(entry instanceof PrivateKeyEntry)) {
    Log.w(TAG, "Not an instance of a PrivateKeyEntry");
    return null;
}
Signature s = Signature.getInstance("SHA256withECDSA");
s.initSign(((PrivateKeyEntry) entry).getPrivateKey());
s.update(data);
byte[] signature = s.sign();
```

Figure 9. Code snippet to sign a data.

3 RESULTS AND DISCUSSION

The proposed methods were implemented using Android 5.0 with Eclipse ADT plugin. Use of proguard and android keystore is very effective. Android keystore require device to be encrypted and password protected. If any device is not encrypted and password protected then android keystore cannot be used on that device. Proguard can be used on any android device to obfuscate the code. It is a very difficult job to find private key without having Mapping.txt file. Mapping.txt file is not stored on the client device. So it is safe to use proguard on all other device where keystore cannot be implemented.

4 CONCLUSION AND FUTURE WORK

In this paper we proposed a method to store sensitive information on android device locally. It has been observed that using keystore and obfuscating the apk make locally stored data more secure. Android for Work can also be explored to make android app more secure for enterprise.

REFERENCES

Android Developer Website http://developer.android.com
Anmol Mishra, A. D. (2008). Android Security: Attacks and Defences. CRC Press.
Kaur S., K. M. (2010). Review Paper on Implementing Security on Android Application. *Journal of Environmental Sciences, Computer Science and Engineering & Technology.*
Makan, K. (2011). Android Security Cookbook. Packt.
Nina, W. (2009). Android On Power Architecture. *ELC, Grenobles.*
Omotosho, V. (2016). Android Apps Security". *ASIN B01C4GS3T2.* Apress.
Powar S., M. B. (2013). Survey on Android Security Framework. *International Journal of Engineering Research and Applications.*

VLSI and embedded system

Communication and Computing Systems – Prasad et al. (Eds)
© 2017 Taylor & Francis Group, London, ISBN 978-1-138-02952-1

Implementation of temperature sensor based on Arduino Uno and LabVIEW 2012

Gaurav Soni, Anshdeep Kaur & Asis Seth
Amritsar College of Engineering and Technology, Punjab, India

ABSTRACT: LabVIEW is a graphical programming language, used in many research fields for designing communication algorithms and simulating traditional as well as real time signals. Arduino kit can be interfaced with LabVIEW for a large number of applications. In this article, a project based on temperature sensor using Arduino Uno and LabVIEW is implemented. We have used LM35 temperature sensor whose output voltage varies in linear proportion to the temperature in centigrade. Thus, providing us an advantage over other sensors to measure the temperature conveniently in degree Celsius. The LabVIEW 2012 provides a simple interface between temperature sensor and Arduino Uno to read output voltage where LabVIEW further plots the measured temperature data obtained in real time in the graphical form.

1 INTRODUCTION

A large number of engineering applications require effective control systems that rely upon several physical quantities like temperature, pressure, humidity, speed, position, current, voltage, etc. Monitoring and control of such parameters being very important for maximizing the performance of systems, it becomes necessary to acquire accurate data corresponding to it. A computer-assisted real time module including suitable data acquisition systems and software can give a proper insight to these physical quantities (Blanco et al. 2014). LabVIEW is one such versatile program development application that can solve our purpose (Bayoumi et al. 2015). It uses graphical programming language. The execution of LabVIEW programs, called Virtual Instruments (VIs), is determined by the structure of graphical block diagrams (Javier et al. 2014). Every VI uses functions to manipulate input from the user and display it. In order to acquaint with its versatility, we have created an application specific VI that runs to acquire data from temperature sensor through Arduino Uno and plots the output in the form of chart. The sequences of steps followed are mentioned through Figure 1.

LM35 temperature sensor is a precision integrated-circuit temperature sensor capable of operating over a temperature range of −55°C to +150°C and has a gradient of 10 mV/°C. Features like linear output, low output impedance and precise calibration make it easier to use (Pradeep et al. 2014). The output voltage that is linearly proportional to the temperature is communicated to the Arduino Uno board. Arduino is an open-source prototyping platform with user-friendly and flexible hardware and software (Gandole et al. 2011). It has a simple microcontroller and an integrated development environment for writing the code. Arduino can be used for numerous purposes like taking input from sensors, controlling variety of lights and motors, or they can be used to communicate with some software running on the computer (Zhu et al. 2012). Here, we have used Arduino UNO that has 14 digital input/output pins, 6 analog inputs, a USB connection, a power jack, a 16 MHz quartz crystal, an ICSP (in-circuit serial programming) header and a reset button. The Arduino UNO takes the voltage corresponding to surrounding temperature from the sensor as its input and then serially communicates this data to the LabVIEW code built in our PC.

2 DESIGN AND IMPLEMENTATION

The hardware setup consists of Arduino UNO connected to LM35 temperature sensor. The technical specifications of Arduino UNO are mentioned in the Table 1. The board operates by providing external power supply through the USB connected to the computer. The first pin of LM35 is connected to the 5V pin on board that provides a regulated 5V supply to operate the sensor. Its second pin that

Figure 1. Basic set-up showing the sequence.

Table 1. Technical specifications of Arduino Uno.

Parameter/Specifications	Value
Microcontroller	ATmega328P
Operating voltage	5V
No. of digital I/O pins	14
No. of analog inputs	6
Input voltage range	6–12V
Flash memory	32 K
Clock frequency	16 MHz
DC current per I/O pin	20 mA

Figure 2. LM35 sending corresponding voltage signal to the Arduino UNO.

provides output voltage is connected to the pin A1 on arduino board. The third pin is connected to the ground pin (GND) on arduino board.

The software required for the whole set-up are as follows:

- Arduino IDE 1.6.6
- NI LabVIEW Version 2012

Arduino being an open source platform, we can download the IDE (Integrated Development Environment) and access the sketch from internet. After compiling and verifying the code, it is uploaded onto the configured board through configured port. Here, we have selected Arduino UNO as board and COM8 as port from the Tools option. The Tx and Rx pins on the arduino UNO glow when the code is successfully uploaded as shown by Figure 2.

Figure 3 illustrates the LabVIEW code built using block diagrams. The VI here uses VISA (Virtual Instrument Software Architecture API) to communicate the data from Arduino to computer. VISA is interface independent and provides easy to use command set for communicating with different instruments. Here, NI-VISA communicates through the USB to connect our computer to the resource (Arduino UNO board). The VISA Configure Serial Port sets up the serial port for communication. VISA Write is used to write data to the USB port. This VI uses the function TO for turn ON and TF for turn OFF. VISA Read is used to read data available at serial port from the arduino UNO. This VI has a while loop that encloses three case structures within it. The first two are for writing a string and the last one is for reading. To activate the reading case, it has to check if the bytes at serial port are greater than 0. If yes, the True Case Structure in VISA Read will return the bytes read.

The Read case structure includes the Frac/Exp String to Number block to convert the string to numeric form. This is further represented in the form of a numerical value and waveform chart form of voltage obtained from sensor. This value is further converted into degree Celsius with readings

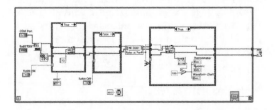

Figure 3. Block diagram view of LabVIEW code.

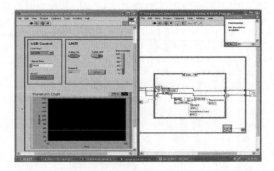

Figure 4. Front panel and block diagram view while running the LabVIEW code.

as shown on thermometer. VISA Close closes the connection established.

3 RESULTS AND DISCUSSION

The observations of temperature as obtained from the implementation of temperature sensor according to the code built in Arduino and LabVIEW are shown by Figure 4.

The numeric value and waveform chart represent the output voltage (in milliVolts) from the temperature sensor and the corresponding temperature value is obtained according to the scaling factor of LM35. The scaling factor of 10 mV/°C used here is as per the datasheet of LM35. The output temperature is represented by the thermometer scale on the front panel which reads 25°C in the above readings. Similarly, we obtained a few more readings by fluctuating the surrounding temperature by means of a match-stick and human hand.

4 CONCLUSION

The results obtained are almost very close to the actual surrounding temperature. This depicts real-time assessment and a low cost implementation of temperature measurement based on Arduino board using LabVIEW as a software platform. We can use the LabVIEW software for finding other parameters like pressure, humidity, soil moisture content etc in a similar way for certain industrial applications where these parameters play a significant role.

REFERENCES

Bayoumi, A., Salem M.A. 2015. A New Checkout-and-Testing-Equipment (CTE) for a Satellite Telemetry using LabVIEW. 36th 2015 IEEE Aerospace Conference, Montana, USA, DOI:10.1109/AERO.2015.7119305.

Blanco, J.R., Ferrero, F.J., Valledor M., and Campo, J.C. 2014. A low-cost open-source data acquisition system. 11th International Multi-Conference on Systems, Signals and Devices, Volume 1–6, 2014.

Gandole, Y.B. 2011. Virtual Instrumentation as an Effective Enhancement to Laboratory Experiment. International Journal of Computer Science and Information Technologies, ISSN-2728-2733, Volume 2, Issue 6, 2011.

Ionel, R., Lascu, M. and Cioabla, A. 2014. Arduino and LabVIEW in educational remote monitoring applications. *2014* IEEE Frontiers in Education Conference (FIE) Proceedings, Madrid, 2014: 1–5.

Javier F., Lara, R.F. and Redel, D.M. 2014. API for communication between Labview and Arduino UNO. *IEEE* Latin America Transactions, vol. 12, no. 6 Sept. 2014. doi: 10.1109/TLA. 2014. 6893988: 971–976.

Pradeep. K.P.J, Reddy, K.S.P., Kumar, D.H., Nagabhushan, K. and Nagaraja, C. 2014. Monitoring of Temperature and Humidity using LIFA. International Journal of IT, Engineering and Applied Sciences Research (IJIEASR), Volume 3, No. 6, June 2014.

Zhu, S. Ji, M., Wang, J. and Lei, Y. 2012. The Design and Implementation of Temperature Acquisition System based on PXI. International Conference on Control Engineering and Communication Technology (ICCECT), 2012: 267–206.

Communication and Computing Systems – Prasad et al. (Eds)
© 2017 Taylor & Francis Group, London, ISBN 978-1-138-02952-1

A relative comparison of FinFET and Tunnel FET at 20 nm and study the performance of clock buffers and inverters using FinFET

S. Ravi, Suprovab Mandal, B. Baskar & Harish M. Kittur
VLSI Design, Department of Micro and Nanoelectronics, VIT University, Tamil Nadu, India

ABSTRACT: FinFET and Tunnel FET (TFET) are the most promising candidate in modern semi-conductor industry for their superior performance into the deep submicron level beyond 45 nm. Where conventional MOSFET tangles with the adverse Short Channel Effects (SCE), both FinFET and TFET provide most desire solutions to these SCE. In this paper, basic performance parameters of the FinFET and TFET has been evaluated and the result shows that a simple D Flip Flop (D FF) made up of FinFET is 2.68 times faster compare to the D FF made up of TFET. Meanwhile, the TFET type D FF consumes 2.79 times less power than the FinFET type D FF.

1 INTRODUCTION

After the decade long regime of the MOSFET in the semiconductor industry, the FinFET and TFET come into the picture due to the irrevocable fatal effects of the MOSFET especially into the deep submicron region. Various SCE degrade the performance of the MOSFET. To overcome this hazardous effects, new devices were very much needed and hence FinFET came into the picture. As compare to MOSFET, FinFET provides a higher drain current due to its 3D fin like structure showing in Figure 1, resulting high speed perform-ance. TFET in the other case provides remarkably low subthreshold swing (for MOSFET, the sub-threshold swing is 60 mv/dec) which is very useful to reduce the static power consumption.

In this paper, the performance characteristics of FinFET and TFET has been studied. This paper also examines the propagation delay and power consumption of some clock buffers and inverters using FinFET.

Also a relative performance comparison is presented with some discrete components like D

FF (sequential) and 1 bit full adder (combina-tional) using 20 nm FinFET, TFET and 45 nm MOSFET.

2 EVICE PRINCIPLE AND SPECIFICATIONS

As there is no built-in models of TFET are avail-able till now so a look-up table based Verilog-A model has been used, provided by the senators of the TCAD Device Simulations. GaSb-InAs hete-ro–junction TFET (hTFET) is used for the circuit simulations. For the FinFET, silicon based Fin-FET model has been availed.

Hemanjaneyulu and Shrivastava (2015).The very early TFET concept (gated p-i-n diode) despite several advancements, such as SiGe source, low-κ drain spacer, high-κ source spacer, low drain doping, highly doped source, abrupt source junc-tion profiles, post silicidation implant, bandgap engineering and double-gate architectures, suf-fered from extremely low ON-currents.

Asra (2011). A modified structure of Tunnel Field Effect Transistor (TFET), is called the Sand-wich Tunnel Barrier FET (STBFET). STBFET has a large tunneling cross-sectional area with a tunneling distance of ~2 nm. STBFET gives a high I_{ON}, exceeding 1 mA/μm at I_{OFF} of 0.1 pA/μm with a subthreshold swing below 40 mV/dec.

Mohata (2012). Staggered tunnel junction (GaAs$_{0.35}$Sb$_{0.65}$/In$_{0.7}$Ga$_{0.3}$ As) can be used to dem-onstrate heterojunction tunnel FET (TFET) with the highest drive current, I_{on}, of 135 $\mu A/\mu m$ and highest I_{on}/I_{off} ratio of 2.7 × 10^4 (V_{ds} = 0.5 V, V_{on} − V_{off} = 1.5 V).

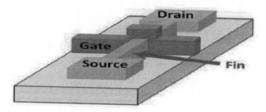

Figure 1. Basic structure of a FinFET.

Table 1. Parameters for the device simulation.

Parameters	Value
Channel Length (L_g)	20 nm
Effective Oxide Thickness	0.7 nm
nHTFET Source (GaSb) Doping	4×10^{19} cm^{-3}
pHTFET Drain (InAs) Doping	2×10^{20} cm^{-3}
Si FinFET Source/Drain Doping	1×10^{20} cm^{-3}
Si FinFET Fin Width (W_{Fin})	10 nm

Khatami and Banerjee (2009), Zhou (2012). Again vertical n-channel Tunnel Field-Effect Transistors (TFETs) with tunneling normal to the gate based on an n$^+$ In$_{x = 0.53 \to 1}$GaAs/p$^+$InP heterojunction have been seen to exhibit simultaneously a high I_{ON}/I_{OFF} ratio of 6×10^5, a minimum subthreshold swing (SS) of 93 mV/dec, and an on-current of 20 $\mu A/\mu m$ at $VDS = 0.5$ V and a gate swing of 1.75 V at 300 K.

Vaddi (2009). The key advantages of operating devices in subthreshold region ($V_{dd} < V_{th}$) compared to superthreshold region ($V_{dd} > V_{th}$) are reduced power due to low V_{dd} and small gate capacitance and also due to exponential I-V characteristics in subthreshold region, the devices will have high trans-conductance gain and thus near ideal Voltage Transfer Characteristics and better noise margins.

TSMC and Synopsys. Double gate FinFETs are also the candidates for subthreshold logic design than equivalent bulk devices, though it is not yet clear at this stage which configuration of DG-FinFETs will be more optimal for subthreshold logic. The device specifications has been listed in Table 1.

3 PERFORMANCE EVALUATION

Table 2 shows the current (I_{on}) variation of the n type FinFET and hTFET with respect to the gate voltage (V_{gs}) for a constant $V_{ds} = 0.6$ V. From this table it can be easily seen that the I_{on} is much higher for low V_{gs} in hTFET as compare to the FinFET indicating faster switching whereas when $V_{gs} = 0.6$ V, both the device gives almost same current.

Table 3 shows the similar comparison as shown in Table 2 between the p type FinFET and hTFET. Here almost the same behavior can be seen by the p type FETs but a dramatically change occurs when $V_{gs} = -0.6$ V. The on current of the p type hTFET becomes lower compared to the FinFET. Further researches are required to find out this type of behavior of the p type hTFET.

Table 4 shows the current (I_{on}) variation of the n type FinFET and hTFET with respect to the drain voltage (V_{ds}) for a constant $V_{gs} = 0.21$ V and 0.602 V.

From Table 4 a very important observation can be made. When V_{gs} is low i.e. 0.21 V, the on current of the hTFET is quite larger than the FinFET

which implies that in the subthreshold region, TFET performs much faster than FinFET. But when V_{gs} is high i.e. 0.602 V, both the devices perform almost equally and even sometimes FinFET overturns TFET.

Now to study the performances of the devices which work in the deep submicron level, a very important parameter is the Subthreshold Swing (SS). As the channel length of the transistor decreases gradually, static power or leakage power becomes the more dominating factor over dynamic power. The SS determines the I_{off} and I_{on} of the device where I_{off} is valuable to evaluate the static power consumption of the device. Table 5 shows the SS of the n type FinFET and hTFET for $V_{gs} = 0.6$ V.

Table 5, we can see that the SS of hTFET is 2.57 times smaller than the FinFET which makes hTFET very attractive for the low power circuit design.

Table 2. On current comparison of nFinFET and nhTFET.

V_{gs} (V)	I_{on} FinFET (μA)	I_{on} of hTFET (μA)
0.3	13.43	112.85
0.4	79.72	233.43
0.5	247.54	387.82
0.6	527.71	544.57

Table 3. On current comparison of pFinFET and phTFET.

V_{gs} (V)	I_{on} FinFET (μA)	I_{on} of hTFET (μA)
−0.3	−13.39	−57.92
−0.4	−79.58	−216.3
−0.5	−247.3	−372.4
−0.6	−527.3	−484.7

Table 4. On current comparison of nFinFET and nhTFET.

V_{gs} (V)	V_{ds} (V)	I_{on} FinFET (μA)	I_{on} of hTFET (μA)
0.21	0.3	0.7666	39.935
	0.4	1.0087	40.07
	0.5	1.2904	40.194
	0.6	1.6146	40.277
0.602	0.3	446.28	304.94
	0.4	481.83	399.6
	0.5	510.08	475.9
	0.6	533.97	546.25

Table 5. Subthreshold swing.

FET	Subthreshold swing
nFinFET	86.36 mV/dec
nhTFET	33.54 mV/dec

4 INVERTER CHARACTERISTICS

4.1 *DC analysis*

In Figures 2–3, the Voltage Transfer Characteristics (VTC) of the FinFET and TFET based inverter have been shown. From fair observation, it can be seen that the VTC of the TFET based inverter is much steeper as compare to the FinFET based inverter.

Vaddi (2013). Benefit from the steep slope switching is the reduced "turn-on" voltage (since V_{th} definition for TFET varies due to the tunneling mechanism, the term, turn-on voltage is used here), which can lead to the improved output voltage. The low turn-on voltage and leakage power trade-off of TFET design is compared with multiple Si FinFET models with different V_{th}. Figure 4 elaborates the VTC of the HTFET and FinFET at a very small gate voltage ($V_{gs} = 0.12$ V). From this it can be clearly seen that the TFET has steeper VTC as compared with the FinFET. Due to the minimum slope, TFET provides better switching or on currents at lower voltages. It also offers better noise margins at lower supply voltages as compared to the FinFET. TFET topology has full voltage swing at small supply voltages in comparison to FinFET one leading to robust digital circuit design with TFETs at small supply voltages.

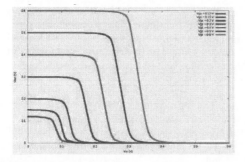

Figure 2.   VTC of FinFET.

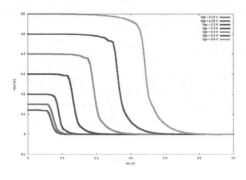

Figure 3.   VTC of TFET.

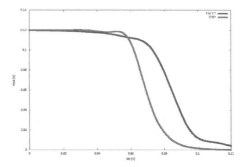

Figure 4.   TFET and FinFET Inverter VTC comparison at Vgs = 0.12 V.

4.2 *Transient analysis*

Figures 5–6 show the transient characteristics of the inverter. In these two graphs, the time required for transition from low voltage to high and vice versa has been shown.

Figure 5 describes the transient response of the inverter when the output is changing from high to low (t_{phl}). It can be seen that the output of the TFET inverter changes more quickly than the FinFET inverter, which also indicates the better switching performance of the device. Here also the slope of the TFET inverter's output is steeper than FinFET.

Figure 6 exhibits the output transition of the inverter from low to high (t_{plh}). In this case also, TFET has better performance than the FinFET because of the steeper slope. Having sub-threshold swing less than 60 mv/dec, TFET offers low propagation delay compared to the FinFET topology. Glitches are present in the TFET plot because of the miller capacitances, but this can be reduced by operating the inverter at higher supply voltages.

In Figures 7–8, the same simulations are performed but at higher supply voltages to eliminate the effect of the glitches. At 0.2 V supply voltage, the glitches are removed and as expected, the performance of the TFET inverter dominates the FinFET.

Now one other important parameter of the inverter is the noise margin. At smaller supply voltages TFET exhibits full voltage swing compare to the FinFET topology under same design environment. Figure 9 represents the noise margin analysis with varying input voltage of the TFET Topology

We can see that when the input voltage is low, then the noise margin i.e. the difference between the NM_H and NM_L is maximum (almost at $V_{in} = 0.1$ V).

5   CLOCK PATH ELEMENTS

In modern day almost every electronic system works with a very high speed clock which has

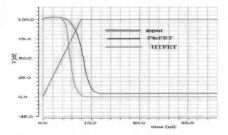

Figure 5. t_{phl} at $V_{dd} = 0.1$ V.

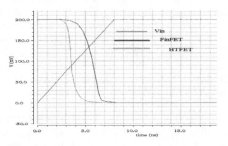

Figure 6. t_{plh} at $V_{dd} = 0.1$V.

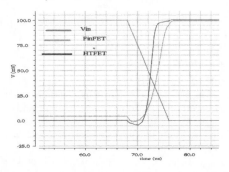

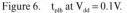

Figure 7. t_{phl} at $V_{dd} = 0.2$V.

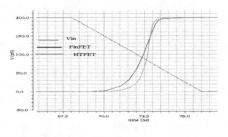

Figure 8. t_{plh} at $V_{dd} = 0.2$V.

frequency in the GHz range. From smart phones to super computers, all of these have a very powerful microprocessor inside them which requires a very fast clock to operate. Now to build a high speed clock system is relatively easy as compare to the distributing of the clocks through various components. Modern PCBs have millions of components and into the deep micron level beyond 32 nm, even small

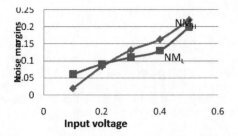

Figure 9. TFET inverter Noise margins with scaling of V_{in}.

interconnect delays make a big difference. As from the beginning of the semiconductor industry, memory management is always a big concern, clock distribution to these memory cells is extremely critical.

In the physical design stage of a chip, the performance of the chip purely depends on the timing. If the chip failures in the timing aspect then it will be of no use. Any extra pessimism in timing analysis not only requires more time to fix the critical paths but could negatively impact other important parameters such as power and area. In the worst case, it might leave no option but to reduce the functional frequency of the design.

On the other hand, silicon failure may occur for the optimism in timing analysis. In Figure 10, a common clock path has shown. From the figure, it can be seen that there are two flops named Launch flop and Capture flop. The reason behind this type of naming is that the launch clock will first get the input and then pass it to the next flop. So basically, the capture flop has to capture the data only after the launch flop passes the data. Now here comes the critical timing analysis.

To get a clear idea about this timing analysis, let's take a real time example. Figure 11 shows a practical clock and assume the clock frequency is 2.0 GHz which gives us the clock period is only 500 ps. To avoid all the timing violations, entire operation i.e. launching and capturing phase has to be done within this one clock cycle. If the cumulative delay of all the flops, data path components and interconnects is less than 500 ps then timing violation could be avoided but in real time this is very hard to achieve. Figure 10 shows the simplest clock path consisting of only two flops and data path elements. But in real time scenario, there will be millions of flops and data path components with their complex interconnections.

So to avoid the timing violations, clock buffers and inverters are used in the clock paths to meet timing requirements. The primary goal of these buffers and inverters is to add some delay in the clock path to slow down the clock frequency and therefore providing some extra time for the proper operation of the components (flops, data path components etc.). With the help of these clock

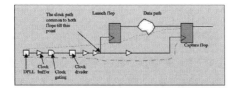

Figure 10.   Typical clock path scenario.

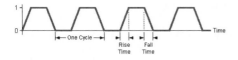

Figure 11.   Clock signal.

path elements, the original clock frequency of the system remains constant which is very much essential for maintaining the high speed performance of the other components such as microprocessor.

Two common timing violations are setup violation and hold violation. In general, setup violations occur when there is a change in input data while the active edge of the clock is also changing, therefore the launch flop does not get enough time to process the data with the relevant clock edge. Meanwhile, the capture flop needs some time to hold the output data from the launch flop. If the output of the launch flop changes before this minimum required hold time, hold violations occur. For a very high speed clock distribution network, clock buffers and inverters in the clock path slow down the original clock for those flops and components which require more time for propagating the input data to output.

For the timing analysis, there are basically 4 types of data paths available. Figure 12 shows the data paths. These are:

1. Input pin to register
2. Register to register
3. Register to output pin
4. Input pin to output pin

Clock buffers and inverters are used to eliminate any timing violations within these paths. To meet the timing constraints, these buffers and inverters are also now redesigned with advance architecture and superior performance.

In this paper, some existed clock buffers and inverters presented in Jilagam (2013), Ravezzi & Partovi (2011), Ravezzi (2013). These clock buffers and inverters have been studied with 20 nm FinFET.

5.1   Clock buffers

5.1.1   Supply noise reduced clock buffer
This is a special type of clock buffer which helps to eliminate the supply voltage variation due to the noise in the supply rails. This buffer lacks with the

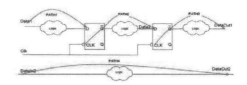

Figure 12.   Data paths for timing analysis.

requirement of high transconductance value of the FETs.

5.1.2   Supply noise reduced clock buffer with gain
This buffer is the modified version of the previous design. Here an extra gain stage has added to provide sufficient gain so that a high transconductance value of the FETs is no longer needed.

5.1.3   Period jitter reduced clock buffer
If there is any jitter in the clock path, this buffer corrects the jitter and make sure that the clock signal is from jitter free.

5.1.4   Tunable delay buffer with capacitive load
This buffer offers variable delay values with the capacitive load connected at its output. In this case there are three capacitive loads and when all these three loads are connected to the output, the buffers gives maximum delay. For three capacitive loads, eight types of different delay values are possible.

5.1.5   Adopted differential buffer
This buffer has pretty common circuitry like the differential amplifier. So when both clock and inverse of clock bar are required, this buffer can be very helpful.

5.1.6   Tunable clock buffer for multiple supply voltages
This buffer offers two modes of operation, full and partial. In the full mode, it acts like a normal buffer while in the partial mode, it acts as a partial buffer, reducing the power consumption

5.2   Clock inverters

5.2.1   Delay adjustable clock inverter
This inverter gives variable delay values by adjusting the input signal at the gate of the FinFETs connected at the bottom of the n type FET in the inverter.

5.2.2   Adjustable inverter for low power operations
Like the tunable delay buffer for multiple supply voltages, this inverter also operates in two modes i.e. full and partial. When the inverter is operated in the partial mode, the power consumption by the inverter becomes significantly less which makes it suitable for low power applications.

Table 6. Performance metrics of clock buffers and inverters.

| Buffer/Inverter | Si FinFET | | | |
	Delay (ps)	Power (uW)	PDP (E-18)	EDP (E-26)
Supply noise reduced clock buffer	22.13	363.1	8040	32200
Supply noise reduced clock buffer with gain stage	38.19	376	14300	57580
Period jitter reduced clock buffer	21.64	35.2	761	3050
Delay adjustable clock inverter	2.42	0.054	0.132	0.5288
Tunable delay buffer with capacitive load	12.3	0.10	1.23	4.94
Adopted differential buffer	60.9	0.090	5.49	21.9
Tunable clock buffer for multiple supply voltages	29.02	0.093	2.70	10.80
Adjustable inverter for low power operations	2.8	0.072	0.199	79.9

Table 7. Relative comparison between discrete components.

FETs	Parameters	D flip flop	1 bit full adder
FinFET	Delay (ps)	6.161	20.365
	Power (μW)	4.076	13.473
	PDP (E–18)	25.11	274.377
TFET	Delay (ps)	16.545	48.555
	Power (μW)	1.46	4.284
	PDP (E–18)	24.155	208.009
MOSFET (45 nm)	Delay (ps)	36	87.48
	Power (μW)	8.567	19.71
	PDP (E–18)	308.412	1724.231

6 SIMULATIONS AND RESULTS

All the circuit simulations have been done in the Cadence® Virtuoso simulator. Chauhan & Hu (2015). The BSIM-CMG verilog-A model files have been used for the FinFET simulations. The results opted from the simulations are presented in Table 6.

7 CONCLUSION

In this paper a relative comparison has been made between the FinFET and TFET. From Table 7,

it can be seen that the PDP for small scale design such as D flip flop is almost equal for both FinFET and TFET but it differs greatly for the large scale complex design such as 1 bit full adder. It can be also understood from Table 7 that the performance of the FinFET as well as the TFET is way ahead as compare to the 45 nm MOSFET.

REFERENCES

Asra R. *et al*. 2011. A Tunnel FET for V_{DD} Scaling Below 0.6 V with a CMOS-Comparable Performance. *IEEE Transactions on Electron Devices*, VOL. 58, NO. 7.

Chauhan Y.S., Hu C., Lu D.D., Vanugopalan S., Khandelwal S., Duarte J.P., Paydavosi N., Niknejad A. 2015. FinFET Modeling for IC Simulation and Design – Using the BSIM-CMG Standard. Academic Press.

Hemanjaneyulu K. & Shrivastava M. 2015. Fin-Enabled-Area-Scaled Tunnel FET, *IEEE Transactions on Electron Devices*. VOL. 62, NO. 10.

Jilagam V.C., Ravi S. & Maillikarjun K.H. 2013. Skew and Power Reduction Using Tunable Clock Buffers and Inverters. *International Conference on Circuits, Power and Computing Technologies (ICCPCT)*.

Khatami Y. & Banerjee K. 2009. Steep Subthreshold Slope n- and p-Type Tunnel-FET Devices for Low-Power and Energy-Efficient Digital Circuits. *IEEE Transactions on Electron Devices*, VOL. 56, NO. 11.

Mohata D.K. *et al*. 2012. Demonstration of Improved Heteroepitaxy, Scaled Gate Stack and Reduced Interface States Enabling Heterojunction Tunnel FETs with High Drive Current and High On-Off Ratio. *Symposium on VLSI Technology Digest of Technical Papers*.

Ravezzi L. & Partovi H. 2011. CMOS Clock buffer with rduced supply noise. *Eletronic Letters* on, Vol. 47 No. 17.

Ravezzi L.2013. Clock buffer with supply noise active compensation for reduced period jitter. *Eletronic Letterrs* on, vol. 49 No. 18.

TSMC & Synopsys. FinFET Technology – Understanding and Productizing a New Transistor. *A joint whitepaper*.

Vaddi R. *et al*. 2009. Investigation of Robustness and Performance Comparisons of 3T-4T DG-FinFETs for Ultra Low Power Subthreshold Logic. *International Conference on Computers and Devices for Communication*.

Vaddi R. *et al*. 2013. Tunnel FET-based Ultra-Low Power, High-Sensitivity UHF RFID Rectifier. *Symposium on Low Power Electronics and Design*.

Zhou G. *et al*. 2012. InGaAs/InP Tunnel FETs With a Subthreshold Swing of 93 mV/dec and I_{ON}/I_{OFF} Ratio Near 10^6. *IEEE Electron Device Letters*, VOL. 33, NO. 6.

Communication and Computing Systems – Prasad et al. (Eds)
© 2017 Taylor & Francis Group, London, ISBN 978-1-138-02952-1

Reduction of reverse leakage current by ULP diode based MTCMOS technique for multiplier

Nancy Bohra Jain & R.H. Talwekar
SSTC(SSGI), Bhilai (C.G.), India

ABSTRACT: The continuously scale down of technology, leakage current is increasing exponentially. Multi-Threshold Complementary Metal-Oxide-Semiconductor (MTCMOS) is one of the most used technique to reduce leakage current in standard circuits. When a conventional MTCMOS circuit operates from the sleep-to-active mode during transaction, the circuit was suffering from ground bounce noise problem. The designed Ultra Low Power (ULP) Diode Based MTCMOS techniques to reduce the ground bounce noise up to 99.4% and the standby leakage current up to 99.75% respectively. There is dependence of ground bounce noise and standby leakage current on voltage variation, temperature and transistor sizing has been shown. Simulation of multiplier with applied technique have been performed on EDA tool CADENCE virtuoso with 90 nm CMOS technology.

Keywords: Multi-Threshold Complementary Metal-Oxide-Semiconductor (MTCMOS): Ground bounce noise

1 INTRODUCTION

One of the important usually used block in VLSI system, to design and implement various blocks of digital CMOS circuit is nothing but Multiplier. The multiplier design and implementation contribute in fields like area, speed and power consumption of computational intensive VLSI systems. With Low-power and high-speed multipliers are in high demand, but to achieve both criteria simultaneously not possible (Meher et al. 2011). As the device size shrinks according to technology, there is occurring of other side-effects with it, so becomes difficult to achieve a good tradeoff by device scaling or sizing the transistors (Bailey et al. 2008).

The basic multiplier has three sequences: parallel, serial and last one is hybrid (parallel- serial) approach. For the high-speed operation and reduces the chip area the parallel approach multiplier is popular (Anuar et al. 2010). The basic operation of parallel multiplier is explained in Fig. 1 (Weste et al. 2011).

And the known generalized equation is shown below in Eq. (1), (2) and (3) [4].

$$P = \sum_{i=1}^{n-1} \sum_{j=1}^{n-1} X_i Y_j 2^{(i+j)} \quad (1)$$

$$X = \sum_{i=1}^{n-1} X_i 2^i \quad (2)$$

$$Y = \sum_{j=1}^{n-1} Y_j 2^j \quad (3)$$

Now with every new technology generation results 30 times increasing in larger gate leakage current and ground bounce noise (Singh et al. 2007). Hence, it extremely important to reduce gate leakage current and ground bounce noise with improved design techniques. A well-known technique is the power gating in which sleep transistor is connected between the actual ground and virtual circuit ground (Kim et al. 2008). At the time of sleep mode, sleep transistor was off so that it cut the leakage path. It shows that power gating technique will reduce the gate leakage with minimum impact on the circuit performance. There is various power gating techniques such as MTCMOS and Transistor Gating are used to reduce gate leakage (leakage current) and ground bounce noise (Jiao & kursun 2011) (Gupta & Nakhate 2012).

			X_3	X_2	X_1	X_0		
		×	Y_3	Y_2	Y_1	Y_0		
					X_3Y_0	X_2Y_0	X_1Y_0	X_0Y_0
				X_3Y_1	X_2Y_1	X_1Y_1	X_0Y_1	
			X_3Y_2	X_2Y_2	X_1Y_2	X_0Y_2		
+		X_3Y_3	X_2Y_3	X_1Y_3	X_0Y_3			
	P_7	P_6	P_5	P_4	P_3	P_2	P_1	P_0

Figure 1. Basic general operation of multiplier.

MTCMOS used to suppress the leakage current and ground bounce noise (Kudithipudi & John 2005). High threshold voltage (High VTH) named sleep transistor is connected to header and footer of circuit. Reduce the leakage current be turned off the sleep transistor. When circuit transition from sleep to active mode, there will be voltage fluctuation which occurs on real power line and real ground line.

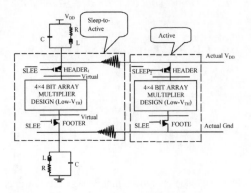

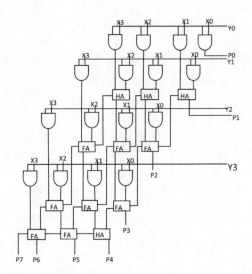

Figure 2. Conventional MTCMOS Technique.

So it is used to cutoff the connection between power supply and ground of the idle Low threshold voltage (Low V_{TH}) circuit blocks (Shi & Howard 2006). These sleep transistors are two types (a) High threshold V_{TH} pMOS transistor or (b) High threshold V_{TH} nMOS transistor. A High threshold V_{TH} pMOS sleep transistor are attached between the real power line i.e. (power supply) and a virtual power line i.e. (Low V_{TH} circuit blocks), as shown in Figure 2. Alternatively in another case a High threshold V_{TH} nMOS sleep transistor is connected between the actual ground (real ground) and virtual ground (Low V_{TH} circuit blocks). To reduce the sub-threshold leakage current in idle standardized circuits, the above mentioned sleep transistors used at header and footer is turned off.

2 PROPOSED DESIGN AND TECHNIQUES

2.1 Multiplier

The array of "n × n" AND gates are used to calculate all the product X_iY_i terms simultaneously. These terms sum by "n(n −2)" number arrays of Full Adders and "n" Half Adder. This is known a Carry-Save Multiplier because the output generated from the Adder are saved and then used in the next stage operation. There is increasing in area cost due to use of additional adder in each stage. However only short wire is use to connect its nearest neighboring cell in it, so can be pipelined easily. As it requires only one critical path rather than compared to the several paths is it's another advantage (Kudithipudi & John 2005). The general structure is shown in Figure 3.

2.2 Ultra low power diode based MTCMOS

In this section the circuit designed with Ultra Low Power (ULP) Diode Based MTCMOS technique Figure 4 (Levacq et al. 2005).

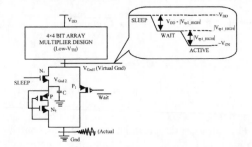

Figure 3. 4 × 4 carry-save multiplier.

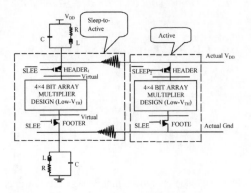

Figure 4. Ultra low power diode based MTCMOS technique.

This technique consists of three parts:

- High threshold transistors (P_2 & N_2) are used as MOS ultra low power diode to reduce ground bounce noise.
- A wait mode is introduced between sleep-to-active mode transitions by high threshold PMOS transistor (P_1).
- Capacitor (C_2) placed between sleep transistor N_1 and P_2 to control the drain current flow through sleep transistor (N_2) in mode transition.

High threshold transistors (P2 & N2) are used as MOS ultra low power diode to reduce circuit ground bounce noise. A wait mode is introduced between sleep to active mode transitions by high threshold PMOS transistor (P1). Between sleep transistor N_1 and P_2, the Capacitor (C_2) placed to control the drain current flow through a sleep transistor (N2) in during mode transition.

In CMOS technology, by connecting drain to gate of MOS transistor it can be used as a diode. When we used diode in reverse biased then gate is connected to source. This reverse leakage current is related to MOS drain current, leakage current increases exponentially for lower threshold voltage. This leakage current is reduced by high threshold transistor. One drawback with high threshold transistor is that the current driving capability in forward biasing mode get reduces.

So, we used Ultra Low Power (ULP) diode its design shown in Figure 5. This Ultra-Low-Power (ULP) diode strongly reduces the reverse leakage current, MOS diode (Adriaensen et al. 2002). The weak inversion drain current for NMOS & PMOS transistor is given below in Eq. (4) & (5) (Levacq et al. 2007).

$$I_{DN2} = I_{SN} e^{\frac{qV_{GSN2}}{nKT}} e^{\frac{\eta q V_{DSN2}}{nKT}} \left(1 - e^{\frac{-qV_{DSN2}}{KT}}\right)$$ (4)

$$I_{DP2} = I_{SP} e^{\frac{-qV_{GSP2}}{nKT}} e^{\frac{\eta q V_{DSP2}}{nKT}} \left(1 - e^{\frac{-qV_{DSP2}}{KT}}\right)$$ (5)

where, V_{GS} = gate to source voltage, V_{DS} = drain to source voltage, V_T = thermal voltage $(V_T = KT/q)$, = Drain Induced Barrier Lowering (DIBL) coefficient. and I_{SN} & I_{SP} = reference currents that correspond to the weak inversion drain current.

When reverse bias voltage is increased more, so increase in ULP diode current due to increase in V_{DS}. The current reaches to peak value, and then strongly decreases with the voltage V_{GS} of transistor. In this way ULP diode decreases the leakage current.

When sleep transistors i.e. (N_1, N_2 and P_1) are turned OFF in standby mode than drain to source potential (V_{DS1}) of sleep transistor N_1 decreases, which causing more body effect due to negative body to source (V_{BS1}) of N_1 and less DIBL. In this way these stacking transistors are used reduced the leakage current.

During sleep to active mode of transition Ground bounce noise is reduced by controlling the current stream flowing through sleep transistors. In mode of transition there are two parts: (1) Sleep to wait transition, (2) wait to active mode transition (Jiao & Kursun 2010). During the first mode (sleep

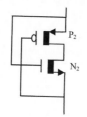

Figure 5. MOS ULP Diode.

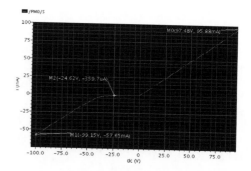

Figure 6. 1-V characteristics of ULP diode.

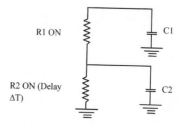

Figure 7. Equivalent Circuit For Stacking Transistor (N1 and N2) During Wait-To-Active Mode Transition.

to wait), transistor P1 turned ON and transistors N1 and N2 are turned OFF. The using of voltage source (V_{BIAS}) we will decrease the threshold voltage of sleep transistor, so more virtual ground can get discharge during wait mode. In second mode (wait to active), transistor P_1 turned OFF and sleep transistor (N_1, N_2) are turned ON. First transistor N_1 turns ON and then after some Delay time (ΔT) transistor N_2 will ON. R_1 is the ON resistance of transistor N_1, C_1 as internal capacitance at virtual ground (V_{GND1}) and C_2 is an external capacitance at intermediate node (V_{GND2}). Capacitor C_1 its voltage V_1 and the capacitance C_2 has its voltage V_2. When $0 < T < ΔT$, then C_1 will start discharge and capacitor C_2 start charging. This process continuously until both capacitors (C_1 and C_2) has the same equal potential (Saxena et al. 2012). So, if control capacitor C_2 and delay (ΔT), intermediate node voltage can be controlled and both transistors i.e. N_1, N_2 are turned ON in region triode and hence the fluctuation in voltage is controlled at ground and will get reduce ground bounce noise.

2.2.1 Calculation of minimum delay

If $ΔT < ΔT_{min}$, transistor N1 turn ON in saturation region and transistor N2 turn ON in triode region. So, when $ΔT ≥ ΔT_{min}$ the ground bounce noise is reduced effectively.

Minimum ΔT condition is (Pattanaik (a) et al. 2010):

$$V_{GS(ST1)} - V_{TH(ST1)} \geq V_{DS(ST1)} \quad (6)$$

When, $0 < t < \Delta T$, Eq. (6) is simplified and written as

$$V_{G(ST1)} - V_{S(ST1)} - V_{TH(ST1)} \geq V_{R1ON(t)}$$
$$V_{DD} - V_{C2(t)} - V_{TH} \geq V_{R1ON(t)} \quad (7)$$

By solving Eq. (7), minimum delay (ΔT) is (Pattanaik et al. 2010):

$$\Delta T_{min} = \tau \ln \left[\dfrac{1}{\left\{ 1 - \left(\dfrac{C_1 + C_2}{C_2} \right) \left(\dfrac{V_{TH}}{V_1} \right) \right\}} \right]$$

where,

$$\tau = R_{1ON} \left(\dfrac{C_1 C_2}{C_1 + C_2} \right)$$

The use of transistor stacking, ULP diode additional and mode between sleep to active mode of transitions will reduce the leakage and ground bounce noise.

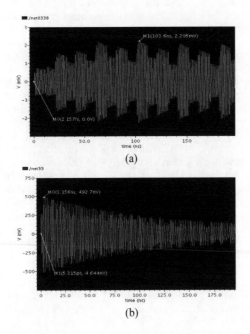

(a)

(b)

Figure 8. Transient response of waveform showing ground bounce noise (a) conventional design, (b) ultra low-power diode based MTCMOS technique.

3 SIMULATED RESUTS

The simulation has been performed with low power multiplier for characterization of ground bounce noise, standby leakage current and active power with ULP Diode Based MTCMOS techniques.

3.1 Ground bounce noise

Ground Bounce Noise for ULP Diode Based MTCMOS techniques has been evaluated in this section.

3.1.1 Voltage scaling effect on ground bounce noise

In MTCMOS technique, Delay (ΔT) are the important parameter to reduce the noise. It observed that ground bounce noise at delay ($\Delta T = 0.6$ ns) is minimum. It has been evaluated that for Bias Voltage (VBIAS = –600 mV) at which ground bounce noise is minimum.

As shown in Table 1, ground bounce noise is minimized for ULP Diode Based MTCMOS technique as compared to Conventional MTCMOS technique. ULP Diode Based MTCMOS techniques reduce ground bounce noise by 99.4% respective as compared to standard conventional circuit.

3.1.2 Effect of transistor sizing on ground bounce noise

Effect of wait transistor size on the ground bounce noise for ULP Diode Based MTCMOS techniques are described in this section.

When size of wait transistor is small, then a discharge of ground voltage in first transition (sleep-to-wait) is less and more ground voltage is discharge in second transition (wait-to-active). Alternatively, for larger wait transistor size, more will be ground voltage is discharge in first period of transition (sleep-to-Wait) and hence, discharge during second transition (wait-to-active) is less.

Ground bounce noise also depends on the size of sleep and wait transistor. If size of wait transistor is greater than sleep transistor, then first peak is dominating of ground bounce noise because more ground voltage will be discharge during first transition. If size of wait transistor is smaller than size of sleep transistor, second peak is dominating. As shown in Table 2, ULP DB MTCMOS techniques increase ground bounce noise by change in transistor size.

Table 1. Voltage scaling on ground bounce noise.

Supply Voltage (V)	0.5	0.6	0.7	0.8	0.9
Conventional (µV)	93.44	610.5	2205	3430	4514
ULP diode based MTCMOS (µV)	0.351	0.412	0.492	0.578	0.652

3.1.3 Temperature effect on ground bounce noise

Effect of temperature scaling on ground bounce noise is evaluated in this section. Ground bounce noise for conventional design, ULP DB MTC-MOS techniques for different temperature (in °C) is shown in Table 3.

At high temperature, carrier mobility reduces and saturation current of sleep transistor also reduce. Virtual ground voltage of MTCMOS technique is reducing which affect the ground bounce noise of the given circuit. Sometimes increased steady state virtual ground voltage and reduced saturation current of sleep transistors with temperature, both can effect in opposite way on Ground Bounce Noise.

So, effect of temperature on ground bounce noise is not monotonic and this can also be seen in Table 4. When temperature varied from 25°C to 110°C than Ground Bounce Noise reduced in ULP DB MTCMOS 99.8% respectively.

3.2 Standby leakage current

Standby leakage current of low power multiplier is evaluated in this section. Table 4 shows standby leakage current comparison of ULP Diode Based MTCMOS techniques with voltage scaling. The Eq. (9)(Roy et al. 2003) shows the basic equation i.e.

$$I_{SUB} = \frac{W}{L}\mu V_{TH}^2 C_{STH} e^{\frac{V_{GS}-V_{TH}+\eta V_{DS}}{\eta V_{TH}}}\left(1-e^{\frac{-V_{DS}}{V_{TH}}}\right) \quad (9)$$

Table 2. Effect of transistor sizing on ground bounce noise.

Wait transistor size (nm)	1	2	3	4	5
Conventional (μV)	4514				
ULP diode based MTCMOS (μV)	0.495	1.012	1.462	1.017	1.288

Table 3. Temperature scaling on ground bounce noise.

Temperature (°C)	25	50	75	100	110
Conventional (μV)	2210	2150	2082	1922	1909
ULP diode based MTCMOS (μV)	0.492	0.501	0.483	0.455	0.449

Table 4. Effect of voltage scaling on leakage current.

Supply Voltage (V)	0.5	0.6	0.7	0.8	0.9
Conventional (μA)	0.3082	1.315	4.765	10.16	18.26
ULP diode based MTCMOS (μA)	0.0247	0.0301	0.0322	0.0413	0.0456

Table 5. Effect of voltage scaling on active power.

Supply Voltage (V)	0.5	0.6	0.7	0.8	0.9
Conven-Tional (μW)	1.658	6.457	19.42	39.2	74.03
ULP diode based MTCMOS (μW)	0.2699	0.4695	0.8053	1.69	13.1

where, W and L = width and length of transistor, μ = carrier mobility, $V_{TH} = (kT/q)$ thermal voltage at temperature T, $C_{STH} = C_{DEP} + C_{IT}$ is the summation of the depletion region and interface trap capacitance both per unit area of MOS gate.

$$\eta = 1 + \left(\frac{C_{STH}}{C_{OX}}\right)$$

where, η = Drain Induced Barrier Lowering (DIBL) coefficient and C_{OX} = gate oxide capacitance.

In Table 4, leakage current for different MTC-MOS techniques are shown. It is clear from tables that if supply voltage increases it also increase the leakage current. It is clear ULP Diode Based MTCMOS techniques reduce standby leakage current by 99.75% respective to standard circuit.

3.3 Active power

This section analysis the active power of multiplier with ULP Diode Based MTCMOS technique. Active power is calculated by giving input vectors and measured the active power dissipation during this time. The active power will include both two i.e. dynamic and static power of the circuit.

The general equation for active power is shown in Eq. (10) (Weste et al. 2011).

$$P_{active} = P_{dynamic} + P_{static} \quad (10)$$

$$= P_{switching} + P_{short-circuit} + P_{leakage} \quad (11)$$

$$= (\alpha_{0\to1} \times C_1 \times V_{dd}^2 \times f_{clk}) + (I_{sc} \times V_{dd}) + (I_{leakage} \times V_{dd}) \quad (12)$$

where, C_L = load capacitance, Fclk = clock frequency, α = probability that a power consuming transition occurs (the activity factor), Isc = short circuit current, $I_{leakage}$ = leakage current.

As shown in Table 5, ULP Diode Based MTCMOS techniques reduce active power by 82.3% respective as compared to standard circuit.

4 CONCLUSIONS

In this paper, ground bounce noise during mode transition and standby leakage current are evaluated in Ultra Low-Power Diode Based MTC-

MOS techniques. We use transistor stacking in ULP DB MTCMOS techniques to reduce ground bounce noise and standby leakage current effectively. Effect of temperature, transistor sizing on ground bounce noise, standby leakage current and active power for MTCMOS techniques are evaluated in this paper. It has been observed that Ultra Low-Power Diode Based MTCMOS technique reduces ground bounce noise by 99.4% and Ultra Low-Power Diode Based MTCMOS reduces standby leakage current by 99.75% respectively. It also has been observed that temperature and transistor sizing has minimum effect on ground bounce noise. Active power of the circuit is reduced by ultra low-power diode based MTCMOS technique up-to 82.3% as compared to conventional multiplier. The proposed multiplier with ULP Diode Based MTCMOS techniques is operated on 45 nm technology.

REFERENCES

Adriaensen S., Dessard, V. & Flandre, D. 2002. Electronics Letters. *25 to 300°C Ultra-Low-Power Voltage Reference Compatible with Standard SOI CMOS Process* 38(19):1103–1104.

Anuar, N., Takahashi, Y. & Sekine, T. 2010. 18th IEEE/IFIP VLSI System on Chip Conference (VLSI-SOC). *4×4-Bit Array Two Phase Clocked Adiabatic Static CMOS Logic Multiplier with New XOR*: 364–368.

Bailey, D., Soenen, E., Gupta, P., Villarrubia, P. & Sang, D. 2008. IEEE/ACM International Conference on Computer-Aided Design (ICCAD).*Challenges at 45 nm and beyond*: 11–18.

Gupta, B. & Nakhate, S. 2012. International Journal of Emerging Technology and Advanced Engineering. *Transistor Gating A Technique for Leakage Power Reduction in CMOS Circuits* 2(4): 321–326.

Jiao, H. (a) & Kursun, V. 2011. Journal of Circuits. *Systems and Computers, Noise—Aware Data Preserving Sequential MTCMOS Circuits with Dynamic Forward Body Bias* 20(1):125–145.

Jiao, H. (b) & Kursun, V. 2010. IEEE Transactions on Circuits & Systems. *Ground-Bouncing-Noise-Aware Combinational MTCMOS Circuits* 57(8): 2053–2065.

Kudithipudi, D. & John E. 2005. Journal of Low Power Electronics. *Implementation of Low Power Digital Multipliers Using 10 Transistor Adder Blocks* 1:1–11.

Levacq, D. et al 2005.IEEE International Symposium on Circuits and Systems. *Ultra-Low Power Flip-Flops for MTCMOS Circuits* 5: 4681–4684.

Levacq, D. et al 2007. IEEE Journal of Solid-State Circuits.*Low Leakage SOI CMOS Static Memory Cell With Ultra-Low Power Diode* 42(3):689–702.

Meher, M.R., Jong, C.C. & Chang Chip-Hong. IEEE Trans. On Very Large Scale Integr. (VLSI) System. *A High Bit Rate Serial-Serial Multiplier With On-the-Fly Accumulation by Asynchronous Counters* 19(10):1733–1745.

Park, J.C. & Mooney III V.J. 2006. IEEE Transactions on Very Large Scale Integration (VLSI) System. *Sleepy Stack Leakage Reduction* 14(11):1250–1263.

Pattanaik, M., Agnihotri, S., Varaprashad, M.V.D.L. & Arasu, T.A. 2010. International Symposium on Electronic System Design (ISED). *Enhanced Ground Bounce Noise Reduction in a Low Leakage 90 nm 1-Volt CMOS Full Adder Cell*:175–180.

Pattanaik, M., Varaprasad, M. V. D. L. & Khan, Fazal Rahim 2010. International Conference on Electronic Devices, Systems and Applications (ICEDSA). *Ground BounceNoise Reduction of Low Leakage 1-bit Nano-CMOS based Full Adder Cells for Mobile Applications*: 31–36.

Roy, K.; Mukhopadhyay S. & Mahmoodi- Meimand H. 2003. Proceedings of the IEEE. *Leakage Current Mechanisms and Leakage Reduction Techniques in Deep-Submicron CMOS Circuits* 91(2):305–327.

Kim, S.; Choi, C.J.; Jeong, Deog-Kyoon; Kosonocky, S. V. & Park, Sung Bae 2008. IEEE Transactions on Electron Devices. *Reducing Ground-Bounce Noise and Stabilizing the Data-Retention Voltage of Power-Gating Structures* 55(1):197–205.

Saxena, C.; Pattanaik, M. & Tiwari R.K. 2012. International Conference on Devices, Circuits and Systems (ICDCS). *Enhanced Power Gating Schemes for Low Leakage Low Ground Bounce Noise in Deep SubmicronCircuits*:239–243.

K. Shi and D. Howard, "Challenges in sleep transistor design and implementation in low-power designs" 43rd ACM/IEEE Design Automation Conference 2006, pp. 113–116.

Singh, H.; Agarwal, K.; Sylvester, D. & Nowka, J.K. 2007. IEEE Transactions on Very Large Scale Integration (VLSI) Systems. *Enhanced Leakage Reduction Techniques Using Intermediate Strength Power Gating* 15(11):1215–1224.

Weste, Neil H.E.; Harris, David & Banerjee, Ayan (3th ed) 2011.*CMOS VLSI Design: A Circuit and System Perspective*:Pearson Education.

Communication and Computing Systems – Prasad et al. (Eds)
© 2017 Taylor & Francis Group, London, ISBN 978-1-138-02952-1

Explicit current output universal filter realization employing CFOA

Soumya Gupta & Tajinder Singh Arora
Maharaja Surajmal Institute of Technology, New Delhi, India

ABSTRACT: Employing current feedback operational amplifiers in current mode and producing explicit currents for obtaining a universal biquad filter are the purpose of this paper. Focusing on the need for the integrated circuit implementation of the configuration only five passive components have been involved, offering very low sensitivities which is beneficial for the performance of the circuit. The paper gives an insight of all the responses of a universal filter, along with the tunability and total harmonic distortion response, which have been generated through PSPICE simulation of the proposed configuration. Comparison between the responses obtained using ideal and non-ideal current feedback operational amplifier is also supported.

1 INTRODUCTION

Analog signal processing, by definition, is the signal processing conducted on continuous analog signals by some analog means. These analog means include a variety of analog devices like operational amplifier (Op-Amp), Current Feedback Operational Amplifiers (CFOA), Operational Transconductance Amplifiers (OTA), Current Differencing Buffer Amplifiers (CDBA) etc (Senani 2013). Considerable efforts have been made in studying the advantageous features of one device over the others, which have been successfully recorded in literature (Deliyannis et al. 1998, Senani 2013, Senani et al. 2015). These studies have initiated the attempts for designing circuits employing different analog devices and with time enhancing the performance of these circuits.

Filters play a key role in signal processing, the defining feature of filter being its ability to completely or partially suppress some output of the signal. By far, there are five basic types of filters namely Low Pass (LP), High Pass (HP), Band Pass (BP), Band Stop (BS) and All Pass (AP). The term universal filter is used while suggesting a circuit that can be used to generate the transfer function of all the five filters. Here, realization of a universal filter is emphasized using a "current feedback operational amplifier" as this device has been rapidly replacing the traditional operational amplifier because of its potential features like "higher bandwidth", "low sensitivity" and "higher slew rate". Moreover, the device has been gaining significant attention as it is available in an integrated circuit, the most commonly used being AD844. The current feedback operational amplifier is a type of electronic amplifier whose inverting input

is sensitive to current rather than to voltage as in conventional voltage feedback operational amplifier. The CFOA is basically a second generation Current Conveyor (CCII) followed by a buffer amplifier. Various attempts have been made by the researchers to design a universal filter using CFOA as an active device. Some of the prominent references are (Abuelma'atti & Al-Shahrani 1996, Abuelma'atti & Al-Zaher 1997, Abuelma'atti & Al-Zaher 1998, Bhaskar & Prasad 2007, Senani 1998, Senani 2013, Senani 1998, Senani 2013, Senani et al. 2015, Sharma & Senani 2003). References (Abuelma'atti & Al-Shahrani 1996, Abuelma'atti & Al-Zaher 1997, Abuelma'atti & Al-Zaher 1998, Senani 1998, Sharma & Senani 2003, Singh & Senani 2005, Soliman 1996) shows the voltage mode universal filter out of which (Abuelma'atti & Al-Zaher 1998, Senani 1998, Singh & Senani 2005, Soliman 1996) are of Single-Input Multiple-Output (SIMO) type of filters. Multi-Input Single Output (MISO) types of filters are available in (Abuelma'atti & Al-Shahrani 1996, Abuelma'atti & Al-Zaher 1997, Senani 2013). References (Bhaskar & Prasad 2007, Senani 1998, Senani 2013) are the Current Modes (CM) or dual mode filters. As per the knowledge of the authors none of the current mode universal filter has been made available in the literature that obtains explicit current output from CFOA.

This manuscript offers one such current-mode universal filter which employs CFOA as an active building block, with fewer passive components. All the five main filter responses can be achieved without doing any change in the hardware configuration. Explicit current output from the high impedance port makes it a better proposition for cascading and designing a higher order filter.

In this manuscript section 2 shows the circuit diagram of the proposed configuration along with the characteristic equations and ideal transfer functions. Section 3 gives sensitivity analysis and non-ideal transfer functions. Simulation results, that support the design configuration, are given in section 4. At last conclusion is provided.

2 PROPOSED CIRCUIT

A Current Feedback Operational Amplifier (CFOA) is symbolically represented as shown in Figure 1.

The characteristic equations for an ideal current feedback operational amplifier are given as:

$$
\begin{aligned}
I_y &= 0 \\
I_z &= I_x \\
V_x &= V_y \\
V_w &= V_z
\end{aligned} \tag{1}
$$

When non-ideal current feedback amplifiers are used for realization of a network, then it is characterized as:

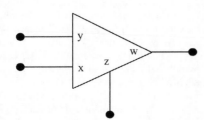

Figure 1.   Symbol of CFOA.

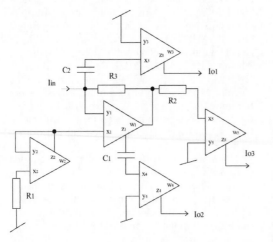

Figure 2.   Realisation of the proposed configuration.

$$
\begin{aligned}
I_y &= 0 \\
I_z &= \alpha I_x \\
V_x &= \beta V_y \\
V_w &= \gamma V_z
\end{aligned} \tag{2}
$$

where α, β, and γ represent non-ideal port transfer ratios of X, Y and Z terminals respectively and ideally $\alpha = \beta = \gamma = 1$.

The connections between the active and passive components, that is, the current feedback operational amplifiers and the resistors and capacitors for implementing a universal filter are shown in Figure 2.

By considering the relationships shown in (1), the transfer functions for the proposed configuration when implemented using ideal device are given in (3)-(7). The natural angular frequency ω_0 and the quality factor Q_0 of the filter are given in (8) and (9) respectively.

$$
\frac{I_{o1}}{I_{in}} = \frac{s^2}{s^2 + s\left(\dfrac{1}{C_2 R_3}\right) + \left(\dfrac{1}{C_1 C_2 R_1 R_3}\right)} \tag{3}
$$

$$
\frac{I_{o2}}{I_{in}} = \frac{\dfrac{-s}{C_2 R_1}}{s^2 + s\left(\dfrac{1}{C_2 R_3}\right) + \left(\dfrac{1}{C_1 C_2 R_1 R_3}\right)} \tag{4}
$$

$$
\frac{I_{o3}}{I_{in}} = \frac{\left(\dfrac{1}{C_1 C_2 R_1 R_2}\right)}{s^2 + s\left(\dfrac{1}{C_2 R_3}\right) + \left(\dfrac{1}{C_1 C_2 R_1 R_3}\right)} \tag{5}
$$

$$
\frac{I_{o1} + I_{o3}}{I_{in}} = \frac{s^2 + \dfrac{1}{C_1 C_2 R_1 R_2}}{s^2 + s\left(\dfrac{1}{C_2 R_3}\right) + \left(\dfrac{1}{C_1 C_2 R_1 R_3}\right)} \tag{6}
$$

$$
\frac{I_{o1} + I_{o2} + I_{o3}}{I_{in}} = \frac{s^2 - \dfrac{s}{C_2 R_1} + \dfrac{1}{C_1 C_2 R_1 R_2}}{s^2 + s\left(\dfrac{1}{C_2 R_3}\right) + \left(\dfrac{1}{C_1 C_2 R_1 R_3}\right)} \tag{7}
$$

$$
\omega_0 = \sqrt{\frac{1}{C_1 C_2 R_1 R_3}} \tag{8}
$$

$$
Q_0 = \sqrt{\frac{C_2 R_3}{C_1 R_1}} \tag{9}
$$

The above equations are the transfer functions of high pass (3), band pass (4), low pass (5), band reject (6) and all pass (7) filter respectively which can be derived easily. This is observed that by summing the functions of high pass and low pass filter, we can synthesize the transfer function of band stop filter. Similarly, a simple current summation of high pass, band pass and low pass response provides the all pass filter function. Simulation results based on the same are given in section 4.

3 NON-IDEAL AND SENSITIVITY ANALYSIS

For getting the non-ideal transfer function of the designed filter, the non-idealities of the port relationship i.e. (2) is considered. The derived non-ideal transfer function of the proposed filter is given in (10)-(14). The angular frequency and quality factor, when focused on the non-idealities of the circuit, are stated in (15) and (16) respectively.

$$\frac{I_{o1}}{I_{in}} = \frac{s^2 \alpha_3}{s^2 + s\left(\dfrac{1}{C_2 R_3}\right) + \left(\dfrac{\alpha_1 \alpha_2 \beta_1 \beta_2 \gamma_1}{C_1 C_2 R_1 R_3}\right)} \tag{10}$$

$$\frac{I_{o2}}{I_{in}} = \frac{\left(\dfrac{-s}{C_2 R_1}\right)(\alpha_1 \alpha_2 \beta_1 \beta_2 \alpha_4)}{s^2 + s\dfrac{1}{(C_2 R_3)} + \left(\dfrac{\alpha_1 \alpha_2 \beta_1 \beta_2 \gamma_1}{C_1 C_2 R_1 R_3}\right)} \tag{11}$$

$$\frac{I_{o3}}{I_{in}} = \frac{\left(\dfrac{1}{C_1 C_2 R_1 R_2}\right)(\alpha_1 \alpha_2 \alpha_5 \beta_1 \beta_2 \gamma_1)}{s^2 + s\dfrac{1}{(C_2 R_3)} + \left(\dfrac{\alpha_1 \alpha_2 \beta_1 \beta_2 \gamma_1}{C_1 C_2 R_1 R_3}\right)} \tag{12}$$

$$\frac{I_{o1} + I_{o3}}{I_{in}} = \frac{s^2 \alpha_3 + \left(\dfrac{1}{C_1 C_2 R_1 R_2}\right)(\alpha_1 \alpha_2 \alpha_5 \beta_1 \beta_2 \gamma_1)}{s^2 + s\left(\dfrac{1}{C_2 R_3}\right) + \left(\dfrac{\alpha_1 \alpha_2 \beta_1 \beta_2 \gamma_1}{C_1 C_2 R_1 R_3}\right)} \tag{13}$$

$$\frac{I_{o1} + I_{o2} + I_{o3}}{I_{in}} = $$

$$\frac{s^2 \alpha_3 - \left(\dfrac{s}{C_2 R_1}\right)(\alpha_1 \alpha_2 \beta_1 \beta_2 \alpha_4) + \left(\dfrac{1}{C_1 C_2 R_1 R_2}\right)(\alpha_1 \alpha_2 \alpha_5 \beta_1 \beta_2 \gamma_1)}{s^2 + s\left(\dfrac{1}{C_2 R_3}\right) + \left(\dfrac{\alpha_1 \alpha_2 \beta_1 \beta_2 \gamma_1}{C_1 C_2 R_1 R_3}\right)} \tag{14}$$

$$\omega_0 = \sqrt{\frac{\alpha_1 \alpha_2 \beta_1 \beta_2 \gamma_1}{C_1 C_2 R_1 R_3}} \tag{15}$$

$$Q_0 = \sqrt{\frac{C_2 R_3 (\alpha_1 \alpha_2 \beta_1 \beta_2 \gamma_1)}{C_1 R_1}} \tag{16}$$

By the sensitivity analysis of the realized network it was found that the proposed configuration has a very low sensitivity for all the active and passive components employed. Therefore, this analysis justified the performance of our network.

$$S^{\omega_0}_{C_2, C_1, R_3, R_1} = \frac{-1}{2}$$

$$S^{\omega_0}_{\alpha_1, \alpha_2, \beta_1, \beta_2, \gamma_1} = \frac{1}{2}$$

$$S^{Q_0}_{C_1, R_1} = \frac{-1}{2}$$

$$S^{Q_0}_{C_2, R_3, \alpha_1, \alpha_2, \beta_1, \beta_2, \gamma_1} = \frac{1}{2}$$

4 SIMULATION RESULTS

To check our proposed and designed filter, the simulations have been performed using ORCAD PSPICE software. The plots generated from simulation depicted that the central frequency of the universal filter was obtained as 1 MHz. Figure 3

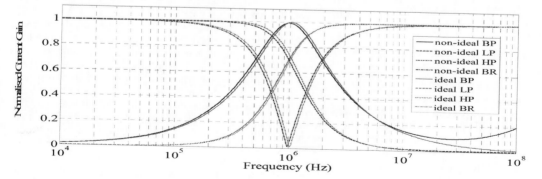

Figure 3. Ideal and non-ideal frequency responses of all the basic filters.

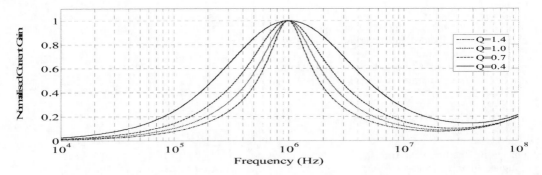

Figure 4. Frequency response of band pass filter for different values of Q_0.

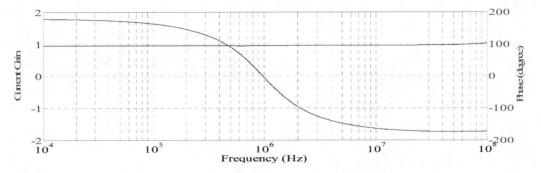

Figure 5. All pass gain and phase response of the suggested biquadratic filter.

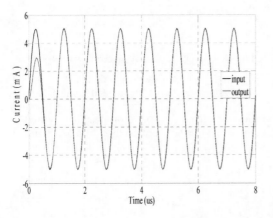

Figure 6. Transient response of the proposed universal filter.

shows the simulation results of the low pass, high pass, band pass and band reject responses of the universal biquadratic filter. Simulations were carried out for the following passive component values $R_1 = R_2 = R_3 = 1$ K and $C_1 = 0.216$nF and $C_2 = 0.11$nF. Plot for both ideal and non-ideal case has been obtained, where, AD844 integrated circuit is used for non-ideal analysis of the circuit

(shown in Figure 1). Figure 3 also shows that there is not much deviation between ideal and non-ideal curves of the designed filter.

Further, keeping in mind the necessity for tuning the circuit, a plot for tunability of band pass filter was generated for which subsequent changes were made in values of passive components. The different plots were made on a single graph to encourage the study of variations obtained by changing value of Q_0, which has been shown in Figure 4. To support the universal filter design, the all pass gain as well as phase response has also been plotted and shown in Figure 5.

The workability of the filter has been validated not only for alternating current input but also for sinusoidal current input. To support this with a simulation result, a plot was obtained showing the transient response of the proposed circuit (shown in Figure 6) along with total harmonic distortion recorded for different values of input current (shown in Figure 7), for better understanding. A sinusoidal input of amplitude 5 mA and frequency 1 MHz is applied and output waveform for band pass filter is recorded. Apparently, keeping the magnitude of sinusoidal input current more than 5 mA distorted the output.

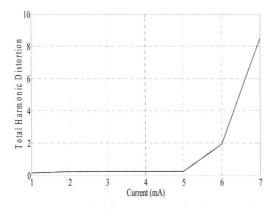

Figure 7. Total harmonic distortion for different magnitudes of input current.

4 CONCLUSION

The paper aims at introducing a new current-mode universal filter, which is capable of giving all five basic filter responses simultaneously, by utilizing the explicit current outputs obtained from a network of current feedback operational amplifiers. Concerning about the practical availability of the proposed circuit, minimum elements have been involved so that integrated circuit implementation can be easier. For better understanding of the configuration, analysis like transient analysis, frequency responses, total harmonic distortion and tunability analysis have also been included in this paper. Explicit current output available at high impedance port makes the circuit worthy of making higher order filters. Suitable comparison between plots of ideal and non-ideal case has also been shown. The simulation results are performed

at suitably high frequency and these are in agreement with the mathematical analysis of the filter.

REFERENCES

Abuelma'atti, M. T., Al-Shahrani, S.M. 1996. New universal filter using two current-feedback amplifiers. *IntJ Electron 80: 753–756.*

Abuelma'atti, M. T., Al-Zaher H. A. 1997. New universal filter using two current-feedback amplifiers. *Active Passive Electron Comp 20: 111–117.*

Abuelma'atti, M. T., Al-Zaher, H.A. 1998. New universal filter with one input and five outputs using current-feedback amplifiers. *Analog Integr Circ Sign Process 16: 239–244.*

Bhaskar, D. R., Prasad, D. 2007. New current mode biquad filter using CFOAs. *J Active Passive Electron Devices 2: 292–298.*

Deliyannis, T., Yichuang, Sun, Fidler, J. K. 1998. Continuous-time active filter design. *CRC Press.*

Senani, R. 1998. Realization of a class of analog signal processing/signal generation circuits: novel configurations using current feedback op-amps. *Frequenz 52:196–206.*

Senani, R. 2013. Current feedback operational amplifiers and their applications. *Springer Sciences and Business Media New York 2013.*

Senani, R., Bhaskar, D.R., Singh, A. K. 2015. Current conveyors, variants, applications and hardware implementations. *Springer Sciences.*

Sharma, R. K., Senani, R. 2003. Multifunction CM/VM biquads realized with a single CFOA and grounded capacitors. *Int J Electron Commun (AEU) 57: 301–308.*

Sharma, R. K., Senani, R. 2004. Universal current mode biquad using a single CFOA. *Int J Electron 91: 175–183.*

Singh, A. K., Senani, R. 2005. CFOA-based state-variable biquad and its high-frequency compensation. *IEICE Electron Express 2: 232–238.*

Soliman, A. M. 1996. Applications of current feedback operational amplifiers. *Analog IntegrCirc Sign Process 11: 265–302.*

Communication and Computing Systems – Prasad et al. (Eds)
© 2017 Taylor & Francis Group, London, ISBN 978-1-138-02952-1

Low power obstacle and skew aware clock tree synthesis

Ravi S., Yash Mittal & Harish M. Kittur
School of Electronics Engineering, VLSI Division, VIT University, Vellore, India

ABSTRACT: Due to continuous scaling down of VLSI based design technology, it has become extremely important to consider the role of Clock Tree Synthesis (CTS) in deciding the performance of complex nanometer chips. Skew minimization with less power dissipation improves efficiency of CTS. This Paper presents Skew aware clock tree routing with obstacles present in the routing path. Thus blockage modeling is incorporated with sorting based partition of points to reduce wire snaking. The routing is based on an efficient algorithm which generates the clock tree, performs merging of points and finally determines the tapping point. Experimental calculations demonstrates the efficiency of our CTS approach with effectively improved skew and Latency by 46% and 57.8%. Moreover our Merge sorting obtains a significant reduction in Power by 75.9% compared to bipartition algorithm. The look up table is built using NGSpice tool to achieve the desired results.

1 INTRODUCTION

An important parameter which holds vital importance in physical design is Clock Tree Synthesis (CTS) and it determines the entire circuits performance. Due to continuous shrinking of VLSI design into nanometer era, CTS with buffer incorporation (Cai yici denz et al. 1992) becomes extremely necessary to minimize the delay and maintain signal integrity. For obtaining a clock tree which delivers low values of skew, delay as well as slew is the main design prospective (Ginneken van L.P.P.P. 1990). Delay models mainly uses Elmore delay to measure the interconnect delay in CTS. But Elmore model limits itself while the measuring the slew values which are of prime importance while measuring the buffer intrinsic delay. Elmore cannot get slew related information in an RC model.

During the process of physical design in VLSI placement process is done before CTS. Therefore the macroblocks, IPs which were placed during the placement process become unavoidable obstacles (Chao T.H. et al. 1992) in the routing phase. Therefore previously work has been on clock tree routing along with obstacle avoiding. In the previous work the obstacles were considered once the zero skew clock routing was done (Chao T.H et al. 1992). Therefore subnets which were in clock proximity with the obstacles need to be rerouted which probably leads to wire snaking and disturbs the skew, capacitance achieved previously. DME algorithm (Edahiro M. 1993) was used to avoid the obstacles (Lillis J. Cheng et al.) which utilized track graph in order to minimize the wire length and cost. Wires in clock tree routing are placed on the metal layers

and obstacles as well as buffers on the poly layers (Rajaram A. et al. 2006). Thus it becomes essential to consider buffers in the obstacle avoiding algorithm. Different Programming methods were introduced to minimize path delay among which dynamic programming was used for a given wire tree. Subsequently in (Chen Ying-yu et al. 2010), various different targets are utilized to insert the buffers, such as minimizing the power, slew as well as area. Previous researchers (Huang H. et al. 2007) adopted a symmetrical buffer insertion technique where they tried to insert buffers on a symmetric tree topology (Liu W.-H. et al. 2010). A hybrid tree structure was adopted where a upper symmetrical tree controls the bottom asymmetrical tree. Tsai et al (Tsay R.S. 1991) presented an exact zero skew algorithm which sizes the wire along with insertion of buffers. This algorithm guarantees minimum skew along with power (Shih X.-W. et al. 2010), delay (Shih X.-W. et al. 2010) and are minimization. In previous published works the buffer was inserted (Ravi S. et al. 2015) on the wire wherever the slew values are violating the constraint which means the delay introduced by the buffer insertion is not carefully analyzed which poses a serious limitation of the previous adopted techniques. In this paper obstacle avoiding algorithm sorts the points in order to minimize the wire length and applies suitable measure to the points near to the obstacles in order to avoid them. Furthermore buffers are inserted in a balanced proportion depending on the capacitance values of the wire with the buffer delays taken into consideration in the Elmore model which guarantees a minimum skew along with no slew constraints violated.

This paper remainder is arranged as follows: section 2 deals with problem formulation, Section 3 presents Merge sorting methodology, section 4 presents obstacle avoiding methodology, Section 5 presents Algorithmic flow, Section 6 presents experimental results and Section 7 presents conclusion.

2 PROBLEM FORMULATION

For provided random clock sink points $SK = \{s1, s2, s3....si\}$, the work of CTS is developing a clock tree network with vertices and edges formulated as $Tr = (V,E)$ and through which every sink is connected to the clock source with one single path.

Here the aim is to develop a clock tree with minimum wirelength, clock skew and power which may vary depending on various applications requirement.

2.1 Zero skew algorithm

This algorithm deals with optimizing the performance in timing sense of various synchronous digital systems. Clock skew is the main deciding factor for the digital system performance, thus optimization of the skew (Chang Y.-C. et al. 2012) can greatly benefit the timing performance by reducing the system's cycle. It follows the bottom approach in a recursive manner. The process if repeated in a bottom up fashion leads to the construction of zero skew clock tree (Khang Andrew B. et al.). Here a assumption is made as that every sub tree has zero skew which implies that delay of signal from the sub tree roots to the leaf nodes is equal. Zero Skew is achieved easily for sub trees with only one leaf nodes. Here the starting sub trees of the network are the leaf nodes.

Now when two sub trees are merged together, the root of the new merged tree is called the Tapping Point. This a Zero Skew point as the delay from this point to the two sub trees is equal. Zero skew tree is constructed for two sub trees i.e. 1 and 2 which is shown in the Figure 1. For each sub tree a lumped delay model is assumed. The interconnection wire is separated into two different halves through the Tapping Point. To get an equal delay from the tapping point to the leaf nodes, the following expression must hold true.

$$r_1(c_{1/2} + C_1) + t_1 = r_2(c_{2/2} + C_2) + t_2 \qquad (1)$$

where r_1 and c_1 denotes the total resistance and capacitance of wire segment 1. Similarly r_2 and C_2 denotes the total resistance and capacitance of wire segment 2. An assumption can be made that the total length of the wire seg-

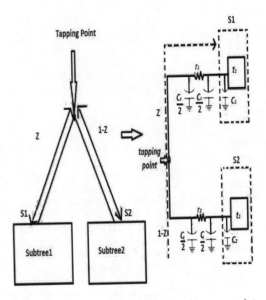

Figure 1. Zero skew algorithm based merge point determination.

ment be l. The wire length starting from the tapping point to the subtree1 is zxl and therefore the wirelength from subtree 2 to the tapping point is (1−z)l Let us consider two constants, α and β where α denotes the resistance per unit length and β denotes the capacitance per unit length. Therefore we have $r_1 = \alpha\,z_1$, $r = \alpha_1$ and $r_2 = (1-z)\alpha_1$. Similarly $c_1 = \beta z_1$, $c = \beta_1$ and $c2 = (1-z)\beta_1$. Hence after simplifying equation (1),we obtain the zero skew condition as

$$z = \frac{(t_2 - t_1) + \alpha l \left(C_2 + \dfrac{\beta l}{2}\right)}{\alpha l (\beta l + C_1 + C_2)} \qquad (2)$$

3 MERGE SORTING ALGORITHM

A clock tree network consists of randomized clock pins which when routed without rearranging them leads to extended wired connections resulting in more clock skew power dissipation and so on. Thus it becomes necessary to sort the clock sinks in order to determine final merge location and thus minimize the wire. Length and clock skew. Two types of sorting techniques are incorporated in zero skew algorithm mainly point sorting and distance sorting.

3.1 Point sorting

Clock sinks are distributed in a randomized fashion in the routing plane comprising of x and y axis respectively. These sink locations are rearranged as based either on there x or y coordinates. In this, Sorting along both planes has been implemented. Post point sorting all the sinks get arranged in the ascending order. This also aids in providing many options as to how many sink locations can a single clock sink be routed.

3.2 Distance sorting

This sorting technique aids in minimizing the wire length. It is applied to the already sorted points and works by calculating the distance between sorted points. Partitioning of sinks based on their minimum routing distance is the key to reduce power dissipation. Minimum wire length guarantees lower capacitance values. Through this sorting, minimum distance between two sinks is calculated and stored in the programming variable and there are many iterations to determine nearest merge point for routing.

4 BLOCKAGE MODELLING

The Obstacles or the Blockage are macro blocks or IPs in the clock routing path and hinder the connections between clock points when lies in between those two points. The given blockage information is stored in the matrix through data structures (Chaturvedi R. et al. 2004). To avoid the blockage we have adopted binary searching technique.

The area around the obstacle is divided into different regions ri where $i = 1, 2...8$ as shown in Figure 2. The nodes located in these regions of obstacle need to be routed in such a manner that their routing path do not collide with the boundaries of the obstacle which are stored already in the matrix of data structures. The clock tree topology after obstacle consideration is comprised of three stages: (1) Blockage information storage (2) binary searching of nearby nodes (3) routing with rearrangement.

In the binary search stage of blockage modeling (Lee D. et al. 2010) the sink locations are searched in the regions of blockage and stored in the matrix. Then boundaries of the obstacle are compared with the sinks routing path and decision is made with the routing distance greater then the obstacle boundary to avoid the overlapping.

Finally the last stage prepares a clock tree with reasonable clock skew based on the Elmore delay model as shown in Figure 3.

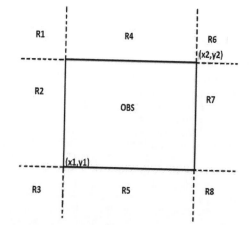

Figure 2. Regions for blockage modelling.

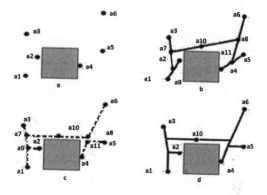

Figure 3. A example of clock tree generation with obstacle consideration in the routing path.

a. It represents a sample input
b. An abstract tree topology generated by the merging sinks around the obstacle.
c. Tapping point determination by merging internal nodes
d. The final constructed tree after routing

5 ALGORITHMIC FLOW

Figure 4 shows the flow of our zero skew algorithm with merge sorting and blockage modeling embedded in to it. Our approach for CTS considers a global watch on obstacles to get the target skew (Cai Yici et al. 1992). The obstacle avoiding clock tree routings begins with the clock sink points as input. A zero skew algorithm (Tsay R.S. 1991) begins routing the points in order to obtain the tapping point or merge point. In this algorithm there are different clock tree stages obtained depending on the number of sinks in each stage. Two clock sinks contribute to a single merge point. This tapping points are merged in the next clock tree stages and contribute to new tapping points. This process repeats until single tapping point is left which is the final stage of clock tree routing. Elmore delay model (Liu W.-H. et al.) is used in this algorithm to calculate the clock latency. At every clock tree generation stage Merge sorting is applied to minimize the routing distance and rearranging the sinks and storing them in the matrix. Manhattan distance is calculated between two sink locations and the minimum one is considered to route through the obstacles.

A check is done for the nodes lying in the region of the obstacles and an appropriate blockage avoiding methodology (Shih X.-W. et al. 2010) help to merge two points without the routing path being overlapped with the rectangular obstacle. Buffers are inserted into the clock tree based on the wire length.

Balanced buffer insertion is done on both the sinks which contribute to the tapping point to balance the final clock skew.

Input: Set of clock pins V, Topology G
Outcome: Zero skew clock tree
In every clock tree stage,

S1. For each clock node V search nearby nodes.
S2. Perform zero skew Merge of two nodes obtained from step1 and obtain a tapping point
S3. Pair up every tapping point .
S4. Repeat steps S1 and S2 until for every clock tree stage only one merge point is left.
S5. Connect the final tapping point to the output node of clock driver.

6 EXPERIMENTAL RESULTS

We executed our design methodology through Matlab tool on a 1.7 Ghz Linux Ubuntu operating workstation with internal memory of 8 GB. We initially generated clock tree for different ISPD'09 benchmarks and validated the design for satisfactory values of skew, slew, latency and power through NGspice simulation. The RC netlist in the form of .cir(circuit) file is given as the input to Ngspice tool which generates a .raw file in the batch mode. The netlist consists lengths from the taping point to respective parent node and the associated capacitances. The raw file contains the information about clock metrices. The same procedure is repeated for five different ISPD benchmarks and the results are shown in Table 1. Figure 5 demonstrates the matlab results of our algorithm and shows the robustness of our design.

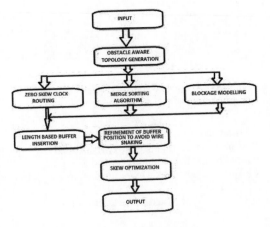

Figure 4.   Flow of obstacle aware algorithm.

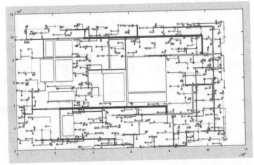

Figure 5.   Developed clock tree for benchmark circuit ispd09S4r3.

Table 1.   Tabulated results.

Benchmarks	No of sinks	No of obstacles	Skew (ps) proposed	Skew (ps)	Worst slew (ps) proposed	Worst slew (ps)	Max latency (ns)	Power (pw) proposed	Power (pw)
f11	121	0	25.8	47.8	83.45	87.4	0.324	42.5	93.51
S2r1	88	31	30.6	31.7	76.3	54.8	0.215	34.08	188
f31	273	88	31.2	35.9	80.8	91.6	0.396	72.93	302.62
S3r1	131	48	40.7	43.5	72.5	86.3	0.238	56.6	276.04
S4r3	623	9	45.4	58	73.4	83.7	0.285	111.4	519.85

REFERENCES

Cai Yici, Deng Chao, Zhou Qiang, Yao Hailong, Niu Feifei and N Cliff 1992, *Clock tree synthesis under aggressive buffer insertion proc DAC june 1992,* 13–18.

Chao T.H., Hsu Y.C. Ho J.M., Bose K.D. and Kahng A.B. 1992, *Zero skew clock routing with minimum wire length IEEE trans.circuits syst November 1992,* vol 39 no. 11,799–814.

Chang Y.-C., Wang C.-K. and Chen H.-M. 2012, On constructing low power and robust clock tree via slew budgeting proc. ISPD, 129–135.

Chaturvedi R. and Hu J. 2004, *Buffered clock tree for high quality IC design proc.* IEEE 5th ISQED, 381–386.

Chen Ying-yu, Dong Chen, Chen Deming 2010, Obstacle avoiding and slew constrained clock tree synthesis with efficient buffer insertion IEEE syst November 2010, vol 39 no. 11, 799–814.

Edahiro M. 1993, *A clustering based optimization algorithm in zero skew routings proc.of the Design Automation Conf. june 1993,* 612–616.

Ginneken van L.P.P.P. 1990, *Buffer placement in distributed RC tree networks for minimal Elmore delay proc. IEEE Int symp.* Circuits Syst. May 1990, 865–868.

Huang H., Luk W.-S., Zhao W. and Zeng X. 2007, *DME based clock routing in the presence of obstacles proc.* 7th AISCON, 1225–1228

Khang Andrew B., Liening, Markov Lgor L. and Hu Jin, *VLSI Physical design From graph partitioning to timing closure.*

Lee D. and Markov I.L. 2010 *CONTAGO integrated optimization of Soc clock network proc.* DAC, 80–85.

Lillis J., Cheng C.K. and Lin T.T., *Optimal wire sizing and buffer insertion for low power and a generalized delay model IEEE J. Solid state circuits,* vol 31, 437–447.

Liu W.-H., Li Y.-L. and Chen H.-C. 2010, *Minimizing clock latency range in robust clock tree synthesis proc. 15th ASPDAC,* 389–394.

Rajaram A. and Pan D.Z. 2006, *Variation tolerant buffered clock network synthesis with cross links in proc. ISPD,*157–164.

Ravi, S. Chandrahasa Reddy, D. Sumanth Kumar, Kittur Harish 2015, *Obstacle avoiding slew aware clock tree synthesis proc. IMAS,* 213–224.

Shih X.-W. and Chang Y.-W. 2010 *Fast timing model independent buffered clock tree synthesis proc. DAC,* 80–85.

Communication and Computing Systems – Prasad et al. (Eds)
© 2017 Taylor & Francis Group, London, ISBN 978-1-138-02952-1

Demand response from residential air conditioning load using smart controller

Aadesh Kumar Arya, Saurabh Chanana & Ashwani Kumar
National Institute of Technology, Kurukshetra, Haryana, India

ABSTRACT: Now-a-days the needs of air conditioning systems are exponential increases, but AC is consumed the major part of energy in residential and commercial buildings. The energy optimization and reducing the consumer's bill is main component of Smart Grid. So Demand response is main component of Smart Grid. With demand response the power consumption is reduced of Residential Air Conditioning. In this paper, the smart controller is designed using fuzzy logic tool box in MATLAB to achieve the condition of Demand Response (DR).

Keywords: Smart Controller, Fuzzy Inference Systems (FIS), Demand Response and Air Conditioning System

1 INTRODUCTION

Now-a-days, the demand of residential and commercial air conditioning system is increasing due to high rate of population and global warming. The production of energy is very low in compare to energy consumption (A. H. Osman et al. 2014). The main target of researchers, scientists and academicians of all over word is to create the smart appliances and reduced energy consumption. The Energy Management Systems (EMS) are best method to reduce the peak energy consumption in the interest of utilities and consumers (K. Le et al. 2007; J. Chen et al. 1995). The Demand Response (DR) systems are the part of energy management systems. For the equilibrium between the utility and the customers in power system, DR is required. The utility can send signals to the customers to indicate a need to reduce demand. On receiving the price signal from utility, customer can take an appropriate action based on his preferences and needs. A variety of DR programs have been explained in (M. H. Albadi and E. F. El-Saadany et al. 2007).

To develop and maintain the modern economy and society, the economical, social and environmental sustainability are required in the energy sector. The smart appliances are the important component of Smart Grid. By using the smart appliances, the electricity consumption can be reduced at the consumer's end. The consumption of energy is very less in smart appliances. The advantages of smart appliances are not limited to making life easier, but very much helpful in energy saving also (Aadesh

Kumar Arya et al. 2013). Now-a-days, Indian are commonly used air conditioning systems in homes, industries and public places for the comfortable zone (H. Nasution, H. Jamaluddin and J. M. Syeriff et al. 2011). There are various techniques to make the smart for the appliances viz. Fuzzy Logic System (FLS) and Artificial Neural Network (ANN). These tools are very important in the interest of engineers, scientist and researcher to develop the smart appliances viz. Air Conditioner (AC), Electric Water Heater (EWH), and microwave oven etc. All peoples are aware from the fuzzy logic system. The main purpose of the fuzzy logic system is to build an intelligent controller, in which human thinking is embedded. By adopting these techniques, the consumers can reduce the electricity bill and researchers can serve to the society. Demand Response (DR) and Demand Side Management (DSM) are very useful to control the energy consumption and reduce the electricity bill. The intelligent controller was fist time implemented based on fuzzy logic system (King et al, 1977). The block diagram represent the process of fuzzy logic controller which shown in Figure 1.

In the modeling of Fuzzy Logic Controller (FLC), the main issue is the tuning of membership function fuzzy rules. The tuning of membership function and fuzzy rules is depending on the human's knowledge and experience.

1.1 TCA thermal model

Home appliances can be classified into three groups: (a) adjustable thermostat devices, (b) controllable

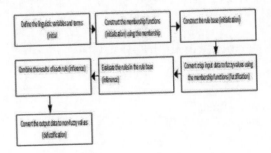

Figure 1.   Process of fuzzy logic controller.

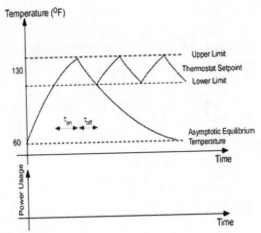

Figure 2.   A typical thermal characteristic curve of an AC load.

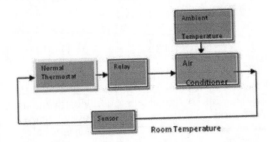

Figure 3.   Block diagram for temperature controller of air conditioner using normal thermostat.

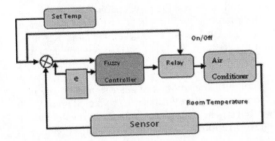

Figure 4.   Block diagram for temperature controlled of air conditioner using fuzzy controller.

non-thermostatically controlled devices (c) and non-controllable devices. Among the adjustable thermostat controlled devices residential HVAC systems, electric water heaters and refrigerators. Among the non-thermostatically controlled appliances washing machines or dryers. Non-controllable appliances cannot be scheduled; therefore, they are modeled by a load profile that is forecast with historical data. Figure 2 shows the thermal dynamic behavior of a Residential Air Conditioner (RAC) unit over time. The rising curves indicate the RAC is "on", and the falling curves represent the RAC is "off". As the RAC unit cycles, the room temperature rises and falls accordingly (Ning Lu et al. 2004). The temperatures limit of the upper and lower are dead band of the thermostat which is used to regulate the power consumption of the TCAs. The exponential growth and the fall in turn, between the upper temperature limit and the lower temperature limit, the curve is almost linear, as shown in Figure 2 (N. Lu et al. 2005). The operations of all TCAs are similar.

Conventionally, the thermostatically controlled loads operate on following logic. Let T_{st} be the thermostat setting. The upper and lower limit for the internal temperature is $T_{st} + \Delta T$ and $Tst - \Delta T$, respectively. Whenever $T > T_{st} + \Delta T$ and thermostat is in OFF state, it would be switched ON; and whenever $T < T_{st} + \Delta T$ and the thermostat is in ON state, it would be switched OFF as illustrated in Figure 2. We term this control as normal thermostat control.

The study of dynamic model for the temperature of a house by a thermostatically driven air conditioning load (Saurabh Chanana et al. 2013) is carried out which is given below. The dynamic model is as follows:

$$\frac{dT}{dt} = \frac{T_f - T - wT_g}{\tau} \tag{1}$$

The symbols which are used in this dynamic mathematical model are following

T = Temperature of room
T_f = Ambient temperature

T_g = Temperature gain of the air conditioner.
w = State of the thermostat (0 – on, 1 – off)
τ = Effective thermal constant of the house

The model is developed of equation (1) with only thermostat relay which shows by block diagram for temperature controlled of air conditioner using normal thermostat in Figure 3. This model gives the information for power consumption and consumed energy. Figure 4 shows the block diagram for temperature controlled of air conditioner with intelligent/fuzzy controller and also provides the same information.

2 METHODOLOGY

2.1 Modeling of Residential Air Conditioning (RAC) load

Figure 4 Shows the Block Diagram for Temperature Controlled of Air Conditioner using Fuzzy Controller.

This block diagram described that the Residential Air Conditioner (RAC) is installed in room for comfort environment. Firstly, the sensor read the room temperature than intelligent controller (Fuzzy Controller) provides the required temperature to the relay of RAC and The RAC worked as per atmosphere. The simulation model is required to obtain the best outcome of RAC with/without implementation of the intelligent controller. To develop the simulation model, dynamic model is required.

2.2 Fuzzy membership function

To design the shapes of membership functions, the membership function editor are required for each membership variable in fuzzy logic tool box (Lotfi A. Zadeh, 1999, 2001). For the input and output variables of smart controller of Residential Air Conditioner (RAC), the membership functions are defined as follows

2.2.1 Input variables

The smart controller having two input variables, temperature error and change in temperature.

2.2.2 Temperature error

The temperature error is described by three membership functions viz. small, medium and large as shown in Table (1).

2.2.3 Change in temperature (ΔT)

The difference between room temperature actually and user temperature, are represented by four membership functions negative, small, medium, large as shown in Table (2).

Table 1. Classification of error temperature.

Input variable	Range	Fuzzy set
Change in temperature (e)	−0.4–0.1	NEGATIVE
	−0.1–0.1	SMALL
	0.85–0.2	MEDIUM
	0.1–0.5	LARGE

Table 2. Classification of change in temperature ΔT.

Input variable	Range	Fuzzy set
Error temperature	0–1	SMALL
	0–9	MEDIUM
	1–10	LARGE

Table 3. Status of RAC.

Output variable	Range	Fuzzy set
Status of AC (ON, OFF & MID)	0–0.5	OFF
	0.5–1.0	ON
	0.1–1.0	MID

2.2.4 Output variables

The smart controller having one output variable status of RAC (ON and OFF). The status of RAC is described by three membership functions viz. ON, OFF and MID point as shown in Table (3).

3 RESULT

In this research, The RAC SIMULINK modeling is obtained by mathematical model. To reduce the peak demands, reduce the power consumption and energy saving. We considered a smart residential building having three unit of RAC of different rating viz. 2 kw, 2.5 kw and 3 kw. The results are carried out by using SIMULINK software in MATLAB. Two different cases are studied, Case 1: RACs with a normal thermostat Case 2: RACs with fuzzy logic model. The specifications of RAC are shown in Table (4).

Case 1: Residential Air Conditioning Load with Normal Thermostat Control
In this case, the SIMULINK model of three RACs is run for 24 hrs with normal thermostat with different duty cycle. The Fig. 10 shows the ambient temperature. The Figures (11)–(13) shows the internal room temperature. The power demand and hourly energy consumption is shown in the Figures 14 and 15 respectively.

Case 2: Residential Air Conditioning Load with RACs with fuzzy logic controller
In this case, the same RACs SIMULINK model is run for same time but with smart controller (fuzzy controller). The membership functions, fuzzy rules and surface views are shown in the Figures (5)–(9) respectively. The Figs. (16)–(18) shows the internal room temperature. The power demand and hourly energy consumption is shown in the Figs. (19) and (20) respectively.

Table 4. Specification of RACs.

	Rating of RAC	Thermal gain T_g	Thermal constant τ
RAC-1	2 kw	20	1000
RAC-2	2.5 kw	19	1100
RAC-3	3 kw	21	900

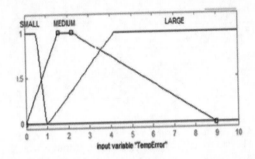

Figure 5. Membership function for error temperature.

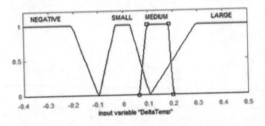

Figure 6. Membership function for change in temperature.

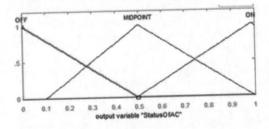

Figure 7. Membership function for status of RAC.

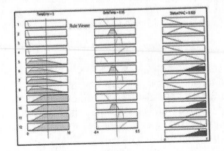

Figure 8. Fuzzy rule base.

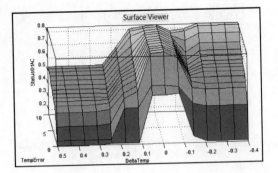

Figure 9. Surface view.

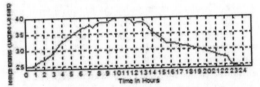

Figure 10. Variation of ambient temperature with time.

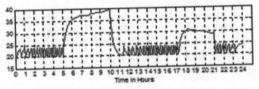

Figure 11. Variation of room temperature with time of AC-1 for case-1.

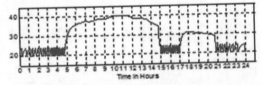

Figure 12. Variation of room temperature with time of AC-2 for case-1.

Figure 13. Variation of room temperature with time of AC-3 for case-1.

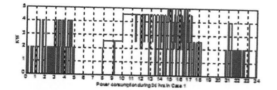

Figure 14.   Variation of power consumption with time in case-1.

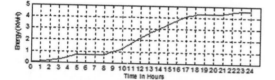

Figure 15.   Variation of energy consumption with time in case-1.

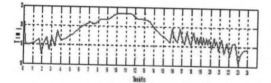

Figure 16.   Variation of room temperature with time of AC-1 for case-2.

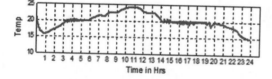

Figure 17.   Variation of room temperature with time of AC-2 for case-2.

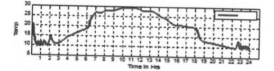

Figure 18.   Variation of room temperature with time of AC-3 for case-2.

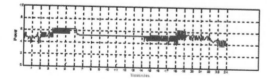

Figure 19.   Variation of power consumption with time in case- 2.

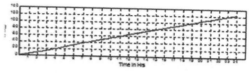

Figure 20.   Variation of energy consumption with time in case-2.

On compare, the power demand and energy consumption in both cases. With smart controller, the RACs consumed less power in 24 hrs over normal thermostat. And also energy is saved in 24 hrs over normal thermostat.

4   CONCLUSIONS

In this paper, the main aim of the research is energy saving and reduced power consumption of Residential Air Conditioning (RAC) with smart controller of RAC using fuzzy logic. Three unit of RAC of different rating and with different set of room temperature are operated by two controller, viz. normal thermostat and smart controller. The Graph Result showed that the power consumption reduced and energy is saved in compared to normal thermostat.

ACKNOWLEDGEMENT

This research paper is made possible through the help and support from College of Engineering Roorkee, Roorkee. We acknowledge to the Administration of College of Engineering Roorkee, Roorkee, Uttarakhand, India for providing the support.

REFERENCES

Aadesh Kumar Arya, Saurabh Chanana and Ashwani Kumar, "Role of Smart Grid to Power System Planning and Operation in India", In Proc. of Int. Conf. on Emerging Trends in Engineering and Technology, GIMT, Kurukshetra, DOI: 03.AETS.2013.3.258, Association of Computer Electronics and Electrical Engineers (ACEEE), U.S.A, p.p., 793–802, 25th–27th October, 2013).

Albadi M. H. and E. F. El-Saadany, "A summary of demand response in electricity market," Electric Power System Research, Vol. 78, No. 11, pp. 1989–1996.

Chen J., F. N. Lee, A. M. Breipohl, and R. Adapa, "Scheduling direct load control to minimize system operational cost," IEEE Trans. on Power Systems, vol. 10, no. 4, pp. 1994–2001, 1995.

King, P.J. and Mamdani, E.H., The Application of Fuzzy Control Systems to Industrial Processes, Automation Vol. 13, No. 3, pp. 235–242, 1977.

Le K., T. Tran-Quoc, J. C. Sabonnadière, C. Kieny, and N. Hadjsaid, "Peak load reduction by using heating regulators: technical impacts of small distributed generation units on LV networks," In *Proc. of the 19th Int. Con. on Electricity Distribution*, Vienna, 2007, pp. 21–24.

Le K., T. Tran-Quoc, J. C. Sabonnadiere, *et al.* "Peak load reduction by using air-conditioning regulators," In *Proc. of the 14th IEEE Mediterranean Electrotechnical Conf.*, Ajaccio, 2008, pp. 713–718.

Lotfi A. (2001), Zadeh. "Fuzzy logic Toolbox".

Lotfi A. Zadeh (1999), "Fuzzy systems", Handbook, Second Edition.

Lu N., D. P. Chassin, and S. E. Widergren, "Modeling Uncertainties in Aggregated Thermostatically Controlled Loads Using a State Queueing Model", *IEEE Transactions on Power Systems*, vol. 20, no. 2, pp. 725–733, May 2005.

Nasution H., H. Jamaluddin, J. M. Syeriff, "Energy analysis for air conditioning system using fuzzy logic controller", TELKOMNIKA, Vol. 9, Issue No. 1, 2011.

Ning Lu and D. P. Chassin, "A State-Queueing Model of Thermostatically Controlled Appliances", *IEEE Transactions on Power Systems*, Vol. 19, Issue 3, pp. 1666–1673, August 2004.

Osman A. H., A. R. Al-Ali, Haroon Khalil, Kiran Prem, Mursalin Haider, and Mishal Eskander, "Distributed Air-Conditioning Energy Management System within the Smart Grid Context", *Journal of Electronic Science and Technology*, Vol. 12, No. 1, March 2014.

Saurabh Chanana, Monika Arora, "Demand Response from Residential Air Conditioning Load Using a Programmable Communication Thermostat", World Academy of Science, Engineering and Technology, *International Journal of Electrical, Electronic Science and Engineering*, Vol. 7, No. 12, 2013.

Communication and Computing Systems – Prasad et al. (Eds)
© 2017 Taylor & Francis Group, London, ISBN 978-1-138-02952-1

Elbow movement classification of a robotic arm using wavelet packet and cubic SVM

Y. Narayan, P. Kumari, Garima & L. Mathew
EE Department, National Institute of Technical Teachers Training and Research, Chandigarh, India

Shallu
ECE Department, National Institute of Technical Teachers Training and Research, Chandigarh, India

ABSTRACT: Now a days, surface Electromyography (sEMG) based robotic devices and vehicles have become a trend for assisting a neuromuscular disordered and amputee person to make their daily life smooth. This offline study focuses only on the de-noising, feature extraction and classification of the sEMG signal with One-Dimensional Wavelet Packet Transform (WPT) with cubic Support Vector Machine (cSVM) classification approach. Five different time domain features were chosen for this study. Selection of Mother Wavelet for this work was random. Wavelet function db2 at level 4 with Shannon entropy cost (logarithm of the squared value of sampled EMG signal) function and un-scale white reduction approaches having tree decomposition have been employed on mother wavelet. In this work, classification accuracy for elbow movement of 94.7% has been achieved with better precision and speed of response.

1   INTRODUCTION

EMG can be defined as the study of muscle functions by means of electrical signals generated from muscular contractions; it may be voluntary or involuntary. EMG signals are based on Motor Unit Action Potentials (MUAPs) in which motor units are building blocks of any neuromuscular system. Analysis of motor units can be used for the study of neuromuscular system or diagnosis of muscles diseases classification. sEMG signals are chosen precisely because of their low amplitude and low Signal-to-Noise Ratio (SNR) and hence become a challenging task in the clinical research areas and industrial applications. The main objective of the study of sEMG signals was to discriminate between various muscle movements into classes by MUAPs during contractions. These signals are detected and recorded using the surface electrodes which are placed adjacent to the skin superimposed on the muscles acupressure point. Signal is generated during the contraction of the electrical activity of muscle fibers. Generated sEMG signals contain some useful feature which can be analyzed through wavelet analysis (Time Frequency Domain (TFD)) for multifunction myoelectric control. Feature extraction is the main criteria which is divided into time domain feature extraction, frequency domain and time-frequency analysis. Feature extraction is done before classification.

A classification technique is meant to distinguish between various movements intention performed by subject. For increasing classification accuracy various classification algorithms can be used. An adaptive certainty-based technique was used which comes under supervised classification scheme (Parsaei & Stashuk, 2012). Another classification technique for EMG was based on the probabilistic neural network in which Log-linearized Gaussian mixture network was used as a classifier (Bu, Okamoto, & Tsuji, 2009). Various other classification methods such as Fuzzy logic (Chan, Yang, Lam, Zhang, & Parker, 2000), (Micera, Sabatini, & Dario, 2001), Autoregressive filters (Graupe, Magnussen, & Beex, 1978), Gaussian mixture matrices (Huang, Englehart, Hudgins, & Chan, 2005), Neural network based (Micera et al., 2001) and nearest-neighbor type approaches have been used. Further, the usage of Linear Discriminant Analysis (LDA) classifier has been seen (Liu & Zhou, 2013), (Li, Schultz, & Kuiken, 2010), and also of the K-Nearest Neighbor (KNN) classifier (Liu & Zhou, 2013). The combination of various time domain and frequency domain features with LDA classifier has also been used (Hartmann, Došen, Amsuess, & Farina, 2015). Classification of sEMG signals is done in 3 steps: (i) Feature extraction, (ii) Dimensionality reduction, (iii) Classification (Kiatpanichagij & Afzulpurkar, 2009). Danies-Bouldin criterion was used for Optimal Wavelet Packet method to classify EMG signals (Wang, Wang, & Chen, 2006). WPT was further classified as Discrete Ordinary Wavelet Packet Transform (DOWPT) and Discrete Harmonic

Wavelet Packet Transform (DHWPT) (Wang, Yan, Hu, Xie, & Wang, 2006).

This paper discusses de-noising, feature extraction and classification of sEMG signal using WPT and cSVM. The paper is divided into four sections. The first section is about introduction which describes work done till date. The second section discusses materials and methods which include data collection from subjects, signal processing and feature extraction by using WPT. The third section covers the results obtained from classifier and the fourth section describes conclusion and future work with scope.

2 MATERIALS AND METHODS

2.1 Data acquisition

Data was acquired from 9 subjects of ages between 23–27 years in National Institute of Technical Teachers Training and Research (NITTTR) laboratory. Data was acquired using four channel recording facility, two channels for Triceps Brachii and remaining two are for Biceps Brachii. Four trials were taken from each subject for each activity, namely Extension and Flexion. The device used for data acquisition is Myotrace 400 with an operating voltage range of 2.6V to 4.2V.

2.2 Signal processing

Raw sEMG possesses non-stationary characteristic which can be used as first objective information. Signal processing techniques was also applied to this raw sEMG signal before feature extraction. sEMG was rectified by full wave rectification facility provided by Myotrace 400 device. The previous study shows that sEMG signal shows random nature, and it cannot be replicated again by its explicit shape. So WPT can effectively be used for such random signal feature extraction. Digital smoothing algorithms and other essential steps were also performed. Data was also normalized in such a way so that it can be rescaled from microvolt to percent of selected reference value. If data was acquired from the left arm of the subject, then we have to nullify the effect of ECG artifacts which are inherently included in the sEMG signal.

2.3 Feature extraction

Feature extraction is done before classification of signal. Feature extraction of the sEMG signal can be done in the Time Domain (TD), Frequency Domain (FD), and Time-Frequency (TFD) domain (wavelet coefficients). Time domain feature can be extracted using wavelet transform. This study focuses on discrete Wavelet Packet Trans-

form (WPT) for de-noising of the myoelectric signal with time domain feature and time scale feature extraction.

Some of time domain features namely Root Mean Square (RMS), Waveform Length (WL), Variance and Slope sign change of EMG were defined mathematically as:

$$RMS = \sqrt{\frac{1}{r}\sum_{r=1}^{R} y_r^2} \qquad (1)$$

$$WL = \sum_{r=1}^{R-1} |y_{r+1} - y_r| \qquad (2)$$

$$VAR = \frac{1}{R-1}\sum_{r=1}^{R} y_r^2 \qquad (3)$$

$$SSC = \sum_{r-2}^{R-1} \left[f\left[(y_r - y_{r-1}) \times (y_r - y_{r+1}) \right] \right] \qquad (4)$$

where r is the number of observation and Y_r is the rth observation of recorded EMG signal.

Data has been acquired from biceps and triceps using Myotrace 400 device and fed into a computer. Further signal processing and all necessary steps were performed on Noraxon (EMG and sensor system). Then the signal was de-noised with the help of wavelet packet. It also extracts time scale feature. Time domain feature and Time scale feature are merged together to form a feature vector that will be input for Cubic SVM classifier. Results of classifier will be used as input signal for controlling external upper limb robotic arm as shown in Figure 1.

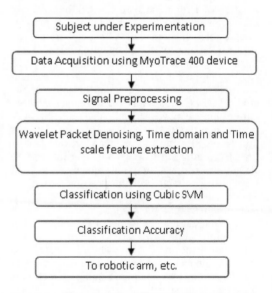

Figure 1. Flow chart for SVM classifier to control a robotic arm.

2.3.1 Wavelet Packet Transform (WPT)

The concept of wavelet packet was brought up by Coifman and Wickerhauser (Hu, Wang, & Ren, 2005). In comparison to time domain analysis, WPT based on time-frequency is much more effectual and efficient (Hu et al., 2005). Decomposition of the signal using wavelet packet depends upon 3 factors: position, scale, and frequency. With the help of wavelet decomposition, wavelet packet tree can be composed in an easy way. Signal can be decomposed into approximation and detail coefficients. Further approximation signals are divided into second level approximation and detail signals. Similarly, detail coefficients are also divided further as shown in Figure 2.

Wavelet packets are built in the form of equations for a sequence of functions as:

$$(U_a(x), a = 0, 1, 2, ...) \tag{5}$$

By

$$U_{2a}(x) = \sqrt{2} \sum_{i=0}^{2A-1} h(i) U_a(2x - i) \tag{6}$$

$$U_{2a+1}(x) = \sqrt{2} \sum_{i=0}^{2A-1} g(i) U_a(2x - i) \tag{7}$$

where $U_0(x) = \varphi(x)$ is scaling function and $U_1(x) = \psi(x)$ is wavelet function.

In WPT, signals were decomposed into subbands of wavelet coefficients with time-frequency resolution. The transformation was carried out by a filter bank composed of two channels, namely a high-pass half-band filter and low-pass half-band filter, and succeeded with down sampling by both. Wavelet Packet method was deprived of translation-invariant feature which goes as due to a slight variation of the signal in time domain, its discrete wavelet packet coefficients change drastically. To gain possession of this property, energy representation was used for wavelet packet method. Optimal Wavelet Packet decomposition was implemented and new feature vectors were constructed (Wang, Wang, et al., 2006). By these feature vectors, the translate-invariance property can be acquired. After which PCA was executed to reduce dimensionality of feature obtained because WPT gives lots of dimensional data. Further clas-

sification was executed with Neural Network (NN) classifier (Wang, Wang, et al., 2006). As WPT coefficients were devoid of time-shift invariance, it could also be taken as noise and effects performance of classifier (Kiatpanichagij & Afzulpurkar, 2009). The upshot of noise in WPT was much less and it yields a lot more useful data than the one acquired from time domain methods in addition to reduced computational time. It gives better outcomes in terms of classification accuracy also.

WPT was further classified as DOWPT and DHWPT (Wang, Yan, et al., 2006). Both of these were deprived of translation-invariant property. DHWT uses numerical implementation consisting of Fast Fourier transform algorithms and disjoint frequency expression. It was applied in various research areas such as signal processing, vibration analysis, and heart beat variance. In feature extraction, energy representation of decomposed signal provides signal features in each subspace. These spaces come with high dimensionality and reduction is a necessity which was further performed by feature selection methods. Dimensionality reduction is preferred to develop strength and design the classifier. A Neural Network (NN) classifier has been used in conjunction with GA to ensure classification performance.

In the biomedical field, signal de-noising and compression should be performed after the data acquisition. Discrete Wavelet Packet Transform (DWPT) can easily fulfill such requirements. DWPT executes adaptive decomposition and with respect to optimization criteria, it builds wavelet tree (L. Brechet, Lucas, Doncarli, & Farina, 2007). In this paper, the optimal basis function for selected mother wavelet (dB7, level 4) was chosen. Mother wavelet was selected such that distortion of the signal was minimized at a given compression ratio. Thereafter optimization was performed and new wavelet coefficients were encrypted by embedded zero-tree algorithm (L. Brechet et al., 2007). This method can also be applied to patterns of EMG and ECG recordings. For EMG signals, fewer efforts were taken in compression as compared to that in ECG signals. In this work, we use the residual of sEMG with time domain feature to form a feature vector for the classifier.

2.4 Classification

SVM technique is used for separation of data into two distinguishable classes, i.e. data of one class was separated from that of other class. SVM approach for multi-class set with conjunction to fast algorithms was also preferred for real-time implementation. SVM is basically used on 3 types of data:

i. Separable data
ii. Non-separable data
iii. Non-linear transformation with Kernels.

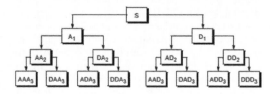

Figure 2. Wavelet packet decomposition of EMG signal.

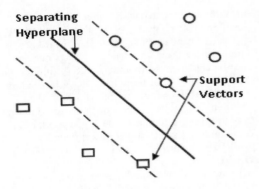

Separating Hyperplane

Support Vectors

Figure 3. Separation of classes using SVM.

For separable data, it was done by means of the best hyperplane which serves as the margin with a maximal width between the two classes as shown in Figure 3. But of all the data points, the ones closest to the separating line or hyperplane are known as support vectors. For non-separable data points, a soft margin was used by SVM to separate as many data points as it can and all the data points could not be divided into classes.

Now, the third type of data classification cannot go on with simple hyperplane separation criteria. A different mathematical approach which was based on the theory of kernel reproduction was applied, and it retains all the basic features of simple SVM hyperplane separation. This approach may be as simple as dot product, to get non-linear classifiers which were hyperspaces in some space say S. Non-linear SVM shows runtime complexity as compared to linear. Non-linear SVM was considered when handling difficult classification work. In general, SVM offers fast training pace.

For binary classification of data sets say $S = \{(p_1, q_1), (p_2, q_2), (p_3, q_3), \ldots, (p_t, q_t)\}$ where q_s is the desired result of respective data point which lies between $\{-1,1\}$, and $p_s \in R^n$ is input data consisting of n attributes and t denotes the number of training sets. The linearly separable hyperplanes are stated as:

$$Z(x) = \sum_{s=1}^{t} v_s x_s + u \qquad (8)$$

where v and u are hyperplane parameters. And for non-linear functions, another function has been defined in terms of kernel function in which a Lagrangian multiplier α_s is defined, and the Kernel function is stated as dot product result of the non-linear functions, as,

$$Z(x) = sign\left(\sum_{s=1}^{t} \propto_s q_s K(x, x_s) + u\right) \qquad (9)$$

$K(x, x_s)$ is the respective kernel function. For Cubic SVM, it can be defined as stated in equation 10 where d could be 3 for cubic SVM.

$$K(x, x_s) = \left(1 + \langle x, x_s \rangle\right)^d \qquad (10)$$

Using SVM involves 3 stages:

i. Training the SVM classifier,
ii. New data to be classified with SVM classifier,
iii. Tuning an SVM classifier.

The important mathematical characteristics of SVM are: (i) Classification of support vectors is used to divide all data points which makes it efficient. (ii) SVM is different from all other methods of pattern recognition like a neural network or nearest neighbor which minimize misclassification errors. SVM works on structural risk minimization. (iii) Even with less no of data points, it works in high dimensional spaces. The basic aim of SVM design is binary classification which may be extended up to multi-class classification. For multiclass, it is done by either combination of many binary classifiers or by taking all data points together, in which optimization task becomes large. It is also cost effective in comparison to the binary problem. A few methods of multi-class classification are, "one-against-all", "one-against-one" and "Directed acyclic graph SVM", and for "Crammer and Singer" method (Hsu & Lin, 2002). For efficiently categorizing aerial images (say on maps), for searching areas from a number of aerial images, or a form of pattern recognition, SVM is an efficient tool with high accuracy for classifying visual characteristics or features in image recognition as in (Zhang, Wang, Hong, Yin, & Li, 2016).

In the present study, Cubic SVM classifier has been used. In comparison to linear SVM which is much easy to perform, Cubic SVM uses cubic kernel functions which are difficult to interpret compared to linear.

3 RESULTS AND DISCUSSION

A classification model for predicting the accurate classes of signal acquired using cubic SVM was proposed in this paper. MyoTrace 400 data acquisition device and Matlab was used for this purpose. For this classification model, the best combination of TD and TFD parameters were derived from the sEMG data. TFD feature was extracted from fourth level approximation coefficient (a_4). EMG signal was de-noised using One-Dimensional Wavelet Packet. Wavelet family selected is dB2 (Daubechies), level 4 and entropy selected was Shannon which is non-normalised. It involves logarithm

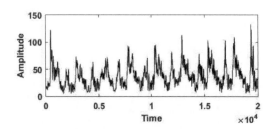

Figure 4.   Raw EMG signal.

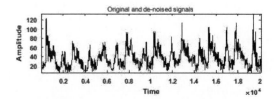

Figure 5.   Denoised EMG signal.

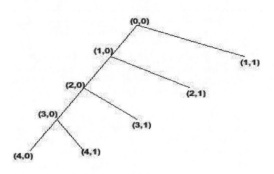

Figure 6.   Best wavelet tree.

of the squared value of a sample of the signal. Figure 4 and Figure 5 shows the original and de-noised signal using wavelet packet method which gives the good white noise rejection capability in a well-defined manner. Figure 6 shows the best decomposition tree using wavelet packet transform. Original signal can be further reconstructed using approximation coefficients and the modified coefficients at level 4. For entropy specified, the optimal wavelet packet tree can be build. Sorted Absolute values of coefficients graph in Figure 7 show a step wise graph for binary classification. Scatter plot classifies different types of data, in which linear or quadratic boundary could not be made, based on colour codes, i.e. different colours and symbols for different group. Scatter plot for Slope Sign Change (SSC) and Standard Deviation (STD) is shown in Figure 8. The Receiver Operating Characteristic (ROC) curve shown in Figure 9 is between true and false positive rate for various threshold values of output. The curve is very near to ideal one, which

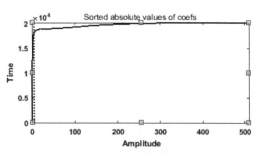

Figure 7.   Sorted absolute value of coefficients.

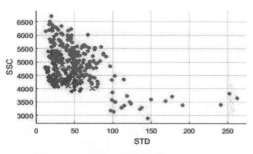

Figure 8.   Scatter plot for SSC & STD.

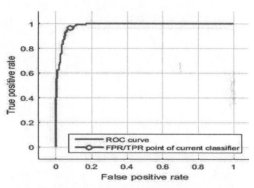

Figure 9.   ROC curve for cubic SVM.

shows good result. The classification accuracy depends on best combinational feature selection, and accuracy increases as the TFD feature was used together with TD features. Scatter plot and ROC curve shows the classification performance of cubic SVM.

4   CONCLUSION AND FUTURE WORK

The method introduced in this paper has a better approach for raw sEMG signal denoising and time scale feature (wavelet coefficient) extraction. Another strong point in this work is good precision and speed of response with an average overall classification accuracy of 94.7%. From the experimental analysis of classifier, it has been observed that training time for SVM classifier was less than 2 seconds with better classification accuracy than another type of SVM. A combination of few time domain (RMS, STD, SSC, WL and Variance) and time scale feature can give better control strategy for various robotic devices. Our future work will be based on a best-selected feature combination (TD, FD, and Wavelet) with Principle Component Analysis (PCA) and optimized SVM classifier with the hybrid series-parallel combination for controlling an exoskeleton robotic arm. It is expected that proposed method in this paper will give good interfacing strategy to an external robotic device with high speed of response that can play a major role in the rehabilitation process of stroke affected persons.

REFERENCES

Brechet L., Lucas, M., Doncarli, C., & Farina, D. (2007). Compression of Biomedical Signals with Mother Wavelet Packet Selection. *IEEE Transactions On Biomedical Engineering, 54*(12), 2186–2192.

Bu, N., Okamoto, M., & Tsuji, T. (2009). A Hybrid Motion Classification Approach for EMG-Based Human—Robot Interfaces Using Bayesian and Neural Networks. *IEEE Transactions On Robotics, 25*(3), 502–511.

Chan, F. H. Y., Yang, Y., Lam, F. K., Zhang, Y., & Parker, P. A. (2000). Fuzzy EMG Classification for Prosthesis Control. *IEEE Transactions on Neural Systems and Rehabilitation Engineering.*

Graupe, D., Magnussen, J., & Beex, A. A. (1978). A Microprocessor System for Multifunctional Control of Upper-Limb Prostheses via Myoelectric Signal Identification. *IEEE Transactions On Automatic Control, 23*(4), 538–544.

Hartmann, C., Došen, S., Amsuess, S., & Farina, D. (2015). Closed-Loop Control of Myoelectric Prostheses with Electrotactile Feedback : Influence of Stimulation Artifact and Blanking. *IEEE Transactions On Neural Systems and Rehabilitation Engineering, 23*(5), 807–816.

Hsu, C.-W., & Lin, C.-J. (2002). A Comparison of Methods for Multiclass Support Vector Machines. *IEEE Transactions On Neural Networks, 13*(2), 415–425.

Hu, X., Wang, Z., & Ren, X. (2005). Classification of surface EMG signal using relative wavelet packet energy. *Computer Methods and Programs in Biomedicine, Elsevier Ireland Ltd., 79*, 189–195. http://doi.org/10.1016/j.cmpb.2005.04.001.

Huang, Y., Englehart, K. B., Hudgins, B., & Chan, A. D. C. (2005). Scheme for Myoelectric Control of Powered Upper Limb Prostheses. *IEEE Transactions On Biomedical Engineering, 52*(11), 1801–1811.

Kiatpanichagij, K., & Afzulpurkar, N. (2009). Biomedical Signal Processing and Control Use of supervised discretization with PCA in wavelet packet transformation-based surface electromyogram classification. *Biomedical Signal Processing and Control, Elsevier Ltd., 4*, 127–138. http://doi.org/10.1016/j.bspc.2009.02.004.

Li, G., Schultz, A. E., & Kuiken, T. A. (2010). Quantifying Pattern Recognition—Based Myoelectric Control of Multifunctional Transradial Prostheses. *IEEE Transactions On Neural Systems and Rehabilitation Engineering, 18*(2), 185–192.

Liu, J., & Zhou, P. (2013). A Novel Myoelectric Pattern Recognition Strategy for Hand Function Restoration After Incomplete Cervical Spinal Cord Injury. *IEEE Transactions On Neural Systems and Rehabilitation Engineering, 21*(1), 96–103.

Micera, S., Sabatini, A. M., & Dario, P. (2001). On automatic identification of upper-limb movements using small-sized training sets of EMG signals. *Medical Engineering & Physics, Elsevier Science Ltd., 22* (2000), 527–533.

Parsaei, H., & Stashuk, D. W. (2012). EMG Signal Decomposition Using Motor Unit Potential Train Validity. *IEEE*, (c).

Wang, G., Wang, Z., & Chen, W. (2006). Classification of surface EMG signals using optimal wavelet packet method based on Davies-Bouldin criterion. *Med. Bio. Eng. Comput., 44*, 865–872. http://doi.org/10.1007/s11517-006-0100-y.

Wang, G., Yan, Z., Hu, X., Xie, H., & Wang, Z. (2006). Classification of surface EMG signals using harmonic. *Physiol. Meas., IOP Publishing Ltd, 1255*, 1255–1267. http://doi.org/10.1088/0967-3334/27/12/001.

Zhang, L., Wang, M., Hong, R., Yin, B., & Li, X. (2016). Large-Scale Aerial Image Categorization Using a Multitask Topological Codebook. *IEEE Transactions On Cybernetics, 46*(2), 535–545.

Communication and Computing Systems – Prasad et al. (Eds)
© 2017 Taylor & Francis Group, London, ISBN 978-1-138-02952-1

Design and analysis of low power, high speed 8 × 8 vedic multiplier using transmission gate

Nitin Dhaka, Mohit Kumar, Rohit Singh Toliya & Ashutosh Pranav
Department of Electronics and Communication Engineering, Graphic Era University, Dehradun, India

ABSTRACT: This paper presents designing and analysis of 8 × 8 Vedic multiplier with the help of transmission gate. The design is implemented on Cadence Virtuoso 6.1.5, 180 nm process technology. Designing of digital circuit is based on transmission gate logic, which provides minimized delay and optimized power consumption. The multiplier architecture is designed by vedic mathematics. Vedic mathematics is a primeval Indian mathematics which has powerful rules to reduce complex calculation. The circuit analysis is done in terms of performance parameters: propagation delay and power consumption. After critically analyzing the design, propagation delay of 39.96 ps and power consumption of 2.92 mW is achieved.

1 INTRODUCTION

In current era high speed, low power consumption and reduced hardware for any system designing is the main area of concern. In signal processing and other applications, system has to perform various arithmetical and logical operations. ALU [2] is used to perform these operations for such systems. For reliable operations, ALU should have high speed blocks with low power consumption. The speed of ALU mainly depends on the performance of multiplier block. Multiplier block is the main source of power consumption and delay generation. Vedic technique is an efficient way for improving the performance of multiplier block. Vedic mathematics was rediscovered from Vedas by swami Bharati Krishna Tirthaji maharaja. Vedic mathematics is mathematics of Vedas, it has 16 principles and sub principles for various arithmetical operations [4]. The Multiplication operation follows the principle of Urdhava Tiryakbhyam sutra. This principle include two words, each having different meaning. The meaning of Urdhava is vertical and of Tiryakbhyam is crosswise, so this method follows two continuous operations, vertical and crosswise multiplication operation. All the performance parameter depends upon the efficiency of algorithm as well as design technique. In the proposed design all the basic digital blocks are designed through transmission gate logic instead of conventional one. Transmission gate has the advantage that it reduces hardware as well as delay time. The proposed architecture has the advantages of an efficient algorithm as well as design approach. So the proposed architecture gives very good result in terms of power consumption and propagation delay.

1.1 Vedic technique for multiplication

The proposed architecture follows Urdhav Tiryakbhyam sutra. First it performs vertical multiplication operation and then crosswise multiplication operation. The advantage of vedic mathematics is that large architectures uses the organized repetition pattern of small architectures. An illustration of vedic Mathematics for multiplication is given in the following steps:

In the Figure 1 vedic multiplication technique for 2 × 2 multiplier is given. Here a1a0 and b1b0 are two input lines and S3S2S1S0 is the 4 bit output lines. The values of output lines are defined as:

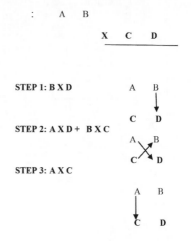

$S0 = a0b0$, $S1 = b0a1 + b1a0$, $S2 = b1a1 + c1$,
$S3 =$ Final carry of half adder

In the design we have used 2 half adder and 4 AND gates. AND gates are used to combined the input lines, which works as primary inputs for half adder.

2 SIMULATION SETUP

Cadence Virtuoso 6.1.5 is used as a simulation tool for simulating vedic multiplier. Advanced features like Cadence Virtuoso Schematic Editor for providing high speed and easy design methods, Cadence Virtuoso Layout Suite for speeding up the physical layout of the design, Cadence Virtuoso Visualization and Analysis for efficiently analyzing the performance of the design and Cadence Assura Physical Verification for reducing overall verification time, because a quick and instinctive debug capability is incorporated within the Virtuoso custom design environment. It easily compares, repair, remove and distinguish errors.

All the transistors used in the designing of proposed design has a channel length of 180 nm and width of 2 μm with finger width of 1.

3 PROPOSED ARCHITECTURE

The hardware architecture of 4×4 and 8×8 high speed multiplier is sown in Figure 2 and Figure 4. In Vedic multiplication technique partial product generation and addition are the two parallel processes that are followed. As a result the speed of vedic multiplier is much greater than any other multiplier. Four 4×4 vedic multiplier are used to design a single 8×8 multiplier and three 8 bit ripple carry adder, same as 4×4 vedic multiplier uses four 2×2 multiplier and three ripple carry adder.

In 4×4 multiplier, b3b2b1b0 and a3a2a1a0 are the two 4 bit input lines and s7 s6 s5 s4 s3 s2 s1 s0

are the output lines. Input a1a0 and b1b0 are provided to the first multiplier, a1a0 and b3b2 to the second, a3a2 and b1b0 to the third and a3a2 and b3b2 to the forth multiplier. Outputs from multiplier 2 and 3 are given as an input to the first ripple carry adder and output from the first ripple carry adder along with 2 MSB output bits from the first multiplier are given to the second ripple carry adder. Finally output of multiplier 4 and second ripple carry adder, along with carry output from the adder 1, serves as an input to the third ripple carry adder. The carry outputs of second and third adders are unused. And so, output s0-s7 will be obtained.

A 4×4 ripple carry adder, shown in Figure 3, consists of three full adders and a single half adder. Inputs a0b0 are given to the half adder which provides a sum output s0 and a carry c0. The carry c0 then goes to the first full adder along with inputs a1b1 thus giving second LSB s1 and carry c1 as the output. Similarly operations with second and third full adder are performed. Finally output s1 s2 s3 s4 are obtained along with final carry c3.

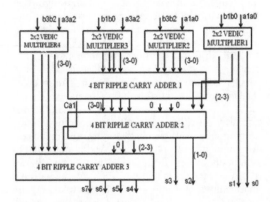

Figure 2. 4×4 vedic multiplier.

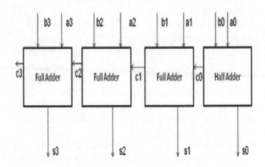

Figure 3. 4 bit ripple carry adder.

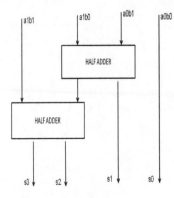

Figure 1. 2×2 vedic multiplier.

4 OPTIMIZED LOGIC DESIGNING USING TRANSMISSION GATE LOGIC

Transmission gate works as a switch or relay. In digital circuit design it uses less number of transistors and provides minimum delay. From delay point of view, transmission gate is very much efficient and due to its optimized designing, it consumes less power as compare to the CMOS logic circuit. Fig. 5 shows AND gate logic with the help of transmission gate logic design.

Similarly Figure 6 shows the logic of OR gate with the help of transmission gate.

Similarly, full adder can be made with the help of two half adder and one OR gate which are formed with the transmission gate.

The transmission gate based half adder and full adder will be used to make a 2×2 vedic multiplier and a 4 bit ripple carry adder. With the help of

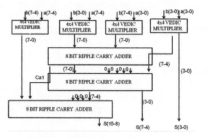

Figure 4. 8×8 vedic multiplier.

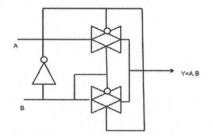

Figure 5.   AND gate using transmission gate logic.

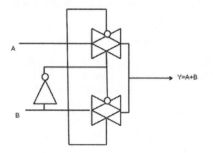

Figure 6.   OR gate logic using transmission gate logic.

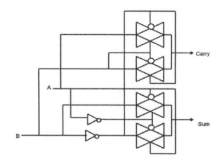

Figure 7.   Half adder using transmission gate logic.

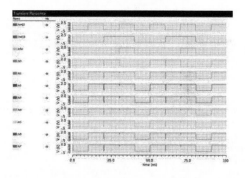

Figure 8.   Transient response of 8 bit ripple carry adder.

these two circuits 4×4 vedic multiplier will be made and it will be completely based on transmission gate logic. Thus, 8×8 vedic multiplier will be created with the help of only transmission gates.

5 RESULTS AND DISCUSSION

The simulated results of 8×8 vedic multiplier which was made through transmission gate are obtained using Cadence Virtuoso 6.5.1.

The transient response of 8 bit ripple carry adder is shown in Figure 8. Net 22 and net 19 are two inputs for 8 bit input lines a (7:0) and b (7:0). Cfnl shows the final carry output and S (7:0) is the output line.

The transient response of 4-bit vedic multiplier is shown in Fig. 9. Net 20 and net 17 are two inputs for 4 bit input lines a (3:0) and b (3:0). P (7:0) is the output lines which show the multiplication result.

Likewise, the transient response of 8 bit vedic multiplier is shown in Fig. 10. Net 22 and net 19 are two inputs for 8 bit input lines a (7:0) and b (7:0). P (15:0) is the output line which shows the multiplication result.

Table 1 shows the comparison between propagation delay of different types of multipliers. The best performance is shown by 8 bit transmission gate

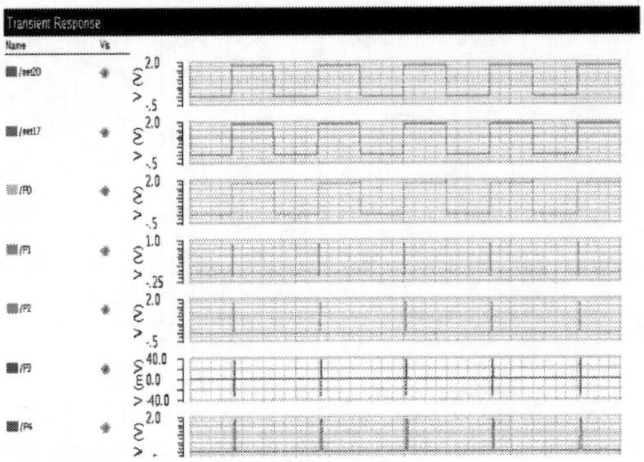

Figure 9. Transient response of 4 bit vedic multiplier.

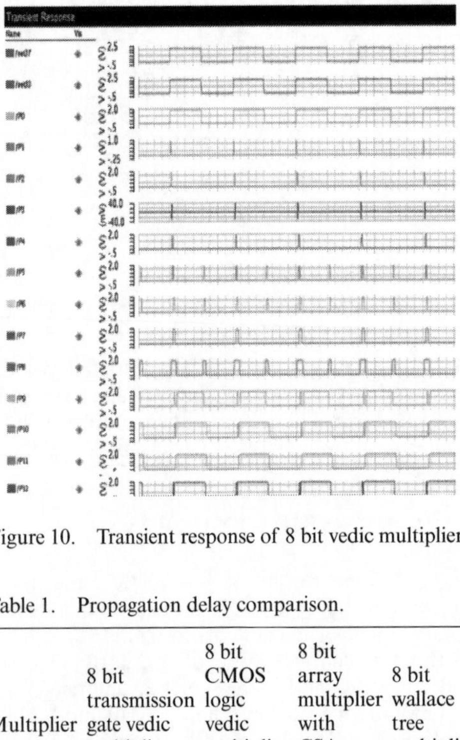

Figure 10. Transient response of 8 bit vedic multiplier.

Table 1. Propagation delay comparison.

Multiplier type	8 bit transmission gate vedic multiplier	8 bit CMOS logic vedic multiplier	8 bit array multiplier with CSA	8 bit wallace tree multiplier
Total delay	39.96 ps	17.430 ns	17.533 ns	15.969 ns

Table 2. Power consumption comparison.

Multiplier type	8 bit transmission gate vedic multiplier	16 bit CMOS logic vedic multiplier	Booth radix-4 multiplier
Power consumption	2.92 mW	29.34 mW	151.34 mW

vedic multiplier with a propagation delay of 39.96 ps. The simulation is operated with v pulse of 0 to 1.8 volt as input and 1.8 volt voltage supply is provided.

Table 2 shows the power consumption of different multipliers with vedic multiplier showing the minimum power consumption of 2.92 mW.

6 CONCLUSION

The vedic multiplier for 8×8 multiplication is designed on cadence virtuoso 6.1.5, 180 nm process technology. Proposed architecture reduces a large power consumption and propagation delay as compared to other multiplication algorithm. Transmission gate logic design uses less number of steps, so it has less computational complexity. Proposed architecture has power consumption of 2.92 mW and propagation delay of 39.96 ps. It has very efficient in terms of power consumption and propagation delay, so it can be used in high speed DSP processors, portable calculating devices which require very small power consumption.

REFERENCES

Anand R. K., Singh K. Verma P. & Thakur A., Design of area and power efficient half adder using transmission gate, *International journal of Research in Engineering and Technology*, vol. 4, no. 4, April 2015.

Bathija R. K., Meena R. S., Sarkar S. & Tinjrit R. S., Low power high speed 16×16 bit multiplier using vedic mathematics, *International Journal of computer Application*, vol. 59, no. 6, December 2012.

Cadence analog and mixed signal labs, Revision 2.0, IC615, Assura 410, Incisive unified simulator 102, Cadence design simulator, Bangalore.

Kang S. M. & Leblibici Y., *CMOS digital integrated circuits: analysis and design*, Tata McGraw Hill, 2003.

Kumar G. G. & Charishma V., Design of high speed vedic multiplier using vedic mathematics techniques, *International Journal of Scientific and Research Publications*, vol. 2, no. 3, March 2012.

Nautiyal P., Madduri P. & Negi S., Implementation of an ALU using modified carry select adder for low power and area-efficient application, *International Conference on Computer and Computational Sciences*, pp. 22–25, 2015.

Nicholas A. P., Williams K. R. & Pickles J., Application of urdhava sutra, *Spiritual Study Group*, Roorkee (India), 1984.

Premananda B. S., Samarth S. P., Shashank B. & Shashank S. B., Design and implementation of 8-bit vedic multiplier, *International Journal of Advanced Research in Electrical, Electronics and Instrumentation engineering*, vol. 2, no. 12, December 2013.

Swami Bharati Krishna Tirthaji Maharaja, vedic mathematics, *Motilal Banarsidass Publishers*, 1965.

Wallace C. S., A suggestion for a fast multiplier, *IEEE Transactions on Electronic. Computers*, vol. 13, no. 1, pp. 14–17, Feb. 1964.

Communication and Computing Systems – Prasad et al. (Eds)
© 2017 Taylor & Francis Group, London, ISBN 978-1-138-02952-1

Design and implementation of a low power and area efficient sequential multiplier

Mohit Kumar, Nitin Dhaka, Rohit Singh Toliya & S.C. Yadav
Department of Electronics and Communication Engineering, Graphic Era University, Dehradun, India

ABSTRACT: A multiplier plays a major role in various digital systems. The area, speed and power consumption of any digital system, which has a multiplier as its component, depend upon the hardware used for the multiplier. In this paper we proposed a sequential multiplier which has a lower area requirement and lower power consumption in comparison of the conventional sequential multiplier. We can use the proposed system where long battery life is required and/or reduced hardware is required. The speed of the proposed multiplier is a little bit slower than the conventional one. So we can use the proposed system where long battery life is required and speed is not the major requirement.

1 INTRODUCTION

Nowadays more and more complex circuits are implemented on a VLSI chip because the scale of integration keeps increasing continuously. These complex circuits need large signal processing systems, hence require significant amount of power. The two major design tools in VLSI system design are area and power consumption. Power expenditure has become a important anxiety in today's VLSI scheme design. The requirement of low power VLSI design arises from two key factors. First one is the stable growth of working frequency and dealing out capability per chip, huge currents have to be delivered and the warmth due to large power consumption must be detached by correct cooling methods. Second one is that the battery life in moveable electronic devices is narrow. Low power design is used to increase the battery backup in these handy devices. Multiplication is an necessary operation in most signal processing operations. Multipliers have bulky area, long intermission and devour substantial amount of power. So low power multiplier design has an vital role in low power VLSI system design. There has been far-reaching effort on low power multiplier design at different levels i.e. technology level, circuit level, physical level, logic levels etc. A multiplier is usually the majority area consuming part of a system. Hence by optimizing the area and power of a multiplier, we can optimize the whole system. This is a key design issue. Region and speed are typically incompatible constraints so that civilizing speed results mostly in bigger areas.

In chapter II the shift and add method is discussed since this method is the base for sequential multiplier design. In chapter III the conventional sequential multiplier is discussed. In chapter IV the proposed sequential multiplier with reduced area and power consumption is explained.

2 SHIFT-AND-ADD METHOD FOR MULTIPLICATION

In scheming of multipliers there is forever a negotiation to be prepared between the speed and the area. One of the simplest multiplication methods is shift-and-add method. Shift-and-add method is sluggish but proficient in use of hardware. This method can be justified by considering guidebook binary multiplication.

To understand the hardware implementation of this method we make some changes to this method. First, as an alternative of having our inspection tip from one bit to another of the operand, we put this operand in a shift register.

Second, as an alternative of lettering one partial product and the subsequently one to its left, when we write a incomplete product, we shift it to the right as we are inscription it and the after that one would not to be placed to the next place.

Lastly, as an alternative of scheming all partial product and adding them up at the conclusion, when a half-done result is intended, we include it to the preceding fractional product and write the recently intended result as the new fractional product.

All these three changes are describes in the example shown in Figure 1 i.e. hardware oriented multiplication process. If the bit being observed of the operand is 0 then there is no need to add

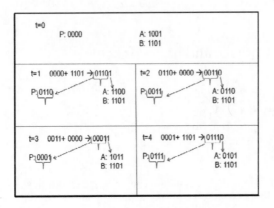

Figure 1. Hardware oriented multiplication process.

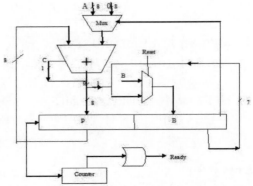

Figure 2. Conventional sequential multiplier.

anything in the previously calculated partial result and only shifting is sufficient. This process of shifting without adding anything is known as "shift". And if the bit being observed of one operand is 1 then another operand is to be summed to the beforehand calculated fractional product and the intended new fractional result should be placed to the right by one place. This process is call "add-and-shift".

3 CONVENTIONAL SEQUENTIAL MULTIPLIER

The architecture of a conventional sequential multiplier by using shift and add method is exposed in Figure 2. There are various key sources of switching activities in this multiplier i.e. shift registers, counter, adder, switching in multiplexer and shifting in the partial product register as shown in Figure 2.

By reducing the switching activity as mentioned above we can design the low power architecture for sequential multiplier. We can reduce these switching activities by using the proposed sequential multiplier architecture.

4 PROPOSED ARCHITECTURE OF SEQUENTIAL MULTIPLIER

If we talk about large digital systems then we have to talk about their data parts and control parts also. To design any large digital system first we have to clarify the design of its datapath and control unit.

The data parts consist of registers, multiplexers, adders and buses interconnecting them. The control unit is a state machine which controls the sequence of operation of the data parts. As shown

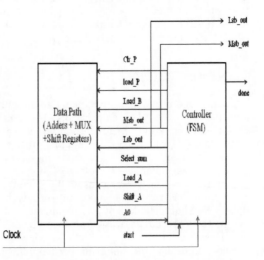

Figure 3. Data and control parts of the proposed architecture.

in Figure 3, the data parts and the control unit both are working on the similar clock pulse. On each increasing edge the regulator (controller) goes into a new state and issues various control signals and the parts of the data path starts reacting according to these signals. The instance for the data parts to complete their individual task is equal to the time from one increasing edge to next increasing frame of the clock pulse. Standards that are applied at the primary inputs of the datapath registers are clocked into these registers at each increasing edging of the clock pulse.

Now let us discuss the functioning of the proposed architecture of the sequential multiplier. As shown in the Figure 4 registers P and B are the 8-bit registers. A multiplexer, an adder and a tri-state buffer amount to the further machinery of the datpath of the proposed architecture.

In the above figure the signals which are shown in bold are the control signals from the control unit. These manage signals manage the register clock, bus coursework and logic element output selection. To load multiplier and multiplicand into the registers A and B, the input databus is used. This data bus is a bidirectional bus and is motivated by the production of P through an authentic tri-state buffer and by the tri-state response of A.

The output of P and B are goes into an adder (8-bit) for the addition of the fractional outcome in P to the B. The production of the adder i.e. P+B, goes to the single face of the multiplexer and the further face of the multiplexer is determined by the P response i.e. P+O. The sel_sum manage input decides which input will pass to the output of the multiplexer i.e. P+B or P+0 as shown in Figure 4. The AND gate selects the carry output from adder or 0 according to the sel_sum manage input. This assessment is concatenated to the immediate left of the multiplexer response and figure a vector of length 9-bits. Following this the least significant bit of this vector come apart and go into the sequential input of the shift register that contain A and further 8-bits goes to the P register. This concatenation of the AND gate production to the left of the multiplexer production and aparting the

right bit from this 9-bit vector produce a shifted outcome that is clocked into P. And finally we get the result in P and A. Msb 4-bits in register P and lsb 4-bits in register A.

5 RESULT AND ANALYSIS

After thoughtful inspection of the architectures of both conservative and proposed architecture of sequential multiplier, the subsequent step was to execute it. In order to achieve this we put pen to paper to write a code in Verilog. This code was synthesized by using Xilinx 14.7 and simulated using Modelsim Simulator. Area utilization, power analysis report and simulation results are shown below.

Table 1. Area utilization of conventional and proposed architecture.

Parameters	Conventional [1]	Proposed
No of slices	68	36
No of 4-input LUTs	179	55
No of bonded inputs	35	13

Table 2. Power analysis of conventional and proposed architecture.

Parameter	Conventional [1]	Proposed
Static Power (mW)	41.70	31.52
Dynamic Power (mW)	8.35	1.30
Total power (mW)	50.12	32.83

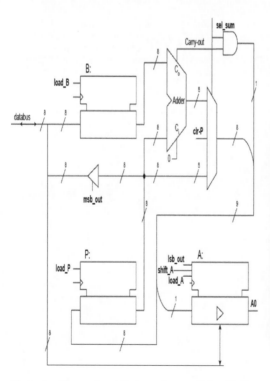

Figure 4. Proposed architecture of sequential multiplier.

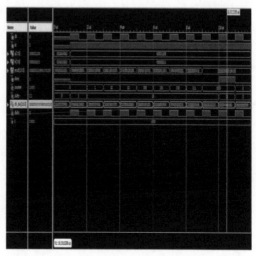

Figure 5. Simulation of 8-bit conventional multiplier.

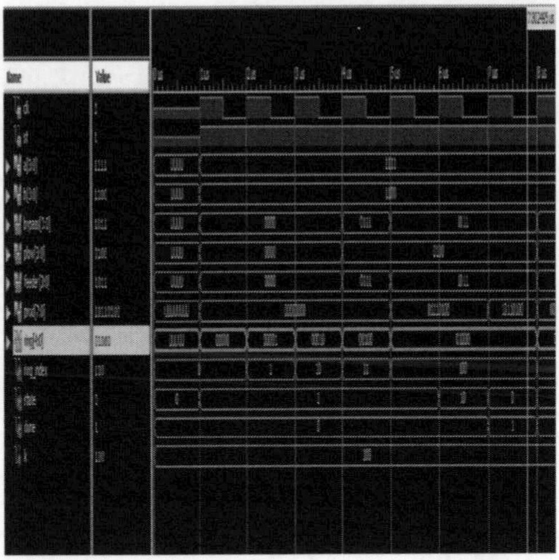

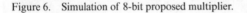

Figure 6.   Simulation of 8-bit proposed multiplier.

6   CONCLUSION

In this paper the architecture for low power and low area sequential multiplier was proposed. Some modifications are done to the conventional multiplier to reduce the switching activities so that the power consumption can be reduced. The results show that there is a reduction of 34% in power consumption of the proposed architecture. We also analyze the area utilization of conventional and proposed sequential multiplier. We can use the proposed architecture in those applications where low power and/or low area is required.

REFERENCES

Chandrakasan A. and R. Brodersen, "Low-Power CMOS Digital Design," IEEE J. Solid-State Circ., vol. 27, no. 4, pp. 473–484, April 1992.

Chen O.T., S. Wang, and Yi-Wen Wu, "Minimization of Switching Activities of Partial Products for Designing Low-Power Multipliers" IEEE Transactions on VLSI Systems, vol. 11, pp. 418–433, June 2003.

Jayaprakash M., M. Peer Mohamed and A. Shanmugam, "Design and Analysis of Low Power and Area Efficient Multiplier," *International Journal of Electrical, Electronics and Mechanical Controls*, vol. 3, no. 1, Jan. 2014.

Marimuthu C. N., Dr. P. Thangaraj and Aswathy Ramesan, "Low Power Shift and Add Multiplier Design" Int. Journal of Computer Science and Info Tech, vol. 2, no. 3, pp. 12–22, June 2010.

Mottaghi-Dastjerdi M., A. Afzali-Kusha and M. Pedram "BZ-FAD: A Low Power Low Area Multiplier Based on Shift and Add Architecture" IEEE Trans. Very Large Scale Integration (VLSI) Syst., vol. 17, no. 2, pp. 302–306, Feb. 2009.

Moshnyaga V. G. and K. Tamaru, "A Comparative Study of Switching Activity Reduction Techniques for Design of Low-Power Multipliers" in Proc. IEEE Int. Symp. Circuits Syst., pp. 1560–1563, April 1995.

Nan-Ying Shen and Oscal T.-C. Chen, "Low-Power Multipliers by Minimizing Switching Activities of Partial Products," in Proc. IEEE International Symposium on Circuits and Systems, vol. 4, pp. 93–96, May 2002.

Prof Prasann, D. Kulkarni, S. P. Deshpandey and G. R. Udupi, "Low Power Add and Shift Multiplier Design BZFAD Architecture," *International Journal of Computer and Electronics Research*, vol. 2, no. 2, pp. 94–107, April 2013.

Communication and Computing Systems – Prasad et al. (Eds)
© 2017 Taylor & Francis Group, London, ISBN 978-1-138-02952-1

Boosters implementation over SCADA architecture

Premkumar Ch, Ravi Prakash, Anuj Awasthi & Sagar Uppal
CIT Department, University of Petroleum and Energy Studies, Dehradun, India

ABSTRACT: Past decades were focused on industrialization. Which lead towards implanting huge machinery at the factories to enhance production rate of goods coming out of industry, which in turn will increase the profit of organization. As the industries expanded their complexity increased and hence their working strategies, which lead to demand of more manpower to operate such system, to overcome such problems SCADA systems were developed which not only helped in reducing manpower demand but also made it simpler to deal with the remote operational units from a centralized control station. The main focus was to develop techniques which help them not only to produce high quality goods but also to hike their profit rates, as the technology advances there is rigorous shift from one technique to other which are surely more efficient and secure than previous ones. Among several advancements in industry, there are some which deals with the remote handling of operational units which is done using SCADA in the industry. Traditional SCADA architecture is being followed over a long period of time, which is obsolete these days. Keeping in mind the computing power of present devices the processing power and response time can be enhanced using Boosters along with traditional architecture. To enhance these systems we can introduce boosters in the traditional architecture for faster and efficient processing of information among the units, we can use boosters along with sensors to monitor their working and to guide them with instructions to act in particular conditions. It will also help in preventing accumulation of faulty data as the working state of sensors is also checked frequently, resulting in enhanced level of precision in units operation, which will eventually increase the profit of organization.

1   INTRODUCTION

The term SCADA stands for 'Supervisory Control and Data Acquisition'. The major use of SCADA is in gathering data from remotely isolated devices such as valves, pumps, transmitters etc. (SCADA Systems 2012) any set of programs that gather data about a system in order to control that system is an SCADA application. Typically, SCADA systems are used to look after those industrial processes where human usage is impractical. i.e. Systems which mainly comprises those dominating factors which require rapid response for proper handling, which can only be achieved if automated systems are used in Industry. (Fresno 2013) But still, human resource was required to keep a check on the sensors which were installed at operational units to monitor their working status. But still SCADA architecture can be implemented to handle any kind of units.

SCADA and Process Control Systems being used by companies were developed many years before even the concepts like mobile computing and distributed computing were a common part of Business Operations. As a result, measures against network security were not anticipated in these systems. (Assessment and Remediation of Vulnerabilities 2005) Now there is urgent need to optimize the traditional architecture as per the demands of present trends, which can be done by introducing Boosters in the architecture.

Boosters are an extra set of control unit introduce in architecture to reduce the complexity in monitoring process. Here boosters act as clusters for sensors and monitors their working and provide them with instructions to operate and control industrial units. Programming of boosters is done at the control end; the response of booster can be manipulated as per the condition demands. They are used to control the functioning of sensors and also to keep an eye on their proper functioning, so that not even a single chunk of faulty data is produced by sensors.

These boosters will then report back to their main control centers with the data and the observations which are accumulated by sensors during the plant operations, basically boosters are additional set of locally installed clusters for sensors. Through these we can monitor sensors around a local area and perform the required maintenance operations for efficient working of plant units.

2 RELATED WORK

The main problem which arises in handling an industry is to monitor various units and their operating state. To overcome these issues many enhancements in the standard operation procedure were introduced like SCADA systems

All the enhancements benefited the industry in a healthy manner, but there is always need for up gradation as per the present trends.

The advantage and convenience that these systems provide in managing plant units have contributed a lot in the advancement of industrial operations. Their robustness and stability enable us to use ICS and SCADA solutions as a tool to enhance security over long periods of time—often in excess of 10 and sometimes 20 years and beyond (SCADA Networks 2015).

Now Cyber-attacks on Industrial Control Systems and critical manufacturing infrastructures is a reality. Power generation facilities, nuclear facilities, country's secret files and factories have become targets of attackers and have been hit recently with an array of Network breaches; Data theft and Denial of service are common these days, so the architecture must be secure enough to protect industry and its crucial data against these attacks (FORTINET).

The traditional SADA system contains set of sensors that accumulate data and report back to units. People have to keep an eye on working of sensors in case they are showing any irregular behavior and maintainers are informed the same.

But identifying those abnormalities may take time and by then it may have pose negative impacts. Introduction of boosters will speed up those responses and will help to prevent any repercussions. The term IOT is yet another buzzword being used to implicate the concepts of SCADA over an extensive environment, despite in IOT various electronics devices are connected over network which in turn communicate with each other and there is flow of data between devices for efficient responses (http://internetofthingsagenda.techtarget.com/definition/Internet-of-Things-IoT).

3 SYSTEM MODEL

Our system implements Boosters which act as clusters for our set of nodes (sensors), each cluster separately monitors a set of sensors. These clusters act as node in the network.

And these nodes are interlinked together to form our network structure.

These clusters act as node which is interlinked together to form our network structure.

These clusters are interconnected among each other through FC-AL loop (Fibre Channel Arbitrated Loop) (http://searchstorage.techtarget.com/definition/Fibre-Channel-Arbitrated-Loop-FC-AL), which allows faster and reliable sharing of data among these clusters and a separate copy of data is maintained at other clusters so that if a Booster is down than other we can get access to its data through other boosters, i.e. multiple copies of data are maintained to deal with the failure conditions. This helps in safe guarding our data and maintains its integrity.

These boosters (clusters) are connected to Base Station through Fabric channel (http://searchstorage.techtarget.com/definition/Fibre-Channel), Fibre Channel is used for transmitting data between end devices at higher speeds up to 4 Gbps which ensures reliability and faster exchange of data. Fibre Channel has begun to replace the Small Computer System Interface (SCSI) as the transmission protocol between servers and clustered storage devices.

P-P (peer to peer) connectivity between base station and a cluster through a switch which ensures data security among peers. In a P2P network, the "peers" are systems connected over a network which enables them to share information among each other. Files and data can be shared directly among systems on the network without the consent of a central server. In other words, each system on a P2P network acts as a client as well as a file server.

Our system not only ensures data reliability, resource optimization, but also keeps in mind fac-

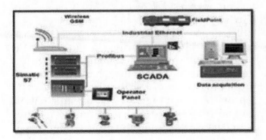

Figure 1.   Traditional SCADA architecture.

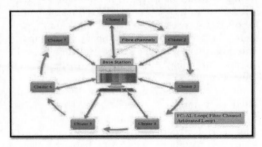

Figure 2.   Connectivity between Base Station and Clusters (boosters).

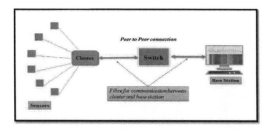

Figure 3. Connectivity between Sensors, Cluster and Base Station.

tors like data security and need for faster access time. Implementing such systems in the industry will reduce the operation time, since time is most delicate recourse for an industry, optimizing that will give industry better outcomes and eventually increase in company profits

4 PROBLEM DEFINITION

The industrial plants implant a complex process at commercial level so the close monitoring of operational units is an essential part to ensure proper working of plant. The failure of even one unit may lead to severe consequences, so the response to abnormalities in working strategies should be rapid. To deal with such type of problems Boosters are the best option to deal with such conditions and generate rapid responses to overcome these conditions. This will also help in reducing down time of plant as higher the down time more loss is faced by industry.

5 CONCLUSION

Today's fast emerging world experiences a new technology every day, these techniques are being developed to reduce human effort and also increase the throughput of overall process. It's very important to understand the implications of these technologies and how it effects lives. By introducing boosters in the SCADA system will not only reduce the complexity which arises in the traditional system but also provide more efficient outcome over traditional methods. The main roots of this paper focuses on enhancing the existing system and make it more secure and easy to maintain and carry out maintenance operations.

REFERENCES

Assessment and Remediation of Vulnerabilities, 2005 Internet Security Systems.
FORTINET – SECURING SCADA INFRA- http://internetofthingsagenda.techtarget.com/definition/Internet-of-Things-IoT
http://searchstorage.techtarget.com/definition/Fibre-Channel:http://techterms.com/definition/p2p
http://searchstorage.techtarget.com/definition/Fibre-Channel-Arbitrated-Loop-FC-AL
Protecting Industrial Control Systems and SCADA Networks, May 1, 2015 Check Point Software Technologies Ltd.
SCADA Systems, March 2012 / White paper by Schneider Electric.
SCADA Tutorial, DPS Telecom, 4955 East Yale Avenue, Fresno, CA 93727 2013 STRUCTURE.

Communication and Computing Systems – Prasad et al. (Eds)
© 2017 Taylor & Francis Group, London, ISBN 978-1-138-02952-1

Reduction of power dissipation in sequential circuit using domino logic

Pragya Srivastava, Priyanka Goyal & Aparna Gangwar
School of ICT, Gautam Buddha University, Greater Noida, India

ABSTRACT: In this paper a low power sequential circuit is designed using domino logic. Since domino logic is one of the power reduction technique to design combinational and sequential circuits. Hybrid latch is designed using domino logic circuitry. The various results of simulation and designs based on with CMOS 180 nm technology on Cadence virtuoso tool is shown in paper. The average power is decreased very much. The layout area is also reduced. Almost 20% of power was saved using this technique as compared to design without domino logic.

1 INTRODUCTION

Flip-flops in digital design are basic and important storage element which used in all digital designs. They can store one bit of information at a time. The difference between latches and flip flop is that latches gives output when enabling signal is asserted on them and flip-flop gives output when clock, is either on rising or on falling edge.

In latches content changes when signal enabled. In flip-flop the content remain unchanged after rising and falling edge. The four types of flip-flops are SR, JK, D, T. All of them have different operations and performances.

Flip-flop performance is measured by three parameters propagation delay, setup time and hold time. Relationship between clock and input data is known as setup time and hold time. Setup time and hold time are basically defines the timing requirements based on input with respect to clock i.e. time taken to sample input and to propagate to output.

Factors that affect the operation of flip-flops are high speed operations, robustness, noise stability, low power consumption, small area, fewer number of transistors, supply voltage scalability, and less internal activity when data activity is low.

Propagation delay (Clock-to-Output) is the time delay is considered to be stable when clock is at active or rising edge of clock.

Propagation delay differs for low-high and high-low transitions. So propagation delay of the flip-flop is by definition maximum value of these two delays in order to function correctly, the edge-triggered flip-flop requires the input to be stable. Setup time-in order to get some active result a time before the rising edge of the clock is known as setup time. It may differ from low to high and high to low transitions. It is the maximum of values which are obtained by high to low and low to high transitions. Hold time-It is the state when signal is holding for a time after the clock edge. At this time signal remains stable.

2 DIFFERENT TYPES OF PULSE TRIGGERED FLIP-FLOPS

2.1 Implicit pulse triggered flip-flop

It consists of AND logic based pulse generator and semi dynamic latch structure which used for storing elements. In this type of design inverters I5 and I6 are used to latch data. Inverters I7 and I8 are used for holding internal node X. The pulse generator is used to generate clock signals.

Two problems occurred firstly, on every rising edge of the clock the NMOS N1 and N2 turn ON which increase the switching power at node X.

Secondly, internal node X controls two MOS transistors known as P2 and N5 which reduces the speed and power.

2.2 Modified hybrid latch flip-flop

Modified hybrid latch has delayed property and having negative setup time. This approach reduces the area of the circuit. In this a sharp pulse is generated on every rising edge of the clock. It's a simple structure, but having maximum number of internal transitions which increase the power consumption of flip-flop.

It's one major drawback is positive hold time, which is generally not in other designs. This approach is used to reduce power consumption in discharging path. Another drawback is that node X

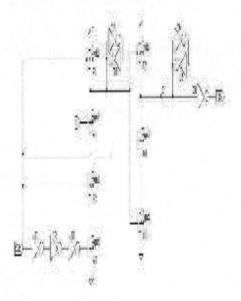

Figure 1.   Design of IPDCO.

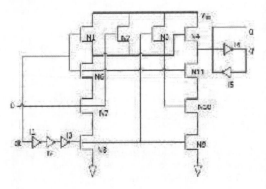

Figure 2.   Modified hybrid latch flip-flop.

becomes floating when input Q and output Q-bar becomes equal.

2.3 *Single ended conditional capturing energy recovery*

Single ended conditional capturing energy recovery using a conditional discharge technique, which is used for low power digital circuits. In this type of inverters I7 and I8 are changed by one PMOS. This PMOS is pull up transistor in support with inverter I2 to reduce load capacitance. In this technique one extra NMOS N3 is placed to reduce the switching power at node X. In this approach N3 is employed to reduce the switching power at path of N1 and N2. N3 is controlled by output Q-bar so there is no discharge if it attains the high value.

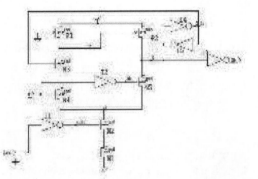

Figure 3.   Single ended hybrid latch flip-flop.

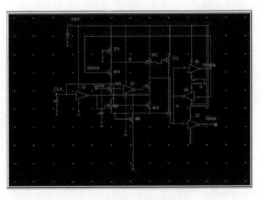

Figure 4.   Static latch CMOS structure.

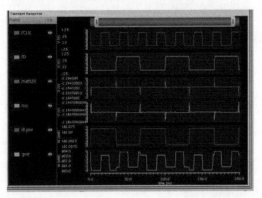

Figure 5.   Static latch waveforms.

3   IMPLEMENTATION AND SIMULATION

Authors designed a static latch structure. The inverter's I5 and I6 are used to latch data and node X is internal node, which is pre-charged periodically by the clock signal. A weak pull up transistor network is controlled by Q-output and it maintains

624

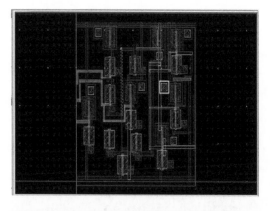

Figure 6. Layout of static CMOS latch structure.

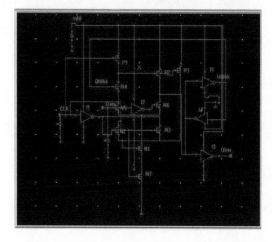

Figure 7. Domino logic static CMOS latch structure.

high level at node X when Q is zero. This design reduced the unwanted power dissipation problem. However, there are few drawbacks are also firstly, when input Q and output Q-bar is zero then node X becomes floating. Secondly, there is more power dissipation at node X which overall degrades the circuit performances. The main function of this design is that pulse input is given to pin D and Q, which is latched by inverters I5 and I6. The design consists of transistor N1 and N2 which forms two input and logic. If both the inputs are zero then means when clock is at falling edge then node Z becomes floating. And at rising edge of the clock inputs N1 and N2 are turned on to pass weak logic high signal. Schematic is shown below.

4 PROPOSED DESIGN

After studying and calculating the results of base paper some new logic is applied to that design by

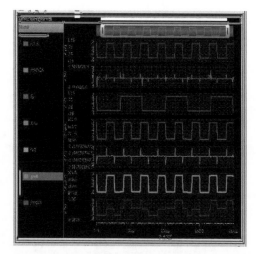

Figure 8. Domino logic static latch waveforms.

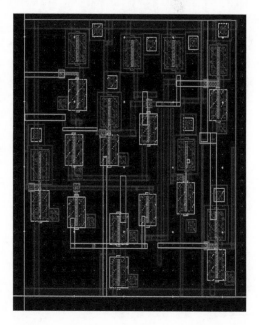

Figure 9. Layout of domino logic static CMOS latch.

Table 1. Comparison between static latch and domino logic static latch.

S. no	Parameters	[6]	Proposed design
1.	No. of transistors	19	20
2.	Power	963.2E-3	920.2E-3
3.	Average power	961.8E-3	917.7E-3
4.	Layout area	20.88 * 20.91 μm	21.28 * 21.36 μm

changing or modified the design to give better results. In our proposed design we made a small change by using domino logic.

We applied a domino logic to the circuit, which is one of the best technique to reduce power consumption. In this logic clock is applied to the pull up PMOS and one NMOS is added in the discharging path for the reduction of power. By adding inverter N1 and N2 there were discharging path created for power dissipation. Domino logic is CMOS based dynamic logic technique based on either PMOS or NMOS transistors. In this logic there is a rail to rail operation to speed up the circuit.

5 CONCLUSIONS

In this paper, the comparison between static flip-flop and domino logic flip flop is made. These are implemented and simulated based on number of transistor, average power and layout area. The simulated results show that the average power for domino logic static CMOS is less than static CMOS latch and a little bit of area is increased due to addition of one NMOS transistor in discharging path. Hence, power is then calculated, which is less than as compared to simple static CMOS latch designs.

REFERENCES

Jin-Fa Lin, Yin-Tsung Hwang and Ming-Hwang Sheun-Novellow complexity Dual moDe pulse generator Designs "IEICE trans. Fun Damenta of Electron., Commun., Comut. Sci., vol. E-91, A, pp. 1812–1815, July 2008.

Kawaguchi H. and T. Sakurai, "A reduced clock-swing flip-flop (RCSFF) for 63% power reduction," IEEE J. Solid–State Circuits vol. 33, no. 5, pp. 807–811, May 1998.

Klass C. Amir, A. Das, K. Aingaran, C. Truong R. Wang, A. Mehta, R. Heald, and G. Yee, "A new family of semi-dynamic and dynamic flip-flops with embedded logic for high performance processors," IEEEJ. Solid-State Circuts, vol. 34, no. 5, pp. 712–716, May 1999.

Kong B., S. Kim, and Y. Jun, "Conditional-capture flip-flop for statistical reduction," IEEE J. Solid-State Circuits, vol. 36, no. 8, pp. 1263–1271, Aug. 2001.

"Low power explicit pulse Triggered flip-flop" by Mr. Suchendranat Popuri in May 2012.

Mahmoodi H., V. Tirumalashetty, M. Cook and K. Roy, "Ultra low power clocking scheme using energy recovery and clock gating," IEEE Trans. Very Large Scale Integr, (VLSI) Syst., vol 17, pp 33–44, Jan. 2009.

Saranya, L., Prof. S. Arumugam, "optimization of power for sequential elements in Pulse Triggered flip-flop using low power topologiesd." IJSTR, March 2013.

Zhao P., T. Darwish and M. Bayoumi, "High Performance and low power conditional Discharge flip-flop," IEEE Trans. Very Large Scale Integr (VLSI) syst, vol. 12, no. 5, pp. 477–484, May 2004.

Communication and Computing Systems – Prasad et al. (Eds)
© *2017 Taylor & Francis Group, London, ISBN 978-1-138-02952-1*

Performance evaluation of 16 nm FinFET based NAND and D flip-flop using BSIM-CMG model

B. Soni, G. Aryan, R. Solanky & A. Patel

U.V. Patel College of Engineering, Ganpat University, Mehsana, India

ABSTRACT: Digital circuits are the heart of any modern micro-processor or micro-controller. Because of the vast advantages over analog integrated circuits, digital circuits are superior in terms of speed, performance and consume less power. Nowadays, digital designs are made up using semiconductor elements like MOSFET. This provides high speed performance. Since the last forty years, integrated circuits have been improving as per the Moore's law. Reduction in size of transistors introduce several problems like SCE. One of the possible solution is FinFET which can mitigate the problems. Using FinFET and its different topologies NAND gate and positive edge-triggered D flip-flop are examined at 16 nm technology. Simulation is done using HSPICE and BSIM-CMG FinFET model. By changing gate geometry and substrate, results are carried out. From the results we can conclude that quadruple gate is better option in terms of delay, average power and current compare to tri-gate and double gate FinFET.

1 INTRODUCTION

As number of transistors per chip are increasing at an exponential rate, in early 1960s, Gorden Moore gave a prediction on the growth rate of chip complexity. This prediction is called as the Moore's law which states that "the number of transistors per chip would quadruple every three years". The semiconductor firms have been following this prediction for the last forty years. It is not every time possible to follow the prediction as limitations and many other problems that won't allow to follow. This law gave birth to the International Technology Roadmap for Semiconductors (ITRS) organization. Every year, ITRS publishes a report for semiconductor companies and this serves as the benchmark for them. Moore's law can be seen graphically in Figure 1.

MOSFET is one of the most important semiconductor device having three terminals which are source, gate and drain. The basic principle of MOSFET is that by using the gate voltage we can control the current between source and drain. The basic structure of MOSFET is shown in Figure 2.

To fulfill the Moore's law, it is required to scale down the MOSFET to integrate more number of transistors on a single chip. It is found that after certain µm and nm scaling range MOSFET performance is degraded. The problems encountered is termed as Short Channel Effects (SCE). There are many problems associated with SCE.

A MOSFET is called a short channel device if the channel length (L_{eff}) is approximately equal to the depletion region thickness (x_d) of the drain

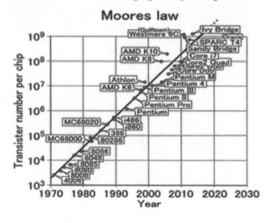

Figure 1. Transistor number per chip over time.

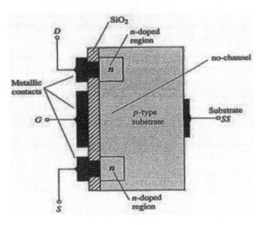

Figure 2. MOSFET structure.

and source. Since, drain and source are doped with n-type and substrate is of p-type, depletion region does exist. So when the channel length between drain and source $L_{eff} \approx x_d$, device is called short channel device.

To counter the problems with MOSFETs, Fin-FET is the possible solution. FinFET structure is shown in Figure 3. It contains vertical fin, which acts as a channel. This vertical fin allows more area of channel to be covered by the gate. Hence, control over the channel is more in the case of FinFET which offers less SCE.

So, as stated earlier that MOSFET is not a good choice at lower technologies (~ < 25 to 30 nm), we can replace it with FinFET. Since many digital circuits are based on the CMOS technology, we can use FinFET to make an inverter.

Based upon the coverage area of gate over the channel there are different FinFETs are available. Figure 3 shows tri-gate FinFET in which gate is covering three sides of channel (fin). Accordingly, there are three gates namely back gate, front gate and top gate. Other combinations are also possible like Double Gate (DG) FinFET, Ω-shaped Fin-FET, cylindrical gate FinFET etc. Figure 4 shows

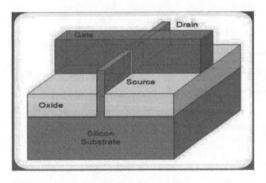

Figure 3.   Tri-gate FinFET.

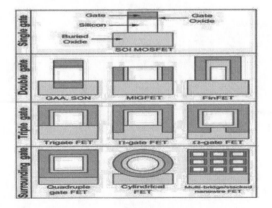

Figure 4.   Different gate structures.

certain combinations. Different configurations give different advantages.

In this paper, NAND gate and D flip-flop are proposed. Since NAND and D flip-flop are one of the most basic elements in digital circuits, it is important to make them work as efficiently as possible in lower technologies.

2   METHODOLOGY

Synopsis HSPICE is taken as a simulation tool, whereas BSIM-CMG model is taken for the MOS-FET model. HSPICE K-2015.06-SP1 and BSIM-CMG 110.0.0 versions are used for HSPICE and MOSFET model respectively. BSIM-CMG contains different model's parameters. Some of them are shown in Table 1. By changing these parameters accordingly, we can improve the performance.

For the NAND gate, gate length is kept at 16 nm. The effective channel width is the function of Number of Fins (NFIN), Thickness of Fin (TFIN) and Height of the Fin (HFIN). The relation can be given as,

$$W\_eff = NFIN \times (TFIN + 2HFIN) \qquad (1)$$

TFIN and HFIN are kept at 12 nm and 26 nm respectively for both nmos and pmos. By changing the NFIN, we can change the W_{eff}. It is important as we need to keep $(W/L)p \approx 2.5(W/L)n$.

By selecting proper W_{eff} of nmos and pmos, we can improve the delay of the circuit. Using logical effort, we can properly size the nmos and pmos. Two input and three input CMOS based NAND gates are shown in Figure 5. Figure 6 shows the positively edge triggered D flip-flop using NAND gate.

In this paper, by selecting the proper NFIN for pmos and nmos, BULKMOD and GEOMOD are changed. BULKMOD = 0 for multigate on SOI substrate and BULKMOD = 1 for multigate on bulk substrate. GEOMOD = 0 for double gate, 1 for triple gate and 2 for quadruple gate. For all these conditions average power, average current, rise and fall time of output and delay are measured for NAND gate and for D flip-flop.

Table 1.   Parameters list.

Parameters	Value
Gate length, L_g	16 nm
Thickness of Fin, TFIN	12 nm
Height of Fin, HFIN	26 nm
Substrate selector, BULKMOD	0 or 1
Structure selector, GEOMOD	0 to 2
Number of Fins, NFIN	Variable
Supply voltage, VDD	0.7 V

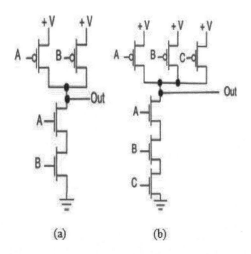

(a)   (b)

Figure 5.   (a) 2-input NAND, (b) 3-input NAND.

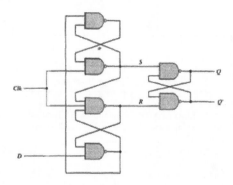

Figure 6.   Positive edge-triggered D flip-flop.

Figure 7.   2-input NAND gate waveforms.

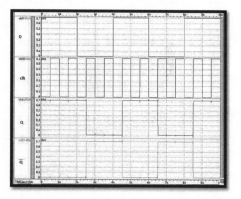

Figure 8.   Positive edge-triggered D flip-flop waveforms.

3 SIMULATION RESULTS

Transient analysis is done for NAND and D flip-flop. The transient times of 1ns to 40 μs and 1ps to 10 ns are applied to NAND gate and D flip-flop respectively. We can use NAND gate to implement D flip-flop. There are total 26 transistors needed to implement D flip-flop as shown in Figure 6. HSPICE allows us to use NAND gate as a sub circuit. D flip-flop is made using sub circuit concept in HSPICE.

The waveforms can be shown using Custom WaveView. The results of NAND gate and D flip-flop are shown in Figure 7 and Figure 8 respectively.

The truth table of the NAND gate can be verified from the waveform. One can verify the same graph for 3-input NAND gate. The simulated results are shown in Table 2 and Table 3. Table 2 consists of 2-input NAND gate results.

As we can see from the graph that at the positive edge of clock, Q changes its state according to input D. Otherwise it retains the input D. $\bar{Q}$ is

the inverted output. Table 3 shows the simulated results of D flip-flop.

As we can see from the Table 2 that for both BULKMOD value i.e. for SOI or Bulk substrate, there are no much differences. The delay is same for both BULKMOD. We can see that GEOMOD has major impact over the circuit. For the quadruple gate (GEOMOD = 2) we are getting the best results. Quadruple gate (All around gate) covers all sides of fin and hence provides better control over the channel. So power dissipation is less because of less leakage current. Similarly, Tri-gate FinFET is better choice over the double gate FinFET.

As D flip-flop contains NAND gates as a sub circuit, more number of transistors are present compare to NAND gate which leads to more current and more power dissipation. Even NFIN for 2-input NAND is different than 3-input NAND gate as per logical effort. We can see from the Table 3 that for quadruple gate on bulk subtract has the least delay compared to others but at the same time provides the highest power dissipation and average current. So there is a tread off between average power, average current and delay.

Table 2. Simulated results for 2-input NAND gate.

BSIM-CMG parameters		NAND				
		Avg. power	Avg. current	Rise time	Fall time	Delay
GEOMOD	BULKMOD	(nW)	(nA)	(ns)	(ns)	(ns)
0	0	12.62	18.03	13.90	33.19	14.53
1		4.55	6.50	25.32	41.56	2.40
2		2.25	3.22	24.93	32.44	2.17
0	1	12.62	18.03	13.90	33.19	14.53
1		4.61	6.58	25.32	41.56	2.40
2		2.28	3.26	24.93	32.44	2.17

Table 3. Simulated results for positive edge-triggered D flip-flop.

BSIM-CMG parameters		D flip-flop				
		Avg. power	Avg. current	Rise time	Fall time	Delay
GEOMOD	BULKMOD	(μW)	(μA)	(ps)	(ps)	(ns)
0	0	0.30	0.44	7.46	5.93	2.52
1		0.33	0.48	7.45	5.72	2.52
2		0.38	0.55	7.33	5.71	2.52
0	1	0.30	0.43	7.46	5.93	2.52
1		0.33	0.47	7.43	5.73	2.52
2		0.41	0.59	7.33	5.74	0.51

4 CONCLUSION

NAND gate and positive edge-triggered D flip-flop are examined. BSIM-CMG model is used as a FinFET model. Different gate topologies and substrates are examined over NAND gate and D flip-flop. 16 nm technology and its related parameters are used for the simulation. We can conclude from the Table 2 and Table 3 that quadruple gate (GEOMOD = 2) is the best choice for NAND gate because of the low power, low average current and less delay. Quadruple gate with bulk substrate is better in terms of delay for D flip-flop but provides more power and current because of more number of transistors. NFIN doesn't provide much impact over delay but it changes the rise time and fall time of the output. Also increase in NFIN, increase the current and ultimately increases the power dissipation. So, double gate FinFET is better in terms of average power and current at the cost of delay in the case of D flip-flop.

ACKNOWLEDGEMENT

We are thankful to BSIM group for providing lower technology based FinFET model.

REFERENCES

Boylestad, 9th ed. Electronic devices and circuit theory.

Bohr, Mark. 2011. Standards 22 nm-3D trigate transistors presentation, *Intel Corporation.*

Binti, A. Tahrim, A. Chin, H. Lim, C. & Tan, M. "Design and Performance Analysis of 1-Bit FinFET Full Adder Cells for Subthreshold Region at 16 nm Process Technology", *Journal of Nanomaterials, Hindawi Publication Corporation*, volume 2015.

"Breaking Moore's law," [Online]. Available: http://betanews.com. [Accessed April 2016].

Colinge, J.P. 2008. FinFET and other multi-gate transistors, Springer.

Debajit, B. & Jha, N.K. "FinFETs: From devices to Architectures", *Hindawi Publication,* volume 2014.

HSPICE® User Guide: Basic Simulation and Analysis, Version K-2015.06, June 2015.

HSPICE® Reference Manual: MOSFET Models, Version K-2015.06, June 2015.

Kang, S.M. Leblebici. CMOS digital integrated circuits Analysis and Design. 3rd edition, Tata McGraw-Hill.

Lim, W. Chin, H. Lim, C. Tan, M., "Performance Evaluation of 14 nm FinFET-Based 6T SRAM cell functionality for DC and Transient Circuit Analysis", *Journal of Nanomaterials, Hindawi Publication Corporation,* volume 2014.

Mano, M.M & Ciletti, M.D. 2007. Digital Design, *Pearson, Prentice Hall.*

Mishra, P. Muttreja, A. & Jha, N.K. "FinFET Circuit Design", *Nano electronic Circuit Design, Springer Publication-2011.*

[Online]. Available: http://electronics.stackexchange.com. [Accessed April 2016].

Patil, N. Martin, C. Oruklu, E. "Performance Evalution of Multi-Gate FETs using the BSIM-CMG Model", *IEEE 2014.*

Communication and Computing Systems – Prasad et al. (Eds)
© 2017 Taylor & Francis Group, London, ISBN 978-1-138-02952-1

IOT based weather monitoring system using Arduino Uno Board

Pankaj Sharma & Vimlesh Kumar
School of Information and Communication Technology, Gautam Buddha University, Greater Noida, UP, India

ABSTRACT: In this paper we present a working model of weather monitoring system. In this our motive is to identify the best result with in minimum cost. Weather plays important role development in farming, defense appliance, industry development and other areas. In different region of India different weather conditions. Different type of sensor used in this work. Before applying the final circuit individual testing of these sensor perform for identify the best result. In this work we are sensing the natural elements like light, temperature, humidity, atmospheric pressure.

1 INTRODUCTION

In India weather play the important role in the life of the people. The farming is the strength of India. They play the important role to development of nation. Now we are facing a big problem is unpredictable behavior of weather. The cause is unbalance in nature. The effect is that we face different problems like unseasonal rain, summer, winter and other natural digester. Now we trying to developed the system that's fully portable. They required two elements power and network. Power required to on my device. And network required is too connected to the server. In this paper we present a low cost IOT based weather monitoring system. We used arduino uno microcontroller and SIM900A GSM module and sensors. The sensors are LM35, DHT11, BMP180, LDR and TEMT6000 Ambient Light Sensor.

2 RELATED WORKED

Many authors present his different model to describe the weather monitoring system. We study many paper and identified the relative work done by him for my study purpose. The authors (Devaraju J.T., Suhas K.R., Mohana H.K., Patil Vijaykumar A. 2015) of this paper present the wireless portable microcontroller based weather monitoring system. In this they used different type of sensor like humidity, temperature, rain, sun radiation, wind speed and wind direction, surface and ambient temperature sensor. The Xbee used to communication and Max-232, 485 used for serial communication. The next paper authors (Srinivasa Dr. K.G., Siddiqui Nabeel, Kumar Abhishek, 2015.) described ParaSense—A Sensor Integrated Cloud based Internet of things Prototype for Real

Time Monitoring Applications. This paper the authors used MIB600 gateway wireless communication and MTS420 cc Sensor board. The next paper authors (Joshi M.A., Jathar M.R., Mehrotra S.C., 2011) described Distributed System for Weather Data Collection through TINI Microcontroller. In this paper they used different sensor 1-Wire temperature sensor, HC3223 humidity sensor, tini microcontroller, wind direction. The next paper authors (Iswanto, Helman muhammad, 2012) described weather monitoring station with remote radio frequency wireless communications. They used sensors LM35 sensor, LDR sensor and wind speed measurement. Atmega16 microcontroller is used and RF communication is used. The next paper authors (Pande Dushyant, Chauhan Jeetender Singh, Parihar Nitin, 2013) described The Real Time Hardware Design to Automatically Monitor and Control Light and Temperature. They used different sensor LM35, LDR sensor, on PIC microcontroller. The next paper authors (Popa Mircea, Iapa Catalin, 2011) described Embedded Weather Station with Remote Wireless Control. They used SEN-08311 USB Weather Board using ATmega328 microcontroller, along with GSM module. SEN-08311 USB weather board contain sensors are SHT15 temperature and humidity, TEMT6000 luminosity sensor, and the SCP1000 pressure sensor.

3 IMPLEMENTATION

To implement IOT based weather monitoring system. In this we used to the communication medium between device and server. We used GSM module because the Wi-Fi connectivity is not available at all place in India. We used GPRS property of GSM module to send the data to the server. We used sensors like LM35, DHT11, LDR, TEMT6000 and BMP180 for different purpose. The BMP180 is used to measure the temperature, DHT11 is used to measure humidity, LDR is used to measure the light intensity and BMP180 is used to calculate atmospheric pressure. DHT11, TEMT6000, LM35 and BMP180 are directly connected to arduino board analog pins. And the LDR is connected with pull-down resistance. We used SIM900A GSM module which have a Frequency range of 900/1800 MHz, the range of frequency accepted by the mobile communication in both CDMA and GSM in India. The final circuit is shown in Figure 1 below.

3.1 *Relative humidity and temperature sensor (RH% and T°C)*

The relative humidity is shows that the moisture in air. That's means they show the water vapor content in air. The relative humidity is the ratio of water vapor content in air, maximum vapor capacity of air. They also measure temperature with the error of ±2%.

$$Relative\ Humidity = \frac{water\ vapor\ content\ in\ air}{Maximum\ vapor\ air} \pm 5\%$$

The range of accuracy is also depend on temperature, if the temperature below 26°C the percentage of error is ±4%.

3.2 *Barometric pressure sensor*

Due to the gravity of earth all the elements pressure down to the earth. The pressure is applied by

the air in per unit volume is called atmospheric pressure. The range of this sensor is between 300 to 1100 millibar. BMP180 is also capable to measure the temperature. They also have I²c Interface property with microcontroller and microprocessor.

3.3 *Light intensity measurement*

3.3.1 *Light Dependent Resistance (LDR)*
The measurement of light intensity is performed by light dependent register. The value of resistance change dynamically with change the intensity of light. In full dark LDR resistance in MΩ. When is fall on its resistance going to its minimum value in Ω. We used LDR with 10 K pull-down resistance. By this two terminal device converted into three terminal device. The generated voltage is result of voltage divider rule. We convert generated voltage into Lux.

$$LUX = \frac{(500 * (5 - Vout))}{10 * Vout}$$

3.3.2 *Ambient light sensor (TEMT6000)*
The other light sensor we used TEMT6000 ambient light sensor. This device consists NPN epiwafer planar phototransistor. In the device of TEMT6000 consist of surface mounted phototransistor with series of 10 kΩ resistance in emitter side. The concept is to provide pull down in the circuit. The sensor packed with anti-moisture transparent packaging. The operating range of this device is –40°C to +85°C. Photoresistor resistance varying of due to change the condition of intensity. Its offer kilo ohms when dark, and close to some ohms with a bright flashlight. The arrangement show in Figure 2 that show low state

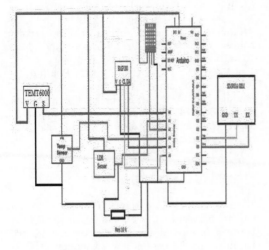

Figure 1. IOT based weather monitor system using microcontroller.

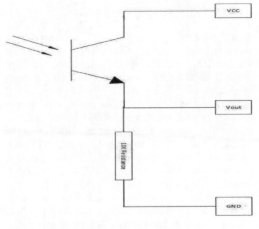

Figure 2. TEMT6000 with series 10 kΩ resistance.

to high state intensity detection. The limitation of this sensor is that its major light intensity maximum 1000 LUX.

3.4 Temperature sensor

The LM35 is the integrated circuit sensor. Which have a property to calculate the temperature with high accuracy range wide range of temperature. LM35 is directly connected to the microcontroller board without extra circuit. The range of accuracy at room temperature is $\pm 0.4°C$ the arrangement shown in Figure 2.

3.5 Communication with server

To provide the communication between server and device model we used SIM900A GSM module. SIM900A is 85.6 kb/s downlink for data. They have a property to directly interface with PC through RS232 interface. Module communicate with different baud rate range from 9600 to 115200 by AT command by any serial communication. We used the GPRS property of GSM modem to send our data to the server link at after every fixed time interval. We are connected to our GSM module two Digital pins of 2 and 3.

3.6 Arduino Uno development board

IOT based weather monitoring system is based on the Arduino Uno development board. This is a widely used platform because open source libraries for additional functionality. Uno board consist AVR microcontroller. In Uno board ATmega328P microcontroller used. To prevent a face a problem of error. We used digital pins to communicate with GSM module. The cause is that programmer is store the data to microcontroller memory through serial communication. When we connected GSM module to Uno Tx and Rx, they also used serial communication so program may be not install successfully in memory. We disconnected GSM module every time when program is burn. To overcome this problem we used digital pins of Uno board to communicate with GSM module. The code must be less than 2 K for good performance. If the size is more nearly 2 K they generate error. Ardunio Uno development board have inbuilt ADC (analog to digital conversion). So need of extra hardware to convert analog data convert into digital data is not required in this. The other property is that if any hardware library not present on board, then we added library simply. We added two library DHT11 and BMP085. BMP180 Library create some I2c error. BMP085 library also exist for BMP180 pressure sensor.

4 WORKING

IOT based weather monitoring system. They contain several sensors for several results. BMP180 is used to measure the Temperature, DHT11 used to calculate humidity and temperature, LDR is used to identify the intensity of light coming from sun and other medium and BMP180 used for calculate biometric pressure pre unit area. The entire sensor is connected to Arduino board 328P. They have 6 analog input pins. We connected the GSM module to digitals pins of arduino board. The working of GSM module is to send the data to the server link. The link can be access any place in the world. The generated analog voltage is converted into digital by divided of 210. So generated voltage is express,

$$Vout = \frac{analog\ reading\ from\ sensor}{1024}$$

$$Vout = 0.0009765625 * analog\ reading\ from\ sensor$$

The working is expressed in term in flow diagram in Figure 3 given below.

5 TESTING AND RESULTS

We are comparison our data through online weather monitoring web sites and comparison our result. We are in Gautam Buddha University. We compare our result online through exist web sites

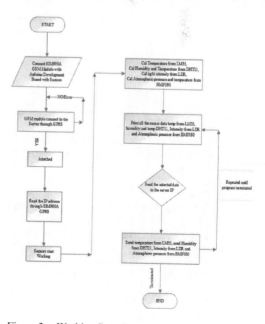

Figure 3. Working flow diagram of IOT based weather monitor system using microcontroller.

Table 1. Temperature in degree centigrade calculate by different sensor at same.

S.no	LM35	DHT11	BMP180
1	29.30	30.00	29.71
2	29.79	30.00	29.76
3	30.27	31.00	29.76
4	29.30	31.00	29.79

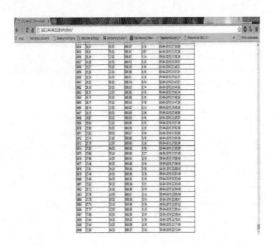

Figure 4. Data store to the server link through SIM900A GSM module.

Table 2. Light intensity in LUX calculate by different light sensor at same time.

S.no	LDR	TEMP6000
1	42.58	40.04
2	44.84	37.11
3	42.58	38.09
4	40.64	36.13

for location in greater Noida, compare all the data and validate the result. In testing phase we see that LM35 is not given constant value the variation in reading is large. We also calculate temperature from DHT sensor, but the temperature accuracy range is ±2°C, and the temperature accuracy range of BMP180 sensor is ±2°C. We calculate temperature from BMP180 sensor. To calculate light intensity we have two sensor LDR and TEMT6000 light sensor. Both connected pull-down with 10 kΩ resistance. The difference between the readings is not much large but the cost difference is large. So make cost-effective our system we used LDR as a light sensor. We show a table below both for temperature and intensity with the difference of reading at a same time. In first table we show that the result of different temperature sensors at a same time and in second table we show different light sensor calculate light intensity at a same time.

After complete all testing, we connected our device to the server and send the data in given IP link. We show the snap shot of web page of our IP. In Figure 4 of IP web page is given below with data.

6 CONCLUSION

The suggest model in this paper is low cost and compact. This is combination of software and hardware. The standard sensor are connected to Arduino Uno development board for monitoring multiple weather variable. The connectivity of IOT is provided of freeness to test in any place town and village or forest where network available. Different GPRS service provider can connected to our device according to network availability. The process for individual service provider to change an APN. The date send to server that is easily access by any place.

REFERENCES

Devaraju J.T., Suhas K.R., Mohana H.K., Patil Vijaykumar A., 2015. "Wireless Portable Microcontroller based Weather Monitoring Station" Esevier, Measurement 76 (2015) 189–200.

Iswanto, Helmanmuhammad, 2012. "Weather Monitoring Station with Remote Radio Frequency Wireless Communications". International Journal of Embedded Systems and Applications (IJESA) Vol.2, No.3, September 2012.

Joshi M.A., Jathar M.R, Mehrotra S.C., 2011. "Distributed System for Weather Data Collection through TINI Microcontroller" International Journal of Environmental Science and Development, Vol.2, No.1, February 2011, ISSN: 2010–0264.

Pande Dushyant, Chauhan Jeetender Singh, Parihar Nitin, 5, May 2013. "The Real Time Hardware Design to Automatically Monitor and Control Light and Temperature". International Journal of Innovative Research in Science, Engineering and Technology, Vol. 2, Issue.

Popa Mircea, Iapa Catalin, 2011. "Embedded Weather Station with Remote Wireless Control". 19th Telecommunications forum TELFOR 2011 Serbia, Belgrade, November 22–24, 2011. 978-1-4577-1500-6/11/$26.00 ©2011 IEEE.

Srinivasa Dr. K.G., Siddiqui Nabeel, Kumar Abhishek, 2015. "ParaSense—A Sensor Integrated Cloud based Internet of things Prototype for Real Time Monitoring Applications", IEEE Region 10 Symposium.

Communication and Computing Systems – Prasad et al. (Eds)
© 2017 Taylor & Francis Group, London, ISBN 978-1-138-02952-1

Four quadrant analog multiplier/divider employing single OTRA

Ujjwal Chadha & Tajinder Singh Arora
Maharaja Surajmal Institute of Technology, Janakpuri, New Delhi, India

ABSTRACT: This paper proposes a new four quadrant analog multiplier/divider cell that uses a single OTRA and six MOSFETs. The circuit can act as an analog multiplier and divider at the same time and uses differential inputs. Applications of the circuit as a squarer and amplitude modulator have been discussed. The multiplication and division operation along with the proposed applications have been simulated in PSPICE to ensure their workability. Frequency characteristics of the circuit have been provided.

1 INTRODUCTION

Operational Transresistance Amplifier (OTRA) has been used by various authors to design applications in analog signal processing (Salama & Soliman 2000, Salama & Soliman 1999, Kılınç, Selçuk, Keskin, & Çam 2007, Çam, UğGur & et al. 2004, Gökçen, Ahmet, Kilinc, & Çam. 2011 and their cited in). The device is thus shown to be highly versatile and flexible.

OTRA is a three terminal device that that provides high transresistance gain and makes the output potential proportional to the input differential current (Salama & Soliman 1999, Mostafa & Soliman 2006). Symbolic representation of OTRA is shown in Figure 1. V_o is the output potential and I_+ and I_- non-inverting and inverting input currents respectively.

The characteristic equations of the device and the input-output relationship is given by (1).

$$\begin{bmatrix} V_+ \\ V_- \\ V_o \end{bmatrix} = \begin{bmatrix} 0 & 0 & 0 \\ 0 & 0 & 0 \\ R_m & -R_m & 0 \end{bmatrix} \begin{bmatrix} I_+ \\ I_- \\ I_o \end{bmatrix} \qquad (1)$$

The equations contained by the above matrix, when written individually give the equations as

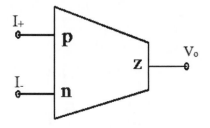

Figure 1. Device symbol of OTRA.

given in (2). The input potentials V_+ and V_- are virtual grounds and the output potential V_o is dependent on the differential input current.

$$V_+ = V_- = 0 \ \& \ V_o = R_m(I_+ - I_-) \qquad (2)$$

Ideally, the transresistance gain R_m approaches infinity and thus the input currents I_+ and I_- are equal.

One of the important non linear operations used in analog signal processing are analog multiplication and division. Various analog multiplier/divider circuits have been proposed in literature but most of them suffer from one or the other problem. Abuelma'atti, & Al-Qahtani (1998), Premont, Abouchi, Grisel & Chante (1999), for example, have proposed analog multiplier / divider circuits using CCCIIs and CCIIs respectively but have the disadvantage of employing two of these active devices. Similarly, the proposed multiplier-divider by Kaewdang, Fongsamut & Surakampontorn (2003), uses three active devices (OTAs). Yuce, E (2008) proposed a multiplier that uses one bipolar CCCII+ but requires current matching in order for the circuit to work as a multiplier. Moreover, the circuit is temperature sensitive. Also Khachab & Ismail (1989) and Khachab (1991) proposed multiplier / divider circuits that employ one op-amp but needs to satisfy a condition for stability of the circuit. This is because of use of both negative and positive feedbacks in the circuit. Liu & Chen (1995) proposed a circuit that uses as many as two CFOAs and 4 MOSFETs to realize only a divider. The four quadrant multiplier by Riewruja & Rerkratn (2011) uses a large number of active devices (5 op-amps). A multiplier circuit employing single OTRA is proposed by Pandey, Pandey, Sriram & Paul (2012), but it does not allow differential inputs.

In this paper, we propose a new four quadrant analog multiplier-divider circuit that employs single OTRA and six MOSFETs. The circuit works on differential input, acts as a multiplier and divider at the same time and has an adjustable gain factor that can be varied by varying the aspect ratio of MOSFETs used. The circuit is simulated in PSPICE software to ensure its workability and all simulation results have been provided. Applications of the circuit as a squarer and amplitude modulator have also been discussed.

The next section gives the circuit of proposed multiplier—divider circuit and the equations defining its behavior. Section III gives the simulation results of the circuit as an (a) analog multiplier, (b) analog divider, (c) squarer and (d) amplitude modulator. It also gives the frequency characteristics of the circuit. Last section gives the concluding remarks.

2 PROPOSED CIRCUIT

The proposed circuit, which acts as a multiplier and a divider simultaneously, is shown in Figure 2. Using the equations in (1), the following relationship between the drain currents can be obtained:

$$I_{d1} + I_{d2} + I_{d5} = I_{d3} + I_{d4} + I_{d6} \quad (3)$$

Now considering the aspect ratios of transistors from M_1 to M_4 to be (W_a/L_a) and that of M_5 and M_6 to be (W_b/L_b) and writing the currents in (3) in terms of the gate and drain voltage of their corresponding NMOS transistors, the output voltage (V_o) can be written as given in (4) when the NMOS transistors are operating in the triode region

$$V_o = \frac{(W_a/L_a)(X_1 - X_2)(Y_1 - Y_2)}{(W_b/L_b)(z_1 - z_2)} \quad (4)$$

The NMOS transistors must operate in linear region. Thus, considering the MOSFET threshold voltage to be V_T, the following conditions must be satisfied:

$$Y_1, Y_2 \leq \min[(X_1 - V_T), (X_2 - V_T)] \text{ for } M_1 \text{ to } M_4 \quad (5)$$

$$\&, V_o \leq \min[(Z_1 - V_T), (Z_2 - V_T)] \text{ for } M_5 \& M_6 \quad (6)$$

3 SIMULATION RESULTS

The proposed circuit shown in Figure 2 was tested for different combinations of input values to ensure its workability and show its operation as a multiplier, divider, squarer and amplitude modulator. All simulations were performed in PSPICE software. OTRA was designed using CMOS technology as shown by Mostafa, & Soliman (2006). The circuit used is redrawn in Figure 3. 0.5 μm process parameters were used for the transistors and their aspect ratios were kept the same as used by Mostafa, & Soliman (2006). The supply voltages of OTRA used were: $+V_{DD} = -V_{SS} = 1.5$ V. The constant bias voltage was −0.5 V.

3.1 Analog multiplier

Equation (4) acts as a multiplier when the potential difference $(Z_1 - Z_2)$ is kept constant. In such a case, V_o is the result of multiplication of potential difference $(X_1 - X_2)$, $(Y_1 - Y_2)$ and some constant gain. For simulation purposes, X_1 and X_2 were taken to be of the form $(X + x)$ and $(X - x)$ respectively, where X is the constant D.C. voltage while x is the varying potential. X was taken to be 3V while x was varied from −100 mV to +100 mV. Thus, the potential difference $(X_1 - X_2)$ was varied from −200 mV to +200 mV. Z_1 and Z_2 were kept constant at 2.8V and 2.7V respectively, maintaining a constant potential difference of 100 mV. All transistor aspect rations were kept same as

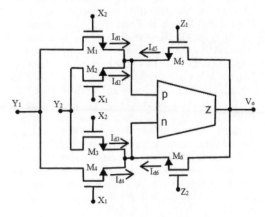

Figure 2. Proposed multiplier/divider circuit.

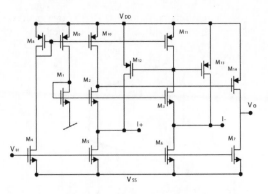

Figure 3. CMOS implementation of OTRA.

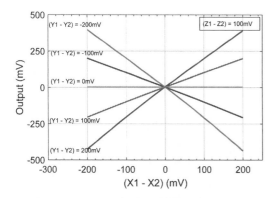

Figure 4. DC sweep result of proposed circuit as a multiplier.

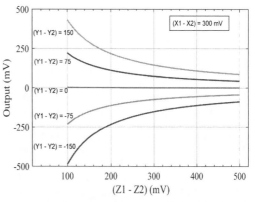

Figure 5. DC sweep result of proposed circuit as a divider.

$W/L = 1.25$ um/9 um (making the aspect ratio gain unity). Now the output potential was plotted for five different values of $(Y_1 - Y_2)$ which varied from (500–300) mV to (500–700) mV. The SPICE plot showing the multiplication output using the above values is given in Figure 4. As can be seen in the figure, the results are in agreement with the theoretical results to a high degree.

3.2 Analog divider

Equation (4) acts as a divider when either one of the potential differences $(X_1 - X_2)$ or $(Y_1 - Y_2)$ is kept constant (let it be $(X_1 - X_2)$ in our case, which acts as the constant gain along with the gain provided by transistor aspect ratios). In such a case, V_o is the result of division of $(Y_1 - Y_2)$ by $(Z_1 - Z_2)$ with some constant gain. For the purpose of simulation, the values taken were: $X_1 = 3.15V$, $X_2 = 2.85V$ (making the potential difference $(X1 - X2) = 300$ mV), $Z_1 = 2.8V$ while Z_2 varied from 2.3V to 2.7V (thus the potential difference $(Z_1 - Z_2)$ varied from 100 mV to 500 mV). All transistor aspect rations were kept same as $W/L = 1.25$ um/9 um (making the aspect ratio gain unity). Now the output potential was plotted for five different values of $(Y_1 - Y_2)$ which varied from (400–250) mV to (400–550) mV. The SPICE plot showing the division output using the above values is given in Figure 5.

3.3 Squarer

In order for the circuit to act as a squarer, the potential difference $(Z_1 - Z_2)$ must be kept constant, while the potential difference $(X_1 - X_2)$ and $(Y_1 - Y_2)$ must be kept same and equal to the required squarer input. During the simulation the values of these potentials were $Z_1 = 2.8V$, $Z_2 = 2.6V$ (thus making the potential difference $(Z_1 - Z_2) = 200$ mV). X_1 and X_2 were taken to be of the form $(X + w)$

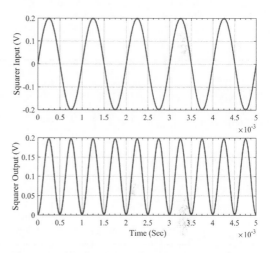

Figure 6. Simulated time response of given circuit as a squarer.

and $(X - w)$ respectively where X was taken as a constant DC voltage (3V) while w was taken to be a sinusoidal wave of 100 mV and 1 KHz frequency. Similarly, Y_1 and Y_2 were taken to be of the form $(Y + w)$ and $(Y - w)$ respectively where Y was taken as a constant DC voltage (500 mV) while w was taken to be a sinusoidal wave of 100 mV amplitude and 1 KHz frequency. Thus, both the potential differences $(X_1 - X_2)$ and $(Y_1 - Y_2)$ were sinusoidal waves of 200 mV amplitude and 1 KHz frequency which acted as the input to squarer. The transistor aspect rations were same as used in previous case (making the gain unity). Since the division factor was a constant DC voltage of 200 mV, the output potential, at all times was square of input potential divided by 200 mV. The simulation result is shown in Figure 6 which resembles the theoretical result to a high degree.

3.4 Amplitude modulator

In order for the circuit to act as an amplitude modulator, the potential difference $(Z_1\text{-}Z_2)$ must be kept constant, while the carrier wave and modulating wave must be applied at $(X_1\text{-}X_2)$ and $(Y_1\text{-}Y_2)$. During the simulation, the values of these potentials were $Z_1 = 2.8V$, $Z_2 = 2.6V$ (thus making the potential difference $(Z_1\text{-}Z_2) = 200$ mV). X_1 and X_2 were taken to be of the form $(X+w)$ and $(X\text{-}w)$ respectively where X was taken as a constant DC voltage (3V) while w was taken to be a sinusoidal wave of 200 mV amplitude and 10 KHz frequency. Thus the potential difference (X1-X2) acted as a carrier wave of 400 mV, 10 KHz frequency. Similarly, Y_1 and Y_2 were taken to be of the form $(Y+m)$ and $(Y\text{-}m)$ respectively where Y was taken as a constant DC voltage (500 mV) while m was taken to be a sinusoidal wave of amplitude 100 mV and 1 KHz frequency. Thus, the potential differences $(Y_1\text{-}Y_2)$ acted as a modulating wave of 200 mV and 1 KHz frequency. The transistor aspect rations were same as used in previous case (making the gain unity). Since the division factor was a constant DC voltage of 200 mV, the output potential, at all times was product of input potentials divided by 200 mV. The simulation result is shown in Figure 7 which match the theoretical results.

3.5 Frequency characteristics

In order to find the frequency response of the proposed circuit, the values of the input potentials were chosen to be: $Z_1 = 2.8V$, $Z_2 = 2.6V$ (making $Z_1 - Z_2 = 200$ mV), $Y_1 = 500$ mV, $Y_2 = 300$ mV (making $Y_1 - Y_2 = 200$ mV). X_1 and X_2 were taken to be of the form $(X + w)$ and $(X - w)$ respectively where X was taken as a constant DC voltage (3V) while w was taken to be a sinusoidal wave of 0.5 mV and varying frequency (making $X_1 - X_2$ to be sinusoidal wave of amplitude 1 mV and varying

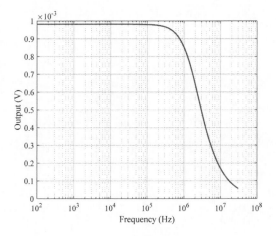

Figure 8. Frequency response of proposed circuit.

frequency). The output potential was the plotted against the frequency to obtain the response shown in Figure 8.

4 CONCLUSION

A new configuration of four quadrant multiplier/divider using single OTRA has been proposed. The circuit uses differential inputs, provides multiplication and division functions simultaneously and uses only six MOSFETs to achieve both operations. The successful PSPICE simulations of the circuit as a simple multiplier/divider as well as squarer and amplitude modulator shows its flexibility to work with different sets and kinds of inputs

REFERENCES

Abuelma'atti, M.T. & Al-Qahtani, M.A (1998.) A current-mode current-controlled current-conveyor based analogue multiplier/divider, International Journal of Electronics, 85(1). 71–77.

Çam, Uğur & et al (2004). Novel two OTRA-based grounded immitance simulator topologies. Analog Integrated Circuits and Signal Processing 39.2: 169–175.

Gökçen, Ahmet, Selcuk Kilinc, & Uğur Çam (2011) Fully integrated universal biquads using operational transresistance amplifiers with MOS-C realization. Turkish Journal of Electrical Engineering & Computer Sciences 19.3: 363–372.

Kaewdang, K., Fongsamut, C. & Surakampontorn W. (2003) A wide band current-mode OTA based. Analog multiplier-divider. Proceedings of the 2003 International Symposium on Circuits and Systems, 1.349–352.

Khachab, N. I. & Ismail, M (1989). MOS multiplier/divider cell for analogue VLSI. Electronics Letters, 25(23), 1550–1551.

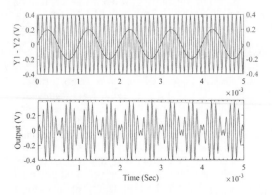

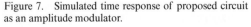

Figure 7. Simulated time response of proposed circuit as an amplitude modulator.

Khachab, N. I. (1991) A nonlinear CMOS analog cell for VLSI signal and information processing, IEEE Journal of Solid State Circuits, 26(11). 1689–1699.

Kılınç, Selçuk, Ali Ümit Keskin, & Uğur Çam. (2007) Cascadable voltage-mode multifunction biquad employing single OTRA. Frequenz 61, no. 3–4: 84–86.

Liu, S. I., & Chen, J.J. (1995). Realization of analogue divider using current feedback amplifier. IEE Proc. Circuits Devices System, 142(1). 45–48.

Mostafa, H., & Soliman, A. M. (2006). A modified CMOS realization of the operational transresistance amplifier (OTRA). Frequenz, 60(3–4), 70–77.

Pandey, R., Pandey, N., Sriram, B. & Paul, S. K (2012). Single OTRA based analog multiplier and its applications. International Scholarly Research Network. Article ID 890615, 7.

Premont, C., Abouchi, N., Grisel, R. & Chante, J.(1999). A BiCMOS current conveyor based four-quadrant analog multiplier. Analog integrated circuits and signal processing, 19. 159–162.

Riewruja, V. & Rerkratn (2011). A Four-quadrant analogue multiplier using operational amplifier. International Journal of Electronics, 98(4). 459–474.

Salama, K. N., & Soliman, A. M. (1999). CMOS operational transresistance amplifier for analog signal processing. Microelectronics Journal 30, no. 3: 235–245.

Salama, K. N., & Soliman, A. M. (2000). Active RC applications of the operational transresistance amplifier. Frequenz, 54(7–8), 171–176.

Yuce, E (2008). Design of simple current-mode multiplier topology using a single CCCII+. IEEE Transaction on Instrumentation and Measurement, 57(3). 631–637.

Communication and Computing Systems – Prasad et al. (Eds)
© 2017 Taylor & Francis Group, London, ISBN 978-1-138-02952-1

Performance analysis of different adaptive algorithms for equalization

P. Aggarwal & S.C. Yadav
Graphic Era Deemed University, Dehradun, Uttrakhand, India

ABSTRACT: The major problems in wireless communication are time dispersion and inter symbol interference. In order to cancel out the effect introduced by the unknown channel and to recover the original signal as from the distorted signal, a channel equalizer is required to compensate the effect of channel distortion, time variation and can adapt it-self to the changes in channel characteristics. The equalizers are expected to have fast convergence rate in communication systems which are difficult to achieve with conventional adaptive algorithms. LMS is widely used because it is simple and robust, but performs poor in terms of convergence rate. NLMS is a improved version of LMS and provides better convergence. RLS exhibit best performance but complex and unstable. In this paper we simulated adaptive algorithms such as LMS, NLMs and RLS algorithms in MATLAB and compared their performance.

1 INTRODUCTION

From past few years, designing of adaptive filter has been an dynamic area of study and realization. An adaptive filter has the quality that automatically adjusts its parameter according to an adaptive algorithm (Haykin S., 2002). Researchers has been studied Adaptive filtering algorithm for many decades because of its usefulness in diverse applications and equalization is one of them. In the field of digital signal processing there has been continuous progress in adaptive algorithm applications and several issues are found to be a focus for everyone's interest together with a large amount of calculation and the difficulty to attain the high-speed in real-time (Proakis J.G, 1995). Figure 1 illustrates the basic building block of adaptive filtering process.

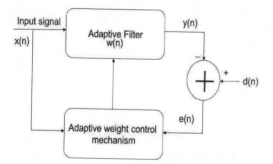

Figure 1. Block diagram of adaptive filtering.

In this paper, to explain the performance of various adaptive algorithms, we took adaptive equalization application. This paper has six sections along with the current introductory Section I. Section II covers brief introduction of channel equalization. Section III discusses adaptive algorithms that are being used for comparison. Section IV shows the MATLAB simulation of both algorithms. Finally, Section V summarize importance conclusion of the research work carried out.

2 CHANNEL EQUALIZATION

When the transmitted signal dispersed in time and amplitude we need an equalizer to nullify those effects. An adaptive equalizer is basically a linear adaptive filter that tries to make the replica of the opposite of transfer function of the channel that signal passes through. This process is also known as De-convolution often in digital signal processing (Ifechor E.C & Jervis B.W, 2002). Figure 2 illustrates the basic process of an equalizer. Here the input u(t) is passing through the unknown channel H(z). The equalization succession changes the impulse response of the unknown channel as $H^{-1}(z)$. The adaptive filtering algorithm calculates the error e(t) which is measured as the variation between the desired signal (delayed version of input signal) and filter output.

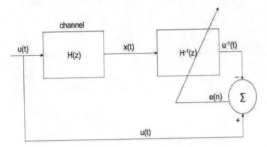

Figure 2.   Basic blocks of an equalizer.

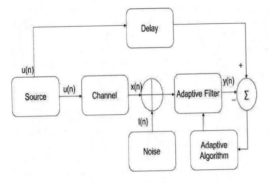

Figure 3.   Transmission system with channel equalizer.

There is another Figure 3 which shows the transmission system with channel equalizer. Additive noise is here represented by the I(n) mixes with the input signal coming from the channel. In digital communication the distortion introduced by the channel is mostly recognized as pulse dispersion effect, which causes inter symbol interference. The receiver performance degrades with the additive noise. Equalizer eliminates the ISI and reduces noise to negligible amount (Sarada C & Amarbabu Y, 2015).

Mathematically the relationship between the unknown channel's transfer function h(n) and transfer function of adaptive filter as:

$$h1(n)h2(n) = 1 \qquad (1)$$

The implementation of an equalizer is generally done in the transversal filter's form. A equalizer initially requires the information of transmitted data symbol or we can say a delayed replica of them. (Haykin S, 1996). This information of data samples will be used for adaptation of equalizer tap weight. Now, equalizer will try to obtain an output closer to transmitted data signal. At the end of the process, equalizer would have converged it tap weight to an optimal value.

3   ADAPTIVE ALGORITHMS

3.1   *Least Mean Square algorithm*

LMS algorithm adjusts the filter coefficients in such a way that produced the least mean squares of the error signal. It is based on stochastic gradient descent method that uses the gradient vector of the filter tap weight to converge of the optimal Wiener solution (Mahmood M et al., 2015). The main advantage of the LMS algorithm is simple computation, easy implementation, fair convergence. The following mathematical model describes the LMS algorithm. The weight updation procedure of LMS algorithm resembles to steepest-descent method except one thing that LMS does not require exact measure of gradient vector but it uses a rough approximation of gradient.

The error of output filter can be expressed as

$$E(n) = d(n) - W^T x(n) \qquad (2)$$

Approximation gradient can be expressed as

$$\zeta = -2E(n)x(n) \qquad (3)$$

After substituting these expression, weight updation equation becomes

$$W(n + 1) = W(n) + 2\mu E(n)x(n) \qquad (4)$$

where W(n) stands for the weight vector, x(n) is the input signal vector, μ is step-size upon which the convergence rate of equalizer depends and e(n) is the error vector. For an N-tap filter, the number of operation get reduce to 2 * N multiplication and N addition per Coefficient update (Reddy B.S & Krishna V.R, 2013). $E^2(n)$ is the mean square error, the difference between the output y(n) and the desired signal which is given by,

$$E^2(n) = [d(n) - y(n)]^2 \qquad (5)$$

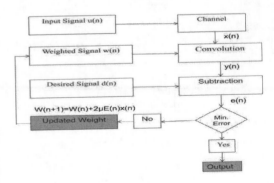

Figure 4.   Flow chart of LMS algorithm.

A large value of step-size is preferred to get high convergence rate and to keep a growing pace ability in the steady state. Large step size causes large steady misadjustment error.

3.2 Recursive Least Square algorithm

There is a least square method upon which RLS is based. The RLS algorithm recursively find out the adaptive filter parameters that minimize the error. In RLS, the input signals are taken deterministic, while for the LMS and NLMS the stochastic input signals are used. In the absence of numerical measure of input signal, it is difficult to adjust the parameter of a filter but RLS can do that (Tato L.M & Miranda H.C). RLS algorithm has the similar procedures as LMS and NLMS except that it provides high convergence rate for fast fading channel, moreover RLS have some stability issues because of covariance updation formula, which is used for automatics change according to the estimated error as follows:

$$W^H(0) = 0 \qquad (6)$$
$$R(0) = \alpha^{-1}I \qquad (7)$$
$$v(n) = R(n-1)x(n) \qquad (8)$$

And gain vector can be expressed as

$$k(n) = \frac{v(n)}{\Delta + x^H(n)v(n)} \qquad (9)$$

$$E(n) = d(n) - W^H(n-1)x(n) \qquad (10)$$
$$W(n) = W(n-1) + k(n)e(n) \qquad (11)$$
$$R(n) = \Delta^{-1}R(n-1) + \Delta^{-1}k(n)x^H(n)R(n-1) \qquad (12)$$

where $W(n-1)$ is the weight value vector that is going to update in present cycle, $W(n)$ is the updated weight value vector, $x(n)$ is the transmitted signal vector for adaptive filter and $E(n)$ is the priori estimation error. The inverse correlation matrix

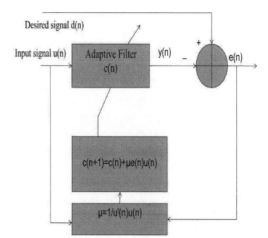

Figure 6. Flow chart of NLMS algorithm.

has to recalculated after the weight updating and process resume with new input values (Eleftheriou E & Falconer D.D).

3.3 Normalized Least Mean algorithm

LMS algorithm has a stagnant step size for every iteration that is one of its disadvantage that can be overcome by using NLMS. The NLMS algorithm is an augmentation of LMS algorithm which uses the greatest step size. It can be said NLMS is an extension of LMS algorithm (Blorisagar K.R & Kulkarni, 2010). The step-size for NLMS calculated by the following formula as:

$$\text{Step-size} = 1/u(n)u^t(n) \qquad (13)$$

where $u(n) = $ input vector

Implementation of NLMS is very much similar to LMS. The steps required for weight updation for NLMS is given as:

$$Y(n) = w^t(n)u(n) \qquad (14)$$

4 IMPLEMENTATION IN MATLAB

The Implementation is done in MATLAB. MATLAB is an interactive environment for numerical computation, design and problem solving. MATLAB is used for the variety of applications including signal processing, communication, control system and computational biology. The MATLAB simulation can be performed either by using MATLAB M-file or by Simulink interface. The experimental work here is done using MATLAB M-file. The result gained are analyzed

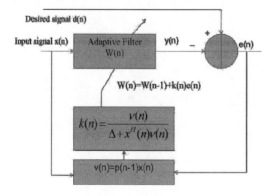

Figure 5. Flow chart of RLS algorithm.

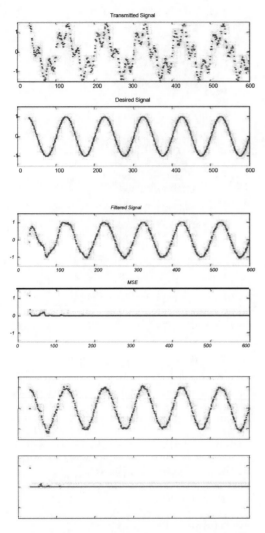

5.2 *Specification of NLMS algorithm in MATLAB*

Number of filter taps = 30
Step size(μ) = 0.02

5.3 *Specification of RLS algorithm in MATLAB*

Forgetting factor(λ) = 0.98
Regularization factor(α) = 1

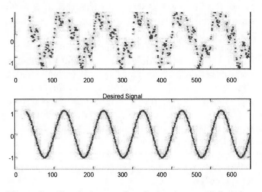

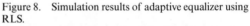

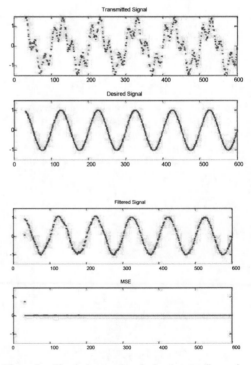

Figure 8. Simulation results of adaptive equalizer using RLS.

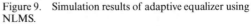

Figure 7. Simulation results of adaptive equalizer using LMS.

and mentioned in the following section of this paper.

$E(n) = d(n) - y(n)$

$\mu = 1/u^t(n)u(n)$

$c(n + 1) = c(n) + \mu e(n)u(n)$

5 SIMULATION RESULTS

5.1 *Specification of LMS algorithm in MATLAB*

Number of filter taps = 30
Step size(μ) = 0.02

Figure 9. Simulation results of adaptive equalizer using NLMS.

6   CONCLUSION

In this paper the comparison of adaptive algorithms is performed on the basis of convergence rate and simulation is done using the software tool MATLAB. The simulation results shows that RLS provides better convergence rate but at the cost of high computational complexity and less stability. LMS filtering is relatively good and it is simple and easy to realize hardware but provides slower convergence rate. Convergence speed can be improved by choosing an appropriate value of step-size but the drawback of LMS algorithm is having a fixed step-size parameter for each iteration, which can be overcome by NLMS which uses the greatest step-size and also provides better convergence speed than LMS.

REFERENCES

"Adaptive Filter Algorithms", IEEE Trans. Acoust. Speech Signal Process., vol. ASSP-34, pp. 1907–1110.

"Algorithms using Real Time Speech Input Signal, Application" IJARECE, vol. 4, pp. 1112–1116, May 2015.

Borisagar K.R., Kulkarni Dr. G.R., "Simulation and Comparative Analysis of LMS and RLS".

"Communication", International Journal of Science and Technology, vol. 2, August 2015.

Eleftheriou E., Falconer D. D., "Tracking the Properties and Steady-State Performance of RLS".

Global Journal of Researches in Engineering, vol. 10, October 2010.

Haykin S., "Adaptive filter theory", 3rd edition, Prentice hall, 2002.

Haykin S., "Analog and Digital communication", Prentice Hall, 1996.

Ifeachor E.C., Jervis B.W, "Digital Signal Processing", A Practical Approach, Prentice Hall, 2002.

Mahmood M. et.al., "Adaptive Filtering Algorithms for Channel Equalization in Wireless".

Proakis JG., "Digital communication", 3rd edition, McGraw hill Inc, 1995.

Reddy B. S, Krishna V. R., "Implementation of Adaptive Filter Based on LMS Algorithm", IJERT, vol. 2, issue 10, October 2013.

Sarada C., Amarbabu Y., "Design and Implementation of Adaptive Filters for Real Time".

Sharma P., Gupta P., "Performance Comparison of ZF, LMS and RLS Algorithms for Linear Adaptive Equalizer" ISSN 2231-1297, vol. 4, pp. 587–592.

Tato L.M., Miranda H.C., "Simulation of an RLS Adaptive equalizer using SIMULINK".

Communication and Computing Systems – Prasad et al. (Eds)
© 2017 Taylor & Francis Group, London, ISBN 978-1-138-02952-1

Implementation and performance analysis of different multipliers

Pooja Karki, Priyanka Aggarwal, Rashmi Bisht & S.C. Yadav
Graphic Era Deemed University, Dehradun, Uttarakhand, India

ABSTRACT: Multipliers are very important and are used in various applications. This paper presents analysis of different multipliers design such as array multiplier, row bypassing multiplier and column bypassing multiplier. The multipliers are implemented using verilog HDL and the simulation is done in Modelsim simulator. The multipliers are compared in terms of delay and area. It is observed that the delay in column bypassing multiplier is 22.48% less than array multiplier and 21.25% less than row bypassing multiplier. It is also observed that the column bypassing multiplier has less area than array multiplier and row bypassing multiplier. Column bypassing multiplier has the least area among array multiplier and row bypassing multiplier.

1   INTRODUCTION

Multiplier is important in digital signal processing and various other applications. With advancement in technology, multiplier which offers high speed, low power consumption, less area etc. are been designed. Digital multipliers are amid the most significant arithmetic functional units in many applications, such as the Fourier transform, discrete cosine transform and digital filtering. Multiplication is most commonly used operations in various applications like math processor and in various scientific applications.

1.1   *Different types of multipliers*

1.1.1   *Array multiplier*
Array multiplier is well known due to its regular structure. Multiplier circuit is basically based on add and shift algorithm. Each partial product is generated by the multiplication of multiplicand with multiplier bit. The partial products are then shifted according to their bit orders and then added respectively. The addition can be performed with carry propagate adder. In multiplication, partial product is the sum of the number of bits of the two operands i.e. multiplicand and multiplier (Devi et al, 2015).

Let us consider the multiplication of 1111 * 0001 bit. In this, we have 4-bit multiplicand bit i.e. 1111 and 4-bit multiplier bit i.e. 0001. The multiplicand bit is multiplied with multiplier bit respectively. The half adder and full adder are used for addition of various bit. Various bits are added respectively and partial product is generated. Here we have a partial product of 8-bit i.e. 00001111.

Figure 1 shows the design of 4 bit array multiplier. Here, various half adder and full adder blocks are used for addition. The multiplication of multiplicand bit and multiplier bit is done and the partial products are generated.

1.1.2   *Row bypassing multiplier*
In bypassing technique, the idle part of the circuit which is not operating is shut down to save the power. In the row bypassing multiplier, if the multiplier bit is zero, the addition operation corresponding to that row is disabled which reduces the switching activity and hence the reduction in power dissipation can be achieved. In this technique, whenever certain multiplier bit bj is zero, then the addition operation in the j-th row is disabled. This technique totally depends on the number of zeroes in the multiplier. In this, the multiplier bits are bj can be used as the selection line of the multiplexer.

Let us consider the multiplication of 1100 * 1010 bit. In this, the multiplier bit consist of two zero bit. During multiplication, the corresponding adders of 1st and 3rd row will be disabled and the previous sum is considered as the current sum. In case of 8*8 row bypassing multiplier, of 16*16 row bypassing multiplier etc, more reduction in power dissipation can be achieved.

Fig. 2 shows the design of 4 bit row bypassing multiplier.

1.1.3   *Column bypassing multiplier*
Column bypassing technique means turning off various columns in the multiplier array each time certain multiplicand bits are zero. Figure 3 shows

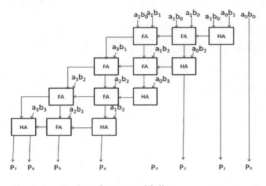

Figure 1. Design of array multiplier.

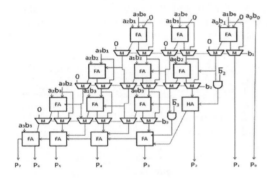

Figure 2. Design of 4*4 row bypassing multiplier.

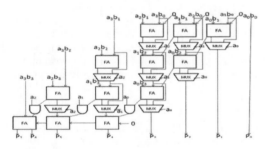

Figure 3. Design of 4*4 column bypassing multiplier.

the design of 4 bit column bypassing multiplier. In column bypassing multiplier technique, some diagonal columns are turned off using the bypassing technique whenever certain multiplicand bits are zero to save the power (Sahu et al, 2013). This technique completely depends on the number of zeroes in the multiplicand bit. In this, the multiplicand bits are ai can be used as the selection line of the multiplexer.

Let us consider the multiplication of 1100 * 1101. In this, the multiplicand bit consist of two zero bit, so the corresponding column of full adder in 1st and 3rd diagonals will be disabled and the

output carry of full adder is 0 and the sum is simply equal to the third bit which is the summation output of its upper full adder (Lin et al, 2014).

2 SIMULATION RESULTS

The design is coded using verilog HDL and the simulation is done using Modelsim simulator. The designs are synthesized and implemented in Xilinx ISE 14.7. Xinilx ISE is a design environment for FPGA product and it is primarily used for circuit synthesis and design. While ModelSin logic simulator is used for system level testing.

2.1 Array multiplier

Figure 4 shows the block diagram of an array multiplier. It is coded in verilog HDL. Here, a bit and b bit is input bit and p bit is partial product (Ramesh & Addanki Purna 2011).

Figure 5 shows the output waveform of an array multiplier in which 1111 bit is multiplied with 0001 and the output is 00001111.

2.2 Row bypassing multiplier

Figure 6 shows the block diagram of a row bypassing multiplier. It is coded in verilog HDL. Here, a bit and b bit is input bit and p bit is partial product.

Figure 7 shows the output waveform of 4*4 row bypassing multiplier (Lonkar et al 2012) in which multiplicand bit 0101 is multiplied with 0011 and the output is 00001111.

2.3 Column bypassing multiplier

Figure 8 shows the block diagram of a 4*4 column bypassing multiplier. It is coded in verilog HDL. Here, a bit and b bit is input bit and p bit is partial product.

Figure 9 shows the output waveform of 4*4 column bypassing multiplier in which multiplicand bit 1111 is multiplied with 0101 and the output is 00000110 (Chirde et al, 2015).

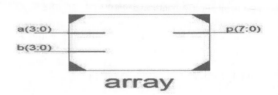

Figure 4. Block diagram of array multiplier.

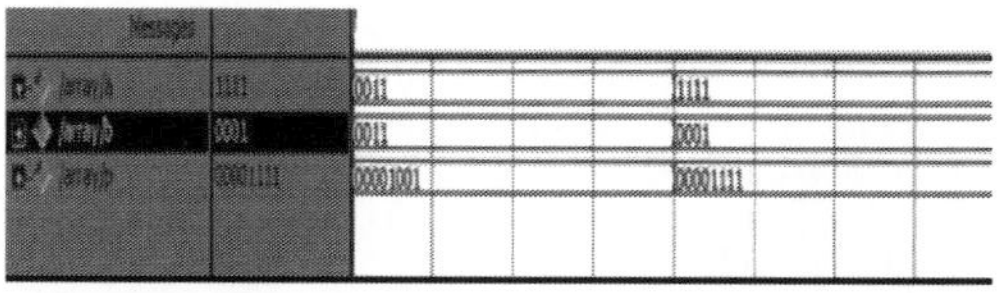

Figure 5. Output waveform of array multiplier.

Figure 6. Block diagram of 4*4 Row bypassing multiplier.

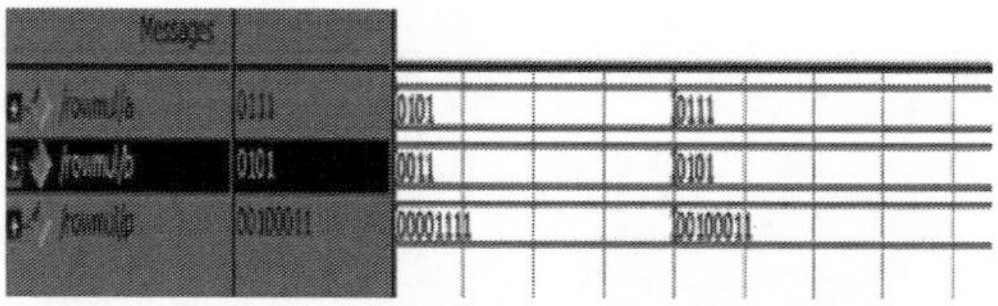

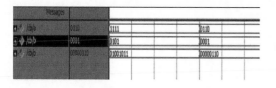

Figure 7. Output waveform of 4*4 Row bypassing multiplier.

a(3:0)
b(3:0)
p(7:0)

cb

Figure 8. Block diagram of 4*4 column bypassing multiplier.

Figure 9. Output waveform of 4*4 Column bypassing multiplier.

3 COMPARISON

The comparison of delay and area of array multiplier, row bypassing multiplier and column bypassing multiplier is done.

3.1 Comparison of time delay

By comparing the time delay obtained in synthesis report, we can decide the high sped multiplier. Table gives the comparison results of array multiplier, row bypassing multiplier and column bypassing multiplier.

From the Table 1, it can be seen that the array multiplier has the time delay of 15.101 ns followed by row bypassing multiplier which has 14.865 ns and column bypassing multiplier that has 11.706 ns. It is observed that the delay in column bypassing multiplier is 22.48% less than array multiplier. Also, it can be seen that column bypassing multiplier has 21.25% less delay than array multiplier.

3.2 Comparison of Area Utilization on the basis of number of slices

The number of slices in column bypassing multiplier is 15 followed by array multiplier that has 17 and then row bypassing multiplier that has 23. It can be observed that column bypassing multiplier has the least area followed by array multiplier and then row bypassing multiplier.

Table 1. Delay and area comparison of different multipliers.

S. No.	Array multiplier	Row by passing multiplier	Column by passing multiplier
1. Time Delay (ns)	15.101	14.865	11.706
2. No. of slices (out of 920)	17 (1%)	23 (2%)	15 (1%)

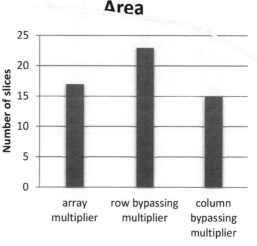

Figure 10. Comparision of area utilization in terms of number of slices of different multipliers.

4 RESULT ANALYSIS

From the table, it is understood that the time delay for column bypassing multiplier is better than any other multiplier. Hence, it can be used for high speed bypassing multiplier. The multipliers are compared in terms of delay and area. It is observed that the delay in column bypassing multiplier is 22.48% less than array multiplier and 21.25% less than row bypassing multiplier. It is also observed that the column bypassing multiplier has less area than array multiplier and row bypassing multiplier.

REFERENCES

Chirde, Vaishali S. Jadav, Usha 2015. Design of Adaptive Hold Logic (AHL) circuit to reduce Aging Effects. National conference on emerging trends in Advanced Communication Technologies (NCETACT): 26–29.

Devi, T. Mathi & Renganayaki 2015. Design of Low Power Vedic Multiplier using Adaptive Hold Logic. International Journal of Emerging Technology in Copmuter Science and Electronics (IJETCSE) 12.

Goyal, C. & Kaur, Gazel Preet 2012. Comparative analysis of low power 4-bit multipliers using 120 nm CMOS technology. International journal of engineering research and applications 2 (4): 551–111.

Lin, I.C., Cho, Y.H. & Yang, Y.M. 2014. Aging aware Reliable Multiplier Design with Adaptive Hold Logic. IEEE Transaction on very large scale integration (VLSI) System.

Lonkar, Rutesh L., Ashtankar, Pravin P. & Shriramwar, S.S. 2012. Analysis of Column Bypass multiplier. The International Journal of Computer Science and Applications (TIJCSA) 1(7).

Nujmun, Roobitha & Cheriyan, Jini 2013. Low Power Variable Latency Multiplier with AH Logic. International Journal of Science and Research (IJSR): 2319–7064.

Ramesh, Addanki Purna 2011. Implementation of Dadda and Array multiplierArchitectures using Tanner tool. International Journal of Computer Science and engineering Technology (IJCSET) 2.

Sahu, Prashant kumar & Meena, Nitin 2013. Comparitive Study of different multiplier architectures. International Journal of engineering and Technology 4(10).

Communication and Computing Systems – Prasad et al. (Eds)
© *2017 Taylor & Francis Group, London, ISBN 978-1-138-02952-1*

Analysis of microwave-tissue interaction using coaxial slot antenna during hyperthermia treatment

Amninder Kaur

Dronacharya College of Engineering, Gurgaon, Haryana, India

ABSTRACT: Microwave ablation is a rising therapy option for the treatment of many cancers and other therapeutic circumstances. In this therapy, Microwaves are directly applied to unhealthy tissues and the same are subject to thermal shock by raising rapid temperature. Heating and temperature increase cause other mechanical and chemical changes to the biological tissues. The engineering design of applicator differs whether the treatment is invasive or non-invasive, also with the consideration for factors such as duration of treatment and the technique. This paper will review the design and operation of a coaxial antenna used during the liver cancer treatment. The goal of hyperthermia treatment Planning is to define the power deposition pattern required to produce the necessary results-raising the temperature of the tumor in liver to therapeutic levels, while keeping the amount of normal tissue damage to acceptable minimum.

Keywords: Electromagnetic (EM), Hyperthermia, microwave ablation (MWA), Specific Absorption Rate (SAR)

1 INTRODUCTION

Biological materials absorb Electromagnetic (EM) energy which becomes heat and this rises the temperature of materials. The rise in temperature causes mechanical and chemical changes to the biological materials. This phenomenon was first noticed by Percy Spencer in 1946. He noticed that when he stood near a microwave radar source a candy bar in his pocket melted. This observation of microwave heating led to the invention of the microwave oven. Thermal therapy is a minimally invasive procedure for destroying tumors using heat (J. Carl et al 2005). Liver tissue of human body is one type of biological material. The basis of this research depends on the effect of microwave heating in liver tissue and the procedure is defined as microwave liver tissue ablation.

2 SPECIFIC ABSORPTION RATE

Biological materials are generally lossy mediums for EM waves with finite electric conductivity. They are usually neither good dielectric materials nor good conductors. Hence as the basic property of EM waves when made propagated though the biological materials, the energy contained in the EM waves is absorbed by the materials. The energy absorption so occurred is defined in terms of SAR i.e Specific Absorption Rate which is the power dissipation rate normalized by material density. It can be shown that:

$$SAR = \frac{1}{\rho} J.E = \frac{\rho}{E} |E|^2 \qquad (1)$$

where σ represents the tissue conductivity [S/m], ρ represents tissue density [kg/m³], and *E* represents the electric field around the tissue [V/m] (Lomako 2000). The equation for SAR shows that the electric field produced by antenna increases the absorption rate by square of its value and is equivalent to the heating source created by the electric field in the tissue. The SAR is a measurement used to quantify the localized deposition of microwave energy. The SAR is a measurement used to quantify and the localized deposition of microwave energy. The SAR units are watts per kilogram (W /kg). SAR denotes the time rate of electromagnetic energy absorption at a given location in the tissue. To determine the energy deposition, point-wise calculated SAR is averaged over a test-volume of tissue. It was also found that the thermal damage to the tumor by elevation of temperature is enhanced by the optimization of the SAR distribution significantly (Salloum et al. 2009).

3 DIELECTRIC PROPERTIES OF BIOLOGICAL TISSUES

Different tissues have differences in their dielectric properties. The material properties of the media(dielectric permittivity and magnetic permeability) effects the transmission of electromagnetic in which the waves propagate (Brace 2010). Tissue dielectric properties have very important roles in microwave tissue ablation. Dielectric properties of tissue directly affect the performance of the ablation probes—the microwave antennas. Liver tumor tissue has higher tissue water content. It has higher permittivity and higher conductivity than normal liver tissue. Stauffer reported that relative permittivity is 12% higher and electric conductivity was 24% higher for human liver tumor tissue than the surrounding normal liver tissue.

The differences of dielectric properties between normal tissues and tumor tissues agree with the relationship between tissue dielectric properties and tissue water content.

3.1 *Frequency dependence*

Tissues dielectric properties change with microwave frequency. Many studies have been conducted to determine the relationship between dielectric properties and frequency for different types of biological tissues. Gabriel summarized most of the previous researches in his 1996 publications. Gabriel suggested the empirical parameterized equation to approximate the measured dielectric properties of different tissues. At lower frequencies (less than 100 MHz) the flow of current to the extracellular regions is restricted by the cell membrane which provides high impedance. Cell membranes have a certain time period for charging. If the time period of the electromagnetic field is larger than this time period, the tissue behaves as a good capacitor. But at high frequencies the field may be much smaller than the charging time period of the cell membranes.

In contrast to the behaviour at lower frequencies the cell membrane at the higher frequencies more than 100 MHz gets short circuited due to increase in the capacitive reactance at such high frequencies. The electrical properties of the tissue are then determined by its water, salt and protein content. It is known that tumor tissues have significantly higher water content than the normal tissue. The reason for the increased water content is not known, but a suggestion has been made that it represents a tendency toward increased protein hydration. Hence the value of permittivity and conductivity of unhealthy tissues makes them suitable to be treated by using microwave frequencies.

3.2 *Temperature dependence*

It is well known that dielectric properties of biological tissues are dependent on temperature changes. Temperature dependence is difficult to measure and related publications are rare. Chin and Sherar measured dielectric properties of bovine liver tissues at 915 MHz in their ex-vivo experiments by heating the liver tissues to different temperatures [www.ncbi]. As per their study the relaxations of two main constituents of tissue i.e water and proteins changes the dielectric properties of liver tissue. This change happens due to heating caused to tissue found to be reversible and changes caused by protein denaturisation were found to be irreversible.

3.3 *Thermal responses of biological tissues during MWA*

Tissue temperature elevates when microwave power is applied. Tissue near the active radiation region of the microwave antenna absorbs more microwave wave energy and has higher temperature than tissue further away from the antenna.

Heat is also transferred from tissue at higher temperature to tissue at lower temperature by thermal conduction and blood perfusion in the liver tissue. The overall effect of the microwave power is to raise tissue temperature in a limited region near the antenna active radiation region. Liver tissue undergoes a few steps of different physical responses to the temperature elevation.

Table 1. Tissue physical response w.r.t temperature elevation.

Temperature	Tissue responses
<48°C	Increase of tissue blood perfusion because of intrinsic response of issue by enlarging blood vessel in order to reduce the tissue Temperature.
48°C	Cell depolarization-Heat-caused pore formation on cell membrane and an increase of membrane fluidity leading an overwhelming number of extracellular ions rushing into the cells and causing cell depolarization
<50°C	Cell physical changes are reversible
>50°C	Heat causes cell trans membrane ion pump activities to stop and eventually cell dies.
>90°C	Tissue water evaporates
>300°C	Tissue charring

4 HYPERTHERMIA

Hyperthermia is defined as the overheating of some selected tissues. Hyperthermia is an ancient but nowadays rapidly developing treatment method in tumor therapy (www.cancerjournal). However, there are some significant technical problems while applying the heat within or outside the body as it a complex issue during the dosing, the selection and the control of heat transfer. The whole process is mandatory for any acceptance of hyperthermia as a modern treatment modality so as to provide reproducibility and standardization. The operation of most of the clinical hyperthermia systems are based upon the exposure of unhealthy tissues to electromagnetic (EM) fields, Infrared radiations, RF waves or ultrasound (US) radiation.

The goal of Hyperthermia treatment planning is to maximize the volume of cancerous tissues raised to therapeutic temperatures while limiting the volume of healthy tissues being damaged to a clinically acceptable value. In most clinical hyperthermia cases heterogeneous temperature distributions are achieved for varying times, so it is difficult to achieve a homogeneous temperature distribution. Simulations presented in this paper for the treatment of cancer tissues present in the liver, are focused on the analysis of control of radiation patterns and power deposited within human tissue. Here a finite element model of a liver tissue is developed using COMSOL Multiphysics, to measure the power absorbed and the distribution of temperature around the coaxial antenna inserted in the tumor. It is a productive method for analyzing the complex structures which provides the flexibility of altering the shape, size, power and temperature distributions of the antenna.

4.1 *Hyperthermia treatment planning*

Hyperthermia Treatment planning includes the elevation of temperature of unhealthy tissues by applying microwave radiations with the insertion of coaxial antenna in the unhealthy part (tumor). Hence the characteristics of unhealthy tissue are vital for the effectiveness of hyperthermia treatment in conjunction with the therapy and duration of (Cabuy 2011). The power of the EM wave is absorbed by the cancerous tissues present in liver and this heats the unhealthy tissues. The tumor (unhealthy tissues) are destroyed when subjected to elevated temperature for a specific duration of time. This whole setup requires basic devices to perform a Microwave Ablation i.e a microwave generator, a microwave applicator (here the coaxial antenna), and a section of flexible coaxial cable to connect the antenna to the microwave generator. For the proper

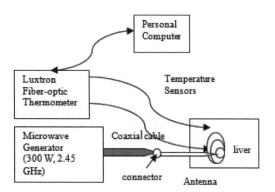

Figure 1. Experimental setup of Microwave Ablation Therapy.

placement of applicator in the tumor, Ultrasound scanners are often used in the MWA procedures. For measuring tissue temperature, Fiber-optic thermometers can be used. MRI scanners can be used to examine lesion size after the procedures. In this procedure, A MWA probe is placed into the tumor with an open surgery or a percutaneous procedure, guided by advanced biological imaging techniques. Microwaves are applied by using the probe which is connected to the Microwave power generator. The heating time for the procedure and power distribution are selected before in accordance to the shape and size of the tumor. A thermal lesion of predicted volume is created by the applied microwave heat to cover the entire tumor with 1 cm margin. Various imaging devices can be used to verify the lesion size and shape before the clinical procedure is finished entirely. The final objective of ablation technology which includes MWA, is to kill the tumor present in the liver while preserving healthy liver tissue effectively.

Figure 1 shows the Schematic of experimental setup of microwave liver tissue ablation. For a clinical procedure, the Luxtron fiber-optic thermometer and the temperature probes are not used.

5 MODEL DESIGN AND SIMULATION

COMSOL Multiphysics is a simulation tool by which all Technical and Engineering complications which are based on Partial Differential equations, can be solved and modelled. The problems based on coupled physics phenomenon can be solved by this software as it provides the provision for easily extending predictable models for unilaterally modeled physics phenomena's into Multiphysics modeled phenomena's. The problem is modeled in the software considering the physical boundary conditions which includes material properties, loads, constraints, sources, and fluxes with the help of built in physics mode present in this tool. This software uses a variety of numerical solvers in which the adaptive meshing and error in the model can be find with the help of finite element analysis (www.comsol).

5.1 Simulation using FEM

Finite element method is deployed for the conversion of physical model to mathematical model of the antenna. The Specific Absorption Rate (SAR) pattern generated after the insertion of selected the microwave coaxial antenna in the tumor is exhaustively studied (Surita 2012). Simulation by using Finite Element Method provides a well-organized method for analyzing the performance of multifaceted structures as it provides the option of altering the shape of antenna. The entire model of task space is distributed into thousands of smaller areas and the local function is used to represent the field in each area. The model is distributed into geometric models of huge number of tetrahedra, where four equilateral triangles form a single tetrahedron.

A Finite Element mesh is created by means of the collection of tetrahedral. The value of a vector field quantity at various points inside it is interpolated from the vertices of the tetrahedron. It stores the components of the field at each vertex that are tangential to the three edges of the tetrahedron. The component of the vector field that is tangential to a face and normal to the edge, its value at the midpoint of selected edges can also be stored. Maxwell equations of electromagnetic can be transformed in the form of matrix by representing field quantities which are solved with the help of traditional numerical methods. FEM models offer manipulators with fast and precise solutions for solving several systems of differential equations.

5.2 Analysis of using FEM—model definition

In Microwave Ablation Therapy, a thin microwave antenna is placed into the tumor during treatment. When the tumor cells are heated with the application of microwaves a coagulated area is produced where the unhealthy tissues are killed. The model is designed to calculate the distribution of temperature, the radiation pattern and the Specific Absorption Rate (SAR) by inserting a thin coaxial slot antenna in the tumor. Bioheat equation is used to compute the distribution of temperature in the tissue (Ito 2005).

$$\rho c \frac{\partial T}{\partial t} = \nabla . k \nabla T + SAR - \rho_{bl} c_{bl} w_{bl} \left(T - T_{bl} \right) \quad (2)$$

where ρ represents the density of tissue (kg/m³), c gives specific heat capacity (J/kg·K), k is thermal conductivity (W/m·K), ρbl represents density of blood (kg/m³), $c$bl gives the blood specific heat capacity (J/kg·K), $w$bl represents blood perfusion rate (kg/m³·s), Tbl represents temperature of blood (K), and SAR is the power distribution of microwaves per unit volume applied during the procedure (W/m³). The antenna geometry of the model consists of a thin coaxial cable which is given in Figure 2. The Antenna geometry contains a ring shaped slot cut which is 1 mm wide on the outer part of conductor and at a distance of 5 mm from the tip which is short-circuited. The antenna is bounded by a sleeve catheter made up of PTFE (polytetra-fluoroethylene) for hygienic purposes. 2.45 GHz frequency is applied to antenna as it is widely used in biomedical coagulation therapy.

5.3 Domain and boundary conditions

A transverse electromagnetic field characterizes the electromagnetic waves propagating in a coaxial cable. Assuming the field of time-harmonics with information of phase having multifaceted amplitudes, the value of electric (E) and magnetic field (H) is given by the equation (3) and (4) (Ortega et al. 2012):

$$E = e_r \frac{C}{r} e^{j(\omega t - kz)} \quad (3)$$

$$H = e_\varphi \frac{C}{Z} e^{j(\omega t - kz)} \quad (4)$$

where z represents the direction of propagation and r and z represents cylindrical coordinates on

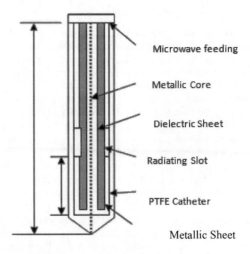

Figure 2. The geometry of the antenna.

Table 2. Boundary conditions for electromagnetism.

Parameters/boundary no	1,3	2,14,18,20,21	8	All other
Boundary condition	Axial Symmetry	Scattering boundary condition	Port	Perfect electric conductor
Wave excitation at this port			Selected	
Pin			1,5,10,15	
Mode specification			Coaxial	
Wave type		Spherical wave		

the central axis of the coaxial cable. P_{av} represents the rate of average power which flows in the coaxial cable. The equation for which is given below:

$$P_{av} = \int_{r_{inner}}^{r_{outer}} \text{Re}\left(\frac{1}{2}E \times H\right)2\pi r = e_z \pi \frac{C^2}{Z}\ln\left(\frac{r_{outer}}{r_{inner}}\right) \quad (5)$$

Here Z gives the value of wave impedance in the dielectric of the coaxial cable. r inner and r outer are The inner and outer radii of this dielectric is denoted by r_{inner} and r_{outer} respectively. ω specifies the angular frequency. The wavelength in the medium is related with the factor of a constant of propagation k given by the equation (6):

$$k = \frac{2\pi}{\lambda} \quad (6)$$

In this model all the external boundaries, boundaries between copper and Teflon and the outer boundaries of the liver tissue are important. Z-axis boundaries are set to be axial-symmetric boundaries. External boundaries of liver tissue and Teflon are set to be low-reflection boundaries except the boundary at the z axis. All copper boundaries are set to be PEC (Perfect Electric Conductor) boundaries. All external boundaries are defined as symmetry/insulation boundaries, including the axial-symmetric boundaries at $r = 0$. The geometry of the model is axial-symmetric which includes only a half portion of the geometry structures of antenna and liver tissue. Rest half portion of the geometry rotates along the z axis at $r = 0$ to form the whole geometry of the antenna immersed in the liver tissue.

6 RESULTS

Figure 3 shows the antenna geometry which includes the coaxial cable having a slot-cut in ring shape on the outer conductor. The tip of antenna is short circuited. The antenna is surrounded by a plastic catheter.

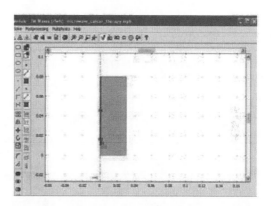

Figure 3. Modeling of antenna in COMSOL.

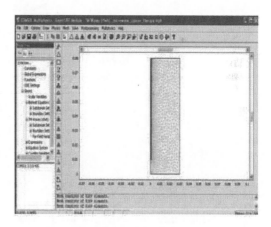

Figure 4. Mesh elements of Coaxial Antenna.

Figure 4 shows the mesh elements present in the model. There are 8169 mesh elements in the model. Figure 5 shows the distribution of temperature in the unhealthy tissues where the co-axial antenna carrying 10 W power, was inserted.

The temperature exhibits highest values near the antenna obeying the circle of influence phenomenon and its value gradually decreases with the increase of distance from the core. The value

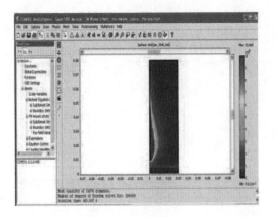

Figure 5. Simulation result: Temperature in the liver tissue for 10 W Power.

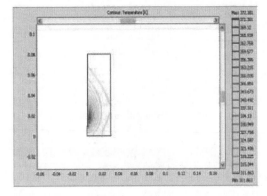

Figure 6. Line contour of temperature distribution.

of temperature at the outer surfaces reaches 310 K closer to the outer boundaries of the computational domain.

7 CONCLUSION

The same procedure can be applied to other body parts prone to cancer. Also the plot for tempera-ture distribution provides a suitable platform to conclude that tumors of different shape and sizes can be cured by varying the values of input power. The comprise between the increase in power should be cautiously balanced with the effects to the surroundings so that there is no damage of healthy tissues, no irritation to the patient and burning of skin will occur.

REFERENCES

Brace L. 2010, Microwave Tissue *Ablation: Biophysics, Technology and Applications,* Crit Rev Biomed Eng Vol 38(1): 65–78.

Cabuy E. 2011, *Hyperthermia in Cancer Treatment, Reliable Cancer Therapies. Energy-based therapies,* Vol 1(2):1–48.

Carl J. Kumaradas1 and Robert H. Kraus 2005:*A Parametric Design Optimization using FEMLAB for a Novel Treatment Method for Cancer,* Proceedings of the COMSOL Multiphysics User's Conference 2005 Boston.

Ito K. 2005: *"Numerical Calcu;lations of Heating Patterns around a Coaxial-Slot Antenna for Microwave Hyperthermia - Aiming at Treatment of Brain Tumor and Bile Duct Carcinoma-",* 2005 IEEE Engineering in Medicine and Biology 27th Annual Conference.

Lomako 2000: *Specific Absorption Rate Testing,* Article Radiofrequency Investigation Limited.

Ortega Palacious,R. A Vera L. Leija 2012 *"Microwave ablation coaxial antenna computational model slot antenna comparison",* 2012 Pan American Health Care Exchanges.

Salloum, M., Ma, R. and Zhu, L.2009, *'Enhancement in treatment planning for magnetic nanoparticle hyperthermia: Optimization of the heat absorption pattern',* International Journal of Hyperthermia, 25: 4, 309–321.

Surita Maini, 2012:*Finite Element Analysis for Optimizing Antenna For Microwave Coagulation Therapy,* Journal of Engineering Science and Technology Vol. 7, No. 4 462–470.

http://www.cancerjournal.net
http://www.comsol.com/showroom/tutorials/
http://www.ncbi.nlm.nih.gov

Communication and Computing Systems – Prasad et al. (Eds)

Organic cylindrical transistor: Analytical modeling and performance parameters extraction

Anshu Singh & Vishal Ramola
Faculty of Technology, Uttarakhand Technical University, Dehradun, India

Poornima Mittal
Department of Electronics and Communication Engineering, School of Engineering and Technology, Graphic Era University, Dehradun, India

Brijesh Kumar
Department of Electronics and Communication Engineering, Madan Mohan Malaviya University of Technology, Gorakhpur (Uttar Pradesh), India

ABSTRACT: Analytical model for surface potential and threshold voltage in Cylindrical Gate Organic Thin Film Transistor (CG-OTFT) is discussed in this paper. The electrical behavior and performance parameter of C-OTFT is extracted using analytical model. Additionally, depth analysis of the CG-OTFT in linear and saturation region is performed in terms of drive current, threshold voltage, mobility. Besides this, a comparative analysis of cylindrical-organic transistor with conventional OTFT device is discussed, concentrating on the measurable extent of the layers forming the device to decide real dissimilarities among the two geometries. At last, the model is validated with simulation results through exploratory outcomes.

1 INTRODUCTION

The organic transistors have risen as a progressive semiconductor innovation in view of their adjustable surface science, low temperature processing and their much straightforward manufacture forms contrasted with their inorganic partner. In addition, the mechanical flexibility of organic materials makes them naturally compatible with plastic, paper substrate making frivolous products. Flexible electronics is a novel technology for construction of electronic circuits by depositing electronic devices on flexible substrate for example synthetic, paper or even fiber (Bonfiglio et al. 2005). Tsumura et al. reported the first organic field-effect transistor in 1986, from that point forward (Tsumura et al. 1996), there has been huge advancement in both the improvement of new creation innovation and materials execution.

The growth of advance electronics has raise to understanding of devices with different geometries for different application. For example, cylindrical geometries are regularly used to get device size lessening without the event of short-channel impacts, like in conventional MOSFET (Auth et al. 1997). Cylindrical geometry has been utilized where long yarn-like structures are required

for adaptable hardware and e-material application. Cylindrical geometry is intended for size lessening with great bowing strength, hysteresis free operation and high pressing thickness, too. This field has as of newly in a solid attention for various purpose that are possibly because of this new innovation: Advance material textile scheme for biomedical observing capacities, man-machine interfaces, and so on Distinctive plan have prompted material transistor whose usefulness is given by the specific geometry and materials of the fiber utilized, or to interlace-designed transistor. Each of the beforehand specified devices were manufactured utilizing organic materials, which have ease of creation and adaptable mechanical properties.

This paper is divided into seven segments, including the present introductory segment I. This is followed by Section II where Organic cylindrical thin film transistor is discussed. This setup is employed for the analyses of the performance of CG-OTFT, fabrication flow of CG-OTFT. It is followed by the brief discussion about analytical modelling and performance parameter extraction of CG-OTFT in segment III. Furthermore, segment IV and V are devoted to the depiction of the model in the two different region (linear and saturation), respectively. In Segment VI, test outcomes for a cylindrical organic thin film transistor are contrasted with the created model. Finally, important conclusions about the recommended work in brief have been discussed in Section VII.

2 ORGANIC CYLINDRICAL THIN FILM TRANSISTOR

Manufacture of CG-OTFT starts with a metallic fiber core of yarn (fiber) that functions as the gate terminal electrode, which is secured by thin insulating (protecting) layer. After that point organic semiconductor layer is kept on the insulator. Source and drain contacts that can be comprised of highly conducting polymer or metal are framed through various procedures (either thermal evaporation or delicate lithography techniques) and the various materials used in organic devices and circuits are discussed in [Mittal et al. 2012, Kumar et al. 2013, 2014a-d]. The CG-OTFT schematic diagram is depicted in Figure 1.

Cylindrical organic thin film transistors can be made either on a solitary substrate of yarn or at the crossing point of two confined yarn. Up-to-date, only a few step have been done towards realization of organic-based e-textile devices.

Cylindrical OTFT is reported based on pentacene with two distinctive polymer gate electrodes. 1) Poly (4-Vinyl Phenol) (PVP) and 2) Poly (Vinyl Cinnamate) (PVCN) with a great bending capability. Author is observed an enhancement in mobility by 2.5 times for transistor with PVCN insulator when contrasted with PVP. Moreover, Maccioni et al. built up the pentacene based Cylindrical OTFT with Poly-3, 4-ethylenedioxythiophene: styrene sulfonic acid (PEDOT: PSS) and gold Source/ Drain electrodes.

When contrasted with gold, the device showed a change in mobility and threshold voltage by 50% and 45%, separately, with PEDOT: PSS contacts [Kumar et al. 2013, Shekar et al. 2004]; the execution parameters of various reported C-Organic Thin Film Transistors are compared in Table 1.

Figure 2 demonstrates the Fabrication flow of cylindrical OTFT in its basic arrangement. The creation of CG-OTFT starts with a metal cylindrical gate electrode; from there on spread it with a thin

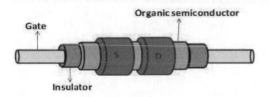

Figure 1. Schematic structure of cylindrical organic transistor.

Table 1. Extracted performance parameters C-OTFT.

Layers	I_{ds} (μA)	μ (cm²/ Vs)	I_{ON}/ I_{OFF}	V_t (v)	Voltage (V) V_{ds}	V_{gs}
OSC: Pentacene, S/D: Gold, I: PVCN, G: Al. Wire, Sub: Al	−10	0.53	4.2 × 10⁻³	7.05	−40	0 to −40
OSC: Pentacene, S, S/D: Gold, I: PVP, G: Al wire, Sub: Al wire	−7	0.24	2.5 × 10³	−4.7	−40	0 to −40
OSC: Pentacene, S, S/D: Gold, G: Polyimide, I: Polyimide, Sub: metallic fiber	−0.7	0.04	7 × 10³	−17.3	−50	0 to −100
OSC: Pentacene, S/D: PED OT: PSS, G: Polyimid I: Polyimide, Sub: metallic fiber	−0.3	0.06	3 × 10³	−9.6	50	0 to −100

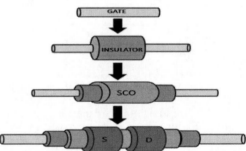

Figure 2. Schematic structure of cylindrical organic thin film transistor.

insulating layer. After this, the Organic semiconductor layers is kept on encasing layer, lastly source and drain contacts either with the metal or conductive polymer are shaped from side to side (warm dissipation or delicate lithography strategy).

3 C-OTFT ANALYTICAL MODELING: ELECTRICAL CHARACTERISTICS AND PERFORMANCE PARAMETER EXTRACTION

The working principal of organic transistors is same as the conventional MOSFETS; notwithstanding, the idea of channel creation is very diverse. Schematic perspective of the C-OTFT is appeared in Fig. 3, A metal cylindrical, with radius

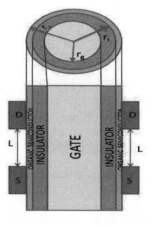

Figure 3. Schematic geometry structure of cylindrical OTFT.

Table 2. Simulation parameters.

Parameter	Symbol	Values
Dielectric permittivity	ε_s	$3\varepsilon_o$
Insulator permittivity	ε_i	ε_o
Density of free carriers	n_o	10^{17} cm^{-3}
Semiconductor doping	N	10^{17} cm^{-3}

r_g, goes about as the gate electrode, and is encompassed by a insulator layer, with thickness d_i and external radius r_i. The semiconductor encompasses the separator with thickness ds and external radius r_s. The S/D electrodes are the two outer rings and their separation, i.e. L is the directing channel length (Kranti et al. 2001).

In this section, we go for demonstrating the impact of the cylindrical geometry on the attributes of device uniqueness. All non-geometrical constraints that are utilized as a part of reenactments are settled to the qualities that are accounted in Table 2.

The above parameters are consider for the analysis of OTFTs. On the other hand, the model is universal and is also suitable for distinct parameters values.

4 ANALYTICAL OF CYLINDRICAL GATE ORGANIC THIN FILM TRANSISTOR (CG-OTFT) IN LINEAR REGIME

In this segment, the I-V equation of the device for the linear region operation will be found out. The key distinction with a conventional (planar) thin film transistor is in the V_{th}, which relies upon the radius of gate rg, as it is connected with the amount of free carriers that are accessible.

Suppose R is the region that controlled by the space of internal radius (r_i) and external radius (r_s)

at any segment of the circular structure. Then area (A) is equivalent to

$$Area\ (A) = \pi\left(r_S^2 - r_i^2\right) \tag{1}$$

The conductivity $\bar{\sigma}$ is given by

$$\bar{\sigma} = \frac{q\mu}{A} \iint_D n(r)\, r\, dr\, d\theta \tag{2}$$

where μ and q is the mobility of carrier and charge, and concentration of carrier at radius r is $n(r)$. Then resistance (dR) is equal to

$$dR = \frac{1}{\sigma} \frac{dz}{A} \tag{3}$$

Total carrier density is defined as

$$n(r) = n_0 + n_a(r) \tag{4}$$

where n_0 is free carriers of the semiconductor and $n_a(r)$ is the carrier that is localized at the boundary between the semiconductor and the insulator, with r_i is the radius. A capacitor with r_g (internal radius) and r_i (outer radius),

$$C_i = \frac{\varepsilon_i}{r_i \ln\left(\dfrac{r_i}{r_g}\right)} \tag{5}$$

here insulators dielectric permittivity is ε_i.

Then n_a is equal to

$$n_a(r) = \frac{C_i V_c}{q}\, \delta(r - r_i) \tag{6}$$

Here δ is define to depiction for the surface sharing of the carriers and applied voltage to the capacitor is V_c and is equal to $V_c = V_g - V_{fb} - V(z)$

By solving Eqs. (6), (4) and (2) leads to

$$\sigma = \frac{q\mu}{A}\left(An_0 + \frac{ZC_i V_c}{q}\right) \tag{7}$$

where Z is the channel width and is equivalent to

$$Z = 2\pi r_i \tag{8}$$

Thus drain current (I_d) is equal to

$$I_d = \frac{Z}{L}\mu C_i\left[\frac{q n_o A}{ZC_i}V_d + \left(V_g - V_{fb}\right)V_d - \frac{V_d^2}{2}\right] \tag{9}$$

Threshold voltage V_{th} is established, which is equal to

$$V_{th} = \pm\frac{q n_0 A}{ZC_i} + V_{fb} = \pm\frac{q n_o (r_s^2 - r_i^2)\ln\left(\dfrac{r_i}{r_g}\right)}{2\varepsilon_i} + V_{fb}. \tag{10}$$

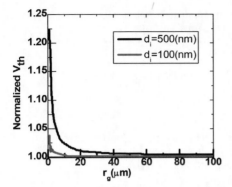

Figure 4. Characteristics curve for Normalized threshold voltage (V_{th}) versus gate radius (r_g) for different insulator thicknesses.

± Sign shows the majority carriers *i.e.* participate either electron or hole. Finally drain current equation is equivalent to

$$I_d = \frac{2\pi}{L} \frac{\varepsilon_i}{\ln\left(\dfrac{r_i}{r_g}\right)} \mu \times \left[\left(V_g - V_{th}\right)V_d - \frac{1}{2}V_d^2 \right] \quad (11)$$

For conditions $[r_g \gg d_i]$ and $[r_g \gg d_s]$, which generally used by several organic transistor, can be cut down to achieve (V_{thp}), in (12) which is the representation of the V_{th} for planar TFT, as illustrate by Horowitz *et al.* [Horowitz et al. 1998].

Fig. 4 demonstrates the V_{th} for a circular gate OTFT as a variable of the r_g for various thicknesses of the insulator (d_i).

$$V_{thp} = \frac{qn_o d_s d_i}{\varepsilon_i} + V_{fb} \quad (12)$$

5 ANALYTICAL MODEL OF CYLINDRICAL GATE ORGANIC THIN FILM TRANSISTOR (CG-OTFT) IN SATURATION REGION

In case of saturation regime, the channel slowly diminishing, so that no free carriers are present in the depletion area and the accumulation layer is restricted to the section of the boundary connecting the semiconductor and the insulator that has not been reduce. Suppose radius of the end point of depletion area is r_d and W is the width, and r_d is equal to

$$r_d = W + r_i \quad (13)$$

To find out the expansion of the depletion area, determine the voltage drop φ_s across the semiconductor. Then Poisson equation is given by

$$\varphi(r) = \frac{qN}{4\varepsilon_s}\left(r^2 - r_d^2\right) - \frac{qNr_d^2}{2\varepsilon_s}\left(\ln r - \ln r_d\right) \quad (14)$$

where N is carrier density, which is equivalent to doping of the semiconductor ($N > n_0$) and dielectric permittivity of semiconductor is ε_s

The equation for the planar device, *i.e.* $\varphi_P(x)$,

$$\varphi_p(x) = \frac{qN}{2\varepsilon_s}(x - W)^2 \quad (15)$$

The potential φs is simplified and equivalent to

$$\varphi_S = \frac{qN}{2\varepsilon_s}W^2 \quad (16)$$

Depletion width can be determined by the following equation

$$-V_g + V_{fb} + V(z) = V_i + \varphi_s \quad (17)$$

$$W_{DR} = \frac{-\dfrac{1}{C_i} + \sqrt{\dfrac{1}{C_i^2} + \left(\dfrac{1}{\varepsilon_s} + \dfrac{1}{r_i C_i}\right)\left(\dfrac{2(-V_g + V_{fb} + V(z))}{qN}\right)}}{\dfrac{1}{\varepsilon_s} + \dfrac{1}{r_i C_i}} \quad (18)$$

Figure 5 illustrates the depletion with versus thickness of oxide t_{ox}. For a cylindrical device, the depletion width is continuously smaller than the planar device width with the identical thicknesses.

Figure 6 shows the depletion regime width (W) is related to the gate radius (r_g). The elemental resistance is equal to

$$dR = \frac{dz}{q\mu n_0 \pi\left(r_S^2 - r_d^2\right)} \quad (19)$$

Thus

$$dV = I_d dR = \frac{I_d dz}{q\mu n_0 \pi\left(r_S^2 - r_d^2\right)} \quad (20)$$

Differentiating W with respect to V gives

$$dW = \frac{dV}{qN\left[\left(\dfrac{1}{\varepsilon_s} + \dfrac{1}{r_i C_i}\right)W + \dfrac{1}{C_i}\right]} \quad (21)$$

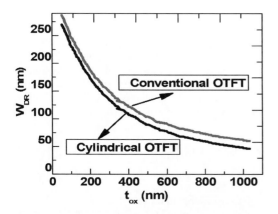

Figure 5. Characteristics curve for variation in dielectric thickness (t_{ox}) and depletion width (W_{DR}) of conventional organic transistor and Cylindrical OTFT.

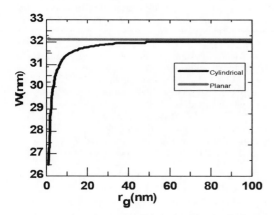

Figure 6. The characteristics for width of the depletion region versus gate radius curve.

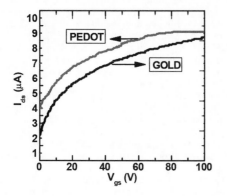

Figure 7. Transfer characteristics curve of Cylindrical OTFT structure with gold and organic electrode (PEDOT: PSS).

The pinch-off voltage V_p, is equal to

$$V_P = \pm qN\left[\frac{d_S^2}{2}\left(\frac{1}{\varepsilon_S} + \frac{1}{r_i C_i}\right) + \frac{d_S}{C_i}\right] \quad (22)$$

The drain saturation current becomes

$$I_{dsat} = \frac{Z}{L}\mu C_i\left(\frac{V_g^2}{2} - V_{th}V_g\right) + \frac{Z}{L}\mu C_i\frac{V_p V_{th}}{2}$$
$$+ \frac{q\mu n_0 \pi}{L}\left[-V_P d_S r_i - V_P\frac{d_S^2}{2} - \frac{qNd_S^3 r_i}{6\varepsilon_S}\right] \quad (23)$$

In condition that $r_i \gg d_s$ and $n_0 = N$ the threshold Voltage is equivalent to the pinch-off voltage ($V_p = V_{th}$), and all expression in the second line of (23) can be abandoned and drain current equation in saturation region is expressed by.

$$I_{dsat} = \left(\frac{Z \times \mu \times C_i}{L}\right)(V_g - V_{th})^2 \quad (24)$$

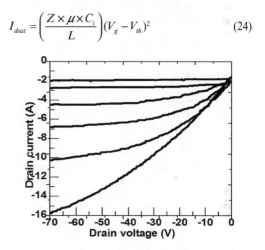

Figure 8a. Id-Vd characteristics curve cylindrical OTFT structure with PEDOT:PSS contacts.

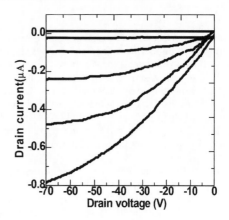

Figure 8b. Id-Vd characteristics curve cylindrical OTFT structure with gold contacts.

661

Table 3. Electrical parameter of simulated analytical model of cylindrical OTFT with PEDOT: PSS electrodes device.

Device	V_{th} (V)	μ (cm²/ vs)	I_{ON}/ I_{OFF}	α	k	R_S
PEDOT: PSS	−9.6	0.06	3×10^3	0.85	0.96×10^{-3}	14
Gold	−17.6	0.04	7×10^{-3}	0.66	1.9×10^{-3}	32

6 RESULTS AND DISCUSSION

The cylindrical OTFT behavior is analyzed using organic module of Atlas Silvaco TCAD simulation. The transfer characteristics analysis is shown in Figure 7.

The output characteristic curves of a cylindrical transistor with PEDOT: PSS contacts and gold contacts are shown in Figure 8(a) and (b). This characteristics curve shows output characteristics of cylindrical OTFT structure extracted by analytical modeling. It is important to note that building up a dependable S/D contacts was especially troublesome because of the low measurements furthermore to the non-planar surface of the wire. Notice that, although above stated problem and exceptionally small value of Z/L such as 1.2 and on-off current ratio is practically high. It is about 10^4. Keeping in mind the end goal to infer a dependable opinion of the electrical parameters, we fixed the I-V curve with a model that makes the cylindrical shape of the transistor.

The results of simulated and analytical modeling of cylindrical OTFT with organic electrode (PEDOT: PSS) and gold electrode are depicted in Table 3.

7 CONCLUSION

In this paper, we have discussed the model for cylindrical gate-Organic Thin Film Transistor (OTFT). The validation is done for proposed analytical model results with simulated results. We observed good agreement between simulation results with analytical results. We have achieved to a shape that portrays the electronic behavior of OTFTs through circular gate shaped. The model turns out to be fundamentally same to its known planar comparable device.

Gate radius is much higher than the insulator and semiconductor thicknesses. The insulator thickness can likewise contain a critical role to extract the value of the threshold voltage (V_{th}). Depletion regime thickness, prompting significant contrast in the behavior of device characteristics. Its structure variable and the utilized matter is completely perfect with a textile process. This transistor has appeared extremely intriguing performance, through distinctive values of the electrical parameters including on-off current ratio, mobility (μ), threshold voltage (V_{th}) exceptionally similar to those of conventional OTFT device. The flexibility of CG-OTFT structure is very much useful for textile industries.

REFERENCES

Auth, C.P. and Plummer, J.D. 1997. Scaling theory for cylindrical, fully depleted, surrounding-gate MOSFETs. IEEE Electron Device Lett. 18(2): 74–76.

Bonfiglio, A., Rossi, D.D., Kirstein, T., Locher, I., Mameli, F. Paradiso, R. and Vozzi, G. 2005. Organic field effect transistor for textile application. IEEE Trans. Inf. Technol biomed 9(3): 319–324.

Horowitz, G., Hajlaoui, R. and Kouki, F. 1998. An analytical model for the organic field-effect transistor in the depletion mode. Application to sexithiophene films and single crystal. Eur. Phys. J. Appl. Phys. 1: 361–367.

Klauk, H., Schmid, G., Radlik, W., Weber, W., Zhou, L., Sheraw, C. D., Nichols, J and Jackson, T. N. 2003. Contact resistance in organic thin films transistors. Solid State Electron. 47(2): 297–301.

Kranti, A., Haldar, S and Gupta, R. 2001. Analytical model for threshold Voltage and I-V characteristics of fully depleted short channel cylindrical surrounding gate MOSFET. Micro electron. Eng. 56(3): 241–259.

Kumar, B., Kaushik, B.K. and Negi, Y.S. 2013. Static and dynamic analysis of organic and hybrid inverter circuits. Journal of Computational Electronics. 12(4): 765–774.

Kumar, B., Kaushik, B.K and Negi, Y.S. 2014a. Organic thin film transistors: Structure, models, materials, fabrication, and applications: A review. Polymer Review. 54(1): 33–111.

Kumar, B., Kaushik, B.K., Negi, Y.S. and Goswami, V. 2014b. Single and dual gate OTFTs based robust organic digital design. Microelectron. Rel. 54(1): 100–109.

Kumar, B., Kaushik, B.K. and Negi, Y.S. 2014c. Design and analysis of noise margin, write ability and read stability of organic and hybrid 6-T SRAM. Microelectron. Rel. 54(12): 2801–2812.

Kumar, B., Kaushik, B.K. and Negi, Y.S. 2014d. Perspectives and challenges for organic thin film transistors: Materials, devices, processes and applications. J. Mater. Sci. Mater. Electron. 25(1):1–30.

Mittal, P., Kumar, B., Kaushik, B.K., Negi, Y. S. and Singh, R. K. 2012. Channel length variation effect on performance parameters of organic field effect transistors. Microelectron. J. 43: 985–994.

Shekar, B.C., Lee, T and Rhee, S.W. 2004. Organic thin films transistor, material, processes and devices. Korean J. Chem. Engg. 21(1): 267–287.

Tsumura, A., Koezuka, H., Ando, T. 1986. Macromolecular electronic Device: Field effect transistor with a polythiophene thin film. Appl. Phys. Lett. 49(18): 1210–1212.

Communication and Computing Systems – Prasad et al. (Eds)
© 2017 Taylor & Francis Group, London, ISBN 978-1-138-02952-1

Novel stress calculation in parallel processor systems using buddy approach with enhanced short term CPU scheduling

Swinky Arora
Department of CSE, LLRIET, Moga, India

Rohit K. Bhullar & Lokesh Pawar
Department of CSE, Chandigarh University, Gharuan Mohali, India

Amit K. Manocha
Department of Electrical Engineering, Maharaja Aggarsain University, Baddi, India

ABSTRACT: Resolute and stable process assignment among processors is a canny effort as such assignments require priori information about the present processor load only then the allocations can be made. First or initial policy structure have no information about processor workload but when the process or task groups arrive and get scheduled for processing, this information acts as a prerequisite for next allotment. Task of workload stability is complicated and tricky when static scheduling strategies are used in allocation, as these strategies need supplementary efforts to balance the load because the techniques designed so far does not follow load stability criteria while performing process or task allocation; instead they follow task adjustment to equalize or balance the load later on that enriches the overhead of context switching of processes among processors. Context switching not only includes the efforts for changing program or process address space, but the position in ready queue is also affected. Work proposed in this paper performs and executes load stability with viable priori information about processor utilization, depending upon and based on this metric value processor space is partitioned or fragmented among different categories. Based on the load status and scenario, processors are categorized and labeled and an appropriate set out of those is figured out that act as buddy or friend for others and handles incoming process queue. Stating in other words stress free "buddy set" handles inbound or incoming job queue in place of their "stress underneath buddy set" processors so as to eliminate the problem and losses of context switching and tasks adjustments after the tasks allocation.

1 INTRODUCTION

Evidently, all computers nowadays are parallel from hardware point of view and multiple systems are configured and interconnected by a network connection to form parallel computing environment. Virtually all the systems whether they are loosely-coupled or tightly-coupled try and intend to achieve parallelism in computing up to a greater extent, which improves the performance of computing. These networks based parallel systems need minimization in recursive bi-section that is, minimizing delay in messages or data communication. Now a days, focus lies on to achieve maximum efficiency from processor inter-connections without performance depreciation. To do that various scheduling strategies are implemented and regularly modified along with an overhead of load stability. The Load stability metric is actually and indeed meant for estimation of stress on processors in the terms of work-load assigned at regular intervals (Kumar 2015) and K. Bhullar 2016). Though some policies considers or follows load stability at earlier stages of load distribution also known as preload balancing technique or mechanism but other considers later on during overloaded system states also known as post load balancing method (Singh 2013). Some policies depend on architecture some are universally changeable to any architecture. With this, some policy structures are based on shared memory inter-connection where the policy environment depends on socket based message passing paradigm. To make consistent system the state interaction among processors should be as less as possible so that processors spend most of the time in doing their own designated work. The quantum of computation assigned or allocated to each processor must be in balanced form so that some processors do not become idle while others are performing heavy and large computation. Basically, on the basis of execution behavior, three categories or classes of job structure may arrive first class of jobs whose computation size is known at the beginning or starting of allocation. Second category of job arrivals whose computation size is available initially but as the computation progresses, the real time needed by the task varies. Third category of job arrivals exists where jobs generate dynamically in the system; also out of order and arbitrary batches of jobs are encountered. So, scheduling strategy and load balancing algorithm must be adaptive in nature. Scheduling and assignment of jobs over processor interconnection is a co-operative process requires supportive interaction with the scheduler. The paper has been divided in to eight sections with section five presenting the proposed work and in section six results has been presented followed by conclusion and future scope.

2 RELATED WORK

Lot of existing literatures has been produced in this area with a view to improve the performance of the parallel processors and maintaining consistent allocation. More than five dynamic scheduling strategies were introduced at very early stages for implementing load sharing and balancing in parallel system. Sender/receiver initiated diffusion called SID/RID in short form is an approach that uses nearest neighbor information to migrate the load evenly among processors b sending load information to each other periodically. High Performance Computing (HPC) will require all these strategies to produce consistency in the system. With this work done leads to the development of more complex system. Similar other approaches like Hierarchical Balancing Method (HBM) that balances the processor load after re-arranging systems' computational components or modules in the hierarchy. Same types of approaches have been based on processor folding into multiple dimensions like Dimension Exchange Method. Strategies proposed had been effective in their approach but have increased complexities of task migration after scheduling. The Gradient Model (GM) which approximates over-loaded and under-loaded processors, again considers task adjustment or task migration (Kumar 2015). The proposed approach in this paper is similar to the gradient method like identifies over-loaded and under-loaded processors for further scheduling of in-bound job arrival. Looking at the other side heterogeneous processor inter-connection produce one new dimensional aspect to the parallel scheduler implementation, lot many of the processor parameters can be taken into consideration e.g. Million Instructions Per Seconds (MIPS) and

Frequency Based load estimations and Scheduling (FBS) and Floating point Operations Per Second (FLOPS) (Arora 2013).

3 SCHEDULING PROCESS CHARACTERIZATION

Generally, for implantation of a parallel scheduler requires lot of intermediate tasks which should be consistently completed on time. The constraint of synchronization among each core activity is required with an increased level of abstraction. Most of the work on parallel processing techniques has synthetic workload experimentation model to explore and find high level attributes of the system. Lots of simulation designs have been the best platform to analyze and check the behavior of any parallel scheduling policy. Though real implementation of simulation may require lot many additional tasks to complete but provides an idea to structure a benchmark or a threshold level. Below are some important points which must be well known beforehand while implementing a parallel scheduler.

3.1 *Processor scheduling*

Scheduling of simultaneous and instant arrival of jobs over parallel environment is a main part of multi-processor scheduling algorithms (Joel 2003). Approaches designed so far tries to place the processes on the processors in such a way that the efficiency of processors is never overlooked and the processors are completed on scheduled time without getting stuck in the processor's queue. Existing approaches that are followed have a keen focus on the optimum tasks placement while other work on task re-allocation and adjustment of the processes so as to balance the load at later stage when the system state gets over-loaded.

3.2 *Thread and process level parallelism*

The structure of the scheduling strategies is designed according and in corroboration to the process behavior. Strategies may consider time sharing and space sharing aspects where the former allocate many processes to the present available processors and later allots multiple processors to a single active job (Arora 2013). This gives a facility to achieve task/process and thread level parallelism. Almost all parallel jobs have many numbers of active threads. So, this needs space sharing problem division. On the reverse side number of concurrent arrivals currently ready to get scheduled makes process level parallelism.

3.3 Dynamic and static scheduling policies

Mechanism of scheduling approaches further requires and involves static and dynamic allocation (Andersson 2001). The static approach allocate the processes to the processors in a non-preemptive way that means the processors will not leave the control while a job is under execution over them. In reverse the dynamic scheme works as a preemptive one in which the jobs currently running can be preempted based on priorities or interrupts (Govardhan 2013). Static scheduling approach considers only the basic features of the configuration like the number of jobs in the queue, processor queue size, number of waiting jobs and the burst cycle requirement of jobs (Joel (2003)). The dynamic approaches or policies in addition to above parameters also takes into account the varying processor load, their computing efficiency, the maximum load that all the processors can handle and many more. Dynamic scheduling is more effective in terms of its flexibility and adaptability to the changing system situations but is also complex to implement (Singh (2009)). Static scheduling on the other different side is quite less complicated and is easy to implement. The dynamic approaches follow task re-scheduling and process context-switching after distribution for load uniformity and balancing.

3.4 Heterogeneous processors and scheduling

The processors architecture connected in a parallel environment can have either homogeneous or heterogeneous configurations. The homogeneous architecture and configurations have all the processors with same clock frequency whereas the heterogeneous systems have all the processors with different clock frequencies (Arora 2013). The scheduling approaches or policies are designed differently for these configurations. The efficiency of the processors is far least compromised if dynamic policies and heterogeneous configuration of processors is pursued. All such policies involve maximum efforts in load distribution as frequency and clock cycles are different and load must be scheduled accordingly. Some time it might be possible that high frequency processor may have less computation intensive workload or tasks and vice versa.

3.5 Context switching and process rescheduling

Despite the complexity involved, context switching of processes plays a critical role among the processors queues with which the load automatically gets normalized and balanced among them but these approaches have an associated overhead associated with them as the processes are rearranged many times to stabilize and normalize the load among the processors. Many a times process waiting time may increases as increased context switches may occur during its execution life span (L.Fet (2007)). The problem of load stability and load balancing may arise in case of static policies and homogeneous configurations where all the processors have same operating speed to run the jobs with different process' cycles' requirements and no preemption is allowed; the jobs once allocated are not changed. In scenario arises the need of following a load balancing procedure so that all the processors have almost same amount of workload to execute and their efficiency will not be affected (Kumar 2015).

3.6 Scheduling queues and their impact

Despite of other aspects, scheduling queues are the major component in a parallel environment. There are many queues in almost every parallel configuration. The top level scheduling is handled by a global queue also known as long term scheduling queue from where the jobs enters in the controller and then controller filters the processes to select most suitable jobs for dispatching them to the processors in the system. One other queue is the local queue also called short term scheduling or scheduler that is available to an individual processor and it's the point from where jobs are picked up by the processor for execution (Arora 2013). Even several other queues are also used and maintained like waiting, priority etc. The usage of all these different queues provides an opportunity to implement various scheduling schemes on different processors configurations. Every policy implementation should be well defined and designed and must achieve performance as per the parameter given.

4 COMPARISONS OF SCHEDULING POLICIES

Contemporarily, task or process scheduling at operating system level involves multi-programming and time sharing systems. Focus has been on to select best possible job for the processor at any given point of time. Typically single controller and single processing unit was available for which the priori task/process selection operation was performed and a ready queue was maintained. In earlier days, process scheduling was not a major issue to be bothered; either a single job was executed or batch processing of jobs were carried out but with the passage of time and parallel computing came into existence, the task of process scheduling became a prime concern while designing of a job controller subsystem. Scheduling jobs on a single processor is bit easy as it only considers which job to execute but on many processors, two dimensions are required i.e. which job to execute and on which processor which complicates the task of scheduling. Despite this, all the scheduling strategies intend to enhance the computational capability and capacity of the systems. Different scheduling strategies are:

FCFS (First Come First Serve Scheduling)
In this Scheme the jobs are dispatched to the processors in the order in which they arrive in the global queue and are executed in non-preemptive manner. This scheduling technique does not consider the parameters like the computation cycles a required by a particular job will take and the processors configuration. FCFS technique is very easy in its implementation but pay little attention to the effective scheduling of jobs to increase processor's efficiency because the average waiting time for the processes is quite large in this case.

MCM (Minimum Computation Mass)
In this policy the job which has weight mass is minimum in the batch is allocated first to the processor. Each and every job has a predefined weight or mass in terms of burst cycle, which is interpreted or calculated prior to the scheduling of jobs among processors. Job with the minimum computation requirement is scheduled first on the processors and so on. This policy is more effective in case of single processor based systems as it enhances the efficiency of processor by improving the average waiting time for jobs. Shortest job selected also increases overall throughput of the system.

HCTS (Highly Critical Tasks Selection)
This is a priority based scheduling technique. In this technique the scheduler selects high critical task from a list of jobs and assigns apriority index to each of them. A priority order is maintained soon after resolving intermediate dependencies among jobs. Such type of jobs which are considered highly critical must be completed with priority according to a predefined sequence. In such real time critical environment, dispatching and scheduling is performed according to this precedence. The loop-hole in this technique is that it causes starvation that means that the low priority jobs may never get a chance to run until a higher priority task releases the CPU. SJF is a type of priority based scheduling in a non-real time system.

FSBP (Fair Share Batch Partitioning)
In this policy scheduler partition's the current batch according to the available processor space. More than one and multiple job may be allocated to each processor. An equivalence division metric is applied during task allocation. These types of methods are static in nature because consideration is on number of jobs in the present batch not on their size, computation requirements etc. At last at each individual CPU level, execution time is allotted among processes in interleaved fashion. On the other facet of the policy the scheduler distributes fair share (equal share) of each batch among available processors in the interconnection so that concurrent and simultaneous operations are performed in an overlapped fashion (Arora 2013).

BBS (Bounded Buffer Schemes)
In this policy the jobs are distributed chronologically to the number of available processors as long as there is a free space in processors buffer queue. Intrinsically this acts as an equivalent partitioning of currently available ready job set. The division is performed chronologically that schedule jobs according to their arrival time. This policy is static and is used in multiple processor systems where each processor implements RR (Round Robin) scheduling in its local queue. LTS (Long Term Scheduling) can either be FCFS or SJF in the global queue. A long term queue is maintained each time a bunch of job arrives; the scheduler performs allocation critically and consistently against these queues; although pre-fetch buffers are also deployed for interaction with processors cache (L. Fet 2007).

ESJE (Effective Single Job Execution Policy)
In this policy short term scheduler of each processor pick up a single process or job from global queue and execute it till its completion. The processors are lightly loaded in this scheme as they have only one job to execute in single unit of time. The scheduler may also use pre-fetch buffers for early selection of next allocation (L. Fet 2007). This strategy is a fixed or static non preemptive in nature.

MJQL (Minimum Job Queue Length)
In this technique selection of the appropriate processor for the next incoming job depends upon the queue size. An in-bound job is assigned to the processor which has minimum number of pending jobs in its local ready queue. The local scheduling is done using traditional RR (Round Robin) scheduling. This particular policy is dynamic in its working and requires prior information about number of jobs in the ready queue of each processor.

LRW (Least Remaining Work)
This strategy is basically a load balancing type of scheduling strategy where the jobs are given to those processors which have minimum pending work to complete or execute. This particular technique works in homogeneous environment and also considers that all processors have same internal configuration. The policy method and structure periodically estimate the present load of each processor and the processor which has minimum quantum of work remaining is the intended processor for the current job (Kumar 2015).

CBS (Cycle Based Scheduling)
This scheduling policy works well in heterogeneous configuration and is has been the improvement

over LRW (Least Remaining Work) policy structure. The in-bound jobs are allocated to the processor which can handle them easily and complete them swiftly. This policy firstly calculates the current load on each processor and then figures out remaining time required to compute that load and incoming job load. Reaming time or left out time is calculated with respect to total processor clock frequency and pending work. After that the processor which will take minimum computation cycle will get that job.

QRA (Quicker Round Robin Access)

In this particular strategy quick task allocation is followed; this policy measures the round robin cycle length of each processor's ready queue. Every cycle length measured from the index of currently executing job till end of corresponding ready queue. This metric provides swift access to next inc-bound job, as the job is allocated to the processor which has shorter round robin cycle. This particular policy structure is different from MJQL policy in which the scheduler determines the number of jobs presently available in the ready queue but QRA determines the round robin cycle length from the index of presently executing job till the end of ready queue.

TAGS (Task Assignment by Guessing Size)

In TAGS strategy, jobs are allocated as soon as they arrive but if it is observed that a job is running for too long in a processor, it is then terminated and restarted either from the scratch on some other different processor or some of its modules may be distributed to other processors for load stability. Such type of tasks must have changeable structures. Stating in other terms, the behavior of the job can be changed as per the current system needs (Kumar 2015). This technique speculates the

size of the current job and based on that guess, it performs the task change and shuffling.

ORBITA (On Demand Restriction For Big Tasks)

ORBITA is policy which is an improvement over TAGS in the way that it collects priori information regarding current ready jobs. In this policy, a processor will run a job only if it will not take too long to execute else it may reject that particular job. All this information regarding the current job is collected from past execution history of same kind of jobs. This particular method is quite good in load balancing of jobs on multiple processors since the processors can somehow decide which work load to accept and which to reject.

5 PROPOSED WORK

BSPI with round robin short term scheduling policy

The policy BSPI (Buddy Set Processor Identification) implemented using first come first serve strategy in the previous discussion is modified in this proposed work where the long term scheduling follows the chronological order as in previous case but the short term scheduling is changed from first come first serve to round robin scheduling. The job controller places the jobs in processors queue according to chronological order that means jobs are placed in a frequent arrival stream. The basic idea here is to identify the processors which can easily handle these jobs without getting into an extra stressing condition. The strategy proposed in previous work focused on completing the jobs in the order in which they arrived where the processors mutually handles each other's stress by becoming the buddies. The existing strategy has been modified in a way that the processors' local queue scheduling has been changed to round robin so that the number of jobs executed per unit time by the overall system is increased as compared to FCFS scheduling. This comparison is done by maintaining a separate thread that count the number of processors who's PUF remains greater than 50% in both scheduling strategies.

Initially all processors interact with the global queue and receives jobs from it. The processors completes the task using round robin scheduling where a time quantum of almost 4 ms is given to each job and the cycle is repeated for all the jobs in queue. Then a measure called, Processor Utilization Factor (PUF) is computed that calculates the number of jobs completed by a single processor in a unit time. The processor unable to complete a good number of jobs is considered under stress and the rest of the processors become its buddies and take the rest of the jobs from the job controller without disturbing the stress underneath proces-

Table 1. Measures for comparison of scheduling policies.

Policy	Preemption of tasks	Migration of tasks	Task walkout	Overall load stability
FCFS	No	No	Yes	No
MCM	Yes	Yes	No	No
HCTS	No	Yes	No	Yes
FSB	Yes	Yes	No	No
BBS	No	Yes	No	Yes
ESJF	No	No	No	Yes
MJQL	No	No	No	Yes
QRA	No	No	No	Yes
LRW	No	No	No	Yes
CBS	No	No	No	Yes
TAGS	Yes	Yes	Yes	No
ORBITA	Yes	Yes	Yes	Yes

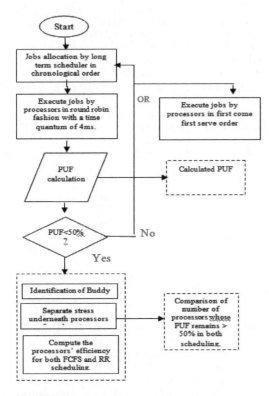

Figure 1.   Flow of work.

sor. By following the round robin fashion instead of first come first serve in CPU queue enhances the overall system efficiency as using round robin reduces the average waiting time for the processes in queue and number of processors under stress decreases.

The scheduling policy is summarized as follows:

Job controller dispatches the jobs queued in a variable length queue to the homogeneous processors in chronological order.

The processors execute the jobs given to them in round robin fashion where each job is given a fixed time quantum by the processor and this cycle continues until the job completes its execution or the queue of processor becomes empty and also in first come first serve order so that a later comparison can be carried out to determine overall efficiency of the system.

A parallel thread calculates the PUF of each processor to estimate the stress on it.

The processors are divided into categories of under stress and stress free processors. The processors whose PUF is less than 50% are considered under stress and the ones whose PUF is more than 50% are considered stress free and thus become the "buddies" of the stress underneath processors.

The stress underneath processor is not assigned any new job by the job controller until its PUF reaches a value greater than 50%.

This whole process is repeated and a thread is maintained parallel with the scheduling thread that computes the overall efficiency of the system when FCFS and RR scheduling is followed as a short term scheduling in processors' queue.

Flow chart

Below figure shows the flow of the proposed work implemented in this dissertation work

Parameters and Metric used

The main parameters here are the calculation of processor utilization factor and overall efficiency of system using comparison between two scheduling policies i.e. FCFS and RR scheduling.

Let "f" be the frequency of processors in Ghz where "f" can be any integral number.

Calculate the following metrics:
Number of processors' cycles per second:

$$Cyc\_Per\_Sec = f \times 10^9 \qquad (1)$$

Seconds consumed for one processor cycle:

$$T_{one_{cycle}} = \frac{1}{f * 10^9} \qquad (2)$$

Stress on processors:

Let β_i specifies queue length at i^{th} processor (i ranges from 1 to 8; there are 8 processors in simulation)

Work assigned to processor in current time unit:

$$Pro_{work_{assigned\,i}} = \sum_{m=1}^{\beta_i}(Job_m\_comp\_cyc\_Pro_i) \qquad (3)$$

The jobs given by job controller to processors may have computation time greater than the computation cycles (cyc_per_sec) of the processor. Thus one cycle of processor is not enough for job completion and incoming cycles are consumed for overall job completion. Thus calculate the pending work of the processor as sum of remaining cycles the crent job require and the computation requirements of the next jobs of the batch.

$$Rem_{job_{cycle\,i}} = curr\_job\_cyc_m - avail\_cyc\_p_i \qquad (4)$$

Pending work:

$$Pro_{work_{pending\,i}} = \sum_{m=1}^{\beta_i}(Job_{mcomp_{cycP_i}} - avail\_cyc\_p_i) \qquad (5)$$

Work Completed:

$$Pro_{work_{completed_i}} = \sum_{m=1}^{\beta_i} (Proc\_cycle_i\_given\_job_m) \quad (6)$$

(Now calculate Processor Utilization Factor (PUF). It considers the number of cycles processor consumed in accordance with the work assigned to it and the work completed by it.

Processor Utilization Factor (PUF):

$$PUF = \sum_{i=1}^{p} \frac{P_{work\_completed_i}}{P_{work\_assigned_i}} \times 100 \quad (7)$$

The PUF is calculated for both FCFS and RR scheduling and a metric is maintained, in common terms a counter is maintained that calculates the number of processors that remained in stress underneath conditions in both scheduling schemes and the efficiency of the system is considered on the basis of the number of stress underneath processors in such a way that the scheme in which number of stress underneath processors are less is considered to be more efficient short term scheduling scheme as in that scheme the number of jobs completed per unit by more stress free processor is more and there is less burden over the job controller to identify the buddy sets.

Algorithm

```
Step1: Initialization
        Set BC:=0
        Set SC:=0
        Set RR_counter:=0
        Set FCFS_counter=0
Step2: For each Processor Pi do where i=0 to n
Calculate Processor Utilization Factor PUF from
equation (7)   [If PUF >=90% then]
                SetFreeState(Pi,1)
                AddBuddySet(Pi)
        BuddySetCount(BC,1)
                Alloc(pi,GlobalQueue)
                [If scheduling is round robin then]
                RR_counter:=RR_counter+1
                Else
                FCFS_counter:=FCFS_counter+1
        [Otherwise Test PUF>= 75% then]
                Stress startup state: still managea-
                ble can be assigned new job; add in
                buddy set
                SetInitiationState(Pi,2)
                AddBuddySet(Pi)
        BuddySetCount(BC,1)
                Alloc(pi,GlobalQueue)
                [If scheduling is round robin then]
                RR_counter:=RR_counter+1
                Else
                FCFS_counter:=FCFS_counter+1
        [Otherwise Test PUF >=50% then]
```

Stress beginning state: can be assigned new job but in next few seconds encounter stress; add in buddy set

```
                AddBuddySet(Pi)
        BuddySetCount(BC,1)
                Alloc(pi,GlobalQueue)
                [If scheduling is round robin then]
                RR_counter:=RR_counter+1
                Else
                FCFS_counter:=FCFS_counter+1
        [Otherwise Test PUF<50% then]
                Stress underneath state: no new
                job assigned
                AddStressUnderneathSet(Pi)
        StressUnderneathCount(SC,1);
        Detach(Pi,GlobalQueue)
                [If scheduling is round robin then]
                RR_counter:=RR_counter+1
                Else
                FCFS_counter:=FCFS_counter+1
        [End of If then Test]
        [End of For construct]
Step-3: [if RR_counter<FCFS_counter then]
        Less number of stress-underneath proces-
        sors in RR scheduling as compare to FCFS
        scheduling and thus RR scheduling better for
        short term scheduling.
Step-4: EXIT
```

6 RESULTS AND DISCUSSIONS

The presented work is implemented in VB6 as a simulation. Below is the figure of the simulation process.

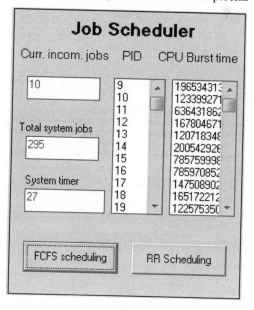

Figure 2.   Simulation of job scheduler.

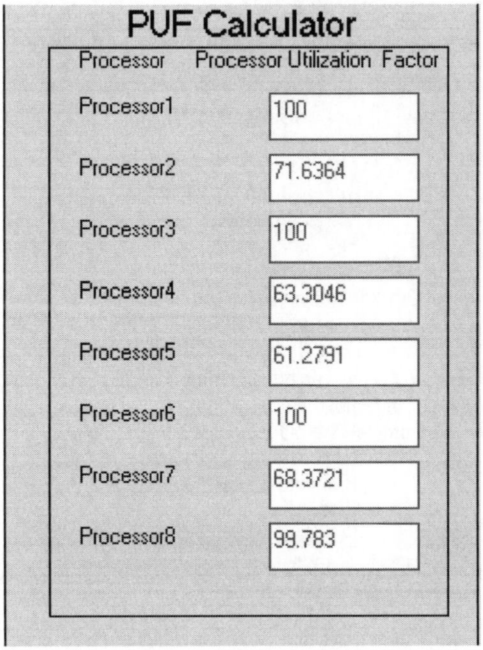

PUF Calculator

Processor	Processor Utilization Factor
Processor1	100
Processor2	71.6364
Processor3	100
Processor4	63.3046
Processor5	61.2791
Processor6	100
Processor7	68.3721
Processor8	99.783

Figure 3.   PUF in RR scheduling.

PUF Calculator

Processor	Processor Utilization Factor
Processor1	22.6633
Processor2	13.9764
Processor3	82.4341
Processor4	25.5885
Processor5	46.9281
Processor6	38.8227
Processor7	10.0866
Processor8	39.4121

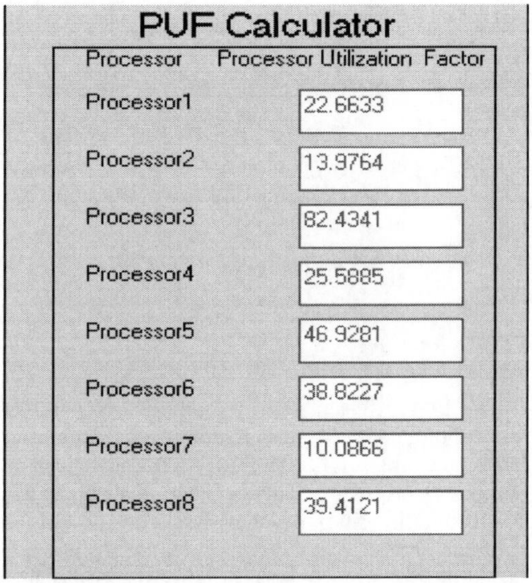

Figure 5.   PUF in FCFS scheduling.

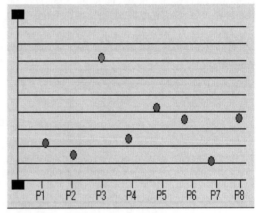

Figure 6.   Processors in FCFS scheduling.

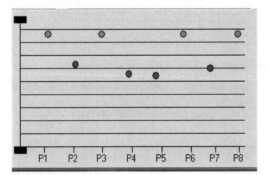

Figure 4.   Processors in RR scheduling.

The algorithm implemented in the presented work gives the result in favor of round robin scheduling.

As depicted from above figure, the processors have PUF of more than 60% and thus are less stressed as due to RR scheduling the jobs are completed in a faster manner.

The green dots represent the processors whose PUF is greater than 90% and the blue dots represent the processors whose PUF is above 60% and thus are in stress initiation stage but due to RR scheduling these processors can easily recover from their stress and can execute the remaining jobs very easily and in a faster manner.

Table 2.   PUF in RR and FCFS.

Processors	PUF in RR	PUF in FCFS
P1	100	22
P2	72	13
P3	100	82
P4	63	25
P5	61	46
P6	100	38
P7	69	10
P8	100	39

This scenario remains for most of the time in RR scheduling. The chances for stress are less in this scheduling.

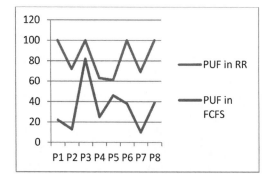

Figure 7.   PUF comparison in RR and FCFS.

Table 3.   Jobs completed in RR and FCFS.

Time slot (ms)	Jobs completed-RR	Jobs completed-FCFS
1	3	2
6	7	5
8	16	12
15	24	21
21	48	37
37	60	58
52	72	64
67	93	82
89	105	99

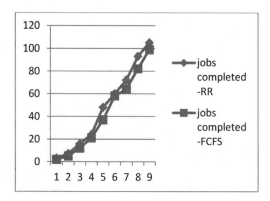

Figure 8.   Jobs completed in RR and FCFS.

In contrast to RR scheduling, FCFS have more number of processors in stressed condition as shown in below figure.

The red dots in above figure depicts the processors whose PUF is less than 50% and thus are under stress condition and there is only processor i.e. processor 3 that acts as buddy for the remaining processors (the one shown with pink dot).

With these statistics, the efficiency of system is improved with RR scheduling as the number of jobs completed with RR scheduling are more as compare to the FCFS scheduling.

The following tables and graphs prove the above stated statement.

7   CONCLUSION AND FUTURE SCOPE

The work proposed in the present paper takes into consideration homogeneous multiprocessor architecture.

Load balancing is key aspect for performance improvement and system utilization so; the workload distribution among the processors should remains in a consistent state with no need of load stability in later stages. The samples taken illustrate that at the outset all the processors are given jobs in sequential order but at very next moment of time; the jobs are allocated depending upon a metric called Processor Utilization Factor such that the load is given to only those processors that are under less stress conditions and the processors that are under high stress conditions are not given any new job until they complete their previously assigned tasks and reach a stress liberated state. The given method of scheduling performs steady and stable load distribution by assigning jobs to only those processors which can handle them without getting into a stress underside stage and with no requirement of supplementary efforts for task adjustments by context switching and load reshuffle. Alongside with highly massive systems; the strategy can also be applied or used in grid systems where homogeneous processors are inter-connected in a network and where resource management is a key concern. Future work may include strategy where processors mutually settle on their buddy sets so that they can formulate a friend list of their buddies and share their load with them without concerning the job controller to decide the load stability criteria throughout job allocation.

REFERENCES

Andersson et al., "A Static-Priority Scheduling on Multiprocessors" RealTime Systems Symposium, 22nd IEEE sponsored conference publication, 2001.

Arora A. et al., "Scheduling Simulations: Novel Experimental Approach to Time-Sharing Multiprocessor Scheduling Schemes" IJCA (0975–8887) Volume 63–No.11, Feb 2013.

Arora S. et al., "Parallel Stress estimation for consistent task scheduling using buddy strategy". IEEE conference on MOOC technologies, Oct 2015.

Arora, H. et al.,"Simulated Heterogeneous Processor Scheduling for Balanced Job Allocation" IJSRCS, Vol. 1, Issue. 4, Nov. 2013.

Cherkasova et al., "A Session-based admission control: A mechanism for peak load management of commercial Web sites". IEEE Requirements on Computers, vol.51, Issue no. 6, 2002.

Fet. L. et al., "Novel Fair Load-Balancing on Parallel System" International conference on parallel processing and systems, 2007.

Govardhan, A et al., "Dynamic Load-Balancing with Central Monitoring of Distributed Job Processing Systems" Foundation of Computing Science New York, 2013.

Joel, B. et al. "Static-priority scheduling of periodic tasks system upon identical multiprocessor platforms" University of North Carolina at Chapel-Hill, 2003.

Kumar, R. et al.,"SPF: Segmented Processor Framework for Energy Efficient Proactive Routing Based Applications in MANET" IEEE conference RAECS pp. 1–7, Dec. 2015.

Paul M. et al., "Stochastic Contention for Single-Chip Heterogeneous Multiprocessor". IEEE conference Volume 59, pp. 1402–1418, Oct 2010.

Rohit K. Bhullar et al., "Cross Platform Application Development for Smartphones: Approaches and Implications" IEEE Conference conf. id. 37465, New Delhi, pp. 2571–2578, 2016.

Rohit Kumar et al., "Specialized hardware Architecture for Smartphones", IJERA, ISSN: 2248–9622 Volume 4, Issue 5, 2014.

Singh at. al., "ASimulated Performance Analysis of Multiprocessor Dynamic Space Sharing Scheduling Method" 2009.

Singh et al., "Efficiency Measurement with Effective Stress Management in Heterogeneous 2-D Mesh Processor" IJCA (0975–8887) Volume 81 – No.12, Nov 2013.

Singh et al., "A Simulated Performance Analysis of Multiprocessor Dynamic Space-Sharing Scheduling policy" IJCA journal, 2009.

Wei-Ming et al., "Task scheduling for multi-processor systems with autonomous performance optimizing control Systems" JISE, San Antonio, 2010.

Communication and Computing Systems – Prasad et al. (Eds)
© 2017 Taylor & Francis Group, London, ISBN 978-1-138-02952-1

Smart farming: IoT based smart sensors agriculture stick for live temperature and moisture monitoring using Arduino, cloud computing & solar technology

Anand Nayyar
Department of Computer Applications and IT KCL Institute of Management and Technology, Jalandhar, Punjab, India

Vikram Puri
G.N.D.U Regional Center, Ladewali Campus, Jalandhar, India

ABSTRACT: Internet of Things (IoT) technology has brought revolution to each and every field of common man's life by making everything smart and intelligent. IoT refers to a network of things which make a self-configuring network. The development of Intelligent Smart Farming IoT based devices is day by day turning the face of agriculture production by not only enhancing it but also making it cost-effective and reducing wastage. The aim/objective of this paper is to propose a Novel Smart IoT based Agriculture Stick assisting farmers in getting Live Data (Temperature, Soil Moisture) for efficient environment monitoring which will enable them to do smart farming and increase their overall yield and quality of products. The Agriculture stick being proposed via this paper is integrated with Arduino Technology, Breadboard mixed with various sensors and live data feed can be obtained online from Thingsspeak.com. The product being proposed is tested on Live Agriculture Fields giving high accuracy over 98% in data feeds.

Keywords: Internet of Things (IoT), Agriculture, Agriculture IoT, Agriculture Precision, Arduino Mega 2560, DS18B20 Temperature Sensor, Smart Farming, Soil Moisture Sensor, Cloud Computing, Solar Technology, ESP8266, Thingspeak.com

1 INTRODUCTION

The next era of Smart Computing will be totally based on Internet of Things (IoT). Internet of Things (IoT), these days is playing a crucial role of transforming "Traditional Technology" from homes to offices to "Next Generation Everywhere Computing". "Internet of Things" (Weber, R.H, 2010) is gaining an important place in research across the nook and corner of this world especially in area of modern wireless communications. The term, Internet of Things (Suo et al, 2012) refers to uniquely identifiable objects, things and their respective virtual representations in Internet like structure which was proposed in year 1998. Internet of Things was discovered by "Kevin Ashton" (Weber, R.H, 2010) in 1999 with regard to supply chain management. These days, the strength and adaptability of IoT has been changed and nowadays it is being used even by normal user. From the point of normal user, IoT (Ashton, 2009) has laid the foundation of development of various products like smart living, e-health services, automation and even smart education. And from commercial point of view, IoT these days is being used in business management, manufacturing, intelligent transportation and even agriculture.

One of main areas where IoT based research is going on and new products are launching on everyday basis to make the activities smarter and efficient towards better production is "Agriculture". Agriculture sector is regarded as the more crucial sector globally for ensuring food security. Talking of India farmers, which are right now in huge trouble and are at disadvantageous position in terms of farm size, technology, trade, government policies, climate conditions etc. No doubt, ICT based techniques have solved some problems but are not well enough for efficient and assured production. Recently, ICT has migrated to IoT which is also known as "Ubiquitous computing" (Patil et al, 2012). Agricultural production requires lots of activities like soil and plant monitoring, environmental monitoring like moisture and temperature, transportation, supply chain management, infrastructure management, control systems management, animal monitoring, pest control etc.

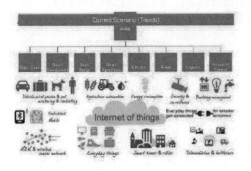

Figure 1. Current scenario of IoT.

IoT based agricultural convergence technology (Lee et al, 2013) creates high value in terms of quality and increased production and also reduces burden on farmers in ample manner. In addition to Agricultural IoT, the future of agriculture is "Precision Agriculture" which is expected to grow at $3.7 billion by 2018. With data generated from GPS and Smart Sensors on agricultural field and integration of smart farming equipment along with Big Data analytics, farmers would be able to improve crop yields and make effective use of water and in turn wastage of any sort would be reduced to a remarkable level.

So, seeing the current scenario of agriculture which is surrounded by tons of issues, it is utmost requirement to have IoT based Smart Farming. In order to implement smart farming in real world, IoT based products are required to be developed and implemented at regular intervals and also at a very fast pace.

The objective of this research paper is to propose IoT Based Smart Stick which will enable farmers to have live data of soil moisture, environment temperature at very low cost so that live monitoring can be done.

The structure of the paper is as follows: Section II will cover over of Overview of IoT Technology & Agriculture—Concept & Definition, IoT Enabling Technologies, IoT Applications in Agriculture, Benefits of IoT in Agriculture and Present and Future Scenario of IoT in Agriculture. Section III elaborates "Novel Proposed IoT Based Smart Farming Agricultural Stick—Overview, Components—Sensors and Modules, Circuit Diagrams and Working. Section IV will highlight live demonstration of IoT Based Smart Stick and live data results. Section V will cover conclusion and future scope.

2   IOT TECHNOLOGY & AGRICULTURE

2.1   *Internet of Things—Concept & Definition*

Internet of Things (IoT) (Atzori et al 2010) (Nayyar, 2016) consists of two words—Internet and Things. The term "Things" in IoT refers to various IoT devices having unique identities and have capabilities to perform remote sensing, actuating and live monitoring of certain sorts of data. IoT devices are also enabled to have live exchange of data with other connected devices and applications either directly or indirectly, or collect data from other devices and process the data and send the data to various servers. The other term "Internet" is defined as Global Communication network connecting trillions of computers across the planet enabling sharing of information.

As forecasted by various researchers, 50 Billion devices based on IoT would be connected all across the planet by year 2020. The Internet of Things (IoT) has been defined as (Smith, 2012):

A Dynamic global network infrastructure with self-configuring capabilities based on standard and interoperable communication protocols where physical and virtual "Things" have identities, physical attributes, and virtual personalities and use intelligent interfaces and are seamlessly integrated into the information network, often communicate data associated with users and their environments".

An ideal IoT device consists of various interfaces for making connectivity to other devices which can either be wired or wireless.

Any IoT based device consists of following components:

- I/O interface for Sensors.
- Interface for connecting to Internet.
- Interface for Memory and Storage.
- Interface for Audio/Video.

IoT devices can be of various forms like wearable sensors, smart watches, IoT smart home monitoring, IoT intelligent transport systems, IoT smart health devices etc.

2.2   *IoT Enabling Technologies*

Internet of Things has a strong backbone of various enabling technologies—Wireless Sensor Networks, Cloud Computing, Big Data, Embedded Systems, Security Protocols and Architectures, Protocols enabling communication, web services, Internet and Search Engines.

Wireless Sensor Network (WSN): It consists of various sensors/nodes which are integrated together to monitor various sorts of data.

Cloud Computing: Cloud Computing also known as on-demand computing is a type of Internet based computing which provides shared processing resources and data to computers and other devices on demand. It can be in various forms like IaaS, PaaS, SaaS, DaaS etc.

Big Data Analytics: Big data analytics is the process of examining large data sets containing various forms of data types—i.e. Big Data—to

Table 1. Various projects and applications are integrated in Agricultural fields leading to efficient management and controlling of various activities.

Application name	Description
Crop Water Management	In order to perform agriculture activities in efficient manner, adequate water is essential. Agriculture IoT is integrated with Web Map Service (WMS) and Sensor Observation Service (SOS) to ensure proper water management for irrigation and in turn reduces water wastage.
Precision Agriculture	High accuracy is required in terms of weather information which reduces the chances of crop damage. Agriculture IoT ensures timely delivery of real time data in terms of weather forecasting, quality of soil, cost of labor and much more to farmers.
Integrated Pest Management or Control (IPM/C)	Agriculture IoT systems assures farmers with accurate environmental data via proper live data monitoring of temperature, moisture, plant growth and level of pests so that proper care can be taken during production.
Food Production & Safety	Agriculture IoT system accurately monitors various parameters like warehouse temperature, shipping transportation management system and also integrates cloud based recording systems.
Other Projects Implemented Till Date	1. The Phenonet Project by Open IoT 2. CLAAS Equipment 3. Precisionhawk's UAV Sensor Platform 4. Cleangrow's Carbon Nanotube Probe 5. Temputech's Wireless Sensor Monitoring.

uncover hidden patterns, unknown correlations, market trends, customer preferences and other useful business information.

Communication Protocols: They form the backbone of IoT systems to enable connectivity and coupling to applications and these protocols facilitate exchange of data over the network as these protocols enable data exchange formats, data encoding and addressing.

Embedded Systems: It is a sort of computer system which consists of both hardware and software to perform specific tasks. It includes microprocessor/microcontroller, RAM/ROM, networking components, I/O units and storage devices.

2.3 IoT applications in agriculture

With the adoption of IoT in various areas like Industry, Homes and even Cities, huge potential is seen to make everything Intelligent and Smart. Even the Agricultural sector is also adopting IoT technology these days and this in turn has led to the development of *"AGRICULTURAL Internet of Things (IoT)"*

2.4 Benefits of IoT in agriculture

The following are the benefits of IoT in Agriculture:

1. IoT enables easy collection and management of tons of data collected from sensors and with integration of cloud computing services like Agriculture fields maps, cloud storage etc., data can be accessed live from anywhere and everywhere enabling live monitoring and end to end connectivity among all the parties concerned.
2. IoT is regarded as key component for Smart Farming as with accurate sensors and smart equipment's, farmers can increase the food production by 70% till year 2050 as depicted by experts.
3. With IoT productions costs can be reduced to a remarkable level which will in turn increase profitability and sustainability.
4. With IoT, efficiency level would be increased in terms of usage of Soil, Water, Fertilizers, Pesticides etc.
5. With IoT, various factors would also lead to the protection of environment.

2.5 IoT and agriculture current scenario and future forecasts

Table 2. Shows the growth of IoT based adoption in Agriculture sector from Year 2000–2016 and Forecasts of year 2035–2050.

Year	Data analysis
2000	525 Million Farms connected to IoT
2016	540 Million Farms till Date are connected to IoT
2035	780 Million Farms would be connected to IoT
2050	2 Billion Farms are likely to be connected to IoT

3 NOVEL PROPOSED IOT BASED SMART AGRICULTURE STICK

In today's era of IoT, lots of new research in terms of Smart IoT based product's development is being carried out to facilitate Smart Farming in terms of Crop Management, Pest Management, Agriculture Precision, Agriculture Fields Monitoring via Sensors and even Drones.

In this section, Smart IoT based Agricultural stick being developed for live monitoring of Temperature, Moisture using Arduino, Cloud Computing and Solar Technology is discussed.

3.1 *Definition—Smart Agriculture IoT Stick*

Smart Farming Based Agriculture IoT Stick is regarded as IoT gadget focusing on Live Monitoring of Environmental data in terms of Temperature, Moisture and other types depending on the sensors integrated with it. Agricultural IoT stick provides the concept of "Plug & Sense" in which farmers can directly implement smart farming by as such putting the stick on the field and getting Live Data feeds on various devices like Smart Phones, Tablets etc. and the data generated via sensors can be easily shared and viewed by agriculture consultants anywhere remotely via Cloud Computing technology integration. IoT stick also enables analysis of various sorts of data via Big Data Analytics from time to time.

3.2 *Components*

In this section, various components i.e. Modules and Sensors being used for Smart IoT Agricultural Stick development is discussed:

3.2.1 *Modles*

3.2.1.1 Arduino Mega 2560
Arduino Mega 2560 is designed for developing Arduino based robots and doing 3D printing technology based research.

Technical Specifications: Arduino Mega 2560 is based on ATmega2560. Consists of 54 digital Input/ Output pins, 16 analog inputs, 4 UART (Universal Asynchronous Receiver and Transmitter). Can simply connect to PC via USB port.

Figure 2.   Arduino mega 2560.

Figure 3.   ESP8266 Wi-Fi module.

3.2.1.2 ESP 8266
ESP8266 Wi-Fi Module is SOC with TCP/IP protocol stack integrated which facilitates any microcontroller to access Wi-Fi network. ESP8266 module is cost effective module and supports APSD for VOIP Applications and Bluetooth co-existence interfaces.

Technical Specifications: 802.11b/g/n; Wi-Fi Direct, 1MB Flash Memory, SDIO 1.1/2.0, SPI, UART, Standby Power Consumption of <1.0 mW.

3.2.1.3 BreadBoard BB400
BreadBoard-400 is a solderless breadboard with 400 connection tie points i.e. 400 Wire insertion points. BB400 has a 300 tie-point IC-circuit area plus four 25-tie point power rails. Housing is made of White ABS plastic, with a printed numbers and letters of rows and columns.

Technical Specifications: 36 Volts, 2 Amps, 400 tie points, 50000 insertions.

3.2.1.4 BreadBoard power supply
Power Module designed for MB102 breadboard.

Technical Specifications: Compatible to 5v or 3.3v, Output Voltage: 5v and 3.3v, Max output current: <700 mA; Suitable for Arduino, AVR, PIC, ARM.

3.2.1.5 Solar plate
6 Watts High-performance solar panel utilizes highly efficient crystalline solar cells to increase light absorption and improve efficiency.

Technical Specifications: 0.53 mA; Voltage: 11.2v.

Figure 4.   BreadBoard.

Figure 5.   BreadBoard power supply.

3.2.1.6 Battery

Li-Ion 11.2V battery is made of 3 A-Grade 18650 cylindrical cells with PCB and poly switch for full protection. It is Light weight and has high energy density.

Technical Specifications: 2200 mAh; 11.2V.

3.2.1 *Sensors*

3.2.2.1 Temperature Sensor-DS18B20

The DS18B20 temperature sensor provides 9-bit to 12-bit Celsius temperature measurements and has alarm function with non-volatile user-programmable upper and lower trigger points. The DS18B20 has 64-bit serial code which allows multiple DS18B20 s to function on same 1-wire bus.

Technical Specifications: Unique 1-Wire Interface; Measures Temperature from −55°C to +125°C; Coverts temperature to 12-bit digital word in 750 ms.

3.2.2.2 Soil Moisture Sensor

Soil Moisture Sensor is used for measuring the moisture in soil and similar materials. The sensor has two large exposed pads which functions as probes for the sensor, together acting as a variable resistor. The moisture level of the soil is detected by this sensor. When the water level is low in the soil, the analog voltage will be low and this analog voltage keeps increasing as the conductivity between the electrodes in the soil changes. This sensor can be used for watering a flower plant or any other plants requires automation.

Technical Specification: 3.3V to 5V; Analog Output; VCC external 3.3 V to 5V.

Figure 6. 6 Watts solar panel.

Figure 7. 11.2 Volts battery.

3.2.3 *Circuit description*

Description:

IoT based Smart Agriculture Stick incorporates Arduino Mega 2560 unit that provides base for live monitoring of temperature and soil moisture and sends the data to the cloud via ESP8266 Wi-Fi module. In this IoT product, 3 values are measured: Environmental Temperature, Soil Moisture and Solar Panel Voltage powering the entire system. DS18B20 Temperature Sensor is relatively accurate digital temperature sensor and uses MAXIM's 1-wire bus protocol for transmitting as well as receiving data in bytes and supports parasite power mode.

Figure 8. DS18B20 waterproof temperature sensor.

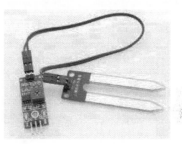

Figure 9. Soil moisture sensor.

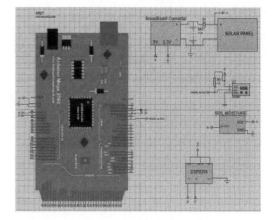

Figure 10. Circuit of "Novel Smart Agriculture IoT Stick for Monitoring Temperature and Soil Moisture"- Designed in proteus software.

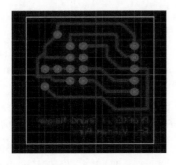

Figure 11.   PCB Design for ESP8266.

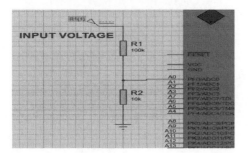

Figure 12.   Implementation of voltage divider.

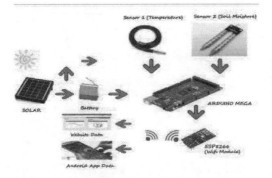

Figure 13.   Overall system working of "Smart IoT based agriculture stick".

The following formula shows to calculate the Temperature:

Temperature = ((HighByte <<8) + LowByte) * 0.0625

Soil Moisture Sensor works on the resistance changing principle. It has two large pads as probes for the Soil Moisture sensing and also acts as a variable resistor. When water level is low in soil, conductivity is less between the pads and resistance is higher. When water level is high in soil, conductivity is high between the pads and resistance is low

Figure 14.   Complete agriculture IoT stick monitoring temperature and humidity.

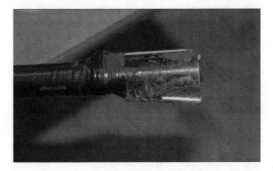

Figure 15.   Soil moisture sensor and temperature sensor mounted on IoT agriculture stick.

Figure 16.   stick mounted with sensors in flowerpot giving live data of temperature and moisture.

and provides higher signal out. ESP8266 is low-cost Serial to Wi-Fi module and easily interfaces with Arduino Mega 2560. ESP8266 is based on AT Commands and fully supports TCP/UDP stack. Arduino is configured as Digital DC Voltmeter to measure the solar voltage. Arduino Mega 2560 is basically measures up to 5V through the analog pins. One diode is used between the solar panel and battery for protection of solar from back current provided by the battery.

R1 = 100 K
R2 = 10 K
VOUT = VIN * (R2 / (R1 + R2))

Figure 17. Complete system with Arduino Board, BreadBoard and laptop giving results using ThingSpeak.com website.

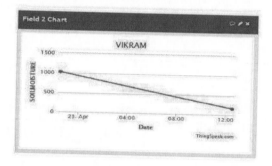

Figure 18. Live data of soil moisture with date and Time from Thingspeak.com.

4 LIVE IMPLEMENATION AND REAL TIME DATA ANALYSIS AND MONITORING

In this section, the overall working of the system is being discussed.

The following diagram shows the Animated View of the Smart IoT Based Agriculture Stick being proposed for Agriculture Temperature and Moisture Monitoring.

The Following Diagram shows the complete details of the system being developed by us: "Smart IoT Agriculture Stick Monitoring—Temperature and Moisture".

5 CONCLUSION

In this Research Paper, a Novel Smart Farming Enabled: IoT Based Agriculture Stick for Live Monitoring of Temperature and Soil Moisture has been proposed using Arduino, Cloud Computing and Solar Technology. The stick has high efficiency and accuracy in fetching the live data

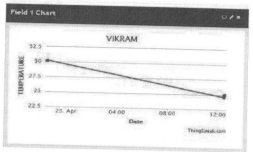

Figure 19. Live data of temperature with date and time from Thingspeak.com.

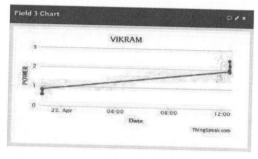

Figure 20. Live data of solar power plates powering the entire agriculture IoT stick.

of temperature and soil moisture. The Agriculture stick being proposed via this paper will assist farmers in increasing the agriculture yield and take efficient care of food production as the stick will always provide helping hand to farmers for getting accurate live feed of environmental temperature and soil moisture with more than 99% accurate results.

6 FUTURE SCOPE

Future work would be focused more on increasing sensors on this stick to fetch more data especially with regard to Pest Control and by also integrating GPS module in this IoT Stick to enhance this Agriculture IoT Technology to full-fledged Agriculture Precision ready product.

REFERENCES

Ashton, K. (2009). That 'internet of things' thing. RFiD Journal, 22(7), 97–114.

Atzori, L., Iera, A., & Morabito, G. (2010). The internet of things: A survey. Computer networks, 54(15), 2787–2805.

Bahga, A., & Madisetti, V. (2014). Internet of Things: A Hands-on Approach. VPT.

http://www.busboard.com/BB400T (Accessed on April 25, 2016)

http://www.cybronyx.com/breadboard-power-suply.html (Accessed on April 25, 2016)

https://www.arduino.cc/en/Main/arduinoBoard-Mega2560 (Accessed on April 25, 2016)

https://www.maximintegrated.com/en/products/analog/sensors-and-sensor-interface/DS18B20.html (Accessed on April 25, 2016)

https://www.sparkfun.com/products/13322 (Accessed on April 25, 2016)

https://www.sparkfun.com/products/13678 (Accessed on April 25, 2016)

Lee, M., Hwang, J., & Yoe, H. (2013, December). Agricultural Production System Based on IoT. In Computational Science and Engineering (CSE), 2013 IEEE 16th International Conference on (pp. 833–837). IEEE.

Nayyar, A. (2016). An Encyclopedia Coverage of Compiler's, Programmer's & Simulator's for 8051, PIC, AVR, ARM, Arduino Embedded Technologies. International Journal of Reconfigurable and Embedded Systems (IJRES), 5(1).

Nayyar, A., & Puri, V. (2016). Data Glove: Internet of Things (IoT) Based Smart Wearable Gadget. British Journal of Mathematics & Computer Science, 15(5).

Patil, V. C., Al-Gaadi, K. A., Biradar, D. P., & Rangaswamy, M. (2012). Internet of things (Iot) and cloud computing for agriculture: An overview. Proceedings of Agro-Informatics and Precision Agriculture (AIPA 2012), India, 292–296.

Smith, I. G. (Ed.). (2012). The Internet of things 2012: new horizons. CASAGRAS2.

Suo, H., Wan, J., Zou, C., & Liu, J. (2012, March). Security in the internet of things: a review. In Computer Science and Electronics Engineering (ICCSEE), 2012 International Conference on (Vol. 3, pp. 648–651). IEEE.

Weber, R. H. (2010). Internet of Things–New security and privacy challenges. Computer Law & Security Review, 26(1), 23–30.

Communication and Computing Systems – Prasad et al. (Eds)
© 2017 Taylor & Francis Group, London, ISBN 978-1-138-02952-1

A hybridized framework to build bias optimized decision tree classifiers

Anuradha & Gaurav Gupta
The Northcap University, Gurgaon, India

ABSTRACT: In the decision tree induction process, it is very crucial to identify core significant attributes which have optimized bias for a given target class. In this paper, a hybridized framework is proposed whose contribution can be stated in two phases. In phase-I, attribute bias is calculated using a modified version of Shannon entropy and in phase II, attribute individual discriminatory power is calculated from Pearson coefficient and fisher score analysis. The hybridization of proposed technique with decision tree induction phase yields better classification accuracy while reducing computational overhead. Four famous decision tree classifiers (HSM, Fuzzy SLIQ, Fuzzy HSM and DCSM) are fused with the proposed method and comparative analysis is done with some existing attribute selection methods (Information Gain, Gain Ratio, CFS and Relief F.

1 INTRODUCTION

Forecasting future events and discovering new knowledge from existing patterns is solely due to advances in data mining technology. Emergence in data analysis and data warehouse technologies had led to exponential dimensional growth in data sets. This undesired dimensional growth increases the size of search space used in induction process. So, to improve the performance of any classification model, it is crucial to identify core significant attributes who alone can give better performance in comparison to whole attribute set. In particular, finding optimal attribute subset is considered as a major performance obstacle in various active research areas of pattern recognition, data mining, statistics and machine learning. Attribute selection process aims at producing optimal or sub optimal results which are relatively significant as compared to that of original data set. The likelihood of irrelevant, redundant and noisy attributes increases with growth in data dimensionality as discussed by (Chang, Verhaegen, & Duflou 2014). Attribute subset selection is a data pre-processing technique to filter out relevant attributes from irrelevant ones so that overall classification results can be improved. The curse of dimensionality is a key parameter to rate the performance of any supervised or unsupervised learning process as highlighted by (Powell 2007a, b, Guyon & Elisseeff 2003). The computational cost of generating training models grow exponentially as the size of attribute set increases, as reported by (Kuo & Sloan 2005). The presence of redundant and irrelevant features makes the induction process more tougher. So, if the data set under consideration have these attributes included,

it will affect accuracy, computational cost and comprehensibility of results as concluded by (Guyon & Elisseeff 2003a, b, Kohavi & John, 1997). As an unseen drawback, attributes could be correlated and have inherited dependency, which should be considered while building classifier models. So, the ultimate motive is to reduce attribute subset to achieve balanced accuracy, efficiency and scalability. Attribute selection techniques can be broadly classified as being 1) filter approach by (Guyon & Elisseeff 2003) 2) Wrapper approach by (Kohavi & John 1997) 3) Hybrid approach by (Huang et al. 2007). Sequential Forward Selection and Backward Elimination are two popular approaches to reduce the traversal space. Sequential Forward Selection follows top down attribute inclusion process by adding attribute at every stage. Bottom up strategy is used in backward elimination which starts with complete attribute set and drops attribute one at a time till stopping condition has not satisfied.

A support vector machine assisted wrapper method was proposed by (Maldonado & Weber 2009), which was able to avoid data over fitting but continuous backward feature selection give rise to computational complexity. (Ladha & Deepa 2011) gave a variant of smart beam search algorithm which starts with k best attributes and then successively adds neighbor attributes till search stops. But, the presence of overlapped attributes in multiple neighborhoods can be misleading yielding increased misclassification rate. (Liu & Yu 2005) used the attribute ranking process through a predefined evaluation function to select the best optimized feature subset. (Hall 2000) proposed CFS (Correlation based feature selection) to rank feature subset instead individual attributes. The attributes having high discernibility for the target class are used to filter out unrelated attributes in the given set. A sequential forward selection method is incorporated by (Miche et al. 2006) to search the best feature subset on some selection criteria and attributes keep added until a feature subset of high accuracy is not identified. (Saeys et al. 2007) identified a major loop whole in filter based selection methods of not interacting with the classifier algorithm. (Xia et al. 2012) gave a wrapper feature selection approach by combining information gain with mutual information. (Jiang et al. 2012) introduced a new attribute selection method based on naïve bayes with aim of improving classifiers accuracy. (Ji-cang et al. 2014) proposed yet another variant of euclidean distance to create filter based feature subset selection. A number of researchers have also explored several combinations of Meta heuristic techniques based on GA, PSO, ACO, simulated annealing and Artificial Bee Colony for effective attribute subset selection.

2 PROPOSED TECHNIQUE

The proposed hybridized framework deals with generating a bias optimized decision tree classifier which aims at increasing classifier accuracy and reducing computational cost by removing irrelevant and redundant attributes. The major processing phases are as follows:

1. Attribute to Class Favor Coefficient (ACFC) is inspired from Shannon entropy and will calculate every attribute bias to every target class. A high value of ACFC is higher bias towards a given target class.
2. A modified fisher score (F-Score), is used to find individual discrimination power of all attributes. A high scoring attribute can give better classification in comparison to others.

A balance of ACFC coefficient and F-Score is used to decide on generic reduced attribute sub set which holds enough knowledge to improve classifier outcome. To perform further pruning over reduced attribute sub set, Pearson correlation coefficient is used. The various steps of the proposed technique are as explained below:

Consider 'D' be a sample data set consisting of two types of attributes:

1. Continuous valued attribute ($Att_c \in A$)
2. Discrete/Dichotomous valued attribute ($Att_d \in A$)

Let 'A' be an attribute vector of length 'm' represented as {$Att_1, Att_2 \dots Att_m$}. The target class set, $C = \{TC_1, TC_2 \dots TC_d\}$ and there are in total 'd' distinct target classes. Consider the data set matrix given below to help understand the working of various steps:

$$
\begin{bmatrix}
Att_1 & Att_2 & & Att_m & C \\
x_{11} & x_{12} & \cdots & x_{1m} & TC_1 \\
x_{21} & x_{22} & & x_{2m} & TC_2 \\
\vdots & & \ddots & & \vdots \\
x_{(n-1)1} & x_{(n-1)2} & \cdots & x_{(n-1)m} & TC_{d-1} \\
x_{n1} & x_{n2} & & x_{nm} & TC_d
\end{bmatrix}
$$

where $i \to$ to m 1

Step 1: Find out Attribute to Class Favor Coefficient (ACFC) for each discrete/dichotomous valued attribute 'Att_d' $\in A$ by Equation 1. Let the count of distinct values attained by 'Att_d' is u_{Att_d} and n_{TC_d} be the instance count belonging to target class 'TC_d'.

$$
ACFC\left(\frac{Att_p}{TC_d}\right) =
$$
$$
\left(1 - \left(\frac{\sum_{k=1}^{u_{Att_d}} \left(-p\left(Att_{dk}/TC_d\right) * log_2\left(p\left(Att_{dk}/TC_d\right)\right)\right)}{log_2(\min\{n_{TC_d}, u_{Att_d}\})} \right) \right)
$$
(1)

The ACFC varies from [0–1], where the higher value of this coefficient indicates attribute favor to the given target class, which needs to be optimized to have a balanced outcome.

Step 2: Find out Attribute to Class Favor Coefficient (ACFC) for each continuous valued attribute 'Att_c' ε A using class distribution histogram. Firstly, sort 'Att_c' in ascending order along with target class 'TC_d' to discretize its input domain. Split points are those where both attribute and target class value changes simultaneously as dictated in Figure 1:

$$
\left\{
\begin{array}{cc}
Att_c & TD_d \\
12 & A \\
21 & A \\
32 & B \\
32 & A \\
42 & A \\
45 & A
\end{array}
\cdots \to \frac{21+32}{2} = 26.5
\right\}
$$

Figure 1. Attribute input space discretization process.

Given below in Table 2 is a two class histogram generated with respect to a split point (SP) and Equation 2 will calculate ACFC count for k split points generated after applying discretization process used in Fig. 1. Here, Lower_Par → Lower data partition with respect to SP and Upper_Par

682

→ Upper data partition with respect to SP. The rest symbols are same as explained in step 1.

$$ACFC\left(\frac{Att_c}{TC_d}\right) = \left(1 - \left(\frac{\sum_{k=1}^{SP}\left(-p\left(Att_{dk}/TC_d\right)*log_2\left(p\left(Att_{dk}/TC_d\right)\right)\right)}{log_2\left(\min\left\{n_{TC_d}, u_{Att_d}\right\}\right)}\right)\right)$$

(2)

Step 3: Next calculate attribute- Target class individual discrimination power using modified F-Score. An attribute ($Att_{i\in c,d}$) having large F_Score will have high weightage to be considered as best attribute for generating training model with respect to a specific target class.

For each attribute $Att_i \in A$, calculate the discernibility power using the following modified fisher similarity index as given below in Equation 3:

$$F\_Score\left(Att_i/TC_k\right) = \frac{\sum_{k=1}^{d} n_k \left(\mu_i^k - \mu_i\right)^2}{\sum_{k=1}^{d} n_k \left(\sigma_i^k\right)^2}$$

(3)

It ranges from [0–1].

Here, Att_i, is the ith attribute, 'd' is the total class count, 'k' is the class variable, n_k is the number of instances under class 'k' in attribute Att_i, μ_i^k statistical mean of ith attribute belonging to class 'k', σ_i^k is statistical variance of ith attribute belonging to class 'k'. The attribute with maximum fisher score is having better discernibility power in comparison to other attributes. $F\_Score\left(Att_i/TC_k\right)$ Value ranges from [0–1].

Step 4: Next, Compute $F\_Score\left(Att_i\right)$ with respect to every target class TC_d present in original data set using Equation (4). The Set 'FS (i)' contains

$$F\_Score\left(F(i)/TC_d\right) = \{F\left(Att_1, TC_d\right), F\left(Att_2/TC_d\right), F\left(Att_3, TC_d\right), ..., F\left(Att_m/TC_d\right)\}$$

(4)

For $Att_i \varepsilon A$ and $TC_{d} \varepsilon C$.

Step 5: Arrange all Att_i in descending order of their ACFC value for every target class $TC_d \in C$ in the pattern depicted below:

For target class $TC_d \in C$ the ordered set is $\{F\left(Att_1, TC_d\right), F\left(Att_2/TC_d\right), F\left(Att_3, TC_d\right), ..., F\left(Att_m/TC_d\right)\}$ with $F\left(Att_1, TC_d\right)$ being the highest score and $F\left(Att_m/TC_d\right)$ being the lowest score.

Step 6: Now, from the two coefficients (ACFC and F_Score) generate an optimized reduced attribute vector for every target class TC_d with the following selection condition given in Equation 5:

$$Reduced_{Attribute\ Set}\left(RA/TC_d\right) = \{\exists_i\ Att_i \epsilon A \mid ACFC\ (Att_i) < 0.5\ and\ F\_Score(Att_i) > 0.5\}$$

(5)

The experimental threshold used to filter out reduced optimized attribute set is 0.5 for both ACFC and F_Score.

Step 7: Finally, remove irrelevant attributes from this Reduced$_{Attribute\ Set}$ (RA/TC$_d$) using modified Pearson correlation coefficient depicted in Equation 6. P_Cof(Att$_c$, Att$_d$) Values close to +1 indicates strong association between observed attributes.

$$P\_Cof\left(Att_c, Att_d\right) = \frac{\sum_{i=1}^{n}\left(Att_c^i - \overline{Att_c}\right)\left(Att_d^i - \overline{Att_d}\right)}{\sqrt{\sum_{i=1}^{n}\left(Att_c^i - \overline{Att_c}\right)^2}\sqrt{\sum_{i=1}^{k}\left(Att_d^i - \overline{Att_d}\right)^2}}$$

(6)

Ranges from [+1 to –1]

where, Att_c^i, Att_d^i is the ith instance value under attribute Att_c and Att_d.

$\overline{Att_c}, \overline{Att_d}$ is the statistical mean value of Att_c and Att_d.

Step 8: So, our final globally optimized reduced attribute set is 'GO$_{AttributeSet}$' $= \{\exists_i Att_i \epsilon A \mid P_{Cof(Att_c, Att_d)} > \delta$ in the set Reduced$_{AttributeSet}$ (RA/TC$_d$). We will discard Att_c in comparison to Att_d if $F\_Score\left(Att_d\right) > F\_Score\left(Att_c\right)$ in the same target class (TC$_d$).$^\delta$ is a user defined threshold to perform further pruning over the Reduced$_{AttributeSet}$ (RA/TC$_d$).$^\delta = 0.5$, is used for the experimental setup.

Step 9: Use this 'GO$_{AttributeSet}$' set to perform decision tree induction.

3 EXPERIMENTAL SETUP AND RESULT DISCUSSION

The proposed method is implemented in Matlab R2009a. Five real data sets are taken from UC Irvine machine learning repository (refer Table 1) and 10 fold cross validation is applied to get averaged outcome of classification results. The fusion of proposed technique is done with four famous decision tree classifiers (Fuzzy SLIQ by B. Chandra et al. (2008), HSM by Venkatesh Babu Kuppli (2011), DCSM Pallath Paul et al. (2010), and Fuzzy HSM by Gaurav Gupta et al. (2015). To perform statistical analysis of results with the proposed variant, four different attribute selection measures (Information Gain (IG), Gain Ratio (GR), Cor-

Table 1. Class histogram for a two class problem.

Att$_c$ Split Point SP	TC1	TC2
Lower_Par	LP1	LP2
Upper_Par	UP1	UP2

relation based Feature Selection (CFS), Relief F (Kira and Rendell in 1992) are used.

Evaluation statistics is collected to record accuracy and computation time. Apart from accuracy and computation time, Specificity (Spec), Sensitivity (Sens) and AUC curve is also measured. Table 2 shows the performance across bank marketing data set on the above mentioned parameters.

Table 2. UCI data set property description.

Data set name	No of data instances	No of attributes
Bank Marketing	45211	17
Credit Card Clients	30000	24
Dermatology	366	33
Breast Cancer	569	32
Cylinder Bands	512	39

Table 3. Evaluation statistics across bank marketing data set.

Decision tree classifier	Accuracy	Sens	Spec	AUC
Fuzzy SLIQ_IG	0.710	0.674	0.729	0.733
Fuzzy SLIQ_GR	0.735	0.693	0.741	0.751
Fuzzy SLIQ_CFS	0.781	0.755	0.801	0.792
Fuzzy SLIQ_Relief F	**0.813**	0.780	0.843	**0.833**
Fuzzy SLIQ_Proposed	0.796	**0.802**	**0.824**	0.827
HSM_IG	0.790	0.766	0.799	0.801
HSM_GR	0.820	0.792	0.815	0.822
HSM_CFS	0.796	0.771	0.781	0.772
HSM_Relief F	0.790	0.787	**0.842**	0.869
HSM_Proposed	**0.803**	**0.832**	0.812	**0.901**
DCSM_IG	0.765	0.742	0.776	0.800
DCSM_GR	0.812	0.823	0.848	0.810
DCSM_CFS	0.795	0.781	0.821	**0.873**
DCSM_Relief F	0.821	0.792	0.842	0.822
DCSM_Proposed	**0.872**	**0.863**	**0.877**	0.862
Fuzzy HSM_IG	0.810	0.788	0.822	0.844
Fuzzy HSM_GR	0.792	0.790	0.810	0.832
Fuzzy HSM_CFS	0.832	0.824	0.830	0.839
FuzzyHSM_Relief F	0.895	0.872	**0.922**	0.887
FuzzyHSM_Proposed	**0.932**	0.891	0.901	**0.926**

The entries highlighted in bold are corresponding to maximum achieved value of a specific performance parameter. It can be observed from Table 2 that proposed attribute selection method gives better accuracy in all decision tree classifiers except SLIQ embedding. Apart from accuracy, In DCSM_Proposed, specificity and sensitivity is also better. AUC is high in two embedding's of proposed variant i.e. HSM_Proposed and FuzzyHSM_Proposed.

Table 4, shows the performance across credit card client data set. The highlighted bold entries show the peak performances across various embedding's. Except HSM_Proposed, all other decision tree classifiers fusion with the proposed technique shows significant improvement in the accuracy. FuzzyHSM_Proposed is better on all performance parameters except sensitivity.

In Table 6, it can be seen that the achieved accuracy is remarkably better in all the fusions of proposed selection method. Here, HSM_Proposed outperformed the others on all the performance ratings.

In breast cancer data set shown in Table 7, FuzzyHSM_Proposed fusion showed significant improvisation in terms of accuracy, sensitivity, specificity and AUC. HSM_Proposed is the only

Table 4. Evaluation statistics across credit card client's data set.

Decision tree classifier	Accuracy	Sens	Spec	AUC
Fuzzy SLIQ_IG	0.792	0.800	0.803	0.810
Fuzzy SLIQ_GR	0.773	0.764	0.795	0.819
Fuzzy SLIQ_CFS	0.752	0.750	**0.810**	**0.854**
Fuzzy SLIQ_Relief F	0.812	0.792	0.800	0.824
Fuzzy SLIQ_Proposed	**0.831**	**0.842**	0.805	0.812
HSM_IG	0.812	0.799	0.805	0.811
HSM_GR	0.823	0.802	0.810	0.823
HSM_CFS	0.837	0.817	0.844	0.842
HSM_Relief F	**0.891**	**0.851**	0.889	0.879
HSM_Proposed	0.870	0.831	**0.904**	**0.895**
DCSM_IG	0.822	0.812	0.821	0.843
DCSM_GR	0.789	0.759	0.792	0.822
DCSM_CFS	0.800	0.777	0.816	0.818
DCSM_Relief F	0.895	**0.846**	0.892	**0.904**
DCSM_Proposed	**0.953**	0.834	**0.902**	0.891
FuzzyHSM_IG	0.804	0.790	0.799	0.823
FuzzyHSM_GR	0.842	0.835	0.832	0.856
FuzzyHSM_CFS	0.882	0.858	0.878	0.874
FuzzyHSM_Relief F	0.927	0.887	0.908	0.902
FuzzyHSM_Proposed	**0.962**	0.893	**0.952**	**0.917**

Table 5. Evaluation statistics across dermatology data set.

Decision tree classifier	Accuracy	Sens	Spec	AUC
Fuzzy SLIQ_IG	0.766	0.780	0.793	0.815
Fuzzy SLIQ_GR	0.783	0.759	0.775	0.809
Fuzzy SLIQ_CFS	0.762	0.742	0.777	**0.874**
Fuzzy SLIQ_Relief F	0.802	0.802	0.820	0.840
Fuzzy SLIQ_Proposed	**0.811**	**0.837**	**0.835**	0.802
HSM_IG	0.802	0.799	0.811	0.811
HSM_GR	0.823	0.802	0.804	0.823
HSM_CFS	0.817	0.797	0.834	0.842
HSM_Relief F	0.859	0.851	0.889	0.879
HSM_Proposed	**0.871**	**0.854**	**0.904**	**0.895**
DCSM_IG	0.822	0.812	0.821	0.843
DCSM_GR	0.789	0.759	0.792	0.822
DCSM_CFS	0.800	0.777	0.816	0.818
DCSM_Relief F	0.895	**0.846**	0.892	**0.904**
DCSM_Proposed	**0.953**	0.834	**0.902**	0.891
Fuzzy HSM_IG	0.795	0.780	0.797	0.832
Fuzzy HSM_GR	0.847	0.835	0.832	**0.871**
Fuzzy HSM_CFS	0.827	**0.868**	0.878	0.857
FuzzyHSM_Relief F	0.947	0.857	**0.908**	0.859
FuzzyHSM_Proposed	**0.952**	0.893	0.892	**0.864**

Table 6. Evaluation statistics across breast cancer data set.

Decision tree classifier	Accuracy	Sens	Spec	AUC
Fuzzy SLIQ_IG	0.720	0.688	0.729	0.737
Fuzzy SLIQ_GR	0.725	0.685	0.741	0.747
Fuzzy SLIQ_CFS	0.771	0.745	0.801	0.772
Fuzzy SLIQ_Relief F	0.823	**0.781**	**0.843**	0.827
Fuzzy SLIQ_Proposed	**0.886**	0.776	0.824	**0.833**
HSM_IG	0.790	0.766	0.791	0.801
HSM_GR	0.816	0.792	0.805	0.812
HSM_CFS	0.796	0.771	0.781	0.772
HSM_Relief F	**0.820**	0.787	**0.822**	0.869
HSM_Proposed	0.803	**0.832**	0.812	**0.901**
DCSM_IG	0.768	0.742	0.776	0.800
DCSM_GR	0.821	**0.823**	0.838	0.810
DCSM_CFS	0.783	0.781	0.821	**0.866**
DCSM_Relief F	0.821	0.802	0.853	0.822
DCSM_Proposed	**0.872**	0.794	**0.862**	0.852
FuzzyHSM_IG	0.807	0.778	0.812	0.834
FuzzyHSM_GR	0.782	0.790	0.810	0.822
FuzzyHSM_CFS	0.838	0.824	0.830	0.849
FuzzyHSM_Relief F	0.885	0.872	0.890	0.878
FuzzyHSM_Proposed	**0.942**	**0.891**	**0.905**	**0.901**

Table 7. Evaluation statistics across cylinder bands data set.

Decision tree classifier	Accuracy	Sens	Spec	AUC
Fuzzy SLIQ_IG	0.799	0.815	0.813	0.808
Fuzzy SLIQ_GR	0.753	0.742	0.785	0.819
Fuzzy SLIQ_CFS	0.728	0.705	**0.815**	0.834
Fuzzy SLIQ_Relief F	**0.832**	0.793	0.800	0.824
Fuzzy SLIQ_Proposed	0.821	**0.845**	0.802	**0.842**
HSM_IG	0.822	0.795	0.835	0.811
HSM_GR	0.832	0.825	0.824	0.823
HSM_CFS	0.852	0.833	0.856	0.842
HSM_Relief F	**0.879**	0.843	0.876	**0.893**
HSM_Proposed	0.870	**0.851**	**0.885**	0.875
DCSM_IG	0.825	0.802	0.821	0.841
DCSM_GR	0.792	0.779	0.792	0.823
DCSM_CFS	0.814	0.800	0.816	0.822
DCSM_Relief F	0.877	0.833	**0.901**	**0.871**
DCSM_Proposed	**0.948**	**0.924**	0.892	0.862
FuzzyHSM_IG	0.814	0.791	0.809	0.803
Fuzzy HSM_GR	0.841	0.855	0.842	0.863
Fuzzy HSM_CFS	0.872	0.863	0.868	0.896
FuzzyHSM_Relief F	0.904	**0.882**	0.908	0.883
FuzzyHSM_Proposed	**0.922**	0.875	**0.952**	**0.903**

variant which is somewhat lacking in terms of accuracy.

In Table 6, a mixed performance response can be observed for sensitivity, specificity and AUC but the three fused variant of proposed method shows increased accuracy. Fig. 3 shows the graphical comparison of accuracy across all data sets.

The graphical analysis in Fig. 4 and Fig. 5 presents computational complexity and attribute set size reduction across various fusions of proposed method in all five data sets.

The experimental analysis performed on various data sets claims the proposed technique is better than the existing state of art methods. The substantial increase in accuracy and reduced computation overhead makes the proposal worthy enough.

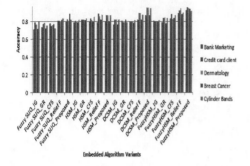

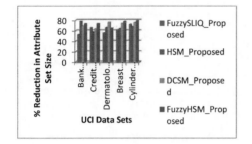

Figure 3. Comparison of accuracy among various data sets.

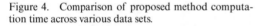

Figure 4. Comparison of proposed method computation time across various data sets.

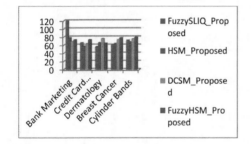

Figure 5. Comparison of proposed method attribute set reduction Across Various Data sets.

3 CONCLUSION

The proposed method fused with decision tree induction process reduces the attribute search space by making it computational effective, which is well suited for any real time system development. The hybridized framework proposed here improves the accuracy of classification model by considering a balance of attribute bias and individual discrimination power, which could be misleading in the presence of noisy attributes. The statistical results shows significant performance improvement using proposed technique in comparison to various other state of art attribute selection techniques (IG, GR, Relief F and CFS).

REFERENCES

Anuradha & Gupta Gaurav (2015), "Fuzzy Decision Tree Construction in Crisp Scenario through fuzzified Trapezoidal Membership Function", in internetworking Indonesia journal, Vol. 7, No. 2, pp. 21–28.

Chandra B., & Kuppili, Venkatanaresh Babu (2011). Heterogeneous Node Split Measure for Decision Tree Construction. Systems, Man, and Cybernetics (SMC). IEEE International Conference, pp. 872–877.

Chandra B., Kothari, R., & Paul, P. (2009). A new node splitting measure for decision tree construction. Pattern Recognition, Vol.43 (8), pp. 2725–2731.

Chang, C., Verhaegen, P. A., & Duflou, J. R. (2014): A comparison of classifiers for intelligent machine usage prediction. Intelligent Environments (IE), pp. 198–201.

Guyon I & Elisseeeff A. (2003): An introduction to variable and feature selection, Journal of Machine Learning, pp. 1157–1182.

Hall M. (2000): Correlation-based feature selection for discrete and numeric class machine learning, in: Proceedings of the 17th International Conference on Machine Learning, 2000, pp. 359–366.

Huang, J., Cai, Y., Xu, X. (2007): A hybrid genetic algorithm for feature selection wrapper based on mutual information. Pattern Recognition Letters 28, 1825–1844.

Ji-cang Lu, Fen-lin Liu, Xiang-yang Luo (2014). Selection of image features for steganalysis based on the Fisher criterion, Digital Investigation 11, pp. 57–66.

Jiang, L., Cai, Z., Zhang, H., and Wang, D. (2012). Not so greedy: Randomly Selected Naive Bayes. Expert Systems with Applications, 2012, 39(12): pp. 11022–11028.

Kohavi, R., & John, G. (1997): Wrappers for Feature Subset Selection. In Artificial Intelligence journal, special issue on relevance, 1997, 97(1–2):273–324.

Ladha, L., & Deepa, T. (2011). Feature selection methods and algorithms. International Journal on Computer Science and Engineering, 3(5), pp. 1787–1797.

Liu, H. & Yu, L. (2005): Toward integrating feature selection algorithms for classification and clustering, IEEE Transactions on Knowledge and Data Engineering, 17(4), pp. 491–592.

Maldonado, S. & Weber, R. (2009). A wrapper method for feature selection using Support Vector Machines, Information Sciences: an International Journal, 179(13), pp. 2208–2217.

Miche, Y., Roue B., Lendasse A. & Bas B. (2006): A Feature Selection Methodology for Steganalysis. Multimedia Content Representation, Classification and Security Lecture Notes in Computer Science, Vol. 4105, 49–56.

Powell, W. B. (2007). Approximate dynamic programming: Solving the curses of dimensionality, Wiley-Interscience, 1st edition.

Saeys, Y., Inza, I. & Larranaga, P. (2007): A review of feature selection techniques in bioinformatics, Bioinformatics, 23(19), pp. 2507–2517.

Xia B.B, Zhao X. F., Feng D. G. (2012): Improved Steganalysis by MWM Feature Selection, Watermarking —Volume 2, Intec, 243–258.

Communication and Computing Systems – Prasad et al. (Eds)
© 2017 Taylor & Francis Group, London, ISBN 978-1-138-02952-1

Real time sentiment analysis of tweets using machine learning and semantic analysis

R. Rajput & A. Solanki

School of ICT, Gautam Buddha University, Greater Noida, India

ABSTRACT: The 21st century has brought in lot of products in form of movies, software, video games, for which users make their opinion or judgments and often express it over internet through use of Social Networking Sites (SNS). This has resulted in rise of user generated content. Sentiment Analysis or Opinion Mining is the computational analysis of public opinions and sentiments towards a particular subject. The social media giants like twitter attract millions of online users sharing their opinions in form of 'tweets'. These tweets can be classified into positive or negative on the basis of their sentimental orientation. Methods used to classify the tweets are semantic and machine learning approaches. These methods are discussed and applied in this research work. A method for classification which is combination of Senti WordNet and Machine Learning algorithms is also proposed. All of these methods are then analyzed on the basis of precision, recall and F-measure. The proposed algorithm gives 86% accuracy as compared to 85% accuracy of Multinomial Naïve Bayes and gives accuracy up to 77% for real time data.

1 INTRODUCTION

The vast and ever-growing popularity of Social Networking Sites (SNS) results in overwhelming amount of information in form of user generated data as in opinions, thoughts and beliefs. Twitter is a well-known social media giant. It has millions of active users. It has become a haven for common people to express their opinion over some product, event, person or news. The reviews given by the users can be used to make important business related decisions as they give insight into product reception and quality. People generally are more likely to have a same sentiment value over a product as being held by their peers. The influence of SNS cannot be overlooked or denied. Presently due to excessive amount of data available online, it is becoming hard for an organization to monitor what sentiment is being held by the general public about some specific product or event. This brings in the urge of Sentiment Analysis automation.

2 RELATED WORK

Research in Sentiment Analysis field ranges from classification on document level (Yessenalina et al. 2010) to classifying words and phrases by learning their polarity (Hatzivassiloglou et al. 1997). It has been reported that events that happen in real life indeed have a significant effect on the public sentiment over SNS. The research work by Connor et al. 2010 predicts poll results and Bollen et al. 2011, predicted stock market using twitter trends supports this report. On the basis of such observations, some other research Mishne et al. 2006, made use of the sentiment orientations in SNS to predict movie sales. The pioneering work of figuring out application and challenges in the field of Sentiment Analysis was presented by Pang et al. 2008 and Liu 2012. They mentioned the techniques used to solve each problem in Sentiment Analysis. Ohana et al. 2009 have used SentiWordNet along with the Machine learning classifiers and Pang et al. 2002 have shown use of machine learning algorithms to classify the documents and comparing results of Naïve Bayes, Maximum Entropy, and SVM with varying n-grams. In another research work, Walia et al. 2012, have used Machine Learning based classification approaches along with the Unsupervised Semantic Orientation based algorithms for sentiment analysis of movie review texts.

3 PROPOSED METHODOLOGY

The proposed system takes in consideration two classification techniques. One being the Machine Learning approach and other being the semantic approach. In subsequent subsections, the changes in the parameters of the algorithms and reasons of doing so are discussed.

3.1 *Machine learning approach*

This approach relies on machine learning algorithms to solve sentiment analysis problem as text classification problem. Most common technique is supervised learning, where labelled dataset is used to train the classifier. The model generated can be used to predict the class of the text. The machine learning approach is explained in subsequent stages.

3.1.1 *Classifier selection*

This paper takes into consideration four types of classifiers namely Naïve Bayes Classifier, Multinomial Naïve Bayes Classifier, Random Forest and Support Vector Machine. These classifiers are on the training dataset and evaluated on the basis of Precision, Recall and F-measure. The results in Table 1 and Graph 1 are over the 10-Fold Cross Validation process on labelled dataset of 50,000 tweets.

As it turns out that Multinomial Naïve Bayes classifier most suits the requirements in terms of Precision, Recall, F-Measure and Accuracy. The other sophisticated algorithms like Random Forest and SVM do somewhat match up in terms of accuracy but Multinomial Naïve Bayes classifier is much more faster than others which will be suited for real-time system, where test-set is fetched in real time.

Multinomial Naïve Bayes is highly studied classifier. It is relatively effective, fast and easy to implement. The earliest description of this classifier can be found in research work of Duda et al.

Table 1. Result of 10-Fold Cross Validation process on labelled dataset.

Classifiers	Precision	Recall	F-Measure
NB	0.771	0.783	0.777
NBM	0.834	0.851	0.837
RF	0.811	0.833	0.778
SVM	0.819	0.841	0.822

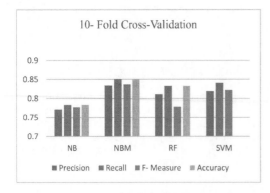

Figure 1. Coherent view of how the above stated classifiers perform on the stated parameters.

1973. Multinomial Naive Bayes is a specialized version of Naive Bayes that is designed more for text documents. The multinomial model captures word frequency information in documents. Research work by McCallullum et al. 1998 is a very coherent explanation of the differences and their application to text classification. For a Feature Vector $x = (x_1, x_2, \ldots x_n)$ where p_i is the probability that event i occurs and x_i is the number of times event I was observed, the multinomial naive Bayes classifier can be expressed as in equation (1):

$$\log p(C_k | x) \propto \log \left(p(C_k) \prod_{i=1}^{n} p_{ki}^{x_i} \right)$$
$$= \log p(C_k) + \sum_{i=1}^{n} x_i \cdot \log p_{ki} \qquad (1)$$

3.1.2 *Tokenizer selection*

Tokenizer is the way to split the data. The unit of measurement of splitted data is 'token'. Finding potentially predictive n-grams such as unigrams, bigrams, and trigrams is an important task especially in the Sentiment Analysis process. The different n-grams gives different results according to the situation. When the data such as 'not heroic' is encountered, the advantages of seeing the two words together is understandable. Table 2 and Graph 2 aims to let user visualize how NBM classifier works with unigram, bigram and trigram model.

From the data in Table 2 and Graph 2, we see that unigram model gives best results thus it is

Table 2. Results of varying n-grams.

Tokenizer	Precision	Recall	F-Measure
Unigram	0.834	0.851	0.837
Bigram	0.824	0.835	0.829
Trigram	0.687	0.829	0.751

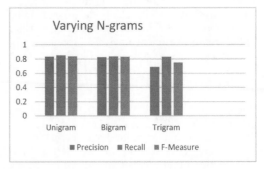

Figure 2. Effect of varying n-grams on precision, recall and F-Measure.

evident that it is best suited for the proposed research. So, the unigram is selected as tokenizer.

3.2 Semantic approach

Semantics is the study of the meanings of words and phrases in language. The semantic classification technique implemented by this paper makes use of dictionary based approach where in lexical resource (SentiWordNet) is used. SentiWordNet is a lexical resource in which each synset of Word-Net is associated to three numerical scores Obj(s), Pos(s) and Neg(s) ranges from 0.0 to 1.0.

3.2.1 Tokenizer selection

Selection of tokenizer in semantic approach plays an important part. Unigram takes one word at a time and lets SentiWordNet to assign sentiment score to it. But in this case, as pointed out by Nicholls et al. 2009, even better approach, where in implicit tokenizer of Stanford Parts Of Speech (POS) tagger is used as for SentiWordNet. The POS tagger reads the text and assigns POS to each word, such as noun, adjective, etc. For the default character encoding of the tagger is UTF-8 (Unicode), so UTF-8 encoding is maintained in the pre-processing. For getting the best results, implicit tokenizer of Stanford POS tags is used.

3.2.2 Classifying using SentiWordNet

The total score is calculated using Word Sense Disambiguation (WSD) and SentiWordNet. As total score remains in between −1 and 1, where 1 being highly positive and −1 being highly negative. This score can be used to our advantage to classify tweets into multiple classes to help understand deeply the sentiment value. Scores used along with their subsequent classes are shown in Table 3.

4 PROCESS FLOW OF THE PROPOSED RESEARCH

The process flow of the proposed system can be visualized by Figure 3.

Table 3. SentiWordNet scores with classification.

Sent score	Classification
Above 0.75	Strong_Positive
Between 0.5 and 0.75	Well_Positive
Between 0.25 and 0.5	Positive
Between 0 and 0.25	Weak_Positive
Between 0 and −0.25	Weak_Negative
Between −0.25 and −0.5	Negative
Between −0.5 and −0.75	Well_Negative
Below −0.75	Strong_Negative

The steps followed in the realization of the proposed system are explained in following steps-:

Step-1: Data Extraction Module
User fires a query through search box. The system makes use of 'Twitter4J' API along with proper credentials to login into the Twitter. The extracted tweets are then stored in Comma Separated Value (CSV) file.

Step-2: Preprocessing Module
Pre-processing is the process of cleaning the data and readying the text for classification. The tweets extracted from the twitter API are leeched with irrelevant details, which will do no good in the text classification task. Pre-processing also speeds up the classification process, thus helping in real time SA. Haddi et al. 2013, have shown that pre-processing the text including data transformations and filtering can significantly improve the performance. The authors propose to eliminate the Uniform Resource Locators (URLs), hashtags, references, special characters and special Twitter Symbols like @, RT etc.

Step-3: Feature Extraction and Selection Module
Feature extraction and Feature Selection is part the of dimensionality reduction. Features in our proposed system are extracted using unigrams in the case of Machine Learning and using POS tags in the case of Semantic Analysis. The 'words' are selected in case of Machine Learning and 'POS tags' are selected in case of Semantic Analysis. Features are not reduced as it will affect the accuracy of the research.

Step-4: Sentiment Classification
Proposed algorithm uses Multinomial Naïve Bayes Algorithm with unigram model and follow up by SentiWordNet algorithm using POS tags.

Proposed Algorithm (pseudo code):
Input: Labelled Training-dataset and search query
    Output: Sentiment polarity

Step 1: Extract the tweets from Twitter API: Test
        set 'ts' // *Extracted tweets becomes test set*
Step 2: Preprocessing
        For tweet t: // *Preprocessing module*
        Preprocessing (String t)
        Removing URLs, special symbols, Non-English Words
        Return t // Return the processed tweet

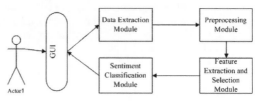

Figure 3.   The process flow of the proposed system.

Step3: Train a classifier 'C' on training dataset 'TS'
    // *Using MNB Classifier*
    Return Classifier model
Step4: Extract Feature Vector list and features
    from //*Feature Extraction and Selection*
    For tweet t:
    StringToWordVector (t) // *filter used*
    Return Feature Vector
    For n in Feature Vector:
    Return Features: Words
Step5: Apply Classifier model Return class (pred)
    // pred is String
Step6: Extract Feature Vector list and features
    //*Feature Extraction and Selection*
    For tweet t:
    POS tagger (t) //Stanford POS Tagger
    Return features: tags
Step 7: Get scores using SentiWordNet
    Classify into classes (sent) // *sent is a String*
Step 8- For tweet t:

If sent = strong_positive or sent = strong_
    negative
Return sent; // *Use results from SentiWordNet*
Else if sent = well_positive and pred = pos
Return sent // *Use results from SentiWordNet*
Else if sent = well_negative and pred = neg
Return sent // *Use results from SentiWordNet*
Else Return pred; // *Use classifier results*

5 IMPLEMENTATION AND WORKING OF SYSTEM

For realisation of the proposed approach, an application is designed which makes use of the java code for implementation. Data is maintained in file system using Comma Separated Value (CSV) files. API used are WEKA, Twitter4J and Stanford POS tagger. WEKA is a popular suite for Machine Learning software, Twitter4j is a library toolkit to integrate java application to the twitter service. Stanford POS tagger is explained above. Main modules that have been designed and implemented are explained as below:

5.1 *Data extraction module*

The application uses 'Twitter4J' API to extract the tweets. This requires user to register for twitter developer profile and get the 'consumer key', 'consumer token' and other credentials. The designed application makes use of these credentials to access the twitter content and extract it. These tweets are saved in CSV format in the file system.

5.2 *Pre-processing module*

Saved Tweets in CSV file is pre-processed according to methods discussed above. The CSV file

is converted to Attribute Relation File Format (ARFF) to match the training dataset. This is now the 'test set'. Changes are made in the test set with help of WEKA filters to match the training dataset. Figure 3 shows the pre-processed tweets saved in ARFF file.

5.3 *Feature extraction and selection module*

Feature are extracted using unigram model in case of Machine Learning Approach and implicit tokenizer of Stanford POS tagger. Then WEKA API is used to train classifier using training dataset. This paper uses the labelled dataset available online on the link given at the end-note. Here n-grams and possible features are chosen. This creates a classifier model.

5.4 *Sentiment classification*

The test set is then fed to the model to find out the class. Using same application, if user wants to get results using Semantic approach. The same CSV file is then used to get sentiment scores using POS tagging and SentiWordNet algorithm. The user can then find the sentiments of the same topic using combined approach as discussed above and get the results.

6 RESULTS AND ANALYSIS

The results are assessed using Precision, Recall and F-Measure using following formulae:

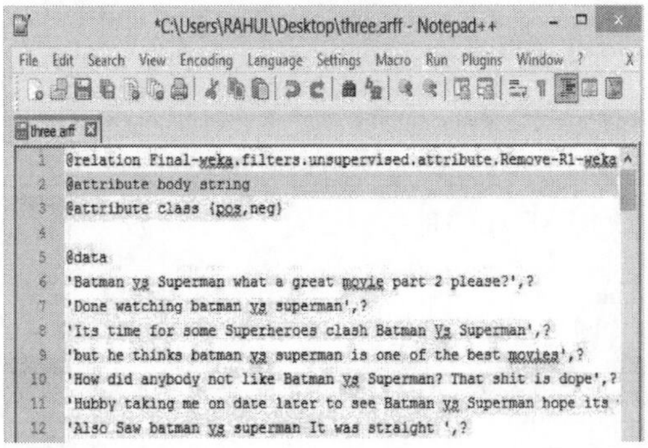

Figure 4.   Pre-processed data in ARFF file.

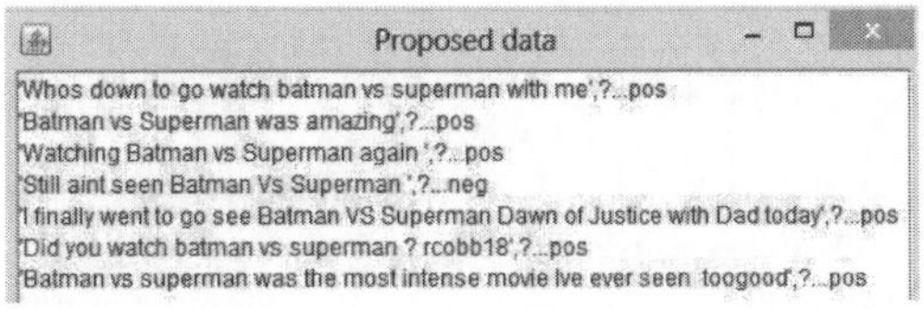

Figure 5.   Snapshot of results by proposed algorithm.

$$Precision = \frac{t_p}{t_p + f_p} \qquad (2)$$

$$Recall = \frac{t_p}{t_p + f_n} \qquad (3)$$

$$F\_Measure = 2 \cdot \frac{Precision \cdot Recall}{Precision + Recall} \qquad (4)$$

$$Accuracy = \frac{t_p + t_n}{t_p + t_n + f_p + f_n} \qquad (5)$$

Here, t_p is True Positive, f_p is False Positive, t_n is True Negative and f_n is False Negative. The results are analysed on two different Test-set. First analysis is done in a conventional way using test-set from a part of training dataset, results of which are shown in Table 4 and Graph 3.

When 100 tweets are taken into consideration, we find out that $T_p = 66$, $T_n = 24$, $F_p = 8$ and $F_n = 4$. Therefore value of precision, recall, f-measure and accuracy is 0.892, 0.942, 0.917 and 0.86 from equations (2), (3), (4) and (5). While taking 200 tweets into consideration, $T_p = 138$, $T_n = 36$, $F_p = 20$ and $F_n = 6$ and value of precision, recall, f-measure and accuracy is 0.873, 0.958, 0.9132 and 0.87 from the same equations.

Second, results are assessed on test set from real-time data, using Precision, Recall and F-Measure as shown in Table 5 and Graph 4:

Table 4. Results of proposed algorithm on conventional test-set.

No. of Tweets	Precision	Recall	F-Measure	Accuracy
100	0.892	0.942	0.917	0.86
200	0.875	0.959	0.915	0.87
500	0.873	0.968	0.9132	0.87

For the real time data, for search query 'Batman vs Superman', number of extracted tweets were 100. After classification and on analysis we find out that Tp = 52, Tn = 24, Fp = 11 and Fn = 13. There value of precision comes out to 0.8254 from equation (2), value of recall is 0.8 from equation (3), value of F-Measure is 0.8123 from equation (4), and accuracy is 0.76 from equation (5).

When going through the results, it can be seen that the accuracy has fallen to 0.76 and 0.75 for BVS and Obama respectively from the results that were seen in (test set) i.e. 0.86, 0.87, 0.87. There are several challenges that were faced during Real Time Sentiment Analysis. Foremost is the tweets that are extracted are spot on, they may not resemble to the training dataset. This affects the performance of the classifier. The major challenge that was faced during sentiment analysis of real time tweets was labelling the tweets with the context. If a user wants to do sentiment analysis of a movie say 'X'. Then it is possible to get tweets which implies different sentiment to 'X', but convey different overall sentiment of tweet. Example, "Everything failed but the movie, including weather, food and the ride. Bad experience" Now it is relevant to stick to either overall sentiment of tweet or just the context of "search query". One more challenge is "blabber" which is insignificant chit chat around the subject, which is more of objective in nature and cannot be put against any class. These can actually be dealt by keeping the 'neutral' class.

Table 5. Results from from real-time data.

Query	Precision	Recall	F-Measure	Accuracy
B vs S	0.8254	0.8	0.8123	0.76
Obama	0.689	0.769	0.727	0.75

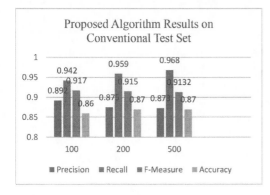

Figure 6. Coherent view of results of proposed algorithm.

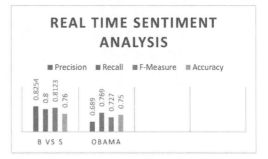

Figure 7. Results of proposed algorithm on real time tweets.

7 CONCLUSION

In this research work, set of Machine Learning algorithms along with semantic approach using POS tagging and SentiWordNet to classify the tweets are analysed. Multinomial Naïve Bayes algorithm is chosen among other classifier as it gives most 85.06% accuracy and is also faster than other sophisticated algorithms. Also unigram model is chosen as it performs best on the available dataset. The MNB classifier is followed with SentiWordNet algorithm which increases the accuracy to 86%. The real time Sentiment analysis is also done, which is the aim of this proposed research. Testing the proposed algorithm on Real Time data, gives up to 77% accuracy. Challenges that were met are also discussed.

REFERENCES

Bollen, J., H. Mao, and X. Zeng, "Twitter mood predicts the stock market," J. Computer Science., vol. 2, no. 1, pp. 1–8, Mar. 2011.

Duda, R.O. and Hart, P.E. (1973), Pattern classification and scene analysis, Wiley, New York.

Haddi, E., Liu, Xi, and Shi, Y., "The Role of Text Pre-processing in Sentiment Analysis" in ITQM2013, Science Direct, 2013.

Hatzivassiloglou, V., McKeown, K., "Predicting the Semantic Orientation of Adjectives", 1997.

Liu, B., "Sentiment analysis and opinion mining" in Synth Lect Human Lang Technol (2012).

McCallullum, A. and Nigam, K., "A Comparison of Event Models for Naive Bayes Text Classification" 1998.

Mishne, G. and N. Glance, "Predicting movie sales from blogger sentiment," in Proc. AAAI-CAAW, Stanford, CA, USA, 2006.

Nicholls, C. and Song, F., "Improving sentiment analysis with Part-of-Speech weighting" in 2009 International Conference on Machine Learning and Cybernetics (Volume: 3)", 2009.

Ohana, B. and Tierney, B., "Sentiment Classification of Reviews Using SentiWordNet" in 9th. IT & T Conference, Dublin Institute of Technology, 2009.

O'Connor, B., R. Balasubramanyan, B.R. Routledge, and N.A. Smith, "From tweets to polls: Linking text sentiment to public opinion time series," in Proc. 4th Int. AAAI Conf. Weblogs Social Media, Washington, DC, USA, 2010.

Pang, B. and Lee, L., "Opinion mining and sentiment analysis" in Found Trends Inform Retriev, 2 (2008), pp. 1–135.

Pang, B. and Lillian, L., "Thumbs up? Sentiment Classification using Machine Learning Techniques" in Proceedings of EMNLP 2002, pp. 79–86.

Waila, P., Marisha, V.K. Singh, and M.K. Singh," Evaluating Machine Learning and Unsupervised Semantic Orientation approaches for sentiment analysis of textual reviews" Computational Intelligence & Computing Research (ICCIC), 2012 IEEE International Conference on 18–20 Dec. 2012.

Yessenalina, A., Yue, Y. and Cardie, C., "Multi-level Structured Models for Document-level Sentiment Classification" in Proceedings of the 2010 Conference on Empirical Methods in Natural Language Processing, pages 1046–1056, MIT, Massachusetts, USA, 9–11 October 2010.

END NOTE

Dataset available at: https://drive.google.com/file/d/0B1 pvkpCwTsiSd1pyTFZkdWVRdEs5Q1NiQW1mRm-F1Zw/view.

Communication and Computing Systems – Prasad et al. (Eds)
© 2017 Taylor & Francis Group, London, ISBN 978-1-138-02952-1

LQG controller design for an industrial boiler turbine

S.K. Sunori
Graphic Era Hill University, Bhimtal, India

P.K. Juneja, M. Chaturvedi & M. Chauhan
Graphic Era University, Dehradun, India

ABSTRACT: In the present work, a boiler turbine industrial process with two manipulated variables, two controlled variables and some dead time has been considered as a case study. In order to eliminate loop interactions its decoupling is done to two separate SISO systems. Finally, an LQG controller has been designed for it and its performance is compared with that of conventional PID controllers based on Ziegler Nichol's tuning and Internal Model Control (IMC) techniques.

1 INTRODUCTION

In boiler turbines, the chemical and thermal energy is transformed to electricity. It is a highly complex, multivariable, time delayed and nonlinear process (S.K. Sunori et al 2016) (Sandeep et al 2015). In a typical boiler turbine plant a header collects all the steam which is generated from number of boilers which is then distributed to several turbines through header. The steam flow is directly proportional to power generation which is the key parameter to be controlled. The other parameter to be controlled is the drum pressure. The ultimate objective is to meet the load demand of electric power. The schematic diagram of a boiler turbine plant is depicted in Figure 1.

The process has two manipulated variables, the Governor Valve position (GV) and the fuel Flow Rate (FR). The variables to be controlled are the Electric Power (EP) and the Steam Pressure (SP). Equation (1) shows the considered model of an industrial boiler turbine process (Zhang Hua & Lilong 2002).

$$
\begin{bmatrix} EP \\ SP \end{bmatrix} = \begin{bmatrix} \dfrac{68.81e^{-2s}}{984s^2 + 94s + 1} & \dfrac{(-23.58s - 2.196)e^{-8s}}{372s^2 + 127s + 1} \\[2ex] \dfrac{e^{-2s}}{6889s^2 + 166s + 1} & \dfrac{2.194e^{-8s}}{6400s^2 + 160s + 1} \end{bmatrix} \begin{bmatrix} GV \\ FR \end{bmatrix} \tag{1}
$$

2 PID CONTROLLER

An industrial PID controller has many extensions over the years that make it a more practical tool for operating a chemical process. Many PID controller tuning methods have been proposed in the literature e.g. Ziegler-Nichols (ZN) tuning, Cohen-Coon tuning, direct synthesis method, Internal Model Control (IMC) etc. In the present work ZN and IMC tuning techniques will be employed. In general, the output of a PID controller is given by equation (2).

$$
u(t) = K_p e(t) + K_i \int_0^t e(\tau)d\tau + K_d \frac{de(t)}{dt} \tag{2}
$$

or

$$
u(t) = K_p \left[e(t) + \frac{1}{T_i} \int_0^t e(\tau)d\tau + T_d \frac{de(t)}{dt} \right] \tag{3}
$$

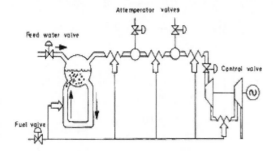

Figure 1. Schematic diagram of the boiler-turbine unit [1].

where K_p, K_i, K_d are proportional gain, integral gain and derivative gain respectively.

T_i, T_d are integral and derivative time constants respectively.

2.1 Ziegler-Nichols (ZN) tuning

In this method, first of all, Ti and Td are set at values ∞ and 0 respectively. Then the proportional gain K is increased from zero to a value at which sustained oscillations are set up at the output. Let this value of gain be Pcr and the corresponding time period be Tcr. Now, the controller tuning is done using the formulas presented in Table 1.

2.2 Internal Model Control (IMC) tuning

Any industrial process control can be successfully achieved by IMC based PID controller as it is highly robust and gives an excellent performance in case of processes with large dead time (Scali et al 1992). An IMC tuning rules table has been developed providing a significant disturbance rejection irrespective of the position at which the disturbance enters (Ian et al. 1996).

An IMC technique has been proposed providing an excellent rejection of the load disturbance in the situation when the desired closed loop dynamics is faster than the process dynamics. Previously done work could not do well in this situation (Scali et al. 1992).

Figures 2 and 3 are depicting the block diagrams of tuned PID controller and the designed IMC controller respectively.

where,

$$q(s) = \frac{c(s)}{1 + p'(s)c(s)} \qquad (4)$$

Table 1. Ziegler-Nichols tuning rules.

Controller	Kp	Ti	Td
P	0.5 P$_{cr}$	⊥	0
PI	0.45 P$_{cr}$	(1/1.2)T$_{cr}$	0
PID	0.6 P$_{cr}$	0.5 T$_{cr}$	0.125T$_{cr}$

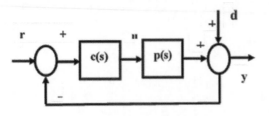

Figure 2. Block diagram of tuned PID controller.

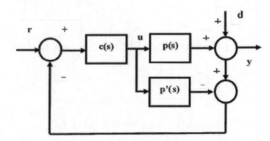

Figure 3. Block diagram of IMC controller.

It consists of an internal model p'(s) and IMC controller q(s). For the perfect internal model p'(s) = p(s). The internal stability of the IMC controller is guaranteed only if both p(s) and q(s) are stable.

3 LQG CONTROLLER

Consider, the state space representation of the plant to be controlled, given as,

$$\dot{x} = Ax + Bu$$
$$y = Cx \qquad (5)$$

The block diagram of LQR (linear quadratic regulator) for this plant is shown in Figure 4.

This regulator minimizes the following objective function,

$$J = \int_0^\infty [x^T(t)Qx(t) + u^T(t)Ru(t)]dt \qquad (6)$$

where, Q and R, are weighting parameters that penalize the states and the control effort respectively. Therefore Q and R represent controller tuning parameters.

$$u(t) = -Kx(t) \qquad (7)$$

K is the gain given as,

$$K = R^{-1}B^T S \qquad (8)$$

where, S is given by the solution of the following equation called Ricatti equation,

$$SA + A^T S + Q - PBR^{-1}B^T S = 0 \qquad (9)$$

Now, if the measurement noise w and the process noise v (assumed to be white Gaussian noise) are also present as shown in the state space representation of the plant in equation (10) then LQR based controller cannot perform well, as in this

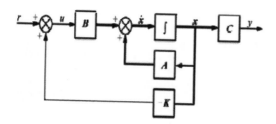

Figure 4.   Block diagram of LQR.

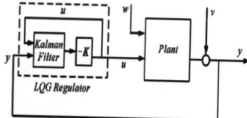

Figure 5.   Simplified block diagram of LQG controller.

case the entire state vector is not available for control at all times. In this case LQG controller is designed.

$$\dot{x} = Ax + Bu + w$$
$$y = Cx + v \tag{10}$$

The simplified and detailed block diagrams of LQG controller are depicted in Figures 5 and 6 respectively showing that the LQG controller combines the LQR with the Kalman filter. The Kalman filter estimates the entire state vector which is required for generation of the optimal control signal u. The estimated error covariance is minimized by this filter.

From this figure, we have,

$$\dot{\hat{x}} = (A - LC - BK)\dot{\hat{x}} + Ly \tag{11}$$

$$u = -K\dot{\hat{x}} \tag{12}$$

here, $\dot{x}$ represents the estimator for the state x and L is called the Kalman gain which is to be determined by the minimization of objective function (13) subject to constraint (14)

$$J = E[(x - \dot{x})^T (x - \dot{x})] \tag{13}$$

$$E[(x - \dot{x})^T y = 0 \tag{14}$$

this gives the following Kalman gain

$$L = S_e C^T R^{-1} \tag{15}$$

where, S_e is given by the solution of the following equation (16)

$$S_e A^T + A S_e + Q - S_e C^T R^{-1} C S_e = 0 \tag{16}$$

where,

$$Q = E(ww^T), R = E(vv^T)$$

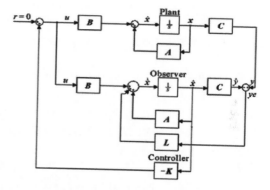

Figure 6.   Detailed block diagram of LQG controller.

4   SIMULATION WORK

Let G(s) be the transfer matrix of a 2 × 2 MIMO system.

$$[G(s)] = \begin{bmatrix} g_{11}(s) & g_{12}(s) \\ g_{21}(s) & g_{22}(s) \end{bmatrix} \tag{17}$$

Then the two respective decoupled SISO systems are given by equations (18) and (19) [3]

$$y_{1(s)} = \left[g_{11}(s) - \frac{g_{12}(s)g_{21}(s)}{g_{22}(s)} \right] v_1(s) \tag{18}$$

$$y_{2(s)} = \left[g_{22}(s) - \frac{g_{12}(s)g_{21}(s)}{g_{11}(s)} \right] v_2(s) \tag{19}$$

Using (17) and (18) we get two independent decoupled SISO systems, G1(s), which represents governor valve-Electric power system (SISO1) and G2(s), which represents Fuel flow rate-Steam pressure system (SISO2). The expressions for G1(s) and G2(s) are given in equations (20) and (21).

Now based on Pade' approximation, controllers are designed here for both decoupled SISO systems using three different techniques namely Ziegler Nichol's (ZN) tuning, Internal Modal Control

(IMC) and Linear Quadratic Gaussian (LQG) technique. The transfer functions of the designed ZN, IMC and LQG controllers for SISO1 are presented as equations (22), (23) and (24) respectively.

The transfer functions of the designed ZN, IMC and LQG controllers for SISO2 are presented as equations (25), (26) and (27) respectively.

corresponding Bode plots are shown in Figure 8. The corresponding performance parameters are specified in Table 2.

The above IMC controller has been designed with dominant closed loop time constant = 23.87.

The set point tracking responses of these controllers are depicted in Figure 9 and the corre-

$$G_{1_1}(s) = \frac{\left(1.485*10^8 s^5 + 4.186*10^8 s^4 + 1.436*10^8 s^3 + 4.334*10^6 s^2 + 4.482*10^4 s + 153.2\right)e^{-2s}}{5.533*10^9 s^6 + 2.551*10^9 s^5 + 2.6*10^8 s^4 + 8.53*10^6 s^3 + 1.248*10^5 s^2 + 849.1s + 2.194} \tag{20}$$

$$G_2(s) = \frac{\left(1.485*10^8 s^5 + 4.186*10^8 s^4 + 1.436*10^8 s^3 + 4.334*10^6 s^2 + 4.482*10^4 s + 153.2\right)e^{-8s}}{1.129*10^{12} s^6 + 4.407*10^{11} s^5 + 2.297*10^{10} s^4 + 5.055*10^8 s^3 + 5.616*10^6 s^2 + 3.117*10^4 s + 68.8} \tag{21}$$

$$C_{ZN} = \frac{0.0053352(1+26s)}{s} \tag{22}$$

$$C_{IMC} = \frac{-1.9865*10^{12}(1+83s)(1+12s)(1+3.6s)(1+0.98s)+(0.54s)^2)}{(1+0.39s)(1-2.6*10^{15} s)(1+3.4s)(1+0.88s+(0.52s)^2)} \tag{23}$$

$$C_{LQG} = \frac{-1.9962*10^{12}(1-0.00011s)(1+0.5s)(1+4.8s)(1+12s)(1+83s)}{(1+0.47s)(1-9.3*10^{14} s)(1+4.5s)(1+1.7s+(1.5s)^2)} \tag{24}$$

$$C_{ZN} = \frac{0.026193(1+91s)}{s} \tag{25}$$

$$C_{IMC} = \frac{8.2346*10^{15}(1+3.5s)(1+85s)(1+75s)\left(1+4s+(2.3s)^2\right)}{(1+0.41s)(1+5.8*10^{17} s)(1+3.3s)\left(1+3s+(2s)^2\right)} \tag{26}$$

$$C_{LQG} = \frac{6.7426*10^{14}(1+0.0004s)(1+75s)(1+85s)\left(1+3.9s+(2.2s)^2\right)}{(1+3*10^{16} s)\left(1+3.3s+(1.8s)^2\right)\left(1+3s+(4.1s)^2\right)} \tag{27}$$

The above IMC controller has been designed with dominant closed loop time constant = 16.9816.

The set point tracking responses of these controllers are depicted in Figure 7 and the

sponding Bode plots are shown in Figure 10. The corresponding performance parameters are specified in Table 3.

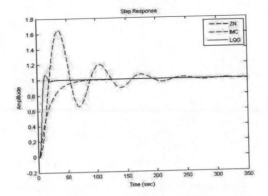

Figure 7. Set point tracking responses of ZN, IMC and LQG controllers for SISO1.

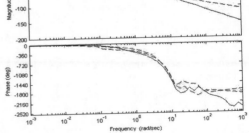

Figure 8. Bode plots for ZN, IMC and LQG controllers for SISO1.

Table 2. Performance parameters of ZN, IMC and LQG controllers for SISO1.

Controller	Rise time (s)	Settling time (s)	Overshoot (%)	Gain margin (dB)
ZN	12	218	65.3	7.24
IMC	37.2	68.2	0	22.8
LQG	5.74	18	.7.05	4.12

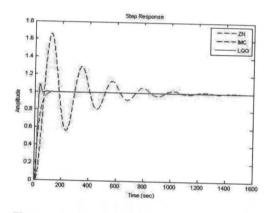

Figure 9. Set point tracking responses of ZN, IMC and LQG controllers for SISO2.

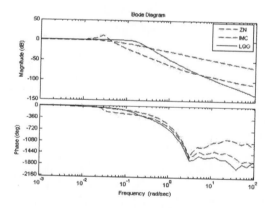

Figure 10. Bode plots for ZN, IMC and LQG controllers for SISO2.

Table 3. Performance parameters of ZN, IMC and LQG controllers for SISO2.

Controller	Rise time (s)	Settling time (s)	Overshoot (%)	Gain margin (dB)
ZN	40	1030	65.5	1.8
IMC	52	99.3	0	14.5
LQG	14.5	50.9	8.67	1.88

5 CONCLUSION

In the present paper a 2×2 MIMO boiler turbine process is taken up. Having done its decoupling to split it into two separate SISO systems, controllers are designed for both using ZN, IMC and LQG techniques and their performance is compared. It has been found that performance of IMC controller is much better than that of ZN controller in terms of both, transient and steady state responses. But, the LQG controller comes up with the best performance amongst all with least settling time and smallest peak overshoot.

REFERENCES

Danial E. Rivera and Melvin E. Flores, "Internal Model Control", *Control Systems, Robotics and Automation-Vol-II.*

Husain Ahmed and Abha Rajoriya, "Performance Assessment of Tuning Methods for PID Controller Parameter used for Position Control of DC motor," *International Journal of u and e Service, Science and Technology,* 2014, pp. 139–150.

Ian G. Horn, Jeffery R. Arulandu, Christopher J. Gombas, Jeremy G. VanAntwerp, Richard D. Braatz, "Improved Filter Design in Internal Model Control," *Ind. Eng. Chem. Res.* 1996, 35, pp. 3437–3441.

Sarailoo, M., B. Rezaie and Z. Rahmani, "MLD Model of Boiler Turbine System based on PWA Linearization approach," *International Journal of Computer Science and Engineering,* 2(4), 2012, pp. 88–92.

Ramneet Singh, Rajni Bala, Bhavi Bhatia, "Internal Model Control (IMC) and IMC based PID Controller," *International Journal of Advanced Research in Computer Science and Software Engineering,* 2014, pp. 915–922.

Ragnar Eide, Per Magne Egelid, Alexander Stamso, Hamid Reza Karimi, "LQG Control Design for Balancing an Inverted Pendulum Mobile Robot", *Intelligent Control and Automation*", 2011, pp. 160–166.

Sunori, S. K., P. K. Juneja, M. Chaturvedi, P. Aswal, S. K. Singh, S. Shree, "GA based optimization of quality of sugar in sugar industry," *Ciencia e Tecnica Vitivinicola journal,* Vol. 31 No 4, 2016.

Sandeep Kumar Sunori, Pradeep Kumar Juneja and Anamika Bhatia Jain, "Model Predictive Control System Design for Boiler Turbine Process," *IJECE,* 2015, pp. 1054–1061.

Sandeep Kumar Sunori and Pradeep Kumar Juneja, "Controller Design for MIMO Boiler Turbine Process," *IJCTA,* 2015, pp. 477–486.

Scali, C., Semino, D., Morari M., "Comparison of Internal Model Control and Linear Quadratic Optimal Control for SISO Systems", *Ind. Eng. Chem. Res.,* 1992.

Zhang Hua Guang and Lilong Cai, "Multivariable Fuzzy Generalized Predictive Control," *Cybernetics and Systems: An International Journal, Taylor & Francis,* 2002, pp 33: 69–99.

Communication and Computing Systems – Prasad et al. (Eds)
© 2017 Taylor & Francis Group, London, ISBN 978-1-138-02952-1

Closed loop compressor control system realization for cryogenic application using PLC

Y. Joshi & H.K. Patel
Institute of Technology, Nirma University, Ahmedabad, Gujarat, India

H. Dave
Institute for Plasma Research, Bhat, Gandhinagar, Gujarat, India

ABSTRACT: As compressors are an integral part of many systems including oil, gas and energy, their control receives attention for better efficiency of the whole plant. This paper presents a realization of a closed loop control scheme for compressor system on Siemens' step7 platform. Here, compressor's performance is simulated and modeled for a cryogenic application where the ultimate goal is to achieve liquid helium at 4.5 K. To implement the scheme on PLC, digitization of controller and process block is required because PLC works on discrete platform and system is continuous. Here, a control loop is implemented for compressor volumetric flow rate control using a recycle valve which is connected between suction and discharge.

Keywords: Step7, PLC, PID controller, Aspen HYSYS

1 INTRODUCTION

With the advent of high speed and reliable digital computers many processes can be easily controlled, manipulated and monitored. In cryogenics to reach up to very low temperature, the fluid is first compressed at high pressure and then expanded in an expansion engine or throttling valve (Barron, 1985; Flynn, 2005). The aim of present study is to simulate compressor performance and implement the control scheme on PLC. In centrifugal compressor system surge is an undesirable and unstable condition, which occurs at specific combination of discharge pressure and volume flow in compressor. In such condition, flow will be reversed and if not controlled, may results in mechanical damage of the compressor. This can be avoided by variable speed control, suction throttling, adjustable inlet guide vanes and bypass throttling. Among which the simplest solution is to recycle the flow from discharge to suction so that compressor can operate within its range of stability (Peter, 2004). The Anti-Surge Control Valve (ASCV) is a fail open solenoid valve. This means that it needs a high signal of 20 mA to close the valve, and a low signal of 4 mA to open the valve. When some failure occurs, the valve will usually receive a low signal and it will open, which is the safe position. When opened, it directs the pressurized fluid from just behind the compressor back to the entrance of the compressor.

This prevents the flow from becoming too low at the compressor inlet.

To design a control strategy for compressor surge is challenging task as compressor has very fast dynamics. We can see compressor steady state as well as dynamic response using Aspen HYSYS process modeling tool. With this tool we can also model and analyze our control strategy before implementation on actual hardware. This paper presents implementation of the above control loop on Siemens Step7 300/400 platform. This is done using Step7-SCL (Structured Control Language) which is a high level language to program PLCs of simatic S7.

2 MODELING COMPRESSOR PERFORMANCE WITH ASPEN HYSYS

Aspen HYSYS is very reliable software for process modeling. Here a centrifugal compressor is modelled and its anti-surge control scheme is also designed using Aspen HYSYS V7.1.

Helium gas is entered at room temperature and above atmospheric pressure for refrigeration system. The gas is compressed up to 14 bar to reach very low temperature of about 4.5 K. In Figure 1, K-100 represents compressor, E-100 represents chiller block and TEE-100 represents splitter block. Compression of gas will be accompanied with high temperature rise (Atrey, n.d., IIT Bombay). Temperature and pressure is related by equation 1:

$$\frac{T2}{T1} = \left(\frac{P2}{P1}\right)^{\frac{gamma-1}{gamma}} \qquad (1)$$

where, gamma = Cp/Cv; Cp = specific heat at constant P; Cv = specific heat at constant V.

Now, gamma = 1.66 for Helium
So, (14)^[(1.66–1)/1.66] = T2/T1 = 2.8
T1 = 27+273 = 300K
So, T2 = 2.8*300 = 840K

Gas will flow from discharge to suction via recycle valve where, makeup stream will mix with it and then it is entered in compressor. The ASCV determines the minimum volumetric flow rate that a compressor should have to prevent from surge phenomena. In Figure 2, volume flow versus discharge pressure is plotted where we can see a constant speed lines on which at a specific flow-head combinations, flow rate minimization is accompanied with lowered discharge pressure and at that point, flow in the compressor will be reversed. Lines connecting this points is called a surge line (Peter, 2004). So, ASCV will take care of this point and will move compressor's performance to surge avoidance

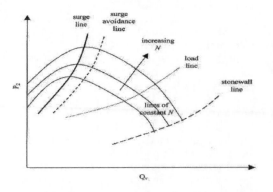

Figure 1.   HYSYS simulation of compressor.

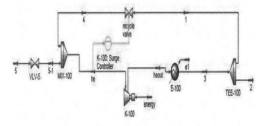

[1]Figure 2.   Schematic of compressor performance map.

---

[1](Peter, 2004).

line which is typically 10% above the surge flow. In HYSYS, anti-surge controller's algorithm is extension of PID algorithm and major difference between it and regular PID controller is that the set point of ASC is calculated and not set. The ASC takes more aggressive action near the surge point.

In HYSYS, the equation governing compressor performance map is given by equation 2.

$$H = A + B*Q + C*Q^2 + D*Q^3 \qquad (2)$$

where, H is head (m) and Q is surge flow (m³/h).

A, B, C, D parameters represents surge curve (Brownrig, n.d., Aspen Technology, Inc.). If one does not have curve in this parameter than using two points of flow-head combinations one can find them with the help of curve fitting. The ASCV is operated in reverse action means when volume flow is decreased, ASCV opening is increased.

3   CONTROL STRUCTURE

The control schematic is shown in Figure 3.

By controlling volume flow rate and hence pressure head, compressor performance within the surge limit can be assured. The ASCV dynamics can be simply modeled as a first order transfer function as the flow rate through valve is linearly proportional to a signal to the valve actuator.

First order system transfer function can be given by equation 3:

$$H(s) = Y(s)/U(s) = Kp/(TpS+1) \qquad (3)$$

where, Kp is system gain and Tp is time constant.

As we are using information from current Process Variable (PV) and it is compared with a setpoint (SP), the system is a feedback control system. The advantage of using feedback is ensuring stability. PLC manipulates the signal given to the ASCV according to the error between PV and SP using PID algorithm. Siemens S7 implements equation 4 in the CONT-C function block which is a continuous PID controller with continuous input and output.

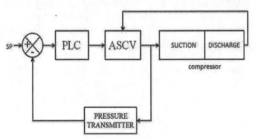

Figure 3.   Control scheme for anti-surge control.

This is a complete software controller with a series of sub-functions that you can activate or deactivate. Using PID assignment window we can modify the parameters of the Function Block (FB) and also monitor and record the curves of interest.

$$CO = Kp(e) + Ki\int_0^t (e)dt + Kd\frac{d(e)}{dt} \qquad (4)$$

where, CO is controller output.

4 EMULATION

Finding the discrete equivalent of a continuous system is desired when the controller is discrete and plant is continuous. Hence, to simulate the whole system as a discrete system, the discrete model of the plant must be obtained. The simplest way is to use numerical integration method. This is called emulation (Nielsen, n. d.).

In S7 CONT_C FB is available already in discrete form. So valve first order dynamics emulation is done as shown below.

Numerical integration can be done using rectangular rule, backward rectangular rule, trapezoid rule (Tustin's method, bilinear transformation), bilinear with pre-wrapping. Among this trapezoid

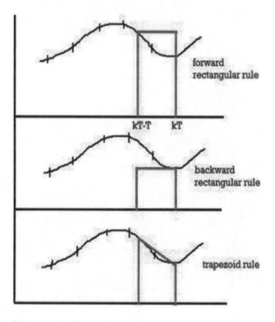

[2]Figure 4. Comparison of methods for numerical integration.

[2](Nielsen, n.d.).

[3]Table 1. Approximation for emulation.

Approximation	Method
S → Z–1/Ts	Forward difference
S → Z–1/TsZ	Backward difference
S →(2/Ts)(Z–1/Z+1)	Tustin's rule

where, Ts is a sampling time.

rule is superior as it approximates the area using average as shown in Figure 4.

First order system can be described using following differential equation,

$$\dot{y} + ay = au$$

Solution of differential equation can be obtained as following,

$$u(t) = \int_0^t (-au(\tau) + ae(\tau))d\tau$$

Choosing time in discrete step t = kT and using Tustin's method for integration, continuous Laplace operator 's' can be replaced according to the following table.

Applying the approximation and using Z transform's time shifting property (Downing, 2002). we can get following difference equation (Vodencarevic & Asim, 2010).

$$y(i) = A\{B(y(i-1)) + C[u(i) + u(i-1)]\}$$

where, A = 1/2Tp+Ts, B = 2Tp–Ts, C = TsKp.

This difference equation can be used to implement valve process function block. It is written in SCL language which comes along with S7 as an optional software package capable to implement complex algorithm (Berger, 2007).

The function block can be called cyclically from organizational block OB35 at exact period of time.

5 IMPLEMENTATION OF CONTROL LOOP ON S7

5.1 Reading analog channel value

To read current PV through analog input MOVE instruction is used. The pressure transmitter is connected with analog input module of PLC. PIW256 (peripheral input word) is the address on

[3](Nielsen, n.d.).

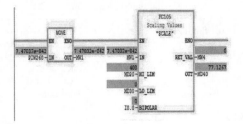

Figure 5.   Reading sensor data.

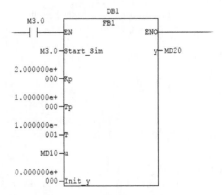

Figure 6.   Valve dynamics block.

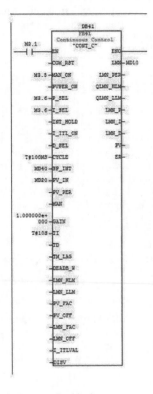

Figure 7.   PID controller block.

which PV is written. MOVE instruction will move PV in PLC's memory word. Transmitters generally have output in the range of 4–20 mA. So, FC105 (function block) will scale this current in real engineering unit and give the result in OUT.

5.2   PID controller and valve dynamics

FB1 represents valve dynamics written in source file and compiled. Then it is called from main program block OB35 as shown below.

FB41 is software PID controller with all parameters which can be activated or deactivated according to application. MD40 contains SP and MD20 is PV. PID controller is set in automatic mode and LMN is a manipulated variable and stored in MD10, which is connected to the valve input u. This completes the loop.

6   SIMULATION RESULTS

We can give PID controller's parameters in assignment window as seen in Figure 8.

As we can see in curve recorder of PID assignment, step input is given to the process and it tracks the input without oscillations because controller is properly tuned.

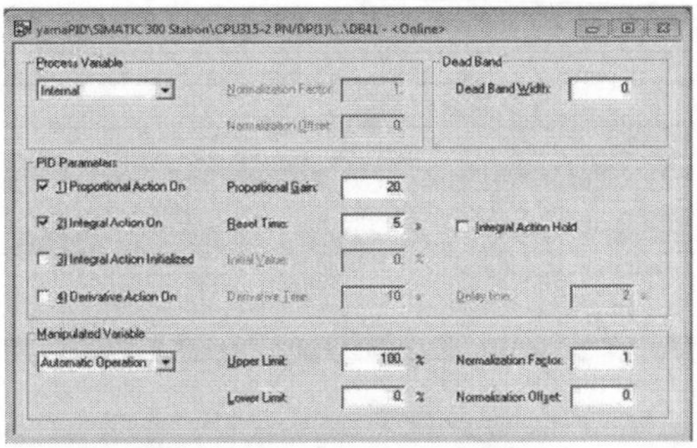

Figure 8.   PID parameter assignment.

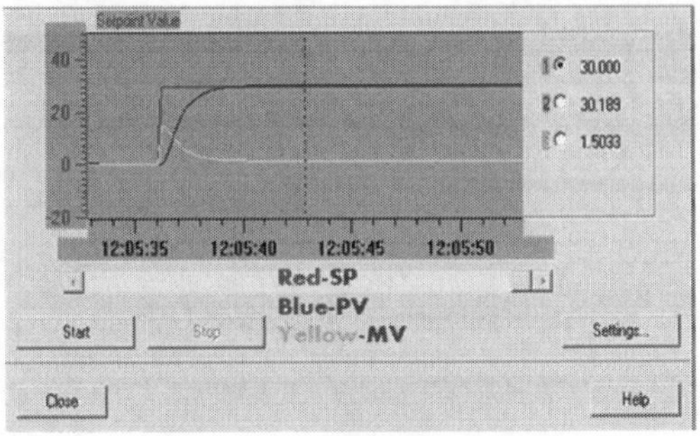

Figure 9.   Response of control loop to step input.

7 CONCLUSION

In this study, PLC based measurement and control system is developed for a compressor system which is a part of a helium liquefier at 4.5 K. Compressors have very fast dynamics means a small disturbance to its steady state will affect its overall performance. So, developing a control system is challenging task for compressor surge phenomena. Various different methods are available. Among them, simple method to recycle a discharge stream at high pressure to the suction side is viable approach. Before implementing any control on actual hardware one can model the system performance with simulation tool like Aspen HYSYS. Here, Compressor surge avoidance process is modeled and tested for cryogenic plant. To implement the process block in PLC, valve dynamics is digitized. This is true for any system implementation using digital computer. Processes involved in actual plants may include large time delays and can have any dynamics. So a method to obtain a closed loop performance using PLC is discussed using Step7 from Siemens.

ACKNOWLEDGEMENT

The authors gratefully acknowledge the support received from department of Large Cryoplant and Cryosystem, Institute for Plasma Research, Bhat, Gandhinagar.

REFERENCES

Atrey, M.D., NPTEL Lectures of Cryogenic Engineering, Department of mechanical engineering, IIT Bombay.

Berger H.: "Automating with STEP 7 in STL and SCL: Programmable Controllers Simatic S7-300/400", Wiley-VCH, 2007

Downing, C.J., T. O'Mahony, 'The Z transform, Digital control' Dept. Electronic Engg., CIT, 2002

Kirsten Mølgaard Nielsen: Analog and digital control (Based on lecture notes by Jesper Sandberg Thomsen).

National Aeronautics and Space Administration, https://www.grc.nasa.gov/www/k12/airplane/compexp.

Nicholas Brownrigg, Jump Start: Compressor Modelling in Aspen HYSYS® Dynamics, A Brief Tutorial (and supplement to training and online documentation), Aspen Technology, Inc.

Randall F. Barron, "Cryogenics Systems", Oxford University Press, New York, 1985.

Thomas M. Flynn, "Cryogenic Engineering", CRC press, New York, 2005.

Tijl, Peter. "Modeling, simulation and evaluation of a centrifugal compressor with surge avoidance control." MD Thesis (2004).

User manual: Standard Software for S7-300 and S7-400 PID Control, step7

Vodencarevic, Asim. "Design of PLC-based Smith predictor for controlling processes with long dead time." *proceeding IMECS, Hongkong* (2010).

Communication and Computing Systems – Prasad et al. (Eds)
© *2017 Taylor & Francis Group, London, ISBN 978-1-138-02952-1*

Effect of delay approximation on set point tracking performance of PID controllers for IPDT process model

Pooja Kholia, M. Chaturvedi & Pradeep K. Juneja
Graphic Era University, Dehradun, India

ABSTRACT: Many industrial systems can be represented as IPDT process model. In the present work, hydraulic control system used in position control application is selected for analysis. The selected process model has inherent delay which is approximated using Pade's first order approximation. PID controllers have been designed via various tuning methods for selected and approximated process model. Various steady-state and transient characteristics have been compared to evaluate the outcome of delay approximation carrying out for PID controllers.

1 INTRODUCTION

Process control is an engineering science to design any system for maintaining the output of a particular process with in a desired operating range. Keeping process variables within a desired operating range is of the greatest importantin industries (Chaturvedi & Juneja 2013).

A mathematical model of the process dynamics often helps us to understand the process behavior under different operational conditions. Hydraulic control system, which is taken under consideration, can be design as IPTD process model. Mathematical model of the IPTD plant is simple, for the reason that it contains only two parameters plant gain and time delay, determined on the basis of the plant step response. There are many physical processes with dead time which is required to control (Hassaan 2014). Integrating plus time delayed process is demanding than any un-delayed stable process. This is simply because of the delayed characteristic of the system is compensated in the analysis by using polynomials in the Laplace operator s which means increasing the order of the closed-loop control system. To overcome this difficulty dead time is compensated using Pade's I order delay approximation technique.

The Pade's approximation for time delay is the ratio of two polynomial, pole zero 1/1 Pade's approximation represented as:

$$e^{-\theta s} = \frac{1-\theta s}{1+\theta s} \qquad (1)$$

In the presence of delay performance of conventional feedback control system limits. PID controllers are extensively used in industrial processes because they have general acceptance in the control industry over the past 60 years, In the process industries 95% of the controller's action PID type by virtue of its modesty, robustness (Seborg et al. 2004; Skogestad 2003).

2 METHODOLOGY

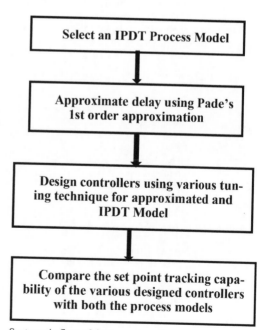

Systematic flow of the present work.

This flow chart shows the systematic flow of the present work. First, IPDT process is taken under consideration and the selected IPDT process is approximated using Pade's I order approximation technique. Different PID controllers designed with various tuning techniques and implemented in a closed loop system with the process models. The effect of delay approximation of the controller is estimated by comparing steady-state and transient characteristics to check the performance of a system.

The selected IPDT system transfer function can be defined as (Zhang et al. 1999):

$$G_1 = \frac{0.2}{s} e^{-7.4s} \tag{2}$$

The above model is approximated using Pade's I order approximation technique as:

$$G_2 = \frac{0.2 - 0.74\ s}{3.7\ s^2 + s} \tag{3}$$

This approximated model shows the second order transfer function. For both the process model approximated and without approximated check the system stability and performance.

3 RESULT AND DISCUSSION

Different PID Controller have been designed to obtain closed loop system with unity feedback, using Ziegler Nicholas, Astrom & hagglund, Chidambaram & Sree, Astrom and Rotach tuning methods. The simulations have been performed to

Table 1. Transient and steady state characteristics of the designed controller.

Tuning methods	Time responses	IPDT Model	Approximated model
Ziegler Nichols	Rise time	13.4	16.4
	Peak time	1.31	1.31
	Settling time	154	147
Astrom & hagglund	Rise time	3.04	6.12
	Peak time	1.63	1.42
	Settling time	72.4	70.7
Chidambaram & sree	Rise time	2.01	4.35
	Peak time	1.91	1.18
	Settling time	81.1	54
Astrom	Rise time	15.8	19.1
	Peak time	1.18	1.18
	Settling time	157	151
Rotach	Rise time	1.8	4.05
	Peak time	2.16	1.49
	Settling time	70.7	54.6

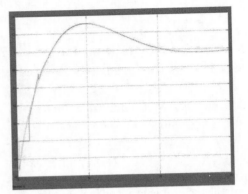

Figure 1. Comparative analysis of G1 and G2 using Ziegler Nichols tuning technique.

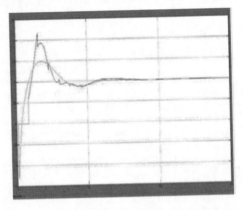

Figure 2. Comparative analysis of G1-G2 process model using Astrom and Hagglund tuning technique.

obtain the desired responses. Table 1 represents the time response characteristics of the PID controller having rise time, peak time and settling time.

Figure 1 represents the comparative closed loop step responses of G1-G2 using Ziegler Nichols tuning. Figure 2 characterizes response for G1-G2 process model is designed using Astrom and Hagglund of PID Controller. Fig. 3 gives the response for G1-G2 using Rotach tuning technique of PID Controller.

Fig. 4 represents comparative Closed loop response for G1-G2 process model is designed using Chidambaram and Sree of PID Controller. Figure 5 gives comparitve response for G1-G2 process model designed for Astrom tuning technique.

The Fig. 6. shows the comparisons for the responses of all the tuning techniques for IPDT process model using PID controller. Here the x-axis represents the time in seconds and y-axis represents the amplitude. The Fig. 7 represents comparative analysis of the controller using all the methods and process approximate using Pade's I method.

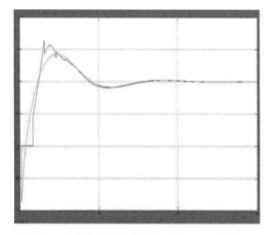

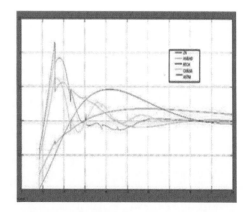

Figure 3. Comparative analysis of G1-G2 using Rotach tuning technique of PID Controller.

Figure 6. Comparative analysis of all controllers designed using various tuning techniques for IPDT process model of PID Controller.

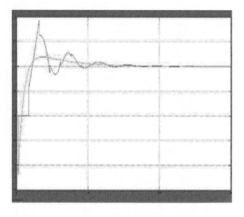

Figure 4. Comparative analysis of G1-G2 using Chidambaram and Sree tuning technique.

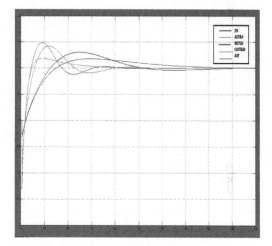

Figure 7. Comparative analysis of controllers designed using various methods and approximated using Pade's I method.

4 CONCLUSION

In the above analysis an IPDT process model is selected and PID controller is designed using various tuning techniques viz. Ziegler Nichols, Astrom & Hagglund, Chidambaram & Sree, Rotach and Astrom. Closed loop step responses are obtained and compared. Following conclusions are made:

Settling time and steady state error indicates the steady state characteristics of the response. Settling time is less in approximated model than the selected process model in case of Chidambaram & Sree tuning method as well as in Rotach tuning method.

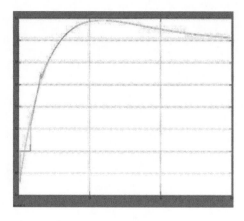

Figure 5. Comparative analysis of G1-G2 process model is designed using Astrom of PID Controller.

Chidambaram & Sree tuning method exhibits optimum set-point tracking for both the process models i.e. selected IPDT model and approximated model.

In case of approximated process model, all characteristics are decreasing except rise time indicating improved overall system performance.

REFERENCES

Chaturvedi, M., Juneja, P.: Effect of dead time approximation on controller performance designed for a second order delayed model. Proceedings of the2013 International Conference on Advanced Electronic Systems, ICAES 2013, Pilani, India, 2013, pp. 313–315.

Hassaan, G.A.: Tuning of a PD-PI Controller Used with an Integrating plus Time Delay Process. International Journal of Scientific & Technology Research Volume 3, Issue 9, September, 2014.

Normey –Rico, J.E., Camcho, E.F.: Control of dead time processes. 2nd edition, springer, 2007.

Panda, R.C., Yu, C., Huang, H.P.: PID tuning Rules for SOPDT systems: Review and some new results. ISA Transactions, Vol. 43, pp. 283–295, 2004.

Seborg, D.E., Edgar, T.F., Mellichamp, D.A.: Process Dynamics and Control. 2nd ed., Wiley, New York, 2004.

Skogestad, S., Simple analytic rules for model reduction and PID controller tuning. Journal of process control 13, 2003, pp. 291–309.

W. Zhang, X. Xu and Y. Sun, "Quantitative performance design for integrating processes with time delay", ibid, Vol.35, 1999, pp.719–723.

Communication and Computing Systems – Prasad et al. (Eds)
© 2017 Taylor & Francis Group, London, ISBN 978-1-138-02952-1

Load current signature analysis of ZSI fed induction motor drive system

Bhawana Negi & Vivek Sharma
Department of Electrical Engineering, Graphic Era University, Dehradun, India

Pawan Negi
Faculty of Technology, UTU Campus Dehradun, Dehradun, India

ABSTRACT: In many of the industrial application step-down and step-up voltage is the key aspect which is to be considering while designing inverter for general purpose motor. The PWM strategies used for the modification and gives the better result. The dissimilar fault detection as the means of change in spectrum and the THD% value is determined using the FFT for the faults condition line to ground, line to line and open circuit. The inductor capacitor arrangement provides boosting of input voltage which is not promising in VSI.

1 INTRODUCTION

Z Source Inverter is a technique having the buck and boost operation in same circuit to find desired output. The Z source inverter system model is capable of used as voltage or current type inverter (Sutar et al. 2012). The MOSFET is used as the switching device to form the inverter module. The PWM is used to provide the frequency as well as the time delay to the MOSFET for the conduction at the specific time period. Z-Source Inverter is being used as the weakness in the VSI and CSI that the limited range of output voltage and undesirable harmonic produces, which cause fall in the efficiency. Harmonics are reduced by introducing this inverter, which helps in giving improved variable speed for speed control of induction motor. The arrangement of capacitor and inductor is to change the output voltage. So the modification in the circuitry due to its X-shaped network can be form. The two inductors and capacitors use in the ZSI circuit work as filter for 2nd order that reduces harmonics as well as the content of ripple which protect the circuit getting damage. Electrical Signature Analysis is the technique to analyzing and detects a variety of faults in the inverter output current & voltage signals (Choi et al. 2015). The output of inverter currents signal selected to analyze or to detect faults before convey to the motor. The selected signal is called a diagnosis media, and the output of the analysis applied to the selected diagnosis media is called a signature. ZSI accomplish the conversion in single stage. This topology suits for the fuel cell and solar cell. The faults can be introduced due to the malfunction in the cir-

cuitry or due to the connected load of the system. The study of these faults is required to protect the circuit and for the smooth operation. There faults are relate the circuitry with power electronics as well as the power system.

Abbreviations and Acronyms
ZSI Impedance Source Inverter
CSI Current Source Inverter
VSI Voltage Source Inverter
PWM Pulse Width Modulation
THD Total Harmonic Distortion
FFT Fast Fourier Transform
MOSFET Metal-Oxide Semiconductor Field-Effect Transistor.

2 METHODOLOGY

Fig. 1 shows methodical flow of the work. The three phase ZSI fed Induction Motor is taken into consideration. Different faults are to be introduced in the inverter (Sharma and Mendiratta 2013).

The effect of faults on the inverter is analyzed by comparing healthy condition with faulty conditions and evaluated by THD% (Benbouzid 2000). As ZSI use the switching devices, so it is important to analyze faults before given to the motor. For the long time if the fault remains undetected can cause the secondary faults in the system and may damage the motor. The different faults introduced in the system which are line to ground fault, line-line fault and open circuit fault (Biswas et al. 2009). In these fault condition the analysis is perform in frequency domain using FFT. The three phase asynchronous motor is connected with the output of

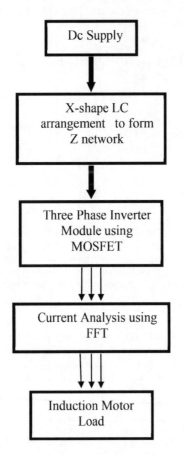

Figure 1. Flow chart for fault analysis of ZSI model.

ZSI and faults analysis carried out between input of motor and output of ZSI.

3 RESULTS AND ANALYSIS

3.1 Three phase ZSI fed Induction motor load

Fig. 2 represents the simulation circuit of the three phase ZSI fed IM drive. Where 6 MOSFET use for the switching with the gate signal provides by the pulse generator.

The input voltage is 120V and the output voltage is 220V. Fig. 3 represents the output of three phase ZSI where faults is not introduced. The current in the healthy condition can be found is 200 A and THD value is 15.14.

3.2 Line to ground fault of Z-source inverter

Fig. 4 shows the simulation circuit based on the line to ground fault, this can be obtain by grounding phase A of the simulation circuit.

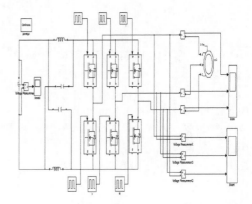

Figure 2. Simulation circuit of the three phase ZSI fed IM.

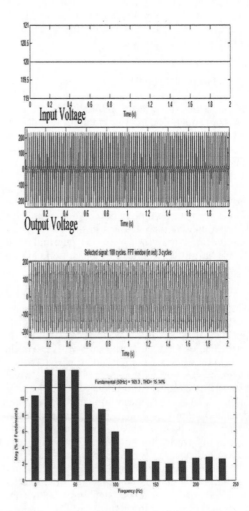

Figure 3. THD% response at healthy condition of Inverter.

Due to the line to ground fault the current goes to the negative and the value of current is –840 A. The THD value with respect to the current is 54.85.

3.3 Line to line fault

This condition is found by connecting phase B and phase C of the Three phase ZSI fed IM Drive.

In the analysis of line to line fault the current is positive and the value of current is 330. The THD% is 28.70.

The responses shown in figure looks similar to the response of the without fault condition and the output current is in the positive as well as negative side.

3.4 Open circuit fault

The open circuit fault on phase A is obtained by removing the supply given to the Drain terminal of

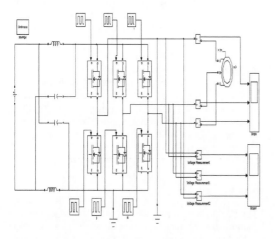

Figure 4.   Simulation circuit of line to ground fault on three phase ZSI fed IM.

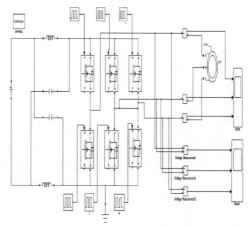

Figure 6.   Simulation circuit of line-line fault on three phase ZSI fed IM.

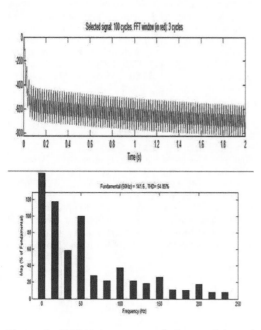

Figure 5.   THD% response at faulty condition of Inverter.

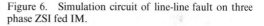

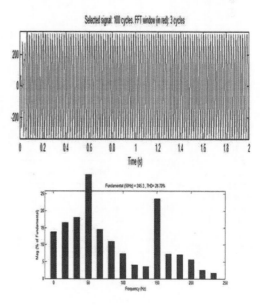

Figure 7.   THD% response at faulty condition of Inverter.

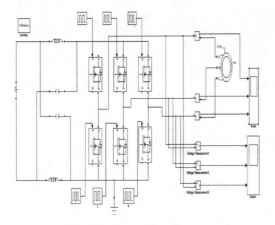

Figure 8. Simulation circuit of open circuit fault on three-phase ZSI fed IM.

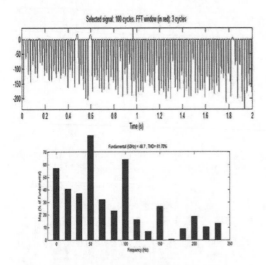

Figure 9. THD% response at faulty condition of Inverter.

Table 1. Analysis of healthy and faulty condition using FFT.

Conditions	THD%	Current (Amp)
Without fault	15.14	200
Line to ground Fault on Phase A	54.85	−840
Line to Line Fault	28.70	330
Open Circuit Fault on Phase A	81.70	−220

the MOSFET. In this condition output is analyze and simulation is performed.

The response of the open circuit fault is presented in Fig. 9. The current is negative and value of current is −220. THD% of this fault is 81.70.

4 CONCLUSION

In the above analysis ZSI model is consider with variety of fault conditions. The THD% response are obtained and compared for the healthy as well as faulty conditions. The current value is consider in all the condition in which we found that the line to line condition is dangerous as the current in phase A is 330 in compared with other faults, in other faults the value is negative. The need of protection from the line to line is required to save the motor from high value of current. The THD% value in case of line-line fault is 28.70. In future the analysis can be form in multilevel inverter for reliable operation.

REFERENCES

Benbouzid, Mohamed El Hachemi 2000. A Review of Induction Motors Signature Analysis as a Medium for Faults Detection. IEEE Transactions on Industrial Electronics, Vol 47 No. 5.

Biswas, B. et al. 2009. Current Harmonics Analysis of Inverter-Fed Induction Motor Drive System under Fault Conditions. Proceedings of the International Multi Conference of Engineers and Computer Scientists, Vol II. ISBN: 978-988-17012-7-5. Hong Kong.

Choi, Jung-Hyun et al. 2015. A Diagnostic Method of Simultaneous Open-Switch Faults in Inverter-Fed Linear Induction Motor Drive for Reliability Enhancement. IEEE Transactions on Industrial Electronics. Vol. 62 No. 7.

Kavitha, N. & Sujitha, N. 2014. Motor current signature analysis of multilevel inverter fed induction motors using wavelets. International Journal of Advanced Information Science and Technology (IJAIST). Vol.22 No.22.

Sharma, Vivek & Mendiratta, Gaurav 2013. Detection of fault in AC to AC Converter fed Induction Motor Driver. International Journal of Scientific & Technology Research. Volume 4 Issue 11.

Sutar, Amol R. & Jagtap, Satyawan R. Tamboli Jakirhusen 2012. Performance Analysis of Z-source Inverter Fed Induction Motor Drive. International Journal of Scientific & Engineering Research Volume 3 Issue 5.

Communication and Computing Systems – Prasad et al. (Eds)
© 2017 Taylor & Francis Group, London, ISBN 978-1-138-02952-1

Uv-Vis studies and quantum analysis of 2,3,5,6-Tetramethyl-1, 4-Benzoquinone using HF and DFT method

Sarvendra Kumar & Surbhi
Amity Institute of Applied Sciences, Amity University, Noida, UP, India

M.K. Yadav
D.N. (P.G.) College Meerut, UP, India

ABSTRACT: The UV-Vis study and quantum analysis of 2,3,5,6-tetramethyl-1, 4-benzoquinone (Duroquinone) has been performed experimentally and theoretically (in Acetonitrile (ACN), chloroform and water) in the range 3500–2300 cm^{-1} in the solution phase. The structural and spectroscopic data of the molecule was obtained by using HF and DFT/B3LYP/6-311++G(d, p) level. The electronic properties such as excitation energy, wavelength corresponding to absorption maxima (λ_{max}), and oscillator strength (f), are calculated by Time-Dependent Density Functional Theory (TD-DFT) using HF/6-311++G(d, p) and B3LYP/6-311++G(d, p) as basis sets. The stability and intra molecular charge transfer have analyzed by the detailed Natural Bond Orbital (NBO) analysis. We have compared our calculated result and the experimentally observed values and found that both are in good agreement with each other.

1 INTRODUCTION

The title compound has been chosen for DFT studies to find out different molecular property has been purchased from Sigma Aldrich of 97% pure quality. The important biological and toxicological roles of quinones could be attributed to their versatile electrophilic and oxidative properties. Quinones are able to undergo Michael addition with cellular thiols such as glutathione and proteins, and promote electron transfer in living systems via redox-cycling. Although the protein adduction of quinones is assumed to be a part of their metabolic fate, the adducts retain the redox-cycling capability of the parent quinones, and thus we can consider the adducts as a type of active metabolite. Herein, the toxicity, reversibility, and selectivity of protein adducts were studied using molecular spectroscopy and molecular modeling (Saleh et al. 2015). The cytotoxic mechanism of many quinones has been correlated to covalent modification of cellular proteins (Saleh et al. 2015). Survey reveals that a set of 25 quinone compounds with anti-trypanocidal activity was studied by using the Density Functional Theory (DFT) method in order to calculate atomic and molecular properties to be correlated with the biological activity. The chemometric methods Principal Component Analysis (PCA), Hierarchical Cluster Analysis (HCA), Stepwise Discriminant Analysis (SDA), Kth nearest neighbor (KNN) and Soft Independent Modeling of Class Analogy (SIMCA) were used to obtain possi-

ble relationships between the calculated descriptors and the biological activity studied and to predict the anti-trypanocidal activity of new quinone compounds from a prediction set (Molfetta et al. 2005). However, literature survey also shows very few ab initio HF/DFT calculations of such compound. Quenching dynamics of excited quinone molecules are given much attention in photochemistry and biochemistry. The study of the viscosity effect on the quenching of triplet excited state of duroquinone (3DQ*) by stable radical, 2,2,6,6-tetramethyl piperidinyloxyl (TEMPO) was done this study measured Chemically Induced Dynamic Electron Polarization (CIDEP) spectra and transient absorptive spectra in various solvents (Xinsheng et al. 2010). The CIDEP spectra of transient radicals during photolysis of the Duroquinone (DQ)/Ethylene Glycol (EG) system in acid, basic, and micellar environments were measured with a highly time-resolved ESR spectrometer indicate that the neutral radical durosemiquonone was formed by proton transfer from EG to 3DQ*, and that DQ•− duroquinone anion radical) was formed by dissociation of neutral durosemiquonone accompanying polarization transfer (Xinsheng et al. 2009).

2 COMPUTATIONAL DETAILS

Calculations of the title compound were carried out with Gaussian03 software program (Frisch et al. 2004) using the HF/6-31+G* and B3LYP/

Figure 1.   Structure of duroquinone.

6-31+G* basis sets. The optimized parameters obtained using DFT approach have been compared with the experimental values and are in close agreement with them. Further we have used the optimized ground state geometry of duroquinone to study the different properties like UV-Vis spectra and NBO analysis. In order to find out the various interactions between the filled and the vacant orbitals, NBO analysis (Glendening & Reed 1998) of the title compound has been done using NBO 3.1 program available in Gaussian 09 package at DFT/6-311++G(2d, 2p) level of theory. The structure of our title compound is presented in Figure 1.

3   RESULT AND DISCUSSION

3.1   UV analysis

The lowest singlet →Singlet spin allowed excited states need to be accounted to investigate the electronic transition (Selvaraj et al. 2001). Acetonitrile (ACN), chloroform, DMSO, methanol and water is used as a solvent to simulate the electronic absorption. Figure 2 represents the computed electronic spectra of duroquinone. The electronic spectra are recorded within a range of 200 nm–800 nm (Diwaker & Gupta 2014). The oscillator strength along with excitation energy for the triplet and the singlet states has also been calculated using TDDFT theory. The different values for excitation energy along with oscillator strength as well as CI expansion coefficients are showed in Table 1. For the title compound the maximum absorption value obtained using TD-DFT/B3LYP/6-311++G(d,p) basis set are 274.40 nm in ACN, 275.27 nm in choloform, and 274.24 nm in water are caused by n→π* transitions, while the smaller intensity bands

are calculated near 299 nm in chloroform and 297 nm in ACN, water phases of the duroquinone are forbidden and so, the oscillator strengths of these phases is nearly zero. The calculated spectra agree with the experimental UV spectra of the title compound is showed in Fig. 2. The theoretical absorption wavelength λ (nm), excitation energies E(eV) and oscillator strengths (f) of the title compound in solvent ACN, chloroform, and water is listed in Table 1 using TD-DFT/B3LYP/6-311++G(d, p) basis set.

4   NBO ANALYSIS

The natural population analysis performed on the electronic structure of title compound clearly describes the distribution of electrons in various sub-shells of their atomic orbitals. The accumulation of charges on the individual atom and accumulation of electrons in the core, valance and Rydberg sub-shells of Duroquinone are presented in Table 2. The most electronegative atoms like C_7, C_{11}, C_{15}, C_{19}, O_{23} and O_{24} have charges –0.72272, –0.73317, –0.72272, –0.73317, –0.53531

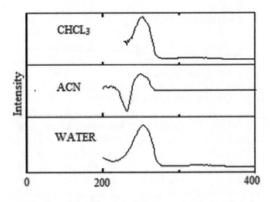

Figure 2.   UV–Vis spectra of Duroquinone (cm⁻¹).

Table 1.   The theoretical absorption wavelength λ (nm), excitation energies E(eV) and oscillation strength (f) of Duroquinone.

Solvent	Excitation sate	λ (nm)	E(ev)	f (a.u)
ACN	Excitation Sate 1	297.2	4.1716	0.0013
	Excitation Sate 2	283.4	4.3747	0.0009
	Excitation Sate 3	274.4	4.5184	0.0393
Chloroform	Excitation Sate 1	299.27	4.1429	0.0003
	Excitation Sate 2	278.45	4.4527	0.0003
	Excitation Sate 3	275.27	4.5042	0.0015
Water	Excitation Sate 1	297	4.1745	0.0017
	Excitation Sate 2	284.06	4.3648	0.0013
	Excitation Sate 3	274.24	4.521	0.089

Table 2. Accumulation of natural charge population of electron in core, valence and Rydberg orbitals of Duroquinone.

Atom	Charge	Natural population			
		Core	Valence	Rydberg	Total
C_1	0.51454	1.99916	3.45269	0.03361	5.48546
C_2	−0.05571	1.99889	4.03993	0.01689	6.05571
C_3	−0.05119	1.99889	4.03598	0.01631	6.05119
C_4	0.51454	1.99916	3.45269	0.03361	5.48546
C_5	−0.05571	1.99889	4.03993	0.01689	6.05571
C_6	−0.05119	1.99889	4.03598	0.01631	6.05119
C_7	−0.72272	1.99945	4.71323	0.01003	6.72272
H_8	0.27041	0	0.72822	0.00137	0.72959
H_9	0.27041	0	0.72822	0.00137	0.72959
H_{10}	0.24806	0	0.75090	0.00104	0.75194
C_{11}	−0.73317	1.99945	4.72467	0.00905	6.73317
H_{12}	0.25731	0	0.74155	0.00115	0.74269
H_{13}	0.25731	0	0.74155	0.00115	0.74269
H_{14}	0.28007	0	0.71820	0.00172	0.71993
C_{15}	−0.72272	1.99945	4.71323	0.01003	6.72272
H_{16}	0.27041	0	0.72822	0.00137	0.72959
H_{17}	0.27040	0	0.72822	0.00137	0.72960
H_{18}	0.24806	0	0.75090	0.00104	0.75194
C_{19}	−0.73317	1.99945	4.72467	0.00905	6.73317
H_{20}	0.25731	0	0.74155	0.00115	0.74269
H_{21}	0.25731	0	0.74155	0.00115	0.74269
H_{22}	0.28007	0	0.71820	0.00172	0.71993
O_{23}	−0.53531	1.99981	6.51452	0.02098	8.53531
O_{24}	−0.53531	1.99981	6.51452	0.02098	8.53531

Core: 23.99130 (99.9637% of 24)
Valence: 63.77933 (99.6552% of 64)
Rydberg: 0.22937 (0.2606% of 88)

and −0.53531 respectively. The most electropositive atom is C_1 and C_4 with same charge 0.51454. From the electrostatic point of view, electronegative atoms have a tendency to donate an electron, whereas the electropositive atoms have a tendency to accept an electron. Further, natural population analysis showed that 88 electrons in the title compound are distributed on the sub-shells as follows:

Core: 23.99130 (99.9637% of 24)
Valence: 63.77933 (99.6552% of 64)
Rydberg: 0.22937 (0.2606% of 88)

5 NBO ANALYSIS

The occupancies and energies of lone pair molecular orbitals (LP) and anti-bonding (BD*) molecular orbitals of the Duroquinone are predicted at HF/6-311++G(d, p) level of theory and is presented in Table 3. The variations in occupancies

Table 3. Second order perturbation theory analysis of Fock matrix in NBO basis (Duroquinone).

Donor NBO	Acceptor NBO	E(2)	E_j-E_i	F(i, j)
C_1-C_2	C_3	2.01	1.83	0.054
	C_6	1.53	1.64	0.045
	O_{24}	0.93	1.52	0.034
	C_1-O_{24}	0.52	1.23	0.023
	C_2-C_3	2.25	1.31	0.048
	C_3-C_7	4.78	1.07	0.064
	C_6-C_{15}	2.06	1.07	0.042
	C_{11}-H_{12}	0.57	1.1	0.023
	C_{11}-H_{13}	0.57	1.1	0.023
C_1-C_6	C_2	2.01	1.58	0.051
	C_5	1.55	1.92	0.049
	C_{15}	0.69	1.42	0.028
	O_{24}	0.75	1.52	0.03
	C_1-O_{24}	0.53	1.23	0.023
	C_2-C_{11}	2.19	1.07	0.043
	C_5-C_6	2.37	1.31	0.05
	C_5-C_{19}	4.65	1.07	0.063
	C_{15}-H_{18}	1.15	1.12	0.032
C_1-O_{24}	C_1	1.62	1.98	0.051
	C_6	0.77	2.04	0.035
	C_1-C_2	0.82	1.5	0.032
	C_1-C_6	0.81	1.5	0.032
	C_2-C_3	0.9	1.71	0.035
	C_5-C_6	1.03	1.71	0.038
	C_2-C_3	5.9	0.42	0.045
	C_5-C_6	6.04	0.42	0.046
C_2-C_3	C_1	0.67	2.05	0.033
	C_4	0.56	2.05	0.031
	C_7	0.53	1.51	0.025
	C_{11}	0.75	1.52	0.03
	C_1-C_2	1.49	1.18	0.038
	C_1-O_{24}	1.93	1.32	0.045
	C_2-C_{11}	2.32	1.16	0.046
	C_3-C_4	1.57	1.18	0.039
	C_3-C_7	2.43	1.16	0.048
	C_4-O_{23}	1.76	1.32	0.043
	C_{11}-H_{14}	0.52	1.23	0.023
	C_7	0.65	1.04	0.024
	C_{11}	0.81	1.01	0.027
	C_1-O_{24}	19.68	0.28	0.067
	C_4-O_{23}	19.4	0.28	0.066
	C_7-H_8	2.31	0.72	0.038
	C_7-H_9	2.31	0.72	0.038
	C_{11}-H_{12}	1.87	0.72	0.034
	C_{11}-H_{13}	1.87	0.72	0.034
C_2-C_{11}	C_1	0.79	1.54	0.031
	C_3	0.88	1.6	0.034
	C_3	1.11	1.8	0.04
	C_1-C_2	0.6	1.06	0.023
	C_1-C_6	2.18	1.06	0.043
	C_2-C_3	3.23	1.27	0.057
	C_3-C_4	3.53	1.06	0.055
	C_{11}-H_{14}	0.58	1.1	0.023

(*Continued*)

Table 3. (*Continued*).　　　　　　　　　　　　Table 3. (*Continued*).

Donor NBO	Acceptor NBO	E(2)	E_j-E_i	F(i, j)
C_3-C_4	C_2	1.55	1.92	0.049
	C_5	2.01	1.58	0.051
	C_7	0.69	1.42	0.028
	C_{23}	0.75	1.52	0.03
	C_2-C_3	2.37	1.31	0.05
	C_2-C_{11}	4.65	1.07	0.063
	C_4-O_{23}	0.53	1.23	0.023
	C_5-C_{19}	2.19	1.07	0.043
	C_7-H_{10}	1.15	1.12	0.032
C_3-C_7	C_2	1.54	1.89	0.048
	C_4	0.93	1.54	0.034
	C_1-C_2	3.42	1.06	0.054
	C_2-C_3	3.28	1.28	0.058
	C_3-C_4	0.65	1.06	0.024
	C_4-C_5	2.36	1.06	0.045
	C_7-H_8	0.51	1.08	0.021
	C_7-H_9	0.51	1.08	0.021
C_4-C_5	C_3	1.53	1.64	0.045
	C_6	2.01	1.83	0.054
	O_{23}	0.93	1.52	0.034
	C_3-C_7	2.06	1.07	0.042
	C_4-O_{23}	0.52	1.23	0.023
	C_5-C_6	2.25	1.31	0.048
	C_6-C_{15}	4.78	1.07	0.064
	C_{19}-H_{20}	0.57	1.1	0.023
	C_{19}-H_{21}	0.57	1.1	0.023
C_4-O_{23}	C_3	0.77	2.04	0.035
	C_4	1.62	1.98	0.051
	C_2-C_3	1.03	1.71	0.038
	C_3-C_4	0.81	1.5	0.032
	C_4-C_5	0.82	1.5	0.032
	C_5-C_6	0.9	1.71	0.035
	C_2-C_3	6.04	0.42	0.046
	C_5-C_6	5.9	0.42	0.045
C_5-C_6	C_1	0.56	2.05	0.031
	C_4	0.67	2.05	0.033
	C_{15}	0.53	1.51	0.025
	C_{19}	0.75	1.52	0.03
	C_1-C_6	1.57	1.18	0.039
	C_1-O_{24}	1.76	1.32	0.043
	C_4-C_5	1.49	1.18	0.038
	C_4-O_{23}	1.93	1.32	0.045
	C_5-C_{19}	2.32	1.16	0.046
	C_6-C_{15}	2.43	1.16	0.048
	C_{19}-H_{22}	0.52	1.23	0.023
	C_{15}	0.65	1.04	0.024
	C_{19}	0.81	1.01	0.027
	C_1-O_{24}	19.4	0.28	0.066
	C_4-O_{23}	19.68	0.28	0.067
	C_{15}-H_{16}	2.31	0.72	0.038
	C_{15}-H_{17}	2.31	0.72	0.038
	C_{19}-H_{20}	1.87	0.72	0.034
	C_{19}-H_{21}	1.87	0.72	0.034
C_5-C_{19}	C_4	0.79	1.54	0.031
	C_6	0.88	1.6	0.034
	C_6	1.11	1.8	0.04
	C_1-C_6	3.53	1.06	0.055
	C_3-C_4	2.18	1.06	0.043
	C_4-C_5	0.6	1.06	0.023
	C_5-C_6	3.23	1.27	0.057
	C_{19}-H_{22}	0.58	1.1	0.023
C_6-C_{15}	C_1	0.93	1.54	0.034
	C_5	1.54	1.89	0.048
	C_1-C_2	2.36	1.06	0.045
	C_1-C_6	0.65	1.06	0.024
	C_4-C_5	3.42	1.06	0.054
	C_5-C_6	3.28	1.28	0.058
	C_{15}-H_{16}	0.51	1.08	0.021
	C_{15}-H_{17}	0.51	1.08	0.021
C_7-H_8	C_3	0.83	1.52	0.032
	C_2-C_3	2.15	1.14	0.044
	C_2-C_3	3.91	0.54	0.042
C_7-H_9	C_3	0.83	1.52	0.032
	C_2-C_3	2.15	1.14	0.044
	C_2-C_3	3.92	0.54	0.042
C_7-H_{10}	C_3-C_4	3.94	0.94	0.055
	C_3-C_7	0.57	0.92	0.021
C_{11}-H_{12}	C_2	0.84	1.44	0.031
	C_1-C_2	1.82	0.93	0.037
	C_2-C_3	3.36	0.54	0.039
	C_2	0.84	1.44	0.031
C_{11}-H_{13}	C_1-C_2	1.82	0.93	0.037
	C_2-C_3	3.36	0.54	0.039
C_{11}-H_{14}	C_2	0.54	1.75	0.028
	C_2-C_3	4.48	1.14	0.064
	C_2-C_{11}	0.59	0.9	0.021
C_{15}-H_{16}	C_6	0.83	1.52	0.032
	C_5-C_6	2.15	1.14	0.044
	C_5-C_6	3.91	0.54	0.042
C_{15}-H_{17}	C_6	0.83	1.52	0.032
	C_5-C_6	2.15	1.14	0.044
	C_5-C_6	3.92	0.54	0.042
C_{15}-H_{18}	C_1-C_6	3.94	0.94	0.055
	C_6-C_{15}	0.57	0.92	0.021
C_{19}-H_{20}	C_5	0.84	1.44	0.031
	C_4-C_5	1.82	0.93	0.037
	C_5-C_6	3.36	0.54	0.039
C_{19}-H_{21}	C_5	0.84	1.44	0.031
	C_4-C_5	1.82	0.93	0.037
	C_5-C_6	3.36	0.54	0.039
C_{19}-H_{22}	C_5	0.54	1.75	0.028
	C_5-C_6	4.48	1.14	0.064
	C_5-C_{19}	0.59	0.9	0.021
LP(1)-O_{23}	C_4	14.68	1.61	0.137
	C_3-C_4	1.43	1.14	0.036
	C_4-C_5	1.73	1.13	0.04

(*Continued*)　　　　　　　　　　　　　　　　　　(*Continued*)

Table 3. (Continued).

Donor NBO	Acceptor NBO	E(2)	E_j-E_i	F(i, j)
LP(2) -O_{23}	C_4	2.15	2.52	0.067
	C_3-$C4$	19.08	0.7	0.105
	C_4-C_5	18.79	0.7	0.104
	C_7-H_{10}	0.57	0.73	0.019
	C_{19}-H_{22}	0.99	0.75	0.025
LP(1)-O_{24}	C_1	14.68	1.61	0.137
	C_1-C_2	1.73	1.13	0.04
	C_1-C_6	1.43	1.14	0.036
LP (2)-O_{24}	C_1	2.15	2.52	0.067
	C_1-C_2	18.79	0.7	0.104
	C_1-C_6	19.08	0.7	0.105
	C_{11}-H_{14}	0.99	0.75	0.025
	C_{15}-H_{18}	0.57	0.73	0.019
C_1-O_{24}	C_1	0.71	2.06	0.114
	C_1	0.67	0.56	0.058
	C_1	0.62	0.61	0.058
	O_{24}	1.44	0.44	0.075
	C_2-C_3	23.53	0.04	0.071
	C_5-C_6	23.93	0.04	0.072
C_4-O_{23}	C_4	0.71	2.06	0.114
	C_4	0.67	0.56	0.058
	C_4	0.62	0.61	0.058
	O_{23}	1.44	0.44	0.075
	C_2-C_3	23.93	0.04	0.072
	C_5-C_6	23.53	0.04	0.071

and energies of the title molecule directly give the evidence for the delocalization of charge upon substitution and this leads to the variation of bond lengths.

6 NBO ANALYSIS

The interactions result in a loss of occupancy from the localized NBO of the idealized Lewis structure into an empty Non-Lewis orbital. NBO analysis of some pharmaceutical compounds has been performed by many spectrosopists (Zhou, Z. R. Hong, L. X. & Zhou, Z. X. 2012, Balachandran, V. Karthick, T. Perumal, S. & Natraraj A. 2013, Balasubramanian, M. & Padma, N. 1963). The loan pair—anti-bonding interaction can be quantitatively described by second-order perturbation interaction (Reed, A. E. & Weinhold, F. 1983, 1985, Reed, A. E. Weinstock, R. B. & Weinhold, F. 1985, Foster, J. P. & Weinhold, F. J. 1980) energy E(2). For each donor (i) and acceptor (j), the stabilization energy E(2) associated with the delocalization i→j is estimated as:

$$E(2) = \Delta E_{ij} = q_i F(ij)^2/\varepsilon_j - \varepsilon_i$$

Where q_i is the donor orbital occupancy, ε_i and ε_j are the diagonal elements and F(i, j) is the off

diagonal NBO Fock matrix element. The NBO analysis provides an efficient method for studying intermolecular and intramolecular bonding. It also provides a convenient basis for Intermolecular Charge Transfer (ICT) or conjugative interactions in molecular system. Table 3 presents the second order perturbation energies (often called as stabilizations energies or interaction energies) of most interaction NBO of Duroquinone. The second order perturbation energies correspond to the hyper conjugative interactions of the title compound such as LP(2) O_{24}→BD*(1)C_1-C_6 and LP(2) O_{23}→BD*(1)C_3-C_4 that are considerably very large with 19.08 kJmol^{-1}, respectively. The interaction such as LP(1) N_1→BD*(1) C_4 and LP(1) N_1→BD*(1) C_1 are little higher than the rest of the interaction as presented in Table 3. These hyper conjugative interactions are most responsible ones for stability of title compound.

7 ELECTRON CONTRIBUTION IN S-TYPE AND P-TYPE SUBSHELLS BO ANALYSIS

NBO analysis of title compound is performed to estimate the delocalization patterns of Electron Density (ED) from the principal occupied

Table 4. Occupancies and energies of one pair orbitals (LP) and anti-bonding (BD*) molecular orbitals of Duroquinone.

Bond	Occupacy	Energy
C_1-C_2	1.97812	-0.75292
C_1-C_6	1.97801	-0.75291
C_1-O_{24}	1.99148	-0.80191
C_1-O_{24}	1.85505	-0.33511
C_1-C_3	1.96990	-0.73671
C_1-C_3	1.73041	-0.28680
C_2-C_{11}	1.98101	-0.61717
C_3-C_4	1.97794	-0.75246
C_3-C_7	1.98098	-0.61668
C_4-C_5	1.97805	-0.75354
C_4-O_{23}	1.99155	-0.80255
C_4-O_{23}	1.85385	-0.33545
C_5-C_6	1.96973	-0.73656
C_5-C_6	1.73199	-0.28681
C_5-C_{19}	1.98111	-0.61764
C_6-C_{15}	1.98096	-0.61679
C_7-H_8	1.97235	-0.52367
C_7-H_9	1.98519	-0.51747
C_7-H_{10}	1.98551	-0.52622
C_{11}-H_{12}	1.97877	-0.52469
C_{11}-H_{13}	1.97568	-0.52568
C_{11}-H_{14}	1.98945	-0.51616

(Continued)

Table 4. (*Continued*).

Bond	Occupancy	Energy
C_{15}-H_{16}	1.97641	−0.52077
C_{15}-H_{17}	1.97667	−0.52065
C_{15}-H_{18}	1.98939	−0.52738
C_{19}-H_{20}	1.97778	−0.52555
C_{19}-H_{21}	1.97769	−0.52557
C_{19}-H_{22}	1.98932	−0.51618
LP(1)-O_{23}	1.98582	−0.81020
LP(2)-O_{23}	1.95089	−0.27779
LP(1)-O_{24}	1.98601	−0.80963
LP(2)-O_{24}	1.95131	−0.27753
C_1-C_2	0.03712	0.54238
C_1-C_6	0.03600	0.54228
C_1-O_{24}	0.02176	0.28077
C_1-O_{24}	0.29797	−0.06613
C_2-C_3	0.02652	0.56079
C_2-C_3	0.16074	0.01219
C_2-C_{11}	0.01686	0.35543
C_3-C_4	0.03641	0.54149
C_3-C_7	0.01676	0.35571
C_4-C_5	0.03690	0.54225
C_4-O_{23}	0.02177	0.28002
C_4-O_{23}	0.029871	−0.06640
C_5-C_6	0.02657	0.56005
C_5-C_6	0.15992	0.01183
C_5-C_{19}	0.01693	0.35497
C_6-C_{15}	0.01672	0.35550
C_7-H_8	0.00665	0.46786
C_7-H_9	0.00583	0.48783
C_7-H_{10}	0.00574	0.47095
C_{11}-H_{12}	0.00570	0.46590
C_{11}-H_{13}	0.00609	0.46488
C_{11}-H_{14}	0.00730	0.49768
C_{15}-H_{16}	0.00606	0.47497
C_{15}-H_{17}	0.00603	0.47530
C_{15}-H_{18}	0.00536	0.47390
C_{19}-H_{20}	0.00560	0.46378
C_{19}-H_{21}	0.00559	0.46372
C_{19}-H_{22}	0.00770	0.49901

Table 5. Natural atomic orbital occupancies of most interacting (lone pair and anti-bonding) NBO's of Duroquinone.

Bond	Hybridization	S%	P%
C_1-C_2	$0.7013sp^{1.72}+0.7129sp^{1.97}$	49.18	50.82
C_1-C_6	$0.7013sp^{1.72}+0.7128sp^{1.97}$	49.19	50.81
C_1-O_{24}	$0.6187sp^{2.80}+0.7856sp^{3.13}$	38.28	61.72
C_1-O_{24}	$0.5803sp^{1.00}+0.8144sp^{1.00}$	33.68	66.32
C_1-C_3	$0.7070sp^{1.87}+0.7072sp^{1.86}$	49.99	50.01
C_1-C_3	$0.07068sp^{1.00}+0.07075sp^{1.00}$	49.95	50.05
C_2-C_{11}	$0.7244sp^{2.18}+0.6894sp^{3.13}$	52.48	47.52

(*Continued*)

Table 5. (*Continued*).

Bond	Hybridization	S%	P%
C_3-C_4	$0.7128sp^{1.97}+0.7014sp^{1.72}$	50.80	49.20
C_3-C_7	$0.7244sp^{2.18}+0.6894sp^{3.13}$	52.47	47.53
C_4-C_5	$0.7013sp^{1.72}+0.7129sp^{1.97}$	49.18	50.82
C_4-O_{23}	$0.6186sp^{2.80}+0.7857sp^{3.12}$	38.26	61.74
C_4-O_{23}	$0.5799sp^{1.00}+0.8147sp^{1.00}$	33.62	66.38
C_5-C_6	$0.7070sp^{1.87}+0.7072sp^{1.86}$	49.98	50.02
C_5-C_6	$0.7075sp^{1.00}+0.7067sp^{1.00}$	50.06	49.94
C_5-C_{19}	$0.7243sp^{2.18}+0.6894sp^{3.13}$	52.47	47.53
C_6-C_{15}	$0.7242sp^{2.19}+0.6896sp^{3.14}$	52.45	47.55
C_7-H_8	$0.7925sp^{3.01}+0.6099sp^{0.00}$	62.80	37.20
C_7-H_9	$0.7986sp^{2.89}+0.6018sp^{0.00}$	63.78	36.22
C_7-H_{10}	$0.7887sp^{2.97}+0.6148sp^{0.00}$	62.20	37.80
C_{11}-H_{12}	$0.7907sp^{3.03}+0.6122sp^{0.00}$	62.52	37.48
C_{11}-H_{13}	$0.7901sp^{3.03}+0.6130sp^{0.00}$	62.42	37.58
C_{11}-H_{14}	$0.8011sp^{2.83}+0.5985sp^{0.00}$	64.18	35.82
C_{15}-H_{16}	$0.7950sp^{2.97}+0.6066sp^{0.00}$	63.20	36.80
C_{15}-H_{17}	$0.7951sp^{2.97}+0.6065sp^{0.00}$	63.22	36.78
C_{15}-H_{18}	$0.7883sp^{2.93}+0.6153sp^{0.00}$	62.14	37.86
C_{19}-H_{20}	$0.7906sp^{3.03}+0.6124sp^{0.00}$	62.50	37.50
C_{19}-H_{21}	$0.7906sp^{3.03}+0.6124sp^{0.00}$	62.50	37.50
C_{19}-H_{22}	$0.8014sp^{2.83}+0.5982sp^{0.00}$	64.22	35.78
LP(1)-O_{23}	$Sp^{0.32}$		
LP(2)-O_{23}	$Sp^{99.99}$		
LP(1)-O_{24}	$Sp^{0.32}$		
LP(2)-O_{24}	$Sp^{99.99}$		
C_1-C_2	$0.7129sp^{1.72}-0.7013sp^{1.97}$	50.82	49.18
C_1-C_6	$0.7128sp^{1.72}-0.7013sp^{1.97}$	50.81	49.19
C_1-O_{24}	$0.7856sp^{2.8}-0.6187sp^{3.13}$	61.72	38.28
C_1-O_{24}	$0.8144sp^{1.00}-0.5803sp^{1.00}$	66.32	33.68
C_2-C_3	$0.7072sp^{1.87}-0.7070sp^{1.86}$	50.01	49.99
C_2-C_3	$0.7075sp^{1.00}-0.7068sp^{1.00}$	50.05	49.95
C_2-C_{11}	$0.6894sp^{2.18}-0.7244sp^{3.13}$	47.52	52.48
C_3-C_4	$0.7014sp^{1.97}-0.7128sp^{1.72}$	49.20	50.80
C_3-C_7	$0.6894sp^{2.18}-0.7244sp^{3.13}$	47.53	52.47
C_4-C_5	$0.7129sp^{1.72}-0.7013sp^{1.97}$	50.82	49.18
C_4-O_{23}	$0.7857sp^{2.80}-0.6186sp^{3.12}$	61.74	38.26
C_4-O_{23}	$0.8147sp^{1.00}-0.5799sp^{1.00}$	66.38	33.62
C_5-C_6	$0.7072sp^{1.87}-0.7070sp^{1.86}$	50.02	49.98
C_5-C_6	$0.7067sp^{1.00}-0.7075sp^{1.00}$	49.94	50.06
C_5-C_{19}	$0.6894sp^{2.18}-0.7243sp^{3.13}$	47.53	52.47
C_6-C_{15}	$0.6896sp^{2.19}-0.7242sp^{3.14}$	47.55	52.45
C_7-H_8	$0.6099sp^{3.01}-0.7925sp^{0.00}$	37.20	62.80
C_7-H_9	$0.6018sp^{2.89}-0.7986sp^{0.00}$	36.22	63.78
C_7-H_{10}	$0.6148sp^{2.97}-0.7887sp^{0.00}$	37.80	62.20
C_{11}-H_{12}	$0.6122sp^{3.03}-0.7907sp^{0.00}$	37.48	62.52
C_{11}-H_{13}	$0.6130sp^{3.03}-0.7901sp^{0.00}$	37.58	62.42
C_{11}-H_{14}	$0.5985sp^{2.83}-0.8011sp^{0.00}$	35.82	64.18
C_{15}-H_{16}	$0.6066sp^{2.97}-0.7950sp^{0.00}$	36.80	63.20
C_{15}-H_{17}	$0.6065sp^{2.97}-0.7951sp^{0.00}$	36.78	63.22
C_{15}-H_{18}	$0.6153sp^{2.93}-0.7883sp^{0.00}$	37.86	62.14
C_{19}-H_{20}	$0.6124sp^{3.03}-0.7906sp^{0.00}$	37.50	62.50
C_{19}-H_{21}	$0.6124sp^{3.03}-0.7906sp^{0.00}$	37.50	62.50
C_{19}-H_{22}	$0.5982sp^{2.83}-0.8014sp^{0.00}$	35.78	64.22

Lewis-type (bond or lone pair) orbitals to occupied non-Lewis (anti-bonding or Rydberg) orbitals. The list of occupancies and energies of most interacting NBOs along with their percentage of hybrid atomic orbitals is listed in Table 4.

The percentage of hybrid atomic orbitals of oxygen lone pair atom (O_{23}) and (O_{24}) shows that oxygen is partially contributed to both s-type and p-type subshells for LP(1), while oxygen is predominantly contributed to p-type subshell for LP(2). In contrast, all the anti-bonding orbitals of title compound shows that oxygen is partially contributed to both s-type and p-type subshell, as stated in Table 5.

8 CONCLUSIONS

A complete electronic analysis has been carried out for Duroquinone using UV-Vis spectroscopy. A good correlation was found between computed and experimental data. The stabilization of the structure has been identified by second order perturbation energy calculations. The UV analysis gives the electronic spectrum of Duroquinone that has revealed the allowed and forbidden transitions with solvent effects.

REFERENCES

Balachandran, V. Karthick, T. Perumal, S. & Natraraj A. 2013. Idian J Pure & Appl Phys 51(2013): 178.

Balasubramanian, M. & Padma, N. 1963. Tetrahedron, 19(1963): 2135.

Diwaker & Gupta, A. K. 2014. Quantum Chemical and Spectroscopic Investigations of (Ethyl 4 hydroxy-3-((E)-(pyren-1-ylimino)methyl)benzoate) by DFT Method International Journal of Spectroscopy 2014 (2014): 15.

Foster, J. P. & Weinhold, F. J. 1980. Am Chem Soc 102(24): 7211–7218.

Frisch, M.J. Trucks, G.W. et al. 2004. Gaussian 03 Revision C. 02, Wallingford, Gaussian.

Glendening, E. D. & Reed, A. E. 1998. NBO Version 3.1, TCL, University of Wisconsin, Madison, Wis, USA.

Molfetta, F. A. 1, Bruni A. T., Honório K. M., da Silva A. B. Apr 2005. A structure-activity relationship study of quinone compounds with trypanocidal activity Eur J Med Chem. 40(4):329–338.

Reed, A. E. & Weinhold, F. 1983. J Chem Phys 78: 4066–4073.

Reed, A. E. & Weinhold, F. 1985. J Chem Phys 83: 1736–1740.

Reed, A. E. Weinstock, R. B. & Weinhold, F. 1985 J. Chem Phys 83: 735–746.

Saleh, Mohamed Elgawish et al. 17 July 2015. Journal of Chromatography A 1403: 96–103.

Saleh, Mohamed et al. 2015. a Molecular modeling and spectroscopic study of quinone–protein adducts: insight into toxicity, selectivity, and reversibility, Toxicol. Res. 4: 843–847.

Selvaraj, K. Narasimhan M. & Mallika, J. 2001. Tansition Metal Chemistry 26: 224.

Xinsheng X. Lixia J. Lei S. & Zhifeng C. 2010. Viscosity Effect Study on Quenching of Photoinduced Excited Triplet Duroquinone by TEMPO Spectroscopy Letters: An International Journal for Rapid Communication 43(4): 310–316.

Xinsheng, X. Lixia, J. Guanglai, Z. & Zhifeng, C. 01/2009. A CIDEP study of photolysis of duroquinone hydrogen donor homogeneous and triton micelle solutions Research on Chemical Intermediates 35(1): 55–61.

Zhou, Z. R. Hong, L. X. & Zhou, Z. X. 2012. Idian J Pure & Appl Phys 50(2014): 719.

Communication and Computing Systems – Prasad et al. (Eds)
© 2017 Taylor & Francis Group, London, ISBN 978-1-138-02952-1

Solar energy harnessing from Delhi metro station rooftops

Rahul Gupta
Department of Electronics and Communication Engineering, Maharaja Agrasen University, Solan, HP, India

Amit Kumar Manocha
Department of Electrical and Electronics Engineering, Maharaja Agrasen University, Solan, HP, India

ABSTRACT: To avoid Energy crisis Government, Corporates and citizens of India are looking for a reliable energy solution, as it is assumed that by next two decades our power supplies will diminish. So, Solar radiation becomes a reliable development resource for India. Solar radiation can provide a lifetime solution for all our energy needs and can also diminish pollution to a large extent. Under the guidance of MNRE, India has taken many measures to go green fuel. One such initiative is taken by Delhi Metro Rail Corporation (DMRC) in their phase 3 project to use renewable energy sources for all their daily power requirements by 2017, which in turn provide a clean transport to Delhi. This paper deals the methods and techniques using simulated experimental results, which can be used to maximize the efficiency of this project and harness energy in large amounts, without wasting precious ground land resource.

1 INTRODUCTION

India is a tropical country with a global annual average solar radiation of 5.31 (kwh/m²/day) (Irwan et al., 2015). Due to insufficient conventional energy resources, our country has decided to move to Renewable Energy Resources (RES) under the governance of Ministry of New and Renewable Energy (MNRE). India is a third largest electricity producing country in the world with a capacity of 1,208,400 (GWh) (BP, 2013). To fulfill the energy needs of its citizens, the country needs to shift to a more reliable energy source for the near future.

Delhi is the National Capital Territory of India. It resides in the northern part of India and has a great population with high land occupancy. To fulfill the needs of citizens, Delhi has an installed electricity generation capacity of 8346.72 MW (as on 30 November 2015) (Central Electricity Authority, 2014). In this number only 10% of electricity is through the Renewable resources i.e. 856.76 MW, all other capacity is through coal, gas i.e. thermal production. Figure 1 [4] Depicts the gateway of harnessing solar energy from the sun using steps to show the areas of working environment.

According to latest reports, Delhi is the most polluted city in the world with annual mean pollutants of PM2.5, ug/m3: 153 (as on 2015)[5]. To tackle the problem of air pollution Ministry of Environment, Forests & Climate Change (MoEFCC) has an Act called—No.14 of 1981, [29/3/1981]—The Air (Prevention and Control of Pollution) Act 1981, amended 1987. Under this act, the ministry makes

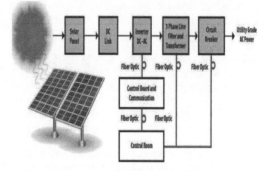

Figure 1. Block diagram for solar energy harnessing.

all its efforts to control the pollution in India. In last few years, Delhi has gained huge media coverage on its alarming pollutants rate. To solve this problem in a productive manner, solar energy gives a reliable solution. By harnessing the solar thermal power and replacing it with the conventional energy sources, we can reduce the pollutants in Delhi in very less duration of time. Figure 2 Shows the amount of pollutants in Delhi in comparison to other states in India and abroad. This figure is a perfect example to visualize the alarm that is ringing to save our environment.

Delhi is well connected by Metro; we have metro stations at every location in Delhi which provides a well-connected grid of transport line to travelers. The network consists of six lines with a total length of 189.63 kilometers (117.83 mi) with 142

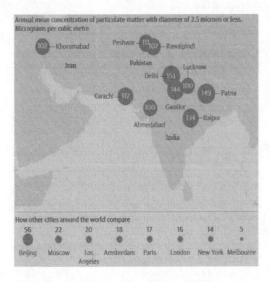

Annual mean concentration of particulate matter with diameter of 2.5 microns or less. Micrograms per cubic metre

Khoramabad 102

Iran

Peshwar 111 — 107 Rawalpindi

Pakistan

Delhi 153 Lucknow

Karachi 117

100

144 100

149 Patna

Gwalior

Ahmedabad

134 Raipur

India

How other cities around the world compare

Beijing	Moscow	Los Angeles	Amsterdam	Paris	London	New York	Melbourne
56	22	20	18	17	16	14	5

Figure 2. Comparison of pollution in Delhi w.r.t to other states in world.

stations, of which more than 100 are elevated (Sharma, Singh, Dhyani, & Gaur, 2014). These elevated stations can act as a source of free, unoccupied land to generate power. If we can supply the electricity loads or consumption of Delhi Metro with clean energy, our task is done. By generating solar power, metro can fulfill all its needs on a daily basis and can contribute to the state. In this case, the metro's stand alone needs can be fulfilled easily and the energy wasted in transmission can also be preserved.

The regular energy requirement of an elevated station is 1,200 units each day whereas it is 3,000 units for an underground metro station. To compare, the monthly power consumption of a household is between 400 and 700 units on an average.

This paper explains the ways to increase the efficiency of solar panels and provide a better model for solar panels placement on metro rooftops. With a better structural design, we can determine the actual energy produced, as by better structure we can to track the sun seasonally and economically. The paper has been divided into six sections; the first section deals with the introduction to solar energy and its possibilities in Delhi, the second section includes the resources used to write this paper, the third section deals with the methodology employed to increase the efficiency, the fourth section divulges result and discussion for implementation method, the fifth method makes comparison with existing methods and the last section includes the conclusion and future scope.

2 RESOURCES

The harnessing of energy at the rooftops need appropriate structures to increase the efficiency of the solar system. To calculate the efficiency and the land available, three softwares used and readily discussed in the paper are:

– PVsyst 6.3.4 SOFTWARE
– HELISCOPE SOFTWARE
– SKETCH UP SOFTWARE

These softwares provided a great platform to study and research for this paper. As they provided me the following data: Area of total land or Rooftop area available; Simulation of various combinations of azimuth and tilt available; Inverter selection and placement; Solar panel selection; Battery selection; Losses occurred in system; Meteorological data according to specific location; Wiring material and number combiners used; Simulate annual yield of power; And choose between panel crystal cell technology, calculate price; Calculate carbon footprint; and determine shading losses. In the visualization of the area on the satellite images, Heliscope software came very handy as it links to Google maps to get the specific location. By this, it becomes easy to visualize the rooftops available for the project and check for any keepouts on the floor. Mechanical and Electrical specifications provided in the Heliscope also allowed me to better understand the topic and motivated me to do relevant research. To better understand the shading losses and 3D structure of rooftop Sketch up software is used. It provided an hourly and yearly variation of sun's position to get the specific shading losses through its toolkits. By studying from it, it becomes easy to place solar panels and calculate correct efficiency. Where the above two softwares helped me to visualize the physical structure, another software called PVsyst helped me to determine the characteristics, efficiency, cost & requirements for the project.

3 IMPLEMENTATION

Looking at all the factors, the main challenge is to find a most appropriable place where solar panels can be planted with minimum or zero shading effects and an area that is totally not utilizable.

All these challenges motivate us to think on rooftops i.e. the rooftops that are just used for shading purposes. The best result we got from the Delhi metro's rooftop. As they are such designed that passengers can have protection from sun and rain. If we can generate a good amount of clean electricity from this system, then it could be fed to the metro itself for its daily standalone needs.

We wanted to make a system on which panels could be such placed that their overall efficiency can be increased. As the energy harnessed by solar panels at plain or global horizontal is less than the energy harnessed at the tilt. Studies show that in Delhi average annual solar yield at global horizontal is 5.4 kwh/m²/day and at global on tilt plane is 6.1 kwh (Irwan et al., 2015). So there is an overall increase of 12.9%, which can be accounted as huge variation in efficiency. To provide tilt, a well-defined structure is needed, which can be resolved by:

- Fixed
- Adjust two seasons
- Adjust four seasons
- Two-axis solar tracking

To implement this, a structure for panel placement has to designed, such that it could be integrated on the top of the metro station rooftop. However, if we use solar tracking to fulfill needs, it would greatly increase the overall cost of the system. So a challenge was to set the panels in such way that if we are even able to change the tilt of the system twice or quarterly, we could generate a good reliable efficiency. So the solution to this to me is to change the tilt quarterly or twice in a year. However, if we are using the quarterly option then again the system structure cost is increased. So we decided to put it on the second option as it could be easily produced using truces or jacks.

To calculate the tilt seasonally we just need to use these formulae's:
Summer:

Tilt (in degrees) = (latitude* 0.93) – 21

Winter:

Tilt (in degrees) = (latitude*0.875) + 19.2

Delhi has an attitude of 28.6139391, So for Delhi, we have a tilt of 5.61° for summer and 44.237° for winters. The date to change the tilt is 30th March for summer and 12th September for winter for optimization [9].

By using scissor lifts the metro rooftops and surface utilization can be increased. With this, we can change the seasonal tilts easily and more efficiently. With these kinds of lifts, the quarterly

Figure 3. Solar energy harnessing Metro plan (Ellis & Torcellini, 2008).

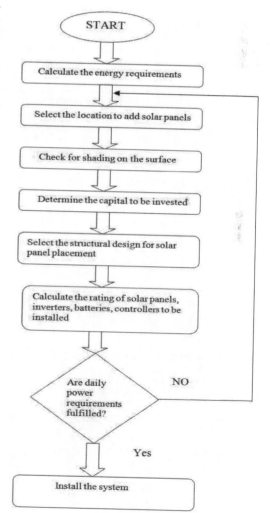

Figure 4. Algorithm to illustrate the solar panel placement strategy.

Table 1. Overall efficiency of system w.r.t type of structure.

SCHEME	Fixed	Adjusted 2 seasons	Adjusted 4 seasons	Solar tracking
% of optimum	71.1%	75.2%	75.7%	100%

cycle of changing tilt is not a big issue. Creating a new surface above the roof of the metro will allow both ventilation to the solar panels and also the flexibility to provide tilt & select azimuth. By this, we are easily able to avoid the shading losses, which are the main energy suckers while placing solar panels.

On an average, every elevated station in Delhi has an area of 5500 m² as total horizontal, which gives an average solar structure platform area of 5527 m².

Since the requirement of elevated metro stations is approximately 1200 units of power. So simulation in PVsyst has been done on that account only.

4 RESULTS AND DISCUSSIONS

The proposed algorithm is simulated in PVsyst and Heliscope for the generation of results, which shows the exact energy produced by the system. Simulating the parameters allowed calculating the losses and cost of the overall system.

These software interfaces provided me the flexibility to add every possible parameter in my research. These results have been generated for a standalone system so that DMRC's all energy needs can be fulfilled by this system only.

Results show that in average annual solar yield at global horizontal is 5.4 kwh/m²/day and at global on tilt plane is 6.1 kwh. So there is an overall increase of 12.9%, which can be accounted as huge variation in efficiency. Since the requirement of elevated metro stations is approximately 1200 units of power. So simulation in PVsyst and Heliscope has been done on that account only. Delhi Metro has already planted solar plants at their rooftops, but the power efficiency w.r.t land invested by their method is very less.

Energy cost of 6.61 INR/kwh gives a very real cost to energy ratio, by keeping in mind that the

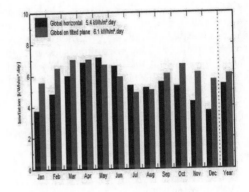

Figure 6.   Global solar radiation in Delhi.

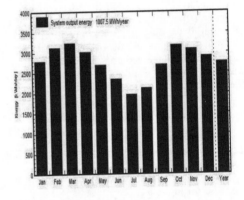

Figure 7.   Monthly yield from system from Pvsyst.

	Gl. horiz. kWh/m².day	Coll. Plane kWh/m².day	System output kWh/day	System output kWh
Jan.	3.77	6.05	4198	130127
Feb.	4.85	6.78	4702	131646
Mar.	6.00	7.00	4855	150520
Apr.	6.84	6.51	4515	135447
May	7.13	5.83	4042	125288
June	6.60	5.08	3525	105748
July	5.35	4.26	2953	91539
Aug.	5.18	4.58	3179	98546
Sep.	5.57	5.85	4055	121652
Oct.	5.26	6.88	4768	147795
Nov.	4.27	6.65	4610	138293
Dec.	3.70	6.27	4346	134721
Year	5.38	5.97	4141	1511323

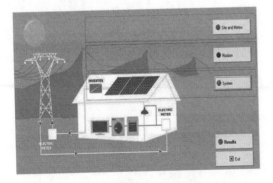

Figure 5.   Software interface of PVsyst for standalone systems.

Figure 8.   System output simulation in PVsyst.

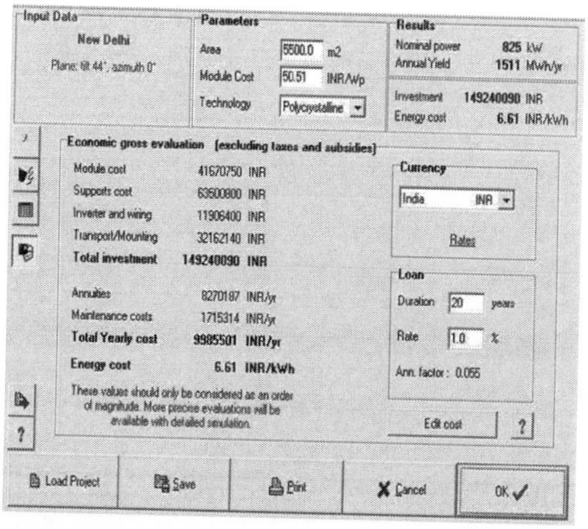

Figure 9. Cost analysis using Polycrystalline technology for cells in Pvsyst.

energy production plant is in the area that is not utilized by the DMRC. The cost of the system can be further decreased if we use concentrators with solar panels so that they can concentrate the sun energy on solar panels and give a better yield. The structural design on which panels reside should not also be totally solid, or I mean that it should have some air ventilation capacity so that solar panels could work at their optimum temperature efficiency.

5 COMPARISON WITH EXISTING MODELS

Delhi Metro Rail Corporation (DMRC) has begun installing solar panels in their phase 3 project to fulfill their need standalone, which is excellent for Delhi to have a clean transport. However, the way of planting solar panels on their rooftops is not that efficient, as they are not utilizing the space available to them (Shi et al., 2012). These solar power plants with total generation capacity of 1,660.4 kWp (Sharma et al., 2014), have been installed, wherein the capital cost has been invested by the Solar Developer. The journey of DMRC started when they collaborated with GIZ in partnership with Delhi Metro Rail Corporation Limited (DMRC). DMRC is looking for a place in Rajasthan to build a 500 MW plant to fulfill all his needs, nut to me this would decrease the efficiency by losing energy in transmission losses. Even if with the proposed model the problem can be fulfilled, then why to look for a place at such great distance (Kepaptsoglou & Karlaftis, 2010).

6 CONCLUSION AND FUTURE SCOPE

By 2017 DMRC has decided to move to total clean energy use at its all stations and by this model, I think, the thought of moving green can be achieved. Since the ridership in DMRC is increasing every year, it is important to build a standalone power source for its use, so that DMRC could grow further (Mann & Banerjee, 2011). Since DMRC have more than 100 elevated metro stations, so the power produced by them can also be used to run the underground metro stations using grid-connected systems (Siemiatycki, 2006). If DMRC could go green fuel, then the carbon footprint in Delhi could also be decreased. In future, DMRC can also upgrade their solar energy system by using: Solar concentrators; Solar tracking; Improved solar cell technology; Solar towers; Using thermocouples with panels; Combining solar energy with the wind; Transparent solar cells on windows; 3d solar panels; By maintaining Temperature of solar panels.

REFERENCES

BP. (2013). *BP Statistical Review of World Energy* Central Electricity Authority. (2014). Government Of India Ministry of Power Central Electricity Authority New Delhi Executive Summary Power Sector, 6. Retrieved from http://www.cea.nic.in/reports/monthly/executive_rep/feb14.pdf.

Ellis, P. G., & Torcellini, P. a. (2008). Energy Design Plugin : An EnergyPlus Plugin for SketchUp. *Conference Paper NREL/CP-550–43569*, 11. http://doi.org/NREL/CP-550–43569.

Irwan, Y. M., Amelia, A. R., Irwanto, M., Leow, W. Z., Gomesh, N., & Safwati, I. (2015). Stand-Alone Photovoltaic (SAPV) System Assessment using PVSYST Software. *Energy Procedia*, 79, 596–603. http://doi.org/10.1016/j.egypro.2015.11.539.

Kepaptsoglou, K., & Karlaftis, M. G. (2010). A model for analyzing metro station platform conditions following a service disruption. In *IEEE Conference on Intelligent Transportation Systems, Proceedings, ITSC* (pp. 1789–1794). http://doi.org/10.1109/ITSC.2010.5624967.

Mann, a., & Banerjee, T. (2011). Institutions and Megaprojects: The Case of Delhi Metro Rail. *Environment and Urbanization Asia*, 2(1), 77–91. http://doi.org/10.1177/097542531000200106.

Sharma, N., Singh, A., Dhyani, R., & Gaur, S. (2014). Emission reduction from MRTS projects— A case study of Delhi metro. *Atmospheric Pollution Research*, 5(4), 721–728. http://doi.org/10.5094/APR.2014.081.

Shi, C., Zhong, M., Nong, X., He, L., Shi, J., & Feng, G. (2012). Modeling and safety strategy of passenger evacuation in a metro station in China. *Safety Science*, 50(5), 1319–1332. http://doi.org/10.1016/j.ssci.2010.07.017.

Siemiatycki, M. (2006). Message in a Metro: Building urban rail infrastructure and image in Delhi, India. *International Journal of Urban and Regional Research*, 30(2), 277–292. http://doi.org/10.1111/j.1468–2427.2006.00664.x

Communication and Computing Systems – Prasad et al. (Eds)
© 2017 Taylor & Francis Group, London, ISBN 978-1-138-02952-1

Induction of radio frequency transmission in Indian railway for smooth running of traffic during fog

Chandra Mukherjee
Christ University, Bangalore, India

Neelam Ruhil
Dronacharya College of Engineering, Gurgaon, India

ABSTRACT: Our railway system drives whole sole based on its electrical signaling but due to poor visibility it becomes impossible to run the traffic smoothly We are suggesting to use radio wave communication technology for running of train when conventional signaling can't be followed due to poor visibility. During winter season, due to heavy fog especially in North India and East India it becomes almost impossible to drive the train on time. Our idea can remove this problem permanently. A dedicated radio frequency band will be used by railway service and a specific frequency will be assigned to all tracks running to a specific direction. All trains will be equipped with a transmitter and a receiver. Train drivers will get notification of received radio frequency within a certain circumference (5 km). So if it receives the same frequency which it is transmitting then the driver will understand another train is there on the same track so signaling room and the driver will also be aware of the fact. Then the control room or the driver can take action considering speed and distance between this two accordingly. If another train will be running on the next track then also it will receive signal but in that case it will run at as usual speed.

Keywords: FM, transmitter, receiver, signal processing

1 INTRODUCTION

A dedicated communication band should be used in this paper for transmission and reception purpose. Secure Radio wave communicators that is Radio frequency transmitter and receiver section will be installed in the trains and it will act as a trans-receiver. Range of the radio transmitters will be kept greater than 5 km to avoid any collision. In this case train drivers and guards will be able to receive all the details of radio frequency existed within 5 km and can communicate with the drivers of another train. Same Trans-receiver will be used in Control Room also. The detector will become active and give a alerting sound if something with that particular frequency (radio waves) comes nearer (say 5 km), in that case drivers will pick up the microphone and confirm the location of other train, whether it lies on up line (example trains going from east to west) or down line (trains going from west to east). For providing more robustness here we will use one emergency contact band also. If due to some circumstances all radio communication gets stopped then with the help of this band personnel will be able to contact each other to avoid any mishappenings. Secure Radio wave

communicators will be installed in the trains and it will act as a transmitting and receiving device, range of the radio transmitters will be >= 5 km. In this case train drivers be able to receive the details of radio frequency existed within 5 km and can communicate with the drivers of another train. The detector will become active and give a alerting sound if something with that particular frequency (radio waves) comes nearer (say 5 km), in that case drivers will pick up the microphone and confirm the location of other train, whether it lies on up line (example trains going from east to west) or down line (trains going from west to east).

In such scenario even if the driver is unable to see anything in fog, he/she will be able to locate the locations of other trains and can keep moving.

Also if trains came very close say (2 km) then detector will give alert sound, then drivers will quickly verify with the other drivers about his location, speed and direction of travel and both can take actions accordingly.

Considering trains move slowly during fog, even if trains are being detected 2 km apart there won't be any real risk of collision.

Also if the driver through conversation came to know that the other train is in down line and he is in up line, because both trains are coming closer so beep sound detected, then he can be sure that he can move as usual.

This system can be modified, we can use separate frequencies for up/down line, although both up/down drivers can listen both up/down frequency.

This technology may be useful in many cases where because of some human error (because drivers jumped the signal) several accident occurs. In this case radio waves will pick up signals and alert driver although he has jumped the signal.

This detector and radio apparatus will be kept both with driver and guard of trains so that closer observation can be made even if someone has missed something.

Since radio apparatus are cheaper so it can be applied on trial basis also for measuring its accuracy and utility. Safety will also increase considerably with this on trains.

2 THEORY AND PRINCIPLE

2.1 Signal transmission and reception

The signal transmission and reception is the most important part of our paper. Radio transmitter will be used for transmission of the signal. A radio transmitter is an electronic device which, when connected to an antenna, produces an electromagnetic signal. The RF module, as the name suggests, operates at Radio Frequency. The corresponding frequency range varies between 30 kHz & 300 GHz. In this RF system, the digital data is represented as variations in the amplitude of carrier wave. This kind of modulation is known as Amplitude Shift Keying (ASK). Transmission through RF is better than IR (infrared) because of many reasons. Firstly, signals through RF can travel through larger distances making it suitable for long range applications. Next, RF transmission is more strong and reliable than IR transmission. RF communication uses a specific frequency unlike IR signals which are affected by other IR emitting sources. As a receiver we can use **tuned radio frequency receiver**, is a type of radio receiver that is composed of one or more tuned Radio Frequency (RF) amplifier stages followed by a detector (demodulator) circuit to extract the audio signal and usually an audio frequency amplifier. Cognitive radio may be defined as a radio that is aware of its environment, and the internal state and with a knowledge of these elements and any stored pre-defined objectives can make and implement decisions about its behavior. As a receiver we can use superheterodyne receiver also. It is one of the most popular forms of receiver in use today in a variety of applications from broad-

cast receivers to two way radio communications links as well as many mobile radio communications systems.

Although initially developed in the early days of radio, or wireless technology, the superhet or superheterodyne receiver offers significant advantages in many applications. Naturally the basic concept has been developed since its early days, and more complicated and sophisticated versions are used, but the basic concept still remains the same.

The latest trans-receiver we can use is "Cognitive radio based system". In general the cognitive radio may be expected to look at parameters such as channel occupancy, free channels, the type of data to be transmitted and the modulation types that may be used. It must also look at the regulatory requirements. In some instances a knowledge of geography and this may alter what it may be allowed to do.

This RF module comprises of an RF Transmitter and an RF Receiver. The transmitter/receiver (Tx/Rx) pair operates at a frequency of 434 MHz. The transmission occurs at the rate of 1 Kbps – 10 Kbps. The transmitted data is received by an RF receiver operating at the same frequency as that of the transmitter

The main feature of this paper is to achieve the technological simplicity, low operation cost, and low manufacturing cost without sacrificing the flexibility.

2.2 Signal processing

Signal processing is main part of any automation system. At first we have to do signal processing for reliable transmission of radio signal. Transmitted radio signal suffers from Path loss (or path attenuation). It is the reduction in power density (attenuation) of an electromagnetic wave as it propagates through space. Path loss may be due to many effects, such as free-space loss, refraction, diffraction, reflection, aperture-medium coupling loss, and absorption. Path loss is also influenced by terrain contours, environment (urban or rural, vegetation and foliage), propagation medium (dry or moist air), the distance between the transmitter and the receiver, and the height and location of antennas. So keeping these things in mind we have to design a system where these effects are minimum. The complex signal processing is done at the receiving end. Here we have to use some filtering methods to eliminate unwanted frequency components from the received signal for proper analysis. Then depending on received signal proper control algorithm has to be developed to drive railway traffic smoothly and some recording system will also be interfaced with the unit so that all communication details can be stored.

2.3 Display

Two display units we have to use here. One is the local display used at the train and one more display unit at the control panel. After the transmission, processing and reception of the radio communication signals received status will be displayed at both of the display for further countable action. We will try to make our display as compact as possible since it will provide us the mobility.

In this paper we can use Distributive Control System panel also for Acquisition, Processing, Analysis and Display purpose.

3 METHODOLOGY

This idea is implemented in three different stages using MATLAB platform. The steps involved for this entire implementation are given stepwise below-

3.1 Signal generation

We have generated 8 bit serial data carrying information about tracks and trains. The bits are specified in following method-

3.1 Signal transmission

For transmitting the signal in RF range Amplitude Shift Keying Technique (ASK) is used. ASK modulation scheme is expressed as a formula, carrier wave C(t) takes the following form.

$$C(t) = Ac.cos(2*pi*Fc*t) \quad (1)$$

$$Vask(t) = m(t)*C(t) \quad (2)$$

$$= m(t)*Ac*cos(2*pi*Fc*t) \quad (3)$$

$$= m(t)*Ac*cos(2*pi*Fc*t) \quad (4)$$

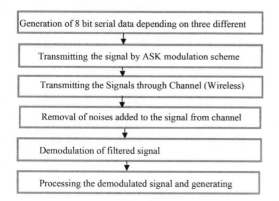

Figure 1. Flow chart of Implemented Algorithm.

Table 1. Coded 8 bit signal for identification.

D_7	D_6 D_5		D_4 D_3 D_2 Distance (Km)		D_1 D_0 Track no	
Future use	Passenger	00	1	001	T_0	00
	Express	01	2	010	T_1	01
	Super fast	10	3	100	T_2	10
	Special	11	4	101	T_3	11

Table 2. Available possibilities to be analyzed by signal processor.

SLNo.	D_7	D_6	D_5	D_4	D_3	D_2	D_1	D_0
1	X	0	0	0	0	1	0	0
2	X	0	1	0	0	1	0	0
3	X	1	0	0	0	1	0	0
4	X	1	1	0	0	1	0	0
5	X	0	0	0	1	0	0	0
6	X	0	1	0	1	0	0	0
7	X	1	0	0	1	0	0	0
8	X	1	1	0	1	0	0	0
9	X	0	0	1	0	0	0	0
10	X	0	1	1	0	0	0	0
11	X	1	0	1	0	0	0	0
12	X	1	1	1	0	0	0	0

the modulating signal $[v_m(t)]$ *is* a normalized binary waveform, where $+1$ V $=$ logic 1 and -1 V $=$ logic 0. Therefore, for a logic 1 input, $v_m(t) = +1$ V, Equation *4* reduces to

$$Vask(t) = A*cos(wc*t) \quad (5)$$

and for a logic 0 input, $v_m(t) = -1$ V, Equation 4 reduces to

$$Vask(t) = A*cos(wc*t) \quad (6)$$

where Ac = Amplitude of carrier; wc = angular frequency of carrier signal; Fc = frequency of carrier, m(t) = message signal; C(t) = Carrier signal; t = sampling interval; and A = amplitude of modulated signal.

In this simulation carrier frequency is taken as 1000 Hz and 8 bit long message signal frequency is taken as 100 Hz.

3.2 Signal reception

While transferring the signal through wireless channel some noises will be added with the signal. So a butterworth Low Pass filter of order 8 is designed at the starting of receiver side. To Design RF signal receiver consists Super Heterodyne Receiver principle has been used.

3.3 *Signal processing*

Now depending on received signal corresponding control logic has to be generated. For a Single Track there are twelve possibilities. The simulation has been done on Track 0. For rest track also this analysis part will remain same. Possible conditions are mentioned in following table-

4 RESULTS

4.1. *Transmitter side*

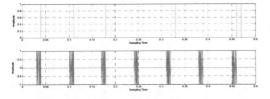

Figure 3.   a. Coded signal based on input parameters from Channel 1 b. Modulated signal of Transmitter side.

4.2. *Receiver side*

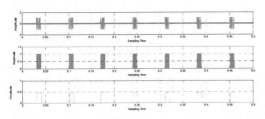

Figure 4.   a. Noise Contaminated signal b. Filtered signal for further processing c. Extracted Message Signal.

5 PROCESSED OUTPUT ON SCREEN

6 CONCLUSIONS

This idea is very much applicable in the winter and monsoon season. Due to low visibility it becomes impossible to drive the train. So our railway communication suffers a lot. Sometimes cargo carrying vehicles also stops these causes financial loss also. If our system is implemented properly then it will be very beneficial to boost up railway commercial service.

REFERENCES

Colletti, Justin (February 4, 2013). "The Science of Sample Rates (When Higher Is Better—And When It Isn't)". Trust Me I'm A Scientist. Retrieved (February 6, 2013).

Lashkarian, N., E. Hemphill, H. Tarn, H. Parekh, and C. Dick, "Reconfigurable digital front-end hardware for wireless base-station transmitters: Analysis, design and fpga implementation," IEEE Trans. Circuits Syst. I: Reg. Papers, vol. 54, pp. 1666–1677, (Aug. 2007). (Pubitemid 47272689).

Martin H. Weik (1996). Communications Standard Dictionary. Springer. ISBN 0412083914.

Platz, J., G. Strasser, K. Feilkas, L. Maurer, and A. Springer, "A direct up-conversion transmitter with integrated prescaler for reconfigurable multi-band/multi-standard base stations," in Proc. IEEE Radio Freq. Integr. Circuits Symp. (RFIC'05), (2005).

Rao, R. Signals and Systems. Prentice-Hall of India Pvt. Limited. (ISBN 9788120338593.)

Shannon, C. E. "Communication in the presence of noise", Proc. Institute of Radio Engineers, vol. 37, no.1, pp. 10–21, Jan. 1949. Reprint as classic paper in: Proc. IEEE, Vol. 86, No. 2, (Feb 1998).

Communication and Computing Systems – Prasad et al. (Eds)
© 2017 Taylor & Francis Group, London, ISBN 978-1-138-02952-1

Smart health care solution—a data mining approach

Kavita
AIM & ACT, Banasthali University, Jaipur, India

Neelam Ruhil
Department of Electronics and Computer Engineering, Dronacharya College of Engineering, Gurgaon, India

ABSTRACT: Due to various environmental changes ailment is a common problem faced by humans these days. Diseases like cold and fever are very common in people these days. Immunity capacity of humans is reduced due to adverse changes in surroundings. In this work we propose a 'Smart Health Care Solution' via data mining application which recommends medicines to the user on the basis of disease and symptoms. In our experimental procedure we have considered 10 common diseases to form the database of symptoms and causes of diseases that are common in today's time like: cold, fever, jaundice etc. We explore the 'Classification' technique on our dataset for our reference. Naïve Bayes algorithm is applied on the dataset. Weka is chosen for implementation of algorithms. We use Explorer application for prediction of Salts for our database on the basis of diseases and their respective symptoms. By applying the significant classification technique on particular dataset, rules set is engender which will help the researchers working in healthcare for intelligent decision making.

1 INTRODUCTION

Due to diverse ecological changes ailment is a common problem faced by humans these days. Bugs present in environment have bad influence on human body. Diseases like cold and fever are very common in people these days. Immunity capacity of humans is reduced due to adverse changes in surroundings. At this time numerous new and rare diseases are exposed in healthcare. Many researchers are currently working in healthcare to discover new medicine or treatment for novel diseases and also to uncover more effective and efficient medicine for diseases which are already exposed and are hazardous for human being. In this work we propose a 'Smart Health Care Solution' via data mining application which recommends medicines to the user on the basis of disease and symptoms.

2 DATA MINING

Data mining process is to analyze facts and figures from diverse perception and abbreviate it into valuable information. The overall goal is to dig out information from a data set and renovate it into comprehensible format. It helps in finding patterns or correlations among various fields in large relational databases which are summarized in the form of feature vectors. Data Mining has attracted great attention from various fields due to wide and large data present in these fields. To convert these large

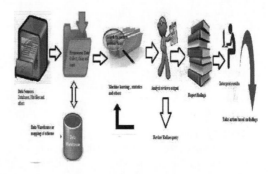

Figure 1.   Data mining process.

data into valuable information and facts data mining is required. The information and knowledge gained by application of data mining techniques can be used in various areas including market analysis, businesses and e-commerce, fraud detection, customer retention, production control, scientific engineering and healthcare etc.

The data mining process is illustrated in Figure 1.

Data mining involves the KDD (Knowledge Discovery in Databases) process which is explained further.

3 KDD PROCESS

Data mining is the analysis step of the "Knowledge Discovery in Databases" process, or KDD. As the name implies knowledge discovery in databases is the process of discovery of knowledge in large databases and to emphasize application of particular data mining method. The main goal of KDD process is to extract useful and meaningful knowledge from large databases. Knowledge is extracted from these large databases by using various data mining methods or techniques. KDD is an iterative process which involves the following steps:

- Selection
- Preprocessing
- Transformation
- Data Mining
- Interpretation/Evaluation
- Knowledge

An outline of the Steps of the KDD Process is given by Figure 2.

Data Selection:
Data selection involves selection of data appropriate to the scrutiny task are salvage from the database.

Data Cleaning and Preprocessing:
Data Cleaning and Preprocessing involves the following steps:

- Elimination of noise or outliers.
- Collecting essential information to model or account for noise.
- Strategies for managing misplaced data fields.

Data Transformation:
In this step data is transformed or merged into the format which is apposite for mining by summing up or amassed process.

Data Mining:
It is fundamental method where intellectual schemes are applied to dig out data prototype.

Interpretation:
Interpretation or Pattern evaluation involves identifying the accurately attention-grabbing patterns which represents knowledge based on various motivating proceedings.

Knowledge Discovery:
Knowledge presentation is the practices in which techniques like visualization and knowledge representation are exploit to make use of mined knowledge to the user.

Figure 2. KDD process.

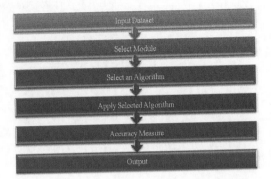

Figure 3. Working of weka.

4 DATABASE AND TOOLS USED IN EXPERIMENT

In our experimental procedure we have considered 10 common diseases to form the database of symptoms and causes of diseases that are common in today's time like: cold, fever, jaundice etc. Common diseases have been selected because genuine information related to these diseases is freely available on various medical sites. There are various data mining techniques and tools that are available and currently used in healthcare sector. We explore the 'Classification' technique on our dataset for our reference.

WEKA Machine learning tool is used for classification tasks. Weka consist a variety of machine learning algorithms for various tasks of data mining. These algorithms are useful in two ways whether user can apply the algorithm directly on the dataset or user is free to call own Java code. Weka includes tools for data pre-processing, classification, clustering, regression, association rules, and visualization. It is also compatible for developing new schemes for machine learning. Weka is free open source software having the GNU General Public License.

5 EXPERIMENT RESULT & ANALYSIS

Experimental work is carried out in Weka. The GUI of Weka includes four buttons each of which represents major application of Weka.

These four buttons refers the following applications:

- **Explorer**: It is an environment used for exploring data with WEKA.
- **Experimenter**: Another environment which perform experiments and accomplish statistical tests among learning schemes.
- **Knowledge Flow**: The environment of Weka allows the user to perform all the same functions which are explored in Explorer along with

a drag-and-drop interface. Additionally it also encourages incremental learning.

- **Simple CLI**: The button provides a interface having simple command-line which promotes execution of WEKA commands (which do not have their own command line interface) directly for operating system.

From the four applications of Weka, Explorer is more suitable for prediction using data mining. Therefore we use Explorer application for prediction of Salts for our database on the basis of diseases and their respective symptoms.

Explorer consists of following Tabs:

1. Preprocess: It allows the user to choose and modify the set of data on which user wants to perform task.
2. Classify: Used for training and testing of learning schemes which perform classification or regression.
3. Cluster: Used for learning clusters for the data.
4. Associate: It allows user to learn association rules for the data.
5. Select attributes: Used for selection of most relevant attributes in the data.
6. Visualize: Used to view an interactive 2D plot of the data.

5.1 Data sets used in Weka

We use two different sets for our experiment one is training set used for training of classifier. Another one is test set used for testing. Both are in ARFF format.

5.2 Training set

Training set contains the following attributes:

Attribute	Type	Other information
Did	Numeric	Refers to disease Id
Disease	Nominal	Name of disease
Cid	Numeric	Cause Id
Sid	Numeric	Symptom Id
SWeight	Numeric	Symptoms Weight
ProposedSalt	Nominal	Salt proposed for disease

5.3 Test set

Test data is used for testing it contains following attributes:

Attribute	Type	Other information
Did	Numeric	Refers to disease Id
Disease	Nominal	Name of disease
Cid	Numeric	Cause Id
Sid	Numeric	Symptom Id
SWeight	Numeric	Symptoms Weight
ProposedSalt	Nominal	Salt proposed for disease

All the attributes of test are same as training set but in test data the attribute 'ProposedSalt' is left undefined because we have to get salt from classification system on the basis of training set.

5.4 Output

The output model saved is again in ARFF format. It contains following attributes:

Attribute	Type	Other information
Did	Numeric	Refers to disease Id
Disease	Nominal	Name of disease
Cid	Numeric	Cause Id
Sid	Numeric	Symptom Id
SWeight	Numeric	Symptoms Weight
Prediction Margin	Numeric	Prediction margin
Predicted ProposedSalt	Nominal	Proposed Salts predicted by the Classifier
ProposedSalt	Nominal	Salt proposed for disease

Prediction ProposedSalt contains all the preferred salts which are predicted by the classification system.

6 SCREENSHOTS GENERATED DURING EXPERIMENT

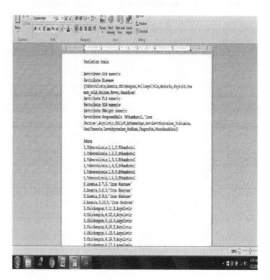

Figure 4. Training set.

Figure 5.   Test set.

Figure 6.   Run information of training set.

Figure 7.   Run information of test set.

Figure 8.   Generate predicted model.

Figure 9.   Predicted output file.

7   CONCLUSION AND FUTURE ENHANCEMENT

In this work data mining application is used to propose appropriate salts to the user on the basis of disease and symptoms. Classification technique is chosen for experimental work in Weka. Naïve Bayes classification algorithm is applied on the dataset to get the salts for respective diseases. In this project classification technique is applied on the dataset but in future other data mining techniques like clustering can also be applied on the dataset to get other interesting and useful facts and knowledge. The dataset and data mining techniques can also be further improved to enhance the overall performance of the system.

The experimental results may vary from data to data depending upon the attributes and their behavior. The results also influenced by classification techniques which can be used differently on different datasets. By knowing the appropriate technique for classification over a dataset rules can be engendered for that particular dataset and these rules will accompaniment in healthcare sector to get better useful knowledge from databases.

REFERENCES

Books:

Jiawei Han and Micheline Kamber-"Data Mining: Concepts and Techniques: Second Edition"; Morgan Kaufmann publications, 2006.

Journals:

Abirami, N. T. Kamalakannan, Dr. A. Muthukumaravel, "A Study on Analysis of Various Datamining Classification Techniques on Healthcare Data", International Journal of Emerging Technology and Advanced Engineering, Volume 3, Issue 7, July 2013.

Boris Milovic, Milan Milovic, "Prediction and Decision Making in Health Care using Data Mining", International Journal of Public Health Science, Vol. 1, No. 2, December 2012, pp. 69–78.

Durairaj, M., V. Ranjani, "*Data Mining Applications in Healthcare Sector: A Study*", International Journal of Scientific & Technology Research, Volume 2, Issue 10, October 2013.

Harleen Kaur and Siri Krishan Wasan, "Empirical Study on Applications of Data Mining Techniques in Healthcare", Journal of Computer Science 2 (2): 194–200, 2006 ISSN 1549-3636.

Hlaudi Daniel Masethe, Mosima Anna Masethe, "Prediction of Heart Disease using Classification Algorithms", Proceedings of the World Congress on Engineering and Computer Science 2014 Vol II, WCECS 2014, 22–24 October, 2014, San Francisco, USA.

Monali Dey, Siddharth Swarup Rautaray, "Study and Analysis of Data mining Algorithms for Healthcare Decision Support System", International Journal of Computer Science and Information Technologies, Vol. 5 (1), 2014, 470–477.

Parvathi I, Siddharth Rautaray, "Survey on Data Mining Techniques for the Diagnosis of Diseases in Medical Domain", International Journal of Computer Science and Information Technologies, Vol. 5 (1), 2014, 838–846.

Parvez Ahmad, Saqib Qamar and Syed Qasim Afser Rizvi, "Techniques of Data Mining In Healthcare: A Review", International Journal of Computer Applications (0975–8887) Volume 120 – No.15, June 2015.

Prakash Mahindrakar, Dr. M. Hanumanthappa, "Data Mining In Healthcare: A Survey of Techniques and Algorithms with Its Limitations and Challenges", Int. Journal of Engineering Research and Applications, Vol. 3, Issue 6, Nov-Dec 2013, pp.937–941.

Shubpreet Kaur and Dr. R.K.Bawa, "Future Trends of Data Mining in Predicting the Various Diseases in Medical Healthcare System", International Journal of Energy, Information and Communications Vol.6, Issue 4 (2015), pp.17–34.

Shukla, D. P. Dr. Shamsher Bahadur Patel, Ashish Kumar Sen, "A Literature Review in Health Informatics Using Data Mining Techniques", International Journal of Software and Hardware Research in Engineering, Volume 2 Issue 2, February 2014.

Shweta Kharya, "Using Data Mining Techniques For Diagnosis And Prognosis Of Cancer Disease", International Journal of Computer Science, Engineering and Information Technology (IJCSEIT), Vol. 2, No. 2, April 2012.

Srinivas, K., B. Kavihta Rani and Dr. A. Govrdhan, "Applications of Data Mining Techniques in Healthcare and Prediction of Heart Attacks", International Journal on Computer Science and Engineering, Vol. 02, No. 02, 2010, 250-255.

World Wide Web

Data Mining [Online] Available https://en.wikipedia.org/wiki/Data_mining.

UCI Machine Learning Repository. [Online] Available: http://archive.ics.uci.edu/ml/.

Weka, Data Mining Machine Learning Software, [Online] Available: http://www.cs.waikato.ac.nz/ml/weka/.

Communication and Computing Systems – Prasad et al. (Eds)
© 2017 Taylor & Francis Group, London, ISBN 978-1-138-02952-1

Non convention source of energy: Push and pop electricity

Sanjeev Kumar

Physics (Hons.) Motilal Nehru College, Delhi University, New Delhi, India

Anurag Vats

Computer Science and Engineering, Dronacharya College of Engineering, Gurgaon, Rohtak, India

Vishal Bharti

Head of Department, Computer Science and Engineering, Dronacharya College of Engineering, Gurgaon, Rohtak, India

ABSTRACT: Conversion of energy from one form to another desired form is both useful and interesting to watch. It becomes even more interesting when you could convert the energy present in your body to electrical and/or magnetic energy by a simple push or a pull. Push/pull can make the bar magnet vibrate, which can induce e.m.f by Faraday's Law of Electromagnetic Induction. The e.m.f produced can be increased sufficiently by using as simple apparatus as spring in our arrangement. The time-varying e.m.f can then be rectified and stored in capacitors and inductors.

1   INTRODUCTION

With the increasing need of energy, we are switching from convention sources of energy to non-conventional sources of energy. Wind energy and solar energy are serving very well. But we require sources of energy which can be used anywhere and are also economical.

Moving and vibrating objects, exercising humans, all posses a great amount of energy and any innovative idea that could convert this energy into electrical energy for our future use would be very interesting. It can be achieved if we attach magnet to these vibrating objects and produce e.m.f using Faraday's law. This e.m.f can then drive the current in our resistors or can be used to fill energy in our capacitors and inductors. The energy produced can be further increased if we put a spring with the vibrating object so that it will not damp easily.

2   THEORETICAL MODELING

2.1   Faraday's law

Consider a bar magnet attached to a rectangular slab with springs at its corners and a circular loop (Figure 1) When a force is applied to the slab, it vibrates along with the magnet. Due to this vibrating magnet, the magnetic flux ϕ through the circular loop changes and thus e.m.f is produced across its terminals.

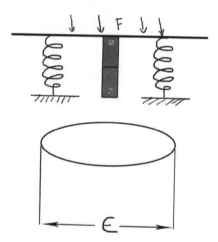

Figure 1.   Faraday's law.

$$\epsilon = -\frac{d\varphi}{dt} \text{ [Faraday's Law]}$$

$$\epsilon = -\frac{d\varphi}{dx} \cdot \frac{dx}{dt} \qquad (1)$$

Where x is the distance between center of magnet and the loop.

$d\varphi/dx$ is the change of flux with x.

dx/dt is the speed of the magnet while it is vibrating.

737

$$B = \frac{\mu m}{4\pi}\left(\frac{1}{z_1^2} - \frac{1}{z_2^2}\right) \tag{2}$$

Where, $z_1^2 = r^2 + (x-l)^2$

$$z_2^2 = r^2 + (x+l)^2 \tag{3}$$

Now, line integral

$$\oint_{s=0}^{s=2\pi r} B.ds = \frac{\mu m}{4\pi}\left(\frac{1}{z_1^2} - \frac{1}{z_2^2}\right)\int_{s=0}^{s=2\pi r} ds$$

$$= \frac{\mu m}{4\pi}\left(\frac{1}{z_1^2} - \frac{1}{z_2^2}\right)2\pi r$$

2.2 Magnet movement

Consider a bar magnet of length **2l** with pole strength **m,** placed at a distance **x** from the circular loop of radius **r** (Fig. 2).

2.3 Magnetic field

Magnetic field at the differential length element **ds**,

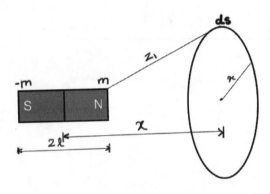

Figure 2. Circular loop/Magnet.

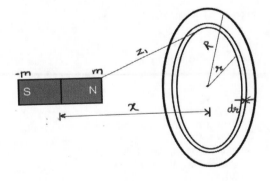

Figure 3. Circular loop/Magnet.

2.4 Calculations

Now, magnetic flux ϕ through the loop of radius R (Fig. 3)

$$\varphi(x) = \int_{r=0}^{r=R}\int_{s=0}^{s=2\pi r} B.ds.dr$$

$$= \int_{r=0}^{r=R}\frac{\mu m}{4\pi}\left(\frac{1}{z_1^2} - \frac{1}{z_2^2}\right).2\pi r.dr \quad \text{(Using 3)}$$

$$= 2\mu mlx\int_{r=0}^{r=R}\frac{1}{\left[r^2 + (x-l)^2\right]}$$

$$.\frac{1}{\left[r^2 + (x+l)^2\right]}.r.dr \quad \text{(Using 2)} \tag{4}$$

$$= 2\mu mlx\left[\frac{1}{8xl}.log\left[\frac{r^2 + (x-l)^2}{r^2 + (x+l)^2}\right]\right]_{r=0}^{r=R}$$

$$\varphi(x) = \frac{\mu m}{4}log\left[\frac{\frac{R^2}{(x-l)^2}+1}{\frac{R^2}{(x+l)^2}+1}\right]$$

From (1), we need $d\varphi/dx$ for e.m.f $\in$.
Thus, $\varphi(x)$ on differentiating w.r.t x gives,

$$\frac{d\varphi}{dx} = \frac{\mu m}{2}\left\{\frac{R^2}{(x+l)^3 + R^2(x+l)} - \frac{R^2}{(x-l)^3 + R^2(x-l)}\right\}$$

So that,

$$-\frac{d\varphi}{dx} = \frac{\mu m}{2}\left\{\frac{R^2}{(x-l)^3 + R^2(x-l)} - \frac{R^2}{(x+l)^3 + R^2(x+l)}\right\} \tag{5}$$

2.5 Plots

e.m.f, $\in = -\dfrac{d\varphi}{dx}.v$, where v is the velocity of magnet.

Let us assume that under the influence of the force F, the slab and the magnet vibrate with a constant velocity v.
And that x is always greater than l.
Now we plot a curve between $\in$ and x(graph1).
Curve between $\in$ and t.

2.6 E.m.f problem rectification

The problem of alternating e.m.f can be solved by using a full wave rectifier (Figure 4).
Here Z is a resistor or a series combination of capacitor and inductor.
The e.m.f $\in$ at the terminals of Z is shown in graph3.
The e.m.f $\in$ becomes n times if there are n circular loops.

2.7 Final plot

In response to our assumption that v remains constant which is the rarest possible case: v will change

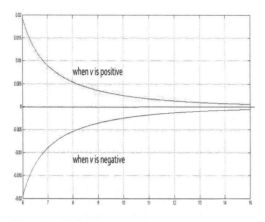

Figure 4. Emf vs x.

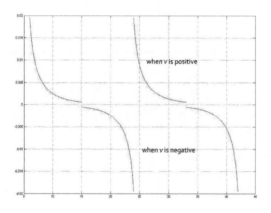

Figure 5. Emf vs t (time).

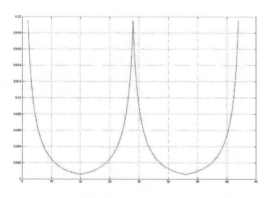

Figure 6. Magnet/Full wave rectifier.

with time depending upon the force F applied to the slab. When v will be more, $\in$ will be more and when v will be less, $\in$ will be less. Then the curve of $\in$ vs t will not be as in graph 3. But we will still get

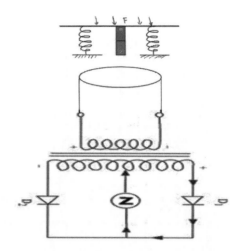

Figure 7. Magnet/Full wave rectifier.

$\in$ in any case and the combination of capacitors and inductors will still store the vibrating energy of slab in the form of electrical and magnetic energy.

3 RELATED WORK

3.1 *Pavegen systems*

Pavegen Systems is a technology company that has developed paving slabs to convert energy from people's footsteps into electrical power. A typical tile is made of recycled polymer, with the top surface made from recycled truck tires.

3.2 *Electric speed-breakers*

Electricity is generated by replacing the traditional speed breakers with some simple mechanism. As vehicles pass over the speed breakers, they spin the rollers which are connected to a generator which in turn generate electricity by EMI.

4 CONCLUSION

Pushing and pulling the slab is a good way of generating the electricity and storing it for future purpose. Since the e.m.f produced is time-varying and the variation depends on velocity of the slab. Thus it is better stored first then using it directly. The production can be increased by varying the variable of the setup. Such setups can be installed in the labs where small amount of electricity can be produced by pushing and popping the slab. Similar arrangements can be installed in toy and floors.

REFERENCES

Bajaj, N.K. The Physics of Waves & Oscillations.

Das, H.K., Dr. Rama Verma. Mathematical Physics.

Elombo, A.I. Renewable energy development, Stellenbosch University, South Africa, IEEEXplore.

Energy Research & Social Science, Science Direct Journals.

Halliday, D., R. Resnick, J. Walker. Fundamentals of Physics, "Faraday's Law of EMI.

http://www.researchgate.net/publication/264706010_ GENERATION_OF_ELECTRICITY_WITH_ THE_USE_OF_SPEED_BREAKERS "PaveGen", Wikipedia.

Kartikeya Singh. Energy for Sustainable Development, International Energy Initiative. Published by Elsevier Inc., Science Direct.

Key, T.S. Future of renewable energy development & deployment, Electric Power Research Institute (EPRI), IEEEXplore.

Leino, J. 2006. An approximative method for calculating. Helsinki University of Technology, TKK, Finland, Article No. 67, ACM New York, NY, USA.

Renewable energy and sustainable development, Associate Professor, Department of Mechanical Engineering, King Fahd University of Petroleum and Minerals, Science Direct.

Renewable energy development in Australia, Sudarshan Dahal; Powerlink Queensland, Virginia, QLD, Australia, IEEEXplore.

Robert D. Strum. Contemporary Linear Systems Using MATLAB, ACM.

ShenJianfei. China's energy development strategy, Training Institute, North China Electric Power University, China, IEEEXplore.

Sustainable energy development, Sila Science and Energy Company, Science Direct.

The Magnet Effect, Jesse Berst, ACM Digital Library.

Verma, H.C. Concepts of physics, "Full Wave Rectifier".

What drives renewable energy development?, L. Alagappan, Energy and Environmental Economics, Inc., Science Direct.

Communication and Computing Systems – Prasad et al. (Eds)
© 2017 Taylor & Francis Group, London, ISBN 978-1-138-02952-1

Fabrication of macroporous silicon and its electrical behaviour study with acetone

S. Haldar, S.S. Mondal & Amit K. Rai
Asansol Engineering College, Asansol, India

Chandra Shekhar Singh
Dronacharya College of Engineering, Gurgaon, India

ABSTRACT: In this paper Macro porous silicon was fabricated by electrochemical etching of p-type silicon with a resistivity range of 0.1–5.0 Ω cm for 60 min in an electrolyte containing hydrofluoric acid (HF), water, and dymethylformamide (DMF). Samples were studied using Scanning Electron Microscopy (SEM). Current–voltage ($I-V$) characteristic was studied for acetone. It is observed that macroporous silicon can be used as a chemical sensor.

1   INTRODUCTION

Various resistive p-type Si wafers were used for macropore formation. It was shown that using the mixture of HF with specific organic solvent, Dimethyle Formamyde (DMF) macropores can also be obtained on low to high resistive p-type Si. Since pore diameter of Porous Silicon (PS) can be changed by controlling the formation parameters like composition of etching solution, resistivity of substrate, current density and etching time (L.T. Canham et al. 1990), the structural advantage, PS layers were used for different chemical and biochemical-sensing applications like detection of bacteria (H. Ouyang et al. 2004), DNA hybridization (M. Archer et al. 2004) and sensing of other biological and organic chemicals (S.J. Kim et al. 2002). In this paper the electrical behaviour of macroporous silicon for acetone was investigated.

2   EXPERIMENT

Boron doped, one-sided polished silicon wafers of {100} crystal orientation and of various resistivity from 0.1 Ω cm to 5 Ω cm were used for the experiments. All wafers were pretreated for 20 min in a mixture of H_2O_2 (98%) and H_2SO_4 (98%) and afterward dipped into a solution of water and few drops of HF (48%) for 30 sec to remove the oxide layer. Anodization of p-type substrates was performed under galvanostatic conditions in a simple O-ring double tank bath with a Pt wire as electrode. Etching solutions were prepared using HF (48%, analytical-grade) and DMF (99.5%,

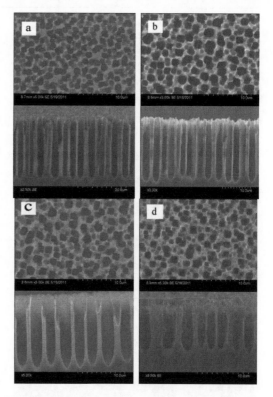

Figure 1.   SEMs of plane and cross-sectional view of p-type silicon substrate for different volumetric ratio of HF and DMF (substrate resistivity = 3 Ω cm, current density = 7 mA/cm² and time of etching = 60 min): (a) HF:DMF = 1:3, (b) HF:DMF = 1:6, (c) HF:DMF = 1:9, (d) HF:DMF = 1:15.

analytical-grade) in different volumetric ratio (e.g. HF: DMF: 1:3, 1:6. etc). KCl was used as electrolytic solution. Anodization process was carried out changing current density and resistivity of the silicon substrate also. A systematic study for macropore formation on p-type Si was carried out changing different parameters of the etching process (S. Haldar et al. 2014). The changing of pore diameter and pore length with different etching parameter (i.e. HF-DMF ratio, resistivity of the sample and current density) are shown in Figs. 1–3.

From the above Fig. (1–3) it is clear that pore density, pore length and pore uniformity depends on the HF, DMF ratio, resistivity of the sample and current density. Since the contact area of the sample with the chemical depends on the pore density, therefore current–voltage $(I - V)$ characteristic also depends on the etching parameter. In this work to study the electrical behaviour of macroporous silicon was fabricated taking HF-DMF ratio as 1:9, resistivity as 3 Ω cm and current density as

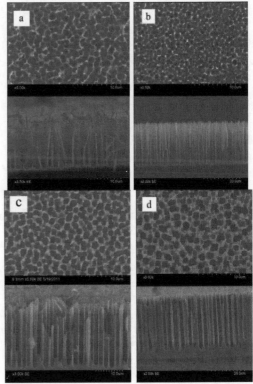

Figure 3. SEMs of plane and cross-sectional view of p-type silicon substrate for different resistive substrate (HF:DMF = 1:9, current density = 7 mA/cm² and time of etching = 60 min): (a) ρ = 0.1 Ω cm, (b) ρ = 1.2 Ω cm, (c) ρ = 3 Ω cm, (d) ρ = 5 Ω cm.

9.5 mA/cm², because these are the optimum value for the etching parameter. The porous layers are rinsed with deionized (DI) water and dried in a oven with temperature 100°C.

The metal contacts of dimensions 4 mm by 1 mm with a separation of around 3 mm on porous layer are then formed by screen printing with commercially available silver-aluminium paste followed by its firing at 700°C for 45 seconds (Fig. 4). Now to study the electrical behaviour organic gas (acetone) was generated by evaporation from organic vapour solution diluted from 1% to 0.5% concentration DI water.

From the above Fig. 5 it is clear that due to the adsorption of organic vapours the current response is changed. The adsorption magnitude of organic vapours at the PS surface is different according to surface state and types of the of organic vapours. Since the oxidation rate of macro PS is very small the conductivity response of macro PS was good for acetone vapours with dipole moment 2.88 debye and vapour pressure 233 tor (Fig. 5).

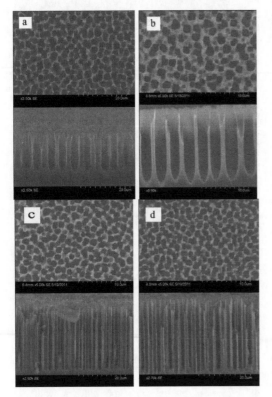

Figure 2. SEMs of plane and cross-sectional view of p-type silicon substrate for different current density (HF:DMF = 1:9, resistivity = 3 Ω cm, time = 60 min) (a) j = 3.5 mA/cm², (b) j = 7 mA/cm², (c) j = 9.5 mA/cm², (d) j = 15.5 mA/cm².

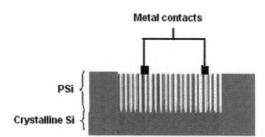

Figure 4. Schematic diagram of Macro PS layer with metal contact.

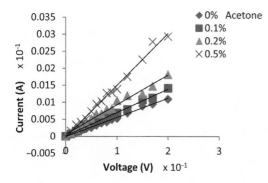

Figure 5. I – V characteristic measured for acetone.

3 CONCLUSION

From this study it can be concluded that conductivity response of the macro PS layer depends on the dipole moment, vapour pressure and adsorption effect of the molecules of the organic chemicals. The adsorption effect of the molecules may depend on the surface structure of the macro PS layer. Now varying the surface structure the conductivity response was not studied in this paper. After all studying the conductivity response of the macro PS layer with different organic chemicals a simple chemical sensor device may be designed.

REFERENCES

Archer, M., M. Christophersen, and P.M. Fauchet, "Macroporous silicon electrical sensor for DNA hybridization detection", *Biomedical Microdevices,* 2004, p. 203.

Canham, L.T. "Silicon quantum wire array fabrication by electrochemical and chemical dissolution of wafers", *Applied Phys. Lett., 1990,* p. 57.

Haldar, S., A. De, S. Chakraborty, S. Ghosh and U. Ghanta, "Effect of dimethylformamide, current density and resistivity on pore geometry in p-type macroporous silicon", *Procedia Materials Science,* vol. 5, 2014, p. 764.

Kim, S.J., S.H. Lee, and C.J. Lee, "Organic vapour sensing by current response of porous silicon layer", *J.Phys.D: Appl. Phys.,* 2002, p. 3505.

Ouyang, H., L.A. DeLouise, M. Christophersen, B.L. Miller, and P.M. Fauchet, *Proc. of SPIE* 5511, 2004, p. 71.

RoyChaudhuri, C., S. Gangopadhyay, R. DevDas, S.K. Datta, and H. Saha, "High performance macroporous silicon chemical sensor with improved phase detection electronics", *International journal on smart sensing and intelligent systems,* vol. 1, 2008, p. 638.

Communication and Computing Systems – Prasad et al. (Eds)
© 2017 Taylor & Francis Group, London, ISBN 978-1-138-02952-1

Performance analysis of 20 nm gate length Fin-FET for different materials and fin-widths

Upendra Kumar Gupta & Vishal Ramola
Faculty of Technology, Uttarakhand Technical University, Dehradun, India

Poornima Mittal
School of Engineering and Technology, Graphic Era University, Dehradun, India

Brijesh Kumar
Madan Mohan Malaviya University of Technology, Gorakhpur, Uttar Pradesh, India

ABSTRACT: Scaling is the most important demand of current scenario as the device technology changes day by day. The channel length of field effect transistors has reduced from micrometers to tens of nanometers. However, lot of limitation and defects are increases with scaling down the device dimensions such as short channel effect, reliability and variability effects issues. To overcome this type of problems related to scaling, new transistor configuration have to be investigated. FinFET is the promising double-gate or tri-gate transistor configuration to extend scaling over planar device. Multiple gates device have better control over the short channel effect. A double gate Fin field effect transistor can reduce Drain Induced Barrier Lowering (DIBL) and improve threshold voltage (short channel effects). In this paper, an important work function geometrical parameter is discussed using Visual TCAD. The transfer characteristics of the FinFET at different fin widths have been acquired at a supply voltage of 0.1V. The comparison is made at different Fin-widths. It is observe that, at larger Fin-widths the drain current is increases in comparison to shorter fin widths.

1 INTRODUCTION

As indicated by Moore's Law, the quantity of transistors in an integrated circuit double in every two year, or we can say the overall processing power of an integrated circuit double in every two year (Scott et al. 2006). A few methodologies have been discussed by different researchers to improve the performance of device and circuits, further downsize and improve the overall performance of the chip and the memory cells through CMOS innovation, SOI innovation, and so on (Mishra et al. 2015). Moreover, the multi-gate transistor is the alternate technique (Hisamoto et al. 2000) that can be utilized to realize the high speed compact circuit.

A MOSFET with more than one gate into a one device is referred to as a multi-gate device or transistor. FinFET, the multi or tri-Gate models can be downsized to deca-nanometer range (Chang et al. 2000). In FinFET, the gate is wrapped around a gate, undoped Si, called a 'fin'; this is from where it determines its name. The sides of the balance are wrapped around by an oxide, which breaks the active part into a few balances and a Gate covers

the channel regions part of the fins. This expands the controlling of the gate over the channel and subsequently high exchanging proportions can be achieved (Hadia et al. 2011).

In a mass Nanoscale MOSFET, the variety of device execution as an after effect of fluctuations in dopant particles is high (Moore et al. 1998).

This can be overcome in a FinFET inferable from its electrostatic control over the channel as specified previously In a MOSFET the control over the SCEs, for example, DIBL, channel length regulation and hot carrier effect and so for is less (Colinge et al. 2007).

This can be enhanced in a FinFET. In a FinFET, we can enhance the control over short channel effect by decreasing the fin width. Yet, decreasing the fin width comes width its own arrangement of disadvantages like, an improved parasitic source to drain resistance, which corrupts the channel current and trans conductance saturation), respectively (Hadia et al. 2011).

As nanometer procedure innovations have multifeature, IC thickness and operational frequency have expanded, making power consuming in battery-operated portable device a major concern.

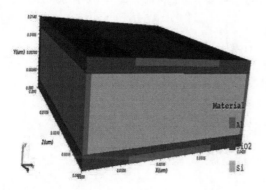

Figure 1. 3D schematic structure of double gate FinFET.

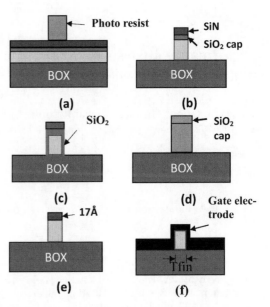

Photo resist

SiN

SiO₂ cap

(a)

(b)

SiO₂

SiO₂ cap

(c)

(d) Gate electrode

17Å

Tfin

(e)

(f)

Figure 2. (a-f) Fabrication flow of FinFET.

Even for non-portable device, power consumption is the most important in point of view increasing packaging and cooling costs and also reliability potential issues. Therefore, the principle plan objective for VLSI manufacturer is to meet performance requirements within a power source device. Subsequently, power efficiency has expected expanded significance. This part investigates how circuits based on FinFETs and developing transistor innovation that is likely to supplant mass CMOS at 20-nm and past, offer interesting delay–power tradeoffs. The main challenges in this regime are twofold: (a) minimize of the leakage current (Sub-threshold + gate leakage), and (b) Reduction in the device-to-device variability to increases yield.

2 STEPS OF DEVICE FABRICATION

In this section, Fig. 2 (a)–(f) demonstrate the schematic structure of the DG FinFET fabrication process. Multiple-fin device were additionally created in this analysis. An important qualification between a Fin field effect transistor and a conventional planar FET is an obviously limited active region (fin). Reducing of the fin width (i.e., body thickness), Tfin, is the most important to the scaling of DG FinFET. What's more, the overall of the gate to the active layer should be successfully controlling to reduce the transistor performance fluctuation.

The FinFET were created on SOI (silicon on transistor) wafers with a changed planar CMOS process. Double doped (n+/p+) poly-Si gate were utilized as gate electrodes. The FinFET CMOS inverters (worked from various fin transistors) were additionally fabricated. A heavily-doped poly-Silicon film wraps around the fin and stable electrical contact to the vertical faces of the fin. The poly-Silicon film greatly reduces the S/D

series resistance and provide a convenient means for local interconnect and making connections to the metal. A gap is etched through the poly-Silicon film to separate the source and drain. A thin sacrificial oxide was shaped and later stripped totally to evacuate the Si surface damage caused during the plasma etching of the fin stack. A thin protecting cap layer is held on top of the Silicon fin.

The circuit schematic of a 2D double gate n type FinFET in TCAD programming is as appeared in the Figure 2. The demonstrating of double gate FinFET has been done in TCAD.

TCAD is a product instrument software tool that models semiconductor fabrications furthermore semiconductor device operation. FinFET has been created on SOI (silicon on insulator) wafer with SiO2 as the covered oxide of thickness 1 nm. The gate length of the FinFET has been 20 nm. All the investigations have been done on this gate length just (Nesamani et al. 2013). The substrate utilized is that of silicon. The gates is ploy silicon and also the source and the drain electrodes have been taken to be made of Al. The work function of both the front and back electrode is 4.5 eV. Both the source and drain have ohmic contacts and the heat transfer coefficient is 1 KW/K/cm².

Utilizing nitride spacers prompts an expansion in the on state current i.e. it enhances the switching ratio. Here we have utilized four nitride spacers i.e. Np1, Np2, Np3 and Np4. We can likewise utilize double material spacer (the spacer with two distinct dielectrics e.g. silicon nitride and hafnium oxide) or triple material spacers (the spacer with

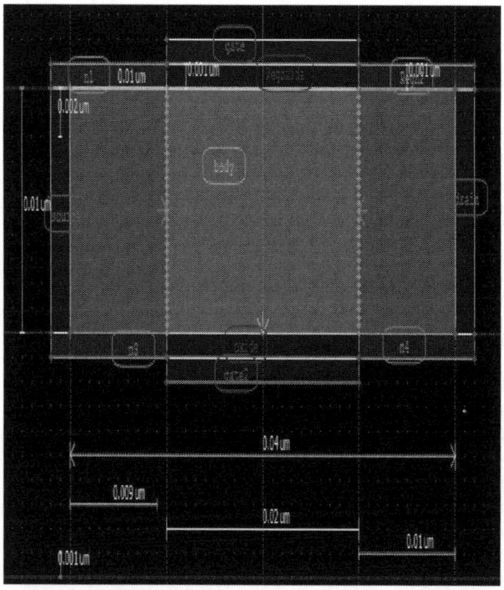

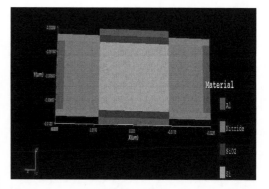

Figure 3. Schematic structure of FinFET in visual TCAD.

Figure 4. FinFET structure with material description.

Table 1. Different parameter at 20 nm technology.

Name of parameters	Dimension
Length of gate (L_g)	20 nm
Spacer width	1 nm
Gate oxide thickness	1 nm
Thickness of fin (Wfin)	Varied 1 to1000 nm
Doping conc. of S/D	1e + 20
Doping conc. of channel	5e + 18
Height of fin	40 nm

three distinct dielectrics e.g. silicon dioxide, silicon nitride and hafnium oxide) to further improve the device performance.

The height of the FinFET is 40 nm. The substrate is doped with acceptor impurity with a doping concentration of 5e+18/cm3, whereas the source and drain are doped with donor type impurity with the doping concentration of 1e+20/cm3. The fin width has been fluctuated from 1 nm to 1000 nm. A major qualification to a FinFET and a customary planar FET is a considerably narrowed active region (fin) Reduction of the fin width (i.e., body thickness), moreover, the overlay of the gate to the active layer should be most effectively controlled to Decrease the transistor performance variation. The device structure is quasiplanar.

3 SIMULATION SETUP AND RESULTS

Device simulations have been performed utilizing the drift diffusion model at room temperature i.e. at 300 K. Here we shifted the fin width of a double gate FinFET from 1 nm to 100 nm. For each fin width an Arrangement of results was acquired and a corresponding plot of the transfer characteristics was gotten. The channel current at various fin widths was then looked at a given gate voltage.

Firstly, we differed the gate voltage (at a steady drain voltage of 0.1V) and got the comparing estimations of channel current (Figures 4 and 5); Secondly we shifted the channel voltage (at a consistent gate voltage of 0.1V) and got the Drain current.

We observe that, as the fin width is expanded, the drain current additionally increments for a given gate voltage. For Higher estimations of fin width, the drain current saturates at a value more than that for the lower estimations of fin width value. The current increases linearly as the gate bias moving from negative to positive and finally attains a saturation value Fin widths the short channel impacts are extremely noticeable. Here likewise, we watch that as the fin width increases the drain current additionally increments however at much lesser short channel effects than that saw in extremely high fin width.

Though when we consider the drain current as a function of drain voltage at a constant value of gate voltage, we observe that, for higher estimations of fin width when the bias is negative i.e. for negative estimations of drain voltage, the drain current is not as much as that for the lower fin widths. But as the gate bias moves from negative to positive the drain current increments for that of higher fin widths, and saturates at a current higher than that for the lower fin widths. In this manner, as we study the drain current as a function of drain voltage, we observe that drain

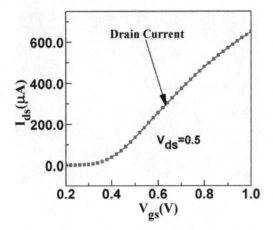

Figure 5. V_{gs} (gate voltage) vs. I_{ds} (drain current) plot.

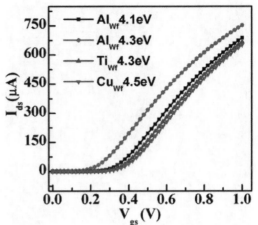

Figure 8. $V_{gs} - I_{ds}$ graph at various work function.

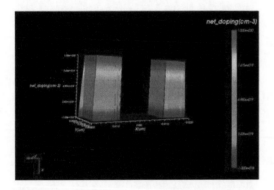

Figure 6.   Double gate FinFET structure view with net doping.

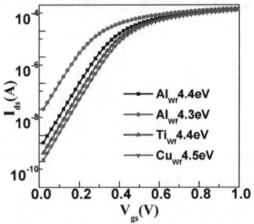

Figure 9. $V_{gs} - I_{ds}$ logarithmic graph at various work functions.

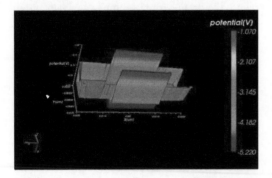

Figure 7.   Double gate FinFET structure view with potential.

this manner greater the fin width, greater is the performance of the FinFET.

4   CONCLUSION

A 20 nm gate length Fin-FET has been fabricated. The performance is examined in 2D utilizing the Visual TCAD simulator at a supply voltage of 0.1 V. The transfer characteristics of the FinFET under various fin widths have been obtained and the comparison of the transfer characteristics at various fin widths has been studied. At shorter fin widths, the control over the short channel impacts is great yet the source to drain resistance

current at first builds directly and afterward achieves a constant value for higher estimations of fin width while, for lower estimations of fin width the drain current remains almost constant at a value lower than that for higher fin width. In increments which prompts a reduction in drain current. Thereafter we presume that at lower value of fin width the drain current is less and expanding the fin width prompts an expansion in drain current and subsequently improve the device performance. Be that as it may, at higher fin widths, the control over the short channel effect is less. Thus, the fin width must be chosen with a specific end goal to type up for both the short channel effect and the drain to source resistance.

REFERENCES

Colinge, J.P. 2007. Multiple-gate SOI MOSFETs. *Microelectronic Engineering*. 84(9–10): 897–905.

Hadia, S. K. Patel, R.R. and Kosta, Y.P. 2011. FinFET architecture analysis and fabrication mechanism. *International Journal of Computer Science*. 8(5): 235–240.

Hisamoto, D. et al. FinFET. A self-aligned double-gate MOSFET scalable to 20 nm. 2000. *IEEE Transactions on Electron Devices*. 47(12): 2320–2325.

Mishra, K. and Sawhney, R. S. 2015. Impact of varying Fin-width in an n-FinFET at 20 nm gate length. *International Journal of Computer Applications*. 122(6): 8–10.

Moore, G.E. 1998. Cramming more components onto integrated circuits. *IEEE Proceedings*. 86(1): 82–85.

Nesamani, I.F.P., Raveendran G. and Prabha, V.L. 2013. Performance improvement of FinFET using nitride spacer. *International Journal of Engineering Trends and Technology*. 4(3): 299–301.

Scott, A. Thompson, E. and Parthasarathy, S. 2006. Moore's law: The future of Silicon microelectronics. *Materials today*, 9(6): 20–25.

Yu, B. Chang, L. et al. 2000. FinFET scaling to 10 nm gate length. *IEEE Transactions on Electron Devices*. 67: 1–4.

Communication and Computing Systems – Prasad et al. (Eds)
© 2017 Taylor & Francis Group, London, ISBN 978-1-138-02952-1

Analysis of optimal performance of PV cell due to variation of irradiance and temperature

Krishna Kumar Bhargav & Vijay Kumar Garg
Department of Electrical Engineering, UIET, Kurukshetra University, Kurukshetra, India

ABSTRACT: The objectives of this paper is to study the dynamic behaviour of the characteristics curve of the PV cell under the consideration of insolation other than uniform type due to variation of irradiance and temperature. This non-uniformity is due to the fact that the position of Sun with respect to certain land mass on earth continuously changes and others due to the environmental conditions. One can say that the overall performance of solar cell primarily depends upon irradiance and temperature. In support of such we have mathematical frame for the electrical physicality of different concern with the phenomenon of same. At the end of the paper we have MATLAB/SIMULINK output curve which show the relative change in the position of knee point of characteristics curve due to the variation of irradiance. According to result, it being found that the irradiance dependence current decreases with decreasing of irradiance.

Keywords: photovoltaic cell, Open-Circuit Voltage (OCV), Standard Test Condition (STC), Short-Circuit Current (SCC), irradiance, temperature, simulation, Maximum Output Power (MOP)

1 INTRODUCTION

During, the last few years, there has been a trend of world energy demand, which is integral part of motivation a lot as investment in alternative energy solution in order to improve energy efficiency and power quality issues. Solar energy has is the most popular renewable energy source is extracted power directly from the Sun using Photovoltaic (PV) cell. The term 'photo' means light (so deduced form of the word photon first coined by G.N. Lewis in 1926) and 'voltaic' means voltage. A Photovoltaic (PV) cell, also known as 'solar cell' is an opto-electronic device which generates electricity when light incident on it. While the Photovoltaic effect was first observed in 1839 by the French scientist "Edmond Becquerel", it was not fully understandable until the development of quantum theory of light and solid state physics in early to middle 1990s. Photovoltaics system is one of the world's fastest growing power generation technologies, which have remarkable applications in the several field, such as solar farms, Distributed Generation (DG) and Micro grids (or μ-grid). All of the different electrical physicality's in concerned with cell like voltage, current, power etc. are the functions of some kind over the phenomenon of light falling and some other of it, as domain. Which can be mathematically expressed in generalized form for all of it as, $E_i = f_i(x,y,...,z,T,\alpha)$ where,

the variables $(x, y, ..., z)$ regarded as independent variables of secondary types but the variables (T,α) here regarded as independent variable of primary types, as they determines the strength of the light falling on it. Thus, we can say that the performance of photovoltaic system is affected primarily by the irradiance and temperature. In order to come with the problem of such, Maximum Power Point Tracking (MPPT) technique must need to be employed whose sole objective is to extract maximum power for the PV panel for the orientation of corresponding. The main reasons for the occurrence of this problem is that the irradiance falling on the surface from the Sun is not constant all the time due to relative rotations and due to fluctuation of environmental physicality's as well. As, a consequence of this I-V curve so achieve changes accordingly, here the role of MPPT starts which tracks the irradiance to meet the demand of knee point in I-V curve efficiently.

2 MATHEMATICAL CALCULATION AND OPERATION

The I-V characteristics curve of PV cell influence the mode of operation of the inverter and the controllers as well so introduce in it. Therefore, it is essential to review the model of the PV cell so designed to obtain the respective I-V and P-V characteristics curves, optimally. When sunlight strikes on a PV cell, the quanta of the observed sunlight, having certain energy, $e = hv$ where h is the Planck's constant having value 6.626×10^{-34} $j\text{-}s$, displace the electrons from the atoms of the cell which have some kinetic in it, which is characterized by the equation, $K.E. = hv - \varphi$ such that $K.E. \geq 0$. The physical process in which a PV cell coverts Sunlight into electricity is known as the "Photovoltaic effect The equivalent circuit of a PV cell is shown in Fig. (1), which is the single diode based model which have a dc current source. It also consists of a diode, a shunt resistance and a resistance in series to overall, so that the current flow in the external circuit is enabled. This effect of photovoltaic, causes the photocurrent, $i.e.$ I_{ph}, to generated by the source which cause the diode to be in forward biased condition, as consequence of which the diode current, $i.e.$ I_d, sets up in parallel with the shunt current I_{sh}, only if, the pathway resistance, $i.e.$ R_{sh}, which may also known as, Parasitic resistance of the cell, holds the value of non-infinitum, other than, these currents through branches of different, there is, the resultant outcome current I_L through series resistance R_s of the cell to the circuit of external.

Now, on applying KCL in the PV circuit cell of Fig. (1), we have,

$$I_{ph} = I_d + I_{sh} + I_L \tag{1}$$

$$where, I_d = I_o\left(e^{\frac{qV_d}{\gamma kT}} - 1\right) \tag{2}$$

Using eq_n (2) on eq_n (1), we get,

$$I_L = I_{ph} - I_o\left(e^{\frac{qV_d}{\gamma kT}} - 1\right) - \left(\frac{V_L + I_L R_s}{R_{sh}}\right) \tag{3}$$

The eq_n (3) so obtained represents the load current in PV cell irrespective of photo current (I_{ph}) and reverse saturation current (I_o) as function of temperature, but it does depend on it, which are given below as,

$$I_0(T) = I_0\left(\frac{T}{T_{nom}}\right)^3 exp.\left[\left(\frac{T}{T_{nom}} - 1\right)\frac{E_g}{\gamma V_t}\right] \tag{4}$$

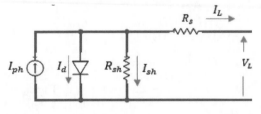

Figure 1. PV cell equivalent circuit.

$$I_{ph}(T) = \left[I_{sc} + K_i(T - 298)\right] \cdot \frac{\alpha}{1000} \tag{5}$$

Again, on applying KVL, we have the required relation for voltage drop across the diode as, $V_d = V_L + R_s I_L$. Using this voltage drop equation on eq_n (1) with the consideration of shunt resistance open-circuited.

$$\frac{I_{ph} - I_L}{I_o} + 1 = e^{q(V_L + R_s I_L)/\gamma kT} \tag{6}$$

Taking algorithm both side of the eq_n (6) which further yields to give the load voltage across the PV circuit as,

$$V_L = \frac{\gamma kT}{q}\ln\left(\frac{I_{ph} - I_L}{I_o} + 1\right) - R_s I_L \tag{7}$$

In case of open load condition $(i.e. I_L = 0)$, we have open circuit voltage relation as,

$$V_{oc} = \frac{\gamma kT}{q}\ln\left(\frac{I_{ph}}{I_o} + 1\right) \tag{8}$$

where,

I_d	Diode current
V_d	Diode voltage
V_{oc}	Open circuit voltage
I_{sc}	Short circuit current
R_{sh}	Shunt resistance
γ	Cascading constant
K	Boltzmann's constant
T	Absolute temperature
q	Charge on an electron
T_{nom}	Nominal temperature
E_g	Band gap energy of the semiconductor
V_t	Thermal voltage
K_i	S.C. current temperature coeff. of PV cell
α	Solar irradiance

The I-V & P-V characteristics curve of PV cell is not linear, due to vary with the level of solar irradiation and temperature, which make the extraction of maximum power a complex task, under the consideration of load variations. It is to be noted that while the I_{sh} is proportional to the insolation V_{oc} does not vary appreciably with it. In fact, for higher level of radiation V_m lies between 0.7 and 0.8 of V_{oc} and I_m is almost equal to 0.8 I_{sc}. The I-V and P-V characteristics of PV cell are shown in Fig. (2). Generally, particular finite data of PV cells/modules at standard test condition the time of testing are provided by the manufacturers, here, the PV cell so design for this paper have following parameters given in Table (1). From eq_n (3), the parameters I_{ph}, I_0, R_s, R_{sh} and γ are needed to

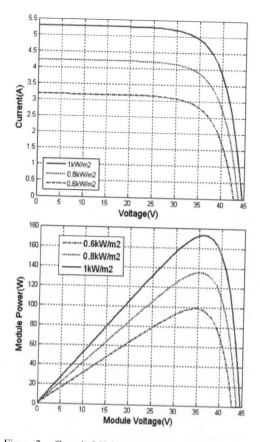

Figure 2. Generic I-V (up) & P-V (down) characteristics curve of solar cell for various irradiance.

Table 1. PV array components parameters.

	PV array parameters	
1	Open circuit voltage	$V_{oc} = 22.2$ V
2	Short circuit current	$I_{sc} = 5.45$ A
3	Voltage at MPP	$V_m = 17.2$ V
4	Current at MPP	$I_m = 4.95$ A
5	Power at MPP	$P_m = 85.14$ W

be determined. Based on the parametric condition of such I-V & P-V characteristics curve for the phenomenon of same is shown in Fig. (2).

Based on the known working points under standard conditions, the following three I-V expressions can be directly obtained for the cases of three, which are, SCC, OCV and MOP are respectively given by the eq_n's (9), (10) and (11) so deduce from the main eq_n (3).

$$I_{ph} - I_0\left[exp\left(\frac{I_{sc}R_s}{\gamma V_t}\right) - 1\right] - \frac{I_{sc}R_s}{R_{sh}} - I_{sc} = 0 \quad (9)$$

$$I_{ph} - I_0\left[exp\left(\frac{V_{oc}}{\gamma V_t}\right) - 1\right] - \frac{V_{oc}}{R_{sh}} = 0 \quad (10)$$

$$I_{ph} - I_0\left[exp\left(\frac{V_m + I_m R_s}{\gamma V_t}\right) - 1\right] - \frac{V_m + I_m R_s}{R_{sh}} - I_m = 0 \quad (11)$$

where $V_t = KT/q$ and V_t is the value at cell temperature $T = 25°C$. The reverse saturation current I_0 is very low, usually 10^{-5} to 10^{-6} times more than I_{sc}, so the equation on short circuit point in eq_n (9) can be simplified as,

$$I_{ph} = I_{sc}\left(1 + R_s / R_{sh}\right) \quad (12)$$

The output power of photovoltaic cells or module is,

$$P = VI_{ph} - VI_0\left[exp\left(\frac{V + IR_s}{\gamma V_t}\right) - 1\right] - \frac{V^2 + PR_s}{R_{sh}} \quad (13)$$

For the maximal value of power i.e. at the maximum power point, the value of power differential equation is zero, that means $dp/dV = 0$, due to the individual peak characteristics of a photovoltaic cell or module shown in Fig. (2). Equation (11) permits to state as follow, where $V = V_m$ and $I = I_m$.

$$\frac{I_0 R_{sh} e^{\frac{I_m R_s + V_m}{\gamma V_t}}\left[I_m V_m R_s - V_m\left(V_m + \gamma V_t\right)\right] + V_m \gamma V_t\left(I_0 R_{sh} + I_{ph} R_{sh} - 2V_m\right)}{V_m\left[I_0 R_{sh} R_s e^{\frac{I_m R_s + V_m}{\gamma V_t}} + \gamma V_t\left(R_s + R_{sh}\right)\right]} = 0 \quad (14)$$

Again, derivative of the power with respect to the voltage is:

$$dP/dV = d(VI)/dV = V.dI/dV + I \quad (15)$$

The derivative of eq_n (3) with respect to the V is

$$\frac{dI}{dV} = -\frac{I_0 R_{sh} exp\left(\frac{V + IR_s}{\gamma V_t}\right) + \gamma V_t}{\gamma V_t\left(R_{sh} + R_s\right) + I_0 R_s R_{sh} exp\left(\frac{V + IR_s}{\gamma V_t}\right)} \quad (16)$$

So, the equation describes the point of maximal can be evaluated by substituting eq_n (16) into eq_n (15).

$$\frac{-I_0 R_{sh} e^{\frac{V_m + I_m R_s}{\gamma V_t}} - \gamma V_t}{\gamma V_t\left(R_s + R_{sh}\right) + I_0 R_{sh} R_s e^{e\frac{V_m + I_m R_s}{\gamma V_t}}} + \frac{I_m}{V_m} = 0 \quad (17)$$

Thus, the five parameters can be extracted by resolving the eq_n (10), (12), (14) & (17) simultaneously, only based on four values in official datasheet such as I_{sc}, V_{oc}, I_m and V_m.

3 MPPT TECHNIQUE

In general, solar panel converts only 30 to 40% of the incident solar radiation into the electrical energy. In order to achieve the maximal efficiency of such MPPT technique as introduce in the earlier section is employed. From the Maximum Power Transfer Theorem (MPTT), we know that the output power of a circuit is maximum only when the Thevenin impedance of the circuit matches with that of its load impedance. In similar way, the objective of the solar irradiance tracking is to meet the point of power maximality in MPPT technique, which must need to be done in such a way that it also matches the impedance of corresponding as stated in MPTT technique. Thus, one can loosely say that MPPT technique based on the problem of impedance matching. For the purpose of such (i.e. matching) in PV grid the element which must need to be introduce is the boost converter at the input side which may regard as Transformer of step-up type for the tracking of optimal one, so that the voltage at the output side get enhanced which can be employed for the application of different types like to drive the motor as load, for lightning etc. By changing the duty cycle of such converter appropriately we can able to match the intrinsic (or Thevenin) impedance with that of the load impedance. If we have variable input, for instance, solar irradiance, the current and voltage will be found to vary correspondingly as shown in Fig. (2). Where, the output power (i.e. simply the product of V & I) is zero at V_{oc} (because $I = 0$) and zero again at I_{sc} (because $V = 0$). In between these two crispy points it rises and then falls, so that there is one point at which the cell delivers maximum power.

4 EVALUATION OF THE PARAMETERS AT DIFFERENT CONDITIONS

The constant output through solar cell is affected as the conversion to electricity of radiation falling on a module not only depends on its tilt but also the intensity of the insolation and its variation over the day. Another factor that affects the module or array output is the temperature fluctuation.

4.1 Photo-generated current I_{ph}

When solar cells exposed to sunlight, then photon from sunlight transfer their energy to some of the electrons in the materials which promoted to the energy of higher level (i.e. conduction band) and it leaves behind holes in the valence band, thus two charge carriers (i.e. electron-hole pair) are generated. The holes, meanwhile, will flow in the opposite direction through the material until they reach another metallic contacts on the bottom of the cell, where they are then 'filled' by electrons entering from the other half of the external circuit. One can loosely speaks that, the photo-current I_{ph} is simply the algebraic sum of electron-hole pair current, i.e. $I_{ph} = I_e + I_h$.

4.2 Revers saturation current I_o

The revers saturation current of diode is generated by thermal excitation which represents the recombination or leakage of minority carriers, which are temperature dependence. The reverse saturation current is expressed approximately by eq_n (4).

4.3 Equivalent shunt and series resistance R_{sh}, R_s

According to practical data from National Institute of Standard and Technology (NIST), the equivalent shunt resistance R_{sh} increases as the absorbed radiation is decreases. Whereas, the series resistance R_s increases with temperature and decreases slightly with solar irradiance.

4.4 The effect of solar radiation

The intensity in turn, depends on the latitude and the atmospheric conditions of the site. Beyond earth's atmosphere (in space), the solar radiation is approximately 1350 w/m². While passing through the atmosphere, the radiation losses some of its power and reduce to 1000 w/m² on the horizontal surface on the earth. This peak value reduces to zero during morning and evening hours and is highest at solar noon when the light has to travel through a thinner atmosphere. The typical variation of incident solar radiation an a horizontal placed surface on earth is in Fig. (3).

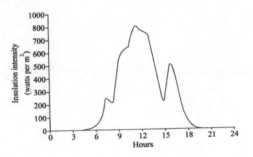

Figure 3. Daily insolation variation on a unit area horizontal surface.

In this case, the Sun rises at about 5 hours (5 A.M.), while the intensity rises from zero to a peak of more than 800 w/m² at noon, then it reduces to zero again at 20 hours in the evening. On integration the area under the curves gives the energy received on the surface from dawn to dusk.

4.5 *The effect of temperature*

The factor which is strongly affect output through the module under the variation of temperature. The data given by the manufacturers are at the Standard Test Condition (STC) but the actual field operation differs in many respects. The ambient temperature, under the Sun, in India varies from almost sub-zero temperatures at higher altitudes. The temperature inside the module is still higher because there is no heat dissipation from the cell summarized there. The Cell Operating Temperature (COT) is normally taken to be 25°C about the ambient temperature.

$$\therefore COT = Ambient\ temp. + 25°C$$

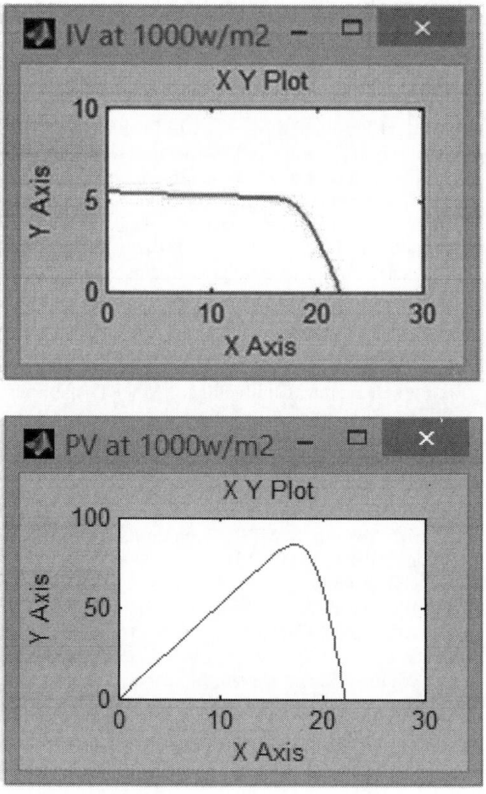

5 RESULT

MATLAB/SIMULINK implementation of PV array with the variation of irradiance along with different parameters. They are, short-circuit current I_{sc}, open-circuit voltage V_{oc}, current I_m and voltage V_m at the Maximum Power Point (MPP), R_s, R_{sh} and γ are the five parameters are also responsible to non-uniform power at maximal point. From Fig. (4), the I-V and P-V characteristics are shown at constant irradiance 1000 W/m². In this situation short circuit current (I_{sc}) is staying almost constant. We know that due to metrological parameters, temperature not constant all day but changes considerably. Thus the results confirm the non-linear characteristics (I-V & P-V) of P-V array. The Simulink results obtained at different irradiance, which are at 1000 w/m², 800 w/m², 600 w/m², 400 w/m² shown in Fig. (4), (5), (6) and (7) respectively, from which it being conclude that as the amount of irradiance available to the solar cell decreases down the performance of cell also drops down which is here shown by the dipping down of its I-V & P-V characteristics curve.

The solar irradiance as a result of the natural changes keeps on changeable but control-mechanism is available that can track this change

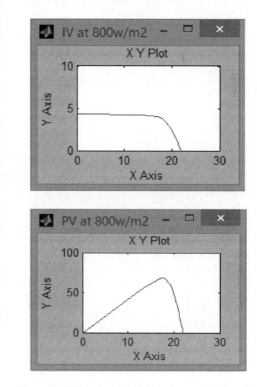

Figure 4.   I-V (up) and P-V (down) curve of PV cell at irradiance of 1000 w/m².

Figure 5.   I-V (up) and P-V (down) curve of PV cell at irradiance of 800 w/m².

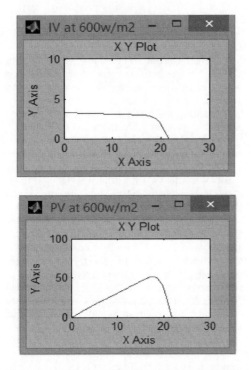

Figure 6.   I-V (up) and P-V (down) curve of PV cell at irradiance of 600 w/m^2.

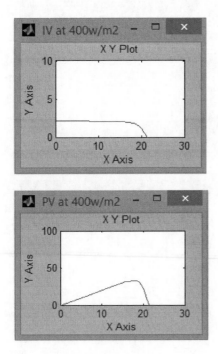

Figure 7.   I-V (up) and P-V (down) curve of PV cell at irradiance of 400 w/m^2.

and can alter the working of solar cell to meet the required load demands.

If solar irradiation is higher, obviously solar input will be high to the solar cell and hence power magnitude would be increase for the same voltage value. From the output curve (I-V) express that the value of short circuit current I_{sc} is decreased different but the value of open circuit voltage is slight different. The P-V curve of Fig. [(4), (5), (6) and (7)] the point of maximal point is varying with variation different irradiation. irradiance of 400 w/m^2. That dictates that if optimal value of power is extracted, according at constant and higher value of irradiation is there.

6   CONCLUSIONS

This paper explains the analysis of giving the best output with the PV panel. The metrological factor which is discussed is very important because they affect the overall performance of the PV panel. If such factors are not considered in installation, it may extremely reduce the power output thus the result loss of capital invested. So suitable techniques should be developed to overcome these problems.

REFERENCES

Arjyadharal Pradhan, S.M. Ali, Chitralekha Jena, 2013 "Analys is of Solar PV cell Perfomance with Changing Irradiance and Temperature", IJECS, pg. 214–220.
Fezzani F, Mahammed Hadi. I, and Said. S, 2014 "MTA-LAB-Based Modeling of Shading Effects in Photovoltaic arrays", STA IEEE, pg. 781–787.
G Michaela. Farr and Joshua S. Stein, 2014 "Spatial Variation in Temperature across a Phptovoltaic Array, IEEE, pg. 1921–1927.
Ioan Viorel BANU, Razwan B, ISTRATE M, 2013 "Comparative Analysis of the Perturb-and-Observe and Incremental Conductance MMPT methods" ATEE, IEEE, pg. 1–4.
Kumar Subhash, Kaur Tarlochan, 2014 "Solar PV Performance-Issues and Challenges", IJIREEICE, pg. 2168–2172.
Moacyr A.G. de Brito, Leonardo P Sampaio, Luigi G. Jr., Guilherme A.e Melo, Carlos A. Canesin, 2011 "Comparative Analysis of MPPT Techniques for PV applications", IEEE, pg. 99–104.
Moballegh. S, Jiang. G, 2011 "Partial Shading Modeling of Photovoltaic System with Experimental Validations", IEEE.
Ovono Aslain Zue, Chandra Ambrish, 2006 "Simulation and Stability Analysis of a 100 KW Grid Connected LCL Photovoltaic Inverter for Industry", IEEE, pg. 1–6.
Shen Ping, Chen Qianhong, Xu Ligang, 2015 "Electrical Characteristics Prediction of Microsatellite Photovoltaic Subsystem in Orbit*", IEEE, pg. 3287–3296.
Shihong Qin, Min Wang, Teng Chen, Xiangling Yao, 2011. "Comparative analysis of Incremental Conductance and Perturb-and-Observation Methods to Implement MPPT in Photovoltaic System", IEEE, pg. 5792–5795.

Communication and Computing Systems – Prasad et al. (Eds)
© 2017 Taylor & Francis Group, London, ISBN 978-1-138-02952-1

Design and analysis of low power 32 bit full adder using CNTFET for DSP applications

Vivek Kumar & Priyanka Goyal
Gautam Buddha University, Greater Noida, India

ABSTRACT: Low power is a changing trend in VLSI, so our main focus is to minimize the power consumption and delay up to a possible extent by implementing 32 bit full adder using CNTFET (Carbon Nano Field Effect Transistor) using cadence tool in 180 nm, 90 nm and 45 nm technology. Firstly, 1 bit full adder using above mentioned gates is implemented and afterwards same is done for 32 bit full adder. Various parameters like layout area, power, power product delay and delay were compared with other designs which are already existing like TGA (Transmission Gate adder), CPL (Complementary pass transistor logic), FA_Hybrid, HPSC, TFA (Transmission Full Adder), C-CMOS etc. When CNTFET as strong transmission gates gets coupled with other CNTFET which acts as weak CMOS inverters, we get very efficient value for delay and power consumption as when power supply is 1.8V, at 180 nm technology, the power consumption and delay was recorded very low. Buffers and capacitors are used after every third stage in 32 bit full adder which gives low delay and also low power consumption. Finally, this 32 bit full adder provides better and significant values for power and delay when compared with all other existing circuits for adder which were mentioned above. Also we find this circuit can be used in ALU, microprocessor, and where we need fast computation and calculation due to the very low delay and low power consumption.

1 INTRODUCTION

Adder is a basic unit and building block for microprocessor and digital signal processing, for address generation in cache memory, floating point unit, ALU and especially where arithmetic operations are needed. As adder simply added two binary bits which is used in many operations like subtraction, division, multiplication etc. So, 1 bit full adder is so crucial for all over system performance, i.e. if performance of 1 bit full adder is anyhow made to get better, then the performance of whole system will going to be improved and we will get fast computation with low delay and low power consumption. Our main focus is to reduce above mentioned parameters by changing its design and by using CNTFET and Comparing with existing 1 bit full adder. Firstly, we will enhance the performance of 1 bit full adder and later on the same can be implemented for 32 bit full adder. Basically, we always want the output of any circuit to be very low Firstly, low power is a changing pattern in VLSI as increased process parameter variability due to aggressive scaling has created problems in yield, reliability and testing (K. Navi et al. 2009).

Secondly, in nanometer technology power has become the most important issue because of increasing transistor counts, higher speed of operations, greater device leakage current. Also packaging and cooling cost is dependent on power dissipation of chip as contemporary high performance processors consume heavy power. Increasing power density of VLSI chips and increased customer demand for hand-held, battery operated devices such as cell phones, PDA, palmtop, laptop etc. are the main reason for requirement of low power. In this paper, we will focus on 32 bit full adder using CNFET for low delay, power, area (A. Khatir et al. 2011) etc. Due to scaling the channel length is continuously decreasing which leads to short-channel effects in nanoscale so to get rid of these limitations, many other alternate devices are proposed. There are wide range of existing technologies for full bit adder which are superior to other, but in some context, they may have some drawbacks also. As in existing designs, the most important logic designs are CPL (complementary pass-transistor logic), CMOS (Standard static complementary metal–oxide–semiconductor), dynamic CMOS logic, TGA (transmission gate full adder) and C-CMOS. There were some other adders also which were made through hybridization of different existing logic designs called hybrid logic design (Chiou-Kou Tung et al. 2007). These hybrid logic circuits enhance the performance of all over circuit, but every style has some pros and cons, so hybrid logic design also does.

The CMOS (standard complementary style) has 28 transistors which have good robustness against transistor sizing and voltage scaling but it needs buffers and also have high input capacitance. Another design like mirror adder also having 28 transistors and same power consumption having much smaller carry propagation delay with respect to CMOS standard full adder. On the other side, CPL having 32 transistors shows better voltage swing restoration, high switching activity, static inverters, but due to increased number of transistor, area gets increased so this logic design can't be used for low power applications. Also major drawback of CPL logic design is voltage degradation which was better in TGA logic design. In TGA number of transistors used are 20 so on-chip area gets reduced as compared to above mentioned logic design. In CPL, as transistors count is large therefore slow speed and also more consumption of power are observed because large on-chip area is directly proportional to high power consumption.

Later on, hybrid adder comes into action as these adders were made of mixing various logic designs. These adders reported to have much better performance like low delay and low power consumption, less area etc. 10T full adder logic (Jin-Fa-Lin 2007) employs only 10 transistors providing less area which finally gives high speed computation but it also has threshold-loss problem. Another logic design namely HPSC circuit, pass transistor simultaneously produced XOR and XNOR functions having only 6 transistors (M. Vesterbacka 1999). Leakage power gets reduced due to decreased transistor count. Although hybrid circuits are better in some context but sometimes when used in cascaded mode, some issues like poor driving capability and degradation in performance occurred.

The parameters like power, delay, PDP etc. were compared with existing designs. The circuit was implemented in cadence virtuoso tool using 180 nm, 90 nm and 45 nm library. By implementing CNTFET's which acts as strong transmission gates with other CNTFET's which acts as weak CMOS inverters. The average power consumption of 1 bit full adder get low for the proposed circuit at 1.8 V supply in 180-nm technology. For 90-nm technology at 1.2-V power supply, the corresponding value gets finer. The design was proved to be better for 32 bit CNTFET full adder.

2 DESIGN OF THE PROPOSED FULL ADDER

The proposed full adder circuit is represented by three modules. Modules 1, 2 are the XNOR

modules that generates the signal 'sum' and module 3 generates the output carry signal 'Cout' as shown in Fig. 1. Each module is evaluated individually such that the entire adder circuit can be optimized in terms of power, delay, and area and finally same can be implemented to 32 bit CNTFET full adder. These modules are discussed below in detail.

2.1 Modified XNOR module

This module drains maximum power consumption from entire circuit. So, this module is designed in such a way that power and delay gets reduced to great extent and minimize the voltage degradation possibly. This can be achieved by using CNFET's in which width of channel being small formed by MP1 and MN1. CNTFET's MP3 and MN3 provides level restoration because output of MP1 and MN2 gets full swing. There are various existing topologies for XOR/XNOR modules. The XOR/XNOR in (J.-M. Wang et al. 1994), (S. Goel et al. 2003) uses only four transistors but at low logic swing. On the other hand, XOR/XNOR uses six transistors, therefore better logic swing was obtained as compared to above design. In this paper, XNOR module also uses six transistors, but due to different arrangement of transistors, parameters like power consumption, high speed, delay gets better and also better logic swing is obtained. Although output is same for all types of

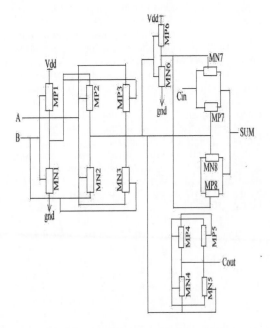

Figure 1. Proposed schematic of 1 bit full adder using CNTFET.

adders, difference occurs due to the arrangement of transistors in the circuit.

3.2 Modified carry generation module

This module can be implemented by MP7, MP8, MN7 and MN8 as shown in Fig. as 'cin' (input carry signal) propagates only through MN7 and MP7, which means through single transmission gates, thus carry propagation path reduces significantly. Also use of the CNTFET's guaranteed further reduction in carry propagation delay (Y. M. Lin et al. 2005).

3 OPERATION OF THE PROPOSED FULL ADDER

Fig. 1 shows the 1 bit full adder using CNTFET in which XNOR modules will give the 'sum'. MP1 and MN1 forms an inverter which generates B', which is used to design the controlled inverter by using MP2 and MN2. Output of controlled inverter is A XNOR B, but some voltage degradation problem arises in it, so to overcome this problem, two CNFET's MP3 and MN3 are used which acted as pass transistors. Insertion of buffers after every third stage of adder is required so that output may not get distorted and also capacitors are required so glitches occurring in output waves may be removed completely. Further, MP4, MP5, MP6, MN4, MN5, MN6 realizes the second stage of XNOR module for complete sum function. By analyzing the truth table of full adder, the expression for cout may be deducted as:

If, $A = B$, then $Cout = B$; else, $Cout = Cin$. Finally for 32 bit, as propagation delay rises due to the large transistor count, so buffer is needed at every third or fourth stage thus, all parameters gets better because of improvement in 1 bit full adder.

4 SIMULATION RESULT AND ANALYSIS

The simulation of 32 bit full adder using CNTFET was carried out using both 180 nm, 90 nm and 45 nm technology and compared with the other existing adder designs reported in [1]–[10]. Our main focus is to optimize delay, power of the circuit and the Power-Delay Product (PDP), i.e., the power consumption has been optimized and reduced to a great extent in the proposed case. It was observed that in this design, the power consumption was minimized by sizing the CNTFET of inverter circuits, while the carry propagation delay can be improved by sizing CNTFET present between Cin to $Cout$. Power consumption, propagation delay, PDP of the 32 bit CNTFET full adder

along with the existing full adders are given in Table 1 for 180 nm, 90 nm and 45 nm technology.

The analysis of 32 bit CNTFET full adder is done in terms of carry propagation delay and power consumption in 180 nm, 90 nm and 45 nm technology,

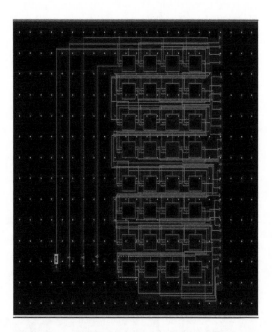

Figure 2. Schematic of 32 bit CNTFET full adder.

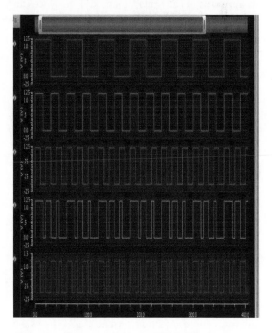

Figure 3. Output of 32 bit CNTFET full adder.

Table 1. Parameters of proposed 32 bit CNTFET full ADDER.

CNTFET	180 nm	90 nm	45 nm
Delay ($\times 10^{-9}$)	4.72	1.767	1.18
Power ($\times 10^{-6}$)	92.33	47.81	29.94
PDP ($\times 10^{-15}$)	435.79	84.48	35.32

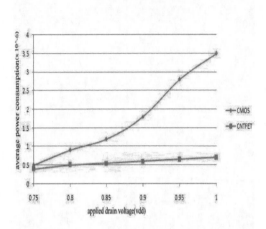

Figure 4. Average power consumed by Cmos and CNT-FET for different Vdd.

i.e., varied power supply. As delay and power consumption gets better so PDP (power product delay) automatically gets enhanced. Finally we can see that using CNTFET, parameters like power, delay and PDP gets better when compared with the CMOS transmission gates as shown in Fig. 4.

5 PERFORMANCE OF 32 BIT FULL ADDER USING CNTFET

A 32-bit CNTFET full adder is implemented by using 1 bit full adder in cascaded mode using capacitors and buffers. It was observed that while using buffer at every third stage (which is experimentally obtained = 3), the parameters like delay and power consumption gets improved. But when no buffer was used, glitches and high propagation delay are obtained. The evaluation of performance of 32-bit adder was carried out in 180 nm, 90 nm and 45 nm. The performance of the 32-bit full adder at 100 MHz in 180 nm, 90 nm and 45 nm technology is carried out. There is a big difference in static and dynamic power consumption in 180 nm technology. But this difference was reduced when implemented in 90 nm due to increase in gate leakage and the sub threshold conduction current. Fig. 3 shows the output of 32 bit full adder using CNTFET. It is

observed that the propagation delay gets increased proportionally with respect to the 1-bit full adder. The simulation was carried out in Cadence Virtuoso tools with 180 nm, 90 nm and 45 nm technology and then compared with other existing logic design like CMOS, TFA, CPL, TGA, and other hybrid logic designs. The simulation results shows that the 32 bit full adder using CNTFET's gives improved PDP as compared with other PDP's of existing logic design. The coupling of CNTFET with weak CNTFET gives high switching speeds. Finally, some glitches are obtained which can be removed by using capacitances and buffers after every third stage of 32 bit CNTFET full adder.

6 CONCLUSION

The simulation was done using Cadence Virtuoso tools in 180 nm, 90 nm and 45 nm technology and then compared with other existing logic design like CPL, CMOS, TFA, TGA[1]–[10] and other hybrid logic designs. The 32 bit full adder using CNTFET provides better and improved PDP as compared with other existing full adder, due to enhancement of these parameters in 1 bit CNTFET full adder. The coupling of CNTFET which acts as strong transmission gates with other CNTFET which acts as a weak CMOS inverter gives fast and high switching speeds. Hence, 32 bit full adder using CNTFET gives efficient results for power consumption and delay and therefore improved PDP. In 90 nm technology, PDP improves when compared to 180 nm. Finally 32 bit adder is implemented in 45 nm and PDP value efficiently reduced. This 32 bit CNTFET adder is also used where high computation is needed with low power consumption and low delay. This technique found to be most promising for DSP applications.

REFERENCES

Chang C. H., Gu J. M., and Zhang M., "A review of 0.18-μm full adder performances for tree structured arithmetic circuits," IEEE Trans. Very Large Scale Integr. (VLSI) Syst., vol. 13, no. 6, pp. 686–695, Jun. 2005.

Chiou-Kou Tung; Yu-Cherng Hung; Shao-Hui Shieh; Guo-ShingHuang. 2009 "A Low-Power High-Speed Hybrid CMOS Full Adder for Embedded Systems" IEEE Design and Diagnostics of Electronic Circuits and Systems, pp. 1–4, Apr. 2007. K. Navi, M. Maeen, V. Foroutan, S. Timarchi, and O. Kavehei, "A novel low-power full-adder cell for low voltage," VLSI J. Integr., vol. 42, no. 4, pp. 457–467, Sep.

Cui Y., Zhong Z., Wang D., Wang W. and Lieber C. M., 2003 "High performance silicon nanowire field effect transistors," Nano Lett., Vol. 3, No. 2, pp. 149–152.

Goel S. Elgamel M., and Bayoumi M. A., "Novel design methodology for high-performance XOR-XNOR circuit design," in Proc. 16th Symp. Integr. Circuits Syst. Design (SBCCI), Sep., pp. 71–76, 2003.

Jin-Fa-Lin, Yin-Tsung Hwang, Ming-Hwa Sheu, and Cheng-Che Ho, 2007 "A Novel High-Speed and Energy Efficient 10-Transistor Full Adder Design", IEEE Trans.on Circuits and Systems I, vol. 54, no. 5, pp. 1050–1059.

Khatir A., Abdolahzadegan S. and Mahmoudi I. High speed multiple valued logic full adder using carbon nano tube field effect transistor. International Journal of VLSI Design & Communication Systems. 2(1):1–9, 2011.

Lin Y. M., Appenzeller J., Knoch J., and Avouris Ph., "High-Performance Carbon Nanotube Field-Effect Transistor with Tunable Polarities," IEEE Trans. on Nanotech. Vol. 4, No. 5, pp. 481–489, 2005.

Shams A. M., Darwish T. K., and Bayoumi M. A., 2002 "Performance analysis of low-power 1-bit CMOS full adder cells," IEEE Trans. Very Large Scale Integr. (VLSI) Syst., vol. 10, no. 1, pp. 20–29.

Tung C.-K., Hung Y.-C., S.-H. Shieh, and Huang G.-S., 2007 "A low-power high-speed hybrid CMOS full adder for embedded system," in Proc. IEEE Conf. Design Diagnostics Electron. Circuits Syst., vol. 13, pp. 1–4.

Vesterbacka M., 1999 "A New six-transistor CMOS XOR Circuits with complementary output," Proc. 42nd Midwest Symp. on Circuits and Systems, Las Cruces, NM.

Wang J.-M., Fang S.-C., and Feng W.-S., 1994 "New efficient designs for XOR and XNOR functions on the transistor level," IEEE J. Solid-State Circuits, vol. 29, no. 7, pp. 780–786.

Zhang M., J. Gu, and Chang C.-H.,2003 "A novel hybrid pass logic withstatic CMOS output drive full-adder cell," in Proc. Int. Symp. Circuits Syst., pp. 317–320.

Communication and Computing Systems – Prasad et al. (Eds)
© 2017 Taylor & Francis Group, London, ISBN 978-1-138-02952-1

Secure architecture for data leakage detection and prevention in fog computing environment

Darakhshinda Parween & Arun Kumar Yadav
ITM University Gwalior, Madhya Pradesh, India

R.S. Rao
Indra Ghandhi National Tribal University, Amarkantak, MP, India

ABSTRACT: Dynamic environment needs on demand location independent computing services. So, cloud computing provides multiple servers for ubiquitous, convenient, and dynamic scalable resources over the internet. But security concern of stored data is important in cloud environment. Data owners need a security gateway because they are unaware where and how data are store in cloud. When a user wants to access information in run time environment, it might be happen that cloud did not provide information at run time. So cloud service provider need to communicate with other service provider which is geographically distributed. If other service provider is agree to provide their Service, then client access information in other cloud server. But storing or accessing sensitive data information via middleware, there is a chance to leaked and access data from unauthorized user. So, this paper presents the comparative analysis for data leakage detection and prevention techniques in cloud computing environment.

1 INTRODUCTION

Cloud computing is a form of computing that shares the computer assets rather than having local server or own devices to handle applications. It is also known as on demand computing. It gives advanced facilities like on demand, pay per use etc mell et al (2009). In cloud computing any information and computer resources from anywhere over the internet can be access with the help of cloud provider Bonomi et al (2012). The facility of sharing information via Internet and cloud computing heedlessly introduce a flourishing complication of data leakage. In the meanwhile, end-users are unfamiliar that their information was leak out or embezzled because many of data is leaked by operations operate in the background. In this implicit and universally distributed network, the mode of transfer sensitive data from the distributor to the trusted third parties ever occurs in modern world. It needs to the security and durability of service based on the demand of users Chaware et al (2012). The concept of modifying the data itself to encounter the leakage is not a new approach. Usually, the sensitive data are leaked by the agents, and the special agent is culpable for the leaked data should always be detected at an early stage. So, detection of data from the distributer to the agents is indispensable. Limitations with cloud computing.

1.1 *Limitation with cloud computing*

- Such that, when methods and procedure of IoT are receiving additional tangled in our life, modern Cloud computing can comparatively satisfies their necessities of mobility sustain locality alertness and low latency?
- Demand of high capacity client access link.
- Advantage of storage space assets, computation restriction and network announcement charge becomes a trouble for latency-sensitive applications.

1.2 *Introduction of fog computing*

We know that fog computing extend the model or pattern of the cloud computing Bonomi et al (2012). Fog computing is a virtualized stage that is commonly placed among final user devices and the cloud data centers hosted within the Internet careersplay et al (2013). Fog contributes information, compute, storage space, and appliance services to end-users as cloud provided. Fog computing combine's computational power, storage capability and network facility at the network edge. The encouragement of Fog computing lies in a sequence of real scenario, like Smart network, smart traffic lights in vehicular networks and software distinct networks Stojmenovice et al (2014).

Fog computing technique provides low down latency, locality awareness, and also improves quality of services (QoI) for streaming and actual time application Armbrust et al (2010). Fog can be different from cloud by its immediacy to end user, the geographical consumer and it's sustain for mobility, Atzori et al (2010), Gupta et al (2014).

But as technology growing more and more risk of data leakage and loss are increases. With the expansion of database production via internet, the data might be insecure after passing through the unsafe network. The customer may scruple to acquire the data facility for the following suspicion. First, the data recipient may expect that the data are mitigated with by unjustified user. Second, they may expect the data received are not formed and hand over by the recognized suppliers. Third, the suppliers and client in fact with unlike interest and should have diverse roles in the database organization. So how to validate and protected the data becomes very essential here. The growth of the internet result in contribution of a spacious scope of web-based application, like as database as a service, digital libraries, e-commerce, online resolution support system etc Ahmad et al (2013). In business, most of time sensitive data should hand over to probably loyal third parties. Likewise, a hospital may provide patient files to researchers who will have preference new treatments. Our objective is to distinguish when the distributor's sensitive data have been leak out through agents, and if feasible to distinguish the agent that leak out the data. We deal with applications where the imaginative sensitive information cannot be bothered. If medical analyst will be treating patients, they may require correct information for the patients. Generally, leakage is managed by watermarking, e.g., an exclusive code is fixed in each dispersed copy. If that information is exposed in the hands of an unrecognized party, the leaker can be analyzed. Watermarks can be extremely valuable in some cases, but it, engage some alteration of the imaginative data Waluni et al (2015). Furthermore, most of time watermarks can be damaged if the data receiver is mischievous.

1.3 Classification of data leakage

Unintentional leakage: It arises as a user unintentionally sends a privileged facts/information to the third party.

- Attached file.
- Zip and copy.
- Copy and pest.

Intentional leakage: It arises while a user attempt to send a privileged manuscript without conscious of company policy and lastly send anyhow.

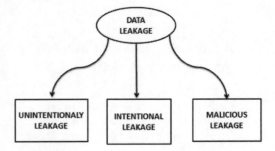

Figure 1. Different classification of data leakage.

- Document Rename.
- Document type change.
- Remove keyword.

Malicious Leakage: It happen at what time a user consciously annoying to slide the private data past the safety policy

- Character encoding.
- Hide data.
- Print screen.

2 LITERATURE SURVEY

Ahmad et al (2013) has proposed a Data Leakage algorithm to represents a major issue for organizations as the quantity of occurrences and the expense to those encountering them keep on expanding Information Leakage is upgraded by the way that dispatch information (both inbound and out-bound), as well as messages, texting, site structures, and record exchanges among others, are to a great extent unfettered and keep under observation on their way to their destinations. The prospective harm and unfavourable results of an information Leakage occurrence can be characterized into two classes: Direct and Indirect Losses. Direct misfortunes allude to substantial harm that is anything but difficult to quantify or to evaluate quantitatively. Aberrant misfortunes, then again, are much harder to evaluate and have a much more extensive effect as far as expense, place, and time. Direct misfortunes incorporate infringement of regulations, (for example, those ensuring client protection) bringing about fines, settlements or client pay expenses prosecution including claims loss of future deals expenses of examination and medicinal or rebuilding charges. Purohit et al (2013) proposed a noteworthy security worries about distributed computing. The significant territories of center are:— Information defence, Virtual Desktop protection, Network protection, and Virtual protection. In today's industry world, numerous associations utilize Information Systems to deal with their touchy and business basic data. The requirement of ensure such a key segment of the association can't be over stressed. Information Loss/Leakage avoidance has been observed to be one of the successful methods for counteracting Data Loss. DLP arrangements distinguish and counteract unapproved endeavours to duplicate or send touchy information, both deliberately or/and unexpectedly, without approval, by individuals why should approved access the delicate data. DLP is intended to recognize potential information rupture occurrences in opportune way and this happens by checking information. Information outflow is an occurrence when the classification of data has been traded off. It alludes to an unapproved broadcast of information from inside of an association to an outer objective. The information that is spilled out can either be confidential in nature and are esteemed classified though Data Loss will be loss of information because of cancellation, system crash and so on.

Panagiotis Papadimitriou et al (2015) proposed an Information Leakage, and how it can affect an association. Since more types of correspondence are being used inside of associations, for example, Instant Messaging; VOIP; and so on, past conventional email, more boulevards for information Leakage have risen. Normal vectors will be investigated, both outer to the association and from inside. The talk will then address a percentage of the suggestions to associations, from lawful and consistence issues to operational issues. Having exhibited the dangers and their related dangers, the paper then analyzes a percentage of the identification and alleviations arrangements accessible. The extension for information Leakage is wide, and not restricted to simply email and web. We are very acquainted with stories of information misfortune from tablet robbery, programmer break-ins, move down tapes being lost or stolen, etc. In what manner would we be able to guard ourselves against the developing risk of information Leakage Attacks by means of informing, social designing, malignant programmers, and that's just the beginning? Numerous makers have items to diminish electronic information Leakage, yet don't address different vectors. It gives an all encompassing talk on information Leakage and its anticipation, and serve as a beginning stage for organizations in their battle against it. Leakage detection has been recommended by Panagiotis Papadimitriou; Hector Garcia Molina which can facilitate us to notice the guilty leaker exclusive of altering the probity of the novel data. Data leakage is the most significant security thread to the organization. In their paper Papadimitriou and Garcia presented a scheme for data leakage detection. In the scheme they address, a distributed distributes a sensitive data to around all agents according to an individual request that is issued for each one of the agent. CRM system is an example of such a scenario in which data owner which client collaborator to call and the client or collaborator facts are dispatched to the third party call agent. If sensitive data is leaked, the data provider would like to be capable to recognize the resource of leakage or slightly to approximate the prospect of each and every agent to have been occupied in the occurrence. Therefore, A guilt mould is projected for estimating the possibility that an agent is tangled in a given data leakage. So that, a data allocation technique that allot data record through the agent based on the agent's requests and optimization scheme are accessible. Papadimitriou and Garcia consider two types of data demand: Explicit request and simple request. An explicit demand holds predefine condition where as in simple request the amount of object to be arbitrarily selected from the whole dataset. Common requirements are not manipulated by the projected algorithms. There are various techniques which are used for data leakage detection some of these are as follows:

- Watermarking the data.
- Steganography.
- Data allocation technique.
- Fake object model or Guilty Agent.

Kumar et al (2014) proposed a technique which focuses on how leakage as well as leakager of dada is identified in cloud based environment and also centralized on how more present work uses the Bell-LaPadula to survey and design for secure computer systems. In his recommended technique the syllabary of "secure state" is explain and it is demonstrate that each state alteration conserve protection by transferring to other secure state, as a outcome inductively demonstrate that system convince security objective model. In Bell-LaPadula representation each subject S has a lattice of right and access rights demonstrated by it are reading down (NRU), writing up (NQD), simple security property, star property, read only, append only, execute only, read write etc.

The primary level of the representation described registration of client to server, after this server conserve database regarding the client which is called as server directory table and some fields of this table are client Id, SHA-512(hash) and (m, n). The secondary level is watermarking of confidential message this method is execute using 2 levels in first the place where the watermarking should be done is calculated:

Row positioning pixel, $m = I\,(1;1) + 2$
Column positioning pixel, $n = I\,(1;2) + 2$

The confidential message, encryption key K, AES-128 are used to calculate cipher text C, for placing of cipher C and authentication code M pixel point of image starting from (m, n) in actual

Table 1. Comparison table for different technique.

Objective	Method	Conclusion	Limitation
Preventing data leakage in distributive strategies by steganography technique	Steganography	Improve the distributer's chances of preventing a leakage	Long document that we want to hide it's hard
A distributed model for data leakage prevention	File distribution model	Distribution model can effectively detect the leak source so as to protect information security	Model cannot distinguish malicious user
Achieving secure and effective data collaboration in cloud computing	Two level HIBE (Hierarchical Identity Based Encryption)	Reducing the computation complexity, communication cost and storage	Achieving more secure data collaboration becoming key challenges in cloud computing
A mantrap inspired, user centric data leakage prevention approach	Kernel space mantrap DLP	It implement its own kernel model to decide which data sending process should be allowed blocked	Pausing alive processes extracting useful information from binary
Taint eraser: Protecting sensitive data leaks using application level taint tracking	Taint eraser	Prevent leak of sensitive data by commercial software	Sensitive data still passed along via shared memory. It improve but not break
Secure cloud computing with RC4 Encryption and attack detection mechanism	RC4 encryption technique	Secure the sharing environment by sharing the data by use of RC4 encryption and decryption mechanism	The roomer client for inter changing the information without losing any type of major security

record is exchanged by cipher C. To perceive client id alter procedure is petition like as first the assignment of cipher C and authentication code M in document is pointed out for this server uses table where the point (m, n) is stored.

3 PROPOSED ARCHITECTURE

After analysing different techniques of data leakage detection and prevention we conclude that many works are done in cloud environment and number of the recommended method present a lot to works in field of the data leakage detection and prevention, but there is a requirement of technique which will allow additional, effective and improved result. Thus proposed plane of future work involves a technique or architecture for data leakage detection and prevention which will work on cloud as well as fog environment In future we want to work on multi-tier security architecture in cloud as well as fog environment. Primarily our architecture has four layer or tier such as client layer, fog layer, cloud layer and middle ware. All clients are geographically distributed to each other. Many cloud service provider does not provide a secure resource/data at run time environment. To overcome the problem of resource allocation or secure data transformation at run time environment we proposed a multi-tier architecture. Different layer of our architecture explained section 3.1. Figure 2 show an expected

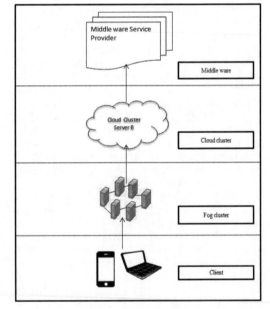

Figure 2. Expected Multi-tier architecture for data leakage detection and prevention.

multi-tier architecture for data leakage detection and prevention. Very ordinary threat today's in geographical distributed cloud frame is data leakage. It is examined as the top threats to user data in

766

cloud environment. Combination of cloud computing and fog computing are such a paradigm which helps in monitoring and identifying the unauthorized access of malicious user. So, it is important to provide secured Multi-tier architecture for data leakage detection and prevention in fog computing environment. As we know that Fog computing is well suited for geographical distributed cloud surroundings to provide better security.

3.1 Components of architecture

3.1.1 Clients
It is an entity. It can be either organization or individual customer, which access a service available by the server. Clients are geographically distributed to each other.

3.1.2 Fog cluster server
Collection of multiple dedicated servers attached with in a server which provide services at the end of the network. Fog cluster server commonly placed between final user devices and the cloud cluster server hosted within the Internet.

3.1.3 Cloud cluster server
It connects several dedicated servers which provide more cost effective ways to achieve performance and availability. It is a valid server that is built, hosted and delivered via a cloud computing paradigm across the Internet. It possesses and display comparable capabilities and functionality to a distinctive server but is accessed slightly from a cloud service contributor.

3.1.4 Middleware services provider
Middleware is generally used to communicate different service provider which are geographically distributed to provide a services. It provide common interface between one server to another server.

Fog computing is well suited for geographical distributed cloud surroundings to provide better security. So, it is important to provide secured Multi-tier technique for data leakage detection and prevention in fog computing environment. In future we ware introduce resource and security algorithm along with the architecture. It is hard to trust the third party service provider. This algorithm has two accepts such as providing a secure resources between two cloud servers which are geographically distributed and ensuring the high security in multiple level.

4 COMPARATIVE ANALYSIS WITH EXISTING APPROACHES

Conventionally, leakage detection is handled by distinguishable technique like as Watermarking, Steganography, file distributed model and so on. By surveying different techniques we conclude that Multi-tier architecture is very applicable as assimilate to the existing approaches. We can deliver secure resources to our customer and even detect if data gets leaked. Multi-tier security architecture is based on providing security in middle ware because most of time our sensitive data was leaked by middle ware or third party. Thus with the help of multi-tier architecture we deliver more secure resources as well as leakage detection.

5 CONCLUSION AND FUTURE SCOPE

This paper concludes the problem of comfortably detecting data leakage and guilty agents in a very large observation database collected by system. Sensitive information can be leaked by the agents mistakenly or maliciously and even if we had to provide sensitive data, in an impeccable world we can used the abstraction of Steganography so that we could add Fake objects. This paper also includes data leakage prevention technique which is handling by MyDLP and taint eraser technique. We have survey number of technique related to data leakage detection and prevention but Data leakage detection and prevention can be implementation more effectively including Multi-tier Security Technique in cloud as well as fog Environment. In future we enhanced the expected architecture and provide a client's authentication algorithm, resource allocation algorithm, security algorithm.

REFERENCES

Ambika Gupta and Dr. Arun Kumar Yadav, "Multi-dimensional Password Generation Algorithm at Multiple Authentication Level in Cloud Computing Environment" published on Second International Conference on Emerging Research in Computing, Information, Communication and Application, August 01–02, 2014.

Armbrust, M., A. Fox, R. Griffith, A.D. Joseph, R. Katz, A. Konwinski, G. Lee, D. Patterson, A. Rabkin, I. Stoica, and M. Zaharia, "A view of cloud computing," Common. ACM, vol. 53, no. 4, pp. 50–58, Apr 2010.

Atzori, L., A. Iera, and G. Morabito, "The internet of things: A survey," Computer. Netw, vol. 54, no. 15, pp. 2787–2805, Oct. 2010.

Bijayalaxmi Purobit, Pawan Singh, "Data leakage analysis on cloud computing", International Journal of Engineering Research and Application, vol. 3, issue 3, May–June 2013, pp. 1311–1316.

Bonomi, F., Connected vehicles, the internet of things, and fog computing. VANET 2011, 2012.

Bonomi, F., R. Milito, J. Zhu, and S. Addepalli, "Fog computing and its role in the Internet of Things," in ACM SIGCOMM Workshop on Mobile cloud Computing, Helsinki, Finland, 2012, pp. 13–16.

Ivan Stojmenovice, Sheng Wen, "The fog computing paradigm: Scenarios and security issue" in Federated Conference on Computer science and Information system vol 2, Issue 7–10 September 2014.

Ma Jun, Wang Zhiying, Ren Jiangchun, Wu Jiangjiang, Cheng Yong and Mei Songzhu, "The Application of Chinese Wall Policy in Data Leakage Prevention" in International Communication Systems and Network Technologies, 2012.

Mell, P. and T. Grance, "The NIST Definition of Cloud Computing," National Institute of Standards and Technology, vol. 53, no. 6, p. 50, 2009. [Online] Available: http://csrc.nist.gov/groups/SNS/cloud-computing/cloud-def v15.doc.

Miss, S.W. Ahmad, Dr G.R. "Bamnote Dept of Computer Sc. & Engg. Data Leakage Detection and Data Prevention using Algorithm" International Journal Of Computer Science And Applications Vol. 6, No. 2, Apr 2013 ISSN: 0974–1011 (Open Access) Available at: www.researchpublications.org

Neeraj Kumar, Vijay Katta, Himanshu Mishra and Hitendra. Garg, "Detection of Data Leakage in Cloud Computing Environment", in 6th International conference on Computational Intelligence and Communication Network, 2014.

Nikhil Chaware, Prachi Bapat, Rituja Kad, Archana Jadhav, S.M. Sangve "Data Leakage Detection" International Journal of Scientific Engineering and technology, vol 1, Issue 6, Dec 2016, pp. 272–273.

Panagiotis Papadimitriou, Hector Garcia Molina, Peter Gordon, "Data Leakage Detection of Guilty Agent" in International journal of Scientific and Engineering Research vol 3, issue 6, june 2015.

Priya Waluni, Priya Tadge, Navnath Kondalkar, Satish Mahamare, "Identification of Data Leakage and Detecting Guilty Agent Using Data Watcher" in International journal of Advanced Research in Computer Science and Software Engineering, vol 5, Issue 2, Feb 2015.

Shobana, V. and M. Shanmuga sundaram, "Data leakage detection using cloud computing," International Journal of Emerging Technology and Advanced Engineering, vol. 3, pp. 111–115, January 2013.

Supriya Singh, "Data Leakage Detection Using RSA Algorithm" in International Journal of Engineering and Management (IJAIEM), vol. 2, issue 5, May, 2013.

www.careersplay.com Seminar "On Data Leakage Detection" Submitted To: www.careersplay.com Submitted By: www.careersplay.com Published on December 16, 2013.

Communication and Computing Systems – Prasad et al.(Eds)
© 2017 Taylor & Francis Group, London, ISBN 978-1-138-02952-1

Gesture recognition: A technique to enhance human-computer interaction using MATLAB

Saurabh Kumar, Sanatan Jha & Vishal Bharti
CSE Department, Dronacharya College of Engineering, Gurgaon, India

ABSTRACT: The main aim of creating hand gesture recognition system is to make a natural interaction between human and computer which is well known as HCI (Human Computer Interaction); it is also known as Man-Machine Interaction (MMI). Human computer interaction moves forward in the field of sign language interpretation especially Indian Sign Language (ISL). As hand plays a very important role for communication for dumb and deaf persons. Gestures are powerful means of communication among humans.

1 INTRODUCTION

The role of computer in today's life has grown very fast in the society. The interaction of a person with the computing devices has been advanced in such a way that it becomes today's life necessity. The aim of HCI is to make a platform where interaction of a human with the computer is as natural as interaction between two humans. Though mouse and keyboard is a good way to HCI but our aim is to make this interaction in 3D. Using hand gesture as a device can help the human to communicate with computers in very effective way. Standard Sign Language (SI) is very helpful for dumb and deaf persons. Sign languages are gestural languages through which a car or robot can be controlled with remote or speech channels. Sign languages vary from country to country even within a country these sign languages it vary from region to region. For example we can say Indian Sign Language (ISL) is language which is used by Indian deaf and dumb community. Now a day's hand gesture is more feasible due to the latest advances in the field of computer and image processing. Now with the wide application of HCI (Human Computer Interaction) it becomes more occlusive for researchers to work on it especially for dumb and deaf persons. Stages: extraction method, feature extraction and classification [5][10].

2 METHOD OF HAND RECOGNITION

For hand gesture recognition first of all we have to decide that the gesture is dynamic or static. The hand gesture in motion or the moving hand is dynamic gesture and the hand gesture which is constant or pointing somewhere is static gesture.

Method of hand recognition can be classified into three steps. The steps are as following:

- Gesture Recognition (Detection).
- Feature Extraction (tracking).
- Classification (actual recognition Phase).

2.1 Extraction method

Extraction method is the first step towards the process of hand gesture recognition. As it is the actual detection of a hand gesture. From extraction we meant the segmentation. Segmentation is the process of dividing the detected image into a particular area separated by boundaries. It basically means that if we have to extract a hand from an image or a video then to divide the hand gesture. Segmentation process depends on the type of gesture if it is dynamic or static. The dynamic picture first has to locate and then tracked. If the input image is static then have segment directly. This process also depends on the skin color. For this process first the hand is to be locating using some bonding box and then the hand is to be tracked. For tracking mainly there are two ways, if we are doing segmentation from a video then the video is divided into

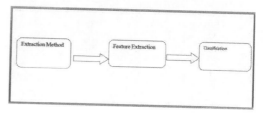

Figure 1. Gesture recognition system steps.

frames and each frame is treated individually as an image and then it is segmented and the second way is using some tracking algorithms like Kalman filter [8][10].

The segmentation of hand gesture is mainly depends on the color of the skin as it is easy to scale, translation and rotation change. Basically there are two model of hand. Skin pixel and non-skin pixel and there is some methods like parametric and non-parametric method for them. Gaussian Model (GM) and Gaussian Mixture Model (GMM) are parametric method and some Histogram based techniques are non-parametric.

In segmentation the problem of illusion and blurred segmentation is always there. There is a method through which some researchers have overcome. This problem can be overcome by using data glove which provide exact information of the color of skin, orientation of fingers and palm, centre of gravity of hand. It has made the task easier. Or for the good segmentation we can use infrared camera but that is quite expensive for us. Fig. 2 shows the segmented image of a hand.

2.2 *Feature extraction*

After the segmentation of hand model it has been noticed that shape is the important visual feature of the hand. The perfect feature extraction is totally based on the good segmentation. Feature extraction means feature vector of segmented image which can be extracted in different ways of application Researcher classification of feature extraction and some techniques based on contour and region based shape representation. Contour based shape representation and description methods are chain code, Polygon, B-spline, perimeter, compactness, Eccentry, shape signature, Hausdoff distance, Elastimatching.

Region based shape representation and description methods are Pseudo-Zernike Moments, Convex Hull, media axis, Euler number, and Geometric moments.

In hand recognition system, shape contour has given more priority than region so; contour based

methods are widely used. But for complex sign as in ISL, region base methods are more suitable. So while choosing features extraction, it must be take care that it should be invariant to translation, rotation and scale. As SL contains a large number of vocabulary so using only one type of technique is not sufficient. Other than contour and region we can utilize fingertips position, palm center created parameters as a feature vector [8][11]. In this we use Self Growing and Self-Organized Neural Gas (SGONG) neural algorithm to capture the shape of the hand. In feature extraction three features are extracted; palm region, palm center, and hand slope. Also the Center Of Gravity (COG) of the segmented hand and the distance from the COG to farthest point in the fingers and extract one binary signal to estimate the number of fingers in the hand region [10]. Dividing the segment image into different blocks size and the each block represent the brightness measurement in the image. Many experiments were applied to decide the rights block size that can achieve good recognition rate. Using Gaussian algorithm to extract geometric central moment as local and global features [1][4]. Fig. 3 shows the feature extraction of the sign languages.

2.3 *Classification*

A classifier and recognition method plays very important role in any gesture recognition system. After it the second step is pattern recognition and machine learning field. The pattern recognition is classified into two methods, either supervised or unsupervised classification. Various supervised classification methods are there such as neighborhood classification with Euclidean distance, Bay's classifier, neural network, linear regression model.

For unsupervised classification methods the algorithms we use are: clustering methods:

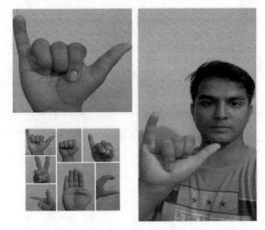

Figure 3. Feature extraction of sign languages.

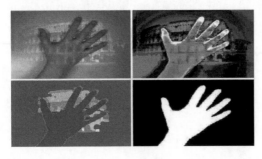

Figure 2. Segmentation process of a hand gesture.

K-mean, Fuzzy K-mean, minimum spanning tree, single link, Mutual Neighborhood, Mixer Decomposition [7][10].

For sign language classification the choice of supervised classification is good. Also in hand

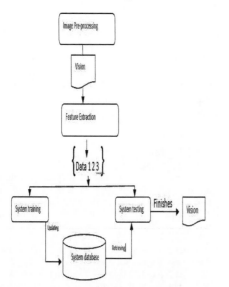

Figure 4.   Architecture of gesture recognition system.

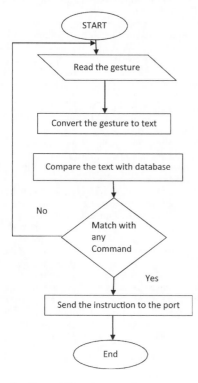

Figure 5.   General flow diagram of system.

gesture recognition, Hidden Markov Model (HMM) is very useful. After the classification as we know that training and testing the system is very important aspect of research work. So in hand movement recognition there are many errors that is ought to have most probable. So there are many error estimation methods available such as redistribution method, Holdout method and leave one out method.

3   RESEARCH STATUS

i. From this literature by D. Zhang and Guojan Lu we got to know many shape representation and description techniques for hand gesture system.

ii. The author T. Maung introduced the hand gesture system to recognize real time gesture in unstrained environment. Efforts are made to make computer. The hand gesture system was done using MATLAB.

iii. The author B. Bauer and Karl Freidrich discussed sign language recognition using linguistic sub units. There were three types of sub units in consideration: learnt from appearance data, inferred from 2D or 3D tracking data.

iv. The author M. M. Hasan and P. K. Mishra includes a new gesture recognition based o image blocking and the gesture is recognized using their suggested brightness factor matching algorithm. They applied two different feature extraction techniques, the one based on features extracted from edge information and other one based on technique for center of mass normalized based on block scaling.

v. In this paper the author A. Malima and Erol Ozgur proposed a fast algorithm for automatically recognizing a limited set of gestures from hand images for a robot control application. They set a fixed set of manual commands, structured environment and developed a simple procedure of hand gesture recognition.

vi. The author Jing Li., Bao-Liang used The Image Euclidean Distance (IMED) which considers the spatial relationship between the pixels of different image in their system. They also an Adaptive Image Euclidean Distance (AIMED) which considered not only the prior spatial knowledge but also the prior gray level knowledge from images.

4   CHALLENGES IN SIGN LANGUAGE INTERPRETATION

Literature survey shows that due to challenges of vision based system, most of the researchers have limited their work, to small subset of full SL. To work on full SL interpretation, close collaboration with SL interpreter and deaf people is required [3].

The researcher reported some difficulties in sign language.

i. Occlusion problem while performing sign.
ii. Signer position may vary in front of camera while performing sign.
iii. Working in 2D camera give loss of depth information.
iv. As each sign varies in time and space, so there may be a change in position and speed with same person or person to person.
v. Co-articulation problem (link between preceding and subsequent sign).

In ISL proper noun is not pronounced as in spoken, but it has been used like pointing to the identity of the particular person with their gender (male/female). In English always a specific structure of sentence is used i.e. SUBJECT, VERB, OBJECT but in ISL not always this same structure is followed. In some places it is used SUBJECT, OBJECT, and VERB [11].

5 SOME OTHER APPLICATIONS OF HAND GESTURE RECOGNITION

a. Number Recognition:
Hand gesture recognition can be also used in the number recognition. In this a real time machine can be created which can understand numerical numbers from 0 to 9 easily.
b. Television Control:
Using hand gesture television can be controlled. Some actions like controlling the volume, turning the TV on or off, muting the television or changing the channel [10].
c. Robot Control:
Controlling a robot using hand gesture system is one of the interesting applications. A robotic system which understand the numeric gesture from 0 to 9 and some other sign signal of hands like pointing the finger or any particular pose of the hand.
d. Virtual Environment (VEs):
Virtual environment is one of the popular applications of the hand gesture system. VEs are especially for communication media system. In this hand gesture is not only used for navigation the VE, but also an interactive device to select and move the virtual objects [6].

6 CONCLUSION

In this paper we discussed various methods for gesture recognition these methods include neural network, HMM, fuzzy c-means clustering, spanning tree, Fourier method besides using orientation histogram for feature representation. For dynamic gesture HMM tools are perfect and have shown its efficiency especially for robot control [4][11]. For feature extraction some methods and algorithms are required even to capture the shape of the hand as in [4][5], [5] applied Gaussian function for fitting the segmented hand which is used to minimize the rotation affection [5]. Data collection for gesture learning is not trivial task. Various algorithms on hand recognition are surveyed in this paper. HMM and its variants can be used I sign language recognition. Due to the complexity of gesture, machine leaning techniques seems perfect for this task. Explanation of gesture recognition issues, detail discussion of recent recognition systems are given as well.

REFERENCES

[1] D. Zhang, C. Lu, "Review of shape representation and description techniques", The journal of the pattern Recognition Society, Elsevier, 2004, pp. 1–19.
[2] T. Maung, "Real-Time Hand Tracking and gesture Recognition System Using Neural Networks", PWASET. Vol. 38, 2009, pp. 470–474.
[3] B. Bauer, Karl-Freidrich, "Towards an Automatic Sign Languages Recognition System Using sub units", LNAI2298, GW-2001, Springer. pp. 34–47.
[4] M.M. Hasan, P. K. Mishra. (2001). "HSV Brightness factor matching for gesture recognition system", International Journal of image Processing (IJIP), Vol. 4(50).
[5] Malima, A., Ozgur, E., Cetin, M. (2006). "A Fast algorithm for vision based hand gesture recognition for robot control", IEEE 14th conference on signal processing and communication applications, pp. 1–4.
[6] J. Li, B. Lu, "an adaptive image Euclidean distance", pattern Recognition Journal, Elsevier, volume 42, 2009, pp. 349–357.
[7] Guo-Dong, A. K. Jain, W. Ma, H. Zhang, "IEEE transaction on neural Networks, Vol.13, No. 4, July 2002, pp. 811–820.
[8] K. K. Wong. R. Cipolla, "Continuous gesture recognition using a sparse Bayesian classifier", International conference on pattern recognition, 2006, pp. 1084–1088.
[9] A. Corradini, "Real time gesture recognition by means of Hybrid Recognizers" GW2001, LNAI2298. Sringer—Verlag Berlin Heidelberg 2002, pp. 34–47.
[10] R. Z. Khan, N. A. Ibrahim, "Hand gesture recognition: a literature review", International Journal of Artificial Intelligence & Applications (IJAIA), Vol. 3, July 2012.
[11] A. S. Ghotkar, Dr. G. K. Kharate, "Study of vision based hand gesture recognition uses Indian sign language" international journal on smart sensing and intelligent systems vol. 7, no. 1, march 2014.
[12] Min B., Yoon, H., Soh, J., Yangc, Y., & Ejima, T. (1997). "Hand Gesture Recognition Using Hidden Markov Models". IEEE International Conference on computational cybernetics and simulation. Vol. 5, Doi: 10.1109/ICSMC.1997.637364.

Communication and Computing Systems – Prasad et al. (Eds)
© 2017 Taylor & Francis Group, London, ISBN 978-1-138-02952-1

Brain wave interfaced electric wheelchair for disabled & paralysed persons

B.M.K. Prasad, Chandra Shekhar Singh & Krishna Kant Singh
Dronacharya College of Engineering, Gurgaon, India

ABSTRACT: For disable and paralysed persons there is always need of assertive robots, electric powered wheelchair is one of them. But the problem with that it operates by conventional inputs such as joysticks, mouse, touchscreen etc. Clinical analysis tells that 50 percent of the disabled and paralysed patients are not able to operate such kind of wheel chair. Persons suffering with Motor Neurone Disease (MND) have only brain and eye functional. In this paper, we discuss the development of brain interfaced electric wheel chair that provides the solution of the problem stated above. The prototype model is a noninvasive type brain interfaced system, where the EEG electrodes are placed on the scalp of the brain, Neuro Sky Mind Wave Brainwave headset is used to record the brain activity and the sensor close to eye detects the Electromyography (EMG) signals generated when eye blinks. The raw data was further processed to obtain the threshold value so that it can be used for decision making. The microcontroller unit gets this data wirelessly and further processes it to perform the desired task. The movement of the electric motor depends upon the interrupts provided by the microcontroller unit.

1 INTRODUCTION

With the recent advancement in the brain wave sensing technology, it is possible to design a small and portable brain wave sensors interfaced electric wheelchair (Luzheng Bi et.al, 2013). For disabled and partially paralysed person there is always need of assertive robots. The previously designed wheel chair works with the help of inputs obtained from keypads, touch screens, Mouse etc. and they are not fully automatic, hence complexity and usage of the system increases and not easy to control. Medical survey also tells that 50% of the disabled patients not able to operate electric wheelchair using conventional inputs. Persons suffering with Motor Neurone Disease (MND) can use only their eyes and brain activity to control (Kanniga et.al, 2013). The prototype model developed by us is fully automatic Electroencephalogram (EEG) based wheelchair and it is useful for disabled as well as paralysed persons. The development of EEG sensing technology opens a new space for the health care industry to design autonomous system for Disable, elderly & paralysed persons (Mahendran R, 2014).

The measurement of electrical activity in the live brain basically depends upon human emotions. EEG is a technique to measure this activity. The First EEG signal obtained by Hans Berger (1924) with the help of single electrode, this wave is known as alpha Wave (Berger's Wave). Now day's more than 20 electrodes are used to detect five fundamentals wave for clinical research, but this device is bulky and very costly (Bi L, Fan et.al, 2013). Technology advancement results in providing inexpensive wireless EEG devices developed by Neurosky, Emotiv EOC etc (NeuroSky BrainWave Signal (EEG) of NeuroSky, Inc). These devices can be easily used to design Brain Interfaced Systems (BCI). BCI provides the interaction of brain activity with the real world applications.

This paper presents a noninvasive type brain interfaced system, where the EEG electrodes are placed on the scalp of the brain. We used NeuroSky Mind Wave Brainwave Headset to record the brain activity and the sensor close to eye detects the Electromyography (EMG) signals generated when eye blinks. The raw data was further processed to obtain the threshold value so that it can be used for

Table 1. Frequency bands of brain waves.

Band	Frequency (Hz)	Activity
Delta	0.4–4	Found during continuous attention tasks
Theta	4–8	Drowsiness in adult
Alpha	8–14	Relaxed
Beta	14–30	Active thinking, high alert case

decision making. The microcontroller unit gets this data wirelessly and further processes it to perform the desired task. The movement of the electric motor depends upon the interrupts provided by the microcontroller unit (Gandhi V et al, 2014).

The paper structured as Section 2, discussed architecture of the system, fundamentals of brain wave and EEG signal processing. Section 3, explains the proposed algorithm. In section 4, discuss the experimental setup. Last section concludes the paper and proposed methodology.

2 SYSTEM ARCHITECTURE

The system architecture for the Brain Wave Interfaced Electric Wheelchair consists of EEG acquisition unit, Wireless interface module and wheelchair control system. The Figure 1 shows the block diagram of the system.

2.1 EEG acquisition

EEG Acquisition is a noninvasive acquisition system to record the brain waves by using multiple sensors (electrodes) mounted on the scalp. Each electrode connected to one of the input of the differential amplifier and other input terminal of differential amplifier connected to reference electrode. The difference of the voltage generated due to the firing of the neurons within the brain due to different activities gets amplified by the amplifier. The output obtained from the amplifier is further filtered and digitalized using analog to digital converter. EEG Signals is a rhythmic activity and generates difference voltage and frequency electrical signal (Mandel C et.al, 2009).

In this proposed model, we use the cheapest single channel Neurosky MindWave MW001 for EEG acquisition. Two EEG dry sensors mounted on the forehead detect the activity of the brain. The Mindwave interfaced with the MATLAB7.0 software to plot raw EEG signal with the maximum sampling frequency of 512 Hz. The FFT and IFFT algorithm is implemented to filter out the information regarding frequency band in the data. Figure 2. Shows the various stages involved in EEG acquisition block.

2.2 Wireless interface module

For Wireless transmission of the data, Bluetooth module HC 06 is used (HC 06 Manual). The EEG processed data further transmitted to the Wheel chair control unit for PWM and direction control of the motor.

2.3 Wheel chair control unit

Wheel chair control unit has the function ability to control the speed and direction of the wheelchair on receiving the EEG processed data from the EEG acquisition block. The main intelligence lies in the microcontroller board (here Arduino board) to process the data and generates interrupts

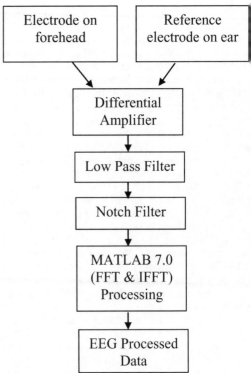

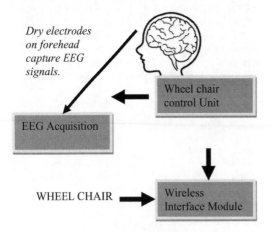

Figure 1. Block diagram of the proposed system.

Figure 2. Stages involved in EEG acquisition.

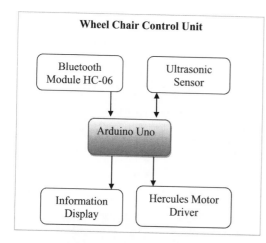

Wheel Chair Control Unit

Figure 3. Wheel chair control unit.

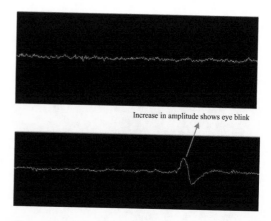

Increase in amplitude shows eye blink

Figure 4. Above) Raw EEG signal without eye blink Below) Raw EEG signal with eye blink.

accordingly to the requirement of the movement. The Arduino output digital pins are connected to Hercules (9V–24V), 15A motor drivers are used to drive high current rating DC motors of the wheelchair. Pulse Width Modulation (PWM) technique is used to control the speed of the wheelchair. Four ultrasonic sensors are also mounted on the four sides of the wheelchair to avoid collision.

3 METHODOLOGY AND PROPOSED ALGORITHM

Thinkgear technology implemented in the NeuroSky Brain wave sensor measures the attention and eye blink values. The attention value increases when the concentration level of the person is high i.e it focusses on a particular point for some time. The attention value varies from 0–100. The average value is to be varies from person to person and in the range of 40–60. The range 60–80 considered as slightly attentive and value above 80 considered as highly attentive. Similarly the eye blinking values also varies from 0–255. (Fischl B et al, 2002).

3.1 Separating eye blinks signals

Eye blinking causes increase the amplitude of Raw EEG signal, but it is very important to distinguish normal eye blink with the rapid signal eye blink. The measurement of the time delay between two eye blinks easily separates this information. When eye blink is detected then counter starts and if the next eye blink occurs in the predefined time then only the blink is considered as a signal blink otherwise considered as normal blink.

Table 2. Logic & outcome of the system.

Interrupt values	Outcome
No of Blinks = 4	Initialize the System
Attention level >65	Forward
Meditation Level >90	left
Meditation Level <90	right
Attention level <65	Stop

3.2 MATLAB processing EEG raw data

MATLAB 7.0 processed the EEG raw data to cancel noise and artifact effects and filter out the signals in the range of Alpha and Beta range. The threshold values are used to classify the signals into attentive and meditation level. The attention level, meditation level and number of detected eye blinks provides logic to drive the motor of the wheelchair to provide forward, left and right movement. The list of logic drive motor and their resulting outcome are listed in Table 2.

4 EXPERIMENTAL SETUP

The Arduino Uno is the main processing unit and provides intelligence to the system. Neurosky brain wave is used, to capture the brain wave signals. The biosensor is placed on the frontal lobe where there is no hair to provide maximum contact. The Thinkgear ASIC technology measures the EEG data and filter out noise and artifacts from packets of data.

The EEG data of the specific user is taken to extract the alpha and beta signal. The User specific average values are calculated in the software

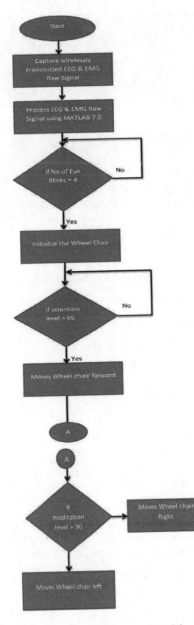

Figure 5. Flow chart of the proposed model.

5 CONCLUSION

In this paper, we developed brain wave interfaced electric wheelchair for disabled & paralysed persons to provide hassle free locomotion. This low cost EEG based wheelchair uses Neuro sky brain waves (attention and meditation level) along with EMG (Eye blinking strength) for controlling the movement. With the further advancement in the sensing and processing technology, the wheelchair can be designed with more control.

REFERENCES

Bi L, Fan X-A, Liu Y. "EEG-based brain-controlled mobile robots: a survey." *IEEE Transactions on Human-Machine Systems, 2013*: 43(2):161–76.

Fischl B, Salat DH, Busa E, Albert M, Dieterich M, Haselgrove C, et al. "Whole brain segmentation: automated labeling of neuroanatomical structures in the human brain." Neuron, 2002: 33(3):341–55.

Gandhi V, Prasad G, Coyle D, Behera L, McGinnity TM. "EEG-based mobile robot control through an adaptive brain-robot interface." *IEEE Transactions on Systems*, Man, and Cybernetics: Systems, 2014: 44(9):1278–85.

Kanniga E, Selvaramarathnam K, Sundararajan M. "Embedded control using mems sensor with voice commands." *Indian Journal of Science and Technology.*, 2013: 6(6), 4796.

Luzheng Bi, Xin-An Fan, Yili Liu. "EEG-Based Brain-Controlled Mobile Robots: A Survey." *IEEE transactions on human-machine systems* 43, no. 2 (2013).

Mahendran R. "EMG signal based control of an intelligent wheelchair." *IEEE International Conference on Communications and Signal Processing (ICCSP).* 2014. 1267–72.

Mandel C, Luth T, Laue T, Rofer T, Graser A, Krieg-Bruckner B. "Navigating a smart wheelchair with a brain-computer interface interpreting steady-state visual evoked potentials." International Conference on Intelligent Robots and Systems, 2009.

Manual, HC 06. "Available from: http://www.tec.reutlingenuniversity.de/uploads/media/DatenblattHC-05_BT-Modul.pdf."

Manual., NeuroSky Mindset Instruction. "Available from: http://developer.neurosky.com/docs/doku.php?id=hinkgear.net_sdk_dev_guide_and_api_reference."

NeuroSky. "Instruction Manual." NeuroSkyInc; Jun 19 2009.

Pires G, Honorio N, Lopes C, Nunes U. "Autonomous wheelchair for disabled people." *Proceedings of the IEEE International Symposium on Industrial Electronics*, 1997.

Rebolledo-Mendez G, Dunwell I, Martinez-Miron EA, Vargas-Cerdan MD, De Freitas S, Liarokapis F, et al. "Assessing neurosky's usability to detect attention levels in an assessment exercise." Human-Computer Interaction New Trends., 2009: 5610:149–5.

to define the threshold levels. Pre signal processing using FFT and IFFT is used to extract the attention and meditation signal. The Arduino board uses these signals to drive the two DC motors of the wheelchair. Ultrasonic Sensors are also mounted on four sides of the wheelchair to avoid obstacles in the path.

Software engineering and emerging technologies

Communication and Computing Systems – Prasad et al. (Eds)
© 2017 Taylor & Francis Group, London, ISBN 978-1-138-02952-1

Software reusability using data mining techniques

A. Jatain, K. Tripath & Abhishek Jain
Amity University, Manesar, Gurgaon, India

ABSTRACT: Software Reusability aims at modifying the existing components into newer version by reducing cost, effort and development time. To extract the existing components for creating he updated version of software we use data mining techniques. In this paper we discussed, Meta level data mining approach for software reusability. We have studied various clustering methods and efficient algorithms for effective storage and retrieval of required high quality software within time and budget.

1 INTRODUCTION

Software reuse is an optimistic approach to achieve software productivity and software quality (Gupta et al. 2013). It leads to decrease in the development time, effort and cost reduction since it operate on modules that are already tested, documented and certified (Smith et al. 1998). Reusability is the extent to which we can reuse software (Prakash et al. 2012). It is better way to provide an efficient way to search and retrieve the asset to attain success on reuse ambition. According to (Murthy et al. 2008), "You first have to find a software module to reuse it".

In today's world, there is improvement in hardware performance, storage capacity, architectures. So computing world can now design and operate the software according to requirements. So there is high pressure for producing high quality software system within limited time and budget. Thus rather developing similar code from scratch, developers reuse the existing software components (Kanungo et al. 2002). (Ramamoorty et al. 2008) discussed that reusability is not only associated with the code, it also include requirement, design, test cases and documents. To reuse e existing knowledge from software scratch such as execution traces, new techniques are required as they contains huge amount of information about software project's status, advancement and growth. Basically software reuse process consists of four steps i.e. identification of software components, knowledge of context, relating software reuse methods, integration and evaluation (Prakash et al. 2012).

Nowadays to extract useful information from already existing software artefacts data mining techniques are used. Data mining is a series of action that produce patterns from historical data. Data mining is a vital drive for transforming data into useful information and it requires KDP (knowledge discovery process). Knowledge discovery is a process of digging and analyzing large amount of data and then extracts the meaning of that data. It mainly consists of six steps: selection of data, cleaning and transformation on data, data mining, data presentation, interpretation and evaluation of data. Data selection is the first step to extract the relevant data for data mining analysis. After that data is stored in database from where data analysis takes place. In data cleaning, transformation, preprocessing phase data is cleaned, prepared, relevant data is converted for preprocessing and desired result is achieved. Meaningful patterns and rules are discovered by analyzing the data using data mining algorithms to produce models. Data mining is the most significant phase of KDP cycle. Various data mining algorithms are used such as clustering, rule association, decision tree, Classification and Regression Tree (CART). The final phase is to interpret and evaluate the results by applying knowledge discovery techniques, which involves making useful decisions making.

The ability to predict needed variability in future assets is an important element for successful reuse. To identify measures of reusability research is needed and methods are needed to validate and estimate the number of potential reuses. So there is a gap between the need of useful data from historical data to software project management practices. To bridge this gap knowledge discovery process is used which can extract useful data effectively and efficiently using various metrics. Finally, project manager uses this knowledge for better management of software process. The remaining paper is organized as follows: Section II describes the related work. Section III provides proposed methodology and section IV gives brief sketch of proposed tool.

2 RELATED WORK

Concept of reuse was given in 1968 and got much attraction in late 1970's which provides new direction of software development in less time and budget (Gomes et al. 2001). (McClure et al. 2001) outlines software reuse "process of bringing software system and software project's modules from existing components". On the other hand (Yu et al. 1991) defined software reuse as "software engineering activities that focus on reconfiguration and conformation for new computing system applications and the identification of reusable software for straight import" and (Feeler. 1993) explains software reuse as "engineering activity that deals on acceptance of generality of system within area. The main focus is to utilize the technology base". According to (Kyo et al. 1992) reuse takes place according to six steps which are performed at each level:

a. Analyzing problem and identification of available solutions to evolve reuse plan.
b. Identification of solution for problem that follows the reuse strategy.
c. Reconfiguration of solutions to improve reuse at next level.
d. Obtaining, initializing and modifying existing reusable modules.
e. Integration of reused and recently developed into products and
f. Evaluation of products.

Many reuse researcher have done a commendable work in this field. (Morisio et al. 2002) mentioned few success and failure factors on running a companywide software reuse that predicted a successful software reuse. In (Shri et al. 2010) proposed a clustering approach to predict reusability of object oriented software system. To generate ranking of reusability software an automatic tool was proposed by Boetticher using soft computing techniques. (Michael. 2002) in used association rules, sequence mining and clustering technique to revel usage patterns of components. A lots of efforts found in literature using concept of reusable components in developing platforms for machine learning and software engineering. Weka (Witten et al. 2005), R-project (Ramamoorty. 2008) and Rapid Miner (Mierswa et al. 2006) are well known open source machine learning platforms. In (Zaki et al. 2008) proposed a DTML (data mining template library) which consists of generic algorithm and containers for pattern mining. Decision tree design comparison is done by (Murthy et al. 2005). These various research direction discussed above acknowledged that reuse is a powerful and essential way against software crisis problems (Ramamoorthy. 1988). Use of software reuse to improve

software quality and productivity is discussed in various studies. For advance searching, matching and modeling tools, software engineers are exploring and adopting many reusability approaches.

3 METHODOLOGY

3.1 Selection of software metrics

To access the quality of software and measure the reusability aspects software metrics and models play in vital role. Software metrics are classified according to procedure and object oriented paradigm. That helps in application of data mining technique for software reuse.

3.2 Data cleaning and data transformation

The quality of software components developed in procedure and object oriented paradigm is measured through software metrics developed. The different metrics that works as input attributes for software components are as follows:

- Weighted Method per Class (WMC)
- Depth of inheritance tree (DIT)
- Number of children (NOC)
- Coupling between Classes (CBO)
- Response for Class (RFC)

There are 167 instances are taken from various open source projects available at www.source.forge.net to measure software components using software metrics.

3.2.1 Precision
To evaluate precision, recall, mean absolute error (MAE), root mean square errors are calculated. Precision for class is calculated by dividing number of true positives by total number of elements belonging to positive class.

$$Precision = TP/ (TP + FP) \qquad (1)$$

3.2.2 Recall
Recall is calculated by dividing total number of true positives by total numbers of elements which actually belong to positive class. Total numbers of elements that actually belong to positive class are the sum of true positives and false negatives, there are the items which should have been labeled as belonging to positive class but are not (Manjas et al., 2010).

$$Recall = TP/ (TP + FN) \qquad (2)$$

3.2.3 Mean Absolute Error
It is average prediction error. First the difference between all the predicted and actual value of all the

best test class is calculated then, there average is MAE.

$$MAE = ((a_1 - c_1)^{2} + (a_2 - c_2)^{2} + ... (a_n - c_n)^{2}) / n \qquad (3)$$

3.2.4 *Root Mean Square Error*

It is the difference between values predicted a values actually observed from the model. The root of Mean Square Error is predicted by given equation.

$$RMSE = ((a_1 - c_1)^{2} + (a_2 - c_2)^{2} + .. (a_n - c_n)^{2}) / n)^{1/2} \qquad (4)$$

3.3 *Basic terminology and proposed framework*

The proposed framework classifies the user's intention and analyzes the collected data. For efficient reuse, similarity index is calculated between already existing software modules expected by user. Then a software system is built that matches with user's expectations and within time and budget. The proposed framework is presented with five main modules i.e. stake holders intentions, classifier, analyzer, cluster and categorizer. Software reusability is predicted stepwise:

Step 1: The graphical user interface is used where the user interacts and uploads intentions. The graphical user interface is web based interface for user interaction with system. After user interaction with GUI classifier then reads the user's intention and select features. This is done to remove unnecessary, irrelevant and redundant information form data specified by user. Feature selection is best done by Genetic algorithms. Removal of irrelevant information helps in improving performance of the system. GA helps in identifying and separating only important and relevant information and keywords from the intention inputted by user. Working of GA involves four steps: a) Selection of attributes b) Fitness function computation c) Fitness evaluation d) reproduction. After separation, pattern matching is performed for those keywords with the existing keywords in repository.

Rabin Karp Algorithm is very efficient string matching algorithm, which can be used to solve the purpose. By ignoring the details like punctuation and cases, this algorithm can easily search for patterns from given source materials. This algorithm practically helps detecting plagiarism. Rabin Karp method produces the most accurate results than any other existing ones. It uses the concept of Situational Method Engineering (Harmsen et al. 1994). According to which in the existing method repository, it retrieve, modify, or assemble other method of interest into new methods then eventually store them again into repository. Expected requirement is expressed in the form of sentence of phrase by the user. Pattern matching classifier helps separating out only important and relevant information by ignoring phrases like "the", "is", "I", "use" etc. Following figure shows how classification results are produced.

Step 2: The data which is obtained from classification and is processed by analyzer. Working of analyzer is done in two stages:

Searching: For this purpose, analyzer searches for relevant keywords from the repository as input from step 1. It then performs various permutations and combinations to obtain appropriate modules for reuse. It works on the basis of maintaining possible synonyms in central repository, then sharing information (synonyms) between proposed framework and central repository.

Matching: In the next stage matching is performed between the proposed intention and existing software components. A minimum criteria is considered; let's say 30% is the criteria for selection. Then matched information and passing criteria is stored in the form of table by analyzer which is available in DB–2.

Step 3: As a result of percentage match, clustering method is performed. For this purpose, we use K-means (Kanungo et al. 2002). Clustering is a process where similar data is groped together. K-means is used because it is fast and efficient even with high number of variables. Also it produces tighter clusters. After clubbing the data of similar type, the next step is computing similarity index count (SIC) (Goldberg et al. 2010). Larger the value of SIC, more the similarity is there with intended information, lower value shows less similarity. For maximum reusability lower SIC are not rejected, but are recommended after higher SIC. This is done because there may be chances that some important functionality is missing from higher SIC, which can then be looked into lower index value modules. Database-3 stores this information in the form of table.

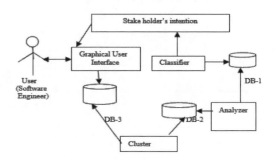

Figure 1.   Framework for software component reusability model.

Table 1. Sample view of analyzer result.

Proposed intention to match	Possible match	Percentage of match
Management	Investment management	75%
	Portfolio management	50%

Table 2. Result computation for cluster formation: sample view.

Clusters	Module numbers	Similarity index count
Cluster 1	M1, M3, M5	75%
....	....	......
Cluster n	M_n, M_{n+2}	@%

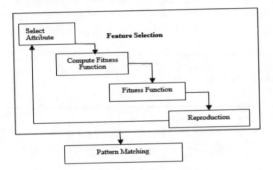

Figure 2. Classifier detailed view.

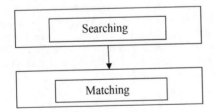

Figure 3. Analyzer detailed view.

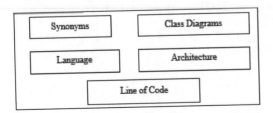

Figure 4. Central software repository: abstract view.

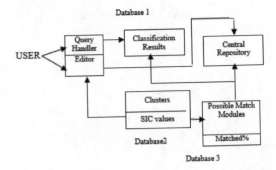

Figure 5. Framework for proposed tool.

4 PROPOSED APPROACH TOOL VIEW

A sketch of proposed framework is made through tool support. It helps describing the nature of proposed framework. Fig. 4 illustrates one main central repository and three small databases i.e. Database-1, Database-2, Database-3.

Database-3 interacts with Database-1 and central repository. It collects the information that is required for searching and matching from Database-1, then interacts with central repository to search the appropriate match modules and their percentage of match.

Database-2 interacts with editor and Database-3. The processed information that is stored in Database-3 is clustered and its similarity index value is calculated, then to analyze the reusability of software modules effectively results are displayed with the help of editor. Editor generally display results in tabular format.

Database-1 interacts with query handler and fetch stakeholders expressed intention for processing. After analysis user is able to pick the desired components. User can also update the central repository with new software component after finishing task with editor interface. We stored this information for further reuse. This stored information can be accessed by software engineers for future analysis and reuse.

5 CONCLUSION

We are able to review the implementation of classifier, analyzer and web based interface. Cluster approach is yet to be implemented. This approach of matching, analyzing and computing the similarity index with software components can be efficient and can really help in reducing the effort of developing similar software modules from scratch. This approach can be helpful in reducing effort, time and cost by identifying the reusable software components effectively. Therefore, data mining techniques are effective in software reuse to improve productivity and quality by reducing development time.

REFERENCES

Feiler P H., 1993. Reengineering: an engineering problem [R]. Software Engineering Institute, Carnegie Mellon University: Special Report.

Frakes W., and Isoda S. 1994. Success factors of systematic reuse. IEEE software, 14–19.

Goldberg M K., Hayvanovych M., Ismail M., 2010. Measuring similarity between sets of overlapping clusters. Second international conference on social computing, 303–308, IEEE.

Gomes P., Bento C., 2001. A case similarity metric for software reuse and design. International journal of artificial intelligence for engineering design, analysis and manufacturing, pp: 21–35.

Gupta C., Rathi M., 2013. A Meta level data mining approach to predict reusability. International Journal of Information Engineering and Electronics Business, 33–39.

Harmsen F., Brinkkemper S., 1994. Situational Method engineering for information system project approaches. International conference on methods and associated tools fo the information systems life cycle, 881–892.

Jatain A., Gaur D., 2012. Estimation of component reusability by identifying quality attributes of component: a fuzzy approach. Second International Conference on Computational Science, Engineering and Information Technology, 738–742.

Kanungo T., Mount D., Netanyahu N. S., Piatko C., Silverman Wu A., 2002. An efficient k-Means clustering algorithm: analysis and implementation. Transaction on pattern analysis and machine intelligence, IEEE, 881–892.

Krishana M., Vasumathi D., 2011. Study of mining software engineering data and software testing. Journal of emerging trends in computing and information sciences, (11): 598–603.

Kyo K C., Shalom C., Robert H., James P., 1992. Reuse based software development methodology. Application of reversible software components project, special report.

Manjas S., Vashisht R., Bhardwaj R., 2010. Framework for evaluating reusability of procedure oriented system using metrics based approach. International journal of computer application, (9).

McClure C., 1997. Software Reuse Techniques. Parentice—Hall, Inc.

Michael A., 2000. Data mining library reuse patterns using generalized association rules. International conference on software engineering, 167–176.

Mierswa I., M. wurst, R. Klinkenberg, M. Scholz and T. Euler, 2006. Yale: Rapid prototyping for complex data mining tasks. International conference on knowledge discovery and data mining, ACM, 935–940.

Morisio, Ezran M. and Tully C., 2002. Success and failure factors in software reuse. Transaction on software engineering, (28): 340–357.

Murthy K., 2008. Automatic construction of decision trees from data: A multidisplinary survey. Journal of data mining and knowledge discovery, 345–38.

Prakash V., Asoka D., Aradhya V., 2012. Application of data mining Techniques for software reuse process. Procedia Technology, 384–389.

Prietman R., Freeman D., 1987. Classifying software for reusability, IEEE software, 6–16.

Ramamoorty R. 2008. R: A language and environment for statistical computing, R foundation for statistical computing, Vienna, Autria.

Ramamoorthy V., 1988. Support for reusability in genesis", IEEE transaction on software engineering. 1145–1154.

Shri A., Sandhu P., Gupta V., 2010. Prediction pf reusability of object oriented software system using clustering approach. World Academy of science, engineering and technology, 853–858.

Smith E., Yasiri A., Merabti M., 1998. Multi-Tiered classification scheme for component retrieval. 24th Euro micro conference, 882–889.

Witten I. and Frank E., 2005. Data mining: practical machine learning tools and techniques. 2nd edition, Morgan Kaufmann, San Francisco.

Yu D., 1991. A view on three R's (3R): reuse reengineering and reverse engineering. Software engineering notes.

Zaki L., Parimi M., Gao N., Phophakdee F., Urban J., Chaoji V., Hasan M. and Salem S., 2005. Towards Generic Pattern Mining. Lecture Notes in Computer Science, 91–97.

Communication and Computing Systems – Prasad et al. (Eds)
© 2017 Taylor & Francis Group, London, ISBN 978-1-138-02952-1

Routing protocol using artificial bee colony for wireless sensor networks

Ashutosh Tripathi
Department of ECE, Amity University Rajasthan, Jaipur, India

Reena Dadhich
Department of CS&I, University of Kota, Kota, Rajasthan, India

Narendra Yadav
Department of CSE, JECRC University, Jaipur, Rajasthan, India

Ravendra Pal Singh
Department of ACEL, Amity University Rajasthan, Jaipur, India

ABSTRACT: The new algorithms proposed with the name of the Predictive Packet Reception (PPRE) with the focus of cluster formation and data transfer used in wireless sensor networks by using soft computing techniques. During the routing, generated information to be collected from all originated nodes and therefore which nodes involved number of times to gathering the information. It leads to the fact that the nearest sensor nodes fail much earlier than remote ones, In the prolong period, traffic from all the sensor nodes is going through one sensor node that is nearest to sink and the former disrupt the work of the rest of WSN. From the point of view of energy saving, big "WSN" with only one sink cannot consume resources effectively. To solve this problem it is necessary to divide the WSN into "clusters". The proposed algorithm would improve the quality of data transfer in a granted way through by increasing the overall packet delivery fraction, throughput, stability of whole networks. The proposed scheme is simulated over NS 2.34 and MATLAB.

1 INTRODUCTION

Wireless sensor networks are deployment of several devices equipped with sensor that performs a collective measurement process. Since 2001 people do start working in field of wireless sensor network due to cost of computing power going down exponentially MEMS technology is going similar and things are going in higher performance and cheaper every year so there are two types of sensor used Wired and Wireless Sensor network. In concerned of wired sensor network, most of the sensor used as wired like Fire sensors, Temperature Sensors, Passive Sensors etc. Even though performance cost is good but installation, connection not cheaper. In July 2012, the major focus about the reliability, Power consumption. So there is a new protocol developed with the name of Time Synchronized MESH Protocol, which is 99% reliable and low power per delivered packet. In this protocol, If "node" A expect to send the data to "node" B then it is expected that "node" B will be in Listen and Vice-Versa. In this protocol "node" A can listen more often than B. And since both are time synchronized, a different a radio frequency can be

used at each wake-up. The time synchronization information transmits in both directions with every packet. So every cost of latency deduction is easy to calculate. The latency v/s Power tradeoff can vary by "node", Time of the day, recent traffic etc. After this protocol Multi Hop routing came into the picture. These protocol usages, global time synchronization to allow sequential ordering of link in a super frame. The measured average latency over the many Hops is $T_{frame}/2$. In time synchronization multi hop protocol having time synchronization on in terms of reliability, Power, Sensor. Reliability mainly considered in terms of frequency diversity, spatial diversity and temporal diversity. Deployed concerned of nodes is easy that's why this technology is grooming day night. There are many stages for researchers to work with wireless sensor networks. In concern of routing algorithms, a protocol can be used to communicate the information from source nodes to sink nodes and there are different stages involved in design a protocol at network layer. Starting from collection of node information to data transfer stage. Either in singe hop or multiple hops the task of routing protocol determine the value of constraints of node in terms of energy,

Power etc. The clustering algorithms designed with the focus of cluster formation and data transfer used in wireless sensor networks. There are many soft computing techniques like Ant colony optimization and Aartificial Bee Colony optimization used for optimization and these are kind of intelligent algorithms based on swarm and its collective behavioural of social insects in self organizing way. Sometimes it is decentralized also. The Ant Colony and Aartificial is widely used in network intelligence technology. This Technology is seemingly expanding human capabilities.

The ACO model is a new paradigm based on swarm techniques it consists three agents to find the route from source to destination nodes and forages will transmit the data packet and evaluate the best way to reaching the data to destination. Forages evaluate the network stability, packet delivery ratio, throughput for wireless sensor network.

A significant point to be noticed of individual nodes used in path while data to be sent from source nodes to cluster and then to sink. So the nodes involved in the process of data sending their resources used more than the other nodes. Apart from cluster such nodes are not establishing any path for data transmission from source to base station/sink. Therefore data delivery is not guaranteed many routing algorithms. The clustering algorithms is used with ACO then not only a path is established but data also can send in guaranteed way to base station. Aim is to solve the problems of more energy consumption and control the number of effective transmission from source to sink nodes. This required scheme need to attempts to provide the guarantee the information to "base station" about the reception of data to and from.

The pattern of the paper is going to organized as follows. In section 2, literature of related technology like ACO family of protocols is given. Section 3, describes the principal statement and proposed solution. Section 4, consists simulation results and performance analysis of PPRP protocol with ABC protocol. Finally, in section 5, The paper will concluded with a discussion on some future work.

2 RELETED WORKS

Helmy Ahmed & Jabed Faruque (2005) & Po-Jen Chuang, (2005) proposed a new idea to find a better packet delivery ratio and "end-to-end delay" with the name of efficient and secure routing protocols for "wireless sensor networks" through SNR based dynamic "clustering" mechanism. Global Mobile Simulator (Glo Mo Sim) 2003 is used for the simulator for the dimension of 1000×1000 meter Sq. flat space for 500 sensor "nodes". The result

compared with RPSDC, LEACH and PEGISIS protocol. But authors not mentioned when energy of the cluster head is lost earlier then other what kind of mechanism used for data transferring. Ramachandran Amuth Subramanian Ganesh et al. 2013 used the cluster formation model with the stable data transfer stage. The simulation platform TOSSIM, OMNET, ATELU is used for 100 sensor "nodes" for dimension of 100×100 meter. This paper explains with different comparisons with number of cluster and simulation time. The major drawback of this paper was "node" destroyed earlier which is used in cluster formation and its affect the overall data transfer from source to destination specially in case of military applications.

M Iqbal et al. proposed a scheme of dynamic aggregation based on the structure of a tree on Directed Acyclic Graph (DAG). This scheme limits the application of the algorithm declares that the event size cannot exceed the size of one grid. The major drawback of this scheme is when energy of the cluster head is lost earlier then other what kind of mechanism used for data transferring.

3 PRINCIPAL AND MODEL APPROACH

3.1 Artificial Bee colony optimization (ABC) model

The ABC algorithm was designed based on the cooperative behavior of natural bees in the swarm. The scout bees evaluate the fitness of the solution (termed nectar amount), and this information is shared with onlooker bees waiting in the hive. After the initial search, all scout bees now become employed bees. The employed bees go to the food sources (solutions) in its memory and determines the neighboring food sources to evaluate the nectar amount. If the neighboring food source contains a better solution, the new position is kept. Otherwise the old position is maintained by Govindan R., D. Estrin, 1999, Li, T. Ellis, M. Iqbal, X. Wang, QoS, 2010.

The info of the new or existing nectar amount then is relayed to awaiting onlookers when the employed bees return. Onlookers bees then select a food source depend on nectar amount relayed. If the nectar amount increases (solution approaching objective), the probability which that food source is selected is higher. The employed bees which carrying high nectar amount will attract onlookers bees toward it food sources position.

After selecting potential food source from employed bees, the onlooker bee goes toward the direction and evaluate the neighboring food source. Same as employed bees, if the neighboring food source contains a better solution, the new position is kept by Govindan R., D. Estrin, 1999, Eberhart

R. C., J. Kennedy, 1997. Otherwise the old position is maintained. The process is repeated between employed and onlooker bees until the food source is finished. Once this happens, scout bees will now be sent to discover new food sources. In ABC, the activation of scout bees is controlled by how many iterations in which no better-quality food sources are discovered by Basturk B., D. Karaboga, 2008.

In order to binarize the ABC algorithm, we follow the concept outlined by Aboutajdine D et al., 2010, by representing the bee positions as "probabilities of change" rather than the actual solution.

3.2 Prediction model

Equation (1), Gives the cost function of pattern i (fi):

$$fi = \frac{1}{N}\sum_{j-1}^{i} d_{max}\left(X_i, P_i(X_i)\right) \tag{1}$$

The location of a food source represents a possible optimization solution [13], and the pheromone volume of a food source corresponds to the quality (fitness) of relevant solution, calculated by Eq. (2):

$$fit = \frac{1}{1+fi}. \tag{2}$$

An ant selects a food source according to the associated probability values. These probability values are expressed in P, calculated by Eq. (3).

$$P_{i(pocket\ reception)} = P_{sd} = \Theta\left(\frac{\sigma 2\Theta z d_{sd}}{g}\right). \tag{3}$$

Where Θ is a hardware related threshold, σ represents the noise power, d denotes the distance between node s and d, and γ is the path loss exponent.

4 RESULT AND ANALYSIS

4.1 Simulation model

Proposed scheme is evaluated the performance with ACO by using "network simulator" NS-2 (V.K,F.K, The network Simulator Ns-2 & Helmy Ahmed, Jabed Faruque, 2005) with incorporation of MIT uAMP project (NS2.34 Extension). Proposed predictive reception based routing protocol guarantees implemented with process that mentioned in Figure 1. Implementation of protocols has been written in "TCL" and C++ programming language and "MATLAB".

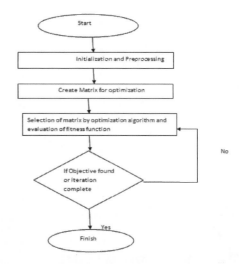

Figure 1. Optimization process.

4.2 Performance measure

4.2.1 Packet delivery ratio
This refers to data rates sent from clustered member to cluster head and further to sink node. So it is measured the total data rates of packets sent from source node to destination node. The ratio of the total number of data packets delivered to the destination to the total number of packets sent is known as packet delivery ratio. The performance of the protocol is high if the value of packet delivery ratio is high.

PDR = Total number of packets received at the destination/total number of packets sent by the source.

4.2.2 Energy
Through this metrics calculate the residual energy when it takes to travel from source nodes to the base station/sink.

4.2.3 Throughput
It is defined as the packets delivered to the destination in total, (S. Li, T. Ellis, M. Iqbal, X. Wang, QoS, 2010) and considered as bits per second (bit/s or bps).

The parameter/size of the network have taken like N = 50.

Analysis of "energy consumption" balance Based on our experiments, we know that using the ACO protocol in the network, the forgoing 20 sec is the network lifetime before the dead nodes have generated.

Figures 2–3. Shows the Data is collected from all the sensor nodes in result the transceivers of the nearest sensor nodes retransmit much more

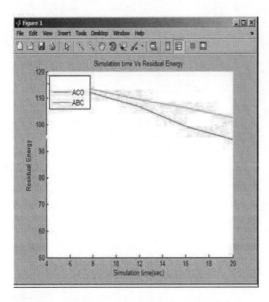

Figure 2. Simulation time and energy.

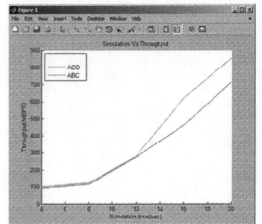

Figure 4. Through put of ACO and ABC.

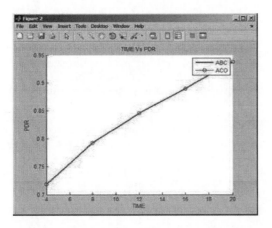

Figure 3. Simulation time and packet delivery ratio.

information, and hence ant colony optimization techniques used to compare the energy of nodes and since sensor nodes are usually all of the same type and have equal energy content, ant colony leads to the fact that the nearest sensor nodes fail much later than the other algorithms, and so the former disrupt the work of the rest of WSN can be improve by ant colony optimization.

Figure 4 shows the packets delivered to the destination in total, (S. Li, T. Ellis, M. Iqbal, X. Wang, QoS, 2010) and considered as bits per second (bit/s or bps) for ACO and ABC.

5 CONCLUSION

Through this paper, soft computing based approach with use of Predictive Packet Reception model (PPRE) has presented. This protocol used to predict the resources of networks in term of Θ is a hardware related threshold, σ represents the noise power, d denotes the distance between node s and d, and γ is the path loss exponent. Initially it starts from message broadcast and then clusters formation and data transferring stage. In this algorithm node degree and their distance is fully considered. In data transfer stage performance considered by using ant colony optimization and ABC protocol design and the basic of considering the packet reception and propagation delay, data can be sent in a precise way to sink node. Through proposed algorithm, it can improve the quality of data transfer in a granted way through by increasing the overall packet delivery fraction, throughput, stability of whole networks.

REFERENCES

Aboutajdine D., B. Elbhiri, S. El Fkihi, R. Saadane, 2010, Clustering in wireless sensor networks based on near optimal bi-partitions, Next Generation Internet (NGI), 2010 6th EURO-NF Conference on, June 2010, pp. 1–6.

Basturk B., D. Karaboga, 2008, "On the performance of artificial bee colony (ABC) algorithm," Applied Soft Computing, vol. 8, pp. 687–697.

Cayirci E., I. Akyildiz, W. Su, Y. Sankarasubramaniam, August 2002, A survey on sensor networks, IEEE Communications Magazine, vol. 40, no. 8, pp. 102–114.

Eberhart R. C., J. Kennedy, 1997, "A discrete binary version of the particle swarm algorithm," in Proc. 1997 Conf. Systems, Man, Cybernetics, Piscataway, NJ, pp. 4104–4108.

Feng W., L. Guangjun, X. Xiaoping, 2011, "Identification of a kind of nonlinear system," 2011 Seventh International Conference on Natural Computation (ICNC), pp. 1730–1733.

Govindan R., D. Estrin, 1999, Next Century Challenges: Scalable Coordination in Sensor Networks, MobiCom 1999, pp. 263–270.

Helmy Ahmed, Jabed Faruque, 2005, "Gradient-Based Routing in Sensor Networks" University of Southern California, Springer-Verlag Berlin Heidelberg.

Hoon-Jae Lee,Triana Mugia Rahayu, Sang-Gon Lee, 2015, A Secure Routing Protocol for Wireless Sensor Networks Considering Secure Data Aggregation" Sensors.

Husain A. R., T. G. Ling, M. F. Rahmat, 2012, System identification and control of an Electro-Hydraulic Actuator system," in IEEE 8th International Colloquium on Signal Processing and its Applications (CSPA), pp. 85–88.

Li S., T. Ellis, M. Iqbal, X. Wang, QoS, 2010, "Scheme for multimedia multicast communications over wireless mesh networks" IET Communication, 2010, Vol. 4, Iss. 11, pp. 1312–1324.

Madria Sanjay, Malik Tubaishat, 2003, Sensor Networks: An Overview, IEEE Potential, pp. 20–23.

Pengfei L., F. Li 2013, "The Research Survey of System Identification Method," in 5th International Conference on Intelligent Human-Machine Systems and Cybernetics (IHMSC), pp. 397–401.

Po-Jen Chuang, 2005, "Efficient Data Gathering Scheme for Wireless Sensor Networks," Springer, pp. 370–379.

Prasanta K. Jana, Suneet K. Gupta, 2015, Energy Efficient "clustering" and Routing Algorithms for Wireless Sensor Networks: GA Based Approach" WPC (2015), Springer.

Ramachandran Amuth Subramanian Ganesh and AUG 2013, Efficient and Secure Routing Protocol for Wireless Sensor Networks through SNR Based Dynamic "clustering" Mechanisms, Journal of Communications and Networks, KICS Vol. 15, No. 4.

V.K., F.K., The network Simulator Ns2," http://www.isi.edu/nsnans.

Willig Andreas, Holger Karl, 2005, "Protocol and Architecture for Wireless Sensor Networks," Wiley Publication.

Ya-nan Zhu De-gan Zhang, Xiang Wang, Xiao-dong Song,Ting Zhang, 2015, A new "clustering" routing method based on PECE for WSN" EURASIP Journal on Wireless Communications and Networking (2015), Springer.

Communication and Computing Systems – Prasad et al. (Eds)
© 2017 Taylor & Francis Group, London, ISBN 978-1-138-02952-1

Fuzzy commitment scheme with cyclic codes

Sonam Chauhan & Ajay Sharma
SRM University, Delhi-NCR, Sonepat, Haryana, India

ABSTRACT: Cyclic Codes ensures the reconstruction of corrupted data using the redundancy added to the data before the transmission or storage. The added parity bits allow the detection and correction of fixed amount of errors depending upon the relationship established by the coding scheme between the actual data and the redundancy. In this paper, data obtained after coding is secured by the Fuzzy Commitment Scheme. Different from the commitment scheme, the fuzzy commitment scheme may utilize a corrupted opening key that is similar to the original opening key. The errors appeared after opening the commitment with a corrupted key may be corrected by the decoding. In the proposed scheme, data is transformed two times before transmission or storage, thus, provide data security at two levels. This paper explores the efficiency of implying fuzzy commitment scheme in conjunction with the cyclic codes.

Keywords: Cyclic Codes, Reed Solomon Codes, Fuzzy Commitment Scheme, Fuzzy Commitment Key

1   INTRODUCTION

To ensure the data to be corruptions free, it is mandatory that the data must be error free or there must be a way to detect and correct the errors. The unidentified errors can be a disaster for the system and may degrade its performance. In 1948, Shannon in his paper (Shannon, 1948) proved that the communication channel could be made more reliable when a fixed fraction of it is used for redundancy. Shannon work results as the starting of Coding theory, which deals with the transmission of messages over noisy channels. The coding techniques that detect and correct the errors in the data have applications ranging from data storage to deep-space telecommunications. The Hamming code was the first known error correction code that was proposed by Richard Hamming in 1950's. In 1959, French Mathematician Alexis Hocquenghem invented a class of error correction code and R.C. Bose and D.K. Ray Chaudhary independently published their work (Bose & Ray-Chaudhari, 1960) on the same error correction code in 1960. Their work is known as the BCH code that comprises the initials of their inventors. I.S. Reed and G. Solomon published a paper in the Journal of the Society for Industrial and Applied Mathematics (Reed I.S. & Solomon, 1960) and introduced RS codes. Gallager developed Low-Density Parity Check Codes (Gallager, 1963) in early 1960's. Gallager's work has been rediscovered by MacKay and Neal in their work (MacKay & Neal, 1996). V.D. Goppa presented the generalization of the BCH codes known as the Goppa codes (Goppa, 1970). Berlekamp summarized Goppa's work in English (Berlekamp, 1973).

Manuel Blum was the first to introduce the Commitment Scheme (Blum, 1981). In Commitment Scheme, the sender encrypts the message of the concealed bit and transmits it. To open the commitment, the receiver requires both key and the committed string. The Commitment Scheme must satisfy the hiding and binding properties. The hiding property confirms that the receiver cannot predict the original message before the open phase, whereas binding property bounds the sender to a single value. The binding property restricts the sender to alter the message after transmission. Claude (Claude, 1997) presented the bit commitment with oblivious transfer based on the Binary Symmetric Channel. A. Juels and M. Wattenberg (Juels & Wattenberg, 1960) proposed the Fuzzy Commitment Scheme and their scheme combines the cryptography with the error correction codes and introduces the Commitment Scheme with fuzziness in it. The introduced fuzziness allows the changed witness that is close to the original witness in some suitable metrics. The FCS also satisfies the hiding and binding properties. This is infeasible for the attacker to learn about the committed value or to reveal the original message with different committed value. Before the open phase, the recipient cannot acquire the committed value. The binding properties restrict the sender for at most one value. In other work (Al-Saggaf & Acharya, 2007), define the FCS that uses the hash functions. The Fuzzy Commitment Scheme with the hash functions suffer the security flaws in hashing. In another paper (Ojha & Sharma, 2010), present their work on the FCS hash functions with the combination of McEliece cryptosystem that suffers the limitation of large key size. The limitations in FCS with MD5 and SHA1 and large key size can be considerably removed by using the Cyclic Error Correction Codes.

1.1 Paper organization

The subsequent sections of this paper are organized as follows: Section 2 describes the Error Correction Codes and cyclic codes. Section 3 and 4 defines the Commitment Scheme and Fuzzy Commitment Scheme and its phases respectively. Section 5 illustrates the proposed scheme and its phases. Section 6 describes the result obtained from the proposed scheme using cyclic RS Codes.

2 ERROR CORRECTION CODES

The retrieval of error free data is the utmost requirement in communication as the errored data may convey an entirely different meaning to the receiver. The correction of errors is confirmed using the coding schemes that transformed the message words into the codewords. The codes can be block or convolutions. In convolution codes, at a particular time instant, k bit information sequence is entered into the encoder, which generates n bit output and the state of the encoder is changed. In block coding, the message sequence of length k is converted into a codeword sequence of length n where $n > k$. For k bit information, there are 2^k valid codewords. The block codes can be either linear or cyclic. Depending upon the number of errors corrected, the code can be single error correcting or burst error correcting code.

2.1 Linear codes

Error Correction Codes for which any linear combination of codewords is also a code word are Linear Block Codes. The Linear Code $C(n,k,t)$ is a linear subspace of dimension k of vector field F_q^n where F_q is a finite field with q elements. The simplest linear block code is hamming codes.

2.2 Cyclic codes

In the Cyclic Codes C having algebraic properties, any cyclic shift of codewords results into a codeword. For every codeword $c \in C$ of length n, and the codeword c' obtained after the circular shift is also a codeword, i.e. $c' \in C$. If codeword $c = (c_1, c_2, ..., c_n)$ belongs to C then the codeword $c' = (c_n, c_1, ..., c_{n-1})$ obtained by right circular shift also belongs to C. The most common cyclic codes include BCH codes, RS codes and Cyclic Hamming codes.

Each cyclic code has different procedures for encoding and decoding. The number of errors corrected by the coding schemes differs from scheme to scheme.

3 COMMITMENT SCHEME

The commitment scheme can be considered similar to the sealed envelope where the sender wants to send a secret message to the receiver and that secret message is kept hidden inside the sealed envelope. The sender encrypts the message of concealed bit and transmits it. The commitment scheme must satisfy the binding and hiding properties. The receiver can recognize the message only after the sender reveal the concealed bit. Mathematically, Commitment Scheme is one to one mapping of the original message into an encoded message using witness value. It is impossible to predict the original value of the message without witness and the value of witness is revealed later by the sender. The mapping function f of original message msg into encoded message en_msg can be represented as

$$f(msg, witness) \rightarrow en\_msg \qquad (1)$$

The above function can be alternatively written as

$$f : msg \times witness \rightarrow en\_msg \qquad (2)$$

To identify the original message, the sender needs the $witness$ value, which is revealed later in the open phase. Thus, the open phase involved the function that evaluates the original message from the encoded message and this can be represented as

$$f : en\_msg \times witness \rightarrow msg \qquad (3)$$

The commitment scheme can be described using the three-phases and three-tuple. The three-tuples of the commitment scheme are $\{msg, E, I\}$ where msg represents the message space, E denotes the event occurring at a specific time and I represents the individual involved in the communication. The three phases of the Commitment Scheme are Setup Phase, Commit Phase, and Open Phase. The initial values are decided and published between the sender and the receiver during the Setup Phase. In the Commit Phase, the message is hidden or encoded along with the witness and send it to the receiver. At the Open Phase, sender reveals the value of witness to the receiver and then receiver open the commitment to obtain the original value of the message.

Many Commitment schemes have been suggested by several researches. The first commitment scheme was introduced by Blum (Blum, 1981). Another scheme was introduced by G. Brassard and Crépeau with the same efficiency parameters as Blum's commitment (Brassard & Crépeau, 1991). Goldwasser, Micali and Rivest (Goldwasser et al., 1988) scheme utilize the concept of collision free permutation-pairs. Pedersen (Pedersen, 1992) presented an efficient and non-interactive scheme that is based on discrete logarithm. Naor (Naor, 1990) proposes a construct that is based on the pseudorandom-generators. His interactive commitment scheme is implemented on the bounded receiver and unbounded sender model and requires two rounds of communication.

4 FUZZY COMMITMENT SCHEME

Fuzzy Commitment Scheme (FCS) is different from the Commitment Scheme with the fact that in the open phase, the receiver can decode the encoded message using slightly different witness value. The fuzziness in the FCS allows small errors in the witness. Thus, the decryption function becomes

$$f(en\_msg, witness') \to msg \qquad (4)$$

where *witness'* may or may not include corruptions.

The FCS can be considered as the three-phases and four-tuple scheme $\{msg, E, I, f\}$ where *msg* represents the message space, *E* describes the event occurring at a specific time and *I* specifies the individuals involved in the communication. Additional tuple *f* represents the error correction function.

During the setup phase, both sender and receiver agree on some values such as decoding method, resilience value, etc. At the commit phase, the sender encodes the message and send it to the receiver. At the open phase, receiver decommit the encoded message and obtain the original message. In this phase, the receiver is involved in the fuzzy decision-making.

5 PROPOSED SCHEME

The existing commitment schemes are based on different concepts, including discrete logarithm, one way permutation, collision free hashing and pseudo-random generators. The scheme in this paper utilizes concept different from the existing constructs. This paper uses the cyclic error correction codes and made it suitable for authentication. Table 1 depicts some of the existing scheme with the technique.

The proposed scheme is the combination of Error Correction Codes with the Fuzzy Commitment Scheme. Similar to the FCS, the proposed scheme is three-phased and four-tuple $\{msg, E, I, Dec\}$ scheme, where *msg* is the Galois array of symbols over $GF(2^m)$, *E* represents the event occurring at any specific time, *I* signifies the individuals involved in the communication and *Dec* describe the decoding algorithm used. The Table 2 defines the various operations performed in each phase of the scheme.

The proposed scheme can be explained using the Fig. 1. In the Setup phase, both sender and

Table 1. Commitment schemes.

Schemes	Complexity assumption
Naor	Pseudo-random Generators
Pederson	Discrete Logarithm
NOVY	One Way Permutation
GMR based	Factoring Blum Integer
Halevi & Micali	Collision Free Hashing
This paper	Cyclic Codes

Table 2. Phases in the proposed scheme.

Phases	Operations performed
Setup Phase	Initialize the Environment and Commitment is made. The sender and receiver agree on some values such as the value of n, k and t.
Commit Phase	Use Cyclic Codes based encoding scheme to encode the message and conceal the encoded message using the Fuzzy Commitment Scheme
Open Phase	The sender reveals the Fuzzy Commitment Key. The receiver first opens the commitment and then applies decoding to obtain the original message.

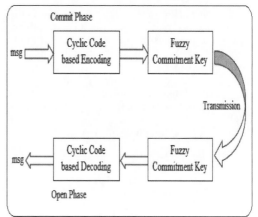

Figure 1. Proposed scheme.

receiver agree on the cyclic code and the decoding technique to be used. During the Commit Phase, the sender first adds the redundancy using the coding scheme. The message along with the parity bits is secured by Fuzzy Commitment Key. Message secured after by FCS is transmitted. The sender later reveals the Fuzzy Commitment Key in the Open Phase. The receiver after receiving the secured message, open the commitment and then apply the decoding procedure to obtain the original message.

The three phases of the proposed scheme and steps involved in these phases can be described as

5.1 Setup or initialization phase

After initializing the environment, both sender and receiver agree on the values of m and t.

5.2 Commit phase

In the open phase, the message is secure at two levels. Initially, the message is encoded using cyclic codes and then is concealed using the fuzzy commitment key. The open phase algorithm can be simply described in two steps

i. Encode msg into C using Cyclic Codes based Encoder.

$$C = CyclicEncoder(msg, n, k)$$

where msg is the Galois array symbols over $GF(2^m)$.

ii. The encoded message C is again secured by Fuzzy Commitment Key, i.e. key_{commit}

$$C_{msg} = FC(C, key_{commit})$$

5.3 Commit phase

At this phase, the fuzzy commitment key is disclosed. This is possible that errors may introduced during the transmission or a corrupted commitment key may be obtained. The commit phase is capable to correct up to t errors. Consider, instead of C_{msg} and key_{commit}, we obtain $t(C_{msg})$ and $t(key_{commit})$ which may or may not involve corruptions. If the combine errors are less than t, then we can successfully correct the errors using the decoder. The steps involved in the Open Phase algorithm are

i. Open the Commitment using the corrupted commitment key.

$$C' = FC(t(C_{msg}), t(key_{commit}))$$

ii. Decode the C' using the decoder

$$msg = CyclicDecoder(C', n, k)$$

6 RESULTS

The proposed scheme is implemented on the MATLAB using RS Codes and all the results are generated using the MATLAB on Intel® Core™ i5-4210U processor with 8.00 GB RAM. The RS Codes are capable of correcting both random and burst errors. The RS codes are non-binary cyclic codes having m bit symbols where m is any positive integer greater than 2. These are based on the polynomial over finite fields. The $RS(n,k)$ code on m bit symbol exists for which $n = 2^m - 1$ where k is the input data to be encoded and n is the number of symbols in the encoded data. Thus, $RS(n,k)$ code can be described as $(n,k) = (2^m - 1, 2^m - 1 - 2t)$ where t is the error correction capability of the code.

The RS cyclic code with burst error correction capability is considered to obtain the results. The random messages and random commitment keys with a random number of errors are considered during execution of the scheme. The time taken by both the Commit Phase and Open Phase code is observed and the mean of twenty-five execution of these phases is evaluated. Initially, different values of n over a fixed error correction capability are taken to observe the effect on the commit and open phase time. Table 3 lists the time taken by both commit and open phase for different values of n with fixed t value. The time for Commit Phase increases rapidly with the value of n as compared with the Open Phase time.

From the results obtained, this can be stated that the Commit phase time depends either on the value of n or k when the proposed scheme uses the RS code as the error correction code. For more precise results, the value of n is kept constant and for different values of k, the commit and open phase time is observed.

The effect of values of k on open and commit phase can be illustrated by the data presented in Table 4. The commit phase time does not depend on the value of k. Thus, this can be concluded that the commit phase time depends largely on the n value but not on k value. The open phase time slightly increases with the increase in the value of k.

The proposed scheme based on the combination of the Fuzzy Commitment Scheme and RS codes provides two level security. It is almost impossible to know the encoded message secured with the fuzzy commitment key without opening the commitment and that commitment key is revealed later to the receiver. If somehow, the commitment key is obtained by an intruder then without the exact value of m, n and k; RS codes cannot be decoded. Thus, for the intruder, it is infeasible to know the message secured by RS code and commitment key.

Table 3. Time taken by commit and open phase for different values of n.

m	n	k	t	Commit phase time (in sec)	Open phase time (in sec)
5	31	7	12	0.1187	0.0985
6	63	39	12	0.1686	0.1018
7	127	103	12	0.2630	0.0996
8	225	231	12	0.4617	0.0985
9	511	487	12	0.8436	0.1005
10	1023	999	12	1.6826	0.1073

Table 4. Time taken by commit and open phase for different values of k.

m	n	k	t	Commit phase time (in sec)	Open phase time (in sec)
4	15	13	1	0.0625	0.0332
4	15	11	2	0.0714	0.0411
4	15	09	3	0.0648	0.0432
4	15	07	4	0.0705	0.0458
4	15	05	5	0.0683	0.0505
4	15	03	6	0.0739	0.0590
4	15	01	7	0.0727	0.0666

In Commit Phase, the message is first encoded using the Error Correction Code and the encoded message is again secured using the Fuzzy Commitment key. The message cannot be found by directly applying decoding without opening the Fuzzy Commitment key and that key is revealed later by the sender in the open phase. The receiver cannot predict the message before the open phase, thus, satisfies the hiding property. The message obtained after the open phase is same as the message send by the sender and the sender cannot deny that the message is not the one that he sent. Thus, satisfies the binding property. The proposed process satisfies both binding and hiding properties.

7 CONCLUSIONS

The proposed scheme provides security at two-level and it is also immune from the intruder thus; it can be applicable for securing data. The proposed scheme ensures both binding and hiding properties. The time taken by the commit phase of the scheme when RS Codes are used depends on the value of n. The fuzziness in our scheme allows it to be used for those applications that can tolerate small errors such as Securing Biometric Templates, etc. Moreover, the proposed scheme satisfies both binding and hiding properties that are essential in case of Commitment Scheme.

REFERENCES

Al-Saggaf & Acharya, H.S. 2007. A Fuzzy Commitment Scheme. *IEEE International Conference on Advances in Computer Vision and Information Technology*: 28–30.

Bennett, C.H. & Brassard, G. 1984. Quantum Cryptography: Public Key Distribution and Coin Tossing. *IEEE International Conference on Computers, Systems, and Signal Processing, IEEE*: 175–179.

Berlekamp, E.R. & Peile, R.E. 1987. The Applications of Error Control to Communications. *IEEE Communications Magazine* 25(4): 44–57.

Berlekamp, E.R. 1968. *Algebraic Coding Theory*. New York: McGraw-Hill.

Berlekamp, E.R. 1973. Goppa Codes. *IEEE Transactions on Information Theory* 19(5): 590–592.

Blum, M. 1981. Coin flipping by telephone. *Advances in Cryptology: A report on CRTPTO'81*: 11–15.

Bose, R.C. & Ray-Chaudhari, D.K. 1960. On a Class of Error Correcting Binary Group Codes. *Information and Control* 3: 68–79.

Brassard, G. & Crépeau, C. 1986. Non-Transitive Transfer of Confidence: A Perfect Zero-Knowledge Interactive Protocol for SAT and Beyond. *Proceedings of 27th IEEE Symp. on Foundations of Computer Science*: 188–195.

Brassard, G. & Crépeau, C. 1991. Quantum bit commitment and coin tossing protocols. *Proceedings of CRYPTO '90*, Lecture Notes in Computer Science, Springer-Verlag, 537: 49–61.

Chauhan, S. & Sharma, A. 2016. Fuzzy Commitment Scheme based on Reed Solomon Codes. *Proceedings of SIN '16*: 96–99.

Claude, C. 1997. Efficient Cryptographic Protocols based on Noisy Channels. *Advances in Cryptology-EURO CRYPT '97, LNCS 1233*: 306–317.

Gallager, R. G. 1963. *Low Density Parity Check Codes*. Cambrigde, MA: MIT Press.

Goldwasser, S., Micali, S. & Rivest, R. 1988. A digital signature scheme secure against adaptive chosen-message attacks. *SIAM Journal of Computing* 17(2): 281–308.

Goppa, V.D. 1970. A new class of linear error correcting codes. *Probl. Peredach. Inform.* 6(3): 24–30.

Halevi, S. 1995. Efficient Commitment with bounded sender and unbounded receiver. *Proceedings of CRYPTO '95, Lecture Notes in Computer Science, Springer-Verlag* 963: 84–96.

Halevi, S., Micali, S. 1996. Practical and Provably-Secure Commitment Schemes from Collision-Free Hashing. *Proceedings of Advances in Cryptology-CRYPTO '96*: 201–215.

Juels, A. & Wattenberg, M. 1960. A Fuzzy Commitment Scheme. *ACM Conference on Computer and Communication Security Vision an Information Technology* 8(2): 28–36.

MacKay, D.J.C. & Neal, R.M. 1996. Near Shannon Limit Performance of Low Density Parity Check Codes. *Electronics letters* 32(18): 1645–1646.

Naor, M. 1990. Bit Commitment using pseudo-randomness. *Proceedings of CRYPTO '89, Lecture Notes in Computer Science, Springer-Verlag* 435:128–137.

Naor, M., Ostrovsky, R., Venkatesan, R. & Yung, M. 1990. Perfect Zero Knowledge arguments for NP can be based on general complexity assumptions. *Proceedings of CRYPTO '92, Lecture Notes in Computer Science, Springer-Verlag* 740:196–214.

Ojha, D.B. & Sharma, A. 2010. A Fuzzy Commitment Scheme with McEliece's cipher. *Surveys in Mathematics and its Application* 5: 73–82.

Pedersen, T.P. 1992. Non-Interactive and Information-Theoretic Secure Verifiable Secret Sharing. *Proceedings of CRYPTO '91, Lecture Notes in Computer Science, Springer-Verlag* 576:129–140.

Reed, I.S. & Solomon, G. 1960. Polynomial Codes over certain Finite Fields. *Journal of the Society for Industrial and Applied Mathematics* 8: 300–304.

Shannon, C.E. 1948. A Mathematical Theory of Communication. *Bell System Technical Journal* 27: 623–656.

Communication and Computing Systems – Prasad et al. (Eds)
© 2017 Taylor & Francis Group, London, ISBN 978-1-138-02952-1

An improved enhanced developed distributed energy efficient clustering algorithm for wireless sensor networks

A.S. Sharma & S. Pathak
School of Information and Communication Technology, Gautam Buddha University, Greater Noida, India

ABSTRACT: Great number of randomly deployed motes of limited battery power together form wireless sensor network. Sensor motes comprised an effective way to sense, process and communicate with Base Station (BS) which dissipates a large amount of energy and affect sensor network lifetime. In WSNs, reducing energy consumption and increase in network lifetime is a great challenge. Clustering is a pivotal approach employed to make an energy efficient WSN. Here, we are proposing a novel Cluster-Head (CH) selection technique: Improved EDDEEC for heterogeneous WSNs which depends on both probability based approach of Enhanced Developed Distributed Energy Efficient Clustering scheme (ED-DEEC) and distance from Base Station. The results from simulation shows that our proposed protocol attains 69.4% longer sensor network lifetime compared to Distributed Energy Efficient Clustering (DEEC), 96% to Developed-DEEC (D-DEEC) and 19% to Enhanced-DEEC (E-DEEC).

Keywords: WSNs; Energy Efficient; Clustering; Heterogeneous; Stability Period; Lifetime

1 INTRODUCTION

Sensor nodes are deployed densely in the physical phenomenon and form wireless sensor network. These tiny sensor nodes (motes) comprised an effective way to sense, process and communicate with its neighboring nodes in a specific network area (Heinzelman et al. 2000). Sensor nodes transmit relatively processed data by carrying out computation locally instead of forwarding raw data. Based on the deployment of sensor nodes WSN is categorized into heterogeneous and, homogenous network. All motes with alike energy level or same battery power level come under homogenous network. There are many clustering algorithms which are only designed for homogenous network such as PEGASIS (Power Efficient Gathering in Sensor Information Systems) and LEACH (Low-Energy Adaptive Clustering Hierarchy) (Qing et al. 2006). In these algorithms for homogenous network all nodes are of same energy level which works poorly in heterogeneous environment. In heterogeneous network all nodes are of different energy level or different battery power. The clustering algorithms which are designed for heterogeneous network are DEEC (Distributed Energy Efficient Clustering), SEP (Selection Aware Protocol).

The rest of the work is assembled as follows. In Section 2, we concisely evaluate related work. Section 3 relates the system model of the proposed algorithm. Section 4 shows the simulation results of proposed algorithm and briefly discuss the results with other DEEC, DDEEC, and EDEEC algorithms. Finally, Section 5 imparts concluding remarks and future work.

2 RELATED WORK

This section outlines related research work in details. Heinzelman et al. 2000, proposed LEACH clustering technique for homogenous WSNs. In this technique nodes randomly elects themselves as CHs. This selection of CH dynamically changes with time to balance the load in network.

Qing et al. 2006, proposed DEEC technique for two—level heterogonous WSNs. In this technique CH election is based on probabilistic approach determined by the ratio of residual energy of sensor nodes (motes) and average energy of sensing field.

Elbhiri et al. 2010, proposed DDEEC technique in which CH election is based on residual energy of sensor nodes. The CH selection criteria dynamically changes in proportion to their residual energy.

Saini et al. 2010, proposed EDEEC clustering technique for three—level heterogonous WSNs. In this technique CH election is based on probabilistic approach determined by the ratio of residual energy of sensor motes and average energy of sensing field.

Javaid et al. 2015, introduced EDDEEC technique for three-level heterogonous WSNs in which CH choice is done in the support of probabilistic approach determined by the ratio of residual energy of sensor nodes and average energy of sensing field. The CH selection criteria dynamically changes in proportion to their residual energy.

Liao et al. 2013, proposed Load Balanced clustering algorithm in which relative distance between node and Base Station takes into consideration for CH selection and Clustering which gives better results.

Our novel clustering idea called improved-EDDEEC (Enhanced Developed Distributed Energy Efficient Clustering) technique is based on EDDEEC with inclusion of distance based approach during CH selection In selecting a tentative cluster-heads phase, a hybrid approach is used for CH selection depends both on distance to Base Station and the ratio of residual energy of sensor nodes (motes) and average energy of sensor network. Thus stability as well as lifetime of the sensor network increases by reducing the overall energy consumption during transmission in the system.

3 SYSTEM MODEL

In this section we discuss our improved—EDDEEC protocol.

3.1 Network scenario

We consider a sensing field of area 100 m × 100 m consisting of three types of N heterogonous sensor nodes randomly deployed over it to continuously monitor the environment. Base Station is location in center of the field as shown in Figure 1.

Three types of nodes are different with each other with respect to their energy levels or battery power. The nodes with low energy i.e. E_o named as normal node, nodes with intermediate energy i.e. $E_o(1 + a)$ is named as advance nodes whereas nodes with high energy i.e. $E_o(1 + b)$ is named as super node. As there are N sensor nodes in the network, so the number of normal, advance and super node in the network is $N(1-m)$, $Nm(1-m_o)$ and Nmm_o. Total energy in the network is calculated as (Javaid et al. 2015):

$$E_T = E_oN(1-m)+E_o(1+a)Nm(1+m_o)+E_o(1+b)Nmm_o$$

$$E_T = NE_o(1+m(a+m_ob)) \qquad (1)$$

Hence, from above calculation we find that three-level heterogonous WSN has $m(a+m_ob)$ times more energy than homogenous network.

3.2 Energy Dissipation Model

We adopt the radio energy model (Heinzelman et al. 2000). This model describes energy dissipation in transmission or reception of one bit message over a distance d. Equation of the model is given as follows:

$$ETX\,(1,d)=l.E_{elec}+l.\varepsilon_{fs}.d^2 \quad \text{if } d<d_o \qquad (2)$$

$$ETX\,(1,d)=l.E_{elec}+l.\varepsilon_{amp}.d^4 \quad \text{if } d\geq d_o$$

$$d_o = \frac{\varepsilon_{fs}}{\varepsilon_{amp}} \qquad (3)$$

ϵ_{amp} and ϵ_{fs} are the energy dissipation models used. d_o is the threshold distance between transmitter and receiver. We use this energy dissipation model information for energy computation in our proposed model. We consider some assumptions about our proposed model:

1. All nodes are stationary, location unaware and assigned with unique Identifier (ID).
2. There is no loss of packet due to collision during transmission.
3. Energy loss due to interference is negligible between signals of heterogeneous nodes.

3.3 An improved—EDDEEC protocol

Here we introduce the detailed explanation of improved EDDEEC protocol. Firstly BS decides whether node is eligible to become a cluster head based on threshold of node which is calculated by proposed percentage and the number of rounds the node has been a cluster-head previously. If the value we get from calculation is less than threshold T(s) value then only node is in the race of CH selection of the existing round. The threshold is set as:

$$T(Si) = \begin{cases} \dfrac{P_i}{1-P_i\left(rmod\dfrac{1}{p_i}\right)} & if\, s \in G \\ 0 \; \text{otherwise} \end{cases} \qquad (4)$$

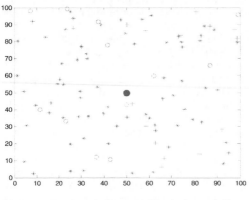

Figure 1. Random deployment of nodes in network.

BS calculates Average energy of network from residual energy of node as:

$$E(r) = \frac{1}{N} E_{total} \left(1 - \frac{r}{R}\right) \tag{5}$$

$$R = \frac{E_{total}}{E_{round}} \tag{6}$$

where E_{round} is the energy dissipated in one round and E_{total} is the total energy in current round.

$$d_{toBS} = 0.765 \frac{M}{2} \tag{7}$$

$$d_{toCH} = \frac{M}{\sqrt{2\pi k}} \tag{8}$$

$$k_{opt} = \frac{\sqrt{N}}{\sqrt{2\pi}} \sqrt{\frac{\in_{fs}}{\in_{mp}} \frac{M}{d_{toBS}^2}} \tag{9}$$

Energy dissipated in single round can be calculated using (7), (8) and (9) as:

$$E_{round} = L(2nE_{elec} + nE_{DA} + k \in_{mp}$$
$$d_{BS}^4 + n \in_{fs} d_{toCH}^2 \tag{10}$$

Average Probability of a node to be cluster-head is expressed as (Javaid et al. 2015):

$$P_i = \begin{cases} \dfrac{P_{opt} E_{i(r)}}{(1 + m(a + m_o b)) E_{(r)}} & \text{for Nml nodes,} \\ & \text{If } Ei(r) \le Tab \\[2mm] \dfrac{P_{opt}(1+a) E_{i(r)}}{(1 + m(a + m_o b)) E_{(r)}} & \text{for adv nodes,} \\ & \text{If } Ei(r) \le Tab \\[2mm] \dfrac{P_{opt}(1+b) E_{i(r)}}{(1 + m(a + m_o b)) E_{(r)}} & \text{for sup nodes,} \\ & \text{If } Ei(r) \le Tab \\[2mm] c\dfrac{P_{opt}(1+b) E_{i(r)}}{(1 + m(a + m_o b)) E_{(r)}} & \text{for Nml,} \\ & \text{Adv, sup;} \\ & \text{If } Ei(r) \le Tab \end{cases} \tag{11}$$

The value of absolute residual energy level, T_{abs}, is written as:

Table 1. Parameters description.

Parameter	Description
pi	Cluster-head desired probability
R	Current Round
E(r)	Average energy of Network
Tsi	Threshold Energy
do	Crossover distance
Eelec	Energy dissipation to start the transmitter or receiver
$\in$ mp	Multipath channel model
$\in$ fs	Free-space channel model
Popt	Optimum probability to be cluster-head
Ei(r)	Residual Energy

$$T_{abs} = 0.7 E_o \tag{12}$$

3.4 Algorithm of our improved protocol

Set up phase
// Begin
// Formation of sensor network
Initialize i, i{s_1, s_2, s_3, ..., s_n}
Set S_{sink}: = s_{n+1} // n+1 node is base station
Cluster-Head formation phase
for i: = 1 to n
// Calculate residual energy of a si, E_{si}
E_{si}: = $E_{initial(si)}$ - $E_{consumed(si)}$
// $E_{consumed(si)}$ is calculated using first radio energy model as defined by [ ]
// $E_{initial(si)}$ is the energy of node in the starting of particular round
// $E_{av(si)}$ is the average energy of network in particular round
If E_{si}< threshold
Use P_i based on node type
end if
for i: = 1 to n
If P_i = high then
set counter CH: = 1
&& CH: = ECH
else if set counter CH: = 0
end if
end else if
end loop
if ECH ≠ ReqCH
then
// Find out distance between node with high P_{si} and Base station

$$|\text{min\_dist}_{n+1}, si|: = \sqrt{(X - x_i)^2 + (Y - y_i)^2}$$

If $|\text{min\_dist}_{n+1,si}|$ = less then
Set counter ECH: = 1
else Set counter ECH: = 0
end if
end else if

799

ECH = ReqCH
If ReqCH = FCH
end if
end if
end loop
Cluster formation phase
for i: = 1 to n
// For non CH nodes s_j, find distance from Final CH

$$|dists_{j,si}| = \sqrt{(x_j - x_i)^2 + (y_j - y_i)^2}$$

If $|dist_{sj,si}| = min\_dis$
set: = 1 form cluster
otherwise
set: = 0 cluster not form
end if
end for

4 SIMULATION RESULTS AND DISCUSSIONS

Compared results are taken between the improved protocol and the compared protocols during simulation in a network field of 100 m × 100 m area. Parameters utilized during simulation are listed in Table 2. In proposed scenario, we located 32% advance nodes and 48% super motes with energy 2 times and 3.5 times more than normal motes (m = 0.8, mo = 0.6, a = 2, b = 3.5).

4.1 Number of alive nodes

Figure 2 gives a comparative statistics of the persisting alive nodes in the network of our proposed algorithm I-EDDEEC with other protocols like DEEC, E-DEEC, D-DEEC. The criteria of more nodes to Alive for further number of rounds, used to find the performance of our proposed protocol with compared protocols regarding Stability Period, and Network Lifetime which are defined as:

- Stability Period: Round at which first node dead during simulation. Figure 2 depicts that for DEEC, E-DEEC, D-DEEC, and I-EDDEEC

Table 2. Simulation parameters.

Parameters Values	Values
N	100
L	4000 bits
ϵ_{elec}	50nJ/bit
ϵ_{fs}	10nJ/bit/m^2
ϵ_{mp}	0.0013p J/bit/m^4
E_o(initial energy of normal nodes)	0.5 J
EDA	5nJ/bit/signal
R (Total Rounds)	12000
do(threshold distance)	70 m
P_{opt}	0.1

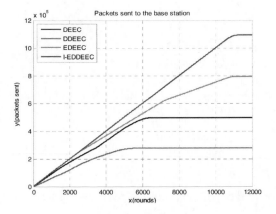

Figure 2. Alive nodes in network.

Figure 3. Packets sent to base station (Sink).

first dies at 4078, 1384, 3846, and 6079 rounds which prove that our improved EDDEEC has more stability period with compared protocols like DEEC, E-DEEC, and D-DEEC.

- Network Lifetime: Round at which all nodes died during network simulation. Also figure 2 depicts that all nodes for DEEC, E-DEEC, D-DEEC, and I-EDDEEC dies at 6635, 11035, 5736, and 11246 rounds which proves that our improved EDDEEC has improved lifetime of network as compared to other protocols like DEEC, EDEEC, DDEEC.

4.2 Packets Sent to Base-Station (BS)

Figure 3 depicts a comparative statistics of numbers of Packets sent to Base Station which is in case of our proposed protocol I-EDDEEC are 877778 which is very high as compared to other chosen protocols. This is due to the more remaining alive nodes present in additional rounds in our proposed algorithm. The Table 3 shows the performance metric of compared protocols.

Table 3. Performance metric of different protocols.

Performance metric of different protocols

Protocol N = 100; L = 4000	Period of stability	Lifetime of network	Packet sent
DEEC	4078	6635	432728
DDEEC	3846	5736	246613
EDEEC	1384	11035	640869
I-EDDEEC	6079	11246	877778

5 CONCLUSION AND FUTURE WORK

An Improved Energy Efficient Clustering Protocol (I-EDDEEC) for heterogeneous network is proposed in this paper. The results clearly shows that, for additional number of rounds the period of stability and network life-time regarding more number of nodes to stay alive and in terms of more packets sent to Base Station is better than compared protocols, DEEC, DDEEC, EDEEC. The percentage of the number of nodes stay alive increased to 69.4%, 96% and 19% in contrast to DEEC, D-DEEC, and E-DEEC.

Future work focuses on the techniques which improves Coverage and connectivity of the network. We can also use different key management techniques to make the communication secure.

In our work we have used stationary sensor nodes and base station, in future we plan to work in mobile environment.

REFERENCES

Elbhiri, B. Rachid, S. Fkihi, S.E. & Aboutajdine, D. 2010. Developed Distributed Energy-Efficient Clustering (DDEEC) for heterogeneous wireless sensor networks. In *Proc. 5th Int. Symposium on I/V Communications and Mobile Network, 2010.*

Heinzelman, W. Chandrakasan, A. & Balakrishnan, H. 2000. Energy-efficient communication protocol for wireless micro sensor networks. In System Science, *Proc. 33rd Annual Hawaii Intern Conf., January 2000.*

Javaid, N. Qureshi, T.N. Khan, A.H. Iqbal, A. Akhtar, E. & Ishfaq, M. 2015. An Energy Efficient Distributed Clustering algorithm for heterogeneous WSNs. *Wireless Communications and Networking* 151: 1–11.

Liao, Y. Qi, H. & Li, W. 2013. Load-balanced clustering algorithm with distributed self-organization for wireless sensor networks. IEEE Sens. 12(5): 1498–1506.

Qing, L. Zhu, Q. & Wang, M. 2006. Design of a distributed energy-efficient clustering algorithm for heterogeneous wireless sensor networks. *Computer Communications* 29: 2230–2237.

Saini, P. & Sharma, A.K. 2010. E-DEEC—Enhanced Distributed Energy Efficient Clustering Scheme for heterogeneous WSN. In *Proc. 1st Int. Conf. Parallel, Distri. and Grid Comput., 2010.*

Communication and Computing Systems – Prasad et al. (Eds)
© 2017 Taylor & Francis Group, London, ISBN 978-1-138-02952-1

Optimized resource allocation for monitoring dengue patients

Sandeep Sharma & Pratiksha Sharma
Gautam Buddha University, Greater Noida, India

ABSTRACT: Dengue is a disease that is caused by mosquito bites. There is not any particular medication for it rather it requires continuous monitoring by doctor. When a person gets affected by dengue it goes under various blood tests to diagnose the level of dengue which may take a little longer to reach the patient who is already having high dengue level, ignorance of which may cause death. In India, during august-september, the dengue, at its peak, causes several deaths due to improper allocation of doctors as well as delay in conducting tests. We provide a solution to scarcity of doctors at one place while plenty at another place by introducing an Optimized algorithm to allocation of doctors. Approach, requires the input data by a web portal consisting spatial information of dengue.

1 INTRODUCTION

Dengue is a disease that is caused by mosquito bites. There is not any particular medication for it rather it requires continuous monitoring by doctor. When a person gets affected by dengue it goes under various blood tests to diagnose the level of dengue which may take a little longer to reach the patient who is already having high dengue level, ignorance of which may cause death. In India, during august-september, the dengue, at its peak, causes several deaths due to improper allocation of doctors as well as delay in conducting tests. We provide a solution to scarcity of doctors at one place while plenty at another place by introducing an Optimized algorithm to allocate of doctors. Approach, requires the input data by a web portal consisting spatial information of dengue in 250–500,000 of patients every year, in southeast Asia. We have provided a solution to the problem that is to apply specific amount of medical facility in a specific region and providing optimal medical resource allocation to them, by continuously monitoring the amount of chronic stages in dengue and outbreak detection. The approach is influenced by Springer published paper which uses a lightweight image processing technique to detect dengue and is done by using a mobile camera. In this approach, dengue is tested using an algorithm that uses a $20USD medical patch that displays the result on a cellular telephone and this data is transmitted further to surveillance agencies for controlling and prevention and additional testing purposes. The transmission of data will enable to access the lists of patients with chronic and non chronic dengue level. This paper also briefly explains the approach of detecting dengue using a mobile phone and a lightweight image processing algorithm. In the end, we summarize with an optimized algorithm for resource allocation where resources are the teams of doctors at various hospitals.

2 DENGUE

Aedes mosquitoes are the primary transmitter of the dengue and dengue hemorrhagic fever. These are the main cause which causes significance illness in the patients if the same are not diagnosed on or treated in time. The vaccine for this disease is currently unavailable. Intensive monitoring of patient and proper fluids are given to the patient as a treatment of this disease. Medical experts use clinical methods to diagnose dengue infection in the patients. Dengue a very old disease is a fever that has re-emerged in past two decades. It was found in the year 1998 that dengue was categorized as a dangerous infectious disease found in tropical areas. Dengue fever caused lives of thousands annually. Out of 5 Lakh cases of dengue fever, 25 Thousand deaths were recorded yearly. The occurrence of dengue is dependent on geographical region. This can be understood using diagnosis of dengue fever in clinics and pathology labs. This way it can be prevented and controlled in that geographical area. Dengue virus is spreading dangerous infection and hence is increasingly recognized as world's major emerging infectious diseases. Many experiments and studies have been carried out to assess dengue fever in the patients. Because of which it is possible to recalculate the severity of dengue level among patients. Aedes mosquitoes have an affinity towards breeding on human blood. Humans are therefore the primary victims which also can

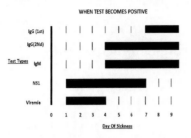

Figure 1. Level of dengue detection.

spread to others. Dengue was one of neglected diseases and now has emerged as a very dangerous disease globally. This infection is now acknowledged as a hazard globally. Aedes Aegypty and also Aedes Elbow Pictus are the mosquitoes through which this disease is transmitted. Both show great adaptability to change of environment in humans. In many tropical areas, dengue is spread at a very fast pace. This is usually found in housing areas which are not properly regulated. The mosquitoes carrying the dengue virus are adapted to modern life environment. These mosquitos breed in clean water lying in the residential areas. Female mosquitos rapidly transmit dengue as they have a natural affinity towards human blood. The Figure 1 illustrates how the dengue level is diagnosed, higher the level, higher are the chances of getting it chronic, which not only depend upon the test type but also the day it has been tested.

3 DENGUE DETECTION

The goal of mobile dengue detection is to improve the quality of life in developing countries through proper disease diagnosis and monitoring. Oscan et al. created an original tool which can do analysis of the blood which helps in detecting malaria. A special lens was used which was based on the concept of Shadow imaging (LUCAS). This lens has special camera on the cell phone. It uses small amount of blood sample of the patient. This blood sample is placed in stacked trays, which is positioned on top of CMOS sensor of the cellar phone. The light from the LED (Light Emitting Diode) filters the wavelength of the light passing through the trays and hence exposes the blood cells. After light is filtered, the blood cells casts shadow images. The shadow images are then uploaded to the server which has the LUCAS processing platform. This then uses template matching algorithm to identify the diseased blood cells. These diseased blood cells can depict dengue virus on the cells. Sometimes also inside the blood cells. However it cannot classify the type of virus strand as good as dengue patch. Due to this

the patch is successful in detecting and diagnosing dengue infection. The test kits which detect dengue fever are based on the PCR or immuno-chromatographic. In India, a version of PCR bioassay methodology was developed for qualitative detection of dengue. This test kit helps in diagnosing IgG and IgM antibodies in human blood called as Erba Den-Go. One of the best solution to track dengue is Erba Den-GO, however such kits are priced very high varying between $200 (Rapid DipStick Test) to $700 (IgG/IgM Card Test Kit).

CMOS color camera module is used along with ARM processors for image processing. Heavy processing power is not required to analyze the patch effectively in the algorithm thus, ARM processor with low processing power is used with camera module for effective dengue detection, also cellular phones devices camera which has limited processing capability can be used for the dengue detection. The research shows simple way to detect dengue. A cellular phone's camera is utilized to process the medical patch that turns different color shades.

The level of disease can be diagnosed by this solution which only utilizes a bio-assay based test along with an image processing algorithm. Other alternate solutions detect dengue by using bio-assays, also with microscopy to detect the virus. Treatment of dengue suspected patient involves initial clinical inspection. Blood test sample is taken for testing purposes. Also the virus containing components are extracted out from the serum of the patient. The test is then administered a lot of times to conclude the diagnosis. These tests can last up to 5 days to diagnose the severity of the infection. The dengue patient is usually then hospitalized where more tests are conducted. The patient is also administered with antibiotics which helps in fighting with the virus. Traditional methodology major drawback is that it takes a lot of time, the level of virus increases each day, and it might get worse for the patient. The new system is described. A suspected patient's blood sample is first taken. Analytes of the patient mix into the chambers of patch. The reactions take places with reagents in each well. After 30 minutes the device is inspected. One can simply tell the abnormalities in the sample just by looking with the naked eye. The medical expert then takes and image of the device, which is processed and color levels in each well are compared with reference randes. This concludes with a quantitative results of the analyte levels in the patient. Depending on results of the above test, patient are given the most appropriate treatment. The treatment usually includes taking in fluids, hospitalization for special care of the patient. This method detects sooner, which allows for faster diagnosis and treatment of same.

4 OPTIMIZED ALGORITHM

The approach to achieve the optimized algorithm is to reach to a zone which has dengue patients based upon chronicity to be treated on prior basis. The zone is segregated and then the total number of chronically affected dengue patients as well as non chronically affected dengue patients are calculated and the area is calculated with the total number of patients thus providing an idea for the calculation of doctors that to be applied for monitoring the dengue patients. The Figure 2 illustrates the area consisting of three regions having chronically affected as well as non-chronically affected dengue patients. The chronically affected patients are indicated with red, while non-chronically affected ones are indicated with orange. The zones are prioritized with the number of increased chronically affected patients not the total affected patients. Hence priority is given to the chronically affected dengue patients. This approachhelps in finding total number of doctors and resources are available

$$C \div N \qquad (1)$$

Where, C is the number of non-chronically affect patients
N is the number of doctors required

Related to the Figure 2, Figure 3 illustrates the algorithm proposed. The algorithm depicts that the region is segregated first and the total number of patients is calculated on the basis of their chronicity. In this algorithm the chronic patients are assigned with one doctor, on the basis of their dengue level, and the non chronic patients are

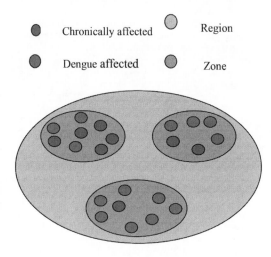

Figure 2.   'Approach to derive an optimized algorithm'.

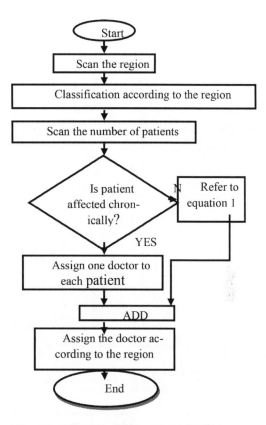

Figure 3.   'Flowchart of proposed algorithm'.

assigned with 'N' number of doctors. This variable 'N' depends upon the total number of doctors available to the number of patients in that zone. After the assignment the total number of doctors required is then calculated to check whether the total number of doctors available is equal to the number of doctors assigned.

4.1   Optimized algorithm in steps

Step 1: Select an area
Step 2: Collect data
    (a) Get total number of Doctors available for that area (N)
    (b) Get Total Chronic Patients for that area (C1)
    (c) Get Total Non Chronic Patients for that area (C2)
Step 3: Calculate total doctors required for chronic patients by dividing C1 with N
Step 4: Similarly, Calculate total doctors required for non-chronic patients by dividing C2 with N
Step 5: Add Total number of doctors required (Q)
Step 6: Calculate the distance of all medical zone in that area

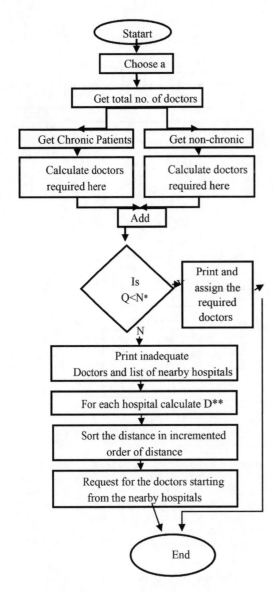

Figure 4. 'Proposed optimized algorithm'.
*RD is total number of required doctors AD is the total available doctors.
**d is the distance from nearby hospital.

Step 7: Satisfy following cases;
   (a) If, Q = N, print data and implement doctors
   (b) If, Q < N, print data and implement doctors
   (c) If, Q > N, generate request to any nearby medical zone
Step 8: If availability of doctors is scarce, then forward request to next nearby hospital and so on until find one to satisfy the step.
Step 9: Implement the algorithm.

5 CONCLUSION

Our optimized algorithm find the total number of doctors requires in that zone to treat an amount of patients and if the scarcity occurs it do have steps to find out further, which nearby zone do have tendency to lend doctors. Thus fulfilling the scarcity of resources in an optimized way, also proving proper allocation of resources

6 FUTURE REFERENCE

To our further enhancement of this algorithm, work on other environments to process the data effectively and try to process the accessing of data. Also, analysis of data to find the zone having high persistency of dengue patients diagnosed and further merging it with the treatment of the factors affecting dengue efficiently.

REFERENCES

Bassil 2012, "A comparitve study on the performance of the top DBMS systems" Journal of computer science and reaserch (JCSCR), Vol.1, pp. 20–31

Bhargava & Chatterjee 2007, "Chikungunya Fever, Falciparum Malaria, Dengue Fever, Japanese Encephalitis...Are We Listening to The Warning Signs For Public Health in INDIA?' Indian Journal of Medical Ethics, Vol. 4, pp. 19–20

Bennet et al. "Epidermics dynamics revealed in dengue evolution", oxford journals, Vol. 27, pp. 811–818

Bernstein 2015, Dengue fever (Online) Available: http://www.m.webmb.com

Datta & Wattal "Dengue NS1 antigen detection: a useful tool in early diagnosis of dengue virus infection" Indian Journal of medical Microbiology 2010, Vol. 28, pp. 107–110

Garg & Amit 2010 "Dengue infection: prevalence study at hospital of western UP India" India journal of applied research Vol. 6, pp. 230–233

Garg, Nagpal, Khairnar, Seneviratne "Economic burden of dengue infection in India" Trans R Soc Trop Med Hyg 2008, Vol. 102, pp. 570–577

Gubler 2006, "Dengue and dengue hemorrhagic "fever" Novartis Found Symp, Vol. 27, pp. 7:3–16

Guzman et al. "Dengue: a continuing global threat" Nature reviews Microbiology 2010, Vol. 496, pp. 504–507

Hahn et al. "Development of a framework evaluating the sustainability of community-based dengue control projects" AM J tropical medical hygiene 2009, Vol. 80, pp. 312–318

Heintze, et al. "What do community-based dengue control programmes achieve? A systematic revie of published evaluations" Trans R soc Trop med hyg 2007, Vol. 101, pp. 317–325

Kroeger et al. "Effective control of dengue vectors with curtains and water container covers treated with insecticide in mexico and venezula: cluster randomized trials" British Medical Journal 2006, Vol. 332, pp. 1247–1252.

Kuberski, Rosen 1977, "A simple technique for the detection of dengue antigen in mosquitoes by immune fluorescence" American Journal of Tropical Medicine and Hygiene, Vol. 26, pp. 533–537

Knox et al. "Critical evaluation of quantative sampling methods for Aedes aegypti immature in water storage containers in Vietnam" J med entomol 2007, Vol. 44, pp. 192–204

Lanciotti et al. 1992, "Rapid detection and typing of dengue viruses from clinical samples by using reverse transcriptase-polymerase chain reaction" J Clin Microbiol, Vol. 30, pp. 545–551

Matthews et al. 2012 "Rapid Dengue and Outbreak Detection with mobile Systems and Social Networks", Springer, Vol. 17, pp. 178–191

Nathan et al.. "Pupal/demographic surveys to inform dengue-vector control" Ann Trop Med Parasitol 2006, Vol. 100, pp. S1–3

Narayan 2015, Dengue crisis exposes negligence, poor medical facilities in India's capita l (Online) Available: http://www.thecitizen.in/index.php/OldNewsPage/?id = 5180

Paddock 2015, Dengue fever: symtoms, treatments, prevention (Online)Avaible:http://www.medicalnewstoday.com/articles/179471.php

Suaya et al. "Cost-effectiveness of annual targeted larviciidng campaigns in Cambodia against the dengue vector Aedes aegypti" Tropical medical international health 2007, Vol. 12, pp. 1026–1036

Suaya et al. "Cost of dengue in eight countries in the America and Asia: a prospective study" AM J Trop Med Hyg 2009, Vol. 80, pp. 846–855

Thivakaran, reviewed by Nithin jayan (2016, May 2) Dengue and dengue fever (Online) Available:http://www.medinida.net/patients/patientnfo/dengue.htm & ei=30qje1p1&Ic=enIN&s=1&m=439&host=www.google.co.in&ts=143334306&sig=APY536yGqERCyvubouiXzjcdUWfJG2 je-Q

Ukey, Bonadade et al. "Study of seroprevalence of dengue fever in central India" Indian j community Med 2010, Vol. 35, pp. 517–519/a-to-z-guides/dengue-fever-reference

Wali et al. 1999, Wig, Diwedi "Dengue Haemorrhagic Fever in Adults: A Prospective Study of 110 Cases" Trop Doct January, Vol. 29, pp. 27–30

Communication and Computing Systems – Prasad et al. (Eds)
© 2017 Taylor & Francis Group, London, ISBN 978-1-138-02952-1

Analyzing the interestingness of association rules extracted from vaccine medical report

M. Kaur & H. Singh
Department of CSE, DAV University, Jalandhar, India

ABSTRACT: Vaccines have been one of the effective public health medications. Although, vaccines are provided to protect from life threatening diseases but these pharmaceutical products can cause adverse effects. This paper presents a method that adopts a text mining system to discover association rules which are interesting and extracted from vaccine medical reports. Text mining technique has been used to extract interesting patterns or knowledge from large text corpus. In this paper medical reports of those patients who suffered from adverse effects of vaccines have been used. The symptoms of patients in medical reports are categorized into background knowledge and target documents. Further the evaluation of interestingness of extracted association rule is analyzed by computing the semantic distance between predecessor and successor of association rule.

1 INTRODUCTION

Every year millions of infants have to undergo vaccination. In year 2010, more than 2000 babies died (Murphy, Xu & Kochanek, 2013) from SIDS (sudden infant death syndrome). Process of vaccination is not 100% safe for everyone; people who are allergic to eggs can't consume flu vaccines because some vaccines contain eggs while manufacturing (Daniel J. DeNoon, 2009). DTap (Diphtheria, Tetanus and acellular Pertussis) vaccine can cause severe allergic reaction. Fever, redness, soreness are common side effects in 1 out of 4 children and in very rare case 1 out of a million dose can cause permanent brain damage (Centers for Disease Control and Prevention [CDC], 2015). This paper presents a methodology for analyzing that the association rules extracted by algorithm are valuable. For measuring the interestingness of the association rules the significant keywords from symptoms of patient's medical reports are mined that act as a background knowledge. Further, keywords are organized into hierarchical form using Probability of Co-Occurrence Analysis (POCA) (Wu & Y. F. B., 2001). It then uses the knowledge extracted from background to extract the association rules from target documents. For extraction of rules Apriori algorithm is implemented. The interestingness factor of Association rules are analyzed by computing the semantic distance between predecessor and successor of rule.

2 RELATED WORK

Extracting interesting and significant patterns from unstructured text document is defined as text mining. The popular conception of text mining is the extension of data mining to textual data (Feldman & Dagan, 1995). A handful of text mining methodologies are available for mining potential knowledge and associations from large medical database. Mahgoub has presented a system called EART (Mahgoub & Rösner, 2006), for automatically extracting association rule from large textual documents. EART integrates XML technology with information retrieval scheme (tf-idf) and for association rule extraction data mining techniques are applied. The knowledge base of biomedical research is expanding. The rapid growth of biomedical text has led to generation of large number of text mining approaches.

As biomedical text contains biological entity such as diseases, drugs, genes and chemical compounds so applying text mining techniques is a strenuous task. Noun phrase extraction is one of the important task which is performed in biomedical text.

Mining Noun phrase extraction consist of two parts, part of speech tagging and noun phrase identification. After assignment of pos tags to text, interested noun phrases can be identified. Association rule mining can generate rules which are not useful and this can surpass the capability of humans to recognize the interesting rules. There are different approaches to analyze interestingness of patterns such as by size of rule, by confidence, by support, and novelty. Piatetsky-Shapiro describes that generally evaluating techniques for interestingness are categorized as subjective and objective measures (Piatetsky-Shapiro 1991). Subjective measures are needed to fulfill the needs of users (Silberschatz & Tuzhilin, 1995) and for objective measures no domain knowledge are needed only features of extracted patterns are considered.

3 CONCEPT HIERARCHY DEVELOPMENT

The methodology and data flow of the system is shown in Figure 1. (TD: target document; BK: background knowledge). For background and target documents retrieval we used VAERS (Vaccine Adverse Event Report System) (Chen, & Robert, 1994) dataset. Rest of the paper is organized as follows: In this section Keywords are extracted from background knowledge, from that keywords concept hierarchy is made. In section 4, from target documents noun phrases are mined and using Apriori algorithm association rules are extracted. Section 5 describes the methodology of analyzing the Interestingness of rules.

3.1 Keywords extraction and indexing

Background information is to be extracted from patient medical report. It consist of symptoms of the patients who consumed vaccine and experienced adverse effect. After extracting keywords, stop words are removed. Stop words are non-functional words which have very little meaning. All keywords are transformed to their core form by eradicating the linguistic variants (e.g. plurals and nouns) (Kantrowitz, Mohit & Mittal, 2000). After removal of stop words and stemming, distinctive keywords are sorted into an integrated list (Salton

& McGill, 1983), where each node represents to the root form and its frequency, as well as pointer to a list of documents where the keywords occur.

3.2 Generating concept hierarchy

The concept hierarchy is defined as a directed acyclic graph which captures the semantic usage of keywords and their relationships in the background knowledge. It organizes data or concepts in hierarchical forms which are used for expressing knowledge in concise way and facilitating knowledge at multiple level of abstraction. In Figure 2, concept hierarchy is shown with heart disease as a root of hierarchy and relationships (heart disease > Arrhythmia > heart block) are shown to demonstrate the hierarchy.

Based on Subsumption rule by the Sanderson and Croft (Sanderson & Croft 1999) and conditional probability, POCA (Probability of Co-Occurrence Analysis) technique is used to develop a concept hierarchy to organize the keywords extracted from documents. POCA (Wu & Y. F. B., 2001) states that in term pair (A, B) A is parent of B If $P(A|B) > P$

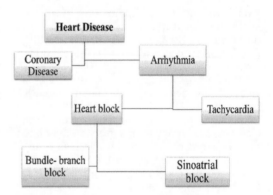

Figure 2. An example of concept hierarchy (Dollah & Aono, 2011).

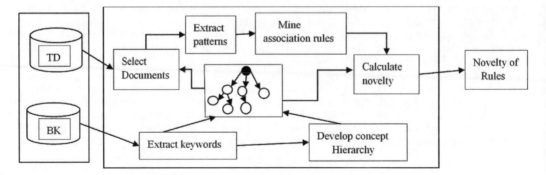

Figure 1. Data flow of system.

(B|A), P (A|B) > = N, where 0 < N < = 1. N is the threshold which affect the number of term pairs derived; larger N results in a smaller number of term pairs. Document Frequency (df) (Forsyth & Rada, 1986) threshold is also applied on the keywords for selecting significant ones. The concept hierarchy is developed from background knowledge.

4 TARGET DOCUMENT KEYWORD DISCOVERY

In the form of association rules, knowledge is to be discovered from the large text corpus. These association rule is to be mined from VAERS database (Chen & Robert, 1994).

4.1 Target document feature extraction

Feature extraction extracts patterns of interest from documents so that structured feature sets are made upon which mining algorithms can be performed. From target documents, noun phrases are to be extracted which consist of causal relations. Causal relation describes the relation between two events, when the occurrence of one event leads to occurrence of another event. The text in Figure 3, is a segment of VAERS dataset which describes medical report of a patient. In this excerpt, causal relation can be identified between vaccine ZOSTAVAX and the adverse event which followed after the patient is vaccinated. Named entity recognition (Leser & Hakenberg, 2005) is used for information extraction, which has overall aim of identifying specific terms such as disease, vaccine and adverse events. After each term is tagged noun phrases can be identified by selecting the sequence of words. After extracting patterns, term frequency-Inverse document frequency (tf-idf) (Salton & Buckley, 1988) weighting scheme is applied to reduce the number of features. Features whose weights are greater than a given threshold are to be selected.

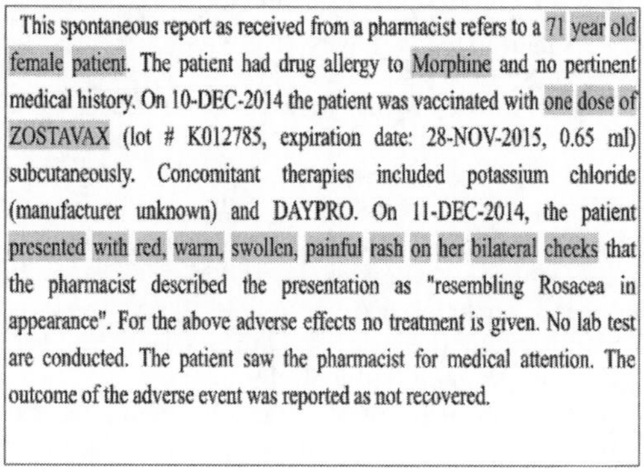

This spontaneous report as received from a pharmacist refers to a 71 year old female patient. The patient had drug allergy to Morphine and no pertinent medical history. On 10-DEC-2014 the patient was vaccinated with one dose of ZOSTAVAX (lot # K012785, expiration date: 28-NOV-2015, 0.65 ml) subcutaneously. Concomitant therapies included potassium chloride (manufacturer unknown) and DAYPRO. On 11-DEC-2014, the patient presented with red, warm, swollen, painful rash on her bilateral cheeks that the pharmacist described the presentation as "resembling Rosacea in appearance". For the above adverse effects no treatment is given. No lab test are conducted. The patient saw the pharmacist for medical attention. The outcome of the adverse event was reported as not recovered.

Figure 3.   An excerpt from VAERS database.

4.2 Mining frequent itemset

After feature extraction, the next step is to find the association rule among the noun phrases extracted in previous step. The job is to extract a set of strong association rules in the form of "$x1^\wedge$... $^\wedge ym => y1^\wedge...^\wedge yn$", where xi (i ∈ {1, m}) and yj (j ∈ {1, n}) are sets of attribute-values.

As described in (Agrawal, Imieliński & Swami, 1993; Agrawal & Srikanth, 1994) association rule are to be extracted using two subtasks, first task is to find all itemsets which are having support value higher than the minimum support value defined by user, and second task is to generate rules using itemsets which are selected in previous step. This two-step process is known as an Apriori Algorithm which makes multiple passes over the data. The vaccine data is an unambiguous text which consists of symptoms of patients who are adversely effected by side effects of vaccines.

As described in (Agrawal & Srikanth, 1994) while extracting association rule from data, the respective data should be in Boolean form. As it is difficult to extract association rules from categorical data so it was necessary to convert vaccine data to Boolean and then use the Apriori Algorithm to extract the association rules. For the transformation process the unique characteristics from the vaccine data is selected and then with the same characteristics partitions are to be generated. As described in the research of Stilou, Bamidis, Maglaveras & Pappas a new partition is created for each unique characteristics of data (Stilou, Bamidis, Maglaveras & Pappas, 2001), so for example in the vaccine data the symptoms of the patients from the side effects of vaccines are stated, so different vaccines could be count as unique characteristics for different patients and could be partitioned. So the coding process could be described as:

ID_OF_Patient_1 = INFLUENZA_1001
ID_OF_Patient_2 = ZOSTER_1002
ID_OF_Patient_3 = HEP B_1003
ID_OF_Patient_4 = VARICELLA_1004

As described above the "001" i.e. the last 3 digits of the code specifies the 1st category of the vaccines and "1" in the code specifies the symptoms regarding the side effects of the particular vaccine. For example "001" would be the INFLUENZA vaccine and in the "1" it will represents the symptoms of the side effects from the vaccine. The user would have to specify the number of ranges. For example, the categorical attribute "age" could be described as follows, the number of categories specified for the "age" is 3, and the 1st group is "0-5", the 2nd "6–15", the 3rd "15 above". Consequently, the code that is generated is the following:

ID_OF_Patient_1 = 7 _ 2002
ID_OF_Patient_2 = 3 _ 2001
ID_OF_Patient_3 = 1 _ 2001
ID_OF_Patient_4 = 24 _ 2003

After the above transformation and coding process the Apriori algorithm can be applied to this data and rules can be extracted. The rules that are generated can be represented in a coded Boolean format:

1001 2002 $\rightarrow$ 1002

which is described as IF Symptoms = "Pyrexia" AND Age = "6–15" THEN, Vaccine = "ZOSTER" with confidence c and support s.

As explained above the association rules are made.

5 INTERESTINGNESS EVALUATION

The concept hierarchy which is developed from the patient's symptoms extracted, is used to calculate the interestingness of association rules. As explained above the background knowledge is collected from patient's medical reports, so the proposed interestingness measure is called user oriented interestingness measure. The interestingness of association rule is stated as the semantic distance among successor and predecessors of the rule (Chen & Wu, 2006). In Figure 2, the concept hierarchy of terms are shown, therefore for calculating novelty of two given term pairs [heart disease, arrhythmia] -> [heart block, bundle branch block], the average distance among all pairs of term i.e. average (D (heart disease, heart block), D (arrhythmia, heart block), D (heart disease, bundle branch block), D (arrhythmia, bundle branch block)) where D(A, B) is semantic distance between item A and B. There are two ways by which semantic distance between two keywords can be calculated—Occurrence Distance and Connection Distance.

5.1 Occurrence distance

Given two keywords A and B, the more often they coexist, the less the occurrence distance. Synonyms or highly interdependent terms are having less occurrence distance. For example in association mining research paper, support and confidence co-occur very frequently. Less the strength of association between two keywords, larger the difference in the occurrences of keywords.

$$D_0(A,B) = \frac{P(A \cup B) - P(AB)}{P(A \cup B)} = 1 - \frac{P(AB)}{P(A \cup B)} \quad (1)$$

In (1) D_o(A, B) is the occurrence distance between A and B, P(AB) is the probability the A and B co-occur, P(AUB) is the probability that A or B occurs. Range of the occurrence distance is from [0, 1]. If A, B occurs together then Do = 1, if not Do = 0.

5.2 Connection distance

Connection distance computes the association between two different keywords considering their associations to other keywords. If two keywords do not co-occur together this doesn't mean that they do not have any relationship, there is still a possibility that they are having certain type of relationship in the concept hierarchy. This type of relationship can be captured by the connection distance. For e.g. horse and fish may not appear in same documents, but if they co-occur with 'animal' concept hierarchy there will be two paths. The common keyword 'animal' suggests that there is a relationship between two keywords. The longer the length of connection path, weaker the connection is. Connection distance is also called hierarchy distance because it measures how far apart two keywords are in the concept hierarchy. Hierarchy distance is standardized by the highest hierarchy distance among two keywords, it is denoted by D_{MAX}. For calculation of hierarchy distance among two keywords X and Y, all the keywords are divided into two categories: background keywords which are derived from the patient's symptoms detail and target keywords which are derived from the VAERS database.

In Figure 4, three areas are defined S1, S2, and S3 which comprises of diverse types of keywords. In S1 and S2 all background keywords are placed. When concept hierarchy is developed, Document Frequency (df) threshold is applied, so keywords whose df is less than the threshold value will be in section S2 and other keywords which satisfy the given threshold are placed In area S1. And target keywords will be placed into any one of the three areas, in Figure 4, grey filled circle corresponds to the target keyword. Black filled circle corresponds to the root of the hierarchy. White ones are for the background keywords, diagonal lines filled circle for both background and target, since there will be many keywords that appear in both background and target documents. H is depth of the hierarchy defined as the number of words in the longest path from root to any leaf. X and Y can both be in concept hierarchy or otherwise. If they are in concept hierarchy the shortest path is selected and its length is assigned as hierarchy distance d_h (X, Y) between keywords X and Y. The reason to select the shortest path is that in concept hierarchy algorithm keywords are placed in multiple positions, so there could be multiple paths between keywords. As root

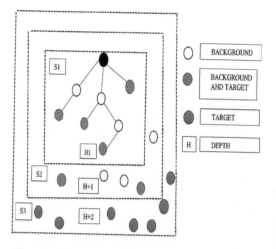

Figure 4. Hierarchy distance calculation.

is not the actual keyword if it is contained in the particular path, +1 should be added in length. If both X and Y do not exists in the hierarchy the D_h will be calculated according to three different cases

- if W (keyword) is in S1 = distance of pathway between keyword and root
- if W appears in S2 = H+1
- if W appears in S3 = H+2

According to the given definition maximum hierarchy distance D_{MAX} is 2(H+2).

5.3 Algorithm for the calculation of hierarchy distance in different conditions (Chen & Wu, 2006)

Input: personal/background knowledge keyword space, PK; concept hierarchy, CH; depth, H; keywords, X and Y; shortestpathlength () function gives length of shortest path between X and Y.
Output: Hierarchy distance, D_h (X, Y);

```
If X∈PK and Y∈PK
   Then Dh(X, Y) = 2*H+2;
Else if X∈(PK-CH)and Y∈PK
   Then Dh(X, Y) = (H+1)+(H+2);
Else if X∈PK and Y∈(PK-CH)
   Then Dh(X, Y) = (H+2)+(H+1);
Else if X∈CH and Y∈PK
   Then
Dh(X, Y) = shortestpathlength(X, r)+(H+2);
Else if X∈PK and Y∈CH
   Then
Dh(X, Y) = (H+2)+shortestpathlength(Y, r);
Else if X∈(PK-CH) and Y∈(PK-CH)
   Then Dh(X, Y) = (H+1)+(H+1);
Else if X∈CH and Y∈(PK-CH)
   Then
```

```
Dh(X, Y) = shortestpathlength(X, r)+(H+1);
Else if X∈(PK-CH) and Y∈CH
   Then
Dh(X, Y) = (H+1)+shortestpathlength(Y, r);
Else if X∈PK and Y∈CH
   Then
Dh(X, Y) = shortestpathlength(X, Y);
End if
```

5.4 Calculation of interestingness

For calculation of interestingness the semantic distance is calculated between predecessor and successor of association rule. In equation (2), semantic distance D(X, Y) (Chen & Wu, 2006), among two keywords X and Y is stated as the square root of the product of their occurrence distance (3) and hierarchy distance (4).

$$D(X,Y)=\sqrt{D_0(X,Y)\cdot D_h(X,Y)} \quad (2)$$

$$D_0(X,Y)=1-P(X,Y)/P(X\cup Y) \quad (3)$$

$$D_h(X,Y)=d_h(X,Y)/2(H+2) \quad (4)$$

6 RESULTS

For evaluation of the performance of user oriented interestingness measure, medical reports of seven patients are extracted from VAERS (Chen, & Robert, 1994) database. Association rules are generated from medical reports of those patients.

As in (Basu, Mooney, Pasupuleti, & Ghosh, 2001) the WordNet lexical database is used to evaluate the interestingness of association rule by computing the semantic distance among the successor and predecessor. One of the drawback of using WordNet database is that it does not consider the background knowledge while making hierarchy. In this methodology, target as well as background knowledge is considered.

While generating hierarchy from background knowledge using POCA (Wu & Y. F. B., 2001), the threshold value N, is set to 0.8. Noun phrases are extracted and after calculating the TF. IDF values, 30% of phrases are eliminated. The support and confidence of extracted association rules are taken as 0.4 and 0.2 respectively. The novelty value lies between 0 and 1. The User oriented novelty prediction is done by analyzing the relationship or connection between interestingness measure and user novelty ratings. The comparison between WordNet novelty (WN) and User-oriented Novelty (UN) are presented in Figure 5. It is clear that the user oriented novelty is always higher than the WordNet novelty.

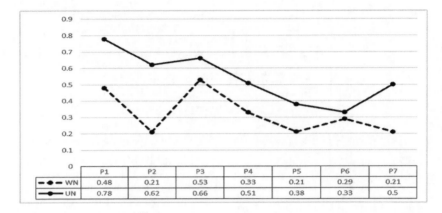

	P1	P2	P3	P4	P5	P6	P7
WN	0.48	0.21	0.53	0.33	0.21	0.29	0.21
UN	0.78	0.62	0.66	0.51	0.38	0.33	0.5

Figure 5.   Comparison between interestingness measures.

7   CONCLUSION

Motivated by the need of extracting interesting rules by association rule mining, this paper has presented a methodology where background knowledge of patient such as vaccine, symptoms and adverse effects are derived and used this information to evaluate the originality of extracted association rules. For future directions, we can test this methodology on other datasets. We can test this methodology using different association rule mining algorithm for fast processing.

REFERENCES

Agrawal, R. & Srikant, R. (1994). Fast Algorithms for Mining Association Rules. Proceedings of the 20th International Conference on Very Large Data Bases: 487–499.

Agrawal, R., Imieliński, T., & Swami, A. (1993). Mining association rules between sets of items in large databases. ACM SIGMOD Record, 22(2): 207–216.

Basu, S., Mooney, R. J., Pasupuleti, K. V., & Ghosh, J. (2001, June). Using lexical knowledge to evaluate the novelty of rules mined from text. In Proceedings of the NAACL work shop and other Lexical Resources: Applications, Extensions and Customizations.

Centers for Disease Control and Prevention (CDC) (July 9, 2015) "Possible side-effects from Vaccines," retrieved from http://www.cdc.gov/vaccines/vac-gen/side-effects.htm #dtap.

Chen, Robert T., et al. "The vaccine adverse event reporting system (VAERS)." Vaccine 12.6 (1994): 542–550.

Chen, X. & Wu, Y. F. (2006, April). Personalized Knowledge Discovery: Mining Novel Association Rules from Text. In SDM: 589–593.

Christiane Fellbaum (1998, ed.) WordNet: An Electronic Lexical Database. Cambridge, MA: MIT Press.

Daniel J. DeNoon. (October 14, 2009). "Is the H1N1 Swine Flu Vaccine Safe?" retrieved from http://www.webmd.com/cold-and-flu/features/is-the-h1n1-swine-flu-vaccine-safe.

Dollah, R. B. & Aono, M. (2011). Ontology based approach for classifying biomedical text abstracts. Int. J. Data Eng, 2(1): 1–15.

Feldman, R. & Dagan, I. (1995, August). Knowledge Discovery in Textual Databases (KDT). In KDD Vol. 95:112–117.

Forsyth, R. & Rada, R. (1986). Machine learning: applications in expert systems and information retrieval. Halsted Press.

Kantrowitz, M., Mohit, B. & Mittal, V. (2000, July). Stemming and its effects on TFIDF ranking (poster session). In Proceedings of the 23rd annual international ACM SIGIR conference on Research and development in information retrieval: 357–359.

Leser, U. & Hakenberg, J. (2005). What makes a gene name? Named entity recognition in the biomedical literature. Briefings in bioinformatics. 6(4):357–369.

Mahgoub, H. & Rösner, D. (2006, August). Mining association rules from unstructured documents. In Proc. 3rd Int. Conf. on Knowledge Mining, ICKM, Prague, Czech Republic: 167–172.

Murphy, S. L., Xu, J. & Kochanek, K. D. (2013). Deaths: final data for 2010. National vital statistics reports: from the Centers for Disease Control and Prevention, National Center for Health Statistics, National Vital Statistics System, 61(4): 1–117.

Piatetsky-Shapiro, G. (1991). Discovery, analysis, and presentation of strong rules. Knowledge discovery in databases: 229–238.

Salton, G. & McGill, M. J. (1983). Introduction to Modern Information Retrieval. McGraw-Hill.

Salton, G., & Buckley, C. (1988). Term-weighting approaches in automatic text retrieval. Information processing & management, 24(5): 513–523.

Sanderson, M., & Croft, B. (1999, August). Deriving concept hierarchies from text. In Proceedings of the 22nd annual international ACM SIGIR conference on Research and development in information retrieval: 206–213.

Silberschatz, A., & Tuzhilin, A. (1995, August). On subjective measures of interestingness in knowledge discovery. In KDD (Vol. 95): 275–281.

Stilou, S., Bamidis, P. D., Maglaveras, N., & Pappas, C. (2001). Mining association rules from clinical databases: an intelligent diagnostic process in healthcare. Studies in health technology and informatics, (2): 1399–1403.

Wu, Y. F. B. (2001). Automatic concept organization: organizing concepts from text through probability of co-occurrence analysis (POCA).

Communication and Computing Systems – Prasad et al. (Eds)
© *2017 Taylor & Francis Group, London, ISBN 978-1-138-02952-1*

Biweight mid variance based energy-efficient virtual machine provisioning

L. Nhapi & A.K. Yadav
Department of Computer Science and Engineering, ITM University Gwalior, MP, India

R.S. Rao
Ambedkar Institute of Advanced Communication Technologies and Research, Geeta Colony, Delhi, India

ABSTRACT: The purpose of this study is to investigate the effect of applying the biweight midvariance for automatically adjusting the server utilization threshold, so as to adapt to workload variation. We consider energy saving and service Quality, (QoS), as key performance parameters of our algorithm. The proposed method is implemented in CloudSim and the outcome proves that it can lower electricity consumed by data centers, and achieve acceptable QoS. Thus it can be made use of, to economically allocate resources and therefore increase profit for cloud data centers.

1 INTRODUCTION

The rapid advancement in cloud services portfolio due to increase in consumer demand in lieu of storage and processing power, has seen cloud providers like Microsoft, Amazon & Google deploying an enormous number of power-hungry planet-scale data centers throughout the globe. These cloud giants are reported to have in excess of 1 million servers individually in their data centers (Meng et al. 2010). Unsurprisingly, these massive data centers gobble a huge volume of electricity so as to keep servers running and cooling systems in operation. A report published in (Nurmi et al. 2009), (Data Center Knowledge report), states that one significant contributor to the Total Cost of Ownership (TCO) variables in managing a data center is electricity. Out of the overall energy usage by data centers, servers and data equipment contribute 55% while cooling equipment contributes a further 30%. Another report by McKinsey (Jung et al. 2009), revealed that world data centers gobble 0.5% of world's electricity and inject more carbon emission than both Argentina and the Netherlands. Clearly, large data centers do not only cost a fortune to maintain, but instead, they also have adverse effects on environment due to carbon emissions.

Among a host of other ways to warrant maximum resource usage inside data centers or server farms is to take advantage of the proficiencies of virtualization technology and accordingly lock horns with the challenge of energy inefficiency (Barham 2003). Cloud service vendors can use the competencies offered by virtualization technology

as it enables multiple Virtual Machine (VMs) instances to be created atop single physical server, thus maximizing the utilization of resources and increasing returns i.e. Returns On Investment (ROI). In order to lower power consumption, the idle nodes can be switched to low-power states (i.e., sleep, hibernation), and as a result eliminate idle electricity use.

This work focuses on energy-efficient provisioning of resources in a cloud environment (e.g. Amazon EC2). We also consider customer satisfaction. The remaining sections of this work are arranged as follows. In section 2 we present a literature review and section 3 outlines the problem statement. The proposed approach is given in section 4 while section 5 presents the system model. In Section 6 and 7 we describe our methodology and give simulation results in that order. Concluding remarks and references are laid out in sections 8 and 9 respectively.

2 LITERATURE SURVEY

Beloglazov & Abawajy (2012) divided the problem of VM provisioning into 4 mini problems: discover overloaded hosts, discover underutilized hosts, identify VMs to migrate and place the identified VMs onto selected physical nodes. For host overload detection, the authors proposed a static or fixed threshold policy (THR). This is based on monitoring CPU usage of a host, if it falls below some set minimum threshold then all the VMs are required to be have to be transferred to another node. The host will then be put to low energy mode or put to sleep so as to conserve power. Conversely, when CPU utilization rises above the maximum threshold, then only some VMs ought to be transferred from this node. If CPU utilization falls under lower threshold, all VMs must be moved from the server to avoid performance degradation. Furthermore, the authors suggested 4 polices for selecting VMs to migrate: Highest Potential Growth (HPG), Single Threshold (ST), Random Choice (RC) and Minimization of Migrations (MM). The suggested algorithms exhibited some flexibility as per simulation results. Nonetheless, using fixed values for the threshold does not fit an environment that has ever changing workloads.

Jung and associates (2008, 2009) proposed the application of gradient search and bin packing techniques in solving the challenges related todt dynamic consolidation of VMs, particularly for multi-tier web-applications. They make use of live migration and in the meantime also considering meeting SLA terms. To model the SLA requirements, the reply time for every single type of transaction particular to the web-application is precomputed resulting in a new VM placement. A utility function is used to account for the SLA fulfillment. The disadvantage of this approach is that it can only be applied to a single web-application setup and, thus, is not adequate for use in a multi-tenant IaaS environment.

In another work, Beloglazov & Buyya (2011) put forward 4 adaptive threshold utilization algorithms for the purpose of approximating CPU utilization and identify the overloaded hosts. These four algorithms make use of insights observed from statistical examination of historical data mined during the lifetime of VMs and they are listed as follows: static Threshold (THR), Median Absolute Deviation (MAD), Robust Local regression (RLL), Inter Quartile Range (IQR) together with three different VM selection policies: the Random Selection (RS), Minimum Migration Time (MMT) and Maximum Correlation (MC). Simulation results point out that the adaptive VM provisioning algorithms outperform the fixed threshold algorithms. The MMT policy was found to be favorable in contrast to the MC and RS policies while dynamic VM provisioning algorithms based on LR performed better than the adaptive-threshold and static-threshold based algorithms.

To formulate an energy-aware placement scenario of applications in a virtualized heterogeneous environment, Verma et al. (2008) modeled it as a continuous optimization process where at every discrete time frame, the current allocation of VMs goes under an optimization process so as to lower energy use and in turn maximize performance. The authors used the bin-packing approach in modeling the problem. Variable sizes of bins were suggested and costs were also taken into consideration. To accomplish a new VM assignment at each discrete

time frame, live migration techniques are used. In contrast to the approach taken by us, we will consider user satisfaction but the algorithms in question do not. The runtime performance of applications can be degraded as the workload varies.

The writers Gandhi et al. (2009) focused on lowering of mean reaction time by investigating the difficulty of allocating an available power budget among servers in a virtualized server farm with heterogeneous resources. They introduced a form of a wait in-line theoretic model which encompasses the forecast of the average reaction time as a function of the power-to-frequency relationship, maximum power budget, arrival rate, etc. The ideal power apportionment for each configuration model of the factors mentioned above is determined by using the proposed model.

As a follow-up research to the challenges arising from setting static usage thresholds, Beloglazov & Buyya (2010) proposed that the system has to dynamically adapt its behavior on the basis of workload trends shown by the applications. They suggested an innovative method for dynamic VM provisioning which is based on automatically adjusting the threshold values on the basis of the statistical analysis of past usage data gathered during the lifespan of VMs. This ensures a significant level of user satisfaction. The total usage of CPU from all VMs running on a node is denoted through a random variable on which the Dynamic Threshold idea (DT) is based. The simulation outcomes showed that the DT's performance is better as compared to other migration-aware policies in regard to the quantity of VMs migrated and user satisfaction measured by SLA violations (i.e. SLAV < 1%). But still, the level of energy consumption remained the same.

In another approach by Zhu et al. (2008), they proposed the usage of three distinct controllers operating at different time scales to tackle the difficulties of automatically allocating resources plus capacity planning. The three controllers are based on long (hours to days), short (minutes) and shorter (seconds) time scales. These are used to allocate similar workloads onto groups of servers, dynamically adapt to non-stationary conditions by reallocating VMs to satisfy the SLAs. A drawback of this method is in setting fixed or static utilization thresholds. Fixed utilization thresholds continue to be inefficient for cloud environments with variable workloads that show volatile resource usage trends.

A unique method of dynamically selecting and allocating VMs so as to decrease power usage and SLAV in cloud server farms was proposed by Cao & Dong (2012). The mean along with the accompanying standard deviation of CPU utilization for VM were employed in deciding which hosts were considered overloaded. Additionally, to select which VMs to transfer from an overloaded host, the positive highest correlation coefficient was used. The proposed process was simulated in CloudSim and the outcome revealed that the suggested overload detection and selection policies perform better, with respect to SLAV reduction, than polices built-in in CloudSim. However, considering the amount of energy consumed, the previous policies suggested by Beloglazov & Buyya (2011) perform marginally better.

3 PROBLEM STATEMENT

Normally, servers in a data center, at any given moment, are in any one of the following four states: overloaded, underloaded, steady and idle. When CPU utilization is 100%, a host is considered as being overloaded. From another perspective, if the overall CPU time requested by VMs on the host surpasses the capability of the host resulting in violation of SLAs (SLAV). If a host is operating with no SLAV and CPU utilization is below minimum threshold, then it is considered as being underloaded. A steady host operates with CPU utilization between minimum and maximum utilization thresholds while an idle host is currently not in use but switched ON.

All together, the process consolidating virtual machines on a single server can be divided into four steps:

1. Determine if a host is overloaded by applying state detection algorithms. If true, then apply VM selection policies choose which VMs to migrate from the host so as to avoid performance degradation.
2. Apply same overload detection algorithms as in step 1 to determine if a host is underutilized. If true, then all the virtual machines have to be transferred by live migration to other hosts and thus the idle host can be switched to a low-power state and eliminate idle power consumption.
3. Place the migrated VMs on other active nodes using VM placement policies.
4. Re-optimize VM placement by repeating step 1 to 3 if required.

According to the work done by Beloglazov et al. (2010), the authors proposed heuristics for deciding the exact moment to migrate VMs from a node based on CPU utilization thresholds. It is based on the idea of maintaining hosts operating in a steady state i.e. keeping CPU utilization between minimum and maximum thresholds. However, fixing values of utilization thresholds is not suitable for an environment with dynamic and random workloads, in which physical resources are shared among different types of applications. An ideal scenario is one in which the system is capable of automatically adjusting its behavior in response to the different workload patterns exhibited by the applications.

4 PROPOSED APPROACH

We therefore, proposition a robust and efficient technique for automatically adjusting the CPU utilization thresholds. The proposed auto-adjustment technique is based on statistical analysis of VMs' footprint i.e. historical data collected during the lifespan of VMs. The chief idea at the back of the proposed dynamic-threshold algorithm is to automatically change the value of the maximum utilization threshold dependent on the deviation strength of the CPU utilization. If the deviation is high, then the upper value of the utilization threshold is low. A higher deviation provides an indication that the CPU utilization will likely reach 100% and cause an SLA violation.

Estimates of central tendency such as the arithmetic mean and measures of dispersion like the standard deviation depend for their understanding on an implied assumption that the data include a random sample from a normal distribution. But in reality, it is known that analytical data usually deviate from that model. They are habitually heavy tailed, meaning that a large proportion contains data that are far from the mean and are also influenced by outliers. Robust statistics provides an efficient and convenient approach of summarizing results in the presence of outliers as compared to classical statistical methods (Huber 1981). The impetus is to produce efficient estimators that are not overly affected by minimal deviations from model assumptions. For Gaussian data, the optimal scale estimator is the standard deviation (or variance). The problem with the standard deviation is that it is not resistant to outliers and therefore it is less successful in providing robustness and efficiency (Wilcox 1993). The Median Absolute Deviation (MAD) is both a robust and resistant scale estimate. But its success in achieving robustness and efficiency is only modest. The biweight midvariance can instead be called upon as an efficient alternative estimator. It provides both resistance and robustness in efficiency. In exploratory work where moderate efficiency is adequate, Mosteller & Tukey recommend using the MAD or interquartile range (Wilcox 1993). When high performance is of paramount importance, the biweight midvariance can be considered.

The biweight midvariance (BiVar) estimate is defined as:

$$\frac{n^* \sum_{i=1}^{n} (x_i - Q)^2 (1 - u_i)^4 \, I(|u_i| < 1)}{\left(\sum_i (1 - u_i^2)\right)(1 - 5u_i^2) I(u_i < 1))^2} \tag{1}$$

where I is the indicator function, Q is the sample median of the x_i, and

$$u_i = \frac{x_i - Q}{9 \times MAD} \tag{2}$$

where MAD is the median absolute deviation.

817

The square root of the BiVar is robust in scale estimation. In the BiVar procedure, as the distance between data points and the median grows, the data points are downweighted resulting in points with greater than 9 MAD units from the median having no influence at all.

We adopt the definition of the upper utilization threshold (T_{High}) by Beloglazov & Buyya (2012), but contrary to them, we apply the biweight midvariance.

$$T_{High} = 1 - m.\sqrt{BiVar} \qquad (3)$$

where $m \in \mathbb{R}^+$ is a parameter of the function that controls how the system aggressively consolidates VMs. In other words, the parameter m allows the adjustment of the safety of the method, the lower value of parameter m, the less the energy consumption, but the higher the level of SLA violations caused by the consolidation.

5 MODEL OF SYSTEM

In this work, the proposed system is for a cloud service provider operating in IaaS environment (Infrastructure as a Service) characterized by a gigantic data center consisting of N heterogeneous physical hosts. The characteristics of each host are listed below.

1. CPU performance, well-defined as Million Instructions Per Second (MIPS).
2. RAM in Megabytes (MB).
3. Network Attached Storage (NAS).

The presence of a NAS storage device allows live migration of VMs. In this type of environment, the processing power of VMs is defined by MIPS, network bandwidth and amount of RAM. Multiple autonomous users submit requests for provisioning of M heterogeneous VMs implying that the resulting workload generated is mixed due to merging multiple VMs on a single physical node.

6 METHODOLOGY AND PERFORMANCE METRICS

6.1 Simulation setup

There is need to assess the proposed policies on large- scale virtualized data center infrastructure. Yet, it is not feasible to implement repeatable experiments on real infrastructure especially at large scale. Instead, we will use CloudSim toolkit 3.0.3 simulator (Calheiros et al. 2010) to evaluate the proposed policy. CloudSim is a popular toolkit commonly used to simulate different aspects of a cloud system. It enables the recreation of real

world set-ups and can be extended to suit a problem being investigated. It also contains built-in power models for data center resources, policies for VMs allocation and selection and provides different types of workloads.

We have simulated a data center with 800 physical nodes using real workload data availed as part of the CoMon project (Park & Pai 2006). The data traces for each day contain a sample of usage traces from over one thousand VMs. Two types of servers are considered: HP ProLiant ML110 G4 and HP ProLiant ML110 G5. For both servers, we use the power models provided by Anon (2016), as shown in Table 3. A summary of the properties of hosts and VMs used in the experiments is given in Tables 1 and 2, in that order.

6.2 Performance metrics

When resources demanded by VMs are fully satisfied, then QoS is guaranteed. We therefore use the forthcoming performance metrics so as to evaluate the proposed policy: SLATAH (SLA violation time per active host), energy consumption, SLAV (SLA violation), ESV (Energy and SLAV), PDM (Performance Degradation due to Migrations) and number of VMs migration.

The main metrics are SLAV and energy consumption by physical nodes. These two metrics are usually antagonistic, meaning that they are negatively correlated as energy consumption can normally be decreased by the cost of the increased level of SLA violations. The objective of dynamic VM provisioning is to lessen both SLA violations and energy consumption. In light of this, a combo metric denoted by ESV, that combines together energy usage and breach of SLAs was proposed.

1. Energy consumption (E)

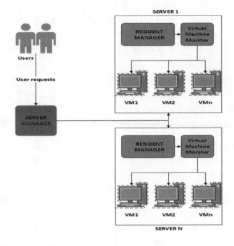

Figure 1. System model.

Table 1. Characteristics of the two types of hosts used in experiments.

| # of Hosts | Type | Ram | | | Storage |
		MB	MIPS	Cores	GB
400	HP Proliant G4	4096	1860	2	100
400	HP Proliant G5	4096	2660	2	100

Table 2. Characteristics of the four types of VMs used in experiments.

| VM Type | RAM | | | Storage |
	MB	MIPS	Cores	GB
1	512	2500	1	5
2	1024	2000	1	5
3	1536	1000	1	5
4	2048	500	1	5

Table 3. Power consumption by two types of hosts at different load level in Kilowatts.

Percentage utilization	HP proliant G5	HP proliant G5
0%	0.0937	0.086
10%	0.097	0.0894
20%	0.101	0.0926
30%	0.105	0.096
40%	0.110	0.0995
50%	0.116	0.102
60%	0.121	0.106
70%	0.125	0.108
80%	0.129	0.112
90%	0.133	0.114
100%	0.135	0.117

This metric represents the total energy consumed by the physical data center resources.

2. SLA Violation (SLAV)

$$SLATAH = \frac{1}{N} \sum_{i=1}^{N} \frac{K_{si}}{K_{ai}}$$ (4)

$$PDM = \frac{1}{V} \sum_{j=1}^{V} \frac{U_{dj}}{U_{rj}}$$ (5)

where N = the number of hosts; V = the number of VMs; K_{ai} = the total time of the host i being in the active state (serving VMs); K_{si} = the total time during which the host i has encountered the utilization of 100% leading to an SLA violation; U_{rj} = the total CPU capacity requested by the VM j during its lifetime; U_{dj} = an estimation of the performance degradation of the VM j caused by migrations.

$$SLAV = SLATAH.PDM$$ (6)

3. Energy and SLA Violations (ESV)
   This combined metric is calculated from both power consumption and SLAV.

$$ESV = E.SLAV$$ (7)

4. Migration Frequency (MF)
   This metric is a measure of the total number of migrated VMs. If the number of VM migrations is low then performance degradation is also minimal else it is high. Therefore, it is imperative to reduce the number of VMs migrated.

7 SIMULATION RESULTS AND ANALYSIS

The proposed overload detection policy, biweight midvariance (BiVar) has been implemented and evaluated against the equally competitive MAD, IQR and THR algorithms. The Minimum Migration Time (MMT) was the preferred VM selection policy over the Random Selection (RS) policy and the Maximum Correlation (MC) policy. The main parameters of interest are Energy consumption and SLAV, however, MF and PDM have also been evaluated. For each algorithm, we have used real workload data. The results have shown that the BiVar policy in conjunction with MMT policy, outperform the aforementioned three algorithms.

Table 4. Simulation results.

Policy	Energy KWh	SLAV x10⁻⁵	PDM %	SLATAH %	MF
THR	191.73	324	0.07	5.84	26634
MAD	184.88	331	0.07	5.03	26292
IQR	188.86	315	0.06	4.96	26476
BIVAR	163.35	326	0.06	5.03	26121

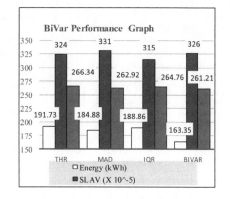

Figure 2. Plot of energy, SLAV & MF.

The overall power consumption was reduced by 20% in contrast with the THR policy. A marginal decrease in the quantity of virtual machine migrations was observed, however, there was a slight increase in the number of SLA violations. A summary of results is shown in Table 4 and Figure 2.

8 CONCLUDING REMARKS AND FUTURE DIRECTIONS

Cloud service providers are continuously being pressured to control the volume of carbon dioxide emissions by their data centers. They are also interested in maximizing resource usage, limit operating expenses and as a result increase their Return on Investment (ROI). By applying energy-efficient VM provisioning techniques, such as adaptive consolidation of VMs and turning idle servers to power-saving modes, Cloud providers can surely improve their profits. In this paper, we have proposed an efficient and robust method for power-aware VM provisioning. We have evaluated the proposed method through simulation in CloudSim. In addition, we have also used workload traces from over a thousand PlanetLab Virtual Machines. Experimental results have shown that the BiVar policy is more efficient than the MAD, IQR and THR policies as it was able to reduce energy consumption and minimize SLA violations.

For future research, we plan on the exploration of Markov chains so as to further improve on results. Apart from monetary benefits associated with this work, this research also contributes to the global climate change initiatives of decreasing carbon dioxide footprints and energy consumption by modern IT infrastructures.

REFERENCES

Anon. 2016. All Published SPEC Specpower_Ssj2008 Results. *Spec. Org.* http://www.spec.org/power_ssj2008/results/power_ssj2008.html.

Barham, P., B. Dragovic, K. Fraser, S. Hand, T. Harris, A. Ho, R. Neugebauer, I. Pratt, and A. Warfield. 2003. Xen And The Art Of Virtualization. *Proceedings Of The Nineteenth ACM Symposium On Operating Systems Principles—SOSP '03.* Association for Computing Machinery (ACM).

Beloglazov, A., and R. Buyya. 2010. Adaptive Threshold-Based Approach For Energy-Efficient Consolidation Of Virtual Machines In Cloud Data Centers. *Proceedings Of The 8Th International Workshop On Middleware For Grids, Clouds And E-Science—MGC '10.* Association for Computing Machinery (ACM).

Beloglazov, A., and R. Buyya. 2011. Optimal Online Deterministic Algorithms And Adaptive Heuristics For Energy And Performance Efficient Dynamic Consolidation Of Virtual Machines In Cloud Data Centers. *Concurrency And Computation: Practice And Experience* 24, no. 13: 1397–1420. Wiley-Blackwell.

Beloglazov, A., J. Abawajy, and R. Buyya. 2012. Energy-Aware Resource Allocation Heuristics For Efficient Management Of Data Centers For Cloud Computing. *Future Generation Computer Systems* 28, no. 5: 755–768. Elsevier BV.

Calheiros, R., R. Ranjan, A. Beloglazov, C. De Rose, and R. Buyya. 2010. Cloudsim: A Toolkit For Modeling And Simulation Of Cloud Computing Environments And Evaluation Of Resource Provisioning Algorithms. *Softw: Pract. Exper.* 41, no. 1: 23–50. Wiley-Blackwell.

Cao, Z., and S. Dong. 2012. Dynamic VM Consolidation For Energy-Aware And SLA Violation Reduction In Cloud Computing. *2012 13Th International Conference On Parallel And Distributed Computing, Applications And Technologies.* Institute of Electrical & Electronics Engineers (IEEE).

Fan, X., W. Weber, and L. Barroso. 2007. Power Provisioning For A Warehouse-Sized Computer. *ACM SIGARCH Computer Architecture News* 35, no. 2: 13. Association for Computing Machinery (ACM).

Gandhi, A., M. Harchol-Balter, R. Das, and C. Lefurgy. 2009. Optimal Power Allocation In Server Farms. *Proceedings Of The Eleventh International Joint Conference On Measurement And Modeling Of Computer Systems—SIGMETRICS '09.* Association for Computing Machinery (ACM).

Huber, P. 1981. Robust Statistics. *Wiley Series In Probability And Statistics.* John Wiley & Sons, Inc.

Jung, G., K. Joshi, M. Hiltunen, R. Schlichting, and C. Pu. 2008. Generating Adaptation Policies For Multi-Tier Applications In Consolidated Server Environments. *2008 International Conference On Autonomic Computing.* Institute of Electrical & Electronics Engineers (IEEE).

Jung, G., K. Joshi, M. Hiltunen, R. Schlichting, and C. Pu. 2009. A Cost-Sensitive Adaptation Engine For Server Consolidation Of Multitier Applications. *Middleware 2009*: 163–183. Springer Science + Business Media.

Meng, X., C. Isci, J. Kephart, L. Zhang, E. Bouillet, and D. Pendarakis. 2010. Efficient Resource Provisioning In Compute Clouds Via VM Multiplexing. *Proceeding Of The 7Th International Conference On Autonomic Computing—ICAC '10.* Association for Computing Machinery (ACM).

Nurmi, D., R. Wolski, C. Grzegorczyk, G. Obertelli, S. Soman, L. Youseff, and D. Zagorodnov. 2009. The Eucalyptus Open-Source Cloud-Computing System. *2009 9Th IEEE/ACM International Symposium On Cluster Computing And The Grid.* Institute of Electrical & Electronics Engineers (IEEE).

Park, K., and V. Pai. 2006. Comon. *ACM SIGOPS Operating Systems Review* 40, no. 1: 65. Association for Computing Machinery (ACM).

Verma, A., P. Ahuja, and A. Neogi. 2016. Pmapper: Power And Migration Cost Aware Application Placement In Virtualized Systems. *Middleware 2008*: 243–264. Springer Science + Business Media.

Wilcox, R. 1993. Comparing The Biweight Midvariances Of Two Independent Groups. *The Statistician* 42, no. 1: 29. JSTOR.

Zhu, X., D. Young, B. Watson, Z. Wang, J. Rolia, S. Singhal, and B. McKee et al. 2008. 1000 Islands: Integrated Capacity And Workload Management For The Next Generation Data Center. *2008 International Conference On Autonomic Computing.* Institute of Electrical & Electronics Engineers (IEEE).

Communication and Computing Systems – Prasad et al. (Eds)
© 2017 Taylor & Francis Group, London, ISBN 978-1-138-02952-1

Securing patient's confidential data in ECG steganography using advanced encryption standards

Suchita Singh & Vimlesh Kumar
School of ICT, Gautam Buddha University, Greater Noida, Uttar Pradesh, India

ABSTRACT: In today's world, the number of elderly cardiac patients is growing tremendously, which arises the need of Point-of-Care (PoC) systems. Therefore, the Body Sensor Networks (BSNs) collect ECG signal from remote patients at their homes only and transmit it along with other physiological readings such as glucose level, temperature, blood pressure, etc. and patient's personal data. All this information is transmitted to hospital servers through internet. As per the HIPPA guidelines, the confidentiality and integrity of the patient's data must be maintained when it is transmitted over internet. In this paper, a combination of cryptography and steganography method is applied to hide patient's secret information. The proposed method uses Advanced Encryption Standards (AES) which encrypts and safeguards the integrity of data for 20–30 years. Image Quality Assessment Techniques such as Peak Signal-to-Noise Ratio and Mean Square Error are used for distortion measurement. It is concluded that the suggested system provides eminent level of safety and privacy for the patient's information and ECG data can be easily diagnosed from the watermarked image as well as from the image obtained after removing the data from the watermarked image.

1 INTRODUCTION

Due to the modern lifestyle and tremendously changing eating habits of people in today's century, there is dynamic increment in the number of cardiac patients around the globe. Cardiovascular diseases demand the availability of fast emergency services due to their unpredictable and sudden occurrence behaviour. Such requirements has led to the establishment of Point-of-Care (PoC) services. In PoC system, patient's biomedical signal and physiological readings are collected from body sensors. The signal along with the readings and patient's confidential information are sent collectively to the patient's PDA device through Bluetooth. Now from the device, all this information is sent to the hospital servers via internet. The patient's biological and personal information is then stored on the hospital server. The doctor then analyses the patient's health status from the received medical report and provides the quick medical treatment accordingly.

In this process of transmitting the patient's confidential information from his/her device to the hospital server, Internet is used as the main channel of communication. With Internet, arises the security and privacy threats to the information which is transmitted. The Health Insurance Portability and Accountability Act mandates that the information, which is transmitted/communicated/transferred over internet, should be safe and preserved from various security threats. The privacy and confidentiality of the information should be maintained throughout the transmission.

Therefore, we need to propose a system that guarantees high security of patient's confidential data and the biological statistics such as glucose reading, blood pressure statistics, temperature, position, etc. collected from the body sensor nodes.

The technique suggested here is a combination of encryption and steganography techniques. The biomedical signal used as the host carrier to hide the restricted data, containing the patient's personal information and physiological readings is the ECG signal because most of the PoCs collect the ECG signal. Additionally, size of the ECG signal is larger than the size of any other data.

In the proposed system, firstly the physiological readings such as blood pressure, glucose level, temperature, ECG signal etc. are collected using BSNs. The BSNs then send the collected readings to the patient's mobile device through Bluetooth. Now, inside that device, patient's secret data and collected readings will be concealed inside his/her ECG signal. All this data of the patient is stored on the hospital server. From the server records, any doctor present in the hospital can see/check the resultant watermarked signal but only an authorized personnel/doctor will be able to extract the information hidden inside the ECG signal.

The proposed technique is designed in order to provide minimum acceptable distortion in the watermarked image so that it can be diagnosed properly and highest security so that no adversary is able to extract the patient's secret information.

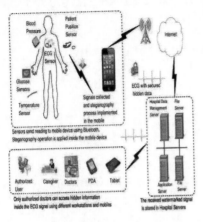

Figure 1. Scenario in Point-of-Care (PoC) systems.

2 METHODOLOGY

The proposed methodology consists of the following five integrated stages which ensures safe hiding of data with lesser distortion in the steganographed signal.

2.1 Stage 1: Encryption

The objective of this stage is to convert the patient's secret data into a format which cannot be understood and read by any unauthorized person. The proposed system uses Advanced Encryption Standards (AES) for this purpose, which is the most advanced technique and ensures the security of data for 20–30 years.

AES is executed according to the following steps:

1. A 128 bit key is used to generate 10 subkeys of 128 bits, before encryption. Each subkey is written in the form of 4×4 matrix.
2. 128 bit block of data is written in the form of 4×4 matrix row-wise, which is called state. Each round generates a new state from the old one and the state after the 10th round contains the cipher text.
3. Every round of AES consists of the following four sequential operations:
4. ByteSub: The 16 bytes of the state matrix (the input data) are substituted by the corresponding values found in the Substitution table i.e. S-box table.

- ShiftRow: The rows of the state matrix are shifted cyclically left.
- MixColumn: The columns of the state matrix are shuffled by a complicated, nevertheless fixed scheme.
- AddRoundKey: The corresponding bytes of the state matrix and the expanded key (also a 4×4 matrix) are XORed.

- Before the 1st round, AddRoundKey transformation is done. In the last round the Mix-Column transformation is left out.

2.2 Stage 2: Wavelet decomposition using Discrete Wavelet Transform (DWT)

A Wavelet is a small wave having its energy concentrated in time that helps in the analysis of transient, non-stationary or time-varying phenomena. Wavelet Transform can be described from the following equation:

$$C(S,P) = \int_{-\infty}^{\infty} f(t) \, \Psi \, (S,P)$$

where

ψ is a wavelet function,
S and P are some positive integers which indicate transform parameters,
C is a coefficient such that it is a function of scale and position parameters.

Wavelet Transform is a dynamic mechanism which combines frequency domain with time domain in a single transformation. A signal is decomposed when wavelet transform is applied to that signal using some band filters. After the band filtering operation, two different outputs will be generated: low frequency and high frequency components of the given signal.

The transform used here is Haar DWT, which is the easiest and effortless DWT. A two-dimensional Haar DWT contains two operations: horizontal and vertical, and hence can be described in the following steps:

Step 1: Scan the pixels going from left to right horizontally. Add and subtract the neighbouring pixels. Store sum on left and difference on right. Same process is iterated until each row is processed. Pixel sum represents low frequency component (L) and difference represents high frequency component (H).

Step 2: Now scan the pixels going from top to bottom vertically. Add and Subtract neighbouring pixels. Store sum on top and difference on bottom. Similar process is iterated each column is operated.

Therefore, 4 sub-bands are generated as a result: LL, HL, LH, and HH respectively. LL sub-band represents the low frequency component. It contains most of the original image. Since, human eyes are more sensitive to low frequencies so any change in the image will be easily noticeable. Therefore we do not embed the secret data in low frequency component. HH contains the edges of the image. So we hide the confidential data in diagonal sub-bands HL and LH.

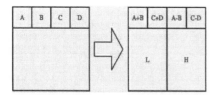

Figure 2.   Horizontal operation on first row.

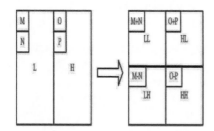

Figure 3.   Vertical operation.

2.3   Stage 3: Embedding encrypted data

Least Significant Bit (LSB) Substitution process is used to embed the secret information in ECG signal. The first bit of the message is embedded in the right-most bit i.e. the LSB bit of the selected sub-band. We substitute the message bits into the LSB bits because they have least weight and hence the substitution process does not affect the original pixel value to a great extent. The next bit of the message is substituted on the next location and so on. The difference between the steganographic ECG image containing the patient's confidential data and the original ECG image is not visually perceptible. Hence, this technique ensures the quality of the water-marked image.

2.4   Stage 4: Inverse wavelet re-composition

The four generated sub-bands, after the embedding operation, are now reconcatenated using inverse wavelet recomposition. The output is a resultant steganographic image that contains the embedded patient's secret information. This inverse process transforms the signal from a combination of time and frequency domain into time domain. Therefore, the resultant steganographed ECG signal resembles with the original ECG signal to a great extent.

2.5   Stage 5: Extraction process

Following is the algorithm for data extraction:
Input: Steganographed Image s.
    Output: Personal and Physiological secret data of patient d.

1. Normalize s.
2. Transform s in to 2 levels using wavelet decomposition.

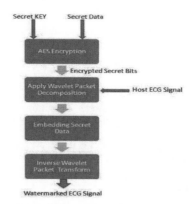

Figure 4.   Block diagram of the sender steganography.

Figure 5.   Block diagram of the receiver steganography.

3. DWT coefficients of cover image c-DWT coefficients of s.
4. Obtain encrypted data.
5. Decrypt the data obtained using AES decryption.
6. Original personal and physiological secret data d is obtained.

3   IMAGE QUALITY ASSESSMENT

Image Quality Assessment is done to measure the difference or resemblance between the modified (steganographed) image and the unmodified (original) image. The popular techniques used for standard quality analysis are: Mean Square Error and Peak Signal-to-Noise Ratio. The major advantages of these two parameters are their easy implementation and low computational complexities.

3.1   Mean Square Error (MSE)

MSE represents the mean squared difference between the original image and steganographed image. Mathematically, MSE is defined as the square of error between cover image and stegoimage. Lower the value of MSE higher the quality of image. MSE can be evaluated as follows:

$$MSE = \left[\frac{1}{M*N}\right]^2 \sum_{i=1}^{M} \sum_{j=1}^{M} \left(X_{ij} - X'_{ij}\right)^2 \qquad (2)$$

where,

Xi: Intensity Value of pixel in cover image
X'ij: Intensity value of pixel in the steganograph image
M*N: Size of an Image

3.2 Peak Signal-to-Noise Ratio (PSNR)

PSNR is the ratio between maximum possible power and corrupting noise that affect representation of image.

Mathematically, it is defined as the ratio of peak square value of pixels by MSE. Higher the value of PSNR, higher the quality of image. It is evaluated as follows:

$$PSNR = 10 \log_{10} \frac{255^2}{MSE} db$$

where 255 is the maximum possible value of the pixel when pixels are represented using 8 bits per sample.

4 EXPERIMENTS AND RESULTS

AES algorithm has the following advantages due to which it is preferred over any other encryption technique:

- Most advanced algorithm used for encryption
- Can resist any kinds of password attacks
- Provides highest data security
- Can be implemented in both hardware and software. In hardware implementation, it has advantage of increased throughput and better security level.
- Suitable for high speed applications in real time.
- Larger key sizes give more security
- Lesser power consumption
- Lesser encryption and decryption time
- Higher simulation speed

To exhibit the performance of the presented system, we obtained 5 abnormal ECG records from the sudden cardiac death holter database of the Physionet database and 5 normal ECG records from the long term ST database of the Physionet database.

Figures 6 to 15 shows the MATLAB implementation of the proposed system. From the comparative study of Tables 1, 2, 3 and 4, it is quite clear that the image quality is maintained, distortion level in water-marked image is lesser and the data is secured with a higher validity of 20–30 years (Varhade & Kasat 2013).

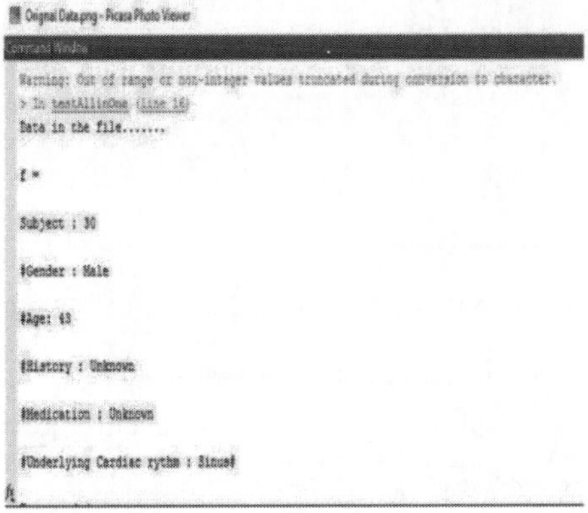

Figure 7. Original data of patient.

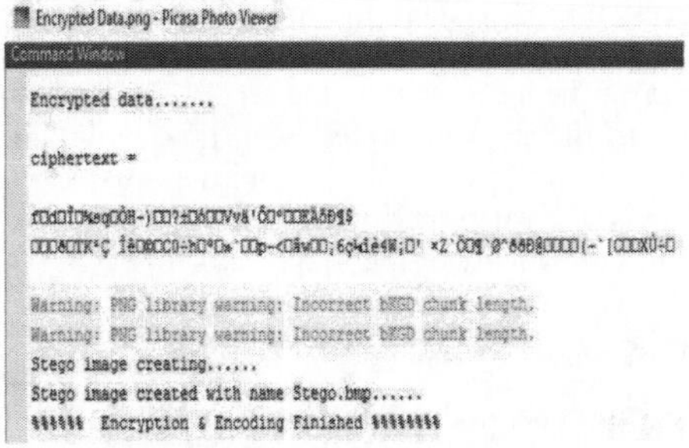

Figure 8. AES encrypted data.

Figure 6. Select the patient data file.

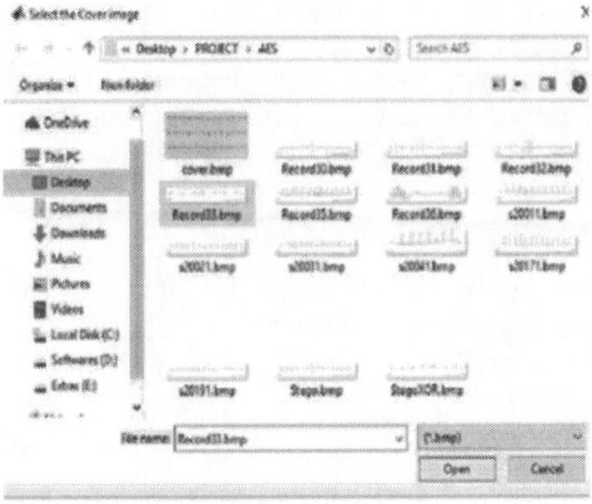

Figure 9. Select the cover image.

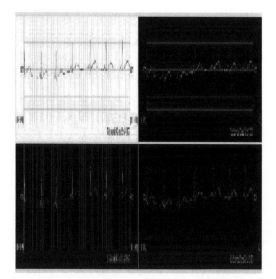

Figure 10. Original ECG signal.

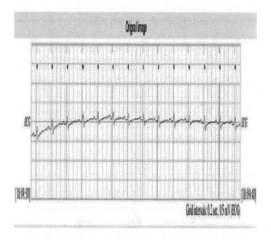

Figure 11. Sub-bands of the original image.

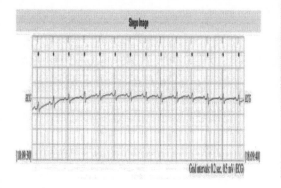

Figure 12. Steganographed image.

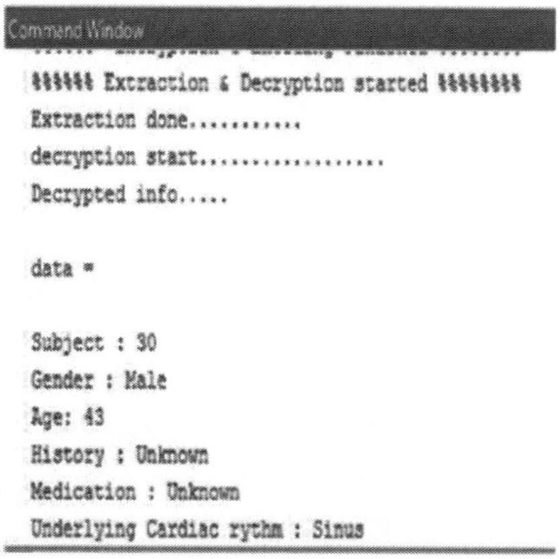

Command Window
%%%%% Extraction & Decryption started %%%%%%%
Extraction done...........
decryption start..................
Decrypted info.....
data =
Subject : 30
Gender : Male
Age: 43
History : Unknown
Medication : Unknown
Underlying Cardiac rythm : Sinus

Figure 13. Decrypted information.

Command Window
Age: 43
History : Unknown
Medication : Unknown
Underlying Cardiac rythm : Sinus
hm : Sinus
Warning: PNG library warning: Incorrect bKGD chunk length.
mse =
0.0320
PSNR_Value =
63.0800

Figure 14. Image quality assessment parameters: MSE and PSNR.

Original image
LL band of image
LH band of image
HL band of image
HH band of image
IDWT of image
Stego image

Figure 15. Integrated results.

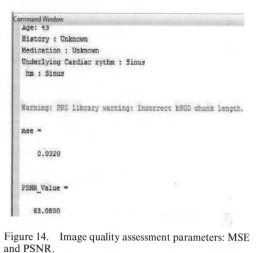

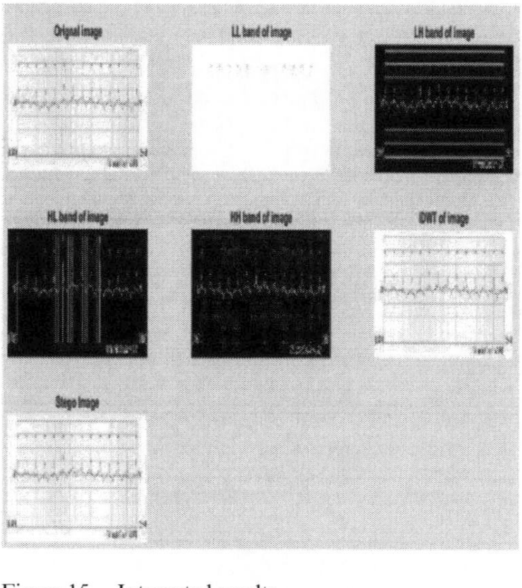

Table 1. MSE and PSNR values for Abnormal ECG Signal (Sudden Cardiac Death Holter Database) when AES encryption is applied.

Record no.	MSE	PSNR
30	0.0320	63.080
31	0.0149	66.3875
32	0.0017	75.954
33	0.0330	62.9404
35	0.0144	66.53908

Table 2. MSE and PSNR values for Abnormal ECG Signal (Sudden Cardiac Death Holter Database) when simple XOR ciphering is applied.

Record no.	MSE	PSNR
30	0.0337	62.8558
31	0.0199	65.1452
32	0.0017	75.7087
33	0.0339	62.8313
35	0.0196	65.2034

Table 3. MSE and PSNR values for Normal ECG Signals (The Long Term Database) when AES encryption is applied.

Record no.	MSE	PSNR
S20011	0.0104	67.9506
S20021	0.0094	68.4052
S20031	0.0108	67.7936
S20041	0.0232	64.4780
S20071	0.0149	66.83871

Table 4. MSE and PSNR values for Normal ECG Signals (The Long Term Database) when simple XOR ciphering is applied.

Record no.	MSE	PSNR
S20011	0.0104	67.9562
S20021	0.0095	68.3671
S20031	0.0111	67.6783
S20041	0.0234	64.4476
S20071	0.0152	66.3044

5 CONCLUSION

The proposed system uses the most advanced encryption standards to provide better security than any other encryption algorithm, when the data is transmitted over the internet (communication channel) so that no attacker will be able to harm the integrity and confidentiality of the patient's data. AES is preferred over XOR because XOR is the most basic encryption technique while AES is the most advanced and hence it can combat many security attacks that XOR cannot. AES makes it difficult for the attacker to extract the confidential data. Therefore, the Health Insurance Portability and Accountability Act (HIPAA) guidelines are also maintained and the data is secured for 20–30 years (Varhade & Kasat 2015). The results of Image Quality Assessment Parameters show that the image quality is maintained and there is less distortion level in the watermarked image. Thus, this system will provide reliable and secure method to transfer the ECG signal, as well as patient's sensitive information, to the responsible doctors in any emergency case. The outcome of the research will improve the performance and security of health-care systems and will save patients lives in many cases.

REFERENCES

Ali Al-Ataby and Fawzi Al-Naima, 2010. A Modified High Capacity Image Steganography Technique Based on Wavelet Transform, The International Arab Journal of Information Technology, Vol. 7, No. 4.

Ankita G. Shirodkar, 2015, Secure Steganography, Compression and Transmission of ECG in Point-of-Care System, IRF International Conference.

Ayman Ibaida, Ibrahim Khalil, 2013, Wavelet based ECG Steganography for Protecting Patient Confidential Information in Point-of-Care system, IEEE transaction on Biomedical Engineering, Vol. 60, No. 12.

Dr. Prerna Mahajan & Abhishek Sachdeva, 2013, A Study of Encryption Algorithms AES, DES and RSA for Security, Global Journal of Computer Science and Technology Network, Web & Security Volume 13 Issue 15 Version 1.0.

Juned Ahmed Mazumder, Kattamanchi Hemachandran, 2013, A High Capacity and Secured Color Image Steganographic Technique Using Discrete Wavelet Transformation, (IJCSIT) International Journal of Computer Science and Information Technologies, Vol. 4 (4), 583–589.

Ms. Pawar Kshetramala Dilip, Prof. V. B. Raskar, 2015, Hiding Patient Confidential Information in ECG Signal Using DWT Technique, International Journal of Advanced Research in Computer Engineering & Technology (IJARCET) Volume 4 Issue 2.

Po-Yueh Chen and Hung-Ju Lin, 2006, A DWT Based Approach for Image Steganography, International Journal of Applied Science and Engineering.

Preeti Motwani and Dimple Chaudhari, 2015, Encrypted Data Concealment in ECG signals using Steganography Technique for Telemedicine Application, International Journal of Advance Foundation And Research In Science & Engineering (IJAFRSE) Volume 1, Special Issue.

Sonali A. Varhade, N. N. Kasat, 2015, A Review on Implementation of AES Algorithm Using FPGA & Its Performance Analysis, International Journal on Recent and Innovation Trends in Computing and Communication, Volume: 3 Issue: 1.

Communication and Computing Systems – Prasad et al. (Eds)
© 2017 Taylor & Francis Group, London, ISBN 978-1-138-02952-1

A novel method for route optimization in inter and intra NEMO

A. Mrinalini, S. Joshi & R.V. Raju
Department of Computer Science and Engineering, Manipal University Jaipur, Jaipur, India

ABSTRACT: With the advancement in technology and the rising need to remain connected to the Internet even while on the move, Network Mobility (NEMO) has come into existence. It is an extension of mobile IPv6 which enables a set of nodes to collectively move from one place to another without disrupting their connection to the Internet. In hierarchical NEMO, due to the high degree of nesting in mobile network, pinball routing problem, inefficient route and end to end delay in packet transmission are quite common. Therefore, considering these problems, we have come up with a solution called Route Optimization for Inter and Intra Network (ROIIN) which will provide one tunnel solution in case of inter-network packet transmission and an efficient route optimization in case of intra-network communications. The proposed solution is simulated in NS2 and the observed results of our simulation shows better efficiency as compared to the existing solutions

Keywords: Network Mobility (NEMO), Mobile Router (MR), Correspondent Node (CN), Mobile Network Node (MNN), Access Router (AR), Top Level Mobile Router (TLMR)

1 INTRODUCTION

The progression in technology is engulfing more and more people to remain connected to the Internet. Even if a person moves around the world, the mobile devices still wants to remain connected to their home network. Additionally, the improvements in wireless communication are allowing people to increase their mobility more rapidly without any fear of getting their mobile devices disconnected from the network.

The Mobile IPv6 is a protocol that assists the nodes to be accessible while moving around in the IPv6 Internet. A node in IPv6 is always acknowledged by its Home Agent (HA). The IPv6 packets that are addressed to a mobile node's Home Address (HoA) are routed transparently to its Care of Address (CoA) thereby maintaining transparency of the mobile node. This protocol binds the mobile node's HoA with its CoA and sends the packet destined for the mobile node directly to its CoA thus compromising with the location of the node in the network. However, with the growing use of IP enabled devices, the demand for mobility support of the entire network of IP enabled devices has increased rapidly. To fulfill this demand, Mobile IP is extended in Network Mobility (NEMO).

The devices on the mobile network are provided uninterrupted Internet access even when they change their point of connection to the internet. Say some people are travelling in a bus and are using

their PDA (devices such as mobile phones, tablets, laptops etc.) which are connected to the network of the bus. As the bus moves, it changes its point of connection to the internet due to the change in its Access Router (AR). Such kind of mobility of the network is called NEMO. In NEMO, every node has—its HoA which is the fixed address through which the node is identified; and its CoA which is the current address of the node in the foreign network.

The basic architecture of the mobile network comprises of Mobile Router (MR), Top Level Mobile Router (TLMR), HA, Correspondent Node (CN), AR and Mobile Network Node (MNN). The MNNs are the PDA devices that are connected wirelessly to the MR which in turn maybe connected to another MR and so on thereby forming a hierarchy. The TLMR is connected to the AR which is connected to the wired network. The mobile nodes register their CoA with their HAs before communicating with the CN. Hence the task of maintaining the CoA of the nodes is done by the HA which resides in the HoA of the node. The architecture of NEMO can be seen in Figure 1. When MR, having its own MNNs, get attached to the TLMR, it forms a nested mobile network.

The NEMO BSP was introduced by IETF working group as an extension of Mobile IPv6 in order to facilitate the movement of the entire IP network as a single unit thereby changing its point of connection to the internet. Though NEMO BSP

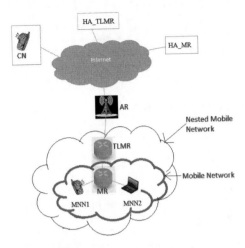

Figure 1.   Basic architecture for NEMO.

provides improved network mobility support as compared to Mobile IPv6, it still has certain limitations in terms of route optimization, end to end packet delay, overheads etc. In case of high degree of nested mobile network, pinball routing problem is quite common.

The MRs send Binding Updates (BU) to their respective HAs in order to register their CoAs with it. The HA_MR provides one or more prefixes to the MR for MNNs under it. It also creates a binding cache to map the HoA of the MR and the prefixes to the CoA of the MR. The HA then sends Binding Acknowledgement (BA) to MR and a bi-directional tunnel is created between MR and the HA_MR for the purpose of mobility transparency.

In Figure 1, when CN wishes to send a packet to MNN1, the packet is first received by the HA_MR because MNN1 is attached to MR and also the HA_MR has provided prefix to MR for MNN1 while maintaining the records of the MNN1's HoA and CoA in its binding cache. The cache states that it is located under MR which is registered with the HA_TLMR. Therefore, the packet is encapsulated by the HA_MR and is forwarded to HA_TLMR. On reaching HA_TLMR the process of looking up the record in the binding cache is repeated and after getting the information required, the packet is encapsulated once more and gets forwarded to TLMR. The TLMR on receiving the packet de-capsulates it and forwards it to the MR. The MR on receiving it, de-capsulates it again and forwards it to MNN1. The number of times that encapsulation and de-capsulation procedure takes place will increase as there will be increase in the nesting of the mobile routers causing an end to end delay in packet transmission and an in-efficient route. This problem is referred to as the pinball routing problem.

The organization of the rest of the paper is as follows—Section 2 contains Literature Review, Section 3 states the Problem Statement, Section 4 gives a detailed introduction about ROIIN, Section 5 shows the Simulation Results and Analysis and finally Section 6 concludes the paper and provides the Future Scope of the work.

2   LITERATURE REVIEW

There are several route optimization techniques that have been proposed in the past years and are still being proposed to overcome the pinball routing problem in case of hierarchical mobile networks. Of those, a few of them are listed below.

2.1   Route Optimization using Tree Information Option

Route Optimization using Tree Information Option (ROTIO) was projected by Hosik Cho, Taekyoung Kwon et al. In this scheme, the TLMR forms the Router Advertisement message or the RA message which has the xTIO (Tree Information Option) which contains CoA of all the MRs contained in the nested mobile network. The xTIO gets appended in each BU message so as to inform an MR of all of its predecessor MRs. Each MR sends two kinds of BUs—a local BU which is sent to the TLMR informing it about the topology of the nested mobile network and a Normal BU which is sent to the HA of MR informing it about the TLMR. The packet from CN is first routed to the nearest MR's HA which has the data about the HA of TLMR through the BU message received from the MR. Hence the packet gets encapsulated and is then forwarded to the HA of TLMR. The HA of TLMR on receiving the packet encapsulates it and sends it to the CoA of TLMR. On receiving the packet, the TLMR de-capsulates it and searches its binding cache to course the packet to the Local Fixed Node (LFN).

Pros: This scheme provides location privacy.
Cons: There are two stages of nested tunnels i.e. one amid the HA of MR and the HA of TLMR and the second amid the HA of TLMR and the CoA of TLMR. This two tunnel solution gives unnecessary delay and overheads in the transmission of packet.

2.2   Mobile Router assisted Route optimization technique

Mobile Router assisted Route optimization Technique (MoRaRo) was projected by Ved P. Kafle, and Eiji Kamioka. In this scheme, the data packets travel along an optimal path from the source to the destination. The MNN performs Route Optimization (RO) with the CN only once at the commencement of the communication. After that, whenever the network moves, the RO is performed by the TLMR on behalf of all the MNNs in the hierarchy. All the MRs in this scheme maintain an MR Binding Cache (MRBC) for all the nested routers below them in the hierarchy. The TLMR maintains another binding cache along with the MRBC called the RO Binding Cache (ROBC) for all the active MNNs which currently are communicating with their respective CNs. The MNN binds the BU message using TLMR's CoA in alternate CoA and sends it to the CN. The CN then creates a binding between the MNN's HoA and the CoA mentioned in the alternate CoA.

Pros: The CN can route the packet directly to the TLMR's CoA without any tunnel.

Cons: It increases the signaling cost due to the large number of BU messages sent to the CNs and it also compromises with the location of the MNN.

2.3 *Route Optimization with Location Privacy*

Route Optimization with Location Privacy (ROLP) was projected by Zial Ul Haq and Adeel Baig. In this scheme, TLMR takes the accountability of the management of the packet transmission from MNN to CN and vice versa. Each MR inside the mobile network has a cache called the MR Cache (MRC) for maintaining records of all the nodes below it in the hierarchy. These cache are exchanged with the TLMR at fixed intermissions. The TLMR maintains another cache besides MRC called the TLMR Cache (TMC) for the management of and connecting the mobile network with the wired network. When the MNN sends a Binding Request (BR) message to TLMR, it makes an entry in its TMC and sends BA signal to the MNN with the CoA of MNN as the HoA of TLMR. Only then can the MNN send the BU message to the CN in alternative CoA for MNN as the HoA of TLMR. The CN then sends the packet to the HoA of TLMR where it gets encapsulated and is forwarded to the CoA of TLMR. The TLMR after checking the information in its TMC gets the CoA of MNN which is inside the mobile network and forwards the packet to it using MRC. In ROLP, the packet transmission between CN and MNN takes place using Routing Header Type 2 (RH2).

Pros: This scheme provides one tunnel solution.

Cons: It does not provide efficient route optimization for packet transfer in case of intra-network communications.

Table 1. Characterization of various route optimization features.

Features / RO techniques	ROIIN	ROTIO	MoRaRo	ROLP
Location privacy	Yes	Yes	No	Yes
One tunnel solution for inter-network communications	Yes	No	No	Yes
Optimized path for intra-network communications	Yes	Yes	Yes	No

3 PROBLEM STATEMENT

As stated in Section II, the ROLP technique does not provide an efficient path for the transfer of packets in case of intra-network communication. However, it does provide an excellent one tunnel solution with efficient route optimization in case of inter-network packet transmission. Therefore, in our research we are making a few modifications in the ROLP technique so as to extend it to work more effectively in case of intra-network communications.

4 PROPOSED SOLUTION

After studying about and analyzing the pros and cons of the route optimization techniques mentioned in Section 2, we are proposing a solution named ROIIN, which will provide one tunnel solution for the transfer of packets from CN to MNN thus reducing end to end delay in packet transfer, as well as will provide a more optimized path for the intra-network communications. In ROIIN, when a specific MNN wants to communicate with a CN, it sends two BUs—one BU will be sent to the HA of the MNN informing it about the CoA of the TLMR and the other BU will be sent to the TLMR informing it that it will be communicating with the CN and thereby provides with the information required for TMC along with the HoA of CN and other authorization data. The TLMR along with the other records in TMC, will also maintain a record of the HoA of CN in it. Therefore, when the packet is sent from CN to the HA of MNN, it encapsulates it and forwards it to the CoA of the TLMR. Now, the TLMR on receiving the packet de-capsulates it and checks it in its TMC for the HoA of MNN, and on finding its location, it forwards the packet to MNN.

Now in case of intra-network, when the MNN wishes to send a packet to CN, it forwards it to

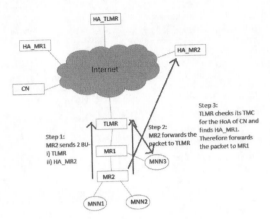

Figure 2.   ROIIN path for inter-network communication.

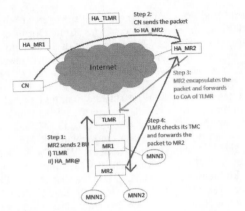

Figure 3.   ROIIN path for intra-network communication.

the TLMR. The TLMR after de-capsulating the packet checks in its TMC for the HoA of MNN, it also checks for the HoA of CN. If the HoA of CN happens to be in the same network, the TLMR forwards the packet to it directly, thus preventing end to end packet delay in case of Intra-Communications. Otherwise, the packet is directed to the HA of MNN from where the packet is sent to the CN.

5 SIMULATION RESULTS AND ANALYSIS

Simulation of our proposed solution ROIIN has been done in NS 2.33 with Mobiwan patch. The network size used for simulation is 1500 m × 1500 m.

The Table 2 shows the end to end delay of the proposed solution ROIIN and the existing solution ROLP.

In case of Inter-network, let the nodes in the NEMO architecture within and outside the network be located at distances say d_{11}, d_{12}, and so on.

Therefore, on comparing the distance a packet has to travel when sent from the CN to an MNN in a nested mobile network of degree 'n' we get,

ROIIN: $d_{IRoiin} = d_{11} + d_{12} + [d_{13} + d_{14}...n + 1 \text{ times}]$
Now, when $n = 2 => d_{IRoiin} = d_{11} + d_{12} + d_{13} + d_{14} + d_{15}$,
When $n = 3 => d_{IRoiin} = d_{11} + d_{12} + d_{13} + d_{14} + d_{15} + d_{16}$,
When, $n = 4 => d_{IRoiin} = d_{11} + d_{12} + d_{13} + d_{14} + d_{15} + d_{16} + d_{17}$ and so on.
ROLP: $d_{IRolp} = d_{11} + d_{12} + [d_{13} + d_{14}...n + 1 \text{ times}]$
Now, when $n = 2 => d_{IRolp} = d_{11} + d_{12} + d_{13} + d_{14} + d_{15}$,
When $n = 3 => d_{IRolp} = d_{11} + d_{12} + d_{13} + d_{14} + d_{15} + d_{16}$,
When, $n = 4 => d_{IRolp} = d_{11} + d_{12} + d_{13} + d_{14} + d_{15} + d_{16} + d_{17}$ and so on.
ROTIO: $d_{IRotio} = d_{11} + d_{12} + d_{13} + [d_{14}...n+1 \text{ times}]$
Now, when $n = 2 => d_{IRotio} = d_{11} + d_{12} + d_{13} + d_{14} + d_{15} + d_{16}$,

Table 2.   End to end delay simulation results.

	Inter-NEMO	Intra-NEMO
Existing solution (ROLP)	72 ms	65 ms
Proposed solution (ROIIN)	70 ms	38 ms

When $n = 3 => d_{IRotio} = d_{11} + d_{12} + d_{13} + d_{14} + d_{15} + d_{16} + d_{17}$,
When, $n = 4 => d_{IRotio} = d_{11} + d_{12} + d_{13} + d_{14} + d_{15} + d_{16} + d_{17} + d_{18}$ and so on.
MoRaRo: $d_{IMoraro} = d_{11} + [d_{12} + d_{13}...n + 1 \text{ times}]$
Now, when $n = 2 => d_{IMoraro} = d_{11} + d_{12} + d_{13} + d_{14}$,
When $n = 3 => d_{IMoraro} = d_{11} + d_{12} + d_{13} + d_{14} + d_{15}$,
When, $n = 4 => d_{IMoraro} = d_{11} + d_{12} + d_{13} + d_{14} + d_{15} + d_{16}$ and so on.

Hence we observe that in case of ROIIN there has been a 16.67% improvement in the path optimization as compared to ROTIO. When compared with ROLP, ROIIN gives a similar result for inter NEMO route optimization as that of it. However, it does not give the result as good as MoRaRo but it does provide location privacy which is absent in MoRaRo.

In the following Figure 4, we see the path followed by packets in case of ROIIN

In the following Figure 5, it shows the analysis of the path followed by the packets in ROIIN, ROLP, ROTIO and MoRaRo in the form of line graphs.

Now, in case of intra NEMO, on comparing the distance a packet has to travel when sent from the MNN to the CN in a nested mobile network of degree 'n' where the depth of CN in the mobile network is 1 and the distances between the nodes are d_{i1}, d_{i2}, d_{i3}, and so on we get,

ROIIN: $d_{iRoiin} = d_{i1} + d_{i2} + [d_{i3}, d_{i4}.....n + 1 \text{ times}]$.
Now, when $n = 2 => d_{iRoiin} = d_{i1} + d_{i2} + d_{i3} + d_{i4} + d_{i5}$

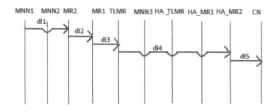

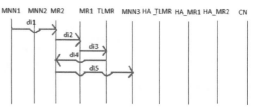

Figure 4. Path followed by a packet when sent from MNN1 to CN in ROIIN in case of Inter NEMO when n = 2.

Figure 6. Path followed by a packet when sent from MNN1 to MNN3 in ROIIN in case of Intra NEMO when n = 2.

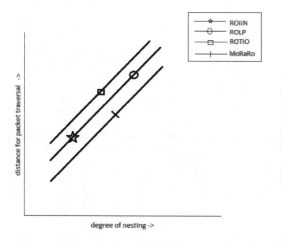

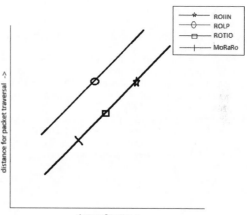

Figure 5. Graph showing inter-network packet traversal time.

Figure 7. Graph showing intra-network packet traversal time.

When $n = 3 => d_{iRoiin} = d_{i1} + d_{i2} + d_{i3} + d_{i4} + d_{i5} + d_{i6}$
When $n = 4 => d_{iRoiin} = d_{i1} + d_{i2} + d_{i3} + d_{i4} + d_{i5} + d_{i6} + d_{i7}$ and so on.

ROLP: $d_{iRolp} = d_{i1} + d_{i2} + d_{i3} + d_{i4} + [d_{i5}, d_{i6}.....n + 1$ times].
Now, when $n = 2 => d_{iRolp} = d_{i1} + d_{i2} + d_{i3} + d_{i4} + d_{i5} + d_{i6} + d_{i7}$
When $n = 3 => d_{iRolp} = d_{i1} + d_{i2} + d_{i3} + d_{i4} + d_{i5} + d_{i6} + d_{i7} + d_{i8}$
When $n = 4 => d_{iRolp} = d_{i1} + d_{i2} + d_{i3} + d_{i4} + d_{i5} + d_{i6} + d_{i7} + d_{i8} + d_{i9}$ and so on.

ROTIO: $d_{iRotio} = d_{i1} + d_{i2} + [d_{i3}, d_{i4}.....n + 1$ times].
Now, when $n = 2 => d_{iRotio} = d_{i1} + d_{i2} + d_{i3} + d_{i4} + d_{i5}$
When $n = 3 => d_{iRotio} = d_{i1} + d_{i2} + d_{i3} + d_{i4} + d_{i5} + d_{i6}$
When $n = 4 => d_{iRotio} = d_{i1} + d_{i2} + d_{i3} + d_{i4} + d_{i5} + d_{i6} + d_{i7}$ and so on.

MoRaRo: $d_{iMoraro} = d_{i1} + d_{i2} + [d_{i3}, d_{i4}.....n + 1$ times].
Now, when $n = 2 => d_{iMoraro} = d_{i1} + d_{i2} + d_{i3} + d_{i4} + d_{i5}$
When $n = 3 => d_{iMoraro} = d_{i1} + d_{i2} + d_{i3} + d_{i4} + d_{i5} + d_{i6}$
When $n = 4 => d_{iMoraro} = d_{i1} + d_{i2} + d_{i3} + d_{i4} + d_{i5} + d_{i6} + d_{i7}$ and so on.

Hence we observe that in case of ROIIN there has been a 28.57% improvement in the path optimization as compared to ROLP. When compared with ROTIO and MoRaRo ROIIN gives a similar result for intra NEMO route optimization as that of them. However, it does provide location privacy which is absent in MoRaRo.

In the following Figure 6, we see the path followed by packets in case of ROIIN.

The following graph Figure 7, shows the analysis of the path followed by the packets in ROIIN, ROLP, ROTIO and MoRaRo in the form of line graphs.

6 CONCLUSION AND FUTURE WORK

In this paper, we have provided a solution named ROIIN with the help of which we are able to achieve 16.67% improvements in result, in case of optimizing the route for the transfer of packets in Inter-NEMO and 28.57% improvements in result in case of optimizing the route for the transfer of packets in Intra-NEMO when compared with the existing solutions. Additionally, we have provided the feature of location privacy in ROIIN as well. In our future works, we would be more focused on designing route optimization technique in case inter and intra NEMO with location privacy in depth.

REFERENCES

Abu Zafar M., Shahriar and Atiquzzamanl, Moham-med Evaluation of the Route Optimization for NEMO in Satellite Networks Journal of Wireless Mobile networks, Ubiquitous Computing, and Dependable Applications, Volume 2 Number 2 pp. 46–66.

Cho, H. Kwon, T. & Choi, Y. 2006 Route Optimization using Tree Information Option for nested mobile networks IEEE J. Sel. Areas Communication, pp. 1717–1724.

Devarapalli, V., Wakikawa, R., Pertruscu, A., & Thubert, P. 2005 Network mobility (NEMO) Basic Support Protocol IETF RFC 3963.

Ernst, T. & Lach, H. 2005 Network Mobility support terminology http://ietfreport.isoc.org/idref/draft-ietf-nemo-terminology.

Goswami, Shubrananda & Das, Chandan Bikash 2014 Performance Evaluation on various Route Optimization techniques in Network Mobility (NEMO) *International Journal of Modern Computer Science (IJMCS)* Volume 2 Issue 6.

Haq, Zia Ul & Baig, Adeel 2012 Route Optimization Technique for Nested Mobile Networks 4th International Conference on Computational Intelligence, Communication Systems and Networks.

IETF Network Mobility (NEMO) Charter. http://www.ietf.org/html.charters/nemo-charter.html

Johnson, D., Perkins, C. & Arkko, J. 2011 Mobility support in IPv6 IETF RFC 6275.

Johnson, D., Perkins, C. & Arkko, J. 2004 Mobility Support in IPv6 *RFC3775.*

Kafle, V. P., Kamioka, E. & Yamada, S. 2006 MoRaRo: Mobile route assisted route optimization for Network Mobility (NEMO) support IEICE Trans. Inf. Syst. pp 158–170.

Kim, M., Lee, J. H. & Chung, T. M. 2009 Route Optimization in Nested NEMO: Classification, Evaluation and Analysis from NEMO Fringe Stub Perspective IEEE Transactions on Mobile Computing, Volume 8 Issue 11.

Shahriar, A. Z. M., Atiquzzaman, M. & Ivancic, W., 2010 Route Optimization in network mobility: Solutions, classification, comparison, and future research directions IEEE Communications Surveys and Tutorials, Volume 12 Issue 1.

Song, Jungwook, Kim, Heemin, Han, Sunyoung & Joo, Bokgyu 2007 LPD Based Route Optimization in Nested Mobile Network *International Federation for Information Processing.*

Thubert, P. & Montavont, N. 2005 Nested NEMO Tree Discovery. http://ietfreport.isoc.org/idref/draft-thubert-tree-discovery.

Communication and Computing Systems – Prasad et al. (Eds)
© 2017 Taylor & Francis Group, London, ISBN 978-1-138-02952-1

Cluster analysis for pollution density

Rahul Kumar, Abhineet Anand & Rajeev Tiwari
UPES, Dehradun, India

ABSTRACT: Clustering is a process in which the objects that fall in one category (by any means of dividing them in those categories) are grouped together and that group has some characteristics different from other groups. In current context, clustering is used to divide the dataset observed for pollution quantities of Uttrakhand area into definable clusters. In this paper the comparative study of two existing algorithms is done and the results derived is used to gather information about the density of pollution particles in a particular area.

1 INTRODUCTION

Basically clustering is defined as the technique that finds significant data underneath random datasets. These data sets can be thus divided into different parts/layers/groups/communities as shown in Figure 1. Algorithms and parameters (e.g. distance parameter etc.) associated with it generally differ according to datasets that are to be computed and the result required. This article emphasizes on locating different clusters in given datasets with the help of two algorithms and thus finding optimized solution for that dataset. To produce dataset, there are a variety of theorems available. This context uses algorithms like K Mean clustering algorithm, Max cut min flow algorithm and uses them to generate clusters on a dataset of Uttrakhand pollution statistics. With those results pollution density can be found out for different areas.

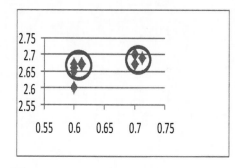

Figure 1. Clustering.

Clustering is not about a particular task but a continuous process to get results based on different scenarios. Clustering has different application in fields of Data mining, information gathering and processing, Bioinformatics, data compression and image processing. E.g. in bioinformatics, Hierarchical clustering is used to enhance protein structure prediction by combining the predictions made by a big number of possible conformation models. IN compression, hierarchical clustering in dendrograms based on fast heuristic for quartet method. These techniques are robust and effective.

2 EXISTING APPROACHES TO FIND CLUSTERS IN A GIVEN DATASET

To define clusters there are many methods that can be deployed with different processes giving off different result sets. Two of them are defined in this document.

2.1 K mean clustering algorithm

K means is a process in which n number of observations are divided into k clusters such that each of the observation belongs to a specific cluster with a nearest mean. Computation of this algorithm is difficult except there are some heuristic techniques that can be employed to get optimum result.

Using k means is similar to using Expectation maximization algorithm except that k means generate clusters that are in a continuous shape while Expectation maximization clusters are of different shapes.

The general form of algorithm is as follows:
1. Given a set of observation m_1, m_2, m_3............ ....m_k.
2. Place these points in the represented space.
3. Allocate a point to a group with nearest centroid.
4. When all the objects are allocated, recalculate the position of the new K centroids.

5. Try above steps until clustering centroid are fixed at a point. This gives a separation of objects into groups
Assignment of observation:

$$S_i^a = \{x^p : \|x^p - m_i^a\|^2 \le \|x_p - m_j^a\|^2 \ \forall j, 1 \le j \le k\},$$

where xp is accorded to exactly one S^a, even if two allotment are possible.

Updation of centroid: $m_i^{a+1} = \dfrac{1}{S_i^a} \displaystyle\sum_{xj \in S_i^a} xj$

There are several variations of k mean clustering algorithm such as Spherical k means, X means clustering algorithm, G means clustering algorithm etc. The Key features of k mean are Euclidian distance, No of clusters that are to be entered as input, Convergence.

The limitation of k mean algorithm is that it cannot accommodate cluster with variable sizes i.e. it's result is Voronoi cells. It cannot have variances and covariance while Expectation maximum algorithm accommodates that.

2.2 Max Flow Min Cut algorithm

This algorithm is a naïve approach to find out the clusters in a directed graph set. It's parts are Minimum cut algorithm and Maximum flow algorithm.

1. Minimum Cut algorithm: A approach that can be applied on both directed and undirected graphs, Min cut divides the datasets into portions that are almost equal. The main motive of using this is to make the sum of weights of edges that are connection two parts of a graph minimum.
2. Maximum Flow algorithm: It works out that the flow from the starting node to the ending node should be maximum (i.e. from source to sink.) The only dominate is that flow should be conserved i.e. the outgoing flow should be equal to incoming flow.

Both Min cut and max flow algorithms can be used together to generate clusters. It was developed by Ford and Fulkerson in 1956. Only a weighed graph can use this kind of approach. It's constraints are:

1. Outgoing flow = incoming flow.
2. Capacity of edge ≥ flow of edge

To calculate the path, a BFS approach is followed and an expanding path line is created. The path line should be free from cycles. This is done until max flow is achieved.

The complexity of this approach is O(ab) where a = no of edges and b = maximum flow. For path finding the complexity is O(a).

The limitations are that flow needs to be integral rational. Also for each iteration residual capacity should be integral.

Max cut min flow can be used in networking data mining as well as many real life scenarios such as Sports team mathematical cutting etc. Many researches are done for the application part eg Bipartite matching and covers, Optimal disclosure in a digraph, Flow feasibility etc.

Some real world scenarios for this algorithm are:

- BaseBall Elimination
- Airline Scheduling
- Car sharing fairness problem

3 FIGURES AND TABLES

3.1 Calculated results from k mean clsutering algorithm

The k means algorithm was applied on the following dataset of Uttrakhand Pollution particles (i.e. concentration levels of No2 and So2 levels) for different areas.

Fig. 2 shows the dataset for the pollution particles present in the air of various places in Dehradun Region.

This is the dataset that is worked upon in this context. It was acquired from http://utrenvis.nic.in/

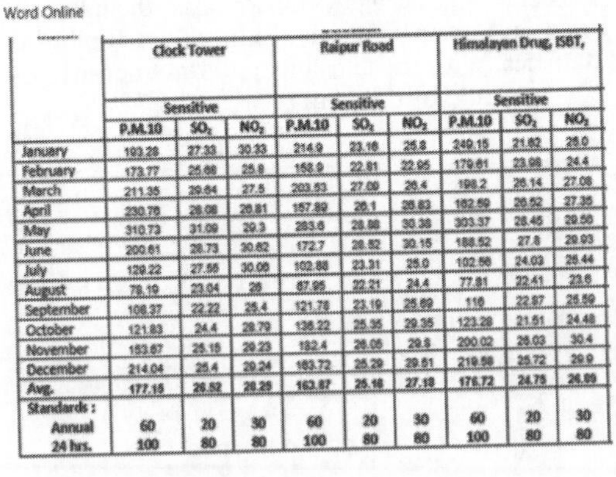

Word Online

	Clock Tower			Raipur Road			Himalayan Drug, ISBT,		
	Sensitive			Sensitive			Sensitive		
	P.M.10	SO₂	NO₂	P.M.10	SO₂	NO₂	P.M.10	SO₂	NO₂
January	193.26	27.33	30.23	214.9	23.16	25.8	249.15	21.82	26.0
February	173.77	25.06	25.6	158.9	22.81	22.05	179.61	23.08	24.4
March	211.35	29.64	27.5	203.53	27.00	26.4	198.2	26.14	27.08
April	230.78	26.08	26.81	157.89	26.1	26.83	182.59	26.52	27.36
May	310.73	31.09	29.3	263.6	28.88	30.38	303.37	28.45	29.56
June	200.61	28.73	30.82	172.7	28.82	30.15	188.52	27.8	29.03
July	129.22	27.56	30.06	102.88	23.31	25.0	192.96	24.03	26.44
August	78.19	23.04	26	87.95	22.21	24.4	77.81	22.41	23.6
September	108.37	22.22	25.4	121.78	23.19	25.69	116	22.87	26.59
October	121.83	24.4	28.79	136.22	25.35	29.35	123.28	21.51	24.48
November	153.67	25.15	29.23	192.4	26.05	29.8	200.02	26.03	30.4
December	214.04	25.4	29.24	183.72	25.29	29.51	176.72	24.75	26.99
Avg.	177.15	26.52	28.29	163.87	25.18	27.18			
Standards :									
Annual	60	20	30	60	20	30	60	20	30
24 hrs.	100	80	80	100	80	80	100	80	80

Figure 2. Data set.

The computation is done in the language R in which the no of clusters = 3.

The Fig. 3 is the result set for the given computation.

Figure 3. R calculations.

Clusters are made such that the higher density pollution particle sets are put together and lower density particle sets were put together.

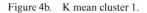

Figure 4a.

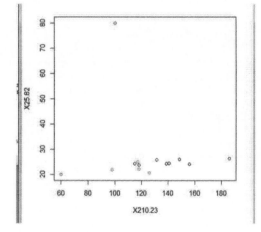

Figure 4b. K mean cluster 1.

Figure 5. K mean cluster 2.

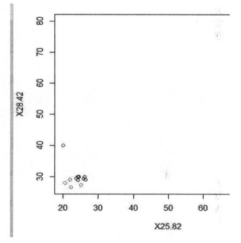

Figure 6. K mean cluster 3.

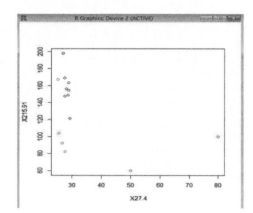

Figure 7. K mean cluster 4.

Cluster sets for different Particle projections are shown in Figures 4,5,6 and 7.

3.2 Calculated results from Min Cut Max Flow clustering algorithm

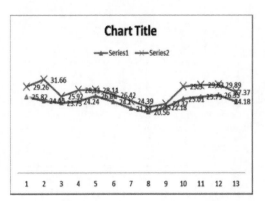

Figure 8.   Chart title.

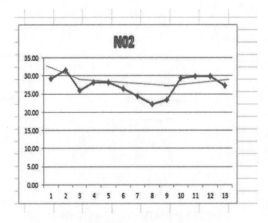

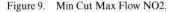

Figure 9.   Min Cut Max Flow NO2.

This result set works on the Graph partition analysis formula.

Using similar cuts, we put the cut sets into a different table to generate clusters.

Cluster 1		Cluster 2	
No2	So2	No2	So2
31.66	26.06	29.26	25.82
29.3	25.79	25.92	24.45
29.83	26.35	28.11	23.75
29.89	25.01	26.42	21.87

The clustering graph is as shown in Figure 11.

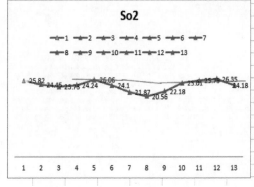

Figure 10.   Min Cut Max Flow SO2.

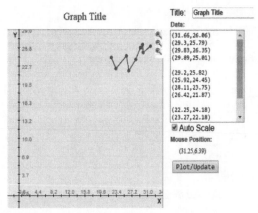

Figure 11.   Cluster through Min Cut Max Flow.

4   CONCLUSION

In this work, an overview of two existing algorithms are provided and they are worked upon a dataset to generate clusters. Both K mean clustering algorithm and Max cut min flow provide us with applicable results and can be used for clustering on similar datasets. But further studies are suggested on these algorithms as these algorithms go not guarantee that clusters formed will be the best possible one. There may be other algorithms that may be better (even if computationally expensive or slow) at providing clusters. Thus the search for such an algorithm can be considered for further study.

REFERENCES

Abhineet Anand, Vikas Kr. Sihag and P S V S Sridhar. Article: Community Structure based on Node Traffic in Networks. *International Journal of Computer Applications* 69(13):15–20, May 2013.

Ashwani Kumar, Surinder Pal Singh and Nitin Arora. Article: A New Technique for Finding Min-cut Tree. *International Journal of Computer Applications* 69(20):1–7, May 2013.

Corrected Version of: IEEE Transactions on Information theory, vol. 51, no 4, April 2005, 1523–154.

Guttmann-Beck, N. and R. Hassin. Approximation algorithms for min-max tree partition *Journal of Algorithms* 24 (1997), 266–286.

Hamann, M., Hartmann, T., Wagner, D.: Complete Hierarchical Cut-Clustering: A Case Study on Expansion and Modularity. In: Bader, D.A., Meyerhenke, H., Sanders, P., Wagner, D. (eds.) Graph Partitioning and Graph Clustering: Tenth DIMACS Implementation Challenge. DIMACS Book, vol. 588, American Mathematical Society (to appear, 2013).

Juhi Gupta and Aakanksha Mahajan. Article: BPSO Optimized K-means Clustering Approach for Data Analysis. *International Journal of Computer Applications* 133(15):9–14, January 2016. Published by Foundation of Computer Science (FCS), NY, USA.

Marghny, M H, Rasha Abd M El-Aziz and Ahmed I Taloba. Article: An Effective Evolutionary Clustering Algorithm: Hepatitis C Case Study. *International Journal of Computer Applications* 34(6):1–6, November 2011.

Mihika Shah and Sindhu Nair. Article: A Survey of Data Mining Clustering Algorithms. *International Journal of Computer Applications* 128(1):1–5, October 2015. Published by Foundation of Computer Science (FCS), NY, USA.

Sadhana Tiwari and Tanu Solanki. Article: An Optimized Approach for k-means Clustering. *IJCA Proceedings on 9th International ICST Conference on Heterogeneous Networking for Quality, Reliability, Security and Robustness 2013* QShine: 5–7, December 2013.

Communication and Computing Systems – Prasad et al. (Eds)
© *2017 Taylor & Francis Group, London, ISBN 978-1-138-02952-1*

Advance approach towards elbow movement classification using discrete wavelet transform and quadratic support vector machine

P. Kumari, Y. Narayan, V. Ahlawat, L. Mathew & Alokdeep
EE Department, National Institute of Technical Teachers Training and Research, Chandigarh, India

ABSTRACT: This paper purposed a method based on Support Vector Machine (SVM) and DWT (Discrete Wavelet Transform) of surface electromyography (sEMG) signal classification of predefined intentional movement rather than imposed standard movement. For this study sEMG signals was acquired from the biceps and triceps muscle of the upper arm. The presented method can be a novel development of dynamic upper extremity, exoskeleton limb for physically weak subject. The overall classification accuracy with few time domain features such as Root Mean Square (RMS), Skewness, Varience (VAR), Waveform Length (WL) and time scale feature (Mean, Median, Standard Deviation (SD), Median Absolute Deviation with Quardatic SVM and DWT was achieved with 93.7%. DWT can simulate the approximation coefficient, detailed coefficient and synthesized signal of raw sEMG signal. For this study we used the fourth level approximation coefficient of sEMG signal as feature vector as an input for classifier.

1 INTRODUCTION

sEMG signal play a vital role in both engineering and medical applications. The sEMG shows the study of muscle function by the analysis of the electrical signal, which is generating by the muscular contraction and flexion, related to tension. The muscular contraction may be two types: Voluntary and Involuntary. EMG signal classification has been done with neural network classifier with DWT. A combination of pattern recognition algorithm with classification accuracy of 97% performed on spinal code injury subject has been implemented (Liu, Zhou, 2013). The previous study shows that residual forearm produce enough signals for myoelectric control of robotic wearable fist wrist (G. Li, Senior, Schultz, & Kuiken, Todd, 2010). An improved and highly accurate algorithm has been demonstrated for heuristic fuzzy logic for EMG pattern recognition for multifunctional prosthesis control (Ajiboye & Weir, 2005). The effect of muscle spasm in long time recording of EMG signal was done by Jeffrey Winslow et al with good accuracy and precision (Winslow, Martinez, & Thomas, 2015). The comparison between Linear Discriminant Analysis (LDA) and SVM has been done and it reached up to 92% accuracy in recognition of trained classes for the collected data from 6 healthy subjects (Z. Li, Wang, Yang, Xie, & Su, 2013). In (Pao, Lan, Yang, & Liao, 2006) presented least square mixed norms vector classifier, which is the combination of 1 norm support vector classifier and 2 norm support vector classifier. It has been studied that SVM has been used in classification problems. An approach has been developed by researcher by using post processor on the SVM classifier. It is generally used in Gaussian radial Kernel function (Lin & Chen, 2008).

2 MATERIALS AND METHODS

2.1 Data acquisition

The data has been acquired from 10 healthy subjects of the age group from 22 to 26 years in National Institute of Technical Teachers Training & Research (NITTTR) Biomedical Instrumentation Lab with the facility of four channels recording. There are two channels for Triceps and rest two are for Biceps Brachii. Six trails have been performed from each subject for each activity such as flexion and extension. The signal was acquired with the MYOTRACE 400 device having the signal processing facility. The data acquisition process is presented in Figure 1.

2.2 Signal processing

Signal processing is done by several processing methods such as: Rectification, Smoothing, Normalization of amplitude and filtering etc by the experimental setup. In rectification process all amplitude values is multiplied by +1, so that amplitude, which is below the zero line become positive.

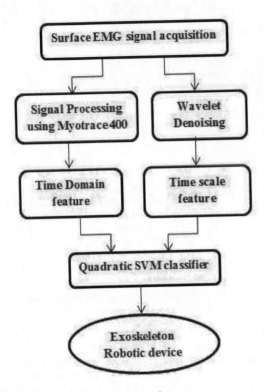

Figure 1. Flow chart for control process.

2.3 Feature selection and extraction

Feature selection is a preliminary and important step before forming feature vector as an input to the classifier. A proper selection of feature can give better result so that the output of classifier can be utilized as a control signal for external wearable robot. The formulation of feature extraction is given in equations (1)–(4).

$$RMS = \sqrt{\frac{1}{N}\sum_{n=1}^{N}x_n^2} \qquad (1)$$

$$WL = \sum_{n=1}^{N-1}|x_{n+1} - x_n| \qquad (2)$$

$$SKW = \frac{1}{N}\sum_{n=1}^{N}(x_n - \bar{x})^3 \Big/ \left(\frac{1}{N}\sum_{n=1}^{N}(x_n - \bar{x})^2\right)^{3/2} \qquad (3)$$

$$VAR = \frac{1}{N-1}\sum_{n=1}^{N}x_n^2 \qquad (4)$$

2.4 Discrete Wavelet Transform (DWT)

Wavelet analysis is an important signal processing tool for the Time-Frequency Domain (TFD).

Wavelet Packet Transform (WPT) and DWT are popular approaches for feature extraction of sEMG signals (Englehart, Hudgins, Parker, & Senior, 2001). Wavelet transform can provide a way for analysing waveform bounded in both frequency and time duration. It allows the use of long time intervals for more precise information of low frequency and shorter regions with high frequency information. Wavelet transform is classified basically in two categories: Continuous Wavelet Transform (CWT) and Discrete Wavelet Transform (DWT).

CWT divides a continuous function in to the wavelets. It offers a good localization of the time and frequency and also has an ability to reconstruct a signal. The computation of CWT is expressed by the integral in equation (5).

$$X_w(a,b) = \frac{1}{|a|^{\frac{1}{2}}}\int_{-\infty}^{\infty}x(t)\overline{\Psi(t)\frac{t-b}{a}}\,dt \qquad (5)$$

where $\Psi(t)$ is a continuous function in both frequency and time domain. It is known as the mother wavelet and the overline represents operation of complex conjugate. The main original signal x(t). a is a scale factor, which either compress or dilates a signal. When scale factor is low, then signal is contracted. When it is high, then signal is stretched. DWT is a transform in which the wavelets are discretely sampled. The key advantage is that it provides sufficient information for both synthesis and analysis of the original signal, by which the computation time is reduced. DWT is such a system of cascading filters. One is known as wavelet filter and other is scaling filter. The wavelet filter operates as a High Pass Filter (HPF) and scaling filter operates as Low Pass Filter (LPF).

In DWT subband algorithm is used, which is shown in Figure 2.

m-level approximation coefficients are –

$$A_m[n] = \sum_{k=-\infty}^{\infty}I_a[K]A_{m-1}[2n-K] \qquad (6)$$

m-level detailed coefficients are –

$$I_m[n] = \sum_{k=-\infty}^{\infty}h_d[K]A_{m-1}[2n-K] \qquad (7)$$

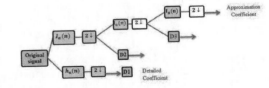

Figure 2. DWT algorithm.

Mother wavelet are mainly two types: Orthogonal and Biorhogonal. Orthogonal wavelet is defined by scaling filter (a low pass finite impulse response filter). In biorthogonal wavelet reconstruction filter and decomposition filter are defined separately. There are seven wavelet families. They have different sub-types. These are given in Table 1.

After selecting the appropriate wavelet function, the first step is denoising. In this paper (procedure), the reconstruction of original sEMG signal is done on the basis of approximation coefficient level upto 4, which is computed by discrete wavelet transformation.

2.5 SVM classifier

Support vector machine is a tool for machine learning and to solve the classification problems. It is based on the statistical learning theory and has been widely used in pattern recognition. SVM is a representation of sample point in space and divided by a hyperplane with a clear gap as wide as possible. It constructs a hyperplane in infinite dimension plane, which is used for classification of samples. The important feature of SVM is the absence of local minima. Vapnik introduced the SVM firstly, to solve the classification problem. In classifica-

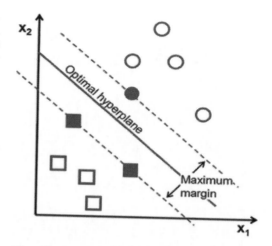

Figure 3. Separation hyper plane.

tion problem, a hyperplane distinct the classes of pattern based on the input space sample vector as shown in Figure 3. For example x_i and y_j are vector in input space $S \in Rn$, for $j = 1, 2, 3, ..., n$. these represent the classes index and taking the value $= 1$ and -1. The Kernel trick is used to transform the data of sample vector (xi) from the input space into the feature space by using non-linear mapping $\emptyset(x)$). A linear function is defined as

$$f(x) = w\emptyset(x) + b \qquad (8)$$

Patterns are classified on the basis of sign in above equation (8). If there is no possibility to draw a hyperplane to distinguish the positive and negative samples, then soft margin method is used to distinct the sample vectors. Some general Kernel functions for different types of SVM are given as:

For Polynomial : $k(x_i x_j) = (x_i x_j + 1)p$

For Radial basis : $k(x_i x_j) = e^{\frac{-x_i - x_j}{2(\sigma^2)}}$

For Sigmoid : $k(x_i x_j) = \tanh(k x_i x_j + c)$

where p, k, c and σ are constants.

3 EXPERIMENTAL RESULTS

Feature extraction is very important for proper classification of sEMG signal. In this study the daubechies wavelet family with wavelet subtype 2 (db2) and level 4 is selected. The waveform of the extracted feature after denoising is shown in Figure 4a and Figure 4b.

Table 1. Wavelet functions.

SN.	Wavelet family	Function in MATLAB	Wavelet subtypes
1	Daubechies	db	db1, db2, bd3 .... db10
2	Haar	haar	haar
3	Symlet	sym	sym2, sym3, sym4, sym5, sym6, sym7, sym8
4	Coiflet	coif	coif1, coif2, coif 3, coif 4, coif 5
5	BiorSplines	bior	bior1.1, bior1.3, bior1.5, bior2.2, bior2.4, bior2.6, bior2.8, bior3.1, bior3.3, bior3.5, bior3.7, bior3.9, bior4.4, bior5.5, bior6.8
6	Reverse Bior	rbio	rbio1.1, rbio1.3, rbio1.5, rbio2.2, rbio2.4, rbio2.6, rbio2.8, rbio3.1, rbio3.3, rbio3.5, rbio3.7, rbio3.9, rbio4.4, rbio5.5, rbio6.8
7	Discrete meyer	dmey	dmey

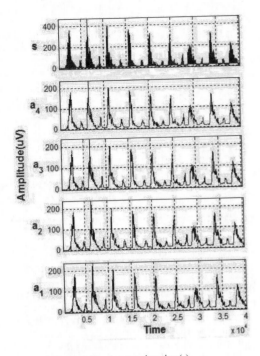

Figure 4a. Signal and approximation(s).

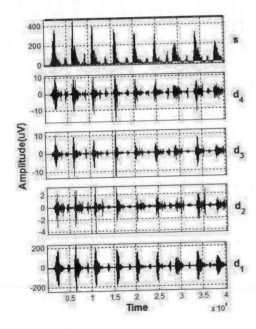

Figure 4b. Signal and detailed(s).

The a_1 and d_1 are the first level approximation coefficient and detailed coefficient. In the same fashion a_2, a_3, a_4 and d_2, d_3, d_4 are approximation coefficients and detailed coefficients respectively.

The approximation coefficients are low frequency signal while detailed coefficients are high frequency signal which also contain noise. So, more accuracy is achieved by the approximation coefficients.

The overall classification accuracy by this proposed method for the elbow movement (flexion and extension) using DWT and Quadratic SVM was 93.7%. The result of classifier has been analysed by confusion matrix, Receiver Operating Curve (ROC) and parallel coordinate plot.

3.1 Classifier analysis by confusion matrix

The confusion matrix shows the overall accuracy for the flexion and extension movements for true class and predicted classes. The True Positive Rates (TPR) and False Negative Rates (FNR) can be obtained mathematically by following formulae:

$$TPR = \frac{TP}{TP + FN} \qquad (9)$$

and

$$FNR = \frac{FN}{TP + FN} \qquad (10)$$

TP = True Positive,
FN = False Negative.

The accuracy for one class (extension) is 96.6% and for another class (flexion) is 92.3% has been obtained separately for per true class as shown in Figure 5.

The overall accuracy of classifier is 93.7% for per predicted class, which is shown as confusion matrix in Figure 6.

The overall accuracy can be evaluated as:

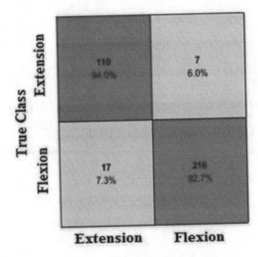

Figure 5. Confusion matrix for Quadratic SVM.

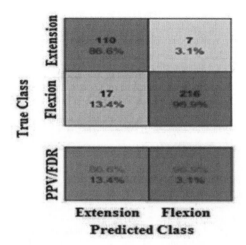

Figure 6.   Confusion matrix for Quadratic SVM.

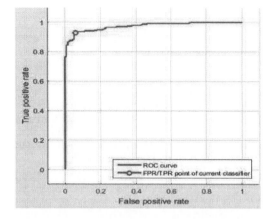

Figure 7b.   ROC curve for flexion movements.

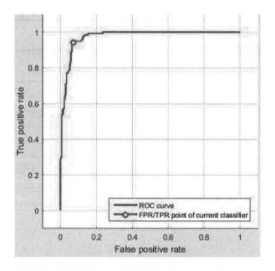

Figure 7a.   ROC curve for extension movements.

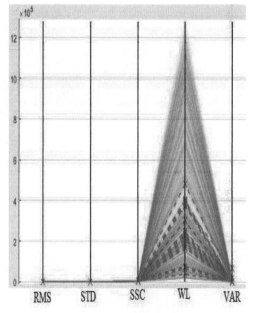

Figure 8.   Parallel coordinator plot.

$$Accuracy = \frac{\sum True\ Positive + \sum True\ Negative}{\sum Total\ Population}$$

(11)

3.2   Classifier analysis through ROC

The ROC curve for flexion and extension classes are shown in Figure 7a and Figure 7b.

The curve shows the relation between TPR and FPR.

The total area under the curve is 0.977917. The area for FPR of current classifier and positive class

is 0.0772532 and that of TPR of current classifier and positive class is 0.965812.

3.3   Classifier analysis through parallel coordinator plot

This plot shows the amplitude of five predictors RMS, STD, SSC, WL & VAR, which is shown by the Figure 8. It shows that the amplitude of WL has more weightage (amplitude) as compared to their successive predictor feature in accordance to their respective significance in classification process.

843

4 CONCLUSION AND FUTURE WORK

This paper presents an advanced approach of classification elbow movements as flexion and extension. This methodology is based on the wavelet transform, which is used to denoise and extract the most significant features of the signals. Quadratic SVM classifier was used to classify the extracted features. The experimental result shows that the proposed method of classification gives better classification accuracy and thus this method is more efficient. In this methodology, the overall accuracy 93.7% is achieved. In future, the classification accuracy can be improved by optimization of classifiers parameters and third level of approximation of sEMG signals.

REFERENCES

Ajiboye, A. B., & Weir, R. F. F. (2005). A heuristic fuzzy logic approach to EMG pattern recognition for multifunctional prosthesis control. *IEEE Transactions on Neural Systems and Rehabilitation Engineering, 13*(3), 280–291. http://doi.org/10.1109/TNSRE.2005.847357

Englehart, K., Hudgins, B., Parker, P. A., & Senior. (2001). A Wavelet-Based Continuous Classification Scheme for Multifunction Myoelectric Control. *IEEE Transactions on Biomedical Engineering, 48*(3), 302–311.

Li, G., Senior, Schultz, A. E., & Kuiken, Todd, S. (2010). Quantifying Pattern Recognition—Based Myoelectric Control of Multifunctional Transradial Prostheses. *IEEE Transactions on Neural Systems and Rehabilitation Engineering, 18*(2), 185–192. http://doi.org/10.1109/TNSRE.2009.2039619

Li, Z., Wang, B., Yang, C., Xie, Q., & Su, C. Y. (2013). Boosting-based EMG patterns classification scheme for robustness enhancement. *IEEE Journal of Biomedical and Health Informatics, 17*(3), 545–552. http://doi.org/10.1109/JBHI.2013.2256920

Lin, K. P., & Chen, M. S. (2008). Releasing the SVM classifier with privacy-preservation. *Proceedings - IEEE International Conference on Data Mining, ICDM,* 899–904. http://doi.org/10.1109/ICDM.2008.19

Liu, J., Zhou, P., & Senior. (2013). A Novel Myoelectric Pattern Recognition Strategy for Hand Function Restoration After Incomplete Cervical Spinal Cord Injury. *IEEE Transactions on Neural Systems and Rehabilitation Engineering, 21*(1), 96–103.

Pao, W., Lan, L., Yang, D., & Liao, S. (2006). The Least-Squares Mixed-Norm Support Vector Classifier. *1-4244-0173-9/06, IEEE,* 375–378.

Winslow, J., Martinez, A., & Thomas, C. K. (2015). Automatic Identification and Classification of Muscle Spasms in Long-Term EMG Recordings. *IEEE Journal of Biomedical and Health Informatics, 19*(2), 464–470. http://doi.org/10.1109/JBHI.2014.2320633

Communication and Computing Systems – Prasad et al. (Eds)
© 2017 Taylor & Francis Group, London, ISBN 978-1-138-02952-1

Automatic creation of NE list for Odia

A. Bhoi, D. Sahoo & R.C. Balabantaray
Department of CSE, IIIT Bhubaneswar, Odisha, India

ABSTRACT: In this paper we proposed a rule base approach for automatic generation of Name Entity list for Odia which is one of the Indo-Aryan languages. NER from Indian languages is a tedious task as they are morphologically rich, words are not capitalized, and uncertainty between common and proper nouns, less resources and also spell variation is there. We have compared our rule based approach with the baseline Moses system and it is found that our system is performing better than the other one.

1 INTRODUCTION

The most weighted words in a document are the name entities. Name Entity Recognition is the process of automatic prediction of the name entities and classifying them to predefined classes such as name of persons, locations, organizations, products, time and date etc from a given text document. Name entity recognition is a very important task in the field of Natural Language Processing (NLP) and Information Extraction. Unlike English not so many Indian languages are highly computerized. Still it a great challenge to do the name entity recognition task with high accuracy due to the complex nature of the sentences in Indian languages. Still among Indian languages Odia is far behind computerization due to lack of resources like a rich Odia corpus. Again whatever work has been done till now for some languages cannot be generalized to other Indian languages due to high diversity among them. So rule based approach is a must for Indian languages, which requires expertise in that language to improve the accuracy. In the literature several approaches have been used for many Indian languages like machine learning based approach, rule based approach and hybrid approach. In section-2 we discuss about the related work. Proposed model for the task has been discussed in section-3. Results and discussions are there in section-4. Finally the conclusion and future work is discussed in section-5.

2 RELATED WORK

Conditional Random Field (CRF) is used to select features which is very essential for name entity recognition in Manipuri (Nongmeikapam et al. 2011). Active learning and Support Vector Machine (SVM) was used to explore different contextual information as well as the orthographic word level features for predicting the name entities from Manipuri text (Singh et al. 2009). A hybridization of statistical Maximum Entropy Model (MaxEnt) and Hidden Markov Model (HMM) was used to develop a two stage language independent NER system for Indian languages (Biswas et al. 2010). Phonetic matching based approach to develop a language independent NER system which requires a set of rules relevant to a particular language (Nayan et al. 2008). A rule based heuristics and Hidden Markov Model was combined to find the NE list from Hindi (Chopra et al. 2012). A hybrid NER system for many Indian languages which use MaxEnt model, rules specific to the languages and gazetteers (Saha et al. 2008). A Max-Ent based NER system has been developed to generate the name entity list from many Indian languages including Odia. They have used some linguistic features for some other Indian languages but not for Odia (Ekbal et al. 2008). A hybridization of Maximum Entropy and Hidden Markov based model to generate the name entity list in Odia language, which is further improvised by certain linguistic rules (Biswas et al. 2010). A system was developed to evaluate the name entity features for Punjabi language. They considered the context word window features, information about digits and words which are infrequent in the text as well as the length of the word as features to evaluate the Punjabi NER system (Kaur et al. 2104). A hybrid method was tried to successfully predict the name entity list in Manipuri language. They considered the Conditional Random and rule base model to define unique word features which helps in accuracy of the task (Jimmy et al. 2013).

3 MODEL LAYOUT

Here we took few Wikipedia pages for crawling using Nutch crawler. Then we applied dumping get the readable text data. We filtered NER list of English from text data using POS and NER annotator in Stanford core NLP tool (Manning et al. 2005). Then we applied the transliteration rules written for Odia language to generate the NE list for Odia. Writing transliteration rules for Odia needs a expertise knowledge in Odia language. The rules are given below in details.

3.1 Steps of the pipeline

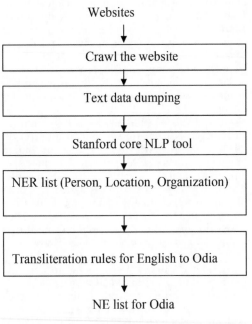

Figure 1. Steps used in Rule based approach to generate NE list for Odia.

3.2 Transliteration rules for English to Odia

If these vowels are immediately followed by a consonant of English alphabet, then it maps to corresponding vowel sign otherwise it will be treated as an independent vowel.

3.2.1 Vowels mapping

Table 1. Vowels mapping.

vowel	odia vowel	vowel sign
a	ଅ(0B05)	Remove virama()(0B3C)
aa	ଆ(0B06)	l(0B3E)
i	ଇ(0B07)	̂(0B3F)
ee	ଈ(0B08)	̄(0B40)
u	ଉ(0B09)	̣(0B41)
uu/oo	ଊ(0B0A)	̣(0B42)
o	ଓ(0B13)	6·l(0B4B)
e	ଏ(0B13)	6(0B47)
ai	ଐ(0B10)	6(0B48)
ou/au	ଔ(0B14)	6·l(0B4C)

3.2.2 Consonants mapping

Table 2. Consonants mapping.

Sl No.	Char-i^{th}	Char-$(i+1)^{th}$	Char-$(i+2)^{th}$	Mapped Char
1	b			ବ(0B2C)
2	b	h		ଭ(0B2D)
3	c			କ(0B15)
4	c	h		ଚ(0B1A)
5	c	h	h	ଛ(0B1B)
6	d			ଡ/ଦ(0B26/0B2)
7	d	h		ଧ(0B27)
8	d	h	h	ଢ(0B22)
9	f			ଫ(0B2B)
10	g			ଗ(0B17)
11	g	h		ଘ(0B18)
12	h			ହ(0B39)
13	j			ଜ(0B1C)
14	j	h		ଝ(0B1D)
15	k			କ(0B15)
16	k	h		ଖ(0B16)
17	l			ଲ/ଳ(0B32/0B3)
18	m			ମ(0B3E)
19	n			ନଣ(0B28/0B23)
20	n	j		ଞ(0B1E)
21	n	k		ଙ(0B19)
22	n	k	h	ଙ(0B19)
23	n	g		ଙ(0B19)
24	n	g	h	ଙ(0B19)
25	p			ପ(0B2A)
26	p	h		ଫ(0B2B)
27	q			କ(0B15)
28	r			ର(0B30)
29	s			ସ(0B38)
30	s	h		ଶ(0B36)
31	s	h	h	ଷ(0B37)
32	t			ଟ/ତ(0B24)
33	t	h		ଥ/ଠ(0B25/0B2)
34	v			ଭ(0B2D)
35	w			□ ଵ(0B71)
36	w	o		□ ଵ(0B35)
37	x			ଜ(0B1C)
38	y			ୟ(0B5F)
39	z			ଯ(0B2F)

3.2.3 Numbers mapping

Table 3. Numbers mapping.

Digits	Digits in Odia	Digits	Digits in Odia
0	୦	6	୬
1	୧	7	୭
2	୨	8	୮
3	୩	9	୯
4	୪	୴	୴
5	୫		

3.2.4 Examples

I.   Tarakasi

T – ଟ , aa - ଆ but as the previous alphabet is a consonant,
aa - ା (vowel sign)   Taa - ଟା   r - ର୍
ra - ର [virama removed]
Taara – ଟାର   k - କ୍   ka - କ
Taaraka – ଟାରକ
s - ସ   i - ଇ But as previous alphabet is a consonant i ି [vowel sign]
si - ସ ି– ସି
Tarakasi - ଟ+ା+ର+କ+ସି – ଟାରକସି

II.   Shaanti

s-h-sh-ଶ୍   sha- ଶ [as 'a' is followed by consonants the virama sign is removed and it will not be map to vowel 'ଅ']
shaa - ଶା   shaan – ଶା ନ୍
shaant – ଶା ନ୍ ତ୍
shaanti – ଶା+ା+ନ୍+ ତ ି [while adding vowel sign ,virama is removed]
shaanti - ଶ ା +ନ୍ +ତ ି - ଶାନ୍ତି

III.   Anka

A - ଅ [as it is not precedeed by any consonants maps to corresponding vowel]
n – ନ୍, Ank- ଅ + ଙ୍ + କ୍
Anka- ଅ + ଙ୍ + କ୍ [virama removed as it is followed by a consonant]
        - ଅ + ଙ୍ + କ - ଅଙ୍କ

IV.   Jagannatha

j-ଜ୍, ja- ଜ୍ - ଜ, jag-ଜ + ଗ୍
jaga –ଜ+ଗ୍ - ଜଗ, jagan –ଜ+ଗ+ନ୍
jagann –ଜ+ଗ+ନ୍+ନ୍ , jaganna – ଜ+ଗ+ନ୍+ନ୍
jagannaa – ଜ+ଗ+ନ୍+ନ୍+ା
jagannath –ଜ+ଗ+ନ୍+ନ+ା+ଥ
jagannaatha – ଜ+ଗ+ନ୍+ନ+ ା+ଥ୍
        - ଜ+ଗ+ନ୍ନ+ ା+ଥ - ଜ+ଗ+ନ୍ନା+ଥ
        -ଜଗନ୍ନାଥ

V.   Antarjyaamee

A – ଅ ,An –ଅ+ନ୍ , Ant –ଅ+ନ୍+ତ୍
Anta –ଅ+ନ୍+ତ୍   Antar –ଅ+ନ୍+ତ+ର
Antarj–ଅ+ନ୍+ତ+ର୍+କ୍ Antarjy–ଅ+ନ୍+ତ+ର୍+ଜ୍+ଯ୍
Antarjya–ଅ+ନ୍+ତ+ର୍+ଜ୍+ଯ୍
Antarjya–ଅ+ନ୍+ତ+ର୍+ଜ୍+ଯ+ା
Antarjyaam –ଅ+ନ୍+ତ+ର୍+ଜ୍+ଯ+ା+ମ୍
Antarjyaamee- ଅ+ନ୍+ତ+ର୍+ଜ୍+ଯ+ା+ମ୍+ୀ
        - ଅ+ନ୍+ତ+ର୍+ଜ୍+ଯ+ା+ମୀ
        -ଅ+ନ୍ତ+ର୍+ଜ୍ଯ+ା+ମୀ
        - ଅ+ନ୍ତ+ର୍ଜ୍ଯ+ା+ମୀ
        - ଅ+ନ୍ତ+ ର୍ଜ୍ଯା+ମ
        - ଅନ୍ତର୍ଜ୍ଯାମୀ

4   RESULTS AND DISCUSSION

To measure the accuracy of this rule based NER system we have taken hundred name entities generated by the Stanford NER for English. To avoid biasness in data list we have taken a list in such a way that it contains NEs that start with all the twenty six alphabets of English. The list contains three categories of NE i.e PERSON, LOCATION and ORGANIZATION.

We tested our system with baseline Moses (Koehn et al. 2007) transliteration system and it is found that rule based system performs better for English to Odia transliteration system.

4.1   Moses training model

For Moses training and language model fifty thousand character split based parallel entries (English-Odia) are created in an automated manner. Sample training data is given below for English Odia pair and for better understanding the equivalent Hindi is also mentioned (Balabantaray et al. 2014).

Table 4.   Results of Moses system.

English	Odia	Hindi
jagadeesh	ଜଗଦୀଶ	जगदीश
suresh	ସୁରେଶ	सुरेश
geeta	ଗୀତା	गीता
ashok	ଅଶୋକ	अशोक
mahesh	ମହେଶ	महेश
aneeta	ଅନୀତା	अनीता
kamala	କମଲା	कमला
sanjay	ସଞ୍ଜୟ	संजय
seema	ସୀମା	सीमा
raju	ରାଜୁ	राजु
ashutosh	ଅଶୁତୋଶ	अशुतोश
deepak	ଦୀପକ	दीपक
rakesh	ରାକେଶ	राकेश

Table 5. Results of Rule based Vs Moses system.

English	NER class	Rulebased System	Moses
Wipro	organization	ଡ଼ିପ୍ରୋ·l	ବି ପ ର 6·l .
Worli	location	ଡ଼ର୍ଲି	ବ 6·l ର ଲି .
Yamuna	location	ଯମୁନ	ମ ୁ ନ
yuvaraj	person	ଯୁଭରଜ	ୟ ୁ ବ ର ଜ
Yunish	person	ଯୁନିଶ	ୟ ୁ ନି ଶ
yugoslavia	location	ଯୁଗୋ·ାସ୍ଲଭିଆ	ୟ ୁ ଗ 6·l ସ ଲ ବ
Yogesh	person	ଯୁଗ·ାଗେଶ	ୟ 6·l ଗ 6 ଶ
zimbabwe	location	ଯିମ୍ବବୁଡ଼6	l ୀ ମ ବ ବ ବ 6
Zafar	person	ଯଫ୍ର	l ଫ ର
Zakir	person	ଯକିର	l କି ର

4.2 Comparison of rule based system with moses

The experimental result shows that the accuracy of rule based system is 72 percent where as that of moses model system gives only 44 percent which has been computed manually.

5 CONCLUSION

In this paper we tried to generate NE list for Odia automatically. For this we extracted NE list for English using Stanford NER tool. Then applied linguistic rule based approach to convert English NE list to Odia NE list. We compared the result of rule based system with Moses statistical model. The experimental result shows that rule based system performs well in comparison to Moses model. In this work we have tried to generate only the unigram NE list for Odia. In our future work we will try to generate multi-gram NE list for Odia and improve the performance of the rule based system by using edit distance to correct few wrongly spelt NEs in Odia with the help of NE corpus.

REFERENCES

Balabantaray R.C. & Sahoo D. 2014. Odia Transliteration engine using Moses, Proceedings of the ICBIM 2014, Durgapur, India: 27–29.

Biswas, S.; Mishara, M. K.; Acharya, S. & Mohanty, S. 2010. A Two Stage Language Independent Named Entity Recognition for Indian Languages. International Journal of Computer Science and Information Technologies. Vol. 1(4): 285–289.

Biswas, S.; Mishra, S. P.; Acharya, S. & Mohanty, S. 2010. A Hybrid Oriya Name Entity Recognition System: Harnessing the Power of Rule. IJAE. Vol.1, Issue.1.

Chopra, D.; Jahan, N. & Morwal, S.2012. Hindi Named Entity Recognition By Aggregating Rule Based Heuristics And Hidden Markov Model. International Journal of Information Sciences and Techniques (IJIST). Vol.2, No.6.

Ekbal, A. & Bandyopadhyay, S. 2008. Named Entity Recognition in Indian Languages Using Maximum Entropy Approach, International Journal of Computer Processing Language, Vol.21, Issue.3.

Finkel, J. R.; Grenager, T. & Manning, C. 2005 Incorporating Non-local Information into Information Extraction Systems by Gibbs Sampling. *Proceedings of the 43rd Annual Meeting of the Association for Computational Linguistics (ACL):* 363–370. http://nlp.stanford.edu/~manning/papers/gibbscrf3. pdf

Jimmy, L. & Kaur, D. 2013. Name Entity Recognition in Manipuri:A Hybrid Approach, GSCL: 104–110.

Kaur, A. & Josan, G. S. 2014. Evaluation of Name Entity Features for Punjabi Language. International Conference on Information and Communication Technologies, 159–166.

Koehn, P. 2007. Moses: Open source toolkit for statistical machine translation. Proceedings of the 45th annual meeting of the ACL on interactive poster and demonstration sessions. Association for Computational Linguistics:177–180.

Nayan, A.; Ravi Kiran Rao, B.; Singh, P.; Sanyal, S. & Sanyal, R. 2008. Named Entity Recognition for Indian Languages. Proceedings of the IJCNLP-08 Workshop on NER for South and South East Asian Languages, Hyderabad, India: 97–104.

Nongmeikapam, K.; Shangkhunem, T.; Chanu, N. M.; Singh, L. N.;Salam, B. & Bandyopadhyay, S. 2011. CRF based Name Entity Recognition (NER) in Manipuri: A highly agglutinative Indian Language. IEEE 0.1109/NCETACS: 92–97.

Saha, S. K.; Chatterji, S.; Dandapat, S.; Sarkar, S. & Mitra, P. 2008. A Hybrid Approach for Name Entity Recognition in Indian Languages. Proceedings of the IJCNLP-08 Workshop on NER for South and South East Asian Languages, Hyderabad, India: 17–24.

Singh, T.D.; Kishorjit, N.; Ekbal, A. & Sivaji, B. 2009. Name Entity Recognition in Manipuri Using SVM. In: Proceedings of Pacific Asia Conference on Language, Information and Computation Hong Kong: 811–818.

Communication and Computing Systems – Prasad et al. (Eds)
© 2017 Taylor & Francis Group, London, ISBN 978-1-138-02952-1

Comparative analysis of travelling salesman problem using metaheuristic algorithms

Ayush Agarwal & Rajendra Bahadur Singh
Gautam Buddha University, Greater Noida, India

ABSTRACT: This paper proposes a solution of Travelling Salesman Problem (TSP) using metaheuristic algorithms. The computation time to solve TSP problem increases exponentially when the number of cities increases. This type of problems demand innovative solutions if they are to be solved within a reasonable amount of time. This paper explores the solution of Travelling Salesman Problem using genetic algorithm, simulated annealing and ant colony optimization algorithms. The object of this paper is to find an efficient solution and comparative study using these algorithms. Travelling salesman problem is one of the most important combinatorial problems. In this paper we present a comparative study of ant colony optimization, simulated annealing and genetic algorithm to study the TSP.

1 INTRODUCTION

In the travelling salesman problem (Yuan et al, 2009), (Zhilong et al 2009), a salesman will travel all the cities taken but the condition followed is, each city can only be visited once. The objective is to minimize the distance of the complete route. The intelligent optimization algorithms to solve the TSP mainly include ACO (Zhilong et al, 2009), (Shang et al, 2007), GA (Geetha et al, 2007), and SA (Li et al, 2011). These algorithms run faster than traditional exact algorithms. Many results on evolutionary optimization for TSP have been published. Our main interest is to estimate the optimal cost in respective of iterations for each of these intelligent systems. By doing this we can comfortably compare and determine the best alternative among the three algorithms when applied to the TSP consequently. The present study has two practical motives. The first is to help beginner programmers to understand and implement the metaheuristics and the second is to exhibit and compare their results.

1.1 Travelling Salesman Problem

Intuitively, TSP is the NP hard problem (Gholamian et al, 2007) in which the salesman starts from his hometown and travels all the given set of customers cities and then back home. The objective is to find the shortest path by following the condition that each city will be visited exactly ones. TSP can be represented by a complete weighted graph $G = (N, C)$ with N being the set of nodes representing the cities and C being the set of arcs. Each arc $(i, j) \in C$ has a length value d_{ij}, which is the distance between cities i and j with i, $j \in N$. For the symmetric TSP, $d_{ij} = d_{ji}$ for all the arcs in C. For asymmetric TSP (Palborn et al, 2004), d_{ij} is dependent on the direction of traversing the arc, that is at least one arc (i, j) for which $d_{ij} \neq d_{ji}$.

2 METAHEURISTIC ALGORITHMS

2.1 Genetic Algorithm (GA)

Genetic Algorithm (Chen et al, 2010) is a search technique. It is used in computing to find approximate solutions as well as exact solutions to optimization and search problem. Using Darwinian principle of reproduction and survival of the fittest and analogs of naturally occurring genetic operation such as crossover and mutation, GA converts a population of individual objects, each with associated fitness value into a new generation of the population. GA has a property of parallelism and it produces optimum solution for noisy environments. The advantage of GA is that it does not require derivative information or other auxiliary knowledge. Also, it can solve multi-dimensional, non-differential, non-continuous and even non-parametrical problems.

2.1.1 Genetic operators
There are mainly three genetic operators:

2.1.1.1 Selection
It equates the survival of the fittest. It randomly chooses members of the population which are based on their fitness to enter a mating pool (Geetha et al,

2009). The members whose fitness value is better, the more likely it is to be selected. Tournament selection, ranking selection and proportional selection are some of the selection methods.

2.1.1.2 Crossover
It represents the mating between individuals. It combines the two parents to produce a new offspring. Some of the crossover techniques are one-point, uniform and half uniform crossover.

2.1.1.3 Mutation
It introduces random modification. It is used to maintain genetic diversity from one generation to next generation. Some of the mutations are flip bit, boundary, uniform, non-uniform etc.

2.2 Simulated Annealing (SA)

Simulated annealing algorithm (Chen et al, 2011) is based on the principle of solid annealing. A material is first heated up to a temperature that allow all its molecules to move freely around and is then cooled down very slowly. SA performs computation that analogous to physical process:

– The energy corresponds to the cost function
– Molecular movement corresponds to a sequence of moves in the set of feasible solution.
– Temperature corresponds to a control parameter T, which control the acceptance probability for a move i.e. a good move.

The main disadvantage of a simulated annealing (Singh, 2016) is that it gives near optimal solution but not stable.

2.3 Ant Colony Optimization (ACO)

ACO is a population based metaheuristic, used to find optimal solution to difficult optimization problem. The ant based metaheuristic consists of three stages i.e. initialization, construction and feedback.
 The primary stage i.e. initialization stage involves the parameters settings such as the number of colonies and the number of ants. The construction stage followed by feedback stage involves the construction of path on the basis of pheromone concentration while the feedback stage deals with the extraction and the reinforcement of ants travelling experiences obtained during the previous searching path. Each ant has the following characteristics:

– It selects the city to go with a probability which is a function of the city distance and the amount of trail present on connecting edge.
– To force the ant to make legal tours. The cities which are already visited are disallowed until a tour is completed.
– When the tour is completed it lays a chemical called pheromone on each edge (i,j)

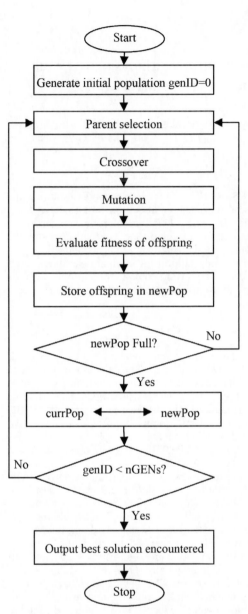

Figure 1. Flowchart of genetic optimization cycle.

 Let the number of cities be N and number of ants be M τ_{ij} (t) be the intensity of pheromone on edge (i, j) at time t

$$\tau_{ij}(t+n) \leftarrow (1-\lambda)\tau_{ij}(t) + \Delta\tau_{ij}(t+n) \qquad (1)$$

where, λ = coefficient of evaporation,

$$\Delta\tau_{ij}(t+n) = \sum_{k=1}^{n} \Delta\tau_{ij}^{k}\Delta\tau_{ij}(t+n) \qquad (2)$$

Now, $\Delta \tau_{ij}^k = \begin{cases} 0; & \textit{if ant doen't travel edge ij} \\ \dfrac{Q}{L_k}; & \textit{Otherwise} \end{cases}$

where, L_k is the cost of tour found by ant k and Q is some constant. So we can see that the amount of pheromone that ants deposit on an edge is inversely proportional to length of tour i.e. L_k. Assume, ant at city i move to city j with probability

$$P \propto \frac{\left[\tau_{ij}\right]^\alpha \left[\eta_{ij}\right]^\beta}{\sum \left[\tau_{ih}\right]^\alpha \left[\eta_{ih}\right]^\beta} \qquad (3)$$

where $\left[\eta_{ij}\right]^\beta$ is the visibility which is inversely proportional to cost (i, j) and 'h' is the set of allowed cities which means those cities are considered only which will not close the loop prematurely.

3   EXPERIMENTAL RESULTS

For the comparative study we have taken total number of cities as twenty and total number of

iterations as two hundred fifty. The simulated work is performed on matlab on Intel core i3 processor.

3.1   Ant Colony Optimization result

In the Figure 3, the yellow dots show the location of coordinates of twenty cities. The way all the twenty cities are connected to each other shows the optimized path.

In the Figure 4, graph is plotted between best cost versus number of iterations. In the graph it is clearly visible that the best cost is 370.7421 after 134 iterations.

3.2   Simulated annealing result

In the Figure 5, the yellow dots show the location of coordinates of twenty cities. The way all the twenty cities are connected to each other shows the optimized path.

In the Figure 6, graph is plotted between best cost versus number of iterations. In the graph it is clearly visible that the best cost is 370.7421 after 168 iterations.

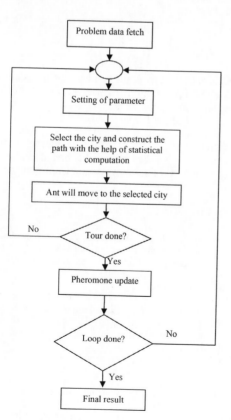

Figure 2.   Ant based metaheuristic algorithm.

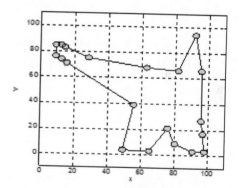

Figure 3.   Result of optimization using ACO.

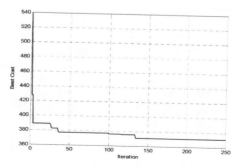

Figure 4.   Best cost (minimum distance) vs number of iteration using ACO.

851

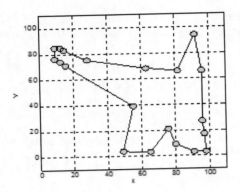

Figure 5.   Result of optimization using SA.

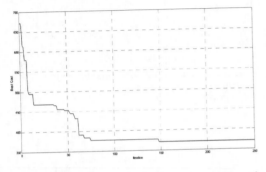

Figure 6.   Best cost (minimum distance) vs number of Iteration using SA.

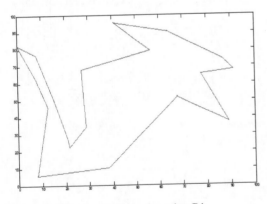

Figure 7.   Result of optimization using GA.

3.3   *Genetic Algorithm result*

In the Figure 7, the way all the twenty cities are connected to each other (shown by red) shows the optimized path.

In the Figure 8, graph is plotted between best cost versus number of iterations. In the graph it is clearly visible that the best cost is 457.3522 after 206 iterations.

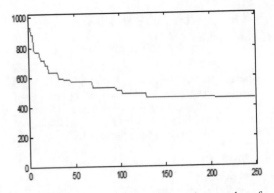

Figure 8.   Best cost (minimum distance) vs number of iteration using GA.

Table 1.   Simulation metrics.

	ACO	SA	GA
Total no. of iterations	250	250	250
Total no. cities	20	20	20
Max. Cost	539.3059	673.4477	960.5434
Best cost	370.7421	370.7421	457.3522
Best cost in no. of iterations	134	148	206

4   CONCLUSION

In the above experimental results, we conclude that ant colony optimization produces the best solution of 370.7421 in only 134 iterations out of 250 iterations. The simulated annealing produces the same result in 148 iterations while the genetic algorithm produces very high cost of 457.3522 in 206 iteration. Therefore ant colony optimization is the best algorithm in respect of best solution versus number of iteration.

REFERENCES

Chen, J., Zhu, W. and Ali, M.M., 2011. A hybrid simulated annealing algorithm for nonslicing VLSI floorplanning. Systems, Man, and Cybernetics, Part C: Applications and Reviews, IEEE Transactions on, 41(4), pp. 544–553.

Chen, J. and Zhu, W., 2010, October. A hybrid genetic algorithm for VLSI floorplanning. In Intelligent Computing and Intelligent Systems (ICIS), 2010 IEEE International Conference on (Vol. 2, pp. 128–132). IEEE.

Geetha, R.R., Bouvanasilan, N. and Seenuvasan, V., 2009, December. A perspective view on Travelling Salesman Problem using genetic algorithm. In Nature & Biologically Inspired Computing, 2009. NaBIC 2009. World Congress on (pp. 356–361). IEEE.

Gholamian, M.R., Ghomi, S.F. and Ghazanfari, M., 2007. A hybrid system for multiobjective problems–A case study in NP-hard problems. Knowledge-Based Systems, 20(4), pp. 426–436.

Gwizdałła, T.M., 2012. The role of different genetic operators in the optimization of magnetic models. Applied Mathematics and Computation, 218(18), pp. 9220–9233.

Hoo, C.S., Jeevan, K., Ganapathy, V. and Ramiah, H., 2013. Variable-order ant system for VLSI multiobjective floorplanning. Applied Soft Computing, 13(7), pp. 3285–3297.

Li, Y., Zhou, A. and Zhang, G., 2011, July. Simulated annealing with probabilistic neighborhood for traveling salesman problems. In Natural Computation (ICNC), 2011 Seventh International Conference on (Vol. 3, pp. 1565–1569). IEEE.

Parsons, S., 2005. Ant Colony Optimization by Marco Dorigo and Thomas Stützle, MIT Press, 305 pp., $40.00, ISBN 0-262-04219-3.

Pasquier, J.L., Balich, I.K., Carr, D.W. and López-Martín, C., 2007, November. A comparative study of three metaheuristics applied to the traveling salesman problem. In Artificial Intelligence-Special Session, 2007. MICAI 2007. Sixth Mexican International Conference on (pp. 243–254). IEEE.

Palbom, A., 2004. Worst case performance of an approximation algorithm for asymmetric TSP. In STACS 2004 (pp. 465–476). Springer Berlin Heidelberg.

Shang, G., Lei, Z., Fengting, Z. and Chunxian, Z., 2007, August. Solving traveling salesman problem by ant colony optimization algorithm with association rule. In Natural Computation, 2007. ICNC 2007. Third International Conference on (Vol. 3, pp. 693–698). IEEE.

Singh, R.B., Baghel, A.S. and Agarwal, A., 2016, March., A Review on VLSI Floorplanning Optimization using Metaheuristic Algorithms. In Electrical, Electronics, and Optimization Techniques (ICEEOT), 2016

Yuan, L., Lu, Y. and Li, M., 2009, December. Genetic Algorithm Based on Good Character Breed for Traveling Salesman Problem. In Information Science and Engineering (ICISE), 2009 1st International Conference on (pp. 234–237). IEEE.

Zhihong, X., Bo, S. and Yanyan, G., 2009, November. Using Simulated Annealing and Ant Colony Hybrid Algorithm to Solve Traveling Salesman Problem. In Intelligent Networks and Intelligent Systems, 2009. ICINIS'09. Second International Conference on (pp. 507–510). IEEE.

Communication and Computing Systems – Prasad et al. (Eds)
© 2017 Taylor & Francis Group, London, ISBN 978-1-138-02952-1

Routing in internet of things: A survey

A. Saini & A. Malik

Department of Computer Science and Engineering, DCRUST Murthal, Sonepat, Haryana, India

ABSTRACT: The IoT enables real world objects to perceive, hear, and contemplate by making them "smart" cooperatively. The Things in IoT are embedded with sensors, actuators and RFID to enable the interaction with each other, with internet and with the people. The IoT transforms these objects from being physical entity to smart by using its underlying technologies such as pervasive and ubiquitous computing. Radio Frequency Identification (RFID) and sensors are the core technology of IoT, which expands the application domain of IoT. Communication will now expand to Human-Machine and Machine-Machine from the conventional Human-Human, hence minimizing the user involvement. Though, there are numerous fields implementing IoT, but as the IoT is developing some concerns related to IoT are also growing like scalability, routing, and privacy and security. For IoT, the routing inflicts an extremely large challenge. This paper surveys a little of the subjects considering routing in Internet of things.

1 INTRODUCTION

The Internet of Things (IoT) extends the concept of Internet wherein the real world things each other and with Internet as well. The things/objects (D. Guisto et al, 2010) are allocated unique identifiers that addresses every single thing uniquely. The Internet of Things (IoT) is a paradigm that is quickly obtaining ground in the scenario of present wireless telecommunications. The main idea is the pervasive presence concerning us of a collection of things or objects—such as Radio-Frequency Identification (RFID) tags (Atzori et al, 2010), sensors, (Gubbi et al, 2013) actuators, mobile phones, etc. —that, across unique addressing schemes, are able to interact with every other thing and cooperate with the neighboring nodes to achieve common goals. The IoT is often considered as an intelligent network that links all things to the Internet for the intention of exchange of data and communication between the devices with some agreed protocols. Routing is a process to determine a path for data transmission between source and destination. In IoT the network layer is mostly used to implement the routing of the data. As IoT consists of huge number of devices, which thus forms multi-hop networks, the intermediate nodes have to relay their packets towards next node in multi-hop networks. While design Routing protocols for IoT, node identification and context awareness is the major concern, as the nodes produce a large amount of data but the challenge is to extract and route semantically relevant information to maximize the overall throughput of the network.

1.1 Technical background

As of now, the Internet is the backbone infrastructure network at the end user's terminals but soon the "smart" interconnected objects will form a pervasive computing environment (Mirondi et al, 2012). This innovation will be achieved by making the things smart by embedding of electronics into the physical things. The things in IoT mainly comprises of sensors, Radio-Frequency Identification Technology (RFID) (Chen et al, 2014) and actuators. RFID is generally considered as the key technology in the Internet of Things, because it has the ability to monitor and track a large number of uniquely identifiable objects alongside the use of Electronic Product Codes (EPC) (Aggarwal et al, 2013). RFID Radio-Frequency Identification (RFID) is a technology which is based on short range communication and consists of RFID tag that uses radio waves to identify the devices globally (Whitemore et al, 2012). For an object, an EPC is a universally unique identifier. Near Field Communication (NFC), a short ranged high frequency wireless technology, is a development of RFID standard and share physical properties that of RFID. NFC is a standard which enables the devices to communicate that are within close proximity (typically 20 cm or less) of one another. NFC's are provided with a tag for unique I identification (Whitemore et al, 2015). With this tag the devices are able to perform tagging and tracking. As the things in Internet of Things are embedded with sensors, they play an important role by collecting surrounding context and environment information by forming a network called Wireless

Sensor Network (WSN). The sensors senses the surrounding environment or the objects and actuators perform the action specified by surrounding information.

1.2 IoT architecture

A typical IoT architecture consists of three layers: Perception layer, Network layer, and Application layer as shown in Figure 1 (Mashal et al, 2015) (Wu et al, 2015).

The Perception layer is the lowest layer, also known as Device Layer (Mashal et al, 2015), encompasses the sensors and the physical world. The main task of this layer is to recognize objects and gather information, such as locations and temperatures. The main technologies in of this layer are RFID for recognizing and detecting things, WSN for sensor communication, and wireless communication protocols for WSNs.

The Network layer is the core of Internet of Things and is often considered as the mind of IoT, its main purpose is to send and process the information (Mashal et al, 2015). The network layer will send and process the data obtained from perception layer. It is accountable for addressing every single object uniquely by using a unique address. Transmission medium and communication protocols like WiFi, Bluetooth, and ZigBee are work on this layer. The topmost layer is the Application

Table 1. Comparison between NFC system, RFID system and Wireless Sensor Network.

	Standard	Processing	Sensing	Range	Network type
NFC	ISO/IEC 18000	No	No	<20 cm	Point-to-point
RFID	ISO/IEC 18000	No	No	Upto 10 m	Point-to-point
WSN	IEEE 802.15.4	Yes	Yes	Upto 100 m	Wireless Mesh-Network

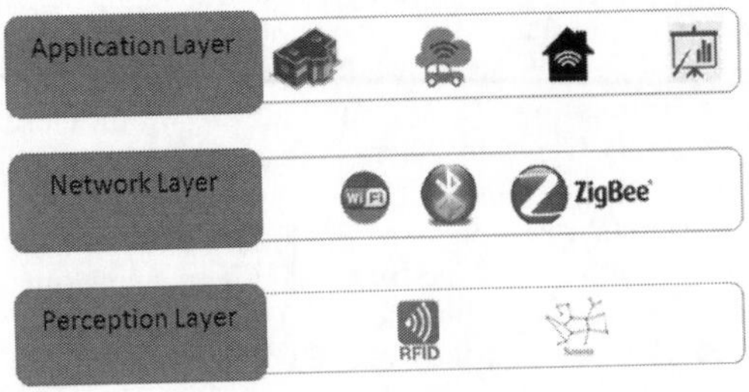

Figure 1. IoT architecture.

Layer that integrates data and deliver the services and applications demanded by users (Wu et al, 2010). For instance, the Application layer can furnish temperature and air humidity measurements to the client who asks for that data. The significance of this layer for the IoT is that it has the skill to furnish high-quality intelligent services to fulfill customers' needs (Al-Fuquaha et al, 2015).

2 ROUTING IN IOT

Routing is an important service in the Internet of Things as routing enables the exchange and transmission of information between things and people, and between the things. The transmission can be single hop or multi-hop transmission when the destination node is not in the close vicinity of the source node. Being expand in large scale and resource constrained, the communication devices in IoT may experience irregular connectivity and dynamic topology changes frequently and impose challenges for the routing on IoT.

2.1 Routing challenges

The basic characteristics of IoT network like resource constraint devices, massive number of things and extreme heterogeneity, give rise to several routing challenges. So it is vital to understand the context while routing the data on the networks. Following are the Routing challenges that the IoT network experiences (Dhumane et al, 2015).

Heterogeneity: The devices in the IoT may not be of same kind, they may vary according to network standards, type of applications they support, resource types because it is highly unlikely that all the devices will have same amount of battery life during the transmission.

Dynamic Routing Topology: There are many reasons of dynamicity in IoT. Being the energy constraint, because the devices are scheduled to be in either sleep or working state to reduce the energy consumption, which results in dynamic routing topology. Another cause may be the unreliable wireless links between nodes, which may cause disconnection and reconfiguration of link.

Fault tolerance: Due to the environmental interferences, deployment mechanisms or power constraints, network is highly prone to failures and affect the overall performance of the network.

Irregular connectivity: Due to battery life constraint, the network topology is not fixed. Routing in this type of dynamic networks is more complex due to frequently changing routing path. Irregular connectivity may also be caused due to high mobility of devices, which may get disconnected when they move.

Scalability: The scale of IoT is quite large, and is likely to be expanded in near future as the number of things becoming "smart" is going to be increased. So, the routing protocol should be flexible enough to work with a huge number of devices as well.

Context awareness: Not all the data that sensors produce is useful, the challenge is to extract semantically relevant information that is context information. The information must be transmitted, according to the context produced and the changes in routing process must be done accordingly.

2.2 *Routing protocols in IoT*

IoT routing protocols must guarantee loop-free routing, connectivity, energy efficiency and QoS between the nodes. As the research of routing protocols of IoT is at its beginning stage (Tian et al, 2010), there is no standardized classification of routing protocols for IoT. One way to classify the IoT routing protocols is by adapting existing routing protocols of ad hoc and wireless network according to the requirements of the IoT (Reina et al, 2013).

The technologies of ad hoc network and Wireless Sensor Network (WSN) are the basic representations for implementing the IoT paradigm. On comparing IoT with Sensor networks and Ad Hoc network, many similarities may be found in node characteristics and node distribution. So it can be find feasible to apply the routing protocols of wireless Sensor networks and Ad Hoc network in future IoT scenarios. Ad hoc networks routing protocols are mainly focused on guaranteeing Quality of Service (QoS) parameters as bandwidth and end-to-end delay (Reina et al, 2013). On the other hand, routing protocols for WSNs are mainly concerned on maximizing network's lifetime by minimizing the energy consumption. Routing protocols for the IoT must guarantee connectivity, fairness and QoS between the nodes both in ad hoc networks and the APs.

Classification based on Ad Hoc Networks: The Traditional ad hoc routing protocols are classified on the basis of their mode of function that is Proactive, Reactive and Hybrid (Dhumane, 2015). Proactive protocols keeps the route information in tabular format before routing begins, reactive protocols maintains on-demand route and hybrid routing use both proactive and reactive routing algorithms.

AODV and AOMDV: The conventional routing protocols for ad hoc networks, Ad Hoc On-Demand Distance Vector (AODV) routing protocol may be modified to satisfy IoT requirements. For example, AOMDV-IOT (Tian et al, 2010) is a modification of Ad Hoc On-demand Multipath Distance Vector routing protocol. The basic AOMDV enables a user to find link-disjoint routes and node-disjoint routes between a source node and a destination node. But, in an IoT perspective, the aim is to find a node connected to the Internet. This is done by implementing an Internet Connecting Table (ICT) in AOMDV-IOT.

OSLR: The Optimized Link-State Routing (OLSR) protocol is a proactive link-state and hop by hop routing protocols. The nodes exchange information frequently with each other and update their own network topology. Some nodes are used as routing nodes and are known as Multi-Point Relay (MPR). MPR nodes are used to broadcast control information. OLSRv2, an extension of OSLR has also been proposed which enables the use of energy aware matrices besides link optimization.

Classification Based on WSN: The existing WSN routing protocols may be classified on the basis of hierarchical topology and negotiation based protocols from the IoT perspective.

RPL: RPL (Routing Protocol for Low power and lossy networks) routing protocol, developed by the IETF in the working group Routing Over Low power and Lossy networks (ROLL), is a distance vector based IPv6 routing protocol. To make the network loop free it makes the topology as Directed Acyclic Graph (DAG) and specifies how to build a Destination Oriented Directed Acyclic Graph (DODAG) (Le Q et al, 2014). The nodes compute their relative rank to root with the help of Objective Function and a set of metrics/constraints. These metrics help in determining the the quality of the paths generated.

SPIN: SPIN (Sensor Protocol for Information via Negotiation) solves the problems of resource wasting like information redundancy and information flooding with the help of negotiating mechanism, by sending the information to the selected nodes only not to the whole network

3 CONTEXT AWARENESS

Context aware system is the core characteristics of ubiquitous and pervasive computing systems. These systems are built on the concept that the

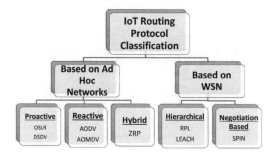

Figure 2.   IoT routing protocol classification.

sensor nodes placed in the environment closely monitors the context parameters, they sense the environment and send the data retrieved for further processing into context. A system is considered as context-aware if it can extract, interpret, fuse and use context information and provides the required services. In the classical Internet, context information is not likely taken into account by the communication services (Kalmar et al, 2013), but for the Internet of things, context- aware is the core feature. The future ubiquitous and emerging technologies provides anytime, anywhere and anywhere services with little or no human interaction with devices. Context is generally considered as any information that characterizes a situation. (Abowed et al., 1999) defined the word context as:

"Context is any information that can be used to characterize the situation of an entity. An entity is a person, place, or object that is considered relevant to the interaction between a user and an application, including the user and applications themselves."

Additionally, (Sanchez et al., 2006) explained the difference between Raw Sensor data and Context Information as follows:

Raw data: The data retrieved directly from its sources, such as, sensors. It is unprocessed data.

Context information: It is the processed data which is gathered from the raw data. It is further checked for consistency and meta data is added.

For example the data produced by sensor nodes is raw sensor data when this data is fused, processed and suitably reasoned according to the application and requirement this data becomes the context information. (Schilit et al., 1994) explained the Context Aware Computing which has the ability to discover and react according to the changes in the environment. Similar to the constant monitoring and the surrounding world information provided by mobile software.

(Perera et al., 2014) proposed four phases of context life cycle, as shown in Figure 3. A data life

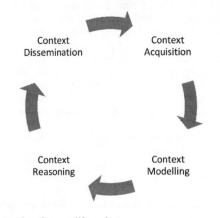

Figure 3. Context life cycle.

Table 2. Comparative analysis of IoT routing protocols.

Protocol	AOMDV-IOT	RPL	OSLR	DSDV	LEACH	SPIN	ZRP
Category	Reactive or On Demand	Proactive or Table Driven	Proactive or Table Driven	Proactive or Table Driven	Cluster Based	Flat Routing	Hybrid
Protocol Type	Distance Vector	Directed Acyclic Graph	Link-State Scheme	Destination Based	Adaptive Cluster based	Data Negotiation and Resource-Adaptive	Link Reversal
Loop-Free	Yes	Yes	Yes	Yes	No	Yes	Yes
Energy-Efficiency	No	Yes	No	No	Yes	Yes	No
Internet Connectivity	Yes	No	No	No	No	No	No
Multipath	Yes	No	Yes	Yes	No	Yes	No
Routing Matrics	Fastest and Shortest Path	RTT (Round Trip Time), ETX (Expected number of Transmissions)	MPR (Multi Point Relay)	Shortest Path	CH (Cluster Head) formation	Meta data Transmission	Hop Count
Scalability	More	Moderate	Low	Low	More	Limited	More
Bandwidth Usage	Low	Moderate	More	More	More	Low	Moderate

cycle depicts how the data transits from one phase to the other phase in software systems (e.g. It particularly explains where the data is generated and where the data is consumed. The context life cycle described in (Perera et al., 2014) consists of four phases.

Context Acquisition: Firstly the data must be collected from data sources that means the context needs to be acquired from various data sources. These sources could be physical or virtual sensors

Context modelling: Then the collected data is to be modelled and context representation needs to be performed according to a meaningful manner.

Context reasoning: The data modelled in previous phase needs to be processed to generate a high-level context information from low-level raw sensor data.

Context dissemination: In the last phase both high-level and low-level context information needs to be distributed to the user.

Context Awareness Challenges

Currently, majority of IoT middleware solutions do not provide context-awareness capabilities (Da Xu et al., 2014). Rather, most of the solutions are highly concerned in device management. There were some middleware solutions in early days which provided the context-aware functionality (Perea et al., 2014), but they did not gratify the necessities that the IoT demands.

- Data Acquisition imposes a major challenge as the sensor nodes produce a huge amount of data periodically.
- To decide which source and technology to use would be a difficult task
- Quality, validation, accuracy and cost and effort of acquisition may be varied based on technology used

4 RELATED WORK

(Tian, Y. et al. 2010) proposed an improved AOMDV (Ad hoc On Demand Multipath Distance Vector) routing protocol for IoT for route designing. This modified of ad-hoc on-demand multipath distance vector aims to find node-disjoint and link-disjoint route and also to create the connection between nodes and internet efficiently. The protocol can select the stable internet transmission path dynamically through regular updating the Internet Connecting Table. The main limitation of this protocol is the increased routing overhead due to requirement of additional ICT table with routing table. RPL (IPV6 Routing Protocol for Low Power and Lossy Network) is designed for networks

similar to IoTs which are resource constraint which provide better routing solutions for low power and lossy networks but in (Le et al., 2014), Le, Q. et al. outlined major limitation of routing protocol RPL, regarding lack of multi path routing. To overcome this limitation Energy Load Balancing-ELB, Fast Local Repair-FLR techniques are proposed. But this scheme do not provide any internet connectivity for the nodes as well as solution to manage large amount of data produced. (Chen, Z. et al., 2012) proposed a routing protocol CASCR (Context-Aware Routing Protocol on Internet of Things Based on Sea Computing Model) for Internet of Things, based on context-awareness. Sea computing model is used in this technique by using the core idea of the context-aware computing into the basic routing protocol. This technique utilizes the context information to define and analyse the work state data of the sensor nodes roundly. (Wang X. et al., 2012) proposed a resource aware clustering based routing protocol in the Internet of Things, RA-AODVjr. The proposed protocol solves the limitations of the constrained memory resources, computing power and energy of nodes. RA-AODVjr protocol is the combination of RA-Cluster and AODVjr routing protocol. The algorithm detects the system resources frequently with a certain time interval. The functionality of the proposed scheme is limited. Some fault tolerant routing schemes are also proposed to make the routing more reliable, Krishna, (Misra et al., 2012) proposed mixed cross-layered and Learning Automata (LA) based fault-tolerant routing protocol for IOTs that ensures successful delivery of packets even when there are faults between source and destination nodes. The LA and cross-layer concepts adopted in the proposed approach is to provide scalability and good performance in a heterogeneous environment. But this protocol lacks in the scalability and usefulness is limited to a few applications domains.

5 CONCLUSION AND FUTURE SCOPE

Route plays a vital role for the communication in Internet of Things as IoT consists of a large number of nodes, so whenever an event is generated or when a user demands, the data must be sent from one node to the other node in multi hop routing. As the nodes in IoT are resource constrained, such as, limited memory, limited battery power, and limited processing capabilities etc., link and node failure are the major cause of disconnection and unreliability in the IoT. Some protocols and schemes have been proposed for routing in IoT but there is demand of protocol that can present better and reliable routing. The protocol or scheme ought to consume less time in relaying data to several nodes, consumes less power.

And also discovers node-disjoint and link-disjoint failures in minimum time and additionally a few nodes ought to be connected to internet that create a network beyond the local network between nodes as well. So, it is of great significance to design a routing protocol that will aid to ascertain limitations present in existing routing protocols.

REFERENCES

Abowd, G. D., Dey, A. K., Brown, P. J., Davies, N., Smith, M., & Steggles, P. (1999, September). Towards a better understanding of context and context-awareness. In *Handheld and ubiquitous computing* (pp. 304–307). Springer Berlin Heidelberg.

Aggarwal, C. C., Ashish, N., &Sheth, A.P. (2013). The Internet of Things: A Survey from the Data-Centric Perspective.

Al-Fuqaha, A., Guizani, M., Mohammadi, M., Aledhari M., & Ayyash, M. (2015). Internet of things: A survey on enabling technologies, protocols, and applications. *Communications Surveys & Tutorials, IEEE, 17*(4), 2347–2376.

Atzori, L., Iera, A., & Morabito, G. (2010). The Internet of Things: A Survey. *Computer networks, 54*(15), 2787–2805.

Balavalad, K. B., Manvi, S. S., & Sutagundar, A. V. (2009, October). Context aware computing in wireless sensor networks. In *Advances in Recent Technologies in Communication and Computing, 2009. ARTCom'09. International Conference on* (pp.514–516). IEEE.

Chen, S., Xu, H., Liu, D., Hu, B., & Wang, H. (2014). A vision of IoT: Applications, challenges, and opportunities with china perspective. *Internet of Things Journal, IEEE, 1*(4), 349–359.

Chen, Z., Wang, H., Liu, Y., Bu, F., & Wei, Z. (2012). A context-aware routing protocol on internet of things based on sea computing model. *Journal of Computers, 7*(1), 96–105.

Da Xu, L., He, W., & Li, S. (2014). Internet of things in industries: a survey. *Industrial Informatics, IEEE Transactions on, 10*(4), 2233–2243.

Dhumane, A., Prasad, J. (2015, March). *Routing Challenges in Internet of Things.* Paper presented at CSI Communications.

Giusto, D., A. Iera, G. Morabito, L. Atzori (Eds.), The Internet of Things, Springer, 2010. ISBN: 987-1-4419-1673-0.

Gubbi, J., Buyya, R., Marusic, S., & Palaniswami, M. (2013). Internet of Things (IoT): A vision, architectural elements, and future directions. *Future Generation Computer Systems, 29*(7), 1645–1660.

Kalmar, A., Vida, R., & Maliosz, M. (2013, December). Context-aware addressing in the Internet of things using bloom filters. In *Cognitive Infocommunications (CogInfoCom), 2013 IEEE 4th International Conference on* (pp. 487–492). IEEE.

Le, Q., Ngo-Quynh, T., & Magedanz, T. (2014, October). RPL-based multipath Routing Protocols for Internet of Things on Wireless Sensor Networks. In *Advanced Technologies for Communications (ATC), 2014 International Conference on* (pp. 424–429). IEEE.

Mashal, I., Alsaryrah, O., Chung, T. Y., Yang, C. Z., Kuo, W. H., & Agarwal, D. P. (2015). Choices for interaction with things on Internet and underlying issues. *Ad Hoc Networks, 28*, 68–90.

Mashal, I., Alsaryrah, O., Chung, T. Y., Yang, C. Z., Kuo, W. H., & Agarwal, D. P. (2015). Choices for interaction with things on Internet and underlying issues. *Ad Hoc Networks, 28*, 68–90.

Miorandi, D., Sicari, S., De Pellegrini, F., & Chlamtac, I. (2012). Internet of Things: VISION, applicatios and research challenge. *Ad Hoc Networks, 10*(7), 1497–1516.

Misra, S., Krishna, P. V., Agarwal, H., Gupta, A., & Obaidat, M. S. (2012, April). An adaptive learning approach for fault-tolerant routing in Internet of things. In *wireless Communications and Networking Conference (WCNC), 2012 IEEE* (PP. 815–819). IEEE.

Perera, C., Zaslavsky, A., Christen, P., & Georgakopoulos, D. (2014). Context aware computing for the internet of things: A survey. *Communications Surveys & Tutorials, IEEE, 16*(1), 414–454.

Reina, D. G., Toral, S. L., Barrero, F., Bessis, N., & Asimakopoulou, E. (2013). The Role of Ad Hoc Networks in the Internet of Things: A Case Scenerio for Smart Environments. In *Internet of things and Inter-Cooperative Computational Technologies for Collective Intelligence* (pp. 89–113). Springer Berlin Heidelberg.

Sanchez, L., Lanza, J., Olsen, R., Bauer, M., & Girod-Genet, M. (2006, July). A generic context management framework for personal networking environments. In *Mobile and Ubiquitous Systems-Workshops, 2006; 3rd Annual International Conference on* (pp. 1–8). IEEE..

Schilit, B. N., Theimer, M. M. (1994). Disseminating active map information to mobile hosts. *Network, IEEE, 8*(5), 22–32.

Tian, Y., & Hou, R. (2010, December). An improved AOMDV routing protocol for internet of things. In *Computational Intelligence and Software Engineering (CiSE), 2010 International Conference on* (pp. 1–4). IEEE.

Wang, X. (2012, August). Resource-aware clustering based routing protocol in the Internet of Things. In *Awareness Science and Technology (iCAST), 2012 4th International Conference on* (pp. 20–25). IEEE.

Weber, R. H. (2010). Internet of Things-New security and privacy challenge. *Computer Law & Security Review, 26*(1), 23–30.

Whitmore, A., Agarwal, A., & Da Xu, L. (2015). The Internet of Things-A survey of topics and trends. *Information Systems Frontier, 17 I*(2), 261–274.

Wu, M., Lu, T. L., Ling, F. Y., Sun, L., & Du, H. Y. (2010, August). Research on the architecture of Internet of things. In *Advanced Computer Theory and Engineering (ICACTE), 2010 3rd International Conference on* (Vol. 5, pp. V5–484). IEEE.

Yang, D. L., F., & Liang, Y. D. (2010, December). A Survey of the internet of Things. In *Proceedings of the 1st International Conference on E-Business Intelligence (ICEBI2010),* Atlantis Press.

Communication and Computing Systems – Prasad et al. (Eds)
© 2017 Taylor & Francis Group, London, ISBN 978-1-138-02952-1

Intelligent traffic light control algorithm at the road intersection

Bharti Sharma
College of Engineering Roorkee, Roorkee, India

Vinod Kumar Katiyar
Department of Mathematics, Indian Institute of Technology, Roorkee, Uttarakhand, India

Arvind Kumar Gupta
Department of Mathematics, Indian Institute of Technology, Ropar, Punjab, India

ABSTRACT: Time allocation in the fixed time intervals operation of traffic signals (green/red lights) do not match with the traffic volume. Congestion can also be measured by speed, travel time, and delay parameters. This paper presents a mathematical model for delay optimization at the road intersection. Optimization of delay is important for congestion on major and minor road networks. Proposed model includes four major phases. In the first phase traffic volume is predicted at the intersection. New mathematical model is applied to calculate new traffic cycle time in the second phase. In the third phase delay model is used to calculate old and new delay considering existing and proposed method respectively. Comparison is done between old and new delay to find the desired output i.e. Traffic cycle time. Performance of the proposed algorithm was verified on the real data sets collected from T-intersection. It was found that the proposed algorithm has good performance and adaptability in case of high traffic flow situation in comparison to current working method i.e. fixed time intervals of traffic light.

1 INTRODUCTION

An intelligent transport system has a critical role in designing different type of traffic control algorithms. ITS systems help to develop the self-regulating algorithms for controlling and managing the road traffic to improve the traffic safety, making the flow of traffic smooth and reduction in fuel consumption on roads. Basically intelligent system is divided into four parts a surveillance system, a communication system, energy efficiency system and a traffic light control system. Modification in existing traffic control systems infrastructure requires involvement of significant human effort and time. Due to increasing number of vehicles day by day, traffic congestion is a critical problem in many urban cities in India. There are many issues related to our daily life due to traffic congestion like high waiting and traveling time and much fuel consumption. These factors lead to a bad impact on the economy of a country as well. Unregulated and heavy traffic volumes are most important factors of the road accidents, which are increasing very rapidly (Garcia et al. 2011; Cheng and Yun 2010). Various methods have been suggested for the implementation of intelligent traffic light control systems, such as Genetic Algorithm (Sumra et al. 2011), Fuzzy Logic Control (Khalid

et al. 2004; Kulkarni and Wainganka 2007), Neural Network (Emad et al. 2008; Srinivasan et al. 2006), Queuing Network (Wu and Miao 2010; Zhou et al. 2010) etc. Most of the existing traffic control systems use a fixed signal interval for adjusting traffic signal cycle at the intersection. However, this system has poor performance and gives less focus on the characteristics of traffic flow and special conditions of traffic. Therefore, there is a great need to design a traffic light control system which can control its traffic cycle interval depending on the traffic flow at the intersections with the objective to minimize average waiting time at the intersections in Indian traffic conditions. In the present study a mathematical model for traffic cycle time is formulated with the objective to minimize the delay considering the traffic volume, average arrival rate and departure rate of vehicles at intersections. Keeping in mind about mixed (i.e. heterogeneous) traffic scenario on Indian roads, the proposed algorithm will be best suited in Indian traffic conditions. The current study is structured as follows. Section II provides a brief summary of literature review. Section III explains sites and data description. Section IV explains the proposed model and Section V defines about results obtained and critical discussion. At the end the conclusion of the study and directions for future work is discussed.

2 LITERATURE REVIEW

An important expansion of traffic controller by traffic lights has been reached as the first traffic control system was implemented in London in 1868. The growth of traffic controller, mainly since 1960, has controlled to overview of other tools in traffic control mechanism, such as computers, telecommunication devices, vehicle detectors, etc. Traffic regulator approaches have also upgraded as the setting up the first traffic controller. The approaches can be classified as fixed-time interval cycle method and the real time method. In the fixed time strategy control, signal plan is measured in advance, using statistical records and the actual data of traffic. Traffic lights, in addition confirming the protection of road trips, may also aid in the reduction of the total time expended by all the vehicles in the intersection, if an optimum regulator approach is applied. The interval for the traffic signal lights are fixed in the way that the traffic travels easily everywhere without bounding the people in the one direction for long period for their turn The traffic cycle intervals are added according to the average levels of traffic at that intersection according to past experience. The signal synchronization problem is to decide the optimal values for signal parameters with respect to a given objective function, meeting the given constraints. There are several algorithms that developed for controlling the traffic lights problem during previous years. A different method to get the optimum traffic signal cycle time strategy at intersection is examined using ant colony optimization method. They measured two dissimilar ant colony optimization methods, that is, the Ant System and the Elitist Ant System. The two methods are used to regulate traffic signals at road junction to do waiting time of vehicles less. Developing algorithm was also applied to develop dynamic traffic light control (David et al. 2009). The effect of signal time modifies as regards the drivers were identified and authors took into account the network design problem to get the optimum time of traffic signal cycles while predicting the reactions of drivers (Teklu 2007). The SATURN package is applied for solving the traffic steadiness problem. The genetic algorithm is applied on the macroscopic traffic flow and got the optimum signal interval cycles, offset, and green signal interval cycles (Vliet et al. 1982). In this study the author used the chromosome grey-code coding in a different way and validity of the study was checked in the town of UK. The researchers also applied genetic algorithm for optimizing the traffic signal cycles of a commercial area of Spain (Sanchez and Rubio 2008). Self-governing cycles at every intersection was proposed using discrete encoding and related grey-code representation was used (Teklu et al. 2007). The genetic algorithm was also applied to increase the efficiency of the traffic light and pedestrians' intersection control in four

way two lane. The method solved the weakness of previous fixed-time control algorithm for passing vehicles and pedestrians (Turky et al. 2009). The authors trained fuzzy logic controller located in each intersection using particle swarm optimization by examining the actual time of green cycle for each period of the traffic lights. The algorithm was tested using network of two basic intersections. In recent research a PSO approach is applied on traffic flow having the microscopic characteristic for the getting the solutions (Xu and Chen 2006). One hypothetical case with a restrictive one-way road with two intersections was taken to examine the particle swarm optimization. In this approach author gave the attention on the dimensions of isolation to keep the variety of the particle swarm optimization population without involving in the problem itself (Peng et al. 2009). A multi objective type of particle swarm optimization was used for improving traffic signal cycles by a predictive mechanism established on a public transport advancement method. In this study, simulation is done using private and public automobiles on a virtual city highway system prepared of 16 junctions and 51 associations (Kachroudi and Bhouri 2009).

3 SITES AND DATA DESCRIPTION

The data sets used in the study were collected from T-intersection at Haridwar, Uttrakhand, India. The three locations Devpura, Rishikul and Shankara Charaya chowk were selected for data collection as shown in Figure 1. Data were collected twice a day from Monday (16/3/15) to Sunday (21/3/2015) between 10:00 am to 11:00 am in morning and 4 pm to 5 pm in the evening. Data were collected from each identified location for six days. All the data records were recorded manually. During data collection study location, date, time, red cycle time, green cycle time and delay parameters were recorded. Total data sets extracted from the recorded data are dependent on total cycle (red cycle time + green cycle time) time.

4 PROPOSED WORK

Block diagram of the proposed algorithm for intelligent traffic light control at road intersections is presented in Figure 2. The abbreviation used in the proposed work is explained in Table 1. The steps of proposed work are described below:

Phase 1: Input traffic volume and traffic signal cycle
The proposed method is based on traffic volume measured at T-intersection during red signal interval. Traffic volume is considered as critical factor

Figure 1. Map for study area.

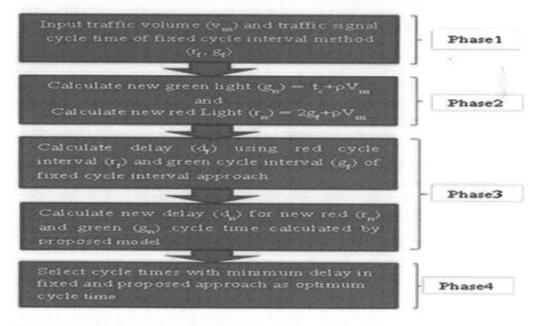

Figure 2. Block diagram of the proposed model.

in calculating the arrival and departure rate of the traffic at intersection. Traffic volume, time to clear queue of vehicles, arrival and departure rates (Webster, F.V. and Cobbe 1966) are used as important factors in the proposed model.

Average arrival time (A_t):

The average time vehicles arrival at the intersection.

$$A_t = \frac{V_m}{r} \qquad (1)$$

Table 1. Variable used in the proposed model.

Variable	Abbreviation
Vm	Traffic Volume
r_f, g_f	Red and green cycle time in fixed cycle time method.
r_n, g_n	Red and green cycle time in the proposed Model
d_f, d_n	Delay calculated according to fixed and proposed method
t_c	Time require to clear the queue
A_t	Average arrival time
D_t	Average departure time
C	Total cycle time
ρ	Constant

Average departure time (Dt):
The average time the departure of the vehicles from the intersection is:

$$D_t = \frac{V_m}{g} \quad (2)$$

Utilization factor (ρ):
Utilization factor is a ration of the traffic jamming of the vehicles at the intersection. The zero value of utilization factor shows small queue at intersection.

$$\rho = \frac{A_t}{D_t} \quad (3)$$

Time required clearing the queue of vehicles (tc):

$$t_c = \frac{\rho V_m}{1-\rho} \quad (4)$$

Phase 2: Mathematical model of the traffic signal time interval
The new mathematical model of calculating the traffic signal time interval is proposed in the current study to calculate the new traffic signal time interval at the intersection considering the traffic volume with the objective of minimizing delay. The equations of the red and green signal interval are given below:
Green light time at T-Intersection

$$g_n = t_c + \rho V_m \quad (5)$$

Red light time

a. For T point junction

$$r_n = 2g_f + \rho V_m \quad (6)$$

Table 2. Proposed algorithm results for Rishikul site's data.

TV	R_{fixed}	G_{fixed}	D_{fixed}	R_{new}	G_{new}	D_{new}
20	86	18	45.7	25	10	15.2
15	86	18	45.4	20	8	12.4
10	86	18	45.1	13	5	8.6
35	86	18	46.2	42	17	24.2
20	86	18	45.7	25	10	15.2
25	86	18	45.9	30	12	17.9
30	86	18	46.1	37	15	21.6
26	86	18	45.9	32	13	18.9
27	86	18	46	32	13	19
23	86	18	45.8	27	11	16.3
31	86	18	46.1	37	15	21.6
22	86	18	45.8	27	11	16.3
21	86	18	45.7	25	10	15.2
19	86	18	45.6	24	10	14.6
18	86	18	45.6	22	9	13.6
20	86	18	45.7	25	10	15.2

Table 3. Proposed algorithm results for Sankara charaya site's data.

TV	R_{fixed}	G_{fixed}	D_{fixed}	R_{new}	G_{new}	D_{new}
21	84	24	44.7	36	15	19.20
22	84	24	44.8	39	16	20.96
10	84	24	44.1	17	7	9.00
15	84	24	44.4	27	11	14.12
20	84	24	44.7	34	14	18.06
26	84	24	44.9	44	18	24.08
30	84	24	45.1	51	21	28.60
26	84	24	44.9	44	18	24.08
27	84	24	45	46	19	25.34
23	84	24	44.8	39	16	21.00
12	84	24	44.2	22	9	11.47
14	84	24	44.4	24	10	12.54
8	84	24	44	15	6	7.98
37	84	24	45.3	63	26	36.99
35	84	24	45.2	58	24	33.43
30	84	24	45.1	51	21	28.60

b. For two way intersection

$$r_n = 3g_f + 2\rho V_m \quad (7)$$

Phase 3: Delay model for the existing system
Delay model is formulated to measure the performance of the proposed model. The formulated equation of the delay model is given below:

$$D_t = \frac{r^2(1+\rho)}{2C} + V_m^{1/3} \quad (8)$$

Using Equation (8) the delay d_f and d_n of the existing and proposed system are calculated.

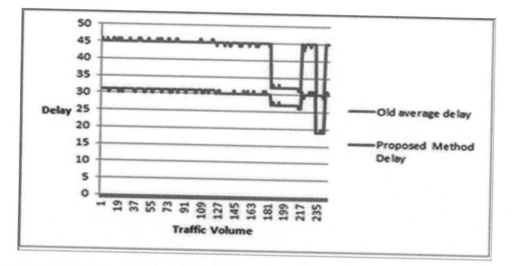

Figure 3. The comparison between current working model and proposed model.

Table 4. Proposed algorithm results for Devpura Chowk site's.

TV	Red_{fixed}	G_{fixed}	D_{fixed}	R_{new}	G_{new}	D_{new}
7	86	18	52.9	10	4	6.9
9	86	18	53.0	10	4	7.0
10	86	18	53.1	12	5	8.1
15	86	18	53.4	17	7	10.9
20	86	18	53.7	22	9	13.7
25	86	18	53.9	27	11	16.4
30	86	18	54.1	34	14	20.1
26	86	18	53.9	30	12	17.9
27	86	18	54	30	12	18
23	86	18	53.8	27	11	16.3
12	86	18	53.2	15	6	9.7
14	86	18	53.4	17	7	10.9
16	86	18	53.5	20	8	12.5
17	86	18	53.5	20	8	12.5
18	86	18	53.6	20	8	12.6
19	86	18	53.6	22	9	13.6

Phase 4: Select optimum traffic signal cycles

Delay model is used to select the optimum traffic signal cycle at the intersection. The delay d_{new} is calculated using equation (8) taking new traffic signal cycles into consideration. The comparison is done between in new delay and old delay. The traffic signal cycles with less delay are taken as optimum traffic signal cycles.

5 EXPERIMENT RESULT

Many algorithms have been proposed by several researchers to control traffic flow at the intersec-

tions in foreign countries although very less work has been done considering the Indian traffic conditions. There is no data set available to test the proposed algorithm in Indian traffic conditions. So, for the current study data is collected from the intersections. T-Intersection is considered for data collection. The effects of proposed study show that the proposed algorithm performs well by minimizing the average delay than current working system i.e. fixed time traffic light method. The average delay helps to reduce various critical factors like the fuel consumption, air pollution and noise pollution. Average delay has been used as the measure of the performance analysis of the proposed model. The average delay helps to reduce various critical factors like the fuel consumption, air pollution and noise pollution. Formulation of delay model is presented in equation (8). Delay model has been formulated by combining queuing theory of various arrivals and departure rates in empirical observations. Experiment results show that red and green cycle times are critical parameters to measure delay at intersections. Average delay reduction has been noticed after implementation of the proposed algorithm in comparison to current working system. i.e. fixed time traffic light control system. Tables 1–3 show the testing of proposed model at three junctions over real data. Testing results show that proposed model performs well and is able to reduce average delay up to 77%.

Figure 3 depicts the comparison between current working model and proposed model. It is clear from graph there is quite good difference between delays calculated by proposed and fixed traffic light cycle.

6 CONCLUSIONS

Traffic flow has been considered as a critical parameter to get the optimal value of red and green light time in the proposed algorithm. Average delay was taken as performance evaluation parameter to find the optimal values of traffic light (Red and Green). Iterative method has been used as an optimization technique in the proposed algorithm to minimize the delay. Proposed algorithm has been tested on real data sets collected from the T-intersection by field study. The results show that proposed algorithm provides an efficient improvement in reducing the average delay time of vehicles at intersections than existing fixed time cycle method.

REFERENCES

Cheng, H. & Yun, W. 2010. A Novel Intelligent Traffic Light Control Scheme, *Ninth International Conference on Grid and Cloud Computing.*

David, R. & Yu, X.H. 2009. Traffic Signal Control with Swarm Intelligence. *IEEE.*

Emad, I.; Abdul, K. & Aman, J. 2008. Intelligent Traffic Light Control Using Neural Network with Multi-Connect Architecture, *National Conference on Information Retrieval and Knowledge Management (CAMP08) in Kuala Lumpur, Malaysia.*

Garcia Nieto, J.; Alba, E. & Olivera, A.C. 2011. Enhancing the urban road traffic with Swarm Intelligence: case study of Crdoba city downtown. *Intelligent Systems Design and Applications (ISDA), 11th International Conference, IEEE.*

Kachroudi, S. & Bhouri, N. 2009. A multimodal traffic responsive strategy using particle swarm optimization, *In Proceedings of the 12th IFAC Symposium on Transportation Systems, Redondo Beach, CA, USA:531–537.*

Khalid, M.; Liang, C.S. & R. Yusof. 2004. Control of a complex traffic junction using fuzzy inference, *5th Asian Control Conference.*

Kulkarni, G. H; Wainganka, P.G. 2007. Fuzzy logic based traffic light controller, *In International Conference on Industrial and Information Systems*

Peng, L.H; Wang, M.; Ping, J.D & Luo, G. 2009. Isolation niches particle swarm optimization applied to traffic lights controlling, *In: Proceedings of the 48th IEEE: Decision and Control Chinese Conference CDC/CCC*: 3318–3322.

Sadaf, M. & Mahmood, F. 2008. *VANET's Communication. IEEE*

Sanchez, J. G & Rubio, E. 2008. Applying a traffic lights evolutionary optimization technique to are alcase: Las Ram blasareain Santa Cruzde enerife, *IEEE Trans. Evol. Compu*, 12(1): 25–40.

Srinivasan, D; Chee, C. M. & Cheu, R.L. 2006. Neural Networks for Real-Time Traffic Signal Control. *Intelligent Transportation Systems, IEEE Transactions* 7(3): 261–272.

Sumra, I. A.; H. Hasbullah, & Manan, J. 2011. A Novel Vehicular SMS System (VSS) Approach for Intelligent Transport System (ITS), *11th International Conference on ITS Telecommunications.*

Teklu, F.; Sumalee, A. & Watling, D. 2007. A genetic algorithm approach for optimizing traffic control signals considering routing. *Comput. Aided Civil Infrastruct. Eng*, 22(3): 31–43.

Turky, A.M.; Ahmad, M.S.; Yuso, M.Z. & Hammad, B.T. 2009. Using genetic algorithm for traffic light control system with a pedestrian crossing, *RSKT 09: Proceedings of the Fourth International Conference on Rough Sets and Knowledge Technology. Berlin, Heidelberg*: 512–519.

Vliet, D.V. 1982. A modern assignment model, *Traffic Eng Control*, 23: 578–581.

Webster, F.V. & Cobbe, B. M. 1966. Traffic Signals. Road Research Technical Paper No. 56. HMSO London UK.

Wu, H.; Miao, Y.C. 2010. Design of intelligent traffic light control system based on traffic flow, *International Conference on Computer and communication Technologies in Agriculture Engineering.*

Xu, J & Chen, L. 2006. Road-junction traffic signal timing optimization by an adaptive particle swarm algorithm. *ICARCV*: 17.

Zhou, B.; Cao, J.; Zeng, X. & Wu, H. 2010. Adaptive traffic light control in wireless sensor network-based intelligent transportation system. *In Vehicular Technology Conference Fall (VTC 2010-Fall), 2010 IEEE 72nd*, 1–5.

Communication and Computing Systems – Prasad et al. (Eds)
© 2017 Taylor & Francis Group, London, ISBN 978-1-138-02952-1

Sound emission based sensor location optimization in fixed axis gearbox using support vector machines

Vanraj, Abhineet Saini, Deepam Goyal, S.S. Dhami & B.S. Pabla
*Department of Mechanical Engineering, National Institute of Technical Teachers'
Training and Research, Chandigarh, India*

ABSTRACT: This paper presents a sound emission-based method for the condition monitoring of fixed axis gearbox to optimize the sensor location. Several sound Emission (SE) statistical parameters, as the monitoring parameters, for the detection of incipient failures of spur gears are reviewed first. The review focuses on the commonly used SE parameters in gearbox, SE signal processing, feature extraction and pattern recognition methods. Experimental work was designed to identify the seeded defect on the spur gear tooth and to optimize the sensor location based on Support Vector Machine (SVM) classifier for detection of defects in the gearbox. The results demonstrated the variation of different SE features at various speeds and loading conditions at different sensor locations. Effectiveness of various SE parameters, for different sound sensor locations based on accuracy of SVM classifier prediction was also determined.

Keywords: acoustic emission; condition monitoring; support vector machines; fixed axis gearbox; signal processing

1 INTRODUCTION

Placement of sensors plays a vital role in building a proficient fault diagnosis system. Reliability and consistency of the captured data entirely depends on location of sensor placement. Recently, there has been a significant increase in the field of condition monitoring for early fault detection of machines. Many researchers presented the utility of vibrations signatures in the anticipation of defects, but only a few emphasized on application of sound emission for the detection of faults. In this paper, an attempt has been made to exploit the sound signal data for fault detection of gearbox using Support Vector Machines (SVMs). Comparison of non-defective and defective gear signals at different locations in combination with different speeds as well as loads makes the detection of faults possible. The seeded faults considered in the present work is with 10% teeth removal in axial direction. Sound emission statistical features like kurtosis, mean, median, and standard deviation etc. were used. The effect of sensor placement on statistical features was studied. Classification was carried out using SVM and subsequently suitable sensor location was proposed based on accuracy of the system. The various vibration monitoring methods and signal processing techniques have been reviewed (Goyal et al. 2015, Goyal et al. 2015).

2 SE BASED CONDITION MONITORING IN GEARBOX

2.1 Characteristics of sound emission

Advanced signal processing techniques employing SE signals have significantly attracted the researchers worldwide for predictive maintenance of rotating machinery (Caesarendra et al. 2016). Although, vibration based monitoring techniques have been applied for fault diagnosis of dynamic machines, SE technology provides some technical advantages over vibration based methods. *Firstly*, SE has the advantage to be placed in either direction as well as periphery of the monitored system whereas vibration monitoring technique requires placement in specified direction for accurate collection of data (Duro et al. 2016, Alkhadafe et al. 2012). *Secondly*, high sensitivity of SE systems paves the way for identifying defects for incipient failures in contrast to other monitoring methods. The various condition monitoring indicators for fault diagnosis of fixed axis gearbox has been reviewed (Goyal et al. 2016).

2.2 SE pattern recognition and clustering methods

The various supervised learning algorithms which are suggested in existing literature are Decision trees, Discriminant analysis, SVM, Naive Bayes, Nearest neighbor, Ensembles.

2.2.1 Support vector machine

Cortes & Vapnik (Cortes & Vapnik 1995) developed a Support Vector Machine (SVM), a learning machine, based on the concept of decision planes for categorization of two group problems. SVMs are termed as supervised learning models in machine learning with associated learning algorithms that was implemented to classify the training data without errors. This technique involves the representation of the mapped data points in space, so that the separate categories could be formed from the data which are divided by a clear gap as wide as possible. Generally, the hyperplane is used for achieving good separation that has largest distance to the nearest training-data point of any category (Saravanan 2010). The Lagrangian multipliers as well as matrix fundamentals are prerequisites in optimizing the data in SVM. This technique is employed for machine fault diagnosis to maximize the distance of the closest point to the boundary curve (Samanta 2003, Guo 2003).

Table 1. Commonly used SE parameters (Goyal et al. 2016).

SE Parameters	Definition
Root Mean Square (RMS)	The square root of the mean of the entirely of the squares of the signal samples
Standard deviation	Measure of the effective energy or power content of the vibration signal.
Variance	Square root of standard deviation
Kurtosis	Flatness or the spikiness of the signal.
Skewness	Degree of asymmetry of a distribution around its mean
Crest Factor	Proportion of maximum highest positive value of the signal
Standard Deviation	Amount of variation or deviation from the mean value

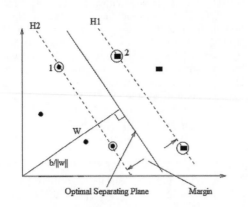

Figure 1. Optimal separating plane and support vectors.

where (w,b) defines the hyperplane separating the two classes of data. 1 and 2 are support vectors, w is normal to the plane, b (scalar constant) is the minimum distance from the origin to the plane, x is a $n \times 1$ vector. In order to make each decision surface (w,b) unique, we normalize the perpendicular distance from the origin to the separating hyperplane by dividing it by $\|w\|$, giving the distance as $\frac{b}{\|w\|}$, as shown in Fig. 1.

3 EXPERIMENTAL SETUP

3.1 Gear dynamic simulator rig

The SE data used was collected from a gear dynamic simulator. The test rig was designed to simulate real working conditions of a gearbox. The rotational speed of the lab test rig could be varied between 1 Hz to 60 Hz by using variable speed drive. For external loading, magnetic particle breaking system was used with loading capacity between 1 lb-inch to 220 lb-inch. The gearbox had single stage fixed axis shafts mounted with spur gears with speed reduction of 3.44:1 as shown in Fig. 2. The shaft carrying gears were supported by two ball bearings. The specifications of gearbox are tabulated in Table 2.

3.2 Sound emission measurement

A G.R.A.S 46 AE type sensors with a frequency range of 5 Hz to 20 kHz was used. The sensor was attached using magnetic mounting onto the top of the system at locations B, C, D in the radial direction at 12.5 cm, 19 cm and 26.5 cm from bottom reference, as shown in Fig. 2 and Fig. 4. The data was acquired at 12.8 kHz sampling rate using NI data acquisition system. The data was acquired for three different speeds corresponding to 15 Hz, 20 Hz and 25 Hz for four different loading conditions viz. no load, 20%, 30% & 50% of maximum loading capacity at all three locations. The above scheme was applied to both non-defective and 10% defective spur gears as shown in Fig. 3.

3.3 Signal processing and feature selection

Sensor data acquired from machines are generally affected by high level of noise and the useful data is masked under some random noise. Therefore, signal processing is required to simplify and extract the useful information which will be utilized for maintenance and decision-making processes. Filtering and amplifying signals are often used to minimize noise and to improve signal-to-noise ratio. The acquired sound amplitude data in time domain contained a lot of random noise

Figure 2.   Experimental Setup of Gear Dynamic Simulator (1: 3 Phase induction motor, 2: Sound Sensor, 3: Magnatic particle break, A: bottom referance of gearbox, B: Location 1, C: Location 2, D: Location 3.

Table 2.   Gearbox specifications.

Specification	Values
Length	27.5 cm
Width	19 cm
Height	26.5 cm
Number of teeth on driver	29
Number of teeth on driven	100
Spur Gear Pressure angle (α_p)	20 (degree)

Figure 3.   Defective gear.

Figure 4.   Top view of gearbox.

which was denoised using Time Synchronous Averaging (TSA) technique. The statistical parameters of TSA signals were then calculated and were analyzed. These calculated parameters were then incorporated to SVM based classifier for categorization of faults. The various SVM models were generated using statistical parameters to optimize the sound sensor location. 70% of the data was used for training and remaining 30% was used for validation purpose followed by testing to find the accuracy of the models.

4   RESULTS AND DISCUSSION

The TSA data for 15 Hz speed without and with defect is shown in Fig. 6 and Fig. 7 respectively. The graphs clearly show the increased amplitude values as the load increased. RMS, kurtosis, skewness, mean, variance, crest factor and standard deviation were then calculated at each location for all combinations of speeds, loads and defect. The scatter plots of all the parameters for location 1 are shown in Fig. 5. The averaged values of all SE statistical parameters for different speeds are listed in Table 3.

The RMS values for both no-defect and defect cases showed similar decreasing trend representing decreasing sound amplitude for farther locations. Hence, the maximum sound amplitude was observed at location 1 for both the cases. The kurtosis value showed an increase in value from location 1 to location 2 and then decreased at location 3. The maximum peak of sound signals was thus obtained at location 2. The crest factor indicated a decreasing trend while moving from location 1 to location 3 similar to RMS value which confirmed the decreasing trend of sound amplitude with increasing displacement from meshing location of gears. The higher value of standard deviation and variance for location 1 in both the cases indicated a more distributed sound amplitude data.

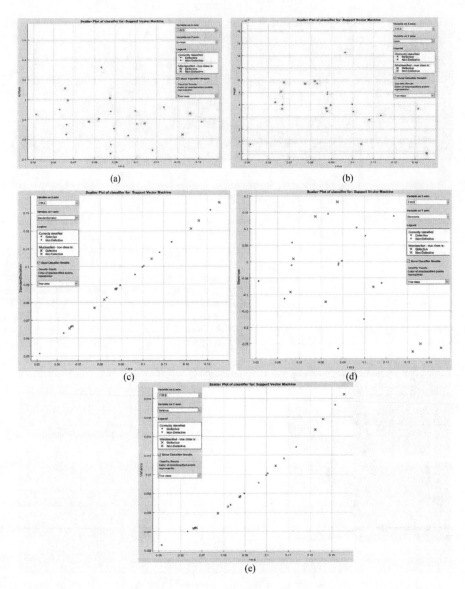

(a)

(b)

(c)

(d)

(e)

Figure 5.   Scatter plots of statistical parameters for location 1.

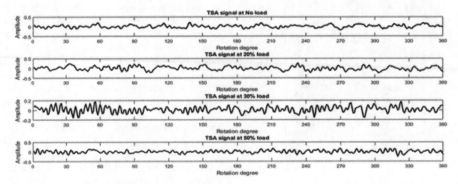

Figure 6.   TSA signal of sound data at 15 Hz with no defect.

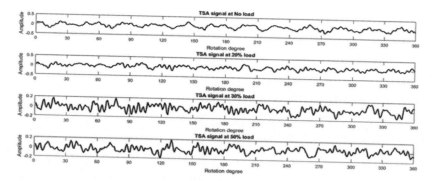

Figure 7. TSA signal of sound data at 15 Hz and with defect.

Table 3. Average values of Statistical parameters for all loading conditions at different sensor locations with and without defect.

	No Defect			Defect		
	L 1	L 2	L 3	L 1	L 2	L 3
RMS	0.096	0.071	0.059	0.088	0.070	0.061
Kurtosis	2.864	3.027	2.950	2.819	2.852	2.750
Skewness	−0.007	−0.022	0.019	−0.071	0.032	−0.006
Mean	0.008	0.005	0.006	0.006	0.007	0.000
Variance	0.009	0.005	0.004	0.008	0.005	0.004
Crest factor	3.360	3.302	3.251	3.157	3.100	3.045
Standard Deviation	0.095	0.070	0.057	0.088	0.070	0.060

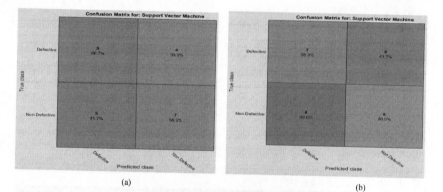

(a)   (b)

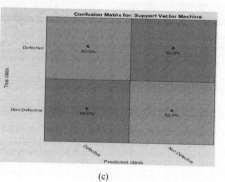

(c)

Figure 8. Confusion matrix of support vector machine classifier (a), (b) and (c) for location 1, 2 & 3 respectively.

871

The confusion matrix for all locations is shown in Fig. 8. SVM model for location 1 (Fig. 8(a)) shows the overall prediction accuracy of 62.5%. The model predicts the defect correctly with an accuracy of 66.7%. and misclassified class at 33.3%. Classification accuracy for non-defective class was around 58.3% whereas it misclassified 41.7%.

SVM model for location 2 (Fig. 8(b)) shows overall prediction accuracy of 56.7% which is less than the location 1 model. SVM model for location 3 (Fig. 8(c)) shows the least prediction accuracy of 50%.

5 CONCLUSIONS

The application of SE for different speed combinations for gearbox has been proposed in this paper. The paper focused on investigation of variation of SE statistical features for different sensor location to find the optimal sensor location based on classification accuracy of proposed SVM model with

SE parameters as input for defective and non-defective spur gears. From the experimental study of spur gears with artificially induced fault, following conclusions were drawn:

1. The investigation of SE parameter results show that the RMS, kurtosis, skewness, variance, mean, crest factor and standard deviation effectively identify significant change in the conditions of spur gear. These changes, however, are too sudden in some parameters which indicated that all the parameters cannot provide adequate information for the classification of defect.
2. The SVM model for location 1 shows the maximum accuracy in predicting the right class of gear. Hence, location 1 can be considered as the optimal location for the sound sensor to collect the necessary data which can be utilized as the input to any classifier.
3. Due to the limitation of the SE parameters in certain cases, some researchers have proposed alternative feature extraction methods e.g. energy ratio, FM4, NA4, and sideband index. Hence more reliable results could be obtained by

using a combination of time domain, frequency domain and time frequency domain features.

REFERENCES

Alkhadafe, H., Al-Habaibeh, A., Daihzong, S., & Lotfi, A. (2012). Optimising sensor location for an enhanced gearbox condition monitoring system. In Journal of Physics: Conference Series, IOP Publishing, Vol. 364(1), p. 012077.

Caesarendra, W., Kosasih, B., Tieu, A. K., Zhu, H., Moodie, C. A., & Zhu, Q. (2016). Acoustic emission-based condition monitoring methods: Review and application for low speed slew bearing, Mechanical Systems and Signal Processing, Vol. 72, pp. 134–159.

Cortes, C., & Vapnik, V. (1995). Support-vector networks. Machine learning, 20(3), 273–297.

Duro, J. A., Padget, J. A., Bowen, C. R., Kim, H. A., & Nassehi, A. (2016). Multi-sensor data fusion framework for CNC machining monitoring, Mechanical Systems and Signal Processing, Vol. 66, pp. 505–520.

Goyal, D., and Pabla, B. S. "Condition based maintenance of machine tools: a review", CIRP Journal of Manufacturing Science and Technology, Vol. 10, 2015, pp. 24–35.

Goyal, D., and Pabla, B. S. "The vibration monitoring methods and signal processing techniques for structural health monitoring: a review", Archives of Computational Methods in Engineering, 2015, doi 10.1007/s11831-015-9145-0.

Goyal, D., Vanraj, Pabla, B.S., & Dhami, S.S. (2016), "Condition Monitoring Parameters for Fault Diagnosis of Fixed Axis Gearbox—A Review", Archives of Computational Methods in Engineering, doi: 10.1007/s11831-016-9176-1.

Guo, M., Xie, L., Wang, S., Zhang, JM. (2003) "Research on an integrated ICA-SVM based framework for fault diagnosis", in: Proceedings of the 2003 IEEE International Conference on Systems, Man and Cybernetics, Vol. 3, Washington, DC, USA, pp. 2710–2715.

Samanta, B., Al-Balushi, K. R., & Al-Araimi, S. A. (2003). Artificial neural networks and support vector machines with genetic algorithm for bearing fault detection, Engineering Applications of Artificial Intelligence, Vol. 16(7), pp. 657–665.

Saravanan, N., Siddabattuni, V. K., & Ramachandran, K. I. (2010). Fault diagnosis of spur bevel gear box using artificial neural network (ANN), and proximal support vector machine (PSVM). Applied Soft Computing, Vol. 10(1), pp. 344–360.

Communication and Computing Systems – Prasad et al. (Eds)
© 2017 Taylor & Francis Group, London, ISBN 978-1-138-02952-1

An R-norm fuzzy entropy of type 'α' on intuitionistic fuzzy sets and its applications in multiple attribute decision making

Rajesh Joshi & Satish Kumar
Department of Mathematics, Maharishi Markandeshwar University, Mullana-Ambala, India

Amit Kumar Manocha
Maharaja Agrasen University, Baddi, Himachal Pradesh, India

ABSTRACT: The theory of Intuitionistic Fuzzy Sets (IFSs) is well suitable to deal with vagueness and hesitancy. In the present communication, a new parametric R-norm fuzzy entropy of type 'α' is proposed with the proof of validity. The intuitionistic fuzzy entropy is useful to represent the decision information in the process of decision making since it is characterized by the degee of satisfiability, degree of non-satisfiability and hesitancy degree. Based on the proposed R-norm fuzzy entropy of type 'α', a new decision-making method of a Multi-Attribute Decision Making problem (MADM) has been introduced. In case of attribute weights, two cases (one with completely unknown attribute weights and other with partially known attribute weights) are discussed. The method is illustrated with the help of suitable examples.

Keywords: Intuitionistic Fuzzy Entropy, R-norm Intuitionistic Fuzzy Entropy of Type α, MADM, TOPSIS, Weighted Hamming Distance

MS Classifications: 94 A15, 94 A24, 26D15.

1 INTRODUCTION

For a multi attribute decision making problem, where there are so many factors which need to be considered simultaneously, it becomes very difficult to arrive at a conclusion. Sometimes the situation is so complex that the crisp data is inadequate or insufficient to handle it. For such type of problems which are uncertain or vague in nature, fuzzy sets play an important role to model the human judgment. The concept of fuzzy sets was first introduced by Zadeh (1965), which assigns each element a membership number. Later in Zadeh (1968), introduced the concept of fuzzy entropy. De Luca and Termini (1972) axiomatized the fuzzy entropy. The concept of fuzzy sets was first generalized by Atanassov (1999) to 'Intuitionistic Fuzzy Sets (IFS)'. The distinguishing fact of the IFS is that it assigns to each element a membership degree, a non-membership degree and a hesitancy degree. Firstly, Burillo and Bustince (2001) defined the entropy on IFS. Szmidt and Kacprzyk (2002) defined the entropy measure on IFS with a different approach based on geomatrical interpretation of IFS. Hung and Yang (2006) used the concept of probability to define the entropy. Vlachos and Sergiadis (2007) proposed a new entropy measure connecting the notion of entropy of a fuzzy set and intuitionistic fuzzy set. Zhang and Jiang (2008) gave another intuitionistic fuzzy entropy by means of intersection and union of membership and non-membership degree IFSs.

In the present paper, we consider an R-norm fuzzy entropy of type 'α', which not only considers the membership and non-membership degree, but also the hesitancy degree of the IF sets. The remainder of the paper is organized as follows: In Section 2, the basic definitions related to fuzzy set theory and intuitionistic fuzzy set theory are briefly given. In Section 3, a new R-norm fuzzy entropy measure of type 'α' is introduced and its axiomatic justification is established. Some mathematical properties of the proposed measure are also studied in this section. Section 4 is devoted to an intuitionistic fuzzy MADM method in which attribute weights are determined using proposed IF entropy measure. For completely unknown weights, an extended ordinary entropy weight method is used and minimum entropy principle is used to determine the weights in case of partially known weights. In Section 5, two methods are explained with the help of two examples. Finally, the paper is concluded with 'Concluding Remarks' in Section 6.

1.1 Preliminaries

In this section, some needed basic concepts and definitions of fuzzy sets and intuitionistic fuzzy sets are introduced.

Definition 2.1. A fuzzy set P in a finite universe of discourse $X = \{z_1, z_2, ..., z_n\}$ is given by

$$P = \{z_i, \mu_P(z_i) / z_i \in X\} \qquad (1)$$

where $\mu_P : X \to [0,1]$ is the membership of function of P. The number $\mu_P(z_i)$ describes the degree of belongingness of $z_i \in X$ in P.

Atanassov (1986, 1999) generalized the idea of fuzzy sets to intuitionistic fuzzy sets as:

Definition 2.2. An Intuitionistic Fuzzy Set P in a finite universe of discourse $X = \{z_1, z_2, ..., z_n\}$ is given as

$$P = \{< z_i, \mu_P(z_i), \nu_P(z_i) > / z_i \in X\}, \qquad (2)$$

where

$$\mu_P : X \to [0,1]; \quad \nu_P : X \to [0,1] \qquad (3)$$

with the condition $0 \le \mu_P(z_i) + \nu_P(z_i) \le 1, \forall z_i \in X$.

In this definition, the numbers $\mu_P(z_i)$ and $\nu_P(z_i)$ respectively denote the degree of membership and degree of non-membership of $z_i \in X$ to the set P.

For each IFS P in X, if $\pi_P(z_i) = 1 - \mu_P(z_i) - \nu_P(z_i), z_i \in X$ then $\pi_P(z_i)$ denotes the degree of hesitancy of $z_i \in X$ to the set P. Also $\pi_P(z_i)$ is called intuitionistic index.

Definition 2.3. Let $IFS(X)$ denote the family of all IFSs in the universe X, and let $P, Q \in IFS(X)$ given by

$$P = \{< z_i, \mu_P(z_i), \nu_P(z_i) > / z_i \in X\} \qquad (4)$$

$$Q = \{< z_i, \mu_Q(z_i), \nu_Q(z_i) > / z_i \in X\} \qquad (5)$$

Then usual set relations and operations Atanassov (1986, 1999) are defined as follows:

1. $P \subseteq Q$ if and only if $\mu_P(z_i) \le \mu_Q(z_i)$, $\nu_P(z_i) \ge \nu_Q(z_i), \forall z_i \in X$;
2. $P = Q$ if and only if $P \subseteq Q$ and $Q \subseteq P$;
3. The complement of the set P denoted as P^c, is $P^c = \{< z_i, \nu_P(z_i), \mu_P(z_i) > / z_i \in X\}$;
4. $P \cap Q = \{\langle \mu_P(z_i) \wedge \mu_Q(z_i) \text{ and } \nu_P(z_i) \vee \nu_Q(z_i) \rangle / z_i \in X\}$
5. $P \cup Q = \{\langle \mu_P(z_i) \vee \mu_Q(z_i) \text{ and } \nu_P(z_i) \wedge \nu_Q(z_i) \rangle / z_i \in X\}$

Szmidt and Kacprzyk (2002) extended the De Luca and Termini (1972) for proposing the entropy measure in the settings of intuitionistic fuzzy sets.

Definition 2.4. An entropy on $IFS(X)$ is a real valued function $E : IFS(X) \to [0,1]$, satisfying the following axioms:

1. (Sharpness): $E(P) = 0$ if and only if P is a crisp set, i.e., $\mu_P(z_i) = 0, \nu_P(z_i) = 1$; or $\mu_P(z_i) = 1, \nu_P(z_i) = 0 \, \forall z_i \in X$;
2. (Maximality): $E(P) = 1$ if and only if $\mu_P(z_i) = \nu_P(z_i) = \pi_P(z_i) = \frac{1}{3}, \forall z_i \in X$;
3. (Symmetry): $E(P) = E(P^c)$;
4. (Resolution): $E(P) \le E(Q)$ if and only if $P \subseteq Q$, i.e. if $\mu_P(z_i) \le \mu_Q(z_i)$ and $\nu_P(z_i) \ge \nu_Q(z_i)$ for $\mu_Q(z_i) \le \nu_Q(z_i)$, or if $\mu_P(z_i) \ge \mu_Q(z_i)$ and $\nu_P(z_i) \le \nu_Q(z_i)$ for $\mu_Q(z_i) \ge \nu_Q(z_i)$ for all $z_i \in X$.

Definition 2.5. Suppose $P = \{< z_i, \mu_P(z_i), \nu_P(z_i) > / z_i \in X\}$ and $Q = \{< z_i, \mu_Q(z_i), \nu_Q(z_i) > / z_i \in X\}$ be two IF sets with the weight of z_i is u_i. Then the Weighted Hamming distance measure of P and Q is defined as follows:

$$s(P,Q) = \frac{1}{2} \sum u_i (|\mu_P(z_i) - \mu_Q(z_i)| + |\nu_P(z_i) - \nu_Q(z_i)| + |\pi_P(z_i) - \pi_Q(z_i)|). \qquad (6)$$

Throughout this paper, $IFS(X)$ will denote the set of all intuitionistic fuzzy sets and $FS(X)$ will represent the set of all the fuzzy sets defined on X.

With these concepts and ideas, we now introduce a new fuzzy information measure called R-Norm Fuzzy Entropy of type 'α' on intuitionistic fuzzy sets with 'α' as a parameter.

1.2 A new R-norm fuzzy entropy of type 'α' on IFS

Now, we introduce a new IF entropy measure as:

$$H_R^\alpha(P) = \frac{R}{R+\alpha-2} \sum_{i=1}^{n} \frac{1}{n} \left\{ (1 - \left(\mu_P^{\frac{R}{2-\alpha}}(z_i) \right. \right.$$
$$\left. \left. + \nu_P^{\frac{R}{2-\alpha}}(z_i) + \pi_P^{\frac{R}{2-\alpha}}(z_i) \right)^{\frac{2-\alpha}{R}} \right\}, \qquad (7)$$

where $R > 0$, $0 < \alpha < 2$ and $R + \alpha \ne 2$ which considers membership degree, non-membership degree and the hesitancy degree of intuitionistic fuzzy sets.

Remark: If $\alpha = 1$ then (7) reduces to R-norm intuitionistic fuzzy information meaure as:

$$H_R(P) =$$
$$\frac{R}{R-1} \sum_{i=1}^{n} \frac{1}{n} \left\{ \left(1 - (\mu_P^R(z_i) + \nu_P^R(z_i) + \pi_P^R(z_i))^{\frac{1}{R}} \right) \right\}$$

Now we prove the following property for proving validity of proposed measure.

Property 7 Under the condition I_3, we have

$$|\mu_P(z_i) - \frac{1}{3}| + |\nu_P(z_i) - \frac{1}{3}| + |\pi_P(z_i) - \frac{1}{3}| \geq$$
$$|\mu_Q(z_i) - \frac{1}{3}| + |\nu_Q(z_i) - \frac{1}{3}| + |\pi_Q(z_i) - \frac{1}{3}| \quad (8)$$

$$\left(\mu_P(z_i) - \frac{1}{3}\right)^2 + \left(\nu_P(z_i) - \frac{1}{3}\right)^2 + \left(\pi_P(z_i) - \frac{1}{3}\right)^2$$
$$\geq \left(\mu_Q(z_i) - \frac{1}{3}\right)^2 + \left(\nu_Q(z_i) - \frac{1}{3}\right)^2 + \left(\pi_Q(z_i) - \frac{1}{3}\right)^2$$
$$(9)$$

Proof. If $\mu_P(z_i) \leq \mu_Q(z_i)$ and $\nu_P(z_i) \leq \nu_Q(z_i)$ with $\max\{\mu_Q(z_i), \nu_Q(z_i)\} \leq \frac{1}{3}$ then $\mu_P(z_i) \leq \mu_Q(z_i) \leq \frac{1}{3}$; $\nu_P(z_i) \leq \nu_Q(z_i) \leq \frac{1}{3}$ and $\pi_P(z_i) \geq \pi_Q(z_i) \geq \frac{1}{3}$ which implies that (8) and (9) hold. Similarly, if $\mu_P(z_i) \geq \mu_Q(z_i)$ and $\nu_P(z_i) \geq \nu_Q(z_i)$ with $\max\{\mu_Q(z_i), \nu_Q(z_i)\} \geq \frac{1}{3}$ then (8) and (9) hold.

Theorem 3.2. Measure (7) is a valid intuitionistic fuzzy information measure.

Proof. To prove (7) is a valid intuitionistic fuzzy information measure, we shall show that it satisfies the properties in Definition (2.4).

1. (Sharpness): If $H_R^\alpha(P) = 0$ then

$$\frac{R}{R + \alpha - 2} \sum_{i=1}^{n} \left\{ 1 - \left(\mu_P^{\frac{R}{2-\alpha}}(z_i) + \nu_P^{\frac{R}{2-\alpha}}(z_i) \right.\right.$$
$$\left.\left. + \pi_P^{\frac{R}{2-\alpha}}(z_i) \right)^{\frac{2-\alpha}{R}} \right\} = 0. \quad (10)$$

Since $R > 0$ and $R + \alpha \neq 2$ therefore this is possible only in the following cases:

1. Either $\mu_P(z_i) = 1$ i.e., $\nu_P(z_i) = \pi_P(z_i) = 0$ or
2. $\nu_P(z_i) = 1$ i.e., $\mu_P(z_i) = \pi_P(z_i) = 0$ or
3. $\pi_P(z_i) = 1$ i.e., $\mu_P(z_i) = \nu_P(z_i) = 0$.

In all the above cases, $H_R^\alpha(P) = 0$ implies that P is a crisp set. Conversely, if P is a crisp set then either $\mu_P(z_i) = 1$ and $\nu_P(z_i) = \pi_P(z_i) = 0$ or $\nu_P(z_i) = 1$ and $\mu_P(z_i) = \pi_P(z_i) = 0$ or $\pi_P(z_i) = 1$ and $\mu_P(z_i) = \nu_P(z_i) = 0$. This implies that

$$1 - \left(\mu_P^{\frac{R}{2-\alpha}}(z_i) + \nu_P^{\frac{R}{2-\alpha}}(z_i) + \pi_P^{\frac{R}{2-\alpha}}(z_i) \right)^{\frac{2-\alpha}{R}} = 0, \quad (11)$$

which gives $H_R^\alpha(P) = 0$. Hence, $H_R^\alpha(P) = 0$ iff P is a crisp set.

2. (Maximality): Since $\mu_P(z_i) + \nu_P(z_i) + \pi_P(z_i) = 1$ therefore to obtain the maximum value of intuitionistic fuzzy entropy $H_R^\alpha(P)$, we write
$$\phi(\mu_P, \nu_P, \pi_P) = \mu_P(z_i) + \nu_P(z_i) + \pi_P(z_i) - 1 \quad \text{and}$$
taking the Lagrange's multiplier λ, we consider

$$\Phi(\mu_P, \nu_P, \pi_P) = H_R^\alpha(\mu_P, \nu_P, \pi_P) + \lambda\phi(\mu_P, \nu_P, \pi_P)$$
$$(12)$$

To find the maximum value of $H_R^\alpha(P)$, we differentiate (12) partially with respect to μ_P, ν_P, π_P and λ and equating them to zero, we get $\mu_P(z_i) = \nu_P(z_i) = \pi_P(z_i) = \frac{1}{3}$. It may be noted that all the first order partial derivatives vanish iff $\mu_P(z_i) = \nu_P(z_i) = \pi_P(z_i) = \frac{1}{3}$. Therefore the stationary point of $H_R^\alpha(P)$ is $\mu_P(z_i) = \nu_P(z_i) = \pi_P(z_i) = \frac{1}{3}$. Next, to prove $H_R^\alpha(P)$ is a concave function of $P \in F(X)$, we calculate its Hessian at the stationary point. The Hessian of $H_R^\alpha(P)$ is given by

$$\hat{H} = 3^{\frac{2-(R+\alpha)}{R}} \times \frac{R}{2-\alpha} \begin{pmatrix} -2 & - & 1 \\ 1 & -2 & 1 \\ 1 & 1 & -2 \end{pmatrix}. \quad (13)$$

For all $R > 0$ and $0 < \alpha < 2$, $\hat{H}$ is a negative semi-definite matrix and hence $H_R^\alpha(P)$ is a concave function having its maximum value at the point $\mu_P(z_i) = \nu_P(z_i) = \pi_P(z_i) = \frac{1}{3}$.

3. (Resolution): Since $H_R^\alpha(P)$ is a concave function of $P \in F(X)$ therefore if $\max\{\mu_P(z_i), \nu_P(z_i)\} \leq \frac{1}{3}$, then $\mu_P(z_i) \leq \mu_Q(z_i)$ and $\nu_P(z_i) \leq \nu_Q(z_i)$ which implies $\pi_P(z_i) \geq \pi_Q(z_i) \geq \frac{1}{3}$. Therefore, by the property (7), we conclude that $H_R^\alpha(P)$ satisfies the condition I_3.

Similarly, if $\min\{\mu_P(z_i), \nu_P(z_i)\} \geq \frac{1}{3}$, then $\mu_P(z_i) \leq \mu_Q(z_i)$ and $\nu_P(z_i) \geq \nu_Q(z_i)$. Therefore, by using property (7), we conclude that $H_R^\alpha(P)$ satisfies condition

4. (Symmetry): It is clear that from the definition that $H_R^\alpha(P) = H_R^\alpha(P^c)$.

Hence, $H_R^\alpha(P)$ satisfies all the properties of intuitionistic fuzzy entropy and therefore, $H_R^\alpha(P)$ is a valid information measure.

The following properties of the proposed measure can be easily proved.

Theorem 3.3. Let P and Q be two intuitionistic fuzzy sets defined in $X = \{z_1, z_2, ..., z_n\}$ where $P = \{\langle z_i, \mu_P(z_i), \nu_P(z_i) / z_i \in X\rangle$ and $Q = \{\langle z_i, \mu_Q(z_i), \nu_Q(z_i) / z_i \in X\rangle\}$, such that for all $z_i \in X$ either $P \subseteq Q$ or $P \supseteq Q$; then

$$H_R^\alpha(P \cup Q) + H_R^\alpha(P \cap Q) = H_R^\alpha(P) + H_R^\alpha(Q) \quad (14)$$

Corollary: For any set $P \in IFS(X)$ and its complement P^c,

$$H_R^\alpha(P) = H_R^\alpha(P^c) = H_R^\alpha(P \cup P^c) = H_R^\alpha(P \cap P^c) \quad (15)$$

1.3 New intuitionistic fuzzy MADM method using proposed entropy measure and hamming distance

Multiple Attribute Decision Making problem can be represented with a fuzzy decision matrix as:

$$X = (\tilde{x}_{ij})_{m \times n} =$$

$$
\begin{array}{c}
 & o_1 & \cdots & o_n \\
A_1 & \begin{pmatrix} (p_{11}, q_{11}) & \cdots & (p_{1n}, q_{1n}) \\ (p_{21}, q_{21}) & \cdots & (p_{2n}, q_{2n}) \\ \vdots & \vdots & \vdots \\ (p_{m1}, q_{m1}) & \cdots & (p_{mn}, q_{nn}) \end{pmatrix} \\
A_2 \\
\vdots \\
A_m
\end{array}
\qquad (16)
$$

where $\tilde{x}_{ij} = (p_{ij}, q_{ij})$ where p_{ij} and q_{ij} are the membership degree and non-membership degree of alternatives A_i's on the attributes o_j satisfying the conditions: $0 \le p_{ij} \le 1$, $0 \le q_{ij} \le 1$ and $0 \le p_{ij} + q_{ij} \le 1$.

Attribute weights play an important role in MADM. For attribute weights $o_j (j = 1, 2, ..., n)$, let $u = (u_1, u_2, ..., u_n)^T$ be the weight vector such that $0 \le u_j \le 1 (j = 1, 2, ..., n)$ and satisfying $\sum_{j=1}^{n} u_j = 1$. Weights of the attributes are not known to us everytime or sometimes partially known to us. In this paper, we have proposed two methods to determine the weights of attributes, first when they are not known, second when they are partially known.

1.3.1 When weights are unknown

Based on the work done by Chen et al. (2010) and Ye (2010), we use (7) to calculate the weights of the attributes when they are completely unknown as:

$$u_j = \frac{1 - e_j}{n - \sum_{j=1}^{n} e_j}, \quad j = 1, 2, ..., n, \qquad (17)$$

where $e_j = \frac{1}{m} \sum_{i=1}^{m} H_R^\alpha(\tilde{x}_{ij})$ and

$$H_R^\alpha(\tilde{x}_{ij}) = \frac{R}{R + \alpha - 2} \sum_{i=1}^{m} \frac{1}{m} \left\{ \left(1 - \left(\mu_P^{\frac{R}{2-\alpha}}(z_i) \right. \right. \right.$$
$$\left. \left. \left. + v_P^{\frac{R}{2-\alpha}}(z_i) + \pi_P^{\frac{R}{2-\alpha}}(z_i) \right)^{\frac{2-\alpha}{R}} \right) \right\}, \qquad (18)$$

where $R > 0$, $0 < \alpha < 2$ and $R + \alpha \ne 2$ which also considers the hesitancy degree alongwith membership and non-membership degrees of the IF sets.

1.3.2 When weights are partially known

In general, there are more constraints for the weight vector $u = (u_1, u_2, ..., u_n)$. The set of known weight information is denoted as H. For MADM

problems with partially known attribute weights under intuitionistic fuzzy enviornment, we use the minimum entropy principle suggested by Wang and Wang (2012) to obtain the weight vector as:

$$E(A_i) = \sum_{j=1}^{n} u_j H_R^\alpha(\tilde{x}_{ij})$$

$$= \frac{R}{R + \alpha - 2} \sum_{j=1}^{n} u_j \left\{ \sum_{i=1}^{m} \frac{1}{m} \left(1 - (\mu_P^{\frac{R}{2-\alpha}}(z_i) \right. \right.$$
$$\left. \left. + v_P^{\frac{R}{2-\alpha}}(z_i) + \pi_P^{\frac{R}{2-\alpha}}(z_i))^{\frac{2-\alpha}{R}} \right) \right\},$$

$$s.t. \quad \sum_{j=1}^{n} u_j = 1, u_j \in H.$$

Since each alternative is made in a fairly competitive enviornment, the weight coefficients corresponding to the same attributes should also be equal; to determine the optimal weight, we construct the following model:

$$\min \quad E = \sum_{i=1}^{m} E(A_i) = \sum_{i=1}^{m} \left[\sum_{j=1}^{n} u_j H_R^\alpha(\tilde{x}_{ij}) \right]$$

$$= \frac{R}{R + \alpha - 2} \sum_{j=1}^{n} u_j \left[\sum_{i=1}^{m} \frac{1}{m} \left(1 - (\mu_P^{\frac{R}{2-\alpha}}(z_i) \right. \right.$$
$$\left. \left. + v_P^{\frac{R}{2-\alpha}}(z_i) + \pi_P^{\frac{R}{2-\alpha}}(z_i))^{\frac{2-\alpha}{R}} \right) \right],$$

$$s.t. \quad \sum_{j=1}^{n} u_j = 1, u_j \in H. \qquad (19)$$

On solving equation (19) with the help of MATLAB software, we get the optimal solution $arg \quad \min \quad E = (u_1, u_2, ..., u_n)^T$.

In summary, the main procedure of the decision making method is listed in the following steps:

1. Determine the attribute weights by using equation (17) and (19).
2. Define Best Solution (Z^+) and Worst Solution (Z^-) as:

$$Z^+ = ((\alpha_1^+, \beta_1^+), (\alpha_2^+, \beta_2^+), ..., (\alpha_n^+, \beta_n^+)), \qquad (20)$$

where $(\alpha_j^+, \beta_j^+) = (1, 0), j = 1, 2, ..., n$.

$$Z^- = ((\alpha_1^-, \beta_1^-), (\alpha_2^-, \beta_2^-), ..., (\alpha_n^-, \beta_n^-)), \qquad (21)$$

where $(\alpha_j^-, \beta_j^-) = (0, 1), j = 1, 2, ..., n$.
3. Using the definition (2.3), the distances of the alternatives Z_i's from Z^+ and Z^- are given as follows:

$$s(Z_i, Z^+) = \frac{1}{2} \sum_{j=1}^{n} u_j (|\alpha_{ij} - \alpha_j^+| + |\beta_{ij} - \beta_j^+|$$
$$+ |\pi_{ij} - \pi_j^+|), \qquad (22)$$

$$s(Z_i, Z^-) = \frac{1}{2}\sum_{j=1}^{n} u_j (|\alpha_{ij} - \alpha_j^-| + |\beta_{ij} - \beta_j^-| + |\pi_{ij} - \pi_j^-|), \quad (23)$$

4. Determine the relative degrees of closeness D_i's as:

$$D_i = \frac{s(Z_i, Z^-)}{s(Z_i, Z^-) + s(Z_i, Z^+)}. \quad (24)$$

5. Rank the alternatives as per the values of D_i's in descending order. The alternative nearest to the Best Solution Z^+ and farthest from the Worst Solution Z^- will be considered as the best alternative.

1.4 Numerical examples

Now, we illustrate the application of MADM method with the help of examples as follows:

Case 1. When weights of attributes are completely unknown.
Consider the MADM proble given in Table 1.
Using the procedure mentioned above for unknown weights, the ranking obtained is $Z_2 \succ Z_4 \succ Z_3 \succ Z_1$ and Z_2 is the desireable alternative.

Case 2. When weights of attributes are partially known
Consider the MADM problem given in Table 2.

Let the attribute weights satisfy the following set:

$$H = \{0.25 \le u_1 \le 0.75, 0.35 \le u_2 \le 0.60, \\ 0.30 \le u_3 \le 0.35\}.$$

Table 1. The IF decision matrix.

Options	Evaluation attributes		
	e_1	e_2	e_3
Z_1	(0.45, 0.35)	(0.50, 0.30)	(0.20, 0.55)
Z_2	(0.65, 0.25)	(0.65, 0.25)	(0.55, 0.15)
Z_3	(0.45, 0.35)	(0.55, 0.35)	(0.55, 0.20)
Z_4	(0.75, 0.15)	(0.65, 0.20)	(0.35, 0.15)

Table 2. The IF decision matrix.

Options	Evaluation attributes		
	e_1	e_2	e_3
Z_1	(0.75, 0.10)	(0.60, 0.25)	(0.80, 0.20)
Z_2	(0.80, 0.15)	(0.68, 0.20)	(0.45, 0.50)
Z_3	(0.40, 0.45)	(0.75, 0.05)	(0.60, 0.30)

Using the equation (19), following programming model can be established:

$$\min E = 0.1273u_1 + 0.1268u_2 + 0.1396u_3,$$

$$s.t. \begin{cases} 0.25 \le u_1 \le 0.75 \\ 0.35 \le u_2 \le 0.60 \\ 0.30 \le u_3 \le 0.35 \\ u_1 + u_2 + u_3 = 1. \end{cases} \quad (25)$$

After solving the above programming model and applying the procedure for partially known weights, we get the following results; $Z_1 \succ Z_2 \succ Z_3$ and Z_1 is the best available option. Thus the best alternative coincides with Li (2005).

1.5 Concluding remarks

Intuitionistic fuzzy sets play an important role in solving multiple attribute decision making problems, i.e., MADM problem. In this paper, we propose an R-norm IF entropy of type α which also considers the hesitancy degree alongwith membership and non-membership degree of the intuitionistic fuzzy sets. A new attribute weight determination method is proposed by using this entropy measure, which is used to solve the MADM problem. Two numerical examples are used to explain the MADM method effectively. The proposed multiple attribute decision making method is also be applicable to many other problems like project installation where there are so many factors to be cosidered, site selection, coding theory and credit evaluation.

REFERENCES

Atanassov, K.T. 1986. Intuitionistic fuzzy sets, Fuzzy Sets Syst. 20: 87–96.
Atanassov, K.T. 1999. Intuitionistic Fuzzy Sets, Springer Verlag, New York, NY, USA.
Bhandari, D., Pal, N.R. 1993. Some new information measures for fuzzy sets, Inf. Sci., 67: 209–228.
Burillo, P., Bustince, H. 2001. Entropy on intuitionistic fuzzy sets and on interval-valued fuzzy sets, Fuzzy Sets Syst., 118: 305–316.
Chen, T., Li, C. 2010. Determining objective weights with intuitionistic fuzzy entropy measures: A comparative analysis, Inf. Sci.,180: 4207–4222.
De Luca, A., Termini, S. 1972. A definition of non-probabilistic entropy in the setting of fuzzy set theory, Inf. Control., 20: 301–312.
Hung, W.L., Yang, M.S. 2006. Fuzzy entropy on intuitionistic fuzzy sets, International Journal of Intelligent Systems, 21: 443–451.
Kaufman, A. 1980. Fuzzy subsets Fundamental Theoretical Elements, Academic Press, New York, 3.
Li, D. 2005. Multiattribute decision-making models and methods using intuitionistic fuzzy sets, J. Comput. Syst. Sci., 70: 73–85.

Szmidt, E., Kacprzyk, J. 2002. Using intuitionistic fuzzy sets in group decision-making, Control Cybern., 31: 1037–1054.

Vlachos, I.K. and Sergiadis, G.D. 2007. Intuitionistic fuzzy information- Applications to pattern recognition, Pattern recognition letters, 28 (2): 197–206.

Wang, J., Wang, P. 2012. Intuitionistic linguistic fuzzy multi-critria deision-making method based on intuitionistic fuzzy entropy, Control. Decis., 27: 1694–1698.

Ye, J. 2010. Fuzzy dcision-making method based on the weighted correlation coefficient under intuitionistic fuzzy enviornment, Eur. J. Oper. Res., 205: 202–204.

Yager, R.R. 1979. On the measure of fuzziness and negation part I: membership in the unit interval, International journal of general systems, 5 (4): 221–229.

Zadeh, L.A. 1965. Fuzzy sets Information and Control, 8: 221–229.

Zadeh, L.A. 1968. Probability measures of fuzzy events, J. Math. Anal. Appl., 23: 421–427.

Zhang, Q., Jiang, S. 2008. A note on information entropy measure for vague sets, Inf. Sci., 178: 4184–4191.

Communication and Computing Systems – Prasad et al. (Eds)

A keypoint effective art network approach for handwritten word recognition

Neha Sahu & Akansha Singh
The NorthCap University, Gurgaon, India

Krishna Kant Singh
EEE, DCE, Gurgaon, India

Nitin Kali Raman
ACEM, Gurgaon, India

ABSTRACT: Patten Recognition is one of the major work areas of image processing. Character recognition is one of the most required pattern recognition applications. Recognizing characters which are handwritten increases the complexity. In this paper, a feature adaptive intelligent method is defined for handwritten word recognition. The work is here is divided in three main layers. In first layer, the noise robustness is improved by applying the Gaussian filter. In second layer, the key point analysis is performed using convolution and mathematical filters. In final stage, the ART network approach is applied to perform the recognition. The experimentation results shows that the presented work has provided the effective recognition rate.

Keywords: handwritten, words, ART, feature points, convolution

1 INTRODUCTION

Numerous automated applications majorly use pattern recognition. These applications provide the automated object identification and classification. The significance of these recognition systems are to provide online and offline object recognition and classification. One of such effective application area is to perform the character or the symbol recognition which includes Optical Character Recognition, Optical Mark Recognition, traffic sign or hand sign recognition etc. In this work, an optimized and intelligent method is presented for handwritten word recognition. In this section, the basic concept of classification model respective to the generalized pattern applications is defined. The section also included a brief study on different stages of a standard classification.

Image processing has its significance in most of the routine applications which are based on the visual system analysis and provide the recognition on different associated objects. Some of these application areas are shown in Figure 1.

Such applications provide human-computer interaction to digitize daily objects and their recognition in an automated way. Because of the vast domain areas and the associated objects availability, it becomes more difficult to perform the

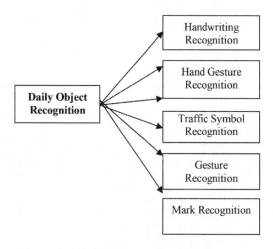

Figure 1. Daily object recognition applications.

recognition. The innovative object recognition can be applied so that the domain specification recognition will be performed. These kind of recognition system also dependent on the associated objects and the object interaction to the application.

Object classification is performed through series of steps; each step is defined as the sub stage for

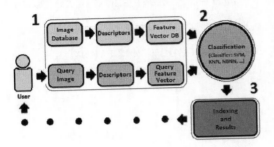

Figure 2. Classification model.

these recognition systems shown in Figure 2. Based on this model, for every raw image which is acquired, its quality is improved using filtration process which includes reduction of image noise or enhancing image features.

As shown in figure, the recognition model requires a filtered and improved image database. This database is considered as the training database on which the recognition process is applied. In same way, the testing image is accepted in the form of query image. If the input image and the databases images are normalized, then the extraction of relative descriptor is obtained. These descriptors are format specific, object specific and application specific. After obtaining such relative information, the feature extraction over the training set and input query image will be performed. After generating the featured training set and featured image, the classification is performed. There are number of classification approaches such as neural network, SVM etc. After this algorithmic approach, the recognized object will be identified. In the final stage, the analysis of work is done in terms of recognition rate analysis.

2 EXISTING WORK

The work defined by earlier researchers is defined on real life object recognition. One of such application area is handwritten character or word recognition. Different researchers provided different methodologies to perform the recognition on such datasets. In this section, the work defined earlier researchers is presented.

M. Egmont-Petersen presented a study on different classification model based on neural network modeling. Author defined the work to improve the recognition based on the relative feature map and the processing model. Author defined a series of process stages including the segmentation, image encoding and feature extraction. Author defined the work on six different models to perform the structural analysis and to perform the object recognition. Anna Bosch provided a work on classification based on real time object recognition. The

shape specific analysis is here performed based on the spatial pyramid based feature generation to extract the effective region and to perform the recognition based on the feature derivation.

S. Mikrut provided a neural adaptive model to improve the recognition process. Author defined the fragmented modeling for neural effective classification. Author used the Hough transformation as the adaptive feature generation so that the recognition over the image will be improved. Giorgio Giacinto has provided an architectural analysis based on the character specific recognition. Author defined the neural adaptive feature level analysis so that the error over the recognition process will be reduced. This kind of analysis is defined to perform the feature level assessment so that the recognition will be improved. Dan C. Ciresan has provided the GPU adaptive classification so that the feature adaptive hand written character recognition will be obtained. Author defined the effective classification and architectural specification so that the relative feature adaptive recognition will be obtained. S. Nagaprasad presented a work on multiple neural network models including the ART network, fuzzy modeling and neural network model to perform recognition on Arabic characters. Author defined the work on holistic method so that the moment specific analysis and object recognition will be obtained.

P Dan Ciresan presented a work on neural adaptive recognition based on the biological features so that the recognition so that the architectural specification will be obtained. Author defined the outlined featured based convolution filter analysis so that the tissue layer based visual feature generation is performed. Author defined the trained neuron adaptive model for object recognition is defined. Yuanqing Maya Lin presented a work on key feature extraction based on the parallel averaging scheme. Author defined the feature adaptive model for feature adaptive recognition. Author provided the SVM modeling to generate the image features and to perform the feature adaptive recognition. Jana Machajdik provided a work on feature sensitive level mapping to generate the color, shape and texture feature analysis. These features are defined based on the rating and later on the adaptive featured results are obtained. Kamal R. Al-Rawi has provided the work on design model to perform the classification based on the ART network model. Author provided the rate and accuracy adaptive system modeling to perform the recognition and classification. Robert L. Harvey provided a work on realistic model so that the effective system design based on the adaptive system features will be obtained. This kind of modeling is proposed and experimented by different researchers on different training and testing sets. Author defined the recognition model using different associated approaches.

3 PROPOSED METHOD

In this paper, a feature Adaptive Resonance Theory network based model is defined for handwriting recognition. ART are capable of developing stable clusters of arbitrary sequences of input patterns by self organizing. ART network are well suited in problems where learning of large and evolving databases is required. Thus in case of handwriting recognition this network will give good results. The flow chart of the proposed model is shown in Figure 3.

3.1 Input data

The input to the system is the grayscale image of the document for which handwriting recognition is to be done.

3.2 Gaussian smoothing

The scanned image is often affected by noise. This noise occurs generally due to the scanner quality or some other artifacts caused during writing. Thus in this paper we use a Gaussian filter to remove the noise. The preprocessed image can then be used for further processing. The Gaussian filter eliminate noise and other illumination effects.

Let the input image be I_D. Then the Gaussian filter be applied using equation 1

$$I_{PD}(x, y) = I_D(x, y) * g(x, y) \tag{1}$$

where * denotes convolution and g(x, y) is a Gaussian function

$$g(x, y) = \frac{1}{2\pi\sigma^2} e^{-\left(\frac{x^2+y^2}{2\sigma^2}\right)} \tag{2}$$

where σ is standard deviation.

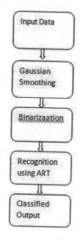

Figure 3.   Flowchart of proposed model.

3.3 Binarization

The preprocessed image is a gray scale image thus before feeding this image to the network we convert it into a binary image using OTSU's thresholding.

$$I_D(x, y) = 255 \text{ if } I_D(x, y) \geq \eta$$

$$I_D(x, y) = 0 \text{ if } I_D(x, y) < \eta$$

where η is the threshold value.

3.4 Recognition using ART network

ART networks are of three types, including:

1. ART-1 1986 that can cluster only binary inputs
2. ART-2 1987 that can handle gray-scale inputs, and
3. ART-3 1989 that can handle analog inputs better by overcoming the limitations of ART-2

The basic structure of ART network comprises of three different layers which include Input layer, Output layer and Reset layer. The processing of all the given inputs is done at the Input processing layer, the output layer deals with the cluster units where as the decision of placing the similar patterns on the same cluster is done by the reset layer through reset mechanism.

The algorithm for the ART network is as follows:

1. Read the number of input units as n, output units as m and learning trials is carried out for all the P input patterns $S_{p=1...P}$. Initialize u_{ij}, d_{ij} and V. u_{ij} being the bottom up weight between input layer I to output layer j, where as d_{ij} being the top down weight between output layer j to input layer i. Set p = 1.
2. Set $O_j = 0$ and $I_i = S_{pi}$, Calculate $S_p = \sum_{i=1}^{n} S_{pi}$
3. $\forall j = 1...m$ if $O_j \neq -1$, then $O_j = \sum_{i=1}^{n} I_i u_{ij}$, where O_j represents jth neuron in output layer, I_i represents ith neuron in Input interface layer and S_{pi} represents input pattern S_p connected to ith neuron of input interface layer.
4. Find J, where $O_J \geq O_j \forall j = 1...m$.
5. If $O_J = -1$ and all other nodes are inhibited, then the pth pattern cannot be clustered.
6. Calculate $I_i = S_{pi} d_{Ji}$ and $I = \sum_{i=1}^{n} I_i$
7. If $\frac{\|I\|}{\|S_p\|} \geq v$, update the weights for unit J as $u_{ij} = \frac{\alpha \times I_i}{\alpha - 1 + I}, d_{Ji} = I_i$, If p = P, end the process otherwise increment p as p = p + 1 and go to step 3, v here indicates vigilance parameter and α indicates learning rate.
8. Otherwise set $O_J = -1$ and move to step 4.

Here Figure 4 is showing the main process model relative to this recognition method.

As shown in the Figure 4, the work is here defined in two main sub stages. In fist stage, the key features from the images are extracted using

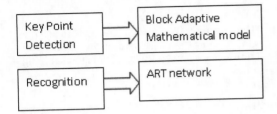

Figure 4.   Process model.

Table 1.   Sample set 1: 5 letter words.

Type of sample set	Handwritten words
Number of characters in each word	5
Sample set	200 words
Completely recognized words	153
Accuracy	76.5%

Table 2.   Sample set 2: 4 letter words.

Type of sample set	Handwritten words
Number of characters in each word	4
Sample set	200 words
Completely recognized words	159
Accuracy	79.5%

Table 3.   Sample set 3: 3 letter words.

Type of sample set	Handwritten words
Number of characters in each word	3
Sample set	200 words
Completely recognized words	182
Accuracy	91%

Table 4.   Sample set 4: 2 letter words.

Type of sample set	Handwritten words
Number of characters in each word	2
Sample set	200 words
Completely recognized words	196
Accuracy	98%

mathematical modeling. For this block adaptive convolution filters are applied. This model identifies the effective key points based on the structural analysis. After generating the key features from the image, the ART network based classification is applied to perform the recognition. The model is here applied on different dataset. The recognition modeling for different experimentations is shown in section IV.

Table 5.   Sample set 5: 1 letter.

Type of sample set	Handwritten letter
Number of characters in each word	1
Sample set	200 words
Completely recognized words	200
Accuracy	100%

4   RESULTS

The stated algorithm is implemented with the help of Neural Network Toolbox and tested over different sets of data in Matlab. Following table shows the result for different datasets.

5   CONCLUSION

In this work, a feature adaptive model is defined to perform the hand written character and word recognition. The presented model is noise robust and used the structural point based recognition. ART network is used as main classifier. The recognition is applied on different datasets. The recognition rate obtained from the work is quite effective. In future the work can be applied on more complex dataset such as overlapped image datasets.

REFERENCES

Anna Bosch, "Image Classification using Random Forests and Ferns".
Chergui Leila, "Neuro Fuzzy for Arabic Handwrittren Recognition System".
Dan Ciresan, "Multi-column Deep Neural Networks for Image Classification".
Dan C. Ciresan, "Flexible, High Performance Convolution Neural Networks for Image Classification", Proceedings of the Twenty-Second International Joint Conference on Artificial Intelligence
Egmont-Petersen M., "Image processing with neural networks—a review".
Giorgio Giacinto, "Design of Effective Neural Network Ensembles for Image Classification Purposes".
Jana Machajdik, "Affective Image Classification using Features Inspired by Psychology and Art Theory".
Kamal R. Al-Rawi, "SUPERVISED ART-II: A New Neural Network Architecture, With Quicker Learning Algorithm, For Learning and Classifying Multivalued Input Patterns", ESANN'1999 proceedings—European Symposium on Artificial Neural Networks.
Mikrut S., "Neural Networks In The Automation".
Nagaprasad S., "Spatial Data Mining Using Novel Neural Networks for Soil Image Classification and Processing", International Journal of Engineering Science and Technology.
Robert L. Harvey, "A Neural Network Architecture for General Image Recognition".
Yuanqing Lin, "Large-scale Image Classification: Fast Feature Extraction and SVM Training".

Communication and Computing Systems – Prasad et al. (Eds)
© 2017 Taylor & Francis Group, London, ISBN 978-1-138-02952-1

A novel technique to isolate and detect byzantine attack in wireless sensor networks

Shivali Goyal, Gurdeep Kaur & Parminder Singh
Department of Information Technology, CEC Landran, Mohali, India

ABSTRACT: The wireless sensor networks are the type of network in which sensor nodes can sense environmental conditions and it is deployed on the far places like a forest, deserts etc. In such type of network, battery consumption is a major issue as on such places creates problem to recharge or change the battery in the network. To reduce battery consumption of these sensor nodes, LEACH protocol is used which is energy efficient. Due to decentralized nature of the network, may malicious nodes join the network which is responsible to trigger an attack and passive type of attacks in this work, LEACH protocol performance is analyzed using misdirection attack. The NS2 simulator is used for simulation and results shows that energy, packet loss due to which byzantine attack is triggered in LEACH protocol.

Keywords: Byzantine attack, AODV protocol, WSNs

1 INTRODUCTION

Wireless sensor network is a collection of tiny particles that are deployed in hostile environment. To analyze the environmental conditions use the independent devices. A WSN system used as a wireless media or connectivity. In this network we extend the size of the network according to the Owner needs. Energy efficient technique applied on the wireless sensor network which improves the performance of the network. It is best suitable in research community of military and civilian (Aditya et al. 2014). In Figure 1 architecture of a clustered wireless sensor. Network represents that cluster, cluster heads, base station and sink. In cluster when sensor nodes are senses the data then it accumulates the data and delivered to the sink or base station directly or through cluster heads.

For working efficiently and good performance of the network used the protocol stack (David et al. 2008).

Sensor/Remote Remote nodes have multiple work in a network. These nodes are sensing such as; data processing and depository; transmitting.

Clusters: It partitioned the wireless sensor network in a efficient manner. For easily communication we break down the clusters which create jamming in the network (G. Padmavathi & Mrs. D. Shanmugapriya, 2009).

Cluster heads: After breaking the cluster these cluster heads are forming in the network which is senses the nodes in the clusters. They can accumulate the data and provide reliable communication schedule of a cluster.

Base Station: It works as a router in the network. It provides the efficient and reliable communication between the sensor nodes and the end user.

End User: In sensor network all information which are used is very important. Because these are used in various applications such as wildlife monitoring, militarily command, intelligent communication etc. [4].

1. Various Challenges of WSN

1. Routing: We can reduce the energy consumption by using various techniques like data aggregation, clustering, data-centric methods, etc. The routing protocols are as follow:

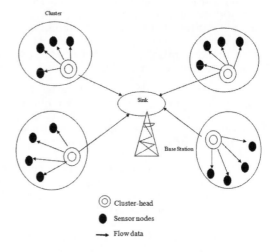

Figure 1. Architecture of a clustered WSN.

- Flat networks: In this network equal nodes are used. Those have flat addressing scheme plays the same role having no logical hierarchy. The example is Routing Information Protocol (RIP).
- Hierarchical networks: In this network nodes are divided into tiny groups called clusters. These clusters have each cluster head which is the coordinator of other nodes. These CHs perform data aggregation so that energy inefficiency may be reduced. It has major advantages of scalability, energy efficiency, efficient bandwidth utilization, reduces channel contention and packet collisions.
- Location-based Routing Protocols: In this type of routing protocols, sensor nodes communicate on the basis of the location of each node with another node. The distance between two neighboring nodes can be estimated by incoming signal strength from the source or using GPS (Hero Modaras et al. 2011).
  Energy Consumption: Sensor nodes are dependent on battery power.
- Energy conservation: Sensor network are placed on hostile environment so replacing the battery is quite impractical. Hence energy conservation and management is a major issue to resolve in wireless sensor network.
- Scalable and flexible architecture: It is required that the network must prevent its scalability. In wireless sensor network we extend the size of the network but we take care about that there is no effect of routing and clustering techniques.

2 SECURITY ATTACK

There is various security mechanisms that helps to prevent the network form the vulnerable attacks. There are various attacks in the network but all are classified into two forms X.800 and RFC 2828, is in terms of Passive attacks and Active attacks. Security is very essential in wireless sensor network. In this type of network alarm has been set when any unwanted actions are performing in the network.

- Passive Attacks: In this attack communication channel is analyzing and sensing by an unauthorized attacker are known as passive attack. In passive attack interchange the data of the node without disturbing the communication from one node to another node.
- Active Attacks: In this attack when an unauthorized attacker converts the data while sensing the communication channel is known as active attack. It can be affirmed that the attack denotes the disturbance of the normal functionality of the network by information interruption and modification etc.

2.1 Active type of attack
There are several active classes of attacks that may ruin the network and the wireless sensor network. These are various attacks that are occurred while transmitting the messages and communication.

2.1.1 Black hole attack
This attack occurs when an attacker catches and attacker re-plan with pair of nodes in the network to block the packets which they receive instead of forwarding towards the base station. Important Event information does not reach the base stations. In the presence of a black hole, attack throughput of the network becomes very low and end-to-end delay increases (Juby joseph & Vinod P, 2014).

2.1.2 Wormhole attack
Attacker records the packets in bits at one location in the network and tunnels those to another location. The tunneling or retransmitting of bits could be done selectively. The attack does not require compromising a sensor in the network rather; it could be performed even at the initial.

2.1.3 Sinkhole attack
In this attack an attacker attracts the traffic as much as possible toward compromised. The main issue of this attack is that it can be used to set up various active attacks on the traffic that is routed through it. These verities of active attacks increase multi-fold especially when these are carried out in collusion. Sinkhole attacks typically work by making a compromised node. In this attack nodes are making fake ID in the network and transform all the data to other nodes in form of packets. This node performs work as a member of the network. (N.ShivKumar & Dr. G. Gunasekaran, 2013).

2.1.4 Gray hole attack
In gray hole attack normal nodes works very unpleasant way which shows itself as a normal node and takes part in the transmission of a packet from packets. This unpleasant node drops the selected packets and only transmits the left packets to the neighbor node.

2.1.5 Byzantine attack
In this attack, a intermediate compromised node carries out attacks such as creating collision forwarding packets on non-optimal paths, routing loops, and dropping packets selectively which result in interruption or dreadful conditions of the routing services. This kind of failures is not easy for identification, since the network seems to be operating very normally in the view of the user. It is an active attack which is performed at Network layer (Ping Yi et al. 2012).

Table 1. Attacks on different layers in wireless sensor network and DOS defenses.

Layer	Attack	Denial-of-service-defense
Transport Layer	Flooding De-synchronization	Client puzzles
Network Layer	Routing attacks like black hole, sink hole, Sybil and wormhole	Authorize and monitoring
Data link Layer	Collision	Error-correction code
Physical Layer	Jamming attack, Tempring	Spread spectrum

If numbers of nodes are compromised then It Would be possible that these nodes can be interacting to gain additional advantages. This allow adversary to triggered more attacks. In Byzantine wormhole two adversaries collude. To create a tunnel to exchange messages and create shortcuts. This tunnel can be built either using existing infrastructure adhoc network or by private communication channel like radios and antennas. The adversaries can send route request and create path in the network and tunnel packets through non-adversaries to execute the path. The adversaries can use the low cost appearance in the wormhole network. It creates the non optimal paths and compromised with the neighboring nodes for passing the information to the other network. Information is passing in the form of packets which deteriorates the network. The byzantine wormhole attack is very dangerous attack that can be performed even if only two nodes have been compromised. It is necessary in this network is that there is only one base station which works as a router in the network.

3 LITERATURE REVIEW

In this paper (Aditya Vempathy et al. 2014) author proposed the design of novel message or technique to detect and isolate byzantine attack in wireless sensor network. Multiple objects Metric (MoM) also defines in this paper. It defines four metrics in the network. When there is availability of MoM in the network then there is changing the position of base station in the network. Due to changing the position of base station in the network creates a non optimal routing and selects forwarding packets. In this paper [2], describes a new form to improvement of confidentially and integrity in multi-hop transmission or code. It gives a protocol and algorithm to integrate confidentially in multi-hop tramission or code. Sometimes it creates jamming and collision in the network which degrades the network performance. This algorithm compares and applies on the network to decrease the multi-hop transmission in the network and improves the efficiency of the network with the various co-efficient. In this paper (G. Padmavathi & D. Shanmugapriya, 2009) describes the analysis performance of Proactive and reactive protocol in the network. Both protocol have work on AODV and DSDV routing protocol. This is based on distance routing which locate the address or calculate the distance between the nodes. In this paper (Hailun Tan et al. 2009) provides the mechanism to defeat many vulnerable attacks and threats using encryption method. To identify the errors alert devices are put in the network so that administrator triggers the attack when alarm raised in the network. In this paper (Hero moda et al. 2009) proposes a algorithm to locate the compromised node in the network by Random Linear Network Coding. This algorithm helps to find the attacker nodes which are compromised in the network. To locate the addresses of these nodes used the shift scheme. This scheme decreases the probability of compromised node in the network. When there are no compromised nodes then there is no loss of information in the packets. In this paper (Jing Deng et al. 2005) describe of the security also depend on the hardware constraints hardware and software are set up on this network and it must be excellent performance of this network. When designing the network it should be focused on the security of WSN. In this paper (Juby Joseph et al. 2014) describes the distributed event detection algorithm. This is a novel algorithm which is based on stastical approach. In this malicious nodes sends the false information to the other nodes which creates a routing loops in the network and becomes the performance of the network is low. For overcome this problem Gaussian approximation is applied on this network. In this paper (N shiv Kumar et al. 2013) describes the issues of designing in FDR-based distributed detection in the presence of byzantine. This was proposed by author varshney in 2011. They analyze the performance of the network by the fusion centre. Byzantine attack is that attack which is compromised in the network and also become malicious node in the network. The malicious node sends wrong information to the fusion centre. Global detection helps to find that there is wrong information which is send fusion centre. So we observed that there is an attacker in the network. The other method is that by fraction values. Fraction values are set in each node while communication. If the fraction value is large then there is a malicious node in the network. Adaptive algorithm is proposed to demonstrate the simulation result before and after the attack in the

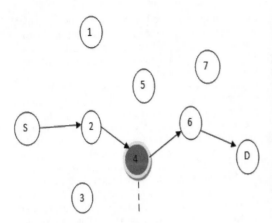

Figure 2. Packet drop by malicious node.

network. Adaptive algorithm also applies on the fusion centre to check the performance of the network. (N Shiv Kumar et al. 2013).

4 PROPOSED METHODOLOGY

The malicious nodes are responsible to trigger various type of security attacks which are broadly classified as active and passive attacks. The major active type of attacks is black hole, wormhole, sinkhole and Sybil attack which reduce network performance at high level. In this work, byzantine attack is detected and malicious nodes are isolated from the network. In this work, novel technique is proposed which is responsible to detect and isolate malicious nodes from the network. The proposed technique is based on trust values. In this techniques adjacent nodes of each node is calculated and nodes which can change its identification its trust value reduced according to time. The nodes which have least trust value is detected as malicious nodes in the network As explained in the flowchart, the sensor network is deployed randomly and it is assumed that some malicious nodes are in the networks. The first step of the technique is to gather in information about the location and adjacent nodes of each node. The second step is the assign trust values to each node. The node which change identification have different adjacent nodes each time, this information will reduce the trust value of the node. The node which have minimum trust value will be detected as the malicious nodes and it will isolated from the network.

As Figure 2 illustrate that there are number of nodes in the network and data is transferred from source to destination. Node 4 becomes malicious node and trigger byzantine attack in the network. This attack is responsible for the packet drop from the network.

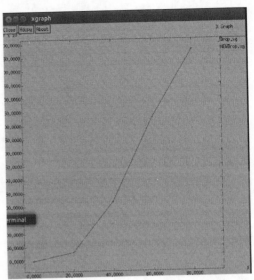

Figure 3. Packet drop graph.

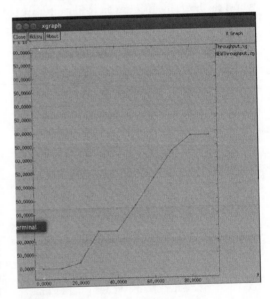

Figure 4. Throughput graph.

4 EXPERIMENTAL RESULTS

The whole scenario is implemented on NS2.

As shown in Figure 3, the packet drop of existing and proposed scenario is shown. It is shown that packet drop in the existing scenario is more as compared proposed scenario due to isolation of byzantine attack in the network

As shown in Figure 4, the throughput of the proposed technique and existing scenario is shown.

Table 2. Simulation table of energy.

Parameter	Proposed approach	Byzantine attack
Minimum Packet loss	10.000	30.000
Average Packet Loss	15.000	50.000
Maximum Packet Loss	20.000	65.000

Table 3. Simulation table of packet loss.

Parameter	Proposed approach	Byzantine attack
Minimum Throughput	30.000	10.000
Average Throughput	50.000	15.000
Maximum Throughput	79.000	20.000

Table 4. Simulation table of throughput.

Parameter	Proposed Approach	Byzantine Attack
Minimum Energy	5	25
Average Energy	9	30
Maximum Energy	11	45

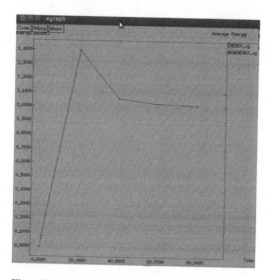

Figure 5. Energy graph.

In the proposed scenario throughput is more due to isolation of byzantine attack in the network

5 CONCLUSION

In this work, it has been concluded that LEACH protocol decreases the energy conservation of wireless sensor networks. Due to decentralized nature of WSN bidirectional attack is possible in the network which is triggered by the malicious node. The simulation results show that energy consumption, packet loss increase due to byzantine attack. The proposed scheme is based upon trust values. The node which has least trust value is considered as a malicious node. In future proposed work will be implemented wormhole attack in the network.

REFERENCES

Aditya Vempaty, Priyadip Ray, Pramod k. Varshney (2014) proposed in their paper, "False Discovery Rate Based Distributed Detection in the Presence of Byzantines", IEEE TRANSACTIONS ON AEROSPACE AND ELECTRONIC SYSTEMS, VOL. 50, NO. 3 JULY 2014.

David R. Raymond, Scott F. Midkiff, "Denial-of-Service in Wireless Sensor Networks: Attacks and Defenses", Published by the IEEE CS no 1536–1268/08/ 2008.

Hailun Tan, Diethelm Ostry, John Zic, Sanjay Jha, "A Confidential and DoS-Resistant Multi-hop Code Dissemination Protocol for Wireless Sensor Network", ACMWiSec09, Zurich, Switzerland, March 16–18, 2009.

Hero Modares, Rosli Salleh, Amirhossein Moravejosharieh, "Overview of Security Issues in Wireless Sensor Networks", IEEE 3rd International Conference on Computational Intelligence, Modelling & Simulation, 2011.

Jing Deng, Richard Han, and Shivakant Mishra, "Defending against Path based DoS Attacks in Wireless Sensor Networks", ACM SASN'05, November 7, 2005.

Juby Joseph, Vinodh P Vijayan," Misdirection Attack in WSN Due to Selfish Nodes; Detection and Suppression using Longer Path Protocol" International Journal of Advanced Research in Computer Science and Software Engineering, Vol 4, pp. 825–829, July 2014.

ShivKumar, N., G. Gunasekaran, "The Quality of Service Support for Wireless Sensor Networks", International Journal of Advanced Research in Computer Science and Software Engineering, pp. 297–302, January – 2013.

Padmavathi, G., Mrs. D. Shanmugapriya, "A Survey of Attacks Security Mechanisms andChallenges in Wireless Sensor Networks", International Journal of Computer Science and Information Security, Vol. 4, pp. 1 & 2, 2009.

Ping Yi, Ting Zhu, Qingquan Zhang YueWu, Jianhua Li, "Green Firewall: An Energy-Efficient IntrusioPrevention Po Wah Yau, Shenglan Hu, Chris 1. Mithell, "Malicious attacks on ad hoc network routing protocols", International Journal of Computer Research Vol 15 Issue 1, 2007. Mechanism in Wireless Sensor Network" IEEE, vol 10, pp.10–19, 2012.

Raj Kumar, Dr.Mukesh Kumar "LEACH: Features, Current Developments, Issues and Analysis" International Journal of Computer Science and Communication Engineering, Vol 1, Oct – 2012.

Roshan Singh Sachan, Mohammad Wazid, Avita Katal, D P Singh, R H Goudar "Misdirection Attack in

WSN: Topological Analysis and an Algorithm for Delay and Throughput Prediction" IEEE, 2012.

Roshan Singh Sachan, Mohammad Wazid, Avita Katal, D P Singh, R H Goudar, "A Cluster-Based Intrusion Detection and Prevention Technique for Misdirection Attack inside WSN", IEEE International conference on Communication and Signal Processing, April 2013.

Ruchita Dhulkar, Ajit Pokharkar, Mrs. Rohini Pise "Survey on different attacks in Wireless Sensor Networks and their prevention system", International Research Journal of Engineering and Technology, Vol. 02, pp. 1067–1072, Oct-2015.

Yi-Ying Zhang, Xiang-zhen LI, Yuan-an LIU, "The detection and defense of DoS attack for wireless sensor network", Elsevier Journal of China Universities of Posts and Telecommunications, Vol 19, pp. 52–56, Oct-2012.

Communication and Computing Systems – Prasad et al. (Eds)

An improved node selection algorithm for routing protocols in VANET

Aruna Sharma & Simar Preet Singh
DAV University, Jalandhar, India

ABSTRACT: Vehicular Ad Hoc Networks (VANET) are an evolution of MANETS and however the routing protocols of MANETs cannot execute on form in VANETs because of the distinct properties. Thus routing protocols need to be modified to the VANET characteristics. This paper addresses the issue of next hope vehicle selection in communication and the flooding of broadcast messages in the network. To deal with these issues, the paper proposes a hybrid scheme of QoS aware node selection algorithm and the Multipoint Relay selection algorithm. The QoS metrics include throughput, average time delay and the total communication time. The simulation results depicts comparative execution analysis of proposed method using simple existing QASA scheme and prove the proposed scheme to be better in terms of considered QoS parameters.

1 INTRODUCTION

The concept of incorporating wireless communication in vehicles is not a new one and dates back to 80 s. But in past few years, due to the strong urge to improve vehicle and road safety, traffic efficiency, and driving convenience, VANETs have gained huge amount of well-deserved attention. VANETs represents the special case of mobile ad hoc networks consisting of vehicles as the communicating entities, and have unfixed or no infrastructure. Over the last few years, many research endeavours have been done to explore different issues related to V2I, V2V, and VRC areas due to their vital part in Intelligent Transportation Systems (ITSs). In VANETs, the vehicles are outfitted by intelligent devices as well as interfaces called On—Board Units (OBUs) which can talk to other on board units plus the road side framework stated by road side units, located at critical points on road, over a single or multi-hops to exchange information about the traffic and vehicle status. The correspondence between OBUs is known as Vehicle-to-Vehicle or Inter vehicle correspondence and the correspondence in the middle of OBU and RSU is termed as Vehicle-to-Infrastructure correspondence.

A vital design aspect of VANET is to build up an efficient, reliable and secure routing protocol. Besides this, it is must to consider the Quality of Service too in ad hoc networks that guarantees to quick plus efficient carting of ITS information and multimedia data. Routing protocols can be considered as the focus of QoS methods. There is a need of framework that not only provides smoother driving in traffic congestion but also increases

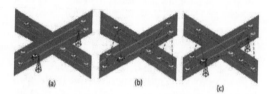

Figure 1. Network architecture in VANETs.

communication capability between vehicles. The main focus lies upon routing and forwarding problems, since searching routes through the required destination in these disconnected areas is called the most compelling problem.

2 RELATED WORK

(Luciana Pelusi et al, 2006) discussed about the opportunistic networking and detailed down a detailed classification about the primary routing as well as forwarding techniques in this tricky environment. The paper presents a survey of some case studies regarding the opportunistic networks. Opportunistic networks enable the mobile nodes to connect with one another devoid of the presence of any route. Opportunistic networks allow the source and destination nodes to exchange messages even without being connected to same network at any same instant. This result into an additional delay in message delivery due to messages being buffered waiting for the path availability through the destination. Hence, routing as well as forwarding issues is main research concerns addressed in the paper. Saleh Yousefi et al in (Saleh et al 2008) offered an analytical method to explain transmission in VAN-ETs. There are many challenges in VANETs that need appropriate attention in every traffic phase where transmission is found reasonable in forced flow state during lowers at weak load as in free-flow phase this can become difficult to send messages to other vehicles due to disconnections. A common model has been used in vehicular traffic scheme that shows some particular point in space depicting cars travelling it divided through time intervals having random time period exponentially scattered. The distance dissipation among cars is done at steady state facilitating the use of techniques originated by the queuing theory. The Laplace Transform involves the probability dissipation having connectivity distance, explicit expressions about estimated connectivity distance, plus probability dissipation and estimation of total count of cars in a platoon. (Ashish Agarwal et al, 2008), addresses the access point placement problem in a hybrid vehicular network that involves multi-hop interaction done with moving automobiles facilitated with access points. This paper considers distinct traffic densities as well as access point partitions beneath the assumption of delay tolerant messaging. This paper involves an extension of the previous infrastructure-less model of message generated in delay tolerant automobile networks providing infrastructure units to vehicles. It clearly demonstrates that beneath delay tolerant networking supposition, less delay and high propagation rates may be achieved for less vehicular traffic densities of 20 vehicles/km while for a route based messaging scheme, same execution could be observed at 40 vehicles/km. (Anna Maria Vegni et al), presented the limitations of V2V and V2I protocols with the resultant challenges in VAN-ETs. Though V2V being a traditional protocol is considered as an utmost appropriate way to deal with low-latency short range vehicular networks, however in some specific situations like unexpected topology change, vehicle speed and in less dense or completely disconnected scenarios, V2V does not prove to be an appropriate scheme. In case of V2I, the main limitations appear in case of some particular requirements of vehicular applications and the performance lays on particular wireless technology being considered for the mentioned RSUs. To address these issues and to use benefits of two of the protocols, a hybrid communication paradigm called Vehicle-to-X (V2X) has been proposed within the paper. V2X involves the combination of the two vehicles V2V and V2I where every vehicle has the ability to shift from V2V to V2I, as well as vice versa, based upon protocol switching decision technique. Ahmad (Mostafa et al, 2014) comes up with a best possible approach to choose the next-hop forwarder node using bridging approach. The paper proposes a quality of service

based node Selection routing protocol for opportunistic networks in VANETs. QASA exceedingly upgrades the QoS measurements for system as it empowers one roadway side vehicles to effectively pick a automobile from the other side. QoS measurements utilized as a part of this paper are all out correspondence time, throughput, plus parcel end-to-end time delay. Shigang Chen et al (Ihn-Han & Stephan, 2013), came up with a distributed QoS routing scheme for ad hoc networks which chooses network route with necessity in a dynamic multi-hop mobile environment. Since many routing techniques rely on precise state information, proposed algorithm mitigates this drawback. The proposed ticket-based scheme succeeds to build up a balance between single-path routing algorithms and the flooding algorithms. It allows multipath routing at the same time avoiding flooding. The proposed scheme intends to achieve the best possible performance with negligible overhead by using a limited tickets and making intelligent hop-by-hop path selection. (Ihn-Han Bae et al, 2013) addresses the idea of the safe message broadcasting in VANETs.

In this regard, paper presents a hybrid intelligent broadcast protocol named Hi-CAST. Proposed protocol includes the use of delay as well as probabilistic broadcast protocols along token protocols to deliver the alert messages effectively. In addition to this, considering the fading factors, four variants of Hi-CAST are also proposed. The performance comparison of the Hi-CAST with other alert message broadcasting algorithms proves Hi-CAST to be superior to other algorithms. Ahmad Mostafa et al [8], comes up with a novel reliable and low-collision packet-forwarding scheme known as Collision-Aware REliable FORwarding (CAREFOR) based on a probabilistic rebroadcasting. The essential thought is to decrease the quantity of rebroadcasts in the system. This minimizes the number of packets in the system, which leads to a lower collision probability and eventually improved throughput. CAREFOR is composed of two phases, a collision probability estimation phase, followed by a reliable forwarding phase. CAREFOR is different from other existing techniques as it accounts for the effect of the next-hop transmission in the rebroadcast decision. Radityo (Anggoro et al, 2011) drew attention towards a significant improvement provided by probabilistic relay for OLSR plus AODV performance beneath realistic VANETs environments. The paper brings out some challenges like broken links and other such circumstances that incur in real time vehicular environments. In order to attain a realistic scenario, VanetMobiSim tool has been used in paper. The routing protocols has been evaluated in terms of Packet Delivery Ratio (PDR) plus Routing Overhead (RO) below different kind of scenarios.

3 ROUTING CHALLENGES IN VANETS

The vehicles on a highway usually move in either direction on a bidirectional roadway. The high mobility of nodes makes the network topology highly dynamic especially while interaction of traffic of opposing lanes. The vehicular traffic density ranges from extremely sparse to dense traffic scenarios on the basis of the time of the day. The VANETs' topology and its dynamic behavior are decided by numerous factors such as the traffic density, the vehicles' speed and the heterogeneous network environment.

Opportunistic network is an interesting evolution in domain of ad hoc networks. Those networks enable mobile nodes to interact to one another even in the absence of any path connecting them. Routes are established dynamically the instant when message transmission is performed between the source and the target as well as each node may conciliatorily be chosen as upcoming hop only when it is liable to convey the information nearest to the target. Hence it comes up with an issue of additional delay since the data are buffered in network while holding for a way to target. Opportunistic networks can be extended in vehicular networks to achieve vehicular communication using V2V to share information.

Bridging technique is used to link partitions of different groups of vehicles travelling along same direction where message dissemination occurs through dynamically occurred links with an issue depicting some vehicle moving in a single direction may choose some other vehicles present on the other side of the roadway.

Considering a starting node 's', a target node 'd', and a delay requirement D, the challenge in delay-constrained routing is identify a feasible route P from sender to target such that delay (P) < D (Ihn-Han, 2013).

The objective is to select the least costing path from multiple feasible paths.

4 PROBLEM IDENTIFICATION AND PROPOSED SCHEME

The numerous protocols designed for MANETs till date somehow are not appropriate for VANET as they are deprived of the high mobility constraints. The main problems faced by these protocols are route instability further leading to packet drops, improved overhead due to route repairs, less delivery ratios as well as more transmission delays. A challenging task in VANETs is defining appropriate set of quality of service parameters. Quality of Service actually is a set of service requirements to be fulfilled with the network while transferring a packet from a sender to its target. It is must to consider QoS requirements in protocols designed for VANETs for safety, emergency and multimedia services. To address these issues and meet all the requirements, a Quality of Service based protocol has been used to facilitate the next hop vehicle selection for communication.

In addition to this, flooding of broadcast messages ultimately leading to traffic overhead is also a potential issue. Hence along with QoS aware node selection approach, MPR algorithm has also been used to deal with the traffic control problem in the network.

4.1 QASA

For quality of service based node selection algorithm, a two lane vehicular network is considered where the vehicles are driven in opposite direction. The vehicular networks possess dynamic topology and the different mobility patterns result into creating vehicles' clusters.

When vehicle V2 is ready for data transmission, there are two candidate recipient vehicles V1 and V3 on W lane within source's transmission range. The most ideal applicant would have the capacity to give the best QoS measurements with best throughput and slightest bundle transmission delay. The probability of a automobile to be a next-hop forwarder is defined as follows:

$$prx = \Pr[d^* \le Ri] \cup \Pr[\Delta T(Ri) > \tau]$$

here d^* [m] represents inter-vehicular distance between next node to the source, $\Delta T(Ri)$ represents the desired time period for the upcoming-hop vehicle in the range of the transmission area Ri taking $i = [1, 2, 3]$), where τ [s] is the mentioned threshold ensuring a greater connectivity time period.

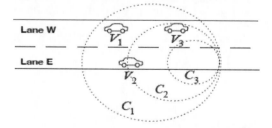

Figure 2. Demonstrates a source vehicle V2 travelling on Lane E and the division of its transmission range into smaller spherical transmission areas (that is taking C1 > C2 > C3) for next-hop selection with R1 as the radius for the circle C1; R2 for C2; and R3 for C3. These transmission spaces are the particular extents from where the starting vehicle may choose an upcoming-hop forwarder vehicle.

QASA recommends that restricting the selection extent to the shortest reach leads to improved throughput but result into unavoidable average time delay. On the other hand, the closer the vehicle gets, it fails to achieve the expected throughput but provides a reduced time delay. Hence it leads to a throughput—time delay trade off.

To manage this exchange off, QASA has to permit the vehicle to pick the farthest conceivable vehicle inside of the transmission range with accomplishing the QoS measurements including throughput of and the normal time delay for bundle transmission.

QASA works through a series of four phases:

- In phase 1, any source vehicle Vs on the east calculates the present status of vehicular thickness in its transmission limit which is further considered to choose the suitable span of the roundabout locale (i.e., C_i with i = 1, 2, 3) where the west path vehicles can react back to Vs;
- In phase 2, Vs shows Request-to-Broadcast (RTB) information to the automobiles accessible within its coverage limit. Request-to broadcast contains points of interest of vehicular thickness at the source hub, the GPS area of Vs, and the sweep processed from phase1;
- In phase 3, after receiving RTB information, all those automobiles evaluate that whether their location matches the needs given in the message. When it does not match, the packet is plunged, in other case the Reliable Forwarding (RF) probability is calculated as:

$$p = \frac{\exp \cdot -p \cdot (z-d)}{c} \cdot \frac{z}{z_i}$$

Here ρ [veh/m] drop speaks to the vehicular thickness, z and zi [m] speak to the transmission scopes of Vs and of the i-th neighboring vehicle individually, and $c \leq 1$ denotes the coefficient that is chosen for the impact of the rebroadcast probability.

Every vehicle choose the bundle retransmission by contrasting the RF likelihood with an impact limit (i.e., Thcoll), as

$$Thcoll = 1 - [\exp(-\rho z_i)]$$

When the RF value is found greater than the threshold, a Clear-to-Broadcast (CTB) information is broadcasted by the automobile. Otherwise, the vehicle does not rebroadcast further.

- In phase 4, on receiving the CTB messages, Vs determines which responding vehicle is the furthest. On deciding that, Vs transmits the bundle whose final location is to the chose vehicle from the west range. If the destination address does not match the IDs of recipient vehicles, in that case they simply drop the packet.

4.2 MPR algorithm

In addition to next hop vehicle selection, traffic control in data forwarding is also a concern. To address the traffic problem, OLSR proactive protocol with Multi Point Relay algorithm is used as the selection of Multipoint Relays provide efficient routing schemes. OLSR is a link-state proactive ad-hoc routing protocol that allows the nodes to identify the network topology and compute routes. It offers shortest-path routes for unicast, minimises the congestion of broadcast messages as well as diminishes drastically the load of control traffic. The goal of Multipoint Relays diminishes the congestion of broadcast messages in the network through curbing the replicated retransmissions locally. Multipoint Relays (MPRs) are subset of neighbours selected by each node to retransmit broadcast packets. This facilitates the neighbours not belonging to MPR aiming that reads the message devoid of retransmitting it, hence controlling the network traffic.

In the OLSR protocol, every node intermittently shows the data about its prompt neighbours which have chosen it as a MPR. Endless supply of this data, every node figures and overhauls its courses to every destination. In MPR broadcasting, a node retransmits a broadcast packet if and only if it was received the first time from a node for which it is an MPR.

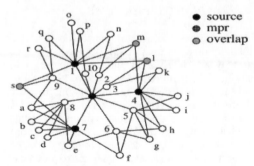

Figure 4. MPR selection.

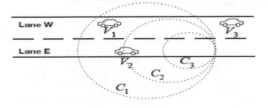

Figure 3. Depicts the settlement performance between the time delay as well as throughput with QASA.

A wireless network is formally stated by bidirectional undirected graph denoted as G (V, E). Any two vehicles say i as well as j allocate a bidirectional link (i, j) only if they can communicate and hear each other. Let N1 (u) be the neighbours of node u. Assume N2 (u) as two-hop neighbours of u (the neighbouring nodes of the neighbours of u which are not already neighbours of u).

The goal of MPR is to choose a littlest conceivable arrangement of nodes in the 1-neighborhood of u (N1(u), which covers the entire 2-neighborhood of u (N2(u)). This is achieved in two steps:

- In step 1, select node of N1(u) which cover confined purposes of N2(u), which are alluded as MPR1 (u).
- In step 2, select the node which covers the most elevated number of purposes of N2(u) among the nodes of N1(u) not chose at the initial step and go ahead till each purposes of N2(u) are secured.

4 RESULTS AND DISCUSSION

The proposed scheme has been simulated and the results have been presented in this section. The results obtained show the QASA performance considering parameters Average Time Delay, Throughput and total communication time versus the Vehicle density. The performance of the proposed scheme involving QASA with MPR has been compared with ordinary QASA with respect to the considered QoS parameters.

The graph clearly shows the average time delay is observed to be more in case of existing QASA scheme while the proposed scheme is showing less delay in the same ranges.

The graph clearly shows that in low ranges also, the proposed scheme is offering more throughput in comparison to the offered by existing one.

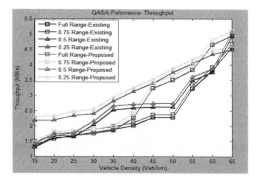

Figure 6. Shows the throughput analysis where high performance is produced having reduced transmission ranges.

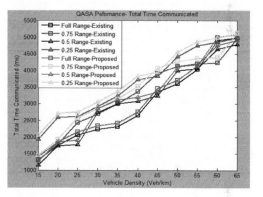

Figure 7. Shows the total time communicated in the period of simulation moves that shows the efficiency of method of restricting the area that lies in coverage.

The graph clearly shows that in low ranges also, the proposed scheme is offering more total communication time in comparison to the offered by existing one.

5 CONCLUSIONS

The paper introduces a hybrid scheme combining the QoS aware node selection algorithm with the Multipoint Relay selection scheme. QASA is the algorithm for forwarding node selection algorithm in opportunistic vehicular networks while MPR provides an effective routing scheme for efficient broadcast and controlling the traffic overhead. The proposed scheme addresses the next hope selection and traffic overhead issue at the same time. Results show comparative execution of proposed scheme against existing QASA scheme in different ranges.

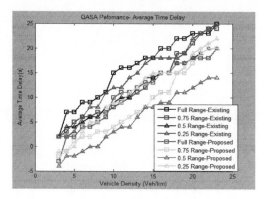

Figure 5. Shows average time delay in case of full range and the ranges with diameter 0.75, 0.5 and 0.25 of starting transmission area.

REFERENCES

Ahmad Mostafa, et al. 2013, "A probabilistic routing by using multi-hop retransmission forecast with packet collision-aware constraints in vehicular networks", 1570–8705.

Ahmad Mostafaa, et al. 2014, "QoS-aware Node Selection Algorithm for Routing Protocols in VANETs", Fourth International Conference on Selected Topics in Mobile & Wireless Networking, Procedia Computer Science 40 (2014) 66–7.

Ali Ouacha, et al. "Intelligent MultiPoint Relays Selection", Wseas Transactions on Communications, E-ISSN: 2224–2864, Volume 14, 2015.

Anna Maria Vegni & Thomas D.C. Little, "Hybrid vehicular communications based on V2V-V2I protocol switching", Int. J. Vehicle Information and Communication Systems, Vol. 2, Nos. 3/4, pp. 213–23.

Ashish Agarwal & Thomas D.C. Little 2008, "Access Point Placement in Vehicular Networking", IEEE Wireless Access for Vehicular Environments (WAVE).

Bernard Mans & Nirisha Shrestha, "Performance Evaluation of Approximation Algorithms for Multipoint Relay Selection", IEEE.

Grossglauser M. & D. N. C. Tse 2002, "Mobility increases the capacity of ad-hoc wireless networks", IEEE/ACM Transactions on Networking, vol. 10, no. 4.

Gupta, P & P. R. Kumar 2000, "The capacity of wireless networks", IEEE Transactions on Information Theory, vol. 46, pp. 388–404.

Ihn-Han Bael & Stephan Olariu, 2013 "Design and Evaluation of Hi-CAST and its Variants for Safety Message Dissemination in VANET", 13th International Conference on Control, Automation and Systems (ICCAS).

Juang, P, et al. 2002, "Energy-efficient computing for wildlife tracking: Design tradeoffs and early experiences with ZebraNet", ACM SIGPLAN Notices, vol. 37, pp. 96–107.

Luciana Pelusi, et al. 2006, "Opportunistic Networking: Data Forwarding in Disconnected Mobile Ad Hoc Networks", IEEE Communications Magazine, 0163–6804.

Pelusi L, et al. 2006, "Beyond MANETs: dissertation on Opportunistic Networking", IIT-CNR Tech. Rep., online available.

Pentland A, et al. 2004, "Rethinking Connectivity in Developing Nations", IEEE Computer, vol. 37, no. 1, pp. 78–83.

Radityo Anggoro1, et al. 2011, "An Evaluation of Routing Protocols with Probabilistic Relay in VANETs", 978-1-4577-0255-6/11© IEEE.

Ruchika Batra, et al. 2015, "Simulation Study of Optimization Techniques of OLSR Protocol in VANETs", International Journal For Advance Research In Engineering And Technology, Volume 3, Issue X, ISSN 2320–6802.

Saleh Yousefi, et al. 2008, "Analytical Model for Connectivity in Vehicular Ad Hoc Networks", IEEE Transactions on Vehicular Technology, Vol. 57, No. 6.

Shigang Chen & Klara Nahrstedt 1999, "Distributed Quality-of-Service Routing in Ad Hoc Networks", IEEE Journal on Selected Areas in Communications, Vol. 17, No. 8.

Communication and Computing Systems – Prasad et al. (Eds)
© 2017 Taylor & Francis Group, London, ISBN 978-1-138-02952-1

Set-point tracking and disturbance rejection capability analysis of controllers designed for SOPDT model

M. Chaturvedi, P.K. Juneja & G. Upreti
Graphic Era University, Dehradun, India

ABSTRACT: Disturbance rejection is one of the key performance benchmark for a controller. For an industrial process, disturbance can be internal or external and it is an important that it can be rejected effectively by an optimal controller design. In the present work, set-point tracking and disturbance rejection capability of the controllers have been evaluated by comparing the closed loop responses for a selected second order delayed model.

1 INTRODUCTION

Controller performance is one of the most significant control objectives of a typical control problem. A control system delivers high performance if it exhibits rapid and smooth responses to disturbances and less variation in set-point. PID controller settings specified for excellent disturbance rejection provides large overshoots for set-point tracking whereas controller settings providing excellent set-point tracking produces sluggish disturbance rejection (Seborg et al. 2004; Zhou et al. 2009).

Many industrial processes, in process control can be analyzed with the help of the model of the process under investigation. The dead time comes naturally in the process model where energy propagates from one place to another (Chaturvedi et al. 2013). This is the reason that the process industry commonly has FOPDT or SOPDT model. A general SOPDT model transfer function is given as:

$$G(s) = \frac{Ke^{-\theta s}}{(1+\tau_1 s)(1+\tau_2 s)} \qquad (1)$$

where K is process gain, θ is time delay, τ_1 is slow and τ_2 is fast time constant (Farkh et al. 2009; Chaturvedi et al. 2014).

Easy implementation and operation of PID controllers makes it extensively applicable in process manufacturing, to control various industrial processes (Visioli 2012). In the literature, lesser PID tuning techniques are available for Second Order plus Delay Time (SOPDT) model compared to First Order plus Delay Time (FOPDT) model (Chaturvedi et al. 2013; Panda et al. 2004). A typical PID controller transfer function is:

$$G_c(s) = K_c \left[1 + \frac{1}{\tau_1 s} + \tau_D s \right] \qquad (2)$$

Most of the PID tuning methods existing in literature lay emphasis on improving set-point tracking performance of the controller rather than disturbance. It is the demand of process control that the tuning methods should be simple and easy to implement which require little information and give moderate performance. A tuning method should be widely applicable (Chen et al. 2002)

For PID controllers, there are various tuning formulae existing in literature for PID controllers, viz. Åström and Häglund (Åström et al. 1995), Ziegler Nichols (ZN) (Ziegler et al. 1942), Cohen Coon (Cohen et al. 1953) and Tyreus Luyben (TL) (Tyreus et al. 1992). Internal Model Control (Morari et al. 1984) and Skogestad (Skogestad 2003) tuning techniques are some other important tuning techniques. Three important tuning formulae, applied in this work are shown in Table 1.

Table 1. PID controller settings for SOPDT process [Seborg et al. 2004].

Tuning parameters	Skogestad (τ_1 (8θ)	Ziegler Nichols	Tyreus Luyben
K_c	$\dfrac{0.5(\tau_1 + \tau_2)}{K\theta}$	$0.6\,K_{cu}$	$0.45\,K_{cu}$
τ_1	$(\tau_1 + \tau_2)$	$\dfrac{P_u}{2}$	$2.2 P_u$
τ_D	$\dfrac{\tau_1 \tau_2}{(\tau_1 + \tau_2)}$	$\dfrac{P_u}{8}$	$\dfrac{P_u}{6.3}$

2 METHODOLOGY

Fig. 1 shows the flowchart for systematic investigation of the presented analysis. Many distributed parameter systems are often modeled as SOPDT model, where a state variable in a process plant is a function of two or more independent variables [Chaturvedi et al. 2013].

The selected SOPDT system transfer function for present analysis is [8]:

$$G(s) = \frac{5e^{-4s}}{(1+20s)(1+8s)} \quad (3)$$

where, Dead Time (θ) = 4 seconds, Process Gain (K) = 5, Slow time constant (τ_1) = 20 seconds and Fast time constant (τ_2) = 8 seconds (Nancy et al. 1995).

3 RESULTS AND DISCUSSION

PID Controllers tuning parameters for the selected SOPDT process are shown in Table 2. Fig. 2 gives the evaluation of the disturbance rejection capability of the controllers designed using Skogestad and ZN tuning techniques. Response obtained in case of PID controller designed using Skogestad tuning technique has less settling time, one of the important steady-state characteristic but more maximum percentage overshoot in comparison to that in case of ZN tuning technique. Although the disturbance response has more peak amplitude but settles faster, in case of PID controller designed using Skogestad tuning technique based controller.

The comparison of responses for the disturbance rejection capability of the controller designed by Skogestad as well as TL tuning techniques is displayed in Fig. 3. Response exhibited by PID controller designed using Skogestad tuning technique has less settling time and maximum percentage peak overshoot in comparison with the response obtained by PID controller using TL tuning technique. The disturbance settles faster in case of Skogestad tuning technique while comparing with PID controller designed using TL tuning technique.

Fig. 4 displays the comparative analysis of the disturbance rejection capability of the controllers designed using ZN as well as TL tuning techniques. The response obtained in case of controller designed using ZN tuning technique has less settling time and maximum percentage peak overshoot while comparing with PID controller designed using TL tuning technique. The disturbance settles

Table 2. PID Controller Settings.

Tuning method	K_c	τ_I	τ_D
Skogestad	0.7	28	5.71
ZN	0.84	16.315	4.09
TL	0.63	71.78	5.18

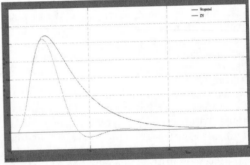

Figure 2. Disturbance rejection capability comparison for Skogestad and ZN techniques.

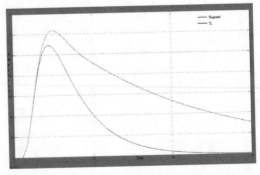

Figure 3. Disturbance rejection capability comparison for Skogestad and TL techniques.

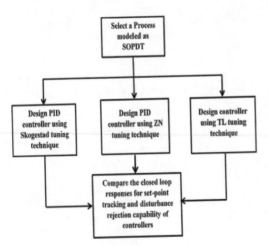

Figure 1. Flowchart for systematic investigation.

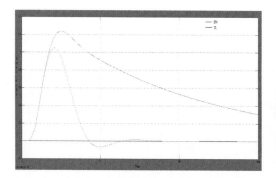

Figure 4. Disturbance rejection capability comparison for ZN and TL techniques.

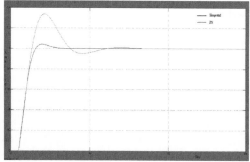

Figure 6. Set-point tracking capability comparison for Skogestad & ZN techniques.

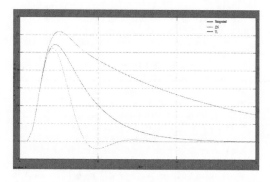

Figure 5. Disturbance rejection capability comparison for Skogestad, ZN and TL techniques.

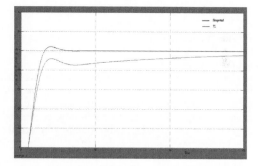

Figure 7. Set-point tracking capability comparison for Skogestad & TL techniques.

Table 3. Relative analysis of significant characteristics.

Characteristics →			
Tuning Methods ↓	Rise time	Settling time	Maximum%age overshoot
Skogestad	7.6	24.2	4.12
ZN	6.51	54.2	33.5
TL	11.6	216	0

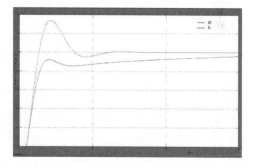

Figure 8. Set point tracking capability comparison for ZN and TL techniques.

faster in case of PID controller designed using ZN tuning method than PID controller designed using TL tuning technique.

Comparison of the disturbance rejection capability of the controller designed for Skogestad, ZN and TL techniques is displayed in Fig. 5. The response exhibits, settling time is least for PID controller designed with Skogestad technique, maximum percentage peak overshoot is least for PID controller designed with ZN technique. The disturbance settles fastest for PID controller designed with Skogestad technique.

Table 3 shows the relative analysis of significant characteristics. Fig. 6 displays the closed loop step

response comparison of PID controllers designed with Skogestad and ZN techniques. The closed loop step response comparison of the controllers designed with Skogestad as well as TL techniques is shown by Fig. 7. Fig. 8 gives the closed loop step response comparison of the PID controllers designed with ZN as well as TL methods. The closed loop step response comparison of the controllers designed with Skogestad, ZN & TL tuning techniques is displayed by Fig. 9.

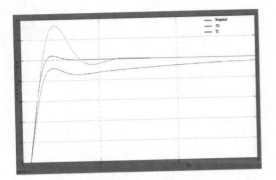

Figure 9. Set point tracking capability comparison for Skogestad, ZN and TL techniques.

4 CONCLUSION

In this paper, Skogestad, Ziegler Nichols and Tyreus Luyben tuning methods were employed for PID controller design for selected SOPDT process model. Disturbance response is best in case of PID controller designed using ZN tuning technique among all the three tuning techniques. PID controller designed using Skogestad tuning technique exhibits best set point tracking capability amongst the three tuning techniques.

REFERENCES

Åstörm, K. J. and Häglund, T. 1995. *PID Controllers: Theory, Design and Tuning.* Second Ed. Research Triangle Park. NC. Instrument Society of America.

Chaturvedi, M. and Juneja, P. 2013. Effect of dead time approximation on controller performance designed for a second order delayed model. *Proceedings of the 2013 International Conference on Advanced Electronic Systems, ICAES 2013.* Pilani: 313–315.

Chaturvedi, M., Juneja, P. K. and Chauhaan, P. 2014. Effect of Implementing Different PID Algorithms on Controllers Designed For SOPDT Process. *Proceedings of the 2014 International Conference on Advances in Computing, Communications and Informatics, ICACCI 2014.* Greater Noida: 853–858, September 2014.

Chen, D. and Seborg, D. E. 2002. PI/PID Controller design based on direct synthesis and disturbance rejection. *Ind. Eng. Chem. Res.* 41: 4807–4822.

Cohen, G. H. and Coon, G. A. 1953. Theoretical considerations of retarded control. *Trans. ASME.* 75: 827–834.

Farkh, R., Laabidi, K. and Ksouri, M. 2009. PI Control for Second Order Delay System with Tuning Parameter Optimization *International Journal of Electrical and Electronics Engineering,* 3(1): 1–7.

Morari, M., Skogestad, S. and Rivera, D. F. 1984. Implications of internal model control for PID controllers. *American Control Conference:* pp. 661–666.

Nancy, J. and Sell, 1995. *Process Control Fundamentals for the pulp and paper industry.* Atlanta, USA. Tappi Press.

Panda, R. C., Yu, C. and Huang, H. P. 2004. PID tuning Rules for SOPDT systems: Review and some new results *ISA Transactions.* 43: 283–295.

Seborg, D. E., Edgar, T. F. and Mellichamp, D. A. 2004. *Process Dynamics and Control.* 2nd Ed. New York. Wiley.

Skogestad, S. 2003. Simple Analytic Rules for Model Reduction and PID Controller Tuning. *J. Process Control.* 13: 291–309.

Tyreus, D. and Luyben, W. L. 1992. Tuning PI controllers for integrator/dead-time process. *Ind. Eng. Chem. Res.* 31: 2625–2628.

Visioli, A. 2012. Research trends for PID controllers. *Acta Polytechnica.* 52(5): 144–150.

Zhou, W., Shao, S. and Gao, Z. 2009. A Stability Study of the Active Disturbance Rejection Control problem by a Singular Perturbation Approach. *Applied mathematical Sciences.* Vol. 3(10): 491–508.

Ziegler, J. G. and Nichols, N. B. 1942. Optimum settings for automatic controllers. *Trans. ASME.* 64: 759–768.

Communication and Computing Systems – Prasad et al. (Eds)
© 2017 Taylor & Francis Group, London, ISBN 978-1-138-02952-1

An implication of multi-objective optimization in test case generation

Kavita Choudhary & Ankit Nahata
JK Lakshmipat University Jaipur, India

Shilpa
ITM University Gurgaon Haryana, India

ABSTRACT: Test case generation is a priority task in software development methodology. Current scenario of testing field is diversified toward automation. The prime objective of paper is automated generation of Boolean-specific test cases; and another objective is optimization of generated test cases using multi-objective genetic algorithm approach for Boolean operator based conditions.

1   INTRODUCTION

This paper presents an approach of Multi Objective Optimization results in test case generation. Traditionally the software's were tested on the basis of specifications or source code implementation. Now-a-days researchers and practitioners are investigating various methods for test case generation. Many areas in engineering domain requires multi-objective optimization framework to solve the problem in efficient and optimized manner. Multi-objective optimization is an ideal approach to explore optimal solution in case of conflicting objectives. The various approaches of multi-objective optimization are Strength Pareto Evolutionary Algorithm (SPEA-2), Simulated Annealing (SA), Multi-Objective Evolutionary Algorithm (MOEA), Particle Swarm Optimization (PSO), Multi-objective Genetic Algorithm (MOGA), and Non-dominated Sorting Genetic Algorithm (NSGA-II). In the paper, we had proposed a technique in the context of test case generation depending on Boolean-operator based conditions. The proposed terminology is performed in two stages. In the first stage, classification tree is generated from Boolean specific conditions (Software under Test) and for every condition, test cases are designed. In the second stage, genetic algorithm based on the multi-objective is introduced on the four Boolean operators. Work related to implication of optimization based on concept of multi-objective in software testing domain is included in Section II. Section III includes the description of formulation of Boolean specific optimization problem. Section IV describes the model proposed for Boolean specific test case generation. Section V includes the simulation results. The testing of simulation results are covered in Section VI. Section VII covers the concluding remarks.

2   LITERATURE SURVEY

(Chang et al. 1996) provided a method for optimization of Boolean logic efficiently. The optimization of Boolean logic is accomplished by adding and removing superfluous wires in a circuit. This defined algorithm is using the scenario of Automatic Test Pattern Generation (ATPG) for detecting efficient redundancy. (Zhang et al. 2000) described the extended version of Boolean Constraints solver. The tool can also take other types of variables such as integers, enumerated type and reds. The approach shows the combination of Boolean logic reasoning and bound propagation with linear programming. With incomplete methods of solving non-linear constraints can be used. (Fen et al. 2009) presented an optimization algorithm of the logic function based on mini-terms by computing on-sets covering to find implement cover and generates those implicates that covers the on-sets without the computation of prime implications. (Sziray et al. 2013) has presented an algorithm for giving the logic conditions, resulted the effects of a cause-effect graph which belongs to given software. The applied algorithm produces a three-valued Boolean algebra depending on the subsequent justification of binary values in a combinational network, here causes are the primary inputs, and effects are the primary outputs. The major benefit of the algorithm is that it decreases the number of decisions made to a high extent by applying don't care values while processing. This paper had presented an algorithm which produces the logic based conditions, resulting in the effects of a cause-effect graph belonging to particular software. The encoding and storing includes the types of gate and the input-output connections of the gates only. The amount of estimations relies on the number of gates with in the network. (Baygiin et al. 1995) presented the methodology to test multiple hypothesis based on random variables which are distributed when fuzzy variables parameterize the hypothesis. This approach presents a Bayesian flavour with the objective of minimizing a fuzzy average decision error probability with help of proper decision regions chosen. The decision rules which are optimal are discussed and also index of the error probability of fuzzy average decision known as TDC is minimized. (Mausa et al. 2012) described how to use the search-based software engineering algorithms and which problem domains can they be used upon. He also explored a potential scope of further research within the area itself. (Harman et al. 2007) had introduced multi-objective based branch coverage and the case study of dynamic memory consumption and branch coverage is presented for real and synthetic both types of programs along with their results using evolutionary algorithms multi objective concepts. (Choudhary et al. 2014) introduced a method of generating automated distribution of test cases uniformly with the use of multi objective genetic algorithm i.e. MOGA. The domain of analysing this method is dominance, MOGA and Pareto optimality.

3 FORMULATION OF BOOLEAN SPECIFIC OPTIMIZATION PROBLEM

The major role of optimization problem is to have two or more conflicting objectives for the software under test and in specified range. The main aim is to perform automation of Boolean specific test case generation and another objective is implication of multi-objective optimization technique such as multi-objective genetic algorithm for Boolean operator based conditions. In this paper, a problem of finding largest among three numbers is considered and accordingly classification tree is generated. Based on classification tree (Figure 1), the Boolean operator based conditions are analysed and for every condition test cases are generated. In the next section, the multi-objective genetic algorithm is applied on the four Boolean operators that are mainly used in software programming practices.

$$if((a >= b) \ \&\& \ (a >= c))$$
$$a -> largest$$
$$if((b >= a) \ \&\& \ (b >= c))$$
$$b -> largest$$
$$if((c >= a) \ \&\& \ (c >= b))$$
$$c -> largest$$

4 PROPOSED MODEL

This section covers the multi-objective platform for Boolean specific test case generation. At first,

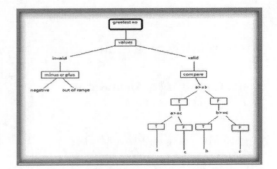

Figure 1. Classification tree.

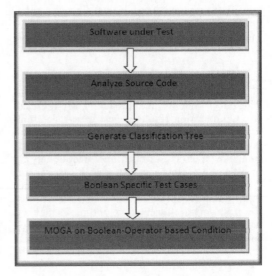

Figure 2. Steps involved in optimization.

test cases are generated on the basis of Boolean operator based conditions from classification tree and then optimization is performed on the four Boolean operators that are mainly used in software programming practices. Boolean algebra is defined as the sub category of algebra which works on only two values 1(true) and 0(false). The Boolean logic is used in software testing on the basis of the conditions. With the help of Boolean algebra, conditions can be represented with each decision and for these conditions a classification tree is generated which shows the resultant.

The four Boolean operators are EQUAL operator, NOT operator, OR operator and AND operator. Every operator is evaluated on the basis of Fitness function. Fig. 2 shows the steps involved in the process of connotation of multi-objective optimization problem in test case formation approach. Based on the optimized output by considering multi-objective genetic algorithm, test cases are reduced on the criteria of Fitness function. Fitness functions are essential part of optimization algorithm as they provide evaluation of individuals which further allows the search to move towards better individuals so as to find the optimal solution. Here fitness functions of Boolean specific operators are discussed and evaluated.

```
Function y = fitness_boolean_op (a, b)
for a=1:100;
for b=1:100;
        if(a==b)   // (EQUAL operator)
        y = (a-b);   // (Fitness Function, (a-b) =0)
        if(a!=b)   // (NOT operator)
        y = (b-a);   // (Fitness Function, (a-b) <0)
        if(a||b)   // (OR operator)
        y = (a||b);   // (Fitness Function, a||b)
        if(a&&b)   // (AND operator)
        y = (a&&b);   // (Fitness Function, a&&b)
```

5 SIMULATION RESULTS

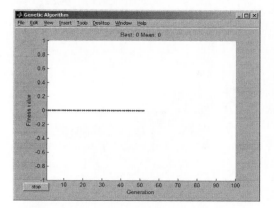

Figure 3.   Equal operator: Best fitness.

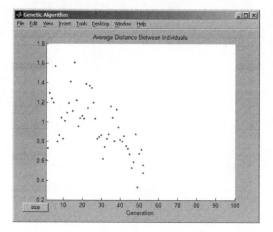

Figure 4.   Equal operator: Best individual.

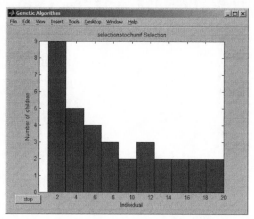

Figure 5.   Equal operator: Distance.

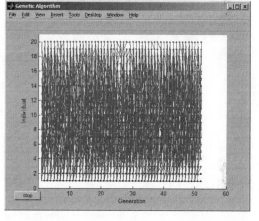

Figure 6.   Equal operator: Genealogy.

Figure 7.   Equal operator: Uniform selection.

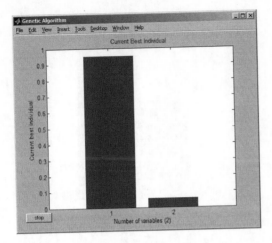

Figure 8.　Not Equal operator: Best individual.

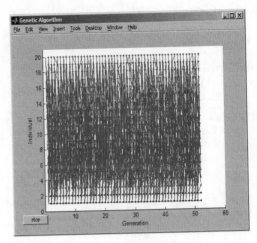

Figure 11.　Not Equal operator: Genealogy.

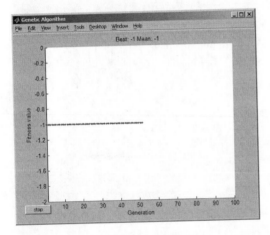

Figure 9.　Not Equal operator: Best fitness.

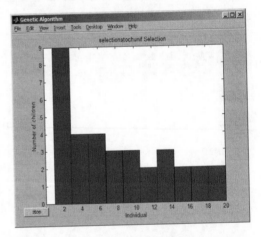

Figure 12.　Not Equal operator: Uniform selection.

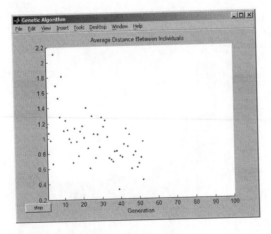

Figure 10.　Not Equal operator: distance.

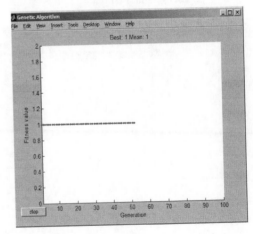

Figure 13.　Or operator: Best fitness.

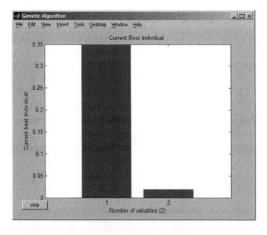

Figure 14.　Or operator: Best individual.

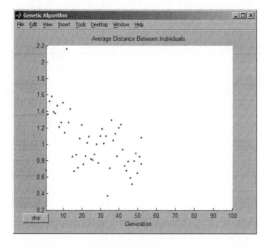

Figure 15.　Or operator: Distance.

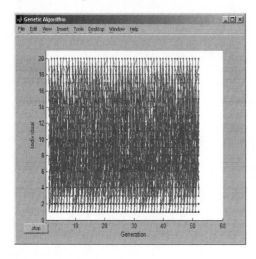

Figure 16.　Or operator: Genealogy.

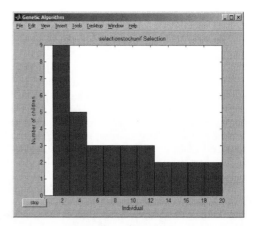

Figure 17.　Or operator: Uniform selection.

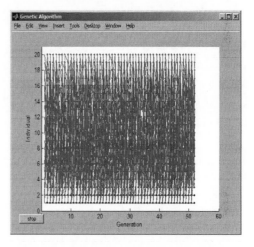

Figure 18.　And operator: Genealogy.

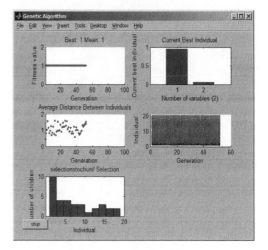

Figure 19.　AND operator: Simulation results.

6 ANALYSIS

Best Fitness function value evaluated for the fitness function of EQUAL operator is shown in Fig. 3. Figs. 4, 5, 6 and 7 shows best individual value, average distance between individuals, genealogy and uniform selection generated values respectively for Equal operator. Figs. 8, 9, 10, 11, 12 represents results in terms of best fitness, best individual value, average distance between individuals, genealogy and uniform selection respectively for NOT equal operator. Figs. 13, 14, 15, 16, and 17 shows results for OR operator. Fig. 18 covers the genealogy for AND operator. Figure 19 shows the collective result for AND operator.

Based on this evaluation, the fitness values for all the operators are adjudged and on the fitness value the number of test cases are reduced to Boolean operator specific conditions.

7 CONCLUSION

It was concluded from the work that test cases are generated on the basis of Boolean operator based conditions. This paper demonstrates the implication of multi-objective optimization in testing domain. Results depict the corresponding fitness values for all the specified Boolean operators which are needed in reduction of test cases. The future work includes the prioritization of Boolean specific test cases.

REFERENCES

Biilent Baygiin 1995. Multiple Fuzzy Hypothesis Testing In IEEE.

Choudhary K, Purohit G.N. 2011. A new testing approach using Cuckoo Search to achieve Multi-Objective Genetic Algorithm In Journal of Computing, USA, Vol. 3 Issue 4.

Choudhary K, Purohit G.N. 2013. Automation of Software Testing Process Based on Multi-objective Optimization, In international Journal of Software Engineering Research & Practices Vol. 3, Issue 2.

Choudhary K, Purohit G.N. 2014. MOGA-UDTG: Automated Uniformly Distributed Testing Approach In IEEE.

Harman M, Kiran Lakhotia, Phil McMinn, 2007. A Multi–Objective Approach To Search–Based Test Data Generation In GECCO'07, July 7–11, ACM 978-1-59593-697-4/07/0007.

Jian Zhang 2000. Specification Analysis and Test Data Generation by Solving Boolean Combinations of Numeric Constraints In IEEE.

József Sziray 2013. Evaluation of Boolean Graphs in Software Testing In IEEE.

Li Fen, Qiu Jian-lin, Chen Jian-ping, Gu Xiang, Ji Dan, Research on Technology of Mini-terms Optimization for Logic Function In IEEE volume 2.

Mausa G, T. Galinac Grbac, B. Dalbelo Basic, 2012. Overview of search-based optimization algorithms used in software engineering In Rijeka. http://www.seiplab.riteh.uniri.hr/wpcontent/uploads/2012/10/Overview-of-search-based-optimization-algorithms-used-in-software-engineering.pdf

Shih-Chieh Chang, Lukas PPP 1996. Fast Boolean Optimization by Rewiring In IEEE.

Communication and Computing Systems – Prasad et al. (Eds)
© *2017 Taylor & Francis Group, London, ISBN 978-1-138-02952-1*

Interval graph and its applications

Divya Srivastava & Suneeta Agarwal

Motilal Nehru National Institute of Technology Allahabad, Allahabad, India

ABSTRACT: Graph theory plays a vital role in Computer Science. They are used to model relations and processes in physical, biological, social, communication and computation flow etc. Interval Graph is a special class of graph with important properties. They are being used to solve large number of problems efficiently that are hard on general graphs. This paper presents Interval Graph: its definition, background, applications in different fields like scheduling, coloring, mystery solution, traffic control signals, order processing sequences etc.

Keywords: Chordal Graphs, Clique, Co-TRO, Graph coloring, Perfect Elimination Ordering, Simplicial Vertex

1 INTRODUCTION

An Interval Graph (Columbic M. 1980) belongs to the family of intervals on the real line, and is also called as an intersection graph. It represents the relations among the interval on the real line. A vertex is used to represent an interval and there is an edge between the two vertices if there is an intersection between the corresponding intervals.

In 1957, Hajos (Columbic M. 1980) a Hungarian mathematician discussed a short problem in German Language mathematical journal. He raised the question "what sort of graphs can be obtained by considering the intersection of intervals?". Thus he introduced Combinatorics in 1957. In 1959 an American biologist Seymour Benzer (Columbic M. 1980) studied the problem of structure of overlapping pieces of genetic material. In his 1959 paper, he hypothesized that these pieces were linear segments of interval on the line and looked at gene strings.

1964, Gilmore and Hoffman (Columbic M. 1980) from IBM gave the first mathematical characterization of interval graph. Finally in 1976, Divya Srivastava (srivastava.d.mnnit@gmail.com) Booth (Columbic M. 1980) and Luker gave the efficient linear time algorithm to recognize this family of graphs. After that interval graph were used in many areas like storage allocation, traffic control signals (Baruah A. et al. 2012), temporal reasoning in artificial intelligence Columbic M. 1980) etc.

Figure 1(a) shows the interval corresponding on a real line, Figure 1(b) shows that the each interval is represented as a vertex and the edges between them represent the intersection of corresponding

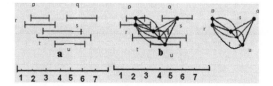

Figure 1. Interval graph.

interval. In Figure 1(c) finally graph is constructed for the set of intervals on the real line. Formally it is an undirected graph formed from family of intervals. Its mathematical definition can be understood as:

let Si represents set of intervals $S_i = \{0, 1, 2, 3, ...\}$, create vertex v_i for each interval and connect v_i and v_j if their interval intersect i.e.:

$$E(G) = \{\{v_i, v_j\} \mid S_i \cap S_j \neq \phi\}.$$

Many problems become polynomial on interval graphs like clique, coloring, independent set problems can be solved in linear time on interval graph.

2 CHARACTERIZATION

Interval graph can be determined in $O(|V+E|)$ time by using an ordering of the maximal cliques of G. The algorithm proposed by Booth (Columbic M. 1980) and Luker (1976) was based on PQ Data structure, but later on in the year 2000, Habib et al. solved the problem by using Lexicographic Breadth-First-Search (Corneil D. 2004). He proved that a graph is an interval graph if and only if it is chordal and its compliment is a comparability graph. Cornein, Olariu and Stewart also used the similar approach by using a 6-sweep LexBFS (Dusart J et al. 2015). For a graph to be an Interval Graph, the following properties are to be satisfied.

1. The Chordal graph property (in case if the graph is undirected). (Fig. 2).
2. The co-TRO (Transitive Orientation) property (in case if the graph is directed). (Fig. 3).

3 PRELIMINARIES

1. Graph Coloring: is the process of labelling / coloring each node of a graph such that no two adjacent nodes shares the same color.

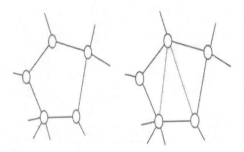

Figure 2.   (a) Not a chordal Graph (b) chordal graph.

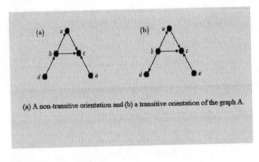

(a) A non-transitive orientation and (b) a transitive orientation of the graph A.

Figure 3.   co-TRO.

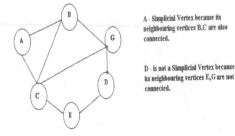

A - Simplicial Vertex because its neighbouring vertices B,C are also connected.

D - is not a Simplicial Vertex because its neighbouring vertices E,G are not connected.

Figure 4.   Simplicial vertex.

2. Clique: a clique is a complete subgraph. i.e it is a subset of vertices of an undirected graph where every two distinct vertices are adjacent.
3. Chordal Graph: A chordal graph 'S' is a graph in which every cycle of length four and greater has a cycle chord Figure 2.
4. Complement Transitive Orientation (co-TRO) Property: Orientation is a diagraph where we have only one direction per edge. A transitive orientation is an orientation which, in addition, obeys the transitive property:

$$if\ (a-> b)\ and\ (b-> c) => (a-> c)$$

5. Simplicial Vertex: A simplicial vertex of a graph G is a vertex V such that the neighbour of V form a clique in G.
6. Perfect Elimination Order (PEO): PEO is an ordering of the vertices such that predecessor of any vertex V(i) is a simplicial vertex. A PEO can be generated by the following algorithm.

Algorithm:

- Choose a simplicial vertex, name it as V1.
- Then delete this vertex from the graph and look for another simplicial vertex in the remaining graph and name it as V2.
- Continue this until the graph is empty and get an Perfect elimination order V1, V2, V3....Vs.
- After finding the order select the first vertex and assign a color C1, to V1. Then take second vertex V2 and check if there exists an edge between V1 and V2 assign different color else assign same color to the corresponding vertex.
- From Figure 5, initially vertex 'a' is selected which is a simplicial vertex and is eliminated. Second vertex 'c' is taken, since it has vertices 'a' and 'b' as its neighbouring vertices and vertex 'a' is already removed so only vertex 'b' is left thus 'c' is also a simplicial vertex and is eliminated next. Similarly by following the mentioned rules obtained PEO is a, c, b, d, e, g, f, h, i, j after which coloring obtained is in Figure 5.

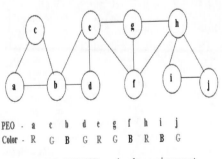

PEO -	a	c	b	d	e	g	f	h	i	j
Color -	R	G	B	G	R	G	B	R	B	G

Figure 5.   PEO ORDER and color assignment.

4 APPLICATIONS OF INTERVAL GRAPH

Some of the primitive applications of interval graphs are:

1. Minimization of order processing.
2. Temporal reasoning in artificial intelligence.

4.1 Minimization of order processing

A manufacturer has a number of orders from the customers to fulfill. Orders requires the making of different products but only one product can be made at a time. Whenever first product of any client is processed, a virtual stack is opened for that client. Stack is closed when all the orders of that client have been served. The objective is to find the sequence in which the products should be manufactured so as to minimize the maximum number of stacks opened simultaneously. This problem is also called as Minimization of Open Stack (MOSP) (Lopes I et al. 2010).

Table 1 shows the clients and the products ordered by them. For example client 1 has ordered the product 1,2 6 and 7.

A graph is constructed by using Table 1. In Figure 6 clients are represented by vertices and there is an edge between two vertices (clients) if the two clients share some common product. For instance, in the Figure 6. there is an edge between 1-3, because client 1 and client 3 share product 7 in common. Similarly there is an edge between 1-5 and 1-4, because they share product 6 and product 2 in common respectively.

The objective is to find such a sequence of the manufacturing products so that the processing of the customer's order is minimized. One possible

Table 1. An instance of the MOSP.

Products:	1	2	3	4	5	6	7	8
Client 1	X	X				X	X	
Client 2				X	X			
Client 3			X				X	
Client 4		X		X				X
Client 5			X			X	X	

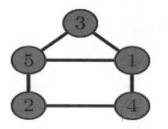

Figure 6. Instance of the graph from Table 1.

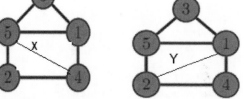

Figure 7. Stacks.

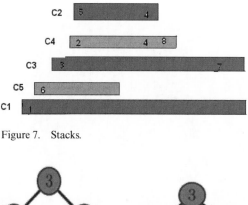

Figure 8. Interval Graph(a) Interval Graph(b).

ordering of the product can be 1-6-3-2-5-4-8-7. This is shown in Figure 7. The maximum number of stacks opened at a time is five. In order to minimize the maximum number of stacks opened simultaneously, Interval graphs can be used.

Such problems can be solved in polynomial time by using interval graph. The graph in Figure 6 is not an interval graph, thus by adding some edge(s) (Lopes I et al. 2010) to it so that it becomes an interval graph. The graph can be converted into interval graph by adding either edge X, or edge Y as in Figure 8(a), 8(b) respectively.

The value of the optimum of the MOSP is equal to the size of the biggest clique in the interval graph. Since interval graphs are chordal graphs and chordal graphs are perfect graphs thus the biggest clique in the interval graph is equal to the chromatic number of the graph. Thus in order to find the chromatic number, first find the PEO and then the chromatic number.

For Figure 8 (a)

PEO:	3	1	5	2	4
Color:	C1	C2	C3	C2	C1

For Figure 8 (b)

PEO:	3	5	1	2	4
Color:	C1	C2	C3	C1	C2

Thus, by taking one of the PEO i.e. 3-1-5-2-4, manufacturing of the products can be done in the following order: 3-6-7-5-4-2-8-1. The maximum number of stacks opened simultaneously in this ordering is three.

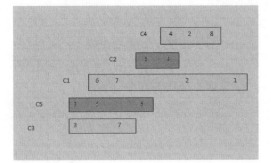

Figure 9.   Stack using PEO.

Thus by using interval graph maximum number of stacks to be opened simultaneously is calculated by using PEO.

4.2   *Temporal reasoning in artificial intelligence*

Six people had been to the library on the same day on which the rare tractate was stolen (Columbic M. 1980). Each had entered once, stayed for some time and then left. If two people were in the library at the same time, then at least one of them saw the other. For example if person X says that he saw person Y, from this statement it can be concluded that person X and person Y had some time interval in common, in which they both were in the library. A detective was appointed to investigate for the same. He questioned each person individually to know who saw whom? And is there any mismatch in their statement as this would help him in his investigation to decide who is lying or who has stated the wrong testimony which can help him to investigate further.

The approach to solve this problem will be:

1. Collect the testimony from each individual separately.
2. Construct a graph from the testimony collected.
3. If in the graph there exists a cycle of four and it does not have a chord, then it is not an interval graph. Thus the detective can conclude that there is some mismatch in the statement made by the people and that can help him to judge that which person has given the wrong testimony which is helpful in further investigation. Cycle of 4 without a chord is not allowed in interval graph this can be seen by the example in Figure 10.

In Figure 10, there are 4 people a, b, c and d. 'a' said he has seen 'b' and 'c', 'b' says he saw and 'd' and 'a', 'c' said he has seen 'a' and 'd' and lastly 'd' said he saw 'b' and 'c'. After constructing the

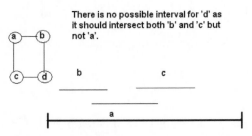

Figure 10.   Chordless cycle of 4 is not possible.

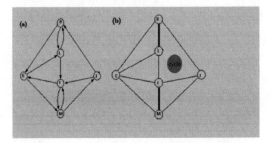

Figure 11.   Graph constructed using testimony.

graph as per their statement and then finding the common interval, it is observed that that there is no possible interval left for person 'd'. Thus such chordless cycle of 4 is not possible in interval graphs.

Steps to solve the problem:

1. Testimony collected from each individual is as follows:
   - Boehner (B) said that he saw Lindsey(L) and Eddie(E).
   - Lindsey reported that he saw Boehner and Ida(I).
   - Michele (M) claimed to have seen John(J) and Ida.
   - John said that he saw Boehner and Ida.
   - Eddie testified to seeing Lindsey and Michele.
   - Ida said that she saw Michele and Eddie.
2. As per the statement made by the people, construct a graph Figure 11 where each person is represented by vertex and the statements made by them is represented by the directed edges. For instance Boehner (B) said that he saw Lindsey(L) and Eddie(E), is shown in the graph as there is an edge from Boehner (B) to Lindsey(L) and Eddie(E). Double arrows can be treated as truth, for example, Lindsey said "I saw Boehner and Ida(I)", and Boehner said "I saw Lindsey and Eddie". Since both Lindsey and Boehner, although interrogated separately

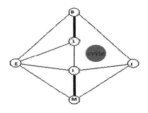

Figure 12.   {B,L,I,J} Cycle remains.

mentioned that they saw each other, such statements can be treated as truth and can be taken that they both were in the library at the same time. Thus in Figure 11(b) double arrows are removed and converted into a single dark line.

3. Now, check whether the constructed graph is an interval graph or not? Since there are three chordless 4-cycles {B,L,I,J}, {B,J,I,E}, {B,E,M,J}, hence it is not an interval graph and thus there is some mismatch in the statements given by the 6 people. Now the detective makes an analysis that who has given a wrong testimony. He sees that Lindsey may not be the liar as he is missing from the second cycle. Similarly since Ida is missing from the third cycle she may not be the liar. Same is with Michele and Eddie as they are missing from second cycle and first cycle respectively. Boehner (B) and John (J) are the two people common in all the three cycles. Thus first taking a look on the statements made by them, if John being truthful and Boehner is being treated as a liar, thus the statements made by him is treated as false. Remove the edges from B except the double arrows which is treated as truth statement. Detective finds that, still there remains a chordless 4 cycle {B, L, I, J} Figure 12. Thus he concludes that even after taking Boehner's statements as false, still there remains a chordless 4-cycle, so it gives an impression that John had lied and now the investigation can be made easier by focusing on John for lying.

5   CONCLUSIONS

Interval Graph theory is an important part of study in the area of structural graph theory. The other applications where interval graphs are much useful, are resource allocation problem in operating system, Graph coloring of the interval graph is done in polynomial time and can be used for assignment problem. There are other variations of interval graphs like Online Interval Graph, Unit Interval Graph, Perfect Interval Graphs that have enhanced the importance of these graphs and have opened the scope for further research.

REFERENCES

Baruah A. & Baruah N. 2012. Signal Groups of Compatible Graph in Traffic Control Problems, Int. J. Advanced Networking and Applications: 04.

Columbic M. 1980: Algorithmic Graph Theory and its Applications. New York: Academic Press.

Corneil D. 2004. Lexicographic Breadth First Search A Surevy. Berlin Heidelberg: LNCS 3353:119.

Dusart J. & Habib M. 2015. A new LBFS based algorithm for cocomparability graph recognition. France: Elsevier: 0166–218.

Hideki H. & Eedsen L. 2010. The Minimization of Open Stacks Problem: a review of some properties and their use in pre-processing operations. European Journal of Operational Research. 203: 559–567.

Kleinberg J. & Eva T. Greedy Algorithms: Chapter 4. Pearson & Addison Wesley.

Kovalyov M. & Cheng E. 2006. Fixed interval scheduling: Models, applications, computational complexity and algorithms. Elsevier.

Lopes I. & Carvalho J 2010. Using Interval Graphs in an Order Processing Optimization Problem. Proceedings of the World Congress on Engineering 03.

Communication and Computing Systems – Prasad et al. (Eds)
© 2017 Taylor & Francis Group, London, ISBN 978-1-138-02952-1

Steady state analysis of compressor and oil removal system with Aspen HYSYS

P. Nema & Himanshu K. Patel
Institute of Technology, Nirma University, Ahmedabad, India

A.K. Sahu
Institute for Plasma Research, Gandhinagar, India

ABSTRACT: For the liquefaction of permanent gas like Helium, there is a need of expansion of it at very low temperature and for the same, compression of Helium gas is necessary. To avoid high heating and to reduce the work requirement in the compression, oil of heavy molecule, having high specific heat is mixed with Helium before compression. Before using this Helium gas for refrigeration, it is necessary to cool it to about room temperature and separate the oil from the mixture. For this purpose, different components like heat exchanger, oil separators tank, oil circulation pump, valves, and filters are needed. This whole system is called as Compressor and Oil Removal System (CORS). This paper introduces the simulation of individual components of CORS system and steady state analysis of each component with Aspen HYSYS simulating tool. All information obtained by performing steady state analysis through Aspen HYSYS is utilized for preparing the platform for dynamic analysis of all these components and process flow diagram of the process as well.

Keywords: CORS system, Simulation, Steady state analysis, Aspen HYSYS, CORS system components

1   INTRODUCTION

CORS System System consists of different components which are described below:

i. Oil removing unit which includes all oil removing components.
ii. Helium gas controlling unit which includes sequential operation of control valves for controlling the desired suction and discharge pressure.

Here, Helium gas is used as fluid for compression and expansion. Thermodynamic properties of Helium fluid play important role in complete plant operations. To produce cooling effect (liquid helium), expansion is necessary and hence compression is required.

Steady state simulating tool is helpful to obtain the optimum perspective and powerful insights into plant behavior. To simulate the actual plant behavior of designed process, steady state analysis is necessary. In a way, it plays vital role for the prediction of process behavior and thereby facilitates in achieving the more efficient and profitable process designs, consistent product quality and

troubleshooting of the process problems. Aspen HYSYS optimizes and automates innovative ideas and promote the best engineering practices.

2   ASPEN HYSYS

Aspen technology is known as Aspen tech's integrated package of software solutions for economic evolution and service provider for engineering design, simulation and optimization. To design and improve the plants and processes and to obtain better operating plant cycles, AspenTech's engineering products are used. Aspen Engineering Suite (AES) is a part of Aspen Technology, which is the Aspen Tech's Process Management (PLM) solution for entire process system, having the capability to integrate rigorous engineering models with data from conceptual engineering through operational performance.

AES is the combination of six integrated product families, one of them is AES simulation and optimization which can optimize plant designs and operations. It is helpful to enable the engineers to evaluate the plant opratatibility, efficiency, safety

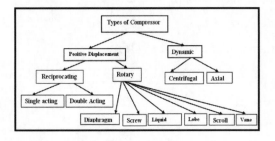

Figure 1. Classification of compressor.

and environmental performance. With the feasibility of steady state, dynamic and batch simulation, it enables the users to optimize the plant operations, capital costs and ensuring the accurate simulation.

3 COMPRESSOR IN CORS SYSTEM

Compressor is a device which is used for increasing the pressure of gas from low to high, using some external energy. Most common use of compressors, is the compression of fluids in a steady state flow process. As shown in Figure 1, compressors are broadly divided into positive displacement and dynamic type. It has dominant influence on cycle efficiency. For CORS system reciprocating type Screw compressor is used. Operation modes of compressor are: Suction, Compression and Discharge. Screw compressor is the most widely used compressor for Helium compression. It provides the isothermal operation by taking the heat from helium gas during compression discharge.

Due to following reasons screw compressor is pr ferred over reciprocating compressor:

- It is having purely moving part which is five times lighter than their reciprocating counterpart of the same capacity and has ten times longer operating life between overhauls.
- It has the capacity to develop pressure around 7.3 times more than a centrifugal compressor operating at same speed, and it is the only rotary unit that operates at tip speeds in excess of 0.12 mach.
- It has the ability to produce much higher pressure ratio.

4 SIMULATION ENVIRONMENT AND ANALYSIS IN ASPEN HYSYS

4.1 Fluid packages and component used

In simulation environment of Aspen HYSYS for steady state analysis of CORS system components, component has taken Helium gas as fluid for all

Single phase flow. For two phase flow there is mixture of Helium gas and Glycol oil is used. For three phase separator components as Helium gas, oil and carbon is used.

Fluid package for Helium Gas is MBWR (Modified Benedict–Webb–Rubin equation) and for mixture of Helium gas and oil Peng-Robin fluid package is used. For Helium gas, Glycol oil and carbon components also Peng-Robin fluid package is used. All fluid packages are used on the basis of applicability of wide ranges of specified temperature and pressure.

4.2 Initial conditions for simulation

Compression Pressure Ratio = 1.05:14 bar
Inlet Temperature = 310K
Inlet Mass Flow Rate = 70 g/s

4.3 Analysis and details

Fig. 2 shows the schematic used for steady state simulation and analysis of compressor, for variation in different parameters values. Purpose of analysis is to observe the thermodynamic properties behavior of compressor regarding the changes made in some parameters.

Fig. 3 shows the changes observed in volumetric efficiency and duty as the inlet temperature increases. With the rise in temperature, efficiency increases whereas the duty increases almost non-linearly. This analysis is helpful to decide the efficiency of compressor for particular temperature. Figs. (4–5) shows that as the mass flow rate of Helium increases, compressor power, speed, capacity and duty requirement increases linearly. All these observations are essential to determine the

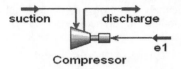

Figure 2. Compressor simulation in aspen.

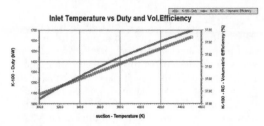

Figure 3. Suction temperature vs Duty and Volumetric efficiency.

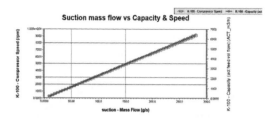

Figure 4. Discharge pressure vs pressure difference.

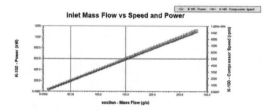

Figure 5. Suction mass flow rate vs speed and power.

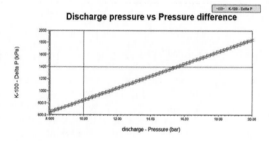

Figure 6. Suction mass flow rate vs capacity and speed.

range of power consumption, speed and duty of compressor for desired requirements. From Fig. 6, it can be seen that if discharge pressure increases, overall pressure difference increases which is the cause of increase in the mass flow rate and outlet temperature beyond desired limit. From Figs. (7–8), it can be analyzed that as the suction pressure increases, volumetric efficiency increases, discharge temperature decreases, compressor speed and feed flow rate decreases nonlinearly which in turn results in the higher pressure ratio, increased leakage rate and decrease the mass flow rate which turns to decrease the volumetric efficiency and adiabatic efficiency. This will finally lead to the successful evolution of entire compressor performance.

5 PUMP

Pumps are, of course, very similar to compressors with one great difference—compressors move gases, while pumps move liquids.

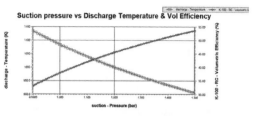

Figure 7. Suction pressure vs discharge temperature and Volumetric efficiency.

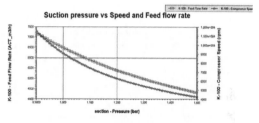

Figure 8. Suction pressure vs speed and feed flow rate.

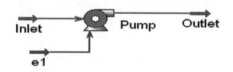

Figure 9. Simulation of pump.

5.1 Oil pump

Oil pump is used to increase the pressure of oil and this pressurized oil is used to support loads due to weight of the rotors and shafts and lubricating side bearing, thrust bearing, mechanical seal. Also pressurized oil is used for slide valve to control capacity of compressors.

Fig. 9 shows steady state simulation of pump, which is used for increasing the inlet pressure. Purpose of analysis of pump is to analysis the thermodynamic properties behavior of pump with respect to variations in different parameters.

5.2 Initial conditions for simulation

Inlet Pressure & Required Outlet Pressure = 13 bar, 30 bar
Inlet Temperature = 310 K
Mass Flow Rate = 70 g/s
Figs. (10-11-12) describe the change in the power requirement for pump for increased outlet pressure, adiabatic efficiency and inlet temperature respectively. It is observed that the pump power is linearly proportinal to outlet pressure and inlet

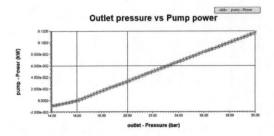

Figure 10. Outlet pressure vs power.

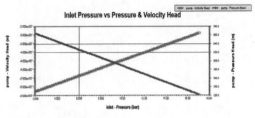

Figure 13. Inlet pressure vs pressure head, velocity head.

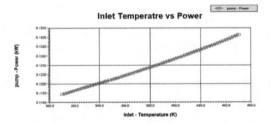

Figure 11. Pump efficiency vs power.

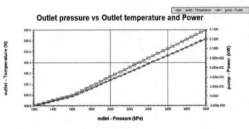

Figure 14. Outlet pressure vs power, temperature.

6 CONTROL VALVE

According to principle of working there are many types of control valves like globe valve, ball valve, butterfly valve, gate valve, plug valve, diaphragm valve. Main quality which all control valve must possess, is excellent throttling characteristics. Globe valve with pneumatic actuator is selected because of excellent throttling characteristics as compared to other types of valves. Initial conditions required for simulation of control valve is shown in Table 1.

6.1 Initial conditions for simulation

6.2 Analysis and detail

Steady state simulation of control valve is shown in Fig. 15. The analysis is intended to review the thermodynamic properties behavior of control valve against the applied changes in some parameters like temperature, pressure and mass flow rate. The main purpose of control valve is to reduce the outlet pressure and change in outlet temperature.

As shown in Fig. 16, as inlet pressure increases, outlet temperature increases. In Fig. 17, it can be observed that, as the inlet temperature increases outlet temperature also increases linearly. Fig. 18 shows that as inlet mass flow rate increases, outlet mass flow rate also increases. Hence from this analysis, one can conclude that if the pressure difference in valve increases then mass flow rate increases and outlet temperature decreases.

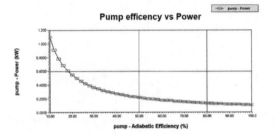

Figure 12. Inlet temperature vs power.

temperature whereas it decreases exponentially with increasing adiabatic efficiency.

5.3 Analysis and details

This analysis is useful to evalute the power requirement for desired operation with respect to changes in different parameters values. It is also observed that (Fig. 13) if inlet pressure increases then pressure head and velocity head increases. From the graph in Fig. 14, it can be diepicted that as inlet mass flow rate increases, heat flow and power consumption also increases. This analysis decides the appropriate relationship between inlet pressure, pressure head and velocity head. It also shows that as flow of fluid increases, there is more heat flow and power required.

Table 1.

Case I	Case II
Inlet Pressure = 14 bar	Inlet Pressure = 14 bar
Inlet Temperature = 300 K	Inlet Mass Flow Rate = 70 g/s
Inlet Mass Flow Rate = 70 g/s	Vapour Fraction of any one stream = 0.6
If Put Pressure Drop get Outlet Pressure and Temperature = 12.95 bar, 300.8 K	If Put the Pressure Drop Get the Outlet Pressure and Temperature = 12.95 bar, 300.8 K

Figure 15.   Simulation of valve.

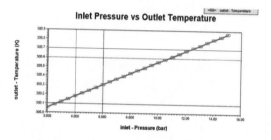

Figure 16.   Inlet pressure vs outlet temperature.

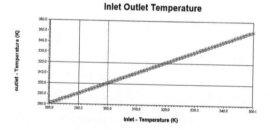

Figure 17.   Inlet temperature vs outlet temperature.

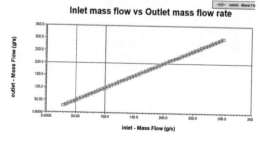

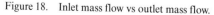

Figure 18.   Inlet mass flow vs outlet mass flow.

This results are helpful to implement the basic thermodynamic cycles and in cryogenics applications.

7 OIL TANK

Oil tank is one type of bulk oil separator in which oil and Helium gas is separated out using the principle of separation based on momentum and gravity settling.

8 SEPARATORS AND FILTERING UNITS

Separator is classified in HYSYS simulation environment as: Tank, Two phase separator, Three phase separator.

The function of two phase separator is to separate out gas and oil from the mixture of gas and oil vapour at specific pressure and temperature. Separator design is based on pressure, temperature, flow rates, and physical properties of the streams as well as the degree of separation required.

In a liquid-vapour separation vessel, there are typically three stages of separation:

The first stage, primary separation, uses an inlet diverter to cause the largest droplets to impinge by momentum and then drop by gravity.

The next stage is gravity separation of smaller droplets as the gas flows through the vapour disengagement section of the separator. The final stage is mist elimination, where the smallest droplets are coalesced on an impingement device.

Coalescer, adsorber Bed are used to reduce the impurity level up to specific PPM level according to their required efficiency. Filters are used at different location for filtering the gas and oil and to remove the solid particle from fluid.

8.1   Initial conditions of separator for simulation

Inlet Pressure = 14 bar
Inlet Temperature = 310 K
Inlet Mass Flow Rate = 70 g/s
Vessel Volume = 20 cm^3
Fraction of Helium and Oil = 0.8:0.2

8.2   Analysis and details

Steady state simulation of separator is shown in Fig. 19. The purpose of steady state analysis of seperator is to evalute the thermodynamic properties behavior of it and how it will affect the whole system performance.

As shown in Fig. 20, if inlet Temperature is increased then separator liquid (oil) mass flow

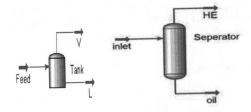

Figure 19.   Simulation of tank and separator.

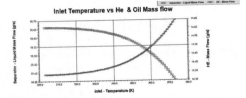

Figure 20.   Inlet temperature vs helium and oil mass flow.

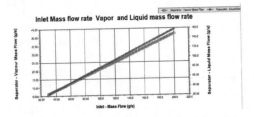

Figure 21.   Inlet mass flow rate vs vapor and liquid mass flow rate.

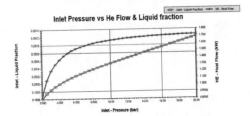

Figure 22.   Inlet pressure vs helium flow and liquid fraction.

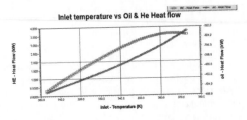

Figure 23.   Inlet temperature vs helium heat flow and oil heat flow.

rate as decreases and gas (helium) mass flow rate increases respectively. As shown in Fig. 21 as inlet mass flow rate increases, vapour mass flow and liquid mass flow increases linearly. As shown in Fig. 22 as inlet pressure increases heat flow and liquid fraction increases. As shown Fig. 23, it can be analyzed that if inlet temperature increases, helium heat flow increases and oil heat flow decreases simultaneously and at a particular point helium heat flow become constant that is try to maintain constant temperature. From this analysis it can observe the thermodynamic properties behaviour of fluids as gas and liquid two different form in separator. In separator there is separation of fluids in two phase, at bottom layer high pressure liquid phase is obtained and on top low pressure gas is collected and on the basis of their amount of fraction characteristics analysis is possible. From these analysis it is concluded that, the amount of liquefaction and gaseous form of fluid is depends on parameter like temperature, pressure, mass flow rate etc.

8.3   *Analysis and detail of coalescer*

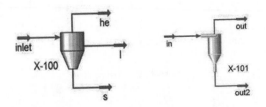

Figure 24.   Simulation of tank and separator.

8.4   *Initial conditions required for filtering units*

Initial conditions required for simulation of filtering units as described for cyclone and solid separator given in Table 2.

Steady state simulation of filtering units like solid separator and cyclone in Aspen HYSYS is shown in Fig. 24. Purpose of analysis of filtering units is to observe the the thermodynamic properties behavior and filtering efficiency of components. efficiency of filters depends on types of filtering elements, their mechanical properties details etc. As shown in Fig. 25, it is observe that, if inlet temperature of filter increases then pressure drop and outlet temperature also increases. From Fig. 26 it is clear that if inlet pressure of filter increases, outlet pressure and pressure drop increases.

Table 2.

Cyclone	Solid separator
Inlet Pressure = 14 bar	Inlet Pressure = 14 bar
Inlet Temperature = 310 K	Inlet Temperature = 323 K
Inlet Mass Flow Rate = 70 g/s	Inlet Mass Flow Rate = 70 g/s
Composition Percentage	Composition Percentage
Efficiency of Particle = 99.9%	Stream Splits Details
Particle Details	

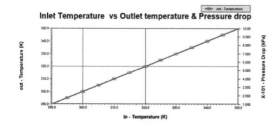

Figure 25.   Inlet vs outlet temperature.

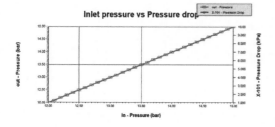

Figure 26.   Inlet pressure vs outlet pressure drop.

9   CONCLUSION

Steady state simulation of different components of CORS system is presented using an excellent simulating tool Aspen HYSYS. Aspen HYSYS has expertise to provide the best engineering practice and optimize solution in various discipline of process control as well as in cryogenic field also.

Nowadays it is used to automate the entire innovation, research, development and modeling the design.

From the steady state analysis results, it is possible to analyze the thermodynamic properties of different parameters of each component with respect to variation in different parameters. it is also possible to determine the operating range and optimum point of particular parameter for required operation.

However, during the steady state analysis, it is not possible to implement the changes in some components. Such as outlet temperature in cooler & heater, and CV of control valve and unable to evaluate the combine impact of all components on system performance. Hence, complete system operation cannot be analyzed. For analyzing the variations in all parameters according to operating condition with time, dynamic analysis is necessary.

ACKNOWLEDGMENT

We are thankful to the Institute for Plasma Research (IPR), INDIA and appreciate their support & efforts to accomplish the Project.

REFERENCES

Randall Barron, "Cryogenic Systems", McGraw-Hill, USA, 1966.
www.aspentech.com application notes, "AES Simulation & Optimization—Steady-State"
en.wikipedia.org application notes, "Aspen Technology"
www.aspentech.com application notes," HYSYS"
www.aspentech.com application notes, "Aspen Engineering Suite™".
Nagam sheshiah, "Experimental and Computational Studies on Oil Injected Twin-Screw Compressor", National institute of technology, rourkela, 2006.
Thomas M. Flynn, "Cryogenic Engineering", Marcel Dekker, New York,2005.
uam.es, application notes, "HYSYS 3.2 operation guide", USA, 2003.
en.wikipedia.org application notes "benedict-webb-rubin equation".
smartdesign.com application notes "aspentechs hysys fluid package-thermodynamics notes".
Department of Energy Fundamentals Handbook, Mechanical science, module 4, valves.

Communication and Computing Systems – Prasad et al. (Eds)
© 2017 Taylor & Francis Group, London, ISBN 978-1-138-02952-1

A deep web search engine for deep pages

Jyoti Singh & Arun Solanki
School of ICT, Gautam Buddha University, Greater Noida, UP, India

ABSTRACT: The information stored on the World Wide Web (WWW) is constantly enlarging, making it continuously more impossible for a single search engine to index the whole web for resources. Web Search engine maintains real-time information by running an algorithm on a web crawler. The data on different websites crawl by a web crawler which systematically browses the WWW. This research proposed a method to collect all the deep pages from different websites for later processing. This paper has proposed and designed the three-tier architecture for deep web search engine. Online Furniture Websites are the proposed domain for deep web search engine. The proposed research has taken this domain for a purpose because there is no such search engine for Online Furniture Websites and this architecture can be used in different domains also.

Keywords: World Wide Web (WWW), web crawler, search engine, deep pages

1 INTRODUCTION

Internet is a network of networks which carries broad range of information resources and services such as the applications and hypertext documents of the World Wide Web (WWW). The WWW has become one of the most important information repositories. However, information in web pages is free from standards in displaying and lacks being standardized in a structured format. It is a challenging work to extract appropriate and useful information from web pages (Muyeba Maybin et al. 2008). The WWW has two parts of web such as: Surface web; Deep web. Surface web is the visible web which is searchable with normal web search engines where as Deep web is the hidden web whose contents are not indexed by normal search engines. The deep web has dark web which refers to web pages that a search engine calculatedly does not index in its database (Kakde Yogesh et al. 2015).

The deep pages are the web pages within a website that is hidden and it can be accessed by n clicks from the home page, it takes too much time and some users can't be able to reach these pages and also very difficult to crawl by conventional search engines. The bulk of the rich and high quality of information is present in deep web and is created from web databases (Yang Jufeng et al. 2007).

A web search engine is designed to search for information on WWW and a program that hunts for and identifies items or products in a database that coincides to keywords or characters specified by the user, especially for finding particular sites on WWW. As the availability of massive amount of information on the internet, makes extraction of data from deep pages difficult. However, due to unstructured web page design of websites, it is hard to retrieve information from deep pages.

In order to solve the problem of retrieving the information from deep pages and displaying the desired result to the user in structured web page design, this paper design and implement a deep web search engine for furniture domains with the help of web crawler. Web crawler is a program that crawl the web in favor of the search engine, and follows all the web pages they visit for later processing by a search engine which lists the downloaded pages (Amrin Andas et al. 2015).

The bulk of web databases server sites has reached to the range from 50,000 to 1, 00,000 and is expanding day after day with rapid very fast rate of growth. The information reserved in the web is predicted about 8000 terabytes and the bulk of the information found in the deep Web is about 550 times of the Surface Web (Bergman Michael K. 2001). Information in deep web databases are usually discriminated whose appearance is unstructured, which is broad and may or may not have subject-oriented content. The extraction, retrieving and mining of the relevant information from the deep web databases are difficult and challenging for various applications. The targeted web can be achieved by inputting keywords with a web search engine. But the problem is that the system is not always able to provide the accurate data and also it is not easy for the system to automatically extract or understand the information contained. Most of the web pages are in Hypertext Markup Language (HTML) format, which is a semi-structured language, and the data are not given in a precise format and changes regularly (Eikvil, L. 1999).

2 RELATED WORK

Umara Noor et al. had explained that a huge part of Deep web constitute of online structured and domain specific databases. These databases are attained by using web query interfaces which is commonly a web page. The information consists of and rescued in these databases is usually related to a particular domain. This highly relevant and useful information is the most apt information for gratifying the needs of the users and for the implementation of large scale deep web integration (Noor Umara et al. 2011).

Nripendra Narayan Das and Ela Kumar have designed the algorithm for extraction of data from deep web. They described that the manual method is not appropriate for huge number of web pages. It is a challenging work and a problem to access and retrieve appropriate and useful information from web pages. Currently, many web mining systems called web wrappers, web crawler have been created. In this paper, some already existing techniques are analyzed and then our new suggested work on web information extraction is explained. This is a fact that some search engines are not able to extract the data from deep web as they lack by some logical issues for extraction of data (Narayan Das Nripendra and Kumar Ela 2014).

Yogesh Kakde et al. have designed a novel technique which is able to build a suitable query which is designed especially for deep web data and a supplementary module which works after building query and extracts data from deep web sources. When data from deep web is achieved we manage the copy of that data in Data Repository with having a number of attributes. They also managed this data repository refreshed by achieving another algorithm. That repository worked as a local cache in their project and makes data retrieval process more efficient. Rahul Choudhari et al. have shown the algorithm to increase the efficiency of search engines. They explained the scheduling problem and given a better solution for crawlers, with the objective in focus is to optimize the resources and improving quality of the indexes for web pages. Also they have divided the web content resources in two portions: Active and Inactive. For the content which is inactive content providers they used crawling agents who continuously crawls the resources of content providers and collect the pattern of updating information of the content providers. They have also proposed an efficient scheduling method which takes advantage on the information taken by the web agents. David Buttler et al. extend this approach in Omini (Object Mining and Extraction System), which uses supplementary heuristics based on HTML tags but no domain knowledge. Andas Amrin et al. used focused web crawler crawling only web pages that are appropriate to the user given keywords or web page link. In this article they reviewed the effective and focused web crawling algorithms that determined the fate of the search system (Amrin Andas et al. 2015).

3 ARCHITECTURE OF PROPOSED SYSTEM

This paper has designed architecture for deep web search engine for Online Furniture Websites. The proposed research has taken this domain for a purpose because there is no such search engine for Online Furniture Websites and this architecture can be used in different domains also.

As shown in Figure 1, the proposed system has three-tier architecture such as:

3.1 Graphical User Interface (GUI)

It allows user to interact with electronic devices through graphical icons and visual indicators, typed command labels or text navigation.

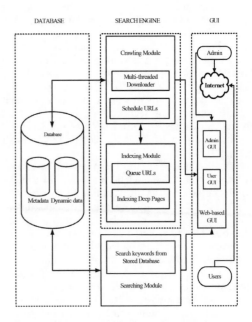

Figure 1. Three-tier architecture of proposed system.

3.2 Search engine

This search engine is designed for extracting the deep pages from deep web and the search results are generally presented in a line of results often referred to as Search Engine Results Pages (SERPs). Search engine also maintain real-time information by running an algorithm on a web crawler. Web search engine has three modules such as:

3.2.1 Web crawling module

The crawling of deep pages starts which visits the lists of URLs to visit. As the crawler reaches these URLs, it searches all the hyperlinks in the pages and adds them to the list of URLs. The crawler has multi-threaded downloader which downloads the pages from the web through crawling process. As the crawler has the option to give the threads count which consists of number of threads downloaded at a single time that these number of threads will be downloaded at a single time because if this is not done then the threads will expire and it will not download the pages efficiently. Then the meta tags are updated so that the links of deep pages store in database. Following steps are applied on crawling module as follows:

Step 1: Pick one by one URL and add in the list.
Step 2: Update the URLs list.
Step 3: Pick one by one URL from the list and internally access all the hyperlinks of the page.
Step 4: Stores the links in database.
Step 5: Move to next URL.
Step 6: Update meta tags.
Step 7: Search for keyword in search engine.
Step 8: Display the result in structured web page.
Step 9: The process is applied for all the URLs and finally result is displayed in single web page.

3.2.2 Indexing module

The downloaded pages are indexed in the database in the form of lists of URLs and it is a technique to efficiently retrieve records from database files based on some attributes or keywords on which the indexing has been done.

3.2.3 Searching module

The user search for the keywords and while searching these keywords the web search engine gives them the lists of those deep pages which were hard to find out from the regular websites but it became easy in our proposed work as it already stored the deep pages in the database and then the user search for any keyword, the proposed search engine gives the single page output in structured web page. Following steps works in the following way:-

Step 1: When keyword is submitted in search box, it matches in database where links are stored of different websites with keywords.

Step 2: Finds the keyword in the lists of URLs or indexed deep pages.
Step 3: If keyword matches with the keyword available in database it collects search result.
Step 4: Display the result in the structured single web page.
Step 5: The result is displayed in structured web page which gives the relevant data from the different web sites on a single web page.

3.3 Database

The metadata and dynamic data is used in the database. Metadata gives the information about the other data which makes searching and working with particular instances of data easier where dynamic data is flexible as the collection of data in the memory which has the flexibility to control the size. The dynamic data refers to an organization or collection of data in memory as the URLs downloaded from the deep web are stored in database which can be updated anytime as the dynamic data is flexible and has the flexibility to control the size.

The URLs are downloaded through the web crawler and then indexed in the database as metadata summarizes the basic information about URLs which makes finding and working with particular instances of URLs easier. The metadata contains descriptions of the page's contents as well as the keywords linked to the content and these are usually expressed in the form of meta tags for the efficient working of search engine so that the result displays the information in structured web page.

4 PROCESS FLOW OF PROPOSED SYSTEM

The proposed system has designed the web search engine through which the user can search the keywords and get the single page output in structured web page as well as the web crawler has designed to run on behalf of the web search engine to store the URLs of deep web pages by crawling the web and indexing those URLs and store data in the database. Figure 2 is shown as process flow diagram for proposed system.

The process of proposed system is discussed step by step as follows:

Step 1: User starts interacting with user interface.
Step 2: User input keyword to search and get the related URLs from the different websites.
Step 3: Crawler starts its working by crawling deep web pages to get the related URLs from the deep web.
Step 4: The crawled deep web has lists of URLs which downloads by the multi-threaded downloader.

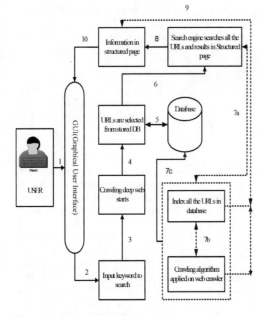

Figure 2. Process flow diagram.

Step 5: The data stored in the database is now being used by selecting the keyword related URLs to give the result for keyword.

Step 6: The selected URLs related to keyword searched by the search engine.

Step 7a: The search engine starts its working by searching the keyword from the database where the data or URLs are stored for later processing.

Step 7b: When the algorithm is applied on crawler the downloaded web pages are indexed by the indexer in the database so that the searching becomes easier.

Step 7c: Finally the indexed data stored in database and ready for searching the keywords from database.

Step 8: Search engine searches all the keyword related URLs from the stored and indexed database and ready to display the result in structured web page.

Step 9: The URLs and data searched from the different websites to give the result on a web search engine with the help of web crawler.

Step 10: Finally the result displayed to the user on single structured web page.

5 IMPLEMENTATION AND WORKING OF PROPOSED SYSTEM

The proposed system is designed in ASP.Net with C#. The ASP.Net is designed for web development to produce dynamic web pages as it is an open source server side web application framework which was developed by Microsoft. ASP.Net web pages is officially known as web forms and these web forms are contained in files with .aspx extension and these files contains static (X)HTML markup. C# is a multi-paradigm programming language which was developed by Microsoft within its. NET initiatives.

Working of system: The working of proposed system is discussing in following steps: -

Step 1: The login page where the user login and then starts searching.

As shown in Figure 3, login page opens up when we start the system as it requires username and password to login the system so that for start searching login is must.

Step 2: Site manager page where websites are added.

As shown in Figure 4, the site manager consists of furniture websites from where the information is retrieved and indexing them on a web search engine. The site manager adds the websites by entering the address and page rank of the website and then submit it. Any website that is already added can be deleted by clicking on the delete button and also it shows the total records of websites.

Step 3: After adding websites in the site manager the crawler starts its working.

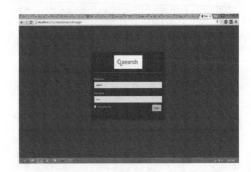

Figure 3. Login page.

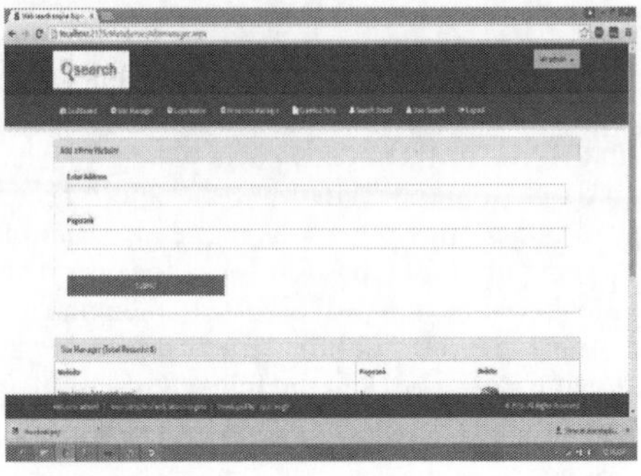

Figure 4. Site manager.

922

As shown in Figure 5, the crawler starts its working by crawling deep web in threads. The crawler crawl the websites one by one and the crawling can be pause by clicking the pause button as well as starts by clicking play button. The number of threads, connection timeout can increase or decrease by setting the MIME types in options. Also it will show the error list and requests list while crawling.

Step 4: Multi-Threaded Downloader downloads the deep pages and starts indexing them in database.

As shown in Figure 6, the multi-threaded downloader downloads crawled deep pages and makes the list of URLs with the keywords, description and title which will show crawled data of different websites and index the deep pages. Indexed data stores in database so that at the time of searching the keywords found and it efficiently gives the output in structured web page.

Step 5: After indexing deep pages, update meta tags.

As shown in Figure 7, the meta tags starts updating so that the data in database is updated and the user can get correct information from the different websites on a single structured page.

Step 6: Input keyword to search.

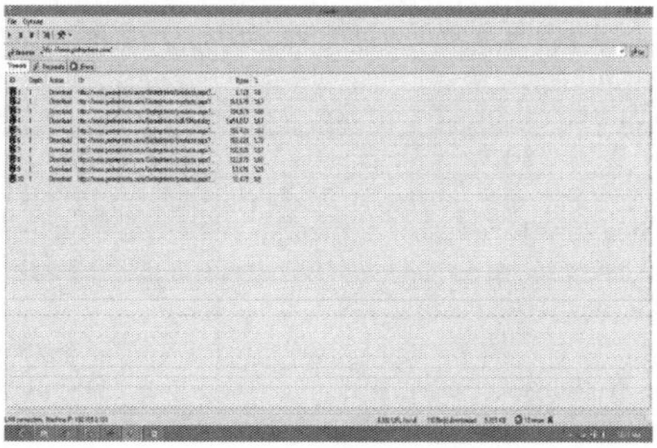

As shown in Figure 8, the user search for keyword as enter the keyword and click the search button and the result will be displayed in structured page containing information or data from different websites i.e. stored in database.

Step 7: Result displayed in structured web page.

As shown in Figure 9, the result is displayed in structured web page containing information from the different websites for a single keyword as in fig keyword searched is "chair" and also if the user wants to see the detailed information about the product click on the open button and it directly opens the web page which contains all the information about the product.

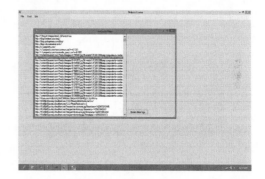

Figure 7.   Update meta tags.

Figure 8.   Search keyword.

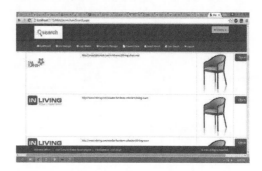

Figure 9.   Result for searched keyword.

Figure 5.   Crawling deep web in threads.

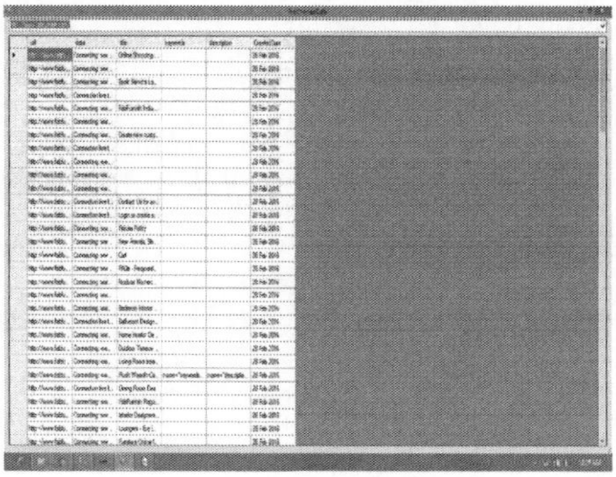

Figure 6.   Indexing deep pages.

6 RESULT AND ANALYSIS

The algorithm mentioned in the process flow and architecture of proposed system are applied to access the deep pages from some Online Furniture Websites containing deep pages in unstructured format. The algorithm of crawling process were applied on web crawler to get the deep pages as resulted in Step 3, 4 & 5 of working of system. The searched result of keyword in single structured web page is resulted in Step 7. In this paper, the web search engine is designed and developed for online furniture websites as there is no such search engine for furniture domain. The search engine and web crawler has executed well for deep pages. The websites which were taken for practical implementation has been shown in Table 1. These websites are added in the site manager and crawler crawl deep pages of these websites and the deep pages from only http linked websites are accessed.

To check the accuracy of getting the deep pages from the sample websites as shown in Table 2 with the help of parameters: precision, recall and f-measure.

By taking a website like fabfurnish.com to show how these parameters (precision, recall and f-measure) are calculated to get the accuracy table. In fabfurnish.com, the unique hits count (URL found) by search engine are 1011 URL found where total hits count are 112 files downloaded and the errors count are 18.

$$(P) = \left(\frac{tp}{tp + fp} \right) \quad (1)$$

In equation (1), P is precision, tp is true positive which is equivalent with hit, fp is false positive which is equivalent with type I error (incorrect rejection of a true null hypothesis). Precision is the retrieved instances that are relevant.

$$P = 112/1011 = 0.11$$

$$(R) = \left(\frac{tp}{tp + fn} \right) \quad (2)$$

In equation (2), R is Recall, tp is true positive which is equivalent with hit, fn is false negative which is equivalent with miss, type II error (failure to reject a false null hypothesis). Recall is the fraction of relevant instances that are retrieved.

$$R = 112/112 + 18 = 0.46$$

$$f - measure = \left(\frac{2(PR)}{P + R} \right) \quad (3)$$

In equation (3), P is precision and R is recall which is a measure that combines precision and recall is the harmonic mean of precision and recall.

F-measure = 0.17

The Table 2 consists of some sample websites with their values of parameters: precision, recall and f-measure where the precision of fabfurnish is 0.11, recall is 0.46 and f-measure is 0.17; precision of inliving is 0.17, recall is 0.34 and f-measure is 0.22; precision of pepperfry is 0.16, recall is 0.48 and f-measure is 0.24; precision of furniture is 0.13, recall is 0.37 and f-measure is 0.19; precision of godrejinterio is 0.18, recall is 0.47 and f-measure is 0.26; precision of woodenstreet is 0.21, recall is 0.47 and f-measure is 0.29.

Graph 1 shows the precision, recall and f-measure for fabfurnish, inliving, pepperfry, furniture, godrejinterio and woodenstreet websites.

As shown in Graph 1, the algorithm used for retrieving the deep pages has returned most of the relevant results as it is shown in the graph that recall is higher and precision measures the quality of deep pages. As the maximum precision from sample websites is of "woodenstreet.com" i.e. 0.21 which means the quality of deep pages of this website is better than other websites as high precision means that an algorithm returned substantially more relevant results than irrelevant. The minimum precision is of "fabfurnish.com" which means among the sample websites this website has lowest quality of deep pages. The highest recall among sample websites is of "pepperfry.com" i.e. 0.48 which means that an algorithm returned most of the relevant results of this website where the lowest recall is of "inliving.com". The f-measure considers both precision and recall of the test to compute the score where f-measure reaches its best value at 1 and worst at 0.

Table 1. Sample websites for retrieval of deep pages.

Name of websites	Deep pages retrieved or not
www.fabfurnish.com	Yes
www.inliving.com	Yes
www.pepperfry.com	Yes
www.furniture.com	Yes
www.godrejinterio.com	Yes
www.woodenstreet.com	Yes
www.urbanladder.com	No (https website)
www.mebelkart.com	No (https website)

Table 2. Accuracy table for retrieval of deep pages.

Websites	Precision	Recall	F-measure
fabfurnish	0.11	0.46	0.17
inliving	0.17	0.34	0.22
pepperfry	0.16	0.48	0.24
furniture	0.13	0.37	0.19
godrejinterio	0.18	0.47	0.26
woodenstreet	0.21	0.47	0.29

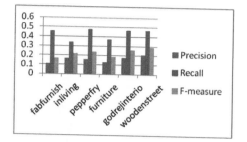

Graph 1.   Precision, recall and f-measure.

7   CONCLUSION

In this paper, an architecture is designed for deep web search engine for Online Furniture Website and this domain is taken for a purpose because there is no such search engine for Online Furniture Websites and this architecture can be used in different domains also. The three-tier architecture is designed which consists of GUI, search engine and database. The search engine consists of three modules i.e. crawling module, indexing module and searching module which explains the process of the system. An algorithm is used for crawling process, crawled the deep pages from different websites and indexing the deep pages which made the web search engine's working possible in efficient manner. The implementation and working of proposed system is explained step by step. Then the result is discussed with the table of sample websites for retrieval of deep pages and to check the accuracy of getting the deep pages from the sample websites with the help of parameters: precision, recall and f—measure. The accuracy table and its graph is discussed in which it explains that the high precision means that algorithm returned substantially more relevant results than irrelevant and high recall means that an algorithm returned most of the relevant results.

REFERENCES

Amrin Andas, Xia Chunlei and Dai Shuguang, May 21, 2015, Focused Web Crawling Algorithms, University of Shanghai for Science and Technology, 516 Junggong Rd., Yangpu District, Shanghai, China. doi: 10.17706/jcp.10.4.245–251

Bergman Michael K., September 24, 2001, The Deep Web: Surfacing Hidden Value, WHITE PAPER, BrightPlanet—Deep Content.

Buttler D., Liu L. and Pu C., 2001, "A Fully Automated Extraction System for the World Wide Web," Proc. 21t Int'l Conf. Distributed Computing Systems (ICDCS 01), IEEE CS Press, pp. 361–370.

Chang C.-H. and Lui S.-L., 2001, "IEPAD: Information Extraction Based on Pattern Discovery," Proc. 10th Int'l Conf. World Wide Web (WWW 01), ACM Press, pp. 681–688.

Choudhari Rahul, Choudhari R.D. and Choudhari Ajay, 2008, Increasing Search Engine Efficiency Using Cooperative Web,, Computer Science and Software Engineering, International Conference on (Volume: 4) on IEEE.

Eikvil, L., (July 1999), Information Extraction from World Wide Web—A Survey, Technical Report 945, Norweigan Computing Center, Oslo, Norway.

Embley D.W., Campbell D.M., Jiang Y.S., Liddle S.W., Lonsdale D.W. and Smith R.D., November 1999, Conceptual-model-based data extraction from multiple-record Web pages, Data & Knowledge Engineering, Volume 31, Issue 3.

Faheem Muhammad, 2012, Intelligent Crawling of Web Applications for Web Archiving, published in WWW companion ACM.

Ferrara Emilio, De Meo Pasquale, Fiumara Giacomo and Baumgartner Robert, November 2014, Web data extraction, applications and techniques: A survey, Knowledge-Based Systems, Volume 70.

He Yeye, Xin Dong, Ganti Venkatesh, Rajaraman Sriram and Shah Nirav, 2013, Crawling, Deep web entity pages. In proceeding of WSDM'.

Kakde Yogesh, Kumar Rawat Manoj and Dangra Jitendra, May 2015, Implementation of Efficient Extraction of Deep Web by Applying Structured Queries and Maintaining Repository Freshness Volume 5, Issue 5, ISSN: 2277 128X.

Kakde Yogesh and Kumar Rawat Manoj, 2 February 2014, Accumulative search engine effectiveness using supportive web, ISSN No: 2347–4890 Volume 2 Issue.

Kausar Md. Abu, Dhaka V. S. and Kumar Singh Sanjeev, February 2013, Web Crawler A Review, International Journal of Computer Applications (0975–8887) Volume 63– No. 2.

Kumar Rahul, Jain Anurag and Agrawal Chetan, September 2014, Survey of web crawling algorithms, Vol. 1, No. 2/3.

Lacroix Zoé, March 2003, Web data retrieval and extraction, Data & Knowledge Engineering, Volume 44, Issue 3.

Lam Man I., Gong Zhiguo and Muyeba Maybin, 2008, A Method for Web Information Extraction, Online ISBN 978-3-540-78849-2.

Narayan Das Nripendra and Kumar Ela, Apr 2014, Automatic extraction of data from deep web page, International Journal of Computer & Mathematical Science ISSN: 2347–8527.

Noor Umara, Rashid Zahid and Rauf Azhar, April 2011, A Survey of Automatic Deep Web Classification technique, International Journal of Computer Applications (0975–8887) Volume 19– No. 6.

Sharma Supriya and Sharma Minakshi, March 2013, Deep Web Data Mining, International Journal of IT, Engineering and Applied Sciences Research (IJIEASR) ISSN: 2319–4413.

Yang Jufeng, Shi Yan Guangshun and Qingren Wang Zheng, 2007, Data Extraction from Deep Web Pages, International Conference on Computational Intelligence and Security, IEEE.

Zheng Qinghua, Zhaohui Wu, Xiaocheng Cheng, Lu Jiang and Jun Liu, September 2013, Learning to crawl deep web, Information Systems, Volume 38, Issue 6.

Communication and Computing Systems – Prasad et al. (Eds)
© *2017 Taylor & Francis Group, London, ISBN 978-1-138-02952-1*

Performance of PI controller for FOPDT process model using Pade's approximation on set point tracking

Pawan Negi & Tushar
Faculty of Technology, UTU, Campus Dehradun, India

Alaknanda Ashok
GBPUAT & UTU, Dehradun, India

ABSTRACT: The industrial application system designed for the control of temperature using the controller or the combinations of controller are designed to meet the requirement. The main requirement of the system is to minimize the error and accelerate the process in the less time duration. The PI controller is designed for the process of heating tank for approximation of time delay is form by the Pade's first order technique. The simulation is performed to find the suitable tuning parameters for the controller using the analysis of characteristic i.e. transient as well as steady state.

1 INTRODUCTION

Among the various approximation techniques the Pade's approximation technique is preferred the most now maximum attraction. The production of many products in the industry the process needsrespective controller. So the identification of tuning for the particular controller carried out and compares the different tuning techniques to find best tuning for that particular task (Hussain et al. 2014). The approximation which is used for the delay approximation in the present work is Pade's first order. The process taken for this work is the heating tank. The process parameter in the case of heating tank which is have to control is temperature, so it may also called temperature control system. This process parameter is controlled by the controller which is to be designed. The PI controller have factor which remove the overshoot. The FOPDT model is consider for the heating tank system. Pade's approximation is used for the process control is

$$e^{-\theta s} = \frac{1 - \theta s/2}{1 + \theta s/2} \qquad (1)$$

Equation 1 shows the mathematical formula to formulate the approximation. Based on which the comparison can be form.

2 METHODOLOGY

In the controller design process the FOPDT (Ling and Baoguo 2013) model in the form of

$$G(s) = \frac{ke^{-s\tau}}{Ts} \qquad (2)$$

The process having the dead time is complex to control. The model for the FOPDT having the parameters are gain (k), dead time (τ) and time constant (T). In the normal system without having any controller shows unwanted results because of non-linearity like dead time, transportation lag (Rajvanshi & Juneja 2013), distance between sensors and most importantly the behaviour of system or model by slight chance in characteristic of different part of system or model. Response of the system is mostly affected by its parts like rise, peak, settling, overshoot time (Rajvanshi & Juneja 2013) so synchronization and setting parameter of parts is required to deliver the required output. Sometimes to avoid hazard or unwanted result additional element may needed in circuit, so the cost and losses of the model is increased. Redundant element which is normally not needed if there is no glitch or unwanted result present in system response. Removal of redundancy will also effects the system frequency as it take some time to get result as system frequency depend upon the turn on and turn off and glitch take some time.

The transfer function of the PI controller is

$$G(s) = K_c\left(1 + \frac{1}{T_i s}\right) \tag{3}$$

The transfer function for the heating tank process is (Darandale et al. 2013)

$$G_1 = \frac{2.2\,e^{-6s}}{40.484s + 1} \tag{4}$$

The given model approximated using the first order Pade's is

$$G_2 = \frac{2.2 - 6.6s}{121.452s^2 + 43.484s + 1} \tag{5}$$

Table 1. Tuning techniques for the PI controller of FOPDT.

Tuning methods	Time responses	IPDT model	Approximated model
Ziegler Nichols	Rise time	4.44	4.3
	Settling time	120	71.6
	Peak time	1.86	1.72
Tyreus Luyben	Rise time	7.51	8.34
	Settling time	43.6	33.4
	Peak time	1.18	1.13
Hagglund & Astrom	Rise time	16.3	17
	Settling time	84.4	84.3
	Peak time	1.2	1.2
Chien, Hrones & Reswich	Rise time	31.4	32.3
	Settling time	101	98.9
	Peak time	0.999	0.998

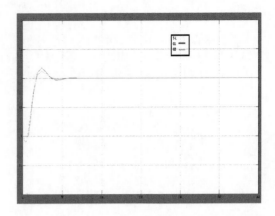

Figure 2. Analysis using G_1 and G_2 using Tyreus Luyben.

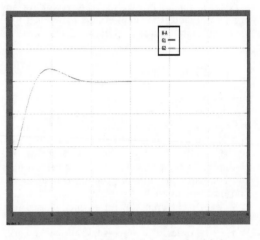

Figure 3. Analysis of G_1 and G_2 by Hagglund & Astrom tech.

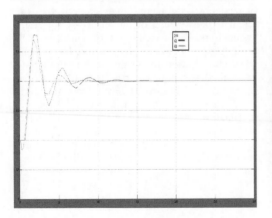

Figure 1. Analysis using dead time G_1 and approximated G_2 model by Ziegler Nichols technique.

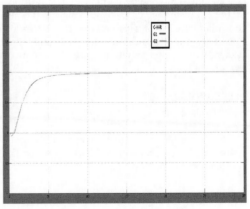

Figure 4. Analysis of G_1 & G_2 by Chien-Hrons & Reswich.

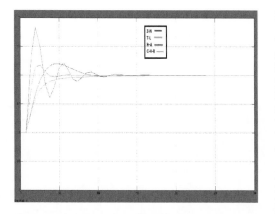

Figure 5. Analysis of transfer function G_1 in all the tuning for the PI controller.

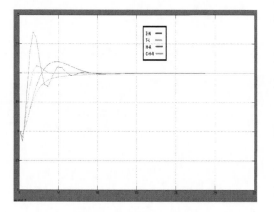

Figure 6. Analysis of transfer function G_2 in all the tuning for the PI controller.

Using the Routh criterion in approximated model one can found the ultimate gain $K_u = 6.588$ and the ultimate period $T_u = 17.59$. Using the value of k, τ, T, K_u and T_u the parameters K_c and T_i for PI controller tuning formulate.

Both response in the figure 5 and figure 6 shows the tuning of PI controller and having the nearby results. The figure 5 is only have the results of the transfer function having dead time. and in figure 6 the results related to the approximation of same transfet function.

3 CONCLUSION

The tabulation and the figures show the performance comparison for the specified model. The curve in the figure having x-axis in the time domain and y-axis defines the amplitude. The output is analyzed and plotted byusing the MATLAB. In case of set point tracking capability the parameter rise time, peak time and settling time is compared for the tuning for both the transfer function i.e. with dead time and approximated.

On the basis of rise time the tuning technique Ziegler Nichols is shown in case of approximated model gives better result due to less rise time and for peak time the minimum time taken by the method Chien, Hrones & Reswich. Based on the settling time the method Tyreus Luyben taking less time to settle down in compare with the other tuning method, and the best response given by the approximated model of Tyreus Luyben.

REFERENCES

Darandale, R.A. et al. 2013. Design of Model Predictive Control for Temperature Process. Proc. of Int. Conf. on Advances in Signal Processing and Communication, Elsevier.

Hussain, K. Mohamed et al. 2014. Comparison of PID Controller Tuning Methods with Genetic Algorithm for FOPTD System. K.M. Hussain et al. Int. Journal of Engineering Research and Applications. ISSN: 2248–9622. Vol. 4, Issue 2 Version 1: pp. 308–314.

Rajvanshi, Saurabh & Juneja, Pradeep 2013. Performance Evaluation of various controller designed for an industrial first order delay process.International Journal of Advanced Research in Electrical, Electronics and Instrumentation Engineering. Vol. 2 Issue 4.

Rajvanshi Saurabh & Juneja, Pradeep K. 2014. Set Point Tracking Capability Analysis of Various Controllers Designed For a First Order plus Dead Time Process. International Journal of Advanced Research in Electrical, Electronics and Instrumentation Engineering. ISSN (Print): 2320–3765 Vol. 3 Issue 5.

Xu, Ling & Xu, Baoguo 2013. A Simple PI/PID Controller Setting for Parallel FOPDT Process via Model Approximation. Journal of Computational Information Systems. 8839–8846.

Communication and Computing Systems – Prasad et al. (Eds)
© 2017 Taylor & Francis Group, London, ISBN 978-1-138-02952-1

Study of backup based issues in inter-intra converged WDM networks

Babandeep Kaur, Gurdeep Kaur & Parminder Singh
CEC Landran, India

ABSTRACT: Traffic is restored at the optical layer through diverse primary and backup light-paths in WDM based Networks. Backup paths share channels in a way that guarantees complete restoration against single event failures. Thus, two backup paths can share a channel only if their corresponding primary paths are diverse. From above problem, we are rectifying to implement 15 routers that may be deployed on different location. Survivability of the overall network to be studied in this paper; it may be inter (public) or intra (inner network). We are implementing decision based graphs to detect the shortest path of the network so that all traffic routed through shortest path.

Keywords: WDM, optical based Networks, Sequence number, NSFNET

1 INTRODUCTION

Wavelength Division multiplexing (WDM) networks, employing wavelength routing have emerged as the dominant technology to satisfy this growing demand for bandwidth. The number request handle by the optical based network to configure each node individually by the server and the modem used to connect the optical fiber through auxiliary port. The number of techniques applies on the network to connect millions of users and it is up to the network administrator which techniques to apply. The network techniques are used by the network administrator like packet switching and circuit switching. An Automatically Switched Optical Network (ASON) applies on the optical based networks to structure the WDM based network test beds. An ASON can be implemented as an AON, using Optical Cross-Connects (OXCs) to switch connections in the optical domain. Although ASONs hold the promise of fast network reconfiguration, carriers are unlikely to deploy an ASON unless it can be shown to have a clear cost benefit over a point-to-point WDM network [4].

An optical network provides circuit-switched end-to-end optical channels or light paths between network nodes and their users, the clients, it is possible for the receiving end to extract the respective lower speed channels again. A light path is made up of a wavelength between two network nodes that can be routed through multiple intermediate nodes and providing almost loss-less transmission over an enormous frequency range, making it capable of carrying enormous levels of traffic. The intermediate nodes direct the wavelengths. The optical

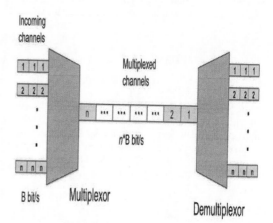

Figure 1. WDM Multiplexor and De-multiplexor [7].

network may thus be thought of as a wavelength-routing network. Light paths are set up and taken down as required by the users of the network. The need for multiplexing arises from the fact that it for most applications is less costly to transmit data at a high bitrate (e.g. Gbit/s) over a single fiber than it is to transmit it at lower rates (e.g. Mbit/s) over multiple fibers. There are two fundamentally different ways of multiplexing the lower bitrates onto a single fiber time division multiplexing (TDM) and wavelength division multiplexing (WDM) [5].

1.1 Characteristics of optical networks

The light paths of the optical network have several important characteristics:

- They are transparent, i.e. they can carry data at various rates, with different protocols etc. This enables the optical layer to support a variety of higher layer protocols concurrently.
- Wavelength and data rate used are set by the terminating nodes.
- Light paths can be set up and taken down on demand, equivalent to the establishment of circuits in a circuit switched network.
- Alternative light paths can be configured and kept in "standby mode" so that in the event of a failure, traffic may be re-routed and the service maintained.
- Wavelengths can be reused. If a light path using a particular wavelength ends in one node, the same wavelength can be re-used in another light path heading in another direction.

The whole concept of WDM and light paths is based on analog optical transmission techniques, making parameters such as dispersion, signal attenuation, optical signal to noise ratio and interference over the whole length of the path important to control.

1.2 Types of WDM

In Figure 2, data from the customers reach the backbone network through local routers. The local routers exchange data using light-paths, which are provided by the backbone optical network. These light paths form the internal traffic of the optical network. Some nodes of the network are directly connected to the local routers. We call these nodes edge nodes. The local routers can also communicate with networks of other carriers, or external networks, through selected nodes in routers.

These all devices worked on the different network types and have been divided into two categories like point to point WDM network and switched optical network that may be automatically. These categories further divided into subcategories i.e. ordinary and links whereas the automatically switched networks divided into all optical, IP over optical and waveband.

2 GAPS IN STUDY

We focus on the design of a backbone network and related to the literature has been studied in this paper.

3 PROBLEM STATEMENT

Carsten Behrens, Ralf Hulsermann, implemented the Metro IP router (MR) to configuring wavelength division multiplex (WDM). MR used to support dense networks as well as to supporting optical fiber based networks. The optimum paths of the fiber connections between metro router (MR) and core nodes have been found based on central offices nearby host nodes. This work analysis solutions that was robust against the inevitable uncertainty with respect to traffic and cost predictions. However, future work may be directed at including the impact of additional criteria increasing number of locations, into the design decisions. Finally, even though a trend to fewer and fewer core router locations is indicated, practical considerations such as survivability and node and link-disjoint routing requirements limit the minimum number of core router locations to approximately 15.

4 PLANNING OF WORK

We focus on the design of a backbone network. When one or more links of a WDM based networks fail due to reasons like a fiber cut, link failure, or IP backbone router failure, the network should be able to recover. For this reason, for each path (known as the primary path or service path) between a source-destination node pair, one or more additional paths (known as secondary paths or restoration paths) are provided in a network. These additional resources are used to ensure the minimum possible recovery time in the event of a failure. Each light-path is then separated into packets and processed by the IP router.

A link is unidirectional, if all light-paths in the link are in one direction; bi-directional, if the light-paths in the link are in two directions.

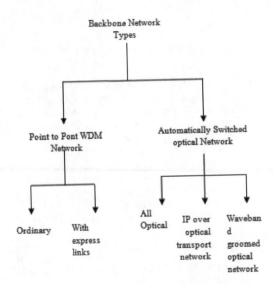

Figure 2. Network Types.

Name of Author	Title	Description	Not to be done in the paper
Marco D. D. et al.	Protection and Minimal Interference in WDM Mesh Networks	This paper works on WDM protection methods such as WDM shared protection and non-protection methods. The WDM shared protection improve the performance by utilizing resources but fail to backup processes because single fibre cut reduce the efficiency of the network. Whereas the non-SRLG protection measured and improve the efficiency of the network but it shared the maximum primary links. This protection method failed to reserve the channels like channel assignment problem is the major issue.	• Channel Assignment • Not create backup paths • Not provide solution if the optical fiber fails
Saraswati Bhakare et al.	Hybrid Approach to Recover from Double Link Failure. Using Backup Path in Ad-Hoc Networks	This paper explained the Mobile AdHoc network (MANETs) and estimates the failure of link if accidentally occurs. The backup process has been formulating and if single failure occurred then the data is safe by employing dual link failures.	It does not discuss wired links. It does not discuss channel assignment Memory cache does not maintained.
Refat Kibria	Multilayer Protection And Survivability In WDM optical Network	This paper works on Multi layer optical networks. It survive the data by using layers like physical, MAC and Network Layer. The Link Layer protection methods has also been discussed in this paper. This paper is related to conceptual not discussed the implementation part. The WDM protection methods was explained in this paper and these methods are: Path based protection, Link based protection and WDM Layer protection. Each methods has been compared and analysed that Link protection methods is best because it uses channelization technique to back-up the data.	It does not discuss the scenario. Implementation skipped Work on wired networks.

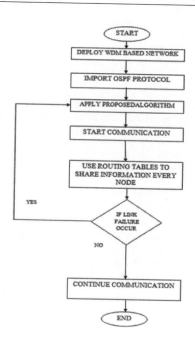

Figure 3.   Flowchart.

5   RESULTS AND DISCUSSIONS

5.1   *Sequence number of received packets*

The time interval from the beginning of the time slot allocated to a particular path until the next recurrence of that particular time slot is commonly referred to as a frame. As a result, timing information is required at both the bit (time slot) and frame levels. The generated packets exactly received by the receiver side and there are some fluctuation and differences on the receiver and sent packets and that could be assumed in dropped packets. The x-axis represents the simulation time of the current network and y-axis refers to the total number of packets and denoted in bytes.

5.2   *Blocking probability*

In this result, we evaluated and compare the three network in terms of network failure rate. We are assuming that if the blocking rate is greater than 30 percent then the chances of packet loss is more. If this blocking rate is less than 30 percent; the network is capable to fix in wide area networks.

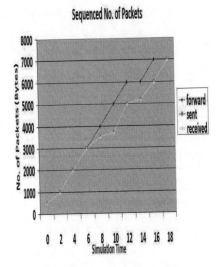

Figure 4. Sequence number of packets.

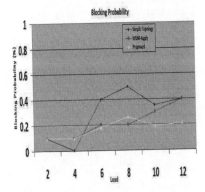

Figure 5. Sequence number of received packets.

Table 1. Blocking probability.

Network topology	Blocking probability (%)
Simple Network	40 (max.)
WDM Network	30 (max.)
Proposed Network	20 (max.)

We are comparing three different topologies, first the network is taken as simple Ethernet based network but congestion rate problems more in the case when TCP send bulk of data and resultant more than 40 percent block rate. Secondly, we take an WDM scenario and this scenario is using fiber based network lies between 10 to 20 percent. If any of failure occur or path diversion is choose as secondary option than blocking probability of the network is 20 to 30

percent. In the third network which is the proposed work and route diversion policy is chooses wisely. Thus the blocking probability of the network.

6 CONCLUSION

It is noticeable that the optimum transmission range obtained by the simulation results for the proposed approach does not exactly match the optimum transmission range obtained previously by numerical analysis. This also concluded that light-path based connection make the WDM network stronger. It also helps to improve the network performance by increasing network throughput and decreasing the delay in the network.

The proposed algorithm defined on the WDM based network and to manage the information cycle on every node. The results described in the section v is concluded and defined in the Table 1.

REFERENCES

Abhishek Bandyopadhyay, Mohtasham Raghib, Uma Bhattacharya, Monish Chatterjee, "Dynamic Survivable Traffic Grooming with Effective Load Balancing in WDM All Optical Mesh Networks", IEEE, 2014, pp. 1388–1394.

Carsten Behrens, Ralf Hulsermann, Monika Jager, Christian Raack, "On the Optimum Topology of a Nation-wide Aggregation and Core Network", IEEE, 2015 http://www.opnet.com/university_program/itguru_academic_edition/.

Farhan Habib, M. Massimo Tornatore, Ferhat Dikbiyik and Biswanath Mukherjee, "Disaster survivability in optical communication networks", Computer Communication, 2013, pp. 630–644.

Mallika and Neeraj Mohan, "Link Failure Recovery in WDM Networks", International Journal of Computer Science and Electronics Engineering, 2013, pp. 599–602.

Paramjeet Singh, Ajay K. Sharma and Shaveta Rani, "Minimum connection count wavelength assignment strategy for WDM optical networks", Elsevier, 2008, pp. 154–59.

Paulo J. S. Junior, Andre C. Drummond, "An Algorithm for Resource Allocation and Partial Protection of Transparent Optical WDM Networks with Service Differentiation", IEEE, 2014, pp. 361–368. Refat Kibria and Md. Aminul Haque Chowdhury, "Multilayer Protection and Survivability in WDM optical Network", Daffodil International University Journal of Science and Technology, 2007, pp. 19–27.

Payman Samadi, Junjie Xu, Ke Wen, Hang Guan, Zhuo Li, Keren Bergma, "Experimental Demonstration of Converged Inter/Intra Data Center Network Architecture", IEEE, 2015, pp. 1–4.

Saraswati Bhakare, C.A. Laulkar and Rashmi Hiraje, "Hybrid Approach to Recover from Double Link Failure Using Backup Path in Ad-Hoc Networks", IRNet Transactions on Computer Science and Engineering, 2011, pp. 47–51.

WDM Reference Guide, FINISAR corp. 2008.

Communication and Computing Systems – Prasad et al. (Eds)
© 2017 Taylor & Francis Group, London, ISBN 978-1-138-02952-1

Design of modified Smith predictor for dead time compensation for SOPDT process

Swati Singh, Mayank Chaturvedi & Pradeep K. Juneja
Graphic Era University, Dehradun, India

ABSTRACT: process industry many time used second order plus dead time process as an industrial process. Time taken between input and output of a phenomena in an industry is called dead time also known as transportation lag. Smith predictor is a dead time compensation technique which is used to compensate dead time. Smith predictor compare process model and actual process of the system until error between them zero, but there are some limitations of smith predictor if actual process and process model is not match. Modified Smith predictor is used to achieve stability and zero steady state error to step load response.

1 INTRODUCTION

In this paper we present a time delay compensation technique which deal with problematic area in process control industries-occurrence of time delay (Seborg). Time delay is also known as dead time, transportation lag. An interval during which an actuating signal produces no response known as dead time (Seborg; Chaturvedi & Juneja 2013). Smith predictor is the dead time compensation technique which is used to compensate the dead time but there are some draw back if actual process and model output is not match (Hang).Here we use modified Smith predictor to compensate the dead time.

The dynamics of most of the industrial processes is modeled by Second Order Plus Dead Time (SOPDT) transfer function. A number of control problems in the process industry are solved using PID controllers because of several reasons-maintenance and operation of PID controllers is simple and also they are robust in nature (Visioli 2012). Tuning rule for PID controller is less for SOPDT as compare to FOPDT (Tan et al. 2006).

In the present analysis second order plus dead time is consider. The representation of second order plus dead time is

$$G(s) = \frac{k}{(\tau_1 + 1)(\tau_2 + 1)} e^{-\theta s} \quad (1)$$

where k is the gain of the process θ is the delay time, τ_1 and τ_2 are the time constant.

The PID controller is probably the most widely-used type of feedback controller. PID stands for Proportional-Integral Derivative, referring to the three terms operating on the error signal to

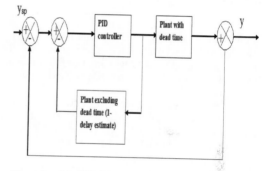

Figure 1. Modified Smith predictor.

produce a control signal. Stability can often be ensured using only the proportional term (Shi & Lee 2004). Proportional-Integral-Derivative (PID) controllers are the most widely-used controllers in process industry, majorly because it is very difficult to improve their cost/advantage ratio (Tan et al. 2006).

PID controllers have simple structure and tuning methods are widely presented (Chaturvedi & Juneja 2013).

$$G_c(s) = K_p \left(1 + \frac{1}{\tau_i s} + \tau_d s \right) \quad (2)$$

2 METHODOLOGY

In this present analysis, a second order plus dead time model is selected (Seborg). After selection of a suitable second order plus dead time process model, controller parameter are calculated to

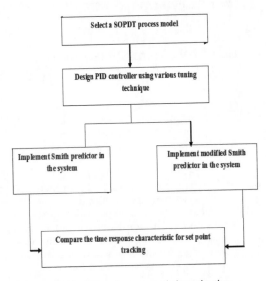

Figure 2. Flow chart for systematic investigation.

design PID controller using various tuning techniques Ziegler Nichols (ZN), Tyreus Luyben (TL), Pettit and Carr (PC), Chau (C), Bucz (B), modified and Damped Oscillation (DO) method is used (Panda).

Implement Smith predictor and modified Smith predictor in the system for dead time compensation. Smith is used PID controller for SOPDT process model (Skogestad 2003). Compare the step responses of PID controller for closed loop system design using various tuning techniques with Smith predictor and with modified Smith predictor. Implement Smith predictor and modified smith in the system for dead time compensation.

SOPDT process model is used for many temperature processes. A temperature control loop is selected from literature (Dwyer 2000).

SOPDT process model is used for many temperature processes (Seborg).

$$G(s) = \frac{2}{(10s+1)(5s+1)}e^{-s} \quad (3)$$

where process gain (k) = 2, Delay time = 1 second, slow time $\tau_1 = 10$ second, fast time $\tau_2 = 5$ second.

3 RESULTS AND DISCUSSION

PID controller has been designed using different tuning techniques and implemented in the closed loop, without Smith Rise time decreases 3.19 second with modified smith predictor.

Comparison of step response of PID controller designed using TL tuning technique without Smith predictor (A_4), with modified Smith predictor, with Smith predictor and with Modified Smith predictor for selected SOPDT process model. Simulation is performed to evaluate time response characteristics without Smith predictor with Smith predictor and Modified Smith predictor for SOPDT model.

Figure 3 exhibit the Comparison of step response of PID controller designed using ZN tuning technique without Smith predictor (A_1), with modified Smith predictor (A_2) and with Smith predictor (A_3) and for selected process model. Settling time is 22 second.

(A_5) and with Smith predictor (A_6) for selected process model show in Figure 4. Figure 5 indicate the Comparison of step response of PID controller designed using PC$_1$ tuning technique without Smith predictor (A_7), with modified Smith pre-

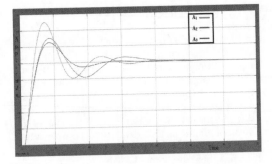

Figure 3. Comparison of step response of PID controller designed using ZN tuning technique without Smith predictor (A_1), with modified Smith predictor (A_2) and with Smith predictor (A_3) and for selected process model.

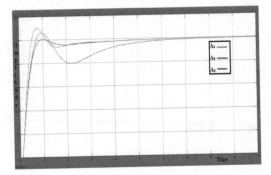

Figure 4. Comparison of step response of PID controller designed using ZN tuning technique without Smith predictor (A_1), with modified Smith predictor (A_2) and with Smith predictor (A_3) and for selected process model.

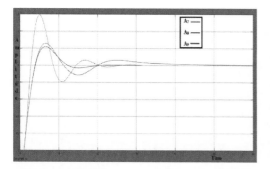

Figure 5. Comparison of step response of PID controller designed using PC_1 tuning technique without Smith predictor (A_7), with modified Smith predictor (A_8) and with Smith predictor (A_9) and for selected process model.

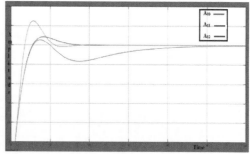

Figure 6. Comparison of step response of PID controller designed using PC_2 tuning technique without Smith predictor (A_{10}), with modified Smith predictor (A_{11}) and with Smith predictor (A_{12}) and for selected process model.

dictor (A_8) and with Smith predictor (A_9) and for selected process model.

Comparison of step response of PID controller designed using PC_2 tuning technique without Smith predictor (A_{10}), with modified Smith predictor (A_{11}) and with Smith predictor (A_{12}) and for selected process model show in Figure 6.

Figure 7 exhibits the Comparison of step response of PID controller designed using PC_3 tuning technique without Smith predictor (A_{13}), with modified Smith predictor (A_{14}) and with Smith predictor (A_{15}) and for selected process model. Comparison of step response of PID controller designed using C_1 tuning technique without Smith predictor (A_{16}), with modified Smith predictor (A_{17}) and with Smith predictor (A_{18}) and for selected process model show in Figure 8.

Comparison of step response of PID controller designed using C_2 tuning technique without Smith predictor (A_{19}), with modified Smith predictor (A_{20}) and with Smith predictor (A_{21}) and for selected process model is given by Figure 9. Figure 10 exhibits the Comparison of step response of PID controller designed using B1 tuning technique without Smith predictor (A_{22}), with modified Smith predictor (A_{23}) and with Smith predictor (A_{24}) and for selected process model.

Comparison of step response of PID controller designed using B_2 tuning technique without Smith predictor (A_{25}), with modified Smith predictor (A_{26}) and with Smith predictor (A_{27}) and for selected process model is given by Figure 11. Fig 12 show the Comparison of step response of PID controller designed using MZ 1 tuning technique without Smith predictor (A_{28}), with modified Smith predictor (A_{29}) and with Smith predictor (A_{30}) and for selected process model.

Comparison of step response of PID controller designed using MZ 2 tuning technique without

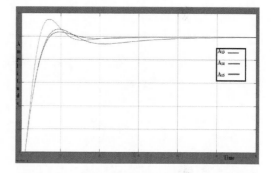

Figure 7. Comparison of step response of PID controller designed using PC_3 tuning technique without Smith predictor (A_{13}), with modified Smith predictor (A_{14}) and with Smith predictor (A_{15}) and for selected process model.

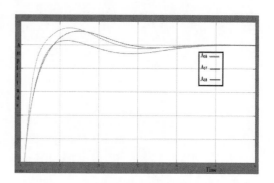

Figure 8. Comparison of step response of PID controller designed using C_1 tuning technique without Smith predictor (A_{16}), with modified Smith predictor (A_{17}) and with Smith predictor (A_{18}) and for selected process model.

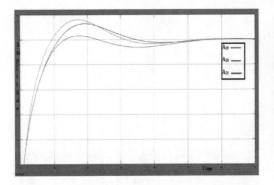

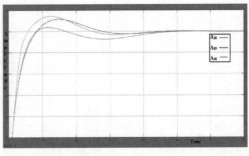

Figure 9. Comparison of step response of PID controller designed using C_2 tuning technique without Smith predictor (A_{19}), with modified Smith predictor (A_{20}) and with Smith predictor (A_{21}) and for selected process model.

Figure 12. Comparison of step response of PID controller designed using MZ_1 tuning technique without Smith predictor (A_{28}), with modified Smith predictor (A_{29}) and with Smith predictor (A_{30}) and for selected process model.

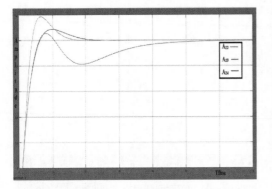

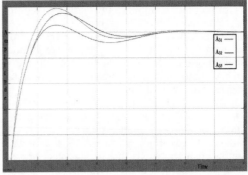

Figure 10. Comparison of step response of PID controller designed using B_1 tuning technique without Smith predictor (A_{22}), with modified Smith predictor (A_{23}) and with Smith predictor (A_{24}) and for selected process model.

Figure 13. Comparison of step response of PID controller designed using MZ_2 tuning technique without Smith predictor (A_{31}), with modified Smith predictor (A_{32}) and with Smith predictor (A_{33}) and for selected process model.

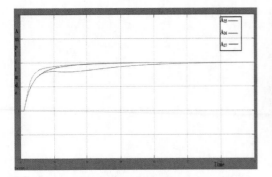

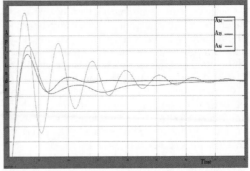

Figure 11. Comparison of step response of PID controller designed using B_2 tuning technique without Smith predictor (A_{25}), with modified Smith predictor (A_{26}) and with Smith predictor (A_{27}) and for selected process model.

Figure 14. Comparison of step response of PID controller designed using DS tuning technique without Smith predictor (A_{34}), with modified Smith predictor (A_{35}) and with Smith predictor (A_{36}) and for selected process model.

Table 1. Comparisons of important time response characteristics.

Tuning technique	Without Smith predictor			With Smith predictor			With Modified Smith predictor		
	Rise time	Settling time	Max.% overshoot	Rise time	Settling time	Max.% overshoot	Rise time	Settling time	Max.% overshoot
ZN	2.26	34	50.1	3.32	14.8	14.9	3.19	22	22.5
TL	3.17	37.2	10.4	5.81	39.4	0	4.85	38.2	.618
PC1 (under damped)	1.41	19.2	59.1	2.39	12.1	11.3	2.29	16.4	21.9
PC2 (critically damped)	1.89	10.4	24.9	3.73	13.5	3.69	3.23	12.9	8.5
PC3 (over damped)	2.94	12.3	11.2	5.5	8.51	.175	4.27	11.4	3.82
C1 (small overshoot)	3.06	36.9	14.4	5.81	25.6	9.43	5.1	36.6	11.7
C2 (without overshoot)	6.04	46.4	16.1	8.35	32	10.4	7.65	45.8	12.9
B1 (over shoot ≤ 20)	2.18	14.3	18.7	4.28	18.4	5.22	3.71	17.1	8.78
	6.8	22.1	0	11.2	25.4	0	9.69	23.6	0
MODIFIED ZN (small overshoot)	3.06	36.9	14.4	5.81	25.7	9.43	5.11	36.6	11.7
MODIFIED ZN (no overshoot)	5.8	45.9	18.8	7.97	44.8	12.1	7.35	46.2	15
Damped Oscillation	1.33	75	90.8	1.97	16.6	19.1	2.02	21.9	35.7

Smith predictor (A_{31}), with modified Smith predictor (A_{32}) and with Smith predictor (A_{33}) and for selected process model is given by Figure 13. Rise time is reduced 7.35 second which is decreases with modified Smith predictor. Figure 14 exhibits the Comparison of step response of PID controller designed using DS tuning technique without Smith predictor (A_{31}), with modified Smith predictor (A_{32}) and with Smith predictor (A_{33}) and for selected process model.

4 CONCLUSION

In present analysis PID controller has been designed using various tuning techniques and implemented in a closed loop, with Smith predictor, with modified smith predictor and without smith predictor for dead time compensation for SOPDT process model. In all of the cases rise time has increased with the implementation of Smith Predictor while Maximum Percentage Overshoot decreased. In most of the cases settling time has also decreased with the use of Smith Predictor.

Modified Smith predictor decreases rise time in most of the cases but settling time and maximum percentage overshoot increases. So effect of dead time is compensated by using modified Smith predictor. But if less settling time is one of the control objective apart from dead time compensation Smith Predictor is more advantageous.

REFERENCES

Abe, N., and K. Yamanaka. "Smith predictor control and internal model control-a tutorial." In SICE 2003 Annual Conference, Vol. 2, pp. 1383–1387. IEEE, 2003.

Chaturvedi, M., P. Juneja, "Effect of dead time approximation on controller performance designed for a second order delayed model, "Proceedings of the 2013 International Conference on Advanced Electronic Systems, ICAES 2013, CEERI, Pilani, Sept. 2013, pp. 313–315.

Chauhan, Prateeksha, Pradeep K. Juneja, and Mayank Chaturvedi. "Controller design and its performance analysis for a delayed process model." In Advances in Computing, Communications and Informatics (ICACCI, 2014 International Conference on, pp. 859–862. IEEE, 2014.

Hang, "Smith Predictor and its Modification," Control Systems, Robotics and Automation, Vol. 2, pp. 119–126.

O'Dwyer, "PI and PID controller tuning rules for time delay processes: a summary Technical Report" AOD-00–01, 1st Edition, Dublin, Ireland, 15 May 2000.

RC Panda "Introduction to PID Controllers—Theory, Tuning and Application To Frontier Areas" InTech publication.

Seborg, D. E., T. F. Edgar, D. A. Mellichamp, F. J. Doyle "Process Dynamics and Control", 3rd Edition, Wiley Publications, United States.

Shi, J., W. S. Lee, "Set point response and disturbance rejection tradeoff for Second-Order plus Dead Time Processes," 5th Asian Conference, pp. 881–887, 2004.

Sigurd Skogestad, "Simple analytic rules for model reduction and PID controller tuning", Journal of Process Control Vol. 13, pp. 291–309, 2003.

Tan, Wen, Jizhen Liu, Tongwen Chen, and Horacio J. Marquez. "Comparison of some well-known PID tuning formulas." Computers & chemical engineering 30, No. 9, pp. 1416–1423, 2006.

Visioli, "Research trends for PID controllers," Acta Polytechnica, Vol. 52, No. 5, pp. 144–150, 2012.

Communication and Computing Systems – Prasad et al. (Eds)
© 2017 Taylor & Francis Group, London, ISBN 978-1-138-02952-1

A preliminary comparison of machine learning algorithms for online news feature extraction and analysis

V.A. Ingle & S.N. Deshmukh
Department of CS and IT, Dr.B.A.M. University, Aurangabad, Maharashtra, India

ABSTRACT: Information from social network sites, blogs and web pages generates massive amount of unstructured data. As this big data is generating at faster rate, analysis part is a challenge to data analysis. To manage and extract useful information from this type of data is a cumbersome job. In this paper, we discuss various machine learning techniques to process and handle this data. The advantages and drawbacks each method is stated based on various metrics such as scalability, real-time processing and data size supported. K-means clustering, SVM, and Genetic algorithms are applied on data to get performance impact of features. A comparison of these machine learning techniques has been performed in order to obtain a reasonable algorithm for detection of useful information from online updating resources such as news and tweets. The best result has been achieved by a k-means clustering.

Keywords: MLE; K-means; Genetic Algorithm; PCA; SVM

1 INTRODUCTION

Machine learning is a core (Rob Schapire et al.) sub domain of artificial intelligence. The other domain such as Statistics, mathematics, physics, theoretical computer science and more in combination with machine learning are used for various research applications. Section 2 briefly describes various machine learning techniques. Section 3 elaborates experimental results with some of machine learning techniques. Further research work is discussed in Section 4.

2 MACHINE LEARNING ALGORITHMS

The various machine learning algorithms are MLE, Decision trees, K-means Clustering, Dimension Reduction, Genetic Algorithms and Support Vector Machines.

2.1 MLE

Maximum Likelihood Estimation [MLE] is a procedure for finding the values (NIST/SEMATECH, 2012) of one or more parameters of a likelihood function which makes that a maximum. Maximum likelihood estimates is defined as the possible values of parameters that produce the maximum sample likely values for the model data.

With the use of MLE, Parameter valuation can be consistently completed. Desired arithmetic and maximum properties can be achieved with MLE methods. The primary disadvantage of MLE is that the likelihood function is required to be designed for a given allocation and estimation problem. For small sample the function is biased as well as optimum values are not applicable. MLE functions are can easily change with alteration of initial values.

2.2 Decision trees

Decision Trees work on the concept of presenting likely output depending on instance of assessment and its corresponding sub consequences. Methods of Decision tree (Murty et al. 2011) use the real values of data attributes to build a model that will deliver possible outcomes.

Decision trees judge the likelihood and outcome (NIST/SEMATECH, 2012) of decisions, using that best possible conclusion can be obtained. If depth of decision tree is small, it is very easy to analyze the results. The cons of decision trees are that, the expectations play a major role in retrieving information from decision trees. Decision trees outcome may change on occurrence of unpredicted events.

2.3 K-means clustering

In K-means Clustering Algorithm, data objects within the same cluster have great similarity, but data objects within the different cluster (Michael et al.) have great non-similarities. A cluster is the spaces which contain the number of data object relatively high dimension cut off by the space which contains the number of data object relatively low dimension.

K-means algorithm is a standard algorithm to define clusters; (StatSoft, 2013) this algorithm is reasonably simple and rapid. The output is relatively flexible and high efficient for large volume of data. It can only process the numerical value, due to limitation of the Euclidean distance.

2.4 Dimension reduction

Principal Component Analysis is a common unsupervised learning technique as dimension reduction for finding (Fodor et al. 2002) patterns in data of high dimension. The process consists of converting a set of observations of possibly correlated variables to linearly uncorrelated variables using orthogonal transformation (Smola et al. 2004) given the orthogonal components.

PCA manages (Karamizadeh et al. 2013) the entire data for the principal components analysis without taking into consideration the fundamental class structure. The drawbacks are, the accuracy of covariance matrix is not guaranteed. The training data must provide explicitly the information to capture small invariance.

2.5 Genetic algorithms

Genetic algorithm (Amsellem, 2014) is used as an optimization algorithm. This means that it is one of the so-called optimization methods for searching optimums (maximum or minima). It can solve every optimization problem. It solves problems with multiple solutions.

With GA it is possible to solve non-differential, multi-dimensional (Safaric et al., 2006), non-continuous, and even non-parametrical problems. Genetic algorithms are easily transferred to existing simulations and models. The shortcoming of GA is that the optimum solution is not guaranteed. Variant Problems are not having solution with GA. It is not feasible to use genetic algorithms for on-line controls in real systems without using them on simulation model.

2.6 Support Vector Machines

Support Vector Machines (SVMs) are used for classification as well as regression. A point in n-dimensional space is data item and coordinate is value of feature particular coordinate. Hyperplane is used to distinguish the two classes.

In general SVM (Ivanciuc, 2007) has multiple continuous and categorical variables. It is very efficient in high dimensional spaces. It uses values

stored in support vectors (Smola et al. 2004), so it saves memory. Customization of kernel as per requirement is possible in SVM. The difficulty arises when samples are very less than features it gives worst performance. Output probability estimates are not directly given it requires calculation with expensive five-fold cross-validation.

3 EXPERIMENTAL RESULTS

There are stages for data processing as shown in (Fig. 1). At stage one we collect online unstructured text data, then for second stage apply basic text mining transformations for data cleaning purpose and third stage data is processed using machine learning algorithms such as k means clustering and genetic algorithm.

3.1 Results with SVM

At first for implementation of Support Vector Machines,

i. The stock market data for a particular company such as 'Tata' as closing and opening values with reference nifty is captured online for a particular time period.
ii. Total data values are in the range of 4500–7000 representing stock values for a particular day approximately 60 days of data is stored in. csv
iii. Data values are plotted in 2 dimensional space as shown in (Fig. 2a)
iv. Line passing through the points is plotted to form linear regression model is also called fitting of line as shown in (Fig. 2b)

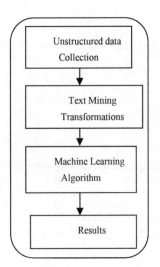

Figure 1.   Data processing stages.

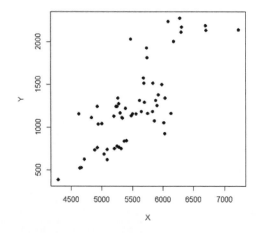

Figure 2a.   Initial plot of data values.

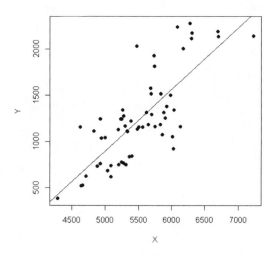

Figure 2b.   Fitting of line for linear regression model.

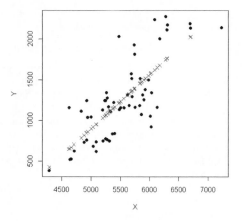

Figure 2c.   Prediction of values with linear regression points in blue.

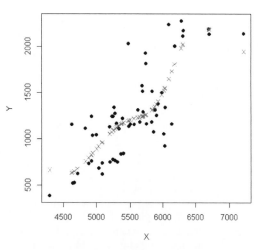

Figure 2d.   Prediction of values with support vector regression cross points in red.

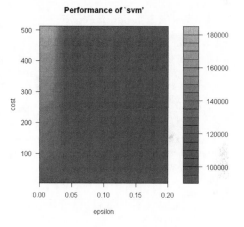

Figure 2e.   Tuning of SVM.

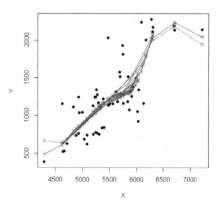

Figure 2f.   Original model in blue and tuned model in red lines.

v. As shown in (Fig. 2c), prediction of values with support vector regression is applied, the points are shown in blue colour.

vi. To find out the best performance of the model we have fine tune the parameters.

vii. (Fig. 2d) shows the model at a glance, fine-tuned model is in blue colour and original model is shown in blue.

viii. With RMSE error taken into account, for the darker region model our error rate decreases. (Fig. 2e) shows the tuning of model.

ix. The value the constant $\in$ is taken between 0 and 0.2. The cots value between 400 and 500 is giving good performance.

x. (Fig. 2f) gives comparison between original model in blue and tuned model in red lines.

xi. The tuned model has a sampling method as 10-fold cross validation and best parameters $\in = 0.2$ and cost = 4 and value of best performance is 103628.6

3.2 Results with K-means clustering

Initially steps for K-means clustering are:

i. Data collection for creation of text corpus is carried out by capturing online news related to stock exchange from news sources such as Google News, Yahoo Finance etc.

ii. The news data is pre-processed by using basic text mining operations.

iii. The data obtained is having large amount of sparse Terms.

iv. The unwanted terms are removed, so that the plot of clustering will not be crowded with words.

v. Hierarchical clustering is used for formation of term clusters.

vi. Hierarchical clustering group data over a variety of scales by creating a cluster tree or dendrogram.

vii. Then the distances between terms are calculated with dist () after scaling.

viii. After that, the terms are clustered with hclust () and the dendrogram is cut into 2 clusters.

ix. The agglomeration method is set to ward, which denotes the increase in variance when two clusters are merged. The output Word clustering dendrogram is shown in (Fig. 3a)

x. Each document score is calculated for positive and negative sentiments using Hu and Liu's "opinion lexicon".

xi. Result analysis of carried out with the histogram of sentiment scores for about 30 news documents as shown in (Fig. 3b)

xii. From this positive sentiments are shown for Tata Company.

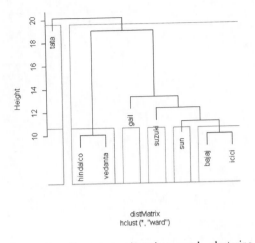

Figure 3a. K means clustering word clustering dendogram.

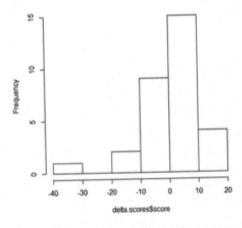

Figure 3b. Histogram of document sentiment scores.

3.3 Results with genetic algorithms

The next part considers use of Genetic algorithms.

i. Daily closing prices with SPY as open high, low, close volume from Google finance is downloaded for sample period from January 2000 to August 2015 and data stored in csv file.

ii. The optimal parameter set is obtained for specific indicators and using this fitness function is optimized over the sample period. Further the specific indicators are measured for values not covered in the sample period.

The SPDR Trust Series I (NYSE: SPY) is an exchange traded fund that tracks the S & P 500 index. [Boehmer et al. 2003]

iii. Long term momentum (Amsellem, 2014) and short term reversal are two main features.

944

Table 1. Values and Parameters results using Genetic algorithm.

Output	Values
Solution Fitness Value:	5.099021e+00
Optimal solution:	X[1]: 3.700000e+01
	X[2]: 5.900000e+01
	X[3]: 3.600000e+01
	X[4]: 7.300000e+01
Solution Found Generation:	12
Number of Generations Run:	28
Optimum Solution:	37 (RSI look-back period)
	59 (RSI threshold)
	36 (Short Term Moving Average)
	73 (Long Term Moving Average)

iv. They are then transformed in terms of moving averages cross over and RSI.

v. Now there are four parameters as
1. Look-back periods for long term moving averages,
2. Look-back periods for short term moving averages,
3. Look-back period for RSI
4. Look-back period for RSI threshold

These are considered as chromosomes. Fitness function is maximum Sharpe ratio and the major functions used are fitness, Trading Statistics, GA engine.

From above three methods k-means clustering algorithm performance measure for feature extraction I is up to 80% compare with other methods.

4 CONCLUSION AND FUTURE WORK

Most machine learning algorithms are designed to learn the most appropriate attributes to use for making their decisions. (Witten et al. 2005) Machine learning algorithms can be used for attribute selection. In this paper experiments are carried out that run selected machine learning algorithms with different parameter settings on a corpus of text data, collect performance statistics.

Future research considers predictive analysis for daily stock prices by using optimum values obtained from genetic algorithms. Investors' sentiments are also valuable for prediction of either low or high next day values. With combination of more than one technique can improve the data interpretation results.

REFERENCES

Amsellem A., Using Genetic Algorithms in Quantitative Trading, http://www.thertrader.com/2014/03/14/using-genetic-algorithms-in-quantitative-trading (Accessed on 15 June, 2016).

Boehmer, B. and Boehmer, E., 2003. Trading your neighbor's ETFs: Competition or fragmentation? Journal of Banking & Finance, 27(9), pp. 1667–1703.

Fodor I., A survey of dimension reduction techniques, 2002, Technical report, https://e-reports-ext.llnl.gov/

Ivanciuc, O., 2007. Applications of support vector machines in chemistry. Reviews in computational chemistry, 23, p. 291.

Karamizadeh S., Abdullah S., Manaf A., 2013, An Overview of Principal Component Analysis, Journal of Signal and Information Processing, 2013, 4, 173–175.

Michael Steinbach, Levent Ertöz, and Vipin Kumar, The Challenges of Clustering High Dimensional Data, [online] available: http://www.users.cs.umn.edu/~kumar/papers/high_dim_clustering_19. Pdf

Murty M., Susheela Devi V., Chapter 6-Decision Trees, Pattern Recognition, An Algorithmic Approach, Springer link Volume 0 2011, ISBN: 978-0-85729-494-4 (Print) 978-0-85729-495-1.

NIST/SEMATECH e-Handbook of Statistical Methods, http://www.itl.nist.gov/div898/handbook/eda/section3/eda3652.htm/ (Accessed on 10 May, 2016).

Rob Schapire, COS 511: Theoretical Machine Learning, http://www.cs.princeton.edu/courses/archive/spr08/cos511/scribe_notes/0204.pdf

Safaric R., Rojko A., 2006, Intelligent Control Techniques in Mechatronics—Genetic algorithm.

Smola, A.J. and Schölkopf, B., 2004. A tutorial on support vector regression. Statistics and computing, 14(3), pp. 199–222.

StatSoft, Inc. (2013). Electronic Statistics Textbook. Tulsa, OK: StatSoft. WEB: http://www.statsoft.com/textbook/

Veeranjaneyulu N., Bhat N. Raghunath A., 2014, Approaches for Managing and Analyzing Unstructured Data, International Journal on Computer Science and Engineering (IJCSE), ISSN: 0975-3397 Vol. 6 No. 01 Jan 2014.

Witten, I.H. and Frank, E., 2005. Data Mining: Practical machine learning tools and techniques. Morgan Kaufmann.

Communication and Computing Systems – Prasad et al. (Eds)
© 2017 Taylor & Francis Group, London, ISBN 978-1-138-02952-1

Performance analysis of cylindrical dual material gate junctionless nanowire transistor

Jaspreet Kaur Sudan & D.S. Gangwar
Faculty of Technology, Uttarakhand Technical University, Dehradun, India

Poornima Mittal
School of Engineering and Technology, Graphic Era University, Dehradun, India

Brijesh Kumar
Madan Mohan Malaviya University of Technology, Gorakhpur, India

ABSTRACT: Continuous scaling of the MOSFET for reduced size, enhanced speed and better performance has tremendously improved the MOS device. However, it has limitations due to some undesirable effects including short channel effects, DIBL, high field effects and many. A lot of research is going on to find suitable solutions to avoid these effects in MOSFETs. Junctionless field effect transistor (JNT) is one of the best alternative to overcome from these problems associated with reduction of channel length in MOSFET. The merits of dual material gate are merged with JNT in Cylindrical Dual Material Gate Junctionless Transistor (CDMG-JNT) structures. In this research paper, electrical behavior of the CDMG-JNT device is analyzed in terms of threshold voltage, drive current, sub-threshold slope and current on-off ratio using state of art industry standard benchmark Atlas three dimensional (3-D) device TCAD simulator. It is observed that as there are no junctions in the JNT device, therefore its fabrication process becomes easy. Besides this, the performance parameters of cylindrical DMG-JNT are extracted analytically for the validation of simulated results. It is observed that there is a good agreement between the simulation and analytical results.

1 INTRODUCTION

VLSI industry has witnessed continuous scaling of conventional MOSFETs for faster circuit speed, higher package density and lower power dissipation which has resulted in improved performance and reduced cost. The International Technology Roadmap for Semiconductors had given prediction of 10-nm gate length with FD-SOI technology in 2015 and a 7-nm gate length with dual gate devices in 2018. However, with the shrinkage of the channel length, many undesirable effects such as short channel effects, high field effects and many more come in effect (Lou et al. 2012). Another problem in short channel devices (channel length less than 10 nm) is in the formation of extremely sharp source and drain junctions that requires very high variation in doping concentration within a few nanometers. This creates challenges for doping techniques and on thermal budget and requires development of costly annealing techniques (Afzalian et al. 2009, Colinge et al. 2012).

To overcome these issues, a new type of Junctionless transistor based on Lilienfeild's first transistor architecture (Lilienfeild et al. 1925) was proposed (Afzalian et al. 2009, Magnus et al. 2009, and Kranti et al. 2010). The Lilienfield's transistor is a unipolar field-effect device which differs from conventional MOSFET as it does not contain any junction. Unlike traditional MOSFET, the doping concentration and type of channel in Lilienfield's device is same as that of the source and drain regions and the gate has a work function of opposite type.

The key to fabricate a Junctionless FET which is also termed as gated resistor is in the creation of very thin and narrow semiconductor layer so that the carriers are fully depleted when device is turned OFF. Another constraint is on the doping concentration which needs to be high so that an acceptable amount of current flows when the device is turned ON. Since the doping concentration gradient is zero in a Junctionless transistor, therefore there is no impurity diffusion in thermal processing and this improves the thermal budget. Also, there is no need for expensive annealing techniques (Golve et al. 2012). Gated resistor exhibits better analog performance and reduced short-channel-effects. It has been shown that junctionless MOSFET have very low leakage current and they have improved Drain Induced Barrier Lowering (DIBL) and sub threshold slope over traditional devices of same dimensions (Doria et al. 2011).

The Dual Material Gate Junctionless Transistor (DMG JNT) structure has a secondary gate of some different material. The functionality of the JNT improves with the addition of a second gate. DMG structure has been demonstrated to enhance the transconductance (gm) and the transport efficiency of carriers in traditional MOS devices (Lou et al. 2012). Therefore, in this research paper, advantages of dual material gate are combined with junctionless transistor to get improved electrical characteristics.

This paper is divided into six sections: Section 1 contains the introduction. The introduction is followed by Section 2 where simulation setup is discussed. This setup analyzes the performance of cylindrical Dual Material Gate FET (DMG-FET). After this, results and characteristics of the device are discussed in Section 3. The modeling of DMG-FET is discussed along with the results in section 4. Finally, conclusions are drawn in section 5.

2 SIMULATION SETUP

The 3-D structure of Dual-Material Gate Junctionless Transistor (DMG-JNT) is shown in Fig. 1. The structure has a cylinder of silicon nanowire for which the doping concentration and the type of doping in the channel region is similar to that of the drain and source regions. There are two metal gates in the structure which are of different materials and equal length. The channel doping is higher in case of JNT devices than in accumulation mode devices due to which current flows in the bulk of the nanowire and no surface accumulation layer formation takes place (Doria et al. 2011).

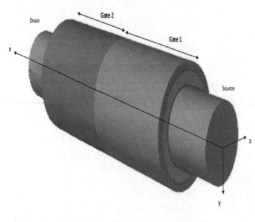

Figure 1. Schematic structure of DMG JNT.

Table 1. Parameters of DMG-JNT and SMG-JNT.

Parameters	SMG-JNT	DMG-JNT	
Channel Doping (cm³)	2×10^{19}	2×10^{19}	
Channel Length (nm)	40	Gate 1	Gate 2
		20	20
Gate Work function (eV)	4.97	4.97	4.27
Radius of the nanowire (nm)	5	5	
Source/drain regions (nm)	10	10	
Gate oxide thickness (nm)	2	2	

Several Dual gate devices have been studied and fabricated in recent years. The DMG structure has been demonstrated to improve the transconductance (gm), the efficiency of the carrier transport and the drain output resistance in MOS structures. The dual material gate structure has resulted in reduction of short channel effects and improvement in transconductance in bulk MOSFETs and SOI (Silicon-On-Insulator) devices (Lou et al. 2012, Shur et al. 1989).

The DMG-JNT device used in this paper has a channel Length (L) of 40 nm and the total silicon nanowire length is 60 nm. The radius of the cylindrical nanowire is 5 nm and the gate oxide thickness is 2 nm. The SMG-JNT FET reported in (Cho et al. 2011) is used for comparison. The two gates of DMG-JNT have length of 20nm each and have work functions of 4.97 eV and 4.27 eV for first gate and second gate respectively. The work function 4.97 eV corresponds to metals like Co and SiO_2 and the work function of 4.27 eV corresponds to metals such as Al and NiAl. The doping concentration is high in junctionless transistors and is taken to be 2×10^{19} cm⁻³. The simulation of Dual Material Gate JNT device is completed using Atlas 3-D device simulator by Silvaco Inc. The performance parameters are illustrated in Table 1. The Auger recombination model and the Shockley-Read-Hall model are used for the generation and recombination regimes in DMG-JNT to account for leakage currents. The drift-diffusion based (without impact ionization) simulation is carried out.

3 RESULTS AND DISCUSSION

The simulated structure and characteristics of the cylindrical Dual Material Gate Junctionless Transistor (DMG-JNT) are presented in this section. Fig. 2 shows the schematic structure of cylindrical DMG-JNT. The transfer characteristics of the

DMG-JNT are shown in Fig. 3 followed by output characteristics in Fig. 4. Fig. 3 demonstrates the variation in the output drain current with increase in the gate bias. While applying bias at the gate in the Dual gate structure, both the gates are tied together and a single potential is applied across both of them. Due to the difference in work function between gate and the channel, the channel is fully depleted initially and at this stage drain current is negligible. When a suitable bias is given at the gate (positive bias for n channel and negative bias for p channel) and a non-zero drain bias is present, the transistor comes out of depletion and current starts flowing (Lilienfeild et al. 1925). In Fig. 4, the output characteristics of the DMG-JNT with plot of drain current versus drain to source voltage for different gate voltage V_{gs} are shown.

Fig. 5 shows the comparison of the DMG JNT and SMG JNT FETS.

The Figure 5 (a) compares the transfer characteristics of both the structures and it can be seen

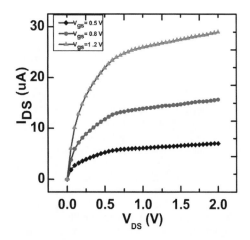

Figure 4.   DMG JNT output characteristics.

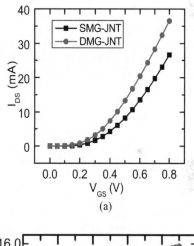

(a)

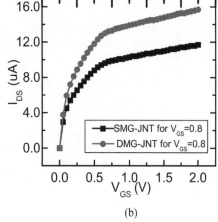

(b)

Figure 5.   Comparison of DMG-JNT and SMG-JNT (a) Comparison of the gate characteristics (b) Comparison of the drain characteristics for $V_{gs} = 0.8V$.

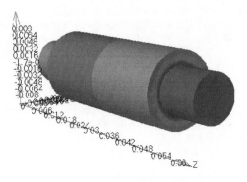

Figure 2.   Schematic structure of DMG JNT.

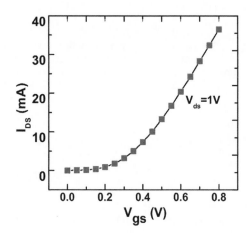

Figure   3.   DMG   JNT   transfer   characteristics.

that DMG JNT provides better ON-state current than SMG-JNT for same channel length and doping profile. The output characteristics are shown in Fig. 5(b) which also shows similar enhancement in the output current.

4 ANALYTICAL MODELING OF DUAL GATE JUNCTIONLESS TRANSISTOR

There are certain rules which govern any device and these are generally developed as equations. The equations once formed can then be used for analysis of the device and to verify its performance. In the modeling of Junctionless cylindrical FET, the expression of the drain current can be obtained by considering the mobile charge density Q_m (per unit length) of the device to have two components so that $Q_m = Q_{dep} + Q_c$. Here Q_{dep} denotes the mobile charge density of the fully semi depleted and depleted regions and Q_c denotes complementary mobile charge added with Q_{dep} (Duarte et al. 2012).

The drain current is given by $I_{DS} = I_{dep} + I_c$ where, I_{dep} denotes the drain current in the fully semi depleted and depleted regions of the device and I_c is the complementary current. These currents are obtained by Pao-Sah integral

$$I_{DS} = -\frac{\mu}{L}\int_0^{V_{DS}}\left(Q_{dep}+Q_c\right)dV \tag{1}$$

$$I_{DS} = -\frac{\mu}{L}\left(\frac{Q_{dep}^2}{2C_{eff}}-v_TQ_{dep}\right)-\frac{\mu}{L}\left(\frac{Q_c^2}{2C_c}-v_TQ_c\right) \tag{2}$$

where μ is the effective mobility and L represents the gate length, C_{eff} is the effective value of gate capacitance in the semi depleted area and V_T is the thermal voltage KT/q.

The currents in all the three regions of working of the device can be determined using the above equations. The drain-to-source current in the subthreshold region is given

$$I_{DS} = 4\pi\varepsilon_{si}v_T^2\frac{\mu}{L}\exp\left[\frac{V_G-V_{TH}}{v_T}\right]\left[1-\exp\left(\frac{-V_{DS}}{v_T}\right)\right] \tag{3}$$

When the gate bias is lower than V_{FB}, the total drain current is governed by the bulk current mechanism and there can be two region of operation. The first region corresponds to the linear region where $V_G-V_{TH}>V_{DS}$, and in this case the drain current model reduces to

$$I_{DS} \approx \left(\frac{\mu C_{eff}}{L}\right)\left(V_G-V_{TH}-\frac{V_{DS}}{2}\right) \tag{4}$$

Second, in the saturation region i.e. $V_G-V_{TH}<V_{DS}$, then the model reduces to

$$I_{DS} \approx \left(\frac{\mu C_{eff}}{2L}\right)\left(V_G-V_{TH}\right)^2 \tag{5}$$

Here V_G-V_{TH} is the drain voltage required to pinch off the channel in the drain region and this voltage determines the saturation region in DMG JNT.

When the device is in flat band condition, and value of V_{DS} is small, then the drain current reduces to

$$I_{DS} \approx \frac{\mu q N_{si}\pi R^2 V_{DS}}{L} \tag{6}$$

This current expression represents a resistor which does not depend on t_{ox}. This differs from the conventional FETs where current is not independent of t_{ox}. When the gate bias is greater than V_{FB}, the total drain current is governed by bulk and surface current mechanisms and three types of behaviors can be found. First, when $V_G-V_{TH}>V_{DS}$ and $V_{FB}>V_{DS}$, the drain current is given by

$$I_{DS} \approx \left(\frac{\mu C_{ox}}{L}\right)\left(V_G-V_{on}-\frac{V_{DS}}{2}\right)V_{DS} \tag{7}$$

At this voltage, the charge accumulation occurs on the channel surface from source to drain region. Second when $V_G-V_{TH}>V_{DS}$ and $V_G-V_{FB}>V_{DS}$, the drain current is given as

$$I_{DS} \approx \left(\frac{\mu C_{ox}}{L}\right)\left(V_G-V_{on}-\frac{V_{DS}}{2}\right)V_{DS} + \left(\frac{\mu C_c}{2L}\right)\left(V_G-V_{FB}\right)^2 \tag{8}$$

At this condition, a semi depletion condition is there in the drain region and accumulation is present near the source region. Third, when $V_G-V_{TH}<V_{DS}$ and $V_G-V_{FB}<V_{DS}$ the drain current is

$$I_{DS} \approx \left(\frac{\mu C_{eff}}{2L}\right)\left(V_G-V_{TH}\right)^2 + \left(\frac{\mu C_c}{2L}\right)\left(V_G-V_{FB}\right)^2 \tag{9}$$

At this bias, the part of channel at the drain side experiences a pinch-off condition and accumulation occurs at the source side. Table 2 gives the comparison of the simulated values and the analytical results. It can be seen from the table that there is a good comparison between both results.

Table 2. Comparisons between simulated and analytical data.

Parameters	Analytical data	Simulated data
Threshold Voltage V_{TH}(V)	0.358473	0.3132
Sub-threshold Slop S.S. (V/Dec.)	0.071988	0.08342
Current Ratio (I_{on}/I_{off})	8459.53	8769.34

5 CONCLUSION

In this paper, an attempt was made to compare the Dual Material Gate with Single Material Gate JNT using 3-D numerical simulations. The drain-to-source current I_{DS} is found to be higher for DMG-JNT than for SMG-JNT for same values of V_{gs}. The graphs are plotted to show this comparison. Furthermore, we also analyzed the transfer characteristics of both the structures, where, we observed that DMG-JNT shown higher values of drive current for same gate voltages applied. The threshold voltage is reduced for the dual material gate structure in compared to single material gate JNT. The analysis is done for all the regions of operation of JNT-FET. The DMG structure has shown improvement in the ON-state current. The introduction of dual gate provided better characteristics by enhancing the channel control ability.

REFERENCES

Afzalian, A., Lee, C.W., Akhavan, N. D., Yan, R., Ferain, I., and Colingea, A.P. 2009. Junctionless multigate field-effect transistor. *Appl. Phy. Lett.* 94(5): 053511.

Cho, S., Kim, K., Park, B. and Kang, I. 2011. RF performance and small-signal parameter extraction of junctionless silicon nanowire MOSFETs. *IEEE Trans. Electron Devices.* 58(5): 1388–1396.

Colinge, J. P., Afzalian, A., Lee, C. W., Akhavan, N. D. and Yan, R. 2010. Nanowire transistors without junctions. *Nature Nanotechnology.* 5: 225–229.

Doria, R. T., Pavanello, M. A., Trevisoli, R. D., Souza, M de, Lee, C. W, Ferain, I., Akhavan, N. D. and Colinge, J. P. 2011. Junctionless multiple-gate transistors for analog applications. *IEEE Trans. on Electron Devices,* 58(8): 2511–2519.

Duarte, J. P., Choi, S. J., Moon, D. I. and Choi, Y. K. 2012. A non-piecewise model for long-channel junctionless cylindrical nanowire FETs. *IEEE Electron Device Lett.* 33(2): 155–157.

Golve, M., Gundapaneni, S. and Kottantharayil, A. 2012. Novel architecture for zinc-oxide junctionless transistor. *International Conference on Emerging Electronics* (ICEE 2012), Mumbai: 1–4.

International technology roadmap for semiconductor, 2015.

Kranti, A., Lee, C. W., Ferain, I., Yu, R., Razavi, P., Akhavan, N. D. and Colinge, J. P. 2010. Junctionless nanowire transistor: Properties and design guidelines. *Proc. ESSDERC* 2010: 357–360.

Lilienfeild, J. E. 1925. Method and apparatus for controlling electric current. *US Patent,* no.: 745175, 1925.

Lou, H., Zhang, L., Zhu, Y., Lin, X., Yang, S., He, J. and Chan, M. 2012. A junctionless nanowire transistor with a dual-material gate. *IEEE Trans. Electron Devices.* 59(7): 1829–1836.

Magnus, W., and Soree, B. 2009. Silicon nanowire pinch-off FET: Basic operation and analytical model. *10th Int. Conf. on Ultimate Integration on Silicon,* 1.4: 246–248.

Shur, M. 1989. Split-gate field-effect transistor. *Appl. Physics Lett,* 54(9):162–164.

Communication and Computing Systems – Prasad et al. (Eds)
© 2017 Taylor & Francis Group, London, ISBN 978-1-138-02952-1

Optimizing unit commitment solution in smart grid environment

Akshita Sharma & Yajvender Pal Verma
Electrical and Electronics Engineering Department, UIET, Panjab University, Chandigarh, India

ABSTRACT: The ever increasing up rise in the energy demand keeping in mind the environmental concerns has encouraged the application of Renewable Energy Sources (RESs) in power systems. Unit Commitment is a basic yet mandatory problem in power system which has gained a significant interest from researchers and industrialists due to increased penetration of RESs and participation of demand side resources. In this paper this unit commitment has been modeled in smart grid environment. Modeling has been carried out in GAMS optimization software using Mixed Integer Non Linear Programming technique on IEEE 14 bus system while considering the power flow and environmental constraints and taking into account the effect of demand side resources comprising of vehicle to grid technology, demand response and distributed generation. Finally, the impact of demand side resources on generation schedule to minimize the overall operation cost and emissions of power system has been analyzed.

1 INTRODUCTION

The salient feature of smart grid in modern power system is a two way digital dialog system. It provides a platform to the system operator for incorporation of Demand Side Resources (DSR) in maintaining power supply demand balance. Now a days the existing grids are becoming smarter in which the concept of Vehicle to Grid (V2G), Demand Response (DR) and Distributed Generation (DG) are gaining significance owing to their environmental friendly nature and the ability to supply power back to grid during peak load hours thereby reducing the burden on the thermal generators and hence minimizing their cost. DSR can readily accommodate the load fluctuations because of the ability to adjust end-use consumption quickly which consequently modifies the load profile of the system.

Unit commitment Problem is a least cost dispatch which determines the ON/OFF status of the available generation resources in order to meet the electrical load demand while satisfying all the technical constraints. Unit commitment problem with various conventional techniques has been solved in (Salam 2007; Cheng & Liu 2000). Due to restructuring of power system in the recent years and integration of demand side resources, the traditional unit commitment problem no longer remains the same and gets modified accordingly. The effect of vehicle to grid technology on the cost and emission of the system was studied in (Saber & Venayagamoorthy 2010;

Lu 2011). Particle Swarm Optimization (PSO) was used to solve unit commitment considering electric vehicles in (Khodayar et al. 2012). The participation of the consumer in striking a balance between supply and demand is promoted by the concept of Demand response. The role of demand response in improving the reliability of power system and reducing the overall cost is studied in (Wang et al. 2013; Ikeda et al. 2012; Du & Lu 2011). Also the incorporation of DG into grid can effectively increase the grid's capacity and reduce carbon dioxide emissions (Moghimi 2013; Safdarian et al. 2013; Doostizadeh & Ghasemi 2013).

Electricity sector is one of the major emitter of greenhouse gases throughout the world. Because of the increasing pollution, global warming and various environmental concerns, it becomes very important to keep a check on these emissions. In this paper, a unit commitment model has been developed and implemented considering vehicle to grid, demand response and distributed generation using GAMS 23.4 optimization software interfaced with MATLAB 7.0.4 (Brooke et al. 1998; Simulink 7 User Guide 2008). The impacts of integration of all these resources on the traditional unit commitment results have been obtained and analyzed.

Rest of the paper is structured as: Nomenclature is described in section II. Problem formulation has been done in section III. Section IV explains the case study followed by results discussion in Section V. The section VI provides the conclusion of the paper.

2 NOMENCLATURE

2.1 *Sets*

l, m   Bus Indexes
j   Generating unit Index
t   Hour Index

2.2 *Variables*

P_{Dgt} Output of DG connected with grid at time t
DG_C Total cost of DG
DR_C Total cost of Demand Response (DR)
P_{DRt} Output of DR at time t
FC_j Fuel cost of unit j
$u_{j,t}$ On/off status of unit j at time t
$Pg_{j,t}$ Output of unit j at time t
$SC_{j,t}$ Start-up cost of unit j at time t
$CH_{j,t}$ State of charge of EV j at time t
$Cost_{Total}$ Total Cost of Unit Commitment (UC)
$V2G_C$ Total Cost of V2G
P_{v2gt} Output of V2G at time t
$D_{ON_{j,t}}$ Duration for which unit j is continuously on at time t
$D_{OFF_{j,t}}$ Duration for which unit j is continuously off at time t
$P_{j,t}$ Net active Power injection at time t
$Q_{j,t}$ Net reactive Power injection at time t
$\delta_{j,t}$ Voltage angle at bus l at time t
$V(l, t)$ Voltage magnitude at bus l at time t
$V(m, t)$ Voltage magnitude at bus m at time t

2.3 *Parameters*

d, e, f   Cost coefficients of DG
x, y, z   Cost coefficients of DR
a, b, c   Fuel Cost coefficients of unit j
p, q, r   Cost coefficients of V2G at time t
C_c Cold start cost of unit j
C_0 Initial charging state of EV
C_f Charging state of EV
t_c Cold start hour of unit j
$P_{Dgt,max}$ Maximum limit of DG at time t
$P_{Dgt,min}$ Minimum limit of DG at time t
$P_{DRt,max}$ Maximum limit of DR at time t
$P_{DRt,min}$ Minimum limit of DR at time t
CH_{max} Maximum state of charge
CH_{min} Minimum state of charge
C_h Hot start cost of unit j
Pd_t Load demand at time t
M_{up_j} Minimum on time of unit j
M_{down_j} Minimum off time of unit j
N   Number of generator units
$Pg_{j,min}$ Minimum output of unit j
$Pg_{j,max}$ Maximum output of unit j
$Loss_t$ Real power loss at time t
ch_{min} Minimum value of state of charge at each hour

$P_{v2g_{max,t}}$ Upper limit of V2G capacity at time t
$\theta_{l,m,}$ Angle of admittance
$Y_{l,m}$ Magnitude of admittance

3 PROBLEM FORMULATION

3.1 *Demand response*

The two way communication between the load and the supply end, and the consumer interaction are the prime features of a smart grid. The upcoming smart meters enables the consumer to participate in the load supply balance during the peak load hours by curtailing their load demand (Zhang et al. 2015) and the incentive based demanresponse encourages the consumer for the same. The load aggregator helps to communicate between the consumers willing to participate in the demand response and the system operator and hence the contracts are signed. The cost of demand response is considered a simple quadratic function mathematically given by equation (1):

$$DR_c\left(P_{DR_t}\right) = x + y * P_{DR_t} + z * P_{DR_t}^2 \qquad (1)$$

3.2 *Distributed generation*

These days the gird is equipped with more DGs which are the cleaner and cheaper source of energy. Like electric vehicles, DGs can also be used to sell power back to the grid especially during peak load hours on assurance of certain incentives. Equation (2) describes the cost function for DG:

$$DG_c\left(P_{Dg_t}\right) = d + e * P_{Dg_t} + f * P_{Dg_t}^2 \qquad (2)$$

3.3 *Vehicle to grid*

In the recent years electric vehicles has gained massive significance because they consume electricity rather than fossil fuel for its working and hence reduce emission. The volatile and intermittent nature of renewable energy soces can be mitigated due to the storage capability of electric vehicles and hence reduce the operation cost. Electric vehicles can also sell their power back to grid i.e. they get discharged during peak load hours once they get charged during off peak hours. Certain incentives are provided to encourage the participation of electric vehicle during peak hours. Cost function for electric vehicle and state of charge has been described by equation (3) and (4) respectively as given below:

$$V2G_c(P_{v2g_t}) = p + q * P_{v2g_t} + r * P_{v2g_t}^2 \qquad (3)$$

$$CH_t = \frac{C_0 + \sum_{u=1}^{t} G2V_u - \sum_{u=1}^{t} P_{v2g_u}}{C_f} \quad (4)$$

$$CH_{min} \leq CH_t \leq CH_{max} \quad (10)$$

$$P_{v2g_t} \leq P_{v2g_{max,t}} \quad (11)$$

3.4 Objective function

The main objective function is to minimise the overall cost which is the sum of fuel cost, demand response cost, vehicle to grid cost, DG cost, start up cost and emission cost. The expressions for the fuel cost and total cost are given below by equation (5) and (6) respectively.

$$FC_j\left(Pg_{j,t}\right) = a + b * Pg_{j,t} + c * Pg_{j,t}^2 \quad (5)$$

$$Cost_{Total} = \sum_{t=1}^{24} \sum_{j=1}^{N} \left[FC_j * Pg_{j,t} * u_{j,t} + u_{j,t} \right.$$
$$\left. * \left(1 - u_{j,t-1}\right) * SC_{j,t} \right] + \sum_{t=1}^{24} \left[DR_C \left(P_{DRt} \right) \right.$$
$$\left. + V2G_c(P_{v2g_t}) + DG_C\left(P_{Dg_t} \right) \right] + Em_{cost} \quad (6)$$

3.5 Operational constraints

In order to meet the load demand the network has to abide certain constraints as described below:

3.5.1 Load balance constraint
The total generation of the system must balance the system demand and losses as described by equation (7)

$$\sum_{j=1}^{N} Pg_{j,t} = Pd_t + Loss_t - P_{DRt} - P_{Dgt} - P_{v2g_t} \quad (7)$$

3.5.2 Generator power constraints
The lower and upper limits for thermal units are described by equation (8).

$$Pg_{j,min} \leq Pg_{j,t} \leq Pg_{j,max} \quad (8)$$

3.5.3 Demand response constraints
There is a limit for load curtailment also. The minimum and maximum limit of demand curtailment at each hour is applied through constraints given below by equation (9).

$$P_{DR_{t,min}} \leq P_{DR_t} \leq P_{DR_{t,max}} \quad (9)$$

3.5.4 Electric vehicle constraints
The state of charge for electric vehicles must be within its minimum and maximum limits given by (10). Equation (11) gives the upper bound for the electric vehicle output at each hour.

3.5.5 DG constraints
The lower and upper limits for thermal units are described by equation (12).

$$P_{Dg_{t,min}} \leq P_{Dg_t} \leq P_{Dg_{t,max}} \quad (12)$$

3.5.6 Power flow constraints
The basic power flow equations are described in (13) and (14)

$$P_j(l,t) = \sum_{m=1}^{1} |V(l,t)| |V(m,t)| |Y_{l,m}|$$
$$\cos\left(\delta(m,t) - \delta(l,t) + \theta_{l,m}\right) \quad (13)$$

$$Q_j(l,t) = -\sum_{m=1}^{1} |V(l,t)| |V(m,t)| |Y_{l,m}|$$
$$\sin\left(\delta(m,t) - \delta(l,t) + \theta_{l,m}\right) \quad (14)$$

4 CASE STUDY

In this paper, IEEE 14 bus stem is taken under consideration with three thermal units at bus no. 1, 2 and 3. The voltage sensitivity analysis was carried out for the system on the basis owhich the position of demand response, electric vehicle and DG were decided at bus number 4, 6 and 8 respectively to improve the voltage profile at these buses. This paper considers three cases which have been explained below. The minimum state of charge for electric vehicle has been assumed zero herefore, it gets charged initially. The upper limit for curtailment of load is taken as 110 kW. The maximum output from DG is 900 kW. Table 1 gives he cost coefficients for DSR.

4.1.1 Case 1 (UC without Demand Side Resources)
In this case, only thermal generators take part in meeting the load requirement of the system while satisfying all the technical consaints. The thermal generators generally have high fuel cost and contribute in greenhouse gas emissions.

Table 1. DSR cost coefficients.

Cost coefficients	V2G	DR	DG
p, x, d ($/h)	8.0	6	2
q, y, e ($/MWh)	2.4	2.4	2.4
r, z, f ($/MWh^2)	0.03	0.05	0.05

4.1.2 *Case 2 (UC with Demand Side Resources)*

In this case, DSR which comprises of demand response, distributed generation and electric vehicle is also taken into account in addition to the conventional generators in unit commitment solution. The combined effect of these resources on various factors like generation, emission and cost for various thermal units is observed. Smart grid environment enables demand side participation to mitigate the high power demand during peak load hours. In DR, the load aggregator helps in signing the contract between customers and system operator. The customer according to the contract can have a definite response in the form of load curtailment thereby reducing the overall demand on the system. Distributed Generation is another way for the demand side resources to take part in power demand fulfillment as they can either sell their generation to the grid or the consumers can directly buy power from DG. Vehicle-to-Grid (V2G) technology has gained high significance in the recent years due to its ability to reduce emission and operation cost by decreasing the dependence on expensive thermal generators. Therefore, the reserve and reliability of the existing system can also be enhanced with the pronounced use of DSR.These demand side resources tend to affect the unit commitment of the conventional generators as well.

5 RESULTS AND DISCUSSIONS

Fig. 1 shows the voltage variation with and without demand side resources. The comparison of the emission and total cost for the cases described above is shown in Fig. 2 and Fig. 3 respectively. Fig. 4 shows the contribution of electric vehicle

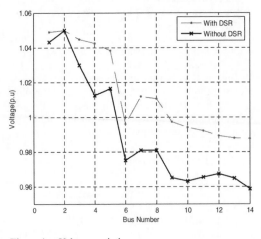

Figure 1.   Voltage variation.

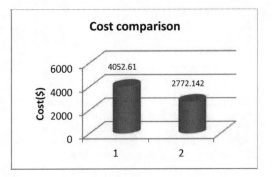

Figure 2.   Total cost comparison in both cases.

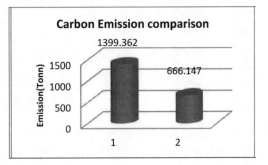

Figure 3.   Carbon emissions comparison.

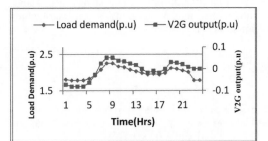

Figure 4.   Charging discharging of V2G in case 2.

Table 2.   Cost and emission comparison in both cases.

Factors	Case 1	Case 2
Carbon emission (ton)	1399.362	666.147
Total cost ($)	4052.61	2772.142

according to the load demand. Table 2 shows the cost and emission comparison under two different cases considered. Table 3 gives the unit commitment schedule for case 1 and case 2.

From Figure 1 we see that the voltage profile is improved at buses 6 and 8 attributed to the

Table 3. Unit commitment schedule.

Time (Hrs)	Case 1			Case 3		
	G_1	G_2	G_3	G_1	G_2	G_3
1	0	1	1	0	0	1
2	0	1	1	0	0	1
3	0	1	1	0	0	1
4	0	1	1	0	0	1
5	0	1	1	0	0	1
6	1	1	1	0	0	1
7	0	1	1	0	0	1
8	1	1	1	0	0	1
9	1	1	1	0	0	1
10	1	1	1	0	0	1
11	0	1	1	0	0	1
12	0	1	1	1	0	1
13	0	1	1	0	0	1
14	0	1	1	0	0	1
15	1	1	1	0	0	1
16	1	1	1	0	0	1
17	1	1	1	0	0	1
18	0	1	1	0	0	1
19	0	1	1	0	0	1
20	0	1	1	0	0	1
21	0	1	1	0	0	1
22	0	1	1	0	0	1
23	1	1	1	0	0	1
24	1	1	1	0	1	1

participation of V2G and DG respectively. Fig. 2 and Fig. 3 depicts that with the integration of DSR, the burden on thermal units and the total emission decreases since DSR are more environmental friendly and comparatively cheaper source of energy. Fig. 4 shows the behavior of V2G with respect to load variation and it was observed that when the load demand is low electric vehicle gets charged and during peak load hours vehicle returns electricity to the grid. The vehicle gets charged in the beginning since the initial state of charge is assumed to be zero. Hence electric vehicle is a suitable option for system operator to compensate the peak load demand. Table 2 shows the cost and emission variation in both cases. In case 1 the generation is solely because of the thermal units and hence the cost is highest. The cost reduction attained in case 2 is due to the involvement of DSR being renewable sources of energy.

Table 3 shows the UC schedule for the case 1 and case3. In case 1 DSR are not available at instants 6, 8, 9 and 10 when the load dand is high, power is supplied by the expensive unit G_1 also in addition to unit G_2 and G_3 which supplies power at all other instants. However in case 2 DSR are included. So more power is supplied from these sources and economically cheaper generator, thereby switch-

ing the status of the thermal units as shown in Table 3.

6 CONCLUSION

The continuously growing demand and global environmental issues invigorate the use of DSR in power systems thereby influencing its operation. This paper formulates a unit commitment solution model considering DSR. The advantages of including DSR in the system are reflected by virtue of decreased dependence and lesser switching of conventional generators. V2G and DG minimizes the overall cost of the system because they are cheaper and green energy sources and promotes the dispatch of the thermal units with relatively lesser emission.

REFERENCES

Brooke, A. et al. 1998. GAMS: a User's Guide. Washington DC: GAMS Development Corporation.
Cheng, C. & Liu 2000. Unit commitment by lagrangian Relaxation and Genetic algorithm. IEEE Transactions On Power Systems 15(2).
Doostizadeh, M. & Ghasemi, H. 2013.Day-ahead scheduling of an active distribution network considering energy and reserve markets. Int. Trans. Elect. Energy Systems 23: 930–945.
Du, P. & Lu, N. 2011. Appliance commitment for household load scheduling, IEEE Trans. Smart Grid 2:411–419.
Ikeda, Y. et al. 2012. A unit commitment model with demand response for the integration of renewable energies. IEEE Power Energy Soc. Gen. Meeting. San Diego, USA.
Khodayar, M.E. et al. 2012. Hourly coordination of electric vehicle operation and volatile wind power generation in SCUC. IEEE Trans. Smart Grid 3:1271–1279.
Lu, L et al. 2011. Unit commitment in power systems with plug-in electric vehicles. Autom. Elect. Power Syst 35:16–20.
Moghimi, H. et al. 2013. Stochastic technoeconomic operation of power systems in the presence of distributed energy resources, Elect. Power Energy Syst 45:477–488.
Saber A.Y. & Venayagamoorthy G.K. 2010. Intelligent unit commitment with vehicle to grid-A cost-emission optimization. J. Power Sources 195:898–911.
Safdarian, A. et al. 2013. A stochastic framework for short-term operation of a distribution company. IEEE Trans. Power Syst 28(4):4712–4721.
Salam, S. 2007. Unit Commitment Solution methods. World Academy of Science Engineering and Technology International Journal of Electrical, Computer, Electronic and Communication Engineering.
Simulink 7 User's Guide: The Math works Inc; Mar, 2008.
Wang, C. et al. 2013. A novel traversal-and-pruning algorithm for household load scheduling. Appl. Energy 102:1430–1438.
Zhang, N. et al. 2015. Unit commitment model in smart grid environment considering carbon emission trading. IEEE transactions on smart grid. 7(1).

Communication and Computing Systems – Prasad et al. (Eds)

An improved data classification technique for data security in cloud computing

Rasmeet Kour & Simar Preet Singh

CSE, DAV University, Jalandhar, India

ABSTRACT: Security of data in cloud computing is still an ultimatum to be achieved. Many techniques came into limelight for data security in cloud. For security in cloud computing, the data encryption is an extensively used technique. This paper introduces a model for the classification of data for data security in cloud computing. Improved bagging classification technique along with the boosting technique is inflected to classify the data on cloud. The objective of using the improved bagging technique is to categorize the data according to its security demands which divides the data into two categories-Non-sensitive data (public) and sensitive data (private). Thus the sensitive data is encrypted using the blowfish algorithm and is stored in the cloud server. The public data is send to server without encryption. Finally to enhance secure cloud system performance, security partitions are created on a single cloud. Also the results show that the improved bagging technique works better than the K-NN classification technique in terms of both accuracy and classification time.

Keywords: Cloud computing, improved bagging, blowfish algorithm, data classification, sensitive/non-sensitive data

1   INTRODUCTION

Cloud computing is a computer technology in which the user can access the resources available on internet without completely controlling them. The companies and the businesses can use the applications available on the internet without even installing them on their computers. Thus cloud computing offers a great deal of computing by consolidating the memory resources, their processing and utilization (UTHSCSA Data Classification report 2006). Cloud computing is emerging as a popular trend in the field of IT and is becoming a dream utility computing. Cloud computing is considered as a regime that has believed to make a sudden change in computer architecture, software development tools, storage and memory usage (Stephen & Yau 2010). The biggest advantage that cloud computing offers is that developers are not bothered of large capital expenditures for making the use of various internet based services. It has a large number of reconfigurable virtual resources that can manage all the demands of the user with extreme usage of resources (Catteddu & Hogben 2009). It represents a model of pay as you use that gives the liability to customers to have the benefit of cloud resources and Service Level Agreement (SLA'S) are provided to guarantee the services of cloud (Rittinghouse & Ransome 2009).

Many business enterprises run various types of cloud services in the cloud now a day's such as customer relationship management, Human resources and other cloud inherited applications (Catteddu & Hogben 2009). Clouds built applications are gaining popularity so they are soon to become the biggest upcoming technologies replacing the traditional applications. Major benefit is the cost saving as it bails out the customers from paying for the resources and services. Thus it is quite clear that the approach of cloud computing is more extensible, secure and steady in comparison to other applications. In addition to this upgrades are provided automatically to customers, and thus they provide security, flexibility and disaster recovery for the applications (Rittinghouse & Ransome 2009).

2   DATA CLASSIFICATION AND DATA SECURITY IN CLOUD COMPUTING

Security is the number one issue when it comes to any upcoming technology and cloud computing is no exception. Cloud computing poses numerous security risks that hinders the customers to use its services. Security of the data is the major risk for the cloud vendors threatening the users to adopt the cloud services (Rittinghouse & Ransome 2009). Cloud server stores the data via two storage methods. The former involves encrypting the data and storing on the server while the latter involves storing the data without encrypting it that is storing the raw data (Zardari et al. 2014). The aforementioned methods are likely to face confidentiality issue. Different data possess different characteristics depending on its type. As the customer's data lie on the on the remote servers and the customer has no idea about its physical location, so there is always a risk of confidentiality exhalation (Rawat et al. 2012). The paper highlights the confidentiality issue in cloud environment. Whenever the data is stored on the cloud server it goes through a security method i.e. encrypting the whole data without even understanding the level of sensitivity of data or data is simply stored on server without securing it (Purushothaman & Abburu 2012). Every data has its own level of sensitivity; therefore it is not an appropriate idea of storing the data not having knowledge of its sensitivity level and security elements (Whitney & Dwyer 1966). Analyzing all the possible elements of security, a data classification technique is proposed for classifying the data based on the security elements and then encrypting the only data which is required to secure using an encryption technique in cloud environment.

Classifying the objects is an indispensable field of research particularly in numerous fields like data mining and cluster analysis, knowledge discovery, pattern analysis and medicine (Zardari et al. 2014). A very intelligent technique to secure the data would be to first classify the data into sensitive and non-sensitive data and then secure the private data only. This will help to reduce the overhead in encrypting the whole data which proves to be very costly in terms of space as well as time (Stephen & Yau 2010). For encrypting the data many encryption techniques can be used and for classifying the data numerous classification algorithms are available in the field of data mining (Wu et al. 2008).

3 RELATED WORK

In the present scenario of cloud computing, where a number of techniques came into picture to provide the security of data in cloud computing, under the cloud computing environment a technique was developed to examine the cloud built application performance. In this scheme a virtual cloud computing environment has been formed then tests are processed that measures the efficiency of cloudlet (Rawat et al. 2012). A crypto co-processor was developed which provides data confidentiality in cloud environment. It provides a mechanism to achieve the utmost security by using the capacity of a cryptographic co-processor (Ram & Sreenivaasan 2010). A technique called DPaas

is formulated that defines a package of security rules incorporating data security and integrity and allows the declaration of security to data proprietors although there exists some malicious software. DPaas ensures examining the outflow of the data, encryption and software updating and execution in cloud environment to restrain from implementing the full-disk encryption algorithm for security of data that restricts the sustainability of low level companies and industries (Song et al. 2012). NCR-neighborhood census rule was proposed to lay out the concepts of optimum thresholds so that unknown samples can be allotted to the presently known classes (Dasarathy 1980). Fuzzy k-Nearest neighbor algorithm was developed which depicts a fuzzy K-NN decision technique as well as fuzzy prototype decision technique embedded with the three techniques that assigns membership numbers to the sample groups. Fuzzy helps in achieving more probabilistic results by assigning an input sample vector, which is of unknown classification to the class of its unknown neighbors. Obviously when more than one neighbor is viewed as, the likelihood that there will be a draw among classes with a most extreme number of neighbors in the gathering of K-nearest neighbor persists (Keller et al. 1985). A technique to prevent the duplication of data in cloud has been proposed with the help of a dekey, an effluent as well as stable convergent key administration technique for secure deduplication. This technique ensures the removal of the repeated information, and is largely employed in cloud computing. Despite of the fact that convergent encryption works well for securing deduplication, the loophole that exist here is the efficient management of a great number of convergent keys (Li et al. 2014). In many areas the classification techniques are used to classify the data first and then securing the private data using encryption technique. K-Nearest Neighbor (K-NN) technique classifies the cases on the basis of some similarity measure. It is the simplest classification technique of supervised learning. The decision rule gives a non-parametric strategy that allocates a tag to input array based on tags of K-Nearest Neighbor of the vector (Zardari et al. 2014).

4 PROPOSED WORK

In the previous models it has been found out that the security is provided up to a large extent but there are still many loopholes that can be resolved and security can be further enhanced in a cloud environment. For classifying the data, more improved techniques can be used. No doubt K-NN provides the confidentiality but the security can still be improved by reducing the classification time (Khan et al. 2001). Furthermore accuracy can be enhanced and the security can be extended so that the data remains confidential that will make a tough job for the hackers which prove beneficial for the end user. In our proposed model we have used the boosting and bagging technique to improve the flaws of K-NN classification technique i.e. the classification time has been reduced and the accuracy has been improved. The improved technique classifies the data into sensitive and non-sensitive in a much lesser time than K-NN. The sensitive data is then secured using a hybrid encryption algorithm that is the blowfish algorithm in this case which is then sent to the cloud. Another task that has been done to improve the security is by creating security partitions on one cloud storage providing a better and more secure cloud performance, thus restricting the data residing on a single cloud.

To separate the public and private data, the improved classification technique i.e. bagging and boosting techniques are used in cloud storage for cloud security. Once the categorization of the data is done, the sensitive data is then shifted to blowfish algorithm to encrypt the sensitive data. Thus the non-sensitive data is immediately allotted a VM and send to the cloud directly, not encrypting it. The VM then operates on the data and disseminate with the cloud servers for storing it. It is a comprehensive idea of determining the security elements of data first and then encrypting it according to the requirements rather than encrypting the whole data (Zardari et al. 2014). This will save the memory resources as well as the time. Fig. 1 depicts the whole scenario to fix the data security concern in cloud environment.

4.1 Bagging and boosting techniques

It is proved by the researchers that bagging and boosting algorithms gives the best results when

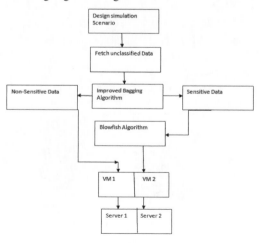

Figure 1. Proposed model.

used together than using any of the single classifier. Bagging gives the best results when the classification approach that is being bagged is little unstable. That means meanwhile little alterations in learning specimen may recurrently leads to detectable variations in prophecy (Bauer & Kohavi 1999). The better performance of bagging depends on the type of classification technique used having very low bias. Boosting is applied along more complicated base classification technique which is unbalanced. Bagging as well as boosting need very small or negligible changes to the present classification techniques. Mostly these techniques act as guide to actual learning technique by accumulating fixed objects of every learning technique as well as its classifier. It is simple to execute in object oriented programming languages. These give the chance to concurrent programming and run the learning technique themselves. Operation of training final classifiers executing fresh objects via learning technique, classification technique takes place currently. A greater level of parallelism persists by differentiating testing and training examples as individual routines (Bauer & Kohavi 1999).

4.1.1 Working of the adaboost algorithm
1. Model Creation:
   Accredit equal weights to every training object.
   For n iterations:
   Impose learning technique to objects that have been assigned weights, keep generated model.
   Calculate error e for objects that have been assigned weights.
   If e = 0 or e>= 0.5
   Stop model creation.

   For every object of the data:
   If the classification of the objects is accurate by the model:
   Update the weight of the tuple by

   $$w \text{ of the tuple} * \frac{Error \ of \ M(i)}{1 - Error \ of \ M(i)}$$

   Perform normalization of the weights for all objects by

   $$w \text{ of the tuple} * \frac{Sum \ of \ old \ weights}{Sum \ of \ new \ weights}$$

2. Classification:
   Accredit weight as 0 to all classes.
   For every n (or less) models
   The model prediction for this class,
   Sum −log e/(1 − e) to the weight of the resulting class.
   Weight of the classifiers vote

$$wi = \log \frac{error \ of \ M(i)}{error \ of \ M(i)}$$

Remark the class that has largest weight.

4.1.2 *Proposed algorithm for data classification*
The present technique generates models (classifiers) of a learning strategy and every model depicts the same weighted prophecy.

Input:
M, a group of m training tuples;
N, a count of models in the group;
The classification strategy Adaboost
Output: A compound model, K*
Procedure:
For i = 1 to T do// generate models;
Generate bootstrap specimen, Gi by sampling G with shuffling;
Employ Gi for generating a model, Ki;
End for:

For employing the compound model on a tuple, Y

1. If classification, then
2. Suppose every N models classify Y and gives back maximum vote;
3. If prediction, then
4. Be every N models predict a number for Y and gives back average predicted number.

4.2 *Cloud simulation environment*

For the simulation purposes of our research, we have used the cloudsim simulator. Fig. 2 shows the detailed description of the methodology used for implementing the data security concern in cloud computing. The first and the foremost step is the creation of the virtual cloud environment that involves formation of brokers, cloudlets, virtual machines, data centers and hosts (Calheiros et al. 2012). Next is to classify the dataset through enhanced bagging technique which automatically diminishes the deviation of the prediction using dataset by incorporating combinations by repetitions to get multiple sets of equal size of dataset. Every multi set involves operation of boosting technique that classifies each instance that in turn creates the model that forms a vote analogous to that model. The outcome of the classifier is the average of all predicted votes, which divides the data into 2 parts: sensitive (private) and non-sensitive (public). Next is encrypting the private data with the help of hybrid encryption algorithm that involves hybridizing the RSA plus blowfish algorithm. For the sensitive data digital signature has been produced with the help of RSA, which is a key for Blowfish algorithm to perform encryption

(Devi & Kumar 2012). In addition to this, the safety is increased by storing the data on different partitions formed on cloud, instead of storing the whole data on an individual cloud. Single cloud is disintegrated into many parts and the whole data is also separated to be stored into these parts. The above procedure secures the encrypted data which remains into cloud. Fig. 2 shows the whole cloud simulation environment. Cloudsim engine is deployed for processing the simulation works. The virtual machine manager on the other hand directs and assigns the VMs to the cloudlets running on the simulator. For all this scenario to take place efficiently it is very necessary to define the number of cloud commodities used for the simulation purposes because the number of cloud items will

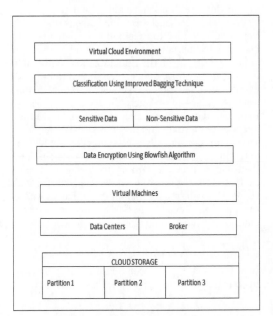

Figure 2. Cloud simulation environment.

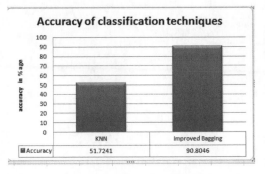

Figure 3. Accuracy of K-NN and improved bagging.

depend on the type of the task to be accomplished on the cloud. In our current scenario of cloud environment, the data is first classified and then the private data being encrypted is transferred to cloud server. So here we have two cloudlets (cloud tasks) that is data classification and data encryption to run in the cloud engine. Figure 3 shows the amount of cloud commodities used for the simulation purposes of our research.

5 RESULTS AND DISCUSSION

This section describes the results of the implemented algorithm i.e. improved bagging and blowfish algorithm to improve the data confidentiality in cloud environment. The algorithms have been tested on a dataset that is employees' record of an organization. Figure 4 shows the increased accuracy using the improved bagging technique in comparison to the K-NN classification algorithm. It is clear that the accuracy of the K-NN algorithm is found out to be 51.7241 while the accuracy of the improved bagging technique is 90.8046 which is approximately half times greater in comparison to K-NN classification technique. Also the time taken by the K-NN for the classification of data is 4773 ms and the improved bagging is 686 ms which proves that the improved bagging technique takes much less time in comparison to K-NN classification algorithm. The graphs show the difference in accuracy and the time taken to classify the data by K-NN and the improved bagging technique.

The time taken by hybrid encryption algorithm to encrypt the sensitive data is 1935 ms. Thus the hybrid encryption algorithm increases the security of the data than by using only the RSA algorithm. No doubt the time taken by the hybrid encryption algorithm is more than the time taken by the RSA algorithm alone but it also enhances the security of the data and is also less vulnerable to attacks than by using the RSA algorithm alone.

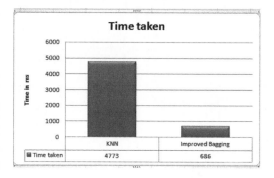

Figure 4.   Time taken by K-NN and improved bagging.

6   CONCLUSION AND FUTURE SCOPE

This paper introduces a model for data security in cloud environment. The focus of the research was to classify the data on the basis of security elements that divides the data into sensitive and non-sensitive. The main augmentation of this security model is data security using a classification algorithm. The sensitive data is encrypted using blowfish algorithm and is stored in the cloud server while the non-sensitive data is stored as it is without being encrypted. Furthermore to enhance the security, the cloud is divided into partitions and the data is stored in chunks on every partition instead of storing the whole data on a single cloud. We have used the cloudsim simulator for the execution of our proposed algorithm. Furthermore in this field a better classifier can be used to increase the accuracy of the classification technique. Time of the classification can also be reduced by using any other classifier. Apart from confidentiality of data, integrity and accountability of data are the key issues to be sorted in future by using an enhanced encryption algorithm to make the cloud system more secure.

REFERENCES

Bauer, Eric, Kohavi, Ron. 1999, "An impirical comparison of voting classification algorithms: Bagging, Boosting and variants", Machine Learning 36, pp. 105–139.

Calheiros R.N., Ranjan R, Beloglazov A.A., Rose F.A., and Buyya R, 2012, "CloudSim: A Toolkit for the Modeling and Simulation of Cloud Resource Management and Application Provisioning Techniques".

Catteddu D, and Hogben G, 2009 "Cloud Computing: Benefits, risks and recommendations for information security", ENISA.

Dasarathy, B.V. 1980, "Nosing around the neighbourhood: A new system structure and classification rule for recognition in partially exposed environments" IEEE Trans. Pattern Anal. Machine Intell. Vol. PAMI-2, pp. 67–71.

Devi, G., Kumar, P. M. 2012, "Cloud computing: A CRM service based on a separate encryption and decryption using Blowfish Algorithm", International Journal of Computer Trends and Technology. Vol. 3 Issue 4, pp. 592–596.

Keller J. M., Gray, M. R. and Givens, J. A. JR. 1985, "A Fussy-K-Nearest Neighbor Algorithm", IEEE Trans. System, Man, and Cybernetics, vol. SMC-15, No. 4, pp. 580–585.

Khan. M, Ding. Q, and Perrizo. W. 2001, "K-Nearest Neighbor Classification on Spatial Data Streams Using P-Trees1, 2", PAKDD LNAI 2336, pp. 517–528.

Li. J, Chen. X, L. Mi, Li. J. 2014, "Secure Deduplication with efficient and reliable convergent key management" IEEE Transactions on parallel and distributed systems, vol. 25, No. 26, pp. 1615–1625.

Purushothaman. D, and S. Abburu. 2012, "An Approach for Data Storage Security in Cloud Computing",

IJCSI International Journal of Computer Science Issues, Vol. 9, Issue 2.

Ram C. P., and Sreenivaasan. G. 2010, "Security as a Service (SasS): Securing user data by coprocessor and distributing the data," Trendz in Information Sciences & Computing (TISC2010), IEEE, pp. 152–155.

Rawat, P.S., Saroha, G.P. and Bartwal, Varun. 2012, "Quality of service Evaluation of SaaS Modeler (Cloudlet) Running on Virtual Cloud Computing Environment using CloudSim", International Journal of Computer Applications (0975–8887), Vol. 53, No. 13, pp. 3538.

Rittinghouse, J.W. and Ransome, J.F. "Cloud Computing: Implementation, Management, and Security", 2010 CRC Press 2009 by Taylor and Francis Group, Journal of high technology law.

Stephen S. Yau Ho G, 2010 "Confidentiality protection in cloud computing sytems", Int J Software Informatics Vol. 4, No. 4.

Song, D., Shi, E., Fischer, I., and Shankar, U. 2012, "Cloud Data Protection for the Masses," IEEE Computer Society, pp. 39–45.

UTHSCSA Data Classification report, 2006 "Protection by Data Classification Security Standard".

Whitney A., and S.J Dwyer, II. 1966, "Performance and implementation of K-nearest neighbor decision rule with incorrectly identified training samples", in Proc. 4th Ann. Allerton Conf. On Circuits Band System Theory.

Wu X., Kumar. V, Quinlan J.R., Ghosh. J, Yang. Q, Motoda, H., McLachlan G.J., 2008 "Top 10 algorithms in data mining", Knowl Inf Syst (2008) 14:1–37.

Zardari, M.A, Jung, L.T, Zakaria. N. 2014, "K-NN classifier for Data confidentiality in cloud computing", IEEE.

Communication and Computing Systems – Prasad et al. (Eds)
© 2017 Taylor & Francis Group, London, ISBN 978-1-138-02952-1

User preference prioritization in multi-criteria collaborative filtering recommenders

Aparna Gangwar & Anurag Singh Baghel
*School of Information and Communication Technology, Gautam Buddha University, Greater Noida,
Uttar Pradesh, India*

ABSTRACT: Recommendation Systems have become a part of every major web service in a relatively short span of time. In today's time, the number of options provided by any website are overwhelming and thus it gets difficult for the user to analyze them all and make informed decisions. As a result, most users end up settling for the best they can find. Most popular recommender systems we know of, primarily work on a single rating provided by the users. Multi-criteria recommender systems based on this Multiple Criteria Decision Making (MCDM), are a fairly new kind of recommenders, popularized by Yahoo! Movies. This research aims to improve multi-criteria recommendation system by adding a preference function which includes adding weights during similarity computation such that priority is given to user's preferences.

1 INTRODUCTION

Recommendation Systems, put simply, are the digital version of sales people, suggesting products according to the customer's preferences. From Google to Flipkart, Recommendation Systems have been employed in almost all major web store or service. Although a little more than two decades of research have been invested in evolving recommender engines, the scope for improvement is immense. The field is constantly going through changes as the core methods used for recommendations are improving in efficiency. Recommender Systems are very quickly taking over the interests of World Wide Web enthusiasts as they are almost a part of every major web service available.

1.1 *Related work*

The first few papers to establish collaborative filtering as a research domain were Burke et al. 2002 and Sarwar et al. 2001. Ever since then, a humongous amount of work has been done in this field. The advent of multi-criteria systems came with Adomavicius et al. 2007. They have presented a study of multi-criteria recommendation systems and the methods to extend single criteria systems to multiple criteria domain. Shambour et al. 2011 have presented an approach to find recommendations in multi-criteria based systems using semantics. They have used tree ontology to classify the items and Jaccard coefficient to eliminate cold-start problem. Palanivel et al. 2010 have proposed

personalization in multi-criteria recommendation systems by employing Fuzzy linguistic approach to populate user-item matrix and fuzzy multi-criteria decision making to make recommendations. Around the same time, Hwang et al. 2010 proposed personalizing multi-criteria systems using genetic algorithms for optimal feature weighing. Lakiotaki et al. 2011 have proposed a framework that clusters users in multi-criteria setting and models their behavior for improved quality of recommendations. Miyahara et al. 2000 have proposed using Naïve Bayes for classification purposes in a single criteria system to improve prediction accuracy.

1.2 *Recommendation systems-an overview*

The process of recommendation involves an active user for whom the system is to make personalized recommendations. For this, some background knowledge about the user and the item space is needed. Based on the type of knowledge used, recommendation systems are divided as follows:

- Collaborative recommendation matches an individual knowledge with similar social knowledge sources and compares the active user's preferences from them.
- Content-based recommendation is entirely focused on the active user, using item features and user opinions to learn a classifier that can predict user preferences on new items.
- Hybrid recommendation systems combine both the techniques-collaborative and content based-using one of the hybridization methods.

A new classification of RS is based on the dimensionality of knowledge used for making recommendations. The RS are hence divided into two types:

- Single-criteria recommendation is when only one criterion is used to make recommendations. For example, MovieLens dataset.
- Multi-criteria recommendation is when more than one criterion is used to make recommendations. For eg. Yahoo! Movies dataset.

Collaborative Filtering based Recommender Systems are classified on the basis of algorithmic type used, into memory and model based systems.

2 MULTI-CRITERIA RECOMMENDATION SYSTEMS

Presently used Recommendation Systems, at large, are single-criteria based systems that represent the usefulness of an item to a user in the two-dimensional *Users × Items* space. The process begins by taking in active users' ratings. An estimation function R is then specified by the recommendation system for each user-item pair that hasn't been rated

$$R: Users \times Items \rightarrow R_0 \qquad (1)$$

R_0 is usually represented by a totally ordered set, i.e., the rating values belong to a pre-known scale. Once we derive the estimation function R, the recommender system easily recommends items using it (or a set of N highest-rated items) for each active user.

Multi-criteria systems incorporate information on more parameters about the user preferences and items and hence are more effective for recommendation making purposes. The rating function for single criteria system as in equation 1 can be extended to multi-criteria domain as follows:

$$R: Users \times Items \rightarrow R_0 \times R_1 \times ... \times R_k \qquad (2)$$

where R_0 represents the overall rating values, and R_i represents the probable rating values for any of the available criterias i ($i = 1, ..., k$), typically on some numeric scale (e.g., from 1 to 5), for any user. A rating in multi-criteria setting $R(u,i) = (r0,r1,...,rk)$ denotes a point in k+1 dimensional space. Therefore, one may calculate user similarities with different existing users.

The very first step for calculating similarity would be to calculate the distance between the two users in question, for the same item, i.e., d_{rating} ($R(a,i)$, $R(u,i)$), where $R(a, i) = (r_0, r_1, ..., r_k)$

and $R(u, i) = (r_0', r_1',...,r_k')$. The following multi-dimensional distance metrics can be used to resolve this purpose.

- Manhattan distance:

$$d(a,u) = \sum\nolimits_{i=0}^{k} |r_i - r_i'| \qquad (3)$$

- Euclidean distance:

$$d(a,u) = \sqrt{\sum\nolimits_{i=0}^{k} |r_i - r_i'|^2} \qquad (4)$$

- Chebyshev (or maximal value) distance:

$$d(a,u) = max_{i=0,...,k} |r_i - r_i'| \qquad (5)$$

Next we find the overall distance between two users a and u using:

$$d_{user}(a,u) = \frac{1}{|I(a,u)|} \sum\nolimits_{i \in I(a,u)} d_{rating}(R(a,i), R(u,i)) \qquad (6)$$

where $I(a, u)$ is the finite set of items that both the users a and u have rated.

Third and the final step is to calculate similarity between the two users, $sim(a,u)$. Distance and similarity are inversely proportional to each other, thus we use a simple conversion between the two as said by (Adomavicius et al. 2007):

$$sim(a,u) = \frac{1}{1 + d_{user}(a,u)} \qquad (7)$$

As distance between the two users increases, similarity $Sim(a,u) \rightarrow 1$ and when distance $d(a,u) \rightarrow 0$, similarity $Sim(a,u) \rightarrow 1$. Similarity is 0 only when the two users haven't rated a single common item. Thus, similarity range is: $Sim(a,u) \in [0,1]$ where $d(a,u) \in [\infty,0]$.

The ultimate goal of a recommendation process is to be able to predict the rating a user a would give to an unseen item i, $R(a,i)$. The following are the two ways in which the rating can be calculated (Adomavicius et al. 2007):

- Weighted average rating:

$$R(a,i) = z \sum\nolimits_{u \in N(u)} sim(a,u) \cdot R(u,i) \qquad (8)$$

- Adjusted weighted average rating:

$$R(a,i) = R_{avg}(a) + z \sum\nolimits_{u \in N(u)} sim(a,u) \\ \cdot (R(u,i) - R_{avg}(u)) \qquad (9)$$

As we can see, the rating for a similar user u for item i, $R(u,i)$ is multiplied by $Sim(a,u)$. This implies, rating for a user that has highest similarity will affect the computed value of rating more than any other user. The multiplier z acts as a normalizing factor and is expressed as:

$$z = 1/\sum_{u \in N(u)} |sim(a,u)| \qquad (10)$$

$R_{avg}(u)$ denotes the mean of all the ratings given by the similar user u and $N(u)$ denotes the set of similar users considered for computing the unknown rating. The size of set $N(u)$ can vary from 1 (only the most similar user is used for computing $R(a, u)$) to all users in the dataset.

Furthermore, due to the analogous nature of users and items in the traditional memory-based collaborative filtering approach, this process can be extended to finding similarity between the two items as well. For instance, equation 10 can be changed as follows, to calculate item-based adjusted weighted mean rating (Adomavicius et al. 2007):

$$R(a,i) = R_{avg}(i) + z \sum_{i \in N(i)} sim(a,u) \cdot \left(R(a,i) - R_{avg}(i) \right) \qquad (11)$$

where z, $R(i)$, $sim(i, i')$, and $N(i)$ are analogous to their user-based counterparts.

3 EXPERIMENTS AND DISCUSSION

3.1 Existing algorithm

The flowchart drawn below represents the most general procedure for developing a memory-based Recommendation System. The flowchart can be modified to incorporate any similarity computation

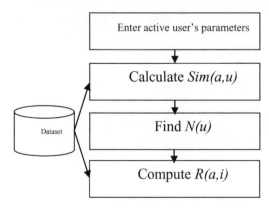

Figure 1. Collaborative filtering recommendation process.

coefficient, rating prediction or different values of k. Also, this flowchart will work for all memory-based techniques, i.e., item-based, user-based or neighborhood based.

3.2 Proposed algorithm

The proposed algorithm implements a memory-based Multi-criteria top-k user-based Collaborative filtering based Recommender System and can be divided into the following steps:

Step 1: Set up system environment and collect data.
Collect data and populate the user rating matrix.
Step 2: Input active user's data.
The active user is asked to input first few ratings before entering the system. This step is necessary to avoid new user cold start problem.
Step 3: Calculate prioritized similarity between active user and existing users.
The distance is then calculated between all active and ith existing user, $d_j(a, u_i)$, separately for each rating j. We have calculated distance using equation 4. The distance is then converted into corresponding similarity value, $Sim_j(a, u_i)$. The similarities are then averaged by multiplying them with corresponding weights.
Thus, average similarity in usual MCRS will be:

$$Sim(a,u_i) = \frac{1}{j}\sum_{n=1}^{j} Sim_n(a,u_i) \qquad (12)$$

And the average similarity of proposed MCRS will be:

$$Sim(a,u_i) = \frac{1}{j}\sum_{n=1}^{j} w_n \cdot Sim_n(a,u_i) \qquad (13)$$

Step 4: Find top-k users.
Top k similar users are derived by sorting the similarity coefficients.
Step 5: Recommend items.
The itemized ratings of the topmost users are then used to find an approximate degree to which the active user may like an item.

3.3 Experimental setup

To compare the results, three recommender systems have been implemented. First is a single criterion based CFRS, named SCRS-E. The second is a multi-criteria based CFRS, named MCRS-E. Third is the proposed system, PMCRS-E, i.e., priority based Multi-criteria CFRS We have used three datasets, extended from MovieLens 100k dataset containing *User X Item* pairs that have rated each

Table 1. Experimental results.

Sparsity									
Algorithms	36.4%			44%			50%		
	K = 3	K = 5	K = 7	K = 3	K = 5	K = 7	K = 3	K = 5	K = 7
SCRS-E	0.4	0.4	0.4	0.4	0.4	0.4	0.267	0.27	0.133
MCRS-E	0.533	0.4	0.4	0.4	0.4	0.53	0.4	0.4	0.4
PMCRS-E	0.67	0.53	0.53	0.67	0.53	0.53	0.53	0.4	0.4

movie on three parameters-general rating, acting, visuals, are used as follows:

Dataset 1: Contains 25 users and 20 movies where each user has rated atleast 10 movies and each movie has been rated atleast 14 times and sparsity is 36.4% ((1−(318/25*20))*100).

Dataset 2: Contains 30 users and 30 movies where each user has rated atleast 15 movies and each movie has been rated atleast 15 times and sparsity is 44%.

Dataset 3: Contains 40 users and 40 movies where each user has rated atleast 20 movies and each movie has been rated atleast 20 times and sparsity is 50%.

These algorithms and datasets have been each tested for three values of top-k users, i.e. k = 3, k = 5 and k = 7.

4 RESULTS AND DISCUSSION

Table 1 shows the F-measure values when we've top-3, 5 and 7 most similar users. Figure 2 shows the graphical representation of the same.

5 CONCLUSION

The proposed system works better than multi-criteria system without preference function in most cases or at least as well as it, but it surely works better than single-criteria system. We notice that data sparsity is inversely proportional to the recommendation accuracy. Also, the relation is similar between the value of "k" in the top-k user approach, i.e., as we consider a larger set of similar users, the prediction accuracy tends to decrease. But this does work to an advantage as the user might get recommended some unconventional that he might like. Although, since the dataset is not real time, thus the results are not an absolute reflection of what might be expected.

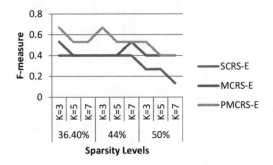

Figure 2. Comparison of experimental results.

Also the system's prediction accuracy keeps improving with the number of parameters. But, since multi-criteria systems involve a much larger amount of data than single-criteria systems, addition of each extra parameter increases the computational cost of the system. Thus, there is a trade-off between the number of parameters used for prediction and the prediction accuracy, which has to be maintained while designing a multi-criteria recommender system.

6 FUTURE SCOPE

The amount of work that can be done in this field is immense. For starters the additions that can be done to this dissertation are, results need to be validated on a real-time database and hence the performance needs to be seen. Further, the system can be converted to a model-based system for working well with large amounts of data. Data mining can be applied to the proposed system to further improve its performance.

We believe that this research work is just a small step in the field of Multi-Criteria Recommendation Systems and that further significant work can be done to improve this issue.

REFERENCES

Adomavicius G., Manouselis N. & Kwon Y., 2011, Multi-Criteria Recommender Systems, *Recommender Systems Handbook Springer:* 796–803.

Adomavicius G. & Kwon Y., 2007, New Recommendation Techniques for Multi-Criteria Rating Systems, *IEEE Intelligent Systems*, Vol. 22, Issue 3: 48–55.

Burke R., 2002, *User modeling and user-adapted interaction, Springer:* 331–370.

Hwang C.S., 2010, Genetic Algorithms for Feature Weighting in Multi-criteria Recommender Systems, *Journal of Convergence Information Technology*, Vol 5, No. 8.

Kumar P.N.V. & Reddy V.R., 2014, A Survey on Recommender Systems (RSS) and Its Applications, *International Journal of Innovative Research in Computer and Communication Engineering*, Vol. 2.

Lakiotaki K., Matsatsinis N.F. & Tsoukias A., 2011, Multi-Criteria User Modeling in Recommender Systems, *IEEE Intelligent Systems*, Vol. 26, No. 2.

Miyahara K. & Pazzani M.J., 2000, Collaborative Filtering with the Simple Bayesian Classifier, *PRICAI 2000 Topics in Artificial Intelligence, Springer*.

Mobasher B., Burke R.,Bhaumik R. & Williams C., 2007, Towards Trustworthy Recommender Systems: An analysis of attack models and algorithm robustness, *ACM Trans. Internet Technology*, Vol. 7.

Palanivel K. & Sivakumar R., 2010, Fuzzy multi-criteria decision making approach for Collaborative recommender systems, *International Journal for Computer Theory and Engineering*, Vol. 2: 1793–2801.

Ram S.K. & Riedl J., 2004, Shilling Recommender Systems for fun and profit, *Proceedings of the 13th international conference on World Wide Web, ACM*: 393–402.

Ricci F., L. Rokach & B. Shapira, 2011, *Recommender Systems Handbook*, Springer.

Sandvig J., Mobasher B. & Burke R., 2007, Robustness of Collaborative Recommendations based on association Rule Mining, *Proceedings of the ACM Conference on Recommender Systems (RecSys'07), ACM*: 285–295.

Sarwar B., Karypis G., Konstan J. & Reidl J., 2001, Item-based collaborative filtering recommendation algorithms, *Proceedings of 10th international conference on World Wide Web, ACM*: 285–295.

Sharma L. & Gera A., 2013, A Survey of Recommendation System: Research Challenges, *International Journal of Engineering Trends and Technology* Vol. 4.

Shambour Q. & Lu J., 2011, A Hybrid Multi-Criteria Semantic-enhanced Collaborative Filtering Approach for Personalized Recommendations, *Web Intelligence and intelligent Agent Technology, IEEE/ WIC/ ACM International Conference*, Vol. 1.

Communication and Computing Systems – Prasad et al. (Eds)
© 2017 Taylor & Francis Group, London, ISBN 978-1-138-02952-1

Monotonic decision trees on rough set theory in machine learning approach

Aarushi Singh & Anurag Singh Baghel
Gautam Buddha University, Greater Noida, Uttar Pradesh, India

ABSTRACT: The traditional theory of Rough Sets states objects by their discrete attributes, and does not account for the ordering of its values. We have introduced the technique of monotone discernibility matrix as well as monotones (object) reduce. Also monotone discrete functions theory, which is developed earlier for representation and computation of decision rules are explained. Feature's values as well as decision values are ordinal in various decision making tasks. Moreover, a monotonic constraint states that the objects having better feature values must not be designated to worse decision class. These types of problems are referred as ordinal classification having monotonicity constraint. A number of learning algorithms have been designed for handling these kinds of tasks. In this paper, we surveyed about the Monotonic Decision Trees.

Keywords: monotonic classification, rough sets, attribute reduction, decision tree, ensemble learning

1 INTRODUCTION

Classification deals with the prediction of an unknown attribute or target attribute from the given known attributes with the assumption that the known attributes contribute to a clear recognition of the target attribute class (Fong & Jahnke 2012). Decision Tree, Naïve Bayes, k-Nearest Neighbors some of the well-known classification algorithms. The simplest of these algorithms is the Decision Tree. Many variations of the decision tree exist such as decision stump, random forest, and fuzzy decision tree and so on.

Some of the well-known clustering algorithms are K-Means, K-Medians, Agglomerative, Divisive, density based, grid based algorithms. Out of these K-Means is the simplest of these. Decision tree algorithm and its hybrids find lot of applications in networking (Dowd et al. 2006; Aggarwal & Yu 2008) for intrusion detection, in wireless networks (Liu et al. 2009), in medicine for detection of diseases (Durairaj & Meena 2011; Kamber et al. 1996; Sweeney 2002), in biometrics (Yellasiri & Rao 2009) and so on.

1.1 Monotonic classification

Monotonic constraints should be taken into account between features and decisions in machine learning and pattern recognition. However, most of the present day techniques do not allow the discovery and the representation of ordinal structures in monotonic type of datasets. Therefore, they are not applicable for monotonic classification. On the other hand, selection for a particular feature has been proven effective in case of improvement of classification performance and avoidance of overfitting. Until now, no particular technique has been designed especially for selecting the features in monotonic classification. We have also introduced a function, which is known as rank mutual information, for evaluating monotonic consistency among various features and decision in such monotonic tasks. Rank mutual information function amalgamates the advantages of rough sets based on dominancy attribute, in reflecting mutual information and ordinal structures on the basis of robustness. Afterwards, the rank mutual information is encapsulated with search strategy of max-relevance and min-redundancy for computing optimal subsets of decisions and features. A set of collection of experiments (numerical) are provided to depict the impact and the effectiveness of the proposed technique.

1.2 Rough sets and data mining

Data Mining and Knowledge Discovery are the two areas application areas of the Rough Set theory. Identification of partial or total dependencies in huge data sets and elimination of redundant data from data sets are the two main applications of rough set theory in data mining. It also grants focus on techniques to void data fields, misplaced data, active data and others. Some of the major ways for amalgamation of rough calculation for the ideas associated for the creation of primordial theory responsible in estimation of multifarious ideas will be constructed, resemblance factors among conceptions, processes for building multifarious ideas using the older concepts. Such tasks can easily be resolved through joining conventional approach to rough set along with up to date additions to this theory.

1.3   Decision tree

They supply sole abilities to enhance, harmonize, and replace customary arithmetical structures of analysis (like manifold linear regression), diversity of data mining tools and methods (like neural networks), lately made multi-dimensional ways of responding and analysis established in for business intelligence.

They are constructed by algorithms which recognize many techniques of splitting and turning in a lateral extension-sort of slice. Such fragments shape up to make the upturned decision tree which starts from the part called the root node which lies at the top of that decision tree. The entity from complete study is revealed from this very first root node as one-dimensional show through tree's edge. The entity (name) of study is shown, next to the movement of ideals which are present in that particular field. Also, Simple and typical decision tree is shown in Figure 1.

All the records, fields, and its field values of data-set which plays a part of entity of study are shown by display of the node. The detection of the decision rule to make segments (branches) below the root is done through a technique that takes out the association amid the entity of study and the fields providing, input ones for producing those lateral extensions. Those values of fields help in guesstimating possible items of the targeted field. That targeted field is referred as a dependent field, outcome, response, or a variable. Splitting rules are put one by one, producing a hierarchy of segments within segments that generates the characteristic upturned decision tree form.

1.4   Ensemble methods

Ensemble learning is learning algorithm which builds a set of base classifiers and then categorize new instances, considering a vote of their guesses

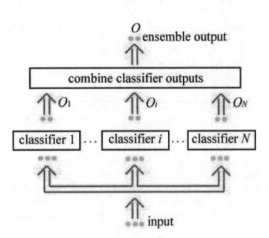

Figure 2.   Framework of ensemble learning.

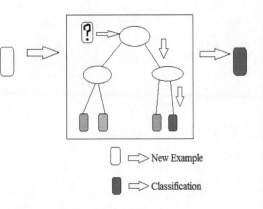

Figure 3.   Decision tree learning.

Figure 1.   Typical diagram of the decision tree.

and it makes one of the principal existing directions in machine learning community.

In the ensemble, classifier 1 to classifier N is first trained with the help of training examples. Then, the guessed output (O_i) of every classifier is amalgamated to create the output O of the ensemble. The basic framework of ensemble learning is illustrated in Figure 2.

Review on numerous traditional rough set data with decision tree is considered. Many algorithms came into existence in the last few decades for performing accurate classification of the machine learning. In addition to the themes identified above, a growing body of study learning addresses other aspects of rough set theory as shown in Figure 3.

P. K. Fong et al. brought confidentiality conserving technique in (Fong & Jahnke 2012), which takes account for decision tree learning suspending any losses for accuracy. It illustrates a technique for the fortification of the privacy of gathered samples of the data in cases when any sort of information is partly lost from these samples. This technique transforms the existing illustrated sets of data and converting in a cluster made of illusory sets, after which all the unusual examples can never reformulate in the absence of entire cluster made of illusory data sets. As shortly as first trial is collected, this technique can be implemented unswervingly to the data storage.

J. Dowd et al. in projected a number of contributions in the direction of privacy-preserving decision tree mining. The new data perturbation technique by means of random substitutions is like the randomization techniques taken into consideration with statistical revelation context but it uses another privacy measure referred as the ρ1-to-ρ2 privacy breaching the γ-diagonal matrix. Some techniques such as C4.5, C5.0, and data mining methods with continuously valued attributes and mixed discretely are also explained. Emphasis on Decision Tree Learning and its classification is done.

With the help of ID3 which uses information gain for deciding what attribute goes into the decision tree. Through Gain, we can measure the amount that training examples are separated from targeted classes on the basis of given attribute. For defining gain, we use entropy. The amount of information in an attribute is called Entropy. It measures the homogeneity of a sample that is calculating the entropy of data sets by this formula.

$$Entropy(S) = \sum_{j=1}^{c} p_j \log_2 p_j$$

Generalization can be done for n > 2 classes and p is the proportion of examples in S that belong to the j class. Measure of the information gain with respect to the attributes,

Gain (A) = E (current set)-∑E (child sets)

Aggarwal et al. 2008, in classifies privacy preserving data taking out techniques, as well as data alteration and cryptographic, statistical, query auditing and perturbation-based strategies. For all intents and purposes, data sets are customized by eradicating and amalgamating infrequent elements. These identical sets behave like patterns for the rest from inside the assembly since these are not supposed to get eminent from the rest.

Building data mining models unswervingly from the disconcerted data without demanding the explanation of the general data distribution restoration as halfway step was brought by Liu et al. 2009, in. Modified version of Privacy preserving decision tree C4.5 (PPDTC4.5) classifier communicating with disconcerted numeric continuous attributes was proposed. The tests state that their proposed methodology can gain an elevated accuracy level when considered for the classification of unusual set of data.

A Cross Prognosis Structure which inculcates Rough Set Theory, along with Artificial Neural Network for prediction and classification of patients' information was brought by Durairaj et al. 2011. It is a new data mining technique and software for providing assistance to cure difficult problems for medical data analysis to produce efficient data analysis along with indicative predictions. In this system, prediction accuracy and reliability is calculated through a comparison between observed and predicted rate. The proposed system is shown here in fig 4.

Kamber et al. 1996 in brought association rule mining for the purpose of Classification. Other techniques like the k-nearest-neighbor classifiers, case-based reasoning, genetic algorithms, rough sets, and fuzzy logic, are also briefly explained. Here, the training tuples are chosen from the database as a part of data analysis task. This data is then presented for the purpose of classification using the classification algorithm. Estimation of test data accuracy based on the classification rules is measured. If it is valid and can be considered as right, the rules are further used for new data tuples classification.

Sweeney et al. 2002, in brought the concept of the K-anonymity which is a utility based data modification approach that uses generalizing attributes for Preserving Privacy, protecting confidential information of the samples. The major drawback of this approach is that it can only be taken into usage after successful completion of data collection.

A new model for classification known as Rough Set Classifier was brought by Yellasiri

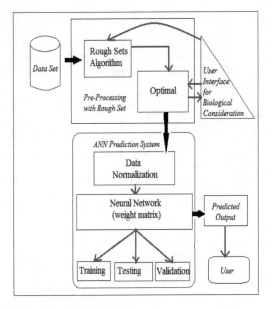

Figure 4. Hybrid IRNNS model.

et al. 2009, for classification of huge amount of protein data which covers data similar to protein's skeletal and serviceable characteristics. The outcome is much faster, precise and efficient. This tool must be considered to be of higher priority, since it is an amalgamated tool that inculcates Progression Calculative approach, Concept Lattice approach with Rough Set Theory, providing 97.7% accuracy, as compared to rest classification tools.

This scales down the field probing area upon 9.8% with no trailing of quality of classification of those proteins. They proposed that the contents of the family should be recognized by individual mathematical techniques and must be taken into account for minimizing the search space (probing area). Such arithmetic conventions are constructed and preserved in the Sequence Arithmetic database.

According to Xiong & Chen 2005, an approach called kernel function, had improved classifier performance for genetic data. Efficiency of this approach was probed, which works on optimizing data -dependent for the kernel model.

Also for interpretation examination, the K-Nearest-Neighbor (KNN) and Support Vector Machine (SVM) could be considered as a classifier. For the healthcare data, classification on the basis of data mining techniques had been developed by Srinivas et al. 2010. Posterior probability depends upon likelihood for conclusion c and observation, x.

2 CHALLENGES IN DECESION TREE

Generally, in decision tree we faces a lots of problem or challenges, here we will display the some challenges:

- The major challenges of rough set-based feature selection are having a constraint that all the data should be discrete.
- A lot of problems in machine learning include high dimensional characterization of input features. Therefore it is not so surprising that much of a research has been carried focusing dimensionality reduction.
- In a classification problem, it is usually sufficient to preserve the ability to discern between classes and not necessarily between individual elements. Here, a minimal attribute set with this property is called a decision reduct or relative reduct.
- Big data refers to problems whose size and complexity render standard machine learning algorithms unable to adequately deal with them (Kamber et al. 1996). It is a hot topic within the machine learning community; as such problems are encountered in many application fields such as bioinformatics, marketing, medicine and so on.
- Concept approximation problem which is an important issue for machine learning and data mining.
- Classification, clustering and association analysis are some of the examples of well known problem which can be taken as concept approximation problem.

3 IMPORTANCE OF ROUGH SET THEORY

Nowadays, machine learning has become increasingly important since the rise of the competitive data mining field.

- There is no need of any preliminary information regarding data—such as statistical probability.
- It offers well-organized techniques, methods and apparatus to locate data patterns (hidden ones).
- It provides reduction of original information, i.e. for finding minimal data sets having same knowledge as of the original data.
- It provides evaluation of the data of its significance.
- It provides automatic way of generation of the data sets through decision rules.
- It is an easier way to understand and implement.
- It provides a straight forward and simpler interpretation of results further obtained.
- It is useful for concurrent (parallel/distributed) processing.

4 DISCUSSION

In this discussion part various Author had studied about the decision tree machine learning technique and so on. In above various literature survey presented by many Authors, we analyze regarding various or many existing research concept in terms of privacy preserving approach, privacy-preserving decision tree mining, Models and Algorithms of data mining, C4.5 decision tree classifier, cross prophecy classification using Rough Set Theory with Neural Network (Artificial) for classification of patients information, k-nearest-neighbor classifiers, K-anonymity, Rough Set Classifier, kernel function and data mining techniques, which are given us to emerging method about Rough set theory operation on the bases of machine learning system that provide consistent accuracy and aware from the time.

5 CONCLUSION

In this paper, we summarize the existing improved algorithms for rough set data. This paper grants an idea for determining and showing the important attributes, according to the strength of an association. The application of Rough Set Theory approach may be used extensively and exclusively for knowledge discovery, data mining or any other field relevant to attribute reduction and feature selection. On the other hand, Decision tree which is one of the most significant and appropriate classification methods used in data mining, its algorithm cannot handle missing data effectively.

REFERENCES

Aggarwal, C. and P. Yu, Privacy-Preserving Data Mining, Models and Algorithms, Springer, 2008.

Dowd, J., S. Xu, and W. Zhang, Privacy Preserving Decision Tree Mining Based on Random Substitions, International Conference on Emerging Trends in Information and Communication Security (ETRICS '06), pp. 145–159, 2006.

Durairaj, M. and K. Meena, A Hybrid Prediction System Using Rough Sets and Artificial Neural Networks, International Journal Of Innovative Technology & Creative Engineering, ISSN: 2045-8711, Vol.1, No.7, July 2011.

Huilin Xiong, and Xue-Wen Chen, Optimized Kernel Machines for Cancer Classification Using gene Expression Data, IEEE Symposium On Computational Intelligence in Bioinformatics and Computational Biology, pp. 1–7, 2005.

Liu, L., M. Kantarcioglu, and B. Thuraisingham, Privacy Preserving Decision Tree Mining from Perturbed Data, 42nd Hawaii International Conference System Sciences (HICSS '09), 2009.

Micheline Kamber, Lara Winstone, Wan Gong, Shang Cheng, and Jiawei Han, Generalization and Decision Tree Induction: Efficient Classification in Data Mining, Canada, V5A IS6, 1996.

Pui K. Fong, and Jens H. Weber Jahnke, Privacy Preserving Decision Tree Learning Using Unrealized Data Sets, IEEE Transactions On Knowledge And Data Engineering, Vol. 24, No. 2, February 2012.

Ramadevi Yellasiri, and C.R. Rao, Rough Set Protein Classifier, Journal of Theoretical and Applied Information Technology, 2009.

Srinivas, K., B. Kavitha Rani, and Dr. A. Govrdhan, Applications of Data Mining Techniques in Healthcare and Prediction of Heart Attacks, International Journal on Computer Science and Engineering, 2010.

Sweeney, L. k-Anonymity: A Model for Protecting Privacy Uncertainty, Fuzziness and Knowledge based Systems, International Journal Of Innovative Technology & Creative Engineering, vol. 10, pp. 557–570, May 2002.

Communication and Computing Systems – Prasad et al. (Eds)
© 2017 Taylor & Francis Group, London, ISBN 978-1-138-02952-1

Performance evaluation of advance leach (A-leach) in wireless sensor network

Rupika Goyal, Shashikant Gupta & Pallavi Khatri
ITM University, Gwalior, Madhya Pradesh, India

ABSTRACT: Wireless Sensor Network (WSN) is a randomized, ad-hoc network, which has gained lot of attention from researchers to work into and bring out better approaches all the time. There are many protocols proposed for the betterment of lifetime and energy of a node but it is also important to know about the characteristics of the topology for the justification that the new techniques are better than their previous counterparts. The improvisation in leach protocol has been carried out earlier and is named as advance leach (A-leach). The main aspect of this work is to generalize various parameters on A-leach. Simulations are carried out on NS-2 (Network Simulator version 2) tool and various parameters are evaluated. Quantitative results clearly exhibit the evaluation of the proposed protocol.

1 INTRODUCTION

Wireless Sensor Networks (WSNs) is networks consisting of various numbers of nodes which are dispersed on huge area. WSN have battery operated nodes, so lifetime of network is main issue to be considered for building up scalable network. Lifetime and energy efficiency is the most demanding task of wireless sensor network as sensors can be deployed only once in their lifetime. These tasks for enhancing the performance of the energy efficient network, the network scenario require high reliability of the sensor networks. To make sensor networks more reliable, the attention to research on wireless sensor networks over different network parameters has been increasing in recent years. Clustering is one of the major techniques used to prolong the lifetime of a entire wireless sensor network by minimizing the energy of the entire network. A sensor network can be made scalable by formation of several clusters in the network for making the network scalable and relevant. A variety of clustering algorithms have been specifically designed to enhance the clustering efficiency and energy improvement for the overall network.

In this paper, multiple scenarios are being created with respect to the Cluster Head (CH) the scenarios consist of different numbers of nodes. The detail of scenarios will be like taking 5 nodes, 10 nodes, 50 nodes, 100 nodes and 150 nodes into exercise. The aim of this paper is to find out the values of each scenario taking various parameters into consideration and collaborating the results into a graph to see the effectiveness of the particular scenarios. The lists of parameters will include Packet Delivery Ratio (PDR), Throughput, Overhead, and Delay.

Structural design of WSN is shown in Figure 1, which have cluster of nodes with one node serving as Cluster Head (CH). The major work of any cluster head is to accumulate the data and perform aggregation and compression on that accumulated data, then it's the duty of CH to bypass compressed information to base station. The base station acts as a medium by which user can access the data via Gateway (for example Internet).

The reminder of this paper is followed by section 2 with literature review; section 3 shows proposed work with section 4 and 5 screening simulation environment and simulation results respectively. Conclusion is cited in section 6 whereas

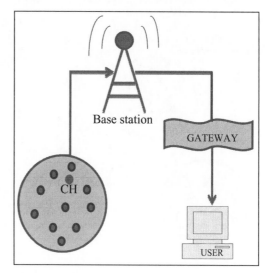

Figure 1. Architecture of WSN.

section 7 and 8 consists of acknowledgement and references.

2 RELATED WORK

Leach is acronym of low energy adaptive clustering hierarchy. Leach is a clustering algorithm and is homogenous in nature. Leach has two phases-Set up phase (creation of cluster head) and Steady state phase (transmission of data between nodes). The main disadvantage of Leach protocol lies in the fact that if the cluster head dies in middle of aggregating the data, whole network becomes dull and transmission becomes inactive. So there was need of alternate method in this case. Therefore substitute of cluster head named secondary cluster head has to be created which will work in case of absence of cluster head and performs the same functioning as cluster head and improves the network lifetime.

Many researchers have given major contribution to the enhancement of energy and lifetime of the scalable WSN network. The Advance leach paper also prioritizes on the enhancement of the energy and lifetime. The proposed methodology of advance leach is discussed below to improve network lifetime. In advance leach, initially the cluster head is elected on the basis of Energy, and when energy of cluster head depletes to fixed threshold value then secondary cluster head is selected.

Figure 2 shows the general description of Advance leach protocol. When the cluster head dies then secondary cluster head is selected based on maximum node degree and minimum distance from base station. The secondary cluster head is then responsible for communication from members' nodes to base station.

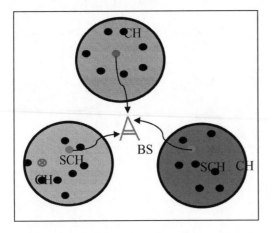

Figure 2. Architecture of advance leach.

Algorithm 1: Pseudo code Cluster setup phase

//Nis the number of nodes that present at any instant of time.
//Ti (n) = Threshold energy

1. for each (node n)
2. n selects random number z between 0 and 1
3. If (z < T (n))
4. n becomes CH
5. Else
6. n becomes a Non CH node
7. Member nodes chooses the CH, this selection is based on the received signal strength
8. Member nodes informs the selected CH and become a member of its cluster
9. End if
10. for each (CH)
11. CH creates TDMA schedule for each cluster member
12. Each cluster member communicates to the CH in its time slot
13. End for

Algorithm 2: Pseudo code for selecting CH and SCH

1. Setup phase
2. Sensor node sends information about its signal strength from base station to sink
3. If sensor node with highest signal strength then
4. Choose CH and broadcast the information to member nodes
5. Else
6. Choose node as member node
7. End if member node sends data to CH
8. CH performs data aggregation
9. Aggregated data is send to base station
10. If (CH energy < threshold value)
11. Then member node will evaluate their data about node density and distance from base station
12. If member node with highest node density and smallest distance from base station then
13. Choose SCH and works alike CH
14. Else
15. Choose node as member node
16. Repeat step 8
17. End.

3 OVERVIEW OF PROPOSED ALGORITHM

Many researchers have given major contribution to the enhancement of energy and lifetime of the scalable WSN network.

The proposed work is on different scenarios taken under NS-2 tool. We are considering various nodes (5 nodes, 10 nodes, 50 nodes, 100 nodes,

150 nodes) and then performing simulations considering parameters like PDR (packet delivery ratio), throughput, overhead, delay.

3.1 *Scalability*

Scalability is the term used when we are increasing the number of nodes in the scenario and the network is still giving the average as well as better performance in respective scenario.

In this paper we are commenting on scalability as respect to the different networks scenario having nodes are respectively 5 nodes, 10 nodes, 50 nodes, 100 nodes, 150 nodes.

Results are varying according to nodes and are elaborated below:

Table 1 elaborates the comparative results of all the respective network parameters.

Network Delay is an important design and performance characteristic. The delay of network specifies how long it takes for a bit of data to transmit across the network.

PDR is one of the vital parameter in case of WSN. Average increment of Packet Delivery Ratio (PDR) makes the network effective. This study gives the average improvement of PDR in respect of 5, 10, 100, 150 scalable nodes.

Throughput of a network is the rate of production or the rate at which something can be processed. Throughput is the rate of successful message delivery over network. The comparative study gives the average network performance over throughput that change inconsistently with respect to the number of nodes. Throughput changes with nodes due to network congestion and connection disputes at several nodes.

Overhead of any network can be define as the arrangement of the excess or indirect computation time, memory, bandwidth, or other resources that are required to attain effective performance enhancement. Overhead is increasing rapidly when the number of node is increases. Table 1 shows the at the node 150 network overhead is high and maximum with compare to the other nodes.

Table 1. Analysis of parameters on different nodes.

Number of nodes	Network delay (ms)	PDR (%)	Throughput (in kbps)	Overhead
5	6000.30	0.8110	654.149	0.006
10	6649.71	0.2114	568.656	0.101
50	6010.20	0.1590	495.450	0.302
100	6012.80	0.3961	606.290	0.226
150	6021.01	0.4010	606.700	0.339

3.2 *Performance*

Network performance is the analysis and review of collective network information, to define the results in networks.

4 SIMULATION SETUP

Network Simulator (NS2) is simulation tool which is used to study behavior of communication networks. We evaluate the performance of our proposed energy efficient leach protocol via simulations in NS2. Performance of different network parameters like PDR, delay, throughput and overhead are calculated and simulation results are plot in graph.

Table 2 shows the overall network environment and the network setup for simulation of the entire network. In simulation performance of the network parameters are evaluated on multiple scalable nodes like 5, 10, 50, 100 and 150. Performance of parameters will also provide energy efficiency and network effectiveness for the entire network.

Figure 3 shows the simulation network scenario of 150 nodes. Simulation is carried out in all respective nodes from 5 to 150. We highlighted only on 150 nodes simulation because of its scalability and network durability. 150 wireless nodes are deployed on random basis and 5, 10, 35 nodes are selected as Cluster Head (CH).

Table 2. Simulation table.

Parameter	Value
Simulation area	100*100
Number of sensor nodes (N)	5, 10, 50, 100, 150
Traffic type	TCP (FTP)
Antenna type	Omni antenna
Cluster head (CH)	5, 10, 35
Rx energy	3.65J
Threshold energy	0.5J
Routing	AODV

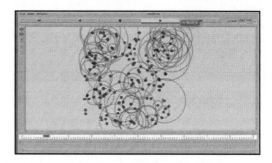

Figure 3. Simulation scenario of 150 nodes.

5 RESULTS AND DISCUSSIONS

This section elaborates the simulation of the entire network. All the network parameters are evaluated with respect to different number of nodes having parametric characteristics of network parameter like PDR, Delay, Throughput and Overhead.

5.1 Packet Delivery Ratio (PDR)

PDR is the ratio of the number of delivered data packet to the destination. This illustrates the level of delivered data to the destination.

$$PDR = \frac{\sum Number\ of\ Packet\ Received}{\sum Number\ of\ Packet\ send} \quad (1)$$

The above figure 4 depicts inclination and declination in the PDR when the different numbers of nodes are taken into action. There are no fixed criteria of increment and decrement because the ratio completely depends on number of packets received by number of packets send.

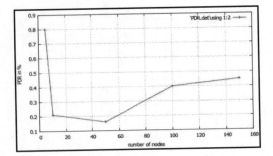

Figure 4.   PDR graph.

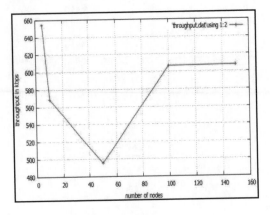

Figure 5.   Throughput graph.

5.2 Throughput

Throughput is a measure of how many units of data and sensor information can process from one node to another node or from source to destination in a given amount of time.

Figure 5 is highlighted on throughput in the network according to evaluation of different nodes. The rate of change of throughput with respect to the number of nodes may vary very rapidly and inconsistently. Some nodes like in 50, 40.10 it is decreasing and increasing at the time when more number of nodes are acting in the network.

5.3 Routing overhead

Routing Protocol overhead refers to metadata and network routing information sent by an application, which uses a portion of the available bandwidth of a communications protocol. This extra data, making up the protocol headers and application-specific information is referred to as overhead.

Figure 6 demonstrate routing overhead graphically. The highest overhead is with 150 nodes that is 0.339 and with 5 numbers of nodes routing overhead sums up at 0.006.

5.4 Network delay

Network delay is depicted by the average time taken by a data packet to succesfully communicate from source to destination. It also includes the delay caused by route discovery process and the queue in data packet transmission. Only the data packets that successfully delivered to destinations that counted.

$$Network\ Delay = \frac{\sum (Arrival\ time - send\ time)}{\sum Number\ of\ connection} \quad (2)$$

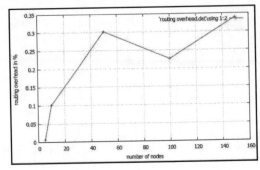

Figure 6.   Routing overhead.

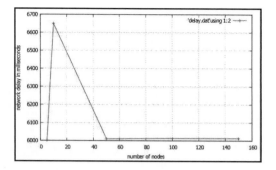

Figure 7. Network delay.

Figure 7 displays the network delay for the entire network consisting with 5, 10, 50, 100 and 150 nodes. The deviation of network delay measured in milliseconds and it deviate very less with respect to node scalability. Delay rate remain almost same for all the nodes from 50 to 150.

6 CONCLUSION

Energy efficiency and prolongation of lifetime over scalable WSN network is the most challenging network phenomenon. This paper proposes an energy efficient protocol over scalable network scenario providing multiple numbers of wireless sensor nodes from to 150 respectively for making the network more scalable. Performance of network parameters like PDR, delay, throughput, normalized network overhead are calculated and evaluate on respective nodes. Parametric graphs of these parameters shows that the network performance are resulting better in almost average cases in compare to the existing protocols and their network performance over scalable WSN. Better improvement of network parameters over large scalable network and residual energy improvisation also invoked in this paper.

REFERENCES

Abbasi, A., Younis, M. "A Survey On Clustering Algorithms For Wireless Sensor Networks". 0140–3664/$—see front matter 2007 Published by Elsevier B.V. DOI=10.1016.

Alhawat, V. Malik, "An Extened Vice-Cluster Selection Approach To Improve V LEACH PROTOCOL IN WSN", Third International Conference on Advanced Computing & Communication Technologies, 2013.

Apoorva Jain, Dr Naveen Hemrajani, "Performance Analysis And Evaluation By Simulation Of Wireless Sensor Network Using Energy Efficient Algorithm", International Journal of Application or Innovation in Engineering & Management Volume 2, Issue 1, January 2013. DOI=2319–4847.

Goyal, R., Gupta, S. "Energy Aware Routing Protocol Over Leach on Wireless Sensor Network", International Conference on Computing, Communication and Automation (ICCCA2016).

Parikha Chawla, Parmender Singh, Taruna Sikka, "Enhance Throughput in Wireless Sensor Network Using Topology Control Approach", International Journal of Soft Computing and Engineering (IJSCE), Volume-2, Issue-3, July 2012. 2231–2307.

Ramesh, K., Dr. K. Somasundaram, "A Comparative Study of Clusterhead Selection Algorithms In Wireless Sensor Networks". International Journal of Computer Science & Engineering Survey (IJCSES) Vol. 2, No. 4, November 2011.

Samar Fakher, Mona Shokair, M.I. Moawad, Karam Sharshar, "The main effective parameters on wireless sensor network performance", International Journal of Scientific and Research Publications, Volume 5, Issue 6, June 2015 DOI=2250–3153.

Shilpi Agarwal, Rajeshwar Lal Dua, "Performance evaluation of parameters Using DSR Routing Protocol in WSNs", International Journal of Advanced Research in Computer Science and Software Engineering, 2012. DOI=2277 128X.

Sindhwani, N., Vaid, R. "V Leach: An Energy Efficient Communication Protocol For Wsn", Vol. 2, No. 2, February–March 2013.

Vidhyapriya, R., Dr P T Vanathi, "Energy Efficient Adaptive Multipath Routing For Wireless Sensor Networks", IAENG International Journal of Computer Science, 34:1, IJCS 3418.2014.

Yadav, L., Sunitha, C. "Low Energy Adaptive Clustering Hierarchy in Wireless Sensor Network (LEACH)", SGT Institute of Engineering & technology, Gurgaon, Haryana-122505, INDIA, International Journal of Computer Science and Information Technologies, Vol. 5 (3), 2014.

Communication and Computing Systems – Prasad et al. (Eds)
© 2017 Taylor & Francis Group, London, ISBN 978-1-138-02952-1

Using Markov model approach to forecast web page caching

Poonam Yadav & Neelam Ruhil
Dronacharya College of Engineering, Gurgaon, India

Vidhi Sharma
California University, California, USA

ABSTRACT: The World Wide Web is a large repository of information. There are lot of user's who regularly accesses this information source, this is simple to invent certain patterns to access resources on web. Web assumption has been implemented in the past for static content. With the increasing Internet (R. Tewari, 1998) traffic and Web content, the Web assumption models are very famous. Data mining methodology categorizes the modules of clients based on their attributes and assumes future activity without allowing instant inferences and interactivity. Here some practices like information retrieval and assumption by partial matching can be used in combination with prediction modelling to increase validity and performance. There is always some scope to improve the web page access based on user requirement. One of the methods given by web is page pre-fetching which means to make available the web page to the user before the user request. In this current work we are presenting an intellectual method created on the history of web page visit used for web page prediction. A three level approach is proposed in which we Markov model is combined with clustered approach and association mining.

1 INTRODUCTION

Prediction model is a very well-known machine learning method and it is differ from the way that data mining does with Data history. Data mining methods categorizes the modules of clients by their attributes and assume future activity without allowing instant inferences and interactivity. Here some practices like information retrieval and assumption by partial matching can be used in combination with prediction modelling to increase validity and performance. Unlike other models, a Web Markov model is mostly challenging because of the several situations that hold the self-motivated Web (Klemm, 1999) in terms of user activities and the content that changes regularly. So to design an assumption model use the prediction probabilistic idea.

1.1 Web cache

A Web cache is a procedure used for the temporary storage and caching (Chen, 1999) of Web documents like HTML pages and images in order to reduce server load, bandwidth usage and perceived lag. It stores the reproduction of documents passing through the web and later it satisfies the request if certain circumstances are met (Dutta, 2007). A Cache is a repository that stores data transparently so that upcoming requests for that data can be served faster.

1.2 Web caches working

1. If the response's headers tell the cache not to keep the information, it won't.
2. If some request is secure or authenticated i.e., HTTPS, it won't be cached.
3. A cached is able to send to a client without checking with the origin server, this representation is considered fresh if:
   a. It contains an expiry time or some age-controlling header set and is still within the fresh period.
   B. If the representation of cache has seen recently and was adapted previously. Fresh demonstrations are aided directly from the cache, without examining the origin server.
4. In case if the representation is old, the source server will be requested to validate it.
5. In circumstances for example, when it is not connected to a network, a cache serves stale responses without examining the origin server using the Markov model which combines the clustered approach and associative mining.

1.2.1 Importance of web cache
1. Reduced Cost of Internet Traffic
2. Reduced latency

1.3 Data mining

Web usage mining is a subset of Web mining operations which itself is a subset of data mining in general. The aim is to use the data and information extracted in Web systems in order to reach knowledge of the system itself. Data mining is a set operations performed on a collection of data or a subset of it so as to extract meaningful patterns on the data. Another definition is "Data mining is the semi-automatic discovery of patterns, associations, changes, anomalies, rules, and statistically significant structures and events in data". That is, data mining attempts to extract knowledge from data. If a subset is to be used, careful and unbiased sampling algorithms should be used to avoid biased result. Data mining is different from information extraction although they are closely related. To better understand the concepts brief definitions of keywords can be given as:

Data: "A class of information objects, made up of units of binary code that are intended to be stored, processed, and transmitted by digital computers".

Information: "It is a set of facts with processing capability added, such as context, relationships to other facts about the same or related objects, implying an increased usefulness. Information provides meaning to data".

Knowledge: "It is the summation of information into independent concepts and rules that can explain relationships or predict outcomes".

2 IMPLEMENTATION

The presented work which is focused on three main concepts called Markov Model, Clustering and the Neural Network. The Markov Model is the basic prediction algorithm, that is been improved by using the C-Means Algorithm as well as the Neural network. The neural network is used as the classification tool to derive the required results. The integration model profits from the decrease the state space complexity of the lower Markov model by using association mining in case of ambiguity. The integration model also provides the complexity of the association rules since the rules are generated only in special cases (M. Junchang, 2006). In brief, the new integration model results in an increase the accuracy and a decrease in state & rule complexity.

2.1 C-mean clustering

It is an important part of cluster analysis is partitioned clustering. Based on various models, several clustering procedures have been implemented and different algorithms remain to act in the work.

The paper's significance can be briefed as:

1. Rendering a new description of the mean, a structure for separated clustering systems, called GCM i.e. General c-means Clustering Model is proposed.
2. By applying confined optimality test, the association between Partitioned clustering and Occam's razor is recognized first time in GCM.
3. A guide for developing and executing clustering algorithm is implemented based on an assumption in partitioned clustering. These decisions are confirmed by numerical experimental results.

3 RESULT

Due to the significant growth in the amount of data the web users are facing the problems of information overload and drowning. The scope of World Wide Web (Greg Barish, 2000) has exceeded 24.39 billion pages in 2010.

The latency by the user is perceived from numerous sources like speed, bandwidth, overhead, accessing the Web page etc. According to "Eight Second Rule", it is detected that Web based latency disturbs the effort made by the user and a lot of work is required to minimize the user latency.

See in Figure 2 the output of Markov Model implementation on a dummy set of Web pages.

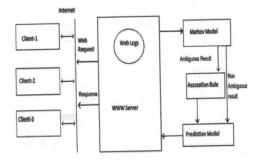

Figure 1. The Integration Markov & Association Model (IMAM).

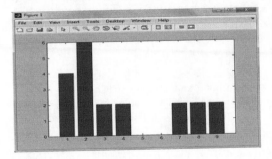

Figure 2. Markov model with web cache.

The x-axis defines the pages and y axis shows the incidence of a page. See here Page 2 is having the highest frequency over all pages.

See the Figure 3 is showing the Markov model with clustering (Xiao, 2001). See as the clustering is applied on Markov model at level 2, it separate the higher frequency pages in separate group. Here page 7, 8, 9 form a group of higher visiting page frequency.

In Figure 4 see the implementation of Markov model on a group of Web pages. To collected the Web page visit information for 5 different sessions.

In Figure 5 see the implementation of Markov model on a group of Web pages. In this output the pages having the lower frequency are pruned from the list. Now the only higher frequency pages in the list. Observe that average page frequency is around 30.

Figure 6 is showing the results of Web page visit with Markov model and implementation of association rule on it. The association model shows the results respective to previous and next visited pages.

Figure 7 is showing the results of clustered Markov model. All pages are divided in 4 clusters (Yan, 2005). High frequency pages are grouped in separate clusters. The cluster 4 is having the low frequency pages.

Figure 8 is showing the results of clustered Markov model. All pages are divided in 4 clusters. High frequency pages are grouped in separate clusters. The cluster 3 is having the low frequency pages. This result is respective to the association model applied on the dataset. Here C-means clustering is applied on it.

Figure 9 is showing the results of clustered Markov model. All pages are divided in 4 clusters. High frequency pages are grouped in separate clusters. The cluster 3 is having the low frequency

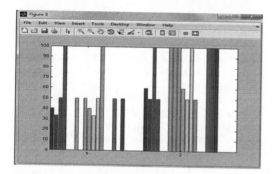

Figure 3. Markov model with clustering.

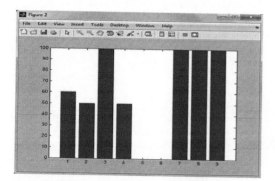

Figure 4. Web pages using markov model.

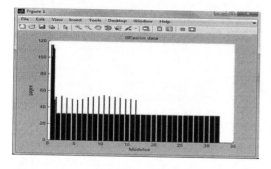

Figure 5. Web prediction model using markov model.

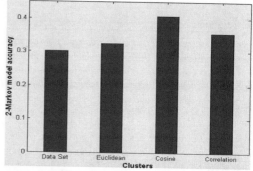

Figure 6. Web page associations using markov model.

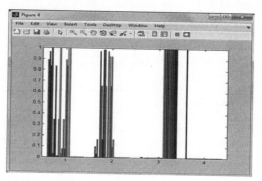

Figure 7. Clustered markov model.

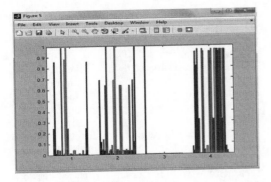

Figure 8. Web page clustering and association using markov model.

Figure 9. C-means clustering using markov model with web page association.

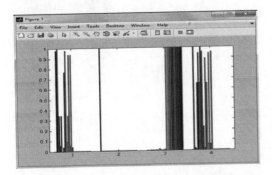

Figure 10. Clustering in C-means with web page associations.

pages. This result is respective to the association model applied on the dataset.

Figure 10 is showing the results of clustered Markov model level 2. All pages are divided in 4 clusters. High frequency pages are grouped in separate clusters. The cluster 2 is having the low frequency pages.

The available Web log dataset is divided in two parts called training dataset and the testing dataset.

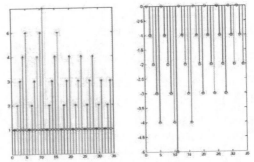

Figure 11. Analysis of training dataset.

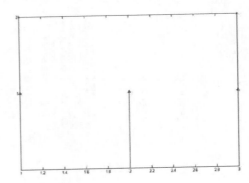

Figure 12. Analysis of training dataset.

The selected 90% data values as the training dataset and 10% as well 30% dataset as the testing dataset. In the above figure, the Web usage analysis based on training dataset is defined. The first image in figure 11 is showing the usage of Web pages (Nanopoulos, 2003) in terms of Web page usage frequency respective to the training dataset.

In the above figure, the Web usage analysis is defined based on testing dataset is defined. For this analysis, just taken the 10% of the available dataset as the training dataset. The figure 12 is showing the usage frequency respective to the testing data's.

4 CONCLUSION

The evolution of Web application, especially in in the field of electronic commerce, thereby needs substantial interest to well understand and analyse Web data usage and imply the information for the betterment of users. Number of issues arises in the area of Web Usage Mining. In the proposed paper, we defined numerous applications of Web Mining recommended by different authors. The paper considers the most stimulating study area that deals with the most of semantics to progress the consequences of Web Tradition Mining applications.

REFERENCES

Chen, X. (1999). Lifetime Behavior and its Impact on Web Caching. WIAPP IEEE.

Davison, D., B. (2001). A Web Caching Primer. IEEE Internet Computing.

Dutta, R. (2007). Offering Memory Efficiency utilizing Cellular Automata for Markov Tree based Web-page Prediction Model. 10th International Conference on Information Technology. IEEE.

Greg Barish, K. O. (2000). World Wide Web Caching: Trends and Techniques. IEEE Communications Magazine.

Gulati, P. (n.d.). Novel Approach for Determining Next Page Access. First International Conference on Emerging Trends in Engineering and Technology. IEEE.

Guo, Y. Z. (2007). Personalized PageRank for Web Page Prediction Based on Access Time-Length and Frequency. International Conference on Web Intelligence. IEEE.

Guo, Y. Z. (2008). Error Correcting Output Coding-based Conditional Random Fields forWeb Page Prediction. International Conference on Web Intelligence and Intelligent Agent Technology. IEEE.

Junchang, M., G. Z. (2006). Finding Shared Fragments in Large Collection of Web Pages for Fragment-based Web Caching. Fifth International Symposium on Network Computing and Applications (NCA'06). IEEE.

Klemm, R. P. (1999). Web Companion: A Friendly Client-Side Web Prefetching Agent. IEEE Transactions.

Nanopoulos, A. (2003). A Data Mining Algorithm for Generalized Web Prefetching. IEEE Transactions on knowledge and data engineering.

Pablo Rodriguez, C. S. (2001). Analysis of Web Caching Architectures: Hierarchical and Distributed Caching. IEEE/ACM Transaction on Networking.

Tewari, R., M. D. (1998). Beyond hierarchies: design considerations for distributed caching on the Internet. 19th International Conference on Distributed Computing Systems (ICDCS).

Xiao, J. (2001). Clustering of Web Users Using Session-based Similarity Measures. IEEE proceedings.

Yu, S., D. C.-R.-Y. (2003). Improving pseudo—relevance feedback in Web information retrieval using Web page segmentation. In Proceedings of the Twelfth International World Wide Web Conference.

Communication and Computing Systems – Prasad et al. (Eds)
© 2017 Taylor & Francis Group, London, ISBN 978-1-138-02952-1

Energy aware task scheduling with adaptive clustering method in cloud computing: A review

Anureet Kaur & Bikrampal Kaur
CGC (Landran), Mohali, Punjab, India

ABSTRACT: Cloud computing is an Internet-based computing that provides on demand data and resources for the users. In Cloud computing there are heterogeneous servers which executes tasks as per the user requirement. The energy consumed by supercomputers and data centers in the cloud is a major environmental issue facing by today's society. In this paper our major focus is to minimize energy consumption in the cloud environment by using task scheduler. In the proposed model, the balanced set based task scheduling will be utilized for better performance of task in the cloud environment to reduce power expenditure in cloud environment. The proposed algorithm uses a method where, tasks are first schedule using an adaptive clustering algorithm into balanced sets and arrange them in a sequence; according to their earliest finish time the tasks are executed by appropriate virtual machines in cloud datacenters to minimize the cost and the consumption of energy in the computing environment.

Keywords: cloud computing; energy consumption, load balancing, task scheduling

1 INTRODUCTION

Cloud computing is the, evolving scientific technology in the IT world. But in the actual cloud computing model, this means that all the data held on a server or a group of servers, and accessing them through the internet. The cloud computing provides on demand resources to users (Zhao et al. 2016). Parallel and distributed computing has been entered into the epoch of cloud computing which provides highly secure and fast IT services through the internet. So it has captured the attention of big IT giant for provide elastic and cost effective economical IT operations (Kokilavani & Amalarethinam 2011). In the past few years, the increased growth of cloud computing also maximizes the number of cloud data center at unrivalled speeds. In the intervening period of time, the energy expenditure by the data centers has kept increasing at higher rates. Therefore, the focus of cloud resource fulfillment and task scheduling has switched from performance to power efficiency. There is an issue called load balancing in cloud computing is an important problem which becomes a major hurdle between the rapid developments of the cloud computing. The large number of clients from all over the world demanding services at rapid rate in every second from cloud service provider (Alahmadi 2015). In cloud computing there are a large number of heterogeneous servers that execute tasks that are assigned to them according to the user's requirements and they consumes a maximum amount of energy. Higher

energy expenditure is not only results in high operational cost but also leads to higher carbon emissions, which causes harmful effect to the environment. Therefore, the issue of energy-awake performance augmentation has attracted remarkable consideration in the cloud environments.

In literature survey, many existing works (Zhao, Qing et al. 2016) discussed some of the issues about energy consumption and resource utilization. (Nidhi Bansal, et al. 2015) also defined issues of task scheduling for balanced resource utilization. The huge volume of power utilizes by supercomputers and cloud computing centers have been a major environmental issue facing today's society. Researchers have shown that a task scheduling optimization is a compelling strategy for the overly performance of the cloud computing environment. The large number of loaded servers is the major reason for lesser resource utilization and high energy consumption. Researchers (Chen et al.) found that even in idle periods, most of servers exhaust up to 50% of their peak power. In cloud data centers, huge amounts of energy exhausted by leaving computing and networking devices—such as computing machines powered on in a low utilization state. Energy conservancy is a major concern in the cloud computing systems because it provides several important benefits such as diminishing operating costs, improves system accuracy, improve resource utilization, and environmental preservation. Meanwhile, energy-awake scheduling approach with adaptive clustering method is an assuring method to accomplish that objective. The energy consumption in a cloud datacenter increasing day by day, while the resources are highly underutilized; these challenges are a hurdle between cloud computing growth that restricts the improvement of cloud computing. The energy consumption in a cloud computing system consists of energy consumed by different kinds of electrical equipment. The resources and computing nodes has to be always available for the incoming jobs or tasks, which are a total waste of energy and time (Cheng 2015). Cloud computing provide so much resources and services to the user on the cloud, this is main reason of development of cloud computing. Energy consumption is a very crucial issue in cloud computing in these days. In cloud data centers the operational cost and energy consumption are increasing day by day which also effect the green computing. So there must be some technique which minimizes energy consumption and deployment cost of cloud datacenters and minimize the environmental impacts (Chaukwale 2015).

The features or advantages provided by the future intelligent energy—aware task scheduling with adaptive clustering method are improved resource utilization, improved overall cost, minimize energy consumption. Researchers have been working for several years to develop a suitable method to minimize energy consumption of cloud environments and improve overall performance. The task scheduling is classified as the internal task in the cloud computing environments, where the tasks or Virtual Machines (VM) share the task data in order to reduce the response time as well as the computational overhead from the selected CPU or primary VM for the tasking of higher number of tasks in the lowest possible time. The wrong clustering mechanism may delay the overall workflow or increase the cost of the workflow execution, which must be used to achieve the best performance at the minimum cost over the given cloud data resources. The intelligent workflow evaluation method will return the type of the workflow, which will help us to select the appropriate clustering method to cluster the task data in the appropriate manner to achieve the best performance. Additionally, the proposed model will be optimized for the energy estimation and cost evaluation for the workflow in focus. The proposed model provides better results with comparison to existing ones.

2 RELATED WORK

Zhao, Qing et al. (2016) worked upon the energy-aware task scheduling technique for data-demanding computing applications or tasks in the cloud. In this method, first, the datasets and tasks are arranged as a binary tree by using a data correlation clustering algorithm, in which both the data correlations developed from the first data-sets and from the common datasets have been examined. So, the extent of universal data transmission can be diminished incredibly, which are helpful to the minimization of SLA violation rates. Second, method is "Tree-to-Tree" task scheduling method based on the evaluation of Task Requirement Degree (TRD) is used, which can boost the energy efficiency of the total cloud system by enhanced the utilization of its computing resources and network bandwidth (Zhao et al. 2016). (Alahmadi et al. 2015) Worked towards the Energy-Awake Task Scheduling schema that mainly uses two energy saving methods, DVFS, VM Reuse, on cloud data centers. Researchers present the scheduling framework and use a specific algorithm, called EATS-FFD, which considers FFD as its base scheduling policy. The proposed technique accomplishes superior energy-efficiency without endure system QoS. The potency of proposed technique is measured under various experimental scenarios using the open source Cloud-Simulator platform (Alahmadi 2015). (Nidhi Bansal, et al. 2015) focused upon the cost of the computing resources (virtual machines) to schedule the given pool of the tasks over the cloud computing model. The cost optimization has been performed over the QoS-task driven task scheduling mechanism, which did not encounter the cost optimization problem earlier. The authors have shown that the earlier QoS-driven task scheduling based studies has been considered the make span, latency and load balancing (Bansal 2015). (Weiwei Chen, et al. 2014) proposed the imbalanced metrics for the optimization of the task clustering on the scientific workflow data executions. They have proposed a horizontal and vertical method for the evaluation of series of task clustering for the widely used scientific workflows. Their proposed model has utilized the in-depth metric values for the real time evaluation of their research model. The authors have proved the efficiency of the proposed model in overcoming the limitations of the existing model under the evaluation analysis (Lin 2014). (Liu & Wang et al. 2012) defined a new task scheduling model in this paper. In the proposed model, author optimizes the task execution time to interpret both the job running time and the resource utilization. So, a PSO-based algorithm is proposed based on the model. This paper works to overcome the load balancing problem in VMs of Cloud data centers in computing environment (Liu 2012). (Li et al. 2011) described in this paper as the cloud computing is the development of network computing and distributed computing. Task organizing is one of the elementary concerns in this environment. It is an NP-hard optimization problem. The main objective of this work is to balance the all system load while trying to minimize the make span of a specific task set (Li et al. 2011). (Panwar et al. 2015) presents a dynamic load management algorithm for the division of the entire incoming request among the virtual machines effectively to balance load between cloud environments. Further, the performance is simulated by using Cloud Analyst simulator based on numerous parameters like data processing time and response time, execution time etc. and compared the result with previous algorithm VM-Assign (Panwar et al. 2015). (Dasgupta et al. 2013) worked upon the load balancing in the cloud computing paradigm, load balancing is one of the challenges, with briskly increased users and their requirement of different services on the cloud computing platform, applicable use of resources in the cloud environment became a demanding issue. The performance measures for load balancing algorithms in cloud environment are response time and waiting time. In the proposed work there are two load balancing algorithms, Min-Min and Max-Min algorithm, which results in better response time of tasks in a cloud environment (Gopinath 2015).

The main objectives of the proposed model are to balance load between virtual machines using an adaptive clustering algorithm to maximize the power management, and minimization of overall cost in cloud datacenters. The proposed model is designed to calculate the CPU time in terms of earliest finish time and it also minimize the overall execution time of tasks in virtual machines which also balances load in the cloud environment. The dynamic threshold is computed on the basis of the communication cost, earliest finish time and CPU cycles, which plays the important role

in taking the offloading decision. The experimental results have proven the potency of the proposed model in comparison with the existing models.

3 SUMMARY OF TECHNIQUES SURVEYED

There are number of techniques which are surveyed to identify the challenges in existing work and to prove that the proposed technique provides better results with comparison to the existing techniques. There are number of problems addressed on which author worked upon with different techniques or algorithms. There are different algorithms which results into different results are surveyed for better results. (Zhao, Qing et al. 2016) discussed some of the issues about energy consumption and resource utilization. (Nidhi Bansal, et al. 2015) also defined issues of task scheduling for balanced resource utilization.

4 CHALLENGES IN EXISTING MODELS

The existing model is designed to solve the problem of energy consumption for data-intensive applications in the cloud environment. The existing model utilizes the task clustering mechanisms to cluster the small tasks into the task group appearing like the single task to process them smoothly and quickly. The task clustering method clusters the task data on the basis of task similarity, dependency, runtime execution, etc. The existing model does not evaluate the energy constraint for the task scheduling. The energy efficient mechanism can help to heal the environment as well as significantly minimizes the cost of electricity

Table 1. The following table represents the number of techniques surveyed in the work.

Authors and Year	Problem addressed	Techniques proposed	Experimental results	Drawbacks
Zhao, Qing et al. (2016)	Resource utilization and energy efficiency.	Data correlation clustering algorithm and tree to tree task scheduling to reduce SLA violation rates.	The proposed algorithm Reduced power consumption and low SLA rates.	Resources are not completely utilized.
A. Alahmadi et al. (2015)	Energy efficiency in the cloud environment.	EATS-FFD algorithm and FFD as its base scheduling algorithm	The proposed algorithm minimizes the energy consumption.	Cost factor is not taken into account
Nidhi Bansal, et al. (2015)	Cost of computing resources in overall cloud.	QoS task driven scheduling mechanism.	The proposed algorithm minimizing the total allocation cost in of resources	Minimize total allocation cost of resources but not energy consumption
Weiwei Chen, et al. (2014)	Divisible task-scheduling problem.	Bandwidth-aware algorithm for divisible task scheduling Algorithm	The proposed mechanism improve overall performance of cloud	Resources are not utilized

expenses. The existing model does not evaluate the adaptive clustering model to cluster the given set of tasks into the balanced sets and the cost-awareness for the scientific workflow executions has not been evaluated under the existing model, which may increase the processing cost due to the wrong data clustering and task scheduling, which may also increase the load of the resources in the given cloud scenario.

The proposed model evaluates the clustering selection problem of the existing models based upon similarity distance factor. Now the proposed model evaluates the overall execution time of the workflow and also reduces the energy consumption in workflows. In the proposed model, the balanced set based task scheduling will be utilized, hence it signifies the current task scheduling to the real-time cloud applications. The proposed model provides better results than existing ones.

5 CONCLUSIONS

In this paper, different techniques are surveyed but they are not still capable to completely minimize energy consumption in the cloud environment. Various different techniques are implemented to minimize energy expenditure in cloud but they are not able to completely offset the energy issue in cloud, So there must be some technique which minimizes energy consumption and deployment cost of cloud datacenters in such a way that resources are highly utilized in cloud and minimize the environmental impacts.

ACKNOWLEDGMENTS

I acknowledge with a deep sense of gratitude and most sincere appreciation, the valuable guidance and unfailing encouragement rendered to me by "Dr. Bikrampal Kaur" Professor for their proficient and enthusiastic guidance, useful encouragement and immense help. I have been a deep sense of admiration for them innate goodness and inexhaustible enthusiasm. The proposed work will be completed under the guidance of the official guide and other experts on the campus of the Chandigarh group of Colleges, Landran, and Mohali.

REFERENCES

Alahmadi, A., Che, D., Khaleel, M., Zhu, M.M. and Ghodous, P., 2015, June. An Innovative Energy-Aware Cloud Task Scheduling Framework. In Cloud Computing (CLOUD), 2015 IEEE 8th International Conference on: 493–500. IEEE.

Bansal, N., Maurya, A., Kumar, T., Singh, M. and Bansal, S., 2015. Cost performance of QoS Driven task scheduling in cloud computing. Procedia Computer Science, 57, pp. 126–130.

Chang, H. and Tang, X., 2010, December. A load-balance based resource-scheduling algorithm under cloud computing environment. In New Horizons in Web-Based Learning-ICWL 2010 Workshops: 85–90. Springer Berlin Heidelberg.

Cheng, C., Li, J. and Wang, Y., 2015. An energy-saving task scheduling strategy based on vacation queuing theory in cloud computing. Tsinghua Science and Technology, 20(1): 28–39.

Chaukwale, R. and Kamath, S.S., 2013, September. A modified ant colony optimization algorithm with load balancing for job shop scheduling. In Advanced Computing Technologies (ICACT), 2013 15th International Conference on: 1–5. IEEE.

Dam, S., Mandal, G., Dasgupta, K. and Dutta, P., 2014. An ant colony based load balancing strategy in cloud computing. In Advanced Computing, Networking and Informatics-Volume 28: 403–413. Springer International Publishing.

Dhurandher, S.K., Obaidat, M.S., Woungang, I., Agarwal, P., Gupta, A. and Gupta, P., 2014, June. A cluster-based load balancing algorithm in cloud computing. In Communications (ICC), 2014 IEEE International Conference on 2921–2925. IEEE.

Gopinath, P.G. and Vasudevan, S.K., 2015. An in-depth analysis and study of Load balancing techniques in the cloud computing environment. Procedia Computer Science, 50: 427–432.

Goyal, S.K. and Singh, M., 2012. Adaptive and dynamic load balancing in grid using ant colony optimization. International Journal of Engineering and Technology, 4(9): 167–174.

Jain, A. and Singh, R., 2014, February. An innovative approach of Ant Colony optimization for load balancing in peer to peer grid environment. In Issues and Challenges in Intelligent Computing Techniques (ICICT), 2014 International Conference on: 1–5. IEEE.

Kokilavani, T. and Amalarethinam, D.D.G., 2011. Load balanced min-min algorithm for static meta-task scheduling in grid computing. International Journal of Computer Applications, 20(2): 43–49.

Lin, W., Liang, C., Wang, J.Z. and Buyya, R., 2014. Bandwidth-aware divisible task scheduling for cloud computing. Software: Practice and Experience, 44(2), pp. 163–174.

Liu, Z. and Wang, X., 2012. A pso-based algorithm for load balancing in virtual machines of cloud computing environment. In Advances in Swarm Intelligence 7331: 142–147. Springer Berlin Heidelberg.

Li, K., Xu, G., Zhao, G., Dong, Y. and Wang, D., 2011, August. Cloud task scheduling based on load balancing ant colony optimization. In Chinagrid Conference (ChinaGrid), 2011 Sixth Annual: 3–9. IEEE.

Panwar, R. and Mallick, B., 2015, October. Load balancing in cloud computing using dynamic load management algorithm. In Green Computing and Internet of Things (ICGCIoT), 2015 International Conference on: 773–778. IEEE.

Xu, G., Pang, J. and Fu, X., 2013. A load balancing model based on cloud partitioning for the public cloud. Tsinghua Science and Technology, 18(1): 34–39.

Zhao, Q., Xiong, C., Yu, C., Zhang, C. and Zhao, X., 2016. A new energy-aware task scheduling method for data-intensive applications in the cloud. Journal of Network and Computer Applications, 59: 14–27.

Communication and Computing Systems – Prasad et al. (Eds)
© 2017 Taylor & Francis Group, London, ISBN 978-1-138-02952-1

Study of big data with medical imaging communication

Yogesh Kumar Gupta & C.K. Jha
Banasthali University, Rajasthan, India

ABSTRACT: The name "Big Data "referred to as the massive amount of huge dataset in Zetabyte or bigger-sized dataset generated in daily life from various sources (social media, healthcare, various sensor etc.) with very high velocity. The big data is characterized by five dimensions such as volume, veracity, velocity, value and variety. This tremendous data is no more time stable in environment; rather it is updated according to time at rapid speed with different format, so that the conventional database management system or algorithms is not able to handle such kind of dataset, thus the big data processing need to describe a very innovative tools and technique (such as Hadoop, Map Reduce, NoSQL database, HPCC and Apache Hive etc.) to acquire, store, distribute, handle and analyze. The process for analyzing the complicated data normally concerned with the disclosing of hidden patterns. In order to healthcare organization, big data analytics tool and techniques are capable to extract the meaningful or helpful information from unattended portion of non-invasive medical imaging modularity (such as X-rays, MRI, and CT-Scan etc.), With different formats (such as DICOM and PACS etc.).

Keywords: big data, storage, dicom, pacs, non-invasive medical imaging modalities

1   INTRODUCTION

As we all know that the tremendous amount of data generated day to day life from various sources such as online transaction, e-mails, research generals and articles, social media sites (Facebook, Twitter, WhatsApp etc.) and forums, different sensor's data composed from various sources such as healthcare science, environmental organizations, meteorological department, business strategically data, trading market, company data etc, in different format such as structured, semi structured and unstructured, With a great Velocity is normally referred to as Big Data (Patel, et al. 2012). Data can be generated on web in various formats like texts, audios, videos, images, texture or social media posts data etc. this tremendous data is no more time stable in environment; rather it is updated according to time at rapid speed. So that put a big number of critical challenges on big data processing and storage. As an outcome, the conventional database computation tools and algorithms as well as data storage and management techniques has not able to deal with these data. Regardless of those challenges, we cannot overlook the prospective and possibilities mendacious in it that can help to support for analytics and identification of hidden texture and information. The tools and techniques used for big data analysis to manage massive amount of huge data are Hadoop, Map Reduce, NoSQL database, HPCC and Apache Hive. The process for analyzing

the complicated data normally concerned with the disclosing of hidden patterns. These technologies manage large volume of huge data in KB, GB, MB, TB, YB, PB, EB and ZB (Patel, et al. 2012).

1.1   *Big data sources and its formats*

The huge data can be classified in three categories:

- Structured Data—the data produced from several research article and generals, Customer Relationship Management (CRM), business applications such as retail, finance, bioinformatics and other such traditional databases in various sources such as RDBMS, OLAP and data warehousing etc.
- Semi-structured Data - such as Rich Site Summary (RSS) or XML formatted data, HTML, CSV and RDF data.

Unstructured Data –These data are created by the users such as trading markets data, web forums, social media sites and user feedback, emails, comments, audios, images, videos etc. or it may be created by machine such as various sensors data, online transactional, web logs *etc.*

1.2   *Characteristics of big data*

The big data have five main characteristics such as Velocity, Volume, Variety, Veracity and Value. Each aspect puts challenges in processing and handling massive amount of huge data to retrieve some relevant information. Such kind of challenges could be in data acquisition, recording, searching, sorting, retrieving, analyzing, and visualizing from a variety of already described key features of the Big Data.

1. Volume: The huge Data is rapidly-increasing day to day in all categories such as KB, MB, GB, TB, PB, YB, EB and ZB of information. So the massive amount of huge data results into large data files (Patel, et al. 2012).
2. Variety: Data generating sources (even from same categories or from different) are excessively heterogeneous. The files are coming from different sources, in different type and in different formats, it may be unstructured, semi-structured or structured [Patel, et al. 2012].On twitter 400 million tweets are sent daily and there are 200 million active users on it (Beakta 2015).
3. Velocity: The massive amount of data comes at high speed at which it is developed and processed For example social media posts (Patel, et al. 2012).
4. Veracity: means accuracy or uncertainty of data. Data is uncertain due to the incompleteness and inconsistency (Beakta 2015). Veracity is the biggest challenge in data analysis when compares to things like velocity and volume.
5. Value: Large amount of huge data has different kinds of values such as statistical data, events, correlations, hypothetical data etc.

1.3 History of big data from various sources

A current study projected that every minute the users of YouTube upload 72 hours of video, Google receives more than 4 million searching queries and users of e-mail send more than 200 million emails, users of Facebook share more than 2 million pieces of substance and more than 350 GB of data is processed, Twitter users produce 277,000 tweets. The approximated figure about the large data is that, it was estimated near about 5 exabytes (EB) till 2003,

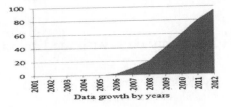
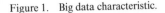

Figure 2.   Data in terabytes for the year 2001–2012.

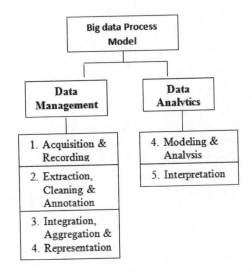

Figure 3.   Big data process model.

2.7 Zettabytes (ZB) till 2012 and it is expected to increase near about 4 times greater till 2016 (Garlasu, et al. 2013). The Figure 2 shows the data in Terabytes (TB) for the year 2001–2012 (Raden, 2012). From Year 2005–2012, it would appear from this graph that the amount of data was exponentially improved within this period due to the significant contribution of Big Data Analytics.

1.4 Big data process model

In this Process model we only focus on Big Data analytics which is use to retrieve the meaningful more information that is relevant to the subject under consideration. The overall process model is separated into 2 Sub-Processes: Data Management and Analytics, which is further divided into 5 stages as shown in the figure 3 (Tiwarkhede and Kakde 2015).

1.5 Big data architecture

The architecture of Big Data is collected with expertise sets that are used for generating a reliable, scalable and completely automated pipelines data in cluster of commodity hardware. The

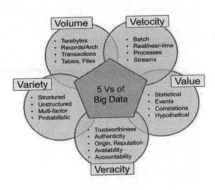

Figure 1.   Big data characteristic.

figure 4 shows the working of completely expertise sets and the architecture of big data.

1. Ingress:- Firstly we have to decide that how to assign the task and data for the particular cluster. For that purpose, we have used two main aspects: batch or batch ingest and event-driven. Batch ingest approach is used for only ingesting the structured data from various sources such as RDBMS. Event ingest aspects is mainly used for real-time events such as transactional data, web logs and it allow us to define redundant agents (Bhardwaj and Balkishan, 2015).
2. Staging:- When the process of incoming data is done then the whole process is give up to the staging only. Normally staging is not only defines that how to store the data but it also gives the full justification to store it in a correct manner and correct place (Bhardwaj and Balkishan, 2015).
3. Access Control:- It is a kind of constraint to access using, entering or we can say consuming something. Another term is authorization which means to permission of accessing something (Bhardwaj and Balkishan, 2015).
4. Hadoop Framework:- Hadoop framework contain some computational and storage tools

such as MapReduce and HDFS respectively. The MapReduce provides computation which ultimately run as map and reduce tasks on the Hadoop cluster node in commodity hardware. And the HDFS is responsible for storing the data in a distributed parallel environment across multiple cluster nodes (Bhardwaj and Balkishan, 2015).

1.6 Paper organisation

The whole paper is divided into four sections that go through the figure 2. The section one introduces the concept of Big Data, heterogeneity in various formats and sources, characteristics, history, process model based on various terminology and architecture. The section two shows the comparison of various existing some well-known tools and techniques based on their limitations and abilities coupled with them. The section Three describes medical imaging communication system. The section four finally concludes this paper with their some positive suggestions and blessing.

2 COMPARISION OF VARIOUS EXISTING COMPUTING AND STORAGE TOOLS & TECHNIQUES

The organization of large Data includes storing, process and analyzing it for a variety of purposes; such as to visualize the infrastructure, to make the hold on Big Data related tasks. Using the detail analysis of different computational and storage tools, we have to get various parameters that help us to compare these tools. The several pros/cons of these tools let us know the correctness of various tools in different kinds of application.

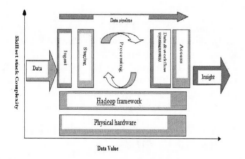

Figure 4. Architecture of big data.

2.1 Comparison of computing tools on various parameters (Prasad and Agarwal 2016)

Computing tools/ parameters	Hadoop & MapReduce	Cloudera impala RTQ	IBM netezza	Apache giraph
Parallel computation	Yes	Yes	Yes—assymatric massively parallel	Yes—bulk parallel processing
Fault tolerance	Vary high	Yes	Yes—use redundant SMP hosts	Yes—by check point
Single point failure	Yes—at master nodes	Yes—if any node fail query execution then whole query process is terminated	At SMP server level	No—multiple master threads running
Data format for analytics	Structured/ unstructured	Structured/unstructured	Structured (RDBMS)	Graph database
Optimized query plan	Not applicable	Yes	Yes	Yes—in terms of graph query/algo

995

2.2 Comparison of storage paradigms/tools on various parameters (Prasad and Agarwal 2016)(Bhosale and Gadekar 2014)

Storage Tools	Hbase	Apache Hive	Neo4j	Cassandra	MongoDB	Redis	Drizzle
Data storage format	Structured i.e. table but not exactly row-oriented relation	Structured/ unstructured	Non-tabular or relational i.e. graph database (schema less)	Structured/semi-structured/ unstructured (schema less)	Documents storage	Key -Value store	Relational DBMS
Failover recovery	More time at node level failure where as at Region Server level	Yes—supports node level recovery	Yes—Select the new master	Yes—for recovery performance it is optimized	Yes	Yes	Yes
Fault tolerance	Yes	Yes—replication method to have synced with meta-store	Yes—by ACID transaction system	Yes—Optional	Yes	Yes	Yes
Access control	Yes	No—built-in security provisions	Yes	Yes—by datastax enterprise	Yes	Yes	Yes
Single point failure	Yes—At region server level	Yes—At master node of underlying hadoop framework	Yes—At master level responsible for write replicas	No-hance high availability	Yes	Yes	Yes

3 MEDICAL IMAGING COMMUNICATION SYSTEM

3.1 Picture Archiving and Communication Systems (PACS)

PACS permits a healthcare association (such as a clinic, radiologic department, hospital) which uses a server to acquire, store, display and share all types of imaging modularity externally and internally over a network. When hosting a PACS, the healthcare association desires to consider the atmosphere in which it will be applicable (ambulatory, inpatient, specialties, emergency) and the other electronic systems with which it will incorporate. In order to healthcare organization, big data analytics tool and techniques are capable to extract the meaningful or helpful information from unattended portion of non-invasive medical imaging modularity. PACS has 4 main components:-

- Medical Imaging modality such as CT-SCAN, X-Ray and MRI etc.
- Secure or strong network for distribution and transmitting patient information.
- Mobile devices or work-stations for interpreting, processing and viewing medical images.
- Archives for retrieval and storage of medical images and related documentation.

Imaging modality of PACS is regularly hosted nearby a Radiology Information System (RIS). An RIS is allowed us to make plan for the appointments of patients and file the patient's testing history or radiology, while PACS focuses on additional such as image retrieval and storage.

3.2 Digital Imaging and Communications in Medicine (DICOM)

It is an application layer network protocol that permits the facility to share medical information and images. DICOM allows PACS, RIS and non-Invasive Imaging modality to connect with and sharing imaging data to other healthcare services. DICOM was originally developed in 1983; the National Electrical Manufacturers Association (NEMA) and American College of Radiology (ACR) formed a joint committee to generate a standard technique for transmitting medical images (such as CAT and MRI) and their associated information. But it is nowadays adopted by DICOM Standards Committee that supports a large range of non-invasive Imaging modality in the fields of cardiology, pathology, radiology and dentistry. DICOM uses TCP/IP as the lower-layer transport protocol. All recent medical Imaging modality (such as X-Rays, USG, CT-SCAN, and MRI) support DICOM and use it

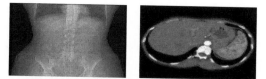

Modality: OT   Modality:CT

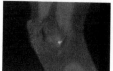

Modality: MR

Figure 5. Examples of medical Imaging modality in DICOM formats.

extensively to holds patient information (name, ID, sex and date of birth etc.). In order to healthcare organization, the vast number of medical images generated in daily life,big data analytics tool and techniques are capable to extract the meaningful or helpful information from attended or unattended portion of non-invasive medical imaging modularity. There are some examples of medical Imaging modality in DICOM formats.

4 CONCLUSION

The main objective of this study is to describe the fundamental concept of Big Data, process model, architecture along with 5Vs dimensions such as volume, veracity, velocity, value and variety of Big Data. The paper also focuses to find and compare some available well known computing and storage tools and techniques basis on some parameters that are being used in current scenarios to address the challenges of Big Data computation. The comparison is made on the most prominent parameters that one looks for before choosing these tools for its application domain to handle Big Data. In order to healthcare organization, big data analytics tool and techniques are capable to extract the meaningful or helpful information from unattended portion of non-invasive medical imaging modularity, with different formats (such as DICOM and PACS etc.).

REFERENCES

Bhardwaj, N. and Balkishan, Kumar, A. 2015. Big Data and Hadoop: A Review. *International Journal of Innovative Research in Science, Engineering and Technology*, An ISO 3297: 2007 Certified Organization, Vol. 4, Issue 6, ISSN-2319-8753.

Bhosale, H. S. and Gadekar, D. P. 2014. A Review Paper on Big Data and Hadoop. *International Journal of Scientific and Research Publications*, Volume 4, Issue 10, ISSN 2250-3153.

Beakta, R. 2015. Big Data And Hadoop: A Review Paper. Volume 2, Spl. Issue 2 e-ISSN: 1694-2329 | p-ISSN: 1694-2345, *RIEECE*.

Dhomse, G., Komal, K., Manali, L., Latika, A. 2015. A Review Approach for Big Data and Hadoop Technology. *International Journal of Modern Trends in Engineering and Research*, www.ijmter.com, e-ISSN No.: 2349-9745.

Garlasu, Sandulescu, D., et al. 17–19 Jan. 2013. A Big Data implementation based on Grid Computing".

Jha, A., Dave, M., Madan, S. 2016. A Review on the Study and Analysis of Big Data using Data Mining. *International journal of latest trades in engineering and technology (IJLTET)* ISSN-2278-621X.

Patel, A. B., Birla, M. and Nair, U. 2012. Addressing Big Data Problem Using Hadoop and Map Reduce. In Nirma University, *International Conference on Engineering (NUiCONE)*.

Prasad, B. R. and Agarwal, S. 2016. Comparative Study of Big Data Computing and Storage Tools: A Review. *International Journal of Database Theory and Application* Vol.9, No.1, pp. 45–66.

Raden, N. 2012. Big Data Analytics Architecture. Hired Brains Inc.

Sriramoju, S. B. 2014. A Review on Processing Big Data. *International Journal of Innovative Research in Computer and Communication Engineering*, ISSN (Online): 2320-9801, (An ISO 3297: 2007 Certified Organization), Vol. 2, Issue 1.

Tiwarkhede, A. S. and Kakde, V. 2015. A Review Paper on Big Data Analytics. *In IJSR*, Volume 4.

http://www.domo.com/blog/2014/04/data-never-sleeps- 2–0/

https://www.emaze.com/@AOTTTQLO/Big-data-Analytics-for-Security-Intelligence.

http://searchhealthit.techtarget.com/definition/picture-archiving-and-communication-system-PACS.

http://www.dicomlibrary.com/DICOM-Samples.

http://dicom.offis.de/dcmintro.php.en/Introduction-to-the-DICOM-Standard.

Communication and Computing Systems – Prasad et al. (Eds)
© 2017 Taylor & Francis Group, London, ISBN 978-1-138-02952-1

Multilateration mechanism in WSN using grid analysis

Monika Yadav, Brijesh Kumar Chaurasia & Shashikant Gupta
ITM University Gwalior, Madhya Pradesh, India

ABSTRACT: In this paper, localization issue addresses in Wireless Sensor ad-hoc Network (WSN). WSN supports data collection and disseminates to the destination to destination. Due to randomly deployment or movement, at the time of operations one of the challenges is computing localization in WSN. In view of this Multilateration mechanism using grid based network proposes. In WSN, anchor nodes having their location and node accuracy in predefined among the sensor nodes. Simulation and results show that proposed localization mechanism may convince the requirement of localization in WSN.

1 INTRODUCTION

WSN consists of huge number of sensor nodes are densely deployed over the region of interest. Nowadays, WSN used in many contexts such as environmental monitoring, health care, traffic control, surveillance systems, anti terrorism operations and localization of services. The important function of a sensor network is to collect and forward data to destination. It is very important to know about the location of collected data. This kind of information can be obtained using localization technique in Wireless Sensor Network (WSN). Localization is a way to determine the location of sensor nodes.

Eventually several types of localization techniques for WSN have been proposed recently. All the localization techniques can be categorized in to range based and range free technique. All the techniques use point-to-point technique as well as broadcasting methodology for calculating location and distance between the neighboring sensors in presence of anchor nodes as well as non-anchor nodes. Different mathematical techniques such as multilateration, trilateration are used by providing respective anchor nodes. Fig. 1 illustrates the different techniques or methods used to identify the location of the nodes. Localization of a sensor node is carried out with the help of neighboring nodes. Several localization techniques are discussed on the predefined diagram.

Known location based localization approach nodes are aware of their location. This is done with the help of manual configuration or GPS devices. Proximity based localization Here network is partitioned into multiple clusters. Every cluster selects a cluster head having a GPS. Angle based localization for identifying the distance which makes use of received single angle or angle of arrival. Range based localization is carried out based on the range.

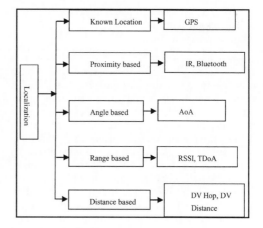

Figure 1. Overview of localization.

The range is estimated by using the Received Signal Strength or Time of Arrival or Time Difference of Arrival. Distance based localization technique uses hop distance among each node to localize the node.

1.1 Multilateration

Multilateration is a mechanism to estimate the coordinates of the unknown nodes, in WSN given the position of some given landmark nodes, known as anchor nodes, whose position are known.

Multilateration localization proposed a tentative algorithm with more anchors that used to reduce the influence of the distance errors on localization results. This technique is used when more than three unknown nodes are calculated for estimate the position of any unknown node. This technique uses multiple reference points with the basis of error estimation and distance based approach.

The distance estimation and system estimation equation are to be elaborated as below.

$$di = di - \epsilon \tag{1}$$

d_i is demonstrated as distance between reference nodes and the unknown nodes.

$$Di - \epsilon = \sqrt{(x - xi)^2 + (y - yi)^2} \tag{2}$$

ϵ is normally considered as the independently normal random variable which has very less mathematical value and which has normally have zero mean estimation value.

This approach and the existing formula uses the equation of multiple values that corresponding with multiple linear equation. The multiple linear equations can be solved by root mean square technique.

i. More than three reference points.
ii. Three known nodes
iii. Known coordinates

The technique that is proposed in this system is the advanced version of trilateration that is basically proposed from multilateration technique. In trilateration technique three reference nodes are take in consideration for estimating the exact position and location of the node. The position of any unknown node is calculated by the help of reference node and estimates the exactness by minimizing the location error. There are so many existing approaches available in the literature. RSSI based multilateration technique for predicting the accuracy and exactness of several mobile sensor nodes estimated according to different algorithm is elaborate by safa hamdoun. Here, RSSI is used for distance measurements between sensor nodes along with the optimal number of the necessary anchors nodes. The comparative study of three models transmit diversity (MISO), receive diversity (SIMO) and the joint transmit-receive diversity (MIMO) is analyzed in this work. The work shows that SISO model is suitable candidature to compute the localization in WSN. However,

the model better work in multilateration algorithm compared with the trilateration.

For estimating the exact location with respect to Selective Iterative multilateration for hop-count Based Localization in WSNs are proposed by Jeffery HS Tay. Iterative multilateration algorithm proposes the improvement in accuracy of multiple sensor nodes for location estimation in hop-count based technique scheme. This paper also supplements the error resolution in cost effective manner. This technique is used for consistently improve the localization accuracy and localization error in both DV-HOP and DHL by taking the cost load low.

Two dimensional location based algorithm approach for distance measurement technique is applied on multilateration is discussed by xinwai wang. This technique used more than three anchor nodes for estimate and analyzes the location of any mobile nodes with reference of those respective nodes. Lateration and angulations techniques for estimating multiple mobile nodes location is discussed in this paper. Obstacle based range-free localization-Error for estimation of WSN architecture is approached by S. Swapna Kumar. The present work focuses on several categories and techniques for sensor topology. In this paper the model was compared with DV-HOP, APIT, ROCRSSI and Amorphous techniques. All this techniques applied for minimize the error of the estimated distance and position of the respective nodes. As the result the localization errors were minimized even in the presence of obstacle where records show that the estimation of exact position is very satisfactory with compare to the existing counterparts' technique. Localization in case of grid based approach as well as error estimation is analyzed by Wei Wang. The monitoring technique uses the whole network as a grid and the network is divided into small random grids and according to their respective position. The objectives position is characterized as an inadequate vector in the discrete space and the quantity of dense and scalable large network. In this work, localization performance is analyzed with different sensing radius, nodes quantity, and targets quantity and measurement noise.

The problem of assessing the security of node localization is demonstrated by Nicola Gatti. The scenario of Verifiable Multilateration (VM) is used to localize nodes and a malicious node. The optimal position for determining exactness according to game theory. In this paper, we propose a grid based multilateration mechanism in WSN. We model it in presence of four anchor nodes and after that five anchor node at intersection point of diagonal of grids. The paper is organized as follows. Section 2 provides a proposed mechanism, Simulation and results show in section 3. Section 4 draws some conclusion.

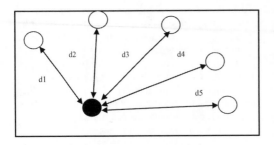

Figure 2. Multilateration mechanism.

2 PROPOSED MECHANISM

The proposed mechanism presents a localization framework considering the presence of obstruction in environment of WSN with random node placement and irregular radio patterns for minimizing estimates of range free localization errors. The mobile anchor nodes are those nodes which are in frequent movement in the wireless network and periodically broadcast beacon message, including their current location approach. The proposed system exhibits the distance based multilateration mechanism for different network scenario with respect to different sensor node architecture and different network category.

The overall description of the proposed work shows on the Fig. 3. The flowchart depicts the overall structure of our proposed mechanism. Initially random nodes are deployed as grid based and without grid based network. The anchor nodes are deployed at different position of the grid. Anchor node is deployed at the diagonal position of the grid. That individual diagonal anchor estimates the position and the location of its neighboring node and broadcast the information of each node. Estimate the location with some respect of error.

2.1 Localization error

Localization error is defined as the average Euclidean distance between the real positions and the recovered positions of the K targets. It is shown as following:

$$Error = \frac{1}{N_t} \sum_{i=1}^{N_t} \sqrt{(x_i - x)^2} + \sqrt{(y_i - y)^2} \qquad (3)$$

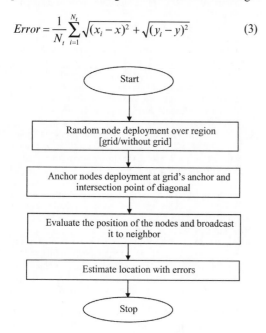

Figure 3. Flowchart representation of proposed mechanism.

Localization error can be finding by the evaluation of Euclidean distance. This paper calculates the euclidean distance for estimation of error in case of distance based estimation as well as position based estimation.

2.2 Root mean square error

Root mean square are use to eliminate the minimum error resolution is case of every error estimation for respective mobile nodes. It can be evaluated by

$$Error = \sqrt{\frac{1}{N} \sum_{i=1}^{N} (x_i - x)^2 + (y_i - y)^2} \qquad (4)$$

3 SIMULATION AND RESULTS

We have simulated multilateration mechanism over grid architecture using MATLAB. We have also considered 100 square meter area with 10–100 sensor nodes and 4–5 anchor nodes. The Fig. 5 and Fig. 6 show that grid architecture with anchor node 4 and anchor node 5 respectively. Sinulation paratementrs show in Table 1.

Fig. 6 shows that the location of sensor nodes estimated when anchor nodes are four. It is observed

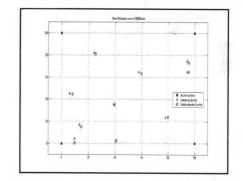

Figure 4. Scenario of 10 nodes with anchor node 40%.

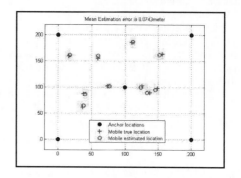

Figure 5. Scenario of 10 nodes with anchor node 50%.

1001

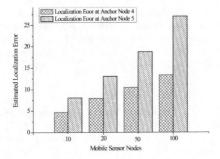

Figure 6. Estimated localization error using anchor nodes.

Table 1. Simulation table.

Parameter	Value
Topology	100*100
Number of Nodes	5,10,50,100
MAC protocol	MAC/802_11
Anchor node	4,6,10,20
Routing protocol	AODV
Antenna type	Omni-Antenna
Network size	100,200

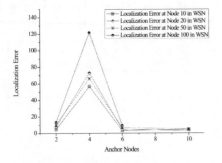

Figure 7. Estimated localization error using anchor nodes at high density WSN.

that estimated error increases exponentially at the density increase respectively in the presence of the anchor nodes it also increases.

We have also measures localization error using multilateration mechanism at 4 to 10 anchor nodes. Fig. 7 shows localization error is more at six anchor nodes and localization error less at anchor nodes 4 and 10 respectively. It is also evidentially shows that more anchor nodes produce less localization error up-to four anchor nodes. However, more anchor nodes incurs complexity and energy issues in WSN.

4 CONCLUSION

The multilateration localization mechanism in WSN using grid architecture was represented. The proposed multilateration mechanism is an effective mechanism for anchor nodes in WSN. The results and simulation is shown that estimated locations accuracy can fulfill the need of logistic applications of WSN. However, more anchor nodes incurs complexity and energy issue in WSN.

REFERENCES

Akyildiz, I. F., W. Su, Y. Sankarasubramanian, and E Cayirci, 2002. "A survey on Wireless sensor network", IEEE Wireless Communications, vol. 40, No. 8, pp. 102–114.

Banerjee C. and A. Das, 2014. "Beaconless Cooperative Localization in Wireless Sensor network," International Conference on Control, Instrumentation, Energy & Communication (CIEC), pp. 550–554.

Boukerche, A., Horacio, A. B. F. Oliveira, E. F. Nakamura and Antonio A. F. Loureiro 2007. "Localization Systems For Wireless Sensor Networks", IEEE Wireless Communications, Dec.

Cheng, L. Chengdong Wu, Yunzhou Zhang, Hao Wu, Mengxin Li and Carten Maple, 2012. "A Survey of Localization in Wireless Sensor Network" Hindawi Publishing Corporation, International Journal of Distributed Sensor Networks, Volume, Article ID 962523.

Gatti, M., M. Morge and S. Sicari, 2012. "A Game Theoritical Analysis of localization Security in wireless sensor networks with adversaties", pp. 1–12.

Hamdom, S., A. Rachedi and A. Benslimane, 2015. "RSSI-based Localization Algorithm using Spatial Diversity in wireless Sensor Networks," International Journal of Ad Hoc and Ubiquitous Computing, Inderscience, Vol. 19, No. 3–4, pp. 157–167.

Harsimran Kaur, Rohit Bajaj, 2015. "Review on Localization Techniques in Wireless Sensor Networks". International Journal of Computer Applications (0975–8887) Volume 116–No.

JHS Tay, V. R. Chandrasekhar, W.K.G. Seah, 2006. "Selective Iterative Multilateration for Hop Count-based localization in Wireless Sensor Networks," 7th IEEE International Conference on Mobile Data Management.

Kumar, S.S., Dr M. Nanda Kumar and Dr V.S Sheeba, 2011. "Obstacle based Range-Free Localization-Error Estimation for WSN" IJCSI International Journal of Computer Science Issues, Vol. 8, Issue 5, No. 2.

Liping Liu, Tingting Cui, Weijie Lv; 2014. "A Range-free Multiple Target Localization Algorithm Using Compressive Sensing Theory in Wireless Sensor Networks". 11th International Conference on Mobile Ad Hoc and Sensor Systems IEEE.

Shahrokhzadeh, M., Abolfazl T. Haghighat, F. Mahmoudi and B. Shahrokhzadeh, 2011. "A Heuristic Method for Wireless Sensor Network Localization" International Symposium on Intelligent Systems Techniques for Ad hoc and Wireless Sensor Networks (IST-AWSN 2011), pp. 812–819.

Wang, X., O. Bischoff, R. Laur and S. Paul, 2009. "Localization in Wireless Ad-hoc Sensor Network using Multilateration with RSSI for Logistic Applications" Pedia Chemistry, Elsevier, Vol. 1, pp. 461–464.

Communication and Computing Systems – Prasad et al. (Eds)
© 2017 Taylor & Francis Group, London, ISBN 978-1-138-02952-1

An estimation of users awareness about features and functionality of search engines

Nidhi Bajpai & Deepak Arora

Department of Computer Science and Engineering, Amity School of Engineering, Amity University, Lucknow, Uttar Pradesh, India

ABSTRACT: The influence of search engines on today's online world is immeasurable and their influence touches every domain. Search engines give access to a vast ocean of information. It is thus important to recognize the power of search engines but more importantly to understand the extent of user ability to exercise control over it. This research work aims at determining users' awareness about Search Engines' features and functionality on different parameters of evaluation. The research work also introduces the concept of misleading factor in search engines. The authors present a method to analyze and evaluate these parameters based on an experiment which was conducted on working professionals employed in various domains like software companies, banks, law firms, educational institutes, Government etc. The sample of the study has 120 working professionals working in different domains.

1 INTRODUCTION

Search Engines are evolving which essentially means addition of new features which are aimed at making the use of search engines easier and the results more accurate. The authors consider users and features of search engine as two faces of the same coin. The accuracy of search results and efficiency of search engines depends as much on the introduction of new features as it does on the understanding of users about the features and functionality of a search engines (Ball, 2013). Awareness about features and functionality of search engines and making use of these features not only saves time and efforts but also provide better search results. There are different search engines available in market today. Most of these search engines provide different advanced features and functionality to help users to retrieve better results from search engines. A study to estimate users' awareness about search engines features is important to understand the gap between search engines and its users in terms of functionality.

The authors have presented different parameters to determine users' awareness about search engines features and functionality. This research work also includes introduction and determination of misleading factor. These parameters are studied and analyzed and their domain wise analysis is performed. Different domains include software companies, law firms, banks, education, government offices and 'others'.

2 IDENTIFICATION OF PARAMETERS OF AWARENESS AND EXPERIMENTAL SET UP

In order to determine the parameters for this research work, the authors conducted an online experiment on working professionals employed in different domains as indicated above. In this experiment, questions related to search engine were presented to the participants. Based on the input received from the participants, the authors have determined parameters to evaluate users' awareness about search engine features. These different parameters are discussed in the headings below:

2.1 Advanced search options usage

In order to facilitate intuitive search for its users, search engines provide advanced search options. Different operators and wildcards can be used for this purpose. They include using quotes " " or using Boolean operators such as AND, OR, NOT or using the wildcard *. Various such options are available in Bing search engine ("Advanced Search Options," 2016) and Google search engines ("Search Operators—Search Help," 2016). Other search engines also have similar kind of operators for advanced search. These options can be specified in different combinations to yield required search results. These options are also provided on a Advanced Search page in Google search engine ("Advanced Search," 2016) and yahoo search

("Advanced Web Search," 2016). With a view to evaluate users' awareness about advanced features, the authors asked the users "Do you use advanced search options?". The response to this question as shown in Table 1. indicates that 38.3% of users working in different domains do not use advanced features. In order to calculate other descriptive statistic for this parameter, variable encoding as shown in Table 1. is done.

Based on variable encoding shown in Table 1, the mean value of advanced search option usage is 1.62 and standard deviation is .488. This is shown in Table 2.

2.2 Awareness about different types of search

As Search engines evolve they add new features to make search engines more useful to the web users. Users can search by keywords. There are also options for search by image, search by audio, search by video. In addition there is an option of semantic search which means that the search engine understands the contextual meaning of terms entered by the user thus improving the search accuracy and making the search more relevant from users' perspective. Table 3 below shows the percentage of awareness among users regarding different types of search.

Table 1. Frequency table for advanced search option usage "Do you use advanced search options ?".

	Frequency	Percent	Cumulative percent	Variable encoding
No	46	38.3	38.3	1
Yes	74	61.7	100.0	2
Total	120	100.0		

Table 2. Descriptive statistics for advanced search option usage.

	N	Minimum	Maximum	Mean	Standard. deviation
Do you use advanced search options?	120	1	2	1.62	0.488

Table 3. Awareness about different types of search.

Type of search	Percent
Keyword based search	100
Image based search	60
Semantic based search	20
Video based search	30
Audio based search	21.67

2.3 Awareness about intelligent search

An intelligent search engine endeavors to be able to understand a query in all its dimensions and provides results which are most appropriate to the intent of the query. This is akin to how a human being understands the context and subtext of any question put to him and tries to frame the answer in the most appropriate terms possible. Hence, instead of returning just a search engine result page containing links to web pages likely to have the answer to the query an intelligent search engine would be able to answer the query directly (Stetzer, 2016). Different search engines like Google and Bing are trying to incorporate different technologies to achieve an intelligent search.The authors analyzed user's responses to know about the extent of awareness about intelligent search. The findings to this aspect are shown in Table 4.

2.4 Awareness about search engine optimization

Search engine optimization is a process by which modifications are done in any particular website in order to increase its primacy on the search engine results page thus increasing the chances of more visitors to the website ("Google Search Engine Optimization Starter Guide," 2016). The authors analyzed the users for awareness about search engine optimization. The findings for this aspect are shown in Table 5 which shows that approximately 40 percent of users are not aware about search engine optimization.

Table 4. Awareness about intelligent search. "Do you think your search engine is an intelligent search engine?".

	Frequency	Percent	Cumulative percent
No	12	10.0	10.8
No Idea about what is intelligent search	21	17.5	28.3
Yes	86	71.7	100.0
Total	120	100.0	

Table 5. Awareness about search engine optimization. "Do you know what is search engine optimization?".

	Frequency	Percent	Valid percent	Cumulative percent
No	47	39.2	39.2	39.2
Yes	73	60.8	60.8	100.0
Total	120	100.0	100.0	

2.5 Misleading factor

Search Engines are generally considered a good source for finding relevant information very quickly. However at times it happens that instead of providing relevant information, the user is mislead by the search engine. The accuracy of search results is effected by auto complete features. It has also been noticed that while searching for particular results a users attention is sought to be diverted to links which are irrelevant to his immediate query. The aforementioned aberrations in the accuracy of search results are collectively known as misleading factor. Misleading factor is determine based on following two aspects

a. The degree to which the auto complete feature in search engine affect the user what they search for. This is termed as Factor-1 (auto complete).
b. The degree to which a users attention is sought to be diverted to links which are irrelevant to his immediate query which may lead him to end up on a link which was completely unconnected to his original query. This is termed as Factor-2 (End up).

The findings to these misleading factors is shown in Table 6 and Table 7. The variable encoding as shown in these tables is done to perform further analysis.

Based on variable encoding shown in Table 5 and Table 6, the mean value of Misleading Factor-1 (Auto complete) is 2.733 and Misleading Factor-2 (End up) is 2.3250 which based on

Table 6. Misleading Factor-1 (Auto complete) "Does the auto complete feature in your search engine affect what you search for ?".

	Frequency	Percent	Variable encoding
Always	4	3.3	5
Never	6	5.0	0
Often	44	36.7	4
Sometimes	66	55.0	2
Total	120	100.0	

Table 7. Misleading Factor-2 (End up) "Does it happen to you that you start searching for something but end up searching something else ?".

	Frequency	Percent	Variable encoding
Always	1	0.8	5
Never	10	8.3	0
Often	28	23.3	4
Sometimes	81	67.5	2
Total	120	100.0	

Table 8. Descriptive statistics for Misleading Factor-1 (Auto Complete).

N	Minimum	Maximum	Mean	Standard deviation
120	0.00	5.00	2.7333	1.20037

Table 9. Descriptive statistics for Misleading Factor-2 (End up).

N	Minimum	Maximum	Mean	Standard deviation
120	0.00	5.00	2.3250	1.11644

variable encoding translates to "sometimes". Hence this happens "sometimes" that search engine tends to mislead the users about what they are searching for.

3 METHODOLOGY USED

3.1 Stage I—qualitative research

In this study, firstly the authors have used qualitative research methodology by doing detailed study about search engines. The exploratory research was done aimed to achieve better understanding of the topic. Literature Review technique was used to gain better understanding of the topic which helped further to frame research questions. This exploratory research step will aid in more powerful analysis of the subject at later stages. This step is important to have better understanding of the topic, users' perspectives and future implications of the topic and provides better insight into the topic.

3.2 Stage II—quantitative research

A Quantitative research was followed which used the online survey methodology. A questionnaire was designed using Google forms and circulated online to working professionals in different domains like Software Company, Government, Banking, Legal firms, Education etc. The responses collected from the working professionals were analyzed using the software GNU PSPP software. s of the practical implications of the results.

4 RESULTS AND DISCUSSION

This research work is based on an experiment conducted online amongst working professionals. The sample consisted of 120 working professionals from different domains which includes 55 professionals working in different software companies, 13 law professionals practicing independently or in legal firms, 13 professionals working in Government services, 12 professionals working in education field, 11 professionals working in different Banks and 16 professionals in 'other' domains. The sample includes 76 male professionals and 44 female professionals. The result for advanced features states that 61.7% of users use advanced search options while 38.3 percent of users do not use advanced search options (Table 1). The domain wise analysis of this parameter shows that the users in education field have maximum usage of advanced search options than any other domain i.e. 1.83 mean value of usage as shown in Figure 1. The value of usage is based on variable encoding shown in Table 1.

There is no considerable difference between male users and female users for this parameter. Male users have mean value of 1.62 while female users have mean value of 1.61.

The analysis for advanced search option parameter amongst different age groups shows that mean value of usage of advanced search option is maximum for age-group between 26–35 years. Thereafter, the mean value of advanced search option usage decreased with increasing age as shown in Figure 2.

The analysis for advanced search option usage parameter amongst users with different experience level shows that mean value of usage of advanced search option is maximum for users with medium level of experience in their domain while its minimum for users belonging to top level positions.

Authors further analyzed users regarding the reason for not using advanced search options. As shown in Figure 4, the major reason of not using advanced search option in education field and software companies is that users don't find it convenient to use while in Banking, Government, Legal and 'others' domain, majority of users opted for option that they don't know about these features.

Analysis of users' awareness about type of search engines shows that 100% of qualified professionals are aware of keyword based search but only 60 percent of qualified professionals are aware of image based search, only 30 percent of users are aware of video based search, only 21.67 percent of users are aware of audio based search and only 20 percent of users are aware about semantic based search.

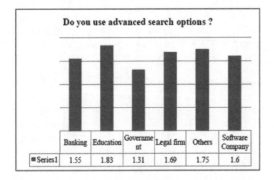

Figure 1. Domain wise analysis of advanced search option based on mean value of parameter.

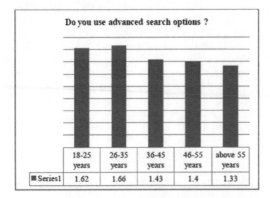

Figure 2. Difference between mean value of advanced search option usage by users in different age groups.

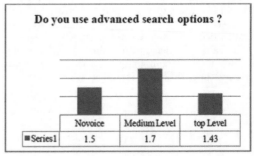

Figure 3. Difference between mean value of advanced search option usage for users belonging to different experience level.

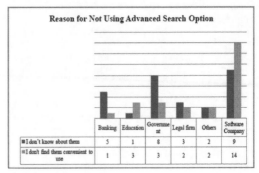

Figure 4. Reason for not using advanced search option in different domains.

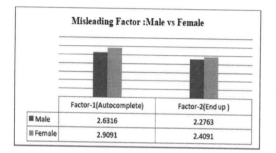

Figure 5. Difference between mean value of Misleading Factor-1 and Misleading Factor-2 between male users and female users.

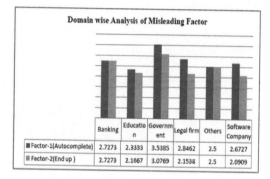

Figure 6. Domain wise analysis of mean value of Misleading Factor-1 and Misleading Factor-2.

The results also shows that 39.2 percent of qualified professionals are not aware about search engine optimization which is an important thing to know to expand any business and its profitability. The authors defined Misleading Factor above. As shown in Figure 5, the mean value of Misleading Factor-1 and Misleading Factor-2 both are greater in female users as compared to male users.

Figure 6 shows domain wise analysis of Misleading Factor. It shows that misleading factor is maximum in users who work in Government domain i.e. the users in Government domain are mislead maximum. Factor-1 is minimum for users belonging to Education field and Factor-2 is minimum for Software companies.

5 CONCLUSION

In this study, authors have presented different parameters to determine users' awareness of search engine's features and functionality based on an experiment performed across a cross section of working professionals. As shown in results

38.3 percent of users are not aware about advanced search options and 39.2 percent of users are not aware about search engine optimization. An almost irrefutable conclusion of the authors analysis of users response for type of search is that users from all the domains are aware about keyword based search engines which is the most conventional form of search engine. The awareness decreases for other types of search .60 percent of users are aware about search by image, 30 percent are aware about search by video, 21.67 users are aware about search by audio and only 20 percent of users are aware about semantic search. Users were also analyzed about their awareness regarding intelligent search.

Authors defined the misleading factor It is noteworthy to add a quick caveat that although the level of awareness about keyword based search engine was 100 percent however a very large number of respondents were not aware at all about different search options like semantic search and search by audio etc. Misleading factor is introduced by the authors which is based on two aspects i.e. the degree to which the auto complete feature in search engine affect the user what they search and the degree to which a users attention is sought to be diverted to links which are irrelevant to his immediate query which may lead him to end up on a link which was completely unconnected to his original query. Misleading factor is greater for female users as compared to male users. Misleading factor was maximum for the users working in Government domain.

The respondents in the authors survey were qualified professionals and it would be safe to conclude the awareness of advanced search options would greatly diminish for other users. It is also concluded by the authors that the introduction of new search features by different search engines are not enough to optimize search results. It is equally important to educate search engine users about the optimum use of advanced search engine features. An increased awareness and understanding of advanced search engine features would increase the usability of search engines. The future scope of this study includes conducting the same experiment on larger sample size by including different domains for further analysis and domain specific understanding.

REFERENCES

"Advanced Search Options". Onlinehelp.microsoft.com. N.p., 2016. Web. 10 May 2016.
"Advanced Search". Google.co.in. N.p., 2016. Web. 10 May.
"Advanced Web Search". search.yahoo.com. N.p., 2016. Web. 10 May 2016.

Agosti, Maristella. *Information Access Through Search Engines And Digital Libraries*. Springer, 2008.

Bajpai, N. and Arora, D., n.d. (in-press) An estimation of user preferences for search engine results and usage. *Advances in Intelligent Systems and computing series Springer*.

Ball, Jon. "Top 10 Search Modifiers: Why They Matter, What They Are & How To Use Them". *searchengineland*. N.p., 2013. Web. 10 May 2016.

"Google Search Engine Optimization Starter Guide". Web. 10 May 2016.

"Search Operators—Search Help". *Support.google.com*. N.p., 2016. Web. 10 May 2016.

Stetzer, Adam "Okay, Now Google'S Artificial Intelligence Division Is Just Showing Off | Search Engine.

Watch". *Searchenginewatch.com*. N.p., 2016. Web. 12 May 2016.

Communication and Computing Systems – Prasad et al. (Eds)
© 2017 Taylor & Francis Group, London, ISBN 978-1-138-02952-1

A comparative review of fast, surf and brisk feature descriptors

Alisha Makkar & Bikrampal Kaur
Department of Computer Science and Engineering, Chandigarh Engineering College, Landran, Mohali, Punjab, India

ABSTRACT: The concept of feature detection is a method to compute abstraction of image information at every point of an image and making local decision at that particular point that there is a feature in an image or not under image processing and computer vision. In this a comparison between three keypoint descriptors have been done and proposed a new combined approach to detect the keypoints present in an image. One of the descriptor is SURF descriptor and other two are new are called BRISK and FREAK. Thus by dividing the time by the number of feature the average time taken for a single feature can be calculated, lesser the number greater is Descriptor in terms of speed and lesser computation power. In the end a flow chart of new proposed methodology is drawn and in future new methodology would be implemented and compare with the other descriptor in my future work.

1 INTRODUCTION

The concept of feature detection is a method to compute abstraction of image information at every point of an image and making local decision at that particular point that there is a feature in an image or not under image processing and computer vision. An image patch near the features found can be produces after successful detection of the features. A high amount of image processing would be needed in this process of extraction whose result is called feature vector also known as feature descriptor. Local histogram and Njets can be mentioned among one of the methods to detect features. The step of feature detection itself can add some additional attributes such as strength of blob and polarity under blob detection and gradient magnitude, edge orientation under edge detection. Features thus extracted are in form of connecting regions, isolated points and continuous curves. Subset of the image domain is resulting features (C. Schaeffer. 2013). A feature definition changes according to the application type, feature of an image can't be bound to an exact or universal definition. An interesting part of an image can be called as feature. Many computer vision algorithms uses feature as starting point thus whole success of these algorithms depends on how good is the feature detector. Repeatability is necessary property for feature detector between two or more different images of same types to detect that have same features or not (Hitul Angrish. 2015).

The focus of this paper is to perform the evaluation between these descriptor so it may help future researcher which tends to exceed in the field of keypoint description, detection and matching. Which is a key area in the computer vision which serves as a base for many of applications such as object reorganization in fields of robotics and other applications, smile detection in cameras for smart phones, face detection in all digital cameras, panorama shots and many more others, The code of all three descriptors that are FAST, SURF and BRISK has been implemented in MATLAB version 2015a.

2 TECHNIQUES

The following techniques have been implemented in this paper. A brief introduction of FAST, SURF and BRISK has been given below.

2.1 Fast

Edward Rosten and Tom Drummond originally developed the FAST detector. Features from Accelerated Segment Test (FAST) FAST is a method to detect corners which is used in many computer vision tasks to extract the feature points and later to track down those features and map the objects. The most advance feature of the FAST detector is its efficiency in computation which seem s to be promising then others. As its name says FAST it is indeed faster than other extraction methods to detect and track features. According to Rosten, it takes about 200 hours on a Pentium 4 at 3 GHz which is 100 repeats of 100,000 iterations to optimize the FAST detector. Like the methods SIFT, Harris, SUSAN and Difference in Gaussian

(DoG). A much better performance can be attained when Machine Learning is applied on FAST which results in better optimized performance and also have lesser strain on the computational resources and gives performance in much less time. For real time processes such as the video processing application this high speed is very beneficial of FAST corner detection (Hitul Angrish 2015).

2.2 SURF

SURF stand for Speeded Up Robust Features first described by Herbert Bay in May 2006. It is basically inspired from SIFT but having some improvements over it is a local feature detection which robust in nature meaning it has some extent of fault tolerance. Hessian blob detector are determined by the use of an integer approximation, with the integral image i.e. three integer operations can be processed very quickly. For detection of features around the point of interest response of Haar wavelet is summed up. For detection of features around the point of interest response of Haar wavelet is summed up. Once again computation of these could be done with the help of an integral image. 3D reconstruction and object recognition are main applications of SURF (Hitul Angrish. 2015).

2.3 BRISK

BRISK stands for Binary Robust Invariant Scalable Keypoints. BRISK is equipped with a mechanism for orientation compensation; by trying to estimate the orientation of the keypoint and rotation the sampling pattern by that orientation, BRISK becomes somewhat invariant to rotation. Preselected pairs are to be used in the BRISK. Thus in the end a binary descriptor which works in the hamming distance in place of Euclidean distances is achieved. It is a 512 bit binary descriptor which calculates the average weighted Gaussian near the keypoints over a select pattern of points. The comparison of the values is done over a specific Gaussian window pairs depending on the greater window value present in the pair it leads either to a 0 or a 1. The BRISK descriptor is different from the descriptors that were explained earlier, BRIEF and ORB, by having a hand-crafted sampling pattern. BRISK sampling pattern is composed out of concentric rings (Hitul Angrish. 2015).

3 RESULTS

In this section the results of techniques FAST, SURF and BRISK is shown with the help of two images named Image One and Image Two respectively.

3.1 Results for image one

This section shows the results after applying the techniques on Original Image One as shown in Figure 1 below.

The Figure 2 shows the results after applying FAST technique to the Original Image One. The Computational Time was 0.076307 seconds and Number of Features detected were 228 and for FAST Average Time per feature is $3.34 \times e^{-4}$.

The Figure 3 shows the results after applying SURF technique to the Original Image One. The Computational Time was 0.21627 seconds and Number of Features detected were 235 and for SURF Average Time per feature is $9.20 \times e^{-4}$.

The Figure 4 shows the results after applying BRISK technique to the Original Image One. The Computational Time was 0.154645 seconds and Number of Features detected were 193 and for BRISK Average Time per feature is $8.012 \times e^{-4}$.

Figure 1. Original image one.

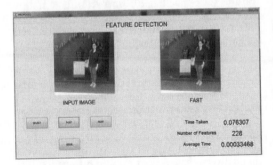

Figure 2. Results of FAST on Image one.

3.2 Results for image two

This section shows the results after applying the techniques on Original Image Hitul as shown in Figure 5 below.

The Figure 6 shows the results after applying FAST technique to the Original Image Two. The Computational Time was 0.0734199 seconds and Number of Features detected were 291 and for FAST Average Time per feature is $2.52 \times e^{-4}$.

The Figure 7 shows the results after applying SURF technique to the Original Image Two. The Computational Time was 0.176498 seconds and Number of Features detected were 444 and for SURF Average Time per feature is $3.97 \times e^{-4}$.

The Figure 8 shows the results after applying BRISK technique to the Original Image Two. The Computational Time was 0.32856 seconds and Number of Features detected were 589 and for BRISK Average Time per feature is $5.578 \times e^{-4}$.

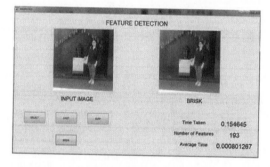

Figure 3.   Results of SURF on image one.

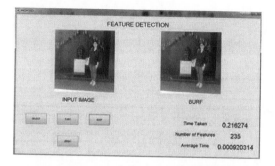

Figure 4.   Results of BRISK on image one.

Figure 5.   Original image two.

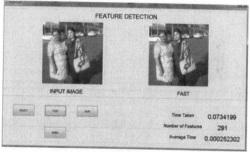

Figure 6.   Results of FAST on image two.

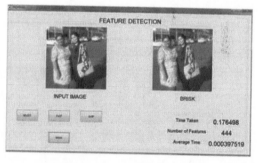

Figure 7.   Results of SURF on image two.

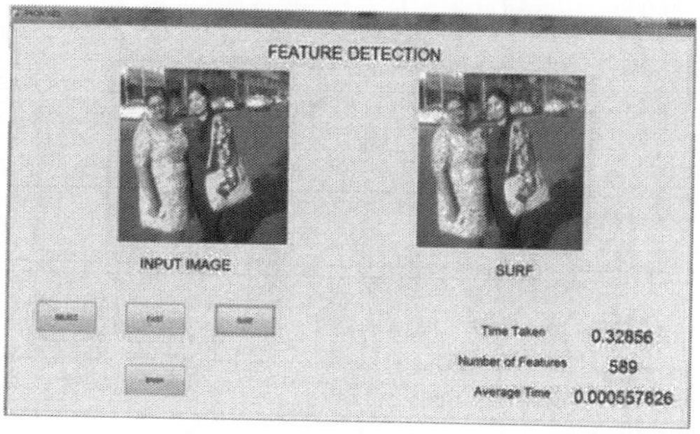

Figure 8.   Results of BRISK on image two.

Figure 9 is bar graph which shows the number of features extracted by FAST, SUR and BRISK on Image One.

Figure 10 is bar graph which shows the time taken to extract features by FAST, SUR and BRISK on Image ONE.

Figure 11 is bar graph which shows the number of features extracted by FAST, SUR and BRISK on Image TWO.

Figure 12 is bar graph which shows the time taken to extract features by FAST, SUR and BRISK on Image TWO.

The table labeled Table 1 above gives a view of all the results of FAST, SURF and BRISK on both

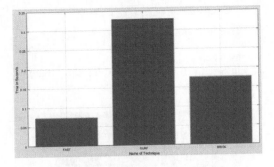

Figure 12.   Bar graph for time taken for image one.

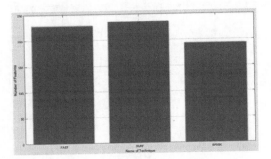

Figure 9.   Bar graph for number of features for image one.

Table 1.   Comparison of results.

Technique	FAST	SURF	BRISK
For image one			
Time	0.076307	0.216274	0.154645
Features	228	235	193
Avg. Time	$2.649 \times e^{-4}$	$9.2031 \times e^{-4}$	$8.0126 \times e^{-4}$
For image two			
Time	0.0734199	0.32856	0.176498
Features	291	589	444
Avg. Time	$2.523 \times e^{-4}$	$5.576 \times e^{-4}$	$3.975 \times e^{-4}$

the Images One and Image Two used for evaluation. From table it is clear FAST is fastest in terms of time while SURF is the slowest among three. SURF payoff its time consumption by finding maximum number of features every time. FAST found second best in Image One case, but least in case of Image Two. BRISK performance increases with increase in number of features in an Image.

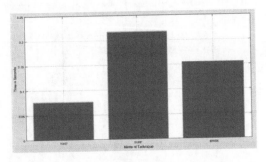

Figure 10.   Bar graph for time taken for image one.

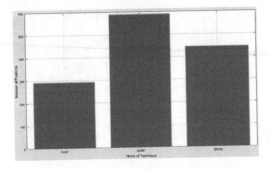

Figure 11.   Bar graph for time taken for image one.

4   NEW PROPOSED TECHNIQUE

A new technique which combines features FAST, SURF and BRISK.

In this new proposed methodology using the features of FAST, BRISK and FREAK combined in the first strep. FAST as its name tells is very fast indeed as compare to other detectors still till this date despite being and old detector. This will provides a certain edge and will not add massive time strains to new method timing. For the time being name of this methodology will be PROPOSED methodology. In the future implementation of the new PROPOSED methodology and the comparison of its results with other techniques will be presented in a paper with results of FAST, SURF and BRISK methods and compare them with PROPOSED methodology.

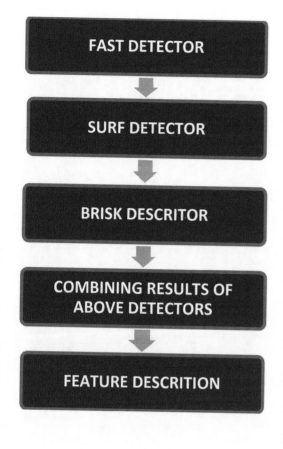

every single time. BRISK also extracts good number of features a little less then SURF but takes lesser time then SURF but it overcomes this drawback by having maximum accuracy. SURF takes maximum time as comparison to for FAST and SURF every single time. FAST is lacking behind in terms of number of features but it also takes much lesser time so this makes it number one in terms of average time taken per feature.

REFERENCES

Herbert Bay, Andreas Ess, Tinne Tuytelaars, Luc Van Gool, 2008. "SURF: Speeded Up Robust Features", Computer Vision and Image Understanding (CVIU), Vol. 110, No. 3, pp. 346–359.

Hitul Angrish, 2015. "A Comparative Review and Optimization of SURF, BRISK and FREAK Feature Descriptors on Images"—International *Journal of Scientific Research and Development*, IISN Online: 2321-0613, PaperID: IJSRDV3I60642, Volume 3, Issue 6, pp. 1368–1372.

Hitul Angrish, 2015. "A Novel Technique to Detect Local Features in Digital Images"—*International Journal of Science Technology and Engineering,* IISN Online: 2349-784X, Paper ID:IJSTEV2I3017, Volume 2, Issue 3, pp. 59–64.

Hitul Angrish, 2015. "A Survey of Different Feature Descriptors Techniques in Image Processing" *International Journal of Science and Research,* IISN Online: 2319-7064, Paper ID: 27041503,Volume 4, Issue 5, pp. 38–42.

Kaur, B., H. Aggarwal, 2010. "Artificial Neural Network based Diagnostic Model For Causes of Success and Failures"—arXiv preprint arXiv: 1005.0965.

Kaur, S., B. Kaur, 2015. "Design of Hybrid Filter with Wavelet denoising and Anisotropic Diffusion Filter for Image Despeckling"—*International Journal of Computer Applications.*

Schaeffer, C. 2013. "A Comparison of Keypoint Descriptors in the Context of Pedestrian Detection: FREAK vs. SURF vs. BRISK".

Stefan Leutenegger, Margarita Chli, Roland Y. Siegwart, 2011. "BRISK: Binary Robust Invariant Scalable Keypoints". ICCV.

5   CONCLUSION

In this paper the main focus was to evaluate the results of three mainstream descriptors under implementation in MATLAB. It has been found that despite of being old FAST is still both more quick and robust in keypoint detection as compare to both SURF and BRISK SURF extracts maximum numbers of features in a very less time

Communication and Computing Systems – Prasad et al. (Eds)
© 2017 Taylor & Francis Group, London, ISBN 978-1-138-02952-1

Anticipation and isolate consequence of DDOS attacks in wireless local area networks

Manveen Kaur
Department of Computer Science, SGTB Khalsa College, Anandpur Sahib, Punjab, India

Baljinder Singh
Department of Computer Applications, CGC Landran, Punjab, India

Parminder Singh
Department of Information Technology, CEC Landran, Punjab, India

ABSTRACT: Many studies have failed to avert the malicious attack on the network. It was observed that security set-up is not robust for even a sensitive network like the base station Node. The attacker Node managed to breach the high security network and carry out the strike in spite of a malicious attack sounded well in advance. This paper dwelled on the fact that a few attacker nodes sneak into the network and extract the meaningful information. The move came following from the recommendation of the Base Station Node to look into the Routing Table related to communication on the network. The base station node is to provide Routing Table and upgrading routing table for nodes of such networks. As a result, there has been a 90% success in the total number of packet send and received in the network environment. We are showing that a higher success rate and yield between 80–90 percent. We assumed that if the success rate fallen by 5%, this would only be the aggressor.

1 INTRODUCTION

Wireless local area networks sometime called as WLAN in which the number of devices so called mobile devices, laptops that able to join the wireless router. These networks follow the IEEE 802.11 standard and cover the local network with data rate up to 1 or 2 Mbps. some standards is providing speed up to 54 Mbps and this is dependent on the device to device or IEEE standard. The wireless standard does not provide any security of the network and thus suspicious user penetrate the security. There are number of attacks that breach the security of the private and public networks and we make standard policies to prevent such types of attacks.

1.1. *Main causes of packet loss in wireless medium*:
1.1.1. The fall in data packets during transmission is mainly because a large number of sent packets from the source node and unable to queue all the packets from the receiver node.
1.1.2. Due to the delay of packets, network was suffering multiple problems like higher congestions, increase overhead and slow payload received by the destination machines.

1.2. *Main Causes Of Attacks In Wireless Medium*:
1.2.1. Interestingly, most attacks on home users where they installed wireless router and setup less security.
1.2.2. The attacks were unable to conclusively determine if the attack was plotted in internal network.
1.2.3. The public has less security than private networks and the attacker seek to steal private information.

While these attacks are growing and becoming more sophisticated, this paper has been taking steps to head this off by developing tighter security and working more closely with wireless local area devices.

2 RELATED WORK

Ahmed A. AlabdelAbass, Mohammad Hajimirsadeghi, Narayan B. Mandayam, Zoran Gajic, "Evolutionary Game Theoretic Analysis of Distributed Denial of Service Attacks in a Wireless Network" (Ahmed A et al. 2016), discussed the distributed denial of service attack with EGT i.e. Evolutionary Game Theory. In this theory it is totally depend on the user whether they transmit the data by measuring the transmission probability method. This method has implemented on Physical Layer where signal strength calculation was done on this research paper. The authors called the attacker as 'Jammers' that launches the attack on the common users and accessible the payload of the data. Here, it is assumed that there are m users and n jammers that correspondingly attack on the network and get benefit from the common users. This paper calculates the Signal to Inference plus Noise Ratio (SINR) and if the value of SINR is low than it was assumes that attack trigger on the network.

Kashif Saghar, HunainaFarid, David Kendall, Ahmed Bourdieu, "Formal Specifications of Denial of Service Attacks in Wireless Sensor Networks" (Kashif Saghar et al. 2016) described the Denial of Service (DoS) that erupted at the wireless node of the network. As a result, other wireless node of the same and different group of the network were affected. This attack was erupted by the group of attacker or single attacker due to access the single channel or compromised the nodes in the network. The increasing number of attacks is not only causing single network damage but also adding to the malfunction or viruses in the network. However, due to weak policies, methodology and other low security devices shrink the security of the network.

A. Kannammal, S. Sujith Roy, "Survey on Secure Routing in Mobile ADVO Networks" (A. Kannammal 2016), has rapped that attack may be active or passive. Passive attack has stolen some useful information that was circulated in between the two node and we assumed that the information was very valuable for both parties. While the Active attack carry on the same network but this attack is less dangerous than passive attack. This attack was disrupting the data during normal operations. The common attacks were performing on the OSI model—Physical Layer, Data Link Layer, Network Layer, Transport Layer and Application Layer. There were also many security solutions but not triggering due to insufficient resources.

Lu Zhou, Mingchao Liao, CaoYuan, Zhongyin Sheng, Haoyu Zhang, "DDOS Attack Detection Using Packet Size Interval" (Lu Zhou et al. 2015), enabled the distributed Denial of Service (DDoS) attack on the Public Network called 'Internet'. This attack has consumed the resources, bandwidth, memory and computing of the nodes—resultant networks down. There were two types of attack detection methods that actively participate on the network i.e. Protocol base Network and Network Flow based method. Protocol based network detect the attack through

Retransmission Time Out (RTO) and Header Information etc. while abnormal flows in the network identify by the Network Flow based methods. The proposed method was suggested by the authors to identify the attack by packet size variations but fails to detect false positive and false negative rate.

3 PROPOSED SOLUTION

In this paper we implement the communication model in section IV, there are N number of devices that stores n number of data and the total number of states is 2 N and

$$\log_2 2^N = N \tag{1}$$
$$\text{then } \log 2M = \log_{\log} M \tag{2}$$

In our proposed theorem, if the path P established between Nodes A and B and Pi packet is allowed in duration t1 then

$$P = \frac{\lim l \to \infty \quad 1.}{\log l} \tag{3}$$

Therefore, number of sequence packets allowed in our proposed experiment is:

$$Pi = P\,(i{-}i1) + P\,(i{-}i2) + P$$
$$(i{-}i3) + \text{------------------} + P\,(i{-}in) \tag{4}$$

From the equation (4), we supposed that P (i–i2) is not received by the destination node and the destination node requested to send the packet again. From the request is receiving by the sender node and generate the same packet again for transmission; if the packet is not received by the target node then it assumes that the packet may be captured by anomaly node. This way, we track down the MAC addresses that are near to the location of destination machine and the base station node verifies from the MAC Table. If any MAC address is encountered as fake or unsuspicious then the node has declared as malicious. This malicious node will be eliminated from the network and the MAC address saves for future reference.

4 SIMULATIONS AND RESULTS

There are 44 nodes has been placed on the network area 1000 meter × 1000 meter. In these nodes there is one base station chosen by the group of node and rest of other nodes adjacent in peer to peer fashion.

The base station node is running the proposed solution and warned against unacceptable information that shortsightedly indulges those who are nurturing this type of false information. However, the nodes are not sharing any information to the neighboring nodes except base station. The attacker node is to creating attack on the network nodes which is also discovering the new nodes that activate on the network (Fig. 2). The attacker node sends false information to the source node and will be attaching to current path. We remain committed to maintaining the proposed solution in our suggested network model (Fig. 1) and additional to reducing the attacks on the experimental network.

As part of the restructuring, the new network node has been entering on the network with permission of base station and the node that refer it. The creation of new node that brings together the existing nodes and security capabilities is in line with the ambitious objective of this paper.

It has almost been confirmed that the detection attacks has failed in its responsibility to

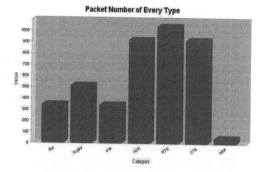

Figure 3.   Packet variation.

Table 1.   Outcomes from the results.

Parameters	Value
Number of nodes	44
Number of receiving nodes	37
Number of generated packets	28861
Number of sent packets	28806
Number of received packets	28702
Number of dropped packets	55
Minimal generated packet size	32
Maximal generated packet size	1040

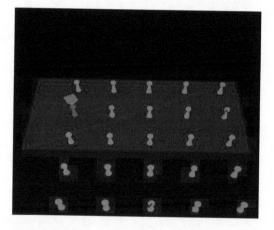

Figure 1.   Network deployment.

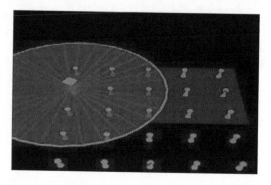

Figure 2.   Attacker node.

ensure the safety of the network. In the results, packet variation raises doubt of possible attacks on TCP between the higher-ups in that protocol. If we consider the attacks on the three layers of OSI model—Physical Layer, MAC Layer, Network Layer and Transport Layer. We assumed that higher attacks on the transport layer as you see the acknowledgment section, the total bytes is 900 bytes whereas TCP protocol uses 300 bytes to send the data on the network and if we calculate the variation then said that packet variation is more in the case. Thus, we said that attack trigger on the network and it is depicted in the Fig. 2. But when we are isolating the attacker node and resultant outcomes show in the Table 1.

From the Table 1, we showed that the drop packets are minimum and we send the higher number of packets to the receiving nodes. These dropped packets have been lost during noise in the communication channel or remains wireless issues. In the Figure 4, we analyze the result of packet loss vs. time interval and we find that the attacker node attack in the duration of 41.0 seconds to 43.0 seconds but when we implement the proposed model then we isolated such types of attacks.

Packet ID

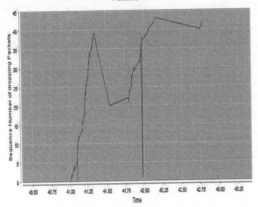

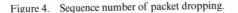
Time

Figure 4.   Sequence number of packet dropping.

5   CONCLUSION AND FUTURE WORK

Earlier lacked specific attacks on the public networks but strongly believes that such a strike was possible, particularly after passive attacks. Still, protocols were not aware of any public network about where or when the attack would occur. The threat level in any public network was raised to its highest but we would neutralize a suspicious node at intranet (own network), and shortly afterwards, a proposed approach gives better result. This research work will be experimenting with new technologies like vehicular Ad Hoc networks so that further research work carried. Deadline The above material should be with the editor before the deadline for submission. Any material received too late will not be published.

REFERENCES

Ahmed A. AlabdelAbass, Mohammad Hajimirsadeghi, Narayan B. Mandayam, Zoran Gajic. 2016, *Evolutionary Game Theoretic Analysis of Distributed Denial of Service Attacks in a Wireless Network*, CISS.

Heng Zhang, Peng Cheng, LingShi, Jiming Chen. 2016 *Optimal DoS Attack Scheduling in Wireless Networked Control System*, IEEE Transactions on Control Systems Technology, pp. 843–852.

Iftakhar Ahmad, kazisinthiakabir, Tanzila Choudhury, A.B.M. Alim AI Islam. 2016, *Enhancing Security in Specialized Use of Mobile IP*, IEEE.

Iman Almomani, Bassam Al-Kasasbeh. 2015, *Performance Analysis of LEACH protocol under Denial of Service Attacks*, ICICS, pp. 292–297.

Jianwul Li, Feng Zebing, Feng Zhiyong, Zhang Ping. 2015, *A Survey of Security Issues in Cognitive Radio Networks*, IEEE, pp. 132–150

Kannammal A., Sujith Roy S. 2016, *Survey on Secure Routing in Mobile AdhocNetworks*, HMI.

Kashif Saghar, Hunaina Farid, David Kendall, Ahmed Bouridane. 2016, *Formal Specifications of Denial of Service Attacks in Wireless Sensor Networks*, IEEE, pp. 324–333.

Simranpreet Kaur, Rupinderdeep kaur, A.K. Verma. 2015 *Jellyfish attack in MANETs: A Review*, IEEE, pp. 1–5.

William Hurst, Nathan Shone, Quentin Monnet. 2015 *Predicting the Effects of DDoS Attacks on a Network of Critical Infrastructures*, IEEE, pp. 1697–1702.

Yuan Yuan, Huanhuan Yuan, Lei Guo, Hongjiu Yang, Shanlin Sun. 2016, *Resilient Control of Networked Control System under DoSAttacks: A Unified Game Approach"*, IEEE Transactions on Industrial Informatics, pp. 1–9.

Zhou Lu., Mingchao Liao, Cao Yuan, Zhongyin Sheng, Haoyu Zhang. 2015, *DDOS Attack Detection Using Packet Size Interval*, IEEE.

Communication and Computing Systems – Prasad et al. (Eds)
© 2017 Taylor & Francis Group, London, ISBN 978-1-138-02952-1

Recent advancements in substation automation systems based on information and communication technologies

Sunil Gupta
Department of Electrical and Electronics Engineering, Maharaja Surajmal Institute of Technology, New Delhi, India

Ikbal Ali, Mini S. Thomas & S.M. Suhail Hussain
Department of Electrical Engineering, Jamia Millia Islamia, New Delhi, India

ABSTRACT: The introduction of communicable Intelligent Electronic Devices (IEDs), advanced digital communication and information technologies in modern substations has offered a wide range of opportunities for utilities to improve and facilitate the effective substation monitoring, control, protection and automation applications that couldn't be realized with their conventional counterparts. This paper discusses the issues and challenges in traditional Substation Automation Systems (SAS), and hence the need of adopting standardized communication in SASs is presented in detail. Further, this paper has presented the evolution of IEC 61850 communication protocol, the standard for '*Communication Networks and Systems in Substations*', as well as its communication approach, key features, benefits, and scope in designing modern digital substation automation systems. Finally, the work examines the huge potential of IEC 61850 standard to impact the design, operation, testing, and performance of Ethernet communication based SAS applications that cannot be achieved with legacy communication protocols.

Keywords: substation automation system, IEC 61850 communication protocol, switched Ethernet, communication and information technology

1 INTRODUCTION

To overcome the shortcomings of conventional substations, there was a need to automate substations that should cover all aspects of intelligence in substation operations. Substation automation is the effective use of smart equipment, communication and networking technologies in substation applications that has the ability to monitor their own functionality. The introduction of communicable Intelligent Electronic Devices (IEDs) and digital communication in substations has offered a wide range of opportunities for utilities to improve and facilitate the effective substation automation applications. Substation IEDs provide communication capabilities with vendor specific communication protocol and hence were incompatible both at the physical layer interface and the communication protocol layer.

Earlier, the proprietary protocols for communication in substation were Profibus, IEC 60870-5 (IEC 60870-5-103, 1996), DNP (Distributed Network Protocol), and Modbus (MODBUS-IDA, 2006). Thus, costly and complicated protocol converters were required to bring the substation

devices onto a common physical network and allow everyone to speak a common application layer protocol. Thus, due to the proliferation of multi-vendor IEDs and communication technologies in substation, there seems to be an immediate need to adopt a standard approach for meeting the critical communication demands of Substation Automation Systems (SAS) and also to be future ready to tackle demand growth and changing scenario due to restructuring and deregulation. The different approaches to achieve this have been discussed by McDonald, 2003. The major challenges encountered were the successful configuration of the multivendor IEDs (using the proprietary configuration tools), their interoperability and performance tests. An effective communication system played a key role to link substation devices within an electric power substation for the high reliability and real-time operation of a SAS (Gungor & Lambert, 2006). Modern substation automation system uses IEC 61850, the standard for "*Communication Networks and Systems in Substation*" for the real time operation of the power system (IEC 61850 communication standard, 2002–2005). Standardized data model, communication approach and the configuration language are some inherent features in IEC 61850 standard that offers various benefits over legacy communication protocols such as Modbus, Modbus Plus, DNP3.0, and IEC 60870-5 (Mackiewicz, 2006).

The paper discusses the issues and challenges in traditional SAS, and hence the need of adopting standardized communication in SASs is presented in detail. Further, it presents the communication approach, key features, benefits, and scope in designing modern digital substation automation systems. Finally, the work examines the huge potential of IEC 61850 standard to impact the design, operation, testing, and performance of Ethernet communication based SAS applications that cannot be achieved with legacy communication protocols.

The rest of the paper is organised as follows: Major shortcomings in traditional SASs are described in section 2 of the paper. Section 3 discusses the communication requirements in designing modern digital substation automation system. Standardization developments and IEC 61850 based SAS is presented in section 4. Finally, concluding remarks are provided in section 5 of the paper.

2 CHALLENGES IN TRADITIONAL SUBSTATION AUTOMATION SYSTEM

Due to expansion in power system over a period of time, substation functionality has been improved to handle power system operations reliably and effectively. So to develop a cost effective and reliable SAS, there was a need to integrate multiple substation automation functions in a single device. To achieve this, Remote Terminal Units (RTUs) at conventional substations are replaced with advanced microprocessor based IEDs which could simultaneously handle multiple substation data from primary equipments and share this information among secondary devices to carry out various substation functions (Kirkman, 2007).

IEDs have started adopting peer to peer and process communication over Ethernet networks with limited data sharing capabilities. However, data sharing is possible only if devices are protocol compatible. Early IEDs provide communication capabilities but followed proprietary protocols such as DNP3.0/MODBUS/MODBUSPLUS/ IEEE 60870-5 etc. of the device manufacturer and produced data in different formats. Hence this presents huge problems to the system integrators. When building the systems, the integrator often had incompatibility problems regarding the physical communication interfaces, the protocol itself, the addressing of information and the data formats.

There exist great benefits to the user, if these IEDs could be integrated and made interoperable in the substation environment (McDonald, 2003). The integration of substation devices not only needs a huge investment in substation engineering and for their maintenance but also faced performance issues due to the insertion of the protocol converters. Thus to reduce the complexity and the amount of cables in traditional SASs, as shown in Figure 1, and to develop an economical and high performance integrated substation automation system, interoperability was desired to enable IEDs from different vendors to communicate without gateways. So the utilities understand the need of a standardized communication protocol to link all these IEDs together in SAS. To achieve this, modern substations are now expected to be equipped with high speed computers, open communication interface and IEDs to carry out substation automation operations reliably.

3 COMMUNICATION REQUIREMENTS IN SUBSTATION AUTOMATION SYSTEMS

Performance requirements for the SAS are special, as the functions performed by SAS are highly time-critical (IEC 61850-5, 2003). Moreover, communication among IEDs, which are part of the SAS, should be highly reliable and fast. Performance requirements for the substation were generally achieved using techniques such as point-to-point and field bus systems, especially of the type MODBUS. Performance of communication aided protection applications is governed by the performance of individual IEDs/communication devices, communication network/device parameter settings, and the communication mechanisms involved in data transmission.

Thus, an effective communication system played a key role to link various protection, control, and monitoring devices within an electric power substation for the high reliability and real-time operation of a substation automation system.

The development of Ethernet technology provides an opportunity to design new and innovative communication systems for power system protection applications. Recently, Switched Ethernet technology, which provides full duplex and collision free communication environment, is suitable for real-time substation automation applications (Skeie et al., 2002). Virtual LANs, RSTP, Simple Network Management Protocol (SNMP) and ElectroMagnetic Interference (EMI) immunity are some special features of switched Ethernet technology, which can be harnessed for enhanced performance of SASs (Decotignie, 2005). Also,

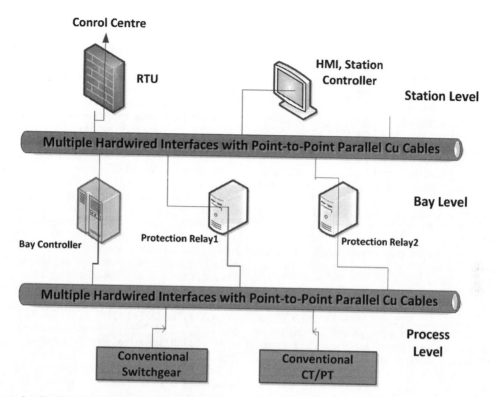

Figure 1. Traditional substation automation system.

substation and its communication architecture play an important role in maintaining high reliability, and availability of the power supply. Thus, a high performance SCN that supports the real time data transfer and others functionalities in SASs is crucial. Ethernet provides high flexibility regarding communication architectures, as well as incorporation of fast growing communication technologies.

Redundancy technologies such as RSTP/Parallel Redundancy Protocol/High-availability Seamless Redundancy/Dual Homing Protocol (PRP/HSR/DHP) facilitate the designing of a suitable substation network topology at various levels in SAS to satisfy different levels of reliability and performance issues.

4 STANDARDIZED DEVELOPMENTS AND IEC 61850 SUBSTATION AUTOMATION SYSTEMS

IEC 61850, the standard for *"Communication Networks and Systems in Substation"*, provides communication interoperability by defining the Object Oriented Data Modeling (OODM) and Substation Configuration Description Language (SCL) approach. The standard specifies object oriented

and hierarchical data modeling approach; defines communication services based on real communication protocols like MMS, TCP/IP, and Ethernet; and allows the automatic information exchange and device configuration based on SCL.

The drawback of most of the proprietary communication standards such as Modbus, Modbus Plus, IEC 60870-5, and DNP3 in SASs is that the data of the functions, the services and the communication protocols are all mixed together (Brunner, 2008, Mohagheghi, 2009). These protocols lacks the standardize representation and organization of data in substation devices in terms of the applications. But the communication standards must be able describe themselves from both a data and services perspective, i.e., the data and its associated information/services must be communicated over the channel.

IEC 61850 resolve this issue by separating the data of the functions, services and communication protocols. The approach in IEC 61850 is based on abstracting the definition of the data items and services i.e. creating data objects and services that are independent of any underlying protocols. With standardized object data models of all possible functions in substations and the related generic services, the standard provides standardized communications in substations. These object data models and its services are mapped to the real time protocols such as Manufacturing Message Specification (MMS), TCP/IP and Ethernet. Client/server communication is based on all 7 layers ISO/OSI stack, only Generic Object Oriented Substation Event (GOOSE) and Sampled Values (SVs) are mapped directly to the link layer of the Ethernet (Brunner, 2008). Since the data services and applications are built above the application layer, when communication technology changes it is necessary only to map the abstract services and standardized data to new protocols. The switched Ethernet technology along with IEC 61850 standards seems to offer cost-effective solutions for the implementations of SAS functions (Skeie *et al.*, 2002, Decotignie, 2005).

Figure 2 shows IEC 61850 based substation automation system architecture which generally consists of three levels.

Station level: It includes Human Machine Interface (HMI) and gateways to communicate with remote control centre and integrate IEDs at the bay level to the substation level. It also performs different process related functions such as implementation of control commands for the process equipment by analyzing data from bay level IEDs.

Bay level: The process level equipments are connected to station bus via IEDs at the bay level that implement monitoring, protection, control and recording functions.

Process Level: It includes switchyard equipments, sensors and actuators. The current and potential transformers are located at the process level to collect system data and send them to bay level devices for automatic control & protection operations which are achieved through circuit breakers and remotely operated switches.

The introduction of serial communication links in IEC 61850 SASs, for all types of substation communication, have not only replaced traditional copper cables carrying binary or analogous information but also allow the time critical information exchange between IEDs. It results in very significant improvements in both cost and performance of electric power system. IEC 61850 based SAS reduces operational and maintenance expenses by integrating multiple functions in a single IED. These functions are distributed among IEDs on the same, or on different levels of the SAS. It enables distributed intelligence in a network for developing various new and improved applications. Author in (Ali *et al.*, 2012 & 2013) analysed the performance of IEC 61850 GOOSE & sampled values messages based substation protection applications.

This improves the functionality, design and construction of modern substations. Hence IEC 61850 communication standard allow the substation

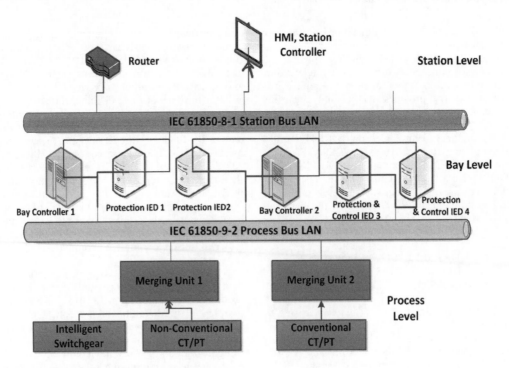

Figure 2.   IEC 61850 substation automation system.

designer to focus more attention on other important issues like intelligence, reliability, availability, security, and efficiency of the power network. Authors in [Ali et al., 2015] proposed fault proof SCN architecture having high reliability and desired real-time performance, for a given substation scenario.

Thus, it has been identified that communication interoperability, object-oriented data modeling, open communication approach, high level SCL based engineering support, and peer-to-peer Generic Object Oriented Substation Event (GOOSE) and Sampled Values (SVs) based communication services are major technical features provided by IEC 61850 protocol over legacy protocols. IEC 61850 standard offers several benefits such as reduction in overall losses, savings in overall life-cycle cost, free configuration of substation functions, new and improved functionality, switched Ethernet networking technology support, and secure and flexible SCN architectures in designing modern SASs. Key Features and Benefits of IEC Substation Automation System include:

- Communication interoperability: Communication interoperability is defined as the ability for the substation IEDs from one or several vendors to exchange information without the need of gateways. The object oriented data modeling approach which specifies data models of all possible functions and devices in SAS and an XML (Extensible Markup Language) based standardized configuration language using Substation Configuration Description Language (SCL) files permits communication interoperability in IEC 61850 substations.
- Free allocation of functions to devices: The standard allows free allocation of functions, i.e., the substation functions can be distributed among IEDs on the same, or on different levels of the SAS. This feature allows not only optimizing the existing applications in substations but also enables distributed intelligence in a network for developing various new and innovative protection and control applications. Thereby, it improves the functionality, design and construction of modern substations.
- Open communications: The standard is based on the idea of separating the object model with its data and services from the ISO/OSI communication stack based MMS, TCP/IP and Ethernet communication protocols. With this, it follows the progress in communication technology and hence safeguards investments in object data models and services.
- Peer-to-Peer communication services: A very high speed peer-to-peer GOOSE messages and SVs based information exchange from

conventional or non-conventional instrument transformers are included in IEC 61850 that are particularly meant for developing fast and reliable control and protection applications. To increase the dependability, real-time and mission-critical messages such as SVs from MUs and GOOSE messages for inter-IED communication are mapped directly to the link layer of the Ethernet.

Although various standard protocols exist for communication outside the substations, the features possessed by IEC 61850 protocols may also be utilized to enhance the performance of applications beyond the substation such as in Distribution Automation (DA) applications and for communication with other substations and control centres.

Although, IEC 61850 standards offers several advantages like overall cost savings, new and improved functionality, switched Ethernet networking technology support, and flexible substation communication architectures in designing modern SASs. However, it was inferred from the literature survey that some of the major technical issues such as configuration, implementation and performance testing of IEC 61850 substation applications; reliability of the process bus based SCN architectures; and their impact on reliability indices of protection functions are left unanswered.

Further, IEC 61850-9-2 process bus has proven to be an attractive and economical technology but it is important to examine the feasibility of implementing 'all-digital' protection schemes, in terms of ETE delay performance of time-critical messages over process bus network. Thus, it is crucial to analyze the performance of process bus based SAS architectures through careful modeling of the network components, communication media, protocols, and the data traffic using LAN simulation tools. Major challenge encountered is the modeling of SCN devices/IEDs that can represent the specific characteristics of the physical IEC 61850 SAS network in the simulation environment.

5 CONCLUSION

The paper has discussed the major shortcomings of traditional SASs and found that they lack in terms of interoperability, and have limited real-time support capability. Further, it presented the evaluation of IEC 61850 communication standard in designing modern digital substation automation systems. The key technical features, major benefits, as well as technical challenges related to the IEC 61850 standards in substation automation are also presented in this paper.

REFERENCES

Ali, I., Thomas, M. S. & Gupta, S. 2013. Integration of PSCAD based power system & IEC 61850 IEDs to test fully digital protection schemes. *IEEE PES Innovative Smart Grid Technologies Conference (IEEE ISGT ASIA 2013)*. Bangalore, India.

Ali, I., Thomas, M. S. & Gupta, S. 2013. Sampled values packet loss impact on IEC 61850 distance relay performance. *IEEE PES Innovative Smart Grid Technologies Conference (IEEE ISGT ASIA 2013)*. Bangalore, India.

Ali, I., Thomas, M. S. & Gupta, S. 2012. Methodology & tools for performance evaluation of IEC 61850 GOOSE based protection schemes. *IEEE Power India International Conference (PICON, 2012)*, DCRUST. Haryana (India).

Ali, I., Thomas, M. S., Gupta, S. & Hussain S. M. 2015. IEC 61850 Substation communication network architecture for efficient energy system automation. *Energy Technology and Policy*, Taylor & Francis, 2(1):82–91. DOI:10.1080/23317000.2015.1043475.

Brunner, C. 2008. IEC 61850 for Power System Communication. *IEEE/PES Transmission and Distribution Conference and Exposition*, 1–6. Chicago.

Decotignie, J.D. 2005. Ethernet-Based Real-Time and Industrial Communications. *Proceedings of the IEEE*, 93(6):1102–1117.

DNP USERS GROUP. Distributed Network Protocol. [Online]. Available: http://www.dnp.org/.

Gungor, V.C. & Lambert, F.C. 2006. A survey on communication networks for electric system automation. *Computer Networks*, 50(7):877–897.

IEC Standard for Telecontrol equipment and system: Transmission Protocols—Companion standard for the informative interface of protection equipment, IEC 60870-5-103, 1996.

IEC 61850: Communications Networks and Systems in Substations. 2002–2005. Available [Online]: http://www.iec.ch./

IEC 61850-5: Communication requirements for functions and device models, IEC INTERNATIONAL STANDARD, July 2003.

Kirkman, R. 2007. Development in substation automation systems. *IEEE International Conference on intelligent systems applications to power systems*, ISAP: 1–6.

Mackiewicz, R.E. 2006. Overview of IEC 61850 and Benefits. *IEEE Power Systems Conference and Exposition, PSCE'06*. Atlanta:623–630.

McDonald, J.D. 2003. Substation automation, IED integration and availability of information. *IEEE Power and Energy Magazine*, 1(2):22–31.

MODBUS, MODBUS-IDA, Modbus application protocol specification, December 2006. [Online]. Available: http://www.modbus-ida.org/.

Mohagheghi, S., Stoupis J. & Wang, Z. 2009. Communication Protocols and Networks for Power Systems- Current Status and Future Trends. Proc. *IEEE PSCE*.

Skeie, T., Johannessen, S., & Brunner, C. 2002. ETHERNET in substation automation. *IEEE, Control Systems Magazine*, 22(3):43–51.

Communication and Computing Systems – Prasad et al. (Eds)
© 2017 Taylor & Francis Group, London, ISBN 978-1-138-02952-1

Malicious URLs detection using random forest classification methodology

Himani Jangra & Chander Diwaker
Computer Engineering Department, UIET, Kurukshetra, India

ABSTRACT: Phishing firms are growing day by day and become more successful to elude mitigation by the blacklists. Due to this detection of malicious URLs is becoming progressively more complex. Phishes can reduce the effectiveness of the blacklists by hosting malicious URLs along with shorter time span. The standard method of detecting malicious URLs use the black-list and examine whether the URLs are listed or not. The highly dynamic behavior of phishing firms needs renewing the models regularly and this creates additional challenges as most of the predictable learning algorithms are also computationally exclusive to maintain. This work uses Random Forest method for detecting malicious URLs that include string values.

Keywords: computer security, url classification, malicious web-page detection, machine learning method, random forest algorithm

1 INTRODUCTION

A Uniform Resource Locator (URL) is used to address a web-page. And there are two types of URLs, one is malicious and other is non-malicious. Malicious URLs are those which contain some malicious data. The term malicious comes from the word "Computer Virus" which is introduced by Cohen, Virus is a piece of code that replicates itself by attaching to any exe file and when that file is executed, the virus is triggered automatically and then infects that system by its malicious activities. As the no. of users on the internet is increasing day by day, the threat of cyber-crime is also increasing rapidly. The internet crime is only through the malicious web sites. As the cyber-crime is increasing, the protecting methods are also developed by the researchers so that the users are eluded from visiting those web sites which are malicious by the nature. For this classical approach is to make blacklist and maintain that blacklist manually. But manual approach was very costly and time consuming. As a result, automated URL classification methods were introduced. These methods were less time consuming. According to a survey made by the McAfee labs, there were 6 million malwares that were created in one month.

2 TYPES OF URLS

There are four types of URL as shown in Figure 1, explained as follows:

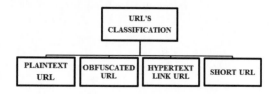

Figure 1. Types of URL.

2.1 Plaintext URL

These are the text-only URLs containing simple text. These URLs are commonly used by the users not having much knowledge about the format of URL.

2.2 Obfuscated URLs

These URL contains some type of code within themselves which is generally contained URL with some symbols and numerical data.

2.3 Hypertext link

These URL contain a hyperlink, e.g.<a href = " ..."> http://3url.org/?67592 </a>.

2.4 Short URLs

Short URLs follow strict length limits e.g. Twitter. Users are familiar to follow a URL that looks like http://bit.ly/lhBa6k, even when the original URL may be http://evil.com/attack?id = 31337. It is very hard to conclude that the URL is malicious or not just by seeing it and become harder in case of short URLs.

3 METHODOLOGY

There are many techniques which are used for URL classification or to detect malicious URLs as shown in Figure 2:

3.1 Binary

This classifier gives result in either 0 or 1, where 0 stands for malicious and 1stands for non-malicious & vice versa.

3.2 Zero-R

It is the straightforward classification method which only predicts the majority class.

3.3 One-R

It generates one rule for each predictor in the dataset that's why it is also known as one-rule classifier.

3.4 Random forest

This method is the collection of many machine learning methods which are used for different type of classifications, regression and other tasks. At training time, it makes many decision trees and generates different classes as output. In the research, it builds decision tree of different blacklists. Afterward it generates ROC (Receiver Operating Characteristics) curve and confusion matrices. Using the generated result, the research classifies the URLs as malicious or non-malicious.

There are lot many methods but in this research the main focus is on Random forest method. This technique is used because it takes string values also. The string values are converted to the numerical values using Term Frequency (TF) and Inverse Document Frequency (IDF) transformations. After the transformation the URLs are classified as malicious or non-malicious.

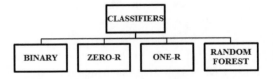

Figure 2. Methodology used in URL classification.

4 RELATED WORK

Hajgude and Ragha described a method of analyzing content of mail and lexical URL. There are two DNS list Blacklist and White list. Phishing mails are detected if link is in blacklist otherwise reasonable DNS. It is analyzed using selection of pattern matching's with the current phishing DNS, contents of mail & the analysis of actual URL only if it is not present in both the list. White list and Blacklist used for avoidance of detection time for legitimate & phishing emails.

Basnet and Sung represented, Production of phishing attacks in latest years that currently referred as an important cyber security area. Authors proposed an innovative scheme that detects phishing URLs automatically using extracting and mining from several Web Services. The scheme is applied on real world data sets and it was validated that the Logistic Regression classifier achieves positive rate of 0.1–1% and false negative rate of 0.7–6.5% with an overall accuracy of 97.2–99.8%, false in non-phishing URLs & detecting phishing.

Akiyama et al. carried out filtering based on blacklist and it is one of the major counter measures against malicious websites. Since malicious URLs have a tendency to be short lived and could be partially changed, blacklist must be updated. Authors also represented a generation method for blacklist URL that identifies neighborhood URLs of a malicious URL using a search engine.

Kandasamy and Koroth described the effects of spam contents over the social media. The proposed algorithm uses spam classification in Twitter as a unified approach. It is a three step process comprises of Natural Language Processing (NLP), URLs analysis & supervised machine learning techniques.

Khare et al. evaluated that internet accessibility was not measured in a proper way. Internet link bandwidths were vulnerable to be misused. The aim of the research was to classify unknown accesses proactively. It will empower campus ISPs to unintended users through denying the access. URLs are classified by the means of short leftovers that are available. The classification accuracy is up to 99.96% on 10 GB of proxy data.

Cao and Caverlee described the influence of social media systems like Facebook and Twitter in providing a global infrastructure for web hyperlinks or sharing information. The research provided new ideas in social media search. The study differentiates click patterns and posting in a sample application domain, the classification of spam URLs. The posting versus clicking behavioral signals provides over-lapping. Different URLs perspectives can be considered in the future designing applications of spam link detection and link sharing.

5 PROBLEM DEFINITION

Phishes are finding out new techniques to add malicious content to the web pages. A URL can it-self describes facts regarding web pages that may contains malicious contents. The experts can extract a lot of data from the URL without expressing the unusual data of the web pages. Relatively, web pages are downloaded first and then examined for containing malicious content because it is easy to classify a URL using different machine learning methods. URLs that contain some extra nasty contents that can be selected and then some process is applied, to facilitate the downloading of malicious data is prohibited.

6 PROPOSED SOLUTION

The working of proposed algorithm is described with help flow chart as shown in Figure 3. The main goal of the research is to classify the URLs as malicious or non-malicious on the basis of segmentation and transformations. The transformations used here are Term Frequency (TF) & Inverse Document Frequency (IDF). These transformations are used to convert string values to the numerical values. At last, the evaluator evaluates the URL and the results are produced.

7 SIMULATION AND RESULT

Receiver Operating Characteristics (ROC) graph is used to organize the classifiers and to see their performance. Use of ROC graph is increasing in the fields of medical and data mining. Using ROC graphs, URL classification become simpler as the results generated by it, are very simple and clean.

In the work, the classifier was tested and trained on analogous number of benign URLs as malicious URLs. The benign-to-malicious URLs ratio does not vary considerably in testing and learning. Such type of conditions arises when the classifier is installed in a dissimilar manner than it was initially trained. For example, presume if a traditional spam filter is used to exclude URLs from doubtful emails with the product advertisements. Such classifier insures the detection of phishing sites that do not exist in list of reputed websites as these sites simply sell the spam-advertised products. Thus, the ratio of benign-to-malicious URLs might be numerous orders of extent lower in testing than the training, for the classifiers organized in this way.

In Figure 4 features of all URLs after Term Frequency (TF) and Inverse Document Frequency (IDF) transformation are shown.

In Figure 5 features of all URLs after Term Frequency (TF) and Inverse Document Frequency (IDF) transformation are shown.

In Figure 6 ROC graph visualize the False Positive Rate (FPR) of the Random Forest classifier.

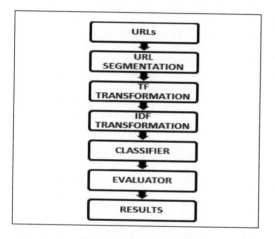

Figure 3.　Process of URL classification.

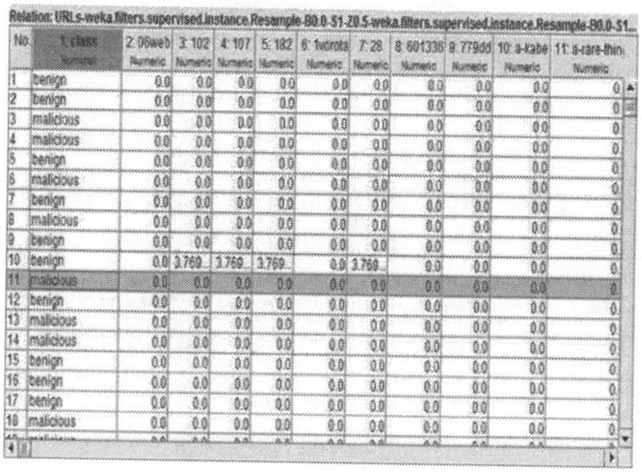

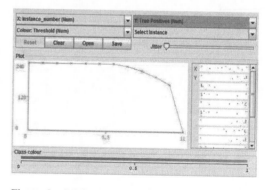

Figure 4.　URL features after TF-IDF transformations.

Figure 5.　ROC analysis of random forest classifier describing TPR.

1027

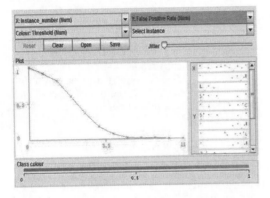

Figure 6. ROC analysis of random forest classifier describing FPR.

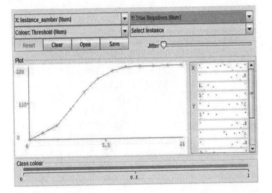

Figure 7. ROC analysis of random forest classifier describing TNR.

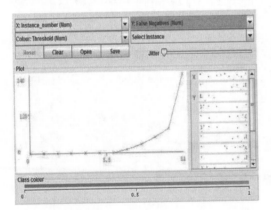

Figure 8. ROC analysis of random forest classifier describing FNR.

In Figure 7 ROC graph visualize the True Negative Rate (TNR) of the Random Forest classifier.

In Figure 8 ROC graph visualize the False Negative Rate of the RF classifier.

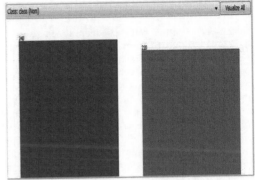

Figure 9. Bar-graph visualization of benign (Blue) and malign (Red) URL.

In Figure 9 the results produced by the evaluator of the classifier are shown here. The bar graph shows the detected ratio of values of malicious and non-malicious URLs.

8 CONCLUSION

Random Forest methodology is used for Malicious URL classification that detects web pages' malicious degree. This supervised technique was applied to a large number of web pages and trained manually which is split into two classes. In nutshell, Random forest based technique shows encouraging performance through accuracies above the range of 91% for the supervised techniques. As a number of web pages amplified the accuracy rates for each type of machine learning model can be improved. An automated URL classification approach is presented to discover host-based and lexical methods for malicious URLs. The string values of the URLs are also taken. This enhances the performance of classifying techniques as both numerical and string values are taken by the random forest method which is purely a feature based classification. Other features are also considered to get the better results.

In Future the research will incorporate bit.ly like URL shortening in dataset for improving the overall classification complexity. Improving performance of classifier can also be another area of research. The work will continue to improvise research for discovering unsafe malicious traffic and clients clustering infected via malware such as malicious worms, bonnets, virus and continuing to incredulous weaknesses. Users can also experiment with a malicious detection systems or primary IDS to appraise the performance of the method.

REFERENCES

Apte, Jitendra & Lima, Marina Roesler. 2007. *Interactive Multimedia Advertising and Electronic Commerce on a Hypertext Network*. U.S. Patent No. 7, 225, 142. 29 May 2007.

Basnet, Ram B. & Sung, Andrew H. *2012. Mining Web to Detect Phishing URLs*. In Proceedings of 11th International Conference on Machine Learning and Applications, vol. 1, pp. 568–573.

Cao, Cheng & Caverlee, James. 2014. *Behavioral Detection of Spam URL Sharing: Posting Patterns versus Click Patterns*. In Proceedings of IEEE/ACM International Conference on Advances in Social Networks Analysis and Mining, pp. 138–141.

Gupta, Neha & Aggarwal, Anupama & Kumaraguru, Ponnurangam. 2014. *bit.ly/malicious: Deep Dive into Short URL based e-Crime Detection*. In Proceedings of IEEE APWG Symposium on Electronic Crime Research, pp. 14–24.

Hajgude, Jayshree & Ragha, Lata. 2012. *Phish Mail Guard: Phishing Mail Detection Technique by using Textual and URL Analysis*. In World Congress on Information and Communication Technologies, pp. 297–302.

Kandasamy, Kamalanathan & Koroth, Preethi. 2014. *An Integrated Approach to Spam Classification on Twitter using URL Analysis, Natural Language Processing and Machine Learning Techniques*. In Proceedings of IEEE Students' Conference on Electrical, Electronics and Computer Science, pp. 1–5.

Khare, Shreya & Bhandari, Akshay & A, Hema Murthy, 2014. *URL Classification using Non Negative Matrix Factorization*. In Proceedings of IEEE 20th National Conference on Communications (NCC), pp. 1–6.

Ma Justin, K Lawrence Saul, Stefan Savage & Geoffrey M. Voelker, 2009. *Beyond Blacklists: Learning to Detect Malicious Web Sites from Suspicious URLs*. in Proceedings of the 15th ACM international conference on Knowledge discovery and data mining.

Mitsuaki, Akiyama & Takeshi, Yagi & Takeo, Hariu. 2013. *Improved Blacklisting: Inspecting the Structural Neighborhood of Malicious URLs*. IT Professional, vol. 15, no. 4, pp. 50–56.

Reddy, Ravula & Ravindar. 2011. *Classification of Malware using Reverse Engineering and Data Mining Techniques*. M.S. Dissertation, University of Akron, CS Department.

Sharma, Priyanka. 2014. *Comparative Analysis of Various Decision Tree Classification Algorithms using WEKA*. In International Journal on Recent Innovation Trends in Computing and Communication, vol. 3, no. 2, pp. 684–690.

Communication and Computing Systems – Prasad et al. (Eds)
© 2017 Taylor & Francis Group, London, ISBN 978-1-138-02952-1

A comparative analysis of cryptographic techniques for data security over cloud

Anshu Kirar & Arun Kumar Yadav
ITM University, Gwalior, Madhya Pradesh, India

R.S. Rao
Indira Gandhi National Tribal University, Amarkantak, Madhya Pradesh, India

ABSTRACT: Today the cloud computing is rising and resourceful technology. The necessity of IT industry is to repository terabyte data generated in every day. For warehouse IT requires many hardware, software and system frameworks. Cloud computing crack this problem in valuable charge way. It also altered the vision completely not only in IT Industries but some other sectors like healthcare, education sector, etc. It has potential to offer servers for wide diversity of resources from research to E commerce. It is very fast because of their features like source potential, network communications, storage space ability, valuable cost, quick access of information. On the other side all the data is virtual and cloud is as open facilities and public network are using for their application and services, which in turn has question on security disputes like verification data loss. This paper review on various ciphering algorithm to improve security files in cloud computing.

1 INTRODUCTION

Cloud computing is solitary of the healthy and leading technical knowledge in present sides. It offers services in tiniest charge way. Than classical tactic user have all services of cloud and it easily share their data. Cloud computing provide services like as Google drive, drop box. Cloud provides their superlative factor but completely all things are available on online network consequently that there are probabilities of data leakage.

We aware here lots of safety worries similar to system and information safety are the extensive sphere. There are many concerns in data security like integrity, Confidentiality, data location and availability authentication but most important and dubious subject is data confirmation. Verification contrivance support to create proof of integrity. The validation system technique certifies that the base of electronic information or record is properly recognized. Cloud service provider always provides unique id and password for file validation and the validation method may be applied in both domain and workgroup. When password is hacked by the intruders then validation is missed and attacks on obtainable data that can be altered, erased. Therefore we attempt to prefer new prototype that can figure out issues of validation so that approved user can access all services. We analysis of numerous encryption algorithms. The Cloud resources

protected the dealings and storage. Security areas of data include three facts: Availability Confidentiality, and Integrity. Confidentiality of records is consummate by cryptography. Cryptography is deliberated arrangement of 3 categories of algorithms. Symmetric key, Asymmetric key both are dashing algorithm. Hashing algorithms establish reliability of data.

Data cryptography is the largely clambering contented of the files like as text, image, audio and video clipping and consequently onward to create the data illegible, obscure or else worthless throughout communication or storage is termed Ciphering. The foremost objective of cryptography is to revenue precaution of data protected from attackers. The inverse process of receiving back the unique data from ciphered data is Decryption, which restores the unique data. In cloud storage we encrypt the data using symmetric and asymmetric key algorithm. Cloud storage space holds huge set of databases for a large databank and symmetric-key algorithm is more relevant as compare to asymmetric-key algorithms.

2 RELATED WORK

Agrawal, Vikas et al (2014) proposed the approach of Ciphering and Decryption are done by using Symmetric key and public key cryptography for secured message. Author studied that in what way the process of Encryption and Decryption be transmit available in terms of Symmetric key and public key cryptography by using AES and DES algorithms and altered RSA algorithm. Color black & white image of any size saved in Tagged Image file Format (TIF) can be encrypted & decrypted using blowfish algorithm. MREA is use to cipher files and transfer cipher files to other end where it is decrypted. Most important feature of this process is that it satisfies the properties of Mix-up and distribution and also has an ideal deduction of encryption key makes decryption unfeasible.

Bakshi, A. et al (2010) introduced a offers an extensive range of profit to small and average enterprises. But security, confidentiality and faith are the major concerns prevent the mass maintained of cloud. A cloud atmosphere provides various services and hosts. Several assets can be secured only by allow genuine clients to access the resources. Therefore strong client verification mechanism narrow illegitimate accesses are the crucial necessity for securing cloud. The surveys on authentication attack in cloud and the equivalent all eviction actions. Elminaam, Abdul. S. D. et al (2009) it deliberate six mutual encryption algorithms AES, DES, 3DES, RC2, Blowfish, and RC6. A valuation has been led for those ciphering algorithms at diverse sets for every algorithm such as different dimensions of data blocks, diverse data kinds, battery-operated power consumption, diverse key length and encryption/decryption speed. Show evaluation of selected symmetric algorithms. Numerous points can be resolved from the simulation results. In First point there is no primary difference when the results are act either in hexadecimal base encrypting or in base 64 encrypting. Second in the situation of shifting packet size, it was established that Blowfish has superior presentation as compare other general encryption algorithms used, followed by RC6. Third in the case of changing data type such as image in its place of text, it was established that RC2, RC6 and also Blowfish has drawback over existing algorithms in case of time consumption. We treasure that 3DES at rest has low performance as compare to DES. Last one when change the key size than higher key size trace to clear change in the battery and time consumption.

Gunasekaran, Dr. S. Professor (2013) introduced a various cipher algorithms to growth the data safety in cloud computing. Cloud Computing gained great approachability from the industry but still here many problems that are obstructing the progress of the cloud. Data security is one of the major concerns which are more difficulty in the implementation of cloud computing. These security issues are avoided by various encryption algorithms. This paper review on RSA, AES encryption algorithms and using the cryptographic techniques to enhancing a data safety.

Guna sundari, T. et al (2014) review on cryptography is the solitary of the core set of computer security it converts info from its standard form then standard form into an illegible form. Author delivers a reasonable comparison between 4 mainly public symmetric key cryptography algorithms: RC2, RC4, RC5, and RC6. The RC6 use a variable number of bits range from 8 to 1024 bits and ciphers the data 16 times. Then it is unfeasible for a hacker to decrypt it.

Gupta, Anjula et al (2014) defined and studied several symmetric algorithms are DES, 3DES, Blowfish, AES and IDEA and asymmetric algorithms are RSA. It analyzed ability to protected data, key size, block size, features. The existing ciphering methods are deliberate and analyzed to stimulate the presentation of the encryption method also to make sure the security actions. To summation of, every one techniques are distinctive in its private way. Throughput rate of BLOWFISH is greater than above further symmetric algorithms. The trial results of existing algorithm in many papers showed that BLOWFISH has superior performance and efficiency than all other block cipher. After that technique extensively uses to protect our information is RSA. RSA algorithm mainly used for data security. It is the more secure & widely used by researchers. And also RSA is used with many techniques like RSA & DES, RSA & AES, by combining cryptography algorithms.

Gupta, Gunjan et al (2012) present a synopsis of diverse block, stream ciphers and algorithms, which are used in cryptography for network security purpose. The cipher's algorithms can generate its private cipher text by making difference into existing cipher text algorithms. Performance evaluation of some ciphers can be done by using the cipher's algorithms.

Hemalatha, N. et al (2014) discussed about the require of validation in cloud. It is description approach of verification by Kerberos and threshold cryptography so that encryption method is more vigorous. Lot of work completed already on security concerns and contests but still there are loop holes. This literature work is exclusive approach since propose scheme minimizes the problem of exchange of key that are usually occurs in symmetric and asymmetric key cryptography.

Kaur, Manpreet et al (2016) reviewed the relative analysis of 8diverse cryptographic algorithms. The intense attention is given to the security jobs correlated with cloud's service models, deployment models, and disputes related to networking capabilities are discussed and studied. Analyzes the problem of security related with cloud. Ciphering is the foremost option for securing the data and author highlights proportional examination of symmetric as well as asymmetric ciphering algorithms for provided that safety in cloud computing systems.

Nithya, B. et al (2016) a review that proportional revises of algorithm is to be study of algorithms taken by several authors. The each algorithm has its personal method, plus point and loss of credit. This paper little narrative about these algorithms DES, TDES, AES, RSA, MD5, RC4 SHA, and ECC. The relative results said that the algorithms AES, RC4, DES, TDES are mainly fast in ciphering time, speed, memory when compared to others. These show symmetric algorithms have best than asymmetric algorithms.

Padmapriya, Dr. A. et al (2013) proposed security of cloud computing mechanisms and presented the relative study of several algorithms. In upcoming we are going to recommend a new plan to resolve security concerns for cloud providers and cloud clients. Author analyses the value of security to cloud. Analyses 3 algorithms are DES, RSA, and Holomorphic encryption for data safety in cloud. They are correlated based on 4 characters; key used scalability, security applied to, and authentication type. In future to propose a backup plan to resolve safety concerns in cloud providers and cloud customers.

Peasant, A. et al mainly focus and discussed on the three basic cloud service prototypes, named as SAAS, PAAS and IAAS, also their major threats and different available assistances. It also analyzed a couple of threats all of such models.

Seth, Mehrotra Shashi et al (2011) deliberate the selected encryption algorithm AES, DES and RSA algorithms are used for performance evaluation. DE Sutilize least encryption time and AES algorithm has lowest amount memory usage while ciphering time. RSA use longest ciphering time and memory usage is also very high but production byte is least. Future scope will include experiments on image,audio data and focus will be to improve ciphering time and less memory usage.

Singh, Preet Simar et al (2106) studied association between the existing algorithms in the data. Authors provide an act contrast among 4 common encryption algorithms: DES, 3DES, Blowfish and AES. The compare has been led by consecutively various encryption sets to procedure diverse sizes of data blocks to evaluate the algorithm's encryption/decryption speed. Major concern is the act of these algorithms beneath different situations, the presented comparison takes into concern the performance of the algorithm when diverse data lots are used. Model use C# language. And the result Blowfish has well performance as compare to generally used encryption algorithms. AES showed

reduced performance as compared to further algorithms, and then it requires more processing power.

Sumitra et al (2013) author use public key and private key. With the facilitate of these keys we cipher and decrypting the records for make secure. Encrypted data is called cipher text and decrypted data s called plaintext. Cryptography is of two types: Symmetric & asymmetric. And two symmetric key security algorithms AES and DES algorithm are related. Symmetric key algorithms have similar key for encryption and decryption. Both algorithms consume different times at different machines. AES is safer as compare to DES.

Yousif, Elfatih Yousif et al (2015) studied various symmetric algorithms are DES, Triple DES, AES and asymmetric algorithms like RSA. Analyzed the act of different encryption method. It can be established that the encryption/decryption speed for DES algorithms is faster than RSA. AES is more protected compared to DES. The output rates for BLOWFISH are greater than all above symmetric algorithms. RC6 uses unpredictable bits range from 8 to 1024 bits and cipher the data 16 times, then it unbreakable for an intruder to decrypt it.

3 COMPARATIVE ANALYSIS

Ciphering is a familiar technology for protective precise data. With the help grouping of public and private Key ciphering to conceal the thoughtful information of users and cipher text recovery.

The each algorithm has its private process advantage and disadvantage. This paper has given short description about the following algorithm.

3.1 BLOWISH

In public domain blowfish is solitary best encryption algorithms. It is deliberate by Bruce Schneider in 1993. Its block size 64 bit and Variable size 32–448 bits. It has 16 rounds or less than 16 rounds. Blowfish is secure cipher and also use for encryption free patents and copyrights. It decent for text ciphering after compare to AES.

3.2 Rc6

RC6 is an imitative of RC5. It is deliberate by Matt Robshaw, Ron Rivest Ray Sidney. RC6used to assemble the necessities of Advance Encryption Standard challenge. It is composed by RSA Security. RC6 attempt superior presentation in standings of security and compatibility. It is based on Feistel Structure and 128 bit plain text use with 20 rounds and a variable Key Size is 128–256 bit.

RC6 mechanism on the standard of RC that can stand a expand variety of key sizes, word-lengths and number of rounds. This algorithm uses unpredictable bits range from 8 to 1024 bits and cipher the data 16 times, so it make hard for an intruder to decrypt it.

3.3   AES

AES also recognized as the Rijndael's algorithm. It is a symmetric block cipher. AES was aware that DES not safe because of evolution in computer processing influence. Cipher data blocks use 128 bits. By default 256 variable key lengths is use. AES may be lying on several platforms like small devices. Ciphering of AES is agile and elastic. AES tested for various security functions. Its encryption consumes minimum encryption time.

3.4   DES

DES was the significant encoding normal to be suggested by NIST. DES is admired set use in symmetric key encryption. Variable key size is 56 and the block length 64. It consist 64 binary digit (0 & 1) of 56 arbitrarily produce and other 8 used for error

finding. DES data processing speed is fast. DES is measured as less security, and this algorithm is not used a large amount since this algorithm has been damaged very easily. The encrypting/decrypting speed of DES algorithms is prior than RSA. This algorithm used to consumes lowest encryption time.

3.5   3DES

3DES set is the subsequent level of DES. It was deliberate to break the attacks that DES meets. To raise the security, it processes DES in three times. 3DES process use 48 rounds and its key size is 168 bits. With use this longer key, it applies to each block and cipher the original text. It runs

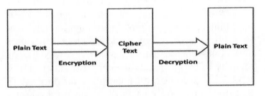

Figure 1.   Encryption to decryption data.

Table 1.   Compression of various encryption algorithms for security in cloud computing.

| Parameters | Algorithms | | | | |
	AES	DES	3DES	RC6	BLOWFISH
Developed	Joan Daeman Vinat Rijmen in 1998	IBM in 1975	IBM in 1978	Ron Rivest, Matt Robshaw in 1998	Bruce Schnier in 1998
Block size	128	64	64	128	64
Key length	128,192,256	56	112	128,192,256	32–448
Security	Considered Secure	Proven Adequate	Considered Secure	Vulnerable	Considered Secure
Data processing speed	Fast	Fast	Moderate	Fast	Very Fast
Algorithm structure	Substitution Permutation Network	Balanced Fiestel Network	Fiestel Network	Fiestel Network	Fiestel Network
Ciphering	High	Medium	Low	High	Very High
Deciphering output	High	Medium	Low	High	Very High
Memory convention	Medium	High	Very High- High	Low	Very Low
Power consumption	Low	Low	High		Very High
Attack	Brute Force	Brute Force	Brute Force, Chosen Plaintext, Known Plaintext		Dictionary Attack
Confidentiality	High	Low	High	High	Very High
Observations	It is Central information processing standard.	It is popular standard use in symmetric key encryption.	It runs 3 times slower than single DES but is considerable more secure.	Very similar to RC5.RC6 use extra multiplication operation	The output rates for BLOWFISH are superior to all symmetric algorithms.

3 times quiet than single DES but is considerable more secure. Triple DES has virtually 1/3 output of DES, in further terms it necessities 3 times slower than DES to development the identical amount of data.

There are three keying options

In keying option 1 is strongest and the 3 keys: K1, K2 and K3 are autonomous.

In keying option 2, the two keys K1 and K2 are self-governing. In keying option 3 K1, K2 and K3 are equal.

Encryption is a technique of adapting "public" to "secret" it secures data from intruders. Here this technique has alternative quota wherever cipher text desires to be decrypted on the last termination to exist implicit.

4 RESULT COMPARISON

Imitation outcomes intended for this concern point are shown Fig. 2 and Table 2 in encryption stage. Overall outcomes show the superiority of Blowfish Algorithm above existing algorithms in standings of the processing time [2]. And also RC6 needs less time as compare to above further algorithms except Blowfish. AES has a lead above further 3DES, DES Utilization and amount produced. 3DES has squat act in standings of power utilization and output as compare with DES for the reason that 3DES always takes more time then DES because of its triple phase Encryption uniqueness comparative analysis for Encryption.

Simulation results intended for this concern point are shown Fig. 3 and Table 3 decryption stage. Blowfish is the superior than further algorithms in output rate and power consumption. RC6 requires less processing times as compare to other algorithms expect BLOWFISH. At last, Triple DES silent requires more time than DES.

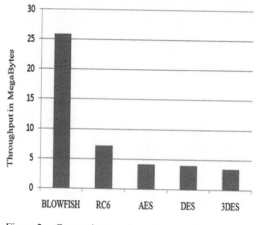

Figure 2. Comparison analysis of encryption.

Table 2. Calculations and proportional study of different algorithms for encryption input data in Kb and CPU execution time for encryption in millisecond.

Input size					
In Kbytes	Blowfish	DES	3DES	AES	RC6
49	36	29	54	56	41
59	36	33	48	38	24
100	37	49	81	90	60
247	45	47	111	112	77
321	45	82	167	164	109
694	46	144	226	210	123
899	64	240	283	258	162
963	66	250	283	208	125
5345.28	122	1296	1466	1237	695
7310.336	107	1695	1786	1366	756
Avg. Time	60.3	389	452	374	217
Throughput (Megabytes/Sec)	25.892	4.01	3.45	4.174	7.19

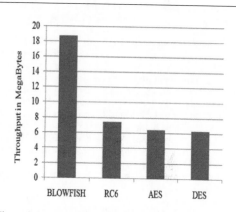

Figure 3. Comparison analysis of decryption.

Table 3. Calculations and proportional study of different algorithms for decryption input data in Kb and CPU execution time for decryption in millisecond.

Input size					
In Kbytes	Blowfish	DES	3DES	AES	RC6
49	38	50	53	63	35
59	26	42	51	58	28
100	52	57	57	60	58
247	66	72	77	76	66
321	92	74	87	149	100
694	89	120	147	142	119
899	102	152	171	171	100
963	80	157	177	164	125
5345.28	149	783	835	655	684
7310.336	140	953	1101	882	745
Avg. Time	83.2	246	275.6	242	210
Throughput (Megabytes/Sec)	18.72	6.347	5.665	6.456	7.43

5 CONCLUSIONS

Cloud computing allows client to accumulation their records in cloud storage space when client required. So cloud is on internet, several security issues are encountered like confidentiality, privacy, authentication. We have deliberated about the necessity of validation. Various cryptographic algorithms used for encryption and decryption of cloud data to get better the security. Security of cloud depends on reliable computing and cryptography. This paper presents an analysis on BLOWFISH, RC6, DES, 3DES, AES ciphering algorithms and using the cryptographic techniques to enhancing a data safety.

REFERENCES

Agrawal, Vikas, Agrawal, Shruti & Deshmukh, Rajesh 2014. Analysis and Review of Encryption and Decryption for Secure Communication, IJSER, Volume 2 Issue 2.

Bakshi, A. & Yogesh, B. 2010. Securing Cloud from DDOS Attacks using Intrusion Detection System in VM, in Proc. IEEE Second Int. Conference on Communication Software and Networks., pp. 260–264.

Elminaam, Abdul, D.S., Abdul Kader, M.H. & Hadhoud, M.M. 2009. Performance Evaluation of Symmetric Encryption Algorithms Communications of the IBIMA Volume 8, ISSN: 1943-7765.

Gunasekaran, Dr. S. & Lavanya, P.M. 2015. A Review on enhancing data security in cloud computing using RSAAND AES algorithms International Journal of Advances in Engineering Research http://www.ijaer.com (IJAER), Vol. No. 9, Issue No. IV, April.

Gunasundari, T. & Elangovan, Dr. K. 2014. A Comparative Survey on Symmetric Key Encryption Algorithms, IJCSMA, Vol. 2, Issue. 2, February.

Gupta, Anjula & Walia, Kaur Navpreet 2014. Cryptography Algorithms: A Review IJEDR1402064 International Journal of Engineering Development and Research (www.ijedr.org) 1668© 2014 IJEDR | Volume 2, Issue 2 | ISSN: 2321-9939.

Gupta, Gunjan & Chawla, Rama 2012. Review on Encryption Ciphers of Cryptography in Network Security", International Journal of Advanced Research in Computer Science and Software Engineering, Volume 2, Issue 7, July.

Hemalatha, N., Jenis, A., Cecil, A. & Arockiam, L. 2014. Encryption Technique and Security Issues in Cloud Computing. International Journal of Computer Application volume 96-No-16 June.

Kaur, Manpreet & Kaur, Kiranbir 2016. A Comparative Review on Data Security Challenges in Cloud Computing. International Research Journal of Engineering and Technology (IRJET), Volume: 03 Issue: 01, January.

Nithya, B. & Sripriya, Dr. P. 2016. A review of cryptographic algorithm in network security International Journal of Engineering and Technology (IJET) Vol 8 No 1 Feburary–March.

Padmapriya, Dr. A. & Subhasri, P. 2013. Cloud Computing: Security Challenges & Encryption Practices. International Journal of Advanced Research in Computer Science and Software Engineering, ISSN: 2277 128X, Volume 3, Issue 3, March, pp. 257. 7.

Prasanth, A., Bajpei, M., Shrivastava, V. & Mishra, G.R. Cloud Computing: A Survey of Associated Services Cloud Computing: A Survey of Associated Services, vol. 13, pp. 1–15.

Seth, MehrotraShashi & Mishra, Rajan 2011. Comparative Analysis of Encryption Algorithms for Data Communication, IJCST Vol. 2, Issue 2, June pp. 292–294.

Singh, PreetSimar & Maini, Raman 2011. Comparison of data encryption algorithm International Journal of Computer Science and Communication Vol. 2, No. 1, January–June 2011, pp. 125–127.

Sumitra 2013. Comparative Analysis of AES and DES security International Journal of Scientific and Research Publications, Volume 3, Issue 1, January.

Communication and Computing Systems – Prasad et al. (Eds)
© 2017 Taylor & Francis Group, London, ISBN 978-1-138-02952-1

Comparative analysis in between the k-means algorithm, k-means using with Gaussian mixture model and fuzzy c means algorithm

Pooja Shrivastava, Kavita & Sarvesh Singh
Jayoti Vidyapeeth Women's University, Jaipur, India

Manoj Shukla
Amity University, Noida, Uttar Pradesh, India

ABSTRACT: K-Means clustering algorithm is partitioning method. Gaussian mixture model is multivariate distribution which is consists of a mixture of one or more multivariate Gaussian distribution component. We are used forest fire data set from UCI machine learning. In this paper we present the comparative analysis of k-means and k-means algorithm using a Gaussian mixture model. Secondly, we presented comparative analysis of k-means algorithm and fuzzy c means algorithm. Firstly, we are taking data points through k-means algorithm. Secondly, we are taking data points through Gaussian mixture model and lastly, we are taking data points through fuzzy c means. In this paper we will improve the accuracy and no of iteration of both algorithms and evaluated the performance of algorithms. The main objective of this research we will found which method is better.

Keywords: data mining, Gaussian mixture model, k-means algorithm, forest fire data set, UCI machine learning, clustering

1 INTRODUCTION

Clustering technique is dividing data elements into different groups. These groups are called as clusters. The elements within a group possess high similarity while they differ from the elements in a different group (Dibya, Jyoti, Bora, 2014 & Dr. Anil kumar Gupta, 2014). K-Means clustering algorithm is partitioning method. In this research firstly, we presented the comparative analysis of k-means algorithm and k-means algorithm with Gaussian mixture model to find No of iterations, Accuracy according to points of misclassified, Accuracy according to sum of distance. Secondly, we presented comparative analysis of k-means algorithm and fuzzy c means algorithm. In this paper we present firstly, we are taking data points through k-means algorithm and allows k-means algorithm. Secondly, we are taking data points through Gaussian mixture model and allows k-means algorithm also lastly, we are taking data point through fuzzy c means algorithm and allows fuzzy c means algorithm. Gaussian mixture model (J.J. Verbeek, 2003 & N. Vlassis, 2003 & B. Krose, 2003) is multivariate distribution. It is consists of a mixture of one or more multivariate Gaussian distribution component. The performance evaluated on the forest fire dataset. We are taking forest fire data set from UCI machine learning. The main objective of this research we will found which method is better.

The rest of the paper is organized as follows. The introduction and review of literature present in section 1. Overview the k-means algorithm present in the section 2. In section 3 Gaussian mixture model, k-means with Gaussian mixture model and fuzzy c means algorithm. The datasets used and interprets the implementation, experimental results and finally conclusion in section 4. The implementation work is done using MATLAB software.

1.1 *Review of literature*

(J.J Verbeek, 2003 & N. Vlassis, 2003 & B. Krose, 2003) presented the Efficient Greedy Learning of Gaussian Mixture Models. Authors compared the result proposed algorithm to other methods on density estimation and texture segmentation are provided in this research. (R. DerSimonian, 1986) worked on Maximum likelihood estimation of a mixing distribution. (N. Vlassis, 2002 & A. Likas, 2002) presented A greedy EM algorithm for Gaussian mixture learning. (W. Fernandez de la Vega, 2003 & Marek Karpinski, 2003 & Claire Kenyon, 2003 & Yuval Rabani, 2003), (Sariel Har-Peled, 2004 & Soham Mazumdar, 2004), (Amit

Kumar, 2004 & Yogish Sabharwal, 2004 & Sandeep Sen, 2004), (Jiri Matousek, 2000) was presented many types of papers competitive algorithms for the k-means problem that are essentially unrelated to Lloyd's method. (David Arthur, 2007 & Sergei Vassilvitskii, 2007) presented k-means++ method is a widely used clustering technique. An experiment showed in this research argument improves both the speed and the accuracy of k-means, often quite dramatically.

2 OVERVIEW THE K-MEANS ALGORITHM

In this section we formally define the k-means algorithm. K-means (Soumi Ghosh, 2013 & Sanjay, kumar, Dubey, 2013) uses the squared Euclidean distance. K-means provides better results to search for lower, local minima. We are given an integer n and a set of k data points. We wish to choose n centers C so as to minimize the local minima.

$$S = \sum_{i=1} min_{x \in c} \|x - c\|^2 \tag{1}$$

K-means assign each observation to the cluster. Which means within-cluster sum of squares (WCSS)? Sum of squares is called the Eulidean distance.

2.1 Steps of basic k-means algorithm

1. Input the value of k.
2. Seletct an initial n centers $C = \{c_1, c_2, \ldots \ldots c_k\}$.
3. Repeat.
4. Re (assign) each objects k to the cluster to which the object is the most similar. It is based on the mean value of the object in the cluster k.
5. Calculate the average of the observation in each cluster to obtain k new centriod location.
6. Until no change.

3 GAUSSIAN MIXTURE MODEL

Gaussian mixture model (J.J. Verbeek, 2003 & N. Vlassis, 2003 & B. Krose, 2003) is multivariate distribution. It is a mixture of one or more multivariate Gaussian distribution component. Gaussian component is defined by a vector of mixing proportions in Gaussian mixture model for each multivariate distribution.

$$d(\varnothing) = \sum_{k=1}^{i} C_k N(u_i, \Sigma k) \tag{2}$$

Where k vector component is characterized by normal distributions with weights C_k, means N and

covariance matrices $\sum k$. Sigma specifies the covariance of each component. Gaussian distribution will have a mean and variance for components.

$$\Delta(x) = \sum_{n=1}^{N} \Delta(x \mid sn, \varnothing_n) \nabla_n \tag{3}$$

Own parameter (u_n, Σ_n)

3.1 K-means algorithm with Gaussian mixture model steps

1-input the data set X from a Gaussian mixture model.

D, Q
$C_i = 1 \ldots \ldots D, C$
$A_i = 1 \ldots \ldots, Q, \varnothing_i = 1 \ldots \ldots N$
$R_i = 1 \ldots \ldots N = $ mean of component
$F_i^2 = 1 \ldots N = $ variance of component
$\Delta_i = 1 \ldots \ldots D \sim$ Categorical (C)
$V_i = 1 \ldots \ldots D \sim \nabla(R_i, F_i^2)$

1. Input the value of k.
2. Select an initial n centers $C = \{c_1, c_2, \ldots \ldots c_k\}$.
3. Repeat.
4. Re (assign) each objects k to the cluster to which the object is the most similar. It is based on the mean value of the object in the cluster k.
5. Calculate the average of the observation in each cluster to obtain k new centriod location.
6. Until no change.

3.2 Overview of fuzzy c means algorithm

Fuzzy C means algorithm is a popular method of clustering. FCM allows one piece of data to belong to two or more clusters. It's method improvement on earlier clusters method. It's method produced by Jim Bezdek in 1981. It's based on following objective function.

$$S_m = \sum_{i=1}^{M} \sum_{j=1}^{N} w_{ij}^n \|x_i - c_j\|^2 \tag{4}$$

Where n is any real number greater than 1, x_i is the ith of d-dimensional measured data and c_j is the d-dimension center of the cluster. S_{ij} Is the degree of membership of x_i in the cluster j.

3.3 Data set

We evaluate the performance of the algorithms on forest fire dataset. We are taking the dataset from UCI Machine Learning Repository. The no of attribute is 13, no of instances is 517, Associated Task is regression, Area is Physical, Attribute Characteristics is real in this dataset. Data set Characteristics is Multivariate.

4 EXPERIMENTAL RESULTS AND DISCUSSION

We are these 3 algorithms using with MATLAB software. First is k-means algorithm second is k-means algorithm with Gaussian mixture model and third is fuzzy c means algorithm. Firstly, we are taking data points through k-means algorithm and allows k-means algorithm. Secondly, we are taking data points through Gaussian mixture model and allows k-means algorithm also lastly, we are taking data point through fuzzy c means algorithm and allows fuzzy c means algorithm. The performance evaluated on the forest fire dataset. We are taking data set from UCI machine learning. Firstly we are compared k-means algorithm and k-means algorithm with Gaussian mixture model.

Forest fire dataset has 13 attributes, 517 instances so we chose value of k = 10,20,30,40 k is the no clusters and X = 517 points.

Experiments are evaluated on random datasets of 517 data points. Results also show that Original k-means algorithm shows different performance measures for same dataset and fixed value of k, k-means algorithm with Gaussian mixture model also shows different performance measures for same dataset and fixed value of k and fuzzy c means algorithm also show different performance measures for same dataset and fixed value k. Table 1 shows the no of iteration, no of points misclassified, sum of distance of k-means algorithm for different clusters, when we used data points through k-means.

Table 2 shows the result of k-means algorithm for different clusters, k-means algorithm determined the accuracy for dataset. First, accuracy is according to No of points misclassified and second accuracy is according to sum of distance for different clusters, when we used data points through k-means.

Table 3 shows the no of iteration, no of points misclassified, sum of distance of k-means algorithm with Gaussian mixture model for different clusters, when we used data points through Gaussian mixture model.

Table 4 shows the result of k-means algorithm with Gaussian mixture model for different clusters, k-means algorithm with Gaussian mixture model determined the accuracy for dataset. First, accuracy is according to No of points misclassified and second accuracy is according to sum of distance for different clusters, when we used data points through k-means with Gaussian mixture model.

In Table 1 we can see that the k-means algorithm increases the no of iteration and reassigns points between clusters and no of points misclassified decreased. Table 3 shows that the k-means algorithm with Gaussian mixture model decreases the no of iterations and no of points misclassified increased.

Table 5 shows the comparison result of k-means and k-means with Gaussian mixture model. In

Table 1. Determine the parameters of k-means algorithm (Comparison between No of clusters k = 10,20,30,40).

No of clusters	No of iterations	No of points misclassified	Sum of distance
10	31	1	3297.12
20	32	1	6678.53
30	28	4	10310.5
40	25	1	13635.8

Table 2. Result of k-means algorithm (Comparison between No of clusters k = 10,20,30,40).

No of clusters	No of points misclassified	Accuracy according to No of points misclassified	Sum of distance	Accuracy according to sum of distance
10	1	99.80%	3297.12	92.63%
20	1	99.80%	6678.53	93.30%
30	4	99.22%	10310.5	94.24%
40	1	99.80%	13635.8	95.89%

Table 3. Determine the parameters of k-means algorithm with Gaussian mixture model (Comparison between No of clusters k = 10,20,30,40).

No of clusters	No of iterations	No of points misclassified	Sum of distance
10	21	3	59833.4
20	19	4	58087.9
30	12	11	55370.2
40	16	1	54628.1

Table 4. Result of k-means algorithm with Gaussian mixture model (Comparison between No of clusters k = 10,20,30,40).

No of clusters	No of points misclassified	Accuracy according to no points of misclassified	Sum of distance	Accuracy according to sum of distance
10	3	99.41%	59833.4	93.47%
20	4	99.41%	58087.9	95.02%
30	11	97.87%	55370.2	93.92%
40	1	99.80%	54628.1	94.18%

Table 5. Comparison result of k-means and k-means with Gaussian mixture model (comparison between No of clusters k = 10,20,30,40).

No of clusters	Accuracy according to No of points misclassified (k-means algorithm)	Accuracy according to sum of distance (k-means algorithm)	Accuracy according to no points of misclassified (k-means algorithm with Gaussian mixture model)	Accuracy according to sum of distance (k-means algorithm with Gaussian mixture model)
10	99.80%	92.63%	99.41%	93.47%
20	99.80%	93.30%	99.41%	95.02%
30	99.22%	94.24%	97.87%	93.92%
40	99.80%	95.89%	99.80%	94.18%

Table 5 we can see that k-means gives highest accuracy according to No of points misclassified, when we used data points through k-means algorithm. When we are taking clusters 10 and 20 then k-means algorithm gives low accuracy according to sum of distance comparison to k-means with Gaussian mixture model but when we used clusters 30 and 40 then k-means provided highest accuracy comparison to k-means with Gaussian mixture model. In Table 1 we can see that k-means algorithm increases the no of iterations and in Table 3 k-means with Gaussian mixture model decreases the no of iteration. So, we can say that when we taken data points through k-means algorithm then we have found better solution comparison to Gaussian mixture model. The performance results of k-means algorithm able to improve accuracy. Now we compared k-means and fuzzy c means algorithm. Table 6 shows the result of k-means and fuzzy c means algorithm. In Table 6 we can see that k-means gives highest accuracy according to sum of distance and increases the no iterations, when we used data points through k-means algorithm. Fuzzy c means algorithm gives low accuracy according to sum of distance and decreases the no of iterations, when we used data points through fuzzy c means. So, we can see that in both case k-means algorithm able to improve accuracy. Table 7 shows that the k-means gives better performance. So, hence prove that when we taken data points through k-means algorithm then we have found better solution comparison to Gaussian mixture model and fuzzy c means algorithm.

Figure 1 shows the No of iterations of k-means, k-means with Gaussian mixture model and fuzzy c means algorithm for different clusters. When, we used data points through k-means then k-means algorithm able to increases the no of iterations for different clusters. Figure 2 shows the accuracy according to No of points misclassified of k-means and k-means with Gaussian mixture model for different clusters. So, we can see that k-means able to improve the accuracy comparison to other method. Figure 3 shows the accuracy

Table 6. Comparison result of k-means and fuzzy c means algorithms (comparison between No of clusters k = 10,20,30,40).

No of clusters	No of iteration (k-means algorithm)	Accuracy according to sum of distance (k-means algorithm)	No of iteration (Fuzzy c means algorithm)	Accuracy according to sum of distance (Fuzzy c means)
10	31	92.63%	13	88.17%
20	32	93.30%	8	85.36%
30	28	94.24%	7	78.18%
40	25	95.89%	7	79.33%

according to sum of distance for different clusters. In this figure we can see that in clusters 10 and 20 k-means gives low accuracy but in clusters 30 and 40 k-means gives the high accuracy comparison to k-means with Gaussian mixture model but in Figure 1 and Figure 2 we can see that k-means gives the better solution for different clusters comparison to k-means with Gaussian mixture model. So, when we used data points through k-means algorithm then k-means gives better solutions, but when we used data points through Gaussian mixture model then k-means not able to improve better performance. K-means is better than k-means with Gaussian mixture model. Figure 4 shows the accuracy according to sum of distance of k-means and Fuzzy c means algorithm for different clusters. So, we can see when we used data points through k-means then k-means able to improve the accuracy comparison to fuzzy c means algorithm. Finally, Figure 5 shows that the comparison result of k-means, k-means with Gaussian mixture model and fuzzy c means algorithms. In Figure 5 we can see that how k-means algorithm gives better result when we used data points through k-means. Hence, it's proved that when we used data points through k-means and allows k-means algorithm then k-means algorithm gives better solution.

Table 7. Comparison result of k-means, k-means with Gaussian mixture model and fuzzy c means algorithm (comparison between No of clusters k = 10,20,30,40).

No of clusters	Accuracy according to sum of distance (k-means algorithm)	Accuracy according to sum of distance (k-means algorithm with Gaussian mixture model)	Accuracy according to sum of distance (Fuzzy c means)
10	92.63%	93.47%	88.17%
20	93.30%	95.02%	85.36%
30	94.24%	93.92%	78.18%
40	95.89%	94.18%	79.33%

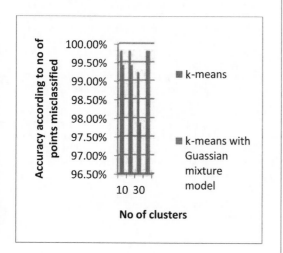

Figure 1. Comparison result of k-means, k-means with Gaussian mixture model and Fuzzy c means algorithms (Comparison between No of clusters k = 10,20,30,40).

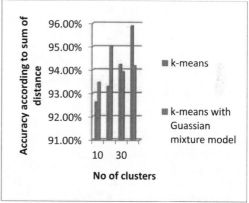

Figure 3. Accuracy according to sum of distance result of k-means and k-means with Gaussian mixture model (Comparison between No of clusters k = 10,20,30,40).

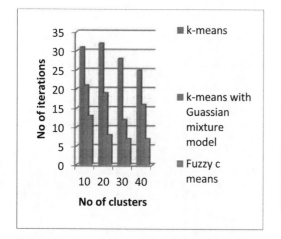

Figure 2. Accuracy according to No of misclassified result of k-means and k-means with Gaussian mixture model (Comparison between No of clusters k = 10,20,30,40).

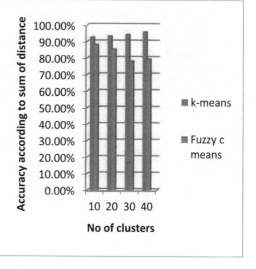

Figure 4. Accuracy according to sum of distance result of k-means and fuzzy c means algorithms (Comparison between No of clusters k = 10,20,30,40).

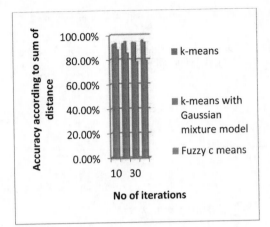

Figure 5. Accuracy according to sum of distance result of k-means, k-means with Gaussian mixture model and Fuzzy c means algorithms (Comparison between No of clusters k = 10,20,30,40).

5 CONCLUSION

In this research firstly, we presented the comparative analysis of k-means algorithm and k-means algorithm with Gaussian mixture model to find No of iterations, Accuracy according to points of misclassified, Accuracy according to sum of distance. Gaussian mixture model improve the quality of the final output. Gaussian mixture model is multivariate distribution. Gaussian mixture model is consists of a mixture of one or more multivariate Gaussian distribution component. K-means gives the better solution for different clusters comparison to k-means with Gaussian mixture model. So, when we used data points through k-means algorithm then k-means gives better solutions, but when we used data points through Gaussian mixture model then k-means not able to improve better performance. K-means is better than k-means with Gaussian mixture model. Secondly we presented comparative analysis of k-means algorithm and fuzzy c means algorithm. So, when we used data points through k-means algorithm then k-means gives better solutions, but when we used data points through fuzzy c means then FCM algorithm is not able to improve better performance. This research is useful to found the quality of k-means algorithm. This algorithm suitable for increases the accuracy in future. K-means will also be modified using the other methods in future.

REFERENCES

Bora, Jyoti, Dibya & Dr. Gupta, Anil, Kumar (2014). A Comparative study Between Fuzzy Clustering Algorithm and Hard Clustering algorithm. International Journal of computer trend and Technology (IJCTT) —volume 10 number 2—Apr, page no 108–113.

DerSimonian, R. (1986). Maximum likelihood estimation of a mixing distribution. J. Roy. Statist. Soc. C, 35:302–309.

Dr. Bhatia, M.P.S. & Khurana, Deepika (2013). Experimental study of Data clustering using k-means and modified algorithms. International Journal of Data Mining & Knowledge Management Process (IJDKP). Vol.3, No.3. Page no 17–30.

Fernandez de la Vega, W., Karpinski, Marek, Kenyon Claire, & Rabani, Yuval (2003). Approximation schemes for clustering problems. In STOC '03: Proceedings of the thirty-fifth annual ACM symposium on Theory of computing, pages 50–58, New York, NY, USA, ACM Press.

Ghosh, Soumi & Dubey, Sanjay, kumar (2013). Comparative Analysis of K-Means and Fuzzy C-Means Algorithms. ((IJACSA) International Journal of Advanced Computer Science and Applications, Vol. 4, No.4. Page no 35–39.

Har-Peled, Sariel & Mazumdar, Soham (2004). On coresets for k-means and k-median clustering. In STOC '04: Proceedings of the thirty-sixth annual ACM symposium on Theory of computing, pages 291–300, New York, NY, USA,. ACM Press.

Kumar, Amit, Sabharwa, Yogish & Sen, Sandeep (2004). A simple linear time (1+_)-approximation algorithm for k-means clustering in any dimensions. In FOCS '04: Proceedings of the 45th Annual IEEE Symposium on Foundations of Computer Science (FOCS'04), pages 454–462, Washington, DC, USA, IEEE Computer Society.

Matousek Jiri (2000). On approximate geometric k-clustering. Discrete & Computational Geometry, 24(1):61–84.

Verbeek, J.J., Vlassis, N. & Krose, B. (2003). Efficient Greedy Learning of Gaussian Mixture Models. Published in Neural Computation, 15(2), pages 469–485.

Vlassis, N. & Likas, A. (2002). A greedy EM algorithm for Gaussian mixture learning. Neural Processing Letters, 15(1):77–87.

Communication and Computing Systems – Prasad et al. (Eds)
© 2017 Taylor & Francis Group, London, ISBN 978-1-138-02952-1

Feasible study of K-mean and K-medoids for analysis of Hadoop mapreduce framework for big data

Subhash Chandra & Deepak Motwani
ITM University Gwalior, Madhya Pradesh, India

ABSTRACT: Big data computing is major research focus for researchers. As information is growing on internet, it is necessity to find the relevant information from huge data storage on multiple servers at different locations. Map reduce frame work is a promising approach to solve the problem. It is used by organizations to process and analyzed the large data in parallel and distributed environment. In recent years, optimizing Hadoop framework gain more importance than others. There are several approaches already proposed for optimizing MapReduce framework such as performance tuning and cluster formation. In this research work we are focusing on clustering approach in Hadoop MapReduce framework. K-means is well studied approach for clustering in Hadoop MapReduce framework. The clustering approaches are more efficiently process and analyze data in less time that will take months for existing approaches. We have proposed improved k-Medoids clustering approach. Our proposed approach performs better than k-means clustering approach.

1 INTRODUCTION

Hadoop includes HDFS as a storage system and MapReduce as a processing engine. Hadoop MapReduce is widely adapted for processing a large datasets on multiple numbers of nodes. In Hadoop, nodes are divided into one master node and multiple slaves. A user program is called job. Job is processed using MapReduce framework. In MapReduce, input data is partitioned by master node into smaller data sets and assigned to slave nodes. Each slave node does their part and passes the result to master node. Big data is emerging area in many engineering and science domain to analyze large data set. Big data is a growing research area for finding useful data by analysing from a large data set. A large data set requires longer process time compare to smaller data set. Big data approaches uses cluster creation for dividing overall large data set to smaller data sets (Jiang et al. 2014).

Hadoop is becoming famous solutions for big data analytics. Hadoop is based on Google's MapReduce (Gu et al. 2014). In Hadoop, data processing is done in distributed and parallel environment. Hadoop performance is depends on many factors such as hardware, algorithm, jobs and input data. Hadoop performance improvement is major focus of research community. Hadoop MapReduce platform is becoming standard for big data analysis. MapReduce frame work

is used by several organizations like advertisement firms, financial companies for processing and data analysis. Internet is becoming a necessity for our life. Every day number of users is increasing. These users are processing large amount of data every day (Shi et al. 2014). This vast amount of data requires large storages and further more processing power for storing and retrieving. Now days, users requests more and more information ranging from education, knowledge, news, finance, health and entertainment. Web search engine are working harder to find out the desired information from large data storages. Hadoop MapReduce make the job more easily compare to existing technologies by providing capability of parallel and distributed environment of data processing from bulk storages of data. MapReduce concept is applied by various organizations. The most widely adopted model is like Apache Hadoop framework (Bansel et al. 2014). MapReduce characteristics are extensive topics for academic study and research to find out the drawback of implementations. Hadoop MapReduce is being developed for last few years. There are multiple extended version of Hadoop MapReduce is implemented proposing improved performance (Gandhi et al. 2014).

In this paper, we evaluate Hadoop MapReduce implementation. The scope of our evaluation is limited to examine Hadoop MapReduce performance with K-Means and K-Medoids clustering algorithms.

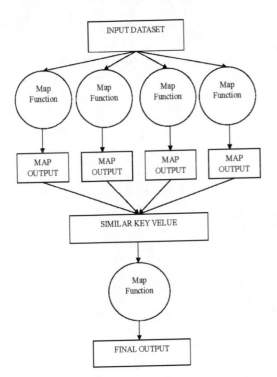

INPUT DATASET

Map Function | Map Function | Map Function | Map Function

MAP OUTPUT | MAP OUTPUT | MAP OUTPUT | MAP OUTPUT

SIMILAR KEY VELUE

Map Function

FINAL OUTPUT

Figure 1. Flow of the MapReduce programming model.

1.1 *Limitation with Hadoop*

The distribution of work to Task Trackers is exceptionally basic. Each Task Tracker has various usable spaces, (for example, "4 openings"). Each dynamic Map or lessen assignment takes up one space. The Job Tracker dispenses work to the tracker closest to the data with an accessible space. There is no thought of the present framework heap of the assigned machine, and henceforth its genuine accessibility. In the event that one Task Tracker is moderate, it can defer the whole Map Reduce work—particularly towards the end of an occupation, where everything can wind up sitting tight for the slowest errand. With theoretical execution empowered, be that as it may, a solitary errand can be executed on different slave node.

- Clients (one or more) submit their work to Hadoop System. Hadoop System accept a Client Request, early it is received by a Master Node.
- Master Node's MapReduce component "Job Tracker" is authoritative for receiving Client Work and divides into manageable autonomous Tasks and assigns them to Task Trackers.
- Slave Node's MapReduce component "Task Tracker" accept those Tasks from "Job Tracker"

and performs those tasks by using MapReduce components.

- Once all Task Trackers accomplished their job, Job Tracker takes those results and incorporate them into final conclusion.
- Finally Hadoop System will send that final result to the Client be very useful in some cases, but it, involve some alteration of the original data (Waluni et al 2015). Furthermore, most of time watermarks can be destroyed if the data recipient is mischievous.

The rest of this paper is organized as follows. Section II discusses literature survey i.e. our mapreduce and k-medoids algorithm methods. Section III we analyze the the existing algorithm in section IV Flow chart of Proposed Algorithm in Section V proposed modify algorithm in Section VI, we provide concluding of this field.

2 LITERATURE SURVEY

Hadoop MapReduce is a flexible model for data processing. It comes with the flexible parameters adjustment according to requirement. The parameters configuration is a challenge in Hadoop MapReduce framework. One must know better combination of MapReduce parameters for best result in timely fashion. Gandhi et al. describes the effect of using several combinations of configuration parameters on Hadoop. Hadoop MapReduce is run for maximum throughput under various conditions by changing configuration parameters. They have analyzed the effect on performance of job execution for parameters in hadoop clusters. Gandhi et al. Concludes that parameters must be effectively selected for every tasks and schedule for achieving maximum performance. A framework is proposed for optimizing hadoop MapReduce for every job. A scheduler is responsible for selecting parameters for particular job and check the performances again and again untill find the best result (Issa et al. 2015).

Hadoop is designed for processing large data set under the large clusters. The applications on hadoop can be different in respect to resources, size and other constrains. Hadoop applications are facing several problems such as ineffective CPU utilizations and memory problems. User should update the hadoop configuration parameters to many criteria such as resource requirement of particular application. Small changes to configuration parameters can produce the huge difference to performance for the same application with same data set (Wantain et al. 2013).

In Hadoop MapReduce big data is processed in distributed and parallel programming model. Jiang et al describes the K-Medoids algorithm HK-

Medoids. HK Medoids is implemented in hadoop MapReduce framework. MapReduce follows strict procedure for every scheduled job. The scheduled job is goes through for many steps. These steps includes map phase, combine phase and reduce phase. In map phase, each input data sample is allocated to one cluster. In combine phase, the centre is being fined for every cluster. In reduce phase, the centre is again calculated. All these phases' repeat themselves until the new centre and old centre doesn't have any variation. Jiang et al. Shows through their experiment that the K-Medoids algorithm has a improved performance (Prabhu et al. 2015).

In Big Data, the data processing and analyzing is very complex job. Apache Hadoop MapReduce provides efficient solution for Big Data analysis. Prabhu et al. describes Hadoop MapReduce performance and find out that it primarily depends on parameters selections. They focus on tuning of hadoop MapReduce parameters for a better result. The tuning of hadoop MapReduce parameters is an effective way to improve performance of job execution in respect to time and disk utilizations. The performance tuning depends on network traffic, memory usage, CPU usage and many more parameters. The several tuning methods have been described for a better result of job execution. In their work they have taken web log as input data and tune the hadoop MapReduce default parameters as baseline. The parameters are changed in several experiments until they have find out the most favorable optimized result for overall system. It is find out that default configuration is of hadoop MapReduce is not favorable to job execution. A more application specific configuration should be preferred for a better result. From the experiments, Prabhu et al shows that application specific parameters tuning will improve hadoop map reduce performance (Kalavri et al. 2013).

The clustering approaches have significant impact in data analysis approach in data mining. In cluster formation, a meaningful data set is being divided into meaningful sub data for more understanding of similar type of data sets. In data mining, clustering approach has been applied to non similar data types. The data sets classification means that they have no predefined category. The clustering approaches have been applied for analysis of pattern recognition, image processing, text mining and many more. (Kalavri et al. 2013) has studied several algorithms of cluster formations. K-means is the widely used approach for data clustering but it is found into he research that in many cases, K-means is not providing better results. (Wu et al. 2015) K-Medoids has better advantages than k-means in all type of cases. K-means calculates means while k-Medoids compute Medoids. Median has more number of advantages to arithmetic means. (Kalavri et al. 2013) has proposed a modified version of k-Medoids algorithm. The experiment result shows that proposed k-Medoids algorithm has better result compare to k-means algorithm in respect to job execution speed and cluster quality. The proposed approach has been compared on the basis of various parameters.

In this paper (Wu et al. 2015) suggest, Apache Hadoop is increasingly being adopted in very large scope of corporation and as a result, Hadoop facility is more beneficial than ever for you and your industries. Using Hadoop, industries can concentrate and evaluate data in ways never before imaginable. Businesses can capture, manage and process data that they used to throw away. You can leverage years of Hadoop participation with practice from www.HadoopExam.com The course/practice is arrange correctly for CEO, CTO to Managers, Software Architect to an personal planner and Testers to improve their scope in very large Data world. You will learn when the use of Hadoop is appropriate, what problems Hadoop addresses, how Hadoop able into your existing environment, and what you need to know about deploying Hadoop. Learn the basics of the Hadoop Distributed File System (HDFS) and MapReduce framework and how to write programs against its API, as well as argue design techniques for larger workflows. This training also covers progressive skills for remove errors MapReduce programs and improve their performance, plus introduces participants to relevant assignment in Distribution for Hadoop such as beehive, Pig, and Oozie. Later completing the training, attendees can leverage our Hadoop Acceptance Exam Simulator for planner as well as Administrator to clear the Hadoop Acceptance Since launch 467+ attendees at present cleaned the exam with the help of our actor.

The challenge with HDFS [Hadoop Distributed File System and Hadoop technique is that, in their initial state, they need a number of hand coding in languages that the average BI professional does not know well, namely Java, R and Hive," said Russom. SAS both stipulation and develop Hadoop. SAS act Hadoop as just another persistent data source, and brings the power of SAS. Storage Analytics and its well-established group to Hadoop implementations. SAS enables users to access and use Hadoop data and processes from within the popular SAS environment for data exploration and analytics. This is analytical; take the skills shortage and the complexity involved with Hadoop. SAS augments Hadoop with world-class data management and partition, which helps ensure that Hadoop will be ready for enterprise expectations. "Hadoop is very important to our customers," said Wayne Thompson, Manager of Data Science Technologies at SAS. It

Table 1. Comparison analysis of different technique.

Author	Objective	Method/simulator	Conclusion
Yuban Jiang, Jiongmin Zhang, IEEE-2014	Parallel K-Medoids Clustering Algorithm Based on Hadoop	K-medoids Algo HK-medoids Design	In this paper the focused on clustering algorithm on Hadoop platform
Dweepna Garg, ICICES-2014	A performance Analysis of MapReduce Applications on Big data in cloud Based Hadoop	HDFS, MapReduce	Analyzed the result of various MapReduce application on big data in cloud based Hadoop cluster
Vasiliki Kalavri, IEEE-2014	MapReduce Limitations, Optimization and Open Issues	HDFS, YARN	Paper concludes the techniques related to optimization of Big Data and their limitations
Swathi Prabhu, IEEE-2015	Performance Enhancement of Hadoop Map reduce Framework for Analysing Big Data	Hadoop Distributed File System MapReduce	Improved the performance of the system in terms of space and CPU time
Mr D Lakshmi Srinivasulu, IJEAS-2015	Improving The Scalability And Efficiency Of K-Medoids By Map Reduce	Mapreduce, k-medoids algorithm	This paper having maximum time complexity once the data size has been increased and also there is a problem to store big amount of data. To solve these kind of problems,
Prakash P S, IJISET-2015	Comparision of k-Means and k-Medoids Clustering Algorithms for Big Data Using MapReduce Techniques	k-mean and k-medoids algorithm	In this paper prakash focused on efficient and better performs compared to the existing algorithms, well effective than existing system

is a very capable way to repository data in a very parallel way to manage not just big data but also critical data. The SAS developer environment, collocating on the Hadoop cluster, enables you to run very advanced, distributed, and statistical and machine learning algorithms Subhashree (2015).

3 PERFORMANCE OPTIMIZATION ANALYSIS OF HADOOP MAPREDUCE

The information on internet is growing in large amount every day. All of this information is usable in some sense. The organizations are very keen to find out the desired analyzed data from vast amount of data stored on various servers in different locations. The clustering approach is very effective to solve the problem of handling so large data set which will take months to final result using existing approaches. So is very obvious that cluster analysis is an important topic for research. There are several clustering approaches is implemented for Hadoop. K-means and k-Medoids are more accepted in academics and scientific researches. A lot of improvement has already be done into k-means algorithm. However these improvement is not sufficient because Hadoop MapReduce frame work still have more scope to improvement. We are proposing a improved version K-Medoids

algorithm. Our proposed algorithm performs better than existing k-means algorithm in Hadoop MapReduce framework Dean et al. (2010).

3.1 Modify Hadoop MapReduce optimization approach

K-means and k-Medoids are more accepted in academics and scientific researches. The main purpose of data partitioned and cluster formation algorithm is to provide data segment of similar types. It is found in experiments that K-means is not effective in many situations. It is not suitable for data partitions in many cases such as Absolute Pearson. K-means select initial Centroids as a first step. This Centroids are selected randomly. This random selection approach sometimes does not provide effective output and result into low quality. Sometime K-means random selection of initial Centroids result into empty and useless clusters. The overall running job on K-means is expensive and requires more time to execute as number of clusters, iterations and data items increases. A lot of improvement has already be done into k-means algorithm (Jiang et al. 2014). However these improvement is not sufficient because hadoop mapreduce frame work still have more scope to improvement. There is lot of work can be done to improve selection of initial centroid selection strategy. We are proposing a improved version K-Medoids algorithm. Our proposed algorithm performs better than existing k-means algorithm in Hadoop MapReduce framework.

4 EXISTING ALGORITHAM (K-MEDOIDS)

Input:
  D = {d1, d2,.......,dn} // set of n data items.
  K // Number of desired clusters.
  Output: A set of k clusters.
  Steps:

1. Initialize: randomly select k of the n data points as the Medoids.
2. Associate each data point to the closest medoids.
3. For each medoids m.
   a. For each non-medoids data point o.
   b. Swap m and o and compute the total cost of the configuration.
4. Select the configuration with the lowest cost. Repeat steps 2 to 4 until there is no change in the medoids.

5 FLOW CHART OF PROPOSED ALGORITHM

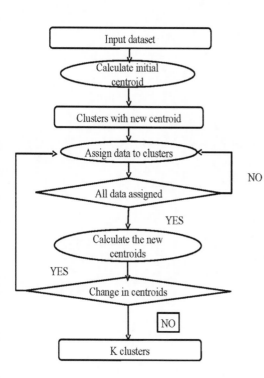

6 PROPOSED ALGORITHM

Input:
  D = {d1, d2,.......,dn} // set of n data items.
  K // Number of desired clusters.

Output:
  A set of k clusters.

Steps:
1. Calculate the initial Centroids.
2. Set the cluster with those Centroids.
3. Initially assign the each data point to the cluster.
4. Calculate the mean value of distance of the all data points of that cluster.
5. Define new Centroids with mean value.
6. Update the Centroids value.
7. Repeat steps 3 & 6 until all data points are assigned to any one of the clusters.
8. Initialize: randomly select k of the n data points as the Medoids.
9. Associate each data point to the closest medoids.
10. For each medoids m.
    a. For each non-medoids data point o.
    b. Swap m and o and compute the total cost of the configuration.
11. Select the configuration with the lowest cost.
12. Repeat steps 9 to 11 until there is no change in the Medoids.

7 CONCLUSION

Apache Hadoop is widely accepted framework for data processing and analysis in parallel and distributed environment. Hadoop is an open source framework which includes Hadoop Distributed File System (HDFS) and the MapReduce. HDFS is responsible for data storages while MapReduce is designed for data processing. Hadoop is developed for storage and processing of very large data set such as web logs. In this research work we have studied Hadoop MapReduce frame work, its working, components and implementations. We have found that clustering approach has vast impact on the hadoop MapReduce performance. K-means and K-Medoids clustering approaches are the subject of research areas. We have proposed improvement into the k-Medoids algorithm. Our proposed algorithm performs better then k-means clustering approach. Our future work includes testing of our approach for different data sets and configuration parameters for hadoop MapReduce framework.

REFERENCES

Dili Wu 2012 "A Self—Tuning System Based on Application profiling and performance analysis for Optimizing Hadoop MapReduce Cluster Configuration" IEEE Conference.

Garvit Bansal, Anshul Gupta, Utkarsh Pyne, Manish Singhal and Subhasis Banerjee 2014 "A Framework for Performance Analysis and Tuning in Hadoop Based Clusters, Workshop on Smarter Planet and Big Data Analytics (SPBDA 2014) held in conjunction with ICDCN.

Gopi Gandhi, Rohit Srivastava "Analysis and Implementation of Modified K-Medoids Algorithm to Increase Scalability and Efficiency for Large dataset" in International Journal of Research in Engineering and Technology, Volume: 03 Issue: 06 | Jun-2014, eISSN: 2319-1163 | pISSN: 2321–7308.

J. Dean and S. Ghemawat 2010. MapReduce: A Flexible Data Processing Tool. CACM, 53(1):72–77.

Ji Wentian, Guo Qingju, Zhong Sheng 2013 "Improved K-medoids Clustering Algorithm under Semantic Web" in Proceedings of the 2nd International Conference on Computer Science and Electronics Engineering.

Josepha. Issa 2015 "Performance Evaluation and Estimation Model Using Regression Method for Hadoop Word Count" in IEEE Access Volume 3, Page(s): 2784–2793, ISSN: 2169-3536,

Juwei Shi, Jia Zou, Jiaheng Lu, Zhao Cao, Shiqiang Li and Chen Wang 2014 "MRTuner: A Toolkit to Enable Holistic Optimization for MapReduce Jobs" in Proceedings of the VLDB Endowment, Vol. 7, No. 13.

Rong Gu, Xiaoliang Yang, Jinshuang Yan, Yuanhao Sun, Bing Wang, Chunfeng Yuan, Yihua Huang 2013 "Hadoop: Improving MapReduce performance by optimizing job execution mechanism in Hadoop clusters" in Journal of Parallel and Distributed Computing Volume 74, Issue 3, Pages 2166–2179.

Subhashree 2015 "Comparison of k-Means and k-Medoids Clustering Algorithms for Big Data Using MapReduce Techniques" IJISET—International Journal of Innovative Science, Engineering & Technology, Vol. 2 Issue 4.

Swathi Prabhu, Anisha P Rodrigues, Guru Prasad M S, Nagesh H R 2015 "Performance Enhancement of Hadoop MapReduce Framework for Analyzing Big-Data" in IEEE International Conference one Electrical, Computer And Communication Technologies, ISBN: 978-1-4799-6084-2, 5–7 March.

Vasiliki Kalavri, Vladimir Vlassov 2013 "MapReduce: Limitations, Optimizations and Open Issues" in 12th IEEE International Conference on Trust, Security and Privacy in Computing and Communications (TrustCom), DOI:10.1109/TrustCom.2013.126, 16-1.

Yaobin Jiang, Jiongmin Zhang "Parallel K-Medoids Clustering Algorithm Based on Hadoop" in 5th IEEE International Conference on Software Engineering and Service Science (ICSESS), 27–29 June 2014 ISSN: 2327-0586.

Communication and Computing Systems – Prasad et al. (Eds)
© 2017 Taylor & Francis Group, London, ISBN 978-1-138-02952-1

Performance analysis of Angular Location Aided Routing (A-LAR) in VANETs

Sachin Kumar Kaushal, Vivek Singh & Jaiprakash Nagar
Gautam Buddha University, Greater Noida, Uttar Pradesh, India

Ram Shringar Raw
Indira Gandhi National Tribal University, Amarkantak, Madhya Pradesh, India

ABSTRACT: Within this paper, we have planned Angular-Location Aided Routing protocols (A-LAR) for VANETs. The performance of protocol is also measured in term of number of hops in request zone, average number of hops and presence of number of nodes arrive within angular zone. Several routing protocols have already been proposed for ad hoc networks. In A-LAR, look for a novel way is done in near a lesser request zone based on angular value. We drive several analytical results based on LAR protocol.

1 INTRODUCTION

Although VANET is an application of MANETs but routing of data packets is more challenging in VANETs (Lee, 2010). This is because of frequent change in network topology due to the highly-dynamic nature of vehicular nodes. VANET address the wireless communication between vehicles (V2V), and vehicles and communications entrée point (V2I). V2V have two types of communication: single hop communication i.e. straight vehicle to vehicle communication and multi hop communication i.e. vehicle relies on additional vehicles to retransmit (Lee, 2010). Here are a lot of problems in front of VANETs systems drawing and performance. In this paper, we contain listening carefully on direction-finding difficulty in V2V. Designing of a routing protocol is a critical problem in VANETs (Verma at el, 2015). Toward job with topology based direction-finding protocols into VANETs, one needs a whole vision of network topology and to have that diffusion of manage packets be completed. This above your head is able to be bargain in location based direction-finding protocols since they think the position in sequence of the vehicular nodes. Location Aided Routing (LAR) (Lee, 2010) be a broadly recognized location based direction-finding protocol, which removes the way finding above your head by utilizing the position in sequence of goal. LAR be an on-demand direction-finding protocol whose action be parallel near Dynamic Source Routing (DSR) (Lee, 2010). Since similar to DSR.

LAR utilizes position in sequence with GPS to boundary the region used for searching a novel way toward a lesser "request zone". The position in sequence is second-hand to remove direction-finding above your head and progress the presentation of direction-finding protocols used for ad hoc networks (Lee, 2010). Here LAR, in its place of flooding the way requests into the whole network, single those nodes into the request zone will forward packets. D-LAR protocol inherits the notion of LAR procedure. Forwarding plan is single of the structure blocks of some location based direction-finding procedure. It dictates how the dispatcher node selects the next reachable forwarder node which be geologically neighboring to the goal (Lee, 2010). There are mainly three types of insatiable promote strategies: 1) mainly promote in Radius (MFR), 2) Adjacent with Promote Development (APD), 3) Directional Scope Direction-finding (DSD). Some researchers also tried to modify the existing greedy forwarding techniques to handle issues, such as dynamic network topology and frequent network detachment (Lee, 2010). D-LAR applied directional greedy forwarding and inherited some features of LAR protocol to complete improved performance in a metropolis traffic picture.

We have a rest of the paper is planned as fallows. The protocol description along with underlying assumptions and notations are given in section 2. The expressions for number of hops and average number of hops are also computed in this section. A numerical illustration to validate the analytical results is shown into fraction 3. Lastly, the conclusion is drawn in section 4.

2 ASSUMPTIONS AND NOTATIONS

In this part, we expand a model to analyze the number of hop count in the expected zone. If N denotes number of nodes located into the round region, adhere Poisson sharing through factor β and R denotes transmission range. Let P(n) is the probability of having N nodes into the expected area.

2.1 Angular-Location Aided Routing (A-LAR) protocol

A-LAR protocol is a position based routing protocol. This protocol is designed by modifying D-LAR. In this procedure, we have increased the advantages of both LAR and D-LAR. In LAR a *request zones* same as a quadrangle, anywhere simply the nodes presence into the quadrangle floods the facts packets. Pursuance near D-LAR protocol, the next-hope node have to exist resides in the request zone for further transmission of facts packets. The next-hop node of S be limited near the single fourth region by the rounded transmission area of S. Into multi-hop network scenarios, every vehicle have to exist drop in the announcement assortment of each additional as shown in Figure 1. These scenarios achieved an uninterrupted connectivity throughout the network.

Here in A-LAR, we have modify the D-LAR, one fourth circular region to angular region with angle (θ) and next-hop node have to exist resides in the request zone used for additional communication of facts packets. The next-hop node by S be limited near the angular value $(\theta/360^0)$ of the rounded transmission area of S. The attendance of average number of nodes in a rounded region count pursuance to equation (1)

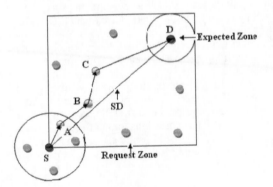

Figure 1. Directional location aided routing.

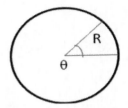

Figure 2. Angular location aided routing.

Number of nodes $(N) = \beta \times$ region of the area

$$= \beta \pi R^2 \left(\frac{\theta}{360^\circ} \right) \tag{1}$$

Equally, division of n nodes in the angular region is

$$n = e^{-\beta \pi R^2} \left(\frac{\theta}{360^\circ} \right) \tag{2}$$

The possibility to here be no. of node into the angular region be

$$P = e^{-n} = e^{-\beta \pi R^2} \left(\frac{\theta}{360^\circ} \right) \tag{3}$$

alike, the possibility to here be on smallest amount single node in the angular region be

$$Q = 1 - P = 1 - e^{-\beta \pi R^2} \left(\frac{\theta}{360^\circ} \right) \tag{4}$$

The possibility to n additional nodes situated into the angular region count pursuance to

$$P(n, A) = \frac{\left(\beta \pi R^2 \left(\frac{\theta}{360^\circ} \right) \right)^n \cdot e^{-\lambda A}}{n!}$$

$$P(n, A) = \frac{\left(\beta \pi R^2 \left(\frac{\theta}{360^\circ} \right) \right)^n \cdot e^{-\beta \pi R^2} \left(\frac{\theta}{360^\circ} \right)}{n!}, \tag{5}$$
$$n = 0, 1, 2, 3, \ldots$$

so, possibility near choose on smallest amount k node in the angular region absent of n count since

$$P_k = 1 - \sum_{n=0}^{k-1} \frac{\left(e^{-\beta \pi R^2} \left(\frac{\theta}{360^\circ} \right) \right)^n \cdot e^{-\beta \pi R^2} \left(\frac{\theta}{360^\circ} \right)}{n!} \tag{6}$$

2.2 Average number of hops calculation

We consider to the purpose of the in cooperation basis and goal nodes be situated within the communication assortment R. If x is the link distance between S and D, we can easily find the PDF used for the space x between S and D as:

$$f(x) = 2\pi\beta x e^{-\pi\beta x^2} \qquad (7)$$

so the possibility of a single-hop count up be able to be probable as adhere:

$$P1 = \int_0^R f(x)d(x) = \int_0^R 2\pi\beta x.e^{-\pi\beta x^2}, \ (0 < x < R)$$
$$P1 = 1 - e^{-\pi\beta R^2} = 1 - e^{-4n} \qquad (8)$$

In case of no direct link, the S and D can communicate via two hops. here be on smallest amount single middle node to have to be in the S and D. so, the possibility of two-hop count up be able to exist predictable while adhere:

$$P2 = \int_R^{2R} \left[2\pi\beta x \cdot e - \pi\beta x^2 dx \right] \times \left[1 - e^{-n} \right], (R < x < 2R)$$
$$P2 = \left[e^{-n} - e^{-16n} \right] \times \left[1 - e^{-n} \right] \qquad (9)$$

as a result, the possibility of multi-hop count up into VANET be able to live general as adhere

$$Pm = [e - 4n(m-1)^2 - e^{-4nm2} \times \left[1 - e^{-n} \right]^{m-1} \qquad (10)$$

By using equations (4) and (8), the average number of hops among basis and goal can be predictable since

$$H_{avg} = \sum_{H=1}^m H.P(H) = p_1 + 2p_2 + 3p_3 + 4p_4 + \cdots mp_m \qquad (11)$$

2.3 Numerical illustration

During this part, we contain obtainable the model job agreed absent inside MATLAB for analyzing the presentation at the A-LAR protocol. The presentation at A-LAR is count up logically as well as numerically together.. The model region be a four-sided figure of 1,0000 m × 1,0000 m. We include count up average number of hops as well as average pathway throughput by changing number at nodes in the show round range of each other, altering number of forwarding nodes, link lifetime, and varying transmission range in the network. The performance of parameter taken for simulation angular values of shaded area $\theta = 65°$ and range 120 to 200.

Figure 3 show rounds round probability of at smallest amount k nodes as of the round area at S. now, we include careful node density 0.0004 and 0.0005 nodes/m² [20]. Figure 5, show round rounds around possibility at choose at least k nodes into the in the shade region. Finally, we watch to, the possibility at vehicle has at least k neighbors here inside the shaded region are relation elevated. His study show round rounds that, in the shade area has fewer number of nodes evaluate to total round area of S.

too, we include watch as of Figure. 5 to the possibility to discover next-hope node in in the shade region is upper evaluate to choose a node from angular round area region. Hop count up is almost certainly the mainly broadly use the way choice technique in

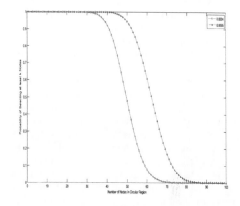

Figure 3. Possibility of K node Vs numbers of nodes in circular region.

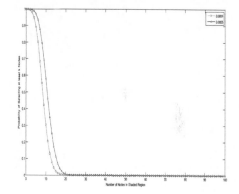

Figure 4. Possibility of K node Vs numbers of nodes in angular circular region.

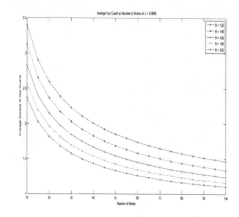

Figure 5. Average number of hops versus number of nodes.

very active network. Figure 6 show round, rounds the achieve at average number of hops used for dissimilar network dimension by dissimilar communication range (120 as well as 200) it explain rounds to the network of nodes have relation huge number of hops. It funds as the number of nodes add to, the average number of hops reduce. It have feature of networks is experiential since in a network to give the connectivity, Nodes should be at additional detachment but they be smaller into numbers as well as they must exist fewer distance but they be bigger in numbers Figure 6 as well shows to while the number of nodes increase, the node density of network also increase. This resources of the performance of A-LAR would raise among the superior density of nodes. It is shown in Figure 7, while communication ranges of the leading node increases, the average number of hops counts decreases. It indicates to the huge communication range used for forwarding packets decrease the hop counts and increases the packet delivery ratio through superior network connectivity.

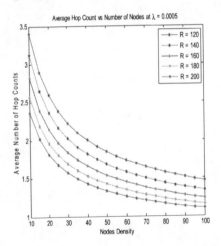

Figure 6. Average number of hops versus nodes density.

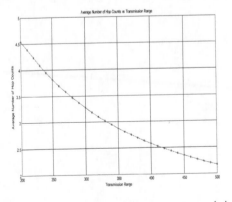

Figure 7. Average number of hops versus transmission range.

3 CONCLUSION

Within this paper, we include examine the act of A-LAR. It protocol be helpful near choose the next-hop node among least angle (θ) all along with the instantly line drawn linking resource and goal node in the request zone parallel near D-LAR protocol. The A-LAR protocol be able to survive simply included into most of the available direction-finding protocols. In future, the act of A-LAR procedure is able to be evaluated among previous QoS metrics e.g., end-to-end delay, packet delivery ratio and road costs. These results additional are able to survive comparison among D-LAR, LAR as well as other direction-finding protocols.

REFERENCES

Boukerche, & Azzedine, et al., 2008 "Vehicular ad hoc networks: A new challenge for localization-based systems." *Computer communications* 31.12: 2838–284.

Johnson, David B., & David A. Maltz, 2010 "Dynamic source routing in ad hoc wireless networks" *Mobile computing* Springer US, 1996. 153–18.

Ko, Young-Bae, & Nitin H. Vaidya, 2000 "Location-Aided Routing (LAR) in mobile ad hoc networks." *Wireless networks* 6.4: 307–321.

Lee, Kevin C., Uichin Lee, & Mario Gerla et al., 2010 "Survey of routing protocols in vehicular ad hoc networks." Advances in vehicular ad-hoc networks: Developments and challenges: 149–170.

Mauve, Martin, Jörg Widmer, & Hannes Hartenstein, 2001 "A survey on position-based routing in mobile ad hoc networks." *Network, IEEE* 15.6: 30–39.

Rao, Meena, & Neeta Singh, 2015 "Routing Issues in Mobile Ad Hoc Networks: A Survey" *International Journal of Computer Science Trends and Technology (IJCST)*, ISSN: 2347-8578, (3) 1.

Raw, Ram Shringar, et al., 2012 "Feasibility evaluation of VANET using directional-location aided routing (D-LAR) protocol." *Ar Xiv preprint arXiv:1212.3216.*

Singh, Harvinder, & Jorge Urrutia, 1999 "Compass Routing on Geometric Networks." *Proc. of the 11th Canadian Conference on Computational Geometry, Vancouver, Canada.*

Verma, Poonam, & Neeta Singh, 2015 "An Analytical Review of the Algorithms Controlling Congestion in Vehicular Networks", *IOSR Journal of Computer Engineering (IOSR-JCE) e-ISSN: 2278-0661, p-ISSN: 2278-8727, Volume 17, Issue 3, Ver. IV: 32–41.*

Wang, Neng-Chung, et al., 2009 "A Greedy Location-Aided Routing Protocol for Mobile Ad Hoc Networks." *WSEAS International Conference. Proceedings. Mathematics and Computers in Science and Engineering.* Ed. Shengyong Chen. No. 8. World Scientific and Engineering Academy and Society.

Yamaguchi, Hirozumi, Weihua Sun, & Teruo Higashino, 2010 "Geographic routing on vehicular ad hoc networks." *IGI Global* 171–178.

Yin, Jijun, et al., 2004 "Performance evaluation of safety applications over DSRC vehicular ad hoc networks" *Proceedings of the 1st ACM international workshop on Vehicular ad hoc networks.* ACM.

Communication and Computing Systems – Prasad et al. (Eds)
© 2017 Taylor & Francis Group, London, ISBN 978-1-138-02952-1

Various advanced applications of bioprocess engineering in the field of biomedical sciences—an insight

Chandra Kant Sharma, Kanika Prasad & Monika Sharma
Department of Bioscience and Biotechnology, Banasthali University, Rajasthan, India

ABSTRACT: Therapeutic application of undifferentiated mass of cells to treat diseases and malfunctions holds great promise. Treatment strategies rely on surgical interventions which require the use of tissue grafts. In-vitro production of cell constructs has direct biomedical application in immunosuppressive therapy, regenerative therapy, and treatment of skeletal defects, osteoarthritis, diabetes mellitus type I and II, acute hepatic failure, Inter Vertebral Disc (IVD) degeneration etc. as well as developmental studies. This review has been categorized into cells of different lineages (Cartilage, Skin, Bone, Ligament/tendon, Liver and trachea) to dis-cuss bioreactor advances in the culturing of each of them.

1 INTRODUCTION

Tissue engineering entails biomedical applications which are involved in the repair of various tissues in the body by tissue grafts. However the generation of tissue grafts by conventional tissue engineering methods poses problems such as necrosis and often the tissue formed is of suboptimal level because of factors such as low mass transfer rates and oxygen diffusion. To avoid this problem, optimal tissue grafts are cultured in closely monitored bioreactors under stringent conditions of temperature, pH, hydrostatic pressure, shear stress, agitation, etc. A variety of novel bioreactors have been developed to culture cells of different lineages such as stem cells, chondrocytes, cardiomyocytes, fibroblasts, hepatocytes, keratinocytes etc. to be applied in the area of tissue engineering of the skin, bone, liver, cartilage and heart (Salehi et al. 2013 & Li et al. 2014).

1.1 Cartilage

Damage to the articular cartilage can be caused by autoimmune diseases or osteoarthritis rendering permanent deterioration among the elderly. Tissue engineering of the cartilage offers a promising treatment strategy which is cell-based (Shahin et al. 2011).

Isolated chondrocytes can be stimulated to develop into cartilage tissue by providing various stimuli through the application of concentric cylinder bioreactors, rotating wall bioreactors and scaffold perfusion bioreactors. Bioreactors allow efficient mixing of nutrients and close monitoring of various parameters that govern the chemical reaction occurring within the system.

Scaffold Perfusion Bioreactors (SPBs), Cylinder Bioreactors (CBs), and well mixed reactors make use of scaffolds and increased nutrient transport through fluid flow. Three dimensional scaffold-based cultures are distinct from conventional cell culture techniques. Seeding the cells inside scaffolds mimics in vivo conditions; guides cell differentiation and stimulates the production of Extra Cellular Matrix (ECM). Shear forces allow increase in the development of ECM. Interactions between cells via ECM allow circulation of nutrients and other metabolic products which are important for eliciting biological responses (Li et al. 2008, Shahin et al. 2011 & Detzel et al. 2011). Seeding of RWBs (rotating wall bioreactors) at densities exceeding 5×105 cells/mL allows spontaneous cell aggregation by providing a low shear, low-gravity environment. Understanding of the interplay among various mechanical forces can unfold the possibilities of developing novel bioreactors.

The novel Continuous Centrifugal Bioreactor (CCBR) is employed for culturing yeast and mammalian cells. For initiating chondrogenesis, both hydrostatic pressure and shear are simultaneously applied alongside maintenance of high-perfusion rates. Cells are immobilized onto a high-density fluidized bed. This is achieved through counter-flow centrifugation in which centrifugal force is balanced with opposing drag and buoyant forces (Detzel et al. 2011).

The use of nanofibrous scaffolds affects distribution and migration of cell mass because of the small size of pores in the fibres. (Li et al. 2008) used electrospun poly (L-lactic) acid (PLLA) as a means to deliver cells into the nanofibrous scaffold thereby modifiying the procedure of cell pelleting. Bovine chondrocytes were mixed with electrospun

PLLA nanofibers followed by centrifugation. The cells were thus seeded into a packed Cell-Nanofiber Composite (CNC). (Li et al. 2008) thus succeeded in resolving the problem of small pore size in the scaffolds, thereby allowing optimal cell distribution within the CNC.

In order to mimic in vivo conditions for enhanced growth of cartilage constructs close monitoring of efficient diffusion and nutrient circulation become necessary. One such novel reactor i.e., the Rotary Wall Vessel (RWV) bioreactor provides a simulated microgravity environment to the cartilage constructs. The rotating medium exerts upward force onto the free falling constructs and through mechanical agitation it allows circulation of nutrients and exchange of oxygen and metabolic wastes. When compared to conventional culture methods involving petri dish and flasks, the RWV bioreactor clearly proves better culture environment conducive to the growth of CNC-based cartilage. The authors substantiated this information by comparing cultures from RVW bioreactor and cultures from conventional method (maintained by pelletizing cells in a 50-ml conical centrifuge tube). To enhance cartilage growth, additional application of growth factors namely Insulin-like Growth Factor-I (IGF-I) as well as Transforming Growth Factor-β1 (TGF-β1) was done. These growth factors have earlier been proven to control chondrocyte differentiation into cartilage (Li et al. 2008).

Poly (a-hydroxy ester)s and their co-polymers viz. Polyglycolic Acid (PGA) and polylactic acid are the ideal choice of scaffold materials owing to their biodegradable nature and safety for surgical use. In PGA scaffolds devoid of alginate, a gradually increasing flow rate is applied during bioreactor culture. Alternatively the scaffolds can be pre-cultured for five days earlier to culture through bioreactor. This method yielded better quality of constructs as compared to the cultures cultivated at elevated flow rate without scaffold pre-culture. Moderate flow rates in scaffold pre-culture prevent ECM from being washed away and permit it to bind to the scaffold and further accumulation of synthesized products. When cultured in the presence of alginate-loaded scaffolds, the cells and the ECM were protected from shear even when operated at the maximum flow rate as well as without scaffold pre-culture (Shahin et al. 2011).

Perfusion flow bioreactors find useful application in cartilage tissue engineering. They apply fluid flow-induced hydrodynamic shear stress and improved material transfer rates to enhance cartilage formation by chondrocytes (Tare et al. 2014).

(Tare et al. 2014) have effectively used a new tissue engineering process, which is a combination of microfluidic perfusion bioreactor technology with uninterrupted use of ultrasound to bioengineer scaffold-free neocartilage grafts. Absence of scaffold makes the graft devoid of any unknown object. The grafts thus obtained are similar to inhabitant hyaline cartilage together histologically as well as biomechanically in addition to possess the capability to mend partial width chondral faults.

The novel method in the Acoustofluidic Perfusion Bioreactor (APB) involves uninterrupted perfusion of culture medium at low-shear rates along with the application of acoustic drive frequencies over the range of 890 to 910 kHz, at a sweep rate of 50 Hz. Mechanical stimulation by flow-induced shear, enhanced the chondrocyte differentiation into cartilage. The chondrocytes exhibited positive response to the fluid convection which aided in enhanced cartilage formation mediated by efficient mass transfer rates.

APBs are constructed with rectangular glass capillaries into which resonant chambers are fabricated and further attached to the transducer. The transducer transmits an ultrasonic standing wave (having half wavelength resonance at 897, 899 and 902 kHz) into the lumen of the glass capillary. A pressure node is created in the centre of the compartment where the aggregation of 3-D multicellular agglomerate occurs. A peristaltic pump is employed to circulate serum free culture medium for chondrocytes to proliferate. The chondrogenic medium is contained within a reservoir for closed loop circulation. Optimal CO_2 concentration is maintained by preconditioning the culture medium for about 24 hours in a standard CO_2 incubator. Also a 5% CO_2 atmosphere is created just above the culture medium by introducing gas through a HEPA filter with 0.22 μm pore size (Tare et al. 2014).

Another innovatory Flow-Chamber Bioreactor (FCBR) offers advantages in steady flow conditions, elimination of poisonous reaction products, elevated cell densities as well as enhanced metabolism. The single flow-channel bioreactor is equipped with counter current movement of gas in addition to medium for tissue-engineered constructs. The closed system is aerated with moisten premixed gas with discretionary composition. Thus it can be hold autonomously from cell culture incubator (Goepfert et al. 2013).

(Portner et al. 2009) designed a single flow-channel bioreactor for the production of 3D cartilage-carrier constructs. The carriers consisted of a bone substitution matter and were enclosed through a 1–2 mm cartilage coating. The reactor was developed for durable cultivation of cartilage-carrier constructs which allows close monitoring of biochemical parameters under constant conditions.

With the intention of, to generate three dimensional engineered cartilage resembling inhabitant articular cartilage. Tarng et al. 2012 synthesized a recirculating flow-perfusion bioreactor to imitate the movement of a native diarthrodial joint by contribution shear stress as well as hydrodynamic pressure all together. The cell constructs obtained from recirculating flow-perfusion bioreactor are comparable to native articular cartilage. The bioreactor provides both stable oscillating laminar stream (maximum shear stress of 250 dyne/cm^2) along with hydrodynamic pressure (amplified from 0 to 15 psi) concomitantly. Shear stress enhances proliferation and differentiation along with lower levels of matrix synthesis. Alternatively hydrodynamic pressure elevates ECM production, and enhances tissue organization.

1.2 Ligament/tendon

Bioreactors used for tendon or ligament manufacturing are distinct from methods that have been employed in tissue engineering of the skin, bone, muscle or liver. Tendon/ligament engineering requires monitoring of the biomechanical and biochemical environment that can be achieved by components such as actuators, culture chamber, media circulation system, etc. The system thus generated is conducive to the differentiation of cells into ligament or tendon.

The actuator system is a vital part of the bioreactor as it provides mechanical agitation to the cell culture. A variety of actuators including pneumatic, linear motors and Step Motorball Screws (SMBSs) can be employed in tendon/ligament engineering. Pneumatic actuator is easy to maintain and offers the advantages of inexpensive, cleanliness as well as elevated power-to-weight proportion. The culture chamber can be designed with stainless steel, polymethylemethacrylate, polyoxymethylene, polycarbonate, glass or silicon. Culture chambers can be integrated or separated. Multiple samples can be cultured in integrated chambers while sharing the same culture medium. On the contrary, separated chambers allow culture of only single sample. A circulating medium easily infiltrates into culture tissue (Wang et al. 2013). This was demonstrated by culturing human umbilical vein in a circulating medium wherein the number of cells were approximately three times greater than those cultured in a quiescent medium (Abousleiman et al. 2009).

Recently there have been recent developments through the introduction of two commercial bioreactor structures have been developed, The Bose® Electro Force® BioDynamic® system as well as the LigaGen system. The LigaGen system is a light weight (< 3 kg) incubator companionable bioreactor. An utmost force of 40N can be applied to the tissue sample as well as simulating compounds. The Bose® ElectroForce® BioDynamic® system contains a load cell and optional laser micrometer which enables it to observe the force or strain curve of engineered tendon or ligament for the duration of the culturing (Wang et al. 2013).

1.3 Bone

Bone transplantation is used to resolve skeletal defects arising due to trauma, injury, infection, genetic predisposition and development. After blood, bone is second most commonly transplanted tissue. The number of transplantations is expected to increase in the upcoming years owing to the aging population. A variety of human cells have the ability to differentiate into osteogenic lineage including osteoblasts, mesenchymal stem cells, amniotic membrane and Adipose-derived Stem Cells (ASCs). Human ASCs are easily available and preferred cell source.

A novel perfusion bioreactor system was developed to culture human ASCs on decellularized bone scaffolds. Approximately 40 ml of culture medium can be accommodated into the bioreactor by distributing it into six channels located at the bottom. These channels distribute the media into individual culture wells. Just above the culture wells, the medium is collected into a reservoir where it is allowed to be equilibrated with respect to oxygen and pH. Circulation of the media is carried out by a multi-channel peristaltic pump. The flow rate through the scaffolds was determined to be approximately 1.8 mL per minute. The cellular constructs obtained however are not very successful after implantation. Proper ways of achieving revascularization of tissues need to be developed (Frohlich et al. 2009).

Several other systems including stirred flasks, rotating bioreactors, and perfusion bioreactors with unique flow patterns have been investigated. Stirred flask reactors are equipped with stirrer around the cell-seeded constructs. These reactors provide elevated levels of cell proliferation and mineralization as compared to static culture systems. Also these reactors are simple to operate and relatively cheaper. However, a stirring system has been negatively associated with appearance of an intense external layer which apparently hinders oxygen as well as nutrient exchange among the cells in the scaffold.

On the contrary, rotating bioreactor systems generate laminar flow when its concentric cylinders rotate horizontally. A rotating wall bioreactor developed by NASA exhibited increased osteoblast performance and mineralization (Barzegari et al. 2012). Maintenance of optimal water density and accurate dimension throughout bioreactor operation are crucial to its performance. Collision of scaffolds in the bioreactor wall is a limitation as they disrupt attached cells.

Another system of perfusion-based bioreactors allows an effective transportation of nutrients in addition to oxygen throughout the whole scaffold and the highest seeding efficiency. The main chamber consists of cell seeded scaffolds and peristaltic pump for delivery of culture medium. The different modes of fluid flow can be steady, oscillating and pulsed and each of these can affect the differentiation capacity of osteoblasts. The deposition of mineralized ECM occurs in the entire scaffold under perfusion based operation of bioreactor. The cell constructs thus obtained have transplantation potential for in vivo conditions.

All the above mentioned bioreactors are promising options to be employed for bone tissue engineering, however then need to comply with GMP guidelines for commercial and clinical applications. Reduction of safety risks and improving efficiency in bioreactor performance are important parameters to be adhered to for developmental strategies (Amini et al. 2012).

A novel perfusion bioreactor developed by (Kleihans et al. 2015) was derived from computation modeling. The bioreactor system houses a medium reservoir that can accommodate 30 ml of growth medium, and a cartridge, which accommodates a scaffold inside silicone housing. A sampling port (Fenwal, Munich, Germany) is supplied to aid in cell seeding and medium replacement. A pressure inlet is provided at the inlet of the cartridge to monitor pressure. The fluid flow is controlled by a peristaltic pump (ISMATEC, Wertheim, Germany). The developers observed homogenous distribution of cells within one week of culture which was devoid of differentiation inducing factors. This bioreactor complied with the culture requirements of bone substitutes. The cell constructs can be applied to treat bone defects.

1.4 Skin

(Helmedag et al. 2015) developed a perfusion based bioreactor that can be operated with sub-merged as well as airlift culture. It requires minimum human intervention and this greatly eliminates the risk of contamination. They investigated a constant flow bioreactor which showed impaired angiogenesis in skin grafts as mediated by agitation of growth medium.

1.5 Trachea

Stenosis, malignancy and injury are some of the conditions that require tracheal tissue replacements for restoration. Recent tracheal replacements have been unsuccessful often leading to inflammation, immunologic rejection, mucous build up and further stenosis. Revascularisation techniques need to be developed to prevent the problem of airway collapse. Bioreactor recellularization enables isolation and selection of specific subpopulations of cells and also evaluation of scaffold prior to transplantation.

The bioreactor is constructed with a translucent, autoclavable polymer, polyphenylsulfone. The body of the bioreactor is a cylindrical culture chamber that houses the tracheal scaffold. Seeding is done on the luminal and exterior surfaces of the scaffold. The scaffold is a rotating one which is equipped with silicone tubing to prevent its twisting. Outflow and inflow ports allow continual replacement of culture medium, collection of living and dead cells, etc.

The perfusion bioreactor produced increased levels of long-segment tracheal scaffolds with higher cell counts and homogeneity scores as compared to static culture methods. Similar conclusions were drawn when dynamic perfusion bioreactors were operated with smaller scaffolds. This method enhanced the reproducibility of the seeding process and has been applied in the expansion of articular chondrocytes, segregation of bone marrow stromal cells along with better spatial organization of different cell types (Haykal et al. 2014).

1.6 Hematopoietic stem cells

When embryonic stem cells are cultured in suspension medium, they tend to differentiate and aggregate into Embryoid Bodies (EBs). These embryoid bodies differentiate into three germinal layers-endoderm, ectoderma and mesoderm. The differentiation pattern is similar to the stages of embryonic development. The mesodermal layers develop into blood tissue as well as Hematopoietic Progenitor Cells (HSPCs). Hematopoietic stem cells have therapeutic applications. They involve transplantation of adult stem cells harvested from blood or bone marrow. (Fridley et al. 2010) investigated suspension cultures of embryonic stem cells in different reactors namely conventional static culture, stirred vessel kind spinner flask structure in addition to the revolving microgravity kind Sythecon system.

The Synthecon rotating vessel bioreactor is reported to enhance EB formation as well as segregation of stem cells into 3 germinal layers. The inoculum was 0.5–0.7×10^6 cells/ml. For differentiation into cardiomyocytes the inoculum was 1×10^5 cells/ml. The increase in the initial cell seeding density is directly proportional to haematopoietic differentiation in the spinner flask system. A seeding density of 7,50,000 cells/ml was shown to maximize EB concentration.

In spinner flask systems, the rotation speed varies between 60 and 100 rpm. (Fridley et al. 2010) reported that 100 rpm is the best possible rotation speed for expansion of undifferentiated stem cells. (Gerecht et al. 2004) demonstrated a rotation speed of 15–20 rpm as for EB formation in the Synthecon system. However EB dimension is not the only governing aspect in hematopoiesis of embryonic stem cells. Other factors like as revolving velocity along with density of cell seeding and important for differentiation (Fridley et al. 2010).

1.7 Pluripotent stem cells

Stirred-Suspension Bioreactors (SSBs) are widely used for proliferation of embryonic stem cells and induced pluripotent stem cells. The advantages include simple operating design, convenient scale up as well as online checking of the culture factors which reduce human intervention and manual work. The cells can be cultured as aggregates, on microcarriers or in scaffolds. As most of the biotechnological industries utilize SSBs, the stem cells developed from them can be easily into commercial usage thereby eliminating the need to develop novel designs.

Stirred suspension culture of mouse embryonic stem cells was conducted by (Magyar et al. 2001). The embryonic bodies were encapsulated in alginate beads. The concentration of alginate in the beads affected cellular organization. This is because cysts were observed in embryoid bodies at 1.1% (w/v) alginate but not at 1.6% (w/v) alginate concentration. Other encapsulation agents that have been used for culture of embryonic stem cells are poly (lactic-co-glycolic acid), poly (l-lactic acid) scaffolds, hydrogels of agarose, synthetic semi-interpenetrating polymers and hyaluronic acid.

The scaffolds used in SSBs can be customized according to requirements by entrenching primordial tissue, minute functional groups or growth factors that can enhance binding of cells. The creation of appropriate microenvironment within the SSB guides the cells towards differentiation into desired lineage and formation of ECM. When the organizations of cell constructs are similar to the native tissue, the chances of immune-rejection are greatly reduced.

Microcarrier bioreactors allow the culturing of cells in macroporous or on microcarrier environment. Macroporous beads direct cells to differentiate into cardiomyocytes. Cells proliferating on the surface of microcarrier beads develop monolayers rather than embryoid bodies. Further, microcarrier cultures have high surface-to-volume ratio which allows higher cell densities compared with static cultures. Culture of human embryonic stem cells on microcarriers at a lower cell density has been shown to may limit the unregulated loss of pluripotency in embryoid bodies. Commercially available beads are coated with collagen (a component of Matrigel) or fibronectin which support human embryonic stem cell proliferation (Kehoe et al. 2010).

1.8 Cardiovascular

Human umbilical cord tissue is a highly preferred cell source for tissue engineering and therapeutic purposes. It is easily acceptable after birth due to lack of regulatory interventions governing medical ethics. Cord arteries derived from human umbilical cord have great potential for cardiovascular tissue engineering. (Reichardt et al. 2013) used two rotating bed bioreactors (non-disposable and disposable) for this purpose. The disposable bioreactor system is gamma sterilized equipped with a culture vessel, tubing system and measuring device with ports for pH and pO_2 sensors. The relatively easy handling and closed system minimizes the instances of contamination.

Perfusion bioreactors have also been used for cardiovascular tissue engineering for enhanced production of ECM.

The culture medium is made to circulate in the reactor via an external loop. The bed of the reactor is slowly rotated along with medium circulation to avoid nutrient gradients. Perfusion bioreactor is operated in continuous mode. The other bioreactors namely stirred and rotary systems operate in batch-feeding modes. (Reihard et al. 2013) concluded through their studies that the rotating bed bioreactor is the ideal system for large scale culture of cord artery cells.

The authors further commented that better manufacturing practices need to be derived for successful clinical applications.

2 CONCLUSION

Bioreactors have become an inevitable of modern day tissue engineering procedures. Every type of tissue (e.g., skin, bone, blood vessel, cartilage, and myocardium) requires a different type, design or operation of bioreactor suited to the needs of the differentiation pattern of the cell. Development of an appropriate bioreactor is possible through proper understanding of the parameters governing the process and the interplay between them. This involves the application of both biological and engineering principles. The commercialization of successful designs of bioreactors requires them to undergo stringent risk assessments to check for compliance with GMP guidelines. Mathematical calculations, modeling techniques and simulation not only check the feasibility of the reactor system but will also allow development of novel systems.

ACKNOWLEDGEMENTS

Authors are thankful to Prof. Aditya Shastri, Vice-Chancellor, Banasthali University, DST, Govt. of India for supporting Banasthali University under its CURIE scheme for providing the necessary facilities.

REFERENCES

Abousleiman, R.I., Reyes, Y., McFetridge, P. & Sikavitsas, V. 2009. Tendon tissue engineering using cell-seeded umbilical veins cultured in a mechanical stimulator. *Tissue Eng Part A* 15(4): 787–795.

Amini, A.R., Laurencin, C.T. & Nukavarapu, S.P. 2012. Bone tissue engineering: Recent advances and challenges. *Crit Rev Biomed Eng* 40(5): 363–408.

Barzegari, A. & Saei AA. 2012. An update to space biomedical research: Tissue engineering in microgravity bioreactors. *BioImpacts* 2(1): 23–32.

Detzel, C.J. & Van Wie, B.J. 2011. Use of a centrifugal bioreactor for cartilaginous tissue formation from isolated chondrocytes. *Biotechnol Prog.* 27(2): 451–459.

Fridley, K.M., Fernandez, I., Li, M.T.A., Kettlewell, R.B. & Roy, K. 2010. Unique differentiation profile of mouse embryonic stem cells in rotary and stirred tank bioreactors. *Tissue Eng Part A* 16(11): 3285–3298.

Frohlich, M., Grayson, W.L., Marolt, D., Gimble, J.M., Velikonja, N.K. & Novakovic, G.V. 2009. Bone grafts engineered from human adipose-derived stem cells in perfusion bioreactor culture. *Tissue Eng Part A* 16(1): 179–189.

Gerecht-Nir, S., Cohen, S. & Itskovitz-Eldor, J. 2004. Bioreactor cultivation enhances the efficiency of human Embryoid Body (hEB) formation and differentiation. *Biotechnol Bioeng* 86: 493.

Goepfert, C., Blume, G., Faschian, R., Meyer, S., Schirmer, C., Wichards, W.M., Müller, J., Fischer, J., Feyerabend, F. & Pörtner, R. 2013. A modular flow-chamber bioreactor concept as a tool for continuous 2D- and 3D-cell culture. *BMC Proceedings* 7(S6): P87.

Haykal, S., Salna, M., Zhou, Y., Marcus, P., Fatehi, M., Frost, G., Machuca, T., Hofer, S.O.P. & Waddell, T.K. 2014. Double-chamber rotating bioreactor for dynamic perfusion cell seeding of large-segment tracheal allografts: Comparison to conventional static methods. *Tissue Eng Part C* 20(8): 681–692.

Helmedag, M.J., Weinandy, S., Marquardt, Y., Baron, J.M., Pallua, N., Suschek, C.V. & Jockenhoevel, S. 2015. The effects of constant flow bioreactor cultivation and keratinocyte seeding densities on prevascularized organotypic skin grafts based on a fibrin scaffold. *Tissue Eng Part A* 21 (1): 343–352.

Kehoe, D.E., Donghui, J., Lock, L.T., & Tzanakakis, E.S. 2010. Scalable stirred-suspension bioreactor culture of human pluripotent stem cells. *Tissue Eng Part A* 16(2): 405–421.

Kleinhans, C., Mohan, R.R., Vacun, G., Schwarz, T., Haller, B., Sun, Y., Kahlig, A., Kluger, P., Wistrand, A.F., Walles, H. & Hansmann, J. 2015. A perfusion bioreactor system efficiently generates cell-loaded bone substitute materials for addressing critical size bone defects. *Biotechnol. J* 10: 1727–1738.

Li, W.J., Jiang, Y.J. & Tuan, R.S. 2008. Cell-nanofiber-based cartilage tissue engineering using improved cell seeding, growth factor and bioreactor technologies. *Tissue Eng Part A* 14(5): 639–648.

Magyar, J.P., Nemir, M., Ehler, E., Suter, N., Perriard, J.C. & Eppenberger, H.M. 2001. Mass production of embryoid bodies in microbeads. *Ann NY Acad Sci* 944: 135–143.

Portner, R., Goepfert, C., Wiegandt, K., Janssen, R., Ilinich, E., Paetzhold, H., Eisenbarth, E. & Morlock, M. 2009. Technical strategies to improve tissue engineering of cartilage carrier constructs—A case study. *Adv Biochem Eng/Biotechnol* 112:145–182.

Reichardt, A., Polchow, B., Shakibaei, M., Henrich, W., Hetzer, R. & Lueders, C. 2013. Large scale expansion of human umbilical cord cells in a rotating bed system bioreactor for cardiovascular tissue engineering applications. *The Open Biomedical Engineering Journal 7:* 50–61.

Reuther, M.S., Briggs, K.K., Wong, V.W., Chang, A.A., Schumacher, B.L., Masuda, K., Sah, R.L. & Watson, D. 2012. Culture of human septal chondrocytes in a rotary bioreactor. *Otolaryngol Head Neck Surg* 147(4): 661–667.

Salehi-Nik, N., Amoabediny, G., Pouran, B., Tabesh, H. et al. 2013. Engineering parameters in bioreactor's design: A critical aspect in tissue engineering. *BioMed Res International* 1–15 http://dx.doi.org/10.1155/2013/762132.

Shahin, K. & Doran, P.M. 2011. Strategies for enhancing the accumulation and retention of extracellular matrix in tissue-engineered cartilage cultured in bioreactors. PLoS ONE 6(8): e23119.

Tare, R.S., Li, S., Jones, P.G., Andriotis, O.G., Ching, K.Y., Jonnalagadda, U.S., Oreffo, R.O.C. & Hill, M. 2014. Application of an acoustofluidic perfusion bioreactor for cartilage tissue engineering. *Lab Chip* 14: 4475–4485.

Tarng, Y.W., Huang, B.F. & Su, F.C. 2012. A novel recirculating flow-perfusion bioreactor for periosteal chondrogenesis. *International Orthopaedics* (SICOT) 36:863–868.

Wang, T., Gardiner, B.S., Lin, Z., Rubenson, J., Kirk, T.B., Wang, A., Xu, J., Smith, D.W., Lloyd, D.G. & Zheng, M.H. 2013. Bioreactor design for tendon/ligament engineering. *Tissue Eng Part B* 19(2): 133–146.

Communication and Computing Systems – Prasad et al. (Eds)
© 2017 Taylor & Francis Group, London, ISBN 978-1-138-02952-1

Review on higher order mutation testing

Neha & Sukhdip Singh
Deenbandhu Chhotu Ram University of Science and Technology, Murthal, Sonepat, Haryana, India

ABSTRACT: Mutation testing is a software testing technique, capable of designing quality test data by inserting faults artificially in system under test. Operators are used to create mutants form original program. Traditionally, mutation testing is done by inserting single fault, which is known as First Order Mutation. But these FOM are easily killed, so Higher Order Mutation testing technique is introduced. This paper reviews the HOM testing techniques.

1 INTRODUCTION

Testing is a process of verification and validation of software programs in Software Development Life Cycle. Its aim is to find and locate the errors. This testing task is very expensive and time consuming as it includes variety of lengthy test cases. On the behalf of test case generation, testing can be classified into two broad categories: White Box Testing and Black Box Testing. White Box Testing technique covers all possible areas which are executed by test cases while Black Box Testing examines performance of program without inquiring the internal areas. Mutation Testing is another most important testing technique which encourages the effect of other white box testing techniques. It was first designed in 1970 by (Millo et al). It is done by injecting faults in the program and then finding those errors. The errors that are seeded are simple faults like syntactical errors, done by programmers while doing programming. By implementing different mutation operators in the program, mutants are originated. Mutation adequacy rule can determine the quality of test set in terms of fault detection. This adequacy criteria is Mutation Score (MS). Location of mutant and type of mutant is chosen carefully to increase the MS. To figure out the quality of test set, these mutants are run against the test cases and results are analysed. If the output of mutated program and original program is same then the mutant is said to be alive else it is killed. If mutant is alive then there can be two possibilities: either mutant is equal to original program or test case is inadequate (Polo et al. 2009). Mutants are divided into: First Order Mutants and Higher Order Mutants. In practical, FOMs cannot detect real faults, hence there is a need for HOMs. HOMs construct high quality mutants by injecting one or more mutants from FOMs into program.

Mutation is one of the most expensive and effective testing technique. For reducing the cost of this technique many researches have been done. The most effective research area is construction of Higher Order Mutants. Mutation Testing have some limitations, like realism, equivalent mutant problem, number of mutants and high execution cost (Nguyen et al. 2014). All these limitations are overcome by HOM Testing technique. Its practical implementation in industry is highly expensive (Kintis et al. 2010).

2 MYTHS ABOUT MUTATION TESTING

Myths are the traditional beliefs or ideas that are listed below (Jia, & Harman, 2009) (Harman et al. 2010).

- Real Fault Representation Myth (RFR): FOM are the errors that a typical programmer can make.
- Unscalability of Mutation Testing Myth (UMT): Number of mutants are created in First Order Mutants, so for large programs Mutation Testing cannot be scaled.
- All Mutants are Equal Myth (AME): All the mutants that are created are given equal efforts to kill them. No biasing should be there. All the mutants that are created and then killed are done by Mutation Testing Tools, and these tools must take some effort to remove equivalent mutants as well. Although in Mutation Sampling Technique and Selective Mutation this AME doesn't hold. In Mutation Sampling, all mutants are equal but we sample only set of mutants. And in Selective Mutants, we focus on mutation operators that generates mutants and then reduces them.
- Global Mutant Operator Myth (GMO): A set of global mutation operators are defined prior

to any program and then applying the same operators to all programs.

- Competent Programmer Hypothesis Myth (CPH): This myth states that programs are the very near to correct or they are within a few keystroke but the make mistakes in the program. Faulty programs may be close but they are not correct.
- Syntactic Semantic Size Myth (SSS): This is an after effect of Competent Programmer Hypothesis. In this approach, small syntactical changes are made.
- Coupling Hypothesis Extension Myth (CHE): This is the extension of coupling effect hypothesis. Coupling effect hypothesis states that the complex faults are paired to simple faults in such a way that test data which detect all simple faults will detect huge number of complex ones also. But (Offutt & J., 1992) does not states that test data that detects all simple faults will also detect all complex faults.

3 MUTATION PROCESS

Traditional Mutation Testing process is shown in Fig 1. (Jia & Harman, 2011).

For original program p, a faulty program p' also known as mutant, is generated using mutation operators. Transformation rule is used that generate faulty programs from original program is known as mutation operator. These mutation operator can be either variable modification, expression replacement, insert operators or delete operators.

In Fig. 1, from original program p mutated program p' is created, then test set T is applied to it. The result is analysed, if output of original program and mutated program is different, then mutant is killed otherwise mutant is alive, these

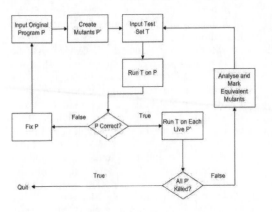

Figure 1.   Traditional mutation testing process.

alive mutants are said to be equivalent mutants. These mutants are optimization of original program. These equivalent mutants are difficult to detect, hence detecting equivalent mutant is an undecidable task. If the program has errors then it must be corrected first and executed thereafter.

Main three steps of mutation testing process are (Polo et al. 2009):

- Mutation Generation: Using mutation operation, mutants are generated.
- Mutant Execution: Test cases are executed.
- Result Analysis: Mutation Score is calculated (Mateo et al. 2010).

Mutation Score is no of killed mutants divided by the total number of non-equivalent mutants (Mateo et al. 2010).

MS (p, t) = $k(m-e)$;
Where:
MS: mutation score
p: program under test
t: test suit
k: no of killed mutants
m: no of mutants
e: no of non-equivalent mutants

4 COST REDUCTION TECHNIQUES

Although the testing phase is costly but the Mutation Testing Technique is much expensive than any other technique (Jia & Harman, 2011).

Some techniques are suggested to reduce the cost of Mutation Testing process. The techniques is divided into three types: Do fewer, Do smarter, Do faster.

But we consider only two: Reduction of generated mutants (do fewer), Reduction of execution cost (do faster and do smarter).

Mutant Reduction Technique: Reducing number of mutants without any loss of test effectiveness can be one possible way to reduce cost. Four techniques are introduce to reduce number mutants.

- Mutant Sampling: This approach chooses a small set of mutants randomly. Then mutation analysis is done on selected mutants while others are discarded.
- Mutant Clustering: Instead of selecting mutants randomly, Mutant clustering chooses a set of mutants by using clustering algorithms. A clustering algorithm is then applied to classify all FOMs which is based on killable or non-killable test cases. Small number of mutants are selected from each cluster for Mutation testing analysis, others in that cluster are discarded.

- Selective Mutants: Reducing number of mutation operators will decrease number of mutants. Small set of mutation operators are used to generate mutants, which will not result in loss of test effectiveness.
- Higher Order Mutation: Jia and Harman (Jia, & Harman, 2008) introduced Higher Order Mutation. Traditionally, First Order Mutant are created by applying mutant operator once while Higher Order Mutants are generated by applying mutant operator more than once.

Execution Reduction Technique: Computational cost can be reduced by refining the mutation execution process. Based on the way, the mutant is killed during the execution process, it is classified into three types:

- Strong: Mutant m of program p is killed if it gives different output from original program p.
- Weak: A program p is made from set of components $C = \{c_1....c_n\}$, mutant m is said to be killed if any execution of c_m gives different result from mutant.
- Firm: This technique is intermediate of both weak and strong mutation.

5 MUTATION STRATEGIES

As Mutation testing is all about inducing faults in the program under test. The faults are syntactical changes which results in mutated versions and these changes are done by set rules also known as mutation operators. There are two types of strategies that are used are: First Order Mutation Testing (FOM) Strategies and Second Order Mutation Testing (SOM) Strategies.

In FOM Testing Strategy, a small portion of code is taken for mutant generation. This small portion is chosen in such a way that significant cost reduction is obtained. Empirical studies shown that 10% mutants will result in 16% overall fault detection loss. In SOM Testing Technique, mutants are created from set of First Order Mutants by pairing them. Every FOM is in atleast one SOM. Hence size of mutant is reduced to half than original First Order Mutant. Different strategies are used to combine First Order Mutants i.e. "RandomMix", "LastToFirst", "DifferentOperators". In "RandomMix", pair of mutants are randomly selected from FOMs using each mutant chosen once. In "LastToFirst", the mutants are paired in the order they are handled by tool. In "DifferentOperators", the pair of mutant is produced by applying different operators. Hence, significant reduction is obtained in number of equivalent mutants.

6 HIGHER ORDER MUTATION CLASSIFICATION

Higher Order Mutation can be classified as COUPLED and SUBSUMING as shown in figure. The central region of Venn diagram represents the domain of all HOM (Jia, & Harman, 2009) (Jia, & Harman, 2008) (Kapoor et al. 2011). And the diagrams around central Venn diagram represent sub-categories. Here, we use Second Order Mutants. There are two First Order Mutants f1 and f2 and one Higher Order Mutant 'h'. The two region of each sub-diagram denotes the test cases that kill First Order f1 and f2. While the shaded region denotes the test cases that kill Higher Order Mutants h.

If both First Order Mutants and Higher Order Mutants are killed by test set then the HOM is known as "coupled HOM". In this figure, sub-diagrams 'a', 'b' and 'f' are "coupled HOM" and the rest ones are "de-coupled HOM". Shaded region from 'c' and 'd' do not overlap the unshaded region, therefore 'c' and 'd' are "de-coupled

Table 1. Mutation testing tools.

S. No	Name	Application	Year	Available
1	PIMS	Fortran	1977	No
2	EXPER	Fortran	1979	No
3	CMS.1	Cobol	1980	No
4	FMS.3	Fortran	1981	No
5	Mothra	Fortran	1987	Yes
6	Proteum 1.4	C	1993	No
7	TUMS	C	1995	No
8	Insure++	C/C++	1998	Commercially
9	Proteum/IM 2.0	C	2001	Yes
10	Jester	Java	2001	Yes
11	Pester	Python	2001	Yes
12	TDS	CORBA IDL	2001	No
13	Nester	C#	2002	Yes
14	JavaMut Nguyen, Q.V., & Madeyski, L. (2014).	Java	2002	Yes
15	MuJava	Java	2004	Yes
16	Plextest	C/C++	2005	Commercially
17	SQLMutation	SQL	2006	Yes
18	Certitude	C/C++	2006	Commercially
19	SEASAME	C, Lustre	2006	No
20	ExMan	C, Java	2006	Yes
21	MUGAMMA	Java	2006	Yes
22	MuClipse	Java	2007	Yes
23	MILU	C	2008	Yes

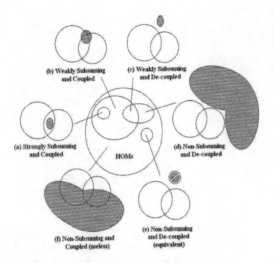

(a) Strongly Subsuming and Coupled
(b) Weakly Subsuming and Coupled
(c) Weakly Subsuming and De-coupled
(d) Non-Subsuming and De-coupled
(e) Non-Subsuming and De-coupled (equivalent)
(f) Non-Subsuming and Coupled (useless)
HOMs

Figure 2. Classification of higher order mutation.

HOM". Sub-diagram 'e' is special case of de-coupled because no test case can kill the HOM, hence, no overlap, hence, this HOM is considered as equivalent mutant.

Subsuming means harder to kill mutants, subsuming HOM are harder to kill than constituent FOMs. Subsuming can be classified into strongly subsuming and weakly subsuming. If test case kill strongly subsuming HOM then it means it kills all its constituent FOM well. Therefore, if shaded region is in the intersection of two FOMs, then it is strongly subsuming HOM i.e. in 'a' otherwise it is weakly subsuming HOM as in 'b' and 'c'. Hence, there are six possible ways:

a. Strongly subsuming and coupled.
b. Weakly subsuming and coupled.
c. Weakly subsuming and decoupled.
d. Non-subsuming and de-coupled.
e. Non-subsuming and de-coupled i.e. equivalent.
f. Non-subsuming and coupled i.e. useless.

7 ADVANTAGES OF HIGHER ORDER MUTATION TESTING TECHNIQUE

As we move from FOMs to HOMs, it brings drastic change (Jia, & Harman, 2009) (Jia, & Harman, 2008). Higher Order Mutants are constructed by combining different First Order Mutants. They are classified as coupled and subsuming. These mutants are very difficult to kill than its constituent FOMs. Three main advantages of HOM are:

Increase Subtlety: Very simple test cases can kill FOMs easily because these are trivial faults. But Higher Order Mutants are more complex faults, so they are difficult to kill. But once killed, then all its constituent First Order Mutants are also killed by set of various test cases.

Reduced Test Effort: As Higher Order Mutants are combination of First Order Mutants, therefore, limited test cases can kill the Higher Order Mutants. The size of mutants is also reduced, as the combination is used instead of single. Hence, test effort is reduced both in terms of number of mutants and test cases.

Reduced Number of Mutants: When First Order Mutants are combined together to form Higher Order Mutants, then the number of mutants are decreased.

8 MUTATION TESTING TOOLS

There is a need of mutation tools to implement the Mutation Testing in actual. 36 different mutation tools have been designed till now. Some of them are Open Source. The author classify the tools into three classes: academic, open source and industrial. Most common tools are: AjMutator, Bacterio, Certitude, JavaMut, Judy, Jumble, MuClipse, muJava, Sesame.

REFERENCES

DeMillo, R.A., Lipton, R.J., & Sayward, F.G. (1978). Hints on test data selection: Help for the practicing programmer. Computer, (4), 34–41.

Howden, W.E. (1982). Weak mutation testing and completeness of test sets. Software Engineering, IEEE Transactions on, (4), 371–379.

Jia, Y., & Harman, M. (2011). An analysis and survey of the development of mutation testing. Software Engineering, IEEE Transactions on, 37(5), 649–678.

Jia, Y., & Harman, M. (2009). Higher order mutation testing. Information and Software Technology, 51(10), 1379–1393.

Jia, Y., & Harman, M. (2008, September). Constructing subtle faults using Higher order mutation testing. In Source Code Analysis and Manipulation, 2008 Eighth IEEE International Working Conference on, 249–258, IEEE.

Kapoor, S., Deptt, C.S.E., & Gni, M. (2011). Test case effectiveness of higher order mutation testing. International Journal of Computer Technology and Applications, 2(5), 1206–1211.

Kintis, M., Papadakis, M., & Malevris, N. (2010, November). Evaluating mutation testing alternatives: A collateral experiment. In Software Engineering Conference (APSEC), 2010 17th Asia Pacific, 300–309, IEEE.

Mateo, P.R., Usaola, M.P., & Offutt, J. (2010, April). Mutation at system and functional levels. In Software Testing, Verification, and Validation Wprkshops (ICSTW), 2010 Third International Conference on, 110–119, IEEE.

Offutt, A.J. (1992). Investigations of the software testing coupling effect. ACM Transactions on Software Engineering and Methodology (TOSEM), 1(1), 5–20.

Papadakis, M., Malevris, N., & Kinitis, M. (2010). Mutation Testing Strategies-A Collateral Approach. In ICSOFT (2), 325–328.

Papadakis, M., & Malevris, N. (2010, April). An empirical evaluation of the first and second order mutation testing strategies. In Third International Conference on Software Testing, Verification, and Validation Workshops, 90–99, IEEE.

Harman, M., Jia, Y., & Langdon, W.B. (2010, April). A manifesto for higher order mutation testing. In Software Testing, Verification, and Validation Workshops (ICSTW), 2010 Third International Conference on, 80–89, IEEE.

Problems of Mutation Testing and Higher Order Mutation Testing. In Advanced Computational Methods for Knowledge Engineering, 157–172, Springer International Publishing.

Polo, M., Piattini, M., & de Guzman, I. G. R. (2009). Decreasing the cost of mutation testing with second-order mutants. Software Testing Verification and Reliability, 19(2), 111–131.

Reales Mateo, P., Polo Usaola, M., & Fernandez Aleman, J.L. (2013). Validating second-order mutation at system level. Software Engineering, IEEE Transactions on, 39(4), 570–587.

Reales, P., Polo, M., Fernandez-Aleman, J.L., Toval, A., & Piattini, M. (2014). Mutation Testing. Software, IEEE, 31(3), 30–35.

Schuler, D., & Zeller, A. (2010, April). (Un-) Covering Equivalent Mutants. In Software Testing, Verification and Validation (ICST), 2010 Third International Conference on, 45–54, IEEE.

Usaola, M.P., & Mateo, P.R. (2010). Mutation testing cost reduction techniques: a survey, IEEE software, 27(3), 80.

Communication and Computing Systems – Prasad et al. (Eds)
© 2017 Taylor & Francis Group, London, ISBN 978-1-138-02952-1

A review paper on cognitive neuroscience

Sarita
CSE Department, Dronacharya College of Engineering, Gurgaon, Haryana, India

Saurabh Mukherjee & Ankur Sharma
Department of Computer Science, Banasthali University, Rajasthan, India

ABSTARCT: Over the past few years, many academicians as well as scholars have depicted a keen interest as well as inclination towards the study of brain and its applications. Their interest in this field even includes study of human cerebral functions in relation to a person's reactions to different stimuli. Cognitive neuroscience has not only made great advancement over the past few years but also has benefitted from the development of this electronic technology. Medical equipment invented for brain studies are also used for brain diagnosis as well as psychological research, which results in many experimental deductions. For conducting experiments with efficient conclusions one require theory, and this field of theoretical neuroscience has advanced drastically through which brains can represent and process information with better results. This theoretical neuroscience is generating new perceptions of the nature representation as well as the computation that can answer questions regarding emotions, consciousness, and creativity. This paper makes an attempt to explore the basic research topics of learning, memory, brain, Dementia which are active fields related to cognitive neuroscience.

1   INTRODUCTION

Understanding the working of a brain is a milestone yet to be achieved in science. Even with the hardest efforts of mankind, computer is the only machine that can be compared to brain; hardware being the brain, software being the mind and algorithm being the thinking procedures of the brain. In the field of cognitive science, it is assumed that the mind has mental representations analogous to that of computer data structures. Scientists and researchers have proposed new ideas using neurons as data structures and algorithms analogous to neuron firing. Since there is no computational model that defines how the mind might work, cognitive science uses the mind, the brain, and the computation as the 3-way complex system.

2   LITERATURE REVIEW

Alzheimer's Disease's (AD) literature has always been noteworthy and has shown incessant deficits in memory and other facets of cognition as the major diagnostic measure of the disease (DSM-IV-TR, American Psychiatric Association 2000; Reisberg et al. 1982; Helkala et al. 1988).

In pursuance of conceptualizing how progressive impairment ensues in the cognitive impairment and illness as it is a significant feature of academic models of similar diseases. Memory loss is a result of this disease which can be seen as semantic knowledge failure, loss of learning power, and repetitiveness in a person. This furthermore leads to poor judgment, planning, impairment in doing the difficult tasks, disinhibiting, poor recognition power, spatial confusion, impaired directed attention of a person. The word "impairment" mostly defines the inability of a person to recall the names of the objects and difficulty in word-finding and poor speaking skills. The earlier recognition of this disease requires memory impairment and impairment in at least one other "cognitive domain".

In accordance to Tappen et al. (2002:63), the inability of a human to communicate properly with others is the most common symptom of the Alzheimer's disease. As, they fail to express their feelings or covey their message to others, it results in loneliness, disturbed behavior, depression, and diminished quality of the person's life. Also, the capacity of emotional concern for people is reduced in the AD patient (Zanetti et al. 1998). This leads to a negative effect on caretakers and marital relationships, particularly in later phases of the disease (de

Vugt et al. 2003). Therefore, this disease exhibits the adverse circumstances of a person's life especially for the people who are in later stage of life.

According to the special issue of Developmental Cognitive Neuroscience on Neuroscience and Education (2012) as researchers have studied the brain mechanisms that underlie learning and memory, and the effects of age, genetics, the environment, emotion and motivation on learning could transform educational strategies and enable us to design programs that optimize learning for people of all ages and of all needs. This research work always helps in changing the experience from infancy onward, along with the neural mechanisms veiled school-based learning and how these could err. Thus, education inculcates the brain, altering it every time a child or an adult gain something new. So, understanding the biological disorders that affect children's educational attainment is a critical step in developing interventions.

As stated by Deak (2011), Cognitive neuroscience of emotion targets on the neural basis of social and emotional processes, and also sturdily contributes to enhanced understanding of biological basis of the emotional processing. Many brain imaging techniques are available such as the functional Magnetic Resonance Imaging (fMRI) which is used to examine the functional connections between perception and emotion, memory, attention, decision making and for localizing particular psychological roles to particular brain areas. This rapidly emerging field has made significant progress in perceiving the neural architecture of human intellect, recognizing a vastly spread network of frontal and parietal regions that upkeep goal-directed and intelligent conduct of a person.

3 UNDERSTANDING THE BRAIN

It is well proven by neuroscientists that the brain is highly robust and the most sensitive to the changes in the environment, a phenomenon called adaptation. Adaptation involves the manipulation of neuronal connections such as creating, strengthening, weakening or eliminating. Type of manipulation depends upon the type of learning for example visual learning forms stronger connections than auditory learning as the visuals are stored for a longer time in the memory. Strength of neural bonds also depends upon time duration of leaning, with infants experiencing extraordinary growth of new synapses. The gist is that the adaptation is the main feature of the brain throughout its age. Many of the environmental factors which improve brain functioning are social interactions, quality of the nutrition, amount of exercise, and both physical as well as mental rest. Without proper understanding

the brain, it would be impossible to compare brain activities associated with experts and novices in their particular task or field. It would be difficult to evaluate how learning declines with age, or why some face difficulties in learning the same thing with the same age group.

The benefits and impact of good diet, physical exertion, and sleep on learning are progressively understood through their effects in the brain. Through cognitive engagement (for example, playing of the ageing brain can be delayed).

The neuroscience by this time has made various contributions towards the diagnosis and identification of 3D's i.e dyslexia, dyscalculia, and dementia.

Dyslexia is a neurobiological reading disorder. Dyslexia is a "difficulty in reading text, phonological processing (the manipulation of sounds), or visual-verbal responding". While the linguistic results of these difficulties are comparatively minor (e.g. confusing words which sound alike), this lack of coordination can be way more vital for mapping lexical sounds to orthographic symbols is the crux of reading in alphabetic languages.

Dyscalculia is a brain-based situation that accounts on the difficulty in understanding the sense of numbers and arithmetic concepts. Some kids suffering from dyscalculia can't understand the logics in basic number. They work hard to learn and memorize basic number facts.

4 MEMORY

Memory is very frequent related with recalling to the mind with whatever has happened earlier. This includes vivid descriptions of this sort resulting a conscious awareness in the person trying to remember or recall whatever has occurred in the past. A hefty amount of memory is obtained from conscious view. Tasks which are complex in nature can even be disrupted when the conscious awareness of the person doing the task is encroached. Memory can also be described in such a way that the ability to recall and use that retained information or knowledge.

Memory is a very crucial component of our existence or being, yet it is hardly understood or worked upon. Various methods of cognitive psychology have led to some useful deductions as well as concluded various theories to understand thoroughly the different types of memory. These methods have hardly given any importance or weight to the biological substrate of memory—the brain. This is due to the complex nature of brain that has obstructed our ability to gain useful observations on memory. In order to understand memory as well as gain knowledge on it various links between process of brain and cognitive theories have come into picture.

Another important aspect which needs to be studied is the division of memory into three different stages involving sensory, short term and long term memory. Sensory memory basically holds the stimuli in a sensory form so that we can pay attention to it. Ichonic sensory memory can be described using the partial report technique. This includes the flash back of word or rather letters in specific, and expecting the participants to name the letters on a particular row. While some participants cannot actually name the whole array and some others can easily name any row immediately. Short Term Memory (STM) is actually a limited or rather a capacity store within certain limits. People suffering from short term memory can retain memory only for a short span of time. The memory trace in case of short term memory fades over the course of few minutes. Long Term Memory (LTM) is the permanent, capacity store with limitless storage capacity. Information is difficult to retain from long term memory. Though, the long term memory consists of all our knowledge of the world and memories of the past.

5 DEMENTIA: A COGNITIVE DISORDER

Dementia is the loss of cognitive functions, which basically affects the decision making capability of a person. A patient suffering from Dementia loses the ability to think, remember or behave normally.

It mainly affects the people lying in the age group of 60+. Dementia is not a specific disease; it consists of a group of conditions that impair judgment. Common symptoms of Dementia are: memory loss, mental decline, confusion, inability to speak or understand, hallucinations, anger, anxiety, loneliness, mood swings, aggression, personality changes, restlessness, depression, paranoia, severe trembling, insomnia and inability to express emotions.

Dementia is caused by damage to the brain cells (neurons). This damage interferes with the neurons to communicate with each other which substantially affects the response time of a patient. The brain consists of almost 100 billion neurons distributed among different parts of the brain to perform different tasks.

5.1 Causes of dementia

Age is one of the factors that causes damage to neurons in our brain, for example when you experience something like dialing a phone number the experience is converted into a pulse of electrical energy that travels along a pattern of neurons, that information first lands in short term memory for a few minutes, it's then transferred to long term memory through areas in brain such as hippocampus and finally to several storage regions across the brain. Neurons inside our brain communicate through dedicated sites called synapses using neurotransmitters but, as we get older synapsis in our brain start decaying affecting how we can recall those memories. Scientists have found that as we age our brain shrinks and loses 5% of its neurons every decade. Depression is also one of the problems people who are depressed are 40% more likely to develop memory problems.

5.2 Alzheimer's Disease

Alzheimer's Disease (AD) is one of the most common forms of dementia among people above the age of 65. It is statistically noted that as many as 5 million Americans aged 65 and older may have AD, and also that these numbers are expected to double in every five years for people beyond age 65. But Alzheimer's is only one of many dementia disorders; an estimated 20 to 40 percent of people with dementia have some other form of the disorder. Age remains the primary risk factor for developing dementia. Due to this, the number of people living with dementia could double in the next 40 years with an increase in the number of Americans who are aged 65 or older—from the 40 million today to more than 88 million in 2050. Notwithstanding the form of dementia, the personal, financial and societal impositions can be devastating.

5.3 Alzheimer's disease vs. dementia

Alzheimer's disease, normal aging and dementia are often confused with each other. Severe memory loss, a defining symptom of AD, is not an effect of normal aging. Healthy aging may incorporate the natural loss of hair, weight and muscle mass. Skin becomes more fragile and bone density is reduced. A loss in hearing and vision may occur, as well as a slowing down of metabolism. It is natural to have a slight decline in memory, including slower recall of information, however cognitive loss that exerts influence on daily life is not a normal sign of healthy aging.

Dementia is observing a significant loss of cognitive abilities severe enough to interfere with routine actions. It is a result from various diseases that cause damage to brain cells. There are many different forms of dementia, all with their own cause and symptoms. For example, vascular dementia is a result of a decrease in blood flow to a part of brain, as caused by a stroke. Dementia may also be present in patients with Parkinson's disease and hydrocephalus.

5.4 *Parkinson's disease*

Parkinson's Disease Dementia (PDD) is a clinical diagnosis related to DLB (Dementia with Lewy bodies) that usually presents in patients with Parkinson's disease. PDD effects memory recall, language, social judgment, or reasoning capabilities. Autopsy studies show that people with PDD often known to have amyloid plaques and tau tangles. Most of people with Parkinson's disease develop dementia, but the time from the arrival of movement symptoms to the beginning of dementia symptoms vary prominently from person to person. Studies suggest that treatment with cholinesterase inhibitors on people with AD might improve cognitive, behavioral, and psychotic symptoms prevalent with Parkinson's disease dementia. The U.S. Food and Drug Administration (FDA) has approved one Alzheimer's drug, rivastigmine, as treatment for cognitive symptoms in PDD.

6 DEMENTIA: THE DIAGNOSIS

Firstly, doctors check if the patient has a more manageable condition such as depression, abnormal thyroid function, drug-induced encephalopathy, normal pressure hydrocephalus, or vitamin B12 deficiency. Early diagnosis of the symptoms play a central role as some causes for the symptoms can be cured. In numerous cases, it is problematic to categorize the type of dementia till the death of the patient.

6.1 *Procedure used in diagnosing dementia*

Brain scans: These tests are able to identify strokes, tumors, and other problems that can result in dementia. The most common scans are Computed Tomographic (CT) scans and Magnetic Resonance Imaging (MRI). CT scans use X-rays to develop images of the brain and other organs. MRI scans use a computer, magnetic fields, and radio waves to produce detailed images of body structures, including tissues, organs, bones, and nerves.

Single-Photon Emission computed Tomography (SPET) is a nuclear medicine tomographic imaging technique with gamma rays. It is similar to conventional nuclear medicine planar imaging with the use of a gamma camera. PET tracers emit positrons that overwhelm with electrons up to a few millimeters away, causing two gamma photons to be emitted in opposite directions. A PET scanner detects these emissions "coincident" in time,

which delivers more radiation event localization information.

Cognitive and neuropsychological tests: These tests measure the memory, language skills, and other mental functioning. For example, people with AD often show deficiency in problem-solving, memory, and the ability to perform once-unconscious tasks. Laboratory tests: These tests are used for quantifying levels of sodium and other electrolytes in the blood, a complete blood count, a blood sugar test, urine analysis, and a check of vitamin B12 levels, cerebrospinal fluid analysis, drug and alcohol tests, and a scrutiny of thyroid function.

6.2 *Neuroimaging techniques*

Imaging: Clinical imaging gives researchers better understanding of changes taking place in the brains of people with dementia, and helps in treating these disorders. Magnetic resonance imaging might reveal structural and functional differences in the brains of individuals with Parkinson's disease and Alzheimer Disease. It identifies small vessel disease. PET scanning uses ligands—radioactive molecules that bind to proteins to show chemical functions of tissues and organs in the body that helps in producing images of brain activity.

6.3 *Neuropsychological examinations can be used to identify cognitive symptoms*

Neuroimaging is a promising aspect of research for detecting AD. There are multiple brain imaging procedures that can be used to identify abnormalities in the brain. Each scan consists of a unique technique and detects specific structures and abnormalities in the brain. Brain imaging is not currently an initial part of AD testing, however current clinical studies have shown promising

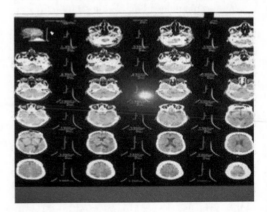

Figure 1. CT scan image.

results that might change the procedure used by physicians to diagnose the disease.

6.4 *PET scans*

Positron Emission Tomography (PET) uses radiation signals to create a three-dimensional color image of the human body. The patient is injected with a radiotracer, composed of a radioactive medicine consisting a naturally occurring chemical. In the study of AD, the chemical usually used is glucose. The radiotracer travels to the organs that use that specific molecule for energy. As the compound is metabolized, positrons are emitted. The energy from these positrons is detected by the PET scan, which converts the input to an image. This image reflects the function of the patient's body by showing how effectively the radiotracer is broken down. The amount of positron energy emitted creates a variety of colors and intensities, which reflects the extent of brain activity. A PET scan has the capacity to detect changes in metabolism, blood flow, and cellular communication processes in the brain.

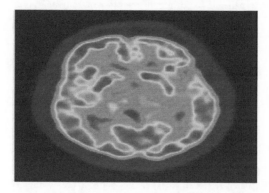

Figure 2.   PET scan of human brain.

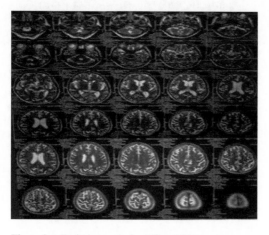

Figure 3.   Brain image derived from MRI scan.

6.5 *CT scan*

- CT scan means a Computed Tomography (CT) scanning which takes a long series of cross-sectional images of the body. With the help of a computer, the individual scans that is the different images are integrated into one detailed image. The CT scan provides the physician with information about the density of tissues in the body usually quantified in Hounsfield unit for attenuation value of the tissue in the image.
- CT scan uses computer-processed combinations of many X-ray images taken from different angles to produce cross-sectional (tomographic) images of specific areas of a scanned object, allowing the researcher to see inside the object without cutting it.
- In Alzheimer's' Disease (AD), CT scan usually shows Shriveling of cortex with prominent cortical sulci and gyri. Shrinkage is more marked in Hippocampus part of brain and ventricles grow large. In PD (Parkinson's disease) image intensity changes are noticed in Basal Ganglia especially in substantia Nigra.

6.6 *MRI scans*

Magnetic Resonance Imaging (MRI) techniques, creates two or three-dimensional images of the body that can be used to diagnose injury and illness. MRI is an imaging technique base on the interaction of radiofrequency induced proton shifts in a strong magnetic environment. Signal intensity is directly proportional to the field strength of the magnets used. The essential component of the MRI system is the superconducting magnet, producing a large and stable magnetic field. These magnets allow scanning of the different parts of the body. The human body is made up of various atoms. However, the hydrogen atoms are altered by the magnetic field. Half of the atoms point towards the patient's head, and half point toward the feet, cancelling each other out. The machine then emits a radio frequency pulse specific to hydrogen, which causes these protons to spin in a different direction. When the spinning ceases, the protons release energy, which is interpreted by the system. Fluid attenuated inversion recovery FLAIR are the mainstay of imaging protocol used in MRI too. Using a contrast dye, each type of tissue responds differently and appears as a unique shade of gray when the image is created.

7   CONCLUSION & FUTURE

Till date, there has been no effective cure found for Dementia, however there has been many steps proposed to reduce Dementia by decrease in smoking, high blood pressure, drinking. Dementia has been accounted to affect 36 million people universally. Greater the life expectancy, dementia becomes a more prone disease. It is beneficial in such times that the patient be given emotional help and care. The most common causes of the Dementia, Alzheimer has had dozens of failed medication program me which influenced the Doctors for its treatment at an early stage, but cases prove that treatment of mild cognitive disease at early stage brings very little success.

Ever since the recognition of Cognitive Neuroscience, we have seen many improvements, for example the MRI technique. The Science of Neurology has yet to grow in understanding of the behaviors, diseases, their causes and cures. Let us work together to contribute our intellect and potential in the progress of Cognitive Neuroscience.

REFERENCES

Barbey, A. K., Colom, R., & Grafman, J. (2014). Distributed neural system for emotional intelligence revealed by lesion mapping. Social cognitive and affective neuroscience, 9(3), 265–272.

Bredeche, N., Shi, Z., & Zucker, J. D. (2006). Perceptual learning and abstraction in machine learning: an application to autonomous robotics. IEEE Transactions on Systems, Man, and Cybernetics, Part C: Applications and Reviews, 36(2), pp–172.

Daw, N. D., & Shohamy, D. (2008). The cognitive neuroscience of motivation and learning. Social Cognition, 26(5), 593.

De Vugt, M. E., Stevens, F., Aalten, P., Lousberg, R., Jaspers, N., Winkens, I. & Verhey, F. R. (2003). Behavioural disturbances in dementia patients and quality of the marital relationship. International journal of geriatric psychiatry, 18(2), 149–154.

Deak, A. (2011). Brain and emotion: Cognitive neuroscience of emotions. Review of Psychology, 18(2), 71–80.

Dehaene, S., & Naccache, L. (2001). Towards a cognitive neuroscience of consciousness: basic evidence and a workspace framework. Cognition, 79(1), 1–37.

DSM-5 American Psychiatric Association. (2013). Diagnostic and statistical manual of mental disorders. Arlington: American Psychiatric Publishing.

Gutiérrez-Rexach, J., & Schatz, S. (2016). Cognitive impairment and pragmatics. SpringerPlus, 5(1), 1–5.

Helkala, E. L., Laulumaa, V., Soininen, H., & Riekkinen, P. J. (1988). Recall and recognition memory in patients with Alzheimer's and Parkinson's diseases. Annals of neurology, 24(2), 214–217.

OCDE. (2007). Understanding the brain: The birth of a learning science.

Reisberg, B., Ferris, S. H., de Leon, M. J., & Crook, T. (1982). The Global Deterioration Scale for assessment of primary degenerative dementia. The American journal of psychiatry.

Shi, Z. (2011). Advanced artificial intelligence (Vol. 1). World Scientific.

Shi, Z., & Shi, J. (2003, August). Perspectives on cognitive informatics. In Cognitive Informatics, 2003. Proceedings. The Second IEEE International Conference on (pp. 129–133). IEEE.

Tampi, R. R. (2013). Clinical Manual of Alzheimer Disease and Other Dementias. Psychiatric Times, 30(3), 30–30.

Tanner, C. M., & Aston, D. A. (2000). Epidemiology of Parkinson's disease and akinetic syndromes. Current opinion in neurology, 13(4), 427–430.

Tappen, R. M., Williams, C. L., Barry, C., & Disesa, D. (2002). Conversation intervention with Alzheimer's patients: Increasing the relevance of communication. Clinical gerontologist, 24(3–4), 63–75.

Weiner, M. F., & Lipton, A. M. (Eds.). (2012). Clinical manual of Alzheimer disease and other dementias. American Psychiatric Pub.

Wray, A. (2014). Formulaic language and threat: the challenge of empathy and compassion in Alzheimer's disease interaction. Dialogue and dementia: cognitive and communicative resources for engagement. Psychology Press, New York, 263–286.

Zanetti, O., Frisoni, G. B., Bianchetti, A., Tamanza, G., Cigoli, V., & Trabucchi, M. (1998). Depressive symptoms of Alzheimer caregivers are mainly due to personal rather than patient factors. International journal of geriatric psychiatry, 13(6), 358–367.

Zheng, R., Kuai, X., Yang, G., & Fu, S. (2012, July). A tri-modal Schema for cognitive neuroscience research. In Intelligent Control and Information Processing (ICICIP), 2012 Third International Conference on (pp. 458–462). IEEE.

Communication and Computing Systems – Prasad et al. (Eds)
© 2017 Taylor & Francis Group, London, ISBN 978-1-138-02952-1

Biomedical applications of green synthesized nanoparticles

Chandra Kant Sharma & Monika Sharma
Department of Bioscience and Biotechnology, Banasthali University, Rajasthan, India

Ambika Panwar
MITS, Gwalior, MP, India

ABSTRACT: Nanoparticles (NPs) can be defined as the small and solid particles from 1–100 nm in diameter, which can't be seen by naked eye. Uniformly and reliably fabrication of nanomaterials at genuine skill and cost are the basic fundamentals of Nanotechnology. NPs carry unique biological, optical and physiochemical properties which are broadly applied in medical, environment, genetic engineering, industrial fields etc. Nanotechnology not only produces new devices but bridges biological, chemical and physical science to yield NPs and Surface Enhanced Raman Scattering. NPs can be synthesized by Chemical, Physical, and Biological methods. Biological methods include wide range of plants and microorganisms for NPs synthesis. Plants used for NPs synthesis have high metal accumulation, reductive potential. The elevated levels of saponins, steroids, alkaloids, flavonoids and phenols act as reducing agents and phyto-constituents as the capping agents providing stability to NPs. Primary and secondary metabolites are used for the NPs production as they are involved in redox reaction. This review paper deals with different biomedical applications of green synthesized nano-particles.

1 INTRODUCTION

Nanoparticles (NPs) can be defined as the small and solid particles from 1–100 nm in diameter, which can't be seen by naked eye (Mody et al. 2010). Apart from this, NPs are also called as the zero dimensional nanomaterials (NMs) (Li et al. 2003). Nanoparticles can be naturally occurring (e.g., volcanic ashes, dust storms etc) or combustional products (e.g., welding, cigarette smoke etc) which are ultrafine particles. Man-made NPs (e.g., fullerenes, quantum dots, metal nanoparticles etc) are desirably produced with specific properties like shape, size and surface area (Ju-Nam & Lead 2008). The study of NPs is known as Nanotechnology, Nanoscience or Nanochemistry. Nanospheres and Nanocapsules are different terms for Nanoparticles. Nanotechnology will be known as Molecular Nanotechnolgy as building machines and mechanisms will be derived at nanoscale dimensions (Saini et al. 2010; Schiek et al. 2008).

Nanotechnology is the cogitation of very tiny or small objects whose size we can alter as per our needs (Sharma et al. 2015; Krukemeyer et al. 2015; Santamaria, 2012; Webster, 2007). It ensures the right designing, production and application of objects at atomic and molecular level for yielding new nanosized materials. NPs show different properties which differ from their bulk materials. Uniformly and reliably fabrication of nanomaterials at genuine skill and cost are the basic fundamentals of Nanotechnology. NPs carry unique biological, optical and physiochemical properties which are broadly applied in medical, environment, genetic engineering, industrial fields etc (Pereira et al. 2015). Nanotechnology not only produces new devices but bridges biological, chemical and physical science to yield NPs and Surface enhanced raman Scattering (Wang et al. 2014).

NPs can be synthesized by Chemical, Physical, and Biological methods. Chemical and Physical methods consist of sol gel, chemical or physical vapor deposition, lithography, electro-deposition (Iravani et al. 2014). These methods demand high energy, toxic solvents and are costly, time consuming and not environment friendly. Biological methods overcome the problems in physical and chemical methods because they are economical, safe, inert, rapid and production at large scale, no need of expertise and equipped lab professionals (Korbekandi et al. 2013).

Biological methods include wide range of plants and microorganisms for NPs synthesis. It can be divided into intracellular and intercellular production of NPs which consist of microbial cells (algae, human cell, bacteria, fungi and yeast) and plant extracts (Song et al. 2008). Microbial cell (metabolites and protein) have the potential to reduce metal ions into atomic metal which gives NPs. There are some limitations with this method like it can't be used at industrial level, not safe and time consuming. To deal these limitations scientists have evolved "Green Nanotechnology" which is sustainable and innovative as well (Makarov et al. 2014). It is based upon fundamental principle of green chemistry which aims at eliminating and minimizing the waste generated from conventional methods.

Plants used for NPs synthesis have high metal accumulation, reductive potential. The elevated levels of saponins, steroids, alkaloids, flavonoids and phenols act as reducing agents and phyto-constituents as the capping agents providing stability to NPs (Elia et al. 2014; Leon et al. 2013). Primary and secondary metabolites are used for the NPs production as they are involved in redox reaction (Rajana et al. 2015).

Investigation as well as analysis will be further efficient—Presently, health investigational studies have been to trial and error. One needs to wait from hours to days for complete state of health (Jain, 2005). By real-time examination of the body's system, it will be achievable to identify unwanted effects previous leading to additionally rapid approach to cure. Diagnosis will be easier and more informative. Then, thousands of diagnostic tests can be built into single and cheap hand held device. It will also eliminate malpractice (Wagner et al. 2006; Boken et al. 2013).

Tiny therapeutic devices can be entrenched everlastingly—Till now, few remedial appliances are permanently fixed. Surgical procedure is less preferable as well as not greatly functionality can be filled into a device to be within a body. Nano—built devices will be extra proficient as well as effective in transplanting. Implantable devices will be constantly sense along with adjust the body's substance stability, in the bloodstream or particular tissue. It will allow even earlier diagnosis of diseases. (Vaddiraju et al. 2010).

Additional health related troubles will be prohibited—numerous medical troubles are avoidable also. Some are attained from the surroundings, including toxins, a number of malignancies in addition to approximately every contagious malady. Extensive examination of wellbeing as well as the environment will permit finding of the basis of such problems before they can harm people. Enhanced infrastructure such as water filtration will also assist to decrease environmentally-attained sickness. Other diseases are associated to way of life. Present way of life recommendation is hard to pursue in addition to be not forever precise. Better study will significantly progress our thoughtfulness of reason as well as consequence, permitting us to survive further healthy ways of life with far fewer attempt. Lastly, some troubles mount up over time as well as untimely recognition

and treatment can rectify the trouble (Chhabra et al. 2015).

Health will improve and life spans increase—Health improvements do not rely directly on molecular manufacturing but it makes accessible to more people. Any treatment can be prescribed to multiple citizens at low-price. Advanced study will pace the progress of treatments for complex troubles for example cancer in addition to aging (Patil et al. 2008).

Various organs will be replaceable—a lot of organs in the body execute easy tasks. Presently, complicated equipment can substitute lung function for hours, heart function for months etc. Molecular manufacturing can build smaller machines than cells which can be used for substitution or expansion like muscles, skin along with sensory purposes (Alberti, 2013).

Systems can be independently enhanced—Our body functions and organs are inter dependent on each other. Small, implanted devices will allow the system to function independently to some extent (Health, 2015).

Facilitation of genetic therapy—Now-a-days, many diseases are incurable which are life threatening like cancer. But by molecular manufacturing, it will be possible to edit the DNA directly in the corpse. The capability to infuse non-cancerous cells would make several kind of genetic remedy greatly safer (Liu et al. 2011).

In recent years, many plants have been used for NPs fabrication which includes *Camellia sinesis*, *Arbutus unedo*, *Moringa oleifera* etc (Sinha et al. 2009; Boken et al. 2014). Plant derived compounds such as flavonoids, terpenoids, steroids etc have received attention due to their diverse pharmacological characteristics like antioxidant and antibacterial activity.

This review paper deals with different biomedical applications of green synthesized nanoparticles:

1.1 Biotherapeutical uses of Silver Nanoparticles (AgNPs)

1.1.1 AgNPs in cancer control

AgNPs have ability to interrupt the mitochondrial respiratory chain, which prompt the genesis of Reactive Oxygen Species (ROS) as well as ATP synthesis, which can cause DNA damage is performing well as cancer therapeutics. Synthesis of silver nanoparticles with *Sesbania grandiflora* leaf extracts were exhibit to be cytotoxic to MCF-7 cancer cells (Sriram et al. 2010). Morphological idiosyncratic, in addition to the splitting of membrane integrity, degrade cell growth, condensation occurs in cytoplasmic fluid and cell clustering, were seen in MCF-7 cells extravagance with AgNPs while control cells continued alive. In addition, features of program cell death, such as shrinking of cell and condensation in nuclear material and disintegration, were also seen in MCF-7 tumor cells after dealing with 20 µg/ml of AgNPs within the duration of 48 h.

S. grandiflora extracts were utilize for the production of silver nanoparticles which activate the formation of free radicals, which consequences in oxidative degradation as well as caspase-mediated apoptosis.

The anticancer action of AgNPs synthesized by immature fruits of *Solanum trilobatum* in opposition to a human breast cancer cell line (MCF-7) was evaluated in vitro using MTT assays, nuclear morphological distinctiveness, RT-PCR as well as western blot investigation (Ahmad et al. 2012). MCF-7 cells treated with either AgNPs or cisplatin demonstrated reduced Bcl-2 expression as well as increased Bax expression, demonstrate the participation of mitochondria in the means of death encouraged by AgNPs. Mitochondria work as significant centers of signaling; their truthfulness can be in conciliation by a variety of regulators of apoptosis. The generation of ROS by AgNPs may also necessitate mitochondria, which may instigate intrinsic caspase-dependent apoptotic pathways leading to cell fatality. Nanoparticles produced with *Rosa indica* extorts have the prospective to be utilized in a broad range of remedial anticancer purposes. AgNPs produced with green petals of *R. indica* work as radical scavengers as well as encourage apoptosis in HCT-15 cells along with the generation of ROS (Umashanker et al. 2011).

1.2 Antidiabetic activity of AgNPs

Control of blood sugar levels were evaluated, utilizing the synthesized AgNPs in the stem extracts of *Tephrosia tinctoria*. AgNPs have ability to remove free radicals, decreases the level of enzymes which take participate ot catalyze the hydrolysis of complex carbohydrates (α-glucosidase and α-amylase), and enhanced the utilization rate of glucose (Rajaram et al. 2015).

Because of the adulteration in the environment causes pollution, water is hugely polluted by the chemical adulteration, Methylene Blue (MB) is one of the chemical which have adverse effects of on the environment, the elimination of MB from wastewater is become an important area of research as well as a prime test for researchers. Aqueous stem extracts of *Salvadora persica* were utilize to synthesized AgNPs, have ability to reduce MB in a light-dependent way; which convert dangerous materials into nonhazardous ones, AgNPs potentially have important uses in the refining of water. MB demonstrates unique absorption peaks

at 663 and 614 nm, which helps to observe the changes occurs during the photo-degradation of MB. Elevated concentrations activate the aggregation of AgNPs, leading to an enlargement in particle dimension along with the reduction in specific surface area as well as surface dynamic sites of particles. The utmost adequate concentration of AgNPs for the photo-deprivation of MB was 8 mg.

1.2.1 Antioxidant action of Silver Nanoparticles (AgNPs)

Leaf extracts of *Leptadenia reticulate* utilize for the synthesis of silver nanoparticles, at a 500 µg/ml concentration, show the best recorded radical scavenging action of 64.81%. Extract of foliages endorse DPPH radical scavenging action of AgNPs, which shows the dependency on dose. Antioxidants have ability to scavenge DPPH radicals is likely attributable to their chemical properties they can efficiently donate hydrogens and efficiently integrate with electrons; the latter is achievable being the existence of host lipophilic radicals. They have property to change their color from purple to yellow, experimented under the 517 nm. The DPPH radical scavenging action of $HAuCl_4$ and $AgNO_3$ was superficially evaluated to nanoparticles, this may happen because of salt surroundings or weaker solubility of metal oxides. A refined response can be experimented amid phenolic compounds as well as phosphotungstic and phosphomolybdic acids in Folin-Ciocalteu reagents. Plant extracts contains different chemical compound such as Phenolic compounds which have elevated antioxidant in addition to reduction actions which are very essential for the production of AgNPs. The greater concentration of total phenolic substance of *Eclipta prostrata* foliage extracts reinforcement the assemblage of silver ions into minute AgNPs, as phenolic compounds efficiently donate the electrons.

1.2.2 Antimicrobial action of AgNPs

The risk posed by the likely onset of antibiotic-resistant microbes is developing worldwide as well as required the initiation in addition to creation of more unique leading platforms for the research as well as development of dynamic antimicrobial agents in opposition to multidrug-resistant strains (Kapil, 2005). Smaller AgNPs have a higher fastening exterior as well as show added bactericidal activity, when compared to larger AgNPs. Membrane organization of gram +ve and −ve bacteria are varies in the compactness and molecular composition which account for the variation in their susceptibility against to AgNPs. Bactericidal action is surely due to alterations in cell wall configuration of bacteria, is a consequence of interactions with deep-seated AgNPs, leading to enhanced membrane permeability and causes death. AgNPs combine with sulfur—as well as phosphorusrich biomaterials, which include intracellular and extracellular components, for example proteins or DNA and membrane proteins respectively (Singhal et al. 2011). These biomaterials control the respiration, splitting up and finally endurance of cells. Beginning with compromising the cell wall of bacteria, silver ions (as part of AgNPs) insert into cells, leading to the chunk of injured DNA and eventually effects synthesis of protein (Boken et al. 2013). AgNPs were produced with *Skimmia laureola* leaf extracts show antibacterial action in opposition to Staphylococcus aureus, followed by *Klebsiella pneumoniae, Pseudomonas aeruginosa* and *Escherichia coli* with utmost inhibition of growth activity (Dar et al. 2012).

Mangrove plant *Avicennia marina* extracts used to AgNPs synthesized, exhibited maximal inhibition action against *E. coli* and lowest against *S. aureus* (Ravikumar et al. 2010). AgNPs, extracts are filled with chemical compounds such as polyphenolic, rupture the cell walls of bacteria, which have greatest specific sensitivity against gram-negative bacteria. Polyphenolic compounds play role for great damage and toxicity in the bacterial cell, have ability to generate free radicals as well as other oxygen-based reactive species, which can persuade abnormality in the cell (Boken et al. 2014). Other damages may occur due to the membranes disruption, leading to the great loss of K^+ ions, consequently reduce in membrane potential. cytoplasmic leakage is a significant result of membrane disruption, which includes the discharge all essential biomolecules such as proteins and lipopolysaccharide required for the survival of life. Lipopolysaccharide is a main constituent of outer membrane of the bacterial cell and is basically asymmetric, at the same time as the inner membrane composed of rigid chains of phospholipids, which is selectively-porous. The interaction system amid AgNPs with bacteria is not exactly understood. AgNPs may act on the cell wall along with thus rupture cell membrane and disturb the permeability and finally cell respiration (Lara et al. 2011). AgNPs can also straightforwardly get into cells since they may attach to cell wall proteins that contain sulfur as well as phosphorus-containing biomolecules for example DNA. As a consequence, silver nanopaticles can effortlessly combine to biomolecules of the bacterial cell furthermore hinder the usual functioning of the cell. An additional probable system is Ag cations are antibacterial in nature which release from the AgNPs (Prakash et al. 2011).

2 CONCLUSION

Applications of nanotechnology are extremely appropriate for biomolecules as a consequence of their exclusive properties. Nanotechnology is a flourishing region of research in the material sciences as well as biological sciences. Researchers are attracted towards the newly emerge research area of silver nanoparticles due to their widespread used in distinct areas like as sensors, integrated circuits, biolabeling, antimicrobial deodorant fibers, filters, cell electrodes, inexpensive paper batteries (silver nano-wires) as well as various antimicrobials activity.

ACKNOWLEDGEMENTS

Authors are thankful to Prof. Aditya Shastri, Vice-Chancellor, Banasthali University, DST, Govt. of India for supporting Banasthali University under its CURIE scheme for providing the necessary facilities.

REFERENCES

Ahmad, N. & Sharma, S. 2012. Green synthesis of silver nanoparticles using extracts of *Ananas comosus*. *Green and Sustainable Chem* 2(4): 1–7.

Alberti, C. 2013. Tissue engineering as innovative chance for organ replacement in radical tumor surgery. *Eur Rev Med Pharmacol Sci* 17(5): 624–31.

Boken, J., Dalela, S., Sharma, C.K. & Kumar, D. 2013. Detection of pathogenic *Escherichia coli* (*E.coli*) strain using robust silver and gold nanoparticles. *Chem Eng Process Technolo* 4(8): 1–6.

Boken, J., Sharma, M., Sharma, C.K. & Kumar, D. 2014. Synthesis of silver nanoparticles by using *Ocimum sanctum* leaf extract and detection of antibacterial activity. In G.C. Mishra (ed.) *Global Sustainability Transitions: Impacts and Innovations* ISBN: 978-93-83083-77-0, excellent publishing house, New Delhi, pp. 278–284.

Chhabra, R., Tosi, G. & Grabrucker, A.M. 2015. Emerging use of nanotechnology in the treatment of neurological disorders. *Curr Pharma Des* 21(22): 3111–3130.

Dar, Y., Zagar, M.I., Agnihotri, V., Qurishi, M.A. & Singh, B. 2013. Chemical composition and antimicrobial activity of the leaf essential oil of *Skimmia laureola* growing wild in Jammu and Kashmir, India. *Nat Product Res* 27(11): 1023–1027.

Elia, P., Zach, R., Hazan, S., Kolusheva, S., Porat, Z. & Zeiri, Y. 2014. Green synthesis of gold nanoparticles using plant extracts as reducing agents. *International Journal of Nanomedicine* 9: 4007–4021.

Health, J.R. 2015. Nanotechnologies for biomedical science and translational medicine. *PNAS* 112(47): 14436–14443.

Iravani, S., Korbekandi, H., Mirmohammadi, S.V. & Zolfaghari, B. 2014. Synthesis of silver nanoparticles: chemical, physical and biological methods. *Res Pharma Sci* 9(6): 385–406.

Jain, K.K. 2005. Nanotechnology in clinical laboratory diagnostics. *Clinica Chimica Acta* 358(1–2): 37–54.

Ju-Nam, Y. & Lead J.R. 2008. Manufactured nanoparticles: An overview of their chemistry, interactions and potential environmental implications. *Sci Total Environ* 400(1–3): 396–414.

Kapil, A. 2005. The challenge of antibiotic resistance: need to contemplate. *Ind J Med Res* 121: 83–91.

Korbekandi, H., Ashari, Z., Iravani, S. & Abbasi, S. 2013. Optimization of biological synthesis of silver nanoparticles using *Fusarium oxysporum*. *Iranian J Pharm Res* 12(3): 289–298.

Krukemeyer, M.G., Krenn, V., Huebner, F., Wagner, W. & Resch, R. 2015. History and possible uses of nanomedicine based on nanoparticles and nanotechnological progress. *J Nanomed Nanotechno* 6(6): 1–7.

Lara, H., Trevino, E.L., Turrent, L.I. & Singh, D.K. 2011. Silver nanoparticles are broad-spectrum bactericidal and virucidal compounds. *J Nanobiotechnol* 9(3): 2–8.

Leon, R., Palomares, I., Navarro, R.E., Urbina, H., Tanori, J. & Maldonado, A. 2013. Synthesis of silver nanoparticles using reducing agents obtained from natural sources (*Rumex hymenosepalus* extracts). *Nanoscale Res Lett* 8(1): 318.

Li, W.H., Yang, C.C., Tsao, F.C. & Lee, K.C. 2003. Quantum size effects on the superconducting parameters of zero-dimensional Pb nanoparticles. *Phys. Rev. B* 68(18): 184507–184512.

Liu, C. & Zhang, N. 2011. Nanoparticles in gene therapy principles, prospects and challenges. *Prog Mol Biol Transl Sci* 104: 509–562.

Makarov, V.V., Love, A.J., Sinitsyna, O.V., Makarov, S.S., Yaminsky, I.V., Taliansky, M.E. & Kalinina, N.O. 2014. Green nanotechnologies: Synthesis of metal nanoparticles using plants. 6(1): 35–44.

Mody, V., Siwale, R., Singh, A. & Mody, H. 2010. Introduction to metallic nanoparticles. *J Pharm Bioall Sci* 2(4): 282–289.

Patil, M., Mehta, D.S. & Guvva, S. 2008. Future impact of nanotechnology on medicine and dentistry. *J Indian Society of Periodontol* 12(2): 34–40.

Pereira, L., Mehboob, F., Stams, A.J., Mota, M.M., Rijnaarts, H.H. & Alves, M.M. 2015. Metallic nanoparticles: microbial synthesis and unique properties for biotechnological applications, bioavailability and biotransformation. *Crit Rev Biotechnol* 35(1): 114–128.

Rajana, R., Chandranb, K., Harperc, S.L. & Kalaichelvana, T. 2015. Plant extract synthesized silver nanoparticles: An ongoing source of novel biocompatible materials. *Industrial Crops and Products* 70: 356–373.

Rajaram, K., Aiswarya, D.C. & Suresh, K.P. 2015. Green synthesis of silver nanoparticle using *Tephrosia tinctoria* and its antidiabetic activity. *Materials letters* 138: 251–254.

Ravikumar, S., Gnanadesigan, M., Suganthi, P. & Ramalaskshmi, A. 2010. Antibacterial potential of chosen mangrove plants against isolated urinary tract infectious bacterial pathogens. *Int J Medi Medical Sci* 2(3): 94–99.

Saini, R., Saini, S. & Sharma, S. 2010. Nanotechnology: The future medicine. *J Cutan Aesthet Surg* 3(1): 32–33.

Santamaria, A. 2012. Historical overview of nanotechnology and nanotoxicology. *Methods mol Bio* 926: 1–2.

Schiek, M., balzer, F., Al-Shamrey, K., Brewer, J.R., Lutzen, A. & Rubahn, H.G. 2008. Organic molecular nanotechnology. *Small* 4(2): 176–181.

Sharma, C.K., Sharma, M., Verma, O. & Sharma, V. 2015. Green synthesis of different nanoparticles and their potential applications in different fields. *Int. J. of Pharma Bio Sciences* 6(3): 555–567.

Singhal, G., Bhavesh, R., Kasariya, K., Sharma, A.R. & Singh, R.P. 2011. Biosynthesis of silver nanoparticles using *Ocimum santum* (Tulsi) leaf extract and screening its antimicrobial activity. *J Nanoparticle Res* 13: 2981–2988.

Sinha, S., Pan, I., Chanda, P. & Sen, S.K. 2009. Nanoparticles fabrication using ambient biological resources. *J Appl Biosci* 19: 1113–1130.

Song, J.Y. & Kim, B.S. 2008. Rapid biological synthesis of silver nanoparticles using plant leaf extracts. *Bioprocess Biosyst Eng.* 32(1): 79–84.

Sriram, M., Kanth, S., Kalishwaralal, K. & Gurunathan, S. 2010. Antitumor activity of silver nanoparticles in Dalton's lymphoma ascites tumor model. *Int J Nanomedicine* 5: 753–762.

Umashanker, M. & Srivastava, S. 2011. Traditional Indian herbal medicine used as antipyretic, antiulcer, antidiabetic and anticancer: A review. *Intl J Res Pharm Chem* 1(4): 1152–1157.

Vaddiraju, S., Tomazos, I., Burgess, D.J., Jain, F. & Papadimitrakopoulos, F. 2010. Emerging synergy between nanotechnology and implantable biosensors: A review. *Biosens Bioelectron* 25(7): 1553–1565.

Wagner, V., Dullaart, A., Bock, A. & Zweck, A. 2006. The emerging nanomedicine landscape. *Nature Biotech* 24: 1211–1217.

Wang, E.C. & Wang, A.Z. 2014. Nanoparticles and their applications in cell and molecular biology. *Integr Biol (Camb)* 6(1): 9–26.

Webster, T.J. 2007. IJN's second year is now a part of nanomedicine. *Int J nanomedicine* 2(1): 1–2.

Communication and Computing Systems – Prasad et al. (Eds)
© 2017 Taylor & Francis Group, London, ISBN 978-1-138-02952-1

Therapeutic efficacy of nanoparticles synthesized by different plants against different hepatotoxicants in the field of medical sciences

Chandra Kant Sharma, Aayushi Rathi & Monika Sharma
Department of Bioscience and Biotechnology, Banasthali University, Rajasthan, India

ABSTRACT: Nanoparticles have various applications. It has been used as disinfecting agent previously and had applications from traditional medicines to cooking items. There are many methods including physical as well as chemical for example electrochemical reduction, chemical reduction, photochemical reduction, heat evaporation etc. which helped in increasing the production of metal nanoparticles. Liver is a vital organ of our body which performs the process of detoxification and is involved in various types of metabolic functions. This is the reason it is exposed to high concentration of toxicants which makes it highly susceptible to injury. These injuries can lead to the alteration in metabolic functions of the organ leading to chronic disorders. Other than silver nanoparticles there are many other metals too which are used for the synthesis of nanoparticles and have therapeutic efficacy.

1 INTRODUCTION

Nanotechnology is the engineering and manufacturing of structures roughly in 1–100 nm size regimes in at least 1 dimension. It is a rapidly progressing field which covers various and diverse arrangement of devices covering fields like engineering, biology, physics and chemistry. In the field of life sciences particularly biomedical devices as well as biotechnology, nanotechnology is making a great sense of anticipation (Amin et al. 2012). On the basis of various specific characteristics like size, distribution and morphology new and improved properties are exhibited by nanoparticles. Metal nanoparticles show distinctive electrical, catalytic and optical properties (Savithramma et al. 2011). For proper tuning in physical, chemical and optical properties, a large spectrum of research has been done to control size as well as shape of nanoparticles. There are many techniques including chemical and physical means such as chemical reduction, electrochemical reduction, photochemical reduction, heat evaporation etc. which helped in increasing the production of metal nanoparticles (Saxena et al. 2014). Surface passivator reagents are used to avoid nanoparticles from aggregation in most of the cases. But there are many organic passivators like thiophenol, thiourea, mercaptoacetate, etc. which are toxic and pollute the environment (Amin et al. 2012).

1.1 History

Before Neolithic revolution, silver was known to be as metal. Greeks used silver in cooking and for keeping water safe. The first reported medicinal use of silver was recorded during 8th century. Previously silver was only known as metal and in nano era it came into existence and then people believed that it can be utilized for the production of nanoparticles. Norio Taniguchi, a researcher at the University of Tokyo, Japan coined the term "nanotechnology" in 1974 while engineering the materials precisely at the nanometer level (Basavaraja et al. 2008).

1.2 Uses

Silver Nanoparticles (SNPs) have various applications. It has been used as disinfecting agent previously and had applications from traditional medicines to cooking items (Boken et al. 2013; Boken et al. 2014; Siddiqui et al. 2015). It is proved that SNPs are non-toxic to humans and are effective against bacteria, virus and various eukaryotic microorganisms at fixed concentrations without any side effects (Sharma et al. 2015). This was tried because many slats of silver and its derivatives are commercially available as antimicrobial agents. Even at some concentrations it is safe for humans and is lethal for microorganisms. SNPs are also having an important application in the field of medical such as ointments for prevention of infection against burn and open wounds (Barathmanikanth et al. 2010). There are various fields like microelectronics, optical devices, catalysis and drug delivery system etc which utilize nanoparticles formed by noble metals and even these show new physiochemical properties which are neither observed in individual molecules nor in bulk metal (Bar et al. 2009).

2 STEPS OF GREEN SYNTHESIS OF SILVER NPS

It basically includes three main steps (Basavaraja et al. 2008).

3 HEPATOTOXICITY

Liver is a vital organ of our body which performs the process of detoxification and is involved in various types of metabolic functions. This is the reason it is exposed to high concentration of toxicants which makes it highly susceptible to injury. These injuries can lead to the alteration in metabolic functions of the organ leading to chronic disorders. These disorders are induced by various types of hepatotoxicants like minerals, microbial metabolites, chemotherapeutic agents and environmental pollutants (Bruchez et al. 1998). These liver problems are not only a great issue for healthcare professionals but even for pharmaceutical industries and drug regulatory agencies. There are many reasons of liver cell injury like:

1. Toxic chemicals like carbon tetra chloride (CCl$_4$), thioacetamide (TAA) etc
2. Some antibiotic compounds
3. Chemotherapeutic agents
4. Excessive alcohol consumption
5. Microbes

Even sometimes the synthetic drugs available in the market for the treatment of liver disinfection can be dangerous for liver and can cause harm to it (Coe et al. 2002). Mainly hepatic injuries show distortion in metabolic processes. The main reason of this ailment is xenobiotic compounds as the toxins after absorption from intestine primarily approaches liver causing health issues related to liver (Coe et al. 2002). The ability to protect liver damage is known as hepatoprotection or antihepatotoxicity and this damage is called hepatotoxicity. Mainly this damage is chemical based. The main function of liver is transmuting and removing chemicals making it more and more susceptible to toxicity. Even some chemicals if taken in high dose as prescribed then they can be hepatotoxic. Other possible ways of toxicity to liver cells are chemicals used in industries and labs, even natural chemicals and sometimes herbal remedies can be toxic to organ. Liver injury causing chemicals are called as hepatotoxins. There are many drugs responsible for damage and according to an estimate there are 900 drugs which cause liver injury and are banned from market. This is the reason that drug screening is done with great precautions as well as with proper testing of the drug on stem cell derived hepatocyte like cells just to have a confirmation of the drug that it is not harmful for the liver. Due to liver injury caused by lung there are 5% of admissions in hospital and 50% of acute liver failures (Kunjiappan et al. 2015).

4 TYPE A

4.1. *Dose response curve is predictable i.e. high concentration is dangerous*

4.1.1. *Toxicity mechanism is well known*

4.1.2. *Directly damages liver tissue*

4.1.3. *Metabolic process is blocked*
Example—Acetaminophen overdose: injury is caused within short duration as it reaches to threshold for toxicity.

5 TYPE B

5.1. *Non predictable dose response curve.*

5.2. *Injury occurs without warning.*

Example—Troglitazone (Rezulin) and Trovafloxacin (Trovan) (Kunjiappan et al. 2015).

6 MECHANISM OF DAMAGE

As our liver is very close to gastrointestinal tract and is very closely related to it, therefore many drugs and other substances have an easy approach towards liver. As well as approximate 75% of blood arrives from gastrointestinal organs to liver and then spleen through portal vein this brings xenobiotics and drugs in approximate undiluted form. Hepatic injury is sometimes due to several mechanisms.

Example—Some chemicals are responsible for disfunctioning of mitochondria which is an intercellular organelle and produces energy which in turn releases excess amount of oxidants and injures hepatic cells.

Some enzyme activates in cytochrome P450 like CYP2E1 which leads oxidative stress leading to liver injury as well as bile acid accumulation take place.

Sometimes defect in non parenchymatous cells like leucocytes, fat storing cells and kupffer cells may also lead to liver injury (Coe et al. 2002).

7 HEPATOTOXICITY INDUCING AGENTS

There are many chemicals which induce hepatotoxicity. Some chemicals are even used to induce experimental hepatotoxicity in to be tested animals like d-Galactosamine/Lipopolysachharide (GalN/LPS), paracetamol, thioacetamide, galactosamine, Carbon tetrachloride (CCl$_4$), antitubercular drugs and arsenic.

7.1   Carbon tetrachloride (CCl$_4$)

First detected liver injury because of CCl$_4$ was in 1965 in rats. CCl$_4$ is synthesized by cytochrome P450 in mitochondria and endoplasmic reticulum forming CCl$_3$O, an oxidative reactive free radical which initiates lipid peroxidaton. A single dose of CCl$_4$ leads to centrilobular necrosis and fatty changes in rat within 24 hr. Within 3 hrs the poison in the liver reaches to its maximum concentration. Slowly the level falls and there is no CCl$_4$ left in the liver by 24 hr. The necrosis development is linked with hepatic enzymes leakage into serum. Hepatotoxicity dosage of CCl$_4$ ranges from 0.1 to 3 ml/kg.

7.2   Thioacetamide

Movement of RNA from nucleus to cytoplasm is hindered by thioacetamide causing membrane injury and one of its metabolite causes hepatic injury. Even it reduces viable hepatocytes number as well as consumption of oxygen decreasing bile volume and its content like deoxycholic acid, bile salts and cholic acid. Dosage is 100 mg/kg subcutaneously (Pattanayak et al. 2015).

7.3   Paracetamol

High dosage of a commonly used analgesic and antipyretic drug i.e. paracetamol can lead to liver damage. It causes centrilobular necrosis characterised by nuclear pyknosis and eosinophilic cytoplasm followed by large excessive hepatic lesion. An oxidative product of paracetamol N-acetyl-P-benzoquinoneimine covalently bounds to sulphydryl groups of protein which results in glutathione peroxidative degradation and hence produces cell necrosis in liver. Dosage is 1 gm/kg post oral (Dauthal & Mukhopadhyay, 2013).

8   PLANTS USED AS HEPATOPROTECTANT

In medical science so many drugs have been synthesized from different plants which are used against liver problems some of these are listed in table 1. (Dauthal & Mukhopadhyay, 2013; Jannu et al. 2012).

Table 1.   Plants and part of plants used as hepatoprotectants.

Name of plant	Plant's part used
Amaranthus caudatus	Whole plant
Anisochilus carnosus	Stems
Asparagus racemosus	Roots
Calotropis procera	Root bark
Cajanus cajan	Leaves
Cajanus scarabaeoides	Whole plant
Clitoria ternatea	Leaves
Cucumis trigonus	Fruits
Ficus religiosa	Stem bark
Garcinia indica	Fruits
Hyptis suaveolens	Leaves
Leucas cilita	Whole plant
Melia azhadirecta	Leaves
Morinda citrifolia	Fruits
Myoporum lactum	Leaves
Myrtus communis	Leaves
Solanum nigrum	Fruits

9   METHOD OF SYNTHESIS

The mainly involved process of biosynthesis of silver nanoparticles is done using AgNO$_3$ with milli Q water. The sample is taken upto 10 gm and mixed with 100 ml. The solution is boiled and is filtered using whatsman filter paper and then 10 ml of filtrate is mixed with 90 ml of AgNO$_3$. The colour changed is observed with time into brownish yellow which indicates the synthesis of AgNPs (Jaeschke et al. 2002).

10   PLANTS USED FOR THE SYNTHESIS OF NANOPARTICLES

There are many plants which have been used as a source for synthesis of silver nanoparticles. One of the examples is Avicennia marina mangrove plant. Then characterization of NPs was done which including in vitro antibacterial activity. The characterization process included FTIR, UV VIS Spectroscopy, XRD, AFM, etc. This proved that plant has high quantity of secondary metabolites like tannins, alkaloids, flavoids and polyphenols. Similarly, there are many more plants which are having hepatoprotection activity and are having property of synthesizing Nanoparticles.

Examples—Cinnamomum camphora (Huang et al. 2007), Delonix regia (Suriyakalaa et al. 2013), Dendrophthoe falcate (Thirumurgan et al. 2014), Terminalia chebula (Vidyanathan et al. 2009) etc.

11 OTHER NANOPARTICLES

Other than silver nanoparticles there are many other metals too which are used for the synthesis of nanoparticles and have therapeutic efficacy.

Gold nanoparticles (AuNPs)

1. *Cinnamomum camphora* trees were cultivated and were harvested. Its leaves were sun dried and using chloroauric acid (HAuCl$_4$) gold nanoparticles were synthesized with proper incubation facilities. Then characterization of AuNPs was done by various techniques like UV-VIS spectra analysis, XRD measurement (X-Ray Diffraction), TEM (Transmission Electron Microscope), SEM (Scanning Electron Microscope) and AFM (Atomic Force Microscope) observation and FTIR analysis (Fourier Transmission Infra Red) (Nandagopal et al. 2014).
2. *Azolla microphylla* commonly known as water fern or fairy moss was used to synthesize gold nanoparticles as it has hepatoprotective and antioxidant properties and was used against acetaminophen induced toxicity.

1 mM HAuCl$_4$.3H$_2$O + Methanolic extract of *Azolla microphylla*

$$\downarrow$$

Then characterization of NPs was done by using techniques like UV-VIS spectra

$$\downarrow$$

Analysis, SPR (Surface Plasmon Resonance) band was used to analyse the shape of NPs, SEM and TEM analysis [17]. After that antioxidant capacity, toxicity studies of AuNPs and its effect on various types of cells including hepatotoxic markers was studied (Pattanayak et al. 2015).

Cerium oxide nanoparticles

Cerium oxide nanoparticles were made and its protection was check against hepatic damage induced by monocrotaline (Sathishkumar et al. 2014).

Solid lipid nanoparticles

Bixin, a carotenoid which is obtained from seeds of achiote tree possessed with properties like antioxidation, anticlastogenicity, antimyeloma, antigenotoxicity. It is studied that methanolic preparation of bixin is having antioxidant property. So this feature of bixin was combined with solid lipid nanoparticles for good results and the nanoparticles produced were characterized and were tested in vitro (Singh et al. 2013).

12 CONCLUSION

As nanotechnology is a progressive field now a day. So this paper is giving a review on how Nanoparticles i.e. particles having size of 1–100 nm can be used for various treatments because of having medicinal value. Moreover it has many other applications too. Liver ailments are basic problem of people now a day due to variety of reasons and some of which have proper treatment while some don't have. So Nanoparticles are used for curing hepatotoxicity caused by various types of Hepatotoxicants. Nanoparticles are synthesized by 'green synthesis' method using plants and its extracts. These synthesized Nanoparticles are used a vehicle to deliver the medicines in human body and resulting in the cure of disease.

ACKNOWLEDGEMENTS

Authors are thankful to Prof. Aditya Shastri, Vice-Chancellor, Banasthali University, DST, Govt. of India for supporting Banasthali University under its CURIE scheme for providing the necessary facilities.

REFERENCES

Amin, K.A., Hassan, M.S., Awad, E.T. & Hashem, K.S. 2012. The protective effects of cerium oxide nanoparticles against hepatic oxidative damage induced by monocrotaline. International Journal of Nanomedicine 6: 143–149.

Bar, H., Bhui, D., Sahoo, G.P., Sarkar, P., De, S.P. & Misra, A. 2009. Green synthesis of nanoparticles using latex of Jatropha curcas Surf. A Physicochem Eng Asp 339: 134–139.

Barathmanikanth, S., Kalishwaralal, K., Sriram, M., Pandian, S.R.K., Youn, HS., Eom, S.H. & Gurunathan, S. 2010. Antioxidant effect of gold nanoparticles restrains hyperglycaemic conditions in diabetic mice. Nanobiotechnology Journal 8: 1–15.

Basavaraja, S., Balaji, S.D., Lagashetty, A., Rajasab, A.H. & Venkataraman, A. 2008. Extracellular biosynthesis of silver nanoparticles using the fungus Fusarium semitectum. Materials Research Bulletin 43: 1164–1170.

Boken, J., Dalela, S., Sharma, C.K. & Kumar, D. 2013. Detection of pathogenic Escherichia coli (E.coli) strain using robust silver and gold nanoparticles. J Chem Eng Process Technology 4(8): 1–6.

Boken, J., Sharma, M., Sharma, C.K. & Kumar, D. 2014. Synthesis of silver nanoparticles by using Ocimum sanctum leaf extract and detection of antibacterial activity. In: Global sustainability transitions: Impacts & innovations (Ed. G. C. Mishra), ISBN: 978-93-83083-77-0, excellent publishing house, New Delhi, 278–284.

Bruchez, M., Moronne, M., Gin, P., Weiss, S. & Alivisatos, A.P. 1998. Semiconductor nanocrystals as fluorescent biological labels. Science 281: 2013–2016.

Coe, S., Woo, W.K., Bawendi, M. & Bulovic, V. 2002. Electroluminescence from single monolayer of nanocrystals in molecular organic devices. Nature 420: 800–803.

Dauthal, P. & Mukhopadhyay, M. 2013. Biosynthesis of palladium nanoparticles using delonix regia leaf extract and its catalytic activity for nitro-aromatics hydrogenation. Industrial & Engineering Chemistry Research 52: 18131–18139.

Gnanadesigan, M., Anand, M., Ravikumar, S., Maruthupandy, M., Syed Ali, M., Vijaykumar, V. & Kumaraguru, A.K. 2012. Antibacterial potential of biosynthesized silver nanoparticles using Avicennia marina mangrove plant. Appl Nanosci 2: 143–147.

Huang, J., Li, Q., Sun, D., Lu, Y., Su, Y., Yang, H., Wang, Y., Shao, W., He, N., Hong, J. & Chen, C. 2007. Biosynthesis of silver and gold nanoparticles by novel sundried Cinnamomum camphora leaf. Nanotechnology 18 (11): 105–104.

Jaeschke, H., Gores, G.J., Cederbaum, A.I., Hinson, J.A., Pessayre, D. & Lemasters, J.J. 2002. Mechanisms of hepatotoxicity. Toxicological Sciences 65: 166–176.

Jannu, V., Baddam, P.G., Boorgula, A.K. & Jambula, S.R. 2012. A review on hepatoprotective plants. International Journal of Drug Development and Research 4(3): 0975–9344.

Kunjiappan, S., Bhattacharjee, C. & Chowdhury, R. 2015. Hepatoprotective and antioxidant effects of Azolla microphylla based gold nanoparticles against acetaminophen induced toxicity in a fresh water common carp fish (Cyprinus caprio). Nanomedicine Journal 2(2): 88–110.

Logeswari, P., Silambarasan, S. & Abraham, J. 2015. Synthesis of silver nanoparticles using plants extract and analysis of their antimicrobial property. Journal of Saudi Chemical Society 19: 311–317.

Nandagopal, S., Ganesh kumar, A., Dhanalakshmi, D.P. & Prakash, P. 2014. Bio-prospecting the antibacterial and anticancer activities of silver nanoparticles synthesized using Terminalia chebula seed extract. Int. J. of Pharmacy and Pharmaceutical Sciences 6(2): 368–373.

Pattanayak, S., Mollick, M.R., Maity, D., Chakraborty, S., Dash, S.K., Chattopadhyay, S., Roy, S., Chattopadhyay, D. & Chakraborty, M. 2015. Butea monosperma bark extract mediated green synthesis of silver nanoparticles: Characterization and biomedical applications. Journal of Saudi Chemical Society http://dx.doi.org/10.1016/j.jscs.2015.11.004

Rao, M.P., Manjunath, K., Bhagawati, S.T. & Thippeswami, B.S. 2014. Bixin loaded solid lipid nanoparticles for enhanced hepatoprotection—Preparation, Characterisation and in vivo evaluation. International Journal of Pharmaceutics 473: 485–492.

Sathishkumar, G., Gobinath, C., Wilson, A. & Sivaramakrishnan, S. 2014. Dendropthoefalacata ettingsh (Neem mistletoe): A potent bioresource to fabricate silver nanoparticles for anticancer effect against human breast cancer cells (MCF-7). Spectrochimica Acta part A: Molecular and Biomolecular Spectroscopy 128: 285–290.

Savithramma, N., Linga rao, M., Rukmini, K. & Suvarnalatha devi, P. 2011. Antimicrobial activity of silver nanoparticles synthesized by using medicinal plants. International Journal of ChemTech Research 0974–4290(3): 1394–1402.

Saxena, A., Tripathi, R.M. & Singh, R.P. 2014. Biological synthesis of silver nanoparticles by using onion extract and their antibacterial activity. Digest journal 5: 483–489.

Sharma, C.K., Sharma, M., Verma, O. & Sharma, V. 2015. Green synthesis of different nanoparticles and their potential applications in different fields. Int. J. of Pharma and Bio Sciences 6(3): 555–567.

Siddiqui, M.H., Al-Whaibi, M.H., Firoz, M. & Al-Khaishany, M.Y. 2015. Role of nanoparticles in plants. M.H. Siddiqui et al. (eds.) Nanotechnology and plant sciences. Springer International Publishing 19–35

Singh, S., Thomas, M.B., Singh, S.P. & Bhowmik, D. 2013. Plants used in hepatoprotective remedies in traditional Indian medicine. Indian Journal of Research in Pharmacy and Biotechnology 1(1): 2320–3471.

Sulaiman, G.M., Mohammed, W.H., Marzoog, T.R., Amir al-amiery, A.A., Kadhum, A.A. & Mohamad, A.B. 2013. Green synthesis, antimicrobial and cytotoxic effects of silver nanoparticles using Eucalyptus chapmaniana leaves extract. Asian Pacific Journal of Tropical Biomedicine 3(1): 58–63.

Suriyakalaa, U., Antony, J.J., Suganya, S., Siva, D., Sukirtha, D., Kamalakkannan, S., Pichiah, P.B.T. & Achiraman, S. 2013. Hepatoprotective activity of biosynthesized silver nanoparticles fabricated using Andrographis paniculata. Colloids Surf B Biointerfaces 102: 189–194.

Thirumurgan, A., Tomy, N.A, Jai Ganesh, R. & Gobikrishnan, S. 2014. Biological reduction of silver nanoparticles using plant leaf extracts and its effect an increased antimicrobial activity against clinically isolated organism. Pharma Chem 2: 279–284.

Vidyanathan, R., Kalishwaralal, K., Gopalram, S. & Gurunathan, S. 2009. Nanosilver—The burgeoning therapeutic molecule and its green synthesis. Elsevier 27: 924–937.

Communication and Computing Systems – Prasad et al. (Eds)
© 2017 Taylor & Francis Group, London, ISBN 978-1-138-02952-1

A vision-based approach for human detection at night-time

Rishav Garg & Gyanendra K. Verma
National Institute of Technology Kurukshetra, Haryana, India

ABSTRACT: This paper presents an effective surveillance system for detecting moving human being in night-time and generating an alarm along with sending SMS. The proposed method identifies human being using image segmentation and pattern analysis techniques. First, local shape features are extracted using three different methods namely HAAR (HAAR wavelet), LBP (Local binary pattern) and HOG (histogram of oriented gradients). The features obtained from above approaches are invariant to shape, orientation, partial occlusion and illumination. The automatic thresholding provides a robust and adaptable detection system that operates well under different night-time conditions. Once, the human detection is conformed, system generates an alarm and send an SMS using SMS module. Our system is tested with a real time video (under home environment), captured from in-house CCTV. The performance of the system is given in terms of TPR (true positive rate), FPR (false positive rate) and RT (response time). Experimental results demonstrate that the proposed surveillance approach is effective for human detection in various night-time environments.

1 INTRODUCTION

People detection and recognition is an essential element of visual surveillance because undesirable activities is increasing persistently due to the support of darker regions. Thus, visual surveillance attracts the keen researchers so that crime rate could be restricted up to a limit. In most of the industrial applications (Garibotto et al. 2013), surveillance systems operate round the clock where the optical cameras are unable to process the images at night because of illumination and shadowing conditions. Under such critical situations, infrared camera is well suited as it work on the principle of sensing the heat radiation and temperature of the object's body. The methods based on local search, contours and colour histogram are used earlier for discriminating the human to other non-human objects. The efficiency of contour based method is directly proportional to the contours models applied to human being. However, a reasonable efficiency can be achieved by these methods if contours are initialized appropriately.

(Shashua et al. 2004) demonstrated an application based study for driving assistance to the drivers using with local histogram in night vision images. (Suard et al. 2006) presents a method for pedestrian detection using supervised classifier and shape based features. The authors found 96% positive results in low quality night time recordings. (Zeng et al. 2007) review the differences in non-urban and urban environments of IR-based human detection and investigate the results for urban conditions. They considered ICA templates with SVM classifier to attain 87% positive detection. (Fang et al. 2004) worked on the segmentation based approach with multi-dimensional shape independent features like brightness, contrast, inertia. They performed horizontal and vertical segmentation of night vision images to improve the results. (Xu et al. 2005) adopted SVM classifier, Kalman filtering and shift mean clustering to monitor, track and predict the behaviour and movements of pedestrian. The authors found 42–90% positive detection and 10% false alarm rate for single infrared sensor. Yet, classification and segmentation based schemes are reliable in order to detect the human being but does not perform well in noisy and blur conditions. Moreover, rigorous training is required to improve the efficiency of algorithm which could be possible while working with a large dataset of night time recordings.

(Dai et al. 2007) proposed a layered based method with EM algorithm with appearance features. The layered techniques contains two passes in which non-moving objects are eliminated in first pass and second pass locate the position of moving pedestrians. Likewise contour based approach, appearance features are also computationally ineffective because feature vector use the notion of contours models. (Li et al. 2010) proposed a technique for monitoring and identifying the presence of pedestrians in night imagery system. The authors employed wavelet transform, entropy function and supervised classifier with non-shape features. (Zhao et al. 2015) presented a high performance system using MSRC classifier with SDH features and draw the results for tracking the movements of pedestrians in images. (Cao et al. 2008) demonstrated an approach for walker tracking using statistical learning with motion features. The authors adopted co-evolutionary algorithms in training process and estimation algorithm in prediction process to achieve 80% of accuracy. (Gilmore III et al. 2009) focused on human detection in night vision mode by using fusion algorithm with optical flow and produced 89% positive and 15% negative detections. The non-shape based features works well for stationary or slowly moving objects in static dynamics but poorly perform under dynamic conditions i.e. great changes in illumination, sharp shadowing effect, complicated background scenes and multiplicity of objects.

Aforementioned discussion reveals that the scene understanding and object localization are very typical task in night vision mode as compared with daylight mode due to lighting conditions. Despite of numerous differences (Wilder et al. 1996; Socolinsky et al. 2001), both types of image share the similarities of texture pattern details and external illumination changes up to some extent. However, fine texture details are getting suppressed with background as the temperature of the body does not change because the intensity value remain constant at that instant. Although, proposed studies are unable to handle the criticalities of day and night mode surveillance along with the accuracy and response time effectively.

Therefore, we present an efficient approach for detecting the presence of human at the restricted places. Our approach also tracked the movement of human being at night time recordings, captured by a single infrared sensor. Our technique is based on supervised classifier scheme using with the shape based features namely HAAR (Viola et al. 2001a), LBP (Ojala et al. 1996a) and HOG (Dalal et al. 2005a). The experiments review the study of three features for different test cases under various background conditions. The performance analysis of tests assessed in terms of True Positive Rate (TPR), False Positive Rate (FPR) and Response Time (RT). The evaluation measures exhibit that the proposed methodology achieves better detection results for real time videos in night vision mode while only daylight mode is popular in processing as well as realizing better results of visual surveillance.

The rest of paper is organised in the following manner: Section II discusses the extraction of illuminated object frames from the video and section III explains about the classification of features. Section IV describes the detection and validation of potential human detection at night-time followed by experimental results in section V. At last, conclusion and future scope is presented in section VI.

2 ILLUMINATED OBJECT EXTRACTION

At first, the real-rime surveillance video which shot at 30 fps in night vision mode by an infrared sensor are loaded into memory along with the classifier which is formed in training phase. Figure 1 shows the typical example of night frames under different background in a house. Further processing, frames are extracted from the input video and then blur and noise are eliminated from the frames so that it can be processed in more explicit way. The noises occurred during acquisition or transmission time are eliminated by applying the filters on each frame of video.

We have applied three filters namely median filter, averaging filter and motion blur filter in order to detect the presence of human being and generate a comparative study for three feature type viz. HAAR, LBP and HOG. As we know that these filter use the notions of correlation or convolution process with a specified mask.

2.1 Median filter

Median filter is a type of non-linear smoothing filter which is based on the ordering of pixels value of image encompassed by filter mask. It can be defined as

$$\hat{f}(x,y) = median\left\{\sum_{-s}^{s}\sum_{-t}^{t} f(x+s, y+t)\right\} \qquad (1)$$

Where $\hat{f}(x,y)$ = filtered image; $f(x,y)$ = input image; $g(x,y)$ = mask of size m × n; s = (m − 1)/2 and t = (n − 1)/2.

The median filter is useful to reduce the random noises effectively with less blurring effect however excellently works to eliminate the salt and pepper noise which might be occurred during transmission of frames from the place where video is shot to server due to analog-to-digital conversion.

2.2 Averaging filter

Averaging Filter is a smoothing linear filter which output is simply the average of pixel values of an image encompassed by the filter mask. It can be defined as

Figure 1.   Sample of night-time scenes of a house (A) video-1 (B) video-2.

$$\hat{f}(x,y) = \frac{1}{\sum g(s,t)} \sum_{-s}^{s} \sum_{-t}^{t} g(s,t) \times f(x+s,y+t) \quad (2)$$

Where $\hat{f}(x,y)$, $f(x,y)$, $g(x,y)$ used usual notions; $s = (m-1)/2$ and $t = (n-1)/2$.

It eliminates the irrelevant or sharp edge details from the image by blurring the sharp transitions in intensities. This filter usually useful to remove the Gaussian noise which might be arose during acquisition of video recording due to poor illumination and high temperature.

2.3 Motion blur filter

Motion blur filter is also known as Gaussian blur filter which is used to eliminate the motion blur from an image. It convolutes original image over a filter mask with a specified angle and produces a blurred image by removing the rapid changes of movements and long exposures that might be arose during the acquisition of video recording. It can be defined as

$$\hat{f}(x,y) = \sum_{x-r}^{x+r} \sum_{y-r}^{y+r} g(i,j) \times f(i,j) \quad (3)$$

Where $\hat{f}(x,y)$, $f(x,y)$, $g(x,y)$ used usual notions and r = radius such that $-r \le x, y \le +r$.

3 FEATURE EXTRACTION

In this paper, we are working on the three different shape based features namely HAAR-like, Local Binary Pattern (LBP) and Histogram Of Gradient (HOG). In order to illustrate the comparative study among the features, we separately computed these three features from the resized frames. However, the study of the results can be made up easily by the pursuer as the size of each and every frame should be same as long as time complexity could be lessened.

3.1 HAAR

(Viola et al. 2001b) inheriting the concept of HAAR wavelet transformation and introduced a powerful feature labelled as HAAR like features. The feature uses the sums of the difference of the pixel intensity with in a region for classifying an image and this target region rolled over each sub-regions of image for computing the differences of each region in whole image. This difference results in generating the feature vector which is used with learned threshold in order to locate an object in detection process. HAAR feature is computed as

$$F_{P|D_0} = \sum_{k=0}^{M-1} \left(\sum s_o - \sum s_n \right) \quad (4)$$

$$f(z) = \begin{cases} F_{P|D_0} & z < 0 \\ 1, & z \ge 0 \end{cases} \quad (5)$$

Where $\sum s_o$ = sum of intensity values of 8-neighbourhood of testing pixel P and $\sum s_n$ = sum of intensity values of equi-circled neighbourhood with diameter D_0 of P.

Although, HAAR feature is a weak learner yet produces marginally better results than arbitrary guessing. However, creation of outsized feature vector may increase the accuracy as well as efficiency. The response time is a typical benefit over others because it took very less time in computation. Moreover, HAAR features are invariant to the position, shape, angle, scale and size of object.

3.2 LBP

(Ojala et al. 1996b) used the concept of binary texture pattern and designed a classification operator known as LBP feature or LBP operator. The operator can be computed by comparing the central pixel to its 8-neighbours with in a region. To create feature vector, this process is repeated until all regions and sub-regions of an image will cover. This feature vector can be used with some classification scheme for getting better results of classification. LBP feature operator is computed as

$$F_{P|D_0} = \sum_{k=0}^{M-1} (s_o - s_n) \quad (6)$$

$$f(z) = \begin{cases} F_{P|D_0}, & z < 0 \\ 1, & z \ge 0 \end{cases} \quad (7)$$

Where s_o = intensity value of 8-neighbourhood testing pixel P and s_n = intensity value equi-circled neighbourhood with diameter D_0 of P.

The local binary pattern feature is a powerful operator due to its computational simplicity and discriminative power. In addition to this, LBP operator is invariant to angle, size, scale and shape of object and plane.

3.3 HOG

Basically, histogram of gradient feature is derived for pedestrian detection by (Dalal et al. 2005b). Feature vector of HOG formed by the combination of two parameters i.e. gradient orientation and gradient magnitude. Gradient magnitude and orientation are calculated for each sub-regions and concatenated the histogram followed by normalisation. In this way, a local descriptor is created with respect to each region which participate in formation of global descriptor. When this process is repeated for each region of an image, local descriptor are merged and form a global one in order to classify the objects in image. HOG feature is computed as

$$F(y,z) = \tan^{-1} \frac{g_z}{g_y} \qquad (8)$$

$$f(y,z) = \sqrt{g_y^2 + g_z^2} \qquad (9)$$

Where g_y = image after applying horizontal mask (i.e. [1 0 −1]) and g_z = image after applying vertical masks (i.e. [1 0 −1]^T).

The histogram of gradient descriptor is a powerful tool for classifying an object due to its formation process. Local object shape and appearance defined by the distribution of gradient values gives it a better discriminative power over others. HOG features are invariant to illumination, geometric transformations and shadowing apart from angle and size of object.

4 NIGHT-TIME HUMAN DETECTION

Aforementioned steps identify the potential detection of human being whenever it will enter into the restricted zone. Since, a single video frame is not enough efficient to obtain the complete information of potential human object however component based tracking is applied to analyse the motion characteristics of a human object which is based on the analysis of consecutive frames of video for potential detection. Then, a message is raised by the system indicating an unusual activity. The confirmation about the activity explicitly depends on the successive detection of individual. Here, we have chosen five successive detection because it does not consume much computation time and hence accuracy is not considerably affected but improves the robustness and reliability of the system. The fundamental steps of algorithm for human detection are shown in Figure 2.

After following the defined procedure, a boundary box of green colour with the smallest area is rendered over those blobs which are ordered as human. After that a warning message is displayed over the screen of administrator indicating that some unusual activity is going on. If warning message is successively raising again and again then it will send an SMS to alert the administrator after

ensuring the confirmation about the activity. In our algorithm, we restrict the successive raising of messages up to five that showing the validation of human being in the frames of video recording.

5 RESULTS AND DISCUSSION

The experiments are brought out on MATLAB R2013a installed in a system with processor of Intel® core™ i3 and 4 GB of main memory. In order to calibrate and accredit our algorithm, we have used our own datasets as there no standard datasets are available for night time recordings. The dataset contains six test cases under different background conditions of a house. All the recordings are of same size i.e. 800 × 600 pixels shot at 30 fps. Our dataset also accommodates a test case of falsely object and multiple objects under each background condition. The training is performed on a dataset which contains 280 positive and 95 negative images of size 1280 × 960 pixels.

In order to identify the presence of human, we consider various measures for evaluating the robustness and efficiency of our framework. These measures are True Positive Rate (TPR), False Positive Rate (FPR) and Response Time (RT). Our approach seems to be economical and outperformer by accounting the greater values of TPR whereas smaller values of FPR and RT.

It is totally infeasible to show all the results in terms of detection frames for each test case. Thus, we show here a few frames as detection results for a test case in different background conditions. The boundary box of green colour rendered over the human in all the frames as shown in Figure 3 and Figure 4 for all three features.

We generate a comparative study of three features, HAAR, LBP and HOG, for four filtering conditions that is without using any filter (no filter) and using techniques namely median filter, averaging filter and motion blur filter on the basis of average of all 6 test cases. The results of corresponding parameters for the features viz. HAAR, LBP and HOG are presented in Table 1, Table 2 and Table 3 respectively using different filtering techniques. The response time, shown in Tables 1–3, is the average of time taken by 250 frames of each test case for processing since we evaluate the performance of our system over 250 frames of each test case.

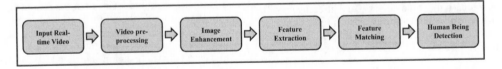

Figure 2. Fundamental steps of algorithm.

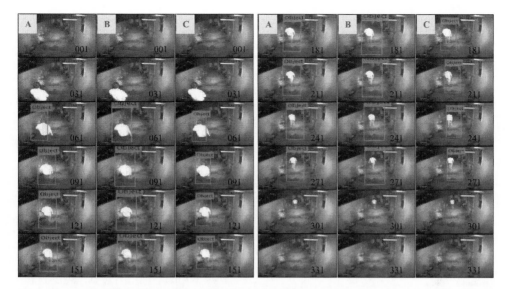

Figure 3.   Sequence of frames with blob using (a) HAAR, (b) LBP, and (c) HOG in test video 1.

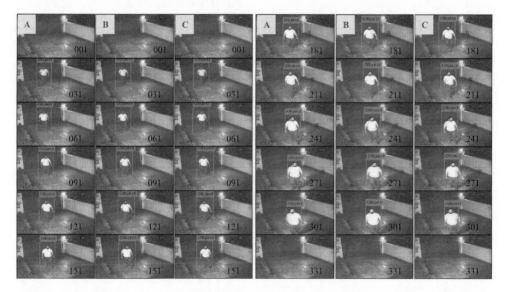

Figure 4.   Sequence of frames with blob using (a) HAAR, (b) LBP, and (c) HOG in test video 2.

Table 1.   Average of TPR, FPR and RT of different environments for HAAR feature.

Filter type	TPR	FPR	RT msec
No filter	0.84	0.13	136
Median filter	0.87	0.11	154
Averaging filter	0.85	0.14	190
Motion blur filter	0.90	0.08	163

Table 2.   Average of TPR, FPR and RT of different environments for LBP feature.

Filter type	TPR	FPR	RT msec
No filter	0.72	0.18	141
Median filter	0.77	0.13	162
Averaging filter	0.75	0.15	196
Motion blur filter	0.79	0.12	168

Table 3. Average of TPR, FPR and RT of different environments for HOG feature.

Filter type	TPR	FPR	RT msec
No filter	0.86	0.12	148
Median filter	0.91	0.08	167
Averaging filter	0.87	0.11	201
Motion blur filter	0.89	0.07	174

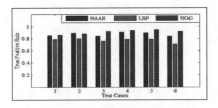

Figure 5. Variation of TPR for different test cases with median filter.

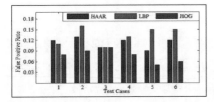

Figure 6. Variation of TPR for different test cases with median filter.

6 CONCLUSION

This paper has proposed an effective night-time human detection with alarm generation and SMS sending system for automatic surveillance. The proposed approach uses an efficient and fast detection and segmentation process based on three detectors i.e. HAAR, LBP and HOG. Actual human being can be efficiently detected from these detectors. The proposed night-time human detection and segmentation approaches are implemented on an in-house real time video of outdoor environment. The experimental results shown that the proposed system is effective and offer advantages for human detection for automatic surveillance in various night-time environments. For further studies, the human detection algorithms can be further improved and extended by integrating some sophisticated machine learning approaches.

REFERENCES

Cao, X. Qiao, H. & Keane, J. 2008. *A Low-Cost Pedestrian-Detection System with a Single Optical Camera*. IEEE Transactions on Intelligent Transportation Systems 9(1): 58–67.

Dai, C. Zheng, Y. & Li, X. 2007. *Pedestrian detection and tracking in infrared imagery using shape and appearance*. Computer Vision and Image Understanding 106(2–3): 288–299.

Dalal, N. & Triggs, B. 2005. *Histograms of oriented gradients for human detection*. Computer Vision and Pattern Recognition 1: 886–893. San Diego, USA: IEEE.

Fang, Y. Yamada, K. Ninomiya, Y. Horn, B.K.P. & Masaki, I. 2004. *A shape-independent method for pedestrian detection with far-infrared images*. IEEE Transactions on Vehicular Technology 53(6): 1679–1697.

Garibotto, G. Murrieri, P. Capra, A. Muro, S.D. Petillo, U. Flammini, F. Esposito, M. Pragliola, C. Leo, G.D. Lengu, R. Mazzino, N. Paolillo, A. D'Urso, M. Vertucci, R. Narducci, F. Ricciardi, S. Casanova, A. Fenu, G. Mizio, M.D. Savastano, M. Capua, M.D. & Ferone A. 2013. *White Paper on Industrial Applications of Computer Vision and Pattern Recognition*. 17th International Conference on Image Analysis and Processing 8157: 721–730. Naples, Italy: Springer.

Gilmore III, E.T. Frazier, P.D. & Chouikha, M.F. 2009. *Improved Human Detection Using Image Fusion*. Proceedings of International Conference on Robotics and Automation: 1–6. Kobe Japan: IEEE.

Ojala, T. Pietikainen, M. & Harwood, D. 1996. *A comparative study of texture measures with classification based on featured distributions*. Pattern Recognition 29(1): 51–59.

Shashua, A. Gdalyahu, Y. & Hayun, G. 2004. *Pedestrian detection for driving assistance systems: single-frame classification and system level performance*. Intelligent Vehicles Symposium: 1–6. Parma, Italy: IEEE.

Suard, F. Rakotomamonjy, A. Bensrhair, A. & Broggi, A. 2006. *Pedestrian Detection using Infrared images and Histograms of Oriented Gradients*. Intelligent Vehicles Symposium: 206–212. Tokyo, Japan: IEEE.

Socolinsky, D.A. Wolff, L.B. Neuheisel, J.D. & Eveland, C.K. 2001. *Illumination invariant face recognition using thermal infrared imagery*. Proceedings of Computer Vision and Pattern Recognition 1: 527–534. Hawaii USA: IEEE.

Viola, P. & Jones, M. 2001. *Rapid object detection using a boosted cascade of simple features*. Proceedings of Computer Vision and Pattern Recognition 1: 511–518, IEEE.

Wilder, J. Phillips, P.J. Jiang, C. & Wiener, S. 1996. *Comparison of visible and infra-red imagery for face recognition*. Proceedings of 2nd International Conference on Automatic Face and Gesture Recognition: 182–187. Vermont USA: IEEE.

Xu, F. Liu, X. & Fujimura, K. 2005. *Pedestrian detection and tracking with night vision*. IEEE Transactions on Intelligent Transportation Systems 6(1): 63–71.

Zeng, J. Sayedelahl, A. Chouikha, M.F. Gilmore, E.T. & Frazier, P.D. 2007. *Human detection in non-urban environment using infrared images*. 6th International Conference on Information, Communications and Signal Processing: 1–6. Singapore: IEEE.

Zhaoa, X. Hea, Z. Zhanga, S. & Liang, D. 2015. *Robust pedestrian detection in thermal infrared imagery using a shape distribution histogram feature and modified sparse representation classification*. Pattern Recognition 48(6): 1947–1960.

Communication and Computing Systems – Prasad et al. (Eds)
© 2017 Taylor & Francis Group, London, ISBN 978-1-138-02952-1

Survey of load balancing algorithms and performance evaluation in cloud computing

Nancy & Amita Malik

Department of Computer Science and Engineering, DCRUST Murthal, Sonepat, Haryana, India

ABSTRACT: Cloud Computing is an emanating technique. It offers brand new framework to provide resources. Cloud Computing systems have many advantages as compared to prevailing conventional service provisions, such as less upfront investment, superior performance with high accessibility of resources and facilities, improved scalability, and high fault-tolerance capability and so on. Conversely, many new companies have emerged with competitive services relaying on Cloud computing systems. Load balancing is a technique for upgrading the capabilities of a parallel as well as distributed system via division of load among the processors or nodes. This Paper surveys few load balancing approaches available in cloud systems, challenges faced by them with their comparative study.

1 INTRODUCTION

With the growth of the Internet, various computing resources have become reasonable as well as more powerful (Singh & Tiwari, 2014). This technological trend lead to the realization and development of a new computing model known as cloud computing, where resources (e.g., CPU and storage) and services are allowed to be used and released by users with the help of Internet in an on-demand basis. In such environment (Singh & Tiwari, 2014), the basic role of service provider is classified into two levels: the infrastructure providers who are responsible to maintain cloud platforms and allocate resources as per to its usage-based and service providers, who rent the resources provided by various infrastructure providers to fulfill the needs of the end users.

Cloud computing (Singh & Tiwari, 2014) is implemented with the help of internet and central remote servers in order to support demands of users and data which they require. Virtualization is one or supplementary physical servers which can be configured and split into countless number of unattached virtual servers and all are working independently and seem to the user as a solitary physical device. Such adjacent servers do not include encompass and can be subsequently elevated in all orders and flaky up or down on the drift lacking changing the finish user.

Cloud computing is a mechanism of distributing the computation that focuses on conferring a expansive scope of users along with the distribution of admission to the virtualized hardware and multimedia groundwork above the internet.

Trusted of cloud computing has attained a lot of attention of users towards parallel based, distributed based as well as virtualization based computing arrangements today. Through virtualization, cloud computing is able to address alongside the comparable physical groundwork a large client center alongside disparate computational needs. The quick progress in the earth of cloud computing additionally increases harsh protection concerns. Lack of protection is the merely hurdle in expansive adoption of cloud computing as well as load balancing (Jonas et al, 2014) is also one of the main and primary concern which hinders in its growth.

1.1 Types of clouds

Public clouds: Public cloud is obtainable for span use alternatively for the enormous industries and is owned by an association vending cloud services. Client has no perceptibility and manipulation of

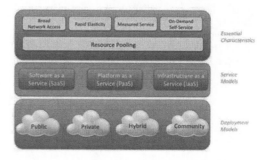

Figure 1.   General architecture of cloud computing.

access where the computing groundwork is hosted. The computing groundwork is area amid every single association.

Private cloud: The cloud infrastructure (Dharmesh & Jaydeep, 2014) is exclusively used by a single organization comprising of multiple consumers. It may be owned or maintained by an organization, a third party, or some combination of both of them, and even it may exist on or off premises. Confidential clouds are extra expensive and extra safeguard in comparison to area clouds. The clouds which are hosted externally i. e the confidential clouds, are merely utilized by one association, but are hosted by third party enumerating in cloud infrastructure. Externally hosted confidential clouds are inexpensive than the On-premise confidential clouds.

Hybrid cloud: Hybrid Cloud merges countless clouds whereas those clouds retain their exceptional individualities, but are attached jointly as a unit. In this, they use their own computational groundwork for area rehearse, but require the cloud for requirements of load [6]. They are engaged together by standardized or proprietary technology which enables the portability of data and various applications.

Community cloud: This cloud is one which has been coordinated to assist a public purpose or general purpose. For example one association or for countless association, but they provide common concerns or responsibilities such as their protection/security, strategies, needs of users and so on.

There are various kinds of Load balancing methods that are meant for cloud computing. Such Load balancing methods are: geographical allocation, static and Dynamic. The geographical allocation of the nodes means a lot in the grouped presentation of each and every single real era cloud computing arrangements and surroundings, specifically when the requests on Twitter, Face book etc come across. A well-distributed composition of nodes in cloud is supportive in acquainting obligation accord and maintaining the efficiency and productivity of the system. Geographical Load Balancing (GLB) can be delineated as a chain of decisions considering all on-stream assignment or migration of Virtual Machines (VMs) or computational tasks to be geographically scattered data centers in order to confront the Service Level Accords (SLAs) or skill deadlines for VMs/tasks and to cut the operative cost of the cloud composition.

2 LOAD BALANCING IN CLOUD COMPUTING

Load balancing is a technique (Dharmesh & Jaydeep, 2014) or method which divides the excess

dynamic load over all the nodes. This technique is used for providing a good service facilities and resource utilization ratio, thereby improving the overall productivity and operation of the system. Incoming tasks come from various locations which are received by the load balancer and further distributed to the data center, for the proper and even load distribution.

2.1 Classification of load balancing algorithms

Static Load Balancing Algorithm: Static Load balancing algorithm does not depend on current state of the system and require the knowledge of the properties of nodes in advance like their capacity, operation, memory etc. The target of static Load balancing is to bring the finished killing era of a synchronous pattern so as to reduce the delays in the links. These algorithms are usually suitable for homogeneous and balanced settings and not suitable for changes during run time. Some of the examples of static Load balancing algorithms are as follows: Round Robin algorithm, Randomized algorithm and Threshold algorithm etc (Dharmesh & Jaydeep, 2014)

Dynamic Load Balancing Algorithm: In Dynamic Load balancing algorithm (Dharmesh & Jaydeep, 2014), the decisions are instituted on the present state of the arrangement, no prior vision is demanded. The main supremacy of dynamic balancing is that if some node fails, it will not halt the arrangement; it will merely change the presentation of the system. These algorithms are supplementary resilient than static algorithms, can facilely change to alteration and furnish larger aftermath in heterogeneous and Dynamic environments. Dynamic Load balancer uses morality for keeping the trail of notified information. There are four strategies for Dynamic Load balancers: selection strategy, transfer strategy, local strategy and data policy. The task of Load balancing is area amid distributed nodes. In a distributed arrangement, Dynamic Load

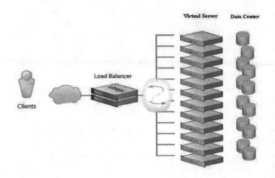

Figure 2.   Load balancing in cloud computing.

balancing can be finished in two disparate ways: distributed and non-distributed.

Distributed Dynamic Load Balancing Algorithm: In such environment, the Dynamic Load balancing algorithm provides all nodes presenting the arrangement and the task of arranging the area amid them. The link between the nodes in Load balancing can grab two forms: cooperative and non-cooperative.

Non-Distributed Load Balancing Algorithm: In the non-distributed or undistributed, the nodes work confidential in order to instate a public goal. Non distributed Dynamic Load balancing algorithms are categorized into two: centralized and semi-centralized.

Semi-distributed Dynamic Load Balancing: In semi-distributed Dynamic Load balancing, the nodes of the arrangement are splits into clusters, whereas the load balancing in every single cluster is of centralized form. A central node is elected in every single cluster by appropriate vote method that seizes care of load balancing inside that cluster. Therefore, the burden of balancing of all arrangement is completed via the central nodes of every single cluster.

Centralized Dynamic Load Balancing: In centralized Dynamic Load balancing (Jonas et al, 2014), the algorithm is merely given by a solitary node in the finished arrangement i.e. central node. This node is immaculately accountable for Load balancing of the finished arrangement and rest of the nodes interacts merely alongside the central node.

3 EXISTING LOAD BALANCING TECHNIQUES IN CLOUD COMPUTING

Round robin based load balancing method

In this method of load balancing, processes are distributed among all the processors. In its working, each and every process is allotted a new processor in round robin fashion. The order, in which the processors are allotted, is kept in cache locally on processors which is independent of allocations from the remote processors. When the load is equal, round robin algorithm performs well. Round Robin and Randomized schemes work effectively and efficiently when number of processes is more than number of processors. Benefits of such an algorithm are that it does not require any inter-process communication. Both of the defined algorithms attain best accomplishment among all load algorithms however in some scenarios Round Robin and Randomized are not expected to attain better performance in general case. It is regarded as one of the simplest scheduling techniques which make the utilization of the principle of time slices

in which the time is distributed into multiple slots and each and every process is allotted a fixed time slot i.e. it utilizes the principle of time scheduling. After the time slot is finished, the next user/process in the queue will arrive for execution. If the completion of the task of the process occurs in between the definite time given to it then the user should not wait else user have to wait for its turn.

Honey bee inspired load balancing

In such algorithm, we perform the computations by modeling the foraging deeds of honey bees. The derivation of such an algorithm is done from the deeds of honey bees that use the technique to search and reap food. In hives of bees, there is a domain of bees which is called as the scout bees and the another domain is known as forager bees. The scout bees, work by foraging for stock of food, and later on they are used to find the food, after which they return to the beehive to inform everyone about this news by retaining a dance known as yelled /waggle/tremble/vibration dance. The motive of this dance is to give the trustiness of the quality and/or number of stock of food and in addition to this also informs about its distance from the beehive. Forager bees subsequently follow the Scout Bees to the local so that they could discover the stock of food and later on onset to reap it. And even they retrace to the beehive and perform a vibration dance for other supplementary bees in the hive in order to inform about the trusted of how distant food is left from their beehive. The tasks on overburdened virtual machines are eliminated as the deeds as Honey Bees. On acceptance to the VM which is not much loaded, will notify about the number of various priority tasks and load of tasks assigned them. This information will be very helpful for supplementary tasks like whenever an elevated priority is to be given to VMs, it must keep a check on the VM's that has only few number of elevated priority tasks, so that the particular task will be provided as soon as possible. All VMs are sorted in a increasing order, the task removed will be given to less loaded VMs. Present workload of all available VMs can be computed on the basis of instituted data perceived from the data center. Benefits of this algorithm are like maximizing the throughput of system; time spent in waiting is minimized and overhead is also reduced. The disadvantage is if supplementary priority instituted queues are there subsequent the lower priority Load can be stay unceasingly in the queue.

Dynamic and adaptive load balancing strategy

It considers distributed architecture for load balancing. The large file systems faces various problems like dynamic file migration and the algorithms which were only based on centralized system. The algorithm known as Self Acting Load Balancing algorithm (SALB), helps in avoiding various such obstacles. In the parallel file ordering the data is shifted amid the recollection and the storage medium so that the data association is an vital deed of the parallel file system. Various challenges were faced across load balancing in the parallel file system are scalability and the possible arrangements, web transmission and the Load migration. In dynamic balancing algorithms, the load in each and every solitary I/O servers is disparate because the workload keeps on changing persistently. Hence there arises a need of slight decision making algorithms in it. In the following decision making ordering, there are most importantly central decision makers, in which the central node acts as a decision maker due to which incase central node fails in its working, the following arrangement becomes down and the reliability lacks. We also choose a cluster decision maker in this arrangement which divides the arrangement into clusters so that the link worth minimizes. This arrangement addressed the load forecast algorithm, effectual load collection mechanism, competent distributed decision maker, migration selection flawless and Dynamic file migration algorithm for a improving Load balancing. The flaws in such algorithm are degradation of such an finished arrangement due to the results of migration side.

Ant colony based load balancing

This technique is one of the static load balancing technique where (Jonas et al, 2014) we search for an optimal path within the source of food and colony which depends on their behavior. This approach optimizes distribution of work load among the node. In Initial stage the start its movement towards the source of food from the head node where Regional Load Balancing Node (RLBN) is selected as Cloud Computing Service Provider (CCSP) as a head node. Ants maintain its data of every node they visits and uses this statistics for decision making accordingly in future. Ant places their pheromones while moving or while they are in motion which informs other ants to choose next node The intensity of pheromones may vary on the bases of certain parameters like distance from food, quality of food etc (Lin et al, 2014). The updation of the pheromones depends upon the completion of their jobs. Whenever an ant is found in under loaded node then it is directed to proceed in forward direction in the path (Patel, 2015). And if it is found in condition of overloaded node, then it is allowed to follow in the backward movement.

Balancing in dynamic structured P2P systems

In this type of peer-to-peer arrangement the unsystematic objects are in the space and the additional load of the node can be adapted constantly due to the various operations like insertion, deletion and supplementary varied operations. In organized arrangements of peer-to-peer systems they impart a Distributed Hash Table (DHT) abstraction which supports in dispatching objects irregularly among the peer nodes. In such methods, a specific identifier is associated with each data item and every node in the system, where the identifier space is segregated among the nodes and thus it helps in forming the Peer-to-Peer (P2P) system. In order to retain all the items, every node is answerable and then they are mapped to an unique identifier in the domain of the space. It is basically drafted to handle the various situations like varying load of node, node capacity, entering and leaving of nodes and also used for various operations like insertion and deletion of the nodes. Major drawback of this algorithm is the reallocation of the adjacent server is difficult.

Load balancing Min-Min

This is a three level dynamic based load balancing algorithm which is termed as Load Balancing Min-Min (LBMM) technique (Patel, 2015), which undertakes a set of new unallocated tasks. Out of every task, firstly we compute the minimum time of completion for all tasks and then out of all these minimum times which we computed, we select the minimal value among all the tasks. Then accordingly, we perform the task. Concurrently when the tasks need to execute or run we can change the time, by adjoining the execution time of the specified task to the execution times of various other tasks on that device. And hence we keep on succeeding the similar method until all the unallocated tasks are located to their specific resources. But most unfavorable reason for not adopting this technique is that it may lead to starvation.

Genetic algorithm for dynamic load balancing

It is a type of algorithm which deploys the principles of natural selection and genetics (George & V, 2013). The exploitation of preceding results is done by GAs by differentiating and combining with the investigation of new domains of the search space. Survival of the fittest techniques are being used in this and then integrate with a structured or the arranged information. A GA can copy various inventive skill of a human search. A generation is termed as assembly of simulated strings where in each and every new generation, strings are produced with the help of statistics from the former ones. In various optimization techniques, we move from one end in the decision area to the other in order to find the next one with the help of some transition or alteration technique. This end-to end method is unsafe in multiple search domains as it may discover wrong peaks or the ends. Working of GA is accomplished with the assistance of a database of points/ends concurrently (a population of strings), climbing various peaks in simultaneously. The possibility of finding a wrong peak is not that much as we do it by differentiating the techniques that go end to end. The approaches of a Genetic algorithm are quite easy and manageable which do not involve much of complications than copying of the strings and then exchanging of partial strings. Importance of Genetic algorithm is clarity of operation and effectiveness of power (George & V, 2013) (Rastogi et al, 2010). The effectiveness relies on suitable combination of investigation and exploitation. There are 3 main operators which are used to attain the better effectiveness in the performance are: selection, crossover, and mutation, where selection is regarded as the origin of exploitation. Other operators like mutation and crossover operators are used for exploration. With the help of mutation the trade off is quite understandable between the exploration and exploitation. Mutation rate is directly proportional to disruptiveness.

4 CHALLENGES AND COMPARISON IN LOAD BALANCING ALGORITHMS

Few metrics which we can enhance for the better performance of our load balancing algorithms:

Fault Tolerance: It is the ability of algorithm to perform load balancing without any non fulfillment of a node. All load balancing algorithms must possess quite high fault tolerance approach.

Migration Time: It is defined as the time taken for shifting a process for execution from one system node to another node. For better results this time taken should be always less (Swamkar et al, 2013).

Response Time: It is defined as the amount of time taken by a particular load balancing technique to respond. This time must be minimized for better results.

Scalability: It is defined as the ability of load balancing technique for a system with any given number of processor and machines. For better results this should be improved (Swamkar et al, 2013).

Throughput: It is defined as the number of given tasks that have completed their execution for a given period of time. Then it is said to have high throughput for better results of the system.

Table 1. Comparison of load balancing algorithms.

Algorithm	Environment	Fault tolerance	Centralized/ decentralized
Round Robin	Static	No	Decentralized
Honey Bee	Dynamic	No	Decentralized
Dynamic N Adaptive	Dynamic	Yes	Centralized
Ant Colony	Dynamic	No	Decentralized
Dynamic Peer To Peer	Dynamic	Yes	Centralized
Min Min	Static	No	Centralized
Genetic Algorithm	Dynamic	Yes	Centralized

5 CONCLUSION AND FUTURE WORK

Load balancing is one of the major challenges in cloud computing framework. One of the solutions for this could be the distributed resolution. As we know, it is not always practical or cost effective to sustain one or more idle services or resources just as to fulfill the demands. Due to which jobs cannot be located to suitable servers and clients individually for effectual load balancing as cloud is a extremely complicated establishment and components are present across a extensive domain. Load balancing algorithms are further classified into static and dynamic algorithms. Static algorithms are basically acceptable for homogeneous and fixed environment and can produce extremely better aftermath in such environments. Though, they are not that much flexible and cannot even compete the dynamic adjustments and settings, to the qualities like killing time. Dynamic algorithms are very much flexible and seized into thought disparate kinds of qualities in the arrangement both prior to and across run-time. A large number parameters and computing techniques can be studied in future.

REFERENCES

Alok Singh, Vikas Kumar Tiwari (2014) "A Survey on load balancing in Cloud Computing using soft computing techniques."In *International Journal of advanced research in computer and communication engineering,* vol 3, Issue 9.

Desai Tushar, Prajapati Jignesh (2013). "A Survey of various load balancing techniques and challenges in cloud computing". In *International Journal of scientific and technology research,* vol 2, issue 11.

Durango Jonas, Dellkrantz Manfred, Maggio Martina et al (2014). "Control-theoretical load-balancing for cloud applications with brownout." In *Conference on Decision and Control (CDC),* IEEE.

Katyal Mayanka, and Mishra Atul (2013). "A Comparative Study of Load Balancing Algorithms in Cloud

Computing Environment." In *International Journal of Distributed and Cloud Computing,* volume 1, issue 2.

Kashyap Dharmesh, Viradiya Jaydeep (2014). "A Survey Of Various Load Balancing Algorithms In Cloud Computing" In *International journal of scientific and technology research*, vol 3, issue 11.

Kaur Rajwinder, Luthra Pawan (2014). "Load Balancing In Cloud Computing" In International Conference on Recent Trends in Information, Telecommunication and Computing ITC, Publication ACEEE.

Kaur Karanpreet, Narang Ashima, and Kaur Kuldeep (2013). "Load balancing techniques of cloud Computing." *International Journal of Mathematics* 1, no. 3.

Lin Weiwei Lin, Liang Chen Liang, Wang James et al (2014). "Bandwidth-aware divisible task scheduling for cloud computing." In *Software: Practice and Experience 44*, no. 2.page no.: 163–174.

Lin Chun-Cheng, Chin Hui-His et al (2014). "Dynamic multiservice load balancing in cloud-based multimedia system." *Systems Journal*, IEEE 8, no. 1.

Mell Peter, Grance Timothy (2011). In *National Institute of Standards and Technology Special Publication* 800–145.

Mittal Patel, ChaitaJani (2015). "A Survey on Heterogeneous load balancing techniques in cloud computing" In *International journal for innovative research in science and technology*, vol. 1, issue 10.

Rajan Rajesh George and Jeyakrishnan. V (2013). "A Survey on Load Balancing in Cloud Computing Environments.". In *International Journal of Advanced Research in Computer and Communication Engineering* 2, no. 12.

Rastogi Divya, Bansal Abhay et al (2010). "Techniques of load balancing in cloud computing: A survey." In *International Conference on Computer Science and Engineering* (CSE), ISBN, pp. 978–993.

Swamkar Namrata, Singh Kumar Atesh et al (2013). "A Survey of Load Balancing Techniques in Cloud Computing." *In International Journal of Engineering Research and Technology*, vol. 2.

Singh Harsukhpal, Singh Mandeep et al (2013). "Cloud based Secure Trust based Middleware for Smartphones for Accessing Enterprise Applications." *International Journal of Computer Applications* 84, no. 7.

Sran Nayandeep and Kaur Navdeep (2013). "Comparative analysis of existing load balancing techniques in cloud computing. "In *International Journal of Engineering Science Invention*, 2 (1).

Saeed Javanmardi, Shojafar Mohammad et al (2014). "Hybrid job scheduling algorithm for cloud computing environment." In *Proceedings of the Fifth International Conference on Innovations in Bio-Inspired Computing and Applications* IBICA 2014, pp. 43–52. Publication: *Springer International.*

Ying Lei, et al. "Stochastic models of load balancing and scheduling in cloud computing clusters."

Zhang Qi, Cheng Lu, Boutaba (2010). "Cloud computing: state-of-the-art and research challenges" In *J Internet ServAppl.*

Zhang Zehua, Zhang Xuejie et al (2010). "A load balancing mechanism based on ant colony and complex network theory in open cloud computing federation." In *Research Gate.*

Communication and Computing Systems – Prasad et al. (Eds)
© 2017 Taylor & Francis Group, London, ISBN 978-1-138-02952-1

QVF: A heuristic for *VM* migration from overloaded servers in cloud computing environment

J.K. Verma & C.P. Katti
School of Computer and Systems Science, Jawaharlal Nehru University, New Delhi, India

ABSTRACT: Increasing demand for high computing power led to the establishment of large-scale datacenters. A datacenter is a collection of millions of servers. These large scale datacenters consume a very huge amount of electrical energy. Managing these servers in autonomic manner for provisioning and deprovisioning of resources in an automatic and efficient way is still a great challenge. The overall problem in this paper is the problem of resource utilization in such a way that help in more and more computation and less wastage of *CPU* cycles. In this paper, we attempt to minimize the power consumption by reducing the number of servers and maximize the resource utilization of the servers that are in use through Virtual Machine (*VM*) consolidation on fewer servers. *VM* consolidation is not so much trivial solution, and therefore the problem is further divided into four modules. We pick *VM* selection to work upon that is one of the module of *VM* consolidation. We propose *QVF* heuristic to offload the overloaded servers for achieving power consumption efficiency while less *SLA* hampering solution of the problem.

1 INTRODUCTION

Cloud computing is one of the most vital and fastest-growing IT model for High Performance Computing (*HPC*). It provides delivery of computing, storage and platform services (Saas, PaaS and IaaS) on metered basis as a public utility. The services are provisioned on-demand and the core idea behind this model is to reduce the burden of processing on the end-user side (Verma & Katti 2014). Users often have a large variety of electronic devices such as laptops, PCs, PDAs, and smart phones and they use them to access various kinds of utility programs, application development platforms, and storage over the Internet using standard protocols as shown in Figure 1. These services are offered ubiquitously in cloud computing environment to ensure accessibility anytime and anywhere with cost saving, high availability and scalability features.

Besides of Ubiquitous accessibility, cloud computing saves energy, reduce datacenter footprint, faster server provisioning, get the client out of hardware lock-in problem and many more (InfoWorld). These benefits are leveraged through virtualization of services and even whole of the underlying hardware. The services and hardware are available in the form of Virtual Machines (*VMs*) in cloud computing environment. VMs are available on pricing model that is *pay—as—you—go* model adopted by cloud providers similar to the public utility like water, electricity etc.

Figure 1. General representation of cloud computing.

Cloud computing is accumulation of millions of servers and they are available on demand basis. It allows users to realize availability of infinite resource (Armbrust et al. 2010) and provision them on-demand. However, cloud computing suffer from lack of full autonomic control over resources that is instant automatic acquisition and release of resource as per the requirement and therefore cloud computing suffer from several flaws among which low resource utilization, energy consumption and high cost are important issues. These issues are very much interconnected and therefore they are very important parameters to go ahead with any cloud provider in terms of cost in the part of client as well as Cloud provider. In this paper, we propose QVF heuristic to minimize power consumption by the servers deployed under cloud computing environment that in turn cut down the cost of running cloud servers and affect the metered cost ultimately. Additionally, it should be noted that we use terminology of host and server interchangeably at several places of this paper to refers the hardware involved.

Rest of the paper is organized as follows. Section 2 present related work followed by Background, proposed work and experimental setup with results in Section 3, Section 4 and Section 5 respectively. Section 6 concludes the paper.

2 RELATED WORK

The proposed heuristic built upon the following considerations. Nathuji and Schwan (Nathuji & Schwan 2007) proposed local and global resource management strategies where VM consolidation is achieved by global policies using live migration. Kusic et al. (Kusic, Kephart, Hanson, Kandasamy, & Jiang 2009) used Limited Lookahead Control (LLC) to address the problem of continuous consolidation in terms of sequential optimization. They used a complex model using simulation-based learning but execution time for their proposed strategy was high enough even for small number of nodes. On the contrary, the proposed algorithm is based on a heuristic that aims high performance for large infrastructure.

Verma et al. (Verma, Ahuja, & Neogi 2008) proposed pMapper for placement of applications in a virtualized system which is power and migration cost-aware based on the idea of bin-packing problem with differently sized bin where servers represent the bins and VMs represent the balls. On the contrary, the proposed algorithm does not ensure the Service Level Agreement (SLA) negotiated due to workload variability.

Gmach et al. (Gmach, Rolia, Cherkasova, & Kemper 2009) and Beloglazov et al. (Beloglazov & Buyya 2010) worked upon the static threshold based approach for dynamic workload consolidation. This approach is not suitable for an Infrastructure-as-a-Service (IaaS) environment serving different application due to the reason that static values of the threshold do not help much for dynamic and unpredictable workload. Beloglazov et al. proposed several adaptive strategies (Beloglazov & Buyya

2012) for energy efficient allocation of resources to overcome the problem of static threshold values. In contrast to the discussion made above, we propose an energy efficient algorithm for reallocation of resources using an adaptive technique, Median Absolute Deviation (MAD) and Inter-Quartile Range (IQR) of setting threshold values dynamically based on the set of VMs instantiated and past historical data of resource usage by the VMs.

3 BACKGROUND

We briefly summarize the modules that are actually involving to solve the problem. The problem of VM consolidation can be divided into four parts in our case. These four parts are as follows:

1. Overloading detection,
2. VM selection,
3. Underloading detection, and
4. VM placement

3.1 Overloading detection

(Beloglazov & Buyya 2012) proposed adaptive and dynamic utilization threshold detection methods like Median Absolute Deviation (MAD), Inter-quartile Range (IQR), Local Regression (LR), and Robust Local Regression (RLR). In this paper, we use MAD and IQR to incorporate with the proposed heuristic.

3.2 VM selection

Once overloaded host is detected the efforts are made to make the host non-overloaded by migrating few VMs from the host. This process of continuous overloading detection for all the host in cloud computing environment goes on till the end of simulation length.

3.3 Underloading detection

Furthermore, host are detected for underloading. In the CloudSim simulation environment minimum utilized host is treated as the underloaded host and therefore all the VMs available on this host shall be migrated.

3.4 VM placement

VM placement module take care of migrated VMs from the overloaded servers and underloaded server. This module find the optimal packing solution for these VMs and pack them on fewer servers using constraint formulated for Bin-Packing problem. Bin packing is non-deterministic polynomial-hard (NP-hard) to solve. This module ensures for packing of *VMs* on server is not more than $\frac{11}{9} \cdot OPT + 1$ where OPT is an optimal variable that provides number of bins/servers to pack *VMs* by optimal solution (Yue 1991).

4 PROPOSED WORK

Power consumption by the resources deployed under cloud computing environment and resource utilization constraints can be presented as shown in (1) and (8).

$$P = \int P(u(t))dt \tag{1}$$

where $u(t)$ is the utilization of servers as a function of time and power consumption is the function of utilization.

Mathematically, objective of this problem is to bring down the utilization level of servers below the utilization threshold of the servers in such a way that there will a minimum gap between threshold and current utilization subject to the constraints of problem formulation as shown in (2), (3), (4), and (5).

$$Min \quad Z = T_u - A_u \tag{2}$$

$$s.t.$$

$$T_u - A_u \geq 0 \tag{3}$$

Algorithm 1: Quartile Position VM First

Require: Supply: *hostList*, Power Model: P_m
1: **for all** *Host* in *hostList* **do**
2:    **if** *HostOverloaded* **then**
3:       $vmList \leftarrow sortByCPUAllocated(vmList)$
4:       $index = \left\lfloor |VM| - \frac{|VM|}{4} - 1 \right\rfloor$
5:    **end if**
6:    $vmsToMigrate \leftarrow vmList.get(index))$
7:    $getNewVmPlacement(vmsToMigrate)$
8: **end for**
9: **for all** *Host* in *hostList* **do**
10:    **if** *HostUnderloaded* **then**
11:       $vmsToMigrate \leftarrow vmList(host)$
12:    **end if**
13:    $getNewVmPlacement(vmsToMigrate)$
14: **end for**

$$T_u, A_u \geq 0 \tag{4}$$

$$and \quad T_u, A_u \leq 1 \tag{5}$$

where T_u is utilization threshold of servers and A_u is the actual or current utilization of the servers.

We take standardized power consumption details of HP Proliant ML110 G4 and HP Proliant ML110 G5 servers on different load levels from *SPECpowerBenchmark* (SPECpower) for our calculations of power consumption by the servers. These details are shown in Table 1. It is quite clear from the benchmark data that even ideal servers consume electrical energy around 50% of a fully loaded server. Ideal servers or underutilized servers causes lots of resource wastage and high metered cost. Therefore, overall emphasis in this paper is to cut down the energy consumption. We solve this problem by the concept of *VM* consolidation where *VMs* running on several servers attempts to consolidate over fewer servers so that energy consumption can be minimized by putting ideal servers into low power consumption mode and when the need arise they are invoked back.

4.1 Random Selection policy (RS)

Random selection criteria selects a *VM* for migration from overloaded host according to uniformly distributed random variable $X \overset{d}{=} U(0, |V_j|)$. The values of X index to the set of *VMs* allocated to the host (Beloglazov & Buyya 2012).

4.2 Quartile position VM First (QVF)

Instead of selecting any random *VM* from the set of *VMs* allocated to the overloaded host, we apply our intuition to select a *VM* from the overloaded host. Our intuition suggest that *VMs* allocated the host are of heterogeneous CPU capacity. If we select a *VM* with highest Million Instruction Per Second (*MIPS*) allocated or the *VM* allocated largest CPU size (in other words), then making an overloaded server as non-overloaded will be much faster and that will save the power consumption. Selecting large CPU size *VM* will cause lesser number *VM* migrations and in practical sense it will save a lot of *VM* transfer cost that may involve bandwidth cost and *QOS* parameters as defined by (Garg, Versteeg, & Buyya 2011).

Table 1. Power consumption by servers in watts.

Server type	0%	10%	20%	30%	40%	50%	60%	70%	80%	90%	100%
HP Proliant G4	86	89.4	92.6	96	99.5	102	106	108	112	114	117
HP Proliant G5	93.7	97	101	105	110	116	121	125	129	133	135

However, selecting a VM for migration with highest CPU allocation will lead to other side-effects as well. On migration of highest CPU allocated VM causes running of a host much below the utilization threshold of the servers and incurs the penalty in terms of more power consumption. Therefore, we propose to select a VM for migration that is available at Upper Quartile position (Doane & Seward 2005) of the sorted list of VMs and do not cause high gap between utilization threshold and current utilization of the servers. We refer this heuristic as Quartile Position VM First (QVF).

Quartiles that is Q_1, Q_2 and Q_3 divides data into quarters (segments of 25% each). The quartile Q_1 corresponds to the 25th percentile, Q_2 to the 50th percentile, and Q_3 to the 75th percentile (Verma & Katti 2015). We define UpperQuartile (i.e. Q_3) in our case as shown in (6).

$$Q_3 = \left\lfloor |VM| - \frac{|VM|}{4} - 1 \right\rfloor \qquad (6)$$

where $|VM|$ represents the cardinality of of set of VMs available on the overloaded host.

We sort VMs available on overloaded servers by CPU size allocated to the VM in ascending order. To select the desired VM, we find the upper quartile position in the sorted list as defined in (6) and select it for migration to offload the host. This scheme is shown pictorially in Figure 2 and in (7).

$$VM_{ij} = \{VM_{i=Q_3} \mid sortVmList(host_j)\} \qquad (7)$$

where VM_i is the VM available at Q_3 position of sorted list of VMs available on host j and VM_{ij} represents i the VM in host j.

Additionally, QVF will save transition cost that will occur during the time when a VM will be put in suspended mode before the migration and the transition cost that will occur during the time after migration of VM on the other server and revoking back the execution.

5 EXPERIMENTAL SETUP AND RESULTS

CloudSim simulator (Calheiros, Ranjan, Beloglazov, De Rose, & Buyya 2011) is used as a test-bed for performing simulations on proposed QVF heuristic and existing RS heuristic. The underlying hardware architecture is *intel i7*, 10 GB RAM, 1 TB storage memory. We used workload traces from PlanetLab servers that were part of the CoMon project (Park & Pai 2006). The workload characteristics are shown in Table 2. We simulated a heterogeneous environment of equal number of HP Proliant ML110 G4 and HP Proliant ML110 G5 servers where total number of hosts are taken as 800. Total number of VMs in the simulated cloud computing environment is dependent on the workload traces as shown in Table 2. Metric used to evaluate SLA violations are *SLAV*, *SLATAH*, and *PDM* as shown in Section 5.1.2, Section 5.1.3, and Section 5.1.4. In our experimentation, we choose *Amazon EC2* instances for VM sizing as shown in Table 3. Results and comparison for RS and QVF using MAD and IQR threshold method are shown in Section 5.1.1, Section 5.1.2, Section 5.1.3, Section 5.1.4, and Section 5.1.5.

5.1 Metrics

We use several metrics for performance evaluation. Power consumption is main focus of this work and

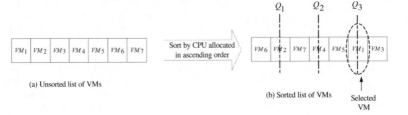

Figure 2. Quartile position VM first.

Table 2. Workload characteristics.

Date	Number of VMs	Mean (%)	St. dev. (%)	Quartile 1 (%)	Median (%)	Quartile 3 (%)
03/03/2011	1052	12.31	17.09	2	6	15
06/03/2011	898	11.44	16.83	2	5	13
09/03/2011	1061	10.70	15.57	2	4	13
22/03/2011	1516	9.26	12.78	2	5	12

Table 3. Amazon EC2 characteristics.

VM type	Medium	Extra large	Small	Micro
MIPS	2500	2000	1000	500
RAM (GB)	0.85	3.75	1.7	613

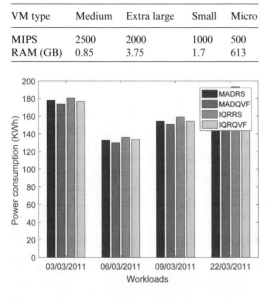

Figure 3. Power consumption metric.

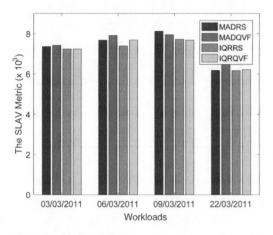

Figure 4. The SLAV metric.

it is calculated as defined in Section 4. We measure the efficiency of the proposed heuristic on power consumption, *SLA* Violations, *SLATAH*, *PDM* and total number of *VM* migration parameters that described in Section 5.1.1, Section 5.1.2, Section 5.1.3, Section 5.1.4, and Section 5.1.5. We perform simulations on workload traces of four days as shown in Table 2. Results are shown in Figure 3, Figure 4, Figure 5, Figure 6, and Figure 7.

5.1.1 *Power consumption*
In this paper, our main focus is to cut down power consumption and therefore the metered cost. In

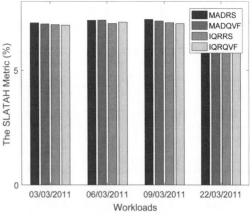

Figure 5. The SLATAH metric.

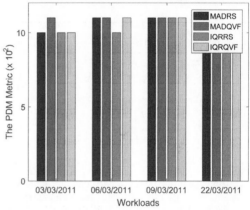

Figure 6. The PDM metric.

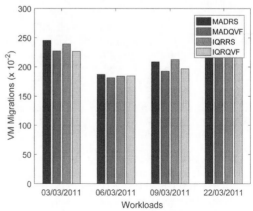

Figure 7. VM migrations.

Table 4. Power consumption in KWh.

	MAD		IQR	
Workload	RS	QVF	RS	QVF
03/03/2011	178.15	173.83	180.53	176.72
06/03/2011	132.99	130.01	136.17	133.87
09/03/2011	154.46	151.04	159.20	154.37
22/03/2011	189.09	187.02	193.53	189.87

Table 5. The SLAV metric.

	MAD		IQR	
Workload	RS	QVF	RS	QVF
03/03/2011	7.36	7.42	7.24	7.24
06/03/2011	7.68	7.91	7.39	7.69
09/03/2011	8.12	7.95	7.72	7.68
22/03/2011	6.17	6.56	6.16	6.21

Table 6. The SLATAH metric (%).

	MAD		IQR	
Workload	RS	QVF	RS	QVF
03/03/2011	7.07	7.02	6.99	6.97
06/03/2011	7.18	7.19	7.04	7.10
09/03/2011	7.22	7.15	7.08	7.04
22/03/2011	7.11	7.28	7.07	7.14

Section 4 we describe the way of calculating power consumption. In a simple way, power consumption can be written as a linear relationship as shown in (8).

$$PC = P_{IdealHost} + P_{ActiveHost}$$
$$= kP_{max} + (1-k)P_{max}(u) \quad (8)$$

where k is the fraction of energy consumed by the ideal host, P_{max} is maximum power consumed by a fully loaded host and u is the level of utilization of host.

5.1.2 SLA violations

Service Level Agreement (SLA) is an important parameter for popularity of any cloud provider. Cloud computing model is a market oriented paradigm and therefore delivery of promised SLAs becomes crucial for reputation of any cloud provider. SLA Violations (SLAV) is a composite parameter capturing SLATAH and PDM together. An algorithm providing higher level of SLA delivery and lower SLA violations are preferable. SLAV are defined as shown in (9).

Table 7. The PDM metric (%).

	MAD		IQR	
Workload	RS	QVF	RS	QVF
03/03/2011	0.10	0.11	0.10	0.10
06/03/2011	0.11	0.11	0.10	0.11
09/03/2011	0.11	0.11	0.11	0.11
22/03/2011	0.09	0.09	0.09	0.09

Table 8. VM migrations.

	MAD		IQR	
Workload	RS	QVF	RS	QVF
03/03/2011	24547	22709	23935	22643
06/03/2011	18708	18137	18399	18453
09/03/2011	20859	19248	21255	19663
22/03/2011	25661	24493	26075	24385

$$SLAV = SLATAH \cdot PDM \quad (9)$$

where $SLATAH$ and PDM are as explained in Section 5.1.3 and Section 5.1.4, respectively.

5.1.3 SLAV time per active host

SLAV Time per Active Host ($SLATAH$) represents the effect on SLAs while host suffered from 100% utilization.

$$SLATAH = \frac{1}{N}\sum_{i=1}^{N}\frac{T_{s_i}}{T_{a_i}} \quad (10)$$

where N is the number hosts, T_{s_i} is the time during which host i experienced 100% utilization and causing SLA violations.

5.1.4 Performance degradation metric

Performance Degradation Metric (PDM) attempts to evaluate the degradation in performance of the VM due to process of migration.

$$PDM = \frac{1}{M}\sum_{j=1}^{N}\frac{C_{d_j}}{C_{r_j}} \quad (11)$$

where M is the number of VMs, C_{d_j} represents degradation in performance due to migration of VM_j and C_{r_j} represents the total CPU capacity requested during the lifetime of VM_j.

5.1.5 VM migrations

The metric VM migrations represents total number of VM migrations occurred due to overloading and underloading detection of hosts during the whole simulation period. In our case we take simulation length as 24 hours.

6 CONCLUSIONS

The experimental study through simulations has shown that QVF performs efficiently as compared to RS on incorporating both of the MAD and IQR for overloading threshold detection techniques. The results of power consumption metric shows that there is a small decrease in power consumption on considering 800 hosts in the cloud computing environment. However, the cloud computing environment has millions of hosts as a participant for resource pooling. Thefore, this small factor of power consumption reduction will show multi-fold effect of QVF on considering large number of hosts.

The results shows that QVF causes lesser number of VM migrations which is a favorable case over RS due to inherent demerits involved with larger number of VM migrations such as non-delivery of promised SLAs.

We look forward to propose more intelligent and more applicable heuristics for minimizing power consumption and delivery of promised SLAs as a future work.

REFERENCES

Armbrust, M., A. Fox, R. Gri_th, A. D. Joseph, R. Katz, A. Konwinski, G. Lee, D. Patterson, A. Rabkin, I. Stoica, et al. (2010). A view of cloud computing. *Communications of the ACM 53* (4), 50–58.

Beloglazov, A. & R. Buyya (2010). Energy efficient allocation of virtual machines in cloud data centers. In *Cluster, Cloud and Grid Computing (CCGrid), 2010 10th IEEE/ACM International Conference on*, pp. 577–578. IEEE.

Beloglazov, A. & R. Buyya (2012). Optimal online deterministic algorithms and adaptive heuristics for energy and performance efficient dynamic consolidation of virtual machines in cloud data centers. *Concurrency and Computation: Practice and Experience 24* (13), 1397–1420.

Calheiros, R. N., R. Ranjan, A. Beloglazov, C. A. De Rose, & R. Buyya (2011). Cloudsim: a toolkit for modeling and simulation of cloud computing environments and evaluation of resource provisioning algorithms. *Software: Practice and Experience 41* (1), 23–50.

Doane, D. P. & L. E. Seward (2005). Applied statistics in business and economics. *USA: Irwin*.

Garg, S. K., S. Versteeg, & R. Buyya (2011). Smicloud: a framework for comparing and ranking cloud services. In *Utility and Cloud Computing (UCC), 2011 Fourth IEEE International Conference on*, pp. 210–218. IEEE.

Gmach, D., J. Rolia, L. Cherkasova, & A. Kemper (2009). Resource pool management: Reactive versus proactive or lets be friends. *Computer Networks 53* (17), 2905–2922.

InfoWorld. Virtualization: Virtualization news, analysis, research, how-to, opinion, and video. http://www.infoworld.com/category/virtualization/. Lasr Accessed: May 02, 2016.

Kusic, D., J. O. Kephart, J. E. Hanson, N. Kandasamy, & G. Jiang (2009). Power and performance management of virtualized computing environments via lookahead control. *Cluster computing 12* (1), 1–15.

Nathuji, R. & K. Schwan (2007). Virtualpower: coordinated power management in virtualized enterprise systems. In *ACM SIGOPS Operating Systems Review*, Volume 41, pp. 265–278. ACM.

Park, K. & V. S. Pai (2006). Comon: a mostly-scalable monitoring system for planetlab. *ACM SIGOPS Operating Systems Review 40* (1), 65–74.

SPECpower. The specpower benchmark. http://www.spec.org/. Lasr Accessed: 21 October, 2016.

Verma, A., P. Ahuja, & A. Neogi (2008). pmapper: power and migration cost aware application placement in virtualized systems. In *Middleware 2008*, pp. 243–264. Springer.

Verma, J. & C. Katti (2014). Study of cloud computing and its issues: A review. *Smart CR 4* (5), 389–411.

Verma, J. & C. Katti (2015). A comparative study into energy efficient techniques for cloud computing. In *Computing for Sustainable Global Development (INDIACom), 2015 2nd International Conference on*, pp. 2062–2067. IEEE.

Yue, M. (1991). A simple proof of the inequality ffd (l) = 11/9 opt (l)+ 1,? 1 for the ffd bin-packing algorithm. *Acta mathematicae applicatae sinica 7* (4), 321–331.

Communication and Computing Systems – Prasad et al. (Eds)
© 2017 Taylor & Francis Group, London, ISBN 978-1-138-02952-1

Structural equation model to analyze the factors affecting agile adoption in Indian software firms

J.K. Bajwa, K. Singh & N. Sharma
Punjabi University, Patiala, Punjab, India

ABSTRACT: The premise of the study is to examine the influencing factors on the resultant agility outcomes. This paper focuses on identifying the relationship between important agile adoption enablers and agility outcomes from the software practitioners in India. Structural equation modeling techniques have been applied to examine the relationships among enablers of interest. Resultant data has been collected from Indian developmental firms to analyze the proposed conceptual framework. The model contains eight hypotheses which are tested using T- test. The results of the study showed that 'organizational culture' has direct positive effect on 'project environment' and 'technical practices'. The indirect positive effect of 'organizational culture' has been observed on agility outcomes through mediating factor 'technical practices'. The analysis indicated that Indian software houses should focus on creating more open organizational environment where project teams and customers could communicate and work jointly. Adoption of agile technical practices should also be given priority to attain agility outcomes.

Keywords: agile software development; agile adoption; partial least square, structural equation modeling

1 INTRODUCTION

A software development methodology is a basic ingredient required to discipline developmental process as well as to quantify developmental efforts. The focus of software organizations and associated research groups has always been to adopt sustainable development methodologies. Recently, agile methodology has grabbed the attention of software organizations because of its abilities to handle turbulent market scenarios with greater efficiency. But its adoption has been reported to be a thorny issue in certain traditional developmental based ecosystems. Primarily, the difficulties in adoption have been attributed to absence of conclusive working environment in such organizations. Agile basically requires a developmental environment where stakeholders can collectively communicate and collaborate to understand various aspects of the concerned project and produce workable software within short time span without compromising on the quality of the product being developed. The agile literature largely reports of adoption studies based in foreign ecosystems but very less information has been retrieved about the agile adoption challenges present in Indian software ecosystems. The primary focus of the current study is to find the relationships between the influencing factors which if properly handled can

enhance agile adoption in Indian software firms. The paper presents partial least square path model for agile adoption and testing the casual relationship between four enabler criteria for the eight proposed hypotheses. The relationship analysis and validation of the model has been done by deploying SPSS and Smart PLS software. The results of the model testing are expected to provide implications regarding adoption of four enablers and their effect on improving the agility outcomes.

2 THEORETICAL BASES

In agile literature there is an unending debate about the successful agile adoption factors. Agile software development, in any kind of organizational setting, not only requires step-by-step implementation of a particular agile methodology but is affected by various variables. Very recently some researchers (Tripp and Armstrong 2014) focused on finding the relationship between motives behind agile adoption and the adopted agile practices. Factors extracted through this study were 'quest to increase quality', 'efficiency and effectiveness' through 'agile practices'. They also advocated tailorability of agile practices to achieve organizational success. One of the studies found 'methodology champion' and 'top management support' as major adoption factors (Senapathi and Srinivasan 2012). Another study extracted four factors of namely 'inefficient Scrum implementation', 'revamping of training mechanism', 'better tool support for requirements management' (Srinivasan and Lundqvist 2009). An empirical study reported 'professional training', 'overzealous teams', 'absence of a pilot project', 'implementation of a new developmental method without an agile master', 'current work pressure', 'upper management' found new transition risky, 'hierarchical / bureaucratic environment' and large documentation requirements' in a governmental system (Hajjdiab and Taleb 2011). Dyba, et al. 2008; identified challenges in agile decision making'. The term 'agility' is defined as an ability to create and respond to change in a turbulent business environment. It is a concept of surviving and gaining profits as well as retaining customer base in highly competitive business world (Steven Goldman et al., 1995).

The stated studies of agile adoption are predominantly covering developmental ecosystems of developed nations. There is dire need to focus on a specific agile adoption model. We tend to take into consideration specific conditions/variables which are specific to Indian ecosystem. In this paper, effort has been made to derive a causal relationship model between adoption factors affecting agility of Indian development organizations.

2.1 Path model and hypothesis development

The proposed model is based on secondary data collected through literature review and empirical survey of 74 software houses operating in India. This model basically propounds that Agile adoption is influenced by four main factors mainly Organizational Culture (OC), Project Environment (PE), Business Approach (BA) and Agile Technical Practices (TP) and their implication on Agility Outcomes (OT). For establishing the SEM model, literature has been reviewed extensively. In this research paper, eight hypotheses have been set up to validate the proposed SEM model. The hypotheses involve 49 attributes for four enablers and six attributes for outcomes.

H1: There is a positive and significant effect of Business environment (BA) on Organizational Culture (OC).

H2: There is a positive and significant effect of Organizational Culture (OC) on agile Technical Practices (TP).

H3: There is a positive and significant effect of Organizational Culture (OC) on Project Environment (PE).

H4: There is a positive and significant effect of Business environment (BA) on agile Technical Practices (TP).

H5: There is a positive and significant effect of Business environment (BA) on Project Environment (PE).

H6: There is a positive and significant effect of Project Environment (PE) on agile Technical Practices (TP).

H7: There is a positive and significant effect on agile Technical Practices (TP) on agility Outcomes (OT).

H8: There is a positive and significant effect of Project Environment (PE) on agility outcomes (OT).

2.1.1 Organizational Culture (OC)
This factor is important to understand as the associated variables focus on the environment within which development is taking place. Twelve attributes are included under this factor-Top Management Support (OC1), Encouragement of team effort (OC2), Organizational policies comply with agile principles (OC3), Communication between offshore and onshore teams (OC4), Senior stakeholder support (OC5), Development supports lean principles (OC6), Adaptable approach (OC7), Mechanisms for continuous deliveries (OC8), Mechanisms for user feedback (OC9), Mechanisms for integrating changing requirements (OC10), Team motivation (OC11), Trust between developers and stakeholders (OC12).

2.1.2 Project Environment (PE)
This factor emphasize on creating an open and flexible work culture. Nine attributes are included under this factor-Trained team members (PE1), Experienced team (PE2), Co-located teams (PE3), Virtually connected offshore teams (PE4), positive perception about agile methods (PE5), Requirements changes allowed during later developmental phases (PE6), Consistent development (PE7), Trust relationship between manager and developer (PE8), Managers as agile coaches (PE 9).

2.1.3 Business Approach (BA)
This is a critical factor during agile adoption because strategies and approaches deployed by business houses require alignment with agile values. Nine attributed are considered under this factor-Business strategies cover stakeholders' risk (BA1), Secure project funding (BA2), Contractual terms support collaboration (BA3), Complete project completion benchmarks (BA4), Terms for incremental progress reviews (BA5), Co-ordination between development and business goals (BA6), Contractual terms align with agile development (BA7), Lifecycle activities align with agile developmental stages (BA8), Consideration of agile approach during project initiation (BA9).

2.1.4 Technical Practices (TP)

This factor highlights the best practices which can be used during agile adoption. Thirteen attributes are considered under this factor-Short iterations (TP1), User stories (TP2), Continuous Integration (TP3), Product backlog (TP4), Product owner (TP5), Pair programming (TP6), Relative estimation (TP7), Self-organizing team (TP8), Daily standups (TP9), Test driven development (TP10), Refactoring (TP11), Retrospectives (TP12), End iteration retrospectives (TP13).

3 RESEARCH METHODOLOGY

3.1 Questionnaire survey

The questionnaire has been used as survey tool for the current study. Seventy four software development houses in Indian IT-hubs have been considered for analysis of agile adoption. For identification of various variables for four adoption factors extensive literature review and expert opinions were done. The outcome factor for assessing agility contains six attributes- Increase in customer base (OT1), Increase in productivity (OT2), Reduction in time to market (OT3), Reduction in product development cost (OT4), Increase in product quality (OT5), Motivated teams (OT6). The questionnaire consisted of two parts. The first part enquired about the demographic profile of respondents and the second part contained 49 statements which enquired about five agile adoption constructs i.e. organizational culture, project environment, business approach, technical practices and outcomes. The responses were recorded on five point Likert scale to access the effect of first four factors on agility outcomes. The attributes gathered from field study assisted the researchers to develop the framework for this study. This framework is tested using PLS-SEM.

3.2 Data analysis

In this study, component-based Partial Least Square (PLS) method has been deployed for SEM analysis. The assessment of conceptual model involves assessment of outer measurement model as well as assessment of inner structural model. The analysis and validity of the model is carried out using smart PLS software. We have used two phased approach for the assessment of proposed PLS-SEM model: first testing the outer measurement model and then testing the inner structural model.

3.2.1 Validity assessment of the outer measurement model

The measurement and testing of outer model involves the estimation of latent constructs and their associated indicators. The outer model is evaluated using convergent validity test and discriminant validity test. The focus of convergent validity is to find the variance between latent variables. For the assessment of convergent validity three measures have been used Cronbach's alpha, Average Variance Extracted (AVE) and Composite reliability score. The statistically accepted values for a reliable model is to have Cronbach's alpha value more than 0.6, AVE value more than 0.5 and composite reliability score more than 0.7. After performing the validity test, the indicators having low values for reliability variances are dropped from the final PLS-SEM model (vide Figure 1).

For determining the discriminant validity test two validity tests are performed mainly cross loading and Fornell-Larcker criterion.

3.2.2 Validity assessment of the inner structural model

The inner structural model involves testing and validating the variance between various endogenous latent variables. The primary test criteria used for evaluation of the structural model are the coefficient of determination (R^2), measures significance of the path coefficients, T-statistic value and effect size.

In the PLS path modeling, the coefficient of determination (R^2) is computed to find the level of variance between key endogenous latent variables. The value of R^2 for each latent variable provides estimation about how precisely the inner model fits the presumptive relationships. The values of R^2 in the present path model lie between 0.29 and 0.74. According to the established statistical rule two of

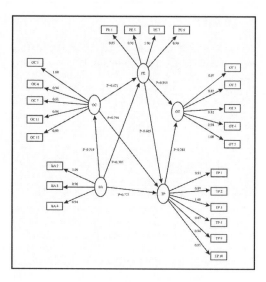

Figure 1. The structural model.

the latent constructs exhibit substantial level of variance.

For assessing the individual path coefficients, bootstrapping procedure has been used. The values of path coefficients are used to check the significance of the proposed hypotheses. We have used 5000 sub-samples in the bootstrapping for checking the significance of structural paths. The standard error value for each path coefficient has been provided by bootstrapping procedure. This value has been used to apply T-test for measuring and testing the significance of eight hypothesized paths.

The relationship between business policies and organizational culture is significant with path coefficient 0.51 and t-value of 12.75. These values indicate that business policies have direct influence on organizational culture. The relationship between business policy and project environment is insignificant with path coefficient value of 0.22 and t-value of 1.57. These values indicate that project environment has direct positive insignificant influence on business & acquisition.

4 MEDIATING EFFECT

In this study the direct and indirect effect of constructs has also been analyzed. In the PLS path model, organizational climate has direct effect (0.31) on project environment and technical practices (0.75). The indirect effect of organizational climate on outcomes (0.47) has been analyzed through mediating construct technical practices.

5 MODEL FIT INDICES

The model fit indices are used as model validating measures as they help in removing erroneous model specifications. The values of GFI indices for this model signify acceptance of model (RMR = 0.097, RMSEA = 0.06, CFI = 0.949 and NFI = 0.926). The values therefore, indicate that the hypothesized PLS model fits nicely with the empirical data.

6 DISCUSSION

This study has examined the casual relationship among four constructs namely organizational climate, business approach, project environment, technical practices and their adoption to achieve agility in Indian software development firms. The aim of this research is to investigate the influencing factors which can improve the

rate of agile adoption in Indian software houses. The results show that technical practices and project-customer environment directly affect the goals for achieving agility. But the technical practices are highly influenced by organizational climate. Here, it is important to see that organizational climate is directly influenced by business approach. The results provide an insight that improving business rules can lead to large scale adoption of agile methodology in Indian software houses.

7 CONCLUSION AND RECOMMENDATIONS

The results of the study demonstrate that software firms in India must focus on the sub variables to create a culture that suits agile methodology and improves the adoption of agile methods and practices.

Organizations should adopt a flexible approach towards accepting a new methodology. Top management (OC1 factor loading = 1) should understand the essence of Agile methodology. They should focus on creation of mutual trust between developers and the customers (OC12 factor loading = 0.9). The collaboration and communication channels between onshore and offshore teams (OC4 factor loading = 0.94) must be open so that bottlenecks between teams can be avoided.

To increase the acceptance of Agile methods, managers should act as agile coaches (PE9 factor loading = 0.99) because trained agile team (PE1 factor loading = 0.86) can consistently develop projects. Moreover, project teams must be trained to use mechanisms like continuous integration (PE7 = 0.99) servers to collaborate off shore and local code integrations. Managements should provide resources such as CI servers, which are helpful in accommodating and integrating requirement changes even in later phases of code development.

Software practices like continuous integration (TP3 factor loading = 1), short iterations (TP1 factor loading = 0.91) (less than two weeks), pair programming (TP6 factor loading = 0.87), daily standup meetings, user stories and test driven development are helpful in achieving organizational agility in Indian software scenario.

As the area of methodological adoption in Agile software engineering is tedious and is influenced by various factors, it is nearly impossible to analyze each influencing variable and factors in a single study. Therefore, future research can focus on different ecosystems. Also larger sample can be analyzed for more accurate results.

REFERENCES

Dybå, T., & Dingsøyr, T. 2008. Empirical studies of agile software development: A systematic review. *Information and software technology, 50*(9): 833–859.

Goldman, S. L. 1995. *Agile competitors and virtual organizations: strategies for enriching the customer. Van Nostrand Reinhold Company.*

Hajjdiab, H., & Taleb, A. S. 2011. *Agile adoption experience: A case study in the UAE. Proc. IEEE 2nd International Conference on Software Engineering and Service Science, 11 July 2011*, 31–34.

Senapathi, M., & Srinivasan, A. 2012. Understanding post-adoptive agile usage: An exploratory cross-case analysis. *Journal of Systems and Software, 85*(6): 1255–1268.

Srinivasan, J., & Lundqvist, K. 2009. Organizational enablers for agile adoption: learning from GameDevCo. *Proc. International Conference on Agile Processes and Extreme Programming in Software Engineering, 5 May 2009:63–72. Springer Berlin Heidelberg.*

Tripp, J. F., & Armstrong, D. J. 2014. Exploring the relationship between organizational adoption motives and the tailoring of agile methods. *Proc. 2014 47th Hawaii International Conference on System Sciences, 6 Jan: 4799–4806. IEEE.*

Wong, K. K.-K. 2013. Partial least squares structural equation modeling (PLS-SEM) techniques using SmartPLS. *Marketing Bulletin, 24*(1): 1–32.

Zain, M., Rose, R. C., Abdullah, I., & Masrom, M. 2005. The relationship between information technology acceptance and organizational agility in Malaysia. *Information & Management, 42*(6): 829–839.

Author index